RADIO
DESIGNER'S
HANDBOOK

RADIO
DESIGNER'S
HANDBOOK

Edited by

F. LANGFORD-SMITH

B.Sc., B.E. (1st class honours)
Senior Member I.R.E. (U.S.A.)
A.M.I.E. (Aust.)

Newnes
An imprint of Butterworth-Heinemann
Linacre House, Jordan Hill, Oxford OX2 8DP
225 Wildwood Avenue, Woburn, MA 01801-2041
A division of Reed Educational and Professional Publishing Ltd

℞ A member of the Reed Elsevier plc group

OXFORD   AUCKLAND   BOSTON
JOHANNESBURG   MELBOURNE   NEW DELHI

First published 1934
Second edition 1935
Third edition 1940
Fourth edition 1953
Revised reprint with addenda 1954, 1955, 1957, 1960, 1963, 1967
Reprinted 1997, 1999

ISBN 0 7506 3635 1

Printed and bound in Great Britain by MPG Books Ltd, Bodmin, Cornwall

FOREWORD

By the Editor of *Wireless World*

The previous edition of this book, published in Australia under the title "Radiotron Designer's Handbook", was first distributed in Great Britain in 1940 from the offices of *Wireless World*. During the second world war it became a widely accepted textbook, and was reprinted many times. The present edition, greatly enlarged, and published by the Amalgamated Wireless Valve Company of Australia under the original Australian title, is printed, bound and distributed in Great Britain by the publishers of *Wireless World*.

London, 1953.

PREFACE

This book has been written as a comprehensive, self-explanatory reference handbook for the benefit of all who have an interest in the design and application of radio receivers or audio amplifiers. Everything outside this field—television, radio transmission, radar, industrial electronics, test equipment and so on—has been excluded to limit the book to a reasonable size.

An effort has been made to produce a handbook which, in its own sphere, is as self-contained as possible. Extensive references to other sources of information have been included for the reader who might require additional detail.

The success of the previous edition, of which over 280,000 copies have been sold throughout the world, encouraged Amalgamated Wireless Valve Company Pty. Ltd., to undertake the compilation of the present up-to-date edition which is more complete, great pains having been taken to fill in many of the gaps in the data published hitherto. This has involved a considerable amount of both experimental and analytical work by the editor and by engineers assisting in the project. Some original work, previously unpublished, has been included.

Although the various chapters have been written by individual authors, all are truly the result of team work, each having been carefully and critically examined by several other engineers specializing in that particular field. In this way the accuracy of mathematical calculations and of individual statements was checked and re-checked to ensure that the reader would not be misinformed.

I wish to express my grateful thanks to Mr. R. Lambie, Manager, Mr. D. M. Sutherland, Works Manager, and to the following collaborating engineers—

J. E. Bailey, Dr. W. G. Baker, D. Barnett, J. D. G. Barrett, Dr. G. Builder, N. V. C. Cansick, W. N. Christiansen, Dr. E. R. Dalziel, K. G. Dean, H. L. Downing, J. Gilchrist, I. C. Hansen, R. Herbert, F. Holloway, I. Hood, D. G. Lindsay, W. S. McGuire, E. J. Packer, J. Pritchard, R. J. Rawlings, B. Sandel, J. Stacey, R. D. Stewart, J. E. Telfer, R. Vine, E. Watkinson.

I also wish to express my thanks to Mr. R. Ainsworth for his invaluable work in sub-editing and assistance in indexing and to Mr. R. H. Aston for compiling data.

F. LANGFORD-SMITH

Amalgamated Wireless Valve Company Pty. Ltd.

NOTES

References to valve types are to the prototype (e.g. 6J7) and include all equivalent types (e.g. 6J7-G, 6J7-GT etc.) unless otherwise indicated.

If any errors are noted, please write to the Editor, Radiotron Designer's Handbook, Amalgamated Wireless Valve Co. Pty. Ltd., 47 York Street, Sydney, Australia.

ACKNOWLEDGEMENTS

The editor wishes to acknowledge his indebtedness to the publishers of the following periodicals for permission to reproduce the diagrams listed below. In some cases the diagrams are based largely on the originals, although modified.

PERIODICALS

Audio Engineering **13.** 50D, 50E. **14.** 1A, B, C, D, E, F, G, H. **15.** 15, 16, 57A, 57B, 60. **16.** 15. **17.** 7, 8, 15A, 15B, 15C, 18, 19A, 19B, 19C, 20, 22, 24B, 24C, 27A, 27B, 27C, 31, 37A. **19.** 6A. **20.** 10, 13, 14.

A.W.A. Technical Review **25.** 33, 34, 35, 36, 37, 38, 39. **26.** 6. **27.** 44.

Communications **7.** 72, 73. **16.** 16. **18.** 17, 18. **23.** 17, 18, 19, 20, 21, 22, 23, 24, 25, 26.

Electronic Engineering **5.** 13E, 13F, 13G, 13H, 13J, 13K, 13L, 13M, 13N, 13P, 13Q, 13R, 13S, 13T, 13U, Table 5. **7.** 51D. **11.** 7, 8, 9, 10, 11, 12, 13. **12.** 3B, 3C, 11A, 11B. **15.** 4, 63. **17.** 21, 26. **18.** 6A, 6B.

Electronic Industries **17.** 6.

Electronics **1.** 8. **4.** 34. **5.** 21, 22. **7.** 51. **9.** 17, 18. **12.** 47A, 47B, 48. **15.** 17, 18A, 18B, 38, 39, 45, 46, 52, 53, 61. **16.** 4, 10, 12, 17, 18, 19, 20, 21, 22, 23, 24, 25, 26, 27, 28. **21.** 9. **35.** 24, 25. **36.** 3, 5.

FM-TV Radio Communication **17.** 35. **20.** 25, 26, 27.

Journal Acoustical Society of America **20.** 28.

Journal British I.R.E. **5.** 13C, 13D.

Proceedings I.R.E. (U.S.A.) **7.** 58, 59. **12.** 7C. **13.** 9B. **14.** 6. **20.** 24. **25.** 1, 2, 3, 4, 5, 6, 7, 8, 9, 10, 11, 12, 13, 14, 15, 16, 17, 18, 19, 20, 21. **30.** 4, 5, 6, 7, 8, 9. **37.** 2, 3, 4.

Proceedings I.E.E. **7.** 60, 65, 69, 70.

Proceedings I.R.E. Australia **35.** 19, 20, 21.

Q.S.T. **16.** 13.

Radio **13.** 50D, 50E. **15.** 47A, 47B.

Radio and Hobbies (Australia) **15.** 14B, 35B, 37B, 37C.

Radio and TV News **15.** 43A, 43B. **16.** 14. **18.** 7A, 7B.

Radio Craft **15.** 42, 44A, 49, 54. **17.** 33.

Radio Electronics **15.** 64.

R.C.A. Review **17.** 37A. **27.** 20, 21, 25, 27.

Tele-Tech **15.** 14A. **36.** 2. **38.** Chart 38.4.

T.V. Engineering **10.** 10.

Wireless Engineer **5.** 13B. **7.** 75, 76, 77, 78, 79, 80, 81, 82, 83. **11.** 1, 6A.

Wireless World **4.** 10, 26. **7.** 51A, 54B, 54C, 55D, 55E, 55F, 59C. **12.** 34, 35, 39. **15.** 5, 13, 41, 50, 51, 55, 56A, 56B, 58A. **16.** 8, 9, 11. **17.** 24A, 28, 32, 35B, 35C, 35D, 35E, 35F, 35G. **18.** 26. **20.** 5, 9, 11, 12, 16, 17. **31.** 6.

Reference numbers are given either in the titles or in the text. Lists of references giving authors' names and details of articles are given at the end of each chapter.

ACKNOWLEDGEMENTS (Continued)

BOOKS AND OTHER PUBLICATIONS

The editor desires to acknowledge his indebtedness to the authors and publishers of the following books and publications for permission to reproduce the diagrams and tables listed below.

Redrawn by permission from ELECTRONIC TRANSFORMERS AND CIRCUITS by R. Lee, published by John Wiley & Sons, Inc., 1947. Fig. 5.23.

Redrawn by permission from General Electric Company from TRANSFORMER ENGINEERING—1st ed. by Blume, Camilli, Boyajian & Montsinger, published by John Wiley & Sons, Inc. 1938. Fig. 5.18B.

Redrawn by permission from ELEMENTS OF ACOUSTICAL ENGINEERING (2nd ed.) by H. F. Olson, copyrighted by D. Van Nostrand Company Inc. Table 19.7. Fig. 20.30.

Redrawn by permission from NETWORK ANALYSIS AND FEEDBACK AMPLIFIER DESIGN by H. W. Bode, Bell Telephone Laboratories Inc., copyrighted 1945 by D. Van Nostrand Company Inc. Figs. 7.55A, 7.55B, 7.55C, 7.56A.

Other acknowledgements are given in the titles.

OTHER PUBLICATIONS

Aerovox Research Worker, Vol. 11, Nos. 1 and 2, January-February 1939, Figs. 5 and 6. Published by permission of Aerovox Corporation. Tables 38. 43, 44. Charts 38. 1, 2.

Allegheny Ludlum Steel Corporation—curves 3.6% silicon steel. Fig. 5. 20.

American Standards Association "American Standard for Noise Measurement" Z24.22-1942. Figs. 19. 7, 8.

Australian Radio Technical Services and Patents Co. Pty. Ltd.—Technical Bulletins, Figs. 15. 22, 23, 24, 25, 26, 27, 43. 31. 3, 4, 5.

Amalgamated Wireless Australasia Ltd. Figs. 18. 7. 28. 2.

Electrical and Musical Industries. Fig. 17. 25.

Jensen Manufacturing Company—Jensen Technical Monograph No. 3. Figs. 14. 2, 3, 4, 5.

P. R. Mallory & Co. Inc. "Fundamental Principles of Vibrator Power Supply" Figs. 32. 1, 2, 3, 4, 5, 6, 7, 8, 9, 10, 11, 12, 13.

National Bureau of Standards Circular C74, Fig. 208 redrawn and published by permission of National Bureau of Standards. Fig. 10. 3.

Standard Transformer Corporation "Engineering a transformer." Figs. 5. 18C, 18D. Table 38. 70.

General Electric Company Ltd. "Radio receiver for use with high fidelity amplifiers." Fig. 15. 62A.

Radio and Television Manufacturers Association (U.S.A.)—Extracts from, or summaries of, the following Standards:

ET-107, ET-109, ET-110, REC-11, REC-101, REC-103, REC-104, REC-105, REC-106-A, REC-107, REC-108, REC-113, REC-114, REC-115-A, REC-116, REC-117, REC-118, REC-119, REC-121, REC-128, S-410, S-416, S-417, S-418, S-504, SE-101-A, SE-103, SE-105, TR-105-A, TR-113-A.

Note:—References in the text to R.M.A. Standards should be taken as having been issued by RTMA.

I.R.E. Standards, Figs. 3. 11B, 15. 37. 1, 2, 3, 4.

Radio Industry Council—Information from standards on resistors and capacitors. Chapter 38 Sections 2 and 3.

Radio Corporation of America. 1. 6. 2. 32A, 33, 34, 35, 47. 3. 16, 17. 10. 6. 13. 26, 30, 48. 17. 5A. 23. 9, 10, 11, 12, 13, 14, 15, 16. 30. 2A, 2B, 2C, 11. 33. 14. 35. 17.

CHAPTER HEADINGS

PART 1 : THE RADIO VALVE

PART 2 : GENERAL THEORY AND COMPONENTS

PART 3 : AUDIO FREQUENCIES

PART 4 : RADIO FREQUENCIES

PART 5 : RECTIFICATION, REGULATION, FILTERING AND HUM

PART 6 : COMPLETE RECEIVERS

PART 7 : SUNDRY DATA

See next page for detailed List of Contents.

CONTENTS

PART 1 : THE RADIO VALVE

CHAPTER 1. INTRODUCTION TO THE RADIO VALVE

CHAPTER 2. VALVE CHARACTERISTICS

CONTENTS

CHAPTER 3. THE TESTING OF OXIDE-COATED CATHODE HIGH-VACUUM RECEIVING VALVES

CONTENTS

CONTENTS

CONTENTS

CONTENTS

CHAPTER 6. MATHEMATICS

CONTENTS

CONTENTS

CHAPTER 13. AUDIO FREQUENCY POWER AMPLIFIERS

(xxi)

CONTENTS

CHAPTER 14. FIDELITY AND DISTORTION

CONTENTS

CHAPTER 15. TONE COMPENSATION AND TONE CONTROL

CONTENTS

CHAPTER 16. VOLUME EXPANSION, COMPRESSION AND LIMITING

CONTENTS

CHAPTER 18. MICROPHONES, PRE-AMPLIFIERS, ATTENUATORS AND MIXERS

CHAPTER 19. UNITS FOR THE MEASUREMENT OF GAIN AND NOISE

CONTENTS

CHAPTER 20. LOUDSPEAKERS

(xxix)

CONTENTS

CHAPTER 21. THE NETWORK BETWEEN THE POWER VALVE AND THE LOUDSPEAKER

PART 4: RADIO FREQUENCIES

CHAPTER 22. AERIALS AND TRANSMISSION LINES

CONTENTS

CHAPTER 23. RADIO FREQUENCY AMPLIFIERS

CONTENTS

CHAPTER 27. DETECTION AND AUTOMATIC VOLUME CONTROL

CONTENTS

CHAPTER 28. REFLEX AMPLIFIERS

CHAPTER 29. LIMITERS AND AUTOMATIC FREQUENCY CONTROL

CONTENTS

PART 5: RECTIFICATION, REGULATION, FILTERING AND HUM

CHAPTER 30. RECTIFICATION

CHAPTER 31. FILTERING AND HUM

CHAPTER 32. VIBRATOR POWER SUPPLIES

CONTENTS

CONTENTS

CONTENTS

CHAPTER 1

INTRODUCTION TO THE RADIO VALVE

by F. LANGFORD-SMITH, B.Sc., B.E.

SECTION 1 : ELECTRICITY AND EMISSION

The proper understanding of the radio valve in its various applications requires some knowledge of the characteristics of the electron and its companion bodies which make up the complete structure of atoms and molecules.

All **matter** is composed of molecules which are the smallest particles preserving the individual characteristics of the substance. For example, water is made up of molecules that are bound together by the forces operating between them. Molecules are composed of atoms that are themselves made up of still smaller particles. According to the usual simplified theory, which is sufficient for this purpose, atoms may be pictured as having a central nucleus around which rotate one or more electrons in much the same manner as the planets move around the sun. In the case of the atom, however, there are frequently several electrons in each orbit. The innermost orbit may have up to 2, the second orbit up to 8, the third orbit up to 18, the fourth orbit up to 32, with decreasing numbers in the outermost orbits (which on y occur with elements of high "atomic numbers"). We do not know the precise shape and positions of the orbits and modern theory speaks of them as "energy levels" or "shells." The electrons forming the innermost shell are closely bound to the nucleus but the forces become progressively less in the outer shells. Moreover, the number of electrons in the outermost shell may be less than the maximum number that this shell is capable of accommodating. In this case, the substance would be chemically active ; examples of such are sodium and potassium.

In a **metal** the various atoms are situated in close proximity to one another, so that the electrons in the outermost shells have forces acting upon them both from their " parent " nucleus and their near neighbour. Some electrons are free to move about throughout the substance and are, therefore, called "free electrons." If an electric potential is applied between two points in the metal, the number of electrons moving from the negative to the positive point will be greater than those moving in the opposite direction. This constitutes an electric current, since each electron carries an electric charge. The charge on the electron is defined as unit negative charge and the accepted direction of current flow is opposite to the net electron movement.

It is interesting to note that the **total current flow**, equivalent to the total movements of all the free electrons, irrespective of their directions, is very much greater than that which occurs under any normal conditions of electric current flow. The directions of movement are such that the external effects of one are generally cancelled by those of another. Thus, in a metal, the oft-quoted picture of a flow of electrons from the negative to the positive terminal is only a partial truth and apt to be misleading. The velocity of the free electrons is very much less than that of the electric current being of the order of only a few centimetres per second. The electron current may be pictured as the successive impacts between one electron and another in the direction of the current. In an **insulator** the number of free electrons is practically zero, so that electric conduction does not take place. In a partial insulator the number of free electrons is quite small.

The **nucleus** is a very complex body, including one or more protons which may be combined with a number of neutrons*. The proton has a positive charge equal and opposite to the charge on an electron but its mass is very much greater than that of an electron. The simplest possible atom consists of one proton forming the nucleus with one electron in an orbit around it—this is the hydrogen atom. Helium consists of two protons and two neutrons in the nucleus, with two electrons rotating in orbits. The neutron has a mass slightly greater than that of a proton, but the neutron has no electric charge. An example of a more complicated atom is that of potassium which has 19 protons and 20 neutrons in the nucleus, thus having a positive charge of 19 units. The number of electrons in the orbits is 19, thus giving zero charge for the atom as a whole, this being the normal condition of any atom. The common form of uranium has 92 protons and 146 neutrons in the nucleus, with 92 electrons rotating in orbits.

Under normal circumstances no electrons leave the surface of a substance since the forces of attraction towards the centre of the body are too great. As the temperature of the substance is raised, the velocity of the free electrons increases and eventually, at a temperature which varies from one substance to another, some of the free electrons leave the surface and may be attracted to a positive electrode in a vacuum. This phenomenon is known as **thermionic emission** since its emission takes place under the influence of heating. There are other types of emission such as **photo emission** that occur when the surface of the substance is influenced by light, or **secondary emission** when the surface is bombarded by electrons.

The radio valve makes use of thermionic emission in conjunction with associated circuits for the purpose of producing amplification or oscillation. The most common types of radio valves have hot cathodes, either in the form of a filament or an indirectly-heated cathode. Many transmitting valves have filaments such as tungsten or thoriated-tungsten, but nearly all receiving valves have what is known as an **oxide** coated **filament or cathode.** The filament, or cathode sleeve, is usually made of nickel or an alloy containing a large percentage of nickel and this is coated with a mixture of barium and strontium carbonates that, during the manufacture of the valve, are turned into oxides. A valve having an oxide-coated cathode has a very high degree of emission as compared with other forms of emitters but requires very great care during manufacture since it is readily poisoned by certain impurities which may be present in the cathode itself or which may be driven out in the form of gas from the bulb or the other electrodes.

Oxide-coated cathodes are generally operated at an average temperature of about 1050° Kelvin (777° C) which looks a dull red. Temperatures much above 1100°K generally cause a short life, while those below 960°K are very susceptible to poisoning of the emission, and require careful attention to maintain a very high vacuum.

The thermionic valve is normally operated with its **anode†** current considerably less than the maximum emission produced by its cathode. In the case of one having a pure tungsten filament no damage is done to the filament if all the electrons emitted

---

*This is in accordance with the theory generally held at the time of writing ; it is, however, subject to later modification.

†The anode (also called the plate) is the positive electrode ; the cathode is the negative electrode.

are drawn away immediately to the anode. This is not so, however, with oxide coated cathodes and these, for a long life and satisfactory service, require a total emission very much greater than that drawn under operating conditions. In such a case a cloud of electrons accumulates a short distance from the surface of the cathode and supplies the electrons that go to the anode. This **space charge** as it is called, is like a reservoir of water that supplies varying requirements but is itself replenished at an average rate. The space charge forms a protection to the cathode coating against bombardment and high electrostatic fields, while it also limits the current which would otherwise be drawn by a positive voltage on the anode. If the electron emission from the cathode is insufficient to build up this " space charge," the cathode coating is called upon to supply high peak currents that may do permanent injury to the coating and in extreme cases may even cause sputtering or arcing.

In multi-grid valves, if one grid has a positive potential and the next succeeding grid (proceeding from cathode to plate) has a negative potential, there tends to be formed an additional space charge. This outer space charge behaves as a source of electrons for the outer electrodes, and is known as a **virtual cathode.**

An oxide-coated cathode, operated under proper conditions, is self-rejuvenating and may have an extremely long **working life.** The life is, therefore, largely governed by the excess emission over the peak current required in normal operation.

A valve having a large cathode area and small cathode current may have, under ideal conditions, a life of the order of 50 000 hours, whereas one having extremely limited surface area, such as a tiny battery valve, may have a working life of less than 1000 hours.

Under normal conditions a valve should be operated with its **filament or heater** at the recommended **voltage** ; in the case of an oxide coated valve it is possible to have fluctuations of the order of 10% up or down without seriously affecting the life or characteristics of the valve [see Chapter 3 Sect. 1(iv)D]. The average voltage should, however, be maintained at the correct value. If the filament or cathode is operated continuously with a higher voltage than that recommended, some of the coating material is evaporated and permanently lost, thus reducing the life of the valve. Moreover, some of this vapour tends to deposit on the grid and give rise to what is known as **grid emission** when the grid itself emits electrons and draws current commonly known as **negative grid current** [for measurement see Chapter 3 Sect. 3 (iv)A].

If the filament or heater is operated for long periods at reduced voltages, the effect is a reduction in emission, but no damage is generally done to the valve unless the cathode currents are sufficient to exhaust the " space charge." Low cathode temperature is, therefore, permissible provided that the anode current is reduced in the proper proportion.

During the working life of the valve, its **emission** usually increases over the early period, reaches a maximum at an age which varies from valve to valve and from one manufacturer to another, and then begins to fall. The user does not generally suffer any detriment until the emission is insufficient to provide peak currents without distortion.

Tests for the measurement of the emission of an oxide-coated cathode are described in Chapter 3 Sect. 3(ii)f.

If a slight amount of **gas** is present some of the electrons will collide with atoms of the gas and may knock off one or more electrons, which will serve to increase the anode current, leaving atoms deficient in electrons. These are known as **positive ions** since they carry a positive charge (brought about by the loss of electrons), and the process is known as ionization. The positive ions are attracted by the negative cathode, and being comparatively massive, they tend to bombard the cathode coating in spite of the protection formed by the space charge.

Some types of rectifiers (e.g. OZ4) have no heaters, and the oxide-coated cathode is initially heated by ion bombardment ; this flow of current is sufficient to raise the cathode temperature so as to enable it to emit electrons in the usual manner. The gas is an inert variety at reduced pressure. Although some types of gaseous thermionic rectifiers will operate (once they have been thoroughly heated) without any filament or heater voltage, this is likely to cause early failure through loss of emission.

Most thermionic valves are **vacuum types** and operate under a very high degree of vacuum. This is produced during manufacture by a combination of vacuum pumps and is made permanent by the flashing of a small amount of " getter " which remains in the bulb ready to combine with any impurities which may be driven off during life. Valves coming through on the production line are all tested for gas by measuring the negative grid current under operating conditions ; methods of testing are described in Chapter 3 Sect. 3(iv)A, where some values of maximum negative grid current are also given. If a valve has been on the shelf for a long time, it frequently shows a higher gas current, but this may usually be reduced to normal by operating the valve under normal conditions, with a low resistance connection between grid and cathode, for a short period. When a valve is slightly gassy, it usually shows a **blue glow** (ionization) between cathode and anode. In extreme cases this may extend outside the ends of the electrodes but a valve in such condition should be regarded with suspicion and tested before being used in any equipment, as it might do serious damage. A slight crack may permit a very small amount of air to enter the bulb, giving rise to a pink/violet glow which may readily be identified by any one familiar with it ; this is a sign of immediate end of life.

The anode current of a thermionic valve is not perfectly steady, since it is brought about by a flow of electrons from the cathode. When a valve is followed by very high gain amplifiers, the rushing noise heard in the loudspeaker is partly caused by the electrons in the valve, and partly by a somewhat similar effect (referred to as the " thermal agitation " or " Johnson noise ") principally in the resistance in the grid circuit of the first valve—see Chapter 4 Sect. 9(i)1, and Chapter 18 Sect. 2(ii). This question of **valve noise** is dealt with in Chapter 18 Sect. 2(ii)c and Chapter 23 Sect. 6.

Some valves show a fluorescence on the inside of the bulb, which may fluctuate when the valve is operating. This is perfectly harmless and may be distinguished from blue glow by its position in the valve. In occasional cases fluorescence may also be observed on the surfaces of the mica supports inside the valve.

SECTION 2 : THE COMPONENT PARTS OF RADIO VALVES

(i) *Filaments, cathodes and heaters* (ii) *Grids* (iii) *Plates* (iv) *Bulbs* (v) *Voltages with valve operation.*

(i) Filaments, Cathodes and Heaters

Cathodes are of two main types—directly heated and indirectly heated. Directly-heated cathodes are in the form of filaments which consists of a core of wire through which the filament current is passed, the wire being coated with the usual emissive coating. Filaments are the most economical form of cathodes so far as concerns the power necessary to heat the cathode. They are, therefore, used in most applications for operation from batteries, particularly dry batteries, and for special applications in which very quick heating is required. Filaments are also used in many types of power rectifiers and power triodes, where the special properties of the filament make it more suitable.

Valves having filaments should preferably be mounted with the filament vertical, but if it is necessary to mount them horizontally, they should be arranged so that the plane of the filament of V or W shaped filaments is vertical ; this reduces the chance of the filament touching the grid.

All filament-type valves having close spacing between filament and grid have a filament tension spring, usually mounted at the top of the valve. Some typical filament arrangements are indicated in Fig. 1.1 where A shows a single " V " shape filament suspended by means of a top-hook at the apex, B shows a " W " shape with two top hooks and C a single strand filament with tension spring as used in 1·4 volt valves.

Indirectly-heated cathodes consist of a cathode sleeve surrounding a heater. The cathode sleeve may have a variety of shapes, including round (D), elliptical (E) and rectangular (F) cross section. They are usually fitted with a light ribbon tag for connection to the lead going to the base pin.

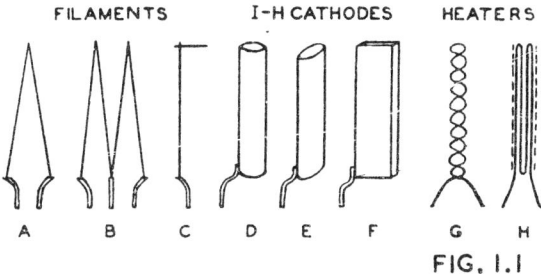

FILAMENTS   I-H CATHODES   HEATERS

A   B   C   D   E   F   G   H

FIG. 1.1

Fig. 1.1.   A, B, C types of filaments ;   D, E, F types of cathodes ;   G, H types of heaters.

In an indirectly-heated valve, the function of the heater is solely to heat the cathode. No emission should take place from the heater and the insulation between heater and cathode should be good. The heater is generally made of tungsten or a tungsten alloy wire coated with a substance capable of providing the necessary insulation at high temperature, such as alundum. In all applications where hum is likely to be troublesome, the heater is preferably of the double helical type, as G in Fig. 1.1. Power valves and other types having elliptical or rectangular cathode sleeves, often employ a folded heater as in H. These are not generally suitable for use in very low level amplifiers whether for radio or audio frequencies.

(ii) Grids

Grids are constructed of very fine wire wound around one, two or four side rods—two being by far the most common. Some valves have two, three, four or five grids inside one another, but all of these are similar in general form although different in dimensions.

In the case of some grids it is necessary to take precautions to limit the grid temperature either to avoid grid emission, in the case of control grids, or to limit the grid temperature to prevent the formation of gas, in the case of screen grids. These may, for better heat radiation, be fitted with copper side rods and blackened radiators either above or below the other electrodes. Grids are numbered in order from the cathode outwards, so that No. 1 grid will be the one closest to the cathode, No. 2 grid the one adjacent to it, and No. 3 the one further out again.

(iii) Plates

The plate of a receiving valve is the anode or positive electrode. It may be in one of a great number of shapes, dependent on the particular application of the valve. The plates of power valves and rectifiers are frequently blackened to increase their heat radiation and thereby reduce their temperature.

(iv) Bulbs

The inside surfaces of glass bulbs are frequently blackened. This has the effects of making them more or less conductive, thereby reducing the tendency to develop static charges, and reducing the tendency towards secondary emission from the bulb.

(v) Voltages with valve operation

All voltages in radio valves are taken with respect to the cathode, in the case of indirectly-heated valves, and the negative end of the filament with directly-heated valves. The cathode is usually earthed or is approximately at earth potential, so that this convention is easy to follow under normal conditions. In some cases, as for example phase splitters or cathode followers, the cathode is at a potential considerably above earth and care should be taken to avoid errors.

Some directly-heated valves may be operated with their filaments on a.c. supply, usually with the centre tap of the filament circuit treated as a cathode. In all such cases the valve data emphasize the fact that the filament is intended for operation on a.c. The plate characteristics are usually drawn with d.c. on the filament and these curves may be applied to a.c. operation by increasing the bias voltage by half the filament voltage.

In cases where resistors or other impedances are connected between the positive electrodes and the supply voltages, the electrode voltages (e.g. E_b, E_{c2}) are the voltages existing between those electrodes and cathode under operating conditions. The supply voltages are distinguished by the symbols E_{bb}, E_{cc2} etc. See the list of symbols in Chapter 38 Sect. 6.

For further information on valve operation see Chapter 3 Sect. 1.

SECTION 3 : TYPES OF RADIO VALVES

(*i*) *Diodes*   (*ii*) *Triodes*   (*iii*) *Tetrodes*   (*iv*) *Pentodes*   (*v*) *Pentode power amplifiers* (*vi*) *Combined valves*   ·(*vii*) *Pentagrid converters.*

(i) Diodes

A diode is the simplest type of radio valve consisting of two electrodes only, the cathode and anode (or plate). The cathode may be either directly or indirectly heated and the valve may be either very small, as for a signal detector, large as for a power rectifier, or any intermediate size. One or two diodes are frequently used in combination with a triode or pentode a-f amplifier as the second detector in receivers ; in most of these cases, a common cathode is used. For some purposes it is necessary to have two diode units with separate cathodes, as in type 6H6. Amplifier types with three diodes, some with a common cathode and others with separate cathodes, have

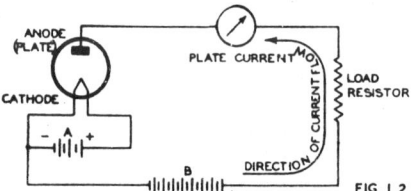

Fig. 1.2. *Fundamental circuit including diode, A and B batteries and load resistor.*

also been manufactured for special purposes. Fig. 1.2 shows the circuit of a diode valve in which battery A is used to heat the filament or heater, and battery B to apply a positive potential to the anode through the load resistor. The plate current is measured by a milliammeter connected as shown, and the direction of current flow is from the positive end of battery B towards the anode, this being the opposite of the electron current flow. It should be noted that the negative end of battery B is returned to the negative end of battery A in accordance with the usual convention. It would be quite permissible to connect the negative end of the battery B to the positive end of battery A so as to get the benefit of the voltage A applied to the anode, but in this case, the total voltage applied to the anode would be A + B. If voltage of battery B is reversed, it will be noted that the plate current is zero, thus indicating the rectification that takes place in a diode. If an alternating voltage is applied, current will only flow during the half-cycles when the anode is positive. This is called a half-wave rectifier since it is only capable of rectifying one half of the cycle.

Full wave* rectifiers are manufactured with two anodes and a common cathode and these are arranged in the circuit so that one diode conducts during one half-cycle and the other during the other half-cycle.

(ii) Triodes

A triode is a three electrode valve, the electrodes being the cathode, grid and anode (or plate). The grid serves to control the plate current flow, and if the grid is made sufficiently negative the plate current is reduced to zero. The voltage on the grid is controlled by battery C in Fig. 1.3, the other part of the circuit being as for the diode in Fig. 1.2. When the grid is negative with respect to the cathode, it does not draw appreciable current ; this is the normal condition as a class A_1 amplifier. Although an indirectly heated cathode has been shown in this instance, a directly heated valve could equally well have been used. The heater in Fig. 1.3 may be supplied either from an a.c. or d.c. source, which should preferably be connected to the cathode or as close as possible to cathode potential.

As the grid is made more negative, so the plate current is decreased and when the grid is made more positive the plate current is increased. A triode is, therefore, capable of converting a voltage change at the grid into a change of power in the load resistor. It may also be used as a voltage amplifier or oscillator.

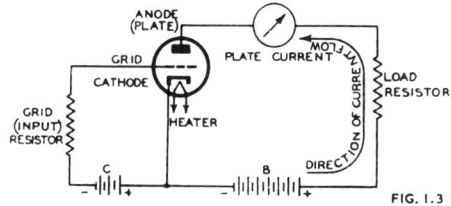

Fig. 1.3. Fundamental circuit including indirectly-heated triode, B and C batteries, and load resistor in plate circuit.

(iii) Tetrodes

The capacitance between the grid and plate can be reduced by mounting an additional electrode, generally called the screen or screen grid, between the grid and plate. The valve thus has four electrodes, hence the name tetrode. The function of the screen is to act as an electrostatic shield between grid and plate, thus reducing the grid-to-plate capacitance. The screen is connected to a positive potential (although less than that of the plate) in order to counteract the blocking effect which it would otherwise have on the plate current—see Fig. 1.4. Owing to the comparatively large spaces between the wires in the screen, most of the electrons from the cathode pass through the screen to the plate. So long as the plate voltage is higher than the screen voltage, the plate current depends primarily on the screen voltage and only to a slight extent on the plate voltage. This construction makes possible a much higher amplification than with a triode, and the lower grid-to-plate capacitance makes the high gain practicable at radio frequencies without instability.

Fig. 1.4. Fundamental circuit including indirectly-heated tetrode, B and C batteries, and load resistor in plate circuit.

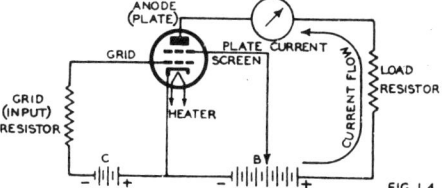

(iv) Pentodes

Electrons striking the plate with sufficient velocities may dislodge other electrons and so cause what is known as " secondary emission." In the case of tetrodes, when

*These are sometimes called biphase half-wave rectifiers.

the plate voltage swings down to a low value under working conditions, the screen may be instantaneously at a higher positive potential than the plate, and hence the secondary electrons are attracted to the screen. This has the effect of lowering the plate current over the region of low plate voltage and thus limits the permissible plate voltage swing. This effect is avoided when a suppressor is inserted between screen and plate. The suppressor is normally connected to the cathode as in Fig. 1.5. Owing to its negative potential with respect to the plate, the suppressor retards the movements of secondary electrons and diverts them back to the plate.

A valve with three grids is known as a pentode because it has five electrodes. Pentodes are commonly used as radio frequency amplifiers and as power amplifiers. Pentode r-f amplifiers are of two main varieties, those having a sharp cut-off* characteristic and those having a remote cut-off*. Valves having sharp cut-off characteristics are generally used as audio frequency voltage amplifiers and anode bend detectors, while remote cut-off amplifiers are used as r-f and i-f amplifiers. The

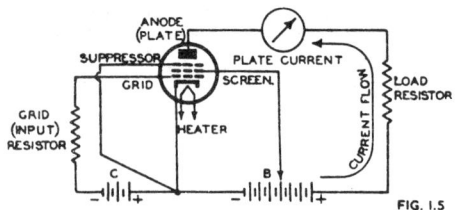

Fig. 1.5. Fundamental circuit including indirectly-heated pentode, B and C batteries, and load resistor in plate circuit,

remote cut-off characteristic permits the application of automatic volume control with a minimum of distortion ; this subject is treated in detail in Chapter 27 Sect. 3.

(v) Pentode power amplifiers

Pentode power amplifiers are commonly used in receiving sets to produce a-f power outputs from about 1 watt up to about 5 watts. They differ from r-f amplifiers in that no particular precautions are made to provide screening, and they are designed for handling higher plate currents and screen voltages. In principle, however, both types are identical and any r-f pentode may be used as a low-power a-f amplifier.

Beam power valves with " aligned " grids do not require a third grid to give characteristics resembling those of a power pentode ; a typical structure is shown in Fig. 1.6. Some " kinkless " tetrodes are also used as r-f and i-f amplifiers. All of these may be treated as being, in most respects, equivalent to pentodes.

(vi) Combined valves

Many combinations of valves have been made. Two triodes are frequently mounted in one envelope to form a " twin triode." One or more diodes are frequently combined with triodes and pentodes to form second detectors. A combination of triode and pentode in one envelope is also fairly common, one application being as a frequency changer. Other combinations are triode-hexodes and triode-heptodes, all of which are primarily intended for use as frequency changers or " converters." In these, the triode grid is generally connected internally to No. 3 grid in the hexode or heptode to provide the necessary mixing of the oscillator and signal voltages. A hexode has four grids while the heptode has five, the outermost of which is a suppressor functioning in the same manner as in a pentode.

In addition to this wide range of combinations, entirely different valves may be combined in one envelope to save space in very small receivers. This is a practice which appears to be dying out, particularly as the envelope size becomes smaller.

(vii) Pentagrid converters

Pentagrids are valves having 5 grids, so that they are really heptodes, but the name pentagrid appears to make a convenient distinction between valves in this group

---

*Sharp cut-off indicates that the plate-current characteristic is as straight as it can be made. A remote cut-off characteristic indicates that the plate current does not become zero until the grid voltage is made very much negative (usually over 30 volts).

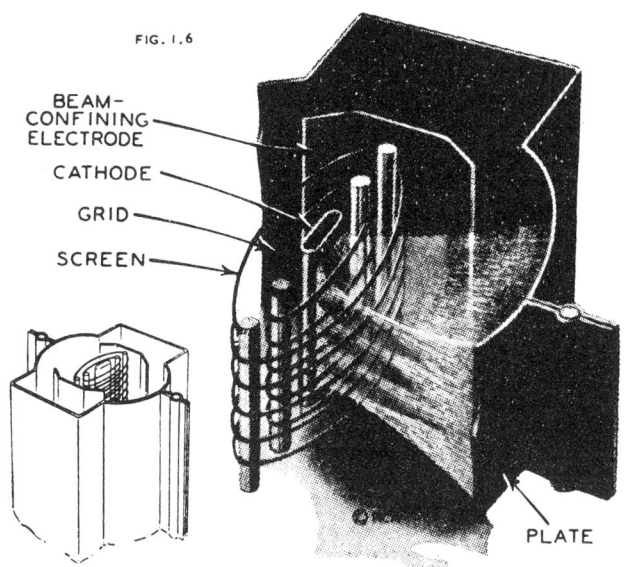

FIG. 1.6

BEAM–
CONFINING
ELECTRODE

CATHODE

GRID

SCREEN

PLATE

ELECTRON BEAM SHEETS FORMED BY GRID WIRES

Fig. 1.6.   *Internal structure of type 6L6 or 807 aligned grid beam power valve (diagram by courtesy of R.C.A.).*

(which do not normally require external oscillators) and those of the hexode or heptode " mixer " type which are used with separate oscillators. Pentagrid converters are of two main groups, the first of these being the 6A8 type of construction which incorporates an oscillator grid and oscillator anode (" anode grid ") as part of the main cathode stream. The other group comprises the 6SA7, 6BE6 and 1R5 type of construction which has no separate oscillator anode, the screen grid serving a dual purpose. The various types of pentagrid converters are described in detail in Chapter 25.

SECTION 4 : MAXIMUM RATINGS AND TOLERANCES

(i) Maximum ratings and their interpretation   (ii) Tolerances.

(i) Maximum ratings and their interpretation

Maximum ratings are of two types—the Absolute Maximum system and the Design Centre system. These are described in detail in Chapter 3 Sect. 1(iv).

(ii) Tolerances

All valves are tested in the factory for a number of characteristics, these usually including plate current, screen current, negative grid current, mutual conductance, noise and microphony, as well as having to pass visual inspection tests for appearance. For methods of testing see Chapter 3. As with any other components such as resistors or capacitors, the characteristics can only be maintained within certain tolerances. For example, a resistor may be bought with a tolerance of plus or minus 10% or 20% ; closer tolerances may be purchased at a higher price.

The subject of tolerances in valve characteristics is covered in detail in Chapter 3 Sect. 2(iii).

Special care should be taken in the screen circuits of beam power amplifiers since in these the screen currents may vary from zero to twice the average figure. Any screen voltage dropping resistor is undesirable with such valves and if the screen is required to be operated at a lower voltage than the plate, it should be supplied from a voltage divider having a bleed current of preferably 5 times the nominal screen current. Alternatively, the screen voltage should be determined for the extreme cases of zero and twice nominal screen current.

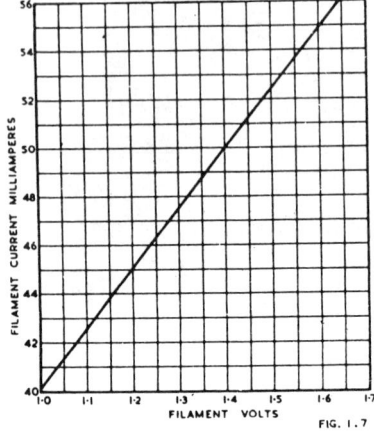

Fig. 1.7. Filament current versus filament voltage for a valve having a 1.4 *volt* 50 *milliampere filament.*

The heater voltage should be maintained at an average voltage equal to the recommended voltage, thus leaving a margin of plus or minus 10% for line fluctuations under normal conditions—see Chapter 3 Sect. 1(iv)D. If any wider variation is required, this will involve decreased maximum grid circuit resistance for a higher heater voltage and decreased plate current for lower heater voltage.

SECTION 5 : FILAMENT AND HEATER VOLTAGE/CURRENT CHARACTERISTICS

A valve filament or heater operates at such a temperature that its resistance when hot is much greater than its resistance when cold. The current/voltage characteristic is curved and does not follow Ohm's Law. Two typical examples are Fig. 1.7 for a battery valve and Fig. 35.14 for an indirectly-heated valve. Approximate curves for general use, on a percentage basis, are given in Fig. 1.8 including also dissipation in watts and temperature (Ref. 7). Filament and heater ratings are covered in Chapter 3 Sect. 1.

SECTION 6 : VALVE NUMBERING SYSTEMS

Receiving valves having the American numbering follow two main systems. The first of these is the numerical system, which is the older, and the second the R.M.A. system. Originally various manufacturers produced the same valve under different type numbers such as 135, 235, 335, 435 etc. This was improved upon by dropping the first figure and using only the two latter figures, e.g. 35.

All the more recent American releases follow the R.M.A. system (Ref. 8) of which a typical example is 6A8-GT. In this system the first figure indicates the approximate filament or heater voltage—6 indicates a voltage between 5·6 and 6·6 volts, while 5 indicates a voltage between 4·6 and 5·6 volts ; 1 indicates a voltage in excess of 0 and including 1·6 volts, while 0 indicates a cold cathode. Lock-in types in the 6·3 volt range are given the first figure 7 (this being the " nominal " voltage), but the normal operating voltages remain at 6·3 volts. In the case of tapped filaments or heaters the first figure indicates the total voltage with both sections in series.

The second symbol is a letter which is allotted in sequence commencing with A, except that I and O are not used ; rectifiers follow the sequence backwards commencing at Z. When all the single letters of a group are exhausted, the system then proceeds with two lettters commencing with AB ; combinations of identical letters are not normally used. The single-ended a.c. range has a first letter S while the second letter may be that of the nearest equivalent in the double-ended range—e.g. type 6SK7 is the nearest single-ended equivalent to type 6K7. Another special case is the first letter L which is used for lock-in types in the battery range.

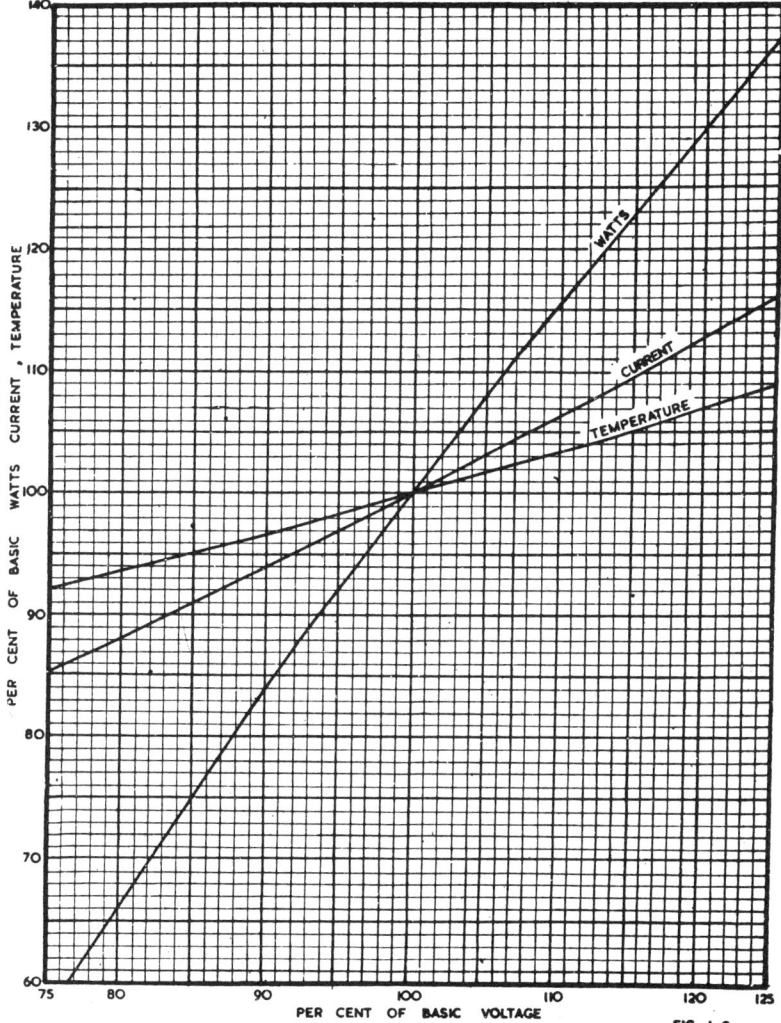

Fig. 1.8. *Filament or heater current, dissipation and temperature plotted against filament or heater voltage, per cent (Ref. 7).*

The final figure denotes the number of "useful elements" brought out to an external connection. The envelope of a metal valve, the metal base of a lock-in valve, and internal shielding having its separate and exclusive terminal(s) are counted as useful elements. A filament or heater counts as one useful element, except that a tapped filament or heater of two or more sections of unequal rated section voltages or currents counts as two useful elements. An octal-based glass valve having n useful elements exclusive of those connected to Pin No. 1 is counted as having $n + 1$ useful elements. Elements connected to terminals identified as "internal connection, do not use" do not count as useful elements. Combinations of one or more elements connected to the same terminal or terminals are counted as one useful element. For example a directly heated triode with a non-octal base is denoted by 3 ; an indirectly-heated triode, with a non-octal base is designated by 4 ; a directly-heated tetrode with a non-octal base is designated by 4. A pentode with the suppressor internally connected to filament or cathode is numbered as a tetrode. A metal envelope or octal-based glass triode with an indirectly-heated cathode is designated by 5, a tetrode (or pentode with the suppressor internally connected) by 6, and a triode-hexode converter usually by 8.

The suffix after the hyphen denotes the type of construction used. In general, metal valves, lock-in types and miniature types have no suffixes, but octal-based glass valves types are given the suffix G for the larger glass bulb or GT for the smaller parallel-sided T9 bulb. The letter M indicates a metal-coated glass envelope and octal base. X indicates a "low loss" base composed of material having a loss factor of 0·035 maximum (determination of loss factor to be in accordance with ASTM Designation D-150-41T). The letter Y indicates an intermediate-loss base composed of material having a loss factor of 0·1 maximum. The letter W indicates a military type. The letters, A,B,C,D,E and F assigned in that order indicate a later and modified version which can be substituted for any previous version but not vice versa.

SECTION 7 : REFERENCES*

(A) GENERAL READING

1. Henderson, F. E. (book) " An Introduction to Valves, including reference to Cathode Ray Tubes " (published by Wireless World, London on behalf of The General Electric Co. of England, 1942).
2. Spreadbury, F. G. (book) " Electronics " (Sir Isaac Pitman & Sons Ltd., London, 1947) Chapters 1-4, 8-10.
3. Reich, H. J. (book) " Theory and Applications of Electron Tubes " (McGraw-Hill Book Co., New York & London, 2nd ed., 1944).
4. Chaffee, E. L. (book) " Theory of Thermionic Vacuum Tubes " (McGraw-Hill Book Co., New York and London, 1933).
5. Emergency British Standard 1106 : 1945 " Code of practice relating to the use of electronic valves other than cathode ray tubes," British Standards Institution, London.
6. " Radio Valve Practice " (The British Radio Valve Manufacturers' Association, London, 1948).
7. Haller, C. E. " Filament and heater characteristics " Elect. 17.7 (July, 1944) 126.
8. " R.M.A.-NEMA Standards for designation system for receiving tubes " (Radio Manufacturers' Association, U.S.A.—Standard ET-110).
9. Couch, W. " Oxide-coated cathodes " Elect. 22.10 (Oct. 1949) 164.
10. Couch, W. " Oxide cathode theory " Elect. 22.10 (Oct. 1949) 190.
11. Koller, L. R. (book) " Physics of Electron Tubes " (McGraw-Hill Book Co. 1934) pp. 41-51.

Additional references will be found in the Supplement commencing on page 1475.

(B) ATOMIC STRUCTURE

12. Stranathan, J. D. " Elementary particles of physics " Elect. 16.8 (Aug. 1943) 122.
13. Darrow, K. K. " Beginnings of nuclear physics " E.E. (Sept. 1945) 315.
14. Stranathan, J. D. (book) " The Particles of Modern Physics " (The Blakiston Co., Philadelphia, U.S.A. 1943).
15. Cork, J. M. (book) " Radioactivity and Nuclear Physics " (D. van Nostrand Co. Inc., New York, 1947) pp. 13-18.
16. Smith, J. J. " The atom and its nucleus " E.E. 66.12 (Dec. 1947) 1165.
17. Shoupp, W. E., and H. Odishaw " The nucleus—its structure and reactions " E.E. 67.2 (Feb. 1948) 125.
18. Lapp, R. E., and H. L. Andrews " Atomic structure " Proc. I.R.E. 36.9 (Sept. 1948) 1068, 1070. Extract from book " Nuclear Radiation Physics " (Prentice-Hall Inc.).
and many others.

---

*For abbreviations of titles of periodicals and references to periodicals see pages 1367-1369.

CHAPTER 2

VALVE CHARACTERISTICS

by F. LANGFORD-SMITH, B.SC., B.E.

SECTION 1 : VALVE COEFFICIENTS

(Otherwise known as Constants, Parameters or Factors)

The triode or multigrid radio valve is a device which allows, under certain operating conditions, an amplified replica of a voltage applied between grid and cathode to appear across an impedance placed between plate and cathode.

A valve, in itself, does not provide amplification of the applied grid-to-cathode voltage. The amplified voltage across the load impedance is due to the action of the valve in controlling the power available from the power supply. The amount of power which can be so controlled is determined by the operating conditions and the characteristics of the valve and of its associated circuits.

The maximum voltage amplification which a valve is capable of giving under ideal conditions is called the amplification factor, generally designated by the Greek symbol μ (mu). This is not truly constant under all conditions (except for an imaginary " ideal valve ") and varies slightly with grid bias and plate voltage in the case of a triode, and is very far from being constant with most multi-electrode valves.

The **amplification factor** (μ) is the ratio of the incremental* change in plate voltage to the incremental change in control grid voltage in the opposite direction, under the conditions that the plate current remains unchanged, and all other electrode voltages are maintained constant.

There are two other principal Valve Coefficients, known as the mutual conductance and the plate resistance (or anode resistance), the values of these also being somewhat dependent upon the applied voltages.

---

*An incremental change of voltage applied to an electrode may be taken as indicating a change so small that the curvature of the characteristics may be neglected. For the mathematical treatment of rate of change, see Chapter 6 Sect. 7(i) and (ii). For treatment of valve coefficients as partial differentials see Chapter 2 Sect. 9(ix).

13

The **mutual conductance** (or grid-plate transconductance) is the incremental change in plate current divided by the incremental change in the control-grid voltage producing it, under the condition that all other voltages remain unchanged.

The **plate resistance**† is the incremental change in plate voltage divided by the incremental change in plate current which it produces, the other voltages remaining constant.

There is a relationship between these three principal valve coefficients, which is exact provided that all have been measured at the same operating point,

$$\mu = g_m \cdot r_p,$$

$$\text{or } g_m = \frac{\mu}{r_p},$$

$$\text{or } r_p = \frac{\mu}{g_m}.$$

The calculation of these " valve coefficients " from the characteristic curves is given in Section 2 of this Chapter, while their direct measurement is described in Chapter 3 Sect. 3. The mathematical derivation of these coefficients and their relationship to one another are given in Section 9 of this Chapter, as is also the representation of valve coefficients in the form of partial differentials.

The reciprocals of two of these coefficients are occasionally used—

$\dfrac{1}{\mu} = D$ where D is called the Durchgriff (or Penetration Factor) and which may be expressed as a percentage.

$\dfrac{1}{r_p} = g_p$ where g_p is called the Plate Conductance (see also below).

Other valve coefficients are described below :—

The **Mu-Factor,** of which the amplification factor is a special case, is the ratio of the incremental change in any one electrode voltage to the incremental change in any other electrode voltage, under the conditions that a specified current remains unchanged and that all other electrode voltages are maintained constant. Examples are

$$\mu_{g1 \cdot g2}, \quad \mu_{g2 \cdot y}.$$

The **Conductance** (g) is the incremental change in current to any electrode divided by the incremental change in voltage to the same electrode, all other voltages remaining unchanged.

Examples are grid conductance (g_g), plate conductance (g_p).

Transconductance is the incremental change in current to any electrode divided by the incremental change in voltage to another electrode, under the condition that all other voltages remain unchanged. A special case is the grid-plate transconductance which is known as the mutual conductance. Another example is the plate-grid transconductance (g_n).

Conversion transconductance (S_c) is associated with mixer (frequency changing) valves, and is the incremental change in intermediate-frequency plate current divided by the incremental change in radio-frequency signal-grid voltage producing it.

The **Resistance** (r) of any electrode is the reciprocal of the conductance ; for example plate resistance is the reciprocal of plate conductance,

$$r_p = 1/g_p.$$

Perveance (G) is the relation between the space-charge-limited cathode current and the three-halves power of the anode voltage. It is independent of the electrode voltages and currents, so long as the three-halves law holds :

$$G = \frac{i_k}{e_b^{3/2}}$$

The measurement of perveance is covered in Chapter 3 Sect. 3(vi)E.

---

†This is strictly the " variational plate resistance " and must be distinguished from the d.c. plate resistance.

SECTION 2 : CHARACTERISTIC CURVES

(i) Plate characteristics (ii) Mutual characteristics (iii) Grid current characteristics (iv) Suppressor characteristics (v) Constant current curves (vi) " G " curves (vii) Drift of characteristics during life (viii) Effect of heater-voltage variation.

It is convenient to set down the measured characteristics of a valve in the form of curves. These are thus a record of the actual currents which flow in a given valve when the specified voltages are applied.

The curves published by the valve manufacturers are those of an " average " valve, and any one valve may differ from them within the limits of the manufacturing tolerances.

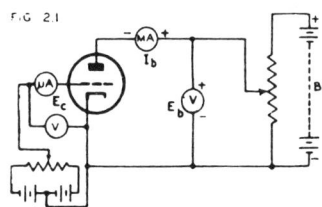

Fig. 2.1. Method of measuring the plate and grid currents of a triode valve.

The method of measuring the plate and grid currents of a triode valve is shown in Fig. 2.1 in which a tapping on the grid bias battery is returned to the cathode so as to permit either positive or negative voltages to be connected to the grid. The grid microammeter and voltmeter should be of the centre-zero type, or provision made for reversal of polarity. For more elaborate testing see Chapter 3 Sect. 3.

(i) Plate characteristics

The Plate Characteristic may be drawn by maintaining the grid at some constant voltage, varying the plate voltage step-by-step from zero up to the maximum available, and noting the plate current for each step of plate voltage. These readings may then be plotted on graph paper with the plate voltage horizontal and plate current vertical. This procedure may be repeated for other values of grid voltage to complete the Plate Characteristic Family.

The Plate Characteristic Family for a typical triode is shown in Fig. 2.2. It is assumed that the plate voltage has been selected as 180 volts, and the grid bias 4 volts. By drawing a vertical line from 180 volts on the E_b axis (point K), the quiescent operating point Q will be determined by its intersection with the " $E_c = -4$ " curve. By referring Q to the vertical scale (I_b) the plate current is found to be 6mA. The plate resistance at the point Q is found by drawing a tangent (EF) to the curve for $E_c = -4$ so that it touches the curve at Q.

The plate resistance (r_p) at the point Q is then EK in volts (65) divided by QK in amperes (6 mA = 0.006 A) or 10 800 ohms.

The amplification factor (μ) is the change of plate voltage divided by the change of grid voltage for constant plate current. Line CD is drawn horizontally through Q, and represents a line of constant plate current. Points C and D represent grid voltages of -2 and -6, and correspond to plate voltages of 142 and 218 respectively. The value of μ^* is therefore (218 − 142) plate volts divided by a change of 4 grid volts, this being 76/4 or 19.

The mutual conductance (g_m) is the change of plate current divided by the change of grid voltage for constant plate voltage. Line AB, which is drawn vertically through Q, represents constant plate voltage. Point A corresponds to 9.6 mA, while point B corresponds to 2.6 mA, giving a difference of 7 mA. Since points A and B also differ by 4 volts grid bias, the mutual conductance* is 7 mA divided by 4 volts, which is 1.75 mA/volt or 1750 micromhos.

---

*The value so determined is not exactly the value which would be obtained with a very small swing, but is sufficiently accurate for most practical purposes.

In these calculations it is important to work with points equidistant on each side of Q to reduce to a minimum errors due to curvature.

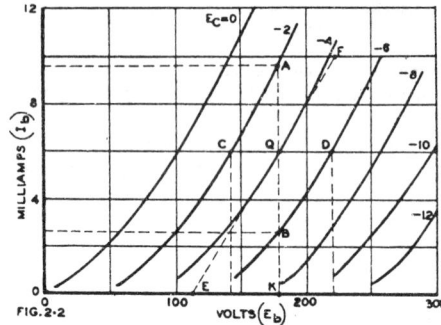

Fig. 2.2. Plate charact-eristic family of curves for a typical triode.

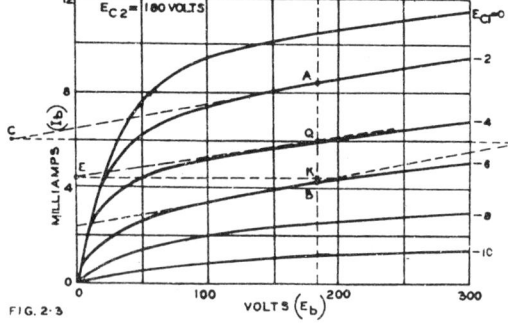

Fig. 2.3. Plate charact-eristics of a pentode, for one fixed screen voltage.

The plate characteristics of a pentode for one fixed screen voltage are shown in Fig. 2.3. Owing to the high plate resistance of a pentode the slope of the portion of the curves above the " knee " is frequently so flat that it is necessary to draw extended tangents to the curves as at A, B and Q. A horizontal line may be drawn through Q to intersect the tangents at A and B at points C and D. As with a triode, points A and B are vertically above and below Q. The mutual conductance is AB (4.1 mA) divided by 4 volts change of grid bias, that is 1.025 mA/V or 1025 micromhos. The amplification factor is the change of plate voltage (CD = 447 volts) divided by the change of grid voltage (4 volts) or 111.7. The plate resistance is EK/QK, i.e. 180/0.001 65 or 109 000 ohms.

The plate characteristics of a beam tetrode are somewhat similar to those of a pentode except that the " knee " tends to be more pronounced at high values of plate current.

The plate characteristics of a screen-grid or tetrode are in the upper portion similar to a pentode, but the " knee " occurs at a plate voltage slightly greater than the screen voltage and operation below the " knee " is normally inadvisable due to instability.

The plate and screen characteristics of a pentode are shown in Fig. 2.4, from which it will be seen that the total cathode (plate + screen) current for any fixed grid bias is nearly constant, except at low plate voltages, and that the plate current increases at the expense of the screen, and vice versa. A pentode is frequently described as a " constant-current device," but the plate current is not so nearly constant as the combination of plate and screen currents, with fixed grid bias and screen voltage.

(ii) Mutual characteristics†

The Mutual Characteristics may be drawn by maintaining the plate voltage constant, and varying the grid from the extreme negative to the extreme positive voltage desired. For any particular plate voltage, there is a negative grid voltage at which the plate current becomes zero ; this is called the point of plate current cut-off, and any increase of grid voltage in the negative direction has no effect on the plate current, which remains zero. If the mutual characteristic were perfectly straight, the point of plate current cut-off would be at a grid voltage of E_b/μ ; in reality, it occurs at a point slightly more negative, owing to the curved foot of the characteristic.

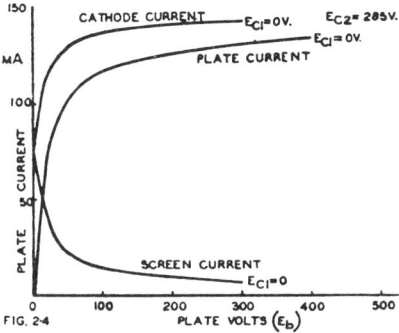

Fig. 2.4. Plate and screen characteristics for a pentode, with
fixed screen and grid voltages,
showing also the cathode current
curve which is the sum of the
plate and screen currents at all
plate voltages.

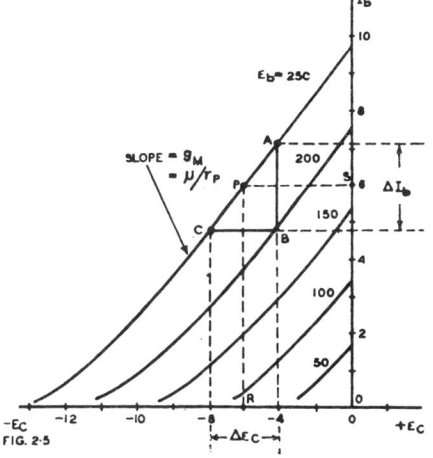

Fig. 2.5. Mutual characteristics
of a triode, with one curve for
each of five fixed plate voltages.

The **Mutual Characteristics of a triode** are shown in Fig. 2.5. Each curve corresponds to a constant plate voltage. Let P be a point on the $E_b = 250$ curve, and let us endeavour to find out what information is available from the curves. The bias corresponding to P is given by R (-6 volts) and the plate current is given by S (6 mA). Let now a triangle ABC be constructed so that AP = PC, AB is vertical, CB is horizontal and point B comes on the $E_b = 200$ curve.

The mutual conductance is given by AB/BC or 2·32 mA/4 volts, which is 0·580 mA/volt or 580 micromhos. Thus the slope of the characteristic is the mutual conductance.**

---

†Also known as Transfer Characteristics.
**This simple construction assumes that A P C is a straight line. In practice it is slightly curved but the construction gives a very close approximation to the slope at point P because the slope of the tangent at P is approximately the slope of the chord joining A and C.

The amplification factor is given by the change of plate voltage divided by the change of grid voltage for constant plate current, that is

$$\mu = \frac{E_{b1} - E_{b2}}{\mathrm{CB}} = \frac{E_{b1} - E_{b2}}{\triangle E_c} = \frac{250 - 200}{4} = 12 \cdot 5.$$

The plate resistance is given by the change of plate voltage divided by the change of plate current for constant grid voltage; that is

$$r_p = \frac{E_{b1} - E_{b2}}{\mathrm{AB}} = \frac{E_{b1} - E_{b2}}{\triangle I_b} = \frac{250 - 200}{2 \cdot 32 \times 10^{-3}} = 21\ 600\ \text{chms.}$$

These curves hold only if there is no series resistance in the plate circuit. They could therefore be used for a transformer-coupled amplifier provided that the primary of the transformer had negligible resistance. In the present form they could not be used to predict the operation under dynamic conditions. The static operation point P may, however, be located by their use.

The mutual characteristics of a pentode, for a fixed screen voltage, are very similar to those of a triode except that each curve applies to a different value of screen (instead of plate) voltage. The plate voltage of pentodes having high plate resistance has only a very minor effect on the plate current, provided that it does not come below the screen voltage.

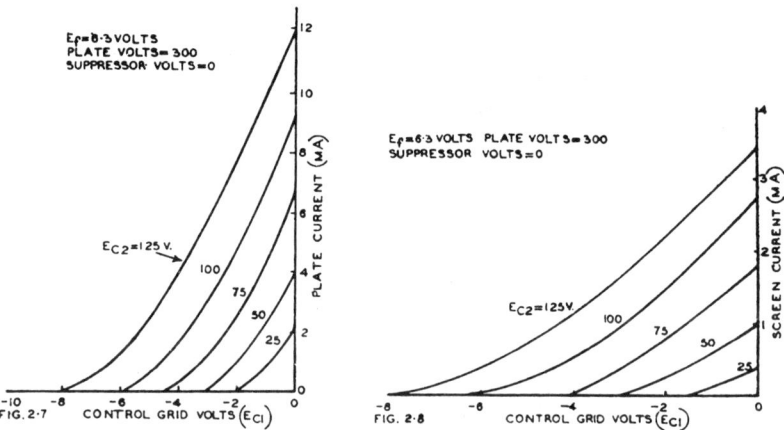

Fig. 2.7. Mutual characteristics of a pentode, with constant plate voltage, and five fixed screen voltages.

Fig. 2.8. Screen current mutual characteristics of a pentode (same as for Fig. 2.7).

The Mutual Characteristic Family for a typical pentode is shown in Fig. 2.7, and the corresponding screen current characteristics in Fig. 2.8.

The resemblance between the shapes of the plate and screen characteristics is very close, and there is an almost constant ratio between the plate and screen currents along each curve.

(iii) Grid current characteristics

Positive grid current in a perfectly hard indirectly-heated valve usually commences to flow when the grid is slightly negative (point Y in Fig. 2.9) and increases rapidly as the grid is made more positive (Curve A). The position of point Y is affected both by the contact potential between grid and cathode and also by the initial electron velocity of emission; the latter is a function of the plate and grid voltages and the amplification factor of the valve, and will therefore vary slightly as the electrode voltages are changed. The grid current commencement point in perfectly hard battery valves is usually slightly positive, so that they may be operated at zero bias with negligible positive grid current (Curve B).

A typical valve at its normal negative bias will have negative (or reverse) grid current which is the sum of gas (ionization) current, leakage current and grid primary emission current. If the two latter are negligibly small, negative grid current (i.e. *gas current*) will be roughly proportional to the plate current, and will increase with the pressure of gas in the valve. If the plate current is maintained constant, the gas current varies approximately as the square of the plate voltage ; reduced cathode temperature has little effect on this relationship (Ref. A12). See Chapter 1 Sect. 1 for general information regarding gas current and Chapter 3 Sect. 3(iv)A for the measurement of reverse grid current.

References to grid current characteristics—A12, H1, H2.

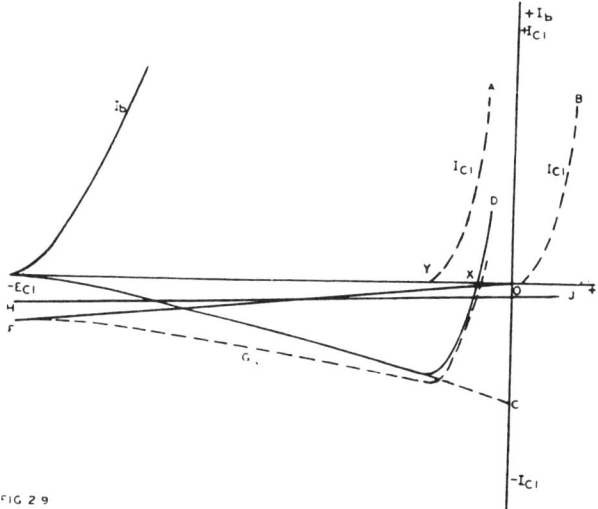

Fig. 2.9. Grid current characteristics of a triode or pentode.

Curve C shows the gas (ionization) current alone, and the solid line D is the combination of curves A and C, this being the grid current characteristic of a typical indirectly heated valve with a slight amount of gas. The maximum negative grid current occurs at a value of grid bias approximately equal to that of the grid current commencement point of the same valve if it could be made perfectly hard (point Y). The point of zero grid current (X) differs from the point of grid current commencement in a perfectly hard valve (Y).

The grid current cross-over point (X) in a new indirectly heated valve is usually between zero and −1·0 volt, and some slight variations in the value are to be expected during life. Change of contact potential between grid and cathode results in a corresponding shift of the mutual characteristics ; a change of contact potential in the direction which makes the grid current cross-over point move in the positive direction during the life of the valve will result in decreased plate current, which may be quite serious in a high-mu triode operating on a low plate supply voltage. This is one reason why grid leak bias (with a grid resistor of about 5 or 10 megohms) is often used with such valves, so that the operating point is maintained in the same relation to the mutual characteristic.

In battery type valves the grid current cross-over point may be either positive or negative and designers should allow for some valves with negative values, particularly in cases of low screen voltage operation.

Contact potential is only one of several effects acting on the grid to change the cross-over point (X)—the others include gas current, grid (primary) emission, leakage, and the internal electron velocity of emission.

The grid variational conductance is equal to the slope of the grid characteristic at the operating point. The conductance increases rapidly as the grid voltage is made less than that corresponding to point Y, irrespective of the value of ionization current, so that input circuit damping due to the flow of electrons from cathode to grid (i.e. the positive component of the grid current) occurs in a typical valve even when the grid current is zero or negative (grid voltages between X and Y in Fig. 2.9). **It is possible for the damping on the positive peaks of applied input voltage to be quite serious even when the microammeter reads zero.** This point is applied in connection with r.c. triodes in Chapter 12 Sect. 2(iv).

If the valve has a leakage path between grid and cathode, the leakage current is given by the line OF, which must be added to the gas current to give the grid current characteristic G. If it has leakage between grid and plate (or screen) the leakage current is given by the line HJ, which intersects the horizontal axis at a positive voltage equal to the plate (or screen) voltage ; this also must be added to the other components to provide the grid current characteristic. The combined leakage currents may be measured by biasing the grid beyond the point of plate current cut-off provided that the grid emission is negligibly small—otherwise see below.

Grid emission with a negative grid is the primary emission of electrons due to grid heating from both cathode and plate (or screen) ; it gradually increases as the valve becomes warmer during operation. It increases the total negative grid current and is included with leakage currents in the total negative grid current indicated by a valve tester. For methods of testing to discriminate between the various components of negative grid current, see Chapter 3 Sect. 3(iv)A.

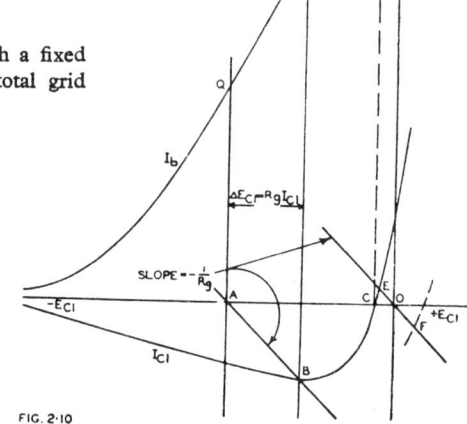

Negative-grid load lines

When a valve is operated with a fixed negative grid bias, but has a total grid circuit resistance R_g, the actual voltage on the grid may differ from the applied bias due to grid current. If negative grid current is present the condition will be as shown in Fig. 2.10 in which OA represents the applied bias. The plate current operating point with no grid current will obviously be Q but if the grid current characteristic is as shown, the grid operating point will be B and the plate operating point Q'. Point

FIG. 2·10

Fig. 2.10.   *Grid current characteristics with grid loadlines.*

B is determined by the intersection of the grid current characteristic and a load line having a slope of $-1/R_g$. The shift in grid bias due to voltage drop across R_g will be $\triangle E_{c1}$ or $R_g.I_{c1}$. The operating point can obviously never be swung beyond the grid-current cross-over point C, so that the static plate current can never go beyond D (Fig. 2.10) due to negative grid current.

If valve is operated with its grid completely open-circuited, the operating point will be at D, since this is the only point corresponding to zero grid current, unless the grid characteristic has a second point of zero grid current at a positive grid voltage (see under grid blocking).

If the valve is operated with zero bias, that is with the grid resistor returned to cathode, the grid static operating point will be at E, the intersection of the grid current characteristic and the grid loadline through O. If the valve is one with positive cross-over point, operating at zero bias, the grid static operating point will occur at F.

In all cases considered above, the operating points are for static conditions, and any large signal voltages applied to the grid may have an effect in shifting the operating point.   If the signal voltage swings the grid sufficiently to draw positive grid current, the operating point will shift as the result of rectified current flowing through R_g.

The effect of negative grid current on the maximum grid circuit resistance and the operation of a-f amplifiers is described in Chapter 12 Sect. 2(iii) and (iv) ; Sect. 3(iv)C and (v) ; also Chapter 13 Sect. 10(i).

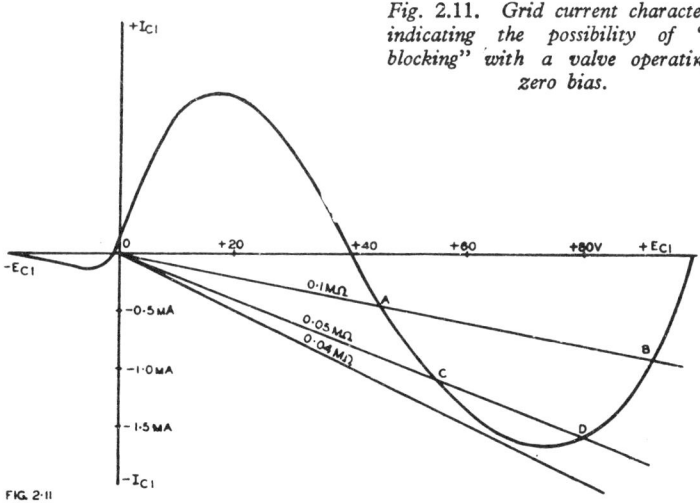

Fig. 2.11. Grid current characteristics indicating the possibility of "grid blocking" with a valve operating at zero bias.

FIG. 2·11

Positive grid voltages and grid blocking

When the grid is made positive, it is bombarded by electrons which cause it to increase in temperature, and it may have both **primary and secondary electron emission.**   This current is in a direction opposite to that of positive grid current flow, and may result in a slight kink in the grid characteristic, or may be severe enough to cause the grid current in this region to become negative.   A typical case of the severe type is shown in Fig. 2.11, in which the greatest negative grid current occurs at a positive grid voltage of 70 or 80 volts.   Such a valve is capable of " grid blocking " if the grid is swung sufficiently positive, and if the grid circuit resistance is high enough.   Grid blocking can only occur if the grid loadline cuts the negative loop of grid current.   In Fig. 2.11 the 0·1 megohm loadline cuts it at points A and B, but point A is unstable and the grid will jump on to point B and remain there until the valve is switched off, or the grid circuit resistance decreased until the grid loadline no longer cuts the curve (e.g. 0.04 megohm in Fig. 2.11).

(iv) Suppressor characteristics

In some pentodes, the suppressor is brought out to a separate pin, and may be used for some special purposes.   Fig. 2.12 shows the mutual characteristics of a pentode suitable for suppressor modulation, although typical of any pentode.   The curves are in the shape of a fan pivoted at the cut-off point, with the slope controlled by the suppressor voltage. The curves of electrode currents versus suppressor voltage are given in Fig. 2.13, and indicate that the plate current curve rises fairly steadily from the point of cut-off at a high negative voltage but flattens out while still at a negative suppressor voltage. The screen current falls as the plate current rises, as would be expected, and the suppressor current commences at a slight positive voltage, although in this case it becomes negative at high voltages due to secondary emission.

The suppressor is occasionally used as a detector in receivers, instead of a diode, but its rectification efficiency is low, since the internal resistance is of the order of 20 000 ohms.

In remote cut-off r-f pentodes the suppressor is sometimes used to provide a more rapid cut-off characteristic. A family of mutual conductance and plate resistance curves for a typical remote cut-off pentode are given in Fig. 2.14.

It will be seen the mutual conductance for any fixed control grid voltage (say $E_{c1} = -3$) may be reduced by making the suppressor voltage negative. This has the additional effect, however of decreasing the plate resistance from 0.8 megohm (at $E_{c3} = 0$) to 35 000 ohms at $E_{c3} = -37$, for $E_{c1} = -3$ volts. The initial rate of reduction is very steep, and occurs with all values of control grid voltage.

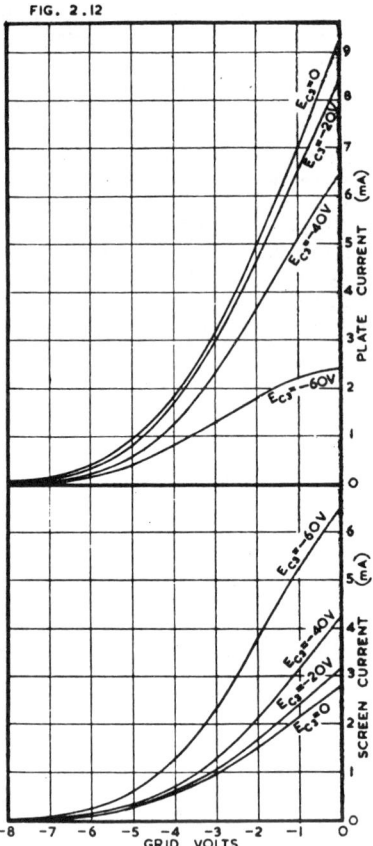

FIG. 2.12

Fig. 2.12. Mutual characteristics of a pentode (6SJ7) for various suppressor voltages.

If the suppressor grid has the same bias control voltage as the control grid, the control characteristic will be as shown by the curve marked "$E_{c1} = E_{c3}$," but in this case the plate resistance, although initially slightly lower for $E_{c3} = -3$ than for $E_{c3} = 0$, rises rapidly as $E_{c1} = E_{c3}$ is made more negative.

(v) Constant current curves

The third principal type of valve characteristic is known as the "Constant Current" Characteristic. A

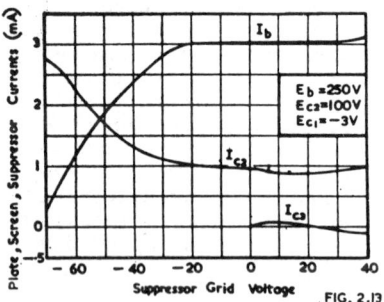

Fig. 2.13. Suppressor characteristics of a pentode (6SJ7) for fixed control grid, screen and plate voltages.

typical family of Constant Current Curves is shown in Fig. 2.15, these being for a typical triode (type 801). The slope of the curves indicates the amplification factor, and the slope of the loadline indicates the stage voltage gain. The operating point is fixed definitely by a knowledge of plate and grid supply voltages, but the loadline is only straight when both plate and grid voltages follow the same law (e.g., both sine wave). Distortion results in curved characteristics, so that this form of representation is not very useful except for tuned-grid tuned-plate or "tank-circuit" coupled r-f amplifiers. Constant Current Curves may be drawn by transferring points from the other published characteristics. For a full treatment the reader is referred to

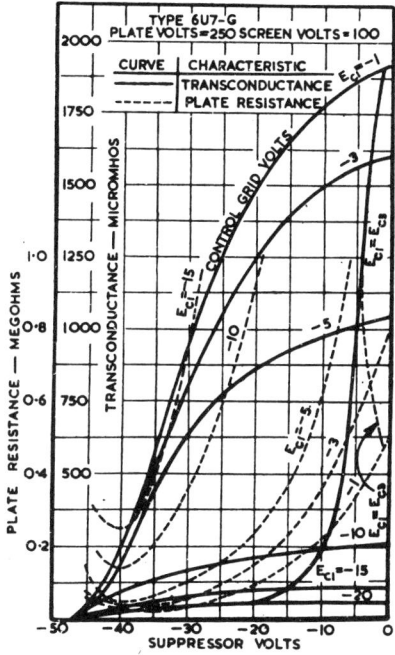

FIG. 2.14

(1) Mouromtseff, I. E., and H. N. Kozanowski " Analysis of the operation of vacuum tubes as Class C Amplifiers " Proc. I.R.E. 23.7 (July, 1935) 752 : also 24.4 (April, 1936) 654.

(2) Everest, F. A., " Making life more simple " Radio 221 (July, 1937) 26.

(3) " Reference Data for Radio Engineers " (2nd edition, Federal Telephone and Radio Corporation, New York, 1946).

(vi) " G " curves

Curves of constant g_m and g_p are plotted for a typical triode in Fig. 13.9B. These are helpful in calculating the voltage gain of resistance-coupled triodes and, to a less extent, pentodes, and in other applications. (Refs. B14, B22, B31, B32).

(vii) Drift of characteristics during life

During the life of a valve there is always a slow drift which is particularly apparent in the plate and screen

Fig. 2.14. Suppressor characteristics of a remote cut-off pentode (6U7-G) for fixed screen and plate voltages.

currents, mutual conductance, negative grid current and the contact potential point. The direction of drift sometimes reverses one or more times during life. It is assumed here that the valve is operated at constant applied voltages throughout its life.

In general, the grid current crossover point (Fig. 2.9) tends to drift in the positive direction during life. The movement of the contact potential point results in a shift of the mutual characteristics which in turn has the effect of reducing the plate current which flows at a fixed grid bias.

Life tests have been carried out (Ref. A12) for a period of 3200 hours on type 6SL7 high-mu twin triodes. The recorded characteristic was the grid voltage to give a plate current of 0·1 mA with a plate voltage of 75 volts. The maximum drift was 0·6 volt (from —1·65 to —1·05 volts), but the majority of the valves did not go outside the limits —1·5 to —1·1 volt (0·4 volt drift). In most cases the drift was generally in a positive direction, but there were two exceptions (out of a total of twelve units) which showed a general tendency to drift in the negative direction for the first hundred hours or so and then to drift in the positive direction, ending up at approximately the same values where they began. However, even those having a positive general direction showed rapid changes in the rate of change, and usually at least one temporary reversal of direction.

It was found that minimum drift occurred for plate currents between 0·1 and 1·0 mA for indirectly-heated types, or between 10 and 100 μA for small filament types.

This drift occurs in diodes and all types of amplifying valves, being particularly noticeable in its effects on high-mu triodes (on account of the short grid base) and on power amplifiers (on account of the decrease in maximum power output). In direct-coupled amplifiers this drift becomes serious, the first stage being the one most affected.

Most of the drift usually occurs during the first hundred hours of operation. If stability is required it is advisable to age the valves for at least 2 days, but in some cases this does not cure the rapid drift. Reference A12, pp. 730-733.

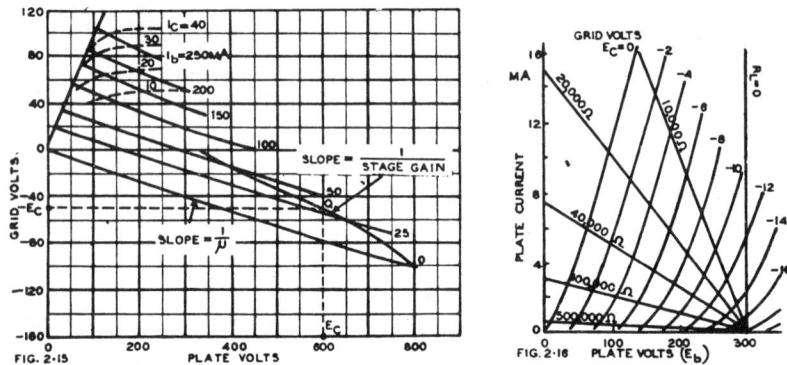

Fig. 2.15. Constant current characteristics for a typical small transmitting triode (801).

Fig. 2.16. Triode plate characteristics with loadlines for five values of load resistance.

(viii) Effect of heater-voltage variation

When a valve is being operated so that the plate current is small compared with the total cathode emission, an increase in heater voltage normally causes an increase in plate current, which may be brought back to its original value by an increased negative bias. With indirectly-heated cathodes the increase in negative bias is approximately 0·2 volt for a 20% increase in heater voltage, whether the valve is a diode, triode or multi-grid valve (Ref. A12, p. 421).

This effect is serious in d-c amplifiers; there are methods for cancelling the effect (Ref. A12, p. 4‒‒).

SECTION 3 : RESISTANCE-LOADED AMPLIFIERS

(i) Triodes   (ii) Pentodes.

(i) Triodes

When there is a resistance load in the plate circuit, the voltage actually on the plate is less than that of the supply voltage by the drop in the load resistor,

$$E_b = E_{bb} - R_L I_b.$$

This equation may be represented by what is known as a **Load Line** on the plate characteristics. Since the load is a pure resistance it will obey Ohm's Law, and the relationship between current and voltage will be a straight line; the loadline will therefore be a straight line.

The static operating point is the intersection of the loadline and the appropriate characteristic curve. Fig. 2.16 shows several loadlines, corresponding to different load resistors, drawn on a plate characteristic family. Zero load resistance is indicated by a vertical loadline, while a horizontal line indicates infinite resistance.

A loadline may be drawn quite independently of the plate characteristics, as in Fig. 2.17. The point B is the plate supply voltage E_{bb} (in this case 300 V); the slope* of the loadline AB $= -1/R_L$ and therefore AO $= E_{bb}/R_L$ (in this case 300/50 000 = 0·006 A = 6 mA). The voltage actually on the plate can only be equal to E_{bb} when the current is zero (point B). At point A the voltage across the valve is zero and the whole supply voltage is across R_L; this is what happens when the valve is short-circuited from plate to cathode. The plate current (E_{bb}/R_L) which flows under these conditions is used as a reference basis for the correct operation of a resistance coupled amplifier (Chapter 12). As the plate voltage, under high level dynamic conditions, must swing about the operating point, the latter must be somewhere in the region of the middle of AB; the plate current would then be in the region of 0·5 E_{bb}/R_L and the plate voltage 0·5 E_{bb}—in other words, the supply voltage is roughly divided equally between the valve and the load resistance. Actually, the operating point may be anywhere within the limits 0·4 and 0·85 times E_{bb}/R_L—see Chapter 12 Sect. 2(vi) and Sect. 3(vi).

In most resistance-loaded amplifiers, the plate is coupled by a capacitor to the grid of the following valve, which has a grid resistor R_g to earth. This resistor acts as a load on the previous valve, but only under dynamic conditions. In Fig. 2.18 the loadline AB is drawn, as in Fig. 2.17, and the operating point Q is fixed by selecting the grid bias (here −6 volts). Through Q is then drawn another line CD having a slope of $-(1/R_L + 1/R_g)$; this is the dynamic loadline, and is used for determining the voltage gain, maximum output voltage and distortion (Chapter 12).

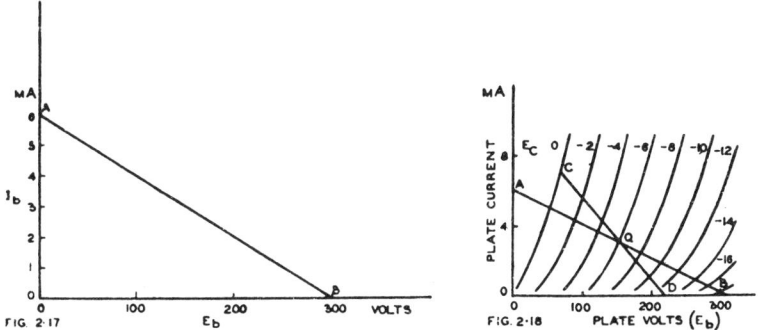

Fig. 2.17. *Loadline is independent of valve curves.*

Fig. 2.18. *Loadlines of resistance loaded triode; AQB is without any following grid resistor, CQD allows for the grid resistor.*

The dynamic characteristic is the effective mutual characteristic when the valve has a resistive load in the plate circuit†. While the slope of the mutual characteristic is g_m or μ/r_p, the slope of the dynamic characteristic is $\mu/(r_p + R_L)$. Owing to $(r_p + R_L)$ being more nearly constant than r_p, the dynamic characteristic is more nearly straight than the mutual characteristic.

---

*The slope of AB is negative since the plate voltage is the difference between the supply voltage and the voltage drop in R_L, and the inverted form $(1/R_L)$ is due to the way in which the valve characteristics are drawn with current vertically and voltage horizontally. The slope of AB is often loosely spoken of as being the resistance of R_L, the negative sign and inverted form being understood.

†It does not make allowance for the following grid resistor, and does not therefore correspond to the dynamic loadline. It is, of course, possible to derive from the dynamic loadline a modified dynamic characteristic which does make allowance for the grid resistor.

A typical dynamic characteristic is shown in Fig. 2.19 applying to a supply voltage of 250 volts and load resistance 0·1 megohm ; the mutual characteristics are shown with dashed lines.

The dynamic characteristic may be drawn by transferring points from along the loadline in the plate characteristic to the mutual characteristic. An alternative method making use of the mutual characteristic is as follows—

When the plate current is zero, the voltage drop in the load resistance is zero, and the plate voltage is equal to the supply voltage (250). For the plate voltage to be 200 volts, there must be a drop of 50 volts in the load resistor (100 000 ohms) and the plate current must therefore be 50/100 000 or 0·5 mA, and so on. A table may be prepared for ease of calculation :

Plate Voltage	Voltage Drop in Load Resistor	Plate Current ($=$ volts drop$/R_L$)
250	0	0
200	50	0·5 mA.
150	100	1·0 mA.
100	150	1·5 mA.
50	200	2·0 mA.

It will be seen that this table is not affected by the shape of the valve characteristics. The dynamic characteristic may then be plotted by taking the intersections of the various plate voltage curves with the plate current values given in the table.

The dynamic characteristic of a triode is very nearly straight along the central portion, with curves at both ends, the " upper bend " being always in the positive grid current region.

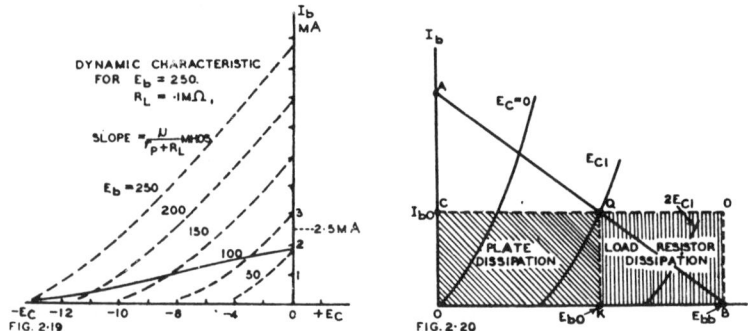

Fig. 2.19. *Triode dynamic characteristic (solid line) for resistance loading.*

Fig. 2.20. *Illustrating power dissipation in a resistance-loaded triode.*

Refer to Chapter 12 for further information on resistance coupled amplifiers.

When a resistance-loaded triode is operated under steady conditions, the **power dissipation** is indicated by Fig. 2.20. The area of the rectangle OCDB represents the total power ($E_{bb}I_{bo}$) drawn from the plate supply. The area of the rectangle OCQK represents the plate dissipation of the valve ($E_{bo}I_{bo}$) and the area of the rectangle KQDB represents the dissipation in the load resistor ($E_{bb} - E_{bo}) I_{bo} = I_{bo}{}^2 R_L$. Under dynamic conditions the plate dissipation decreases by the amount of power output, and the load resistor dissipation increases by the same amount, provided that there is no a.c. shunt load and that there is no distortion. The case of transformer-coupled loads is treated in Chapter 13.

(ii) Pentodes

Pentodes with resistive loads are treated in the same manner as triodes, the only complication being the screen voltage which must be selected at some suitable value and maintained constant (Fig. 2.21). The operating point as an amplifier will normally, as with a triode, be in the region of the middle of the loadline so that the voltage across the valve and that across R_L will be approximately the same. The only special case is with very low values of R_L (e.g. 20 000 ohms) where grid current occurs at approximately $E_{c1} = 0$, thereby limiting the useful part of the loadline.

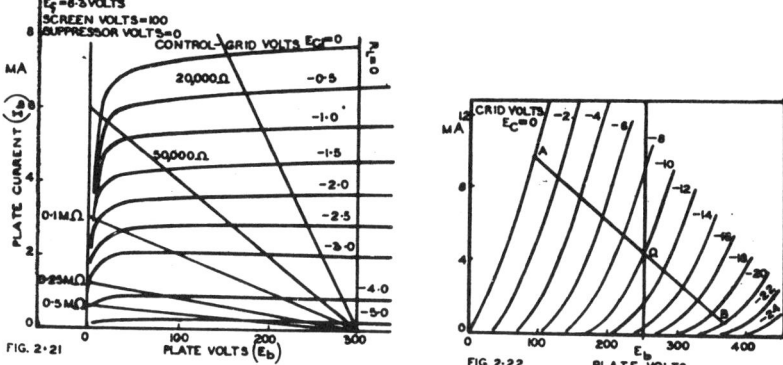

Fig. 2.21.   *Loadlines of resistance-loaded pentode.*

Fig. 2.22.   *Triode plate characteristics and loadline with transformer-coupled load.*

With any value of screen voltage, and any value of load resistance, it is possible to select a grid bias voltage which will give normal operation as an amplifier. With load resistance of 0·1 megohm and above, pentodes give dynamic characteristics which closely resemble the shape of triode dynamic characteristics with slightly greater curvature at the lower end ; at the upper end, provided that the screen voltage is not too low, the pentode has a curved portion where the triode runs into grid current. The top bend of the pentode dynamic characteristic is often used in preference to the bottom bend for plate detection—see Chapter 27 Sect. 1(ii)C.

For further information on resistance coupled pentode amplifiers, reference should be made to Chapter 12 Sect. 3.

SECTION 4.: TRANSFORMER-COUPLED AMPLIFIERS

(i) With resistive load.   *(ii) Effect of primary resistance*   *(iii) With i-f voltage amplifiers* *(iv) R-F amplifiers with sliding screen*   *(v) Cathode loadlines*   *(vi) With reactive loads.*

(i) With resistive load

When the load resistance is coupled to the valve by an ideal transformer, there is no direct voltage drop between the supply voltage and the plate. The slope cf the loadline, as before, is $-1/R_L$ but the loadline must be lifted so that it passes through the operating point. Fig. 2.22 shows a typical triode with $E_b = 250$ volts, and $E_c = -10$ volts, thus determining the static operating point Q. The loadline AQB is then drawn through Q with a slope corresponding to a resistance of 30 000 ohms. It is not taken beyond point A ($E_c = 0$) because in this case it is intended to be a Class A amplifier, operating without grid current. It is not taken beyond B because this is the limit of swing in the downward direction corresponding to A in the upward direction and having twice the bias of point Q (i.e. −20 volts). Of course, AB could be projected upwards and downwards if it were desired to increase the grid swing without regard to grid current or distortion.

(ii) Effect of primary resistance

If the primary circuit includes resistance, the point Q must be determined by drawing through E_b a straight line with a slope of $-1/R'$, where R' includes all resistances in the primary circuit other than the plate resistance of the valve. R' will include the d.c. resistance of the transformer primary winding and any equivalent internal resistance of the plate supply source. Fig. 2.23 is a typical example, with $R' = 1500$ ohms, from which point Q can be determined as previously. The total a.c. load on the valve is then $(R_L + R')$, in this case 31 500 ohms, which will give the slope of AQB. In these examples it is assumed that fixed bias is used, and that the negative side of the supply voltage is applied directly to the cathode of the valve.

(iii) With i-f voltage amplifiers

I-F amplifiers, when correctly tuned, operate with the valve working into practically a resistive load. I-F and r-f amplifier valves are in two principal groups—sharp cut-off and remote cut-off.

Sharp cut-off r-f pentodes operate in much the same manner as a-f pentodes, and the tuned transformer in the plate circuit reduces any distortion which might occur through non-linearity of the characteristics. The d.c. resistance of the transformer is usually so small that it may be neglected and the loadline drawn through Q with a slope corresponding to the dynamic load resistance of the transformer (including its secondary load, if any, referred to the primary).

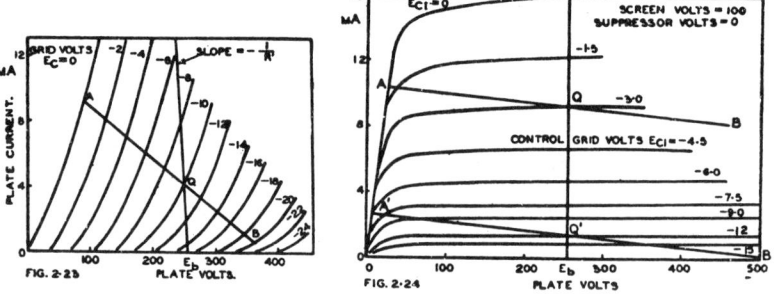

Fig. 2.23. Triode plate characteristics and loadlines with transformer-coupled load, allowing for the resistance of the primary winding.

Fig. 2.24. Plate characteristics of typical remote cut-off pentode with fixed screen and suppressor voltages.

Remote cut-off r-f pentodes are similar, except that the mutual characteristics are curved, and the distortion is greater. Fig. 2.24 shows the plate characteristics of a typical remote cut-off pentode, with $E_b = 250$ volts. Two loadlines (AQB, A' Q' B') have been drawn for grid bias voltages of -3 and -12 volts respectively, with a slope corresponding to a load resistance of 200 000 ohms, as for an i-f amplifier. This application of the loadline is not entirely valid, although it gives some useful information, since the tuned plate circuit acts as a " flywheel " to improve the linearity and reduce the distortion. This is a case in which constant current curves could be used with advantage. However, the ordinary plate characteristics at least indicate the importance of a high Q (high dynamic resistance) second i-f transformer if it is desired to obtain high output voltages at even moderately high negative bias voltages ; a steeper loadline would reach plate current cut-off at the high voltage peak.

(iv) R-F Amplifiers with sliding screen

Remote cut-off pentodes may have their cut-off points made even more remote by supplying the screen from a higher voltage (generally the plate supply) through a

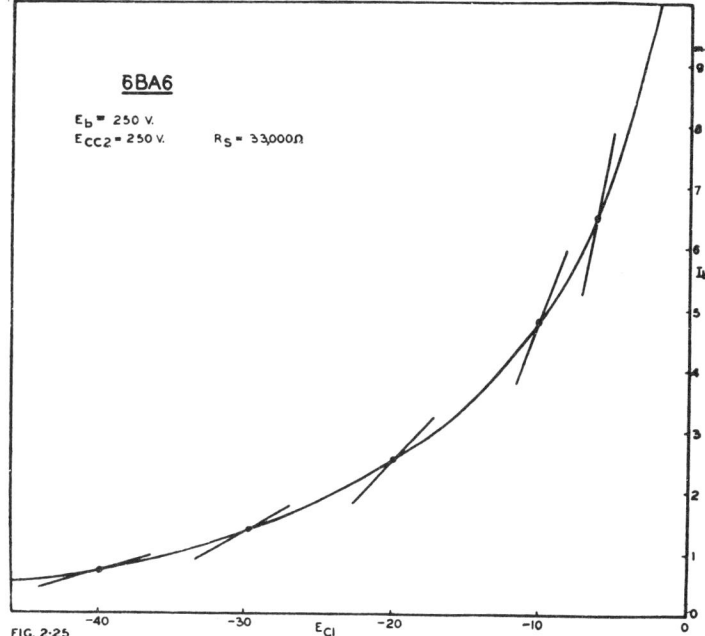

FIG. 2·25

*Fig. 2.25.   Plate current characteristic of remote cut-off pentode with " sliding screen."
The straight lines indicate the mutual conductances at several points.*

resistor designed to provide the correct screen voltage for the normal (minimum bias) operating condition.   The screen requires to be by-passed to the cathode.

The same method may be used with a sharp cut-off pentode to provide a longer grid base.   This does not make it possible to obtain the same results as with a properly designed remote cut-off pentode, although it does increase the maximum input voltage which can be handled with a limited distortion.   It is important to remember that the extended plate current characteristic curve obtained by this method cannot be used to determine the dynamic slope, since the latter is higher than would be calculated from the characteristic.   This is demonstrated in Fig. 2.25 which shows the " sliding screen " plate current characteristic, with straight lines drawn to indicate the mutual conductance at several points.

The procedure for deriving the " sliding screen " plate current characteristic from the fixed voltage data is as follows—

Let plate and screen current curves be available for screen voltages of 50, 75, 100 and 125 volts (Fig. 2.26) and take the case with a series screen resistor (R_s) of 250 000 ohms from a supply voltage of 300.

E_{c2}	E_{drop}*	I_{c2}**	Point	E_{c1}†	Point	I_b‡
50 V	250 V	1·0 mA	A	−0·1	E	3·7 mA
75	225	0·9	B	−1·7	F	3·15
100	200	0·8	C	−3·3	G	2·6
125	175	0·7	D	−5·2	H	2·1

*The voltage drop in the screen resistance $= 300 - E_{c2}$.
**$I_{c2} = E_{drop}/R_s$.
†Derived from the screen characteristics and transferred to the plate characteristics.
‡Derived from the plate characteristics.

(v) Cathode loadlines

The static operating point with cathode self bias may be determined graphically by the use of the mutual characteristic.   The mutual characteristic of a triode shown in Fig. 2.27 applies to the voltage between plate and cathode—the total supply voltage will be greater by the drop in the cathode resistor R_k.

Through O should be drawn a straight line OD, having a slope of $-1/R_k$ ohms. The point P where OD intersects the curve corresponding to the plate-to-cathode voltage (here 250 V) will be the static operating point, with a bias $-E_{c1}$ and plate current I_{b1}.

In the case of pentodes, with equal plate and screen voltages, the " triode " mutual characteristic should be used, if available. With the plate voltage higher than the screen voltage, the triode mutual characteristic may be used as a fairly close approximation, provided that the triode curve selected is for a voltage the same as the screen voltage.

Alternatively, pentodes may be treated as for triodes, except that the slope of OD should be

$$-\frac{1}{R_k} \cdot \frac{I_b}{I_b + I_{c2}}$$

where I_b and I_{c2} may be taken to a sufficient degree of accuracy as being the values under published conditions. The plate current $(I_b{}')$ may then be read from the curve, and the screen current calculated from the ratio of screen to plate currents.

For the use of cathode loadlines with resistance coupled triodes and pentodes, refer to Chapter 12.

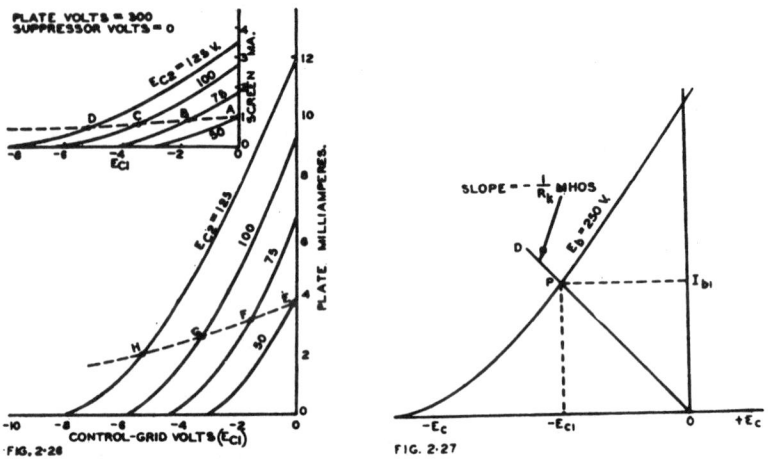

Fig. 2.26. *Plate and screen current characteristics of pentode illustrating procedure for deriving " sliding screen " characteristics.*

Fig. 2.27. *Triode mutual characteristics with cathode bias loadline OD.*

(vi) With reactive loads

When the load on the secondary of the transformer is not purely resistive, the loadline is normally in the form of an ellipse instead of a straight line. Fig. 2.28 shows three different examples of elliptical loadlines for purely reactive loads. A purely capacitive load has exactly the same shape of loadline as a purely inductive one, but the direction of rotation of the point is opposite, as indicated by the arrows. Curve A is for a high reactance, curve B for an intermediate value of reactance, and curve C for a low reactance. In each case the maximum current is E_0/X_0 where E_0 is the peak voltage across the reactance and $X_0 = \omega L$ for the inductive case, and $X_0 = 1/\omega C$ for the capacitive case. The voltage E_0 is shown as negative to the right of O, so as to be suitable for applying directly to the plate characteristics of the valve.

For convenience in application, the horizontal and vertical scales should be the same as in the valve characteristics to which the loadline is to be applied. For example, if on the plate characteristics one square represents 1 mA in the vertical direction and 25 V in the horizontal direction, the same proportion should be maintained for the elliptical loadline. Having drawn the ellipse for any convenient value of E_o, it may be expanded or contracted in size, without changing its shape (that is the ratio of the major to the minor axis when both are measured in inches).

(a) **Resistance and inductance in series**

The load is more commonly a combination of resistance and reactance. When the load is a resistance R_L in series with an inductive reactance ωL, the maximum current through both will be I_o and the procedure is to draw both the straight resistive loadline for R_L (AB in Fig. 2.29) and the elliptical loadline for ωL, and then to combine them in series. It will be seen that in Fig. 2.29 the peak current of the ellipse and of the resistive loadline are identical (I_o).

To combine these in series, it is necessary to consider the phase relations. When the current is a maximum (OE), the voltage drop across R_L is a maximum (AE) and that across L is zero, because there is 90° phase difference between the voltage and current : the total voltage drop across R_L

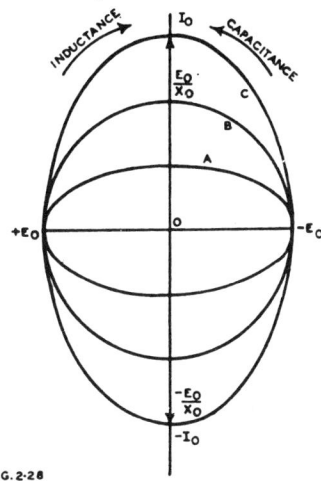

FIG. 2-28

Fig. 2.28.   Three examples of elliptical loadlines for purely reactive loads.

and L in series is therefore AE and point A is on the desired loadline. When the current is zero, the voltage drop across R_L is zero, and that across L is OC ; the total voltage drop is therefore OC, and point C is on the desired loadline. At any intermediate point (OF) with current increasing the voltage drop across R_L is FG, and that across L is FH, so that the total drop is

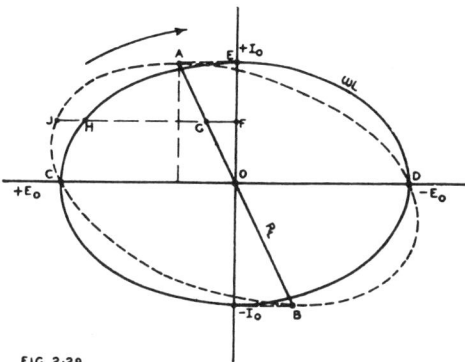

FIG. 2·29

Fig. 2.29.   Resistive loadline (R_L); inductive loadline (ωL); and elliptical resultant for R_L in series with ωL (dashed curve).

$FJ = FG + FH$. With similar procedure in the other three quadrants, the combined loadline is shown to be an ellipse CJADB which is tilted, or rotated in the clockwise direction as compared with the original ellipse. The maximum voltage drop is greater than that across either R_L or L alone, as would be expected.

If an elliptical loadline is known, as for example the dashed ellipse of Fig. 2.29, its series components may readily be determined. Mark points A and B where the ellipse reaches its maximum and minimum current values, then draw the line AB ; the slope of AB gives R_L. Mark O as the centre of the line AB ; draw COD horizontally to cut the ellipse at points C and D.

Then $\omega L = E_0/I_0$ ohms,

where E_0 = voltage corresponding to length OD

and I_0 = current (in amperes) corresponding to max. vertical height of ellipse above line COD.

Alternatively

$$\omega L = \frac{\text{Length of horizontal chord of ellipse through O, in volts}}{\text{Maximum vertical extent of ellipse, in amperes}}$$

(b) Resistance and inductance in parallel

When the load is a resistance R_L in parallel with an inductive reactance ωL, the maximum voltage across both will be E_0, and the resistive loadline and reactive ellipse may be drawn as for the series connection. In this case, however, the currents have to be added. In Fig. 2.30 the maximum current through R_L is CK (corresponding to $+E_0$), while the maximum current through L is OE. When the voltage is zero and increasing, the current through R_L is zero, and that through L is the minimum value OP ; point P is therefore on the desired loadline. When the voltage is its positive maximum (OC), the current through R_L is CK and that through L is zero ; point K is therefore on the desired loadline. Similarly with points E and M. At an intermediate value, when the voltage is negative and approaching zero (OR), the current through R_L is RS, and that through L is RT ; the total current is therefore RT + RS = RW, and W is on the desired loadline. The loadline is therefore the ellipse PKEMW.

Fig. 2.30. Resistive loadline (R_L); inductive loadline (solid ellipse) ; and elliptical resultant for R_L in parallel with ω_L (dashed curve).

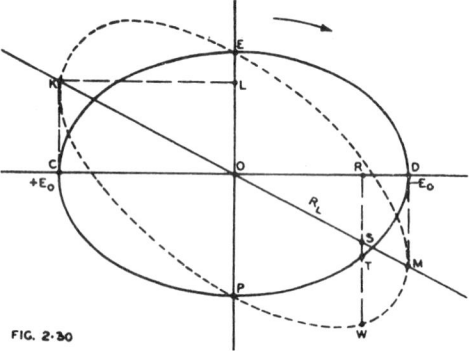

FIG. 2·30

If an elliptical loadline is known, as for example the dashed ellipse of Fig. 2.30, its parallel components may readily be determined. Mark points K and M where the ellipse reaches its maximum and minimum voltage values, then draw the line KM ; the slope of KM gives R_L. Mark O as the centre of the line KM ; draw EOP vertically to cut the ellipse at points E and P.

Then $\omega L = E_0/I_0$ ohms,

where E_0 = voltage difference between points O and K,

and I_0 = current corresponding to length OE, in amperes.

Alternatively

$$\omega L = \frac{\text{Maximum horizontal length of ellipse, in volts}}{\text{Length of vertical chord of ellipse through O, in amperes}}$$

(c) Resistance and capacitance

A similar shape of loadline is obtained when the inductance is replaced by a capacitance of equal reactance, except that the direction of rotation is opposite.

(d) Applying elliptical loadlines to characteristics

The elliptical loadlines derived by the methods described above may be applied to the plate characteristics of a valve, but it is first necessary to enlarge or reduce their size until they just fit between grid voltage curves corresponding to extreme swing

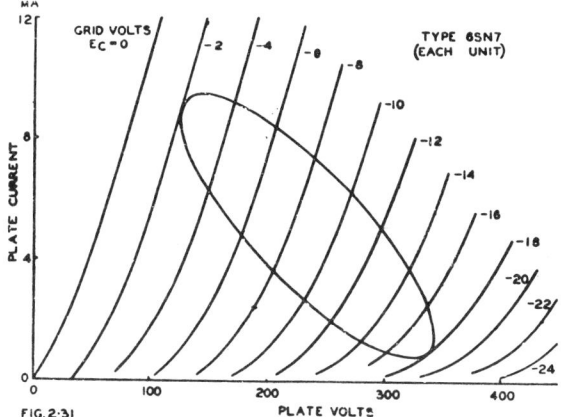

FIG. 2·31

Fig. 2.31. Triode plate characteristics with elliptical loadlines corresponding to resistance 25 000 ohms in series with reactance of 18 000 ohms.

in each direction. The examples taken have all been based on an arbitrary current (I_o) or voltage (E_o), which may be made larger or smaller as desired. In Fig. 2.31 there is shown the elliptical loadline corresponding to a resistance of 25 000 ohms in series with a reactance of 18 000 ohms, on triode plate characteristics with $E_b = 250$ volts, $E_c = -10$ volts and peak grid amplitude $E_{gm} = 8$ volts.

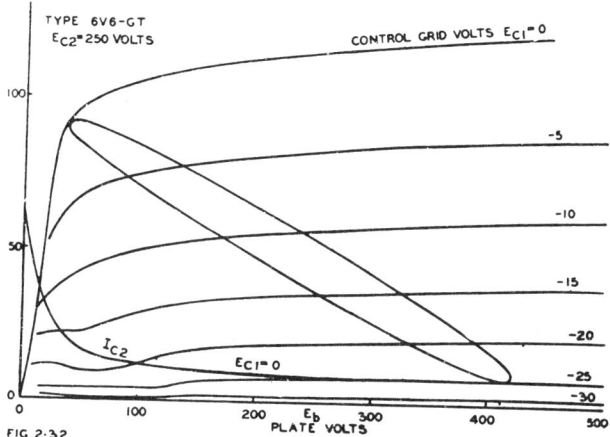

FIG. 2·32

Fig. 2.32. Beam power amplifier plate characteristics with elliptical loadline corresponding to a resistance of 4750 ohms in parallel with a reactance of 23 000 ohms.

Fig. 2.32 shows a typical beam power amplifier with an elliptical loadline with a resistive load of 4750 ohms shunted by a reactance of 23 000 ohms. The plate voltage is 250 volts, grid bias $-12\cdot5$ volts, and grid swing from 0 to -25 volts.

In all applications of elliptical loadlines to characteristics, the shape of the ellipse (i.e. the ratio of its major to its minor axis) and the slope of the major axis are determined solely by the nature of the load. The ellipse can be imagined as being slowly blown up, like a balloon, until it just touches without cutting the two curves of extreme voltage swing. If there is no distortion, the centre of the ellipse will coincide with the quiescent working point, but in the general case the centre of the ellipse will be slightly displaced.

SECTION 5 : TRIODE OPERATION OF PENTODES

(i) *Triode operation of pentodes*   (ii) *Examples of transconductance calculation* (iii) *Triode amplification factor*   (iv) *Plate resistance*   (v) *Connection of suppressor grid.*

(i) Triode operation of pentodes

Any pentode may be operated as a triode, provided that none of the maximum ratings is exceeded, and the characteristics may readily be calculated if not otherwise available.

When the cathode current of a valve is shared by two collecting electrodes (e.g. plate and screen) the mutual conductance of the whole cathode stream (i.e. the " triode g_m ") is shared in the same proportion as is the current.

Let I_k = cathode current
I_{c2} = screen current
I_b = plate current
g_m = pentode transconductance (to the plate)
g_t = triode transconductance (with screen and plate tied together)
and g_s = screen transconductance (with pentode operation).

Then $I_k = I_{c2} + I_b$ (1)
$g_t = g_m + g_s$ (by definition) (2)
and $g_m/g_t = I_b/I_k$ (3)

If it is desired to find the screen transconductance, this can be derived from the expression

$g_s/g_m = I_{c2}/I_b$ (4)
or $g_s/g_t = I_{c2}/I_k$ (5)

(ii) Examples of transconductance calculation

Example 1 : Type 6J7-G as a pentode with 100 volts on both screen and plate, and with a grid bias of. −3 volts, has the following characteristics :—

Transconductance   1185 micromhos
Plate Current   2·0 mA
Screen Current   0·5 mA

It is readily seen that the cathode current (see equation 1 above) is given by
$$I_k = 0·5 + 2·0 = 2·5 \text{ mA.}$$
The triode transconductance is calculated by inverting equation (3) above,
$$g_t/g_m = I_k/I_b .$$
Therefore $g_t/1185 = 2·5/2·0$
and g_t = 1482 micromhos.

The example selected was purposely chosen so as to have equal plate and screen voltages. Under these conditions the method is exact, and the calculated triode mutual conductance applies to the same conditions of plate and grid voltages as for the pentode operation (in this example 100 volts and −3 volts respectively).

Example 2 : Type 6J7-G as a pentode with 250 volts on the plate, 100 volts on the screen, and −3 volts grid bias.

In this case a similar method may be used, but it is necessary to make an assumption which is only approximately correct. Its accuracy is generally good enough for most purposes, the error being within about 5% for most conditions.

The assumption (or approximation) which must be made is—*That the plate current of a pentode valve does not change as the plate voltage is increased from the same voltage as that of the screen up to the voltage for pentode operation.*

This assumption means, in essence, that the plate resistance is considered to be infinite—a reasonable approximation for most r-f pentodes, and not seriously in error for power pentodes and beam power valves.

In this typical example we can take the published characteristics, and assume that the plate current and transconductance are the same for 100 as for 250 volts on the plate. From then on the procedure is exactly as in the previous example. It is important to note that the calculated triode characteristics only apply for a triode plate voltage of 100 volts and a grid bias of −3 volts.

Example 3 : To find the screen transconductance under the conditions of Example 1.

From eqn. (2) we may derive the expression—

$$g_s = g_t - g_m = 1482 - 1185$$
$$= 297 \text{ micromhos.}$$

This could equally well have been derived from eqn. (4) or (5).

(iii) Triode amplification factor

The triode amplification factor (if not available from any other source) may be calculated by the following approximate method.

Let μ_t = triode amplification factor
        E_{co} = negative grid voltage at which the plate current just cuts off
and E_{c2} = screen voltage.
Then μ_t $= E_{c2}/E_{co}$ approx.   (6)

For example, with type 6J7-G having a screen voltage of 100 volts, the grid bias for cut-off is indicated on the data sheet as being −7 volts approx. This is the normal grid bias for complete plate current cut-off, but it is not very suitable for our purpose since equation (6) is based on the assumption that the characteristic is straight, whereas it is severely curved as it approaches cut-off. The preferable procedure is to refer to the plate current-grid voltage characteristic, and to draw a straight line making a tangent to the curve at the working point—in this case with a screen voltage of 100 volts and grid bias −3 volts. When this is done, it will be seen that the tangent cuts the zero plate current line at about −5 volts grid bias. If this figure is used, as being much more accurate than the previous value of −7 volts, the triode amplification factor will be

$$\mu_t = 100/5 = 20.$$

Alternatively, if only the plate characteristics are available, much the same result may be obtained by observing the grid bias for the lowest curve, which is generally very close to plate current cut-off.

In the case of remote cut-off characteristics it is essential to adopt the tangent method, and the result will only apply to the particular point of operation, since the triode amplification factor varies along the curve.

The amplification factor of the screen grid in a pentode valve with respect to the control grid is almost exactly the same as the triode amplification factor.

The amplification factor of the plate of a pentode valve with respect to its screen grid may be calculated from the expression—

$$\mu_{g1 \cdot p} = \mu_{g1 \cdot g2}\, \mu_{g2 \cdot p}$$ (7)

where $\mu_{g1 \cdot p}$ = pentode amplification factor
and $\mu_{g2 \cdot p}$ = screen grid-plate mu factor.

This expression can only be used when the pentode amplification factor is known. If this is not published, it may be determined from a knowledge of the plate resistance and mutual conductance. If the former is not published, it may be derived graphically ; this derivation is only very approximate in the case of sharp cut-off r-f pentodes, since the characteristics are nearly horizontal straight lines.

For example type 6AU6 has the following published values—

$$r_p = 1.5 \text{ megohms}$$
and $$g_m = 4450 \text{ } \mu\text{mhos}$$ $\Big\}$ at $E_b = 250, E_{c2} = 125, \quad E_{c1} = -1 \text{ V}$

from which $\mu = 6675$.
But $\mu_{g1 \cdot g2} = 36$ approx.
Therefore $\mu_{g2 \cdot p} = 6675/36 = 185$ approx.

(iv) Plate resistance

The "plate resistance" of each electrode (plate or screen) in the case of pentode operation, and the "triode plate resistance" when plate and screen are tied together, may be calculated from the corresponding values of μ and g_m.

(v) Connection of suppressor grid

The suppressor may be connected either to cathode or to the screen and plate, with negligible effect on the usual static characteristics. Some valves have the suppressor internally connected to the cathode, so that there is no alternative. In other cases, connection to cathode slightly increases the output capacitance. In low level amplifiers, connection of the suppressor to cathode may give lower noise in certain cases if there is a high resistance leakage path from suppressor to cathode ; similarly its connection to screen and plate will give lower noise if there is leakage to the latter electrodes.

SECTION 6 : CONVERSION FACTORS, AND THE CALCULATION OF CHARACTERISTICS OTHER THAN THOSE PUBLISHED

(*i*) *The basis of valve conversion factors* (*ii*) *The use of valve conversion factors* (*iii*) *The calculation of valve characteristics other than those published* (*iv*) *The effect of changes in operating conditions.*

Conversion Factors provide a simple approximate means of calculating the principal valve characteristics when all the voltages are changed by the same factor. It is possible to make certain additional calculations so as to allow for the voltage of one electrode differing from this strict proportionality.

(i) The basis of valve conversion factors

Valve Conversion Factors are based on the well-known mathematical expression of valve characteristics

$$I_b = A(E_b - \mu E_c)^x \tag{1}$$

where I_b = plate current
E_b = plate voltage
E_c = grid voltage
A = a constant depending upon the type of valve
μ = amplification factor
and x = an exponent, with a value of approximately 1.5 over the nearly straight portion of the characteristics.

If we are concerned merely with changes in the voltages and currents, then we can reduce the expression to the form

$$I_b \propto (E_b - \mu E_c)^x. \tag{2}$$

Now if we agree to change the grid voltage in the same proportion as the plate voltage, we obtain the very simple form

$$I_b \propto E_b^x. \tag{3}$$

Finally, if we take x as 1.5 or 3/2, we have the approximation

$$I_b \propto E_b^{3/2}. \tag{4}$$

Put into words, this means that the plate current of a valve varies approximately as the three-halves power of the plate voltage, provided that the grid voltage is varied in the same proportion as the plate voltage.

The same result may be obtained with pentodes, provided that both the grid and screen voltages are varied in the same proportion as the plate voltage. This result is the basis of Valve Conversion Factors, so that we must always remember that their use is restricted to cases in which all the electrode voltages are changed in the same proportion.

Let F_e be the factor by which all the voltages are changed (i.e. grid, screen, and plate), and let I_b' be the new plate current.

Then $I_b' \propto (F_e.E_b)^{3/2}$.   (5)

But $I_b' = F_i.I_b$

where F_i is the factor by which the plate current is changed.

Therefore $F_i.I_b \propto (F_e.E_b)^{3/2}$.   (6)

From the combination of (4) and (6) it will be seen that

$$F_i = F_e^{3/2}$$ (7)

Now the power output is proportional to the product of plate voltage and plate current so that

$$P_o \propto E_b.I_b$$ (8)

and $P_o' \propto (F_e.E_b)(F_i.I_b)$ (9)

so that $P_o' \propto F_e.F_i(E_b.I_b)$ (10)

$\propto F_e.F_i(P_o)$.   (11)

We may therefore say that the power conversion factor F_p is given by the expression

$$F_p = F_e.F_i.$$ (12)

Therefore $F_p = F_e^{5/2}$.   (13)

The mutual conductance is given by

$$g_m = \frac{\text{change of plate current}}{\text{change of grid voltage}}$$

Therefore $F_{gm} = F_i/F_e = F_e^{3/2}/F_e = F_e^{\frac{1}{2}}$.   (14)

The Plate Resistance is given by

$$r_p = \frac{\text{change of plate voltage}}{\text{change of plate current}}$$

Therefore $F_r = F_e/F_i = F_e/F_e^{3/2} = F_e^{-\frac{1}{2}}$.   (15)

This also applies similarly to the load resistance and cathode bias resistance.

We may therefore summarize our results so far :—

$$F_i = F_e^{3/2}$$ (7)

$$F_p = F_e^{5/2}$$ (13)

$$F_{gm} = F_e^{\frac{1}{2}}$$ (14)

$$F_r = F_e^{-\frac{1}{2}}$$ (15)

These are shown in graphical form on the Conversion Factor Chart (Fig. 2.32A).

(ii) The use of valve conversion factors

It is important to remember that the conversion factors may only be used when all the voltages (grid, screen and plate) are changed simultaneously by the same factor. If it is required to make any other adjustments, these may be carried out before or after using conversion factors, by following the method given under (iii) below.

Conversion factors may be used on any type of valve whether triode, pentode or beam tetrode, and in any class of operation whether class A, class AB1, class AB2 or class C.

The use of conversion factors is necessarily an approximation, so that errors will occur which become progressively greater as the voltage factor becomes greater. In general it may be taken that voltage conversion factors down to about 0·7 and up to about 1·5 times will be approximately correct. When the voltage factors are extended beyond these limits down to 0·5 and up to 2·0, the accuracy becomes considerably less, and any further extension becomes only a rough indication.

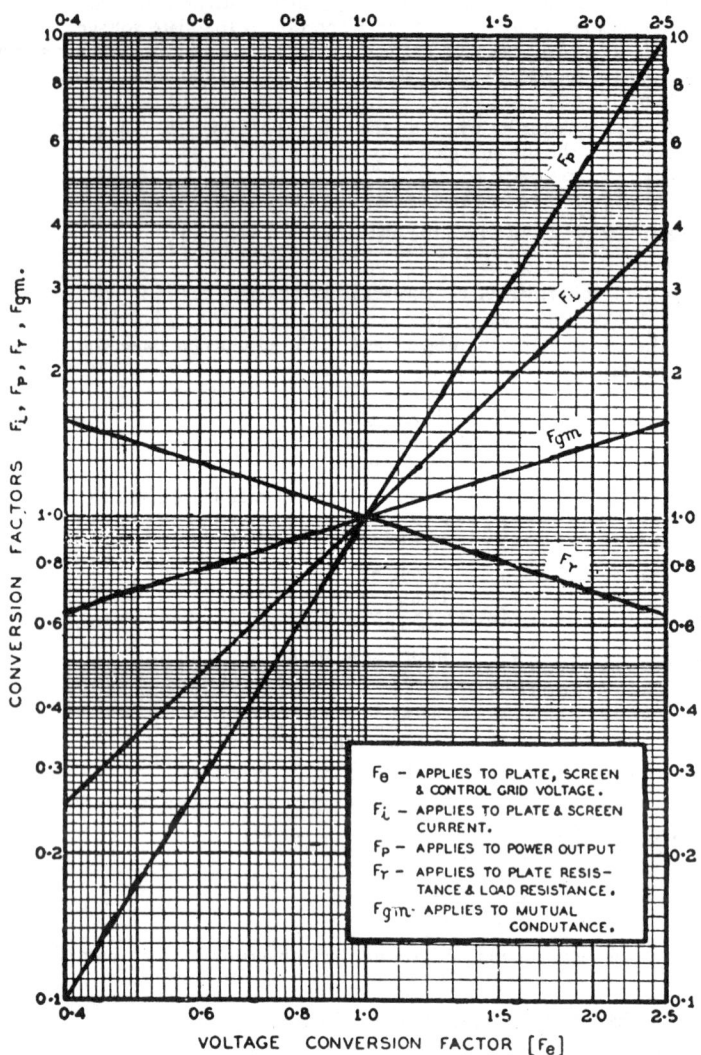

Fig. 2.32A.   Conversion factor chart (by courtesy of R.C.A.).

The example given below is a straightforward case of a pentode valve whose characteristics are given for certain voltages and which it is desired to operate at a lower plate voltage.

Plate and screen voltage	250 volts
Control grid voltage	−15 volts
Plate current	30 mA

Screen current	6 mA
Mutual conductance	2,000 μmhos
Power Output	2·5 watts.

It is required to determine the optimum operating conditions for a plate voltage of 200 volts.

The Voltage Conversion Factor $(F_e) = 200/250 = 0·8$.

The new screen voltage will be $0·8 \times 250 = 200$ volts.

The new control grid voltage will be $-(0·8 \times 15) = -12$ volts.

Reference to the chart then gives the following :

Current Conversion Factor (F_i)	0·72
Mutual Conductance Conversion Factor (F_{gm})	0·89
Power Output Conversion Factor (F_p)	0·57

The new plate current will be $0·72 \times 30 = 21·6$ mA.

The new screen current will be $0·72 \times 6 = 4·3$ mA.

The new mutual conductance will be $0·89 \times 2000 = 1780$ μmhos.

The new power output will be $0·57 \times 2·5 = 1·42$ watts.

There are two effects not taken into account by conversion factors. The first is contact potential, but its effects only become serious for small grid bias voltages. The second is secondary emission, which occurs with the old type of tetrode at low plate voltages ; in such a case the use of conversion factors should be limited to regions of the plate characteristic in which the plate voltage is greater than the screen voltage. With beam power amplifiers the region of both low plate currents and low plate voltages should also be avoided for similar reasons.

The application of conversion factors to resistance-capacitance-coupled triodes and pentodes is covered in Chapter 12 Sect. 2(x) and Sect. 3(x) respectively.

FIG. 2 .33

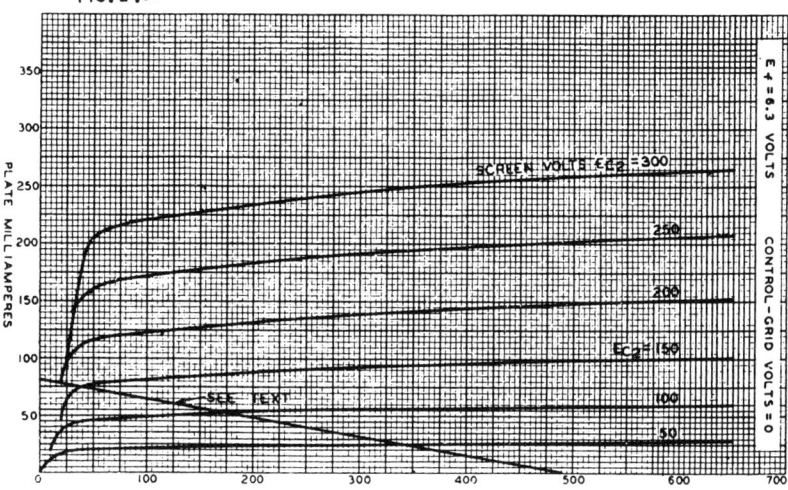

Fig. 2.33. *Zero bias plate characteristics for type* 807 *beam power amplifier with six values of screen voltage (Ref. E2).*

Greater accuracy in the use of conversion factors over a wide range of screen voltages may be obtained, if curves are available for zero bias at a number of different screen voltages as in Fig. 2.33 (Ref. E2).

When the plate, screen, and grid voltages of a pentode or beam power amplifier are multiplied by the same voltage conversion factor, the ratio of the plate current at a given grid bias to that at zero bias does not change. In order to convert a given family of plate characteristics to a new screen voltage condition, it is therefore only necessary to have a zero-bias plate characteristic for the screen voltage of interest.

Example

Suppose that the family of plate characteristics shown in Fig. 2.34, which obtains for a screen voltage of 250 volts, is to be converted for a screen voltage of 300 volts. The zero-bias plate characteristic for $E_{c2} = 300$ volts, which is shown in Fig. 2.33, is replotted as the upper curve in Fig. 2.35.

Since all bias values shown in Fig. 2.34 must be multiplied by $300/250 = 1.2$, corresponding plate characteristics for the new family obtain for bias values that are 20 per cent. higher than those shown in Fig. 2.34. Consider the conversion of -10-volt characteristic of Fig. 2.34. At a plate voltage (E_b) of 250 volts in Fig. 2.34, $AB/AC = 100/187 = 0.535$. On the new characteristic in Fig. 2.35 which corresponds to a bias of -12 volts, $A'B'/A'C'$ must also equal 0.535 at $E_b = 300$ volts. Therefore, $A'B' = 0.535 \times A'C'$. From the given zero-bias characteristic of Fig. 2.35, $A'C' = 244$ at $E_b = 300$ volts; hence $A'B' = 131$ milliamperes. At $E_b = 200$ volts in Fig. 2.34 $DE/DF = 98/183 = 0.535$. Therefore, at $E_b = 200 \times 1.2 = 240$ volts in Fig. 2.35, $D'E' = 0.535 \times 238 = 127$ milliamperes. This process is repeated for a number of plate voltages and a smooth curve is drawn through the points on the new characteristic.

FIG. 2.34

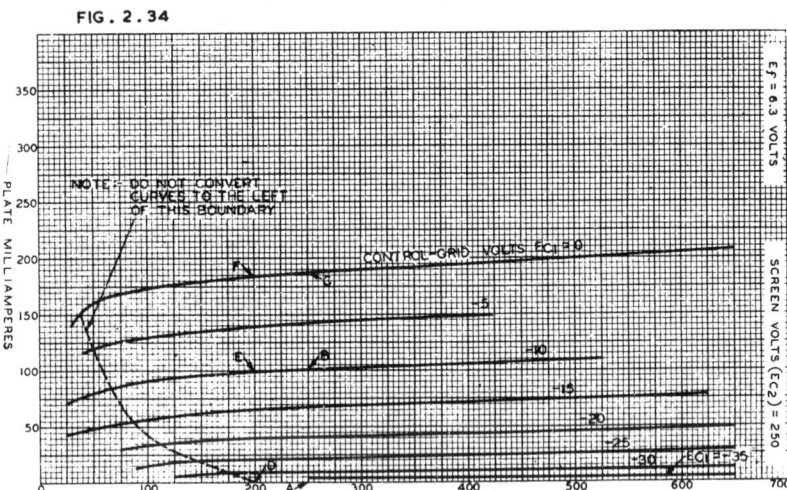

Fig. 2.34. *Plate characteristics for type* 807 *with fixed screen voltage and eight values of grid voltage* (Ref. E2).

The factor 0.535 can be used for the -10-volt characteristic at plate voltages greater than that at which the knee on the zero-bias characteristic of Fig. 2.34 occurs; for plate voltages in the immediate region of the knee, a new factor should be determined for each point. The plate characteristics of Fig. 2.34 should not be converted to the left of the dashed line of Fig. 2.34 because of space-charge effects. This limitation is not a serious one, however, because the region over which the valve usually operates can be converted with sufficient accuracy for most applications. The converted plate characteristic of Fig. 2.35 for $E_{c1} = -30$ volts was obtained in a similar manner to that for $E_{c1} = -12$ volts.

The curves of Fig. 2.35 were checked under dynamic conditions by means of a cathode-ray tube and the dotted portions show regions where measured results departed from calculated results.

(iii) The calculation of valve characteristics other than those published

It is frequently desired to make minor modifications in the operating conditions of a valve, such as by a slight increase or decrease of the plate voltage, change in grid bias or load resistance. It is proposed to describe the effects which these changes will have on the other characteristics of the valve.

The procedure to be adopted is summarized below :—

FIG . 2 . 35

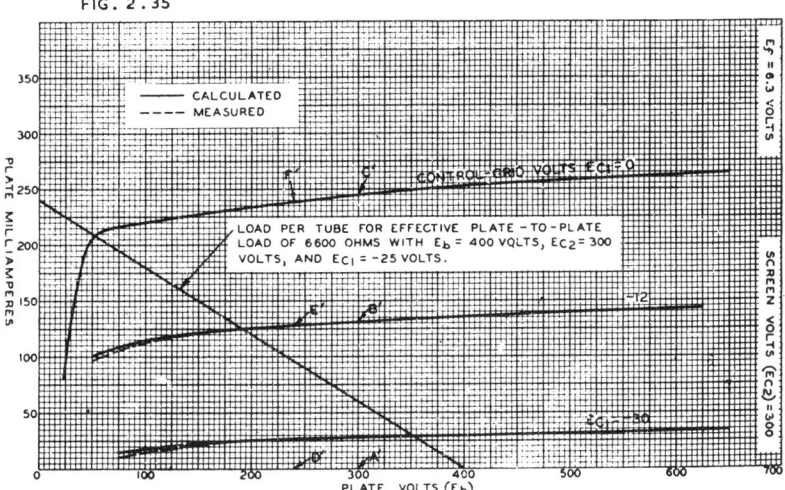

Fig. 2.35.   Derived plate characteristics for type 807 *with different screen voltage, making use of Figs.* 2.33 *and* 2.34 *with conversion factors* (*Ref. E2*).

(a) In the absence of valve curves

Triode—Use conversion factors to adjust the plate voltage to its new value, and apply the correct conversion factors to all other characteristics ; then adjust the grid bias to its desired new value by the method given below, and finally adjust the load resistance.

Pentode or beam power amplifier—Use conversion factors to adjust the screen voltage to its new value, and apply the correct conversion factors to all other characteristics ; then adjust the plate voltage to the desired new value by the method given below ; then adjust the grid bias to its desired new value, and finally adjust the load resistance.

(b) When valve curves are available

Triode with no d.c. load resistance in the plate circuit : Refer to the published characteristics to find the maximum plate dissipation ; calculate the maximum plate current which can be permitted at the desired new plate voltage ; select a suitable plate current for the particular application (which must not exceed the maximum) ; and refer to the curves to find the grid bias to give the desired plate current.

If the valve is a power amplifier, the load resistance may be determined by one of the methods described in Chapter 13 [e.g. triodes Sect. 2(iii) ; pentodes Sect. 3(iii)A].

Triode with resistor in plate circuit : Use conversion factors, with adjustments as required in accordance with the method given in (iv) below.

Pentode or beam power amplifier : If curves are available for the published value of screen voltage, use the method in (iv) below to obtain the characteristics for a plate voltage such that, when conversion factors are applied, the plate voltage is the desired value. For example, if curves and characteristics are available for plate and screen voltages of 250 volts, and it is desired to determine the characteristics for a plate voltage of 360 volts and screen voltage of 300 volts : firstly determine the characteristics for a plate voltage of 300 and screen voltage of 250 ; then apply voltage conversion factors of 1·2 to the plate, screen and grid voltages so as to provide the desired conditions.

If curves are available for the new value of screen voltage, use conversion factors to bring the screen voltage to the desired value, then apply the method below to adjust the plate voltage, load resistance and grid bias.

(iv) The effect of changes in operating conditions

(A) Effect of Change of Plate Voltages of Pentodes and Beam Power Amplifiers

(a) On plate current

The plate current of a pentode or beam power valve is approximately constant over a wide range of plate voltages, provided that the plate voltage is maintained above the " knee " of the curve. The increase of plate current caused by an increase in plate voltage from E_{b1} to E_{b2} is given by the expression

$$\triangle I_b = \frac{\triangle E_b}{r_p} = \frac{E_{b2} - E_{b1}}{r_p}. \tag{16}$$

In many cases the plate characteristic curves are available, and the change in plate current may be read from the curves.

(b) On screen current

In the case of both pentodes and beam power valves the total cathode current (i.e., plate plus screen currents) is approximately constant over a wide range of plate voltages (see Fig. 2.4). The increase in plate current from E_{b1} to E_{b2} is approximately equal to the decrease in screen current over the same range.

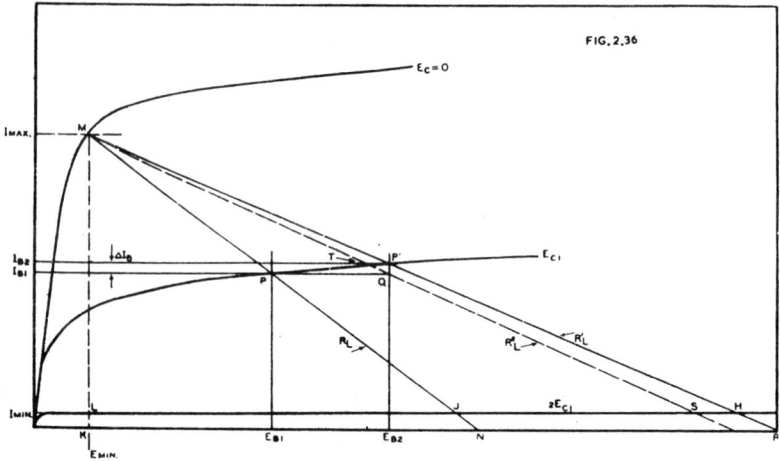

Fig. 2.36.   Plate characteristics of power pentode illustrating effect of change of plate voltage.

(c) On load resistance and power output

The plate characteristics of a typical power pentode are shown in Fig. 2.36 in which I_{b1} is the " published " plate current at plate voltage E_{b1} and grid bias $-E_{c1}$. The loadline MPJ swings up to I_{max} at $E_c = 0$ and down to I_{min} at $2E_{c1}$, the assumption being made that the $2E_{c1}$ curve is straight and horizontal over the range of plate voltages in which we are interested.

If the plate voltage is increased to E_{b2}, the new loadline will be MP′H, the point M being common to both, since it is at the knee of the characteristic. The quiescent operating point P′ is at a higher plate current than P, the difference being $\triangle I_b$.

Since the power output is proportional to the area of the triangle under the loadline, it is also proportional to the value of the load resistance, all triangles having ML as a common side. It may readily be shown that

$$R_L = \frac{E_{b1} - E_{min}}{I_{max} - I_{b1}}$$

and

$$R_L' = \frac{E_{b2} - E_{min}}{I_{max} - I_{b2}}.$$

Therefore

$$\frac{R_L'}{R_L} = \frac{E_{b2} - E_{min}}{E_{b1} - E_{min}} \cdot \frac{I_{max} - I_{b1}}{I_{max} - I_{b2}} \qquad (17)$$

which is also the ratio of the output powers. If $I_{b2} = I_{b1}$ or the rise of plate current is neglected as an approximation, then

$$\frac{R_L'}{R_L} = \frac{E_{b2} - E_{min}}{E_{b1} - E_{min}}. \qquad (18)$$

As an example, apply this to type 6V6-GT under the following conditions—

	Published Condition	Desired Condition
Plate voltage	250	300 V
Screen voltage	250	250 V
Grid voltage	-12.5	-12.5 V
Load resistance	5000	(see below) ohms
Plate current (I_{b1})	47	48* mA
Peak plate current (I_{max})	90*	90* mA
Min. plate current (I_{min})	8*	8* mA
Min. plate voltage (E_{min})	35	35 V
Power output	4.5	(see below) W

*From curve.

Using equation (17)—

$$\frac{R_L'}{R_L} = \frac{300 - 35}{250 - 35} \cdot \frac{90 - 47}{90 - 48} = \frac{265}{215} \cdot \frac{43}{42} = 1.26$$

whence $R_L' = 1.26 \times 5000 = 6300$ ohms.

The increase of power output is in proportion to the increase in load resistance. i.e. $P_o = 4.5 \times 1.26 = 5.66$ watts.

This method is remarkably accurate when there is very small rectification in the plate circuit, as is usually the case with power pentodes. With beam power amplifiers of the 6L6 and 807 class, in which the rectification is considerable (strong second harmonic component), the "corrected" loadline should be used as a basis, and the values of I_{max}, I_{b1} and E_{min} should be those corresponding to the corrected loadline.

If the rise in plate current ($\triangle I_b$) is considerable, the point P' will be above the centre point of the loadline MH; and there will be an appreciable amount of second harmonic distortion ; this may be reduced to zero (if desired) by increasing the load resistance slightly.

(B) Effect of change of load resistance

In a r.c.c. triode the effect of a change in R_L on stage gain is very slight, provided that $R_L \geqslant 5r_p$. In any case where the change cannot be neglected, eqn. (7) of Chapter 12 Sect. 2 may be used to calculate stage gain.

In a r.c.c. pentode the effect of a change in R_L on stage gain is given by eqn. (7) of Chapter 12 Sect. 3, bearing in mind that the mutual conductance at the operating plate current is increased when R_L is decreased. As a rough approximation, the voltage gain is proportional to the load resistance. If optimum operating conditions are to be obtained, conversion factors should be applied to the whole amplifier—see Chapter 12 Sect. 3(x)C.

(C) Effect of change of grid bias

In any valve which is being operated with fixed voltages on all electrodes and without any resistance in any of the electrode circuits, a change of grid bias will result in a change of plate current as given by the expression

$$\triangle I_b = \triangle E_c \times g_m \tag{19}$$

where $\triangle I_b$ = increase of plate current,
    $\triangle E_c$ = change of grid bias in the positive direction,
and g_m = mutual conductance of valve at the operating plate current.

In most practical cases, however, the valve is being operated with an impedance in the plate circuit and in some cases also in the screen circuit.   The effect of a change in grid bias is therefore treated separately for each practical case.

(a) On resistance-coupled triodes

In this case a plate load resistor is used, resulting in a considerable voltage drop and a decrease in the effective slope of the valve.

The change in plate voltage brought about by a change in grid bias is given by the expression

$$\triangle I_b = \triangle E_c \times \mu/(r_p + R_L) \tag{20}$$

where μ = amplification factor of valve at the operating point,
    r_p = plate resistance of valve at the operating point,
and R_L = resistance of plate load resistor.

(b) On resistance-coupled pentodes

The change of plate current with grid bias is given by the expression.

$$\triangle I_b = \triangle E_g \times g_d \tag{21}$$

where g_d = dynamic transconductance at the operating point,
    = slope of dynamic characteristic at the operating point.

The change of screen current (with fairly low screen voltages) is approximately proportional to the plate current up to plate currents of $0.6\ E_{bb}/R_L$ and the change in screen current is given by the expression

$$\triangle I_{c2} = \triangle I_b\ (I_{c2}/I_b)\ \text{approx.} \tag{22}$$

where I_{c2} = screen current
and E_{bb} = plate supply voltage.

For further information on resistance coupled valves, see Chapter 12, Sects. 2 and 3.

(c) On i-f or r-f amplifier

In this case there is no d.c. load resistor and the full supply voltage reaches the plate of the valve.   The change of plate current is given by eqn. (19) while the change in screen current may be calculated from the ratio of screen and plate currents, which remains approximately constant.   The voltage gain is proportional to the mutual conductance* of the valve, and is therefore a maximum for the highest plate current at the minimum bias.   A decrease in bias will therefore normally result in increased gain, while increased bias will result in decreased gain.   The limit to increased gain is set by the plate or screen dissipation of the valve, by positive grid current, and, in some circuits, by instability. In most cases the mutual conductance curves are published so as to enable the change of gain to be calculated.

(d) On power valves

This subject is covered in detail in Chapter 13.

---

*The voltage gain is also affected by the plate resistance, but this is quite a secondary effect unless the plate resistance is less than 0.5 megohm.   In most remote cut-off pentodes the plate resistance falls rapidly as the bias is decreased towards the minimum bias, but this is more than counterbalanced by the rise in mutual conductance.

SECTION 7 : VALVE EQUIVALENT CIRCUITS AND VECTORS

(i) Constant voltage equivalent circuit (ii) Constant current equivalent circuit (iii) Valve vectors.

Much useful information can be derived from an equivalent circuit of a valve, even though this may only be valid under limited conditions. The equivalent circuit is only a convenient fiction, and it must be remembered that it is the plate supply which, in reality, supplies the power—the valve merely controls the current by its varying d.c. plate resistance. The equivalent circuit is merely a device to produce in the load the same a.c. currents and voltages which are produced by the valve when alternating voltages are applied to its grid.

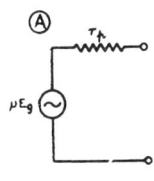
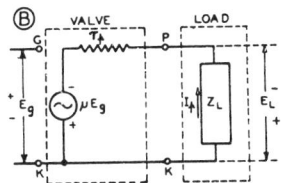

Fig. 2.37.   (A) Equivalent circuit of valve using constant voltage generator (B) Equivalent circuit of valve and load.

FIG. 2·37

(i) Constant voltage equivalent circuit
The simplest equivalent circuit treats the valve as an a.c. generator of constant r.m.s. voltage μE_g, which is applied through an internal generator resistance r_p (Fig. 2.37A). This is valid for small alternating voltages (under which conditions the characteristics are practically uniform) but is of no assistance in determining direct currents or voltages, phase angles or operating conditions. It is also limited to frequencies at which the effects of capacitances are negligible.

This may be elaborated, as in Fig. 2.37B, with the inclusion of the input circuit GK and the load Z_L. The input voltage E_g is shown by the $\pm$ signs to be such that the grid is instantaneously positive, and the plate negative (with respect to the cathode) at the same instant. It is assumed that the grid is biased sufficiently to prevent grid current flow.

The current I_p flowing through the load Z_L produces across the load a voltage E_L which is of opposite sign to E_c. It will be noted that the " fictitious " voltage μE_g is opposite in sign to E_g, although μ is positive ; this apparent inversion is a consequence of treating the valve as an a.c. generator.

In the simplest case, Z_L is a resistance R_L. We can then derive the following relationships—

$$\mu E_g \doteq (r_p + R_L) I_p \tag{1}$$

$$E_L = -I_p R_L = -\mu E_g \frac{R_L}{r_p + R_L} \tag{2}$$

and
$$\frac{E_L}{E_g} = \text{voltage gain} = -\frac{\mu R_L}{r_p + R_L} \tag{3}$$

If the load is made up of a resistor R_L and an inductor X_L in series—

Complex Values	Scalar Values	
$Z_L = R_L + jX_L$	$\sqrt{R_L{}^2 + X_L{}^2}$	(4)
$\mu E_g = (r_p + R_L + jX_L) I_p$	$\sqrt{(r_p + R_L)^2 + X_L{}^2}.I_p$	(5)
$\dfrac{E_L}{E_g} = -\dfrac{\mu(R_L + jX_L)}{r_p + R_L + jX_L}$	$= -\dfrac{\mu\sqrt{R_L{}^2 + X_L{}^2}}{\sqrt{(r_1 + R_L)^2 + X_L{}^2}}$	(6)

and similarly for any other type of load.

The **interelectrode capacitances** are shown in the equivalent circuit of Fig. 2.38, and may be taken as including the stray circuit capacitances.   This circuit may be applied at frequencies up to nearly 10 Mc/s, beyond which the inductances of the leads and electrodes become appreciable.   It may also be applied to a screen grid (tetrode) or pentode, provided that the screen is completely by-passed to the cathode ; in this case C_{gk} becomes the input capacitance $(C_{g1.k} + C_{g1.g2})$ and C_{pk} becomes the output capacitance (C_p to all other electrodes).

(ii) Constant current equivalent circuit

An alternative form of representation is the constant current generator equivalent circuit (Fig. 2.39), this being more generally convenient for pentodes, in which the plate resistance is very high.   Either circuit is equally valid for both triodes and pentodes.

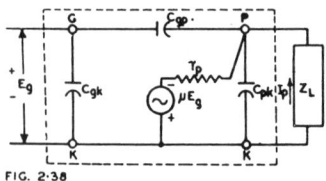

FIG. 2·38

Fig. 2.38.   Equivalent circuit of valve on load, with interelectrode capacitances.

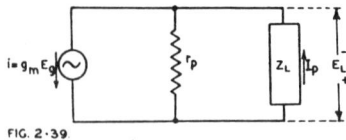

FIG. 2·39

Fig. 2.39.   Equivalent circuit of valve on load, using constant current generator.

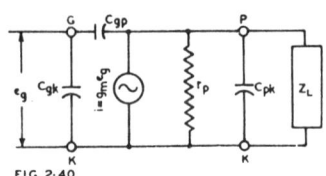

FIG. 2·40

Fig. 2.40.   Equivalent circuit of valve on load, using constant current generator, with interelectrode capacitances.

In the constant voltage generator equivalent circuit, the current varies with load impedance and plate resistance ; in the constant current equivalent circuit, the voltage across the load and plate resistance varies with load impedance and plate resistance.

A constant current generator equivalent circuit, in which account is taken of capacitances, is shown in Fig. 2.40.   This circuit may be applied at frequencies up to nearly 10 Mc/s, beyond which the inductances of the leads become appreciable. It will be seen that C_{pk} (which may be taken to include all capacitances from plate to cathode, and the output capacitance of a pentode) is shunted across both r_p and Z_L. In the case of a resistance-capacitance coupled stage, Z_L would be the resultant of R_L and R_g (following grid resistor) in parallel.

Maximum power output is obtained when the valve works into a load resistance equal to its plate resistance provided that the valve is linear and completely distortion-less over the whole range of its working, and also that it is unlimited by maximum electrode dissipations or grid current.   In practice, of course, these conditions do not hold and the load resistance is made greater than the plate resistance.

At frequencies of 10 Mc/s and above, the effects of the inductance of connecting leads (both internal and external to the valve) become appreciable.   Although it is possible to draw an equivalent circuit for frequencies up to 100 Mc/s, in which each capacitance is split into an electrode part and a circuit part (Ref. B21 Fig. 38) the circuit is too complicated for analysis, and the new circuit elements that have been introduced cannot be measured directly from the external terminals alone.

At frequencies above about 50 Mc/s, transit time effects also become appreciable. The circuit which is commonly used for frequencies above 50 Mc/s is Fig. 2.47 in which the valve is treated as a four terminal network with two input and two output terminals.   This is described in Sect. 8(iii)e.

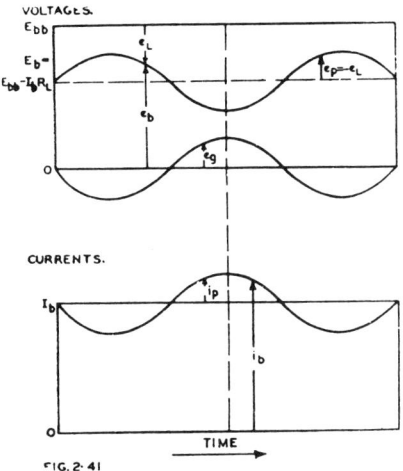

Fig. 2.41.   Voltage and current relationships in a resistance-loaded valve.

(iii) Valve vectors

Vectors [see Chapter 6 Sect. 5(iv)] may be used to illustrate the voltage and current relationships in a valve, but great care must be taken on account of the special conditions.   Vectors are normally restricted to the representation of the a.c. voltages and currents when the grid is excited with a sine-wave voltage limited to such a value that the operation is linear.   The grid and output voltages (with respect to the cathode) are normally of opposite polarity when the load is resistive ; under the conditions noted above this is almost the same as being 180° out of phase except there is no half-cycle time lag between them.

The voltage and current relationships for a resistance loaded valve are shown in Fig. 2.41 ; peak total plate current occurs with peak positive grid voltage and results in maximum voltage across the load (e_L) and minimum voltage from plate to cathode (e_b).   It will be seen that $e_b + e_L = E_{bb}$ (the supply voltage) under all conditions, and that e_L is naturally measured in the downward direction from E_{bb}.   If only alternating components are considered, a negative peak e_g corresponds to a positive peak e_p and a negative peak e_L.   If the supply voltage E_{bb} is omitted from the equivalent circuit, we are left with $e_p = -e_L$.

Each case must be considered individually and the vectors drawn to accord with the conditions.   The only general rule is that E_g and μE_g are always either in phase or of opposite phase.

Fig. 2.42 shows the vector diagram (drawn with respect to the cathode) of an amplifying valve with a resistance load and a.c. grid excitation.   Commencing with the grid-to-cathode voltage E_g, the vector μE_g is drawn in the same direction but is μ times as large.   The output voltage E_L is also in the same direction as μE_g, but smaller by the value $I_p r_p$.   All of these voltages are with respect to the cathode and the centre-point of the vector diagram has accordingly been marked K.   The a.c. component of the plate current (I_p) is in phase with E_L, since E_L is the voltage drop which it produces in R_L.   The grid-to-plate voltage E_c is the sum of E_L and E_g owing to the phase reversal between grid and plate.

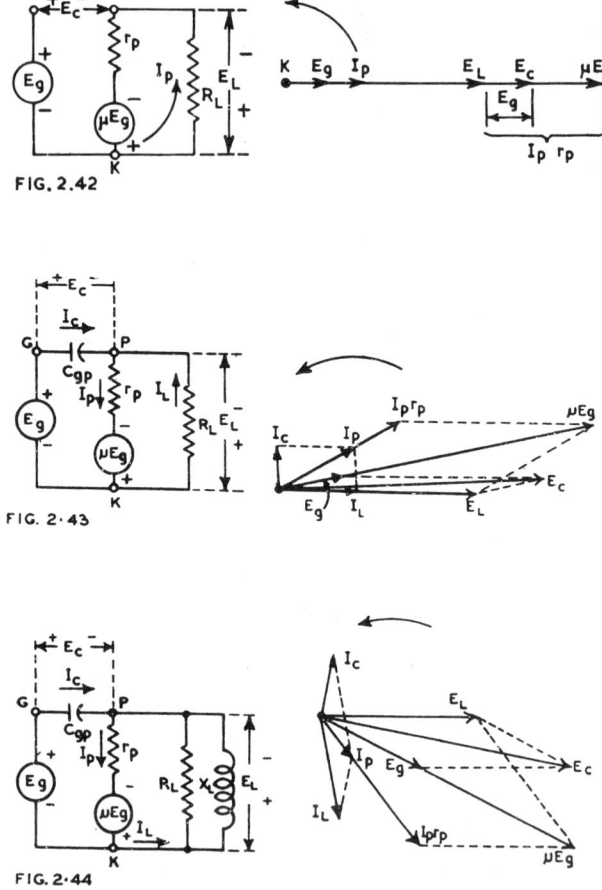

FIG. 2.42

FIG. 2·43

FIG. 2·44

Fig. 2.42. *Equivalent circuit and vector diagram of resistance-loaded valve.*
Fig. 2.43. *Vector diagram of valve with resistance load and capacitance from grid to plate.*
Fig. 2.44. *Vector diagram of valve with partially inductive load and capacitance from grid to plate.*

When the equivalent circuit includes more than one mesh, it is usual to proceed around each mesh in turn, using some impedance, common to both, as the link between each pair of meshes. For example Fig. 2.43 shows a valve with a capacitor C_{gp} from grid to plate, and a resistive load. Firstly, set down E_L in any convenient direction (here taken horizontally to the right) and I_L in the same direction; then draw I_c leading by approximately 90° (actually I_c leads E_c by 90°) and complete the parellelogram to find the resultant current I_p; then draw $I_p r_p$ in the same direction as I_p and complete the parallelogram to find the resultant μE_g—this completes the first mesh. Finally take E_g along μE_g and complete the parallelogram to find the resultant of E_g and E_L, which will be E_c.

If the load is partially inductive (Fig. 2.44) the plate current and $I_p r_p$ lag behind E_L and the resultant μE_g is determined by the parallelogram ; E_g and E_L combine to give the resultant E_c ; I_c leads E_c by 90°, and I_L is determined by completing the parallelogram of which I_c is one side and I_p the resultant.

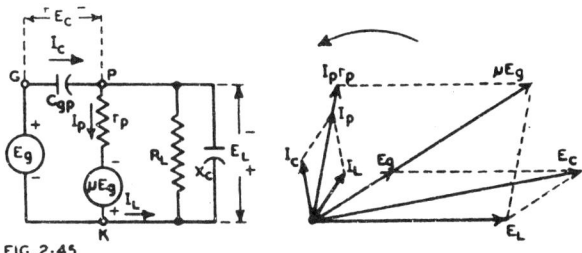

FIG. 2·45

Fig. 2.45. Vector diagram of valve with partially capacitive load and capacitance from grid to plate.

With a partially capacitive load (Fig. 2.45) the plate current and $I_p r_p$ lead E_L, and the resultant μE_g is determined by the parallelogram ; E_g and E_L combine to give the resultant E_c ; I_c leads E_c by 90°, and I_L is determined by completing the parallelogram of which I_c is one side and I_p the resultant.

SECTION 8 : VALVE ADMITTANCES

(i) Grid input impedance and admittance (ii) Admittance coefficients (iii) The components of grid admittance—Input resistance—Input capacitance—Grid input admittance (a) with plate-grid capacitance coupling ; (b) with both plate-grid and grid-cathode capacitance coupling ; (c) with grid-screen capacitance coupling ; (d) with electron transit time ; (e) equivalent circuit based on admittances (iv) Typical values of short-circuit input conductance (v) Change of short-circuit-input capacitance with transconductance (vi) Grid-cathode capacitance (vii) Input capacitances of pentodes (published values) (viii) Grid-plate capacitance.

(i) **Grid input impedance and admittance**

When a valve is used at low audio frequencies, it is sometimes assumed that the grid input impedance is infinite. In most cases, however, this assumption leads to serious error, and careful attention is desirable to both its static and dynamic impedances.

As with any other impedance (see Chapter 4 Sect. 6) it may be divided into its various components :—

Component	Normal	Reciprocal
Resistive	Grid input resistance (r_g)	conductance (g_g)
Reactive	Grid input reactance (X_g)	susceptance (B_g)
Resultant	Grid input impedance (Z_g)	admittance (Y_g)

Normal values are measured in ohms, while reciprocal values are measured in reciprocal ohms (mhos). It is interesting to note that

a resistance of		a conductance of
a reactance of	is equivalent to	a susceptance of
an impedance of		an admittance of
1 megohm		1 micromho
0·1 megohm		10 micromhos
10 000 ohms		100 micromhos
1000 ohms		1000 micromhos = 1 mA/volt

The following relationships hold :

$$g_g = \frac{r_g}{r_g{}^2 + X_g{}^2}, \qquad r_g = \frac{g_g}{g_g{}^2 + B_g{}^2} \qquad (1)$$

$$B_g = \frac{X_g}{r_g{}^2 + X_g{}^2}, \qquad X_g = \frac{B_g}{g_g{}^2 + B_g{}^2} \qquad (2)$$

$$|Y_g| = \sqrt{g_g{}^2 + B_g{}^2} = 1/|Z_g|, \qquad Y_g = g_g + jB_g = 1/Z_g \qquad (3)$$

Similar relationships hold for other electrodes.

It is usual to carry out calculations with admittances, even though the resultant may then have to be changed to the form of an impedance. With a number of conductances (or susceptances) in parallel, the total conductance (or susceptance) is found by adding all together, with due regard to positive and negative quantities :—

$$\text{e.g. } g_g = g_1 + g_2 + g_3 + \ldots + g_k \qquad (4)$$
$$B_g = B_1 + B_2 + B_3 + \ldots + B_k \qquad (5)$$

Inductive reactance is regarded as positive.
Capacitive reactance is regarded as negative.
Inductive susceptance is regarded as positive.
Capacitive susceptance is regarded as negative.

With Complex Notation (see Chapter 6 Sect. 6) we have

$$Z_g = R_g + jX_g, \qquad Y_g = g_g - jB_g. \qquad (6)$$

(The " j " merely indicates a vector at 90° which must be added vectorially.)

(ii) Admittance coefficients

The operation of a valve may be expressed by the two equations

$$i_p = A e_g + B e_p \qquad (7)$$
$$i_g = C e_g + D e_p \qquad (8)$$

where A, B, C and D are complex values determined by the valve characteristics, being in the form of admittances and known as the Admittance Coefficients. The effect of these Admittance Coefficients may be understood more easily by considering two special cases, one with a short-circuited output (i.e. short-circuited from plate to cathode) and the other with a short-circuited input (i.e. short-circuited from grid to cathode).

Case 1 : Short-circuited output $(e_p = 0)$.
From equation (7), $i_p = A e_g$
From equation (8), $i_g = C e_g$
where   A is defined as the short-circuit forward admittance,
and   C is defined as the short-circuit input admittance.

Case 2 : Short-circuited input $(e_g = 0)$.
From equation (7), $i_p = B e_p$
From equation (8), $i_g = D e_p$
where B is defined as the short-circuit output admittance,
and   D is defined as the short-circuit feedback admittance.

At frequencies up to about 10 Mc/s, the Admittance Coefficients are given approximately by :

Short-circuit forward admittance $(A) = g_m - j\omega C_{gp} \approx g_m$ $\qquad (9)$
Short-circuit output admittance $(B) = 1/r_p + j\omega(C_{pk} + C_{gp})$ $\qquad (10)$
Short-circuit input admittance $(C) = 1/r_g + j\omega(C_{gk} + C_{gp})$ $\qquad (11)$
Short-circuit feedback admittance $(D) = j\omega C_{gp}$ $\qquad (12)$

If the grid is negatively biased to prevent the flow of positive grid current, the grid resistance r_g becomes very high, and $1/r_g$ may be negligible in the expression for C.

At frequencies above 10 Mc/s the Admittance Coefficients are somewhat modified, the capacitances and admittances containing a term which is proportional to the square of the frequency.

The short-circuit forward admittance (A) is affected by the transit time of electrons and the inductance of the cathode lead, thus causing a phase shift between anode current and grid voltage. This is treated in detail in Chapter 23 Sect. 5.

The short-circuit output admittance (B) is affected by the reduction in r_p which occurs with increasing frequency due to the capacitances and inductances of the electrodes. The capacitance term is practically constant.

The short-circuit input admittance (C) is affected by the transit time of electrons, the inductances of the electrodes (particularly the cathode) and the capacitance between grid and cathode. The capacitance term is practically constant.

The short-circuit feedback admittance (D) remains purely reactive even at very high frequencies, although it changes from capacitive at low frequencies, through zero, to inductive at high frequencies. This can cause instability in certain circumstances.

(iii) The components of grid admittance

Input resistance may be due to several causes :
1. Leakage between the grid and other electrodes.
2. Negative grid current (caused by gas or grid emission).
3. Positive grid current (may be avoided by negative grid bias).
4. Coupling between the grid and any other electrode presenting an impedance to the input frequency (e.g. $C_{g p}$).
5. Transit time of the electrons between cathode and grid (at very high frequencies only).

Input capacitance (C_{in}) is dependent on several factors :
1. The static (cold) capacitance (C_i) from the grid to all other electrodes, except the plate.

$$\text{For a pentode,} \quad C_i = C_{g1 \cdot k} + C_{g1 \cdot g2} \tag{13}$$
$$\text{For a triode,} \quad C_i = C_{g k}. \tag{14}$$

2. The very slight increase in capacitance caused by thermal expansion of the cathode (0·1 to 0·6 $\mu\mu$F for the majority of r-f pentodes).
3. The increase in capacitance caused by the space charge and by conduction (0·5 to 2·4 $\mu\mu$F for r-f pentodes).
4. Coupling between the grid and any other electrode presenting an impedance to the input frequency ; this holds both with capacitive and inductive reactance (Miller Effect— see below).
5. Transit time of the electrons between cathode and grid (at very high frequencies only).

References to change of input capacitance : B13, B15, B16, B17, C1, C4, C5.

The measurement of interelectrode capacitances is covered in Chapter 3 Sect. 3(ii)g, together with some general comments and a list of references to their significance and measurement.

Grid Input Admittance
(a) With plate-grid capacitance coupling

In the circuit of Fig. 2.46A, in which $C_{g1 \cdot k}$ and $C_{g1 \cdot g2}$ are not considered, it may be shown* that

$$r_g = \frac{1}{g_g} = \frac{(g_p + G_L)^2 + (B_L + B_{g p})^2}{B_{g p}[g_m \cdot B_L + B_{g p}(g_p + G_L + g_m)]} \tag{15}$$

and

$$C_g = \frac{B_g}{\omega} = \frac{C_{g p}[(g_p + G_L + g_m)(g_p + G_L) + B_L(B_L + B_{g p})]}{(g_p + G_L)^2 + (B_L + B_{g p})^2} \tag{16}$$

where $g_p = 1/r_p$, $Y_L = G_L + jB_L = 1/Z_L$, $B_{g p} = 1/X_{C g p} = 1/2\pi f C_{g p}$.
As an approximation, if $g_p \ll G_L$, and $B_{g p} \ll B_L$,

$$r_g = \frac{1}{g_g} = -\frac{1}{g_m \omega^2 L_g C_{g p}} \quad \text{when the load is inductive} \tag{17}$$

$$= \frac{C_L}{g_m C_{g p}} \quad \text{when the load is capacitive} \tag{18}$$

*Sturley, K. R. " Radio Receiver Design, Part 1 " (Chapman & Hall, London, 1943) p. 37 et seq

$$C_g = C_{gp} \left[1 + \frac{g_m G_L}{G_L{}^2 + B_L{}^2} \right] \tag{19}$$

When B_L is infinite, i.e. when $Z_L = 0$, R_g is infinite and $C_g = C_{gp}$.

When the load is inductive, the input resistance is usually negative, thus tending to become regenerative, although for values of B_L between 0 and

$$-B_{gp}\,(g_p + G_L + g_m)/g_m,$$

the input resistance is positive.

When the load is capacitive, the input resistance is always positive, thus causing degeneration.

As an approximation, if $B_{gp} \ll g_m$ and $B_{gp} \ll (g_p + G_L)$, the positive and negative minimum values of r_g are given by

$$r_g \, (min) \approx \pm \frac{2(g_p + G_L)}{g_m B_{gp}} \tag{20}$$

and these occur at $B_L = \pm (g_p + G_L)$.

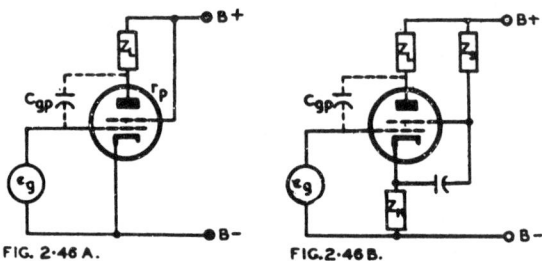

FIG. 2·46 A.   FIG. 2·46 B.

Fig. 2.46. Conditions for deriving input admittance (A) with plate-grid capacitance coupling (B) general case including cathode circuit impedance.

Similarly, the maximum value of the input capacitance is given by

$$C_g \, (max) = C_{gp} \left[1 + \frac{g_m r_p R_L}{r_p + R_L} \right] = C_{gp} \left[1 + \frac{\mu R_L}{r_p + R_L} \right] \tag{21}$$

which occurs at $B_L = - B_{gp}$. This is the well known " Miller Effect " [see Chapter 12 Sect. 2(xi) for a-f amplifiers]. The effect on the tuning of r-f amplifiers is treated in Chapter 23 Sect. 5, and on i-f amplifiers in Chapter 26 Sect. 7.

In the circuit of Fig. 2.46B, which includes an impedance Z_k in the cathode circuit, with the screen decoupled to the cathode, the input resistance is given approximately by

$$r_g = \frac{1}{g_g} \approx \frac{G_L{}^2 + B_L{}^2}{g_m B_{gp} B_L} \left[1 + \frac{g_m(g_m + 2G_k)}{G_k{}^2 + B_k{}^2} \right] \tag{22}$$

where B_{gp} and g_p are neglected in comparison with the other components, and $(B_L G_k - B_k G_L)$ is very small. Thus the reflected resistance is increased, and the damping decreased, as the result of the insertion of Z_k.

The input capacitance under these conditions is given by

$$C_g = C_{gp} \left[\frac{(G_L + g_m)G_L + B_L{}^2}{G_L{}^2 + B_L{}^2} - \frac{g_m[g_m{}^2 G_L + g_m(G_k G_L - B_k B_L)]}{(G_L{}^2 + B_L{}^2)[(G_k + g_m)^2 + B_k{}^2]} \right] \tag{23}$$

which is less than with $Z_k = 0$.

If the screen is by-passed to the cathode,

$$r_g = \frac{1}{g_g} = \frac{G_L{}^2 + B_L{}^2}{g_m B_{gp} B_L} \left[1 + \frac{g_i (g_i + 2G_k)}{G_k{}^2 + B_k{}^2} \right] \tag{24}$$

$$C_g = C_{gp} \left[\frac{(G_L + g_m) G_L + B_L{}^2}{G_L{}^2 + B_L{}^2} - \frac{g_m[g_i{}^2 G_L + g_i(G_k G_L - B_k B_L)]}{(G_L{}^2 + B_L{}^2)[(G_k + g_i)^2 + B_k{}^2]} \right] \tag{25}$$

where g_t = triode g_m (whole cathode current)

$$= g_m (I_p + I_{g2})/I_p.$$

(b) **With both plate-grid and grid-cathode capacitance coupling**

The circuit is as Fig. 2.46B with the addition of a capacitance C_{gk} between grid and cathode. The input resistance is given by

$$r_g = \frac{[(G_k + g_m)^2 + B_k{}^2][G_L{}^2 + B_L{}^2]}{g_m[B_{gp}B_L(G_k{}^2 + B_k{}^2) - B_{gk}B_k(G_L{}^2 + B_L{}^2)]} \tag{26}$$

This becomes infinite when $B_{gp}B_L(G_k{}^2 + B_k{}^2) = B_{gk}B_k(G_L{}^2 + B_L{}^2)$, which is the condition for input resistance neutralization (see Chapter 26 Sect. 8 for i-f amplifiers). This condition may be put into the form

$$\frac{B_{gp}}{B_{gk}} = \frac{C_{gp}}{C_{gk}} = \frac{L_k}{L_p} \tag{27}$$

Thus, by including an inductance $L_p = L_k C_{gk}/C_{gp}$ between the load and the plate, the input resistance may be increased to a very high value. The same effect may also be achieved by means of an inductance in the screen circuit.

The input capacitance under the conditions of Fig. 2.46B is given approximately by

$$C_g \approx C_{gp} + C_{gk} + g_m\left[\frac{C_{gp}\,G_L}{G_L{}^2 + B_L{}^2} - \frac{C_{gk}(G_k + g_m)}{(G_k + g_m)^2 + B_k{}^2}\right] \tag{28}$$

If $g_m \ll G_k$ it is possible to prevent change of input capacitance when g_m is varied (for example with a.v.c.), by making

$$\frac{C_{gp}}{C_{gk}} = \frac{R_k}{R_L}.$$

(c) **With grid-screen capacitance coupling**

Conditions as in Fig. 2.46B, but with screen by-pass capacitor.

$$r_g = \frac{(g_{g2} + G_s)^2 + (B_s + B_{g1 \cdot g2})^2}{B_{g1 \cdot g2}[g_{g1 \cdot g2}B_sB_{g1 \cdot g2}(g_{g2} + G_s + g_{g1 \cdot g2})]} \tag{29}$$

$$C_g = C_{g1 \cdot g2}\left[\frac{(g_{g2} + G_s + g_{g1 \cdot g2})(g_{g2} + G_s) + B_s(B_s + B_{g1 \cdot g2})}{(g_{g2} + G_s)^2 + (B_s + B_{g1 \cdot g2})^2}\right] \tag{30}$$

where g_{g2} = screen conductance,

$g_{g1 \cdot g2}$ = grid-screen transconductance,

and $B_{g1 \cdot g2}$ = susceptance due to capacitance from $g1$ to $g2$.

In an r-f amplifier, $B_{g1 \cdot g2}$ and $(g_{g2} + G_s)$ may usually be neglected in comparison with B_s, and thus

$$r_g \approx -\frac{1}{g_{g1 \cdot g2}\omega^2 L_s C_{g1 \cdot g2}} \quad \text{when } B_s \text{ is inductive} \tag{31}$$

$$r_g \approx \frac{C_1}{g_{g1 \cdot g2}\,C_{g1 \cdot g2}} \quad \text{when } B_s \text{ is capacitive} \tag{32}$$

The input resistance may be made infinite by making

$$C_{gi \cdot g2}/C_{gk} = L_k/L_s \tag{33}$$

(d) **With electron transit time**

This subject is treated fully in Chapter 23 Sect. 5.

(e) **Equivalent circuit based on admittances**

In determining valve admittances at frequencies higher than approximately 10 Mc/s, it is not practicable to introduce voltages or measure them directly at the electrodes of a valve. The lead inductances and interelectrode capacitances form a network too complex for exact analysis. The most practical method of avoiding such difficulties is to consider the valve, the socket, and the associated by-pass or filter circuits as a unit, and to select a pair of accessible input terminals and a pair of accessible output terminals as points of reference for measurements. When such a unit is considered as a linear amplifier, it is possible to calculate performance in terms of four admittance coefficients. These are :

Y_{in} = short-circuit input admittance
     = admittance measured between input terminals when the output terminals are short-circuited for the signal frequency.

Y_{for} = short-circuit forward admittance
     = value of current at output terminals divided by the voltage between the input terminals, when the output terminals are short-circuited for the signal frequency.

Y_{out} = short-circuit output admittance
     = admittance measured between output terminals when the input terminals are short-circuited for the signal frequency.

Y_{fb} = short-circuit feedback admittance
     = value of current at the input terminals divided by the voltage between the output terminals, when the input terminals are short-circuited for the signal frequency.

Each of these admittances can be considered as the sum of a real conductance component and an imaginary susceptance component. In the cases of the input and output admittances, the susceptance components are nearly always positive (unless the valve is used above its resonant frequency) and it is, therefore, common practice to present the susceptance data in terms of equivalent capacitance values. The short-circuit input capacitance is the value of the short-circuit input susceptance divided by 2π times the frequency. The capacitance values are more convenient to work with than the susceptance values because they vary less rapidly with frequency and because they are directly additive to the capacitances used in the circuits ordinarily connected to the input and output terminals. However, when frequencies higher than 200 Mc/s and resonant lines used as tuning elements are involved, the use of susceptance values may be preferable.

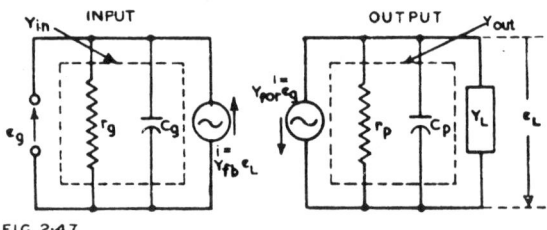

FIG. 2-47

Fig. 2.47. Alternative form of equivalent circuit for deriving input admittance

In Fig. 2.47 the short-circuit input admittance is represented by a resistor r_g and a capacitor C_g in parallel across the input terminals. The value of r_g is equal to the reciprocal of the short-circuit input conductance and the value of C_g is equal to the short-circuit input capacitance. The short-circuit output admittance is represented by a similar combination of r_p and C_p across the output terminals.

Since the input and output circuits are separated, allowance may be made for their interaction by an additional constant current generator in each. A constant current generator is shown at the output terminals producing a current equal to the product of the short-circuit forward admittance and the input voltage. A similar generator is shown at the input terminals producing a current equal to the product of the short-circuit feedback admittance and the output voltage.

The principal differences in the performance of receiving valves at high and low frequencies can be attributed to the variations of the short-circuit input conductance with frequency. The other short-circuit admittance coefficients, however, contribute to the input admittance actually observed in an operating circuit as follows :

$$\text{Voltage gain } (A) = \frac{Y_{for}}{Y_{out} + Y_L} \approx \frac{g_m}{Y_{out} + Y_L} \tag{34}$$

Added current at input terminals due to presence of load $= e_g A Y_{fb}$. (35)

Phase angle of added component = phase angle of voltage gain + phase angle
of feedback admittance.   (36)

Grid input admittance $(Y_g) = Y_{in} + AY_{fb}$

$$= Y_{in} + \frac{Y_{for} Y_{fb}}{Y_{out} + Y_L} \qquad (37)$$

See also Ref. B21.

The measurement of the four short-circuit admittances is covered in Chapter 3
Sect. 3(vi) A, B, C and D and also Refs. B17, B21.

(iv) Typical values of short-circuit input conductance

Pentodes tested under typical operating conditions (Ref. B17).

Type	Input conductance approx. (micromhos)						Mutual Conductance μmhos
	$f = 50$	60	80	100	120	150 Mc/s	
6AB7	200	310	600	980			5000
6AC7	380	600	1200	1970			9000
6AG5	100	145	280	326	480		5000
6AK5	40	57	92	134	185		5100
6AU6	180	280	490	759	1100		5200
6BA6	150	230	410	603	950		4400
6CB6	125	170	300	460	(Ref. B20)		6200
6BJ6				275	(Ref. B19)		3800
6SG7	190	270	430	604	670		4700
6SH7	200	300	470	632	880		4900
6SJ7		260	380	528			1650
6SK7	138	190	320	503	660		2000
9001			44	60	96	141	1400
9003			48	66	100	145	1800
Z77	110 at 45 Mc/s (Data from M.O.V.)						7500

(v) Change of short-circuit input capacitance with transconductance

($f = 100$ Mc/s). Ref. B17 unless otherwise indicated.

Increase in capacitance ($\mu\mu F$) from cut-off

Type	to $g_m = 1000$	2000	4000	Typical operation	
				$\mu\mu F$	μmhos
6AB7	0·55	1·0	1·7	1·8	5000
6AC7	0·65	1·2	1·8	2·4	9000
6AG5	0·5	0·8	1·25	1·4	5000
6AK5	0·3	0·6	1·0	1·1	5100
6AU6	0·6	1·1	2·0	2·5	5200
6BA6	0·75	1·4	2·2	2·2	4400
6BH6 (Ref. B18)				1·8	4600
6CB6 (Ref. B20)				1·54	6200
6BJ6 (Ref. B19)				1·6	3800
6SG7	0.8	1·5	2·2	2·3	4700
6SH7	0·75	1·3	2·05	2·3	4900
6SJ7	0·8	—	—	1·0	1650
6SK7	0·65	1·18	—	1·2	2000
9001	0·35	—	—	0·5	1400
9003	0·39	—	—	0·5	1800
Z77 (M.O.V.)				2·2	7500
Limits	0·3-0·8	0·6-1·5	1·0-2·2	0·43-2·38	

Value of unbypassed cathode resistor needed for complete compensation of input capacitance change with bias change (Ref. B18)

Valve type	Interelectrode capacitances			Unbypassed cathode resistor	Gain factor*
	C_{in}	C_{out}	C_{gp}		
6BA6	5·5 $\mu\mu$F	5·0 $\mu\mu$F	0·0035 $\mu\mu$F	100 ohms	0·62
6AU6	5·5	5·0	0·0035	85	0·61
6AG5	6·5	1·8	0·025	50	0·75
6AK5	4·0	2·8	0·02	50	0·75
6BJ6	4·5	5·5	0·0035	135	0·59
6BH6	5·4	4·4	0·0035	110	0·59
Z77	7·4	3·1	0·009	60	0·64

*degeneration due to unbypassed cathode resistor (see below).

$$\text{Gain factor} = \frac{\text{gain with cathode unbypassed}}{\text{gain with cathode by-passed}} \tag{38}$$

$$= \frac{1}{1 + R_k g_m (I_b + I_{c2})/I_b} \tag{39}$$

where R_k = cathode resistor for complete compensation of input capacitance change with bias

$$\approx \frac{\triangle C}{C_{gk} g_m (I_b + I_{c2})/I_b}$$

$\triangle C$ = change in input capacitance in farads from normal operating condition to cut-off,

C_{gk} = grid-to-cathode capacitance in farads measured with valve cold,

g_m = mutual conductance in mhos at normal operating condition,

I_b = direct plate current in amperes

and I_{c2} = direct screen current in amperes.

(vi) Grid-cathode capacitance

The mathematical treatment of the effects of grid-cathode capacitance has been given above. Methods of neutralization are described in Chapter 26 Sect. 8.

The published grid-plate capacitances are usually in the form of a maximum value, without any indication of the minimum or average value. In some cases the average is fairly close to the maximum, while in others it may be considerably less. The average value is likely to vary from one batch to another, and from one manufacturer to another. Equipment should be designed to avoid instability with the maximum value, although fixed neutralization should be adjusted on an average value, determined by a test on a representative quantity of valves.

Effect of electrode voltages on grid-cathode capacitance—see Ref. B23.

(vii) Input capacitances of pentodes (published values)

1. Indirectly heated	$\mu\mu$F
High slope r-f (metal)	8 to 11
Ordinary metal r-f All glass and miniature r-f }	4·3 to 7
Small power amplifiers	5 to 6
Ordinary power amplifiers	6·5 to 10
Large power amplifiers	10 to 15
Pentode section of diode-pentodes :	
Metal	5·5 to 6·5
Glass	3 to 5·5
2. Directly heated	
2 volt r-f pentodes	5 to 6
1·4 volt r-f pentodes	2·2 to 3·6
2 volt power pentodes	8
1·4 volt power pentodes	4·5 to 5·5

(viii) Grid-plate capacitance

The grid-plate capacitance decreases with increasing plate current. Eventually the rate of change becomes very small and even tends to become positive. The total change in triodes does not usually exceed 0.06 $\mu\mu F$ for high-mu types, or 0.13 $\mu\mu F$ for low-mu voltage amplifiers, although it may exceed 2 $\mu\mu F$ in the case of triode power amplifiers (Ref. B13).

SECTION 9 : MATHEMATICAL RELATIONSHIPS

(i) General (ii) Resistance load (iii) Power and efficiency (iv) Series expansion ; resistance load (v) Series expansion ; general case (vi) The equivalent plate circuit theorem (vii) Dynamic load line—general case (viii) Valve networks—general case (ix) Valve coefficients as partial differentials (x) Valve characteristics at low plate currents.

(i) General

Valve characteristics may be represented mathematically as well as graphically (see Chapter 6 for mathematical theory).

The plate (or space) current is a function (F) of the plate and grid voltages and may be expressed exactly as

$$i_b = F(e_b + \mu e_c + e_1) \qquad (1)$$

where e_1 is the equivalent voltage which would produce the same effect on the plate current as the combined effects of the initial electron velocity of emission together with the contact potentials. The amplification factor μ is not necessarily constant. There will be a small current flow due to e_1 when e_b and e_c are both zero.

As an approximation, when e_b and μe_c are large, e_1 may be neglected. The function in eqn. (1) may also be expressed approximately in the form

$$i_b \approx K(e_b + \mu e_c)^n \qquad (2)$$

in which K is a constant. The value of n varies from about 1.5 to 2.5 over the usual operating range of electrode voltages, but is often assumed to be 1.5 (e.g. Conversion Factors) over the region of nearly-straight characteristics, and 2.0 in the region of the bottom bend (e.g. detection). We may take the total differential* of eqn. (2),

$$di_b = \frac{\partial i_b}{\partial e_b} de_b + \frac{\partial i_b}{\partial e_c} de_c \qquad (3)$$

which expresses the change in i_b which occurs when e_b and e_c change simultaneously. Now—see (ix) below—provided that the valve is being operated entirely in the region in which μ, g_m and r_p are constant,

$$\frac{\partial i_b}{\partial e_b} = \frac{1}{r_p} \text{ and } \frac{\partial i_b}{\partial e_c} = g_m,$$

so that

$$di_b = \frac{1}{r_p} de_b + g_m de_c. \qquad (4)$$

If i_b is held constant,

$$\frac{1}{r_p} de_b + g_m de_c = 0,$$

and thus

$$g_m r_p = -\frac{de_b}{de_c} \quad (i_b \text{ constant}) \qquad (5)$$

whence $g_m r_p = \mu$ [see (ix) below]. $\qquad (6)$

The treatment so far has been on the basis of the total instantaneous voltages and currents, e_b, e_c, i_b ; it is now necessary to distinguish more precisely between the steady (d.c.) and varying (signal) voltages and currents.

---

*For total differentiation see Chapter 6 Sect. 7(ii).

R.D.H.—3

Instantaneous total values e_b e_c i_b
Steady (d.c.) values E_{bo} E_{cc} I_{bo}
Instantaneous varying values e_p e_g i_p
Varying (a.c.) values (r.m.s.) E_p E_g I_p
Supply voltages E_{bb} E_{cc}

For definitions of symbols refer to the list in Chapter 38 Sect. 6.

In normal operation each of the voltages and currents is made up of a steady and a varying component:

$$\left. \begin{array}{l} i_b = I_{bo} + i_p \\ e_b = E_{bo} + e_p \\ e_c = E_{cc} + e_g \end{array} \right\} \quad (7)$$

Eqn. (4) may therefore be extended in the form

$$d(I_{bo} + i_p) = \frac{1}{r_p} d(E_{bo} + e_p) + g_m d(E_{cc} + e_g).$$

But the differentials of constants are zero, and the relation between the varying components may be expressed in the form

$$i_p = \frac{e_p}{r_p} + g_m e_g \quad (8)$$

or

$$i_p = \frac{e_p + \mu e_g}{r_p}. \quad (9)$$

This only holds under the condition that μ, g_m and r_p are constant over the operating region.

(ii) Resistance load

If there is a resistance load (R_L) in the plate circuit,

$$e_b + e_L' = E_{bb} \quad (10)$$

where e_L' is the instantaneous total voltage across R_L.
Breaking down into steady and varying components,

$$(E_b + e_p) + (E_L + e_L) = E_{bb} \quad (11)$$

where e_L is the instantaneous varying voltage across R_L, and E_L is the steady (d.c.) voltage across R_L.

Under **steady conditions** $e_p = e_L = 0$, and therefore

$$E_b + E_L = E_{bb}.$$

Now, by Ohm's Law, $E_L = I_b R_L$

Therefore

$$I_b = \frac{E_{bb} - E_b}{R_L}. \quad (12)$$

Eqn. (12) represents a straight line on the plate characteristics, passing through the points

$$E_b = E_{bb}, I_b = 0 \text{ and } E_b = 0, I_b = E_{bb}/R_L,$$

in other words, **the loadline.**

The quiescent operating point must satisfy both the equation for the valve characteristics (1) and that for the loadline (12), therefore

$$F(e_b + \mu e_c + e_1) = \frac{E_{bb} - E_b}{R_L}. \quad (13)$$

Under **varying conditions**, neglecting steady components, we may derive from equation (11) the relation

$$e_p + e_L = 0$$

i.e.

$$e_p = -e_L. \quad (14)$$

Also $e_L = i_p R_L$ by Ohm's Law,
Therefore

$$e_p = -i_p R_L. \quad (15)$$

Substituting this value of e_p in equation (9) we obtain

$$i_p^- = \frac{-i_p R_L + \mu e_g}{r_p}$$

i.e.

$$i_p = \frac{\mu e_g}{r_p + R_L} \qquad (16)$$

which is a fundamentally important relation but which holds only in the region where μ, g_m and r_p are constant.

Substituting $- e_L$ (eqn. 14) in place of e_p in equation (9) we obtain, **for the valve alone,**

$$i_p^{\blacktriangledown} = \frac{-e_L + \mu e_g}{r_p}$$

i.e.

$$e_L = \mu e_g - i_p r_p \qquad (17)$$

which is **the basis of the constant voltage generator equivalent circuit** as in Sect. 7(i).

Eqn. (17) may be put into the form

$$e_L = r_p (g_m e_g - i_p). \qquad (18)$$

This voltage e_L across the valve and the load can be developed by means of a current $g_m e_g$, passed through r_p in the opposite direction to i_p, so that the total current through r_p is $(g_m e_g - i_p)$. Eqn. (18) is **the basis of the constant current equivalent circuit** as in Sect. 7(ii).

The **voltage gain** (A) of an amplifying stage with a load resistance R_L is

$$A = \left| \frac{e_L}{e_g} \right| = \left| \frac{i_p R_L}{e_g} \right| = \left| \frac{\mu R_L}{r_p + R_L} \right|. \qquad (19)$$

When the load is an impedance Z_L, the voltage gain may be shown to be

$$A = \left| \frac{\mu Z_L}{r_p + Z_L} \right| = \left| \frac{\mu}{1 + r_p/Z_L} \right| \qquad (20)$$

where r_p and Z are complex values (see Chapter 6).

If $Z_L = R_L + jX_L$, the scalar value of A is given by

$$A = \mu \frac{\sqrt{R_L^2 + X_L^2}}{\sqrt{(r_p + R_L)^2 + X_L^2}} \qquad (21)$$

The voltage gain may also be put into the alternative form

$$A = \left| g_m \frac{r_p Z_L}{r_p + Z_L} \right| \qquad (22)$$

$$\approx |g_m Z_L| \text{ if } r_p \gg Z_L. \qquad (23)$$

(iii) Power and efficiency

When the operation of a valve as a Class A_1 amplifier is perfectly linear we may derive* the following :—

Zero-Signal :

Plate current	$= I_{bo}$.	
Power input from plate supply	$P_{bb} = E_{bb} I_{bo}$.	
D.C. power absorbed in load	$P_{dc} = I_{bo}^2 R_L = E_{Lc} I_{bo}$.	
Quiescent plate dissipation	$P_{po} = E_{bo} I_{bo}$.	
But	$E_{bb} = E_{bo} + E_{Lo}$.	
Therefore	$P_{bb} = E_{bo} I_{bo} + E_{Lo} I_{bo}$.	
	$= P_{po} + P_{dc}$.	(24)

Signal Condition :

Average value of total input $= P_{bb} = E_{bb} I_{bo}$ (25)

which is constant irrespective of the signal voltage.

*After book by M.I.T. Staff " Applied Electronics " (John Wiley & Sons Inc. New York, 1943) pp. 419-425.

Power absorbed by load $= P_L$

$$P_L = \frac{1}{2\pi} \int_0^{2\pi} i_b^2 R_L \, d(\omega t) \tag{26}$$

$$= \frac{1}{2\pi} \int_0^{2\pi} (I_{bo} + i_p)^2 R_L d(\omega t)$$

$$= \frac{1}{2\pi} \int_0^{2\pi} I_{bo}^2 R_L d(\omega t) + \frac{1}{2\pi} \int_0^{2\pi} 2I_{bo} i_p R_L d(\omega t) + \frac{1}{2\pi} \int_0^{2\pi} i_p^2 R_L d(\omega t) \tag{27}$$

$$= I_{bo}^2 R_L + 0 + P_{ac}$$

i.e. $P_L = P_{dc} + P_{ac} \tag{28}$

where $P_{ac} = \dfrac{1}{2\pi} \displaystyle\int_0^{2\pi} i_p^2 R_L d(\omega t) \tag{29}$

Plate dissipation $P_p = \dfrac{1}{2\pi} \displaystyle\int_0^{2\pi} e_b i_b d(\omega t) \tag{30}$

From eqns. (28) and (29) it will be seen that the power absorbed by the load increases when the signal voltage increases, but the power input remains steady ; the plate dissipation therefore decreases as the power output increases,

$$\begin{aligned}
\text{i.e. } P_p &= P_{bb} - P_L \tag{31}\\
\text{from (26), (28)} &= E_{bb}I_{bo} - E_{Lo}I_{bo} - P_{ac} \tag{32}\\
&= E_{bo}I_{bo} - P_{ac}\\
&= P_{po} - P_{ac} \tag{33}
\end{aligned}$$

where $P_{po} = E_{bo}I_{bo}$.

That is to say, the plate dissipation (P_p) is equal to the apparent d.c. power input to the valve (P_{po}) minus the a.c. power output.

The plate efficiency $\eta_p = \dfrac{\text{power output}}{\text{d.c. power input}}$

$$= \frac{P_{ac}}{E_{bo}I_{bo}}. \tag{34}$$

$$\left[\text{For non-linear operation} \qquad \eta_p = \frac{P_{ac}}{E_b I_b} \right] \tag{35}$$

With sinusoidal grid excitation, linear Class A_1 valve operation and resistive load,
$$P_{ac} = E_p I_p = I_p^2 R_L.$$
Applying equation (16),

$$P_{ac} = \frac{\mu^2 E_g^2 R_L}{(r_p + R_L)^2}. \tag{36}$$

Differentiating with respect to R_L and equating to zero in order to find the condition for maximum power output,

$$\frac{dP_{ac}}{dR_L} = \mu^2 E_g^2 \left[\frac{(r_p + R_L)^2 - 2R_L(r_p + R_L)}{(r_p + R_L)^4} \right] = 0$$

i.e. when $(r_p + R_L)^2 - 2R_L(r_p + R_L) = 0$
or when $R_L = r_p$ \hfill (37)
and the maximum power output is

$$P_{acm} = \frac{\mu^2 E_g{}^2}{4r_p} = \frac{E_g{}^2}{4} \mu g_m. \tag{38}$$

The factor μg_m is a figure of merit for power triodes.

If the load is an impedance $(Z_L = R_L + jX_L)$ the condition for maximum power output is when

$$R_L = \sqrt{r_p{}^2 + X_L{}^2}. \tag{39}$$

In the general case, with a resistive load, the power output is given by eqn. (36) which may be put into the form

$$P_{ac} = \frac{\mu^2 E_g{}^2}{r_p} \cdot \frac{1}{\frac{r_p}{R_L} + 2 + \frac{R_L}{r_p}}. \tag{40}$$

If $R_L/r_p = 2$, the loss of power below the maximum is only 11%, while if $R_L/r_p = 4$ the loss of power is 36%, so that " matching " of the load is not at all critical.

The treatment above is correct for both triodes and pentodes provided that both are operated completely within the linear region, that is with limited grid swing. A pentode is normally operated with a load resistance much less than the plate resistance on account of the flattening of the output voltage characteristic which would otherwise occur at low plate voltages.

This subject is considered further in Chapter 13, under practical instead of under ideal conditions.

(iv) Series expansion ; resistance load

Except in eqn. (1), which is perfectly general, certain assumptions have been made regarding linearity and the constancy of μ, g_m and r_p which restrict the use of the equations. If it is desired to consider the effects of non-linearity in causing distortion in amplifiers and in producing detection or demodulation, it is necessary to adopt a different approach.

The varying component of the plate current of a valve may be expressed in the form of a series expansion :

$$i_p = a_1 e + a_2 e^2 + a_3 e^3 + a_4 e^4 + \ldots \tag{41}$$

This form may be derived* from eqn. (1), and it may be shown that

$$a_1 = \frac{\mu}{r_p + R_L} \tag{42}$$

$$a_2 = -\frac{\mu^2 r_p}{2(r_p + R_L)^3} \cdot \frac{\partial r_p}{\partial e_b} \tag{43}$$

$$a_3 = \frac{\mu^3 r_p}{6(r_p + R_L)^5}\left[(2r_p - R_L)\left(\frac{\partial r_p}{\partial e_b}\right)^2 - r_p(r_p + R_L)\frac{\partial^2 r_p}{\partial e_b{}^2}\right] \tag{44}$$

If μ is assumed to be constant (this is only approximately true for triodes and not for pentodes)

$$e = e_g + \frac{v_p}{\mu}$$

where v_p = instantaneous value of plate excitation voltage. (Normally for an amplifier $v_p = 0$ and $e = e_g$).

The value of $\partial r_p/\partial e_b$ may be determined by plotting a curve of r_p versus e_b for the given operating bias, and drawing a tangent at the point of operating plate voltage. The value of $\partial^2 r_p/\partial e_b{}^2$ may be determined by plotting a curve of ∂r_p versus ∂e_b and treating in a similar manner.

The higher terms in the series expansion (41) diminish in value fairly rapidly, so that a reasonably high accuracy is obtained with three terms if the valve is being used as an amplifier under normal conditions with low distortion.

---

*Reich, H. J. " Theory and Applications of Electron Tubes " (2nd edit.). McGraw-Hill, New York and London, 1944), pp. 74-77.

The first term $a_1e = \mu e/(r_p + R_L)$ is similar to eqn. (16) above, which was regarded as approximately correct for small voltage inputs ; that is to say for negligible distortion.

The first and second terms

$$i_p = a_1e + a_2e^2$$

express the plate current of a " square law detector " which is closely approached by a triode operating as a grid or plate ("anode-bend") detector with limited excitation voltage.

The second and higher terms are associated with the production of components of alternating plate current having frequencies differing from that of the applied signal—i.e. harmonics and (if more than one signal frequency is applied) intermodulation frequencies.

For example, with a single frequency input,

$$e = E_m \sin \omega t \tag{45}$$

Therefore

$$e^2 = E_m^2 \sin^2 \omega t = \tfrac{1}{2}E_m^2 - \tfrac{1}{2}E_m^2 \cos 2\omega t \tag{46}$$

and

$$e^3 = \tfrac{3}{4}E_m^3 \sin \omega t - \tfrac{1}{4}E_m^3 \sin 3\omega t. \tag{47}$$

The second term (e^2) includes a d.c. component ($\tfrac{1}{2}E_m^2$) and a second harmonic component. The third term includes a fundamental frequency component ($\tfrac{3}{4}E_m^3 \sin \omega t$) and a third harmonic component.

If the input voltage contains two frequencies (f_1 and f_2) it may be shown that the second term of eqn. (41) produces
  a d.c. component
  a fundamental f_1 component
  a second harmonic of f_1
  a fundamental f_2 component
  a second harmonic of f_2
  a difference frequency component ($f_1 - f_2$)
  a sum frequency component ($f_1 + f_2$)

The third term of equation (41) produces
  a fundamental f_1 component
  a third harmonic of f_1
  a fundamental f_2 component
  a third harmonic of f_2
  a difference frequency component ($2f_1 - f_2$)
  a difference frequency component ($2f_2 - f_1$)
  a sum frequency component ($2f_1 + f_2$)
  a sum frequency component ($2f_2 + f_1$)

In the case of an A-M mixer valve, f_2 may be the signal frequency and f_1 the oscillator frequency. The normal i-f output frequency is ($f_1 - f_2$) while there are spurious output frequencies of ($f_1 + f_2$), ($2f_1 + f_2$), ($2f_2 + f_1$), ($2f_1 - f_2$) and ($2f_2 - f_1$). Even though no oscillator harmonics are injected into the mixer, components with frequencies ($2f_1 + f_2$) and ($2f_1 - f_2$) are present in the output, thus demonstrating mixing at a harmonic of the oscillator frequency.

If the input voltage contains more than two frequencies, or if the terms higher than the third are appreciable, there will be greater numbers of frequencies in the output. The effect of this on distortion is treated in Chapter 14.

It may be shown that the effect of the load resistance, particularly when it is greater than the plate resistance, is to decrease the ratio of the harmonics and of the intermodulation components to the fundamental. This confirms the graphical treatment in Chapter 12.

(v) Series expansion : general case

The more general case of a series expansion for an impedance load and variable μ has been developed by Llewellyn* and the most important results are given in most text books.†

(vi) The equivalent plate circuit theorem

It was shown above (eqn. 41) that the plate current may be expressed in the form of a series expansion. If the distortion is very low, as may be achieved with low input voltage and high load impedance, sufficient accuracy may be obtained by making use of only the first term in the equation, i.e.

$$i_p = \frac{\mu e}{r_p + Z_L} \tag{48}$$

where $e = e_g$ (for amplifier use)
and $Z_L =$ impedance of the plate load at the frequency of the applied voltage.
For amplifier use this may be put into the form

$$I_p = \frac{\mu E_g}{r_p + Z_L} \tag{49}$$

This is the same as eqn. (16), except that R_L has been replaced by Z_L.

This is the basis of the Equivalent Plate Circuit Theorem which states‡ 'that the a.c. components of the currents and voltages in the plate (load) circuit of a valve may be determined from an equivalent plate circuit in one of two forms—

(1) a fictitious constant-voltage generator (μE_g) in series with the plate resistance of the valve, or
(2) a fictitious constant-current generator ($I = g_m E_g$) in parallel with the plate resistance of the valve.

These are applied in Sect. 7 of this chapter.

If a distortionless Class A amplifier or its equivalent circuit is excited with an alternating grid voltage, the a.c. power in the load resistor R_L (i.e. the output power) is $I_p{}^2 R_L$.

The d.c. input from the plate supply to the valve and load (in the actual case) is $P_{bb} = I_{bo} E_{bb}$. Under ideal Class A_1 conditions the d.c. current I_{bo} remains constant, since the a.c. current is symmetrical and has no d.c. component.

Now the a.c. power input from the generator is

$$P_g = \mu E_g I_p \tag{50}$$

But

$$I_p = \frac{\mu E_g}{(r_p + R_L)}$$

Therefore

$$\mu E_g = r_p I_p + R_L I_p$$

and

$$P_g = \mu E_g I_p = r_p I_p{}^2 + R_L I_p{}^2. \tag{51}$$

In this equation

P_g = a.c. power input from generator
$r_p I_p{}^2$ = a.c. power heating plate
$R_L I_p{}^2$ = a.c. power output = P_{ac}.

The a.c. power P_g can only come from the d.c. power P_{po} dissipated in the valve, which decreases to the lower value P_p when the grid is excited.
The total plate dissipation (P_p) is therefore

$$P_p = P_{po} - P_g + r_p I_p{}^2 \tag{52}$$

where $P_{po} =$ d.c. plate dissipation.
This may be put into the form

$$P_p = P_{po} - (P_g - r_p I_p{}^2)$$
$$= P_{po} - P_{ac} \tag{53}$$

where

$$P_{ac} = R_L I_p{}^2 = (P_g - r_p I_p{}^2) = \text{a.c. power output.}$$

---

*Llewellyn, F. B. Bell System Technical Journal, 5 (1926) 433.
†such as Reich, H. J. " Theory and Application of Electron Tubes," p. 75.
‡for a completely general definition see Reich, H. J. (letter) " The equivalent plate circuit theorem," Proc. I.R.E. 33.2 (Feb., 1945) 136.

The statement may therefore be made, that the plate dissipation is equal to the d.c. plate dissipation minus the a.c. power output.

A more general statement covering all types of valve amplifiers and oscillators is that the plate input power is equal to the plate dissipation plus the power output.

This analysis, based on the equivalent plate circuit, reaches a conclusion in eqn. (46) which is identical with eqn. (33) derived from a direct mathematical approach. It is, however, helpful in clarifying the conditions of operation of a distortionless Class A_1 amplifier.

The preceding treatment only applies to amplifiers ($e = e_g$ in eqn. 41), but it may be extended to cover cases where the load impedance contains other e.m.f's, by using the principle of superposition—see Chapter 4 Sect. 7(viii).

It is possible to adopt a somewhat similar procedure to develop the **Equivalent Grid Circuit,** or that for any other electrode in a multi-electrode valve.

(vii) Dynamic load line—general case

If the a.c. plate current is sinusoidal,
$$i_p = I_{pm} \sin \omega t$$
and
$$e_p = -I_{pm}|Z_L| \sin (\omega t + \theta)$$
where
$$\theta = \tan^{-1} X_L/R_L.$$
From this it is possible to derive[*].
$$e_p{}^2 + 2e_p i_p R_L + i_p{}^2|Z_L{}^2| = I_{pm}{}^2 X_L{}^2 \tag{54}$$
which is the equation of an ellipse with its centre at the operating point, this being the dynamic path of operation.

(viii) Valve networks ; general case

The ordinary treatment of a valve and its circuit—the Equivalent Plate Circuit Theorem in particular—is a fairly satisfactory approximation for triodes or even pentodes up to frequencies at which transit-time effects become appreciable. If it is desired to calculate, to a higher degree of precision, the operation of a valve in a circuit, particularly at high frequencies, a very satisfactory approach is the preparation of an equivalent network which takes into account all the known characteristics. This method has been described[†] in considerable detail, and those who are interested are referred to the original article.

(ix) Valve coefficients as partial differentials

Valve coefficients, as well as other allied characteristics, may be expressed as partial differential coefficients—see Chapter 6 Sect. 7(ii).

Partial differential coefficients, designated in the form $\dfrac{\partial y}{\partial x}$ are used in considering the relationship between two of the variables in systems of three variables, when the third is held constant.

" $\dfrac{\partial y}{\partial x}$ " is equivalent to " $\dfrac{dy}{dx}$ (z constant) " when there are three variables, x, y and z. Partial differentials are therefore particularly valuable in representing valve coefficients.

Let e_p = a.c. component of plate voltage,
e_g = a.c. component of grid voltage,
and i_p = a.c. component of plate current.
(These may also be used with screen-grid or pentode valves provided that the screen voltage is maintained constant, and is completely by-passed for a.c.).

[*]Reich, H. J. " Theory and application of electron tubes " (2nd edit.), p. 99.
[†]Llewellyn, F. B. and L. C. Peterson " Vacuum tube networks," Proc. I.R.E. 32.3 (March, 1944), 144.

Then $\mu = -\dfrac{\partial e_p}{\partial e_g}$ $(i_p = \text{constant})$,   (55)

or more completely $\div \dfrac{\partial i_p/\partial e_g}{\partial i_p/\partial e_p}$,

$g_m = +\dfrac{\partial i_p}{\partial e_g}$ $(e_p = \text{constant})$.   (56)

$r_p = \div \dfrac{\partial e_p}{\partial i_p}$ $(e_g = \text{constant})$,   (57)

or more correctly* $\div \dfrac{1}{\partial i_v/\partial e_p}$.

In a corresponding manner the gain (A) and load resistance (R_L) of a resistance-loaded amplifier may be given in the form of total differentials—

$$\left| A \right| = \left| \frac{de_p}{de_g} \right| \dagger$$ (58)

and $$R_L = -\frac{de_p}{di_p}.$$ (59)

Particular care should be taken with the signs in all cases, since otherwise serious errors may be introduced in certain calculations.

(x) Valve characteristics at low plate currents

In the case of diodes and diode-connected triodes at very low plate currents (from 1 to about 100 microamperes) an increment of plate voltage of about 0·21 volt produces a 10-fold increase of plate current.   If the $\log_{10}$ of current is plotted against plate voltage, the result should approximate to a straight line with a slope of $1/0·21$.

In the case of triodes operating as triodes the relationship of plate current to grid-cathode voltage is still approximately logarithmic, up to a value of plate current which varies from type to type, but the slope of the curve is decreased by the plate-grid voltage.   The decrease in slope is approximately proportional to the grid bias and therefore to $1/\mu$ times the plate voltage.   The curve at a given plate voltage is, in general, steeper for a high-mu than for a low-mu triode.   Over the region in which the logarithmic relationship holds, the mutual conductance is proportional to the plate current.   For a given plate voltage and plate current, the g_m in the low-current region is greater for high-mu than for low-mu triodes, regardless of ratings.   Also, for a given triode at a given plate current, g_m is greater than at lower plate voltages. Maximum voltage gain in a d-c amplifier is obtained if the valve is operated at as low a plate voltage as possible, and at a plate current corresponding to the top of the straight portion of the characteristic when $\log_{10}$ of current is plotted against the grid voltage.

With pentodes at low plate currents, the maximum gain is obtained when the screen voltage is as low as is permissible without resulting in the flow of positive grid current.

Reference A12 pp. 414-418.

---

*The simple inversion of partial differentials cannot always be justified.

†A is a complex quantity which represents not only the numerical value of the stage gain but also the phase angle between the input and output voltages.   The vertical bars situated one on each side of A and its equivalent indicate that the numerical value only is being considered.

SECTION 10 : REFERENCES

(A) The following text books will be found helpful for general reading :—
A1. Reich, H. J. "Theory and Applications of Electron Tubes." (McGraw-Hill Book Company, New York and London, 2nd edit. 1944) Chapters 1, 2, 3, 4 ; Appendix pp. 671-673.
A2. Massachusetts Institute of Technology "Applied Electronics" (John Wiley and Sons, New York ; Chapman and Hall, London, 1943) Chapters 4, 8, 12.
A3. Sturley, K. R. "Radio Receiver Design" (Chapman and Hall, London, 1943 and 1945) Part 1, Chapter 2 ; Part 2, Chapters 9, 10.
A4. Everitt, W. L. "Communication Engineering" (McGraw-Hill Book Company, New York and London, 2nd edit. 1937) Chapters 12, 13, 14, 15.
A5. Chaffee, E. L. "Theory of Thermionic Vacuum Tubes" (McGraw-Hill Book Company, New York and London, 1933) Chapters 4, 5, 6, 7, 8, 10, 11, 12, 15, 19, 20, 21, 23.
A6. Terman, F. E. "Radio Engineering" (McGraw-Hill Book Company, New York and London 3rd edit. 1947).
A7. Terman, F. E. "Radio Engineers' Handbook" (McGraw-Hill Book Company, New York and London, 1943) Sections 4, 5, 7.
A8. Everitt, W. L. (Editor) "Fundamentals of Radio" (Prentice-Hall Inc. New York, 1943) Chapters 4, 7.
A9. Henney, K. (Editor) "The Radio Engineering Handbook" (McGraw-Hill Book Company, New York and London 3rd edit. 1941) Section 8.
A10. Albert, A. L. "Fundamental Electronics and Vacuum Tubes" (Macmillan Company, New York, 1938) Chapters 4, 5, 6, 12.
A11. Preisman, A. "Graphical Constructions for Vacuum Tube Circuits" (McGraw-Hill Book Company, New York and London, 1943).
A12. Valley, G. E., & H. Wallman. "Vacuum Tube Amplifiers" (M.I.T. Radiation Laboratory Series, McGraw-Hill Book Co., New York and London, 1948).

(B) GENERAL
B1. Chaffee, E. L. "Variational characteristics of triodes" (Book "Theory of Thermionic Vacuum Tubes" p. 164 Sect. 79). .
B2. R.C.A. "Application Note on receiver design—Battery operated receivers" (grid blocking) No. 75 (May 28, 1937).
B3. Jonker, J. L. H. "Pentode and tetrode output valves" Philips Tec. Com. 75 (July, 1940).
B4. "Application of sliding screen-grid voltage to variable-mu tubes" Philips Tec. Com. 81 (April 1941).
B5. Chaffee, E. L. "Characteristic Curves of Triodes" Proc. I.R.E. 30.8 (Aug. 1942) 383—gives method for obtaining static characteristic curves from one experimental curve.
B6. Thompson, B. J. "Space current flow in vacuum tube structures" Proc. I.R.E. 31.9 (Sept. 1943) 485.
B7. Pockman, L. T. "The dependence of inter-electrode capacitance on shielding" Proc. I.R.E. 32.2 (Feb. 1944) 91.
B8. Zabel, L. W. "Grid-current characteristics of typical tubes" Elect. 17.10 (Oct. 1944) 236. (Curves of 6J5, 6SJ7-GT, 6K6-G triode, 6V6-GT triode, 38).
B9. Thurston, J. N. "Determination of the quiescent operating point of amplifiers employing cathode bias" Proc. I.R.E. 33.2 (Feb. 1945) 135.
B10. Radio Design Worksheet No. 49 "Perveance" Radio (June 1946) 29.
B11. Haefner, S. J. "Dynamic characteristics of pentodes" Comm. 26.7 (July 1946) 14.
B12. Warner, J. C., & A. V. Loughren. "The output characteristics of amplifier tubes" Proc. I.R.E. 14.6 (Dec. 1926) 735.
B13. Zepler, E. E. "Triode interelectrode capacitances" W.E. 26.305 (Feb. 1949) 53.
B14. Pullen, K. A. "The use of G curves in the analysis of electron-tube circuits" Proc. I.R.E. 37.2 (Feb. 1949) 210.
Pullen, K. A. "G curves in tube circuit design" Tele-Tech. 8.7 (July 1949) 34 ; 8.8 (Aug. 1949) 33.
B15. Jones, T. I. "The dependence of the inter-electrode capacitances of valves upon the working conditions" Jour. I.E.E. 81 (1937) 658.
B16. Humphreys, B. L., & B. G. James. "Interelectrode capacitance of valves—change with operating conditions" W.F. 26.304 (Jan. 1949) 26.
B17. R.C.A. "Application Note on input admittance of receiving tubes" No. 118 (April 15, 1947).
B18. R.C.A. Application Note "Use of miniature tubes in stagger-tuned video intermediate frequency systems" No. 126 (Dec. 15, 1947). Reprinted in Radiotronics No. 134.
B19. R.C.A. Application Note "A tube complement for ac/dc AM/FM receivers" (Jan. 2, 1948).
B20. R.C.A. Application Note "Use of sharp cut-off miniature pentode RCA-6CB6 in television receivers" No. 143 (March 31, 1950)
B21. "Standards on electron tubes—methods of testing 1950" Proc. I.R.E. Part 1 38.8 (Aug. 1950) ; Part 2 38.9 (Sept. 1950).
B22. Pullen, K. A. "Use of conductance, or G, curves for pentode circuit design" Tele-Tech. 9.11 (Nov. 1950) 38.
Additional references will be found in the Supplement commencing on page 1475.

(C) INPUT IMPEDANCE ; HIGH FREQUENCY OPERATION
C1. North, D. O. "Analysis of the effects of space-charge on grid impedance" Proc. I.R.F. 24.1 (Jan. 1936) 108.
C2. Bakker, C. J. "Some characteristics of receiving valves in short-wave reception" Philips Tec. Rev. 1.6 (June 1936) 171.
C3. Strutt, M. J. O., & A. Van Der Ziel. "The behaviour of amplifier valves at very high frequencies" Philips Tec. Rev. 3.4 (Apr. 1938) 103.
C4. R.C.A. "Application Note on input loading of receiving tubes at radio frequencies" No. 101 (Jan. 25, 1939).
C5. Hudson, P. K. "Input admittance of vacuum tubess" Comm. 23.8 (Aug. 1943) 54.
C6. R.C.A. "Application Note on use of miniature tubes in stagger-tuned video intermediate-frequency systems" No. 126 (Dec. 15, 1947).
C7. Zepler, E. E., & S. S. Srivastava "Interelectrode impedances in triodes and pentodes" W.E. 28.332 (May 1951) 146.
See also B17, B18, B19, B20, B21 and Chapter 23 Sect. 5.

(D) NON-LINEAR OPERATION

D1. Preisman, A. " Graphics of non-linear circuits " R.C.A. Rev. (1) 2.1 (July, 1937) 124 : (2) 2.2 (Oct. 1937) 240 (gives loadlines with inductive load and dynatron).
D2. Shelton, E. E. " Some applications of non-linear current potential characteristics " Electronic Eng. 15.179 (Jan. 1943) 339.

(E) CONVERSION OF CHARACTERISTICS

E1. R.C.A. " Application Note on the operation of the 6L6 " No. 60 (June 10, 1936).
E2. R.C.A. " Application Note on the conversion of a 6L6 plate family to new screen voltage conditions " No. 61 (June 25, 1936).

(F) VALVE EQUIVALENT CIRCUITS AND NETWORKS

F1. Chaffee, E. L. " Equivalent-plate-circuit theorem " (Book " Theory of Thermionic Vacuum Tubes " p. 192 Sect. 83).
F2. Chaffee, E. L. " Equivalent grid circuit theorem " (Book " Theory of Thermionic Vacuum Tubes " p. 196 Sect. 84).
F3. Chaffee, E. L. " Equivalent circuits of the triode " (Book " Theory of Thermionic Vacuum Tubes " p. 197 Sect. 85).
F4. Chaffee, E. L. " Alternative equivalent circuits of a triode with constant current generator " (Book " Theory of Thermionic Vacuum Tubes " p. 199 Sect. 86).
F5. Terman, F. E. " Equivalent circuit of the vacuum-tube amplifier " (Book " Radio Engineering," 2nd edit. pp. 172-178).
F6. Terman, F. E. " Equivalent amplifier circuits " (Book " Radio Engineers Handbook " p. 354).
F7. Terman, F. E. " Exact equivalent circuit of Class A amplifier and application to analysis of distortion and cross modulation " (Book " Radio Engineers' Handbook " p. 462).
F8. Stockman, H. " Signs of voltages and currents in vacuum tube circuits " Comm. 24.2 (Feb. 1944) 32.
F9. Llewellyn, F. B., & L. C. Peterson. " Vacuum tube networks " Proc. I.R.E. 32.3 (March, 1944) 144.
F10. Butler, F. " Cathode coupled oscillators " W.E. 21.254 (Nov., 1944) 521 (includes equivalent circuits all major arrangements).
F11. Wheeler, H. A. " Equivalent triode networks " Elect. 18.3 (Mar., 1945) 304.
F12. Walker, G. B. " Theory of the equivalent diode " W.E. 24.280 (Jan. 1947) 5 ; Letter W. E. Benham 24.281 (Feb. 1947) 62.
F13. Jonker, J. L. H., & B. D. H. Tellegen. " The current to a positive grid in electron tubes " (i) The current resulting from electrons flowing directly from cathode to grid ; (ii) returning electrons, Philips Research Reports 1.1 (Oct. 1945) 13.
F14. Salzberg, B. (letter) " Valve equivalent circuit " W.E. 24.283 (April 1947) 124 ; also G.W.O.H. (editorial) p. 97.
F15. Keen, A. W. (letter) " Valve equivalent circuit," W.E. 24.286 (July 1947) 217.
F16. Salzberg, B. (letter) " Valve equivalent circuit," W.E. 24.286 (July 1947) 218.
See also (H) below.

(G) VALIDITY OF EQUIVALENT PLATE-CIRCUIT THEOREM FOR POWER CALCULATIONS—DISCUSSION

G1. Stockman, H. " The validity of the equivalent plate-circuit theorem for power calculations " (letter) Proc. I.R.E. 32.6 (June, 1944) 373.
G2. Preisman, A., & H. Stockman. " The validity of the equivalent plate-circuit theorem for power calculations " (letter) Proc. I.R.E. 32.10 (Oct. 1944) 642.
G3. Reich, H. J. " The validity of the equivalent plate-circuit theorem for power calculations " (letter) Proc. I.R.E. 33.2 (Feb. 1945) 136.
G4. Stockman, H. " The validity of the equivalent plate-circuit theorem for power calculations " (letter) Proc. I.R.E. 33.5 (May 1945) 344.

(H) GRID CURRENT CHARACTERISTICS

H1. Nielsen, C. E. " Measurement of small currents : Characteristics of types 38, 954 and 959 as reduced grid current tubes " Rev. Sci. Instr. 18.1 (Jan. 1947) 18.
H2. Crawford, K. D. E. " H.F. Pentodes in electrometer circuits " Electronic Eng. 20.245 (July 1948) 227.

CHAPTER 3

THE TESTING OF OXIDE-COATED CATHODE HIGH VACUUM RECEIVING VALVES

by N. V. C. CANSICK, B.SC.*
and F. LANGFORD-SMITH, B.SC., B.E.

SECTION 1 : BASIS OF TESTING PRACTICE

(i) Fundamental physical properties.
(ii) Básic functional characteristics.
(iii) Fundamental characteristic tests.
(iv) Valve ratings and their limiting effect on operation :
  (A) Limiting ratings.
  (B) Characteristics usually rated.
  (C) Rating systems.
  (D) Interpretation of maximum ratings.
  (E) Operating conditions.
(v) Recommended practice and operation :
  (a) Mounting.
  (b) Ventilation.
  (c) Heater-cathode insulation.
  (d) Control grid circuit resistance.
  (e) Operation at low screen voltages.
  (f) Microphony.
  (g) Hum.
  (h) Stand-by operation.

---

*Senior Engineer, in charge of Quality Control and Valve Application Laboratory, Amalgamated Wireless Valve Co. Pty. Ltd.

(i) Fundamental physical properties

The characteristics and operation of a high vacuum oxide-coated cathode- or filament-type valve under normal conditions of operation, initially and throughout life, are determined primarily by the following fundamental general physical properties—

(a) the **vacuum** within the envelope surrounding the cathode and electrode structure.

(b) the total available **cathode emission** and **uniformity of activation of the cathode surface.**

(c) the **geometrical configuration of the electrode system.**

(d) the **electrode contact potentials** with respect to the cathode.

(e) **primary and secondary emission** from electrodes, other than the cathode.

(f) the **interelectrode admittances.**

(g) the **stability of the electrical characteristics.**

(h) the **stability, robustness and durability of the mechanical construction.**

(j) the **external size and shape,** and the **system of electrode connections.**

The maintenance of a satisfactory **vacuum** under maximum rating conditions and continued operation, together with the availability of an adequate cathode emission under all prescribed electrode conditions and normal variations of heater/filament power, are essential to the fundamental operation of a valve, since an activated cathode surface requires a certain degree of vacuum to be maintained for its emissive properties to remain unimpaired, while the emission available from the cathode must always be sufficient to supply the total peak and average space currents required.

At the same time, it is necessary that the cathode surface be uniformly activated over the regions from which emission current is supplied, since the parameter transconductance is directly proportional to the activated cathode area.

The **electrode geometry** determines the electrical static and dynamic characteristics, within the limitations imposed by the available cathode emission, to the extent to which the emission is incompletely space charge limited and the capacity of the gettering to maintain a satisfactory vacuum.

Electrode contact potentials and electrode **primary and secondary emission** depend upon the electro-chemical condition of the electrode surfaces. This is determined usually by the extent to which contamination of these surfaces has occurred during manufacture or operation, due to the deposition of active material from the cathode or getter. **Contact potential** is of importance in the case of a low- or zero-voltage electrode since it may contribute a sufficiently large fraction of the total electrode voltage to modify significantly the characteristics determined by the electrode geometry and applied electrode voltages alone.

Primary and secondary electrode emission may occur if electrode temperatures become sufficiently high and the deposited films of active material in combination with the electrode material have a low work function. The effect on operation of primary or secondary electrode emission depends on the value of resistance in series with the emitting electrode.

The interelectrode short-circuit admittances are determined both by the geometry of the electrode construction, the d.c. and r-f properties of the insulation supporting the electrode system and electrode connections, and the frequency and conditions of operation. The susceptive components are dependent mainly on the interelectrode and lead capacitances and inductances. The conductive components are due to transit-time effects and the interelectrode resistances which are determined mainly by the d.c. resistance of the interelectrode insulation, but may also include a frequency-dependent component. The d.c. insulation resistance is determined by the surface condition of the electrode supports (mica, glass, ceramic), the extent to which contamination of these surfaces has occurred, and also by leakage between electrode leads within the base or between base pins due to leakage over or through the base material. A high value of interelectrode resistance is required throughout the life of a valve in order to avoid—

(1) **uncontrolled changes in electrode voltages** supplied through series resistive circuits from the electrode voltage source, due to the flow of leakage conduction currents through or across the interelectrode insulation from higher to lower voltage electrodes and thence through the externally connected circuits to the negative terminal of the electrode voltage supply. This characteristic is particularly important, e.g., in the case of negatively biased control grids which obtain their bias through a high value of grid resistor, and power dissipating electrodes which obtain their voltages through high resistance series dropping resistors.

(2) **noisy operation** due to intermittent conduction under the conditions described in (1).

(3) **lowering the resistive component** of the interelectrode admittances and thus increasing the damping by the valve on resonant circuits connected between electrodes, either directly or as a result of increased " Miller Effect ".

The **interelectrode r-f resistance** must also be high throughout the life of the valve up to all normal frequencies of operation, in order to avoid damping high impedance resonant circuits connected between electrodes. In general, it is determined by the r-f properties of—

(a) the interelectrode insulation,
(b) films of active material deposited on the interelectrode insulation,
(c) the base material.

The **stability of the electrical characteristics** depends on the goodness of the vacuum and its maintenance, stable emission from a properly and uniformly activated cathode surface, stable electro-chemical surface conditions, high interelectrode resistance, and the mechanical stability of the electrode structure. Variations of any of these characteristics throughout life produce changes in the valve parameters and, through them, changes in circuit performance.

Two types of instability of electrical characteristics occur ; initial instability which is sometimes experienced early in life due to incomplete or unsatisfactory processing, and the normal gradual deterioration of characteristics dependent on emission, which continues until the valve is no longer serviceable. Generally, the ultimate electrical life of a valve which has been properly exhausted, which has a properly activated cathode and which has been operated within its maximum ratings, depends mainly on the cathode current at which it has been operated and the ability of the getter to prevent deterioration of the vacuum and consequent " poisoning " of the cathode emission, by absorbing the gases produced in the cathode coating during operation, and also the gases released from the electrodes, the walls of the envelope and other parts of the internal construction during storage, and throughout life.

The **stability, robustness and durability of the mechanical construction** in general determine the ultimate reliability and effective life of a valve, also the extent to which electrical performance is affected by conditions of vibration or mechanical shock under which it may be transported or operated and the ability of the construction to withstand such conditions without mechanical failure.

(ii) Basic functional characteristics

Space current—When the cathode surface is raised to its normal emitting temperature, by supplying the required power to the filament or heater, and suitable voltages are applied to the other electrodes, an electron space current flows from the cathode emitting surface to the electrode system, its value depending on the electrode geometry and the combined effect of the electrode voltages acting in the plane(s) of the control electrode(s).

For electrode systems consisting of an inner control electrode adjacent to the cathode and one or more outer electrodes, providing that a virtual cathode is not formed between the control electrode and the outermost electrode, the **total space current from the cathode** may be expressed, generally, to a good approximation by the following equation—

$$i_s \approx G \left[(e_{c1} + \epsilon_1) + \frac{e_{c2}}{\mu_{1-2}} + \frac{e_{c3}}{\mu_{1-3}} + \ldots + \frac{e_b}{\mu_{1-b}} \right]^{3/2} \tag{1}$$

where i_s = total cathode space current,
  e_{c1} = voltage of the first electrode,
   ϵ_1 = contact potential of the first electrode,
$e_{c2}, e_{c3},$
  etc. = voltages of the successive outer electrodes,
   e_b = voltage of the outer plate current collecting electrode,
μ_{1-2} = amplification factor of the first electrode with respect to the second electrode,
μ_{1-3} = amplification factor of the first electrode with respect to the third electrode,
μ_{1-b} = amplification factor of the first electrode with respect to the outermost electrode,
and G = perveance of the cathode-control-electrode region

$$\approx 2.33 \times 10^{-6} \times \frac{\text{cathode area}}{\text{(cathode-to-first-electrode spacing)}^2}$$

In eqn. (1), the terms within the brackets represent the effect of the various electrode voltages considered as acting in the plane of the first electrode, so that the sum of the terms combined represents an " equivalent electrode voltage ". For purposes of space current calculation, the factors depending on the geometry of the electrode system and on the effect of the electrode voltages may thus be reduced to the perveance of the cathode-to-first-electrode space and the equivalent electrode voltage acting in the plane of the first electrode, so that the space current equation becomes—·

$$i_s \approx G \text{ [equivalent electrode voltage] }^{3/2} \qquad (2)$$

The equations for the total cathode current in diodes, triodes, tetrodes and pentodes follow directly from eqn. (1).

. In the case of electrode systems having two control electrodes and which operate with a virtual cathode before the second control electrode, as in converter and mixer valves, the space current cannot be expressed in a simple form. Such systems may be regarded as consisting of two separate but related systems, and reduced to equivalent diodes, with the space current of the outer equivalent diode supplied from, and controlled by, the space current of the equivalent diode adjacent to the cathode.

Generally, it can be stated that whatever the electrode system, the **total space current** from the cathode depends primarily upon the total activated cathode surface area, the total available emission, the cathode temperature, the extent of temperature and space charge limiting of emission, the geometry of the electrode system and the electrode voltages.

The **distribution of the space current** to the various electrodes depends only upon electrode geometry, space charge effects and the electrode voltages and, in general, cannot be expressed in a simple form.

In addition to the electron space current from the cathode, there are always some positive ions present due initially to the presence of the infinitesimal traces of residual gases remaining after the exhausting and gettering processes and, during life, to the release of absorbed gases from the surfaces of the electrode structure and envelope walls, and the gases produced as a result of physical-chemical changes occurring in the cathode coating during the emission processes.   The positive ions, so produced, flow to the negative voltage electrodes and are prevented from bombarding the cathode surface during operation, when space current is flowing, by collision effects and the presence of the space charge surrounding the cathode.

A primary or secondary electron emission current may also flow from the surface of an electrode and represents a negative electrode current.   Small positive or negative leakage conduction currents, due to conducting or semi-conducting leakage paths over or through the interelectrode insulation, may also contribute to the electrode currents ; their values depend on the voltages of the electrodes concerned, and the conducting properties of the contaminated surfaces or insulating material, and often on their temperature.   When the effects above are present, the total current of an electrode is the algebraic sum of the positive and negative components.

Static Characteristics—The various electrical characteristics, which result from the electrode geometry and the application of steady **direct voltages** only to the electrodes, with the cathode emitting, are termed " static characteristics ". These characteristics consist of—

(a) the voltage-current relationship for each electrode, when constant voltages are applied to all remaining electrodes.

e.g., Diode, $— I_b — E_b$,

Triode, $— I_b — E_b$, E_{c1} constant

$I_{c1} — E_{c1}$, E_b constant

and similarly in the case of other types.

(b) the mutual voltage-current relationships between electrodes, when constant voltages are applied to all electrodes except the voltage-varying electrode.

e.g., Triode, $— I_b — E_{c1}$ E_b constant

Pentode, $— I_b — E_{c1}$ E_b, E_{c3}, E_{c2} constant

$I_{c3} — E_{c1}$ E_b, E_{c3}, E_{c2} constant

$I_{c2} — E_{c1}$ E_b, E_{c3}, E_{c2} constant

and similarly in the case of other types.

Derived Characteristic Parameters—By considering infinitesimal changes of voltage and current of the static characteristics, described under (a) and (b) above, the characteristic parameters—amplification factor, transconductance and variational plate resistance are obtained, which are related as follows,

$$\text{Transconductance} = \frac{\text{Amplification Factor}}{\text{Plate Resistance}}$$

These characteristic parameters are derived by the following relations—

$$\text{transconductance} = \frac{\delta I_b}{\delta E_{c1}} ; \quad E_b, E_{c3}, E_{c2} \text{ constant} \tag{3}$$

$$\text{amplification factor} = \frac{\delta E_b}{\delta E_{c1}} ; \quad E_{c3}, E_{c2} \text{ constant} \tag{4}$$

$$\text{plate resistance} = \frac{\delta E_b}{\delta I_b} ; \quad E_{c3}, E_{c2}, E_{c1} \text{ constant} \tag{5}$$

The amplification factor is a function of the electrode geometry only, but becomes also dependent on the control grid voltage as the cut-off condition is approached. The transconductance, as defined by eqn. (3) may be derived from the general space current eqn. (1), and is a function of the electrode geometry, the total cathode space current and uniformity of activation of the cathode surface.

The characteristic parameters normally used to describe the electrode geometry, and in terms of which performance is interpreted, are—the amplification factor, transconductance of the control grid with respect to the plate and the (a.c.) plate to cathode resistance.

Dynamic Characteristics—By superimposing alternating voltages on the direct voltage(s) of the control electrode(s), with suitable impedances in series with the output electrode(s), dynamic characteristics are obtained which depend directly on the static characteristics and characteristic parameters. These dynamic characteristics include rectification, frequency conversion, voltage and power gain, oscillation, and impedance transformation characteristics.

The static characteristics, characteristic parameters and the fundamental properties previously described comprise the basic mechanical and electrical characteristics, which determine the serviceability and application of a valve and its performance under given conditions.

In order to appreciate the significance and limitations of the various tests, which normally are applied to determine the condition of a valve and the probability that it will continue to operate satisfactorily, the performances and individual tests must be properly interpreted in terms of all the relevant characteristics and corresponding fundamental physical properties. The manner and extent of the dependence and interdependence of the various characteristics and fundamental physical properties must also be recognized and understood.

The important basic functional characteristics, together with the fundamental physical properties on which they depend, are shown in the following
tabulation :

GENERAL PHYSICAL PROPERTIES	RELATED FUNCTIONAL CHARACTERISTICS
1. Mechanical	
Size, shape and material of external construction	Physical dimensions
	Type of envelope
	Type of base
	Type of top cap
System of electrode connections	Base pin and top cap connections
Stability, robustness and durability of mechanical construction	Ruggedness of envelope, basing and internal electrode structure
	Microphony
	Noise
2. Electrical	
Interelectrode resistance	D.C. interelectrode insulation resistance
	R-F interelectrode insulation resistance
	Noise
Vacuum	Gas pressure within the envelope*
Cathode emission	Total available peak cathode current*
	Total available average cathode current*
Uniformity of cathode activation	Transconductance or dynamic performance characteristic at reduced heater filament voltage*
Electrode geometry	Electrode currents*
	Interelectrode transconductances*
	Interelectrode mu factors*
	Interelectrode variational resistances*
	Interelectrode capacitances
Electrode contact potentials	⎫ Modify characteristics dependent upon
Primary and secondary emission	⎭ electrode geometry (see above)
Stability of electrical characteristics	Affect all electrical characteristics.

(iii) Fundamental characteristic tests

The characteristic tests which are of fundamental importance in specifying the
performance and determining the condition, acceptability and usability of an oxide-
coated cathode- or filament-type valve, are as follows :

(a) **Tests common to all valve types**
Visual inspection of internal and external construction, and finish.
Maximum overall length.
Maximum diameter.
Interelectrode short-circuits.
Electrode continuity (open-circuits).
Interelectrode insulation resistance (d.c.).*
Heater/filament power.*
Heater/cathode leakage.*
Emission or emission-dependent dynamic test.*
Interelectrode capacitances.*

(b) **Tests common to all types except diodes***
Reverse grid current (gas, leakage, grid emission).
⎰ Control grid current commencement potential, or
⎱ Positive control grid current.
Electrode currents (d.c.).
Electrode current (d.c.) cut-off.

---

*Under specified operating conditions.

Transconductance (see Ref. 47).
Amplification factor.
Plate resistance.
Noise (a-f and/or r-f).
Microphony.

(c) **Tests common to diodes only***
Signal diodes   Plate current commencement or zero signal plate current.
Power diodes   Back emission.
(d) **Dynamic performance tests***
Signal diodes   Rectification (operation)
Power diodes   Rectification (operation)
A-F amplifiers   A.C. amplification
R-F and I-F amplifiers   Stage gain.
Converters   Conversion stage gain or conversion transconductance
                                           and
                                           Oscillator transconductance or oscillator grid current.
Power output types   Power output.

(e) **Dynamic performance tests at reduced heater/filament voltage***

Dynamic performance tests are also performed at reduced filament/heater voltages during manufacture, in order to control initial characteristics and performance, and also to provide a manufacturing process control of the uniformity of activation of the emitting surface. In some cases, with the reduction of filament/cathode temperature, changes occur in contact potential and/or reverse grid current, which make the normal-voltage tests ineffective or unsatisfactory as a control of activation, depending on the conditions of operation. In such cases a transconductance test is normally used instead.

Notwithstanding the control exercised by such tests over initial characteristics and performance, it is a very difficult manufacturing problem to avoid the wide variations and deterioration of characteristics which often occur under conditions of reduced filament/heater operation throughout life although the characteristics at normal filament/heater voltage may be satisfactory. For this reason, and also as performance may be critically dependent on circuit design and electrode supply voltages, particularly the operation of the oscillator circuits of a converter valve and the power output and distortion of a power output valve, it is not in general normal practice for valve specifications to specify any minimum requirements for characteristics or performance at reduced filament/heater voltage during life. Reasonable performance at reduced filament/heater and electrode voltage conditions is achieved by most manufacturers, however, as a result of the pressure of competition combined with the user's demand for acceptable performance under slump voltage conditions.

Under American practice the reduced heater/filament voltages at which specified dynamic performance tests are normally performed during manufacture are as follows :

0.625	volt types	0.55 volt
1.25	volt types	1.1 volts
1.4	volt types	1.1 volts
2	volt types	1.6 or 1.7 volts
2.5	volt types	2.2 volts
6.3	volt types	5.5 volts
7.5	volt types	6.0 volts
12.6	volt types	11.0 volts
19	volt types	16.5 volts
25	volt types	22 volts
26.5	volt types	23.5 volts

*Under specified operating conditions.

32	volt types	28.5 volts
35	volt types	31.0 volts
45	volt types	40 volts
50	volt types	44 volts
117	volt types	100 volts

(iv) Valve ratings and their limiting effect on operation

(A) Limiting ratings

In order to assist the designer to obtain the maximum service from a given valve type, within the limits of safe operation, in the various applications in which it may be used, the limiting conditions under which the type can be operated without impairing its performance and normal life are usually specified by the valve manufacturer in the form of maximum (or minimum) ratings for relevant basic characteristics which may only be exceeded at the user's risk.   These ratings, unless otherwise stated, apply to valves having bogie values of characteristics.   It is therefore the responsibility of the equipment designer to see that a bogie valve will not exceed any of its ratings.

Maximum ratings are established on the basis of life tests and operating performance. It is a matter of experience that when one or more ratings are exceeded for any appreciable time, depending on the period and the extent of the overload, the particular rating exceeded and the capability of the type to withstand such overload, the serviceability of the particular valve may be impaired, its life may be shortened or its performance may be unsatisfactory.

Maximum ratings usually are specified only for those characteristics which normally determine the limits of safe operation in the intended applications for which a particular type has been designed.   Generally, the limits of safe operation for any application may be prescribed completely by specifying the maximum ratings for the electrode voltages, peak inverse voltage in the case of rectifiers, electrode dissipations and the maximum peak and average total cathode current.

In the case of high transconductance and power output types it is necessary, in addition, to limit the maximum value of resistance which may be used between the negative control grid and cathode, in order to avoid either excessive variation of circuit performance, excessive space current or excessive electrode dissipations due to reverse grid current.   Maximum ratings, in general, can only be determined satisfactorily by extensive life tests.

(B) Characteristics usually rated

The characteristics of individual types for which ratings are required to specify performance are as follows :

(a) General mechanical ratings (common to all types)

Maximum overall length.
Maximum seated height.
Maximum diameter.
Dimensions and locations of top caps, bases and base pins.
Materials and design of external construction.

(b) General electrical ratings common to all types

	Nature of rating
Heater/filament voltage (a.c. or d.c.)	nominal
Heater/filament current (a.c. or d.c.)	nominal
Heater to cathode voltage (d.c.)	maximum
Control grid circuit resistance	maximum
Interelectrode capacitances—	
control grid to plate capacitance	maximum
all other capacitances (single unit)	nominal
capacitances between electrodes of multiple units	nominal or maximum

(c) **Specific additional electrical ratings applied to particular types**

Type	Application	Fundamental rating	Nature of rating
Diodes	Detector	Average (d.c.) plate current per plate	Maximum
		Peak plate current per plate	Maximum
	Power Rectifier	Peak inverse plate voltage	Maximum
		Average (d.c.) plate current per plate	Maximum
		Peak plate current per plate	Maximum
Triodes	Voltage amplifier	Plate Voltage	Maximum
		Plate dissipation	Maximum
		Grid voltage—negative	Maximum
		positive	Maximum
		Cathode current (d.c.)	Maximum
	Power amplifier	Plate voltage	Maximum
		Plate dissipation	Maximum
		Grid voltage—negative	Maximum
		positive	Maximum
		Grid dissipation**	Maximum
		Cathode current (d.c.)	Maximum
		Cathode current (peak)*	Maximum
Pentodes	Voltage amplifier	Plate voltage	Maximum
		Screen voltage	Maximum
		Screen supply voltage	Maximum
		Grid voltage—negative	Maximum
		positive	Maximum
		Plate dissipation	Maximum
		Screen dissipation	Maximum
		Cathode current (d.c.)	Maximum
	Power amplifier	Plate voltage	Maximum
		Screen voltage	Maximum
		Grid voltage—negative	Maximum
		positive	Maximum
		Plate dissipation	Maximum
		Screen dissipation	Maximum
		Grid dissipation**	Maximum
		Cathode current (d.c.)	Maximum
		Cathode current (peak)*	Maximum
Converters	Frequency converter	Mixer plate voltage	Maximum
		Mixer screen voltage	Maximum
		Mixer screen supply voltage	Maximum
		Oscillator plate voltage	Maximum
		Oscillator plate supply voltage	Maximum
		Mixer signal grid voltage—	
		positive	Maximum
		negative	Maximum
		Mixer plate dissipation	Maximum
		Mixer screen dissipation	Maximum
		Oscillator plate dissipation	Maximum
		Cathode current (d.c.)†—	
		mixer	Maximum
		oscillator	Maximum

*Power amplifier types only, for other than Class A operation.
**For operation in the positive grid current region.
†In converters of the 6SA7, 6BE6 class having no separate oscillator positive electrode, the total cathode current is the fundamental rating.

Notwithstanding the ratings shown against each type, the actual ratings of an individual type depend entirely on its intended applications and may not include all those listed.   The absence of a rating for a particular characteristic may be taken to indicate either that it has not previously been necessary to specify a rating for this characteristic, or that the type in question was not intended originally for this application and has not been processed accordingly.   In the latter case, operation may not be satisfactory, and reference should be made to the valve manufacturer for specific information.

(C) **Rating systems**

Valves are rated by either of two systems—the " absolute maximum " system, or the " design-centre maximum " system.   The absolute maximum system originated in the early days of valve development and was based on the voltage characteristics of battery supplies.   Battery voltages could fall below their nominal values but seldom appreciably exceeded them, so that valve maximum ratings set on the basis of specified battery voltages were absolute maximum ratings.   This system is still widely used by British and European valve manufacturers and is the system of ratings used in the British Services' Electronic Valve Specification K1001 and the U.S.A. Services' Specification for Electron Tubes, JAN-1A (Ref. S2).   With the introduction of power line and car-radio operated receivers and the tendency of many designers to interpret absolute maximum ratings as nominal values, it became necessary to re-rate valves according to a system in which allowance was made in the ratings for the variations which occur under both conditions of operation.   Accordingly, the design-centre system was adopted in U.S.A. by the Radio Manufacturers Association in 1939 for the rating of receiving valves and since then has become the standard system for rating most receiver types of American design, manufactured both in U.S.A. and elsewhere.

With either system, each maximum rating for a given valve type must be considered in relation to all other maximum ratings for that type, so that no one maximum rating will be exceeded in utilizing any other maximum rating.

Thus it will often happen that one rating alone will determine the limiting operating conditions while other characteristics are below their maximum ratings.

(D) **Interpretation of maximum ratings**

In the absolute maximum or " absolute " system (except for filament or heater voltage) the maximum ratings are limiting values above which the serviceability of the valve may be impaired from the viewpoint of life and satisfactory performance. Therefore, in order not to exceed these absolute ratings, the equipment designer has the responsibility of determining an average design value for each rating below the absolute value of that rating by an amount such that the absolute values will never be exceeded under any usual condition of supply-voltage variation, load variation, or manufacturing variation in the equipment itself.

The equipment should be designed to operate the **filament or heater** of each valve type at rated normal value for full-load operating conditions under average voltage-supply conditions.   Variations from this normal value due to voltage-supply fluctuation or other causes, should not exceed $\pm$ 5 per cent. unless otherwise specified by the valve manufacturer.

Under the " British Standard Code of Practice " B.S.1106, 1943 (Ref. 49) and the British Radio Valve Manufacturers' Association's publication " Radio Valve Practice," August, 1948 (Ref. 48), recommended British practice is that in general it is not permissible that the heater voltage should vary more than 7 per cent. from the rated value.

It is a matter of experience, however, that the heaters and filaments of most modern receiving valve oxide-coated cathode and filament types may be operated at voltages whose maximum fluctuations do not exceed $\pm$ 10 per cent. from their rated values, without serious effect on life or marked reduction in performance, provided that the maximum ratings of the other electrodes are not exceeded.   In cases where the heater or filament voltage variations exceed, or are likely to exceed, $\pm$ 10 per cent., the maximum ratings should be reduced and recommendations obtained from the valve manufacturer as to the maximum ratings permissible under the particular conditions.

In the design-centre maximum or " design-centre " system, the maximum ratings are working design-centre maxima.   The basic purpose underlying this system is to provide satisfactory average performance in the greatest number of equipments on the assumption that they will not be adjusted to local power-supply conditions at the time of installation. Under the design-centre system, ratings are based on the normal voltage variations which are representative of those experienced with the three important types of power supply commonly in use, namely a.c. and d.c. power lines, storage battery with connected charger, and dry batteries, so that satisfactory performance from valves so rated will ordinarily be obtained in equipment which is so designed that the design-centre maximum ratings are not exceeded at the supply design-centre voltage.

The following interpretation of receiving valve design-centre ratings is based partly on R.M.A. (U.S.A.) Standard M8-210 entitled " Tube Ratings " (Ref. S11), and partly on established design practice recommended by valve manufacturers.

1. **Cathode**
The heater or filament voltage is rated as a normal value, unless otherwise stated, so that transformers or resistances in the heater or filament circuit should be designed to operate the heater or filament at the rated value at the supply design-centre voltages, indicated hereunder. Where the permissible heater or filament voltage variations are exceeded, or are liable to be exceeded, maximum ratings should be reduced in accordance with recommendations obtained from the valve manufacturer.

2. **Indirectly-heated and a.c. filament types**
(2.1) **A.C. or d.c. power line operation**—Maximum ratings have been chosen so that valves will give satisfactory performance at these maximum ratings in equipment operated from power line supplies, the normal voltage of which, including normal variations, fall within ± 10 per cent. of a specified value. Heaters or filaments as well as positive and negative voltage electrodes, unless otherwise specified, may therefore be operated at voltages up to ± 10 per cent. from their rated values, provided that at the specified line design-centre voltage the heater or filament is operated at its rated voltage and the maximum ratings of plate voltages, screen-supply voltages, electrode dissipations, total cathode current and rectifier output currents are not exceeded.

The prescribed power line voltage variation of ± 10 per cent. is based on surveys made in the U.S.A., which have shown that the line voltages delivered fall within 10 per cent. of 117 volts, which is taken as the line design-centre voltage in that country. In using design-centre ratings with other power line systems elsewhere, it is usually satisfactory to regard ± 10 per cent. as being representative of the line voltage variation likely to be experienced under normal conditions, so that it is then only necessary to determine the line design-centre voltage. In extreme cases, where power line variations exceed ± 10 per cent., as for example when operation is in remote areas supplied by long lines subject to variable and heavy peak loading, adjustment to the equipment should be made locally.

(2.2) **Storage battery operation**—When storage battery equipment is operated without a charger, it should be designed so that the heaters or filaments are operated at their normal rated values and published maximum values of plate voltages, screen-supply voltages, electrode dissipations, total cathode currents and rectifier output currents are never exceeded for a terminal voltage at the battery source of 2.0 volts per cell. When storage battery equipment is operated with a charger or similar supplies, the normal battery fluctuation may be as much as 35 per cent. or more, which imposes severe operating conditions on valves. Under these conditions, the equipment should be designed so that the effect of high heater or filament voltages on valve life and performance (due mainly to excessive reverse grid current) is reduced to a minimum, and so that 90 per cent. of the above maximum ratings is never exceeded for a terminal voltage at the battery source of 2.2 volts. In both classes of operation, progressively reduced and unreliable performance is to be expected as the heater or filament voltage falls below 90 per cent. of its rated value and approaches the " slump " value specified for the reduced performance tests of the valve test specification.

(2.3) **" B " battery operation**—The design-centre voltage for " B " batteries supplying positive voltage electrodes is the normal voltage rating of the battery block, such as 45 volts, 90 volts, etc. Equipment should be designed so that under no condition of battery voltage will the plate voltages, screen-supply voltages, electrode dissipations, or total cathode currents ever exceed the recommended respective maximum values specified for each valve type by more than 10 per cent.

(3) 2.0 Volt battery filament types

(3.1) **Filament**—The 2.0 volt battery filament types are designed to be operated with 2.0 volts across the filament. In all cases the operating voltage range should be maintained within $\pm$ 10 per cent. of the rated filament voltage, i.e., within the limits of 1.8 to 2.2 volts.

(3.2) **Positive and negative electrodes**—The electrode voltage supplies may be obtained from dry-cell batteries, storage batteries or from a power line and should be chosen so that the maximum ratings of plate voltages, screen-supply voltages, electrode dissipations and total cathode current are not exceeded at the supply design-centre voltages specified under 2.3, 2.2 and 2.1.

(4) 1.4 volt battery valve types

(4.1) **Filament**—The filament power supply may be obtained from dry-cell batteries, from storage batteries, or from a power line.

(4.11) **Dry-cell battery supply**—The filament may be connected either directly across a battery rated at a terminal potential of 1.5 volts, or in series with the filaments of similar valves across a power supply consisting of dry cells in series. In either case, the voltage across each 1.4 volt section of filament should not exceed 1.6 volts. In order to meet the recommended conditions for operating filaments in series it may be necessary to use shunting resistors across the individual 1.4 volt sections of filament.

(4.12) **Storage battery supply**—The filament may be operated either singly or in series with the filaments of similar valves. For such operation in either case, design adjustments should be made so that, with valves of rated characteristics, operating with all electrode voltages applied on a normal storage-battery voltage of 2.0 volts per cell (without a charger) or 2.2 volts per cell (with a charger), the voltage drop across each 1.4 volt section of filament will be maintained within a range of 1.25 to 1.4 volts with a nominal centre of 1.3 volts. In order to meet the recommended conditions for operating filaments in series it may be necessary to use shunting resistors across the individual 1.4 volt sections of filament.

(4.13) **Power line supply**—The filament may be operated either singly or in series with the filaments of similar valves. For such operation, in either case, design adjustments should be made so that, with valves having rated characteristics operating with all electrode voltages applied and on a line voltage equal to the design-centre voltage, the voltage drop across each 1.4 volt section of filament will be maintained within a range of 1.25 to 1.4 volts with a nominal centre of 1.3 volts. In order to meet the recommended conditions for operating filaments in series, it may be necessary to use shunting resistors across the individual 1.4 volt sections of filament.

(4.2) **Positive and negative voltage electrodes**—The electrode voltage supplies may be obtained from dry-cell batteries, storage batteries, or from a power line. For such operation the electrode voltages should be chosen so that the maximum ratings of plate voltages, screen-supply voltages, electrode dissipations and total cathode currents are not exceeded at the supply design-voltages specified under 2.3, 2.2 and 2.1 respectively.

(5) General (all types)

(5.1) **Screen voltage supply**—When the screen voltage is supplied through a screen-dropping resistor, the maximum screen voltage rating may be exceeded provided :

(a) The screen supply voltage does not exceed the maximum plate voltage rating.

(b) At any signal condition, the average screen dissipation does not exceed the maximum rating.

(c) At the signal condition which results in maximum screen current, the screen voltage does not exceed the maximum rating.

(5.2) **Grid voltage limits**—Where a grid voltage is specified as " positive bias value 0 maximum " or " never positive ", this indicates that grid dissipation is not permitted.   In such cases it may be inferred that the grid has not been suitably treated to permit dissipation.

(E) Operating conditions

Typical operation—In addition to maximum ratings, information is published on typical operating conditions for most of the various types, when used in particular applications.   These typical operating conditions are intended to provide guiding information for the use of each type.   They must not be considered as ratings, because each type can, in general, be used under any suitable conditions within its rating limitations.

Datum point for electrode potentials—In published data, it is standard practice for the values of grid bias and positive-potential-electrode voltages to be given with reference to a specified datum point, as follows :—For types having filaments heated with direct current, the negative filament terminal is taken as the datum point to which other electrode voltages are referred.   For types having filaments heated with alternating current, the mid-point (i.e., the centre tap on the filament transformer secondary, or the mid-point on a resistor shunting the filament) is taken as the datum point.   For types having equi-potential cathodes (indirectly heated) the cathode is taken as the datum point.

Grid bias for a.c. or d.c. filament excitation—If the filament of any type whose data are given for a d.c. filament voltage is to be operated from an a.c. supply, the grid bias given for d.c. filament operation should be increased by an amount approximately equal to one half the rated filament voltage and be referred to the filament mid-point.   Conversely, if it is required to use d.c. filament excitation on any filament type whose data are given for an a.c. filament voltage, the value of grid bias given should be decreased by an amount approximately equal to one-half the rated filament voltage, and be referred to the negative filament terminal, instead of the mid-point as in a.c. operation.   This rule is only approximate and does not, in general, provide identical currents for both types of filament excitation.

(v) Recommended practice and operation

The following additional limitations on valve practice and operation are based partly on the recommendations of the British Standard Code of Practice (Ref. 49) also on the B.V.A. Radio Valve Practice (Ref. 48), and on established design practice.

(a) **Mounting**—(1) Unless otherwise stated it is desirable that valves should be mounted base down and in a vertical position.   Where it is necessary to depart from vertical mounting, the plane of the filament of directly heated valves should be vertical.   Similarly, the plane of the grid side rods (or major axis of the control grid) of indirectly heated valves having a high transconductance and/or a long unsupported cathode, should be vertical.   This reduces the possibility of filament- and cathode-to-grid short circuits and microphony in filament valves.

It is particularly undesirable that valves having high plate dissipation ratings should be mounted base upwards without agreement from the valve manufacturer, as this method of mounting seriously affects the flow of air around the bulb and may result in the limiting temperature being exceeded.

Depending on the distribution of the bulb temperature, gas may be released from the getter deposit under these conditions and the vacuum and emission affected.

(2) It is particularly important that the connections to floating contacts of sockets for glass based valves should be as flexible as possible and that the contacts themselves should float properly and not become rigidly locked in position.   The use of a wiring jig, having the nominal specified dimensions for the base type, inserted in the valve socket during wiring, is desirable in order to locate the socket contacts correctly, so that strain on the base pin seals is minimized when the valve is inserted.   Prior to insertion, the base pins of miniature valves should be straightened by means of a pin-straightener, if misaligned.   The pin-straightener may consist simply of a steel block drilled with countersunk holes of the correct diameter and location, to slightly larger tolerances than those specified for the pins of the base type.   For specific information and design, reference should be made to the valve manufacturer.

(3) It is undesirable to use socket contacts as connecting tags in circuit wiring because of circumstances which may arise if the valve is subsequently replaced by another

having different or additional pin connections. In particular, contact No. 1 of octal sockets should not be used in this connection, as frequently internal base shielding is brought out to pin No. 1 of octal-based valves, which should be connected to chassis. Furthermore, in glass-based valves the above-mentioned practice may adversely affect the valve characteristics by the application of a voltage to pins which are not connected to any valve element, but which project into the envelope.

(4) Valves with rigid pins in glass bases and valves without bases which have short rigid lead-out wires are normally intended to be mounted in sockets, and it is recommended that such valves should not be soldered directly into the wiring, as such connections can impose sufficient strain to endanger the seals.

(5) If valves are to be subjected to continuous vibration, means should be employed to damp out such vibration by the use of cushioned valve socket mountings.

(b) **Ventilation**

(1) The layout and design of equipment should afford sufficient ventilation to ensure a safe bulb temperature under all conditions. As a general guide, the maximum temperature of the hottest part of the bulb under operating conditions in the equipment should not exceed by more than 20°C. that temperature which would be attained if the valve were operated at its maximum ratings under conditions of free air circulation in an ambient temperature of 20°C. Where exceptional increases of temperature may occur (e.g., when valves are used in screening cans or in equipment working in tropical conditions) the valve manufacturer should be consulted on each type concerned.

The present trend in valve design is to reduce dimensions with the object of saving space and of improving efficiency at high frequencies, and the extent to which the reduction can be made is usually limited by the amount of heat which can be dissipated from the exterior surface of the bulb.

For valves of present day sizes operating at normal temperatures, about half the heat is dissipated by convection and half by radiation. It is therefore necessary to allow free convection of reasonably cool air past the bulb and free radiation from the surface of the bulb to cooler surroundings.

The increase in the temperature of air in convective cooling is very small and it is therefore more important that the flow of air should be copious and unimpeded than that it should be particularly cold. No great risk is incurred if the air is slightly warmed by passing near other components if this allows it to flow through a less obstructed path.

In order to improve radiation from a valve, surrounding surfaces should not be polished but should be kept as cool as possible. The temperature of a valve surrounded by a plated shield can, or by components at about its own temperature, may rise seriously.

(2) When valves are mounted in other than upright vertical positions, greater care should be taken to ensure that adequate ventilation is provided.

(3) Adequate ventilation is particularly important in the case of output valves and rectifiers.

(c) **Heater-cathode insulation**

(1) It is generally desirable to avoid a large potential difference between heater and cathode. This potential should not normally exceed 100 volts except in the case of certain indirectly-heated rectifier valves and valves specially designed for a.c./d.c. operation. Where a design necessitates higher heater-cathode potentials, a recommendation of the maximum permissible value should be obtained from the valve manufacturer. For convenience, the maximum heater-cathode voltage rating is usually given as a d.c. value.

(2) The insulation resistance between the heater and the cathode should not be included in r-f circuits where frequency stability is required or in a-f circuits followed by a high gain amplifier.

The leakage currents make themselves apparent as noise or hum, which may assume serious proportions if the valve in which they originate is followed by a large degree of amplification. Moreover, if the heater-cathode insulation is included in a tuned circuit, any alteration to the physical or electrical properties of the insulation will alter the frequency to which the circuit is tuned, and if both r-f and mains frequency voltages exist across the insulation, there is a risk of modulation hum, particularly in cathode-coupled oscillators and the like.

(d) **Control grid circuit resistance**

In all cases where published recommendations are available from the valve manufacturer, they should be followed. The maximum value of grid circuit resistance in the case of a particular type of valve cannot be specified without knowledge of the valve characteristics and conditions of operation. In no circumstances should valves be operated without a d.c. connection between each electrode and cathode.

The maximum value of control grid circuit resistance which may be used under any condition of operation depends initially upon the stability of performance required but ultimately upon the transconductance under that condition, the increase in cathode current and electrode dissipations which may occur due to the cumulative effects of reverse grid current (gas, grid emission and leakage) permitted by the specification, the maximum ratings for these characteristics, and the limiting effect of resistances in series with the cathode and in the electrode circuits in series with the various electrode supply voltages. The following formulae, expressing the relationships between the various resistances, direct currents and electrode parameters in typical grid controlled triodes, tetrodes and pentode circuits hold providing the control grid current is negligible in comparison with the total cathode current.

For pentodes and tetrodes (Ref. 41)

$$R_{g1} = \frac{\Delta I_k}{\Delta I_{c1}} \left[\frac{1}{g_k} + R_k \left(1 + \frac{1}{\mu_{g1 \cdot g2}} \right) + P \frac{R_{g2}}{\mu_{g1 \cdot g2}} \right] \tag{6}$$

For triodes (Ref. 41)

$$R_{g1} = \frac{\Delta I_b}{\Delta I_{c1}} \left[\frac{1}{g_m} + R_k \left(1 + \frac{1}{\mu} \right) + \frac{R_L}{\mu} \right] \tag{7}$$

where ΔI_k = change in cathode current permitted by maximum ratings of plate dissipation and/or cathode current; or change in performance which can be tolerated.

ΔI_b = change in plate current permitted by maximum ratings of plate dissipation and/or cathode current; or change in performance which can be tolerated.

ΔI_{c1} = change in control grid (No. 1) current which is likely to occur or is permitted by the valve specification.

R_{g1} = grid resistor (ohms).

R_{g2} = series screen resistor (ohms).

R_k = cathode resistor (ohms).

R_L = series plate load resistor (ohms).

$\mu_{g1 \cdot g2}$ = triode amplification factor of pentode or tetrode.

μ = amplification factor of triode.

P = ratio of screen current (I_{c2}) to cathode current

$$= \frac{I_{c2}}{I_{c2} + I_b}$$

g_m = grid-plate transconductance (mhos).

g_k = grid-cathode transconductance (mhos)

$$= g_m \frac{I_k}{I_b} \cdot$$

To determine the limiting value of grid resistor which may be used, ΔI_{c1} should be taken as the maximum value of reverse grid current permitted by the specification, under maximum electrode dissipation and cathode current conditions; ΔI_b and ΔI_k are the changes in plate and cathode currents which result in maximum rated plate and/or screen dissipations and/or maximum cathode current.

The maximum rating which is reached first determines the limiting value of the grid resistor.

If the specified value of ΔI_{c1} is not available, the highest typical value of reverse grid current given in Sect. 3(iv)A for each major group of valve types may be used with discretion. In cases where the valve is being operated below maximum rated conditions and at a total cathode current less than that for the grid current test conditions, the value of ΔI_{c1} may be taken as varying approximately as the cathode current.*

It should be clearly understood that equations (6) and (7) are fundamental, and, in cases where no maximum grid circuit resistance values are published, may be used to calculate maximum safe values. These equations are based on the margin (ΔI_b) between the operating plate current and that value of plate current which gives maximum plate dissipation or maximum cathode current or which gives the maximum tolerable change in performance, the limiting condition of operation being determined by the rating which is reached first. It may happen that the published typical operating conditions give maximum rated plate dissipation—in this case ΔI_b or ΔI_k is zero, and the equations cannot therefore be applied to calculate R_{g1} directly. In such a case the procedure is to assume reasonable values of R_{g1} and ΔI_{c1}, and then to use the equations to derive ΔI_b or ΔI_k. The operating plate or cathode current would then be made less than the permissible maximum by the margin ΔI_b or ΔI_k. Alternatively a lower value of R_{g1} may be chosen and the conditions of operation recalculated.

If a maximum value of R_{c1} is published for fixed bias operation, then the value for cathode bias may be determined from equations (8) to (13) :

For pentodes and tetrodes—general case

$$\frac{R_{g1} \text{ for cathode bias}}{R_{g1} \text{ for fixed bias}} = 1 + g_k \left[R_k \left(1 + \frac{1}{\mu_{g1 \cdot g2}} \right) + \frac{PR_{g2}}{\mu_{g1 \cdot g2}} \right] \qquad (8)$$

$$\text{or approximately} \approx 1 + g_k \left[R_k + \frac{PR_{g2}}{\mu_{g1 \cdot g2}} \right] \qquad (9)$$

$$(\text{or when } R_{g2} = 0) \approx 1 + g_k R_k \qquad (10)$$

For triodes—general case

$$\frac{R_{g1} \text{ for cathode bias}}{R_{g1} \text{ for fixed bias}} = 1 + g_m \left[R_k \left(1 + \frac{1}{\mu} \right) + \frac{R_L}{\mu} \right] \qquad (11)$$

$$\text{or approximately} \approx 1 + g_m \left[R_k + \frac{R_L}{\mu} \right] \qquad (12)$$

$$(\text{or when } R_L = 0) \approx 1 + g_m R_k \qquad (13)$$

In the case of resistance-loaded triodes and pentodes with series screen resistors, the values of R_{g1} as derived from eqn. (6) or (7) respectively will be very high. In such cases the limiting factor is the effect of the reverse grid current on the operating bias. This effect is considered in detail in Chapter 12, Sect. 2(iv) for triodes with fixed bias and Sect. 2(iii) with cathode bias also Sect. 3(v) for pentodes with fixed bias, and Sect. 3(iv)C with cathode bias.

A high value of resistance between the control grid and the cathode should be avoided wherever possible. In B.S.1106, 1943 (Ref. 49) it is recommended that, in general, 1 megohm with self-bias and 0.5 megohm with fixed bias are suitable maxima and that with receiving valves having an anode dissipation exceeding 10 watts the grid-circuit resistance should not exceed 0.5 megohm when self-bias is used and 0.1 megohm with fixed bias. These values must be reduced when one resistor is common to more than one control grid circuit.

The maximum values of grid circuit resistance with r-f pentodes controlled by v.c. is covered in Chapter 27 Sect. 3(iv)b.

Some valve types have an inherent negative resistance region in the positive portion of the grid characteristic, due to secondary emission from the grid as a result of the deposition or evaporation of active material from the cathode or deposition from the getter during manufacture. In such cases the resultant grid current may change sign from positive to negative as the grid voltage increases from zero in a positive direction, ultimately becoming positive again. The maximum value of grid circuit resistance

---

*This assumes that the grid leakage current is small compared with the ionization current, and that the latter does not increase appreciably during operation.

which may be used under these conditions without **grid blocking** occurring, is that value of resistance represented by the line drawn from the operating bias point on the abscissa, tangential to the cross-over loop of the grid current characteristic below the abscissa as shown in Fig. 2.11.

If the grid circuit resistance is such that the line cuts the loop of the characteristic below the abscissa, blocking will occur if the instantaneous values of grid voltage exceed the voltage of the point at which the line crosses the grid characteristic nearer to the origin.   See also Chapter 2 Sect. 2(iii).

Because of the difficulty of controlling the grid current commencement point of oxide-coated cathode and filament valves during manufacture, due mainly to **contact potential variations** caused by variable control grid surface conditions, it is desirable that operating conditions should be chosen, such that the control grid bias is always sufficiently large in relation to the contact potential, so that variations of the latter produce only minor effects on performance.   In general, the grid current commencement voltage of indirectly heated cathode types is negative and may have a value up to −1.0 volt but as a rule varies during life.   The grid current commencement voltage of filament types is usually positive and seldom exceeds about +0.5 volt, but usually becomes less positive during life and may even become negative.   The plate current commencement voltage of diodes due to contact potential behaves in a similar manner.

(e) Operation at low screen voltages

As the grid current commencement voltage may in certain cases be dependent on the value of screen voltage and may become more negative as the screen voltage is reduced, it is in general undesirable to operate valves having low or zero control grid-bias at low screen voltages.   In the case of zero-bias operated filament types, operation at low screen voltages may result in the grid current commencement voltage becoming negative and in the flow of positive current, causing either a change in operating conditions due to additional negative bias developed across the grid resistor or damping of tuned grid circuits due to lowered input resistance.   In either case undesirable and often wide variations in performance may result.

(f) Microphony

Small variations of electrode spacing cause corresponding variations in the output of the valve, and it is desirable to ensure that little or no vibration reaches the valve. Such vibration may reach the valve by way of the valve socket or acoustically, and it should be noted that microphonic trouble may originate in the converter or i-f stages through modulation of the signal or i-f carrier at audio frequencies.   It is recommended that, if possible, the position of the valve in relation to the source of vibration be so chosen that microphony effects will be at a minimum.   Methods which may be used to minimize microphony effects are described in Chapter 35 Sect. 3(iv).

(g) Hum

Hum due to conditions within the valves is analysed in Chapter 31 Sect. 4(i).   See also Ref. 91 on heater-cathode leakage as a source of hum.

(h) Stand-by operation

Where it is necessary to maintain cathodes and filaments at normal operating temperatures during stand-by periods, better life in general will be obtained when the equipment is so designed that **some,** rather than no, cathode current flows during such periods.   It is also essential that the heater or filament voltage be maintained as close to the rated value as possible.   Continuous operation of heaters or filaments at voltages exceeding their rated values by more than 10 per cent., without cathode current flowing, will result, generally, in progressively short life being obtained.

In equipments which are designed so that no cathode current flows during stand-by periods for periods of less than 15 minutes, the filament voltage of quick-heating filament types should be reduced to 80 per cent. of normal, while the heater voltage of indirectly heated cathode types should be maintained at normal rated value.   For longer periods, both filament and heater power should be turned off.

SECTION 2 :   CONTROL OF CHARACTERISTICS DURING MANUFACTURE

(i) Importance of control over characteristics   (ii) Basic manufacturing test specification   (iii) Systematic testing   (iv) Tolerances on characteristics

(i) Importance of control over characteristics

The successful design and continued satisfactory operation of electronic equipment require that the valves used have certain prescribed characteristics, the initial values of which do not vary from valve to valve sufficiently to affect operation seriously and which, throughout life, remain within a prescribed range with only gradual change until the end of useful life is reached   Owing however to the highly complex nature of the physical-chemical processes involved in manufacture and the difficulty of controlling the variations of many of these processes under mass production conditions, the ultimate degree of uniformity obtainable in the final product, both in initial characteristics and their variation throughout life, must always be a compromise between the performance required by the equipment designer and user and the manufacturing cost in obtaining that performance.   In any individual case, however, the degree to which variation of one or more characteristics affects operation, either initially or during life, depends entirely on the particular application and the critical features of the circuit design.   For good circuit performance, it therefore is essential to take into consideration the range of variation of the various characteristics on which operation depends and to ensure that operation, so far as is practicable, is independent of the variation of characteristics, particularly of critical characteristics and characteristics over which limited or no control is exercised in manufacture by the valve test specification.

(ii) Basic manufacturing test specification

Owing to the relatively limited range of variation of mechanical and electrical characteristics for which usability and performance of the various types of valves in their intended applications may be satisfactory, it is necessary to apply tolerances, or " limits " to control the range of characteristics obtained under mass production conditions.   For commercial applications these tolerances are chosen to be both satisfactory to the equipment designer and sufficiently wide to be economical and thus enable valves to be manufactured in large quantities at a reasonable cost.   **These requirements are embodied in the complete manufacturing test specification, which normally comprises—**

(1) Maximum permissible conditions of operation or **ratings,** as determined by emission capability, goodness of vacuum maintenance and interelectrode insulation.

(2) Nominal values for the principal physical dimensions and electrical **characteristics,** relevant to intended applications under specified operating conditions.

(3) A schedule of mechanical and electrical **tests** (including tolerances and sequence) sufficiently comprehensive to ensure that the prescribed dimensions, and electrical characteristics are maintained in production, and sufficiently severe to indicate likely failure, during life, when operated under maximum ratings.

(4) A **life test** schedule sufficiently severe in conditions of operation and permissible deterioration of the main functional characteristics, to indicate reliability and performance of the product on life, under both normal recommended and maximum rating conditions.

As most valve types are designed to give a certain performance and life in a specific application or limited range of applications, it is important to note that each type is processed accordingly and that, in general, the test specification for a particular type normally includes only those tests which are necessary for adequate and economic control of the characteristics, and are required for satisfactory performance and life in the intended applications.   Unintended applications not covered by the maximum ratings or test specification should be referred to the valve manufacturer, as performance and life in many cases can often be decided only by laboratory investigation and special life tests.

Reference to quality control : Ref. 45.   Reference to life testing : Ref. 97.

(iii) Systematic testing

For the purpose of systematic testing, valve characteristics may be divided into the following three categories.

(a) **Primary characteristics,** which are fundamental to the basic operation and life of the valve as a thermionic device and its functional operation and usability. These include the mechanical stability of the envelope and electrode structure, external physical dimensions and construction, continuity of electrode connections, characteristics indicating the goodness of the vacuum and emission and the principal functional characteristics dependent on the electrode geometry.

(b) **Secondary characteristics,** which are not fundamental to the operation of the valve, but are, in general, dependent on the primary functional characteristics and essentially determine the degree of performance obtainable in intended applications of the particular type.

(c) **Tertiary characteristics,** which are not fundamental to the operation of the valve, but are dependent on, and are in general controlled by the primary and/or secondary characteristics and either affect performance in a minor degree only, seldom vary sufficiently to affect performance or are important only in certain applications or for purposes of valve design.

In order to control the above-mentioned characteristics after manufacture in accordance with their specification, the acceptance testing procedure is usually organized into the following group of tests, the functions of which are described in detail in the following subsections.

(A)   Production Tests.
(B)   Design Tests.
(C)   Recorded Readings.
(D)   Life Tests.
(E)   Warehouse Tests.

(A) Production Tests

(1) First Testing

All primary characteristics, except those which are subject only to minor variation from valve to valve, are usually tested 100 per cent. in the factory after manufacture Tests performed on this basis are designated Production or Factory Tests and failure of any valve to meet the prescribed test limits for any production test constitutes cause for total rejection of that particular valve. When characteristics are uniform and quality satisfactory, production tests are usually limited to primary characteristics only and particular characteristics which cannot be otherwise controlled satisfactorily. The normal production tests are performed in the sequence prescribed by the specification and for all types the tests normally include interelectrode shorts and continuity, reverse grid current (gas, grid emission, leakage), cathode current, series-resistance-supplied electrode currents, transconductance, emission or equivalent dynamic characteristic test, noise and microphony.

(2) Holding period

After the schedule of production tests has been completed, it is usual for all valves to be held in storage for periods varying from 24 hours upwards to allow any deterioration of mechanical or electrical characteristics, due to unsatisfactory manufacture, to develop.

(3) Second testing

Following the holding period, all, or the most important production tests are usually repeated either on all valves 100 per cent. or on a representative sample using statistical sampling procedures which have been established as reliable, economical methods of determining either that the maximum percentage of defectives in a given quantity shall not exceed a certain value, or that the average percentage of defectives in the outgoing product shall not exceed a certain level.

When statistical sampling is used, if the product fails to meet the sampling acceptance criteria in one or more tests, or totally, the batch in question is then re-tested 100 per cent. for those tests in which failure occurred, and all defectives screened out. In addition to repeating first-production tests, tests are often included at the second production-sampling test to control characteristics, 100 per cent. testing of which may not be warranted or necessary.

Subject to the product having passed the design and life test criteria to be described, it is now usually transferred to the warehouse for shipping and distribution to manufacturers and wholesalers.

(B) Design tests^t

All important secondary characteristics are controlled by regularly testing and recording the characteristics of a relatively small sample selected at random, either from each production batch, or each day's production in the case of continuous production, after all the first-production tests have been performed and the defectives for these tests screened out. The sample size is determined by the production batch size on the daily production level, provided characteristics are reasonably uniform. Typical sample sizes used are 1 per cent. (minimum 5 valves) of each batch, or a fixed sample size of 5 to 10 valves per type per day for types in continuous production.

The tests performed on the characteristics included in this group, on the above basis, are called Design Tests and the usual procedure is that failure of more than a stipulated percentage of the valves in the sample to meet the prescribed test limits for a particular characteristic, or failure of more than a stipulated percentage of all the characteristics to meet their prescribed limits, constitute cause, initially for retesting a larger sample— usually 100 valves—for the characteristic(s) failed, and in the event of the failure being confirmed, for 100 per cent. testing of the complete batch for those characteristics in which failure occurred. Failure of any valve of the batch to meet the specification for the characteristic(s) in question is then cause for rejection.

In the case of continuous production, the usual practice is for the failed design tests to be made production tests temporarily until satisfactory control of the characteristics in question has been re-established and the design test criteria for these characteristics can again be met consistently.

Typical design test acceptance criteria require that not more than 10 per cent. of the valves of the sample fail for any one design test or that all the design tests failed do not exceed 20 per cent.

In addition to essential secondary characteristics, other characteristics which are often controlled on a design test basis include those characteristics which are not independent parameters but have their values determined by characteristics which are tested 100 per cent., also characteristics which require to be controlled for particular applications.

Design tests are performed only on valves which meet the prescribed production tests. To this end, the latter normally are repeated and recorded when design tests are performed.

(C) Recorded readings

Tertiary characteristics are controlled, in general, on the basis of criticism and correction from recorded readings of all important characteristics obtained from the design test samples, taken either daily, weekly or monthly, depending on the importance of the characteristic and its liability to variation.

Failure of characteristics tested on this basis to meet the requirements of the specification (or, where not specified, to meet the average range of variation usually maintained) in respect of the design test criteria is cause for criticism only, and not rejection ; individual valves having the characteristics indicated are not screened from the product, unless the effect of their variation on performance is likely to be serious.

Where a recorded reading test indicates continuing unsatisfactory control of any particular characteristic, it is usual for that recorded reading to be made a design test temporarily, if considered warranted, until satisfactory control has been re-established.

(D) Life tests^t

Due to the varying extent to which the mechanical and electrical characteristics may be affected by undetected and consequently uncontrolled variations of manufacturing processes, it is not possible to predict reliably by any schedule of instantaneous tests, however comprehensive, the performance obtainable from any particular valve under operating conditions, and it is necessary to rely on recorded observations of characteristic variation during continuous or intermittent operation under controlled test conditions, to determine ultimate performance. Such tests are known as Life Tests.

Provided maximum ratings have not been exceeded, the life of most oxide-coated cathode high vacuum valves, assuming that the initial emission and electrode insulation are satisfactory, is determined almost entirely by the extent to which the initial vacuum is maintained during storage and during periods of operation and non-operation, as in general the supply of emission-producing material available on the average filament or cathode greatly exceeds that required for the lives normally obtained in practice.

The average rate and uniformity of deterioration of the vacuum- and emission-dependent characteristics which normally occur slowly and uniformly in all valves during operation, and which vary from valve to valve, and any excessive deterioration of mechanical and electrical characteristics due to defective manufacture are controlled by regular life tests of small samples of each type. These life tests are operated normally under maximum rating conditions, for a specified period during which the performance determining characteristic(s) may not deteriorate beyond prescribed values, or for the period (which may not be less than a prescribed minimum) required for such deterioration to occur.

Owing to limitations of equipment and cost, regular control life tests are run usually for periods of 500 to 1000 hours on small samples of the order of 5 valves per type per week for types in continuous production. As variations in characteristics are most liable to occur early in life, readings are usually spaced at increasing intervals to give an indication of the complete life characteristic over the control periods.

A typical life test acceptance criterion of satisfactory life is that the average life obtained per valve, considering all valves of the sample, must be not less than 80 per cent. of the specified duration of the life test. The valves used must also be selected at random and comply with the production and design test acceptance criteria.

In the event of the completed product failing to comply with the life test criteria, delivery of the product is then withheld from the warehouse, until satisfactory life has been re-established.

(E) Warehouse tests

When quality is uniform, the controls normally imposed prior to packing are retests of electrode mechanical stability and continuity, and maintenance of the vacuum, by means of a suitable interelectrode hot " shorts and continuity " test during which each valve is tapped lightly with a specified mallet. Each valve is also re-inspected for any deterioration of the envelope in the case of glass valves, cementing of phenolic bases and top caps, dry or badly soldered pin and top cap connections, type designation and general mechanical finish. Pins of miniature valves are also straightened, if necessary.

(iv) Tolerances on characteristics
(a) Initial characteristics

The tolerances on types of American origin to-day are substantially those published for these types in the American JAN-1A Specification which is based on common valve manufacturing practice in that country and was compiled by a committee which included the major valve manufacturers (Ref. S2 ; for tabulated characteristics see Ref. 65).

(b) Life test end points

Apart from the information published in the relevant American and British Service Specifications, no information is generally available concerning accepted life test end points. In R.M.A. Standard Specification ET-107 (Ref. S7) there is published a list of life test end points which may be regarded as typical of the practice followed by American valve manufacturers. A valve is considered to have reached its life test end point when, at rated filament or heater voltage and specified electrode voltages, the following values have been reached :

1. 65% of rated transconductance for r-f and i-f amplifiers.
2. 50% of rated conversion transconductance and 65% of rated oscillator grid current for converter and mixer types.
3. 50% of rated transconductance for general purpose triode types.
4. 50% of rated power output for power output types.
5. 40% of minimum rated direct current for diode types.
6. 80% of rated current or voltage for rectifier types.
7. 70% of normal alternating output voltage for resistance-coupled amplifier types.

Note.—Rated values are those referred to on R.M.A. Electron Tube Characteristic Sheets under maximum typical operating conditions.

SECTION 3: METHODS OF TESTING CHARACTERISTICS*

(i) General conventions

(ii) General characteristics
    (a) Physical dimensions
    (b) Shorts and continuity
    (c) Heater (or filament) current
    (d) Heater-to-cathode leakage
    (e) Inter-electrode insulation
    (f) Emission
    (g) Direct interelectrode capacitances

(iii) Specific diode characteristics
    (a) Rectification test
    (b) Sputter and arcing
    (c) Back emission
    (d) Zero signal or standing diode current

(iv) Specific triode, pentode and beam tetrode characteristics
    (A) Reverse grid current
    (B) Grid current commencement voltage
    (C) Positive grid current
    (D) Positive voltage electrode currents
    (E) Transconductance or mutual conductance
    (F) Amplification factor
    (G) Plate resistance
    (H) A.C. amplification
    (I) Power Output
    (J) Distortion
    (K) Microphony
    (L) Audio frequency noise
    (M) Radio frequency noise
    (N) Blocking
    (O) Stage gain testing
    (P) Electrode dissipation

(v) Specific converter characteristics
    (A) Methods of operation including oscillator excitation
        (1) Oscillator self-excited
        (2) Oscillator driven
        (3) Static operation

*The assistance of A. H. Wardale, Member I.R.E. (U.S.A.) and D. H. Connolly, A.S.T.C., is acknowledged.

(B) Specific characteristics
    (a) Reverse signal-grid current
    (b) Signal-grid current commencement
    (c) Mixer positive voltage electrode currents
    (d) Mixer conversion transconductance
    (e) Mixer plate resistance
    (f) Mixer transconductance
    (g) Oscillator grid current
    (h) Oscillator plate current
    (i) Oscillator transconductance
    (j) Oscillator amplification factor
    (k) Oscillator plate resistance
    (l) Signal-grid blocking
    (m) Microphony
    (n) R-F noise
(vi) Tests for special characteristics
    (A) Short-circuit input admittance
    (B) Short-circuit feedback admittance
    (C) Short-circuit output admittance
    (D) Short-circuit forward admittance
    (E) Perveance
(vii) Characteristics by pulse methods—point by point
(viii) Characteristics by curve tracer methods

(i) General conventions

The methods described in this section are typical of good practice and should only be taken as a guide of good practice. They represent, in general, the main operational tests which are used to control the performance of various types of receiving valves.

The valve under test should have its heater (or filament) operated at the specified voltage for constant voltage types or at the specified current for constant current types. Direct current is generally specified for all characteristic testing, although a.c. may be used for indirectly heated types and directly-heated a.c. power types, but it is essential that the filaments of all directly-heated battery types be operated from direct current.

The positive electrodes (e.g., plate, screen) should be supplied from suitable direct-voltage sources of which can be adjusted to the values specified. Good regulation (i.e., good voltage stability) is highly desirable for characteristic testing but is expensive to attain with valves drawing high cathode currents. The recommended source is an electronic voltage regulator such as that of Fig. 33.10, the output terminals being shunted by a r-f by-pass capacitor (e.g., 0.01 μF mica). A large capacitance should not be used since this results in a tendency for the voltage regulator to " hunt ". In other cases all supply voltages should be adequately by-passed for a.c. components.

In the case of emission testing, a special low voltage high current source of low internal resistance is essential. Where operation is required from a.c. mains, a selenium rectifier supply is usually the most suitable and inexpensive for high current emission testing.

The negative electrodes (e.g., control grid, suppressor) do not normally draw much current, and the voltage regulation of the current source is therefore not very important, but the voltage applied to the terminals of voltage divider supplies, as normally used, must be well stabilized. Either an electronic voltage regulator or a gas type voltage regulator tube, or both, may be used. If the characteristics are to be tested with the grid positive, the " screen " source may be used for triode grids ; otherwise an additional electronic voltage regulator (of the same type as for the plate and screen) should be used. All direct electrode voltages are to be measured with respect to the cathode. In dynamic tests the voltages are to be measured under operating conditions.

In all methods of testing, depending upon the particular characteristic, well filtered supplies both positive and negative should be used.

The basic circuit for testing electrode currents with variable applied voltages is indicated in Fig. 3.1 for use with diode, triode or pentode valves and negative grid voltages. If instability is experienced with valves having high transconductance, a screen stopper of 50 to 100 ohms with or without a control grid stopper of say 500 ohms should be connected directly to the electrodes concerned. It is very important that these resistors be non-inductive and, in the case of the screen stopper, of sufficient rated dissipation. The screen stopper should be by-passed directly from the supply side to the cathode using a mica capacitor of, say, 0.01 μF capacitance.

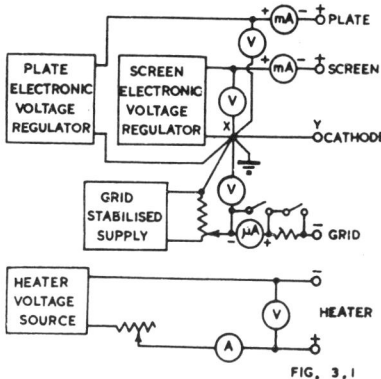

FIG. 3.1

Fig. 3.1.   Basic circuit for testing electrode currents.

Pre-heating

Pre-heating with heater voltage applied to the heater only is generally adequate for all ordinary purposes where it is necessary to conserve time, except where full electrode dissipation pre-heating is required by particular valve specifications under acceptance testing conditions, or when testing for reverse grid current (gas, grid emission and hot leakage). When full electrode dissipation is applied, the time of pre-heating is normally 2 minutes (JAN-1A) to 5 minutes (R.M.A.). Maximum valve ratings should not be exceeded during pre-heating.
See References S2, S7.

(ii) General characteristics

(a) **Physical dimensions**—Valves may be checked for overall length and maximum diameter either by the use of " go/not go " gauges or by an adjustable length gauge and outside calipers respectively.
Standard ring gauges for checking maximum and minimum base and sleeve diameters, also pin alignment gauges, are specified by R.M.A. Standard ET-106 (Ref. S6).

(b) **Shorts and continuity**—It is important to ensure that electrodes such as control and signal grids and suppressors, which do not normally dissipate any power, should not be subjected to any appreciable power dissipation during the shorts and continuity test, otherwise the valve may be damaged as the degassing treatment of these grids during manufacture is normally much lighter than that of electrodes which are required to dissipate power. In addition, the cathode current drawn, particularly in filament types, should not be excessive. It is also important that shorts be checked between the various electrode pin connections and pins designated as no connection, also that the continuity of internal jumper connections between base pins be checked.
The circuit arrangement accepted as the most generally satisfactory for both shorts and continuity with regard to sensitivity and speed of testing is that known as the hexaphase shorts and continuity test. This circuit is now widely used, particularly by valve manufacturers (e.g., Ref. 1, Fig. 54). A modification is shown in Fig. 3.2 which uses 6 110 volt ½ watt neon lamps and is entirely satisfactory for valves not requiring more than 6 independent electrodes. The values shown for this circuit have been

chosen to suit the majority of receiving type valves in that the dissipations of the normally negative electrodes are kept low for reasons previously stated—see (b) above—and
may not be satisfactory for all valve types.   In the case of a semi-universal tester it is
essential that the requirements above be observed, and also that all electrode dissipations and peak and average currents be kept within their maximum ratings, otherwise
valves may be damaged. **It is also essential as in any shorts tester that the cathode-
to-grid voltage, and battery-type filament-to-grid voltage particularly, be
kept to a value not exceeding approximately 100 volts,** owing to the small
spacing between these electrodes resulting in unreliable " flick " indications of the
neon lamps in normally good valves.   The test is carried out with the cathode hot, and
the resistor R_1 is for the purpose of limiting the peak cathode current to a safe value
(3 mA per 50 milliamp. filament strand for 1.4 volt battery valves)   The tester uses
split anode neon lamps and continuity of each valve electrode is indicated by the
glowing of one half of the split-anode neon lamp connected to that electrode.   In the
case of a short-circuit to cathode, one lamp will light on both electrodes, while in the
case of a short-circuit between two or more other electrodes, two or more lamps will
so light.   There is an optimum arrangement of connections from the terminals
A, B, C, D and E to give fairly low voltages between cathode and control grid or other
low-dissipation electrodes and to give uniform illumination on all lamps.   In some cases,
experimenting may be necessary in changing over connections to obtain fairly uniform
illumination on all lamps.   It is necessary that the insulation of the sockets and
wiring be sufficiently good to avoid residual glows on the neon lamps when a valve is not
in the socket.   This is particularly important in humid climates.

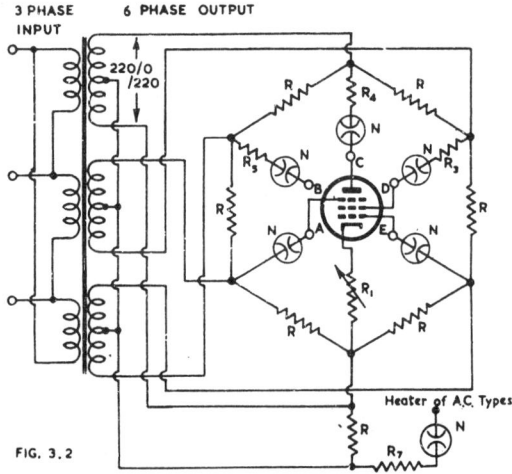

Fig. 3.2.   Hexaphase shorts and continuity tester.   Suitable values are $R = 5000$ ohms,
$R_3 = R_5 = 7500$ ohms, $R_4 = 1500$ ohms, $R_7 = 5000$ ohms, $R_1 = 50\,000$ ohms for battery
valves other than power output, 33 000 ohms for normal a.c. types, or 15 000 ohms for
power output valves.   Lamps (N) are 110 volt $\frac{1}{2}$ watt split-anode neon.

If more than six lamps are required, additional lamps may be incorporated at suitable
points in the " ring " network, but all such arrangements have the limitation that one
or more conditions exist where no shorts indication is possible on one or more pairs
of lamps.

In addition to the neon shorts and continuity test, filament type lamps may be used
to indicate continuity from shield to cathode or along a " jumper " linking two pins.

Continuity of heaters and filaments is indicated by means of a current meter of
suitable scale in series with the heater or filament, an open-circuit being indicated by

zero current. In the absence of a current meter and where a rheostat is used to drop the filament/heater source voltage, the voltmeter reading may be used to indicate open-circuited filaments or heaters since the voltmeter reading will then be much higher than normal.

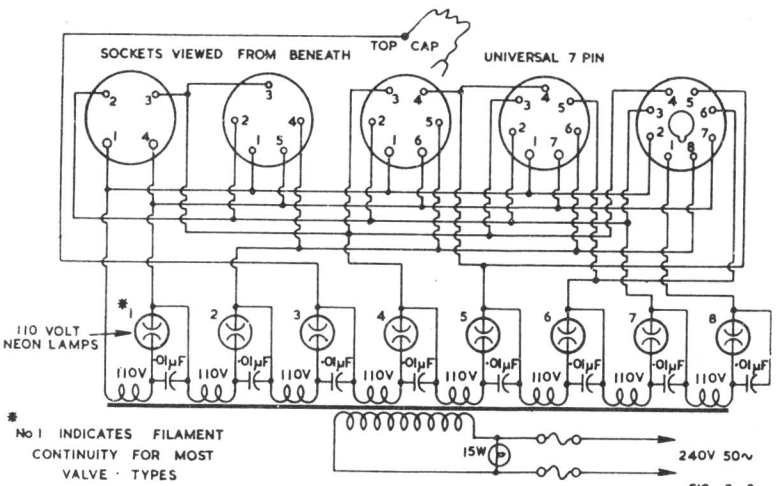

Fig. 3.3.   *Older type of shorts and continuity tester.*

The older type of shorts tester is shown in Fig. 3.3 and operates on single phase and requires a total voltage of 880 volts for 8 lamps. This may be used as a hot or cold cathode shorts tester, but has the disadvantages that it only indicates electrode continuity of those electrodes which draw sufficient current to operate lamps to which they are connected. It also has the severe disadvantage that the voltages cannot be arranged conveniently with regard to the electrode configuration for semi-universal testing, as all voltages are in phase, and quite high voltages may be developed between adjacent electrodes in the absence of complicated switching.

When using this type of tester with either hot or cold filament (or cathode), it is essential to connect grid No. 1 to the lowest voltage in the chain with respect to earth, otherwise damage may be done to the valve, or else good valves may indicate as having a short circuit. The other points in the chain should be connected in the same sequence as the grids and plate.

A fairly satisfactory form of single lamp shorts and continuity tester commonly used in service testers is shown in Fig. 3.4. The electrode switch *S* is rotated with the cathode heated. If the neon lamp glows on one side only the particular electrode is connected ; if on both sides a short-circuit is indicated. If only one electrode indicates short-circuit, the short is to heater (filament) ; if two or more electrodes indicate a short-circuit, they are shorted together. The continuity test only works well for the electrodes close to the cathode, the sensitivity decreasing towards the anode.

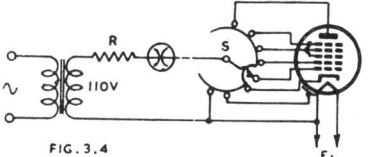

Fig. 3.4.   *Simpler type of shorts and continuity tester, employing a single lamp.*

The same circuit Fig. 3.4 may also be used for a cold cathode shorts test.

(c) **Heater (or filament) current**—This is measured with a d.c. heater source, using the heater circuit as in Fig. 3.1. The heater voltmeter current passes through the ammeter *A*, so that it is necessary for greatest accuracy to subtract the voltmeter current from the current reading. However, with a 1000 ohms-per-volt meter, the current is less than 1 mA and the error is quite small with heater currents of the order of 0.3 A. When measuring filament currents of small battery valves (e.g., 50 mA), the ammeter zero may be set to read zero when the voltmeter is at its nominal reading (e.g., 1.4 volts). The current is normally measured with no connections to any other electrodes. See also Ref. S12, Sect. 2.1.

(d) **Heater-to-cathode leakage**—This may be tested by applying the maximum rated direct voltage between one heater terminal and cathode and measuring the current with a microammeter (say 0-500 μA) with a safety resistance in series (not greater than 100 000 ohms). The leakage current should be not greater than 20 μA for ordinary 6.3 volt 0.3 ampere valves, or 50 to 100 μA for power valves, with 100 volts applied.

The heater may be operated with either a.c. or d.c. supply if below 35 volts ; at 35 volts and above it is usual to operate it on a.c. supply. The voltage between heater and cathode should be applied with both polarities, except in the case of rectifiers where the heater must be negative with respect to the cathode.

(e) **Inter-electrode insulation**—This is not normally a regular commercial test, except to meet service specifications. The test voltage specified is normally 300 volts for small valves or 500 volts for those which are able to stand the higher voltage. The insulation resistance may be measured by any ordinary type of insulation tester, but preferably by an electronic megohm-meter. In this test, the cathode should always be positive. The test is carried out on a valve which has reached a stable temperature under normal operating conditions. The minimum permissible insulation resistance usually specified is 10 megohms (Ref. S2).

This test is not used between heater and cathode.

(f) **Emission**—The purpose of the emission test is to ensure that the cathode emission is adequate to provide the peak and average space currents for the particular application of the type by a margin which has been found to be satisfactory for good life and performance.

Emission is normally tested by applying a suitable direct positive voltage to all grids connected together with the plate, and measuring the total cathode current.

Because of the very high value of the emission current normally drawn from the cathode, the resulting excessive dissipation of the inner grids and the gas produced as a result of this dissipation and also as a result of ion bombardment of the cathode, the emission test is a damaging test and should be performed as seldom as possible and then always very carefully.

In order to avoid damage to the cathode by this test, the applied voltage must be sufficiently large to approach current saturation without drawing sufficient current to damage the valve due to excessive dissipation of the inner grids. At the same time the test must be of sufficient duration to enable the stability of the emission to be indicated without being so long as to give risk of poisoning the emission. A safe value generally accepted for the duration of the emission test is 3 seconds. During this period the emission should not fall below a value recommended by the valve manufacturer. In general, emission is a minimum reading. As most of the cathode current goes to the No. 1 grid, which is usually wound with very fine wire, and treated lightly during manufacture, the applied voltage must be restricted to a value such that the emission current drawn does not result in damage to the valve. The voltages normally used in manufacture are listed below.

Emission testing with applied alternating voltage is also used, but in this case the test becomes one intermediate between a peak emission test and a high current trans-conductance test. In practice, there is no exact and simple correlation between a.c. and d.c. emission tests due to the fact that the a.c. emission current depends in part on the emission capability of the valve as well as on its geometry. Experience has shown that mains frequency a.c. emission tests tend to be unreliable owing to the fact that the emission " gets a rest " between cycles and, in a practical case, the rating of peak to average emission current drawn cannot be made sufficiently high to be effective.

Voltages given below are based on JAN-1A (Ref. S2) ; values of voltages and minimum emission currents are given by Ref. 65.

Signal diodes are normally tested for emission with an applied direct voltage of 10 volts, except that 20 volts are normally applied to type 6H6.

Power rectifiers are normally tested with applied voltages as listed below (suffixes of valve types have been omitted) :
20 volts—1V, 12A7, 32L7.
25 volts—84/6Z4, 70L7.
30 volts—5V4, 6ZY5, 7Y4, 12Z3, 25A7, 25Z5, 25Z6, 35Z3, 35Z4, 35Z5, 45Z3, 45Z5, 50Y6, 83V, 117N7, 117Z4, 117Z6.
50 volts—5Z4, 6W5, 6X5, 7Z4, 28Z5.
75 volts—5R4, 5T4, 5U4, 5W4, 5X3, 5X4, 5Y3, 5Y4, 5Z3.

Amplifier valves are normally tested for emission with an applied direct voltage of 30 volts. Some of the exceptions are listed below :
10 volts—1AB5, 6AC7, 6AG5, 7E5, 7F8, 7G8, 7V7, 7W7, 26A7.
15 volts—6BE6.
20 volts—1D8, 1J6, 2A3, 3A8, 6AB7, 6AG7, 6SD7, 6SQ7, 6SH7, 7G7, 7H7, 7J7, 7L7, 7Q7, 7R7, 7S7, 19, 28D7, 35L6, 70L7.
25 volts—117N7.
50 volts—6A6, 6B5, 6L6, 6N6, 6N7, 12A5, 24A, 53, 59, 79.
All the voltages listed above are subject to variation by different valve manufacturers.

(g) **Direct interelectrode capacitances**—The following information is based on recognised engineering practice by the manufacturers of American-type valves, and is in line with the I.R.E. Standard (Ref. S1, S12), the R.M.A.-NEMA Standard (Ref. S8) and with JAN-1A (Ref. S2).

The capacitances which are measured are direct interelectrode capacitances, and not the total (self) capacitances which are the sum of two or more direct capacitances. The measured value of inter-electrode capacitance of a valve is dependent upon the valve shield or, in the absence of a shield, upon the geometry of the external environment. It is therefore necessary for the test to be made under strictly specified conditions of valve shielding and of electrode and internal valve screen connections.

Published values of interelectrode capacitances are stated either for the condition where the valve is shielded with a specified standard valve shield (see below) or is unshielded. Interelectrode capacitances are normally measured with the cathode cold and with no direct voltage applied to the electrodes. The base pins and leads are shielded from each other and from other elements of the valve so that they and their connections do not form part of the capacitance being measured. All external metal parts integral with the valve should be connected to the cathode ; such parts include lock-in valve bases, metal base shells, and pins with no connection. When capacitance measurements are made on indirectly-heated valves the heater should be connected to the cathode unless in special cases the measurement is between the cathode only and other elements, or the heater and other elements. For valves having elements other than a control grid, a plate, and a cathode, the additional elements of the active section including internal shields should be connected to the filament or cathode by the shortest possible connections. For a multiple-unit valve structure, all elements of the other section(s) should be earthed except when reading inter-section coupling capacitances.

The test sockets should be Standard Capacitance Sockets as specified by R.M.A.-NEMA (Fig. 3.4A). Valves with top caps should be tested with the R.M.A.-NEMA Standard Cap Connectors (Fig. 3.4B).

The connections to be made to the electrodes while testing for interelectrode capacitances are given below :

Input capacitance is tested by measuring the capacitance between grid and cathode, the latter being connected to heater, screen and shields ; the plate is earthed.

Output capacitance is tested by measuring the capacitance between plate and cathode, the latter being connected to heater, screen and shields ; the grid is earthed.

Capacitance from grid to plate for all types of valves is measured with all other electrodes earthed.

In multiple unit valves, the capacitance between the grid of one section and the plate of the other is measured with all other electrodes earthed ; similarly from one plate to the other.

The input capacitance of a diode is measured between its plate and cathode, the latter being connected to heater and shields ; other sections are earthed.

The capacitance between diode-plate and grid or plate of other sections is measured with all other electrodes earthed.

In a converter, the r-f input capacitance is measured between the signal grid and all other electrodes connected together.

The mixer output capacitance is measured between the mixer plate and all other electrodes connected together.

The capacitance from oscillator grid to oscillator plate is measured with all other electrodes earthed.

The oscillator input capacitance is measured between oscillator grid and cathode, the latter being connected to heater and shields ; the oscillator plate and all electrodes of the other section being earthed.

The oscillator output capacitance is measured between oscillator plate and cathode, the latter being connected to heater and shields ; the oscillator grid and all electrodes of the other section being earthed.

The capacitance between oscillator grid and signal grid, or oscillator plate and signal grid, is measured with all other electrodes earthed.

In converters in which there is usually a r-f voltage between cathode and earth (e.g., 6BE6), the oscillator output capacitance is measured between cathode and heater, the latter being connected to screen and shields ; the oscillator grid being earthed.

The oscillator input capacitance is measured between oscillator grid and all other electrodes connected together. The capacitance between oscillator grid and cathode is measured with all other electrodes earthed.

Standard sockets, cap connectors and shields for use in the measurement of valve capacitances (R.M.A.-NEMA, Ref. S8).

(1) **Sockets**

The construction and shielding of capacitance sockets and leads shall be such that when the holes for the insertion of the base pins are covered with a grounded, flat metal plate, the capacitance between any one socket terminal and all other socket terminals tied together does not exceed 0.000 10 $\mu\mu$F for receiving valves.

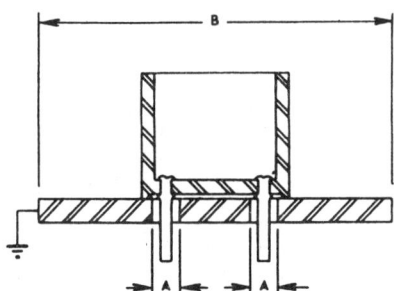

Fig. 3.4A.   RMA-NEMA Standard Capacitance Sockets (Ref. S8).

The hole for the accommodation of the locating lug of octal and lock-in bases shall be less than 0.500 inch diameter.

The diameter of holes for the insertion of the base pins (see Fig. 3.4A) shall be limited to the values shown in Table 1. The socket face plate shall be flat and shall have a minimum diameter as shown in Table 1. Any structure above the face plate shall have negligible effect on the capacitance being measured.

A thin insulating film may be permanently attached to the face plates of capacitance sockets to provide insulation for ungrounded shielding members.

The socket shall be so constructed that the base of the valve under test will seat on the face plate.

Table 1.

Base Designation	Max. diameter (A)*	Max. diameter (B)*
Standard 4, 5, 6 or 7-pin	0.250 in.	3 in.
Octal	0.175 in.	3 in.
Lock-in	0.093 in.	3 in.
7-pin miniature	0.075 in.	$2\frac{1}{2}$ in.
9-pin miniature	0.075 in.	$2\frac{3}{4}$ in.

*see Fig. 3.4A.

(2) **Cap connectors.**

Standard Cap Connectors shall be made as shown in Fig. 3.4B. A thin insulating film may be placed on surface X in Fig. 3.4B. Dimensions in inches shall be as tabulated below.

Designation	A	B	C
Medium	$\frac{21}{32} \pm \frac{1}{64}$	0.556	0.850
Small	$\frac{17}{64} \pm \frac{1}{64}$	0.352	0.750
Miniature	$\frac{11}{64} \pm \frac{1}{64}$	0.242	0.750

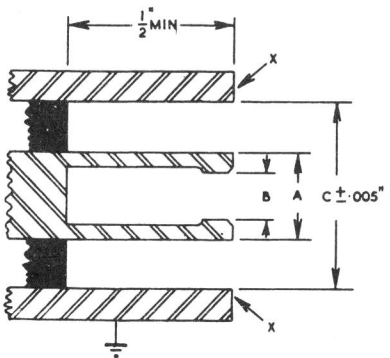

Fig. 3.4B. RMA-NEMA Standard Cap Connectors for use in measurement of valve capacitances (Ref. S8).

(3) **Shields.**

Standard shields are shown in Fig. 3.4C. It is recommended that these shields be used as indicated below, or as specified by the valve manufacturer.

Shield No.		Max. seated height with top cap	without top cap
308	GT glass types with T9 bulbs*	3	$2\frac{7}{8}$ in.
309	G glass types with T9 bulbs*	$3\frac{3}{4}$	$3\frac{9}{16}$ in.
311	G glass types with ST12 bulbs*	$4\frac{5}{16}$	$3\frac{11}{16}$ in
312	{ G glass types with ST14 bulbs*	$4\frac{13}{16}$	$4\frac{1}{16}$ in.
	{ G glass types with ST16 bulbs*	$5\frac{1}{8}$	$4\frac{3}{4}$ in.
315	Miniature types with T6½ bulbs*	—	$2\frac{3}{8}$ in.
316	Miniature types with T5½ bulbs*	—	$2\frac{3}{8}$ in.

(4) **Methods of measuring inter-electrode capacitances** are described in the I.R.E. Standard (Ref. S12). The R.M.A.-NEMA Standard (Ref. S8) states that the r-f bridge method (Refs. 77, S8, S12) and the transmission method (Refs. S8, S12) shall be the standard methods of measuring interelectrode capacitances ; both methods are described. These measurements are normally made only in valve laboratories.

*The maximum outside diameters of these bulbs are : T9 1-3/16 in. ; ST12 1-9/16 in. ; ST14 1-13/16 in. ; ST16 2-1/16 in. ; T5½ 3/4 in. ; T6½ 7/8 in.

RMA-NEMA STANDARD SHIELDS FOR USE IN MEASUREMENT OF VALVE CAPACITANCES (Ref. S8)

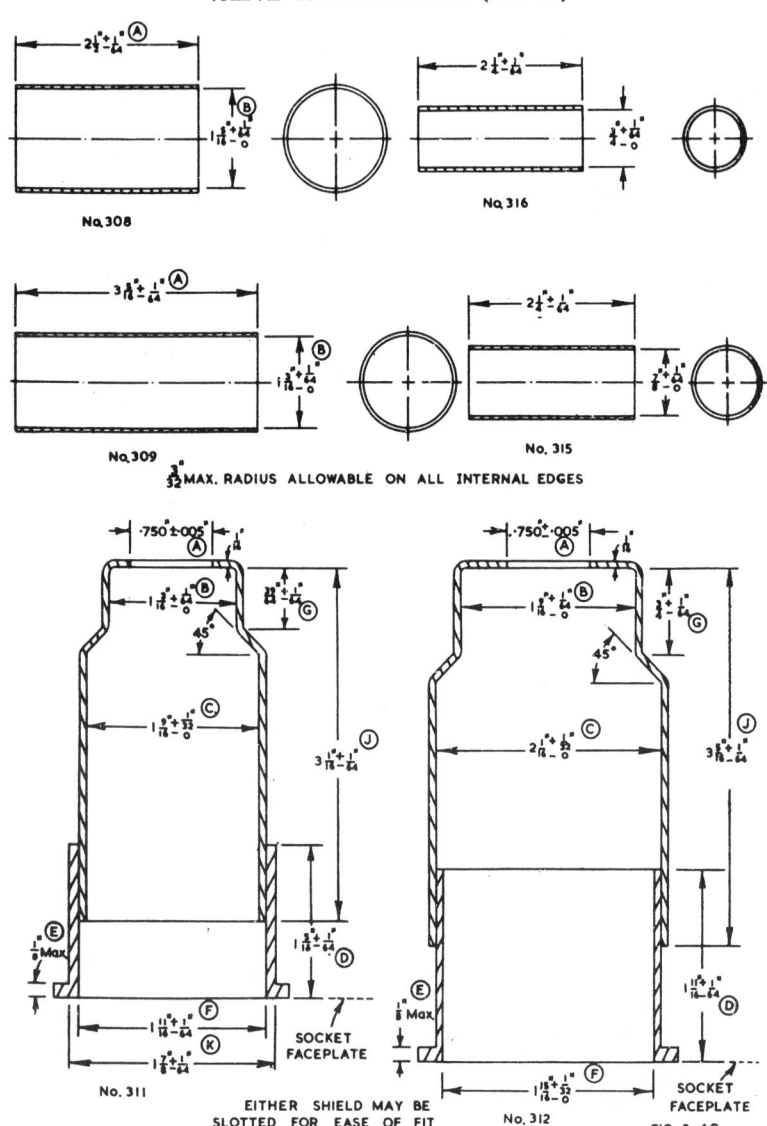

FIG. 3.4C

The significance of interelectrode capacitances and their measurement are covered by many text books and some articles including Refs. 3, 4, 5, 8, 9, 10, 11, 12, 13, 16, 18, 21, 23, 24, 25, 26, 27, 28, 29, 30, 31, 32, 74, 75, 76, 77.

The special low range capacitance bridge circuit used in the inter-electrode capacitance meter of Ref. 3 is capable of measuring directly the capacitances, with a range from

0.0001 to 2 $\mu\mu$F, between any two electrodes of a valve with all the other electrodes earthed, at a frequency of 400 Kc/s.

Capacitances above 2 $\mu\mu$F are usually measured on a simple capacitance bridge at 1000 c/s. Commercial bridges having a logarithmic scale are available for this purpose, reading from 0 to 30 $\mu\mu$F. However, audio-frequency bridges are out of the question for very low capacitances, because the insulation resistance between electrodes may be lower than the reactance of a 0.0001 $\mu\mu$F capacitance. Even at 1000 Kc/s the reactance of this value of capacitance is 1500 megohms.

The capacitance measured with an audio-frequency capacitance bridge is frequently higher than that measured with a capacitance bridge operating at radio frequency. This effect is associated with the presence of resistive films on the insulators supporting the valve electrodes. Such films may be caused by deposition of getter or by the evaporation of metal from the cathode either during the manufacture or during the life of the valve. The value of the resistance may be as high as 1000 megohms or more, but is capable of modifying the measured value of the capacitance and prevents the prediction of the performance of the valve at frequencies higher than that of the measurement (Ref. 23).

The actual capacitance of a valve depends on its operating conditions and the impedances in the electrode circuits. For purposes of specification testing the valve is tested with the cathode cold and without any voltages applied to the electrodes. In any particular case under specified operating conditions the interelectrode capacitances will differ from the static capacitances measured in this way. For further information see Chapter 2 Sect. 8(iii) and the references there listed.

(iii) Specific diode characteristics

(a) **Rectification test.** **Power diodes**—The rectification test is an operation test as a rectifier at the maximum ratings for applied r.m.s. voltage with condenser input filter, and with the maximum rated values of average and peak currents. The basic circuit is shown in Fig. 3.5 in which T_1 is the plate transformer and T_2 is the filament transformer. The average current is measured with the d.c. milliammeter, while the voltage across R_L may either be calculated from the resistance and current or may be measured by the direct voltmeter V. The condenser C_L should be an impregnated paper type.

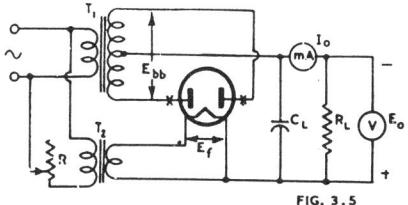

Fig. 3.5. Rectification test.

FIG. 3.5

The plate supply impedance is that of transformer T_1 which should be one having good regulation and low winding resistance.

In a transformer the plate supply impedance per plate is given approximately by

$$R_S \approx N^2 R_{pri} + R_{sec} \qquad (1)$$

where N = voltage ratio of transformer at no load (primary to half secondary in the case of full-wave rectification)

$$= \frac{\text{(half) secondary voltage}}{\text{primary voltage}} \text{ at no load}$$

R_{pri} = resistance of primary winding in ohms

and R_{sec} = resistance of secondary winding in ohms (or half-secondary in the case of full-wave rectification).

If the transformer plate supply impedance per plate is less than the required value, two equal resistances should be added at points XX in Fig. 3.5 to make up the deficiency. If the transformer plate supply impedance per plate is greater than the required value, the test will not be so severe since it will reduce the peak plate current and also reduce the output voltage or current.

In view of this limitation, a rectification test which may be used and which should be satisfactory for all practical purposes is to make use of the maximum published conditions with a condenser input filter. These conditions include the maximum r.m.s. applied voltage, the maximum d.c. load current, the minimum plate supply impedance per plate and the maximum filter capacitance for these conditions, if published. In cases where the latter information is not available, it is usually safe to test with the maximum value of filter capacitance used in the published curves or typical operation data, provided that the plate supply impedance per plate is equal to the minimum rated value or to the value used for the curves or typical operation data, whichever is the greater. In all cases it is essential that the maximum peak plate current rating per plate be not exceeded.

The rectification test may be used for the purpose of checking the output voltage and so determining end of life. The circuit constants for the rectification test may be obtained by the following procedure. The value of R_L is given by E_0/I_0 where E_0 and I_0 are the direct voltage and current respectively across the load resistance under maximum ratings. The value of E_0 is obtainable from the usual operational curves published by valve manufacturers, and is the direct voltage for the maximum r.m.s. applied voltage at the maximum value of load current (I_0). If curves are published for more than one value of filter capacitance, the curve corresponding to the highest capacitance should be selected. The rectification test should be carried out with the value of C_L equal to that used for the derivation of E_0, while the total effective plate supply impedance per plate should be as specified above. Under these conditions an average new valve should give a voltage E_0 and a current I_0 approximately equal to the published values, and individual valves will have values either higher or lower than the average as permitted by the manufacturing tolerances. The end-of-life point is commonly taken as 80 per cent. of the value obtained with a valve having bogie characteristics (R.M.A., Ref. S7).

The rectification test may be carried out either with full-wave rectification as in Fig. 3.5 or with half-wave rectification, each unit being tested separately with half the total average load current. The test is usually performed by inserting a cold valve and waiting until it has attained normal temperature.

In all cases, rectifier valves are required to operate satisfactorily in the rectification test without arcing or sputtering.

Where valve failure has been due to arcing or sputtering, the conditions under which this occurred may be reproduced for the purposes of testing either by

(1) allowing the valve to heat up to normal temperature from cold, the heater and plate supply voltages being applied simultaneously, or

(2) operating the valve at normal temperatures with heater and plate voltages applied, and then " keying " the plate voltage.

(b) Sputter and arcing

Sputter and arcing are closely akin. When an indirectly-heated close-spaced rectifier is warming up with plate voltage applied, there is no space charge and the electrostatic field gradient at any sharp peaks on the cathode surface may be high enough to vapourize the coating material and lead to the formation of an arc. Even when there are no sharp points on either cathode or plate, a discharge of gas leading to an arc may occur when there is no space charge, and the peak current required is greater than the available emission. Sputter does not usually occur with directly-heated rectifiers.

Signal diodes may be tested in a rectification test as an alternative or addition to the emission test. Diodes in diode-amplifier valves may be tested as half-wave rectifiers with $R_L = 0.25$ megohm, $C_L = 2\,\mu$F and an applied voltage of 50 volts r.m.s.

(c) **Back emission**—Back emission is emission from the plate to the cathode during the half-cycle when the cathode is positive with respect to the plate. Any appreciable amount of back emission results in severe bombardment of the cathode and ultimate plate-to-cathode arcing. Back emission may be tested by the circuit of Fig. 3.6. Switch S_1 is normally closed, and is only opened momentarily while taking a reading. The combined effect of contact potentials and initial electron velocities in the 6X5 and 6H6 rectifiers is to cause a residual current of perhaps 400 μA to pass through the back emission meter while S_1 is closed. The back emission current is then taken as the increase in the meter reading when S_1 is opened.

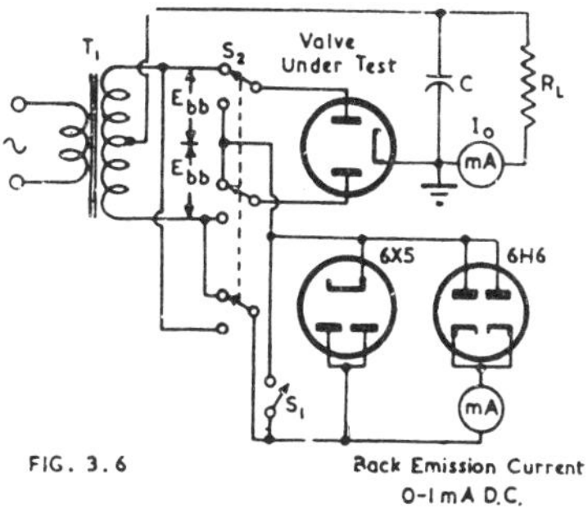

FIG. 3.6   Back Emission Current 0-1 mA D.C.

Fig. 3.6. Back emission test.

(d) **Zero signal or standing diode current**—This is usually read by measuring the current through, say, 0.25 megohm connected between diode plate and cathode with no signal applied ; the standing diode current is likely to vary during life. It should be noted that the diode direct plate voltage required to make the plate current just zero should never be positive if loss of detector sensitivity at very low signal levels is to be avoided.

(iv) Specific triode, pentode and beam tetrode characteristics
(A) Reverse grid current

Reverse grid current is the sum of gas current, leakage current and grid (primary) emission current. The test is normally carried out after pre-heating—see Sect. 3(i) above—since this is the condition which obtains under normal operation in equipment. Reverse grid current is measured by inserting a microammeter* in the grid circuit in series with a resistor, with maximum plate and screen voltages and dissipations and maximum cathode current. The circuit used is shown in Fig. 3.1 in which the microammeter is protected by a shorting key which is only opened when the reading is to be taken. The maximum value of the grid circuit resistance during the test as specified by JAN-1A should not exceed 100 000 ohms, although values up to 500 000 ohms are commonly used. The reading is normally taken after the reverse grid current has reached a steady value.

Alternatively, a suitable value of grid resistor (R_c) is connected in the grid circuit, with a switch for short-circuiting the resistor when desired. With the switch closed, the grid and plate voltages are adjusted to the desired values and the plate current is read. The switch is then opened and the grid bias readjusted so that the plate current returns to its former value   The reverse grid current may be computed from

$$I_c = \Delta E_c / R_c \qquad (2)$$

where ΔE_c =change in grid voltage to maintain constant plate current.
When the mutual conductance is known (or as an approximation using the nominal value) the reverse grid current may be computed from

Grid current in microamperes $= 1000 \; \Delta I_b / g_m R_c$

where ΔI_b =change in plate current in milliamps.
     g_m =mutual conductance in micromhos
and R_c =resistance of grid resistor in megohms.

Typical values of maximum reverse grid current in new valves, based on American JAN-1A Specifications (Ref. S2).

These values apply to types commonly used in radio receivers. They are only useful as a general guide, since commercial valves may be tested to slightly different specifications

---

*Alternatively, an electronic microammeter may be used, incorporating a less sensitive indicating instrument. One possible form uses a 6SN7-GT twin triode, cathode coupled ; the maximum resistance in the grid circuit is 50 000 ohms for a 0-3 μA range. This arrangement has the advantage that the instrument can be made self-protecting so that the indicating meter is not damaged by excessive gas current.

High-mu triode ($\mu \nless 65$)	0.5 to 0.6 μA
Medium-mu triode	1.0 to 2.0 μA
Twin triode—medium mu	1.0 to 1.5 μA each section
high mu ($\mu > 65$)	0.5 to 1.0 μA each section
R-F pentode	
sharp cut-off (generally)	0.5 to 1.0 μA*
remote cut-off	1.0 to 2.0 μA
Power output	2.0 to 4.0 μA
Converter	1.0 to 2.0 μA

*Exceptions are : 6AK5 & 713A 0.1 μA, 6AS6 0.2 μA, 6AJ5 0.25 μA, 1231 & 7G7/1232 1.5 μA.

These reverse grid currents apply rigidly only when measured under the specified electrode conditions (for values see Ref. 65, Sec. 14). These are, in general, identical with the published typical operating condition or, if there is more than one, at the typical operating condition having the greatest plate dissipation.

The maximum value of grid current which may be used in any particular application without seriously affecting the performance, is dependent upon the actual value of grid circuit resistance—see Sect. 1(v)d.

(a) **Gas current (ionization current)**—Gas current cannot be readily measured directly† It may be determined approximately by measuring the total reverse grid current under conditions of maximum dissipation (see above), and then subtracting the hot leakage current and grid emission current under the same conditions.

(b) **Cold and hot leakage**—Leakage currents may be measured with a voltage applied between the grid and each electrode in turn, with the other electrodes floating. Measurements may be made either with the valve cold (" cold leakage "), or immediately after the filament (or heater) has been switched off after testing for reverse grid current (" hot leakage "). The grid to cathode hot leakage current, with the grid biased as for the negative grid current test, may be too small to measure on the microammeter ; in this case it may be regarded as zero in deriving the gas current. If any grid emission is present, the procedure will be as indicated in (c) below.

(c) **Grid (primary) emission**—Grid primary emission current may be derived by operating the valve under maximum rated voltages and dissipations for, say, 5 minutes, or until the characteristics have reached stable values, then by increasing the grid bias to the point of plate current cut-off without any other change in electrode voltages, and by measuring the negative grid current at this point. This negative grid current is then the sum of hot leakage and grid emission currents. There is no simple method of distinguishing between real grid emission current and those hot leakage currents which fall during the first few seconds as the valve cools. However, it is generally possible to distinguish hot leakage and grid emission by plotting the cooling curve of the reverse grid current versus time characteristic (this test cannot be applied to valves of the pentagrid type).

Grid current characteristics are described in Chapter 2 Sect. 2(iii).

(d) **Grid secondary emission**—The amount of secondary emission from the grid of a valve cannot be measured directly. Valves of the same type, however, may be compared as to the relative amount of secondary emission present by plotting, or tracing on a C.R.O., the positive grid characteristic. If the amount of secondary grid emission is sufficient for the characteristic to cut the axis, blocking may occur—see Sect. 3(iv)N.

(B) **Grid current commencement voltage**

The circuit of Fig. 3.1 is used except that the polarity of the grid microammeter is reversed. The grid voltage is then measured for which the grid current is positive and of the smallest value discernible on the microammeter (e.g., 0.2 μA).

(C) **Positive grid curren**

This test applies usually to zero biased r-f pentodes, particularly filament types, to ensure that conduction does not occur during service when low screen voltages are used. The test is normally performed under low screen voltage conditions, with other

---

†Gas current may be measured by the method which converts the desired ion current to an alternating current by modulation of the ionizing electron stream while leaving undesired stray currents (e.g., leakage) unmodulated. See Ref. 40.

electrode conditions normal. It is a matter of experience that the grid current commencement voltage moves in a negative direction in some filament type pentodes as the screen voltage is reduced.

(D) **Positive voltage electrode currents**
The currents of the positive electrodes may be measured at the electrode voltages specified under typical (or any other) operating conditions, by using the circuit of Fig. 3.1. The valve may be damaged if it is operated, even momentarily, at electrode currents and dissipations in excess of the maximum ratings.

The electrode currents may tend to drift, particularly when the valve is operated at

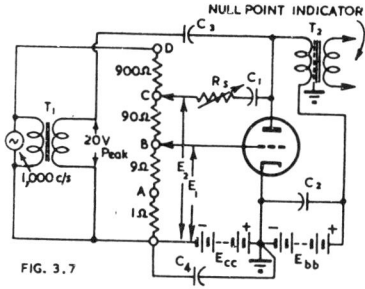

FIG. 3.7

Fig. 3.7. Dynamic voltage-ratio method of measuring mutual conductance (or transconductance).

other than the cathode current at which it has been stabilized. If a series of curves is to be plotted, it is desirable to check one of the earlier readings at intervals during the test, to see whether the valve has drifted.

(E) **Transconductance or mutual conductance***
Transconductance may be determined from the slope of a characteristic which has been plotted carefully. An approximation to the grid-plate transconductance, which generally errs on the high side, is to measure the plate current at equal grid-voltage increments on either side of the desired voltage, and to calculate the slope—see Chapter 2 Sect. 2(ii).

Grid-plate transconductance and mutual conductance may be measured by many methods, but the most satisfactory and most generally used laboratory method is the dynamic voltage ratio method developed by Tuttle (Ref. 2).

A simple adaptation which is satisfactory for all practical purposes is described below. In Fig. 3.7 a 1000 c/s generator is connected to a voltage divider and voltages E_2 and E_1 are applied to the plate (through R_s and C_1) and grid respectively of the valve under test. Resistor R_s is then adjusted to give a null point on the indicator. Condenser C_3 is provided to balance out reactive currents arising from stray capacitances : the secondary of transformer T_1 should be connected in the direction which allows capacitive balancing.

At balance, the value of mutual conductance is given by

$$g_m = \frac{E_2}{E_1} \cdot \frac{1}{R_s}. \tag{3}$$

One convenient form is shown in Fig. 3.7 where E_1 and E_2 are adjustable in steps of 10, while R_s is variable from 5000 minimum to 100 000 ohms maximum.

Range	Grid tap	Plate tap	E_2/E_1	R_s min. ohms	R_s max. ohms
1-10 μmhos	C	B	1/10	10 000	100 000
10-100	B	B	1	10 000	100 000
100-1000	B	C	10	10 000	100 000
1000-20 000	A	C	100	5 000	100 000

Values of the other components may be :
$C_1=0.5$ μF, $C_2=1$ μF, both 400 V working
C_3=one gang condenser section
T_1=step up transformer, 1 : 2 turns ratio
T_2=electrostatically shielded output transformer, with primary resistance not greater than 50 ohms.

The reactances have been selected to give an error in mutual conductance not greater than 1 per cent. in each case.

*For definitions, see Chapter 2 Sect. 1.

Pentodes may be tested if a suitable screen supply is added to Fig. 3.7.

Further information on methods of measuring mutual or transconductance is given by Refs. 2, 15, 16, 17, 18, 21, 22, 59, 78, 79, 80, 81, 82, 83, 84, S12.

(F) **Amplification factor**

The amplification factor is usually tested by some form of bridge circuit as in Fig. 3.8. This is suitable for measurements of μ up to 1000. Suggested values of components are :

$R_1 = 10$ ohms, $R_2 = 90$ ohms, $R_3 = 900$ ohms, $R_4 = 10\,000$ ohms max., $L_1 = 5$ henrys min., $L_2 = 20$ henrys min. at max. plate current. $C_1 = 0.5$ μF, $C_2 = 0.5$ μF, C_3 may be two paralleled sections of a gang condenser.

Choke L_1 should be of very low resistance, so that the voltage drop caused by any reverse or positive grid current does not affect the point of operation. Transformer T_1 should be electrostatically shielded.

The amplification factor is given by

$$\mu = \frac{R_4}{R}$$

where $R =$ effective total grid resistance in grid arm of voltage divider (i.e., either R_1 or $R_1 + R_2$ or $R_1 + R_2 + R_3$).

With R_4 adjustable from 1000 to 10 000 ohms, position 1 of S_1 will give a range of μ from 100 to 1000, position 2 from 10 to 100, and position 3 from 1 to 10. A satisfactory value of signal voltage is about 10 volts peak. Too high a value will result in operation over a non-linear portion of the characteristic, while also introducing the possibility of positive grid current with some types of valves operated at low grid bias.

Capacitance currents may be balanced out by adjusting the value of C_3 and by moving switch S_2 to the position giving capacitance balance.

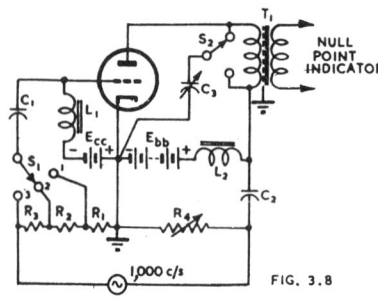

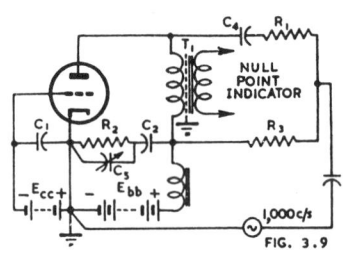

Fig. 3.8. *Bridge test for measuring amplification factor.*

Fig. 3.9. *Method of testing for plate resistance.*

The amplification factor test of Fig. 3.8 may be used for pentodes with the addition of a suitable screen supply. For values of amplification factor over 1000 a higher ratio of the plate and grid arms of the bridge is required, that is R_4/R. This will result in some loss in accuracy at the higher values of μ due to capacitance effects.

References to measurement of mu factor and amplification factor : Refs. 2, 15, 16, 17, 18, 21, 22, 78, 79, 80, 81, 82, 84, S12.

(G) **Plate resistance**

Plate resistance may be measured conveniently by the dynamic voltage ratio method developed by Tuttle (Ref. 2) and an adaptation is shown in Fig. 3.9. The value of plate resistance is given by

$$r_p = \frac{R_1 R_2}{R_3} \tag{4}$$

where R_1, R_2 and R_3 may have any convenient values of resistance.

One possible combination of resistances is :

Range of r_p	R_1 (fixed) ohms	R_2 (fixed) ohms	R_3 (variable) ohms
1000 to 10 000 ohms	10 000	1000	10 000 to 1000
10 000 to 100 000 ohms	100 000	1000	10 000 to 1000
0.1 to 1.0 megohm	100 000	10 000	10 000 to 1000
1 to 10 megohms*	100 000	100 000	10 000 to 1000

*On this range the accuracy is poor.

This circuit may be used for pentodes with the addition of a screen supply. Under these conditions C_2 should have a minimum capacitance of 1 μF and C_4 a minimum capacitance of 0.1 μF (paper dielectric, 400 V working) for a maximum error of 1% in each case, while choke L_1 should have an inductance of at least 12.5 henrys for a maximum error of 1% (except on the maximum range), for a frequency of 1000 c/s.

On the maximum range, the inductance of L_1 must be at least 125 henrys for a maximum error of 1 per cent., but the plate current through L_1 will normally be only a few milliamperes on this range. Thus two separate chokes will be required if the maximum range is to be used. Transformer T_1 should be electrostatically shielded, and its primary should have a resistance less than 50 ohms. Condenser C_3 is merely a blocking condenser of any convenient value, connected in series with the 1000 c/s source ; it does not affect the readings.

Condenser C_5 is for the purpose of balancing out capacitance currents ; it may be a single section of a gang condenser.

The combination of mutual conductance, amplification factor and plate resistance into a single unit introduces serious design difficulties and necessitates difficult compromises. A more elaborate combined bridge using Tuttle's method has been marketed by the General Radio Company and the theory is published in Ref. 2.

Plate resistance may also be measured by the well known bridge circuit (e.g., see Refs. 8, 9, 18, 21, 22, S12). The quadrature current may be balanced out by a small capacitor in the arm of the bridge adjacent to the arm containing the unknown.

References to the measurement of plate resistance : Refs. 2, 15, 16, 17, 18, 21, 22, 33, 78, 79, 80, 81, 82, 84.

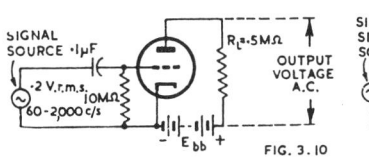

Fig. 3.10. Method of testing for a.c. amplification (after RMA).

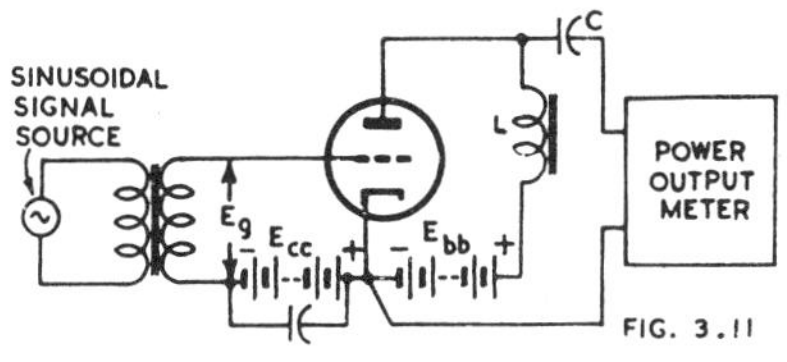

Fig. 3.11. Method for testing for power output.

(H) A.C. amplification

The standard R.M.A. test for a.c amplification (Ref. S7) is shown in Fig. 3.10. This is applied to audio frequency voltage amplifier pentodes, and to a-f voltage amplifier high-mu triodes   This is a resistance-coupled amplifier test normally performed at zero bias with a 10 megohm grid resistor. The internal impedance of the signal source should not exceed 2500 ohms. The test is made at rated heater or filament voltage and normally with a plate load resistor of 0.5 megohm, and an appropriately by-passed screen resistor in the case of a pentode. In performing this test, it is important to specify whether the output voltage is to be read across the plate load resistor alone (as in the R.M.A. test of Fig. 3.10) or across a capacitively-coupled following grid resistor of specified value. It is also important, if accurate results are to be obtained, that the valve voltmeter used be true r.m.s. reading—that is, one having a square-law characteristic.

(I) Power output

Power output from the plate may be measured under the conditions indicated in Fig. 3.11. The peak value of the sinusoidal voltage E_g applied to the grid is normally equal to the bias voltage E_{cc}, unless otherwise specified. The signal frequency is preferably either 400 or 1000 c/s, in order to reduce the size of L and C. The impedance of the choke L at the signal frequency should be at least, say, 7 times the impedance of the load, while the reactance of C should be less than, say, one seventh of the load impedance.

The power output meter for approximate and routine testing may be of the rectifier/d.c. meter type. In this case the load resistance will be incorporated in the power output meter as drawn in Fig. 3.11.

For accurate measurements, however, it is essential to measure either the true r.m.s. voltage across, or the current through, a load resistor of specified value and unity power factor. Generally the latter measurement is the more convenient as the current through the load may be measured by means of a thermo-couple ammeter.

The power output of a Class B amplifier may be measured either with push-pull operation or on each section separately. In the latter case the load resistance is one quarter that of the plate-to-plate value for push-pull operation, and the unit not under test is tied to the cathode. The power output from one section, owing to shifting of the loadline, is greater than half the power output with push-pull operation (Ref. 58), but the relationship must be established experimentally in each case. Alternatively, a resistive load may be connected directly in the plate circuit to give a closer approach to normal push-pull operation. The value of the resistance is arranged to be one quarter of the plate-to-plate load resistance with push-pull operation, and the plate supply voltage equal to the specified plate voltage for normal push-pull operation. Under these conditions the power output is given by (Ref. 58) :

$$P_o \approx 2.47 \ (I_b - 0.25 I_{bo})^2 R_L \qquad (4a)$$

where P_o = power output from one section, in watts

I_b = direct plate current in amperes, as measured by a d.c. meter*

I_{bo} = plate current with no signal, in amperes,

R_L = load resistance per section, in ohms.

The value of I_{bo} may be taken, with sufficient accuracy for many purposes, as the publ shed value.

In ᵗall cases with Class B amplifiers the specified value of input voltage is applied to the grid circuit in series with a specified value of resistance. The plate voltage source should have good voltage stability, an electronic voltage regulator being satisfactory. If a plate source impedance is specified, a series resistor of the appropriate value must be inserted in the plate supply circuit.

The relevance of a particular output measured under specified conditions is significant only in relation to the distortion present under those conditions.

(J) Distortion

Distortion in the power output test may be measured by one of several suitable methods, provided that the signal source is truly sinusoidal. The effect of harmonics from the signal source may be reduced by connecting a low-pass filter between the signal source and the grid.

The distortion may be measured by a harmonic analyser, of which several types have been described in the literature. When merely the value of the total harmonic distortion is desired, as in determining the undistorted output, those analysers which measure the root-mean-square value of all harmonics present are preferable to those which measure the separate harmonics.

The method of Suits (Ref. 85) is a particularly good example of the type of analyser which measures the harmonics separately. The Suits method requires only the simplest apparatus, and where laboratory facilities are limited this advantage may outweigh the disadvantages involved in the computation of the total harmonic distortion.

The Belfils analyser (Refs. 86, 87) utilizes an alternating-current Wheatstone-bridge balance for the suppression of the fundamental, and is particularly useful for direct measurement of the total harmonic distortion. For maximum convenience, the frequency of the audio-frequency source should be very stable. This instrument can be operated so that it is direct reading, by maintaining a constant input voltage.

In the McCurdy-Blye analyser (Ref. 88), low- and high-pass filters are used to separate the harmonics from the fundamental. This instrument is superior to the Belfils type in that the frequency of the source may vary somewhat without necessitating readjustment.

A differential analyser especially designed for power-output work has been described by Ballantine and Cobb (Ref. 89).

If an iron-cored choke is employed for shunt feed in the plate circuit (Fig. 3.11), care should be exercised in its selection or design to avoid the generation of harmonics in it due to the non-linear and hysteretic behaviour of the iron.

For measurement of distortion see Chapter 14 Sects. 2, 3 ; Chapter 37 Sect. 3.

(K) **Microphony**

There is no published standard test for microphony outside of Service or manufacturers' specifications. However an indication of the microphony of valves may be obtained using the same conditions as the R.M.A. standard audio frequency noise test (see below). It is reasonable to adjust the amplifier gain to be of the same order as the gain of the subsequent portion of the amplifier in which the valve is to operate. Nevertheless, valves unless specifically designed for low-level high-gain pre-amplifier use, should not be expected to be non-microphonic and free from noise at low levels when tested under such conditions.

In radio receiver factories the valves are usually tested in the chassis in which they are to operate, the set being tuned to a signal generator with internal modulation, and the volume control set at maximum. Any microphonic sounds which quickly die away are not considered as cause for removal of the valve, but a sustained howl is so considered. Such a valve may not be microphonic in another position in the chassis or in another model of receiver. Actions to be taken to reduce microphony have been described in Chapter 35 Sect. 3(iv). Notwithstanding the considerations above, all valves cannot be expected to be non-microphonic when placed directly in the acoustic field of, and close to, the loudspeaker in a receiver, particularly in small cabinets.

Ref. S2.

(L) **Audio frequency noise**

The R.M.A. standard for audio frequency noise test is with normal voltages applied to all electrodes, and the plate of the valve under test is coupled to the input of an amplifier with a frequency range of 100 to 2000 c/s ($\pm$5 db of response at 400 c/s). A dynamic loudspeaker with a rating of at least 2.5 watts is connected to the output of the amplifier. The minimum external grid resistance is 100 000 ohms. The gain of the amplifier and the plate load resistance of the valve under test are to be as specified. The valve under test may be tapped lightly with a felt or cork mallet weighing not over $\frac{1}{2}$ ounce. Any objectionable noise heard in the loudspeaker, is cause for rejection (Ref. S7). See also Ref. S2.

(M) **Radio frequency noise**

The R.M.A. standard test for radio frequency noise is for normal voltages applied to all electrodes of the valve under test ; the plate is coupled to the input of a r-f amplifier (of a radio receiver) at a frequency within the range of 50 to 1600 Kc/s. The minimum band-width is to be 5 Kc/s at 10 times signal. The audio frequency response is to be within $\pm$5 db of the response at 400 c/s over the frequency range from 100 to 2000 c/s. A suitable tuned transformer giving a resonant impedance of 50 000 to 200 000 ohms is to be used to couple the valve under test to the amplifier. The external grid circuit impedance of the valve under test is to be from 5000 to 25 000 ohms. Other conditions of test are as for the audio frequency noise test (Ref. S7). See also Ref. S2.

(N) **Blocking**

Tests on grid blocking are normally applied only to certain types subject to this characteristic. A valve may be tested for blocking by either of two methods. In the first method a high positive grid voltage (say 100 volts) is applied instantaneously with normal electrode voltages and the maximum value of grid resistance. The plate current should return to its initial value on the removal of the applied voltage.

The second method (" cross-over ") is to apply a gradually increasing positive grid voltage with substantially zero grid circuit resistance, and to plot the grid current characteristic for positive values of grid voltage up to the point at which the grid current changes from positive to negative (" cross-over " point). See Chapter 2 Sect. 2(iii).

(O) **Stage gain testing**
A circuit such as that shown in Fig. 3 11A can be used to measure the gain of a r-f pentode operating under typical i-f amplifier conditions. Stage gain testing has the advantage of evaluating at the one time and under operating conditions at the appropriate frequency the effects of all valve characteristics which influence gain, such as mutual conductance, output impedance, input resistance, reverse grid current and even the effect of high screen or plate currents in increasing control grid bias or decreasing screen grid voltages

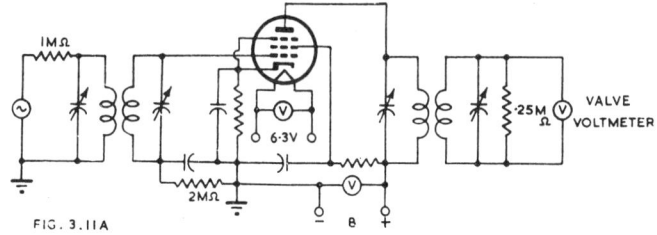

FIG. 3.11A

Fig. 3.11A. Method of testing for i-f stage gain.

In Fig. 3.11A the input and output circuits are typical high-impedance i-f transformers with secondary tuning of the first and primary tuning of the second, adjustable from the front panel of the tester to allow accurate alignment with varying capacitances in the valves under test. The 1 megohm resistor represents the plate impedance of a preceding converter stage and the 0.25 megohm resistor simulates the impedance of a diode detector. Screen and bias resistors are designed to give the required voltages with a valve of bogie characteristics, and the high impedance in the control grid circuit allows the effect of reverse grid current to be evaluated.

The output meter is calibrated to show stage gain in db for a given input at 455 Kc/s and limits are marked on the scale at a suitable number of db above and below the stage gain of a bogie valve.

(P) **Electrode dissipation**
The dissipation of an electrode is determined by subtracting the power output from that electrode —see Sect. 3(iv)I—from the d.c. power input to that electrode. Where the power output from an electrode is zero, the electrode dissipation is then equal to the d.c. power input.

(v) **Specific converter characteristics**
(A) **Methods of operation including oscillator excitation**
A converter valve essentially includes two sections, an oscillator and a mixer, which may be either separate units supplied by independent cathode streams as in a triode-hexode or combined to use a single cathode stream as in the pentagrid types. The characteristics of a converter depend both on the electrode voltages of the mixer and oscillator and on the excitation of the mixer by the oscillator.

For purposes of testing, the oscillator may be either self-excited or driven by an external source of alternating voltage of suitable amplitude ; alternatively a limited range of characteristics may be determined under purely static conditions with direct voltages only applied to the electrodes. In general, the latter is not recommended for routine testing outside of valve factories, owing to the practical difficulties of correlating characteristics with published characteristics obtained under oscillating conditions.

(1) **Oscillator self-excited**
This is the normal method of operation in most radio receivers, and although it has much to commend it as a simultaneous test of both mixer and oscillator sections, the

oscillator circuit characteristics require to be specified exactly and closely controlled in order to obtain consistent results. For general service testing, less critical methods are therefore to be preferred. This method of testing, with a self-excited oscillator, requires a completely specified oscillator coil, particularly as regards the resonant impedances presented by the circuit to the valve which, together with the electrode voltages, determine the peak and average plate currents of the various electrodes. In practice, it is usually convenient to standardize the circuit by using a valve having specified characteristics under particular conditions (transconductance, amplification factor, plate current) and adjusting the resonant impedance of the tuned circuit by means of the shunting resistor to give a prescribed grid current.

The oscillator frequency is usually either an audio frequency or 1 Mc/s, the latter being more typical of normal operation

(2) **Oscillator driven**

This is a fairly commonly used test, because it has less dependence of the mixer on the oscillator characteristics. It suffers from the disadvantage, however, that inasmuch as the driving voltage is applied to the oscillator grid, it does not provide a satisfactory test of oscillator characteristics. With this method of testing, if grid current is regarded as a criterion of oscillator characteristics, the oscillator emission needs to be adequate only to supply the rectified current resulting from the excitation but may be quite inadequate to supply the high peak oscillator plate currents which may occur in normal operation, the values of which depend upon the impedance in the oscillator plate circuit.

For this reason, it is generally more satisfactory in service testing to use the driven method with prescribed excitation to test the mixer characteristics and to test the oscillator by means of either a transconductance test at a high plate current or a prescribed self-excited oscillator circuit test which calls for a reasonably high peak plate current, with rectified grid current as the criterion. In the latter case provided the oscillator circuit characteristics, that is the resonant impedances presented to the valve, are precisely specified the test is not critical as regards grid current.

(3) **Static operation**

With this method of testing, static voltages are applied to all electrodes and the electrode currents are measured.

(B) **Specific characteristics**

(a) **Reverse signal grid current**

This test is performed under the same general conditions as for triodes and pentodes, except that the method of operation may be with the oscillator self-excited, or with the oscillator driven, or with static operation.

(b) **Signal grid current commencement**

In general this test may be performed as for triodes and pentodes, except that the oscillator may be either self-excited or driven.

(c) **Mixer positive voltage electrode currents**

These may be measured, in general, as for triodes and pentodes, except that the oscillator may be either self-excited or driven.

(d) **Mixer conversion transconductance***

Mixer conversion transconductance is determined from measurements of the magnitude of a beat-frequency component, $(f_1 - f_2)$ or $(f_1 + f_2)$, of the output current, and the magnitude of the input voltage of frequency f_1 where f_2 is the oscillator frequency. The method of measurement which has been standardized by JAN-1A (Ref. S2) and R.M.A. (Ref. S7), and which uses a driven oscillator, is to apply voltages of identical phase and frequency — usually mains frequency for convenience — to the specified electrodes (that is the signal and oscillator grids) with provision for a phase reversal of 180° of one with respect to the other. The change in direct plate current due to the phase reversal represents the difference-frequency plate-current component.

---

*Also referred to as conversion conductance. Mixer conversion transconductance is defined as the quotient of the magnitude of a single beat-frequency component $(f_1 + f_2)$ or $(f_1 - f_2)$ of the output electrode current by the magnitude of the control-electrode voltage of frequency f_1, under the conditions that all direct electrode voltages and the magnitude of the electrode alternating voltage f_2 remain constant and that no impedances at the frequencies f_1 of f_2 are present in the output circuit, and that the magnitude of the signal voltage f_1 is very small.

This value divided by twice the peak value of the sinusoidal voltage applied to the signal input electrode is taken as the mixer conversion transconductance. Specified voltages are applied to the valve electrodes under specified circuit conditions and the oscillator excitation voltage is adjusted to the specified value or to give the specified oscillator grid current. Other methods of testing are acceptable if correlated with this method.

The relationship may be expressed in the form

$$g_c = \frac{I_{b1} - I_{b2}}{2E_{1m}} \qquad (4b)$$

where g_c = conversion conductance in mhos

    I_{b1} = plate current for the in-phase condition of signal and oscillator voltages E_1 and E_2 in amperes

    I_{b2} = plate current for the out-of-phase condition of E_1 and E_2, in amperes

and E_{1m} = peak value of the signal voltage.

It should be clearly understood that the value of conversion transconductance measured by this method is dependent on the signal grid voltage and on the curvature of the mixer characteristics and consequently only approaches a true " differential " transconductance at the point of operation, as distinct from an average value, if the signal grid voltage amplitude is sufficiently small. In most cases, if the signal grid voltage does not exceed say 25 mV, exact correlation can be obtained for all types. At higher values, up to 100 mV, the error may be acceptable, depending upon the type and the condition of operation.

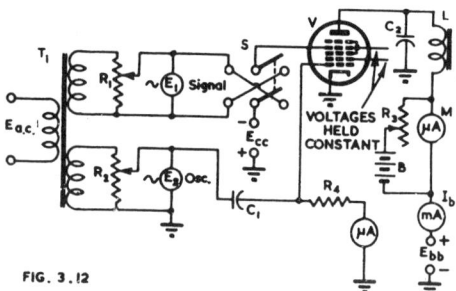

Fig. 3.12. Phase reversal method of measuring conversion transconductance.

FIG. 3.12

The circuit is shown in Fig. 3.12 in which one double-wound transformer, T_1 with two secondary windings provides the signal and oscillator voltages, which are controlled by means of potentiometers R_1 and R_2. A switch S permits phase reversal of the signal voltage. The grid coupling capacitor C_1 and grid resistor R_4 are provided, with a meter to measure oscillator grid current through R_4. In the plate circuit of the valve under test there is a by-pass capacitor C_2 and iron-cored choke L to pass the direct current I_b from the plate voltage source E_{bb}. The increase in plate current is measured by the low resistance microammeter M after the standing current I_{b2} has been " bucked out " by the bucking battery B and rheostat R_3 which should have a minimum resistance at least 100 times that of the microammeter.

In order to obtain exact results, as indicated previously, very small signal voltages are required, which produce only small incremental readings of plate current and demand stabilized electrode voltage sources. In any particular case, the signal voltage may be increased until the error due to the large signal voltage becomes measurable, and tests should then be carried out at a somewhat lower signal voltage.

If a signal voltage of $1/2\sqrt{2}$ volt (0.5 volt peak) is used, the incremental plate current in milliamps then reads conversion transconductance in milliamps per volt. It seems that values of transconductance frequently quoted in published data are obtained using the above-mentioned value of signal voltage, as in many cases the true conversion transconductance under the conditions published differs from this value. As a result, it is frequently difficult to calculate the exact performance of converter stages from published data of the converter used.

The shape of the curve of conversion transconductance plotted against oscillator grid current is very much influenced by the signal voltage used in the tests. The actual operating conversion transconductance in a practical case is a function of the input signal level.

The phase-shift method of testing conversion transconductance is the only **absolute** experimental method of measuring this characteristic, and accurate results can be obtained provided all electrode voltages, including signal and oscillator excitation voltages, are well stabilized, the electrode supply voltages free from hum, the excitation voltages sinusoidal, of low frequency and correct phase and the signal grid voltage sufficiently small.

Side-band method

The side-band method of measuring conversion transconductance is shown by the circuit of Fig. 3.13. The oscillator may be either self-excited at any convenient frequency, say 1 Mc/s as in Fig. 3.13, or may be driven. A signal voltage E_1 at audio frequency (say 1000 c/s) is applied to the control grid. The oscillator is adjusted to give the required grid current through the specified grid resistor R_2. In the plate circuit there is a resonant circuit $L_1 C_1$, tuned to the oscillator frequency and shunted by a resistance R_1, the value of which should be as low as possible and which should not exceed 10 000 ohms.

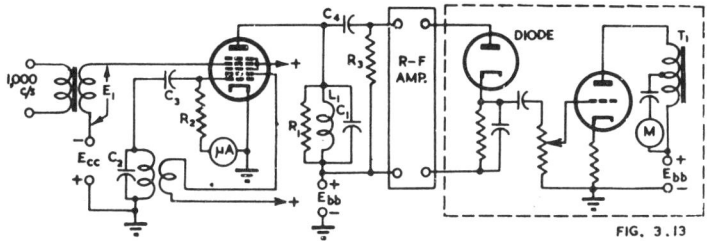

Fig. 3.13. Side-band method of measuring conversion transconductance.

The voltage across the resonant circuit comprising the oscillator frequency, the signal frequency and the two resulting sidebands are applied to a radio frequency amplifier followed by a diode detector across whose diode load is connected a valve voltmeter to indicate the amplitude of the audio frequency components. If an r-f amplifier is not used, the region of non-linearity of the diode characteristic results in the calibration of the system being dependent on the voltage developed across the tuned circuit and would thus be unsatisfactory for valves of low conversion conductance. However, the r-f amplifier may be omitted provided the signal applied to the diode is sufficiently high to ensure linear rectification. Where this is not so, the simplified circuit may not be suitable for measurements of low values of conversion conductance.

If the r-f amplifier is omitted, the signal voltage E_1 may be about 0.5 volt peak, and R_1 may be 10 000 ohms.

The tester must be calibrated against the standard phase reversal method and the circuit may be adjusted so that the meter is direct reading.

(e) **Mixer plate resistance**

This may be measured either with a self-excited or driven oscillator, the value of plate resistance being not affected by the method used. Measurement of plate resistance is performed as for a pentode, except for the excitation of the oscillator.

(f) **Mixer transconductance**

This may be measured as for a pentode valve, except that the oscillator should be either self-excited or driven.

(g) **Oscillator grid current**

The oscillator performance of a converter valve depends upon several valve and circuit parameters that are not simply defined.

The oscillator grid current may be used as the criterion of oscillator performance in a self-excited oscillator, operating under specified conditions. The circuit should be designed to draw typical peak and average total space currents. The Boonton Converter Oscillator test circuit is widely used to check oscillator performance, owing to its simplicity of design and ease of adjustment and control and, while not giving an *exact* criterion of oscillator performance, does provide a dynamic test of the usual parameters under conditions that simulate average receiver operation.

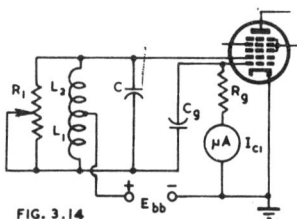

Fig. 3.14. *Self-excited oscillator, using the Hartley circuit, for testing oscillator performance.*

The circuit (Fig. 3.14) is a series-fed Hartley having a fixed feedback ratio and having the tank circuit impedance variable so that the magnitude of oscillation may be adjusted. Normal voltages are applied to all electrodes.

It is necessary to specify the oscillator plate voltage, oscillator plate series resistor (if used), oscillator grid resistor (R_g) the inductance of each section of the coil (L_1, L_2), the mutual inductance between the two sections (M), the capacitance shunted across the whole coil (C), the grid coupling capacitance (C_g), and the resonant impedance of the tuned circuit (R_d). Normally a coil having a higher reactance than the specified value of resonant impedance is used and is shunted by a variable resistance (R_1) to give the specified resonant impedance. Bias is obtained by means of a capacitor and a wire-wound gridleak. The wire-wound grid resistor is used because it has considerably higher impedance to radio frequencies than to direct current. The high radio-frequency impedance diminishes the shunting effect of the gridleak on the tuned circuit. The rectified grid current is read on a microammeter (suitably by-passed) in series with the grid resistor. Before the test oscillator is used, the relationship between the tuned impedance of the tank circuit and the setting of the variable resistor must be determined

The R.M.A. standard (Ref. S7) and JAN-1A (Ref. S2, sheet dated 19 July, 1944) specify the component and coil design values as being : $C=100$ $\mu\mu$F (mica) ; $C_g=200$ $\mu\mu$F (mica) ; $R_1=50\,000$ ohms potentiometer ; $R_g=50\,000$ ohms wire-wound resistor ; $L_1=83$ μH ; $L_2=48$ μH : $M=23.3$ μH ; coil diameter$=1.25$ in. ; winding length$=59/64$ in. ; wire No. 30 a.w.g. enamelled copper ; turns$=83$; tap at 33 turns from anode end. See also Ref. S12.

The constants quoted above have been selected to simulate average circuit conditions in broadcast receivers for composite converters having transconductances less than 1500 μmhos at zero bias. For valves having higher transconductances or for circuits above broadcast frequencies, the circuit with the constants given may not be satisfactory, as spurious oscillations may make it impossible to obtain correlation between test readings and receiver performance.

The ability of a valve to oscillate when the shunt impedance of a tank circuit is low is one criterion of the value of a valve as an oscillator. The oscillation test is made by applying the desired electrode potentials to the valve under test and reading the rectified grid current at some known setting of tank-circuit impedance. A valve that will not oscillate with the tank-circuit impedance below about 9000 ohms is considered to be a weak oscillator. A valve that will oscillate with a tank-circuit impedance as low as 3000 ohms is considered to be a strong oscillator. The minimum tuned impedance at which oscillation will start or cease as indicated by the rectified grid current will depend on the oscillator characteristics of the particular type and will vary from type to type.

(h) **Oscillator plate current**
   The oscillator plate current may be read while the valve is being tested for grid current as a self-excited oscillator, or alternatively at zero grid voltage and specified plate voltage (usually 100 volts).

(i) **Oscillator transconductance**
   The oscillator transconductance is usually measured at zero grid voltage and specified plate voltage (usually 100 volts).

(j) **Oscillator amplification factor**
   The oscillator amplification factor is measured as for any triode, usually with a plate voltage of 100 volts and zero grid voltage.

(k) **Oscillator plate resistance**
   The oscillator plate resistance is measured as for any triode, usually with a plate voltage of 100 volts and zero grid voltage

(l) **Signal grid blocking**
   Signal grid blocking may be treated as for the general case of grid blocking in Sect. 3(iv)N.

(m) **Microphony**
   The general remarks in Sect. 3(iv)K regarding microphony apply also to converters.

(n) **R-F noise**
   Converters may be tested for r-f noise under published electrode voltage and typical receiver circuit conditions, following the relevant recommendations of the R.M.A. r-f noise test — see Sect. 3(iv)M.

(vi) **Tests for special characteristics**
   The following tests are not normally applied to receiving type valves and the undermentioned information is provided to assist designers in obtaining data of these characteristics for particular purposes.

(A) **Short-circuit input admittance** (y_{11})
   At frequencies up to 10 Mc/s it is possible to use conventional bridge methods for the measurement of the four admittances, but at higher frequencies some version of the susceptance-variation or resistance-substitution method is necessary.

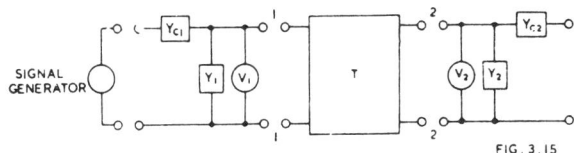

FIG. 3.15

Fig. 3.15. Semi-schematic diagram of equipment for measuring admittances by the susceptance-variation method (Ref. S12).
T is the 4 terminal transducer to be measured, Y_1 and Y_2 are calibrated variable-admittance elements which may be of various forms, such as coils or capacitors. V_1 and V_2 are signal-frequency voltage-measuring devices ; these may be simply crystal or diode voltmeters or heterodyne receivers. Variable admittances Y_{c1} and Y_{c2} are used for coupling the input or output circuits to the signal oscillators.

(1) **Susceptance-variation method of measurement** (Refs. 51, S12)
   In this method, the circuit is detuned with either the capacitor or the inductor adjustment to a point giving half the power output observed at resonance. The increment in susceptance, determined from the capacitance calibration curves, is then equal to the circuit conductance. In practice, the mean value obtained by using the half-power points on each side of resonance is used.
   Using the circuit of Fig. 3.15 (Ref. S12) —
   (a) Short-circuit the output terminals 2-2. This may be done either by detuning Y_2 sufficiently or by placing a suitable by-pass capacitor directly across terminals 2-2.
   (b) Excite the input circuit by coupling the signal oscillator loosely through Y_{c1} to Y_1.

(c) Adjust Y_1 for resonance as indicated by a maximum reading of V_1. In order to insure that the coupling to the oscillator is sufficiently small, reduce the coupling until further reduction does not change the setting of Y_1 for resonance. Record the calibrated values of G_1 and B_1 for this setting.

(d) Vary Y_1 on either side of resonance until the voltage V_1 is reduced by a factor $1/\sqrt{2}$. Record the calibrated values of this total variation of Y_1 between half-power points as ΔG_1 and ΔB_1. In order to insure that the oscillator and detector are not loading the circuit, reduce the coupling until further reduction does not change the susceptance variation ΔB_1. The short-circuit input susceptance is then given by the relation

$$B_{11} = -B_1 \tag{5}$$

and the short-circuit input conductance by the relation

$$y_{11} = \tfrac{1}{2}\Delta B_1[(1+2\eta^2)^{\frac{1}{2}}+\eta]-G_1 \tag{6}$$

$$\text{or } y_{11} \approx \tfrac{1}{2}\Delta B_1[1+\eta+\eta^2]-G_1 \tag{7}$$

or even further by the relation

$$y_{11} \approx \tfrac{1}{2}\Delta B_1 - G_1 \tag{8}$$

if η is negligible.

FIG. 3.16

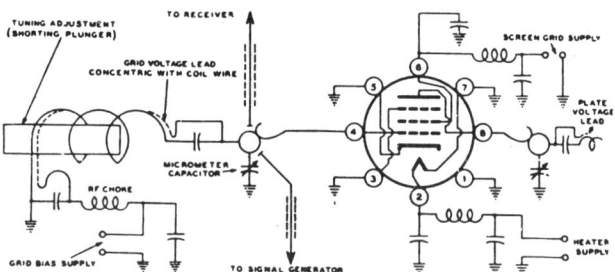

Fig. 3.16. *Circuit diagram of equipment used for measurement of short-circuit input admittance (Ref. 51).*

A practical form (Ref. 51) for the frequency range 50 to 150 Mc/s is shown in Figs 3.16 and 3.17. The valve under test is used as a part of a resonant circuit which includes a continuously-variable inductor and a small concentric-cylinder capacitor built on a micrometer head. The high-potential end of the inductor is connected to the high-potential electrode of the micrometer capacitor inside a cylindrical cavity open at the top. Fig. 3.17 also shows the positions of some of the by-pass capacitors used with the octal socket. These are button-type, silver-mica capacitors of approximately 500 $\mu\mu$F capacitance. The socket is of the moulded phenolic type. Terminals 1, 3, 5 and 7 are connected directly to the mounting plate at a point directly below the terminal in each instance. Terminals 2 (heater) and 6 (screen grid) are by-passed to ground, and a lead is brought from each of these terminals through the mounting plate to a small r-f choke and a second by-pass capacitor. Terminal 8 (plate) is by-passed and fitted with a spring contacting the terminal for one of the circuits. Terminal 4 (grid) has only the contact spring. A similar arrangement is used with miniature valves. In this case, the socket is of the wafer type with mica-filled rubber insulation. Terminals 2, 3 and 7 are grounded and terminals 4, 5 and 6 are by-passed to ground. Terminals 1 (grid) and 5 (plate) have contact springs. The by-pass capacitors are closer to the mounting plate than in the case of the octal socket ; the capacitors at terminals 4 and 6 overlap the capacitor to terminal 5. A 10 ohm resistor, mounted inside a cylindrical shield to minimize lead inductance, is connected between socket terminal 5 (plate) and the by-pass capacitor. This component was added to suppress a parasitic oscillation observed with certain type 6AK5 valves. The resistor was found to have no

FIG. 3.17

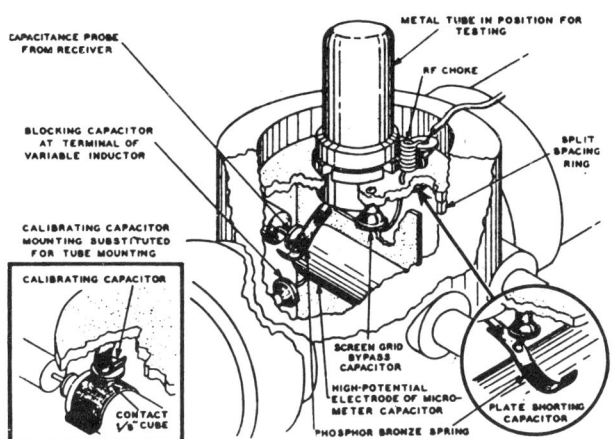

Fig. 3.17. Physical arrangement of circuit elements of Fig. 3.16 (Ref. 51).

measurable effect on input admittance readings obtained with valves, either of the 6AK5 type or of other types, not subject to the parasitic oscillation.

In order to obtain susceptance values, the circuit must first be calibrated for the capacitance required for resonance at each test frequency. The circuit is calibrated by determining the inductor settings for resonance with each of a number of small, disc-shaped, calibrating capacitors substituted for the valve. The insert in the lower left corner of Fig. 3.17 shows a cutaway view of the cavity with one of the calibrating capacitors in place. The length of the phosphor-bronze contact spring used with the calibrating capacitors is approximately the same as that used with the valve. Thus, the inductance of this lead is accounted for in the calibrating procedure. The reference terminals for the valve are the socket plate and the grid terminal of the socket or, possibly, a point on the grid terminal a little inside the body of the socket. The calculated inductance of the contact spring is 4.5 milli-microhenries per centimeter of length within about ±25 per cent., but the difference in effective lengths of the springs for the socket connection and the calibrating connection is not more than 2 or 3 millimeters.

The range of the micrometer capacitor is sufficient for measurements of the circuit with the calibrating capacitors, with most cold valves, and with some valves under operating conditions. For other cases, adjustment of the inductor is required.

If the conductance being measured is small enough to be measured by the available range of the micrometer capacitor, the susceptance and conductance variations between $1/\sqrt{2}$ voltage points are given by

$$\Delta B = \omega \Delta C$$

and $\Delta G = 0$

where ω is the angular frequency of measurement and ΔC is the micrometer capacitance variation between the half-power points.

The calibration curves used at each test frequency are :
(1) Capacitance for resonance
(2) Slope of the capacitance curve
(3) Conductance at resonance of the circuit with the calibrating capacitors.

These three quantities are plotted against the inductor adjustment readings. Since the conductance values for the calibrating capacitors themselves are too small to affect the calibration appreciably, the conductance curve corresponds, essentially, to the equipment.

(2) Resistance-substitution method of measurement

The resistance-substitution method applies only to the measurement of the conductive component of the admittance. The susceptive component must be measured by means of a calibrated susceptance element, as for the susceptance-variation method of measurement. Ideally, the resistance-substitution method involves, in the case of a two-terminal admittance, the removal of the electron-tube transducer from the calibrated admittance element and its replacement by a standard pure resistance of such a value that the voltage reading between the two terminals is the same as that obtained with the transducer in place. The calibrated susceptance element must be adjusted to resonance both before and after substitution of the resistance. If the measurement is of y_{11}, the following relations are obtained :

$$B_{11} = -B_1 \qquad (9)$$

as in (1) above, and

$$G_{11} = 1/R \qquad (10)$$

where R is the standard resistance of a value that satisfies (10).

There are practical difficulties in obtaining standard resistors having negligible reactance at frequencies of the order of 100 Mc/s or higher. Wire-wound resistors are not usable at such frequencies. The most satisfactory types available are the metallized-glass or ceramic-rod resistors of relatively small physical size, having low-inductance terminals and very little distributed capacitance. A further difficulty arises from the fact that such resistors are obtainable only in discrete values of resistance. It would not be practicable to obtain the very large number of resistors needed to match the resistance of any electron-tube transducer. Hence, it is necessary to utilize a transformation property of the admittance-measuring equipment in order to match any arbitrary admittance with some one of a reasonably small set of standard resistors. A suitable resistance-substitution set consists of a transmission line of length l short-circuited at one end, having a characteristic admittance Y_o and a propagation constant γ. If a known admittance Y is placed across the line at a distance x from the short-circuited end, the admittance Y_t at the open end of the line is given by the relation

$$Y_t = Y \left[\frac{\sinh \gamma x}{\sinh \gamma l} \right]^2 \frac{1}{1 + \dfrac{Y \sinh \gamma x}{Y_o \sinh \gamma l} \sinh \gamma (l-x)} \qquad (11)$$

Conversely, the admittance Y_t is the admittance that would have to be placed at the open end of the line to produce the same effect there as the known admittance Y at the position x. Equation (11) then represents the property of the transmission line of converting admittance Y at position x into admittance Y_t at position l. This expression can be simplified for a low-loss line having a characteristic admittance Y_o large compared with the bridging admittance Y. Thus if $Y_o \gg Y$,

$$\left| \frac{Y}{Y_o} \frac{\sinh \gamma x}{\sinh \gamma l} \sinh \gamma (l-x) \right| \ll 1, \qquad (12)$$

and the real part of the propagation constant γ of the line is small,

$$\left[\frac{\sinh \gamma x}{\sinh \gamma l} \right]^2 \approx \left[\frac{\sinh \beta x}{\sinh \beta l} \right]^2 \qquad (13)$$

where $\beta = 2\pi/\lambda$ Equation (11) then simplifies to

$$Y_t = Y \left[\frac{\sin \beta x}{\sin \beta l} \right]^2 \qquad (14)$$

If Y is a pure conductance of value $1/R$, then Y_t is the pure conductance

$$G_T = \frac{1}{R} \left[\frac{\sin \beta x}{\sin \beta l} \right]^2 \qquad (15)$$

In the measurement of the short-circuit input admittance Y_{11}, a low-loss transmission line of large characteristic admittance is coupled loosely to an oscillator near the short-circuited end. At the open end are a voltage-detecting device and a calibrated capacitor, by means of which B_{11} is obtained from (9). With the unknown transducer across the open end and the capacitor adjusted to obtain resonance, a voltage reading is taken. The transducer is removed and one of the standard resistors placed across the line bridging the two conductors. The position of this resistor along the line is then adjusted and the system is readjusted for resonance with the calibrated capacitor until the voltage, as measured at the end of the line, is the same as before. By (15) we have

$$G_{11} = G_T = \frac{1}{R} \left[\frac{\sin \beta x}{\sin \beta l} \right]^2 \qquad (16)$$

where R is the resistance of the standard placed x centimeters from the short-circuited end of the line. It is evident that a resistor must be selected having a resistance value near to but not larger than the reciprocal of G_{11}.

Since the transmission line should have low loss and low characteristic impedance, a coaxial line is desirable. The line will require a longitudinal opening or slot in order to permit one of the standard resistors to bridge the line at an adjustable position to satisfy the required voltage condition. A valve voltmeter is capacitively coupled to the open end of the line across which the electron-tube transducer may be attached. Socket and filter arrangements for wire-lead valves can be attached to this line. Radiation difficulties arising from the longitudinal opening in the line, together with the increasing difficulty in obtaining resistance standards at frequencies much above 300 Mc/s, appear to make this type of measuring equipment impracticable for measurements at higher frequencies on surface-lead valves. Ref. S12.

(B) Short-circuit feedback admittance (y_{12})
Susceptance-variation method of measurement
See (A)1 above for general description.

(a) With the input termination still set at the value for resonance obtained in step (c) in (A) above, excite the output circuit through Y_{C2}. In the event that oscillation difficulties are encountered, detune the output circuit Y_2 or load it until oscillation stops.

(b) Record the voltmeter readings V_1 and V_2.

The magnitude of the feedback admittance is then given by the relation

$$|y_{12}| = \left| \frac{V_1}{V_2} \right| \frac{\Delta B_1}{2} \left[(1 + 2\eta^2)^{1/2} + \eta \right] \qquad (17)$$

or

$$|y_{12}| \approx \left| \frac{V_1}{V_2} \right| \frac{\Delta B_1}{2} \left[1 + \eta + \eta^2 \right] \qquad (18)$$

or

$$|y_{12}| \approx \frac{\Delta B_1}{2} \left| \frac{V_1}{V_2} \right| \qquad (19)$$

where ΔG_1 and ΔB_1 are the values obtained in the preceding measurement of y_{11}.
Ref. S12

(C) Short-circuit output admittance (y_{22})
The short-circuit output admittance may be measured by following the procedure outlined for the short-circuit input admittance, the signal being coupled through Y_{C2}. If the subscripts 1 and 2 are interchanged, all of the above formulae concerning y_{11} may be used to relate y_{22} to the measured data.
Ref. S12.

(D) Short-circuit forward admittance (y_{21})
The magnitude of the forward admittance may be measured by following the procedure outlined previously for the measurement of the magnitude of y_{12}. If the subscripts 1 and 2 are interchanged, all of the formulae concerning y_{12} may be used to relate y_{21} to the measured data.
Ref. S12.

(E) Perveance
(1) Perveance of a diode
The perveance of a diode may be derived by plotting the current against the three-halves power of the voltage, when the slope of the curve will give the value of G. Voltages should be chosen sufficiently high so that such effects as those produced by

contact potential and initial electron velocity are unimportant. If this is not possible, a correction voltage ϵ should be added to the diode voltage, where

$$\epsilon = (3/2)I_b r_b - e_b$$

where r_b = anode resistance of diode.

Ref. S12.

(2) Perveance of a triode

The perveance of a triode may be derived as for a diode where the diode anode voltage is taken as the composite controlling voltage

$$e' = \frac{e_c + (e_b/\mu)}{1 + (1/\mu)} \tag{20}$$

Multi-grid valves may be considered, with sufficient accuracy for most purposes, as triodes with the screen grid as the anode and the screen grid voltage as the anode voltage.

Low voltage correction—The effects of initial electron velocity and contact potential may be represented by an internal correction voltage ϵ that is added to the composite controlling voltage, where

$$\epsilon = \frac{3I_b \mu}{2g_m(\mu+1)} - \frac{\mu e_c + e_b}{\mu+1} \tag{21}$$

The value of g_m to be used is that obtained with all the electrodes except the cathode and control grid tied together to form an anode which is held at the voltage ordinarily used for the screen grid.

Ref. S12.

(vii) Characteristics by pulse methods—point by point

The characteristics of valves in the region where electrode dissipations or currents exceed safe values may be obtained by pulse methods, in which the valve is allowed to pass current only for short intervals of such duration and recurrence frequency that it is not damaged.

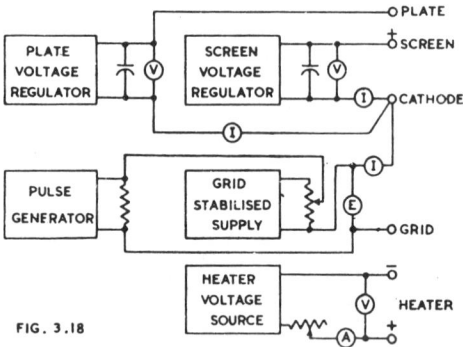

FIG. 3.18

Fig. 3.18. *Circuit for measuring electrode currents using single pulse generator method (I=peak-reading current indicator, E=peak-reading voltage indicator, V=moving coil voltmeter). If high accuracy is required, correction should be made for the voltage drop across the current indicators. The regulation of the electrode supplies is not important for point by point testing, provided that the shunt capacitors are sufficiently large.*

(a) Single generator method

The valve is operated with the desired screen and plate voltages, but with the grid biased beyond cut-off, in the absence of pulses (Fig. 3.18). If secondary emission effects produce a negative impedance, the pulse generator must be shunted by a non-inductive load resistor of such value as to maintain an overall positive impedance at its terminals.

Refs. 38, S12.

(b) **Multiple generator method**

With this method, no direct voltages are applied but each electrode has its own pulse generator, all generators having their pulses synchronized. This method is not commonly employed with receiving type valves. Refs. 68, 72, S12.

References to characteristics by pulse methods : Refs. 38, 68, 72, 93, S12.

(c) **Pulse generators**

Pulse generators are of several types. In the capacitor discharge type a condenser is switched from charge to discharge through the valve under test. If there is no series impedance, the peak voltage applied to the valve is equal to the charging voltage. However, because of the very short duration of the pulse, an accurate current indicator is difficult to design.

Refs. 68, 69, S12.

A rectangular pulse shape is obtainable from a direct supply source in series with a high-vacuum triode (or several in parallel) whose grid is normally biased beyond cut-off. A rectangular control pulse, obtainable from a conventional source, is applied in the grid circuit to permit the triode to conduct only during the duration of the pulse. (Ref. S12). Alternatively an ignitron may be used (Ref. 38).

A half sine-wave, or portion of a half sine-wave, pulse is obtainable from an alternating source with gas triode or other control to provide the desired gap between pulses. Alternatively a d.c. source may be used with a gas triode and LC network. Ref. S12.

(d) **Peak-reading voltage and current indicators**

The input circuit must have a time constant that is large with respect to the time interval between successive pulses. The greatest accuracy is obtained with a rectangular waveform, good accuracy may be obtained from a half sine-wave, while the sharply peaked capacitor discharge method is the least accurate.

A peak-reading voltage indicator may be employed to indicate peak current by using it to measure the peak voltage drop across a non-inductive resistor of known resistance.

One form of indicator is a diode type valve voltmeter with a high resistance load shunted by a large capacitance. The use of this indicator for current-measuring purposes may result in serious errors if the electrode characteristic of the valve under test exhibits any negative slope.

A cathode ray oscillograph may be used directly to measure peak voltage, or peak voltage drop across a non-inductive resistor of known resistance, provided that there is no negative slope in the electrode characteristics. If negative slope is likely to occur, a linear horizontal sweep voltage, synchronised to the pulse generator frequency, may be used to spread out the current trace. In this manner the detection of possible errors is simplified. The vertical deflection corresponding to the maximum horizontal deflection is the one required ; for characteristics having only positive slopes, this vertical deflection will also be the maximum.

Amplifiers may be used provided that they are designed with sufficient bandwidth and sufficiently linear phase-frequency response for the high-frequency components of the pulse. The attenuator should be capacitively balanced, making the ratio of capacitance to resistance of all sections alike.

Reference to peak-reading indicators : Ref. S12.

(viii) Characteristics by curve tracer methods

A characteristic curve may be traced on the screen of a C.R. tube, using a suitable type of pulse generator for one electrode and fixed voltages for the other electrodes. The simplest characteristic for this purpose is the grid voltage versus plate current characteristic, either with zero plate load resistance or with a specified load resistor (" dynamic " characteristic).

A triangular or half sine-wave pulse generator characteristic is satisfactory. The grid is normally biased beyond cut-off, but is swung over the useful range of voltages by the pulse.

The plate voltage versus plate current characteristic may be traced by maintaining the grid (and screen, if any) at constant voltage and by applying the pulse voltage to the plate. This requires a peak voltage of several hundred volts for receiving valves, together with high peak currents. The grid voltage must be maintained constant, the regulation requirements being severe. The regulation requirements of the other electrodes are not important, if large capacitors are used.

A method of showing the two axes for any single characteristic is described in Ref. 73 which makes use of a standard vibrator.

References to curve tracer methods : 37, 39, 69, 70, 71, 72.

Display of valve characteristics

If it is desired to view a family of curves, the voltage of one electrode must be changed in steps for successive pulses. This may be accomplished by means of synchronous contactors (Refs. 34, 35, 36) or by electronic means (Ref. 39). Good regulation (high stability) for the electrode voltage supplies is very important.

See also Refs. 37, 90, 94, 96, S12.

SECTION 4 : ACCEPTANCE TESTING

(i) *Relevant characteristics* (ii) *Valve specifications* (iii) *Testing procedure.*

(i) Relevant characteristics

The relevant characteristics for acceptance testing are those characteristics that are subject to deterioration subsequent to manufacture, it being assumed that the valves have satisfactorily passed the valve manufacturing test specification. These relevant characteristics for acceptance testing may be divided into mechanical and electrical characteristics.

Mechanical characteristics that require checking for faults include loose bases and top caps, dry joints in connections, cracks in envelopes caused by glass strain, damage resulting from inherent strain or careless handling, damage to internal structure resulting in misalignment of electrodes, shorts and open circuits, loss of vacuum, noise and microphony from excessive handling causing damage to the mount and micas.

Electrical characteristics that require checking for faults include gas, a relevant dynamic test, emission and noise.

(ii) Valve specifications

The most widely known of the official specifications are those of the British (K1001) and American (JAN-1A) authorities covering types used in their respective areas.

(iii) Testing procedure

The testing procedure in each case is as laid down in the prescribed specifications. Where, however, valves do not require to be accepted to a prescribed specification, it is satisfactory to adopt the following procedure :

(a) Visually inspect all valves 100 per cent. for mechanical defects.

(b) Test all valves 100 per cent. for shorts and open circuits.

(c) Select at random from all the valves a representative sample (or samples) and test for a suitable dynamic characteristic using established statistical acceptance procedure. The test selected should give a satisfactory indication of the overall performance of the type in its normal application.

References to sampling inspection : Refs. 42, 43, 44.

SECTION 5: SERVICE TESTING AND SERVICE TESTER PRACTICE*

(i)   Purpose and scope of service testing and discussion of associated problems.
(ii)   Fundamental characteristics which should be tested.
(iii) Types of commercial testers.
(iv)   Methods of testing characteristics in commercial service testers
   (a) Shorts testing
   (b) Continuity testing
   (c) Heater to cathode leakage
   (d) Emission testing
   (e) Mutual conductance testing
   (f) Plate conductance testing
   (g) Reverse grid current testing
   (h) Power output testing
   (i) Conversion conductance testing
   (j) Oscillator mutual conductance testing
   (k) Noise testing
   (l) A.C. amplification testing.
(v)   A.C. versus d.c. electrode voltages in testers.
(vi)   Pre-heating.
(vii) Testing procedure.

(i) Purpose and scope of service testing and discussion of associated problems

Service testing is only carried out on valves which have previously been factory tested and which have also operated satisfactorily in a new receiver. The faults to be expected are those which may occur after manufacture.

Service testing is for the purpose of determining either

(1) whether or not a valve in working condition has reached the point in life when it should be replaced, or (2) whether a valve with unknown history is faulty or not.

The determination of the point of life when a valve should be replaced is a very complex problem. This is a function of the application of the valve and of the standard of performance expected by the user. For example, a valve which still operates reasonably well as a resistance-coupled amplifier may yet give low gain as an r-f amplifier, and a valve which will still operate as an amplifier may fail to oscillate when used in the oscillator position. A receiver with old valves in the r-f and i-f stages may have very much less gain that it would have with new valves, but the effect may not be noticed by the user unless the set is used for the reception of weak signals.

A list of the generally accepted life test end-points is given in Sect. 2(iv)b.

The radio service-man is usually called upon to test a number of valves, frequently the whole valve complement of a receiver, to determine which (if any) is faulty. If one is obviously faulty, for example due to a short-circuit, open-circuit, gas, or extremely low emission, the problem is simple. In other cases it is helpful to test the valves in the original receiver, with a second complement of new valves for comparison, to make certain that the receiver itself is not faulty   If the original receiver is not accessible, the valves may be plugged into another receiver (preferably of the same model) for test.

Microphony should be checked in the original receiver. Motor-boating may be due to gas or grid emission, but it may also be caused by a faulty component in the receiver. Distortion is generally due to the power valve, but may also be caused by a defective filter by-pass capacitance in the receiver and in some cases by the signal diode or a defective resistor. Hum is usually caused by heater-to-cathode leakage or breakdown in the valve or a faulty filter capacitor in the circuit.

*The assistance of A. H Wardale, Member I.R.E. (U.S.A.) and D. H. Connolly, A.S.T.C. is acknowledged.

(ii) Fundamental characteristics which should be tested

The fundamental characteristics which should be capable of being tested in a service tester are

(a) Shorts and continuity.

Testing for short circuits is essential.
Testing for continuity of the electrodes is highly desirable.

(b) Dynamic tests.

The dynamic tests which should be carried out in a service tester are :

Class of valve	Preferred test	Less desirable alternative
Rectifier (power diode)	Emission	Rectification test
Signal diode	Emission	—
Triode, low mu	Mutual conductance	—
Triode, high mu	A.C. amplification	—
Pentode, r-f	Mutual conductance	—
Pentode, battery a-f	A.C. amplification	—
Pentode, power	Power output	—
Converter	Conversion conductance	Mixer g_m
	Oscillator grid current	Oscillator g_m

See Sect. 3 for methods of testing these characteristics.

The dynamic test should be one in which the peaks and average plate currents approximate to the maximum current in normal operation.

(c) Other tests

Other tests which should be carried out in a service tester are :
Gas (amplifier types only).
Noise (amplifier types only).
Heater to cathode leakage (indirectly-heated types only).

(iii) Types of commercial testers

Commercial service testers are necessarily a compromise. They must be fairly simple to use, flexible with regard to future valve developments, as free as possible from damage to indicating instruments through faulty valves or misuse, speedy in operation and true in their indications.

The possible combinations of characteristics tested are :

Type of valve	Essential tests	Desirable tests
All	Shorts	Continuity
		Heater to cathode leakage
Vacuum types	At least one dynamic test	{ Emission Mutual conductance A.C. amplification Power output
	or for converters	{ Conversion conductance or Mixer g_m and Oscillator g_m
	together with	Noise
		Gas

Out of a representative group of 22 modern service testers examined, 17 incorporated some form of mutual conductance test, 7 incorporated an emission test other than for diodes and all 22 incorporated a diode emission test. One provided a plate conductance test. With regard to the remaining tests, 8 incorporated a heater to cathode leakage test, with 7 having a gas test and 7 having a noise test.

It seems to be generally admitted that an emission test alone is not a very satisfactory dynamic test for all amplifying valves. The better types of testers have both mutual conductance and diode emission ; they are deficient in that they do not provide a power output test for power valves, or a.c. amplification for high-mu triodes and a-f pentodes. The method of testing for mutual conductance with a high grid swing is not the equivalent of a power output test. Neither mutual conductance nor emission testing is a satisfactory substitute for a.c. amplification testing of high-mu triodes or a-f pentodes.

It is obvious, therefore, that any ordinary commercial service tester does not give a true indication of the condition of all types of valves, and its readings should be used with discretion.

(iv) Methods of testing characteristics in commercial service testers

(a) Shorts testing

Testing for shorts should be carried out at mains frequency or on d.c., and the voltage applied to the neon lamp circuit should not exceed 110 volts r.m.s. The use of higher voltages results in some good valves being classed as filament-to-grid short circuits.

With a.c. supply, it is usual to design the circuit so that resistances over 0.5 megohm do not indicate as shorts ; this is to avoid capacitance effects. With d.c. supplies there is no similar limitation, but different models vary considerably in their shorts testing sensitivity.

The most satisfactory switching arrangement is one which tests for shorts from any pin to all other pins tied together. The circuit of Fig. 3.4 may be used, except that the cathode need not be heated.

(b) Continuity testing

Continuity testing may be carried out in a manner closely resembling shorts testing, but with the cathode hot. Positive or alternating voltage is applied through the neon lamp to each electrode in turn ; a glow indicates continuity.

The circuit of Fig. 3.4 may be used in a simple service tester, but a more elaborate tester such as the hexaphase (Fig. 3.2) is much more satisfactory for general use.

For further information on continuity testing see Sect. 3(ii)b.

(c) Heater to cathode leakage

The maximum leakage resistance from heater to cathode in new valves varies from 1 or 2 megohms for power valves to 5 megohms for ordinary 6.3 volt 0.3 ampere voltage amplifier valves and even higher for rectifier types having separate heaters and cathodes. A neon lamp will normally glow slightly when used in a shorts tester and applied between heater and cathode. A neon lamp may be de-sensitized either by shunting by a resistor (of the order of 0.1 to 0.5 megohm) or by a mica condenser (a value of 0.01 μF gives a sensitivity of about 5 megohms—this is only useful with a.c. supply).

(d) Emission testing

Of the 7 commercial testers which included emission testing of amplifying valves, 5 used a.c. testing, 1 used half-wave rectified a.c. testing, and 1 used d.c. testing. Owing to the poor regulation obtainable from most service valve testers with d.c. supplies, other than batteries, it is doubtful whether this has any practical advantages over a.c. testing. With a.c. supply, the voltage in most service testers is 30 volts r.m.s. ; this is only a compromise which is far from the optimum for all cases. For laboratory testing of emission see Sect. 3(ii)f.

(e) Mutual conductance testing

Of the 17 commercial testers having a mutual conductance test, 7 used some form of a.c. signal voltage applied to the control grid, and 6 used some grid shift method, the methods used by the other 4 being unknown.

All the American mutual conductance testers examined apply an a.c. signal to the grid, and measure the signal current in the plate circuit ; all but one of these use a mains frequency signal, the exception being one with a 5 Kc/s signal and a tuned signal plate current meter.

In most cases this is not an accurate g_m test for all types of valves because the signal voltage on the grid is fairly high—between 4 and 5 volts r.m.s. in some cases—sufficient to run beyond grid cut-off in one direction and well on to the " flat top " in the other, with short grid-base valves. Even those testers with reduced signal voltages have quite appreciable errors with some types of valves.

On the other hand, all but one of the examined English mutual conductance testers use a grid shift method. This introduces some errors due to plate supply regulation and curvature of the characteristics, but these can be allowed for in the calibration.

(f) Plate conductance testing

This is used by one American tester only as a reasonably close equivalent of the mutual conductance test, the amplification factor being assumed to be constant. Pentodes are triode-connected. A moderately high signal voltage is applied, this being possible with less error than with direct g_m measurement owing to the slight compensating change of μ with plate current.

(g) Reverse grid current testing

Of the post-war testers examined, all those of English design incorporate reverse grid current testing (often referred to as " gas testing ") but this test was incorporated in only two of American design. The usual methods of testing are either by grid current meter or by a shorted grid resistor. For further details see Sect. 3(iv)A.

A valve which passes the reverse grid current test on a service tester may not operate satisfactorily in a receiver. The usable maximum value of negative grid current depends upon the application, the mutual conductance of the valve, the resistance in the grid circuit, the type of bias, the bias voltages, the total dissipation of the valve and its ventilation. The valve may be checked in the actual receiver after running for a period sufficiently long for the valve to reach its stable operating temperature ; it may be tested by variation in the plate current of the valve under test when its grid resistor is shorted.

(h) Power output testing

This is not incorporated into any of the commercial service testers examined, but is the most satisfactory test for all power valves. Triode testing of pentode and beam power valves is quite satisfactory. Methods of testing are described in Sect. 3(iv)I.

(i) Conversion conductance

Only one of the testers examined provided a test for conversion conductance, the method being by phase reversal. For further details see Sect. 3(v)Bd.

A satisfactory alternative is to measure the transconductance of the mixer, this being the method commonly employed.

(j) Oscillator mutual conductance

None of the testers provided a test for oscillator grid current under self-oscillating conditions, although this is the most satisfactory form of test. For details see Sect. 3(v)Bg.

The method commonly employed is to measure the oscillator mutual conductance.

The measurement of oscillator grid current in a suitable receiver is a simple way of checking the oscillator section of a converter.

(k) Noise

The usual form of noise test is to connect the two neon lamp terminals to the aerial and earth terminals of a receiver, during the shorts test. For other forms of noise test see Sect. 3(iv)L and M.

(l) A.C. amplification

This test should be carried out with zero bias and a high value of grid resistor, as described in Sect. 3(iv)H. No test with fixed or self-bias is able to determine how a valve will operate with grid resistor bias.

(v) A.C. versus d.c. electrode voltages in testers

All but four of the commercial testers examined, which made provision for some form of g_m test, used d.c. electrode supplies. The general trend seems to be in the direction of using d.c. supply to all electrodes, and the additional circuit complications appear to be justified.

For comparison between laboratory testing of emission by a.c. or d.c. methods, see Sect. 3(ii)f.

(vi) Pre-heating

When more than one valve is to be tested, a pre-heating socket should be provided to save time. If not incorporated into the tester itself, the pre-heater may be a separate unit with one of each type of socket and a filament voltage switch and transformer. No voltages need normally be applied to the other electrodes except in cases when the valve is initially satisfactory but becomes defective during operation.

(vii) Testing procedure

Always test for shorts before carrying out any other test. When testing for emission, allow ample time for the valve to heat thoroughly before pressing the emission switch ; then press only long enough to give a reading. Do not leave the valve for more than three seconds on the emission test as otherwise the valve may be permanently damaged. The reading may be regarded as satisfactory provided that it is not below the limit and that it does not fall rapidly to the limit in the three second period.

When testing for mutual conductance, the switch may be pressed for any desired length of time, as no damage will be done to the valve provided that the valve is operated within its maximum ratings.

If a valve is gassy, the readings of emission and mutual conductance will be affected (usually increased for a slight amount of gas and decreased by excessive gas). If a valve is both gassy and of low emission, the true fault is " gas ".

If a valve indicates " no emission " while the heater is continuous, the fault may be excessive gas, air in the bulb, cathode coating " poisoned " by gas, or open-circuited cathode. Air in the bulb is indicated by a continuous heater, a cathode not visibly red, and a bulb which gradually becomes warm.

References to service testing : Refs. 55, 56, 57, 60, 61, 62, 63, 64.

SECTION 6 : REFERENCES

1. Benjamin, M., C. W. Cosgrove and G. W. Warren. " Modern receiving valves : design and manufacture " Jour. I.E.E. 80.484 (April 1937) 401.
2. Tuttle, W. N. " Dynamic measurement of electron tube coefficients " Proc. I.R.F. 21.6 (June 1933) 844.
3. Lehany, F. J., and W. S. McGuire. " A radio-frequency interelectrode-capacitance meter " A.W.A. Tec. Rev. 7.3 (April 1947) 271. See also Ref. 5.
4. Pockman, L. T. " The dependence of interelectrode capacitance on shielding " Proc. I.R.F. 32.2 (Feb. 1944) 91.
5. Dobbie, L. G., and R. M. Huey. " A.W.A. interelectrode capacity meter Type R1782 for screen grid valves " A.W.A. Tec. Rev. 2.4 (Oct. 1936) 107.
6. Van der Bijl (book). " The Thermionic Vacuum Tube " (McGraw-Hill Book Co. 1920).
7. Pidduck, F. B. (book). " A Treatise on Electricity " (Cambridge University Press, 1925).
8. Hague, B. (book). " Alternating Current Bridge Methods " (Sir Isaac Pitman & Sons Ltd., London, 5th edit., 1943).
9. Hartshorn, L. (book). " Radio Frequency Measurements by Bridge and Resonance Methods " (Chapman & Hall Ltd., London, 1942).
10. Thomson, J. J. (book). " Elements of Electricity and Magnetism " (Cambridge University Press, 5th edit., 1921).
11. Jeans, J. (book). " The Mathematical Theory of Electricity and Magnetism " (Cambridge University Press, 5th edit., 1943).
12. Smythe, W. R. (book). " Static and Dynamic Electricity " (McGraw-Hill Book Co., 2nd ed. 1950).
13. Maxwell, J. C. (book). " Electricity and Magnetism " (published 1892).
14. Valley, G. E., and H. Wallman (editors). " Vacuum Tube Amplifiers " Radiation Laboratory Series, McGraw-Hill Book Co., 1948.
15. Reich, H. J. (book). " Theory and Applications of Electron Tubes " (McGraw-Hill Book Co., 1939).
16. Chaffee, E. L. (book). " Theory of Thermionic Vacuum Tubes " (McGraw-Hill Book Co., 1933).
17. Scroggie, M. G. (book). " Radio Laboratory Handbook " (Wireless World, London).
18. Terman, F. E. (book). " Measurements in Radio Engineering " (McGraw-Hill Book Co., 1935).
19. Wireless Telegraphy Board Specification K1001 – radio valves.
20. Hund, A. (book). " High Frequency Measurements " (McGraw-Hill Book Co., 1933).
21. Miller, J. M. " A dynamic method for determining the characteristics of three-electrode vacuum tubes " Proc. I.R.E. 6 (June, 1918) 141.
23. James, E. G., and B. L. Humphreys. " Resistive films in valves – effect on interelectrode capacitance " W.E. 26.306 (March 1949) 93.
24. Booth, R. H. (letter). " Triode interelectrode capacitances " W.E. 26.309 (June 1949) 211.
25. Walsh, L. " A direct capacity bridge for vacuum-tube measurements " Proc. I.R.E. 16.4 (April 1928) 482.
26. Wheeler, H. A. " Measurement of vacuum-tube capacities by means of a transformer balance " Proc. I.R.E. 16.4 (April 1928) 476.

27. Hoch, E. T. "A bridge method for the measurement of interelectrode admittance in vacuum tubes " Proc. I.R.E. 16.4 (April 1928) 487.
28. Loughren, A. V., and H. W. Parker. " The measurement of direct interelectrode capacitance of vacuum tubes " Proc. I.R.E. 17.6 (June 1929) 957.
29. Jones, T. I. " The measurement of the grid-anode capacitance of screen-grid valves " Jour. I.E.E. 74.450 (June 1934) 589.
30. Astbury, N. F., and T. I. Jones. " A capacitance attenuator and its application to the measurement of very small capacitances " Jour. Sci. Instr. 13.12 (Dec. 1936) 407.
31. Barco, A. A. " An improved interelectrode capacitance meter " R.C.A. Rev. 6.4 (April 1942) 434.
32. Tillman, J. R., and A. C. Lynch. " Apparatus for the measurement of small three-terminal capacitances " Jour. Sci. Instr. 19.8 (Aug. 1942) 122.
33. Proctor, R. F., and E. G. James. " A radio-frequency and conductance bridge " Jour. I.E.E. 92 Part III. 20 (Dec. 1945) 287.
34. van Suchtelen, H. " Applications of cathode ray tubes : iv—the recording of diagrams " Philips Tec. Rev. 3.11 (Nov. 1938) 339.
35. Douma, Tj., and P. Ziilstra. " Recording the characteristics of transmitting valves " Philips Tec. Rev. 4.2 (Feb. 1939) 56.
36. van der Ven, A. J. H. " Testing amplifier output valves by means of the cathode ray tube " Philips Tec. Rev. 5.3 (March 1940) 61.
37. Millman, J., and S. Moskowitz. " Tracing tube characteristics on a cathode ray oscilloscope " Elect. 14.3 (March 1941) 36.
38. Easton, E. C., and E. L. Chaffee. " Pulse-type tester for high-power tubes " Elect. 20.2 (Feb. 1947) 97.
39. Webking, H. E. " Producing tube curves on an oscilloscope " Elect. 20.11 (Nov. 1947) 128.
40. Herold, E. W. " An improved method of testing for residual gas in electron tubes and vacuum systems " R.C.A. Rev. 10.3 (Sept. 1949) 430.
41. Kauzmann, A. P. " New television amplifier receiving tubes " R.C.A. Rev. 3.3 (Jan. 1939) 271.
42. Dodge, H. F., and H. G. Romig. " Single sampling and double sampling inspection tables " B.S.T.J. Jan. 1941 20.1 (Jan. 1941) 1.
43. Dodge, H. F., and H. G. Romig (book). " Sampling Inspection Tables " (John Wiley & Sons Inc., New York, Chapman & Hall, London, 1944).
44. Keeling, D. B., and L. E. Cisne. " Using double sampling inspection in a manufacturing plant " B.S.T.J. 21.1 (June 1942) 37.
45. Davies, J A. " Quality control in radio tube manufacture " Proc. I.R.E. 37.5 (May 1949) 548.
46. Harris, J. A. " On the co-ordination of circuit requirements, valve characteristics and electrode design " J. Brit. I.R.E. 9.4 (April 1949) 125.
47. Slonczewski, T. " Transconductance as a criterion of electron tube performance " B.S.T.J. 28.2 (April 1949) 315.
48. British Radio Valve Manufacturers' Association (booklet). " Radio Valve Practice " (August 1948).
49. British Standard Code of Practice BS1106 (Feb. 1943). " The use of radio valves in equipment."
51. Application Note. " Input admittance of receiving tubes " (AN-118, April 15, 1947) Radio Corporation of America.
52. " Tube Handbook " HB3, Radio Corporation of America.
53. " Radio Components Handbook " (1st ed. 1948, Technical Advertising Associates, Cheltenham, Penn., U.S.A.).
55. Hartkopf, R. E. " Simple valve tester—checking insulation, mutual conductance and emission " W.W. 42.12 (Dec. 1946) 386.
56. Everett Edgcumbe Service Valve Tester, W.W. 42.12 (Dec. 1946) 416.
57. " Dynamic mutual conductance tube tester " (Hickok design) Service 16.3 (March 1947) 2.
58. Heacock, D. P. " Power measurement of Class B audio amplifier tubes " R.C.A. Rev. 8.1 (March 1947) 147.
59. Walker, H. S., C. G. Mayo and J. Tomlinson " Portable sub-standard of mutual conductance " B.B.C. Research Dept. Report H.009 (27 Oct. 1938).
60. " A universal tube checker " Radio Craft (Feb. 1944) 292.
61. Dewar, J. A. " Dynamic tube tester " Radio Craft (Nov. 1944) 94.
62. Planer, F. E. " Valve Testing—rapid determination of amplification factor and anode impedance under operating conditions " W.W. 51.6 (June 1945) 20.
63. Simpson. " Dynamic tube tester " Elect. 19.2 (Feb. 1946) 338.
64. " Hickok, Tube Tester Models 532C and 532P " Elect. 19.3 (Mar. 1946) 282.
65. Blackburn, J. F. (editor). " Components Handbook " (M.I.T. Radiation Laboratory Series, No. 17, 1st ed. 1949, McGraw-Hill Book Co.).
66. Laws, F. A. (book). " Electrical Measurements " (McGraw-Hill Book Co. 2nd ed. 1938).
67. Appleton, E. V. (book). " Thermionic Vacuum Tubes " (Methuen & Co. Ltd., London).
68. Livingston, O. W. " Oscillographic method of measuring positive-grid characteristics " Proc. I.R.E. 28.6 (June 1940) 267.
69. Kozanowski, H. N., and I. E. Mouromtseff. " Vacuum tube characteristics in the positive grid region by an oscillographic method " Proc. I.R.E. 21.8 (April 1933) 1082.
70. Chaffee, E. L. " Oscillographic study of electron tube characteristics " Proc. I.R.E. 10.6 (Dec. 1922) 440.
71. Schneider, W. A. " Use of an oscillograph for recording vacuum tube characteristics " Proc. I.R.E. 16.5 (May 1928) 674.
72. Chaffee, E. L. " Power tube characteristics " Elect. 11.6 (June 1938) 34.
73. Walker, A. H. B. " Cathode-ray curve tracer " W.W. 50.9 (Sept. 1944) 266.
74. Zepler, E. E. " Triode interelectrode capacitances " W.E. 26.305 (Feb. 1949) 53.
75. Jones, T. I. " The dependence of the inter-electrode capacitances of valves upon the working conditions " Jour. I.E.E. 81 (1937) 658.
76. Humphreys, B. L., and B. G. James. " Interelectrode capacitance of valves—change with operating conditions " W.E. 26.304 (Jan. 1949) 26.
77. Young, C. H. " Measuring interelectrode capacitances " Tele-Tech 6.2 (Feb. 1947) 68.
78. Terman, F. E. " Radio Engineers' Handbook " (McGraw-Hill Book Co., 1943).
79. Dawes, C. L. (book). " A Course in Electrical Engineering " Vol. 2 (McGraw-Hill Book Co. 3rd edit., 1934).
80. Hickman, R., and F. Hunt. " The exact measurement of electron tube coefficients " Rev. Sci. Instr. 6 (Sept. 1935) 268.
81. Eastman, A. V. (book). " Fundamentals of Vacuum Tubes " (McGraw-Hill Book Co. 1937).
82. Albert, A. L. (book). " Fundamental Electronics and Vacuum Tubes " (The Macmillan Co., New York, 1938).

83. Aiken, C. B., and J. F. Bell. " A mutual conductance meter " Comm. 18.9 (Sept. 1938) 19.
84. Millman, J., and S. Seely (book). " Electronics " (McGraw-Hill Book Co. 1941).
85. Suits, C. G. " A thermionic voltmeter method for the harmonic analysis of electrical waves " Proc. I.R.E. 18 (Jan. 1930) 178.
86. Belfils, G. " Mesur du ' residu ' des courbes de tension par la methode du pont filtrant " Rev. Gen. d'Elect. 19 (April 1926) 526.
87. Wolff, Irving. " The alternating current bridge as a harmonic analyzer " Jour. Opt. Soc. Amer. 15 (Sept. 1927) 163.
88. McCurdy, R. G., and P. W. Blye. " Electrical wave analyzers for power and telephone systems " Jour. A.I.E.E. 48 (June 1929) 461.
89. Ballantine, S., and H. L. Cobb. " Power output characteristics of the pentode " Proc. I.R.E. 18 (March 1930) 450.
90. " Curve generator for tubes " developed by M. L. Kuder of N.B.S., Elect. 23.7 (July 1950) 144.
91. " Heater-cathode leakage as a source of hum " Elect. 13.2 (Feb. 1940) 48.
92. Zepler, E. E. and S. S. Srivastava " Interelectrode impedances in triodes and pentodes " W.E. 28.332 (May 1951) 146.
93. Wagner, H. M. " Tube characteristic tracer using pulse techniques " Elect. 24.4 (Apr. 1951) 110.
Additional references will be found in the Supplement commencing on page 1475.

Standards Specifications.
S1. Standards on electronics (1938) The Institute of Radio Engineers. U.S.A. See also S12.
S2. Joint Army-Navy Specification JAN-1A (U.S.A.).
S3. Joint Army-Navy Specification JAN-S-28 Sockets, electron tube, miniature.
S4. RMA-NEMA* Standards for electron tube bases caps and terminals (Feb. 1949) – RMA Standard ET-103A.
S5. RMA-NEMA* Dimensional characteristics of electron tubes (Jan. 1949) – RMA Standard ET-105A.
S6. RMA-NEMA-JETEC Standard ET-106A (Feb. 1952) Gauges for electron tube bases.
S7. RMA Standard ET-107 (Dec. 1946) Test methods and procedures for radio receiving tubes.
S8. RMA-NEMA* (Dec. 1947) Standards for measurement of direct interelectrode capacitances – RMA Standard ET-109.
S9. RMA-NEMA* (Feb. 1948) Standard for designation system for receiving tubes – RMA Standard ET-110.
S10. RMA-NEMA* (Feb. 1949) Designation system for electron tube shells – RMA Standard ET-112.
S11. " Tube Ratings " – RMA Standard M8-210 (Jan. 8, 1940).
S12. " Standards on electron tubes : methods of testing 1950 " Proc. I.R.E., Part 1, 38.8 (Aug. 1950) 917 ; Part 2, 38.9 (Sept. 1950) 1079.

CHAPTER 4

THEORY OF NETWORKS

by F. LANGFORD-SMITH, B.SC., B.E.

For ease of reference

SECTION 1 : CURRENT AND VOLTAGE

(i) Direct current (ii) Alternating current (iii) Indications of polarity and current flow.

(i) Direct current

We speak of the flow of an electric current in more or less the same way that we speak of the flow of water, but we should remember that the conventional direction of current is opposite to the actual flow of electrons. In most electrical circuit theory however, it is sufficient to consider only the conventional direction of current flow. In Fig. 4.1 there is a battery connected to a load ; the current flows from the positive (+) terminal, through the load, to the negative (−) terminal, and then through the battery to the positive terminal.

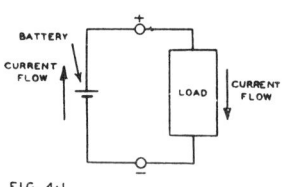

FIG. 4·1

Fig. 4.1. *Flow of current with battery and load.*

Batteries (or cells) may be connected in series as in Fig. 4.2 and the total voltage is then equal to the sum of the voltages of the individual batteries (or cells). When calculating the voltage of any intermediate point with respect to (say) the negative terminal, count the number of cells passed through from the negative terminal to the tapping point, and multiply by the voltage per cell. When batteries are connected in series, each has to supply the full load current.

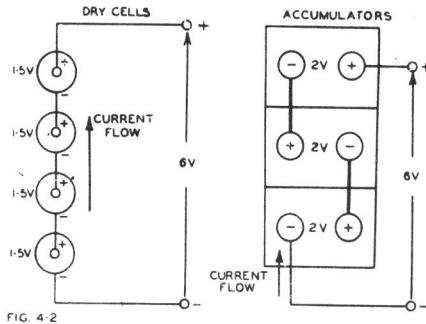

FIG. 4·2

Fig. 4.2. *Cells in series.*

Batteries (or cells) are occasionally connected in parallel as in Fig. 4.3. In this case the terminal voltage is the same as the voltage per battery (or cell). The current does not necessarily divide uniformly between the cells, unless these all have identical voltages and internal resistances.

Direct current may also be obtained from a d.c. generator, or from rectified and filtered a.c. supply. In all such cases there is a certain degree of ripple or hum which prevents it from being pure d.c.; when the a.c. component is appreciable, the supply may be spoken of as " d.c. with superimposed ripple (or hum)." and must be treated as having the characteristics of both d.c. and a.c. When we speak of d.c. in a theoretical treatise, it is intended to imply pure, steady d.c.

(ii) Alternating current

The ordinary form of d.c. generator actually generates a.c., which is converted to d.c. by the commutator. If a loop of wire is rotated about its axis in a uniform magnetic field, an alternating voltage is generated across its terminals. Thus a.c. is just as fundamental as d.c. The usual power-house generates 3 phase a.c., but in radio receivers we are only concerned with one of these phases. A " sine wave "

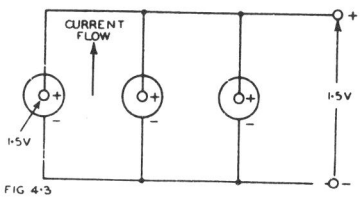

Fig. 4.3. *Cells in parallel.*

alternating current is illustrated in Fig. 4.4, where the vertical scale may represent voltage or current, and the horizontal scale represents time.* A **cycle** is the alternation from *A* to *E*, or from *B* to *F*, or from *C* to *G*.

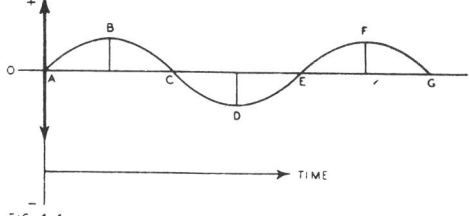

FIG 4·4

Fig. 4.4. *Form of sine-wave alternating current.*

Most power supplies have **frequencies** of either 50 or 60 cycles per second (c/s). The **period** is the time taken by one cycle, which is 1/50 or 1/60 second, in these two cases.

*For mathematical treatment of periodic phenomena see Chapter 6 Sect. 4.

The precise shape of the wave is very important, and the sine wave has been adopted as the standard a.c. **waveform,** since this is the only one which always has the current waveform of the same shape as the voltage, when applied to a resistance, inductance or capacitance.† In practice we have to deal with various waveforms, some of which may be considered as imperfect ("distorted") sine waves, while others are of special shapes such as square waves, saw-tooth, or pulse types, or rectified sine waves. However, when we speak of a.c. in any theoretical treatise, it is intended to imply a distortionless sine wave. Other waveforms may be resolved by Fourier Analysis into a fundamental sine wave and a number of harmonic frequency sine waves, the latter having frequencies which are multiples of the fundamental frequency. This subject is treated mathematically in Chapter 6 Sect. 8.

When deriving the characteristics of any circuit, amplifier or network, it is usual to assume the application of a pure sine-wave voltage to the input terminals, then to calculate the currents and voltages in the circuit. In the case of a valve amplifier (or any other non-linear component) the distortion may be either calculated, or measured at the output terminals. If the device is to operate with a special input waveform (e.g. square wave), it is usual to resolve this into its fundamental and harmonic frequencies, and then to calculate the performance with the lowest (fundamental) frequency, an approximate middle** frequency, and the highest harmonic frequency—all these being sine waves.

(iii) Indications of polarity and current flow

In circuit diagrams the polarity of any battery or other d.c. voltage source is usually indicated by + and − signs ; alternatively it may be indicated by an arrow, the head of the arrow indicating positive potential (e.g. Fig. 4.2). A similar convention may be used for the voltage between any points in the circuit (e.g. Fig. 4.14A). The direction of d.c. current flow is indicated by an arrow.

In the case of a.c. circuits a similar convention may be used, except that an arbitrary instantaneous condition is represented (Fig. 4.18A).

SECTION 2 : RESISTANCE

(i) Ohm's Law for d.c.   (ii) Ohm's Law for a.c.   (iii) Resistances in series   (iv) Resistances in parallel   (v) Conductance in resistive circuits.

(i) Ohm's Law for direct current

All substances offer some obstruction to the flow of electric current. Ohm's Law states that the current which flows is proportional to the applied voltage, in accordance with the equation

$$I = E/R \tag{1}$$

where R is the total **resistance** of the circuit. For example in Fig. 4.5 an ideal battery, having zero internal resistance, and giving a constant voltage E under all conditions, is connected across a resistance R. The current which flows is given by eqn. (1) above, provided that

I is expressed in amperes,

E is expressed in volts,

and R is expressed in ohms.*

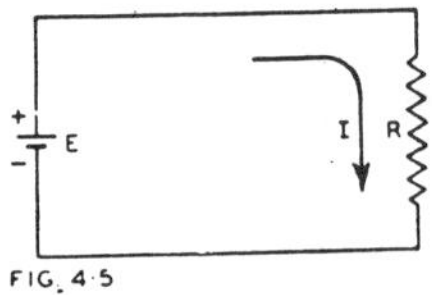

Ohm's Law may also be arranged, for convenience, in the alternative forms

$$E = IR \tag{2}$$
$$\text{and } R = E/I \tag{3}$$

FIG. 4·5

Fig. 4.5. Circuit illustrating Ohm's Law for d.c.

---

†A sinewave has its derivative and integral of the same form as itself.

**Preferably the geometrical mean frequency which is given by $\sqrt{f_1 f_2}$ where f_1 and f_2 are the lowest and highest frequencies.

*It is assumed that the resistance remains constant under the conditions of operation. For other cases see Sect. 7(i).

In a circuit containing more than one battery (or other source of direct voltage), the effective voltage is determined by adding together those voltages which are in the same direction as the current, and subtracting any opposing voltages.

Ohm's Law also holds for any single resistor or combination of resistances. The voltage drop across any resistance R_1, no matter what the external circuit may be, is given by

$$\text{Voltage drop} = IR_1 \qquad (4)$$

where I is the current flowing through R_1.

(ii) Ohm's Law for alternating current

Ohm's Law holds also for alternating voltages and currents, except that in this case the voltage (E) and the current (I) must be expressed in their **effective** or **root-mean-square** values of volts and amperes.

(iii) Resistances in series

When two or more resistors are connected so that the current through one is compelled to flow through the others, they are said to be in series, and the total resistance is the sum of their individual resistances. For example, in Fig. 4.6 the total resistance of the circuit is given by

$$R = R_1 + R_2 + R_3 \qquad (5)$$

and the current is given by

$$I = E/R = E/(R_1 + R_2 + R_3).$$

It is interesting to note that R_1, R_2 and R_3 form a **voltage divider**, across the battery E. Using eqn. (4) :

$$\text{voltage drop across } R_1 = IR_1 = [R_1/(R_1 + R_2 + R_3)] \times E$$
$$\text{voltage drop across } R_2 = IR_2 = [R_2/(R_1 + R_2 + R_3)] \times E$$
$$\text{voltage drop across } R_3 = IR_3 = [R_3/(R_1 + R_2 + R_3)] \times E$$
$$\text{total voltage drop} = IR_1 + IR_2 + IR_3$$
$$= I(R_1 + R_2 + R_3) = E.$$

For example, if $E = 6$ volts, $R_1 = 10$ ohms, $R_2 = 10$ ohms and $R_3 = 10$ ohms, then

$$I = E/(R_1 + R_2 + R_3) = 6/(10 + 10 + 10)$$
$$= 6/30 = 0.2 \text{ ampere.}$$

$$\text{Voltage between points } C \text{ and } D = 10 \times 0.2 = 2 \text{ volts}$$
$$B \text{ and } D = 20 \times 0.2 = 4 \text{ volts}$$
$$A \text{ and } D = 30 \times 0.2 = 6 \text{ volts.}$$

The voltage across any section of the voltage divider is proportional to its resistance (it is assumed that no current is drawn from the tapping points B or C).

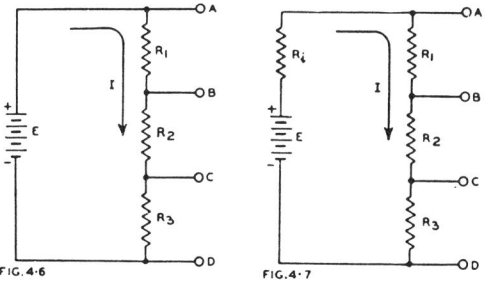

Fig. 4.6.   *Resistances in series.*

Fig. 4.7.   *Resistances in series, allowing for internal resistance of battery.*

If the battery has any appreciable internal resistance the circuit must be modified to the form of Fig. 4.7 where R_i is the equivalent internal resistance. Here we have four resistances effectively in series and $I = E/(R_i + R_1 + R_2 + R_3)$. The voltages between any of the points A, B, C or D will be less than in the corresponding case for zero internal resistance, the actual values being

$$\frac{R_1 + R_2 + R_3}{R_i + R_1 + R_2 + R_3} \times \text{voltage for } R_i = 0.$$

If R_i is less than 1% of $(R_1 + R_2 + R_3)$, then its effect on voltages is less than 1%.

(iv) Resistances in parallel

When two resistances are in parallel (Fig. 4.8) the effective total resistance is given by

$$R = \frac{R_1 R_2}{R_1 + R_2} \tag{6}$$

When $R_1 = R_2$, $R = R_1/2 = R_2/2$.
When any number of resistances are in parallel (Fig. 4.9) the effective total resistance is given by

$$\frac{1}{R} = \frac{1}{R_1} + \frac{1}{R_2} + \frac{1}{R_3} + \ldots . \tag{7}$$

When two or more resistors are in parallel, the total effective resistance may be determined by the graphical method of Fig. 4.10.   This method* only requires a piece of ordinary graph paper (or alternatively a scale and set square).   As an example, to find the total resistance of two resistors, 50 000 and 30 000 ohms, in parallel take any convenient base AB with verticals AC and BD at the two ends.   Take 50 000 ohms on BD and draw the straight line AD ; take 30 000 ohms on AC and draw CB ; draw the line XY from their junction perpendicular to AB.   The height of XY gives the required result, on the same scale.

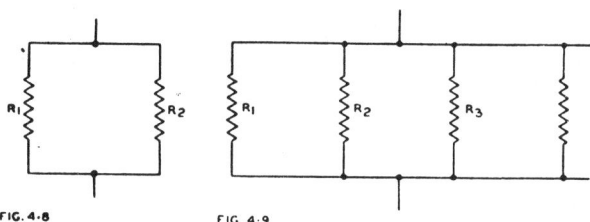

FIG. 4·8   FIG. 4·9

Fig. 4.8.   *Two resistances in parallel.*   *Fig.* 4.9.   *Several resistors in parallel.*

If it is required to determine the resistance of three resistors in parallel, the third being say 20 000 ohms, proceed further to join points E and Y, and the desired result is given by the height PQ.   This may be continued indefinitely.

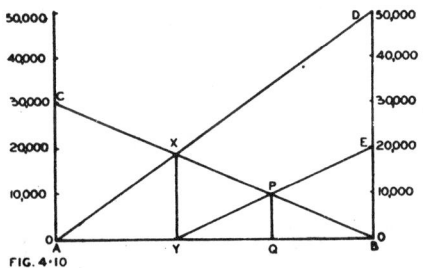

FIG. 4·10

Fig. 4.10.   *Graphical method for the determination of the effective resistance of two or more resistors in parallel (after Wireless World).*

The same method may be used to determine suitable values of two resistors to be connected in parallel to give a specified total resistance.   In this case, select one value (C) arbitrarily, mark X at the correct height and then find D ; if not a suitable value, move C to the next available value and repeat the process until satisfactory.

*"Resistances in Parallel—Capacitances in Series," W.W. 48.9 (Sept. 1942) 205.
" Diallist " " Series C and Parallel R," W.W. 51.4 (April 1945) 126.

When a network includes a number of resistors, some in series and some in parallel, firstly convert all groups in parallel to their effective total resistances, then proceed with the series chain.

(v) Conductance in resistive circuits

The conductance (G) of any resistor is its ability to conduct current, and this is obviously the reciprocal of the resistance—

$$G = 1/R \tag{8}$$

Applying Ohm's Law, we derive

$$I = EG \tag{9}$$

The unit of conductance is the mho (i.e. the reciprocal ohm).

When resistances are in parallel, their effective total conductance is the sum of their individual conductances—

$$G = G_1 + G_2 + G_3 + \ldots \tag{10}$$

When a number of resistors are in parallel, the current through each is proportional to its conductance. Also,

$$\frac{I_1}{I_{total}} = \frac{G_1}{G_{total}} \tag{11}$$

SECTION 3 : POWER

(i) Power in d.c. circuits   (ii) Power in resistive a.c. circuits.

(i) Power in d.c. circuits

The power converted into heat in a resistance is directly proportional to the product of the voltage and the current—

$$P = E \times I \tag{1}$$

where P is expressed in watts, E in volts and I in amperes. This equation may be rearranged, by using Ohm's Law, into the alternative forms—

$$P = I^2 R = E^2/R \tag{2}$$

where R is expressed in ohms.

The total **energy** developed is the product of the power and the time. Units of energy are

(1) the **watt-second** or **joule**
(2) the **kilowatt-hour** (i.e. 1000 watts for 1 hour).

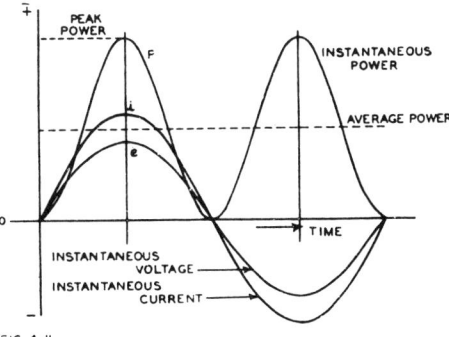

Fig. 4.11. One cycle of sine-wave voltage (e) and current (i) with zero phase angle. The instantaneous power (P) is always positive ; the average power is half the peak power.

FIG. 4·11

(ii) Power in resistive a.c. circuits

The same general principles hold as for d.c., except that the voltage, current and power are varying. Fig. 4.11 shows one cycle of a sinewave voltage and current. The instantaneous power is equal to the product of the instantaneous voltage and instantaneous current at any point in the cycle. The curve of instantaneous power (P) may be plotted point by point, and is always positive.

The heating of a resistor is obviously the result of the **average** or **effective power,** which is exactly half the peak power.

$$P_{peak} = E_{max} I_{max} \qquad (3)$$

$$P_{av} = \frac{E_{max} I_{max}}{2} = \frac{E_{max}}{\sqrt{2}} \times \frac{I_{max}}{\sqrt{2}} \qquad (4)$$

$E_{max}/\sqrt{2}$ is called the **effective** or **r.m.s. voltage**
$I_{max}/\sqrt{2}$ is called the **effective** or **r.m s. current.**

The effective values of voltage and current are the values which have the same heating effect as with d.c. The initials r.m.s. stand for root mean square, indicating that it is the square root of the average of the squares over the cycle.

In a.c. practice, any reference to voltages or currents without specifying which value is intended, should always be interpreted as being r.m s. (or effective) values. Measuring instruments are usually calibrated in r.m.s. values of currents and voltages.

The **form factor** is the ratio of the r.m.s. to the average value of the positive half-cycle. The following table summarizes the principal characteristics of several waveforms, over the positive half-cycle in each case (see Chapter 30 for rectified waveforms) :

	Sine wave	Square wave	Triangular wave (isosceles)
Form factor (= r.m.s./average)	$\pi/2\sqrt{2} = 1.11$	1.00	$2/\sqrt{3} = 1.15$
Peak/r.m.s.	$\sqrt{2} = 1.414$	1.00	$\sqrt{3} = 1.73$
R.M.S./peak	$1/\sqrt{2} = 0.707$	1.00	$1/\sqrt{3} = 0.58$
Peak/average	$\pi/2 = 1.57$	1.00	2.0
Average/Peak	$2/\pi = 0.64$	1.00	0.5

SECTION 4 : CAPACITANCE

(i) *Introduction to capacitance* (ii) *Condensers in parallel and series* (iii) *Calculation of capacitance* (iv) *Condensers in d.c. circuits* (v) *Condensers in a.c. circuits.*

(i) Introduction to capacitance

A capacitor* (or condenser) in its simplest form, consists of two plates separated by an insulator (dielectric).

Any such condenser has a characteristic known as capacitance† whereby it is able to hold an electric charge. When a voltage difference (E) is applied between the plates, current flows instantaneously through the leads connecting the battery to the condenser (Fig. 4.12) until the latter has built up its charge, the current dropping gradually to zero. If the battery is removed, the condenser will hold its charge indefinitely (in practice there is a gradual loss of charge through leakage). If a conducting path is connected across the condenser plates, a current will flow through the conductor but will gradually fall to zero as the condenser loses its charge.

It is found that the charge (i.e. the amount of electricity) which a condenser will hold is proportional to the applied voltage and to the capacitance.

This may be put into the form of an equation :

$$Q = CE \qquad (1)$$

where Q = quantity of electricity (the charge) in coulombs,
C = capacitance in farads,
and E = applied voltage.

*The American standard term is " Capacitor."
†It is assumed here that the condenser is ideal, without series resistance, leakage, or dielectric lag.

The unit of capacitance—the Farad (F)—is too large for convenience, so it is usual to specify capacitance as so many microfarads (μF) or micro-microfarads* ($\mu\mu$F). Any capacitance must be converted into its equivalent value in farads before being used in any fundamental equation such as (1) above

$$1\,\mu\text{F} = 1 \times 10^{-6} \quad \text{farad}$$
$$1\,\mu\mu\text{F} = 1 \times 10^{-12} \quad \text{farad}$$

Note : *The abbreviations mF or mmF should not be used under any circumstances to indicate microfarads or micro-microfarads, because mF is the symbol for milli-farads (1×10^{-3} farad). Some reasonable latitude is allowable with most symbols, but here there is danger of serious error and misunderstanding.*

The **energy stored** in placing a charge on a condenser is

$$W = \tfrac{1}{2}(QE) = \tfrac{1}{2}(CE^2) = Q^2/2C \tag{2}$$

where W = energy, expressed in joules (watt-seconds)
      Q = charge in coulombs
      C = capacitance in farads
and E = applied voltage.

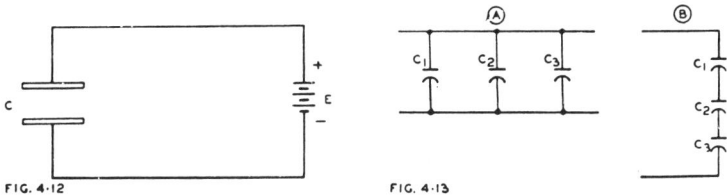

Fig. 4.12. *Condenser connected to a battery.*
Fig. 4.13. *(A) Condensers in parallel. (B) Condensers in series.*

(ii) Condensers in parallel and series

When two or more condensers are connected in parallel (Fig. 4.13A) the total capacitance is the sum of their individual capacitances :

$$C = C_1 + C_2 + C_3 + \cdots \tag{3}$$

When two or more condensers are connected in series (Fig. 4.13B) the total capacitance is given by :

$$\frac{1}{C} = \frac{1}{C_1} + \frac{1}{C_2} + \frac{1}{C_3} + \cdots \tag{4}$$

When only two condensers are connected in series :

$$C = \frac{C_1 C_2}{C_1 + C_2} . \tag{5}$$

When two or more equal condensers (C_1) are connected in series, the total capacitance is

$$C = C_1/2 \text{ for 2 condensers}$$
and $C = C_1/n$ for n condensers.

Note : The curved plate of the symbol used for a condenser indicates the earthed (outer) plate of an electrolytic or circular paper condenser ; when this is not applicable the curved plate is regarded as the one more nearly at earth potential.

(iii) Calculation of capacitance
Parallel plate condenser
When there are two plates, close together, the capacitance is approximately :

$$C = \frac{AK}{11.31d} \quad \mu\mu\text{F when dimensions are in centimetres}$$

---

*The name picofarad (*pF*) is also used as an alternative.

or $C = \dfrac{AK}{4.45d}\ \mu\mu\text{F}$ when dimensions are in inches

where A = useful area of one plate in square centimetres (or inches). The useful area is approximately equal to the area of the smaller plate when the square root of the area is large compared with the gap.

K = dielectric constant (for values of common materials see Chapter 38 Sect. 8. For air, $K = 1$.

d = gap between plates in centimetres (or inches).

Capacitance with air dielectric, plates 1 mm. apart
  $C = 0.884\ \mu\mu\text{F}$ per sq. cm. area of one plate.

Capacitance with air dielectric, plates 0.10 inch apart
  $C = 2.244\ \mu\mu\text{F}$ per sq. inch area of one plate.

When there are more than two plates, the " useful area " should be interpreted as the total useful area.

Cylindrical condenser (concentric cable)
  $C = \dfrac{7.354K}{\log_{10} D/d}\ \mu\mu\text{F}$ per foot length

where D = inside diameter of outside cylinder (inches)
  d = outside diameter of inner cylinder (inches)
and K = dielectric constant of material in gap.

(iv) Condensers in d.c. circuits

An **ideal condenser** is one which has no resistance, no leakage, and no inductance. In practice, every condenser has some resistance, leakage and inductance, although these may be neglected under certain conditions of operation.

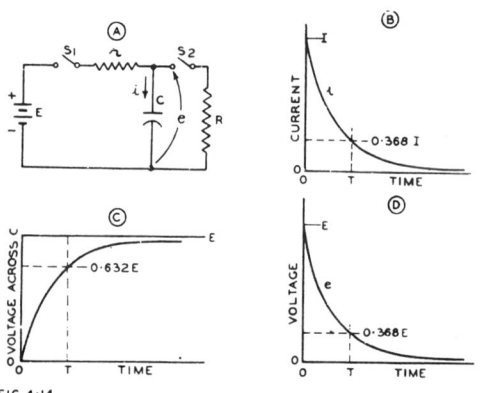

FIG. 4·14

Fig. 4.14. Condenser charge and discharge (A) Circuit (B) Discharge current characteristic (C) Charge voltage characteristic (D) Discharge voltage characteristic.

In Fig. 4.14A, C is an ideal condenser which may be charged by closing switch S_1; r represents the combined internal resistance of the battery E and the resistance of the leads in the circuit. When S_1 is closed, current (i) will flow as indicated in diagram B, the peak current being $I = E/r$ at time $t = 0$. The time for C to become fully charged is infinite—in other words the current never quite reaches zero, although it comes very close to zero after a short period. The equation for the current is of logarithmic form :

$$i = \frac{E}{r}\,\epsilon^{-t/rC} \tag{6}$$

where ϵ = base of natural logarithms ($\approx$ 2.718)
      t = time in seconds after closing switch S_1
      E = battery voltage
      r = resistance in ohms
and C = capacitance in farads.

The voltage (e) across the condenser is

$$e = E - ri = E\,(1 - \epsilon^{-t/rC}) \tag{7a}$$

and the curve (diagram C) is of the same shape as the current curve except that it is upside down.   The voltage never quite reaches the value E, although it approaches it very closely.

The charge on the condenser is given by

$$q = Q\,(1 - \epsilon^{-t/rC}) \tag{7b}$$

where q = instantaneous charge on condenser
and Q = EC = final charge on condenser
which follows the same law as the voltage (eqn. 7a).

If we now assume that the condenser C is fully charged, switch S_1 is opened, and switch S_2 is closed, the **discharge characteristic** will be given by

$$i = -\frac{E}{R}\,\epsilon^{-t/RC} \tag{8}$$

which is of the same form as diagram B, except that the current is in the opposite direction.

The curve of voltage (and also charge) against time for a discharging condenser is in diagram D, and is of the same shape as for current, since $e = Ri$.   These charge and discharge characteristics are called **transients**.

In order to make a convenient measure of the time taken to discharge a condenser, we adopt the **time constant** which is the time taken to discharge a condenser on the assumption that the current remains constant throughout the process at its initial value.   In practice, as explained above, the discharge current steadily falls with time, and under these conditions the time constant is the time taken to discharge the condenser to the point where the voltage or charge drops to $1/\epsilon$ or 36.8% of its initial value.   The same applies also to the time taken by a condenser in process of being charged, to reach a voltage or charge of $(1 - 1/\epsilon)$ or 63.2% of its final value.

The time constant (T) is equal to

$$T = RC \tag{9}$$

where T is the time constant in seconds,
      R is the total resistance in the circuit, either for charge or discharge, in ohms,
and C is the capacitance in farads.
This also holds when R is in megohms and C in microfarads.

(v) Condensers in a.c. circuits

When a condenser is connected to an a.c. line, current flows in the circuit, as may be checked by inserting an a.c. ammeter in series with the condenser.   This does not mean that electrons flow through the condenser from one plate to the other ; they are insulated from one another.

Suppose that a condenser of capacitance C farads is connected directly across an a.c. line, the voltage of which is given by the equation $e = E_m \sin \omega t$.   The condenser will take sufficient charge to make the potential difference of its plates at every instant equal to the voltage of the line.   As the impressed voltage continually varies in magnitude and direction, electrons must be continually passing in and out of the condenser to maintain its plates at the correct potential difference.   This continual charging and discharging of the condenser constitutes the current read by the ammeter.

At any instant, $q = Ce$, where q is the instantaneous charge on the condenser.   The current (i) is the rate of change (or differential* with respect to time) of the charge,

---

*See Chapter 6, Section 7.

i.e., $i = dq/dt$
But $q = Ce = CE_m \sin \omega t$

Therefore $i = \dfrac{d}{dt}(CE_m \sin \omega t)$

Therefore $i = \omega CE_m \cos \omega t$ (10)

Eqn. (10) is the equation of the current flowing through the condenser, from which we may derive the following facts :

1. It has a peak value of ωCE_m ; the current is therefore proportional to the applied voltage, also to the capacitance and to the frequency (since $f = \omega/2\pi$).

2. It has the same angular velocity (ω) and hence the same frequency as the applied voltage.

3. It follows a cosine waveform whereas the applied voltage has a sine waveform. This is the same as a sinewave advanced 90° in phase—we say that the current leads the voltage by 90° (Fig. 4.15).

Considering only the magnitude of the condenser charging current,
$$I_m = \omega CE_m \quad \text{(peak values)}$$
Therefore $I_{rms} = \omega CE_{rms}$ (effective values)
Where I and E occur in a.c. theory, they should be understood as being the same as I_{rms} and E_{rms}.

This should be compared with the equivalent expression when the condenser is replaced by a resistance (R) :

$$I_{rms} = \frac{E_{rms}}{R}.$$

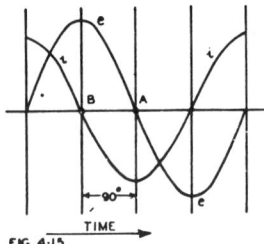

FIG. 4·15

Fig. 4.15. Alternating current through an ideal condenser.

It will be seen that R in the resistance case, and $(1/\omega C)$ in the capacitance case, have a similar effect in limiting the current. We call $(-1/\omega C)$ the **capacitive reactance**† (X_c) of the condenser, since it has the additional effect of advancing the phase of the current. We here adopt the convention of making the capacitive reactance negative, and the inductive reactance positive ; the two types of reactance are vectorially 180° out of phase.

The relationships between the various voltages and currents are well illustrated by a **vector diagram***. Fig. 4.16A shows a circuit with R and C in series across an a.c. line with a voltage $e = E_m \sin \omega t$. The instantaneous** values of voltage (e) and current (i) are shown with arrows to indicate the convention of positive direction. It is quite clear that the same current which passes through R must also pass through C. This causes a voltage drop RI across R and $(I/\omega C)$ across C where I is the r.m.s. value of the current. The relative phase relationships are given by the angular displacements in the vector diagram. These are determined trigonometrically by the peak values I_m, RI_m and $I_m/\omega C$; for convenience, the lengths of the vectors are marked in Fig. 4.16B according to the effective values I, RI and $I/\omega C$. The current vector is distinguished by a solid arrowhead ; it may be to any convenient scale since there is no connection between the voltage and current scales. Since R is purely resistive, RI must be in phase with I, but the current through C must lead the voltage drop across C by 90°. This is shown on the vector diagram 4.16B where the direction of I is taken as the starting point ; RI is drawn in phase with I and of length equal to the voltage drop on any convenient scale ; $I/\omega C$ is drawn so that I leads it by 90° ; and the resultant (E) of RI and $I/\omega C$ is determined by completing the parallelogram. The resultant E is the vector sum of the voltage drops across R and C, which must therefore be the applied (line) voltage. It will be seen that the current I leads the voltage E by an angle ϕ where

$$\tan \phi = \frac{I}{\omega C} \cdot \frac{1}{RI} = \frac{1}{\omega CR} = \frac{|X_c|}{R} \qquad (10a)$$

If $R = 0$, then $\tan \phi = \infty$, and $\phi = 90°$.

Since ϕ is the phase angle of the current with respect to the voltage, the angle ϕ in the circuit of Fig. 4.16A is positive.

The instantaneous current flowing through the circuit of Fig. 4.16A is therefore given by

$$i = I_m \sin (\omega t + \phi) \qquad (10b)$$

where $\phi = \tan^{-1} (1/\omega CR)$.

The value of I_m in eqn. (10b) is given by

$$I_m = E_m/\mathbf{Z} \qquad (10c)$$

where $\mathbf{Z}$ is called the impedance.

Similarly in terms of effective values,

$$I = E/\mathbf{Z} \qquad (10d)$$

Obviously $\mathbf{Z}$ is a vector (or complex) quantity having phase relationship as well as magnitude, and is printed in bold face to indicate this fact. This may be developed further by the use of the j notation.

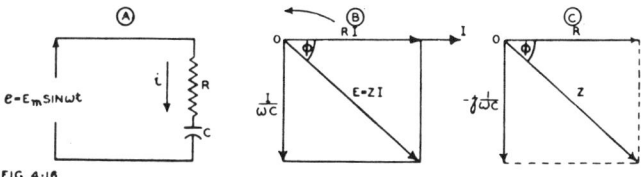

FIG. 4·16

Fig. 4.16. (A) Resistance and capacitance across a.c. line (B) Vector diagram of voltage relationships (C) Vector diagram of impedance (Z) with its real component (R) and reactive component $-j$ *(1/ωC).*

Using the j Notation

The operator j^* signifies a positive vector rotation of 90°, while $-j$ signifies a negative rotation of 90°. Instead of working out a detailed vector diagram, it is possible to treat a circuit problem very much more simply by using the j notation. For example in Fig. 4.16, we may equate the applied voltage E to the sum of the voltage drops across R and C:

$$E = RI - j\frac{I}{\omega C} \qquad (11)$$

the $-j$ indicating 90° vector rotation in a negative direction, which is exactly what we have in diagram B.

It is sometimes more convenient to put $-jI/\omega C$ into the alternative form $+I/j\omega C$ which may be derived by multiplying both numerator and denominator by j (since $j^2 = -1$). Thus

$$E = RI + I/j\omega C \qquad (12)$$

From (11) we can derive:

$$\mathbf{Z} = E/I = R - j(1/\omega C) \qquad (13)$$

For example, in Fig. 4.16A let $R = 100$ ohms and $C = 10\,\mu F$, both connected in series, and let the frequency be 1000 c/s.

Then $\omega = 2\pi \times 1000 = 6280$

$1/\omega C = 1/(6280 \times 10 \times 10^{-6}) = 15.9$ ohms.

and $\mathbf{Z} = R - j(1/\omega C)$

$= 100 - j\,15.9 \qquad (13a)$

Eqn. (13a) indicates at a glance a resistance of 100 ohms in series with a negative reactance (i.e. a capacitive reactance) of 15.9 ohms. Values of capacitive reactances

---

*See Chapter 6 Sect. 5(iv).

for selected capacitances and frequencies are given in Chapter 38 Sect. 9 Table 42. Even if $R = 0$, we still write the impedance in the same form,

$$\mathbf{Z} = 0 - j\,15.9 \tag{13b}$$

Thus $\mathbf{Z}$ is a complex* quantity, that is to say it has a " real " part (R) and an " imaginary " part $(1/\omega C)$ at $90°$ to R, as shown in Fig. 4.16C. The absolute magnitude (modulus) of $\mathbf{Z}$ is :

$$|Z| = \sqrt{R^2 + (1/\omega C)^2} \tag{14}$$

and its phase angle ϕ is given by

$$\tan \phi = -\frac{1}{\omega CR} \tag{15}$$

which is the same as we derived above from the vector diagram, except for the sign. The negative sign in eqn. (15) is because ϕ is the phase shift of $\mathbf{Z}$ with respect to R.

In a practical condenser, we may regard R in Fig. 4.16A as the equivalent series loss resistance of the condenser itself.

It is obvious from Fig. 4.16B that

$$\cos \phi = R/\mathbf{Z} \tag{16a}$$

where $\mathbf{Z}$ is given by equation (13) above.
Cos ϕ is called the " Power factor."

If ϕ is nearly $90°$, X is nearly equal to Z, apart from sign, and so

$$\cos \phi \approx R/|X| \tag{16b}$$

where $|X|$ indicates that the value of the reactance is taken apart from sign. With this approximation we obtain

Power factor $= \cos \phi \approx \omega CR$ (17)

Note that the power factor of any resistive component is always positive. A negative power factor indicates generation of power.

The total power dissipated in the circuit (Fig. 4.16A) will be

$$P = EI \cos \phi = EI \times \text{power factor} \tag{17a}$$

$$= I^2R = \frac{E^2}{R[1 + X_c^2/R^2]} \tag{17b}$$

where P = power in watts
$X_c = 1/\omega C$
and cos ϕ is defined by eqn. (16a) or the approximation (16b).

The **Q factor** of a condenser is the ratio of its reactance to its resistance—

$$Q = 1/\omega CR = \tan \phi \approx 1/(\text{power factor}) \tag{18}$$

SECTION 5 : INDUCTANCE

(i) Introduction to inductance (ii) Inductances in d.c. circuits (iii) Inductances in series and parallel (iv) Mutual inductance (v) Inductances in a.c. circuits (vi) Power in inductive circuits.

(i) Introduction to inductance

An inductor, in its simplest form, consists of a coil of wire with an air core as commonly used in r-f tuning circuits. Any inductor has a characteristic known as inductance whereby it sets up an electro-magnetic field when a current is passed through it. When the current is varied, the strength of the field varies ; as a result, an electromotive force is induced in the coil. This may be expressed by the equation :

$$e = -N\frac{d\phi}{dt} \times 10^{-8} \text{ volts} \tag{1}$$

where e = e.m.f. induced at any instant
N = number of turns in the coil
ϕ = flux through the coil

*See Chapter 6, Sect. 6(i).

and $\dfrac{d\phi}{dt}$ = rate of change (differential with respect to time) of the flux.

The direction of the induced e.m.f. is always such as to oppose the change of current which is producing the induced voltage. In other words, the effect of the induced e.m.f. is to assist in maintaining constant both current and field.

We may also express the relationship in the form :

$$e = -L\,\frac{di}{dt} \tag{2}$$

where all the values are expressed in practical units—

   e = e.m.f. induced at any instant, in volts
   L = inductance, in henrys
   i = current in amperes

and $\dfrac{di}{dt}$ = rate of change of current, in amperes per second.

The inductance L varies approximately* with the square of the number of turns in the coil, and may be increased considerably by using iron cores** (for low frequencies) or powdered iron (for radio frequencies). With iron cores the value of L is not constant, so that eqn. (2) cannot be used accurately in such cases.

The **energy stored in a magnetic field** is
$$\text{energy} = \tfrac{1}{2}\,LI^2 \tag{2a}$$
where energy is measured in joules (or watt seconds)
   L is measured in henrys
and I is measured in amperes.

(ii) Inductances in d.c. circuits

When an inductance (L henrys) with a total circuit resistance (R ohms) is connected to a d.c. source of voltage (E), the current rises gradually to the steady value $I = E/R$. During the gradual rise, the current follows the logarithmic law

$$i = \frac{E}{R}\,(1 - \epsilon^{-Rt/L}) \tag{3}$$

where i = current in amperes at time t,
   ϵ = base of natural logarithms $\approx$ 2.718,
and t = time in seconds after switch is closed.

The **time constant** (T) is the time in seconds from the time that the switch is closed until the current has risen to $(1 - 1/\epsilon)$ or 63.2% of its final value :
$$T = L/R \tag{4}$$
where L = inductance in henrys
and R = resistance in ohms.

The decay of current follows the law
$$i = \frac{E}{R}\,(\epsilon^{-Rt/L}) \tag{5}$$

The rise and decay of current are called **transients**.

(iii) Inductances in series and parallel

Inductances in series

The total inductance is equal to the sum of the individual inductances, provided that there is no coupling between them :
$$L = L_1 + L_2 + L_3 + \ldots \tag{6}$$

Inductances in parallel

The total inductance is given by eqn. (7), provided that there is no coupling between them :

$$\frac{1}{L} = \frac{1}{L_1} + \frac{1}{L_2} + \frac{1}{L_3} + \ldots \tag{7}$$

*See Chapter 10 for formulae for calculating the inductance of coils.
**See Chapter 5 for iron cored inductances.

(iv) Mutual inductance

When two coils are placed near to one another, there tends to be coupling between them, which reaches a maximum when they are placed co-axially and with their centres as close together as possible.

If one such coil is supplied with varying current, it will set up a varying magnetic field, which in turn will induce an e.m.f. in the second coil. This induced e.m.f. in the secondary is proportional to the rate of current change in the first coil (primary) and to the mutual inductance of the two coils :

$$e_2 = -M \frac{di_1}{dt} \tag{8}$$

where e_2 = voltage induced in the secondary,

$\dfrac{di_1}{dt}$ = rate of change of current in the primary in amperes per second,

and M = mutual inductance of the two coils, in henrys.
(Compare Equations 2 and 8.)

M may be either positive or negative, depending on the rotation of, or connections to, the secondary. M is regarded as positive if the secondary voltage (e_2) has the same polarity as the induced voltage in a single coil.

The maximum possible (theoretical) value of M is when $M = \sqrt{L_1L_2}$, being the condition of unity coupling, but in practice this cannot be achieved. The coefficient of coupling (k) is given by

$$k = M/\sqrt{L_1L_2} \tag{9}$$

so that k is always less than unity.

If the secondary is loaded by a resistance R_2 (Fig. 4.17), current will flow through the secondary circuit.

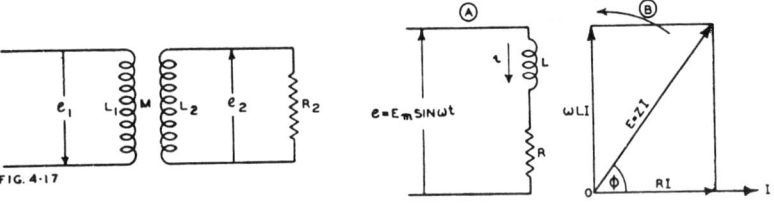

FIG. 4·17   FIG. 4·18

Fig. 4.17.   *Two inductances coupled by mutual inductance (M) with the secondary loaded by a resistance.*
Fig. 4.18.   *(A) Equivalent circuit diagram of practical inductance (B) Vector diagram of voltage relationships.*

(v) Inductance in a.c. circuits

If an ideal inductance (L henrys) is connected across an a.c. line, the voltage of which is given by the equation $e = E_m \sin \omega t$, a current will flow having the same waveform as the line voltage, but the current will lag 90° behind the voltage. The inductance is said to have an **inductive reactance*** (X_L) equal to ωL, and the current will be

$$I_{rms} = E_{rms}/\omega L \tag{10}$$

We may helpfully compare this with the case of a condenser :

	Current	Reactance	Phase shift†
Capacitance :	$I_{rms} = \omega C E_{rms}$	$X_c = -1/\omega C$	+90°
Inductance :	$I_{rms} = E_{rms}/\omega L$	$X_L = \omega L$	−90°

Every practical inductance has appreciable resistance, and we may draw the equivalent circuit of any normal inductor as an ideal inductance in series with a resistance

---

*A table of inductive reactances is given in Chapter 38 Sect. 9 Table 41.
†Of current with respect to voltage.

(Fig. 4.18A). If there is any other resistance in the circuit, it may be added to the resistance of the inductor to give the total resistance R.

If an alternating voltage is applied across L and R in series (Fig. 4.18A) the vector diagram may be drawn as in (B). The current vector I is first drawn to any convenient scale ; the vector of voltage drop across R is then drawn as RI in phase with I ; the vector of voltage drop across L is then drawn as ωLI so that I will lag behind it by 90°—hence ωLI is drawn as shown ; the parallelogram is then completed to give the resultant $E = ZI$ with a phase angle such that $\tan \phi = \omega L/R$.

Using the j notation we may write :
$$E = RI + j\omega LI \qquad (11)$$
where $+j$ indicates 90° vector rotation in a positive direction.

From (11) we can derive :
$$\mathbf{Z} = E/I = R + j\omega L \qquad (12)$$
For example, if $R = 150$ ohms, $L = 20$ henrys and $f = 1000$ c/s, then
$$\omega = 2\pi \times 1000 = 6280$$
$$\omega L = 6280 \times 20 = 125\,600 \text{ ohms}$$
and
$$\mathbf{Z} = 150 + j\,125\,600 \qquad (12a)$$
Eqn. (12a) indicates at a glance a resistance of 150 ohms in series with a positive reactance (i.e. an inductive reactance) of 125 600 ohms. Values of inductive reactances are given in Chapter 38 Sect. 9 Table 41.

The magnitude (modulus) of $\mathbf{Z}$ is :
$$|Z| = \sqrt{R^2 + (\omega L)^2} \qquad (13)$$
and its phase angle is given by
$$\tan \phi = \omega L/R \qquad (14)$$
as also shown by the vector diagram.

(vi) Power in inductive circuits

The power drawn from the line in Fig. 4.18A is the integral over one cycle of the instantaneous values of $e \times i$. As shown by Fig. 4.19, during parts of each cycle energy is being taken by the circuit, while during other parts of the cycle energy is being given back by the circuit. The latter may be regarded as negative power being taken by the circuit, and is so drawn. The expression for the power is
$$P = E_m \sin \omega t \times I_m \sin (\omega t - \phi)$$
$$= E_m I_m \sin \omega t \cdot \sin (\omega t - \phi)$$
$$= E_m I_m \sin \omega t (\sin \omega t \cos \phi - \cos \omega t \sin \phi)$$
$$= E_m I_m (\sin^2 \omega t \cos \phi - \sin \omega t \cos \omega t \sin \phi).$$
Now the average value of $(\sin \omega t \cos \omega t)$ over one cycle is zero*,

Therefore $P = E_{m} I_m (\sin^2 \omega t \cos \phi)$
$$= \tfrac{1}{2} E_m I_m (1 - \cos 2\omega t)(\cos \phi)$$
$$= E_{rms} I_{rms} \cos \phi \qquad (15)$$
(since the average value of $\displaystyle\int_0^{2\pi} \cos 2\omega t = 0*$)

where P is expressed in watts, E and I in volts and amperes.

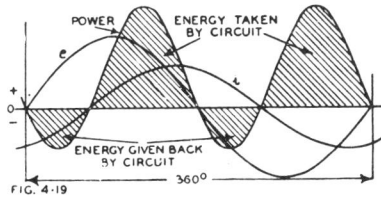

Fig. 4.19.   Power in an inductive circuit   with   applied   sine-wave voltage (e).

From eqn. (15) the power is equal to the product of the effective voltage and current multiplied by cos ϕ, which is called the **Power Factor**, its value being given by cos $\phi = R/\sqrt{R^2 + (\omega L)^2}$.

When the load is purely resistive, $L = 0$ and the power factor $= 1$; the power is therefore $P = E_{rms} I_{rms}$.

From (15) we may derive for the general case :

$$P = EI \cos \phi = EI . R/\sqrt{R^2 + (\omega L)^2}$$

Therefore $P = I . I\sqrt{R^2 + (\omega L)^2} \times R/\sqrt{R^2 + (\omega L)^2} = I^2 R$ (16a)

$$\text{or } P = \frac{E^2}{R[1 + (\omega L)^2/R^2]} \tag{16b}$$

Therefore $R = P/I^2$ (17)

where $P =$ power in watts

          $\omega L =$ inductive reactance in ohms

          $R =$ resistance in ohms

and $I =$ current in amperes.

In other words, the total power taken from the line in the circuit of Fig. 4.18A is the power dissipated by the effective total resistance R. There is no loss of power in an ideal inductance with zero resistance, although it draws current from an a.c. line, because the power factor is zero ($\phi = 90°$, therefore cos $\phi = 0$). The product of $I \times E$ in this case is called the **wattless power** or **reactive power**, or more correctly the **reactive volt-amperes**.

The power factor at any given frequency gives the ratio of the resistance of a coil to its impedance and may be used as a figure of merit for the coil. A good coil should have a very small power factor.

The power factor is almost identical with the inverse of the coil magnification factor Q (Chapter 9), and the error is less than 1% for values of Q greater than 7 :

$$Q = \omega L/R = \tan \phi$$
$$\text{Power factor } = \cos \phi = R/Z$$
$$\tan \phi \approx 1/\cos \phi \text{ (error } < 1\% \text{ for } \phi > 82°)$$

Therefore Power Factor $\approx 1/Q$ for $Q > 7$.

SECTION 6 : IMPEDANCE AND ADMITTANCE

(i) Impedance a complex quantity   (ii) Series circuits with L, C and R   (iii) Parallel combinations of L, C and R   (iv) Series-parallel combinations of L, C and R   (v) Conductance, susceptance and admittance   (vi) Conversion from series to parallel impedance.

(i) Impedance, a complex quantity

Impedance has already been introduced in Sections 4 and 5, in connection with series circuits of C and R or L and R. Impedance is a complex quantity, having both a resistive and a reactive component. We are therefore concerned, not only with its magnitude, but also with its phase angle.

(ii) Series circuits with L, C and R

When a resistance, an inductance and a capacitance are connected in series across an a.c. line (Fig. 4.20A), the same current will flow through each.

In using the j notation, remember that j simply means 90° positive vector rotation (the voltage drop for an inductance) and $-j$ means 90° negative vector rotation (the voltage drop for a capacitance).

In terms of the effective values

where $E =$ applied voltage

and $I =$ current through circuit, we have :

Voltage drop through $R = RI$

    „   „   „ $L = j\omega LI$

    „   „   „ $C = (-j/\omega C)I$

The total voltage drop is equal to the applied voltage,

therefore $E = [R + j(\omega L - 1/\omega C)]\,I$

Hence $Z = R + j(\omega L - 1/\omega C)$ (1)

Here R is the resistive component of **Z**, while $(\omega L - 1/\omega C)$ is the reactive component.

It may therefore be written as

$$Z = R + jX \text{ where } X = (\omega L - 1/\omega C).$$

For example, let $R = 500$ ohms, $L = 20$ henrys, and $C = 1\ \mu F$, all connected in series across an a.c. line with a frequency of 50 c/s.

Then $\omega = 2\pi \times 50 = 314$

$1/\omega C = 1/(314 \times 10^{-6}) = 3180$ ohms

$X = \omega L - 1/\omega C = 6280 - 3180 = 3100$ ohms

(the positive sign indicates that this is inductively reactive) and

$$Z = R + jX = 500 + j\,3100.$$

If L had been 5 henrys, then X would have been

$$1570 - 3180 = -1610 \text{ ohms}$$

which is capacitively reactive.

The phase angle is given by $\tan \phi = X/R$. The magnitude of the impedance is given by $|Z| = \sqrt{R^2 + (\omega L - 1/\omega C)^2}$ (1a)

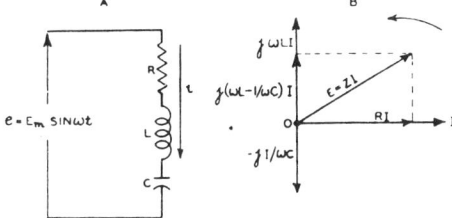

Fig. 4.20. (A) Circuit of R, L and C in series across a.c. line (B) Vector diagram of voltage relationships.

This may be illustrated by means of a vector diagram (Fig. 4.20 B) where I is drawn to any convenient scale. The voltage-drop vectors are then drawn, to the selected voltage scale, RI in phase with I, $j\omega LI$ at 90° in advance of I and $-jI/\omega C$ lagging 90° behind I. The simplest method of combining the three vectors is to take either ωLI or $I/\omega C$, whichever is the greater, and then to subtract the other from it, since these two are exactly in phase opposition. In diagram B, ωLI is the greater, so that $I/\omega C$ is subtracted from ωLI to give $(\omega L - 1/\omega C)I$. The two remaining vectors, RI and $j(\omega L - 1/\omega C)I$ are then combined by completing the parallelogram, to give the resultant ZI which, of course, must be equal to the applied voltage E.

A special case arises when $\omega L = 1/\omega C$ in eqn. (1) ; the reactive component becomes zero and $Z = R$. This is the phenomenon of **series resonance** which is considered in greater detail in Chapter 9.

Special cases of equation (1)

R and L only : Put $1/\omega C = 0$, $Z = R + j\omega L$

R and C only : Put $\omega L = 0$, $Z = R - j/\omega C = R + 1/j\omega C$

L and C only : Put $R = 0$, $Z = j(\omega L - 1/\omega C)$

R only : $Z = R + j0$

L only : $Z = 0 + j\omega L$

C only : $Z = 0 - j/\omega C = 1/j\omega C$

Note that $Q = |\tan \phi| = |X|/R$.

A table of inductive reactances is given in Chapter 38 Sect. 9(i) Table 41.

A table of capacitive reactances is given in Chapter 38 Sect. 9(ii) Table 42.

A table and two charts to find X, R or Z, when a reactance and resistance are connected in series, any two values being known, is given in Chap. 38 Sect 9(iv) Table **44**.

Series Combinations of L, C and R

Series combination	Impedance $Z = R + jX$ ohms	Magnitude of impedance $\lvert Z \rvert = \sqrt{R^2 + X^2}$ ohms	Phase angle $\phi = \tan^{-1}(X/R)$ radians	Admittance* $Y = 1/Z$ mhos
R	R	R	0	$1/R$
L	$+j\omega L$	ωL	$+\pi/2$	$-j(1/\omega L)\cdot$
C	$-j(1/\omega C)$	$1/\omega C.$	$-\pi/2$	$j\omega C$
$R_1 + R_2$	$R_1 + R_2$	$R_1 + R_2$	0	$1/(R_1 + R_2)$
$L_1(M)L_2$	$+j\omega(L_1 + L_2 \pm 2M)$	$\omega(L_1 + L_2 \pm 2M)$	$+\pi/2$	$-j/\omega(L_1 + L_2 \pm 2M)$
$C_1 + C_2$	$-j\dfrac{1}{\omega}\left(\dfrac{C_1 + C_2}{C_1 C_2}\right)$	$\dfrac{1}{\omega}\left(\dfrac{C_1 + C_2}{C_1 C_2}\right)$	$-\dfrac{\pi}{2}$	$j\omega\left(\dfrac{C_1 C_2}{C_1 + C_2}\right)$
$R + L$	$R + j\omega L$	$\sqrt{R^2 + \omega^2 L^2}$	$\tan^{-1}\dfrac{\omega L}{R}$	$\dfrac{R - j\omega L}{R^2 + \omega^2 L^2}$
$R + C$	$R - j\dfrac{1}{\omega C}$	$\sqrt{\dfrac{\omega^2 C^2 R^2 + 1}{\omega^2 C^2}}$	$-\tan^{-1}\dfrac{1}{\omega RC}$	$\dfrac{\omega^2 C^2 R + j\omega C}{\omega^2 C^2 R^2 + 1}$
$L + C$	$+j\left(\omega L - \dfrac{1}{\omega C}\right)$	$\left(\omega L - \dfrac{1}{\omega C}\right)$	$\pm\dfrac{\pi}{2}$	$-\dfrac{j\omega C}{\omega^2 LC - 1}$
$R + L + C$	$R + j\left(\omega L - \dfrac{1}{\omega C}\right)$	$\sqrt{R^2 + \left(\omega L - \dfrac{1}{\omega C}\right)^2}\cdot$	$\tan^{-1}\left(\dfrac{\omega L - 1/\omega C}{R}\right)$	$\dfrac{R - j(\omega L - 1/\omega C)}{R^2 + (\omega L - 1/\omega C)^2}$

*See Sect. 6(v) below.

General case with a number of arms connected in series, each arm being of the form $R + X$

Arm (1) : $Z_1 = R_1 + jX_1$ (i.e. R_1 in series with X_1)
Arm (2) : $Z_2 = R_2 + jX_2$
Arm (3) : $Z_3 = R_3 + jX_3$ etc.

then the total impedance is given by

$$Z = (R_1 + R_2 + R_3 + \ldots) + j(X_1 + X_2 + X_3 + \ldots)$$

(iii) Parallel combinations of L, C and R

When a number of resistance and reactive elements are connected in parallel across an a.c. line, the same voltage is applied across each. For example in Fig. 4.20C there are three parallel paths across the a.c. line, and the current through each may be determined separately.

Let E_o = r.m.s. value of line voltage.

Then r.m.s. current through $L = I_1 = E_o/\omega L$

„   „   „   „ $C = I_2 = E_o\omega C$

„   „   „   „ $R = I_3 = E_o/R$.

The phase relationships between these currents are shown in Fig. 4.20D. There is 180° phase angle between the current through L and that through C, so that the resultant of these two currents is $(I_2 - I_1)$. The vector resultant of $(I_2 - I_1)$ and I_3 is given by I_o in Fig. 4.20D. Thus the current through L and C may be considerably greater than the total line current I_o.

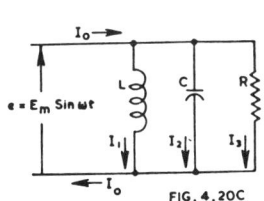
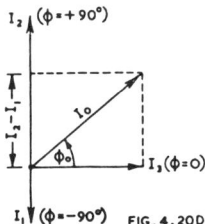

Fig. 4.20.*C   Circuit of R, L and C in parallel across a.c. line.*
Fig. 4.20.*D   Vector diagram of voltage relationships.*

The impedance of the parallel combination may be derived by considering L and C as being replaced by a single reactive element having a reactance of $1/(\omega C - 1/\omega L_j$. Note that with parallel connection, the convention is that the phase of the capacitive element is taken as positive. Thus

$$I_o = I_3 + j(I_2 - I_1)$$
$$= \frac{E_o}{R} + jE_o\left(\omega C - \frac{1}{\omega L}\right)$$

Therefore $\mathbf{Z} = \dfrac{E_o}{I_o} = \dfrac{1}{\dfrac{1}{R} + j\left(\omega C - \dfrac{1}{\omega L}\right)} \times \dfrac{\dfrac{1}{R} - j\left(\omega C - \dfrac{1}{\omega L}\right)}{\dfrac{1}{R} - j\left(\omega C - \dfrac{1}{\omega L}\right)}$

$$= \frac{\dfrac{1}{R} - j\left(\omega C - \dfrac{1}{\omega L}\right)}{\left(\dfrac{1}{R}\right)^2 + \left(\omega C - \dfrac{1}{\omega L}\right)^2}$$

2

Parallel combinations of L, C and R

Parallel combination	Impedance $Z = R + jX$	Magnitude of impedance $\lvert Z \rvert = \sqrt{R^2 + X^2}$	Phase angle $\phi = \tan^{-1}(X/R)$	Admittance* $Y = 1/Z$
R_1, R_2	ohms $\dfrac{R_1 R_2}{R_1 + R_2}$	ohms $\dfrac{R_1 R_2}{R_1 + R_2}$	radians 0	mhos $\dfrac{R_1 + R_2}{R_1 R_2}$
C_1, C_2	$-j\dfrac{1}{\omega(C_1 + C_2)}$	$\dfrac{1}{\omega(C_1 + C_2)}$	$-\dfrac{\pi}{2}$	$+j\omega(C_1 + C_2)$
L, R	$\dfrac{\omega^2 L^2 R + j\omega L R^2}{\omega^2 L^2 + R^2}$	$\dfrac{\omega L R}{\sqrt{\omega^2 L^2 + R^2}}$	$\tan^{-1}\dfrac{R}{\omega L}$	$\dfrac{1}{R} - \dfrac{j}{\omega L}$
R, C	$\dfrac{R - j\omega R^2 C}{1 + \omega^2 R^2 C^2}$	$\dfrac{R}{\sqrt{1 + \omega^2 R^2 C^2}}$	$\tan^{-1}(-\omega R C)$	$\dfrac{1}{R} + j\omega C$
L, C	$+j\dfrac{\omega L}{1 - \omega^2 L C}$	$\dfrac{\omega L}{1 - \omega^2 L C}$	$\pm\dfrac{\pi}{2}$	$j\left(\omega C - \dfrac{1}{\omega L}\right)$
$L_1(M)L_2$	$+j\omega\dfrac{L_1 L_2 - M^2}{L_1 + L_2 \mp 2M}$	$\omega\dfrac{L_1 L_2 - M^2}{L_1 + L_2 \mp 2M}$	$\pm\dfrac{\pi}{2}$	$-j\dfrac{1}{\omega}\left(\dfrac{L_1 + L_2 \mp 2M}{L_1 L_2 - M^2}\right)$
L, C, R	$\dfrac{\dfrac{1}{R} - j\left(\omega C - \dfrac{1}{\omega L}\right)}{\left(\dfrac{1}{R}\right)^2 + \left(\omega C - \dfrac{1}{\omega L}\right)^2}$	$\dfrac{R}{\sqrt{1 + R^2\left(\omega C - \dfrac{1}{\omega L}\right)^2}}$	$\tan^{-1} - R\left(\omega C - \dfrac{1}{\omega L}\right)$	$\dfrac{1}{R} + j\left(\omega C - \dfrac{1}{\omega L}\right)$

Note that $Q = \lvert \tan \phi \rvert$.

*See Sect. 6(v) below.

The magnitude of the impedance of a reactance X in parallel with a resistance R is given by

$$|Z| = \frac{RX}{\sqrt{X^2 + R^2}} \tag{3a}$$

$$= \frac{\omega LR}{\sqrt{\omega^2 L^2 + R^2}} \quad \text{when } X = \omega L \tag{3b}$$

$$= \frac{R}{\sqrt{1 + \omega^2 R^2 C^2}} \quad \text{when } X = 1/\omega C \tag{3c}$$

$$= \frac{R}{\sqrt{1 + R^2 (\omega C - 1/\omega L)^2}} \quad \left\{ \begin{array}{l} \text{when } L \text{ and } C \\ \text{are in parallel} \end{array} \right\} \tag{3d}$$

A table and a chart to find X, R or Z when a reactance and resistance are connected in parallel, is given in Chapter 38 Sect. 9(iii) Table 43.

A simple graphical method for determining the resultant impedance of a reactance and a resistance in parallel is given in Fig. 4.20E where OA and OB represent R and X respectively and OD is drawn at right angles to the hypotenuse. The length OD represents the scalar value of the resultant impedance, while the angle DOA is equal to the phase angle ϕ between the applied voltage and the resultant current.

For other graphical methods for impedances in parallel see Reed, C.R.G. (letter) W.E. 28.328 (Jan. 1951) 32 ; Benson, F. A. (letter) W.E. 28.331 (April 1951) 128.

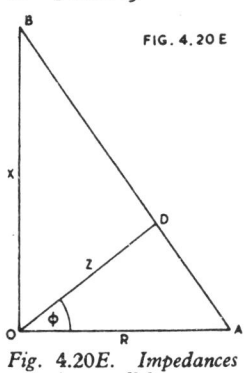

FIG. 4.20E

Fig. 4.20E.   Impedances in parallel.

(iv) Series-parallel combinations of L, C and R

Some simple combinations of L, C and R are described in (A) to (E) below. The general procedure for other combinations is described in (F) below.

(A) Parallel tuned circuit with losses in both L and C (Fig. 4.21A)

Impedance of branch (1) : $Z_1 = R_1 + j\omega L$ (4a)

 ,, ,, ,, (2) : $Z_2 = R_2 - j/\omega C$ (4b)

Let Z = total impedance of Z_1 and Z_2 in parallel.

Then $Z = \dfrac{Z_1 Z_2}{Z_1 + Z_2} = \dfrac{(R_1 + j\omega L)(R_2 - j/\omega C)}{(R_1 + R_2) + j(\omega L - 1/\omega C)}$ (4c)

Multiplying both numerator and denominator by $(R_1 + R_2) - j(\omega L - 1/\omega C)$, we have

$$Z = \frac{(R_1 R_2 + j\omega L R_2 - jR_1/\omega C + L/C)[R_1 + R_2 - j(\omega L - 1/\omega C)]}{(R_1 + R_2)^2 + (\omega L - 1/\omega C)^2} \tag{5}$$

Let $Z = \dfrac{A + jB}{(R_1 + R_2)^2 + (\omega L - 1/\omega C)^2}$ (6)

and let us now determine the values of A and B :

Numerator $= R_1^2 R_2 + jR_1 R_2 \omega L - jR_1^2/\omega C + R_1 L/C + R_1 R_2^2 + jR_2^2 \omega L$
$- jR_1 R_2/\omega C + R_2 L/C - jR_1 R_2 \omega L + R_2 \omega^2 L^2 - R_1 L/C$
$- j\omega L^2/C + jR_1 R_2/\omega C - R_2 L/C + R_1/\omega^2 C^2 + jL/\omega C^2$
$= R_1^2 R_2 + R_1 R_2^2 + R_2 \omega^2 L^2 + R_1/\omega^2 C^2 + j[R_2^2 \omega L - R_1^2/\omega C$
$- \omega L^2/C + L/\omega C^2]$.

Therefore $A = R_1 R_2 (R_1 + R_2) + R_2 \omega^2 L^2 + R_1/\omega^2 C^2$ (7)

and $B = R_2^2 \omega L - R_1^2/\omega C - (L/C)(\omega L - 1/\omega C)$ (8)

which values should be used in eqn. (6).

Note that A divided by the denominator (eqn. 6) is the effective resistance of the total circuit, while B divided by the denominator is the effective reactance.

The magnitude of the impedance may be obtained most readily from eqn. (4c) by replacing Z_1 and Z_2 in the numerator only by $Z_1 \angle \phi_1$ and $Z_2 \angle \phi_2$ respectively.

$$Z = \frac{(Z_1 \angle \phi_1)(Z_2 \angle \phi_2)}{(R_1 + R_2) + j(\omega L - 1/\omega C)} = \frac{Z_1 Z_2 \angle (\phi_1 + \phi_2)}{(R_1 + R_2) + j(\omega L - 1/\omega C)} \qquad (9)$$

where $\quad Z_1 = \sqrt{R_1{}^2 + \omega^2 L^2}$

and $\quad\quad Z_2 = \sqrt{R_2{}^2 + 1/\omega^2 C^2}$

whence $\quad |Z| = \left[\dfrac{(R_1{}^2 + \omega^2 L^2)(R_2{}^2 + 1/\omega^2 C^2)}{(R_1 + R_2)^2 + (\omega L - 1/\omega C)^2} \right]^{\frac{1}{2}} \qquad (10)$

The phase angle is given by $\tan^{-1} B/A$

i.e. $\quad \phi = \tan^{-1}\left[\dfrac{R_2{}^2 \omega L - R_1{}^2/\omega C - (L/C)(\omega L - 1/\omega C)}{R_1 R_2 (R_1 + R_2) + R_2 \omega^2 L^2 + R_1/\omega^2 C^2} \right] \qquad (11)$

When $LC\omega^2 = 1$, eqn. (10) may be reduced to

$$|Z| = \frac{L}{C} \cdot \frac{[1 + (C/L)(R_1{}^2 + R_2{}^2) + (C^2/L^2) R_1{}^2 R_2{}^2]^{\frac{1}{2}}}{R_1 + R_2} \qquad (12)$$

or $\quad |Z| \approx (L/C)(R_1 + R_2) \qquad\qquad\qquad (13)$

with an error $< 1\%$ provided $Q > 10$

where $Q = \omega L/(R_1 + R_2)$.

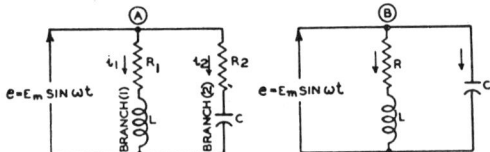

Fig. 4.21. *(A) Network incorporating 4 elements for impedance calculations (B) Simplified network with 3 elements.*

(B) Parallel tuned circuit with loss in L only (Fig. 4.21B)

This is a special case of (A) in which $R_2 = 0$.

It may be shown that eqn. (5) becomes

$$Z = \frac{R - j\omega [CR^2 + L(\omega^2 LC - 1)]}{\omega^2 C^2 R^2 + (\omega^2 LC - 1)^2} \qquad (14)$$

The magnitude of the impedance derived from eqn. (10) becomes

$$|Z| = \left[\frac{R^2 + \omega^2 L^2}{\omega^2 C^2 R^2 + (\omega^2 LC - 1)^2} \right]^{\frac{1}{2}} \qquad (15)$$

The phase angle derived from eqn. (11) becomes

$$\phi = \tan^{-1} \frac{\omega [L(1 - \omega^2 LC) - CR^2]}{R} \qquad (16)$$

(a) The effective reactance of eqn. (14) is zero (i.e. the power factor is unity) when

$$CR^2 = L(1 - \omega^2 LC)$$

i.e. when $\omega^2 L^2 C = L - CR^2$

i.e. when $\quad \omega = \sqrt{\dfrac{1}{LC} - \dfrac{R^2}{L^2}} \qquad (17)$

which can be written in the form

when $\quad \omega = \dfrac{1}{\sqrt{LC}} \sqrt{1 - \dfrac{CR^2}{L}} = \omega_0 \sqrt{1 - \dfrac{CR^2}{L}} \qquad (18)$

where $\quad \omega_0 = 1/\sqrt{LC}$ is the value of ω when the resistance R is zero.

This condition, namely that the effective reactance is zero, which may also be expressed as the condition of unity power factor, is one of several possible definitions of **parallel resonance**. This is the definition used in Chapter 9. It should be noted that the expression giving the value of ω is not independent of the resistance in the circuit as is the case with series resonance.

If $LC\omega^2 = 1$, which is the condition for resonance if the resistance is zero, the impedance is given very closely, when R is small, by

$$Z \approx L/CR \tag{19}$$

The remaining possible definitions of parallel resonance are conditions to give maximum impedance. The condition of maximum impedance is sometimes called *anti-resonance*. Maximum impedance occurs under slightly different conditions for the variables C, L and ω^*.

(b) Condition of maximum total impedance when C is the variable

The maximum value of Z which can be obtained is

$$|Z|_{max} = \frac{R_1^2 + \omega^2 L^2}{R_1} = R_1(1 + Q^2) \tag{20}$$

where $Q = \omega L/R_1$

which occurs when $C = \dfrac{L}{R_1^2 + \omega^2 L^2}$ \hfill (21)

This is also the condition giving unity power factor.

(c) Condition of maximum total impedance when L is the variable

The maximum value of Z which can be obtained is

$$|Z|_{max} = \frac{2R_1}{\sqrt{1 + 4\omega^2 C^2 R_1^2} - 1} \tag{22}$$

which occurs when $L = \dfrac{1 + \sqrt{1 + 4\omega^2 C^2 R_1^2}}{2\omega^2 C}$ \hfill (23)

or when $C = \dfrac{L}{\omega^2 L^2 - R_1^2}$ \hfill (24)

If L is the variable and Q is maintained constant, the maximum value of Z which can be obtained is

$$|Z|_{max} = Q/\omega C \tag{25}$$

which occurs when $L = \dfrac{Q^2}{\omega^2 C(1 + Q^2)}$ \hfill (26)

(d) Condition of maximum total impedance when the applied frequency is the variable

The maximum value of Z which can be obtained is

$$|Z|_{max} = \frac{L}{C\sqrt{R_1^2 - \dfrac{L}{C}\left(\sqrt{2R_1^2\dfrac{C}{L} + 1} - 1\right)^2}} \tag{27}$$

which occurs when

$$\omega = 2\pi f = \sqrt{\frac{\sqrt{(2R_1^2 C/L) + 1}}{LC} - \frac{R_1^2}{L^2}} \tag{28}$$

In practice, for all normal values of $Q(= \omega L/R)$ as used in tuned circuits, these four cases are almost identical and the frequency of parallel resonance is approximately the same as that for series resonance.

We may summarise the resonance frequencies for the various conditions given above :

Series resonance $\quad LC\omega^2 = 1$
Parallel resonance (a) $LC\omega^2 = 1 - CR^2/L$
(b) $LC\omega^2 = 1 - CR^2/L$
(c) $LC\omega^2 = 1 + CR^2/L$
(d) $LC\omega^2 = \sqrt{1 + 2CR^2/L} - CR^2/L$

*See R. S. Glasgow (Book) "Principles of Radio Engineering" (McGraw-Hill Book Co., New York and London, 1936) pp. 35-44.

Note that $CR^2/L = 1/Q^2$ under conditions (a) and (b)
       $CR^2/L \approx 1/Q^2$ under conditions (c) and (d).
For further information on tuned circuits see Chapter 9.

(C) Circuit of Fig. 4.21C
This is a special case of Fig. 4.21A in which
$$(1/\omega C) = 0. \quad \text{From eqn. (6)}$$
$$Z = \frac{R_1 R_2 (R_1 + R_2) + R_2 \omega^2 L^2 + j\omega L R_2^2}{(R_1 + R_2)^2 + \omega^2 L^2} \tag{29}$$

From eqn. (10),
$$|Z| = R_2 \left[\frac{R_1^2 + \omega^2 L^2}{(R_1 + R_2)^2 + \omega^2 L^2} \right]^{\frac{1}{2}} \tag{30}$$

From eqn. (11)
$$\phi = \tan^{-1} \frac{\omega L R_2}{R_1(R_1 + R_2) + \omega^2 L^2} \tag{31}$$

(D) Circuit of Fig. 4.21D
Impedance of arm (1) $= Z_1 = R_1 + j(\omega L - 1/\omega C_1)$
       „   „   „   (2) $= Z_2 = R_2 + j(\omega L_2 - 1/\omega C_2)$
Let $X_1 = \omega L_1 - 1/\omega C_1$
and $X_2 = \omega L_2 - 1/\omega C_2$
then $Z_1 = R_1 + jX_1$; $Z_2 = R_2 + jX_2$.
$$Z = \frac{Z_1 Z_2}{Z_1 + Z_2} = \frac{(R_1 + jX_1)(R_2 + jX_2)}{(R_1 + R_2) + j(X_1 + X_2)}$$
$$= \frac{(R_1 R_2 - X_1 X_2) + j(R_1 X_2 + R_2 X_1)}{(R_1 + R_2) + j(X_1 + X_2)}$$
which may be put into the form
$$Z = \frac{R_1(R_2^2 + X_2^2) + R_2(R_1^2 + X_1^2) + j[X_1(R_2^2 + X_2^2) + X_2(R_1^2 + X_1^2)]}{(R_1 + R_2)^2 + (X_1 + X_2)^2} \tag{32}$$
The magnitude of the impedance may be derived by the method used for deriving
eqn. (10)—
$$|Z| = \left[\frac{(R_1^2 + X_1^2)(R_2^2 + X_2^2)}{(R_1 + R_2)^2 + (X_1 + X_2)^2} \right]^{\frac{1}{2}} \tag{33}$$
The phase angle may be derived from eqn. (32)—
$$\phi = \tan^{-1} \frac{X_1(R_2^2 + X_2^2) + X_2(R_1^2 + X_1^2)}{R_1(R_2^2 + X_2^2) + R_2(R_1^2 + X_1^2)} \tag{34}$$

(E) Circuit of Fig. 4.21E
1. Determine the impedance (Z_1) of L_1 C_1 and R_1 in series, using eqn. (1)—
$$Z_1 = R_1 + j(\omega L_1 - 1/\omega C_1) \tag{35}$$
2. Determine the impedance (Z_2) of $L_2 C_2$ and R_2 in parallel using eqn. (2) but
separating the resistive and reactive components—
$$Z_2 = \frac{\dfrac{1}{R_2}}{\left(\dfrac{1}{R_2}\right)^2 + \left(\omega C_2 - \dfrac{1}{\omega L_2}\right)^2} - j\frac{\left(\omega C_2 - \dfrac{1}{\omega L_2}\right)}{\left(\dfrac{1}{R_2}\right)^2 + \left(\omega C_2 - \dfrac{1}{\omega L_2}\right)^2} \tag{36}$$
3. Determine the combined impedance (Z) by adding the resistive and reactive
components of (14) and (15)—
$$Z = A + jB \tag{37}$$
where $A = R_1 + \dfrac{\dfrac{1}{R_2}}{\left(\dfrac{1}{R_2}\right)^2 + \left(\omega C_2 - \dfrac{1}{\omega L_2}\right)^2}$

and $\quad B = \omega L_1 - \dfrac{1}{\omega C_1} - \dfrac{\omega C_2 - \dfrac{1}{\omega L_2}}{\left(\dfrac{1}{R_2}\right)^2 + \left(\omega C_2 - \dfrac{1}{\omega L_2}\right)^2}$.

The magnitude of the impedance is given by

$$|Z| = \sqrt{A^2 + B^2} \tag{38}$$

The phase angle ϕ is given by

$$\phi = \tan^{-1}(B/A) \tag{39}$$

where A and B have the same values as for eqn. (37).

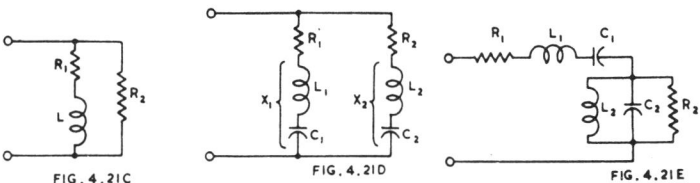

FIG.4.21C   FIG.4.21D   FIG.4.21E

Fig. 4.21.   (C) (D) (E) *Series-parallel networks for impedance calculations.*

(F) General procedure to find the impedance of a two-terminal network

1. If possible, divide the network into two or more parallel branches, each of which is connected to the two terminals but has no other connection with any other branch.

2. Find the impedance of each branch, using the methods described in A, B and C above.

3. Determine the impedance (Z) of the network from the relation

$$\frac{1}{Z} = \frac{1}{Z_1} + \frac{1}{Z_2} + \frac{1}{Z_3} + \cdots \tag{40}$$

Note : If there are more than two parallel branches it is more convenient to work in terms of admittance—see Sect. 6(v) below.

Alternatively, if the circuit cannot be divided into parallel branches, treat it as a series circuit, firstly determining the resistive and reactive components of each section, and then adding all the resistive and all the reactive values separately, as in (E).

(v) Conductance, susceptance and admittance

In an arm* containing both reactance and resistance in series, the **conductance** (G_1) of the arm is given by

$$G_1 = \frac{R_1}{R_1^2 + X_1^2} \text{ mhos} \tag{41}$$

where $X_1 = (\omega L_1 - 1/\omega C_1)$.

When there are a number of such arms in parallel, the resultant conductance is the sum of their separate conductances, that is

$$G = G_1 + G_2 + G_3 + \cdots$$

The **susceptance** (B_1) of the arm under similar conditions, is given by

$$B_1 = \frac{X_1}{R_1^2 + X_1^2} \text{ mhos} \tag{42}$$

where $X_1 = (\omega L_1 - 1/\omega C_1)$.

[Inductive susceptance is regarded as positive.   Capacitive susceptance is regarded as negative.]

When there are a number of such arms in parallel, the resultant susceptance is the sum of their separate susceptances,

$$B = B_1 + B_2 + B_3 + \cdots \tag{43}$$

When any arm includes only resistance, the conductance of the arm is $1/R_1$ and the susceptance zero,

i.e. $G_1 = 1/R_1$ $\qquad\qquad\qquad B_1 = 0.$

---

*An arm is a distinct set of elements in a network, electrically isolated from all other conductors except at two points.

When any arm has only inductance, the conductance of the arm is zero and the susceptance is $1/\omega L_1$,

i.e. $G_1 = 0$ $\qquad\qquad\qquad$ $B_1 = 1/\omega L_1$

When any arm has only capacitance, the conductance of the arm is zero, and the susceptance is $-\omega C$,

i.e. $G_1 = 0$ $\qquad\qquad\qquad$ $B_1 = -\omega C$

The following relationships hold between R, X, G **and** B

$$R = \frac{G}{G^2 + B^2}\ ;\ X = \frac{B}{G^2 + B^2} \tag{44}$$

The **admittance** (Y_1) **of any arm containing resistance and reactance in series** is given by

$$\mathbf{Y_1} = G_1 - jB_1 \tag{45}$$

i.e. $\mathbf{Y_1} = G_1 - jB_1 = \dfrac{R_1 - jX_1}{R_1{}^2 + X_1{}^2} = \dfrac{1}{R_1 + jX_1} = \dfrac{1}{\mathbf{Z_1}},$ $\qquad$ (46)

indicating that **the admittance is the reciprocal of the impedance.**

Thus the value of the admittance may always be derived from the impedance—

$$\mathbf{Y} = \frac{1}{\mathbf{Z}} = \frac{1}{R + jX} \cdot \frac{R - jX}{R - jX} = \frac{R - jX}{R^2 + X^2} \tag{46a}$$

Similarly $\mathbf{Z} = \dfrac{1}{\mathbf{Y}} = \dfrac{1}{G - jB} \cdot \dfrac{G + jB}{G + jB} = \dfrac{G + jB}{G^2 + B^2}$ $\qquad$ (46b)

The negative sign in front of jB in eqn. (45) deserves special attention—
Admittance of inductive arm (R and L in series) :

$$\mathbf{Y} = G - jB = \frac{R - j\omega L}{R^2 + \omega^2 L^2} \tag{47}$$

Admittance of capacitive arm (R and C in series) :

$$\mathbf{Y} = G + jB = \frac{R + j/\omega C}{R^2 + 1/\omega^2 C^2} = \frac{\omega^2 C^2 R + j\omega C}{\omega^2 C^2 R^2 + 1} \tag{48,}$$

The admittance of any arm containing resistance, capacitance and inductance in series (Fig. 4.20A) is given by

$$\mathbf{Y} = G - jB \tag{49}$$

where $\quad G = \dfrac{R}{R^2 + X^2}$

$\qquad\quad X = (\omega L - 1/\omega C)$

and $\quad B = \dfrac{X}{R^2 + X^2} = \dfrac{(\omega L - 1/\omega C)}{R^2 + (\omega L - 1/\omega C)^2}$

therefore $\mathbf{Y} = \dfrac{R - j(\omega L - 1/\omega C)}{R^2 + (\omega L - 1/\omega C)^2}$ $\qquad$ (50)

Values of admittance for various series combinations are included in the table in Sect 6(iii).

Admittance of parallel-connected impedances

L and R in parallel : $\quad G = G_1 + G_2 = 0 + 1/R = 1/R$

$\qquad\qquad\qquad\qquad B = B_1 + B_2 = 1/\omega L + 0 = 1/\omega L$

$\qquad\qquad$ therefore $\mathbf{Y} = G - jB = (1/R) - j(1/\omega L)$

Similarly

C and R in parallel : $\quad \mathbf{Y} = G - jB = (1/R) + j\omega C$

L and C in parallel : $\quad \mathbf{Y} = G - jB = j(\omega C - 1/\omega L)$

L, C and R in parallel : $\mathbf{Y} = G - jB = (1/R) + j(\omega C - 1/\omega L)$

See also table of Parallel Combinations of L, C and R—Column 5, in Sect. 6(iii).

Admittance of series-parallel-connected impedances

When there are a number of arms in parallel, each arm including resistance and reactance in series,

Arm (1) : $Z_1 = R_1 + jX_1$ or $Y_1 = G_1 - jB_1$
Arm (2) : $Z_2 = R_2 + jX_2$ or $Z_2 = G_2 - jB_2$
Arm (3) : $Z_3 = R_3 + jX_3$ or $Y_3 = G_3 - jB_3$ etc.
where $X_1 = (\omega L_1 - 1/\omega C_1)$ etc.
then the total admittance is given by the vector sum

$$\begin{aligned} Y &= Y_1 + Y_2 + Y_3 + \ldots \\ &= (G_1 + G_2 + G_3 + \ldots) - j(B_1 + B_2 + B_3 + \ldots) \end{aligned} \tag{51}$$

The following examples may alternatively be handled by the use of eqn. (46a) provided that the value of Z is known.

Example (A) : Fig. 4.21A

Arm (1) is inductive : $Y_1 = G_1 - jB_1$

$$G_1 = \frac{R_1}{R_1{}^2 + X_1{}^2} \; ; \; B_1 = \frac{X_1}{R_1{}^2 + X_1{}^2} \qquad (X_1 = \omega L)$$

Arm (2) is capacitive · $Y_2 = G_2 + jB_2$

$$G_2 = \frac{R_2}{R_2{}^2 + X_2{}^2} \; ; \; B_2 = \frac{X_2}{R_2{}^2 + X_2{}^2}, \qquad (X_2 = -1/\omega C)$$

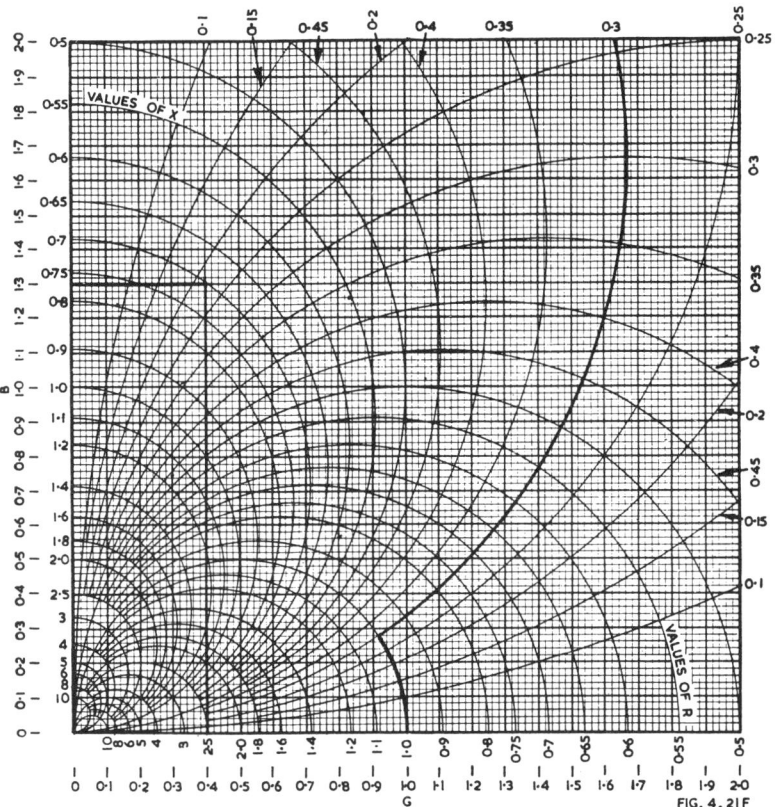

Fig. 4.21F.   *Chart for conversion between resistance and reactance, and conductance and susceptance*

$$Z = (R \pm jX) \, 10^n, \quad Y = (G \mp jB) \, 10^{-n}.$$

On collecting the terms we get

$$\begin{aligned} \mathbf{Y} &= (G_1 + G_2) - j(B_1 + B_2) \\ &= \frac{R_1 + \omega^2 C^2 R_1 R_2 (R_1 + R_2) + \omega^4 L^2 C^2 R_2}{(R_1{}^2 + \omega^2 L^2)(1 + \omega^2 C^2 R_2{}^2)} \\ &\quad + j\omega \left[\frac{CR_1{}^2 - L + \omega^2 LC (L - CR_2{}^2)}{(R_1{}^2 + \omega^2 L^2)(1 + \omega^2 C^2 R_2{}^2)} \right] \end{aligned} \tag{52}$$

Example (B) : Fig. 4.21B.

This is a special case of (A) in which $R_2 = 0$. From eqn. (52)

$$\mathbf{Y} = \frac{R - j\omega[L(1 - \omega^2 LC) - CR^2]}{R^2 + \omega^2 L^2} \tag{53}$$

Example (C) : Fig. 4.21C.

This is a special case of Example (A) in which $X_2 = (1/\omega C) = 0$.

$$\mathbf{Y} = \frac{R_1 (R_1 + R_2) + \omega^2 L^2 - j\omega LR_2}{R_2 (R_1{}^2 + \omega^2 L^2)} \tag{54}$$

Example (D) : Fig. 4.21D.

$$\begin{aligned} \mathbf{Y}_1 &= G_1 - jB_1 = \frac{R_1}{R_1{}^2 + X_1{}^2} - j\,\frac{X_1}{R_1{}^2 + X_1{}^2} \\ \mathbf{Y}_2 &= G_2 - jB_2 = \frac{R_2}{R_2{}^2 + X_2{}^2} - j\,\frac{X_2}{R_2{}^2 + X_2{}^2} \\ \mathbf{Y} &= \frac{R_1}{R_1{}^2 + X_1{}^2} + \frac{R_2}{R_2{}^2 + X_2{}^2} - j\left[\frac{X_1}{R_1{}^2 + X_1{}^2} + \frac{X_2}{R_2{}^2 + X_2{}^2} \right] \\ &= \frac{R_1 R_2 (R_1+R_2) + R_1 X_2{}^2 + R_2 X_1{}^2 - j[R_1{}^2 X_2 + R_2{}^2 X_1 + X_1 X_2 (X_1 + X_2)]}{(R_1{}^2 + X_1{}^2)(R_2{}^2 + X_2{}^2)} \end{aligned} \tag{55}$$

Chart for conversion between resistance and reactance, conductance and susceptance.

Fig. 4.21F can be used in conversions between resistance and reactance, and conductance and susceptance.

Example 1. The impedance of a circuit is $1 + j\,0.3$; determine its admittance.

Method : $\mathbf{Z} = 1:0 + j\,0.3$. Enter the chart on the semi-circle $R = 1.0$ and follow it until it meets the arc $X = 0.3$. The corresponding values of G and B are 0.917 and 0.275. Thus $\mathbf{Y} = 0.917 - j\,0.275$.

Example 2. The admittance of a circuit is $0.000\,004 + j\,0.000\,013$; determine its impedance.

Method : $\mathbf{Y} = 0.000\,004 + j\,0.000\,013 = (0.4 + j\,1.3) \times 10^5$.

Enter the chart on the lines $G = 0.4$ and $B = 1.3$.

The intersection is at $R = 0.22$, $X = 0.7$. Thus

$$\mathbf{Z} = (0.22 - j\,0.7)10^5 = 22\,000 - j\,70\,000.$$

At series resonance, $\omega L = 1/\omega C$ and $X = 0$, so that $Y = G = 1/R$.

The admittance at any frequency is given graphically by the **Admittance Circle Diagram** (Fig. 4.22) in which the vector **OY** represents the admittance, where Y is any point on the circle. The diameter of the circle is equal to $1/R$ and the admittance at series resonance, when the frequency is f_0, is represented by OA.

As the frequency of the voltage applied to the series tuned circuit (Fig. 4.20A) is increased from zero to in-

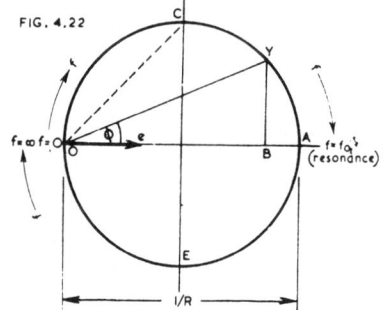

Fig. 4.22. *Admittance circle diagram.*

finity, so the point Y moves from O (at zero frequency) through C to A (at resonance) and thence through E to O again (at an infinite frequency). The angle ϕ is the angle by which the current i leads the applied voltage e ; when ϕ is negative the current lags behind the voltage. Thus when OY lies in the upper portion of the circle the circuit is capacitive, and when it lies in the lower portion of the circle the circuit is inductive.

Fig. 4.22.A   Series form of imped-ance ; (B) Equivalent parallel form.

FIG. 4.22A   FIG. 4.22B

When Y is at the point C, the admittance will be represented by OC and ϕ will be 45° ; the reactive and resistive components of the impedance will therefore be equal. At zero frequency ϕ will be +90° and the admittance will be zero : at infinite frequency ϕ will be −90° and the admittance will also be zero.

(vi) Conversion from series to parallel impedance

It is possible to convert from the series form $\mathbf{Z} = R_s + jX_s$ as shown in Fig. 4.22A

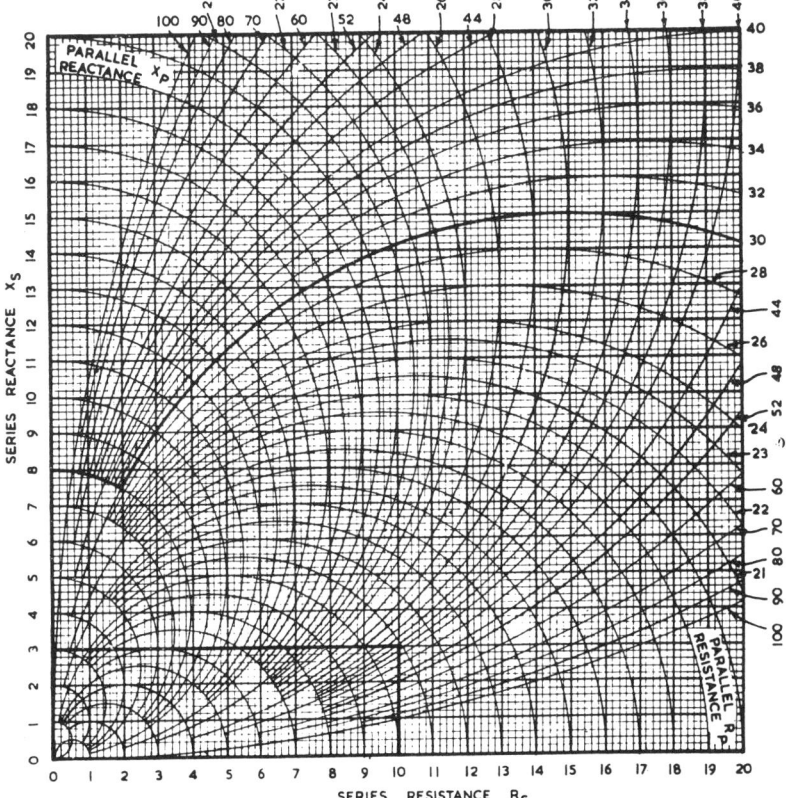

Fig. 4.22C.   Chart for conversion between series resistance and reactance and equivalent parallel resistance and reactance.

to the equivalent parallel form (B) and vice versa.

$$R_s = \frac{R_p X_p{}^2}{R_p{}^2 + X_p{}^2} = \frac{X_p}{\dfrac{X_p}{R_p} + \dfrac{R_p}{X_p}} \tag{56}$$

$$X_s = \frac{X_p R_p{}^2}{R_p{}^2 + X_p{}^2} = \frac{R_p}{\dfrac{X_p}{R_p} + \dfrac{R_p}{X_p}} \tag{57}$$

The following approximations hold with an error within 1%:

(1) If $R_p > 10X_p$,   or if $R_s < 0.1\ X_s$,
    then $R_s \approx X_p{}^2/R_p$; $X_s \approx X_p$;   (58)
    and $R_p \approx X_s{}^2/R_s$; $X_p \approx X_s$.   (59)

(2) If $R_p < 0.1\ X_p$,   or if $R_s > 10X_s$,
    then $R_s \approx R_p$; $X_s \approx R_p{}^2/X_p$;   (60)
    and $R_p \approx R_s$; $X_p \approx R_s{}^2/X_s$;   (61)

Fig. 4.22C can be used to make this conversion.

Example 1. Find the parallel circuit equivalent to a series connection of a 10 ohm resistor (R_s) and an inductor with a reactance of 3 ohms (X_s).

Method: Enter the chart vertically from $R_s = 10$ and horizontally from $X_s = 3$. These lines intersect at $R_p = 10.9$ and $X_p = 36$, so the equivalent parallel connection requires a 10.9 ohm resistor and an inductor with a reactance of 36 ohms.

Example 2. Find the series circuit equivalent to a parallel connection of a 30 000 ohm resistor and a capacitor with a reactance of -8000 ohms.

Method: Enter the chart on the arcs $R_p = 30$ and $X_p = 8$. The intersection is at $R_s = 2$ and $X_s = 7.5$, so the appropriate series elements are $R_s = 2000$ ohms and $X_s = -7500$ ohms.

SECTION 7 : NETWORKS

(i) Introduction to networks (ii) Kirchhoff's Laws (iii) Potential Dividers (iv) Thevenin's Thereom (v) Norton's Theorem (vi) Maximum Power Transfer Theorem (vii) Reciprocity Theorem (viii) Superposition Theorem (ix) Compensation Theorem (x) Four-terminal networks (xi) Multi-mesh networks (xii) Non-linear components in networks (xiii) Phase-shift networks (xiv) Transients in networks (xv) References to networks.

(i) Introduction to networks

A network is any combination of impedances (" elements ")—whether resistances, inductances, mutual inductances or capacitances. Ohm's law may be applied either to the voltage drop in any element, or in any branch, or to the voltage applied to the whole network, involving the total network current and total network impedance, provided that the impedances of the elements are constant. Other laws and theorems which may be used for the solution of network problems are described below.

In network analysis it is assumed that the impedances of elements remain constant under all conditions ; that is that the elements are **linear devices.** Some types of resistors and capacitors and all air-cored inductors are linear, but iron-cored inductors and amplifying valves are **non-linear.** Other non-linear devices include granule-type microphones, electrolytic condensers, glow lamps, barretters (ballast tubes), electric lamp filaments, temperature-controlled resistances such as thermistors*, and thyrite.† However, it is usually found that satisfactory results may be obtained by applying the average characteristics of the non-linear devices under their operating conditions. Further consideration of non-linear components is given in Sect. 7(xii).

---

*Thermistors are resistors with a high negative temperature coefficient. See Sect. 9(i)n.
†Thyrite is a conductor whose resistance falls in the ratio 12.6 : 1 every time the voltage is doubled, over a current ratio 10 000 000 to 1. See K. B. McEachron " Thyrite, a new material for lightning arresters " General Electric Review (U.S.A.) 33.2 (Feb. 1930) 92.

Rectifiers, whether thermionic or otherwise, are non-linear devices ; they are frequently represented by an equivalent circuit having a fixed series resistance in the conducting direction and infinitely high resistance in the other.   This approximation is very inaccurate at low levels where they have effective resistance varying as a function of the applied voltage, while some rectifiers pass appreciable current in the reverse direction.

Most elements (resistors, capacitors and inductors) transmit energy equally in either direction and are referred to as " bilateral", but thermionic valves operate only in one direction (" unilateral ") ; when the latter form part of a network it is necessary to exercise care, particularly if they are represented by equivalent circuits.

Hints on the solution of network problems

A complete solution of a network involves the determination of the current through every element (or around every mesh).   With simple networks the normal procedure is to apply Kirchhoff's Laws—Sect. 7(ii)—until all the currents and their directions have been determined.   The voltage drop across any element may then be derived from a knowledge of the impedance of the element and the current through it.

As a first stage it is important to simplify the circuit, and to draw an equivalent circuit diagram for analysis.

If in any arm there are two or more resistors connected in series, the equivalent circuit diagram should be drawn with

$$R = R_1 + R_2 + \dots$$
Similarly with inductance $L = L_1 + L_2 + \dots$
and with capacitance
$$C = \cfrac{1}{\cfrac{1}{C_1} + \cfrac{1}{C_2} + \dots}.$$

If in any part of the circuit there are two or more elements of the same kind in parallel (whether R, L, or C) the resultant should be determined and applied to the equivalent circuit diagram.

An exception to this rule is where it is merely required to calculate the output voltage from a passive resistive 4-terminal network.   In this case it is sometimes helpful to arrange the network in the form of a potential divider, or sequence of dividers, and to use the method of Sect. 7(iii).

It is very important to mark on the equivalent circuit diagram the directions or polarities of the applied voltages (whether direct or alternating) and the assumed directions of the currents ; if any one of the latter is incorrect, this will be shown by a negative sign in the calculated value.   A clockwise direction for the flow of current around any mesh is conventional.

In some cases it may be found simpler to reduce a passive 4-terminal network to an equivalent T or π section—see Sect. 7(x)—than to analyse it by means of Kirchhoff's Laws.

Definitions

An **element** is the smallest entity (i.e. a distinct unit) which may be connected in a network—e.g. L, C or R.

An **arm** is a distinct set of elements, electrically isolated from all other conductors except at two points.

A **series arm** conducts the main current in the direction of propagation.

A **shunt arm** diverts a part of the main current.

A **branch** is one of several parallel paths.

A **mesh** is a combination of elements forming a closed path.

A **two-terminal network** is one which has only two terminals for the application of a source of power or connection to another network.

A **four-terminal network** is one which has four terminals for the application of a source of power or connections to other networks.   The common form of four-terminal network has two input and two output terminals ; this term is used even when one input terminal is directly connected to one output terminal, or both earthed.

A **passive network** is one containing no source of power.

An **active network** is one containing one or more sources of power (e.g. batteries, generators, amplifiers).

The **input circuit** of a network is that from which the network derives power.

The **output circuit** of a network is that into which the network delivers power.

Impedance matching—two impedances are said to be matched when they have the same magnitude and the same phase angle.

Reference may also be made to I.R.E. Standard 50IRE4.S1 published in Proc. I.R.E.39.1 (Jan. 1951) 27.

Examples

An amplifier is a four-terminal active network.

An attenuator is a four-terminal passive network.

A conventional tone control is a two-terminal passive network.

Differentiating and Integrating Networks

Based on the fundamental mathematical analysis of the circuit, the following terms are sometimes used in connection with 4-terminal networks.

Differentiating Networks—(1) Series resistance and shunt inductance
                                      or (2) Series capacitance and shunt resistance
Integrating Networks   —(1) Series inductance and shunt resistance
                                      or (2) Series resistance and shunt capacitance.

(ii) Kirchhoff's Laws

(1) The algebraic sum of all the instantaneous values of all currents flowing towards any junction point in a circuit is zero at every instant.

This is illustrated for d.c. in Fig. 4.23. It will be seen that all junctions and corner points are lettered for reference. The polarities of the two batteries and their voltages are marked. The currents are marked in the obvious directions or, if this is not clear, then arbitrarily in either direction (clockwise around each loop is preferred). Both currents and voltages are referred to by the point lettering,

e.g. i_{ab} is the current flowing from a to b

i_{ba} is the current flowing from b to a

e_{bc} is the potential of point b with respect to point c.

Applying Kirchhoff's first law, at point b

$i_{ab} + i_{fb} - i_{bc} = 0$

or $i_{ab} + i_{fb} \qquad = i_{bc}$ which is really obvious.

Positive current is taken as flowing towards the junction point ; negative current as flowing away from it.

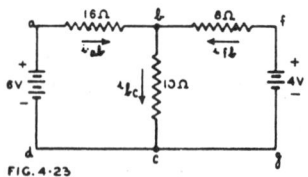 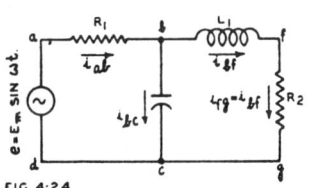

FIG. 4·23   FIG. 4·24

Fig. 4.23.   Network incorporating 3 elements and 2 d.c. voltage sources.

Fig. 4.24.   Network incorporating 4 elements and one a.c. voltage source.

This law is illustrated for a.c. in Fig. 4.24 which follows the same general rules as for d.c. The instantaneous generator voltage e is shown in an arbitrary direction and the directions of the currents are then determined.

At point b,

$$i_{ab} - i_{bf} - i_{bc} = 0$$

following the general procedure as in the d.c. case.

(2) **The total rise or fall of potential at any instant in going around any closed circuit is zero.**

This is illustrated for d.c. in Fig. 4.23, taking each closed circuit in turn and proceeding clockwise in each case. Any voltage source is here regarded as positive if it assists in sending current clockwise around the closed circuit (i.e. voltage rise). The voltage across any impedance is regarded as negative if the current arrow is in the same direction as the direction of travel around the closed circuit (i.e. voltage drop).

Circuit $d\ a\ b\ c\ d$: $\qquad +6 - 16\,i_{ab} - 10\,i_{bc} = 0$ (1)

Circuit $d\ a\ b\ f\ g\ c\ d$: $+6 - 16\,i_{ab} + 8\,i_{fb} - 4 = 0$ (2)

Circuit $f\ b\ c\ g\ f$: $\qquad +4 - 8\,i_{fb} - 10\,i_{bc} = 0$ (3)

Applying Kirchhoff's first law to point b,

$$i_{ab} + i_{fb} = i_{bc} \qquad (4)$$

To find the values of the three unknown currents, it is necessary to apply three suitable equations.

[*Similarly for all other cases—the total number of equations must be equal to the number of unknowns. The number of equations based on Kirchhoff's first law should be one less than the number of junction points ; those based on his second law should equal the number of* **independent** *closed paths.*]

Equations (1), (3) and (4) would be sufficient, since (2) merely duplicates parts of (1) and (3).

From (4) : $\quad i_{ab} = i_{bc} - i_{fb}$

Applying in (1) : $\ +6 - 16(i_{bc} - i_{fb}) - 10\,i_{bc} = 0$

Therefore $\qquad +6 - 16\,i_{bc} + 16\,i_{fb} - 10\,i_{bc} = 0$

Therefore $\qquad +6 - 26\,i_{bc} + 16\,i_{fb} = 0$ (5)

Adding twice (3): $+8 - 20\,i_{bc} - 16\,i_{fb} = 0$

Therefore $\qquad +14 - 46\,i_{bc} = 0$

Therefore $\quad i_{bc} = 14/46$ ampere.

The other currents may be found by applying this value in (5) and then in (4). Kirchhoff's Second Law is illustrated for a.c. in Fig. 4.24. Here again, as in all cases, it is assumed that we move around each loop of the network in a clockwise direction.

The voltage across—

a resistance is $\quad -Ri$ ⎫ where i is in the clockwise direction around
an inductance is $\ -j\omega Li$ ⎬ the loop.
a capacitance is $\ +j(1/\omega C)i$ ⎭

Circuit $d\ a\ b\ c\ d$: $e - R_1 i_{ab} + j\,(1/\omega C_1)i_{bc} = 0$

Therefore $\quad R_1 i_{ab} - j(1/\omega C_1)i_{bc} = e$ (6)

Circuit $b\ f\ g\ c\ b$: $\ -j\omega L_1 i_{bf} - R_2 i_{bf} - j(1/\omega C_1)i_{bc} = 0$

Therefore $\quad (j\omega L_1 + R_2)i_{bf} + j(1/\omega C_1)i_{bc} = 0$ (7)

Applying Kirchhoff's first law to junction b :

$i_{ab} - i_{bf} - i_{bc} = 0$

Therefore $\quad i_{ab} = i_{bf} + i_{bc}$ (8)

Adding (6) and (7)

$R_1 i_{ab} + (j\omega L_1 + R_2)i_{bf} = e$

Applying (8), $\quad R_1 i_{bf} + R_1 i_{bc} + (j\omega L_1 + R_2)i_{bf} = e$

Therefore $\quad R_1 i_{bc} + (R_1 + R_2 + j\omega L_1)i_{bf} = e$ (9)

Multiplying (7) by $(-jR_1\omega C_1)$, remembering that $j^2 = -1$,

$R_1 i_{bc} + (R_1\omega^2 L_1 C_1 - jR_1 R_2\omega C_1)i_{bf} = 0$ (10)

Subtracting (10) from (9),

$[(R_1 + R_2 - R_1\omega^2 L_1 C_1) + j(\omega L_1 + R_1 R_2\omega C_1)]\,i_{bf} = e$ (11)

which gives the value of i_{bf} when e is known.

The value of i_{bc} may be found by substituting this value of i_{bf} in eqn. (7) ; i_{ab} may then be determined by eqn. (8).

(iii) Potential Dividers

The fundamental form of potential divider (also known as voltage divider or potentiometer) is shown in Fig. 4.25. Here a direct line voltage E is " divided " into two voltages E_1 and E_2, where $E = E_1 + E_2$. If no current is drawn from the junction (or tap) B, the voltage across BC is given by

$$E_2 = \left(\frac{R_2}{R_1 + R_2}\right)E, \tag{1}$$

and $I_1 = I_2 = E/(R_1 + R_2)$ \hfill (2)

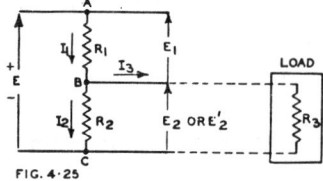

FIG. 4·25

Fig. 4.25. *Potential divider across a d.c. line.*

On load

When a current I_3 is drawn from B, the simplest analysis is to consider the effective load resistance R_3 which will draw a current I_3 at a voltage E_2', i.e. $R_3 = E_2'/I_3$. We now have resistances R_2 and R_3 in parallel, and their total effective resistance is therefore

$$R' = R_2 R_3/(R_2 + R_3).$$

In this case the voltage divider is composed of R_1 and R' in series, and the voltage at the point B is given by

$$E_2' = \left(\frac{R'}{R_1 + R'}\right)E$$

$$= \left(\frac{R_2 R_3}{R_1 R_2 + R_1 R_3 + R_2 R_3}\right)E$$

$$E_2' = \left(\frac{R_2}{R_1 + R_2}\right)E - \left(\frac{R_1 R_2}{R_1 + R_2}\right)I_3 \tag{3}$$

The first term on the right hand side is the no-load voltage E_2 ; the second term is the further reduction in voltage due to I_3—this being a linear equation.

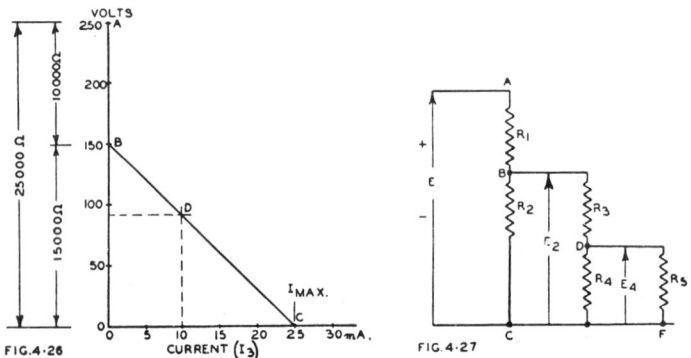

FIG.4·26

FIG.4·27

Fig. 4.26. *Graphical method for determining the output voltage from a potential divider.*

Fig. 4.27. *A potential divider with a load in the form of another potential divider.*

Eqn. (3) may be plotted* as in Fig. 4.26 where, as a typical example, $E = 250$ V, $R_1 = 10\,000$ and $R_2 = 15\,000$ ohms, so that $(R_1 + R_2) = 25\,000$ ohms. E_2, for

*Cundy, P. F. " Potential Divider Design," W.W. 50.5 (May 1944) 154.

no load, is obviously 150 volts (point B) while point C is for the condition of maximum current (I_{max}) and zero voltage. The latter may be determined by putting $E_2' = 0$ in eqn. (3) which gives

$$I_{max} = E/R_1 \qquad (4)$$

In this condition we have, effectively, a resistance R_1 only, in series with a load of zero resistance.

If points B and C in Fig. 4.26 are joined by a straight line, we may then determine the voltage for any value of load current (I_3) between zero and maximum. For example, with a load current of 10 mA, the voltage will be 90 V (point D).

It is sometimes useful to calculate on the basis of a drop in voltage (based on the no-load voltage E_2) of so many volts per milliamp of load current; in this case the drop is 150 volts for 25 mA, or 6 volts per milliamp. This rate of voltage drop is actually the negative slope of the line BC.

The equivalent series source resistance (R_s) is given by

$$R_s = \frac{E_2}{I_{max}} = \frac{150}{25 \times 10^{-3}} = 6000 \text{ ohms.}$$

This may be put into the alternative form

$$R_s = \frac{R_1 R_2}{R_1 + R_2}. \qquad (5)$$

" **Regulation** " is defined differently in American and British practice.

American definition : The percentage voltage regulation is the difference between the full-load and no-load voltages, divided by the *full-load* voltage and multiplied by 100

British definition : The percentage voltage regulation is the difference between the full-load and no-load voltages, divided by the *no-load* voltage and multiplied by 100.

In the example above for a current of 10 mA,

$$\text{Regulation} = \frac{150 - 90}{150} \times 100 = 40\% \text{ by British definition.}$$

The line BD in Fig. 4.26 may be called the " **regulation characteristic** " for the conditions specified above.

To find the load current corresponding to a specified value of E_2', eqn. (3) may be re-arranged in the form

$$I_3 = \frac{E}{R_1} - E_2'\left(\frac{R_1 + R_2}{R_1 R_2}\right) \qquad (6)$$

$$\text{or } I_3 = I_{max} - E_2'\left(\frac{R_1 + R_2}{R_1 R_2}\right) \qquad (7)$$

Special Case 1

If it is known that the voltage drops from E_2 at no load to E_2' for a load current I_3, then the voltage E_x corresponding to a load current I_x is given by

$$E_x = E_2 - (E_2 - E_2')(I_x/I_3) \qquad (8)$$

Special case 2

If the voltages across the load (E_x, E_y) for two different values of load current (I_x, I_y) are known, the voltage at zero load current is given by

$$E_2 = \frac{E_x I_y - E_y I_x}{I_y - I_x} \qquad (9)$$

This is often useful for determining the no load voltage when the voltmeter draws appreciable current. A method of applying this with a two range voltmeter has been described*, and the true voltage is given by

---

*Lafferty, R. E. " A correction formula for voltmeter loading " (letter) Proc. I.R.E. 34.6 (June 1946) 358.

$$E_2 = \frac{(S - 1) E_x}{S - (E_x/E_y)} \tag{10}$$

where S = ratio of the two voltmeter scales used for the two readings E_x and E_y,
          E_x = voltmeter reading on the higher voltage scale,
and E_y = voltmeter reading on the lower voltage scale.
If $S = 2$, then

$$E_2 = \frac{E_x}{2 - (E_x/E_y)}. \tag{11}$$

Complicated divider network

When a voltage divider has another voltage divider as its load (Fig. 4.27) the best procedure is to work throughout in resistances and voltages, and to leave currents until after the voltages have been determined. The final equivalent load resistance (R_5) must be determined before commencing calculations, then proceed—

R_4 in parallel with R_5 : $R' = R_4R_5/(R_4 + R_5)$
R_3 in series with R' : $R'' = R_3 + R'$
R'' in parallel with R_2 : $R''' = R_2R''/(R_2 + R'')$

Then $E_2 = \left(\dfrac{R'''}{R_1 + R'''}\right)E$ \hfill (12)

and $E_4 = \left(\dfrac{R'}{R_3 + R'}\right)E_2$ \hfill (13)

A somewhat similar procedure may be adopted in any divider network.

(iv) Thévenin's Theorem (pronounced " tay-venin's ")

This theorem* may be expressed in various ways, of which one is :

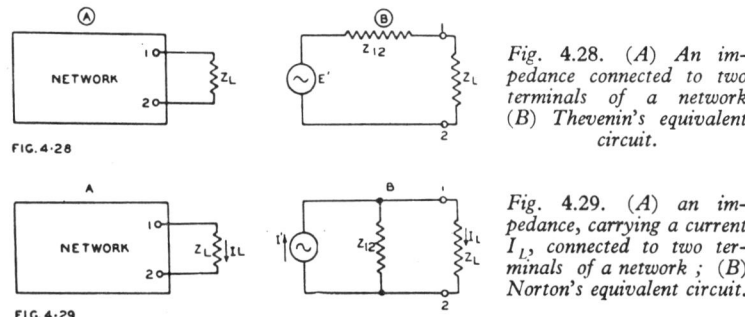

Fig. 4.28. (A) An impedance connected to two terminals of a network (B) Thevenin's equivalent circuit.

Fig. 4.29. (A) an impedance, carrying a current I_L, connected to two terminals of a network ; (B) Norton's equivalent circuit.

The current in any impedance, $\dot{Z}_L$, connected to two terminals of a network consisting of any number of impedances and generators (or voltage sources) is the same as though Z_L were connected to a simple generator, whose generated voltage is the open-circuited voltage at the terminals in question, and whose impedance is the impedance of the network looking back from the terminals, with all generators replaced by impedances equal to the internal impedances of these generators.

In Fig. 4.28, (A) shows an impedance Z_L whose two ends are connected to the terminals 1, 2 of any network. Diagram (B) shows Thevenin's equivalent circuit, with a generator E' and series impedance Z_{12} where :

E' is the voltage measured at the terminals 1, 2, with Z_L removed,
and Z_{12} is the impedance of the network measured across the terminals 1, 2, when looking backwards into the network, with all generators out of operation and each replaced by an impedance equal to its internal impedance.

---

*For the proofs of this and subsequent theorems, see W. L. Everitt, Ref. 1 pp. 47-57.

(v) Norton's Theorem

This is similar in many ways to Thevenin's Theorem, but provides a constant current generator and a shunt impedance.

The current in any impedance Z_R, connected to two terminals of a network, is the same as though Z_R were connected to a constant current generator whose generated current is equal to the current which flows through the two terminals when these terminals are short-circuited, the constant-current generator being in shunt with an impedance equal to the impedance of the network looking back from the terminals in question.

In Fig. 4.29, (A) shows an impedance Z_L through which flows a current I_L, connected to a network. Diagram (B) shows Norton's equivalent circuit, with a constant current generator delivering a current I' to an impedance Z_{12} in shunt with Z_L. As in diagram A, the current through Z_L is I_L.

Here $I' = E'/Z_{12}$
where E' and Z_{12} are the same as in Fig. 4.28B (Thevenin's Theorem).

(vi) Maximum Power Transfer Theorem

The maximum power will be absorbed by one network from another joined to it at two terminals, when the impedance of the receiving network is varied, if the impedances (looking into the two networks at the junction) are conjugates* of each other.

This is illustrated in Fig. 4.30 where E is the generated voltage, Z_g the generator internal impedance and Z_L the load impedance. In the special case where Z_g and Z_L are pure resistances,

Z_g will become R_g
Z_L will become R_L
and maximum power transfer will occur when $R_L = R_g$.

In the general case, $Z_g = R_g + jX_g$ and $Z_L = R_L + jX_L$ while for maximum power transfer $R_L = R_g$ and $X_L = -X_g$.

In other words, if Z_g is inductive, Z_L should be capacitive, and vice versa.

If the magnitude of the load impedance may be varied, but not the phase angle, then the maximum power will be absorbed from a generator when the absolute value of the load impedance is equal to the absolute value of the impedance of the supply network.

See Ref 3 (References to networks). Sect. 7(xv).

(vii) Reciprocity Theorem

In any system composed of linear bilateral impedances, if an electromotive force E is applied between any two terminals and the current I is measured in any branch, their ratio (called the " transfer impedance ") will be equal to the ratio obtained if the positions of E and I are interchanged.

In Fig. 4.31 a generator supplies a voltage E to a network, and an ammeter A reads the current I_2. The transfer impedance is E/I_2. If now E and A are reversed, the new transfer impedance will have the same value as previously. In other words, E being unchanged, the ammeter reading in the new position will be the same as previously.

This theorem proves that a network of bilateral impedances transmits with equal effectiveness in both directions, when generator and load have the same impedance.

(viii) Superposition Theorem

In any network consisting of generators and linear impedances, the current flowing at any point is the sum of the currents which would flow if each generator were considered separately, all other generators being replaced at the time by impedances equal to their internal impedances.

This theorem considerably simplifies the analysis of any network containing more than one generator. It is important to note the linearity requirements, as the theorem

---

*Two impedances are conjugates of each other when their resistive components are equal, and their reactive components are equal in magnitude but opposite in sign.

breaks down under other conditions. It is therefore only applicable to valves when
these are being operated to give negligible distortion.

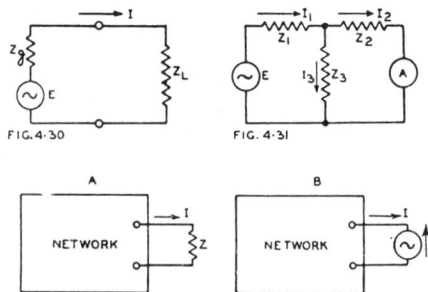

*Fig. 4.30. A generator, with
internal impedance Z_g, connected
to a load Z_L, to illustrate the
maximum power transfer.*

FIG. 4·30   FIG. 4·31

*Fig. 4.31. A generator E supply-
ing voltage to a network, with
an ammeter A to read the current
I_2. The reciprocity theorem
states that, when A and E are
reversed, the transfer impedance
E/I_2 will be unchanged.*

FIG. 4·32

*Fig. 4.32. (A) An impedance Z
in a network with current I and voltage drop ZI (B) Equivalent circuit having
identical results so far as current and voltage drop are concerned, with a generator
developing a voltage E = ZI. This illustrates the Compensation Theorem.*

(ix) Compensation Theorem

**An impedance in a network may be replaced by a generator of zero internal
impedance, whose generated voltage at any instant is equal to the instantaneous potential difference produced across the replaced impedance by the
current flowing through it.**

This is illustrated in Fig. 4.32 where in (A) a current I is flowing through an impedance Z with a voltage drop IZ. This is equivalent to an identical network, as
in (B), where Z has been replaced by a generator of zero internal impedance, whose
generated voltage (E) is equal in magnitude to IZ, and is in a direction opposing the
flow of current.

(x) Four-terminal networks

The most common fundamental types of four terminal networks are illustrated
in Fig. 4.33, where (A) is a T section, (B) is a Π section and (C) a Lattice section.
Both A and B are called 3 element networks, and C a 4 element network, from the
number of arms containing impedances. In conventional operation, the left-hand
terminals 1, 2, are regarded as the input terminals, to which is connected some generator, or other network containing a generator. Terminals 3, 4, are normally regarded
as the output terminals, across which is connected a load impedance Z_L.

If we are concerned only with the observable impedances between terminals, it is
possible—by selecting suitable values—to convert a T section to a Π section, and
vice versa, but only for one particular frequency. This equivalence is independent
of the character of the generator or load.

Equivalent T section   *Equivalent Π section*

$$Z_1 = \frac{Z_A Z_B}{Z_A + Z_B + Z_C}$$

$$Z_A = \frac{Z_1 Z_2 + Z_2 Z_3 + Z_1 Z_3}{Z_2}$$

$$Z_2 = \frac{Z_B Z_C}{Z_A + Z_B + Z_C}$$

$$Z_B = \frac{Z_1 Z_2 + Z_2 Z_3 + Z_1 Z_3}{Z_3}$$

$$Z_3 = \frac{Z_A Z_C}{Z_A + Z_B + Z_C}$$

$$Z_C = \frac{Z_1 Z_2 + Z_2 Z_3 + Z_1 Z_3}{Z_1}$$

$$(14)$$

The bridged T section of Fig. 4.33E may be reduced to the equivalent T section
of Fig. 4.33F, or vice versa, by using the same equivalent values as for equivalent T
and Π sections (above).

Any complex four-terminal network can be reduced, at a single frequency, to a single T or Π section. The procedure is to measure (or calculate) the three constants :

Z_{o1} = the impedance at the input terminals when the output terminals are open-circuited.

Z_{o2} = the impedance at the output end looking back into the network with the input terminals open-circuited.

Z_{s1} = the impedance at the input end with the output terminals short-circuited.

When these are known, the constants for the equivalent T and Π sections are :

Equivalent T section   *Equivalent Π section*

$$Z_1 = Z_{o1} - \sqrt{Z_{o2}(Z_{o1} - Z_{s1})}$$

$$Z_A = \frac{Z_{s1}Z_{o2}}{Z_{o2} - \sqrt{Z_{o2}(Z_{o1} - Z_{s1})}}$$

$$Z_2 = Z_{o2} - \sqrt{Z_{o2}(Z_{o1} - Z_{s1})}$$

$$Z_B = \frac{Z_{s1}Z_{o2}}{\sqrt{Z_{o2}(Z_{o1} - Z_{s1})}}$$

$$Z_3 = \sqrt{Z_{o2}(Z_{o1} - Z_{s1})}$$

$$Z_C = \frac{Z_{s1}Z_{o2}}{Z_{o1} - \sqrt{(Z_{o1} - Z_{s1})Z_{o2}}}$$

(15)

It would also be possible to derive expressions for the equivalent sections, using Z_{s2} in place of Z_{s1}, where Z_{s2} = impedance at the output end with the input terminals short-circuited.

A lattice network may be reduced to an equivalent T or Π network (see any suitable textbook) but it is interesting to note that it is essentially a " bridge " circuit. Fig. 4.33 (C) may be re-drawn as in (D) without any change being made.

Four-terminal networks are considered further in Sect. 8 ; they may also be treated as multi-mesh networks as in Sect. 7(xi) below.

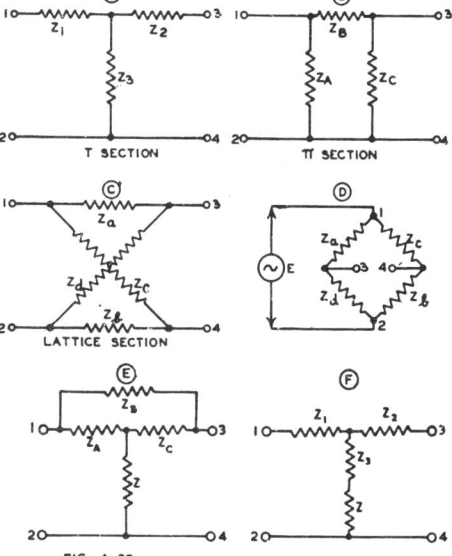

FIG. 4.33

Fig. 4.33. Four Terminal Networks (A) T Section (B) Π Section (C) Lattice Section (D) Lattice Section redrawn in the form of a bridge (E) Bridged T Section (F) T Section equivalent to Bridged T.

(xi) Multi-Mesh Networks

A typical flat multi-mesh network is shown in Fig. 4.33G. A " flat " network is defined as one which can be flattened out without having any lines crossing over each other ; the treatment in this handbook is limited to flat networks. Junctions at which the current can divide are called branch points (A, B, C, D, E). In Fig. 4.33G there are 5 meshes, and the circulating mesh current of each is marked (I_1, I_2, etc.) in the conventional clockwise direction. The simplest form of solution is by means of the **Mesh Equations.** The basis for the use of these equations is given in Ref. **4,** Sect. 7(xv). Impedances in the network are numbered Z_{10}, Z_{20}, etc., when they form part of one mesh only (mesh 1, mesh 2, etc.). Impedances which are common to two meshes are numbered Z_{12}, Z_{13}, etc., and are called **mutual impedances,** the suffixes indicating the meshes to which they are common. The applied alternating voltages are also numbered E_1, E_2, etc. where the suffix indicates the mesh number. If there is more than one voltage source in any mesh, E_1 etc. will indicate the vector sum of these voltages.

There must be the same number of mesh equations as there are meshes in the network. The mesh equations are written in the general form for n meshes as

$$
\left.
\begin{aligned}
Z_{11}I_1 + Z_{12}I_2 + Z_{13}I_3 + \quad &\ldots Z_{1n}I_n = E_1 \\
Z_{21}I_1 + Z_{22}I_2 + Z_{23}I_3 + \quad &\ldots Z_{2n}I_n = E_2 \\
&\ \ \vdots \\
Z_{n1}I_1 + Z_{n2}I_2 + Z_{n3}I_3 + \quad &\ldots Z_{nn}I_n = E_n
\end{aligned}
\right\}
\tag{16}
$$

where each Z is of the form $R + j(\omega L - 1/\omega C)$ and Z_{11}, Z_{22}, etc. is called the **self-impedance** of the individual mesh, i.e., the impedance round the mesh if all other branches of the network other than those included in the mesh in question were open-circuited.

For example, in Fig. 4.33G—

$$
\begin{aligned}
Z_{11} &= Z_{10} + Z_{13} + Z_{12} \\
&= R_{10} + jX_{10} + R_{13} + jX_{13} + R_{12} + jX_{12} \\
&= R_{10} + j(\omega L_{10} - 1/\omega C_{10}) + R_{13} + j(\omega L_{13} - 1/\omega C_{13}) \\
&\quad + R_{12} + j(\omega L_{12} - 1/\omega C_{12})
\end{aligned}
$$

i.e. $\quad Z_{11} = (R_{10} + R_{13} + R_{12}) + j((\omega L_{10} + \omega L_{13} + \omega L_{12})$
$$
- (1/\omega C_{10} + 1/\omega C_{13} + 1/\omega C_{12})]
\tag{17}
$$

Note that Z_{21} is the same as Z_{12} etc. and that the signs of the mutual impedances may be positive or negative (see below).

An impedance that is common to two branches is considered to be a positive **mutual impedance** when the arrows representing the corresponding mesh currents pass through the impedance in the same direction ; or a negative mutual impedance if the arrows pass through the impedance in opposite directions.

Thus in Fig. 4.33G the arrows representing the corresponding mesh currents pass through the impedances Z_{12}, Z_{13}, Z_{24} and Z_{34} in opposite directions, so that these constitute negative mutual impedances, and have negative signs in eqn. (16).

A **mutual inductance** may be defined as positive or negative according to whether it acts with a polarity the same as, or opposite to, that of a corresponding common inductance*.

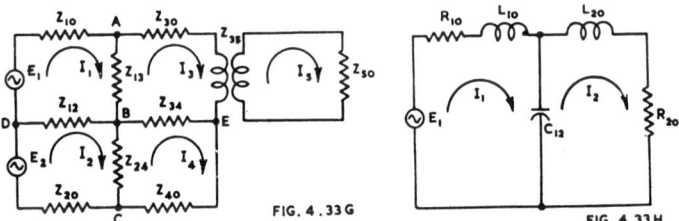

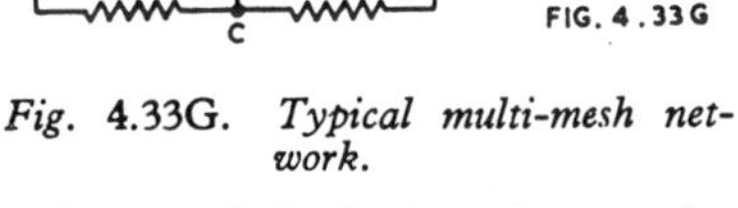

FIG. 4.33 G

FIG. 4.33 H

Fig. 4.33G. *Typical multi-mesh network.* 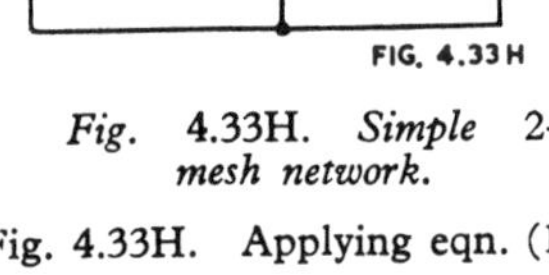
       Fig. 4.33H. *Simple 2-mesh network.*

A simple example is the 2-mesh network of Fig. 4.33H. Applying eqn. (16) we have

$$Z_{11} = R_{10} + j[\omega L_{10} - 1/\omega C_{12}] \tag{18}$$
$$Z_{22} = R_{20} + j[\omega L_{20} - 1/\omega C_{12}] \tag{19}$$
$$Z_{12} = Z_{21} = j(1/\omega C_{12}) \tag{20}$$

Since there are two meshes, there will be two mesh equations—

$$
\left.
\begin{aligned}
Z_{11}I_1 + Z_{12}I_2 &= E_1 \\
Z_{21}I_1 + Z_{22}I_2 &= 0
\end{aligned}
\right\}
\tag{21}
$$

---

*The opposite definition is also used.

These linear simultaneous equations may be solved by elimination, but any more complicated network would have to be solved by the use of Determinants, for which see any suitable mathematical textbook.

The total current in any common branch may be determined by the difference between the two mesh currents.

The method of handling **mutual inductance** is illustrated by the 2-mesh network of Fig. 4.33I. Here

$$Z_{11} = (R_{10} + R_{12}) + j(\omega L_{10} + \omega L_{12} - 1/\omega C_{10}) \tag{22}$$
$$Z_{22} = (R_{12} + R_{20}) + j(\omega L_{12} + \omega L_{20} - 1/\omega C_{20}) \tag{23}$$
$$Z_{12} = Z_{21} = R_{12} + j(\omega L_{12} \pm \omega M_{12}) \tag{24}$$

and the two mesh equations will be as eqn. (21).

The polarity of M_{12} in eqn. (24) must be determined in accordance with the accepted convention, as described in connection with eqn. (16).

A solution of eqn. (16) by means of Determinants shows that in the general case the Determinant D is given by

$$D = \begin{vmatrix} Z_{11}Z_{12} & \cdot\cdot & Z_{1n} \\ Z_{21}Z_{22} & \cdot\cdot\cdot & Z_{2n} \\ \cdot\cdot & \cdot\cdot & \cdot \\ Z_{n1}Z_{n2} & \cdot\cdot\cdot & Z_{nn} \end{vmatrix} \tag{25}$$

and the current I_k in the kth mesh that flows as the result of the voltage E, acting in the jth mesh is

$$I_k = E_j \frac{B_{jk}}{D} \tag{26}$$

where B_{jk} is the principal minor of D multiplied by $(-1)^{j+k}$. This minor is

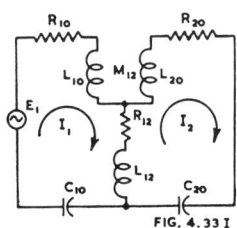

formed by cancelling the jth row and the kth column and then moving the remainder together to form a new determinant with one less row and column than D, where D is the determinant defined by eqn. (25). The row cancelled corresponds to the mesh containing the input voltage, the column cancelled to the mesh containing the required current.

The **input impedance** of a passive network with a single applied voltage (Fig. 4.33J) is given by

Input impedance $= E_1/I_1 = D/B_{11}$ (27)

Fig. 4.33 I. Two-mesh network incorporating mutual inductance.

where E_1 = voltage applied to input terminals
I_1 = input current
D = determinant defined by eqn. (25)
and B_{11} is the minor of D obtained by cancelling the first row and column.

The **transfer impedance** of a 4-terminal network (Fig. 4.33K) is defined as the ratio of the voltage applied to the input terminals to the resulting current through the load impedance connected to the output terminals (i.e., the nth mesh).

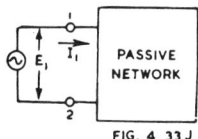

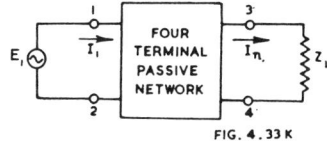

FIG. 4.33 J   FIG. 4.33 K

Fig. 4.33J. Illustrating input impedance of network.
Fig. 4.33K. Four terminal network illustrating input and transfer impedances.

Transfer impedance $= E_1/I_L = D/B_{1n}$ (28)
where D = determinant defined by eqn. (25)

and B_{1n} is the minor of D obtained by cancelling the first row and n^{th} column, and applying the factor $(-1)^{n+1}$.

References to multi-mesh networks : Refs. 1, 2, 4, 5, 6, 7. Sect. 7(xv).

(xii) Non-linear components in networks

In radio engineering the principal non-linear components in networks are valves, although certain resistors and iron-cored inductances are also non-linear. The non-linearity can only be neglected when the voltage swing across any non-linear component is so limited that the characteristic is substantially constant over the range of operation.

In a non-linear component the impedance is not constant, but varies with the applied voltage*. Each such impedance must be treated as having an impedance which is a function of voltage,
i.e. $$Z = F(e).$$

The mathematical theory of circuits containing non-linear components is complicated and, in the general form, outside the scope of this handbook. Those who are interested are referred to the list at the end of this subsection.

The treatment of non-linearity in valve characteristics is covered in Chapter 27 for detection and Chapter 25 for frequency conversion. Distortion through curvature of valve characteristics is covered in Chapter 2 Sect. 9 and Chapter 13.

References to non-linearity
(other than those covered elsewhere)

Chaffee, E. L. " Theory of Thermionic Vacuum Tubes " (McGraw-Hill Book Company, New York and London, 1st edit. 1933) Chapter 21.

Llewellyn, F. B., and L. C. Peterson. " Vacuum tube networks " Proc. I.R.E. 32.3 (March 1944) 144.

(xiii) Phase–shift networks

The bridge type phase-shift network of Fig. 4.34 has the advantage of providing full range phase shifting from 0 to 180°, with constant attenuation (6db) for all degrees of phase shift†.

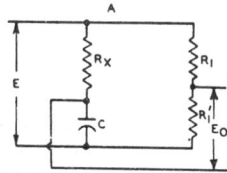 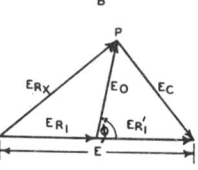

FIG.4·34

Fig. 4.34 *(A) A phase-shift network providing full range phase shifting from 0 to 180° with 6 db attenuation for all degrees of phase shift (B) vector diagram of voltage relationships.*

It may be shown, when $R_1 = R_1' = 1/\omega C$, that
$$\left|\frac{E_o}{E}\right| = \tfrac{1}{2} \qquad \text{and } \phi = \tan^{-1}\left(\frac{2R_1R_x}{R_x^2 - R_1^2}\right)$$
which is illustrated by the vector diagram Fig. 4.34B, where E_{R1} and E_{R1}' are equal, and both equal to E_o ; E_{Rx} and E_c are at right angles, but their vector sum is E. Point P is therefore on the circumference of a semi-circle.

*It is assumed here that the impedance is not a function of time ; this occurs in the case of barretters and lamp filaments.
†Lafferty, R. E. " Phase-shifter nomograph," Elect. 19.5 (May 1946) 158.

(xiv) Transients in Networks

The treatment of networks earlier in this Section has been on the basis of a steady applied direct voltage, or a steady alternating voltage, or a combination of the two. It is also important to know the transient currents which may flow during the period from the application of a voltage until the steady state has been reached or during the period from the disconnection of the voltage until a steady state has been reached, or due to some sudden change in operating conditions after the steady state has been reached.

The simple cases of a capacitance and inductance each in series with a resistance, on direct charge and discharge, have been covered in Sects. 4 and 5. These charge and discharge characteristics are of a logarithmic form, and non-oscillatory. However, in any network including L, C and R, the transient characteristics tend to be oscillatory. These oscillatory transients may be more or less heavily damped, and normally become of negligible value after a short period of time. On the other hand, they may continue as sustained oscillations.

Amplifiers, particularly those with feedback, should have a sufficiently damped oscillatory transient in the output circuit when the input is increased instantaneously from zero to some predetermined steady voltage (this is called "unit step" input). In practice it is more convenient to use a periodic rectangular wave input, with the time period of the " flat top " sufficiently long to allow for the decay of damped oscillatory transients. A C.R.O. is commonly used for observation of such a waveform.

Alternatively, an impulse having very short time duration may be applied to the network or amplifier. This differs from unit step input in that the input voltage returns to zero almost instantaneously.

The complete mathematical analysis of all but very simple networks is very complicated and specialized, and outside the scope of this Handbook.

For further information see Refs. 4, 7 and 8.

(xv) References to Networks

1. Everitt, W. L. (book) " Communication Engineering." (McGraw-Hill Book Co. Inc., New York and London, 1937).

2. Shea, T. E. (book) " Transmission Networks and Wave Filters " (D. Van Nostrand Co. Inc., New York, 1943).

3. Ellithorn, H. E. " Conditions for transfer of maximum power " Comm. 26.10 (Oct. 1946) 26.

4. Guillemin, E. A. (book) " Communication Networks " Vol. 1. (John Wiley and Sons Inc. New York ; Chapman and Hall Ltd., London, 1931).

5. Terman, F. E. (book) " Radio Engineers' Handbook " (McGraw-Hill Book Co., New York and London, 1943).

6. Johnson, K. S. (book) " Transmission Circuits for Telephonic Communication " (D. Van Nostrand Co. Inc. New York, 1931).

7. Valley, G. E., and H Wallman (book) " Vacuum Tube Amplifiers " (McGraw-Hill Book Co. New York and London, 1948).

8. Gardner, M. F., and J. L. Barnes (book) " Transients in Linear Systems " (John Wiley and Sons, Inc. New York, 1942).

See Supplement for additional references.

SECTION 8 : FILTERS

(i) Introduction to filters   (ii) Resistance-capacitance filters, high-pass and low-pass (iii) Special types of resistance-capacitance filters   (iv) Iterative impedances of four-terminal networks   (v) Image impedances and image transfer constant of four-terminal networks   (vi) Symmetrical networks   (vii) " Constant k " filters   (viii) M Derived filters   (ix) Practical filters   (x) Frequency dividing networks   (xi) References to filters.

(i) Introduction to filters

A filter is any passive* network which discriminates between different frequencies, that is to say it provides substantially constant " transmission " over any desired range of frequencies and a high degree of attenuation for all other frequencies.

Filters are conveniently grouped as under :

Low pass filters—transmission band from zero (or some very low) frequency to a specified frequency.   Attenuation for all higher frequencies.

High pass filters—transmission band from some specified frequency to very high frequencies.   Attenuation for all lower frequencies.

Band pass filters—transmission band from one to another specified frequency. Attenuation for all lower and higher frequencies.

Band elimination filters—" traps."

Practical filters, particularly of the simple variety, have only a gradual change in attenuation.   The sharper the required change in attenuation, the more complicated becomes the filter.

Some very simple filters are :

1. The grid coupling condenser and grid resistor of an amplifier (high pass resistance-capacitance filter).

2. The series condenser and variable resistor of a conventional tone control (low pass resistance-capacitance, with adjustable attenuation).

3. The smoothing filter of a power supply, including one or two inductances and two or three capacitances (low pass filter).

4. An overcoupled i-f transformer (tuned band pass filter).

5. A tuned aerial coil or r-f transformer (tuned narrow band pass filter).

See Chapter 6 for mathematics.

(ii) Resistance—capacitance filters

Fig. 4.35A shows a r.c. high pass filter as for grid coupling in an amplifier.   This is essentially a voltage divider in which C forms a reactive, and R a resistive, arm.

If the generator has zero resistance and if there is no loading on the output, the ratio of output to input voltages is given by

$$\frac{E_o}{E_i} = \frac{R}{R + jX_c} \quad \text{where} \quad X_c = -1/2\pi fC.$$

Therefore $\left|\dfrac{E_o}{E_i}\right| = \dfrac{R}{\sqrt{R^2 + X^2}} = \dfrac{1}{\sqrt{1 + (X/R)^2}}$ \hfill (1)

If we select a frequency (f_1) at which $|X_c| = R$, then $|E_o/E_i| = 0.707$ which is practically equivalent to an attenuation of 3 db. This frequency is the reference point used for design ; it is called the theoretical cut-off frequency.   Its value is given by†

$$f_1 = 1/(2\pi RC) \text{ c/s} \tag{2}$$

where R and C are in ohms and farads (or in megohms and microfarads).   The value RC is called the **time constant** and is measured in seconds (see Sect. 4(iv)) so that

---

*i.e. not including a valve or generator.
†A nomogram to determine the value of f_1 is given by E. Frank " Resistance Capacitance Filter Chart " Elect. 18.11 (Nov. 1945) 164.

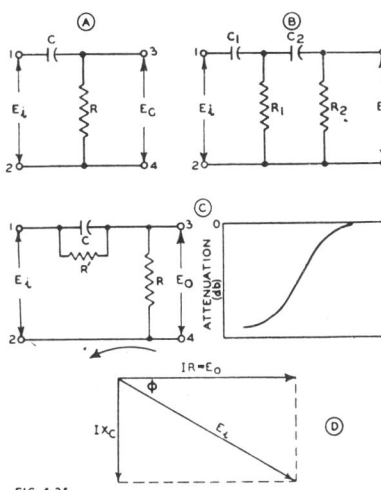

FIG. 4·35

Fig. 4.35. (A) is a resistance capacitance high-pass filter ; (B) is a two section filter ; (C) is a modified form of A, providing attenuation within top and bottom limits ; (D) is the vector diagram of A.

the low frequency response of an amplifier is sometimes defined by specifying a time constant of so many microseconds. If the time constant is, say 3000 microseconds,

$$RC = 3000 \times 10^{-6} \text{ seconds,}$$

so that

$$f_1 = 1/(2\pi \times 3000 \times 10^{-6}) = 53 \text{ c/s.}$$

Fig. 4.36 shows the attenuation in db plotted against frequency for $R = 1$ megohm and selected values of C. It will be seen that the slope of the attenuation characteristics approaches 6 db per octave (i.e. the attenuation increases by 6 db every time the frequency is halved), and is very close indeed to this value for attenuations beyond 10 db. Each of these characteristics is exactly the same shape as the other, only moved bodily sideways.

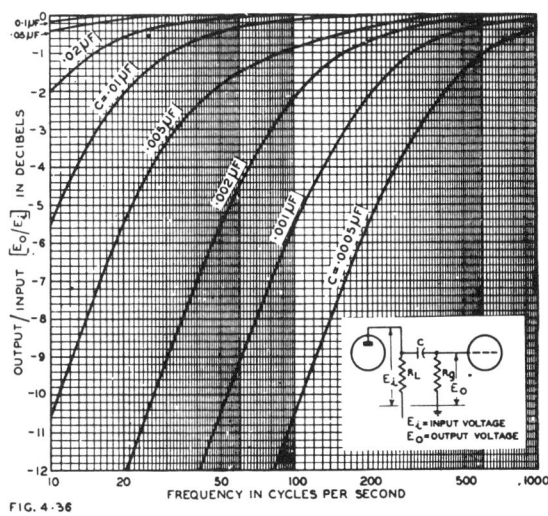

FIG. 4·36

Fig. 4.36. Attenuation in decibels versus frequency for a resistance-capacitance coupling or filter (Fig. 4.35A) in which the total resistance (R) = 1 megohm. If applied to a resistance-coupled valve amplifier, $R = R_g + r_p R_L/(r_p + R_L)$. If r_p is less than 10 000 ohms, the error in neglecting the second term is less than 1%. If the valve is a pentode, R may be taken as $(R_g + R_L)$ with a sufficient accuracy for most purposes.
If $R = 0.5$ megohm, multiply values of C by 2 and similarly in proportion for other resistances.

One of these attenuation characteristics is shown in Fig. 4.36A together with the straight line AB which is the tangent to the curve and has a constant slope of 6 db/octave (or 20 db/decade). The point of intersection between AB and the zero db line is point A which corresponds to the theoretical cut-off frequency f_1. For ease in calculation, the " straight-line " approximate characteristic CAB is sometimes used in calculations in place of the actual attenuation characteristic, the maximum' error being 3 db.

In Fig. 4.35(B) there are two such filters in cascade and further sections may also be added. A two-section filter in which $R_1 = R_2$ and $C_1 = C_2$ will have somewhat more than twice the attenuation in decibels of a single section filter, and the ultimate slope of the attenuation characteristic will approach 12 db/octave.

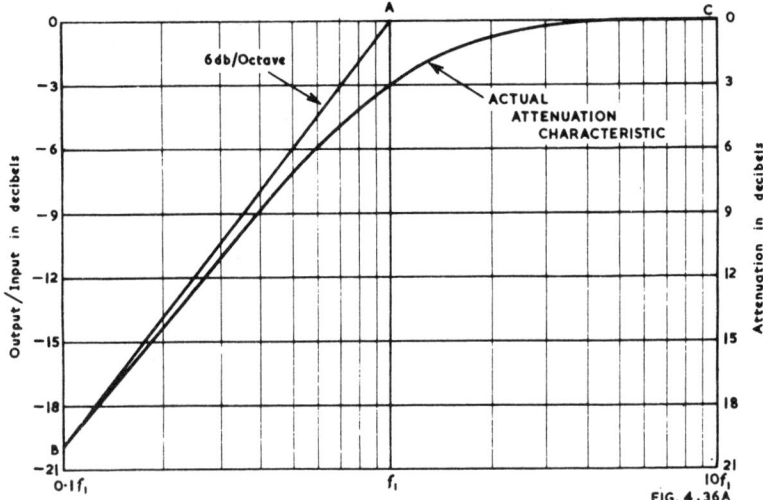

Fig. 4.36A. Actual and approximate attenuation characteristics, using the theoretical cut-off frequency f_1 as the reference frequency.

If C in diagram (A) is shunted by a resistance R', as in Fig. 4.35(C), the effect is to limit the attenuation to that given by R' and R as a voltage divider, i.e., $E_o/E_i = R/(R + R')$. The shape of the attenuation characteristic at low values of attenuation is very little affected by R'.

The reactance of C in Fig. 4.35A causes a phase difference between E_o and E_i, as shown by the vector diagram (D). The phase angle is given by

$$\cos \phi = E_o/E_i \qquad (3)$$

For an attenuation of 3 db, $E_o/E_i = 0.707$ and $\phi = 45°$; thus E_o leads E_i by 45°.

Fig. 4.37 shows a typical **r.c. low-pass filter,** as used for tone control, decoupling in multistage amplifiers, or smoothing filters for power supplies when a large voltage drop in the filter is permissible. It is readily seen that this is the same as Fig. 4.35A except that R and C have been interchanged. The theoretical cut-off frequency f_1, at which $X_c = R$, is therefore unchanged and equal to $1/(2\pi RC)$. Provided that the generator impedance is zero, and that there is no load across the output terminals, the ratio of output to input voltages is given by

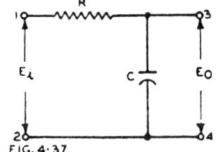

Fig. 4.37. Is a low pass resistance-capacitance filter

$$\frac{E_o}{E_i} = \frac{jX_c}{R + jX_c} \text{ where } X_c = -1/2\pi fC$$

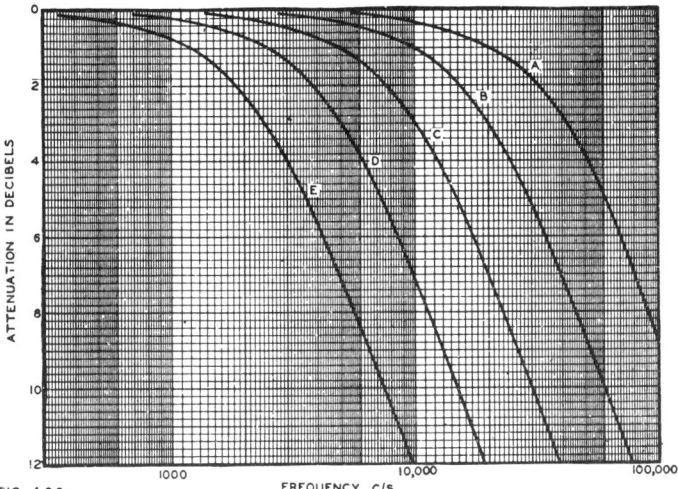

FIG. 4·38

Fig. 4.38.   *Shows the attenuation characteristics for the low-pass filter of Fig.* 4.37
for the values of R *and* C *shown below:*

R (ohms)	C ($\mu\mu F$)				
	Curve A	B	C	D	E
400,000	10	20	40	80	160
200,000	20	40	80	160	320
100,000	40	80	160	320	640
50,000	80	160	320	640	1280
20,000	200	400	800	1600	3200
10,000	400	800	1600	3200	6400

Therefore
$$\left|\frac{E_o}{E_i}\right| = \frac{X_c}{\sqrt{R^2 + X_c{}^2}} = \frac{1}{\sqrt{1 + (R/X_c)^2}} \qquad (4)$$

If the generator has a resistance R_g, then R in the eqn. (4) should include R_g.

Attenuation characteristics derived from this equation are given in Fig. 4.38 in a form capable of adaptation to most problems involving limited attenuation.

The rate of attenuation approaches 6 db/octave, and the shape of the curve is a mirror-image of that for the high-pass filter (Figs. 4.35A and 4.36).

If the filter is used as a smoothing filter, the ratio E_o/E_i is frequently 0.1 or less and under these conditions (with a maximum error of 1%)—

$$\left|\frac{E_o}{E_i}\right| \approx \frac{1}{2\pi fRC} \qquad (5)$$

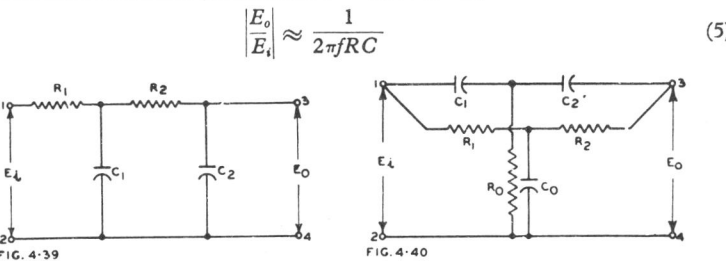

Fig. 4.39 *is a two section low pass resistance-capacitance filter.*
Fig. 4.40 *is a Parallel T Network for the complete elimination of a particular input frequency from the output voltage.*

If additional filtering is required, it is possible to use two or more filter sections (Fig. 4.39). Provided that $R_1 > 10X_{c1}$; $R_2 > 10X_{c1}$ and $R_2 > 10X_{c2}$, the voltage ratio is given (with a maximum error of 3%) by

$$\left|\frac{E_o}{E_i}\right| \approx \frac{1}{40f^2 \times R_1 R_2 C_1 C_2} \tag{6}$$

(iii) Special types of resistance-capacitance filters

An important method of eliminating a particular input frequency from the output voltage is the **Parallel T Network** as in Fig. 4.40. In the usual symmetrical form*,
$$R_1 = R_2 = 2R_o; \quad C_1 = C_2 = \tfrac{1}{2}C_o; \quad \text{and} \quad R_1 = 1/(2\pi f C_1);$$
infinite attenuation is obtained at a frequency f where
$$f = 1/(2\pi R_1 C_1) \tag{7}$$
provided that the output is unloaded.

The mathematical analysis of the general case has been published†, showing that the unsymmetrical case provides an improvement in the discrimination.

The case with finite generator resistance and terminated into a resistive load has also been analysed‡.

(iv) Iterative impedances of four-terminal networks

A four-terminal network or filter is shown in Fig. 4.41 with a generator E_g, having a series generator impedance Z_g, applied to the input terminals, 1, 2, and a load impedance Z_L across the output terminals 3, 4. The input impedance under these conditions, looking into the network from terminals 1, 2 is shown as Z_{k1}. The output impedance, looking into the network from terminals 3, 4 is shown as Z_{k2}. When Z_L is adjusted to be equal to Z_{k1} we have a special case which is of interest when similar filter sections are to be connected together in a chain so that each section is a load on the one preceding it. Under those conditions we call Z_{k1} an **iterative impedance** of the network

We must also adjust Z_g to be equal to Z_{k2}, this being the second iterative impedance. If the network is a single T section, as in Fig. 4.42, and if $Z_1 = Z_2$, then
$$Z_{k1} = Z_{k2} = \sqrt{Z_1(Z_1 + 2Z_3)} \tag{8}$$
and the two iterative impedances are equal.

If the network is a single L section (as in Fig. 4.42 but with $Z_2 = 0$) then the iterative impedances are given by

$$Z_{k1} = \sqrt{\frac{Z_1^2}{4} + Z_1 Z_3} + \frac{Z_1}{2} \tag{8a}$$

$$Z_{k2} = \sqrt{\frac{Z_1^2}{4} + Z_1 Z_3} - \frac{Z_1}{2} \tag{8b}$$

If we now define a quantity P, called the iterative transfer constant, such that
$$P = \log_\epsilon (I_1/I_2) \tag{9a}$$
where I_1 = input current
and I_2 = output current
the formulae for the L section may be written in the alternative forms

$$Z_{k1} = \frac{Z_1 \epsilon^P}{\epsilon^P - 1} = Z_3(\epsilon^P - 1) \tag{9b}$$

$$Z_{k2} = Z_{k1}\epsilon^{-P}$$

where $\cosh P = 1 + \frac{Z_1}{2Z_3} = \frac{Z_1 + 2Z_3}{2Z_3}$.

(See Chapter 38 Sect. 21 Table 73 for hyperbolic sines, cosines and tangents.)

*Scott, H. H., " A new type of selective circuit and some applications." Proc. I.R.E. 26.2 (Feb. 1938) 226.
†Wolf, A. " Note on a parallel-T resistance-capacitance network," Proc. I.R.E. 34.9 (Sept. 1946) 659. See also Hastings, A. E. " Analysis of a resistance-capacitance parallel-T network and applications," Proc. I.R.E. 34.3 (March 1946) 126P ; McGaughan, H. S. " Variation of an RC parallel-T null network," Tele-Tech. 6.8 (Aug. 1947) 48.
‡Cowles, L. C. " The parallel-T resistance-capacitance network " Proc. I.R.E. 40.12 (Dec. 1952) 1712.

For a T network with iterative impedances,

$$P = \cosh^{-1}[(Z_1 + Z_2 + 2Z_3)/2Z_3] \qquad (10)$$

If a number of networks are connected in a chain, with each section having the same two iterative impedances Z_{k1} and Z_{k2} as Fig. 4.43, we may regard the combined networks as equivalent to a single network having iterative impedances Z_{k1} and Z_{k2} respectively. The iterative transfer constant (P) is then given by

$$P = (A_1 + A_2 + A_3) + j(B_1 + B_2 + B_3) \qquad (11)$$

where P_1 = propagation constant of the first section of the network
$\qquad = A_1 + jB_1$
$\qquad P_2 = A_2 + jB_2$ etc.
$\qquad A_1$ = attenuation constant (nepers)
and $\qquad B_1$ = phase constant (radians).

(v) Image impedances and image transfer constant of four-terminal networks

An alternative way of expressing the constants of a network is by the use of image impedances. A T network is shown in Fig. 4.44 with generator and load impedances Z_{I1} and Z_{I2} respectively. It is possible to adjust Z_A, Z_B and Z_C to give an input impedance (looking into the network from terminals 1, 2 with load connected) equal

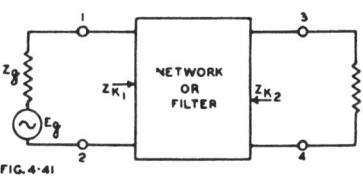

Fig. 4.41 shows a four terminal network with generator and load.

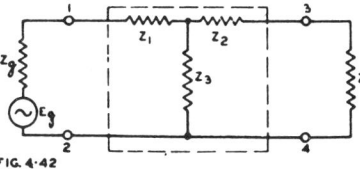

Fig. 4.42 is a single T section network applicable to Fig. 4.41.

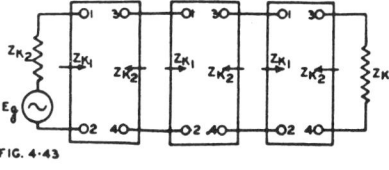

Fig. 4.43 is a combination of three networks in cascade, with iterative impedance relationships.

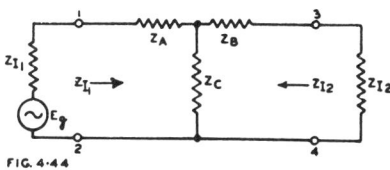

Fig. 4.44 is a single T section network terminated in its image impedances.

to Z_{I1}, and at the same time to give an output impedance (looking from terminals 3, 4 with generator connected) equal to Z_{I2}. Under these conditions the impedance on each side of terminals 1, 2 is an " image " of the other (since they are both identical) and similarly with the impedances on each side of terminals 3, 4. The two image impedances are given by

$$Z_{I1} = \sqrt{\frac{(Z_A + Z_C)(Z_A Z_B + Z_A Z_C + Z_B Z_C)}{(Z_B + Z_C)}} \quad (12a)$$

$$Z_{I2} = \sqrt{\frac{(Z_B + Z_C)(Z_A Z_B + Z_A Z_C + Z_B Z_C)}{(Z_A + Z_C)}} \quad (12b)$$

Similarly a Π network as shown in Fig. 4.44A may have the values of the elements adjusted so that the network is terminated in its image impedances. The two image impedances are then given by

$$Z_{I1} = Z_1 \sqrt{\frac{(Z_2 + Z_3)Z_3}{(Z_1 + Z_3)(Z_1 + Z_2 + Z_3)}} \quad (13a)$$

$$Z_{I2} = Z_2 \sqrt{\frac{(Z_1 + Z_3)Z_3}{(Z_2 + Z_3)(Z_1 + Z_2 + Z_3)}} \quad (13b)$$

The two image impedances may also be expressed in terms of the open-circuit and short-circuit impedances (Sect.7(ix))

$$Z_{I1} = \sqrt{Z_{01}Z_{s1}} \quad (14)$$

$$Z_{I2} = \sqrt{Z_{02}Z_{s2}} \quad (15)$$

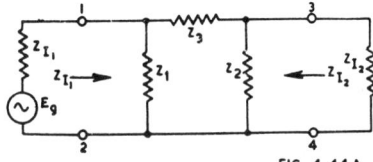 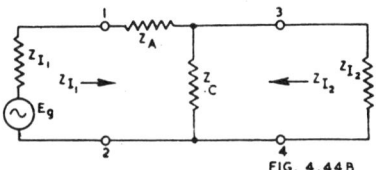

FIG. 4.44A   FIG. 4.44B

Fig. 4.44A is a single Π section network terminated in its image impedances.
Fig. 4.44B is a single L section network terminated in its image impedances.

The transfer of power is indicated by the **image transfer constant** θ whose value is given by

$$\theta = \tfrac{1}{2} \log_\epsilon \frac{E_1 I_1}{E_2 I_2} = \log_\epsilon \frac{I_1}{I_2}\sqrt{\frac{Z_{I1}}{Z_{I2}}} = \log_\epsilon \sqrt{\frac{I_1}{I_2} \cdot \frac{I_2'}{I_1'}} \quad (16)$$

provided that the network is terminated in its image impedances,
where E_1 and I_1 are voltage and current at terminals 1, 2
          E_2 and I_2 are voltage and current at terminals 3, 4
          I_1' and I_2' are currents at terminals 1, 2 and 3, 4
respectively with transmission in the reversed direction.

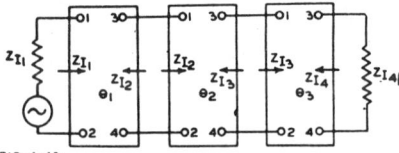

FIG. 4·45

Fig. 4.45 is a combination of three networks in cascade, with image impedance relationships.

When every section of a filter is working between its image impedances, there are no reflection effects. Fig. 4.45 shows a three section group connected on an image impedance basis. This is equivalent to a single network having image impedances Z_{I1} and Z_{I4} respectively,
and $\theta = (\alpha_1 + \alpha_2 + \alpha_3) + j(\beta_1 + \beta_2 + \beta_3) = \alpha + j\beta \quad (16a)$
where $\theta_1 = \alpha_1 + j\beta_1$, etc.

The real part $(\alpha_1 + \alpha_2 + \alpha_3 = \alpha)$ of the image transfer constant θ is called the **image attenuation constant,** and the imaginary part $(\beta_1 + \beta_2 + \beta_3 = \beta)$ is called **the image phase constant.**

Where the values of the elements of the T and Π networks are known, the value of the image transfer constant is given by
T Network (Fig. 4.44):

$$\tanh \theta = \sqrt{\frac{(Z_A Z_B + Z_A Z_C + Z_B Z_C)}{(Z_A + Z_C)(Z_B + Z_C)}} \tag{16b}$$

Π Network (Fig. 4.44A):

$$\tanh \theta = \sqrt{\frac{Z_3(Z_1 + Z_2 + Z_3)}{(Z_1 + Z_3)(Z_2 + Z_3)}} \tag{16c}$$

If the image impedances and transfer constant are known, the impedances of T, Π and L networks are given by
For T Section (Fig. 4.44):

$$Z_C = \frac{\sqrt{Z_{I1} Z_{I2}}}{\sinh \theta} \tag{17a}$$

$$Z_B = \frac{Z_{I2}}{\tanh \theta} - Z_C \tag{17b}$$

$$Z_A = \frac{Z_{I1}}{\tanh \theta} - Z_C \tag{17c}$$

For a Π Section (Fig. 4.44A):

$$Z_3 = \sqrt{Z_{I1} Z_{I2}} \sinh \theta \tag{17d}$$

$$Z_2 = \frac{1}{\dfrac{1}{Z_{I2} \tanh \theta} - \dfrac{1}{Z_3}} \tag{17e}$$

$$Z_1 = \frac{1}{\dfrac{1}{Z_{I1} \tanh \theta} - \dfrac{1}{Z_3}} \tag{17f}$$

For an L Section (Fig. 4.44B):

$$Z_A = \sqrt{Z_{I1}(Z_{I1} - Z_{I2})} \tag{17g}$$

$$Z_C = Z_{I2}\sqrt{\frac{Z_{I1}}{Z_{I1} - Z_{I2}}} \tag{17h}$$

$$\cosh \theta = \sqrt{(Z_{I1}/Z_{I2})} \tag{17i}$$

(vi) Symmetrical networks

When a network is symmetrical, that is when it may be reversed in the circuit with respect to the direction of propagation without alterations in the voltages and currents external to the network, the two iterative impedances become equal to each other and to the two image impedances:

$$Z_K = Z_{K1} = Z_{K2} = Z_I = Z_{I1} = Z_{I2} \tag{18}$$

(this is sometimes called the characteristic impedance)
Also $P = \theta = \alpha + j\beta$ \hfill (19)
where α = image attenuation constant
and β = image phase constant.

(vii) " Constant k " filters

A " constant k " filter is one in which

$$Z_1 Z_2 = k^2 \tag{20}$$

where Z_1 and Z_2 are the two arms of a filter section, and k is a constant, independent of frequency. Fig. 4.46 shows a symmetrical T type section terminated in its image impedances, which is a " constant k " filter provided that $Z_1 Z_2 = k^2$. The constant k^2 has the dimensions of a resistance squared, so that we replace k^2 by R^2 in the following analysis. This requirement is fulfilled when Z_1 and Z_2 are reciprocal reactances; the simplest case is when Z_1 is a capacitance with zero resistance and Z_2 an inductance with zero resistance or vice versa. Some popular combinations are given below :

Original (Z_1)	Reciprocal (Z_2)
L	$C = L/R^2$
$L + r$ in series	$C = L/R^2$ in shunt with R^2/r
$C + r$ in series	$L = R^2C$ in shunt with R^2/r
$L + C$ in series	$C' = L/R^2$ in shunt with $L' = R^2C$
$r + L + C$ in series	R^2/r in shunt with $C' = L/R^2$
	and with $L' = R^2C$

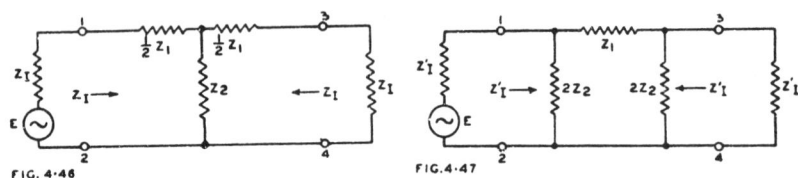

F IG. 4·46   FIG. 4·47

Fig. 4.46 is a symmetrical T section terminated in its image impedances.
Fig. 4.47 is a symmetrical Π section terminated in its image impedances.

Z_I is called the **mid-series image impedance** in a symmetrical T section (Fig. 4.46) while Z_I' is called the **mid-shunt image impedance** in a symmetrical Π section (Fig. 4.47) :

T section : mid-series image impedance

$$Z_I = \sqrt{Z_1Z_2 + (Z_1^2/4)} = R\sqrt{1 + (Z_1/2R)^2} \tag{21}$$

Π section : mid-shunt image impedance

$$Z_I' = \sqrt{\frac{Z_1Z_2}{1 + Z_1/4Z_2}} = \frac{R}{\sqrt{1 + (Z_1/2R)^2}} \tag{22}$$

Therefore $Z_IZ_I' = R^2 = Z_1Z_2$ (23)

If Z_1 is a pure reactance (X_1),

then $Z_I = R\sqrt{1 - (X_1/2R)^2}$ (24)

and $Z_I' = \dfrac{R}{\sqrt{1 - (X_1/2R)^2}}$ (25)

The image transfer constant θ for either T or Π sections is given by

cosh $\theta = 1 + (Z_1/2Z_2)$ (25a)

Half sections have exactly half the image transfer constant of a full section.

In the pass band : $\alpha = 0$ (25b)

cos $\beta = 1 + (Z_1/2Z_2)$ (25c)

In the stop band : cosh $\alpha = | 1 + (Z_1/2Z_2) |$ (25d)

Phase shift $= 0$ or $\pm 180°$ (25e)

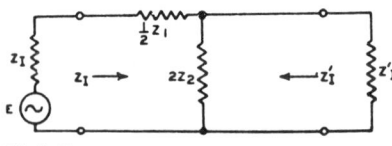

Fig. 4.48 is a half section terminated
in its image impedances.

F IG. 4·48

Fig. 4.48 is a half-section terminated in its image impedances, which in this case are unequal. Two such half-sections, with the second one reversed left-to-right, are equivalent to a single T section.

$$Z_I = \sqrt{Z_1Z_2 + (Z_1^2/4)} \tag{25f}$$

$$Z_I' = \sqrt{\frac{Z_1Z_2}{1 + (Z_1/4Z_2)}} \tag{25g}$$

An ideal filter in which the reactances have zero loss has zero attenuation for all frequencies that make $(Z_1/4Z_2)$ between 0 and -1 ; this range of frequencies is called the **pass band**. All other frequencies are attenuated and are said to lie in the **stop band** (or attenuation band) of the filter.

Low pass filter—constant k type

Fig. 4.49 shows three forms of simple low-pass filters of the constant k type. In each case

$$Z_1 = j\omega L$$

and $Z_2 = 1/j\omega C$.

Now $Z_1 Z_2 = R^2$, therefore $L/C = R^2$, where R may have any convenient value. The mid-series image impedance is

$$Z_I = R\sqrt{1 - (\omega L/2R)^2} = R\sqrt{1 - (f/f_0)^2} \tag{26}$$

where $f_0 = 1/\pi\sqrt{LC}$, $f_0 =$ cut-off frequency.

When $f = f_0$, $Z_I = 0$ and there is infinite attenuation if L and C have no resistance. The mid-shunt image impedance is

$$Z_I' = \frac{R}{\sqrt{(1 - (\omega L/2R)^2}} = \frac{R}{\sqrt{1 - (f/f_0)^2}} \tag{27}$$

When $f = f_0$, Z_I' becomes infinite.

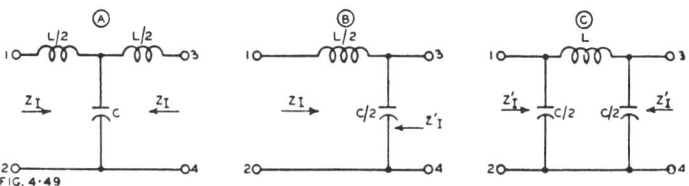

Fig. 4.49. *Three varieties of low-pass constant k filters*: (A) T section (B) Half sèction (C) Π section.

With both T and Π arrangements, the ideal filter has zero attenuation for frequencies less than f_0, a sharp cut-off at f_0, and a very rapid attenuation immediately above f_0. However the rate of attenuation gradually falls as the frequency is increased, and approaches 12 db/octave for the single section at frequencies much greater than f_0.

Both image impedance characteristics are purely resistive below f_0 and purely reactive at higher frequencies.

The **phase shift** varies from zero at zero frequency to $180°$ at f_0, but is constant at $180°$ for all frequencies higher than f_0.

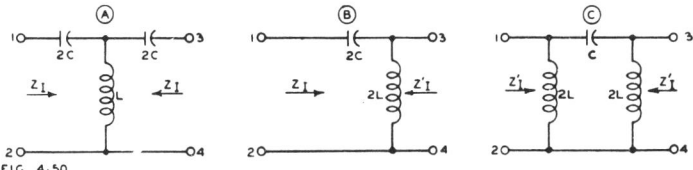

Fig. 4.50. *Three varieties of high-pass constant k filters* (A) T section (B) Half section (C) Π section.

High-pass filter—constant k type

Fig. 4.50 shows three forms of simple high-pass filters of the constant k type In each case

$$Z_1 = 1/j\omega C$$
$$Z_2 = j\omega L.$$

Now $Z_1 Z_2 = R^2$, therefore $L/C = R^2$, where R may have any convenient value. The mid-series image impedance is

$$Z_I = R\sqrt{1 - (1/2R\omega C)^2} = R\sqrt{1 - (f_0/f)^2} \tag{28}$$

where $f_0 = 1/(4\pi\sqrt{LC})$, $f_0 =$ cut-off frequency.

The mid-shunt image impedance is

$$Z_I = \frac{R}{\sqrt{1 - (1/2R\omega C)^2}} = \frac{R}{\sqrt{1 - (f_0/f)^2}} \tag{29}$$

With both T and Π arrangements, the ideal filter has zero attenuation for frequencies greater than f_0, a sharp cut-off at f_0, and a very rapid attenuation immediately below f_0. However, the rate of attenuation gradually falls as the frequency is decreased, and approaches 12 db/octave for the single section for frequencies much less than f_0.

The **phase shift** varies from nearly zero at very high frequencies to 180° at f_0, but is constant at 180° for all frequencies lower than f_0.

Both image impedance characteristics are purely resistive above f_0, and purely reactive at lower frequencies.

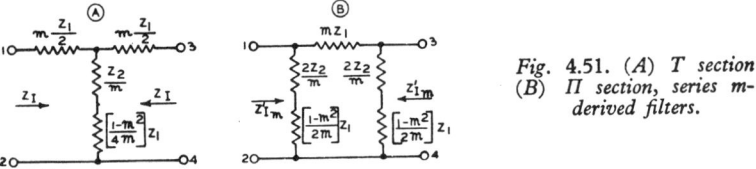

FIG. 4.51

Fig. 4.51. (A) T section (B) Π section, series m-derived filters.

(viii) M-derived filters

This is a modified form of the constant k filter. Fig. 4.51A shows a T section series m-derived filter, which has the same value of Z_I as its prototype (Fig. 4.46 and eqn. 21). In fact, when $m = 1$, it becomes the prototype. In general, m may have any value between 0 and 1. It may be joined, at either end, to a constant k or m-derived section, or half-section, having an image impedance equal to Z_I. This derived filter has the same pass band and cut-off frequency f_0 as the prototype, but different attenuation characteristics with sharper cut-off and infinite attenuation at the resonant frequency of the shunt arm, provided that the elements have zero loss.

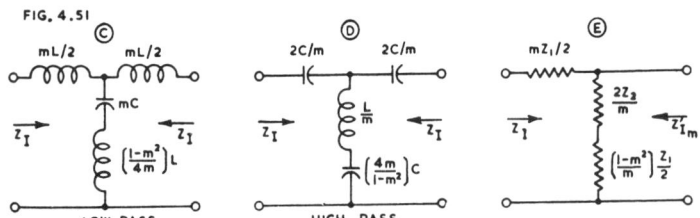

Fig. 4.51. (C) Low pass (D) High pass T section m-derived filters (E) m-derived half-section for matching purposes.

Values of Z_1 and Z_2 are as for the prototype (constant k) filter. A low-pass T section series m-derived filter is shown in Fig. 4.51C, and an equivalent high-pass section in Fig. 4.51D. In both cases the shunt arm becomes resonant at a frequency $f\infty$ given by

$$\text{Low pass} \quad f\infty = \frac{1}{\pi\sqrt{(1 - m^2)LC}} \tag{30}$$

$$\text{High pass} \quad f\infty = \frac{\sqrt{1 - m^2}}{4\pi\sqrt{LC}} \tag{31}$$

In the theoretical case when the reactances have zero loss, the shunt arm will have zero impedance and therefore infinite attenuation at frequency $f\infty$.

The cut-off frequency f_0 is given by

$$f_0 = 1/(4\pi\sqrt{LC}) \tag{32}$$

and the following relationships hold.

Low pass   **High pass**

$$f\infty = f_0/(\sqrt{1 - m^2}) \qquad\qquad f\infty = f_0\sqrt{1 - m^2} \qquad\qquad (33)$$

$$m = \sqrt{1 - (f_0/f\infty)^2} \qquad m = \sqrt{1 - (f\infty/f_0)^2} \qquad (34)$$

The frequency of "infinite" attenuation (sometimes called the peak attenuation frequency) may be controlled by varying the value of m, which variation does not affect the image impedances.

It is generally desirable, for good attenuation characteristics, that the ratio of the cut-off frequency to the frequency of peak attenuation in a high-pass filter should be as high as possible, and not less than 1.25, and in a low pass filter it should be as low as possible and not greater than 0.8. The value of m as given by eqn. (34) is 0.6 when the ratio is 1.25 or 0.8 respectively.

The equivalent Π section series m-derived filter is shown in Fig. 4.51B but here the mid-shunt image impedance is different, and this section cannot be connected at either end to a constant k or m-derived T section except through a half-section to match the respective image impedances.

$$Z_{Im}' = Z_I' \left[1 + (1 - m^2)(Z_1/2R)^2\right] \qquad (35)$$

In other respects (B) has the same characteristics as (A). An m-type half-section for matching sections having different image impedances is Fig. 4.51E, the values shown being for matching a constant k section on the left and a series m-derived Π section on the right. The value of m should be approximately 0.6 to provide the most nearly constant value of image impedance in the pass band.

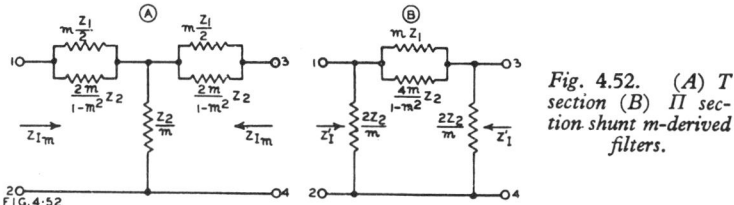

Fig. 4.52. (A) T section (B) Π section shunt m-derived filters.

FIG.4·52

Two forms of shunt m-derived filters are shown in Fig. 4.52. The Π section (B) may be joined at either end to a constant k section or half-section of mid-shunt image impedance Z_I'. The T section requires the medium of a half-section to match the impedances before being so connected.

(A) $Z_{Im} = \dfrac{Z_I}{1 + (1 - m^2)(Z_1/4Z_2)}$ (36)

(B) Z_I' is as for Fig. 4.47 (Eqn. 22)

Therefore $Z_{Im}Z_{Im}' = Z_IZ_I' = R^2$ (37)

•The design of multiple-section filters

A multiple-section filter is made up from any desired number of intermediate sections (either T or Π), and usually of the m-derived type, together with a terminal half-section at each end. All sections in the filter are matched at each junction on an image-impedance basis. The intermediate sections usually have different values of m such that frequencies which are only slightly attenuated by one section are strongly attenuated by another. The image impedances of these sections are far from constant over the pass band of the filter ; hence the necessity for using suitable terminal half-sections.

The terminal half-sections should be designed with a value of m approximately equal to 0.6 to provide the most nearly constant image impedance characteristics in the pass band ; with this value of m the image impedance is held constant within 4% over 90% of the pass band. Each of the terminal end sections should, however, be so designed that its frequency of peak attenuation is staggered with respect to the other, and the intermediate sections should then be staggered for the best overall attenuation characteristic.

The total attenuation or phase shift of the combined multi-section filter would be given by the sum of the attenuations or phase shifts respectively of the individual sections or half-sections.

When the design has been completed for all the sections in the filter, the practical form will be obtained by neglecting the junction points between sections and by adding together the values of the inductors in series in each arm, and those of the capacitors in parallel in each arm. If capacitors are in series, or inductors in parallel, the total effective value should be calculated and used in the practical form of the filter.

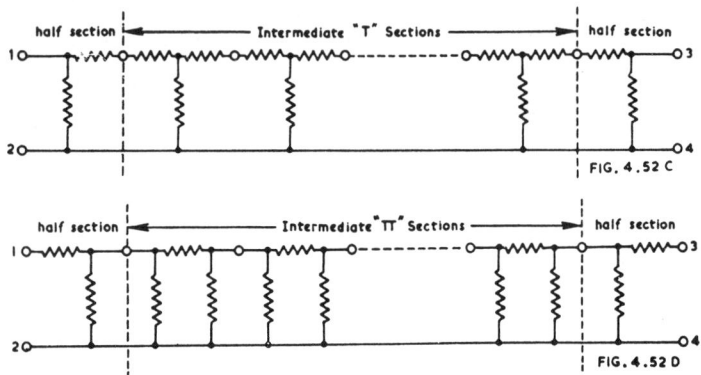

Fig. 4.52.   (C and D) *Method of building up a multi-section filter using* T *or* Π *intermediate sections.*

For further information on the design of m-derived filters, reference may be made to Shea (Ref. 1) pp. 244-285 ; Guillemin (Ref. 4) Vol. 2, pp. 324-338 ; Johnson (Ref. 6) pp. 192-195, 204-228, 293-303 ; Terman (Ref. 2) pp. 226-238 ; Everitt (Ref. 3) pp. 194-214 ; (Ref 6) pp. 130-152 ; (Ref. 7) pp. 6-33 to 6-62.   References are listed in Sect. 8(xi).

(ix) Practical filters

In practice a filter network is usually terminated, not by its image impedance which is a function of frequency, but by a resistance of fixed value R where

$$R = \sqrt{Z_1 Z_2} \qquad\qquad (38)$$

As a result, the impedance mismatching causes some attenuation in the pass-band, which is increased further by the unavoidable losses in the inductors. Moreover, the attenuation in the attenuation-band is less than for the ideal case, and of course never reaches infinity except in the case of a null network.

(x) Frequency dividing networks

Frequency dividing networks are of two types, the filter type (which has only approximately constant input impedance) and the constant-resistance type.

With either type, the **cross-over frequency** (f_c) is the frequency at which the power delivered to the two loads is equal. This occurs with an attenuation of 3 db for each load, with an ideal dividing network having no loss.

The nominal **attenuation per octave** beyond the crossover frequency may be :

6db : available with constant-resistance, but attenuation not sufficiently rapid for general use with loudspeaker dividing networks (Fig. 4.53A)

$$L_0 = R_0/(2\pi f_c) \qquad\qquad C_0 = 1/(2\pi f_c R_0)$$

12db : available with either type, and very suitable for general use with loudspeaker dividing networks.   Fig. 4.53B shows the constant-resistance type.   This is a very popular arrangement.

$$L_1 = R_0/(2\sqrt{2}\pi f_c) \qquad\qquad L_2 = R_0/(\sqrt{2}\pi f_c)$$
$$C_1 = 1/(\sqrt{2}\pi f_c R_0) \qquad\qquad C_2 = 1/(2\sqrt{2}\pi f_c R_0)$$

The filter type has identical circuit connections but the condensers and inductors are unequal (see Chapter 21 Sect. 3).

18db : available with filter type.   This is the maximum rate of attenuation normally used with loudspeaker dividing networks (Fig. 4.53C).

$$L_4 = R_0/(2\pi f_c) \qquad\qquad L_3 = (1 + m)R_0/(2\pi f_c)$$
$$L_6 = 2R_0/(2\pi f_c) \qquad\qquad L_4 = R_0/(2\pi f_c)$$
$$L_7 = R_0/(1 + m)(2\pi f_c) \qquad L_5 = R_0/(4\pi f_c)$$
$$C_5 = 1/(2\pi f_c R_0) \qquad\qquad C_3 = 1/(\pi f_c R_0)$$
$$C_6 = 1/(4\pi f_c R_0) \qquad\qquad C_4 = 1/(1 + m)(2\pi f_c R_0)$$
$$C_7 = (1 + m)/(2\pi f_c R_0) \qquad C_5 = 1/(2\pi f_c R_0)$$
$$(L \text{ in henrys} ; \ C \text{ in farads})$$

See also Chapter 21 Sect. 4.

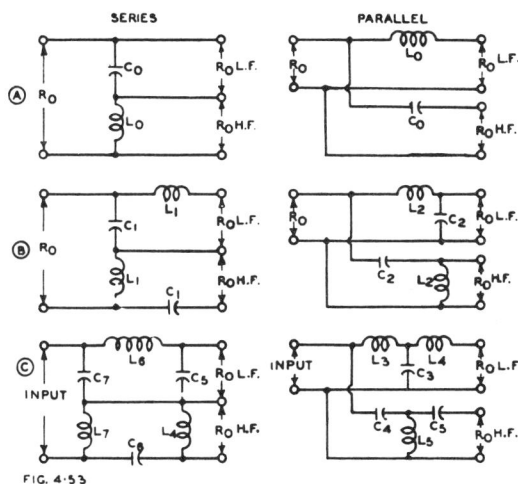

FIG. 4·53

Fig. 4.53. Frequency dividing networks (A) 6db (B) 12db (C) 18db for the octave beyond the cross-over frequency. L.F. indicates low frequency, H.F. indicates high frequency speakers.

(xi) References to filters

Many textbooks including—

1. Shea, T. E. " Transmission Networks and Wave Filters " (D. van Nostrand Company, New York, 1943).

2. Terman, F. E. " Radio Engineers' Handbook " (McGraw-Hill Book Company, New York and London, 1943).

3. Everitt, W. L. " Communication Engineering " (McGraw-Hill Book Co. Inc. New York and London, 1937).

4. Guillemin, E. A. " Communication Networks " Vols. 1 and 2 (John Wiley and Sons Inc. New York ; Chapman and Hall Ltd. London, 1931).

5. Johnson, K. S. " Transmission Circuits for Telephonic Communication " (D. van Nostrand Co. Inc. New York, 1931).

6. " Reference Data for Radio Engineers " (Federal Telephone and Radio Corporation, 3rd ed. 1949).

7. Pender, H., and K. McIlwain " Electrical Engineers' Handbook : Electric Communication and Electronics " (John Wiley and Sons, Inc. New York ; Chapman and Hall Ltd., London, 1950).

Charts for the prediction of audio-frequency response :

Crowhurst, N. H. " The prediction of audio-frequency response " Electronic Eng. No. 1—Circuits with single reactance element, 23.285 (Nov. 1951) 440.   No. 2—Circuits with two reactive elements, 23.286 (Dec. 1951) 483 ; 24.287 (Jan. 1952) 33 ; 24.288 (Feb. 1952) 82.   No. 3—Single complex impedance in resistive network, 24.291 (May 1952) 241.   No. 4—Step circuits 24.293 (July 1952) 337.

See Supplement for additional references.

SECTION 9 : PRACTICAL RESISTORS, CONDENSERS AND INDUCTORS

(i) Practical resistors (ii) Practical condensers (iii) Combination units (iv) Practical inductors (v) References to practical resistors and condensers.

(i) Practical resistors

Resistors are in two main groups, wire-wound and carbon, although each group is subdivided. Wire-wound resistors are available with ordinary and non-inductive windings. Nichrome wire is suitable for operation at high temperatures but has a large temperature coefficient ; it is suitable for use with $\pm 5\%$ or 10% tolerances. Advance (or constantan, or eureka) is limited in its operating temperature, but is used for tolerances of about $\pm 1\%$. Wire-wound resistors are also graded by the type of coating material.

Carbon resistors are divided into insulated and non-insulated types, while the resistance material may be composition or cracked carbon (high stability). The resistance material may be either solid (e.g. rod) or in the form of a film.

For American and English standard specifications on resistors see Chapter 38 Sect. 3(i) and (ii), including standard resistance values. For Colour Codes see Chapter 38 Sect. 2.

(a) Tolerances

Every resistor has tolerances in resistance, and the price increases as the percentage tolerance is made smaller. Composition resistors are usually obtainable with the following tolerances :

$\pm$ 5% For critical positions only

$\pm$ 10% Desirable for semi-critical use in radio receivers and amplifiers (e.g., plate, screen and bias resistors)

$\pm$ 20% For non-critical positions only (e.g. grid resistors).

" High stability " carbon resistors are available with resistance tolerances of $\pm 5\%$, $\pm 2\%$ and $\pm 1\%$ (Ref. A26).

Wire-wound resistors are available with almost any desired tolerances in resistance ($\pm 5\%$, 10% are usual values in radio receivers).

Comment on tolerances in components

When a manufacturer of resistors or capacitors selects simultaneously for large quantities of each of three tolerances, $\pm 5\%$, $\pm 10\%$ and $\pm 20\%$, there is a distinct possibility that the $\pm 10\%$ tolerance group may be nearly all outside the $\pm 5\%$ tolerances, and therefore in two " channels " differing by more than 10%. Similarly with $\pm 10\%$ and $\pm 20\%$ tolerances. It is therefore good engineering practice to design on the expectation of a large percentage of components lying close to the two limits.

(b) Stability

The resistance of carbon resistors tends to drift with time. Ordinary composition resistors may drift as much as $\pm 2\%$ during storage for 3 months at 70°C and normal humidity (Ref. A27). See also Ref. A35.

Some high stability carbon resistors are limited to a maximum change in resistance of $\pm 0.5\%$ after 3 months' storage at 70°C (Ref. A26).

(c) Dissipation

Composition resistors are usually available with nominal dissipation ratings of $\frac{1}{3}$, $\frac{1}{2}$, 1, 2, 4 and 5 watts (JAN-R-11). The English RIC/113 standard (Ref. A27) includes ratings of 1/10, $\frac{1}{4}$, $\frac{1}{2}$, $\frac{3}{4}$, 1 and 2.5 watts. Other manufacturers produce $\frac{1}{2}$, $\frac{1}{2}$, $1\frac{1}{2}$ and 3 watt ratings.

English **high stability resistors** are available with 1/10, $\frac{1}{8}$, $\frac{1}{4}$, $\frac{1}{2}$, $\frac{3}{4}$, 1, $1\frac{1}{2}$ and 2 watt ratings (Ref. A26).

In accordance with the American R.M.A. Standard and JAN-R-11 characteristics *A*, *B*, *C* and *D* [see Chapter 38 Sect. 3(i)], composition resistors may be used at their maximum ratings with **ambient temperatures** up to 40°C.

The JAN-R-11 characteristic *G* and some English resistors (Refs. A3, A26, A27) may be used with maximum ratings up to an ambient temperature of 70°C. In one case (Ref. A27) higher ratings than the nominal value are permitted below 70°C ambient temperature (see below).

At ambient temperatures greater than the maximum rating, resistors may only be used with reduced dissipation and maximum voltage, in accordance with **derating curves.**

The following is a typical power dissipation derating curve for the temperature limits 40°C and 110°C (JAN-R-11 types, *A*, *B*, *C*, *D*):

Temperature	40°	50°	60°	70°	80°	90°	100°	110°
Dissipation	100%	83%	66%	50%	33%	17%	5%	0

The following is the English derating curve for high stability Grade 1, RIC/112 (Ref. A26):

Temperature	70°	80°	90°	100°	110°	120°	130°	140°	150°
Dissipation	100%	87.5%	75%	62.5%	50%	37.5%	25%	12.5%	0

The following is the English rating and derating curve for Grade 2, RIC/113 (Ref. A27):

Temperature	40°	50°	60°	70°	80°	90°	100°	110°
Dissipation	175%	150%	125%	100%	75%	50%	25%	0

A typical **voltage de-rating curve** is the American R.M.A., see Chapter 38 Sect. 3(i).

It is good engineering practice to select a composition resistor, for any particular application, such that the actual dissipation is about 60% of the resistor rated dissipation, after making allowance for any de-rating due to high ambient temperature.

Wirewound resistors are usually available with nominal dissipation ratings of ¼, ½, 1, 2, 3, 5, 8, 10, 12, 16, 20 watts and upwards. These are also normally available with tappings.

Wirewound resistors, vitreous enamelled, are usually available in dissipation ratings of 5 or 10 watts and upwards. These are also normally available with tappings.

Adjustable voltage dividers are manufactured in 8 or 10 watt dissipation ratings, together with larger sizes.

" **Radio** " **voltage dividers** usually have resistances of 25,000 and 15,000 ohms suitable for connection across 250 volts. The maximum dissipation is something like 5 watts.

(d) Voltage ratings

With low resistances, the applied voltage is always limited by the permissible dissipation. With high resistances there is an additional condition to be met in the maximum voltage rating. In general the maximum voltage is between 250 and 500 volts for ¼ to 1 watt ratings, although there are a few below 250 volts, some (½ watt and over) with maximum voltages of 700 volts or more, and some (¾ watt and over) with maximum voltages of 1000 volts or more.

(e) Temperature rise

A temperature rise of from 40°C to 62°C (with 40°C ambient temperature) is to be expected with ordinary composition resistors at maximum ratings. English resistors which are rated at an ambient temperature of 70°C have a maximum surface temperature not greater than 120°C (Grade 2) or 150°C (Grade 1, high stability) (Refs. A26, A27).

A typical small wire-wound resistor may have a maximum temperature of 110°C for 1 watt or 135°C for 2 watts dissipation.

A typical large bare or organic-coating wire-wound resistor has a maximum surface temperature from 170°C to 220°C.

A typical lacquered wire-wound resistor has a maximum surface temperature of about 130°C.

A typical vitreous-enamelled resistor normally has a maximum surface temperature from 200°C to 270°C, but even higher temperatures are sometimes used.

High dissipation resistors should be mounted vertically to allow good air circulation, and should be spaced well away from other resistors or components. When mounted in a confined space they should be used at about half the rated maximum dissipation.

(f) Effect of temperature on resistance

The resistance of composition resistors always tends to rise as the temperature is reduced below 20° or 25°C, but at higher temperatures the resistance may either rise or fall, or may fall and then rise (for curves see Ref. A14).

For a temperature rise from 25° to 80°C, the change of resistance is usually under $\pm$ 7% for low values of resistance.

For the same rise of temperature, high stability resistors have a change of resistance not greater than $+ 0 - 2.2\%$ for resistances up to 2 megohms (1 and 2 watts, Ref. A26). See also Ref. A36.

The resistance of composition resistors is also affected by a temporary severe change in ambient temperature, followed by return to normal (Ref. A35).

The resistance of a wire-wound resistor generally increases as the temperature is increased, the change being usually not more than 0.025% per °C in the case of low temperature units.

See also Chapter 38 Sect. 3(i) and (ii) for standard specifications.

The effect of soldering

Small ($\frac{1}{4}$ watt) composition resistors are subject to as much as 3% change in resistance, approximately half of which may be permanent, due to soldering ; larger dissipation resistors are usually not affected more than 1% (Ref. A24). The maximum permissible change in resistance due to soldering is $\pm$ 2% (Class 2) or $\pm$ 0.3% (Class 1, high stability) for resistors coming under the English RIC Specifications (Refs. A26, A27).

(g) Effect of voltage on resistance

The resistance of a composition resistor decreases when the voltage applied across it increases. The percentage change increases as the resistance increases. For a 1 megohm resistor, a typical percentage fall in resistance is given by (Erie Resistor Co.) :

Size	$\frac{1}{4}$	$\frac{1}{2}$	1	2	watts
Voltage from zero to	200	350	500	500	volts
Fall in resistance	2.1	2.5	1.3	1.5	%

The voltage coefficient is defined as (Ref. A27) :

$$\text{Voltage coefficient} = \frac{100 (R_1 - R_2)}{R_2 (E_1 - E_2)}$$

where R_1 and E_1 are the resistance and voltage respectively at the normal maximum rating, and R_2 and E_2 are the values at one-tenth of E_1. Limiting values of the voltage coefficient are (Ref. A27) :

0.025% per volt for values below 1 megohm
0.05% per volt for values above 1 megohm.

This effect is much reduced by the use of " high stability " resistors, a typical value being 0.4% fall in resistance for 1 megohm, with voltage change from zero to 500 volts (Dubilier). The limiting value of the voltage coefficient for Specification RIC/112 (Ref. A26) is 0.002% per volt.

Some applications require a resistor whose resistance falls as the applied voltage is increased ; a wide range of characteristics is available (e.g. Carborundum " Globar " ceramic resistors).

See Chapter 38 Sect. 3(i) for standard specifications.

(h) Effect of humidity on resistance

The effect of humidity is to increase the resistance by up to about 3% under normal conditions. Extreme tropical humidity may cause an increase in resistance generally less than 10%. Some insulated resistors have less than 1% change in humidity due to humidity tests (e.g., I.R.C. type BTA). See Chapter 38 Sect. 3(i) for humidity tests and limits.

(i) Capacitance of resistors

Every resistor has a capacitance which, at the lower radio frequencies, may be considered as a capacitance between the two ends of the resistor (usually between 0.1 and 1.0 $\mu\mu$F for composition resistors). This capacitance may usually be neglected in normal applications in radio receivers. At higher frequencies it is necessary to consider the capacitance as being distributed along the resistance element. This leads to a reduction in resistance which, unlike the end-to-end capacitance, is not removable by tuning [see (k) below]. References A5, A6, A7, A8, A9, A14.

(j) Inductance of resistors

Every resistor has an inductance, partly due to the inductance of the resistor itself and partly due to the leads. However, experience shows that at high frequencies the effect of the inductance is negligible compared with the effects of capacitance.

Typical values of inductance of composition resistors (Dubilier) are given below :

Resistance		100 ohms	1 megohm	
Inductance	$\frac{1}{4}$ watt	0.0007	0.06	μH
	1 watt	0.017	2.0	μH

Wire-wound resistors have much greater inductance than composition types, but where they must be used it is possible to adopt a " non-inductive " winding which reduces the inductance to a low value.

(k) Effect of frequency on resistance

Largely as a result of the distributed capacitance effects, the effective resistance of a rod type carbon or composition resistor falls as the frequency is increased. With resistances up to approximately 1 megohm, the theoretical curve of effective resistance plotted against frequency is given in Ref. A29 (Fig. 5). The effective resistance drops to 90% of the d.c. value

when $f = \dfrac{0.3}{C_d R_{d\,c}}$ cycles per second where C_d = total distributed capacitance of rod ;

or as an approximation, assuming $C' \approx C_d/3$, when $f \approx 1/10 C' R_{d\,c}$ cycles per second where C' = equivalent shunt capacitance at low frequencies (this is usually the published capacitance of the resistor).

If C' is measured with the resistor in its operating position in relation to other components and metal parts, the proximity effects will be included.

C' is constant below 4 Mc/s (Ref. A29).

References A4, A5, A6, A7, A8, A9, A14, A29.

We give below some experimental values :

Ratio of effective resistance to d.c. resistance

R (megohms d.c.) $\times f$ (Mc/s)=	0.1	.5	1.0	5	10	20
I.R.C. type BTR ($\frac{1}{2}$W)	0.98	0.93	0.89	0.62	0.46	0.30
I.R.C. type BTA (1W)	0.95	.80	.71	.48	.37	.24
I.R.C. type BT-2 (2W)	0.80	.53	.40	.19	.14	.11
I.R.C. type BTS	1.00	.89	.79	.61	.57	—
I.R.C. type F (lower limit)	—	—	—	.84	.80	.75
Allen Bradley GB-1	0.85	.60	.48	.24	.17	.12
Allen Bradley EB$\frac{1}{2}$	0.90	.68	.57	.46	.23	.15
Speer SCT $\frac{1}{2}$	0.92	.70	.60	.35	.27	.20

Reference A14.

(l) Noise of resistors

All resistors have an inherent minimum noise voltage due to **thermal agitation** (" Johnson noise ") which is given at 30°C (80°F) by

$$e \ = \ 1.29 \ \times \ 10^{-10} \sqrt{\triangle f \times R}$$

where e = r.m.s. noise voltage

$\triangle f$ = bandwidth in c/s of the noise measuring instrument

and R = resistance in ohms.

The thermal agitation noise of ideal resistors at 30°C is tabulated below for a bandwidth of 5 000 c/s.

Resistance	1000	10 000	100 000	1 Meg	8 Meg	ohms
Noise	0.29	0.91	2.9	9.1	25.7	μV

When a current flows through the resistor, there is a small increase in the magnitude of the thermal agitation noise.

In addition to the thermal agitation noise, carbon and metallized resistors also have a noise voltage which is approximately proportional to the direct voltage applied across the resistance.   This has been called **current noise** or **resistance fluctuation noise** (Ref. A25).   The frequency distribution of this additional noise component, unlike thermal agitation noise, is not uniform but its value decreases with increasing frequency from 30 c/s upwards.   The amount of noise varies according to the material and construction of the resistor, and even varies considerably from one resistor to another equivalent type.   The " current noise " voltages of some typical English composition resistors (Ref. A3) are given below :

Resistance	1000	10 000	100 000	1 Meg	8 Meg	ohms
Average noise	0.03	0.18	0.35	0.5	0.6	μV/V
Max. noise	0.13	0.62	1.3	1.8	2.2	μV/V

Noise values up to 20 μV/V have been measured in commercial radio resistors (Ref. A25).   The English Specification RIC/113 (Grade 2) issued June 1950, gives the limit of noise as $\log_{10} R\,\mu$V/V ; this is equivalent to $6\,\mu$V/V for 1 megohm.

Lower " current noise " voltages are produced by high stability cracked carbon resistors (see below) and also by palladium film resistors (Ref. A21).

In addition to the steady " current noise " fluctuations, all carbon composition resistors show abnormal **fluctuations** which do not appear to bear any simple relationship to the steady " current noise " (Ref. A21).

References A2, A3, A4, A13, A14, A15, A16, A17, A18, A19, A20, A21, A25, A32.

(m) High stability cracked carbon resistors (Refs. A10, A26, A28)

These not only have high short and long period stability and close tolerances (up to $\pm$ 1%) but also have low noise, low voltage coefficient and low temperature coefficient.   They have practically no non-linearity of resistance and the inductance is very low except for the higher resistance values.   They are manufactured in England with resistances from 10 ohms to 10 megohms and dissipations from ⅛ to 2 watts. The extremely high resistance values are only obtainable in the higher wattage ratings. See also (b), (e), (f) and (g) above, and Chapter 38 Sect. 3(i) for standards.

The historical development, constructions and special features of cracked carbon resistors, with a very extensive bibliography, are given in Ref. A28.

These resistors are particularly suited for use in low-level high-gain a-f amplifiers, and in r-f applications up to 100 Mc/s.   The inductance varies from 0.001 μH for a small 100 ohm resistor to 2μH for a large (spiral element) 1 megohm resistor.

The I.R.C. deposited carbon resistors have tolerances of $\pm$ 1%, $\pm$ 2% and $\pm$ 5%. The voltage coefficient is approx. 10 parts per million per volt.   The temperature coefficient varies linearly from $-$ 0.025 to $-$ 0.05 (10 MΩ type DCH) or $-$ 0.065 (10 MΩ type DCF).   Maximum dissipations for high stability are ¼ and ½ watt ; when high stability is not essential the values are ½ and 1 watt.

(n) Negative temperature coefficient resistors (Thermistors)

Negative temperature coefficient resistors are sometimes used in a.c./d.c. receivers to safeguard the dial lamps when the set is first switched on.   For example, one such NTC resistor has a resistance of 3 000 ohms cold and only 200 ohms when the heater current of 0.1 ampere is passing through it.   By its use the initial surge on a 230 volt supply may be limited to 0.12 ampere.   See also page 1267.

They are also known as " Varistors."   Refs. A22, A28, A31, A34.

(o) Variable composition resistors ("potentiometers")

Standard variable composition resistors are described in Chapter 38 Sect. 3(viii). In radio receivers and amplifiers these are commonly used as diode load resistances and as volume controls (attenuators). In general it is desirable to reduce to a minimum or eliminate entirely any direct voltage across them, and any current drawn by the moving contact. Whether or not a current is passed through the terminations, noise voltages appear across any two of the three terminations and, so long as the rotor is stationary, the noise does not differ from that of a fixed resistor of equal resistance. However, when the rotor is turned, additional noise is produced which is of the order of 1 or 2 millivolts per volt applied across the extreme terminations for a speed of rotation of one full rotation per second. The noise produced at any point of the track is approximately proportional to the voltage gradient at this point; consequently the rotation-noise is greater over that portion of a logarithmic resistance characteristic where the most rapid change in resistance occurs, than for the other end of the same characteristic or for a linear characteristic. When a logarithmic characteristic is used for volume control, the rotation noise will therefore be much lower at settings for low volume where noise would be most noticeable (Ref. A30).

Some receiver manufacturers avoid the increased rotation noise caused by direct diode current in a volume control by using a fixed diode load resistor with capacitance coupling to a separate volume control having at least four times the resistance of the diode load resistor.

Variable composition resistors are available with a choice of up to 6 tapers; tappings may be provided at 38% and 62% effective rotation (Mallory). They are also available, if desired, with a switch.

(ii) Practical condensers

(a) Summary of characteristics

A condenser has tolerances in the value of its capacitance. For most radio receiver applications, tolerances of ± 20% or even higher may be used. The closest tolerances available with paper dielectric condensers are ± 10% [see Chapter 38 Sect. 3(iii)]. Where closer tolerances are required it is necessary to adopt mica dielectric condensers in which the tolerance may be any standard value between ± 20% and ± 2% [Chapter 38 Sect. 3(v)].

The capacitance changes with frequency, with temperature, and with age.

A condenser has inductance, so that at some high frequency it becomes series-resonant.

A condenser has a.c. resistance, and dissipates energy in the form of heat; the loss is approximately proportional to the square of the frequency and is also affected by the temperature. This energy loss is partly dielectric loss (which predominates at low frequencies) and partly electrode and lead losses (which predominate at higher frequencies). It may be replaced for the purpose of calculation by the "equivalent series resistance."

A condenser thus has a complex impedance, with resistive, capacitive and inductive components. The impedance may be capacitive over one range of frequencies, inductive over another, and resistive at one or more frequencies.

A condenser with a solid or liquid dielectric takes a longer time to charge than an ideal condenser having the same capacitance; this effect is due to "dielectric absorption." When such a condenser is short-circuited it fails to discharge instantaneously —a second discharge may be obtained a few seconds later. As a consequence the capacitance of such a condenser is a function of the duration of the applied direct voltage; when it is used in an a.c. circuit, the capacitance decreases as the frequency rises. This effect is pronounced with paper dielectric, but is very small with mica dielectric condensers.

A condenser has d.c. leakage, and behaves as though it were an ideal condenser with a high resistance shunted across it.

A condenser with a solid dielectric tends to deteriorate during service, and may break down even when it is being operated within its maximum limits.

(b) **The service life of a condenser**

Condensers under voltage may be subject to gradual deterioration and possible breakdown, due to the solid dielectric (if any) and other insulation. This deterioration is much more rapid with some dielectrics than with others, and also varies considerably between different batches from the same factory.

Except for electrolytic condensers, there should be no deterioration with age except while under voltage ; the service life may therefore be taken as the time of operation under voltage.

The maximum temperature of a condenser has a pronounced effect on the service life. In some types, every 10°C rise in temperature causes 50% decrease in life ; other types are less sensitive to temperature.

The safe working voltage of a condenser at a given temperature is much less than the " ultimate dielectric strength "—as low as one tenth of this value in some cases. The " factory test " voltage is intermediate between these two values, but it is no guide to the safe working voltage. The only satisfactory procedure, if long life is essential, is to obtain from the manufacturer the safe working voltage at the proposed temperature of operation.

Condensers for operation on a.c. should have maximum a.c. ratings for safe working voltage, frequency and temperature. The peak operating voltage, whether a.c. or pulse, should be within the maximum voltage rating. The maximum surge voltage (usually during " warming up ") should be within the surge rating, if quoted, or alternatively should not exceed the maximum working voltage by more than 15%.

(c) **Electrolytic condensers***

An electrolytic condenser provides more capacitance in a given space and at a lower cost per microfarad than any other. It is usually manufactured with a capacitance of $4 \mu F$ or more. The capacitance **tolerances** may be -20%, $+100\%$, or -20%, $+50\%$, while JAN-C-62 permits $+250\%$. Unlike other types, electrolytic condensers may only be used on substantially direct voltage, and they must be correctly connected with regard to polarity. Electrolytic condensers are generally used in radio receivers and amplifiers with a steady direct voltage plus an a.c. ripple. The highest rated voltage rarely exceeds 500 V peak, even for use under the most favourable conditions.

The capacitance increases somewhat with **increase of temperature,** and decreases rapidly with temperatures below $-5°C$. The capacitance also decreases with **age**—one dry type shows a 5% decrease in capacitance after 7000 hours operation at 20°C ambient temperature, and a 20% decrease at 40°C.

The **capacitance** at 10 000 c/s is usually less than that at low frequencies. Typical wet types at 10 000 c/s have only from 30% to 50% of the capacitance at 50 c/s. Typical dry types are better in this regard, having capacitances at 10 000 c/s from 42% to 85% of that at 50 c/s, at 20°C. However, the **temperature** has a marked effect on the capacitance versus frequency characteristic. One etched-foil type (Ref. B13) has 42% of its nominal capacitance at 20°C, 95% at 33°C, and 107% at 50°C.

The **series-resistance** of a new condenser at ordinary working temperature is fairly low (not more than 25 ohms for $8 \mu F$ at 20°C, 450 V working) but it rises rapidly at higher temperatures and temperatures below 10°C. The series resistance rises considerably during life and eventually may be the cause of unsatisfactory operation of a receiver.

Among dry electrolytics, those with etched foil anodes are much inferior to those with plain foil electrodes when used for a-f by-passing, owing to their high impedance particularly at the higher frequencies. For example, at 10 000 c/s an ideal 8 μF capacitor has an impedance of approximately 2 ohms, whereas typical plain-foil electrolytics have impedances from 3.5 to 6.5 ohms and those with etched-foil electrodes have impedances from 8 to 22 ohms (Ref. B13).

---

*The following remarks apply to aluminium electrodes. However tantalum electrodes are also used (Ref. B17).

The **dissipation factor** is a function of both frequency and temperature One 10 μF etched-foil capacitor has a dissipation factor of 10% at 100 c/s, 68% at 1000 c/s and 92% at 10 000 c/s, at 20°C. The dissipation factor decreases rapidly with increase of temperature, and at 1000 c/s is 30% at 33°C and 13% at 50°C for the same capacitor (Ref. B13).

When electrolytic condensers are required to be operated across **voltages of more than 450 volts**, two or more condensers may be connected in series but the effective total capacitance will then be half (or less) that of the single unit.   In such a case it is advisable to connect a resistor, say 0.25 megohm, across each capacitor.

Electrolytic condensers have **self-healing properties**—after a momentary surge of over-voltage, resulting in break-down of the dielectric, the electrolytic condenser is more likely to recover than a non-electrolytic type.   Wet electrolytic condensers are very good in this respect.

Electrolytic condensers have an appreciable **leakage current** ; this may be from 0.002 to 0.25 mA per microfarad and varies considerably with the type of condenser and the " working voltage," being higher for higher values of working voltage.   The maximum leakage current in milliamperes permitted by JAN-C-62 is (0.04 × capacitance in microfarads) + 0.3.   This is equivalent to a shunt resistor of 0.7 megohm for an 8 μF condenser, 450 V rating.   At voltages lower than the working voltage, the leakage current falls, but at higher voltages it increases very rapidly.   The leakage current also increases rapidly at higher temperatures.

If an electrolytic condenser is **left idle** for some days, the initial leakage may be quite substantial, but it tends to become normal after a few minutes.   If the condenser has been left idle for several months, the time of recovery is longer.

The **power factor** of an electrolytic condenser may be between 2% and 3% for the best condensers and is usually between 5% and 10%   Some of the older wet types had power factors of 25% or greater.

Electrolytic condensers should not be used in positions where the **ambient temperature** is high or the alternating current component is excessive, otherwise the service life will be short.   Ambient temperatures up to 50°C are always satisfactory while 60°, 65°C or 70°C is permissible for many types and some may be used at higher temperatures (e.g. 85°C).

Special types are available for very low temperatures (Ref. B18).

Electrolytic condensers are in two major groups.

Wet electrolytic condensers have vents, and must be mounted vertically with the vent unobstructed.   They are valuable as first filter condensers in a rectifier system.   Some wet types are used as voltage regulating devices, to limit the peak voltage during the warming-up period.   All wet types have rather greater leakage currents than dry types. **Dry electrolytic condensers** are very widely used as filter and by-pass condensers.   They are inferior to the wet type as regards frequent and severe voltage surges and short period overloads, when they are liable to fail permanently, but are preferable to the wet type in most other respects.   They are manufactured in several forms—plain foil, etched, sprayed or fabricated foil.   Most modern compact units have etched or fabricated foil, but the plain foil type has a lower impedance at radio frequencies.   " Surgeproof " types are available with a safe operating voltage of 450 volts but which have heavy leakage current when the voltage exceeds 500 volts.   This type is able to handle very heavy ripple currents without deterioration.

Reversible dry electrolytic condensers are manufactured, but they have higher leakage currents than standard types.

Electrolytic condensers when used as **first filter condensers** in condenser-input filters require careful consideration.   The d.c. voltage plus the peak value of the ripple voltage must not exceed the rated voltage of the capacitor while the ripple current must not exceed the ripple current rating.   The ripple current may be measured by a low-resistance moving iron, or thermal, meter ; a moving-coil rectifier type of instrument is not suitable.   Alternatively the ripple current may be calculated —see Chapter 30 Sect. 2.

Some typical **ripple current ratings** are given below (T.C.C.).   Plain foil types have a higher ripple current rating than equivalent etched foil types.

Ambient temperature	20°C	40°C	60°C	70°C
8 μF 350 V " micropack " plain foil	148	125	85	32 mA
16 μF 350 V " micropack " plain foil	250	200	110	50 mA
8 μF 450 V etched foil	88	67	33	10 mA
16 μF 450 V etched foil	162	122	62	20 mA
16 μF 450 V plain foil	300	260	160	85 mA
32 μF 450 V plain foil	500	405	230	100 mA

With multiple capacitor units, only one of the units is normally intended for use as the first filter condenser ; see catalogues for identification.

See Chapter 38 Sect. 3(x) for standard ratings.

(d) Paper dielectric condensers

Impregnated paper forms a very useful dielectric, being intermediate between electrolytic and mica condensers as regards cost, size and leakage for a given capacitance. It is usually manufactured in units from 0.001 to 0.5 μF, larger values being built up from several smaller units in parallel in one container. The impregnating material may be resin, wax, oil or a synthetic compound. Some impregnating materials enable condensers to withstand extremely wide temperature ranges (e.g. $-50°$ to $+125°$C—Sprague " Prokar " with plastic impregnant). Waxes may be used for moderate voltages and temperatures, as in radio receivers (from $-30°$C to $+65°$C for R.M.A. Class W). Other impregnants are used for higher temperatures (e.g., 85°C, as R.M.A. Class M ; 100°C as T.C.C. " metalpack " and " metalmite "). The permissible insulation resistance at 25°C is not less than 5000 megohms for capacitances up to 0.15 μF, falling to 1000 megohms for 1 μF, but this falls rapidly at higher temperatures, being 35% of these values at 40°C.

A typical 1 microfarad wax paper condenser designed for audio frequency applications has the following characteristics :

Frequency	1000 c/s	10 000 c/s	100	360	Kc/s
Inductance	0.2	0.2	0.2	0.2	μH
Resistance (effective)	1.1	0.43	0.3	0.25	ohms
Reactance	$-159*$	$-15.9*$	-1.50	0**	ohms
Q	145	37	5	0	
Power factor	0.007	0.027	0.19	1	
Percentage power factor	0.7	2.7	19	100	%

* equal to ideal   ** resonance

Some paper dielectric capacitors are impregnated with a high permittivity wax to reduce the dimensions of the capacitor (Ref. B13). These capacitors have less desirable electrical characteristics than those with normal waxes. At 10 000 c/s the capacitance falls to 89% of its value at 100 c/s ; the capacitance varies from -19% to $+6\%$ as the temperature is varied from $-30°$ to $+70°$C.

Wax-impregnated paper dielectric condensers are sometimes used for grid coupling purposes from a preceding plate at high potential, but plastic-impregnated or mica dielectric is to be preferred on account of leakage. As an example, take a paper condenser with capacitance = 0.01 $\mu\mu$F at 40°C. The minimum insulation resistance will be 5000 $\times$ 0.35 = 1750 megohms. If the grid resistor has a resistance of 1 megohm and the preceding plate voltage is 175 volts, there will be a voltage of 0.1 volt on the grid as the direct result of leakage.

In most other applications the leakage may be neglected entirely.

Plastic (polystyrene) impregnated paper dielectric condensers have a very high insulation resistance, of the order of 500 000 megohms per microfarad, and are much more suitable for use as grid coupling condensers (e.g. T.C.C. Plastapacks, 50 to 5 000 $\mu\mu$F). A test after 9 months' handling under bad climatic conditions showed insulation resistances of 24 000 to 100 000 megohms (Aerovox Duranite, 0.01 to 0.22 μF). These condensers have power factors as low as those of mica condensers, while the temperature coefficient of capacitance is from -100 to -160 parts in a million per °C. For maximum stability they should not be operated above 60°C, but the insulation resistance remains very high even up to 75°C. These are available in capacitances from 100 to 10 000 $\mu\mu$F in tubular form and from 0.02 to 4 μF in rectangular metal boxes (Ref. B11).

Mineral oil is used as an impregnant for working voltages from 1000 to 25 000 volts and operating temperatures from $-30°$ to $+71°$C (T.C.C. " Cathodray "). The insulation resistance of a mineral oil impregnated capacitor is greater than that with petroleum jelly impregnation, in the ratio of 2.5 to 1 at 0°C, rising to 12.5 to 1 at 70°C (Ref. B13).

Paper dielectric condensers are made in two forms—inductive; and non-inductive. The former is limited to a-f applications, while the latter may be used at radio frequencies.

Ordinary paper dielectric capacitors should not be subjected to high a.c. potentials. Special types are produced by some manufacturers for use under these conditions, for example with vibrator power packs and line filters.

For Standard Specifications see Chapter 38 Sect. 3(iii) and (iv).

Metallized paper dielectric condensers

This type utilizes a metal-sprayed or metal-evaporated paper dielectric instead of the more conventional metal foil and paper construction. This construction results in considerable reduction in size, while it also has a partial self-healing property in the case of breakdown. The insulation resistance of unlacquered condensers is quite low—of the order of 100 megohm microfarads—but some of those with a lacquered film have an average insulation resistance as high as 8000 megohm microfarads at 25°C (Refs. B14, B15).

There is a gradual reduction in the insulation resistance due to the self-healing property, the degree depending on the number of punctures. It is desirable for the total circuit resistance to be not less than 500 or 1000 ohms, to reduce the carbonising effect of the arc. However if the circuit resistance is high, there may be insufficient current to clear completely any breakdown, and the insulation resistance may fall. Consequently, this type of condenser should not be used in high impedance circuits without seeking the advice of the manufacturer.

A metallized paper dielectric unit should be used with discretion as the first filter condenser following a thermonic rectifier, since the high peak breakdown current may damage the rectifier unless the circuit resistance is sufficiently high.

The ratio of reactance to resistance (Q) of one lacquered unit with a capacitance of 2 μF is 200 at 0.5 Kc/s, 140 at 2 Kc/s and 60 at 10 Mc/s ; a 0.1 μF unit has $Q = 98$ at 10 Mc/s (Ref. B15).

The inductance may be made very low, and these condensers are very useful for a-f and r-f by-passing. The usual (English) temperature limit is 71°C for d.c. operation, 60°C for a.c. In tubular form (wax-coated) these are available from 0.0001 to 2 μF with voltage ratings 150, 250, 350 and 500 V d.c. (RIC/136). Larger sizes are available with capacitances up to 20 μF (400 V d.c. or 250 V a.c.) and voltages up to 550 V d.c. (4 μF).

Voltage-temperature derating curves for Astron (U.S.A.) are 100% up to 86°C, linearly down to 38% at 120°C for units up to 1 μF ; larger units are 100% up to 76°C, down to 22% at 120°C (Ref. B16).

The effect of 5000 hours' operating life on capacitance is negligible up to 65°C and 8% at 100°C. The effect of the same operation is to increase the power factor from an initial value of 0.5% to 0.6% at 65°C, or 0.8% at 100°C (Ref. B16).

The paper is usually impregnated with wax, although mineral oil has also been used. Mineral wax impregnated units are generally preferred because of their higher breakdown voltage, although their capacitance falls about 10% as the temperature is increased from 50°C to 85°C. Mineral oil impregnated units have more constant capacitance with temperature change.

These condensers are damaged by moisture and the unit is therefore well dried initially and hermetically sealed to prevent the ingress of moisture.

References B10, B12, B14, B15, B16.

Standard Specifications—Chapter 38 Sect. 3(ix).

(e) Mica dielectric condensers

Mica has very high electrical stability and very low a.c. loss. It also permits the manufacture of condensers with close tolerances in capacitance, and low leakage. It is used in the manufacture of condensers with capacitances from 5 $\mu\mu$F to 0.047 μF for radio receiver applications, with voltage ratings from 300 to 2500 V. The insulation resistance is in excess of 3000 for the cheaper grade (Class A) and 6000 megohms for other classes (American R.M.A. REC-115). The value of Q is over 1000 for a typical capacitance of 200 $\mu\mu$F at a frequency of 1 Mc/s ; the maximum value of Q occurs at frequencies about 100 Kc/s.

Mica condensers are available in metal, moulded and ceramic casings [see Chapter 38 Sect. 3(v)].

" **Silvered mica** " condensers are used when very high precision is required. Such a condenser with a capacitance of 1000 $\mu\mu$F exhibits a capacitance change of less than 0.1% over a frequency range from low frequencies to 2 Mc/s. The effect of temperature on capacitance is a change of less than 60 parts in a million for 1°C temperature change (RIC/137).

A typical 0.001 microfarad silvered mica condenser has the following characteristics :

Frequency	1000 c/s	10 000 c/s	100	500	Kc/s
Resistance (effective)	0.024	0.024	0.024	0.024	ohm
Q	3400	5500	7000	5800	
Power factor	0.00029	0.00018	0.00014	0.00017	
Percentage power factor	0.029	0.018	0.014	0.017	%

Silvered mica condensers are normally available with capacitances from 5 to 20 000 $\mu\mu$F with tolerances $\pm$ 1%, $\pm$ 2%, $\pm$ 5%, $\pm$ 10% and $\pm$ 20% (subject to minimum tolerance $\pm$ 1 $\mu\mu$F) (RIC/137). The average temperature coefficient is + 25 × 10^{-6} per °C, with limits from + 5 × 10^{-6} to + 50 × 10^{-6} per °C(U.I.C.).

Standard Specifications—Chapter 38 Sect. 3(v).

(f) Ceramic dielectric condensers

Ceramic dielectric condensers may be grouped under five heads :

1. Types intended primarily for temperature compensation, having a series of negative and positive temperature coefficients with close tolerances on the coefficients.

2. Types having temperature coefficients nearly zero.

3. General purpose types with a broad spread of temperature coefficients. This may be further divided into two groups, those having positive and negative temperature coefficients.

4. Types with temperature coefficients not specified. These are available with capacitances from 0.5 $\mu\mu$F upwards.

5. High-K types having relatively poor power factors and indeterminate temperature coefficients.

The first group is intended for use in the tuned circuits of radio receivers, in which their special temperature versus capacitance characteristics are used to reduce frequency drift during warming-up and running. A condenser having a negative temperature coefficient may be used to compensate the positive temperature coefficient of the tuned circuit alone. For standard specifications see Chapter 38 Sect. 3(vi).

The second group is intended for use in tuned circuits requiring nearly constant capacitance, such as in i-f transformers.

All except High-K types have high stability and values of Q from 335 to 1000, except for capacitances less than 30 $\mu\mu$F. The insulation resistance is not less than 7 500 megohms (R.M.A.).

The High-K types are not suitable for compensation purposes, but are used for by-passing and other non-critical applications. They are manufactured with capacitances up to 15 000 $\mu\mu$F, with capacitance tolerances of $\pm$ 20%. The value of Q is from 30 to 100.

(g) Gang condensers

Gang condensers usually have air dielectric, and are available in 1, 2 and 3 (occasionally 4) gang units. Some are fitted with trimmer condensers—see (h) below—while others are not. The shape of the plates may be designed to provide any desired capacitance characteristic (Refs. B5a, B5b) among which are :

(1) **Straight line capacitance** : Each degree of rotation should contribute an equal increment in capacitance.

(2) **Straight line frequency** : Each degree of rotation should contribute an equal increment in frequency.

(3) **Logarithmic Law** : Each degree of rotation should contribute a constant percentage change of frequency.

(4) **Square Law** : The variation in capacitance should be proportional to the square of the angle of rotation.

Some gang condensers have all sections identical, while others have the oscillator section with specially shaped plates to give correct tracking without the use of a padder condenser. For standard specifications, see Chapter 38 Sect. 3(vii).

The normal construction incorporates an earthed rotor, but condensers with insulated rotors are also available. Condensers with split stators and either earthed or insulated rotors are available for special applications.

(h) Trimmer condensers (" compensators ")

These are available in innumerable forms, and can only be briefly mentioned.

The compression mica type is the least satisfactory of all since it tends to suffer from drift in capacitance, and is far from being linear in its characteristic. It is used in receivers for the medium frequency broadcast band and in the less expensive dual-wave receivers.

The concentric or vane air-dielectric types are more expensive but have greater stability, are easier to adjust, and the better types are more satisfactory under tropical conditions. Ceramic trimmers are also obtainable.

Trimmers are available in a wide range of capacitances, but for ordinary use with gang condensers should preferably have a minimum capacitance not greater than 2 $\mu\mu$F, and a capacitance change not less than 15 $\mu\mu$F (R.M.A. REC-101, REC-106-A).

See also Chapter 38 Sect. 3(vii and xi) for standard specifications.

(iii) Combination units

Combinations of one or more capacitors with one or more resistors are becoming common, and are very convenient. Some popular combinations are

(a) Diode filters incorporating one resistor and two capacitors with a common earth return.

(b) Cathode bias units incorporating one resistor shunted by a capacitor.

(c) Plate and grid decoupling units incorporating one capacitor and one resistor.

(d) Audio frequency coupling unit incorporating a plate load resistor, coupling capacitor and grid resistor with also (in one example) a grid stopper resistor and capacitor.

(iv) Practical inductors

Iron-cored inductors are covered in Chapter 5. Radio frequency inductors, both air-cored and iron-dust cored, are covered in Chapter 11. The calculation of inductance of air-cored inductors at all frequencies is covered in Chapter 10.

(v) References to practical resistors and condensers

(A) REFERENCES TO PRACTICAL RESISTORS

A1. Pender, H., & K. McIlwain (book) " Electrical Engineers' Handbook—Electrical Communication and Electronics " (John Wiley & Sons, New York ; Chapman & Hall Ltd., London, 4th edit. 1950) Section 3.
A2. American R.M.A. Standards (see Chapter 38 Sect. 3(i) and (ii)).
A3. Spratt, H. G. M. " Resistor ratings," W.W. 54.11 (Nov. 1948) 419.
A4. British Standard BS/RC.G/110 " Guide on fixed resistors " (Issue 1) Aug. 1944. To be superseded by RC.G./110—see Chapter 38 Sect. 3.
A5. G.W.O.H. " Behaviour of high resistances at high-frequencies," W.E. 12.141 (June 1935) 291.
A6. Puckle, O. S. " Behaviour of high resistance at high-frequencies " W.E. 12.141 (June 1935) 303.
A7. G.W.O.H. " A further note on high resistance at high-frequency " W.E. 12.143 (Aug. 1935) 413.
A8. Hartshorn, " The behaviour of resistances at high-frequency " W.E. 15.178 (July 1938) 363.
A9. G.W.O.H. " The behaviour of resistances at high-frequencies " W.E. 17.206 (Nov. 1940) 470.
A10. Wilton, R. W. " Cracked carbon resistors " FM & T. 9.2 (Feb. 1949) 29.
A12. Cooper, W. H., & R. A. Seymour " Temperature dependent resistors " W.E. 24.289 (Oct. 1947) 298.
A13. Fixed composition resistors (draft specification not complete) B.R.M.F. Bulletin 1.7 (July 1948) 8-10. Superseded by Refs. A26, A27.
A14. Blackburn, J. F. (Editor) " Components Handbook " (1st edit. 1949, McGraw-Hill Book Co., New York & London, for Massachusetts Institute of Technology). Also gives extensive bibliography.
A15. American JAN-R-11 Specification with Amendment No. 3 (see Chapter 38 Sect. 3 for details).
A16. Catalogues and reports from resistor manufacturers.
A17. Valley, G. E., & H. Wallman (Editors) " Vacuum tube amplifiers " (McGraw-Hill Book Co., New York & London, 1948).
A18. Data Sheet XIX " Circuit noise due to thermal agitation " Electronic Eng. 14.167 (Jan. 1942) 591.
A19. Johnson, J. B. " Thermal agitation of electricity in conductors " Phys. Rev. 32.1 (July 1928) 97.
A20. Nyquist, H. " Thermal agitation of electric charge in conductors " Phys. Rev. 32.1 (July 1928) 110.
A21. Campbell, R. H., & R. A. Chipman " Noise from current-carrying resistors 20 to 500 Kc " Proc. I.R.E. 37.8 (Aug. 1949) 938.
A22. " Properties and uses of negative coefficient resistors—thermistors " W.W. 55.10 (Oct. 1949) 405.
A23. " Deposited-carbon resistors " Elect. 22.10 (Oct. 1949) 182.
A24. " Radio Components Handbook " (1st ed., 1948, Technical Advertising Associates, Cheltenham, Pennsylvania, U.S.A.).
A25. Oakes, F. " Noise in fixed resistors " Electronic Eng. 22.264 (Feb. 1950) 57 ; letters 22.267 (May 1950) 207.
A26. The Radio Industry Council, London, Specification No. RIC/112 (Issue 1, May 1950) Resistors, fixed, composition, Grade 1.
A27. The Radio Industry Council, London, Specification No. RIC/113 (Issue 1, June 1950) Resistors, fixed, composition, Grade 2.
A28. Coursey, P. R. " Fixed resistors for use in Communication equipment, with special reference to high stability resistors " Proc. I.E.E. 96. Part 3.41 (May 1949) 169. Discussion 96. Part 3.44 (Nov. 1949) 482.
A29. Arthur, G. R., & S. E. Church " Behaviour of resistors at high frequencies " TV Eng. 1.6 (June 1950) 4.
A30. Oakes, F. " Noise in variable resistors and potentiometers " Electronic Eng. 22.269 (July 1950) 269.
A31. Rosenberg, W. " Thermistors " Electronic Eng. 19.232 (June 1947) 185.
A32. Oakes, F. " The measurement of noise in resistors " Electronic Eng. 22.273 (Nov. 1950) 464. See also References Chapter 18 Sect. 2(ii)b.
A33. Pavlasek, T. J. F., & F. S. Howes " Resistors at radio frequency—characteristics of composition type " W.E. 29.341 (Feb. 1952) 31.

(B) REFERENCES TO PRACTICAL CONDENSERS (books)

B1. Brotherton, M. " Capacitors—their use in electronic circuits " (D. Van Nostrand Co. Inc. New York, 1946).
B2. Coursey, P. R. " Electrolytic Condensers " (Chapman & Hall Ltd. London, 2nd edit. 1939).
B3. Deeley, P. McK. " Electrolytic Capacitors " (Cornell-Dubilier Electric Corp., South Plainfield, N.J. 1938).
B4. Georgiev, A. M. " The Electrolytic Capacitor " (Murray Hill Books Inc. New York and Toronto, 1945).
B5a. Blakey, R. E. " The Radio and Telecommunications Design Manual " (Sir Isaac Pitman & Sons Ltd., London, 1938) Section 3.
B5b. Henney, K. " Radio Engineering Handbook " (McGraw-Hill Book Co. New York & London, 4th ed. 1950).

OTHER REFERENCES

B6. Brotherton, M. " Paper capacitors under direct voltages," Proc. I.R.E. 32.3 (March 1944) 139.
B7. American and English Standard specifications—see Chapter 38 Sect. 3.
B8. Roberts, W. G. " Ceramic capacitors " Jour. Brit. I.R.E. 9.5 (May 1949) 184.
B9. Gough, Kathleen A. " Choosing capacitors " W.W. 55.6 (June 1949) S.5.
B10. Cornell, J. I. " Metallized paper capacitors " Comm. (Jan. 1947) 22.
B11. Cozens, J. H. " Plastic film capacitors " W.W. 55.6 (June 1949) S. 11.
B12. Trade catalogues and data.
B13. Bennett, A. E., & K. A. Gough " The influence of operating conditions on the construction of electrical capacitors " Proc. I.E.E. Part III 97.48 (July 1950) 231.
B14. McLean, D. A. " Metallized paper for capacitors " Proc. I.R.E. 38.9 (Sept. 1950) 1010.
B15. Weeks, J. R. " Metallized paper capacitors " Proc. I.R.E. 38.9 (Sept. 1950) 1015.
B16. Fisher, J. H. " Metallized paper capacitors " Elect. 23.10 (Oct. 1950) 122.
B17. Whitehead, M. " New electrolytic capacitors—use of tantalum for electrodes " FM-TV 11.2 (Feb. 1951) 26.

Additional references will be found in the Supplement commencing on page 1475.

CHAPTER 5

TRANSFORMERS AND IRON-CORED INDUCTORS

by G. Builder, B.Sc., Ph.D., F.Inst.P.,
I. C. Hansen, Member I.R.E. (U.S.A.), and
F. Langford-Smith, B.Sc., B.E.

SECTION 1 : IDEAL TRANSFORMERS

(i) Definitions (ii) Impedance calculations—single load (iii) Impedance calculations—multiple loads.

(i) Definitions

An **ideal transformer** is a transformer in which the winding reactances are infinite, and in which winding resistances, core loss, leakage inductances and winding capacitances are all zero. In such a transformer the voltage ratio between any two windings is equal to the turns ratio of the windings, under all conditions of loading, as illustrated in Fig. 5.1. Also, in such a transformer the currents in any two windings are inversely proportional to the ratio of turns in the windings under all load conditions

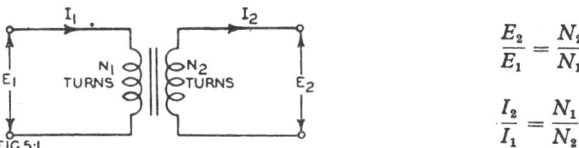

$$\frac{E_2}{E_1} = \frac{N_2}{N_1}$$

$$\frac{I_2}{I_1} = \frac{N_1}{N_2}$$

Fig. 5.1. Ideal two-winding transformer. E_1 and E_2 are alternating voltages (r.m.s.). I_1 and I_2 are alternating currents (r.m.s.) but the indicated directions of current flow are at a chosen instant and correspond to the direction of voltage at that instant. Similar remarks apply to Fig. 5.2.

Modern iron-cored transformers often approach so closely to perfection for their particular purposes that their analysis on the basis of ideal transformer theory may give useful practical approximations for design purposes.

A **double-wound transformer** is one in which, as illustrated in Fig. 5.1, separate primary and secondary windings are used to permit isolation of the primary and secondary circuits except through mutual inductive coupling.

199

Auto-transformers may be used with economy in some cases : a single winding is tapped to give the required turns ratio, which may be greater or less than unity, between primary and secondary. An ideal step-up auto-transformer is shown in Fig. 5.2.

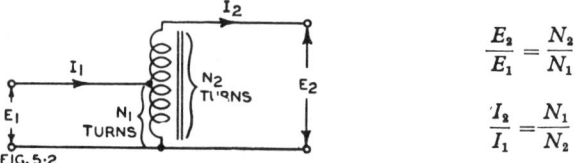

$$\frac{E_2}{E_1} = \frac{N_2}{N_1}$$

$$\frac{I_2}{I_1} = \frac{N_1}{N_2}$$

Fig. 5.2. Ideal step-up auto-transformer.

An auto-transformer is always more economical to construct than a two winding transformer as indicated in the following tabulation :

Type of Winding	Ratios	Total Volt-ampere rating of windings	Transformer Output V.A.
Auto	10 : 9 or 9 : 10	20	100
Auto	2 : 1 or 1 : 2	100	100
Auto	3 : 1 or 1 : 3	133	100
Auto	5 : 1 or 1 : 5	160	100
Auto	10 : 1 or 1 : 10	180	100
Double	any ratio	200	100

The currents in the primary and secondary sections are exactly 180° out of phase, and the resultant current flowing through the common portion of the winding is the difference between the two. When the ratio is 1 : 2 or 2 : 1, the currents in the two sections of the winding are equal.

In ideal transformer theory there is no distinction between auto- and two-winding transformers and they need not therefore be considered separately in this section.

(ii) Impedance calculations—single load

When the secondary of a simple two-winding ideal transformer is loaded with a resistance R_2 as shown in figure 5.3, the equivalent or **transformed load** R_1 as measured between the primary terminals is

$$R_1 = (N_1/N_2)^2 . R_2 \qquad (1)$$

and since the voltage ratio between primary and secondary is equal to the turns ratio, this may be written as

$$R_1 = (E_1/E_2)^2 . R_2 \qquad (2)$$

A transformer therefore merely transforms a load imposed on its secondary. Its primary does not in the ideal case impose any load unless a load is applied to the secondary. It is the turns ratio between primary and secondary, and not the number of turns in the primary, that governs the transformed or reflected load impedance.

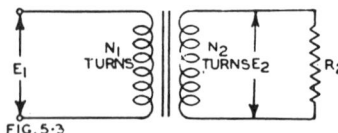

Fig. 5.3. Ideal two-winding transformer with loaded secondary.

If however a transformer is to approximate to the ideal, the number of turns in the primary must be sufficient to make the reactance of the primary high compared with the transformed value of the load impedance as measured across the primary terminals.

For an ideal transformer with a centre-tapped primary as shown in Fig. 5.4, the transformed load measured across the whole primary (between terminals, P, P) is equal to the transformed value R_1.   If however only one half of the primary is used (between either of terminals P and terminal C.T.) the transformed load presented is $\frac{1}{4} R_1$.

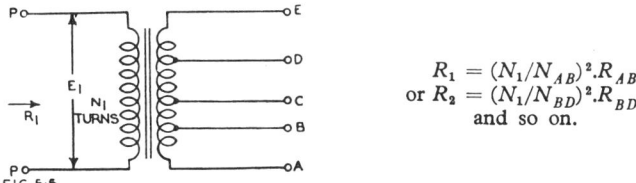

$$\frac{E_1}{E_2} = \frac{N_1}{N_2}$$
$$R_1 = (N_1/N_2)^2 . R_2$$
$$= (E_1/E_2)^2 . R_2$$

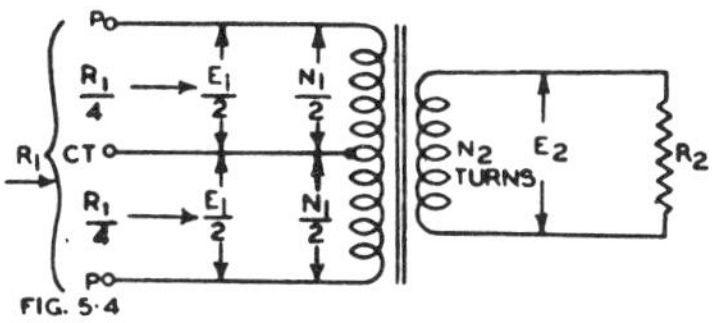

FIG. 5·4

Fig. 5.4.   Ideal transformer with primary centre-tap and loaded secondary.

As a practical example of the primary centre tap, consider the use of the transformer of Fig. 5.4 to feed a 500 ohm line ($R_2 = 500$ ohms).   If the transformer has an impedance ratio $(N_1/N_2)^2$ equal to 10 : 1, the transformed load across the whole of the primary, e.g., when the primary is fed by a push-pull amplifier, is $10 \times 500 = 5000$ ohms. If, however, only one half of the primary were used for connection to a single-ended amplifier, the load presented to the amplifier would be 1250 ohms.

For an ideal transformer with a winding tapped for load matching, as shown in Fig. 5.5, the calculation of the tap to be selected for any particular load follows from the application of eqn. (1).

$$R_1 = (N_1/N_{AB})^2 . R_{AB}$$
$$\text{or } R_2 = (N_1/N_{BD})^2 . R_{BD}$$
$$\text{and so on.}$$

FIG. 5·5

Fig. 5.5.   Ideal transformer with secondary tapped for load matching.

In these equations a load connected across terminals A and B is denoted by R_{AB} and the number of turns in the secondary between these two terminals is given by N_{AB}, and corresponding designations apply for any pair of terminals across which the load is connected.

If, for example, the terminals A, B provide a match with a 10 ohm secondary load with a total of N_{AB} secondary turns, the number of turns N_{AD} required between terminals A and D to provide a similar match to a 500 ohm line is given simply by

$$(N_{AD}/N_{AB})^2 = 500/10 = 50, \text{ so that}$$
$$N_{AD}/N_{AB} = \sqrt{50} = 7.07.$$

The number of turns in the 10 ohm winding is approximately one-seventh of the number of turns for matching to 500 ohms.   It is, of course, permissible to use any pair of secondary terminals such as B, C or C, D and so on, so that a wide range of transformation ratios is available from a transformer arrangement such as that shown in Fig. 5.5.

(iii) Impedance calculations—multiple loads

Where two or more loads are connected simultaneously to the windings of a transformer, the conditions for matching may be determined readily by the following methods.

Consider an ideal transformer having two secondaries of N_2 and N_3 turns connected to loads R_2 and R_3 respectively as shown in Fig. 5.6. It follows immediately from eqn. (2) that

$$E_2/E_1 = N_2/N_1 \; ; \; E_3/E_1 = N_3/N_1 \; ; \; E_3/E_2 = N_3/N_2' \qquad (3)$$

and, since the transformer is an ideal one, these relations hold irrespective of the relative values of the loads.

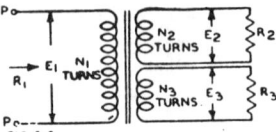

Fig. 5.6.   Ideal transformer
with multiple loads.

It is sometimes convenient to draw the diagram as in Fig. 5.7, which is equivalent in every way to Fig. 5.6. If, as a special case, $N_2/N_3 = R_2/R_3$, then the voltages of the two points A and B will be the same, and the link AB may be omitted without any effect. The currents in both sections of the transformer and in R_2 and R_3 will then be the same.

The value of the reflected or transformed load R_1 measured across the primary terminals can be obtained by considering the power in relation to the voltages specified in eqn. (3).

Fig. 5.7   Ideal transformer with
multiple loads as in Fig. 5.6, but
with the secondaries connected at
points A, B.

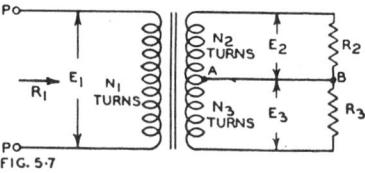

Let W_1 = total input watts to the primary = $E_1{}^2/R_1$
$\quad W_2$ = watts in the load R_2 $\qquad\qquad = E_2{}^2/R_2$
$\quad W_3$ = watts in the load R_3 $\qquad\qquad = E_3{}^2/R_3$

Then $W_1 = W_2 + W_3$ (since there are no transformer losses)
so that $\dfrac{E_1{}^2}{R_1} = \dfrac{E_2{}^2}{R_2} + \dfrac{E_3{}^2}{R_3}$

$$= \frac{E_1{}^2}{R_2}\left(\frac{N_2}{N_1}\right)^2 + \frac{E_1{}^2}{R_3}\left(\frac{N_3}{N_1}\right)^2$$

We therefore have

$$\frac{1}{R_1} = \frac{1}{R_2(N_1/N_2)^2} + \frac{1}{R_3(N_1/N_3)^2} \qquad (4)$$

so that the total load R_1 presented by the primary is equal to the parallel combination of the two transformed loads $R_2(N_1/N_2)^2$ and $R_3(N_1/N_3)^2$.

If an additional winding of N_4 turns is connected to a load R_4 we obviously have in the same way

$$\frac{1}{R_1} = \frac{1}{R_2(N_1/N_2)^2} + \frac{1}{R_3(N_1/N_3)^2} + \frac{1}{R_4(N_1/N_4)^2} \qquad (5)$$

and so on, for any number of loads. Such expressions are equally applicable whether the secondary windings used are separate windings or whether they form part of a single tapped secondary. For example if, in the transformer shown in Fig. 5.5, we have loads R_{AB}, R_{AC} and R_{BD} connected to terminals AB, AC, and BD respectively, we will have

$$\frac{1}{R_1} = \frac{1}{R_{AB}(N_1/N_{AB})^2} + \frac{1}{R_{AC}(N_1/N_{AC})^2} + \frac{1}{R_{BD}(N_1/N_{BD})^2}$$

and so on.

A typical practical case is one in which a known power output W_1 from an amplifier is fed into a known reflected impedance R_1 with two secondaries feeding loads R_2

and R_3 (such as two loudspeakers) which are required to operate with power inputs of W_2 and W_3 respectively, so that $W_1 = W_2 + W_3$. The required transformer turns ratios are then given by

$$R_1 W_1 = E_1{}^2 \; ; \; R_2 W_2 = E_2{}^2 \; ; \; R_3 W_3 = E_3{}^2 \qquad (6a)$$

so that $R_2 W_2 / R_1 W_1 = E_2{}^2 / E_1{}^2 = N_2{}^2 / N_1{}^2 \qquad (6b)$

and $\quad R_3 W_3 / R_1 W_1 = E_3{}^2 / E_1{}^2 = N_3{}^2 / N_1{}^2 \qquad (6c)$

For example if $W_2 = 3$ watts $\quad R_2 = 500$ ohms

$\quad\quad\quad\quad\quad\quad W_3 = 4$ watts $\quad R_3 = 600$ ohms

and $\quad\quad\quad\quad W_1 = 7$ watts $\quad R_1 = 7{,}000$ ohms

we have

$$\frac{N_2{}^2}{N_1{}^2} = \frac{500 \times 3}{7000 \times 7} = \frac{1}{32.7} \; . \; \text{Therefore} \; \frac{N_2}{N_1} = \frac{1}{5.7}$$

and similarly $N_3 / N_1 = 1/4.5$.

Expressions such as (6) may be written in the more general form

$$N_n / N_1 = \sqrt{(R_n / R_1)(W_n / W_1)} \qquad (7)$$

where N_n = number of turns on secondary n,

$\quad\quad\quad N_1$ = number of turns on primary,

$\quad\quad\quad R_n$ = load applied to secondary n,

$\quad\quad\quad R_1$ = transformed total primary load,

$\quad\quad\quad W_n$ = watts in load R_n,

and $\quad W_1$ = total watts input to primary,

but it must be noted that these relations hold only when all loads are connected so that the specified input conditions to the primary do exist. Eqn. (7) is also applicable to determine the turns ratios of two or more separate transformers feeding two or more loads from a common amplifier which delivers W_1 watts into a total load of R_1 ohms.

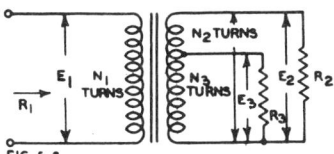

FIG. 5·8

Fig. 5.8. Ideal transformer with multiple loads and tapped secondary, which is effectively identical with Fig 5.6.

If a transformer is supplying power to two or more loads, such as loudspeakers, and one of these is switched out of circuit, the impedance reflected on to the primary will change due to the reduction of loading on the secondary. In order to avoid the resultant mismatching it is advisable to switch in a resistive load, having a resistance equal to the nominal (400 c/s) impedance of the loudspeaker, so as to take the place of the loudspeaker which has been cut out of circuit. In this case the resistance should be capable of dissipating the full maximum power input to the loudspeaker. Such an arrangement will also have the result that the volume from the remaining speakers will be unchanged

Alternatively if it is desired to switch off one loudspeaker and to apply the whole power output to a single speaker, it will be necessary to change the number of secondary turns so as to give correct matching. This change may generally be arranged quite satisfactorily by the use of a tapped secondary winding. In this case the loudspeaker would be used on the intermediate tap when both speakers are in use, and on the whole winding for single speaker operation.

It does not matter whether two or more secondary windings or a single tapped winding is employed. The arrangement shown in Fig. 5.8 is effectively identical with that of Fig. 5.6.

SECTION 2 : PRACTICAL TRANSFORMERS

(i) *General considerations*   (ii) *Effects of losses.*

(i) General considerations

The treatment in section 1 based on ideal transformer theory is an extremely useful first approximation in design problems, particularly if the transformers to be used are so liberally designed that their general characteristics approximate to the ideal.

In practice it is usually necessary to take into account :

(a) The resistance of each winding
(b) The core loss
(c) The inductances of the windings
(d) The leakage inductances
(e) The capacitances between windings and between each winding and ground, and the self capacitance of each winding.

A useful equivalent circuit of a practical transformer is shown in Fig. 5.9.

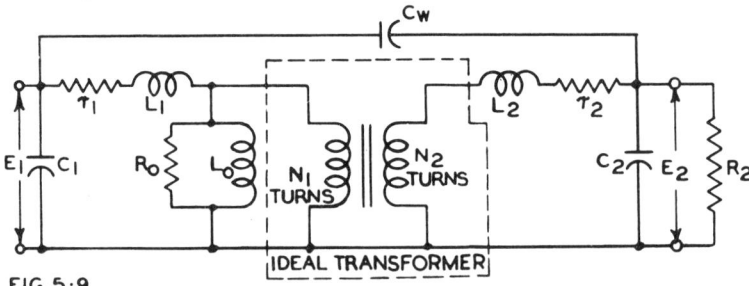

FIG. 5·9

Fig. 5.9.   Equivalent circuit of a practical transformer.

In this equivalent circuit we have an ideal transformer with a turns ratio N_2/N_1 (equal to the turns ratio of the actual transformer), with the incidental characteristics of the actual transformer represented by separate reactances and resistances.

r_1 = the resistance of the primary winding
L_1 = the equivalent primary leakage inductance
r_2 = the resistance of the secondary winding
L_2 = the equivalent secondary leakage inductance
R_0 = the equivalent core-loss resistance (including both hysteresis and eddy current losses)
L_0 = the inductance of the primary winding
C_1, C_2 = the primary and secondary equivalent lumped capacitances
C_w = the equivalent lumped capacitance between windings
R_1 = input resistance of transformer on load
and R_2 = the load resistance across the secondary.

Such an equivalent circuit is capable of representing a practical design with considerable accuracy, but actual calculations would be tedious and in some cases very difficult.

The reactances and resistances shown therein have varying effects on the input-output voltage ratio according to the frequency of the signal which the transformer is handling (Ref. A10). In general, the equivalent circuit can be presented in three distinct simplified forms for use when considering the transformer operating at low, medium and high frequencies respectively (Figs. 5.10B,C,D).

Audio transformers can be conveniently dealt with in this manner, whereas power transformers operating over a very limited frequency range can be more simply designed on the basis of Fig. 5.10A.

The whole of the equivalent circuit of Fig. 5.10A can be referred to the primary, as in Fig. 5.11 where the ideal transformer has been omitted and r_2 and R_2 multiplied by the square of the turns ratio.   This is often a convenient way of making calculations.

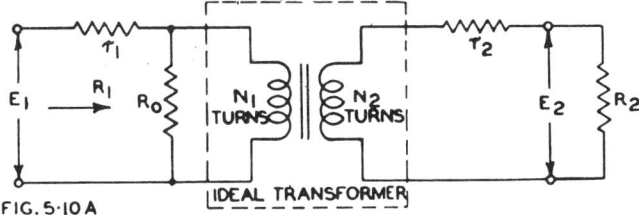

FIG. 5·10 A

Fig. 5.10A. Simplified equivalent circuit for calculating the effect of losses.

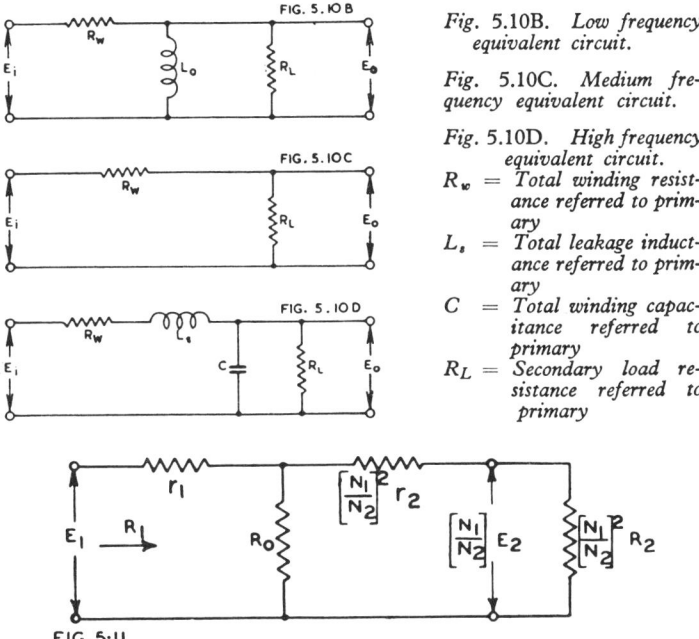

FIG. 5·11

Fig. 5.10B. Low frequency equivalent circuit.

Fig. 5.10C. Medium frequency equivalent circuit.

Fig. 5.10D. High frequency equivalent circuit.

R_w = Total winding resistance referred to primary

L_s = Total leakage inductance referred to primary

C = Total winding capacitance referred to primary

R_L = Secondary load resistance referred to primary

Fig. 5.11. Simplified equivalent circuit referred to the primary.

(ii) The effects of losses

If a transformer is designed to deliver power into a resistance load, such as valve heaters, the primary and secondary copper losses are usually designed to be of the same order.

A particular transformer reaches its maximum efficiency when the copper losses have become equal to the iron losses (proof given in Ref. A10), although this does not necessarily occur at full load unless the transformer is so designed.

The **efficiency** is the ratio of the output power to the output power plus the losses. For audio and power transformers for radio purposes typical efficiencies range from 70 to 95 per cent, with the majority of power transformers falling between 80 and 90 per cent.

The **regulation** of a transformer is defined differently in American and British practice, the primary voltage being held constant throughout.

American definition. The percentage voltage regulation is the difference between the full-load and the no-load secondary voltages, divided by the **full-load** voltage and multiplied by 100.

British definition. The percentage voltage regulation is the difference between the full-load and the no-load secondary voltages, divided by the **no-load** secondary voltage and multiplied by 100 (B.S.205 : 1943).

However, the difference between the two definitions is quite small for small percentages of voltage regulation.

The regulation of audio transformers when operating over a limited frequency range (say 200-2000 c/s), that is the ratio of E_2/E_1, is affected mainly by the copper losses.

Thus $E_2/E_1 \approx \eta T$ and $I_2/I_1 \approx 1/T$,   (1)
where η = efficiency
and T = turns ratio.

As a further consequence the impedance ratio is changed, and, making the same assumptions as above,

$$R_1 \approx r_1 + r_2/T^2 + R_2/T^2 \tag{2}$$

At low frequencies, the reflected impedance is altered by the shunt effect of the primary inductance, while at high frequencies, a similar change is caused by the leakage inductance and winding capacitances. This is covered in Sect. 3(iii).

The damping of a loudspeaker, connected as a load to the secondary of an output transformer, is also affected to some degree by the losses. Where an accurate indication of the damping factor is required, these losses should be taken into account. Refer to Chapter 21 Sect. 3.

SECTION 3 : AUDIO-FREQUENCY TRANSFORMERS

(*i*) *General considerations* (*ii*) *Core materials* (*iii*) *Frequency response and distortion*—(*a*) *Interstage transformers* (*b*) *Low level transformers* (*c*) *Output transformers* (*iv*) *Designing for low leakage inductance* (*v*) *Winding capacitance* (*vi*) *Tests on output transformers* (*vii*) *Specifications for a-f transformers.*

(i) General considerations (References A6 and A10)

Audio frequency transformers can be divided into three major categories,
     (a) Low level input
     (b) Medium level interstage
     (c) Output.

For design purposes it is necessary to know

    (1) Operating level, usually expressed in db above or below a reference level of 1 milliwatt (i.e. dbm),
    (2) Frequency response, with permitted deviation, quoting reference level at which measurements are made,
    (3) Permissible distortion, at specified operating levels and frequencies,
    (4) Impedance, phase angle, and nature of source and load between which transformer is to be connected,
    (5) Value of d.c. (if any) flowing through any winding,
    (6) Hum reduction requirements,
    (7) Phase shift permitted.

Operating level
This restricts the choice of core materials and determines the physical size. Suitable core alloys include *MUMETAL* and *RADIOMETAL* together with silicon

steels in various grades. As *MUMETAL* saturates at a comparatively low flux density, it is only suitable for low level transformers. Economic factors will probably dictate which material is finally used.

Frequency response

At low frequencies the response falls off due to the finite value of primary inductance. At high frequencies, the winding capacitance and leakage inductance are responsible for the response limitations.

Distortion (a) Low frequency

This is mainly dependent on the maximum operating flux density at the lowest frequency of interest. Distortion due to this cause falls off rapidly with increase in frequency. Other sources of distortion are observed when transformers operate in valve plate circuits. The drop in valve load impedance due to the shunting effect of the primary inductance may cause the valve to distort. Further, the load impedance will also change to a partially reactive one at low frequencies. The valve will therefore operate with an elliptical loadline and will introduce additional distortion unless care is taken.

A simple method of measuring the harmonic distortion in the cores of a-f transformers is described in Ref. C33.

Distortion (b) High frequency

At this end of the audio spectrum the load impedance changes again in magnitude and sign, thus causing an associated valve to generate distortion.

Source and load

Before the design of a transformer can be proceeded with, something must be known about the impedances between which the transformer is required to operate. Assuming that it is a low level unit, for example, it may be intended to operate from a ribbon microphone, a low impedance line, a gramophone pickup or the plate of a valve. It may have to feed the grid of a pentode or triode, a line or a mixer circuit. The secondary may be shunted by a resistance or a frequency correcting network. Unless these factors concerning the external circuits are known it is not possible to predict with any degree of accuracy, the ultimate performance of the transformer.

D.C. polarization

If an unbalanced d.c. component is present in one of the windings, this will cause a reduction in inductance over that attainable without d.c. This would require a larger transformer to meet the same performance specifications. In some cases where the d.c. magnetizing force is high, the use of high permeability alloys is not feasible. As far as practicable, unbalanced d.c. should be avoided in transformers, either by using a push-pull connection or shunt feed. For calculation of primary inductance with a d.c. component present, refer to Section 6.

Hum reduction

When the operating level is very low, it may be found desirable to shield the transformer to decrease the hum level to a suitable magnitude. This can be achieved in several ways, *MUMETAL* shields up to three in number being particularly effective. An outer case of sheet metal or cast iron is normally employed. The use of a balanced structure such as a core type instead of a shell type lamination will assist in reducing the effects of extraneous a.c. fields.

Phase shift

In certain applications it is desirable to apply negative feedback over an amplifier incorporating one or two transformers. To achieve stability with the desired amount of gain reduction, it is necessary to exercise careful control of phase shift over a frequency range very much wider than the nominal frequency range of the transformer (see Chapter 7 Sect. 3).

(ii) Core materials

High permeability alloys are now produced by several manufacturers under a variety of trade names. These are listed below, with silicon steel shown for comparison.

See also Ref. C34.

NICKEL IRON AND OTHER ALLOYS

Material	ρ	μ_0 d.c.	μ_{max} d.c.	(B-H) Sat. Gauss	Lines/sq.in.
Mumetal	62	30 000	130 000	8 500	55 000
Permalloy C	60	16 000	75 000	8 000	52 000
Radiometal	55	2 200	22 000	16 000	103 000
Permalloy B	45	2 000	15 000	16 000	103 000
Permalloy A	20	12 000	90 000	11 000	71 000
Cr-Permalloy	65	12 000	60 000	8 000	52 000
Mo-Permalloy	55	20 000	75 000	8.500	55 000
1040	56	40 000	100 000	6 000	39 000
Megaperm	97	3 300	68 000	9 300	60 000
Hipernick	46	3 000	70 000	15 500	100 000
45 Permalloy	45	2 700	.23 000	16 000	103 000
Rhometal*	95	250—2 000	1 200—8 500	12 000	78 000
4% Silicon Steel	55	450	8 000	19 500	125 000

ρ = resistivity in microhm cm ; μ_0 = initial permeability
μ_{max} = maximum permeability obtainable.

For audio transformer work the first four are frequently used. *RHOMETAL* has a special field of application, namely for transformers handling ultrasonic and radio frequencies up to several megacycles.

Flux densities of the order of 22 000 lines per square inch can be used with *MUMETAL* and approximately double this value with *RADIOMETAL,* the upper limit being set by the permissible distortion. For further information on this point see Sect. 3(iii).

For higher power **output transformers,** high silicon content (up to $4\frac{1}{2}\%$) sheet steel is in general use. To retain high permeability at low flux densities, the strips or laminations should be annealed after shearing and punching. Spiral cores of grain-oriented silicon steel are of considerable use in this application.

As a general rule, the output transformer should have the largest core which is practicable or permissible having regard to cost or other factors. A large core of ordinary silicon steel laminations is usually better than a small core of special low-loss steel.

The weight of steel in the core is a function of the minimum frequency, the permissible distortion, the core material, and the maximum power output. As a rough guide, subject to considerable variation in practice, the core may be taken as having

Weight in lbs. = 0.17 × watts output
Volume in cu. ins. = 0.7 × watts output.

These are for normal typical conditions, and may be decreased for less extended low frequency response or for a higher permissible distortion. For good fidelity, an increase in core size to double these values is desirable.

Several new **core materials** are now available including *CASLAM* and *FERROX-CUBE.*

CASLAM is a soft magnetic core material with finely laminated structure for use at frequencies from 50 c/s up to at least 10 Kc/s. It is composed of flake iron particles pressed into a compact mass of the desired shape in such a way as to produce innumerable thin magnetic layers aligned in the plane of the flux. By virtue of its dense compacted structure many of the assembly and fixing problems associated ·with the older stacking method are eliminated. Grade 1 is a low density material with a maximum permeability of 860. Grade 2 is a denser material with a higher maximum permeability of 1000. Grade 3 is similar to grade 1 in magnetic characteristics but has better strength and machining qualities.

For choke cores a pair of E's can be butted together but where minimum gap is required the block can be broken and then rejoined after the coil is positioned. Because of the fibrous laminated structure exposed by ·breaking, microscopic interleaving occurs when the join is remade in the correct manner.

*Properties depend upon different heat treatment deliberately given.

Rigid clamping is less important since there are no free laminations to vibrate under load. For this reason also, combined with the somewhat discontinuous nature of the material, the acoustic noise emitted by the block is considerably reduced, especially at higher frequencies.

PHYSICAL PROPERTIES OF CASLAM

Property	Caslam 1	Caslam 2	Caslam 3
Maximum permeability	860 at 4 Kg	1000 at 4 Kg	
Effec. permeability	500 at 10 Kg	830 at 10 Kg	Similar
Hys. loss for $B_{max} = 10$ Kg. at $50\sim$	7000	7000 ergs/cc/cycle	to
Coercive force $B_{max} = 10$ Kg. at $50\sim$	2.4	2.4 oersteds	grade 1
Sat. flux density	18 Kg	18 Kg	
Total a.c. loss, $B_{max} = 10$ Kg. at $50\sim$	2.5	2.75 watts/lb.	
Density	7.0	7.4 gms/cc	
Max. oper. temperature	110°C	110°C	150°C
Resistivity (ohm-centimetre)			
(a) Normal to plane of laminae	0.04		0.03
(b) In plane of laminae	0.003		0.002

The a.c. loss is almost entirely hysteresis, the eddy current loss being less than 10% of the total. For this reason, the a.c. permeability at $50\sim$ is approximately equal to the d.c. figure and at higher acoustic frequencies, blocks of *CASLAM* compare favourably in magnetic properties with stacks built up from normal silicon iron sheet.

Dust cores generally suffer from low permeability and, to reduce eddy currents, particle size has to be reduced ; this causes further reduction in permeability.

Other new ferromagnetic materials such as the ferromagnetic spinels and *FERROX-CUBE* are described in Chapter 11 Sect. 3(v)E, and find their applications principally at frequencies above the audio range.

(iii) Frequency response and distortion

(a) Interstage transformers—Class A and B

At the mid-frequency* the amplification is very nearly equal to the amplification factor of the valve multiplied by the turns ratio of the transformer, where the secondary is unloaded.

At low frequencies the gain falls off due to the decrease in primary reactance. The ratio of the amplification at a low frequency A_{lf} compared with that at the mid-frequency A_{mf} can be expressed thus—

$$\frac{A_{lf}}{A_{mf}} = \frac{1}{\sqrt{1 + \left(\frac{R}{\omega L_0}\right)^2}} \tag{1}$$

where L_0 = primary inductance
and R = plate resistance plus primary resistance.

The response will fall off 3 db at a frequency such that $\omega L_0 = R$. At a frequency such that

$$\omega L_0 = 2R \tag{2}$$

the response will be down approximately 1 db from the mid-frequency level.

At high frequencies the leakage reactance and shunt capacitance, in conjunction with the plate and winding resistances, form a low Q series resonant circuit. Above this resonant frequency the gain will fall off rapidly. In the neighbourhood of resonance the change in gain will depend on Q_0, the Q of the resonant circuit. This factor can be varied by adding external resistance or by winding the secondary partly with resistance wire (Ref. C3). The resonant frequency can be varied by changing the value of the total leakage inductance, L_t or the interwinding capacitances. These are both functions of the transformer structure. See also page 518.

By careful choice of core material, lamination dimensions and method of sectionalizing the winding, it is possible to achieve a frequency response extending beyond the normal audio range (Refs. C2, C4, C5, C6 and Figs. 5.12 and 5.13A).

---

*The mid-frequency is the frequency at which maximum gain is obtained.

As the leakage inductance L_t is proportional to the square of the turns, N, it is possible to extend the high frequency response by reducing N. This, of course, reduces L_0 in the same proportion as L_t, but this effect can be overcome by the use of a high permeability alloy.

For low level working it is usual to assume an initial a.c. permeability of 350 for silicon steel. If a *RADIOMETAL* core with an initial a.c. permeability of say, 1600, and of similar dimensions is substituted, N could be reduced in the ratio of $\sqrt{1600/350}$ or 2.14/1 and L_t approximately 4.6/1 over that for the silicon steel transformer. Similarly if a *MUMETAL* core having an initial a.c. permeability of 10 000 and of similar dimensions is substituted, N could be reduced in the ratio of $\sqrt{(10\,000/350)}$ or 5.34/1 and L_t approximately 28.5/1.

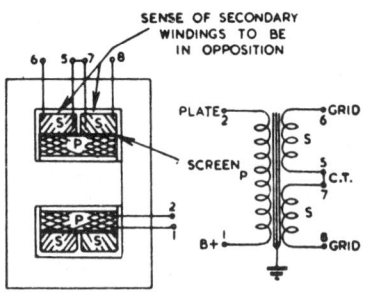

Fig. 5.12. *Transformer wound with balanced secondaries.*

In those examples, the primary inductance has remained constant, but the leakage inductance has decreased over 4 times and 28 times respectively without any sectionalizing of the winding or interleaving. This indicates the improvement possible with the use of high permeability alloys.

Care must be taken, with the reduction in turns, that the flux density in the core does not exceed safe limits from the point of view of distortion (see Refs. C7, C8, C9, C10).

In the case of transformers working at low levels and hence, generally, a low flux density, the distortion may not be a consideration. It is possible to achieve an improved high frequency response, while keeping the distortion at the same level at low frequencies, by increasing the core cross section and proportionally reducing the turns. This however, increases the mean length of turn and also L_t, so the nett reduction of L_t is not as great as would be anticipated. In addition, the cost of the transformer increases, particularly if a high permeability alloy is used. Class B input (driver) transformers are usually called upon to handle appreciable amounts of power during part of the audio frequency cycle, but are designed on the basis of open circuit working when considering primary inductance. Winding resistance and leakage inductance must be kept low to avoid distortion (Refs. C21, C22).

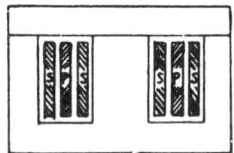 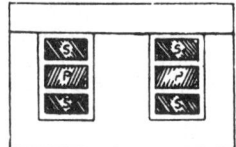

Fig. 5.13A. *Typical coil arrangements in audio transformers.*

(iii) (b) Low level transformers

Transformers working about or below a zero level of 1 milliwatt (0 dbm) usually employ a *MUMETAL* core. In most instances hum shielding is necessary to attain the desired signal to hum ratio. This is generally required to be in excess of 60 db. Satisfactory shielding can be obtained by using around the transformer one or more shields made of high permeability alloy. Where a.c. fields are strong, the outer shield is normally designed to be made of an alloy which has a high saturation flux density. Thus the inner shields of, say, *MUMETAL*, are working at maximum efficiency in a low a.c. field where their permeability is a maximum. Sometimes sheet steel or thick cast iron outer cases are used. More often a special alloy, such as Telcon 2129, is used. This material has magnetic properties similar to *RADIO-METAL*, but is suitable for deep drawing. Inner shields are often of copper to give shielding from electric fields (Refs. C11, C12). Some improvement is possible if a core type structure is used in place of a shell type, owing to the cancellation of stray voltages induced in the winding by external fields.

As an example of the use of high permeability alloys, the following design problem is presented. Calculate the primary turns for a 50 ohm to 50 000 ohm transformer working at zero level (0 dbm). The frequency response must not fall more than 1db below mid-frequency response, at 50 c/s Distortion must not exceed 1% at zero level at 50 c/s. The source impedance is 50 ohms resistive and the secondary is unloaded. Core material to be used is *MUMETAL*.

1st Step. Calculate primary inductance.
For 1 db attenuation, $\omega L_0 = 2R$ (Eqn. 2).
$L_0 = 2R/\omega = 2 \times 50/2 \times \pi \times 50 = 0.32$ henry.

2nd Step. Calculate primary turns.
Assume square stack of Magnetic and Elec. Alloys No. 35 lamination. Length of magnetic path, $l = 4.5$ inches.
Cross sectional area = 0.56 square inches.

$$L_0 = \frac{3.2A \; \mu N^2}{10^8 \times l} \qquad \text{Therefore } N^2 = \frac{L_0 \times 10^8 \times l}{3.2 \times A \times \mu}$$

$$N^2 = \frac{0.32 \times 10^8 \times 4.5}{3.2 \times 0.56 \times 10^4} \qquad \text{Therefore } N = 90 \text{ turns.}$$

3rd Step. Calculate working flux density.
Primary voltage $E = \sqrt{W.R}$
where W = input power
Therefore $E = \sqrt{1 \times 10^{-3} \times 50} = 0.224$ V
and $\qquad B = \dfrac{0.224 \times 10^8}{4.44 \times 50 \times 90 \times 0.56}$
$\qquad\qquad$ = 2000 lines per square inch
$\qquad\qquad$ = 310 Gauss.

4th Step. Determine percentage distortion.

Referring to Fig. 5.13B and using curve for $\omega L_0 = 2R$, it will be noted that the percentage distortion is approximately 0.75%. This assumes that the permeability value, μ, is still 10 000 at the operating flux density. In practice μ may exceed this figure and thus the distortion as calculated above may be larger than would be measured in a finished transformer. In this problem, for simplification, no account has been taken of the stacking factor, which would modify the result slightly.

When the secondary of the transformer is loaded, R in eqn. 1 then becomes R_A as in eqn. 3 and R in Fig. 5.13B is read as R_A. The calculation for distortion then follows in a similar manner to the unloaded secondary example worked earlier.

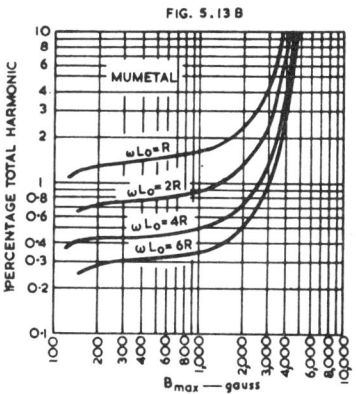

FIG. 5.13B

Fig. 5.13B. Total harmonic distortion plotted against flux density for different ratios of primary reactance to R where R = plate resistance + primary resistance (Ref. C9).

(iii) (c) Output transformers

The factors affecting the frequency response of output transformers (Ref. C13) are similar to those affecting interstage transformers. Refer to Figs. 5.10B, C and D.

The response falls off from the mid-frequency gain by 3 db at a low frequency such that

$$\omega L_0 = R_A \tag{3}$$

Attenuation is 1 db when

$$\omega L_0 = 2R_A \tag{4}$$

where $\quad R_A = \dfrac{(r_p + R_W)R_L}{r_p + R_W + R_L}$

$\qquad r_p$ = plate resistance of valve

$\qquad R_L$ = load resistance referred to the primary

and $\qquad R_W$ = total winding resistance referred to the primary.

The response falls off from the mid-frequency gain by 3 db at a high frequency such that

$$\omega L_S = R_B \tag{5}$$

Attenuation is 1 db when

$$\omega L_S = 0.5\, R_B \tag{6}$$

where $\quad R_B = r_p + R_W + R_L$

and $\qquad L_S$ = total leakage inductance referred to primary.

The gain at the mid-frequency $= \dfrac{E_0}{E_i} = \dfrac{\mu R_L}{R_B}$ $\tag{7}$

where E_i = voltage input to grid of output valve

$\qquad E_0$ = voltage output across R_L

and $\qquad \mu$ = amplification factor of output valve.

It is thus possible to specify the primary and leakage inductances permissible when the frequency response requirements are known. An example will illustrate this.

Determine the minimum primary inductance and maximum leakage inductance permitted in an output transformer designed to match a pair of Class A 2A3 triode valves with a 5 000 ohm load. The response is to be within 1 db from 50 to 10 000 c/s. The plate resistance of each valve is 800 ohms. Neglect R_W.

Make all calculations from plate to plate.

For a fall of 1 db at 50 cycles per second,

$$\omega L_0 = 2R_A.$$

Now $\quad R_A = \dfrac{5000 \times 1600}{5000 + 1600} = 1200$ ohms approx.

$\quad \omega L_0 = 1200 \times 2 = 2400$

$\quad L_0 = 2400/2 \times \pi \times 50$

$\qquad = 7.6$ henrys approx.

This is the value that would be measured on a bridge at low induction.

For a fall of 1 db at 10 000 cycles per second,

$$\omega L_S = 0.5\, R_B$$

Now $R_B = 5000 + 1600 = 6600$ ohms.

$\quad \omega L_S = 0.5 \times 6600 = 3300$

$\quad L_S = 3300/2 \times \pi \times 10\,000$

$\qquad = .052$ henry approx. $= 52$ millihenrys.

Note particularly that distortion requirements may necessitate an increase in L_0 and a decrease in L_S.

The following table indicates the relationship between the low frequency attenuation and the ratio of $\omega L_0/R_A$:

TABLE 1

Loss	Relative amplification	$\omega L_0/R_A$
0.5 db	0.94	2.76
1.0 db	0.89	1.94
2.0 db	0.79	1.30
3.0 db	0.71	1.00
6.0 db	0.50	0.58

The inductances required for various values of R_A, for a bass response loss of 1 decibel are as follows, correct to two significant figures :

TABLE 2

Value of R_A ohms	Bass response down 1 db at			
	150 c/s	100 c/s	50 c/s	30 c/s
800	1.7 H	2.6 H	5.1 H	8.5 H
1200	2.6 H	3.8 H	7.6 H	13 H
1500	3.2 H	4.8 H	9.5 H	16 H
2000	4.3 H	6.4 H	13 H	21 H
3000	6.4 H	9.5 H	19 H	32 H
4000	8.5 H	13 H	26 H	42 H
5000	11 H	16 H	32 H	53 H
7500	16 H	24 H	48 H	80 H
10 000	21 H	32 H	64 H	110 H
15 000	32 H	48 H	95 H	160 H
20 000	43 H	64 H	130 H	210 H
50 000	110 H	160 H	320 H	530 H

where R_A is approximately equal to the load resistance R_L in parallel with the effective plate resistance of the valve ; however see comments below. When the plate resistance is very high, as for a pentode without feedback, R_A may be taken as being approximately equal to R_L.

For loss in bass response to be reduced to 0.5 decibel these inductance values should be increased by a factor of 1.9 times. For a reduction of 2 db the factor becomes 0.67.

Since the permeability of the core material varies with induction, the frequency response will vary with signal level—the limiting low frequency will usually extend lower as the signal level is increased. It is therefore desirable that the inductance values tabulated above are calculated or measured at low signal levels. In the case of feedback from the secondary, this effect should be taken into account.

Table 2 gives the value of inductance to provide a nearly constant output voltage, but this is only one of several requirements to be satisfied. The **core distortion** is a function of the ratio $\omega L_0/R_A$ (this may be derived from equations 9 and 10) and for low distortion a high ratio of inductive reactance to R_A is required, and this is equivalent to having a low value of bass attenuation (see Table 1). The bass attenuation in Table 2 (1 db at specified minimum frequencies) is based on $\omega L_0/R_A = 1.94$, resulting in fairly low core distortion. Still lower core distortion would be achieved by increasing the inductance and hence, incidentally, decreasing the bass attenuation.

The other important secondary effect resulting from a finite value of inductance is the **phase angle of the load** presented to the output valve(s). If the value of inductance from Table 2 is used for bass response down 1 db at a specified frequency, the phase angle of the load presented to the output valves will be between 45° and 90° for a triode either without feedback or with negative voltage feedback. This will cause a pronounced elliptical load-line, normally resulting in severe valve distortion at this frequency, at full power output.

If R_A in Table 2 is taken as being equal to R_L, the maximum phase angle of the load will be less than 28° for bass response 1 db down at the specified frequencies. If the factor 0.67, for bass response down 2 db, is applied to Table 2, the maximum phase angle of the load will be less than 38°. If the factor 2.0, for bass response down about 0.5 db, is applied to Table 2, the maximum phase angle of the load will be about 15°. This appears to be a reasonable value for good fidelity.

Table 2 may therefore be used as a general guide to the choice of inductance values where more exact calculations are not required—see below.

Summary of general application of Table 2

To give low core distortion, nearly constant output voltage and a total load impedance effective on the valve which is not too reactive :—

1. Take $R_A = R_L$.
2. Apply Table 2 as printed for ordinary use.
3. Multiply inductance values by a factor of 2 for good fidelity.
4. The specified frequencies in Table 2 are to be interpreted as the minimum frequencies of operation for the transformer.

Where the source impedance is high, the high frequency fall off is determined largely by L_S and the winding capacitances as in the case of interstage transformers.

Similar devices to that employed in the construction of interstage transformers can be used to extend the range of output transformers. The use of *RADIOMETAL*, specially annealed high silicon content steel (such as *SUPERSILCOR*), and grain oriented steels, are all common in better quality output transformers (Refs. C14, 15).

It is feasible to " build-out " a transformer into a half section filter and thus maintain the impedance, viewed either by the source or the load, constant over a wide frequency range. In addition the phase angle variation is reduced towards the extremes of the range. This reduces distortion and maintains full power output to a greater degree than otherwise possible.

This idea can be applied to interstage transformers and output transformers quite successfully (Refs. C19, C20). Even large modulation transformers and class B driver transformers for broadcast equipment have provided improved performance when treated in this way (Refs. C16, 17, 18).

In radio receivers and record players advantage can be taken of this " building-out " procedure, to limit the high frequency response to any given point, say 6000 c/s, with rapid attenuation thereafter. This usually involves only one extra component ; a condenser across the secondary winding of the output transformer. This is quite an effective " top " limit, more so than the normal tone control. The output transformer is designed to have the necessary amount of leakage inductance for the network to function as intended.

The **winding resistances** are not of major importance in interstage transformers, but assume greater significance in output transformers. An appreciable amount of power may be lost unless the resistances of both primary and secondary are kept to reasonable proportions. In the normal good quality transformer the total resistance reflected into the primary side is approximately double the measured d.c. resistance of the primary. The total winding resistance (referred to the primary) will vary between 10 and 20 per cent of the load resistance, which means an insertion copper loss of 0.5 to 1 db. This extra resistance must be considered when choosing the turns ratio to reflect the correct load (see Eqn. 2, Sect. 2), otherwise an impedance error of 10 to 20 percent will occur. Core losses will not materially affect the calculation as these losses do not reach their maximum except at full power at the lowest audio frequency of interest.

Distortion in output transformers

When a transformer has its primary connected to an audio frequency source of zero impedance, the waveform of the voltage on the secondary will be the same as that of the source—in other words there is no distortion.

When a transformer is connected in the plate circuit of a valve, the latter is equivalent to a resistance r_p in series with the source and the transformer primary. If the secondary of the transformer is also loaded by a resistance R_2, this is equivalent (as regards its effect on distortion) to a total primary series resistance R,

where $$\frac{1}{R} = \frac{1}{r_p} + \frac{1}{R_2(N_1/N_2)^2}$$

N_1 = primary turns

and N_2 = secondary turns.

In the following treatment the symbol R is used to indicate the total effective primary series resistance, whether it is caused by r_p alone or by a combination of this and secondary loading.

The resistance R in series with the primary causes a voltage drop proportional to the current flowing through it, which is the magnetizing current. Now the form of the magnetizing current is far from being sine-wave, since it is distorted by the non-linear *B-H* characteristic of the core material.

This distorted current waveform has no bad effect when R is zero, but results in distortion of the voltage waveform which becomes progressively greater as R is increased, for any one fixed value of B_{max}.

The resulting harmonic distortion with silicon steel has been calculated by Dr. N. Partridge, and his results are embodied in the formula which follows (Refs. C24, C25, C26, C27).

$$\frac{V_h}{V_f} = S_H \frac{10^9}{8\pi^2} \frac{l}{N^2 A} \cdot \frac{R_A}{f}\left(1 - \frac{R_A}{4Z_f}\right) \tag{8}$$

This formula can be modified to include the core stacking factor, 90%, and to use inch units instead of centimetres. It then becomes

$$\frac{V_h}{V_f} = \frac{5.54 S_H l R_A}{N^2 A f}\left(1 - \frac{R_A}{4Z_f}\right) \tag{9}$$

where V_h = the harmonic voltage appearing across the primary,

$\quad\quad V_f$ = the fundamental voltage across the primary,

$\quad\quad S_H$ = the distortion coefficient of the magnetic material,

$\quad\quad l$ = the length of the magnetic path,

$\quad\quad N$ = number of primary turns,

$\quad\quad R_A$ = resistance (or equivalent resistance) in series with the primary, (refer under eqn. 4_f Sect. 3),

$\quad\quad A$ = cross-sectional area of core,

$\quad\quad f$ = frequency of fundamental in cycles per second,

$\quad\quad Z_f$ = impedance of primary at fundamental frequency $\approx 2\pi f L$

and $\quad L$ = inductance of primary in henrys at chosen flux density.

In most cases the final term $(1 - R_A/4Z_f)$ can be omitted with a further simplification.

The right hand side of this equation gives the value of the fractional harmonic distortion ; the percentage harmonic distortion may be obtained by multiplying this value by 100. The formula holds only for values of R/Z_f between 0 and 1 ; this limits its application to output circuits having a maximum attenuation of 3 db.

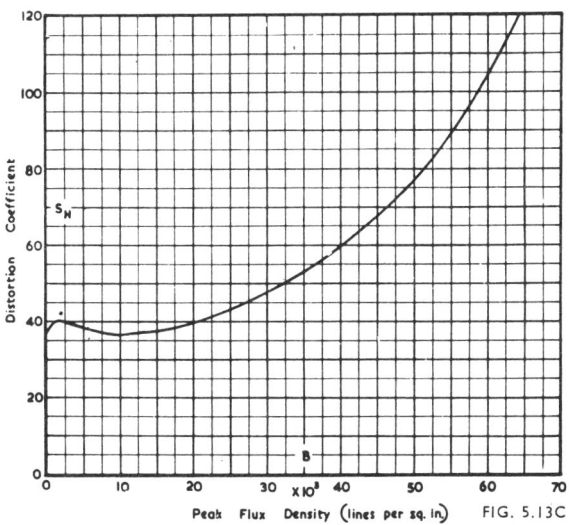

Peak Flux Density (lines per sq. in.)   FIG. 5.13C

Fig. 5.13C.   *The distortion coefficient of Silcor 2 as a function of B* (Ref. C25).

Values of S_H for Silcor 2 can be obtained from Fig. 5.13C. Similarly Fig. 5.13D can be used to determine the permeability of Silcor 2 and hence the inductance of the primary from the formula

$$L = \frac{2.88 N^2 A \mu}{10^8 l}\text{henrys} \tag{10}$$

where $\quad \mu$ = permeability at operating flux density

and core stacking factor is 90%.

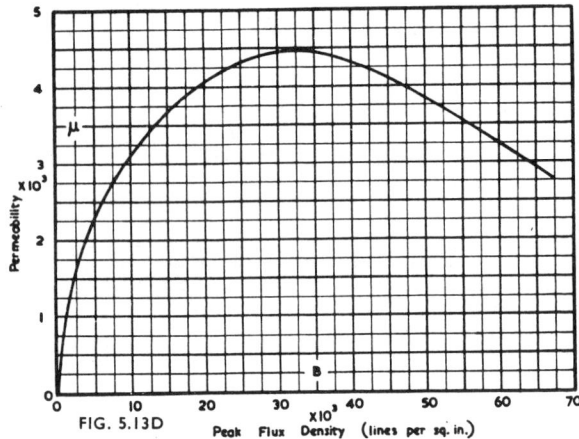

FIG. 5.13D

Fig. 5.13D. Variation of permeability with B at 50 c/s of Silcor 2 laminations, 0.014 in. thick (Ref. C25).

Both of these figures have been adapted from those published by Partridge (Ref. C25).

An example is quoted to demonstrate the use of these formulae.

Determine the transformer distortion produced when a pair of KT66 valves (**very** similar to type 6L6-G) connected as triodes are operated in conjunction with a transformer having the following characteristics.

Lowest frequency of operation	50 c/s.
Maximum flux density	40 000 lines/sq. in.
Core stack of M.E.A. 78 Pattern lams. Silcor 2	2 ins.
(Refer to Sect. 5 and also Fig. 5.18C for lamination data).	

Operating conditions of KT66 valves

Plate voltage	400 volts
Plate resistance (per valve)	1450 ohms
Load resistance (total)	10 000 ohms
Power output	14 watts

1st step. Calculate primary voltage

$$E = \sqrt{WR} = \sqrt{14 \times 10\,000} = 374 \text{ V.}$$

2nd step. Calculate primary turns

$$N = \frac{E \times 10^8}{4 \times fBA} \quad \text{(allowing 90\% core stacking factor)}$$

$$= \frac{374 \times 10^8}{4 \times 50 \times 4 \times 10^4 \times 1.25 \times 2} = 1870 \text{ turns.}$$

3rd step. Determine μ from Fig. 5.13D

$$\mu = 4300.$$

4th step. Calculate Z_f ($\approx 2\pi fL$)

$$Z_f = \frac{2\pi f\, 2.88\, N^2 A \mu}{10^8 l}$$

$$= \frac{2 \times \pi \times 50 \times 2.88 \times 1.87 \times 1.87 \times 10^6 \times 1.25 \times 2 \times 4300}{10^8 \times 7.5}$$

$$= 45\,300 \text{ ohms.}$$

5th step. Calculate R_A (parallel resistance of plate and load resistance).

$$R_A = \frac{(r_p + R_W) R_L}{r_p + R_W + R_L} \quad \text{(see eqn. 4, Sect. 3)}$$

assuming $R_W = 400$ ohms

then $R_A = \dfrac{(2900 + 400)\ 10\ 000}{13\ 300} = 2480$ ohms.

6th step.   Calculate $1 - (R_A/4Z_f)$
      $1 - (R_A/4Z_f) = 1 - (2480/4 \times 45\ 300) = 0.986.$
Thus this factor can be neglected without serious error.

7th step.   Determine S_H from Fig. 5.13C
      $S_H = 60.$

8th step.   Calculate fractional distortion
$$\frac{V_h}{V_f} = \frac{5.54\ S_H lR_A}{N^2 Af} = \frac{5.54 \times 60 \times 7.5 \times 2480}{1.87 \times 1.87 \times 10^6 \times 2.5 \times 50}$$
$$= .014.$$

Thus percentage distortion is $.014 \times 100$ or 1.4%.

Note : High fidelity output transformers may be designed with distortion less than 0.05% at maximum power output at 50 c/s.

The value of the distortion coefficient S_H is constant for any given material operating at any one value of B_{max}.   The value of S_H will be different for each harmonic, but its value for the third harmonic (S_3) is very close to the r.m.s. sum of all harmonics (S_{rms}) when there is no direct current component.

It will be noticed that the curve in Fig. 5.13C reaches a minimum at about $B_{max} = 10\ 000$ lines/sq. in., and that the distortion coefficient rises at both lower and higher values of B_{max}, although the lower rise is only slight.   Actually the lower part of the curve (below the knee) drops away rapidly and eventually reaches zero at $B_{max} = 0$. Some alloy core materials have appreciable values of the distortion coefficient even when approaching $B_{max} = 0$ (Ref. C9).

When there is a direct polarizing field, even as well as odd harmonic distortion are both evident.   Under these conditions, below 65 000 lines/sq. in., the r.m.s. sum of the second and third harmonic currents S_{2+3} approximates to the r.m.s. sum of all the harmonic currents.

It will be seen from this analysis that it is desirable to use a low impedance source, (e.g., triodes), and a high inductance primary for best results at low frequencies.

Distortion at high frequencies

At high frequencies distortion is produced, apart from normal valve distortion due to non-linearity of characteristics, by the leakage inductance and winding capacitances which change the magnitude and phase angle of the load impedance.

The load on the secondary, if a loudspeaker, also complicates this trouble.   Here again, a low impedance source is desirable ; a high impedance source will accentuate the distortion due to this effect.   With Class B amplifiers, it is essential that the leakage inductance between each half of the primary be as small as possible, otherwise there will not be proper cancellation of even order harmonics, and higher order harmonics will be generated (Ref. C23).   A static shield between primary and secondary will prevent any stray even harmonics being fed into the secondary by capacitive coupling.   This shield will alter the winding capacitances and increase the leakage inductance, hence it must be employed judiciously.

In Class B output transformers, high leakage inductance and winding capacitances cause distortion and decrease in power output.   To a considerable degree, these remarks on distortion at high frequencies also apply to Class AB transformers.

(iv) Designing for low leakage inductance*

Assuming that the turns and winding dimensions are kept constant, leakage inductance can be progressively reduced by interleaving the winding structure until the limit is reached when
$$a/3N^2 < c \tag{11}$$
where a = total thickness of all winding sections
      N = number of leakage flux areas
and c = thickness of each insulation section.

───

. *This treatment follows Crowhurst, N. H. (Ref. C28).

From eqn. (11) it will be seen that the insulation between the sections is the limiting factor. Fig. 5.13E shows that the largest value of N^2 for a given number of sections is achieved when there is a half section at the end of each winding structure. Although series connections are shown, similar results can be obtained by paralleling the sections. In this case all the turns in each paralleled section must be equal. An example of the use of Fig. 5.13H to determine leakage inductance follows :

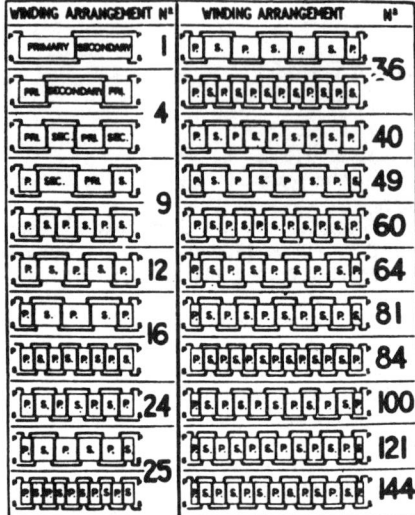

FIG. 5.13E

Fig. 5.13E. Table of winding arrangements (Ref. C28).

A push-pull output transformer is to be wound on a former 1.875 inches long and have a winding height of 0.4 inch, allowing for clearance. The mean length of turn is 9.5 inches. The primary and secondary turns are 2700 and 120 respectively. Insulation between sections is 0.015 inch. Assuming the seventh winding arrangement in left hand column of Fig. 5.13E, it is required to determine the leakage inductance between the whole primary and secondary. The total primary winding height is 0.24 inch and the total secondary winding height is 0.16 inch.

$$a = 0.4 \text{ inch.} \qquad N^2 = 16.$$

Referring to Fig. 5.13H, intercepts of a and N^2 give 0.008.

Adding $c = 0.015$ gives 0.023.

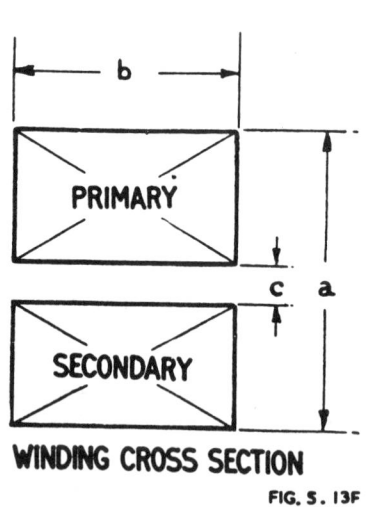

WINDING CROSS SECTION

FIG. 5.13F

Fig. 5.13F. Dimensions used in chart (Fig. 5.13H).

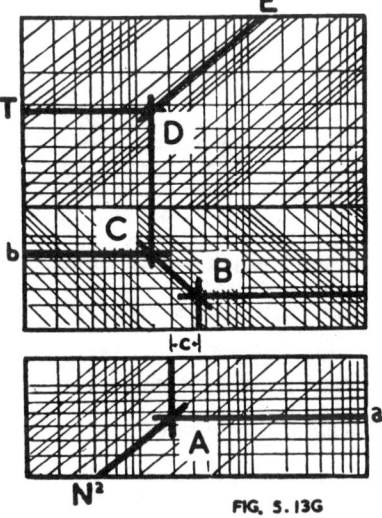

FIG. 5.13G

Fig. 5.13G. Showing use of chart (Fig. 5.13H).

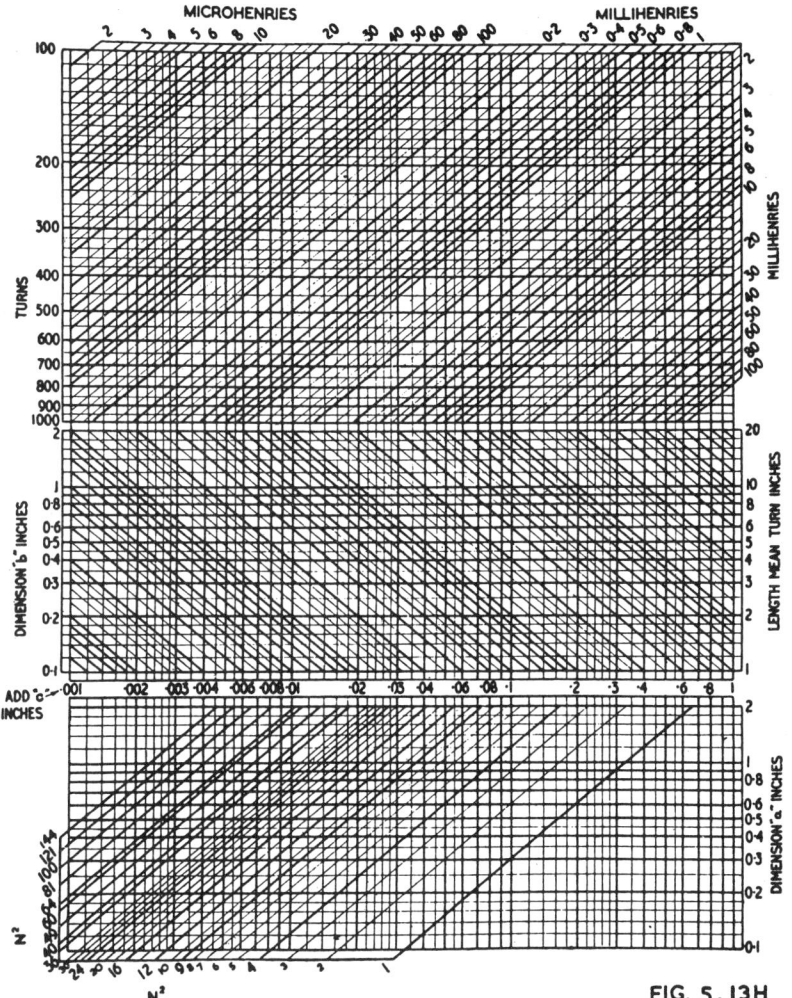

Fig. 5.13H.   Leakage inductance chart (Ref. C28).

Intercepts with the mean length of turn, 9.5 inches and winding width, *b*, of 1.875 inches gives a vertical of 0.012.

Assuming 270 turns, leakage inductance = 0.14 millihenry, or 14 millihenrys for 2700 turns.

(v) Winding ·capacitance*

This information assists in computing the various winding capacitances of multi-layer windings used in the construction of audio-frequency transformers, chokes and other equipment.   They apply to windings in which turns are wound on layer by layer, either interleaved or random wound, so that the P.D. between adjacent turns belonging to consecutive layers will be much greater than that between adjacent turns

---

*Reprint of an article by Crowhurst, N. H., in Electronic Eng.   (Ref. C29).

in the same layer.   The capacitance effect between adjacent turns in the same layer
is neglected, only that between layers being considered.   Capacitance between winding and core, and electrostatic screens, if used, must also receive attention.

Fig. 5.13J illustrates a cross section of a piece of winding in which layer interleaving is used.   It is seen that the dielectric between adjacent conductors in consecutive layers is complex both in shape and material.   The turns on the top layer shown
fall so that each turn drops in the space between two turns on the second layer, corru-
gating the interleaving material with a slight resulting increase in capacitance compared with that between the middle and bottom layers shown,   Due to the spiral
form of each layer, the position of turns in consecutive layers to one another will
change at different points round the direction of winding, thus the capacitance between any pair of layers will automatically take up the average value.   The composite
dielectric is made up of conductor insulation, most commonly enamel, interleaving
material and the triangular shaped spaces left between adjacent turns and the interleaving material.   These spaces will be filled with dry air if the windings are dried
out and hermetically sealed, or with impregnating compound if the windings are
vacuum impregnated.   The latter procedure will give rise to a somewhat higher
capacitance.

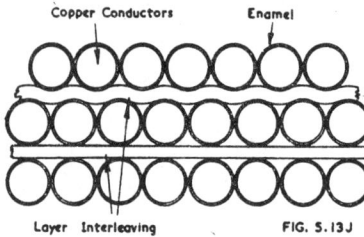

Fig. 5.13J. Section through layer interleaved winding (Ref. C29).

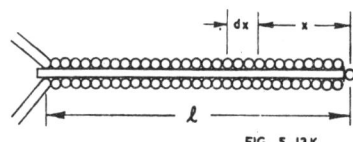

Fig. 5.13K. Effective capacitance of a pair of layers (Ref. C29).

In practice the major controlling factor determining the total capacitance between
two adjacent layers of winding is the thickness of the interleaving material, thus distributed capacitance may conveniently be estimated in terms of the thickness of interleaving material, giving this material a value of dielectric constant empirically obtained, allowing for the average effect of the other dielectrics in the composite arrangement.

Effective layer to layer capacitance

Take a winding having only two adjacent layers, its turns distributed uniformly
throughout the two layers as shown in section at Fig. 5.13K.   Consider an element
dx at a distance x from the end of the winding where the conductor steps up from one
layer to the other.   In the complete winding the elemental capacitance due to the
section dx will be transformed so that it can be represented as an equivalent value
across the whole winding.   If the length of the whole layer is l and the capacitance
per unit layer length C, then the effective capacitance of the element referred to the
whole winding will be $(x/l)^2 C.dx$.   The capacitance due to the whole winding will be

$$\int_0^1 (x/l)^2 C.dx = \frac{1}{3} l.C.$$

Thus the effective capacitance of such a two-layer winding is one-third of the capacitance between two layers measured when their far ends are unconnected.

Take now a winding consisting of n whole layers : there will be $(n-1)$ adjacent
pairs of layers throughout the winding and the effective capacitance of each pair of
layers, referred to the whole winding, will be a capacitance of $(2/n)^2$ times their capacit-

ance referred to the high potential end and considered as a pair. Thus the capacitance of n whole layers referred to the whole winding becomes

$$\frac{4}{3} \cdot \frac{(n-1)}{n^2} \cdot Cl$$

For large values of n the capacitance becomes inversely proportional to the number of layers. Fig. 5.13Q illustrates a typical winding shape together with the dimensions as used in the related diagrams. The capacitance per layer, given in the foregoing formula as Cl, is proportional to the product of the length per mean turn L_{mt}, and the length of layer L_w.

Vertical sectionalizing

It is sometimes of advantage to sectionalize the winding as shown at Fig. 5.13L, each vertical space being filled completely before proceeding to fill the next one.

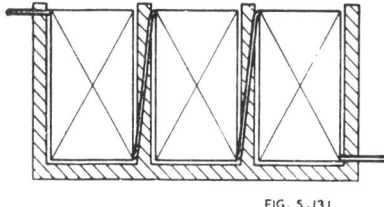

Fig. 5.13L. Sectionalizing to reduce capacitance (Ref. C29).

FIG. 5.13 L

Although physically the winding will have the same overall cross-sectional dimensions its self-capacitance will be equivalent to that of a winding having $1/N \times$ layer length and $N \times n$ layers, where N is the number of vertical sections. The distributed capacitance of the winding due to such sectionalizing thus is reduced by the factor $1/N^2$. Note that this rule applies only to referred interlayer capacitance and does not apply to capacitance between the top and bottom of the winding and adjacent windings or screens. The various reduction factors for vertical sectionalizing are given in Table 4.

TABLE 4

Number of vertical sections	Distributed capacitance component	Winding to screen capacitance arrangement as Figure 5.13M		
		(a) One side earthy	(a) Centre point earthy	(b) One side earthy
1	1	1	0.5	0
2	0.25	0.75	0.25	0.125
3	0.111	0.704	0.185	0.185
4	0.0625	0.6875	0.1875	0.219
5	0.04	0.68	0.168	0.24
6	0.0278	0.676	0.176	0.255
∞	0*	0.667	0.167	0.333

*The method of winding is here changed so that for the purposes of this column, the dimensions L_w and T_w will change places.

Effect of mixing windings

In the design of a transformer it is often necessary to mix the primary and secondary windings in order to reduce leakage inductance. This arrangement will generally be a disadvantage as regards minimizing winding capacitance, since it exposes greater surface area of winding in proximity to either the other winding or an earthed screen. If the ratio of the transformer is fairly high, then from the high impedance winding the whole of the low impedance winding appears at common audio potential, usually earthy. But if the ratio of the transformer is not very high, capacitance between points at differing audio potentials in the two windings may have serious effects, and it is generally best to arrange the windings so as to avoid such capacitance.

Fig. 5.13N shows a cross-section suitable for an inter-valve transformer designed to operate two valves in push-pull from a single valve on the primary side. The H.T. end of the primary is earthy and is therefore diagrammatically earthed. The high potential end of the primary is adjacent to the earthy end of one of the half secondaries so that the capacitance between windings at this point is effectively from anode to earth. The two high potential ends of the secondary are remote from the primary and so minimize the possibility of unbalanced capacitance transfer from primary to one half secondary.

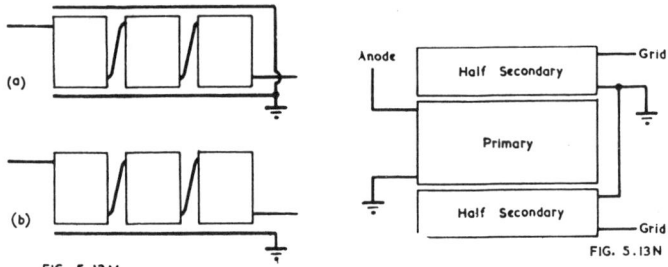

FIG. 5.13M   FIG. 5.13N

Fig. 5.13M. *Arrangements using vertical sectionalizing. See Table 1 (Ref. C29).*
Fig. 5.13N. *Push-pull secondary intervalve transformer secondary arrangements*
                  (Ref. C29).

Another problem which often arises is in the design of push-pull output transformers, particularly for Class AB or Class B circuits, where it is essential that each half of the primary be well coupled to the whole secondary. From the viewpoint of leakage inductance and winding resistance, it is unimportant whether the secondary sections are connected in series or parallel. Fig. 5.13P illustrates three arrangements for a transformer of this type, each of which may be best suited under different circumstances. At (a) is an arrangement which gives minimum primary capacitance, but suffers from the defect that leakage inductance and winding resistance are unequal for the two primary halves. For Class A operation using valves requiring an optimum load of high impedance this arrangement is sometimes the best. At (b) is an arrangement intended to equalize winding resistance and leakage inductance from each half-primary to the whole secondary as well as primary self-capacitance. This arrangement is particularly suited to circuits employing low loading Class AB or Class B operation. The alternative arrangement shown at (c) results in a slightly lower referred capacitance across one half only of the primary. In general this unbalance is not desirable, but if leakage inductance is adequately low, the coupling between all the windings may be so good that the reduction in capacitance may be apparent across the whole primary.

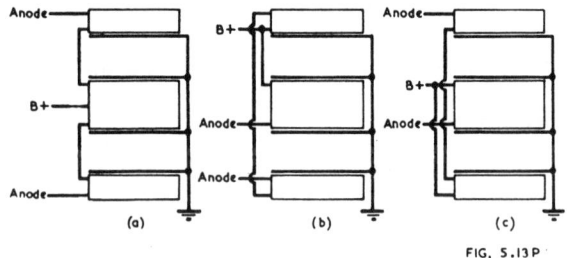

                              FIG. 5.13P

Fig. 5.13P. *Arrangements for push-pull output transformer (Ref. C29).*

TABLE 5.

WINDING & SCREEN ARRANGEMENT & CONNECTION	CAPACITANCE FACTOR		WINDING & SCREEN ARRANGEMENT & CONNECTION	CAPACITANCE FACTOR	
	One side earthy	Centre point earthy		One side earthy	Centre point earthy
(diagram)	—	·25	*(diagram)*	1·25	·25
(diagram)	1	·5	*(diagram)*	1·11	·11
(diagram)	·22	·055	*(diagram)*	2·11	·611
(diagram)	·5	—	*(diagram)*	* 1	1
(diagram)	1·5	·5	*(diagram)*	3	1·5
(diagram)	* —	·5	*(diagram)*	1·94	·44
(diagram)	2	1	*(diagram)*	1·75	·25

Table 5 gives a pictorial representation of various ways in which high impedance windings may be arranged in relation to earthy points shown as screens. The table is equally applicable if these points are earthy low impedance windings. The capacitance factors for alternative connexions of the windings, with either one side or the centre point at earth potential, are given relative to the average capacitance between one end layer of the winding and one screen. The two arrangements marked with an asterisk indicate that it is necessary to reverse the direction of winding in order to achieve the capacitance factor shown.

Random winding

In what is known as random winding, no interleaving material is used. For ideal

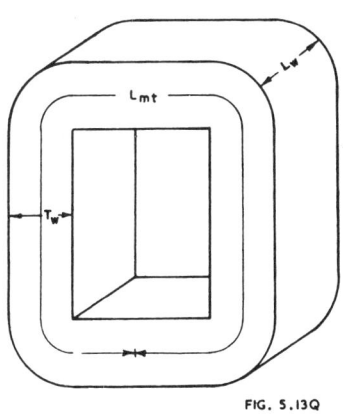

FIG. 5.13Q

Fig. 5.13Q. Dimensions used in figures 5.13R, 5.13S, 5.13T, 5.13U (Ref. C29).

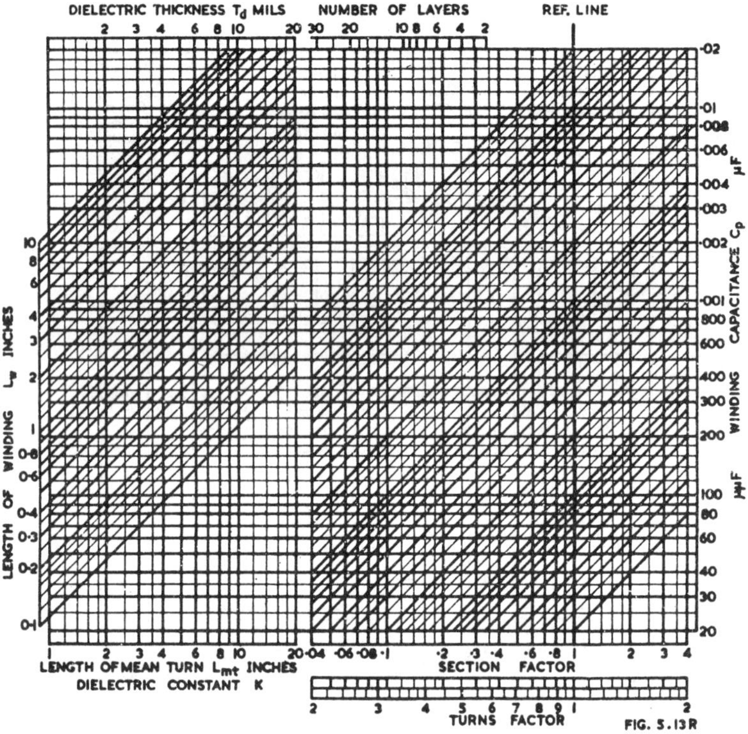

DIELECTRIC THICKNESS T$_d$ MILS   NUMBER OF LAYERS   REF. LINE

Fig. 5.13R and 5.13S. For distributed capacitance due to layer winding (Ref. C29).
Refer length of winding L$_w$, on their respective scales, to intercept at A then along the
horizontal reference lines to intercept with the thickness of dielectric material T$_d$,
at B. From this point refer along the slanting reference lines to intercept with the
empirically determined value of dielectric constant k, at C, then along the horizontal
reference lines to the unity reference vertical at D. From this point refer down the
slanting reference lines to intercept with the number of layers n, at E, whence the
referred capacitance is read off on the scale at the right.

For interwinding, or winding to screen capacitance: As above, for the winding
dimensions and dielectric thickness to a point corresponding to D, whence, for winding

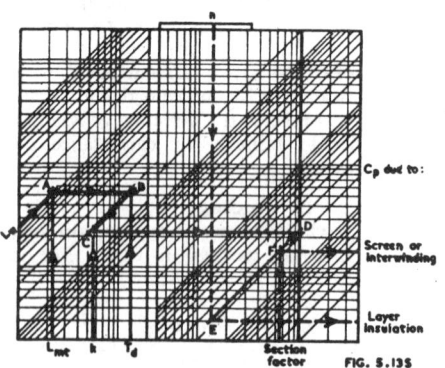

to screen capacitance, or interwinding capacitance when the
other winding is all at low potential, refer down the slanting reference lines to intercept with the
section capacitance factor obtained from Table 4. For interwinding capacitance where there
is appreciable potential in the
adjacent portions of both windings, individual attention will
be necessary for each interwinding
space, and the turns factor scale
will assist here.

random winding the layers should be built up so that at all times during winding the top surface is level. Failure to do this will not greatly affect the self-capacitance, but will result in increased danger of breakdown due to electrical or mechanical stresses. As the number of layers is always large, it is convenient to reduce the calculation to simple terms of the winding dimensions.

The essential variables are : length of winding L_w, length of mean turn L_{mt}, winding thickness T_w (see Fig. 5.13Q), and number of turns, T. Considering variation of each of these quantities in turn, the others being taken as constant : variation of L_w will vary the referred capacitance per layer as before, and additionally the effective number of layers will vary inversely as $L_w^{\frac{1}{2}}$; variation of L_{mt} simply varies the

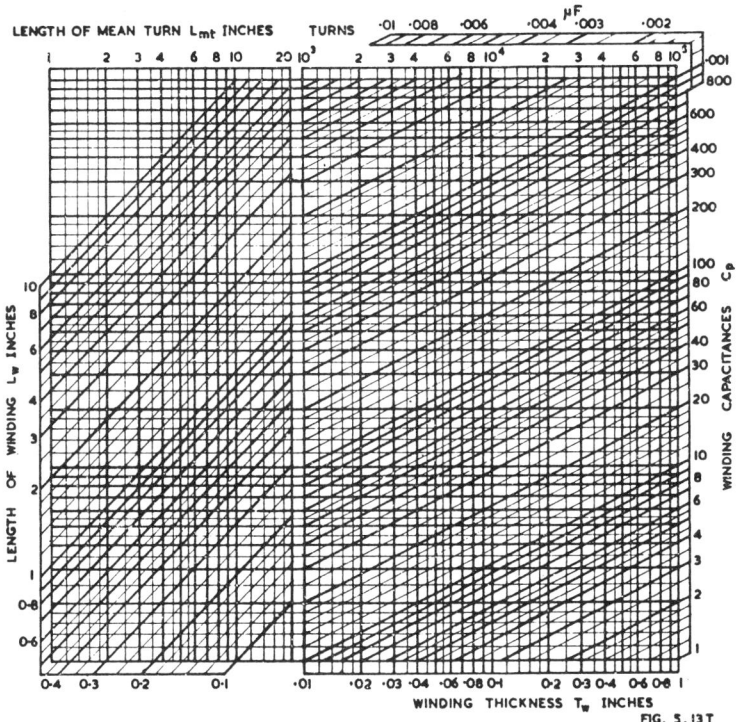

FIG. 5.13 T

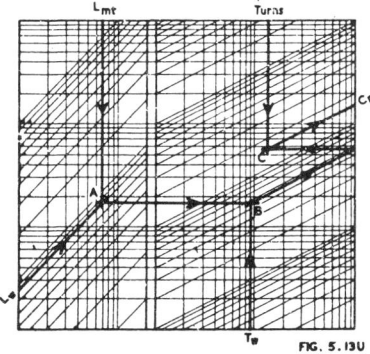

FIG. 5.13U

Fig. 5.13T and 5.13U (Ref. C29). For distributed capacitance due to random winding. Refer length of winding L_w, and length of mean turn L_{mt}, on their respective scales, to intercept at A then along the horizontal reference lines to intercept with the vertical line for thickness of winding T_w, at B. From here refer up the slanting reference lines to the right-hand edge of the data sheet, and then back along the horizontal lines to intercept with the vertical lines for the number of turns at C from which referred capacitance is read off along the slanting lines at C_p.

layer size, as before ; variation of both T_w and T varies the number of layers in direct proportion to $T_w^{\frac{1}{4}}$ or T. Thus the whole expression for variation of capacitance can be written,

$$C_p \propto \frac{L_{mt} L_w^{3/2}}{T_w^{\frac{1}{4}} T^{\frac{1}{2}}}$$

Figures 5.13T and 5.13U are based on this relation and empirical values obtained from average results with random windings.

Example 1

A push-pull output transformer is arranged as at Fig. 5.13N : primary winding has a total of 12 layers, $T_d = 3$, $k = 2$: insulation between primary and screen, $T_d = 15$, $k = 5$; main dimensions, $L_w = 2$ in., $L_{mt} = 8$ in.

Distributed capacitance, using Figs. 5.13R and 5.13S, $L_w = 2$ in., $L_{mt} = 8$ in., $T_d = 3$, $k = 2$ and $n = 12$, is $C_p = 240$ $\mu\mu$F.

Capacitance from each primary to screen, substituting $T_d = 15$, $k = 5$, is $C_p = 1200$ $\mu\mu$F. The capacitance factors for the arrangement of Fig. 5.13N are : (a) 0.25 ; (b) 0.5 ; (c) 0.375. Thus the total capacitance referred to the whole primary for each method of connexion is,

    (a) 300 $\mu\mu$F + 240 $\mu\mu$F = 540 $\mu\mu$F.
    (b) 600 $\mu\mu$F + 240 $\mu\mu$F = 840 $\mu\mu$F.
    (c) 450 $\mu\mu$F + 240 $\mu\mu$F = 690 $\mu\mu$F.

Example 2

An intervalve transformer, to operate from push-pull to push-pull, uses a simple arrangement having both windings all in one section : $L_w = 0.6$ in, $L_{mt} = 2.5$ in, T_w (each winding) = 0.1 in. ; insulation between windings, T_d, $k = 3$; Turns, 4 000 c.t./12 000 c.t.

Distributed capacitance, using Figs. 5.13T and 13U $L_w = 0.6$ in., $L_{mt} = 2.5$ in., $T_w = 0.1$ in., T (primary) = 4 000, is $C_p = 58$ $\mu\mu$F. For secondary, $T = 12\,000$, $C_p = 34$ $\mu\mu$F.

Capacitance coupling between one-half primary and one-half secondary, using Figs. 5.13R and 13S, actual capacitance, $L_w = 0.6$ in., $L_{mt} = 2.5$ in., $T_d = 10$, $k = 3$, is $C_p = 100$ $\mu\mu$F. Both windings wound in same direction, turns factor across this capacitance referred to whole primary is $\frac{1}{2} + 1\frac{1}{2} = 2$, so referred capacitance is 400 $\mu\mu$F. Windings wound opposite directions, turns factor referred to primary is $1\frac{1}{2} - \frac{1}{2} = 1$, so referred capacitance is reduced to 100 $\mu\mu$F. Referring these two values to secondary, turns factors are $1/2 + 1/6 = 2/3$ or $1/2 - 1/6 = 1/3$, giving capacitance values referred to whole secondary of 45 $\mu\mu$F or 11 $\mu\mu$F respectively.

A complete analysis would need to consider leakage inductance between each primary and each half secondary, and separate source and load impedances applied to each half. For this purpose, the primary and secondary shunt capacitances across each half would be 116 $\mu\mu$F and 68 $\mu\mu$F respectively, while the capacitance coupling would be 1,600 $\mu\mu$F or 400 $\mu\mu$F referred to half primary. The secondary shunt capacitances referred to the half primaries would be 610 $\mu\mu$F.

Example 3

A direct coupled inter-valve transformer is arranged as at Fig. 5.13M : $L_w = 1.5$ in., $L_{mt} = 6$ in., $T_d = 20$, $k = 1$ (air spaced) ; T_w (each whole winding) = 0.2 in. ; Turns 4 000/12 000 c.t.

Distributed capacitance, using Figures 5.13T and 5.13U, $L_w = 1.5$ in., $L_{mt} = 6$ in., $T_w = 0.2$ in., T (primary) = 4 000, gives $C_p = 400$ $\mu\mu$F. Secondary, $T = 12\,000$ gives $C_p = 225$ $\mu\mu$F (or 450 $\mu\mu$F per half secondary).

Interwinding capacitance, using Figs. 5.13R and 5.13S, actual capacitance, $L_w = 1.5$ in., $L_{mt} = 6$ in., $T_d = 20$, $k = 1$, give RC_p just over 90 $\mu\mu$F say 100 $\mu\mu$F.

Vertical sectionalizing will reduce the distributed component in each case, but will also vary the interwinding capacitance. Using the information in Table 4 the results may be presented as in Table 6.

In practice three sections for the primary and four or five sections for each half-secondary will be best, remembering capacitance reduction is more important in the secondary. By making the earthy end of primary and secondary at opposite ends of the vertical groups, interwinding capacitance coupling effects are minimized.

TABLE 6

Number of vertical sections	Primary capacitance			Half secondary capacitance		
	Dis-tributed	Inter-winding	Total	Dis-tributed	Inter-winding	Total
1	400	100	500	450	—	450
2	100	75	175	112.5	12.5	125
3	44.4	70.4	115	50	18.5	68.5
4	25	69	94	28	22	50
5	16	68	84	18	24	42
6	—	—	—	12.5	25.5	38

(vi) Tests on output transformers
Summary of R.M.A. Standard SE-106A (Sound systems)

The distortion shall be measured with a zero-impedance source of voltage in series with a pure resistance R_{TG} of value 0.4 ($\pm$ 5%) times the square of the standard distribution voltage, V_{TR}, from which the tap is designed to work, divided by the manufacturer's rating, W_{TR}, for the power drawn by that tap at that distribution voltage :

$$R_{TG} = (0.4\ V_{TR}^2/W_{TR})(1 \pm 0.05).$$

Measurements shall be made at the lowest frequency of the rated frequency response or 100 c/s, whichever is the higher ; at 400 c/s and at 5000 c/s if within the rated frequency response.

The power-handling capacity of a speaker matching transformer is the maximum r.m.s. power drawn by the transformer at which the specified distortion (which shall be not more than 2%) is not exceeded. The power drawn by the transformer, W_T, shall be determined by dividing the square of the actual voltage measured across the primary terminals, V_{TP}, by the square of the standard distribution line voltage, V_{TR}, from which the primary tap is designed to work, and multiplying this quotient by the manufacturer's rating for the power, W_{TR}, drawn by that tap at that distribution voltage :

$$W_T = W_{TR}\ (V_{TP}/V_{TR})^2$$

The rating shall be determined as follows : The power at the stated distortion at 100 c/s shall be multiplied by 2. This figure shall be compared with the other measurements made at the other test frequencies, and the lowest power figure shall be taken as the power-handling capacity. In case the transformer is provided with more than one primary tap, the rating shall be given for the tap drawing the highest power at the rated distribution voltage. If there is more than one secondary tap, the power rating shall be given for the tap with the lowest measured power rating when properly terminated.

If the transformer is to be used in a system employing emphasized bass, a transformer must be chosen which has a rating higher than the nominal power to be handled in proportion to the bass emphasis employed. Likewise, transformers to handle organ music must have a rating at least four times the nominal power to be handled.

The frequency response of a speaker matching transformer is the variation of output voltage as a function of frequency, with a constant source voltage in series with a known impedance connected to the primary, expressed as a variation in db relative to the output voltage at 400 c/s.

For measurement, the transformer shall be connected as for distortion measurement (see above). The frequency response shall be measured using a constant source voltage, which will deliver one-half rated input power, W_{TR}, to the transformer at 400 c/s

The loss of a speaker matching transformer is the inverse ratio of the power delivered by the secondary of the transformer to a pure resistance equivalent to the rated load impedance, to the power delivered by the same source if the transformer is replaced by an ideal transformer of the same impedance ratio, expressed in db. For measurement, the transformer shall be connected as for distortion measurement (see above).

The impedance presented by the ideal transformer to the source shall be taken as
$$R_{T1} = V_{TR}{}^2 / W_{TR}.$$

The power delivered to the secondary load is $(V_{TS}{}^2 / R_{TL})$ where V_{TS} is the voltage across the load resistance R_{TL}.

The power delivered to the ideal transformer is $(V_{T1}{}^2 / R_{T1})$ where V_{T1} is the voltage across the load resistance R_{T1}.

The loss is given then by
$$\text{Loss} = 10 \log_{10} \frac{V_{T1}{}^2 / R_{T1}}{V_{TS}{}^2 / R_{TL}}.$$

The loss shall be measured at 400 c/s and at a value of source voltage at which rated power is delivered to the ideal transformer. The loss shall not exceed 2 db (equivalent to 63% minimum efficiency).

(vii) Specifications for a-f transformers

The following details are suggested for forming part of a specification for a transformer. See also Ref. C32.

In all cases, it is desirable to submit a circuit showing the transformer application when writing a specification.

(1) Input transformers
   (a) Operating level ; this should be quoted in db above or below specified reference level, usually 1 milliwatt.
   (b) Frequency response with permissible variation in db from a reference frequency, generally 1000 c/s.
       Conditions of measurement must be specified—usually normal operating conditions.
   (c) Impedance ratio or turns ratio.
   (d) Positions of any taps should be stated.
   (e) Source and source impedance.
   (f) Load and load impedance. This may be the grid of a following amplifier valve. The secondary winding may also be shunted by a frequency-correcting network ; if so, full details should be given.
   (g) Total r.m.s. harmonic distortion ; this should be measured at max. output at the lowest frequency of interest.
   (h) Minimum resonant frequency.
   (i) Insertion loss in db—frequently quoted at 400 c/s.
   (j) Permissible phase characteristics at lowest and highest frequencies of interest.
   (k) Direct currents in windings.
   (l) Magnetic and electrostatic shielding.

(2) Interstage transformers
In general, as for input transformers, with the addition that the type of valves used in the preceding stage, together with their operating conditions, should be specified. Where push-pull input is intended, the maximum out-of-balance current should be stated.

(3) Output transformers
As (2) above. Where multiple secondary windings are employed the power to be delivered to each should be stated.

SECTION 4 : MAGNETIC CIRCUIT THEORY

(i) Fundamental magnetic relationships (ii) The magnetic circuit (iii) Magnetic units and conversion factors.

(i) Fundamental magnetic relationships

Just as we have an electrical circuit, so the core of a transformer can be regarded as a magnetic circuit through which a flux passes, its value depending on the magneto-motive force producing it and on the nature of the magnetic circuit. These are related by an equation resembling Ohm's Law for electrical circuits :

$$F = \phi R \text{ or } \phi = F/R \text{ or } R = F/\phi \tag{1}$$

where F = magnetomotive force (m.m.f.) in **gilberts** (c.g.s. electromagnetic units) analogous to electromotive force (e.m.f.) in electrical circuits,

ϕ = total flux = the total number of lines of flux (or maxwells ; 1 maxwell = 1 line of flux),

R = reluctance (equivalent to resistance in electrical circuits). The c.g.s. electromagnetic unit of reluctance is the reluctance of 1 cubic centimetre of vacuum, which is very closely that of 1 cubic centimetre of air. Reluctances are combined in series or parallel like resistances ; when in series they are added.

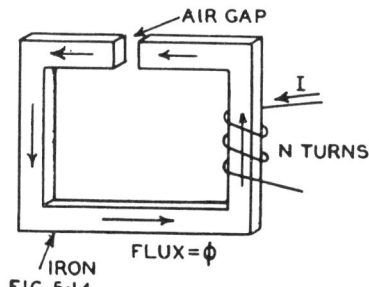

Fig. 5.14. *Typical magnetic circuit with iron core and air-gap.*

A typical magnetic circuit is shown in Fig. 5.14 where we have the greater part of the magnetic circuit through iron, and a short length (the air gap) through air. In this case the total reluctance is the sum of the reluctances of the iron-circuit and the air gap. The flux ϕ is caused by N turns of wire carrying a current I amperes and producing a magnetomotive force given by

$$F = (4\pi/10) NI \approx 1.257 NI \tag{2}$$

where F = magnetomotive force in gilberts,
N = number of turns,
and I = current in amperes.
Thus 1 ampere turn $\approx$ 1.257 gilberts. $\tag{3}$

Instead of considering the total flux, it is often more convenient to speak of the **flux density**, that is the number of lines (maxwells), per square inch or per square centimetre (c.g.s. electromagnetic unit = 1 gauss = 1 maxwell per square centimetre). The symbol for flux density is B. Thus

$$\phi = BA \tag{4}$$

where ϕ = total flux,
B = flux density, either in lines per square inch or in gauss (maxwells per square centimetre),
and A = cross sectional area of magnetic path (practically equal to the cross sectional area of the core in the iron section), in square inches or square centimetres respectively.

The **magnetizing force** (also known as the magnetic potential gradient or the magnetic field intensity) is defined as the magnetomotive force per unit length of path :

$$H = F/l \qquad (5)$$

where H = magnetizing force in oersteds (or gilberts/centimetre),
$\quad F$ = magnetomotive force in gilberts,
and $\quad l$ = length of path in centimetres.

Alternatively, if F is expressed in ampere-turns, H may be expressed in ampere-turns per inch or per centimetre.

The **permeability** (μ) is defined by the relationship ;

$$\mu = B/H \qquad (6)$$

where B = flux density in gauss (maxwells per square centimetre),
$\quad \mu$ = permeability*,
and $\quad H$ = magnetizing force in oersteds.
In air, H is numerically equal to the flux density (B).

Permeability is the equivalent of conductivity in electrical circuits. Permeability in iron cores is not constant, but varies when the flux is varied. The relationship between B, H and μ is shown by the " BH characteristics " of the iron, as shown for example in Fig. 5.15. The value of μ at any point is the value of B divided by the value of H at that point.

For example, the permeability at point C is equal to OE/OD, where OE represents the flux density at point C, and OD represents the magnetizing force at point C. The permeability is therefore the slope of the line OC. A curve may be drawn indicating the value of μ for any value of H, and this curve may be plotted on the same graph with, of course, the addition of a μ scale (dashed curve in Fig. 5.15).

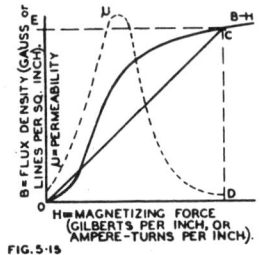

FIG.5·15

Fig. 5.15. BH characteristics of a typical transformer steel. The dashed curve is the permeability μ where $\mu = B/H$.

The **incremental permeability** is the permeability when an alternating magnetizing force is superimposed on a direct magnetizing force.

The **initial permeability** is the permeability at values of H approaching zero.

The **hysteresis loop** of a typical magnetic material is shown in Fig. 5.16. When the magnetizing force increases from zero (demagnetized condition) to the positive peak, the B-H characteristic is followed from O to A, where A represents the maximum (peak) values of H and B. As the value of H decreases to zero and then increases in the opposite direction, the path followed is along the curve ACDE, where E is the negative equivalent to A, occurring half a cycle later than A. From E, the path followed is along the curve EFGA†. The area of the curve ACDEFGA represents power loss, known as hysteresis loss, in the magnetic material.

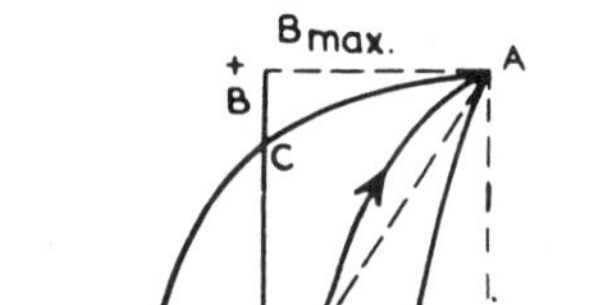

Point C is the value of B for zero magnetizing force (i.e., $H = 0$) and it represents the residual magnetism ; this value of B is called the **remanence**, or **remanent flux density**‡.

FIG. 5·16

Fig. 5.16. Hysteresis loop of a typical magnetic material.

---

*Strictly speaking, μ is measured in gauss per oersted, but it is common practice in engineering work to speak of the permeability as a pure number which is the ratio between μ and μ_0 where μ_0 is the permeability of vacuum, and has a value of unity.

†Actually this " cyclic condition " is not reached until after a number of cycles have occurred. In the early cycles the position of A falls slightly each cycle.

‡This strictly applies only to the initial cycle ; the term residual flux density is used for symmetrical cyclically magnetized conditions. The latter term is also sometimes used when it is not desired to distinguish between the initial and cyclic conditions.

Point D is that at which the applied negative magnetizing force brings the value of the residual magnetism to zero.

The length OC is called the **coercive force** (strictly this applies only for symmetrical cyclically magnetized conditions).

The **average permeability** is the slope of the straight line EOA (shown dashed in Fig. 5.16).

The locus of the extremities (A or E) of the normal hysteresis loops of a material is called its **normal magnetization curve**; this is the same as the *B-H* curve of Fig. 5.15.

When alternating current is passed through the winding, the iron will pass through the hysteresis curve ACDEFGA (Fig. 5.16) each cycle. The maximum value of H is called H_{max}, and the corresponding value of B is called B_{max}, the curve being normally symmetrical in the positive and negative directions in the absence of a direct current component.

When dealing with alternating currents, it is usual to quote ampere-turns per inch in r.m.s. values; the corresponding values of flux density may be quoted either in terms of B or B_{max}. It is obvious that $H_{max} = \sqrt{2}H$, but B_{max} is normally greater than $\sqrt{2}B$.

(ii) The magnetic circuit

A typical magnetic circuit is shown in Fig. 5.14. Certain assumptions are generally made for simple theoretical treatment, including—

1. That the flux confines itself entirely to the iron over the whole length of the iron path (in practice there is always some leakage flux, which is more serious when there is an air gap).

2. That the flux is uniformly distributed over the cross-sectional area of the iron.

In Fig. 5.14 we therefore have:

Total magnetomotive force $\qquad F \approx 1.26\ NI$ from (2)

Total reluctance $\qquad R = R_{iron} + R_{air}$

where R_{iron} = reluctance of iron path
and R_{air} = reluctance of air path.

Total flux $\phi \quad = F/R = F/(R_{iron} + R_{air})$ from (1).

The reluctance of the air gap is given by
$$R_{air} = F/\phi = Hl/BA = kl/A \qquad (7)$$
where l = length of air gap,
A = equivalent area of air gap, allowing for " fringing "
$\approx (a + l)(b + l)$
and a, b = actual dimensions of pole faces.
Values of k are given by the table below:

H	oersteds	ampere turns/cm	ampere turns/inch
B	gauss	gauss	lines/sq. inch
l	centimetres	centimetres	inches
A	sq. centimetres	sq. centimetres	square inches
k	1	0.796	0.313

The reluctance of the iron path is not constant, so that the best approach is graphical.

The magnetic **potential difference** (U) is the equivalent of potential difference in electrical circuits. The sum of the potential differences around any magnetic circuit is equal to the applied magnetomotive force.

Applying this to Fig. 5.14,
$$F = U_{iron} + U_{air} \qquad (8)$$

where F = total magnetomotive force,

U_{iron} = magnetic potential difference along the whole length of iron,

and U_{air} = magnetic potential difference across the air gap.

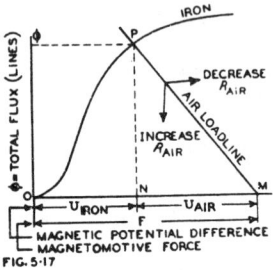

FIG. 5·17

Fig. 5.17. *Total flux versus magnetic potential difference, with air loadline.*

This is applied in Fig. 5.17 where OM represents the applied magnetomotive force ($F \approx 1.26\ NI$), and the curve shows the total flux (ϕ) plotted against F for iron only. The shape of the curve is the same as that of the B-H characteristic of Fig. 5.15, but the vertical scale $\phi = Ba$, and the horizontal scale $F = lH$.

The " loadline " through M represents the effect of the air-gap ; it follows the equation

$$U = F - \phi R_{air} \qquad (9)$$

and its slope is $-(1/R_{air})$. The intersection of the " loadline " and the curve at point P gives the operating point. Therefore ON represents the magnetic potential difference along the iron path, while NM represents the magnetic potential difference across the air gap.

It will thus be seen that variation of the air gap merely changes the slope of the " loadline " PM and moves point P, without changing the base line OM.

In order to be more generally applicable, the scales of Fig. 5.17 may be changed—

vertical scale :   from ϕ to B (Note $\phi = Ba$)

horizontal scale :   from F to H (Note $F = lH$)

    from U to U/l (i.e., magnetic potential difference per inch)

as in Fig. 5.18A.

(iii) Magnetic units and conversion factors

The basic units generally adopted are the c.g.s. electromagnetic units such as the gauss, the oersted and the gilbert, with the centimetre as the unit of length. Practical units such as the lines per square inch and ampere-turns per inch are widely used in engineering design. More recently, the Giorgi M.K.S. system, with its webers and webers per square metre, has achieved considerable popularity. The full range of these various systems of units is given in Chapter 38 Sect. 1.

The following table of conversion factors will be helpful in converting from one system to another.

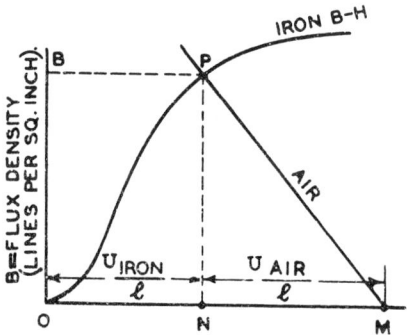

H= MAGNETIZING FORCE.
(GILBERTS PER INCH, OR AMPERE-TURNS PER INCH).
FIG. 5·18A

Fig. 5.18A. *Flux density versus magnetizing force, with air loadline.*

Magnetic Units—Conversion Factors

Multiply	by	to obtain
F in ampere-turns	$0.4\pi = 1.257$	F in gilberts
F in gilberts	$1/0.4\pi = 0.796$	F in ampere-turns
F in pragilberts*	0.1	F in gilberts
F in gilberts	10	F in pragilberts*
F in ampere-turns	$4\pi = 12.57$	F in pragilberts*
F in pragilberts*	$1/4\pi = 0.0796$	F in ampere-turns
H in ampere-turns/in.	$0.4\pi/2.54 = 0.495$	H in oersteds
H in oersteds	$2.54/0.4\pi = 2.02$	H in ampere-turns/in.
H in praoersteds*	10^{-3}	H in oersteds
H in oersteds	10^3	H in praoersteds*
H in ampere-turns/in.	495	H in praoersteds*
H in praoersteds*	0.00202	H in ampere turns/in.
B in maxwells/sq. in.	$1/6.45 = 0.155$	B in gauss
B in gauss B in maxwells/sq. cm.	6.45	B in maxwells/sq. in.
B in webers/sq. metre*	10^4	B in gauss
B in gauss	10^{-4}	B in webers/sq. metre*
B in maxwell/sq. in.	$10^{-4}/6.45 = 0.155 \times 10^{-4}$	B in webers/sq. metre*
B in webers/sq. metre*	6.45×10^4	B in maxwells/sq. in.
ϕ in maxwells ϕ in lines of flux	10^{-8}	ϕ in webers*
ϕ in webers*	10^8	ϕ in maxwells

SECTION 5 : POWER TRANSFORMERS

(i) General (ii) Core material and laminations (iii) Primary and secondary turns (iv) Currents in windings (v) Temperature rise (vi) Typical design (vii) Specifications for power transformers.

(i) General

The general design principles of power transformers have been dealt with in detail elsewhere (Refs. D1, D2, D3, D4) but an outline is given below of the design procedure for small power transformers for use in radio and electronic equipment. For transformers of this kind, efficiencies ranging from 80 to 90 per cent are common. The ratio of copper to iron losses is usually about 2 to 1. Winding capacitances have little effect on circuit operation and are usually neglected. For radio receivers it is common practice to provide some form of electrostatic screening between the primary and the other windings. This can be achieved in one of several ways (Ref. D5):

(a) By using a turn of shim copper or brass of full winding width between primary and secondary, taking care to insulate the ends to avoid a shorted turn. This shield is then earthed.

(b) By winding the earthed low voltage filament between primary and secondary.

(c) By winding the high voltage secondary in two separate halves, one on top of the other, over the primary. The innermost and outermost leads are joined and become the centre tap which is then earthed. The two leads from the middle of the winding then become the high tension outers, and are connected to the rectifier plates. In this way the capacitance from primary to secondary is made very small. It is important in this method of construction to ensure that adequate insulation is used between the two plate leads within the winding, as the whole of the potential difference of the high tension winding appears between them.

With the normal method of construction the leakage inductance between either half of the secondary and the primary is unequal. In large transformers this becomes

*M.K.S. unit : See Chapter 38 Sect. 1.

important and, to maintain balance, the primary is wound between the halves of the secondary or the latter are wound side by side over the primary.

(ii) Core material and size

Various grades of core material are available all differing in silicon content from 0.25 to 4.5 per cent (Refs. D6, 7). They feature lower loss, increasing cost and brittleness with increase in silicon. Several manufacturers make somewhat similar grades and these are listed hereunder (Ref. D8).

TRANSFORMER STEELS

Silicon Content*	M. and E.A.	Baldwin	Sankey	Allegheny	Armco
4%	Silcor 1	Quality 5	Super Stalloy	Transf. C	Trancor 2
3½%	Silcor 2	Quality 4	Stalloy	Transf. D	Trancor 1
2¾%	Silcor 3	Quality 3B	42 Quality	Electrical	Spec. Elec.
1%	Silcor 4	Quality 1	Lohys	Armature	Armature

*This applies exactly to Silcor (Magnetic and Elec. Alloys Ltd.).

The core losses for the various Silcor grades measured at 50 c/s. with 0.014 inch sheet, at two different values of flux density are shown below.

B_{max}	Watts lost per pound			
	Silcor 1	Silcor 2	Silcor 3	Silcor 4
10 Kilogauss	0.59	0.63	0.89	1.32
13 Kilogauss	1.04	1.07	1.51	2.24

It will be observed that for small changes in flux density the core loss varies as the square of the flux density. For radio power transformer work, core materials similar to Silcor 2 are commonly used. Measurements of losses can be made with a low power factor wattmeter, or by the three ammeter method. A system suitable for mass production testing has been described recently (Ref. D9).

Cold rolled, grain oriented 2.7 per cent silicon steels are becoming of increasing importance. Typical trade names for this material are " Hipersil " and " Crystalloy " (Ref. 10). Flux densities in excess of 110 000 lines per square inch (17 kilogauss), can be employed without high core loss. Owing to low core losses at such high flux densities, the application of this material to small transformers for electronic equipment is increasing. It can be used in strip form, being wound around the winding in one method of assembly. In another method, the strip core is sawn in half and the two halves clamped together around the winding. Either method results in a considerable saving in material and labour over the present method of punching laminations and hand stacking the winding with them.

The most popular laminations in current use are the E and I " scrapless " variety (Refs. D11, 12, 13, 14). These are so dimensioned that the I is punched from the window of the E thus avoiding wastage. The usual ratio of dimensions of these laminations are as follows : Window height 1, tongue 2, window width 3, I-length 6, magnetic path 12.

Similar laminations are stamped by several different firms :

Standard Lamination Sizes
Pattern Number

Tongue	M.E.A.	Baldwins	Sankey	Allegheny	Chicago
9/16 in.	18	—	—	EI-56	—
⅝	145	392	—	EI-625	F000
¾	35	217	70	EI-75	D000
⅞	147	—	—	EI-11	B000
1	29	430	111	EI-12	1000
1⅛	—	—	158	EI-112	13000
1¼	78	420	133	EI-125	14000
1⅜	152	362	—	EI-138	—
1½	120	—	149	EI-13	—

Laminations for small power transformers are generally 0.014 inch thick ; thicker laminations up to 0.025 inch are occasionally used, but result in increased losses and shortened lamination die life.

To determine the core size, it is first necessary to estimate the power requirements. Then we may apply the empirical relation

$$A = \frac{\sqrt{V.A}}{5.58} \qquad (1)$$

where A = cross-sectional area in square inches
and VA = voltamps output.

For ease of production an approximately square stack is desirable. Thus having determined A, the tongue size can be estimated from $\sqrt{A}$. Where high voltage windings, or windings operating above ground, or at a high potential between other windings, are used, it becomes necessary to employ a lamination which has a different ratio of dimensions from those quoted earlier, the window height being increased to allow room for extra insulation. The input current taken by a transformer on no-load is commonly called the magnetizing current. In fact, it consists of two components in phase quadrature. The in-phase component, usually small, is the iron loss plus a very small copper loss. The quadrature portion is the true magnetizing current. Their vector sum does not normally exceed about a third of the full load current.

(iii) Primary and secondary turns

The primary turns required can be determined from the fundamental transformer equation.

$$N = \frac{E \times 10^8}{4.44\, fBA} \qquad (2)$$

where N = primary turns, f = frequency in c/s
    B = max. flux density
and A = cross-sectional area of core in sq. ins.

Assuming an average stacking factor of 90 per cent, the factor 4.44 becomes 4 in the denominator. The stacking factor reduces the apparent height of the stack, the reduction being caused by insulation, scale and burr due to die wear. With very thin laminations, the stacking factor decreases to 80 per cent approximately. For a flux density of 64 500 lines per square inch (10 kilogauss) eqn. (2) becomes :

For 240 volts 50 c/s   230 volts 50 c/s   117 volts 60 c/s

$$N = \frac{1860}{A} \qquad N = \frac{1780}{A} \qquad N = \frac{755}{A}$$

where A = gross cross-sectional area of core in sq. ins.

The flux density employed depends on the application, the power rating, the core material and the frequency. For oscilloscopes and pre-amplifiers, densities of 40 000 to 50 000. lines per square inch (about 7 000 gauss) are used. For small transformers below about 50 watts, densities up to 90 000 lines per square inch (14 000 gauss) are used, gradually decreasing to about 65 000 lines (10 000 gauss) as the transformer size increases to several hundred watts.

Having chosen a suitable flux density, the turns required for each secondary winding may be calculated.

It can be shown that, approximately,

$$N_2 = \frac{E_2 . N_1}{E_1 . \sqrt{\eta}} \qquad \text{where } \eta = \text{transformer efficiency.}$$

As a first approximation it may be assumed that $\eta = 0.85$ and $\sqrt{\eta} = 0.92$.

The values thus obtained for the secondary turns may be checked by more detailed calculations once wire gauges have been chosen and winding resistances calculated. For this purpose the equivalent circuit of Figure 5.11 may be used.

If a secondary is to feed a rectifier and the d.c. output of the valve is specified, reference should be made to Chapter 30 or a valve data book to determine the required secondary voltage. Allowance must be made for any voltage drops due to the d.c. resistance of the rectifier filter.

(iv) Currents in windings

To enable the wire gauges to be chosen and to assess the copper losses in the windings, it is necessary to estimate the current in each winding. Where windings are used to supply valve heaters and resistance loads, the winding current is the same as the load current. In a secondary winding feeding a rectifier the winding current must be estimated from a knowledge of the type of rectifier and its associated filter and their characteristics. For normally-loaded full wave rectifiers the following values of secondary current may be used as a fairly close guide for design purposes.

Condenser input filter : The r.m.s. current in each half of the transformer secondary may be taken approximately as 1.1 times the direct current to the load. For further details see Chapter 30 Sect. 2.

Choke input filter : The r.m.s. current in each half of the transformer secondary may be taken approximately as 0.75 times the direct current to the load. For further information, see Chapter 30, Sect. 4 (also Sect. 3). With half-wave rectification there will be a d.c. component of the current which will affect the transformer design if the total d.c. ampere turns are considerable. For half-wave battery chargers using bulb rectifiers, this must be taken into consideration when selecting core size and wire gauges (Ref. D15).

In the case of transformers supplying full-wave rectifiers, the full load primary current can be estimated by calculating the total secondary loading in voltamps, allowing an efficiency of 85% as a first approximation. This is then the primary input in voltamps which, when divided by the primary voltage, will give the desired current. Where secondaries feed a resistive load, the loading is the product of the voltage and current. Where a secondary feeds a full-wave rectifier the load is the product of the direct current and the direct voltage output from the rectifier plus the power lost in the rectifier. This latter can be calculated, for a condenser input filter, from data presented in Chapter 30, Sect. 2. It should be noted that indirectly heated, close-spaced rectifiers such as the 6X4, are more efficient than types such as the 5Y3-GT, with its heavier filament power and lower plate efficiency. This is one reason why the former are almost exclusively used in the majority of small a.c. radio receivers.

The primary input current as calculated above should be accurate to within 10% for small transformers.

Wire gauges and copper losses

For the usual type of radio receiver power transformer it will be safe to choose wire gauges on the basis of 450-800 circular mils* per ampere (i.e. 2830-1590 amperes per square inch), but values up to 1000 circular mils per ampere (i.e. 1270 amperes per square inch) may be desirable in larger units, or if a high flux density is used with high loss lamination steel. The latter figure is easy to remember as 1 circular mil per milliamp.

In practice, wire gauges may be chosen arbitrarily on a basis of (say) 700 circular mils per ampere (i.e. approximately 1800 amperes per square inch) and check calculations should then be made to see that

(a) the build of the winding (i.e. winding height) is satisfactory for the window space available,

(b) the copper loss is not so high as to cause excessive temperature rise,

(c) the voltage regulation is satisfactory.

Wire tables in this Handbook will simplify these calculations (Chapter 38 Sect. 19).

(v) Temperature rise

This is dependent upon the cooling area, the total loss and the ratio of iron to copper loss. To avoid deterioration of the insulating materials, it is necessary to limit the working temperature to 105°C for Class A insulation which includes paper, cotton,

---

*A circular mil is the area of a circle 1 mil (1/1000 inch) in diameter.

silk, varnish and wire enamel (Refs. D16, 17). The temperature rise in the winding, as measured by the change of resistance method, will be about 10°C lower than the maximum (hot spot) temperature. Thus with an ambient temperature of 40°C (104°F) plus a margin of 10°C for the difference between measured and hot spot temperature, it will be seen that the maximum permissible rise is 55°C, as measured by the change of resistance.

It is common practice to allow 10°C margin for change in line voltage, frequency, or operation in situations with restricted ventilation. Thus 45°C is generally accepted as the maximum permitted rise above ambient when measured by resistance change.

The temperature difference between winding and core varies between 10° and 20°C according to the distribution of losses. This means that even with an ambient temperature of 25°C (77°F), the core temperature may be 60° (140°F). This will feel quite hot to the touch, although the internal temperature may be well under the permitted maximum. Measurement of the core temperature may be made with a spirit thermometer if good thermal contact is maintained between the core and the thermometer bulb.

The winding temperature rise can be calculated by measuring the cold resistance, R_0 at an observed temperature T_1. After a heat run at full load the hot resistance R is measured and the ambient temperature T_2 is again measured.

Taking the temperature coefficient of resistivity (a) of copper as 0.003 93, the temperature rise T is found from the formula

$$T = \frac{R - R_0}{R_0 a}.$$

To correct for any changes in ambient temperature during the heat run it is necessary to subtract the difference $T_2 - T_1$ from T to find the actual rise. An example will illustrate this.

The primary resistance of a transformer measured when cold was 30 ohms at an ambient temperature of 20°C. After an 8 hour full load run, the resistance was 36 ohms, while the ambient temperature was then 18°C. It is required to find the winding temperature rise.

$$T = \frac{36 - 30}{30 \times 0.00393} = \frac{6}{30 \times 0.00393} = 51°C.$$

Now $T_2 - T_1 = (18 - 20)°C = -2°C.$

Winding temperature rise = $[51° - (-2)]°C = 53°C.$

Standard methods of testing small radio receiver power transformers have been published (Refs. D18, D22). The cooling area of a transformer for the " scrapless " type laminations with standard ratio of dimensions is

$$A = T(7.71\,T + 11S)$$

where A = cooling area

　　　　　T = width of tongue

and S = height of stack.

For a square stack $S = T$ and $A = 18.71\,T^2$ (Ref. 11).

Having calculated the iron and copper losses and the total cooling area, it is possible to estimate the temperature rise before a sample transformer is wound. The watts lost per square inch of cooling area is first calculated. Reference to Fig. 5.18B will then show the temperature rise to be expected with an accuracy of approximately ± 10%.

(vi) Typical design

(Refs. D19, D20, D21, D23).

Specifications :

　Primary　　　　240 V 50 c/s

　Secondary (i)　6.3 V 2.6 A

　Secondary (ii) 300 V + 300 V r.m.s. for full wave 6X4 rectifier with condenser
　　　　　　　　　　input filter to deliver 60 mA d.c.

Secondary loading

Secondary (i)	6.3 V 2.6 A	or 16.5 watts
D.C. output	343 V 0.06 A	or 20.5 watts

Rectifier loss   (see Chap. 30) $\underline{2.0}$ watts approx.

Total loading   39.0 watts

Primary input (assuming an efficiency of 85%) 46.0 watts.

Thus the input current is 46/240 or 0.19 A.   Core cross-section : $A = \sqrt{39}/5.58$ = 1.12 square inches.   Choosing pattern EI-112 lamination of Silcor 2 material, a suitable stack for calculated cross section is 1.125 inches, that is, a square stack.

Primary turns :

With a flux density of 13 kilogauss (84 000 lines per square inch) and a stacking factor of 0.9,

$$N_p = 240 \times 10^8/4 \times 50 \times 84\,000 \times 1.125 \times 1.125$$
$$= 1130.$$

Secondary (ii) turns :

$$N_1 = E_1 N_p/E_p\sqrt{\eta} = 300 \times 1130/240\sqrt{0.85}$$
$$= 1530 + 1530.$$

Turns per volt $= N_1/E_1 = 1530/300 = 5.1$.

Secondary (i) turns $\quad = (6.3 \times 5.1) = 32$.

Wire gauges :

Assume a current density of 600-700 circular mils per ampere.   Referring to wire tables, a suitable primary gauge is 29 A.W.G. enam.

Turns per layer $= 95$. Refer Figs. 5.18C and wire tables in Chapter 38 Sect. 19.

No. of layers $= 1130/95 = 12$.

R.M.S. current in secondary (ii) is $(0.06 \times 1.1) = 0.066$ A.

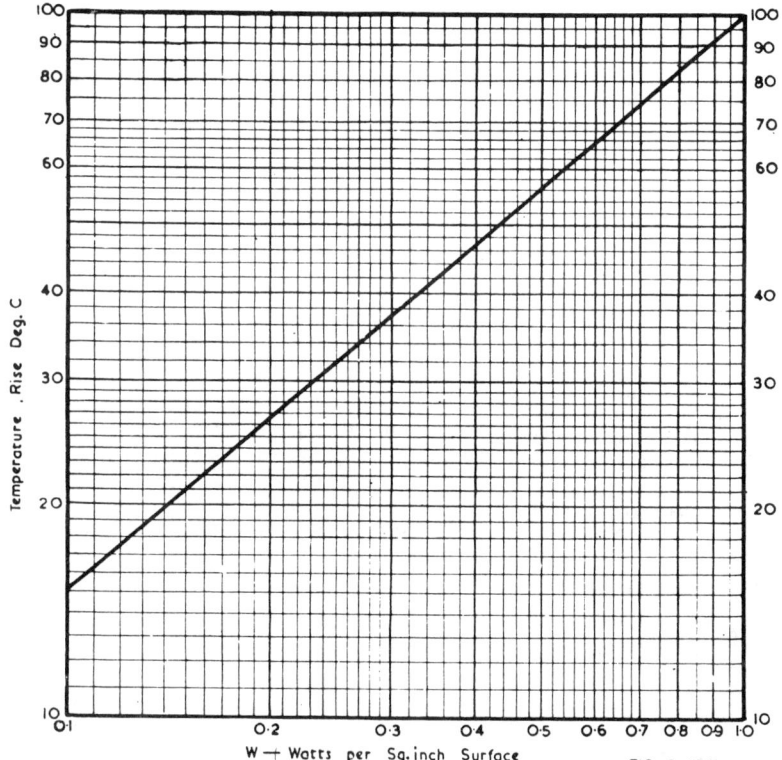

$W \longrightarrow$ Watts per Sq. inch Surface

FIG. 5.18B

Fig. 5.18B. *Temperature rise versus loss per square inch of cooling surface for ambient temperature* 25°C (*Ref. D.1.*)

Suitable wire gauge is 34 A.W.G. enam.
Turns per layer = 167.
No. of layers = 3060/167 = 19.
Suitable secondary (i) gauge is twin 20 A.W.G. enam.
Turns per layer of twin wire = 16.
No. of layers = 32/16 = 2.
A twin wire is used in preference to a single wire of 17 or 18 A.W.G. in order to save winding height.

Winding Build :

Primary build	= No. of layers × (enam. wire diam. + interlayer insulation)
	= 12 × (12.2 + 2) × 10^{-3} inches
	= 0.170 inch.
Secondary (ii) build	= 19 × (6.9 + 1) × 10^{-3} inch
	= 0.150 inch.
Secondary (i) build	= 2 × (32.4 + 5) × 10^{-3} inch
	= 0.075 inch.

Allowing 50 mil former thickness and 10 mil insulation between windings and over outer winding, total build is (0.170 + 0.150 + 0.075 + 0.050 + 0.040) inch
= 0.485 inch.
Winding height   = 0.562 inch.
Build expressed as a percentage of window height
= 0.485/0.562 × 100 = 86.5%.

In this particular design, the heater winding is wound between primary and high tension windings to serve as a static shield.

Mean length of turn calculations—see Fig. 5.18D.

A = Former build plus insulation	= 0.050 + 0.010	= 0.060 inch
B = Primary build plus insulation	= 0.170 + 0.010	= 0.180 inch
C = Secondary (i) build plus insulation	= 0.075 + 0.010	= 0.085 inch
D = Secondary (ii) build plus insulation	= 0.150 + 0.010	= 0.160 inch
$4S$ = 2 × (stack + tongue)	= 4.5 inches	
Primary mean length of turn	= 5.44 inches	
Secondary (i) mean length of turn	= 6.27 inches	
Secondary (ii) mean length of turn	= 7.04 inches.	

Winding resistance :

If each value for the mean length of turn is multiplied by the number of turns in its own winding and then divided by twelve, the resulting quantity will be the number of feet of wire in each winding. By referring to the wire tables, the resistance in ohms per thousand feet for any particular gauge can be found. Dividing the wire length by a thousand and multiplying by the resistance per thousand feet, will determine the winding resistance.

Primary resistance	= 5.44 × 1130 × 81.8/12 × 1000	= 42 ohms
Secondary (i) resistance	= 6.27 × 32 × 12/2 × 12 × 1000	= 0.09 ohms
Secondary (ii) resistance	= 7.04 × 3060 × 261/12 × 1000	= 470 ohms

Copper loss :

Primary copper loss	= $(0.19)^2$ × 42	= 1.52 watts
Secondary (i) copper loss	= $(2.6)^2$ × 0.09	= 0.61 watts
Secondary (ii) copper loss	= $(0.066)^2$ × 470	= 2.05 watts
Total copper loss		= 4.18 watts

Iron loss :

Loss per pound of Silcor 2 at flux density of 13 kilogauss = 1.07 watts.
Weight of core = 2.17 lbs.
Therefore total iron loss = (1.07 × 2.17) = 2.32 watts.

Temperature rise :

Total of iron and copper loss = (2.32 + 4.18) = 6.5 watts.
Cooling area of square stack = 18.71 × $(1.125)^2$ = 23.6 sq. inches.
Therefore watts lost per sq. inch = 6.5/23.6 = 0.275 watts.
From Fig. 5.18B temperature rise = 35°C.

Efficiency :

Efficiency = Power output/(power output plus losses)
= 39/(39 + 6.5) × 100 = 86%.

Regulation, calculated from British definition :

Primary voltage drop = 0.19 × 42 = 8 V

Regulation due to resistance of primary = 8 × 100/240 = 3.3%.

FIG. 5.18C

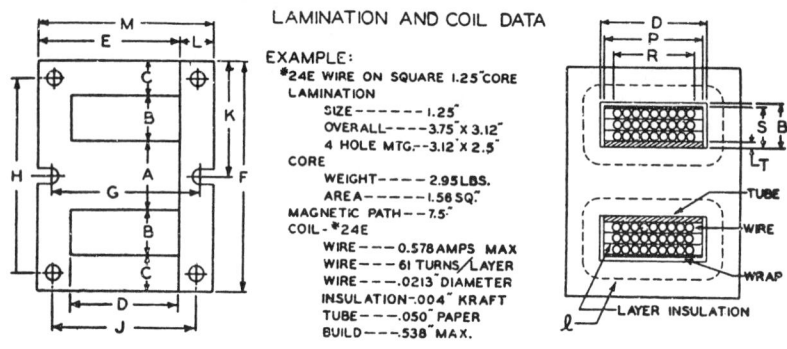

LAMINATION AND COIL DATA

EXAMPLE:
*24E WIRE ON SQUARE 1.25"CORE
LAMINATION
  SIZE ------ 1.25"
  OVERALL ---- 3.75"X 3.12"
  4 HOLE MTG.--3.12"X 2.5"
CORE
  WEIGHT ---- 2.95LBS.
  AREA ----- 1.56 SQ."
MAGNETIC PATH -- 7.5"
COIL - *24E
  WIRE --- 0.578 AMPS MAX
  WIRE --- 61 TURNS/LAYER
  WIRE --- .0213"DIAMETER
  INSULATION-.004" KRAFT
  TUBE -- -.050"PAPER
  BUILD --- -.538"MAX.

Lamination Size	Width Center Leg, In.	A	0.500	0.625	0.750	0.875	1.000	1.125	1.250	1.375	1.500
Window Dimensions	Area, Sq. In.	B × D	0.250	0.292	0.422	0.573	0.750	0.950	1.17	1.42	1.69
	Length, In.	D	0.812	0.937	1.125	1.31	1.50	1.69	1.87	2.06	2.25
	Width, In.	B	0.312	0.312	0.375	0.437	0.500	0.562	0.625	0.687	0.750
Coil Dimensions	Build, Max. In.	S	0.256	0.256	0.315	0.367	0.430	0.483	0.538	0.590	0.646
	Length Overall, In.	P	0.750	0.875	1.06	1.25	1.44	1.56	1.75	1.94	2.12
	Length Winding, In.	R	0.500	0.625	0.750	1.00	1.12	1.25	1.37	1.56	1.75
Core Tube	Thickness, In.	T	0.035	0.035	0.050	0.050	0.050	0.050	0.050	0.050	0.060
"E" & "I" Piece Overall Dimensions	Length, In.	F	1.62	1.87	2.25	2.62	3.00	3.37	3.75	4.12	4.50
	Width, In.	M	1.31	1.62	1.87	2.19	2.50	2.81	3.12	3.44	3.75
4 Hole Mounting Dimensions	Length, In.	H	...	....	....	....	2.50	2.81	3.12	3.44	3.75
	Width, In.	J	....	....	....	....	2.00	2.25	2.50	2.75	3.00
	Diameter Hole, In.	...	....	....	....	....	0.219	0.219	0.219	0.219	0.281
2 Slot Mounting Dimensions	Width, In.	G	....	....	1.62	1.95	2.16	2.44	2.78	2.94	3.12
	From End, In.	K	....	....	1.12	1.31	1.50	1.69	1.87	2.06	2.25
"E" Piece Dimensions	Width, Overall, In.	E	1.06	1.25	1.50	1.75	2.00	2.25	2.50	2.75	3.00
	Width, Outer Leg, In.	C	0.250	0.312	0.375	0.437	0.500	0.562	0.625	0.687	0.750
"I" Piece	Width, In.	L	0.250	0.312	0.375	0.437	0.500	0.562	0.625	0.687	0.750
For Square Stack Center Leg	Volume, Cu. Inches	Area × h	0.810	1.46	2.54	4.03	6.00	8.55	11.70	15.60	20.20
	Weight, Pounds	Vol. × Density	0.21	0.37	0.64	1.05	1.50	2.14	2.95	3.90	5.06
	Area, Sq. Inches	A × A	0.250	0.390	0.562	0.766	1.000	1.27	1.56	1.90	2.25
Magnetic Path	Length, In.	l	3.25	3.75	4.50	5.26	6.00	6.75	7.50	8.24	9.00

Fig. 5.18C. *Lamination and coil data (Ref. D20).*

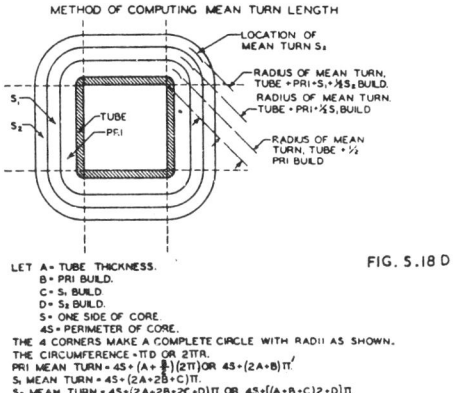

METHOD OF COMPUTING MEAN TURN LENGTH

LET A = TUBE THICKNESS.
B = PRI BUILD.
C = S_1 BUILD.
D = S_2 BUILD.
S = ONE SIDE OF CORE.
4S = PERIMETER OF CORE.
THE 4 CORNERS MAKE A COMPLETE CIRCLE WITH RADII AS SHOWN.
THE CIRCUMFERENCE = πD OR 2πR.
PRI MEAN TURN = 4S + (A + $\frac{B}{2}$)(2π) OR 4S + (2A + B)π
S_1 MEAN TURN = 4S + (2A + 2B + C)π
S_2 MEAN TURN = 4S + (2A + 2B + 2C + D)π OR 4S + [(A + B + C)2 + D]π.

Fig. 5.18D. *Method of computing mean length of turn* (*Ref.* D20).

The percentage regulation of any secondary winding is calculated by dividing the full-load voltage drop of the winding by the open-circuit winding voltage and multiplying by 100. By adding the percentage regulation of the primary winding, the overall regulation for the secondary winding under examination is found.

Secondary (i) voltage drop	$= 2.6 \times 0.09 = 0.234$ V
Regulation, overall (i)	$= (0.234 \times 100/6.8) + 3.3 = 6.7\%$
Secondary (ii) voltage drop	$= 0.066 \times 470 = 31$ V
Regulation, overall (ii)	$= (31 \times 100/650) + 3.3 = 8.1\%$
Full load secondary (i) voltage	$= 6.35$ V
and full load secondary (ii) voltage	$= 300$ V $+ 300$ V.

This should be a satisfactory design.

As the temperature rise is low, it might be feasible to effect economies by re-design. For example, a higher loss core material could be used. Alternatively the flux density could be increased by decreasing the turns or the size of the stack. A new design could also be tried on the next smaller size lamination.

(vii) Specifications for power transformers

The following may be incorporated into a specification for a power transformer.

(a) Input voltage and frequency ; should these quantities vary, the range of variation must be stated.

(b) Secondary full load voltages and currents. If tapped windings are required this should be indicated. Tolerances on voltages should be stated.

(c) Regulation of secondary voltages other than for rectifier plate circuit.

(d) Power factor of each secondary load.

(e) Rectifier system, if d.c. output is required ; e.g. full wave, half wave, etc.

(f) Type of rectifier valve to be used.

(g) Type of filter circuit ; e.g. choke or condenser input, etc.

(h) Capacitance of input condenser, or inductance and resistance of input choke.

(i) D.C. full load current and voltage at filter input.

(j) D.C. regulation.

(k) External voltages between windings, or between windings and ground. Should transformer windings be interconnected externally, this should be indicated.

(l) Static shields. Number and position of each should be indicated.

(m) Any limiting dimensions together with mounting and lead terminations.

(n) Ambient temperature in which the transformer is to operate.

A circuit should be supplied showing intended use of transformer.

In all cases it is desirable for the design of the whole rectifier system, comprising transformer, rectifier and filter system, to be carried out by the one engineer.

SECTION 6 : IRON-CORED INDUCTORS

(i) General   (ii) Calculations-general   (iii) Effective permeability   (iv) Design with no d.c. flux   (v) Design of high Q inductors   (vi) Design with d.c. flux   (vii) Design by Hanna's method   (viii) Design of inductors for choke-input filters   (ix) Measurements   (x) Iron-cored inductors in resonant circuits.

(i) General

Iron cored inductors fall into several different categories depending upon the circuit requirements. In some applications these inductors may have to carry a.c. only, in others, both a.c. and d.c. They may have to work over a wide range of frequencies, or at any single frequency up to the ultrasonic range. Iron-cored inductors are employed as smoothing and swinging chokes in power supplies, as equalizer elements in audio frequency equipment, as modulation chokes and as filter elements in carrier equipment, to mention a few varied applications.

To design such an inductor it is therefore necessary to know some or all of the following specifications—

(a) inductance, or range of inductance if variable,
(b) alternating voltage across the coil,
(c) direct current through the coil,
(d) frequency of operation,
(e) maximum shunt capacitance,
(f) minimum frequency of self-resonance,
(g) minimum Q over frequency range and/or
(h) frequency of maximum Q,
(i) d.c. resistance,
(j) shielding,
(k) temperature rise,
(l) size and weight limitations,
(m) insulation requirements.

(ii) Calculations—general

The inductance L of an iron-cored coil may be calculated using the relation

$$L = \frac{3.2 \times N^2 \times \mu \times a}{10^8 \times l} \text{ henrys} \tag{1}$$

where N = number of turns
a = effective cross sectional area of coil in square inches
l = length of magnetic circuit in inches
and μ = effective permeability.*

The effective permeability depends on the type of steel used, on the a.c. and d.c.

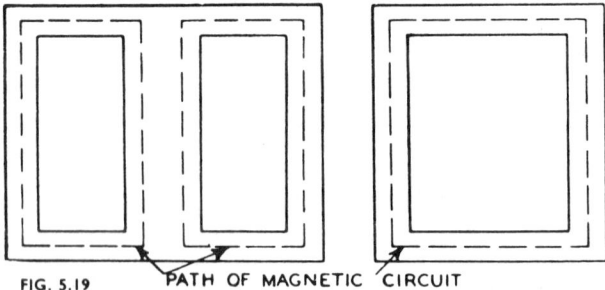

FIG. 5.19   PATH OF MAGNETIC CIRCUIT

Fig. 5.19.   Magnetic cores, showing method for calculation of length of magnetic path.

---

*A more precise term is "inductance ratio."

flux densities in the core, and on whether the core laminations are interleaved or whether there is an air gap in the magnetic circuit.

The length of the magnetic circuit is the length measured around the core at the centre of cross section of each magnetic path. Referring to Fig. 5.19, the path taken will be along the centre of each leg except where there are two windows, when each path through the centre leg will be along a line one-quarter of the way across the leg. In this latter case only a single path round the window is considered in calculating the magnetic circuit length.

Eqn. (1) is properly applicable only when the cross-sectional area of the core is uniform throughout the magnetic circuit. Where the cross section is non-uniform, a conservative value of L will usually be obtained by using for a the minimum value of the cross-sectional area. For more accurate calculations in such cases reference should be made to a suitable text-book (e.g. Ref. E1).

(iii) Effective permeability

For any given sample of lamination steel the effective permeability depends primarily on the a.c. and d.c. flux densities in the core.

Irrespective of the presence of air gaps in the magnetic circuit, the maximum a.c. flux density B_{max} in the core is determined by the cross section of the core, the number of turns in the coil, the alternating voltage across the coil and the frequency, and may be calculated directly from the relation derived from eqn. (2) in Sect. 5,

$$B_{max} = \frac{E \times 10^8}{4.44\,fNa} \tag{2}$$

The d.c. magnetizing force in the core depends on the total number of d.c. ampere-turns, less the number of ampere-turns absorbed in any air gap, divided by the length of the magnetic circuit in inches (see Sect. 4).

Fig. 5.20 shows the variation of effective permeability with variation of a.c. flux density and d.c. magnetizing force for typical electrical sheet steel. It will be observed that the effective permeability increases up to a maximum as the a.c. flux density increases, and then drops rapidly due to saturation of the core. For any particular value of a.c. flux density the effective permeability decreases as the d.c. magnetizing force is increased.

Where the a.c. flux in the core is much less than the d.c. flux, it is convenient to refer to the effective permeability as the **incremental permeability** since it depends on the characteristics of the core material in relation to small changes in the total flux. This term is therefore generally used in dealing with filter chokes and other inductance coils having a relatively large number of turns and carrying a relatively large direct current.

(iv) Design with no d.c. flux

If there is no d.c. flux in the core the calculation of inductance is straightforward because the effective permeability depends only on the number of turns, the frequency, the core cross section, and the applied voltage. This case arises when the windings of a coil or transformer do not carry direct current or when the windings carrying direct current are so arranged that the d.c. flux in the core is zero as, for example, with transformers used in balanced push pull amplifiers.

Design of a coil to obtain a required value of inductance is therefore also quite straightforward, and the following steps may be followed—

(1) Choose an available core and calculate from eqn. (1) the number of turns required to give the necessary inductance assuming a value of effective permeability of between 1 000 and 5 000 depending on a very rough estimate of the a.c. flux density in the core.

(2) Calculate the a.c. flux density for this number of turns in relation to the known frequency and voltage across the coil.

(3) Correct the estimate of the number of turns using a revised value of permeability for the calculated flux. If the first estimate was far out, repeat steps (2) and (3).

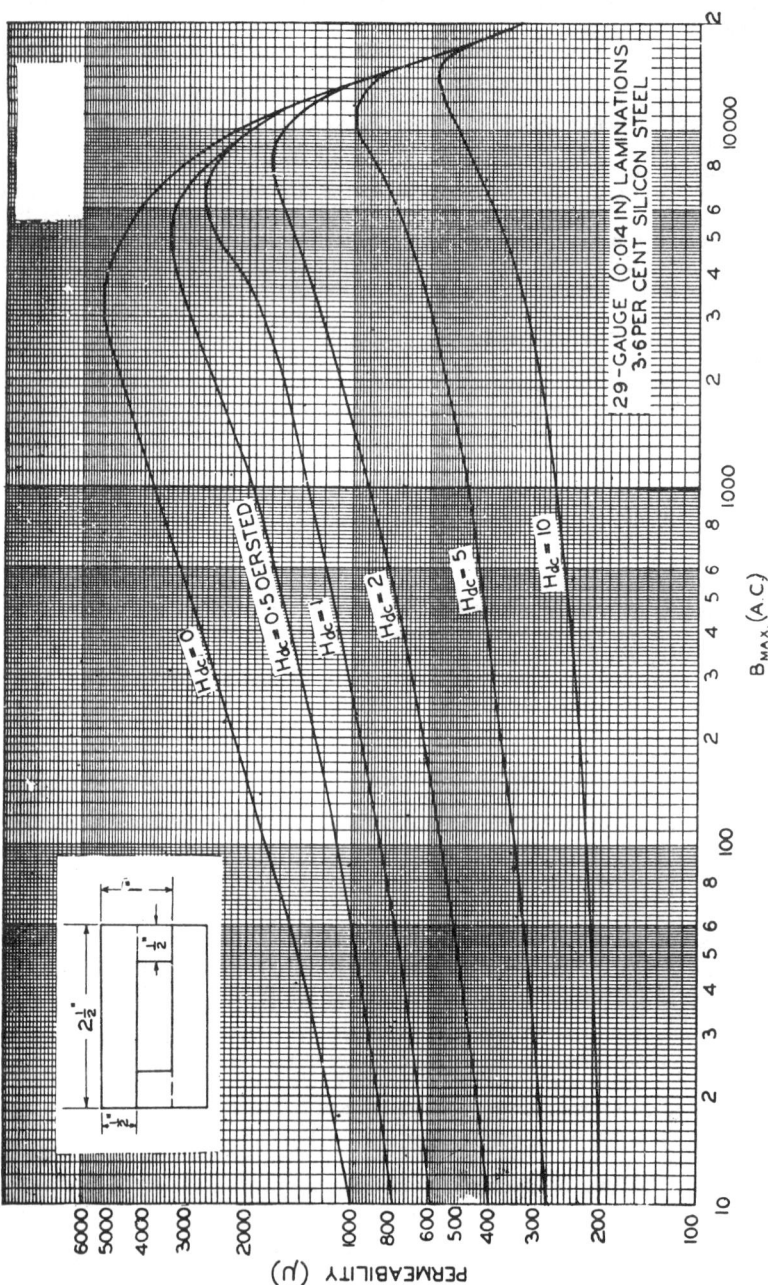

Fig. 5.20.   *Effective permeability (μ) of silicon steel versus a.c. maximum flux density, for various values of d.c. magnetising force H_0 (by courtesy of Allegheny Ludlum Steel Corporation).*

(4) If the flux density is excessive (i.e. above about 50 000 lines per square inch) increase the number of turns until the flux density is satisfactory.

In case step 4 is necessary, the magnetic circuit must be broken by an air gap and the gap adjusted until the required value of inductance is obtained ; but if the larger value of inductance is satisfactory or useful the air gap may be omitted.

(5) Calculate the wire gauge required, and the copper and iron losses as in the design of a power transformer. Check also that the d.c. resistance of the winding is not excessive for the purpose for which the coil is to be used.

If an air gap of length α inches is used in the core, the value of inductance becomes

$$L = \frac{3.2N^2a}{10^8.l} \times \frac{1}{(1/\mu + \alpha/l)} \tag{3}$$

and this expression may be used for calculations. Alternatively, it will be clear that if the air-gap is large, so that $1/\mu \ll \alpha/l$, eqn. (3) approximates to

$$L = 3.2N^2 . a/10^8\alpha \tag{4}$$

FIG. 5.21.

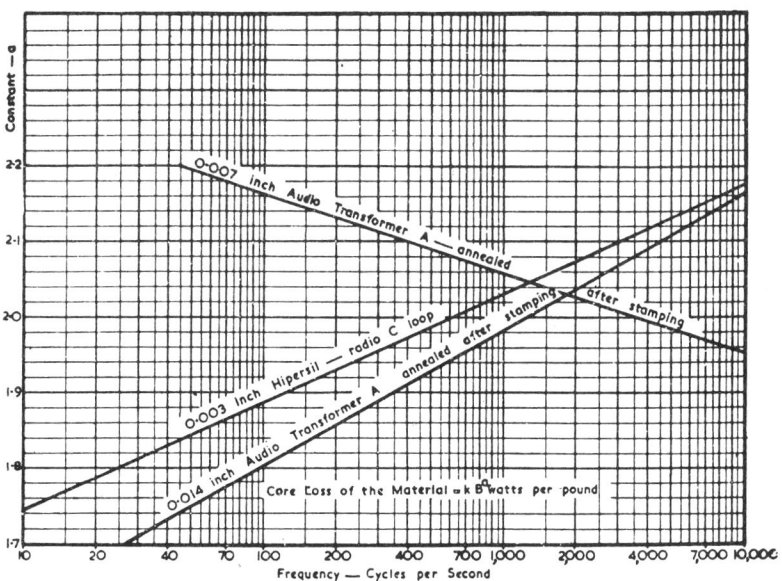

Fig. 5.21. *Experimentally determined curve showing values of " a " as a function of frequency for several magnetic materials (Ref. E2).*

(v) Design of high Q inductors

This sub-section follows S. L. Javna (Ref. E2) and is set out in a step-by-step form for ease of working. The following table defines the symbols used by Javna in the formulae presented here.

A = Effective cross-sectional area of magnetic-flux path, in sq. in.

A_g = Gross cross-sectional area of magnetic-flux path, in sq. in.

a = Empirical constant, see Fig. 5.21.

B = Maximum flux density in the core, in lines per sq. in.

F = Fraction of core-window area occupied by copper wire of coil.

f = Frequency, in cycles per second.

g = Actual gap length, in inches.

k = Empirical constant, see Fig. 5.22.

L = Inductance of iron-cored coil, in henrys.

l = Mean length of magnetic path, in inches.

m = Mean length of a turn of the coil, in inches.

$\mu_\triangle$ = Incremental permeability of core with respect to air at operating frequency.
N = Number of turns in coil.
n = A.W.G. (B. and S.) wire gauge number of conductor.
$R_{a.c.}$ = Apparent a.c. coil resistance caused by core loss, in ohms.
$R_{d.c.}$ = Copper loss resistance, in ohms.
s = Total lamination window area, in sq. in.
V = Voltage across coil, in volts.
w = Weight of core, in pounds.

A typical problem will be solved to illustrate the design procedure.

A 5 henry inductor is to be designed on a one inch square stack of Allegheny pattern EI-12 audio transformer A silicon steel annealed laminations. The voltage across the coil will be 10 V a.c. at 1000 c/s. It is required to determine the turns and gauge of wire, the gap width and Q at the operating frequency. Assume a stacking factor of 0.9.

Tabulation :
l = 6.0 ins.
w = 1.5 lbs.
A_g = 1.0 sq. in.   } from Fig. 5.18C
A = 0.9 sq. in.
s = 0.75 sq. in.
m = 5.5 ins. from Fig. 5.18D
$F \approx 0.3$, a typical value for this lamination
k = 1.3×10^{-8} from Fig. 5.22.
a = 1.987 from Fig. 5.21.
$\mu_\triangle$ is found from Fig. 5.20.

Calculation :

$$B = \left(\frac{1.74mV^4 10^7}{akwf^4 L^2 A^2 sF} \right)^{1/(a+2)} \qquad = 48.6 \text{ lines/sq. in.}$$

$$R_{a.c.} = 39.5kB^a wf^2 L^2/V^2 \qquad = 430 \text{ ohms}$$

$$R_{d.c.} = \frac{3.44mV^2 10^8}{sFB^2 A^2 f^2} \qquad = 440 \text{ ohms}$$

$$Q = \frac{2\pi fL}{R_{a.c.} + R_{d.c.}} \qquad = 36$$

$$N = \frac{V 10^8}{4.44 fBA} \qquad = 5150 \text{ turns.}$$

$$n = 49.8 + 9.96 \log (1.2R_{d.c.}/mN) \qquad = 33 \text{ A.W.G.}$$

$$g = \left(\frac{1.59N^2 A_g}{L 10^8} - \frac{lA_g}{2\mu_\triangle A} \right) \qquad = 0.08 \text{ in.}$$

Coil build check :
Turns/layer using 33 A.W.G. enam. = 133 from winding data chart, thus number of layers = 5150/133 = 39.
Interlayer insulation from winding data chart = 0.0015 in., thus winding build = 39 (0.0078 + 0.0015) = 0.363 in.
Former thickness = 0.05 in.
Insulation under and over winding = 0.02 in., thus total coil build = (0.363 + 0.05 + 0.02) = 0.433 in.
Max. build from Fig. 5.18C = 0.43 in., thus the winding should fit the window satisfactorily.

It should be noted that $R_{a.c.}$ and $R_{d.c.}$ are usually very nearly equal and where a square stack of scrapless laminations is used, m will approximate l. Due to the large air gap the inductance will remain practically independent of voltage and frequency. The basic design equations may be used and applied to almost any magnetic material. The constants a and k for other magnetic materials can be obtained from graphs drawn from measurements made in accordance with the method mentioned in the

FIG. 5.22

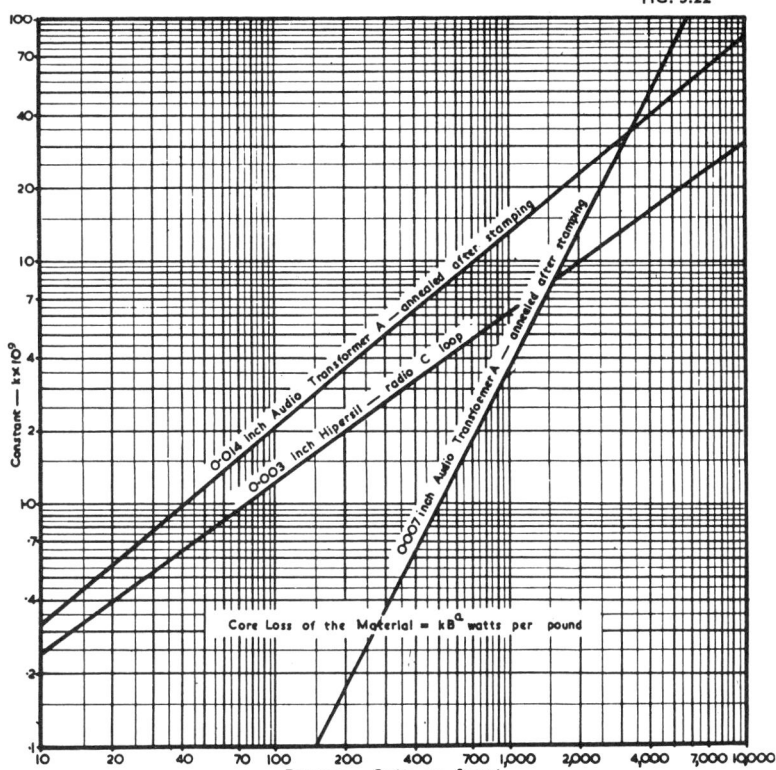

Fig. 5.22. *Experimentally determined curve showing values of " k " as a function of frequency for several magnetic materials (Ref. E2).*

original reference. Further information on iron-cored inductor design is contained in Refs. E3, E4, E5, E6, E7, E8, E9.

(vi) Design with d.c. flux
(Ref. E10).

When a d.c. flux is set up in the core by unbalanced direct currents in the windings, the effective permeability is decreased, as is shown in the curves of Fig. 5.20.

The amount of d.c. flux set up in the core depends on the applied d.c. magnetizing force (i.e. the total unbalanced ampere-turns in the windings), on the length of the magnetic circuit, and on the length of any air gap in the magnetic circuit.

The effective permeability depends on
   (i) the d.c. flux density in the core
   (ii) the a.c. flux density in the core
and (iii) the length of air gap in the magnetic circuit.

There is an optimum air gap giving a maximum value of the effective permeability for any particular value of total d.c. magnetizing force and a.c. flux density (Ref. E11).

For very low values of a.c. flux density, the following table gives the variation of incremental permeability with d.c. magnetizing force (A.T./inch of total length of magnetic circuit) and the ratio of the length of air gap to the total length of the magnetic circuit.

Incremental Permeability of High Silicon Steel
Total d.c. magnetizing force

Gap Ratio	1.0 AT/in.	2.0 AT/in.	5 AT/in.	10 AT/in.	20 AT/in.	30 AT/in.
0	1000	820	490	340	250	140
0.0005	720	700	680	560	360	170
0.0010	530	520	510	490	410	250
0.0015	—	425	410	400	370	270

Such data may be obtained using the methods given in "Magnetic Circuits and Transformers" (Ref. A3)   Ch. 7, which sets out the general methods suitable for calculation of inductance with d.c. flux in the core.

The method developed by Crowhurst (Ref. E18) makes a considerable saving in time by the use of charts, particularly when a number of inductors are to be designed.

However, when the a.c. flux density is very small, as for example in many filter chokes, the method developed by Hanna may be used.

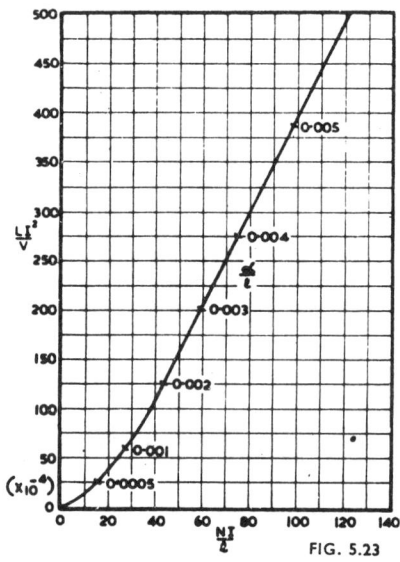

Fig. 5.23. LI^2/V plotted against NI/l with various gap ratio intercepts (Ref. E13).

(vii) Design by Hanna's Method
(Ref. E12)

Strictly, the method is applicable only when the core to be used is of constant cross section throughout the length of the magnetic path, but it gives a useful first estimate in cases where the cross section is not uniform.

Let N = number of turns in coil
    I = direct current through coil
    L = Inductance in henrys (at low a.c. flux density)
    l = length of magnetic gap (inches)
    α = length of air gap (inches)
    α/l = air-gap ratio
    a = cross-sectional area of core (square inches)
and V = $l.a$ = Volume of core (cubic inches).

Then, assuming in a typical case that L is 12 henrys and I is 80 mA proceed thus—

(a) choose lamination size and stack	e.g.	EI-11 $\frac{7}{8}$ in.
(b) find V from Fig. 5.18C		4.03 cub. in.
(c) calculate LI^2/V		190×10^{-4}
(d) find NI/l from Fig. 5.23		58
(e) find l from Fig. 5.18C		5.26 in.
(f) calculate N		3800 turns
(g) assume suitable wire gauge		33 A.W.G.
(h) find turns per layer from winding data chart		118
(i) calculate number of layers		33
(j) calculate coil build using 0.0015 in. insulation		0.31 in.
(k) calculate winding build using 0.05 in. former		0.36 in.
(l) find maximum build from Fig. 5.18C so that winding will fit window satisfactorily		0.367 in.
(m) calculate mean length of turn from Fig. 5.18D		4.8 ins.
(n) calculate coil resistance		315 ohms
(o) calculate power lost in coil		2 watts
(p) calculate cooling area (see Sect. 5)		14.3 sq. ins
(q) calculate dissipation in watts/sq. in.		0.14 watts/sq. in.
(r) find temperature rise from Fig. 5.18B		20°C.
(s) find gap ratio α/l from Fig. 5.23		0.0029
(t) calculate air gap $(\alpha/l) \times l$		0.0075 in.

With any particular core, the highest inductance is obtained with the largest possible number of turns, as limited by the window space available, the permissible value of d.c. resistance, and heating of the winding with the specified current.

(viii) Design of inductors for choke-input filters
(Ref. E13)

(a) **The input choke**

Chapter 30 Sect. 3 Eqn 1, 2 or 3 gives the minimum value required for the inductance of the input choke of a choke-input filter for any particular values of load resistance and frequency.

The required inductance for any particular value of load resistance may be achieved by designing the choke by Hanna's method as given above. It is only necessary to check that the value of the a.c. flux density in the core is small, so that Hanna's method will be applicable. For a full-wave rectifier the a.c. voltage applied to the choke is of twice the supply frequency and its peak amplitude is approximately two-thirds of the direct voltage at the input to the filter.

If, however, good regulation is required over a wide range of loads, it is necessary for the inductance of the input choke to be sufficiently high at the highest value of load resistance, that is, at the minimum value of load current. Some minimum load current greater than zero must be provided because the inductance would need to be infinite to maintain regulation down to zero load.

For a wide range of load variation (say 10 : 1) it would be very uneconomical to design a choke to have a constant inductance at all load currents as large as that required at minimum load. The choke is therefore designed so that its inductance will vary with the direct current through it, in such a way that the inductance is sufficiently high at all values of the load current. This method is discussed fully by Dunham (Ref. E14) and Crowhurst (Ref. E19) and reference should be made to their papers for detailed design considerations.

For many purposes, the following simple design procedure is adequate.

(1) From Eqn. (1) (2) or (3) of Chapter 30 Sect. 3, determine the required values of the choke inductance at minimum and maximum d.c. load currents.

(2) Select an interleaved core and, assuming a permeability of (say) 1000, design a choke to give the inductance required at minimum d.c. load current. For this purpose use the procedure of Sect. 6(iv) above but correct the assumed value of permeability using Fig. 5.20 and taking into account the number of d.c. ampere-turns

per inch of core path length. The peak a.c. voltage, of double mains frequency, across the choke, may be taken to be about two thirds of the d.c. voltage at the input to the filter.

(3) Using Fig. 5.20 note the change in permeability μ when maximum d.c. load current flows through the choke and check that the value of inductance under these conditions is sufficient. If it is not sufficient an increase in turns may be satisfactory and this can be checked by repeating the procedure above. If this is not satisfactory a small air gap may be necessary, but calculations then become complex.

A choke so designed is commonly referred to as a **swinging choke** because its inductance varies with the direct current through it.

(b) The second choke
The design of the second choke of a choke input filter is straightforward, and Hanna's method may be used after determining the required value of inductance by the methods set out in Chapter 31 Sect. 1.

(ix) Measurements
Measurements of inductances of iron-cored coils must be made under conditions similar to those under which the coils are to be used, because the value of inductance depends to a marked degree on the a.c. flux density in the core and also on any d.c. flux set up in the core by direct currents in the windings.

Bridge methods are desirable for accurate measurements of inductance but must be arranged to simulate the operating conditions of the coil being measured. Owen and Hay bridges are widely used when there is a large flux due to direct current (Refs. E15, E16).

For many purposes it is satisfactory to determine the effective inductance (or more accurately, the impedance) of a coil by measuring the current through the coil when the rated a.c. voltage is applied to it ; but this method is not usually feasible when a d.c. component is also present.

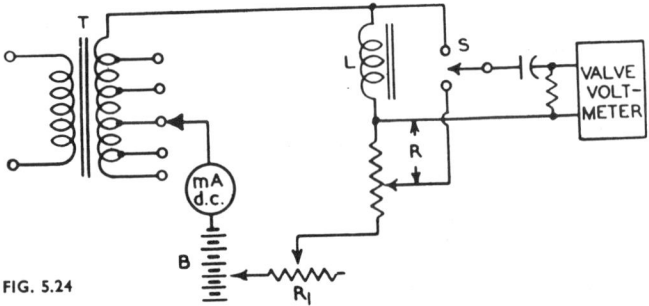

FIG. 5.24

Fig. 5.24. Determination of the impedance of an inductor L carrying direct current.

When a d.c. flux must be produced in the core to simulate operating conditions, the circuit arrangement of Fig. 5.24 may be used to determine the a.c. impedance of the coil. A valve voltmeter is used to adjust the a.c. voltage drop across the known variable resistance R to equality with the a.c. voltage drop across the inductance L. The required value of direct current through the coil is obtained by adjusting the tapping on battery B and the rheostat R_1 (or by adjusting any other variable direct current source), the current being measured by a d.c. ammeter or milliammeter. The required alternating voltage across the choke coil is then adjusted by varying the tapping on transformer T, the voltage across the coil being read on the valve voltmeter ; this adjustment is not usually critical. The resistance R is then varied until the same reading is obtained on the valve voltmeter when it is connected across R

by operating the switch S. If the valve voltmeter responds to direct voltages, it must be connected to the circuit through a blocking condenser and grid leak as shown in the diagram.

The value Z obtained for the impedance of the coil will differ from its reactance owing to coil losses, but for most purposes it will be satisfactory to assume that it is equal to the reactance. If the frequency used is f, the value of inductance is given approximately by

$$L \approx Z/2\pi f.$$
or
$$L \approx Z/314 \text{ Henrys, for } f = 50 \text{ c/s.}$$
$$\approx Z/376 \text{ Henrys for } f = 60 \text{ c/s.}$$

where Z is numerically equal to the value of R for balance.

For choke coils carrying direct current, and operating with a high a.c. flux density, reference should be made to the method of approximate measurement given by F. E. Terman (Ref. E15) Fig. 40 and pp 57-58.

(x) Iron-cored inductors in resonant circuits

The performance of iron-cored inductors in resonant circuits cannot be calculated mathematically owing to the immense complexity. While most work is carried out empirically, it is helpful to have a general grasp of the problem, and this may perhaps best be carried out graphically (see Ref. E17).

SECTION 7 : REFERENCES

(A) GENERAL

A1. Still, A. (book) " Principles of Transformer Design " (John Wiley and Sons, New York, 1926).
A2. Still, A. (book) " Elements of Electrical Design " (2nd ed. McGraw-Hill Book Co. 1932).
A3. Massachusetts Institute of Technology (book) " Magnetic Circuits and Transformers " (John Wiley and Sons, New York ; Chapman and Hall, London, 1943).
A4. Blume, L. F. and others (book) " Transformer Engineering " (John Wiley and Sons, New York, 1938).
A5. Terman, F. E. (book) " Radio Engineers' Handbook " (McGraw-Hill Book Co. 1943).
A6. Lee, R. (book) " Electronic Transformers and Circuits " (John Wiley and Sons 1947).
A7. Beatty, R. T. " Radio Data Charts " (Iliffe and Sons Ltd.).
A8. Roters, H. C. (book) " Electromagnetic Devices " (John Wiley and Sons 1941).
A9. Miner, D. F. (book) " Insulation of Electrical Apparatus " (McGraw-Hill Book Co. 1941).
A10. Connelly, F. C. (book) " Transformers " (Sir Isaac Pitman and Sons Ltd. London, 1950).
A11. Crowhurst, N. H. " Winding space determination " Electronic Eng. 23.282 (Aug. 1951) 302.

Additional references will be found in the Supplement commencing on page 1475.

(B) REFERENCES TO PRACTICAL TRANSFORMERS

B1. Honnell, P. M. " Note on the measurement of transformer turns ratio " Proc. I.R.E. 33.11 (Nov. 1945) 808. See also Refs. A1 to A10 inclusive.

(C) REFERENCES TO AUDIO FREQUENCY TRANSFORMERS

C1. Harrison, E. B. " Notes on transformer design " Elect. 17.2 (Feb. 1944) 106.
C2. Koehler, G. " Audio-frequency amplifiers " included in Henney's " Radio Engineering Handbook " Chapter 8, p. 355, 4th ed., 1950 (McGraw-Hill Book Company).
C3. Webb, E. K. " Response of a.f. transformers at high frequencies " Proc. I.R.E. Aust. 5.3 (Jan. 1944) 3.
C4. Klipsch, P. W. " Design of a.f. amplifier circuits using transformers " Proc. I.R.E. 24.2 (Feb. 1936) 219.
C5. Rohner, A. J. (Chapter) " Ferrous-cored inductors " (" Elec. Eng. Handbook " Edit. by Pender and McIlwain 4th ed. 1950) Communication and Electronics, Sect. 3-42.
C6. Wrathall, E. T. " Audio-frequency transformers " W.E. 14.6, 7, 8 (June, July, August 1937).
C7. Sowter, G. A. V. " The properties of nickel iron alloys and their application to electronic devices " J. Brit. I.R.E. 2.4 (Dec. 1941—Feb. 1942) 100.
C8. Peterson, E. " Harmonic production in ferromagnetic materials at low frequencies and low flux densities." B.S.T.J. 7.4 (Oct. 1928) 762.
C9. Story, J. G. " Design of audio-frequency input and inter-valve transformers " W.E. 15.173 (Feb. 1938) 69.
C10. Roddam, T. " Distortion in audio-frequency transformers " W.W. 51.7 (July 1945) 202.
C11. Terman, F. E. " Radio Engineers' Handbook " 1st edit. (1943) p. 131.
C12. " Telcon Metals " Booklet Pub. by Telegraph Construction and Maintenance Co. Ltd., Greenwich, London 1948.
C13. Terman, F. E. (book) " Radio Engineering " (3rd edit. 1947) p. 291.
C14. Baxandall, P. J. " High quality amplifier design " W.W. 54.1 (Jan. 1948) 2.
C15. Lee, R. " Recent transformer developments " Proc. I.R.E. 33.4 (April 1945) 240.
C16. Strong, C. E. " Final stage class B modulation " Elect. Comm. 16.4 (April 1938) 321.
C17. McLean, T. " An analysis of distortion in class B amplifiers " Proc. I.R.E. 24.3 (March 1936) 487.
C18. Lee, R. (book) " Electronic transformers and circuits " 1st edit. (1947) p. 163.
C19. Terman, F. E. " Radio Engineers' Handbook " (1st edit. 1943) p. 392.
C20. Mortley, W. S. " The design of low frequency transformer coupled amplifiers " Marconi Review No. 45 (Nov. Dec. 1933) 25.
C21. Preisman, A. (book) " Graphical Constructions for Vacuum Tube Circuits " (1st edit. 1943 McGraw-Hill) p. 195.
C22. McLean, T. " An analysis of distortion in class B audio amplifiers " Proc. I.R.E. 24.3 (March 1936) 487.
C23. Sah, A. P-T. " Quasi-transients in Class B audio frequency push-pull amplifiers " Proc. I.R.E. 24.11 (Nov. 1936) 1522.
C24. Partridge, N. G. R. " Distortion in transformer cores " W.W. 44.25 (June 22, 1939) 572 ; 44.26 (June 29, 1939) 597 ; 45.1 (July 6, 1939) 8 ; 45.2 (July 13, 1939) 30.
C25. Partridge, N. G. R. " An introduction to the study of harmonic distortion in audio frequency transformers " J. Brit. I.R.E. 3.1 (June-July, 1942) 6.
C26. Partridge, N. G. R. " Harmonic distortion in audio frequency transformers " W.E. 19.228 (Sept. 1942) 394 ; 19.229 (Oct. 1942) 451 ; 19.230 (Nov. 1942) 503.
C27. Partridge, N. G. R. " Transformer distortion " W.W. 48.8 (Aug. 1942) 178.
C28. Crowhurst, N. H. " Leakage inductance " Electronic Eng. 21.254 (April 1949) 129.
C29. Crowhurst, N. H. " Winding capacitance " Electronic Eng. 21.261 (Nov. 1949) 417.
C30. Harrison, E. B. " Test methods for high quality audio transformers " Tele.-Tech. 9.3 (March 1950) 40.
C31. " Reference Data for Radio Engineers " Chap. 11 (Federal Telephone and Radio Corp. N.Y. 3rd edit. 1949).
C32. Lord, H. W. " Design of broad-band transformers for linear electronic circuits " E.E. 69.11 (Nov. 1950) 1020.
C33. Williams, T., and R. H. Eastop " Harmonic distortion in iron-core transformers " Audio Eng. 35.4 (April 1951) 18.
C34. Crowhurst, N. H. " Transformer iron losses " Electronic Eng. 23.284 (Oct. 1951) 396.

Additional references will be found in the Supplement commencing on page 1475.

(D) REFERENCES TO POWER TRANSFORMER DESIGN

D1. Blume, L. F. (book) " Transformer Engineering " (John Wiley and Sons, Inc. New York, 1938).
D2. Stigant, S. A. and H. M. Lacey " The J. and P. Transformer Book " (Johnson and Phillips Ltd., London).
D3. M.I.T. Staff (book) " Magnetic Circuits and Transformers " (John Wiley and Sons, Inc. New York, 1943).

D4. Reed, E. G. (book) "Essentials of Transformer Practice" (McGraw-Hill Book Co. Inc. New York 2nd edit. 1927).
D5. Ledward, T. A. "Transformer screening" W.W. 50.1. (Jan. 1944) 15.
D6. "British Standard Specification for Steel Sheets for Transformers for Power and Lighting" B.S. 601-1935 (British Standards Institution, London).
D7. "Standard Methods of Test for Magnetic Properties of Iron and Steel" A-34 (1940). American Society of Testing Materials, N.Y.
D8. Partridge, N. G. R. "Harmonic distortion in audio frequency transformers" W.E. 19.8 (Aug. 1942) 180.
D9. Medina, L. "A device for the measurement of no-load losses in small power transformers" Proc. I.R.E. Aust. 7.9 (Sept. 1946) 13.
D10. Horstman, C. C. "Rolled steel cores for radio transformers" Elect. 16.6 (June 1943) 110.
D11. Partridge, N. G. R. "Lamination design—influence of shape on the cost and weight of transformers and chokes" W.W. 48.12 (Dec. 1942) 286.
D12. Mawson, R. "The design of stampings for L.F. transformers" Electronic Eng. 16.195 (May 1944) 514.
D13. "Dry Type Power Transformers for Radio Transmitters" R.M.A. Standard TR-102 Nov. 1947.
D14. "Power Transformers for Radio Broadcast Receivers—Core Laminations, Vertical and Horizontal Channel Frames." R.M.A. Standard REC-120 Oct. 1948.
D15. Prince, D. C., and P. B. Vogdes (book) "Mercury Arc Rectifiers" (McGraw-Hill Book Co. Inc., New York, 1927).
D16. "British Standard Specification for Electrical Performance of Transformers for Power and Lighting" B.S. 171—1936 ; Amendment No. 3, Jan. 1948 (British Standards Institution, London).
D17. American Standards
     C57.10—1948 Terminology for transformers, regulators and reactors.
     C57.11—1948 General requirements for transformers, regulators and reactors.
     C57.12—1949 Distribution, power and regulating transformers other than current limiting.
D18. "Power and Audio Transformers and Reactors—Home Receiver Replacement Type" C.16.9 (1943) (American Standards Association, New York).
D19. "Reference Data for Radio Engineers" Chapter 11. (Federal Telephone and Radio Corp. N.Y. 3rd edit. 1949).
D20. "Engineering a transformer." Booklet (Standard Transformer Corp. Chicago).
D21. Roesche, C. "Practical transformer design and construction" Radio News 37.6 (June 1947) 60 ; 38.1 (July 1947) 58 ; 38.2 (Aug. 1947) 60.
D22. Radio Industry Council, London, Specification "Transformers, Power, up to 2 KVA Rating" No. RIC/214 (Issue No. 1—Feb. 1951).
D23. "Components Handbook" Vol. 17 of M.I.T. Radiation Lab. Series. Chapter 4 "Iron-Core Inductors" (McGraw-Hill Book Co. Inc. New York, 1949).
D24. Hamaker, H. C., and Th. Hehenkamp "Minimum cost transformers and chokes" Philips Res. Rep. 5.5 (Oct. 1950) 357 ; 6.2 (April 1951) 105.
D24. R.C.S.C. Standard RC.L/191 Cores, magnetic.
D25. R.C.S.C. Standard RC.L/193 Cores, magnetic, strip wound.
D26. R.C.S.C. Standard RC.S/214 Specification for transformers, power.
also A10.

(E) REFERENCES TO IRON-CORED INDUCTORS

E1. M.I.T. Staff (book) "Magnetic Circuits and Transformers" (John Wiley and Sons N.Y. 1943).
E2. Javna, S. L. "High Q iron cored inductor calculations" Elect. 18.8 (Aug. 1945) 119.
E3. Campbell, C. A. "High Q audio reactors—design and production" Comm. 24.3 (March 1944) 37.
E4. Arguimbau, L. B. "Losses in audio-frequency coils" General Radio Exp. 11.6 (Nov. 1936) 1.
E5. McElroy, P. K. and R. F. Field, "How good is an iron-cored coil?" General Radio Exp. 16.10 (Mar. 1942) 1.
E6. McElroy, P. K. "Those iron-cored coils again" G.R. Exp. 21.7 (Dec. 1946) ; 21.8 (Jan. 1947).
E7. Payne-Scott, R. "A note on the design of iron-cored coils at audio-frequencies" A.W.A. Tec. Rev. 6.2 (1943) 91.
E8. Ryder, J. D. "Ferroinductance as a variable electric-circuit element" Trans. A.I.E.E. 64.10 (Oct. 1945) 671.
E9. Lamson, H. W. "Alternating-current measurement of magnetic properties" Proc. I.R.E. 36.2 (Feb. 1948) 266.
E10. Roters, H. C. (book) "Electromagnetic Devices" (John Wiley and Sons New York 1942).
E11. Legg, V. E. "Optimum air gap for various magnetic materials in cores of coils subject to superposed direct current" Trans. A.I.E.E. 64.10 (Oct. 1945) 709. Discussion in Supplement to Trans. 64.12 (Dec. 1945) 969.
E12. Hanna, C. R. "Design of reactances and transformers which carry direct current" Trans. A.I.E.E. 46 (1927) 155.
E13. Lee, R. (book) "Electronic Transformers and Circuits" (John Wiley and Sons N.Y. 1947).
E14. Dunham, C. R. "Some considerations in the design of hot cathode mercury vapour rectifier circuits" Jour. I.E.E. 75.453 (Sept. 1934) 278.
E15. Terman, F. E. (book) "Measurements in Radio Engineering" (McGraw-Hill Book Co. N.Y. 1935) pp. 51-57.
E16. "Magnetic Materials for use under combined D.C. and A.C. Magnetisation" British Standard Specification 933-1941 ; Amendment No. 1, April 1944.
E17. Drake, B. "Ferro-resonance" Electronic Eng. 21.254 (April 1949) 135.
E18. Crowhurst, N. H. "Design of iron-cored inductances carrying d.c." Electronic Eng. 22.274 (Dec. 1950) 516.
E19. Crowhurst, N. H. "Design of input (regulation control) chokes" Electronic Eng. 23.279 (May 1951) 179.

CHAPTER 6

MATHEMATICS

By F. Langford-Smith, B.Sc., B.E.

For ease in reference

Mathematical signs   see Chapter 38 Sect. 6(vii)
Greek letters   see Chapter 38 Sect. 14
Definitions   see Chapter 38 Sect. 15
Decimal equivalents of fractions   see Chapter 38 Sect. 16
Numerical values   see Chapter 38 Sect. 18
Logarithm tables   see Chapter 38 Sect. 20
Trigonometrical and hyperbolic tables   see Chapter 38 Sect. 21
Log. scales and log. scale interpolator   see Chapter 38 Sect. 22

Mathematics, to the radio engineer, is merely a tool to be used in his design work. For this reason it is often used in a slovenly manner or with insufficient precision or understanding.

There are normally three stages in the solution of a problem—

1. Transferring the mechanical or electrical conditions into a mathematical form.

2. Solving the mathematics.

3. Interpreting and applying the mathematical solution.

The first stage is dealt with in Chapter 4 ; the second stage is the subject of this chapter, while the third stage requires careful consideration of all the relevant conditions. A solution only applies under the conditions assumed in stage one, which may involve some approximations and limits. In all cases the solution should be checked either experimentally or theoretically to prove that it is a true solution.

This chapter is not a textbook on mathematics, although it is in such an easy form that anyone with the minimum of mathematical knowledge should be able to follow it. It has been written primarily for those who require assistance in " brushing up " their knowledge, and for the clarification of points which may be imperfectly understood. It is " basic " rather than elementary in its introduction, and could therefore be read with advantage by all.

Sufficient ground is covered for all normal usage in radio receiver design, except for that required by specialists in network and filter design.

Reference data have been included for use by all grades

SECTION 1 : ARITHMETIC AND THE SLIDE RULE

(*i*) *Figures* (*ii*) *Powers and roots* (*iii*) *Logarithms* (*iv*) *The slide rule* (*v*) *Short cuts in arithmetic*

(i) Figures

A figure (e.g. 5) indicates a certain number of a particular object—e.g. five resistors or five radio receivers ; or else five units of a particular scale—e.g. five inches, five microfarads or five ohms. No matter what we may do with adding, subtracting, multiplying or dividing, if we begin with ohms we must finish with ohms, and so on with any other unit. We cannot add together dissimilar objects without identifying each type, and similarly with any other mathematical process.

(ii) Powers and roots

As figures are often too large, or too small, to be shown completely in the ordinary form, there is a Scientific Notation commonly used—

Numbers above unity	Numbers below unity
$10^1 = 10$	$10^{-1} = 0.1$
$10^2 = 100$	$10^{-2} = 0.01$
$10^3 = 1\ 000$	$10^{-3} = 0.001$
$10^4 = 10\ 000$	$10^{-4} = 0.000\ 1$
$10^5 = 100\ 000$	$10^{-5} = 0.000\ 01$
$10^6 = 1\ 000\ 000$	$10^{-6} = 0.000\ 001$
	and so on

$$10^0 = 1$$

Note that 10^2 means $10 \times 10 = 100$; 10^3 means $10 \times 10 \times 10 = 1000$ and so on. The " 2 " and the " 3 " are called exponents, or indices (plural of index).

10^2 is called " 10 squared "
10^3 is called " 10 cubed "
10^4 is called " 10 to the fourth " etc.
10^{-1} is called " 10 to the minus 1 " etc.

Examples : 2 750 000 is written 2.75×10^6
0.000 025 is written 2.5×10^{-5}.

Multiplying is carried out as in the examples :
$$(1.5 \times 10^6) \times (4 \times 10^{-3}) = 6 \times 10^3$$
$$(4.5 \times 10^3) \times (2 \times 10^2) = 9 \times 10^5$$

Reference should be made to the table of multiples and sub-multiples in Chapter 38 Sect. 17, Table 59.

The same procedure may be applied, not only to 10, but to any figure, e.g. 3 ;

$3^0 = 1$
$3^1 = 3$
$3^2 = 3 \times 3 = 9 \qquad\qquad 3^{-2} = 1/3^2 = 1/9$
$3^3 = 3 \times 3 \times 3 = 27 \qquad 3^{-3} = 1/3^3 = 1/27$

Roots : The expression $9^{\frac{1}{2}}$ may be written $\sqrt{9}$, where the sign $\sqrt{}$ is called the square root. Similarly $27^{1/3}$ may be written $\sqrt[3]{27}$ where the sign $\sqrt[3]{}$ is called the cube root.

The whole question is dealt with more fully, and more generally under " Algebra " in Sect. 2.

(iii) Logarithms

We may write $100 = 10^2$
or we may express this in different language as
$$2 = \log_{10} 100$$
which is spoken of as " log 100 to the base 10." Here 2 is the logarithm of 100 to the base 10.

Tables of logarithms to the base 10 are given in Chapter 38 Sect. 20, Table 71. They are useful in multiplying and dividing numbers which are too large to handle conveniently by the ordinary procedure.

A typical logarithm is 2.4785. Here the figure 2 to the left of the decimal point is called the **index**, and the figures 4785 to the right of the decimal point are called the **mantissa**.

(1) The **index** of the logarithm of a number greater than unity is the number which is less by one than the number of digits (figures) in the integral* part of the given number ; for example the index of the logarithm of

57 640	is 4
5 764	is 3
576.4	is 2
57.64	is 1
5.764	is 0

If the number is less than unity, the index is negative, and is a higher number by one than the number of zeros that follow the decimal point of the given number ; for example the index of the logarithm of

.576 4	is -1
.005 764	is -3.

To denote that the index only is negative, the minus sign is usually written above it ;
e.g. $\bar{1}, \bar{3}$.

(2) The **mantissa** of the logarithm is found from the tables. Proceed to find the first two figures in the left hand column of the table, then pass along the horizontal line to the vertical column headed by the third figure. To this number add the number in the difference column under the fourth figure of the given number. The mantissa is the result obtained by this process with a decimal point before it.

For example, to find $\log_{10} 5764$.

The index is	3
The mantissa is	$.7604 + .0003 = .7607$

Therefore $\log_{10} 5764 = 3.7607$.

Similarly, $\log_{10} 0.5764 = \bar{1}.7607$ (note that the index is negative, but the mantissa positive).

To find the number whose logarithm is given, it is possible to use either antilog. tables (if these are available) or to use the log. tables in the reverse manner. In either case, only the mantissa (to the right of the decimal point) should be applied to the tables.

The procedure with log. tables is firstly to find the logarithm, on the principal part of the table, which is next lower than the given logarithm, then to calculate the difference, then to refer to the difference columns to find the number—exactly the reverse of the previous procedure.

For example, to find the number whose logarithm is 2.5712. The mantissa is .5712 and the nearest lower logarithm on the tables is 5705, the difference being 7 (in the fourth figure). The number whose log = 5705 is 3720, and to this must be added the figure 6 which corresponds to the difference of 7 in the fourth column ; the number is therefore 3720 + 6 = 3726. The decimal point must be placed so as to give 2 + 1 = 3 digits to the left of the point, i.e. 372.6.

The application of logarithms

If two numbers are to be **multiplied** together, the answer may be found by adding their logarithms, and then finding the number whose logarithm is equal to their sum. For example, suppose that it is desired to multiply 371.6×58.24,

$$\log 371.6 = 2.5701$$
$$\log 58.24 = 1.7652$$
$$\text{sum} = 4.3353$$

The number whose log = .3353 is 2164.

The decimal point should be placed so as to give 4 + 1 = 5 digits to the left of the point ; i.e. 21 640.

---

*To the left of the decimal point.

Therefore $371.6 \times 58.24 = 21\,640$.

It should be remembered that the last figure of four figure logs. is correct to the nearest unit, and that slight errors creep into the calculations through additions and other manipulations.   The first three digits of the answer will be exact, and the fourth only approximate.

If one number is to be **divided** by a second number, the answer may be found by subtracting the logarithm of the second from the logarithm of the first, then finding the number whose logarithm is equal to their difference.

Logarithms may also be used to find the **powers of numbers**   For example, to find the value of $(3.762)^3$:

$$\log (3.762)^3 = 3 \log 3.762 = 3 \times .5754$$
$$= 1.7262$$

Therefore $(3.762)^3 =$ antilog $1.7262 = 53.23$.

For other applications of logarithms see Sect. 2(xvii)

(iv) The Slide Rule

The Slide Rule is a mechanical device to permit the addition and subtraction of logarithms so as to effect multiplication and division of the numbers.   The small size of the normal slide rule does not give as high a degree of accuracy as four figure log. tables, but is sufficiently accurate for many calculations.

The usual 10 inch slide rule has four scales A, B, C, D of which B and C are on the slide.   The scales C and D (the lower pair) are normally used for multiplication and division, and each covers from 1 to 10.   The upper scales A and B cover from 1 to 100. The **square** of any number from 1 to 10 is found by adjusting the line on the cursor (runner) to fall on the number on the D scale, and reading the answer where the line cuts the A scale.   The **cube** of a number from 1 to 10 may be found by squaring, and then multiplying the result by the original number on the B scale, reading off the answer on the A scale.   If the number is not between 1 and 10, firstly break it up into factors, one of which should be a multiple of 10, and the other a number between 1 and 10, then proceed as before.   For example
$$300^2 = (3 \times 100)^2 = (3)^2 \times (100)^2 = 3^2 \times 10^4.$$
The value of 3^2 is found in the normal way to be 9 ; this is then multiplied by 10^4 to give the answer $9 \times 10^4 = 90\,000$.

Square roots may be found by the reverse procedure.   Firstly reduce the number to factors, one of which should be a multiple of 100 and the other between 1 and 100, then apply the cursor to the number on the A scale and read the answer on the D scale to be multiplied by the square root of the 100 factor.   For example, to find the square root of 1600 : $\sqrt{1600} = \sqrt{16 \times 100} = \sqrt{16} \times \sqrt{100} = \sqrt{16} \times 10.$
The value of 4 on the D scale is then multiplied by 10 to give the answer 40.

Cube roots of numbers between 1 and 100 may be determined by setting the cursor to the number on the A scale, then moving the slide until the B scale cursor reading is the same as the D scale reading below 1 on the C scale.

Slide rules which have log/log scales may be used to determine any power of a number 1.1 or greater (up to a maximum value of 100 000).   Set the cursor to the number on the upper log/log scale, then set 1 on the C scale to the same cursor line. Move the cursor to the required power on the C scale and read the answer on the log/log scale.   If the number is too high to be on the upper log/log scale, carry out the same procedure on the lower log/log scale.   If the number is found on the upper scale, but the answer is beyond the limits of this scale, set the mark* (e.g. W) on the slide immediately below the number on the upper scale, and read the answer on the lower scale, immediately below the power on the C scale.

If several figures are to be multiplied and divided, carry out multiplication and division alternately, e.g.
$$\frac{75 \times 23 \times 5}{41 \times 59 \times 36}$$
should be handled as $75 \div 41 \times 23 \div 59 \times 5 \div 36$.

---

*With slide rules having no special mark, use 10 on the C scale as the " mark."

In a complicated calculation, especially with very large and very small numbers, it is highly desirable to arrange the numerator and denominator in powers of 10. For example

$$\frac{75\,000 \times 0.0036 \times 5900}{160\,000 \times 0.000\,001\,7} = \frac{7.5 \times 10^4 \times 3.6 \times 10^{-3} \times 5.9 \times 10^3}{1.6 \times 10^5 \times 1.7 \times 10^{-6}}$$

$$= \frac{7.5 \times 3.6 \times 5.9 \times 10^5}{1.6 \times 1.7}$$

The slide rule does not indicate the position of the decimal point, and it is necessary to determine the latter by some method such as inspection ; this is much easier when the individual numbers are all between 1 and 10 as in the example above. It is also possible to keep track of the decimal point by noting how often the manipulation passes from end to end of the rule.

To find the logarithm of a given number, move the 1 on C scale to the number on the D scale, then turn the rule over and read the logarithm on the L scale against the mark (this will be a number between 0 and 1).

To find the decibels corresponding to a ratio, proceed as for the logarithm, but multiply by 10 for a power ratio or 20 for a voltage ratio.

To find the sine or tangent of an angle, first set the angle on the S or T scale to the mark, then read the value on the B scale, below 1 on the A scale, and divide by 100.

There are countless special types of slide rules, and in all such cases the detailed instructions provided by the manufacturers should be studied.

Hints on special calculations on the slide rule

(1) $Z = \sqrt{R^2 + X^2} = R\sqrt{1 + (X^2/R^2)}$

Procedure : For example if $X = 3$ and $R = 2$ set cursor to 3 on D scale, move slide to give 2 on C scale. The value of $(X/R)^2$ is given by the value on A scale opposite 1 on B scale—in this case 2.25. Move the slide up to 3.25 ($= 2.25 + 1$) and then move the cursor to 2 on C scale, reading 3.61 on D scale as the answer.

(2) If a large number of figures is to be divided by one figure, divide unity or 10 (D scale) by the divisor (C scale) and then, with fixed slide, move the cursor to each dividend in turn on the C scale, reading the answer on the D scale.

(v) Short cuts in arithmetic

(a) Approximations involving π

π^2 may be taken as 10 with an error less than 1.4%

π may be taken as 25/8 with an error less than 0.6%

$1/\pi$ may be taken as 8/25 with an error less than 0.6%

2π may be taken as 25/4 with an error less than 0.6%

$1/2\pi$ may be taken as 4/25 with an error less than 0.6%

$(2\pi)^2 = 39.5$ with an error less than 0.06%

(b) Approximations with powers and roots

General relation : $(x \pm \delta)^n \approx x^n \pm nx^{n-1}\,\delta$

where δ is small compared with x. Examples are given below.

Squares $(n = 2)$

$10.1^2 = (10 + 0.1)^2 \approx 10^2 + 2 \times 10 \times 0.1 \approx 102$ (error $= 0.01\%$)

$10.2^2 = (10 + 0.2)^2 \approx 10^2 + 2 \times 10 \times 0.2 \approx 104$ (error $= 0.04\%$)

$9.9^2 = (10 - 0.1)^2 \approx 10^2 - 2 \times 10 \times 0.1 \approx 98$ (error $= 0.01\%$)

Square roots $[n = 0.5 \,; (n - 1) = -0.5]$

$$\sqrt{x \pm \delta} \approx \sqrt{x} \pm \frac{0.5\delta}{\sqrt{x}}$$

$$\sqrt{101} = \sqrt{100 + 1} \approx \sqrt{100} + \frac{0.5 \times 1}{\sqrt{100}} \approx 10.05 \text{ (error} = 0.001\%)$$

$$\sqrt{110} = \sqrt{100 + 10} \approx \sqrt{100} + \frac{0.5 \times 10}{\sqrt{100}} \approx 10.5 \text{ (error} = 0.12\%)$$

$$\sqrt{50} = \sqrt{49+1} \approx \sqrt{49} + \frac{0.5 \times 1}{\sqrt{49}} \approx 7.0714 \text{ (error} = 0.004\%)$$

Cubes $(n = 3)$
$(10.2)^3 = (10 + 0.2)^3 \approx 10^3 + 3 \times 10^2 \times 0.2 \approx 1060 \text{ (error} = 0.1\%)$
Cube roots $[n = 0.333 ; (n - 1) = -0.667]$

$$\sqrt[3]{x \pm \delta} \approx \sqrt[3]{x} \pm \frac{\delta}{3(\sqrt[3]{x})^2}$$

$$\sqrt[3]{66} = \sqrt[3]{64+2} \approx \sqrt[3]{64} + \frac{2}{3 \times 4^2} \approx 4.042$$

For more accurate approximations see Sect. 2 eqns. (82) and (83).

(c) Approximations in multiplication
$a \times b \approx \frac{1}{4}(a + b)^2$ where a and b are close together
e.g. $49 \times 51 \approx \frac{1}{4}(49 + 51)^2 \approx \frac{1}{4}(100)^2 \approx 2500$ (error 0.04%).
This may be put into the alternative form :
$a \times b \approx$ (arithmetical mean between a and $b)^2$
e.g. $68 \times 72 \approx (70)^2 \approx 4900$ (error 0.08%).

(d) Exact multiplication
$a \times b = \frac{1}{4}\{(a + b)^2 - (a - b)^2\}$ (no error)
$\quad = $ (arithmetical mean between a and $b)^2 - \frac{1}{4}(a - b)^2$.
When $(a - b) = 1$, the second term in this expression becomes $\frac{1}{4}$ and we have the
exact application :
$$3\tfrac{1}{2} \times 4\tfrac{1}{2} = 16 - \tfrac{1}{4} = 15\tfrac{3}{4} \text{ (exact)}.$$
Another application is illustrated by the example
$$98 \times 102 = 100^2 - \tfrac{1}{4}(4)^2 = 10\,000 - 4 = 9996.$$

(e) To multiply by 11
To multiply a number by 11, write down the last figure, add the last and last but
one and write down the result, carrying over any tens to the next operation, add the
last but one and the last but two and so on, finishing by writing down the first
$$\text{e.g. } 11 \times 42\,736 = 470096 \text{ (no error)}$$
(f) For approximations based on the Binomial Theorem see Sect. 2(xviii).
(g) For general approximations see Sect. 2(xx).

SECTION 2 : ALGEBRA

(*i*) *Addition* (*ii*) *Subtraction* (*iii*) *Multiplication* (*iv*) *Division* (*v*) *Powers*
(*vi*) *Roots* (*vii*) *Brackets and simple manipulations* (*viii*) *Factoring* (*ix*) *Proportion*
(*x*) *Variation* (*xi*) *Inequalities* (*xii*) *Functions* (*xiii*) *Equations* (*xiv*) *Formulae
or laws* (*xv*) *Continuity and limits* (*xvi*) *Progressions, sequences and series* (*xvii*)
Logarithmic and exponential functions (*xviii*) *Infinite series* (*xix*) *Hyperbolic
functions* (*xx*) *General approximations.*

See Section 6 for Complex Algebra.

Algebra is really only arithmetic, except that we use alphabetical symbols to stand
for figures. It is frequently more convenient to put an expression into an algebraic
form for general use, and then to apply it to a particular case by writing figures in place
of the letters. All algebraic expressions are capable of being converted into arithmetical ones, and the fundamental mathematical processes of algebra may be used
in arithmetic.

(i) Addition

If a, b, and c are all values of the one unit (e.g. all resistances in ohms) we can add them together to find the sum d, where d will also be in the same unit,

$$d = a + b + c$$

For example, if $a = 5$, $b = 10$, $c = 15$ ohms,
then $d = 5 + 10 + 15 = 30$ ohms.

(ii) Subtraction

Subtraction is the opposite of addition, or negative addition, and can only be applied when the quantity to be subtracted is in the same unit as the quantity from which it is to be taken. For example let

$$a = b - c$$

where a, b and c are all voltages.

If $b = 6$ volts and $c = 2$ volts, then

$$a = 6 - 2 = 4 \text{ volts.}$$

As another example, let a, b and c be readings of a thermometer in degrees—say $b = 10°C$ and $c = 20°C$, then $a = 10° - 20° = -10°C$. This is commonly described as " 10 degrees below zero " or " a temperature of minus 10 degrees." Thus a negative temperature has a definite value and is readily understood. Its magnitude is given by the figure, while its direction above or below zero is given by the positive or negative sign.

Similarly a negative current is one with a magnitude as indicated by the figure but with a direction opposite to that of a positive current. In most cases the direction of a positive current is arbitrarily fixed, and if the answer comes out negative it merely indicates the actual direction of current flow is the opposite of the direction assumed. The same principle holds in all cases.

(iii) Multiplication

Multiplication is continued addition—

$$4 \times a = a + a + a + a$$

and is commonly written as $4a$. In the general sense we can write ba where b is any number ; this has the same value as ab,

i.e. $ab = ba$ or $a.b = b.a.$

It should be noted that

$$4 \times (-a) = -4a$$
$$\text{and } (-1) \times (-1) = +1.$$

(iv) Division

Division is the breaking up of a number of things into a given number of groups, e.g.

$$6a \div 3 = (2a + 2a + 2a) \div 3 = 2a.$$

We may write this in the alternative forms

$$\frac{6a}{3} = 2a, \text{ or } 6a/3 = 2a.$$

(v) Powers

Powers are continued multiplication,

$$\text{e.g. } a^3 = aaa = a \times a \times a$$
$$a^m = a \times a \times a \times a \dots (m \text{ factors})$$
$$a^n = a \times a \times a \times a \dots (n \text{ factors})$$

Therefore $a^m \times a^n = a^{(m+n)}$ (1)

We can write, as a convenience,

$$\frac{1}{a^n} \text{ in the form } a^{-n}$$

where the $-n$ is not a true index (or exponential) but merely a way of writing $1/a^n$.

Therefore $\dfrac{a^m}{a^n} = a^m \times a^{-n} = a^{m-n}$ (2)

which indicates that the $-n$ can be treated as though it were a true index.

The following can also be derived :

$$\frac{a^m}{a^m} = a^{(m-m)} = a^0 \tag{3}$$

but $\dfrac{a^m}{a^m} = 1$

Therefore $a^0 = 1$ (4)

$$(a^m)^n = a^{m \times n} = a^{mn} \tag{5}$$

$$(ab)^n = a^n b^n \tag{7}$$

$$\left(\frac{b}{a}\right)^{-n} = \left(\frac{a}{b}\right)^n = \frac{a^n}{b^n} \tag{8}$$

$$(-a)^n = (-1)^n \times a^n \tag{9}$$

$$= +a^n \text{ if } n \text{ is even} \tag{9a}$$

$$\text{or} = -a^n \text{ if } n \text{ is odd} \tag{9b}$$

These identities hold even when m and n are negative or fractions.

(vi) Roots

$$\sqrt{a \times a} = a \text{ or } \sqrt{a^2} = a \tag{10}$$

$$\sqrt[3]{a \times a \times a} = a \text{ or } \sqrt[3]{a^3} = a \tag{11}$$

We may adopt as a convenience the form

$$\sqrt{a} = a^{\frac{1}{2}} \tag{12}$$

$$\sqrt[3]{a} = a^{1/3} \tag{13}$$

$$\sqrt[n]{a} = a^{1/n} \tag{14}$$

This may be extended to include

$$\sqrt[n]{a^m} = (a^m)^{1/n} = a^{m/n} \tag{15}$$

so that $a^{m/n}$ is the nth root of a^m.

Note that $1/\sqrt[3]{a} = \sqrt[3]{1/a}$; $1/\sqrt[n]{a} = \sqrt[n]{1/a}$ (16)

(vii) Brackets and simple manipulations

$$a(a + b) = a \times a + a \times b = a^2 + ab \tag{17}$$

$$x(a + b - c) = xa + xb - xc \tag{18}$$

$$-x(a + b) = -xa - xb \tag{19}$$

$$-x(a - b) = -xa + xb = x(b - a) \tag{20}$$

$$-[(a - b) - (c + d)] = -(a - b) + (c + d) \tag{21}$$

$$= -a + b + c + d \tag{21a}$$

$$= (b + c + d) - a \tag{21b}$$

$$\frac{ax + bx}{cx + dx} = \frac{(a + b)x}{(c + d)x} = \frac{a + b}{c + d} \times \frac{x}{x} = \frac{a + b}{c + d} \tag{22}$$

$$(a + b)^2 = (a + b)(a + b) = a(a + b) + b(a + b) \tag{23}$$

$$= a^2 + 2ab + b^2 \tag{23a}$$

$$(a - b)^2 = (a - b)(a - b) = a(a - b) - b(a - b) \tag{24}$$

$$= a^2 - 2ab + b^2 \tag{24a}$$

$$(a + b)(a - b) = a(a - b) + b(a - b) = a^2 - b^2 \tag{25}$$

$$(a + b)(x + y) = ax + ay + bx + by \tag{26}$$

$$(a + b + c)^2 = a^2 + b^2 + c^2 + 2ab + 2bc + 2ca \tag{27}$$

$$\frac{a}{-b} = -\frac{a}{b}; \quad \frac{-12ac}{4a} = -3c; \quad \frac{a^5}{a^2} = a^3 \tag{28}$$

$$+\frac{a}{b} = \frac{-a}{-b} = -\frac{-a}{b} = -\frac{a}{-b} \tag{29}$$

$$a \times 0 = 0; \quad \frac{a}{0} = \text{infinity*} = \infty \tag{30}$$

(note that it is not possible to divide by 0 in algebraic computations).

*Infinity may be described as a quantity large without limit.

$$\frac{a}{c} \pm \frac{b}{d} = \frac{ad \pm bc}{cd} \; ; \qquad \frac{a}{c} \pm \frac{b}{c} = \frac{a \pm b}{c} \; ; \qquad \frac{a}{c} \pm \frac{a}{d} = \frac{a(d \pm c)}{cd} \tag{31}$$

The sign $\pm$ means either plus or minus. When $\pm$ and/or $\mp$ signs are used on both sides of an equation, the upper signs in both cases are to be taken in conjunction as one case, while the lower signs are to be taken as the other case.

$$\frac{a}{b} \times \frac{c}{d} = \frac{ac}{bd} \; ; \qquad \frac{a}{b} = \frac{ac}{bc} \tag{32}$$

$$\frac{a}{b} \div \frac{c}{d} = \frac{a}{b} \times \frac{d}{c} = \frac{ad}{bc} \tag{33}$$

$$\frac{x}{y} - \frac{a+b}{c+d} = \frac{x(c+d)}{y(c+d)} - \frac{y}{y}\left(\frac{a+b}{c+d}\right) \tag{34a}$$

$$= \frac{x(c+d) - y(a+b)}{y(c+d)} \tag{34b}$$

(viii) Factoring—Examples

$$6ab + 3ac = 3a(2b + c) \tag{35}$$
$$x^2 - 7xy + 12y^2 = (x - 4y)(x - 3y) \tag{36}$$
$$2x^2 + 7x + 6 = (2x + 3)(x + 2) \tag{37}$$
$$x^3y - 4y^3 = y(x^2 - 4y^2) = y(x + 2y)(x - 2y) \tag{38}$$

(ix) Proportion

(1) If $\quad \dfrac{a}{b} = \dfrac{c}{d}$ $\qquad\qquad$ then $\quad \dfrac{a}{c} = \dfrac{b}{d}$ $\tag{39}$

also $\quad \dfrac{a}{b} - \dfrac{c}{d} = 0$ $\qquad\qquad$ therefore $\dfrac{ad - bc}{bd} = 0$ $\tag{40}$

from which $ad - bc = 0$ $\qquad\qquad$ and thus $ad = bc$ $\tag{41}$

(2) If $\quad \dfrac{a}{b} = \dfrac{c}{d}$ and also $\dfrac{e}{f} = \dfrac{g}{h}$,

then $\quad \dfrac{ae}{bf} = \dfrac{cg}{dh}$ $\tag{42}$

(x) Variation

If $y = kx$, then $y \propto x$
$\qquad\qquad$ i.e. y is directly proportional to x.

If $y = \dfrac{k}{x}$, then $y \propto \dfrac{1}{x}$
$\qquad\qquad$ i.e. y is inversely proportional to x.

If $y = kxz$, then y varies jointly as x and z.

If $y = k\dfrac{x}{z}$, then y varies directly as x and inversely as z

(xi) Inequalities

The letter symbols below are positive and finite.

If $a > b$ then $\qquad a + c > b + c, \qquad b < a$ $\tag{43a}$

$\qquad\qquad\qquad a - c > b - c, c - a < c - b$ $\tag{43b}$

$\qquad\qquad\qquad ac > bc, \qquad\qquad bc < ac$ $\tag{43c}$

$$\frac{a}{c} > \frac{b}{c}, \qquad\qquad \frac{c}{a} < \frac{c}{b} \tag{43d}$$

If $a - c > b$ then $a > b + c$ (44)

If $a > b$ and $c > d$

then $a + c > b + d$, and $ac > bd$ (45)

(xii) Functions

We may describe $3x + 4$ as " a function of x " because its value depends upon the value of x. This is usually written as
$$F(x) = 3x + 4.$$
Other typical functions of x are
$$2x^2 + 3x + 5 ; \quad x(x^2 + 3x) ;$$
$$\cos x ; \quad \log x ; \quad 1/x.$$

In such functions, x is called the " independent variable." It is usual to write $F(a)$ as meaning " $F(x)$ where $x = a$."

(xiii) Equations

An equation is a statement of conditional equality between two expressions containing one or more symbols representing unknown quantities. The process of determining values of the unknowns which will satisfy the equation is called solving the equation.

An **Identical Equation** is one which holds for all values of its letter symbols.

A **Linear Equation** is one in which, after getting rid of fractions, the independent variable only occurs in the first degree (e.g. x).

Example : $y = 5x + 3$.

A **Quadratic Equation** is one in which, after getting rid of fractions, the independent variable occurs in the second degree (e.g. x^2) but not in higher degree.

Example : $y = 4x^2 + 5x + 3$.

A quadratic equation in one unknown has two roots, although both may be complex (i.e. with an imaginary term).

A **Cubic Equation** is one in which, after getting rid of fractions, the independent variable occurs in the third degree (e.g. x^3) but not in higher degree.

Example : $y = 3x^3 + 4x^2 + 2x + 5$.

Note : x and y are usually taken as unknowns ; a, b, c and d as known constants.

Rules for solution of equations

1. The same quantity may be added to (or subtracted from) both sides.
   Example : If $x + 3 = 2$; then $x + 5 = 4$ and $x + 1 = 0$.

2. A term can be moved from one side to the other provided that its sign is changed.
   Example : If $a = b$; then $a - b = 0$.

3. All signs in the equation may be changed together.
   Example : If $x - a = y - b$; then $a - x = b - y$.
   This is equivalent to multiplication throughout by -1.

4. Both sides can be multiplied (or divided) by the same quantity.
   Example : If $x + 2 = 5$; then $2x + 4 = 10$.

5. The reciprocal of one side is equal to the reciprocal of the other.
   Example : If $x = a$; then $\dfrac{1}{x} = \dfrac{1}{a}$.

6. Terms can be replaced by terms that are equal in value.

7. Both sides can be raised to the same power.
   Example : If $x = a$; then $x^2 = a^2$.

8. Both sides can be replaced by the same root of the original.
   Example : If $x = a$; then $\sqrt[3]{x} = \sqrt[3]{a}$, and

in general, $\sqrt[n]{x} = \sqrt[n]{a}$ if n is odd, but if n is even we must write
$$\sqrt[n]{x} = \pm \sqrt[n]{a}$$

which means that
$$\text{either } \sqrt[n]{x} = \sqrt[n]{a} \text{ or } \sqrt[n]{x} = -\sqrt[n]{a}.$$
In such cases two roots are obtained, and **both** should be tested in the original equation.

Warning
If both sides of an equation are squared, or if both sides are multiplied by a term containing the unknown, a new root may be introduced.

Solution of equations
   (1) **Linear equations with one unknown**
   Example : $ax + b = 0$.
   Solution : $x = -b/a$. (46)
   (2) **Linear equations with two unknowns**
   Any linear relation between two variables, x and y, can be written in the general form
$$ax + by + c = 0 \tag{47}$$
or (provided that b is not zero) in the alternative form
$$y = mx + n \tag{48a}$$
   This type of equation is not limited to one or two solutions, but has a corresponding value of y for every possible value of x. It is very helpful to plot the value of y against the value of x on squared paper—see Sect. 5(i). With any equation of this type, the graph will be a straight line, and it is only necessary to determine
      (1) one point on the line
      (2) the slope of the line at any point.
The most convenient point is usually $x = 0$, and in eqn. (48a)
this will give $y = n$
or in eqn. (47) this will give $y = -c/b$
   The slope of the line is given by the difference of the y values of two points, divided by the difference of their x values.
   In eqn. (47) the slope is $-a/b$
while in eqn. (48a) the slope is m.
   A particular form of eqn. (47) is
$$\frac{x}{a} + \frac{y}{b} = 1 \tag{48b}$$
and in this case the line cuts the x axis at $x = a$ and cuts the y axis at $y = b$. The slope is equal to $-b/a$.
   An equation of the form
$$\frac{a}{x} + \frac{b}{y} = c \tag{49}$$
may be solved by regarding $1/x$ and $1/y$ as the unknowns, then following a similar procedure as for an equation in x and y, and solving for $1/x$ and $1/y$.

   (3) **Simultaneous linear equations (two unknowns)**
$$\left.\begin{array}{l} a_1x + b_1y = c_1 \\ a_2x + b_2y = c_2 \end{array}\right\}$$
$$x = \frac{c_1b_2 - c_2b_1}{a_1b_2 - a_2b_1} \tag{50a}$$
$$y = \frac{a_1c_2 - a_2c_1}{a_1b_2 - a_2b_1} \tag{50b}$$
provided that $(a_1b_2 - a_2b_1)$ is not zero.
   Alternatively the solution may be derived by determining x in terms of y from the first equation, and then substituting in the second.
   Checking solutions :
   After any solution has been found, particularly with more than one solution, it is highly desirable to check these in the original equation.

   (4) **Three simultaneous equations (three unknowns)**
   Given $\begin{cases} ax + by + cz + d = 0. \\ a_1x + b_1y + c_1z + d_1 = 0 \\ a_2x + b_2y + c_2z + d_2 = 0 \end{cases}$

Then

$$x = \frac{d(b_2c_1 - b_1c_2) + d_1(bc_2 - b_2c) + d_2(b_1c - bc_1)}{a(b_1c_2 - b_2c_1) + a_1(b_2c - bc_2) + a_2(bc_1 - b_1c)}$$
$$y = \frac{d(a_1c_2 - a_2c_1) + d_1(a_2c - ac_2) + d_2(ac_1 - a_1c)}{a(b_1c_2 - b_2c_1) + a_1(b_2c - bc_2) + a_2(bc_1 - b_1c)}$$
$$z = \frac{d(a_2b_1 - a_1b_2) + d_1(ab_2 - a_2b) + d_2(a_1b - ab_1)}{a(b_1c_2 - b_2c_1) + a_1(b_2c - bc_2) + a_2(bc_1 - b_1c)}$$

(51)

It will be noticed that the three denominators are identical.

(5) Quadratic equations

$(x - a)(x + b) = 0$; $x = a$ or $x = -b$ (52)

$ax^2 + bx + c = 0$; $x = \dfrac{-b \pm \sqrt{b^2 - 4ac}}{2a}$ (53)

Note that in eqn. (53)
when $b^2 = 4ac$, the two roots are equal
when $b^2 - 4ac$ is positive, the two roots are real
when $b^2 - 4ac$ is negative, the roots are imaginary.

(6) Quadratic equations with two variables
Example : $y = ax^2 + bx + c$.
This type of equation is not limited to one or two solutions, but has a corresponding value of y for every possible value of x. It is helpful to plot part of the curve on squared paper—see Sect. 5(i). The curve may cut the x axis at two points, or it may touch at one point, or it may not touch it at all. Let $y = 0$, then
$$ax^2 + bx + c = 0$$
and the points at which the curve cuts the x axis will be
$$x = \frac{-b \pm \sqrt{b^2 - 4ac}}{2a}$$

(54)

If $b^2 > 4ac$, the curve will cut at two points.
If $b^2 = 4ac$, the curve will touch at one point.
If $b^2 < 4ac$, the curve will not cut the x axis.

(xiv) Formulae or laws

A formula is a law, or rule, generally in connection with some scientific relationship, expressed as an equation by means of letter symbols (variables) and constants.
For example, Ohm's Law states that $E = RI$ where each of the letter symbols has a precise meaning. If we know any two of the variables, we can determine the third,
$$R = \frac{E}{I} \text{ and } I = \frac{E}{R}$$

Another example is
$$X_c = \frac{1}{2\pi fC}$$

which gives X_c for any desired values of f and C. Note that 2π is a constant.
All formulae or laws may be rearranged in accordance with the rules for the solution of equations, so as to give the value of any variable in terms of the others.

(xv) Continuity and limits

Some functions are " continuous," that is to say they are smooth and unbroken when plotted as curves. Other functions are said to be " discontinuous " if at some value of x the value of y is indeterminate or infinite, or there is a sharp angle in the plotted value of $y = F(x)$. Examples of points of discontinuity are :
(1) $y = 1/x$ is discontinuous at $x = 0$.
(2) $y = 10^{1/(x-1)}$ is discontinuous at $x = 1$.
Even when a function is discontinuous at one or more points, it may be described as continuous within certain limits.

It frequently happens that a function approaches very closely to a limiting value, although it never quite reaches it for any finite values of the independent variable. For example, the voltage gain of a resistance coupled amplifier is given by

$$A = \mu R_L/(R_L + r_p)$$

and it is required to find the limiting value of A when R_L is made very great.

The formula may be put in the form

$$A = \mu \cfrac{1}{1 + \cfrac{r_p}{R_L}}. \tag{55}$$

and as R_L is made very much greater than r_p, the value of r_p/R_L becomes very small although it never actually reaches zero. We may express this in the form

$$\underset{R_L \to \infty}{Lt} \left(\frac{r_p}{R_L}\right) = 0 \tag{56}$$

which may be stated "the limit of (r_p/R_L), as R_L approaches infinity, is zero." The limiting value of A, as R_L approaches infinity, is therefore

$$\underset{R_L \to \infty}{Lt} A = \mu \tag{57}$$

(xvi) Progressions, sequences and series

A **Sequence** is a succession of terms so related that each may be derived from one or more of the preceding terms in accordance with some fixed law.

A **Series** is the sum of terms of a sequence.

Arithmetical Progression is a sequence, each term of which (except the first) is derived from the preceding term by the addition of a constant number.

General form : a, $(a + d)$, $(a + 2d)$, $(a + 3d)$

Example :   2,   5,   8,   11,   etc.

(here $a = 2$ and $d = 3$).

The nth term $= a + (n - 1)d$ (58)

The sum of n terms is $S = \frac{1}{2}n[2a + (n - 1)d]$ (59)

When three numbers are in Arithmetical Progression, the middle number is called the "arithmetical mean." The **arithmetical mean** between a and b is $\frac{1}{2}(a + b)$.

Geometrical Progression is a sequence, each term of which (except the first) is derived from the preceding term by multiplying it by a constant ratio (r).

General form : a, ar, ar^2 ar^3,

Examples :   3,   6,   12,   24, $(r = 2)$

             4, -2, $+1$, $-\frac{1}{2}$ $(r = -\frac{1}{2})$

With the general form above,

the nth term $= ar^{n-1}$ (60)

and the sum of the first n terms is

$$S = a\left(\frac{r^n - 1}{r - 1}\right) \tag{61}$$

When three numbers are in Geometrical Progression, the middle number is called the Geometrical Mean. The **Geometrical Mean** of two numbers a and b is $\sqrt{ab}$.

If the ratio r is less than unity, the terms become progressively smaller, and the sum of a very large number of terms approaches

$$\underset{n \to \infty}{S} = \frac{a}{1 - r} \tag{62}$$

Harmonic Progression : The terms a, b, c, etc. form a harmonic sequence if their reciprocals

$$\frac{1}{a}, \ \frac{1}{b}, \ \frac{1}{c}, \text{ etc.}$$

form an arithmetical sequence.

Example : 1, $\frac{1}{2}$, 1/3, $\frac{1}{4}$, 1/5 is a Harmonic Progression

because   1, 2, 3 , 4, 5   is an Arithmetical Progression

The **Harmonic Mean** between a and b is $\dfrac{2ab}{a + b}$.

Note that the Arithmetical Mean between two numbers is greater than the Geometrical Mean, which in turn is greater than the Harmonic Mean.

With any form of progression or sequence, we are often concerned with the sum of a number of terms of which the general term is·given. It is possible to write this sum in a shortened form, for example

$$1 + 2 + 3 + 4 + \ldots + k + \ldots + n = \sum_{k=1}^{k=n} k$$

where the Greek letter capital sigma is used to indicate the sum of a number of terms ; k is merely the general term ; and the values of k are to be taken from $k = 1$ (beneath sigma) to $k = n$ (written above sigma).

(xvii) Logarithmic and exponential functions

If $a^x = N$

then x is the logarithm to the base a of the number N. This may be put in the form

$$x = \log_a N$$

where the base a may be any positive number except 1 or 0.

The two principal systems of logarithms are

(1) The Naperian (or natural) system, using the base $e = 2.718\ 28 \ldots$ (preferably written with the Greek ϵ), and

(2) The Briggs (or common) system, using the base 10.

Only one set of tables is required, for it is possible to convert a logarithm to one base (b) into a logarithm to any other base (a) :

$$\log_a y = \log_b y \times \log_a b$$

If it is required to find the logarithm to the base ϵ, given the logarithm to the base 10,

$$\log_\epsilon y = \log_{10} y \times \log_\epsilon 10 \tag{63a}$$
$$= \log_{10} y \times 2.3026 \tag{63b}$$
$$\log_{10} y = \log_\epsilon y \times 0.4343 \tag{63c}$$

Some properties of ϵ

The value of ϵ is given by the right hand side of eqn. (86).
Values shown below in brackets are to four decimal places.

$\epsilon = 2.71828 \ (= 2.7183)$

$1/\epsilon = 0.367879 \ (= 0.3679)$

$\log_{10} \epsilon = 0.43429 \ (= 0.4343)$

$\log_\epsilon 10 = 2.30259 \ (= 2.3026)$

$\log_\epsilon 10 = 1/\log_{10} \epsilon$

$\log_{10} \epsilon^n = n \times 0.43429 \ (= n \times 0.4343)$

Some manipulations with logarithmic functions

$$\log a/b = \log a - \log b \tag{64}$$
$$\log 1/a = -\log a \tag{65}$$
$$\log y^n = n \times \log y \tag{66}$$
$$\log y^{-1} = -\log y \tag{67}$$
$$\log y^{m'n} = (m/n) \times \log y \tag{68}$$
$$\log \sqrt{y} = \log y^{1/2} = (1/2) \log y \tag{69}$$
$$\log \sqrt[3]{y} = \log y^{1/3} = (1/3) \log y \tag{70)}$$

To find the cube root of 125—

$$\sqrt[3]{125} = (125)^{1/3}$$

Therefore $\log (125)^{1/3} = (1/3) \log 125 = (1/3) (2.0969) = 0.699$.

Then antilog $0.699 = 5.00$ (from tables).

$$\log \sqrt[4]{y^3} = \tfrac{3}{4}\log y \tag{71}$$
$$\log abc = \log a + \log b + \log c \tag{72}$$
$$\log (ab/cd) = \log a + \log b - \log c - \log d \tag{73}$$
$$\log (a^m b^n c^p) = m.\log a + n.\log b + p.\log c \tag{74}$$
$$\log (ab^m/c^n) = \log a + m.\log b - n.\log c \tag{75}$$
$$\log (a^2 - b^2) = \log [(a + b)(a - b)] = \log (a + b) + \log (a - b) \tag{76}$$

$$\log \sqrt{a^2 - b^2} = \tfrac{1}{2} \log (a + b) + \tfrac{1}{2} \log (a - b) \tag{77}$$

Logarithmic Functions are closely related to **Exponential Functions,** and any equation in one form may be put into the other form. If the curves are plotted, the two will be the same.

Example : Exponential form $y = r^x$
            Logarithmic form $x = \log_r y$

Numerical example :
If $r = 10$ and $x = 3$
Then $y = 10^3 = 1000$.
This may be handled by the logarithmic form of the equation,
$$x = \log_r y = \log_{10} 1000 = 3 \text{ as before.}$$

Logarithmic decrement : If the equation is of the form
$$y = a \epsilon^{-bx}$$
where $(- b)$ is negative, the value of y decreases as x is increased, and $(- b)$ is called the Logarithmic Decrement.

(xviii) Infinite series

It was noted, when dealing with Geometrical Progression, that it is possible to take the limit of the sum of a very large number of terms, as the number approaches infinity, provided that the terms become progressively smaller by a constant ratio. Such a series is called " convergent" and is defined as having a finite limit to the sum to infinity. Infinite series which do not comply with this definition may be " divergent " (these are not considered any further) or else they may be " oscillating."

(a) Binomial series :
$$1 + mx + \frac{m(m - 1)}{1.2} x^2 + \frac{m(m - 1)(m - 2)}{1.2.3} x^3 + \text{etc.} \tag{78}$$

The nth term is
$$a_n = \frac{m(m - 1)(m - 2) \ldots (m - n + 2)}{1.2.3 \ldots (n - 1)} x^{n - 1} \tag{79}$$

The denominator is usually written in the form
$$(n - 1) !$$
which is called " factorial $(n - 1)$."

This Binomial Series is convergent, provided that x is numerically less than 1.

(b) Binomial theorem :
Case 1 :
$$(1 \pm x)^m = 1 \pm mx + \frac{m(m - 1) x^2}{2 !} \pm \frac{m(m - 1)(m - 2)x^3}{3 !} + \ldots$$
$$\pm \frac{m(m - 1) \ldots (m - n + 2)x^{n - 1}}{(n - 1) !} + \ldots \tag{80}$$

which holds for all values of x if m is a positive integer, and for all values of m provided that x is numerically less than 1.

The Binomial Theorem is useful in certain approximate calculations. If x is small compared with 1, and m is reasonably small,

$$\left. \begin{array}{l} (1 + x)^m \approx 1 + mx \\ (1 - x)^m \approx 1 - mx \\ (1 + x)^{-m} \approx 1 - mx \\ (1 - x)^{-n} \approx 1 + mx. \end{array} \right\} \tag{81}$$

To a closer approximation (taking three terms),
$$\left. \begin{array}{l} (1 + x)^m \approx 1 + mx + \tfrac{1}{2} m(m - 1) x^2 \\ (1 - x)^m \approx 1 - mx + \tfrac{1}{2} m(m - 1) x^2 \\ (1 + x)^{-m} \approx 1 - mx + \tfrac{1}{2} m(m + 1) x^2 \\ (1 - x)^{-m} \approx 1 + mx + \tfrac{1}{2} m(m + 1) x^2 \end{array} \right\} \tag{82}$$

Numerical example : To find the cube root of 220.

$$\sqrt[3]{220} = (216 + 4)^{1/3} = \left\{216\left(1 + \frac{4}{216}\right)\right\}^{1/3} = 6\left(1 + \frac{1}{54}\right)^{1/3}$$

Applying the approximation from the Binomial Theorem,
$$(1 + x)^m \approx 1 + mx$$
to the evaluation of the quantity above, we may make $x = 1/54$ and $m = 1/3$, from which

$$\left(1 + \frac{1}{54}\right)^{1/3} \approx 1 + \frac{1}{3} \cdot \frac{1}{54} \approx 1 + \frac{1}{162} \approx 1.006\ 17$$

Therefore $\sqrt[3]{220} \approx 6 \times 1.006\ 17 \approx 6.037$.

Case 2 :

We can also expand $(a + x)^m$, which is convergent when x is numerically less than a :

$$(a \pm x)^m = a^m \pm ma^{m-1}x + \frac{m(m-1)}{2!}a^{m-2}x^2 \pm \frac{m(m-1)(m-2)}{3!}a^{m-3}x^3 + \ldots \tag{83}$$

Approximation :

$$\frac{1}{a+1} = (a+1)^{-1} \approx \frac{1}{a}\left(1 - \frac{1}{a}\right) \text{ when } a > 1 \tag{84}$$

(c) Exponential series

From the Binomial Theorem, putting $x = 1/n$ and $m = nx$, we may derive

$$\left(1 + \frac{1}{n}\right)^{nx} = 1 + x + \frac{x^2}{2!} + \frac{x^3}{3!} + \frac{x^4}{4!} + \ldots + \frac{x^{n-1}}{(n-1)!} + \ldots \tag{85}$$

When $x = 1$ this becomes

$$\left(1 + \frac{1}{n}\right)^n = 1 + 1 + \frac{1}{2!} + \frac{1}{3!} + \frac{1}{4!} + \ldots + \frac{1}{(n-1)!} + \ldots \tag{86}$$

The right hand side of this equation is the value ϵ, which is equal to 2.71828 (to five places of decimals). Taking the xth power of each side of this equation,

$$\left(1 + \frac{1}{n}\right)^{nx} = \epsilon^x$$

Therefore $\epsilon^x = \left(1 + \frac{1}{n}\right)^{nx}$

and $\quad \epsilon^x = 1 + x + \frac{x^2}{2!} + \frac{x^3}{3!} + \frac{x^4}{4!} + \ldots + \frac{x^{n-1}}{(n-1)!} + \ldots \tag{87}$

which is called the Exponential Series.

(d) Logarithmic series

The logarithmic series is the expansion of $\log_\epsilon(1 + x)$ in ascending powers of x :

$$\log_\epsilon(1 + x) = x - \frac{x^2}{2} + \frac{x^3}{3} - \frac{x^4}{4} + \ldots \tag{88}$$

which is convergent if x is numerically less than 1.

(e) Trigonometrical series

$$\sin x = x - \frac{x^3}{3!} + \frac{x^5}{5!} - \frac{x^7}{7!} + \ldots \tag{89}$$

$$\cos x = 1 - \frac{x^2}{2!} + \frac{x^4}{4!} - \frac{x^6}{6!} + \ldots \tag{90}$$

$$\tan x = x + \frac{x^3}{3} + \frac{2x^5}{15} + \frac{17x^7}{315} + \frac{62x^9}{2835} + \ldots (|x| < \pi/2) \tag{91}$$

For derivation of eqns. (89) and (90) see Sect. 6, eqns. (17) and (18).

(xix) Hyperbolic functions

These are combinations of the sum and difference of two exponential functions.

$\dfrac{\epsilon^x - \epsilon^{-x}}{2}$ is called the hyperbolic sine of x, designated by sinh x

$\dfrac{\epsilon^x + \epsilon^{-x}}{2}$ is called the hyperbolic cosine of x, designated by cosh x,

$\dfrac{\epsilon^x - \epsilon^{-x}}{\epsilon^x + \epsilon^{-x}}$ is called the hyperbolic tangent of x, designated by tanh x

and similarly with the inverses

cosech x = 1/sinh x
sech x = 1/cosh x
coth x = 1/tanh x

Note $\epsilon = 1 + \dfrac{1}{1!} + \dfrac{1}{2!} + \dfrac{1}{3!} + \dfrac{1}{4!} + .. \approx 2.718\,28$ (92)

The following may be derived :

$\cosh^2 x - \sinh^2 x = 1$ (93)

$\operatorname{sech}^2 x + \tanh^2 x = 1$ (94)

$\coth^2 x - \operatorname{cosech}^2 x = 1$ (95)

$\sinh(-x) = -\sinh x\,;\ \cosh(-x) = \cosh x\,;\ \tanh(-x) = -\tanh x$ (96)

$\sinh x = \dfrac{\tanh x}{\sqrt{1 - \tanh^2 x}} = \sqrt{\cosh^2 x - 1}$ (97)

$\cosh x = \dfrac{1}{\sqrt{1 - \tanh^2 x}} = \sqrt{\sinh^2 x + 1}$ (98)

$\tanh x = \dfrac{\sinh x}{\cosh x} = \sqrt{1 - \operatorname{sech}^2 x}$ (99)

$\sinh x + \sinh y = 2 \sinh \dfrac{x + y}{2} \cosh \dfrac{x - y}{2}$ (100)

$\sinh x - \sinh y = 2 \sinh \dfrac{x - y}{2} \cosh \dfrac{x + y}{2}$ (101)

$\cosh x + \cosh y = 2 \cosh \dfrac{x + y}{2} \cosh \dfrac{x - y}{2}$ (102)

$\cosh x - \cosh y = 2 \sinh \dfrac{x + y}{2} \sinh \dfrac{x - y}{2}$ (103)

$\tanh x \pm \tanh y = \dfrac{\sinh(x \pm y)}{\cosh x \cosh y}$ (104)

$(\sinh x + \cosh x)^n = \cosh nx + \sinh nx$ (105)

$\sinh^{-1} x = \log_\epsilon(x + \sqrt{x^2 + 1}) = \cosh^{-1}\sqrt{x^2 + 1}$ (106)

$\cosh^{-1} x = \log_\epsilon(x + \sqrt{x^2 - 1}) = \sinh^{-1}\sqrt{x^2 - 1}$ (107)

$\tanh^{-1} x = \tfrac{1}{2} \log_\epsilon \dfrac{1 + x}{1 - x}$ (108)

$\qquad = \cosh^{-1} \dfrac{1}{\sqrt{1 - x^2}} = \sinh^{-1} \dfrac{x}{\sqrt{1 - x^2}}$ (109)

$\sinh(x \pm y) = \dfrac{\epsilon^{x \pm y} - \epsilon^{-(x \pm y)}}{2} = \sinh x \cosh y \pm \cosh x \sinh y$ (110a)

$\cosh(x \pm y) = \dfrac{\epsilon^{x \pm y} + \epsilon^{-(x \pm y)}}{2} = \cosh x \cosh y \pm \sinh x \sinh y$ (110b)

$\tanh(x \pm y) = \dfrac{\epsilon^{x \pm y} - \epsilon^{-(x \pm y)}}{\epsilon^{x \pm y} + \epsilon^{-(x \pm y)}} = \dfrac{\tanh x \pm \tanh y}{1 \pm \tanh x \tanh y}$ (111)

$\sinh 2x = 2 \sinh x \cosh x = \dfrac{2 \tanh x}{1 - \tanh^2 x}$ (112)

$\cosh 2x = \cosh^2 x + \sinh^2 x = \dfrac{1 + \tanh^2 x}{1 - \tanh^2 x}$ (113)

$\tanh 2x = \dfrac{2 \tanh x}{1 + \tanh^2 x}$ (114)

$\cosh x + \sinh x = \epsilon^x$ (115a)

$\cosh x - \sinh x = \epsilon^{-x}$ (115b)

$\sinh x = x + \dfrac{x^3}{3!} + \dfrac{x^5}{5!} + \cdots$ (115c)

$$\cosh x = 1 + \frac{x^2}{2!} + \frac{x^4}{4!} + \frac{x^6}{6!} + \ldots \tag{115d}$$

(these are convergent for all real values of x).

Also in complex form (see Section 6)

$$\sinh jx = j \sin x \; ; \; \cosh jx = \cos x \; ; \; \tanh jx = j \tan x \tag{116}$$
$$\sin jx = j \sinh x \; ; \; \cos jx = \cosh x \tag{117}$$
$$\sinh (x \pm jy) = \sinh x \cos y \pm j \cosh x \sin y \tag{118}$$

Note that x and y in $\sin x$, $\cos x$, $\tan x$ etc. in eqns. 116 to 119 must be expressed in radians.

$$\cosh (x \pm jy) = \cosh x \cos y \pm j \sinh x \sin y \tag{119}$$
$$\sinh \left(\frac{x}{2}\right) = \sqrt{\frac{\cosh x - 1}{2}} \; ; \quad \cosh \left(\frac{x}{2}\right) = \sqrt{\frac{\cosh x + 1}{2}} \tag{120}$$
$$\tanh \left(\frac{x}{2}\right) = \frac{\cosh x - 1}{\sinh x} = \frac{\sinh x}{\cosh x + 1} \tag{121}$$

(xx) General approximations

Let δ be an extremely small quantity and x be a quantity very large compared with δ, then

$$\frac{1}{1 - \delta} \approx 1 + \delta \qquad\qquad \frac{1}{1 + \delta} \approx 1 - \delta \tag{122}$$
$$\frac{1 + \delta_1}{1 + \delta_2} \approx 1 + \delta_1 - \delta_2 \tag{123}$$
$$(1 \pm \delta)^n \approx 1 \pm n\delta \tag{124}$$
$$\frac{1}{(1 \pm \delta)^n} \approx 1 \mp n\delta \qquad \left.\begin{array}{l}\text{where } n \text{ may be integral,} \\ \text{fractional or negative}\end{array}\right\} \tag{125}$$
$$\sqrt{1 + \delta} \approx 1 + \tfrac{1}{2}\delta \qquad\qquad \sqrt{1 - \delta} \approx 1 - \tfrac{1}{2}\delta \tag{126}$$
$$\frac{1}{\sqrt{1 + \delta}} \approx 1 - \tfrac{1}{2}\delta \qquad\qquad \frac{1}{\sqrt{1 - \delta}} \approx 1 + \tfrac{1}{2}\delta \tag{127}$$
$$(1 + \delta)^2 \approx 1 + 2\delta \qquad\qquad (1 - \delta)^2 \approx 1 - 2\delta \tag{128}$$
$$(x \pm \delta)^n \approx x^n \pm nx^{n-1}\delta \tag{129}$$

Eqn. (129) is used in Sect. 1(v)b for approximations with powers and roots.

$$\sqrt{x(x + \delta)} \approx x + \tfrac{1}{2}\delta \qquad\qquad \sqrt{x(x - \delta)} \approx x - \tfrac{1}{2}\delta \tag{130}$$
$$(1 + \delta_1)(1 \pm \delta_2) \approx 1 + \delta_1 \pm \delta_2 \tag{131}$$
$$(1 + \delta_1)(1 + \delta_2)(1 + \delta_3) \approx 1 + \delta_1 + \delta_2 + \delta_3 \tag{132}$$

where δ_1, δ_2 and δ_3 are all extremely small quantities.

$$\epsilon^\delta \approx 1 + \delta \qquad\qquad \epsilon^{-\delta} \approx 1 - \delta \tag{133}$$
$$\epsilon^{t/RC} \approx 1 + t/RC \qquad\qquad \epsilon^{-t/RC} \approx 1 - t/RC \tag{134}$$

Eqn. (134) has an error less than 0.6% when t/RC does not exceed 0.1.

$$\log_\epsilon (x \pm \delta) \approx \log_\epsilon x \pm \frac{\delta}{x} - \tfrac{1}{2}\left(\frac{\delta}{x}\right)^2 \tag{135}$$
$$\log_\epsilon (1 \pm \delta) \approx \delta - \tfrac{1}{2}\delta^2 \tag{136}$$
$$\sinh \delta \approx \delta \qquad\qquad \cosh \delta \approx 1 \qquad\qquad \tanh \delta \approx \delta \tag{137}$$
$$\sinh^{-1}\delta \approx \delta \qquad\qquad \tanh^{-1}\delta \approx \delta \tag{138}$$
$$\sinh (x + \delta) \approx \sinh x + \delta \cosh x \; ; \; \sinh (x - \delta) \approx \sinh x - \delta \cosh x \tag{139}$$
$$\cosh (x + \delta) \approx \cosh x + \delta \sinh x \; ; \; \cosh (x - \delta) \approx \cosh x - \delta \sinh x \tag{140}$$
$$\tanh (x + \delta) \approx \tanh x + \delta \operatorname{sech}^2 x \; ; \; \tanh (x - \delta) \approx \tanh x - \delta \operatorname{sech}^2 x \tag{141}$$

When L is a large quantity

$$\sinh L \approx \tfrac{1}{2}\epsilon^L \qquad\qquad \cosh L \approx \tfrac{1}{2}\epsilon^L \qquad\qquad \tanh L \approx 1 \tag{142}$$

Trigonometrical relationships

When δ is an extremely small quantity, so that an angle of δ radians is a very small angle, and x is an angle very large compared with δ,

$$\sin \delta \approx \delta \qquad \cos \delta \approx 1 \qquad \tan \delta \approx \delta \qquad (143)$$
$$\sin^{-1} \delta \approx \delta \qquad \cos^{-1} \delta \approx \tfrac{1}{2}\pi(4K - 1) + \delta \qquad \tan^{-1} \delta \approx \delta \qquad (144)$$

where K is any integer. See Sect. 3(iii) for inverse functions.

$$\sin (x + \delta) \approx \sin x + \delta \cos x \qquad \sin (x - \delta) \approx \sin x - \delta \cos x \qquad (145)$$
$$\cos (x + \delta) \approx \cos x - \delta \sin x \qquad \cos (x - \delta) \approx \cos x + \delta \sin x \qquad (146)$$
$$\tan (x + \delta) \approx \tan x + \delta/\cos^2 x \qquad \tan (x - \delta) \approx \tan x - \delta/\cos^2 x \qquad (147)$$

SECTION 3 : GEOMETRY AND TRIGONOMETRY

(i) *Plane figures* (ii) *Surfaces and volumes of solids* (iii) *Trigonometrical relationships.*

(i) Plane figures
Angles
Two angles are complementary when their sum is equal to a right angle (90°).
Two angles are supplementary when their sum is equal to two right angles (180°).
The three angles of a triangle are together equal to two right angles (180°)

$$2\pi \text{ radians} = 360°$$
$$\pi \text{ radians} = 180°$$
$$1 \text{ radian} \approx 57.29578°$$
$$1° \approx 0.0174533 \text{ radian.}$$

When an angle is measured in radians, and incorporates the sign π, it is usual to omit the word " radians " as being understood—e.g. π, 2π.

Right Angle Triangles (Fig. 6.1)

Sine :* $\dfrac{a}{c} = \sin A$ $\qquad\qquad$ $a = c \sin A$ $\qquad\qquad$ (1)

Tangent : $\dfrac{a}{b} = \tan A$ $\qquad\qquad$ $a = b \tan A$ $\qquad\qquad$ (2)

Cosine : $\dfrac{b}{c} = \cos A$ $\qquad\qquad$ $b = c \cos A$ $\qquad\qquad$ (3)

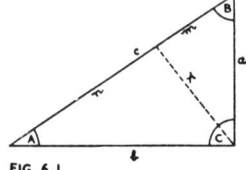

Cosecant : $\dfrac{c}{a} = \operatorname{cosec} A = 1/\sin A$ $\qquad$ (4)

Secant : $\dfrac{c}{b} = \sec A = 1/\cos A$ $\qquad$ (5)

Cotangent : $\dfrac{b}{a} = \cot A = 1/\tan A$ $\qquad$ (6)

c is called the hypotenuse.

FIG. 6.1

$$a^2 + b^2 = c^2 \qquad\qquad (7)$$
$$a = \sqrt{(c + b)(c - b)} = \sqrt{mc} \qquad\qquad (8)$$
$$a = c \sin A = b \tan A \qquad\qquad (9)$$
$$b = \sqrt{(c + a)(c - a)} = \sqrt{nc} \qquad\qquad (10)$$
$$b = c \cos A = a \cot A = a/\tan A \qquad\qquad (11)$$
$$c = \sqrt{a^2 + b^2} = m + n \qquad\qquad (12)$$
$$c = a \operatorname{cosec} A = a/\sin A = b \sec A = b/\cos A \qquad\qquad (13)$$
$$\text{Area} = \tfrac{1}{2} ab = \tfrac{1}{2} a^2 \cot A = \tfrac{1}{2} b^2 \tan A \qquad\qquad (14)$$
$$= \tfrac{1}{4} c^2 \sin 2A = \tfrac{1}{2} bc \sin A = \tfrac{1}{2} ac \sin B \qquad\qquad (15)$$

The perpendicular (p) to the hypotenuse is the mean proportional (or mean geometrical progression) between the segments of the hypotenuse,

*See Sect. 3(iii) for trigonometrical relationships.

$$\frac{m}{p} = \frac{p}{n}, \quad p = \sqrt{mn} \tag{16}$$

also $\quad \dfrac{m}{a} = \dfrac{a}{c}, \ \dfrac{n}{b} = \dfrac{b}{c}.$ (17)

Any triangle inscribed in a semicircle, with the diameter forming one side, is a right angle triangle.

Equilateral Triangle (Fig. 6.2a)

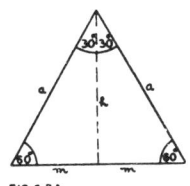

FIG. 6·2A

Each side $= a$

Each angle $= 60°$

$m = \frac{1}{2} a$

$h = \sqrt{3}\, m = (\sqrt{3}/2)a \approx 0.866a$ (18)

area $= \frac{1}{2} ah = (\sqrt{3}/4)a^2 \approx 0.433a^2$ (19)

Any triangle (Fig. 6.2b)

Area $= \frac{1}{2} bh = \sqrt{s(s-a)(s-b)(s-c)}$ (20)

where $b =$ base, $h =$ height, $s = \frac{1}{2}(a + b + c)$.

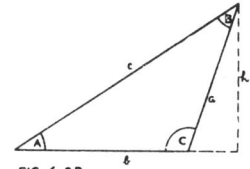

FIG. 6.2B

$$\frac{a}{\sin A} = \frac{b}{\sin B} = \frac{c}{\sin C} \tag{21}$$

$a^2 = b^2 + c^2 - 2bc \cos A$ (22)
$b^2 = c^2 + a^2 - 2ca \cos B$
$c^2 = a^2 + b^2 - 2ab \cos C$ (23)
$a = b \cos C + c \cos B$ (24)
$b = c \cos A + a \cos C$
$c = a \cos B + b \cos A.$ (25)

Rectangle (Fig. 6.3)

Area $= ab$ (26)

$d =$ diagonal $= \sqrt{a^2 + b^2}$ (27)

Parallelogram (Fig. 6.4)

Area $= bh = ab \sin C; \ a = c; \ b = d.$ (28)

Angle $A =$ angle $C;$ angle $B =$ angle $D.$ (29)

Trapezoid (Fig. 6.5).

Side d is parallel to side b.

Area $= \frac{1}{2}(b + d)h.$ (30)

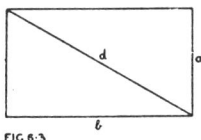

FIG. 6·3

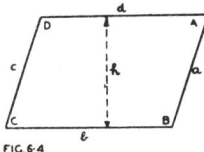

FIG. 6·4

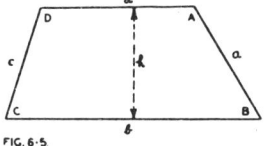

FIG. 6·5

Polygons and Quadrilaterals

To find area, divide into triangles, calculate the area of each, and add.

Circle

Circumference $= \pi \times$ diameter $\approx 3.1416 \times$ diameter (31)
$= 2\pi \times$ radius $\approx 6.2832 \times$ radius (32)

Area $= \pi \times (\text{radius})^2 = \frac{1}{2}(\text{circumference} \times \text{radius})$ (33)
$= (\pi/4) \times (\text{diameter})^2 \approx 0.7854 (\text{diameter})^2$ (34)

Sector (Fig. 6.6)

$A =$ angle subtended at centre $= s/r$ radians (35)

$c = 2\sqrt{r^2 - a^2} = 2\,r \sin(A/2) = 2a \tan(A/2) = 2\sqrt{2hr - h^2}$ (36)

$a = \frac{1}{2}\sqrt{4r^2 - c^2} = \frac{1}{2}\sqrt{d^2 - c^2} = r \cos(A/2) = \frac{1}{2} c \cot(A/2)$ (37)

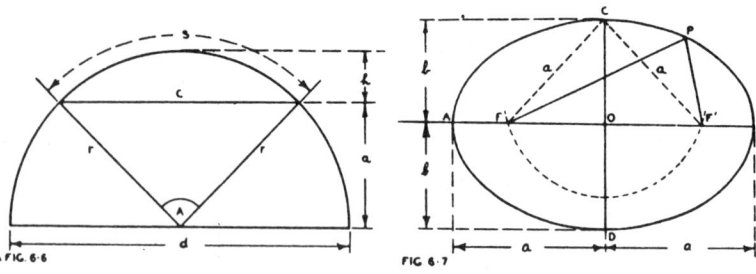

FIG. 6·6   FIG 6·7

$$s = \text{length of arc} = rA \qquad (A \text{ in radians}) \tag{38}$$
$$= \pi rA/180 \qquad (A \text{ in degrees}) \tag{39}$$
$$h = r - a = r(1 - \cos A/2) \tag{40}$$
$$\text{Area of sector} = \tfrac{1}{2} rs = \tfrac{1}{2} r^2 A \qquad (A \text{ in radians}) \tag{41}$$
$$= \pi r^2 A/360 \qquad (A \text{ in degrees}) \tag{42}$$
Area of segment (bounded by chord c, curve s)
$$= \tfrac{1}{2} r^2 (A - \sin A) \qquad (A \text{ in radians}) \tag{43}$$
$$= \tfrac{1}{2} r(s - r \sin s/r) \qquad (s/r \text{ in radians}) \tag{44}$$
$$= \tfrac{1}{2} [r(s - c) + ch] \text{ for segments less than half a circle.} \tag{45}$$

Ellipse (Fig. 6.7)

The ellipse has two foci, F and F', and for any point P on the perimeter,
FP + PF' is constant = FB + BF' = FA + AF' (46)
Major axis = $AB = 2a$; minor axis = $CD = 2b$ (47)
Area of ellipse = $\pi ab \approx 0.7854$ major axis × minor axis (48)
Perimeter $\approx a(4 + 1.1m + 1.2m^2)$, where $m = b/a$ (49)

An ellipse may be drawn by putting a pin into the paper at each focus (F, F'), tying the ends of a short length of cotton thread around the pins leaving a slack portion in the middle, and running a pencil point around in the loop of the thread.

To find the foci, draw an arc with centre at C and radius a to intersect the X axis (AB) at F and F'.

Parabola (Fig. 6.8)

The parabola has a focus (F) and a directrix (MN) and for any point P on the parabola,
FP = PM where PM is the perpendicular to the directrix (50)
Area of segment cut off by chord $PP' = (2/3) ch$. (51)

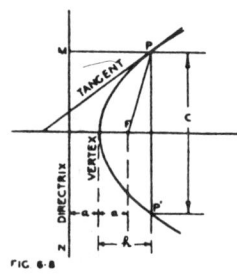

FIG 6·8

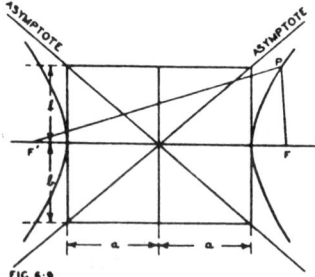

FIG 6·9

Hyperbola (Fig. 6.9)

The hyperbola has two foci (F, F') and two asymptotes, and for any point P on either curve,
$$F'P - FP = \text{constant} \tag{52}$$

General rules for areas

Areas bounded by straight sides may be calculated by dividing the area into triangles, calculating the area of each and adding.

Areas bounded by irregular curves may be divided into parallel strips and the area calculated by one of the following approximations :

Trapezoid rule
Area $= d(\tfrac{1}{2}y_1 + y_2 + y_3 + \cdots\cdots + y_{n-1} + \tfrac{1}{2}y_n)$ $\qquad$ (53)
where $d =$ width of each strip
and $y_1, y_2, y_3 \ldots y_n$ are measured lengths of each of the equidistant parallel chords.
Note that the first (y_1) and the last (y_n) do not cut the area, and may be zero if the
surface is sharply curved.

Simpson's Rule
Area $= \dfrac{d}{3}(y_1 + 4y_2 + 2y_3 + 4y_4 + 2y_5 \ldots + 2y_{n-2} + 4y_{n-1} + y_n)$ $\qquad$ (54)
where n must be odd
$\qquad d =$ width of each strip
and $\quad y_1 \ldots y_n$ are measured lengths of equidistant parallel chords.

(ii) Surfaces and volumes of solids
Cube (length of side $= a$)
Volume $= a^3$ $\qquad$ (55)
Surface area $= 6a^2$ $\qquad$ (56)
Length of diagonal $= a\sqrt{3}$ $\qquad$ (57)

Rectangular prism
(length $= l$, breadth $= b$, height $= h$)
Volume $= lbh$ $\qquad$ (58)
Surface area $= 2(lb + lh + bh)$ $\qquad$ (59)
Diagonal $= \sqrt{b^2 + l^2 + h^2}$ $\qquad$ (60)

Cylinder, solid right circular
(length l, radius r)
Volume $= \pi r^2 l \approx 0.7854\,d^2 l$ $\qquad$ (61)
Area of curved portion $= 2\pi rl = \pi dl$ $\qquad$ (62)
Area of each end $= \pi r^2$ $\qquad$ (63)
Total surface area $= 2\pi r(l + r)$ $\qquad$ (64)

Hollow cylinder, right circular
(length l, outer radius R, inside radius r)
Volume $= \pi l(R^2 - r^2)$ $\qquad$ (65)

Any pyramid or cone
Volume $= 1/3$ (area of base $\times$ distance from vertex to plane of base) $\qquad$ (66)

Sphere
Volume $= \dfrac{4}{3}\pi r^3 = \pi d^3/6 \approx 4.1888\,r^3 \approx 0.5236\,d^3$ $\qquad$ (67)
Surface area $= 4\pi r^2 = \pi d^2$ $\qquad$ (68)

(iii) Trigonometrical relationships
We have already introduced the sine, cosine and tangent of an angle, and their
inverses, under the subject Angles (Fig. 6.1). The following table may readily be
derived with the assistance of Fig. 6.10 :

Angle	Sine	Cosine	Tangent
0	0	1	0
30°	1/2	$\sqrt{3}/2$	$1/\sqrt{3}$
45°	$1/\sqrt{2}$	$1/\sqrt{2}$	1
60°	$\sqrt{3}/2$	1/2	$\sqrt{3}$
90°	1	0	∞ *

*Approaches infinity as angle approaches 90°.
The values for any angle between 0° and 90° may be found from Table 72, Trigo-
nometrical Relationships, in Chapter 38 Sect. 21, or from any book of Mathematical
Tables.

Angles of any magnitude

If the line OX (Fig. 6.11) revolves about O to a new position OP, the amount of rotation is the angle XOP between its original position OX and its new position OP. Such a counter-clockwise rotation is called positive, while the opposite direction of rotation is called negative.

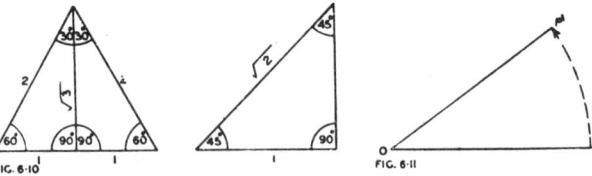

Examples of angles in all four sectors are shown in Fig. 6.12. It will be seen that, for any angle A, the position of OP is the same for a movement of angle A in a positive direction, or for a negative movement of $(360° - A)$; for example,

$$+ 330° = - (360° - 330°) = - 30°$$

In the case of angles greater than 360° we are generally only concerned with the final position of OP, so that for these cases we may subtract 360°, or any multiple of 360°, from the angle so as to give a value less than 360°. For example $390° = 360° + 30°$; $800° = 720° + 80°$; $1125° = 1080° + 45°$.

In trigonometry it is also necessary to define the polarity of the three sides of the triangles from which we derive the sine, cosine and tangent. The hypotenuse (OP_1) is always positive (see Fig. 6.13). The base (OX_1) is positive when X is to the right hand side of O, and negative when X is to the left of O (e.g. OX_2). The perpendicular is positive when P is above X (e.g. $P_1 X_1$) and negative when P is below X (e.g. $= P_4$).

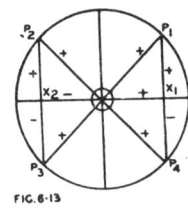

1st Quadrant : All sides positive $(OX_1 P_1)$.
                 Sine, cosine, tangent all positive.
2nd Quadrant : OX_2 negative, other sides positive.
                 Sine $= X_2 P_2 / OP_2$ which is positive.
                 Cosine $= OX_2 / OP_2$ which is negative.
                 Tangent $= X_2 P_2 / OX_2$ which is negative.
3rd Quadrant : OX_2, $X_2 P_3$ negative ; OP_3 positive.
                 Sine $= X_2 P_3 / OP_3$ which is negative.
                 Cosine $= OX_2 / OP_3$ which is negative.
                 Tangent $= X_2 P_3 / OX_2$ which is positive.
4th Quadrant : OX_1, OP_4 positive, $X_1 P_4$ negative.
                 Sine $= X_1 P_4 / OP_4$ which is negative.
                 Cosine $= OX_1 / OP_4$ which is positive.
                 Tangent $= X_1 P_4 / OX_1$ which is negative.

A convenient method for determining graphically the value of the sine, cosine and tangent of an angle is shown in Fig. 6.14, where the circle is drawn with radius 1 (to any convenient scale, say 1 inch). To the same scale, PX gives the value of the sine, OX the cosine and AT the tangent.

" Inverse " functions

If sin $\theta = n$, we may describe θ as the angle whose sine is n. This is conventionally written in the form

$$\theta = \sin^{-1} n,$$

where the " $\sin^{-1}$ " is to be regarded purely as an abbreviation for " the angle whose sine is."

The same system is used with all trigonometrical functions—$\cos^{-1}$, $\tan^{-1}$, $\operatorname{cosec}^{-1}$, $\sec^{-1}$, $\cot^{-1}$

These are occasionally written as arc sin, arc cos, arc tan, etc.

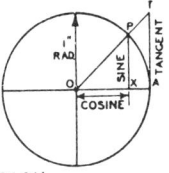

FIG. 6·14

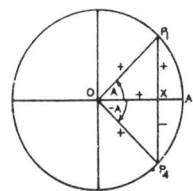

FIG. 6·15

Summary of trigonometrical relationships

1. $\operatorname{cosec} A = 1/\sin A = \cot A/\cos A$ (69)
   $\sec A = 1/\cos A = \tan A/\sin A$ (70)
   $\cot A = 1/\tan A = \cos A/\sin A$ (71)

2. $\tan A = (\sin A)/(\cos A) = \sin A \sec A$ (72)

3. $\sin A = \cos (90° - A) = \sin (180° - A)$ (73)
   $\cos A = \sin (90° - A) = - \cos (180° - A)$ (74)
   $\tan A = \cot (90° - A) = - \tan (180° - A)$ (75)
   $\operatorname{cosec} A = \sec (90° - A) = \operatorname{cosec} (180° - A)$ (76)
   $\sec A = \operatorname{cosec} (90° - A) = - \sec (180° - A)$ (77)
   $\cot A = \tan (90° - A) = - \cot (180° - A)$ (78)

4. $\sin^2 A + \cos^2 A = 1$; $\sec^2 A - \tan^2 A = 1$ (79)
   (Note : $\sin^2 A$ is the square of $\sin A$)

   $$\sin A = \pm \sqrt{1 - \cos^2 A} = \pm \frac{\tan A}{\sqrt{1 + \tan^2 A}} = \pm \frac{1}{\sqrt{1 + \cot^2 A}}$$ (80)

   $$\cos A = \pm \sqrt{1 - \sin^2 A} = \pm \frac{1}{\sqrt{1 + \tan^2 A}} = \pm \frac{\cot A}{\sqrt{1 + \cot^2 A}}$$ (81)

   the choice of signs being determined by the quadrant.

5. $\sec^2 A = 1/\cos^2 A = 1 + \tan^2 A$ (82)
   $\operatorname{cosec}^2 A = 1/\sin^2 A = 1 + \cot^2 A$ (83)

6. $\sin (A + B) = \sin A \cos B + \cos A \sin B$ (84)
   $\cos (A + B) = \cos A \cos B - \sin A \sin B$ (85)

   $$\tan (A + B) = \frac{\tan A + \tan B}{1 - \tan A \tan B}$$ (86)

7. $\sin (A - B) = \sin A \cos B - \cos A \sin B$ (87)
   $\cos (A - B) = \cos A \cos B + \sin A \sin B$ (88)

   $$\tan (A - B) = \frac{\tan A - \tan B}{1 + \tan A \tan B}$$ (89)

8. Negative angles (Fig. 6.15)
   $\sin (- A) = XP_4/OP_4 = - (XP_1/OP_1) = - \sin A$ (90)
   $\cos (- A) = OX/OP_4 = OX/OP_1 = \cos A$ (91)
   $\tan (- A) = XP_4/OX = - (XP_1/OX) = - \tan A$ (92)
   $\operatorname{cosec} (- A) = OP_4/XP_4 = - (OP_1/XP_1) = - \operatorname{cosec} A$ (93)
   $\sec (- A) = OP_4/OX = OP_1/OX = \sec A$ (94)
   $\cot (- A) = OX/XP_4 = - (OX/XP_1) = - \cot A$ (95)

9. $\sin 2A = 2 \sin A \cos A = \dfrac{2 \tan A}{1 + \tan^2 A}$ (96)

   $\cos 2A = \cos^2 A - \sin^2 A = 1 - 2 \sin^2 A = 2 \cos^2 A - 1$ (97)

   $\tan 2A = \dfrac{2 \tan A}{1 - \tan^2 A}$; $\cot 2A = \dfrac{\cot^2 A - 1}{2 \cot A}$ (98)

10. $\sin \tfrac{1}{2}A = \pm \sqrt{\tfrac{1}{2}(1 - \cos A)} = \pm (\tfrac{1}{2}\sqrt{1 + \sin A} - \tfrac{1}{2}\sqrt{1 - \sin A})$ (99)

   $\cos \tfrac{1}{2}A = \pm \sqrt{\tfrac{1}{2}(1 + \cos A)} = \pm (\tfrac{1}{2}\sqrt{1 + \sin A} + \tfrac{1}{2}\sqrt{1 - \sin A})$ (100)

   $\tan \tfrac{1}{2}A = \dfrac{1 - \cos A}{\sin A} = \dfrac{\sin A}{1 + \cos A}$ (101)

11. $\sin 3A = 3 \sin A - 4 \sin^3 A$
    $\cos 3A = 4 \cos^3 A - 3 \cos A$ $\tan 3A = \dfrac{3 \tan A - \tan^3 A}{1 - 3 \tan^2 A}$ (102)

12. $\sin^2 A = \frac{1}{2}(1 - \cos 2A)$ (103)
    $\cos^2 A = \frac{1}{2}(1 + \cos 2A)$ (104)

13. $\sin^3 A = \frac{1}{4}(3 \sin A - \sin 3A)$ (105)
    $\cos^3 A = \frac{1}{4}(\cos 3A + 3 \cos A)$ (106)

14. Approximations for small angles :
    where A is measured in radians
    $\sin A \approx A - (A^3)/6$ error for 30° is 0.06% (107)
    $\cos A \approx 1 - (A^2)/2$ error for 30° is 0.35% (108)
    $\tan A \approx A + (A^3)/3$ error for 30° is 1.03% (109)
    and for very small angles $\sin A \approx A$ (110)
    $\cos A \approx 1$ (111)
    $\tan A \approx A.$ (112)

    See also Sect. 2(xx).

15. $\sin A = \dfrac{\epsilon^{jA} - \epsilon^{-jA}}{2j}$ [Sect. 6, eqn. (19)] (113)

    $\cos A = \dfrac{\epsilon^{jA} + \epsilon^{-jA}}{2}$ [Sect. 6, eqn. (20)] (114)

16. versine $A = 1 - \cos A = 2 \sin^2 (A/2)$ (115)

SECTION 4 : PERIODIC PHENOMENA

The rotation of a wheel and a train of waves are two examples of periodic phenomena, that is to say the same action takes place repeatedly, each of such phenomena being called a " cycle." The number of cycles which occur in 1 second is called the frequency, and is expressed in one of the forms

cycles per second   c/s
kilocycles per second   Kc/s
megacycles per second   Mc/s.

If we take a point (P in Fig. 6.16) which is rotating with uniform angular velocity* about the point O, we can plot the height PX against the angle of rotation (θ). When P is at A (which we may regard as the zero point, since here $\theta = 0$), PX = O, and we mark point B. When P is at angle θ, the perpendicular PX gives point C, and CK = PX. When $\theta = 90°$, P will be at the top of the circle, giving D on the curve. When $\theta = 180°$, P will be at the extreme left hand side of the circle, and the height above OA will be zero, thus giving point E. When $\theta = 270°$, P will be at the bottom of the circle, giving point F on the curve. Lastly, P will return to point A (the zero point) and the height will be zero, point G. If the process is continued, the curve (GH etc.) will repeat the shape of the first cycle (BD etc.) and so on indefinitely. Thus BCDEFG represents one cycle.

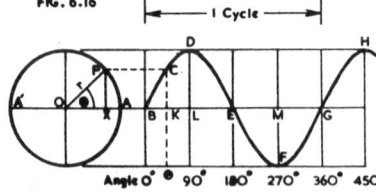

FIG. 6.16

Angle 0° 90° 180° 270° 360° 450°

The length PX = $r \sin \theta$, so that its projected height CK is proportional to the sine of θ. The curve is therefore called a " sine curve " or " sine wave." A cosine curve has exactly the same shape, except that it begins at D (since $\cos 0° = 1$) and the cycle ends at H.

The motion of the point X, as it oscillates about O between the extremes A and A is called **Simple Harmonic Motion.**

Angular velocity (usually represented by the small Greek letter omega—ω) is the number of radians per second through which the point P travels. In each revolution

*i.e. uniform rate of rotation.

(360°) it will pass through 2π radians, and if it makes f revolutions per second, then the angular velocity will be

$$\omega = 2\pi f \text{ radians per second}$$

where f = frequency in cycles per second.

In most mathematical work it is more convenient to write ω than to write $2\pi f$.

SECTION 5 : GRAPHICAL REPRESENTATION AND j NOTATION

(i) Graphs   (ii) Finding the equation to a curve   (iii) Three variables   (iv) Vectors and j notation.

(i) Graphs

Graphs are a convenient representation of the relationships between functions. For example, Ohm's Law

$$E = RI$$

may be represented by a graph (Fig. 6.17) in which I is plotted horizontally (on the X axis) and E vertically (on the Y axis) for a constant value of R.

Any point P on the plotted " curve " has its position fixed by coordinates. The horizontal, or x—coordinate (OQ) is called the abscissa, while the vertical, or y—coordinate (QP) is called the ordinate. The position of the point is written as a, b, thus indicating that $x = a$ and $y = b$, where $a = $ OQ and $b = $ QP.

Any function of the form

$$y = mx + n \quad (m \text{ and } n \text{ being constants})$$

is a linear equation, since the plotted curve is a straight line. It only passes through the origin (O) if $n = 0$

In the general case, the axes extend in both directions about the origin (Fig. 6.18) forming four quadrants and allowing for negative values of both x and y. These are known as " Cartesian Coordinates."

When the two variable quantities in an equation can be separated into

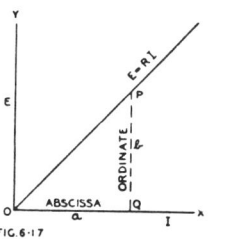

FIG. 6·17

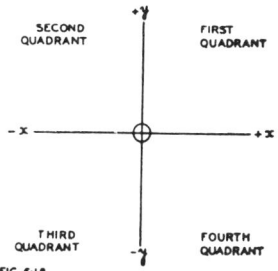

FIG. 6·18

" cause " and " effect," the " cause " (known as the independent variable) is plotted horizontally on the x axis, and the " effect " (known as the dependent variable) vertically on the y axis. In other cases the choice of axes is optional.

Any convenient scales may be used, and the x and y scales may differ.

The procedure to be adopted to plot a typical equation

$$y = 2x^2 + 4x - 5$$

is as follows. Select suitable values of x (which will be regarded as the independent variable) and calculate the value of y for each :

$x = 3$	$y = 18 + 12 - 5 = + 25$
$x = 2$	$y = 8 + 8 - 5 = + 11$
$x = 1$	$y = 2 + 4 - 5 = + 1$
$x = 0$	$y = 0 + 0 - 5 = - 5$
$x = -1$	$y = 2 - 4 - 5 = - 7$
$x = -2$	$y = 8 - 8 - 5 = - 5$
$x = -3$	$y = 18 - 12 - 5 = + 1$

Then plot these points, as in Fig. 6.19A and draw a smooth curve through them.

The tangent at any point (e.g. P in Fig. 6.19A) is a straight line which is drawn so as to touch the curve at the point. The slope of the tangent, which is the same as that of the curve at the point, is defined as the tangent of the angle θ which it makes with the X axis. Between points B and C on the curve, θ is positive, therefore $\tan \theta$ is positive and the slope is called positive. Between points B and A the angle θ is negative, and the slope negative. At point B, $\theta = 0$ and the slope is zero. It is important to remember that the curve normally extends in both directions indefinitely unless it has limits, or turns back on itself. It is therefore advisable, when plotting an unknown function, to take very large positive and negative values of x and calculate the corresponding values of y, even though the points cannot be plotted on the graph paper. This will indicate the general trend of the curve beyond the limits of the graph paper.

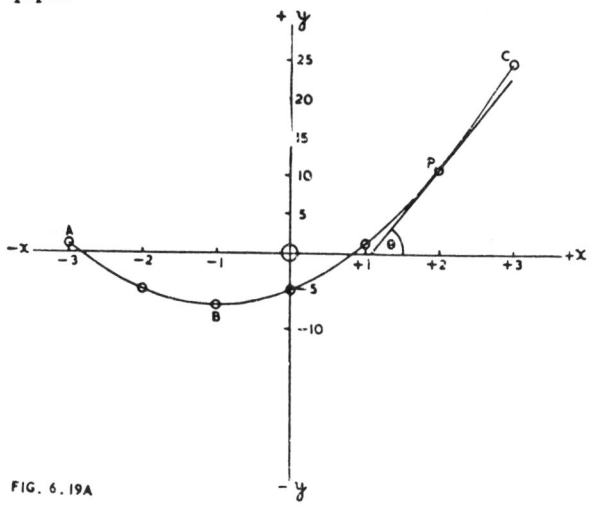

FIG. 6.19A

The equations of some common curves are :

Straight line through origin	$y = mx$	(1)
Straight line not through origin	$y = mx + n$	(2)
Circle with centre at origin	$y^2 = r^2 - x^2$	(3)
Circle with centre at (h, k)	$(x - h)^2 + (y - k)^2 = r^2$	(4)
General equation of circle	$x^2 + y^2 + dx + ey + f = 0$	(5)
Ellipse	$\dfrac{x^2}{a^2} + \dfrac{y^2}{b^2} = 1$	(6)
Hyperbola	$\dfrac{x^2}{a^2} - \dfrac{y^2}{b^2} = 1$	(7)
Parabola (origin at vertex)	$y^2 = 2px$	(8)
Focus is at	$x = p/2, y = 0$	(9)

Areas and average heights

The average height of a curve (i.e. the average length of the ordinates) may be determined by dividing the area beneath the curve into strips of equal width, and then using either the Trapezoid Rule or Simpson's Rule [see Sect. 3 eqn. (54)] to determine the area, and dividing the area by the length (abscissa).

Logarithmic paper

Logarithmic ruled paper is frequently used in the plotting of curves, particularly when the x or y coordinates cover a range of 100 : 1 or more. Single cycle* paper accommodates a range of 10 : 1, and may be drawn by hand, using the whole of the C scale, or half the B scale on a slide rule. Two " cycle " log paper accommodates a range of 100 : 1, and may be drawn with the whole B scale on a slide rule. Each

*Strictly this should be called single decade.

of the " cycles " has the same linear length. Additional " cycles " may be added as desired (for examples see a.v.c. characteristics in Chapter 27).

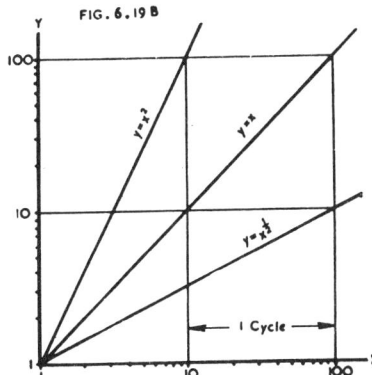

FIG. 6.19 B

Occasionally it is desirable to use log. log. ruled paper ; this is ruled logarithmically on both X and Y axes. An important feature of this form of representation is that a curve of the type

$$y = ax^n$$

is shown as a straight line with a slope of n where the slope is measured as the number of " cycles " on the Y axis per " cycle " on the X axis (see Fig. 6.19B).

In plotting readings (say for a valve I_b, E_b characteristic) which are likely to follow a power law, it is often helpful to use log. log. paper. A straight line indicates a true power law, and its slope gives the value of the exponential—usually not an integer. A slightly curving line indicates a close approach to a power law, and a tangent or chord may be drawn to give the slope at a point or the average over a region.

(ii) Finding the equation to a curve

A method which may be used to find the equation to a curve is given by K. R. Sturley in his book " Radio Receiver Design—Part 1 " (Chapman and Hall, London, 1943) pages 419-421.

(iii) Three variables

The plate current of a triode is given approximately by

$$I_b = K(\mu E_c + E_b)^{3/2}.$$

In one typical case $K = 10 \times 10^{-6}$ mhos (the perveance)

and $\mu = 20$.

Here there are three variables, E_c, E_b and I_b. We can select a suitable value of E_b, calculate the curve, and then repeat the process for other values of E_b. Here E_b is called the **parameter**.

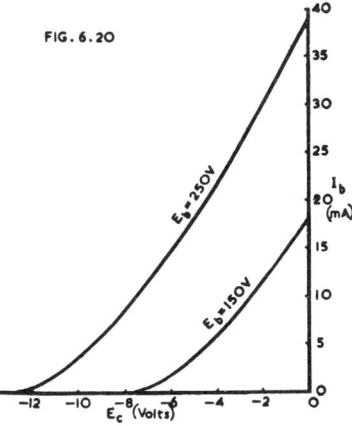

FIG. 6.20

If $E_b = 250$ volts, the plate current in milliamperes will be

$$I_b = 10^{-2} (20 \times E_c + 250)^{3/2} \text{ mA}$$

If $E_c = 0$, $I_b = 10^{-2} (0 + 250)^{3/2} = 10^{-2} (250)^{3/2} \approx 39$ mA

$E_c = -4$, $I_b = 10^{-2} (-80 + 250)^{3/2} = 10^{-2} (170)^{3/2} \approx 22$ mA

$E_c = -8$, $I_b = 10^{-2} (-160 + 250)^{3/2} = 10^{-2} (90)^{3/2} \approx 8.5$ mA

$E_c = -12$, $I_b = 10^{-2} (-240 + 250)^{3/2} = 10^{-2} (10)^{3/2} \approx 0.3$ mA

These points have been plotted in Fig. 6.20, and a curve has been drawn through them marked $E_b = 250$ V.

Similarly, for $E_b = 150$ V,

If $E_c = 0$, $I_b = 10^{-2} (0 + 150)^{3/2} = 10^{-2} (150)^{3/2} \approx 18$ mA

$E_c = -2$, $I_b = 10^{-2} (-40 + 150)^{3/2} = 10^{-2} (110)^{3/2} \approx 11.5$ mA

$E_c = -4$, $I_b = 10^{-2} (-80 + 150)^{3/2} = 10^{-2} (70)^{3/2} \approx 5.8$ mA

$E_c = -6$, $I_b = 10^{-2} (-120 + 150)^{3/2} = 10^{-2} (30)^{3/2} \approx 1.6$ mA

Cut off occurs at $E_c = -(150/20) = -7.5$ volts.

These points have been plotted and a smooth curve drawn through them. Similar curves could be drawn for any other plate voltage, thus forming a " family " of curves. This is actually a three-dimensional graphical diagram reduced to a form suitable for a flat surface.

(iv) Vectors and *j* notation

Any physical quantity which possesses both magnitude and direction is called a vector. Vectors may be represented on paper by means of straight lines with arrow-heads. The length of the line indicates (to some arbitrary scale) the magnitude of the quantity, and the direction of the line and arrow-head indicates the direction in which the vector is operating. The position of the line on the paper is of no consequence.

Addition of vectors

Vectors may be added by drawing them in tandem, and taking the resultant from the beginning of the first one to the end of the last one. Fig. 6.21A shows two vectors, **a** and **b**, which are added together to give the resultant **c**. Exactly the same result is obtained by placing **a** and **b** together, as in Fig. 6.21B, completing the parallelogram, and taking **c** as the diagonal. Vectors are generally printed in bold face type, to distinguish them from **scalar** values, which have no direction, although they have magnitude and sign (i.e. positive or negative).

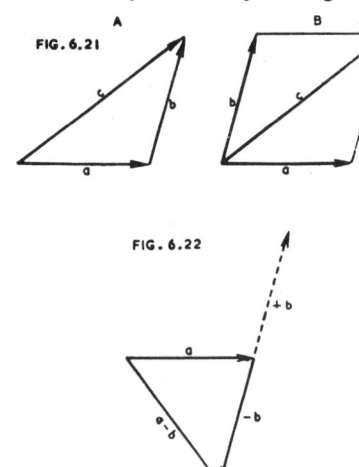

FIG. 6.21

FIG. 6.22

Vector negative sign

A vector (− **a**) has the same magnitude as a vector (**a**) but its direction is reversed.

Subtraction of vectors

The vector to be subtracted is reversed in direction, and then the vectors are added.

In Fig. 6.22, to find **a** − **b**, the direction of **b** is reversed to give (− **b**) and then **a** and (− **b**) are added to give the resultant (**a** − **b**).

Multiplication of a vector by a number (*n*)

The resultant vector has the same direction, but its length is increased *n* times

$$\text{e.g. } \mathbf{a} \times n = n\mathbf{a}$$

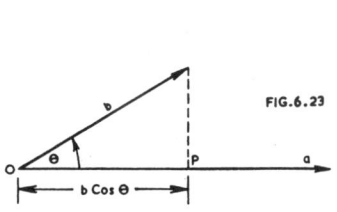

FIG.6.23

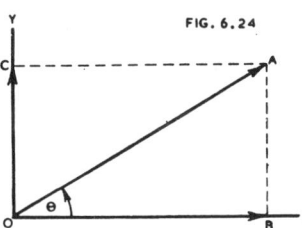

FIG. 6.24

The scalar product of two vectors

The scalar product of two vectors (**a** and **b** in Fig. 6.23) is $ab \cos \theta$, where θ is the angle between them. This may be written

$$\mathbf{a} . \mathbf{b} = ab \cos \theta$$

where **a** . **b** indicates the multiplication of two vectors. From Fig. 6.23 it will be seen that the scalar product is the product of the magnitude of one vector and the " projection " of the other on it.

Components of a vector

Any vector can be resolved into two component vectors in any two desired directions. For example, in Fig. 6.21 the vector **c** can be resolved into the component vectors **a** and **b**. If the component vectors are at right angles to one another they are called rectangular components ; in such a case they are usually taken horizontally (along the X axis) and vertically (along the Y axis). The vector OA in Fig. 6.24 can be resolved into two rectangular components OB and OC where

$$|\mathbf{OB}| = |\mathbf{OA}| \cos \theta, \ |\mathbf{OC}| = |\mathbf{OA}| \sin \theta.$$

Polar coordinates

An alternative form of defining a vector OP is

$$\mathbf{OP} = r \ \angle \ \theta$$

where r is the magnitude of OP, and $\angle \theta$ indicates that there is an angle θ between it and OX (Fig. 6.25).

A graphical device has been described* for the conversion from complex to polar forms.

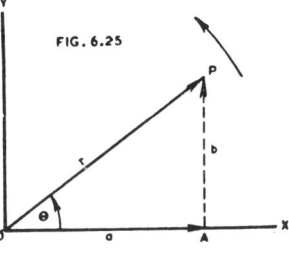

FIG. 6.25

Rotating vectors** (j notation)

If a vector **X** (OA in Fig. 6.26) is rotated 90° in a positive direction to a position OB, the new vector is called j**X**, and j **is described as an " operator " which rotates a vector by 90° without changing its magnitude.**

If the vector j**X** (OB in Fig. 6.26) is operated upon by j, it will be rotated 90° to the position OC, where it is called j^2**X**, the j^2 indicating that it has been rotated $2 \times 90° = 180°$ from its original position OA.

If the vector j^2**X** (OC in Fig. 6.26) is operated upon by j, it will be rotated 90° to the position OD, where it is called j^3**X**, the j^3 indicating that it has been rotated $3 \times 90° = 270°$ from its original position OA.

If the vector j^3**X** (OD in Fig. 6.26) is operated upon by j, it will be rotated 90° to the position OA where it would be called j^4**X**, the j^4 indicating that it has been rotated $4 \times 90° = 360°$ from its original position OA.

There is an important deduction which is immediately obvious. j^2 **indicates a reversal of direction which is the same as a change of sign. The operator j^2 is therefore equivalent to multiplication by -1.**

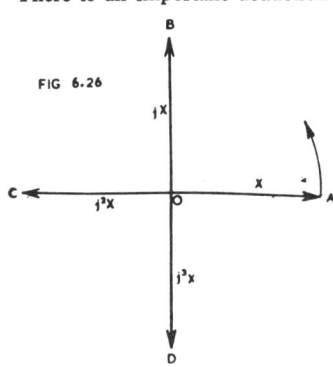

FIG 6.26

If the operator j^2 is applied twice in succession, the result should be equivalent to multiplication by -1 twice in succession, i.e. $(-1) \times (-1) = +1$. This is so, because the operator j^4 brings the vector back to its original direction.

Since the operator j^2 is equivalent to multiplication by -1, we may deduce that the operator j is equivalent to multiplication by $\sqrt{-1}$, even though this in itself does not mean anything.

Operator	Equivalent to multiplication by
j	$\sqrt{-1}$
j^2	-1
j^3	$-\sqrt{-1}$
j^4	$+1$
$-j$	$-\sqrt{-1}$ (same as operator j^3)

*Snowdon, C. " A vector calculating device," Electronic Eng. 17.199 (Sept. 1944) 146.
**Also called " radius vectors." In pure mathematics " i " is used in place of " j."

Operator	Equivalent to multiplication by
$\dfrac{1}{j}$	$-\sqrt{-1}$ (same as operator $-j$)
$\dfrac{1}{-j} = -\dfrac{1}{j}$	$\sqrt{-1}$ (same as operator j)

See Chapter 4 Sects. 4(v), 5(v) and 6 for the application of the *j* notation to a.c. circuits.

The direction of any vector can be defined in terms of the *j* notation. In Fig. 6.25, OX represents the axis of reference, **OP** the vector, and θ the angle of rotation of OP from OX. The perpendicular PA may be drawn from P to OX and then, by the simple theory of vectors, **OP** is the sum of the vectors **OA** and **AP**. Using *j* notation, we may say that

$$\mathbf{OP} = a + jb$$

which means that the vector **OP** is the vector sum of a and b, where a is in the direction of the reference axis OX and b is rotated 90° from OX. In other words, a is the component of OP in the direction OX, and b is the component of OP in the direction OY. The values of a and b are given by

$$a = OP \cos \theta, \quad b = OP \sin \theta,$$

where $\theta = \tan^{-1} b/a$.

This is sometimes called the Argand Diagram.

In electrical a.c. circuit theory (Chapter 4) it is well known that the current through an inductance lags behind the applied voltage, while the current through a capacitance leads the voltage. This is most clearly shown by rotating vectors. It should be borne in mind that all the vectors are rotating at the same angular velocity—one revolution per cycle of the applied voltage—and that the pictorial representation is for any one instant. Fig. 6.27 shows the vector diagram for peak a.c. current (I) flowing through a resistance (R) and an inductance (L) in series. The current vector (**I**) may be placed in any convenient direction—say horizontally (Fig. 6.27); a solid arrow head is used to distinguish it from voltage vectors. The peak voltage drop (**RI**) in the resistance must be " in phase " with the current, and is so shown. The scale to which the voltage vectors are drawn has no connection with the scale to which the current (**I**) is drawn—in fact we are here only concerned with the direction of **I**. The peak voltage drop across L is $\omega L \mathbf{I} = \mathbf{X}_L \mathbf{I}$, and this vector is drawn vertically so as to lead the current **I** by 90°. The total peak voltage drop (**E**) is found by " completing the parallelogram of vectors " as previously. It will be seen that **I** lags behind **E** by the angle θ.

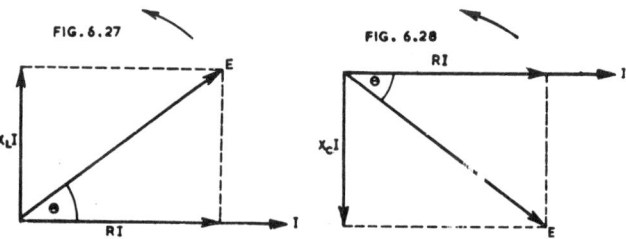

A similar procedure applies with a capacitance instead of an inductance (Fig. 6.28) except that the capacitive reactance is X_c instead of X_L and the vector of peak voltage drop across C ($\mathbf{X}_c\mathbf{I}$) is drawn vertically downwards, since it must be opposite to $\mathbf{X}_L\mathbf{I}$. The current (**I**) here leads the voltage (**E**) by the angle θ.

SECTION 6 : COMPLEX ALGEBRA AND DE MOIVRE'S THEOREM

(i) Complex algebra with rectangular coordinates   (ii) Complex algebra with polar coordinates   (iii) De Moivre's Theorem.

(i) Complex algebra with rectangular coordinates

Complex algebra should preferably be called " the algebra of complex quantities," and it is really quite simple to understand for anyone who has even a limited knowledge of mathematics.

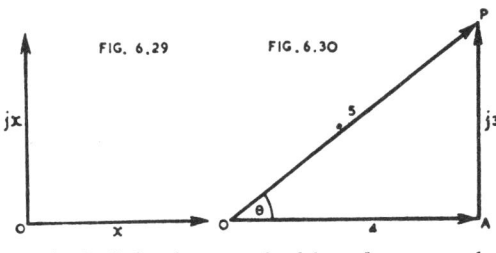

It was explained in Section 5 that the letter j* in front of a vector indicated that it had been rotated 90° from the reference, or positive X, axis (Fig. 6.29). We can make use of this procedure to indicate the magnitude and direction of any vector. In Fig. 6.30 the vector **OP**, with magnitude 5, has been resolved into the rectangular components—

$$\mathbf{OA} = 4 \text{ in the } + X \text{ direction}$$
$$\text{and } \mathbf{AP} = 3 \text{ in the } jX \text{ direction (90° to } X).$$

The X axis is taken horizontally through O, and is sometimes called the " real " axis, while the jX direction is called " imaginary." It is therefore possible to describe **OP** in both magnitude and direction by the expression

$$4 + j3.$$

Its magnitude is the vector sum of 4 and 3, which is $\sqrt{4^2 + 3^2} = \sqrt{25} = 5$. Its direction is given by the angle θ from the X axis, where $\cos \theta = 4/5$.

Any other vector, such as **a** in Fig. 6.31, can similarly be resolved into its components :

$$b = \mathbf{a} \cos \theta \text{ in the } X \text{ direction}$$
$$\text{and } c = \mathbf{a} \sin \theta \text{ in the } jX \text{ direction}$$

and be written as

$$b + jc$$

where

$$b = \mathbf{a} \cos \theta$$
$$\text{and } c = \mathbf{a} \sin \theta.$$

FIG. 6.31

The expression $b + jc$ is called a complex quantity, the word " complex " indicating that the addition is not to be made algebraically, but by vector addition.

This use of a complex quantity such as $b + jc$ is not limited to true vectors, but is found useful in many applications in electrical engineering. Its application to alternating currents is covered in Chapter 4, and the following treatment is a general introduction to the methods of handling complex quantities.

Modulus

The modulus is the magnitude of the original vector (**a** in Fig. 6.31) and is numerically equal to the square root of the sum of the squares of the magnitudes of the two components,

$$|a| = \sqrt{b^2 + c^2} \tag{1}$$

Addition and subtraction of complex quantities

The primary rule is to place all " real " numbers together in one group and all " imaginary " numbers in another. These two groups must be kept entirely separate and distinct throughout. The final form should be

$$(\ldots) + j(\ldots).$$

---

*Or i in pure mathematics.

Example : Add $A + jB$, $C + jD$, $E - jF$.
Total is $A + C + E + jB + jD - jF$
$$= (A + C + E) + j (B + D - F).$$

Multiplication of complex quantities

The multiplication is performed in accordance with normal algebraic laws for real numbers, j^2 being treated as -1.
$$(x_1 + jy_1)(x_2 + jy_2) = x_1x_2 + jy_1x_2 + jy_2x_1 + j^2y_1y_2$$
$$= (x_1x_2 - y_1y_2) + j (x_1y_2 + x_2y_1) \tag{2}$$

Division of complex quantities

The denominator should be made a real number by multiplying both the numerator and the denominator by the conjugate of the denominator (i.e. the denominator with the opposite sign in front of the j term).
$$\frac{x_1 + jy_1}{x_2 + jy_2} = \frac{(x_1 + jy_1)(x_2 - jy_2)}{(x_2 + jy_2)(x_2 - jy_2)} = \frac{x_1x_2 + jy_1x_2 - jx_1y_2 + y_1y_2}{x_2^2 + y_2^2}$$
$$= \left(\frac{x_1x_2 + y_1y_2}{x_2^2 + y_2^2}\right) - j\left(\frac{x_1y_2 - y_1x_2}{x_2^2 + y_2^2}\right) \tag{3}$$

Similarly
$$\frac{1}{x + jy} = \frac{(1)(x - jy)}{(x + jy)(x - jy)} = \left(\frac{x}{x^2 + y^2}\right) - j\left(\frac{y}{x^2 + y^2}\right) \tag{4}$$

Square root of complex quantities

To find the square root of $(a + jb)$, assume that the square root is $(x + jy)$ and then proceed to find x and y.
$$(x + jy)^2 = (x + jy)(x + jy) = x^2 - y^2 + 2jxy$$
Therefore $\quad a + jb = (x^2 - y^2) + 2jxy = (x^2 - y^2) + j(2xy)$
Therefore $(x^2 - y^2) = a$
and $\qquad\qquad 2xy = b$.

Modulus $\qquad = r = \sqrt{a^2 + b^2}$
Square of modulus of $(x + jy) = x^2 + y^2$.
Therefore $x^2 + y^2 = r$
But $\qquad\quad x^2 - y^2 = a$ (see above)
Therefore $\qquad x^2 = \frac{1}{2}(r + a)$ and $y^2 = \frac{1}{2}(r - a)$
Therefore $\qquad x = \pm \sqrt{\frac{1}{2}(r + a)}$ and $y = \pm \sqrt{\frac{1}{2}(r - a)}$ $\tag{5}$
The signs should be checked to see which are applicable.

(ii) Complex algebra with polar coordinates

In complex numbers we can write (Fig. 6.31)
$$a = b + jc$$
but $b = |a| \cos \theta$
and $c = |a| \sin \theta$
Therefore $\qquad a = |a| \cos \theta + j |a| \sin \theta$
Therefore $\qquad a = |a| (\cos \theta + j \sin \theta) \tag{6}$
Here $|a|$ is the magnitude of the vector and $(\cos \theta + j \sin \theta)$ may be called the **trigonometrical operator** which rotates the vector through the angle θ in a positive (counter-clockwise) direction from the x axis.
As explained in Sect. 5(iv), a vector may be defined by
$$r \angle \theta$$
where r is the magnitude and $\angle \theta$ the angle between the vector and the reference axis. It will therefore be seen that the trigonometrical operator $(\cos \theta + j \sin \theta)$ is effectively the same as $\angle \theta$ with polar coordinates.
i.e. $\angle \theta = \cos \theta + j \sin \theta \tag{7}$

Pure mathematical polar form

By the use of the Exponential Series we can express ϵ^x in the form (Sect. 2, eqn. 87)
$$\epsilon^x = 1 + x + \frac{x^2}{2!} + \frac{x^3}{3!} + \frac{x^4}{4!} + \ldots\ldots$$

Putting $j\theta$ for x we obtain

$$\epsilon^{j\theta} = 1 + j\theta + \frac{(j\theta)^2}{2!} + \frac{(j\theta)^3}{3!} + \frac{(j\theta)^4}{4!} + \frac{(j\theta)^5}{5!} + \cdots$$

$$= 1 + j\theta - \frac{\theta^2}{2!} - j\frac{\theta^3}{3!} + \frac{\theta^4}{4!} + j\frac{\theta^5}{5!} - \cdots$$

Grouping the j terms,

$$\epsilon^{j\theta} = (1 - \frac{\theta^2}{2!} + \frac{\theta^4}{4!} - \ldots) + j(\theta - \frac{\theta^3}{3!} + \frac{\theta^5}{5!} - \ldots)$$

But $\cos \theta = 1 - \dfrac{\theta^2}{2!} + \dfrac{\theta^4}{4!} - \ldots$ from eqn. (17) below

and $\sin \theta = \theta - \dfrac{\theta^3}{3!} + \dfrac{\theta^5}{5!} - \ldots$ from eqn. (18) below.

Therefore $\epsilon^{j\theta} = \cos \theta + j \sin \theta$ \hfill (8)

Thus the pure mathematical polar form is

$$|a|\ \epsilon^{j\theta}$$

and is called the Exponential Form.

The vector can thus be written in the various forms

$$a = b + jc \tag{9}$$
$$a = |a| \angle \theta \tag{10}$$
$$a = |a|\ \epsilon^{j\theta} \tag{11}$$
$$a = |a| (\cos \theta + j \sin \theta) \tag{12}$$
$$a = |a| \cos \theta + j|a| \sin \theta \tag{13}$$

A graphical method for converting from the complex form $(b + jc)$ to the polar form and vice versa has been described.* It will therefore be seen that there is a connection with the operator j :

j turns a vector through a right angle

$(\cos \theta + j \sin \theta)$ or $\epsilon^{j\theta}$ } turns a vector through an angle θ.

Addition and subtraction in polar form

Addition and subtraction may be done either graphically, or by expressing each in rectangular components and proceeding as for rectangular coordinates.

Multiplication of polar vectors

The product is found by multiplying their magnitudes and adding their angles.

Division of polar vectors

The quotient is found by dividing their magnitudes and subtracting the angle of the divisor from the angle of the dividend.

Square root of polar vectors

The root is found by taking the square root of the magnitude and half the angle.

(iii) De Moivre's Theorem

De Moivre's Theorem states that

$$(\cos \theta + j \sin \theta)^n = \cos n\theta + j \sin n\theta \tag{14}$$

where n may be positive or negative, fractional or integral.

It was explained in (ii) above that $(\cos \theta + j \sin \theta)$ may be regarded as a trigonometrical operator which rotates the vector through an angle θ. If this is applied twice in succession, the trigonometrical operator becomes

$$(\cos \theta + j \sin \theta)(\cos \theta + j \sin \theta) = (\cos \theta + j \sin \theta)^2$$

giving a total rotation of an angle 2θ.

Similarly if this is applied three times in succession, the trigonometrical operator becomes

$$(\cos \theta + j \sin \theta)(\cos \theta + j \sin \theta)(\cos \theta + j \sin \theta) = (\cos \theta + j \sin \theta)^3$$

giving a total rotation of an angle 3θ.

---

*Snowdon, C. " A vector calculating device " Electronic Eng. 17.199 (Sept. 1944) 146.

Thus, in the general case, the trigonometrical operator $(\cos \theta + j \sin \theta)^n$ gives a rotation of an angle $n\theta$ which, as explained above, is equivalent to a trigonometrical operator

$$(\cos n\theta + j \sin n\theta).$$

We have, in this way, proved De Moivre's Theorem for the case when n is a positive integer, and indicated the significance of the Theorem.

Application of De Moivre's Theorem :

1. To express $\cos n\theta$ and $\sin n\theta$ in terms of $\cos \theta$ and $\sin \theta$, where n is a positive integer.

$$\cos n\theta + j \sin n\theta = (\cos \theta + j \sin \theta)^n$$

$$= \cos^n \theta + nj \cos^{n-1} \theta . \sin \theta + \frac{n(n-1)}{2} j^2 . \cos^{n-2} \theta . \sin^2 \theta + \ldots$$

(as in Sect. 2, eqn. 83).

Then equate the real and imaginary parts of the equation, giving

$$\cos n\theta = \cos^n \theta - \frac{n(n-1)}{2} \cos^{n-2} \theta . \sin^2 \theta + \frac{n(n-1)(n-2)(n-3)}{4!} \cos^{n-4} \theta .$$

$$\sin^4 \theta + \ldots \tag{15}$$

$$\sin n\theta = n \cos^{n-1} \theta . \sin \theta - \frac{n(n-1)(n-2)}{3!} \cos^{n-3} \theta . \sin^3 \theta +$$

$$\frac{n(n-1)(n-2)(n-3)(n-4)}{5!} \cos^{n-5} \theta . \sin^5 \theta + \ldots \tag{16}$$

2. To express $\cos \theta$ and $\sin \theta$ in terms of θ, write θ in the form $n(\theta/n)$ and expand the sine and cosine as in eqns. (15) and (16). When n becomes large, $\cos (\theta/n)$ may be taken as unity and $\sin (\theta/n)$ as (θ/n) itself. In the limit as n tends to infinity it can then be shown that

$$\cos \theta = 1 - \frac{\theta^2}{2!} + \frac{\theta^4}{4!} - \frac{\theta^6}{6!} + \ldots \tag{17}$$

$$\sin \theta = \theta - \frac{\theta^3}{3!} + \frac{\theta^5}{5!} - \frac{\theta^7}{7!} + \ldots \tag{18}$$

where θ is expressed in radians.

3. To express $\sin \theta$ and $\cos \theta$ in terms of $\epsilon^{j\theta}$.

By substituting $j\theta$ in place of x in the Exponential Series (Sect. 2, eqn. 87), and by using the relationship

$$\epsilon^{j\theta} = \cos \theta + j \sin \theta$$

and equations (17) and (18) we may obtain

$$\sin \theta = \frac{\epsilon^{j\theta} - \epsilon^{-j\theta}}{2j} \tag{19}$$

$$\text{and } \cos \theta = \frac{\epsilon^{j\theta} + \epsilon^{-j\theta}}{2} \tag{20}$$

SECTION 7: DIFFERENTIAL AND INTEGRAL CALCULUS

(*i*) *Slope and rate of change* (*ii*) *Differentiation* (*iii*) *Integration* (*iv*) *Taylor's Series* (*v*) *Maclaurin's Series*.

(i) Slope and rate of change

The **slope** of any straight line is the ratio of the lengths of the vertical and horizontal projections of any segment of the line. In Fig. 6.32 the line QP has its vertical projection MN and horizontal projection AB ; its slope is therefore MN/AB. Its slope could equally well be based on the projections of its segment CP, and its slope BP/CB. In both cases the result is the same, and is equal to the tangent of the angle of inclination which the line makes with the horizontal axis—

$$\text{slope} = MN/AB = BP/CB = \tan \theta.$$

The lengths of the projections must be measured in terms of the scales to which the line is drawn. Thus horizontal projections such as CB must be measured, not in inches, but in the equivalent number of units corresponding to the length CB on the X axis. Similarly with the vertical projections on the Y axis, which usually has quite a different scale from that for the X axis.

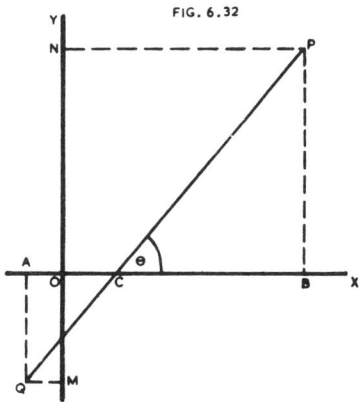

FIG. 6.32

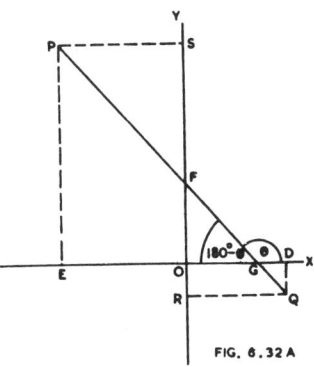

FIG. 6.32 A

If the line QP in Fig 6.32 is rotated approximately 90° in the counter-clockwise direction, the result will be as shown in Fig. 6.32A, the angle θ being more than 90°. The vertical projection is RS and the horizontal projection is DE. The slope is therefore RS/DE or OF/GO, the latter applying to the segment FG. It is important in all this work to consider the directions as well as the magnitudes of the lines : OF is in a positive direction but GO is in a negative direction. We can therefore replace GO by −OG ; the slope of QP will then be given by −(OF/OG) which is described

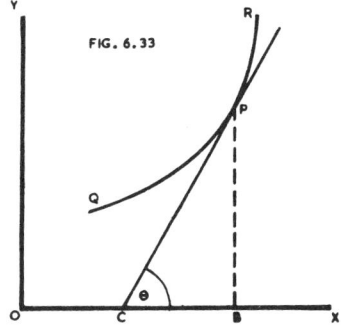

FIG. 6.33

as a negative slope. The loadlines on valve plate characteristics are examples of lines with negative slope (e.g. Fig. 2.22).

When a line is curved, its slope varies from point to point. The slope at any point is given by the slope of the tangent at that point. In Fig. 6.33 the curve is QPR, and the slope at P is given by BP/CB or $\tan \theta$.

Rate of change

One of the most important relationships between the variables in any law, formula or equation is the rate of change of the whole function with its independent variable.

Definition

The rate of change is the amount of change in the function, per unit change in the value of the independent variable. The rate of change of the function is therefore the ratio of change in the function to the change in the variable which produces it.

Consider the equation for a straight line

$$y = ax + b$$

where a and b are constants. Let us take two values of x, one equal to x_1 and the other $(x_1 + \Delta x)$, where Δx is a small increment of x.

Point 1: $x = x_1$	$\therefore y_1 = ax_1 + b$	(1)
Point 2: $x = (x_1 + \Delta x)$	$\therefore y_2 = a(x_1 + \Delta x) + b$	(2)
Subtracting (1) from (2),	$(y_2 - y_1) = a.\Delta x$	
Putting $(y_2 - y_1) = \Delta y$,	$\Delta y = a.\Delta x$	
Dividing both sides by Δx,	$\dfrac{\Delta y}{\Delta x} = a$	(3)

Here Δy is the amount of change in the function for a change Δx in the independent variable. The " rate of change " is defined as the amount of change in the function per unit change in the independent variable, that is

$$\text{rate of change} = \Delta y/\Delta x \qquad (4)$$

Referring to Fig. 6.34, we have a graph of $y = ax + b$ which is, of course, a straight line cutting the Y axis at a height b above the origin. The first point (x_1, y_1) is at P, and the second point $(x_1 + \Delta x, y_1 + \Delta y)$ is at Q. In the preceding argument we found that $\Delta y/\Delta x = a$. In Fig. 6.34 $\Delta x = $ PR and $\Delta y = $ RQ, so that $\Delta y/\Delta x = $ RQ/PR, which is the slope of the line $y = ax + b$. Thus

$$\text{slope} = \frac{\text{RQ}}{\text{PR}} = \frac{\Delta y}{\Delta x} = a \qquad (5)$$

Equations (4) and (5) prove that **the rate of change is the same as the slope for a straight line.**

We now proceed to consider the general case when the function is not linear. In Fig. 6.35 there is plotted a function which may be of the form

$$y = ax^2 + bx + c$$

where a, b and c are constants. As before, the independent variable (x) is plotted horizontally, while the dependent variable (y) is plotted vertically. The vertical coordinate of any point P (i.e. AP in Fig. 6.35) represents, to its proper scale, the value of $ax^2 + bx + c$ which is a function of x, while the horizontal coordinate (OA) represents, to its own scale, the value of x.

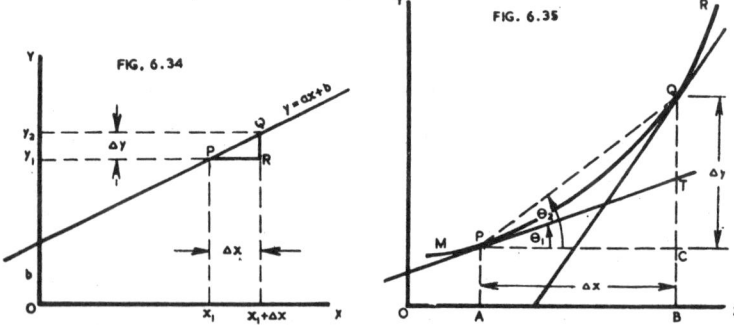

FIG. 6.34   FIG. 6.35

As with the simpler case of the straight line, take a second point (Q) with its x coordinate increased by Δx. Also, as before, call the increment in the y coordinate Δy. It will therefore be seen that, commencing at point P, an increment ($\Delta x = $ PC) in the value of x results in an increment ($\Delta y = $ CQ) in the value of y.

The **average rate of change** over this increment is, as before, $\Delta y/\Delta x$ which is the slope of the chord PQ and the tangent of the angle θ_2 which PQ makes with the horizontal. If now, leaving point P unchanged, we gradually move Q along the curve towards P we will see that, as Q approaches P, θ_2 approaches θ_1 until in the limiting case the slope of the chord approaches the slope of the tangent (PT). The tangent to a curve at any point shows the instantaneous slope of the curve at that point and therefore the instantaneous rate of increase of the function at that point.

(ii) Differentiation

We may express the foregoing argument in the mathematical form of limits—

$$\underset{\Delta x \to 0}{Lt} \quad (\Delta y/\Delta x) = \tan \theta_1 \tag{6}$$

which says that the limit (as Δx is made smaller and approaches zero) of $\Delta y/\Delta x$ is $\tan \theta_1$ or the slope of the tangent PT, which is the instantaneous rate of increase at point P.

This is given the symbol dy/dx which is " the **differential coefficient** (or derivative) of y with regard to x," and is spoken of as " dee y by dee x."

It should be noted that dy/dx is a single symbol, not a fraction, and is merely a short way of writing

$$\underset{\Delta x \to 0}{Lt} \quad (\Delta y/\Delta x).$$

Differentiation is the process of finding the differential coefficient (or derivative). Some examples are given below and in each case the result may be obtained by considering the increase in the function which results from an increase Δx in the independent variable.

Note : u and v are functions of x ; a, b and c are constants.

1. Derivative of a constant $[y = c]$ $dy/dx = 0$ (7)
2. Derivative of a variable with respect to itself
   $[y = x]$: $dy/dx = dx/dx = 1$ (8)
3. Derivative of a variable multiplied by a constant
   $[y = cx]$: $dy/dx = c$ (9)
4. Derivative of powers of a variable
   $[y = x^2]$: $dy/dx = 2x$ (10)
   $[y = x^3]$: $dy/dx = 3x^2$ (11)
   $[y = x^4]$: $dy/dx = 4x^3$ (12)
   $[y = x^n]$: $dy/dx = nx^{n-1}$ (13)
   This applies for n negative as well as positive.
5. Derivative of a constant times a function of a variable
   $[y = cx^2]$: $dy/dx = 2cx$ (14)
   $[y = c.u]$: $dy/dx = c.du/dx$ (15)
6. Derivative of fractional powers of a variable
   $[y = x^{\frac{1}{2}}]$: $dy/dx = \frac{1}{2}x^{-\frac{1}{2}}$ (16)
   $[y = x^{1/n}]$: $dy/dx = (1/n)x^{(1/n)-1}$ (17)
7. Derivative of a sum or difference
   $[y = u \pm v]$: $dy/dx = du/dx \pm dv/dx$ (18)
   $[y = ax^3 + bx^2 - cx]$: $dy/dx = 3ax^2 + 2bx - c$ (19)
8. Derivative of a product of two functions
   $[y = u.v]$: $dy/dx = u\dfrac{dv}{dx} + v\dfrac{du}{dx}$ (20)
   $[y = (x + 1) x^2]$: $dy/dx = (x + 1).2x + x^2.1$
       $= 3x^2 + 2x$ (21)
9. Derivative of a quotient of two functions
   $[y = u/v]$: $\dfrac{dy}{dx} = \dfrac{v\dfrac{du}{dx} - u\dfrac{dv}{dx}}{v^2}$ (22)
10. Differentiation of a function of a function
    $[y = F(u)$ where $u = F(x)]$: $\dfrac{dy}{dx} = \dfrac{dy}{du} \cdot \dfrac{du}{dx}$ (23)
    $[y = u^2$ where $u = ax^2 + b]$: $dy/dx = 2(ax^2 + b)(2ax)$ (24)

11.	$[y = \epsilon^x]$:	$dy/dx = \epsilon^x$	(25)
	$[y = a.\epsilon^{mx}]$:	$dy/dx = a.m.\epsilon^{mx}$	(26)
	$[y = \epsilon^u]$:	$dy/dx = \epsilon^u . du/dx$	(27)
12.	$[y = \log_\epsilon x]$:	$dy/dx = 1/x$	(28)
	$[y = \log_\epsilon u]$:	$dy/dx = (1/u)(du/dx)$	(29)
	$[y = \log_{10} u]$:	$dy/dx = (1/u)(du/dx) \log_{10} \epsilon$	(30)
		$= (0.4343)(1/u)(du/dx)$	(31)
13.	$[y = \sin x]$:	$dy/dx = \cos x$	(32)
	$[y = \cos x]$:	$dy/dx = - \sin x$	(33)
	$[y = \tan x]$:	$dy/dx = \sec^2 x$	(34)
	$[y = \cot x]$:	$dy/dx = - \operatorname{cosec}^2 x$	(35)
	$[y = \sec x]$:	$dy/dx = \sec x . \tan x$	(36)
	$[y = \operatorname{cosec} x]$:	$dy/dx = - \operatorname{cosec} x . \cot x$	(37)
14.	$[y = \sin^{-1} x]$:	$dy/dx = 1/\sqrt{1 - x^2}$*‡	(38)
	$[y = \cos^{-1} x]$:	$dy/dx = - 1/\sqrt{1 - x^2}$†‡	(39)
	$[y = \tan^{-1} x]$:	$dy/dx = 1/(1 + x^2)$‡	(40)
15.	$[y = \sinh x]$:	$dy/dx = \cosh x$	(41)
	$[y = \cosh x]$:	$dy/dx = \sinh x$	(42)
	$[y = \tanh x]$:	$dy/dx = \operatorname{sech}^2 x.$	(43)

Successive differentiation

If y is a function of x, then dy/dx will also be a function of x and can therefore be differentiated with respect to x, giving

$$\frac{d}{dx}\left(\frac{dy}{dx}\right) \text{ which is written as } \frac{d^2y}{dx^2}$$

and spoken of as " dee two y by dee x squared." Here again this is only a symbol which must be handled in accordance with its true meaning. The same procedure may be applied again and again.

Example : $y = x^n$ . . . . function $F(x)$

$$\frac{dy}{dx} = nx^{n-1} \dots \dots \dots \text{ derivative } F'(x)$$

$$\frac{d^2y}{dx^2} = n(n - 1)x^{n-2} \dots \dots \text{ second derivative } F''(x)$$

$$\frac{d^3y}{dx^3} = n(n - 1)(n - 2)x^{n-3} \dots \text{ third derivative } F'''(x)$$

Application of differentiation

The plate current versus grid voltage characteristic of a triode valve (for constant plate voltage) is a function of the grid voltage, and follows approximately the law

$$I_p = K(\mu E_g + E_p)^{3/2}$$

where K, μ and E_p are constants. The derivative with regard to E_g is dI_p/dE_g, which is the mutual conductance. The second derivative is the rate of change of the mutual conductance with regard to E_g, and is useful when we want to find the conditions for maximum or minimum mutual conductance.

In Fig. 6.36 there is a curve with a **maximum** at point M and a **minimum** at point N. It will be seen that the instantaneous slope of the curve at both points M and N is zero, that is $dy/dx = 0$.

Part of curve :	P to M	M	M to N	N	N to Q
Slope (dy/dx) :	$+$ve	O	$-$ve	O	$+$ ve
d^2y/dx^2 :		$-$ ve		$+$ ve	

A maximum is indicated by : $\begin{cases} dy/dx = 0 \\ d^2y/dx^2 \quad \text{negative} \end{cases}$

A minimum is indicated by : $\begin{cases} dy/dx = 0 \\ d^2y/dx^2 \quad \text{positive} \end{cases}$

‡ x measured in radians.
*Positive sign if $\sin^{-1} x$ lies in first or fourth quadrant, negative sign if $\sin^{-1} x$ lies in second or third quadrant.
†Negative sign if $\cos^{-1} x$ lies in first or second quadrant, positive sign if $\cos^{-1} x$ lies in third or fourth quadrant.

A point of inflection* is indicated by :
$$d^2y/dx^2 = 0$$
Curve concave upwards indicated by : d^2y/dx^2 positive
Curve concave downwards indicated by : d^2y/dx^2 negative

Examples

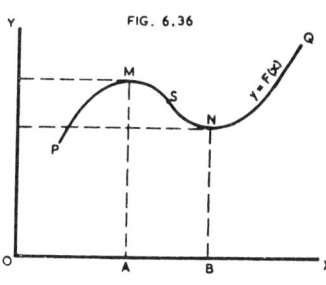

FIG. 6.36

(1) To find the maximum value of
$$y = 2x - x^2 + 4$$
$$dy/dx = 2 - 2x$$

For a maximum $dy/dx = 0$, therefore $2 - 2x = 0$, and $x = 1$.

This is the value of x at which a maximum, or a minimum, or a point of inflection occurs.

To see which it is, take the second derivative—
$$d^2y/dx^2 = -2 \text{ which is negative.}$$

Therefore the point is a maximum.
To find the value of y at this point, put the value
($x = 1$) into $y = 2x - x^2 + 4$.
Therefore $y = 2 - 1 + 4 = 5$.
(2) To find the points of inflection in the curve
$$y = x^4 - 6x^2 - x + 16.$$
$$dy/dx = 4x^3 - 12x - 1$$
$$d^2y/dx^2 = 12x^2 - 12.$$
For points of inflection, $d^2y/dx^2 = 0$, therefore $12x^2 - 12 = 0$
Therefore $12x^2 = 12$, Thus $x = \pm 1$.
There are thus two points of inflection, one at $x = +1$, the other at -1. The values of y at these points are given by substituting these values of x in the function :
$$x = +1: y = 1 - 6 - 1 + 16 = +10.$$
$$x = -1: y = 1 - 6 + 1 + 16 = +12.$$
It is always wise to make a rough plot of the curve to see its general shape. Some curves have more than one value of maximum and minimum.

Partial differentiation

Partial differential coefficients, designated in the form $\partial y/\partial x$ (the symbol ∂ may be pronounced " der " to distinguish from " d " in dy/dx) are used in considering the relationship between two of the variables in systems of more than two variables such as the volume of an enclosure having rectangular faces, the sides being of length x, y and z respectively :
$$v = x\,y\,z \tag{44}$$
Thus, the rate of change of volume with the change in length of the side x, while the sides y and z remain constant, is
$$\partial v/\partial x = y\,z \tag{45}$$
Similarly $\partial v/\partial y = z\,x$, where z and x are constant $\tag{46}$
And $\partial v/\partial z = xy$ where x and y are constant $\tag{47}$
In three-dimensional differential geometry, the equation representing a surface may be represented generally in the form
$$y = F(x, z) \tag{48}$$
In this case, the partial differential coefficient $\partial y/\partial x$ represents the slope at the point (x, y, z) of the tangent to the curve of intersection of the surface with a plane parallel to the plane passing through the x and y axes and separated by a fixed distance z from the latter.
Thus $\partial y/\partial x$ represents the slope of a tangent to a cross section of a three-dimensional solid, the partial derivative reducing the three-dimensional body to a form suitable for two-dimensional consideration. " $\partial y/\partial x$ " is equivalent to " dy/dx (z constant)" when there are three variables, x, y and z.

*A point of inflection is one at which the curvature changes from one direction to the other (e.g. S in Fig. 6.36). It is necessarily a point of maximum or minimum slope.

Partial differentials are therefore particularly valuable in representing Valve Co-efficients [see Chapter 2 Sect. 9(ix)].

Total differentiation

When there are three independent variables (x, y, z) which are varying simultaneously but independently of each other,

$$u = F(x,y,z) \tag{49}$$

the total differential is

$$du = \frac{\partial u}{\partial x} dx + \frac{\partial u}{\partial y} dy + \frac{\partial u}{\partial z} dz \tag{50}$$

and similarly for two, or any larger number of independent variables.

When the independent variables are functions of a single independent variable (t) the total differential with respect to t is

$$\frac{du}{dt} = \frac{\partial u}{\partial x} \cdot \frac{dx}{dt} + \frac{\partial u}{\partial y} \cdot \frac{dy}{dt} + \frac{\partial u}{\partial z} \cdot \frac{dz}{dt} \tag{51}$$

(iii) Integration

Integration is merely the inverse of differentiation. For example—

Differentiation— $\dfrac{d}{dx}(4x^3 + 2) = 12x^2$

Integration— $\int 12x^2\, dx = 4x^3 + C$

The sign $\int$ (called " integral ") before a quantity indicates that the operation of integration is to be performed on the expression which follows.

The dx which follows the expression is merely a short way of writing " with respect to x." Just as the constant 2, in the function above, disappeared during the process of differentiation, so it is necessary to replace it in the inverse procedure of integration. But when we are given the integral alone, we do not know what was the value of the constant, so we add an unknown constant C, the value of which may be determined in some cases from other information available.

Useful rules for integrals

a, b, c = constants ; C = constant of integration ; u and v are functions of x.

1. $\int a.F(x)\, dx = a\int F(x)\, dx$; $\int a\, dx = ax + C$ $\hspace{2em}$ (52)

2. $\int (u \pm v)\, dx = \int u\, dx \pm \int v\, dx$ (similarly for more than two) $\hspace{1em}$ (53)

3. $\int x^n\, dx = \dfrac{1}{n+1} x^{n+1} + C$ $\hspace{3em}$ $(n \neq -1)$ $\hspace{2em}$ (54)

4. $\int a^x\, dx = (a^x/\log_\epsilon a) + C$ $\hspace{2em}$ (55)

5. $\int \epsilon^x\, dx = \epsilon^x + C$; $\int \epsilon^{ax}\, dx = (1/a)\epsilon^{ax} + C$ $\hspace{2em}$ (56)

$\hspace{1.5em}\int x\,\epsilon^x dx = \epsilon^x(x-1) + C$; $\int x^m \epsilon^x dx = x^m \epsilon^x - m \int x^{m-1}\epsilon^x dx + C^*$ $\hspace{1em}$ (57)

6. $\int u\dfrac{dv}{dx}dx = uv - \int v\dfrac{du}{dx}dx$ (integration by parts) $\hspace{2em}$ (58)

7. $\int \dfrac{dx}{x} = \int \dfrac{1}{x}\, dx = \log_\epsilon x + C = \log_\epsilon c\,x$ $\hspace{2em}$ (59)

8. $\int a^x \log_\epsilon a\, dx = a^x + C$; $\int \log_a x\, dx = x \log_a(x/\epsilon) + C$ $\hspace{2em}$ (60)

9. $\int (ax+b)^n\, dx = \dfrac{(ax+b)^{n+1}}{a(n+1)} + C$ $\hspace{2em}$ $(n \neq -1)$ $\hspace{2em}$ (61)

___
* $m > 0.$

10. $\int \dfrac{dx}{ax + b} = \int \dfrac{1}{ax + b} dx = (1/a) \log_\epsilon (ax + b) + C$ (62)

11. $\int \dfrac{x\, dx}{ax + b} = \int \dfrac{x}{ax + b}\, dx = (1/a^2) [ax + b - b \log_\epsilon (ax + b)] + C$ (63)

12. $\int \dfrac{x\, dx}{(ax + b)^2} = \int \dfrac{x}{(ax + b)^2}\, dx = \dfrac{1}{a^2}\left[\dfrac{b}{ax + b} + \log_\epsilon (ax + b)\right] + C$ (64)

13. $\int \dfrac{x^2\, dx}{ax + b} = \int \dfrac{x^2}{ax + b}\, dx = \dfrac{1}{a^3}\left[\dfrac{(ax + b)^2}{2} - 2b(ax + b) + \right.$

$\left. b^2\log_\epsilon (ax + b)\right] + C$ (65)

14. $\int \dfrac{dx}{x^2 + a^2} = \int \dfrac{1}{x^2 + a^2}\, dx = \dfrac{1}{a} \tan^{-1} \dfrac{x}{a} + C = -\dfrac{1}{a} \cot^{-1} \dfrac{x}{a} + C$ (66)

15. $\int \dfrac{dx}{x^2 - a^2} = \int \dfrac{1}{x^2 - a^2} dx = \dfrac{1}{2a} \log \dfrac{x - a}{x + a} + C = \dfrac{1}{2a} \log_\epsilon \dfrac{a - x}{a + x} + C$ (67)

16. $\int \dfrac{dx}{\sqrt{a^2 - x^2}} = \int \dfrac{1}{\sqrt{a^2 - x^2}}\, dx = \sin^{-1} (x/a) + C = -\cos^{-1} (x/a) + C$ (68)

17. $\int \dfrac{dx}{\sqrt{x^2 \pm a^2}} = \int \dfrac{1}{\sqrt{x^2 \pm a^2}}\, dx = \log_\epsilon (x + \sqrt{x^2 \pm a^2}) + C$ (69)

18. $\int \sin ax\, dx = -(1/a) \cos ax + C$ (70)

19. $\int \cos ax\, dx = (1/a) \sin ax + C$ (71)

20. $\int \tan ax\, dx = -(1/a) \log_\epsilon \cos ax + C = (1/a) \log_\epsilon \sec ax + C$ (72)

21. $\int \mathrm{cosec}\, ax\, dx = (1/a) \log_\epsilon (\mathrm{cosec}\, ax - \cot ax) + C$ (73)

$= (1/a) \log_\epsilon \tan (ax/2) + C$ (74)

22. $\int \sec ax\, dx = (1/a) \log_\epsilon (\sec ax + \tan ax) + C$ (75)

$= (1/a) \log_\epsilon \tan [(ax/2) + \pi/4] + C$ (76)

23. $\int \cot ax\, dx = (1/a) \log_\epsilon \sin ax + C = -(1/a) \log_\epsilon \mathrm{cosec}\, ax + C$ (77)

24. $\int \sin^2 ax\, dx = x/2 - (1/2a) \sin ax \cos ax + C$ (78)

$= x/2 - (1/4a) \sin 2ax + C$ (79)

25. $\int \cos^2 ax\, dx = x/2 + (1/4a) \sin 2ax + C$ (80)

26. $\int \sinh x\, dx = \cosh x$ (81)

27. $\int \cosh x\, dx = \sinh x$ (82)

Rules to assist integration

1. If the function is the sum of several terms, or can be put into this form, integrate term by term.
2. If the function is in the form of a product or a power, it is usually helpful to multiply out or expand before integrating.
3. Fractions may be either divided out, or written as negative powers.
4. Roots should be treated as fractional powers.
5. A function of x may be replaced by u, and then

$$\int F(x)\, dx = \int F(x) \dfrac{dx}{du}\, du.$$

Areas by integration

If the area under a given portion of a curve is A (Fig. 6.37), then a small increase Δx on the horizontal axis causes an increase ΔA in area, where

$$\Delta A = \Delta x(y + \tfrac{1}{2}\Delta y).$$

As Δx and Δy are made smaller, in the limiting case as Δx approaches zero, the value of $(y + \tfrac{1}{2}\Delta y)$ approaches y,

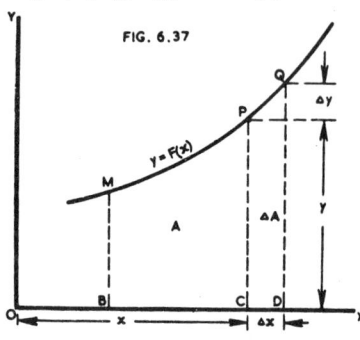

FIG. 6.37

i.e.

$$\underset{\Delta x \to 0}{Lt}\left[\frac{\Delta A}{\Delta x}\right] = y$$

Therefore $\dfrac{dA}{dx} = y = F(x)$

Therefore $dA = y.dx = F(x)\,dx$

Thus $A = \displaystyle\int y\,dx = \int F(x)\,dx$ (83)

The area is therefore given by the integral of the function, over any desired range of values of x.

Example

To find the area under the curve $y = 3x^2$ from $x = 1$ to $x = 4$.

$$A = \int 3x^2\,dx = x^3 + C$$

when $x = 1$, $A = 0$, therefore $x^3 + C = 0$, therefore $C = -1$
when $x = 4$, $A = x^3 - 1 = 4^3 - 1 = 63$.

Definite integrals

When it is desired to indicate the limits in the value of x between which the integral is desired, the integral is written, as for the example above,

$$\int_{x=1}^{x=4} 3x^2\,dx \text{ or } \int_{1}^{4} 3x^2\,dx.$$

These limits are called the **limits of integration**, and the integral is called the **definite integral**. For distinction, the unlimited integral is called the **indefinite integral**.

The definite integral is the difference between the values of the integral for $x = b$ and $x = a$ —

$$\int_{a}^{b} f(x)\,dx = \left[F(x)\right]_{a}^{b} = F(x = b) - F(x = a) \qquad (84)$$

Owing to the subtraction, the constant of integration does not appear in definite integrals.

Special properties of definite integrals

$$\int_{a}^{b} f(x)\,dx = -\int_{b}^{a} f(x)\,dx \qquad (85)$$

$$\int_{a}^{c} f(x)\,dx = \int_{a}^{b} f(x)\,dx + \int_{b}^{c} f(x)\,dx \qquad (86)$$

Examples :

$$\int_{0}^{\pi/2} \sin\theta\,d\theta = \left[-\cos\theta\right]_{0}^{\pi/2} = [-\cos\pi/2 + \cos 0] = 0 + 1 = 1 \qquad (87)$$

$$\int_0^{\pi/2} \cos\theta \, d\theta = \left[\sin\theta \right]_0^{\pi/2} = [\sin \pi/2 - \sin 0] = 1 - 0 = 1 \tag{88}$$

$$\int_0^{\pi} \sin\theta \, d\theta = \left[-\cos\theta \right]_0^{\pi} = [-\cos \pi + \cos 0] = 1 + 1 = 2 \tag{89}$$

$$\int_0^{\pi} \cos\theta \, d\theta = \left[\sin\theta \right]_0^{\pi} = [\sin \pi - \sin 0] = 0 - 0 = 0 \tag{90}$$

$$\int_0^{2\pi} \sin\theta \, d\theta = \left[-\cos\theta \right]_0^{2\pi} = [-\cos 2\pi + \cos 0] = -1 + 1 = 0 \tag{91}$$

$$\int_0^{2\pi} \cos\theta \, d\theta = \left[\sin\theta \right]_0^{2\pi} = [\sin 2\pi - \sin 0] = 0 - 0 = 0 \tag{92}$$

$$\int_0^{\pi} \sin^2 n\theta \, d\theta = \tfrac{1}{2}\left[\theta - (1/2n) \sin 2n\theta \right]_0^{\pi} = \pi/2 \tag{93}$$

$$\int_0^{\pi} \cos^2 n\theta . d\theta = \int_0^{\pi} (1 - \sin^2 n\theta \, d\theta) \, d\theta = \left[\theta \right]_0^{\pi} - \pi/2 = \pi/2 \tag{94}$$

The following may also be derived* where $m \neq n$

$$\int_0^{2\pi} \sin n\theta \, d\theta = 0 \qquad\qquad \int_0^{2\pi} \cos n\theta \, d\theta = 0 \tag{95}$$

$$\int_0^{2\pi} \sin^2 n\theta \, d\theta = \pi \qquad\qquad \int_0^{2\pi} \cos^2 n\theta \, d\theta = \pi \tag{96}$$

$$\int_0^{2\pi} \sin m\theta \cos n\theta \, d\theta = 0 \qquad\qquad \int_0^{2\pi} \sin m\theta \sin n\theta \, d\theta = 0 \tag{97}$$

$$\int_0^{2\pi} \cos m\theta \cos n\theta \, d\theta = 0 \qquad\qquad \int_0^{2\pi} d\theta = 2\pi \tag{98}$$

$$\int_0^{2\pi} \sin n\theta \cos n\theta \, d\theta = 0 \tag{99}$$

Average values by definite integral

The value of the definite integral is the area under the curve between the limits on the horizontal (x) axis. The average value of the height is determined by dividing the area by the length, or in other words the average value of y is determined by dividing the definite integral by the difference in the limiting values of x.

Examples :

1. $y = 3x^2$ from $x = 1$ to $x = 4$.

$$y_{av} = \frac{1}{4-1} \int_1^4 3x^2 \, dx = \frac{1}{3}\left[x^3 \right]_1^4 = \frac{1}{3}\left[64 - 1 \right] = \frac{63}{3} = 21.$$

2. $e = E_m \sin\theta$ from $\theta = 0$ to $\theta = \pi$.

$$E_{av} = E_m \frac{1}{\pi} \int_0^{\pi} \sin\theta \, d\theta = E_m \frac{1}{\pi}\left[-\cos\theta \right]_0^{\pi} = \frac{2}{\pi} E_m.$$

---

*This also applies with limits from k to $(k + 2\pi)$.

This is the average value of a sine wave voltage. It was taken from 0 to π since this is the range over which it is positive. The other half cycle is similar but negative, so that the average over the whole cycle is zero.

3. $(I_{rms})^2 = \dfrac{1}{2\pi} \displaystyle\int_0^{2\pi} (I_m \sin \theta)^2 \, d\theta$

FIG. 6.38

$= \dfrac{I_m^2}{2\pi} \left[\dfrac{\theta}{2} - \dfrac{\sin 2\theta}{4} \right]_0^{2\pi}$

$= \dfrac{I_m^2}{2\pi} [\pi - 0 - 0 + 0]$

$= \dfrac{I_m^2}{2\pi} \cdot \pi = \dfrac{I_m^2}{2}.$

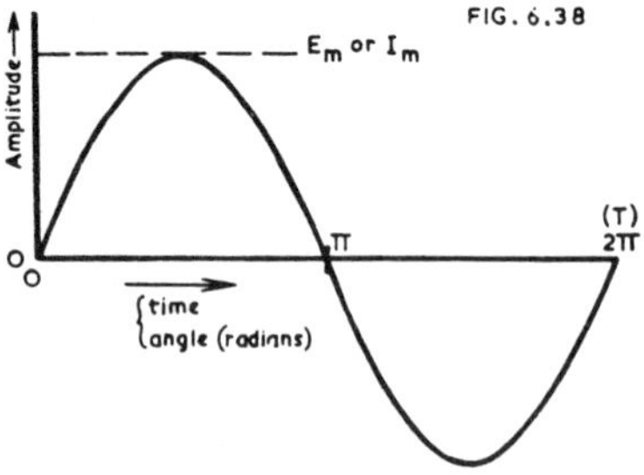

Therefore $I_{rms} = I_m/\sqrt{2} \approx 0.707 \, I_m.$

This is the root mean square value of the current with sine waveform.

(iv) Taylor's Series

If $f(x)$ be a function with first derivative $f'(x)$, second derivative $f''(x)$, third derivative $f'''(x)$, etc., and if the function and its first n derivatives are finite and continuous from $x = a$ to $x = b$, then the following expansion holds true in the interval from $x = a$ to $x = b$:

$$f(x) = f(a) + \frac{x - a}{1!} f'(a) + \frac{(x - a)^2}{2!} f''(a) + \frac{(x - a)^3}{3!} f'''(a)$$

$$+ \ldots + \frac{(x - a)^{n-1}}{(n - 1)!} f^{n-1}(a) + \frac{(x - a)^n}{n!} f^n(x_n) \tag{100}$$

where $a < x_n < b.$

The final term is called the **remainder** ; if this can be made as small as desired by making n sufficiently large, the series becomes a convergent infinite series, converging to the value $f(x).$

Another form of Taylor's Series is :

$$f(a + h) = f(a) + \frac{h}{1!} f'(a) + \frac{h^2}{2!} f''(a) + \frac{h^3}{3!} f'''(a) + \ldots$$

$$\ldots + \frac{h^{n-1}}{(n - 1)!} f^{n-1}(a) + \frac{h^n}{n!} f^n(a) + \ldots \tag{101}$$

The sum of the first few terms of Taylor's Series gives a good approximation to $f(x)$ for values of x near $x = a.$

Examples of the use of Taylor's Series :

(1) To expand $\sin (a + h)$ in powers of $h.$
    $f(a) = \sin a$ and $f(h) = \sin h$

Differentiating in successive steps we get
    $f'(a) \quad = \cos a$
    $f''(a) \quad = - \sin a$
    $f'''(a) \quad = - \cos a$
    $f''''(a) = \sin a$

Applying the alternative form of Taylor's Series,

$$\sin (a + h) = \sin a + h \cos a - \frac{h^2}{2!} \sin a - \frac{h^3}{3!} \cos a + \frac{h^4}{4!} \sin a + \ldots \tag{102}$$

(2) Similarly

$$\cos (a + h) = \cos h - h \sin a - \frac{h^2}{2!} \cos a + \frac{h^3}{3!} \sin a + \frac{h^4}{4!} \cos a + \ldots \tag{103}$$

(v) Maclaurin's Series

Maclaurin's Series is a special case of Taylor's Series where $a = 0$.

$$f(x) = f(0) + \frac{f'(0)x}{1!} + \frac{f''(0)x^2}{2!} + \frac{f'''(0)x^3}{3!} + \ldots \qquad (104)$$

The sum of the first few terms of Maclaurin's Series gives a good approximation to $f(x)$ for values of x near $x = 0$.

Example of the use of Maclaurin's Series :

$$f(x) = \cos x$$

then $\cos x = f(0) + \dfrac{f'(0)x}{1!} + \dfrac{f''(0)x^2}{2!} + \dfrac{f'''(0)x^3}{3!} + \dfrac{f''''(0)x^4}{4!}$

where $\quad f(0) = \cos 0 = 1$
$f'(0) = -\sin 0 = 0$
$f''(0) = -\cos 0 = -1$
$f'''(0) = \sin 0 = 0$
$f''''(0) = \cos 0 = 1$

The series may then be written down as

$$\cos x = 1 - \frac{x^2}{2!} + \frac{x^4}{4!} - \frac{x^6}{6!} + \frac{x^8}{8!} - \ldots \qquad (105)$$

Similarly

$$\sin x = x - \frac{x^3}{3!} + \frac{x^5}{5!} - \frac{x^7}{7!} + \frac{x^9}{9!} - \ldots \qquad (106)$$

$$j \sin x = j(x - \frac{x^3}{3!} + \frac{x^5}{5!} - \ldots)$$

$$\cos x + j \sin x = \left(1 - \frac{x^2}{2!} + \frac{x^4}{4!} - \frac{x^6}{6!} + \ldots\right) + j\left(x - \frac{x^3}{3!} + \frac{x^5}{5!} - \frac{x^7}{7!} + \ldots\right)$$

$$= 1 + jx + \frac{j^2 x^2}{2!} + \frac{j^3 x^3}{3!} + \frac{j^4 x^4}{4!} + \ldots \qquad (107)$$

Also it may be proved that

$$\log_\epsilon (1 + x) = x - \frac{x^2}{2} + \frac{x^3}{3} - \frac{x^4}{4} + \ldots \qquad (108)$$

and $\qquad \epsilon^{jx} = 1 + jx + \dfrac{j^2 x^2}{2!} + \dfrac{j^3 x^3}{3!} + \dfrac{j^4 x^4}{4!} + \ldots$

$$= \cos x + j \sin x \text{ (see eqn. 107)} \qquad (109)$$

SECTION 8 : FOURIER SERIES AND HARMONICS

(i) *Periodic waves and the Fourier Series*   (ii) *Other applications of the Fourier Series*   (iii) *Graphical Harmonic Analysis.*

(i) Periodic waves and the Fourier Series

The equation for any periodic wave can be written by substituting the correct values in Fourier's Series :

$$y = F(\theta) = B_0/2 + A_1 \sin \theta + A_2 \sin 2\theta + A_3 \sin 3\theta + \ldots$$
$$+ B_1 \cos \theta + B_2 \cos 2\theta + B_3 \cos 3\theta + \ldots \qquad (1)$$

where $B_0/2$ is a constant which is zero if the wave is balanced about the x axis ; its value is the average value of y over one cycle and may be determined by putting $n = 0$ in the expression for B_n below,

$$A_n = \frac{1}{\pi} \int_0^{2\pi} F(\theta) \sin n\theta \, d\theta \text{ where } n = 1,2,3, \text{ etc.}$$

$$= 2 \times \text{average value of } F(\theta) \sin n\theta \text{ taken over 1 cycle,}$$

$$B_n = \frac{1}{\pi} \int_0^{2\pi} F(\theta) \cos n\theta \, d\theta \text{ where } n = 0,1,2,3, \text{ etc.}$$

$$= 2 \times \text{average value of } F(\theta) \cos n\theta \text{ taken over 1 cycle,}$$

and
$$\begin{aligned} \theta &= \omega t = 2\pi f t & : \quad f = \text{fundamental frequency} \\ 2\theta &= 2\omega t = 2\pi(2f)t & : \quad (2f) = \text{second harmonic frequency} \\ 3\theta &= 3\omega t = 2\pi(3f)t & : \quad (3f) = \text{third harmonic frequency} \\ & \quad\quad \text{etc.} \end{aligned}$$

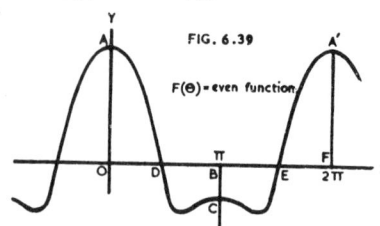

FIG. 6.39

$F(\theta)$ = even function

Special cases
(1) $F(\theta)$ is an even function

If the waveform is symmetrical about the y axis (e.g. Fig. 6.39), $F(\theta)$ is called an even function and $A_n = 0$, giving the simplified form

$$F(\theta) = B_o/2 + B_1 \cos\theta + B_2 \cos 2\theta + B_3 \cos 3\theta + \ldots \tag{2}$$

This is the equation which applies to all types of distortion introduced by valves.

(2) $F(\theta)$ is an odd function

If the waveform is such that the value of y is equal in magnitude but opposite in sign for plus and minus values of x (e.g. Fig. 6.40), $F(\theta)$ is called an odd function and $B_n = 0$, giving the simplified form :

$$F(\theta) = A_1 \sin\theta + A_2 \sin 2\theta + A_3 \sin 3\theta + \ldots \tag{3}$$

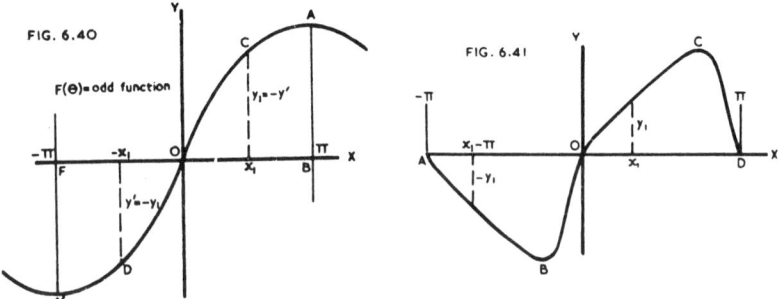

FIG. 6.40

$F(\theta)$ = odd function

FIG. 6.41

This is the equation for the condition when the fundamental and all the harmonics commence together at zero.

(3) $F(\theta) = -F(\theta \pm \pi)$

If the waveform (Fig. 6.41) is such that the value of y is equal in magnitude but opposite in sign for $x = x_1$ and $x = (x_1 \pm \pi)$, the expansion contains only odd harmonics :

$$\begin{aligned} F(\theta) = A_1 \sin\theta &+ A_3 \sin 3\theta + A_5 \sin 5\theta + \ldots \\ &+ B_1 \cos\theta + B_3 \cos 3\theta + B_5 \cos 5\theta + \ldots \end{aligned} \tag{4}$$

(4) $F(\theta)$ is an even function, with the positive and negative portions identical and symmetrical (Fig. 6.42) :

$$F(\theta) = B_1 \cos\theta + B_3 \cos 3\theta + B_5 \cos 5\theta + \ldots \tag{5}$$

or if the origin is taken at A,

$$F(\theta) = B_1 \sin\theta - B_3 \sin 3\theta + B_5 \sin 5\theta - \ldots \tag{6}$$

This is the equation for a balanced push-pull amplifier.

The general equation (1) can also be expressed in either of the alternative forms :

$$F(\theta) = B_o/2 + C_1 \sin(\theta + \phi_1) + C_2 \sin(2\theta + \phi_2) + \ldots \tag{7}$$

$$F(\theta) = B_o/2 + C_1 \cos (\theta - \phi_1') + C_2 \cos (2\theta - \phi_2') + \ldots \tag{8}$$

where $C_n = \sqrt{A_n^2 + B_n^2}$

and $\tan \phi_n = B_n/A_n$; $\tan \phi_n' = A_n/B_n$.

The angles $\phi_1, \phi_2, \ldots$ in eqn. (7) are· the angles of lead between the harmonics of the sine series and the corresponding sine components in eqn. (1). The angles $\phi_1', \phi_2', \ldots$ in eqn. (8) are the angles of lag between the harmonics of the cosine series and the corresponding cosine components in eqn. (1). All the angles ϕ and ϕ' in equations (7) and (8) are measured on the scales of angles for the harmonics

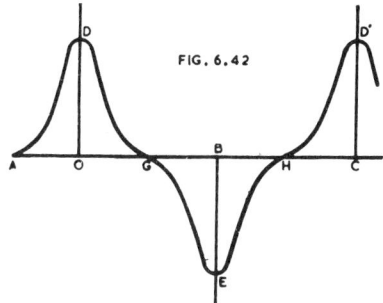

FIG. 6.42

Harmonic composition of some common periodic waves (Fig. 6.43)

Square wave (A)

$$y = \frac{4E}{\pi}\Big(\cos \theta - \frac{\cos 3\theta}{3} + \frac{\cos 5\theta}{5} - \frac{\cos 7\theta}{7} + \ldots\Big) \tag{9}$$

Triangular wave (B)

$$y = \frac{8E}{\pi^2}\Big(\cos \theta + \frac{\cos 3\theta}{9} + \frac{\cos 5\theta}{25} + \ldots\Big) \tag{10}$$

Sawtooth wave (C)

$$y = \frac{2E}{\pi}\Big(\sin \theta - \frac{\sin 2\theta}{2} + \frac{\sin 3\theta}{3} - \frac{\sin 4\theta}{4} + \ldots\Big) \tag{11}$$

FIG. 6.43

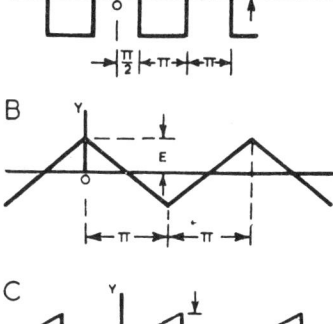

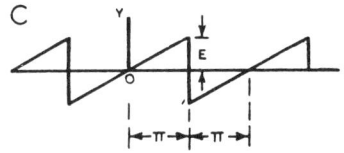

Short rectangular pulse (D)

$$y = E\left[k + \frac{2}{\pi}\left(\sin k\pi \cos \theta + \frac{\sin 2k\pi \cos 2\theta}{2} + \ldots \right. \right.$$
$$\left.\left. + \frac{\sin nk\pi \cos n\theta}{n} + \ldots\right)\right] \qquad (12)$$

Half-wave rectifier output (E)

$$y = \frac{E}{\pi}\left(1 + \frac{\pi \cos \theta}{2} + \frac{2 \cos 2\theta}{3} - \frac{2 \cos 4\theta}{15} + \frac{2 \cos 6\theta}{35} - \ldots \right.$$
$$\left. \ldots (-1)^{n/2+1} \frac{2}{n^2 - 1} \cos n\theta \ldots \right) (n \text{ even}) \qquad (13)$$

Full wave rectifier output (F)

$$y = \frac{2E}{\pi}\left(1 + \frac{2 \cos 2\theta}{3} - \frac{2 \cos 4\theta}{15} + \frac{2 \cos 6\theta}{35} - \ldots \right.$$
$$\left. \ldots (-1)^{n/2+1} \frac{2 \cos n\theta}{n^2 - 1} \ldots \right) (n \text{ even}) \qquad (14)$$

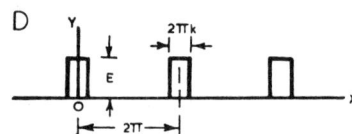

Fig. 6.43

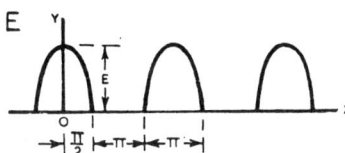

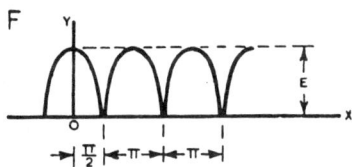

(ii) Other applications of the Fourier Series

The Fourier Series is particularly useful in that it may be applied to functions having a finite number of discontinuities within the period, such as rectangular and saw-tooth periodic pulses.

The Fourier Series may be put into the exponential form, this being useful when the function lacks any special symmetries.

The Fourier Series may also be applied to non-periodic functions.

For information on these applications, see the list of references—Sect. 9(B).

(iii) Graphical Harmonic Analysis

Any irregular waveform may be analysed to determine its harmonic content, and the general method is to divide the period along the X axis into a suitable number of divisions (e.g. Fig. 6.44 with 24 ordinates), the accuracy increasing with the number of divisions.

Ordinates are drawn at each point on the X axis and the height of each ordinate is measured. The minimum number of ordinates over the cycle must be at least twice the power of the highest harmonic which it is desired to calculate. Various methods for carrying out the calculations have been described. Some are based on equal divisions of time (or angle) while others are on equal divisions of voltage.

In the harmonic analysis of the distortion introduced by valves on resistive loads, it is possible to make use of certain properties which simplify the calculations :

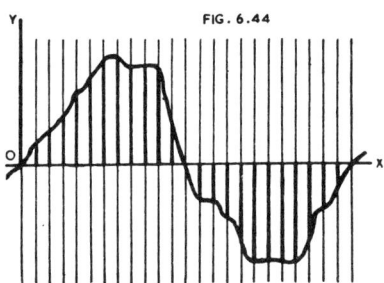

FIG. 6.44

(1) All such distortion gives a waveform which is symmetrical on either side of the vertical lines (ordinates) at the positive and negative peaks.

(2) It is therefore only necessary to analyse over half the cycle, from one positive peak to the following negative peak, or *vice versa*.

(3) Even harmonic distortion results in positive and negative half cycles of different shape and area, thus causing a steady ("rectified") component.

(4) Odd harmonic distortion results in distorted waveform, but with the positive and negative half cycles similar in shape.

(5) Even harmonics are in phase with the positive fundamental peak, and out of phase with the negative peak, or *vice versa*; they are always maxima when the fundamental is zero.

(6) Odd harmonics are always exactly in phase or 180° out of phase with both positive and negative fundamental peaks, and are zero when the fundamental is zero.

The relative phases of the fundamental (H_1) and the harmonics (up to H_5) are shown in Fig. 6.45. The fundamental and third, fifth and higher order odd harmonics have zero amplitude at 0°, 180° and 360° on the fundamental scale. The second,

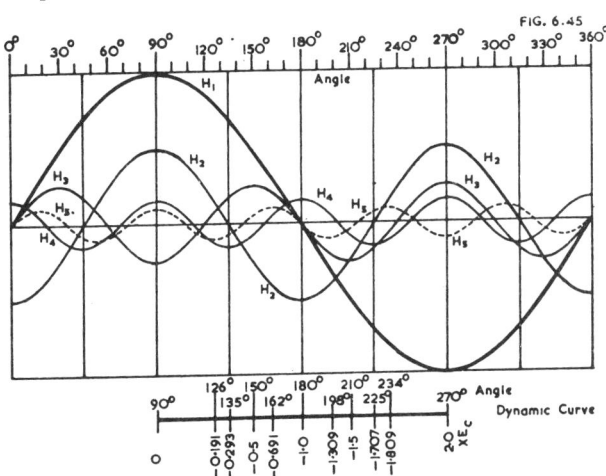

FIG. 6.45

fourth, and higher order even harmonics reach their maximum values (either positive or negative) at 0°, 90°, 180°, 270° and 360° on the fundamental scale.

The amplitudes of the harmonics as drawn in Fig. 6.45 have been exaggerated for convenience in drawing, while their relative magnitudes are quite arbitrary. Their relative phases are, however, quite definite.

In proceeding with Graphical Harmonic Analysis it may be shown that it is possible to select **thirteen points** on the X axis which will enable the exact values of the first, second, third and fourth harmonics to be calculated (Ref. C9) on the assumption that there are no harmonics of higher order than the fifth, or that these are negligibly small. These points are limited to the range from 90° to 270° on the fundamental scale, as in Fig. 6.45. They may be expressed in terms of the grid voltage E_c, the static operating point being $-E_c$ and the operating point swinging from $E_c = 0$ on the one side to $2E_c$ on the other side.

It is only necessary to determine the plate currents at the specified grid voltages, to insert these into the formulae given in the article and to calculate the values of the harmonics.

The preceding exact method has been approximated by R.C.A. (Ref. C10) to give greater ease in handling. In the approximation there are **eleven specified points** in place of 13 in the exact form, the values of the grid voltages being (see Fig. 6.45) : 0 ; $-0.191E_c$; $-0.293E_c$; $-0.5E_c$; $-0.691E_c$; $-E_c$; $-1.309E_c$; $-1.5E_c$; $-1.707E_c$; $-1.809E_c$; $-2.0E_c$.

These have been approximated by R.C.A. to the nearest decimal point, and the approximate values have been used in the " eleven selected ordinate method " of Chapter 13 Sect. 3(iv)D and Fig. 13.24.

The equation giving the second harmonic distortion—eqn. (28) in Chapter 13 Sect. 3(iv)—only requires the values of plate current at three points. This is an exact form and is used for triodes in Chapter 13 Sect. 2(i) eqns. (6) to (7b) inclusive and Fig. 13.2.

The " five selected ordinate method," described in Chapter 13 Sect. 3(iv)A and used for calculating second and third harmonic distortion in pentodes, is exact provided that there is no harmonic higher than the third. It is, however, a very close approximation under all normal conditions. The same remarks also apply to the **simple method for calculating third harmonic distortion** in balanced push-pull amplifiers, described in Chapter 13 Sect. 5(iii) eqn. (23) and Fig. 13.37.

An alternative method, based on **equal grid voltage divisions**, has been devised by Espley (Ref C6). This gives harmonics up to one less than the number of voltage points. Two applications are described in Chapter 13 Sect. 3(iv)—**five ordinates** giving second, third and fourth harmonic distortion, and **seven ordinates** giving up to sixth harmonic distortion.

When the loadline is a closed loop, as occurs with a partially reactive load, these conditions and equations do not apply, or are only approximated.

SECTION 9 : REFERENCES

(A) HELPFUL TEXTBOOKS ON MATHEMATICS FOR RADIO
Elementary :
Sawyer, W. W. " Mathematician's Delight " (a Pelican Book, A121, published by Penguin Books, England and U.S.A. 1943). Possibly the best introduction to mathematics.
Cooke, N. M. and J. B. Orleans, " Mathematics essential to electricity and radio " (McGraw-Hill, New York and London, 1943). Highly recommended for general use. 418 pages.
Everitt, W. L. (Editor) " Fundamentals of radio " (Prentice-Hall Inc. New York, 1943). Chapter 1 only.
" Radio Handbook Supplement " (The Incorporated Radio Society of Great Britain, London, 2nd ed., 1942). Chapters 2, 9.
" Radio Handbook " (10th edit., Editors and Engineers, Los Angeles, California). Chapter 28. (Not 11th edition.)
Basic :
Colebrook, F. M. " Basic Mathematics for Radio Students " (Iliffe and Sons Ltd., London, 1946).
Complete textbooks, commencing from elementary level :
Dull, R. W. " Mathematics for Engineers " (McGraw-Hill Book Co., New York and London, 2nd edit. 1941). Covers whole ground. 780 pages.
Cooke, N. M. " Mathematics for Electricians and Radiomen " (McGraw-Hill Book Coy., New York and London, 1942). Generally at lower level than R. W. Dull and less comprehensive. Useful for those with limited mathematical background. 604 pages.

Smith, Carl E. " Applied Mathematics for Radio and Communication Engineers " (McGraw-Hill Book Coy., New York and London, 1945). Excellent treatment in limited space. 336 pages.
Wang, T. J. " Mathematics of Radio Communications " (D. van Nostrand Coy., New York, 1943). 371 pages.
Rose, W. N. " Mathematics for Engineers " Parts 1 and 2 (Chapman and Hall Ltd., London, 2nd edition 1920).
Textbooks commencing at higher level :
Warren, A. G. " Mathematics Applied to Electrical Engineering " (Chapman and Hall, London, 1939). 384 pages.
Sokolnikoff, I. S. and E. S. " Higher Mathematics for Engineers and Physicists " (McGraw-Hill Book Coy., New York and London, 2nd edit., 1941). 587 pages.
Toft, L., and A. D. D. McKay " Practical Mathematics " (Sir Isaac Pitman and Sons Ltd., London, 2nd edit., 1942). 612 pages.
Jaeger, J. C. " An Introduction to Applied Mathematics " (Oxford University Press).
Limited application :
Sturley, K. R. " Radio Receiver Design " Part 1 (Chapman and Hall, London, 1943). Appendix 1A— j notation ; 2A—Fourier Series.
Golding, E. W., " Electrical Measurements and Measuring Instruments " (Sir Isaac Pitman and Sons Ltd., London, 3rd edit., 1944). Chapter 15—Wave forms and their determination.
Lawrence, R. W. " Principles of Alternating Currents " (McGraw-Hill Book Coy., New York and London, 2nd ed., 1935). Chapter 4—Non-sinusoidal waves.
Eshbach, O. W. " Handbook of Engineering Fundamentals " (John Wiley and Sons, New York ; Chapman and Hall, London, 1944). Section 1—Mathematical and Physical Tables : Section 2— Mathematics.
Also other handbooks.

(B) REFERENCES TO FOURIER ANALYSIS
[see also references under (C) Graphical Harmonic Analysis]
B1. Hallman, L. B. " A Fourier analysis of radio-frequency power amplifier wave forms," Proc. I.R.E. 20.10 (Oct. 1932) 1640.
B2. Lockhart, C. E. " Television waveforms—an analysis of saw-tooth and rectangular waveforms encountered in television and cathode ray tube practice, " Electronic Eng. 15.172 (June, 1942) 19.
B3. Williams, H. P. " Fourier analysis by geometrical methods," W.E. 21.246 (March, 1944) 108.
B4. de Holzer, R. C. " The harmonic analysis of distorted sine waves," Electronic Eng. 17.208 (June, 1945) 556 ; 17.209 (July, 1945) 606.
B5. Moss, H. " Complex waveforms," Electronic Eng. The harmonic synthesiser, 18.218 (April, 1946) 113 ; Analysis of complex waves, 18.220 (June, 1946) 179 ; 18.222 (Aug., 1946) 243.
B6. Furst, U. R. " Harmonic analysis of overbiased amplifiers," Elect. 17.3 (March, 1944) 143. Gives curves of harmonics of idealized straight characteristic and sharp angle cut-off.
B7. Espley, D. C., " Harmonic Analysis by the method of central differences " Phil. Mag. 28.188 (Sept., 1939) 338.

(C) REFERENCES TO GRAPHICAL HARMONIC ANALYSIS
Books
C1. Manley, R. G. " Waveform analysis " (Chapman and Hall, London, 1945) and most radio and electrical engineering text books.
Articles
Equal time divisions
C2. Kemp, P. " Harmonic analysis of waves containing odd and even harmonics," Electronic Eng. 15.172 (June, 1942) 13 (12 ordinates per cycle).
C3. Denman, R. P G. " 36 and 72 ordinate schedules for general harmonic analysis," Elect. 15.9 (Sept., 1942) 44. Correction 16.4 (April, 1943) 214.
C4. Cole, L. S. " Graphical analysis of complex waves," Elect. 18.10 (Oct., 1945) 142. (6, 8 and 12 points per cycle.)
C5. Levy, M. M. " Fourier Series," Jour. Brit. I.R.E. 6.2 (March-May, 1946) 64. (Calculations up to 160 harmonics)
and many other references.

Equal voltage divisions
C6. Espley, D. C. " The calculation of harmonic production in thermionic valves with resistive loads," Proc. I.R.E. 21.10 (Oct., 1933) 1439.

Selected ordinates
C7. Hutcheson, J.-A. " Graphical Harmonic Analysis," Elect. 9.1 (Jan., 1936) 16 (odd and even harmonics in amplifiers).
C8. Chaffee, E. L. " A simplified harmonic analysis," Review of Scientific Instruments, 7 (Oct., 1936) 384 (Gives 5, 7, 9, 11 and 13 point analysis).
C9. Mouromtseff, I. E., and H. N. Kozanowski," A short-cut method for calculation of harmonic distortion in wave modulation," Proc. I.R.E. 22.9 (Sept., 1934) 1090.
C10. R.C.A. Application Note " Use of the plate family in vacuum tube power output calculations " No. 78 (July, 1937).

CHAPTER 7

NEGATIVE FEEDBACK

By F. Langford-Smith, B.Sc., B.E.

SECTION 1 : FUNDAMENTAL TYPES OF FEEDBACK

(i) Feedback, positive and negative (ii) Negative voltage feedback at the mid-frequency (iii) Negative current feedback at the mid-frequency (iv) Bridge negative feedback at the mid-frequency (v) Combined positive and negative feedback at the mid-frequency (vi) Comparison between different fundamental types at the mid-frequency.

(i) Feedback, positive and negative

Feedback may be applied to any amplifier, at any frequency, and may be either positive or negative. The application is illustrated in Fig. 7.1 where an amplifier, with voltage amplification A, develops a voltage E_0 across its load. Portion of the voltage across the load (βE_0) is fed back in series with the input terminals, so that the input voltage E_i' differs from the voltage (E_i) across the input terminals of the amplifier itself. It is obvious that

$$E_0 = AE_i \text{ and that } E_i' = E_i \pm \beta E_0,$$

the sign in front of βE_0 taking account of phase relationships. The quantity β is sometimes called the transfer coefficient.

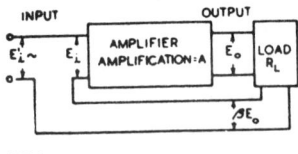

Fig. 7.1. *Block diagram of an amplifier with feedback.*

FIG. 7·1

The effective overall amplification with feedback is therefore

$$A' = \frac{E_0}{E_i'} = \frac{E_0}{E_i \pm \beta E_0}.$$

If the effect of feedback is to increase the gain, the feedback is positive* ; if it decreases the gain, the feedback is negative. Positive feedback is used to convert

---

*If βE_0 is much greater than E_i, then the gain will theoretically be reduced irrespective of the sign in front of βE_0.

an amplifying valve into an oscillator (see Chapter 24).   Negative feedback is used mainly in amplifiers, both at radio and audio frequencies, although this Handbook covers only low frequency applications.

The voltage βE_0 may be proportional to E_0, in which case it is described as **voltage feedback** ; or it may be proportional to the current through the load, when it is called **current feedback**.   These two are, of course, identical if the load is a constant resistance, since the voltage and current are then proportional.

Bridge feedback is a combination of voltage and current feedback.

A network may be inserted in the feedback circuit to change the phase of the feedback voltage, or to change its magnitude and/or phase so as to discriminate between different frequencies.

Use of symbols

Black (Ref. A1) and others use the symbols $\mu\beta$ to indicate the same quantity which is here indicated by βA.   Some authorities use the symbols $A\beta$, but in certain equations (e.g those involving effective plate resistance) it is necessary to introduce $\beta\mu$ or $\mu\beta$ where μ is the amplification factor of the final valve.   The latter form might be confused with Black's $\mu\beta$ which has an entirely different significance.   Hence the use of βA in this chapter.   A further, but minor, advantage from the use of βA is that the beginner, in considering operation at the mid-frequency, is able to regard the magnitude for β as being simply a fraction.

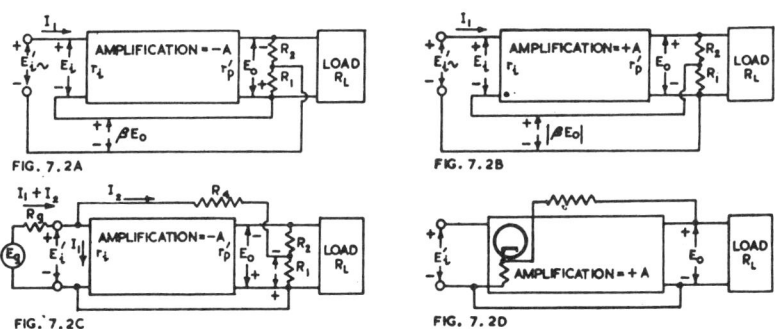

Fig. 7.2.   Block diagrams of amplifiers with negative feedback.   It is assumed that $(R_1 + R_2)$ has a resistance very much greater than the load impedance ; (A) and (B) have feedback external to the amplifier connected in series with the input voltage (C) has feedback connected in shunt with the input voltage (D) has the feedback connected to a cathode in the amplifier.

(ii) Negative voltage feedback at the mid-frequency

Block diagrams of two conventional forms for applying negative voltage feedback are shown in Figs. 7.2A and B.   The polarities marked on the diagrams are instantaneous values at an arbitrary time.   In each case the feedback voltage is derived from a voltage divider $R_1 R_2$ across the output and is **applied in series with the input voltage** so as to give negative feedback.   In Fig. 7.2A the polarity of the output voltage is opposite that of the input voltage—this is regarded in network analysis as being equivalent to an amplification of $-A$.   The amplifier of Fig. 7.2A may consist, for example, of one, or any other odd number of resistance-coupled stages.

In Fig. 7.2B the polarity of the output voltage is the same as that of the input voltage—this being regarded as equivalent to an amplification of $+A$.   In Fig. 7.2B the feedback voltage is marked $|\beta E_0|$ thus indicating the magnitude of the voltage and leaving its polarity to be indicated by the positive and negative signs.

In this section the treatment is limited to the mid-frequency at which the amplifier has maximum gain and zero phase shift.

The combined amplifier and feedback circuit, from input back to input, is called the **feedback loop**.

The magnitude of β is equal to $R_1/(R_1 + R_2)$. In all cases, for negative feedback, βA must be negative and the feedback network must be arranged to give this result.

An alternative method of applying the feedback voltage to produce degeneration (Fig. 7.2C) connects the **feedback voltage in shunt with the input voltage**. In this case the amplification must be negative, that is to say there must be an odd number of stages unless a transformer is incorporated (Ref. A17), and β must therefore be positive.

A still further variation is **when the feedback voltage is applied to the cathode of one of the stages in the amplifier** (Fig. 7.2D). In this case the amplification must be positive, for example an even number of r.c.c. stages with the feedback returned to the cathode of the first stage, which must have some impedance between cathode and earth. Alternatively the feedback may be returned to the cathode of the penultimate stage of a multi-stage r.c.c. amplifier. The value of β should be considered as negative on account of the method of connection of the feedback loop.

The following treatment is based on the circuit of Fig. 7.2B but the results may also be applied to any amplifier with the feedback applied externally to the amplifier and in series with the input voltage. It is understood that in all cases the polarity of the feedback voltage is arranged to give negative feedback.

(A) Gain without feedback
$$A = E_0/E_i \tag{1}$$
and if feedback is applied and the input voltage increased to E_i' to give the same output voltage E_0 as without feedback, then follows

(B) Gain with feedback
$$A' = \frac{E_0}{E_i'} = \frac{E_0}{E_i - \beta E_0} \tag{2a}$$
Combining (1) and (2a),
$$A' = \frac{A}{1 - \beta A} = \frac{1}{(1/A) - \beta} = \frac{A}{1 + [AR_1/(R_1 + R_2)]} \tag{2b}$$
(for chart showing relationship between A' and A see Ref. A21), and if $|\beta A| \gg 1$
then $A' \approx 1/(-\beta)$ \hfill (2c)

The value of β is given by
$$\beta = - R_1/(R_1 + R_2) \tag{3a}$$
The effect of feedback on gain is therefore
$$\frac{A}{A'} = \frac{E_0/E_i}{E_0/(E_i - \beta E_0)} = 1 - \frac{\beta E_0}{E_i} = 1 - \beta A \tag{3b}$$

The quantity βA is negative for negative feedback ; $A/A' = 1 - \beta A$ is always greater than unity.

The quantity $(1 - \beta A)$ is called the **feedback factor***. The degree of feedback is indicated by the reduction in gain. For example, an amplifier with " 20 db feedback " is one in which gain has been reduced 20 db by feedback.

In the general case, when the phase angle displacement is not restricted,
$$A' = \frac{A}{\sqrt{1 + |\beta A|^2 - 2|\beta A| \cos \phi}} \tag{3c}$$
[which becomes $A' = A/(1 - \beta A)$ when $\phi = 180°$]
where ϕ = phase angle displacement of amplifier and feedback circuit loop.

(C) Effect of feedback on harmonic distortion†
Let the input voltage E_i' be expressed in the form $E_{im}' \cos \omega t$.
Then the output voltage (E_0) will be given by
$$E_{0m} \cos \omega t + E_{2m} \cos 2\omega t + E_{3m} \cos 3\omega t \tag{4}$$

---

*Some authorities call βA the feedback factor, and care should therefore be taken in the use of this term.

†Treatment adapted from K. R. Sturley.

where E_{0m} = peak value of fundamental frequency,

$\quad E_{2m}$ = peak value of second harmonic,

and $\quad E_{3m}$ = peak value of third harmonic.

The feedback voltage $(-\beta E_0)$ will therefore be

$\quad -\beta(E_{0m} \cos \omega t + E_{2m} \cos 2\omega t + E_{3m} \cos 3\omega t)$.

The voltage applied to the input of the amplifier will be

$E_i' + \beta E_0$

$= E_{im}' \cos \omega t + \beta(E_{0m} \cos \omega t + E_{2m} \cos 2\omega t + E_{3m} \cos 3\omega t)$

$= (E_{im}' + \beta E_{0m}) \cos \omega t + \beta E_{2m} \cos 2\omega t + \beta E_{3m} \cos 3\omega t$.

The output voltage will therefore be

AE_i + harmonic distortion

$= A[(E_{im}' + \beta E_{0m}) \cos \omega t + \beta E_{2m} \cos 2\omega t + \beta E_{3m} \cos 3\omega t]$

$\quad + A(E_{im}' + \beta E_{0m})(H_2 \cos 2\omega t + H_3 \cos 3\omega t)$ ⟨(5)⟩

where H_2 = ratio second harmonic to fundamental voltages in amplifier without feedback,

and $\quad H_3$ = ratio third harmonic to fundamental voltages in amplifier without feedback.

But we already have the output voltage in (4) above.

Equating fundamental components in (4) and (5),

$E_{0m} = A(E_{im}' + \beta E_{0m})$

$E_{im}' = (1 - \beta A)E_{0m}/A$ ⟨(6)⟩

Equating the second harmonic components in (4) and (5),

$E_{2m} = \beta AE_{2m} + A(E_{im}' + \beta E_{0m})H_2$.

Inserting the value of E_{im}' from (6) above,

$E_{2m} = \beta AE_{2m} + [(1 - \beta A) E_{0m} + \beta AE_{0m}]H_2$

Therefore $E_{2m} (1 - \beta A) = E_{0m}H_2$

Therefore $\dfrac{E_{2m}}{E_{0m}} = \dfrac{H_2}{1 - \beta A}$ ⟨(7)⟩

Similarly for the third or any higher harmonic,

$\dfrac{E_{3m}}{E_{0m}} = \dfrac{H_3}{1 - \beta A}$ ⟨(8)⟩

That is, the magnitudes of all the harmonics (and of course the intermodulation products) introduced by the amplifier are reduced by negative voltage feedback in the same proportion that the gain is reduced.

This result is based on the assumptions :

(1) that E_0 is exactly 180° out of phase, or exactly in phase, with E_i

(2) that the amplification A is the same for the harmonics as for the fundamental

(3) that the intermodulation voltages are negligibly small. The presence of intermodulation* has no effect on the second harmonic, but in certain cases the higher order harmonics can increase owing to a small amount of negative feedback ;· when the feedback is raised sufficiently these harmonics will decrease again. When $H_2 = 0$, H_3 with feedback will be as indicated by eqn. (8). In any case, the effect is small if the amplifier distortion is initially fairly small.

(D) Effect of feedback on hum or noise introduced by the amplifier

Following a similar method—

Input voltage $\quad = E_i' = E_{im}' \cos \omega t$

Output voltage $\quad = E_0 = E_{0m} \cos \omega t + E_{nm}' \cos \omega_n t$ ⟨(9)⟩

$\quad$ where E_{nm}' = peak value of noise or hum output voltage introduced by the amplifier with feedback,

$\quad$ and ω_n = $2\pi \times$ frequency of noise or hum voltage.

---

*Tellegen, B. D. H., and V. C. Henriquez " Inverse feed-back," W.E. 14.167 (Aug. 1937) 409— quoting R. Feldtkeller ; also correspondence by R. W. Sloane and J. Frommer " Distortion in negative feedback amplifiers " W.E. 14.164 (May 1937) 259 ; 14.166 (July 1937) 369 ; 14.170 (Nov. 1937) 607 ; 15.172 (Jan. 1938) 20.

The feedback voltage is therefore
$$- \beta E_0 = - \beta(E_{0m} \cos \omega t + E_{nm}' \cos \omega_n t).$$
The voltage applied to the input of the amplifier is
$$E_i = E_i' + \beta E_0 = E_{im}' \cos \omega t + \beta(E_{0m} \cos \omega t + E_{nm}' \cos \omega_n t)$$
$$= (E_{im}' + \beta E_{0m}) \cos \omega t + \beta E_{nm}' \cos \omega_n t.$$
The output voltage is therefore
$$E_0 = AE_i + \text{noise voltage without feedback}$$
$$= A(E_{im}' + \beta E_{0m}) \cos \omega t + \beta A E_{nm}' \cos \omega_n t + E_{nm} \cos \omega_n t \qquad (10)$$
Equating the fundamental components in (9) and (10),
$$E_{0m} = A(E_{im}' + E_{0m}).$$
Therefore $E_{im}' = (1 - \beta A) E_{0m}/A$ \qquad (11)
Equating the noise components in (9) and (10),
$$E_{nm}'(1 - \beta A) = E_{nm}$$
$$\frac{E_{nm}'}{E_{nm}} = \frac{1}{1 - \beta A} \qquad (12)$$

Thus the feedback reduces noise or hum voltages introduced by the amplifier in the same proportion that the gain is reduced, provided the following conditions are fulfilled :

(1) That the signal input voltage is increased with feedback, to maintain the signal output voltage constant.
(2) That β and A have the same values for the hum (or noise) frequency as for the signal frequency.
(3) The voltage fed back does not include any voltages other than the fraction of the amplifier output voltage. In many circuit arrangements this condition is not satisfied ; the most important of these are covered in Sect. 2.

When all noise is considered as originating at the input (e.g. thermal noise), the signal-to-noise ratio is unchanged by the feedback.

(E) Effect of voltage feedback on the output resistance of the amplifier

Every linear network which has a pair of output terminals may, as regards its external effects, be replaced by its equivalent open-circuit voltage in series with its equivalent resistance.

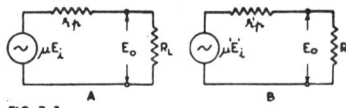

FIG. 7·3

Fig. 7.3. Equivalent circuit diagrams of amplifier and load resistance (A) without feedback (B) with feedback.

The following treatment is based on a single stage, but may be applied to any amplifier provided that the values of μ and r_p are as defined below.

Fig. 7.3A shows the equivalent circuit of the amplifier **without feedback** :
$$E_0 = \mu E_i \left(\frac{R_L}{r_p + R_L} \right) \qquad (13)$$
$$A = \frac{E_0}{E_i} = \frac{\mu R_L}{r_p + R_L} \qquad (14a)$$

where μ = amplification factor (i.e. amplification with an infinite load resistance)
and r_p = plate resistance of final stage.

If the amplifier has more than one stage μ must be defined by
$$\mu = A_1\mu_2 \qquad (14b)$$
where A_1 = amplification between input and grid of final stage (with due regard to phase reversal)
and μ_2 = amplification factor of final stage.

Fig. 7.3B shows the equivalent circuit of the amplifier **with voltage feedback** :[**]
$$E_0 = \mu'E_i' \left(\frac{R_L}{r_p' + R_L} \right) \qquad (15)$$

**Based on B. D. H. Tellegen " Inverse feed-back " Philips Tec. Rev. 2.10 (Oct. 1937) 289.

But $E_0 = A'E_i' = \dfrac{E_i'}{(1/A) - \beta}$ from eqn. (2b).

Applying the value of A from (14a),

$$E_0 = \frac{\mu R_L E_i'}{r_p + R_L (1 - \beta\mu)} \tag{16}$$

This may be arranged in the form

$$E_0 = \left(\frac{\mu.E_i'}{1 - \beta\mu}\right)\left(\frac{R_L}{r_p/(1 - \beta\mu) + R_L}\right) \tag{17}$$

which is the same as eqn. (15) except that

$$\mu' = \frac{\mu}{1 - \beta\mu} \tag{18}$$

and $\quad r_p' = \dfrac{r_p}{1 - \beta\mu}$. $\tag{19}$

Equations (18) and (19) give the effective amplification factor and plate resistance respectively with feedback. It will be seen that these are each equal to the corresponding value without feedback divided by $(1 - \beta\mu)$, whereas the gain, internally produced hum or noise, and distortion are divided by $(1 - \beta A)$. If the amplifier has more than one stage, the value of μ must be as defined by eqn. (14b), while the plate resistance r_p will be that of the final stage.

However, it is important to remember that the actual valve characteristics are not changed by feedback, since feedback is external to the valve ; while the optimum value of load resistance is also unchanged by feedback except under certain very special conditions.

It may be shown that negative voltage feedback as a method for reducing the effective plate resistance, when compared with a transformer giving the same reduction of gain, is more effective than the transformer with low feedback factors, is equally effective with a particular feedback factor, and is less effective than the transformer with feedback factors greater than this value.

Example : 6V6-GT with $\mu = 218$, and $A = 17$.

Gain	$|\beta A|$	$|\beta\mu|$	r_p/r_p'	Transformer impedance ratio
1/2	1	12.8	1/13.8	1/4
1/10	9	115	1/116	1/100
1/100	99	1270	1/1271	1/10000

The " looking-backwards " output terminal impedance of the amplifier is equal to r_p' in parallel with R_L,

$$R_0' = \frac{r_p' R_L}{r_p' + R_L} = \frac{r_p R_L}{r_p + R_L(1 - \beta\mu)}. \tag{19a}$$

If r_p' is very much less than R_L, eqn. (17) becomes

$$E_0 \approx \frac{\mu E_i'}{1 - \beta\mu} \tag{20}$$

that is to say, the output voltage is independent of the value of R_L. This is an indication of good regulation.

(F) Effect of voltage feedback on the input resistance

The effect of feedback on the input resistance depends on the method of application of the feedback and not on whether it is voltage or current feedback. For example, consider the amplifier of Fig. 7.2B. Let the input resistance of the amplifier without feedback be r_i.

Then $r_i = E_i/I_1$.

With feedback, the same current will flow, but the input voltage will be
$$E_i' = E_i(1 - \beta A)$$
Therefore $r_i' = E_i'/I_1 = (E_i/I_1)(1 - \beta A)$
Therefore $r_i' = r_i(1 - \beta A)$. $\qquad$ (21)

Thus in this case the input resistance is increased by feedback in the same proportion that the gain is decreased. This applies to all cases in which the feedback voltage is applied in series with the input voltage, including the current feedback case of Fig. 7.4, as will be shown later.

On the other hand, if the feedback voltage is applied in shunt with the input voltage the input resistance will be decreased as proved below for Fig. 7.2C—

The input resistance of the amplifier without feedback is
$$r_i = E_i/I_1.$$

With feedback, the current I_1 will be unchanged, but there will be an additional current I_2 through R_4 and of direction such as to increase the current from the source, which will therefore become $(I_1 + I_2)$. The input voltage applied to the input terminals of the amplifier will be the same as without feedback

Therefore $r_i' = r_i \cdot \dfrac{I_1}{I_1 + I_2}$

which is less than r_i.

Other special cases are covered in Sect. 2.

(iii) Negative current feedback at the mid-frequency

The block diagram of an amplifier with negative current feedback is shown in Fig. 7.4. In this case the feedback voltage is derived from the voltage drop across a resistor R_3 in series with the load impedance. The resistance of R_3 should be very small compared with the load impedance. This diagram also is for the case with the polarity of the output voltage opposite that of the input voltage.

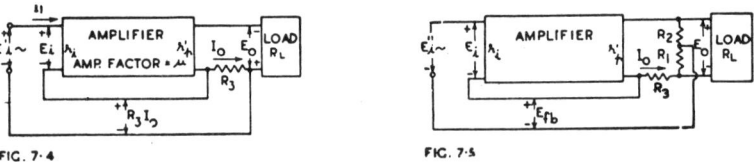

FIG. 7·4 $\qquad\qquad\qquad\qquad$ FIC. 7·5

Fig. 7.4. *Block diagram of an amplifier with negative current feedback. The resistance R_3 is very much smaller than the load impedance.*
Fig. 7.5. *Block diagram of an amplifier with bridge feedback.*

Without feedback but with R_3 in position (Fig. 7.4)
$$A = \frac{E_0}{E_i} = \frac{\mu R_L}{r_p + R_3 + R_L} \qquad (22)$$
With current feedback,
$$A' = \frac{E_0}{E_i'} = \frac{E_0}{E_i + R_3 I_0} = \frac{E_0}{E_i + (R_3/R_L)E_0} = \frac{1}{(E_i/E_0) + R_3/R_L}$$
Therefore $A' = \dfrac{\mu R_L}{(\mu + 1)R_3 + r_p + R_L} \qquad (23)$

Therefore $A' = \dfrac{A}{1 + AR_3/R_L} = \dfrac{A}{1 + \mu R_3/(r_p + R_3 + R_L)} \qquad (24a)$

It is sometimes helpful to substitute
$$-\gamma = R_3/R_L$$
where γ has an application similar to that of β in the case of voltage feedback. Eqn. (24a) may thus be put into the form
$$A' = A/(1 - \gamma A). \qquad (24b)$$
Equation (23) indicates that for a single stage amplifier

(A) **the amplification factor with current feedback** is the same (μ) as without feedback

(B) **the effective plate resistance with current feedback** is
$$r_p' = r_p + (\mu + 1)R_3 \tag{25a}$$
and $r_p'/r_p = 1 + (\mu + 1)R_3/r_p$ (25b)

The increase in plate resistance (eqn. 25b) is proportionally greater than the decrease in gain (eqn. 24a).

When the voltage is fed back over more than one stage, the effective plate resistance with feedback is given by
$$r_p' = r_p + A_1 (\mu_2 + 1)R_3, \tag{25c}$$
where A_1 = amplification between input and grid of final stage (with due regard to phase reversal)

and μ_2 = amplification factor of final stage.

The "**looking-backwards**" **output terminal impedance** of the amplifier is equal to r_p' in parallel with R_L,
$$R_0' = \frac{r_p' R_L}{r_p' + R_L} = \frac{[r_p + (\mu + 1)R_3]R_L}{r_p + (\mu + 1)R_3 + R_L} \text{ for single stage}$$

or $R_0' = \dfrac{[r_p + A_1 (\mu_2 + 1) R_3] K_L}{r_p + A_1 (\mu_2 + 1) R_3 + R_L}$ for more than one stage.

(C) **The ratio of amplification with and without feedback** is given by
$$\frac{A}{A'} = \frac{(\mu + 1)R_3 + r_p + R_L}{R_3 + r_p + R_L} \tag{26}$$
which may be put into the form
$$A/A' = 1 + A(R_3/R_L) = 1 - \gamma A. \tag{27}$$

(D) **Approximations when μ is very large**

If μ is very large,
$$A' \approx R_L/R_3 \approx 1/(-\gamma) \tag{28}$$
which is independent of the amplifier characteristics.

Also, if μ is very large,
$$I_0 \approx E_0/R_L \approx E_i'/R_3 \tag{29}$$
which gives an output current which is constant, irrespective of R_L.

(E) **Input resistance with negative current feedback**

In the circuit of Fig. 7.4—

Without feedback
$$r_i = E_i/I_1$$

With feedback
$$r_i' = \frac{E_i'}{I_1} = \frac{E_i + (R_3/R_L)E_0}{I_1} = \frac{E_i + (R_3/R_L)AE_i}{I_1}$$

Therefore $r_i' = r_i(1 + AR_3/R_L) = r_i(1 - \gamma A)$ (30)

which indicates that with this circuit the input resistance is increased with negative current feedback in the same proportion that the gain is decreased (eqn. 27) and the effect is therefore the same as with Fig. 7.3.

(F) **Harmonic distortion with negative current feedback**

It may readily be shown that, as with negative voltage feedback, the harmonic distortion is reduced in the same proportion that the gain is reduced, subject to the conditions enumerated above (following eqn. 8) :
$$\text{e.g., } \frac{E_{2m}}{E_{0m}} = \frac{H_2}{1 + AR_3/R_L} = \frac{H_2}{1 - \gamma A} \tag{31}$$

(iv) Bridge negative feedback at the mid-frequency

Bridge feedback is a combination of voltage and current feedback, as shown in Fig. 7.5. It is assumed that $(R_1 + R_2) \gg R_L$ and that $R_3 \ll R_L$.

Without feedback : $A = E_0/E_i = \mu R_L/(r_p + R_3 + R_L)$.

With feedback :

Feedback voltage $= R_1 E_0/(R_1 + R_2) + I_0 R_3$

Therefore $E_i' = E_i + R_1 E_0/(R_1 + R_2) + I_0 R_3$

Therefore $A' = \dfrac{AE_i}{E_i'} = \dfrac{AE_i}{E_i + AE_iR_1/(R_1 + R_2) + R_3AE_i/R_L}$

Therefore $A' = \dfrac{A}{1 + AR_1/(R_1 + R_2) + AR_3/R_L}$ (32)

Comparing this with (2b) and (24a), it will be seen that the denominator in (32) includes the second terms of the denominators of both (2b) and (24a). Equation (32) may be put into the alternative forms

$$A' = \frac{\mu R_L}{r_p + (\mu + 1)R_3 + R_L(1 + \mu R_1)/(R_1 + R_2)}$$ (33)

and $A' = \dfrac{\mu}{1 + \mu R_1/(R_1 + R_2)} \cdot \dfrac{R_L}{\dfrac{r_p + (\mu + 1)R_3}{1 + \mu R_1/(R_1 + R_2)} + R_L}$ (34)

Equation (34) indicates that—

(1) the amplification factor is reduced by the factor

$\dfrac{1}{1 + \mu R_1/(R_1 + R_2)} = \dfrac{1}{1 - \beta\mu}$ which is the same as for voltage feedback

alone (equation 18).

(2) the effective plate resistance is given by

$$r_p' = \frac{r_p + (\mu + 1)R_3}{1 + \mu R_1/(R_1 + R_2)} = \frac{r_p - (\mu + 1)(\gamma R_L)}{1 - \mu\beta}$$ (35a)

Compare this with equation (25a) for current feedback alone and (19) for voltage feedback alone.

If the amplifier has more than one stage, μ must be defined as

$\mu = A_1\mu_2$ (35b)

where $A_1 =$ amplification between input and grid of final stage (with due regard to phase reversal)

and $\mu_2 =$ amplification factor of final stage, thus giving the expression for effective plate resistance

$$r_p' = \frac{r_p + (A_1\mu_2 + 1)R_3}{1 - A_1\mu_2\beta} = \frac{r_p - (A_1\mu_2 + 1)(\gamma R_L)}{1 - A_1\mu_2\beta}$$ (35c)

where $\gamma = -R_3/R_L$.

Special applications of bridge feedback (Ref. A9)

(A) The output resistance may be adjusted so as to equal the load resistance.

If the voltage and current feedback resistors are adjusted so that

$R_1/(R_1 + R_2) = R_3/R_L$, that is $\beta = \gamma$,

then $r_p' \approx R_L$

provided that $(-\beta\mu) \gg 1$

and that $(-\beta\mu R_L) \gg r_p$.

(B) Negative voltage feedback may be combined with positive current feedback to decrease the plate resistance r_p' to zero or even to make it negative. This is little used in amplifiers because the positive current feedback increases the harmonic distortion. However, it is possible to combine negative feedback in the output stage with positive feedback in an earlier stage to give very useful results—see Sect. 2(xi). This somewhat resembles one form of the balanced feedback in (v) below, except that the design is less restricted.

(C) Negative current feedback may be combined with positive voltage feedback to give very high effective plate resistance.

(v) Combined positive and negative feedback at the mid-frequency

The distortion in a two or three stage amplifier is mainly in the output stage, and the distortion in a well designed first stage will be relatively quite small. It is practicable to apply positive voltage feedback to the first stage only, and then to apply negative voltage feedback over two or three stages in order to secure very low distortion and low output resistance.

This arrangement is shown in the simplified block diagram of Fig. 7.6. The effect of positive feedback on A_1 is to increase its gain from A_1 without feedback to A' with positive feedback. The effect on the whole amplifier, so far as effective plate resistance, distortion in the final stage and gain are concerned, is the same as though A_1 with positive feedback were replaced by another amplifier without feedback but with gain A'. The positive feedback in A_1 will increase the distortion of this stage, but

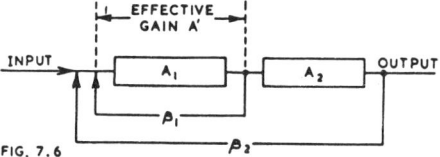

Fig. 7.6. *Simplified   block diagram illustrating one possible arrangement of combined positive and negative feedback.*

FIG. 7.6

will decrease the distortion of the whole amplifier provided that the initial distortion in A_1 is small in comparison with that in A_2. A general purpose triode may be used in the first stage, and yet have an effective stage gain greater than that of a pentode with a plate load resistor of 0.25 megohm.

Practical amplifiers using this principle are described in Sect. 2(xi).

Balanced feedback amplifiers

A balanced feedback amplifier is one using both positive and negative feedback in such proportions that the overall gain with feedback is equal to that without any feedback (Ref. A20). There does not seem to be any real advantage in achieving exact balance between the two feedback systems and the designer will generally prefer complete liberty on this point.

(vi) Comparison between different fundamental types at the mid-frequency

Negative feedback :

Voltage Feedback	Current Feedback	Effect
decreases	decreases	gain
decreases	decreases	effect of variations of valves and other components on gain
decreases	increases	effective plate resistance
increases or decreases	increases or decreases	input resistance*
decreases	decreases	harmonic distortion
decreases	increases	effect of load impedance on output voltage
increases	decreases	effect of load impedance on output current
increases	decreases	damping on loudspeaker
increases	increases	frequency response (band width)
decreases	decreases	hum and unwanted voltages introduced in the amplifier (subject to certain conditions).

Positive feedback :

Voltage Feedback	Current Feedback	Effect
increases	increases	gain
increases	decreases	effective plate resistance
increases	increases	harmonic distortion

*The effect on the input resistance depends on the method of applying the input voltage.

Characteristics of amplifiers with negative feedback

Characteristic	No feedback	Voltage feedback	Current feedback†
Voltage gain	$\dfrac{\mu R_L}{r_p + R_L}$	$\dfrac{\mu R_L}{r_p + R_L(1 - \beta\mu)}$	$\dfrac{\mu R_L}{r_p + R_L[1 - \gamma(\mu+1)]}$
or	A	$\dfrac{A}{1 - \beta A}$	$\dfrac{A}{1 - \gamma A}$
Effective amplification factor‡	μ	$\dfrac{\mu}{1 - \beta\mu}$	μ
Effective plate resistance of final stage	r_p ‡	$\dfrac{r_p}{1 - \beta\mu}$	$\begin{cases} r_p + (\mu + 1)R_3 \\ = r_p - \gamma(\mu + 1)R_L \end{cases}$
Input resistance	r_i	$r_i(1 - \beta A)$*	$r_i(1 - \gamma A)$**
Second harmonic distortion	H_2	$\dfrac{H_2}{1 - \beta A}$	$\dfrac{H_2}{1 - \gamma A}$
Hum or noise introduced by amplifier	E_n	$\dfrac{E_n}{1 - \beta A}$	$\dfrac{E_n}{1 - \gamma A}$

Note : β is negative throughout.
γ is defined by $-\gamma = R_3/R_L$.

SECTION 2 : PRACTICAL FEEDBACK CIRCUITS

(i) *The cathode follower* (ii) *The cathode degenerative amplifier and phase splitter* (iii) *Voltage feedback from secondary of output transformer* (iv) *Voltage feedback from plate—transformer input* (v) *Voltage feedback from plate—r.c.c. input* (vi) *Voltage feedback over two stages* (vii) *Voltage feedback over three stages* (viii) *Cathode coupled phase inverters and amplifiers* (ix) *Hum* (x) *Some special features of feedback amplifiers* (xi) *Combined positive and negative feedback* (xii) *Choke-coupled phase inverter.*

The idealized conditions assumed in Section 1 do not always hold. At very low and very high audio (or ultrasonic) frequencies, the reactances cause phase angle displacements tending to nullify the "negative" feedback which, in extreme cases, may even become positive. The resulting effects may be evident as poor response to transients (damped oscillations), parasitic oscillations extending over a portion of a cycle of sine waveform or, in extreme cases, oscillation at a very low or very high frequency. This aspect is considered in further detail in Section 3 for those who wish to have anything beyond the most elementary understanding of the problem.

Other complications arise from (a) certain circuit connections affecting the amount of hum, and (b) the application of feedback in parallel with an impedance within the amplifier.

(i) The cathode follower

With a cathode follower the load impedance is connected between cathode and earth instead of between plate and earth in the more conventional arrangement. It is there-

---

*Applies only to circuits such as Fig. 7.2A or 7.2B in which the feedback voltage is applied in series with the input voltage.

†Over single stage.

‡If the amplifier has more than one stage, the amplification factor must be defined by $\mu = A_1\mu_2$ where A_1 = amplification between point at which feedback is introduced and grid of final stage and μ_2 = amplification factor of final stage.

**Applies only to circuits such as Fig. 7.4 in which the feedback voltage is applied in series with the input voltage.

fore frequently referred to as " cathode loading " in distinction to the conventional
" plate loading." As a result of 100 per cent. negative voltage feedback inherent in
a cathode follower, both the distortion and the output impedance may be very low.

The basic circuit of a triode cathode follower is shown in Fig. 7.7. It is understood
that the input circuit must be conductive to provide a grid return path, and that a
suitable grid bias value is achieved by inserting an appropriate battery at, say, point D.
The input voltage is applied between grid and earth while the load resistance (R_k)
is connected between cathode and earth. The voltage developed across R_k is the output voltage, and the maximum output voltage is slightly greater than for the conventional case with R_k as the plate load resistor. The stage gain, however, is necessarily less than 1 since the input voltage is equal to the grid-to-cathode voltage plus
the output voltage (with a medium-mu valve the voltage gain normally approximates
0.9). Since the stage gain is always less than unity, the **input voltage** is always
greater than the output voltage. Thus the input voltage may be very much greater
than the grid bias, but grid current will not flow until the instantaneous difference
between the input and the output voltage exceeds the bias.* In some cases the input
voltage will be very high as, for example, when a high power output is required from
a cathode-loaded low-mu triode operated with a high plate voltage. In extreme
cases there may be difficulty in obtaining sufficient voltage in the preceding stage to
excite fully the cathode follower.

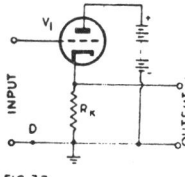

FIC 7.7

· *Fig. 7.7. Basic circuit of
a cathode follower.*

Since a cathode follower does not amplify the input voltage it might, at first sight,
be thought that the arrangement fulfilled no useful purpose. Its principal usefulness,
however, lies in its· impedance characteristics. The input impedance is high while
the output impedance (R_0') is low, and the whole device may therefore be regarded
as a kind of Impedance Transformer see (F1) below.

The low output impedance of the cathode follower makes it particularly useful as a
driver stage for Class AB_2 or Class B amplifiers, or for pulse techniques where a low
impedance source is required.

With the cathode follower, the grid and cathode are in phase with one another, so
that there is no reversal of polarity as with plate-loaded amplifiers this characteristic
is sometimes useful in amplifiers, and particularly with pulse techniques.

A cathode follower will work into practically any value of load resistance, although
there is a broad optimum value for any particular application.

The effect of interelectrode and wiring capacitances on the gain is generally negligible up to a frequency of 1 Mc/s.

It is important to remember that all electrode voltages must be stated with respect
to the cathode ; these will normally have the same values as for plate loading.

The graphical treatment is given on pages 390-394.

(A) Voltage gain at low frequencies
    The voltage gain without feedback would be

$$A = \frac{\mu R_k}{R_k + r_p}.$$

· The value of the transfer coefficient β will be β 1.

---

*More precisely, the difference between the bias and the voltage of the grid current cross-over point.

The **gain with feedback** will therefore be

$$A' = \frac{A}{1 - \beta A} = \frac{A}{1 + A}$$

Therefore $A' = \dfrac{\mu R_k}{(\mu + 1)R_k + r_p}$.   (1a)

which may be written in the form

$$A' = \frac{\mu}{\mu + 1} \cdot \frac{R_k}{R_k + r_p/(\mu + 1)} \tag{1b}$$

Equation (1a) may also be written in the form

$$A' = \frac{R_k}{\left(\dfrac{\mu + 1}{\mu}\right)R_k + \dfrac{1}{g_m}} \tag{1c}$$

where g_m = mutual conductance in mhos.

If $\mu \gg 1$, the gain with feedback will be approximately

$$A' \approx \frac{g_m R_k}{1 + g_m R_k} \tag{2}$$

R_k is the resultant of all resistances between cathode and earth, whether internal or external to the amplifier itself. There is very little increase in gain through making R_k greater than about twice r_p; provided that μ is not less than 10, then the increase of gain through any further increase in R_k is always less than 5 per cent. (It is assumed that the input voltage is small enough to avoid grid current and/or plate current cut-off).

Charts giving gain
Voltage gain (exact) based on μ, R_k and r_p—Ref. C19.
Voltage gain (approx.) based on g_m and R_k for pentodes—Ref. C21.
Gain in decibels (approx.)—Ref. C24.

(B) Effective plate resistance (r_p') at low frequencies
The effective plate resistance in the arrangement of Fig. 7.7, as with any case of voltage feedback, is equal to r_p divided by $(1 - \beta\mu)$, i.e.

$$r_p' = \frac{r_p}{1 - \beta\mu} = \frac{r_p}{1 + \mu} = \frac{1}{g_m} \cdot \frac{\mu}{\mu + 1} \tag{3}$$

If μ is very much greater than 1,
then $r_p' \approx 1/g_m$ (4a)
with an error not exceeding 5 per cent. if μ is not less than 20.

As with the gain, so too with the effective plate resistance, there is very little change brought about by an increase in R_k beyond $2r_p$, provided that μ is not less than 20. Chart of r_p' (approx.)—Ref. C24.

(C) Equivalent valve characteristics
The results obtained with a cathode follower (e.g. eqn. 1b) are equivalent to those which would be obtained from an equivalent plate-loaded triode* having :
    plate resistance $= r_p' = r_p/(\mu + 1)$
    amplification factor $= \mu/(\mu + 1)$
    mutual conductance $= g_m$ (unchanged)
    and working into a load resistance $R_L = R_k$ (unchanged).

Example 1 : type 6J7 (triode connection)
For 250 volts between plate and cathode, $\mu = 20$, $g_m = 1900$ micromhos and $r_p = 10\,500$ ohms. The value of R_k for greatest power output may be assumed to be approximately equal to $2r_p$, so that a value of 20 000 ohms may be adopted (see Sect. 5 for graphical treatment). The effective plate resistance is therefore
    $r_p' = r_p/(\mu + 1) = 10\,500/21 = 500$ ohms,
and the gain will be

---
*See Sect. 5(i) for graphical treatment.

$$A' = \frac{20\,000}{(21/20)(20\,000) + 525} = 0.93.$$

Example 2 : type 6V6 (triode connection)

With 250 volts between plate and cathode, $\mu = 9.6$, $g_m = 4000$ micromhos and $r_p = 2400$ ohms. If R_k is made equal to 5000 ohms we have

$$r_p' = 2400/(9.6 + 1) = 2400/10.6 = 226 \text{ ohms, and the gain will be}$$

$$A' = \frac{5000}{(10.6/9.6)5000 + 250} = 0.866.$$

(D) Distortion

The distortion, within the limits of plate-current cut-off on one hand, and grid current on the other, is reduced by the same factor that the gain is reduced, namely $1/(1 + A)$ where A is the gain without feedback. In a practical case, if $\mu = 20$ and $R_k = 2r_p$, then the distortion is reduced to 3/43 of the distortion with plate loading.

The **maximum input voltage** which may be handled without distortion may be calculated graphically (see Sect. 5).

(E) Calculated operating conditions

The following treatment is based on ideal valve characteristics with straight lines instead of curves, and is therefore only approximate (Ref. C31). It is assumed that the bias is adjusted to allow for maximum voltage swing over the whole loadline.

Operating point : $I_b = \frac{1}{2}E_{bb}/(r_p + R_k)$ (4b)

$$E_k = \frac{1}{2}E_{bb}R_k/(r_p + R_k) \quad\quad\quad (4c)$$

$$E_c = -E_{bb}/(2\mu) \quad\quad\quad\quad\quad\quad (4d)$$

where E_k = voltage across R_k.

If the circuit of Fig. 7.10 is used to provide grid bias :

$$R_2 = (r_p + R_k)/\mu = R_k/A = (1 + R_k/r_p)g_m \quad\quad (4e)$$

where $A = \mu R_k/(r_p + R_k)$

        = amplification of valve with plate loading.

Example : Type 6J7 (triode connection)

$$I_b = 125/30\,500 = 4.1 \text{ mA}$$
$$E_k = 125 \times 20\,000/30\,500 = 82 \text{ V}$$
$$E_c = -250/40 = -6.25 \text{ V}$$
$$A = 20 \times 20\,000/30\,500 = 13.1$$
$$R_2 = 30\,500/20 = 1525 \text{ ohms.}$$

In the general case, use may be made of the following table, based on eqn. (4e) :

when $R_k =$	$0.1/r_p$	$1.0r_p$	$2r_p$	$3r_p$	$4r_p$	$5r_p$
then $R_2 =$	$1.1/g_m$	$2/g_m$	$3/g_m$	$4/g_m$	$5/g_m$	$6/g_m$

The peak amplitude of the voltage E_k is given by.

$$E_{km} = \frac{1}{2}E_{bb}R_k/(r_p + R_k) \quad\quad\quad (4f)$$

Special case (1)

If it is desired to use the minimum value of R_k to give a specified peak output voltage E_{km}, the value of R_k is given by

$$R_k = 2E_{km}r_p/(E_{bb} - 2E_{km}) \quad\quad\quad (4g)$$

and the optimum value of bias resistor (Fig. 7.10) is given by

$$R_2 = \frac{E_{bb}}{g_m(E_{bb} - 2E_{km})} \quad\quad\quad (4h)$$

If E_{km} is small compared with E_{bb}, R_2 is thus approximately equal to $1/g_m$.

Special case (2)

If R_k is greater than $1/g_m$ and the whole of R_k is used to provide grid bias, the maximum output voltage is thereby reduced,

$$E_{km} = \frac{E_{bb}R_k}{r_p + (\mu + 1)R_k} \quad\quad\quad (4i)$$

Special case (3)

If R_k is made equal to $1/g_m$ the cathode follower works into a load equal to its own effective plate resistance (r_p'). The value of peak output voltage is given by

$$E_{gm} = \tfrac{1}{2}E_{bb}/(\mu + 1) \tag{4j}$$

and the optimum value of bias resistor R_2 is given by

$$R_2 = \frac{r_p}{\mu}\left(\frac{\mu + 1}{\mu}\right) \approx R_k \tag{4k}$$

so that the valve may be considered perfectly biased by the whole of its cathode resistor R_k.

Special case (4)

When the value of cathode resistance required to provide bias is small compared with $1/g_m$, the arrangement of Fig. 7.11 may be used in which $R_1 = R_k = 1/g_m$, and R_2 is calculated to provide the correct bias. Since R_1 is small compared with $1/g_m = r_p/\mu$, the maximum peak output voltage will be given approximately by

$$E_{km} \approx \tfrac{1}{2}E_{bb}R_1/r_p \tag{4l}$$

Special case (5)

If R_k is small compared with $1/g_m$ and R_k alone used to provide bias, the bias will not be optimum and the maximum peak output voltage will be given by

$$E_{km} = \frac{\mu R_k{}^2 E_{bb}}{r_p[r_p + (\mu + 1)R_k]} = \frac{\mu n^2 E_{bb}}{(\mu + 1)n + 1} \tag{4m}$$

where R_k is neglected in comparison with r_p and $n = R_k/r_p$.

The ratio of the output voltage with correct bias (eqn. 4l) and that with self bias from cathode resistor (eqn. 4m) is given by

$$\text{Ratio} = \frac{2}{1 + 1/\mu n} \tag{4n}$$

(F1) The " looking-backwards " output terminal impedance (R_0')

The impedance, seen when looking backwards from the output terminals of Fig. 7.7 into the amplifier, is equal to r_p' in parallel with R_k. This corresponds to the plate-loaded case where we have to consider r_p and R_L in parallel.

$$R_0' = \frac{r_p'R_k}{r_p' + R_k} = \frac{R_k}{(r_p + R_k)/r_p + g_m R_k} = \frac{1}{g_m + 1/R_k + 1/r_p} \tag{5a}$$

Therefore $R_0' = \dfrac{r_p R_k}{(\mu + 1)R_k + r_p}$ \hfill (5b)

If $r_p \gg R_k$, $R_0' \approx \dfrac{R_k}{1 + g_m R_k}$ \hfill (6)

In special cases, where a very low value of R_0' is required, it may be convenient to obtain it by decreasing the value of R_k, even though this results in a serious reduction in maximum power output. For example, with type 6V6 (triode connection) under the conditions given above, where $r_p' = 226$ ohms,

when R_k = 2400,   1000,   500,   200,   100,   50 ohms.
then R_0' = 206,   184,   156,   105,   69,   41 ohms.

Charts :

R_0' (approx.) based on R_k and g_m—Ref. C23.

R_0' (approx.) but with correction factor to give exact value—Ref. C22.

(F2) Resistance-capacitance coupled cathode follower

An alternative form of load is illustrated in Fig. 7.8 where R_k is the d.c. load resistance, C_c the coupling condenser, and R_g the a.c. shunt load resistance. In cal-

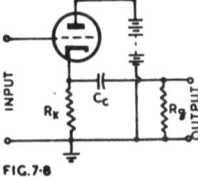

INPUT

R_k C_c R_g OUTPUT

FIG. 7·8

Fig. 7.8. Cathode follower with d.c. load resistance R_k, capacitance coupled to a shunt load resistance R_g.

culations regarding grid bias, the resistance of R_k should be used ; in calculations regarding gain the resultant $R = R_k R_g/(R_k + R_g)$ should be used. As for the similar case with plate-loading, it is usually desirable to have R_g considerably greater than R_k. R_g may take the form of a volume control.

Alternatively, R_k may be selected to suit the grid bias, while R_g is then calculated to provide the desired total a.c. load.

When the circuit of Fig. 7.8 is used as a power amplifier stage, it may be shown (Ref. C32) that maximum power is dissipated in R_g when $R_g = r_p$ and $R_k = \sqrt{2}\,r_p$, and that when these conditions are fulfilled the power dissipated in R_g is given by

$$P_{max} = E_1{}^2/93 r_p$$

where E_1 = plate supply voltage E_{bb} less the voltage where the tangent to the
          $E_c = 0$ characteristic cuts the axis (point B in Fig. 7.8A)

and r_p = plate resistance of valve at junction of d.c. loadline with $E_c = 0$
          characteristic (point D in Fig. 7.8A).

If in Fig. 7.8 the load impedance is removed to a distance and the leads between the cathode follower and the output terminals are lengthened, the " link " between the amplifier and the output may be used as a low impedance loaded a-f line, although it is loaded unequally at the two ends.

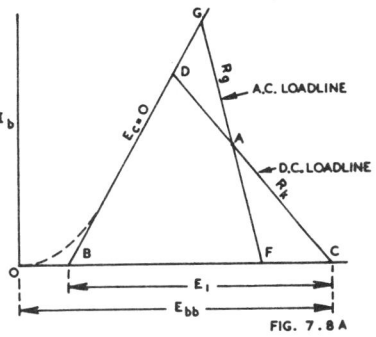

The term " link " is used to describe a line which is so short that its characteristics at the frequency of operation may be neglected in comparison with the loading. It should be carefully distinguished from a " transmission line," which has a characteristic impedance determined by the self inductance and capacitance per unit length

*Fig. 7.8A. Loadlines of circuit Fig. 7.8
used as a power amplifier.*

and which, for the suppression of standing waves, requires to be terminated by an impedance equal to the characteristic impedance.

(G) Transformer coupled cathode follower

With the resistance-coupled case of Fig. 7.7 the plate supply voltage must be higher than the plate-to-cathode voltage and the whole of the power output plus the d.c. losses are dissipated in the resistance R_k. It is possible to replace R_k by the primary of a transformer, and the secondary of the transformer may then be loaded, e.g. by driving the voice coil of a loudspeaker or other electrical network (Fig. 7.9). In this case the full supply voltage, except for the small resistance loss in the transformer primary, is applied between plate and cathode.

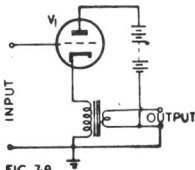

*Fig. 7.9. Cathode follower having the primary
of a transformer in its cathode circuit, the secondary
of the transformer being loaded.*

The maximum power output with the same load resistance as for plate loading will be approximately the same. The distortion and the output resistance will, however, be very much lower than for plate loading, and the damping on the loudspeaker will be greater.

A special case of some importance is when V_1 is the driver valve of a Class B power amplifier (see Chapter 13 Sect. 7).

The optimum load resistance may best be calculated graphically (see Sect. 5).

The resistance of the transformer primary will provide at least part of the voltage for grid bias ; the balance may be supplied by a suitable resistance (shunted by the usual by-pass capacitance) between the cathode and the upper transformer primary terminal.

Alternatively, the primary of the transformer may be connected in place of R_g in Fig. 7.8, and C_c may be adjusted to tune the primary to a suitable low frequency to extend the frequency band to lower frequencies or to give bass boost (Ref. C14).

(H) **Grid circuit arrangements and their effect on input resistance and R_0'**

If the voltage drop in R_k is greater than the desired grid bias, then a convenient arrangement is to return the grid resistor (R_g) to a tapping point (X) on the cathode load resistor, the position of X being adjusted for optimum operating conditions (Fig. 7.10) ; the bias will be

$$E_{c1} = R_2 I_b \qquad (7)$$

where E_{c1} is the bias in volts, R_2 is expressed in ohms and I_b the plate current in amperes. The total load resistance is $(R_1 + R_2)$ which corresponds to R_k in Fig. 7.7. The correct value of C should always be selected on the basis of the input resistance (R_i') and not on the basis of R_g.

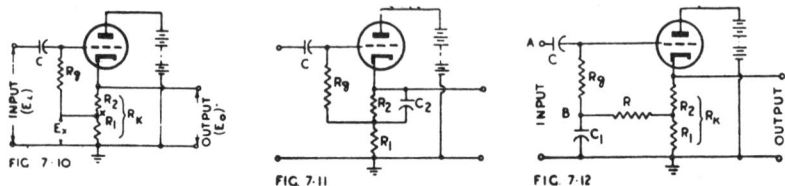

FIG. 7.10   FIG. 7.11   FIG. 7.12

Fig. 7.10. *Cathode follower with grid bias supplied by a tapping point (X) on the load resistance.*
Fig. 7.11. *Cathode follower with separate resistances for bias (R_2 by-passed by C_2) and load (R_1).*
Fig. 7.12. *Cathode follower with alternative form of grid bias circuit.*

The circuit of Fig. 7.10 does not apply the full degree of feedback (as in Fig. 7.8) unless the output resistance of the preceding stage is very much smaller than R_g. In other cases the feedback is reduced by a factor $R_0/(R_g + R_0)$ where $R_0 = r_p R_L/(r_p + R_L)$ for the preceding stage.

The **input resistance** (R_i') due to the grid resistor (R_g in Fig. 7.10) is greater than R_g because the voltage across R_g is less than E_i.

Thus $R_i' = E_i/i$
where $E_i =$ the input voltage
and $i =$ the current through R_g (Fig. 7.10).
Now $i = (E_i - E_x)/R_g$
and $E_x = A'E_i R_1/(R_1 + R_2)$.

Therefore $R_i' = \dfrac{E_i}{i} = \dfrac{E_i R_g}{E_i - E_x} = \dfrac{E_i R_g}{E_i[1 - A'R_1/(R_1 + R_2)]}$

Therefore $R_i' = \dfrac{R_g}{1 - A'R_1/(R_1 + R_2)}$ \hfill (8)

Special cases :

If $R_2 = 0$, i.e. R_g returned to cathode : $R_i' = R_g/(1 - A')$ \hfill (9a)
If $R_1 = 0$, i.e. R_g returned to earth : $R_i' = R_g$ \hfill (9b)

As a practical example take type 6J7 (triode connection) with the same conditions as previously (Example 1). The cathode load resistance $(R_1 + R_2)$ is therefore 20 000 ohms. The cathode current (from the published data) is 6.5 mA while the grid bias is -8 volts. The resistance of R_2 is therefore 1230 ohms and R_1 is 18 770 ohms. If R_g is taken as 1 megohm the input resistance will be

$$R_i' = \frac{1}{1 - 0.93 \times 18\,770/20\,000} = \frac{20\,000}{20\,000 - 17\,500} = 8 \text{ megohms.}$$

The "**looking-backwards**" **output terminal impedance** with the circuit of Fig. 7.10 becomes (Ref. C28)

$$R_0' = \cfrac{1}{g_m\left[1 - \cfrac{R_s}{R_s + R_g}\cdot\cfrac{R_1}{R_1 + R_2}\right] + \cfrac{1}{R_1 + R_2} + \cfrac{1}{r_p}} \qquad (10)$$

where R_s = source impedance of input voltage.

[Compare eqn. (10) with eqn. (5a) which applies to the circuit of Fig. 7.7.]

If R_s is not greater than one fifth of R_g, the effect on the output terminal impedance will usually be negligible.

If R_2 in Fig. 7.11 is by-passed, the a.c. load is lower than the static load resistance, which is usually undesirable unless R_2 is very much smaller than R_k. For example, if R_1 is at least 10 times the value of R_2, the effect is usually so small as to be negligible. The input resistance is

$R_i' = R_g/(1 - A')$ which is the same as (9a).

The same remarks apply as for Fig. 7.10 with regard to the output resistance of the preceding stage.

An alternative circuit is shown in Fig. 7.12 in which the reactance of C_1, at the lowest frequency to be amplified, should be small compared with the resistance of R. The input resistance is approximately equal to R_g, provided that $(R + R_1)$ is much larger than the reactance of C_1. With this circuit, the full degree of feedback is obtained irrespective of the output resistance of the preceding stage.

Another method of securing the correct grid bias voltage is to connect a positive grid bias to counteract the too-great negative bias due to R_k, either at the point D in Fig. 7.7 or at the low potential end of the grid resistor R_g. One method of obtaining the negative bias without requiring a separate bias voltage source is shown in Fig. 7.13 where R_3 and R_g form a voltage divider across the plate voltage source. The input resistance here is equal to R_g and R_3 in parallel.

Direct coupling to a cathode-follower is shown in Fig. 7.14. If the first valve has low plate resistance, it would be practicable to drive the cathode follower into the grid current region. The voltage drop in R_{L1} must equal the grid-plate voltage of the cathode follower.

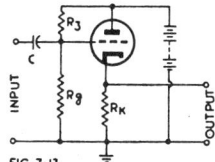 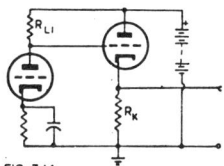 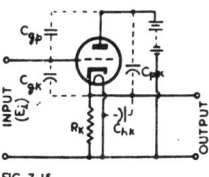

FIG. 7·13   FIG. 7·14   FIG. 7·15

Fig. 7.13. Cathode follower with positive grid bias from voltage divider across $B+$, in order to counteract the too-great negative bias from R_k.
Fig. 7.14. Direct coupling to the grid of a cathode follower.
Fig. 7.15. Triode cathode follower showing capacitances.

(I) Input capacitance (purely resistive load)

The capacitance from grid to plate (C_{gp}) is effectively across the input (Fig. 7.15). The effect of the capacitance from grid to cathode is reduced by the ratio of the voltage from grid to cathode to the voltage from grid to earth. Thus

$$C_i' = C_{gp} + (1 - A')C_{gk} \qquad (11)$$
or $\quad C_i' = C_{gp} + C_{gk}/(1 + A) \qquad (12)$

where A has the same meaning as with plate loading, being the stage gain without feedback, and A' is the actual stage gain, being always less than 1 (eqns. 1a and 1b). Note the reversed Miller Effect whereby the C_{gk} is reduced by the factor $1/(1 + A)$.

In a typical case, using type 6J7-GT (triode connection) :—

$A' = 0.93$; $C_{gp} = 1.8$ $\mu\mu$F ; and $C_{gk} = 2.6$ $\mu\mu$F.
$C_i' = 1.8 + 0.07 \times 2.6 = 1.8 + 0.18 = 1.98$ $\mu\mu$F.

This may be compared with an input capacitance of 22.4 $\mu\mu$F under equivalent plate loaded conditions.

(J) Output capacitance

The capacitance from plate to cathode (C_{pk}) and also the capacitance from heater to cathode (C_{hk}) are across the output (Fig. 7.15). The voltage across the capacitance from grid to cathode (C_{gk}) is $(1 - A')e_i$ and the current through C_{gk} is, therefore, $(1 - A') e_i.j\omega C_{gk}$. The current which would flow through C_{gk} if connected directly across R_k is $A'e_i.j\omega C_{gk}$. Therefore, the effect of C_{gk} connected between grid and cathode is the same as that of a capacitance of $[(1 - A')/A']$ $C_{gk} = C_{gk}/A$ connected across R_k. It is assumed that the source of input voltage has a resistance and a reactance each negligibly small in comparison with the reactance of C_{gk}. With this assumption the effect of the capacitance from grid to plate (C_{gp}) is zero, since it merely shunts the input voltage.

Thus, the output capacitance (C_0') under the assumed conditions is given by

$$C_0' = C_{pk} + C_{hk} + C_{gk}/A \qquad (13)$$

This capacitance is effectively shunted across R_0' (see eqn. 5).

In the case of type 6J7-GT (triode connection), $C_{pk} = 17.0$ $\mu\mu$F, $C_{hk} = 10.5$ $\mu\mu$F (approx.), $C_{gk} = 2.6$ $\mu\mu$F, and $A' = 0.93$ (see Example 1).

Therefore $C_0' = 17.0 + 10.5 + 2.6 (0.07/0.93) = 27.7$ $\mu\mu$F.

This is effectively in shunt with the output resistance (R_0') of 487 ohms. In the plate-loaded case there would be a capacitance of 17.0 $\mu\mu$F shunted across 6900 ohms.

In all practical cases, both for input and output capacitances, it is necessary to make allowances for stray capacitances in addition to the valve capacitances.

(K) Special considerations with pentodes

The discussion so far has been principally confined to triodes, or pentodes connected as triodes. It is readily seen that if a pentode is operated with the screen volt-

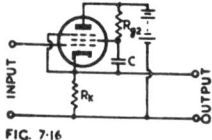

FIG. 7·16

Fig. 7.16. Pentode cathode follower with screen by-passed to cathode.

age equal to the plate voltage, without special precautions to isolate the screen and plate, the result is equivalent to triode operation. The reason is that there is, in the cathode circuit, an impedance which is common to both plate and screen currents, so that the signal voltage drop across it affects the voltages of both plate and screen. Under these conditions the valve ceases to behave like a pentode.

If it is desired to obtain pentode operation, the voltage between screen and cathode must be maintained steady and free from signal frequency fluctuations. This may be done by connecting an impedance (either a voltage dropping resistor or a choke) between the plate and the screen, and by-passing the screen to the cathode by means of a sufficiently large capacitance (Fig. 7.16).

The screen dropping resistor R_{g2} is effectively in parallel with R_k so far as signal voltages are concerned. Unless R_{g2} is more than 20 times greater than R_k, allowance should be made for its effect.

If the by-pass condenser (C) is returned to earth instead of to the cathode the result is, again, effectively triode operation except that the screen may be operated at a

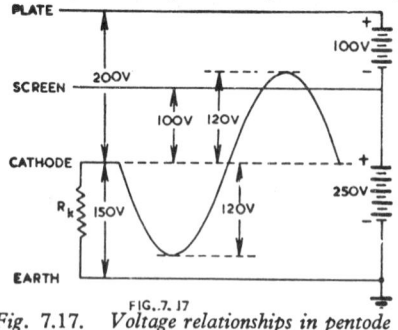

Fig. 7.17. Voltage relationships in pentode cathode follower.

voltage lower by a fixed amount than that of the plate. For example, if the static voltage from cathode to plate is 200 volts, and from cathode to screen 100 volts, then the voltage from screen to plate is 100 volts (Fig. 7.17). But, when a signal voltage is being amplified, the cathode voltage has a signal-frequency component which may reach a peak value of, say, 120 volts (see diagram). This makes the cathode 20 volts more positive than the screen, or this may be expressed the other way around as the screen being 20 volts negative to the cathode. The result is plate-current cut-off and severe distortion, which can only be avoided by a reduced output voltage. In the case of pentode operation, with the screen by-pass condenser returned to the cathode (Fig. 7.16) this type of overloading cannot occur.

When valves having high mutual conductance are connected as triodes, it may be found that parasitic oscillation occurs at a very high frequency. This may be prevented by connecting a 100 to 500 ohm non-inductive resistor in the link between plate and screen, using short leads.

(L) **Voltage gain with pentodes**

Owing to the very high values of μ with most pentodes, eqn. (2) may be taken as being almost exact :

$$A' = \frac{g_m R_k}{1 + g_m R_k} \tag{2}$$

where g_m = pentode mutual conductance under operating conditions in mhos.

In Example 1, the pentode g_m may be taken as being $(I_b/I_k) \times$ triode g_m ; i.e. $0.8 \times 1900 = 1520$ micromhos.

With $R_k = 20\,000$ ohms,

$$A' = \frac{1520 \times 20\,000 \times 10^{-6}}{1 + 1520 \times 20\,000 \times 10^{-6}} = 0.97.$$

Values of A' up to about 0.99 are practicable with pentodes, using higher values of load resistance.

(M) **Input capacitance with pentodes, screen by-passed to cathode** (Fig. 7.16)

The input capacitance with pentode operation is different from that with triode operation of the same valve. The capacitance from grid to plate may be neglected, and the total input capacitance is approximately given by

$$C_i' = (1 - A') \times \text{pentode input capacitance} \tag{14}$$

In the case of type 6J7-GT the pentode input capacitance is 4.6 $\mu\mu$F and with a stage gain (A') of 0.97 the input capacitance will be $0.03 \times 4.6 = 0.14$ $\mu\mu$F. This is less than one fourteenth the input capacitance of the same valve with triode connection under similar conditions (previously shown to be 1.98 $\mu\mu$F).

If the screen is by-passed to earth the input capacitance will be the same as for triode operation.

(N) **Output capacitance with pentodes (screen by-passed to cathode)**

With pentode operation the screen and suppressor are effectively at cathode potential at signal frequency (Fig. 7.16). The " pentode output capacitance " is the capacitance from the plate to all other electrodes except the signal grid and will therefore take the place of the plate-to-cathode capacitance in a triode (eqn. 13). The output capacitance is therefore given by

$$C_0' = \text{pentode output capacitance} + C_{hk}$$
$$+ (1/A) \times \text{pentode input capacitance} \tag{15}$$

In the case of type 6J7-GT with a stage gain of 0.97, the pentode output capacitance is 12 $\mu\mu$F, and the pentode input capacitance 4.6 $\mu\mu$F, so that C_0' is $12 + 10.5 + 0.14 = 22.6$ $\mu\mu$F. This may be compared with 27.7 $\mu\mu$F for triode operation under similar conditions.

(O) **Circuit to avoid screen current through the load resistance**

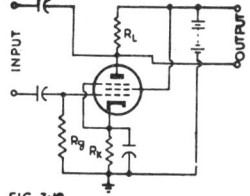

With the normal pentode cathode follower, the screen current passes through the load resistance with resulting complications (see also Sect. 5). One way of avoiding these is shown in Fig. 7.18 which produces results identical with those of the cathode follower. The input is applied between grid and plate, the load resistance is in the plate circuit, and the cathode bias resistor has the same value as for normal operation. It has the disadvantage that neither input terminal is earthed, and transformer input is necessary.

FIG. 7·18

Fig. 7.18. Alternative to the cathode follower, giving similar results, with input applied between grid and plate.

(P) **Circuit to make the screen current constant**
One other method of avoiding the troubles caused by the screen current involves the use of an additional valve to apply signal voltage of the correct value and phase to the screen, and thereby maintain the screen current constant (Ref. C25).

(Q) **Screen-coupled cathode follower**
The cathode of a cathode follower may be directly coupled to the screen of the following stage, this being particularly valuable in d.c. amplifiers. See Ref. C26, also Chapter 12 Sect. 9(vi).

(R) **" Infinite impedance detector "**
The circuit arrangement of the cathode follower is the same as for the so-called " infinite impedance detector " or " reflex detector," but in the latter case it is essential for the plate current to cut off for part of every r-f cycle.
For further information see Chapter 27 Sect. 1(ii)D.

(S) **Amplification at all frequencies—complex load**
It is impossible in practice to have a purely resistive load owing to the unavoidable effect of the interelectrode capacitances plus stray wiring capacitances.
When the load is a resistance in parallel with a capacitance, the input resistance (and conductance) may become negative, thus leading to oscillation under certain conditions. This may be prevented, where necessary, by the addition of a non-inductive grid-stopper of 100 to 2000 ohms resistance ; in resistance-coupled a-f applications a grid stopper is unnecessary. When an inductance is coupled to the grid circuit (e.g. the secondary of an a-f transformer) it may be necessary to shunt it by means of a fairly low value of resistance to avoid instability.
When the load is inductive the input resistance and conductance are always positive.
The **input reactance** is always capacitive.
References C14, C20.

(T) **Gain with capacitive load**
When the load impedance is a resistance R_k (as in Fig. 7.16) shunted by a capacitance C_k, the gain is given approximately by (Ref. C6),

$$A' \approx \frac{g_m R_k}{1 + g_m R_k} \sqrt{\frac{1}{1 + \left(\dfrac{f/f_0}{1 + g_m R_k}\right)^2}} \cdot \underline{/\theta} \qquad (16)$$

where $\mu > 20$
   f = frequency at which the gain is calculated,
   $f_0 = 1/2\pi R_k C_k$
and $\theta = \tan^{-1} -\left(\dfrac{f/f_0}{1 + g_m R_k}\right).$

The expression within the square root sign in eqn. (16) is known as the **relative gain,** and has been plotted graphically (Ref. C6b, Data Sheet 39). The **phase angle** θ has also been plotted (Data Sheet 41).
References C6, C9 ; Reich (Ref. A30) pp. 166-168 equations 6-34 to 6-34F.
The transient performance under pulse conditions is given in Refs. J1, J5, J6.

(U) **Input impedance with capacitive load**
Input resistance

$$R_i' \approx \frac{(1 + g_m R_k)^2 + (f/f_0)^2}{\omega C_{gk} R_k g_m (f/f_0)} \qquad \text{when } \mu \gg 1. \qquad (17)$$

Input capacitance

$$C_i' \approx C_{vp} + C_{gk} \frac{1 + g_m R_k + (f/f_0)^2}{(1 + g_m R_k)^2 + (f/f_0)^2} \qquad \text{when } \mu \gg 1 \qquad (18)$$

References C6, C9, C20, J7 ; Sturley (Ref. C34) Vol. 2, pp. 119-120.

(V) Plate resistance with capacitive load

$$r_p' = r_p\left[1 + \frac{g_m R_k}{1 + j(f//f_0)} \cdot \frac{\mu + 1}{\mu}\right] \tag{19}$$

(W) Effect of impedance of input voltage source
This is covered partly by Lockhart (Ref. C6) and more completely by Jeffery (Ref. C9)

(X) Wide band amplifiers
The cathode follower is a valuable method of coupling in wide band amplifiers (Refs. C2, C10, C14).

(Y) Cut-off effect and overloading with capacitive load
When a cathode follower with a capacitive load has a quickly-changing voltage applied to its input, the plate current tends to cut off and positive grid current tends to flow even with input voltages less than those practicable with low frequency sine-wave input.   At low input voltages it behaves quite normally, and the time constant of its load impedance (R_k and C_k in parallel) is
$R_k C_k r_p' / (r_p' + R_k)$.
When the plate current has been cut off, the time constant becomes $R_k C_k$, which is more than μ times greater (it is assumed that R_k is not less than r_p), and the advantages of the cathode follower are lost.   It is therefore essential to ensure that plate-current cut-off never occurs.   This effect occurs with a rectangular waveform, a steep saw-tooth waveform or with a high frequency sinewave input.
References C13, C15, J3, J4, J5, J6, J9

(Z) Noise in cathode followers
A triode cathode follower has the same reflected noise voltage in the grid circuit, and the same reflected equivalent noise resistance as when used with plate loading. The high input resistance has the effect of improving the signal/noise ratio.
A pentode cathode follower always has a higher equivalent noise resistance than a triode, but the noise resistance is greater when the screen is by-passed to earth than to the cathode.
Refs. B1, B2, C6a (Part 3-A), J12.

(ii) The cathode–degenerative amplifier and phase splitter
With conventional plate loading, the whole load is in the plate circuit ; with the cathode follower the whole load is in the cathode circuit.   We now consider the case when part of the load is in the plate circuit and part in the cathode circuit.
Fig. 7.19 shows the general form of the cathode degenerative amplifier with plate load Z_L and cathode load Z_k.   The output may be taken from K and earth, or from P and earth, or from both.

(A) Unbypassed cathode resistor
Fig. 7.20 is a particular case commonly known as the unbypassed cathode resistor.[*]
Although not shown in the diagram, there are valve and stray capacitances which can only be neglected at low audio frequencies ; this condition will be assumed as a first approach.   Here R_L is the load resistance, R_k is the unbypassed cathode resistor and R_y the grid resistor.   In this application R_k usually has a much lower resistance than R_L.   If the signal current through the valve and through R_L is called I_p, then the signal voltage drop across R_k is $R_k I_p$, which is proportional to the signal current through the load ; hence the current feedback.   The formulae used below are taken from Sect. 1(iii), putting R_k in place of R_3.
Gain (equation 23, Section 1) :

$$A' = \frac{\mu R_L}{(\mu + 1)R_k + r_p + R_L} \tag{20a}$$

For pentodes

---
*For graphical treatment see Sect. 5(v) and (vi).

$$A' \approx \frac{g_m R_L}{g_m R_k + (r_p + R_L)/r_p} \tag{20b}$$

Ratio of gain with and without feedback

From equation (24a) Section 1,

$$\frac{A'}{A} = \frac{1}{1 + \mu R_k/(r_p + R_k + R_L)} \tag{21}$$

or as a rough approximation for triodes

$$A'/A = R_L/(R_L + \mu R_k) \tag{22}$$

An alternative exact form is

$$\frac{A'}{A} = \frac{1}{1 + g_m R_k r_p/(r_p + R_k + R_L)} \tag{23a}$$

or as a rough approximation for pentodes

$$A'/A \approx 1/(1 + g_m R_k) \tag{23b}$$

From equations (24a) and (24b) of Section 1 :

$$\frac{A'}{A} = \frac{1}{1 - \gamma A} = \frac{1}{1 - \dfrac{\mu \gamma R_L}{r_p + R_L(1 - \gamma)}} \tag{24}$$

where $- \gamma = R_k/R_L$.

Example (1) : 6J5 with $\mu = 20$, $r_p = 20\,000$ ohms (under resistance-coupled conditions), $R_L = 100\,000$ ohms, $R_k = 2700$ ohms.

$A'/A = 1/[1 + 20 \times 2700/(20\,000 + 2700 + 100\,000)] = 0.695$.

Using the rough approximation, $A'/A \approx 0.65$.

Example (2) : Pentode with $g_m = 2000$ micromhos and
$R_k = 2000$ ohms, $A'/A \approx 1/(1 + 4) \approx 1/5$.

With a pentode, the gain is approximately halved when $R_k = 1/g_m$ (see eqn. 23b). Thus when $g_m = 2000$ micromhos, a value of $R_k = 500$ ohms will halve the gain and halve the distortion.

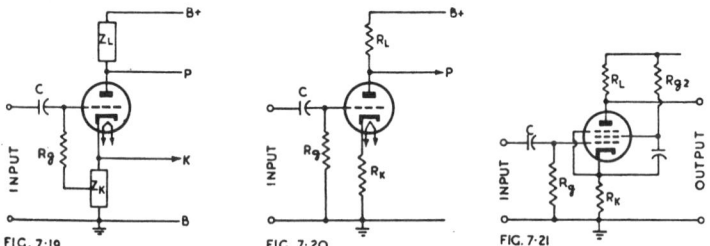

FIC. 7·19   FIG. 7·20   FIG. 7·21

Fig. 7.19. General circuit of a cathode-degenerative amplifier.
Fig. 7.20. Resistance coupled amplifier with unbypassed cathode bias resistor,
giving cathode degeneration with current feedback.
Fig. 7.21. Resistance coupled pentode with unbypassed cathode bias resistor.

Input resistance : As the grid resistor R_g (Fig. 7.20) is returned to earth, the input resistance is R_g. If it had been returned to cathode, eqn. (30) Sect. 1 could then be applied, putting R_g in place of r_i :

Input resistance with R_g returned to cathode—

$$r_i' = R_g(1 + AR_k/R_L) = R_g[1 + \mu R_k/(R_L + R_k + r_p)] \tag{25}$$

Effective plate resistance (equation 25a, Sect. 1)

$$r_p' = r_p + \mu R_k \tag{26}$$

and $r_p'/r_p = 1 + \mu R_k/r_p$ $\tag{27}$

With the arrangement of Fig. 7.20, the total load on the valve is $R_L + R_k$ and this is the value which should be used in drawing the loadlines on the characteristics. The value of R_k is, however, regarded as fixed, while R_L is regarded as capable of being varied—otherwise the " output resistance " would be meaningless. In practice,

R_L may be modified by the connection to point P of a coupling capacitance and following grid resistor, thus providing an a.c. shunt load.

If the valve is a pentode, the suppressor is normally connected to the cathode, although it may be connected to earth if the voltage drop across R_k is very small ; the screen should be by-passed to the cathode (Fig. 7.21). In calculating the grid bias, allowance should be made for the d.c. screen current through R_k.

If a less degree of feedback is desired, portion of R_k may be by-passed. In Fig. 7.22, the value of R_k for calculations of gain, r_i' and r_p' should be R_2, but for grid bias should be $(R_1 + R_2)$. Alternatively, R_2 could be by-passed, instead of R_1.

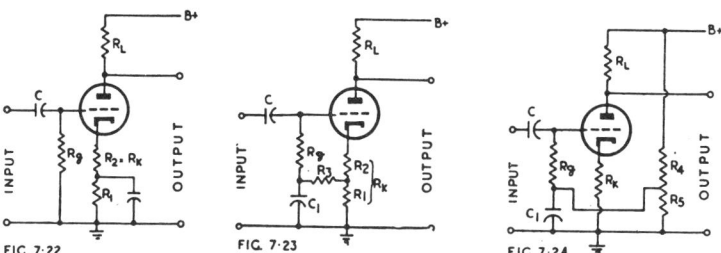

FIG. 7·22   FIG. 7·23   FIG 7·24

Fig. 7.22.   Cathode degenerative amplifier with a small degree of current feedback.
Fig. 7.23.   Cathode degenerative amplifier with a large degree of current feedback.
Fig. 7.24.   Cathode degenerative amplifier with an alternative method of obtaining grid bias.

If a higher degree of feedback is desired there are several alternatives. In Fig. 7.23 $(R_1 + R_2) = R_k$ for signal frequencies, but only R_2 is effective in producing grid bias. C_1 should have a reactance much lower than $(R_3 + R_1)$. As a slight modification, C_1 may be returned to any tapping point along R_1 and R_2, to give any desired degree of feedback between zero and maximum. An alternative modification is to by-pass R_2, and to return C_1 to any tapping point along R_1. Finally, the arrangement of Fig. 7.24 provides a positive voltage from the voltage divider R_4, R_5 which may be adjusted to give the correct bias voltage from grid to cathode.

When R_L and R_k are not purely resistive, they may be replaced by Z_L and Z_k in the expressions for gain which then become vectors.

Current feedback is undesirable in transformer loaded amplifiers for it tends to stabilize the output transformer's magnetizing current (i.e. make it sinusoidal) and thus produce a distorted output voltage.

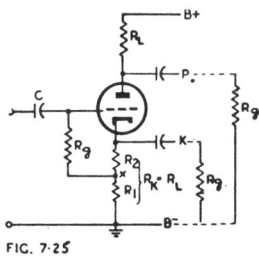

FIG. 7·25

Fig. 7.25.   Conventional phase splitter.

(B) Phase splitter

If R_k is made equal to R_L, we have the well known " phase splitter " (Fig. 7.25), in which the output from P to earth is equal and opposite in phase to that from K to earth. Obviously $(R_1 + R_2) = R_k = R_L$, and the same signal current (I_p) passes through R_L, r_p, R_2 and R_1. The grid resistor R_g is returned to a point on R_k to give the desired grid bias $(E_c = - R_2 I_b)$.

The gain to either output channel is then

$$A' = \frac{\mu R_L}{(\mu + 2)R_L + r_p} \qquad (28)$$

which is always less than 1. In a typical example where $\mu = 20$ and $R_L = 4r_p$, $A' = 0.9$.

The input resistance is

$$r_i' = \frac{R_g}{1 - A'R_1/(R_1 + R_2)} \qquad (29)$$

If $A' = 0.9$ and $R_1 = 9R_2$, then $r_i' = 5.25R_g$.

If R_g is returned to earth, or to a voltage divider, or to a separate bias supply, the input resistance is R_g.

If R_g is returned to cathode, or if R_2 is by-passed, then $r_i' = R_g/(1 - A') \approx 10R_g$.

If R_2 is by-passed, then R_1 must be increased to equal R_L and channel K should be taken from point X. If channel K is taken from point X, and $R_1 = R_L$, then the by-passing of R_2 is unimportant, except that the maximum output voltage will be reduced slightly ; this is of greater importance with low mu valves.

The effective output resistance is different for the two output channels, since P operates with current feedback and K with voltage feedback.

Channel P : $r_p' = r_p + (\mu + 1)R_k$ where $R_k = R_L$ (30)
(from equation 25a, Sect. 1)

Channel K : $r_p' = (r_p + R_L)/(\mu + 1)$ (31)
(from equation 3)

but this does not affect the balance at either low or high frequencies when the total effective impedance of channel P is equal to that of channel K. The same signal plate current which flows through one impedance Z_P also flows through the other impedance Z_K, and if $Z_P = Z_K$ then the two output voltages are equal. The ratio of the output voltages is $R_L : R_K$ at low frequencies and $Z_P : Z_K$ at any high frequency. The capacitive component of Z_P is the sum of the input capacitance of the following stage, wiring and stray capacitances, and the output capacitance of P channel (eqn. 33), and similarly with Z_K (eqn. 34). The output capacitances of the phase splitter normally differ by only 1 or 2 $\mu\mu$F, having a negligible effect on the balance at 10 000 c/s— see also pages 522-523.

Input capacitance (Fig. 7.25)
$C_i' = C_{gk}/(1 - A') + C_{gp}(1 + A')$ (32)

Output capacitance
P channel : $C_0' = 2C_{pk} + C_{gp}(1 + 1/A')$ (33)
K channel : $C_0' = 2C_{pk} + C_{gk}(1/A' - 1) + C_{hk}'$ (34)

The equivalent source impedance that determines the high frequency attenuation for each channel due to shunt capacitance is given by

$$R_0 = \frac{r_p R_L}{r_p + R_L (\mu + 2)}$$ (34a)

and is of the order of 1000 ohms. Hence the excellent high frequency response characteristic (Ref. A33).

(C) **Tone control with cathode degenerative amplifier**
If R_k (Fig. 7.20) is shunted by a capacitor C_k, it will cause a rise in gain which will reach the maximum of a normal non-degenerative amplifier at frequencies where the reactance of C_k is small compared with R_k. C_k may also be connected across portion of R_k.

If R_k is shunted by an inductor L_k, it will cause a similar rise of gain at low frequencies.

(D) **Degenerative cathode impedance** $(R_k + L_k)$
A degenerative cathode impedance with R_k in series with L_k may be used with r-f or i-f amplifiers.

(F) **Other forms of cathode degenerative amplifiers**
Cathode-coupling is described in Section 2(viii).

A cathode-degenerative amplifier may be used as the first or second stage of an amplifier having an overall feedback loop, to provide improved stability and a flatter frequency response characteristic—see Sect. 3(v)E and 3(vii)C.

(iii) **Voltage feedback from secondary of output transformer**
Voltage feedback* may be taken from the secondary of the output transformer and applied to the grid or any other suitable electrode in the power amplifier valve or any earlier stage. Fig. 7.26 shows one form with feedback to the grid of the power amplifier valve. This is limited to transformer input, and there is a further limitation

*For graphical treatment see Sect. 5(iii) and (iv).

in that the voltage across the voice coil may not be great enough to provide sufficient degeneration. It is important to avoid accidental reversal of the secondary terminals. This, and all its modifications, may be treated as pure voltage feedback following the formulae of Sect. 1, at least in the middle frequency range where β is equal to T_2/T_1, the transformer turns or voltage ratio. It has the properties of maintaining constant the voltage across the secondary, thereby avoiding the effects of transformer leakage inductance at high frequencies, and of reducing the transformer distortion. Unfortunately, at very low and very high audio frequencies, the phase angle introduced by the output transformer tends to cause instability, the tendency being more pronounced as the feedback is applied over 2 or 3 stages. This subject is treated in detail in Section 3.

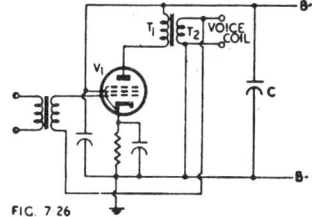

Fig. 7.26. Voltage feedback from secondary of output transformer, over one stage.

FIG. 7.26

The push-pull form is merely a mirror image of the single-ended variety, and involves two separate secondary windings on the input transformer, and a centre-tap on the secondary of the output transformer, which is earthed.

In either case, if the feedback voltage is greater than desired, the connection may be taken from a voltage divider across the voice coil (or each half of the transformer secondary).

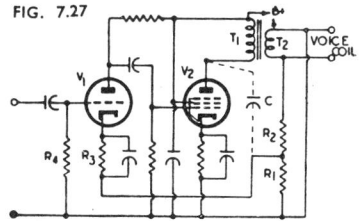
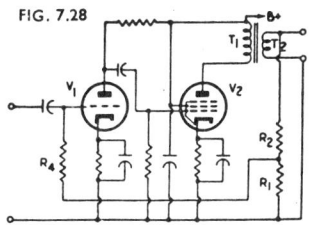

Fig. 7.27. Voltage feedback from secondary of output transformer to cathode of preceding stage.
Fig. 7.28. Voltage feedback from secondary of output transformer to grid of preceding stage.

A form suitable for feedback over two stages is shown in Fig. 7.27 where the feedback voltage is reduced by the voltage divider R_1R_2 and applied to the cathode of V_1. The voltage divider total resistance $(R_1 + R_2)$ should be at least 20 times the voice coil impedance, while R_1 should preferably be less than one tenth of R_3, otherwise there will be an appreciable amount of negative current feedback* in V_1. Here

A = voltage gain from grid of V_1 to plate of V_2,

$$\beta = \left(\frac{R_1}{R_1 + R_2}\right)\left(\frac{T_2}{T_1}\right) \tag{35}$$

and $r_i' = R_4$ at mid frequencies (36)

Portion of the plate current of V_1 will flow through the secondary winding. It is sometimes found that a very small condenser C connected from the plate of V_2 to the cathode circuit of V_1 reduces the tendency to instability.

A modified form providing feedback to the grid circuit of V_1 is shown in Fig. 7.28 (note the reversed connections to the secondary of the output transformer). This

---

*In practice, this may be desirable.

avoids d.c. through the secondary winding, but the impedance of the preceding stage (R_0) affects the value of β, A being unchanged,

$$\beta = \left(\frac{R_1}{R_1 + R_2}\right)\left(\frac{T_2}{T_1}\right)\left(\frac{R_0}{R_0 + R_4}\right). \tag{37a}$$

In this case the input resistance is decreased because the feedback voltage is applied in shunt from grid to cathode,

$$r_i' = R_4/(1 - \beta A). \tag{37b}$$

This principle is also incorporated in Fig. 28.3 where feedback is taken from the secondary to a tapping on the volume control.

A further modification, in which the amount of feedback decreases as the volume control is increased, is described in Chapter 35 (Fig. 35.3).

See Sect. 2(vi) below for an analysis of a two stage amplifier with feedback from the secondary of the output transformer.

(iv) Voltage feedback from plate—transformer-input

A typical power output stage with negative voltage feedback from the plate to the bottom end of the secondary of the input transformer is shown in Fig. 7.29. The reactance of C is normally small compared with the total resistance of the voltage divider ($R_1 + R_2$). The condenser C is placed in this position instead of between R_2 and plate since in the latter position the circuit may oscillate owing to the capacitance of C to grid. If there is a tendency to oscillate due to the leakage inductance of the transformer, a small condenser or a resistance, or both, may be connected across the input transformer secondary.

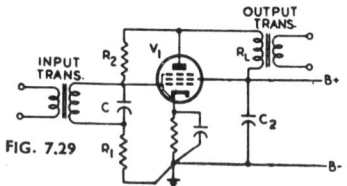

Fig. 7.29. Voltage feedback from the plate to the grid circuit, using transformer-coupled input.

This arrangement reduces the amplitude distortion resulting from saturation of the output transformer core at low frequencies, and improves the low-frequency response. It does not, however, counteract the effect of leakage reactance at high frequencies.

A = voltage gain of valve from grid to plate

$\beta = R_1/(R_1 + R_2)$

and r_i' is extremely high, being unchanged by the feedback (valve input impedance alone).

The same method may be applied to a push-pull stage, but a transformer is required having two separate secondaries, and each valve must have its own feedback network. This circuit is not suitable for use with resistance coupling.

(v) Voltage feedback from plate—r.c.c. input

If it is desired to employ negative feedback from the plate to a resistance-capacitance input coupling, it is impossible to apply the feedback voltage in series opposition to the signal voltage. The only manner of applying it is, so to speak, in shunt with the input signal.

The simplest circuit is Fig. 7.30 in which V_1 is r.c. coupled to V_2, and a feedback path $R_1 C_1$ is provided from plate to grid of V_2. For normal applications C_1 is merely a blocking condenser to isolate grid and plate for d.c. The resistance R_1 acts as a shunt path for signal frequencies in much the same way as the Miller Effect; the circuit behaves as though an additional shunt resistance equal to $R_1/(A_2 + 1)$ were connected from grid to earth, A_2 being the numerical voltage gain of V_2. This decreased input resistance causes a reduction in overall amplification. The amplification of V_2 is not affected, but V_1 now has a heavy a.c. shunt load, leading to increased distortion and possible overloading. As an example, for the gain to be reduced to half, the additional a.c. shunt resistance across R_g will be equal to the total resistance of r_{p1}, R_L and R_g in parallel. A larger than normal value of coupling condenser (C) will obviously be required.

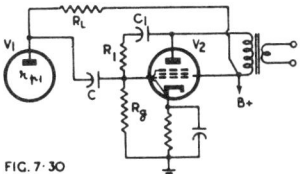

Fig. 7.30.   Voltage feedback from plate to grid, with resistance capacitance coupling. The feedback makes the a.c. shunt load on V_1 considerably less than R_g.

FIG. 7·30

As a modification of Fig. 7.30, R_1 may be made zero and C_1 may have a capacitance of about 10 to 100 $\mu\mu$F to provide degeneration principally at high audio frequencies. This has been used very successfully in small receivers with very little margin of gain, but is only fully effective when no plate by-pass capacitor is used.

An improved circuit is Fig. 7.31 (Refs. E1, E3) in which R_1 and R_2 form a voltage divider across the primary of the output transformer, R_L being connected to the junction (X). If ($R_1 + R_2$) is at least 10 times the load resistance of V_2, the divider will have little effect on V_2. It is obvious that the degree of feedback can be adjusted from zero to maximum by moving the tapping point X on the voltage divider from the B+ end to the plate end. If $R_s + R_L \gg R_1$, where $R_s = r_{p1}R_g/(r_{p1} + R_g)$,

then $\beta \approx -\dfrac{R_s}{R_s + R_L} \cdot \dfrac{R_1}{R_1 + R_2}$ (38)

An electrically equivalent circuit is Fig. 7.32 (Ref. E2) in which the two resistances R_3 and R_4 take the place of the three resistances R_L, R_1 and R_2 of Fig. 7.31. For identical operating conditions :

$$R_3 = \frac{(R_1 + R_2)R_L}{R_2} \text{ and } R_4 = \frac{R_3 R_L}{R_3 - R_L} \cdot$$ (39)

The value of β in Fig. 7.32 is given by

$$\beta = -\frac{R}{R_4 + R} \text{ where } \frac{1}{R} = \frac{1}{r_{p1}} + \frac{1}{R_g} + \frac{1}{R_3} \cdot$$ (40)

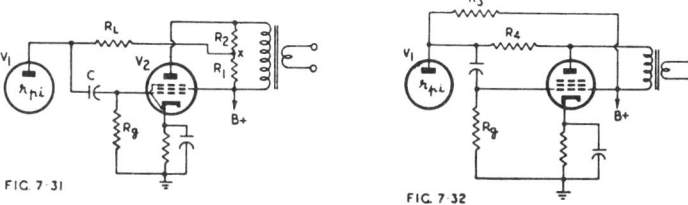

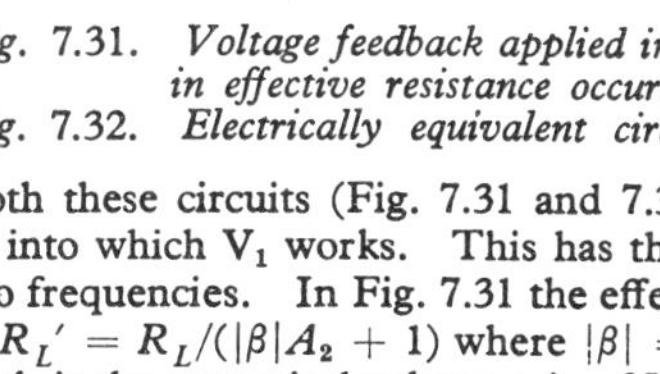

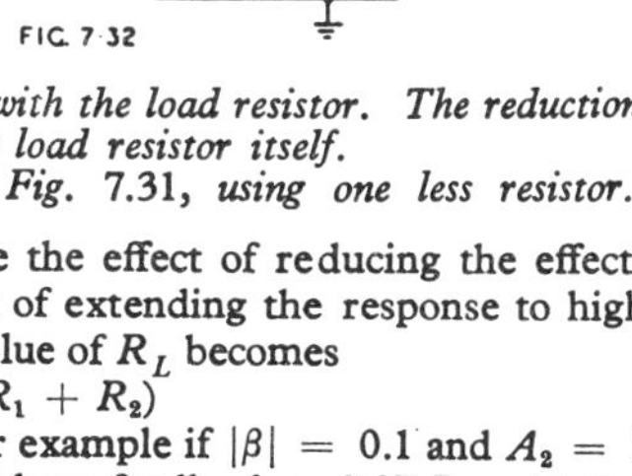

FIG. 7·31

FIG. 7·32

Fig. 7.31.   Voltage feedback applied in series with the load resistor. The reduction in effective resistance occurs in the load resistor itself.
Fig. 7.32.   Electrically equivalent circuit to Fig. 7.31, using one less resistor.

Both these circuits (Fig. 7.31 and 7.32) have the effect of reducing the effective load into which V_1 works. This has the merit of extending the response to higher audio frequencies. In Fig. 7.31 the effective value of R_L becomes

$R_L' = R_L/(|\beta|A_2 + 1)$ where $|\beta| = R_1/(R_1 + R_2)$

and A_2 is the numerical voltage gain of V_2. For example if $|\beta| = 0.1$ and $A_2 = 17$, the effective load resistance changes from R_L without feedback to 0.37 R_L with feedback. The load R_L' into which V_1 works is, however, not constant because A_2 varies due to distortion. This circuit is usually limited to values of $|\beta|$ not greater than say 0.05 to 0.1 for typical applications.

A similar effect occurs with the equivalent circuit Fig. 7.32 in which the feedback causes the effective value of R_4 to change to $R_4/(A_2 + 1)$. In both circuits V_1 is preferably a pentode, although a high-mu triode may be used with less effectiveness.

Mezger (Ref. E4) has shown that it is equally valid to regard the plate resistance of V_1 (and consequently its amplification factor) as being reduced by feedback, leaving the external resistors unchanged.

(vi) Voltage feedback over two stages

Voltage feedback over two stages permits a wide choice of circuits, without the limitations which frequently arise with a single stage.

(A) One of the deservedly most popular circuits is Fig. 7.33 in which feedback is taken from the plate of V_2 to the cathode of V_1. The most serious limitation to its use is that it cannot be applied without complications* to combined second detector and amplifier valves of the conventional type. It actually involves two types of feedback—the primary voltage feedback, and the subsidiary current feedback caused by the unbypassed cathode resistor in V_1.

This is a form of Duerdoth's multiple feedback—see Sect. 3(v)E—and the two feedback voltages must be added together to determine the performance. The subsidiary feedback increases the stability by reducing the slope of the βA characteristic over a wide range of very low and a wide range of very high frequencies.

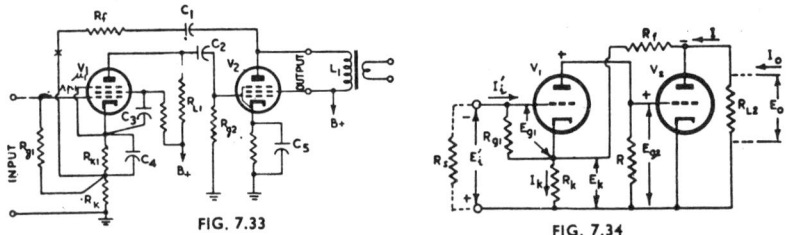

FIG. 7.33   FIG. 7.34

Fig. 7.33. Amplifier with voltage feedback from the plate of V_2 to the cathode of V_1, also incidentally incorporating negative current feedback due to the unbypassed resistor R_k.

Fig. 7.34. Simplified a.c. equivalent of Fig. 7.33 for calculations.

The following theoretical treatment† is based on the simplified a.c. diagram Fig. 7.34. Here V_1 has constants μ_1, g_{m1}, r_{p1} while V_2 has μ_2, g_{m2} and r_{p2}. Resistor R represents R_{L1} and R_{g2} in parallel. It is assumed that, unless otherwise specified, the impedance of the input voltage source (R_s) is zero.

The amplification of stage V_1 without negative current feedback is defined as A_1 where

$$A_1 = E_{g2}/E_{g1} = \mu_1 R/(R + R_k + r_{p1}) \tag{41a}$$

The amplification of stage V_2 is given by

$$A_2 = E_0/E_{g2} = \mu_2 R_{L2}/(R_{L2} + r_{p2}) \tag{41b}$$

Also

$$\beta \approx R_k/(R_f + R_k) \tag{41c}$$

Now

$$E_0 = A_1 A_2 E_{g1} \tag{41d}$$

$$E_i' = E_{g1} + E_k = E_{g1} + R_k I_k \tag{41e}$$

and

$$I_k = [\mu_1 E_{g1}/(R + R_k + r_{p1})] + \beta E_0/R_k \tag{41f}$$

$$= E_{g1}[\mu_1/(R + R_k + r_{p1}) + \beta A_1 A_2/R_k] \tag{41g}$$

(neglecting $E_i'/(R_{g1} + R_k)$ as being very small).

Applying these values in equation (41e)

$$E_i' = E_{g1}\left[1 + \frac{\mu_1 R_k}{R + R_k + r_{p1}} + \beta A_1 A_2\right] \tag{41h}$$

Also

$$E_0 = A_1 A_2 E_{g1} = \frac{\mu_1 R}{R + R_k + r_{p1}} \cdot A_2 E_{g1}$$

*See Fig. 35.3 for one possible arrangement.

†The Editor is indebted to Mr. E. Watkinson for this method. This method assumes that $(r_{p1} + R)$ is large compared with R_k, and that R_f is large compared with R_{L2}.

$$= \left(\frac{\mu_1 R}{R + R_k + r_{p1}}\right)\left(\frac{E_i'}{1 + \beta A_1 A_2 + \mu_1 R_k/(R + R_k + r_{p1})}\right)A_2$$

$$= \left(\frac{\mu_1 R}{R + R_k + r_{p1}}\right)\left(\frac{A_2 E_i'(R + R_k + r_{p1})}{R + R_k + r_{p1} + \mu_1 R_k + \beta A_1 A_2(R + R_k + r_{p1})}\right)$$

$$= E_i'\left(\frac{A_2 \mu_1 R}{R + (\mu_1 + 1)R_k + r_{p1} + \beta A_1 A_2(R + R_k + r_{p1})}\right)$$

Therefore $E_0 = E_i' \dfrac{A_2 \mu_1 R}{\left[R_k\left(\dfrac{1 + \mu_1 + \beta A_1 A_2}{1 + \beta A_1 A_2}\right) + R + r_{p1}\right]\left[1 + \beta A_1 A_2\right]}$ (41i)

Also when $R_k = 0$ there is no current feedback:

$$E_0 = E_i \frac{A_2 \mu_1 R}{R + r_{p1}} \tag{41j}$$

The overall amplification with feedback is given by
$$A' = E_0/E_i' \tag{41k}$$
while that without feedback is given by
$$A = E_0/E_i = A_2 \mu_1 R/(R + r_{p1}) \tag{41l}$$

Therefore $\dfrac{A'}{A} = \left[\dfrac{A_2 \mu_1 R}{\left[R_k\left(\dfrac{1 + \mu_1 + \beta A_1 A_2}{1 + \beta A_1 A_2}\right) + R + r_{p1}\right]\left[1 + \beta A_1 A_2\right]}\right]\left(\dfrac{R + r_{p1}}{A_2 \mu_1 R}\right)$

Therefore $\dfrac{A'}{A} = \dfrac{R + r_{p1}}{\left[R_k\left(\dfrac{1 + \mu_1 + \beta A_1 A_2}{1 + \beta A_1 A_2}\right) + R + r_{p1}\right]\left[1 + \beta A_1 A_2\right]}$ (41m)

Output impedance

The effect of the unbypassed cathode resistor depends on the resistance of the source (R_s).
Let $\delta = R_s/(R_s + R_v)$ (42a)
Let voltage E_0 be applied across R_{L2} from an infinite impedance source with no input voltage applied to the amplifier.

Then $I_0 = \dfrac{E_0}{R_{L2}} + \dfrac{E_0}{r_{p2}} + \dfrac{E_0}{R_f + R_k} + \dfrac{E_0 \beta \delta A_1 \mu_2}{r_{p2} + R_{L2}}$ (42b)

The final term in eqn. (42b) is the effect of the voltage E_0, a fraction of which is fed back across R_k, and amplified by V_1 and applied to the grid of V_2. The effect in the plate circuit of V_2 is the same as though a voltage $E_0 \beta \delta A_1 \mu_2$ were applied to a series connection of r_{p2} and R_{L2}.

Thus the "looking backwards" output terminal impedance is given by

$$R_0' = \frac{E_0}{I_0} = \frac{1}{\dfrac{1}{R_{L2}} + \dfrac{1}{r_{r2}} + \dfrac{1}{R_f + R_k} + \dfrac{\beta \delta A_1 \mu_2}{r_{v2} + R_{L2}}} \tag{42c}$$

i.e. the effect of the feedback is to add another impedance,
$$\frac{r_{v2} + R_{L2}}{\beta \delta A_1 \mu_2},$$
in parallel with the output, as shown in Fig. 7.41.

Input impedance

$$R_i' = \frac{E_i'}{I_i'} = \frac{E_{g1}\left[1 + \dfrac{\mu_1 R_k}{R + R_k + r_{p1}} + \beta A_1 A_2\right]}{E_{g1}/R_{g1}} \tag{42d}$$

Therefore $R_i' = R_{g1}\left[1 + \dfrac{\mu_1 R_k}{R + R_k + r_{p1}} + \beta A_1 A_2\right]$ (42e)

Therefore $\dfrac{R_i}{R_i} = 1 + \dfrac{\mu_1 R_k}{R + R_k + r_{p1}} + \beta A_1 A_2$ (42f)

where R_i is defined as the input resistance when $R_k = 0$.

Conclusions

Eqn. (41m) shows that the effect of feedback in reducing gain in this circuit (Figs. 7.33 and 7.34) is not given simply by the product of the current feedback gain reduction and the voltage feedback gain reduction, each considered separately

Eqn. (41h) indicates that the two feedback voltages are effectively added so far as their effect on gain is concerned. This is in line with Duerdoth's multiple feedback theory in Sect. 3(v)E.

Eqn. (42c) shows that the effective plate resistance of V_2 differs from that of an equivalent amplifier without current feedback only on account of δ. When the source impedance is zero, δ has the value unity, and the effective plate resistance of V_2 is identical with that of an equivalent amplifier with voltage feedback only.

It is obvious that current feedback in V_1 cannot directly affect the plate resistance of V_2. The only indirect effect (through δ) is the result of partial application of the feedback voltage to the grid of V_1, which must also affect the gain. In other words, the impedance of the input voltage source affects the gain, the effective plate resistance and the output resistance R_0'.

If the source is reactive, some non-linear frequency characteristic will occur— for example, if the source is a crystal pickup there will be some measure of bass boosting.

The circuit of Fig. 7.33 tends to produce peaks at low and high frequencies in the response characteristic unless the feedback factor is low. An analysis of the peaks has been made by Everest and Johnston (Ref. H5) based on the ideal condition of zero-impedance screen circuits and unbypassed cathode resistors ; it also omits any complications arising from an output transformer. Methods for reducing or eliminating these peaks are described in Sect. 3(vi) and (vii).

The condenser C_1 in the feedback network of Fig. 7.33 is normally only a blocking condenser ; it may, however, be designed to produce a fixed degree of bass boost. Alternatively, C_1 may remain as a blocking condenser and a second condenser inserted at point X to provide bass boosting ; the latter may be shunted by a variable resistance to form a continuously variable tone control (see Chapter 15 Sect. 2 and Fig. 15.11).

It will be demonstrated in Sect. 3 that feedback over two stages of r.c. coupling is normally stable ; this does not necessarily hold if cathode or screen by-pass condensers are used or if an iron-cored transformer forms any part of the feedback loop.*

(B) Feedback from secondary of output transformer

See Sect. 2(iii) and Figs. 7.27 and 7.28 for the general description of such an amplifier.

An illustration of the effects of negative feedback on a simple 2 stage a-f amplifier can be obtained by calculating† the low frequency response of the circuit of Fig. 7.35. In this simplified circuit batteries are used as voltage sources to minimize calculation, but the effect of cathode by-passes (or of screen by-passes) could be covered without difficulty by using Figs. 12.11 A and B and 12.3 A and C.

The values of components used in interstage couplings have also been chosen for simplicity. Referring to Table 42 in Chapter 38 it will be seen that the reactance of a 0.01 μF condenser (the coupling between the 6AV6 and 6AQ5) at 50 c/s is 318 000 ohms. It is assumed that the series impedance of R_7, and the parallel combination of R_6 and the 6AV6 plate resistance [see Chapter 12 Sect. 2(xiii) Eqn. 15] is also 318 000 ohms, so that at 50 c/s the reactance and resistance of the interstage coupling network are equal.

---

*The feedback loop is the complete path, commencing from the point to which the feedback is returned, through the amplifier to the point from where the feedback is taken, back through the feedback network to the starting point.

†The Editor is indebted to Mr. E. Watkinson for the calculation of this example.

Similarly the reactance of the output transformer primary at 50 c/s is equal to the plate resistance of the 6AQ5 in parallel with the reflected load and winding resistances (coupling 3).

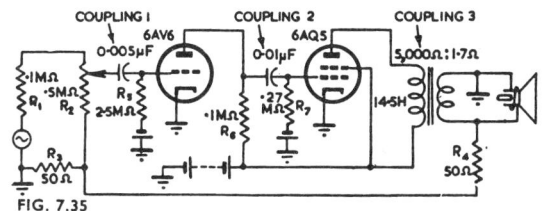

Fig. 7.35. Simple feedback amplifier in which setting of volume control alters feedback.

FIG. 7.35

Movement of the input volume control is assumed to have no effect on the frequency ($12\frac{1}{2}$ c/s) at which the reactance of the input condenser is equal to the resistance in series with it, say 2.55 megohms (coupling 1).

The amplification factor, μ, of the 6AV6 is 100 and its plate resistance, r_p, 80 000 ohms, so the mid-frequency 6AV6 gain, A_1 [see Chapter 12 Sect. 2(vii)] is

$$A_1 = \frac{100 \times 73\,000}{80\,000 + 73\,000} = 48$$

where 73 000 ohms is the value of R_6 and R_7 in parallel.

The 6AQ5 μ is 210 and its r_p 52 000 ohms so that the 6AQ5 gain

$$A_2 = \frac{210 \times 5000}{52\,000 + 5000} = 18.4.$$

The output transformer impedance ratio is 5000/1.7 so that its gain

$$A_3 = \sqrt{1.7/5000} = 1/54.$$

Total gain from 6AV6 grid to voice coil :

$$A = A_1 \times A_2 \times A_3 = 48 \times 18.4/54 = 16 \text{ times} = 24.1 \text{ dbvg.*}$$

When the volume control is turned to its minimum setting, one half of the output voltage is applied to the amplifier input, i.e. $\beta = -\frac{1}{2}$ and the factor

$$\frac{1}{1 - \beta A} = \frac{1}{1 + 16 \times \frac{1}{2}} = \frac{1}{9}.$$

As the volume control is turned up, a smaller amount of the negative feedback voltage is applied to the 6AV6 control grid, and when the control is 50 000 ohms from its maximum setting only one quarter of the previously used feedback voltage is effective (ignoring the shunting effect of R_5). In this case

$$\frac{1}{1 - \beta A} = \frac{1}{3}.$$

The low frequency response curves will be calculated for each of these volume control settings, but it is first necessary to obtain the response of the amplifier without feedback. This is done in Table 1. Lines 1 and 2 of Table 1 are obtained from Figs. 12.9A and B with $12\frac{1}{2}$ c/s taken as the frequency for 70.7 per cent. voltage gain (when the reactance is numerically equal to the resistance) in line 1, and 50 c/s in line 2. Thus for the 25 c/s response in line 1

$$\frac{\text{Actual frequency}}{\text{Frequency for 70.7 per cent. frequency response}} = 2$$

so that attenuation = 1 db and phase shift = 27°, and so on for each of the other frequencies in the two lines.

Line 3 is obtained by adding twice the attenuation and phase shift of line 2 (because there are two 50 c/s couplings) to line 1.

Line 4 is the result of adding the mid-frequency gain of the amplifier (24.1 db) to the attenuation and phase shift at each of the tabulated frequencies. This gives the frequency response as plotted at A in Fig. 7.36.

In line 5 the gain A is expressed numerically, giving the actual gain at each frequency.

*Gain expressed in decibels of voltage gain—see Chapter 19 Sect. 1(vi)A.

Table 2 sets out the calculations necessary to obtain the modification of the frequency response by the application of feedback when $\beta = -\frac{1}{2}$. Line 1 results from converting the attenuation expressed in db in line 3 of Table 1 to fractions, and line 2 is obtained by multiplying the mid-frequency gain, the attenuation and phase (for the frequency concerned) by β. The result is βA which is plotted in Fig. 7.37.

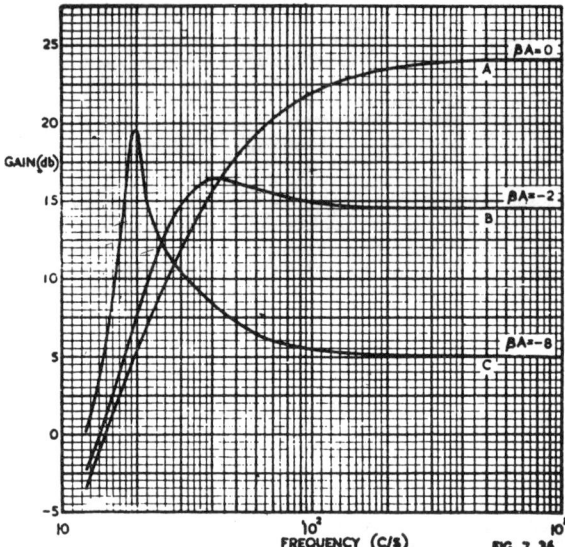

Fig. 7.36.   *Frequency response of amplifier of Fig. 7.35 with zero feedback and two selected values of feedback.*

For the method of plotting this curve and for its significance see Sect. 3 of this Chapter. It should be noted that once the βA curve has been obtained the effects of feedback on the gain and phase shift of the amplifier can readily be obtained graphically as discussed below.   Lines 3 to 9 of Table 2 are the mathematical equivalents of the measurements mentioned below (1 and 5) which give the degree of degeneration or regeneration and the phase shift with feedback between output and input voltages at a given frequency.

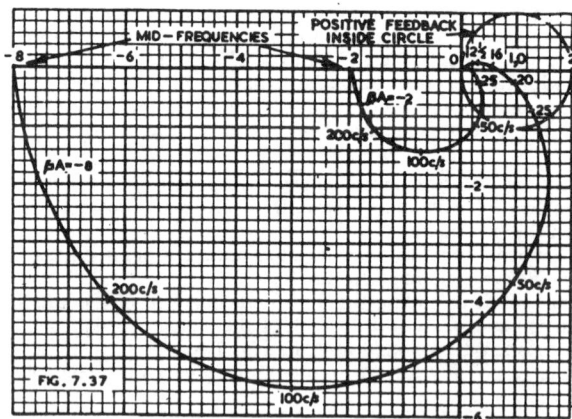

Fig. 7.37.   *Polar diagram for two selected values of feedback.*

TABLE 1.

	Mid-frequency	200 c/s	100 c/s	50 c/s	37½ c/s	25 c/s	20 c/s	18¾ c/s	16 c/s	12½ c/s
(1) Coupling 1 Attenuation (db)	0∠0°	0∠4°	0.10∠7°	0.3∠14°	0.5∠18°	1.0∠27°	1.4∠32°	1.5∠34°	2.1∠38°	3∠45°
(2) Coupling 2 or 3 Attenuation (db)	0∠0°	0.3∠14°	1.0∠27°	3.0∠45°	4.5∠53°	7.0∠63°	8.5∠68°	9.1∠69.5°	10.3∠72°	12.3∠76°
(3) Couplings 1 + 2 + 3. Attenuation (db)	0∠0°	0.6∠32°	2.1∠61°	6.3∠104°	9.5∠124°	15.0∠153°	18.4∠168°	19.7∠173°	22.7∠182°	27.6∠197°
(4) Gain $A = A_1 \times A_2 \times A_3$ (db)	24.1∠0°	23.5∠32°	22.0∠61°	17.8∠104°	14.6∠124°	9.1∠153°	5.7∠168°	4.4∠173°	1.4∠182°	−3.5∠197°
(5) Gain A	16∠0°	14.9∠32°	12.6∠61°	7.7∠104°	5.3∠124°	2.9∠153°	1.9∠168°	1.66∠173°	1.17∠182°	0.67∠197°

Lines 3 and 4 are the values for the sine and cosine of the angle of phase shift [see Chapter 38 Table 72 for magnitude ; Chapter 6, Sect. 3(iii) for sign]. Line 5 expresses $1 - \beta A$ in the form $a + jb$. Multiplying the magnitude of βA in line 2 by the sine of the associated angle gives the j term and multiplying by the cosine and adding 1 (because of the 1 in $1 - \beta A$) gives the real term. For example, at 100 c/s, $1 + 6.3 \times 0.4848 = 4.05$ and $6.3 \times 0.8746 = 5.5$ so that $1 - \beta A = 4.05 + j\,5.5$.

Lines 6 and 7 convert $1 - \beta A$ to polar co-ordinates again (the change to rectangular co-ordinates was necessary to add 1 to βA), line 6 being expressed in the form of the square root of the sum of the squares of the two terms in line 5 at the angle whose tangent is the imaginary term divided by the real term, e.g. at 100 c/s the angle is $\tan^{-1} 5.5/4.05 = \tan^{-1} 1.36 = 54°$ to the nearest degree. Where negative terms occur the appropriate quadrant is determined from the knowledge of the sign of the sine, cosine and tangent, all of which are known at this stage.

Line 8 is the reciprocal of line 7, so that magnitudes are divided into unity, and angles are reversed in sign. In line 9 the magnitudes are expressed in db, and in line 10 the original gain A at each frequency is multiplied by $1/(1 - \beta A)$ to give the response of the amplifier with feedback. Since lines 9 and 4 are both expressed in db, the multiplication is carried out by adding the values in decibels. Line 10 is plotted in curve C of Fig. 7.36.

Table 3 gives similar calculations when $\beta = -1/8$ and the result is plotted in curve B of Fig. 7.36, while the smaller βA polar diagram is also plotted in Fig. 7.37.

Several interesting aspects which are common to all feedback amplifiers are brought out by these calculations and curves.

TABLE 2

	Mid-frequency	200 c/s	100 c/s	50 c/s	25 c/s	20 c/s	18¼ c/s	16 c/s	12¼ c/s
(1) Couplings 1+2+3. Gain	1.0	0.93	0.79	0.48	0.18	0.12	0.104	0.073	0.042
(2) $\beta A(\beta=-\frac{1}{3})$	$-8\angle 0°$	$-7.45\angle 32°$	$-6.3\angle 61°$	$-3.8\angle 104°$	$-1.44\angle 153°$	$-0.96\angle 168°$	$-0.83\angle 173°$	$-0.58\angle 182°$	$-0.34\angle 197°$
(3) Sine	0	0.5299	0.8746	0.9703	0.4540	0.2079	0.1219	-0.0349	-0.2924
(4) Cosine	1.0	0.8480	0.4848	-0.2419	-0.8910	-0.9781	-0.9925	-0.9994	-0.9563
(5) $1-\beta A$	$9+j0$	$7.3+j3.9$	$4.05+j5.5$	$0.08+j3.7$	$-0.28+j0.65$	$0.06+j0.20$	$0.18+j0.1$	$0.42-j0.02$	$0.67-j0.1$
(6) $1-\beta A$	$\sqrt{9^2+0^2}\angle 0°$	$\sqrt{53.3+15.2}\angle 28°$	$\sqrt{16.4+30.3}\angle 54°$		$\sqrt{0.078+0.42}\angle 113°$	$\sqrt{0.004+0.04}\angle 73°$			
(7) $1-\beta A$	$9\angle 0°$	$8.3\angle 28°$	$6.8\angle 54°$	$3.7\angle 87°$	$0.71\angle 113°$	$0.21\angle 73°$	$0.21\angle 29°$	$0.42\angle -3°$	$0.67\angle -8°$
(8) $\dfrac{1}{1-\beta A}$	$0.11\angle 0°$	$0.12\angle -28°$	$0.147\angle -54°$	$0.27\angle -87°$	$1.4\angle -113°$	$4.76\angle -73°$	$4.76\angle -29°$	$2.4\angle 3°$	$1.5\angle 8°$
(9) $\dfrac{1}{1-\beta A}$ (db)	$-19.2\angle 0°$	$-18.4\angle -28°$	$-16.6\angle -54°$	$-11.4\angle -87°$	$+3.0\angle -113°$	$13.6\angle -73°$	$13.6\angle -29°$	$7.6\angle 3°$	$3.6\angle 8°$
(10) $\dfrac{A}{1-\beta A}$ (db)	$4.9\angle 0°$	$5.1\angle 4°$	$5.4\angle 7°$	$6.4\angle 17°$	$12.1\angle 40°$	$19.3\angle 95°$	$18\angle 144°$	$9.0\angle 185°$	$0.1\angle 205°$

TABLE 3

	Mid-frequency	200 c/s	100 c/s	50 c/s	37½ c/s	25 c/s	20 c/s	16 c/s	12½ c/s
(1) $1+2+3$ Gain	$1\angle 0°$	$0.93\angle 32°$	$0.79\angle 61°$	$0.48\angle 104°$	$0.33\angle 124°$	$0.18\angle 153°$	$0.12\angle 168°$	$0.073\angle 182°$	$0.042\angle 197°$
(2) βA ; $\beta=-\tfrac{1}{8}$	$-2\angle 0°$	$-1.86\angle 32°$	$-1.58\angle 61°$	$-0.96\angle 104°$	$-0.66\angle 124°$	$-0.36\angle 153°$	$-0.24\angle 168°$	$-0.146\angle 182°$	$-0.082\angle 197°$
(3) Sine	0	0.5299	0.8746	0.9703	0.8290	0.4540	0.2079	-0.0349	-0.2924
(4) Cosine	1	0.8480	0.4848	-0.2419	-0.5592	-0.8910	-0.9781	-0.9994	-0.9563
(5) $1-\beta A$	$3+j0$	$2.58+j0.98$	$1.77+j1.4$	$0.77+j0.93$	$0.63+j0.55$	$0.68+j0.16$	$0.755+j0.05$	$0.855-j0.005$	$0.92-j0.024$
(6) $1-\beta A$	$\sqrt{3^2+0^2}\angle 0°$	$\sqrt{6.65+0.96}\angle 21°$	$\sqrt{3.13+1.96}\angle 38°$	$\sqrt{0.59+0.86}\angle 50°$	$\sqrt{0.40+0.30}\angle 41°$	$\sqrt{0.46+0.02}\angle 13°$			
(7) $1-\beta A$	$3\angle 0°$	$2.76\angle 21°$	$2.26\angle 38°$	$1.21\angle 50°$	$0.84\angle 41°$	$0.69\angle 13°$	$0.76\angle 4°$	$0.86\angle 0°$	$0.92\angle -2°$
(8) $\dfrac{1}{1-\beta A}$	$0.33\angle 0°$	$0.36\angle -21°$	$0.44\angle -38°$	$0.82\angle -50°$	$1.19\angle -41°$	$1.45\angle -13°$	$1.3\angle -4°$	$1.16\angle 0°$	$1.09\angle 2°$
(9) $\dfrac{1}{1-\beta A}$ (db)	$-9.6\angle 0°$	$-8.9\angle -21°$	$-7.1\angle -38°$	$-1.7\angle -50°$	$1.6\angle -41°$	$3.2\angle -13°$	$2.3\angle -4°$	$1.3\angle 0°$	$0.7\angle 2°$
(10) $\dfrac{A}{1-\beta A}$ (db)	$14.5\angle 0°$	$14.6\angle 11°$	$14.9\angle 23°$	$16.1\angle 54°$	$16.2\angle 85°$	$12.3\angle 140°$	$8.0\angle 164°$	$2.7\angle 182°$	$-2.8\angle 199°$

(1) The degree of degeneration (or regeneration) at any frequency marked on the polar diagram Fig. 7.37 can be obtained by measuring the distance at the appropriate frequency from the curve to the point 1,0. The same scale is used for the measurement as for plotting βA and the distance gives the amount of degeneration, e.g. the 50 c/s point on the $\beta A = -8$ curve is $3.66\beta A$ units from 1,0, so gain reduction $= 3.66$ times $= 11.3$ db (compare 11.4 db Table 2 line 9). Again at 20 c/s, distance $= 0.2$ units so gain reduction $= 0.2$ times, i.e. gain increase $= 5$ times $= 14$ db (compare 14.1 db Table 2 line 9).

It follows that at whatever frequency the βA curve crosses the circle with centre 1,0 and radius 1, the gain of the amplifier is unchanged by feedback. For example, with $\beta A = -8$, the locus cuts the circle with radius 1 unit at a frequency of 28 c/s, which is therefore the frequency at which the gain of the amplifier is unchanged by feedback.

The same result may be obtained from Fig. 7.36 where the intersection of curves $\beta A = 0$ and $\beta A = -8$ is at 28 c/s. At lower frequencies the feedback is positive.

(2) When $\beta A = -8$ the low frequency peak is relatively much larger than when $\beta A = -2$ (14 db when $\beta A = -8$ and $2\frac{1}{2}$ db when $\beta A = -2$) and it occurs at a lower frequency. The reason can be seen from the polar diagram. When βA is large the curve is increased in size so that lower frequencies on the curve come closer to the 1,0 point.

(3) When $\beta A = -8$, the angle of βA varies from $0°$ to $197°$ but the angle of $1 - \beta A$ varies from $0°$ to $113°$ to $-8°$ (Table 2). The reason can be seen from the polar diagram. Note the rapid $1 - \beta A$ phase shift near the regenerative peak and the reason from the polar diagram.

(4) From the calculations it will be seen that, close to the mid-frequency, phase shift correction is approximately equal to $1/(1 - \beta A)$ e.g. when $\beta A = -8$ at 200 c/s in Table 2, a phase shift of $32°$ is reduced to $4°$.

(5) Although the feedback is regenerative as soon as the βA curve cuts the circle with centre 1,0 and radius 1, there is phase correction until the βA curve cuts the horizontal axis (between $18\frac{3}{4}$ and 16 c/s when $\beta A = -8$).

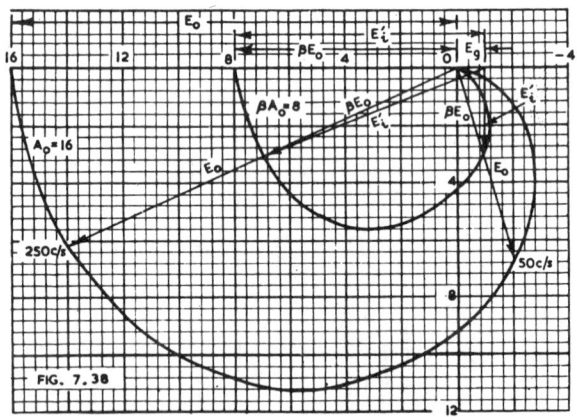

Fig. 7.38. Construction to demonstrate correction of phase shift by negative feedback.

The phase shift correction due to negative feedback is not directly obtainable from Fig. 7.37, but a construction to illustrate this effect is given in Fig. 7.38. In the diagram the larger curve ($A_0 = 16$) is the polar diagram of the gain of an amplifier, plotted as before, while the smaller curve ($\beta A_0 = 8$) is drawn for $\beta = \frac{1}{2}$. (For convenience, β is here taken as positive).

E_g represents the voltage applied between grid and cathode of the input valve, and since it is drawn 1 unit in length, the amplification curve also represents the output voltage E_0 ($= 16$ at the mid-frequency). Similarly the βA_0 curve represents the feedback voltage, and the vector sum of the grid to cathode voltage and the feedback voltage gives the required input voltage with feedback,

i.e. $E_g + \beta E_0 = E_i'$.

Thus at the mid-frequency, 1 volt of input gives 16 volts of output without feedback, but 9 volts of input are required for the same output voltage with feedback. Since there is no phase shift, there is no phase correction.

However at 250 c/s a phase shift of 23° has occurred and E_0 has the magnitude and phase shown in the diagram. E_i' is still the vector sum of E_g and βE_0 as shown on the diagram, and the phase angle between feedback input voltage E_i' and output voltage E_0 is only about 3°.

This checks as closely as the angles can be measured with Eqn. (51) Sect. 3(viii) of this chapter, taking into account the fact that E_0 has a smaller value at 250 c/s than at the mid-frequency.

At 50 c/s a phase shift of 106° between the voltage between grid and cathode of the first valve and the output voltage is reduced to about 17° between input voltage with feedback and output voltage, which is also in agreement with eqn. (12).

As the phase shift in the amplifier proper approaches 180°, phase shift reduction from inverse feedback becomes smaller, disappearing at 180° and becoming an increase at greater angles. This can be visualized from Fig. 7.38 or demonstrated on a polar diagram on which the βA curve cuts the X axis between 0 and 1 and returns to zero after passing through angles greater than 180°.

(C) Feedback from plate, with transformer coupling

The circuit is conventional, but there is danger of instability. See Sect. 3.

(D) Multiple feedback

The theory of multiple feedback for the purpose of improving stability is covered in Sect. 3(v)E. Fig. 7.33 is an example of dual feedback, with a principal feedback loop and subsidiary feedback through the unbypassed cathode resistor in the circuit

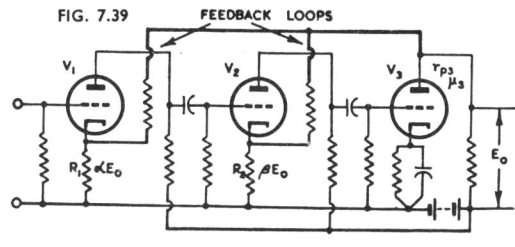

Fig. 7.39. *Amplifier with interconnected positive and negative feedback loops in which the negative loop is completely contained within the positive.*

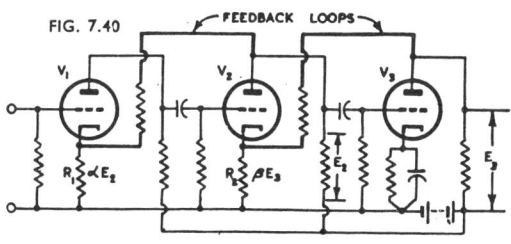

Fig. 7.40. *Amplifier with interconnected feedback loops in which the loops overlap.*

of V_1. In the case of a 3 stage amplifier, the unbypassed cathode resistor could be in the first or second stage. It is highly desirable, in general, for some such form of subsidiary feedback to be used. It is normally desirable that the degeneration due to the subsidiary feedback should not exceed say 10 per cent. of that due to the principal feedback loop, since the subsidiary feedback increases the distortion in the final stage. For further details and more complicated circuits see Sect. 3(v)E.

There are two general cases of interconnected feedback loops, the first in which one loop is completely contained within the other (Fig. 7.39) and the second in which the loops overlap as in Fig. 7.40. In each of the circuits the gain of the valves V_1, V_2 and V_3 is A_1, A_2 and A_3 respectively, and the size of the resistors R is such as to introduce negligible current feedback.

For the circuit of 7.39 the gain reduction factor is

$$\frac{1}{1 - \alpha A_1 A_2 A_3 - \beta A_2 A_3}$$

(with α subject to the same convention as β) and the effective plate resistance of V_3 becomes

$$r_{p3}' = \frac{r_{p3}}{1 - \alpha A_1 A_2 \mu_3 - \beta A_3 \mu_3}$$

For the circuit of 7.40 the gain reduction factor is

$$\frac{1}{1 - \alpha A_1 A_2 - \beta A_2 A_3}$$

and the effective plate resistance of V_3 becomes

$$\frac{r_{p3}}{1 - \dfrac{\beta A_2 \mu_3}{1 - \alpha A_1 A_2}} .$$

(vii) Voltage feedback over three stages

When the feedback is taken over three stages, there is a strong likelihood of experiencing instability at extremely low and high audio (or ultrasonic) frequencies (see Sect. 3). There are several popular circuits :

FIG. 7.42

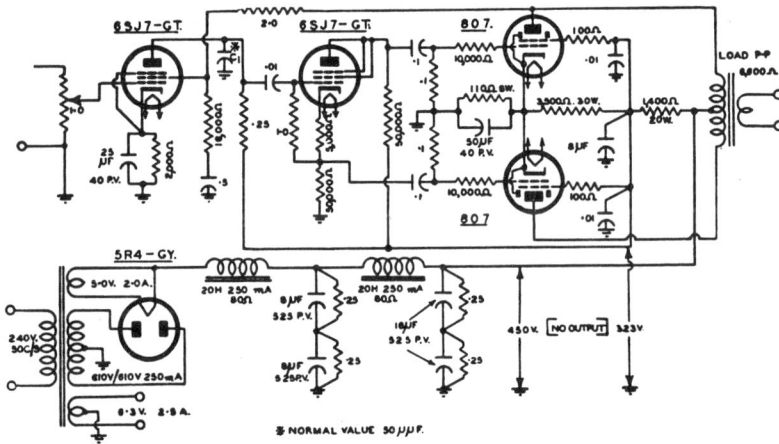

Fig. 7.42. 30 watt push-pull amplifier with voltage feedback from one output plate to the screen of the first stage. Stability is ensured by adjustment of condenser C_1.

(A) Feedback from plate of V_3 to screen of V_1 (Fig. 7.42)

This circuit is very satisfactory for push-pull amplifiers. As the feedback is taken from one plate only, the output transformer should have very tight coupling between the two halves of the primary (i.e. very low leakage reactance). The amplifier and feedback network have been designed to give a flat response over a wide band of frequencies, with the intention of introducing additional attenuation at one point only. No instability has been experienced, and practically no rise in response, at low audio frequencies. On a resistive load, the high frequency response has a slight peak at about 40 000 c/s, while on a loudspeaker load the peak is much larger and occurs at a lower frequency. Attenuation is provided by the condenser C_1 from the plate of the first valve to earth, with the results on frequency response shown in Fig. 7.43, for resistive loading.

FIG. 7.43

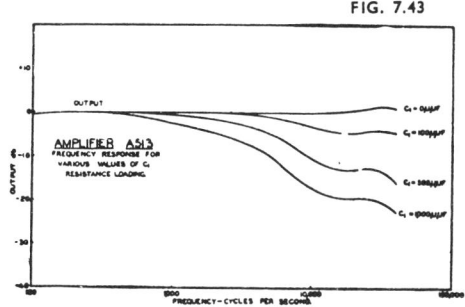

Fig. 7.43. *Frequency response curves for amplifier of Fig. 7.42 on resistive load with several values of C_1.*

In the absence of C_1 and on a loudspeaker load, parasitic oscillations may occur at a fairly high output level.   These are largely independent of the input signal frequency but are found to occupy a small part of the cycle, with a parasitic frequency in the region of 30 000 to 40 000 c/s.   The effect on music is not very apparent, but a single input frequency shows a slight " breaking up " effect when the parasitics appear.   The best check is by the use of a C.R.O.

The feedback is equivalent to overall voltage feedback, and follows the usual formulae, except that it is necessary to allow for μ_{g1g2} of the first valve in the calculations for β and A.   In this case :

$\beta = 16\ 000/(16\ 000 + 2\ 000\ 000) = 0.008$

A = amplification from first grid to final plate divided by μ_{g1g2}

$= 2730/20 = 136.5$

Calculated gain reduction 1/2.1

Measured gain reduction 1/2.2

Input voltage for 30 watts output = 0.2 volt r.m.s.

Distortion at 30 watts output : $H_2 = 0.7\%$, $H_3 = 0.9\%$, $H_4 = 0.1\%$, $H_5 = 0.2\%$.   Total 1.16% r.m.s.

Intermodulation distortion (Amplifier A513) :

Conditions of test—input frequencies 60 c/s and 2000 c/s (voltage ratio 4:1)

Power output (r.m.s.)*	4	10	14	20	24 watts
Equivalent power**	5.9	14.7	20.6	29.4	35.4 watts
Intermodulation	2.9	7.7	10	17.5	42%

*r.m.s. sum of two output frequencies

**r.m.s. sum × 25/17 to give the single frequency power having the same peak voltage swing (see Chapter 14 Sect. 3).

Any further increase in the degree of feedback is likely to prove difficult (see Section 3).

When a circuit is used which applies **negative feedback to the screen of a pentode,** the input capacitance is increased by

$\mu_t C_{g1g2} |\beta| A$

where μ_t is the " triode amplification factor "

C_{g1g2} is the capacitance from control grid to screen

$|\beta|$ is numerical value of β

and A is voltage gain from grid to the point from which the feedback is returned.

(B) Feedback from secondary of output transformer

This introduces additional phase shift in the amplifier and requires very careful design of the output transformer for all but small degrees of feedback.   The feedback voltage is usually taken to the cathode of the first valve, but if there is transformer input to the first stage then it may be returned to the transformer secondary (as Fig. 7.26).   This method may be applied equally to both single-ended and push-pull amplifiers.

Fig. 7.44 is the circuit of such an amplifier using push-pull triodes, with **extraordinarily** low distortion. It is based on the design of Williamson (Refs. F4, F6). The first triode (V_1) is direct-coupled to the grid of a phase splitter (V_2) which in turn is r.c. coupled to push-pull triodes (V_3 and V_4) and thence to the output stage. No by-pass condensers are used and the only reactances to cause phase shift at low frequencies are the grid coupling condensers in two stages. The circuit has a number of refinements which are described in the references.

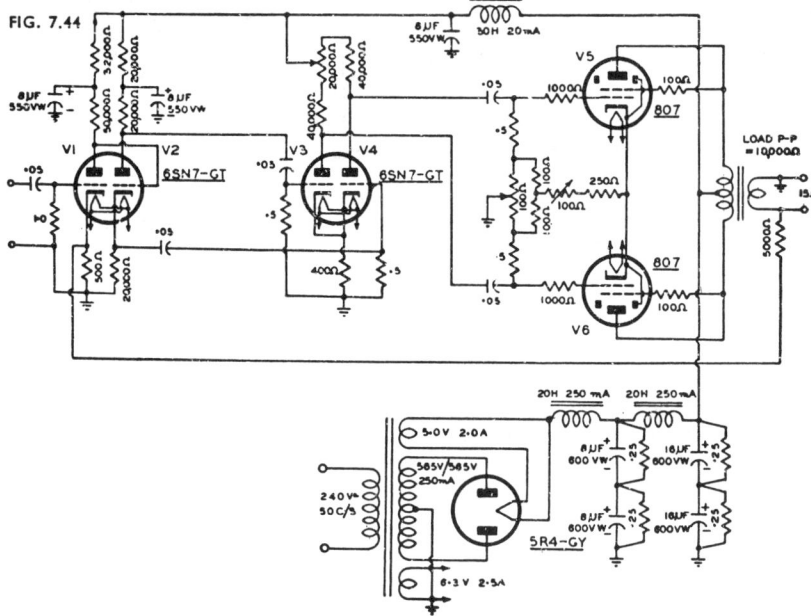

FIG. 7.44

Fig. 7.44. Amplifier employing voltage feedback from the secondary of the output transformer, with push-pull triodes. The amplifier is virtually distortionless up to an output of 11 watts, and has a smooth overload up to 16 watts.

The original version used type L63 valves in place of each half of the 6SN7-GT, and type KT66 as triodes in the output stage. The circuit shown (A515) gives almost identical results, but type 807 valves are used in place of type KT66.

The specifications laid down by Williamson for the **output transformer** are as follows :

Primary load impedance	10 000 ohms, centre-tapped
Primary inductance	100 henrys, min.
Leakage inductance, series	
(whole prim. to whole sec.)	30 millihenrys, max.
Primary resistance	250 ohms, approx.

Commercial transformers wound to the original winding data supplied by Williamson perform quite satisfactorily, although some difficulty may be experienced where the winding structure differs from that recommended. This is brought about by the increase in interwinding capacitances which may result.

Several manufacturers have recently marketed some wide range output transformers which are a considerable improvement on that specified above.

In one particular transformer, the Partridge CFB, the series leakage inductance, the winding resistances and weight have been reduced by a third, but the **power handling** capabilities have been increased four times. The core distortion at a level of 16 watts has been reduced ten times and the important leakage inductance existing between the halves of the primary has been decreased thirty times.

These improvements mean that this transformer, when used in the original Williamson circuit, will give a greater stability margin or alternatively will permit the use of a greater amount of feedback.

Distortion will be appreciably reduced in any amplifier by the use of a well designed transformer of such a type.

Distortion with amplifier A515

Power Output	H_2	H_3	H_4	H_5	Total r.m.s.
11 watts	0.01	0.04	0.01	0.015	0.045%
14 watts	0.075	0.35	0.01	0.26	0.44%
16 watts	0.04	0.92	0.11	0.45	(overload)

The smooth overload is a particularly valuable feature.

Intermodulation distortion

Conditions of test— input frequencies 60 c/s and 2000 c's; higher frequency 12 db lower than 60 c/s level.

Power output (r.m.s.)*	4	6	8	10	12 watts
Equivalent power**	5.9	8.8	11.8	14.7	17.6 watts
Intermodulation	0.17	0.27	0.72	3.7	8.8%

*r.m.s. sum of two output frequencies
**r.m.s. sum $\times$ 25/17 to give the single frequency power having the same peak voltage swing (see Chapter 14 Sect. 3).
Note :
In the circuit of Fig. 7.44, a capacitance may be inserted in series with the 5000 ohm resistor in the feedback circuit to provide bass boosting for equalizing purposes in record reproduction ; see Chapter 15 Sect. 9(ii)B.

A modification of this circuit, which provides for bass boosting, is given in Fig. 15.58A.

A new version of the Williamson amplifier, together with pre-amplifier and tone control (Ref. F9) is given in Figs. 17.35B,C,D,E,F,G. See also Chapter 13 Refs. F4, F5, H15.

Modified forms of the Williamson amplifier are given in Refs. F6, F8 and Chapter 13 Refs. H5, H6, H8, H10, H12.

(C) Bridge circuits

Amplifiers can be designed with bridge circuits in both input and output circuits, which prevent any modification of the input and output impedances and which also eliminate any phase shift round the feedback loop caused by reactances introduced through the input or output circuits (Refs. A1, A2, A11).

(viii) Cathode-coupled phase inverters and amplifiers

(A) Cathode coupled phase inverters

The fundamental circuit is Fig. 7.45 and requires two triode units with both cathodes linked, a common cathode resistor R_k and separate plate resistors R_{l1} and R_{l2}. Three tappings are required on the battery or potential divider across the B supply. The input voltage is applied between the two grids (Ref. G1).

If both V_1 and V_2 have identical characteristics and $R_{l1} = R_{l2} = R_L$, it may be shown (Ref. C7) that

$$E_{AB} = \mu E_s R_L/(R_L + r_p) \qquad (43)$$

where E_s = signal input voltage applied between the two grids
and E_{AB} = signal output voltage between points A and B.

Similarly, where E_A and E_B are the voltages from these points to earth,

$$E_A = \tfrac{1}{2}\left(\frac{\mu E_s R_L}{R_L + r_p}\right)\left(1 + \frac{R_L + r_p}{R_L + r_p + 2(\mu + 1)R_k}\right) \qquad (44)$$

$$\text{and } E_B = -\tfrac{1}{2}\left(\frac{\mu E_s R_L}{R_L + r_p}\right)\left(1 - \frac{R_L + r_p}{R_L + r_p + 2(\mu + 1)R_k}\right) \qquad (45)$$

The two voltages thus differ by a small amount, which can be reduced by using a large value of R_k and/or by using high mu valves. If the remaining out-of-balance is still serious, it may be eliminated entirely by making R_{L1} slightly smaller than R_{L2} so as to give exact balance.

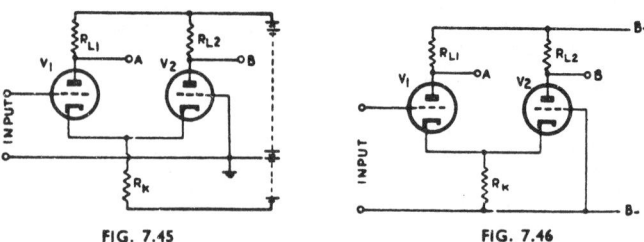

FIG. 7.45   FIG. 7.46

Fig. 7.45. Fundamental circuit of a cathode coupled phase inverter.
Fig. 7.46. Fundamental circuit of a cathode-coupled amplifier.

One stage of a cathode-coupled phase inverter may be coupled to a second (Ref. C7) and the output of the latter will be almost exactly balanced.

The same principle may be applied to the power output stage (Ref. G6).

The circuit of Fig. 7.46 may also be used as a phase-inverter.

See also Chapter 12, Sect. 6(vi).

(B) Cathode-coupled amplifiers

The fundamental circuit of a cathode-coupled amplifier is shown in Fig. 7.46 in which V_1 and V_2 are twin triodes with a common cathode resistor R_k. The output may be taken from either A or B, A being 180° out of phase with the input, while B is in phase with the input voltage.

Output from A : Gain $A_1' = \dfrac{\mu R_{L1}}{r_p + R_{L1} + r_p R_k(\mu + 1)/[r_p + R_k(\mu + 1)]}$ (46)

(Note : R_{L2} may be shorted-out.)

The plate resistance is increased by r_p and $R_k(\mu + 1)$ in parallel. The minimum value of A' is half the gain without feedback (when $R_k = 0$). (Ref. G2).

Output from B : (Note : R_{L1} may be shorted out) (Refs. G2, G4)

Gain $A_2' = - \dfrac{\mu R_{L2}}{2r_p + R_{L2} + r_p(r_p + R_{L2})/R_k(\mu + 1)}$ (47)

If $\mu R_k \gg r_p$, then $A_2' \approx - \mu R_{L2}/(2r_p + R_{L2})$ (48)

In reality, the values of μ and r_p are not constant and not equal for the two units in the twin triode. A rigorous analysis (Ref. G17) provides a straight forward method for the accurate derivation of the amplification when the output is taken from B.

Outputs from A and B (special case of phase inverter).

The gain is somewhat less than with Fig. 7.45, but the circuit is often more convenient.

There are many applications of this circuit which are described in References (G) also C7, D6 and J2 ; see also Chapter 12 Sect. 9(iii).

(ix) Hum

The hum in a practical amplifier is not always reduced by the factor $1/(1 - \beta A)$ as in eqn. (12), Sect. 1 but sometimes is decreased according to a different law, and sometimes is even increased by the application of feedback. All hum arising from sources within the amplifier and which is independent of the signal level, may be represented by an equivalent hum voltage in series with the input voltage ; this is reduced in all cases by the factor $1/(1 - \beta A)$.

(A) **Hum originating in the grid circuit** of a single stage amplifier will be reduced by $1/(1 - \beta A)$.

(B) **Hum caused by heater-cathode leakage** is reduced by $1/(1 - \beta A)$ with voltage feedback. With feedback from an unbypassed cathode resistor, however, the cathode-to-earth impedance is often very high, and the hum is thereby increased.

With the circuit of Fig. 7.20 the hum voltage in the output is

$$\frac{\frac{1}{2}E_h R_k}{R_k + R_{hk}} \cdot \frac{\mu R_L}{(\mu + 1)R_k + r_p + R_L} \tag{49}$$

where E_h = heater voltage (r.m.s.)

and R_{hk} = leakage resistance from heater to cathode.

(C) Hum originating in the plate supply voltage

[*The examples below are based on the following power amplifier valves under typical operating conditions*:

Type	Triode 2A3	Pentode 6V6-GT
μ	4.2	215
μ (*triode connection of pentode*)		9.6
r_p	800	52 000 *ohms*
R_L	2500	5000 *ohms*
A	3.2	18.8 *times*

Voltage feedback is taken as 20% (*i.e.* $\beta = -0.2$).
Current feedback is based on $R_k = 250$ *ohms.*]

Case (1) Transformer-coupled output with voltage feedback from the secondary (Fig. 7.26)

In this case the hum is multiplied by the factor $1/(1 - \beta A)$ [= 0.61 *for type 2A3* and 0.21 *for type 6V6-GT*]. The result is that the feedback reduces the hum slightly in the case of a triode and very considerably in the case of a pentode. If a hum-bucking coil is used, the number of turns will require readjustment for negative feedback.

Case (2) Transformer-coupled output with voltage feedback from the plate (Fig. 7.29)

In this case the feedback voltage is a fraction of the output voltage plus the hum voltage across the condenser C_2.

If V_1 is a **triode**, the hum voltage **without feedback** is

$$E_{h0} = E_h R_L/(R_L + r_p) \tag{50}$$

[*Example* : $E_{h0} = 0.76 E_h$ *for type 2A3*.]

while the hum voltage **with feedback** is

$$E_{h0}' = E_h R_L/(R_L + r_p') \tag{51}$$

[*Example* : $E_{h0}' = 0.85 E_h$ *for type 2A3*]

where E_{h0} = hum voltage output without feedback

E_{h0}' = hum voltage output with feedback

and E_h = hum voltage across condenser C_2.

Since $r_p' = r_p/(1 - \beta\mu)$ is less than r_p, the hum with feedback is greater than the hum without feedback :

$$\frac{E_{h0}'}{E_{h0}} = \frac{R_L + r_p}{R_L + r_p'} = \frac{1 - \beta\mu}{1 - \beta A} \tag{52}$$

[*Example* : $E_{h0}'/E_{h0} = 1.12$ *for type 2A3*]

If V_1 is a **pentode**, the hum voltage **without feedback** is

$$E_{h0} = E_h [R_L/(R_L + r_p) + R_L \mu_{g2p}/(R_L + r_p)] \tag{53}$$

[*Example* : $E_{h0} = (0.09 + 1.96) E_h = 2.05 E_h$ *for type 6V6-GT*]

while the hum voltage **with feedback** is

$$E_{h0}' = E_h[R_L/(R_L + r_p') + R_L \mu_{g2p}/(R_L + r_p - \beta\mu R_L)] \tag{54}$$

[*Example* : $E_{h0}' = (0.81 + 0.41)E_h = 1.22 E_h$ *for type 6V6-GT*]

where μ_{g2p} = mu factor from screen to plate without feedback

$\approx \mu_{g1p}/\mu_{g1g2}$ [= 215/9.6 = 22.4 *for type 6V6-GT*.]

It will be noticed that the screen and plate effects are additive ; it is only with parallel feed (see Case 3 below) that they oppose one another. The " screen " component of E_{h0}' is equal to that without feedback multiplied by $1/(1 - \beta A)$. This is not a good circuit for low hum, either with or without feedback ; these remarks also apply to Figs. 7.30, 7.31, 7.32, 7.33 and 7.42.

If the screen is perfectly filtered, the second term in eqn. (54) becomes zero. [*Example* : $E_{h0}' = 0.81E_h$ *for type* 6V6-GT.] The effect of feedback on the hum is then given by eqn. (52) with the substitution of the correct values of μ and A [*Example* : $E_{h0}'/E_{h0} = 9.2$ *for type* 6V6-GT].

Case (3) Parallel feed with voltage feedback (Fig. 7.47).

When V_1 is a triode, this is an excellent circuit for low hum, both with and without feedback, provided that L_1 has a reactance, at the hum frequency (i.e. twice supply frequency for full wave rectification) at least several times the impedance of r_p' and R_L in parallel.

When V_1 is a pentode, the hum due to the plate circuit is low but that due to the screen is high.   Screen filtering (L_2C_2) is required for low total hum ; this is improved by feedback so that the screen filtering may be omitted in some applications.

Without feedback, when V_1 is a **triode,** the hum voltage is applied to a voltage divider with ωL_1 in one arm and R in the other, where $R = r_p R_L/(r_p + R_L)$.   For ease in calculation it may be expressed in the form

$$E_{h0} = \cos \theta . E_h \text{ where } \theta = \tan^{-1}L_1/R \tag{55}$$

[*If* $\omega L_1 = 25\,000$ *ohms, and* $R = 606$ *ohms as for type* 2A3, then $\theta = \tan^{-1}41.2 = 88.6°$.   *Thus* cos $\theta = 0.024$ *and* $E_{h0} = 0.024E_h$].

With feedback, $E_{h0}' = E_{h0}/(1 - \beta A)$ \hfill (56)

[*Example* : $E_{h0}' = 0.024/1.64 = 0.015$ *for type* 2A3].

When V_1 is a **pentode without feedback**

$$E_{h0} \approx E_h[\cos \theta - R_L\mu_{g2p}/(R_L + r_p)] \tag{57}$$

where $\theta = \tan^{-1} \omega L_1/R$ as for the triode case
and $R = r_p R_L/(r_p + R_L)$.

The second term in eqn. (57) is only approximate, since it is assumed that the shunting effect of ωL_1 on R_L has no effect on the gain ; this is very nearly true when ωL_1 is at least equal to $10R_L$.

[*Example* : If $\omega L_1 = 25\,000$ *ohms, and* $R = 4560$ *ohms as for type* 6V6-GT, *then* $\theta = \tan^{-1} 5.48 = 79.7°$.   *Thus* cos $\theta = 0.179$ *and* $E_{h0} = (0.179 - 1.96)E_h$ $= -1.78E_h$.   *With screen filtering, this would be considerably improved.*]

With feedback, $E_{h0}' = E_{h0}/(1 - \beta A)$. \hfill (58)

[*Example* : $E_{h0}' = (-1.78/4.8)E_h = -0.37E_h$ *for type* 6V6-GT.]

If L_1 is replaced by a resistor R, the necessary substitution for $j\omega L$ may be made in the equations for hum.   Similarly, the transformer-coupled load (R_L) may be replaced by a following grid resistor R_g.

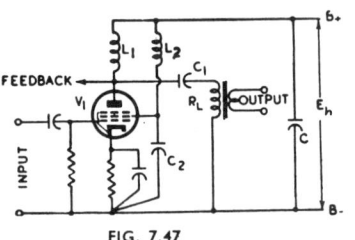

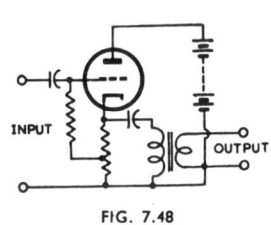

FIG. 7.47   FIG. 7.48

Fig. 7.47.   Circuit of an amplifier with parallel-fed transformer output and negative feedback supplied from the plate.

Fig. 7.48.   Cathode follower with parallel feed.

With this type of amplifier, hum balancing arrangements inside the amplifier are unaffected by the feedback.   An adjustment of the hum bucking coil (if used) in the loudspeaker will be necessary.   Any type of feedback circuit may be used, either from the plate, or from the primary or secondary of the output transformer.

Case (4) Cathode follower with parallel feed (Fig. 7.48)

In this case the plate resistance is reduced by the factor $1/(\mu + 1)$, and the hum output voltage is decreased by the factor $1/(\mu + 1)$.   If a pentode is used, the screen capacitance by-pass to cathode (C in Fig. 7.16) may be adjusted to neutralize hum (Ref. C14).

Case (5) Transformer-coupled output with feedback from an unbypassed cathode resistor (Fig. 7.49)

The hum arising from the plate and screen circuits is the same as for Case (2) except that r_p' is now greater than r_p (Eqn. 51): $r_p' = r_p + (\mu + 1)R_k$. With a triode the hum is less with current feedback than without feedback. [*Example type 2A3* : $E_{ho}' = 0.54E_h$]. With a pentode the total hum from plate and screen is always decreased by current feedback

$$E_{ho}' = E_h\left[\frac{R_L}{R_L + r_p + R_k(\mu + 1)} + \frac{\mu_{g2p}R_L}{R_L + r_p + (\mu + \mu_{g2p})R_k}\right] \qquad (59)$$

where the first term is for the plate, and the second for the screen circuit.

[*Example* : $E_{ho}' = (0.04 + 0.97)E_h = 1.01E_h$ *for type* 6V6-GT.]

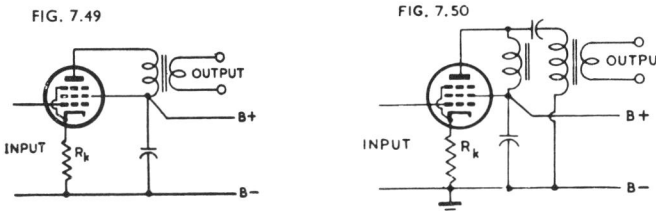

FIG. 7.49   FIG. 7.50

Fig. 7.49. Transformer-coupled output with feedback from an unbypassed cathode resistor.

Fig. 7.50. Parallel-fed transformer-coupled output with feedback from an unbypassed cathode resistor.

Case (6) Parallel feed, with feedback from an unbypassed cathode resistor (Fig. 7.50)

With a pentode, the " plate " hum is increased by feedback from an unbypassed cathode resistor, while the " screen " hum is the same as for transformer-coupling (Eqn. 59) except for the sign.

$$E_{ho}' \approx E_h\left[\frac{r_p + R_k(\mu + 1)}{R_L + r_p + R_k(\mu + 1)} - \frac{\mu_{g2p}R_L}{R_L + r_p + (\mu + \mu_{g2p})R_k}\right] \qquad (60)$$

The first term also applies to a triode. The approximation is from regarding R_k as the load, instead of R_k in parallel with ωL.

[*Example of triode* : $E_{ho}' = 0.46E_h$ *for type* 2A3.

Example of pentode : $E_{ho}' = (0.95 - 0.97)E_h = -0.02E_h$.

This gives almost exact hum neutralization.]

References to hum : B3, B4, B5, B7.

Summary—hum originating in the plate and screen supply voltages

With triodes, the output hum voltage is always less than the hum voltage from the plate supply. In the examples given above, it varies from 46% to 85% of the plate supply hum voltage with one exception. The exception is transformer-coupled output with parallel-feed, feedback being taken from the plate, for which the output hum is only 1.5% of the plate supply hum voltage (Case 3).

With pentodes or beam power amplifiers the output hum may exceed the plate supply hum voltage, owing to the effect of the screen. This occurs with Case (2) which is the conventional transformer-coupled output both without feedback and with voltage feedback from the plate. The only circuit giving very low output hum is Case (6) which applies to transformer-coupled output with parallel feed and with feedback from an unbypassed cathode resistor. The second best is Case (1) with feedback from the secondary, while the third best is Case (3) with parallel feed, having feedback from the plate.

The examples given above have been tabulated for ease in comparison, including some cases also without feedback, and some with the screen circuit perfectly filtered.

Case	Fig.	Relative hum voltage with 2A3	6V6-GT	Conditions
1.	7.26	0.61	0.21	with feedback
2.	7.29	0.76	2.05	without feedback
		0.85	1.22	with feedback
		—	0.09	{ without feedback and with screen perfectly filtered
		—	0.81	{ with feedback and with screen perfectly filtered
3.	7.47	0.024*	1.78	without feedback
		0.015*	0.37	with feedback
		—	0.179	{ without feedback but with screen perfectly filtered
		—	0.037	{ with feedback and with screen perfectly filtered
5.	7.49	0.54	1.01	with feedback
6.	7.50	0.46	0.02*	with feedback

*Very low output hum.

(x) Some special features of feedback amplifiers

(A) It is important to ensure that the frequency range of the input signal applied to a feedback amplifier does not extend in either direction beyond the " flat " frequency range of the amplifier. The reason for this is explained below.

The voltage applied to the grid of the first valve in a multi-stage amplifier with overall feedback is equal to the difference between the input signal and the fedback voltage. For example, assuming 20 db of feedback, the signal voltage may be 1 volt, the fed-back voltage 0.9 volt and the voltage applied to the first stage will then be $1 - 0.9 = 0.1$ volt. At some frequency well outside the " flat " frequency range of the amplifier, where the amplifier output is 20 db down, the fed-back voltage will be 0.09 volt for a 1 volt signal, leaving a difference of 0.91 volt applied to the grid of the first valve—i.e. nine times its normal input. This may not overload the first stage, but it will probably overload the subsequent stages and cause serious distortion, even though the frequency is itself inaudible.

Even at a frequency which is only 1 db down, the signal voltage may be 1 volt, the fed-back voltage will then be 0.8 volt and the difference of 0.2 volt will be applied to the grid of the first valve ; twice normal. It is therefore essential to restrict the input range of frequencies, and it is also advisable to place the coupling having the highest attenuation as close as possible to the input of the amplifier (Ref. F7).

(B) **Damping on transients**

See Sect. 3(v) including " Design Tests."

(C) **Overload test**

See Sect. 3(v) " Design Tests."

(xi) Combined positive and negative feedback

The principle of combined positive and negative feedback has been outlined in Sect. 1(v).

A practical version of such an amplifier with push-pull output is shown in Fig. 7.51 (Ref. F10). The overall negative feedback is obtained from the secondary of the output transformer T_1 and is fed through R_6 to the cathode of V_2. Shunt capacitor C_5 affords some feedback phase correction at very high frequencies. The feedback gain reduction is 9 db, and becomes 11 db with the positive feedback disconnected.

The positive feedback is obtained from the grid of V_4 and is fed through R_1 and C_1 to the grid of V_2. The positive feedback voltage is developed primarily across R_2

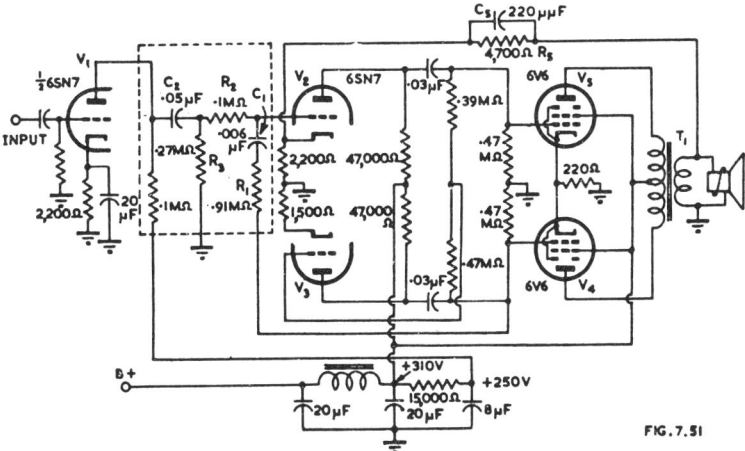

Fig. 7.51.   *Two stage amplifier using combination of positive and negative feedback*
(*Ref.* F10). T_1 *ratio is* 10,000 : 15 *ohms.*

and C_2 since the plate resistance of V_1 is relatively small, and the input resistance of
the grid of V_2 is high.   The increase in gain due to positive feedback is about 26 db
at 400 c/s.   About one tenth of the voltage on the grid of V_4 is fed back to the grid of
V_2.   The resistance of R_1 is therefore made about nine times that of R_2, and C_2 has
about nine times the capacitance of C_1.   Because of the highly degenerative nature
of the phase inverter, the balance is not appreciably affected by the additional load
of the positive feedback network.

Some phase shift in the positive feedback is obtained at extreme frequencies in
the stages V_2 and V_3 due to electrode and stray capacitances, and due to the blocking
capacitors.   The input capacitance of the grid of V_2 causes a further phase shift, so
that the polarity of the product $\beta_1 A_1$ reverses from positive to negative at extremely
high frequencies, where β_1 and A_1 apply to stage V_2.   The input capacitance of V_2
is primarily Miller Effect due to feedback through its grid-plate capacitance at very
high frequencies where the overall feedback is positive or small.

In some designs it may be necessary to connect a small capacitor from the grid of
V_2 to earth, or to use a more elaborate phase shift network to obtain a sufficiently
rapid phase turnover in the local feedback.

At extremely low frequencies most of the local feedback current flows through R_3
instead of through C_2, so that a phase shift is obtained, which together with the phase-
shifting action of the 0.03μF blocking capacitors in stages V_2 and V_3, is sufficient to
cause the desired phase reversal.   In practice, the phase reversal frequencies are
placed as far outside the desired pass band as good stability permits.

The output transformer is quite small, the core area being only $\frac{3}{4}$ in. by $\frac{3}{4}$ in.

The maximum third harmonic distortion at 400 c/s is 0.24%, higher harmonics
being relatively small, at 8 watts output, measured at the secondary.   At 100 c/s
the highest harmonic is the second, with 0.12% for 8 watts output.   At 50 c/s, with
5 watts output, third harmonic distortion is 0.88%, while at 2000 c/s with 4 watts
output the third harmonic is 0.23%.   The intermodulation distortion is 40% with
no feedback, 8% with negative feedback alone and 1.9% with combined positive and
negative feedback under the following conditions—output 8 watts, 4:1 ratio with
frequencies 60 and 7000 c/s.   With frequencies of 100 and 7000 c/s the intermodulation distortion is only 0.84%.   The output circuit regulation is quoted as 0.1 db at
400 c/s as compared with 2.7 db for positive feedback disconnected, or 19 db with
no feedback (Ref. F10).

It is possible to apply overall negative feedback together with positive feedback from a tapping on the cathode resistor of a phase-splitter through a coupling resistor to the unbypassed cathode of the preceding r.c. pentode, and thus increase the gain. The increased gain may require additional filtering (Ref. F16). This may also be used with direct coupling (Fig. 7.51A and Ref. F11).

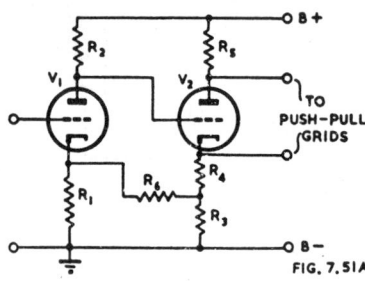

Fig. 7.51A. *Direct-coupled triode amplifier V_1 and phase splitter V_2, with positive feedback from the cathode circuit of V_2 to the cathode of V_1 (Ref. F11).*

FIG. 7.51A

It is possible to omit the by-pass condensers from the cathode bias resistors in the final and penultimate stages of an amplifier or receiver (or any other two successive stages) and to bring the gain back to normal by means of positive feedback, using only one resistor coupling the two cathodes as in Fig. 7.51B.

Fig. 7.51B. *Positive feedback from the cathode of V_2 through R_4 to the cathode of V_1 to offset the loss of gain by the omission of both by-pass condensers.*

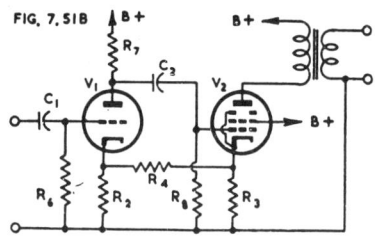

FIG. 7.51B

This circuit (Fig. 7.51B) may also be used with any desired degree of feedback. The effective plate resistance of V_2 may be increased or decreased by the positive feedback as shown by the following analysis and Fig. 7.51C. Here R_1 replaces R_7 and R_8 in parallel, no input signal is applied but a signal generator E is inserted in the plate circuit. It is assumed that R_4 is very much greater than R_3. We thus obtain

$$I = \frac{E}{R_A} - \frac{IR_3\mu_2}{R_A} + \frac{IR_3|\beta|A_1\mu_2}{R_B} \qquad (61)$$

where $R_A = R_5 + r_{p2} + R_3$
$\qquad R_B = R_5 + r_{p2} + R_3(\mu_2 + 1)$

$$A_1 = \frac{\mu_1 R_1}{R_1 + r_{p1} + R_2}$$

and $\quad |\beta| = \dfrac{R_2}{R_2 + R_4}$

From (61) we may derive

$$I = \frac{E}{R_B - R_3|\beta|A_1\mu_2 R_A/R_B}$$
$$= \frac{E}{R_5 + r_{p2} + R_3[\mu_2 + 1 - (|\beta|A_1\mu_2 R_A)/R_B]}$$

and hence $r_{p2}' = r_{p2} + R_3[\mu_2 - (|\beta|A_1\mu_2 R_A)/R_B] \qquad (62)$

where $|\beta|$ and A have the values defined below eqn. (61). The second term on the right hand side of eqn. (62) is the increase in effective plate resistance due to the un-bypassed cathode resistor of V_2, while the third term is the decrease in effective plate resistance due to positive feedback. It is obvious that, by a suitable choice of

βA_1, it is possible to obtain any desired value of effective plate resistance from very high values to very low, even zero or negative values. If the resistor R_4 is made variable, the effective plate resistance may be varied over a wide range. The addition of overall negative feedback will make the gain nearly constant for all settings of R_4, and will also decrease the distortion and reduce the effective plate resistance to any values desired (Ref. F17). See also Ref. J19.

The addition of positive feedback will increase the phase shift and may increase the tendency towards instability at very low and very high frequencies ; care should be taken to reduce to a minimum the phase shift within the positive feedback loop. The general principles of designing for stability are the same as for negative feedback —see Sect. 3.

It is normally inadvisable to return the positive and negative feedback loops to the same point in the input circuit. If both are taken to the same electrode, a bridge network may be used for isolating their effects. Each 6 db increase in gain due to positive feedback will double the distortion in this stage ; for example 24 db increase in gain will increase the distortion in this stage by 16 times. This will then be reduced by the negative overall feedback in accordance with the usual relationship. References to positive and negative feedback—Refs. F10, F11, F12, F13, F15, F16, F17 ; Sect. 3(v)E.

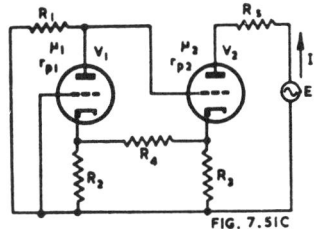 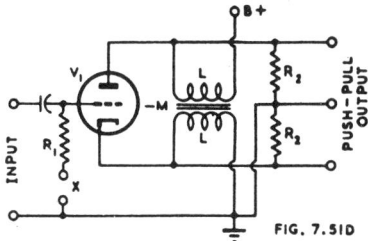

FIG. 7.51C FIG. 7.51D

Fig. 7.51C. *Analysis of circuit of Fig.* 7.51B *to determine the effective plate resistance of* V_2.

Fig. 7.51D. *Choke coupled phase inverter (Ref.* F14).

(xii) Choke-coupled phase inverter (Ref. F14)

The choke-coupled phase inverter of Fig. 7.51D has been designed to give higher output voltages than the conventional phase splitter, to be practically unaffected by the flow of grid current in V_1 and to have small phase shift. The two identical chokes L in the plate and cathode circuits are tightly coupled so that this duty is performed by a 1 : 1 a-f transformer. This transformer applies negative voltage feedback to the input circuit, and thus causes a low effective plate resistance. Design must aim at a high total inductance to keep phase errors small, and a low leakage inductance if it is desired to keep the effect of grid current small. A fixed bias voltage is applied between terminals X, having polarity and magnitude to give, in conjunction with the voltage drop in the cathode choke, the correct grid bias. The complete analysis is given in Ref. F14.

The voltage gain is given approximately by

$$A' \approx \frac{\mu R_2}{(\mu + 2)R_2 + r_p} \text{ where } X \gg R_2$$

and the phase shift between input and output is given exactly by

$$\tan \phi = \frac{r_p}{(\mu + 2)R_2 + r_p} \cdot \frac{R}{X}$$

where $X = 2\omega(L + M)$.

X may be taken to include any self-capacitance or other stray capacitance.

As a result of the small phase shift, this circuit may readily be used in feedback amplifiers.

SECTION 3 : STABILITY, PHASE SHIFT AND FREQUENCY RESPONSE

(i) Stability and instability (ii) Conditions for stability (iii) Relationship between phase shift and attenuation (iv) Design of 1 and 2 stage amplifiers (v) Design of multi-stage amplifiers (vi) Effect of feedback on frequency response (vii) Design of amplifiers with flat frequency response (viii) Constancy of characteristics with feedback (ix) Effect of feedback on phase shift.

(i) Stability and instability

Equation (2b) of Sect. 1 gives the amplification with negative feedback, at the mid-frequency, in the form

$A' = A/(1 - \beta A)$.

It is obvious that both A and βA must be vector quantities having different magnitudes and phase angles at very low and very high frequencies from those they have at the mid-frequency.   If βA is real and negative, we have negative feedback with a decrease in amplification.   If βA is real, positive, and less than 1 in absolute value, we have positive feedback with an increase in amplification.   If $\beta A = 1$ the amplification becomes theoretically infinite, and the system is unstable.   If βA is real, positive, and greater than 1 in absolute value, the system will be only " conditionally stable " and is likely to oscillate during the warming-up period.

The quantity βA is called the **loop amplification.**

The conditions for stability described in (ii) below are based on the well-known Nyquist criterion of stability.   There is, however, an alternative known as the Routh-Hurwitz stability criterion which is useful when the expression is in analytical form. See Ref. H16.

(ii) Conditions for stability

Before proceeding with any particular amplifier, it is necessary to calculate or measure its amplification and phase angle over a very wide band of frequencies. The frequency and phase angle characteristics of an imaginary amplifier are shown in Fig. 7.52.   The frequency f_0 at which the phase angle is zero is called the mid-frequency ; f_1 and f_2 are the frequencies at which the phase angle displacement is a lead and lag of $90°$ respectively ; frequencies f_3 and f_4 are those giving a lead and lag of $180°$ respectively.

Taking first the case of a single stage with r.c. coupling, it is possible to plot the locus of the values of βA from zero frequency to infinity with polar coordinates (Fig. 7.53A).   This is commonly known as a " Nyquist diagram."   If A_0 is the amplification at the mid-frequency f_0, this will normally be its maximum value and the phase angle displacement will be zero.   It will, of course, be negative for negative feedback, thus giving the vector βA_0 as drawn.   At any lower frequency f_1, the value of A may be taken from the curve for this amplifier ; let us call it A_1.   The value of β may be regarded as a fixed negative fraction less than unity.   We thus have the value of βA_1, and the phase angle displacement ϕ_1 may also be determined from the phase angle curve.   This may then be plotted as a radial vector from 0 to βA_1, with a leading phase angle displacement ϕ_1, which implies that the phase of the feedback voltage leads the input voltage by the angle ϕ_1.

Similarly for any higher frequency f_2 ; we can plot βA_2 with a length proportional to the amplification at this frequency, and a lagging phase angle displacement ϕ_2. As the frequency increases still further, we have a smaller βA_3 with a larger phase displacement, while in the extreme limit $\beta A = 0$ with $\phi = 90°$ lagging.   At the other frequency extreme $(f = 0)$ we have $\beta A = 0$ with $\phi = 90°$ leading.   In this case the shape of the locus is a circle which is all in the negative region ; this amplifier is therefore always stable.

Any amplifier may have its βA locus plotted by this method ; a typical case is Fig. 7.53B for a 2 stage r.c.c. amplifier.   Here βA_1 still has appreciable length with ϕ_1 greater than $90°$, but $\beta A = 0$ when $\phi = 180°$ either leading or lagging.

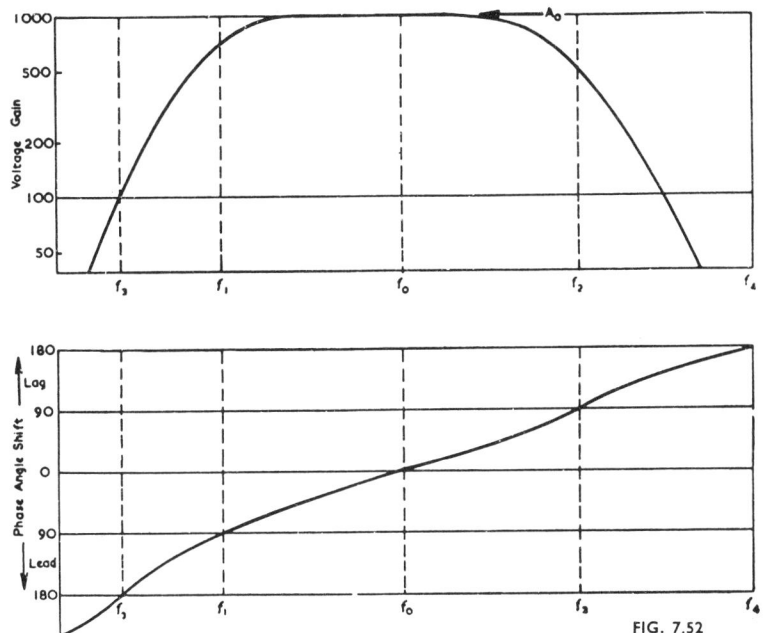

Fig. 7.52. Voltage gain and phase angle shift of an imaginary amplifier. The mid-frequency is f_o and the maximum voltage gain A_o.

The distance from any point on the locus to point K is equal to $(1 - \beta A)$ where β is negative for negative feedback (see Fig. 7.53B).

If it is desired to find the magnitude and phase angle of βA at any frequency, it is only necessary to measure the gain without and with feedback, to calculate the ratio A_o/A' between them, to draw a circle with centre K and radius A_o/A' units, then to draw a second circle with centre 0 and radius equal to the measured fed-back voltage $|\beta A|$. The line from the origin to the point of intersection between the two circles is the complex number βA.

For any amplifier to be stable, the βA locus must not include the point K (1,0). It is also desirable for the locus not to cut the X axis beyond the point K,

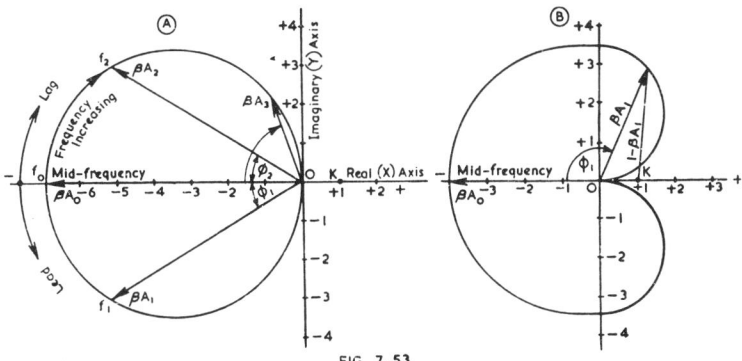

FIG. 7.53

Fig. 7.53. (A) Locus of βA vectors (" Nyquist diagram ") for a single stage resistance capacitance coupled triode amplifier with fixed bias (B) Similar locus for a 2 stage amplifier.

otherwise oscillation may occur during the warming-up period ; such an amplifier is said to be conditionally stable.   In the case of more complicated circuits, particularly those with peculiarly shaped βA loci, it is advisable to plot loci for successively increasing values of β up to and somewhat beyond the desired value of β.   The criterion of stability should be applied to each—that the locus should not enclose the point K (1,0).

The following rule is also helpful as it covers most cases and avoids having to draw a βA locus.   **If βA is less than unity at $\phi = +180°$ and at $\phi = -180°$ the amplifier is stable.**

Another useful rule is that **the maximum permissible value of βA is equal to the maximum value of A (i.e. A_0) divided by the value of A at $\phi = \pm180°$ (whichever is the larger).**

Also, **if the phase shift is not more than 180° at zero and at infinite frequencies, the amplifier is always stable with any value of β.**

The " attenuation " is the attenuation with respect to the maximum value of A (i.e. A_0).   Both the attenuation and the loop amplification βA may be expressed in decibels.

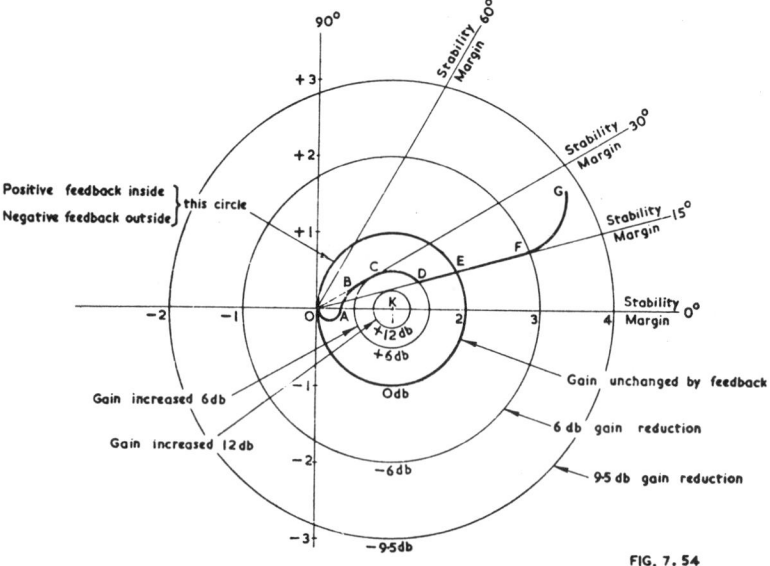

FIG. 7.54

Fig. 7.54.   Additional information furnished by a " Nyquist diagram " (see also Fig. 7.53 A and B.)

Certain aspects of the polar diagram for amplifiers with feedback are shown in Fig. 7.54.   Circles are shown with centres at point K (1,0) with radii of 0.25, 0.5, 1.0, 2 and 3 units.   The circumference of the circle with a radius of 1 unit indicates the locus of points at which the gain is unchanged by feedback.   At all points outside this circle, the feedback is negative and degenerative.   For example, the circumference of the circle with a radius of 2 units is the locus of points having a gain reduction of 6 db.   At all points inside the circumference of the circle with a radius of 1 unit, the feedback is positive and regenerative although not necessarily unstable. For example, the circumference of the circle with a radius of 0.5 unit is the locus of points having a gain increase of 6 db.

Instability occurs when the point K is included by the βA locus.   Straight lines are drawn radiating from the origin with stability margins of 15°, 30°, 60°, and 90°. The 30° line is tangential to the circle with a radius of 0.5 unit, so that the point of tangency has 30° stability margin and 6 db gain increase.   At other points along this line the gain increase will be less than 6 db.

An example of the use of the polar diagram in connection with the design of negative feedback amplifiers is given in Sect. 2(vi)B and Figs. 7.37 and 7.38.

It would be permissible for an amplifier to have a βA locus such as the line OABCDEFG in Fig. 7.54 (see Ref. A29).   Here the maximum increase of gain due to positive feedback is 6 db, which occurs between C and D.   The feedback is negative over the path EFG.   The distance OA should not exceed say 0.3 unit, so as to allow for a possible increase in amplifier gain of about 3 db without further increasing the peak of 6 db.

Feedback amplifiers which are designed with a small stability margin should have narrow tolerances on components that have a direct effect on gain or frequency characteristics.   It is desirable for βA to be measured and its locus plotted for the pilot model of each design for three conditions, with normal (bogie) valves and with valves at the upper and lower limits for mutual conductance.   This is only required in the region of the low and high frequency peaks of response, which are normally beyond the working frequency range.   Alternatively and more simply, the height of the low and high frequency peaks may be measured in relation to the response without feedback ; in accordance with good practice, the rise with feedback should not exceed 6 db, and lower values are desirable.

A special stability problem occurs when an amplifier may not always be connected to its correct load.   One way of overcoming this problem is to design a high-and-low pass filter pair for connection between the amplifier and the load, the high-pass filter being terminated by a resistance (Ref. A29).

If an amplifier is stable with the output terminals open-circuited, short-circuited, or operated into its rated resistive load, then it is stable under all load conditions.[*]   The measurements of stability on open-circuit and under rated load resistance can be made with an oscilloscope across the output terminals to check for the presence of ultrasonic oscillations.   A check on the short-circuit conditions can best be made using a high-frequency ammeter across the output terminals.   An oscilloscope of very high input impedance might also be used by connecting it across the input of one of the stages in the feedback loop (Ref. A28).

(iii) Relationship between phase angle displacement and attenuation

The phase angle displacement of an amplifier is normally a function of the attenuation characteristic.   Because of this fact, it is possible to design a feedback amplifier on the basis of either the phase angle displacement or the attenuation characteristic. If the attenuation characteristic[†] is a straight line with a slope of 6 decibels per octave, then the ultimate phase angle is 90° and so on in proportion ; 12 db/octave gives 180° phase angle and 18 db/octave gives 270°.   The condition which must be satisfied is that the amplifier is a " minimum phase shift network."   This condition is satisfied by most amplifiers, the exceptions being (1) when it contains a transmission line or equivalent circuit with distributed constants, and (2) when it includes an allpass section, either as an individual structure or in a combination which can be replaced by an all-pass filter section plus some other physical structure (Refs. H6, H10, H11).

Fig. 7.54A shows ultimate slopes of 6 and 12 db/octave and higher slopes.   The 12 db/octave attenuation characteristic is the limiting value for stability, being on the verge of instability.   A practical amplifier requires a safety margin between the slope of its design characteristic and the limiting value of 12 db/octave.   A typical design slope is 10 db/octave, giving an angular safety margin of 30°.

Any single reactive element such as a grid coupling condenser, a shunt capacitance or a shunt inductance in conjunction with a resistance provides an attenuation characteristic with an ultimate slope of 6 db/octave.   Such a combination is known as a single time constant circuit.   Attenuation characteristics are given in Fig. 4.36 (grid coupling condenser or shunt inductance) and Fig. 4.38 (shunt capacitance), for a single time constant in each case.

---

[*]This applies to a resistive load only.   With a capacitive load, instability is likely to occur.
[†] Plotted with logarithmic frequency scale.

If a parallel-fed transformer is used, the coupling condenser resonates with the primary inductance at some low frequency and the ultimate slope of the stage is approximately 12 db/octave below the frequency of resonance. If the secondary of a transformer is unloaded, the leakage inductance resonates with the distributed capacitance, giving a slope for the stage of approximately 12 db/octave above the frequency of resonance. Each of these combinations is equivalent to two time constants.

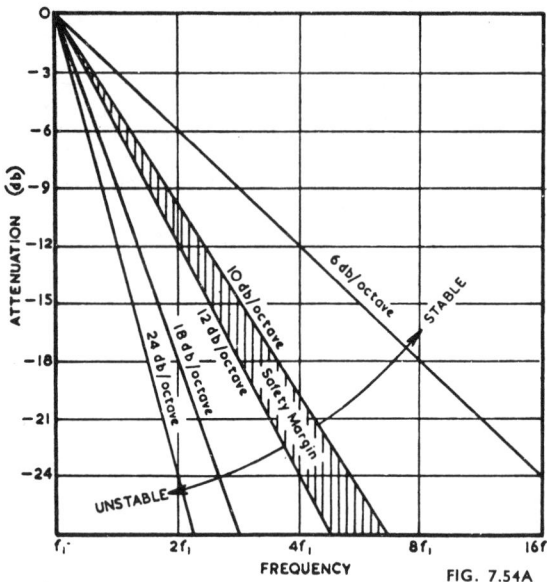

FIG. 7.54A

Fig. 7.54A. *Attenuation characteristics for* 6, 12, 18 *and* 24 *decibels per octave as given by the " asymptotic " or ultimate characteristics of the amplifiers in their simplest form. The region beyond 12 db/octave is unstable ; that below 10 db octave is stable, and the region between 10 and 12 db/octave is a safety margin to allow for discrepancies between the calculated design and the finished amplifier.*

Cathode and screen by-passing

In the case of cathode and screen by-passing, the attenuation characteristics are noticeably curved and the phase angle displacement cannot be calculated as for a straight line, but may be obtained from Figs. 12.3C and 12.11B respectively. In both cases the phase angle displacement reaches a maximum value at some frequency and gradually decreases at lower and higher frequencies.

Cathode by-passing

The maximum slope of the attenuation characteristic and maximum phase angle displacement are given belcw

B	=	2	3	5	10	
Max. loss	=	6	9.5	14	20	db
Max. slope	=	2.25	2.6	3.6	4.8	db/octave
Max. angle	=	20°	30°	42°	55°	

where $B = 1 + \dfrac{R_k(\mu + 1)}{r_p + R}$ and $R = \dfrac{R_L R_{g2}}{R_L + R_{g2}}$.

For a pentode $B \approx 1 + g_m R_k$.

For typical r.c. triodes and pentodes, B is usually less than 3.

Screen by-passing

This has an effect similar to that of cathode by-passing

B	=	2.5	5	10	20	
Max. loss	=	8	14	20	26	db
Max. slope	=	2.4	4	4.5	5.4	db/octave
Max. angle	=	25°	42°	54.5°	64.5°	

where $B = 1 + \dfrac{R_s g_m}{m\mu_t(1 + R_L/r_p)}$

$\quad\quad R_s$ = series screen resistor in ohms

$\quad\quad g_m$ = mutual conductance at operating point, in mhos

$\quad\quad m$ = ratio of plate to screen currents

$\quad\quad \mu_t$ = triode amplification factor

$\quad\quad R_L$ = plate load resistance in ohms

and $\quad r_p$ = plate resistance in ohms.

Phase angle characteristics of "step circuit"

The step circuit of Fig. 7.59A has a maximum phase angle displacement at the frequency of the half-attenuation point, and approaches zero at lower and higher frequencies. The curves for a particular case are plotted in Fig. 7.54B, and the value of the maximum phase angle as a function of the attenuation in Fig. 7.54C.

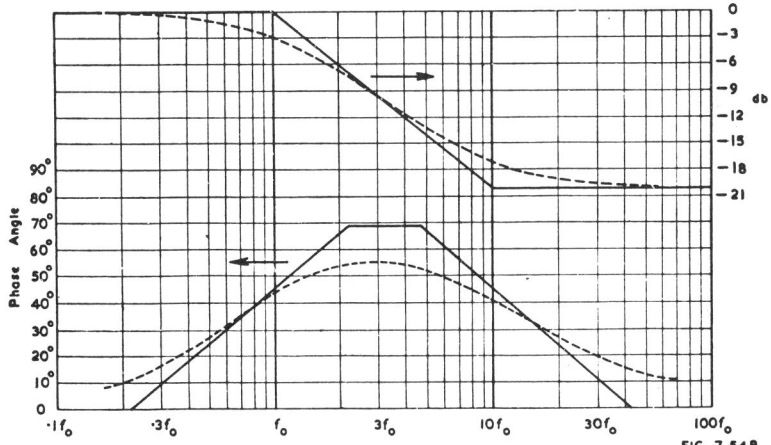

Fig. 7.54B. *Attenuation and phase characteristics of " step circuit " Fig. 7.59A for particular case with step attenuation 20db (Ref. H18).*

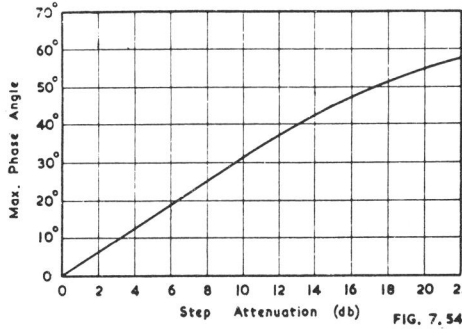

Fig. 7.54C. *Maximum phase shift as a function of attenuation in the circuit of Fig. 7.59A (Ref. H18).*

Total slope of attenuation characteristic

The total slope of the attenuation characteristic of an amplifier in db/octave, at any frequency, is the sum of the slopes of the attenuation characteristics of all the reactive elements in the amplifier at the same frequency, provided that the slopes are constant.

In practice, the best procedure is to plot the individual attenuation characteristics with a linear db scale, then to add the decibel ordinates at various frequencies to determine the total attenuation characteristic. The slope of the latter may then be determined graphically.

Determination of phase angle

When the attenuation characteristic is a straight line forming a sharp angle at the cut-off frequency f_0, as in Fig. 7.55A, the phase angle displacement is $k \times 45°$ at f_0, where k is one sixth of the slope in db/octave, and the phase angle characteristic is asymptotic to $0°$ and $k \times 90°$.

When the attenuation characteristic is that of a normal r.c. amplifier, $k \times 3$ db down at f_0 (as broken curve in Fig. 7.55D) the phase angle displacement is also $k \times 45°$ at f_0, but at other frequencies the phase angle characteristic differs slightly from that of Fig. 7.55A.

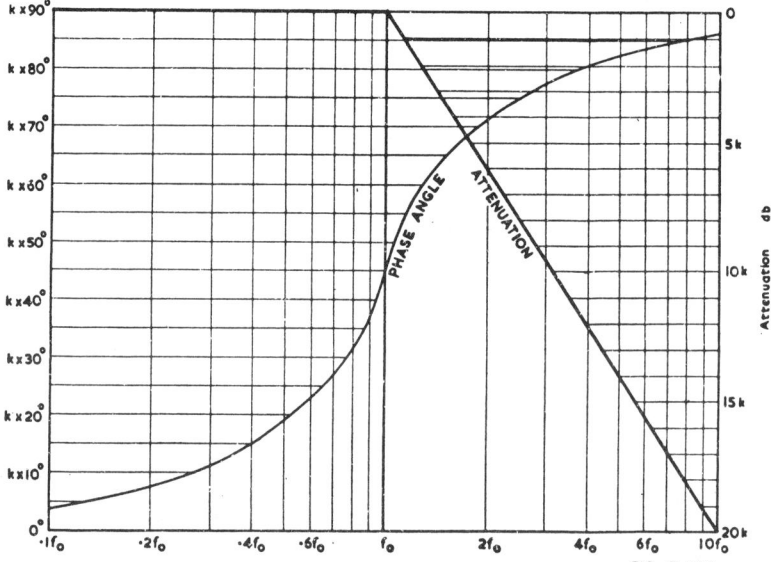

Fig. 7.55A. *Phase angle for attenuation characteristic as shown ; $k = 1$ for 6db/octave, $k = 2$ for 12db/octave etc. (Based on Bode, Refs. H6, H11).*

On account of this fairly close agreement between the phase angle characteristics for the two cases, it is possible to replace the actual attenuation characteristic by an approximate " straight-line " equivalent. Where the attenuation characteristic does not have constant slope throughout, it may be replaced by a succession of intersecting straight lines having slopes of zero or any multiple of 6 db/octave, either positive or negative (e.g. Fig. 7.55B). These, in turn, may be resolved into " semi-infinite "* lines of constant slope, as in Fig. 7.55C. The phase characteristics may be derived by the use of Fig. 7.55A for each junction. The resultant phase angle characteristic may be determined by adding these individual characteristics.

---

*A " semi-infinite " straight line is one commencing from a definite point and proceeding to infinity in one direction only.

More accurate results may be obtained in some cases by the use of phase angle curves for attenuation characteristics having a sloping portion and a horizontal portion, as given by Bode (Ref. H11, Chapter 15).

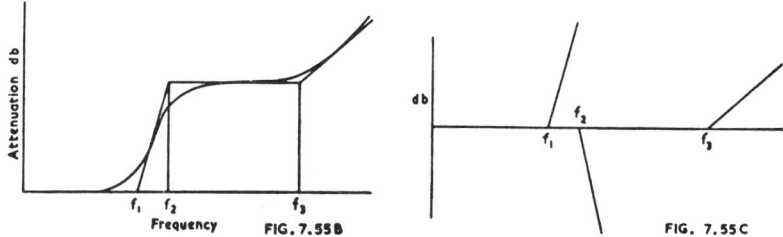

Fig. 7.55B.   Attenuation characteristic with straight line approximation (based on Bode, Ref. H11).

Fig. 7.55C.   Semi-infinite lines of constant slope corresponding to straight line approximations in Fig. 7.55B (based on Bode, Ref. H11).

More accurately again, **when the slope of the attenuation characteristic is varying,** the phase angle at any frequency may be determined from a measurement of the amplitude characteristic over a wide frequency range.   This procedure is facilitated by the use of special graph paper, plotting the slope of the amplitude characteristic in db/octave against a function of frequency, then measuring the area under the curve and so enabling the slope at one point to be determined (see Ref. A29).

When the slope of the attenuation characteristic reaches a maximum value over a limited frequency range and is less than this value at lower and higher frequencies, the phase angle at the point of maximum slope is less than that indicated by the slope, while at frequencies below and above the region of higher slope the phase angle is greater than that indicated by the slope.

Simple method for determination of phase angle (Ref. H18)

A very simple approximate method, which gives the total phase angle, is based on the straight-line approximate characteristics (Fig. 7.55D).   The phase angle displacement at the cut-off frequency f_0 is $45°$;   the approximate straight-line phase characteristic is taken as a tangent to the actual curve at the point of greatest slope, with sharp bends at the intersections with $0°$ and $90°$.   The maximum error is about $10°$.

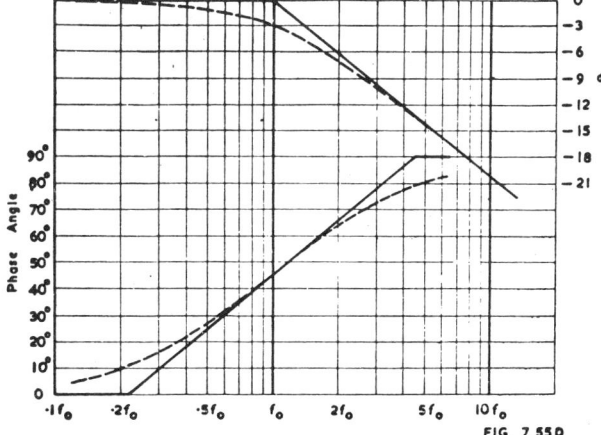

Fig. 7.55D " Straight line " approximate characteristics with an attenuation slope of 6db/octave (actual characteristics shown with broken lines) and corresponding phase angle characteristics (Ref. H18).

A typical attenuation characteristic is shown in Fig. 7.55E with three component " semi-infinite " slopes.

The corresponding phase angle characteristics are shown in Fig. 7.55F, where 1 and 2 are to be added, and 3 is to be subtracted to give the resultant. In each case the position of the component phase characteristic is determined by the relationship that the cut-off frequency on the attenuation characteristic gives 45° phase angle.

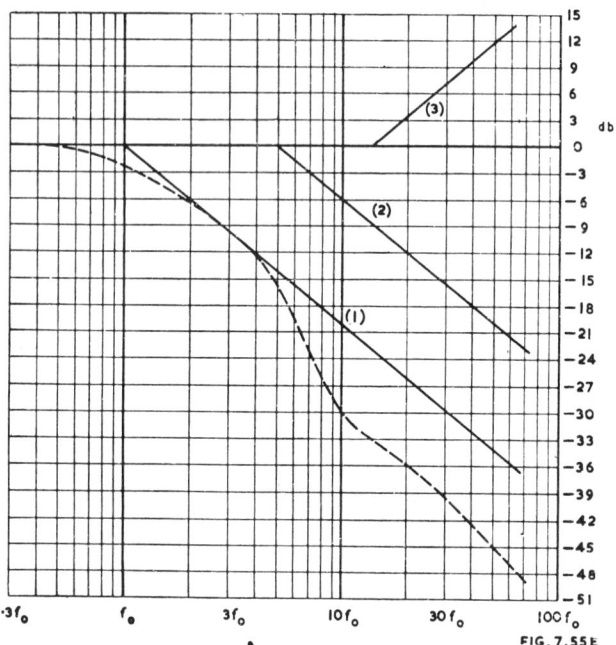

Fig. 7.55E. *Typical attenuation characteristic of amplifier with feedback (broken curve) and component " semi-infinite " slopes (1, 2 and 3) from which Fig. 7.55A is constructed (Ref. H18).*

The broken line in Fig. 7.55F is the actual phase characteristic, indicating reasonable accuracy for the method. In all cases the actual characteristic tends to " round the corners " of the approximate characteristic.

The same curves (Figs. 7.55D, E and F) may be used for the low frequency attenuation characteristics by inverting the frequency scales.

(iv) Design of 1 and 2 stage amplifiers

A **single stage** resistance capacitance coupled amplifier has + 90° phase angle displacement at zero frequency due to the grid coupling condenser. The screen and cathode by-pass condensers may cause phase angle displacements up to about + 65° and + 30° respectively (see Figs. 12.11B and 12.3C) but these come back to zero again at very low frequencies. Consequently the total phase angle displacement is normally less than 180° at any frequency. At the high frequency end the maximum phase angle displacement is − 90°. Such an amplifier is stable under all conditions.

A **two stage** r.c.c. amplifier has twice the phase angle displacement of a single stage amplifier, if both stages are identical. It may therefore produce peaks of frequency response due to positive feedback at low and high frequencies, and care is necessary in design—see Sect. 3(vi) below. It is desirable to " stagger " the low frequency attenuation characteristics by choice of widely different time constants for the grid condensers and following grid resistors ; the larger the ratio, the larger will be the permissible feedback factor.   One or both cathode by-pass condensers may be omitted, or the by-passed section of R_k may be reduced in resistance.   One or both screen dropping resistors may be reduced in resistance, or replaced by a voltage divider with other consequential adjustments.   The values of the cathode and screen by-pass capacitors may be adjusted so that their frequencies of maximum phase angle are " staggered," preferably in the ratio of at least 20 to 1.

The high frequency peak may be reduced by staggering the high frequency response of the two stages, for example by shunting a capacitor from one plate to earth and increasing the capacitance experimentally until the peak is sufficiently reduced. For an exact design method see Sect. 3(vii) below.

The phase angle displacement is increased by the inclusion of an iron-cored trans-former within the feedback loop, also by any resonance effects, and such cases should be treated by the method described for multi-stage amplifiers.

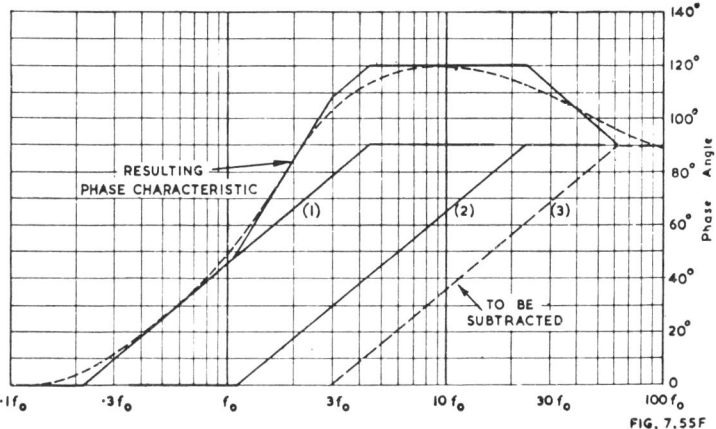

FIG. 7.55F

Fig. 7.55F.   Phase angle characteristic derived from Fig. 7.55E (Ref. H18).

(v) Design of multi-stage amplifiers

In the design of multi-stage amplifiers, Bode's method (A) is most commonly used or its simplification by Learned (B).   This method is intended primarily for use with amplifiers having only one feedback path.

The addition of subsidiary feedback to a single feedback loop leads to Duerdoth's multiple feedback system (E) which is based on a modified stability margin (C) and the theory of the summation of attenuation characteristics and subsidiary voltages (D).   Duerdoth's system is capable of providing a greater feedback factor than Bode's method, with the same minimum value of stability margin in each case.   Some de-lightfully simple applications of Duerdoth's method are possible and it seems that any amplifier with a single feedback loop can be improved as regards stability by the addition of one or more subsidiary feedback paths.   The latter, in their simplest form, may be merely unbypassed cathode resistors.

With any multi-stage amplifier, an important feature is the **degree of damping on transients.**   Insufficient experimental work has been carried out to indicate what degree of damping is desirable in a-f amplifiers.   A reasonably safe inference is that damping heavier than critical damping is undesirable, because it results in sluggish uptake.   Whether critical damping—see Sect. 3(vii)C—is desirable, or some lighter degree of damping is preferable, is an unsolved problem which can only be tackled on an experimental basis.   It is stated by those who advocate a light degree of damping that this gives very rapid uptake, and the overshoot which occurs is at an ultrasonic frequency which is certainly inaudible, and in any case would be very seriously attenuated by the loudspeaker.   On the other hand, it is well known that some forms of ultrasonic parasitic oscillation give rise to objectionable reproduction.

Design tests

It is highly desirable to measure and plot both the frequency response and phase angle over a sufficiently wide frequency range ; both curves may be plotted on the same sheet of logarithmic graph paper. The required frequency range is the useful frequency range of the amplifier plus (at each end) one octave for each 10 db of feedback plus at least one octave.

When it is desired to achieve the maximum degree of feedback it is essential to plot the polar (Nyquist) diagram as in Figs. 7.53B and 7.54. For a method of measuring the phase angle see Refs. H7, H19, H20.

An amplifier may be tested for transient response by applying a rectangular waveform to the input and observing the waveform of the output with a C.R.O. having good frequency and phase characteristics up to the maximum test frequency. If there is any oscillatory response or " overshoot " the feedback may be reduced or some modification made to the feedback loop to provide the desired degree of damping.

Another useful design test is to overload the amplifier with input voltages of various frequencies, and to determine the level to which the input must be reduced to return to normal linear operation (Ref. A29 discussion).

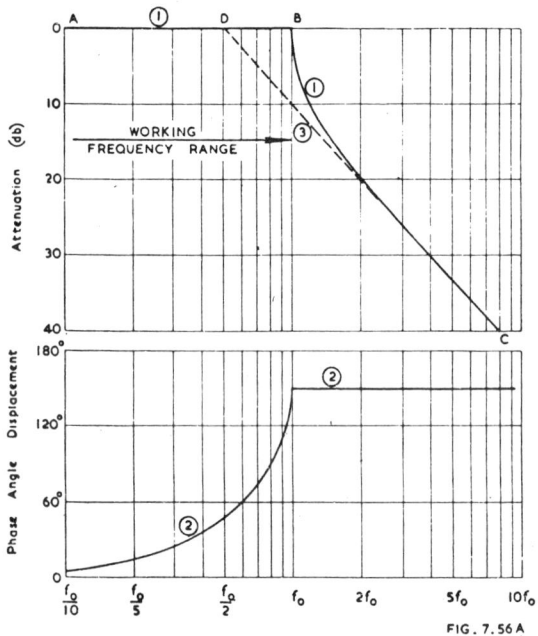

FIG. 7.56A

Fig. 7.56A. *Bode's method—Curve (1) attentuation characteristic and (2) phase angle displacement characteristic for constant 150° phase angle displacement above upper working frequency f_o ; curve (3) constant slope 10 db/octave (Ref. H11).*

Alternative methods of design

(a) One method sometimes used is the " cut and try " method of constructing an amplifier and then applying feedback with the feedback factor increasing in steps until instability occurs, and finally decreasing the feedback to provide a safe margin.

If instability occurs with only a small degree of feedback, there are devices which may be experimented with, such as

1. Shunting the primary of the output transformer by a condenser.
2. Connecting a very small condenser from the plate of one of the earlier stages to earth.

3. Shunting the feedback resistor by a condenser, or the more elaborate network of Fig. 7.59C.

(b) A preferable alternative is to measure and plot the frequency response curve with and without feedback, over a frequency range sufficiently wide to include both low and high frequency peaks. The feedback factor may be increased until the response with feedback is greater than that without feedback by a predetermined amount (say from 1 to 6 db) at any frequency.

(c) Alternatively the amplifier may be designed with " staggered " frequency response. In a 2-stage amplifier, there should be one stage with wide, and the other with a narrow frequency response ; in a 3-stage amplifier there may be two stages with wide and one with narrow frequency response. The use of " maximal flatness " —as described in Sect. 3(vii)A and B—results in oscillatory response to transients, and the feedback factor should be reduced to half the specified value if it is desired to produce critical damping. A preferable arrangement is " compensated " critical damping as described in Sect. 3(vii)C.

(A) **Bode's method** (Refs. H6, H11)

In accordance with Bode's method, the usual procedure is to adopt a constant phase angle displacement, expressed in the form of an attenuation characteristic, as the safety margin. For example, Fig. 7.56A shows the attenuation characteristic BC of Curve 1 for a constant phase angle displacement of 150°, thus giving a constant safety margin of 30°. The ultimate slope of this characteristic is 10 db/octave, which is extended upwards in Curve 3 to point D. The characteristic ABC gives an additional octave of working frequency range as compared with characteristic ADC and is therefore to be preferred if the maximum possible feedback is to be used, together with an adequate safety margin.

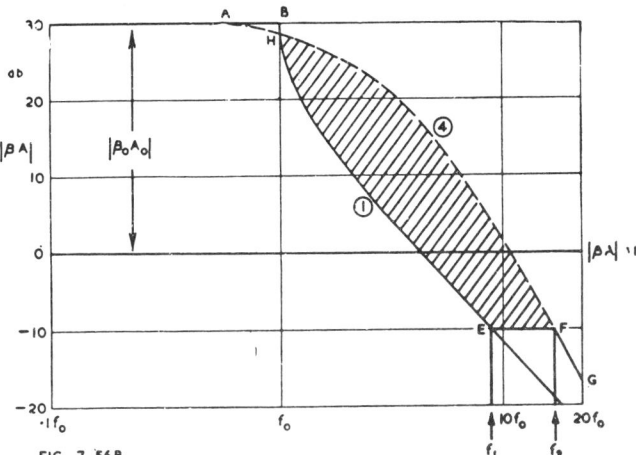

FIG. 7.56B

Fig. 7.56B. *Bode's method—Curve* (1) *as in Fig.* 7.56A ; (2) *normal attenuation characteristic of typical amplifier. Shaded portion indicates additional attenuation required in the feedback loop.*

For convenience in feedback design, Curve 1 has been redrawn in Fig. 7.56B with 0 db corresponding to $|\beta A| = 1$, for 30 db of feedback at the mid-frequency. Curve 4 is the normal attenuation characteristic of a typical amplifier, so that the shaded portion indicates the additional attenuation required in the feedback loop to produce the desired curve 1. The shaded portion is limited at the lower end by the line EF which provides a safety margin (here 10 db) below $|\beta A| = 1$. The desired characteristic is therefore BEFG, where FG is portion of the normal attenuation characteristic of the amplifier. The safety margin is therefore a constant angle of 30° from B to E, and a constant attenuation of 10 db from E to F   The effect of the horizontal " step " EF is to limit the phase angle displacement in this region, to a value less than 180°, which will hold provided that

$$\frac{f_2}{f_1} = \frac{\text{ultimate slope of actual characteristic (Curve 4) in db/octave}}{\text{ultimate slope of desired characteristic (Curve 1) in db/octave}}$$

Since point F is fixed by the actual attenuation characteristic, the position of point E, and hence Curve 1, will be determined by the ratio f_2/f_1.

Methods for providing the " step " in the characteristic are described in (B) below.

The following relations hold, provided that the ultimate slope of curve 1 is 10 db/ octave and the safety margin of the " step " below $|\beta A| = 1$ is 10 db.

Let n = ultimate slope of actual attenuation characteristic, in db/octave
$|\beta_0 A_0|$ = desired value of $|\beta A|$ at the mid-frequency
and f_2 = frequency at which the actual attenuation is equal to $(|\beta_0 A_0| + 10)$ db, then $f_1 = f_2 (10/n)$
number of octaves between f_0 and f_1 = $|\beta_0 A_0|/10$
 ,,   ,,   ,,   ,, f_1 and f_2 = 3.32 log$(n/10)$.

The frequency ratio may be derived from the number of octaves by the relation—
frequency ratio = antilog $(0.301 \times$ number of octaves)
or by the use of the table below :

Frequency Ratio, Octaves and Decades

Frequency Ratio	No. of Octaves	No. of Decades	Frequency Ratio	No. of Octaves	No. of Decades
1.07	0.1	0.03	5.28	2.4	0.72
1.15	0.2	0.06	6.07	2.6	0.78
1.23	0.3	0.09	6.97	2.8	0.84
1.32	0.4	0.12	8.00	3.0	0.90
1.42	0.5	0.15	9.18	3.2	0.96
1.52	0.6	0.18	10.0	3.32	1.00
1.63	0.7	0.21	10.6	3.4	1.02
1.74	0.8	0.24	12.1	3.6	1.08
1.87	0.9	0.27	13.9	3.8	1.14
2.00	1.0	0.30	16.0	4.0	1.20
2.30	1.2	0.36	18.4	4.2	1.26
2.64	1.4	0.42	21.2	4.4	1.33
3.03	1.6	0.48	24.3	4.6	1.39
3.48	1.8	0.54	27.9	4.8	1.45
4.00	2.0	0.60	32.0	5.0	1.51
4.60	2.2	0.66			

Frequency ratio   = antilog (number of decades)
                     = antilog $(0.301 \times$ number of octaves)
Number of decades = log (frequency ratio)
                     = 0.301 $\times$ number of octaves
Number of octaves = 3.32 $\times$ log (frequency ratio)
                     = 3.32 $\times$ number of decades.

Relationship between db/octave and db/decade

db/octave	3	6	9	10	12	15	18
db/decade	10	20	30	33.3	40	50	60

In addition to the attenuation of the shaded area in Fig. 7.56B, it is also necessary to boost the gain near the limits of the working frequency band to provide nearly constant gain (area AHB) and thereby achieve almost zero phase angle at the cut-off frequency f_0. This may be accomplished by the circuit of Fig. 7.56C in which the Q of the $L_1 C_1$ circuit should be unity, as given by

$$L_1 = 1/\omega^2 C_1 \text{ and } r = 1/\omega_0 C_1$$

where r = series resistance of L_1. More complicated but more satisfactory networks (low-pass or band-pass " Wheeler " networks) are described in Ref. A29. These have an ultimate attenuation of 6 db/octave and a phase angle displacement of approximately 90° at all frequencies outside the working band. For other methods see Ref. H11.

The low frequency end may be treated in a similar manner throughout.

Means for producing the desired attenuation characteristics are described by References H6, H10, H11. See also (B) below.

This method of design requires the amplifier to have a very much wider frequency range before feedback is applied than it would need without feedback. The extension is approximately one octave in each direction for each 10 db reduction of gain by feedback plus a sufficient margin of safety (say 1 octave at each end). Thus an amplifier to cover from 40 to 15 000 c/s with 30 db reduction of gain by feedback would require to have a frequency response without feedback (at 40 db down) from 2.5 to 240 000 c/s. This actually applies to the transmission characteristics of the whole feedback loop (amplifier and feedback network together).

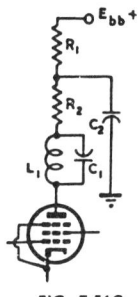

FIG. 7.56C

Fig. 7.56C. Simple network in plate circuit to give more nearly constant gain over the working frequency range (based on Ref. A29).

(B) **Simplified treatment by Learned** (Ref. H9)

The following treatment is based on the simplified assumption that the attenuation characteristics are straight lines. This is approximately correct except in the octave immediately above and below the useful frequency band of the amplifier, where the method gives a wider safety margin than is necessary. The method does not, therefore, give the same degree of feedback obtainable with the precise method of Ref. H6. At the same time it is quite suitable for general use.

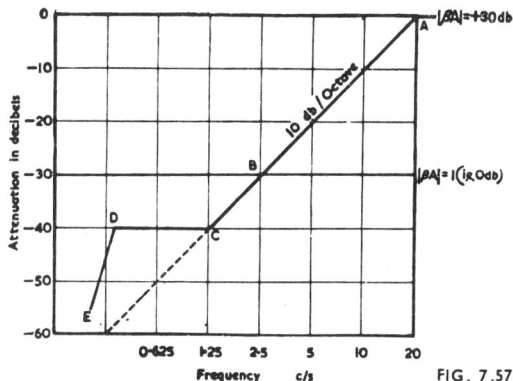

Fig. 7.57. Low frequency attenuation characteristics of multi-stage amplifier with feedback (simplified treatment).

Low frequency attenuation characteristics

As an example for illustrating the general principle, take an amplifier with nominally flat frequency response down to 20 c/s (actually it will be − 2 or − 3 db at this frequency). Let the loop amplification $|\beta A|$ be 30 db at useful frequencies.

Draw the attenuation characteristic with a slope of 10 db/octave from 20 c/s downwards (line ABC in Fig. 7.57). Point B is where the value of the loop amplification becomes unity. Extend the line AB to C which is at a level 10 db below $|\beta A| = 1$. At C, insert a horizontal " step " CD, the length of which will be given later. Beyond D the attenuation will fall at a rate of 12, 18, 24, 30 db or more depending on the circuit ; each single time constant or non-resonant transformer contributes 6 db, while each resonant circuit contributes 12 db. The frequency ratio between points

C and D is made equal to the ratio of the slopes of *DE* and *AC*, which is equal to the slope of *DE* divided by 10.   The line *ABCDE*, thus determined, is to be used as a guide to the low frequency attenuation characteristic of the amplifier.   In reality, of course, the sharp corners will be rounded—the important features are that the slope of the *AC* region should not exceed 10 db/octave, and that the step should be sufficiently long.

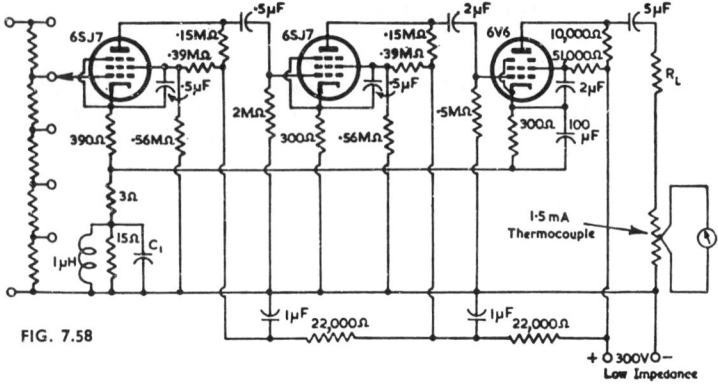

FIG. 7.58

Fig. 7.58. *Three stage amplifier with negative feedback illustrating design method* (*Ref. H*9).

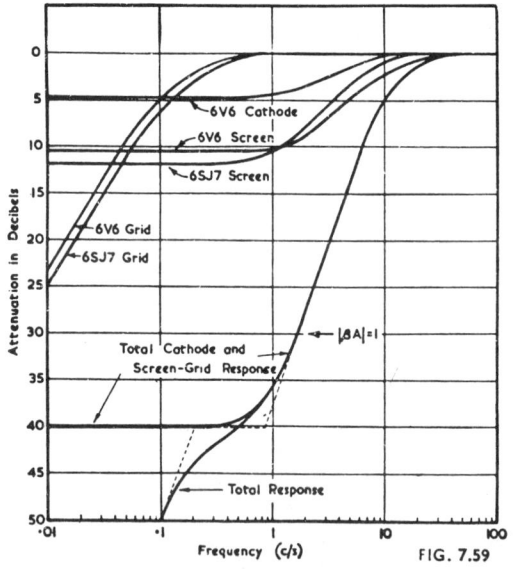

FIG. 7.59

Fig. 7.59. *Low frequency attenuation characteristics of Fig.* 7.58 (*Ref. H*9).

Example

An example of an amplifier designed in accordance with this method is Fig. 7.58 (Ref. H9).   The low frequency attenuation characteristics are given in Fig. 7.59. The 6V6 cathode, 6V6 screen and 6SJ7 screen characteristics are designed to provide a very close approach to a total slope of 10 db/octave with a limiting attenuation of 40 db, so providing the basis of a step.   The two grid condensers are designed to give effective attenuation beyond the limit of the step.   It is evident that it is possible to design an amplifier which provides the necessary low frequency attenuation characteristics, including the step, without using any resonant circuits. The tolerances on the frequency-dependent components (R and C) must be small.

The feedback loop does not include the plate circuit of the third stage, so that the slope of the high-frequency attenuation characteristic is nominally 12 db/octave. The resonant circuit (incorporating C_1) in the feedback path is employed to stabilize the frequencies in this region by providing a step in the high-frequency attenuation characteristic.

Corrective networks

The design of simple corrective networks is well summarized in Ref. H9. More complicated designs incorporate LCR 2-terminal networks as the plate load impedances and (some) cathode bias impedances (Ref. H11). See also Ref. J16.

One of the simplest and most effective methods of providing the required step in the high frequency attenuation characteristic is the RC network shunting the plate load resistor of the pentode V_1 in Fig. 7.59A (based on Ref. H9). The ultimate attenuation of the step in decibels is given approximately by

$$\text{attenuation} \approx 20 \, \log \, (R_1 + R)/R.$$

The cut-off frequency is given by

$$f_1 = 1/2\pi R_1 C$$

and the " flattening-out " frequency by

$$f_2 \approx 1/2\pi RC$$

where $R_1 = R_L R_g/(R_L + R_g)$

and the " flattening-out " frequency is defined as the frequency at which the attenuation is 3 db less than the ultimate attenuation of the step.

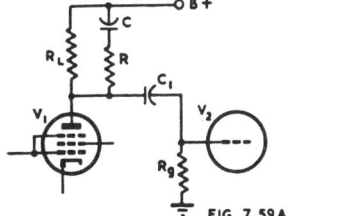 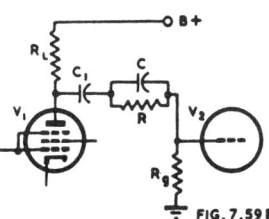

FIG. 7.59A   FIG. 7.59 B

Fig. 7.59A. Circuit to provide step in high frequency attenuation characteristic (based on Ref. H9).

Fig. 7.59B. - Circuit to provide step in low frequency attenuation characteristic (based on Ref. H9).

The equivalent to Fig. 7.59A for the low frequency end is provided by Fig. 7.59B :

Step attenuation $\approx 20 \, \log \, (R + R_g)/R_g$

Cut-off frequency $f_1' = 1/2\pi C(R_L + R_g)$

Flattening-out frequency $f_2' \approx 1/2\pi RC$

The value of C_1 is selected to give negligible attenuation at f_2' ; say

$$C_1 \approx 5/[\pi f_2'(R + R_L + R_g)].$$

With these step circuits it is desirable to have as many separate circuits as possible, each with a small step.

Fig. 7.59C. Feedback path including CR network to give a step in the attenuation characteristic, and thereby to improve stability (Ref. H18).

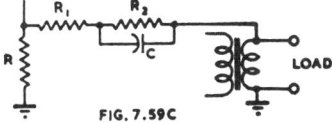

FIG. 7.59C

A step in the attenuation characteristic may also be provided by an LCR circuit in the feedback path, or more simply by a CR circuit as in Fig. 7.59C. Here R may be the cathode resistor of the first valve, $R_1 + R_2$ take the place of the usual feedback resistor, and C effectively increases the value of β at high frequencies. This is a most useful device to improve the stability of a feedback amplifier. A simpler form, omitting R_1, is sometimes sufficient ; this is similar to Flood's " compensated feedback " Sect. 3(vii)C.

(C) **Duerdoth's stability margin** (Ref. A29)

Duerdoth adopts a pre-determined value (e.g. 6 db) of the rise of gain with feed-back at high and low frequencies as a stability margin over values of βA from 0.3 to some value less than 2.0. At higher values of βA he adopts an angular stability margin (e.g. 15°) while below $\beta A = 0.3$ there is no danger of instability. An am-plifier designed in accordance with his method might have a high-frequency charac-teristic such as OABCDEFG in Fig. 7.54. See discussion on stability and tolerances in components with this characteristic in Sect. 3(ii) above. One method for achieving such a shape of characteristic is given in Sect. 3(v)E below.

(D) **Summation of attenuation characteristics and subsidiary voltages**

When two voltages have to be added, as for example with a loop feedback voltage and a subsidiary voltage connected in series, the attenuation characteristic can only be derived by adding the voltages at selected frequencies, with due regard to the phase angle between them.

The phase angle between the two voltages is a function of the difference in slope when plotted as attenuation characteristics, provided that the slopes are constant :

Slope difference	2	4	6	8	10	12	14	16	18 db/octave
Phase angle diff.	30	60	90	120	150	180	210	240	270 degrees

For example : If one characteristic has a slope of 4 db/octave and the second 12 db/octave, the slope difference is 8 db/octave, and the phase angle between the volt-ages is therefore 120°.

Take the intersection of the two characteristics as the origin (as in Fig. 7.60) and assume that 0 db = (say) 1 volt—both assumptions being for convenience.

Let x be the voltage contributed by the lower slope characteristic at a certain fre-quency. It may readily be shown that the voltage contributed by the high slope characteristic at this frequency is x^a, where a is the ratio of slopes in db/octave. These two voltages, x and x^a must then be added vectorially, the angle between them being as given by the table above.

The magnitude of the combined characteristic may then be plotted as in Fig. 7.60, and the values of the phase angle may be marked along it.

The method of deriving the magnitude and phase angle of the resultant is illustrated by Fig. 7.61 for the condition where the higher slope characteristic has a slope of 12 db/octave, corresponding to 180°. Vector OA, having magnitude x and phase angle

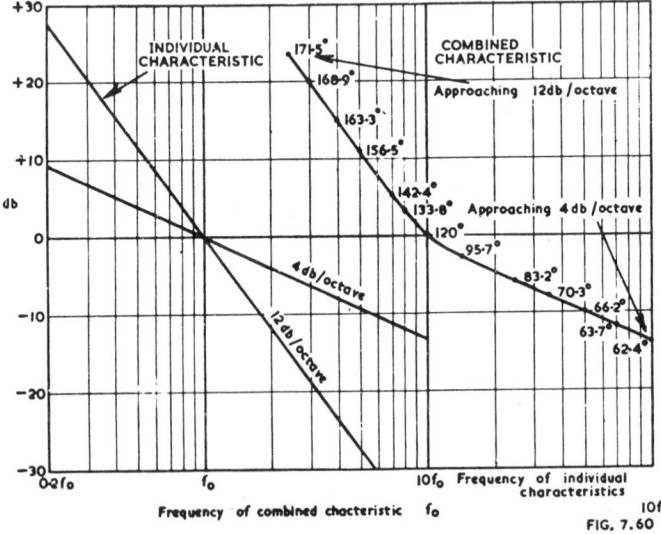

Fig. 7.60. *Illustrating summation of attenuation characteristics (based on Ref. A29).*

ϕ indicates the lower slope characteristic, the value of ϕ in degrees being given by 15 × slope in db/octave.   Vector OB, having magnitude x^a and phase angle 180°, indicates the higher slope characteristic with a slope of 12 db/octave.   For example, when the slopes are 4 and 12 db/octave, $a = 3$ and $\phi = 60°$.

The magnitude of the resultant is given by

$$\text{Resultant} = \sqrt{(x \sin \phi)^2 + (x^a - x \cos \phi)^2}.$$

The phase angle θ of the resultant is given by

$$\theta = \tan^{-1} \frac{x \sin \phi}{x \cos \phi - x^a} \, .$$

For ease in calculations the following table has been derived for use in all normal cases when the phase angle between the two voltages is a multiple of 30° and less than 180° and when the higher slope is 12 db/octave.

Difference in phase angle*	Resultant	Phase Angle θ of Resultant
30°	$x\sqrt{1 + 1.73x^{0.2} + x^{0.4}}$	$\tan^{-1} - \dfrac{1}{2x^{0.2} + 1.73}$
60°	$x\sqrt{1 + x^{0.5} + x}$	$\tan^{-1} - \dfrac{1.73}{2x^{0.5} + 1}$
90°	$x\sqrt{1 + x^2}$	$\tan^{-1} - 1/x$
120°	$x\sqrt{1 - x^2 + x^4}$	$\tan^{-1} \dfrac{1.73}{1 - 2x^2}$
150°	$x\sqrt{1 - 1.73x^5 + x^{10}}$	$\tan^{-1} \dfrac{1}{1.73 - 2x^5}$

*between the two voltages, i.e. 180° − ϕ.

A similar procedure can, of course, be applied to any other value of maximum slope.

Procedure when lower slope is zero

The foregoing procedure cannot be used when one of the two attenuation characteristics has a slope of zero, that is when one is a fixed voltage.   In this case the procedure is as follows.

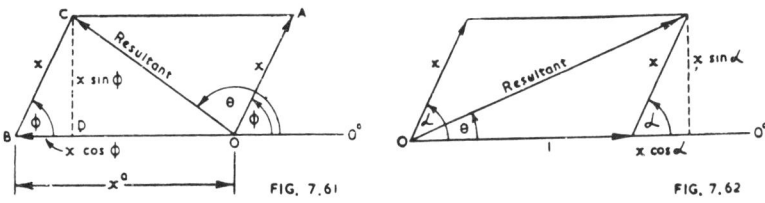

FIG. 7.61   FIG. 7.62

Fig. 7.61.   Vector relationships in summation of attenuation characteristics.
Fig. 7.62.   Vector relationships when one characteristic has zero slope.

As previously, the origin is the point of intersection of the two characteristics, and its level is taken as 0 db equal to 1 volt.   The vector relationships are shown in Fig. 7.62.

$$\text{Resultant} = \sqrt{x^2 + 1 + 2x \cos \alpha}$$

$$\text{Phase angle of resultant} = \theta = \tan^{-1} \frac{x \sin \alpha}{1 + x \cos \alpha}$$

where x = voltage at a selected frequency, on the sloping characteristic
and α = phase angle of sloping characteristic (i.e. angle in degrees = 15 × slope in db/octave).

Angle α	Resultant	Phase angle θ of Resultant
30°	$\sqrt{x^2 + 1.73x + 1}$	$\tan^{-1} \dfrac{x}{2 + 1.73x}$

Angle x	Resultant	Phase angle θ of Resultant
60°	$\sqrt{x^2 + x + 1}$	$\tan^{-1}\dfrac{1.73x}{2 + x}$
90°	$\sqrt{x^2 + 1}$	$\tan^{-1} x$
120°	$\sqrt{x^2 - x + 1}$	$\tan^{-1}\dfrac{1.73x}{2 - x}$
150°	$\sqrt{x^2 - 1.73x + 1}$	$\tan^{-1}\dfrac{x}{2 - 1.73x}$
165°	$\sqrt{x^2 - 1.93x + 1}$	$\tan^{-1}\dfrac{0.26x}{1 - 0.97x}$

An example of the summation of two attenuation characteristics, one having zero slope and the other having a slope of 10 db/octave, is given in Fig. 7.63. The resultant is asymptotic to 10 db/octave and to the horizontal characteristic, but reaches a minimum value at the frequency of the point of intersection O. The phase angle is marked at points along the resultant, and at all frequencies the phase angle is less than that of the 10 db/octave characteristic, the effect becoming more pronounced

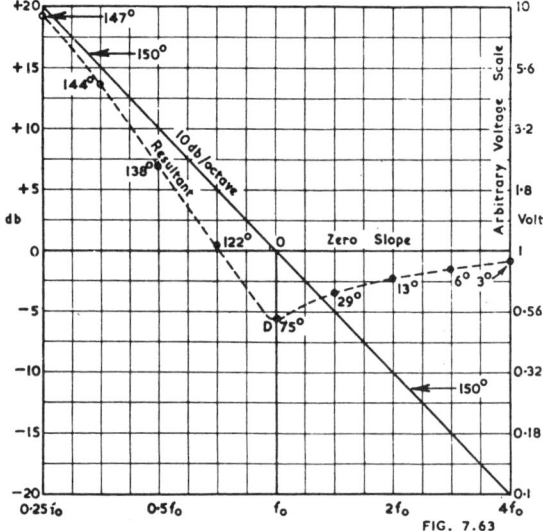

Fig. 7.63. *Summation of two attenuation characteristics having slopes of zero and 10 db/octave.*

at higher frequencies, particularly beyond point D. The fact that the slope of the resultant in Fig. 7.63 exceeds 12 db/octave to the left of point D, has no effect whatever on the stability as indicated by the angles. It is therefore obvious that the relationship between phase angle and slope of the attenuation characteristic which applies to minimum phase shift networks, does not apply here.

A similar calculation has been made for a slope of 11 db/octave, Fig. 7.64.

The position of point D on the resultant characteristic is a function of the slope of the attenuation characteristic.

slope in db/octave	2	4	6	8	10	11
height of point D	+ 5.7	+ 4.7	+ 3	0	− 5.6	− 11.6 db.

(E) Duerdoth's method employing multiple feedback paths (Ref. A29)

In accordance with this method, one or more additional subsidiary feedback voltages are introduced into the feedback loop so as to decrease the phase angle over the

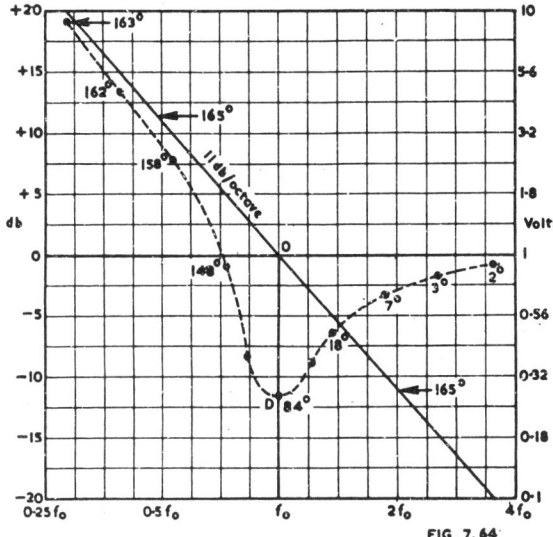

Fig. 7.64. *Summation of two attenuation characteristics having slopes of zero and 11 db/octave.*

attenuation portion of the amplitude characteristic without appreciably affecting the characteristics within the working frequency range.

With subsidiary feedback applied over the first stage of a 2-stage amplifier and loop feedback over the whole amplifier, the ratio of gain with and without feedback is given by

$$\frac{A'}{A} = \frac{1}{1 - \beta_1 A_1 - \beta_2 A} \tag{1}$$

where A = overall amplification without feedback
   A' = overall amplification with feedback
   A_1 = amplification of first stage without feedback
   β_1 = value of β for subsidiary feedback over first stage only
and β_2 = value of β for loop feedback over whole amplifier.

It is evident that when subsidiary feedback is used, the several feedback voltages must be added at their common point. This holds even when the subsidiary voltage is entirely inside the feedback loop.

Under the same conditions, the ratio of distortion in the final stage with and without feedback is given by

$$\frac{D'}{D} = \frac{1 - \beta_1 A_1}{1 - \beta_1 A_1 - \beta_2 A} = \frac{A'}{A}(1 - \beta_1 A_1) \tag{2}$$

Thus the application of negative subsidiary feedback to the first stage increases the distortion in the final stage in the same proportion that the gain of the first stage is reduced, provided that the overall feedback loop is unchanged.

If, however, the subsidiary feedback is positive, the gain of the first stage will be increased and the distortion in the final stage will be reduced further. The use of positive subsidiary feedback need not present any stability problems, since the stability criterion is dependent on $(\beta_1 A_1 + \beta_2 A)$. See Sect. 2(xi) and Refs. F11, F12.

With multiple loop amplifiers, the usual Nyquist stability criterion applies to any loop, provided that the amplifier remains stable when the particular loop is broken. Even with amplifiers which are always stable, the Nyquist diagrams obtained by measurements of the several loops in turn may have different shapes and the definition of stability margin becomes meaningless. However, if a mesh of the amplifier can be found which, if broken, simultaneously breaks all the loops, then there is no possibility of oscillation due to the disconnection and the definition of stability margin remains applicable (Ref. A29).

When employing multiple feedback paths with summation of attenuation characteristics and subsidiary voltages, the crossing point of the characteristics should be controlled with considerable accuracy. The crossing point will be modified when the gains of the various stages change owing to changes in the valve characteristics with age, or to overloading caused by an excessive input voltage. The latter may be avoided or reduced in more elaborate amplifiers by the addition of a cathode follower stage, operating as a limiter, as part of the first stage. Precautions to be taken in design are described in Ref. A29.

Some applications employing multiple feedback paths are described below. See also Sect. 2(vi)A.

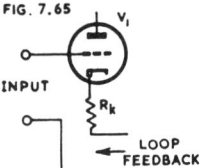

FIG. 7.65

Fig. 7.65. Part of first stage of feedback amplifier showing one method of applying subsidiary feedback (Ref. A29).

(1) One simple application is the circuit of Fig. 7.65 where the signal voltage across the unbypassed cathode resistor R_k is added to the loop feedback voltage. This is an application of the principle of summation of attenuation characteristics as shown in Figs. 7.63 and 7.64. The voltage across R_k is practically constant over the critical region of the high frequency attenuation characteristic, so that its characteristic has zero slope. This method may only be used when the phase angle of the βA locus is always less than 180°, i.e. the slope of the βA attenuation characteristic is less than 12 db/octave. This method does not reduce the distortion in the final stage to the same level as without subsidiary feedback, or to that using one of the alternative methods below. For this reason the subsidiary voltage should be considerably smaller than the loop feedback voltage. However, this method will improve the stability margin of an amplifier and reduce the amplitude of the high frequency peak.

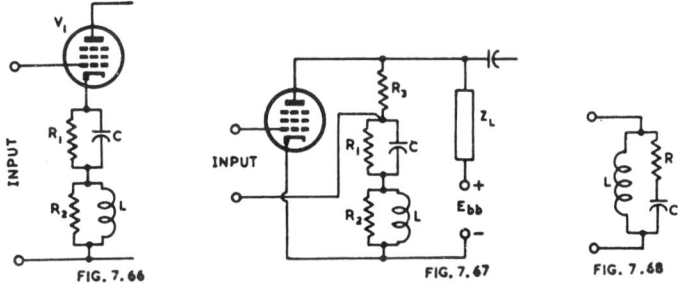

Fig. 7.66. One form of subsidiary feedback (after Ref. A7).
Fig. 7.67. Another form of subsidiary feedback (after Ref. A7).
Fig. 7.68. Two terminal network as cathode load impedance.

(2) One modification of this method, which has only small degeneration in the working range, is shown in Fig. 7.66. This appears to have been first described by Farren (Ref. A7) and applied to one of the stages inside the feedback loop. The phase shift reaches maxima at low and high frequencies outside the working range, beyond which it approaches zero in both directions. The degeneration increases gradually in both directions to the limiting design values provided by R_1 and R_2. This arrangement does not modify the fundamental phase shift between the input and output voltages of the stage due to the phase angle of the plate load impedance.

(3) A further modification by Farren (Ref. A7) is shown in Fig. 7.67 which has the advantage at very high and very low frequencies that it reduces the phase shift between the input and output voltages of the stage due to the phase angle of the plate load impedance which would be characteristic of the stage if there were no subsidiary feedback.

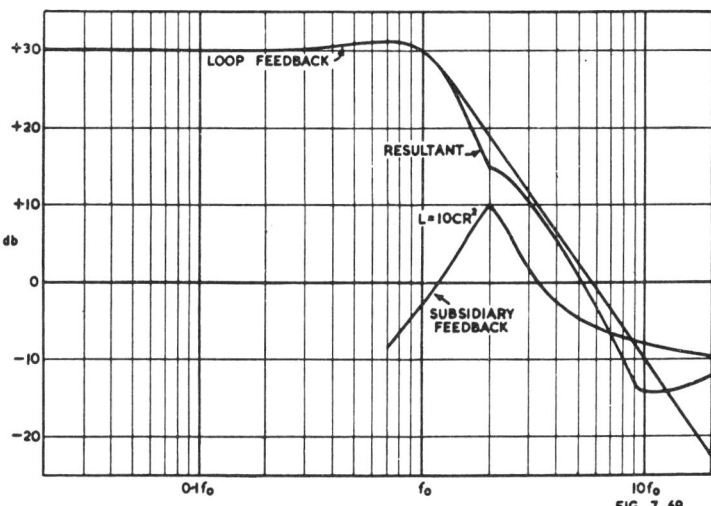

Fig. 7.69.   *Summation of loop and subsidiary feedback voltages to provide improved stability (Ref. A29).*

(4) Two terminal networks of the form of Fig. 7.68 are described by Duerdoth as cathode impedances for improving the stability of feedback amplifiers with an ultimate attenuation of 12 db/octave.   A value of $L = 10CR^2$ may be used as a first trial and maximum feedback and zero phase angle will then occur at a frequency of approximately

$$f = 0.048/CR$$

where C and R are in farads and ohms respectively.   An example of the application of subsidiary feedback on an amplifier with an ultimate attenuation slope of 12 db/octave is given in Fig. 7.69.   In this example the peak of subsidiary feedback occurs at twice the maximum frequency of the working band.   The stability margin in the critical region has been increased from $10°$ to $36°$ with subsidiary feedback, as shown by the Nyquist diagram (Ref. A29).

It often happens that the addition of the subsidiary voltage results in an increase in phase angle just above the working frequency band where the magnitude of the

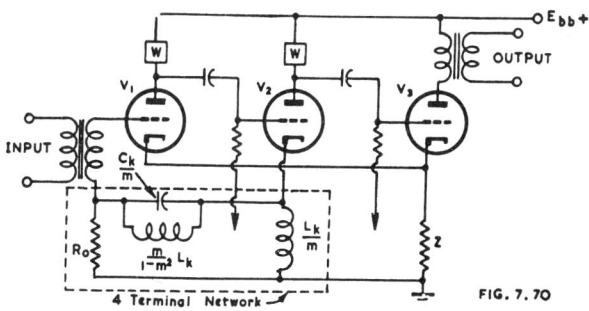

Fig. 7.70.   *Three stage amplifier incorporating subsidiary feedback in the form of a 4 terminal network. m = 0.87. W = " Wheeler " network (based on Ref. A29).*

subsidiary voltage is increasing with frequency, although a decrease in phase angle occurs, where the magnitude of the subsidiary voltage is steady or decreasing with frequency. This increase in phase angle will reduce the angular margin over a range of frequencies and, in the extreme case when the frequency of the peak subsidiary voltage is greater than 3 or 4 times the maximum frequency of the working band, may approach " conditional stability."

A similar technique may also be applied to low frequency attenuation characteristics.

(5) Four terminal networks of the form included in Fig. 7.70 are described by Duerdoth for providing subsidiary feedback in amplifiers with an ultimate attenuation slightly over 12 db/octave. Fig. 7.70 has loop feedback from the cathode of V_3 to the cathode of V_1, main subsidiary feedback from the cathode of V_2 to the grid circuit of V_1, and minor subsidiary feedback due to the unbypassed cathodes of V_1 and V_3. The four terminal network is a high-pass m-derived half-section filter terminated in R_o. The cut-off frequency of the filter is made about twice the maximum frequency of the working band (Ref. A29).

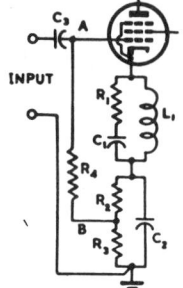

FIG. 7.71

Fig. 7.71. Combination of high and low frequency subsidiary feedback.

(6) A combination of both high and low frequency subsidiary feedback is shown in Fig. 7.71 where $R_1C_1L_1$ is the high-frequency network (as in Fig. 7.68) and $R_2R_3C_2$ is the low frequency network. The latter provides a subsidiary negative feedback at low frequencies only ; its value will be affected by the source impedance of the input voltage. If transformer coupling is used, the secondary should be connected to points A and B.

(7) In a three stage amplifier with an ultimate slope of about 18 db/octave, the second stage being a phase splitter which introduces little phase shift, some reduction in the total loop phase angle may be achieved by means of a phase shifting network in the feedback line. A subsidiary feedback path is added from the plate circuit of the phase splitter through a high-pass filter to the cathode of the first stage. This voltage is suppressed in the working band by the filter, but above the working band the loss of the network falls and the subsidiary voltage becomes the predominant factor, thus reducing the ultimate slope to 6 db/octave. Complete stability is thus achieved (Ref. A29 Fig. 26).

(8) A third feedback voltage may be added, if desired, to give two subsidiary feedback paths.

(vi) Effect of feedback on frequency response

In the case of a single stage r.c.c. amplifier (assuming perfect cathode and screen by-passing) the feedback merely widens the frequency range without changing the shape of the voltage gain characteristics (Fig. 7.72).

With two such identical stages, the frequency characteristics are as Fig. 7.73. The peaks, which occur as the value of βA is increased, are due to the reduction in effective negative feedback and the development of positive feedback through phase angle shift, which approaches $180°$ at very low and very high frequencies. F_1 and F_2 are the frequencies giving 0.707 relative voltage gain for each single stage without feedback. Ref. H4.

These peaks may be reduced or eliminated entirely by designing the amplifier with one stage having a much wider frequency range than the other. Design methods to produce " maximal flatness " are described in Sect. 3(vii) and these may be modified to produce critical damping, if desired.

Alternatively a multiple feedback system may be adopted—for example one overall feedback loop in conjunction with an unbypassed cathode resistor in the first stage (Duerdoth's method) as described in Sect. 3(v)E.

See also Sect. 3(iv) and Sect. 2(vi) for the design of two-stage amplifiers.

A special application to a particular two-stage amplifier is Ref. H5.

See also Refs. J8, J11.

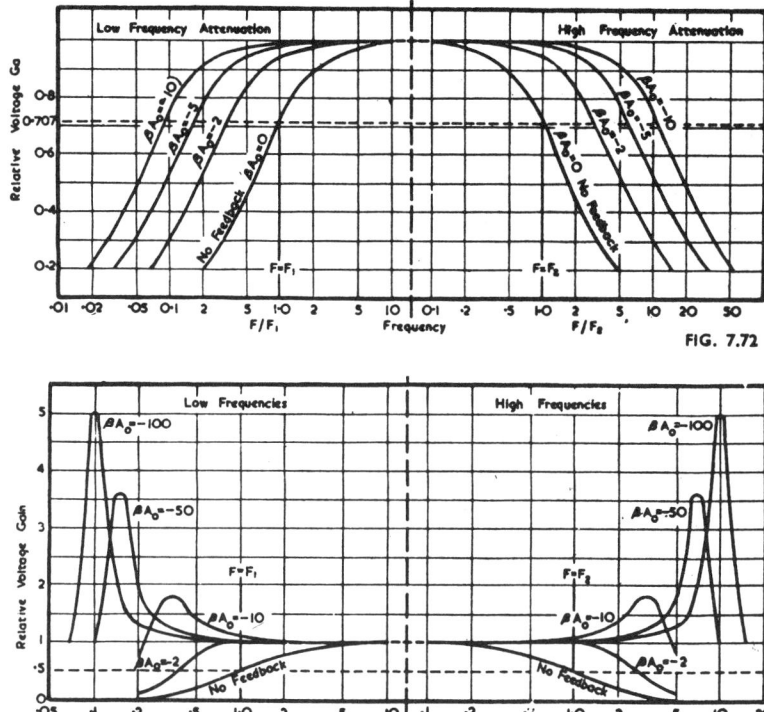

Fig. 7.72. Relative voltage gain of a single stage r.c.c. amplifier without feedback and with three selected values of feedback (Ref. H4).

Fig. 7.73. Relative voltage gain of two identical r.c.c. amplifier stages, without feedback and with four selected values of feedback (Ref. H4).

Graphical method for two-stage r.c.c. amplifier

An ingenious graphical method has been developed by Barter (Ref. H17), and may be used for determining the R and C values to give any desired height of peaks at low and high frequencies.

(vii) Design of amplifiers with flat frequency response

(A) **Method of H. Mayr** (Ref. H13). See also Ref. J13.

A resistance-capacitance-coupled amplifier Fig. 7.74 may be regarded, so far as frequency response is concerned, as a tuned circuit with a Q less than 0.5. The value of this Q is given by

$$Q = \frac{1}{b(d + k) + (1/b)(p + k)} \tag{3}$$

where $b = \sqrt{R_g/R_L}$; $d = C_g/C$; $k = C_c/C$;
$p = C_p/C$ and $C = \sqrt{C_gC_c + C_cC_p + C_cC_p}$.

This method may be applied to multi-stage amplifiers with feedback, it being assumed that all stages have the same resonant frequency ($\omega_0/2\pi$). The conditions for maximum flat response are then given by:

1 stage $|(A_0/A)| = \sqrt{1 + x^2}$ (4)
$\phi = -\tan^{-1}x$ (5)

$$\text{where } x = \frac{Q}{n}\left(\frac{\omega}{\omega_0} - \frac{\omega_0}{\omega}\right) \tag{6}$$

N.B. n must be greater than 2.

2 stages $|(A_0/A)| = \sqrt{1 + x^4}$ $\tag{7}$

$$\phi = -\tan^{-1}\sqrt{2}x/(1 - x^2) \tag{8}$$

$$\text{where } x = Q_2\sqrt{\frac{Q_1}{nQ_2}}\left(\frac{\omega}{\omega_0} - \frac{\omega_0}{\omega}\right) \tag{9}$$

and $\qquad Q_1/Q_2 = (n - 1) + \sqrt{n(n - 2)}$ $\tag{10}$

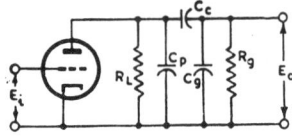

Fig. 7.74. *Resistance-capacitance coupled stage.*

FIG. 7.74

3 stages $|(A_0/A)| = \sqrt{1 + a_2x^4 + x^6}$ $\tag{11}$

$$\phi = -\tan^{-1}(b_1x - x^3)/(1 - b_2x^2) \tag{12}$$

$$\text{where } x = Q_2\sqrt[3]{\frac{Q_1}{nQ_2}}\left(\frac{\omega}{\omega_0} - \frac{\omega_0}{\omega}\right) \tag{13}$$

$$a_2 = \frac{1}{\sqrt[3]{n^2}}\left[\sqrt[3]{\frac{Q_2^4}{Q_1^4}} + 2\sqrt[3]{\frac{Q_1^2}{Q_2^2}}\right] \tag{14}$$

$$b_1 = \frac{1}{\sqrt[3]{n^2}}\left[\sqrt[3]{\frac{Q_1^2}{Q_2^2}} + 2\sqrt[3]{\frac{Q_2}{Q_1}}\right] \tag{15}$$

$$b_2 = \frac{1}{\sqrt[3]{n}}\left[\sqrt[3]{\frac{Q_2^2}{Q_1^2}} + 2\sqrt[3]{\frac{Q_1}{Q_2}}\right] \tag{16}$$

$$Q_1/Q_2 = 2(n - 1) + \sqrt{2n(2n - 3)} \tag{17}$$

and $\qquad Q_2 = Q_3.$ $\tag{18}$

N.B. n must be greater than 1.5.

This has also been extended to four stages (Ref. H13).

Symbols used :

A_0 = amplification with feedback at mid-frequency ($\omega_0/2\pi$)

A = amplification with feedback at any frequency ($\omega/2\pi$)

α_0 = amplification without feedback at mid-frequency ($\omega_0/2\pi$)

ϕ = phase angle

n = $1 - \alpha_0\beta_0$ = feedback factor

β_0 = transfer coefficient (negative for negative feedback)

ω_0 = $1/(C\sqrt{R_L R_g})$.

These equations have been plotted in Fig. 7.75 from which it will be seen that the feedback factor has no effect on the shape of the curves for 1 or 2 stages, and only a very slight effect for 3 or 4 stages (curves for limiting values of n are given). Although the curves are shown only for the high frequency limit of the flat top, they may be applied to the low frequency side by changing the sign of the abscissae and, in the case of phase angle, also changing the sign of the ordinates.

This method of design provides the maximum flat frequency response, but the maximum degree of feedback only occurs at the mid-frequency and it falls off rapidly outside the frequency band of the " selective " stage. The feedback becomes zero near the knees of the flat top, and positive at still higher and lower frequencies.

This design ensures stability at all frequencies from zero to infinity.

This method also makes it possible to design for any desired response curve, within the limits of the amplifier. The procedure is to commence with the desired response curve and to determine the response (either modulus or phase angle) corresponding

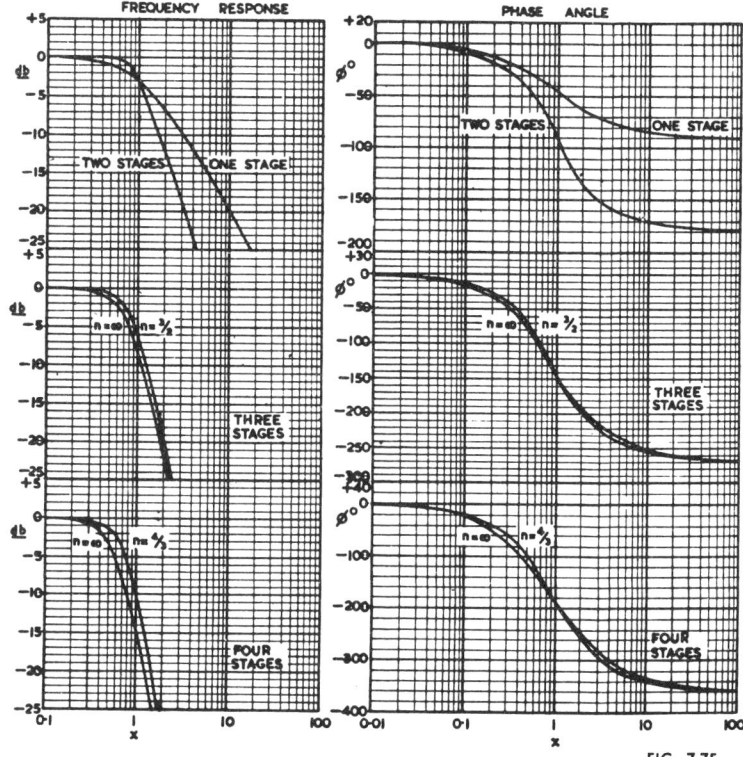

FIG. 7.75

Fig. 7.75. Frequency response and phase angle of frequency response for 1,2,3, and 4 r.c.c. stages designed for maximum flat response by Mayr's method (Ref. H13).

to two arbitrary frequencies ($\omega_1/2\pi$) and ($\omega_2/2\pi$), one near the lower and the other near the upper frequency limit. Then we read from the curves the values of x_1 and x_2 corresponding to the two chosen response values ; the value of x_1 corresponding to the lower frequency is, of course, negative.

The mid-frequency may be computed from

$$\omega_0{}^2 = \omega_1\omega_2 \frac{x_2\omega_1 - x_1\omega_2}{x_2\omega_2 - x_1\omega_1} \tag{19}$$

and the ratio Q_1/Q_2 from equation (10), or (17) for 2 or 3 stages or

$$(Q_1/Q_2) = 3(n - 1) + \sqrt{3(3n - 4)} \text{ for 4 stages.}$$

The Q's of the various stages are then given by :

1 stage : $Q = \dfrac{nx_1}{\dfrac{\omega_1}{\omega_0} - \dfrac{\omega_0}{\omega_1}}$ (20)

2 stages : $Q_2 = \sqrt{\dfrac{nQ_2}{Q_1} \cdot \dfrac{x_1}{\dfrac{\omega_1}{\omega_0} - \dfrac{\omega_0}{\omega_1}}}$; $Q_1 = Q_2 \cdot \dfrac{Q_1}{Q_2}$ (21)

3 stages : $Q_2 = Q_3 = \sqrt[3]{\dfrac{nQ_2}{Q_1} \cdot \dfrac{x_1}{\dfrac{\omega_1}{\omega_0} - \dfrac{\omega_0}{\omega_1}}}$; $Q_1 = Q_2 \cdot \dfrac{Q_1}{Q_2}$ (22)

4 stages : $\quad Q_2 = Q_3 = Q_4 = \sqrt[4]{\dfrac{nQ_2}{Q_1} \cdot \dfrac{x_1}{\dfrac{\omega_1}{\omega_0} - \dfrac{\omega_0}{\omega_1}}} \; ; \quad Q_1 = Q_2 \cdot \dfrac{Q_1}{Q_2}$ \hfill (23)

These amplifiers are made up of one rather selective stage, corresponding to Q_1, and a number of equal broadly-tuned stages, corresponding to $Q_2, Q_3 \ldots$ With increasing feedback the selectivity of Q_1 must be increased, while the selectivity of Q_2 etc. approaches a limiting value which is of the order of the Q of a single stage amplifier without feedback, having the same bandwidth as the complete amplifier with feedback.

The amplifier is first designed, neglecting all capacitances, to have a gain approximately n times the final value. The feedback is designed n with a purely resistive feedback network to decrease the gain n times. The value of C_g is then determined at the minimum practicable value, that is the input capacitance of the following stage plus an allowance for strays. Then the values of the other capacitances are given by

$$C_g = \frac{p}{\omega_0 \sqrt{R_L R_g}} \; ; \quad C_c = \frac{k}{\omega_0 \sqrt{R_L R_g}} \tag{24}$$

where $\quad p \;\; = \dfrac{b}{2Q}\left[1 - \sqrt{1 + 4Q_b^d - 4Q^2\left(1 + \dfrac{1}{b^2} + d^2\right)}\right]$

$\quad\quad k \;\; = \dfrac{1 - pd}{p + d} \; ; \; b = \sqrt{R_g/R_L}$ and $d = C_g \omega_0 \sqrt{R_L R_g}.$

The amplification without feedback at the mid-frequency is

$$\alpha_0 = kQ \cdot g_m \sqrt{R_L R_g} \tag{25}$$

There are limits to the physically realizable values of C_p and C_c which are quoted in the article (Ref. H13).

It is shown in (C) below that the condition of maximal flatness gives an oscillatory transient response ; for instance when the input voltage is a unit step, the two-stage amplifier has a response which overshoots 4.3%, while the three-stage amplifier overshoots 8% (Ref. H15).

A maximal flatness 2 stage amplifier has a small rise in the response at some high frequency due to positive feedback, compared with the response without feedback. For example, as may be shown by drawing a Nyquist diagram, a 2-stage amplifier with $\alpha_2/\alpha_1 = 10$, $A_0 = 316$, $\beta = 0.016$ has a rise of about 1.4 db due to positive feedback at some high frequency.

(B) Method of C.F. Brockelsby (Ref. H12)

A different approach to the same goal is made by Brockelsby who provides " staggering " by changing the values of load resistance and thereby obtaining different values of gain and frequency response. This method was published earlier than that of Mayr but is less comprehensive, although providing useful additional information. A summary of this method is given below. See also Ref. J13.

Two-stage amplifiers and maximal flatness

The widest frequency range possible without any peaks whatever is known as the condition of " maximal flatness." With a 2 stage amplifier in which both stages have identical gain and frequency response, the condition of maximal flatness is that $\beta A_0 = -1$, corresponding to a gain reduction of 2 times. On the other hand if two r.c.c. pentode stages have " staggered " gain and high frequency response (by using different values of load resistance) the condition of maximal flatness at the high frequency end of the range is that

$\quad \frac{1}{4}(b + 1/b)^2 = 1 - \beta A_0$ \hfill (26)

where A_0 = amplification at mid-frequency without feedback,

$\quad\quad b$ = factor by which the normal value of load resistance is multiplied for one stage,

$1/b$ = factor by which the normal value of load resistance is multiplied for the other stage,

and β = fraction of the output voltage fed back to the input.

Equation (26) may be expressed in the alternative form

$$F = 2S^2 \tag{26a}$$

where $F = 1 - \beta A_0$ = feedback factor

and $S = \frac{1}{2}(b + 1/b)$ = " staggering coefficient."

In a two stage amplifier with a high frequency peak, the required staggering to produce a peak of known value is given by

$$S^2 = (F/2)(1 - \sqrt{1 - \alpha^2}) \tag{27}$$

where α = ratio of voltage gain at middle frequencies to that at the peak (α is less than 1).

The results are summarized in the following table :

$b =$	1	1.5	2	3	4	5
$S =$	1	1.08	1.25	1.7	2.1	2.6

Feedback factor (F) for maximal flatness (Eqn. 26a) :

$F =$	2	2.3	3 1	5.8	8.8	13.5

For 1 db rise (Eqn. 27) :—

$F =$	3.7	4.3	5.7	10.6	16	25

For 3 db rise (Eqn. 27) :—

$F =$	6.8	8.0	10.7	19.7	30	46

Fig. 7.76 shows the frequency characteristics obtained with 2 stage amplifiers with maximal flatness and for several other conditions. Here

$$x = \omega CR$$

and $F = 1 - \beta A_0$ = feedback factor.

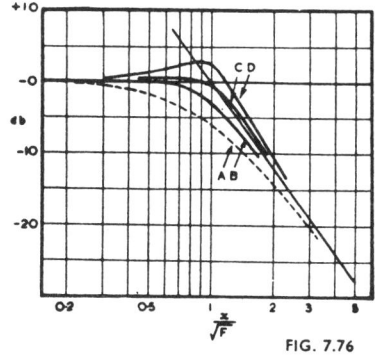   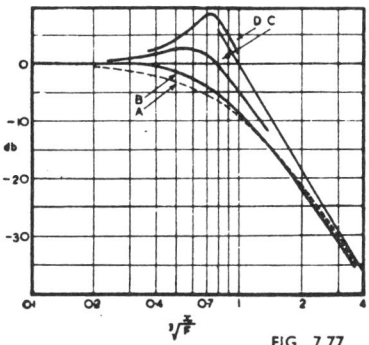

FIG. 7.76   FIG. 7.77

Fig. 7.76. Frequency characteristics of two-stage amplifier. Curve A is without feedback and curves B, C and D respectively for maximal flatness, a 1 db peak and a 3 db peak ($x = \omega CR$; C = normal shunt capacitance of each stage ; R = total normal a.c. load resistance for each stage ; pentode valves (Ref. H12).

Fig. 7.77. Frequency characteristics of three-stage amplifier. Curve A is without feedback and curves B, C, D are respectively for maximal flatness and with feedback equal to two and four times the m.f. value (Ref. H12).

If the condition of maximal flatness is to hold at the low as well as at the high frequency end of the band, the centre-frequencies of both stages should be identical (Ref. H13).

Three-stage amplifiers and maximal flatness

It has been shown (Ref. H12) that if a substantially flat response over the maximum possible frequency band is required, the optimum arrangement for a three-stage amplifier is to have two stages with wide frequency response and one with narrow.

This finding is based on the assumption that the amplifier has maximal flatness ; it is not necessarily the most desirable arrangement under all conditions.*

The condition for maximal flatness is that

$$F = \tfrac{1}{2} \cdot \frac{(B^3 + 2)^2}{2B^3 + 1} \qquad (28)$$

where F = feedback factor

B = staggering factor = $1/b$

and $\quad b$ = factor by which the normal value of load resistance is multiplied for the one (narrow) stage.

[B is always equal to or greater than unity.]

Equation (28) may be approximated within 1.4% when $B \geq 2$ by

$$F \approx 1 + (B^3/4) \qquad (29)$$

For example when $F = 10$ (i.e. 20 db feedback), $B \approx 3.3$ and the ratio of bandwidths is approximately 11.

As with the two-stage amplifier, an increased amount of feedback beyond that to give maximal flatness gives high frequency peaks (Fig. 7.77).

The critical amount of feedback to provide self-oscillation is

$$F_c = 5 + 2B^3 + 2/B^3 = 8F, -3 \qquad (30)$$

where F_c = feedback factor to give critical feedback

and $\quad F_f$ = feedback factor to give maximal flatness.

The stability margin of a three-stage maximal-flatness amplifier is $(F_c - F_f)$ and in a typical example

$\qquad F_f = 3.16$ (i.e. 10 db reduction in gain) for maximal flatness

then $\quad F_c = 8 \times 3.16 - 3 = 22.2$ (i.e. 27 db approx.) to give self-oscillation. The stability margin is therefore $27 - 10 = 17$ db approximately.

If the feedback is made very large, and the staggering is adjusted so as always to give maximal flatness, $F_c \approx 8F_f$ which for this example becomes $8 \times 3.16 = 25.2$ (i.e. 28 db approx.) giving a stability margin of 18 db approximately.

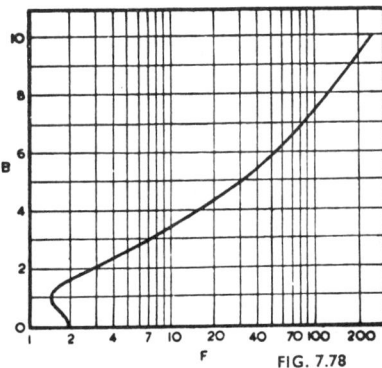

Fig. 7.78. *Relation between the staggering factor B and the gain reduction factor F for maximal flatness in a three-stage amplifier. Values of B less than unity apply to the " two narrow, one wide " condition ; values of B greater than unity apply to the " two wide, one narrow " condition (Ref. H12).*

FIG. 7.78

For both two- and three-stage maximal-flatness amplifiers (high frequency response)

1. The feedback which is fully effective in reducing distortion begins to fall at a frequency about F times lower than the knee.

2. The effective gain-reduction factor is about unity, indicating no feedback, at the knee.

3. The knee of the curve for the two-stage amplifier (Fig. 7.76) is at $x = \omega CR = \sqrt{F}$; at this point the attenuation is 3 db. In the three-stage amplifier (Fig. 7.77) the knee of the curve is at $x = \omega CR = \sqrt[3]{F}$; the attenuation is 3 db at $x = (2/3)^3 \sqrt{F}$.

*The arrangement to provide the greatest value of the product (effective feedback) × (bandwidth) is that due to Duerdoth described in Sect. 3(v)E.

Low frequency response

The equations for the high frequency end may be applied to the low frequency end of the response characteristic, in the simple case when the only frequency variable is the grid coupling condenser, by giving x the value

$$x = 1/\omega C_g R_g \tag{31}$$

where C_g = grid coupling capacitance

and R_g = grid resistor plus the effective source impedance of the preceding stage (approximately R_L in the case of a pentode).

(C) Conditions for critical damping

Critical damping may be defined as that value which gives the most rapid transient response which is possible without overshoot.

The maximal flatness amplifier—see (A) and (B) above—has been examined by J. E. Flood (Ref. H15) in the light of critical damping ; the following summary is based on his work. It has been shown that a two-stage maximally-flat amplifier has a response which overshoots 4.3% while the three-stage amplifier overshoots 8% when a step waveform is applied to the grid circuit.

Although this article is concerned mainly with video-frequency or pulse amplification, it is possible that critical damping is a desirable feature in a-f amplifiers and that a condition of under-damping with oscillatory response to transients is to be avoided. See general comments in Sect. 3(v), also Ref. J8.

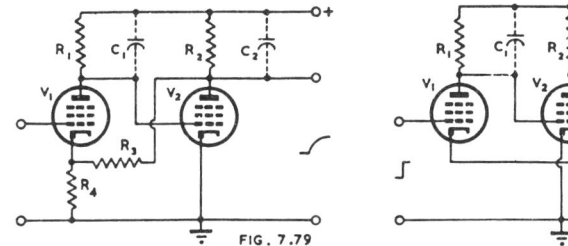

Fig. 7.79. *Two-stage uncompensated amplifier with feedback (Ref. H15).*
Fig. 7.80. *Three-stage uncompensated amplifier with two time constants within the feedback loop (Ref. H15).*

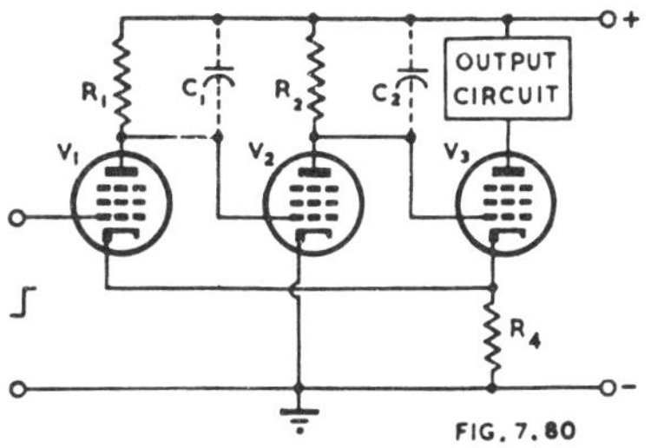

Two stage amplifier with constant value of β (Fig. 7.79)

The following may also be applied to the three stage amplifier of Fig. 7.80 which has only two time constants within the feedback loop.

It may be shown that the condition for critical damping is that

$$1 - \beta A_0 = \frac{(\alpha_1 + \alpha_2)^2}{4\alpha_1\alpha_2} \tag{32}$$

where A_0 = amplification at the mid-frequency without feedback

β is negative and independent of frequency

$\alpha_1 = 1/R_1C_1$ = inverse time constant of V_1

and $\alpha_2 = 1/R_2C_2$ = inverse time constant of V_2.

When a higher feedback factor than that indicated by eqn. (32) is used, the transient response will be oscillatory ; when a lower feedback factor is used, the transient response will be over-damped.

Equation (32) may be put into the form

$$1 - \beta A_0 = (1 + K)^2/4K = K/4 + 1/2 + 1/4K \tag{33}$$

where $K = \alpha_2/\alpha_1$ = ratio of inverse time constants.

This is plotted in the lower curve of Fig. 7.81.

If a fairly large amount of feedback is to be used, the ratio of time constants must be large.

The resultant inverse time constant of the amplifier is given by

$$\alpha = \tfrac{1}{2}(\alpha_1 + \alpha_2) = \tfrac{1}{2}\alpha_1(1 + K) \tag{34}$$

This relationship is plotted in the upper curve of Fig. 7.81.

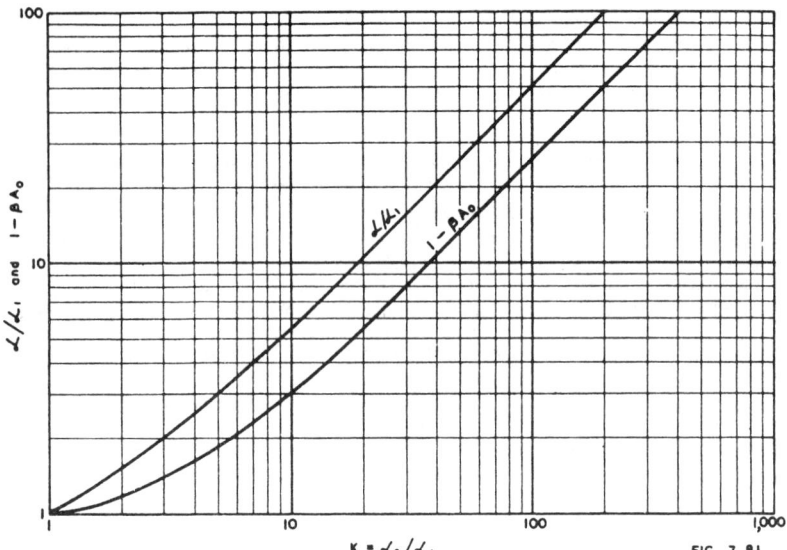

Fig. 7.81. Conditions for critical damping of two-stage uncompensated amplifier
(Ref. H15).

Note : The " staggering coefficient " S as used by Brockelsby (Ref. H12) is related to α_1 and α_2 by the equation

$$S = \frac{\alpha_1 + \alpha_2}{2\sqrt{\alpha_1 \alpha_2}} = \tfrac{1}{2}\left(\sqrt{K} + \frac{1}{\sqrt{K}}\right) \tag{35}$$

The condition for maximal flatness is therefore given by

$$1 - \beta A_0 = \frac{(\alpha_1 + \alpha_2)^2}{2\alpha_1 \alpha_2} \tag{36}$$

Comparing equations (32) and (36), it will be seen that for any particular ratio of α_2/α_1, the condition of critical damping permits a feedback factor of only one half that required for maximal flatness of the frequency characteristic. The **transient response of the maximally-flat amplifier is therefore oscillatory**.

A 2-stage critically damped amplifier will normally have a small amount of positive feedback at high frequencies, although the rise in response in a typical case will be less than 1 db above the zero-feedback curve.

Two-stage compensated amplifier

If resistor R_3 in Fig. 7.79 is shunted by a capacitor C_3 and if R_3 is very much greater than R_4 then critical damping is obtained when the inverse time constant of the feedback path is given by

$$\alpha_3 = \frac{-\alpha_1\alpha_2\beta A_0}{2\sqrt{\alpha_1\alpha_2(1 - \beta A_0)} - (\alpha_1 + \alpha_2)} \tag{37}$$

where $\alpha_1 = 1/R_1C_1$; $\alpha_2 = 1/R_2C_2$; $\alpha_3 = 1/R_3C_3$
and β is negative for negative feedback.

Similarly if an inductance L is connected in series with R_4 in Fig. 7.80, and if g_mR_4 is very much less than 1, then critical damping is obtained as indicated by eqn. (37) where $\alpha_3 = R_4/L$.

The inverse time constant for the amplifier is given by

$$\alpha = \sqrt{\alpha_1\alpha_2(1 - \beta A_0)} \tag{38}$$

Using the value of α_3 given by eqn. (37) for the inverse time constant of the amplifier with feedback, critical damping is obtained and the response to a step waveform of the feedback amplifier is made identical with that of an amplifier without feedback having two stages, each having an inverse time constant equal to the geometrical mean of the inverse time constants of the two stages multiplied by the square root of the feedback factor (eqn. 38).

A particularly simple special case is obtained when

$$\alpha_2 = \alpha_1(1 - \beta A_0) \ ;$$

then $\alpha = \alpha_3 = \alpha_2 = \alpha_1(1 - \beta A_0)$.

Values of α_3 for critical damping may be derived from the curves of Fig. 7.82 and values of α from Fig. 7.83.

Three-stage amplifier

Critical damping cannot be obtained in a three-stage amplifier with constant β.

Three-stage amplifier with single time constant in feedback path

In this case there is only one value of feedback factor as given by eqn. (39) for which critical damping can be obtained.

$$1 - \beta A_0 = \frac{(\alpha_1 + \alpha_2 + \alpha_3)^2}{27\alpha_1\alpha_2\alpha_3} \tag{39}$$

$$\alpha = (\alpha_1 + \alpha_2 + \alpha_3)/3 \tag{40}$$

$$= [(\alpha_1\alpha_2\alpha_3(1 - \beta A_0)]^{1/3} \tag{41}$$

$$\alpha_4 = \frac{- 3\alpha_1\alpha_2\alpha_3\beta A_0}{\alpha_1^2 + \alpha_2^2 + \alpha_3^2 - (\alpha_1\alpha_2 + \alpha_2\alpha_3 + \alpha_3\alpha_1)} \tag{42}$$

where α_3 = inverse time constant of third stage

α_4 = inverse time constant of feedback path

and α = inverse time constant of amplifier.

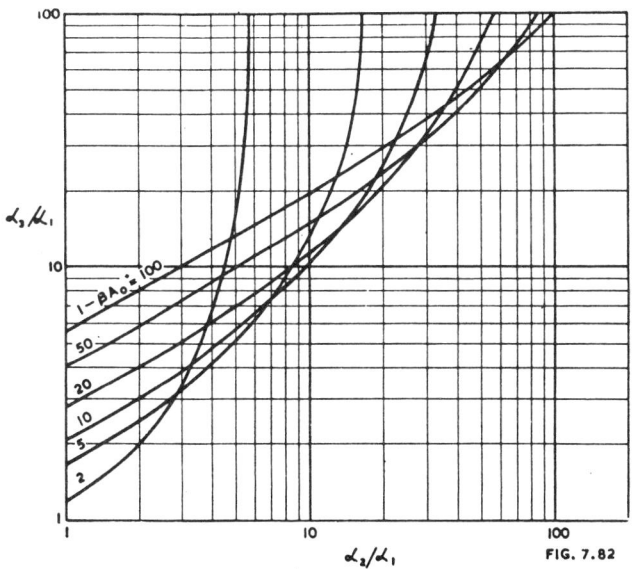

Fig. 7.82. *Curves for determining α_3 for critical damping of two-stage compensated amplifier (Ref. H15).*

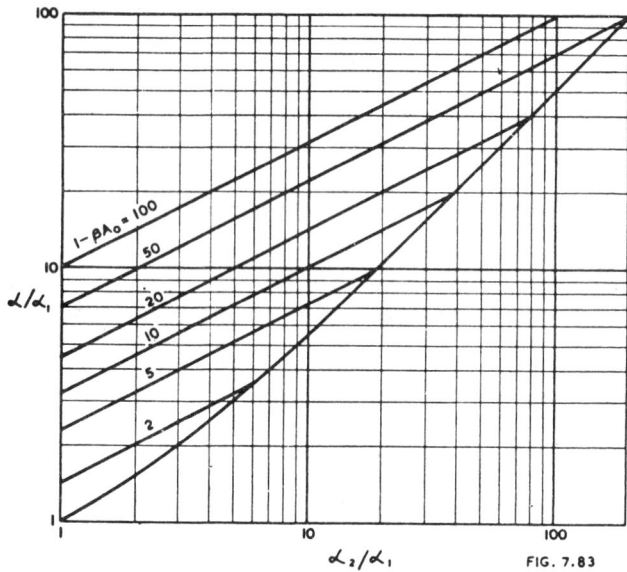

Fig. 7.83. Curves for determining α for critical damping of two-stage compensated amplifier (Ref. H15).

Three-stage amplifier with two time constants in feedback path

Critical damping can be obtained provided that

$$1 - \beta A_0 > \frac{(\alpha_1 + \alpha_2 + \alpha_3)^3}{27\alpha_1\alpha_2\alpha_3} \tag{43}$$

Under this condition,

$$\alpha_4 = \alpha_5 = (b \mp \sqrt{b^2 - 4ac})/2a \tag{44}$$
$$\alpha^3 = \alpha_1\alpha_2\alpha_3(1 - \beta A_0) \tag{45}$$

where α = inverse time constant of amplifier

$\alpha_1, \alpha_2, \alpha_3$ = inverse time constants of three stages

α_4, α_5 = inverse time constants in feedback path

$a = 3\alpha - (\alpha_1 + \alpha_2 + \alpha_3)$

$b = 3\alpha^2 - (\alpha_1\alpha_2 + \alpha_2\alpha_3 + \alpha_3\alpha_1)$

and $c = \alpha^3 - \alpha_1\alpha_2\alpha_3$.

When $1 - \beta A_0$ is less than the right hand side of eqn. (43) the transient response is over damped and less rapid.

A particularly simple case is obtained when

$\alpha_2 = \alpha_3 = \alpha_1(1 - \beta A_0)$

then $\alpha = \alpha_5 = \alpha_4 = \alpha_3 = \alpha_2 = \alpha_1(1 - \beta A_0)$ \hfill (46)

(viii) Constancy of characteristics with feedback

This is sometimes called " stability " (Ref. H8) but refers to the more or less complete independence of the gain of an amplifier with feedback (inside the amplifier frequency range) with regard to changes in valve characteristics and supply voltages. Becker defines the " stability factor " as

$$SF = 1 + \beta A_0 X \tag{47}$$

where $A_0 X$ = lowest gain which the amplifier ever has under the worst operating conditions, without feedback

and SF = stability factor with negative feedback.

(ix) Effect of feedback on phase shift*

The amplification without feedback may be expressed in the complex form

$$A = a + jb = |A| \angle \theta \qquad (48)$$

where $|A|$ is the magnitude of the voltage amplification and θ is the phase angle of the output voltage of the amplifier relative to the input voltage, without feedback. It is obvious that $\theta = \tan^{-1}(b/a)$ and $|A| = \sqrt{a^2 + b^2}$.

The amplification with feedback may be expressed in the form

$$A' = \frac{A}{1 - \beta A} = \frac{a + jb}{1 - (a + jb)\beta} = \frac{a - (a^2 + b^2)\beta + jb}{(1 - a\beta)^2 + b^2\beta^2} \qquad (49)$$

provided that there is no phase shift in the feedback network. The phase angle of the output voltage relative to the input voltage, with feedback, is given by

$$\alpha = \tan^{-1} \frac{b}{a - (a^2 + b^2)\beta} = \tan^{-1} \frac{b/a}{1 - \beta A\sqrt{1 + b^2/a^2}} \qquad (50)$$

$$= \tan^{-1} \frac{\tan \theta}{1 - \beta A\sqrt{1 + \tan^2 \theta}}. \qquad (51)$$

When β is negative, as is the case when the feedback is negative, it is evident from (51) that the angle α is less than the angle θ. Hence the phase shift of the output voltage relative to the input voltage is reduced by negative feedback.

SECTION 4 : SPECIAL APPLICATIONS OF FEEDBACK

There are so many applications of feedback, many of which are outside the subject of radio receiver and amplifier design, that it is impossible even to list them here. The following selected applications have some interesting features.

(1) Electronic attenuators

A cathode follower may be used with various cathode load resistors selected by rotary switch, to give the coarse steps, and then the screen voltage may be adjusted to give a fine adjustment of attenuation.

Smith, F. W., and M. C. Thienpont " Electronic attenuators " Comm. 27.5 (May 1947) 20.

(2) Stable admittance neutralization

Two valves in cascade with negative voltage feedback may be used to neutralize the input admittance of an amplifying valve (mainly the result of its input capacitance) and thereby enable the use of a load resistance of 23 megohms with a phototube for a cut-off frequency of 20 000 c/s.

" Stable admittance neutralization " Electronic Eng. 14.167 (Jan. 1942) 594.

(3) An amplifier without phase distortion

O. H. Schade has developed an amplifier without cathode by-pass condensers, screen by-pass condensers, decoupling condensers or output condenser in B-supply filter, which has negligble phase distortion from 10 to 30 000 c/s.

" An amplifier without phase distortion," Elect. 10.6 (June 1937) 26.

(4) Stabilized negative impedances

Negative impedance is produced from a positive impedance of the desired type by positive feedback through an amplifier stabilized with negative feedback.

Ginzton, E. L. " Stabilized negative impedances " Elect. (1) 18.7 (July 1945) 140 ; (2) 18.8 (Aug. 1945) 138 ; (3) 18.9 (Sept. 1945) 140.

(5) Tone Control

This is covered in detail in Chapter 15.

---

*Method after H. J. Reich " Theory and application of electron tubes."

SECTION 5 : VALVE CHARACTERISTICS AND FEEDBACK

(i) Triode cathode follower   (ii) Pentode cathode follower   (iii) Triode with voltage feedback   (iv) Pentode with voltage feedback, transformer-coupled   (v) Cathode degenerative triode   (vi) Cathode degenerative pentode   (vii) Cathode-coupled triodes (viii) Feedback over two stages.

Although it is important to bear in mind that the application of feedback does not alter the characteristics of a valve, yet there are certain features which can best be visualized by the graphical use of effective valve characteristics with feedback.

(i) Triode cathode follower

The cathode-follower characteristics (Fig. 7.85) may be drawn very readily from the published triode plate characteristics (Ref. D2). The $E_c = 0$ curve remains as the boundary of the grid current region. If the grid circuit impedance is high, it is not practicable to go beyond the $E_c = 0$ curve, just as with plate-loading.

FIG. 7.84

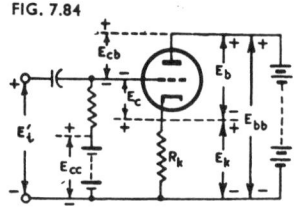

Fig. 7.84. Basic circuit of cathode follower.

The basic circuit is shown in Fig. 7.84 from which it is evident that with zero input voltage $(E_i' = 0)$

$$E_{cb} = E_b - E_c \qquad (1)$$

The plate is the most convenient electrode to regard as the common basis for electrode voltages, since it is the only one directly connected to the voltage source. Valve characteristics may be plotted by the method described below so that each curve, instead of applying to constant grid-to-cathode voltage applies to constant grid-to-plate voltage (E_{cb}). In effect, the valve and its load resistor are inverted as compared with normal practice and the cathode here takes the place of the plate.

Applying eqn. (1) for the case where E_{cb} is constant,
$$E_b = E_{cb} + E_c \qquad (2)$$
For example, take the case when $E_{cb} = 150$ volts. We may calculate the points along the curve using eqn. (2) :

$E_{cb} =$	150	150	150	150	150	volts
$E_c =$	0	−2	−4	−6	−8	volts
$E_b =$	150	148	146	144	142	volts

These points may be plotted to give the $E_{cb} = 150$ volt curve on the existing "plate characteristics," as has been done in Fig. 7.85.

(A) Resistance loaded triode cathode follower

The cathode-follower characteristics may then be used as any ordinary resistance-loaded triode for the determination of power output, voltage gain and distortion. Maximum power output will generally be achieved when R_k is between r_p and $3r_p$ with slightly less distortion with the higher values of R_k. In Fig. 7.85, R_k is taken as 15 000 ohms, with $E_{bb} = 200$ volts. If it were desired to obtain the highest possible voltage swing, a much higher value of load resistance would be selected (e.g. 50 000 ohms). In each case the operating point should be selected so as to avoid both grid current and plate-current cut-off with the highest input voltage. For maximum output, a point slightly above half-way along the loadline is desirable. In Fig. 7.85, a suitable operating point (O) is the intersection of the $E_{cb} = 150$ volt curve with the loadline which occurs at a plate-to-cathode voltage of 145 volts, thus indicating that $E_c = -5$ volts. Alternatively, the bias could have been read from intersection of the $E_c = -5$ (plate characteristic) with the loadline.

When the operating point (O) has been established, it is then necessary to work entirely with the E_{cb} curves, neglecting the E_c curves. In Fig. 7.85 the operating point is at $E_{cb} = 150$ volts, and point A is $E_{cb} = 100$, so that the peak input voltage

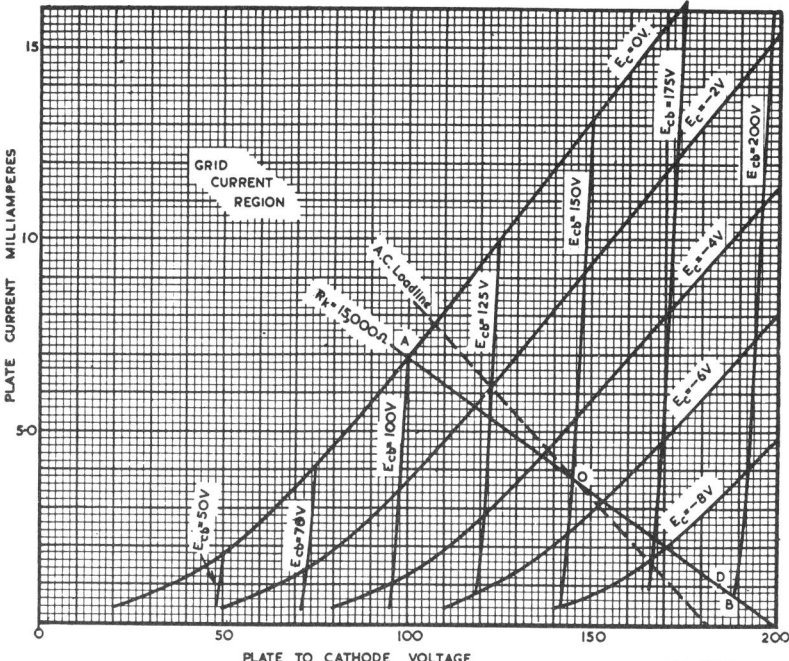

Fig. 7.85. *Cathode follower characteristics of small general purpose triode. The solid loadline is for $R_k = 15\,000$ ohms, the dashed loadline is for the addition of an a.c. shunt load.*

amplitude is 50 volts.   It is obvious that the opposite swing will extend to $E_{cb} = 200$ volts (point B) thus giving a peak-to-peak swing of $2 \times 50 = 100$ volts. It will be seen that OA corresponds to $E_c = -5$ volts, while OB corresponds to -5.7 volts, but this is due to the improved linearity with the cathode follower.   With plate loading the loadline would be AD, so that cathode loading causes a longer loadline by the amount DB, thus resulting in a slightly greater power output.

The voltage gain is given by

$$A' = \frac{E_B - E_A}{E_{cb(B)} - E_{cb(A)}} \tag{3}$$

where subscripts A and B indicate the voltages at points A and B on the loadline.

If the cathode is coupled through a coupling condenser C_c to a following grid resistor R_g (Fig. 7.8), the a.c. loadline will be drawn through the operating point O with a slope of $-1/R'_L$ where $R_L = R_k R_g / (R_k + R_g)$.

The mutual characteristics of cathode followers may be drawn and interpreted by the method of Ref. C29.

(B) Transformer-coupled triode cathode follower

The same form of cathode-follower characteristic may be used when the load is coupled through a transformer so as to reflect a resistance R_k across the primary. In most cases the load resistance may be taken (at least as a first approximation) to have the same value as for plate loading.   In Fig. 7.86 the loadline AOB extends from A (the junction of the $E_c = 0$ and the $E_{cb} = 100$ V curves) to B, on the $E_{cb} = 500$ V curve.   The operating point O is the intersection of the loadline with the $E_{cb} = 300$ V curve, thus giving a peak input voltage of 200 volts in both directions.   The power output and distortion may be calculated as for plate loading.

If the load resistance R_k is increased, the loadline will rotate about O to the position (say) CD with constant input voltage.   This will usually result in a decrease in both power output and distortion—there is the same advantage (as with plate loading) in keeping the loadline out of the region of bottom curvature. It would be possible, with the higher resistance load, to increase the input voltage so as to extend the loadline from OC to meet the $E_c = 0$ curve.

If the load resistance R_k is decreased, the loadline will rotate about O to the position (say) EF. If the input voltage is unaltered, the plate current will cut off at F and the valve will run into grid current at E, both resulting in distortion. In this, the cathode follower differs from a plate loaded triode, and it is important to ensure that the load resistance does not fall appreciably below the designed value. Fortunately, a loudspeaker has an impedance characteristic which, although it rises considerably above its nominal (400 c/s) value, does not drop appreciably below it. A cathode follower with a loudspeaker load is therefore a good combination from the loading point of view.

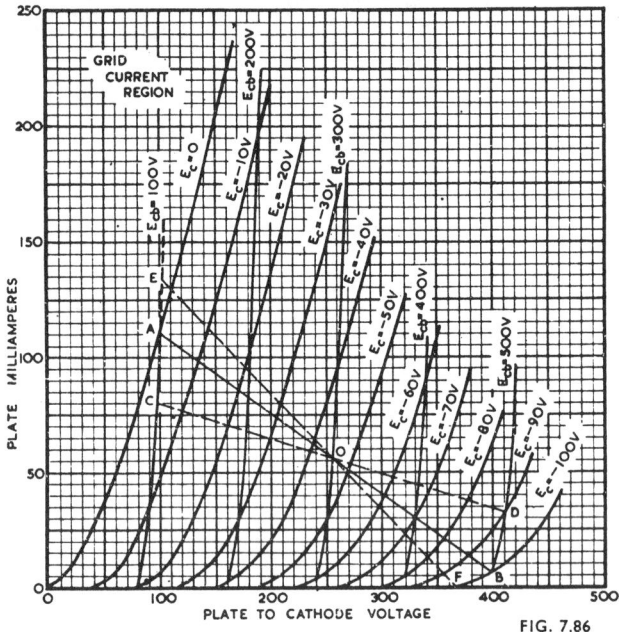

Fig. 7.86. *Cathode follower characteristics of power triode (type 2A3) with transformer-coupled load. Loadline AB is normal ; CD is high resistance load line ; EF is low resistance loadline illustrating grid current and plate current cut-off.*

If a low impedance driver is used, it is possible to drive a cathode follower into the grid current region with considerably greater power output and efficiency. In the ideal limiting case, the conditions are :

Plate voltage	E_b volts
Plate current	I_b amperes
Load resistance	E_b/I_b ohms
Power output	$\frac{1}{2}E_b I_b$ watts
Plate efficiency	50%

If the cathode follower is to be used as a high-input-impedance voltage amplifier, the conditions vary with the output level. At low output voltages, the operating conditions are of no consequence (within reasonable limits) so far as voltage gain is concerned, and R_k may be equal to r_p. For very high output voltages, the load resistance should be as high as practicable (assuming a purely resistive load), say $R_k = 5r_p$, and the operating current should be as low as practicable without causing plate-current cut-off. A somewhat lower load resistance may be used for fairly high output voltages.

The transformer primary inductance should have the same value as for plate loading under similar conditions, if full power output is desired at low frequencies. A transformer with low primary inductance will give uniform gain at low level output, but at high output it will cause plate-current cut-off and grid current as for loadline EF in Fig. 7.86.

References C14, C16, C17, C29, D1, D2, D3, D8, D9, D10, D11.

(ii) Pentode cathode follower

A pentode (or tetrode) may be connected in several ways arising from the screen supply and by-pass—

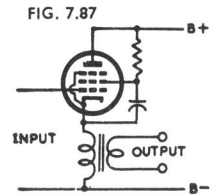

FIG. 7.87

Fig. 7.87. Pentode cathode follower, trans-former-coupled, screen by-passed to cathode.

(A) **Triode connection (screen tied to plate)**

The published "triode" characteristics may generally be used. If no triode characteristics are available, ascertain the characteristics for the screen voltage corresponding to the desired conditions (the plate voltage may be equal or higher).

Then $I_k \approx I_b + I_{c2}$ (4)

$$g_m \approx (\text{pentode } g_m) \times (I_k/I_b) \qquad (5)$$

and $\mu \approx E_{c2}/E_{c1}$ (cut-off).   (6)

(B) **Screen by-passed to earth**

There is no exact method using published characteristics. An approximate method has been described by Shapiro (Ref. D2).

(C) **Screen by-passed to cathode—Transformer-coupled** (Fig. 7.87)

The cathode follower characteristics may be drawn by the same method as for a triode (Fig. 7.88). The procedure for calculating the power output and distortion is :

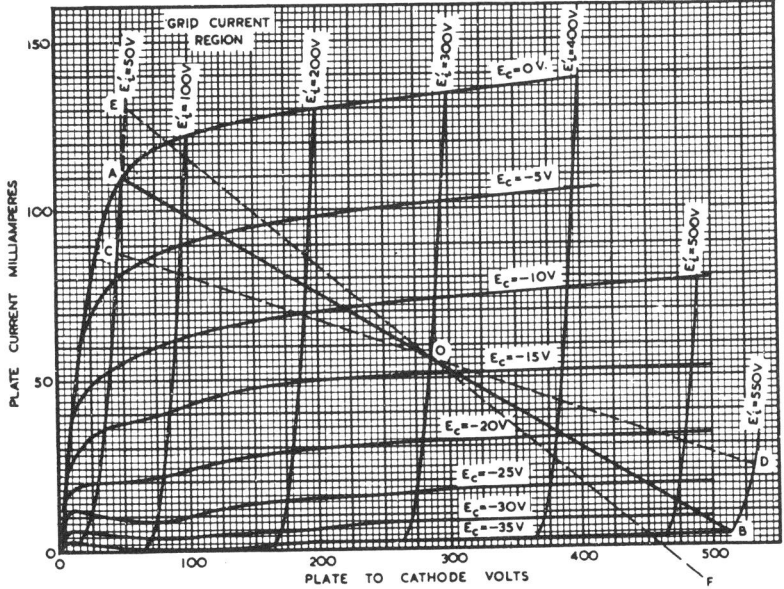

FIG. 7.88

Fig. 7.88. Cathode follower characteristics of beam power tetrode and typical of all pentodes. Loadline AOB is normal ; CD high resistance ; EF low resistance illustrating grid current and plate current cut-off.

1. To determine the maximum plate voltage and current for the operating point. Here $E_b = 285$ V and happens to coincide with the $E_{cb} = 300$ V curve. If desired, mark the value of E_i' corresponding to point O.

2. Try several values of load resistance (i.e. slope of loadline) until an optimum position for point A is determined, so as to make OA as long as possible, and at the same time give OB = OA with B slightly above plate-current cut-off.

3. Determine E_b for point A (in this case 50 V)—A is then automatically on the $E_{cb} = 50$ V curve.

4. Draw the other limiting E_{cb} curve (in this case $E_{cb} = 550$ V, giving a peak amplitude of 250 V each way).

5. Calculate power output and distortion as for a power triode, using the loadline AOB and the E_{cb} curves.

When a higher resistance load is used with the same input voltage, the loadline rotates about O to (say) COD, where the power output is lower, the distortion slightly less, and point C is well negative with respect to $E_c = 0$. When a lower resistance load is used with the same input voltage, the new loadline may be (say) EOF where E is well into the grid-current region and F is beyond plate-current cut-off. In this respect the pentode is similar to the triode, but the pentode is rather more critical regarding low load resistances ; it is, however, quite satisfactory with a speaker load.

If cathode bias is used, the procedure is as for a plate-loaded power pentode.

(D) Screen by-passed to cathode—Resistance-loaded (Fig. 7.89)

The curves are generally the same as for transformer coupling, and the treatment is similar to that for resistance-loaded triodes, except that it is necessary to allow for the d.c. screen current flowing through R_k. The screen current may be taken from the published data for typical operating conditions, or estimated from the ratio of plate to screen currents. The voltage drop through R_k additional to that due to the plate current will be $R_k I_{c2}$. The procedure is to take as the effective plate supply voltage the value $(E_{bb} - R_k I_{c2})$ and then to proceed normally, as for a triode.

(E) Screen voltage from separate supply (Fig. 7.90)

If the screen voltage is obtained from a separate supply, such as a battery, returned to the cathode, then the voltage from cathode to screen may be maintained constant without smoothing out the screen current variations. In this case the screen current does not flow through the cathode resistor R_k. The pentode characteristics for a constant screen voltage may be used without any adjustment for the effects of screen current.

References D1, D2, D3, D11.

FIG. 7.89   FIG. 7.90

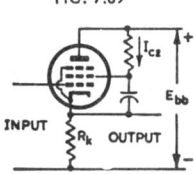

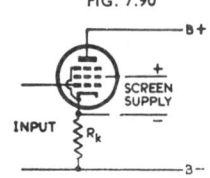

Fig. 7.89.   Pentode cathode follower, resistance-loaded, screen by-passed to cathode.

Fig. 7.90.   Pentode cathode follower, resistance-loaded screen from separate supply.

(iii) Triode with voltage feedback

The basic circuit is Fig. 7.91 in which we can consider the conditions with direct applied voltages (E_i' and E_b). A fraction of the plate voltage is applied degeneratively to the grid so that

$$|\beta| = R_1/(R_1 + R_2) \tag{7}$$

where $|\beta|$ is the numerical value of β without regard to its sign.

Consider first the condition for E_c to be zero. It is evident that the condition is that

$$E_i' = |\beta| E_{b0} \tag{8}$$

where E_{b0} = value of E_b when $E_c = 0$.

Now consider the general condition when the grid is negative with respect to the cathode. If the input voltage E_i' is kept constant, then the grid-to-cathode voltage E_c will change when E_b is changed. The relationship is

$$\Delta E_c = |\beta|(\Delta E_b) \quad \text{when } E_i' \text{ is constant.}$$

Therefore $\Delta E_b = \Delta E_c/|\beta|$

Therefore $E_b = E_{b0} - \Delta E_b = E_{b0} - \Delta E_c/|\beta|$ \hfill (9)

Eqn. (9) may be used for the calculation of points on the constant E_i' characteristics, as for example :

E_{b0}	=	100	100	100	100	100 volts
$|\beta|$	=	0.2	0.2	0.2	0.2	0.2
E_i'	=	20	20	20	20	20 volts
$-\Delta E_c$	=	0	-1	-2	-3	-4 volts
$-\Delta E_c/|\beta|$	=	0	-5	-10	-15	-20 volts
E_b	=	100	95	90	85	80 volts

These have been plotted in Fig. 7.92. It should be noted that E_{b0} is the plate voltage corresponding to zero grid voltage.

(iv) Pentode with voltage feedback—Transformer coupled

Pentodes with a substantial degree of negative voltage feedback (say $|\beta| \geqq 0.2$)

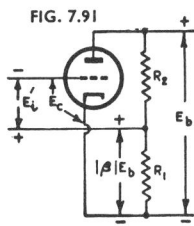

behave more or less like a cathode follower. Pentodes with a limited degree of feedback, or none, exhibit some special peculiarities. Fig. 7.93 shows input voltage curves for three different values of β ; the curvature on the $\beta = -0.1$ curve continues along its whole length while its slope is gradual and bottom curvature extensive.

Fig. 7.91. Basic circuit of triode with negative voltage feedback.

The full effect of a low degree of feedback is illustrated in Fig. 7.94 where $\beta = -0.1$. The curves are drawn as in the triode case except for the change in β. A suitable loadline AOB has been selected to give about the same power output as without feedback. Provided that the load resistance remains constant, the performance is satisfactory, but the value is critical. With constant input voltage (40 volts peak)

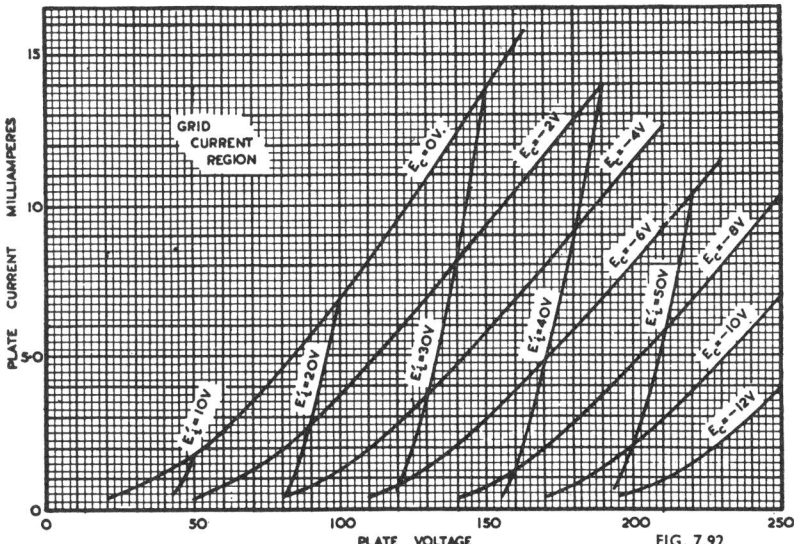

Fig. 7.92. Feedback characteristics of small general purpose triode with 20% negative voltage feedback ($\beta = -0.2$).

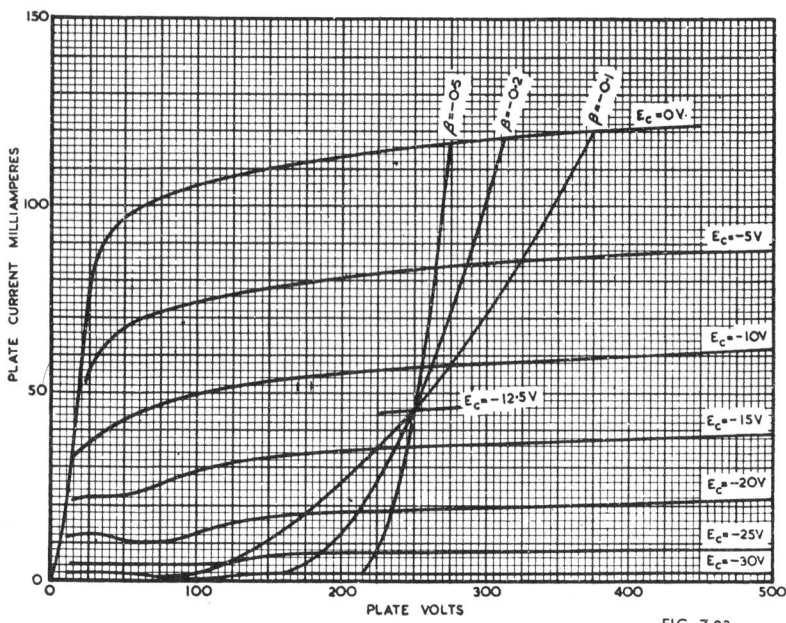

FIG. 7.93

Fig. 7.93. Single input voltage curve for each of three values of voltage feedback for beam power tetrode (6V6-GT with $E_{c2} = 250$ V), to illustrate effect of feedback.

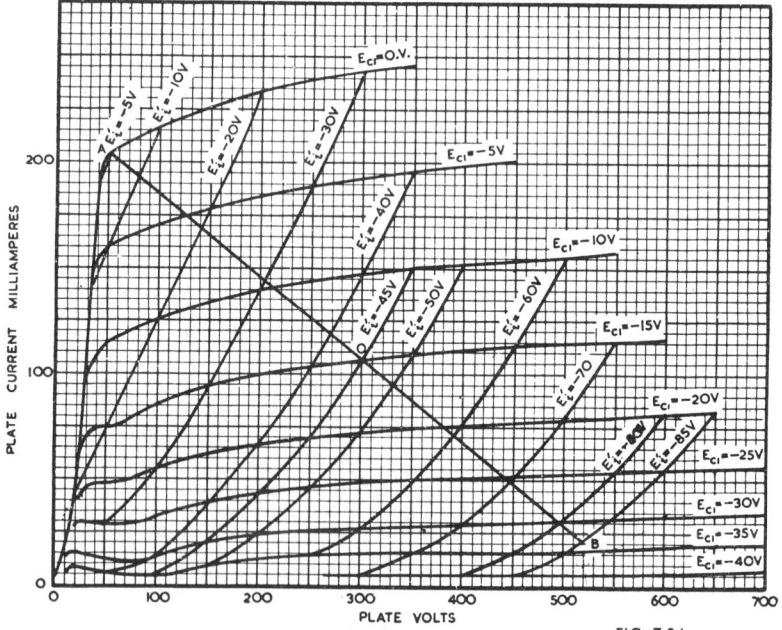

FIG. 7.94

Fig. 7.94. Feedback characteristics of beam power tetrode (6L6 or 807 with $E_{c2} = 300$ V) and 10% voltage feedback ($\beta = -0.1$).

the valve will run into grid current as soon as the load resistance either increases or decreases.   It is therefore not suitable for use on a loudspeaker load unless the input voltage is reduced. FIG. 7.95 A 10% reduction in input voltage (19% in power output) would be some improvement, but greater reduction is desirable, say 20% to 30% of the input voltage, depending on the operating conditions.   Even with an input reduction giving a power output of only 50% maximum, the pentode is still less flexible with regard to load resistance than an ordinary triode.   This position improves as the amount of feedback is increased.

See References D4, D12.

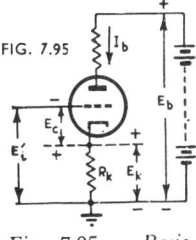

Fig. 7.95.   Basic circuit of cathode degenerative triode.

(v) Cathode degenerative triode

The circuit diagram of a cathode degenerative triode is shown in Fig. 7.95 from which it is evident that

$$E_c = E_i{}' - E_k \tag{10}$$

The curves of a cathode degenerative triode may be drawn by the procedure outlined below, although a special set of characteristics is required for each value of R_k. Fig. 7.96 shows the curves for a typical general purpose triode with $R_k = 1000$ ohms. The input voltage curves are straighter than those without feedback, and the two curves coincide only at $I_b = 0$.   The $E_i{}'$ curves have a lower slope, indicating a higher plate resistance, than the E_c curves.

Take the $E_i{}' = -4$ V curve as an example of the calculations.   We know that $R_k = 1000$ ohms and $E_i{}' = -4$ V.   If $I_b = 2$ mA, then $E_k = 2$ V and $E_c = E_i{}' - E_k = -4 - 2 = -6$ V.

Refer to the plate characteristics to find the plate voltage which will give a plate current of 2 mA at a bias of -6 V —the value is $E_{pk} = 144$ volts.

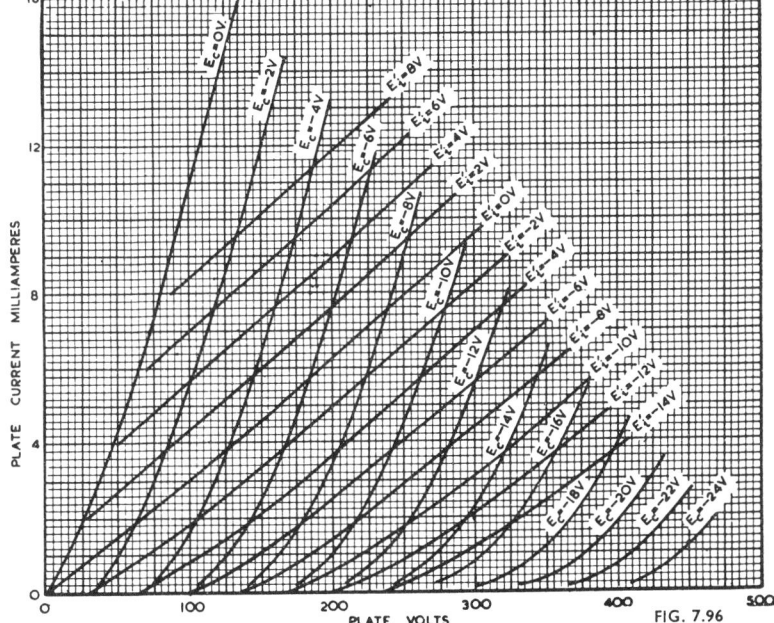

FIG. 7.96

Fig. 7.96.   Current feedback characteristics of general purpose triode (6SN7-GT single unit) with $R_k = 1000$ ohms.

The total voltage (E_b) from plate to earth is
$$E_b = E_{pk} + E_k = 144 + 2 = 146 \text{ V}.$$

Therefore the point ($E_b = 146$ V ; $I_b = 2$ mA) is on the $E_i' = -4$ V curve. Repeat this procedure for $I_b = 0$, 4, and 6 mA to give the whole curve ; then perform a similar operation for $E_i' = -6$ V, -8 V and so on.

The loadline may then be drawn in the normal manner, except that its slope will be $-1/(R_k + R_L)$. The E_i' curves are to be used for calculating gain and distortion.

The method generally employed with cathode degenerative triodes makes use of the published characteristics. There are several methods including those of Middleton, McIlroy, Huber, Lonsdale and Main ; the following treatment is based on Krauss, and may also be applied to cathode followers.

As an example take the characteristics of Fig. 7.97 with $R_k = 8\,000$ ohms, $R_L = 32\,000$ ohms and $E_{bb} = 400$ volts. Draw the loadline corresponding to $R_k + R_L = 40\,000$ ohms, as shown. Add an E_k scale below the E_b scale, based on the equation $E_k = R_k I_b$, commencing from E_{bb}. The value of E_k at any point on the loadline may then be found by projecting downwards to the E_k scale.

The input voltage E_i' is not proportional to any scale on the diagram, because the amplification is not constant. The value of E_i' at any point of intersection along the loadline may be found by drawing up the following table where each value of E_c is taken in turn. The input voltage is actually ($E_k + E_c$), but for convenience the point ($E_c = -8$) has been selected as the operating point, and so shown in the E_i' column.

E_c	E_k	$(E_k + E_c)$	E_i'	
0	64	64	34.7	Point A
−2	57	55	25.7	
−4	50.2	46.2	16.9	
−6	43.6	37.6	8.3	
−8	37.3	29.3	0	Operating point (O)
−10	31.2	21.2	−8.1	
−12	25.1	13.1	−16.2	
−14	19.6	5.6	−23.7	
−16	14.7	−1.3	−30.6	Point B at $E_k = 11.9$
−18	10.1	−7.9	−37.2	and $E_i' = -34.7$
−20	6.2	−13.8	−43.1	

The values of E_i' may be plotted against E_k and therefore also against E_b to give the dynamic characteristic, but this is usually unnecessary. Once the maximum input voltage has been selected it is only necessary to mark the extremities on the loadline. For example, if it is desired to swing to $E_c = 0$, then the peak E_i' will be 34.7 volts in each direction. The peak in one direction will be A where $E_c = 0$, and the other will be B, determined by interpolation :

$$E_c = -16 \qquad E_k = 14.7 \qquad E_i' = -30.6$$
$$E_c = -18 \qquad E_k = 10.1 \qquad E_i' = -37.2$$

$$\text{Diff.} = \quad\quad\quad\quad\quad\quad 6.6$$

Also the difference between the desired value (-34.7) and -30.6 is 4.1. The ratio is therefore 4.1/6.6.

Similarly with E_k : $14.7 - 10.1 = 4.6$.

The value of E_k for point B is therefore $14.7 - 4.6\,(4.1/6.6) = 14.7 - 2.8 = 11.9$.

The voltage gain and power output may be calculated from $E_b = 5E_k$, using the values of E_k at $E_i' = 34.7$ (point A), $E_i' = 0$ (point O) and $E_i' = -34.7$ (point B). The second harmonic distortion may be found from the ratio AO/OB measured in volts (E_k) ; i.e. $(64 - 37.3)/(37.3 - 11.9) = 26.7/25.4 = 1.05$. Therefore $H_2 \approx 1\%$—see Chapter 13 Sect. 2(i).

References D1, D3, D5, D8, D9, D10, D11.

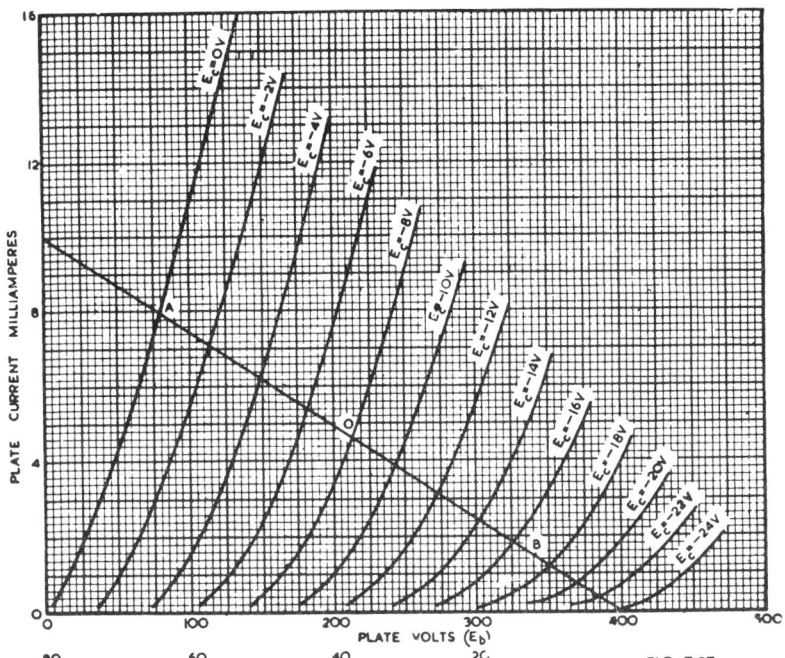

Fig. 7.97. Method of calculating performance of cathode degenerative triode without drawing special characteristics.

(vi) Cathode degenerative pentode

The general procedure is the same as for triodes, except that allowance must be made for the d.c. screen current flowing through R_k. This is done by taking $(E_{bb} - R_k I_{c2})$ as the effective plate supply voltage, and then by carrying on as for the triode case.

For valves with a high plate resistance this method is not very satisfactory, and a more practical method has been described by Pratt.

References D1, D11.

(vii) Cathode-coupled triodes

It is possible to make a graphical analysis of the cathode-coupled amplifier, using ordinary published valve characteristics (Ref. D6).

(viii) Feedback over two stages

It is possible to draw the equivalent characteristics of two stages in cascade with feedback over both stages, following the method of Pratt (Ref. D1).

Alternatively, it is possible to use the published valve characteristics to obtain a graphical analysis (Ref. D7).

SECTION 6 : REFERENCES TO FEEDBACK

A) GENERAL REFERENCES TO THEORY OF FEEDBACK
A1. Black, H. S. " Stabilised feedback amplifiers " E.E. 53 (Jan. 1934) 114 ; Also B.S.T.J. 13.1 (Jan. 1934) 1.
A2. Black, H. S. " Feedback amplifiers " Bell. Lab. Rec. 12 (June 1934) 290.
A3. Editorial Review " Feedback amplifiers " Elect. 9.7 (July 1936) 30.
A4. Edit. " Negative feedback amplifiers—a new development " W.W. 39.19 (6th Nov. 1936) 475.
A5. Tellegen, B. D. H., and V. C. Henriquez, " Inverse feedback " W.E. 14.167 (Aug. 1937) 409.

A6. Tellegen, B. D. H. " Inverse feedback " Philips Tec. Rev. 2.10 (Oct. 1937) 289 (General treatment and stability).
A7. Farren, L. I. " Some properties of negative feedback amplifiers " W.E. 15.172 (Jan. 1938) 23.
A8. Frommer, J., and M. Marinesco (letters) " Distortion in negative feedback amplifiers " W.E. 15.172 (Jan. 1938) 20.
A9. Mayer, H. F. " Control of the effective internal impedance of amplifiers by means of negative feedback " Proc. I.R.E. 27.3 (March 1939) 213.
A10. Bartels, H. " Input resistance of feedback amplifiers " (abstract) Elec. 13.1 (Jan. 1940) 74 (Eleck-trische Nachrichten Technik, June, 1939).
A11. Sandeman, E. K. " Feedback " W.E. 17.203 (Aug. 1940) 342.
A12. Fairweather, A. " Equivalent circuits of feedback amplifier " W.E. 18.211 (April 1941) 151.
A13. Schulz, E. H. " Comparison of voltage and current feedback amplifiers " Proc. I.R.E. 31.1 (Jan. 1943) 25.
A14. Erhorn, P. C. " Notes on inverse feedback " Q.S.T. 27.6 (June 1943) 13 ; Correction 28.2 (Feb. 1944) 106.
A15. Mezger, G. R. " Feedback amplifier for C-R oscilloscopes " Elect. 17.4 (April 1944) 126.
A16. Crane, R. W. " Influence of feedback on source impedance " Elect. 17.8 (Sept. 1944) 122 ; Corres. 18.1 (Jan. 1945) 382.
A17. Builder, G. " Negative voltage feedback " Proc. I.R.E. Aust. 6.2 (Aug. 1945) 3.
A18. Winternitz, T. W. " A variation on the gain formulae for feedback amplifiers for a certain driving-impedance configuration " Proc. I.R.E. 34.9 (Sept. 1946) 639.
A19. Bode, H. W. (book) " Network analysis and feedback amplifier design " (D. Van Nostrand Co. New York 1945).
A20. Ginzton, E. L. " Balanced feed-back amplifiers " Proc. I.R.E. 26.11 (Nov. 1938) 1367.
A23. James, E. J. " Negative feedback calculations " W.W. 54.9 (Sep. 1948) 326.
A24. Madler, M., and V. Toma (letter) " Degenerative feedback " Proc. I.R.E. 37.1 (Jan. 1949) 60.
A25. Sturley, K. R. " Combined current and voltage feedback " Electronic Eng. 21.255 (May 1949) 159.
A27. Winternitz, T. W. " Equivalent circuit method for feedback amplifier analysis " Comm. 29.11 (Nov. 1949) 14.
A32. Burwen, R. S. " Equivalent circuits to simplify feedback design " Audio Eng. 35.10 (Oct. 1951) 11.
A33. Jones, G. E. " An analysis of the split load phase inverter " Audio Eng. 35.12 (Dec. 1951) 16.

Charts
A21. " Graphic aid for the design of degenerative amplifiers " Elect. 12.11 (Nov. 1939) 64.   Gives A' for values of β from -0.01 to -0.95.
A22. Korn, T. E. " Stabilizing gain " Elect. 22.2 (Feb. 1949) 122.
A26. Felker, J. H. " Calculator and chart for feedback problems " Proc. I.R.E. 37.10 (Oct. 1949) 1204.
A28. Keroes, H. I. " Considerations in the design of feedback amplifiers " Audio Eng. 34.5 (May 1950) 15 ; 34.6 (June 1950) 17.
A29. Duerdoth, W. T. " Some considerations in the design of negative-feedback amplifiers." Proc. I.E.E. Part III. 97.47 (May 1950) 138.

Books
    This subject is covered more or less adequately by most radio text books. The following are particularly helpful :
A30. Reich, H. J. " Theory and Application of Electron Tubes " (McGraw-Hill Book Co., 1939).
A31. Valley, G. E., and H. Wallman (Editors) " Vacuum Tube Amplifiers " (McGraw-Hill Book Co., 1948).
    See also A19, H10, H11.

(B) REFERENCES SPECIALIZING ON HUM AND NOISE
B1. Bell, D. A. (letter) " Signal/noise ratio of cathode follower " W.E. 19.227 (Aug. 1942) 360.
B2. Burgess, R. E. (letter) " Signal/noise ratio of the cathode-follower " W.E. 19.229 (Oct. 1942) 450.
B3. Macklen, F. S. " Feedback " (short note only on " improvement " in signal noise ratio using feed-back) Elec. 16.4 (April 1943) 211 ; also letters J. T. Pratt and W. Palmer Elect. 16.6 (June 1943) 336.
B4. " Cathode Ray," " Negative feedback and hum " W.W. 52.5 (May 1946) 142.
B5. Builder, G. " The effect of negative voltage feedback on power-supply hum in audio-frequency amplifiers " Proc. I.R.E. 34.3 (Mar. 1946) 140W.
B6. Burgess, R. E. " Aerial-to-line couplings—cathode-follower and constant-resistance network " W.E. 23.275 (Aug. 1946) 217.
B7. " The effect of negative feedback on hum in the output of a-f amplifiers," A.R.T.S. and P. Bulletin No. 111 (4 August 1941) 1.

(C) REFERENCES TO CATHODE FOLLOWERS.
C1. Preisman, A. " Some notes on video-amplifier design " R.C.A. Rev. 2.4 (April 1938) 421.
C2. Barco, A. A. " An iconoscope preamplifier " R.C.A. Review 4.1 (July 1939) 89 (especially appendix).
C3. Williams, E. " The cathode follower stage " W.W. 47.7 (July 1941) 176.
C4. Hanney, E. A. " Cathode follower again " W.W. (July 1942) 164.
C5. Nordica, C. F. " Cathode-coupled amplifiers " Radio, No. 271 (Aug. 1942) 28.
C6 (a). Lockhart, C. E. " The cathode follower " Electronic Eng. 15.178 (Dec. 1942) 287 ; 15.180 (Feb. 1943) 375 ; 16.184 (June 1943) 21.
C6 (b). Data Sheets 39, 40 and 41 " The performance of the cathode follower circuit " Electronic Eng. 15.178 (Dec. 1942) 290.
    Data Sheets 42, 43 and 44 " The cathode follower—Part 2 " Electronic Eng. 15.180 (Feb. 1943) 380.
C7. Richter, W. " Cathode follower circuits " Elect. 16.11 (Nov. 1943) 112.
C8. Mitchell, C. J. " Cathode follower output stage " W.W. 50.4 (April 1944) 108 ; also letters 50.6 (June 1944) 188.
C9. Jeffery, C. N. " An analysis of the high-frequency operation of the cathode follower " A.W.A. Tech. Rev. 6.6 (1945) 311.
C10. Moskowitz, S. " Cathode followers and low-impedance plate-loaded amplifiers " Comm. 25.3 (March 1945) 51.
C11. Greenwood, H. M. " Cathode follower circuits " Q.S.T. (June 1945) 11.
C12. Cathode Ray " The cathode follower—what it is and what it does " W.W. 51.11 (Nov. 1945) 322.
C13. Goldberg, H. " Some considerations concerning the internal impedance of the cathode follower " Proc. I.R.E. 33.11 (Nov. 1945) 778 ; Discussion Proc. I.R.E. 35.2 (Feb. 1947) 168.

C14. Schlesinger, K. " Cathode-follower circuits " Proc. I.R.E. 33.12 (Dec. 1945) 843.
C15. Cocking, W. T. " Cathode follower dangers—output circuit capacitance " W.W. 52.3 (March 1946) 79.
C16. McIlroy, M. S. " The cathode follower driven by a rectangular voltage wave " Proc. I.R.E. 34.11 (Nov. 1946) 848. Letter H. L. Krauss, Proc. I.R.E. 35.7 (July 1947) 694. '
C17. Rifkin, M. S. " A graphical analysis of the cathode-çoupled amplifier " Comm. 26.12 (Dec.1946) 16.
C18. Smith, F. W., and M. C. Thienpont " Electronic attenuators " Comm. 27.5 (May 1947) 20..
C19. Kline, M. B. " Cathode follower nomograph " Elect. 20.5 (May 1947) 136.
C20. Reich, H. J. " Input admittance of cathode-follower amplifiers " Proc. I.R.E. 35.6 (June 1947) 573.
C21. Kline, M. B. " Cathode follower nomograph for pentodes " Elect. 20.6 (June 1947) 136. Letter P. M. Hackett 20.8 (Aug. 1947) 250.
C22. Audio design notes, " The cathode follower " (output impedance graph and power output) Audio Eng. 31.5 (June 1947) 39.
C23. Kline, M. B. " Cathode follower impedance nomograph " Elect. 20.7 (July 1947) 130.
C24. Houck, G. " Gain chart for cathode followers " Tele-tech. 6.8 (Aug. 1947) 54.
C25. E.M.I. Laboratories " Cathode-follower circuit using screen-grid valves," Electronic Eng. 19.229 (March 1947) 97.
C26. Yu, Y. P. " Cathode-follower couplings in d-c amplifiers " Elect. 19.8 (Aug. 1946) 99.
C27. Diamond, J. M. " Circle diagrams for cathode followers " Proc. I.R.E. 36.3 (March 1948) 416.
C28. Sowerby, J. McG. " Notes on the cathode follower " W.W. 54.9 (Sept. 1948) 321.
C29. Parker, E. " The cathode-follower " Electronic Eng. (1) 20.239 (Jan. 1948) 12 ; (2) 20.240 (Feb. 1948) 55 ; (3) 20.241 (Mar. 1948) 92 ; (4) 20.242 (April 1948) 126.
C30. Diamond, J. M. " Circle diagrams for cathode followers " Proc. I.R.E. 36.3 (March 1948) 416.
C31. Amos, S. W. " Valves with resistive loads " W.E. 26.307 (April 1949) 119.
C32. Scroggie, M. G. " RC coupled power stage " W.E. 27.318 (March 1950) 81.

Books

C33. Reich, H. J. " Theory and Application of Electron Tubes " pp. 164-174.
C34. Sturley, K. R. " Radio Receiver Design " Vol. 2 pp. 118-120.

(D) REFERENCES TO VALVE CHARACTERISTICS WITH FEEDBACK

D1. Pratt, J. H. " The equivalent characteristics of vacuum tubes operating in feedback circuits" R.C.A. Rev. 6.1 (July 1941) 102.
D2. Shapiro, D. L. " The graphical design of cathode-output amplifiers " Proc. I.R.E. 32.5 (May 1944) 263. Correction 32.8 (Aug. 1944) 482.
D3. Schlesinger, K. " Cathode-follower circuits " Proc. I.R.E. 33.12 (Dec. 1945) 843.
D4. " Cathode Ray," " Negative feedback—(2) its effect on optimum load and distortion," W.W. 52.3 (March 1946) 76.
D5. Middleton, R. G. " Graphical analysis of degenerative amplifiers " Radio, 30.3 (March 1946) 23.
D6. Radio Design Worksheet No. 51 " Graphical analysis of the cathode-coupled amplifier " Radio 30.8 (Aug. 1946) 20.
D7. Radio Design Worksheet No. 53 " Graphics of negative feedback in cascade," Radio, 30.10 (Oct. 1946) 17.
D8. McIlroy, M. S. " The cathode follower driven by a rectangular voltage wave " Proc. I.R.E. 34.11 (Nov. 1946) 848. Letter H. L. Krauss, 35.7 (July 1947) 694.
D9. Krauss, H. L. " Graphical solutions for cathode followers " Elect. 20.1 (Jan. 1947) 116 (with bibliography).
D10. Huber, W. A. " Graphical analysis of cathode-biased degenerative amplifiers " Proc. I.R.E. 35.3 (March 1947) 265.
D11. Lonsdale, E. M., and W. F. Main " A method of graphically analyzing cathode-degenerated amplifier stages " Proc. I.R.E. 35.9 (Sept. 1947) 981.
D12. Langford-Smith, F. " The Design of a high-fidelity amplifier, (2) Negative feedback beam power amplifiers and the loudspeaker " Radiotronics No. 125 (May/June 1947) 53.
D13. Schade, O. H. " Beam power tubes " Proc. I.R.E. 26.2 (Feb. 1938) 137.
See also C31.

Books

Sturley " Radio Receiver Design " Vol. 2 pp. 126-128.

(E) REFERENCES TO VOLTAGE FEEDBACK FROM PLATE WITH R.C.C. INPUT

E1. Laboratory staff of Amalgamated Wireless Valve Company, " Inverse feedback," Radio Review of Australia, 5.3 (March 1937) 64.
E2. R.C.A. " Application Note on an inverse feedback circuit for resistance-coupled amplifiers " No. 93 (June 8, 1938).
E3. Laboratory staff of Amalgamated Wireless Valve Company, " Negative feedback in RC amplifiers " W.W. 43.20 (Nov. 17, 1938) 437.
E4. Mezger, G. R. " Feedback amplifier for C-R oscilloscopes " Elect. 17.4 (April 1944) 126.
E5. Winternitz, T. W. " A variation on the gain formula for feedback amplifiers for a certain driving-impedance configuration " Proc. I.R.E. 34.9 (Sept. 1946) 639.

(F) REFERENCES TO PRACTICAL FEEDBACK AMPLIFIERS
(additional to those listed in A)

F1. Terman, F. E. " Feedback amplifier design " Elect. 10.1 (Jan. 1937) 12.
F2. Day, J. R., and J. B. Russell, " Practical feedback amplifiers " Elect. 10.4 (April 1937) 16.
F3. Reference Chart " Methods of applying negative feedback " Electronic Eng. 17.213 (Nov. 1945) 770
F4. Williamson, D. T. N. " Design of a high quality amplifier " W.W. 53.4 (April 1947) 118 ; 53.5. (May 1947) 161.
F5. Aston, R. H. " Radiotron 30 watt amplifier A513 " Radiotronics No. 124 (March/April 1947) 19.
F6. Langford-Smith, F., and R. H. Aston " The design of a high fidelity amplifier—(3) A design using push-pull triodes with negative feedback," Radiotronics No. 128 (Nov./Dec. 1947) 99.
F7. Booth, H. (letter) " Amplifiers with negative feedback " W.W. 54.6 (June 1948) 233.
F8. Sarser, D., and M. C. Sprinkle " Musician's Amplifier " (Williamson circuit) Audio Eng. 33.11 (Nov. 1949) 11.

F9. Williamson, D. T. N. "High-quality amplifier—new version" W.W. 55.8 (Aug. 1949) 282 ; 55.10 (Oct. 1949) 365 ; 55.11 (Nov. 1949) 423.
F10. Miller, J. M. "Combining positive and negative feedback" Elect. 23.3 (March 1950) 106.
F11. Roddam, T. "More about positive feedback—applied locally it can improve negative-feedback amplifier performance" W.W. 46.7 (July 1950) 242.
F12. Llewellyn, F. B. U.S. Patent 2 245 598.
F13. Shepard, F. H. U.S. Patents 2 313 096, 2 313 097, 2 313 098.
F14. Seymour, R. A., and D. G. Tucker "A choke-coupled phase-invertor of high accuracy" Electronic Eng. 23.276 (Feb. 1951) 64.
F15. Banthorpe, C. H. "Positive feedback in a.f. amplifiers" Electronic Eng. 22.273 (Nov. 1950) 473 ; correspondence 23.276 (Feb. 1951) 70.
F16. Cross, R. M. "Positive feedback in a.f. amplifiers" Electronic Eng. 23.276 (Feb. 1951) 70.
F17. Roddam, T. "Output impedance control" W.W. 56.2 (Feb. 1950) 48.
F18. Griffiths, E. "Negative feedback—its effect on input impedance and distortion" W.W. 56.3 (March 1950) 111.
F19. Mitchell, R. M. "Audio amplifier damping" Elect. 24.9 (Sept. 1951) 128.

(G) REFERENCES TO CATHODE COUPLED AMPLIFIERS

G1. Schmitt, O. H. "Cathode phase inversion" Jour. Sci. Instr. 15 (March 1938) 100.
G2. Williams, E. "The cathode coupled double-triode stage" Electronic Eng. 16.195 (May 1944) 509.
G3. Noltingk, B. E. (letter) "The cathode coupled double-triode stage" Electronic Eng. 16.196 (June 1944) 34.
G4. Butler, F. "Cathode coupled oscillators" W.E. 21.254 (Nov. 1944) 521.
G5. Sziklai, G. E., and A. C. Schroeder, "Cathode-coupled wide-band amplifiers" Proc. I.R.E. 33.10 (Oct. 1945) 701.
G6. Amos, S. W. "Push-pull circuit analysis : Cathode coupled output stage" W.E. 23.269 (Feb. 1946) 43.
G7. Pullen, K. A. "The cathode-coupled amplifier" Proc. I.R.E. 34.6 (June, 1946) 402. Discussion 35.12 (Dec. 1947) 1510.
G8. Le Bel, C. J. "Graphical characteristics of cathode-coupled triode amplifiers" Audio Eng. 31.3 (May 1947) 40.
G9. Wheeler, M. S. "An analysis of three self-balancing phase inverters" Proc. I.R.E. 34.2 (Feb. 1946) 67P.
G10. Bird, E. K. M. "A note on the phase-splitting amplifier circuit" Electronic Eng. 17.198 (Aug. 1944) 103 (Schmitt circuit).
G11. Schmitt, O. H. (letter) "Re—an original phase inverter" Radio Craft, 13.2 (Aug. 1941) 68.
G12. Cocking, W. T. "Phase splitting in push-pull amplifiers" W.W. 44.15 (April 13th, 1939) 340.
G13. Korman, N. I. (letter) "Cathode-coupled triode amplifiers" Proc. I.R.E. 35.1 (Jan. 1947) 48.
G14. Clare, J. D. "The twin triode phase-splitting amplifier" Electronic Eng. 19.228 (Feb. 1947) 62.
G15. Campbell, N. R., V. J. Francis and E. G. James "Common-cathode amplifiers" W.E. 25.297 (June 1948) 180.
G16. Rifkin, M. S. "A graphical analysis of the cathode-coupled amplifier" Comm. 26.12 (Dec. 1946) 16.
G17. Ross, S. G. F. "Design of cathode-coupled amplifiers" W.E. 27.322 (July 1950).
G18. Lyddiard, J. A. "Cathode coupled amplifier—analysis and design" W.E. 29.342 (March 1952) 63.

(H) STABILITY AND MAXIMAL FLATNESS

H1. Nyquist, H. "Regeneration theory," B.S.T.J. 11.1 (Jan. 1932) 126.
H2. Peterson, E., J. G. Kreer and L. A. Ware, "Regeneration theory and experiment" Proc. I.R.E. 22.10 (Oct. 1934) 1191.
H3. Reid, D. G. "The necessary conditions for instability (or self-oscillation) of electrical circuits" W.E. 14.170 (Nov. 1937) 588.
H4. Terman, F. E. and Wen-Yuan Pan "Frequency response characteristic of amplifiers employing negative feedback" Comm. 19.3 (March 1939) 5.
H5. Everest, F. A., and H. R. Johnston "The application of feedback to wide-band output amplifiers" Proc. I.R.E. 28.2 (Feb. 1940) 71.
H6. Bode, H. W. "Relations between attenuation and phase in feedback amplifier design" B.S.T.J. 19.3 (July 1940) 421. Also U.S.A. Patent, 2 123 178.
H7. Watton, A. "Modulated beam cathode-ray phase meter" Proc. I.R.E. 32.5 (May 1944) 268.
H8. Becker, S. "The stability factor of negative feedback in amplifiers" Proc. I.R.E. 32.6 (June 1944) 351.
H9. Learned, V. "Corrective networks for feedback circuits" Proc. I.R.E. 32.7 (July 1944) 403.
H10. Terman, F. E. "Radio Engineers Handbook" (McGraw-Hill, 1943) pp. 218-226 ; 396-402.
H11. Bode, H. W. (book) "Network Analysis and Feedback Amplifier Design" (D. Van Nostrand Co. Inc. New York, 1945).
H12. Brockelsby, C. F. "Negative feedback amplifiers—conditions for maximal flatness" W.E. 26.305 (Feb. 1949) 43.
     Letters T. S. McLeod, W.E. 26.308 (May 1949) 176 ; 26.312 (Sept. 1949) 312 ; C. F. Brockelsby 26.310 (July 1949) 247 ; 26.314 (Nov. 1949) 380.
H13. Mayr, H. "Feedback amplifier design—conditions for flat response" W.E. 26.312 (Sept. 1949) 297.
H14. West, J. C. "The Nyquist criterion of stability" Electronic Eng. 22.267 (May 1950) 169.
H15. Flood, J. E. "Negative feedback amplifiers—conditions for critical damping" W.E. 27.322 (July 1950) 201.
H16. Bothwell, F. E. "Nyquist diagrams and the Routh-Hurwitz stability criterion" Proc. I.R.E. 38.11 (Nov. 1950) 1345.
H17. Barter, L. D. "Graphical solution for feedback amplifiers" Elect. 23.11 (Nov. 1950) 204.
H18. Roddam, T. "Stabilizing feedback amplifiers" W.W. 57.3 (March 1951) 112.
H19. Ragazzini, J. R., and L. A. Zadeh "A wide band audio phase meter" Rev. Sci. Inst. 21.2 (Feb. 1950) 145.
H20. Baldwin, T., and J. H. Littlewood "A null method measuring the gain and phase shift of comparatively low frequency amplifiers" Electronic Eng. 23.276 (Feb. 1951) 65.
     Also A1, A6, A7, A11.
H21. Lynch, W. A. "The stability problem in feedback amplifiers" Proc. I.R.E. 39.9 (Sept. 1951) 1000.
     Additional references will be found in the Supplement commencing on page 1475.

CHAPTER 8

WAVE MOTION AND THE THEORY OF MODULATION

By I. C. Hansen, Member I.R.E. (U.S.A.)

SECTION 1 : INTRODUCTION TO ELECTROMAGNETIC WAVES

(i) Wave motion (ii) Electromagnetic spectrum (iii) Wave propagation.

(i) Wave motion

Before considering the various methods of radio reception, it is desirable to have an understanding of the nature of radio waves and of how they are used to transmit intelligence.

The theory of electromagnetic radiation (Refs. 1, 2, 3) is essentially a mathematical one and any attempt to reduce the theory to non-mathematical terms requires certain assumptions to be made which are not strictly true.

It should be noted that sound waves in air, waves in water and radio waves are propagated in entirely different ways.

In the case of sound, the waves are called " longitudinal " and consist of alternate compressions and expansions of air.

The layers of air as shown in Fig. 8.1 do not move steadily forward with the wave but move to and fro through a limited path in the direction of motion of the wave.

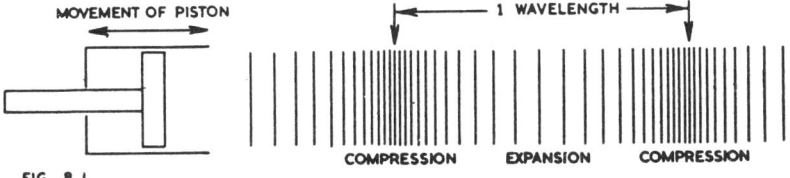

FIG. 8.1

Fig. 8.1. Representation of a sound wave in air.

Referring to Fig. 8.2 and assuming this to be a representation of a water wave, it will be seen that as the wave progresses, a particle (such as a cork) as shown in the figure, moves in a vertical direction which is at right angles to the direction of motion of the wave. This form of wave motion is known as " transverse."

The distance between two successive crests is a measure of the length of the wave, i.e. the wavelength, and the number of complete waves (or cycles) passing any point in one second denotes the frequency in cycles per second.

The medium in which the wave is transmitted determines the speed or velocity of propagation. In the case of sound waves in air, the velocity is approximately 3.4×10^2 metres per second. The velocity of sound in some common media is given in Chapter 20 Sect.. 8(iii).

Electromagnetic waves are also transverse. However, radiated electromagnetic energy consists of two component fields, respectively magnetic and electric, both in phase, existing at right angles to each other and to the direction of motion of the wave. In contrast to sound waves in air, electromagnetic waves in free space have a velocity of 3×10^8 metres per second.

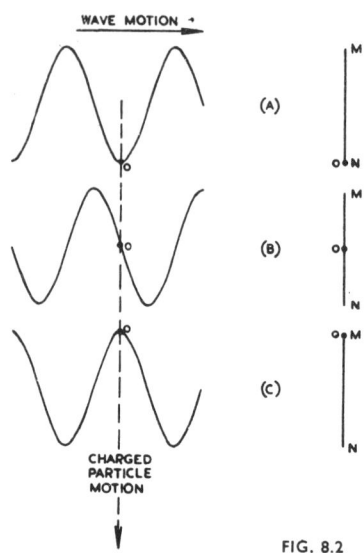

FIG. 8.2

Fig. 8.2. Representation of a water wave.

(ii) Electromagnetic frequency spectrum (Ref. 4)

Type of radiation	Wavelength
Hertzian Waves (used for radio communication)	30 000 M — 0.01 cm
Infra Red	0.04 cm — 0.000 07 cm
Visible Light*	7000 A° — 4000 A°
Ultra Violet	4000 A° — 120 A°
X-Rays	120 A° — 0.06 A°
Gamma Rays	1.4 A° — 0.01 A°
Cosmic Radiation	about 0.0001 A°

1 Angstrom unit $A° = 1 \times 10^{-8}$ cm.

(iii) Wave propagation

Assuming a simple non-directional aerial, the electromagnetic waves are radiated in straight lines at all angles. That portion of the radiation at a large angle to the horizontal is called the " sky " wave. Depending upon the frequency and the angle of radiation, this wave may be reflected by an ionized layer in the earth's upper atmosphere and thus return to the earth at a considerable distance from its origin. By making use of this phenomenon, long distance short wave communication can be accomplished.

In recent years much work has been done in forecasting frequencies that will give best results for communication between any two points at any given time. These predictions, based on a study of the ionosphere, are published regularly in different parts of the world and are an exceedingly valuable aid to reliable communication services. A handbook has been published explaining the basis of the predictions and the methods used in forecasting (Refs. 5, 6, 7, 8, 9, 10). This subject is covered more comprehensively in Chapter 22, Sect. 6.

For the effect of the ionosphere on the reception of radio signals see Chapter 22, Sect. 6.

*For further details, see Chapter 38 Section 13 and Fig. 38.8.

SECTION 2 : TRANSMISSION OF INTELLIGENCE

(i) Introduction. (ii) Radio telegraphy (iii) Radio telephony.

(i) Introduction

There are several methods of conveying information using the radiated waves. By switching on and off the transmitter in accordance with a prearranged code, audible sounds will be heard in a suitable receiver. This is known as radio telegraphy. Alternatively the intelligence in the form of speech or music can be superimposed on the transmitted radio frequency by a process of " modulation." This is referred to as radio telephony. There are four major types of this latter system of transmission, known as amplitude, frequency, phase and pulse modulation (Refs. 17, 18, 33).

(ii) Radio telegraphy

This is used primarily for communication purposes over long distances. By means of automatic transmission and reception it is possible to convey messages at a rapid rate half way round the world. Such circuits are less susceptible to interference and mutilation of the message text than corresponding circuits using radio telephony. A radio telegraph (C.W.) transmitter usually consists of a crystal oscillator followed by several r-f power amplifier stages. An automatic or manual switch, operating in a power circuit supplying current to one of the r-f amplifier electrodes, serves to turn the r-f power on or off as required by the coded messages (Ref. 11).

(iii) Radiotelephony

(A) Amplitude Modulation (A-M)

This is commonly used in local medium frequency and long distance shortwave broadcast transmitters. The amplitude of the r-f radiated wave (the carrier) is varied at an audio frequency rate according to similar variations of the intelligence which it is required to transmit. This is usually accomplished by varying the plate voltage of a r-f amplifier by the audio signal. The r-f amplifier may be in the final transmitter stage, in which case a large audio power is required for complete modulation—50% of the modulated r-f stage plate input power. This is called " high-level " modulation. As an alternative, modulation can be employed at an early stage in the transmitter, and linear r-f amplifiers used to raise the power level to that required. Much smaller powers are necessary for this " low-level " modulation system, but the Class B linear r-f amplifiers are less efficient than the Class C counterparts used in the " high-level " method of modulation. Both systems are used extensively, but most high power transmitters employ " high-level " modulation. With amplitude modulation, the power output of the transmitter varies, but the phase and frequency of the carrier remain unaffected (Refs. 12, 13).

(B) Frequency Modulation (F-M)

This is now coming into its own in the mobile equipment field. While it is being used to supplement medium frequency A-M broadcasting, F-M may also be used for the audio channel of television (TV) transmitters. In this system the carrier frequency is varied at audio frequency above and below the unmodulated carrier frequency. This can be achieved by using a reactance valve to alter the frequency of the transmitter master oscillator, the reactance valve being controlled by the audio frequency signal.

In contrast to A-M, the power output from an F-M transmitter does not change, but the frequency of the carrier varies, the deviation in frequency being proportional to the amplitude of the modulating signal (Refs. 14, 15, 16, 24, 32, 33).

(C) Phase Modulation

In this system the audio signal is used to shift the phase of the carrier, the change in phase angle being proportional to the instantaneous amplitude of the modulating signal. The carrier frequency also changes but the power output remains constant.

P-M is not yet in general use and will therefore not be considered further (Refs. 19, 20, 21, 22, 23, 24, 31; 33).

(D) Pulse Modulation

The numerous variations of this system are finding their chief application in the very high frequency communication field. Radar and related distance measuring equipments rely largely on pulse modulation for their successful operation. This modulation system is outside the scope of the Handbook (Refs. 25, 26, 27, 28, 29, 30, 33).

SECTION 3 : REFERENCES

1. Dow, W. G. (Book) " Fundamentals of Engineering Electronics " Chapter 17. John Wiley and Sons Inc. N.Y. 1937.
2. Everitt, W. L. (Editor) " Fundamentals of Radio " Chapter 9. Prentice-Hall, Inc. N.Y. 1943.
3. Skilling, H. H. (Book) " Fundamentals of Electric Waves " 2nd edit. John Wiley and Sons Inc. N.Y. 1948.
4. Spectrum Chart " Electronics Buyers Guide " McGraw-Hill Book Co. N.Y. 19B.6B (June 1946) 80.
5. Bennington, T. W. (Book) " Short-Wave Radio and the Ionosphere ". Iliffe and Sons Ltd. London 1950.
6. " CRPL-D Basic Radio Propagation Predictions " Sup. of Docs. U.S. Govt. Printing Office Wash. D.C.
7. " I.R.P.L. Radio Propagation Handbook " Part 1, 1943.
8. " Radio Propagation Bulletins " I.P.S. Commonwealth Observatory, Canberra, Australia.
9. " Handbook for use with Radio Propagation Bulletins " Radio Research Board C.S.I.R., Australia, 1945.
10. Foley, W. R. " Forecasting Long Distance Transmission " Q.S.T. 30.2 (Feb. 1946) 36.
11. Nelson and Hornung (Book) " Practical Radio Communication " 2nd Edit. McGraw-Hill Book Co. 1943.
12. Sandeman, E. K. (Book) " Radio Engineering " Vol. 1, 1st edit. Chapman and Hall London 1947.
13. Ladner and Stoner (Book) " Short Wave Wireless Communication."
14. Tibbs, C. E. (Book) " Frequency Modulation Engineering " Chapman and Hall, London, 1947.
15. Rider, J. F. (Book) " Frequency Modulation " Rider Publisher Inc. N.Y. 1940.
16. Hund, A. (Book) " Frequency Modulation " McGraw-Hill Book Co. N.Y. 1942.
17. Hund, A. " Amplitude frequency and phase modulation relations " Elect. (Sept. 1942) 48.
18. Smith, E. J. " Theoretical signal-to noise ratios "(comparison between various modulation systems) Elect. 19.6 (June 1946) 150.
19. Tibbs, C. " Phase and frequency modulation " W.W. 48.9 (Sept. 1942) 210.
20. Farren, L. I. " Phase detectors—some theoretical and practical aspects " W.E. 23.279 (Dec. 1946) 330.
21. Hund, A. " Frequency and phase modulation " Proc. I.R.E. 32.9 (Sept. 1944) 572.
22. Stockman, H., and G. Hok, " Frequency and phase modulation " Proc. I.R.E. 33.1 (Jan. 1945) 66.
23. Jaffe, D. L., D. Pollack, A. Hund and B. E. Montgomery (letter) " Frequency and phase modulation " Proc. I.R.E. 33.3 (March 1945) 200 ; 33.7 (July 1945) 487. See Refs. 21, 22, 31.
24. Crosby, M. G. " Exalted-carrier amplitude and phase-modulation reception " Proc. I.R.E. 33.9 (Sept. 1945) 581.
25. Deloraine, E. M., and E. Labin, "Pulse-time modulation " Elect. 18.1 (Jan. 1945) 100.
26. " Post war prospects for modulation pulse transmission " Elect. 18.7 (July 1945) 244.
27. " Pulse-width modulation—its basic principles described " W.W. 51.12 (Dec. 1945) 361.
28. " Pulse-time modulation—an explanation of the principle " W.W. 52.2 (Feb. 1946) 45.
29. Roberts, F. F., and J. C. Simmonds " Multichannel communication systems " preliminary investigation of systems based upon modulated pulses—W.E. 22.266 (Nov. 1945) 538 ; 22.267 (Nov. 1945) 576. See (30) below.
30. Fitch, E., and E. R. Kretzmer, "Pulse modulation " (Ref. 29 above) letters W.E. 23.275 (Aug. 1946) 231 ; 23.277 (Oct. 1946) 288.
31. Stockman, H., and G. Hok " A note on frequency modulation terminology " letter to Proc. I.R.E. 32.3 (March 1944) 181. See Refs. 21 and 22 above.
32. Armstrong, E. H. " A method of reducing disturbances in radio signalling by a system of frequency modulation " Proc. I.R.E. 24.5 (May 1936) 689.
33. "Reference data for radio engineers " (Federal Telephone and Radio Corporation, 3rd ed. 1949)

CHAPTER 9

TUNED CIRCUITS

based on original chapter in previous edition written by L. G. Dobbie, M.E.
Revised by G. Builder, B.Sc., Ph.D., F. Inst. P.

SECTION 1 : INTRODUCTION

When a violin is tuned, the tensions of its strings are adjusted to permit vibration at particular frequencies. In radio, when an arrangement of L, C, and R responds to particular frequencies, it is called a " tuned " circuit.

In principle, the tuned circuit is similar to a pendulum or violin string, tuning fork, etc.—it has the property of storing energy in an oscillating (vibrating) state, regularly changing from kinetic form (magnetic field when current flows through the coil) to potential form (electric field, when the condenser is charged) and back again at a frequency called the **natural resonant frequency.**

FIG. 9.1

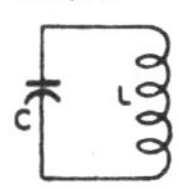

In Fig. 9.1 let the condenser C be charged. It will discharge its energy through the inductance L, causing the current to increase all the while, until it reaches the maximum when there is no potential across C. At that instant the energy is all magnetic, and the current continues, fed by the magnetic field, to build a voltage of reversed polarity across C. When all the energy has been transferred from L to C, the voltage across C has its original value, but is reversed in sign, and the current is diminished to zero. The process then reverses, and repeats itself indefinitely, cycle after cycle.

407

SECTION 2 : DAMPED OSCILLATIONS

Were there no loss of energy, each cycle would be the same, but this is never the case in practice, as the coil must have resistance. The rate at which the energy decreases is proportional to the energy (remaining) in the circuit.

For a detailed account of damped oscillations, as these periodic changes of decreasing amplitude are called, reference should be made to standard textbooks relating to the theory of radio circuits. A list of several such books is given in the accompanying bibliography.

Here we consider briefly the circuit shown in Fig. 9.2a. Let I_o be the maximum value of the current during a given cycle of the oscillation, and let the time t be measured from the instant of this maximum ; then the value of the current i at any subsequent time is given by the equation

$$i = I_o \epsilon^{-\alpha t} \cos \omega_n t, \tag{1}$$

where $\alpha = r/2L$ = damping factor,

and $\omega_n = \sqrt{\dfrac{1}{LC} - \dfrac{r^2}{4L^2}}$.

The quantity $f_n = \omega_n/2\pi$ is called the **natural resonant frequency**. In the above formulae L is in henrys, C in farads, r in ohms, ω_n in radians per second and f_n in cycles per second.

The **resonant frequency** f_o (corresponding to $r = 0$) of the circuit is defined by

$$2\pi f_o = \omega_o = \sqrt{1/LC} \tag{2}$$

The ratio of the natural resonant frequency to the resonant frequency is

$$f_n/f_o = \sqrt{1 - 1/(4Q^2)} \tag{3}$$

where $Q = \dfrac{\omega_o L}{r} = \dfrac{2\pi f_o L}{r}$

$$= \frac{\text{reactance of the coil at resonant frequency}}{\text{coil resistance}}.$$

This relationship shows that in practice there is little difference between these two frequencies. Q must, for example, be less than four to make f_n differ by 1% from f_o. Q normally exceeds fifty, for which value the two frequencies differ by about one part in twenty thousand.

If now, in the equation for the current, t is increased by an amount $2\pi/\omega_n$, the period of one cycle, we arrive at the corresponding point in the next cycle. Let i' denote the current at this point, so that

$$i' = I_o \epsilon^{-\alpha(t+2\pi/\omega_n)} . \cos \omega_n(t + 2\pi/\omega_n)$$

$$= i\epsilon^{-2\pi\alpha/\omega_n}$$

or, $i'/i = \epsilon^{-\pi r/\omega_n L}$, since $\alpha = r/2L$ \hfill (4)

This gives the ratio of the amplitude of one cycle to that immediately preceding.

Logarithmic decrement, δ, is defined by

$$\delta = \pi r/\omega_n L = \log_\epsilon (i/i') \tag{5}$$

and is thus the naperian logarithm of the ratio of the amplitudes of two successive cycles.

See also Sect. 11 Summary of Formulae.

SECTION 3 : SERIES RESONANCE

The **series impedance** z of the circuit in Fig. 9.2a at any frequency ($f = \omega/2\pi$) is given by the expression

$$z = \sqrt{r^2 + [\omega L - (1/\omega C)]^2} \tag{6}$$

Thus if an alternating voltage of frequency f be applied in series with the circuit, when the value of r is fixed, we see that z is least, and hence the current reaches its maximum, when $\omega L - 1/\omega C = 0$, i.e. when $\omega = 1/\sqrt{LC} = \omega_o$, the resonant frequency.

It is perhaps surprising that the frequency for maximum current is independent of the circuit resistance r and that maximum current does not occur when the frequency of the applied voltage is equal to the natural frequency f_n. The state of maximum current flow is called **series resonance**.

If E be the r.m.s. value of the applied alternating voltage, the r.m.s. value I of the current produced in the circuit is clearly

$$I = \frac{E}{z} = \frac{E}{\sqrt{r^2 + [\omega L - (1/\omega C)]^2}} \tag{7}$$

$$= E/r, \text{ when } \omega = 1/\sqrt{LC} = \omega_o \tag{8}$$

The voltages across the several parts of the circuit under this condition of series resonance are (Fig. 9.2a) :

$$rI = E \text{ across the resistance} \tag{9}$$
$$\omega_o LI \text{ across the inductance} \tag{10}$$
and $- (1/\omega_o C)I$ across the capacitance. $\tag{11}$

The voltages across the inductance and capacitance are equal and opposite in sign, thus cancelling each other, and usually they are large compared with the voltage across the resistance. The voltage across the inductance may be expressed as $(\omega_o L/r)E$, and therefore its ratio to the voltage applied in series with the circuit is $\omega_o L/r$. **This ratio, usually denoted by Q**, is the ratio of the reactance of the coil at resonance to the resistance in series with it, and is called the **magnification factor** or **quality factor**. Thus,

$$Q = \frac{\omega_o L}{r} = \frac{1}{r}\sqrt{\frac{L}{C}} = \frac{1}{\omega_o Cr} \tag{12}$$

Q may also be called the **energy factor** and is then defined by

$$Q = 2\pi \frac{\text{peak energy storage}}{\text{energy dissipated per cycle}}$$

$$= \omega \frac{\text{peak energy storage}}{\text{average power loss}}$$

That this is equivalent to $\omega_o L/r$ is shown by

$$Q = \omega_o \frac{\tfrac{1}{2}LI^2}{\tfrac{1}{2}I^2 r} = \frac{\omega_o L}{r}.$$

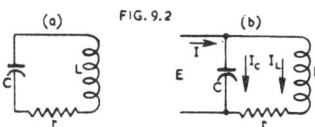

FIG. 9.2

(a)   (b)

It is shown in Chapter 4 Sect. 5(vi) that the **power factor** is approximately equal to $1/Q$; when Q is greater than 7 the error is less than 1%.

See also Sect. 11 Summary of Formulae.

The general treatment of series circuits with L, C and R is given in Chapter 4 Sect. 6(ii).

SECTION 4 : PARALLEL RESONANCE

Let us now examine the result of applying an alternating voltage E across the condenser C in the circuit shown in Fig. 9.2b. The current divides between the two branches, and we find that the (r.m.s.) value of the current in L and r is given by

$$I_L = \frac{E}{\sqrt{r^2 + \omega^2 L^2}} \tag{13}$$

and the r.m.s. value of the current in C by

$$I_c = \omega C E \tag{14}$$

Adding the two currents, with due regard to their phase relation, the r.m.s. value of the total current I is found to be

$$I = E \sqrt{\left(\omega C - \frac{\omega L}{r^2 + \omega^2 L^2}\right)^2 + \left(\frac{r}{r^2 + \omega^2 L^2}\right)^2} \tag{15}$$

When $\omega C = \omega L/(r^2 + \omega^2 L^2)$, the total current I is in phase with the applied voltage E, and has its minimum value

$$I = E \frac{r}{r^2 + \omega^2 L^2} \tag{16}$$

This condition is termed " parallel resonance* ", as distinct from the " series resonance " considered earlier. At resonance, the currents in the condenser and coil are large compared with the current in the external circuit and they are very nearly opposite in phase.

The value of ω at which parallel resonance occurs can be determined from the equation

$$\omega_o = \frac{1}{\sqrt{LC}} \cdot \frac{1}{\sqrt{1 + (r^2/\omega_o^2 L^2)}} \tag{17}$$

It was pointed out in the previous section that $\omega L/r$, the ratio of the reactance of the coil to the resistance in series with it, is usually greater than 50. Thus, with an error of about one part in 5000 at $Q = \omega L/r = 50$ and correspondingly smaller errors at larger values of Q, we find that the parallel resonant frequency f_o is given by

$$\omega_o = 2\pi f_o \approx 1/\sqrt{LC}.$$

This is the same result, to the accuracy indicated above, as that obtained in the series resonance case.

With a similarly small degree of error we may write the following simple relations for the **currents at resonance**. The current in the external circuit (Fig. 9.26) is given by

$$I \approx E \frac{r}{\omega_o^2 L^2} \approx \frac{E.Cr}{L} \approx E.\omega_o^2 C^2 r,$$
$$\approx \frac{E}{\omega_o L} \cdot \frac{1}{Q} \approx \frac{E.\omega_o C}{Q} \tag{18}$$

where $Q = \omega_o L/r$ as before. **The current in the inductance and resistance** is very closely equal in value to that in the capacitance,

$$I_L \approx -I_c \approx E/\omega_o L \approx -\omega_o CE \tag{19}$$

and **is Q times larger than the current in the external circuit**

The formulae given so far hold only when the condenser loss is negligible. In order to generalize our expressions in a simple manner let us first consider the two circuits shown in Fig. 9.3a and Fig.*9.3b. At (parallel) resonance for Fig. 9.3a we have found that the current I in the external circuit is

$$I \approx ECr/L \approx E/Q\omega_o L \text{ etc. };$$

---

*Parallel resonance may be defined either as
  (a) the frequency at which the parallel impedance of the circuit is a maximum, or
  (b) the frequency at which the equivalent reactance of the complete parallel circuit becomes zero
     (i.e. when the impedance has unity power factor and acts as though it were a pure resistance at
     the resonant frequency). This can also be expressed by saying that the parallel circuit has
     zero susceptance at the resonant frequency.
For further details see Chapter 4 Sect. 6(iii) and (iv).
Definition (b) is used in this chapter.

and for Fig. 9.3b, to the same approximation, it is evident that
$$I \approx E/R_e.$$
These two circuits are equivalent at resonance provided we set
$$R_e = L/Cr = Q\omega_o L = Q/\omega_o C = Q\sqrt{L/C} \qquad (20)$$
Note that $Q = R_e\sqrt{C/L}$, and that R_e is the parallel resistance (across C) equivalent to the series coil resistance r.

It has been shown that the circuit Fig. 9.3a with a condenser C having no losses and an inductance L having series coil resistance r may be replaced by the equivalent circuit Fig. 9.3b having an ideal tuned circuit LC, without losses, shunted by the resistor R_e having a value given by equation (20). It is obvious that the impedance of the parallel combination LCR_e in Fig. 9.3b at resonance is R_e, this being the " resonant impedance " of the circuit. At frequencies other than the resonant frequency, the impedance will be less than the " resonant impedance."

The values of Q in terms of series coil resistance r and equivalent parallel resistance R_e are grouped below for convenience.

In terms of r : $Q = \dfrac{\omega_o L}{r} = \dfrac{1}{\omega_o Cr} = \dfrac{1}{r}\sqrt{\dfrac{L}{C}}$

In terms of R_e : $Q = \dfrac{R_e}{\omega_o L} = \omega_o CR_e = R_e\sqrt{\dfrac{C}{L}}$

We can now consider the important practical case of the circuit shown in Fig. 9.4, in which a resistance R appears in shunt with the condenser C. R represents the effect of all insulation losses in condenser, coil, wiring, switches and valves, together with the plate or input resistance of the valves.

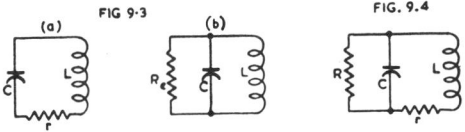

FIG 9·3   FIG. 9.4

Fig. 9.3(a). *Parallel resonance with series loss resistance.*
Fig. 9.3(b). *Parallel resonance with parallel loss resistance.*
Fig. 9.4. *Parallel resonance with both series and parallel loss resistances.*

In the present case we have R in parallel with our equivalent parallel coil resistance R_e of Fig. 9.3b. The resultant parallel resistance at resonance, which we will call the **resonant impedance** is, of course,
$$R_D = \frac{1}{1/R + Cr/L} \qquad (21)$$
Therefore, the **resultant value of** Q **is**
$$Q = \sqrt{\frac{C}{L}} \cdot R_D = \frac{1}{(1/R)\sqrt{L/C} + r\sqrt{C/L}} = \frac{1}{(\omega_o L/R) + (r/\omega_o L)} \qquad (22)$$
Note that at the resonant frequency the expression $\sqrt{L/C}$ is equal to the reactance of the inductance and also that of the condenser, i.e.
$$\omega_o L = \sqrt{L/C} = 1/\omega_o C. \qquad (23)$$
See also Sect. 11 Summary of Formulae.

SECTION 5 : GENERAL CASE OF SERIES RESONANCE

Fig. 9.5 shows the general type of series resonant circuit. It is often convenient to express the effect of the two resistances at resonance as a resultant equivalent series resistance, r' say. In the circuit of Fig. 9.3 we saw that the effect of a resistance r in series with L was equivalent at resonance to that of a resistance L/Cr shunted across C. By similar reasoning it may be shown in the present case (Fig. 9.5) that

FIG. 9.5

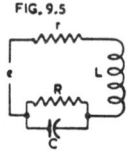

Fig. 9.5.   Series resonance with losses in inductive and capacitive elements.

the effect of R at resonance is equivalent to that of a resistance of value L/CR in series with L. The resultant equivalent series resistance r' is thus equal to $r + (LC/R)$, and the resultant value of Q is $\omega_o L/r'$.

See also Sect. 11 Summary of Formulae.

SECTION 6 : SELECTIVITY AND GAIN

(i) *Single tuned circuit* (ii) *Coupled circuits—tuned secondary* (iii) *Coupled circuits—tuned primary, tuned secondary* (iv) *Coupled circuits of equal Q* (v) *Coupled circuits of unequal Q.*

(i) Single tuned circuit

The currents and voltages, and hence the gain, of single tuned circuits at resonance are determined by the equivalent series resistance (r' as defined in Sect. 5) and the resonant impedance (R_D as defined in eqn. 21).

At frequencies other than the resonant frequency, the reactances of the coil and the condenser no longer balance. In the series circuit the resistance r' becomes an impedance z which is greater than r'. In the parallel circuit the resonant impedance R_D becomes an impedance Z which is less than R_D.

The appropriate ratios of these quantities determine the **selectivity** or response of the circuit, and they are related to Q and f by the following expression :

$$\frac{A_o}{A} = \frac{z}{r'} = \frac{R_D}{Z} = \sqrt{1 + Q^2\left(\frac{f}{f_o} - \frac{f_o}{f}\right)^2} \qquad (24)$$

where A_o is the voltage gain at the resonant frequency f_o and A the gain at frequency f.

Note that A_o/A is the ratio of current at resonance to that at frequency f in the series case, and the ratio of **total** current at frequency f to that at the resonant frequency in the parallel case (see below).

The phase angle between the applied voltage and the total current is such that

$$\tan \phi = \pm Q\left(\frac{f}{f_o} - \frac{f_o}{f}\right) \qquad (25)$$

the positive sign pertaining to the series case and the negative sign to the parallel case.

The equation (24) leads to a simple method for determining Q from the response curve. We see that when $Q[(f/f_o) - (f_o/f)] = \pm 1$ the total current will be decreased by the factor $\sqrt{2}$ in the series case (i.e. decreased to 70.7% of the resonance value), and increased by the same factor in the parallel case. It will be observed that at either of the frequencies satisfying this condition the phase angle ϕ is numerically equal to 45°, as $\tan \phi = \pm 1$, and that the resistance r' or R_D is equal to the reactance.

The condition
$$Q(f/f_o - f_o/f) = \pm 1 \qquad (26)$$
may be written
$$f/f_o - f_o/f = \pm 1/Q ; \qquad (27)$$
and for values of Q not too low'(> 50, say) we have, very closely,
$$2\Delta f_o/f_o \approx \pm 1/Q \qquad (28)$$
where $\Delta f_o = f - f_o$.

Thus $\Delta f_o \approx \pm \dfrac{f_o}{2Q}$ \qquad (29)

and the current is decreased, or increased, by the factor $\sqrt{2}$ at two frequencies f_1 and f_2, one on each side of the resonant frequency f_o, determined by

$$f_1 = f_o - \frac{f_o}{2Q} ; f_2 = f_o + \frac{f_o}{2Q} \qquad (30)$$
Hence, $Q \approx f_o/(f_2 - f_1)$ \qquad (31)
The frequencies f_o, f_1 and f_2 may be found experimentally, and hence Q may be calculated.

At frequencies very different from f_o, so that A_o/A is greater than about 10, the equation giving the response is very approximately
$$A_o/A \approx Q(f/f_o - f_o/f). \qquad (32)$$
Also, the expression for the phase angle may be written in the alternative forms
$$\tan \phi = \pm Q(f/f_o - f_o/f) \qquad \text{from equation (25)}$$
$$= \pm Q \frac{\Delta f_o}{f_o} \cdot \frac{2 + (\Delta f_o/f_o)}{1 + (\Delta f_o/f_o)} \qquad (33)$$
Under these conditions we see that **in the series case**
(a) Across the coil :
$$\frac{\text{Voltage at resonance}}{\text{Actual voltage when well off resonance}} = \frac{L\omega_o i_o}{L\omega i} = \frac{Lf_o i_o}{Lfi} = Q\left(1 - \frac{f_o^2}{f^2}\right) \quad (34)$$
and (b) Across the condenser :
$$\frac{\text{Voltage at resonance}}{\text{Actual voltage when well off resonance}} = \frac{C\omega i}{C\omega_o i} = Q\left(\frac{f^2}{f_o^2} - 1\right) \quad (35)$$
Similarly, **in the parallel case,** we have
$$\frac{\text{Impedance at resonance}}{\text{Impedance at frequencies well off resonance}} = Q\left(\frac{f}{f_o} - \frac{f_o}{f}\right) \quad (36)$$
from which it can be shown that
(a) In the coil :
$$\frac{\text{Current at resonance}}{\text{Current at frequency } f} = \frac{f}{f_o} \qquad (37)$$
and (b) in the condenser :
$$\frac{\text{Current at resonance}}{\text{Current at frequency } f} = \frac{f_o^2}{f} \qquad (38)$$

(ii) Coupled circuits—tuned secondary

We consider briefly now two examples of coupled circuits. The first example, shown in Fig. 9.6 illustrates a typical case of a **high frequency transformer with tuned secondary** in a radio receiver.

The symbols to be used are set out below :
g_m = mutual conductance of the valve in mhos,
r_p = plate resistance of the valve in ohms,
L_1 = primary inductance in henrys,
L_2 = secondary inductance in henrys,
M = mutual inductance between L_1 and L_2 in henrys,
k = $M/\sqrt{L_1 L_2}$ = coupling factor,
Q_2 = $L_2\omega_o/r = 1/rC\omega_o$,
ω_o = $2\pi \times$ resonant frequency of secondary in cycles per second,

$R_D = \omega_o L_2 Q_2$ = resonant impedance of secondary in ohms,
e_i = input voltage,
e_o = output voltage,
and A = stage voltage gain (amplification).

When the secondary is tuned, its impedance (at ω_o) is simply r, and it reflects into the primary a resistance equal to $M^2 \omega_o{}^2/r$.   The primary signal current I_p is, therefore,

$$I_p = \frac{g_m \cdot r_p \cdot e_i}{\sqrt{X_p{}^2 + (r_p + M^2 \omega_o{}^2/r)^2}} \tag{39}$$

where X_p is the reactance of the primary.

When the conditions are such that the primary reactance can be neglected, we have

$$I_p \approx g_m \cdot r_p \cdot e_i/(r_p + M^2 \omega_o{}^2/r) \tag{40}$$

It follows that the secondary current I_s is given by

$$I_s = \frac{M\omega_o I_p}{r} \approx \frac{g_m \cdot e_i}{(r/M\omega_o) + (M\omega_o/r_p)} \tag{41}$$

Hence the induced voltage across L_2 (and C), that is e_o, is

$$e_o = \frac{g_m \cdot e_i \cdot L_2 \omega_o}{(r/M\omega_o) + (M\omega_o/r_p)} \tag{42}$$

or $$\frac{e_o}{e_i} = A_o = \frac{g_m}{(1/M\omega_o Q_2) + (M/r_p L_2)} \tag{43}$$

$$= \frac{g_m k R_D \sqrt{L_1/L_2}}{1 + k^2(R_D/r_p) \cdot (L_1/L_2)} \tag{44}$$

When r_p is very much greater than $\omega_o{}^2 M^2/r [= k^2 R_D(L_1/L_2)]$ we have simply

$$I_p \approx g_m \cdot e_i, \tag{45}$$
$$I_s \approx g_m \cdot e_i \cdot M\omega_o/r, \tag{46}$$

and $e_o/e_i = A_o \approx g_m \omega_o M Q_2 = g_m \cdot k R_D \sqrt{L_1/L_2}$ (47)

If the inductances L_1 and L_2 have the same ratio of diameter to length, or form factor, and the turns are N_1 and N_2 respectively, then $\sqrt{L_1/L_2}$ in the above formulae may be replaced by N_1/N_2.   The plate resistance r_p in parallel with the primary is reflected as a series resistance $M^2 \omega_o{}^2/r_p$ into the secondary.   If the value of this reflected resistance is greater than say 5% of r, its effect should be taken into account when computing the selectivity of the secondary.   This selectivity, then, is determined by means of the formula

$$z/r' = \sqrt{1 + Q_2'^2[f/f_o - f_o/f]^2} \tag{48}$$

where $r' = r + M^2 \omega_o{}^2/r_p$,

and $Q_2' = \dfrac{\omega_o L_2}{r'} = \dfrac{1}{(1/Q_2) + (k^2 L_1 \omega_o/r_p)}$.

Note that Q_2' and r' must not be used when calculating the gain A_o.

FIG. 9.6   FIG. 9.7

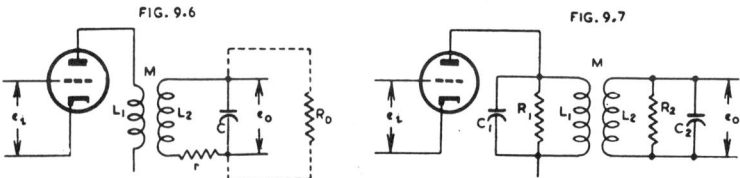

Fig. 9.6. *Amplifier stage using a high frequency transformer with a tuned secondary.*
Fig. 9.7. *Amplifier stage using a double tuned high frequency transformer.*

(iii) Coupled circuits—tuned primary, tuned secondary

The second example of coupled circuits is shown in Fig. 9.7.   This is a typical high frequency transformer with tuned primary and tuned secondary.   Intermediate frequency transformers in super-heterodyne receivers are usually of this type.   Very thorough discussions of such transformers have been given from the theoretical point of view by Aiken.(Ref. C2). For the design procedure see Chapter 26 Sect. 4. In this case the resistance in parallel with the primary at resonance is

$L_1/C_1 r_p + L_1/C_1 R_1 = L_1/C_1 R'$, where $1/R' = 1/r_p + 1/R_1$;

and the resistance of the secondary is $L_2/C_2 R_2$, so that the resistance reflected into the primary is $\omega_o{}^2 M^2 C_2 R_2/L_2.'$ The primary current I_p is, therefore,

$$I_p = \frac{g_m e_i}{\omega_o C_1} \cdot \frac{1}{L_1/C_1 R' + \omega_o{}^2 M^2 C_2 R_2/L_2} \tag{49}$$

The secondary current I_s can readily be shown to be

$$I_s = (M \omega_o C_2 R_2/L_2) I_p; \tag{50}$$

and the induced voltage e_o across L_2 and C_2 to be

$$e_o = L_2 \omega_o I_s = (M C_2 R_2 \omega_o{}^2) I_p.$$

Therefore the voltage gain at the resonant frequency, A_o, is given by

$$\frac{e_o}{e_i} = A_o = \frac{g_m M \omega_o}{L_1/R' R_2 C_2 + \omega_o{}^2 M^2 C_1/L_2} . \tag{51}$$

This relation for the gain may also be expressed as

$$A_o = \frac{g_m k \omega_o \sqrt{L_1 L_2}}{k^2 + 1/Q' Q_2} \tag{52}$$

or, $\quad A_o = \dfrac{g_m \sqrt{R' R_2}}{k\sqrt{Q' Q_2} + 1/(k\sqrt{Q' Q_2}}$, \tag{53}

where, as before, $k = M/\sqrt{L_1 L_2}$, and Q', Q_2 have their usual meaning, i.e. $R'/L_1 \omega_o$ and $R_2/L_2 \omega_o$ respectively.

The expression for calculating the selectivity is lengthy and complicated and a graphical treatment described later is preferable (Sect. 7).

As k is increased from low values, the gain increases until $k\sqrt{Q' Q_2} = 1$, after which it decreases. This value of $k = 1/\sqrt{Q' Q_2}$ is known as the **critical coupling factor** (k_c).

(iv) Coupled circuits of equal Q

When the primary and secondary circuits are identical, and the coupling factor is equal to k_c, and the plate and grid return resistances are very high, we see that the voltage gain obtained is exactly half that with a single tuned circuit. The critical coupling factor in this case, of course, is

$k_c = 1/Q$ where $Q = Q' = Q_2$.

For values of k less than k_c, the response curve (gain versus frequency) has a single maximum at f_o, the resonant frequency of each of the circuits. When k exceeds the critical value, however, the amplification curve becomes double-humped, i.e. there are two frequencies of maximum response, and these are separated by equal amounts above and below f_o. The distance between these two peaks increases with k, and very approximately we find that

$$(f_2 - f_1)/f_o \approx \sqrt{k^2 - 1/Q^2} \tag{54}$$
$$\approx \sqrt{k^2 - k_c{}^2}, \tag{55}$$

when f_1 and f_2 are the frequencies for maximum response, i.e. $(f_2 - f_1)$ is **the band width between peaks**. The amplitude of these two peaks is substantially the same as the maximum possible gain $g_m R/2$, where $R = R_1 = R_2$.

Frequently, an approximate formula for band-width is used :

$$(f_2 - f_1)/f_o \approx k \tag{56}$$

While k largely determines the band-width, the depth of the valley at f_o, and hence the uniformity of the response in the pass band of frequencies, is determined by the relation of Q to k. For a constant value of k (above critical coupling) the dip becomes more pronounced as Q is increased, while the frequency separation between peaks becomes greater ; conversely as Q is decreased the dip becomes less pronounced and the frequency separation between peaks becomes less. The ratio of the response at f_o to that at the two peaks is found to be $2.b/(1 + b^2)$, where $b = k/k_c$.

A value for k in the order of 1.5 times critical, i.e. $kQ = 1.5$, is often used for i-f amplifiers requiring band pass characteristics. However, the exact value chosen for kQ depends of course, on the bandwidth requirements.

Further points on the resonance curve can be obtained from the result that the frequency band width between the points on either flank of the resonance curve, at which the response is equal to the minimum in the " valley " between the two peaks, is $\sqrt{2}$ times the peak separation. It can be shown also that, in general, the gain at any frequency f is given by

$$\frac{A_o}{A} = \sqrt{\left[1 - \frac{Q^2 Y^2}{1 + k^2 Q^2}\right]^2 + \left[\frac{2QY}{1 + k^2 Q^2}\right]^2} \tag{57}$$

where $Y = f/f_o - f_o/f$.

This expression may well be solved graphically according to a procedure developed by Beatty (Ref. C7) ; this procedure will be described below.

(v) Coupled circuits of unequal Q

In the general case where Q' and Q_2 are unequal, the two peaks of maximum response do not appear immediately k exceeds the critical value k_c. The value of k at which these two peaks just appear has been defined as the **transitional coupling factor** by Aiken (Ref. C2). The value of this coupling factor k_t is

$$k_t = \sqrt{\tfrac{1}{2}(1/Q'^2 + 1/Q_2{}^2)}. \tag{58}$$

The band width between peaks is found to be

$$(f_2 - f_1)/f_o = \sqrt{k^2 - k_t{}^2} \tag{59}$$

This useful result is discussed in an editorial by G. W. O. Howe (Ref. C5). Further it has been shown that here, as in the symmetrical case, the band width between the points on the flanks, level with the minimum response in the " valley " between the peaks, is $\sqrt{2}$ times the peak separation.

In this case, too, the selectivity curve remains symmetrical as k increases (above k_t). The amplitude of the peaks decreases, however, as k increases, and also as the ratio of R_1/R_2 (or R_2/R_1) increases. Aiken (Ref. C2) gives selectivity curves for the three cases (i) $R_1 = 10R_2$, (ii) $R_1 = 50R_2$ and (iii) $R_1 = 200R_2$ with $L_1 = L_2$ and $C_1 = C_2$. Some idea of the magnitude of this decrease in peak amplitude may be obtained from the following figures taken from Aiken's curves :

$R_1/R_2 = 10$		$R_1/R_2 = 50$		$R_1/R_2 = 200$	
k/k_c	A/A_{opt}	k/k_c	A/A_{opt}	k/k_c	A/A_{opt}
3	0.67	7	0.33	15	0.17
6	0.62	15	0.28	20	0.15

In this table A_{opt} is the optimum value of the gain, i.e. the gain at resonant frequency with critical coupling $(= g_m\sqrt{R_1 R_2}/2)$.

The gain at f_o when $k > k_t$ is given by $[2b/(1 + b^2)]A_{opt}$, where $b = k/k_c$, as in the case of equal Q's ; as the gain at the peaks is less than A_{opt} however, the response curve is flatter in the present case.

See also Sect. 11 Summary of Formulae.

SECTION 7 : SELECTIVITY—GRAPHICAL METHODS

(i) Single tuned circuit   (ii) Two identical coupled tuned circuits.

(i) Single tuned circuit

A single tuned circuit in the plate load of a valve has the well-known frequency response shown in Fig. 9.8. At the resonant frequency, where the reactance is zero there occurs the maximum value of the response, and the gain falls away on both sides. This curve may be computed from eqn. (24), namely

$$A_o/A = \sqrt{1 + Q^2 Y^2},$$

and $\tan \phi = \pm QY$,   from eqn. (25)

where $Y = (f/f_o - f_o/f)$.

These formulae, however, lend themselves to a simple graphical treatment as indicated in Fig. 9.9.

The ratio of the gain A_o at resonance to the gain A at any other frequency may be plotted as a vector quantity, OP in Fig. 9.9, having both magnitude and phase. At resonance, when the frequency is f_o, it becomes OP$_o$ in Fig. 9.9, where OP$_o$ is of unit length since A_o/A is then equal to unity. At any other frequency f, the ratio A_o/A is then given by OP where the point P is fixed by the relation

length $P_oP = Q(f/f_o - f_o/f)$.

The phase angle is the angle P$_o$OP.

Near resonance, when f is nearly equal to f_o, this may be approximated by

length $P_oP \approx 2Q\Delta f/f_o$ where $\Delta f = f - f_o$

so that Δf measures the amount of detuning. Thus, near resonance, the length P$_o$P is nearly proportional to the amount of detuning.

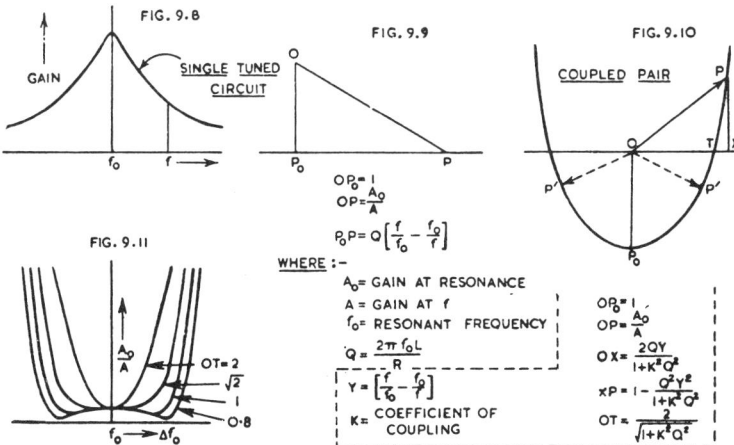

FIG. 9.8

GAIN

SINGLE TUNED CIRCUIT

FIG. 9.9

FIG. 9.10

COUPLED PAIR

FIG. 9.11

$$OP_o = 1$$
$$OP = \frac{A_o}{A}$$
$$P_oP = Q\left[\frac{f}{f_o} - \frac{f_o}{f}\right]$$

WHERE :—

A_o = GAIN AT RESONANCE
A = GAIN AT f
f_o = RESONANT FREQUENCY
$Q = \dfrac{2\pi f_o L}{R}$
$Y = \left[\dfrac{f}{f_o} - \dfrac{f_o}{f}\right]$
K = COEFFICIENT OF COUPLING

$OP_o = 1$
$OP = \dfrac{A_o}{A}$
$OX = \dfrac{2QY}{1+K^2Q^2}$
$XP = 1 - \dfrac{Q^2Y^2}{1+K^2Q^2}$
$OT = \dfrac{2}{\sqrt{1+K^2Q^2}}$

(ii) Two identical coupled tuned circuits

It is found that when two identical tuned circuits are coupled, either by some common reactance in the circuit or by mutual inductance, the locus of the point P is the parabola

$$x^2 = (y + 1)4/(1 + k^2Q^2).$$

This parabola is shown in Fig. 9.10 where OP$_o$ again represents A_o. To use this graph to determine the gain, we first compute the quantity $OX = 2QY/(1 + k^2Q^2)$, then from X draw a line perpendicular to OX to cut the curve at P. The line OP represents the gain (A_o/A) while P$_o$OP is the phase angle.

The form of the parabola depends upon the magnitude of kQ. It is found that when OT $< \sqrt{2}$, corresponding to $kQ > 1$, there are two frequencies of maximum gain as shown by the two vectors marked OP$'$ in Fig. 9.10. When the attenuation is plotted against Δf_o, as in Fig. 9.11, it becomes clear that a much flatter top may be obtained by using coupled pairs of circuits than by using single tuned circuits. Fig. 9.11 serves also to show the variation in band width with variations of OT (i.e. changes of kQ). It will be seen that the shape of the skirt of the curve is practically independent of the value of kQ.

See also Sect. 11 Summary of Formulae.

SECTION 8 : COUPLING OF CIRCUITS

(i) Mutual inductive coupling   (ii) Miscellaneous methods of coupling   (iii) Complex coupling.

(i) Mutual inductive coupling

As already emphasized, when the mutual inductive coupling between two tuned circuits is increased above a critical value, k_t, two peaks appear in the response curve, symmetrically situated with regard to the resonant frequency f_o. No other types of coupling possess this useful property. Where optimum gain and selectivity are required it can be shown that these will be obtained with a coupling about 80 per cent of the critical value—(Ref. C45). Greater selectivity can be achieved with less coupling than this value while increased gain will result from tighter coupling.

When the highest possible selectivity without serious loss of gain is desired from a pair of tuned coupled circuits, a practical compromise is to reduce coupling to 0.5 k_c at which value the gain is 0.8 times the optimum. The selectivity then approaches that which would be obtained by separating the two circuits with a valve (assuming this be done without altering Q and Q_2). For other relationships between gain and selectivity, refer to Reed (Ref. C1) or to Aiken (Ref. C2).

(ii) Miscellaneous methods of coupling

There are other types of coupling which may be used between tuned circuits as alternatives to mutual inductance. Four such circuits are shown in Figs. 9.12, 9.13, 9.14 and 9.15. **High impedance or " top " coupling** is used in the circuits shown in Figures 9.12 and 9.13 and low impedance, or " bottom " coupling, is used in the circuits of Figures 9.14 and 9.15

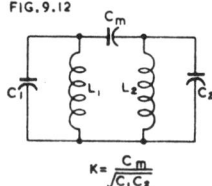

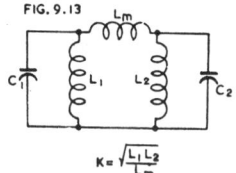

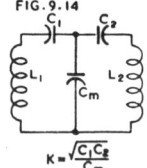

Fig. 9.12. High impedance capacitive coupling.

Fig. 9.13. High impedance inductive coupling.

Fig. 9.14. Low impedance capacitive coupling.

A fifth type is **link coupling** shown in Fig. 9.16A, in which a relatively small coupling inductance L_1' is coupled to L_1 and similarly L_2' to L_2 and L_1' is connected directly in series with L_2'. The behaviour of this circuit is the same as that to be described for Fig. 9.15.

" The coupling between two circuits, from a general point of view, is the relation between the possible rate of transfer of energy and the stored energy of the circuits " (Ref. C5, Sept. 1932).

From this definition it follows that for **low impedance coupling**

$$k = \frac{X_m}{\sqrt{(X_1 + X_m)(X_2 + X_m)}} \approx \frac{X_m}{\sqrt{X_1 X_2}} \text{ when } X_1 \text{ and } X_2 \gg X_m$$

and that for **high impedance coupling**

$$k = \sqrt{\frac{X_1 X_2}{(X_1 + X_m)(X_2 + X_m)}} \approx \frac{\sqrt{X_1 X_2}}{X_m} \text{ when } X_1 \text{ and } X_2 \ll X_m$$

where X_m is the coupling reactance and X_1 and X_2 are the effective reactances of either the coils or the condensers* with which the circuits are tuned.

---

*X_1 and X_2 must be of the same " kind " (i.e. either inductive or capacitive) as X_m.

The effective reactances X_1 and X_2 in the high impedance case are calculated by regarding the actual tuning reactances ($L_1\omega_o$ and $L_2\omega_o$ or $1/C_1\omega_o$ and $1/C_2\omega_o$) as being in parallel with the coupling reactance X_m ; while in the low impedance case, X_1 and X_2 are calculated by taking X_m to be in series with the actual tuning reactances.

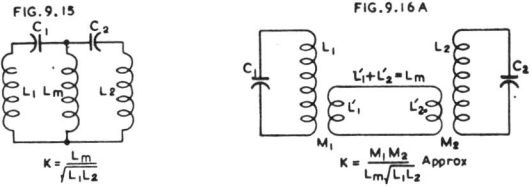

FIG. 9.15

$K = \dfrac{L_m}{\sqrt{L_1 L_2}}$

Fig. 9.15. Low impedance inductive coupling.

FIG. 9.16A

$L_1' + L_2' = L_m$

$K = \dfrac{M_1 M_2}{L_m \sqrt{L_1 L_2}}$ Approx

Fig. 9.16A. Link coupling.

Mutual inductive coupling belongs to the low impedance coupling group ; here $X_m = M\omega_o$, $X_1 = L_1\omega_o$, $X_2 = L_2\omega_o$; so that $k = M/\sqrt{L_1 L_2}$, in agreement with our previous definition in this particular case.

For a general analysis of the calculation of coupling coefficients see Chapter 26 Sect. 4(vii).

Application of the formulae given above to the circuits shown in Figures 9.12, 9.13, 9.14, 9.15 and 9.16A give the following results for the coefficient of coupling k :

Circuit	k (exact)	k (approximate)
Fig. 9.12	$\dfrac{C_m}{\sqrt{(C_1 + C_m)(C_2 + C_m)}}$	$\dfrac{C_m}{\sqrt{C_1 C_2}}$, when $C_m \ll (C_1, C_2)$
Fig. 9.13	$\sqrt{\dfrac{L_1 L_2}{(L_1 + L_m)(L_2 + L_m)}}$	$\dfrac{\sqrt{L_1 L_2}}{L_m}$, when $L_m \gg (L_1, L_2)$
Fig. 9.14	$\sqrt{\dfrac{C_1 C_2}{(C_1 + C_m)(C_2 + C_m)}}$	$\dfrac{\sqrt{C_1 C_2}}{C_m}$, when $C_m \gg (C_1, C_2)$
Fig. 9.15 ·	$\dfrac{L_m}{\sqrt{(L_1 + L_m)(L_2 + L_m)}}$	$\dfrac{L_m}{\sqrt{L_1 L_2}}$, when $L_m \ll (L_1, L_2)$
Fig. 9.16A	$\dfrac{M_1 M_2}{L_m \sqrt{\left(L_1 - \dfrac{M_1^2}{L_m}\right)\left(L_2 - \dfrac{M_2^2}{L_m}\right)}}$ or $\dfrac{k_1 k_2}{\sqrt{(1 - k_1^2)(1 - k_2^2)}}$	$\dfrac{M_1 M_2}{L_m \sqrt{L_1 L_2}}$, when individual couplings are small. $k_1 k_2$, when individual couplings are small.

where $L_m = L_1' + L_2'$,

$$k_1 = \frac{M_1}{\sqrt{L_1 L_m}}, \text{ and } k_2 = \frac{M_2}{\sqrt{L_2 L_m}}.$$

When the coupling increases above k_t for the tuned circuits shown in Figs. 9.12, 9.13, 9.14, 9.15 and 9.16A, the two peaks in the response curve of the secondary move at unequal rates from the original resonant frequency (determined by $L_1 C_1 \omega_o^2 = L_2 C_2 \omega_o^2 = 1$). For the first four of these examples in the special case $L_1 = L_2$, $C_1 = C_2$, and for the fifth generally, one peak remains approximately stationary (at ω_o), while the other peak moves to one side : the shift of the semi-stationary peak depends upon the series resistances of the two circuits and decreases with them, being zero in the ideal case $r_1 = r_2 = 0$: the second peak is lower in frequency in Figs. 9.12 and 9.15, but higher in Figs. 9.13, 9.14 and 9.16A. The selectivity and bandwidth (between peaks) may be calculated from the formulae (57) and (55) respectively—already quoted for transformer coupling—provided the appropriate value of k is used. It is theoretically possible, although seldom convenient, to combine two types of coupling in equal amounts to give symmetrical separation of the two peaks. Normally, when this is required simple mutual inductive coupling is used.

It may be shown that for all types of coupling the centre frequency is determined by the effective reactances obtained by taking the coupling reactance into account. Thus, for example, for Fig. 9.15, $(L_2 + L_m)C_2\omega_c{}^2 = 1$, where ω_c corresponds to the frequency of the minimum between the peaks ; while for Fig. 9.16A, $(L_2 - M_2{}^2/L_m)C_2\omega_c{}^2 = 1$.

(iii) Complex coupling

With any single type of coupling the gain and the band width vary with the frequency. Clearly, then, a single type of coupling cannot give satisfactory performance in the tuned radio frequency stages of a receiver where the frequency range is two or three to one. From the formulae already given it can be seen that for transformer coupling both the gain and the band width are approximately proportional to the frequency (assuming that Q_1 and Q_2 do not vary greatly) ; for other types of simple coupling k depends upon the square of the frequency, and hence band width and gain are functions of frequency.

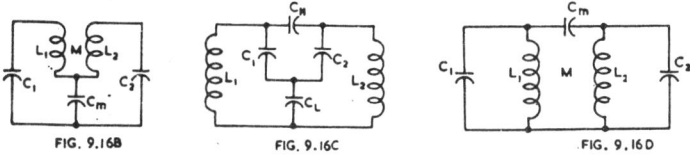

FIG. 9.16B   FIG. 9.16C   FIG. 9.16D

Fig. 9.16B. *Constant bandwidth using inductive and capacitive coupling.*
Fig. 9.16C. *Constant bandwidth using capacitive coupling.*
Fig. 9.16D. *Complex coupling with mutual inductive and top capacitive coupling.*

In practice, a reasonably constant band width over the tuning range can be obtained by a suitable combination of the types of coupling already described. Two common arrangements are shown in Figs. 9.16B and 9.16C.

For the circuit shown in Fig. 9.16B the coupling reactance $X_m = \omega M + 1/\omega C_m$, while for that shown in the Fig. 9.16C it is

$$\frac{1}{\omega C_L} + \frac{C_H}{\omega C_1 C_2}.$$

The corresponding coupling factors are respectively

$$k \approx \frac{M + 1/\omega^2 C_m}{\sqrt{L_1 L_2}} \qquad (k < 0.05),$$

and $k \approx \dfrac{C_H}{\sqrt{C_1 C_2}} + \dfrac{\sqrt{C_1 C_2}}{C_L} \qquad (k < 0.05).$

Aiken (Ref. C2) gives a practical design procedure for obtaining the values of the components which give the best average results over the whole tuning range.

When k has been determined, the band width and selectivity may be calculated from eqns. (55) and (57) given earlier. Also, for circuits with the same values of Q, a suitable value of k is given by $kQ = 0.5$—as in the case of transformer coupling—when the greatest possible selectivity without notable loss of gain is desired.

In Fig. 9.16D there is a combination of mutual inductive coupling and top capacitive coupling, as commonly used in aerial and r-f coils, and, effectively, in i-f transformers. The analysis is given in Chapter 26 Sect. 4 (vii).

See also Sect. 11 Summary of Formulae.

SECTION 9 : RESPONSE OF IDENTICAL AMPLIFIER STAGES IN CASCADE

When two or more amplifier stages having identical circuits and values are connected in cascade, the overall gain is the product of the gains, and the resultant selectivity is the product of the selectivities of all the stages. For n stages, therefore, the **total gain** is $A_o{}^n$; the **selectivity for single tuned circuits is**

$$(A/A_o)^n = (1 + Q^2 Y^2)^{-n/2},$$

while **for coupled pairs**

$$\left(\frac{A}{A_o}\right)^n = \left[\left(1 - \frac{Q^2 Y^2}{1 + k^2 Q^2}\right)^2 + \left(\frac{2QY}{1 + k^2 Q^2}\right)^2\right]^{-n/2}$$

When $k^2 Q^2$ is very small, the selectivity of n coupled pairs is almost the same as that of $2n$ single tuned circuits. Thus there is a limit to the improvement of selectivity obtained by reduction of the coupling of coupled pairs of tuned circuits. When it is possible to increase Q, there is a corresponding improvement in selectivity. The tendency with several stages of single tuned circuits is to produce a very sharp peak at the centre frequency, which may seriously attenuate the higher audio frequencies of a modulated signal. Conditions are much better with coupled pairs, because two peaks with small separation appear as Q is increased, if the coupling is not too close. Difficulties occur when Q is increased so much that a deep " valley " or trough occurs between the peaks. The practical limit is usually a ratio of 1 : 1.5 overall gain between the response at the bottom of the valley and that at the two peaks. It is then good practice to add another stage employing a single tuned circuit which substantially removes the " valley " of the preceding circuits. The procedure by which the best results may be obtained is described by Ho-Shou Loh (Ref. C13). In this manner a nearly flat response may be obtained over a range of frequencies 10 Kc/s to 20 Kc/s wide, with very sharp discrimination against frequencies 20 Kc/s or more away from the centre frequency, 450 to 460 Kc/s.

See also Sect. 11 Summary of Formulae.

SECTION 10 : UNIVERSAL SELECTIVITY CURVES

In Figs. 9.17 and 9.18 are shown universal selectivity curves taken from Maynard's data (Ref. D2). These curves apply to a pair of coupled tuned circuits, and are not restricted to circuits of equal Q. Fig. 9.17 gives the gain at various frequencies off the centre frequency in terms of the gain at the centre frequency, for various coefficients of coupling ; the ordinate scale D is proportional to $Q \varDelta f_o / f_o$. The Q shown in Figs. 9.17 and 9.18 is Q_2 for all expressions containing a and b.

The phase difference between the currents in the primary and secondary circuits can readily be obtained from Fig. 9.18. There we have plotted an angle θ as a function of D for various coefficients of coupling, and the phase shift is $\theta \pm 90°$, the positive sign being taken when the coupling is negative, and the negative sign when k is positive.

The parameter b, which is a measure of the coupling, becomes simply k/k_c when the tuned circuits have equal values of Q ; also, in this case, the variable D becomes simply $2Q \varDelta f_o / f_o$. For convenience, scales have been added to give A_o / A in terms of $Q \varDelta f_o$ for a number of values of f_o for two identical coupled circuits.

In deriving the curves of Figs. 9.17 and 9.18 it was assumed that Q and k do not vary appreciably over the range considered, thus giving symmetrical selectivity curves ;

and that Q is reasonably high (> 25 say). Very low values of Q require a different curve for each value, but the effect is only to alter slightly the skirts of the curves without altering appreciably the main portions.

As an illustration of the use of Fig. 9.17 consider the example $f_0 = 1000$ Kc/s, $Q_1 = Q_2 = 200$, $k/k_c = 2 = b$: we see that the peak occurs at $Q\Delta f_0 \approx 800$, i.e. $\Delta f_0 = 4$ Kc/s, and that the gain in the valley compared with the gain at the peak $= 0.8$; these results agree well with those calculated from the formulae already given, namely

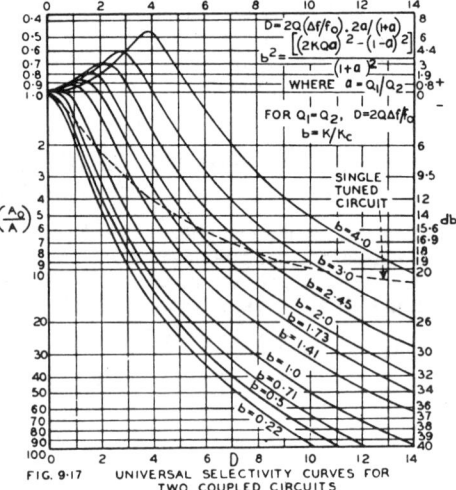

FIG. 9·17   UNIVERSAL SELECTIVITY CURVES FOR TWO COUPLED CIRCUITS

Band width is : $f_0\sqrt{k^2 - k_c^2} \approx 8.6$ Kc/s.
Ratio of gain in valley to that at peak is :
$$\frac{2k/k_c}{1 + (k/k_c)^2} = \frac{2b}{1 + b^2} = \frac{4}{5}.$$

Phase change at $\Delta f_0 = (+) 4$ Kc/s : (from Fig. 9.18) phase change $\approx 20 \pm 90°$.

In conclusion, we give the selectivity curves of Fig. 9.19 to illustrate some of our remarks in preceding sections. These curves have been derived from those of Fig. 9.17, but here the abscissa is A_{opt}/A, where A_{opt} is the gain at the centre frequency where $k = k_c$.

These curves are for the case of two identical coupled circuits ; they show how the maximum gain, band width and depth of the valley between peaks vary with the ratio k/k_c.

In all expressions using a, the Q mentioned is the secondary Q, namely Q_2.

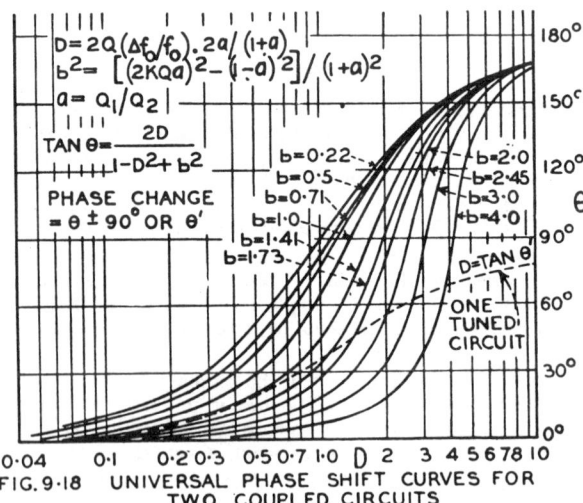

FIG. 9·18   UNIVERSAL PHASE SHIFT CURVES FOR TWO COUPLED CIRCUITS

SECTION 11 : SUMMARY OF FORMULAE FOR TUNED CIRCUITS

(1) NOMENCLATURE

L = inductance (in henrys unless otherwise stated)

C = capacitance (in farads unless otherwise stated)

f_o = resonant frequency (in cycles per second unless otherwise stated)

f_n = natural resonant frequency (in cycles per second unless otherwise stated)

f = frequency (in cycles/sec. unless otherwise stated)

Δf_o = $f - f_o$ (in cycles/sec.)

Y = $(f/f_o - f_o/f)$

π = 3.1416 approximately

k = coefficient of coupling

ω = $2\pi f$, $\omega_o = 2\pi f_o$, $\omega_n = 2\pi f_n$ radians per second

r = series resistance (in ohms)

r' = resistance of a series resonant circuit at resonance (in ohms)

R = shunt resistance (in ohms)

R_e = effective shunt resistance of a parallel resonant circuit at resonance when $R = \infty$ (in ohms)

R_D = resonant impedance (in ohms)

e = voltage across the circuit at a time t

E = initial voltage of charged condenser

ϵ = 2.718 (ϵ is the base of Naperian Logarithms)

t = time (in seconds)

α = damping factor

δ = logarithmic decrement

λ = wavelength in metres

i = current at frequency f (in amperes)

i_o = current at resonant frequency f_o

A = gain at frequency f

A_o = gain at frequency f_o

Q = magnification factor.

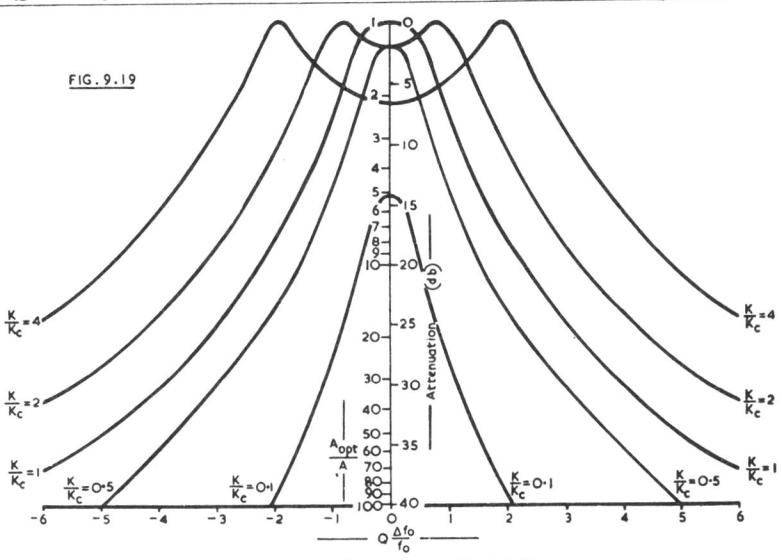

FIG. 9.19

Selectivity Curves for Two Identical Coupled Circuits,
Showing the Variation in Maximum Gain,
Band Width and Depth of Valley with K/K_c

(2) NATURAL RESONANT FREQUENCY (f_n)

Exact formula

$$f_n = \frac{1}{2\pi}\sqrt{\frac{1}{LC} - \frac{r^2}{4L^2}} \text{ cycles per second} \tag{1}$$

Approximate formula for use when r is small compared with $2\sqrt{L/C}$:—

$$f_n \approx f_o = \frac{1}{2\pi\sqrt{LC}} \text{c/s.} \tag{2}$$

For numerical use this may be put in the form

$$f_n \approx f_o = \frac{159\,200}{\sqrt{LC}} \text{ c/s, where } L \text{ is in microhenrys} \tag{3}$$
$$\text{and} \quad C \text{ is in microfarads ;}$$

or $\quad f_n \approx f_o = \dfrac{159\,200}{\sqrt{LC}}$ Kc/s, where L is in microhenrys $\qquad$ (4)
$$\text{and} \quad C \text{ is in micromicrofarads.}$$

(3) WAVELENGTH (λ)

Wavelength (in metres) $= 1884\sqrt{LC}$, $\qquad\qquad$ (5)
where L is in microhenrys and C is in microfarads.

Wavelength $\times$ frequency $= 2.9979 \times 10^8$ metres per second $\qquad$ (6)
$$\approx 3 \times 10^8 \text{ metres per second.}$$

Note : Equations (5) and (6) are based on the velocity of electromagnetic radiation in a vacuum as determined by Dr. Essen and others in 1951 (see Ref. E1).

$$\text{Wavelength} \approx \frac{300\,000}{\text{frequency in Kc/s}} \approx \frac{300}{\text{frequency in Mc/s}} \tag{7}$$

(4) DAMPED OSCILLATIONS

$e = E\epsilon^{-\alpha t} \cos \omega_n t$, where $\alpha = r/2L$ (damping factor) $\qquad$ (8)
$\delta = r/2f_n L$ (logarithmic decrement) $\qquad\qquad$ (9)

(5) SERIES RESONANCE

L, C, and ω_o :
For resonance
$LC\omega_o^2 = 1$
$\qquad \omega_o L = 1/\omega_o C$ ohms

$\qquad \omega_o = 1/\sqrt{LC}$ radians/second

$\qquad \omega_o L = \sqrt{L/C}$ ohms
$\qquad \omega_o C = \sqrt{C/L}$ mhos.

L, C and f_o :
For resonance $\qquad\qquad$ (10)
$2\pi f_o L = 1/2\pi f_o C$ ohms

$$LC = \frac{1}{39.48 f_o^2} \quad \text{(henrys} \times \text{farads)}$$

$$\text{or } LC = \frac{2.533 \times 10^{10}}{f_o^2} \ \mu H \times \mu F$$

where L is in microhenrys
and $\quad C$ is in microfarads.

$$Q = \omega_o L/r' = 1/r'C\omega_o, \tag{11}$$
where r' is the effective series resistance at resonance,
i.e. $\quad r' = r + LC/R$
where r is the series resistance
and R the parallel resistance.

$$\text{Therefore } Q = \frac{\omega_o L_1}{r + LC/R} = \frac{1}{r/\omega_o L + 1/RC\omega_o} \tag{12}$$

$$= \frac{1}{\omega_o Cr + 1/RC\omega_o} = \frac{1}{r\sqrt{C/L} + \sqrt{L/C}/R} \tag{13}$$

When $L/CR \ll r$: $Q \approx \omega_o L/r \approx 1/\omega_o Cr$. $\qquad$ (14)

Magnification factor (Q) : Ratio of the voltage across either reactance to the voltage across the circuit.

(6) PARALLEL RESONANCE

At resonance, $\omega_o = \dfrac{1}{\sqrt{LC}} \cdot \dfrac{1}{\sqrt{1 + r^2/\omega_o^2 L^2}} \approx \dfrac{1}{\sqrt{LC}}.$ $\qquad$ (15)

$$Q = R_D/\omega_o L = R_D \omega_o C, \tag{16}$$

where R_D, the resonant impedance, is the effective shunt resistance of the circuit at resonance,
and $1/R_D = 1/R + Cr/L$.

Therefore $Q = \dfrac{1}{r\sqrt{C/L} + \sqrt{L/C}/R}$ etc., as in (12), (13) above.

When R is infinite (i.e. no shunt resistance),

$$R_D = R_e = L/Cr = Q/\omega_o C = \omega_o LQ = Q^2 r. \tag{17}$$

Magnification factor

Ratio of total circulating current to input current $= Q$ $\qquad(18)$

(7) SELECTIVITY
(a) Series Resonant Circuit

$$i_o/i = A_o/A = \sqrt{1 + Q^2(f/f_o - f_o/f)^2} = \sqrt{1 + Q^2 Y^2}. \tag{19}$$

$$\tan\phi = QY = Q\Delta f_o/f_o \cdot \frac{2 + \Delta f_o/f_o}{1 + \Delta f_o/f_o}; \tag{20}$$

i lags behind i_o when $f > f_o$, and leads i_o when $f < f_o$.

When $\Delta f_o/f_o$ is small,

$$i_o/i \approx \sqrt{1 + 4Q^2(\Delta f_o/f_o)^2}, \tag{21}$$

and $\tan\phi \approx 2Q\Delta f_o/f_o$. $\qquad(22)$

When i_o/i is large, $i_o/i \approx QY$. $\qquad(23)$

(b) Parallel Resonant Circuit

$$A_o/A = R_D/Z = i/i_o = \sqrt{1 + Q^2 Y^2} \tag{24}$$

$$\tan\phi = -QY; \tag{25}$$

where i and i_o are the total currents;
i leads i_o when $f > f_o$ and lags behind i_o when $f < f_o$.

When $\Delta f_o/f_o \ll 1$, $i/i_o \approx \sqrt{1 + 4Q^2 \Delta f_o^2/f_o^2}$, $\qquad(26)$

and $\tan\phi \approx -2Q\Delta f_o/f_o$. $\qquad(27)$

When $i/i_o \gg 1$, $i/i_o \approx QY$. $\qquad(28)$

(8) R-F TRANSFORMER, UNTUNED PRIMARY, TUNED SECONDARY

When the primary impedance can be neglected,

$$\text{Gain } A_o = \frac{g_m}{\dfrac{M}{r_p L_2} + \dfrac{1}{\omega_o M Q_2}} = \frac{g_m}{\dfrac{1}{kR_D}\sqrt{\dfrac{L_2}{L_1}} + \dfrac{k}{r_p}\sqrt{\dfrac{L_1}{L_2}}} \tag{29}$$

The gain may be expressed in a number of alternative forms, for example,

$$A_o = \frac{\mu\omega_o M Q_2}{r_p + \dfrac{(\omega_o M)^2 Q_2}{\omega_o L_2}} = \frac{\mu\omega_o M Q_2}{r_p + \dfrac{\omega_o^2 M^2}{r}} \tag{30}$$

where r is the series resistance of the secondary.

In the special case where $\omega_o^2 M^2/r \ll r_p$, we have

$$A_o \approx g_m \omega_o M Q_2. \tag{31}$$

Selectivity

To determine selectivity, the effect of the resistance reflected into the secondary should be taken into account when its value $\omega_o^2 M^2/r_p > 5\%$ of r. The effective series resistance r' is $r + \omega_o^2 M^2/r_p$; so that the effective value of Q_2, Q_2' say, is

$$Q_2' = \omega_o L_2/r' = \frac{1}{\dfrac{1}{Q_2} + \dfrac{k^2}{r_p} \cdot \omega_o L_1}. \tag{32}$$

The selectivity is then obtained from

$$A_o/A = \sqrt{1 + Q_2'^2 Y^2}. \tag{33}$$

(9) R.F. TRANSFORMER, TUNED PRIMARY, TUNED SECONDARY

Gain

The gain at resonance (A_o) is

$$A_o = \frac{g_m \sqrt{R'R_2}}{k\sqrt{Q'Q_2} + \dfrac{1}{k\sqrt{Q'Q_2}}}, \tag{34}$$

where $1/R' = 1/r_p + 1/R_1$ and $1/Q' = L_1\omega_o/R' = \omega_o L_1/r_p + 1/Q_1$.

Maximum gain occurs when $k = k_c = 1/\sqrt{Q'Q_2}$, the **critical coupling coefficient,** and is given by

$$A_o \text{ (max.)} = g_m\sqrt{R'R_2/2}. \tag{35}$$

Also, when $k = k_c/2$, the value of A_o is approximately 0.8 of A_o (max.).

Identical circuits $(L_1 = L_2, \ C_1 = C_2, \ R' = R_2 = R, \ Q' = Q_2 = Q)$

Critical Coupling : $k_c = 1/Q.$ (36)

Maximum gain (at resonance) $= g_m R/2.$ (37)

= half gain of a single tuned circuit.

Band width between peaks $(f_1$ and $f_2)$,

$$(f_2 - f_1)/f_o = \sqrt{k^2 - k_c^2}. \tag{38}$$

Selectivity

$$A_o/A = \sqrt{\left[1 - \frac{Q^2 Y^2}{1 + k^2 Q^2}\right]^2 + \left[\frac{2QY}{1 + k^2 Q^2}\right]^2} \tag{39}$$

Gain at peaks, f_2 and f_1, is very closely equal to the optimum value A_o (max.).

Circuits of unequal Q

Transitional coupling factor, k_t, is

$$k_t = \sqrt{\tfrac{1}{2}(1/Q'^2 + 1/Q_2^2)}. \tag{40}$$

Band width between peaks is

$$(f_2 - f_1)/f_o = \sqrt{k^2 - k_t^2}. \tag{41}$$

Selectivity

$$\frac{A_o}{A} \approx \sqrt{\left[1 - \frac{Q'Q_2 Y^2}{1 + k^2 Q'Q_2}\right]^2 + \left[\frac{2Y\sqrt{Q'Q_2}}{1 + k^2 Q'Q_2}\right]^2} \tag{42}$$

When $k^2 Q^2 \gg 1$ and $\Delta f_o/f_o \ll 1$,

$$A_o/A \approx \sqrt{1 - 8\Delta f_o^2/f_o^2 k^2}. \tag{43}$$

(10) COUPLING COEFFICIENTS

High impedance coupling*

$$k \approx \frac{C_m}{\sqrt{C_1 C_2}} \text{ for capacitive coupling (Fig. 9.12)} \tag{44}$$

$$= \frac{\sqrt{L_1 L_2}}{L_m} \text{ for inductive coupling (Fig. 9.13)} \tag{45}$$

where C_m = coupling capacitance and L_m = coupling inductance.

Low impedance coupling*

$$k \approx \frac{\sqrt{C_1 C_2}}{C_m} \text{ for capacitive coupling (Fig. 9.14)} \tag{46}$$

$$\approx \frac{L_m}{\sqrt{L_1 L_2}} \text{ for inductive coupling (Fig. 9.15)} \tag{47}$$

where C_m = coupling capacitance and L_m = coupling inductance.

Link coupling*

$$k \approx \frac{M_1 M_2}{L_m\sqrt{L_1 L_2}} \text{ (Fig. 9.16A),} \tag{48}$$

where M_1 = mutual inductance between L_1 and L_m,
and M_2 = mutual inductance between L_2 and L_m.

---

* For exact values see table in Section 8.

SECTION 12 : REFERENCES

(A) BOOKS DEALING WITH RADIO TUNED CIRCUIT THEORY

A1. McIlwain, K., and J. G. Brainerd, "High Frequency Alternating Currents" (John Wiley and Sons Inc., New York, Chapman and Hall Ltd., London, 1931).
A2. Harnwell, G. P. "Principles of Electricity and Magnetism" (McGraw-Hill, New York and London, 1938).
A3. Terman, F. E. "Radio Engineering" (McGraw-Hill Book Company Inc., New York and London, 3rd ed. 1947).
A4. Everitt, W. L. "Communication Engineering" (McGraw-Hill Book Company Inc., New York and London 2nd ed. 1937).
A5. Nilson, A. R., and J. L. Hornung, "Practical Radio Communication" (McGraw-Hill Book Company Inc., New York and London, 2nd ed. 1943).
A6. Henney, K. "Principles of Radio" (John Wiley and Sons Inc., New York, 4th edit. 1942).
A7. Glasgow, R. S. "Principles of Radio Engineering" (McGraw-Hill Book Co., New York and London, 1936).
A8. Henney, K. "Radio Engineering Handbook" (McGraw-Hill Book Company Inc., New York and London, 4th ed. 1950).
A9. "Admiralty Handbook of Wireless Telegraphy" (His Majesty's Stationery Office, London).
A10. Sturley, K. R. "Radio Receiver Design" (Chapman and Hall Ltd., London, 1943) Part 1.
A11. Terman, F. E. "Radio Engineers' Handbook" (McGraw-Hill Book Company, New York and London, 1943).
A12. Reich, H. J. "Theory and Applications of Electron Tubes" (McGraw-Hill Book Company, New York and London, 1944).
A13. Sandeman, E. K. "Radio Engineering" (Chapman and Hall Ltd., London, 1947).
A14. Welsby, V. G. "The Theory and Design of Inductance Coils" (Macdonald and Co., London, 1950).
A15. "Reference Data for Radio Engineers" (Federal Telephone and Radio Corp. 3rd ed. 1949) pp. 114-129.

(B) REFERENCES TO THE THEORY OF RADIO FREQUENCY SINGLE TUNED CIRCUITS AND COUPLINGS

B1. Reed, M. "The design of high frequency transformers" E.W. and W.E. 8.94 (July 1931) 349.
B2. Wheeler, H. A., and W. A. MacDonald, "The theory and operation of tuned radio frequency coupling systems" Proc. I.R.E. 19.5 (May 1931) 738 and discussion by L. A. Hazeltine 19.5 (May 1931) 804.
B3. Wheeler, H. A. "Image suppression in superheterodyne receivers" Proc. I.R.E. 23.6 (June 1935) 569.
B4. Bayly, B. de F. "Selectivity a simplified mathematical treatment" Proc. I.R.E. 19.5 (May 1931) 873.
B5. Purington, E. S. "Single- and coupled-circuit systems" Proc. I.R.E. 18.6 (June 1930) 983.
B6. Smith, V. G. "A mathematical study of radio frequency amplification" Proc. I.R.E. 15.6 (June 1927) 525.
B7. Callendar, M. V. "Problems in selective reception" Proc. I.R.E. 20.9 (Sept. 1932) 1427.
B8. Takamura Satoru, "Radio receiver characteristics related to the sideband coefficient of the resonance circuit" Proc. I.R.E. 20.11 (Nov. 1932) 1774.
B9. Sandeman, E. K. "Generalised characteristics of linear networks" W.E. 8.159 (Dec. 1936) 637.
B10. Everitt, W. L. "Output networks for radio frequency power amplifiers" Proc. I.R.E. 19.5 (May 1931) 725.
B11. Walker, L. E. Q. "A note on the design of series and parallel resonant circuits" Marconi Review No. 63 (Dec. 1936) 7.
B12. Hughes, D. H. "The design of band-spread tuned circuits for broadcast receivers" Jour. I.E.E. 93, Part III (March 1946) 87.
B13. Najork, J. "Simple L and C Calculations" Q.S.T. 31.9 (Sept. 1947) 31.
B14. Design Data (15) "Link coupling," W.W. 53.8 (Aug. 1947) 291.
B15. Hudson, A. C. "Efficiency of inductive coupling" Elect. 20.12 (Dec. 1947) 138.
B16. "Cathode Ray," "Transformers, obvious and otherwise," W.W. 53.10 (Oct. 1947) 388.
B17. Rehfisch, T. J. "Resonance : an experimental demonstration of series and parallel resonant circuits for radio training classes," Electronic Eng. 17.197 (July 1944) 76.
B18. McComb, C. T., and A. P. Green, "Single inductor coupling networks" Elect. 17.9 (Sept. 1944) 132.
B19. Sabaroff, S. "Impulse excitation of a cascade of series tuned circuits" Proc. I.R.E. 32.12 (Dec. 1944) 758.
B20. Amos, S. W. "Wavetraps—modern applications of a well-tried device" W.W. 51.2 (Feb. 1945) 43.
B21. Haworth, J. E. "The tapped inductor circuit" Electronic Eng. 18.223 (Sept. 1946) 284.
B22. Reed, M. B. "Frequency response of parallel resonant circuit" Elect. 14.8 (Aug. 1941) 43.
B23. Brunetti, C., and E. Weiss "Theory and application of resistance tuning" Proc. I.R.E. 29.6 (June 1941) 333.
B24. Schade, O. H. "Radio-frequency-operated high-voltage supplies for cathode-ray tubes" (gives summary of theory of coupled circuits) Proc. I.R.E. 31.4 (April 1943) 158.
B25. Blow, T. C. "Mutual inductance of concentric coils" Elect. 19.11 (Nov. 1946) 138.
B26. Tucker, D. G. "The transient response of a tuned circuit" Electronic Eng. 18.226 (Dec. 1946) 379.
B27. Builder, G. "The graphical solution of simple parallel-tuned circuits" Radio Eng. 16.8 (Aug. 1944) 20.

(C) REFERENCES TO THE THEORY OF TUNED COUPLED CIRCUITS

C1. Reed, M. "The design of high frequency transformers" E.W. and W.E. 8.94 (July 1931) 349.
C2. Aiken, C. B. "Two-mesh tuned coupled circuit filters" Proc. I.R.E. 25.2 (Feb. 1937) 230 and errata 26.6 (June 1937) 672.
C3. Scheer, F. H. "Notes on intermediate-frequency transformer design" Proc. I.R.E. 23.12 (Dec. 1935) 1483.
C4. Bligh, N. R. "The design of the band pass filter" W.E. and E.W. 9.101 (Feb. 1932) 61.
C5. Howe, G. W. O., Editorials W.E. 14.165 (June 1937) 289 ; 14.166 (July 1937) 348 ; 9.108 (Sept. 1932) 486.

C6. Buffery, G. H. " Resistance in band pass filters " W.E. 9.108 (Sept. 1932) 504.
C7. Beatty, R. T. " Two element band pass filters " W.E. 9.109 (Oct. 1932) 546.
C8. Oatley, C. W. " The theory of band-pass filters for radio receivers " W.E. 9.110 (Nov. 1932) 608.
C9. Wheeler, H. A., and J. K. Johnson " High fidelity receivers with expanding selectors " Proc. I.R.E. 23.6 (June 1935) 595.
C10. Cocking, W. T. " Variable selectivity and the i.f. amplifier " W.E. 13.150 (March 1936) 119 ; 13.151 (April 1936) 179 ; 13.152 (May 1936) 237.
C11. Reed, M. " The analysis and design of a chain of resonant circuits " W.E. 9.104 (May 1932) 259 ; 9.105 (June 1932) 320.
C12. Baranovsky, C., and A. Jenkins, " A graphical design of an intermediate frequency transformer with variable selectivity " Proc. I.R.E. 25.3 (March 1937) 340.
C13. Loh, Ho-Shou, " On single and coupled circuits having constant response band characteristics " Proc. I.R.E. 26.4 (April 1938) 469 and errata 12 (Dec. 1938) 1430.
C14. Christopher, A. J. " Transformer coupling circuits for high frequency amplifiers " B.S.T.J. 11 (Oct. 1932) 608.
C15. " Tuned impedance of i.f. transformers " (nomogram) Comm. 18.2 (Feb. 1938) 12.
C16. Roberts, W. Van B. " Variable link coupling " Q.S.T. 21.5 (May 1937) 27.
C17. " A generalised theory of coupled circuits " A.R.T.S. and P. No. 73 (May 1939).
C18. Landon, V. D. " The band-pass—low-pass analogy " Proc. I.R.E. 24.12 (Dec. 1936) 1582.
C19. Nachod, C. P. " Nomograms for the design of band-pass r-f circuits " Radio Eng. 16.12 (Dec. 1936) 13.
C20. Erickson, C. V. " A graphical presentation of band-pass characteristics " Radio Eng. 17.3 (March 1937) 12 ; 17.4 (April 1937) 13. Also see 17.6 (June 1937) 19.
C21. Dudley, H. " A simplified theory of filter selectivity " Comm. 17.10 (Oct. 1937) 12.
C22. Everitt, W. L. " Coupling networks " Comm. 18.9 (Sept. 1938) 12 ; 18.10 (Oct. 1948) 12.
C23. Clifford, F. G. " The design of tuned transformers " Electronic Eng. Part 1, 19.229 (March 1947) 83 ; Part 2, 19.230 (April 1947) 117.
C24. Dishal, M. " Exact design and analysis of double and triple-tuned band-pass amplifiers," Proc. I.R.E. 35.6 (June 1947) 606.
C25. Tellegen, B. D. H. " Coupled circuits " Philips Research Reports 2.1 (Feb. 1947) 1.
C26. Rudd, J. B. " Theory and design of radio-frequency transformers " A.W.A. Tec. Rev. 6.4 (1944) 193 with bibliography.
C27. " Theory and design of radio frequency transformers," A.R.T.S. and P. Bulletin (1) No. 134 (1st May, 1944) ; (2) No. 135 (12th June, 1944).
C28. Editorial " Coupled circuits " W.E. 21.245 (Feb. 1944) 53.
C29. Sturley, K. R. " Expression for the voltages across the primary and secondary of two tuned circuits coupled by mutual inductance," Appendix 1 to article " The phase discriminator " W.E. 21.245 (Feb. 1944) 72.
C30. Sandeman, E. K. " Coupling circuits as band-pass filters " W.E. Part (i) 18.216 (Sept. 1941) 361 ; Part (ii) 18.217 (Oct. 1941) 406 ; Part (iii) 18.218 (Nov. 1941) 450 ; Part (iv) 18.219 (Dec. 1941) 492.
C31. Korman, N. I. " Coupled resonant circuits for transmitters," Proc. I.R.E. 31.1 (Jan. 1943) 28.
C32. Sturley, K. R. " Coupled circuit filters "—generalised selectivity, phase shift and trough and peak transfer impedance curves—W.E. (1) 20.240 (Sept. 1943) 426 ; (2) 20.241 (Oct. 1943) 473.
C33. Beatty, R. T. " Two-element band pass filters " W.E. 9.109 (Oct. 1932) 546.
C34. Editorial " Effect of stray capacitance on coupling coefficient " W.E. 21.251 (Aug. 1944) 357.
C35. Editorial " Coupling coefficient of tuned circuits " W.E. 22.256 (Jan. 1945) 1.
C36. Espy, D. " Know your coupled circuits " Q.S.T. 29.10 (Oct. 1945) 76.
C37. Williams, H. P. " H.F. Band-Pass Filters " Electronic Eng. (1) General Properties 18.215 (Jan. 1946) 24 ; (2) Similar Circuits 18.216 (Feb. 1946) 51 ; (3) Dissimilar circuits and miscellaneous properties 18.217 (Mar. 1946) 89 ; (4) Practical example in band-pass design 18.219 (May 1946) 158.
C38. Duerdoth, W. T. " Equivalent capacitances of transformer windings " W.E. 23.273 (June 1946) 161.
C39. Tucker, D. G. " Transient response of tuned-circuit cascades " W.E. 23.276 (Sept. 1946) 250.
C40. Richards, P. I. " Universal optimum-response curves for arbitrarily coupled resonators " Proc. I.R.E. 34.9 (Sept. 1946) 624.
C41. Spangenberg, K. R. " The universal characteristics of triple-resonant-circuit band-pass filters " Proc. I.R.E. 34.9 (Sept. 1946) 629.
C42. Sherman, J. B. " Some aspects of coupled and resonant circuits " Proc. I.R.E. 30.11 (Nov. 1942) 505.
C43. Varrall, J. E. " Variable selectivity IF amplifiers " W.W. 48.9 (Sept. 1942) 202.
C44. Ferris, W. R. " Some notes on coupled circuits " R.C.A. Rev. 5.2 (Oct. 1940) 226.
C45. Adams, J. J. " Undercoupling in tuned coupled circuits to realize optimum gain and selectivity " Proc. I.R.E. 29.5 (May 1941) 277.
C46. Editorial " Natural and resonant frequencies of coupled circuits " W.E. 18.213 (June 1941) 221.
C47. Mather, N. W. " An analysis of triple-tuned coupled circuits " Proc. I.R.E. 38.7 (July 1950) 813.

(D) REFERENCES TO UNIVERSAL SELECTIVITY CURVES

D1. Terman, F. E. (book) " Radio Engineering."
D2. Maynard, J. E. " Universal performance curves for tuned transformers " Elect. (Feb. 1937) 15.
D3. Budenbom, H. T. " Some methods for making resonant circuit response and impedance calculations " Radio Eng. 15.8 (Aug. 1935) 7.

(E) MISCELLANEOUS REFERENCES

E1. Editorial " The velocity of light " W.E. 27.331 (April 1951) 99.

CHAPTER 10

CALCULATION OF INDUCTANCE

based on original chapter in the previous edition by L. G. Dobbie, M.E.

Revised by G. Builder, B.Sc., Ph.D., F. Inst. P. and E. Watkinson, A.S.T.C., A.M.I.E. (Aust.), S.M.I.R.E. (Aust.).

SECTION 1 : SINGLE LAYER COILS OR SOLENOIDS

(i) Current-sheet inductance   (ii) Solenoid wound with spaced round wires   (iii) Approximate formulae   (iv) Design of single layer solenoids   (v) Magnitude of the difference between L_s and L_0 (vi) Curves for determination of the " current sheet " inductance   (vii) Effect of concentric, non-magnetic screen.

(i) Current-sheet inductance

For the ideal case of a very long solenoid wound with extremely thin tape having turns separated by infinitely thin insulation we have the well-known formula for the low frequency inductance which is called the " current-sheet inductance " L_s :

$$L_s = 4\pi N^2 A'/l \text{ electromagnetic units} \tag{1}$$

where N is the total number of turns, A' is the cross-sectional area in cm.^{2} and l is the length in cm. This result for L_s may be expressed also as

$$L_s = 0.100\ 28a^2 N^2/l \text{ microhenrys} \tag{2}$$

when l is measured in inches, and a is the radius of coil, also in inches.

For solenoids of moderate length—more precisely, for those for which a/l is not small compared with unity—there is an end-correction, and we find that

$$L_s = (0.100\ 28a^2 N^2/l)K, \tag{3}$$

where K is a function of a/l, which approaches unity as a/l tends to zero.   Values of K, computed by Nagaoka (Bulletin, Bureau of Standards, 8, p. 224, 1912), are shown by the curves in Fig. 10.1.

The concept of current sheet inductance is introduced because such (theoretical) inductances can be calculated with high precision, and the formulae used in practical cases can be derived from these results by making approximate allowances for the deviations from the ideal case.   In many cases these deviations are less than 1%.

(ii) Solenoid wound with spaced, round wires

The low frequency inductance, L_o, of an actual solenoid wound with round wire is obtained from the equivalent cylindrical current-sheet inductance L_s by introducing two correction terms, thus :

$$L_o = L_s - 0.0319aN(A + B) \text{ microhenrys,} \qquad (4)$$

where a is the radius of the coil, in inches, measured to the centre of the wire, A is a constant taking into account the difference in self-inductance of a turn of wire from that of a turn of the current sheet, and B depends on the difference in mutual inductance of the turns of the coil from that of the turns of the current sheet. The quantity A is a function of the ratio of wire diameter to pitch, and B depends only on the total number of turns, N. Values of A and B are shown by means of curves in Fig. 10.2.

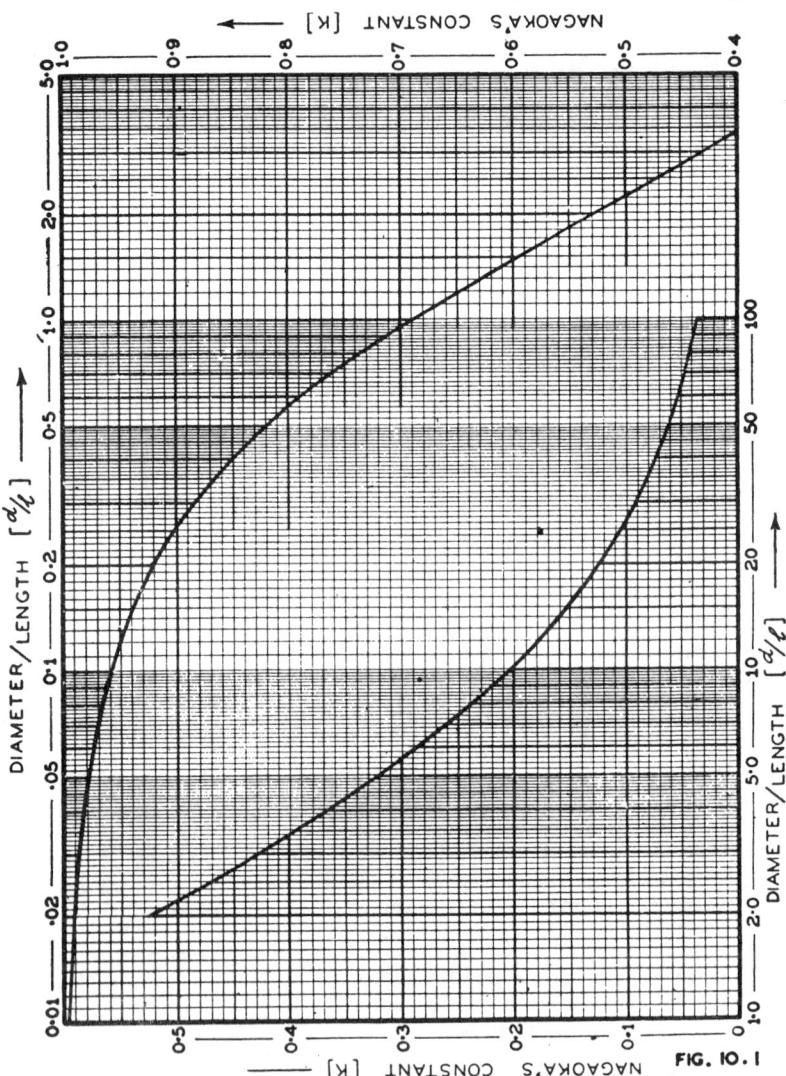

Fig. 10.1. Nagaoka's Constant (K) for a wide range of d/l.

This formula for L_o, together with the values of A and B, have been taken from the above quoted Bulletin of the Bureau of Standards.   The value thus obtained for L_o is given as correct to one part in a thousand.

The equation given above for L_o can be expressed in the alternative forms :

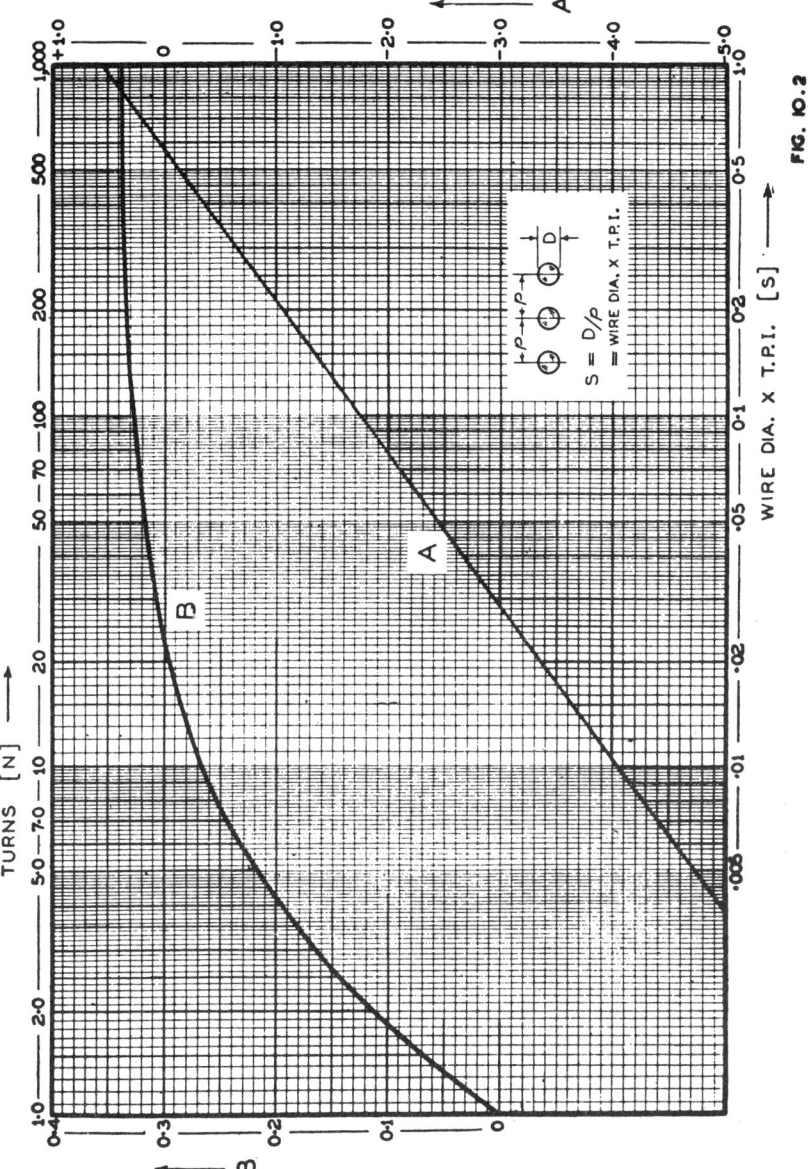

Fig. 10.2. *Constants A and B as used in the formula for the correction of " current sheet " formulae for application to round wire with spaced turns (eqn. 4).*

$$L_o = L_s\left[1 - \frac{l(A+B)}{\pi aNK}\right] = L_s\left[1 - \frac{P}{\pi aK}(A+B)\right] \qquad (5)$$

where P, the pitch of the winding in inches, is equal to l/N.

As an example of the use of these formulae and curves, let us consider the following case : a coil of 400 turns of round wire of bare diameter $D = 0.05$ inch, is wound with a pitch of 10 turns per inch, on a form of such a diameter that the mean radius to the centre of the wire is 10 inches. Then,

$$a = 10, l = NP = 40, N = 400, P = 0.1, D/P = 0.5.$$

The value of K corresponding to $2a/l = 0.5$ is 0.8181 from Fig. 10.1. Therefore,

$$L_s = 0.100\,28\,(400)^2 \times (100/40) \times 0.8181$$
$$= 32\,815 \text{ microhenrys.}$$

From Fig. 10.2 with $D/P = 0.5$, $N = 400$,

$$A = -0.136$$
$$B = 0.335$$

Therefore $A + B = 0.199$.

The correction is, therefore, $0.0319 \times 10 \times 400\,(0.199)$
$$= 25.4 \text{ microhenrys.}$$

The total inductance $L_o = 32\,815 - 25 = 32\,790$ microhenrys, and the error which would be introduced by calculating as a current sheet inductance is less than 0.08%.

(iii) Approximate formulae

In many instances it is not necessary to calculate L_o to the accuracy given by the expression above. There are available a number of approximate formulae suitable for slide-rule computation. For example :

$$A \approx 2.3 \log_{10} 1.7\,S \qquad (6)$$

where $S = D/P$, i.e. the ratio of wire diameter to pitch, with an accuracy of 1% for all values of S.

$$B \approx 0.336\,[1 - (2.5/N + 3.8/N^2)] \qquad (7)$$

accurate to 1% when N exceeds four turns.

(a) Wheeler's Formula

$$L_s = \frac{a^2N^2}{9a + 10\,l} \qquad (8)$$

This expression is accurate to 1% for all values of $2a/l$ less than 3. Wheeler's formula gives a result about 4% low when $2a/l = 5$.

(b) Approximate expression based on a value of K given by **Esnault-Pelterie**

$$L_s \approx 0.1008\,\frac{a^2N^2}{l + 0.92a} \qquad (9)$$

This expression is accurate to 0.1% for all values of $2a/l$ between 0.2 and 1.5.

(c) For solenoids whose length is small compared with the diameter

$$L_s = \frac{a^2N^2}{[9 - (a/5l)]\,a + 10l} \qquad (10)$$

which is accurate to 2% for all values of $2a/l$ up to 20. The error approaches $+ 2\%$ when $d/l = 2.0$ to 3.5 and at $d/l = 20$. The error approaches $- 2\%$ in the range $d/l = 10$ to 12.

Let us apply Wheeler's and Esnault-Pelterie's formulae to the solenoid already considered, namely :

$$a = 10 \text{ in.}, l = 40 \text{ in.}, N = 400.$$

Wheeler's formula gives $L_s = \dfrac{10^2 \times 400^2}{90 + 400} = 32\,650$ microhenrys.

Esnault-Pelterie's formula yields

$$L_s = 0.1008\,\frac{10^2 \times 400^2}{40 + 9.2} = 32\,790 \text{ microhenrys.}$$

These results agree, within the stated limits, with the value 32 815 obtained previously.

(iv) Design of single layer solenoids

The difference between L_s and L_o is usually less than 1%.   Design formulae are based on L_s, as this is easier to compute, and then the correction is estimated.

Two of the many possible formulae are given in the Bureau of Standards Circular 74 (Ref. 2).

(A) **Where it is required to design a coil which shall have a certain induct-ance with a given length of wire, the dimensions of the winding and the kind of wire being unrestricted within broad limits.**   This design problem includes a consideration of the question as to what shape of coil will give the required inductance with the minimum resistance.

We have the relations :

$L_s = 0.100\,28 a^2 N^2 K/l$ from (3),

$l = NP$, $\lambda = 2\pi aN$,

where λ is the length of wire.   Eliminating N gives :

$$L_s = 0.004\,50 \, \frac{\lambda^{3/2}}{\sqrt{P}} \left(K \sqrt{\frac{2a}{l}} \right) \text{ microhenrys} \tag{11}$$

This gives the inductance in terms of the length of wire, the pitch P, and the shape $2a/l$, as K is a function of $2a/l$.

From the graph of the quantity $K\sqrt{2a/l}$ against $2a/l$, its maximum value is found to occur at $2a/l = 2.46$.   **Thus, for a given length of wire, wound with a given pitch, that coil has the greatest inductance, which has a shape $d/l = 2.46$ approximately** ; or to obtain a coil of a required inductance, with a minimum re-sistance, this relation should be realized.   Further, the inductance diminishes rapidly for coils longer than this optimum value, but decreases only slowly for shorter coils.

The optimum value of d/l can also be obtained roughly from the approximate expression for K :

$K = 1/(1 + 0.45 \, d/l)$.

Therefore $(d/l)_{opt} = (1/0.45) = 2.2$ approximately   (12)

(B) **Given the diameter of the coil, the pitch and inductance, to determine the length of the coil.**

A suitable form of the equation is obtained by substituting for N its value l/P in the formula :

$L_s = 0.100\,28 \, a^2 N^2 \, K/l$ from (2) ;

we find :

$$L_s = 0.200\,56 \, (a^3/P^2)(Kl/2a) \tag{13}$$
$$\approx 0.200 \, (a^3/P^2) f \tag{14}$$

where $f = K(l/2a)$.   The quantity f, clearly a function of $2a/l$, is shown by the curve in Fig. 10.3.

For example, supposing we require a coil of 200 microhenrys to be wound on a former of 1 inch diameter; let us determine the length of the coil for a pitch of 0.02 inch.

From the equation,

$$f = \frac{200 \times (0.02)^2}{0.200 \times (0.5)^3} = 3.2.$$

Then, from Fig. 10.3, the value of $2a/l$ corresponding to $f = 3.2$ is 0.28.   Hence the length of the coil is

$l = 2a/0.28 = 1/0.28 = 3.6$ inches.

The number of turns required would be :

$N = l/P = 3.6/0.02 = 180$ turns.

The length of wire needed would be :

$\lambda = 2\pi aN = \pi 180 = 565.5$ inches.

When the calculated length is too long, the turns per inch should be increased. Let us consider the effect of using 100 T.P.I. in the above example :

$$f = \frac{200 \times (0.01)^2}{0.200 \times (0.5)^3} = 0.8.$$

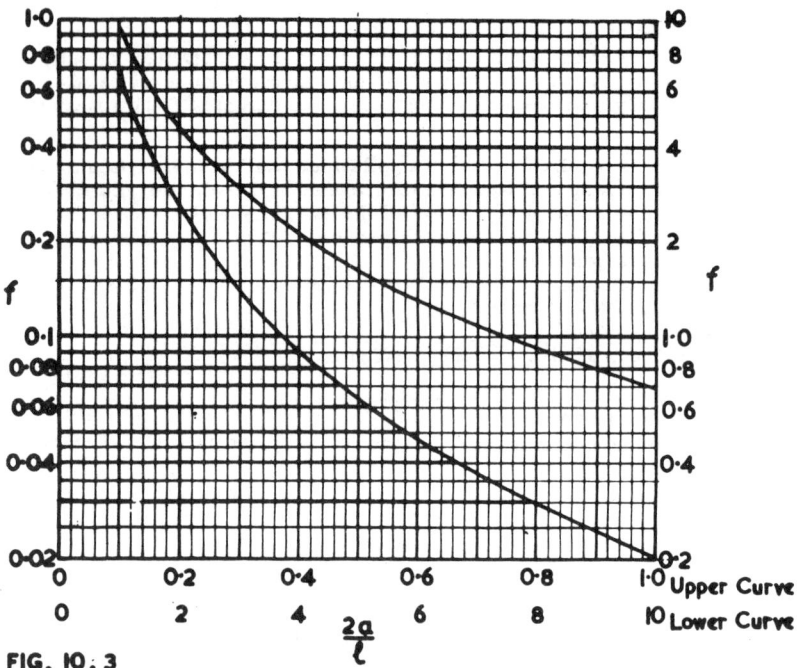

FIG. 10 : 3

Fig. 10.3. Variation of f with (2 a/l). Curves derived from U.S. Dept. of Commerce Circular C74 " Radio instruments and measurements ".

From Fig. 10.3, we see that the value of $2a/l$ is 0.9, and hence
$l = 1/0.9 = 1.1$ inches.

Formula for determining the coil length when the coil diameter and winding pitch are known for coils such that $d < 3l$.

As an alternative to the use of Fig. 10.3, we derive a simple result for solenoids such that $2a/l < 3$. Here we have

$$K \approx 1/[1 + 0.45\,(2a/l)] \qquad (15)$$

Hence, $f = K \cdot \dfrac{l}{2a} = \dfrac{l}{2a} \cdot \dfrac{1}{1 + 0.45\,(2a/l)} \qquad (16)$

Therefore, $0.9(2a/l) = -1 + \sqrt{1 + 1.8/f} \qquad (17)$

Thus, in the first case above, where $f = 3.2$,
$0.9(2a/l) = -1 + \sqrt{1 + 1.8/3.2} = 0.25,$
and $2a/l = 0.25/0.9 = 0.28$;

for the second case, where $f = 0.8$,
$0.9(2a/l) = \sqrt{1 + (1.8/0.8)} - 1 = 0.803,$
and $2a/l = 0.803/0.9 = 0.89.$

These values agree well with those above.

The limitation $2a/l < 3$ is equivalent to $f \not< 0.14$.

Eqn. (14) for L_s is suitable too when the former diameter and coil length have been decided. For example, suppose we require a coil of inductance 500 microhenrys on a former of diameter 2 inches, and having a coil length of 2 inches.

Then, from equation (14)

$$L_s = 0.200(a^3/P^2) \cdot f, \qquad (18)$$

we have

$$P = \sqrt{(0.200a^3 \cdot f)/L_s} \qquad (19)$$

Here $a = 1, L_s = 500, f = Kl/2a = 0.688.$

Therefore,

$$P = \sqrt{0.200 \times 0.688/500} = 0.0167 \text{ inch,}$$

equivalent to 60 T.P.I.

Simple procedure for computing the length and total turns when the coil diameter and pitch of winding are known for coils such that $d < 3l$.

Substituting in equation (17),

$$0.9(2a/l) = -1 + \sqrt{1 + 1.8/f},$$

where $f = 5L_s P^2/a^3 = 5L_s/n^2 a^3$, n being T.P.I.

Putting $F = 1.8/f$ and $y = 0.9(2a/l)$,

we obtain

$$y = -1 + \sqrt{1 + F}, \tag{20}$$

where $F = 0.36 n^2 a^3/L_s$,

and $l = 1.8a/y$.

As an example, let us consider a coil of 2 inches diameter, wound with 33 T.P.I. of required inductance 380 microhenrys.

(a) $F = 0.36 n^2 a^3/L_s = 0.36 \times 33^2 \times 1^3/380 = 1.031$

(b) $y = -1 + \sqrt{1 + F} = -1 + \sqrt{2.031} = 0.426$

(c) $l = 1.8a/y = 1.8/0.426 = 4.23$ inches

(d) $N = nl = 33 \times 4.23 = 139\frac{1}{2}$ turns.

Another method applicable to the same type of problem is due to Hayman.

From Wheeler's formula (equation 8) :

$$L_s \approx \frac{a^2 N^2}{9a + 10l},$$

the length l is eliminated by the substitution $l = N/n$, n being the turns per inch ; and then the resulting expression is solved for N, yielding :

$$N = \frac{5L_s}{na^2}\left[1 + \sqrt{1 + \frac{0.36 a^3 n^2}{L_s}}\right] \tag{21}$$

For convenience in computation, the quantity $x = 5/na^2$ is introduced, so that finally

$$N = xL_s[1 + \sqrt{1 + (9/ax^2 L_s)}] \tag{22}$$

As an example, we take the same problem as above, namely, $a = 2$ inches, $n = 33$, $L_s = 380$ microhenrys. The procedure is as follows :

(a) $x = 5/na^2 = 5/(33 \times 1) = 0.151$,

(b) $x^2 = 0.0227$,

(c) $9/ax^2 L_s = 9/(1 \times 0.0227 \times 380 = 1.042$,

(d) $N = 380 \times 0.151 \times (\sqrt{2.042} + 1) = 139$ turns,

(e) $l = N/n = 139/33 = 4.2$ inches.

(C) Given the diameter and length of coil, value of capacitance and frequency of resonance, to determine the number of turns.

See Chapter 38 Sect. 9(v)B.

(v) Magnitude of the difference between L_s and L_o

In Sect. 1(iv) it was stated that in many practical cases the difference between L_s and L_o is less than 1 per cent. Here, we examine briefly the necessary conditions for this difference to be small.

A formula given in Sect. 1(ii) for L_o, namely :

$$L_o = L_s[1 - P(A + B)/\pi aK] \text{ from equation (5)}$$

is in a useful form ; clearly the ratio of $P(A + B)/\pi aK$ to unity determines the order of the difference between L_s and L_o. From the curves of Fig. 10.2, $A = +0.5$ for $P/D = 0.95$ and decreases steadily to -0.7 for $P/D = 0.25$, while $B = 0.114$ at $N = 2$ and 0.336 at $N = 1000$, and remains approximately constant for higher values of N. Hence, for coils having P/D between 0.95 and 0.25—as most have— the value of $(A + B)$ is less than 0.83 ; i.e.

$$P(A + B)/\pi aK < (P/a) \times (0.26/K).$$

Thus, when $(2a/l) = d/l < \frac{1}{2}$, $K > 0.82$, so that $P(A + B)/\pi aK < 0.32\, P/a$.

The correction is less than 1% when $P/a < 0.03$.

Similarly, for $2a/l < 1$, $K > 0.69$, and the correction is less than 1% when $P/a < 0.026$.   Corresponding results for $2a/l < 5$ and $2a/l < 10$ are $P/a < 0.012$ and $P/a < 0.0078$.

In the table opposite are given some examples of the minimum turns per inch, n, for a number of coil diameters required to make the difference between L_s and L_o less than 1% :

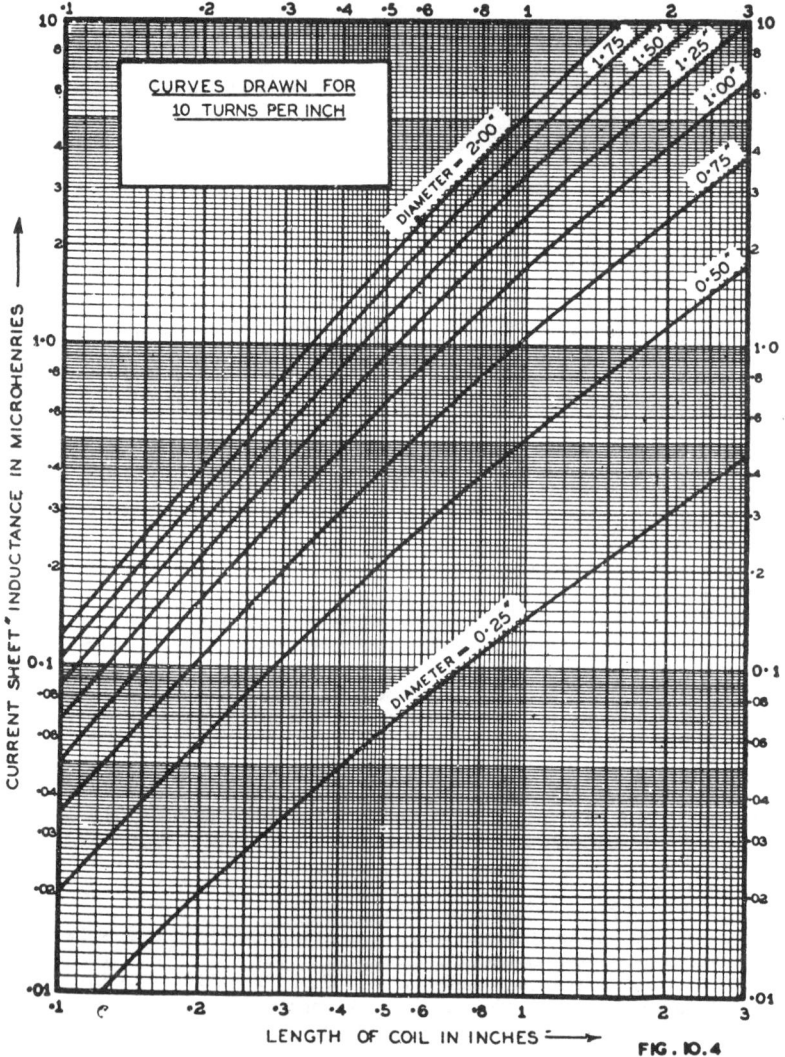

CURVES DRAWN FOR
10 TURNS PER INCH

CURRENT SHEET "INDUCTANCE IN MICROHENRIES ——→

DIAMETER = 2·00"

1·75" 1·50" 1·25" 1·00"

0·75"

0·50"

DIAMETER = 0·25"

LENGTH OF COIL IN INCHES ——→   FIG. 10.4

Fig. 10.4.   Curves for the determination of Current Sheet Inductance (L_s) for small solenoids, plotted against length of coil (l). For other pitches refer to Fig. 10.5.

Minimum turns per inch, n for

$\left(\dfrac{\text{Coil diameter}}{\text{length}}\right)$ not exceeding	$d = 5$ inches	$d = 2$ inches	$d = 1$ inch	$d = \frac{1}{2}$ inch
0.5	13	33	66	132
1	16	40	80	160
5	32	80	160	320
10	52	130	260	520

In many practical cases the ratio of pitch to wire diameter lies between 0.8 and 0.95 in which range $A = 0.4 + 0.1$; and N, the number of turns, exceeds 10, so that $B = 0.3 \pm 0.035$. In such cases $A + B = 0.7 \pm 0.135$. Hence, from eqn. (4) :

$$L_o = L_s - 0.0319aN(A + B),$$

we have, under such conditions,

$$L_o \approx L_s - 0.0223aN, \tag{23}$$

with an error not exceeding 20% of $0.0223aN$; and thus in many cases not exceeding 0.2% of L_o.

(vi) Curves for determination of the " current sheet " inductance
(A) Method of using the curves

Figure 10.4 applies to a winding pitch of 10 turns per inch only ; for any other pitch the inductance scale must be multiplied by a factor, which is easily determined from Fig. 10.5. The diameter of a coil is considered to be twice the distance from coil axis to centre of the wire.

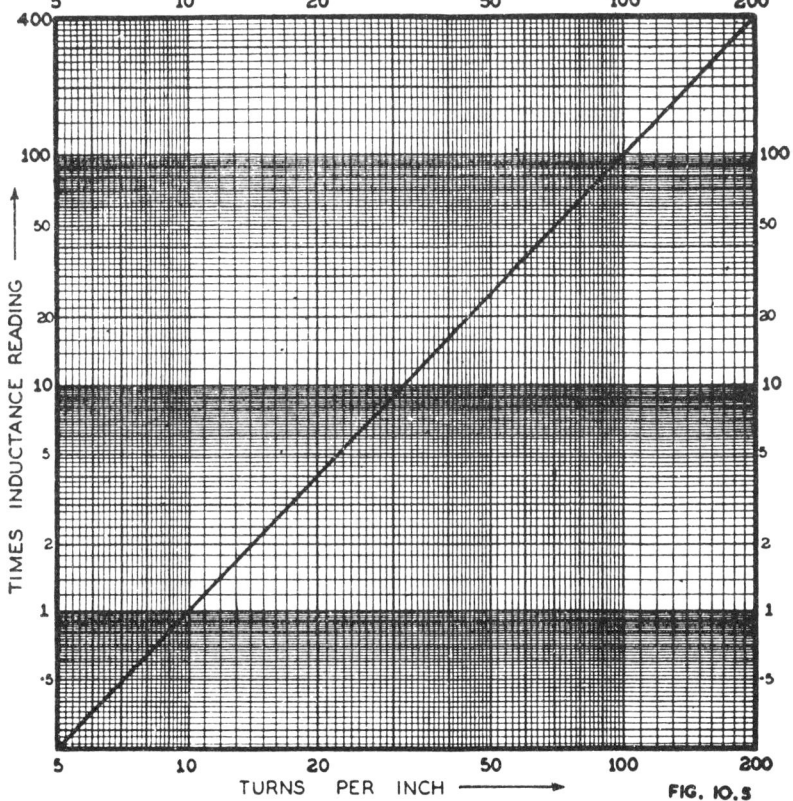

Fig. 10.5. Winding pitch correction for Fig. 10.4.

(B) To design a coil having required " current sheet " inductance

Determine a suitable diameter and length, and from Fig. 10.4 read off the " current sheet " inductance for a pitch of 10 T.P.I. The required inductance may then be obtained by varying the number of turns per inch. The correct number of turns may be found by calculating the ratio of the required inductance to that read from Fig. 10.4 and referring it to Fig. 10.5 which will give the required turns per inch.

Alternatively if the wire is to be wound with a certain pitch, a conversion factor for that pitch may first be obtained from Fig. 10.5, and the required inductance divided by that factor. The resultant figure of inductance is then applied to Fig. 10.4 and suitable values of diameter and length determined.

(C) To find the " current sheet " inductance of a coil of known dimensions

Knowing the diameter and length, determine from Fig. 10.4 the inductance for a pitch of 10 T.P.I. Then from Fig. 10.5 determine the factor for the particular pitch used, and multiply the previously determined value of inductance by this factor.

(vii) Effect of concentric, non-magnetic screen

A shield surrounding a coil acts as a short-circuited turn coupled to the coil and reflects an impedance into the coil. The value of this reflected impedance is given by the expression $M^2 \omega^2 / Z_t$, where M is the mutual inductance between the coil and the shield, and Z_t is the impedance of the shield. In practice, the resistance of the screen may be neglected, so that the effective impedance of the coil Z_c', is

$$Z_c' = r_o + j\omega L_o + M^2 \omega^2 / j\omega L_t, \tag{24}$$

where r_o and L_o are the resistance and the inductance of the coil in the absence of the shield, and L_t is the inductance of the shield. The expression for Z_c' may be put in the forms :

$$Z_c' = r_o + j\omega(L_o - M^2/L_t), \tag{25a}$$
$$= r_o + j\omega(1 - M^2/L_o L_t), \tag{25b}$$
$$= r_o + j\omega L_o', \tag{25c}$$
$$= r_o + j\omega L_o(1 - k^2) : \tag{25d}$$

where L_o' is the effective inductance of the enclosed coil, and k is the coefficient of coupling between the coil and the can. The presence of the shield thus lowers the effective inductance, and we have

$$L_o' = L_o(1 - k^2), \tag{26a}$$
$$\text{or} \quad (L_o - L_o')/L_o = k^2 \tag{26b}$$

It has not proved possible to obtain a simple, accurate formula for the apparent decreases in inductance of a shielded coil, but various estimates have been made by Hayman, Kaden, Davidson and Simmonds, Bogle and others (see bibliography). Here we review briefly some of the published work.

(A) H. Kaden (Electrishe Nachrichten Tecknik, July, 1933, p. 277)

Kaden first showed that the shape of the shield is not important. Then, in his theoretical treatment, he replaced the actual solenoid in the concentric cylindrical shield by a magnetic dipole placed at the centre of a spherical screen ; the dipole having the same magnetic moment as the solenoid and the spherical screen a radius equal to the geometrical mean of the three dimensions of the cylinder. In this way he obtained an expression for the relative decrease in inductance of the solenoid.

$$\frac{L_o - L_o'}{L_o} = \frac{2}{3}\frac{V_c}{V_t}\frac{\alpha}{K}, \tag{27}$$

where V_c is the volume of the coil, V_t is the volume of the shield, K is Nagaoka's constant, and α is a constant which depends upon the permeability and the shield and the dimensions. When the shielding is effective and the shield is non-magnetic, α is approximately 1 and then :

$$\frac{L_o - L_o'}{L_o} \approx \frac{2}{3}\frac{V_c}{V_t}\frac{1}{K} \tag{28}$$

(B) C. F. Davidson and J. C. Simmonds (Ref. 41)

Following Kaden, these authors also considered the case of a spherical screen. Here the solenoid was taken to consist of a small number of closely spaced circular turns placed at the centre of the sphere. The change in inductance was found to be

$$L_o - L_o{'} = 0.1 \ aN^2(a/b)^3 . f(a/b), \tag{29}$$

where a is the coil radius, N is the number of turns, b is the shield radius and $f(a/b)$ is a function of (a/b). This function f is approximately 0.5 for values of $(a/b) \not> 0.5$, and it increases to 0.73 at $(a/b) = 0.9$. Thus, for values of $(a/b) < 0.5$, we have :

$$L_o - L_o{'} \approx (0.05 \ aN^2)(a/b)^3, \tag{30}$$

in agreement with Kaden's result. The authors point out that their formula holds only for very short solenoids; they indicate how the change in inductance of long solenoids in spherical screens may be calculated, but it is clear that no simple formula can be thus obtained.

(C) R.C.A. Application Note No. 48 (June, 1935)

In this Application Note (Ref. 48) curves are given for k^2 in the case of a concentric cylindrical shield whose length exceeds the coil length by at least the radius of the coil. These curves (Fig. 10.6) show k^2 as a function of the ratio coil length to coil diameter $(l/2a)$ for various values of the ratio coil radius to shield radius ranging from 0.2 to 0.9. It is stated that these values of k^2 have been calculated and also verified experimentally. The shields are not closed at the ends, but see (D) below.

It is stated that these curves may also be used for cans of square cross-section by taking A as 0.6 of one side of the can.

(D) A. G. Bogle (Ref. 45)

Bogle obtains an approximate theoretical formula for the effective change in inductance, valid in a restricted range, and then, from measurements of $L_o{'}$, he derives an empirical expression which is of the same form as the theoretical formula, but which gives good results over a wide range.

For a coil inside a concentric cylindrical shield, Bogle's expression is :

$$\frac{L_o - L_o{'}}{L_o} = \frac{1}{1 + 1.55g/l} \left(\frac{a}{b}\right)^2 \tag{31}$$

where b is the shield radius, and $g = b - a$, the distance between the coil and the shield. It is assumed that the shield length l_t is not less than $l + 2g$. An accuracy of 2% over a wide practical range is claimed for this simple result.

An investigation of this expression shows that it is useful in almost all practical cases, the only restriction being the non-stringent one that the length of the coil must not be very much less than the distance between the coil and the shield, i.e. $g/l \not> 1$.

For very long coils this formula shows that

$$(L_o - L_o{'})/L_o = a^2/b^2$$

in agreement with theoretical expectations. It is in accord, too, with the R.C.A. curves for k^2 in that the proportional change in inductance depends upon the two ratios a/b and l/a, as can be seen by writing the formula in the form :

$$\frac{L_o - L_o{'}}{L_o} = \frac{1}{1 + 1.55\frac{a}{l}\left(\frac{b}{a} - 1\right)} \left(\frac{a}{b}\right)^2 .$$

Bogle showed also that when $l_t > l + 2g$ the effect on $L_o{'}$ of closing the ends of the shield is negligible. He investigated the effect of eccentricity of the coil, and ellipticity of the screening tube, and of small axial displacements in a closed screen ; and found that in practice these would not need to be taken into account.

Sowerby has published data charts (Ref. 44) calculated from Bogle's formula.

(E) W. G. Hayman (Ref. 51)

Hayman gives an empirical formula for coils in concentric cylindrical cans :

$$(L_o - L_o{'})/L_o = (a/b)^3, \tag{32}$$

provided $l_t \gg l$. For coils of length $l > \frac{1}{2}l_t$ there is an end-correction factor, and he gives :

$$(L_o - L_o')/L_o = (1 - l/2l_t)^2 (a/b)^3 \qquad (33)$$

It is true that there is fair agreement between Hayman's calculated and measured values of L_o' for a number of coils ; but, nevertheless, the useful range of eqn. (32) must be somewhat restricted.

(F) Where the coil diameter and length are equal, the curves of Fig. 11.13 can be used for a shield can with ends.

It is recommended that either the R.C.A. curves for k^2, reproduced in Fig. 10.6, or Bogle's formula be used. Comparison of the results obtained from these curves and from the formula will be found to be in good agreement except for very

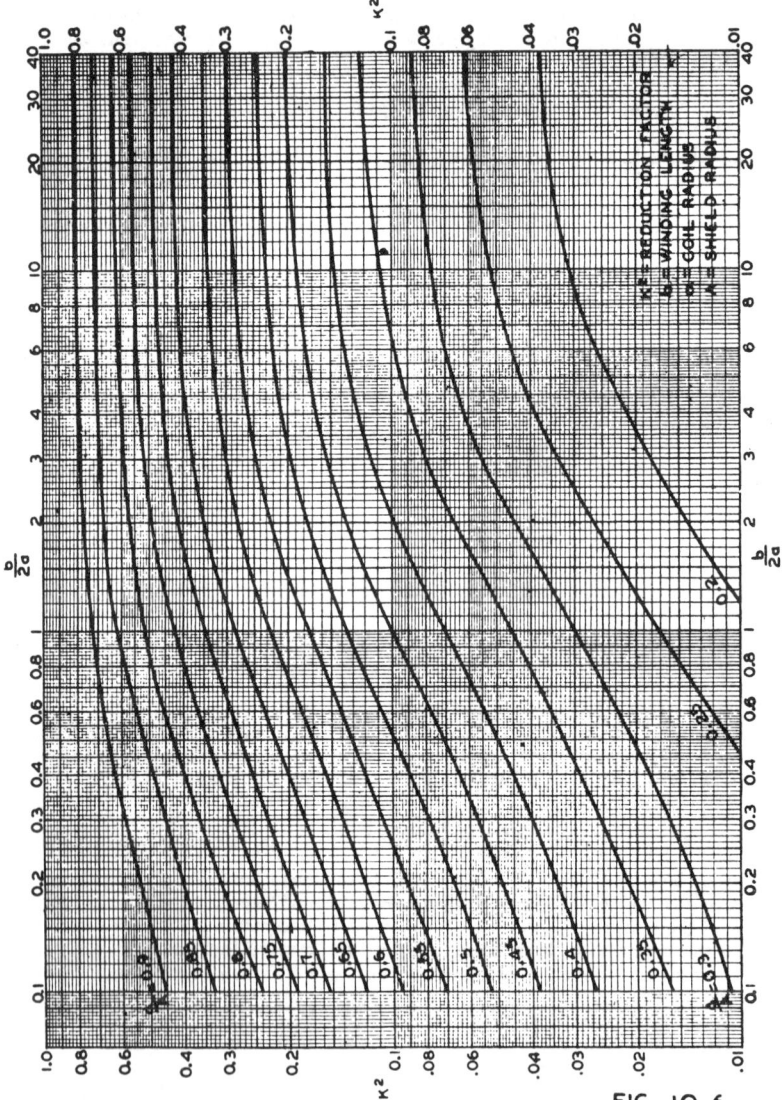

FIG. 10.6

Fig. 10.6.   Curves for determination of decrease in inductance produced by a coil shield (reprinted by kind permission of Radio Corporation of America).

small values of l/g where Bogle's formula should not be used.   In the tables below comparisons are made in a number of typical cases :

Table 1 : Comparison between estimates of the percentage reduction in inductance caused by a cylindrical screen for various coil shapes when the ratio of coil diameter to screen diameter = 0.7.

l/d	$(L_o - L_o')/L_o$		g/l
	R.C.A.	Bogle	
20	49%	48.2%	0.0107
10	48%	47.3%	0.0215
5	46%	46.0%	0.0430
1	35%	36.8%	0.215
0.2	20%	18.3%	1.075
0.1	15%	11.3%	2.15

where l = coil length, d = coil diameter, g = distance between coil and shield.

Table 2 : Comparison between estimates of the percentage reduction in inductance caused by a cylindrical screen for various ratios of coil diameter to screen diameter.

a/b	$(L_o - L_o')/L_o$					
	$l/d = 10$		$l/d = 1$		$l/d = 0.2$	
	R.C.A.	Bogle	R.C.A.	Bogle	R.C.A.	Bogle
0.9	80%	80.6%	.72%	74.5%	54%	56.5%
0.7	48%	47.3%	35%	36.8%	20%	18.3%
0.5	23%	23.2%	13.6%	14.0%	6.7%	5.7%
0.3	7.8%	7.6%	3.0%	3.2%	1.4%	0.9%

where a = coil radius, b = shield radius.

SECTION 2 : MULTILAYER SOLENOIDS

(i) Formulae for current sheet inductance   (ii) Correction for insulation thickness (iii) Approximate formulae   (iv) Design of multilayer coils   (v) Effect of a concentric screen.

(i) Formulae for current sheet inductance
(A) Long coils of a few layers
The Bureau of Standards Circular No. 74 gives a simple formula for the current sheet inductance, correct to 0.5% for long solenoids of a few layers ; it is

$$L_s = L_s' - \frac{0.0319N^2ac}{l}(0.693 + B_s), \qquad (1)$$

where N = total number of turns,
   c = radial depth of winding,
   l = length of winding
   B_s = a tabulated function of l/c
   a = radius of coil to centre of winding,
and $L_s' = (0.100\,28a^2N^2/l)K$.
This result may be expressed in the form :

$$L_s = L_s'\left[1 - \frac{c(0.693 + B_s)}{\pi aK}\right] \qquad (2)$$

The quantity $B_s = 0$ for $l/c = 1$; it rises to 0.279 for $l/c = 10$, and then increases very slowly, being 0.322 at $l/c = 30$.

(B) Short coils of rectangular cross-section
The solution of the problem for short coils is based on that for the ideal case of a circular coil of rectangular cross section.   Such a coil would be closely realized by a winding of wire of rectangular cross section, arranged in several layers, with negligible insulating space between adjacent wires.

R.D.H.—15.

When the dimensions l and c are small in comparison with a (see Fig. 10.7) the inductance is given closely by **Stephan's formula**, which, for $l > c$, takes the form :

$$L_s = 0.031\,93aN^2\left[2.303\left(1 + \frac{l^2}{32a^2} + \frac{c^2}{96a^2}\right)\log_{10}\frac{8a}{g'} - y_1 + \frac{l^2y_2}{16a^2}\right] \tag{3}$$

while for $l < c$, that is for pan-cake coils, the formula becomes :

$$L_s = 0.031\,93aN^2\left[2.303\left(1 + \frac{l^2}{32a^2} + \frac{c^2}{96a^2}\right)\log_{10}\frac{8a}{g'} - y_1 + \frac{c^2y_3}{16a^2}\right] \tag{4}$$

where $g' = \sqrt{c^2 + l^2}$ and y_1, y_2 and y_3 are functions of c/l which are tabulated in the Bureau of Standards Circular No. 74 (p. 285).

The formulae of this section have been put in the form of a family of curves by Maynard (Ref. 38).

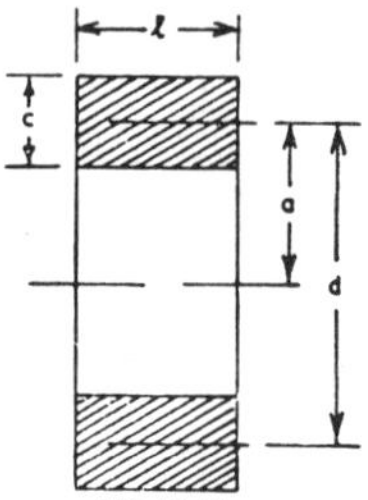

FIG. 10.7

(ii) Correction for insulation thickness

Unless the percentage of the cross section occupied by the insulation is large the formulae given in Sect. 2(i) apply very well to an actual coil. When the spacing P is appreciably greater than the diameter of the wire D a correction term should be added to L_s to give L_o. Thus :

$$L_o = L_s + 0.073\,35aN\left[\log_{10}(P/D) + 0.0675\right] \tag{5}$$

(iii) Approximate formulae

(A) Long coils of a few layers

For long coils of a few layers $l/c > 10$ and $B \approx 0.3$, so that we obtain from equation (1) ;

$$L_s \approx \frac{0.1003a^2N^2K}{l}\left[1 - \frac{c}{\pi aK}\right] \tag{6}$$

$$\approx \frac{a^2N^2}{9a + 10l}\left[1 - \frac{c}{10\pi al}(9a + 10l)\right] ; \tag{7}$$

$$\text{or}\quad L_s \approx \frac{a^2N^2}{9a + 10l} - \frac{caN^2}{10\pi l} \tag{8}$$

(B) Short coils

For short coils such that both l and c are much less than a, it follows from Stephan's formula that :

$$L_s = 0.073\,35aN^2\left[\log_{10}\frac{8a}{\sqrt{l^2 + c^2}} - 0.4343y_1\right] \tag{9}$$

$$= \frac{aN^2}{13.5}\log_{10}\frac{3.6a}{\sqrt{l^2 + c^2}} \quad \text{when } \frac{l}{c} \text{ or } \frac{c}{l} \text{ lies between 0.35 and 1} \tag{10}$$

$$= \frac{aN^2}{13.5}\log_{10}\frac{4.02a}{\sqrt{l^2 + c^2}} \quad \text{when } \frac{l}{c} \text{ or } \frac{c}{l} \text{ lies between 0.15 and 0.35} \tag{11}$$

$$= \frac{aN^2}{13.5}\log_{10}\frac{4.55a}{\sqrt{l^2 + c^2}} \quad \text{when } \frac{l}{c} \text{ or } \frac{c}{l} \text{ lies between 0 and 0.15} \tag{12}$$

These results are accurate to 5% as l and c approach a, and are increasingly accurate as l and c decrease compared with a. When l/a and c/a are both **very** small, it is sufficient to use the approximation :

$$L_s \approx \frac{aN^2}{13.5}\log_{10}\frac{4}{\sqrt{l^2 + c^2}}$$

for all values of the ratio l/c.

(C) Bunet's formula

Bunet (Ref. 24) gives the following approximate formula applicable to coils whose diameters are less than three times their length :

$$L_s = \frac{a^2N^2}{9a + 10l + 8.4c + 3.2cl/a} \tag{13}$$

The range of usefulness of this formula can be best judged from its accuracy for various coil dimensions given below :

Accuracy

For $c/a = 1/20$ 1% for $(2a/l)$ up to 3 4% for $(2a/l) = 5$
For $c/a = 1/5$ 1% for $(2a/l)$ up to 5 2% for $(2a/l) = 10$
For $c/a = 1/2$ 1% for $(2a/l)$ up to 2 3% for $(2a/l) = 5$
For $c/a = 1$ 1% for $(2a/l)$ up to 1.5 5% for $(2a/l) = 5$

It is easy to verify that for long coils of a few layers Bunet's formula is approximately :

$$L_s \approx \frac{a^2N^2}{9a + 10l} - \frac{a^2N^2(3.2cl/a)}{(9a + 10l)^2} \tag{14}$$

$$\approx \frac{a^2N^2}{9a + 10l} - \frac{CaN^2}{10\pi l} \tag{15}$$

in accord with the result given previously for this type of coil.

(D) Wheeler's formula for short coils

Wheeler gives a formula for short coils having l and c less than a (Ref. 22). This result, which he states was obtained theoretically, is :

$$L_s = \frac{aN^2}{13.5} \log_{10} \frac{4.9a}{l + c} \tag{16}$$

The accuracy is given as 3% when $(l + c) = a$, and is stated to improve as $(l + c)/a$ decreases towards zero.

This formula is slightly different in form from those given in (B) of this section. Nevertheless, the numerical agreement is quite good, as can be seen from the following considerations :

For the case $l = c$,

$$\frac{3.6a}{\sqrt{l^2 + c^2}} = \frac{3.6a}{0.707(l + c)} = \frac{5.1a}{l + c} ;$$

when $l = 3c$ or $c = 3l$,

$$\frac{4.02a}{\sqrt{l^2 + c^2}} = \frac{4.02a}{0.792(l + c)} = \frac{5.1a}{l + c} ;$$

when $l = 10c$ or $c = 10l$,

$$\frac{4.55a}{\sqrt{l^2 + c^2}} = \frac{4.55a}{0.91(l + c)} = \frac{5.0a}{l + c} :$$

Thus when $\log 4.9/(l + c)$ is compared with the logarithms of these quantities the percentage difference is found to be small.

Another formula due to Wheeler (Ref. 22) which covers the shape of many universal windings, is

$$L_s = \frac{0.8a^2N^2}{6a + 9l + 10c}$$

accurate to within about 1% when the three terms in the denominator are about equal.

(iv) Design of multilayer coils

(A) Short coils

The Bureau of Standards Circular No. 74 gives two of the possible approaches to the design of short multilayer coils.

(1) The two forms of Stephan's equation given in eqns. (9) to (12) may be expressed as :

$$L_s = \frac{\lambda^{5/3}}{p^{2/3}} G, \tag{17}$$

where λ is the wire length, P is the distance between centres of adjacent wires and G is a function of the shape of cross section (l/c) and of the shape ratio of the coil (c/a). The quantity G is represented by means of curves. This equation is used, in cases where the ratios l/c and c/a have been decided, to determine the necessary wire-length for a given pitch, or vice versa. The mean radius of the winding is then obtained from

$$a = \sqrt[3]{\frac{\lambda}{2\pi}\frac{c}{l}\frac{P^2}{(c/a)^2}};\tag{18}$$

and the total number of turns is given by $N = \lambda/2\pi a$ or lc/P^2.

It can be shown that, for a given resistance and coil shape, the square cross section $(l/c = 1)$ gives a greater inductance than any other form ; and, further, for a square cross section, the inductance for a given length of wire is a maximum for $c/a = 0.662$.

(2) Stephan's formulae may also be expressed in the form :
$$L_s = 0.031\,93aN^2g,\tag{19}$$
where g is a known function of l/c and c/a.

This form is useful when a, c and l have been decided to give N, and thence P. We have
$$P = \frac{lc}{N}$$
and $\lambda = 2\pi aN$.

(B) Universal coils

The usual problem is the calculation of the inductance of a given number of turns before winding the coil, with only former size and wire size known. It is stated in Ref. 20 that the inductance of universal coils is about 10% greater than that of normal multilayer windings of the same external dimensions.

A suitable procedure is to determine the gear ratio from the formula,
$$\text{gear ratio} = \tfrac{1}{2}n(P + 1)/P\tag{20}$$
as described in Chapter 11, Sect. 3(iv). The radial depth of winding, c, can then be obtained from the formula
$$c = Nqw'/(P + 1)\tag{21}$$
where N = number of turns in coil,
$\quad\quad q$ = number of crossovers per winding cycle
$\quad\quad\quad$ [Refer Table 1 in Chapter 11, Section 3(iv)]
$\quad\quad w'$ = diameter of wire plus insulation,
and $\quad P$ = an integer defined in Chapter 11, Section 3(iv).
Since the length of the winding, l, is equal to the sum of the cam throw and the wire diameter, the dimensions of the coil are then known and the inductance can be calculated from the methods previously described.

Example : A coil of 500 turns of 38 A.W.G. enamelled wire (0.0044 inch) is to be wound on a 0.5 inch diameter former with a 0.1 inch cam. From Chapter 11 Sect. 3(iv), $P = 43$ and $q = 2$. Therefore
$\quad\quad c = 500 \times 2 \times 0.0044/44$ inch = 0.1 inch
$\quad\quad l = 0.1$ inch + 0.0044 inch = 0.1044 inch
and $a = 0.25$ inch + $c/2 = 0.3$ inch.

(v) Effect of a concentric screen

No formulae have apparently been given for the change in inductance produced by a concentric shield. It is clear that the percentage change in inductance of a multi-layer coil will be less than that of a single layer coil of equal outside dimensions ; and that the greater the winding depth c the less will be the effect of the shield. For solenoids of a few layers the percentage change in inductance will be very closely the same as for a single layer solenoid of corresponding dimensions.

SECTION 3 : TOROIDAL COILS

(i) Toroidal coil of circular section with single layer winding (ii) Toroidal coil of rectangular section with single layer winding (iii) Toroidal coil of rectangular section with multilayer winding.

(i) Toroidal coil of circular section with single layer winding

The current sheet inductance of a single layer coil wound on a torus, that is a ring of circular cross-section, is given by

$$L_s = 0.031\ 93N^2\ [R - \sqrt{R^2 - a^2}]\ \text{microhenrys,} \tag{1}$$

where R is the distance in inches from the axis to the centre of cross-section of the winding, a is the radius of the turns of the winding and N is the total number of turns.

(ii) Toroidal coil of rectangular section with single layer winding

For this type of coil the current sheet inductance is readily shown to be :

$$L_s = 0.011\ 70N^2h\ \log_{10} \frac{r_2}{r_1}\ \text{microhenrys,} \tag{2}$$

where r_1 is the inner radius of the winding, r_2 is the outer radius of the winding and h is the axial depth of the winding.

The difference between L_s and L_o for single layer toroidal coils is usually small. Where high precision is required the value of L_o may be obtained from L_s by using the corrections shown in Bulletin, Bureau of Standards, 8 (1912) p. 125.

(iii) Toroidal coil of rectangular section with multilayer winding

Dwight (Ref. 11) has obtained an expression for the current sheet inductance of this type of coil. His expression contains only simple terms, but it is too long to reproduce here. In an example quoted by the author the calculated and measured values of an inductance agreed within 1%.

Richter (E.E. Supplement. Dec. 1945, p. 999) stated that in practice fair results are obtained by use of the empirical formula :

$$L = 0.0117N^2h'\ \log_{10} (r_2'/r_1') \tag{3}$$

where h', r_2' and r_1' are determined as follows :

Let t_1 be the thickness of winding on the inside face,

$\quad\quad t_2$ be the thickness of winding on the outside face,

$\quad\quad t_3$ be the thickness of winding on the inside edge of the top and bottom faces, and

$\quad\quad t_4$ be the thickness of winding on the outside edge of the top and bottom faces,

then

$\quad\quad r_2' = r_2 + t_2/3$; $r_1' = r_1 - t_1/3$; $h' = h + (t_3 + t_4)/3$;

h, r_1 and r_2 having the same meanings as in Section 3(ii).

SECTION 4 : FLAT SPIRALS

(i) Accurate formulae (ii) Approximate formulae.

(i) Accurate formulae

The current sheet inductance of flat spirals can be obtained from Stephan's formula for short multilayer coils [see Sect. 2(i)].

(a) Wire of rectangular cross section

For spirals wound with metal ribbon or with thicker rectangular wire the procedure is as follows :

Use Stephan's formula with these conventions :

for l put the width of the wire ;

for c put NP, where P is the distance between the centre of cross section of a turn and the corresponding point of the next turn ;

for a put $a_o + (N - l)P/2$, where $2a_o$ is the distance (Fig. 10.8) across the centre of the innermost end of the spiral.

The correction for cross section is given in the Bureau of Standards Circular No. 74 in the form :
$$L_0 = L_s - 0.0319aN(A_1 + B_1),$$
where A_1 and B_1 are tabulated functions.   The accuracy obtained for L_0 is 1%.

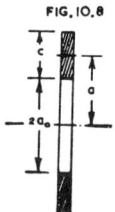

FIG. 10.8

(b) Round wire

For spirals wound with round wire the same conventions are adopted to obtain a and c, but a simplified form of Stephan's equation, obtained by setting $l = 0$, is used.   It is

$$L_s = 0.3193aN^2\left[2.303 \log_{10} \frac{8a}{c} - \tfrac{1}{2} + \frac{c^2}{96a^2}\left(2.303 \log_{10} \frac{8a}{c} + \frac{43}{12}\right)\right]$$

The inductance L_o of a spiral is given more closely by :
$$L_o = L_s - 0.0319aN(A + B),$$
where A and B are shown in Fig. 10.2.

(ii) Approximate formulae

For **round wires** it follows immediately that

$$L_s \approx 0.031\,93 \times 2.303aN^2\left[\log \frac{8a}{c} - \frac{0.5}{2.303}\right]$$

$$\approx \frac{aN^2}{13.5} \log \frac{4.9a}{c}.$$

Wheeler has given an approximate formula for spirals where $c > 0.2a$.   It is :
$$L_s = \frac{a^2N^2}{8a + 11c}.$$

The first of these formulae is the more accurate as the ratio c/a decreases, while the second is the more accurate the closer c/a approaches unity.   The agreement between them is quite fair :

For $c/a = 0.2$, we have from the first result,
$$L_s \approx (aN^2/13.5) \log 24.5 = 0.103aN^2,$$
while from Wheeler's formula
$$L_s \approx aN^2/(8 + 2.2) = 0.098aN^2.$$
For $c/a = 0.5$, we obtain, respectively,
$$L_s \approx (aN^2/13.5) \log 9.4 = 0.97aN^2/13.5 \text{ and}$$
$$L_s \approx aN^2/(8 + 5.5) = aN^2/13.5.$$

SECTION 5 : MUTUAL INDUCTANCE*

(*i*) *Accurate methods*   (*ii*) *Approximate methods.*

(i) Accurate methods

Accurate methods for the calculation of mutual inductance between coils of many different shapes and relative dispositions are given in the Bureau of Standards Circular No. 74, and in the Bureau of Standards Scientific Paper No. 169.   The book by Grover (Ref. 15) covers the subject comprehensively in Chapters 12, 15 and 20.   The possible accuracy of these methods is always better than one part in one thousand.

The mutual inductance of coaxial single-layer coils, with tables to facilitate the calculation, is covered in Ref. 15 Chapter 15 (see also Ref. 5).

*By the Editor.

There are few simple formulae which can be used for the more common practical cases, such as are possible with self inductance.   The following exact method may be used for two windings on the same former with a space between them, both windings being similar in pitch and wire diameter (Fig. 10.9).

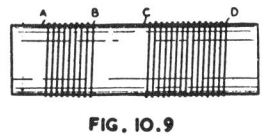

FIG. 10.9

Assume that the space between the windings is wound as a continuation of the windings on the two ends, to form a continuous inductor from A to D with tappings at points B and C.   Then the required mutual inductance, M, between the two original windings is given by

$$M = \tfrac{1}{2}(L_{AD} + L_{BC} - L_{AC} - L_{BD}) \tag{1}$$

where L_{AD} is the inductance between points A and D, and similarly for other terms. These inductances may be calculated from the formulae given in earlier sections of this chapter.

References to accurate methods : Refs. 2, 3, 5, 13, 14, 15, 18.

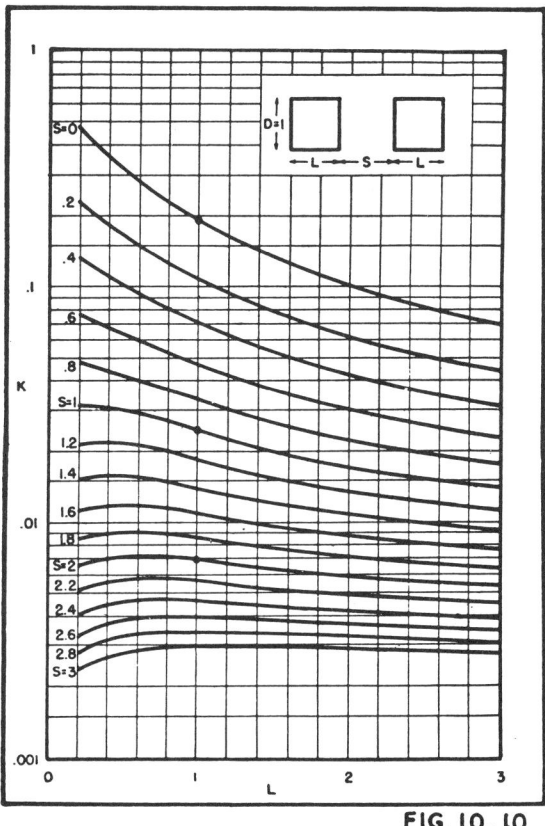

FIG. 10.10

Fig. 10.10.   *Chart giving coefficient of coupling for a specified spacing between two coaxial solenoids having identical dimensions.   The quantities S and L are measured in terms of coil diameters.*   (Ref. 56.)

(ii) **Approximate methods**

The coefficient of coupling for a specified spacing between two coaxial solenoids having identical dimensions, but not necessarily identical numbers of turns, is given by Fig. 10.10 (Ref. 56). The accuracy should be better than 5%.

SECTION 6 : LIST OF SYMBOLS

Note : Inductances are given in microhenrys and dimensions in inches.

L_s = current sheet inductance
L_o = low frequency inductance
L_o' = effective inductance of a shielded coil
N = total turns in coil
n = turns per unit length
l = coil length
P = pitch of winding = $1/n$
a = coil radius (from axis to middle of winding)
$d = 2a$ = coil diameter
A' = cross sectional area of coil
D = (bare) wire diameter
$S = D/P$
λ = total length of wire
K = Nagaoka's constant—see Fig. 10.1
A, B = correction terms—see Fig. 10.2
$f = Kl/2a$—see Fig. 10.3
$F = 1.8/f$
$y = 1.8/l$
b = radius of concentric cylindrical screen
l_t = length of concentric cylindrical screen
$g = b - a$, i.e. coil to screen spacing
k = coefficient of coupling between coil and concentric can
c = radial depth of winding
$g' = \sqrt{l^2 + c^2}$
B_s = correction term, function of l/c
y_1, y_2, y_3 = functions of l/c in Stephan's formula
q = number of crossovers per winding cycle
R = mean radius of a toroidal coil of circular cross section
r_1 = inner radius of a toroidal coil of rectangular cross section
w' = diameter of wire plus insulation
r_2 = outer radius of a toroidal coil of rectangular cross section
h = axial depth of a toroidal coil of rectangular cross section
M = mutual inductance

SECTION 7 : REFERENCES

EXACT METHODS OF CALCULATING SELF AND MUTUAL INDUCTANCE
1. de Holzer, R. C. '' Calculating mutual inductance ; co-axial and co-planar multi-layer coils " W.E. 25.300 (Sept. 1948) 286.
2. Bureau of Standards Circular No. C74 (1937).
3. Bureau of Standards Scientific Paper No. 169 (1912).
4. Nagaoka, H. " The inductance coefficients of solenoids " Journal of the College of Science, Tokyo, Vol. 27, Art. 6 (August 15, 1909) p. 1.
5. Grover, F. W. " Tables for the calculation of the mutual inductance of any two coaxial single layer coils," Proc. I.R.E. 21.7 (July 1933) 1039 (includes further bibliography).
6a. Reber, G. " Optimum Design of Toroidal Inductances " Proc. I.R.E. 23.9 (Sept. 1935) 1056.
6b. Moullin, E. B. " The use of Bessel Functions for Calculating the self-inductance of single-layer solenoids " Proc. I.E.E. 96. Part 3.40 (March 1949) 133. See also Ref. 55.
Also Refs. 15, 25b.

APPROXIMATE FORMULAE FOR SELF AND MUTUAL INDUCTANCE
7. Simon, A. W. " Calculating the inductance of universal-wound coils " Radio, 30.9 (Sept. 1946) 18.
8. Ricks, J. B. " A useful formula for solenoid inductor design—determining coil dimensions with known capacitance," Q.S.T. 31.5 (May 1947) 71.
9. A. I. Forbes Simpson " The design of small single-layer coils " Electronic Eng. 19.237 (Nov. 1947) 353.

10. Simon, A. W. " Wire length of universal coils " Elect. 19.3 (Mar. 1946) 162.
11. Dwight, H. B. " Self inductance of a toroidal coil without iron " A.I.E.E. Trans. (Nov. 1945) 805 ;
    discussion supplement A.I.E.E. Trans. (Dec. 1945) 999.
12. Blow, T. C. " Solenoid inductance calculations " Elect. 15.5 (May 1942) 63.
13. Di Toro, M. J. " Computing mutual inductance " Elect. 18.6 (June 1945) 144.
14. Maddock, A. J. " Mutual inductance : simplified calculations for concentric solenoids " W.E.
    22.263 (Aug. 1945) 373.
15. Grover, F. W. (book) " Inductance calculations : Working Formulae and Tables " (New York,
    D. Van Nostrand Co., Inc., 1946).   Includes extensive bibliography.
16. Dwight, H. B. (book) " Electrical coils and conductors : Their Electrical Characteristics and
    Theory " (McGraw-Hill Book Co., Inc. New York, 1945).
17. Amos, S. W. " Inductance calculations as applied to air-cored coils " W.W. 47.4 (April 1941) 108.
18. Turney, T. H. " Mutual inductance : a simple method of calculation for single-layer coils on the
    same former " W.W. 48.3 (March 1942) 72.
19. Everett, F. C. " Short wave inductance chart " Electronics, 13.3 (March 1940) 33.
20. Zepler, E. E. (book) " The Technique of Radio Design " Chapman and Hall, 1943.
21. Jeffery, C. N. " Design charts for coaxial transformers—co-axial windings of common diameter
    and pitch " A.W.A. Tech. Rev. 8.2 (April 1949) 167.
22. Wheeler, H. A. " Simple inductance formulas for radio coils " Proc. I.R.E. 16.10 (Oct. 1928) 1398
    and discussion Proc. I.R.E. 17.3 (March 1929) 580.
23. Esnault-Pelterie, M. R. " On the coefficient of self inductance of a solenoid " Comptes Rendus,
    Tome 205 No. 18 (Nov. 3, 1937) 762 and No. 20 (Nov. 15, 1937) 885.
24. Bunet, P. " On the Self Inductance of Circular Cylindrical Coils " Revue General de l'Electricite,
    Tome xliii, No. 4 (Jan. 22, 1938) 99.
25a. Hayman, W. G. " Approximate Formulae for the Inductance of Solenoids and Astatic Coils "
    E.W. and W.E. 8.95 (Aug. 1931) 422.
25b. Welsby, V. G. (book) " The Theory and Design of Inductance Coils " (Macdonald & Co. Ltd.,
    London, 1950).

NOMOGRAMS AND CHARTS FOR SELF INDUCTANCE
26. Sabaroff, S. (letter) " Nomogram for Rosa inductance correction " Proc. I.R.E. 35.8 (Aug. 1947) 793.
27. Pepperberg, L. E. " Coupling coefficient chart " Elect. 18.1 (Jan. 1945) 144.
28. Data Sheets 32 and 33 " The inductance of single layer solenoids on square and rectangular formers "
    Electronic Eng. 15.173 (July 1942) 65.
29. Data Sheets 26, 27 " Skin effect " Electronic Eng. 14.170 (April 1942) 715.
30. Data Sheets 23-25, 28 " The shunt loaded tuned circuit " Electronic Eng. 14.169 (March 1942)
    677, and 14.170 (April 1942) 715.
31. " Inductance chart for single layer solenoids " G.R. Exp. 15.3 (Aug. 1940) 6.
32. Purington, E. S. " Simplified inductance chart " (inductance of multi-layer coils) Elect. 15.9 (Sept.
    1942) 61.
33. " Instructions for use of Federal Band Pass Nomographs A to C and 1 to 3," Federal Telephone
    and Radio Corporation.
34. " A convenient inductance chart for single layer solenoids " G.R. Exp. (based on Circular 74, U.S.
    Bureau of Standards) 15.3 (Aug. 1940) 6.
35. Data Sheets 12-17 " Inductance of single layer solenoids on circular formers " Electronic Eng.
    14.164 (Oct. 1941) 447 and 14.165 (Nov. 1941) 495.
36. Shiepe, E. M. " The Inductance Authority " published by Herman Bernard, New York, 1933
    (contains 50 large sized charts based on formulae and tables of Bureau of Standards Circular No. 74).
37. Seki, H. " A new Abac for single layer coils " Wireless Engineer, 10.112 (Jan. 1933) 12 (nomo-
    gram based upon Hayman's formula).
38. Maynard, J. E. " Multilayer coil inductance chart " Elect. (Jan. 1939) 33.
39. Beatty, R. T. " Radio Data Charts " published by Iliffe and Sons, London.
40. Nachod, C. P. " Nomogram for coil calculations " Elect. (Jan. 1937) 27-28.
40a. Sabaroff, S. " Multi-layer coil inductance chart " Comm. 29.12 (Dec. 1949) 18.
40b. Sulzer, P. G. " R.F. coil design using charts " Comm. 29.5 (May 1949) 10.
40c. " Reference Data for Radio Engineers," Federal Telephone and Radio Corp., 3rd ed. 1949, p. 75.
40d. Wheeler, H. A. " Inductance chart for solenoid coil " Proc. I.R.E. 38.12 (Dec. 1950) 1398.

THE EFFECTS OF METAL SCREENS ON COILS
41. Davidson, C. F. and J. C. Simmonds " Effect of a spherical screen on an inductor " W E. 22.256
    (Jan. 1945) 2.
42. Anderson, A. R. " Cylindrical shielding and its measurement at radio frequencies " Proc. I.R.E.
    34.5 (May 1946) 312.
43. " Effect of shield can on inductance of coil " Radiotronics No. 108 p. 67.
44. Sowerby, J. McG. " Effect of a screening can on the inductance and resistance of a coil " Radio
    Data Charts (2) W.W. 48.11 (Nov. 1942) 254.
45. Bogle, A. G. " The effective inductance and resistance of screened coils " Jour. I.E.E. 87 (1940) 299.
46. Reed, M. " Inter-circuit screening—survey of the general principles " W.W. 48.6 (June 1942) 135.
47. Davidson, C. F., Looser, R. C. and Simmonds, J. C. " Power loss in electromagnetic screens "
    W.E. 23.268 (Jan. 1946) 8.
48. R.C.A. Radiotron Division Application Note No. 48 (June 12, 1935) reprinted in Radio Engineering
    (July 1935) 11.
49. Sowerby, A. L. M. " The modern screened coil " W.W. Sep. 23, 30 and Oct. 7, 14 ; 1931.
50. Howe, G. W. O. editorials, W.E. 11.126 (March 1934) 115 and 11.130 (July 1934) 347.
51. Hayman, W. G. " Inductance of solenoids in cylindrical screen boxes " W.E. 11.127 (April 1934) 189.
52. " Formulas for shielded coils " Radio Eng. 16.12 (Dec. 1936) 17.
53. Levy, S. " Electro-magnetic shielding effect of an infinite plane conducting sheet placed between
    circular coaxial coils " Proc. I.R.E. 24.6 (June 1936) 923 (contains further bibliography).
54. Moeller, F. " Magnetic screening with a plane sheet at audio frequencies " Elektrische Nachrichten
    Technik, Band 16, Heft 2 (Feb. 1939), 48.
55. Phillips, F. M. " A note on the inductance of screened single-layer solenoids " Proc. I.E.E. 96. Part
    III, 40 (March 1949) 138 ; R. G. Medhurst, discussion, 98.   Part III, 53 (May 1951) 248.   Sequel
    to Ref. 6b.

CHARTS FOR COEFFICIENT OF COUPLING
56. Sulzer, P. G. " Coupling Chart for solenoid coils " TV Eng. 1.6 (June 1950) 20.

CHAPTER 11

DESIGN OF RADIO FREQUENCY INDUCTORS

BY E. WATKINSON, A.S.T.C., A.M.I.E. (AUST.), S.M.I.R.E. (AUST.)

SECTION 1 : INTRODUCTION

The space available for coils in a radio receiver is invariably limited, and part of the design work on each coil is the obtaining of maximum Q in a minimum volume. In the case of i-f transformers the volume is clearly defined by the shield used, and the same applies to r-f coils in cans, but where shielding is not used with r-f coils the increased Q obtainable from larger diameter formers is offset against the increased damping from components and magnetic materials, such as the receiver chassis, adjacent to the coil.

The resulting form factor for each individual winding requiring maximum Q and minimum self-capacitance is usually such that the length of the winding is very approximately equal to its diameter. To obtain this shape for i-f, broadcast and short-wave coils the winding method is varied.

I-F transformers use universal windings, and where the Q and distributed capacitance requirements are not severe a single coil wound with a large cam (perhaps $\frac{3}{8}$ in.) will be satisfactory, even without an iron core, if litz wire [see Sect. 5(i)] is used. However higher Q values are usually required than are obtainable in this way and in such cases each winding consists of two or more pies, perhaps with an iron core.

The same types of coils are satisfactory in the broadcast band, although in the case of signal frequency coils which are tuned over a ratio of more than three to one it is essential to use very narrow windings or two or more pies to reduce the distributed capacitance of the winding. Litz wire and iron cores are commonly used to give a secondary Q of 100 or more.

Another method of decreasing distributed capacitance is the use of progressive universal windings in which the winding finger travels along the former in addition to moving to and fro. Such coils are rarely built up to a greater height than five wire thicknesses, and are equally useful with or without iron cores. Solenoids are also used either with iron cores or on comparatively large diameter formers if an iron core is not used.

At frequencies between 2 and 6 Mc/s (which are used in receivers providing continuous coverage between the normal short-wave and broadcast bands, or for oscillators giving second harmonic mixing for the short wave band) the progressive universal winding is particularly useful because the winding length of solenoids may be too great with the usual range of former diameters, and the number of turns is insufficient to allow them to be split into many sections.

A set of worm driving gears which spreads the appropriate number of turns over a length approximately equal to the coil diameter is required. This usually necessitates a faster worm drive than is used for broadcast coils. Litz wire with a large number of strands is useful in obtaining maximum Q.

For the 6 to 18 Mc/s band the solenoid is used, almost always with solid wire, and this type of coil remains useful at least up to 100 Mc/s. Increased wire diameters and spacings are used at higher frequencies to obtain form factors similar to those used at low frequencies.

SECTION 2 : SELF-CAPACITANCE OF COILS

(i) *Effects of self-capacitance* (ii) *Calculation of self-capacitance of single-layer solenoids* (iii) *Measurement of self-capacitance.*

(i) Effects of self-capacitance

The self-capacitance of a coil is due to the electrostatic coupling between individual turns and between the turns and earth. When the self-capacitance is between uninsulated turns in air its Q may be high, but the greater the amount of dielectric in the field of the coil the greater will be the losses.

Short wave coils of enamelled wire on a solid former do not have serious dielectric losses if good quality materials are used, but at broadcast and intermediate frequencies the universal windings used have comparatively high-loss dielectrics and unless self-capacitances are kept to a low value the reduction in Q may be appreciable.

In the case of coils which are to be tuned over a range of frequencies, comparatively small values of shunt capacitance can have a large effect on the possible tuning range.

In all cases the self capacitance of a coil has an apparent effect on its resistance, inductance and Q, and at frequencies considerably below the self resonant frequency of the coil—

$$\text{apparent inductance} = L\left(1 + \frac{C_0}{C}\right)$$

$$\text{apparent resistance} = R\left(1 + \frac{C_0}{C}\right)^2$$

$$\text{and apparent } Q = Q/\left(1 + \frac{C_0}{C}\right)$$

where C_0 = self-capacitance of coil
C = external capacitance required to tune L to resonance
L = true inductance of coil
R = true resistance of coil
and Q = true Q of coil.

(ii) Calculation of self-capacitance of single-layer solenoids

Until recently the work of Palermo (Ref. C3) had been taken as the standard on the self-capacitance of single-layer solenoids. However Medhurst (Ref. C2) has disputed the theoretical grounds on which Palermo's work is based. As the result of a careful analysis and a large number of measurements he states that the self-capacitance, C_0, of single-layer solenoids with one end earthed, and without leads, is

$C_0 = HD$ μμF   (1)
where D = diameter of coil in cm.

and H depends only on the length/diameter ratio of the coil. A table of values of H is given below and when used in conjunction with a capacitance correction for the " live " lead, Medhurst states that the accuracy should be within 5%.

TABLE 1

$\left(\dfrac{\text{Length}}{\text{Diameter}}\right)$	H	$\left(\dfrac{\text{Length}}{\text{Diameter}}\right)$	H	$\left(\dfrac{\text{Length}}{\text{Diameter}}\right)$	H
50	5.8	5.0	0.81	0.70	0.47
40	4.6	4.5	0.77	0.60	0.48
30	3.4	4.0	0.72	0.50	0.50
25	2.9	3.5	0.67	0.45	0.52
20	2.36	3.0	0.61	0.40	0.54
15	1.86	2.5	0.56	0.35	0.57
10	1.32	2.0	0.50	0.30	0.60
9.0	1.22	1.5	0.47	0.25	0.64
8.0	1.12	1.0	0.46	0.20	0.70
7.0	1.01	0.90	0.46	0.15	0.79
6.0	0.92	0.80	0.46	0.10	0.96

Lead capacitance can be determined separately and added to the coil self-capacitance. Fig. 11.1 (from Medhurst) can be used for this purpose.

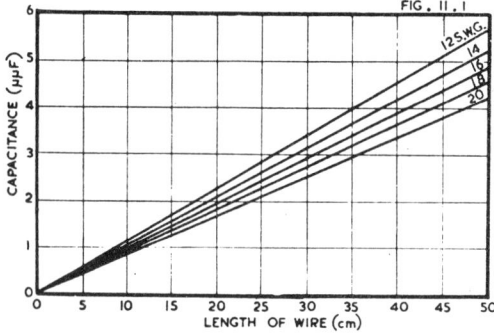

Fig. 11.1. *Variation of capacitance with wire length for vertical copper wires of various gauges (Ref. D6).*

It is interesting to compare Medhurst's formula with another by Forbes Simpson [Ref. G17 and Sect. 5(i) of this chapter]. Forbes Simpson's formula is applicable to coils with a length/diameter ratio of unity, a pitch of 1.5 times the wire diameter and with leads at each end equal in length to the diameter of the coil. His formula is

$$C_0 = D(0.47 + a) \ \mu\mu\text{F} \tag{2}$$

where $D =$ diameter of coil in cm.

and $a =$ a constant depending on the gauge of wire, lying between 0.065 for 42 S.W.G. (38 A.W.G. approx.) and 0.11 for 12 S.W.G. (10 A.W.G. approx.) where one lead of the coil is connected to chassis.

In eqn. (2) there is no equivalent of the H in eqn. (1) because the length diameter ratio is a constant. The length of " live " lead taken into account in eqn. (2) is equal to D, and it is possible to determine its effect from Fig. 11.1. For 12 S.W.G. wire the capacitance is 0.11 $\mu\mu$F per cm. so that eqn. (1) (with lead correction) could be written

$$C_0 = 0.46D + 0.11D = 0.57D$$

for the conditions of eqn. (2) and using 12 S.W.G. wire. Similarly, using the value of a given for 12 S.W.G. wire in eqn. (2), eqn. (2) could be written

$$C_0 = 0.47D + 0.11D = 0.58D.$$

The data in Fig. 11.1 are not sufficient to compare the equations using the lower limit of a given by Forbes Simpson but for 20 S.W.G. eqn. (1) becomes

$$C_0 = 0.55D$$

for the conditions of eqn. (2), while eqn. (2) for 42 S.W.G. becomes

$$C_0 = 0.535D.$$

(iii) Measurement of self-capacitance

A graphical method of self-capacitance determination due to Howe is shown in Fig. 11.2. Values of external capacitance in $\mu\mu$F required to tune a coil to resonance are plotted against $1/f^2$ where f is the resonant frequency in Mc/s. The self-capacitance is the negative intercept of the straight line with the capacitance axis.

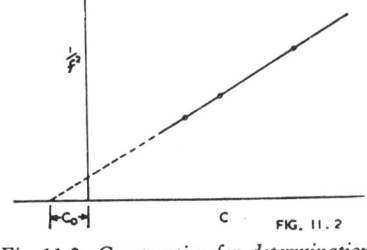

Fig. 11.2. Construction for determination of self-capacitance of coil.

An alternative method is to determine the external capacitances necessary to tune a coil to frequencies of f and $2f$. If the two capacitances are C_1 and C_2 respectively

$$\text{self-capacitance} = \frac{C_1 - 4C_2}{3}.$$

SECTION 3 : INTERMEDIATE-FREQUENCY WINDINGS

(*i*) *Air-cored coils*   (*ii*) *Iron-cored coils*   (*iii*) *Expanding selectivity i-f transformer*
(*iv*) *Calculation of gear ratios for universal coils*   (*v*) *Miscellaneous considerations*

(i) Air-cored coils

Most commercial receivers have two i-f transformers with four circuits tuned to about 455 Kc/s. To obtain adequate selectivity in such a case the required Q for each winding (mounted on the chassis but not connected in the circuit) is 100 or more.

This Q can readily be obtained with air cored coils of about 1 mH inductance provided that litz wire and comparatively large coil cans are used. Without litz it is difficult to exceed a Q of 50, but even this may be sufficient when more i-f stages than usual are used.

When a single pie is used, Q is to some extent dependent on coil shape but if five times the winding depth plus three times the winding width equals the external diameter, Q will be close to the maximum for the wire and type of winding.

There are three methods of increasing the Q, and when all are used to practical commercial limits, production Q figures of 150 can be maintained. The first requirement is litz wire, probably nine strands for a Q of the order of 100, and twenty or more strands for a Q of 150. Large coil cans are necessary, up to two inches in diameter, with formers of such a size as to make the outside coil diameter little more than half the can diameter. Lastly, self-capacitance must be reduced to a minimum, because its Q is always low and in the case of a single pie winding it may amount to say 25 $\mu\mu$F, a large percentage of the total capacitance. Splitting the winding into pies reduces self-capacitance, at the same time improving the form factor, and a suitable compromise between Q and economy is obtained by winding with three pies. Self-capacitance is also reduced by winding narrow pies, although a limit is set by the difficulty of winding litz wires with a cam of less than 0.1 inch, or perhaps even ⅛ inch. In addition, as the cam is reduced the height of the coil increases and larger losses from damping of the coil by the can may more than offset the reduced losses in a smaller self-capacitance.

When the windings have more than one pie, the inductance is dependent on pie spacing. For this reason, and to increase the speed of winding, multi-section coils are wound with the former located by a gate, a different slot in the gate being used for each individual pie. Spacing between primary and secondary windings is kept constant by means of a double winding finger which winds primary and secondary at the same time.

A disadvantage of air cored i-f coils is that a trimmer capacitance is needed to re-sonate the tuned circuit.   The cheapest types of trimmers usually have poor stability with respect to time, vibration, humidity and temperature, and satisfactory types may cost more than a variable iron core with provision for adjusting it.

(ii) Iron-cored coils

The advantages of iron cores used in i-f transformer windings are that
(a) they increase the inductance of the winding, thereby giving a saving in wire cost and winding time for a given inductance,
(b) they increase the Q of the winding, thereby allowing cheaper litz (or solid wire in an extreme case) to be used,
(c) they restrict the field of each winding, to an extent depending on the type of core used, and thus allow a smaller, cheaper coil can to be used without excessive damping,
(d) they provide a satisfactory method of adjusting the tuned circuit to resonance. Because of this the total tuning capacitance can be of a stable high Q type (e.g. a silvered-mica capacitor), and there is no trouble from capacitance between primary and secondary trimmers giving an asymmetrical resonance curve, or from capacitance between say first i-f and second i-f trimmers leading to re-generation.

To obtain the greatest benefit from an iron core it is necessary to have all of the turns close to the core.   In particular cases, especially at high frequencies, Q may be decreased due to increased distributed capacitance if the coil is too close (e.g. wound on the core) but in normal i-f applications in which a former comes between coil and core the thinner the former can be made, the better will be the Q.

Also because of this effect it sometimes happens that by winding a coil with thinner litz (fewer strands), and so bringing the top turns closer to the iron and further from the can, an increase in Q is obtained.

Another benefit from bringing all of the turns as close to the iron core as possible is that the range of inductance adjustment provided by the core is increased.   If fixed capacitors with $\pm\ 10\%$ tolerance are used to tune the i-f windings, it is necessary to provide at least $\pm\ 15\%$ inductance adjustment from the iron core to allow for winding variations and changes in stray capacitances due to valves and wiring.   It is necessary to split the i-f winding into pies to obtain a variation of this order with normal cheap i-f cores.

An additional advantage of splitting the winding is that at least half of the turns in the coil are brought much closer to the core so that Q is increased above the amount to be expected from the reduction in distributed capacitance.

To obtain the previously mentioned Q of 100 at an inductance of about 1 millihenry with a small iron core, a coil can somewhat larger than one inch diameter would pro-bably be needed with a coil wound in three pies of 5 strand 41 A.W.G. litz.

Much higher Q's are obtainable with special core shapes and materials, and a 455 Kc/s i-f transformer with a Q of 260 in the can is mentioned in Ref. A15.   However difficulty is usually experienced from instability even with valves of comparatively low slope, such as the 6SK7-GT, if i-f transformers with a Q above about 150 are used.

A possibility of increasing Q with iron-cored coils which does not usually arise with air-cored coils is that if the coil is wound directly on to a non-adjustable iron core or on to a small individual former, it can be mounted in different planes inside the coil can.   By mounting the coil with its axis at right angles to the axis of the can it is possible to obtain an increase in Q of the order of 10% (depending on the diameters of coil and can) and this method also has the advantage of making both cores adjust-able from above the chassis.   The chief disadvantage is that coupling between primary and secondary takes place mainly between adjacent edges of the two windings and variations in the height of the windings, due to variations in wire thickness, notice-ably affect the coupling.   In addition, trouble may be experienced from regeneration because with this method of mounting the field of the bottom winding extends further outside the coil can.   Unless a good joint is made between can and chassis this trouble will be further aggravated.

(iii) Expanding selectivity i-f transformers

The simplest method of expanding i-f transformer response to give peaks symmetrically spaced about the intermediate frequency is by switching a tertiary winding. Two other possibilities* of obtaining the desired result are the simultaneous switching of series and shunt coupling capacitors between the two windings, and the switching of capacitors to alter the coupling in a Π or T network (Ref. G13). In each of these cases, however, it is necessary to switch components in high potential sections of the circuit whereas tertiary switching can be carried out in low potential circuits.

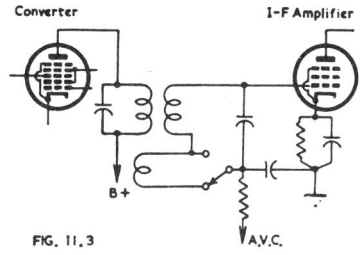

B +

FIG. 11.3   A.V.C.

Fig. 11.3. *Expanding selectivity with switched tertiary coil.*

Receivers have been manufactured in which movement of a complete winding altered the coupling, and whilst this eliminates switching problems, the additional mechanical problems are at least as troublesome. Moreover. it is difficult to keep constant the damping effect of the coil can and the mechanical coupling of the moving coil so that the whole object of the variable coupling (i.e. symmetrical expansion of the pass-band) is liable to be lost.

Although symmetrical expansion becomes more difficult as the Q of the transformer increases, it is possible with tertiary switching to maintain satisfactory symmetry in production with a Q of 150 and with simple alignment procedures.   In a normal receiver with a single i-f amplifier it is only necessary to expand the selectivity of the first i-f transformer because the diode damping on the secondary (and perhaps primary) of the second i-f reduces its selectivity considerably.

Fig. 11.3 gives a typical circuit.   It will be seen that the inductance of the tertiary winding is switched into the tuned circuit in the expanded position.   This does not lead to appreciable detuning because the inductance of the secondary is some thousands of times larger than that of the tertiary.

Suitable methods of winding the tertiary coil vary with the method of primary winding but the object is always to obtain maximum coupling with minimum tertiary inductance.   When the primary is pie-wound the tertiary should be wound between two pies as close to the middle of the coil as possible.   From three to five tertiary turns are required (assuming approximately critical coupling without the tertiary and three hundred or more primary turns in a 455 Kc/s i-f transformer).   The type of wire is not important except for its covering, and solid wire of the finest gauge which will give no trouble with handling or cleaning is probably most suitable.

It is essential that the insulation resistance between primary and tertiary be very high.   In a typical case a leakage of 100 megohms between primary and tertiary would give the control grid of the following i-f amplifier a positive voltage of five volts. Because of this the tertiary winding should be double-covered wire and the completed coil must be thoroughly dried out and then immediately impregnated in some moisture resisting compound.

If the primary consists of a single winding the tertiary is best wound in solenoid form on top of it.   Similar results are obtainable for the same number of coupling turns.

To align such a transformer the switch should be turned to the " narrow " position and the receiver aligned normally for maximum gain.   When the switch is turned to " broad " the output should drop, but when the signal generator is detuned, equal peaks should be found symmetrically spaced about the intermediate frequency.

*See also Chapter 9 Sect. 8 ; Chapter 26 Sect. 5.

If the peaks are not symmetrical, undesired couplings are probably responsible. All traces of regeneration must be removed and, because the stability requirements are more severe than usual, unusual effects are liable to be uncovered. For example, regeneration may occur due to coupling between a loop formed by the generator input leads and the output of the i-f amplifier (twisting generator leads will cure this) or coupling may occur between first and second i-f transformers within the steel chassis. When this happens, rotating one or both transformers will probably give cancellation but leave production receivers susceptible to the trouble. Additional spacing between transformers is advisable with each primary and secondary wired for minimum regenerative coupling.

The switch used for the tertiary winding should preferably be of the " break before make " variety. A " make before break " switch momentarily short circuits the tertiary winding during switching, giving a sudden reduction and increase in sensitivity which can be heard as a click.

Even after a satisfactory i-f selectivity curve has been obtained, the over-all curve may be too narrow if a r-f stage is used in the receiver. In such cases the r-f stage should also be expanded [Chapter 35 Sect. 5(iii)].

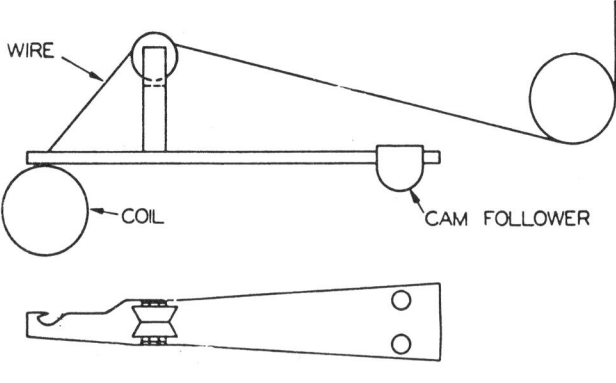

FIG. 11.4

Fig. **11.4.** *Type of finger recommended for universal winding.*

(iv) Calculation of gear ratios for universal coils

To obtain electrical consistency between universal coils in production runs it is essential to wind the coils with good mechanical stability. For this, machines must be adjusted correctly, but a suitable gear ratio is of equal importance.

The method of gear ratio calculation given below has been used successfully for production coils, and gives a straight pattern on the side of each coil.

The instructions make provision for a spacing between centres of adjacent wires of 8/7 of the wire diameter. This diameter should be measured, and the required spacing depends on the type of winding finger used. The most satisfactory types, and the ones for which the spacing factor will be found suitable, are those in which the wire passes to the bottom side of the finger before being placed in position on the coil. Fig. 11.4 shows such a finger, and it will be seen that the tension on the wire pulls the finger down to the face of the coil.

Fingers in which the wire passes through a groove on the top and then takes up its own position on the surface of the coil may require more spacing than that specified. Other factors may also affect the spacing required, but a small amount of experiment will decide this. It will be noticed that some types of litz wire tend to spread when wound, and so need more spacing than would be expected from measurement.

Symbols

d = former diameter (inches)
c = cam throw (inches)
n = nominal number of crossovers per turn (see Table 1)
q = number of crossovers per winding cycle, i.e. before wire lies alongside preceding wire (see Table 1).
v = nominal number of turns per winding cycle (see Table 1)
R = gear ratio = former gear/cam gear
w = modified wire diameter (inches)—see Note below
P = $qc/(w + x)$ = an integer
x = smallest amount necessary to make $(P + 1)/v$ an integer (inches).

Note. For fabric covered wire, w = (measured diameter of covered wire) × 8/7. If the wire is enamelled only, the same formula is used but the bare wire diameter is multiplied by 8/7.

Procedure

(A) From $n \not> (2d/3c)$ determine the largest convenient value for n. Do not use values of n less than 2 for bare enamelled wire. Obtain values of q and v from Table 1 below for the value of n chosen.

TABLE 1

n	4	2	1	2/3	$\frac{1}{2}$	1/3	$\frac{1}{4}$
q	4	2	2	2	2	2	2
v	1	1	2	3	4	6	8

(B) Determine w from information given in the note above.

(C) Calculate P from $P = qc/(w + x)$.

(D) Obtain R from $R = \frac{1}{2}n(P + 1)/P$.

Example 1

Given $d = \frac{1}{2}$ in. and $c = 0.1$ in. determine the gears to wind a coil with 42 S.W.G. enamelled wire.

(A) $(2d/3c) = 1.0/0.3$. Take $n = 2$, giving $q = 2$ and $v = 1$ from the table.

(B) The diameter of bare 42 S.W.G. wire is 0.004 in., so $w = 0.004\,57$ in.

(C) $P = qc/(w + x)$ and $(P + 1)/v$ must be an integer, i.e. P must be an integer since $v = 1$.
$qc/w = 200/4.57 = 43.7$.
But $P = qc/(w + x)$ = an integer.
Therefore $P = 43$.

(D) $R = \frac{1}{2}n(P + 1)/P = \frac{1}{2} \times 2(43 + 1)/43 = 44/43$.

Example 2

Given $d = \frac{1}{4}$ in. and $c = \frac{1}{4}$ in. determine the gears to wind a coil with 0.016 in. litz wire.

(A) $2d/3c = 2/3$. Take $n = 2/3$, giving $q = 2$ and $v = 3$.

(B) $w = 0.016$ in. × 8/7 = 0.0183 in.

(C) $qc/w = 500/18.3 = 27.3$.
But $(P + 1)/v$ = an integer.
Therefore $P = 26$.

(D) $R = \frac{1}{2}n(P + 1)/P = 1/3 \times 27/26$.
To obtain suitable gears :—
$R = 2/3 \times \frac{1}{2} \times 27/26 = (28/42) \times (27/52)$.

When it is known that n will be 2, as is the case with the majority of coils, the method reduces to dividing the modified wire diameter into twice the cam throw (ignoring any fractions in the answer). This gives P and the required ratio is $(P + 1)/P$.

For further information on universal coil winding see Refs. I 1 to I 8.

(v) Miscellaneous considerations

(A) Direction of windings

Although the coupling between primary and secondary is assumed to be due to mutual inductance, the capacitive coupling is appreciable. Depending upon the direction of connection of the windings, the capacitive coupling can aid or oppose the inductive coupling. When the two types are in opposition and of the same order a slight change in one—the capacitive coupling is particularly liable to random variation —gives a much larger percentage variation in the effective coupling. In a bad case, production sensitivity variations from this cause may be quite uncontrollable.

To avoid the trouble, i-f transformers are usually connected for aiding capacitive and inductive coupling. This is done by connecting the i-f amplifier plate to the beginning of one winding and the following grid to the end of the other winding when the two coils are wound in the same direction. Other connections to give the same winding sense will give the same result.

Even with aiding couplings it is desirable to reduce capacitive coupling to a minimum to obtain a symmetrical response curve and care should be taken in the placing of tuning capacitors and with details such as keeping the grid wire of the secondary winding well away from the plate wire or the plate side of the primary winding.

(B) Amount of coupling

An undesirable feature, from a production point of view, of i-f transformers in which the coupling is less than critical, is that receiver sensitivity becomes more dependent on i-f coil spacing. For a transformer with approximately critical coupling a spacing difference of 1/32 inch makes no appreciable difference to sensitivity in a typical case. However if the transformer were under-coupled the sensitivity change would be noticeable.

Transformers which are slightly over-coupled can be aligned for symmetrical response curves without undue trouble by detuning each winding (say with an additional capacitor of the same size as the tuning capacitor) while the other winding of the transformer is aligned for maximum sensitivity. Symmetry is of course dependent on the absence of regeneration.

(C) Losses

Unless care is taken, the sum of a number of apparently negligible losses may result in appreciable reduction in Q. Many artificial coverings have greater losses than silk, but this is usually obvious when a sample coil is wound. Less obvious losses may occur in details such as the material used to seal the end of a winding or in the placing of shunt capacitors or even coil lugs close to the actual i-f winding. An assembly of straight parallel wires between top and bottom of the i-f can may cause appreciable decrease in Q, particularly if large blobs of solder are placed close to the windings.

To eliminate such losses it is advisable to check the Q of a winding, sealed with low-loss material and baked to remove all moisture, when suspended well away from any substances which will introduce losses. The mechanical structure of the complete i-f assembly can then be added, one section at a time, and the Q read at each stage, a final reading being taken when the coil is mounted in the receiver.

(D) F-M i-f transformers

For details of F-M i-f construction and other i-f information of practical interest see Chapter 26 Sect. 4(vi).

(E) Other ferromagnetic materials

The development of a non-metallic ferromagnetic material named **Ferroxcube** has been announced by Philips (Refs. A28, A29, A30, A31, A32, A33).

Several grades of Ferroxcube are manufactured. They have in common a high specific resistance of 10^2 to 10^8 ohm cm. and a high initial permeability of from 50 to 3000 depending on the type. Ferroxcube IV, which is useful to 40 Mc/s, has a permeability of 50, and Ferroxcube III with an upper frequency limit of about 0.5 Mc/s has a minimum permeability of 800.

Because of the closeness of the Curie point to room temperature (the Curie point of Ferroxcube III is 110°—160°C) some change of permeability occurs with changing temperature. For instance the permeability of Ferroxcube III can be almost halved by an increase in temperature from 20°C to 80°C. However, the permeability of Ferroxcube V is decreased less than 10% by a similar temperature change under the same conditions. Between 10° and 40°C the change in permeability averages 0.15% per 1°C for the various types of Ferroxcube.

In cases in which the magnetic circuit is normally completely enclosed, the high permeability of Ferroxcube can be used to minimize losses by means of an air gap. If the gap is such as to reduce the effective permeability to one tenth of its original value (which could still be high), losses and the effect of heat on effective permeability will also be reduced in the same proportion.

The properties of Ferroxcube allow considerable reductions to be made in the size of such items as i-f transformers or carrier-frequency filters, and it is an excellent material for magnetic screens or for permeability tuning. However its saturation point is rather low and it is not used for power transformer or output transformer cores.

Ref. A27 describes research on **ferromagnetic spinels** by the Radio Corporation of America. These spinels are ceramic-like ferromagnetic materials characterized by high permeability (up to greater than 1200), high electric resistivity (up to 10^8 ohm cm), and low losses at radio frequencies. Wide ranges in these and other properties are obtainable by varying the component ingredients and methods of synthesis.

Ferrospinels are being used increasingly in electronic equipment operating in the frequency range of 10 to 5000 Kc/s. At power and low audio frequencies the ferrospinels are not competitive with laminated ferromagnetic materials, and at very high frequencies the losses in ferrospinels are excessive when high permeability is required. It is possible however to produce a ferrospinel, with low permeability, useful at frequencies in the order of 100 Mc/s. The application of ferrospinels as core bodies in the deflection yoke, horizontal deflection transformer and high voltage transformer for television receivers is now finding wide acceptance. The ferrospinels are especially suited to television video frequencies as their use in these components results in improved performance at lower cost and with smaller space.

In the standard broadcast receiver, the ferrospinels are expected to be used in the radio frequency circuits as " trimmers " and as permeability tuning cores. With a properly designed coil it is possible to tune a circuit, by the movement of a ferrospinel rod, from 500 to 3000 Kc/s, or to cover the standard American broadcast band (540 to 1730 Kc/s) with only three eighths of an inch movement of the rod.

By using a ferrospinel with a high electric resistivity as the core body for radio frequency inductances, the wire body may be placed on the ferrospinel without additional insulation. In fact, the conductor may be affixed by the printed circuit technique for some applications.

SECTION 4 : MEDIUM WAVE-BAND COILS

(i) Air-cored coils   (ii) Iron-cored coils   (iii) Permeability tuning   (iv) Matching.

(i) Air-cored coils

With large diameter formers reasonable Q can be obtained on the broadcast band with air-cored solenoids. However in limited spaces, coils are wound with two or more pies or by progressive universal winding when high Q is required. The progressive winding has the advantage of being less susceptible to inductance variation through careless handling or winding, and from the production point of view it is desirable because it can be wound without stopping.

Although high Q is not necessary for broadcast band oscillator coils, it is desirable. Low Q tuned circuits need larger reaction windings which in turn give increased phase shift in the voltage fed back from the plate into the grid circuit. Consequently an even larger reaction winding is required and the result may be reactance reflected into the secondary circuit which makes it impossible to tune the required range—if the police band (U.S.A.) is to be included. In any case the reflected reactance, which is a variable throughout the tuning range, complicates the tracking problem.

A method of decreasing the coupling inductance required, which is applicable to progressive but not to ordinary universal windings, is to thread both secondary and primary wires together through the winding finger, wind the required number of primary turns, terminate the primary and proceed with the secondary winding until completed. This method is possible with progressive windings because of the gaps which occur throughout the winding, whereas with a normal universal winding there is no room for a primary between turns of the secondary.

For r-f and particularly for aerial coils the need for high Q is greater. The sensitivity, signal-to-noise ratio and image ratio of a receiver are all governed by the aerial coil Q.

Since the amount of wire used in an aerial coil secondary is so much less than in the i-f transformers it is often possible to use more expensive litz.

Probably the most satisfactory method of coupling the aerial to the grid of the first valve (or the plate of the first valve to the grid of the second) is by means of a high impedance primary. This can take the form of a universal winding of the finest wire the machines will conveniently handle, wound at an appropriate distance from the secondary to give about 20% coupling (depending on the Q of the secondary, the amount of mistracking that can be tolerated, and the types of aerial expected to be used). In the case of progressive universal secondaries it sometimes happens that the primary must be wound as closely as possible to the secondary to obtain the required degree of coupling.

The primary should be resonated by its own self-capacitance, or by an additional capacitor, below the intermediate frequency so that no normal aerial can tune the primary to the intermediate frequency and so cause instability. A typical value of primary inductance is 1 mH and a 150 $\mu\mu$F capacitor would be suitable for use with it.

To increase the aerial coil gain at the high frequency end of the broadcast band a capacitor of the order of 4 $\mu\mu$F might be used between aerial and grid of the first valve. Since the capacitor decreases image ratio at the same time as it increases sensitivity and signal-to-noise ratio, some requirement other than flat sensitivity across the broadcast band might be desirable.

Similar considerations apply to the coupling of the r-f valve except that primary inductances are usually larger (say 4 mH) and their tuning capacitors may be increased in size until the gain is reduced to the required figure.

The direction of connection for high impedance coils and oscillator coils is the same as that already detailed for i-f transformers ; the grid is connected to the end of the secondary further from the primary, to reduce capacitance coupling, and then, considering the secondary and primary to be one tapped coil wound in the one direction, the aerial (or plate) is connected to the other extreme of the winding.

(ii) Iron-cored coils

The use of suitable iron cores greatly simplifies the problem of obtaining high Q windings. In fact, for any reasonable Q requirement it is only a matter of obtaining a suitable core.

Whilst for a given Q and coil size a coil wound on an iron slug will be less affected by its surroundings than an air-cored coil, owing to the concentration of the flux in the core, high Q secondaries (e.g. 250) will be found to require very careful placement if serious Q losses are to be avoided, particularly at the high frequency end of the broadcast band. Even an open circuited primary at the appropriate distance for correct coupling may noticeably decrease the Q.

The primary should be wound on the opposite side of the secondary from the iron core, which will not pass completely through the secondary because of the necessity for leaving room for adjustment. Under these conditions movement of the core to compensate for differences between individual receivers will not seriously affect the coupling between primary and secondary.

To avoid losses due to adjacent components with high Q coils, iron cores can be used which completely enclose the winding. These cores give higher Q's than the more usual slugs with a similar winding and reduce the external field of the coil to such an extent that in extreme cases it may be difficult to obtain sufficient coupling between the secondary and an external high impedance winding.

An advantage of using an iron core in the oscillator coil is that, apart from giving increased Q, it can greatly increase the coupling between primary and secondary, thereby decreasing the primary reactance for a given amount of coupling.

(iii) Permeability tuning

The increasing use of permeability tuning is an indication of the extent to which the design and production difficulties associated with this type of tuning are being overcome.

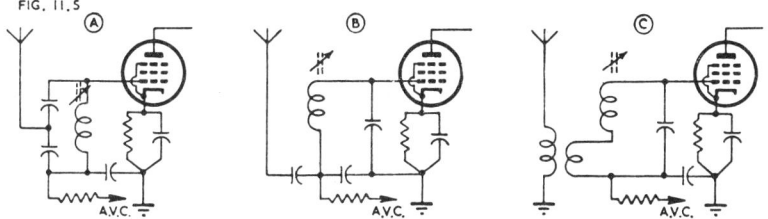

Fig. 11.5. Aerial coupling circuits used with permeability tuning.

No gang condenser is needed, so that even with the added complication in the coil assembly a cheap broadcast band tuning unit is possible. Gang microphony is automatically eliminated, and a value of tuning capacitor can be chosen which will make the effects of, for instance, capacitance variations due to valve replacement negligible over the whole tuning range. This capacitor can be a low-loss type and it may be given a suitable temperature coefficient if desired. The space saving owing to the elimination of the tuning capacitor is appreciable, even taking into account the additional space required for the iron cores and their driving mechanism.

By a suitable choice of coil and core constants the oscillator grid current can be maintained at or very close to the desired figure at all points in the broadcast band. Aerial coil gain can also be kept constant.

One of the most serious troubles in the design of a permeability tuning unit is in obtaining satisfactory coupling between the aerial and the permeability-tuned aerial coil. If a normal primary is used the coupling variations with changing core position, as the coil is tuned over the broadcast band, are excessive. Similarly a tapped inductor can not be used because the core alters the coupling between the two sections of the winding too much. Fig. 11.5A, B and C show three methods which have been used.

The circuit of Fig. 11.5A couples the aerial straight into the tuned circuit and gives good aerial coil gain when correctly aligned. However image rejection is poor and the detuning effect when different aerials are connected to the receiver is considerable.

Improved image rejection is obtained from the circuit of Fig. 11.5B but sensitivity is lost because much of the voltage developed across the tuned circuit is not applied between grid and cathode of the valve.

Fig. 11.5C gives a more satisfactory solution (Ref. A24). There is no inductive coupling between primary and permeability tuned secondary and the coupling between primary and coupling coil is unaffected by the core position. A constant gain of about three times over the band can be obtained from a coil with average effective Q of 100, with satisfactory image ratios across the band and reasonable freedom from detuning effects with different aerials.

The maintenance of correct calibrations in production receivers is a problem which concerns the designer of the iron dust core, but cores have been produced with permeability consistent enough for station names for the whole broadcast band to be marked on the dial and used.

Iron cores are available with sufficient variation in permeability for the complete broadcast band to be tuned. However if the police band is required and space for mounting the tuning unit is restricted (resulting in appreciable damping on the coil) the permeability requirements for the core are severe.

The Colpitts oscillator circuit is suitable for use in permeability tuning un ts, and padding can be obtained by using a suitably shaped core for the oscillator coil. Alternatively the oscillator coil can be wound in two series-connected sections, one with fixed inductance (or pre-tuned for alignment purposes) and the other with inductance varied by the moving core. Many receivers have been built with complete tracking i.e. a series inductor corresponding to the usual (parallel) trimmer, and a parallel inductor corresponding to the usual (series) padder.

The rate of change of inductance with core movement does not give a linear relationship between tuning control and the frequency tuned unless special precautions are taken. Linearity, if required, can be obtained in two ways The drive between tuning spindle and cores can include some type of worm or cam arranged to vary the speed of the drive in suitable parts of the tuning range, or the coils themselves can be progressive universal windings to give the same effect. Progressive universal coils lend themselves very well to this form of winding because the progression is provided by a screw thread and this can be cut with the pitch varied over the length of the winding to give the required tuning rate.

An important drawback to permeability tuning is the difficulty involved in providing coverage of a number of wavebands without complications which outweigh the saving of a tuning condenser.

One simple method of adding short wave ranges to a broadcast band permeability tuner is to connect the broadcast coil in parallel with a short wave coil which is to be tuned. This gives satisfactory tuning but restricted bands. It is suitable for a band-spreading circuit but when every international short wave broadcasting band is to be covered, a large number of ranges is required. By designing a special short wave permeability tuner it is possible to cover the 6 to 18 Mc/s band in two ranges (Ref. A24).

For more restricted ranges such as the F-M broadcasting band, permeability tuning is satisfactory provided that a suitable type of iron is available. Cores have been made which can be used in tuned circuits at 150 Mc/s without introducing undue loss.

References A11, A16, A20, A22, A23, A24, A34.

(iv) Matching

Methods of coil matching depend on the type of coil and to a certain extent upon the subsequent alignment procedure.

Air-cored solenoids can be wound with a spacing towards the end of the winding equivalent to one turn. The inductance is then increased by sliding turns across the gap from the small end section to the main body of the coil, or decreased by sliding turns in the opposite direction.

Air-cored pie-wound coils can be matched by pushing pies together to increase inductance, but the method has disadvantages. Firstly, the coils are liable to be damaged, even to the extent of collapsing completely, and secondly coils which are rigid enough to withstand the pushing without damage may have enough elasticity to return towards their original positions over a period of time, thus destroying the matching. If coils matched in this way are to be impregnated it is necessary to re-check (and readjust in many cases) after impregnation, and recheck again after flash dipping, if a reasonable degree of matching accuracy is to be maintained.

Another method which has been used with air-cored coils is to wind each coil with slightly too much inductance, bake, impregnate and flash-dip the coil and then seal it in an aluminium can. A groove is then run in the can above the appropriate coil and the depth of the groove is increased until the inductance of the coil inside the can is reduced to that of a standard. All production inductance variations can be absorbed in the matching process with insignificant Q loss.

When iron cores are used, matching problems are much simplified. Even when the core is not threaded it can be pushed through a winding for inductance adjustment. With threaded cores the adjustment is so simple that in many cases the coils are not matched. Given correct calibrations and close tolerance padders in a complete receiver, the oscillator coil trimming and padding can be adjusted to suit the calibration and the aerial coil trimming and padding to suit the oscillator with results which may even be (from the user's point of view) superior to those obtained from more complicated matching and alignment methods.

When matching individual windings on formers containing two or more coils it is possible for errors to occur due to variations in windings other than the one being matched. For example an aerial coil secondary may be matched to the same inductance as a standard secondary but may be coupled to a primary with an inductance different from that used on the standard coil. Under operating conditions in a receiver the two effective secondary inductances would be different.

This difficulty can be overcome by suitable connection of associated coils during matching. Normally the best procedure is to short-circuit and earth coils which are not being matched, but on occasion it may be better to earth the cold end only.

SECTION 5 : SHORT-WAVE COILS

(*i*) *Design* (*ii*) *Miscellaneous features.*

(i) Design*

Much has been written on the subject of designing coils suitable for use on short waves. A number of references will be found at the end of this chapter. The work by Pollack, Harris and Siemens, and Barden and Grimes is very complete from the practical design viewpoint. The papers by Butterworth, Palermo and Grover, and Terman are basically theoretical. Austin has provided an excellent summary and practical interpretation of Butterworth's four papers. Medhurst's paper gives the results of measurements which in some cases disagree with Butterworth's theoretical values. For coils whose turns are widely spaced the measurements of high-frequency resistance are in good agreement but for closely-spaced coils the measured values are very considerably below those of Butterworth. Theoretical reasons are given for these differences.

(A) **Pollack** (Ref. G4) summarizes the procedure for the optimum design of coils for frequencies from 4 to 25 Mc/s as follows :—

1. Coil diameter and length of winding : Make as large as is consistent with the shield being used. The shield diameter should be twice the coil diameter, and the ends of the coil should not come within one diameter of the ends of the shield.

2. A bakelite coil form with a shallow groove for the wire, and enamelled wire may be used with little loss in Q. The groove should not be any deeper than is necessary to give the requisite rigidity. The use of special coil form constructions and special materials does not appear to be justified (except for the reduction of frequency drift due to temperature changes).

3. Number of turns : Calculate from
$$N = \sqrt{L(102S + 45)/D}$$
where S = ratio of length to diameter of coil,
D = diameter of coil in centimetres,
and L = inductance in microhenrys.

4. Wire size : Calculate from
$$d_0 = b/N\sqrt{2} = \text{optimum diameter in centimetres.}$$
where b = winding length in centimetres.

---

*This is a revision and expansion of the chapter on this subject in the previous edition by L. G. Dobbie.

That is, the optimum wire diameter is $1/\sqrt{2}$ times the winding pitch, measured from centre to centre of adjacent turns.

(B) Barden and Grimes (Ref. G16) recommend for coils working near 15 Mc/s. that No. 14 or No. 16 A.W.G. enamelled wire on a form not less than one inch diameter at a winding pitch equal to twice the wire diameter is desirable. The screen diameter should be not less than twice the coil diameter. A comparison of coils of equal inductance on 0.5 in. and 1 in. forms in screens double the coil diameter indicates that the value of Q is twice as great for the larger diameter coil. No. 24 A.W.G. wire was used for the small diameter coil.

(C) Harris and Siemens (Ref. G21) quote the following conclusions :—
(1) Q increases with coil diameter.
(2) Q increases with coil length, rapidly when the ratio of length to diameter is small, and very slowly when the length is equal to or greater than the diameter.
(3) Optimum ratio of wire diameter to pitch is approximately 0.6 for any coil shape. Variation of Q with wire diameter is small in the vicinity of the optimum ratio ; hence, selection of the nearest standard gauge is satisfactory for practical purposes.

(D) The shape of a coil necessary for minimum copper loss (from Butterworth's paper) is stated by **Austin** (Ref. D7) as follows :—
(1) Single layer solenoids : Winding length equal to one-third of the diameter.
(2) Single layer discs (pancake) : Winding depth equal to one-quarter of the external diameter.

(E) Butterworth's paper (Ref. D1) deals with the copper loss resistance only, and insulation losses must be taken into account separately. **Insulation losses** are minimized by winding coils on low loss forms, using a form or shape factor which provides the smallest possible self-capacitance with the lowest losses. Thus air is the best separating medium for the individual turns, and the form should provide only the very minimum of mechanical support. Multilayer windings in one pie have high self-capacitance due to proximity of the high and low potential ends of the winding. The same inductance obtained by several pies close together in series greatly reduces the self-capacitance and associated insulation losses. Heavy coatings of poor quality wax of high dielectric constant may introduce considerable losses.

(F) In the section of **Medhurst's paper** (Ref. D6) dealing with h.f. resistance he states that so long as the wire diameter is less than one half of the distance between wire centres, Butterworth's values are applicable. An increasing error occurs as the ratio of wire diameter to distance between centres increases, the values being 190% too high when the ratio is 0.9.

A graph is presented giving variation in optimum wire spacing with variation in the ratio of coil length to coil diameter (Fig. 11.6A).

It is shown that a good approximation to the high frequency Q of coils of the type measured is given by the simple expression

$$Q = 0.15 R \psi \sqrt{f}$$

where R = mean radius of coil (cm),
$\quad f$ = frequency (c/s)
and ψ depends on the length/diameter and spacing ratios.
A comprehensive table of ψ values is presented.

(G) M. V. Callender (Ref. D9) points out that Medhurst's formula for Q can be approximated within a few per cent by the equation

$$Q = \sqrt{f}/(6.9/R + 5.4/l)$$

where R = radius of coil (cm).
and l = length of coil (cm).
An even simpler expression

$$Q = 0.15\sqrt{f}/(1/R + 1/l)$$

follows the data to a few per cent provided $l > R$.

The range of conditions under which this formula applies is the same as that to which Medhurst's data refer : in particular—

(a) The ratio of wire diameter to wire spacing must approximate to the optimum shown in Fig. 11.6A.

(b) The formulae apply only for very high frequencies because of skin effect considerations. However the following table, giving the thinnest wire for which the formula applies within $\pm 10\%$, shows that most practical solenoids will be covered—

f (Mc/s)	1.0	4.0	16
wire	22 S.W.G.	28 S.W.G.	37 S.W.G.

(c) The formulae do not hold for coils of very few turns (or extremely short coils).

(d) Dielectric loss is not allowed for. This is unlikely to be serious except where the coil has a rather poor dielectric (bakelite or worse) and is used in a circuit having a low parallel tuning capacitance.

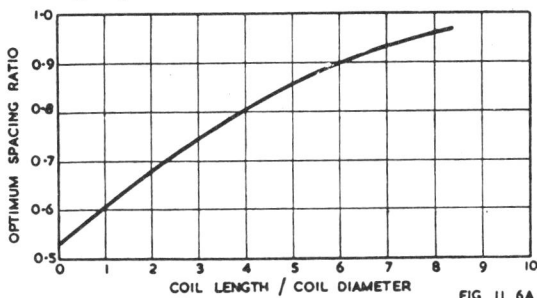

Fig. 11.6A. Variation of optimum spacing ratio with length/diameter (Ref. D6).

(H) **Meyerson** (Refs. G3 and G11) gives the results of measurements on a large number of coils tuned over a range between 25 and 60 Mc/s. The following information is obtained from his work :—

(a) Q at any frequency within the band, and the frequency for maximum Q, both increase with an increase in wire size for a given coil diameter. 10 A.W.G. wire (the largest used) was nearly 10% better over the range than 12 A.W.G. wire on a one inch diameter former.

(b) Maximum Q increases, and the frequency for maximum Q decreases, with an increase in coil diameter for a given wire size, number of turns and number of turns per inch.

(c) The coil diameter required for highest Q throughout the tuning range decreases with increasing frequency for a given wire size, number of turns and number of turns per inch.

(d) Maximum Q is obtained with a spacing between adjacent wires which is slightly greater than the bare wire diameter.

(e) No variation in Q was detected between coils wound with bare wire, enamelled wire or silver plated wire.

On the other hand, cotton covering decreased Q by as much as 5% at 50 Mc/s and annealing of the copper wire increased Q by less than one per cent.

(I) In Refs. G10 and G12 **Meyerson** gives details of work with frequency ranges between 60 and 120 Mc/s on the Q and frequency stability of inductors made of wire, tubing and strip in various shapes e.g. disc, hairpin, folded lines, etc.

(J) Other points taken from various references are given below. **Dielectric losses** present in the self capacitance of the coil are reduced by altering the shape to separate the high potential end from all low potential parts of the circuit. These losses become relatively more important the higher the frequency Thus, the shape of a solenoid for minimum total losses may need to be increased beyond what would otherwise be the optimum length.

The high frequency alternating field of a coil produces **eddy currents** in the metal of the wire, which are superimposed upon the desired flow of current. The first effect is for the current to concentrate at the outside surface of the conductor, leaving the interior relatively idle. In a coil where there are numbers of adjacent turns carrying current, each has a further influence upon its neighbour.

In turns near the centre of a solenoid the current concentrates on the surface of each turn where it is in contact with the form, i.e. at the minimum diameter. In turns at either end of a solenoid the maximum current density occurs near the minimum diameter of the conductor, but is displaced away from the centre of the coil.

Thus most of the conductor is going to waste. Multi-strand or litz (litzendraht) wires have been developed to meet this difficulty. Several strands (5, 7, 9, 15 being common) are woven together, each being of small cross section and completely insulated by enamel and silk covering from its neighbours. Owing to the weaving of the strands, each wire carries a nearly similar share of the total current, which is now forced to flow through a larger effective cross section of copper. The former tendency towards concentration at one side of a solid conductor is decreased and the copper losses are correspondingly reduced.

Litz wire is most effective at frequencies between 0.3 and 3 Mc/s. Outside of this range comparable results are usually possible with round wire of solid section, because at low frequencies " skin effect" steadily disappears while at high frequencies it is large even in the fine strands forming the litz wire, and is augmented by the use of strands having increased diameter.

Screens placed around coils of all types at radio frequencies should be of non-magnetic good-conducting material to introduce the least losses. In other words, the Q of the screen considered as a single turn coil should be as high as possible. In addition, the coupling to the coil inside it should be low to minimize the screen losses reflected into the tuned circuit. For this reason the screen diameter should, if possible, be at least double the outside diameter of the coil. A ratio smaller than 1.6 to 1 causes a large increase in losses due to the presence of the screen.

The design of coils for use with iron core materials depends mainly upon the type of core material and the shape of the magnetic circuit proposed. Nearly closed core systems are sometimes used with high permeability low loss material. More commonly, however, the core is in the form of a small cylindrical plug which may be moved by screw action along the axis of the coil and fills the space within the inside diameter of the form. The main function of the core in the latter case may be only to provide a means of tuning the circuit rather than of improving its Q. When improvement in Q is possible with a suitable material, the maximum benefit is obtained by ensuring that the largest possible percentage of the total magnetic flux links with the core over as much of its path as possible ; the ultimate limit in this direction is of course the closed core.

An excellent series of charts for the design of single-layer solenoids for a required inductance and Q is presented by A. I. Forbes Simpson in Ref. G17. These charts and the instructions for their use are reprinted by permission of " Electronic Engineering " and the author (Figs. 11.7 to 11.13 inclusive).

The use of the charts is simple and gives direct answers. As so many unascertainable factors govern the final inductance of a coil in position, no attempt has been made to achieve an accuracy better than 1 per cent. The inductances indicated by the charts assume that the leads to the coil are of the same wire as the coil and perpendicular to it, and are each a coil diameter long.

Coils of this form have a low self-capacitance which is largely independent of all save the coil diameter and to a lesser extent the wire gauge.

The capacitance may be expressed as

$C = D(0.47 + a) \ \mu\mu F$

where D is coil former diameter in cms. and a is a constant depending on the gauge of wire, lying between 0.065 for 42 S.W.G. and 0.11 for 12 S.W.G. wire, where one lead of the coil is connected to chassis.

Use of charts

The charts shown may be used as follows :

If the desired Q is known, then reference to Fig. 11.8 will suggest a suitable diameter of former.

If, however, as great a Q as possible is required, then the largest possible former should be used unless the coil is to be sited in a can, in which case the diameter of the former should be less than one half of the internal diameter of the can and spaced from the end by the same amount or more.

When the diameter of the former is determined, reference should be made to the appropriate chart, and as shown in Fig. 11.7 the inductance intercept A corresponds to C which gives the wire gauge (D) which gives the number of turns per inch for that gauge, and E the number of turns required.

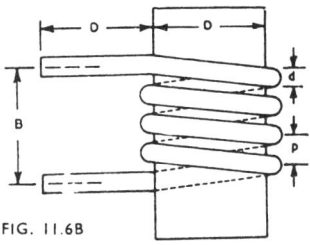

FIG. 11.6B

Fig. 11.6B. Dimensions of coil referred to in text.

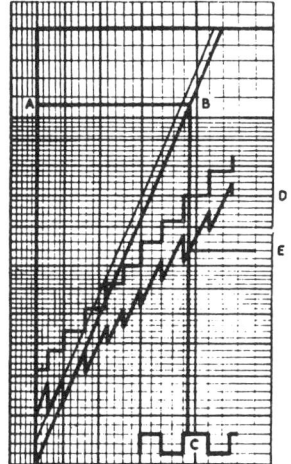

Diagram to illustrate use of Charts

FIG. 11.7

Fig. 11.7. Diagram to illustrate use of charts (Ref. G17).

It will be seen that " even " gauges of wire have been used. If it is desired to use an " odd " gauge, a little simple geometry will soon give the result.

The Q in air of a coil wound to this information, when tuned by capacitances between 45 and 500 $\mu\mu$F may be read directly from Fig. 11.8.

The Q values shown in the chart have been corrected for coil capacitance and are somewhat higher than those indicated by the usual Q meter. The accuracy of these curves is rather less than that of the inductance charts, but various values have been checked by several methods which have given substantial agreement.

Example

As an example* of the use of the curves, suppose we have a variable condenser of 450 $\mu\mu$F swing and desire to find a coil with which it will tune from 12 Mc/s to 4 Mc/s. If the total capacitance at 12 Mc/s. be C $\mu\mu$F then

$$(C + 450)/C = 12^2/4^2 = 9 \text{ and } C = 56.25 \ \mu\mu F$$
$$L = 1/4\pi^2f^2C$$
$$= \frac{10^{12} \cdot 10^6}{4\pi^2 \times 12^2 \times 10^{12} \times 56.25} \ \mu H = 3.15 \ \mu H.$$

Inspecting Fig. 11.8 we see that the Q at 4 Mc/s. will lie between 116 for $\frac{1}{2}$ in. and 168 for a $\frac{7}{8}$ in. former, while that at 12Mc/s. will lie between 182 for a $\frac{1}{2}$ in. and 265 for $\frac{7}{8}$ in. former.

If we choose a $\frac{5}{8}$ in. former we find that the line corresponding to this in Fig. 11.10 cuts the L line at a wire diameter of 0.025 in. This line intersects T.P.I. at 23.8 and the " No. of turns " is 18.5 with a wire gauge of 22 S.W.G.

Referring to Fig. 11.8—we find for $L = 3.15 \ \mu H$:

C $\mu\mu F$	45	100	200	400	500
Q	212	195	175	150	142

Z_o is readily obtained by writing $Z_o = Q\sqrt{L/C}.$

---

*This example is presented in a different form from that in the original article.

If, however, we had used the ½ in. former we would find that 18 turns of 26 S.W.G. at 37 T.P.I. would give

C μμF	45	100	200	400	500
Q	181	165	147	123	117

(ii) Miscellaneous features
(A) Matching
Short wave coils are frequently matched by altering the position of the last few turns on the former. This sometimes leads to damage to the coil, particularly if there is an interwound primary.

An alternative method, which is applicable to coils terminated by passing the wire diametrically through the former, is to push sideways this terminating wire in the

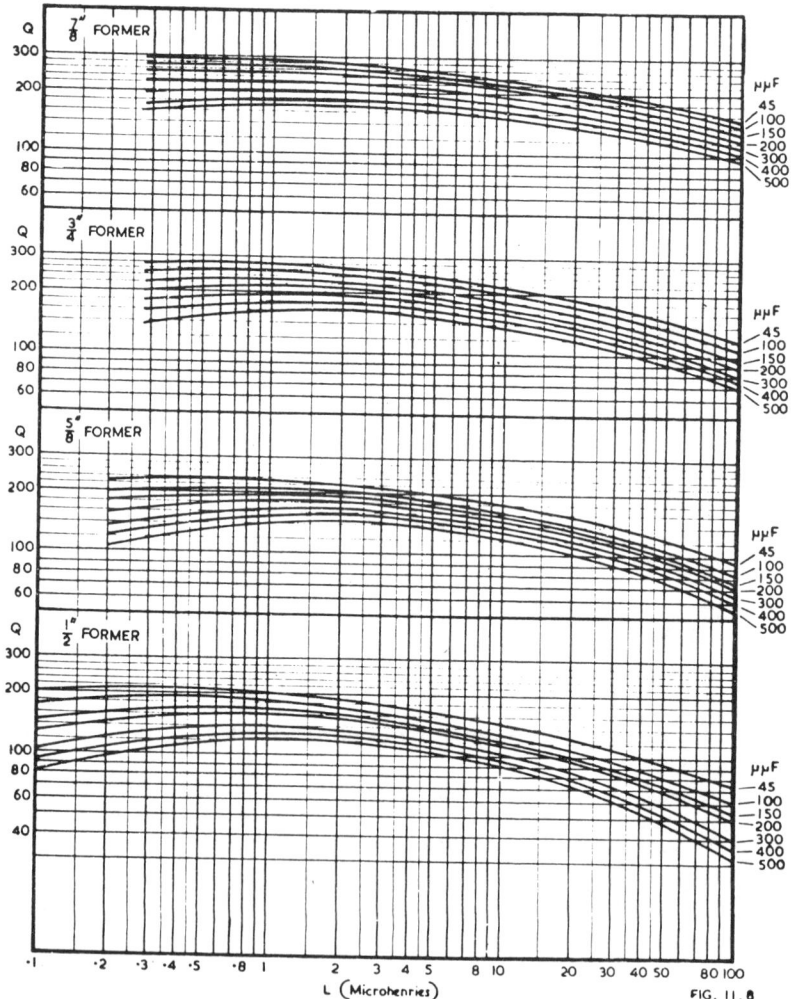

Fig. 11.8. *Q/L chart for obtaining suitable size of former (Ref. G17).*

middle of the coil. In this way the last half turn of the coil can be increased almost to a complete turn or decreased to a small fraction of a turn. Both the first and last half turns on the coil can be treated in this way, thereby providing a possible inductance adjustment equivalent to approximately one complete turn of the winding. When matching short wave coils it is essential that associated primaries, and any other coils on the same former, be connected in a similar manner to their ultimate connection in the receiver, i.e. earthed ends of coils should be earthed, and hot ends either earthed or left open depending on which is more suitable.

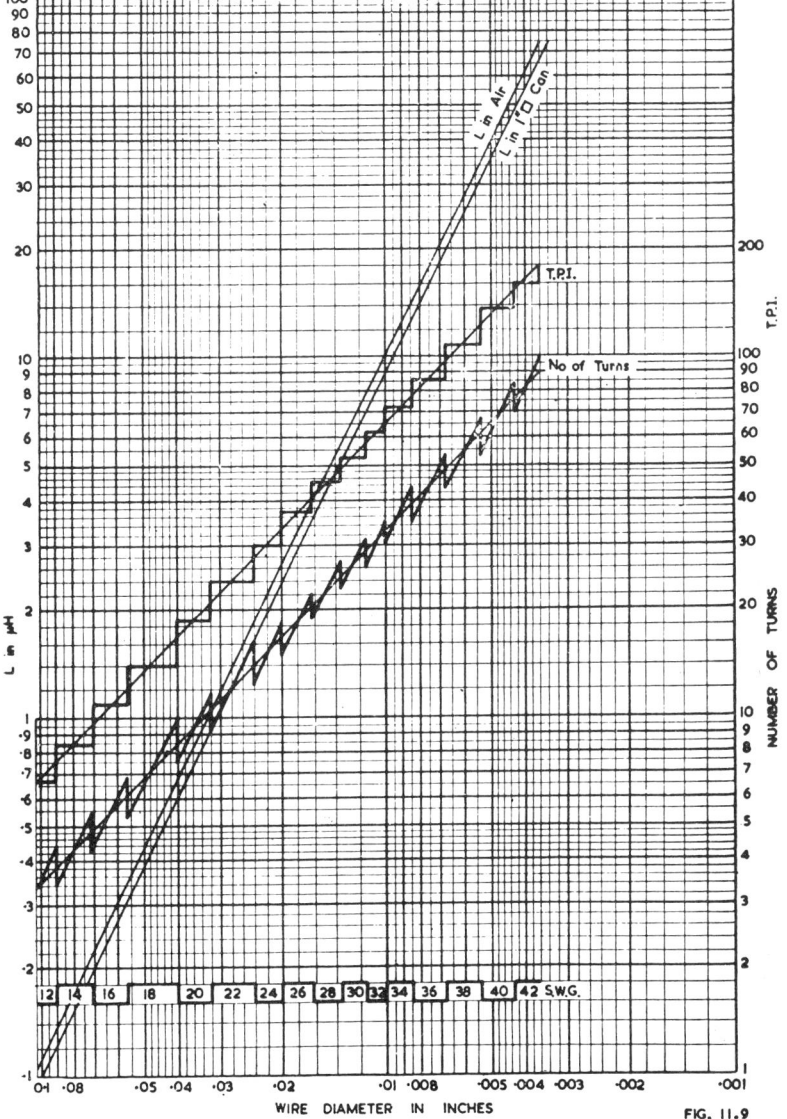

Fig. 11.9. Optimum Q, ½ in. former (Ref. G17).

Iron cores are widely used for adjusting the inductance of short wave coils. Because at higher frequencies increased losses in cores offset the advantage of increased permeability, the cores in some cases do not increase the Q of a coil, but provide a convenient method of inductance adjustment.

In shortwave oscillator coils an iron core also has the advantage of increasing the coupling between primary and secondary for a given winding. Thus increased grid current may be obtained although Q is not increased.

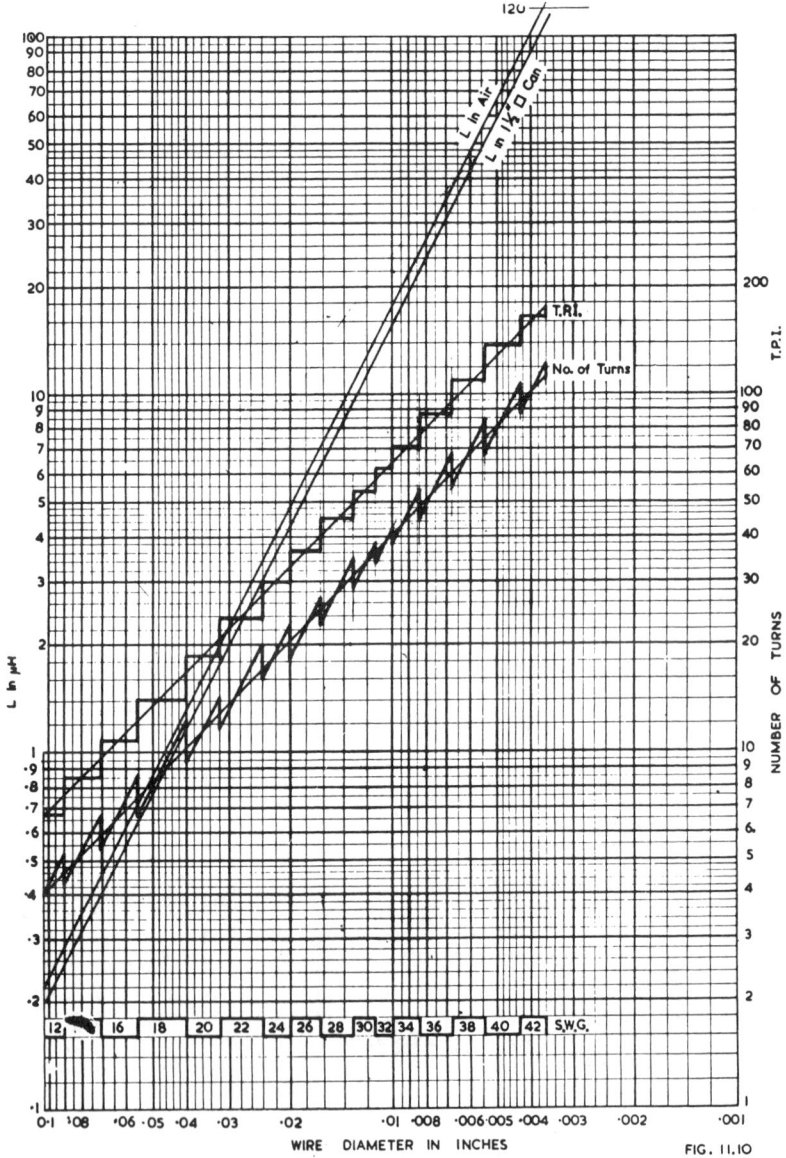

Fig. 11.10. *Optimum Q, $\frac{5}{8}$ in. former (Ref. G17).*

At frequencies of the order of 100 Mc/s the difficulties of manufacturing suitable iron cores are greater although they are made and used (Ref. A22) but it is comparatively simple to wind air cored coils with a Q of say 300. Because of this, inductance adjustment can also be carried out by means of copper slugs which are adjusted in the field of the coil in the same way as are iron cores at lower frequencies. The slugs are sometimes silver plated to minimize losses, but on the other hand brass slugs are also used, apparently without undue losses.

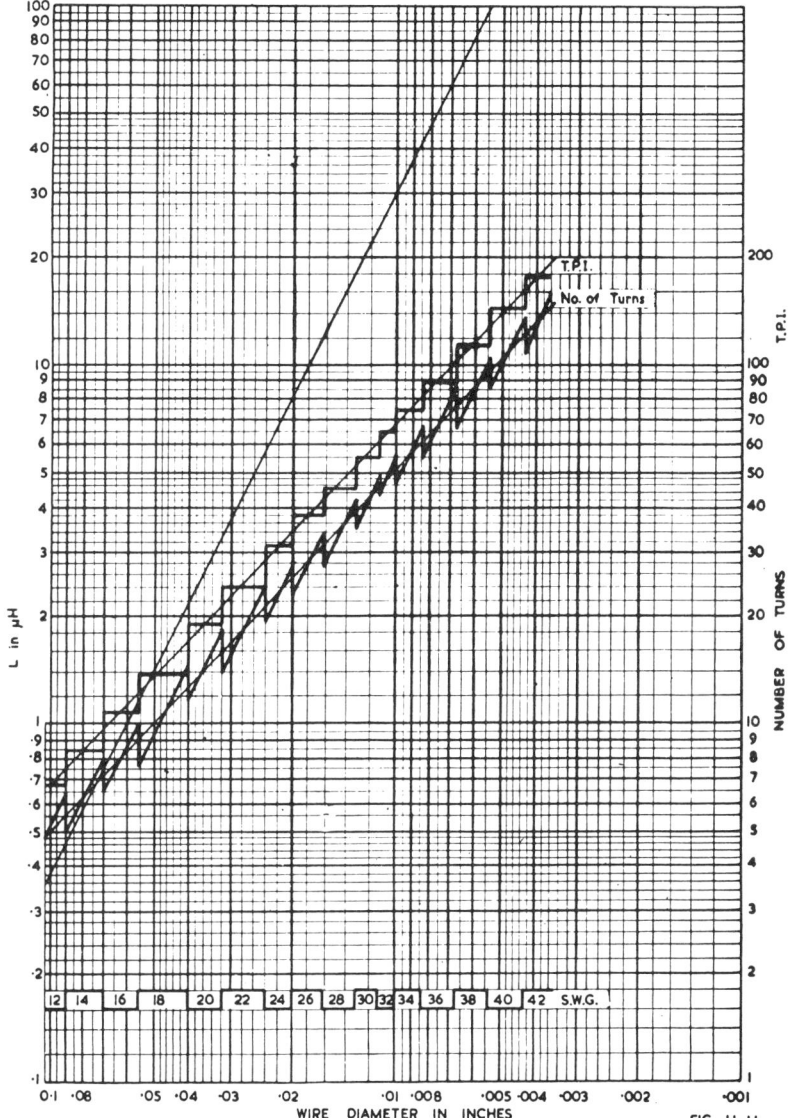

Fig. 11.11. *Optimum Q, ¾ in. former (Ref. G*17).

(B) Self-capacitance

To reduce capacitances in a coil with an interwound primary winding without noticeably decreasing the coupling, it is possible—if the spacing factor of the secondary winding allows it—to wind the primary against the wire on the low potential side of each space between secondary wires. A definite advantage is gained from this method of winding and at the same time the cause of appreciable random deviations in coil capacitance is removed.

It is usually advisable to start an interwound primary winding just outside a tuned secondary winding for a desired coupling with minimum secondary capacitance. Best results will probably be obtained with the primary started between one half

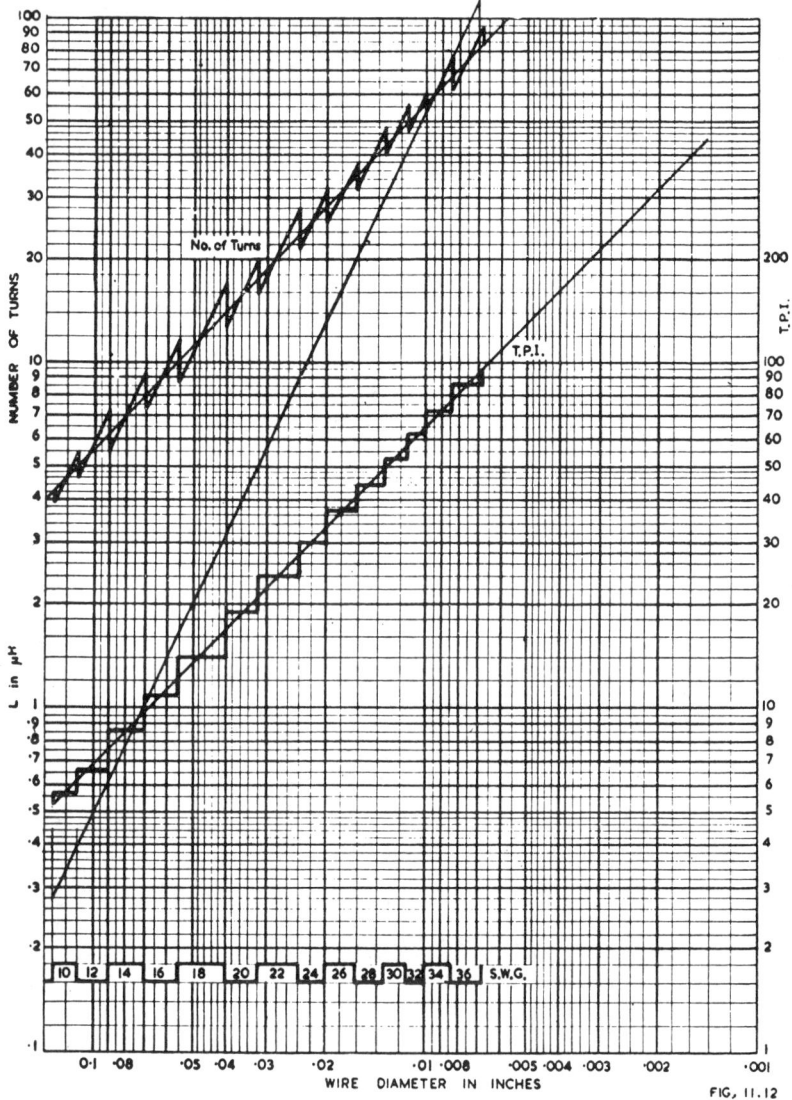

FIG. 11.12

Fig. 11.12.   *Optimum* Q, $\frac{7}{8}$ *in. former (Ref. G17).*

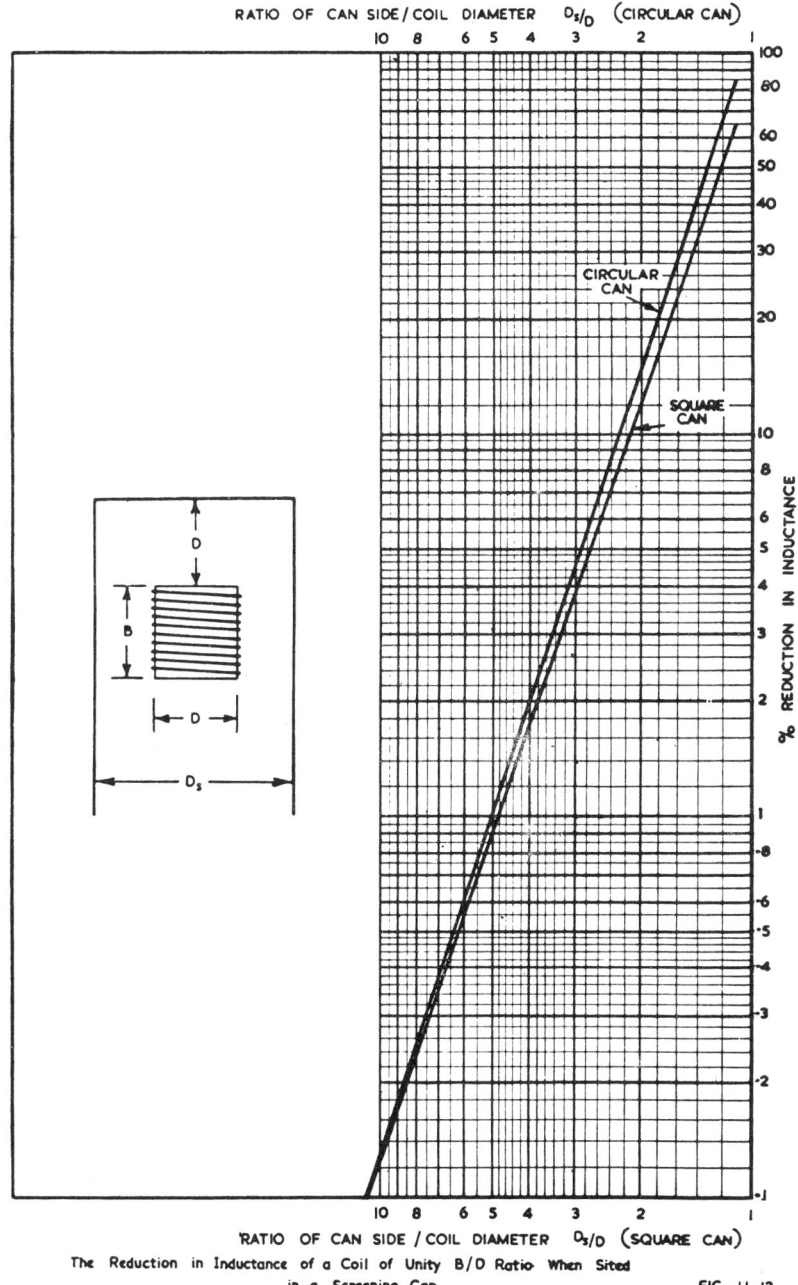

The Reduction in Inductance of a Coil of Unity B/D Ratio When Sited in a Screening Can

FIG. 11.13

Fig. 11.13. *Effect of circular and square screening cans on inductance (Ref. G17).*

and one and one half turns outside the cold end of the secondary, close spacing being used for parts of the primary outside the secondary and the external primary turns being wound as close to the last secondary turn as possible

(C) Aerial primary windings
On the 6 to 18 Mc/s short wave bands both interwound and external high imped-ance primaries are used.   External primaries have the advantage of reducing the capac-itive coupling between the aerial and the grid of the first valve with the result that oscillator re-radiation from superheterodyne receivers is noticeably reduced.   In addition, impulse interference is apparently decreased.   The gain obtainable over a band of frequencies is at least as good as that with low impedance interwound primaries and the tracking troubles experienced when a particular aerial resonates an interwound primary within the band are not so likely to occur.

A satisfactory method of winding for an external primary is to use a universal winding (perhaps 36 or 38 A.W.G. S.S.C.) with a small cam (1/10 inch or 1/16 inch) and space the primary as closely as possible to the cold end of the secondary.   About 25 turns on a 3/8 inch former would be suitable for the range from 6 to 18 Mc/s but final details must be determined in a receiver. Additional capacitance coupling will sometimes provide increased gain at the high frequency end of the band, although with the loss of some of the advantages outlined above.

(D) Direction of windings
For interwound oscillator coils the method of connection is the same as that pre-viously described—considering primary and secondary as one continuous tapped winding in the one direction, the grid is connected to the end of the secondary further from the primary (for a tuned grid oscillator) and the plate is connected to the other extreme, i.e. the end of the primary further from the grid connection.

Interwound aerial and r-f coils are connected in the opposite sense, i.e. grid to end of secondary further from primary, and plate (or aerial) to end of primary closer to grid.

External primaries are connected in the same sense as oscillator coils i.e. aerial and grid at extremes of the two windings when considered as one tapped winding in the same direction.   Care is needed with short wave coil connections because second-ary and external primary are wound on different types of machines and may be wound in opposite directions.   This necessitates reversing the connections to the primary winding.   Examination of a coil will show that if the primary is moved from one end of the secondary to the other it will also be necessary to reverse the connections to the primary.

SECTION 6 : RADIO-FREQUENCY CHOKES

(i) Pie-wound chokes   (ii) Other types.

(i) Pie-wound chokes
Radio frequency chokes are of two main types, those used above their main self-resonant frequency and those used below it. · In either case the choke is normally required to have, over a range of frequencies, an impedance higher than some minimum value and a high reactance which does not have sudden changes in its value.   The need for high impedance is obvious, and a constant value of reactance is necessary if the choke is in parallel with a ganged tuned circuit.

At frequencies below its natural resonance the impedance of a choke is due to its inductive reactance, modified by the self-capacitance as shown in Sect. 2(i) of this chapter, and so increases with frequency until self-resonance is reached.   The choke may have one or more resonances, series as well as parallel, as explained below, but above some high frequency its impedance is approximately equal to the reactance of its self-capacitance, i.e. the impedance is inversely proportional to frequency and eventually reaches values low enough to render the choke useless.

Thus to obtain high impedance over a wide band of frequencies a choke requires maximum inductance and minimum self-capacitance. For lower frequencies universal windings are used, usually split into two or more spaced pies with each pie as thin as possible for minimum self-capacitance.

To keep the impedance high over a band of frequencies it is necessary to avoid any tuning of sections of the choke by other sections. A typical case of such tuning would occur in a r-f-c consisting of two pies of different sizes. The impedance-frequency curve of such a choke would show two maxima at the self-resonant frequencies of the two sections, but between these peaks there would be a serious drop in impedance at the frequency at which the inductive reactance of one pie became series resonant with the capacitive reactance of the other. At such a frequency there would also be a sudden change in the effective reactance of the choke which would give tracking errors.

Series resonance of the type described occurs whenever pies with different self-resonant frequencies are connected together. To eliminate them it is not sufficient to wind several pies with the same numbers of turns because the effects of mutual inductance in different sections of the coil still give different self-resonant frequencies to adjacent pies. To avoid this, Miller (Ref. H6) recommends winding progressively smaller pies towards the centre of a multi-pie r-f-c to make the self resonant frequency of each pie the same. He gives figures which show a considerable increase in uniformity of impedance over a range of frequencies when this method is adopted.

In addition to the need for individual pies to be resonant at the same frequency, Wheeler (Ref. H5) gives the requirement (for two-pie chokes) that the pies should be connected so that mutual inductance opposes the self-inductance of the pies. This almost entirely removes minor resonances which occur at approximately harmonically related frequencies, and gives a very smooth curve of apparent capacitance for the r-f-c. The spacing between pies and the number of turns on the pies should be experimentally adjusted for best results with the choke located in the position in which it is to be used.

(ii) Other types

At high frequencies (above 10 Mc/s) self resonant chokes can be wound to give a high impedance over a reasonably wide band of frequencies. The winding length of such chokes for a given former diameter and wire size can be obtained from Ref. H1 for frequencies between 10 and 100 Mc/s.

In other cases it is necessary to design chokes for maximum impedance, i.e. maximum inductance and minimum self-capacitance with a minimum of resistance. Such an application occurs in the battery input circuit of a vibrator unit. Solenoids have been used for this purpose but higher impedance can be obtained with pyramid winding (say eight turns on the first layer, seven on the second and finishing with a top layer of one turn).

Even better results are obtainable by using an iron core as a former, although the winding must be arranged in such a way that the iron core does not increase the self-capacitance unduly.

Another application requiring a minimum-resistance r-f-c is a mains filter. Universal windings of, say, 26 A.W.G. wire are used in some cases, but where better performance is required a wooden former of perhaps 2 inches diameter is used, and slots in the former are filled with jumble windings of heavy wire.

SECTION 7 : TROPIC PROOFING

(*i*) *General considerations* (*ii*) *Baking* (*iii*) *Impregnation* (*iv*) *Flash dipping* (*v*) *Materials.*

(i) General considerations

The use of a suitable wax or varnish is not the only way in which the moisture resisting properties of a coil can be improved. The higher the impedance of a circuit, the greater the effect of a certain resistance in parallel with it. Thus a leakage between coil terminals (or other appropriate parts of the circuit) due to moisture will have a greater effect on a high Q circuit than on a low Q one. Depending upon the importance attached to maintenance of performance in humid conditions, it may be desirable to use low impedance (low Q and low L) circuits even at the expense of an additional stage in a receiver.

Another aspect of the question is that the moisture absorption of a coil is governed by its surface area. High Q coils usually have a larger surface (due to the use of pie-winding for instance) and thus are more liable to deteriorate rapidly under humid conditions.

This has been demonstrated in a test in which two sets of coils were used. The first set consisted of 1 mH coils wound in a single pie with a 3/8 in. cam, giving an initial Q of 75, the second set of 1 mH coils were wound with three pies and a 1/10 in. cam giving a Q of 145. Both sets of coils were baked, impregnated and flash dipped and then exposed to conditions of high humidity. At the end of one week the single pie windings had an average Q of 65, and the three pie windings an average Q of about 60. As the test continued the improved performance of the single pie windings became even more marked. In addition the variations between individual coils were much less in the single pie windings, which may have been due to the more variable nature of the flash-dip on the three-pie windings.

No coating or impregnating treatment for coils prevents moisture penetration if the exposure to high humidity is prolonged. On the other hand, some treatments will delay the ingress of moisture much longer than others—for instance the Q of a coil with a single flash-dip will fall much more rapidly than that of a coil which has been vacuum impregnated and flash-dipped twice. The corollary should not be overlooked ; after prolonged exposure to humidity the coil with a single flash dip will return to its original performance when the humidity is decreased more rapidly than will the other. Conditions of use determine requirements, but it cannot be taken for granted that the treatment which gives the slowest deterioration in Q is the best for all types of operation.

Apart from maintaining Q, moisture proofing is also required to maintain insulation resistance between different windings or between windings and earth. This is particularly important in the case of interwound coils (e.g. an oscillator coil) in which one winding is earthed and the other is of fine wire with a d.c. potential applied to it. Leakage between the two windings in the presence of moisture can produce electrolysis and a consequent open-circuit in the fine wire.

(ii) Baking

Irrespective of the moisture proofing treatment which is to be applied to a coil, the first requirement is to remove all moisture. This is done most conveniently by baking at a temperature above the boiling point of water. Materials used in coil winding are liable to be damaged at temperatures greatly in excess of 100°C and a satisfactory oven temperature is from 105° to 110°C.

The time required depends on the components being treated, and it is essential for all parts of a coil to be raised to oven temperature and maintained at that temperature at least for a short time. Baking for a quarter of an hour is a minimum for simple coils, and half an hour is a more satisfactory time.

One detail of oven design which must not be overlooked is that provision must be made for removing the water vapour as it is expelled from the coils. A current of fresh air passing through the oven is satisfactory for this purpose.

Whether the baking is followed by impregnation or flash-dipping only, the next process should begin before the coils have cooled.

(iii) Impregnation

Although vacuum impregnation gives the best penetration of the impregnant and best removal of moisture vapour and gas, a " soaking " treatment is often used and gives satisfactory results for commercial requirements.

A typical soaking treatment calls for the coils to be immersed in wax which is maintained at 105° to 110°C for a period of from one quarter to one half hour, depending on the type of winding, e.g. a thick universal winding needs more time than a spaced solenoid. In any case the coil should be left in the wax until no more bubbles are given off.

The temperature of the wax-tank needs accurate adjustment as apart from damage to coils which may result if it is too high, the characteristics of the wax may be altered by excessive heating.

Soaking in varnish (air-drying or oven-drying) is also possible but it is necessary to keep a close check on the viscosity.

(iv) Flash-dipping

Flash-dipping is commonly practised without impregnation and properly carried out it can be a satisfactory commercial treatment. The flash-dip should be of the same type of material as normally used for impregnation.

A wax-tank for flash-dipping can conveniently be maintained at a lower temperature than one for soaking, and a suitable temperature is one at which a skin of wax just develops on the surface of the tank on exposure to air. By keeping the top of the tank covered when not in use the skin will not be troublesome and it can readily be cleared away when necessary.

With the wax temperature low, a thicker coating is obtained on the coil when dipped and the tendency for " blow-holes " to develop is minimized. This is partly due to the thicker coating and partly because the coil itself is heated less by the wax and so has less tendency to expel air through the coating of wax as it is setting.

Once a hole has developed it is desirable to seal it with a hot instrument. To attempt to seal it with a second flash-dip usually results in failure as air is expelled through the hole in the first coat as the coil is warmed up and this punctures the second coating in the same place.

If flash-dipping alone is used for moisture proofing a minimum treatment might be, first flash-dip, seal all blow-holes with wax, then second flash dip.

Similar troubles are experienced with holes when varnish sealing is used.

(v) Materials

The first requirement for an impregnating compound is good moisture resistance. It is also important that its losses be low at the frequencies concerned, and it is quite possible to obtain impregnants which can be used without noticeably affecting the Q of the impregnated winding.

Because it is usually desirable to have minimum self-capacitance in a coil, the dielectric constant should be low, and the temperature coefficient should also be known. Waxes can be obtained with dielectric constants having a negative temperature coefficient and this is useful in offsetting the temperature coefficient of the coil itself, which is positive in all usual cases.

Apart from the resistance of a moisture proofing compound to moisture absorption, its surface properties are important. A compound which does not wet easily is desirable and tests on moisture proofing should be designed to separate the effects of surface leakage from those of absorption.

Where flash-dipping only is carried out, some of the properties mentioned above are not of such importance but they cannot be ignored completely.

Other points which must be considered when choosing a moisture-proofing compound are that it must be capable of maintaining its properties over the range of temperatures to which it is liable to be subjected. Melting of waxes is the trouble most likely to be encountered, but crazing (and the consequent admission of moisture) may be experienced at low temperatures.

Waxes and varnishes do not themselves support **fungus growth**, but they are liable to collect coatings of dust which may do so. Suitable precautions to avoid this may be desirable. More important is the avoidance in coils as far as possible of materials which are subject to fungus attack. Some types of flexible tubing fall in this category and should not be used, and the same applies to some cheap coil former materials. The fabric covering of wires also may support fungi and should be completely coated with the wax or varnish to avoid this occurrence.

SECTION 8 : REFERENCES

(A) REFERENCES TO IRON CORES
A1. Nottenbrock, H., and A. Weis " Sirufer 4 " Radio Review of Australia 4.4 (April 1936) 5.
A2. Austin, C., and A. L. Oliver " Some notes on iron-dust cored coils at radio frequencies " Marconi Review, No. 70 (July 1938) 17.
A3. Editorial " Iron powder compound cores for coils " W.E. 10.112 (Jan. 1933) 1.
A4. Editorial " Iron core tuning coils " W.E. 10.117 (June 1933) 293.
A5. Editorial " Iron powder cores " W.E. 10.120 (Sept. 1933) 467.
A6. Friedlaender, E. R. " Iron powder cores : their use in modern receiving sets " W.E. 15.180 (Sept. 1938) 473.
A7. Editorial " Distribution of magnetic flux in an iron powder core " W.E. 15.180 (Sept. 1938) 471.
A8. Welsby, V. G. " Dust cored coils " Electronic Eng. (1) 16.186 (Aug. 1943) 96 ; (2) 16.187 (Sept. 1943) 149 ; (3) 16.188 (Oct. 1943) 191 ; (4) 16.189 (Nov. 1943) 230 ; (5) 16.190 (Dec. 1943) 281. Correspondence by E. R. Friedlaender and reply by V. G. Welsby Electronic Eng. 16.192 (Feb. 1944) 388.
A9. Cobine, J. D., J. R. Curry, C. J. Gallagher and S. Ruthberg " High frequency excitation of iron cores " Proc. I.R.E. (Oct. 1947) 1060.
A10. Buckley, S. E. " Nickel-iron alloy dust cores " Elect. Comm. 25.2 (June 1948) 126.
A11. Polydoroff, W. J. " Permeability tuning " Elect. 18.11 (Nov. 1945) 155.
A12. Bushby, T. R. W. " A note on inductance variation in r-f iron cored coils " A.W.A. Tec. Rev. 6.5 (Aug. 1944) 285.
A13. White, S. Y. " A study of iron cores " Comm. 22.6 (June 1943) 42.
A14. Friedlaender, E. R. " Permeability of dust cores " Correspondence, W.E. 24.285 (June 1947) 187.
A15. Fetherston, N., and L. W. Cranch " Powder metallurgy and its application to radio engineering " Proc. I.R.E. Aust. 5.9 (May 1945) 3.
A16. Tucker, J. P. " A permeability tuned push button system " Elect. 11.5 (May 1938) 12.
A17. " Powdered-iron cores and tuning units " Comm. 21.9 (Sept. 1941) 26.
A18. Friedlaender, E. R. " Magnetic dust cores " J. Brit. I.R.E. 5.3 (May 1945) 106.
A19. Foster, D. E., and A. E. Newlon " Measurement of iron cores at radio frequencies " Proc. I.R.E. 29.5 (May 1941) 266.
A20. Jacob, F. N. " Permeability-tuned push-button systems " Comm. 18.4 (Apr.l 1938) 15.
A21. Fetherston, N. " Magnetic iron and alloy dust cores " R. and E. Retailer (1) 22.2 (28 Sept. 1944) 18 · (2) 22.3 (12 Oct. 1944) 20.
A22. Polydoroff, W. J. " Coaxial coils for FM permeability tuners " Radio 31.1 (Jan. 1947) 9.
A23. Polydoroff, W. J. " Ferro-inductors and permeability tuning " Proc. I.R.E. 21.5 (May 1933) 690.
A24. Information supplied by Mr. L. W. Cranch of Telecomponents Pty. Ltd., Petersham, N.S.W.
A25. Samson, H. W. " Permeability of dust cores " W.E. 24.288 (Sept. 1947) 267.
A26. Bardell, P. R. " Permeability of dust cores " Elect. 18.6 (Feb. 1947) 63.
A27. Harvey, R. L., I. J. Hegyi and H. W. Leverenz " Ferromagnetic spinels for radio frequencies " R.C.A. Rev. 11.3 (Sept. 1950) 321.
A28. Strutt, M. J. O. " Ferromagnetic materials and ferrites " W.E. 27.327 (Dec. 1950) 277.
A29. Snoek, J. L. " Non-metallic magnetic material for high frequencies " Philips Tec. Rev. 8.12 (Dec. 1946) 353.
A30. " Ferroxcube " Philips Industrial Engineering Bulletin, Australia, 1.5 (Jan. 1951) 18.
Additional references will be found in the Supplement commencing on page 1475.

(B) REFERENCES TO INDUCTANCE CALCULATION
B1. Parington, E. S. " Simplified inductance chart " Elect. 15.9 (Sept. 1942) 61.
B2. Maddock, A. J. " Mutual inductance : simplified calculations for concentric solenoids " W.E. 22.263 (Aug. 1945) 373.
B4. " The inductance of single layer solenoids on square and rectangular formers " Data Sheet, Electronic Eng. 15.173 (July 1942) 65.
B5. Amos, S. W. " Inductance calculations " W.W. 47.4 (April 1941) 108.
B6. Everett, F. C. " Short wave inductance chart " Elect. 13.3 (Mar. 1940) 33.
B7. Turney, T. H. " Mutual inductance : a simple method of calculation for single-layer coils on the same former " W.W. 48.3 (March 1942) 72.
B8. " The inductance of single layer solenoids " Data Sheets Electronic Eng. (1) 14.164 (Oct. 1941) 447 ; (2) 14.165 (Nov. 1941) 495.
B9. Blow, T. C. " Design chart for single layer inductance coils " Elect. 16.2 (Feb. 1943) 95.
B10. Grover, F. W. (book) " Inductance Calculations " (D. van Nostrand Co. Inc. 1946).
B11. Blow, T. C. " Solenoid inductance calculations " Elect. 15.5 (May 1942) 63.

(C) REFERENCES TO SELF-CAPACITANCE OF COILS
C1. Morecroft, J. H. " Resistance and capacity of coils at radio frequencies " Proc. I.R.E. 10.4 (Aug. 1922) 261.
C2. Medhurst, R. G. " H.F. resistance and self-capacitance of single-layer solenoids " (1) W.E. 24.281 (Feb. 1947) 35 ; (2) W.E. 24.282 (March 1947) 80.
C3. Palermo, A. J. " Distributed capacity of single-layer coils " Proc. I.R.E. 22.7 (July 1934) 897.
C4. " The self capacity of coils : its effect and calculation " Data Sheets, Electronic Eng. 14.7 (Jan. 1942) 589.

(D) REFERENCES TO LOSSES IN COILS
D1. Butterworth, S. " Effective resistance of inductance coils at radio frequency " (1) W.E. 3.31 (April 1926) 203 ; (2) W.E. 3.32 (May 1926) 309 ; (3) W.E. 3.34 (July 1926) 417 ; (4) W.E. 3.35 (Aug. 1926) 483.
D2. Terman, F. E. " Some possibilities for low loss coils " Proc. I.R.E. 23.9 (Sept. 1935) 1069.
D3. Reber, G. " Optimum design of toroidal inductances " Proc. I.R.E. 23.9 (Sept. 1935) 1056.
D4. Palermo, A. J., and F. W. Grover (1) " A study of the high frequency resistance of single layer coils " Proc. I.R.E. 18.12 (Dec. 1930) 2041 ; (2) " Supplementary note to the ' Study of the high frequency resistance of single layer coils ' " Proc. I.R.E. 19.7 (July 1931) 1278.
D5. Morecroft, J. H. " Resistance and capacity of coils at radio frequencies " Proc. I.R.E. 10.4 (Aug. 1922) 261.
D6. Medhurst, R. G. " H.F. resistance and self-capacitance of single-layer solenoids " (1) W.E. 24.281 (Feb. 1947) 35 ; (2) W.E. 24.282 (March 1947) 80.
D7. Austin, B. B. " The effective resistance of inductance coils at radio frequency " W.E. 11.124 (Jan. 1934) 12.
D8. Mitchel, P. C. " The factor-of-merit of short-wave coils " G. E. Review, 40.10 (Oct. 1937) 476.
D9. Callendar, M. V. (letter) " Q of solenoid coils " W.E. 24.285 (June 1947) 185.
D10. Editorial " The Q-factor of single-layer coils " W.E. 26.309 (June 1949) 179.

(E) REFERENCES TO THE EFFECT OF SCREENS
E1. Davidson, C. F., and J. C. Simmonds " Effect of a spherical screen upon an inductor " W.E. 22.256 (Jan. 1945) 2.
E2. Sowerby, J. McG. " Effect of a screening can on the inductance and resistance of a coil " Data Charts W.W. 48.11 (Nov. 1942) 254.
E3. Bogle, A. G. " The effective inductance and resistance of screened coils " Jour. I.E.E. 87 (Sept. 1940) 299.
E4. " The effect of a shield can on the inductance of a coil " Radiotronics 108 (Dec. 1940) 67.
E5. Editorial " The Q-factor of single-layer coils " W.E. 26.309 (June 1949) 179.

(F) REFERENCES TO SKIN EFFECT
F1. Whinnery, J. R. " Skin effect formulas " Elect. 15.2 (Feb. 1942) 44.
F2. " Depth of current penetration in conductors " and " The a.c. resistance of a round wire " Data Sheets, Electronic Eng. 14.170 (April 1942) 715.
F3. Wheeler, H. A. " Formulas for skin effect " Proc. I.R.E. 30.9 (Sept. 1942) 412.
F4. Shepperd, W. B. " Skin effect in round conductors " Comm. 25.8 (Aug. 1945) 56.

(G) REFERENCES TO COIL DESIGN
G1. " Coil design factors " Radio Review of Australia (1) 4.3 (March 1936) 32 ; (2) 4.5 (May 1936) 16.
G2. Scheer, F. H. " Notes on intermediate-frequency transformer design " Proc. I.R.E. 23.12 (Dec. 1935) 1483.
G3. Meyerson, A. H. " V-H-F coil construction " Comm. 24.4 (April 1944) 29.
G4. Pollack, D. " The design of inductances for frequencies between 4 and 25 megacycles " R.C.A. Rev. 2.2 (Oct. 1937) 184 and E.E. 56.9 (Sept. 1937) 1169.
G5. Rudd, J. B. " Theory and design of radio frequency transformers " A.W.A. Tech. Rev. 6.4 (March 1944) 193.
G6. Jeffery, C. N. " Design charts for air cored transformers " A.W.A. Tec. Rev. 8.2 (April 1949) 167.
G7. Maynard, J. E. " Coupled circuit design " Comm. 25.1 (Jan. 1945) 38.
G8. Meyerson, A. H. " U-H-F design factors " Comm. 22.6 (June 1943) 20.
G9. Everett, F. C. " Tuned circuits for the u-h-f and s-h-f bands " Comm. 26.6 (June 1946) 19.
G10. Meyerson, A. H. " V-H-F coil design " Comm. 26.6 (June 1946) 46.
G11. Meyerson, A. H. " Coil Q factors at v-h-f " Comm. 24.5 (May 1944) 36.
G12. Meyerson, A. H. " Coil design for v-h-f " Comm. 25.9 (Sept. 1945) 50.
G13. Varrall, J. E. " Variable selectivity IF amplifiers " W.W. 48.9 (Sept. 1942) 202.
G14. " Theory and design of radio frequency transformers " A.R.T.S. and P. (1) 134 (May 1944) 1 ; (2) 135 (June 1944) 1.
G15. Adams, J. J. " Undercoupling in tuned coupled circuits to realize optimum gain and selectivity " Proc. I.R.E. 29.5 (May 1941) 277.
G16. Barden, W. S. and D. Grimes, " Coil design for short-wave receivers " Elect. 7.6 (June 1934) 174.
G17. Forbes Simpson, A. I. " The design of small single-layer coils " Electronic Eng. 19.237 (Nov. 1947) 353.
G18. Sulzer, P. G. " R.F. coil design using charts " Comm. 29.5 (May 1949) 10.
G19. Simon, A. W. " Winding universal coils—short cut procedures to obtaining exact self-inductance, mutual-inductance and centre-taps " Elect. 18.11 (Nov. 1945) 170.
G20. Welsby, V. G. (book) " The Theory and Design of Inductance Coils " (Macdonald and Co. Ltd., London, 1950).
G21. Harris, W. A., and R. H. Siemens " Superheterodyne oscillator design considerations " R.C.A. Radiotron Division Publication No. ST41 (Nov. 1935).

(H) REFERENCES TO RADIO-FREQUENCY-CHOKES
H1. " Resonant R.F. chokes " W.W. 53.7 (July 1947) 246.
H2. Cooper, V. J. " The design of H.F. chokes " Marconi Review 8.3 (July 1945) 108.
H3. Scroggie, M. G. " H.F. chokes : construction and performance " W.W. (1) 36.20 (May 1935) 486 ; (2) 36.21 (May 1935) 529.

H4. Hartshorn, L., and W. H. Ward " The properties of chokes, condensers and resistors at very high frequencies " Jour. Sci. Instr. 14.4 (April 1937) 132.
H5. Wheeler, H. A. " The design of radio-frequency choke coils " Proc. I.R.E. 24.6 (June 1936) 850.
H6. Miller, H. P. " Multi-band r-f choke coil design " Elect. 8.3 (Aug. 1935) 254.

(I) **REFERENCES TO UNIVERSAL WINDINGS**
I1. Simon, A. W. " Winding the universal coil " Elect. 9.10 (Oct. 1936) 22.
I2. Simon, A. W. " On the winding of the universal coil " Proc. I.R.E. 33.1 (Jan. 1945) 35.
I3. Kantor, M. " Theory and design of progressive and ordinary universal windings " Proc. I.R.E. 35.12 (Dec. 1947) 1563.
I4. Simon, A. W. " On the theory of the progressive universal winding " Proc. I.R.E. 33.12 Pt 1 (Dec. 1945) 868.
I5. Simon, A. W. " Wire length of universal coils " Elect. 19.3 (Mar. 1946) 162.
I6. Hershey, L. M. " The design of the universal winding " Proc. I.R.E. 29.8 (Aug. 1941) 442.
I7. Joyner, A. A., and V. D. Landon " Theory and design of progressive universal coils " Comm. 18.9 (Sept. 1938) 5.
I8. Simon, A. W. " Universal coil design " Radio 31.2 (Feb. 1947) 16.
I9. Simon, A. W. " Winding universal coils—short cut procedures to obtaining exact self-inductance, mutual-inductance and centre-taps " Elect. 18.11 (Nov. 1945) 170.
I10. Arany, D., and M. Macomber, " Universal coil-winding graph " Comm. 29.10 (Oct. 1949) 28.
I11. Simon, A. W. " The theory and design of ordinary progressive and universal windings " Proc. I.R.E. 37.9 (Sept. 1949) 1029.
I12. Watkinson, E. " Universal coil winding " Proc. I.R.E. Aust. 11.7 (July 1950) 179. Reprinted Jour. Brit. I.R.E. 11.2 (Feb. 1951) 61.

(¶) **GENERAL REFERENCES**
J1. Soucy, C. I. " Temperature coefficient effects of r.f. coil finishes " Tele-Tech. (1) 6.12 (Dec 1947) 52 ; (2) 7.1 (Jan. 1948) 42.
J2. Bureau of Standards Circular, C74.
J3. Much useful u-h-f information and a good bibliography are to be found in a series of five articles in Elect. 15.4 (April 1942) 37. The articles are (a) Kandoian, A. G. " Radiating systems and wave propagation " (b) Mouromtseff, I. E., R. C. Retherford and J. H. Findley, " Generators for u.h.f. waves " (c) Dudley, B. " U.H.F. reception and receivers " (d) Jaffe, D. L. " Wide band amplifiers and frequency multiplication " (e) Lewis, R. F. " Measurements in the u-h-f spectrum."
J4. Orr, H. " Corrosion in multiple layer wound coils " Comm. (1) 29.1 (Jan. 1949) 22 ; (2) 29.7 (July 1949) 18 ; (3) 29.8 (Aug. 1949) 22.

CHAPTER 12

AUDIO FREQUENCY VOLTAGE AMPLIFIERS

By F. LANGFORD-SMITH, B.Sc., B.E.

SECTION 1 : INTRODUCTION

(i) Voltage amplifiers

A voltage amplifier is one in which the voltage gain is the criterion of performance. To be strictly correct it is not possible to have voltage without power since infinite impedance does not exist in amplifiers, but for all ordinary purposes a " voltage amplifier " is one in which a " voltage " output is required. Voltage amplifiers generally work into high impedances of the order of 1 megohm, but in certain cases lower load impedances are used and there is no sharp demarcation between " voltage " and " power " amplifiers. In cases where transformer coupling is used between stages, the secondary of the transformer may be loaded only by the grid input impedance of the following stage and the numerical value of the impedance may not be known. In such cases the transformer is usually designed to operate into an infinite impedance, and the effect of normal grid input impedances on the transformer is very slight compared with the primary loading.

It is important to bear in mind the reversal in polarity which occurs in any valve when used as an amplifier with a load in the plate circuit. As a consequence, the a-f voltage from grid to plate is

$$E_{gp} = E_{gk} + E_{kp} = E_{gk}(1 + A) \tag{1}$$

where A is the voltage gain from grid to plate.

SECTION 2 : RESISTANCE-CAPACITANCE COUPLED TRIODES

(i) Choice of operating conditions (ii) Coupling condenser (iii) Cathode bias (iv) Fixed bias (v) Grid leak bias (vi) Plate voltage and current (vii) Gain and distortion at the mid-frequency (viii) Dynamic characteristics (ix) Maximum voltage output and distortion (x) Conversion factors with r.c.c. triodes (xi) Input impedance and Miller Effect (xii) Equivalent circuit of r.c.c. triode (xiii) Voltage gain and phase shift (xiv) Comments on tabulated characteristics of resistance-coupled triodes.

(i) Choice of operating conditions* (Fig. 12.1)

Any triode may be used as a r.c.c. amplifier, but for most purposes the valves specially suitable for this application may be grouped :

1. General purpose triodes with μ from 15 to 50, and plate resistance from 6000 to 10 000 ohms (with battery types somewhat inferior). These are also called " medium mu " triodes.

2. High-mu triodes with μ from 50 to 100, and plate resistances say from 50 000 to 100 000 ohms.

The **load resistance** (R_L) may be any value from a few ohms to many megohms, but for normal operation R_L should never be less than twice the plate resistance at the operating point, with a higher value preferred. The load resistance should never be greater than the following grid resistance (R_{g2}) and should preferably be not more than one quarter of R_{g2}. The following table is a good general guide, but capable of modification in special circumstances.

Valve type—	General purpose		High-mu		
Following grid resistor—	0.22 to 1	0.22	0.47	1	MΩ
Load resistance	0.1	0.22	0.22	0.22	MΩ
				or 0.47	or 0.47 MΩ

The optimum combination of R_L and R_{g2} for maximum output voltage is covered in Sect. 2(ix).

The **plate supply voltage** (E_{bb}) should generally be as high as practicable provided that the maximum ratings are not exceeded. Plate supply voltages up to 300 volts are safe for use with all types of indirectly-heated valves unless otherwise stated. Somewhat higher supply voltages may be used with triodes provided that the designer ensures that, other than momentarily when switching on, the maximum plate voltage rating is not exceeded under any possible conditions.

The **input grid resistor** (R_{g1}) should not normally exceed 1 megohm with in-

FIG. 12.1

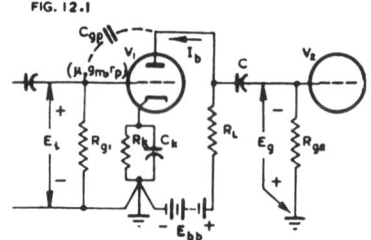

Fig. 12.1. Triode valve (V_1) resistance-capacitance coupled to V_2.

directly-heated valves, although higher resistances are satisfactory with 1.4 volt battery valves. Higher grid resistances will not do damage to a r.c.c. triode, provided that the plate load resistance is not less than, say, 0.1 megohm with 300 volts plate supply, but the reverse grid current may be sufficient to shift the operating point into the region of distortion and lower gain. For this reason cathode bias is to be preferred to fixed bias and the input grid resistor should not be higher than necessary, particularly with high-mu triodes—see (iv) below. For example, little is gained by using a grid resistor having more than four times the resistance of the plate load resistor of the preceding stage.

Values of input grid resistor in excess of 1 megohm may only be used satisfactorily in low-level pre-amplifier operation where the valve is one specially manufactured or tested for this class of service, under a specification which ensures that the reverse grid current is very low. With a maximum reverse grid current of 0.2 μA and with

*See Chapter 3 Sect. 1(iv) for valve ratings and their limiting effect on operation ; also (v) Recommended practice and operation.

μ not greater than 40, a grid resistor up to 5 megohms may be used, provided that the total resistance of the plate load resistor and the plate decoupling resistor is not less than 0.1 megohm.

If grid-leak bias is used, R_g may be from 5 to 10 megohms.

The output grid resistor (R_{g2}) may be the **maximum recommended** for the following stage by the valve manufacturers—usually 0.5 megohm for power valves with cathode bias—or as determined by eqn. (6) or (7) in Chapter 3 Sect. 1(v)d. Lower values are desirable to reduce the effects of reverse grid current in the following stage, and there is no appreciable advantage in either gain or distortion through the use of a resistance greater than $4\,R_L$. Calculation of the maximum grid resistance for use with power valves is covered in Chapter 13 Sect. 10(i).

(ii) Coupling condenser

The coupling condenser (C) may be selected to give the desired low frequency response. The loss of voltage due to C may be calculated by the use of a vector diagram or by the following equation, which applies to the circuit of Fig. 12.1 provided that the input resistance of V_2 is very high.

$$E_g/E = R/|Z| \tag{1a}$$

where E_g = signal voltage on grid of V_2
          E = signal voltage across R_L
          $R = R_{g2} + r_p R_L/(r_p + R_L)$
          or as an approximation, $R \approx R_{g2}$ if the plate resistance of V_1 is small compared with R_{g2},
      $|Z|$ = magnitude of series impedance of R and C
          $= \sqrt{R^2 + X_c^2}$

and $X_c = 1/\omega C = 1/2\pi f C$.

For example, if $R = 1$ megohm and $f = 50$ c/s, the following results will be obtained—

db loss	E_g/E	X_c/R	X_c	C
1	0.891	0.51	0.51 megohm	0.006 24 μF
2	0.794	0.76	0.76 megohm	0.004 19 μF
3	0.708	1.00	1.0 megohm	0.003 18 μF

The phase angle shift is given by

$$\phi = \tan^{-1} 1/\omega CR = \tan^{-1} 1/2\pi f CR \tag{1b}$$

In certain cases a low value of C is adopted intentionally to reduce the response to hum arising from preceding stages. However, the use of a low coupling capacitance, when the following grid resistor is 0.5 megohm or more, increases the hum contributed by the following valve through the a.c. operation of its heater. A low coupling capacitance should therefore not be used on a low-level stage. In high fidelity amplifiers a fairly large value of C is generally adopted, thus not only improving the low frequency response but also reducing phase shift and possibly also improving the response to transients. However, excessively large values of C are undesirable.

The following table gives the approximate values of C for certain selected conditions. Note that R must be as defined below eqn. (1a).

COUPLING CONDENSER

Attenuation 1 db at	12.5	25	50	100	200 c/s
2 db at	8.5	17	34	67	134 c/s
3 db at	6.5	13	26	51	102 c/s
$R =$ 10 000 ohms	2.5	1.25	0.62	0.31	0.15 μF
50 000 ohms	0.5	0.25	0.12	0.06	0.03 μF
100 000 ohms	0.25	0.12	0.06	0.03	0.015 μF
0.25 megohm	0.1	0.05	0.025	0.012	0.006 μF
0.5 megohm	0.05	0.025	0.012	0.006	0.003 μF
1.0 megohm	0.025	0.012	0.006	0.003	0.001 5 μF
2 megohms	0.012	0.006	0.003	0.001 5	0.000 8 μF
5 megohms	0.005	0.002 5	0.001 2	0.000 6	0.000 3 μF

The **effect** of selected values of capacitance for $R = 1$ megohm is shown graphically in Fig. 4.36. These curves have an ultimate slope of 6 db/octave.

A general curve of attenuation and phase shift is given in sub-section (xiii) below and Fig. 12.9. A Nomogram of transmission factor and phase shift is given by Ref. A2.

(iii) Cathode bias

(A) Cathode bias (Fig. 12.1) is generally preferable to fixed bias as it is largely self-compensating. The plate current flowing through R_k produces a voltage drop which is smoothed out by C_k and applied through R_{g1} to the grid. The full voltage on the grid (in the absence of grid current) is

$$E_c = -I_b R_k \qquad (1c)$$

where I_b = plate current in amperes

and R_k = resistance of cathode resistor in ohms.

If negative* grid current (e.g. gas or grid emission current) **is flowing**, the bias will be decreased by $I_g R_{g1}$

where I_g = direct grid current in microamperes (taken as positive)

and R_{g1} = resistance of grid resistor in megohms.

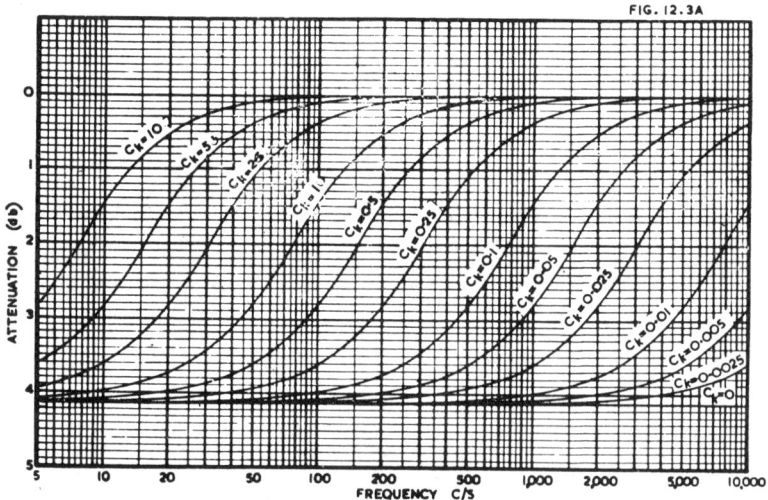

Fig. **12.3A.** *Frequency characteristics due to cathode by-pass condenser with general purpose triode having* $\mu = 20$, $r_p = 10,000$ *ohms,* $R_L = 0.1$ *megohm,* $R_{g2} = 0.5$ *megohm,* $R_k = 2700$ *ohms.*

If positive grid current is flowing the bias will be increased by $I_g R_{g1}$.

The cathode by-pass condenser C_k is only fully effective at high frequencies, and it becomes increasingly ineffective as the frequency is decreased—under these conditions there is degeneration (negative current feedback—see Chapter 7). **The effect of R_k and C_k on the voltage gain** at any frequency is given by the equation

$$\left|\frac{A'}{A}\right| = \sqrt{\frac{1 + (\omega C_k R_k)^2}{\left[1 + \frac{(\mu + 1)R_k}{R' + r_p}\right]^2 + (\omega C_k R_k)^2}} \qquad (2)$$

where A' = stage voltage gain at frequency f with self bias resistor R_k by-passed by condenser C_k

A = stage voltage gain with R_k completely by-passed

= mid-frequency voltage gain

$\omega = 2\pi f = 2\pi \times$ frequency of input signal

---

*Also known as reverse grid current.

C_k = capacitance of by-pass condenser in farads
R_k = resistance of self bias resistor in ohms
μ = valve amplification factor at the operating point
r_p = valve plate resistance in ohms at the operating point
R_L = resistance of plate' load resistor in ohms
R_{g2} = resistance of following grid resistor in ohms
and $R' = R_L R_{g2}/(R_L + R_{g2})$.

The derivation of eqn. (2) is given in Ref. B11 ; see also Refs. A11, A13.

The **attenuation characteristics** of a typical general-purpose triode **with cathode bias** are given in graphical form in Fig. 12.3A. It will be seen that all curves have the same shape, but are shifted bodily sideways depending on the value of C_k. The maximum slope of the curves in Fig. 12.3A is 1.4 db/octave ; the slope does not normally exceed 3 db/octave when R_k is the optimum value to provide bias.

The limiting loss of gain at zero frequency due to R_k is given by

$$\left|\frac{A'}{A}\right| = \frac{R' + r_p}{R' + r_p + (\mu + 1)R_k} \tag{3}$$

Examples of limiting loss of gain at zero frequency (based on equation 3)

Type	R_L	R_{g2}	R	R_k	μ*	r_p*	A'/A	loss
6J5	0.1MΩ	0.5MΩ	0.08MΩ	2700	18	17 000	0.65	3.7 db
6Q7	0.25MΩ	0.5MΩ	0.17MΩ	3000	68	70 000	0.53	5.5 db
6SQ7	0.25MΩ	0.5MΩ	0.17MΩ	3900	100	100 000	0.40	8.0 db

*At operating point.

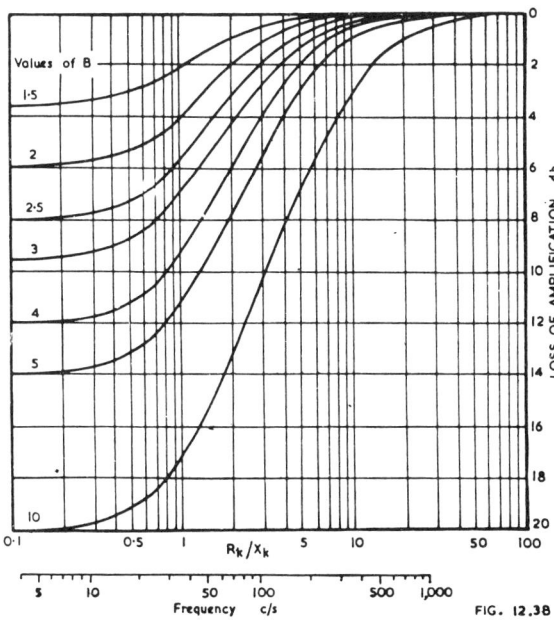

FIG. 12.3B

Fig. 12.3B. *Universal attenuation curves with cathode bias* (*Ref. A*11 *Part* 4).

Universal attenuation curves are given in Fig. 12.3B (Ref. A11 Part 4) in which

$$B = 1 + \frac{R_k(\mu + 1)}{r_p + R'}$$

$$R' = R_L R_{g2}/(R_L + R_{g2})$$

and $X_k = 1/(\omega C_k)$.

(B) The phase angle shift is a maximum at the frequency where the slope of the attenuation characteristic is a maximum, and drops towards zero at zero frequency and at higher audio frequencies. The maximum value for any normal r.c.c. amplifier does not exceed about 30°.

Universal curves showing phase angle shift are given in Fig. 12.3C in which the symbols have the same meaning as in Fig. 12.3B.

(C) Choice of cathode bias resistor

The best method is to determine the grid voltage as for fixed bias—see (iv) below—and then to calculate R_k. A fairly satisfactory approximation is to select R_k by the following table.

μ	R_L MΩ	Low level	Inter-mediate level	High level $R_g = R_L$	$R_g = 2R_L$
12 to 25	any	$R_k = 0.25R_L/\mu$	$0.5R_L/\mu$	$0.6R_L/\mu$	$0.8R_L/\mu$
30 to 50	any	$R_k = 0.5R_L/\mu$	$0.7R_L/\mu$	$0.7R_L/\mu$	$0.9R_L/\mu$
70 to 100	0.1	$R_k = 1.0R_L/\mu$	$1.2R_L/\mu$	$1.2R_L/\mu$	$1.6R_L/\mu$
	0.22	$R_k = 0.8R_L/\mu$	$1.0R_L/\mu$	$0.9R_L/\mu$	$1.2R_L/\mu$
	0.47	$R_k = 0.65R_L/\mu$	$0.8R_L/\mu$	$0.75R_L/\mu$	$1.0R_L/\mu$

Conditions : Plate supply voltage 200 to 300 volts. Values of μ are the published values.

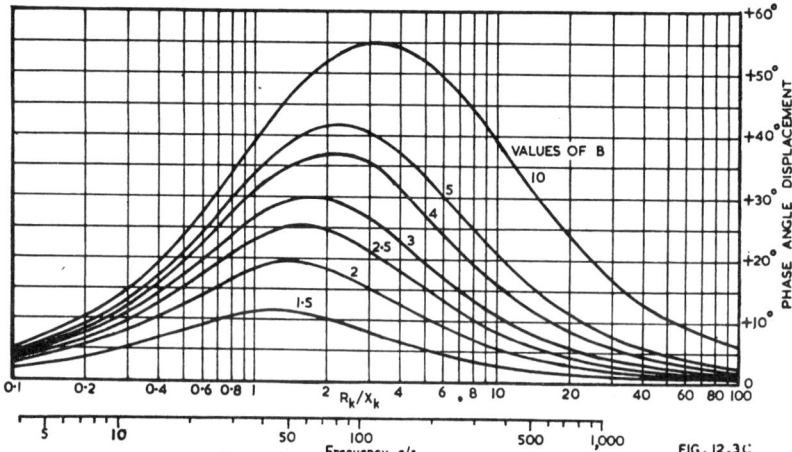

Fig. 12.3C. *Universal phase angle shift curves with cathode bias (Ref. A11 Part 4).*

(D) Cathode bias loadlines

Cathode bias loadlines may be drawn on the plate characteristics as on the mutual characteristics [Chapter 2 Sect. 4(v) and Fig. 2.27] but they will no longer be straight lines. Fig. 12.4 shows the loadline and plate characteristics for a high-mu triode. The cathode bias loadline may very easily be plotted for any selected value of R_k, e.g. 10 000 ohms. In Fig. 12.4 each curve applies to increments of 0.5 volt in E_c, so that (with $R_k = 10\,000$ ohms) the successive increments of I_b are 0.05 mA—

$E_c = 0$ -0.5 -1 -1.5 -2 -2.5 -3 volts
$I_b = 0$ 0.05   0.1   0.15   0.2   0.25   0.3 mA

The intersection of the selected R_k loadline with the R_L loadline gives the operating point (Fig. 12.4).

Alternatively, the intersection of the R_L loadline with each grid curve may be marked with the corresponding value of R_k ; e.g. in Fig. 12.4 at point A, $E_c = -1.5$ V and $I_b = 0.47$ mA, therefore $R_k = (1.5/0.47) \times 1000 = 3200$ ohms. Similarly at point B, $R_k = 7800$ ohms. If greater accuracy is required in determining **an in-**

dividual point, the values of R_k may be plotted against E_c, and the desired value may be selected.

For greater accuracy, particularly with relatively high values of cathode resistance, the slope of the loadline in Fig. 12.4 should be that corresponding to $(R_L + R_k)$.

(E) Maximum grid resistance with cathode bias

It may be shown that the maximum permissible grid resistance with cathode bias is greater than that with fixed bias, as indicated by the approximation :

$$\frac{R_g \text{ with cathode bias}}{R_g \text{ with fixed bias}} \approx 1 + g_{md}R_k \tag{3a}$$

where g_{md} = slope of dynamic characteristic at operating point

$$\approx \frac{g_m r_p}{(r_p + R_L)} \approx \frac{\mu}{(r_p + R_L)}$$

g_m = mutual conductance at operating point
μ = amplification factor at operating point
r_p = plate resistance at operating point

and R_k = cathode bias resistor.

This ratio (eqn. 3a) is usually between 1.2 and 2.

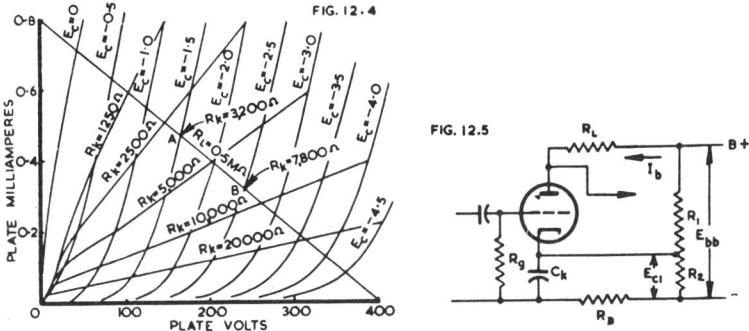

Fig. 12.4. *Plate characteristics of high-mu triode (6SQ7) with $E_{bb} = 400$ volts and $R_L = 0.5$ megohm. Cathode bias loadlines for selected values of cathode bias resistance (R_k) have been drawn in.*

Fig. 12.5. *Resistance-capacitance coupled triode with fixed bias from voltage divider.*

(iv) Fixed bias

(A) It is sometimes desired to operate the valve with fixed bias, either from a separate bias supply (battery or rectifier/filter combination) or from a voltage divider across the plate supply. Fig. 12.5 is an example of the latter, with R_1 and R_2 forming a voltage divider. Here

$$E_{c1} = \frac{R_2 E_{bb} + R_1 R_2 I_b}{R_1 + R_2} \text{ and } R_2 = \frac{E_{c1}}{I_b + (E_{bb} - E_{c1})/R_1} \tag{4}$$

Condenser C_k is for by-passing R_2, but as R_2 may have a fairly low value, thus necessitating a large capacitance for C_k, it is frequently economical to add R_3 and so permit a smaller value of C_k. In addition, $C_k R_3$ forms a useful hum filter for the bias voltage.

(B) **The optimum grid bias** $(-E_c)$ is a function of the input voltage ; the bias should be the minimum (k_1) which can be used without running into damping due to positive grid current, with a margin for differences between valves, together with a further margin (k_2) to allow for the effects of reverse grid current in the grid resistor.

If the input voltage is known,

$$E_c = 1.41 E_i + k_1 + k_2 \text{ volts} \tag{5a}$$

If the output voltage is known,

$$E_c = (1.41 E_0/A) + k_1 + k_2 \text{ volts} \tag{5b}$$

where E_i = r.m.s. input voltage

E_0 = r.m.s. output voltage, from plate to cathode

A = voltage gain of stage

k_1 = 0.75 to 1.0 for high-mu indirectly heated valves*

= 0.6 (approx.) for general purpose indirectly heated valves*

= 0.25 (approx.) for battery valves*

$k_2 \approx \dfrac{7.5\ R_{g1}}{100\ R_L}$ for high-mu valves

$\approx \dfrac{3\ R_{g1}}{100\ R_L}$ for general purpose valves.

R_{g1} = resistance of grid resistor (upper limit)

and R_L = resistance of plate load resistor (lower limit).

These approximate values of k_2 are based on typical American manufacturing specifications for maximum grid current (limiting values), and on the assumption that the reverse grid current is proportional† to the plate current.

Values of k_2 are tabulated below for three typical conditions for a plate supply voltage of 250 volts.

Case 1: R_{g1} = 1 megohm

	General purpose triode	High-mu triode
R_L = 0.1 megohm	k_2 = 0.31	k_2 = 0.75 volt
R_L = 0.22 megohm	k_2 = 0.14	k_2 = 0.34 volt
R_L = 0.47 megohm	k_2 = 0.07	k_2 = 0.16 volt

The additional bias (k_2) provided to allow for the effect of reverse grid current is of no great consequence with general purpose triodes, but it is very serious with high-mu triodes

Case 2: $R_{g1} = 2R_L$,

k_2 = 0.15 volt for high-mu triodes

k_2 = 0.06 volt for general purpose triodes.

Case 3: $R_{g1} = 4R_L$

k_2 = 0.3 volt for high-mu triodes

k_2 = 0.12 volt for general purpose triodes.

Thus with **general purpose triodes operated as fixed bias r.c.c. amplifiers,** a grid resistor having a resistance of 1 megohm or even higher may be used without serious effects. However, it is always desirable to keep the resistance as low as practicable—not more than four times the resistance of the preceding plate load resistor, and not exceeding 1 megohm.

If the effect of reverse grid current on the operating point is to be kept small in **fixed-bias high-mu triodes,** it is essential to use a grid resistor R_{g1} (that is, its own grid resistor) having a resistance not more than twice that of its plate load resistor. This will restrict the maximum change of bias to 0.15 volt (i.e. k_2 = 0.15 volt). For example a load resistance of 0.47 megohm will permit the use of a grid resistor of 1 megohm, while a load resistance of 0.22 megohm will permit the use of a grid resistor of 0.47 megohm.

(C) **Cathode bias is preferable to fixed bias** in that the shift of operating point due to reverse grid current is minimized. With general purpose triodes employing cathode bias there is no advantage in making any adjustment to the bias voltage as calculated for fixed bias (eqns. 5a, 5b) but with high-mu triodes employing cathode bias the value of k_2 may be taken as approximately

$k_2 \approx \dfrac{5\ R_{g1}}{100\ R_L}$ for high-mu valves with cathode bias.

Consequently, with high-mu triodes, the resistance of the grid resistor R_{g1} may be made 1.5 times that with fixed bias for the same effect on the operating-point.

---

*These values of k_1 are only typical, and are likely to be exceeded by some valves. In each case, a value should be determined for the valves being used.

†This is approximately true when the reverse grid current is nearly all due to ionization—see Ref. G1, Fig. 1.

(D) Damping due to positive grid current

It is shown in Chapter 2 Sect. 2(iii) that damping on the positive peaks of grid input voltage may be quite serious, even when the peak grid voltage does not reach the grid current " cross-over point." This damping on positive peaks introduces an objecttionable form of distortion which is particularly important when the preceding stage has high effective impedance (looking backwards from the grid of the stage being considered) as, for example, with a high-mu triode or pentode. The damping at the positive peak of the input voltage is proportional to the grid conductance at this point, which value increases rapidly as soon as electrons commence to flow from cathode to grid, and even while the resultant grid current is still negative.

If grid current damping is to be avoided, the grid bias must be increased sufficiently to avoid this region.

(v) Grid leak bias

If a grid resistor of 5 or 10 megohms is used, it is possible to obtain the grid bias for a high-mu triode by means of the voltage drop in R_g [see Chapter 2 Sect 2(iii)]. This imposes damping on the input circuit due to the grid current, and the average input resistance is approximately $R_g/2$. The distortion is approximately the same as with the optimum fixed bias, but the great advantage is that it accommodates itself to variations from valve to valve, while fixed bias is critical.

It may also be used with general purpose triodes operating with input voltages not exceeding, say, 1 volt peak.

Grid leak bias is not very suitable for use with low-level (pre-amplifier) stages owing to hum.

(vi) Plate voltage and current

As discussed in Chapter 2 Sect. 3(i), the voltage on the plate is less than the supply voltage by the voltage drop in the load resistor. This is illustrated in Fig. 12.6 in which the loadline is fixed by the plate supply voltage E_{bb} and the load resistance R_L. The quiescent operating point Q is fixed by the intersection of the loadline and the grid curve for E_{c1}, the bias voltage having been previously determined. Alternatively, the cathode bias loadline may be drawn by the method described above, thus determining the operating point.

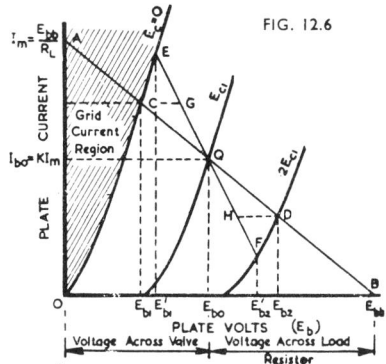

FIG. 12.6

Fig. 12.6. Plate characteristics of r.c.c. triode with normal loadline AB, operating point Q, and working loadline EQF due to grid resistor shunting.

Having fixed Q, the quiescent plate current I_{b0} and plate voltage E_{b0} will automatically be fixed. If

$I_{b0} = KI_m$, where K is a constant less than 1 and $I_m = E_{bb}/R_L$, then $E_{b0} = (1 - K)E_{bb}$. If $K = 0.5$, for example, then $I_{b0} = \tfrac{1}{2}I_m = \tfrac{1}{2}E_{bb}/R_L$ and $E_{b0} = \tfrac{1}{2}E_{bb}$; in this case the voltage across the valve is equal to the voltage across the load resistor. In practice, K may have values between about 0.25 and 0.85.

Typical values of K are tabulated below

	Low level	Intermediate level	High level*
General purpose triodes	0.75 — 0.85	0.6 — 0.75	0.5 — 0.6
High-mu triodes	0.6 — 0.65	0.5 — 0.6	0.45 — 0.55($R_L = 0.5M\Omega$)
			0.4 — 0.5 ($R_L = 0.25M\Omega$)

*If R_{g2} is less than $2R_L$, then the values of K for high level operation should be decreased.

Relation between K and cathode bias resistor
Knowing K, it is possible to calculate R_k or vice-versa :

$$R_k \approx \frac{R_L - K(R_L + r_p)}{\mu K} \qquad (6a)$$

$$K \approx \frac{R_L}{R_L + r_p + \mu R_k} \qquad (6b)$$

Relation between K and grid voltage

$$E_c \approx [R_L - K(R_L + r_p)][(E_{bb}/\mu R_L). \qquad (6c)$$

Eqns. (6a), (6b) and (6c) are exact for linear characteristics, but only approximate in practice.

Example : $\mu = 20$, $r_p = 10\,000$ ohms, $R_L = 0.1$ megohm, $E_{bb} = 250$ volts

K =	0	0.1	0.2	0.3	0.4	0.5	0.6	0.7	0.8	0.9	1.0
E_c =	−12.5	−11.1	−9.7	−8.4	−7.0	−5.6	−4.3	−2.9	−1.5	−0.1	+1.25

(vii) Gain and distortion at the mid-frequency

The voltage gain at the mid-frequency is

$$A_0 = \frac{\mu R_L}{R_L + r_p} = \frac{g_m}{(1/r_p) + (1/R_L)} \qquad (6d)$$

where there is no a.c. shunt load, and where μ, g_m and r_p are the values at the operating point, that is at the operating plate current. Alternatively this may be calculated from the loadline (Fig. 12.6) :

$$A_0 = \frac{E_{b2} - E_{b1}}{2E_{c1}} \qquad (6e)$$

where E_{b2} is plate voltage corresponding to negative peak grid signal voltage

E_{b1} is plate voltage corresponding to positive peak grid signal voltage
and E_{c1} is peak grid signal voltage.

When there is a following grid resistor R_{g2}, the loadline must be rotated about Q to a new position EQF with a slope of $-(1/R_L + 1/R_{g2})$. The voltage gain for the same input voltage is then

$$A' = \frac{E_{b2}' - E_{b1}'}{2E_{c1}} \qquad (6f)$$

which may be shown by the equivalent circuit [see (xii) below] to be

$$A' = \frac{g_m}{1/r_p + 1/R_L + 1/R_{g2}} \qquad (7)$$

also $A'/A = R_{g2}/(R + R_{g2})$ (8)
where $R = R_L r_p/(R_L + r_p)$
and g_m and r_p are the values at the operating point.

In practice, the new loadline EQF extends downwards with point F in the region of increasing distortion, and in the extreme case plate current cut-off may occur. This effect may be avoided by reducing the input voltage so that the output voltage will be

$$R_{g2}/(R_L + R_{g2}) \times \text{output voltage without shunting} \qquad (9)$$

Gain in terms of g_m and g_p

Equation (6d) may be put into the alternative form

$$A_0 = \frac{g_m R_L}{1 + g_p R_L} \qquad (9a)$$

where g_p = plate conductance at the operating point = $1/r_p$.
When there is a following grid resistor R_{g2} the gain is given by

$$A_0 = \frac{g_m R}{1 + g_p R} \qquad (9b)$$

where $R = R_L R_{g2}/(R_L + R_{g2})$.

The values of g_m and g_p may be derived from " G " curves (e.g. see Fig. 13.9B). By this means the actual values of g_m and g_p at the operating point may be determined with a good degree of accuracy without any assumptions or manipulations (Ref. **A15**).

Distortion

The percentage second harmonic distortion—see Chapter 13 Sect. 2(i)—is given by

$$H_2\% = \frac{EQ - QF}{2(EQ + QF)} \times 100 \tag{10}$$

(viii) Dynamic characteristics

The dynamic characteristic of a resistance-loaded triode is described in Chapter 2 Sect. 3(i) and Fig. 2.19. In shape it closely resembles the dynamic characteristic of a resistance-loaded pentode, and a comparison between the two is made in Sect. 3(viii) below and Fig. 12.15. With the triode dynamic characteristic the greater part or the whole of the " top bend " is in the grid current region and therefore cannot be used. The most nearly straight portion is therefore that of minimum bias.

An " ideal " (linear) dynamic characteristic of a general purpose triode is shown in Fig. 12.7A. It is limited at the lower end by plate-current cut-off at A, while the other end is in the grid current region. The usable part extends from A to G, but in reality the lower part is curved, and is avoided as far as possible. The upper end (GB) is also curved, but as this is in the grid current region it cannot be used in any case. The curves are to scale for $\mu = 20$, $r_p = 10\ 000$ ohms, $R_L = 0.1$ megohm and $E_{bb} = 250$ volts.

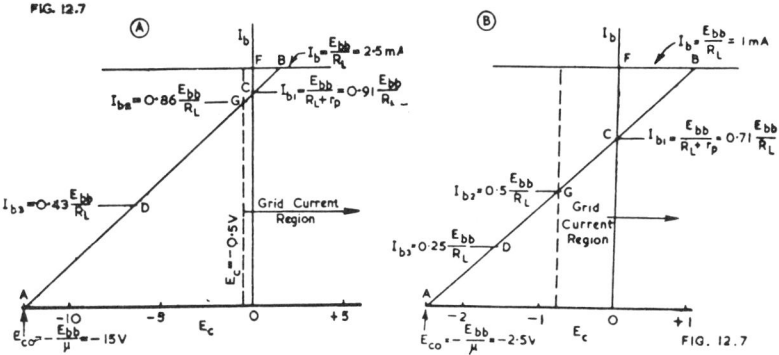

FIG. 12.7

Fig. 12.7. " Ideal " linear dynamic characteristics (A) for general purpose triode with $\mu = 20$, $r_p = 10\ 000$ ohms, $R_L = 0.1$ megohm and $E_{bb} = 250$ volts ; (B) for high-mu triode with $\mu = 100$, $r_p = 0.1$ megohm, $R_L = 0.25$ megohm and $E_{bb} = 250$ volts.

Point C at $E_c = 0$ has a plate current $I_{b1} = E_{bb}/(R_L + r_p)$ which in this case is $0.91\ E_{bb}/R_L$. This point and point A may be used as the two basic points for plotting the loadline. Alternatively, point B may be plotted, since $FB = E_{bb}/g_m R_L$. Point G is the commencement of grid current ($E_c = -0.5$ volt) and in this case its $I_{b2} = 0.86\ E_{bb}/R_L$. The highest operating point cannot exceed $0.85\ E_{bb}/R_L$, and this is only possible for extremely small input voltages. The lowest useful operating point (for high-level operation) is D, which is the mid point of AG, with $I_{b3} = 0.43\ E_{bb}/R_L$. Thus the operating point must be within the limits 0.43 and $0.85 \times E_{bb}/R_L$.

The dynamic characteristic in Fig. 12.7B applies to a high mu triode with $\mu = 100$, $r_p = 0.1$ megohm, $R_L = 0.25$ megohm and $E_{bb} = 250$ volts. The grid current is taken as commencing at $E_c = -0.75$ volt. The usable part of the characteristic extends from A to G, and the operating point must be within the limits 0.25 and 0.5 multiplied by E_{bb}/R_L. The upper limit would be somewhat extended if the load resistance were increased to 0.5 megohm.

(ix) Maximum output voltage and distortion

It is difficult to lay down any limit to the maximum voltage output, since overloading occurs very gradually. It is assumed that in all cases the grid bias is sufficient to avoid positive grid current.

An approximate method for determining the conditions for maximum output voltage are given by Diamond (Ref. A16). In Fig. 12.7C, the tangent FCA at $E_c = 0$ cuts the voltage axis at F where $E_b = e_1$. The dynamic loadline AQD corresponds to R_L in parallel with the following grid resistor R_{g2}. Point B is predetermined by the minimum plate current permissible on account of distortion (I_{min}).

The optimum value of R_L is given by

$$R_L \approx \frac{r_p}{\delta + \sqrt{(\delta + \gamma/x)}}$$

and the maximum value of peak-to-peak voltage swing is then

$$E_0 \approx (E_{bb} - e_1)\frac{1 - \delta}{1 + (1/x) + 2(\delta + \sqrt{\delta + \gamma/x})}$$

where r_p = plate resistance at point C (as given by slope of FC)

$\delta = I_{min}r_p/(E_{bb} - e_1)$

$x = R_{g2}/r_p$

$\gamma = \dfrac{\text{`` positive '' swing}}{\text{total swing}} = \dfrac{GK}{GH}$

E_{bb} = plate supply voltage

E_0 = peak-to-peak total voltage swing = GH

and e_1 = plate voltage corresponding to intersection of tangent FC with voltage axis.

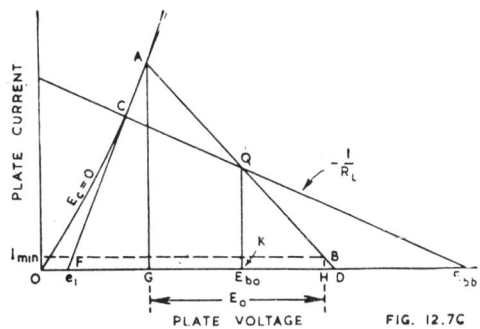

Fig. 12.7C. *Approximate method for determining conditions for maximum output voltage (Ref. A16).*

The value of γ for low distortion in a single valve is 0.5 ; values appreciably higher than 0.5 are possible with push-pull operation. The value of e_1 is a function of the valve characteristics, the load resistance and the plate supply voltage ; it is usually less than 10 volts for R_L not less than 0.1 megohm and E_{bb} not greater than 300 volts.

Example : Type 6J5, r_p = 10 000 ohms, $e_1 = 5$ volts, E_{bb} = 250 volts, R_{g2} = 0.5 megohm, I_{min} = 0.2 mA, γ = 0.5. We obtain x = 50, δ = 0.008 17, γ/x = 0.01, and from the equations, $R_L \approx$ 70 000 ohms optimum ; E_0 = 186 volts peak-to-peak.

Output voltage and distortion

The maximum output voltage for a general purpose triode (e.g. type 6J5) for 14% intermodulation distortion* and E_{bb} from 180 to 300 volts is given approximately by :

Load resistance	Following grid resistance	Maximum output voltage r.m.s.
0.1 megohm	0.1 megohm	0.155 × E_{bb}
	0.2 megohm	0.19 × E_{bb}
	0.4 megohm	0.215 × E_{bb}
0.25 megohm	0.25 megohm	0.20 × E_{bb}
	0.5 megohm	0.24 × E_{bb}
	1.0 megohm	0.27 × E_{bb}

*For details of intermodulation distortion see Chapter 14 Sect. 3.

For lower values of I.M. distortion, the preceding values of maximum output voltage should be multiplied by :

I.M.	$R_L = 0.1$ megohm	$R_L = 0.25$ megohm
10%	Factor $= 0.8$	Factor $= 0.72$
5%	0.48	0.36
2.5%	0.26	0.18

With a high-mu triode the voltage output for 14% intermodulation distortion is only about $0.17\,E_{bb}$ to $0.21\,E_{bb}$ for $R_L = 0.25$ and $R_{g2} = 0.5$ megohm.

Note that with intermodulation distortion the ouput voltage is taken as the arithmetical sum of the component voltages—see Chapter 14 Sect. 3(ii).

The intermodulation distortion of a triode and a pentode are compared in Fig. 12.16A [Sect. 3(ix)]. At low output voltages the pentode gives less distortion while at high output voltages the triode gives less distortion. The triode is, however, less critical than the pentode, and the distortion increases at a lower rate than with the pentode when the bias is made more negative than the optimum value.

(x) Conversion factors with r.c.c. triodes

Some valve manufacturers publish values of μ, g_m and r_p plotted against plate current. These may be used fairly accurately under any conditions of plate or grid voltage.

If these are not available :

1. The **amplification factor** is nearly constant. In most high-mu triodes it drops about 5% to 10% (15% for types 6SL7-GT and 5691) as I_b is reduced to 0.25 mA, while in general purpose triodes it drops about 15% as I_b is reduced to 1 mA.

2. The **mutual conductance** is largely a function of the plate current and is given by the approximate equation

$$g_m \approx F g_{m0} \tag{11a}$$

where g_m = desired mutual conductance at any operating point with plate current I_b

g_{m0} = published mutual conductance at plate current I_{b0}

and F is given by the following table based on the equation

$$F = \sqrt[3]{F_i} = \sqrt[3]{I_b/I_{b0}}.$$

I_b/I_{b0}	1.0	0.9	0.8	0.7	0.6	0.5	0.4	0.3	0.2	0.1
F	1.0	0.97	0.93	0.89	0.84	0.79	0.74	0.67	0.58	0.46

Alternatively, if the gain with resistance-coupled operation is known, the mutual conductance at the operating point may be calculated from

$$g_m = \frac{\mu \times \text{voltage gain}}{R(\mu - \text{voltage gain})} \tag{11b}$$

where $R = R_L R_g/(R_L + R_g)$,

and μ = amplification factor at the operating point.

3. The **plate resistance** may be calculated from $r_p = \mu/g_m$. It is approximately equal to $r_{p0}\sqrt[3]{I_{b0}/I_b}$.

(xi) Input impedance and Miller Effect

In the amplifier of Fig. 12 1 the grid to plate capacitance C_{gp} has impressed across it a voltage $(A + 1)E_i$ where A is the voltage gain of V_1. The current flowing through C_{gp} is therefore $(A + 1)$ times the current which would flow through the same capacitance when connected from grid to cathode. This is one example of the Miller Effect which occurs in all amplifiers. For example take type 6SQ7 high-mu triode with $C_{gp} = 1.6$ $\mu\mu$F and $A = 62$. The effective input capacitance due to C_{gp} alone is therefore $63 \times 1.6 = 101$ $\mu\mu$F.

The total input capacitance also includes the capacitance from grid to cathode plus $(A + 1)$ times any stray capacitance from grid to plate.

If the plate load is partially reactive the input impedance is equivalent to a capacitance C' and a resistance R' in parallel from grid to cathode,

where $C' = C_{gk} + (1 + A \cos \phi) C_{gp}$, \hfill (12)

$R' = -1/(2\pi f C_{gp} A \sin \phi)$ \hfill (13)

and ϕ = angle by which the voltage across the load impedance leads the equivalent voltage acting in the plate circuit (ϕ will be positive for an inductive load and negative for a capacitive load)

When the load is capacitive, as usually with a r.c.c. amplifier, R' is positive and there is some slight additional loading of the input circuit. When the load is inductive, R' is negative and self oscillation may occur in an extreme case.

(xii) Equivalent circuit of r.c.c. triode

The exact a.c. equivalent circuit of a r.c.c. triode is given in Fig. 12.8 where the valve V_1 is replaced by a generator of voltage μE_i in series with r_p. The plate load resistor R_L is shunted by C_0, which includes the valve output capacitance plus stray capacitance. The grid resistor of V_2 is shunted by C_i which includes the input capacitance of V_2 (including the Miller Effect capacitance from the plate) plus stray capacitance. Any additional condenser connected from plate or grid to earth should be added to C_0 or C_i respectively.

FIG. 12.8

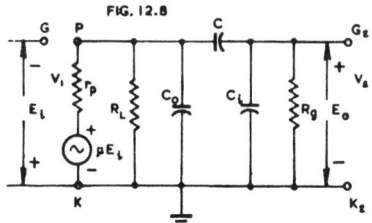

Fig. 12.8. *Exact equivalent circuit of r.c.c. triode.*

The cathode bias resistor is here assumed to be adequately by-passed at all signal frequencies.

This equivalent circuit is the basis of the calculations of gain and phase angle shift at all frequencies.

(xiii) Voltage gain and phase shift

It has been shown by Luck and others that the voltage gain/frequency characteristic of a r.c.c. amplifier has the same mathematical form as a tuned circuit except that the Q is very low (never greater than 0.5). There is a mid-frequency at which the gain is a maximum (A_0) and the phase shift zero. At both lower and higher frequencies the gain falls off, and when the gain is $A_0/\sqrt{2}$ the absolute values of the resistive and reactive components are equal. These two reference frequencies, which correspond approximately to 3 db attenuation, are the basic points on the attenuation characteristics, and, with the mid-frequency gain A_0, are sufficient to determine the whole frequency characteristic. The value of Q is given by

$$Q = f_0/(f_b - f_a) \qquad (14)$$

where f_0 = mid-frequency (at which voltage gain is A_0)

f_b = (high) reference frequency at which voltage gain is 0.707 A_0

and f_a = (low) reference frequency at which voltage gain is 0.707 A_0.

The low reference frequency f_a is given by

$$f_a = 1/2\pi CR' \qquad (15)$$

where $R' = R_g + r_p R_L/(r_p + R_L)$

and the constants are as in Fig. 12.8.

The high reference frequency (f_b) is given by

$$f_b = 1/2\pi(C_0 + C_i)R'' \qquad (16)$$

where $R'' = r_p R_L R_g/(r_p R_L + r_p R_g + R_L R_g)$.

When f_a and f_b have been determined, the attenuation at any other frequency may be found by reference to Fig. 12.9A.

The phase shift at any frequency may be determined from Fig. 12.9B ; the phase shift at f_a and f_b is 45° leading and lagging respectively.

Alternatively, if the low reference frequency is known, the **gain at any frequency** f is given by (Ref. F1):

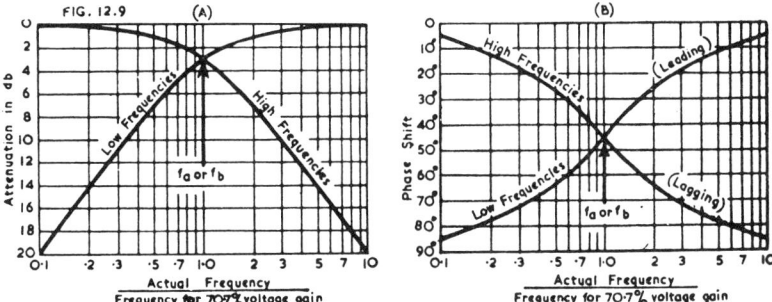

Fig. 12.9(A). *Attenuation in decibels for r.c.c. triode at low and high frequencies*
(B) *Phase shift at low and high frequencies. Values of phase shift for frequencies beyond
the limits of the curves may be estimated with reasonable accuracy by taking the angle
as proportional or inversely proportional to the frequency.*

$$\frac{\text{gain at frequency } f}{\text{gain at mid-frequency}} = \cos \tan^{-1} \frac{f_a}{f} \tag{17}$$

while the **phase angle shift** is given by

$$\tan \phi = f_a/f \tag{18}$$

where f_a = low reference frequency (3 db attenuation).

Similarly if the high reference frequency is known, the **gain at any frequency** f
is given by

$$\frac{\text{gain at frequency } f}{\text{gain at mid-frequency}} = \cos \tan^{-1} \frac{f}{f_b} \tag{19}$$

while the **phase angle shift** is given by

$$\tan \phi = f/f_b \tag{20}$$

where f_b = high reference frequency (3 db attenuation).

(xiv) Comments on tabulated characteristics of resistance-coupled triodes

Published operating characteristics of resistance-coupled triodes (unless the test
conditions are specified) appear to be taken under test conditions with a low source
impedance, a low d.c. resistance between grid and cathode, and a small value of positive grid current. When this is so, they cannot be applied directly to the normal practical case where there is a high resistance d.c. path between grid and cathode, and
where the source impedance may be high. The usual value of total harmonic distortion, unless otherwise specified, appears to be about 5% under the conditions of test
(Ref. A17).

In a practical resistance-coupled amplifier the maximum output voltage for 5% total
harmonic distortion will be less than the tabulated value, depending on the source
impedance and the d.c. resistance from grid to cathode, in the presence of grid-circuit
damping. This damping is caused, not only by positive grid current, but also by
the grid input conductance which, on the extreme tip of the positive peaks, may be
appreciable. See page 20 under Grid variational conductance and page 489 under
Damping due to positive grid current.

Thus, for any practical amplifier, the tabulated maximum output voltage is misleading, and a lower value should be used for design purposes. In addition, there
will be a slight effect on the gain.

SECTION 3 : RESISTANCE-CAPACITANCE COUPLED PENTODES

(i) Choice of operating conditions (ii) Coupling condenser (iii) Screen by-pass (iv) Cathode bias (v) Fixed bias (vi) Dynamic characteristics of pentodes (vii) Gain at the mid-frequency (viii) Dynamic characteristics of pentodes and comparison with triodes (ix) Maximum voltage output and distortion (x) Conversion factors with r.c.c. pentodes (xi) Equivalent circuit of r.c.c. pentode (xii) Voltage gain and phase shift (xiii) Screen loadlines (xiv) Combined screen and cathode load-lines and the effect of tolerances (xv) Remote cut-off pentodes as r.c.c. amplifiers (xvi) Multigrid valves as r.c.c. amplifiers (xvii) Special applications (xviii) Comments on tabulated characteristics of resistance-coupled pentodes.

(i) Choice of operating conditions

A r.c.c. pentode may be treated as a special case of a r.c.c. triode, and many features are common to both provided that the necessary adjustments are made for the differences in μ, g_m and r_p.

The **load resistance** (R_L) may be any value from a few ohms to many megohms, but for normal operation it is generally from 0.1 to 0.5 megohm. Lower values are used when a reduction in gain or an extended high frequency response is desirable, cr when some form of tone correction is intended. Higher values are occasionally used for special applications, but the frequency response is seriously limited unless negative voltage feedback is used. The load resistance should never be greater than the following grid resistor (R_{g2}) and should preferably be not more than one quarter of R_{g2}. The choice of load resistance has only a small effect on the distortion, lower values of R_L (of the order of 0.1 megohm) being somewhat better in this respect.

Plate supply voltage : As for triodes.

Screen supply : A series resistor from the plate supply voltage is generally preferred. Valves in which the screen current has wide tolerances (e.g. some remote cut-off types and most tetrodes) should be supplied from a voltage divider.

Following grid resistor : As for triodes.

Input grid resistor : As for triodes.

(ii) Coupling condenser

The triode formulae and curves in Sect. 2(ii) may be used except that R may be taken as being approximately equal to $R_{g2} + R_L$. See Sect. 3(x) below for more accurate calculations involving r_p.

(iii) Screen by-pass

For normal performance it is necessary for the screen to be adequately by-passed to the cathode. In practice this is usually obtained by by-passing from screen to earth and also from earth to cathode, the value of C_k being normally much larger than C_s (Fig. 12.10).

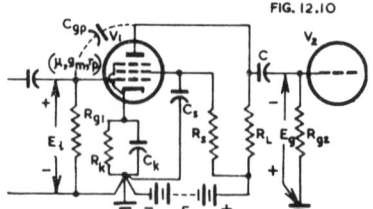

FIG. 12.10

Fig. 12.10. Circuit of resistance-capacitance coupled pentode with screen dropping resistor and cathode bias.

The effect of incomplete screen by-passing on gain is given by eqn. (1) on the assumption of complete cathode by-passing (Fig. 12.10), the derivation being given in Ref. B3 (see also Refs. A11, A12, A13, B2).

$$\left|\frac{A'}{A}\right| \approx \sqrt{\frac{1 + R_s{}^2\omega^2 C_s{}^2}{B^2 + R_s{}^2\omega^2 C_s{}^2}} \tag{1}$$

where A' = stage voltage gain at frequency f with screen resistor R_s by-passed by condenser C_s

A = stage voltage gain with R_s completely by-passed
   = mid-frequency voltage gain

ω = $2\pi f$

$$B \approx 1 + \frac{R_s g_m}{m\mu_t(1 + R_L/r_p)} = 1 + \frac{R_s g_{md}}{m\mu_t} \tag{1a}$$

m = I_b/I_{c2} (assumed constant)

g_m = mutual conductance at operating plate current [see Sect. 3(vii)]

g_{md} = slope of dynamic characteristic at operating plate current

and μ_t = triode mu (with screen tied to plate).

If the screen voltage is obtained from a voltage divider (R_1, R_2) then R_s should be a ken as $R_1 R_2/(R_1 + R_2)$.

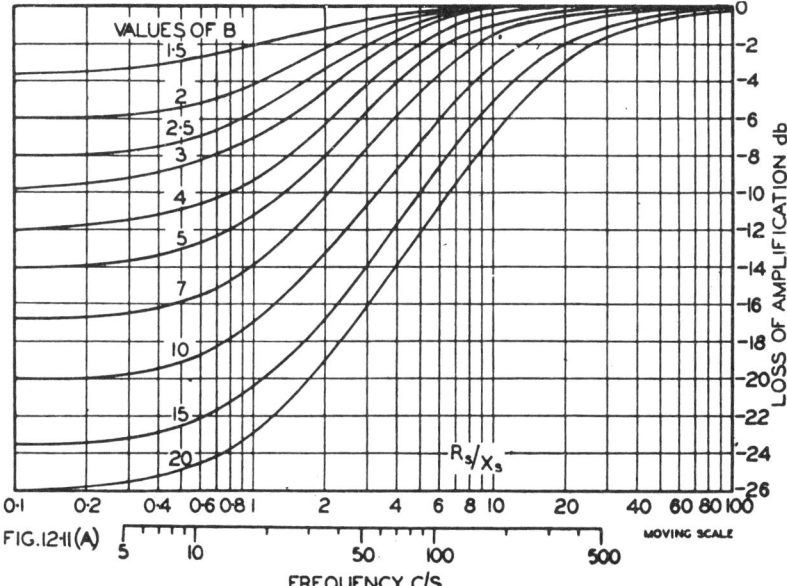

FIG.12-11(A)

FREQUENCY C/S

Fig. 12.11A. Curves for the frequency response of a resistance-coupled pentode with screen dropping resistor R_s and screen by-pass reactance X_s ($= 1/\omega C_s$). The value of B is given by eqn. (1a). The frequency for $R_s/X_s = 1$ is given by $f = 1/2\pi R_s C_s$. The frequency scale should be traced and the tracing moved horizontally until f corresponds to $R_s/X_s = 1$ (method after Sturley).

Eqn. (1) is plotted in Fig. 12.11A for selected values of B versus R_s/X_s, which is equal to $2\pi f R_s C_s$ and therefore proportional to the frequency.

For most purposes it is sufficient to use the **approximation**, for a loss not exceeding 1 db at a frequency f :

$$C_s \approx B/\pi f R_s \tag{2}$$

where $B \approx 1 + R_s g_m/m\mu_t$
and $m = I_b/I_{c2}$.

For example, type 6J7 with $R_L = 0.25$, $R_s = 1.5$ megohms, $g_m = 850$ μmhos, $\mu_t = 20$ and $m = 4$ has $B \approx 16.5$. For 1 db attenuation at $f = 50$ c/s, $C_s \approx 0.07$ μF.

The **limiting loss of voltage gain** at very low frequencies is $1/B$; in the example above this is 0.06 (i.e. 24.3 db). If the screen had been supplied from a voltage divider, the loss of gain at very low frequencies would have been much less.

An unbypassed screen resistor of $g_m r_s$ ohms gives the same degree of degeneration as a 1 ohm cathode resistor

where r_s = dynamic screen resistance $(\partial e_s / \partial i_s)$

　　　　$\approx$ triode plate resistance.

For example, when g_m = 2000 micromhos and r_s = 10 000 ohms, an unbypassed

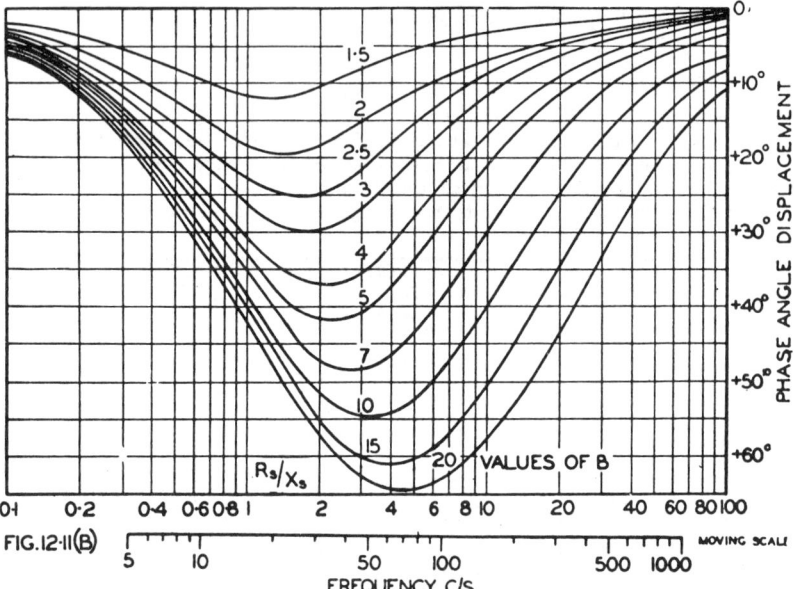

FIG.12·11(B)

Fig. 12.11B.　*Phase angle displacement corresponding to Fig. 12.11A.*

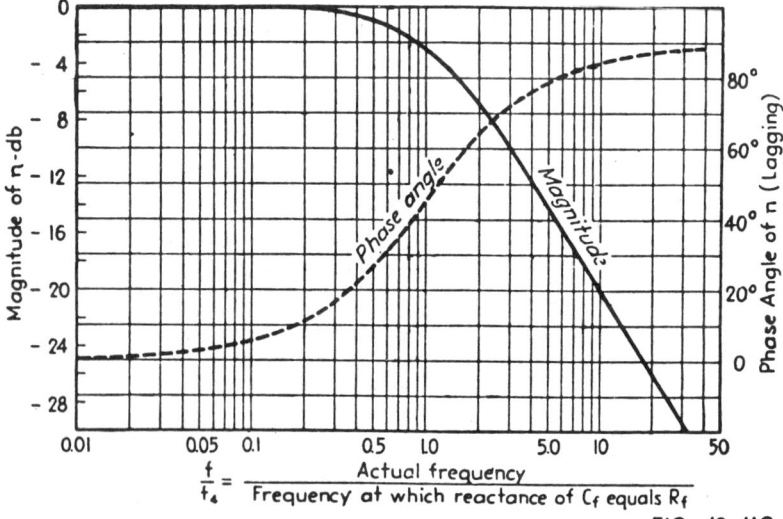

FIG. 12. IIC

Fig. 12.11C.　*Magnitude and phase of factor for use in Fig. 12.11D (Ref. A12). Reprinted by permission from RADIO ENGINEERS' HANDBOOK by F. E. Terman, copyrighted in 1943, McGraw-Hill Book Company Inc.*

screen resistor of 20 ohms will give the same degree of degeneration as a 1 ohm cathode resistor.

The effect of the screen by-pass capacitance on **phase shift** is indicated by Fig. 12.11B. Under normal conditions with a high resistance dropping resistor, the angle does not exceed 65°. This angle may be considerably reduced by reducing the effective value of R_s, as for example with a voltage divider. Such action is normally only necessary with negative feedback.

(iv) Cathode bias
(A) Effect of incomplete by-passing on gain

Provided that the screen is adequately by-passed at all frequencies of operation, the procedure is as for triodes (Sect. 2) except that the d.c. current flowing through

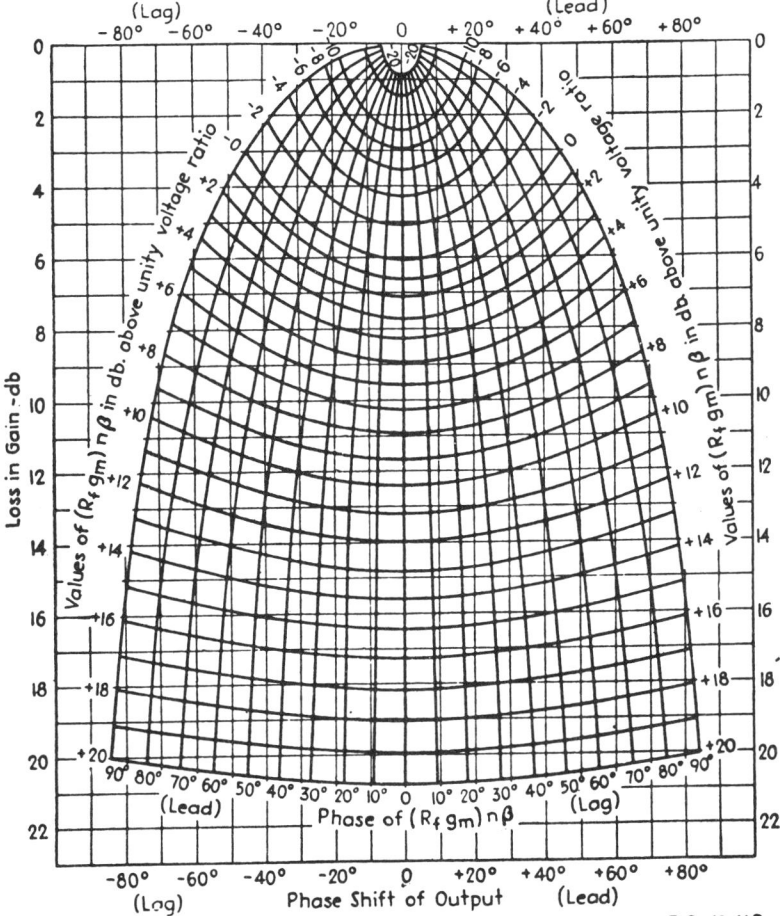

FIG. 12.11D

Fig. 12.11D. Curves from which the loss in gain and phase shift of output at low frequencies resulting from resistance-condenser bias impedance R_fC_f can be calculated in a resistance-coupled amplifier for the general case where both screen and bias impedances are of importance (Ref. A12). Reprinted by permission from RADIO ENGINEERS' HANDBOOK by F. E. Terman, copyrighted in 1943, McGraw-Hill Book Company Inc.

R_k is now $I_k = I_b + I_{c2}$. The effect of R_k and C_k on the voltage gain at any frequency is given by the approximation (derived from Sect. 2 Eqn. 2)

$$\left|\frac{A'}{A}\right| \approx \sqrt{\frac{1 + (\omega C_k R_k)^2}{(1 + g_m R_k)^2 + (\omega C_k R_k)^2}} \tag{3a}$$

on the assumption that the screen is completely by-passed.

When the screen is not completely by-passed, so that both cathode and screen circuits are attenuating simultaneously, the cathode attenuation characteristic is affected by the screen circuit. Under these circumstances it is possible to use the method due to Terman (Refs. A12, B2) whose curves are reproduced in Figs. 12.11C and 12.11D.

Terman expresses the relationship in the form

$$\frac{\text{Actual output voltage}}{\text{Output voltage with zero bias impedance}} = \frac{1}{1 + R_f g_m \eta \beta} \tag{3b}$$

where g_m = mutual conductance of valve in mhos, at operating point
  R_f = cathode bias resistance in ohms

$$\eta = \frac{1}{1 + j(f/f_4)}$$

  f = actual frequency in c/s
  $f_4 = 1/2\pi C_f R_f$ = frequency at which the reactance of C_f equals bias resistance R_f

and $\beta = \left|\dfrac{A'}{A}\right|$ contributed by the screen circuit at frequency f as given by eqn.

(1). (Note that $\beta = 1$ for complete by-passing).

The procedure in a practical case is
1. Determine $\beta = |A'/A|$ from equation (1).
2. Knowing f and f_4, calculate f/f_4.
3. Apply this value of f/f_4 to Fig. 12.11C, thus determining the magnitude of η in db and also its phase angle.
4. Calculate the value of $R_f g_m \eta \beta$, in db above unity.
5. Apply the value of $R_f g_m \eta \beta$ and the phase angle of η to Fig. 12.11D to determine the loss in gain (in decibels).

It will be seen that incomplete screen by-passing results in a smaller value of attenuation by the cathode impedance than would occur with complete screen by-passing.

Attenuation curves for a typical r.c.c. pentode are given in Fig. 12.12 for complete screen by-passing.

As a useful **rule of thumb,** sufficient for most design purposes other than for amplifiers incorporating negative feedback over 2 or 3 stages,

$$fC_k \approx 0.55 g_m \text{ for 1 db attenuation} \tag{4a}$$
$$fC_k \approx 0.35 \, g_m \text{ for 2 db attenuation} \tag{4b}$$

where C_k is in microfarads,
and g_m = mutual conductance in micromhos **at the operating plate current.**

The effect of incomplete cathode by-passing on phase angle displacement is approximately the same as for a triode, and the curves of Fig. 12.3C may be used, but the value of B may be taken as roughly

$$B \approx 1 + g_m R_k$$

on the assumption that μ is very large and that R' is small compared with r_p, where $R' = R_L R_{g2}/(R_L + R_{g2})$.

The combined effect of incomplete cathode and screen by-passing and of a grid coupling condenser on gain and on phase angle displacement may be determined by adding the attenuations in decibels and the phase angles in degrees, provided that eqn. (3b) is used for calculating the attenuation due to the cathode impedance.

(B) Cathode bias resistance

The cathode bias resistance should preferably be the smallest value which can be used without any danger of grid current on the maximum signal. Fortunately, however, the resistance is not critical and a higher value has very little deleterious

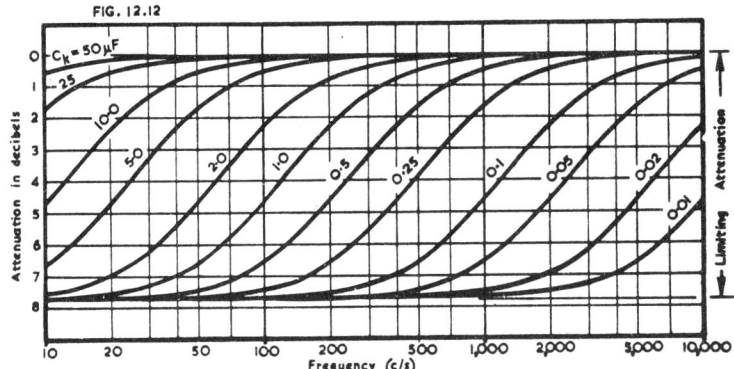

Fig. 12.12. *Attenuation characteristics for selected values of cathode by-pass condenser with valve type* 6J7 *and* $E_{bb} = 250$ *volts,* $R_L = 0.25$, $R_g = 0.5$, $R_s = 1.5$ *megohms,* $R_k = 2000$ *ohms : screen adequately by-passed.*

effect under normal operation, provided that the screen voltage is adjusted to give the correct operating plate current.

Optimum values may be calculated for any specific case by first finding the fixed bias and then calculating $R_k = E_{c1}/(I_b + I_{c2})$.

Alternatively, the value of R_k may be determined by eqn. (5f) below.

The procedure for the calculation of the cathode bias resistor when a series screen resistor is also used, is given in Sect. 3(vi)C, while the graphical method is described in Sect. 3(xiii) and (xiv).

(C) Effect of reverse grid current with cathode bias

When the screen voltage is maintained constant from a low impedance source, the following approximate relationship may be derived from Chapter 3 Sect. 1 eqn (6) :

$$\frac{R_{g1} \text{ for cathode bias}}{R_{g1} \text{ for fixed bias}} \approx 1 + R_k g_k \approx 1 + R_k g_m I_k / I_b \qquad (4c)$$

For a typical pentode, type 6J7, the ratio is approximately 1.7, 2.1 and 2.6 for $R_L = 0.1$, 0.22 and 0.47 megohm respectively. Now the value of R_{g1} for fixed bias is derived in (v)a below for a reverse grid current of 1 μA, so that for type 6J7 the maximum grid resistor under this condition with cathode bias and fixed screen voltage is approximately

$$0.56 \text{ megohm for } R_L = 0.1 \text{ megohm,}$$
$$0.91 \text{ megohm for } R_L = 0.22 \text{ megohm,}$$
$$.1.5 \text{ megohms for } R_L = 0.47 \text{ megohm.}$$

With a series screen resistor the following approximate relationship may be derived from Chapter 3 Sect. 1 eqn. (6).

$$\frac{R_{g1} \text{ for cathode bias}}{R_{g1} \text{ for fixed bias}} \approx 1 + R_k g_k + \frac{P g_k R_{g2}}{\mu_{g1g2}} \qquad (4d)$$

The first two terms are the same as in eqn. (4c), while the third term is the same as the second term of eqn. (5b) below, so that the ratio in eqn. (4d) will normally exceed 7 : 1, for $R_L = 0.1$ megohm and will be considerably greater than this value for $R_L = 0.22$ megohm or higher.

However, it is generally advisable to limit the grid resistor to 2.2 megohms maximum for amplifiers having a reduced frequency range and 1 megohm or less for amplifiers having a maximum frequency of 10 000 c/s or more.

(v) Fixed bias

The optimum bias is the smallest which can be used without danger of positive grid current, provided that the screen voltage is adjusted to give the correct operating plate current.

If the input voltage is known,

$$E_{c1} = 1.41 E_i + k_1 + k_2 \qquad (4e)$$

If the output voltage is known,

$$E_{c1} = (1.41E_0/A) + k_1 + k_2 \qquad (4f)$$

where E_i = r.m.s. input voltage

E_0 = r.m.s. output voltage from plate to cathode

A = voltage gain of stage

k_1 = bias voltage to avoid damping due to positive grid current—a value of 1.0 volt maximum would cover all normal indirectly heated valves*

and k_2 = increment of bias voltage to allow for the effects of reverse grid current in the grid resistor (see below).

If a fixed screen voltage is used, and the grid resistor has the maximum value determined in (a) below, then the value of k_2 may be taken as 0.1 volt for high slope valves (e.g. 6AU6) with R_L not less than 0.1 megohm, and the same value for low slope valves with R_L not less than 0.22 megohm. The value of k_2 may be taken as 0.2 volt for low slope valves having R_L less than 0.22 but not less than 0.1 megohm.

If a series screen resistor is used, and if the grid resistor does not exceed 1 megohm, k_2 may be taken as 0.1 volt maximum.

With some valves, as the screen voltage is decreased, the grid current commencement (or cross-over) point may tend to move to a more negative value in indirectly-heated types, and to a less positive value (possibly even negative) in directly-heated battery pentodes. **Low screen voltages should therefore be avoided with zero-bias operation of battery pentodes, particularly when a high plate load resistance is used.**

The effect of reverse grid current with fixed grid bias
(a) With fixed screen voltage

The position of the operating point in pentodes, for minimum distortion, is fairly critical. Thus when the correct screen and grid bias voltages have been applied, any reverse grid current that may flow through the grid resistor R_{g1} will cause a change in bias and hence a change in plate current to a value less than the optimum. The writer considers a change of 0.1 E_{bb}/R_L as being the maximum permissible change in plate current due to the flow of grid current. On this basis, the maximum permissible grid resistance with fixed bias and fixed screen voltage can be shown to be given approximately by

$$R_{g1} \approx K_p \left(\frac{R_L}{1000} \right)^{0.38} \qquad (5a)$$

where R_{g1} = maximum grid resistance in ohms,

R_L = load resistance in ohms, ·

$K_p = \dfrac{8I_{b0}}{g_{m0}E_{bb}I_{c1}}$

I_{b0} = plate current in amperes at which both g_{m0} and I_{c1} are measured,

g_{m0} = mutual conductance in mhos at plate current I_{b0},

E_{bb} = plate supply voltage,

and I_{c1} = maximum rated reverse grid current in amperes at plate current I_{b0}.

In this calculation the assumption was made that the ionization current is proportional to the cathode current, which relationship only holds approximately and then only when the grid leakage current is small compared with the ionization current and provided that the latter does not increase during operation.

Values of K_p have been derived for three valve types—

Type 6J7 K_p = 5.2 × 10⁴ ⎫ for E_{bb} = 250 volts
Type 6SJ7 K_p = 5.8 × 10⁴ ⎬ and I_{c1} = 1 μA.
Type 6AU6 K_p = 5.7 × 10⁴ ⎭

Values of the function of R_L are given below—

---

*Higher values may occur in a few cases, but are not typical. See also comments on triodes in Sect. 2(iv)B. For grid damping see Sect. 2(iv)D.

R_L	$\left(\dfrac{R_L}{1000}\right)^{0.38}$
100 000	5.75
220 000	7.8
470 000	10.4

From which the following values may be calculated :

Type				
6J7	$R_L = 0.1$	0.22	0.47 megohm	
6SJ7	$R_{g1} = 0.3$ max.	0.4 max.	0.54 max. megohm	
6AU6	$R_{g1} = 0.33$ max.	0.45 max.	0.6 max. megohm	
	$R_{g1} = 0.33$ max.	0.44 max.	0.59 max. megohm	

We may therefore conclude that, for a maximum reverse grid current of 1 μA, the value of R_{g1} with fixed bias and fixed screen voltage should not exceed

0.33 megohm for $R_L = 0.1$ megohm,
0.43 megohm for $R_L = 0.22$ megohm,
0.56 megohm for $R_L = 0.47$ megohm.

Where the variations of reverse grid currents are such that the large majority of valves would have values below half the maximum value—i.e. in this case below 0.5 μA—double these values of R_{g1} would be satisfactory. This would also apply in any cases where the maximum value of reverse grid current is specified as 0.5 μA.

(b) With series screen resistor

When fixed bias is used in conjunction with a series screen resistor supplied from the plate voltage source, it may be shown from eqn. (9) of Chapter 3 Sect. 1 that

$$\frac{R_{g1} \text{ for series screen resistor}}{R_{g1} \text{ for fixed screen voltage}} \approx 1 + \frac{P g_k R_{g2}}{\mu_{g1g2}} \tag{5b}$$

where $P = I_{c2}/I_k$,
  $g_k = g_m(I_k/I_b)$ at the operating point,
  R_{g2} = resistance of series screen resistor in ohms,
and μ_{g1g2} = " triode " amplification factor.

Ratios calculated from eqn. (4c) for typical valves, with a load resistance of 0.1 megohm or more, exceed 6 times, so that when a high-resistance series screen resistor is used, the grid resistor may be at least 6 times the value quoted above for fixed screen voltage.

However, it is generally advisable to limit the grid resistor to 2.2 megohms maximum for amplifiers having a reduced frequency range and 1 megohm or less for amplifiers having a maximum frequency of 10 000 c/s or more.

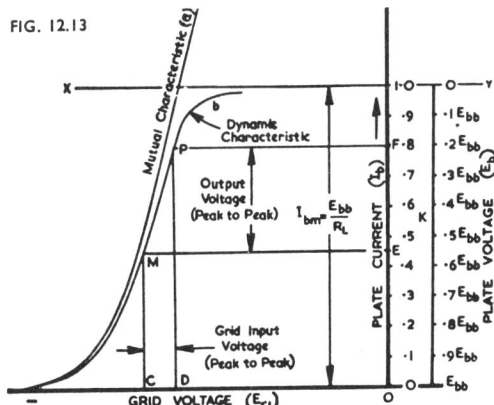

FIG. 12.13

Fig. 12.13. Dynamic characteristic of r.c.c. pentode for fixed screen voltage (b) compared with mutual characteristic (a).

(vi) Dynamic characteristics of pentodes
(A) General description

A single dynamic characteristic of a pentode, together with the equivalent mutual characteristic, are shown in Fig. 12.13. The slope of the dynamic characteristic at any point is given by

$$g_{md} = g_m \frac{r_p}{r_p + R_L} = \frac{g_m}{1 + (R_L/r_p)} \tag{5c}$$

The values of g_m and g_{md} are normally within 10% provided that the operating plate current does not exceed $0.8\ E_{bb}/R_L$. Values of g_m and g_{md} for valve types 6J7 and 6SJ7 are given in (vii) below.

Pentodes differ from triodes in that there is an unlimited number of dynamic characteristics for any selected plate voltage and load resistance, there being one characteristic for each value of screen voltage. Some typical dynamic characteristics are shown in Fig. 12.14A from which it will be seen that these have much the same general form, but that they are moved bodily sideways. Careful examination will show that the curves for lower screen voltages are less curved than those for higher screen voltages. As a general rule, the screen voltage should therefore be as low as possible, provided that the correct operating point can be maintained and that positive grid current does not flow.

FIG. 12.14A

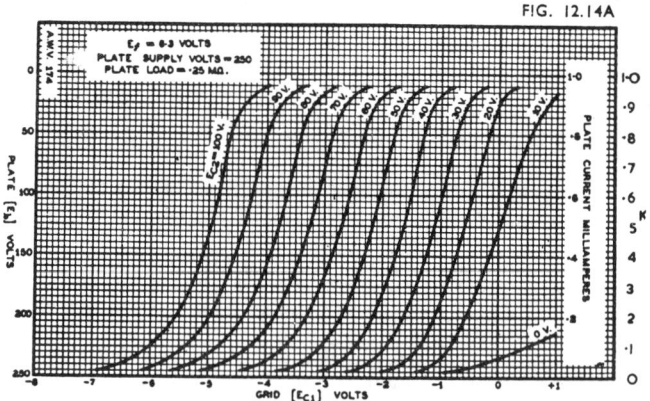

Fig. 12.14A. Family of dynamic characteristics for typical r.c.c. pentode (type 6J7 with $E_{bb} = 250$ volts, $R_L = 0.25$ megohm).

(B) Optimum operating conditions

A very complete investigation (Ref. B8) has shown that all pentodes normally used with r.c. coupling may be adjusted for **maximum gain and minimum non-linear distortion** merely by ensuring that the plate current is a certain fraction (K) of E_{bb}/R_L. This plate current may be achieved either by fixing the grid bias (or cathode bias resistor) and varying the screen voltage (or series screen resistor) or vice versa. The effect of reverse grid current in the grid resistor is to decrease the effective grid bias by an amount which may be appreciable with fixed screen voltage, but may usually be neglected when a high-resistance series screen resistor is used. In the former case, with fixed screen voltage, it is suggested that the design be based on a value of half the maximum specified reverse grid current flowing through the grid resistor.

The **optimum value of** K is not appreciably affected by either the following grid resistor or the electrode voltages, provided only that the screen voltage is kept reasonably low. The optimum value of K for minimum distortion is, however, a function of the output voltage as indicated below.

Output voltage (r.m.s.)	=	$0.04E_{bb}$	$0.08E_{bb}$	$0.16E_{bb}$	$0.24E_{bb}$
Output voltage (r.m.s.)*	=	10	20	40	60
Optimum K	=	0.78	0.76	0.70	0.62

*For E_{bb} = 250 volts.
where $K = I_b/I_{bm}$ and $I_{bm} = E_{bb}/R_L$, and $R_g = 2R_L$.
   For example, with E_{bb} = 250 volts and R_L = 0.25 megohm, I_{bm} = 1 mA.
   The **voltage between plate and cathode** is a function of K : $E_b = (1 - K) E_{bb}$.

When K =	0.78	0.76	0.70	0.62
Then E_b =	$0.22E_{bb}$	$0.24E_{bb}$	$0.3E_{bb}$	$0.38E_{bb}$

(C) Determination of series screen and cathode bias resistors when plate and screen current curves are available
If curves of both plate and screen currents versus grid voltage under resistance loaded conditions (e.g. Figs. 12.14A and 12.18) are available, the procedure is—
   1. Determine E_{c1}—see (v) above.
   2. Determine the optimum value of K—see (vi)B above.
   3. From the I_b curves, determine E_{c2} to give the required values of E_{c1} and K ; also determine I_b at the operating point.
   4. From the I_{c2} curves, determine I_{c2} at the operating point.
   5. Then $R_k = E_{c1}/(I_b + I_{c2})$ (5d)
and $R_s = (E_{cc2} - E_{c2})/I_{c2}$ (5e)

(D) Calculation of series screen and cathode bias resistors
The procedure for calculating R_s and R_k is—
   1. Determine E_{c1} as for fixed bias—see (v) above.
   2. Determine the optimum value of K—see (vi)B above.
   3. Then $R_k \approx \dfrac{E_{c1}R_Lm}{KE_{bb}(m + 1)}$ (5f)

where E_{c1} = control grid voltage (taken as positive)
and $m = I_b/I_{c2}$.

   Typical values of m at normal voltages are :

Type	6J7	6SJ7	6SF7	6SH7	6AC7	6AU6	1S5	1U5
m	4.0	3.75	3.75	2.63	4.0	2.54	4.0	4.0

Equation (5f) is perfectly general, and may be used with any screen voltage source. The only approximation is the assumption that m is constant.

   4. Then $E_{c2} \approx E_{c1}\mu_t + E_2 - \dfrac{KE_{bb}\mu_t}{R_{L1}g_{md}}\left(1 - \dfrac{R_{L1}}{R_L}\right)$ (5g)

where E_{c2} = screen voltage
             μ_t = triode mu
             E_2 = screen voltage at which I_b = 2 mA for $E_{c1} = 0$
                $\approx$ 22 (for 6AU6), 25 (for 6SJ7), 39 (for 6J7)
             R_{L1} = 0.1 MΩ
and g_{md} = max. slope of dynamic characteristic with R_L = 0.1 MΩ.
Equation (5g) may also be used with fixed bias E_{c1}.

   5. Then $R_s \approx \dfrac{mR_L}{KE_{bb}}(E_{bb} - E_{c2})$ (5h)

   Note : This procedure is based on several approximations (e.g. that m, g_{md} and E_2 are constant) but is sufficiently accurate for all practical purposes for any pentode.
   It will be seen that a single operating condition is incapable of giving optimum performance for both low and high levels.   For low level operation, E_{c1} may be taken as − 1.3 volts, while for high level operation E_0 may arbitrarily be taken as $0.24E_{bb}$ volts r.m.s. unless it is desired to determine some intermediate condition.
   It will be seen from eqn. (5g) that the optimum screen voltage is a function, not only of E_{c1}, but also of R_L—a higher value of load resistance permits a lower screen voltage.

(E) Screen supply from voltage divider

Although a series screen resistor is normally preferable, the screen may be supplied in certain applications from a voltage divider. The procedure is straight forward, but the voltage should be adjusted manually or else the two resistors should have $\pm 5\%$ tolerances.

When a screen voltage divider is used, the equivalent series screen resistance is given by the resultant of the two sections of the voltage divider in parallel. This value may be used as R_{g2} in eqn. (6) of Chapter 3 Sect. 1, although if R_{g2} is less than 50 000 ohms, its effect is small.

If the equivalent series screen resistance is less than 50 000 ohms, the screen voltage may be considered as fixed, and it is then necessary to increase the grid bias to allow for the **effect of reverse grid current.**

The increase in grid bias is

$$\Delta E'_{c1} = R_{g1} I_{c1} I_b / I_{b0} \tag{6a}$$

where I_{b0} = plate current at which the reverse grid current is measured, and I_{c1} may be taken as half the maximum specified reverse grid current (say 1 μA) ; i.e. (say) $I_{c1} = 0.5 \ \mu$A.

The increased value of cathode bias resistor is given by

$$R_k' = R_k + \Delta R_k \tag{6b}$$

where $\Delta R_k = \dfrac{\Delta E_{c1}}{I_b + I_{c2}} = \dfrac{\Delta E_{c1}}{I_b} \cdot \dfrac{m}{m+1}$

and E_{c1} is given by equation (6a).

Using the suggested value of I_{c1}, we have

$$R_k' = R_k + \frac{R_{g1} \times 0.5 \times 10^{-6}}{I_{b0}} \cdot \frac{m}{m+1} \tag{6c}$$

For example, if $R_{g1} = 1$ megohm, $I_{b0} = 2$ mA, and $m = 4$, then

$$R_k' = R_k + 200 \text{ ohms.}$$

(vii) Gain at the mid-frequency

(A) The voltage gain at the mid frequency (for small input voltages) is given by

$$A_0 = \frac{g_m}{(1/r_p) + (1/R_L) + (1/R_g)} \tag{7}$$

where g_m = mutual conductance at operating plate current,

$\quad r_p$ = plate resistance at operating point,

and $\quad R_g$ = resistance of following grid resistor.

The mutual conductance at the working plate current is often an unknown factor, and methods for determining it are given below (E).

(B) Alternatively, the gain may be calculated from the slope of the dynamic characteristic :

$$A_0 = g_{md} R\left(\frac{r_p + R_L}{r_p + R}\right) \approx g_{md} R \tag{8}$$

where g_{md} = slope of dynamic characteristic at the operating point—see (E) below, and $\quad R = R_L R_g / (R_L + R_g)$.

(C) **The gain may also be calculated graphically from the dynamic characteristic** (Fig. 12.13). If the peak-to-peak grid input voltage is CD, then the instantaneous plate current will swing between the extreme limits M and P, giving a peak-to-peak output voltage EF measured on the plate voltage scale. To allow for the effect of R_g, the output voltage must be multiplied by the factor $R_g / (R_g + R')$ where $R' = r_p R_L / (r_p + R_L) \approx R_L$.

Typical voltage gains are :

R_L	R_g	6J7	6SJ7	6AU6
0.1 megohm	0.5 megohm	94	104	168
0.25 megohm	0.5 megohm	140	167	230*
0.5 megohm	2.0 megohms	230	263	371**

$\quad$ *$R_L = 0.22$ megohm. $\quad$ **$R_L = 0.47$ megohm.

If the resistance of the following grid resistor R_g is limited, the **maximum gain on low input voltages** is obtained when the load resistance is approximately equal to the following grid resistance, i.e. $R_L \approx R_g$.

(D) Gain in terms of g_m and g_p

Eqn. (7) may be put into the alternative form

$$A_0 = \frac{g_m R}{1 + g_p R} \tag{9a}$$

where $R = R_L R_g / (R_L + R_g)$
and g_p = plate conductance = $1/r_p$.

Unfortunately " G," curves for pentodes hold only for a fixed screen voltage and are not flexible (g_m curves for type 7E7 are shown in Ref. A15).

(E) The determination of g_m, g_{md}, and r_p

The values of g_m, g_{md} and r_p for use in equations (7), (8) and (9) are difficult to calculate with any precision, but may be measured on an average valve.

Curves of constant g_{md} could, with advantage, be plotted on the family of dynamic characteristics such as Fig. 12.14A, although this has not yet been done.

Mutual conductance

For the general case see (x)A below. Fig. 12.14B gives an approximation to the maximum value of g_m with any load resistance for two typical pentodes. This value is believed to be accurate within $\pm 10\%$ for an average valve.

Slope of dynamic characteristic

The value of the slope of the dynamic characteristic at any point is given by

$$g_{md} = g_m / (1 + R_L / r_p) \tag{9b}$$

where g_{md} = slope of dynamic characteristic at operating point
g_m = mutual conductance at operating point
and r_p = plate resistance at operating point.

FIG. 12.14B

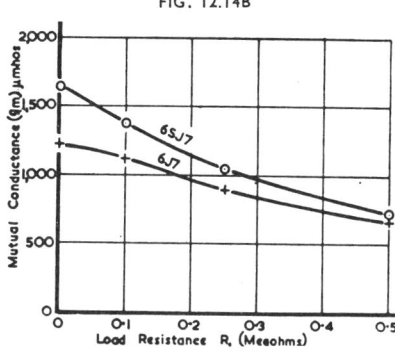

Fig. 12.14B. Mutual conductance of resistance-capacitance coupled pentodes types 6J7 and 6SJ7 plotted against load resistance. In each case the value of K is taken as being adjusted to give maximum mutual conductance.

Plate resistance

The value of plate resistance r_p at the operating point may be estimated from Sect. 3(x) below.

Data for types 6J7 and 6SJ7

Detailed values of mutual conductance (g_m), slope of the dynamic characteristic (g_{md}) and plate resistance (r_p) for types 6J7 and 6SJ7 are tabulated below. It is emphasized that there are considerable variations between valves, particularly with the plate resistance.

Valve type Plate supply voltage Published g_m*		6J7 250 1225			6SJ7 250 1650			volts μmhos
R_L MΩ	K	g_m μmhos	g_{md} μmhos	r_p MΩ	g_m μmhos	g_{md} μmhos	r_p MΩ	
0.1	0.78	1065	980	0.6	1390	1280	0.7	
	0.76	1080	1010	0.67	1380	1290	0.9	
	0.70	1130	1080	1.1	1370	1310	1.3	
	0.62	1120	1080	1.5	1340	1290	1.4	
	0.55	1100	1060	2.1	1335	1280	1.7	
0.25	0.78	896	834	1.6	1050	990	1.7	
	0.76	880	836	1.9	1040	985	1.8	
	0.70	860	830	3.0	1000	960	2.4	
	0.62	820	790	3.3	940	900	3.3	
	0.55	790	765	3.7	920	845	3.4	
0.5	0.78	670	640	3.0	720	660	3.0	
	0.76	660	630	3.6	710	660	3.0	
	0.70	647	620	4.5	650	648	3.7	
	0.62	595	575	5.2	590	580	5.0	
	0.55	550	540	6.0	519	495	5.4	

*With $E_b = 250$ volts, $E_{c2} = 100$ volts, $E_{c1} = -3$ volts.

(viii) Dynamic characteristics of pentodes and comparison with triodes

A family of dynamic characteristics for a typical pentode is shown in Fig. 12.14A. There is one curve for every possible value of screen voltage, but it is obvious that those for lower screen voltages (e.g. 30 volts) are less curved than those for higher voltages. In actual operation the quiescent operating point is fixed, and the signal voltage on the grid causes the plate current to swing along the dynamic curve. This only holds strictly when there is no following grid resistor or other shunt load, but it is sufficient for design purposes since allowance can readily be made for the effect of R_g in reducing both the voltage gain (eqn. 8) and the output voltage.

The voltage drop in R_L is proportional to the plate current, and therefore a plate voltage scale may be added to Fig. 12.14A. The general treatment and the calculation of gain are identical with those for ideal linear dynamic characteristics as applied to triodes in Sect. 2(viii). In the case of pentodes the operation can be kept outside the grid current region, without any other loss, by selecting a suitable screen voltage.

A single pentode dynamic characteristic (that is one for a fixed screen voltage) is drawn as curve b in Fig. 12.13. For comparison, curve a has also been included, this being the ordinary mutual characteristic for the same screen voltage. At low plate currents the two are practically identical but they diverge steadily up to the point of inflexion* P, beyond which the dynamic characteristic forms the " top bend."

The line XY is at a plate current $I_{bm} = E_{bb}/R_L$ at which the factor $K = 1$. This curve is typical of the shape of all pentode dynamic characteristics which differ mainly in the slope and the horizontal displacement of the curve. The effects of changes in R_L are largely overcome by the use of the factor K in place of the actual plate current.

Comparison between triode and pentode dynamic characteristics

Both triode and pentode dynamic characteristics are shown in Fig. 12.15 in such a way as to enable a comparison to be made between them. Over the region from $K = 0.4$ to $K = 0.6$ they appear to be very similar, but in the region from $K = 0.15$ to 0.4 the pentode characteristic appears less curved than the triode, while in the region from $K = 0.6$ to 0.8 the triode characteristic appears less curved than the pentode.

*The point of inflexion is the point at which the curvature changes from one direction to the other, and is the point of greatest slope.

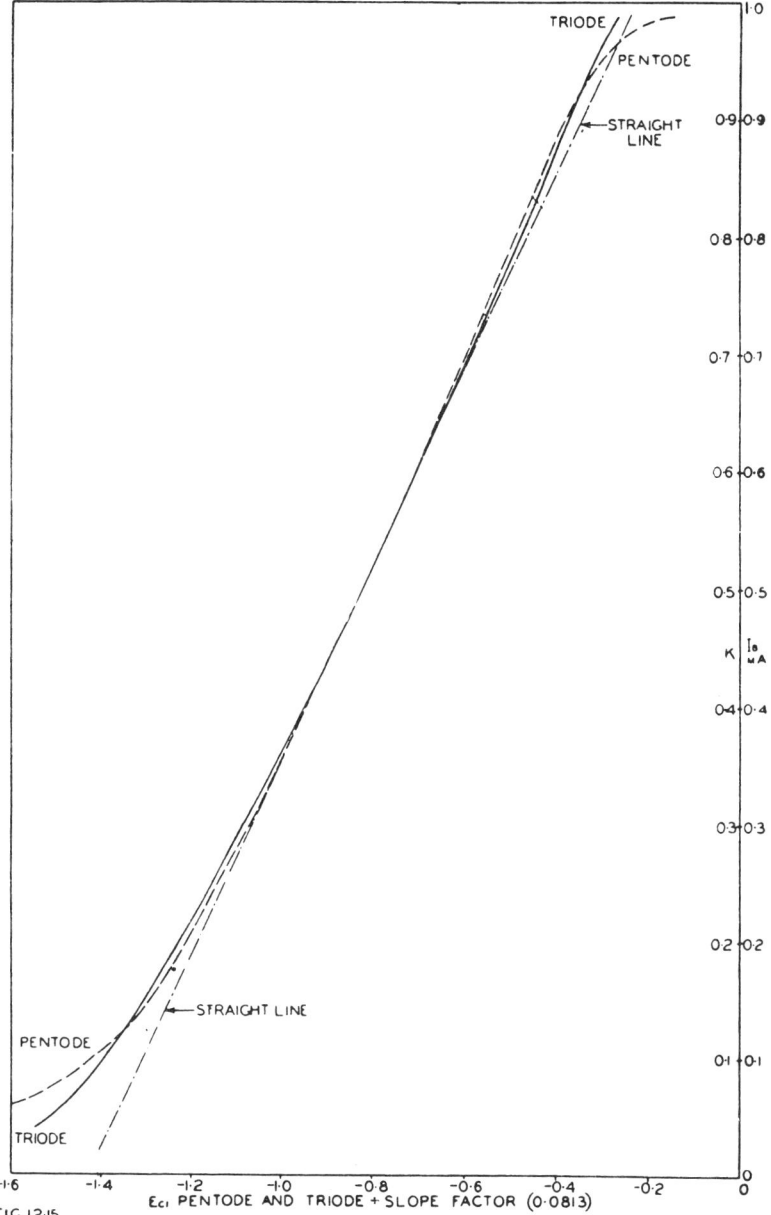

Fig. 12.15. *Comparison between pentode and triode dynamic characteristics of the same valve (6SJ7 with $E_{bb} = 250$ volts, $R_L = 0.25$ megohm, $E_{c2} = 22.5$ volts for pentode, E_{c2} connected to plate for triode characteristics ; triode characteristics divided by slope factor 0.0813 and superimposed so that points for $I_b = 0.5$ mA coincide).*

(ix) Maximum voltage output and distortion

The maximum voltage output, with any specified limit of distortion, is obtained when R_g is much greater than R_L. If R_g is limited, the maximum voltage output is obtained when R_L is much less than R_g. The following table is typical of r.c.c. pentodes such as types 6J7 and 6SJ7, and also enables a comparison to be made with triodes. The optimum value of K has been used throughout.

Intermodulation distortion* with $E_{bb} = 250$ volts

Output voltage†		10	19	37	63
Valve type 6J7			Intermodulation distortion %		
$R_L = 0.1$ MΩ $R_g = 0.1$ MΩ		0.35	2.2	8.6	—
	0.2	0.3	1.5	4.6	14.0
	0.4	0.26	0.85	3.9	12.5
$R_L = 0.25$ MΩ $R_g = 0.25$ MΩ		0.65	1.9	7.4	37
	0.5	0.5	1.6	4.4	17
	1	0.35	1.4	3.5	15
$R_L = 0.5$ MΩ $R_g = 0.5$ MΩ		1.0	3.7	14.0	54
	1	0.8	2.6	9.5	32
	2	0.5	2.0	6.5	19
Valve type 6SJ7					
$R_L = 0.1$ MΩ $R_g = 0.1$ MΩ		1.0	2.9	12.0	—
	0.2	0.9	2.2	8.3	24
	0.4	0.8	1.5	5.9	20
$R_L = 0.25$ MΩ $R_g = 0.25$ MΩ		2.1	6.0	16.0	67
	0.5	1.1	3.8	13.0	38
	1	1.0	3.0	11.0	23
$R_L = 0.5$ MΩ $R_g = 0.5$ MΩ		3.0	8.3	18.0	76
	1	2.1	6.6	15.0	42
	2	1.5	4.4	12.5	28

The maximum output voltage for limited intermodulation distortion (r.m.s. sum) is indicated below for type 6SJ7 pentode with $R_g = 4R_L$.

R_L	$E_{0(r.m.s.)}/E_{bb}$			
	I.M. = 2.5%	5%	10%	20%
0.1 MΩ	0.09	0.13	0.18	0.25
0.25	0.08	0.11	0.16	0.22
0.5	0.07	0.09 •	0.13	0.20

When $R_g = 2R_L$, the factors above should be multiplied by 0.87. When $R_g = R_L$, the factors above should be multiplied by 0.7. The values are for $E_{bb} = 250$ volts, but hold closely over the range from 200 to 300 volts. Optimum operating conditions are assumed.

Harmonic distortion

Type 6SJ7 with $E_{bb} = 250$ volts, $R_L = 0.25$ MΩ, $R_g = 0.5$ MΩ

$K =$	0.8	0.76	0.72	0.65	0.6	0.56
$E_{0(r.m.s.)}$ volts	8	16	30	52	62	78
$E_{0(r.m.s.)}/E_{bb}$	0.03	0.06	0.12	0.21	0.25	0.31
H_2 %	0.3	0.9	2.6	3.5	5.4	4.4
H_3 %	0.24	0.19	1.25	4.2	5.0	12.0
H_4 %	0.14	0.11	0.19	1.9	2.2	2.85
H_5 %	0.01	0.02	0.18	0.67	0.28	0.17

*Modulation method—r.m.s. sum. For details and for relation between intermodulation and harmonic distortion see Chapter 14 Sect. 3.
†The arithmetical sum of the r.m.s. values of the two component waves.

Comments

1. Type 6J7 (with published $g_m = 1225$ μmhos) has less distortion than type 6SJ7 (with published $g_m = 1650$ μmhos) under the same conditions.

The distortion for a given output voltage increases when a valve is replaced by another having higher mutual conductance, although there are also differences between valve types having approximately the same mutual conductance.

2. Load resistance $R_L = 0.1$ megohm provides lower distortion than higher values of load resistance.

3. The distortion under any given conditions decreases when the resistance of the following grid resistor is increased.

Comparison between triode and pentode

[Refer Sect. 2(ix)]

The comparison is based on intermodulation distortion with type 6SJ7 as both triode and pentode, having $R_L = 0.25$ and $R_g = 1.0$ megohm (Fig. 12.16A). Generally similar results are obtained with type 6J7 and with other load resistances (Ref. B8).

1. At the level used in the first a-f stage in a typical receiver ($E_0 = 10$ volts r.m.s.) the pentode gives only about one eighth of the intermodulation distortion given by a triode, when both are adjusted for minimum distortion.

2. The two curves in Fig. 12.16A cross, and the intermodulation distortion is therefore the same for both triode and pentode, at about 31 r.m.s. volts output.

3. At higher output voltages the pentode gives the greater intermodulation distortion, the ratio being 2.3 : 1 at 63 r.m.s. volts output.

4. The pentode, to give the minimum value of distortion, requires fairly critical adjustment. Under the working conditions recommended in this section, however, the distortion with a pentode is likely to be less than with a general-purpose triode at output voltages up to about 20 volts r.m.s.

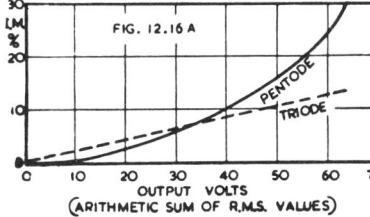

Fig. 12.16A. Intermodulation distortion for type 6SJ7 with $E_{bb} = 250$ volts, $R_L = 0.25$ $M\Omega$, $R_g = 1.0$ $M\Omega$, in both triode and pentode operation.

(x) Conversion factors with r.c.c. pentodes

(A) **The mutual conductance** may be derived from published data (generally plotted against E_{c1}) or if these are not available then it may be calculated as for triodes, Sect. 2(x). See also values quoted for 6J7 and 6SJ7 in (vii) above.

(B) **The plate resistance** may be estimated by the following method from published data. The usual data are for either equal plate and screen voltages (e.g. 100 V) or for 250 V and 100 V respectively. Unfortunately there is no general rule for calculating the effect of a change of plate voltage on the plate resistance, and the accuracy obtainable by graphical means is very poor. However, as some sort of guide, the plate resistance of type 6SJ7 is increased about twice, and of type 6J7 about 2.75 times, for a change from 100 to 250 volts on the plate, with 100 volts on the screen and -3 volts grid bias.

The plate resistance of a r.c.c. pentode may vary from slightly below the published value at $E_b = E_{c2} = 100$ volts, to 6 or 8 times this value, depending on the load resistance and value of K. Typical values for types 6J7 and 6SJ7 are given below [see also table in (vii) above] :

$R_L = 0.1$ megohm	$r_p = 0.6$ to 2.1 megohms
$R_L = 0.25$ megohm	$r_p = 1.6$ to 3.7 megohms
$R_L = 0.5$ megohm	$r_p = 3.0$ to 6.0 megohms
	for $K = 0.78$ to 0.55 respectively.

There are considerable variations between different valves, and some may have plate resistances less than half these values, while others may have higher plate resistances.

(C) **Conversion factors applied to whole amplifiers**

If one set of operating conditions is available, it is possible to calculate others.

Given :	E_{bb}	R_L	R_k	R_s	R_g	A	E_0
	300 V	0.1 MΩ	450 Ω	0.5 MΩ	0.25 MΩ	82	81 V

Example :

(1) To calculate conditions for $R_L = 0.25$ megohm,
$E_{bb} = 300$ volts :—

$F_r = 2.5$ (i.e. resistance conversion factor).

$R_k = 2.5 \times 450 = 1130$ ohms.

$R_s = 2.5 \times 0.5 = 1.25$ megohms.

$R_g = 2.5 \times 0.25 = 0.75$ megohm.

E_0 will be approximately the same for the same distortion.

A is affected both by the load and the mutual conductance.

The load resistance factor is 2.5 ; the mutual conductance factor is $\sqrt[3]{I_{b2}/I_{b1}} = \sqrt[3]{1/2.5} = 0.74$. Therefore $A = 82 \times 2.5 \times 0.74 = 152$.

Note : If E_{bb} had been altered, the mutual conductance factor would have been $\sqrt[3]{E_{bb2}R_{L1}/E_{bb}R_{L2}}$.

(2) If all resistors are multiplied by a factor F_r, leaving E_{bb} unchanged, the voltage gain will be increased approximately as tabulated :

F_r	0.4	0.5	0.75	1.0	1.5	2	2.5
Voltage gain	0.54A	0.63A	0.83A	A	1.43A	1.58A	1.85A approx.

In practice these may vary ± 10% or even more.

(xi) Equivalent circuit of r.c.c. pentode

The exact a.c. equivalent plate circuit of a r.c.c. pentode is given in Fig. 12.16B where the " constant current generator " circuit has been adopted. This could equally be applied to the triode case (Fig. 12.8) or vice versa. In other respects the triode and pentode equivalent circuits are identical [see Sect. 2(xii)]. Both the cathode and screen circuits are assumed to be adequately by-passed at all signal frequencies, or the effects to be separately calculated.

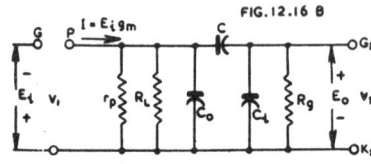

FIG. 12.16 B

Fig. 12.16B. *Exact a.c. equivalent plate circuit of a r.c.c. pentode.*

(xii) Voltage gain and phase shift

Provided that the cathode and screen circuits are adequately by-passed at all signal frequencies, or that their voltages are obtained from low impedance sources, the voltage gain and phase shift will be the same as for the triode case [Sect. 2(xiii)]. In the case of a pentode it is usually possible to neglect the " Miller Effect " capacitance from the plate circuit.

The **low frequency response** [Sect. 2(xiii) ; eqn. (15)] will be slightly higher than for a triode with the same values of C_1, R_L and R_g owing to the higher value of r_p.

The **high frequency response** [Sect. 2(xiii), eqn. (16)] will normally be lower than that for a triode, owing to the higher value of R'' due again to the higher plate resistance. The valve output capacitance is usually greater for a pentode than for a triode, this being a further contributing factor. The high frequency response may be extended to higher frequencies by reducing the resistance of the plate load resistor, although at the cost of gain. A similar result may be achieved by the use of negative voltage feedback.

The effect of the screen and cathode by-passing has been covered in (iii) and (iv) above.

(xiii) Screen loadlines

(A) Exact method using I_{c2} versus E_{c2} characteristics

These characteristics (Fig. 12.17) only apply to a single value of supply voltage and plate load resistance. The screen loadline is a straight line drawn from A to B.
Point A : $E_{c2} = E_{cc2}$ (here 250 volts), $I_{c2} = 0$
Point B : $E_{c2} = 0$, $I_{c2} = E_{cc2}/R_s$ (here $250/1.5 = 167$ μA).

It is then necessary to transfer the values of E_{c2}, at the intersections of the E_{c1} curves and the loadline, to a second curve in which E_{c1} is plotted against E_{c2}. From the latter, the values of E_{c1} corresponding to the screen voltages of the dynamic characteristics (Fig. 12.14A) are determined, and transferred directly to the plate dynamic characteristics, and the screen loadline drawn as a smooth curve (Fig. 12.19).

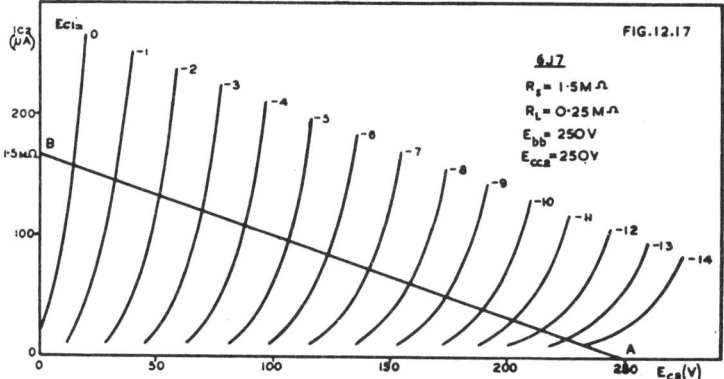

Fig. 12.17. Screen I_{c2} versus E_{c2} characteristic for type 6J7 ; $E_{bb} = 250$ volts, $R_L = 0.25$ MΩ. A screen loadline is shown.

(B) Exact method using I_{c2} versus E_{c1} characteristics

This is an alternative method illustrated by Fig. 12.18 on which the screen loadline (which in this case is slightly curved) may be plotted. On each E_{c2} curve plot the corresponding value of I_{c2} derived from the equation

$$I_{c2} = (E_{cc2} - E_{c2})/R_s \qquad (10)$$

Then draw a smooth curve (CD in Fig. 12.18) through the plotted points. The values of E_{c1} corresponding to the points of intersection are then transferred directly to the plate dynamic characteristic, and the loadline drawn as a smooth curve (Fig. 12.19).

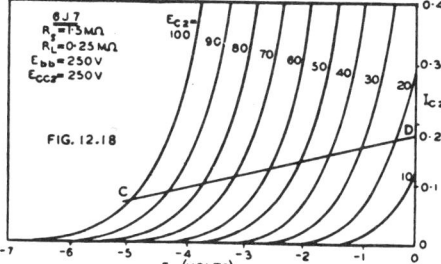

Fig. 12.18. Screen I_{c2} versus E_{c1} characteristics for type 6J7 ; static curves are for $E_{bb} = 250$ volts, $R_L = 0.25$ MΩ. A screen loadline is shown.

(C) Approximate method*

This is based on the assumption that the ratio of plate and screen currents is constant, such being only a rough approximation ; it is, however, good enough for many purposes. Using the conditions for Fig. 12.19 and assuming $I_b/I_{c2} = 4.0$ (as for type 6J7) and $R_s = 1$ megohm, we can draw up the table :

E_{c2}	$(E_{cc2} - E_{c2})$	I_{c2}	$I_b = 4I_{c2}$
100 V	150 V	0.15 mA	0.60 mA
80	170	0.17	0.68
60	190	0.19	0.76
40	210	0.21	0.84
20	230	0.23	0.92

The values of I_b are then plotted on the characteristics corresponding to the respective screen voltages (E_{c2}). The screen loadline (Fig. 12.19) is almost a straight line which may be extended to cut the horizontal axis at approximately

$$E_{c1} \approx -E_{cc2}/\mu_t \tag{11}$$

and which may be shown to have a slope of approximately

$$m\mu_t g_{md}/(R_s g_{md} + m\mu_t) \tag{12}$$

where μ_t = valve triode mu,

       g_{md} = slope of plate dynamic characteristic at point of interest,

and m = ratio of plate to screen currents.

Alternatively, the point of intersection of the screen loadline with the horizontal axis may be determined by the point of cathode current cut-off on the "triode" characteristics (if available) where $E_b = E_{cc2}$.

The plate current at $E_{c1} = 0$ is given approximately by

$$I_{b0} \approx mg_{md}E_{cc2}/R_s g_{md} + m\mu_t \tag{13}$$

The value of screen resistor to provide a plate current $I_b = KE_{bb}/R_L$ at a fixed grid voltage E_{c1} when $E_{cc2} = E_{bb}$ is given approximately by

$$R_s \approx \frac{m}{K}\left\{ R_L - \frac{\mu_t R_L E_{c1}}{E_{bb}} - \frac{K\mu_t}{g_{md}} \right\} \tag{14}$$

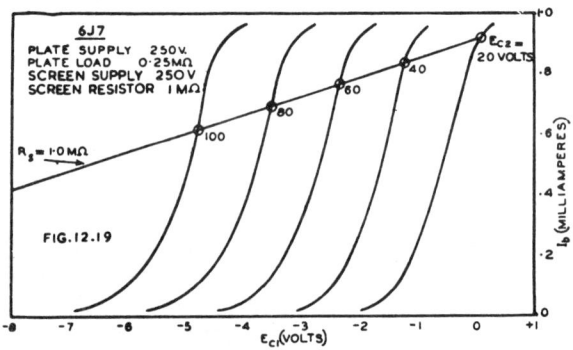

Fig. 12.19. Plate dynamic characteristics for type 6J7 with screen loadline.

Variations in ratio between plate and screen currents

Although considerable variations exist, particularly in the region of high plate current, there is a comparatively large region within which the ratio is constant within ± 5%, this being fortunately in the region most generally useful for amplification (e.g. shaded area in Fig. 12.20).

The variations which occur are mainly caused by the variations in the ratio of plate to screen voltages. It is clear that they will be serious for E_b less than E_{c2} (above broken line in Fig. 12.20) but the variations become greater as E_b approaches E_{c2}. If it is desired to make accurate calculations involving an assumed constant value of m (= I_b/I_{c2}) it is desirable to maintain E_b/E_{c2} nearly constant. The use

*A further approximate method is described in Ref. B12.

of a series screen resistor with a resistance equal to mR_s (or slightly above this value)
assists in maintaining the constant current ratio.

(xiv) Combined screen and cathode loadlines and the effect of tolerances

(A) Cathode loadlines

The method normally adopted is an approximate one but very convenient, since
it may be used directly with the plate dynamic characteristics. Provided that $I_b/I_{c2} = m = $ constant, the current through R_k will be $(m + 1) I_b$, so that it is necessary to
use a " conversion factor " of $m/(m + 1)$ in respect to both the current and slope of
the loadline. For example if $m = 4$, the loadline for a cathode bias resistor of 2000
ohms would have a slope of $- 1/2500$ mhos. The effect of any error in the value of
m is minimized through the screen current being only a small fraction of the cathode
current.

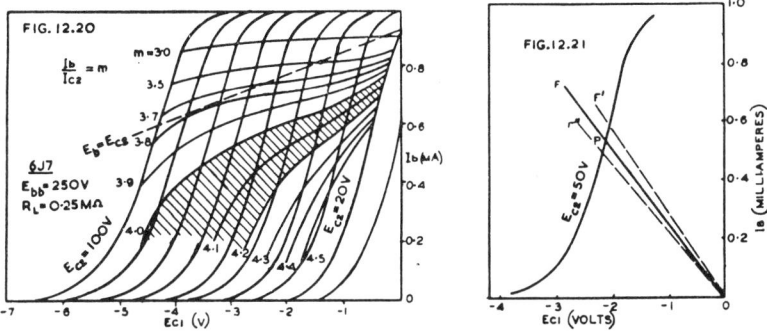

Fig. 12.20. *Plate dynamic characteristics for type 6J7 with curves of constant
plate to screen current ratio. The shaded area has m within ± 5%.*

Fig. 12.21. *Cathode loadline (OF) with single plate dynamic characteristic for
type 6J7. The broken lines show the limits of ± 10% tolerances in R_k.*

The method is illustrated in Fig. 12.21 where only one plate dynamic characteristic
is shown, with a cathode bias loadline OF having a slope of $- m/(m + 1)R_k$. The
static operating point is at P, the intersection of OF with the dynamic characteristic.
Owing to the very large plate load resistors commonly employed, there is very little
rise in plate current due to rectification effects, so that P may be regarded as the
dynamic operating point.

The effect of $\pm 10\%$ variation in R_k is illustrated by lines OF'' and OF' respectively in Fig. 12.21.

It may readily be shown geometrically that $\pm 10\%$ variation in R_k has less effect
on I_b than $\pm 10\%$ variation in E_{c1}.

(B) Cathode and screen loadlines and tolerances

When both cathode and screen resistors are used, the point of operation is the
intersection of the two loadlines (Fig. 12.22). In this diagram it has been assumed
that the screen loadline is a straight line, this being closely correct except at very low
and high values of I_b. However, any practicable method for determining the screen
loadline may be used—see (xiii) above.

The effect of $\pm 10\%$ variation in the resistance of R_s is indicated by the broken
lines, and the combined effect with $\pm 10\%$ variation of both R_s and R_k is indicated
by the region shown shaded. It is on account of these inevitable variations, together
with valve variations, that it is inadvisable to operate with too low a nominal screen
voltage and consequently very close to the grid current point.

(C) Tolerances in general

All three resistors R_L, R_s and R_k have partially self-compensating effects which
enable fairly satisfactory results to be obtained with ordinary 10% tolerances in the
resistors. If operation is required for minimum harmonic distortion at low level or

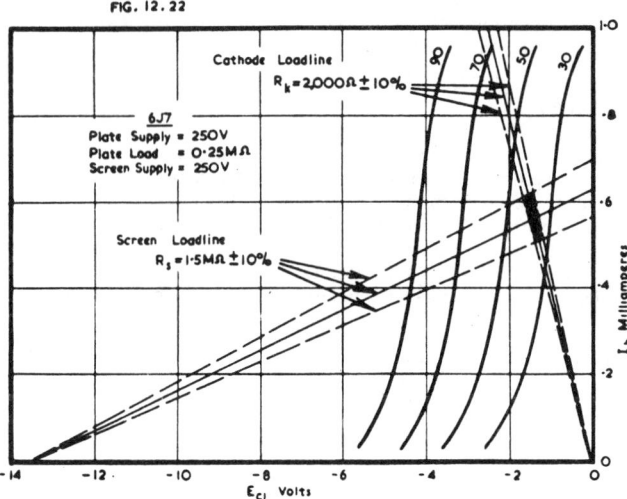

FIG. 12.22

Fig. 12.22.   *Plate dynamic characteristics of type* 6J7 *with screen and cathode loadlines, showing the effect of* $\pm$ 10% *tolerances in both* R_s *and* R_k.

for a high output voltage, the resistors may beneficially have closer tolerances or be selected so that high R_L, high R_s and high R_k go together, and similarly with low values.

The supply voltage may be varied by a ratio up to 2 : 1 in either direction without very serious effects in normal operation and without necessarily making any change in R_L, R_s or R_k.

(xv) Remote cut-off pentodes as r.c.c. amplifiers

Remote cut-off (" variable-mu ") pentodes may be also used satisfactorily as r.c.c. amplifiers, although they give higher distortion for the same voltage output, and should therefore be restricted to low level operation. However, they are quite satisfactory in ordinary radio receivers for the first a-f stage driving an output pentode. Operation is slightly more critical than with sharp cut-off pentodes, but a value of $K = 0.78$ is reasonably close for low level operation.

As a typical example, intermodulation distortion on type 6SF7 is tabulated for $E_{bb} = 250$ volts.

R_L	R_s	$E_0 = 10$	19	37	63 volts r.m.s.
0.25 MΩ	0.25 MΩ	1.8%	5.5%	17%	—
	0.5	1.5%	4.0%	13.5%	52%
	1.0	1.2%	3.0%	10.5%	34%
	Optimum value of K				
0.25 MΩ		0.8	0.78	0.73	0.63

(xvi) Multigrid valves as r.c.c. amplifiers

All types of multigrid valves may be used as r.c.c. amplifiers, and behave effectively the same as pentodes provided that the unused grids are connected to suitable fixed voltages. Operating conditions, including suitable values of K, are the same as for pentodes.

(xvii) Special applications

1. An article giving useful information on 28 volt operation is Ref. B4.
2. An article describing a special pentode operating as two triodes in cascade, giving gains up to 500 with a 45 volt plate supply, is Ref. B6.

(xviii) Comments on tabulated characteristics of resistance-coupled pentodes

Comments are as for triodes Sect. 2(xiv). See Sect. 3(v), (vi)B and E for optimum operating conditions.

SECTION 4 : TRANSFORMER-COUPLED VOLTAGE AMPLIFIERS

(i) Introduction (ii) Gain at the mid-frequency (iii) Gain at low frequencies (iv) Desirable valve characteristics (v) Equivalent circuits (vi) Gain and phase shift at all frequencies (vii) Transformer characteristics (viii) Fidelity (ix) Valve load-lines (x) Maximum peak output voltage (xi) Transformer loading (xii) Parallel feed (xiii) Auto-transformer coupling (xiv) Applications · (xv) Special applications.

(i) Introduction

Transformer-coupled voltage amplifiers usually employ general purpose triode valves with plate resistances about 6 000 to 10 000 ohms. Valves having higher plate resistance require excessively large transformer inductances, while valves having lower plate resistances are only used in special applications, with transformers designed to handle the higher plate currents.

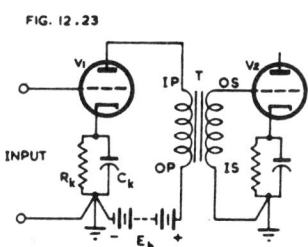

FIG. 12.23

A typical transformer-coupled amplifier stage is shown in Fig. 12.23 where V_1 is the valve under consideration, with transformer T coupling it to the grid of the following valve (V_2). The stage gain is the voltage gain from the grid of V_1 to the grid of V_2. Push-pull transformers are covered in Section 6. Read also Chapter 5 Sections 1, 2 and 3.

Fig. 12.23. Circuit diagram of transformer-coupled voltage amplifier

(ii) Gain at the mid-frequency

An unloaded transformer with high inductance and low losses has a very high impedance at the mid-frequency ; an input impedance of 1 megohm at 1000 c/s is not uncommon.

With the circuit of Fig. 12.23, the voltage gain at the mid-frequency is very nearly

$$A_0 = \mu T \qquad (1)$$

where μ = amplification factor of valve
T = turns ratio = N_2/N_1
N_1 = primary turns
and N_2 = secondary turns.

The gain with loaded transformers is dealt with in (xi) below.

(iii) Gain at low frequencies

At low frequencies the gain is reduced by the shunting effect of the primary inductance. The gain relative to that at the mid-frequency (A_0) is given in Chapter 5 Sect. 3(iii)a particularly eqn. (1).

(iv) Desirable valve characteristics

(a) Low plate resistance—gives better bass response.
(b) High amplification factor —gives higher gain.

(c) Low operating plate current—reduces the direct current through the transformer primary and thus increases the effective inductance compared with a valve having a higher plate current. If necessary, any valve may be overbiased so as to bring the plate current to a suitable value—5 mA is quite usual.

(d) High maximum plate voltage—only desirable if a high input voltage is required by the following stage (e.g. push-pull low mu triodes, or cathode follower).

(v) Equivalent circuits

See Chapter 5 Sect. 2, particularly Fig. 5.9, the input terminals of which are to be understood as being connected to a generator μe_g in series with r_p.

(vi) Gain and phase shift at all frequencies

A transformer tends to produce a peak in the gain/frequency characteristic due to the series resonance between the total leakage inductance and the distributed capacitance of the secondary. This peak may be made negligible by ensuring that

$$Q_s = \omega_1 L'/[r_1 + (r_2/T^2) + r_p] \leq 0.8 \qquad (2)$$

where Q_s = Q of resonant circuit in secondary

ω_1 = $2\pi \times$ frequency of resonance = $\sqrt{1/L'C_2T^2}$

L' = leakage inductance referred to primary

C_2 = secondary equivalent shunt capacitance

r_1 = d.c. resistance of primary

r_2 = d.c. resistance of secondary

and T = turns ratio = N_2/N_1.

Irrespective of the value of Q_s, there is a 90° phase angle shift (lagging) at the frequency of resonance which extends asymptotically to 180° at infinite frequency. The rate of change is more gradual for lower values of Q_s. It is on account of this phase shift that transformer coupling is avoided as far as possible with negative feedback, and that (when unavoidable) the secondary is heavily damped to give a low Q_s.

At low frequencies a transformer-coupled amplifier has the same attenuation and phase angle shift (leading) characteristics as a r.c.c. amplifier due to its grid coupling condenser alone (see Fig. 12.9).

See Refs. A12 (pp. 366-371 and curves Fig. 13) ; A13 (pp. 28-38).

(vii) Transformer characteristics

See Chapter 5 Section 3.

(viii) Fidelity

Distortion caused by the valve is usually very low in the vicinity of the mid-frequency with unloaded transformers of high impedance—about 1% to 2% second harmonic with maximum grid swing—but it increases as the transformer is loaded, up to about 5% second harmonic and 2% or 3% third harmonic. At frequencies below 1000 c/s the impedance steadily falls, while the loadline opens out into a broad ellipse at low frequencies. The valve distortion may be considerable at high output levels under these conditions [see Chapter 2 Sect. 4(vi)].

The transformer (core) distortion is usually quite appreciable and may cause serious intermodulation distortion [see Chapter 5 Sect. 3(iii)].

(ix) Valve loadlines

The loadline for the mid-frequency approximates to a resistive loadline [see Chapter 2 Sect. 4(i) and (ii)]. The procedure is identical to that for Fig. 2.23 except that the loadline AQB is at a smaller angle to the horizontal.

(x) Maximum peak output voltage

To determine the maximum peak output voltage for the mid-frequency, refer to the published curves (say type 6J5), determine I_b (say 5 mA) and E_b (say 250 V) ; mark Q at $E_b = 250$, $I_b = 5$ mA ; draw a horizontal loadline through Q ; determine E_{bo} for the grid current point ($E_c = -0.5$ V) on the loadline—here $E_{bo} = 65$ volts.

The maximum peak output voltage in this example is $(E_b - E_{b0})T = (250 - 65)$
$T = 185\ T$ volts $= 0.74\ E_b T$. This is reduced as the load becomes less, and a value
of about $0.65\ E_b T$ is fairly typical.

(xi) Transformer loading

A resistor shunted across the secondary reduces the value of Q_s, thereby giving more
uniform high frequency response, and also extends the low frequency response. A
resistor shunted across the primary extends the low frequency response. If f_1 is the
low frequency giving a specified attenuation with an unloaded transformer, then the
same attenuation is reached at $0.5 f_1$ when the total shunt load effective across the primary is equal to r_p, and at $(2/3) f_1$ with a shunt load of $2r_p$.

Loading results in higher valve distortion and secondary loading also results in
higher transformer distortion. Loading is generally undesirable, although unavoidable in some applications ; it reduces the maximum peak output voltage and the gain.

The effective load on the valve is given by $R_I = R_1 + R_2/T^2$
where R_1 = resistance shunted across the primary
$\quad\quad R_2$ = resistance shunted across the secondary
and $\quad T = N_2/N_1 =$ turns ratio.

(xii) Parallel feed

Parallel feed (Fig. 12.24) avoids the direct plate current passing through the transformer primary and thereby increases its effective inductance, and decreases transformer distortion. Owing to the drop in R_L the average plate voltage will be con-

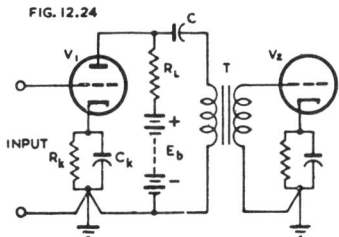

FIG. 12.24

Fig. 12.24. Circuit diagram of parallel-fed transformer-coupled volt-age amplifier.

siderably less than the supply voltage and this may restrict the maximum output
voltage. If a high output voltage is required, the supply voltage may be increased
up to the maximum rating. It is generally desirable to make R_L a resistance of 3 or 4
times the plate resistance of the valve. Higher values of R_L result in increased distortion at low frequencies due to the elliptical load-line. Lower values of R_L result
in lower gain and increased distortion at all frequencies.

The optimum value of C is dependent upon the transformer primary inductance,
and the following values are suggested—

$L =$	10	20	30	50	100	150 henrys
$C =$	4.0	2.0	2.0	1.0	0.5	0.5 μF.

These values of capacitance are sufficiently high to avoid resonance at an audible
frequency. Use is sometimes made of the resonance between C and the inductance
of the primary to give a certain degree of bass boosting. By this means a transformer
may be enabled to give uniform response down to a lower frequency than would otherwise be the case. It should be noted that the plate resistance of the valve, in parallel
with R_I, forms a series resistance in the resonant circuit. The lower the plate resistance, the more pronounced should be the effect [see Chapter 15 Sect. 2(iii)C].

It is frequently so arranged that the resonant frequency is sufficiently low to produce a peak which is approximately level with the response at middle frequencies,
thereby avoiding any obvious bass boosting while extending the frequency range to a
maximum.

For mathematical analysis see Ref. A13 pp. 38-41.

In all cases when making use of any resonance effects involving the inductance of
the transformer primary, it is important to remember that this is a variable quantity.

Not only are there considerable variations from one transformer to another, but there are large variations of inductance caused by the a.c. input voltage (see Chapter 5).

The series resonant circuit presents a low impedance to the valve at the resonant frequency, thus tending to cause serious distortion, particularly when the valve is being operated at a fairly high level. Fot these reasons the resonance method is not used in good design.

(xiii) Auto-transformer coupling

An " auto-transformer " is a single tapped inductance which is used in place of a transformer. Fig. 12.25 shows a parallel-fed auto-transformer coupled amplifier. The auto-transformer may be treated as a double-wound transformer (i.e. with separate primary and secondary) having primary turns equal to those between the tap and earth, and secondary turns equal to the total turns on the inductance. A step-up or step-down ratio may thus be arranged. An ordinary double-wound transformer may be connected with primary and secondary in series (with sections aiding) and used as an auto-transformer, but capacitance effects between windings may affect the high-frequency response of certain types of windings.

The inductance between the tapping point and earth should be the same as for a normal transformer primary. With the parallel-fed arrangement of Fig. 12.25 the plate current does not flow through the inductance, but an alternative arrangement is to omit the parallel-feed and to add a grid coupling condenser and grid resistor for V_2.

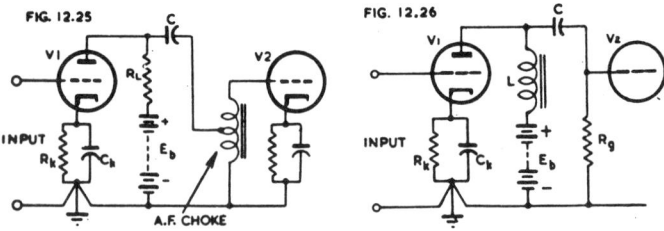

Fig. 12.25. *Circuit diagram of parallel-fed auto-transformer coupled voltage amplifier.*
Fig. 12.26. *Choke-capacitance coupled voltage amplifier.*

(xiv) Applications

The cost of a transformer having linear response over a wide frequency range is considerable and, since equally good response may generally be obtained by a very simple resistance-coupled amplifier, the transformer is only used under circumstances where its particular advantages are of value. Some of these are—

(1) High output voltage for limited supply voltage,

(2) Stepping up from, or down to, low-impedance lines,

(3) When used with split or centre-tapped secondary for the operation of a push-pull stage, and

(4) When a low d.c. resistance is essential in the grid circuit of the following stage.

(xv) Special applications

(a) **Cathode loading** : V_1 in Fig. 12.25 may be connected as a cathode follower either with the transformer primary in the cathode circuit or with parallel feed (see Chapter 7 Sect. 2(i)G).

(b) **28 volt operation** : see Reference B4.

SECTION 5 : CHOKE-COUPLED AMPLIFIERS

(i) Performance   (ii) Application.

(i) Performance

A typical choke-capacitance coupled amplifier is shown in Fig. 12.26. The operation and design are similar to those of a transformer-coupled amplifier (Fig. 12.23) with a transformer ratio 1 : 1 except that C must be designed as in a r.c.c. amplifier to avoid additional low frequency attenuation. An amplifier of this type produces a higher maximum output voltage than a r.c.c. amplifier but less than that with a step-up transformer.

(ii) Application

It is occasionally used with valves having rather high plate current, for which a suitable transformer may not be available. It was also used with tetrode valves of old design, the inductance being several hundred henries, shunted by a resistance of about 0.25 megohm.

SECTION 6 : METHODS OF EXCITING PUSH-PULL AMPLIFIERS

(i) Methods involving iron-cored inductors   (ii) Phase splitter   (iii) Phase inverter (iv) Self-balancing phase inverter   (v) Self-balancing paraphase inverter   (vi) Common cathode impedance self-balancing inverters   (vii) Balanced output amplifiers with highly accurate balance   (viii) Cross coupled phase inverter.

Normally we begin with a single-sided amplifier, and then at some suitable level a stage may be inserted having a single input and a push-pull output. In a radio receiver such a stage usually immediately precedes the output stage, but in more ambitious amplifiers there may be several intermediate push-pull stages. In this section we consider the methods of exciting push-pull amplifiers.

(i) Methods involving iron-cored inductors

(A) **Tapped secondary transformer** (Fig. 12.27A)

This needs little explanation except that the input transformer step-up ratio (primary to half secondary) does not usually exceed 1 : 2. The fidelity is largely dependent upon the quality of the transformer. This method may be used with almost any type of amplifier, and the arrangement illustrated is merely typical. For example, fixed bias operation, or operation with triodes in place of pentodes could equally well be adopted.

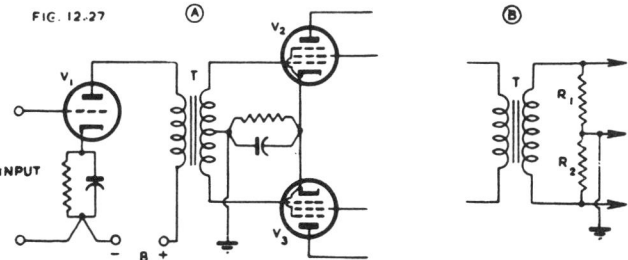

Fig. 12.27. *(A) Triode valve (V₁) followed by transformer (T) having centre-tapped secondary and exciting the grids of the push-pull stage (V₂, V₃). (B) Modified form with non-centre-tapped transformer, using two resistors to provide an equivalent centre-tap.*

(B) **Centre-tapped resistor across secondary** (Fig. 12.27B)

This is an alternative arrangement which does not require centre-tapping of the transformer secondary. In this case an ordinary transformer with a single secondary winding is used, and a centre-tapped resistor ($R_1 = R_2$) is connected across the secondary. The resistors cause a load to be reflected into the plate circuit of V_1 which is equal to $(R_1 + R_2)/T^2$ where $T = N_2/N_1$. For example if $R_1 = R_2 = 0.1$ megohm, and $T = 3$, the load reflected into the plate circuit will be 200 000/9 = 22 000 ohms. This load is lower than some triodes are capable of handling without noticeable distortion, and it might be necessary to increase R_1 and R_2, the limit being set by the maximum grid circuit resistance permitted with valves V_2 and V_3. In addition, this arrangement gives greater distortion than the centre-tapped transformer when the valves are slightly over-driven and pass grid current.

(C) **Centre-tapped choke** (Fig. 12.28)

This is an alternative method sometimes used, involving parallel feed. The effective voltage gain from primary to half secondary is 1/0.5, indicating a loss as compared with a transformer. It is difficult to obtain perfect balance between the two sides.

(D) **Choke-coupled phase inverter**—see page 355.

(ii) **Phase splitter**

(A) This is an excellent method which is also self-balancing. Its principal characteristics are given in Chapter 7 Sect. 2(ii)B and Fig. 7.25. Perfect balance* is obtained at low frequencies if the plate and cathode load resistors are equal, while commercial tolerances in the following grid resistors have only a small effect on the balance. General purpose triode valves are normally used, and the gain is about 0.9 to each side of the output. The input resistance is of the order of 10 megohms, and the harmonic distortion is extremely low.

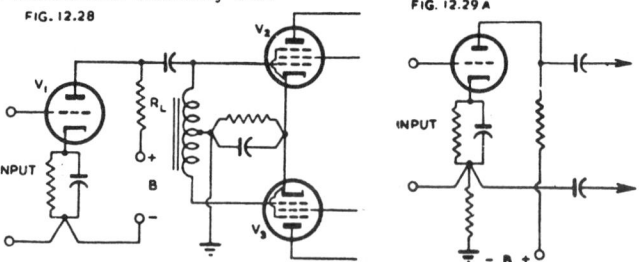

Fig. 12.28. *Triode valve* (V_1) *with parallel feed and centre-tapped choke exciting the grids of the push-pull stage.*

Fig. 12.29A. *Modified form of phase splitter giving full gain without degeneration. The input is from grid to cathode and cannot be earthed. The cathode resistor is by-passed to provide full gain.*

An analysis of the balance with equal load resistors and shunt capacitors is given in Ref. C27, indicating that under these conditions the balance is perfect at all frequencies. However, in practice, the total shunt capacitances across the two channels differ slightly and there is a slight (and generally negligible) unbalance at high frequencies ; see also page 330.

If a high gain amplifier is placed between the phase splitter and the output stage, hum may be troublesome. Part of the hum is due to the difference of potential between the heater and cathode. This may be reduced by operating the heater of the phase splitter from a separate transformer winding which may be connected to a suitable point in the circuit at a potential approximating that of the cathode.

It is normal practice to assume a maximum r.m.s. output voltage (grid-to-grid)

*In an experimental test it is important to reverse the phase of the input from the B.F.O. when the valve voltmeter is moved from one output to the other.

of $0.18E_{bb}$ for less than 2% total harmonic distortion, equivalent to $0.25 E_{bb}$ peak-to-peak output. The distortion drops rapidly as the output voltage is decreased. With a plate supply voltage of 400 volts to the phase splitter, the output is sufficient to excite push-pull Class A 2A3 valves operating with 250 volts on their plates and -45 volts bias.

A complete circuit with 3 stages incorporating a phase splitter and negative feedback is shown in Fig. 7.42.

(B) **A modified form of phase splitter** (Fig. 12.29A) gives the full gain without degeneration, but the input is floating and cannot be earthed. For this reason it cannot generally be used with a pickup, although it may be applied to a radio receiver. In the latter application, the valve may be a duo-diode triode performing detection and 1st a-f stage amplification ; a.v.c. may be operated with some complication. This circuit is particularly prone to suffer from hum, owing to the high impedance from cathode to earth and the high gain. The hum may be minimized by adjusting the potential on the heater to approximately that of the cathode. References (C) 1, 3, 7, 11, 12.

(C) **It is possible to apply positive feedback** from a tapping on the cathode resistor, through a coupling resistor to the unbypassed cathode of the preceding r.c. pentode, and thus increase the gain. A direct-coupled version is shown in Fig. 7.51A. See Chapter 7 Sect. 2(xi).

(D) **Another modification** (Fig. 12.29B) also gives high gain not from the phase splitter itself but from the preceding stage. It makes use of the high input resistance of the phase splitter as the dynamic load on V_1, thereby increasing its gain. The full analysis of its operation is given in Reference C18 and the following is a summary

The effective cathode load on V_2 (apart from the cathode resistor R_8) is R_2 and R_5 in parallel, i.e. 20 000 ohms, which is the same as R_4. The cathode resistor R_3 is by-passed in order to increase the input resistance.

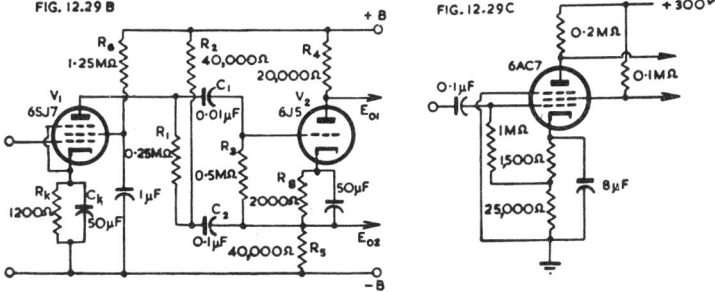

Fig. 12.29B. *Phase splitter* (V_2) *using high input impedance to increase the gain of the preceding stage* (V_1) *by about six times.*
Fig. 12.29C. *Phase splitter using pentode with unbypassed screen and suppressor grid.*

The d.c. load resistance in the plate circuit of V_1 is $R_1 + R_2 = 0.29$ megohm, while the dynamic load is the input resistance of V_2, i.e. [see Chapter 7 Sect. 2(ii)B]

$$r_i' = R_g/(1 - A')$$

where $R_g = R_1R_3/(R_1 + R_3) = 167\,000$ ohms
and $A' \approx 0.9$,
so that in this example, $r_i' = 1.67$ megohms.
The gain of V_1 is given by

$$A_1 = \frac{g_m}{(1/r_p) + (1/r_i')} = \frac{950}{1/4 + 1/1.67} = 1120,$$

which is about 6 times the gain under normal conditions.
The circuit is nearly balanced if $R_4 = R_c$ and

$$\frac{E_{01}}{E_{02}} = \frac{\mu_2 R_g - R_c}{\mu_2 R_g + r_{p2} + R_c}$$

where $R_c = R_2R_5/(R_2 + R_5)$.

The out-of-balance, being about 1.2% in this example, is negligible.

(E) **A further type makes use of a heptode (mixer) valve** in which unequal load resistors are placed in both plate and screen circuits, the push-pull output being taken from plate and screen. The input is taken to grid No. 3.

Ref. C20.

(F) **A further modification makes use of a pentode with an unbypassed screen.** The two output voltages are taken from plate and screen, and the suppressor grid is maintained at a negative potential with respect to the cathode (Fig. 12.29C and Ref. C23).

(iii) Phase inverter (Fig. 12.30A)

This is a popular arrangement with twin triode valves, either general purpose or high-mu. It is not self-balancing, and requires individual adjustment for accurate balance both during manufacture and after the valve has been replaced. It is slightly out of balance at very low frequencies owing to the two coupling condensers operating in the lower channel, but C_2 may be made larger than C_1 if desired. It gives a gain (to each channel) equal to the normal gain of one valve.

If it is preferred to avoid individual balancing, the value of R_2 is given by $R_2 = (R_1 + R_2)/A$ where A is the voltage gain of valve V_2. If R_1 and R_2 both have $\pm 10\%$ tolerances, the maximum possible out of balance will be nearly 20% due to the resistors alone, plus valve voltage gain tolerances.

Separate cathode resistors, each by-passed, are helpful in reducing valve gain tolerances, but require independent cathodes. If a common cathode resistor is used, it may be unbypassed, thus introducing negative feedback for out-of-balance voltages. The hum level is quite low.

N.B. This circuit was originally named Paraphase, but the latter name covers a large number of different circuit arrangements and cannot therefore be used to distinguish one from another.

References C1, C3, C7, C11, C14.

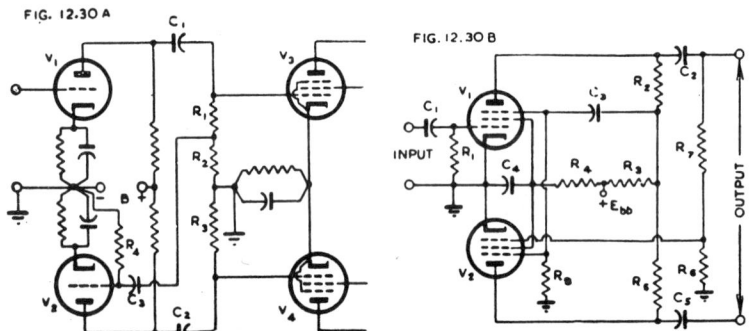

Fig. 12.30A. *Conventional form of phase inverter in which V_1 excites V_3, and V_2 excites V_4, the grid of V_2 being connected to a tapping on the grid resistor of V_3.*
Fig. 12.30B. *Phase inverter with pentodes using the suppressor grids for self-balancing* (*Ref. C19*).

(iv) Self-balancing phase inverter (Fig. 12.30B)

In this circuit V_1 and V_2 are pentodes, and any unbalanced voltage appears across the common plate resistance R_3 and is fed to both suppressors through the blocking condenser C_3, thus causing degeneration in the valve producing the larger signal output, and regeneration in the other. Ref. C19.

(v) Self-balancing paraphase inverter
(A) **Floating paraphase** (Fig. 12.31)

This circuit is, to a considerable extent, self-balancing thereby avoiding any necessity for individual adjustment except in cases where a very high accuracy in balancing is required.

In order to visualize the operation of this circuit consider firstly the situation with V_2 removed. Resistors R_5 and R_9 in series form the load on valve V_1, and the voltage at the point X will be in proportion to the voltage at the grid of V_3. When V_2 is replaced, the voltage initially at point X will cause an amplified opposing voltage to be applied to resistors R_7 and R_9. If resistor R_7 is slightly greater than R_5, it will be found that the point X is nearly at earth potential. If the amplification of V_2 is high, then R_7 may be made equal to R_5 and point X will still be nearly at earth potential. The point X is therefore floating, and the circuit a true Paraphase ; the derivation of the name " Floating Paraphase " is obvious.

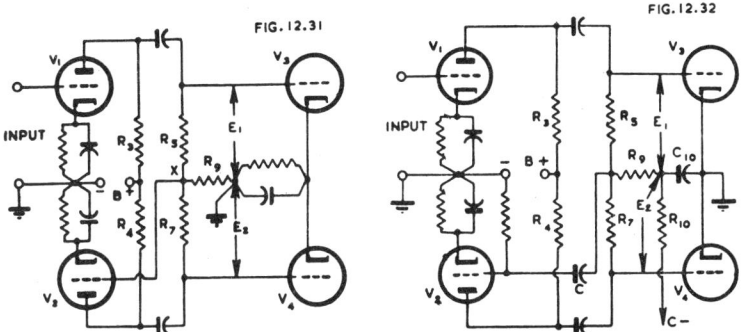

Fig. 12.31. *The Floating Paraphase self-balancing phase inverter with cathode bias.*
Fig. 12.32. *The Floating Paraphase circuit with fixed bias in the following stage.*

The degree of balance is given by

$$\frac{E_1}{E_2} \approx \frac{R_5}{R_7} + \frac{1}{A_2}\left(1 + \frac{R_5}{R_7} + \frac{R_5}{R_9}\right) \tag{1}$$

where $A_2 =$ voltage gain of V_2 into plate load resistor R_4 and following grid resistor R_7.
If $R_5 = R_7 = R_9$, then $E_1/E_2 = 1 + 3/A_2$ (2)
If V_2 is type 6J5 (or half type 6SN7-GT) $R_4 = 0.1, R_5 = R_7 = R_9 = 0.25$ megohm, then $A_2 = 14$ and $E_1/E_2 = 1.21$ which is too high to be acceptable. In such a case R_7 may be increased to, say, 0.3 megohm giving $E_1/E_2 = 1.03$.
If V_2 is type 6SQ7 with $R_4 = R_5 = R_7 = R_9 = 0.25$ megohm, then $A_2 = 48$ and $E_1/E_2 = 1.06$ which is generally acceptable.
If V_2 is a pentode (e.g. type 6J7) with $R_4 = R_5 = R_7 = R_9 = 0.25$ megohm, then $A_2 = 104$ and $E_1/E_2 = 1.03$, which is very close.
The gain from the grid of V_1 to the grid of V_3 is slightly greater than the gain with $R_9 = 0$.

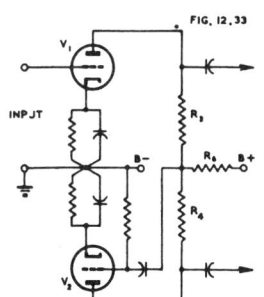

Fig. 12.33. *Common plate impedance self-balancing phase inverter.*

If fixed or partially-fixed bias is employed, it is necessary to couple the grid of V_2 to point X through a suitable condenser (C in Fig. 12.32). In addition, a hum filter (R_{10}, C_{10}) may be required, because most partially-fixed bias sources contain appreciable hum voltage ; any hum voltage appearing across the grid resistor of V_2 is amplified by V_2 and V_4. References C12, C14, C16.

(B) Common plate impedance (Fig. 12.33)

This follows the same principle as the Floating Paraphase, except that the common impedance is in the d.c. plate circuit instead of in the shunt a.c. (following grid) circuit. Here similarly

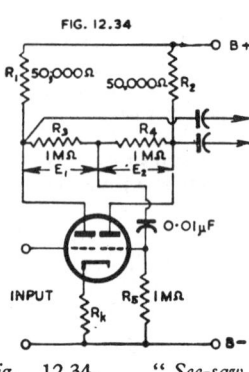

FIG. 12.34

$$\frac{E_1}{E_2} = \frac{R_3}{R_4} + \frac{1}{A_2}\left(1 + \frac{R_3}{R_4} + \frac{R_4}{R_6}\right) \qquad (3)$$

while if $R_3 = R_4 = R_6$, then $E_1/E_2 = 1 + 3/A_2$ (4) where A_2 = voltage gain of V_2 into plate load resistance R_l. Ref. C12.

(C) See-saw self-balancing phase inverter (Fig. 12.34)

This is merely another form of the common plate impedance circuit, with two separate resistors.
When $R_3 = R_4 = R_6$,
    then $E_1/E_2 = 1 + 3/(A_2 + 3)$ (5)
For perfect balance,
    $R_4 = R_3 [1 + 6/(2A_2 - 3)]$. (6)

Fig. 12.34. "See-saw" self-balancing phase inverter (Ref. C9).

It is desirable for A_2 to be greater than 40, thereby giving an out-of-balance less than 7%. References C9, C10, C17.

(D) Modified see-saw self-balancing phase inverter (Fig. 12.35)

This is very similar to the original form, except that it eliminates the loading of the grid resistor R_5 (Fig. 12.34). As a result the balance is improved :
When $R_3 = R_4$, then
    $E_1/E_2 = 1 + 2/(M + 2)$ (7)
while for perfect balance, $R_4 = R_3 [1 + 2/(A_2 - 1)]$ (8)
References C9, C10.

(vi) Common cathode impedance self-balancing inverters
(A) The Schmitt phase inverter

This circuit has been described in Chapter 7 Sect. 2(viii)A. It is very useful with twin triode valves, and exact balance may be obtained by suitably proportioning the two load resistors. A slightly modified practical form is shown in Fig. 12.36. The gain with this type of circuit is only about half the normal amplification. References C7, C12.

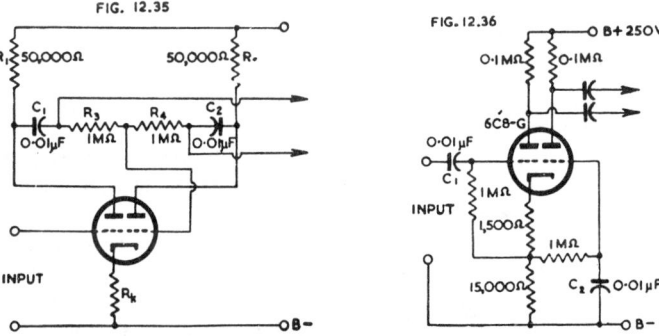

Fig. 12.35. Modified "see-saw" self-balancing phase inverter (Ref. C10).
Fig. 12.36. Practical "Schmitt" common cathode impedance self-balancing inverter using type 6C8-G twin triode ($\mu = 36$). C_1 and C_2 should have high insulation resistance.

(B) Two stage circuit (Fig. 12.37)

This circuit includes a normal push-pull triode stage V_1V_2 with the input applied to V_1 grid only. The grid of V_2 is excited from the common cathodes of the following stage (V_3V_4) through blocking condenser C. Thus the out-of-balance voltage across R excites V_2. No analysis appears to have been published.

(vii) Balanced output amplifiers with highly accurate balance

For certain specialized applications it is necessary to have highly stable and accurate balance. This may be achieved in various ways including

(a) the use of two or more inverters in cascade, the later ones operating with push-pull input.

(b) the use of an additional valve to amplify the out-of-balance voltage before applying it for correction (Ref. C13).

(c) the use of the so-called "phase compressor" following the inverter—see Sect. 7(v).

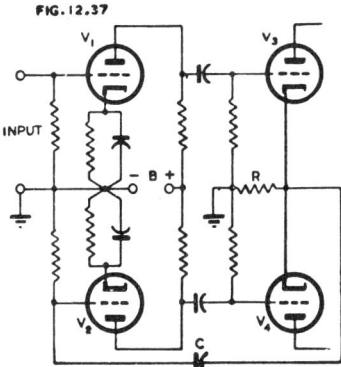

FIG. 12.37

Fig. 12.37. *Two stage self-balancing phase inverter with common cathode impedance.*

(viii) Cross coupled phase inverter

The cross coupled phase inverter (Ref. C25) employs two twin triodes, the circuit being that of the second and third stages of Fig. 15.43A. In this application the single input may be connected between either of the two grids and earth, the other being connected to earth. The hum level from the plate supply is low because hum voltages balance out.

SECTION 7 : PUSH-PULL VOLTAGE AMPLIFIERS

(i) Introduction   (ii) Cathode resistors   (iii) Output circuit   (iv) Push-pull impedance-coupled amplifiers—mathematical treatment   (v) Phase compressor.

(i) Introduction

A push-pull voltage amplifier stage is one having push-pull (3 terminal) input and push-pull output. Two separate valves (or one twin valve) are required. The two valves are each treated as for a single-ended amplifier, whether resistance- or transformer-coupled.

(ii) Cathode resistors

A common cathode resistor may generally be used and should not normally be by-passed. Separate cathode resistors are beneficial in that they aid in the correct adjustment of the operating point when this is at all critical, such as

(a) r.c.c. pentodes when extremely low distortion is required*,

(b) any r.c.c. valves operating near to the overload point,

(c) transformer-coupled triodes when it is desired to reduce to a minimum the out-of-balance direct current.

(iii) Output circuit

The voltage from plate to plate does not include any appreciable even harmonic distortion owing to the cancellation between the two valves, but it does include odd harmonics. The voltage from either plate to earth, however, includes even harmonics. If such a stage is r.c. coupled directly to the grids of a following similar push-pull stage, there is no benefit obtained in the form of reduced distortion. For the latter achievement it is necessary to use either a push-pull transformer with a

---

*Alternatively a similar result may be achieved by adjusting the screen resistors.

centre-tapped primary, or the so-called " phase compressor "—see (v) below. The same also holds true when feeding into a single-ended stage.

Many push-pull transformers are designed to operate with an out-of-balance plate current not exceeding about 1 or 2 mA, thus necessitating matched valves, or adjusted bias, or parallel feed.

(iv) Push-pull impedance-coupled amplifiers—mathematical treatment (Fig. 12.38)

Such a 6 terminal amplifier involves four gain factors

$$A = e_2/e_1 \qquad \text{for} \qquad e_1' = 0 \qquad (1)$$
$$A' = e_2'/e_1' \qquad \text{for} \qquad e_1 = 0 \qquad (2)$$
$$\gamma = -e_2'/e_1 \qquad \text{for} \qquad e_1' = 0 \qquad (3)$$
$$\gamma' = -e_2'/e_1' \qquad \text{for} \qquad e_1 = 0 \qquad (4)$$

In a perfectly balanced amplifier $A = A'$ and $\gamma = \gamma'$.

FIG. 12.38

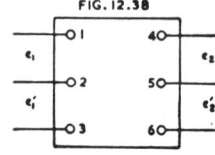

Fig. 12.38. 6 terminal push-pull amplifier.

If there is no cross-coupling (e.g. from a cathode resistor common to both sides) $\gamma = \gamma' = 0$.

In a linear amplifier—

$$e_2 = Ae_1 - \gamma'e_1' \qquad (5)$$
$$e_2' = A'e_1' - \gamma e_1 \qquad (6)$$

Differential gain $= (e_2 - e_2')/(e_1 - e_1') \qquad (7)$
$\qquad\qquad = \tfrac{1}{2}(A + A' + \gamma + \gamma') \qquad (8)$

In-phase gain $= (e_2 + e_2')/(e_1 + e_1') \qquad (9)$
$\qquad\qquad = \tfrac{1}{2}(A + A' - \gamma - \gamma') \qquad (10)$
Inversion gain $= (e_2 - e_2')/\tfrac{1}{2}(e_1 + e_1') \qquad (11)$
$\qquad\qquad = A - A' + \gamma - \gamma' \qquad (12)$
Differential unbalance $= (e_2 + e_2')/(e_1 - e_1') \qquad (13)$
$\qquad\qquad = \tfrac{1}{2}(A - A' - \gamma + \gamma') \qquad (14)$

In an amplifier without cross-coupling, $\gamma = \gamma' = 0$:

Differential gain $= \tfrac{1}{2}(A + A') =$ average gain of two sides.

In-phase gain $=$ differential gain.

Inversion gain $= A - A' =$ difference between gains of two sides.

Differential unbalance $= \tfrac{1}{2}(A - A')$.

See also pages 573-574 for the theory of push-pull amplification based on the expansion of the valve characteristic into an infinite series.

(v) Phase compressor

The so-called " phase compressor " is rather a misnomer since its function is to eliminate the in-phase components (e.g. even harmonics) from a nominally push-pull output. In this application it operates in very much the same manner as a transformer. The circuit is given in Fig. 12.39 and is a push-pull phase splitter with the addition of capacitances C_1 and C_2 from each plate to the opposite output terminal. It is these condensers which attenuate in-phase components. The gain for push-pull input is approximately 0.9 from input to output.

Even if the input signal is imperfectly balanced, the output is still balanced. This circuit, unlike a transformer, does not remove hum from the B+ line, but reduces it to about half at the output terminals. Ref. C8.

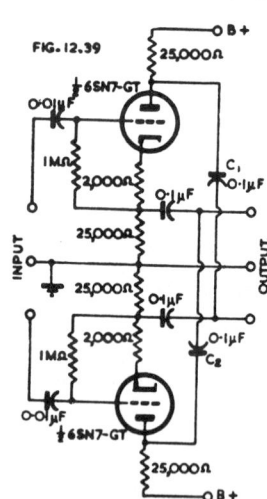

FIG. 12.39

Fig. 12.39. Circuit of the " phase compressor " for eliminating in-phase components and passing on a pure push-pull output voltage (Ref. C8).

SECTION 8 : IN-PHASE AMPLIFIERS

(i) Cathode-coupled amplifiers   (ii) Grounded-grid amplifiers   (iii) Inverted input amplifiers   (iv) Other forms of in-phase amplifiers.

(i) Cathode-coupled amplifiers
These are described in Chapter 7 Sect. 2(viii)B and have many forms for special applications.   In Fig. 7.46 the output from terminal B is in phase with the input. See references Chapter 7 Sect. 6(G).

(ii) Grounded-grid amplifiers
In these, the input voltage is applied to the cathode, the grid is earthed, and the output is taken from the plate, being in phase with the input.   Driving power is required, so that it is not strictly a voltage amplifier.

(iii) Inverted input amplifiers
In Fig. 7.18 the input is from grid to plate and the output voltage is in phase with the input.

(iv) Other forms of in-phase amplifiers
There are many other forms too numerous to mention.

SECTION 9 : DIRECT-COUPLED AMPLIFIERS

(i) Elementary d-c amplifiers   (ii) Bridge circuit   (iii) Cathode-coupled   (iv) Cathode follower   (v) Phase inverter   (vi) Screen coupled   (vii) Gas tube coupled   (viii) Modulation systems   (ix) Compensated d.c. amplifiers   (x) Bridge-balanced direct current amplifiers   (xi) Cascode amplifiers.

(i) Elementary d-c* amplifiers
A direct-coupled amplifier is one in which the plate of one stage is connected to the grid of the next stage directly, or through a biasing battery or equivalent.   It usually receives the plate, screen and grid voltages from sources which do not include any reactances such as filter or by-pass condensers.   If this condition is fulfilled, it may amplify down to zero frequency without attenuation or phase shift ; it may also be used to amplify direct voltages.

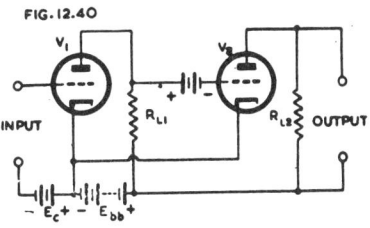
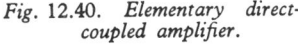

FIG. 12.40

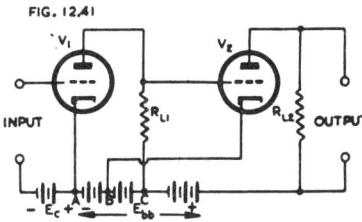

FIG. 12.41

Fig. 12.40.   Elementary direct-coupled amplifier.

Fig. 12.41. Direct-coupled amplifiers without " hot" grid bias battery.

The most elementary form is Fig. 12.40 in which the plate of V_1 is coupled to the grid of V_2 through a bias battery to provide the correct grid bias.   As it stands it is of little practical use, owing to the " hot " bias battery.

A more practical circuit is Fig. 12.41 in which all the batteries may be at earth potential.   Tapping point B has to be adjusted to give the correct bias between grid

*The abbreviation d-c amplifier is used in this Handbook to indicate a direct-coupled amplifier.

and cathode of V_2. The voltage drop across R_{L1} will normally be more than half the voltage between A and C.

A circuit requiring only a single source of voltage is Fig. 12.42 ; the two dividers are desirable to avoid interaction (degeneration) between the two stages unless the bleed current is very high. The output terminal is returned to a point on the voltage divider having the same potential.

Another circuit requiring only a single source of voltage is Fig. 12.43 ; this has only a single voltage divider. The voltage drop across R_{k2} must equal the plate voltage of V_1 minus the grid bias of V_2. There will be degeneration caused by the unbypassed cathode resistors, which may be avoided by a push-pull arrangement with common cathode resistors for both stages.

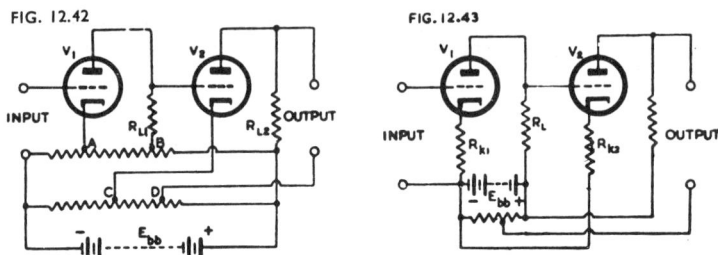

Fig. 12.42.   *Direct-coupled amplifier with only one source of voltage* (E_{bb}).
Fig. 12.43.   *Direct-coupled amplifier with one source of voltage and one voltage divider.*

Pentodes may be used, if desired, in all these circuits by making suitable provision for the screen voltages. In Fig. 12.43 the screen of V_1 may be taken to a tap on R_{k2}, provided that either the screen is by-passed to cathode, or R_{k2} is by-passed ; the amplifier would then be limited to audio frequencies only (Ref. D40).

When the first amplifier stage is a pentode, its load resistor may be increased to values much greater than those conventionally used, provided that sufficient negative voltage feedback is applied to secure an acceptable high frequency response (e.g. Ref. D40).

A pentode may also be used with another pentode as its plate resistor. By this means a gain of several thousand times may be obtained, but this is only useful in electronic measuring instruments.

Such circuits (Figs. 12.40—12.43) are generally limited to two stages. If any increase is made, there is distinct danger of slow drift occurring in the direct plate current, due to variations in battery voltages and valve characteristics. These may be minimized by voltage regulators and controlled heater voltage or current, or may be avoided by one of the special methods described below (viii to x). See also Refs. D36, D39.

Negative feedback may be applied to any d-c amplifier in the normal manner.

(ii) **Bridge circuit**

The bridge circuit (Fig. 12.44) may be used with any number of stages in cascade from a single B supply. The basic design equations are :

$$R_2 = R_1(E_b + E_3)/[E_1 - E_b(1 + R_1/R_p)] \qquad (1)$$
$$R_3 = R_1(E_2 - E_3)/[E_1 - E_b(1 + R_1/R_p)] \qquad (2)$$

where R_p = the d.c. plate resistance of $V_1 = E_b/I_b$
and E_b = the direct plate voltage.

Typical operation with 6SJ7 pentode as V_1

E_{bb} = 400 volts, $E_1 = E_2$ = 200 volts, E_b = 100 volts, $E_3 = 0$, R_1/R_p = 0.5. Amplification is 71.4% of that as a r.c.c. amplifier. A two stage amplifier using type 6J7 pentodes has an amplification of 100 per stage with fixed voltages for bias and screen supplies, or with push-pull connection.

Negative feedback may be applied to this type of circuit.

References D8, D15.

FIG. 12.44

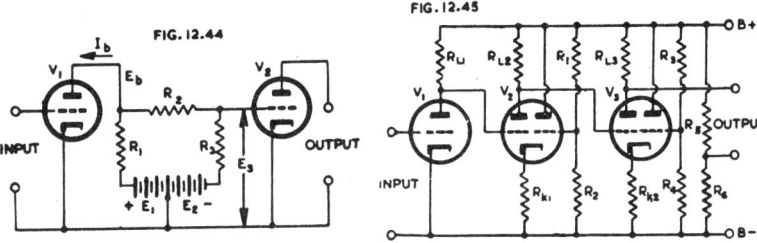

Fig. 12.44. *"Bridge" circuit, direct-coupled amplifier.*

FIG. 12.45

Fig. 12.45. *Three stage d-c amplifier with V_2 and V_3 as cathode-coupled twin triodes.*

(iii) Cathode-coupled

The fundamental form of a cathode-coupled amplifier is covered in Chapter 7 Sect. 2(viii)B.

Fig. 12.45 shows a conventional d-c single triode (V_1) followed by two twin triodes as d-c cathode-coupled amplifier stages. For example, V_2 has one triode as a cathode follower with its grid at a fixed voltage from the voltage divider R_1R_2, while the other triode operates as an amplifier but sharing the common cathode resistor R_{k1}.

FIG. 12.46

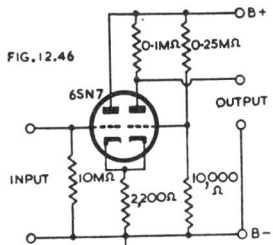

Fig. 12.46. *Direct-coupled amplifier incorporating twin triode with relative positions opposite to those in Fig.* 12.45 *(extracted from Ref.* D19).

Fig. 12.46 is an alternative arrangement in which the positions of the two triodes are reversed. This input circuit was primarily for use with a phototube (Ref. D19) but could be adapted to any other requirement.

In the form shown, the output circuit has a direct potential difference, being intended for coupling directly to the screen of the 6V6 following.

See also Section 6(vi).

References D1, D19 ; also Chapter 7 Refs. (G).

(iv) Cathode follower

The cathode follower may also be used as a d-c amplifier. One circuit is Fig. 12.47A in which the total cathode load is $R_k + R_6R_7/(R_6 + R_7)$. The value of R_7 is equal to E_c/I_b so as to eliminate the undesired direct voltage across R_6.

References D15, D36.

FIG. 12.47 A

FIG. 12.47 B

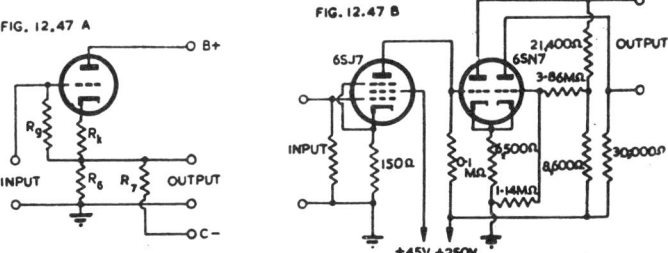

Fig. 12.47A. *Cathode follower used as d.c. amplifier (Ref.* D15).
Fig. 12.47B. *Direct-coupled amplifier with r-c pentode exciting phase inverter (Ref. D14).*

(v) **Phase inverter**

A direct coupled amplifier having in the first stage a resistance coupled pentode, and in the second stage a twin triode phase inverter, is shown in Fig. 12.47B. The voltage gain is 67 db, with uniform gain up to 12 000 c/s.

References D14, D15.

(vi) **Screen-coupled**

The preceding stage may be directly-coupled to the screen of a cathode follower pentode (Fig. 12.48). This circuit has a voltage gain of 30 db with 0.5% total harmonic distortion at 0.85 volt peak output. Output terminal A has a d.c. potential of -1.5 volts, which may be used as bias for the following stage.

Screen-coupled cathode followers are stable, with a wide-band frequency response, but the distortion is higher than with normal operation owing to the non-constant ratio of plate to screen currents. About 85% of this distortion can be cancelled by a push-pull arrangement.

References D14, D19.

(vii) **Gas tube coupled**

Fig. 12.49 shows the simplest form of gas tube coupling in which a gas tube (GT) provides the desired voltage drop from the plate of V_1 to the grid of V_2. The gas tube here must be a glow tube or neon lamp, a voltage regulator tube being unsuitable because its d.c. plate resistance is of the same order of magnitude as the plate load resistor. The values of R_{L1} and R_g must be carefully selected to meet the various limitations.

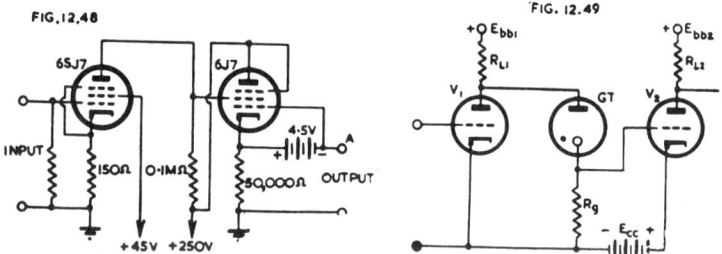

Fig. 12.48.   *Two stage amplifier with the plate of the first stage directly-coupled to the screen of the second stage (Ref. D14).*

Fig. 12.49.   *Two stage amplifier with gas tube coupling from the plate of the first stage to the grid of the second stage.*

Fig. 12.50 is an improved circuit in which an additional valve (6J5) is used as a cathode follower with its cathode impedance composed of the voltage regulator tube and the resistor R_g. With this arrangement there is no d.c. load on the first amplifier and the input impedance of the cathode follower is so high that it does not affect the gain of the first stage. Almost all the signal voltage drop occurs across R_g. The design of the first stage is independent of the d.c resistance of the V.R. tube, but R_g must be much larger than the dynamic impedance of the V.R. tube. Negative feedback for improved stability is provided by R_k in the second stage.

The gas tube introduces noise, hence should not be used in low-level amplifiers. For design, see Reference D13.

(viii) **Modulation systems**

Although not strictly direct-coupled amplifiers, they may be used in many applications. The d.c. signal to be amplified modulates a carrier wave, and after sufficient amplification the modulated wave is detected to obtain the amplified signal. In a modified arrangement, the input signal is interrupted or " chopped."

References D2, D8, D9, D11, D20, D21, D22, D28.

FIG. 12.50

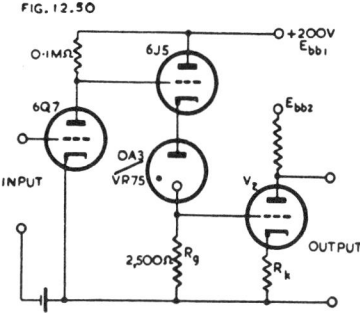

Fig. 12.50. Improved two stage amplifier with gas tube coupling, in which a cathode follower is introduced for better performance.

(ix) Compensated direct current amplifiers

In these, some variable characteristic of the amplifying valve is balanced against the same variations in another valve, or against a different characteristic of the same valve.

(A) Cathode compensation

A typical circuit is Fig. 12.51A and makes use of a twin triode with common cathode resistance. This largely compensates for contact potential drift, and provides a stable amplifier provided that an accurately regulated power source is used. Valves with common cathodes are also used (e.g. 6SC7).

A diode-triode or diode-pentode valve with a common cathode may also be used (Refs. D36, D39).

Both cathode and B supply compensation may be obtained by returning the lower input terminal of Fig. 12.51A to a tapping point ($+ E_b/A$) on the voltage divider (broken line) instead of to earth (Ref. D8).

Other circuits used are series balance, and cascode series balance (Ref. D39); also cathode coupled phase splitter with single ended output (Refs. D1, D39).

References D1, D4, D8, D15, D23, D24, D36, D39.

(B) Compensation for filament and plate voltages

This is used in the electrometer tube circuit, and has low drift but cannot be cascaded. Ref. D8.

(C) Compensation for emission

This can be obtained by a circuit using a pentagrid valve. Refs. D4, D8.

(D) Push-pull operation

A degree of compensation is provided by any push-pull amplifier. An alternative form is a push-pull circuit in which one half only of each stage is used as an amplifier, and the other half as a dummy to reduce drift (Ref. D25).

FIG. 12.51A

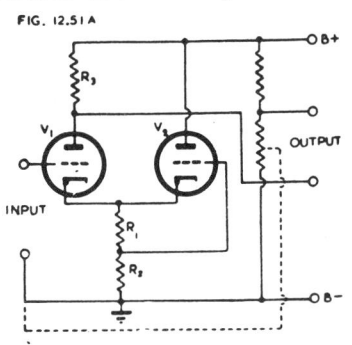

Fig. 12.51A. Cathode-coupled twin-triode used as d.c. amplifier with cathode compensation.

(x) Bridge-balanced direct current amplifiers

With this type of direct current amplifier the regulation of plate and filament supplies usually becomes unnecessary. These are normally used only in laboratory instruments. Ref. D8.

(xi) Cascode amplifiers

The cascode amplifier fundamentally consists of two triodes connected in series (Fig. 12.51B). The usual arrangement in practice is to provide a fixed positive voltage for the grid of V_1.

A cascode amplifier may be considered as a single valve having the characteristics μ', g_m', r_p'. The load into which V_2 works is given by

$$\frac{r_p + R_L}{\mu + 1}$$

where μ, g_m and r_p are the characteristics of both V_1 and V_2.

The amplification of V_2 is therefore given by

$$A' = \frac{\mu R_L}{r_p + (r_p + R_L)/(\mu + 1)} \tag{3}$$

$$= \frac{1}{\dfrac{\mu + 2}{g_m(\mu + 1)R_L} + \dfrac{1}{\mu(\mu + 1)}} \tag{4}$$

Eqn. (4) may be compared with the ordinary form for expressing amplification, namely

$$A = \frac{1}{\dfrac{1}{g_m R_L} + \dfrac{1}{\mu}}$$

and it will be seen therefore that from equation (4)

$$\mu' = \mu(\mu + 1) \tag{5}$$

$$g_m' = g_m(\mu + 1)/(\mu + 2) \tag{6}$$

and therefore

$$r_p' = \frac{\mu'}{g_m'} = (\mu + 2)r_p \tag{7}$$

It is preferable to make R_L very much greater than r_p in order to avoid distortion in V_2 due to the low load resistance into which it works ; a value of $2\mu r_p$ is satisfactory

For example, consider a twin triode with $\mu = 20$ and $r_p = 10\,000$ ohms under resistance-coupled conditions. A suitable value for the load resistance is $2\mu \times 10\,000 = 400\,000$ ohms—say 0.5 megohm.

From eqn. (5) : $\mu' = 20 \times 21 \times 500\,000 \div 510\,000 = 412$

From eqn. (6) : $g_m' = \mu/r_p = 20/10\,000 = 2000$ μmhos.

From eqn. (7) : $r_p' = 21 \times 10\,000 \times 500\,000 \div 510\,000 = 206\,000$ ohms.

A high-mu triode would show even higher values of amplification factor and plate resistance, resembling those of a sharp cut-off r-f pentode.

Curves have been drawn for some typical twin triodes operating as cascode amplifiers ; they resemble the curves of pentodes except that the rounded knee has been replaced by a nearly straight, sloping line (Ref. D35).

One special application is as a low-noise r-f amplifier (Refs. D33, D35).

Another application is as a voltage stabilizer (Refs. D32, D34).

The cascode amplifier has been used as a direct current amplifier responding to zero frequency (Ref. D35).

Two cascode amplifiers have been used in a " floating paraphase " push-pull amplifier operating with single-ended input, to deliver a balanced output. In this application, two high-mu twin-triode type 6SL7-GT valves were used each as a cascode amplifier, to deliver an output of about 30 volts peak, each side. Only one coupling capacitor was used in the whole stage, thus simplifying the design of the feedback circuit (Ref. D41).

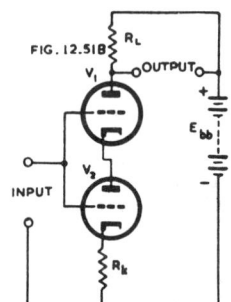

FIG. 12.51B

Fig. 12.51B. *Fundamental circuit of cascode amplifier.*

References to cascode amplifiers : D32, D33, D35, D36, D41.

References to direct-coupled amplifiers (general) Refs. (D).

Ref. D36 is particularly valuable as it gives a detailed examination of the whole subject, including all causes of " drift." See also Chapter 2 Sect. 2(vii) Drift of characteristics during life, and (viii) Effect of heater voltage variation.

SECTION 10 : STABILITY, DECOUPLING AND HUM

(i) Effect of common impedance in power supply   (ii) Plate supply by-passing   (iii) Plate circuit decoupling   (iv) Screen circuit decoupling   (v) Grid circuit decoupling (vi) Hum in voltage amplifiers.

(i) Effect of common impedance in power supply

Every form of power supply has some impedance—even a dry battery has appreciable internal resistance, particularly when partially discharged. This is represented by the resistance R_4 in Fig. 12.52.

When two circuits operating at the same frequency have an impedance common to both there is coupling between them, and the phase relationships may be such that the coupling is either regenerative or degenerative. In the former case instability may result.

A two-stage resistance coupled a-f amplifier has degenerative coupling through the common power supply since the plate currents are out of phase.

A three stage r.c.c. amplifier (Fig. 12.52) has the signal plate currents of the first and third stages in phase, and the total signal current through R_4 is $I_{p1} + I_{p3} - I_{p2}$. Since $(I_{p1} + I_{p3})$ is greater than I_{p3}, the resultant signal current through R_4 will be in the direction of I_{p3}, thus causing across R_4 a signal frequency voltage drop E_4. As a result, the signal voltage applied to the grid of V_2 through R_{L1} and C_1 will be in phase with the normal signal on the grid of V_2, thus giving positive feedback. If the gain of the amplifier is high, there may be sufficient positive feedback to cause oscillation. This effect may be prevented by the use of decoupling.

On the subject of stability, see Chapter 7 Sect. 3 (stability with feedback), also Ref. E1 (requires a high standard of mathematics).

FIG. 12.52

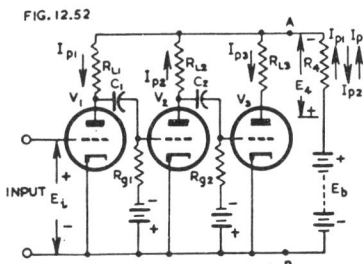

FIG. 12.53

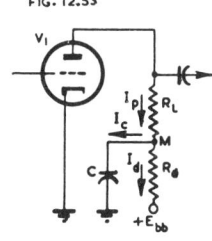

Fig. 12.52. *Three stage amplifier demonstrating the effects of impedance* (R_4) *in the power supply.*

Fig. 12.53. *Plate circuit decoupling.*

(ii) Plate supply by-passing

The simplest form of decoupling, but one having limited usefulness, is a large capacitance across the power supply (points A and B in Fig. 12.52). This reduces the effective power supply impedance to a low value except at very low frequencies, but it may not be sufficient to prevent low frequency instability (" motor-boating ") and frequently requires to be supplemented by plate circuit decoupling.

This by-pass capacitance also fulfils a useful purpose in that it completes the circuit for signal frequencies without appreciable signal currents passing through the B battery or power supply.

Electrolytic condensers are generally used ; if paper condensers are used they should be of a type having low inductance. Either type of condenser may need, in certain rare cases, to be shunted by an additional condenser with a small value of capacitance.

(iii) Plate circuit decoupling

The most popular method is illustrated in Fig. 12.53 where R_d is the decoupling resistance and C the decoupling capacitance. The signal plate current I_p divides between the path through C and the path through the B supply in the ratio

$$\frac{I_c}{I_d} = \frac{R_d + R_4}{X_c} \tag{1}$$

where R_4 = resistance of B supply
and $X_c = 1/\omega C = 1/(2\pi f C)$.

If X_c is very much less than $R_d + R_4$ then almost the whole of the signal current will pass through C, thus reducing the coupling through the B supply. If the value of R_4 is unknown, it may be neglected as an approximation, since R_d is usually much greater than R_4.

The decoupling circuit comprising C and R_d is actually a resistance-capacitance filter [for theory see Chapter 4 Sect. 8(ii)]. The frequency at which the current divides equally between the two paths is given by

$$f_1 \approx 1/2\pi R_d C \tag{2}$$

A normal minimum value of $R_d C$ is $10\,000 \times 10^{-6}$, giving a time constant 0.01 second and $f_1 \approx 16$ c/s. Typical combinations for this value of RC are :

R_d	5000	10 000	20 000	40 000	100 000	ohms
C	2	1	0.5	0.25	0.1	μF

Higher values of C may, of course, be used if desired ; these higher capacitances are often necessary in pre-amplifiers where several stages operate from the same power supply.

The total d.c. load resistance in the plate circuit is $R_L + R_d\,(+ R_4$ if desired), and the quiescent operating conditions should be based on this value. On the other hand, the dynamic (a.c.) load is approximately R_L.

The dynamic operating conditions (gain, voltage output etc.) may be determined by referring to the published data for a supply voltage the same as that for point **M** (Fig. 12.53). The voltage from M to earth is given by

$$E_M = E_{bb} - I_b R_d \tag{3}$$

where $I_b = K E_{bb}/R_L$
and the value of K (for low level operation) may be taken as approximately

$\quad K = 0.75$ for general purpose triodes and pentodes

$\quad K = 0.6$ for high-mu triodes

[for more exact values see Sects. 2 (vi) for triodes and 3 (vi) for pentodes].

A good general value of R_d for most purposes is one fifth of R_L. If the stage is operating at low level, R_d may be increased up to about the same resistance as R_L to give better decoupling. If the stage is operating at high level, R_d should be reduced as much as possible provided that sufficient decoupling can be maintained. This may be assisted by increasing C, but the cost and size of paper condensers may set a limitation. Two stages of decoupling can sometimes be used to advantage. Electrolytic condensers should be used with caution, since their leakage currents are appreciable and they tend to cause noise if used in low level stages. They should never be used with values of R_d above 50 000 ohms, and very much lower values are desirable.

The plate decoupling circuit of Fig. 12.53 has the effect of increasing the bass response, since at extremely low frequencies the total plate load becomes very nearly $(R_L + R_d)$. If V_1 is a triode, and R_L is greater than 5 r_p, then the effect is slight. If V_1 is a pentode, or a triode with R_L less than 5 r_p, the effect may be appreciable. The increase of gain and the phase angle are both identical in form (except for the sign) with the loss of gain and phase angle caused by the cathode bias resistor and by-pass condenser—see Section 2 (iii) for triodes and 3 (iv) for pentodes ; also Ref. E3 for curves.

The limiting increase in gain is given by

$$\left|\frac{A'}{A}\right| = 1 + \frac{1}{(R_L{}^2/R_d)(1/r_p + 1/R_g) + R_L(1/r_p + 1/R_g + 1/R_d)} \tag{4}$$

or if $r_p \gg R_L$ as for pentodes,

$$\left|\frac{A'}{A}\right| \approx 1 + \frac{R_g R_d}{R_L(R_L + R_d + R_g)} \tag{5}$$

or if the effect of R_g may be neglected,

$$\left|\frac{A'}{A}\right| \approx 1 + \frac{r_p R_d}{R_L(R_L + R_d + r_p)} \tag{6}$$

Example : Pentode with $R_L = 0.25$, $R_d = 0.05$, $R_g = 1$ megohm.
  $A'/A \approx 1.15$.

This increase of gain at low frequencies may be exactly cancelled by a suitable choice of cathode bias resistor and by-pass condenser (Refs. E3, E5) provided that R_L and R_d are small compared with r_p and R_g : for triodes—

$$R_k = R_d/R_L g_m \text{ and } C_k R_k = CR_d \tag{7}$$

and then $A = g_m r_p R_L/(r_p + R_L)$. \hfill (8)

This generally gives a value of R_d greater than R_L, and is of limited usefulness.

A special circuit giving **plate decoupling without decoupling condensers,** suitable for use at low frequencies, is shown in Fig. 12.54 (Ref. E4).   The plate currents of the last two valves, which are in phase opposition, flow through separate impedances Z_1, Z_2, of such value that the back e.m.f.'s developed across them and the common impedance Z_3 of the plate supply balance out at the junction points J1, J2. The screens and plates of the preceding valves are supplied from J2 and J1 through pairs of resistances R, R and R_1, R_1 in order to balance out undesirable ripple.

The need for decoupling is less when **push-pull operation** is used, although there is still the possibility of motor-boating arising from the effect on earlier stages of plate supply voltage variations caused by the final stage.   The latter are small when pure Class A operation is employed.

Chokes may be employed instead of resistors for decoupling, and are frequently used when the plate currents are large—e.g. power amplifiers.

It is desirable in all cases to reduce the internal resistance of the power supply, so that even at very low frequencies there may be no tendency towards the production of relaxation oscillations (" motor-boating "). Power supplies having good regulation have low effective internal impedance ; thermionic valve type voltage regulators are highly desirable for special applications and have the feature of retaining low internal impedance characteristics down to the lowest frequencies, if correctly designed.

Motor-boating in transformer-coupled amplifiers may frequently be cured by reversing the transformer connections on either primary or secondary.   This has the effect of providing degeneration instead of regeneration, and may have an adverse effect on the frequency response (Ref. E3).

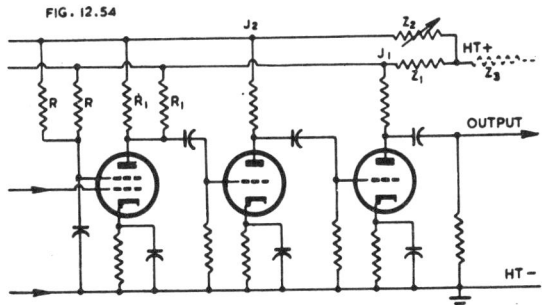

Fig. 12.54.   *Special circuit giving plate decoupling without decoupling condensers (Ref. E4).*

(iv) Screen circuit decoupling

Resistance-coupled pentodes in multistage amplifiers should always have their screens separately decoupled from B+.   By this means any impedance in the screen supply may be prevented from causing instability, and the hum is much reduced.

(v) Grid circuit decoupling

Fig. 12.55 is a method of applying grid-circuit decoupling with transformer input and cathode bias. The condenser C acts as a by-pass to resistors $R + R_k$ and if its reactance is small at the lowest frequencies to be amplified, a negligible portion of the signal voltage will occur across C at all signal frequencies. This method requires a smaller value of C than for cathode bias in the ratio $R_k : (R + R_k)$. This circuit is only effective with transformer coupling.

A similar arrangement for fixed bias is shown in Fig. 12.56, which may be used with any form of coupling.

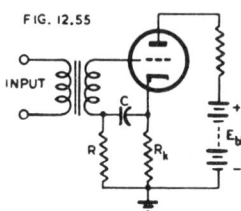

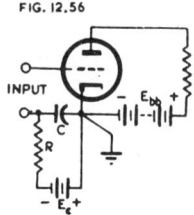

Fig. 12.55. Grid cir-
cuit decoupling with
transformer input and
cathode bias.

Fig. 12.56. Grid
circuit decoupling
with fixed bias.

(vi) Hum in voltage amplifiers

(A) Hum from plate supply

The plate supply is generally filtered sufficiently for the final stage, but additional filtering is usually necessary for the earlier stages. This is usually provided by the plate decoupling circuits, which may be made more effective than demanded by stability in order to reduce the hum.

In a normal r.c.c. amplifier, such as Fig. 12.1, the hum voltage passed on to the following stage is given by

$$E_h' = E_h r_p / (R + r_p) \qquad (9)$$

where E_h = hum voltage across the plate power supply

E_h' = hum voltage across R_{g2}

r_p = effective plate resistance of V_1, taking into account any feedback

and $R = R_L R_{g2} / (R_L + R_{g2})$.

If r_p is very much less than R, the hum voltage passed on to the following stage is much less than E_h. On the other hand, if r_p is greater than R_1 then E_h' may approach E_h.

With a transformer-coupled amplifier (Fig. 12.23) the corresponding expression becomes, in vector form,

$$E_h' = E_h \mathbf{Z} / (\mathbf{Z} + r_p) \qquad (10)$$

where E_h' = hum voltage across transformer primary

and $\mathbf{Z}$ = input impedance of transformer at hum frequency.

Normally $\mathbf{Z} \gg r_p$, and E_h' approaches E_h.

In the case of parallel-feed (Fig. 12.24) the expression becomes, in vector form,

$$E_h' = E_h \mathbf{Z}' (\mathbf{Z}' + R_L) \qquad (11)$$

where $\mathbf{Z}' = r_p \mathbf{Z} / (r_p + \mathbf{Z}) \approx r_p$.

The hum passed on to the following stage is normally only a small fraction of E_h.

The hum voltages of successive stages are usually out of phase, thus resulting in some cancellation, but this is very slight if the same degree of filtering is provided for all stages. In such a case the hum of the first stage is the only one to be considered.

For amplifiers using **negative feedback,** see Chapter 7 Sect. 2(ix).

(B) **Hum from the screens** may also be reduced by screen decoupling. The hum from the screen is out of phase with the hum from the plate, but the screen hum predominates. For amplifiers using negative feedback, see Chapter 7 Sect. 2(ix).

(C) **Hum from the grid bias supply** may be reduced in some cases by grid decoupling. This hum voltage may be either in phase, or out of phase, with the plate circuit hum, depending on the source.

(D) **Hum neutralization**

In a r.c.c. pentode the hum voltages from the plate and screen are out of phase, so that it is possible to neutralize the hum by a circuit such as Fig. 12.57. Here the screen is fed from a series resistance R_{g2}, and the hum voltage is applied to the screen by the capacitance divider C_1C_2. If V_1 is type 6J7 with $R_L = 0.25$ and $R_{g2} = 1.5$ megohms, minimum hum is obtained when $C_1 = 0.05$ and $C_2 = 0.5$ μF, the hum then being about 14% of that without neutralization. C_1 and C_2 may have $\pm 10\%$ tolerances in quantity production. For perfect neutralization it is necessary to balance both resistive and capacitive elements.

An alternative form (Ref. E2) is shown in Fig. 12.58 in which, for perfect neutralization at all frequencies,

$$\frac{C_k}{C_1} = \frac{R_{L1}}{R_g}\left(1 + \frac{R_g}{r_{p1}}\right) \text{ and also } \frac{C_3}{C_1} = \frac{R_k}{R_y}\left(1 + \frac{R_{L1}}{r_{p1}}\right) \tag{12}$$

where r_{p1} = plate resistance of V_1 (preceding stage) and the other values are marked on the diagram. This circuit gives neutralization both for hum and for low frequency regeneration.

There are other methods which may also be used for hum neutralization or reduction (see Chapter 31 Sect. 5).

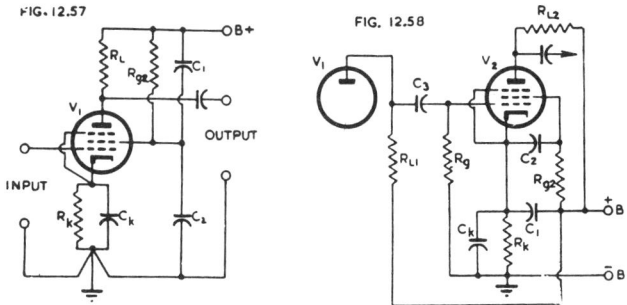

Fig. 12.57.   Hum neutralization in a r.c.c. pentode.
Fig. 12.58.   Alternative circuit giving perfect neutralization for hum and for low frequency regeneration at all frequencies (Ref. E2).

(E) **Hum caused by inductive coupling**

The most common cause is an a-f transformer or choke, which may be placed in an electromagnetic field. This effect may be minimized by altering the position of the transformer or choke, but cannot usually be eliminated entirely. The best procedure is to avoid using the transformer, to remove the cause of the electromagnetic field, or to employ a properly shielded transformer.

Some valves exhibit hum when they are placed in a strong field — the cure is obvious.

See also Chapter 18 Sect. 2(iii) on pre-amplifiers, and Chapter 31 Sect. 4 on hum.

(F) **Hum caused by electrostatic coupling**

This may be almost entirely eliminated by suitable placing of the mains and rectifier leads, and by electrostatic shielding of all low-level wiring (especially the first grid lead) and by shielding the valves and other susceptible components. See also Chapter 18 Sect. 2(iii) on pre-amplifiers.

(G) Hum due to heater-cathode leakage or emission

When a valve is operated at a fairly high input level, it is often sufficient to earth one side of the heater supply and to use a large by-pass condenser (40 μF or more) with cathode bias, or an earthed cathode with fixed bias.

At a somewhat lower input level, hum due to heater-cathode leakage or emission may be reduced as required, by one or more of the following devices—

1. Centre-tapped heater supply, with tap earthed.

2. Centre-tapped heater supply, with tap connected to a fixed positive or negative voltage (say 15 or 20 volts bias).

3. Potentiometer across heater supply, with moving arm connected to earth and adjusted for minimum hum.

4. Potentiometer across heater supply, with moving arm connected to a positive or negative voltage (not more than 50 volts maximum) selected experimentally to give minimum hum, in conjunction with adjustment of the potentiometer.

For further information see Chapter 18 Sect. 2 (pre-amplifiers) and Chapter 31 Sect. 4 (hum).

SECTION 11 : TRANSIENTS AND PULSES IN AUDIO FREQUENCY AMPLIFIERS

(i) Transient distortion   (ii) Rectangular pulses.

(i) Transient distortion

A transient is a complex wave which does not repeat periodically ; it may be analysed into a fundamental and harmonic frequencies (see Chapter 6 Sect. 8). If an a-f amplifier contains a tuned circuit, e.g. in a compensated wide band amplifier, or in a tone control stage, distortion of transients can occur if the damping of the tuned circuit is too slight. The effect is actually shock excitation of the tuned circuit ; this may be reduced to acceptable proportion if Q does not exceed 0.7 (Ref. A12 p. 428).

(ii) Rectangular pulses

The design of special pulse amplifiers is beyond the scope of this handbook, but amplifiers are frequently tested with a rectangular input voltage (" square wave ") and their performance under these conditions is of interest.

FIG. 12.59

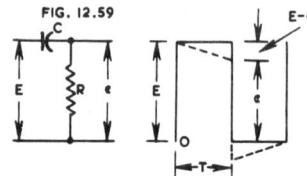

Fig. 12.59. A single resistance-capacitance coupling and its response to a rectangular waveform.

In the case of a single r.c. coupling (Fig. 12.59), the voltage across R is given by
$$e = E \epsilon^{-t/RC} \tag{1}$$
where E = amplitude of pulse voltage input
$$R = R_g + r_p R_L/(r_p + R_L) \approx R_g$$
and t = time measured from commencement of pulse, in seconds.

If t is small compared with RC, we may write the equation as the approximation (Ref. F1)
$$e \approx (1 - t/RC)E \tag{2}$$
or $E/(E - e) \approx RC/t =$ (say) X \hfill (3)
the error being negligible if t is not greater than 0.1 RC.

The value of the time constant RC to give any desired value of X at the end of a pulse is given by
$$RC \approx TX \tag{4}$$
where T = time length of pulse,
and $X = E/(E - e)$,
provided that X is not less than 10.

For example, if X is required to be 20 (i.e. the amplitude of the square top falls by 5% at the end of the pulse) and if $T = 0.01$ second, then $RC \approx 0.2$ second. If $R = 0.5$ megohm, then $C = 0.4$ μF.

If the pulses are repeated periodically with the length of pulse equal to the time between pulses, and if the applied voltage is zero during the period between pulses (as Fig. 12.59),

then $RC \approx X/2f_1$ (5)

where f_1 = frequency in cycles per second.

With sine-wave input, the frequency (f_0) at which the response of a single-stage r.c.c. amplifier falls by 3 db is given by

$f_0 = 1/(2\pi RC)$ (6)

But with " square-wave " input, $RC \approx X/2f_1$ (7)

where f_1 = frequency of square wave in cycles per second.

Therefore $f_0 = f_1/\pi X$ (8)

By using a square wave input, and noting on a C.R.O. the frequency f_1 at which there is (say) 10% drop at the end of the pulse (i.e. $X = 10$), it is possible to calculate the frequency for 3 db attenuation with sine-wave input :

$f_0 \approx f_1/10\pi$ [for $X = 10$]   (9)

Vice versa, by measuring the frequency f_0 at which the sine-wave response is 3 db below that at the mid-frequency, it is possible to calculate the frequency f_1 at which the square wave shows a specified drop at the end of the pulse :

$f_1 = X\pi f_0$ (10)

where $X = E/(E - e)$.

Equations (9) and (10) may also be used in connection with multi-stage amplifiers.

SECTION 12 : MULTISTAGE VOLTAGE AMPLIFIERS

(i) Single-channel amplifiers   (ii) Multi-channel amplifiers.

(i) Single channel amplifiers

Almost any desired number of single r.c.c. stages may be connected in cascade, if adequate provision is made for decoupling. A practical limit is reached when the noise and hum from the first stage become excessive (see Chapter 18 Sect. 2—pre-amplifiers).

The total voltage gain of the amplifier (A) is given by

$A = A_1 \times A_2 \times A_3 \times \ldots$

where A_1 = voltage gain of first stage, etc.

For decibel calculations see Chapter 19 Sect. 1.

The attenuation in decibels below the mid-frequency gain (A_0) is given, for any frequency f, by the sum of the attenuations in decibels of the individual stages at the same frequency.

It is normal practice to design such an amplifier with a flat, or nearly flat, response over the desired frequency range, and to introduce one or more stages for either fixed tone compensation or manual tone control, or both (see Chapter 15). These stages usually give very limited gain at the mid-frequency.

(ii) Multi-channel amplifiers

Amplifiers having 2 or 3 channels are sometimes used. For example, a 3 channel amplifier may be preceded by a frequency dividing network such that each channel only amplifies a limited range of frequencies—low, middle and high. Each channel may have its own attenuator, with an additional attenuator for the whole amplifier (see Chapter 18). The three outputs may feed three separate power amplifiers, or they may be recombined to form a tone compensating amplifier (see Chapter 15).

SECTION 13 : REFERENCES

(A) **REFERENCES TO RESISTANCE-CAPACITANCE-COUPLED TRIODES**
(including general articles)
A1. Mitchell, C. J. " Miller effect simplified " Electronic Eng. 17.196 (June 1944) 19.
A2. Sowerby, J. McG. " Radio Data Charts—18 : Transmission and phase shift of rc couplings "
W.W. 51.3 (Mar. 1945) 84.
A3. Sturley, K. R. " The frequency response of r.c. coupled amplifiers—Data Sheet " Electronic Eng.
17.209 (July 1945) 593.
A4. Design Data 1 " Cathode Bias—effect on frequency response " W.W. 52.1 (Jan. 1946) 21.
A5. Luck, D. G. C. " A simplified general method for resistance-capacity coupled amplifier design "
Proc. I.R.E. 20.8 (Aug. 1932) 1401.
A6. Cowles, L. G. " The resistance-coupled amplifier " Supplement Trans. A.I.E.E. (June 1945) 359.
A7. Seletzky, A. C. " Amplification loci of resistance coupled amplifiers " Trans. A.I.E.E. 55 (Dec.
1936) 1364 ; discussion 56 (July 1937) 877.
A8. Sowerby, J. McG. " Radio Data Charts (16) Voltage gain of resistance coupled amplifiers " W.W.
50.7 (July 1944) 209.
A9. Thurston, J. N. " Determination of the quiescent operating point of amplifiers employing cathode
bias " Proc. I.R.E. 33.2 (Feb. 1945) 135.
A10. Data Sheets 45 and 46 " Performance of resistance-capacity coupled amplifiers " Electronic Eng.
15.181 (March 1943) 421.
A11. Sturley, K. R. " Low frequency amplification " Electronic Eng. (1) 17.201 (Nov. 1944) 236 ;
(2) 17.202 (Dec. 1944) 290 ; (3) Frequency response for low and high frequency 17.203 (Jan. 1945)
335 ; (4) Cathode bias 17.204 (Feb. 1945) 378 ; (5) Anode decoupling 17.205 (Mar. 1945) 429 ;
(6) Screen decoupling 17.206 (Apr. 1945) 470 ; (7) Increasing l.f. response 17.207 (April 1945) 510.
A12. Terman, F. E. (book) " Radio Engineers Handbook."
A13. Sturley, K. R. (book) " Radio Receiver Design " Part 2.
A14. Roorda, J. " Improved analysis of the r-c amplifier " Radio, 30.10 (Oct. 1946) 15.
A15. Pullen, K. A. " Using G curves in tube circuit design " Tele-Tech 8.7 (July 1949) 35.
A16. Diamond, J. M. " Maximum output from a resistance-coupled triode voltage amplifier " Proc.
I.R.E. 39.4 (April 1951) 433.
Additional references will be found in the Supplement commencing on page 1475.

(B) **REFERENCES TO RESISTANCE-CAPACITANCE-COUPLED PENTODES**
B1. Staff of Amalgamated Wireless Valve Company Ltd. " Resistance-coupled pentodes " W.W. 41.13
(Sept. 24, 1937) 308.
B2. Terman, F. E., W. R. Hewlett, C. W. Palmer and Wen-Yuan Pan, " Calculation and design of
resistance coupled amplifiers using pentode tubes " Trans. A.I.E.E. 59 (1940) 879.
B3. Baker, W. G., and D. H. Connolly " Note on the effect of the screen by-pass capacity on the res-
ponse of a single stage " A.W.A. Tec. Rev. 4.2 (Oct. 1939) 85.
B4. Hammond, C. R., E. Kohler, and W. J. Lattin " 28 volt operation of receiving tubes " Elect. 17.8
(Aug. 1944) 116.
B5. Haefner, S. J. " Dynamic characteristics of pentodes " Comm. 26.7 (July 1946) 14.
B6. Adler, R. " Reentrant pentode a-f amplifier " (using CK511X valve) Elect. 19.6 (June 1946) 123.
B7. Terlecki, R., and J. W. Whitehead " 28 volts H.T. and L.T. ? " Electronic Eng. 19.231 (May 1947)
157.
B8. Langford-Smith, F. " The choice of operating conditions for resistance-capacitance-coupled
pentodes " Radiotronics No. 132 (July/Aug. 1948) 63.
B9. Edwards, G. W., and E. C. Cherry " Amplifier characteristics at low frequencies, with particular
reference to a new method of frequency compensation of single stages " Jour. I.E.E. 87.524 (Aug.
1940) 178.
B10. Crawford, K. D. E. " H.F. pentodes in electrometer circuits " Electronic Eng. 20.245 (July 1948)
227.
B11. " Cathode by-passing—derivation of mathematical formula " Radiotronics No. 113 (June 1941) 39.
B12. Shimmins, A. J. " The determination of quiescent voltages in pentode amplifiers." Electronic
Eng. 22.271 (Sept. 1950) 386.
See also A6, A11 (Parts 1, 2, 5, 6) A12, A15.

(C) **REFERENCES TO PHASE INVERTERS ETC.**
C1. McProud, C. G. and R. T. Wildermuth, " Phase inverter circuits" Elect. 13.10 (Oct. 1940) 50.
C3. Paro, H. W. " Phase Inversion " Radio Eng. 16.10 (Oct. 1936) 13.
C7. Cocking, W. T. " Phase splitting in push-pull amplifiers " W.W. 44.15 (April 13th, 1939) 340.
C8. Parnum, D. H. " The phase compressor "—a resistance-capacity output circuit complementary
to the phase splitter—W.W. 51.1 (Jan. 1945) 19. Correction 51.2 (Feb. 1945) 38.
C9. Scroggie, M.G. " The See-saw circuit : a self-balancing phase splitter " W.W. 51.7 (July 1945)
194.
C10. Carpenter, R. E. H. " See-saw or paraphase "—origin of the circuit—W.W. 51.8 (Sept. 1945)
263, with reply from M. G. Scroggie.
C11. Saunders, L. A. (letter) " Phase splitter " Electronic Eng. 18.216 (Feb. 1946) 63. Reference
Chart " Methods of driving push-pull amplifiers " 17.214 (Dec. 1945) 816.
C12. Wheeler, M. S. " An analysis of three self-balancing phase inverters " Proc. I.R.E. 34.2 (Feb.
1946) 67P.
C13. E.M.I. Laboratories " Balanced output amplifiers of highly stable and accurate balance " Elec-
tronic Eng. 18.220 (June 1946) 189.
C14. R.C.A. " Application Note on a self-balancing phase-inverter circuit " No. 97 (Sept. 28, 1938).
C16. " Resistance-coupled push pull—Floating paraphase circuit " Radiotronics No. 83 (January 1938)
97.
C17. E.M.I. Laboratories " Balanced amplifier circuits " Electronic Eng. 17.209 (July 1945) 610.
C18. Jeffery, E. " Push-pull phase splitter—a new high gain circuit "—W.W. 53.8 (Aug. 1947) 274.
C19. Crawley, J. B. " Self-balancing phase inverter " Elect. 20.3 (March 1947) 212. U.S. Patent
2,383,846 Aug. 28, 1945.
C20. Beard, E. G. " A new high gain phase splitting circuit " Philips Tec. Com. No. 8 (Sept. 1947) 10.
C21. Cocking, W. T. " Push-pull input circuits," W.W. (1) General principles W.W. 54.1 (Jan. 1948)
7 ; (2) Cathode follower phase splitter 54.2 (Feb. 1948) 62 ; (3) Phase reversers, 54.3 (Mar. 1948)
85 ; (4) The anode follower, 54.4 (April, 1948) 126 ; (5) Cathode-coupled stage, 54.5 (May 1948)
183.

C22. Johnson, E. " Directly-coupled phase inverter," Elect. 21.3 (March 1948) 188.
C23. Sulzer, P. G. " Applications of screen-grid supply impedance in pentodes." Comm. 28.8 (Aug. 1948) 10.
C24. Sowerby, J. McG. " The see-saw circuit again " W.W. 54.12 (Dec. 1948) 447.
C25. Van Scoyoc, J. N. " A cross-coupled input and phase inverter circuit " Radio News 40.5 (Nov. 1948) 6.
C26. E. E. Carpentier, U.S. Patent 2,510,683 described by R. H. Dorf " Audio Patents—Phase inverter improvement " Audio Eng. 35.2 (Feb. 1951) 2.
C27. Jones, G. E. " An analysis of the split-load phase inverter " Audio Eng. 35.12 (Dec. 1951) 16.

(D) REFERENCES TO DIRECT COUPLED AMPLIFIERS
D1. Miller, S. E. " Sensitive d.c. amplifier with a.c. operation " Elect. 14.11 (Nov. 1941) 27.   " Stable dc amplification " W.W. 48.5 (May 1942) 111.
D2. " A novel dc amplifier " Electronic Eng. 14.170 (April 1942) 727.
D3. Hay, G. A. " High sensitivity dc amplifier—another application of the C.R. tuning indicator" W.W. 49.1 (Jan. 1943) 9.
D5. Mezger, G. R. " A stable direct-coupled amplifier " Elect. 17.7 (July 1944) 106.
D6. Lawson, D. I. " An analysis of a d.c. galvanometer amplifier " Electronic Eng. 17.198 (Aug. 1944) 114.
D7. Goldberg, H. " Bioelectric-research apparatus " Proc. I.R.E. 32.6 (June 1944) 330.
D8. Artzt, M. " Survey of D-C Amplifiers " Elect. 25.8 (Aug. 1945) 112, with extensive bibliography.
D9. Williams, J. A. " Crystal-driven modulator for d-c amplifiers " Elect. 18.12 (Dec. 1945) 128.
D10. Middleton, R. G. " An analysis of cascode coupling " Radio 30.6 (June 1946) 19.
D11. Lampitt, R. A. " A d.c. amplifier using a modulated carrier system " Electronic Eng.18.225 (Nov. 1946) 347.
D12. Noltingk, B. E. " D.C. amplifiers " (letter with reply from R. A. Lampitt) Electronic Eng. 18.226 (Dec. 1946) 389.
D13. Iannone, F., and H. Baller " Gas tube coupling for d-c amplifiers " Elect. 19.10 (Oct. 1946) 106.
D14. Yu, Y. P. " Cathode follower coupling in d-c amplifiers " Elect. 19.8 (Aug. 1946) 99.
D15. Ginzton, E. L. " D.C. amplifier design technique " Elect. 17.3 (March 1944) 98.
D16. R.C.A. " Application Note on special applications of the type 79 tube" No. 28 (Nov. 9th, 1933) —Fig. 5, d-c amplifier.
D18. Mezger, G. R. " Feedback amplifier for C-R oscilloscopes " Elect. 17.4 (April 1944) 126.
D19. Scully, J. F. " A phototube amplifier " Elect. 18.10 (Oct. 1945) 168.
D20. " Balanced modulator," U.S. Patent, 1,988,472 (Jan. 1935).
D21. Whitaker and Artzt " Development of facsimile scanning heads—Radio facsimile " (R.C.A. Institutes Technical Press, Oct. 1938).
D22. Black, L. J., and H. J. Scott " A direct-current and audio-frequency amplifier," Proc. I.R.E. 28.6 (June 1940) 269.
D23. Goldberg, H. " A high-gain d-c amplifier for bioelectric recording " E.E. 59 (Jan. 1940) 60.
D24. Richter, W. " Cathode follower circuits " Elect. 16.11 (Nov. 1943) 112.
D25. Shepard, W. G. " High-gain d-c amplifier " Elect. 20.10 (Oct. 1947) 138.
D26. Aiken, C. B., and W. C. Welz " D-C amplifier for low-level signals " Elect. 20.10 (Oct. 1947) 124.
D27. Anker, H. S. " Stabilized d-c amplifier with high sensitivity " Elect. 20.6 (June 1947) 138.
D28. Gall, D. C. " A direct-current amplifier and its application to industrial measurements and control " Jour. I.E.E. 89 Part 2, 11 (Oct. 1942) 434.
D29. Offner, F. F. " Balanced amplifiers " Proc. I.R.E. 35.3 (March 1947) 306.
D30. Lash, J. F. " Feedback improves response of d-c amplifier " Elect. 22.2 (Feb. 1949) 109.
D31. Bishop, P. O. " A note on interstage coupling for d.c. amplifiers " Electronic Eng. 21.252 (Feb. 1949) 61.
D32. Hunt, F. V., and R. W. Hickman " On electronic voltage stabilizers " Rev. of Sci. Instr. 10.1 (Jan. 1939) 6.
D33. Wallman, H., A. B. Macnee and C. P. Gadsden, " A low-noise amplifier " Proc. I.R.E. 36.6 (June 1948) 700.
D34. Sowerby, J. McG. " The cascode amplifier " W.W. 54.7 (July 1948) 249.
D35. Sowerby, J. McG. " The cascode again " W.W. 55.2 (Feb. 1949) 50.
D36. Valley, G. E., and H. Wallman (book) " Vacuum tube amplifiers " (M.I.T. Radiation Laboratory Series, McGraw-Hill Book Co. New York and London, 1948).
D38. Harris, E. J., and P. O. Bishop " The design and limitations of d.c. amplifiers " Electronic Eng. 21.259 (Sept. 1949) 332 ; 21.260 (Oct. 1949) 355.
D39. Sowerby, J. McG. " Reducing drift in d.c. amplifiers " W.W. Part 1 : 56.8 (Aug. 1950) 293 ; Part 2 : 56.10 (Oct. 1950) 350.
D40. Volkers, W. K. " Direct-coupled amplifier starvation circuits " Elect. 24.3 (Mar. 1951) 126.
D41. Passman, B., and J. Ward " A new theatre sound system " Jour. S.M.P.T.E. 56.5 (May 1951) 527.
Additional references will be found in the Supplement commencing on page 1475.

(E) REFERENCES TO STABILITY, DECOUPLING AND HUM
E1. En-Lung Chu " Notes on the stability of linear networks " Proc. I.R.E. 32.10 (Oct. 1944) 630.
E2. Wen-Yuan Pan " Circuit for neutralizing low frequency regeneration and power supply hum " Proc. I.R.E. 30.9 (Sept. 1942) 411.
E3. Sturley, K. R. " Low frequency amplification—Part 5, The anode decoupling circuit " Electronic Eng. 17.205 (March 1945) 429.
E4. British Patent 567021, Furzehill Laboratories Ltd., 27/7/43 " Decoupling Circuits " Review W.W. 51.9 (Sept. 1945) 288.
E5. Design Data (1) " Cathode bias—effect on frequency response " W.W. 52.1 (Jan. 1946) 21.
E6. Zepler, E. E. (book) " The technique of radio design " (1943) pp. 206-219, 241.
E7. Reich, H. J. (book) " Theory and applications of electron tubes " (1944) pp. 132, 209-211.
E8. Zakarias, I. " Reducing hum in pentodes " Elect. 21.11 (Nov. 1948) 170.

(F) REFERENCES TO PULSE AMPLIFIERS
F1. Moskowitz, S. " Pulse amplifier coupling," Comm. 25.10 (Oct. 1945) 58.

(G) GRID CIRCUIT RESISTANCE
G1. Crawford, K. D. E. " H.F. pentodes in electrometer circuits " Electronic Eng. 20.245 (July 1948) 227—also gives additional references.
See also Chapter 2 Ref. H1, Chapter 3 Ref. 41.

CHAPTER 13

AUDIO FREQUENCY POWER AMPLIFIERS

By F. Langford-Smith, B.Sc., B.E.

SECTION 1 : INTRODUCTION

(i) Types of a-f power amplifiers (ii) Classes of operation (iii) Some characteristics of power amplifiers (iv) Effect of power supply on power amplifiers.

(i) Types of a-f power amplifiers

An interesting analysis of possible types of a-f power amplifiers is given in Ref. G1. However, for practical applications we may sub-divide the arrangements as under :

On basis of form of loading—
 (a) Plate loaded (the normal arrangement).
 (b) Cathode loaded (i.e. cathode follower).
 (c) Combined plate and cathode loaded (e.g. the McIntosh amplifier—see Fig. 13.50D and the Acoustical QUAD amplifier, Refs. H4, H6).

On basis of coupling to load—
 (a) Transformer coupled.
 (b) Choke-capacitance coupled (shunt feed).

On basis of load connection—
 (a) Single-ended output (i.e. one valve).
 (b) Push-pull output.
 (c) Parallel output.
 (d) Push-pull parallel output.
 (e) Single-ended push-pull (Refs. E32, H1, H2).

On basis of excitation—
 (a) Single input to grid.
 (b) Push-pull input to grids.
 (c) Single input to one grid, with other valve excited from common plate or cathode circuit—see Sect. 6(viii).
 (d) Grounded-grid single input.*
 (e) Grounded-grid push-pull input*.

---

*Not normally used with a-f amplifiers.

On basis of input coupling—
(a) Resistance-capacitance.
(b) Choke-capacitance.
(c) Transformer.
(d) Direct coupling.

On basis of type of valve—
(a) Triode.
(b) Pentode.
(c) Beam power amplifier.

On basis of use of feedback (see Chapter 7)—
(a) Without feedback.
(b) Negative voltage feedback.
(c) Negative current feedback.
(d) Cathode follower.
(e) Combined positive and negative feedback.

On basis of Class of Operation—see (ii) below.

On basis of input power required—
(a) No grid input power.
(b) With grid input power.

(ii) Classes of operation

Class A operation* is the normal condition of operation for a single valve, and indicates that the plate current is not cut off for any portion of the cycle.

Limiting Class A push-pull operation* is operation such that one valve just reaches plate current cut-off when the other reaches zero bias.

Class AB operation* indicates overbiased conditions, and is used only in push-pull to balance out the even harmonics.

Class B operation* indicates that the valves (which are necessarily in push-pull) are biased almost to the point of plate current cut-off.

The numeral " 1 " following A or AB indicates that no grid current flows during any part of the cycle, while " 2 " indicates that grid current flows for at least part of the cycle. With Class B operation the " 2 " is usually omitted since operation with grid current is the normal condition.

(iii) Some characteristics of power amplifiers

Amplifiers incorporating **negative feedback** are covered in Chapter 7. The design of **output transformers** is covered in Chapter 5 Sect. 3.

For a limited supply voltage Class A_1 gives the lowest **power output** with given valves, while Class AB_1 and Class AB_2 give successively higher outputs. Pentodes and beam power amplifiers give greater power output than triodes under the same conditions. Negative feedback does not affect the maximum power output.

The **power plate efficiency** is the ratio of the audio frequency power output to the d.c. plate and screen power input. It is least for Class A_1, and increases progressively with Class AB_1, AB_2 and B. It is less for Class A_1 triodes than for Class A_1 pentodes or beam power amplifiers.

The **sensitivity**† is normally taken as the ratio of milliwatts output to the square of the r.m.s. grid voltage. Pentodes and beam power amplifiers have considerably greater sensitivity than triodes. Class AB_1 or any push-pull operation decreases the sensitivity. Amplifiers with grid current require power in the grid circuit ; sensitivity cannot be quoted for such types except for the whole section including the driver valve.

---

*See also definitions in Sect. 5(i)B.
†The writer favours the use of the alternative form

$$\text{Sensitivity} = \frac{\backslash \text{ power output in milliwatts}}{\text{grid voltage r.m.s.}}$$

which gives values from slightly over 1 (e.g. type 45) to nearly 50. This form has the advantage that, for the same power output, the value is proportional to the voltage gain. This value is the square root of the " sensitivity " as normally measured.
Another alternative form is the ratio of milli-watts output to the square of the peak grid voltage.

The **effective plate resistance** (or output resistance) of a power amplifier is an important characteristic when the load is a loudspeaker. The optimum value of plate resistance depends upon the loudspeaker, but in the majority of cases the optimum is about one fifth of the load resistance for the best frequency response—lower values give heavier loudspeaker damping but a loss of bass response. Feedback amplifiers with very high feedback factors normally have a very low value of output resistance as the result of negative voltage feedback ; in some cases the output resistance is purposely raised to a more suitable value by some device such as bridge feedback—see Chapter 7 Sect. 1(iv). The general question of the optimum plate resistance is covered in Chapter 21 Sect. 1(ii), while loudspeaker damping is covered in Chapter 21 Sect. 3.

Critical load resistance—Pentodes are much more critical than triodes as regards the effects of variation from the optimum value of load resistance—this holds with or without feedback in both cases. See Sect. 2(iv), Sect. 3(viii), Sect. 5(ii), Sect. 6(ii), and summary Chapter 21 Sect. 1(iii).

Distortion—Single Class A_1 triodes are usually operated with 5% second harmonic distortion at maximum output, while the third and higher order harmonics are small under the same conditions. All published data for such valves are based on 5% second harmonic unless otherwise specified. With push-pull Class A_1 triodes the even harmonics are cancelled and only small third and higher order odd harmonics remain. Push-pull class A_1 triode operation is regarded as providing the best fidelity obtainable without the use of feedback.

As the bias is increased towards Class AB_1 operation the odd harmonic distortion increases only slightly until cut-off is just reached during the cycle (i.e. up to Limiting Class A operation), beyond which point a kink appears in the linearity (transfer) characteristic, and the distortion is more displeasing to the listener than is indicated by the harmonic distortion.

Power pentodes operated under Class A_1 conditions on a resistive load may have very slight second harmonic distortion but from 7% to 13% total distortion. This is largely third harmonic with appreciable higher order harmonics. When operated into a loudspeaker load the harmonic distortion is much more severe at low and high frequencies due to the variation of loudspeaker impedance with frequency. Negative feedback may be used to reduce distortion at all frequencies.

With a load of varying impedance, such as a loudspeaker, there is a selective effect on the harmonic distortion. For example, if the impedance of the load is greater to a harmonic than to the fundamental, the harmonic percentage will be greater than with constant load resistance equal to that presented to the fundamental. See Sect. 11(iii) and Fig. 13.54, also Sect. 2(iv) for triodes, Sect. 3(viii) for pentodes.

Owing to the fact that the dominant harmonic with power pentodes is the third, there is very little reduction of distortion due merely to push-pull operation. If, however, the load resistance per valve is decreased, the effect is to increase the second harmonic per valve (which is cancelled out in push-pull) and to decrease the third harmonic, and thus to improve the fidelity.

Some of the distortion occurring with Class AB_2 or Class B operation is due to the effect of grid current on the input circuit. The design of such amplifiers is treated more fully later in Sections 7 and 8.

Normally the harmonic distortion is stated for full power output, but the **rate of increase** is also of importance. Second harmonic distortion (Class A_1 triodes or beam power tetrodes) increases more or less linearly from zero to full power. Third harmonic distortion in pentodes increases less rapidly at first, and then more rapidly as full output is approached. Higher order odd harmonics show this effect even more markedly.

Beam power amplifiers in Class A_1 have considerable second harmonic, but less third and higher order harmonics. When operated in push-pull the second harmonic is cancelled, and the total harmonic distortion on a constant resistive load is small. On a loudspeaker load, however, the same objections apply as for pentodes, and negative feedback is necessary in all cases where good fidelity is required.

For further information on fidelity and distortion see Chapter 14.

In calculating the **frequency response** of an output stage, it should be noted that the inductance of the output transformer varies with the applied signal voltage. When the d.c. plate current remains nearly constant at all output levels the **regulation of the power supply** is not important as regards the output power, provided that it is adequately by-passed—see (iv) below.   With Class AB$_1$ operation there is a greater variation in current drain from zero to maximum signal, and improved regulation is required in the power supply in order to avoid loss of power and increased distortion.   Class AB$_2$, and particularly Class B amplifiers, require extremely good power supply regulation owing to the large variations in current drain.

The use of **self bias** (cathode bias) reduces the variation of plate current due to change of signal level, and frequently enables less expensive rectifier and filter systems to be used, although in some cases the output may be slightly reduced and the distortion slightly increased as a result.   Self bias cannot be used with Class AB$_2$ or Class B operation.

Parasitic oscillation in the power stage is sometimes encountered, either of a continuous nature or only under certain signal conditions.   High-mutual-conductance valves are particularly liable to this trouble, which may be prevented by one or more of the expedients listed below.   Class AB$_2$ or Class B amplifiers sometimes suffer from a negative slope on portion of the grid characteristic ; this may sometimes be recognised by a " rattle " in the loudspeaker.   Improvement in most cases may be secured by the use of one or more of the following expedients :

A small condenser from each plate to earth.

A condenser from each grid to earth (with transformer input only).

Series stopping resistors in grid, screen and plate circuits, arranged as close as possible to the valve.

Improved layout with short leads.

Input and output transformers with less leakage inductance.

In addition to these expedients, it is usually helpful to apply negative feedback from the plate of the output valve to its grid circuit or to the cathode of the preceding stage.

A Class AB$_2$ or Class B amplifier requires a **driver stage** and (usually) coupling transformer in addition to the final stage.   These, together with the additional cost due to the good regulation power supply, should all be considered in calculating the total cost.   It is desirable to consider the whole combination of driver valve, driver transformer and push-pull power stage as forming the power amplifier, and the input voltage to the driver will generally be comparable with that required by a single power pentode.

When fixed bias is required for a class AB$_1$, AB$_2$ or B amplifier, this may be obtained from a battery or from a separate rectifier and filter.   In order to reduce the cost, back-bias with the addition of a heavy bleeder resistance is often used.   Some variation in bias is inevitable with this arrangement, and a loss of power output and an increase of distortion will result.   The additional cost of the power supply and filter needed to handle the total current of valves and bleeder must also be considered.

Pentodes and beam power amplifiers may be used as triodes by connecting plate and screen.   If the suppressor is brought out to a separate pin it may be connected to the plate.   Care should be taken to avoid operating such a valve at a plate voltage higher than the maximum rated screen voltage unless special " triode " maximum ratings are available.

The load impedance with triode power amplifiers may be open-circuited (if necessary) without any serious effects ; it should never be short-circuited.   The load impedance with pentodes may be short-circuited only if the valve is being operated considerably below its maximum plate dissipation ; it should never be open-circuited unless an appreciable degree of negative voltage feedback is being used.

(iv) Effect of power supply on power amplifiers

The usual by-pass capacitor connected across the power supply is large enough to act as a reservoir and maintain practically constant voltage, except at low frequencies.

The analysis of a power supply shunted by a condenser and supplying sine-wave signal current does not appear to have been published.

An analysis has been made (Ref. B6) for the simpler case where the signal current is of rectangular form. The minimum frequency of rectangular waveform for a specified variation of supply voltage over the half-cycle is given by

$$f_r = \frac{K(R_L/R_s) + 1}{4.606KR_LC \log_{10} A} \tag{1}$$

where $K = I_{b0}/(I_{b0} + I_p)$
    R_s = resistance of voltage source
    $R_L = E_{b0}/I_{b0}$
    C = shunt capacitance

$$A = \frac{(1 - K)}{(1 - K) - [K(R_L/R_s) + 1]x/100}$$

E_{b0} = direct voltage across C
    I_{b0} = direct plate current of valve
    I_p = peak signal current through valve plate circuit

and x = fall of supply voltage between beginning and end of positive half-cycle, expressed as a percentage of E_{b0}.

The only approximation in eqn. (1) is that the voltage across the condenser at the commencement of the positive half-cycle is taken as being E_{b0}.

Eqn. (1) has been applied to the following example—
    $E_{b0} = 250$ V, $I_{b0} = 40$ mA, $I_p = 30$ mA, $C = 16\mu$F, $x = 2$, to give these results :

When R_L/R_s =	5	10	20	30	35
then f_r =	170	157	127	96	49 c/s

This example is a severe one since a rectangular waveform is more severe than a sine waveform with the same peak current, while the value of x only allows 2% drop in voltage over the half-cycle. If 5% voltage drop is permissible, f_r becomes 39 c/s for $R_L/R_s = 10$ The importance of good power-supply regulation for satisfactory performance at low frequencies is demonstrated.

This effect does not occur to any appreciable extent with push-pull Class A amplifiers, and is not so pronounced with Class AB amplifiers as it is with single-ended Class A or with Class B.

Amplifiers should always be tested for frequency response and distortion at maximum power output, as well as for frequency response at a lower level.

SECTION 2 : CLASS A SINGLE TRIODES

(i) *Simplified graphical conditions, power output and distortion*   (ii) *General graphical case, power output and distortion*   (iii) *Optimum operating conditions*   (iv) *Loudspeaker load*   (v) *Plate circuit efficiency and power dissipation*   (vi) *Power sensitivity* (vii) *Choke-coupled amplifier*   (viii) *Effect of a.c. filament supply*   (ix) *Overloading* (x) *Regulation and by-passing of power supply.*

(i) Simplified graphical conditions, power output and distortion

The basic circuit of a single Class A$_1$ triode is Fig. 13.1.   The load resistance (R_2) is normally connected to the secondary of a transformer (T) whose primary is connected in the plate circuit of the valve.   The load resistance (R_1) presented to the valve is given (see Chapter 5 Sect. 1) by

$$R_1 = (N_1/N_2)^2R_2 \tag{1}$$

where N_1/N_2 = transformer turns ratio, primary to secondary.   In the ideal case, the primary and secondary windings of T may be assumed to have zero resistance, and the d.c. plate current I_b to remain constant under all operating conditions.

The valve characteristics are shown by Fig. 2.22 in which the loadline AQB has a slope $- 1/R_L$ and passes through the operating point Q.   In this example, $E_b = 250$ volts and $E_c = - 10$ volts, and Q is vertically above $E_b = 250$ volts, because the full E_b voltage is applied between plate and cathode.

The maximum grid swing which can be used is 20 volts, that is from A ($E_c = 0$) to B ($E_c = -20$ volts) since the $E_c = 0$ curve is the border of the grid current region. Actually a slight grid current usually flows at zero bias, but this is generally neglected.

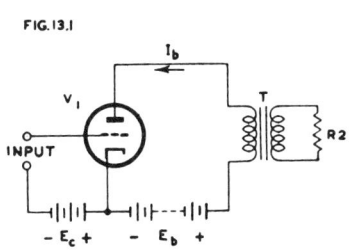

FIG.13.1

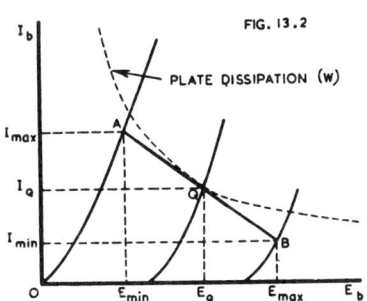

FIG. 13.2

Fig. 13.1. *Basic circuit of a Class A_1 triode with transformer-coupled load.*

Fig. 13.2. *Loadline applied to triode plate characteristics to determine power output and distortion.*

The **power output** may be calculated from a knowledge of the maximum and minimum voltages and currents along the loadline, assuming a sinewave input voltage (Fig. 13.2):

$$\text{Power Output} \approx \tfrac{1}{8}(E_{max} - E_{min})(I_{max} - I_{min}) \tag{2}$$
$$\approx \tfrac{1}{8}R_L(I_{max} - I_{min})^2 \tag{3}$$
$$\approx \tfrac{1}{8}(E_{max} - E_{min})^2/R_L \tag{4a}$$
$$\approx \tfrac{1}{2}E_b I_Q(1 - E_{min}/E_b)(1 - I_{min}/I_b) \tag{4b}$$

Equations 2, 3 and 4 give the exact fundamental frequency power output when the odd harmonic distortion is negligible. The power output of 5% second harmonic frequency is only 1/400 of the fundamental power output, which is negligible.

The load resistance corresponding to the loadline is

$$R_L = (E_{max} - E_{min})/(I_{max} - I_{min}) \tag{5a}$$
$$= (E_b/I_b)(1 - E_{min}/E_b) \tag{5b}$$

The percentage second harmonic distortion

$$= \frac{\tfrac{1}{2}(I_{max} + I_{min}) - I_Q}{I_{max} - I_{min}} \times 100 \tag{6}$$

$$= \frac{AQ - QB}{2(AQ + QB)} \times 100 \tag{7a}$$

$$= \frac{AQ/QB - 1}{2(AQ/QB + 1)} \times 100 \tag{7b}$$

The equation (7b) has been plotted graphically in Fig. 13.3.

The **voltage gain** of a power amplifier is given by

$$M = (E_{max} - E_{min})/(2E_{c1})$$
$$= 2.82\sqrt{P_0 R_L} \tag{8}$$

where P_0 = power output in watts.

The peak values of the fundamental and second harmonic components of the signal plate current are:

Fundamental $\quad I_{h1} = \tfrac{1}{2}(I_{max} - I_{min})$ (9)

Second harmonic $I_{h2} = \tfrac{1}{4}(I_{max} + I_{min} - 2I_Q)$ (10)

If $AQ/QB = 11/9 = 1.22$, then the second harmonic distortion will be 5%. This

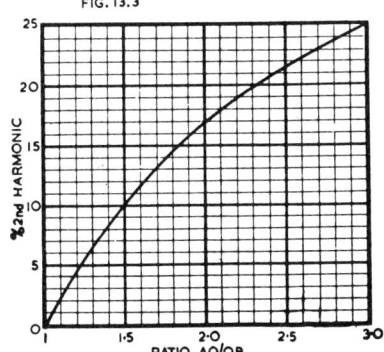

FIG. 13.3

Fig. 13.3. *Curve of second harmonic distortion plotted against the ratio AQ/QB for the two portions of the loadline (Fig. 13.2).*

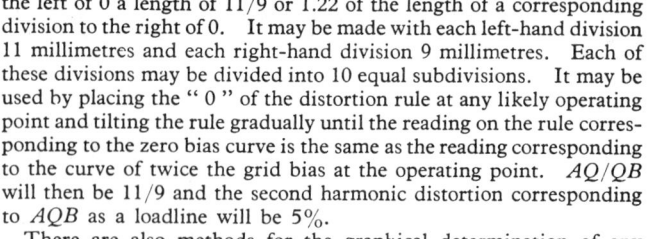

is the principle of " **5% distortion rule** " reproduced in Fig. 13.4. The rule has each division to the left of 0 a length of 11/9 or 1.22 of the length of a corresponding division to the right of 0. It may be made with each left-hand division 11 millimetres and each right-hand division 9 millimetres. Each of these divisions may be divided into 10 equal subdivisions. It may be used by placing the " 0 " of the distortion rule at any likely operating point and tilting the rule gradually until the reading on the rule corresponding to the zero bias curve is the same as the reading corresponding to the curve of twice the grid bias at the operating point. AQ/QB will then be 11/9 and the second harmonic distortion corresponding to AQB as a loadline will be 5%.

There are also methods for the graphical determination of any degree of second harmonic distortion. One of these was originated by Espley and Farren (Ref. B1) and is illustrated in Fig. 13.5 ; it must be drawn on transparent material. It is applied to the curves in a manner similar to the 5% distortion rule, except that a loadline is first drawn in, and then the Harmonic Scale is moved so that DBC is parallel with the loadline. For example, with the loadline EFG the distortion is 20% while with E′FG the distortion is 10%. The scale may be constructed on the basis of :

2nd Harmonic Distortion %	0	5	10	15	20
DB/BC	1.0	1.22	1.5	1.86	2.33

If the valve is to be operated with a known plate voltage, the usual procedure is to take as the operating point (Q) the intersection of the vertical line through E_b with one of the E_c curves such that Q is either on, or below, the " maximum plate dissipation curve " (Fig. 13.2). If the latter is not included in the published curves, it may readily be plotted over a small range. This dissipation curve only affects the operating point Q, and there is no harm if the loadline cuts the curve. In general, if the plate voltage is fixed, a triode gives greatest output when it is operated at the maximum permissible plate current, with the limit of $I_b = \frac{1}{4}E_b/r_p$ (see eqn. 22).

If there is no predetermined plate voltage, a triode gives increasing power output as the plate voltage is increased, even though the plate current is limited in each case by the dissipation.

(ii) General graphical case, power output and distortion

In practice there are several additional factors which should be taken into account.

(A) Resistance of transformer primary

Provision may be made for this resistance by giving the line QE_b a slope of $-1/R'$ where R' is the resistance of the transformer primary or choke (Fig. 13.6). By this means the plate voltage of point Q is less than E_b by the value $R'I_Q$. This does not affect the slope of the loadline, which is determined solely by the effective impedance of the load reflected across the primary of the transformer. It is assumed that the final filter condenser has sufficient capacitance to supply the varying current over each cycle without appreciable change in voltage.

(B) Effect of primary inductance

The reactance of the transformer primary gives an elliptical loadline [see Chapter 2 Sect. 4(vi)]. This may be reduced to any desired extent by increasing the inductance. Obviously this effect is only serious at the lowest signal frequencies ; suitable values of inductance are given in Chapter 5 Sect. 3(iii)c.

Fig. 13.4.   5% distortion rule for use in calculating the power output of triodes.

(C) Effect of regulation of power supply

This has only a slight effect on Class A triodes, provided that the final filter condenser has sufficient capacitance to supply the varying current over each cycle without appreciable change in voltage.

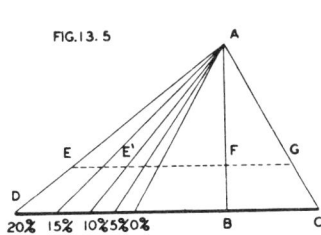

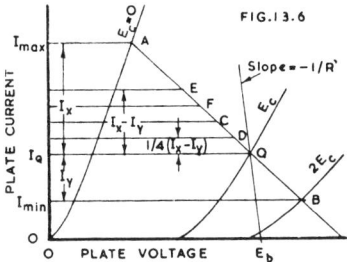

Fig. 13.5. *Alternative form of distortion rule for application to triode loadlines (Ref. B1).*
Fig. 13.6. *Plate characteristics of triode exaggerated to show rectification effects.*

(D) Rectification effects

Owing to the second harmonic component of the signal frequency plate current, the average plate current under operating conditions is greater than with zero signal. The increase in plate current is given by

$$\Delta I_b = \tfrac{1}{4}(I_{max} + I_{min} - 2I_Q) \tag{11}$$

which is also equal to the peak value of the second harmonic current (I_{h2}) as shown by eqn. (10). This may be put into the form

$$\Delta I_b = \tfrac{1}{4}(I_A - I_B) \tag{12}$$

where $I_A = I_{max} - I_Q$ and $I_B = (I_Q - I_{min})$.

This is illustrated in Fig. 13.6 where AE = QB and EQ = AQ — QB. Obviously AQ/QB $\propto I_A/I_B$. The line EQ is then divided into four equal parts; C is the centre point of AB and D is the centre point of CQ. The plate current of point D is greater than that of point Q by the amount $\tfrac{1}{4}(I_A - I_B)$ which is equal to ΔI_b. Point D is the only point on the loadline which fulfils the conditions regarding current, while point Q is the only point which similarly fulfils the conditions regarding voltage. The condition is therefore an impossible one, and the loadline must shift upwards until the point of average current lies on the line QE_b.

Unfortunately, as the loadline moves upwards, the relationship between I_A and I_B changes, and the simplest procedure is to draw a second loadline A'B' (Fig. 13.7) parallel to AB, then to determine its points E'F'C'D' as for the original loadline.

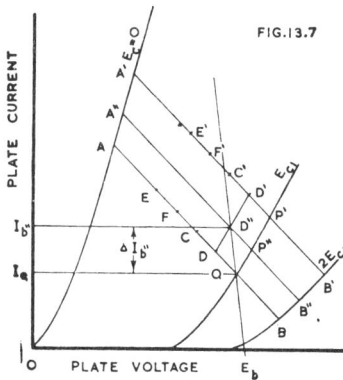

Fig. 13.7. *Method for graphical determination of " shifted " loadline due to rectification.*

Points D and D' are joined by a straight line DD', and the intersection of this with QE_b gives point D'' which must be on the working loadline. Through D'' draw a loadline parallel to AQB, and A''D''B'' is then the maximum signal dynamic loadline. The only error is through assuming that DD' is a straight line, whereas it is actually the locus of point D ; the error is small if DD' is short.

It is evident that the change from the no-signal quiescent point Q to the maximum signal dynamic loadline A''B'' will be a gradual process. As the signal increases, so will the loadline move from AQB to A''P''B''. At intermediate signal voltages the loadline will be intermediate between the two limits. The average plate current for zero signal will be I_Q but this will rise to I_b'' at maximum signal.

It will be observed that D″ does not correspond to the intersection of the loadline and the static bias curve. Point P″ is not the quiescent point (which is Q) but may be described as " the point of instantaneous zero signal voltage on the dynamic load-line." Point P″ must therefore be used in the calculations for harmonic distortion at maximum signal. The loadline A″D″P″B″ provides the data necessary for the calculation of power output, second harmonic distortion, and average direct current. All these will, in the general case, differ from those indicated by the loadline AQB.
Summary :

1. Power output is calculated from the loadline A″B″ in the usual manner.

2. Second harmonic distortion (per cent) at maximum signal = (2D″P″/A″B″) × 100.

3. Average d.c. current at maximum signal = $I_b″ = I_Q + \varDelta I_b″$ = current for point D″.

The graphical method above is accurate, within the limits of graphical construction, but rather slow. Approximate results may be calculated from the original loadline by the equations :

$$I_b″ \approx I_Q + \tfrac{1}{4}(I_{max} + I_{min} - 2I_Q)(1 + R_L/r_p) \qquad (13)$$
$$\varDelta I_b″ \approx \tfrac{1}{4}(I_{max} + I_{min} - 2I_Q)(1 + R_L/r_p) \qquad (14)$$

Alternatively the rise in current may be calculated from the second harmonic distortion (H_2) and power output :

$$\varDelta I_b″ \approx \sqrt{2}H_2\sqrt{P_0/R_L}(1 + R_L/r_p) \qquad (15)$$

In another form, the rise in current may be calculated from the harmonic distortion, I_Q and I_{min} :

$$\varDelta I_b″ \approx (I_Q - I_{min})(1 + R_L/r_p)[H_2/(1 - 2H_2)] \qquad (16)$$

or less accurately by

$$\varDelta I_b″ \approx (0.78\,I_Q)(1 + R_L/r_p)[H_2/(1 - 2H_2)] \qquad (17)$$

(on the assumption that $I_{min} = 0.22\,I_Q$).
Eqn. (17) is interesting, since it does not involve any data beyond those normally published. If $H_2 = 0.05$ it may be reduced to

$$\varDelta I_b″ \approx 0.043.\,I_Q(1 + R_L/r_p) \qquad (18)$$

Eqns. (13) to (18) give results slightly lower than the graphical method.
For rectification effects with cathode bias see below.

(E) Cathode bias

Cathode bias loadlines may be drawn on the mutual characteristics [Chapter 2 Sect. 4(v)] but the position is complicated by the rise in plate current caused by rectification. The simplest approach is to assume a voltage between plate and cathode (E_b), determine the plate current under maximum signal conditions as for fixed bias

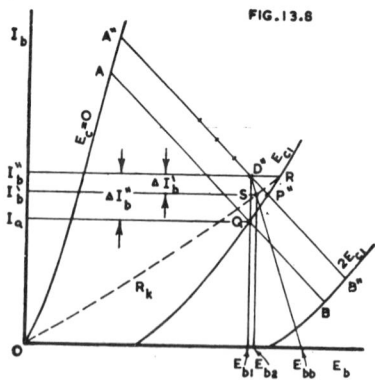

($I_b″$) and then to calculate the cathode bias resistor (R_k) from the equation $R_k = E_c/I_b″$. It is then necessary to check for plate dissipation at no signal ; this may be done by the method of Fig. 2.27, drawing a cathode loadline having the correct slope, and noting the plate current (I_{b1}) at which the cathode loadline intersects the mutual characteristic for the desired value of E_b. The dissipation with no signal input is then $E_b I_{b1}$. There is a slight error through assuming that E_b remains constant from zero to maximum signal, but this is usually negligible. Alternatively, the maximum value of I_{b1} may be calculated from the maximum dissipation, and marked on the corresponding E_b curve ; through this point a cathode loadline may be drawn whose slope will indicate the minimum permissible value of cathode bias resistor.

FIG.13.8

Fig. 13.8 *Determining change of plate current with fixed and cathode bias, having operating conditions identical at maximum signal.*

Alternatively, cathode loadlines may be drawn on the plate characteristics, but they will be slightly curved [for method see Chapter 12 Sect. 2(iii)].

Rectification effects with cathode bias may be determined entirely from the plate characteristics (Fig. 13.8). The loadline corresponding to maximum signal conditions is $A''B''$ (as for Fig. 13.7) while point P'' is the point of instantaneous zero signal voltage. The maximum signal plate current is I_b'' as for fixed bias. For the sake of comparison, the zero signal loadline AQB for fixed bias has also been added. From D'' a straight line is then drawn to E_{bb} where E_{bb} is the total supply voltage across both valve and R_k; this line will have a slope of $-1/R_k$.

The cathode bias loadline must pass through the point R on the E_{c1} curve at which the plate current is I_b'', because $R_k = E_{c1}/I_b''$. The R_k loadline also passes through O as shown; it intersects the sloping line through E_{bb} at point S with plate current I_b'. This point S is the quiescent operating point with cathode bias and has a plate voltage E_{b2} which is slightly greater than E_{b1}.

The change in average plate current from no signal to maximum signal is given by $\Delta I_b'$ for cathode bias and $\Delta I_b''$ for fixed bias. It is obvious that $\Delta I_b'$ is always less than $\Delta I_b''$. More specifically,

$$\frac{\Delta I_b'}{\Delta I_b''} \approx \frac{r_p}{R_k} \cdot \frac{E_{c1}}{E_{b1}} \tag{19A}$$

As a typical example, for type 2A3, $E_b = 250$ volts, $E_{c1} = -45$ volts, $r_p = 800$ ohms, $R_k = 750$ ohms; $\Delta I_b'/\Delta I_b'' \approx 0.2$.

In words, the change of plate current with fixed bias is five times that with cathode bias.

If the value of the cathode bias resistor and the total supply voltage $(E_{bb} = E_b + E_c)$ are known, the following procedure may be adopted (Fig. 13.9A).

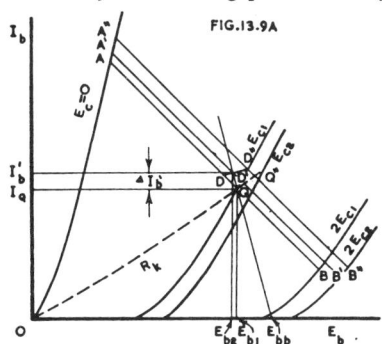

Fig. 13.9A. *Determining position of loadline and rise in plate current, given plate supply voltage $(E_{bb} = E_{b1} + E_{c1})$ and cathode bias resistor only.*

1. Draw the cathode bias loadline (R_k) on the plate characteristics.
2. Draw a straight line through E_{bb} with a slope of $-1/R_k$.
3. Take the point of intersection (Q) and draw a vertical line to the horizontal axis. This will give the quiescent plate-to-cathode voltage E_{b1}, and Q will be the quiescent operating point, with grid bias $-E_{c1}$.
4. Through Q draw a loadline with the desired slope from $E_c = 0$ to $E_c = -2E_{c1}$ —this will be AQB. Mark the point of average plate current (D).
5. Select the adjacent characteristic curve $(-E_{c2})$ and arbitrarily select point Q'' having slightly higher plate current than Q. Through Q'' draw a parallel loadline $A''Q''B''$ from $E_c = 0$ to $E_c = -2E_{c2}$. On this loadline mark the point of average plate current (D'').
6. Join DD'', and mark its point of intersection (D') with the sloping line through E_{bb}. Through D' draw a parallel loadline $A'B'$ which will be the maximum signal loadline. Obviously B' will be intermediate between B and B'', and its position may be fixed approximately by joining BB'' and making B' the point of intersection with the new loadline. The rise in plate current from the quiescent condition (I_Q) to the maximum signal condition (I_b') is given by $\Delta I_b'$.

If it is desired to make allowance for the transformer primary resistance, the slope of $D''E_{bb}$ in Fig. 13.8 should be $-1/(R' + R_k)$ where R' = primary resistance.

In general, any single Class A triode may be operated either with fixed or cathode bias as desired. The maximum value of grid circuit resistance frequently depends on the source of bias. When it is permissible to use a value of cathode bias resistor which provides the same grid bias as required for fixed bias, the load resistance, distortion, and power output will be identical. When it is necessary, on account of plate dissipation, to use a higher value of R_k, then the conditions in the two cases will be different.

Cathode by-passing—It is important for the by-pass condenser to be sufficiently large to maintain the bias voltage constant over each cycle—any fluctuation leads to increased distortion and loss of power output. When operating at low levels it is usually sufficient to ensure that the reactance of the by-pass condenser, at the lowest frequency to be amplified, does not exceed one tenth of the resistance of the cathode bias resistor.

At high operating levels it is necessary to ensure that the direct voltage from B + to cathode remains substantially constant over each half-cycle. This may be treated in the same manner as the by-passing of the power supply—see Sect. 1(iv) and Eqn. (1).

The network comprising the resistive elements R_s and R_k and the capacitive elements C and C_k may be replaced by its equivalent network comprising C_1 in parallel with R_1, connected from B+ to cathode. In the special case when $C/C_k = R_k/R_s$ we have $R_1 = R_s + R_k$ and $C_1 = CC_k/(C + C_k)$, but the general case is more involved. Eqn. (1) of Sect. 1 may then be applied directly.

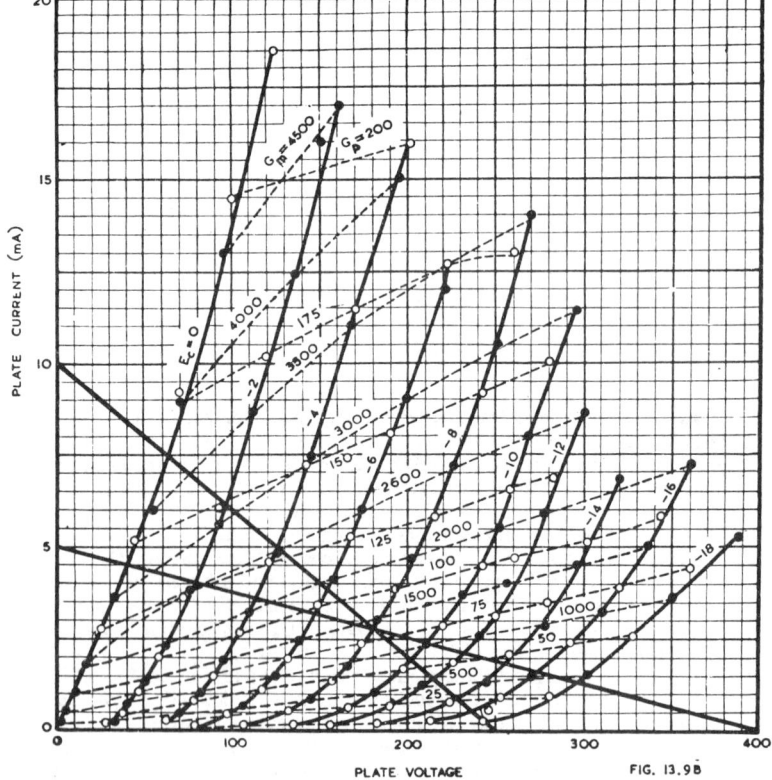

Fig. 13.9B. *Plate characteristics type 6J5 with g $_m$ and g $_p$ curves and loadlines (Ref. B5).*

(F) Alternative method using g_m and g_p curves

$$A = -g_m R_L/(1 + g_p R_L) \qquad\qquad (19B)$$

where A = amplification with small input signal
         g_m = mutual conductance
and g_p = plate conductance = $1/r_p$.

For predominant second harmonic distortion :

$$A' = \tfrac{1}{2}(A_1 + A_2) \qquad\qquad (19C)$$

Second harmonic distortion per cent = $25(A_1 - A_2)/(A_1 + A_2)$

where A' = amplification with maximum input signal
         A_1 = amplification at maximum positive excursion
and A_2 = amplification at maximum negative excursion.

The values of A_1 and A_2 may be calculated, using eqn. (19B), from the values of g_m and g_p corresponding to the relevant points on the loadline which may be derived from Fig. 13.9B for type 6J5. Alternatively, the values of g_m and g_p may be derived by measurement or graphically at the points of maximum positive and negative excursion on the loadline. By this means it is possible to calculate amplification and distortion with high signal inputs from measurements made at low inputs.

This method may be used for cathode followers by using eqn. (19D) in place of (19B) :

$$A = g_m R_k/[1 + (g_m + g_p)R_k] \qquad\qquad (19D)$$

It may also be used for degenerative amplifiers by using eqn. (19E) :

$$A = -g_m R_L/[1 + (g_m + g_p)R_k + g_p R_L] \qquad\qquad (19E)$$

References B5, B7, B8, H17.

(iii) Optimum operating conditions

The preceding graphical treatment enables the optimum load resistance to be determined to give maximum power output for limited distortion, provided that valve curves are available. It is frequently desirable to be able to make an approximate calculation without going to so much trouble. The optimum load resistance is a function of the operating conditions (see Refs. A). The following treatment is based on Ref. A14, and relates to " ideal " (linear) valve characteristics.

Case 1 : Grid current and distortion zero, fixed signal input voltage, no limitations on plate current or voltage

See loadline AQB in Fig. 13.10 in which the characteristics are parallel and equidistant straight lines. The loadline may be placed anywhere between the $E_c = 0$ curve and the E_b axis. Maximum power output will be obtained when $R_L = r_p$, that is when the slope of AQB is equal in magnitude to the slope of the characteristic curves. The power output with any load resistance is given by

$$P_0 = \frac{(\mu E_g)^2 R_L}{(r_p + R_L)^2} \qquad\qquad (20)$$

where E_g = r.m.s. grid input voltage.

Since there is no advantage in having the loadline in the position shown, it may be transferred to the position A′Q′B′ where it has the lowest possible grid bias and plate voltages. Under these conditions the " plate circuit efficiency "* is 16.6% when $R_L = r_p$.

Case 2 : Grid current and distortion zero, fixed plate voltage, no limitations on plate current or signal input voltage

Let E_{b1} be the fixed plate voltage. The loadline may be CPD in Fig. 13.10 where C is the intersection with the $E_c = 0$ curve which is the boundary of the grid current region. PD must equal CP for zero distortion. Maximum power output is obtained when the loadline is adjusted so that its lower extremity touches the E_b axis, as with EFG, and when $R_L = 2r_p$. Under these conditions the plate circuit efficiency* is 25%.

The maximum power output with any load resistance is given by

$$P_0 = \frac{\tfrac{1}{2}E_{b1}{}^2 R_L}{(2r_p + R_L)^2} \qquad\qquad (21)$$

See also equations (31) to (34) inclusive.

*See Section 2(v).

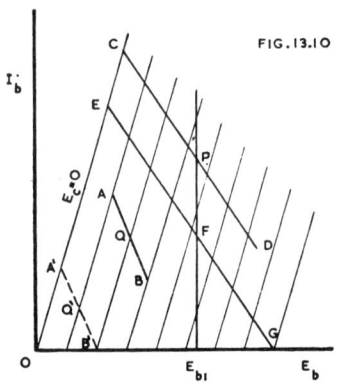
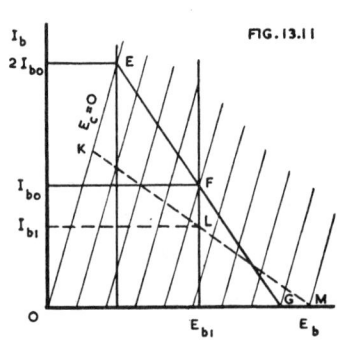

Fig. 13.10.   *" Ideal " plate characteristics of triode to illustrate choice of optimum loadline.*

Fig. 13.11.   *" Ideal " plate characteristics of triode with fixed plate voltage to illustrate choice of optimum loadline.*

Case 3 : Grid current and distortion zero, fixed plate voltage and maximum plate current, no limitation on signal input voltage

Let E_{b1} = fixed plate voltage

and I_{b1} = maximum plate current.

It is always advantageous to operate the valve with the maximum plate voltage, which is here regarded as being fixed.   There is, however, an optimum value of plate current, and any increase or decrease in plate current results in lower power output. The optimum plate current is given by

$$I_{b0} = \tfrac{1}{4}E_{b1}/r_p \tag{22}$$

where $R_L = 2r_p$

as illustrated in Fig. 13.11 with loadline EFG.

If the maximum plate current is higher than I_{b0}, then it is preferable to operate the valve at I_{b0} so as to avoid loss of power output and waste of current.

If the maximum plate current is less than I_{b0}, it is necessary to increase the grid bias to reduce the plate current to I_{b1} (point L).   One possible loadline is KLM where KL = LM and for this condition

$$R_L = (E_{b1}/I_{b1}) - 2r_p \tag{23}$$
$$P_0 = \tfrac{1}{2}(E_{b1} - 2I_{b1}r_p)I_{b1} \tag{24}$$

and the plate circuit efficiency is given by

$$\eta = \tfrac{1}{2}(1 - 2r_pI_{b1}/E_{b1}) \tag{25}$$

The load resistance for this condition is always greater then $2r_p$.

Case 4 : Grid current and distortion zero, fixed plate dissipation, no limitation on plate voltage, plate current or signal input voltage

The power output under these conditions is given by eqn. (24) which may be put into the form

$$P_0 = \tfrac{1}{2}(P_{pm} - 2I_{b1}{}^2r_p) \tag{26}$$

where P_{pm} = maximum plate dissipation.

Obviously the power output continues to increase as E_{b1} is increased.   The plate circuit efficiency approaches 50% as E_{b1} is made very large.   The value of load resistance is given by eqn. (23).

Case 5 : Grid current and distortion zero, fixed plate dissipation, fixed minimum instantaneous plate current

Owing to the existence of " bottom bend " curvature in actual valve characteristics, it is necessary to fix a minimum value of instantaneous plate current if distortion is to be reasonable (Fig. 13.12).

It is here assumed that the portions of the characteristics above I_{bmin} are straight, and that operation is restricted to the linear region. E_{bmin} is the plate voltage at which the extended straight portion of the characteristic cuts the axis.

Applying this procedure to each of the previous cases, we have :

(1) No change—$R_L = r_p$.

(2) No change—$R_L = 2r_p$.

(3) $P_0 = \frac{1}{2}[E_{b1} - (2I_{b1} - I_{b\,min})r_p - E_{b\,min}][I_{b1} - I_{b\,min}]$ (27)

The optimum value of load resistance is given by

$R_L = (E_{b1}/I_{b1}) - 2r_p + (E_{b1}/I_{b1} - r_p - E_{b\,min}/I_{b\,min})I_{b\,min}/(I_{b1} - I_{b\,min})$ (28)

The plate circuit efficiency is given by

$\eta = \frac{1}{2}(1 - 2r_pI_{b1}/E_{b1} + r_pI_{b\,min}/E_{b1} - E_{b\,min}/E_{b1})(1 - I_{b\,min}/I_{b1})$ (29)

(4) The optimum value of load resistance for maximum power output is given by

$R_L = (4I_{b1}/I_{b\,min} - 1)r_p + E_{b\,min}/I_{b\,min}$ (30)

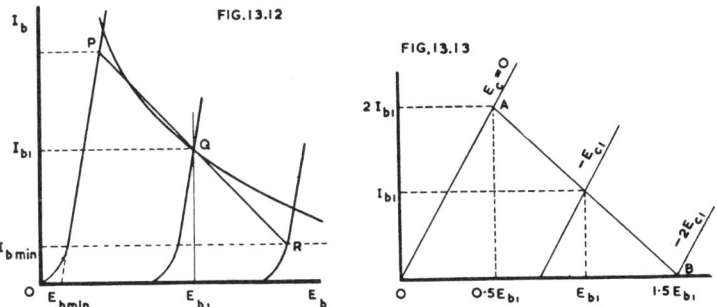

Fig. 13.12. *Plate characteristics of imaginary triode with "ideal" characteristics above $I_{b\,min}$ and curved characteristics below this line.*
Fig. 13.13. *Ideal triode characteristics with $R_L = 2r_p$.*

Summary of operating conditions

(A) **Ideal characteristics**

If there is no limitation on plate current, the optimum load $R_L = 2r_p$ and the load-line is as Fig. 13.13.

Grid bias :	$E_{c1} = 0.75E_{b1}/\mu$	(31)
Plate current :	$I_{b1} = E_{b1}/4r_p$	(32)
Max. power output :	$P_0 = E_{b1}{}^2/16r_p$	(33)
Plate circuit efficiency : η	$= 25\%$	(34)

(B) **Practical characteristics**

The optimum plate current for ideal characteristics is given by

$I_{b0} = \frac{1}{4}E_{b1}/r_p$

and in practice is slightly higher.

Grid bias for optimum plate current :

$E_{c1} = 0.75(E_b - E_t)/\mu$ (35)

where E_t = plate voltage at intersection of E_b axis and tangent to $E_c = 0$ curve at current I_{max},

and E_{c1} = grid bias for d.c. filament operation (add $E_f/2$ for a.c. operation).

Usual values of E_t are between 30 and 45 volts. Examples are :

Type 2A3 E_t = 35 volts approx.

Type 45 E_t = 40 volts approx.

Note : At voltages above about 180 volts, the grid bias will have to be increased to reduce the plate dissipation.

When a triode is being operated at the optimum plate current, the load resistance should be approximately $2r_p$. Fig. 13.14 shows the variation in power output, and second harmonic distortion, of a typical small power triode indicating that maximum power output occurs at slightly less than $2r_p$. The percentage **third harmonic distortion** is usually between one third **and** one tenth of the second harmonic distortion up to 5% of the latter ; the ratio **decreases** as the load resistance is increased.

When a triode is being operated at a plate current which is less than the optimum, then R_L will be greater than $2r_p$ for maximum power output. Refer to Case (3), eqns. 23, 24 and 25.

(iv) Loudspeaker load

A loudspeaker (see Chapters 20 and 21) presents a load impedance which is neither purely resistive nor constant.   At most frequencies it causes an elliptical loadline [see Chapter 2 Sect. 4(vi)] with the shape of the ellipse varying widely over the audio frequency range.   All that can be said here is that the elliptical loadline results in higher distortion and lower power output than a purely resistive load.

The variation in impedance is almost entirely an increase above the nominal (400 c/s) impedance.   As a result, with constant signal voltage applied to the grid, the distortion and the power output decrease as the load resistance is increased (see Fig. 13.14).

A triode applies nearly constant voltage across the load impedance.   This is a standard condition of test for a loudspeaker, and some models of loudspeakers are designed to operate under these conditions (see Chapters 20 and 21).   A triode is almost the ideal output stage for a loudspeaker load when looked at from the load point of view, with or without feedback.

· See also Sect. 3(viii) for pentodes, Sect. 11(iii) and Fig. 13.54, and Chapter 21 Sect. 1.

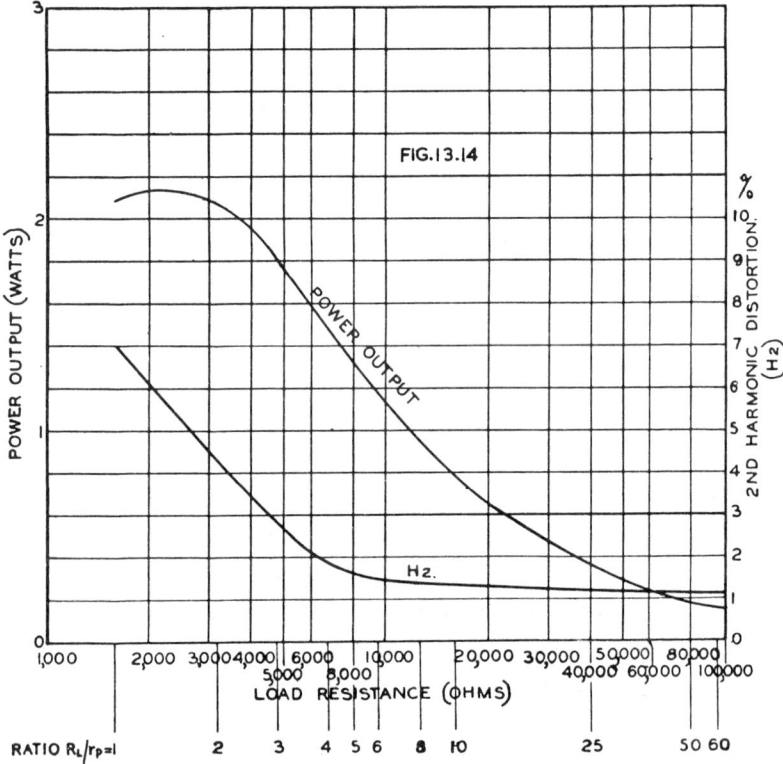

Fig. 13.14.   *Power output and distortion of type 45 triode plotted against load resistance. The curves were derived graphically, with allowance for shifting loadline.   The operating conditions are* : $E_b = 250$ *volts,* $E_c = -50$ *volts, peak grid signal voltage 50 volts,* $I_b = 34$ *mA (zero signal),* $r_p = 1610$ *ohms.*

(v) Plate circuit efficiency and power dissipation

The plate circuit efficiency is defined as the ratio of the maximum signal frequency power output to the d.c. power input under these conditions. (Owing to rectification, the d.c. power input under maximum signal conditions is usually greater than with zero signal input).

The plate circuit efficiency cannot exceed 50% in any Class A amplifier, and usually does not exceed 25% with Class A triodes. The plate circuit efficiency in an ideal Class A triode is given by

$$\eta = R_L/(2R_L + 4r_p) \tag{36}$$

The power relationships in a Class A triode are shown in Fig. 13.15 on the assumption that no second harmonic or rectification effects are present, and that fixed bias is employed. Allowance is made for the power loss in the d.c. resistance of the transformer primary. The plate-to-cathode voltage is E_b, but the total supply voltage is E_{bb}.

The plate dissipation in a Class A triode is greatest with zero input signal. If there are no rectification effects, the power input remains constant and the power output is simply power transferred from heating the plate to useful output. With normal rise of plate current for 5% second harmonic distortion, the plate dissipation is more nearly constant, but it is still sufficient to base the design on the zero-signal condition.

The quiescent operating point may be limited by one or more of the following :

1. Maximum plate dissipation.
2. Maximum plate voltage.
3. Maximum plate current.

Fig. 13.16 is an example including all three. The dissipation curve may be plotted from the relationship $I_b = P_0/E_b$. The horizontal line AB is the maximum plate current, while the vertical line CD is the maximum plate voltage. The quiescent operating point must not be outside the area ABCD.

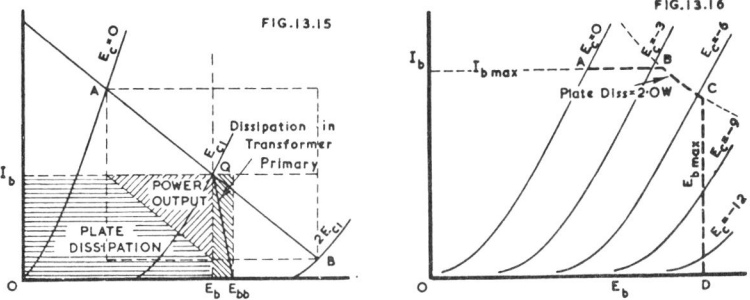

Fig. 13.15. *Power output and dissipation relationships in a Class A triode.*
Fig. 13.16. *Plate characteristics of small battery operated power triode with limits of maximum plate dissipation, maximum plate voltage and maximum plate current.*

(vi) Power sensitivity

A triode suffers from low power sensitivity when compared with pentodes and beam power amplifiers, but the difference becomes smaller when a higher degree of negative feedback is applied to the latter. In any case, it is usually a simple matter to provide sufficient gain in the voltage amplifier.

Maximum power sensitivity is obtained when $R_L = r_p$, although this limits the maximum power output for a fixed value of distortion.

See Sect. 1(iii) for the form in which the sensitivity is expressed.

(vii) Choke-coupled amplifier

The choke-coupled amplifier of Fig. 13.17 may be used as an alternative to the transformer-coupled arrangement of Fig. 13.1, and the foregoing discussion may be applied exactly on the understanding that

1. the choke inductance and resistance correspond to the transformer primary inductance and resistance,

2. the condenser C offers negligible impedance to signal frequencies compared with R_1.

(viii) Effect of a.c. filament supply

The curves of power triodes are usually drawn for the condition with d.c. filament supply and with the negative filament terminal regarded as the cathode. When a.c. is applied to the filament, roughly the same performance is achieved by increasing the grid bias by $E_f/2$. The increase is taken as 1.5 volts for a 2.5 volt filament.

FIG.13.17   FIG.13.18

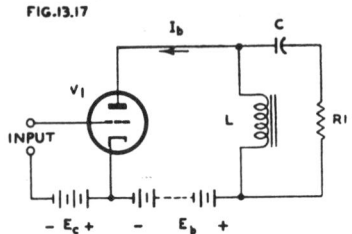
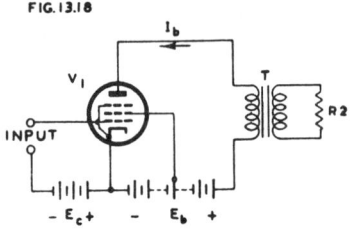

Fig. 13.17. *Choke-coupled ampli-*   Fig. 13.18.   *Basic   circuit   of*
fier.   *power pentode.*

(ix) Overloading

A triode has no " cushioning " effect such as occurs with a pentode when the peak signal voltage approaches the bias voltage. There is therefore no warning of the impending distortion which occurs when grid current flows. The distortion may be minimized by the use of a low-plate-resistance triode in the preceding stage, a low grid circuit resistance or a grid stopper resistor (10 000 ohms is normally satisfactory, but values from 1000 to 50 000 ohms have been used).

(x) Regulation and by-passing of power supply

The regulation and by-passing of the power supply affect the minimum frequency which can be handled satisfactorily at full power output—see Sect. 1(iv).

SECTION 3 : CLASS A MULTI-GRID VALVES

(*i*) *Introduction* (*ii*) *Ideal pentodes* (*iii*) *Practical pentodes—operating conditions* (*iv*) *Graphical analysis—power output and distortion* (*v*) *Rectification effects* (*vi*) *Cathode bias* (*vii*) *Resistance and inductance of transformer primary* (*viii*) *Loudspeaker load* (*ix*) *Effects of plate and screen regulation* (*x*) *Beam power valves* (*xi*) *Space charge tetrodes* (*xii*) *Partial triode operation of pentodes.*

(i) Introduction

Multi-grid valves include pentodes, beam power amplifiers and similar types, and space-charge valves. All of these have a family resemblance, in that they have higher plate circuit efficiencies and greater power sensitivities than triodes, but on the other hand they have higher distortion, particularly odd harmonic distortion. Their advantages over triodes generally outweigh their disadvantages, particularly as the distortion may be minimized by negative feedback, and they are almost exclusively used in ordinary commercial radio receivers. Reference should be made to Chapter 7 for amplifiers incorporating negative feedback.

The basic circuit of a power pentode with transformer-coupled load is Fig. 13.18. A choke-capacitance coupled load may also be used, as for a triode (Fig. 13.17). The only essential difference from a triode is that provision must also be made for a constant voltage to be applied to the screen. In some cases the screen is operated at the same voltage as the plate, so as to avoid a separate screen voltage supply.

In a practical circuit the screen, the plate return and the grid return should each be by-passed to the cathode.

(ii) Ideal pentodes

An ideal pentode is one having infinite plate resistance, a 90° angular knee and equally spaced characteristic curves (Fig. 13.19). The zero bias characteristic is OAC, the $-E_{c1}$ characteristic is ODE, while the $-2E_{c1}$ characteristic is OB. The operating point is Q, the optimum loadline is AQB where AQ = QB and the distortion is zero.

The following may readily be derived—

$$\text{Optimum } R_L = E_b/I_b \tag{1}$$

Then for optimum load resistance :

$$P_0 = \tfrac{1}{2}E_b I_b \tag{2}$$

$$\text{D.C. power input} = E_b I_b \tag{3}$$

$$\text{Plate circuit efficiency} = \tfrac{1}{2}E_b I_b/E_b I_b = 50\% \tag{4}$$

$$E_{c1} = E_{c2}/2\mu_t \tag{5}$$

where E_{c2} = screen voltage
and μ_t = triode amplification factor = μ_{g1g2}.

For any value of load resistance :

$$P_0 = \tfrac{1}{2}I_b{}^2 R_L \text{ when } R_L \text{ is less than } E_b/I_b \tag{6}$$

$$P_0 = \tfrac{1}{2}E_b{}^2/R_L \text{ when } R_L \text{ is greater than } E_b/I_b \tag{7}$$

(It is assumed here that the signal voltage is reduced for other than optimum R_L in order to avoid distortion.)

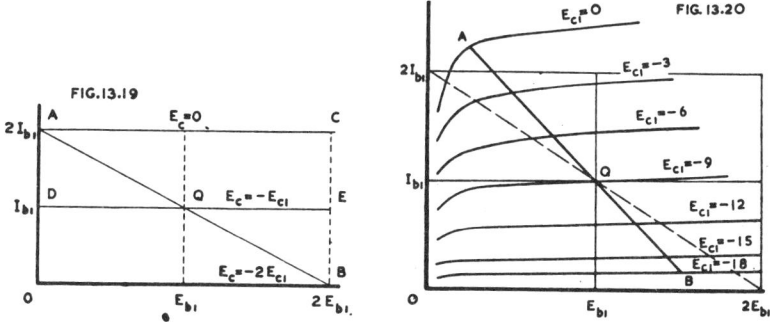

Fig. 13.19. *Plate characteristics of ideal pentode.*

Fig. 13.20. *Plate characteristics of typical pentode compared with ideal characteristics.*

(iii) Practical pentodes—operating conditions

A practical pentode has characteristics of the form of Fig. 13.20 on which has been superimposed the " ideal characteristics " for the same plate voltage and current.

(A) The **load resistance** corresponding to the loadline AQB is slightly lower than the ideal value E_b/I_b. A good general rule for all pentodes is

$$R_L \approx 0.9 \ E_b/I_b \tag{8}$$

which is usually correct within $\pm 10\%$ for maximum power output. Eqn. (8) is safe to use under all circumstances, although an adjustment may be made experimentally or graphically to secure the best compromise between distortion and power output.

(B) The **power output** is always less than $\tfrac{1}{2}E_b I_b$, and may be calculated graphically [see (iv) below].

(C) Owing to rectification effects, the direct **plate current** may rise or fall from zero signal to maximum signal [see (v) below].

(D) The **plate circuit efficiency** is usually between 28% and 43% for total harmonic distortion less than 10%.   Of course, the screen current is wasted, so that the total plate + screen circuit-efficiency is lower.   The plate dissipation is equal to the d.c. watts input less the signal power output.   The screen dissipation is equal to the d.c. screen input.

(E) **Screen current** : In a pentode, if the control grid bias is kept constant and only the plate voltage varied, the total cathode current (plate + screen) will remain nearly constant, decreasing slightly as the plate voltage is reduced down to the knee of the curve.   Below this plate voltage the screen current increases more rapidly until zero plate voltage is reached, at which point the screen current is a maximum (Fig. 13.21).   It is evident therefore that if a dynamic loadline cuts the zero bias curve below the knee, the screen current will rise rapidly and the screen dissipation may be exceeded.   The average maximum-signal screen current may be calculated from the approximation

$$I_{c2 \, av} \approx \tfrac{1}{4}I_A + \tfrac{1}{2}I_Q \qquad\qquad (9)$$

where I_A = screen current at minimum plate voltage swing and zero bias (point A), and I_Q = screen current at no signal and normal bias.

The screen dissipation is therefore P_{g2} where

$$P_{g2} = E_{c2}(\tfrac{1}{4}I_A + \tfrac{1}{2}I_Q) \qquad\qquad (10)$$

The variation of screen current with change of control grid voltage is such that the ratio between plate and screen currents remains approximately constant provided that the plate voltage is considerably higher than the knee of the curve.   This ratio may be determined from the published characteristics.

A pentode, or beam power tetrode may be used as an amplifier with the plate voltage in the region of the knee of the curve of the $E_c = 0$ characteristic, that is somewhere about one fifth of the screen voltage, provided that care is taken to keep both plate and screen dissipations within their ratings.   This gives operation very similar to that of a triode with a non-critical and low value of load resistance, which may be useful in some special applications (Ref. C6).   This method of operation is not here described in detail.

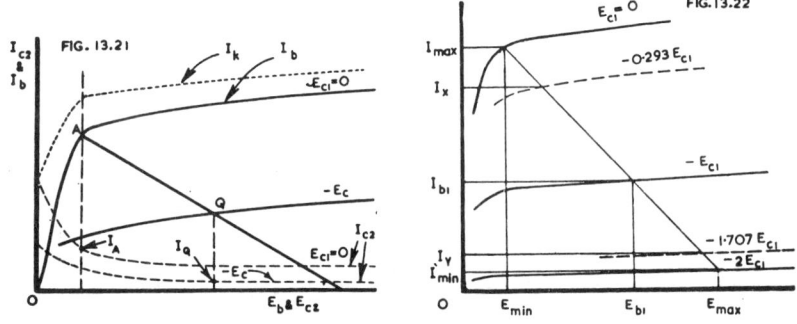

Fig. 13.21. Plate, screen and cathode current characteristics of pentode.
Fig. 13.22. Calculation of pentode power output and distortion using the five selected ordinate method.

(F) The screen should be supplied from a voltage source of good regulation ; a voltage divider, whose " bleed " current is very much higher than the maximum screen current under dynamic conditions, is the most common source when the screen voltage is lower than the plate voltage.

(G) The **voltage gain** is determined as for a triode [Section 2(i) eqn. 8].

(H) If **oscillation or parasitics** are experienced, a grid stopper resistance of about 500 ohms may be used.   A small mica condenser from the plate terminal to earth is often beneficial.   The bias resistor and its electrolytic by-pass condenser may be by-passed by a 0.001 μF mica condenser.   The shells of metal valves should be earthed directly.   The length of leads should be as short as possible, particularly between the valve terminals and their by-pass condensers.

With type 807, a stopper resistor of 100 ohms (non-inductive) may be connected directly to the screen terminal of the valve, and a by-pass condenser (0.01 μF mica) taken from the end of the resistance remote from the screen, directly to earth.

(I) **Overload characteristics**—Owing to the crowding together of the characteristics at and below the knee of the curves, a pentode tends to give a smoother overload characteristic than a triode. There is no point at which the distortion begins sharply, but rather a gradual flattening effect on the top, or top and bottom, of the signal current wave. Pentodes or beam power amplifiers with negative feedback lose most of this " cushioning effect," and more closely resemble triodes as regards overload. The distortion due to grid current may be minimized by a low impedance (triode) preceding stage and a fairly low value of grid resistor ; this will then leave the distortion due to the characteristics as the predominant feature, and provide a reasonably smooth overload.

(iv) Graphical analysis—power output and distortion

The choice of loadline for a pentode is usually made by firstly selecting the screen voltage and then selecting a convenient plate voltage. The third step is to note the grid voltage which allows a small plate current to flow, and to divide this voltage by 2 to obtain the grid voltage for the working point—thus determining the quiescent operating point Q. A scale is then swung around Q (a small pin is helpful as a pivot) until the two parts of the loadline are equal. If the scale is calibrated in millimetres, the 10 cm. calibration may be held at Q and the two parts compared directly by eye. When the two parts are equal, the second harmonic distortion will be zero, and the loadline is a first approximation for maximum power output.

In the case of some pentodes, and particularly beam power amplifiers of the 6L6 class, the loadline for zero second harmonic is obviously far from the knee of the curve. In such a case the loadline should be taken to the knee of the curve, if sharply defined.

To determine the exact loadline for maximum power output, it is necessary to take several angles, and to calculate the power output for each. The final choice of loadline is usually a compromise between power output and distortion ; in such a case the load resistance is always less than the one giving maximum power output.

Load resistances higher than that giving maximum power output are to be avoided because they give greater harmonic distortion, less power output and higher screen current. It is better to err on the low side, particularly with nominal loudspeaker loads.

Owing to the presence of appreciable percentages of third and higher order odd harmonics, the formulae for triodes are not suitable for pentodes. There are several methods for pentodes, the choice depending on whether third harmonic only is required, or third and higher harmonics. Reference should be made to Chapter 6 Sect. 8(iii) " Graphical Harmonic Analysis."

The treatment here is based on a working loadline. If the amount of second harmonic is very small, the rise of plate current due to rectification may be neglected. If there is an appreciable amount of second harmonic, it is necessary to use the method of Sect. 3(v) for drawing the corrected loadline ; the formulae below should then be applied to the corrected loadline.

(A) **Five selected ordinate method** (For second and third harmonics only)
This method makes use of the plate currents corresponding to grid voltages $-$ 0.293 and $-$ 1.707 E_{c1} in addition to the three currents used with Class A$_1$ triodes. If these are all available, the loadline of Fig. 13.22 is all that is required—if not, it will be necessary to plot the dynamic characteristic as in Fig. 13.24.

$$P_0 = [I_{max} - I_{min} + 1.41 (I_X - I_Y)]^2 R_L/32 \tag{11}$$

$$R_L = (E_{max} - E_{min})/(I_{max} - I_{min}) \tag{12}$$

$$\% \text{ 2nd harmonic} = \frac{I_{max} + I_{min} - 2I_{b1}}{I_{max} - I_{min} + 1.41 (I_X - I_Y)} \times 100 \tag{13}$$

$$\% \text{ 3rd harmonic} = \frac{I_{max} - I_{min} - 1.41 (I_X - I_Y)}{I_{max} - I_{min} + 1.41 (I_X - I_Y)} \times 100 \tag{14}$$

where I_X = plate current at $- 0.293 \, E_{c1}$
       I_Y = plate current at $- 1.707 \, E_{c1}$.

Eqns. (11) to (14) are exact provided that there is no harmonic higher than the third. The power output is that for fundamental frequency only, but the power output for 10% harmonic distortion is only 1% of the fundamental power output.

The presence of third harmonic distortion has a flattening effect on both the positive and negative peaks, thus increasing the power output for a limited value of $I_{max} - I_{min}$. For example, 10% third harmonic reduces the value of I_{max} by 10% while it increases the total power output by 1%. If there were no distortion, and the value of I_{max} were maintained at 90%, the power output would be only 81%. Thus the ratio of power output with and without 10% third harmonic distortion is $101/81 = 1.25$, on condition that $I_{max} - I_{min}$ is kept constant. In simple language, for the same plate current swing, 10% third harmonic distortion increases the total power output by 25%. This will be modified by the presence of fifth and higher harmonics [see iv(D) below ; Sect. 7(iii)].

(B) **Five equal-voltage ordinate method** (Espley)

For second, third and fourth harmonics only.

The previous method requires characteristics for $- 0.293$ and $- 1.707 \, E_{c1}$, which are often not directly available. The following method is usually more convenient and is exact provided that all harmonics above H_4 are zero (see Fig. 13.23 for symbols).

$$P_0 = (I_0 - I_{2.0} + I_{0.5} - I_{1.5})^2 R_L / 18 \tag{15}$$

$$\% \text{ 2nd harmonic} = \frac{3(I_0 - 2I_{1.0} + I_{2.0})}{4(I_0 + I_{0.5} - I_{1.5} - I_{2.0})} \times 100 \tag{16}$$

$$\% \text{ 3rd harmonic} = \frac{(I_0 - 2I_{0.5} + 2I_{1.5} - I_{2.0})}{2(I_0 + I_{0.5} - I_{1.5} - I_{2.0})} \times 100 \tag{17}$$

$$\% \text{ 4th harmonic} = \frac{(I_0 - 4I_{0.5} + 6I_{1.0} - 4I_{1.5} + I_{2.0})}{4(I_0 + I_{0.5} - I_{1.5} - I_{2.0})} \times 100 \tag{18}$$

$$\text{D.C. plate current} = \frac{1}{6}(I_0 + 2I_{0.5} + 2I_{1.5} + I_{2.0}) \tag{19}$$

A third harmonic scale may be prepared for reading the third harmonic percentage (Ref. A14 pp. 71-72).

(C) **Seven equal-voltage ordinate method** (Espley)

This is exact for harmonics up to the sixth, provided that higher harmonics are zero. It is sometimes more convenient than the five ordinate method, when there are no $- 0.5$ and $- 1.5 \, E_{c1}$ curves, even when the higher harmonics are of no interest.

The symbols have the same significance as in Fig. 13.23 with the subscript indicating the grid voltage.

$$P_0 = (167I_0 + 252I_{0.33} - 45I_{0.67} + 45I_{1.33} - 252I_{1.67} - 167I_{2.0})^2 \, R_L / 819\,200 \tag{20}$$

$$H_2\% = 25(559I_0 + 486I_{0.33} - 1215I_{0.67} + 340I_{1.0} - 1215I_{1.33} + 486I_{1.67} + 559I_{2.0})/I \tag{21}$$

$$H_3\% = 250(45I_0 - 36I_{0.33} - 63I_{0.67} + 63I_{1.33} + 36I_{1.67} - 45I_{2.0})/I \tag{22}$$

$$H_4\% = 450(17I_0 - 42I_{0.33} + 15I_{0.67} + 20I_{1.0} + 15I_{1.33} - 42I_{1.67} + 17I_{2.0})/I \tag{23}$$

$$H_5\% = 4050(I_0 - 4I_{0.33} + 5I_{0.67} - 5I_{1.33} + 4I_{1.67} - I_{2.0})/I \tag{24}$$

$$H_6\% = 2025(I_0 - 6I_{0.33} + 15I_{0.67} - 20I_{1.0} + 15I_{1.33} - 6I_{1.67} + I_{2.0})/I \tag{25}$$

where $I = 167I_0 + 252I_{0.33} - 45I_{0.67} + 45I_{1.33} - 252I_{1.67} - 167I_{2.0}$.

(D) **Eleven selected ordinate method**

To use this method, it is first necessary to plot the loadline on the plate characteristics, and then to transfer it to a dynamic mutual characteristic (Fig. 13.24) from which the required values of plate current may be derived.

$$P_0 = \tfrac{1}{4}(0.5I_0 - 0.5I_{2.0} + I_{hm3} - I_{hm5})^2 R_L \tag{26}$$

$$I_{h1} = 0.5I_0 - 0.5I_{2.0} + I_{hm3} - I_{hm5} = \text{fundamental} \tag{27}$$

$$I_{h2} = \tfrac{1}{4}(I_0 + I_{2.0} - 2I_{1.0}) \tag{28}$$

$$I_{h3} = 0.167(2I_{0.5} + I_{2.0} - I_0 - 2I_{1.5}) \tag{29}$$

$$I_{h4} = \tfrac{1}{8}(I_0 + 2I_{1.0} + I_{2.0} - 2I_{0.3} - 2I_{1.7}) \tag{30}$$

$$I_{h5} = 0.1(2I_{0.7} + I_0 + 2I_{1.8} - 2I_{0.2} - 2I_{1.3} - I_2) \tag{31}$$

Percentage second harmonic $= (I_{hm2}/I_{hm1}) \times 100$

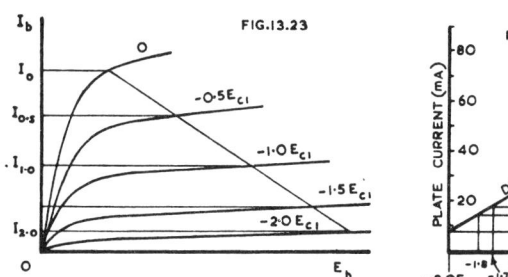
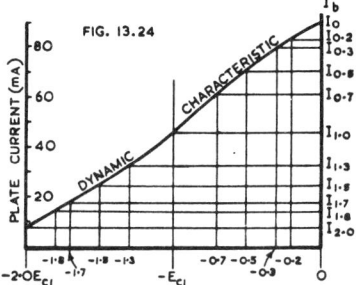

Fig. 13.23.   *Calculation of pentode power output and distortion using the five equal-voltage ordinate method.*

Fig. 13.24.   *Calculation of pentode power output and distortion using the eleven selected ordinate method.*

Percentage third harmonic $= (I_{h\,m3}/I_{h\,m1}) \times 100$ etc.

Note that all currents, including $I_{h\,m1}$, $I_{h\,m2}$ etc., are peak values. The exact values of grid voltage are 0, 0.191, 0.293, 0.5, 0.691, 1.0, 1.309, 1.5, 1.707, 1.809 and 2.0. The approximate values are, however, sufficiently accurate for most purposes.

Eqn. (26) shows that third harmonic distortion adds to the power output, while fifth harmonic subtracts from the power output as calculated for distortionless conditions. See also (iv)A above ; Sect. 7(iii).

For theoretical basis see Chapter 6 Sect. 8(iii).

(v) Rectification effects

The general effects are the same as for a triode except that it is sometimes possible with a pentode to have zero second harmonic and no loadline shift. In all other cases it is necessary to determine the corrected loadline before applying harmonic analysis or calculating power output. The method of deriving the corrected loadline is the same for pentodes as for triodes [Sect. 2(ii)D]. For example see Fig. 13.28.

With a triode, the loadline always shifts upwards into such a position that the distortion is less than it would otherwise be. With a pentode, the loadline shifts upwards when the load resistance is lower than a certain value, and shifts downwards when the load resistance is higher than this value. When a pentode loadline shifts downwards it causes increased distortion or decreased power output.

(vi) Cathode bias

The operation of cathode bias is the same as with triodes [Section 2(ii)E] except that the total current flowing through the cathode bias resistor is the sum of the plate and screen currents [Chapter 2 Section 4(v)]. If the screen is supplied from a voltage divider which is returned to the cathode, the bleed current must be added to the cathode current in calculating R_k. This tends to stabilize the bias voltage. Even in cases where there is no rise of plate current at maximum signal, there will always be a rise in screen current [Section 3(iii)E].

When a tentative value of R_k has been determined for maximum signal conditions, it is necessary to check both plate and screen dissipations at zero signal. If the plate and screen voltages are equal, the simplest method is the use of the " triode " mutual characteristics, if available [Chapter 2 Sect. 4(v)].

If there is a " bleed " current through R_k, the procedure is shown in Fig. 13.25 where the triode mutual characteristic is shown in the upper part, with OA representing the bleed current in the lower part. The cathode loadline is drawn from A (instead of from 0) with a slope of $-1/R_k$ where R_k is the value determined for maximum signal conditions.

The no-signal grid bias is $-E_{c1}$ and the cathode current $(I_b + I_{c2})$ is I_{k1}. The dissipation on plate and screen is given by

$$P_p = I_{k1}E_b m/(m+1) \; ; \; P_{g2} = I_{k1}E_{c2}/(m+1) \qquad (32)$$

where $m =$ ratio of plate to screen currents $= I_b/I_{c2}$.

The rectification effects with cathode bias may, like triodes (Fig. 13.8) be determined from the plate characteristics but an approximation is involved since $D''E_{bb}$ must have a slope of $-m/R_k(m+1)$ where $m = I_b/I_{c2}$. The value of I_{c2} may be taken as that for maximum signal, and a slight error will then be introduced when deriving the zero signal condition. The cathode bias loadline should, ideally, be drawn as two loadlines—one for zero signal, and the other for maximum signal. For most purposes, however, the maximum signal cathode loadline may be used for both conditions. The zero bias lines are, in each case, above the lines for maximum signal, and the difference in slope is of the order of 5% to 10%.

In general, any Class A₁ pentode may be operated either with fixed or cathode bias, as desired, except for the limitations on the maximum grid circuit resistance [see Sect. 10(i)]. It occasionally happens that a condition is permissible only with fixed bias, owing to the rise in plate dissipation at zero signal.

For back bias, further details regarding fixed bias, and grid circuit resistance see Sect. 10.

For cathode by-passing see Sect. 2(ii).

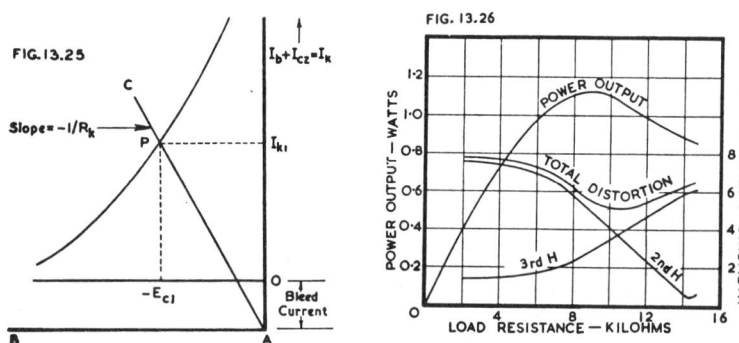

Fig. 13.25. *Cathode loadline of pentode with additional bleed current passing through the cathode resistor, plotted on mutual characteristics.*
Fig. 13.26. *Power output and harmonic distortion of typical pentode plotted against load resistance. Valve type 6AK6, $E_b = E_{c2} = 180\ V$, $E_{c1} = -9$, peak signal = 9 volts.*

(vii) Resistance and inductance of transformer primary

The **resistance** of the transformer primary may be allowed for in the same way as with triodes [Sect. 2(ii)A]. If R' is the resistance of the primary, and R_k is the cathode bias resistance, then the slope of the line QE_b in Fig.·13.6 should be $-1/(R' + R_k)$.

The **transformer primary inductance** may be based on Chapter 5 Sect. 3(iii)c.

For the same load impedance and the same high frequency attenuation, pentodes may have higher transformer **leakage inductance** than triodes if frequency response is the only criterion. However, owing to the distortion with reactive loads at high output levels, it is very desirable to maintain the leakage inductance as low as practicable, particularly in push-pull amplifiers.

(viii) Loudspeaker load

A pentode is, unfortunately, critical in its load resistance for both maximum power output and distortion. At low operating levels the output power rises steadily as the load resistance is increased up to the value $R_L = r_p$ provided that the grid input voltage is maintained constant (the low operating level is to avoid overloading under any conditions). A loudspeaker has pronounced impedance peaks at the bass resonant frequency and at high audio frequencies. When a loudspeaker is operated at a low level, the acoustical output is accentuated at the bass resonant frequency and at high audio frequencies. The latter may be reduced to any desired degree by a

shunt filter—a resistance R in series with a capacitance C. If R is variable, the combination is the simplest form of tone control. If R is fixed, typical values are :
$R = 1.3R_L$; $C = 0.025\mu F$ for $R_L = 5000$ ohms.
This does not affect the rise of impedance at the bass resonant frequency, which is a function of the loudspeaker design and the type of baffle—see Chapter 20 Sect. 2(iv) and Sect. 3.

At maximum signal voltage the conditions are somewhat different (Fig. 13.26). The power output in this case reaches a maximum at $R_L = 9000$ ohms. The second harmonic reaches a minimum (practically zero) at about $R_L = 14\,000$ ohms, and then rises steadily ; actually it undergoes a change of phase near $R_L = 14\,000$ ohms. The third harmonic rises all the way from zero to the limit of the graph. Minimum " total distortion " occurs at $R_L = 10\,000$ ohms, which is the published typical load, being a close approach to maximum power output. In this particular case the load resistance for zero second harmonic is not that for maximum power output.

In the case of a loudspeaker load, the load resistance may rise from the nominal value to (say) 6 or 8 times this value ; all the variation is in the upwards direction. If full signal voltage is maintained for all frequencies, the distortion will be very severe and the maximum power output will be reduced at low and high frequencies. The only methods of minimizing the trouble are the use of a loudspeaker and baffle with less prominent impedance peaks, and the use of negative voltage feedback (see Chapter 7) or reduced signal voltage on the grid. A pentode, operating well below its nominal power output, is capable of giving reasonable fidelity even on an ordinary loudspeaker load. In a normal radio receiver, a power pentode with a nominal maximum power output of 4 or 5 watts can give reasonable fidelity up to somewhat over 1 watt, but it has the advantage of being capable of delivering its full power output when the distortion can be tolerated.

The effect of these high impedance loads, which are here assumed to be purely resistive for the purpose of illustration, is shown by the beam power amplifier plate characteristics in Fig. 13.27. The effect may be minimized by reducing R_L slightly below the optimum value.

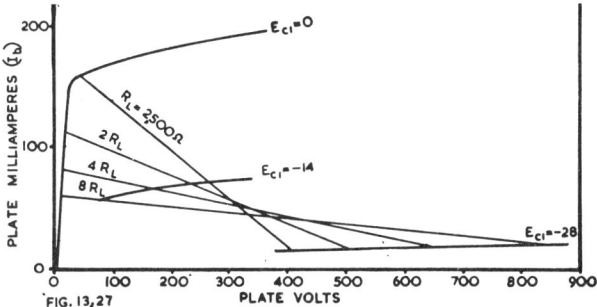

Fig. 13.27. *Plate characteristics of 6L6 or 807 beam tetrode with loadlines of optimum resistance also twice, four times and eight times optimum. The loadlines have been corrected for the rectification effect.*

The impedance of a loudspeaker is, however, far from being resistive (see Chapters 20 and 21), having a reactive component varying with frequency, which must be combined with the shunt reactance of the transformer primary at low frequencies and that of the shunt capacitance from plate to earth at high frequencies. The combined reactive components increase the distortion and reduce the power output [see also Chapter 5 Sect. 3(iii)c and Chapter 2 Sect. 4(vi) ; Ref. C4].

The published values of power output apply to highly efficient output transformers. The available power output from the secondary of a normal power transformer is equal to η times the published value, where η is the efficiency [see Chapter 5 Sect. 2(ii) and Sect. 3(vi)]. Typical efficiencies are from 70% to 95% (depending on the price class) for well-designed transformers.

(ix) Effects of plate and screen regulation

The internal resistances of plate and screen supply sources cause a reduction in power output.   If the regulation of the power supply is such that the rise of plate and screen currents causes a decrease of 1% in all the electrode voltages, the decrease in power output (by the use of conversion factors) will be approximately 2.5%.   This will be modified by the shape of the characteristics and rectification effects, and the only accurate method is the graphical one outlined below.   See also remarks on push-pull—Sect. 6(iii).

In addition, the regulation and by-passing of the power supply also affect the minimum frequency which can be handled satisfactorily at full power output—see Sect. 1(iv).

Graphical method (Ref. B4).

Let R_1 = internal resistance of plate supply source,
R_2 = internal resistance of screen supply source,
ΔI_b = increase in plate current from zero to maximum signal, with constant electrode voltages,
ΔI_{c2} = increase in screen current, with constant electrode voltages,
ΔE_b = change in plate voltage,
ΔE_{c1} = change in screen voltage,
E_b' = plate voltage at maximum signal,
and E_{c2}' = screen voltage at maximum signal.

Then $E_b' = E_b - \Delta E_b$ (33) $E_{c2}' = E_{c2} - \Delta E_{c2}$ (34)
where $\Delta E_b = R_1 \, \Delta I_b$ (35) $\Delta E_{c2} = R_2 \Delta I_{c2}$ (36)

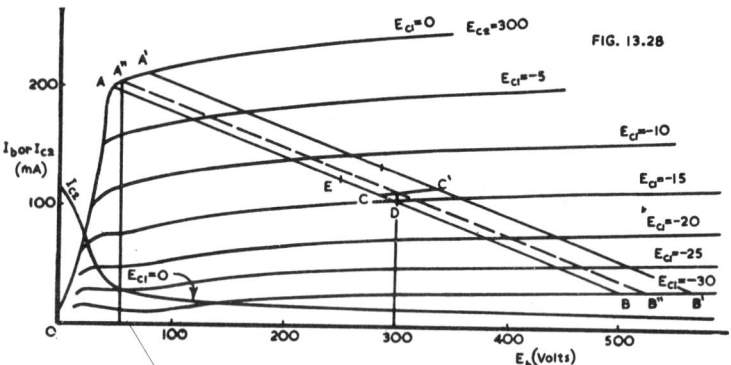

Fig. 13.28.   *Plate characteristics of 6L6 or 807 illustrating method of correcting the loadline for rectification.* *The corrected loadline is A″ B″.*

On the plate characteristics (Fig. 13.28) AB is the uncorrected loadline with C as its average current point, A′B′ is a parallel loadline with C′ as its average current point, A″B″ is the corrected loadline.   The average current may be determined from the point on A″B″ which has a plate voltage 300 volts ; from this the rise of current ΔI_b may be calculated.   The value of ΔI_{c2} may be determined from eqn. (9), ΔE_{c2} may be calculated by eqn. (36) and E_{c2}' by eqn. (34).   In the example of Fig. 13.28 let it be assumed that the plate, screen and grid voltages are all reduced proportionally, F_e* being 0.9 :

$E_b = E_{c2} = 300$ volts $E_b' = E_{c2}' = 270$ volts
$E_{c1} = -15$ volts $E_{c1}' = -13.5$ volts

The procedure is :

1. Plot a new curve for $E_{c1} = 0$, $E_{c2}' = 270$ volts, by the use of conversion factors*. This is drawn on Fig. 13.29.   For $F_e = 0.9$, $F_i = 0.86$.
2. Plot a curve for $E_{c1}' = -13.5$, $E_{c2}' = 270$ volts similarly.   This will be the -15 volt curve (for $E_{c2} = 300$ volts) with the current ordinates multiplied by the factor 0.86.

*For conversion factors see Chapter 2 Sect. 6 and Fig. 2.32A.

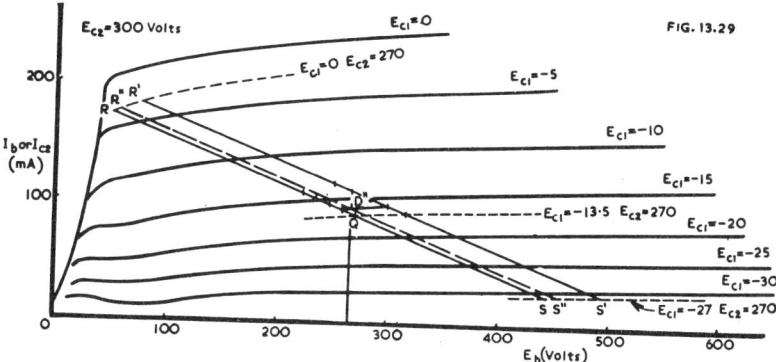

Fig. 13.29.   *Plate characteristics of 6L6 or 807 giving graphical method for deriving the effect of internal resistance in the voltage supply source.*

3. Plot a curve for $E_{c1}' = -27$, $E_{c2}' = 270$ volts. This will be the -30 volt curve (for $E_{c2}' = 300$ volts) with the current ordinates multiplied by 0.86.
4. Mark the point Q at $E_b = 270$, $E_{c2} = 270$, $E_{c1} = -13.5$ volts.
5. Through Q draw the uncorrected loadline RQS.
6. Draw the corrected loadline R″D″S″ by the same method as previously (Fig. 13.28).
7. Determine the rise of plate current from Q to D″ (intersection of R″S″ with $E_b = 270$ volts). This should be approximately the same as derived in the first case (ΔI_b). If there is an appreciable error, some adjustment should be made.

(x) Beam power valves

Beam power amplifier valves, otherwise known as output tetrodes, are in two principal classes. The first class includes most of the smaller valves, which have characteristics so similar to pentodes that they may be treated as pentodes in all respects. The second class includes types such as 6L6, 807 and KT66, which differ from pentodes principally in having sharper " knees " to their plate characteristics, more second harmonic but less third harmonic distortion. The optimum load resistance is more critical than with ordinary power pentodes.

The screen currents of many types of beam power valves, due to variations in grid alignment, may have considerably greater tolerances than in pentodes—up to $\pm$ 100% in some instances—and screen dropping resistances should not be used unless recommended by the valve manufacturer. The screen supply should normally be obtained from a voltage divider.

The distortion and power output of type 6L6 beam power valve are plotted against load resistance in Fig. 13.30. The second harmonic is 9.6% at the rated load resistance, the third harmonic only 2.4% and all higher harmonics negligible. At lower load resistances the second harmonic rises, although not seriously, the third harmonic decreases steadily, and all higher harmonics are negligible—the overall effect being quite satisfactory. At higher load resistances the performance is not good, and the overall effect is roughly the same as with a pentode. See Chapter 7 for the effect of negative feedback.

Reference C3.

(xi) Space charge tetrodes

A space charge tetrode is actually a triode operating on a virtual cathode provided by the thermionic cathode and the inner (space charge) grid. It is capable of low distortion, even lower than a triode, but the power output for a given d.c. power input is less than that of a triode, owing to the power taken by the space-charge grid.

See Refs. C1, C2.

(xii) Partial triode operation of pentodes ("ultra-linear" operation)

When the screen and plate of a pentode are being operated at the same voltage, pentode operation is obtained when the screen is connected to the B+ end of the output transformer primary, while triode operation is obtained when the screen is connected to the plate end of the primary. Any desired intermediate condition can be obtained by connecting the screen to a suitable tap on the primary. In such intermediate condition the valve operates as a pentode having negative feedback applied to the screen, with a section of the load impedance common to both electrodes, and minimum high level distortion with push-pull operation is usually obtained when the tapping point is about 43% of the total primary turns (Refs. C7, H5, H6).

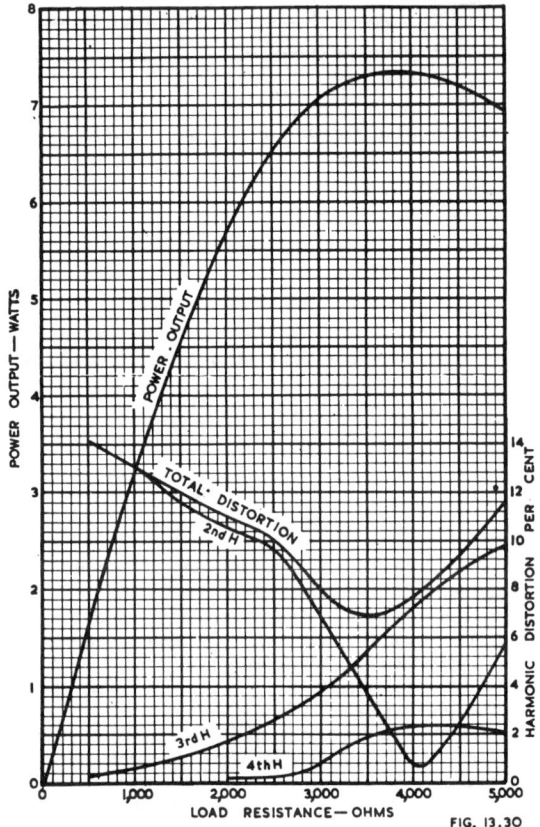

FIG. 13.30

Fig. 13.30. Power output and distortion of type 6L6 beam power amplifier plotted against load resistance, for $E_b = E_{c2} = 250$ volts, $E_{c1} = -14$ volts, peak signal 14 volts.

SECTION 4 : PARALLEL CLASS A AMPLIFIERS

Any two Class A amplifier valves may be connected in parallel, with suitable provision for their correct operation, to provide double the power output. It is assumed that these have identical characteristics (the normal manufacturing tolerances have only a very slight effect). The load resistance will be half that for one valve. The distortion will be the same as for one valve. The input voltage will be the same as for one valve. The total plate current will be twice that for one valve. The effective plate resistance will be half that for one valve.

Parasitics are likely to occur, particularly with high-slope valves. **Precautions to be taken include the following :**

1. The two valves should be placed closely together, with very short leads between grids and plates (and screens in the case of pentodes).
2. A grid stopper should be connected directly to one or both grids. It is usually cheaper—and just as effective—to have one grid stopper of, say, 200 ohms than two stoppers each of 100 ohms.
3. Screen stoppers (50 or 100 ohms for each screen) are very helpful, particularly with types 6L6 or 807 [see Sect. 3(iii)H].

N.B. Plate stoppers are less helpful, are wasteful of power, and are generally unnecessary.

The advantages of parallel operation lie principally in the elimination of the phase-splitter or input transformer required with push-pull operation. The disadvantages are :

1. The necessity for handling the heavy direct plate current. This necessitates either a special output transformer or a choke (say 20 henrys) with parallel feed to the output transformer.
2. The higher distortion—this is not serious if negative feedback is used, and in any case is no worse than that of a single valve.
3. The attenuation of lower frequencies at maximum power output due to the limited size of the by-pass capacitor. This effect is also a function of the plate supply regulation—see Sect. 1(iv).

Parallel operation may be used with a cathode-follower stage, permitting the use of two smaller valves with lower plate voltage in place of one valve with higher plate voltage, and thereby reducing the difficulties of grid excitation.

SECTION 5 : PUSH-PULL TRIODES CLASS A, AB$_1$

(i) Introduction (ii) Theory of push-pull amplification (iii) Power output and distortion (iv) Average plate current (v) Matching and the effects of mismatching (vi) Cathode bias (vii) Parasitics.

(i) Introduction

(A) Fundamental principles of push-pull operation

The fundamental circuit of a push-pull power amplifier is Fig. 13.31. A balanced (push-pull) input voltage must be applied to the three input terminals, and a balanced (push-pull primary) output transformer must be connected to the two plates with its centre-tap connected to E_{bb} +. For the best results the input voltage must be exactly balanced, the valves must have identical characteristics and the output transformer must be exactly balanced between the two sections of the primary, with perfect coupling between them. Under these conditions, any even harmonics introduced by the valves will be cancelled, but the odd harmonics will not be affected ; the flux in the core due to the d.c. plate currents would be zero.

FIG. 13.31

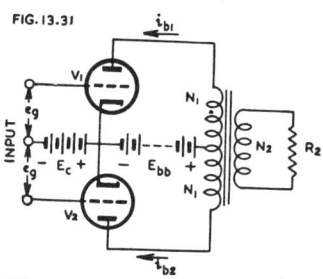

Fig. 13.31. Fundamental circuit of push-pull power amplifier.

The load resistance R_2 is connected across the secondary, and the reflected resistance* across the whole primary is

$$R_L = 4R_2(N_1{}^2/N_2{}^2)$$

*See Chapter 5 Sect. 1(ii).

and the reflected resistance across half the primary is

$$R_L{}'' = \tfrac{1}{4}R_L = 2R_2(N_1{}^2/N_2{}^2)$$

where N_1 = turns in half primary winding

and N_2 = turns in secondary winding.

If valve V_2 in Fig. 13.31 were removed from its socket, the load resistance effective*
on V_1 would then be

$$R_L{}' = \tfrac{1}{4}R_L = R_2(N_1{}^2/N_2{}^2)$$

which is half the load resistance on V_1 under push-pull conditions. This is the condition which occurs when one of the valves reaches plate current cut-off.

If the output transformer were replaced by two separate transformers, one from
each plate to E_{bb} +, the even harmonics would be cancelled but some advantages of push-pull operation would be lost. The whole principle of push-pull operation is based on the assumption that the two plates are always exactly 180° out of
phase with one another. This is achieved by an " ideal " output transformer (see
Chapter 5 Sect. 1) even if one of the valves is not operating.

If two valves are to be operated in push-pull, the plate and grid voltages may be
the same as for single valve operation, the plate-to-plate load resistance may be twice
and the plate currents twice the respective values for a single valve ; under these
conditions the power output will be exactly twice that for a single valve.   On the other
hand, higher power output is obtainable, with some increase in third harmonic distortion, merely by decreasing the plate-to-plate load resistance without any change
in grid bias (Ref. E7).   Still higher power output is obtainable by increasing the grid
bias and the signal input voltage, also by increasing the plate voltage to its maximum
value.

(B) Classes of operation

A **Class A amplifier** is an amplifier in which the grid bias and alternating grid
voltages are such that the plate current of the output valve or valves flows at all times.
The suffix 1 indicates that grid current does not flow during any part of the input
cycle.

A **Class AB₁ amplifier** is an amplifier in which the grid bias and alternating
grid voltages are such that the plate current in any specific valve flows for appreciably
more than half, but less than the entire, input cycle.   The suffix 1 has the same
meaning as with Class A.

A very useful operating condition is the borderline case between Class A and Class
AB₁, that is when the plate current just reaches the point of cut-off—this is called
Limiting Class A₁ operation.

Class A₁ triodes may be operated from poor regulation power supplies without
serious loss of power output owing to the comparatively small rise of current at maximum signal.   Class AB₁ triodes require good regulation of the power supply.   Class
AB₁ also requires very tight coupling between the two halves of the transformer
primary.

Automatic bias control has been developed (Refs. A24, A25) whereby the operation is pure Class A₁ with small input voltages, with the bias automatically increasing
with the input voltage to provide firstly Class AB₁ and finally Class AB₂ operation.
See also Sect. 11(ii).

(C) Important features

Class A₁ triodes may be operated in push-pull from power supplies having poor
regulation without any serious loss of power output, owing to the comparatively small
rise of current at maximum signal.   In addition, with normal by-passing, the regulation of the power supply does not affect the minimum frequency that can be handled
satisfactorily at full power output.

Good regulation of the power supply is necessary with Class AB operation, on both
these counts.

With Class AB operation, the current in one half of the output transformer is zero
for part of the cycle—this causes a very rapid rate of change of current at the cut-off
point, tending to cause parasitics due to the leakage inductance of the transformer.
This effect may be minimized by the use of a transformer with low leakage induct-

---

*See Chapter 5 Sect. 1(ii).

ance—see Chapter 5 Sect. 3(iii)c—or by reducing the rate of change of current. The latter is accomplished in limiting Class A operation owing to the avoidance of the sharp bend or " discontinuity " in the characteristic at the cut-off point. Valves of the 6L6 or 807 class, when connected as triodes, have a slower rate of cut-off than normal triodes, and are therefore particularly adapted to Class AB_1 operation, which merges closely into limiting Class A.

The use of a resistive network to pass a steady current through the primary of the transformer does nothing to reduce the rate of change of current, although it may help in damping out any parasitics which may occur.

(ii) Theory of push-pull amplification

It is assumed that the input voltage is sinusoidal, that the operation is Class A_1, that the output transformer is ideal having no resistance or leakage reactance, and that the input and output voltages are balanced. The valves are assumed to be perfectly matched. For circuit and conditions see Fig. 13.31.

(A) The plate-current grid-voltage characteristic of any valve may be expanded into an infinite series—

$$i_{b1} = a_0 + a_1e_g + a_2e_g{}^2 + a_3e_g{}^3 + a_4e_g{}^4 + a_5e_g{}^5 + \dots \dots \quad (1)$$

where i_{b1} = instantaneous plate current,

e_g = instantaneous grid signal input voltage,

a_0 = plate current at zero signal,

and a_1, a_2, a_3, etc. are coefficients.

The instantaneous fundamental power output is given by

$$P_0 = (a_1e_g)^2R_L \quad (2)$$

where R_L = load resistance on one valve.

In a push-pull amplifier the second valve V_2 has a grid voltage opposite in polarity to V_1 :

$$i_{b2} = a_0 - a_1e_g + a_2e_g{}^2 - a_3e_g{}^3 + a_4e_g{}^4 - a_5e_g{}^5 + \dots \dots \quad (3)$$

where i_{b2} = instantaneous plate current of V_2,

a_0 = plate current of V_2 at zero signal,

and a_1, a_2, a_3 etc. have the same meanings as in eqn. (1).

Eqn. (3) has been derived from eqn. (1) merely by making e_g negative, so that $e_g{}^2$ becomes positive and $e_g{}^3$ becomes negative.

As the plate currents are in phase opposition to the output load, the net flux-producing current is

$$i_d = i_{b1} - i_{b2}$$
$$= 2a_1e_g + 2a_3e_g{}^3 + 2a_5e_g{}^5 + \dots \dots \quad (4)$$

where i_d = net flux-producing current in N_1 turns, i.e. one half the primary winding,

and all currents are instantaneous values.

The d.c. components and all the even harmonic terms are seen to have been cancelled, only the fundamental and odd harmonic terms remaining.

The total plate current from E_{bb} is given by

$$i_t = i_{b1} + i_{b2}$$
$$= 2a_0 + 2a_2e_g{}^2 + 2a_4e_g{}^4 + \dots \dots \quad (5)$$

from which it will be seen that the d.c. components plus the even harmonic terms are present. Thus the total supply current will only remain constant when there is no even harmonic distortion.

(B) **The effect of hum in the plate and grid supply voltages**

If the plate supply voltage E_{bb} in a distortionless Class A amplifier is changed slightly, both plate currents will change together, and the net flux-producing current (i_d) will be unchanged. Hum from the supply voltage would not therefore appear in the output. The effect of hum from the grid bias supply will be similar.

In a practical amplifier, owing to the curvature of the valve characteristics, the effect may be analysed as follows, commencing with hum in the grid bias supply voltage. Assume that there is a hum voltage $E_h \cos pt$ in series with E_c, and that the input signal is $E_g \cos qt$.

The input voltage to V_1 is $E_h \cos pt + E_g \cos qt$.

The input voltage to V_2 is $E_h \cos pt - E_g \cos qt$.

Substituting these values in equation (1) we have

$i_{b1} = a_0 + a_1(E_h \cos pt + E_g \cos qt) + a_2(E_h \cos pt + E_g \cos qt)^2 + \ldots\ldots$

$= a_0 + a_1 E_h \cos pt + a_1 E_g \cos qt + a_2 E_h{}^2 \cos^2 pt + a_2 E_g{}^2 \cos^2 qt +$
$2a_2 E_h E_g \cos qt \cos pt + \ldots\ldots$

Similarly

$i_{b2} = a_0 + a_1(E_h \cos pt - E_g \cos qt) + a_2(E_h \cos pt - E_g \cos qt)^2 + \ldots\ldots$

$= a_0 + a_1 E_h \cos pt - a_1 E_g \cos qt + a_2 E_h{}^2 \cos^2 pt + a_2 E_g{}^2 \cos^2 qt -$
$2a_2 E_h E_g \cos qt \cos pt + \ldots$

Therefore $i_d = i_{b1} - i_{b2} = 2a_1 E_g \cos qt + 4a_2 E_h E_g \cos qt \cos pt$

$\qquad\qquad = 2a_1 E_g \cos qt[1 + 2(a_2/a_1)E_h \cos pt]$ \hfill (6)

Eqn. (6) has the form of a carrier $(2a_1 E_g \cos qt)$ modulated to the depth $2(a_2/a_1)E_h$ by a hum frequency $\cos pt$.

The effect of hum in the plate supply voltage is similar to its effect on the grid bias supply, that is to say the hum frequency modulates the signal frequency. It should be noted that these modulation components appear in single-ended amplifiers to the same extent, but in combination with the fundamental and harmonics of the ripple, which latter are absent with push-pull operation.

Summary—Push-pull operation tends always to reduce the effects of hum in either the grid bias or plate supply voltage.

(C) **Effects of common impedance**

The fact that no fundamental component is found in the total plate current i_t (eqn. 5) prevents any fundamental signal voltage from being fed back to earlier stages as the result of a common impedance in the plate voltage source. As the even harmonics are fed back, each will again result in higher order harmonics, so that there can be no instability caused by feedback around a push-pull stage.

General deductions

1. Because the d.c. components of the plate currents cancel each other, no steady flux is maintained in the core of the output transformer.

2. Because the even harmonics are zero, the limit placed on single-ended amplifiers no longer applies. It is usual to design push-pull amplifiers for maximum power output without primarily considering the odd harmonics ; when distortion is objectionable, this may be reduced—at the expense of power output—by increasing the load resistance.

The effects of the regulation and by-passing of the power supply are covered in Sect. 5(i).

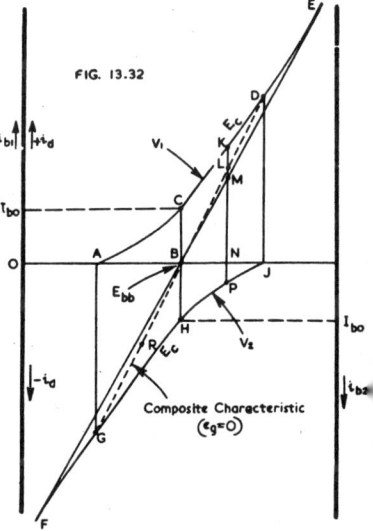

FIG. 13.32

(D) **Application to characteristic curves—Composite characteristics**

From eqn. (4) the net flux-producing current in one half of the primary winding is

$i_d = i_{b1} - i_{b2}$ \hfill (7)

Thus a " composite " characteristic may be drawn for the fixed grid voltage $- E_c$ by subtracting the currents in the two valves. At the quiescent plate voltage E_{bb}, both valves draw the same plate current and therefore $i_d = 0$. The composite characteristic must therefore pass through the point $E_b = E_{bb}$, $i_d = 0$. At other plate voltages the value of i_d is given by $i_{b1} - i_{b2}$ when the plate voltage of one is increased

Composite Characteristic
$(\epsilon_g = 0)$

Fig. 13.32. The derivation of the composite characteristic for matched push-pull valves.

by the same voltage that the other is decreased. This may be applied graphically
as shown in Fig. 13 32 ; the upper half includes the E_c characteristic for V_1 (ACDE)
while the lower half includes the E_c characteristic for V_2 but inverted and placed left
to right (FGHJ). Point C on the V_1 characteristic is the quiescent operating point,
and point B corresponds to the plate voltage E_{bb}. The V_2 characteristic is placed
so that H, the quiescent operating point, comes below C ; then since BC = BH = I_{bb}
the point B is on the composite characteristic. At any other plate voltage N, the point
L on the composite characteristic is found by subtracting PN from KN, giving LN
as an ordinate. Valve V_2 cuts off at point J, so that the amount to be subtracted from
the V_1 ordinate is zero, giving D (and its opposite number G) as points on the composite characteristic. The composite characteristic is therefore FGRBLDE which
may be compared with the straight line FBE. Sudden bends occur at D and G, but
the portion between D and G is fairly straight ; the latter includes the whole Class A
operating region. It is obvious that the non-linearity of the composite characteristic
becomes worse as the quiescent operating point is moved towards the foot of the characteristic—that is as I_{bo} becomes less. Thus we have the Class A condition (including
limiting Class A) with nearly straight composite characteristics, the Class AB$_1$ condition which includes the kinks at the points of plate-current cut-off, and finally the
Class B condition with quite considerable non-linearity.

The composite characteristic of Fig. 13.32 is that for zero signal input voltage
($e_g = 0$). Other composite characteristics may be drawn by a somewhat similar
method, except that the ($E_c + e_g$) characteristic of V_1 must be combined with the
($E_c - e_g$) characteristic of V_2 to give the $+ e_g$ composite characteristic. For example,
if $E_c = - 60$ volts, we may take e_g in increments of 10 volts, giving :

e_g	0	+10	+20	+30	+40	+50	+60	volts
$E_c + e_g$	−60	−50	−40	−30	−20	−10	0	volts
$E_c - e_g$	−60	−70	−80	−90	−100	−110	−120	volts

A family of composite characteristics is shown in Fig. 13.33 in which $E_{bb} = 300$
volts and $E_c = - 60$ volts, with values of e_g in accordance with the table above. The

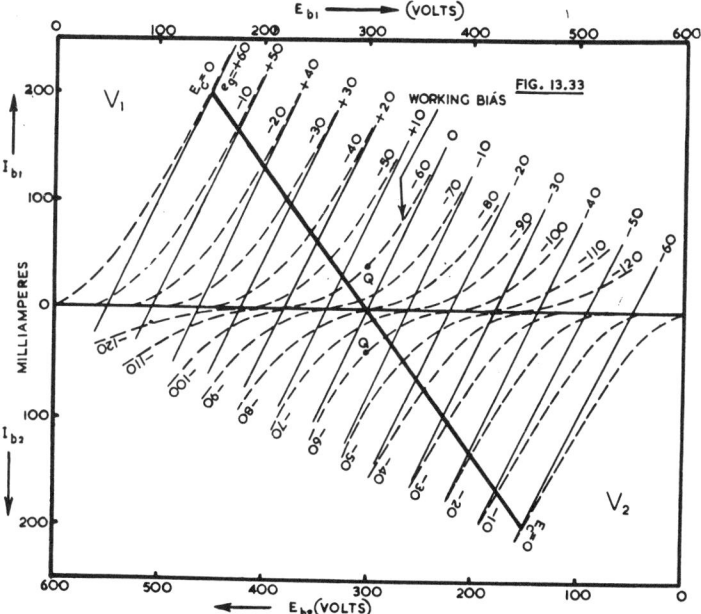

Fig. 13.33.   *Family of composite characteristics for two type 2A3 triodes.*

composite operating point is $e_g = 0$ and $E_b = E_{bb} = 300$ volts. The composite loadline is a straight line through this point, with a slope corresponding to $R_L' = \frac{1}{4}R_L$, where R_I = load resistance plate to plate. We may therefore imagine a composite valve, taking the place of both V_1 and V_2, working into half the primary winding with the other half open-circuited. This composite valve will have a plate resistance (r_d) as indicated by the slope of the composite characteristic, which value is approximately half that of one valve at the quiescent operating point (r_{po}). For this reason the slope of the composite characteristic changes slightly with the grid bias.

Maximum power output is obtained from the composite valve when its load resistance is equal to its plate resistance
i.e. when $R_L' = R_L/4 = r_{p0}/2$ or $R_L = 2r_{p0}$ (8)
On the composite characteristics therefore, maximum power output is obtained when the slope of the loadline is the negative of the slope of the composite characteristics.

Owing to the good linearity of the composite characteristics for Class A₁ operation, and the freedom from limitations in the vertical direction, **elliptical loadlines** may be accommodated with less distortion than with any other method. **Negative voltage feedback** makes such an amplifier practically distortionless for any type of load, resistive or reactive, of any value of impedance ; the only limitation is regarding grid current.

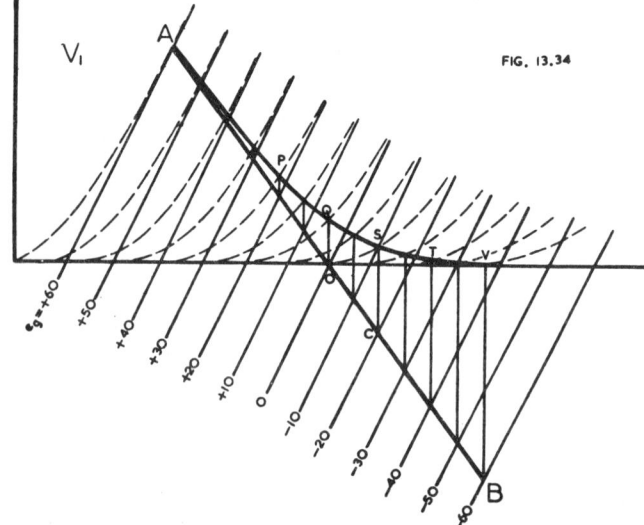

Fig. 13.34. *Method of deriving the loadline on an individual valve (AQV) from the composite characteristic AOB.*

The loadline on one valve (e.g. V_1) is determined as Fig. 13.34 in which the composite characteristics, the composite loadline AB, and the individual V_1 characteristics are the same as in Fig. 13.33. At every intersection of a composite characteristic with the composite loadline (e.g. point C) draw a vertical line (e.g. CS) to cut the corresponding V_1 characteristic. Then join these points APQSTV etc. with a smooth curve, which is the loadline on a single valve, Q being the quiescent operating point. It is obvious that this loadline is curved, although it is less curved with Class A than with Class AB₁ operation.

Equivalent circuit for push-pull amplifier
    There are various forms which an equivalent circuit might take, but the one adopted here (due to Krauss) has some special advantages for the purpose (Fig. 13.35). The two valves are assumed to have equal constant amplification factor (μ), and plate resistances $(r_{p1}$ and $r_{p2})$ which are functions of the plate current.

From equation (4) we have
$$i_d = i_{b1} - i_{b2}.$$
We may write
$$i_{b1} = I_{b0} + \Delta i_{b1} \text{ and } i_{b2} = I_{b0} - \Delta i_{b2} \tag{9}$$
where I_{b0} = quiescent plate current of either valve
  Δi_{b1} = change of plate current in valve V_1
and Δi_{b2} = change of plate current in valve V_2.
Combining (4) and (9),
$$i_d = \Delta i_{b1} + \Delta i_{b2} \tag{10}$$
in which the varying components of i_{b1} and i_{b2} add so far as i_d is concerned. Let the varying components be i_{p1} and i_{p2}, thus leading to the equivalent circuit of Fig. 13.35 in which two generators each developing a voltage μe_g through their internal resistances r_{p1} and r_{p2}, are effectively in parallel to supply the load current i_d through the load resistance R_L'. All quantities except R_L' and μ are instantaneous values.

FIG. 13.35

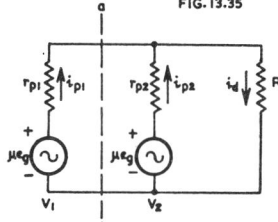

Fig. 13.35. Equivalent circuit which may be used for deriving certain impedance and current relationships.

The impedance seen by V_1 is the impedance of the circuit to the right of the line *ab*, which is called r_{ab}.
It may be shown (Ref. E7) that
$$r_{ab} = R_L'(1 + r_{p1}/r_{p2}) \tag{11}$$
The dynamic plate resistance of the composite valve may be expressed
$$r_d = \Delta e_{b1}/i_d \tag{12}$$
From (10) $i_d = \Delta i_{b1} + \Delta i_{b2}$.
Now $\Delta i_{b1} = \Delta e_{b1}/r_{p1}$ (13)
and $\Delta i_{b2} = \Delta e_{b2}/r_{p2}$ (14)
From (12), (13), (14), $r_d = r_{p1}r_{p2}/(r_{p1} + r_{p2})$ (15)
This indicates that the plate resistance of the composite valve at any instant is equal to the parallel combination of the individual plate resistances of V_1 and V_2. For Class A operation r_d is very nearly constant, so that
$$r_d \approx r_{p0}/2 \tag{16}$$
where r_{p0} = plate resistance of V_1 or V_2 at the quiescent operating point.
Now if $R_L' = r_d$ (the condition for maximum power output)
$$R_L' = r_d \approx r_{p0}/2 \tag{17}$$
The load impedance seen by a single valve (V_1 or V_2) in Class A operation is given by
[From (11), (15), (17)], $r_{ab} \approx r_{p1}$ (18)
so that each valve at any instant is working into a load resistance approximately equal to its own plate resistance.

(iii) Power output and distortion

It has been shown in the preceding subsection that maximum power output is obtained from two matched valves in push-pull when the load resistance from plate to plate (R_L) is equal to 4 times the plate resistance of the imaginary composite valve (r_d) or approximately twice the plate resistance of one of the valves at the quiescent operating point (r_{p0}). This value of R_L may be regarded as the minimum value, since any decrease would cause loss of power output, increased distortion, and high peak currents. In some circumstances it is found desirable to increase R_L—even though this reduces the power output—thereby reducing the odd harmonic distortion, the peak currents and plate dissipation.

The value of load resistance to provide maximum power output may be determined approximately from the plate characteristics of one valve (Fig. 13.36). Since the $E_c = 0$ characteristic approximately follows the 3/2 power law, it may be shown (Ref. E5) that maximum power output occurs when the loadline intersects the $E_c = 0$ curve at $0.6E_{bb}$. The plate current at this point is I_{bm} and the other values are :

$$R_L = 1.6E_{bb}/I_{bm} \text{ plate to plate} \qquad (19)$$
$$P_0 = 0.2E_{bb} I_{bm} \text{ for 2 valves} \qquad (20)$$

If it is desired to determine the optimum value of R_L in a particular case, several half-loadlines may be drawn as in Fig. 13.36 but radiating from B. The power output for each may be calculated from the expression

$$P_0 = \tfrac{1}{8}I_{bm}{}^2 R_L \qquad (21)$$
$$\text{or} \quad P_0 = \tfrac{1}{2}I_{bm}(E_{bb} - E_{min}). \qquad (22)$$

Eqns. (21) and (22) may be used with reasonable accuracy for Class AB_1, since the third harmonic distortion is usually less than 3%. A useful rule is to multiply the power output, as indicated by these equations, by the factor $10\,000/(100 - H_3\%)^2$ to obtain a close approximation to the actual power output. Values of this factor for various third harmonic percentages are given below :

H_3	1%	2%	3%	5%	7%	10%
Factor	1.02	1.04	1.06	1.11	1.15	1.23

It is here assumed that fifth and higher order odd harmonics are negligible.

The loadline slope is unaffected by the grid bias, but the two extremities are slightly affected—as the grid bias is decreased, the point A (which is really the intersection of the loadline with the composite characteristic) moves slightly towards B. Thus with Class A_1 the output will be slightly less than indicated by eqns. (20), (21) and (22).

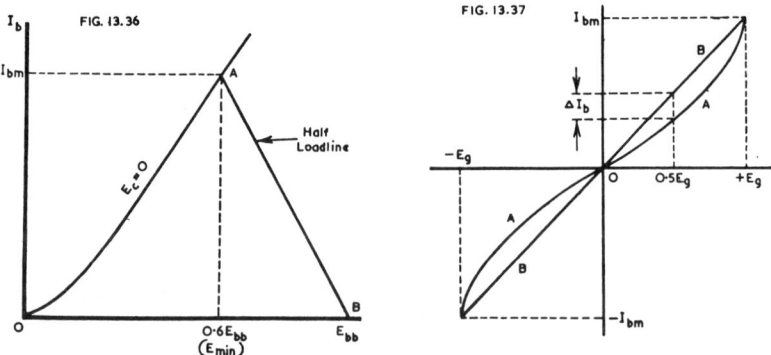

Fig. 13.36. *Method of deriving the approximate power output and load resistance of a Class AB_1 amplifier from the characteristics of a single valve. The method may also be used with poorer accuracy for Class A_1.*

Fig. 13.37. *Calculating third harmonic distortion with balanced push-pull. Curve A is the dynamic characteristic.*

The dissipation at maximum signal may be calculated from the product of the plate supply voltage E_{bb} and the average plate current [see (iv) below], minus the power output. The normal procedure is firstly to select a grid voltage such that the plate dissipation is slightly below the maximum rating at zero signal, then to adjust the loadline so that it approaches as closely as possible the value for maximum power output, without exceeding the plate dissipation limit at maximum signal.

Simple method for calculating third harmonic distortion

This method is only accurate in the absence of all distortion other than third harmonic ; it is a close approximation under normal conditions provided that all even harmonics are zero.

The procedure is to draw the loadline on the plate characteristics, then to transfer this to the mutual characteristics in the form of a dynamic characteristic (Fig. 13.37).

Curve A is the dynamic characteristic while B is a straight line joining the two ends of A and passing through O. ΔI_b is the difference in plate current between curve A and line B at one half of the peak grid voltage. The percentage of third harmonic is given by

$$\left(\tfrac{2}{3}\Delta I_b \times 100\right)\Big/\left(I_{bm} - \tfrac{2}{3}\Delta I_b\right) \tag{23}$$

or $\left(\tfrac{2}{3}\Delta I_b \times 100\right)/I_{bm}$ approximately. (24

Calculating up to fifth harmonic distortion

The usual method is to transfer from the loadline to a dynamic characteristic, then to proceed using the " eleven selected ordinate method " (Sect. 3(iv)D). This gives the power output and harmonics up to the fifth. If the balance is good, only the third and fifth harmonics are likely to be required ; the fifth is very much less than the third harmonic for Class A operation.

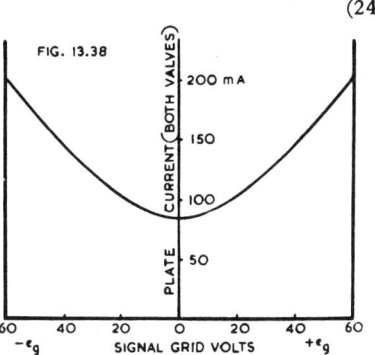

FIG. 13.38

Fig 13.38. Plate· current (both valves) plotted as a function of the signal grid voltage.

(iv) Average plate current

The average plate current with maximum signal input is always greater than under quiescent conditions—slightly greater for " single valve " conditions, more so for limiting Class A_1, and considerably greater for Class AB_1.

The average plate current may be calculated approximately by the expression

$$I_b \approx I_{b0} + \tfrac{1}{4}(I_{max} + I_{min} - 2I_{b0}) \tag{25}$$

provided that the plate current does not actually cut off (Ref. E1).

More generally, and more accurately, the average plate current may be determined by adding the plate currents of the two valves instead of subtracting them as for the composite characteristics. The total plate current may then be plotted as in Fig. 13.38 as a function of the signal grid voltage. In order to find the average current it is generally most convenient to take equal angle increments over the cycle, for example, every 10˙ as shown in Fig. 13.39. The plate current should then be noted at each point corresponding to 10 increase in angle over the whole 360 . The average plate current is then the average of these individual values.

In order to reduce the amount of work involved in this calculation, use may be made of the fact that each quadrant (90˙) is similar. It is very easy to make an error in this calculation and the following method of obtaining the average from the plate current curve over one quadrant is therefore given.

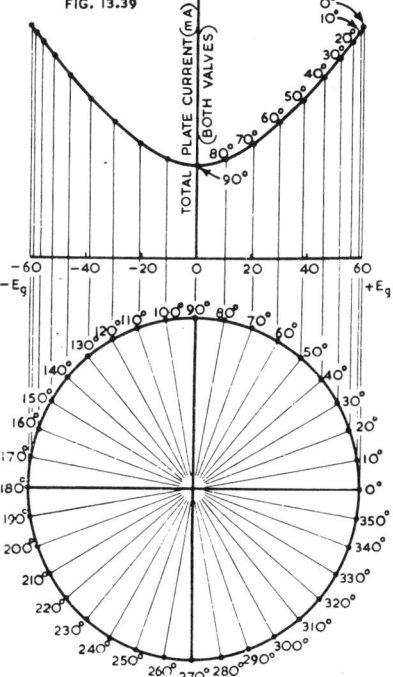

FIG. 13.39

Fig. 13.39. Plate current (both valves) with 10˙ angular increments.

This method is illustrated in Fig. 13.40 which shows the total plate current for both valves over one-quarter of a cycle.   Grid voltages are shown as fractions of the peak grid voltage.   The plate currents corresponding to grid voltages of 0, 0.17, 0.34, 0.5, 0.64, 0.77, 0.87, 0.94, 0.98 and 1.0 times the peak voltage are shown as $I_0, I_{0.17}$ etc. The average plate current (I_{av}) is then given by

$$I_{av} = \tfrac{1}{9}(\tfrac{1}{2}I_0 + I_{0.17} + I_{0.34} + I_{0.5} + I_{0.64} + I_{0.77} + I_{0.87} + I_{0.94} + I_{0.98} + \tfrac{1}{2}I_{1.0}). \qquad (26)$$

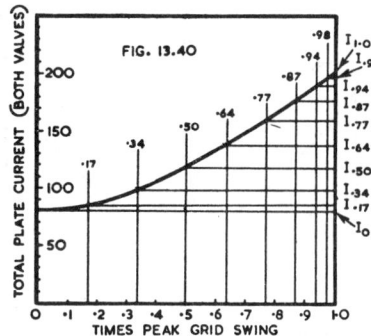

Fig. 13.40.   *Plate current (both valves) plotted over one quadrant for the calculation of average plate current.*

(v) Matching and the effects of mismatching

Matching is the process of selecting valves for satisfactory push-pull operation. Most valve manufacturers are prepared to supply, at some additional cost, valves which have been stabilized and matched for the application required.

It is important to specify the conditions of operation when ordering matched valves, as unless the valves are stabilized and matched under conditions similar to those in which the valves are to be used, they will drift apart during life or may even be mismatched initially if the matching conditions are unsuitable.

All valves intended for matching should be operated for at least 50 hours under similar conditions to those under which the valves are intended to be operated in the amplifier.

The matching technique varies with the class of operation for which the valves are intended.

Matching valves for class A_1 service is not as critical as for valves intended for class AB_1, class AB_2 or class B.   For class A_1 service it is usually sufficient to match for zero signal plate current only.

Triodes for class AB_1, AB_2 or class B service should be checked at a number of points on the plate current grid bias curve.   The points usually taken are (a) zero signal condition (b) a bias corresponding to the maximum permissible plate dissipation.   The plate currents so measured should agree at all points within 2%. Triodes intended for class AB_2 or class B service should also be matched for amplification factor.

When matching tetrodes and pentodes it is usually sufficient to match for zero signal plate current and power output.

Even with perfect initial matching, valves are likely to drift apart during life and in critical applications it is desirable to provide some means of balancing the plate currents of the valves in the equipment.   This may take the form of separate bias resistors in the case of self-bias or adjustable bias supplies in the case of fixed-bias applications.   Pentodes and tetrodes may also be balanced by adjusting the screen voltages.

It is important that matched valves should never be run, even momentarily, at dissipations or ratings in excess of those recommended by the valve manufacturers as such treatment will render the valves unstable and destroy the matching.

Effects of mismatching with Class A_1 triodes

In Class A_1 push-pull triodes, a considerable degree of mismatching between the valves is permissible without serious effects, provided that the valves are being operated under single valve conditions as regards grid bias.   There will be only a slight effect on the maximum power output or the odd harmonic distortion, but there will be some second harmonic distortion and some out-of-balance flux-producing current in the transformer.   The second harmonic distortion will normally be small, particularly if the plate-to-plate load resistance is not much less than $4r_{p0}$, and for most purposes it may be neglected with valves of ordinary tolerances.   The maximum out-of-balance plate current should be provided for in the design of the output transformer and this additional cost should be compared with the alternative additional cost of using valves which have been stabilized and matched for quiescent plate current only.   There is normally no real need for matching for any other characteristic.

In order to demonstrate the effects of abnormal mismatching, composite characteristics have been drawn in Fig. 13.41 for two valves of entirely different types.   It has been shown by Sturley (Ref. E27) that the method of deriving composite characteristics also holds with mismatching.

Type	2A3	45
Amplification factor	4.2	3.5
Mutual conductance	5250	2175 μmhos
Plate resistance	800	1610 ohms

There is a difference of 12% in amplification factor, while there is a ratio exceeding 2 : 1 for the other characteristics.   The selected operating conditions are : $E_b = 250$ volts, $E_c = -50$ volts, $R_L = 5800$ ohms (plate-to-plate).   It will be seen that the composite characteristics (dashed lines) are not quite parallel, although they are very nearly straight.   On account of the unmatched condition, rectification occurs, leading to a shift of the loadline, and the corrected loadline may be derived by the method of Sect. 2(ii)D and Fig. 13.7.   The second harmonic distortion on the corrected loadline is only 5%.

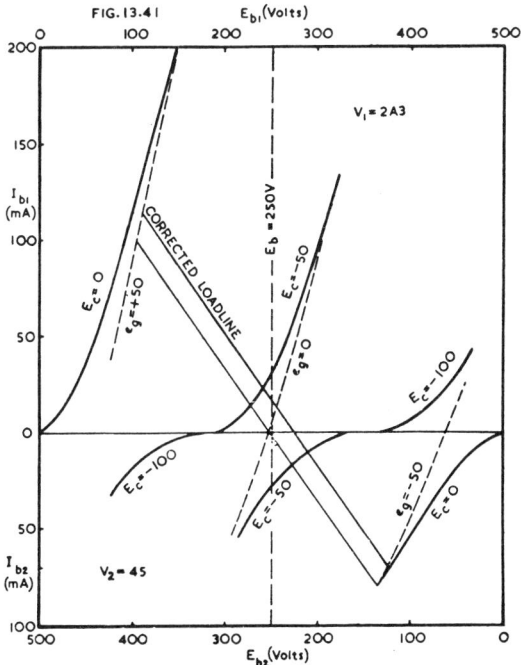

Fig. 13.41.   *Composite characteristics for two different valves (2A3 and 45) in Class A_1 push-pull. $E_{bb} = 250$ volts, $E_c = -50$ volts, $R_L = 5800$ ohms plate-to-plate.*

The fact that two valves so different in characteristics can be used in Class A_1 push-pull to give what is generally classed as good quality (5% second harmonic) indicates the wide latitude permissible, provided that the bias is retained at the value for a single valve and also that the load resistance is not much less than the sum of the single valve loads (or 4 times the average plate resistance).

For Class A_1 triodes it has been suggested (Ref. E8) that the following are reasonably satisfactory limits in conjunction with ordinary commercial tolerances for valves :

	Max. unbalance
Signal input voltages on the grids of V_1 and V_2 :	5%
Phase unbalance at high and low frequencies :	
Quadrature component	= 3%

These are easily achieved by attention to the phase splitter or other source (see Chapter 12 Sect. 6). When testing for balance with a C.R.O. on the grids of V_1 and V_2, it is important to reverse the connections of the B.F.O. when changing from one grid to the other.

(vi) Cathode bias

In an accurately balanced push-pull Class A_1 amplifier there is no point in by-passing the common cathode bias resistor, since there is no fundamental signal current flowing through it. In Class A_1 amplifiers which are not accurately balanced, there will be some degeneration, and it is usual to by-pass the cathode bias resistor, although this is not essential.

In Class AB_1 amplifiers it is essential to by-pass the cathode bias resistor.

Provision may be made, with cathode bias, for balancing the plate currents provided that they do not differ too seriously. One excellent arrangement is incorporated in Fig. 7.44 which may be used, with the necessary adjustments in the values of the resistances, for triodes, pentodes or beam power amplifiers.

The value of cathode bias resistance may be determined, in the same way as for a single valve, on the basis of the maximum signal total plate current and desired bias voltage. This will give the same performance as fixed bias, but it is necessary to check for plate dissipation at zero signal [for method see Sect. 2(ii)E]. If the dissipation at zero signal is too great, it will be necessary to increase the bias resistance. This will introduce a tendency to change from Class A_1 to AB_1, which may be undesirable ; it may be minimized by the use of a bleed resistor to pass current from E_{bb} to the cathode, and thence through the bias resistor, thus giving an approach towards fixed bias. Alternatively, the load resistance may be increased, thereby reducing the maximum signal plate current and grid bias ; this will also reduce the power output.

Cathode bias causes a smaller change in average plate current from no signal to maximum signal than fixed bias. This permits a poorer regulation power supply than may be used with fixed bias. However, the regulation and by-passing of the power supply also affect the minimum frequency which can be handled satisfactorily at full power output—see Sect. 1(iv).

Changes in effective gain occur in Class AB_1 amplifiers employing cathode bias, during heavy low frequency **transients**, which add to the distortion measured under steady conditions.

(vii) **Parasitics**

Parasitic oscillations in the plate circuit may occur with Class AB_1 operation when the plate current is cut off for an appreciable part of the cycle, as a result of the transformer leakage inductance and the rapid rate of change of current at the cut-off point—see Sect. 5(i)C. They may usually be cured by the use of a RC network shunted across each half of the primary of the output transformer—see Sect. 7(i)—and, if necessary, by the use of a transformer with lower leakage inductance—see Chapter 5 Sect. 3(iii)c.

Parasitics in the grid circuit are not usually troublesome except when the valves are driven to the point of grid current flow. Grid stoppers up to 50 000 ohms are often used with both Class A and AB_1 operation to give a smoother overload without parasitics.

SECTION 6 : PUSH-PULL PENTODES AND BEAM POWER AMPLIFIERS, CLASS A, AB_1

(i) Introduction   (ii) Power output and distortion   (iii) The effect of power supply regulation   (iv) Mismatching   (v) Average plate and screen currents   (vi) Cathode bias   (vii) Parasitics   (viii) Phase inversion in the power stage   (ix) Extended Class A   (x) Partial triode (" ultra-linear ") operation.

(i) Introduction

Push-pull pentodes follow the same general principles as triodes (see Sect. 5) although there are some special features which will here be examined.   Provision must be made for the supply of voltage to the screens, and for it to be maintained constant with respect to the cathode.   For economy, the screens are often operated at the same voltage as the plates.   If the screens are to be operated at a lower voltage, the alternatives are the use of a separate power supply, or a low resistance voltage divider. The McIntosh Amplifier may also be used for Class A or AB_1 operation—see Sect. 8. Quiescent push-pull pentodes are covered in Sect. 7(vii).

(ii) Power output and distortion

Composite characteristics may be drawn as for triodes, on the assumption that the screen voltage is maintained constant, but they will not be straight (Fig. 13 42).   The loadline will pass through the point $(E_b = E_{bb}, I_b = 0)$ and the two knees of the characteristics $(e_g = \pm E_{c1})$.

In practice it is not necessary to draw the composite characteristics, if we are only interested in power output.   In limiting Class A_1 or in AB_1 one valve reaches cut-off, so that the $E_{c1} = 0$ characteristic is the same as the e_g characteristic.   Even with ordinary Class A_1 the error due to the approximation is small.   The loadline may therefore be drawn (AB in Fig. 13.43) and the plate-to-plate load resistance will be 4 times that indicated by the slope of AB.   With valves of the 6L6 class, the third

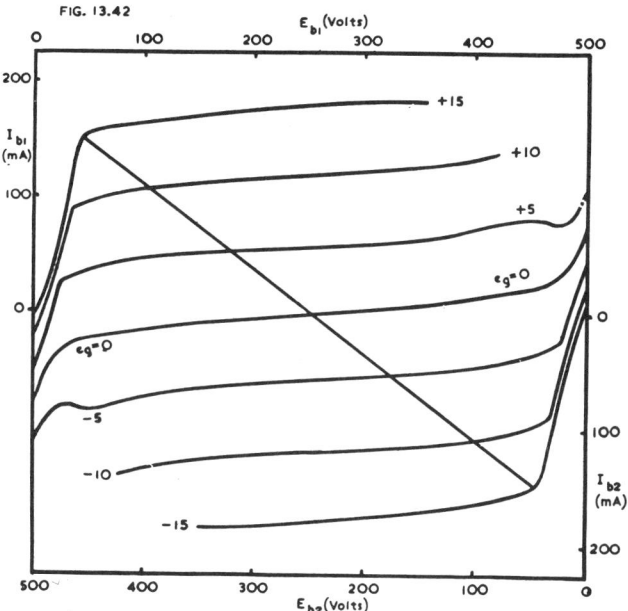

FIG. 13.42

Fig. 13.42.   *Composite characteristics for push-pull Class A_1 beam power amplifiers type* 6L6, $E_{bb} = 250$ *volts,* $E_{c2} = 250$ *volts,* $E_{c1} = -15$ *volts.*

harmonic is so small that its effect on the power output may be neglected, so that
$$P_0 = \tfrac{1}{2} I_{bm}(E_{bb} - E_{min}) \qquad (1)$$
which is the same as for triodes. With pentodes the power output will be somewhat higher than indicated by eqn. (1) owing to the third harmonic distortion. In general,

the third harmonic distortion is slightly less than half that with a single valve, owing to the effect of the lower load resistance. If it is desired to calculate the harmonic distortion, it will be necessary to plot at least portion of the composite characteristics.

The effect of a higher load resistance is to increase rapidly the odd harmonic distortion, while the effect of a lower load resistance is to decrease the power output. It is therefore advisable, with a loudspeaker load, to adopt a nominal impedance rather less than the value for maximum power output.

The power output of Class A push-pull pentodes is only slightly greater than twice that for a single valve.

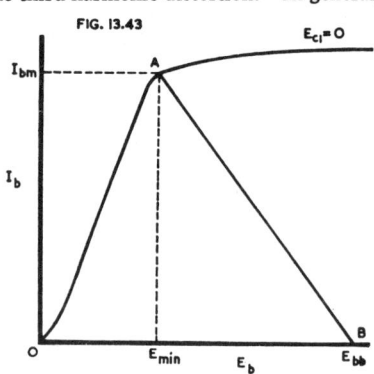

FIG. 13.43

Fig. 13.43.   Calculation of power output using $E_{c1} = 0$ characteristic of one valve only.

(iii) The effect of power supply regulation

As with the case of a single valve the screen is more sensitive to voltage changes than the plate. If neither screen nor plate is being operated at its maximum rating, the simplest procedure is to adjust the voltages as desired for maximum signal and allow them to rise with no signal. Alternatively, if it is desired to obtain maximum power output, both screen and plate may be adjusted to their maximum ratings with no signal, and allowed to fall with increasing signal. If it is desired to calculate the maximum power output, the procedure for a single valve [Sect. 3(ix)] may be followed, except that in this case there is no loadline shift caused by rectification.

The regulation of a common plate and screen power supply will not have a serious effect on the minimum frequency which can be handled satisfactorily at full power output with Class A push-pull operation, provided that a reasonably large by-pass capacitor is used. If separate plate and screen supplies are used the screen supply regulation is very much more important than the plate supply regulation, with Class A operation.

With Class AB_1 operation, the regulation of both plate and screen supplies should be good.

(iv) Matching and the effects of mismatching

Matching is covered generally in Sect. 5(v). It is important to match the valves under the operating conditions in the amplifier.

The effects of mismatching with pentodes are more serious than with triodes. If no care is taken in matching, or in the design of the output transformer, it is possible for the distortion to be higher than with two valves in parallel. The advantages of push-pull operation will only be obtained in proportion to the care taken to achieve correct balance, particularly with regard to the quiescent plate currents and the signal input voltages. As with triodes, Class AB_1 is more sensitive to mismatching than Class A_1.

(v) Average plate and screen currents

The average plate current for Class A_1 may be calculated approximately as for triodes [Sect. 5(iv), Eqn. 25]. In all other cases, the composite characteristics are required, following the same method as for triodes.

The average screen current may be calculated by the same method as for single pentodes [Sect. 3(iii)E].

(vi) Cathode bias

See Sect. 5(vi) as for push-pull triodes, except that the screen dissipation must also be checked.

(vii) Parasitics

See Sect. 5(vii) as for push-pull triodes, also Sect. 3(iii)H.

(viii) Phase inversion in the power stage

In the interests of economy, push-pull is sometimes used in the output stage without a prior phase inverter. All such methods—except the Cathamplifier—have inherently high distortion, and some have serious unbalance between the two input voltages.

(A) Phase inverter principle (Fig. 13.44)

The grid of V_2 is excited from the voltage divider $R_3 R_4$ across the output of V_1. $R_3 + R_4$ must be very much greater than the load resistance (say 50 000 ohms). R_5 and R_6 are grid stoppers. All other components are normal. R_k may be by-passed if desired.

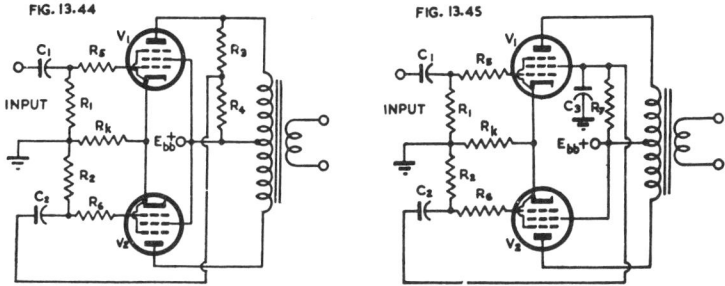

Fig. 13.44. Push-pull circuit using phase inversion in the power stage.
Fig. 13.45. Push-pull circuit using screen resistance coupling from V_1 to the grid of V_2.

The signal voltage on the grid of V_2 must first pass through V_1 where it is distorted, then through V_2 where it will be distorted again. Thus the second harmonic will be the same as for a single valve, and the third harmonic will be approximately twice the value with balanced push-pull. The balance, if adjusted for maximum signal, will not be correct for low volume, owing to the third harmonic " flattening."

(B) Screen resistance coupling (Fig. 13.45)

This is a modification of (A) being an attempt to obtain from the screen a more linear relationship than from the plate. No comparative measurements have been published. R_7 may be about 1500 ohms for type 6V6-GT or 2500 for type 6F6-G, with $E_b = E_{c2} = 250$ volts—the exact value should be found experimentally ; C_3 may be 0.002 μF. For better balance an equal screen resistor might be added for V_2. Ref. E10.

(C) Common cathode impedance (Fig. 13.46)

R_1 and R_2 in series provide a common cathode coupling impedance [see Chapter 12 Sect. 6(vi)]. R_2 may have a value of, say, 1000 ohms to give an approach towards balance, but necessarily must carry the plate currents of both valves—say, 70 or 80 mA—and will have a voltage of, say, 70 to 80 with a dissipation around 6 watts. Care should be taken to avoid exceeding the maximum heater-cathode voltage rating.

See Reference E23.

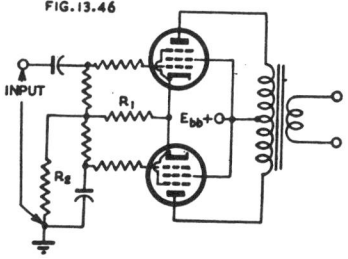

Fig. 13.46. Push-pull circuit with common cathode impedance coupling.

(D) The Parry " Cathamplifier "

The basic circuit is Fig. 13.46A, the two cathodes being coupled by a centre-tapped a-f transformer, whose secondary winding excites the grid of V_2. A theoretical analysis is given in Ref. E29, while some practical designs are in Ref. E30.

For balance, $\dfrac{N_3}{N_1 + N_2} = \dfrac{1 + g_m R/2}{g_m R}$ and $N_1 = N_2$

where g_m = mutual conductance of V_2.

Distortion is reduced by the factor $T(2T - 1)$
where $T = N_3/(N_1 + N_2)$.

Note that T should normally be slightly greater than 1.

Gain is reduced by the factor $T(T - 0.5)$.

The common cathode resistor R_0 helps to reduce unbalance.

In practice, R is made variable (say 100 ohms total) so as to permit the amplifier to be balanced experimentally. One method is to connect a valve voltmeter across R_0, and to adjust R for minimum reading.

Instability may occur if R is too small.

A modified circuit is Fig. 13.46B in which the centre-tapped primary of T_1 is not necessary.

Fig. 13.46C permits both a.c. and d.c. balancing.

Fig. 13.46D keeps the circulating screen current out of the cathode circuit and so maintains the ratio between plate and screen currents at the negative voltage peak swing. Resistors R_1 are to prevent coupling from cathode to cathode through the screen by-pass condensers ; their values should be low—say 100 to 250 ohms each.

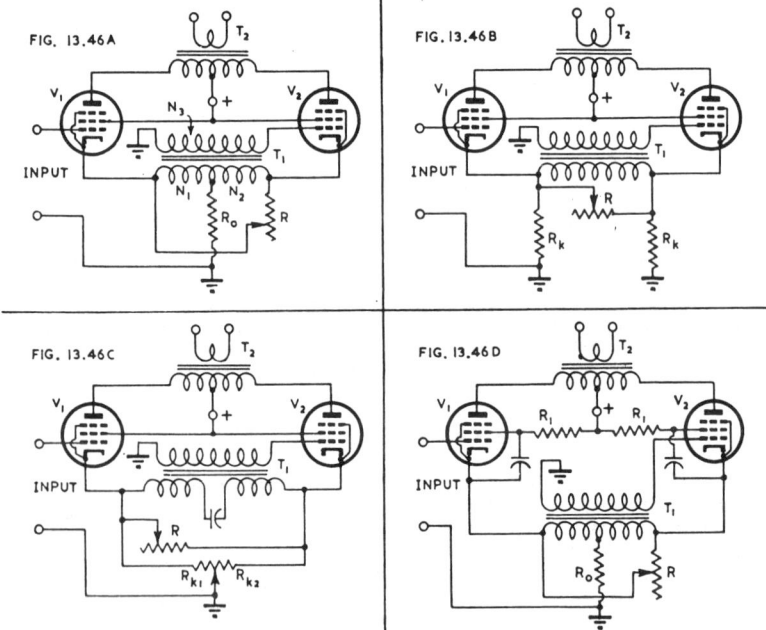

Fig. 13.46A. Basic circuit of Parry Cathamplifier, (B) Modified circuit, (C) With both a.c. and d.c. balancing, (D) Keeps circulating screen current out of cathode circuit (Ref. E30).

(ix) Extended Class A

Extended Class A is the name given to a push-pull amplifier using a triode and a pentode in parallel on each side. The amplifier operates entirely on the push-pull triodes at low levels, with the pentodes (or beam power amplifiers) cut off ; at high levels the output is mainly from the pentodes. Consequently there is some curvature in the linearity (" transfer ") characteristic at the transition point. The total dissipation is, however, only about one third of that of a Class A amplifier with the same maximum power output (Ref. E31).

This principle might also possibly be applied to any Class AB or Class B amplifier merely to avoid current cut-off in the transformer and its resultant parasitics (Ref. E13) unless one with very low leakage inductance is used.

(x) Partial triode (" ultra-linear ") operation

See page 570.

SECTION 7 : CLASS B AMPLIFIERS AND DRIVERS

(i) Introduction (ii) Power output and distortion—ideal conditions—Class B_1 (iii) Power output and distortion—practical conditions—Class B_2 (iv) Grid driving conditions (v) Design procedure for Class B_2 amplifiers (vi) Earthed-grid cathode coupled amplifiers (vii) Class B_1 amplifiers—quiescent push-pull.

(i) Introduction

A Class B amplifier is an amplifier in which the grid bias is approximately equal to the cut-off value, so that the plate current is approximately zero when no signal voltage is applied, and so that the plate current in a specific valve flows for approximately one half of each cycle when an alternating signal voltage is applied.

Class B amplifiers are in two main groups—firstly Class B_1 (otherwise known as quiescent push-pull, see below) in which no grid current is permitted to flow, secondly Class B_2 (generally abbreviated to Class B) in which grid current flows for at least part of the cycle.

Class B_2 amplifiers have inherently high odd harmonic distortion, even when the utmost care is taken in design and adjustment. This distortion frequently has a maximum value at quite a low power output, making this type of amplifier unsuitable for many applications. They are also comparatively expensive in that a driver valve and transformer together with two output valves form an integral part of the stage, and together give only the same order of sensitivity as a pentode.

The one outstanding advantage of a Class B_2 amplifier is in the very high plate-circuit efficiency, although the current drawn by the driver stage should be included. The principal applications are in battery-operated amplifiers, public address systems and the like

The grid bias must be fixed (either battery or separate power supply) and special high-mu triodes have been produced to permit operation at zero bias to avoid the necessity for a bias supply. In the smaller sizes, twin triodes are commonly used. For the best results, accurate matching of the two valves is essential. If they are being operated at a negative bias it is possible to match their quiescent plate currents by adjusting the bias voltages separately. If the dynamic characteristics are not matched, a difference of 10% in the two plate currents, measured by d.c. milliammeters under operating conditions, will produce roughly 5% second harmonic distortion (Ref. E18). The matching of valves is covered in Sect. 5(v).

Well regulated plate, screen (if any) and bias supplies (if any) are essential.

The rate of change of current in each half of the output transformer at the plate current cut-off point is considerable, often resulting in parasitics

Parasitics in the grid circuit may be eliminated by the use of a driver transformer having low leakage inductance [Chapter 5 Sect. 3(iii)a] and, if necessary, by connecting a small fixed condenser from each grid to cathode—a typical value is 0.0005 μF—see (iv) below.

Parasitics in the plate circuit may be eliminated by the use of an output transformer with low leakage inductance—see Chapter 5 Sect. 3(iii)c—together with a series resistance-capacitance network connected across each half of the primary or alternatively from plate to plate.

Typical values for connection across each half of primary :

Load resistance (p-p)	6000	10 000	12 000	ohms
R	3300	5600	6800	ohms
C	0.05	0.05	0.05	μF

Typical values for connection from plate to plate :

Load resistance (p-p)	6000	10 000	12 000	ohms
R	6800	12 000	15 000	ohms
C	0.03	0.02	0.02	μF

Note—The McIntosh Amplifier may also be used for Class B operation—see Sect. 8(iii).

(ii) Power output and distortion—ideal conditions—Class B_2

(Circuit Fig. 13.31 ; characteristics Fig. 13.36.)

If the input voltage is sinusoidal, the output transformer ideal, the plate characteristics equidistant straight lines, and the valves biased exactly to cut-off, the operating conditions will be—

Power output (total) : $P_0 = \frac{1}{2}I_{bm}(E_{bb} - E_{min})$ (1)

or $P_0 = \frac{1}{8}I_{bm}^2 R_L = 2(E_{bb} - E_{min})^2/R_L$ (2)

where I_{bm} = maximum (peak) plate current of either valve

        E_{min} = minimum plate-voltage of either valve

and R_L = load resistance plate-to-plate.

Load resistance (plate-to-plate) : $R_L = 4(E_{bb} - E_{min})/I_{bm}$ (3)

Load resistance per valve : $R_L' = R_L/4 = (E_{bb} - E_{min})/I_{bm}$ (4)

Maximum power output is obtained when $R_L = 4r_p$

                 i.e. when $E_{min} = 0.5E_{bb}$

where r_p = plate resistance of one valve.

Average plate current (each valve) : $I_b = I_{bm}/\pi \approx 0.318I_{bm}$ (5)

Power input from plate-supply : $P_b = 2E_{bb}I_b \approx 0.637E_{bb}I_{bm}$ (6)

Plate circuit efficiency : $\eta = (1 - E_{min}/E_{bb}) \times 0.785$ (7)

Plate dissipation : P_p = d.c. power input − power output

             = $I_{bm}(0.137E_{bb} + 0.5E_{min})$ (8)

Power output in terms of plate dissipation and plate circuit efficiency :

             $P_0 = P_p\eta/(1 - \eta)$ (9)

The current in the secondary of the transformer will be sinusoidal, there being no distortion. The plate current in each valve will have the form of a rectified sine wave.

The slope of the composite characteristic is half that for Class A operation, so that the slope of the composite plate resistance will be twice that for Class A.

(iii) Power output and distortion—practical conditions—Class B_2

The practical treatment is based on the use of the composite characteristics [Sect. 5(ii)D]. If the valves are biased completely to cut-off, each half of the composite characteristic is identical with the individual valve characteristic, and a considerable degree of non-linearity occurs in the middle region (Fig. 13.47A). As a result, the greatest distortion of the loadline occurs with fairly small signals. For any particular characteristic there is one value of grid bias beyond which the distortion increases rapidly with increase of bias—there is a small quiescent plate current at this point (Fig. 13.47B). Even in this case the whole of the composite characteristic, except the small middle portion, is identical with the single valve characteristic.

It is therefore practicable to calculate the **power output** from the characteristics of a single valve (e.g. Fig. 13.48). The power output as calculated by eqns. (1) and (2) is modified by the presence of harmonics and should be multiplied by the factor

      $F = (1 + H_3 - H_5 + H_7 - H_9 + H_{11})^2$ (10)

where H_3 = third harmonic distortion (i.e. I_3/I_1),

and H_5 = fifth harmonic distortion (i.e. I_5/I_1), etc.

The various harmonics are unpredictable, but at maximum power output H_3 usually predominates, and a purely arbitrary approximation is to take

$F = (1 + 0.6H_3)^2$.

Usual values of H_3 at maximum power output vary from 0.05 to 0.10, so that F varies from 1.06 to 1.12, a reasonable design value being 1.08. Thus, from eqn. (2),

$$P_0 \approx 0.135I_{bm}{}^2R_L \qquad (11)$$

Alternatively, the total power output may be calculated from a knowledge of the average and quiescent currents of one valve, with an error not exceeding 9% (see eqns. 12 and 13 below)—

$$P_0 \approx 2.47(I_b - 0.25I_{b0})^2R_L \qquad (12)$$

The more accurate method is to use the composite characteristics (which are the same as the single valve characteristics) of Fig. 13.48 over the range concerned) and to apply the " eleven selected ordinate method " [Sect. 3(iv)D] for the determination of power output and distortion. Although harmonics higher than the fifth are appreciable, it is difficult to calculate them graphically with any degree of accuracy.

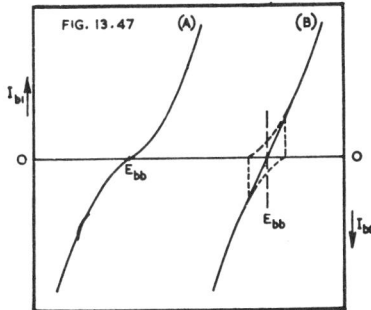
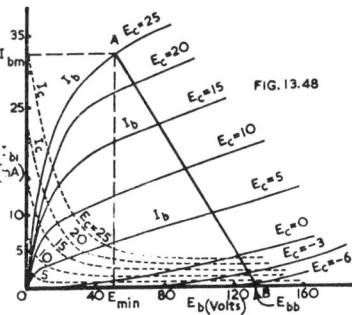

Fig. 13.47. *Composite characteristics of Class B amplifiers (A) biased to cut-off (B) biased to the point of minimum distortion.*

Fig. 13.48. *Plate and grid characteristics of a typical small battery-operated Class B triode (1J6–G).*

The **load resistance** may be calculated by eqn. (3) or (4).

Maximum power output is usually achieved when the slope of the loadline is numerically equal to the slope of the individual valve characteristic corresponding to the peak grid voltage, at its point of intersection with the loadline. A different loadline is usually necessary for each value of peak grid voltage. Each loadline should be checked for peak current, average current, plate dissipation and grid driving power.

When any loadline has been established as permissible, a higher load resistance may be used with perfect safety without any further checking. This will require less grid driving power, and will draw less plate current, but will give a lower power output; the distortion will be roughly unchanged, and the plate circuit efficiency will be higher.

The **average plate current** may be calculated by the accurate method of Sect. 5(iv) and eqn. (26). Alternatively, with an accuracy within 3% for third harmonic not exceeding 10% (Ref. E11)—

(for each valve) $I_b \approx 0.318I_{bm} + 0.25I_{b0}$ $\qquad (13)$

Eqn. (13) may also be put into the form—

$I_{bm} \approx 3.14I_b - 0.785I_{b0}$ $\qquad (14)$

The average power input from the plate supply is given by

$P_b = 2E_{bb}I_b \approx E_{bb}(0.637I_{bm} + 0.5I_{b0})$. $\qquad (15)$

Plate circuit efficiency : $\eta = P_0/2E_{bb}I_b$ $\qquad (16)$

[Usual values of plate circuit efficiency are from 50% to 60%.]

Plate dissipation : $P_p \approx E_{bb}(0.637I_{bm} + 0.5I_{b0}) - P_0$ $\qquad (17)$

With a plate circuit efficiency of 60%, the power output is 1.5 times the total plate dissipation, whereas with Class A operation and a plate circuit efficiency of 25%, the power output is 0.25 times the plate dissipation. Thus, if in both cases the plate dissipation is the only limiting factor, six times more output may be obtained from the same valves in Class B than in Class A.

(iv) Grid driving conditions

Grid current characteristics are provided for use with valves suitable for Class B amplification (e.g. Fig. 13.48). It is usual practice to select a likely peak positive grid characteristic and loadline, then to calculate the grid driving power and power output. This may be repeated for several other loadlines, and the final choice will be made after considering all the relevant features.

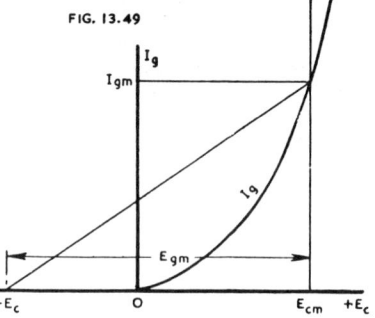

FIG. 13.49

In the case of a particular loadline such as AB in Fig. 13.48, the peak grid current may be determined by noting the intersection of the vertical through A and the grid current curve (shown with a broken line) corresponding to the E_c characteristic which passes through A. This is shown in Fig. 13.49 where $- E_c$ indicates the fixed grid bias and the positive grid current commences to flow at approximately $E_c = 0$. The peak grid current is I_{gm} and the peak signal grid voltage is E_{gm}, corresponding to

Fig. 13.49. Typical grid mutual characteristic illustrating Class B operating conditions.

a positive grid bias voltage E_{cm}. The peak grid input power is then given by

$$P_{gm} = E_{gm}I_{gm} \qquad (18)$$

and the minimum grid input resistance is given by

$$r_{gmin} = E_{gm}/I_{gm} \qquad (19)$$

[Note : The minimum variational grid resistance is derived from the slope of the I_g curve—it is always less than r_{gmin}.]

The driver valve has to supply this peak power, plus transformer losses, into a half-secondary load varying from infinity at low input levels to r_{gmin} at the maximum input. The basic driver circuit is shown in Fig. 13.50A, where V_1 is the driver valve and T_1 the step-down transformer with a primary to half-secondary turns ratio N_1/N_2. In practice, the transformer has losses, and it may be represented by the equivalent circuit Fig. 5.9 (omitting C_H).

The efficiency is usually calculated (or measured) at 400 c/s where r_1, r_2 and R_0 are the principal causes of loss. An efficiency of from 70 to 80% at peak power is typical of good practice. The voltage applied to the grid of V_2, provided that the iron losses are small, is approximately [Chapter 5 Sect. 2(ii)]

$$E_{g2} \approx E_p \eta(N_2/N_1) \qquad (20)$$

where η = percentage efficiency $\div$ 100,
and E_p = voltage applied across primary.

The load presented to V_1 is roughly as calculated for an ideal transformer [Chapter 5 Sect. 2(ii)] driving the same grid.

At high audio frequencies, the effects of L_1 and L_2 (Fig. 5.9) become appreciable, causing an additional loss of voltage during the time of grid current flow, and a tendency towards instability. The instability is brought about by a negative input resistance which some valves possess over a portion of the grid characteristic [Chapter 2 Sect. 2(iii)]. It is therefore important to reduce the leakage inductance of the transformer to the lowest possible value. Fortunately the bad effects of any remaining leakage inductance may be minimized by connecting a small condenser across each half of the secondary so as to resonate at a frequency about 1.5 times the highest frequency to be amplified (Ref. E18). The condenser and leakage inductance form a half-section of a simple two-element constant k type low-pass filter [Chapter 4 Sect. 8(vii)].

At low audio frequencies the primary inductance L_0 causes loss of gain as with any a-f transformer. It should be designed for an unloaded secondary (Chapter 5 Sect. 3) because it is unloaded for part of the cycle. Secondary loading resistances should be avoided with Class A triode drivers (Refs. E17, E18).

In order to avoid excessive distortion due to the non-linear grid current characteristic, the effective impedance looking backwards from the grid-cathode terminals towards the driver, must be small compared with the effective input impedance of the valve. The " looking backwards " impedance has as its components (Fig. 5.9) :

" Driver resistance " $R' = r_2 + (N_2/N_1)^2(r_{p1} + r_1)$ (21)

" Driver inductance " $L' = L_2 + (N_2/N_1)^2 L_1$ (22)

It is usual to restrict the driver resistance R' to a value less than $0.2 \, r_{g\,min}$ for the most favourable conditions with high-mu valves and with limited grid drive, where $r_{g\,min}$ does not differ seriously from the minimum variational grid resistance ; in other cases it should be less than one fifth of the minimum variational grid resistance (Refs. E2, E12). For minimum distortion, R' should be made as low as practicable. Hence the transformer usually has a step-down ratio from primary to half-secondary.

The " driver inductance," if large and uncompensated, produces an effect like a faint high pitched hiss or scratch that rises and falls with the signal (Ref. E18).

The effect of the varying load resistance on the driver valve is shown by the curved loadline in Fig. 13.50B. The horizontal portion corresponds to the conditions without grid current flow, while the slope at any point corresponds to the variational grid resistance at that point. The broken line joining the two extremities has a slope corresponding to a resistance of $r_{g\,min}(N_1/N_2)^2$; this should not be less than $4r_{p1}$. The driver valve should preferably be operated near its maximum rated plate voltage and dissipation.

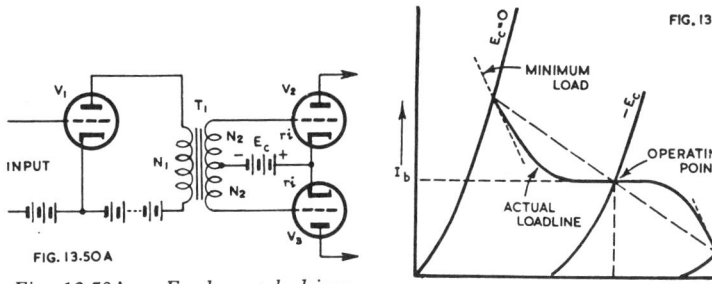

Fig. 13.50A. Fundamental driver and Class B (or AB) amplifier circuit.

Fig. 13.50B. Loadline on a typical triode driver.

As the output valves are driven harder, the driver valve is called upon to provide increased power into a decreased minimum grid resistance ; this necessitates a greater step-down ratio in the transformer. The extreme limit is when the " diode line " is reached. As a guide which may be used for a first trial, the minimum plate voltage may be taken as twice the " diode line " voltage at the peak current level. The diode line is shown in many characteristic curves, and is the envelope of the characteristics—it is the line where the grid loses control, in other words the minimum plate voltage for any specified plate current no matter how positive the grid may be.

The cathode follower makes an excellent driver for all forms of amplifiers drawing grid current, having very low effective plate resistance and low distortion—see Chapter 7 Sect. 2(i).

A cathode follower has been applied as a transformerless driver, one 6SN7-GT twin triode being used as a push-pull driver to the grids of two 6L6 valves giving an output of 47 watts (Ref. E22). However, this is not usually the most efficient arrangement—see Chapter 7 Sect. 1(ii).

(v) Design procedure for Class B_2 amplifiers

The major difficulty is in keeping the electrode voltages stabilized.

The following procedure applies to the case where the voltages are stabilized and the plate voltage and desired power output are known.

1. Assume a value of plate-to-plate load resistance R_L. Draw a loadline with a slope corresponding to $R_L/4$ on the characteristics for one valve, through the point $E_b = E_{bb}$, $I_b = 0$.

2. Select a value of grid bias to give the minimum distortion of the composite characteristic (Fig. 13.47).

3. Determine the quiescent plate current I_{b0} of one valve.

4. Calculate the peak plate current from the approximation :

$$I_{bm} \approx \sqrt{7.4 P_0 / R_L} \qquad (23)$$

5. Calculate the average power input from the plate supply, using eqn. (15).

6. Calculate the plate dissipation from equation (17).

7. Check the plate dissipation and the peak plate current to see that they do not exceed the maximum ratings.

8. Determine the maximum positive grid voltage E_{cm} from the loadline and I_{bm}.

9. Determine the peak grid current I_{gm} corresponding to extreme point of the loadline.

10. Calculate the peak signal grid voltage : $E_{gm} = E_{cm} + E_c$.

11. Calculate the peak grid input power : $P_{gm} = E_{gm} I_{gm}$.

12. Calculate the minimum input resistance : $r_{gmin} = E_{gm}/I_{gm}$.

13. Repeat steps 1 to 12 for several other values of R_L and select what appears to be the best compromise, with a view to the lowest driving power

14. Assume a reasonable peak power transformer efficiency—say 70%. This value is used in the following step.

15. Select a driver valve with a maximum power output at least $0.9 P_{gm}$ under typical operating conditions (this makes allowance for the higher load resistance).

16. Select a driver load resistance twice the typical value or at least four times the plate resistance.

17. Draw the assumed (straight) driver loadline on the driver plate characteristics and determine the maximum available peak signal voltage E_{pm}.

18. Determine the transformer ratio from primary to half-secondary :
$N_1/N_2 = \eta E_{pm}/E_{gm}$.

19. Calculate the " driver resistance " from eqn. (21) using values for r_1 and r_2 indicated* from similar transformers. These resistances are usually less than 1900 and 350 ohms respectively for 10 watt amplifiers, or 2700 and 500 ohms respectively for low power battery amplifiers.

20. Check the " driver resistance " to see that it is less than $0.2 r_{gmin}$ for high-mu valves or $0.1 r_{gmin}$ for other cases (this assumes that r_{gmin} is approximately twice the minimum variational grid resistance).

*Alternatively the driver transformer may be tentatively assumed to have 10% primary and 10% secondary copper losses, and 10% iron losses. On this basis, $r_2 = r_{gmin}/7$ and $r_1 \approx r_{gmin}(N_1^2/N_2^2)/9$.

(vi) Earthed-grid cathode-coupled amplifiers

A cathode-coupled output stage, with earthed grids, has the advantages of a much more nearly constant input resistance than a Class B stage, but it requires more driver power for the same output. It is more suited to high-power amplifiers or modulators than to conventional a-f amplifiers (Ref. E15). The increase in plate resistance due to feedback is disadvantageous.

(vii) Class B_1 amplifiers—Quiescent push-pull

Class B_1 amplifiers do not draw any grid current whatever, but in other respects resemble Class B_2 amplifiers. Either triodes or pentodes may be used, the latter being more popular—they are used as " quiescent push-pull pentodes " in battery receivers. The two valves (or units) must be very accurately matched, and an adjustable screen voltage is desirable.   A high-ratio step-up input transformer is used to supply the high peak signal voltage, nearly equal to the cut-off bias.   Both the input and output transformers should have low capacitance and leakage inductance.   The output transformer should have its half-primary inductance, and its leakage inductance, accurately balanced.   Condensers (0.002 to 0.005 μF) are usually connected across each half primary (Ref. E4).

SECTION 8 : CLASS AB₂ AMPLIFIERS

(*i*) *Introduction*   (*ii*) *Bias and screen stabilized Class AB₂ amplifier*   (*iii*) *McIntosh amplifier.*

(i) Introduction

Class AB₂ amplifiers closely resemble Class B₂ amplifiers, but the valves are biased as for Class AB₁ operation.   Consequently they are less critical with regard to matching, and the distortion which occurs in the plate circuit is much the same as for Class AB₁.   The plate circuit efficiency is intermediate between Class AB₁ and B₂ operation —typical values for triodes are from 40 to 48%.   The variation in plate current from zero to maximum signal is less than with Class B operation, being of the order of 1 : 2.

The matching of valves is covered in Sect. 5(v).

Pentodes and beam power amplifiers may be used quite successfully in Class AB₂ operation.

Fixed bias is essential.

Type 6L6 or 807 beam power amplifiers may be used in Class AB₂ with plate circuit efficiencies of about 65% (or 61% including screen losses).   The peak grid input power does not exceed 0.27 watt for power outputs from 30 to 80 watts.   The total harmonic distortion under ideal laboratory conditions is 2% at maximum signal and less than 2% at all lower output levels (this assumes a low-distortion driver stage).

Well regulated plate, screen and bias power supplies are essential.

Parasitics in plate and grid circuits—see Sect. 7(i).

References E20, E21.

(ii) Bias and screen stabilized Class AB₂ amplifier (Ref. E26)

Fig. 13.50C gives a circuit with 807 valves operated in Class AB₂ with stabilization of the bias and screen supply provided by the driver valve.   Two high tension sources are used, obtained from separate windings on the power transformer, and neither needs good regulation, so that condenser input filters can be used.

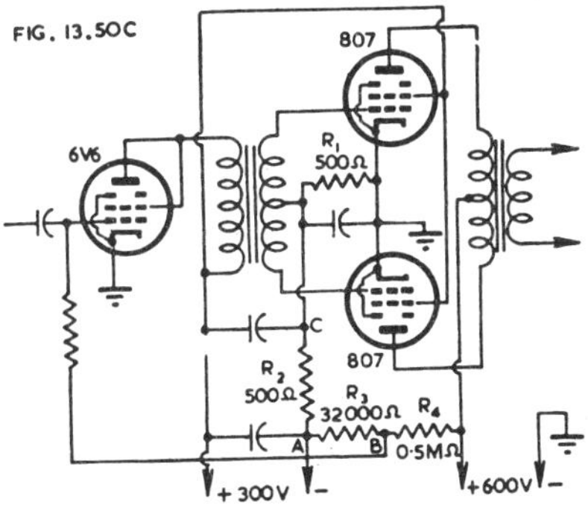

The circuit offers three types of compensation, the combined effect being to allow the output valves to draw a minimum of plate current in the no signal condition, although still being capable of delivering an output of 60 watts for small distortion.   The three types of compensation are

(a) As the 807 plate voltage falls, due to increasing plate current with signal input, the screen voltage is increased and the bias reduced.

(b) When the 807's are driven into grid current their bias is increased due to the current in the 500 ohm resistor

Fig. 13.50C.   Class AB₂ amplifier with stabilized bias and screen voltages (Ref. E26).

R_1. This is minimized by a d.c. negative feedback circuit giving several times reduction of the effect, and is offset by a corresponding increase in 807 screen voltage.

(c) Any tendency for the 807 screen voltage to fall and for the bias voltage to increase with rising 807 screen current due to signal input is minimized by the same negative feedback circuit.

Referring to Fig. 13.50C, the voltage divider from the positive side of the 600 volt supply to the negative side of the 300 volt supply has values to provide a suitable bias for the triode-operated 6V6 so that it draws about 45 mA with a standing plate voltage of 250 V. A decrease in the 807 plate voltage causes the 6V6 grid to become more negative, decreasing its plate current which also flows through R_1 and so decreases the 807 bias. The low voltage power supply is deliberately given poor regulation with resistive filtering so that the decrease in 6V6 plate current appreciably increases its plate voltage and·thus the 807 screen voltage.

The 807 grid current at large output levels increases the negative potential at C and also at A and B and thus increases the 6V6 bias. This reduces the 6V6 plate current which makes point C less negative and minimizes the increase in 807 bias, and at the same time increases the 807 screen voltage thus off-setting the bias increase.

The 807 screen current rises with increasing signal input and this tends to decrease the screen voltage and increase the 807 bias. However the increased bias is also ·applied to the 6V6, reducing its plate current and thus tending to restore the 807 bias and screen voltages to their original values.

Between no output and full output the 6V6 plate current falls from 45 mA to 25 mA but the valve is still well able to provide the small power required to drive the 807's.

An additional feature of the circuit is that a very large input reduces the ability of the driver to overload the output valves, and after full output is reached very effective limiting is provided.

(iii) McIntosh Amplifier (Ref. E28)

With any push-pull amplifier in which each valve is cut-off during portion of the cycle, some form of quasi-transient distortion tends to occur at a point in each cycle at the higher audio frequencies (Ref. E13). This distortion is caused by the leakage reactance of the primary of the output transformer, which cannot be reduced sufficiently by conventional transformer design technique. A completely new approach to the problem is made by the McIntosh amplifier which incorporates special types of driver transformer and output transformer, together with many other novel features. The basic principles are indicated in Fig. 13.50D in which both driver transformer T_1 and output transformer T_2 have two windings wound together in a bifilar manner so that the coupling between them is almost unity. It is claimed that it is practicable to wind coils with a ratio of primary inductance to leakage inductance better than 200 000 to 1, whereas conventional transformers do not nearly reach the minimum requirement (for low distortion) of 80 000 to 1. This type of transformer is cheaper to wind than a sectionalized winding as used in conventional high quality transformers.

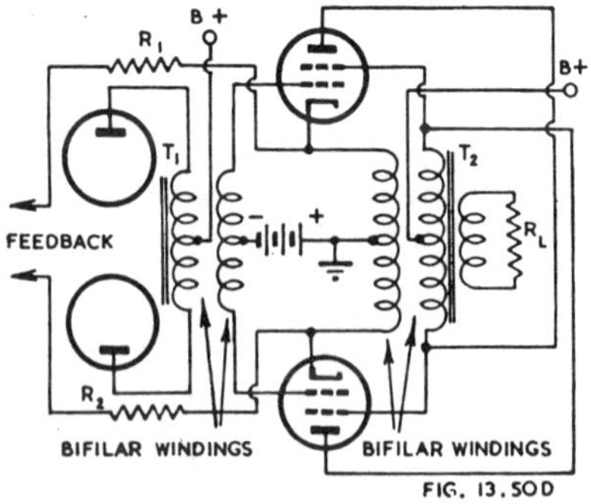

Fig. 13.50D. *Basic principles of McIntosh Amplifier (Ref. E28).*

Each output valve works into two primary sections, one in its plate circuit and the other in its cathode circuit, but these have practically unity coupling. The effective number of primary turns for each valve is equal to the total turns for each of the bifilar windings. The output transformer should therefore be designed to have a total impedance on each of its primaries equal to one quarter of the plate-to-plate load impedance.

In the case of the amplifier of Fig. 13.50E the plate-to-plate load imped-

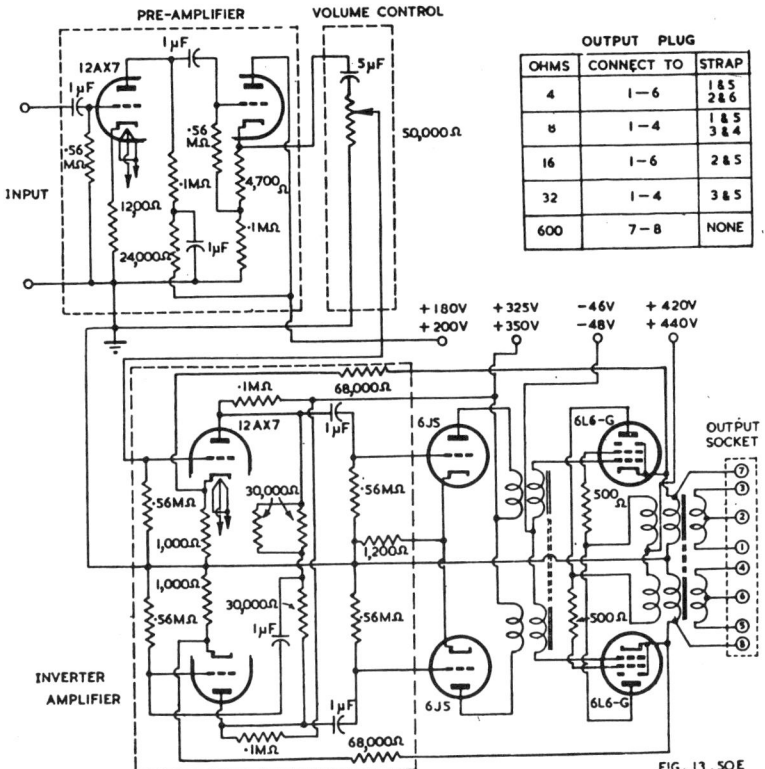

OUTPUT	PLUG	
OHMS	CONNECT TO	STRAP
4	1 – 6	1 & 5 2 & 6
8	1 – 4	1 & 5 3 & 4
16	1 – 6	2 & 5
32	1 – 4	3 & 5
600	7 – 8	NONE

Fig. 13.50E. *Circuit diagram of complete 50 watt McIntosh Amplifier (Ref. E28).*

ance is 4000 ohms and the total impedance of each primary winding is 1000 ohms ; the impedance from each cathode to earth is only 250 ohms. These low impedances reduce the effects of capacitive shunting and thereby improve the high frequency performance.

The voltage from each screen to cathode is maintained constant by the unity coupling between the two halves of the bifilar windings, no screen by-pass capacitor being required. This arrangement, however, has the limitation that it can only be used for equal plate and screen voltages.

The driver transformer T_1 makes use of the same bifilar winding method adopted in the output transformer. The primary impedance in this design is above 100 000 ohms from 20 to 30 000 c/s, while the response of the whole transformer is within 0.1 db from 18 to 30 000 c/s. These high performances are made necessary by the inclusion of this transformer in the second feedback path of the whole amplifier.

The method of loading the output stage, with half the load in the plate and half in the cathode circuit, provides negative feedback as a half-way step towards a cathode follower. Additional feedback is achieved by connecting suitable resistors between the cathodes of the output valves and the cathodes of the phase inverter stage. The complete amplifier (Fig. 13.50E) has a typical harmonic distortion of 0.2% from 50 to 10 000 c/s, rising to 0.5% at 20 c/s and 0.35% at 20 000 c/s, at an output level of 50 watts. The frequency response under the same conditions, measured on the secondary of the transformer, is level from 20 to 30 000 c/s, − 0.4 db at 10 c/s and − 0.3 db at 50 000 c/s. The phase shift is zero from 50 to 20 000 c/s, − 10° at 20 c/s and + 4° at 50 000 c/s.

The output resistance is one tenth of the load resistance, thus giving good damping and regulation

These same principles may be applied to any type of push-pull output whether triode or pentode, with any class of operation, A, AB_1, AB_2 or B.

The high power output of the McIntosh amplifier as described above is due to the operation of the screens at voltages greater than 400 volts, which is very considerably in excess of the maximum rating of 270 volts (design centre) for type 6L6 or 300 volts (absolute) for type 807. It is unfortunate that the McIntosh amplifier is limited to operation with equal plate and screen voltages, but these should always be within the maximum ratings for the particular valve type.

See also Refs. E31, G1.

The principle of combined plate and cathode loading is applied in some high-fidelity Class A amplifiers, e.g. Acoustical QUAD, Refs. H4, H6.

SECTION 9 : CATHODE-FOLLOWER POWER AMPLIFIERS

The principles of cathode followers have been covered in Chapter 7 Sect. 2(i). A cathode follower may be used either as driver for a Class B or AB_2 stage, or as the output stage itself.

A cathode follower forms almost an ideal driver stage, having very low plate resistance and distortion, although it requires a high input voltage. It is commonly used, either singly or in push-pull, in high power a-f amplifiers where the distortion must be reduced as much as possible. If parallel-feed is used, the hum is reduced by the factor $1/(\mu + 1)$; see Chapter 7 Sect. 2(ix) Case 4. A cathode follower driver stabilizes the a-f signal voltage, but does not stabilize the grid bias.

Cathode follower output stages introduce serious problems, and are not suitable for general use, even though their low plate resistance and low distortion appear attractive. The difficulty is in the high input voltage which is beyond the capabilities of a resistance-coupled stage operating on the same plate supply voltage. Two methods are practicable, either a step-up transformer in the plate circuit of a general purpose triode, or a resistance-coupled amplifier with a plate supply voltage about 3 times the plate-cathode voltage of the cathode follower. In order to take advantage of the low distortion of the cathode follower, the preceding stage should also have low distortion. A general purpose triode is to be preferred to a pentode or high-mu triode with resistance coupling, and it may have an unbypassed cathode resistor.

One practical amplifier which has been described in the literature (Ref. F3) uses 700 volts supply voltage to the 6SN7 penultimate stage and eight 6V6-GT valves in push-pull parallel operation in the cathode follower output stage. Negative feedback is used from the secondary of the output transformer, and the damping on the loudspeaker is as high as practicable. However, the total harmonic distortion at 50 c/s is over 1% at 8 watts output, and 1.7% at 20 watts. The high output voltage which must be delivered by the resistance-coupled penultimate stage thus shows its effect on the distortion, even though the plate supply voltage has been increased to a dangerously high value.

SECTION 10 : SPECIAL FEATURES

(i) Grid circuit resistance (ii) Grid bias sources (iii) Miller Effect (iv) 26 volt operation (v) Hum.

(i) Grid circuit resistance

A maximum value of grid circuit resistance is usually specified by the valve manufacturer, a higher value being usually permitted with cathode (self) bias than with fixed bias. The reason for the latter is that cathode bias provides increased bias as the plate current rises, and so gives a degree of protection against "creeping" plate current. This effect is due to the combined gas and grid emission currents which flow through the grid-cathode circuit—see Chapter 2 Sect. 2(iii), also Chapter 3 Sect. 1(v)d, and Chapter 12 Sect. 2(iii)E, (iv)B, (iv)C, (iv)D ; Sect. 3(iv)C, (v). Typical values (e.g. type 6V6-GT) are :

    With cathode bias   0.5 megohm (maximum)
    With fixed bias   0.1 megohm (maximum).

If the heater is operated, even for limited periods, more than 10% above its average rated voltage, the grid circuit resistance must be reduced considerably ; this holds in automobile receivers.

If partial self-bias operation (" back bias ") is used, the maximum value of grid circuit resistance may be found from the relation :

$$R_{gm} = R_{gf} + P(R_{gs} - R_{gf})\qquad(1)$$

where R_{gf} = max. grid circuit resistance (fixed bias)
       R_{gs} = max. grid circuit resistance (self bias)
and P = ratio of cathode current in output valve to the total current flowing through the bias resistor.

In cases where the maximum value of grid resistance for a particular application is not specified, the procedure of Chapter 3 Sect. 1(v)d should be followed, using eqn. (6) for pentodes or tetrodes, or eqn. (7) for triodes. The value of maximum reverse grid current (ΔI_{c1}) should be obtained from the specifications or from the valve manufacturer ; failing this, a value of $2\mu A$ may be tentatively assumed for valves of the 6F6, 6V6 class, and $4\mu A$ for valves having higher cathode currents.

If either the value of R_{g1} for cathode bias, or that for fixed bias is specified, the value for the other may be calculated from Chapter 3 Sect. 1 eqns. (8) to (13).

(ii) Grid bias sources

Fixed bias is normally obtained from a separate power source with rectifier, filter and load resistance. A typical circuit is Fig. 13.51 in which V_1 is the usual rectifier, V_2 is the bias rectifier and R_4 the bias load resistance. V_2 may be any half-wave indirectly-heated rectifier ; if the V_2 heater is operated from a common heater winding, V_2 should be a type capable of withstanding a high voltage between heater and cathode. Alternatively a shunt diode may be used (see Chapter 30 Sect. 6). If the output stage draws positive grid current, as in Class AB_2 or overloaded Class A operation, the grid current flowing through the bias load resistance increases the bias voltage. It is therefore advisable to design this to be as low as practicable. The effect of negative grid current on the bias is usually negligible.

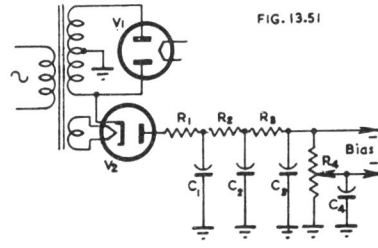

FIG. 13.51

Fig. 13.51. Method of obtaining fixed bias using half of the transformer secondary (plate) winding. Typical values of resistors are : $R_1 = 2500$, $R_2 = 25\,000$, $R_3 = 15\,000$, $R_4 = 3000$ ohms ; $C_1 = 8\mu F$, $C_2 = 16\mu F$, $C_3 = C_4 = 50\mu F$. Alternatively R_2 may be replaced by a choke.

Voltage stabilized grid bias supplies are sometimes used (see Chapter 33).

It is sometimes convenient to have two plate voltage supplies, one to provide for the plates of the output stage only, the other for the earlier stages together with the screens of the output stage and grid bias. The second supply is loaded by a heavy current voltage divider (say 150 mA total drain) tapped near the negative end and the tapping connected to the cathodes of the output stage ; one section provides grid bias, and the other section provides the positive potentials.

Back bias is intermediate between cathode bias and fixed bias, so far as its constancy is concerned. With this arrangement, the whole return current of a receiver or amplifier is passed through a resistor between the cathodes (which are generally earthed) and $- E_{bb}$ (Fig. 13.52). The value of R_1 is given by E_{c1}/I where E_{c1} is

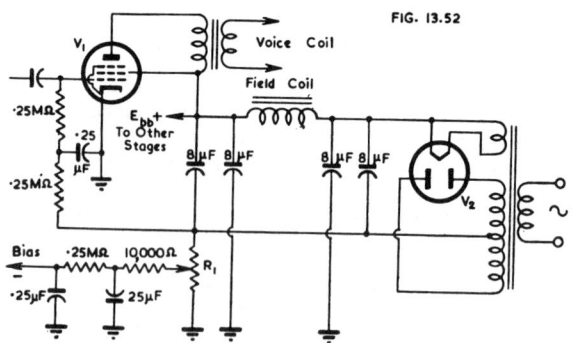

Fig. 13.52. *Circuit of portion of a receiver using back bias. V_1 is the power amplifier, V_2 is the rectifier.*

the bias required by V_1 and I is the total current passed through R_1. Bias for earlier stages is obtainable from a tapping or tappings on R_1. A hum filter is necessary for each separate bias system. [Note that the curved plate of an electrolytic condenser represents the negative terminal ; while on a paper condenser the curved plate represents the outside electrode.] Back bias works best in large receivers where the total current is much greater than the cathode current of the power amplifier.

An alternative form having fewer components is shown in Fig. 13.53. Here the field coil is in the place of R_1 in Fig. 13.52. The bias on V_1 is equal to the voltage drop across the field coil multiplied by $R_2/(R_2 + 0.5$ megohm). The filtering in this circuit is not so complete as in Fig. 13.52.

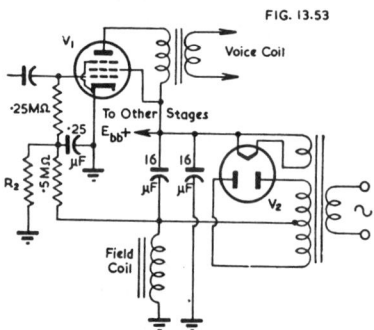

Fig. 13.53. *Alternative form of back bias, with the speaker field coil in the negative lead.*

(iii) Miller Effect

The Miller Effect has been introduced in Chapter 12 Sect. 2(xi). If the output valve is type 6V6-GT under typical operating conditions, $A = 17$ and $C_{gp} = 0.7 \ \mu\mu F$, then the additional input capacitance, due to C_{gp} alone, with a resistive load is $18 \times 0.7 = 12.6 \ \mu\mu F$. If the valve is a triode the effect is more pronounced. For example, with type 2A3 under typical operating conditions $A = 2.9$ and $C_{gp} = 16.5 \ \mu\mu F$; the additional input capacitance is $3.9 \times 16.5 = 64 \ \mu\mu F$.

(iv) 26 volt operation

Standard power valves give very limited power output with plate and screen both at 26 or 28 volts. Special types (e.g. 28D7) have been developed to give higher power output under these conditions.

(v) Hum from plate and screen supplies

The hum of amplifiers with and without feedback is covered in Chapter 7 Sect. 2(ix)C.

Power amplifiers not using feedback are here summarized briefly, using the conditions for types 2A3 and 6V6 as in Chapter 7.

Triode with (a) conventional output transformer—
Relative hum voltage $= 0.85$
(b) parallel feed (series inductor)—
Relative hum voltage $= 0.024$

Pentode with conventional output transformer
and (a) common plate and screen supplies—
Relative hum voltage $= 2.05$
(b) screen perfectly filtered—
Relative hum voltage $= 0.09$

Pentode with parallel feed (series inductor)
(a) common plate and screen supplies
Relative hum voltage $= 1.78$
(b) screen perfectly filtered
Relative hum voltage $= 0.179$

See also Chapter 31 Sect. 4(ii)—Effect of the output valve on hum originating in the plate supply voltage.

SECTION 11 : COMPLETE AMPLIFIERS

(i) Introduction   (ii) Design procedure and examples   (iii) Loudspeaker load.

(i) Introduction

While every amplifier must be built up of individual stages, the purpose of the design is to produce a complete amplifier having specified characteristics in regard to overall gain, maximum power output, frequency range, output resistance, distortion, hum, noise level and special features such as overload characteristics and tone control. A designer's job is to produce the required results at the minimum cost within any limitations imposed by space or by the availability of valve types.

(ii) Design procedure and examples

The correct procedure is to commence with the loudspeaker, and then to work backwards through the amplifier. For example, after investigation of the loudspeaker and speaker transformer efficiency (see Chapter 20 Sect. 6) it should be possible to estimate the maximum power output required from the output valve(s). Then the output valve(s) should be selected allowing for the possible use of feedback. Then the preceding stage should be designed to work into the known following grid resistor and effective input capacitance with a satisfactory frequency response and distortion level. Finally, the input stage should be designed to give the required hum and noise performance. If, after a first calculation, the overall gain is too high, it may be reduced by one of many expedients such as increasing (or adding) negative feedback on the output stage, removing a cathode by-pass, changing a voltage amplifier stage from pentode to triode operation, or changing one or more valves to a lower gain type. In large amplifiers a pre-set second volume control may be used, to be adjusted in the laboratory.

The following information is a general guide in the design of typical amplifiers
A $=$ public address (speech)
B $=$ typical radio receiver (a-f amplifier only)
C $=$ good quality radio receiver or amplifier
D $=$ high fidelity receiver or amplifier

Performance applies to amplifier only (resistance load).

Category	A	B	C	D	
Frequency range	200-5 000	100-5 000	50-10 000	40-15 000	c/s
Attenuation at limits	3	3	2	1	db
Total harmonic distortion*	10	10	2	<1	%
Output resistance	—	—	$<R_L/2$	$<R_L/10\ddagger$	
Noise level†	—	−25	−25	−30	dbm
Hum level (120 c/s) **	—	−15 to −30	−35 to −40	−40 to −45	dbm
Intermodulation distortion*—					
r.m.s. sum (voltage ratio 4 : 1)	40	40	8	<4	%

*at maximum power output.

**For 3% loudspeaker efficiency. In all cases the hum should not be audible under listening conditions. See Chapter 31 Sect. 4(iii) for hum.

†For room volume 3000 cubic feet, 3% loudspeaker efficiency—see Chapter 14 Sect. 7(v). For measurement of noise see Chapter 19 Sect. 6.

‡Or as required to give the desired degree of damping at the bass resonant frequency.

See also Chapter 14 Sect. 12(i) for high fidelity.

With careful design and the use of negative feedback it is possible to make the total harmonic distortion of the amplifier less than 0.1% and the intermodulation distortion less than 0.5%. This has been achieved in the circuit of Fig. 7.44 and in the Williamson amplifier on which it is based (the new version of the Williamson amplifier is given in Figs. 17.35 B, C, D, E, F, G), also in the Leak amplifier. See Refs. F1, F4, F5, H4, H15, H16.

An example of category C is given in Figs. 7.42 and 7.43.

Undoubtedly one of the best and most versatile single-ended amplifiers incorporating negative feedback (and, if desired, also tone control) is the circuit Fig. 7.33. The feedback should be as high as permissible, being usually limited by the sensitivity required by a pickup. The only serious limitation is that V_1 cannot be a combined second detector and amplifier.

The **Lincoln Walsh amplifier** with automatic bias control (E24, E25) is quoted as having intermodulation distortion of 0.2% at 5 watts, 0.6% at 10 watts and 1.7% at 25 watts, the power in each case being the low frequency power only (50 c/s). The valves used in this amplifier are type 6B4-G (similar to 2A3).

The **McIntosh 50 watt Class AB₂ amplifier** (Fig. 13.50E) has harmonic distortion less than 0.5% from 20 to 30 000 c/s, with exceptionally flat frequency response and low phase shift—see Sect. 8(iii).

References to complete amplifiers : Refs. (F) and (H).

(iii) Loudspeaker load

Most amplifiers are designed to work into a constant resistive load, whereas a loudspeaker presents a far from constant impedance which is largely inductive or capacitive at almost all frequencies within its range. This feature is covered from the loudspeaker angle in Chapter 20 Sects. 1, 2 and 7. See also Sect. 2(iv) for triodes and Sect. 3(viii) for pentodes.

A simple 2 stage amplifier with a triode output stage is shown in Fig. 13.54, together with the fundamental, second, third and fourth harmonics (see Chapter 20 for comments). An obvious conclusion is that all amplifiers should be tested for output and distortion while delivering full power into a loudspeaker load. While highly desirable, this requires a large sound-proof room for all except low levels. An alternative which is highly recommended is to construct a dummy load which has the same impedance as the loudspeaker at 400 c/s, at 10 000 c/s, and at the bass resonant frequency, based on the circuit of Fig. 20.3. The value of $(R_0 + R_1)$ in Fig. 20.3 should be equal to the measured value of the working impedance of the loudspeaker at the bass resonant frequency as described in Chapter 20 Sect. 2(iv).

FIG. 13.54

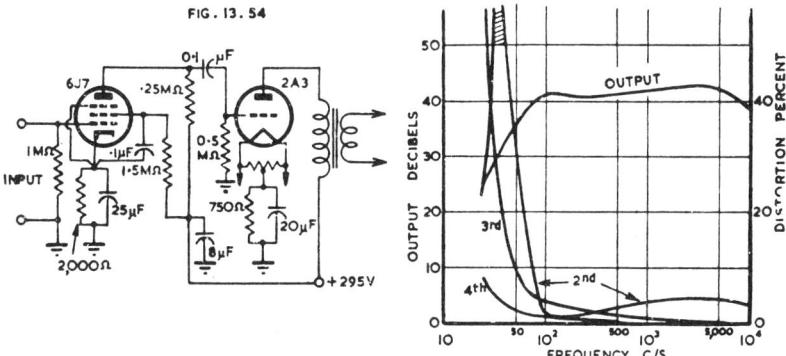

Fig. 13.54 *Harmonic distortion of amplifier having type* 6J7 *pentode coupled to* 2A3 *grid resistor* 0.5 *megohm triode loaded by a loudspeaker on a flat baffle. The distortion was measured by a wave analyser connected across the voice coil. The bass resonance frequency is* 70 *c/s.*

SECTION 12 : REFERENCES

(A) REFERENCES TO OPTIMUM LOAD RESISTANCE AND DISTORTION
A1. Brain, B. C. " Output characteristics of thermionic amplifiers " W.E. 6.66 (March 1929) 119.
A2. Notthingham, W. B. " Optimum conditions for maximum power in Class A amplifiers " Proc. I.R.E. 29 12 (Dec. 1941) 620.
A3. Editorial " Optimum conditions in Class A amplifiers " W.E. 20.233 (Feb. 1943) 53.
A4. Hadfield, B. M. (letter) " Optimum conditions in Class A amplifiers," W.E. 20.235 (April 1943) 181.
A5. Sturley, K. R. " Optimum load, Ra or 2Ra " ? W.W. 50.5 (May 1944) 150.
A6. Sturley, K. R. (letter) " Optimum conditions in Class A amplifiers " W.E. 20.235 (April 1943) 181.
A7. Bradshaw, E. (letter) " Optimum conditions in Class A amplifiers " W.E. 20.237 (June 1943) 303.
A8. Benham, W. E. (letter) " Optimum conditions in Class A amplifiers " W.E. 20.237 (June 1943) 302.
A9. Gladwin, A. S. (letter) " Optimum conditions in Class A amplifiers " W.E. 20.240 (Sept. 1943) 436.
A10. Date, W. H. (letter) " Optimum load," W.W. 50.7 (July 1944) 222.
A11. Good, E. F. (letter) " Optimum load " W.W. 50.8 (Aug. 1944) 252.
A12. Hughes, E. " Optimum valve load—unified treatment for different operating conditions " W.W. 51.8 (Aug. 1945) 246.
A13. Foster, H. G. " Load conditions in Class A triode amplifiers " Electronic Eng. 19.227 (Jan. 1947) 11.
A14. Sturley, K. R. (book) " Radio Receiver Design " Part II pp. 56-64 (Chapman and Hall, London, 1945).
A15. Warner, J. C., and A. V. Loughren " The output characteristics of amplifier tubes " Proc. I.R.E. 14.6 (Dec. 1926) 735.
A16. Kilgour, C. E. " Graphical analysis of output tube performance " Proc. I.R.E. 19.1 (Jan. 1931) 42.
A17. Lucas, G. S. C. " Distortion in valve characteristics " W.E. 8.98 (Nov. 1931) 595.   Letter P. K. Turner 8.99 (Dec. 1931) 660.
A18. Nelson, J. R. " Calculation of output and distortion in symmetrical output systems," Proc. I.R.E. 20.11 (Nov. 1932) 1763.
A19. Ferris, W. R. " Graphical harmonic analysis for determining modulation distortion in amplifier tubes " Proc. I.R.E. 23.5 (May 1935) 510.
A20. Mouromtseff, I. E., and H. N. Kozanowski " A short-cut method for calculation of harmonic distortion in wave modulation " Proc. I.R.E. 22.9 (Sept. 1934) 1090.
A21. Hutcheson, J. A. " Graphical harmonic analysis " Elect. 9.1 (Jan. 1936) 16.
A22. Harries, J. H. O. " Amplitude distortion " W.E. 14.161 (Sept. 1937) 63.
A23. Kemp, P. " Harmonic analysis of waves up to eleventh harmonic " Electronic Eng. 23.284 (Oct. 1951) 390.
See also References to Graphical Harmonic Analysis Chapter 6 Sect. 9(C).

(B) GENERAL REFERENCES TO CLASS A AMPLIFIERS
B1. Espley, D. C., and L. I. Farren, " Direct reading harmonic scales," W.E. (April 1934) 183.
B2. R.C.A. " Application Note on design of audio systems employing type 2A3 power amplifier triodes " No. 29 (Dec. 29, 1933).
B3. R.C.A " Application Note on short-cut method for determining operating conditions of power output triodes " No. 42 (Sept. 5, 1934).
B4. R.C.A. " Application Note on use of the plate family in vacuum tube power output calculations " No. 78 (July 28, 1937)—Reprinted Radiotronics 80 (Oct. 11, 1937) 72.
B5. Pullen, K. A. " The use of G curves in the analysis of electron-tube circuits " Proc. I.R.E. 37.2 (Feb. 1949) 210.
B6. Langford-Smith, F., and W. J. Steuart " Effect of power supply on power amplifiers "—unpublished analysis.
B7. Pullen, K. A. " G curves in tube circuit design " Tele-Tech 8.7 (July 1949) 34 ; 8.8 (Aug. 1949) 33.
B8. Pullen, K. A. " Use of conductance, or G, curves for pentode circuit design " Tele-Tech 9.11 (Nov. 1950) 38.

(C) REFERENCES TO MULTI-GRID AMPLIFIERS

C1. Brian, W. S. " Experimental audio output tetrode " Elect. 20.8 (Aug. 1947) 121.
C2. Pickering, N. " Space-charge tetrode amplifiers " Elect. 21.3 (Mar. 1948) 96.
C3. Schade, O. H. " Beam power tubes " Proc. I.R.E. 26.2 (Feb. 1938) 137.
C4. Jonker, J. L. H. " Pentode and tetrode output valves " Philips Tec. Com. No. 75 (July 1940) 1.
C5. " Advantages of space-charge-grid output tubes " (Review of paper by N. C. Pickering) Audio Eng. 31.9 (Oct. 1947) 20.
Co. Hadfield, B. M. " Amplifier load impedance reduction," Tele-Tech 7.5 (May 1948) 33. See also B5.
C7. Hafler, D., & H. I. Keroes "An ultra-linear amplifier" Audio Eng. 35.11 (Nov. 1951) 15.

(D) REFERENCES TO PARALLEL AMPLIFIERS

D1. Jones, F. C. " Parallel tube high fidelity amplifiers " Radio 29.10 (Oct. 1945) 27.
D2. Jones, F. C. " Additional notes on the parallel tube amplifier " Radio (June 1946) 26.

(E) REFERENCES TO PUSH-PULL AMPLIFIERS

E1. Thompson, B. J. " Graphical determination of performance of push-pull audio amplifiers " Proc. I.R.E. 21.4 (April 1933) 591.
E2. Barton, L. E. " High audio output from relatively small tubes " Proc. I.R.E. 19.7 (July 1931) 1131.
E3. M.I.T. Staff (book) " Applied Electronics " (John Wiley and Sons, New York ; Chapman and Hall, London, 1943) pp. 433-448, 548-558.
E4. Sturley, K. R. " Radio Receiver Design " Part 2 (Chapman and Hall, London, 1945) pp. 84-97.
E5. " Push-pull triode amplifiers " Radiotronics No. 79 (Sept. 1937) 64 ; No. 80 (October 1937) 78.
E6. Adorjan, P. " Power amplifier design," Radio Eng. 16.6 (June 1936) 12.
E7. Krauss, H. L. " Class A push-pull amplifier theory " Proc. I.R.E. 36.1 (Jan. 1948) 50.
E8. Cocking, W. T. " Push-pull balance " W.W. 53.11 (Nov. 1947) 408.
E9. R.C.A. Application Note No. 54 "Class AB operation of type 6F6 tubes connected as pentodes," (Dec. 1935).
E10. Williams, W. N. " Novel 10 watt amplifier for P.A." Radio and Hobbies, Australia (April 1942) 37.
E11. Heacock, D. P. " Power measurement of Class B audio amplifier tubes " R.C.A. Rev. 8.1 (Mar. 1947) 147.
E12. Barton, L. E. " Application of the Class B audio amplifier to a-c operated receivers " Proc. I.R.E. 20.7 (July 1932) 1085.
E13. Sah, A. Pen-Tung " Quasi transients in Class B audio-frequency push-pull amplifiers," Proc. I.R.E. 24.11 (Nov. 1936) 1522.
E14. Nelson, J. R. " Class B amplifiers considered from the conventional Class A standpoint " Proc. I.R.E. 21.6 (June 1933) 858.
E15. Butler, F. " Class B audio-frequency amplifiers " W.E. 24.280 (Jan. 1947) 14.
E16. R.C.A. Application Notes Nos. 5, 14, 18, 33 and 54 giving operating data on types 79, 53, 19, 800, 54 respectively.
E17. Barton, L. E. " Recent developments of the Class B audio- and radio-frequency amplifiers " Proc. I.R.E. 24.7 (July 1936) 985.
E18. McLean, T. " An analysis of distortion in Class B audio amplifiers " Proc. I.R.E. 24.3 (March 1936) 487.
E19. Strafford, F. R. W. " Join-up distortion in Class B amplifiers," W.E. 12.145 (Oct. 1935) 539.
E20. R.C.A. Application Note No. 40 " High power output from type 45 tubes " (June 1934).
E21. Reich, H. J. " Theory and applications of electron tubes " (McGraw-Hill, 2nd edit. 1944).
E22. Greenwood, H. M. " Cathode-follower circuits " Q.S.T. 29.6 (June 1945) 11, especially Fig. 20.
E23. Amos, S. W. " Push-pull circuit analysis—cathode-coupled output stage " W.E. 23.269 (Feb. 1946) 43.
E24. Edinger, A. " High-quality audio amplifier with automatic bias control " Audio Eng. 31.5 (June 1947) 7. Based on patent by Lincoln Walsh.
E25. Minter, J. " Audio distortion and its causes " Chapter 6. " Standard FM Handbook " (FM Company, Great Barrington, Mass., U.S.A. 1st edit. 1946)—gives intermodulation test results on Lincoln Walsh amplifier.
E26. Patent application, inventors C. G. Smith, W. Storm, E. Watkinson, assignee Philips Electrical Industries of Aust. Pty. Ltd.
E27. Sturley, K. R. " Push-pull A.F. amplifiers—load curves for Classes A, B and C conditions " W.E. 26.313 (Oct. 1949) 338.
E28. McIntosh, F. H., and G. J. Gow " Description and analysis of a new 50-watt amplifier circuit " Audio Eng. 33.14 (Dec. 1949) 9.
E29. Parry, C. A. " The cathamplifier " Proc. I.R.E. Aust. 11.8 (Aug. 1950) 199.
     Parry, C. A. " The cathamplifier " Proc. I.R.E. 40.4 (April 1952) 460.
E30. " The Cathamplifier Handbook " (Mingay Publishing Co., Sydney, Australia).
E31. Sterling, H. T. " Extended Class A audio " Elect. 24.5 (May 1951) 101. See also A13, A18, A20, B5, C3.
E32. Peterson, A., and D. B. Sinclair "A single-ended push-pull audio amplifier" Proc. I.R.E. 40.1 (Jan. 1952) 7.

(F) REFERENCES TO COMPLETE AMPLIFIERS

F1. Williamson, D. T. N. " Design for a high-quality amplifier " (1) Basic requirements : alternative specifications, W.W. 53.4 (April 1947) 118 ; (2) Details of final circuit and its performance, W.W. 53.5 (May 1947) 161.
F2. Baxandall, P. J. " High-quality amplifier design—advantages of tetrodes in the output stage " W.W. 54.1 (Jan. 1948) 2 ; Correction 54.2 (Feb. 1948) 71.
F3. Gibson, W. F., and R. Pavlat " A practical cathode-follower audio amplifier " Audio Eng. 33.5 (May 1949) 9.
F4. Williamson, D. T. N. " High-quality amplifier—new version " W.W. 55.8 (Aug. 1949) 282 ; 55.10 (Oct. 1949) 365 ; 55.11 (Nov. 1949) 423.
F5. " The Williamson Amplifier," Reprint of articles published in the Wireless World (Iliffe and Sons Ltd., London).

(G) GENERAL REFERENCES

G1. Sulzer, P. G. "A survey of audio-frequency power-amplifier circuits" Audio Eng. 35.5 (May 1951) 15.

Additional references H will be found in the Supplement commencing on page 1475.

CHAPTER 14

FIDELITY AND DISTORTION

By F. LANGFORD-SMITH, B.Sc., B.E.

SECTION 1 : INTRODUCTION

(i) Fidelity   *(ii) Types of distortion*   *(iii) Imagery for describing reproduced sound.*

(i) Fidelity

True fidelity is perfect reproduction of the original. In the case of an amplifier, true fidelity means that the output waveform is in all respects an amplified replica of the input waveform.

In acoustical reproduction, true fidelity is achieved if the listener has the same aural sensations that he would have if present among the audience in the studio or concert hall.

In practice, true fidelity can only be regarded as an ideal to be aimed at.

The concept of " hole in the wall " listening put forward by Voigt (e.g. Ref. A49) is a useful device. The listener, surrounded by his listening room, is imagined as being in the concert hall and able to hear directly, through his open window, the sound coming from the stage, together with echoes. Echoes from his own room will come from all directions, but echoes from the concert hall only come through the open window. This is probably the nearest approach to true fidelity which there is any hope of approaching with single channel transmission. See also Ref. A50.

The purpose of high fidelity reproduction of music is to satisfy a particular listener, who is primarily interested in the emotions arising from what he hears. The complete process involves sensations and emotions which cannot be treated objectively and must bring in personal preferences and differences of opinion.

It is manifestly impossible to reproduce at the two ears of the listener an exact equivalent of the sounds which he would hear in the concert hall. The greatest deficiencies are probably the single point sound reproducer (the loudspeaker) and the volume level. The use of stereophonic reproduction is briefly mentioned in Chapter 20 Sect. 6(v)—it is impracticable with the existing radio and disc recording techniques. In the home it is rarely that orchestral reproduction is heard at the same maximum volume level as in the concert hall. This is the real dilemma from which there is no satisfactory escape. If we listen at a lower volume level we suffer attenuation of the low and high frequencies. If we apply tone correction in the form of bass and treble boost, it is difficult to gauge the correct amount, and most correcting circuits only give a rough approximation to the ideal. If we leave the adjustment of tone controls to the listener, he will adjust them to please himself and the result is usually far from a true reproduction of the original.

Refs. A10, A20.

(ii) Types of distortion

Distortion is lack of fidelity. In the case of an amplifier, distortion occurs when the output waveform differs in some respect from the input waveform. The purpose of the amplifier, and also of the whole equipment, is to reproduce the input waveform, not a band of frequencies. The latter may be necessary in order to achieve the former, but is only a means to an end.

Distortion may be grouped into six main classes.

1. Non-linear distortion (also known as amplitude distortion), and resulting in
1a. Harmonic distortion and
1b. Intermodulation distortion.
2. Frequency distortion (unequal amplification of all frequencies).
3. Phase distortion.
4. Transient distortion.
5. Scale distortion (or volume distortion).
6. Frequency modulation distortion.

There are other features in reproduction which are not normally classed as distortion, although they affect the listener as being untrue to the original. These include background noise and the use of a single point source of sound. Background noise includes needle scratch, hum and adjacent channel whistles. There are also some types of distortion peculiar to loudspeakers—see Chapter 20 Sect. 7.

(iii) Imagery for describing reproduced sound

The following is based largely on Ref. A48, and is merely a proposal which has not been generally accepted. It is included for general interest.

(a) Frequency Range Notation

Extreme Lows	Below 100 c/s
Lows	100— 300 c/s
Lower Middles	300— 800 c/s
Upper Middles	800—1500 c/s
Lower Highs	1500—4000 c/s
Highs	4000—8000 c/s
Extreme Highs	Above 8000 c/s

(b) Distortion

General : Dirty, non-linear distortion.
Overload : Hash-up, mush-up.
Thump (sudden rectification when signal hits bottom).
Sound often becomes strident if harmonic energy peaks in the lower highs.
Records : Fuzz or lace. Inability of stylus to track at high groove curvatures.
Crackle : Same as fuzz, but occurring principally on high-amplitude peaks.
Rattles or buzz, rub or wheeze.
Swish—scratch periodic with rotation of record.
Carbon Microphones : Frying, popping, sizzle.
Sub-harmonics : Breakup, birdies, tweets.
Intermodulation : Harsh, rough.

Cross-over distortion : Marbles, garble.
Transient distortion : Hang-over.
Attack—good or slurred.
Intermodulation with peak in the high-frequency region :
Violins sound wiry.
Male voices have kazoo.
Brass instruments show jamming in upper octaves.

(c) General Terms

Position presence : Localization, mass of sound ; advance, come forward, stand out ; distant, dead, recede, lost.

Intimacy presence : Intimate, rapport.

Detail presence : Transparent, translucent, clear, opaque, acoustic fog, veiled, muddy.

Source size : Live ; broad, volume, floods out, big tones, well-focused tones ; dead and flat, compressed, from a hole in the wall, out of a barrel.

Realism : Presence, natural, life-like, pleasing ; canned music.

Reproduction : Realistic, perfect, adequate.

	Lows	Lower Middles	Upper Middles	Lower Highs	Highs	Extreme Highs
Excess	Grunt	Sock Muddy Solid Dead, dull or thick		Tinkly* Shrill Brassy	Harsh Hard	
				Metallic		
	Boom	Flat-sounding Body Mellow	Masculine Baritone	Bright Brilliant Crisp	Brittle	Brittle
Deficiency		Lean Thin Tinny	Warm Soprano		Soft	Soft

*If confined to upper part of region.

SECTION 2 : NON-LINEAR DISTORTION AND HARMONICS

(*i*) *Non-linearity* (*ii*) *Harmonics* (*iii*) *Permissible harmonic distortion* (*iv*) *Total harmonic distortion* (*v*) *Weighted distortion factor* (*vi*) *The search for a true criterion of non-linearity.*

(i) Non-linearity

A distortionless amplifier has an input voltage versus output voltage characteristic that is a straight line passing through the origin (OA in Fig. 14.1), the slope of the line indicating the constant voltage gain. When non-linearity occurs, as may happen with curvature of the valve characteristics, the input-output characteristic becomes curved as in OB, thereby indicating that the gain of the amplifier is a variable quantity.

(ii) **Harmonics**

One effect of this non-linearity is the production of harmonic frequencies in the output when a pure sine-wave input voltage is applied, hence the name harmonic distortion. For example, if the input voltage is a pure sine wave of frequency 100 c/s, the output may consist of a fundamental frequency 100 c/s, a second harmonic of 200 c/s, a third harmonic of 300 c/s, and so on. Only the fundamental frequency is present in the input, and the harmonics are products of the non-linearity (for mathematical treatment see Chapter 6 Sect. 8).

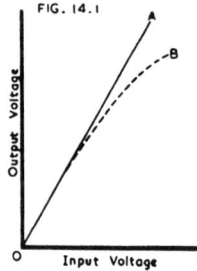

FIG. 14.1

Output Voltage

Input Voltage

Fig. 14.1. *Linearity characteristic* (*or transfer characteristic*) *— OA for distortionless amplifier ; OB for non-linear amplifier.*

Harmonics themselves are not necessarily displeasing, since all musical instruments and voices produce complex sounds having many harmonics (or overtones). But all sounds have certain relationships between the fundamental and harmonic frequencies, and it is such relationships that give the sound its particular quality*. If certain harmonics are unduly stressed or suppressed in the reproduction, the character of the sound (i.e. the tone) will be changed. For example, it is possible for a displeasing voice to be reproduced, after passing through a suitable filter, so as to be more pleasing.

The critical ability of the human ear to distinguish harmonic distortion depends upon the frequency range being reproduced and also upon the volume. Thus, with wide frequency range, the limit of harmonic distortion which can be tolerated is noticeably lower than in the case of limited frequency range, for the same volume level.

If the amplifier does not amplify the harmonics to the same extent as the fundamental, the effective harmonic distortion will be changed. Bass boosting reduces the harmonic distortion of bass frequencies, while treble boosting increases the harmonic distortion of fundamental frequencies whose harmonics are in the frequency range affected by the boosting.

The shape of the audibility curves for the ear, which has maximum sensitivity at about 3000 c/s (Fig. 19.7), indicates that there is some effective boosting of harmonics compared with the fundamental for harmonics up to about 3000 c/s, and an opposite effect for fundamental frequencies above 3000 c/s. This effect, however, is quite small at frequencies up to 1500 c/s, at the usual maximum listening levels (say 65 to 80 db on a sound level meter). At lower listening levels the effect will be appreciable, but the distortion is normally fairly low at these levels in any case. With fundamental frequencies above 3000 c/s there is an appreciable reduction of harmonic distortion at all levels (this effect has been referred to by Ladner, Ref. E14). For example, with a fundamental frequency of 3000 c/s, at a loudness level of 80 phons, the second harmonic will be attenuated by 9 db and the third harmonic by 15 db.

Harmonic voltages or currents may be expressed in the form of a percentage of the fundamental voltage or current. For example, if the fundamental voltage is 100 volts, and there is a second harmonic voltage of 5 volts, the second harmonic percentage is 5%.

Some harmonics are dissonant with the fundamental and (unless at a very low level) are distinctly unpleasing to the listener as regards their direct effect, quite apart from

---

*These remarks apply to sounds which have slow attack and recovery times. In the case of most orchestral instruments this effect is largely overshadowed by the more prominent effects of transients. Many sounds (e.g. vowels) have an inharmonic content in addition to the harmonic content (Ref. E14).

the secondary effect on intermodulation. The following table gives the relevant details when the fundamental frequency (C) is taken for convenience as 250 c/s.

Harmonic	Harmonic frequency	Musical scale*	Effect
2nd	500	C^i	
3rd	750	G	
4th	1000	C^{ii}	
5th	1250	E	
6th	1500	G	
7th	1750	—	dissonant
8th	2000	C^{iii}	
9th	2250	D	dissonant
10th	2500	E	
11th	2750	—	dissonant
12th	3000	G	
13th	3250	—	dissonant
14th	3500	—	dissonant
15th	3750	B	dissonant
16th	4000	C^{iv}	
17th	4250	—	dissonant
18th	4500	D	dissonant
19th	4750	—	dissonant
20th	5000	E	
21st	5250	—	dissonant
22nd	5500	—	dissonant
23rd	5750	—	dissonant
24th	6000	G	
25th	6250	G #	dissonant

*Natural (just) scale.   See Chapter 20 Sect. 8(iv).

(iii) Permissible harmonic distortion

The effect of non-linear distortion usually first becomes apparent to the listener through the production of inharmonic frequencies by intermodulation when two or more frequencies are present in the input (see Sect. 3). In practice, therefore, the percentages of the various harmonics which can be tolerated are fixed by their indirect effects rather than by their direct effect.

Under ideal theoretical conditions†, the magnitudes of the individual harmonics to produce equal intermodulation distortion are inversely proportional to the order of the harmonic. For example, 5% second harmonic produces the same intermodulation distortion as 3.3% third harmonic or 0.77% thirteenth harmonic—see below (v) Weighted distortion factor.

In order to make true comparisons between different sound systems, it is desirable to specify the frequency ranges, and the amplitudes of all the harmonics up to, say, the thirteenth. In the special case of single Class A triodes (below the overload point) the effect of the harmonics higher than the third may be neglected.

For a typical single Class A triode (type 2A3, Ref. E3) the harmonics, as derived from Olson, are :

Harmonic	2nd	3rd	4th	5th	
level	− 30	− 50	− 70	—	db
percentage	3.16	0.3	0.03	‡	%

With push-pull operation, the third harmonic becomes the dominant harmonic and, if the grid excitation is increased to give the same value of total harmonic distortion as with the single valve, the higher harmonics become more significant. With Class AB₁ operation, the higher harmonics are still further increased and, when weighted in proportion to the order of the harmonics, odd harmonics up to the thirteenth are prominent, and harmonics up to the twenty fifth may be appreciable.

†These conditions require the distortion to be similar to that produced by a Class A triode with not more than about 5% second harmonic distortion and without running into positive grid current.
‡Less than 0.01%.

With a **single pentode** the weighted values of the higher order harmonics are quite appreciable. The values of distortion—columns 1 and 2—are derived from Olson† for type 6F6 (Ref. E3).

Harmonic	level (db)	percentage	weighting factor‡	weighted distortion‡
2nd	−20	10	1	10%
3rd	−25	5.6	1.5	8.4%
4th	−35	1.8	2	3.6%
5th	−45	0.56	2.5	1.4%
6th	−62	0.08	3	0.24%
7th	−62	0.08	3.5	0.28%
8th	−63	0.07	4	0.28%
9th	−65	0.06	4.5	0.27%
10th	−73	0.022	5	0.11%
11th	−74	0.02	5.5	0.11%
12th	−73	0.022	6	0.13%
13th	−67.5	0.04	6.5	0.26%
14th	−74	0.02	7	0.14%

‡These are based on the ideal theoretical condition for equal intermodulation distortion ; they may differ considerably from the true effect on the listener.

With any system of amplification other than Class B, the percentage of total harmonic distortion, as defined in (iv) below, decreases as the power output level is reduced. Moreover, the percentages of the highest order harmonics decrease more rapidly than those of the lower order harmonics, as the power level is reduced.

The use of **negative feedback*** merely reduces all harmonics in the same proportion and does not affect their relative importance, except when the overload point is approached.

Comparative tests have been carried out by Olson (Ref. E3) which indicate
1. That slightly greater distortion is permissible with speech than with music.
2. That a higher value of total harmonic distortion is permissible with a single triode than with a single pentode, for the same effect on the listener.
3. That the permissible distortion decreases as the cut-off frequency is increased. However the rate of change varies considerably with the three categories of distortion, as given below.
4. **Perceptible distortion** (for definition see below).
Distortion becomes perceptible when the measured total harmonic distortion reaches the level of 0.7% for music, and 0.9% for speech, with a frequency range of 15 000 c/s and a pentode valve. Even with the very limited frequency range of 3750 c/s a total harmonic distortion of 1.1% on music, and 1.5% on speech, is perceptible with a pentode valve. The difference between pentode and triode is negligibly small.
5. **Tolerable distortion** (for definition see below).
The measured total harmonic distortion to give tolerable distortion on music, with either triode or pentode, increases about four times when the frequency range is reduced from 15 000 to 3750 c/s. The value of tolerable distortion with a pentode is 1.35% total harmonic distortion with a frequency range of 15 000 c/s, and 5.6% with a frequency range of 3750 c/s on music, and 1.9% and 8.8% respectively on speech. The amount of permissible total harmonic distortion with a triode is greater than that with a pentode for the same level of tolerable distortion, in the approximate ratio of 4 to 3.
The effect of frequency range on tolerable distortion is much greater than on perceptible distortion.
6. **Objectionable distortion** (for definition see below).
Objectionable distortion, on music, with either triode or pentode, increases about 5.5 times when the frequency range is reduced from 15 000 c/s to 3750 c/s ; the

---

†By kind permission of the author and of the publishers and copyright holders, Messrs. D. Van Nostrand Company Inc.
*It is assumed that the feedback network is not frequency-selective.

effect of frequency range on objectionable distortion with music is thus greater than its effect on tolerable distortion and much greater than on perceptible distortion. The values of total harmonic distortion to provide objectionable distortion are 2% with a frequency range of 15 000 c/s and 10.8% with a frequency range of 3750 c/s for music, and 3% and 12.8% respectively for speech, with a pentode. The value of total harmonic distortion with a triode to give the same level of objectionable distortion as a pentode is approximately in the ratio 5/4.

Definitions—Perceptible distortion is defined as the amount of distortion in the distorting amplifier which is just discernible when compared with a reference system of very low distortion. Tolerable distortion is the amount of distortion which could be allowed in low-grade commercial sound reproduction. Objectionable distortion is the amount of distortion which would be definitely unsatisfactory for the reproduction of sound in phonograph and radio reproduction. The two latter are dependent upon personal opinion.

Test conditions—Information derived from these tests has been tabulated below, the conditions being

Triode : single 2A3, power output 3 watts.
Pentode : single 6F6, power output 3 watts.

The sound was reproduced in a room with acoustics similar to a typical living room with a noise level of about 25 db. The tests were performed with a limited number of critical observers.

Test Results (Ref. E3)

Cut-off frequency	3750	5000	7500	10 000	15 000 c/s
		Objectionable distortion			
Music Triode	14.0	8.8	4.8	3.4	2.5%
Pentode	10.8	6.0	4.0	2.8	2.0%
Speech Triode	14.4	10.8	6.8	5.6	4.4%
Pentode	12.8	8.8	6.4	4.4	3.0%
		Tolerable distortion			
Music Triode	6.8	5.6	4.4	3.4	1.8%
Pentode	5.6	4.0	3.2	2.3	1.35%
Speech Triode	8.8	7.2	4.8	3.6	2.8%
Pentode	8.8	5.2	4.0	3.0	1.9%
		Perceptible distortion			
Music Triode	1.2	—	0.95	—	0.75%
Pentode	1.1	—	0.95	—	0.7%
Speech Triode	1.4	—	1.15	—	0.9%
Pentode	1.5	—	1.2	—	0.9%

(iv) Total harmonic distortion

The **distortion factor** of a voltage wave is the ratio of the total r.m.s. voltage of all harmonics to the total r.m.s. voltage.

The **percentage of total harmonic distortion** is the distortion factor multiplied by 100 :

$$D = \frac{\sqrt{E_2^2 + E_3^2 + E_4^2 + \dots}}{\sqrt{E_1^2 + E_2^2 + E_3^2 + E_4^2 + \dots}} \times 100 \qquad (1)$$

where D = percentage of total harmonic distortion
E_1 = amplitude of fundamental voltage
and E_2 = amplitude of second harmonic voltage, etc.

If the distortion is small, the percentage of total harmonic distortion is given approximately by

$$D \approx \sqrt{(H_2\%)^2 + (H_3\%)^2 + \dots} \qquad (2)$$

where $(H_2\%)$ = second harmonic percentage, etc.
with an error not exceeding 1% if D does not exceed 10%.

The measurement of total harmonic distortion is covered in Chapter 37 Sect. 3(ii).

(v) **Weighted distortion factor**

Some engineers use a weighted distortion factor in which the harmonics are weighted in proportion to their harmonic relationship :

$$W.D.F. = \frac{1}{2}\sqrt{\frac{(2E_2)^2 + (3E_3)^2 + \ldots}{E_1{}^2 + E_2{}^2 + E_3{}^2 + \ldots}} \qquad (3)$$

For example, if $E_1 = 100$, $E_2 = 5$, $E_3 = 8$ volts,
i.e. second harmonic $= 5\%$
and third harmonic $= 8\%$
then $W.D.F. = 0.129$ (or 12.9%)
and Total distortion factor $= 0.094$ (or 9.4%).

When all the distortion is second harmonic, this weighted distortion factor is the same as the total distortion factor.

With this weighted distortion factor it is desirable to measure harmonics up to a high order (say the thirteenth) at frequencies where the high order harmonics come within the frequency band of the amplifier.

(vi) **The search for a true criterion of non-linearity**

Tests carried out by Olson—see (iii) above—indicate that, under the test conditions, the value of total harmonic distortion is a fairly accurate indication of " perceptible distortion ". Total harmonic distortion may thus be regarded as a true criterion of the subjective effects of distortion in an amplifier, provided that the value is of the order of 1% or less, and provided that overloading or " kinks " in the linearity characteristic do not occur. It has yet to be proved whether or not total harmonic distortion of the order of 1% is a true measure of the subjective effects of distortion under all conditions.

It is generally admitted that the value of total harmonic distortion does not provide a true criterion of non-linearity between different types of amplifiers under all possible conditions, even though the measurements may be made over the whole audible frequency range, and even though the frequency ranges may be identical.

It is the writer's opinion that the conventional weighted distortion factor—see (v) above—also fails to provide a true criterion of non-linearity.

It has been found that when measuring intermodulation distortion, the r.m.s. sum method is less effective than the arithmetical peak sum method in comparing different types of amplifiers—see Sect. 3. Since the principle of intermodulation testing inherently gives harmonic weighting, an obvious step would be to adopt the arithmetical sum of the weighted harmonics, using the conventional weighting factor.

One interesting investigation by Shorter (Ref. A45) has indicated that a very much more drastic weighting, using the square of the conventional weighting factor and including all harmonics with at least 0.03% amplitude, showed distortion values in the correct subjective sequence. The results obtained by Shorter are summarized in the table on page 611.

One inference which may reasonably be drawn is that any sharp kinks in the linearity curve, as usually occur in any Class AB_1 or AB_2 amplifier, have a far more serious subjective effect than is indicated by any of the standard methods of measuring distortion—whether total harmonic distortion, conventional weighted distortion factor or the standard form of intermodulation testing.

The excellent work by Olson—as described in (iii) above—has provided the correlation between frequency range and harmonic distortion for either music or speech, triode or pentode, for three levels of subjective distortion. What still remains to be done is for a comprehensive investigation to be made into the equivalent subjective effects (preferably following the definitions laid down by Olson) of all types of amplifiers—both good and bad design—measuring the individual harmonics up to, say, the twenty-fifth. The tests need only be taken with a single frequency range, since the effect of frequency has been adequately covered by Olson.

The test on each amplifier should be made at maximum rated power output into a resistive load, and also, if desired, at a specified level below maximum rated power

Results obtained by Shorter

System	Subjective Classification of Distortion	R.M.S. Sum of Harmonics	Weighted* Distortion	More drastic† Weighted Distortion	Linearity Curve‡
A	Bad	2.3%	5.1%	19.4%	Pronounced kink
D	Bad	3.7%	6.7%	16.5%	Pronounced kink
B	Perceptible	3.3%	5.1%	8.6%	
E	Just perceptible	0.62%	1.3%	4.5%	Slight kink
C	Just perceptible	2.6%	2.8%	3.3%	
F	Not perceptible	0.41%	0.8%	2.2%	

*As (v) above.
†With square of weighting factor of (v) above.
‡Indicated on a C.R.O. (input versus output voltage).

output.   Tests should also be made on the same amplifiers for intermodulation distortion, using both r.m.s. and peak sum methods, with suitable test frequencies.
In all cases, any distortion measurements should be supplemented by
(1) An oscillographic inspection of the waveform, and
(2) An oscillographic inspection of the linearity (input versus output) characteristics for sharp " kinks."
From these results it should be possible to derive a system of harmonic weighting such that the weighted harmonic distortion is a true criterion of the subjective effects of the distortion.

SECTION 3 : INTERMODULATION DISTORTION

(i) Introduction   (ii) Modulation method of measurement—r.m.s. sum   (iii) Difference frequency intermodulation method   (iv) Individual side-band method   (v) Modulation method of measurement—peak sum   (vi) Le Bel's oscillographic method   (vii) Comparison between different methods   (viii) Synthetic bass.

(i) Introduction

Intermodulation distortion is one of the effects of non-linearity when more than one input frequency is applied.   It is evident to the listener in two forms, amplitude-modulation of one frequency by another, and the production of sum and difference frequencies.   See also Chapter 2 Sect. 9(iv).
An example of amplitude-modulation is the effect of a non-linear amplifier on the reproduction of a choir with a heavy organ bass accompaniment.   The choir is amplitude-modulated by the organ—a displeasing effect to the listener.   This effect is negligibly small in amplifiers if the total harmonic distortion is less than say 2%, but it is apparent with some loudspeakers (see Chapter 20 Sect. 7).
The formation of sum and difference frequencies is the second and more serious form of intermodulation distortion.   These frequencies are normally in-harmonic and are the principal cause of the distorted reproduction noticed by any listener—sometimes described as harsh, buzz, rough or unpleasant.
In actual operation there are very many input frequencies applied simultaneously, but it is possible to make comparative tests using only two frequencies.   Tests for

intermodulation distortion may be made in accordance with any of the methods described below. For measurement see also Chapter 37 Sect. 3(ii)A.

The indicated value of intermodulation differs significantly with the method used, and the conditions of testing. It is important, in all cases, to specify both the method and the conditions of testing.

In all cases the two input voltages must be mixed, without intermodulation, before reaching the input terminals of the amplifier—bridge networks or hybrid coils are commonly used.

(ii) Modulation method of measurement—r.m.s. sum

In accordance with this method, one of the two input frequencies is preferably weaker than the other (a ratio of 4 : 1 in voltage or 12 db is used generally throughout this handbook unless otherwise indicated), and the stronger is usually lower in frequency than the other*. The sum and difference frequencies are usually expressed in the form of the percentage modulation of the weaker (high frequency) fundamental voltage. For one pair of sidebands the modulation percentage is given by

2 × sideband amplitude × 100/fundamental amplitude.

When testing, in order to make a fair comparison with single-frequency conditions, the input should be adjusted so as to give the same peak output voltage as under single frequency conditions. **With the 4 : 1 ratio of the two input voltages, the equivalent single-frequency power output is 25/17 or 1.47 times the indicated power output under I.M. conditions** (Ref. B14). The indicated power output is that calculated from the r.m.s. voltage across the load resistance.

Since there are many of these sum and difference frequencies, plus harmonics of both applied frequencies, it is necessary to take account of their magnitudes. This may be accomplished, in accordance with this method, in special equipment for the measurement of total intermodulation distortion by reading the r.m.s. sum of all extraneous frequencies (Refs. B5, B10). For measurement see Chapter 37 Sect. 3(ii).

There is no simple **relationship between harmonic and intermodulation distortion.** For example, it has been shown that in record manufacture the excessive polishing of masters greatly increases intermodulation distortion but does not much affect harmonic distortion (Ref. B6). See also Refs. A45 (correspondence), B7, B11, B19, B20.

If all harmonics are within the frequency range of the amplifier it may be shown that—

1. If only second harmonic is present (a condition which never occurs in practice), the ratio of total intermodulation distortion to second harmonic distortion is 3.2 (Refs. B7, B11).
2. If only third harmonic is present (another condition which never occurs in practice), the ratio of total intermodulation distortion to third harmonic distortion is approximately 3.84 at low values of distortion (Refs. B7, B11).
3. If the distortion is small, the intermodulation sidebands are approximately given by (Ref. B8) :

    Modulation percentage of first ⎫ ≈2 × second harmonic distortion percent-
    intermodulation sideband   ⎭   age.
    Modulation percentage of second ⎫ ≈ 3 × third harmonic distortion percent-
    intermodulation sideband   ⎭   age.

Thus intermodulation distortion is automatically weighted by the order of the distortion.

As a very rough approximation, the ratio I.M./H.D. may be taken as :

   3.2 for single-ended triodes

and   3.8 for push-pull (triodes or pentodes)

where I.M. = total intermodulation distortion (r.m.s. sum)

and   H.D. = total harmonic distortion,

---

*With a-f systems with peaked response in the middle-to-high frequency range, such as high efficiency speech systems and hearing aids, better results are obtained by having the higher input frequency stronger than the lower frequency (Ref. B13). The two methods of measurement will, of course, give different results.

provided that the operation is restricted to the normal low-distortion region (Refs B7, B11). The ratio tends to increase as the distortion increases. These ratios may increase suddenly as the amplifier reaches the overload point. They are also affected if a lower I.M. frequency is selected which gives appreciable attenuation. The ratio is increased if there is any second harmonic cancellation in successive stages in the amplifier or any attenuation of harmonics. In practice therefore, with many disturbing factors, the ratio may vary from less than 1 to more than 6—see summary Ref. B19.

The usual test frequencies are 40, 60, 100, 150 or 400 c/s and 1000, 2000, 4000, 7000 or 12 000 c/s. It is helpful to make tests at two low frequencies—one of these (40 or 60 c/s) should approximate to the low frequency limit of the amplifier ; the other may be 100 or 150 c/s. The distortion at the lower of these two frequencies is largely influenced by any iron-cored transformers, while that at the higher of these two frequencies gives a more normal overall value. The upper frequency may approximate to half the upper frequency limit of the amplifier—this does not, however, give a stringent test of the distortion at high frequencies.

Intermodulation distortion may be visually observed with a C.R.O. (Refs. B9, B19) —see also Sect. 3(vi) below.

Reliable intermodulation measurements may be made in the presence of considerable noise since the latter is excluded by filters to a greater degree than with harmonic distortion. It may also be successfully applied in the case of restricted frequency range where the harmonics would be outside the range of the amplifier. In such a case it is important to remember that the distortion is not zero merely because the harmonics are not reproduced.

Permissible intermodulation distortion

The following is an arbitrary grouping which may be useful (Modulation method— r.m.s. sum) :

Extremely high fidelity a-f amplifier (40 c/s)	I.M. less than 2%
High fidelity a-f amplifier (40 c/s)	I.M. less than 4%
Good fidelity a-f amplifier (60 c/s)	I.M. less than 8%
Fairly good fidelity a-f amplifier (60 c/s)	I.M. less than 20%
Typical radio receiver—a-f amplifier only (150 c/s)	I.M. less than 40%

The values quoted are only a rough guide, since so many factors are involved. They apply to the equivalent single frequency maximum power output, and the lower test frequency is given in brackets.

See also comments in (vii) below.

Tests with 1 : 1 voltage ratio

More stringent tests at the higher frequencies may be made by the use of a 1 : 1 voltage ratio.

(iii) Difference-frequency intermodulation method

In accordance with this method, two frequencies f_1 and f_2, of equal amplitude, are applied to the input and the relative amplitude of the intermodulation component at the difference frequency $(f_2 - f_1)$ is then considered a measure of the inter-modulation distortion, which may be expressed as a percentage of either output voltage. A wave analyser may be used to measure the difference frequency. Typical lower frequencies are 1000, 5000 and 9000 c/s, while difference frequencies range from 50 to 500 c/s (Ref. A51) or even higher (Ref. B22). For the precise interpretation of results, a fairly low difference frequency is desirable.

Equipment has been designed so that the frequencies of the two input voltages can be varied over the a-f range with a constant difference frequency maintained between them (Refs. B3, B22).

See also comments in (vii) below.

References A51, B3, B12, B13, B22.

(iv) Individual sideband method

In accordance with this method the individual sidebands are measured separately, and some attempt made to select the most important ones (Ref. B1).

If all significant sidebands are measured, the distortion may be recorded in accordance with the peak sum modulation method—see (v) below.

(v) Modulation method of measurement—peak sum

This newer method of defining and measuring intermodulation distortion (Ref. B16) has outstanding advantages over the older method based on the r.m.s. sum. This peak sum method will measure the arithmetical sum of the amplitudes of the modulation products involved, with no discrimination against the weaker modulation products.

In accordance with this method, the percentage intermodulation is defined as the arithmetical sum of the amplitudes of the " in phase " modulation products divided by the amplitude of the high frequency carrier,

i.e. percentage intermodulation $= \dfrac{A_1 + A_{-1} + A_2 + A_{-2} + \cdots}{A} \times 100$

Where A_1 and A_{-1} are the peak amplitudes of the modulation products of frequencies $\omega_2 \pm \omega_1$,

A_2 and A_{-2} are the peak amplitudes of the modulation products of frequencies $\omega_2 \pm 2\omega_1$, etc.,

A is the peak amplitude of the high frequency carrier,

ω_1 is the angular velocity of the low frequency input,

ω_2 is the angular velocity of the high frequency input.

The input amplitude of ω_2 is 12 db below that of ω_1.

In order to measure the sum of these voltages accurately, it is necessary to use a peak-reading voltmeter. Ref. B16 describes such an analyser primarily for use with frequencies of 400 and 4000 c/s for testing pickups, but also capable of use with input frequencies less than 400 or greater than 4000 c/s.

See also comments in (vii) below.

(vi) Le Bel's oscillographic method

The use of an oscilloscope to give a qualitative indication of intermodulation distortion is well known—a good description is given in Ref. B9. A quantitative method has subsequently been developed by Le Bel (Ref. B19). The two voltages of different frequencies are applied to the input of the amplifier in the usual way, the output is passed through a high-pass filter to an oscilloscope with the sweep adjusted to cover one cycle of the low frequency. If there is no intermodulation, the high frequency wave has constant peak amplitude and the envelope is rectangular as in Fig. 14.1A.

FIG. 14.1A   FIG. 14.1B

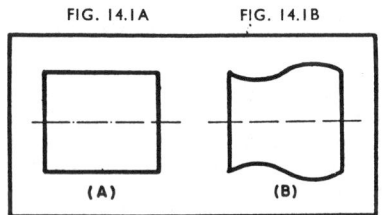

(A)   (B)

FIG. 14.1C

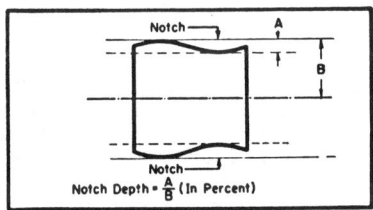

Notch

A

B

Notch

Notch Depth = $\frac{A}{B}$ (In Percent)

Fig. 14.1A & B. Envelope of oscilloscope images without intermodulation (A), and with intermodulation (B). Ref. B19.
Fig. 14.1C. Definition of notch depth (Ref. B19).

Intermodulation results in an envelope such as Fig. 14.1B. Normally, the intermodulation causes a " notch " or notches in the envelope and the notch depth is defined by Fig. 14.1C. The experimental relationship which has been determined between notch depth and intermodulation (using the normal modulation method of measurement—r.m.s. sum—as in Ref. B2) is shown in Fig. 14.1D. This curve is practically linear below 50% notch depth (i.e. 10% intermodulation), so that a scale may be used directly on the screen for measuring the value corresponding to each notch. If there is more than one notch, Le Bel defines total notch depth as the arithmetical

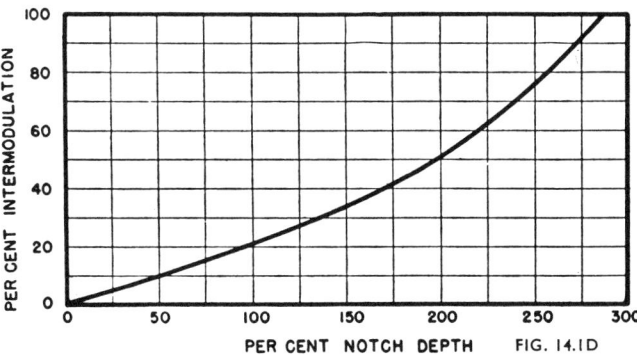

Fig. 14.1D. Relation between notch depth and per cent. intermodulation (Ref. B19).

FIG. 14.1E   FIG. 14.1H

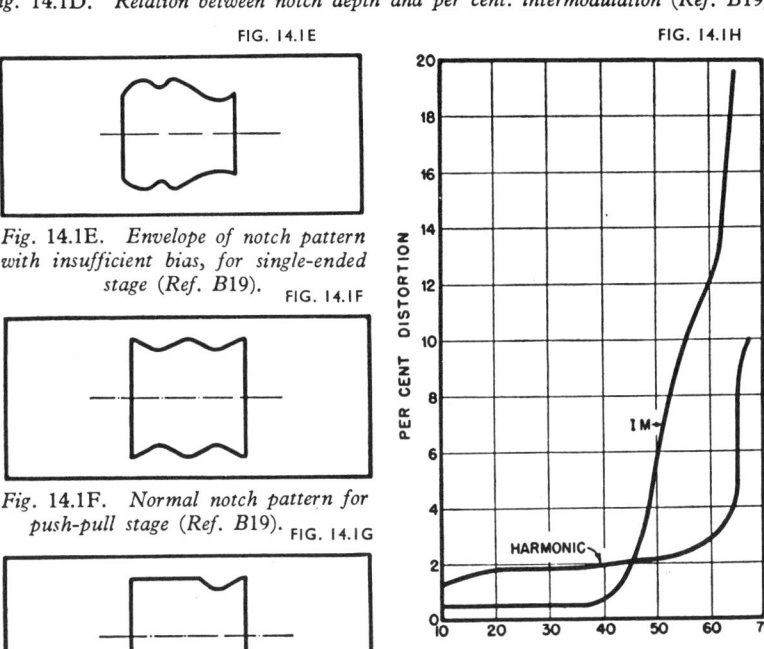

Fig. 14.1E. Envelope of notch pattern with insufficient bias, for single-ended stage (Ref. B19).   FIG. 14.1F

Fig. 14.1F. Normal notch pattern for push-pull stage (Ref. B19).   FIG. 14.1G

Fig. 14.1G. Push-pull output stage with single-ended driver stage showing effect of driver overload (Ref. B19).

Fig. 14.1H. Intermodulation (IM) and harmonic distortion characteristics of a push-pull amplifier showing that the ratio of the two parameters changes (Ref B19).

sum of the individual notch depths. Each notch in the low frequency cycle on top and/or bottom of the envelope is counted. Some typical envelope patterns are shown in Figs. 14.1E, F and G. If a peak instead of a notch occurs at any point, this indicates regeneration.

The harmonic and intermodulation* distortion characteristics of a push-pull amplifier are shown in Fig. 14.1H. At low output levels the intermodulation is less than the harmonic distortion, while at high output levels the reverse is true. This

---

*In accordance with Le Bel's oscillographic method.

indicates that the method is a much more sensitive indication of the approach towards the overload point than is total harmonic distortion.

See also comments in (vii) below.

(vii) Comparison between different methods

In recent years the modulation method of measurement based on the r.m.s. sum has been widely used. However this method has a serious shortcoming in that the lower amplitude modulation products have a negligibly small effect on the reading, which is almost entirely controlled by a few modulation products of high amplitude. It has been clearly demonstrated that this method does not give an indication proportional to the subjective effect on the listener, particularly when discontinuities (" kinks ") occur in the valve characteristics as with Class AB or Class B operation, or when running into grid current with a high impedance source.

This defect is at least partially overcome by the use of the peak reading analyser described in (v) above, which is distinctly preferable as giving an indication more in line with the results from listening tests.

It seems that Le Bel's method described in (vi) above is an oscillographic equivalent of the peak sum method, and close correlation would be expected between them.

It remains to be proved whether or not the peak sum method gives readings proportional to the subjective effect on the listener under all conditions. It may be that some " weighting " method is required to give even greater prominence to the higher-order products that arise from sharp kinks in the linearity characteristic.

All modulation methods with the lower frequency at the higher level are primarily tests on low frequency distortion and, as such, fulfil a useful purpose. It is essential, for design purposes, to have separate measurements of distortion both at low and high frequencies. Insufficient investigation has been made into the general use of modulation methods with the higher frequency at the higher level, for testing high frequency distortion (however see Refs. B13, B19).

The difference-frequency method is very suitable for testing distortion at middle frequencies, and also at high frequencies where conventional* modulation methods are deficient. Unfortunately there are no published data to enable a comparison to be made with the peak sum modulation method.

The distortion indicated by the several methods may differ appreciably. For example, consider high-frequency pre-emphasis as used in recording and F-M. If distortion takes place after pre-emphasis and before de-emphasis, and if the distortion is measured after de-emphasis, the percentage harmonic distortion is lowered by the de-emphasis ; the modulation method of measuring intermodulation is not significantly altered by de-emphasis ; while the difference-frequency method gives a higher value as the result of de-emphasis (Ref. B13).

See also Ref. A43 and Chapter 37 Sect. 3(ii)A.

(viii) Synthetic bass

(A) When two or more input frequencies are applied to a non-linear amplifier the output will include sum and difference frequencies located about each of the higher input frequencies. For example with input frequencies of 50 and 150 c/s, the output will include frequencies of 50, 100, 150 and 200 c/s. Even if the lowest frequency is very much attenuated by the amplifier, the sum and difference frequencies tend to create the acoustical impression of bass. With more than two input frequencies the effect is even greater, so that fairly high distortion has the effect of apparently accentuating the bass.

(B) Owing to the peculiar properties of the ear, a single tone with harmonics may be amplified, the fundamental frequency may be completely suppressed, and yet the listener hears the missing fundamental (Ref. A18).

These two effects assist in producing " synthetic bass "—Chapter 15 Sect. 12(ii)— when the natural bass is weak or entirely lacking. It should be emphasized that this is not the same as true bass, and does not constitute fidelity.

---

*With the lower frequency having the greater amplitude.

SECTION 4 : FREQUENCY DISTORTION

(i) Frequency range   (ii) Tonal balance   (iii) Minimum audible change in frequency range   (iv) Sharp peaks.

(i) Frequency range

Frequency distortion in an amplifier is the variation of amplification with the frequency of the input signal. A high fidelity a-f amplifier should have nearly constant amplification over the whole range audible to the most critical listener—from say 30 to 20 000 c/s (see Sect. 7). There is no serious difficulty in designing such an amplifier, but it can only be usefully applied when it is fed from a wide-range source and excites a wide-range loudspeaker system.

In practice there are limitations imposed by both the source and the loudspeaker, but it is good practice to design the amplifier for full-power-handling capacity with negligible non-linear distortion over a wider frequency range than it will normally be required to handle. The limitation in frequency range (if not inherent in the source) should preferably be applied either by a filter between the source and the input terminals, or in an early stage in the amplifier—see Chapter 15 Sect. 1(iii).

The full frequency range is only audible at very high levels as is demonstrated in Sect. 7. If a particular equipment is to be operated always at comparatively low levels (e.g. dinner music) or in noisy locations, reduction of the frequency range is quite correct and may even be beneficial.

Amplifiers for other than wide-range high-fidelity are usually designed to meet the requirements of the average listener, with a frequency range (at maximum orchestral level) of about 45 to 15 000 c/s, which may be still further reduced to about 70-13 000 c/s at typical levels for home listening—see Sect. 7(ii).

In commercial quantity-produced equipment the frequency range is often restricted to the extreme. By this means non-linear distortion and hum are reduced to a bearable level without expense. Unfortunately, frequency range can only be extended, while remaining free from obvious distortion, at a cost which rises at a rapidly increasing rate—largely due to the loudspeaker. It is important to reduce the distortion before widening the frequency range—see Sect. 12(ii).

(ii) Tonal balance

It has been found that good **tonal balance** between high and low frequencies is obtained when the product of the limiting frequencies is about 500 000* (Refs. A8, A17, A38, C1, C10).

For example :
1. 25-20 000 c/s. Very wide frequency range,
2. 33-15 000 c/s. Wide frequency range,
3. 50-10 000 c/s. Fairly wide frequency range,
4. 60- 8 000 c/s. Medium frequency range,
5. 100- 5 000 c/s. Restricted frequency range,
6. 150- 3 300 c/s. Very restricted frequency range,

but these should not be taken as more than a general guide. (5) is typical of a medium quality console, and (6) a mantel model receiver with " mellow " tone. These frequency ranges should include both amplifier and loudspeaker. Wide frequency range is only comfortable to the listener so long as other forms of distortion are imperceptible.

(iii) Minimum audible change in frequency range

Just as with sound level there is a minimum audible change in level, so there is also a minimum audible change in frequency range. The limen has been proposed (Ref. A25) as being the minimum difference in band width that is detectable by half the observers. Tests were made on a limited number of observers resulting in the following list of frequencies in steps of 1 limen : 15 000, 11 060, 8000, 6400, 5300,

---

*Various authorities place this figure from 400 000 to 640 000.

4400 c/s.   It would therefore be possible to reduce the frequency range of an amplifier from 15 000 c s (say the limit of hearing in a particular case) to 11 000 c/s without a non-critical listener noticing the restriction.   The same principle also holds in reverse, when widening the frequency range.   (See also Ref. A9).

(iv) Sharp peaks

One particularly objectionable form of frequency distortion is that due to sharp peaks in the output, as may be caused by loudspeaker cone resonances, especially in the 2000-3500 c s range.   Sharp troughs are relatively unimportant through their direct effects, although they may be accompanied by poor transient response.

The effect of restricted frequency range on articulation is covered in Sect. 11.

SECTION 5 : PHASE DISTORTION

Phase distortion is the alteration by the amplifier of the phase angle between the fundamental and any one of its harmonics or between any two component frequencies of a complex wave.   Phase distortion causes the output waveform to differ from the waveform of the input voltage.

A fixed phase shift of $180°$ (or any multiple of $180°$) at all frequencies does not constitute phase distortion.   A phase shift which is proportional to the frequency also does not cause phase distortion.

It has been demonstrated in monaural listening (Ref. A23) that the tonal quality of a complex steady-state waveform depends not only upon the component frequencies and their relative amplitudes, but also—over certain ranges of frequency and level—upon the envelope shape, which is controlled by the phase angles.   Large changes in aural perception can occur, through changes in phase only, among frequencies which are related harmonically or otherwise.   These effects vary from a raucous to a smooth-sounding quality, depending on the relative phases.   The frequency range, over which these effects were observed, increases as the acoustical level increases, at least up to 60 phons.

The authors state that " in the case of very complex, but continually varying sounds, such as voice or a musical instrument with many harmonics, phase effects of this type are probably not noticeable under usual listening conditions because the patterns are coming in and out non-simultaneously in various parts of the frequency range."

Phase shift has a serious effect on transients (see Sect. 6).

In the light of the known effects, and with insufficient data to determine what degree of phase shift is permissible for high fidelity reproduction, it is wise to reduce the phase shift in amplifiers to the lowest practicable value.

With normal circuits, the phase shift is determined by the overall attenuation at the extreme low and high frequencies of the sound system.   If the frequency characteristic is practically flat over the whole useful frequency range, the phase shift may normally be neglected.   However, when a high-pass or low-pass filter is used to limit the frequency range, or when some form of tone control is used, the phase angle may be quite large over a limited frequency range.

Refs. A15, A23, A28, A40, A42, A57, E11.

SECTION 6 : TRANSIENT DISTORTION

(i) General survey (ii) Testing for transient response.

(i) General survey

It has been demonstrated by Prof. Richardson that it is the attack and decay times of sounds that largely determine their tonal quality, rather than their harmonic content (Ref. A41).

An amplifier which gives fairly good reproduction of steady tones may give serious distortion with transients. The distortionless reproduction of a short pulse requires

1. Very wide frequency response—possibly higher than the limit of audibility.
2. No phase distortion.
3. No " hang-over." The duration should not be greater than the original pulse.

" Hang-over " effects are caused by insufficiently damped LC circuits such as may occur with a-f transformers, tone correction circuits and filters—see Chapter 15 Sect. 1(ix). Loudspeakers are particularly prone to this effect, and it is desirable for them to have heavy acoustical and electro-magnetic damping, with reduction of mass of moving parts.

Some measurements of the phase shift introduced by microphones and loudspeakers are given in Ref. A40.

References : A40, A41, E1.

(ii) Testing for transient response

Amplifiers and loudspeakers may be tested by applying to the input terminals a waveform with a sharp discontinuity—e.g. square wave or saw-tooth.

One of the most serious problems with certain kinds of music is the occurrence of extremely high transient peaks which may possibly reach a level 20 db above the reading of the volume indicator. This effect has been observed with choral music, orchestral string passages and similar highly complex sounds (Ref. E15—see also A3). These peaks cause overloading and distortion in the amplifier and loudspeaker, but particularly in recording. Some form of peak limiting is usually adopted in recording and broadcasting in order to avoid blasting or " buzz." However, for true fidelity, these transient peaks should be reproduced in their original form.

One interesting approach is the use of **white-noise** as a test signal. White-noise is random fluctuation noise, but the construction of a white-noise signal generator whose output power is uniformly distributed over the a-f region of the spectrum is difficult. One white-noise generator, which heterodynes a selected portion of the r-f noise spectrum to produce nearly uniform voltage over the a-f range, is described in Ref. D18. A simpler method is described in Ref. D17 which employs a disc recording of white-noise (40 to 20 000 c/s) together with bands having high frequency cut-off at 7000, 9000 and 12 000 c/s and low frequency cut-off at 80 and 150 c/s. This is recorded on a constant velocity basis. A flat noise voltage characteristic may be obtained by playing back with a velocity pickup of the dynamic or magnetic type, without any equalizing network. The recording is corrected for translation loss— 4 db at 20 000 c/s. The following comments are based on Ref. D17.

Loudspeakers may be tested for undamped resonances by applying white-noise signal, picking up the sound by a suitable microphone and applying the output to the vertical plates of an oscilloscope. The sweep frequency is then varied to pick up the resonances. Cross modulation may be recognised by a periodic thinning out of the fine-grain high frequency noise. It is stated that a miniature condenser microphone is particularly suitable for this test.

When an amplifier is overloaded, the peaks on the oscilloscope appear beaded.

The elliptical pattern, obtained by applying the white-noise to both pairs of plates with approximately $90°$ phase difference, may be used as a sensitive indication of overloading. The circuit is arranged to give larger ellipses with higher frequencies.

SECTION 7 : DYNAMIC RANGE AND ITS LIMITATIONS

(i) Volume range and hearing (ii) Effect of volume level on frequency range (iii) Acoustical power and preferred listening levels (iv) Volume range in musical reproduction (v) The effect of noise.

(i) Volume range and hearing

Every source of sound, for example speech or music, has a variation in sound level from its minimum to its maximum.   The loudness level is measured in phons (see Chapter 19 Sect. 5) while the maximum variation in level may be expressed in decibels —this is known as the **volume range,** and is measured by a standard sound-level meter or volume indicator.   As this instrument does not indicate short sharp peaks, the **peak dynamic range** is usually 10 or occasionally up to 20 db greater than that indicated by the instrument.*

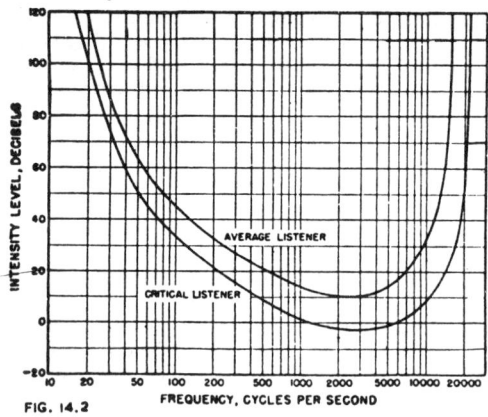

FIG. 14.2

Fig. 14.2.   Threshold of audibility curves for average and very critical (5% most acute) listeners in the absence of noise. Reference level 0 db = 0.0002 dyne per sq. cm. Curves by courtesy of Jensen Radio Manufacturing Company, based on Fletcher (Ref. A3).

A symphony orchestra, with a peak dynamic range of up to 70 db provides the most difficult problem for the reproducing equipment.   This range is equivalent to a power ratio of 10,000,000 : 1.

Before proceeding further, it is necessary to consider some of the characteristics of hearing.   Fig. 14.2 shows the hearing characteristics of average and very critical listeners in the absence of noise—at any one frequency, a listener is only able to hear sound intensity levels above the curve.

Fig. 14.3.   Masking levels for noise in average and very quiet residences. Curves by courtesy of Jensen Radio Manufacturing Company, based on Refs. A3, D6, D7, E9.

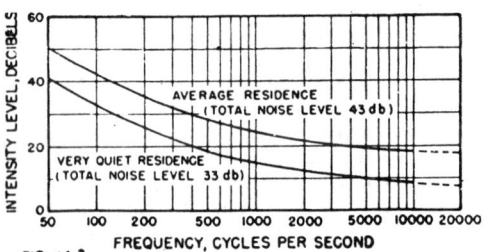

FIG. 14.3

The masking† effect of room noise is shown by Fig. 14.3 for average and for very quiet residences.   These curves may be applied directly to the hearing characteristics of Fig. 14.2.   The result is shown in Fig. 14.4 for average conditions and in Fig. 14.5

*The 10 db ratio is given in Ref. A51, quoting Refs. E7, E16.   The 20 db ratio is mentioned in Sect. 6(ii), based on Refs. E15 and A3.
†The masking level is the level of pure tones which can just be perceived in the presence of noise. The masking level is higher than the noise (spectrum) level at the same frequency by a margin of 15db up to 1000 c/s, increasing to 28 db at 10 000 c/s (Ref. A53).

for a very critical listener and low room noise level (these latter are the extreme conditions for high fidelity). In each case the room noise reduces the effective hearing over a frequency range from about 150 to 6000 or 9000 c/s.

The effect of noises other than room noise is covered in (v) below.

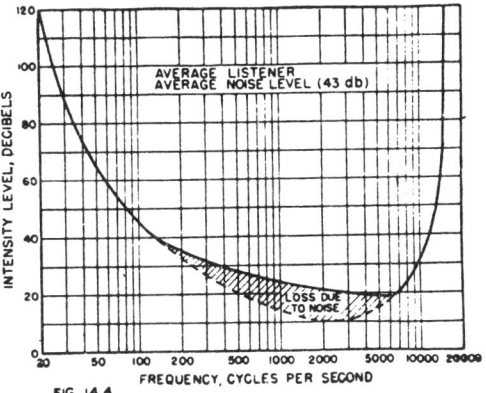

Fig. 14.4. Effective hearing characteristic for average listener with average noise level. Curves by courtesy of Jensen Manufacturing Company.

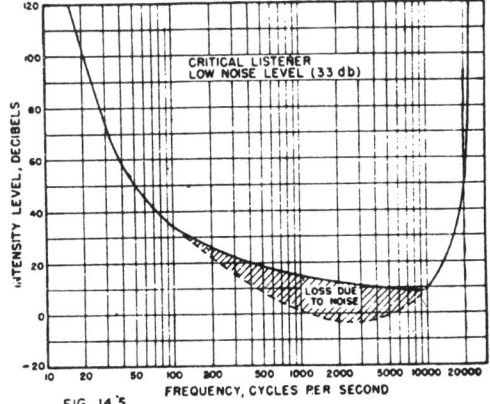

Fig. 14.5. Effective hearing characteristic for very critical listener with low room noise level. Curves by courtesy of Jensen Manufacturing Company.

(ii) Frequency range

When listening to the rapidly varying intensity levels which occur with speech or music, the ear appears to integrate these varying sounds over about 1/4 second intervals. When considering frequency range, the measurements of the sound levels created by the various types of musical instruments should therefore be reduced to their **equivalent sound levels in 1/4 second intervals.**

If we compare the loudness of a narrow band of thermal noise with that of a pure tone having the same intensity, the two will be judged to have equal loudness if the width of the transmitted frequency band of noise is limited to a critical value called the **critical bandwidth** (Ref. A3). For this reason, measurements on the sound levels of musical instruments should be reduced to intensity levels which would have been obtained if the frequency bandwidths used in the filters had been equal to the critical bandwidths (Ref. A3).

The integrated sound energy in a critical band over a 1/4 second interval will sound as loud as a pure tone in the same frequency band which produces the same sound energy in each 1/4 second interval (Ref. A3).

Fig. 14.6 gives the maximum r.m.s. intensity levels in 1/4 second intervals in critical frequency bands for certain musical instruments and an orchestra at a distance of 20 feet from the sound source, together with the threshold of audibility curve for average listeners, indicating that a frequency range from 40 to 15 000 c/s meets all normal requirements for an average listener to an orchestra.*

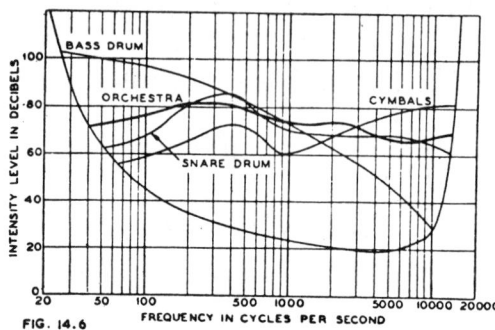

Fig. 14.6. Maximum r.m.s. intensity levels in 1/4 second intervals in critical frequency bands for whole orchestra and certain instruments. After Fletcher (Ref. A3).

Fig. 14.7 (highest curve) shows the same " orchestra " curve as Fig. 14.6, but combined with the threshold of audibility curve for a very critical listener, indicating that the maximum possible frequency range is from 32 to 21 000 c/s. These are the most extreme conditions to be taken into account for high fidelity reproduction. If this orchestra is reproduced at lower levels as indicated by the lower " orchestra " curves in Fig. 14.7, the frequency range is more restricted, particularly at the lower end. The values for both average and critical listeners are tabulated below :

Peak value	100	90	80	70 db
Average listener				
Lower frequency	45	50	70	100 c/s
Upper frequency	15 000	14 000	13 000	12 000 c/s
Critical listener				
Lower frequency	32	40	50	70 c/s
Upper frequency	21 000	20 000	19 000	18 000 c/s

Fig. 14.7. Maximum r.m.s. intensity levels in 1/4 second intervals in critical frequency bands for whole orchestra, with threshold of audibility curve for very critical listener. Also illustrating effect on frequency range of reproduction at lower levels.

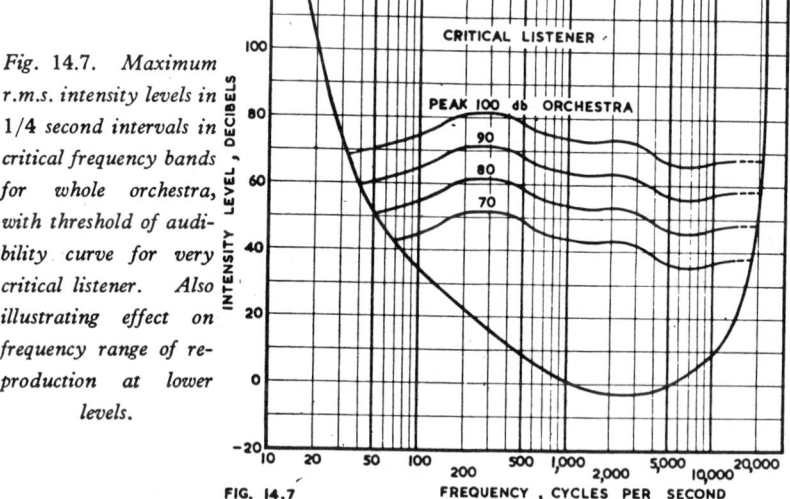

*The bass drum, played solo and very loudly, and the loud organ have frequencies in the vicinity of 30 c/s audible to an average listener.

The curves of Fig. 14.6 and the orchestra curves used in Fig. 14.7 are only suitable for determining the audible frequency ranges of the various sound sources—they cannot be used for calculating the peak power.

(iii) Acoustical power and preferred listening levels

The r.m.s. peak sound level at a desirable seat in a concert hall for " full orchestra " is not likely to exceed a value of 100 db (Refs. A51, E13).

If we accept the value of 100 db as being " maximum orchestral level," then a maximum r.m.s. peak acoustical power of about 0.4 watt will be required for reproduction at the same level in a fairly large living room—see Chapter 20 Sect. 6(ii). If the loudspeaker is 3% efficient over the frequency band, then the maximum power from the amplifier will need to be 13 watts—with a more efficient loudspeaker, the power from the amplifier would be less.

Peaks above 100 db may occur occasionally with a large orchestra or choir, but they are usually " peak limited " before being broadcast or recorded. However in the design of the amplifier it is desirable to allow a margin to provide for possible peaks above this level.

Since sound levels are normally measured with a sound-level meter, it is necessary—when considering acoustical reproduction—to make allowance for the margin of 10 db or more between this level and the r.m.s. peak level.

Preferred listening levels

Tests conducted by the B.B.C. (Ref. C12) on listeners give the following preferred maximum sound level, as measured with a sound level meter :

	Public		Music-ians	Programme Engineers		En-gineers
	Men	Women		Men	Women	
Symphonic music	78	78	88	90	87	88 db
Light music	75	74	79	89	84	84 db
Dance music	75	73	79	89	83	84 db
Speech	71	71	74	84	77	80 db

Individual variations varied from 60 to 97 db for symphonic music, but in all cases 50% of the subjects were within ± 4 db of the mean. Increasing age showed a preference for lower listening levels.

The following table gives an approximate guide for home listening :

Loudness	Sound-level meter reading	Normal r.m.s. peak
Very loud	80 db	90 db
Loud (serious listening)	70 db	80 db
Medium (as a background)	55-65 db	65-75 db

(iv) Volume range in musical reproduction
(A) Orchestral reproduction

If the full peak orchestral dynamic range of 70 db is to be reproduced, the receiver (if any) and amplifier must have a peak dynamic range of the same amount. This may be accomplished with the peak sound intensity in the room either equal to that in the concert hall, or at a lower level. However, if the reproduced dynamic range is to be effective it must all be above the noise masking level.

Fig. 14.8 shows the extreme conditions for high fidelity reproduction, with the threshold of audibility curve for a very critical listener, combined with the masking

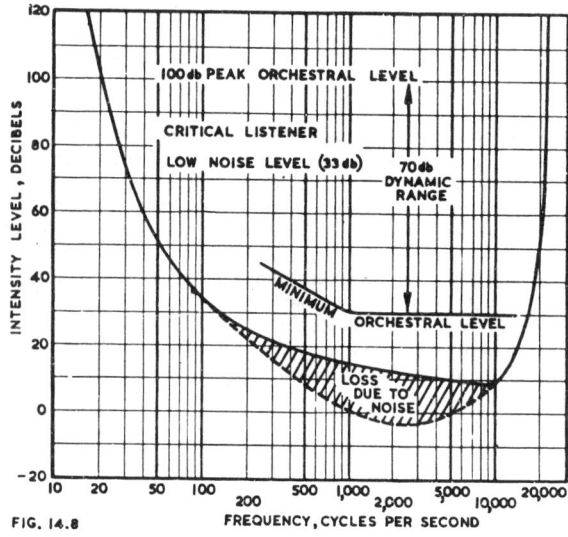

FIG. 14.8

Fig. 14.8. *Effective hearing characteristic for very critical listener with low room noise level, showing* 70 db dynamic range for orchestra in a concert hall. Minimum orchestral level based on Ref. E13.

effect of low room noise.   The 70 db dynamic range of an orchestra in a concert hall is shown, with the minimum orchestral level of 30 db well above the masking level of the noise.   In this case it would be possible to reproduce the full dynamic range at a level 15 db lower than the level in the concert hall, without loss due to noise. With an average listener and average noise level, however, the reproduction could not be more than 5 db below the level in the concert hall without loss due to noise.   Any further reduction in level will result in the masking of the softest passages by the noise.

(B) Volume range broadcast or recorded
   The volume range actually broadcast or recorded is frequently less than that of the original.

Source	Volume range
F-M broadcast (direct)	at least 60 db (F.C.C.)
A-M broadcast (direct)	at least 50 db (F.C.C.)
Transmission over telephone lines	40-50 db*
ditto (special lines)	50-75 db*
Lateral-cut disc records :	
shellac	30-45 db*
microgroove, say	40-56 db*
transcription (orthacoustic)	45-58 db*
Vertical-cut (hill and dale)	45-50 db*

   To reduce the volume range, it is usual to employ automatic peak limiters together with volume compressors, although some manual adjustment is also made on occasions. To restore the volume range, it is sometimes possible to use volume expansion (see Chapter 16).

(v) The effect of noise
   Noise as heard by the listener is contributed by many sources—for example by background noise in the studio, by the microphone and pre-amplifier, the transmitter, the receiver, the a-f amplifier and room noise.   If the sound source is a record, there will be additional noise due to the recording.   Under conditions for good fidelity, noise due to the transmitter and receiver (if any) can be made negligibly small.   The remaining noise sources may be grouped as

---
*Maximum signal to noise ratio.   The maximum dynamic range may be somewhat higher—see Sect. 7(v).

1. Noise from the sound source.
2. Noise from the amplifier.
3. Room noise.

It is assumed here that hum from the amplifier has been made inaudible, and then the total* noise arising from amplifier and room noise may be determined by the nomogram Fig. 19.6A when both component values are expressed in decibels. By following a similar procedure, this value may then be combined with noise from the sound source to give the total noise from all sources.

The value of room noise is 43 db for the average level (Fig. 14.4) and 33 db for the low noise level (Fig. 14.5).

It is usually simpler to combine these noises electrically than acoustically, a convenient point being the primary of the output transformer. The equivalent electrical value of the room noise for a typical living room of 3000 cubic feet is given approximately by†

Room noise	Electrical noise power (milliwatts)
33 db	$0.009/\eta$
43 db	$0.09/\eta$

Where η = loudspeaker efficiency percentage.

If the loudspeaker efficiency is 3%, the equivalent electrical value of the room noise is approximately — 25 dbm for low room noise, or — 15 dbm for average room noise.

The masking value of the total noise can only be determined accurately by calculating or measuring the total noise spectrum and applying the masking curves of Fletcher and Munson (Ref. E9 Fig. 15 ; also reproduced in Olson Ref. E3 Fig. 12.30). However for the simple case where room noise predominates, it may be taken as a reasonable approximation that the curves of Fig. 14.3 may be moved vertically as required and that the masking ordinate at 1000 c/s is 18 db below the total noise level.

Thus the masking effect at 1000 c/s of room noise alone is approximately — 43 dbm for low room noise, or — 33 dbm for average room noise, for 3% loudspeaker efficiency and a room volume of 3000 cubic feet. If the total noise from other sources is equal to the room noise, these values would be increased by 3 db.

For high fidelity reproduction, the amplifier and pre-amplifier noise should be kept below the lowest room noise level. Under the preceding conditions, with room volume 3000 cubic feet and loudspeaker efficiency 3%, a reasonable value for total amplifier noise‡ appears to be about — 30 dbm, giving a masking level 5 db below that of the lower value of room noise. For values of loudspeaker efficiency other than 3%, the value of total amplifier noise is given by

$$\text{amplifier noise power} = 0.003/\eta \ \text{mW} \qquad (1)$$
$$\text{or amplifier noise in dbm} = 10 \log (0.003/\eta) \qquad (2)$$

where η = loudspeaker efficiency percentage.

See also Chapter 18 Sect. 2(ii) pages 782-783 for noise in pre-amplifiers ; Chapter 19 Sect. 6(iii) page 829 for the measurement of noise in amplifiers.

SECTION 8 : SCALE DISTORTION

When music or speech is reproduced in a room with flat frequency response and at a level such that the listener experiences the same sensation of loudness as he would in the concert hall or studio, the tonal balance will be the same as in the original (assuming that his seat occupies the same position as the microphone). If the level of reproduction is less than the original, the bass will be weak in comparison with the middle frequency range from 600 to 1000 c/s, while the frequencies above 1000 c/s will also be attenuated, although only slightly. This is illustrated in Fig. 14.9 for one selected condition with 90 phons average original level, reproduced at an average level of 70 phons ; the loss at 50 c/s is over 10 db.

---

*It is assumed that the room noise can be treated as random noise. Random noises are additive.
†Based on Chapter 20 Sect. 6(ii)B.
‡For measurement of amplifier noise see Chapter 19 Sect. 6.

FIG. 14.9

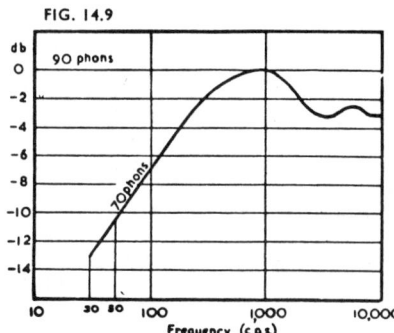

Fig. 14.9. Loss of bass and high audio frequencies caused by reproduction at a loudness of 70 phons as compared with the original at 90 phons (based on Fig. 19.7).

On the other hand, if the reproduction is louder than the original, the bass (and to some extent the treble) will be accentuated. This commonly occurs with the human voice when reproduced at a high level, unless some correction is applied in the studio amplifier.

It is for this reason that orchestral or organ music, when reproduced at normal room volume, sounds weak and uninteresting. The application of the correct amount of bass boosting assists considerably in maintaining realism and tonal balance. This application of bass boosting may be made automatically with the adjustment of the volume control, provided that maximum setting of the control corresponds to maximum volume from the loudspeaker on musical peaks (Chapter 15 Fig. 15.59). This provision makes a second (semi-fixed) volume control desirable in reproduction from records—it calls for an exceptionally flat a.v.c. characteristic in a receiver, in addition to the double volume control. If the automatic arrangement is adopted without these precautions being taken, it is only partially effective, and many users prefer a manually operated bass-boost. The automatic arrangement is only satisfactory when the tonal balance is correct at the maximum setting of the volume control ; it merley boosts the bass when the setting of the control is lowered.

Refs. A7, A29, A30, A33, A34, A56.

SECTION 9 : OTHER FORMS OF DISTORTION

(i) *Frequency-modulation distortion*   (ii) *Variation of frequency response with output level*   (iii) *Listener fatigue.*

(i) Frequency-modulation distortion
Frequency-modulation distortion (or " wow ") tends to occur in recording, and is a cyclic variation of a higher frequency by a lower frequency. Frequency-modulation distortion may also occur in loudspeakers—see Chapter 20 Sect. 7(ii).

(ii) Variation of frequency response with output level
The frequency response of an amplifier should be substantially the same at all working levels. Amplifiers should be tested for frequency response at maximum output and at 30 and 60 db below maximum output. Output transformers are a frequent cause of variations in frequency response at different levels.

(iii) Listener fatigue
There is no doubt that listening to reproduced music causes listener fatigue much more rapidly than listening to the original. This listener fatigue is caused by the necessity for mental processes arising from the unnatural effects in the hearing system.

Probably the creation of synthetic bass, intermodulation distortion and transient distortion play important parts in producing this fatigue.

An excessively high background noise level, whether caused by hum, record scratch or any form of extraneous noise, is also an important factor contributing to listener fatigue.

Ref. C7.

SECTION 10 : FREQUENCY RANGE PREFERENCES

(i) Tests by Chinn and Eisenberg   (ii) Tests by Olson   (iii) Single-channel versus dual-channel tests   (iv) Summing up.

(i) Tests by Chinn and Eisenberg (Ref. C1)

This was the first large scale series of tests on a number of listeners in order to determine their preferences in the frequency range of reproduced music. The ranges were (3 db attenuation) :

Wide   35 to 10 000 c/s.
Medium   70 to   6 500 c/s.
Narrow   150 to   4 000 c/s.

The equipment included a single channel amplifier and loudspeaker system.

Only 12% of the listeners preferred the wide range on classical music, and 21% on male speech ; the remainder had no preference or were divided between narrow and medium. Tests were made at indicated sound intensities of 50, 60 and 70 db, the preference being between 60 and 70 db, varying with the type of music ; a somewhat higher sound intensity was preferred for speech than for music.

These tests gave rise to considerable discussion—see also (iv) below.

Refs. C1, C2, C3, C4, C5, C6, C7, C8, C9, C14, A12, A34.

(ii) Tests by Olson (Ref. C10)

In this case preferences were made on the frequency range for orchestral music with the listeners in the same room as the orchestra. No amplifying equipment was used ; the restriction of frequency range was accomplished by an acoustical filter. Listeners were given a choice between a high frequency limit of 5 000 c/s and full frequency range.

From 66% to 69% of the listeners preferred the full frequency range on music, and there was a majority preference for the full range on speech in " all acoustical " tests.

Refs. C10, C11, C7.

(iii) Single-channel versus dual-channel tests

Webster and McPeak (A19) carried out tests with single channel and dual channel (audio perspective) transmission of a live programme, compared with transcribed music. The majority preferred the transmission of a live programme to transcribed music. The majority also preferred the dual-channel to the single-channel transmission. The results, however, have been seriously challenged regarding their interpretation (C7, A12). The tests gave no indications of the preferred frequency range.

Hilliard (A27) expressed the opinion that the entertainment value of a two channel system reproducing frequencies up to 3500 c/s is better than a single channel system with 15 000 c/s response.

It seems to be generally agreed that any satisfactory form of stereophonic reproduction is very much preferable to single channel—see Chapter 20 Sect. 6(v).

(iv) Summing up

The tests by Chinn and Eisenberg indicated that, under their conditions of test, the majority preferred a restricted frequency range.   The tests by Olson indicated that, under his " all acoustical " listening conditions the majority preferred an unrestricted frequency range.   Obviously the conditions of the tests are of vital importance—see discussion (i) above, also summing-up by Olson, C10.

Subsequent amplifier tests by Olson have indicated that an audience was able to perceive a value of 0.7% total harmonic distortion (see Sect. 2 and Ref. E3).   A reasonable inference is that the reduction in frequency range preferred by listeners in the Chinn and Eisenberg tests was caused by distortion present in the sound system.

It has been pointed out (Ref. C14) in connection with the Chinn and Eisenberg tests, that the choice between " wide " and " medium " (at 60 and 70 db) and between " medium " and " narrow " (at 50 and 60 db) was complicated by inadequate discriminability.

SECTION 11 : SPEECH REPRODUCTION

(*i*) The characteristics of speech   (*ii*) Articulation   (*iii*) Masking of speech by noise   (*iv*) Distortion in speech reproduction   (*v*) Frequency ranges for speech.

(i) The characteristics of speech

It has been shown (Ref. A3) that conversational speech audible at $2\frac{1}{2}$ feet occupies a frequency band from 62 to 8000 c/s.   The average conversational speech power is about 10† microwatts, loud shouting power is about 5 milliwatts, while the softest conversational level is about 0.01 microwatt (Ref. E5).   The peak dynamic range is about 40 db for any one speaker in ordinary conversation, but may reach 56 db as an extreme limit in the general case (Ref. A3).

The peak power in conversational speech may reach 100 or even 200 times the average speech power.   The octave with maximum energy in men's voices is from 300 to 600 c/s, and for women's voices from 555 to 1110 c/s, the energy in the octave being 31% of the total for both cases (Ref. E5).

The characteristics of a man's conversational speech at a distance of $2\frac{1}{2}$ feet have been given as (Ref. A3) :

Level of speech above threshold ($+$ 5 db)	63 db
Long root-mean-square intensity level	68 db
Peak level in 1/8 second intervals exceeded 5% of the time	85 db
Root-mean-square level in 1/8 second intervals exceeded only 1% of the time	78 db

Refs. A3, D5, D9, D20, E4, E5, E6.

(ii) Articulation

The articulation of speech is defined as the percentage of syllables interpreted correctly (Ref. D8).   The results for an average listener receiving syllables at optimum intensity are given below (Ref. A3) :

Articulation	Minimum frequency	or	Maximum frequency
98%	40 c/s		15 000 c/s
98	100		12 000
98	250		7000
96	570		5000
94	720		3900
90	960		3100
80	1500		2300
70	1920		1970
60	2300		1700
50	2600		1500

†A more recent test gave 34 microwatts for men and 18 microwatts for women (Ref. D5).

Columns 2 and 3 are the results of separate tests with a high-pass filter and a low-pass filter respectively. They cannot be combined to give a pass-band with the limiting frequencies as tabulated, except as a rough approximation for high percentages of articulation ($> 90\%$) ; in this case the two articulation values should be multiplied together (e.g. $96\% \times 96\% = 92\%$ articulation approximately for a pass band from 570 to 5000 c/s).

There is no loss in articulation with a frequency range (pass band) from 250 to 7000 c/s, while the loss is quite small with a range from 570 to 5000 c/s. With continuous speech the loss in articulation is noticeably less than with syllables.

Owing to the unnaturally high-pitched tone and obvious lack of balance, these frequency ranges are extended towards the bass for all communication purposes.

An articulation index has been proposed (Ref. D3) which may be calculated for any known conditions, including noise. A value of 1.0 is the maximum (for perfect articulation) while a value of 0.5 or over is entirely satisfactory.

Those who are interested in the intelligibility of speech either with or without noise interference should read Refs. D10, D11, and D12.

Experiments have shown that intelligibility is impaired surprisingly little by the type of amplitude distortion known as **peak clipping**. Conversation is possible even over a system that introduces " infinite " peak clipping, i.e., that reduces speech to a succession of rectangular waves in which the discontinuities correspond to the crossings of the time axis in the original speech signal. The intelligibility of the rectangular speech waves depends critically upon the frequency-response characteristics of the speech transmission circuits used in conjunction with the " infinite clipper." A word articulation value of 97 was obtained using firstly a filter providing a frequency-response characteristic rising 6 db/octave, followed by the infinite peak clipper, followed in turn by a filter with a characteristic falling at the rate of 6 db/octave (Ref. D13).

References to articulation : A3, D3, D8, D10, D11, D12, D13, D19.

(iii) Masking of speech by noise

Noise has a similar masking effect with both speech and music. If the noise level is reasonably low there is very little effect on the reproduction of speech, since the dynamic range is not excessive. It is sometimes required to operate loudspeakers in very noisy locations, and in such cases it is helpful to provide the amplifier with a bass cut-off at 500 c/s, since the maximum intensity levels of speech (particularly declamatory* speech) occur below this frequency, and the effect of the lower frequencies on the articulation is negligible. This permits the middle frequencies from say 1000 to 3500 c/s to be boosted to the threshold of pain at 120 db, thus bringing them above the noise. An alternative arrangement is to give a treble boost of 6 db per octave over the whole useful frequency band, up to at least 3000 c/s—this will automatically reduce the bass frequencies as desired (D2, D4). This latter method requires much less acoustical power—19% of that of a flat system with 500 c/s cut-off —but the effect sounds unnatural.

Refs. A3, D2, D3, D4, D14, D15, D16, E4.

(iv) Distortion in speech reproduction

A considerable degree of distortion is possible before speech becomes objectionably distorted—about 15% second harmonic with a high frequency cut-off at 3750 c/s. This may be put in the form of about 48% intermodulation distortion (r.m.s. sum) with a similar cut-off frequency.

If the frequencies below 500 c/s are attenuated, the distortion for a given loudness will be much less than for a wide frequency range.

---

*Very loud public speaking.

(v) Frequency ranges for speech

Application	Articulation	Frequency range
High fidelity reproduction	98%	62-8000 c/s
Good fidelity	98%	150-7000 c/s
Fair fidelity (Public address)	96%	200-5000 c/s
Restricted bass*. Unbalanced	96%	500-5000** c/s
Restricted bass and treble*	95%	500-4000** c/s
Very restricted*	90%	500-3000** c/s
Telephone		300-3400** c/s

*For noisy locations.   **Response may be peaked.

SECTION 12 : HIGH FIDELITY REPRODUCTION

(i) The target of high fidelity   (ii) Practicable high fidelity   (iii) The ear' as a judge of fidelity.
(General references A3, A9.)

(i) The target of high fidelity

High fidelity reproduction is essentially reproduction such that the most critical person can listen intently to it without any apparent distortion and without any appreciable fatigue, other than any effects due to the single point source of sound. With the present limited state of knowledge we have no definite measurable limits which can be set for each form of distortion to ensure high fidelity.

However, it is generally agreed that a high fidelity amplifier should comply with the following specification. Values in brackets are rather extreme.

Frequency range :   40-15 000   (30-20 000) c/s.

Variation in output $\pm$ 1 db ($\pm$ 0.5 db) at three levels.

Total harmonic distortion not more than 1% (0.7%).

Intermodulation distortion not more than 3% (2%), measured at 40 c/s and, say, 7000 c/s—r.m.s. sum.

Power output sufficient to ensure that overloading does not occur.

Phase angle as small as practicable.

The desired degree of damping on loudspeaker at bass resonant frequency.

Hum inaudible.

Noise* to give specified dynamic range 70 (80) db :

Total noise† 52 (62) db below max. r.m.s. peak signal.

Noise* to be inaudible in low noise room :

For room volume 3000 cubic feet—

Loudspeaker efficiency	3%	10%	30%	45%
Total noise	−30	−35	−40	−42 dbm

Conditions of test—Including microphone (or other source) and loudspeaker.

This specification does not include some very important loudspeaker characteristics, such as frequency-modulation distortion, time delay for specified increase in sound pressure, transient decay characteristics, and sub-harmonics (see Chapter 20 Sect. 7). These have been omitted owing to difficulties in testing and the lack of standardized procedure.

(ii) Practicable high fidelity

There is no great difficulty in meeting the specification given in (i) above so far as a microphone and amplifier are concerned. However, when a loudspeaker is added, to say nothing of the additional distortion contributed by transmission and reception, or reproduction from disc records, it is impossible to meet this specification.

---

*See Sect. 7(v).
†Based on total noise 18 db above masking level of noise at 1000 c/s.   For measurement of noise see Chapter 19 Sect. 6.

This raises the question whether such high standards are essential. The principal characteristics in question are variation in output, distortion, and frequency range.

Variation in output over the frequency range

A tolerance of the order of $\pm 5\%$ appears to be generally acceptable for the overall electro-acoustical performance of a sound system.

Distortion and frequency range

The permissible non-linear distortion in a high fidelity amplifier is dependent on the frequency range, the type of programme, the critical character of the listener and the sound level. It has been observed that the sensitivity of the ear to distortion in music appears to be a maximum for sound levels in the vicinity of 70 to 80 db (Ref. A51). Hence for sound levels of the order of 90 db (sound level meter indication), somewhat higher values of distortion than those specified in (i) above would be permissible—possibly of the order of 2% total harmonic distortion with a cut-off frequency of 10 000 c/s, and 2.5% with a cut-off frequency of 7500 c/s.

Most direct radiator loudspeakers exceed 2% total harmonic distortion over certain frequency bands when operated to give a sound level (meter reading) of 90 db in a fairly large living room. However, the problem is complicated by the fact that many types of music have maximum intensity levels in the 200 to 500 c/s frequency band where loudspeakers often have distortion values well below their maximum values.

When the distortion in any individual case is distressing to a listener, the high cut-off frequency should be reduced until he is relieved from the discomfort. It is better to have freedom from discomfort than to have a wide frequency response, particularly when listening for a sustained period. This leads to the conclusion that " the true. measure of the quality of an electro-acoustical system is the maximum bandwidth which the public finds acceptable " (Ref. C7). Thus all high fidelity amplifiers should have a choice of high roll-off* frequencies.

In addition, it is necessary to provide the listener with tone controls on treble and bass, with the choice of boosting or attenuation on each (see Chapter 15). He can then please himself on the choice of tonal balance and frequency range.

In the writer's opinion, nearly distortionless reproduction of bass frequencies up to say 400 c/s is far more important than reproduction of frequencies above 10 000 c/s.

The importance of **improved bass response** in reproduced music is gradually gaining recognition. Various methods are being adopted to produce acceptable extreme bass response without requiring an excessively large loudspeaker. Fortunately the extension of the bass range, with or without bass boosting, does not directly result in any increase in the subjective effects of the distortion such as occurs with extension of the treble range.

High fidelity reproduction with a level frequency response, when compared with a distorting amplifier, seems to lack bass. If a fair comparison is to be made between them, when operating at the same sound levels, the high fidelity amplifier will require bass boosting.

A direct radiator loudspeaker, or a conventional horn type, cannot give high fidelity reproduction at frequencies lower than the bass resonance, because the distortion rises very rapidly. Thus the minimum frequency for high fidelity depends upon the loudspeaker and any extension to lower frequencies will require another, and usually larger and more expensive, loudspeaker.

The output transformer should, for high fidelity, have low distortion at the lowest frequency which the loudspeaker can handle. A transformer to handle high power at very low frequencies and low distortion is both heavy and expensive, but is not otherwise difficult either in design or manufacture.

If the output transformer and loudspeaker are incapable of handling frequencies below a critical value without serious distortion, it is wise to filter out these lower frequencies and so rid the amplifier of their bad effects. It is better to have a clean bass limited to say 80 or 100 c/s, than to have a distorted condition resulting in intermodulation effects over a wide frequency range.

*Sharp cut-off characteristics are undesirable for fidelity. However a roll-off and sharp cut-off may be combined to give an acceptable result.

(iii) The ear as a judge of fidelity

It is common practice to regard the ear as the final judge of fidelity, but this can only give a true judgment when the listener has acute hearing, a keen ear for distortion, and is not in the habit of listening to distorted music. A listener with a keen ear for distortion can only cultivate this faculty by making frequent direct comparisons with the original music in the concert hall.

Non-linear distortion in any good quality **amplifier** should be so low as to be inaudible to the most critical listener. This distortion can therefore only be checked by measurements.

The ear is the only final judge of fidelity with **loudspeakers**, although it should be supplemented by measurements of harmonic distortion, frequency response, frequency-modulation distortion, damping of bass resonance, time delay for 60 db increase in sound pressure, transient decay characteristics (Shorter's method), and subharmonics—see Chapter 20 Sect. 7.

The ear is the only judge of **tonal balance**.

In any acoustical test, the **sound level** should be that of normal loudspeaker operation.

SECTION 13 : REFERENCES

(A) REFERENCES TO DISTORTION AND FIDELITY—GENERAL
A1. Massa, F. " Permissible amplitude distortion of speech in an audio reproducing system " Proc. I.R.E. 21.5 (May 1933) 682.
A2. Wheeler, H. A. " High fidelity problems " Hazeltine Corporation Laboratory Report, reprinted in A.R.T.S. and P. Bulletin No. 7.
A3. Fletcher, H. " Hearing, the determining factor for high-fidelity transmission " Proc. I.R.E. 30.6 (June 1942) 266.
A4. Ebel, A. J. " Characteristics of high fidelity systems " Comm. (1) 23.4 (April 1943) 38 ; (2) 23.5 (May 1943) 24.
A5. Hartley, H. A. " Aesthetics of sound reproduction—high fidelity or judicious distortion ?" W.W. 50.7 (July 1944) 198 ; W.W. 50.8 (Aug. 1944) 236.
A6. " Aesthetics of sound reproduction—replies to queries " W.W. 50.10 (Oct. 1944) 318.
A7. Stevenson, P. " Scale distortion and visual analogies " Electronic Eng. 17.200 (Oct. 1944) 207.
A8. Hanson, O. B. " Comments on high fidelity " Elect. 17.8 (Aug. 1944) 130.
A9. " Frequency range and power considerations in music reproduction " Technical Monograph No. 3, Jensen Radio Mfg. Co., Nov. 1944.
A10. Discussion " High fidelity reproduction of music " J. Brit. I.R.E. 5.5 (Oct.-Dec. 1945) 190.
A11. Toth, E. " High fidelity reproduction of music " J. (June 1947) 108.
A12. Haynes, N. M. (letter) " Listening tests " Elect. 20.6 (June 1947) 268.
A14. Williamson, D. T. N. " Design for a high-quality amplifier " W.W. 53.4 (April 1947) 118.
A15. Goodell, J. D., and B. M. H. Michel " Auditory perception " (with bibliog.) Elect. 19.7 (July 1946) 142.
A16. Amos, S. W. " Distortion in radio receivers " Electronic Eng. 14.169 (March 1942) 686.
A17. Hanson, O. B. " Down to earth on high fidelity " Radio 28.10 (Oct. 1944) 37.
A18. Van der Ven, A. J. " Output stage distortion " W.E. 16.192 (Sept., 1939) 444.
A19. Webster, N. D., and F. C. McPeak " Experiments in listening " Elect. 20.4 (April 1947) 90.
A20. " What is good reproduction ?—Discussion by British Sound Recording Association," W.W. 54.1 (Jan. 1948) 20.
A21. " Supersonic high fidelity ? " W.W. 54.1 (Jan. 1948) 23 ; Comments by " Diallist " p. 36.
A22. Morgan, R. L. " Noise measurements," Comm. 28.2 (Feb. 1948) 28 ; 28.7 (July 1948) 21.
A23. Mathes, R. C., and R. L. Miller " Phase effects in monaural perception " J. Acous. Soc. Am. 19.5 (Sept. 1947) 780.
A24. Minter, J. " Audio distortion and its causes " F.M. Radio Handbook (1946) 94.
A25. Gannett, D. K., and I. Kerney " The discernibility of changes in program band width " B.S.T.J. 23 (Jan. 1944) 1.
A26. Glover, R. P. " The problem of frequency range in speech and music reproduction " (review) J. Acous. Soc. Am. 17.1 (July 1945) 103.
A27. Hilliard, J. K. " Audio quality—intermodulation tests " (review of I.R.E. paper) Elect. 19.4 (April 1946) 218.
A28. Lloyd, M. G. and P. G. Agnew " Effect of phase of harmonics upon acoustic quality " Nat. Bureau of Standards Sci. Paper No. 127, 6 (1909) 255.
A29. " Cathode Ray " ' Scale distortion " W.W. 41.13 (Sept. 24, 1937) 318
A30. " Cathode Ray " " Loudspeaker versus orchestra," W.W. 42.10 (March 10, 1938) 210.
A31. Hawkings, J. N. A. " Notes on wide-range reproduction " Audio Eng. 31.11 (Dec. 1947) 19.
A32. Camras, M. " How high is high fidelity ? " Tele-Tech 7.7 (July 1948) 26.
A33. " Cathode Ray " " Scale distortion again " W.W. 54.11 (Nov. 1948) 392.
A34. Le Bel, C. J. " Psycho-acoustic aspects of higher quality reproduction " Audio Eng. 33.1 (Jan. 1949) 9.
A35. " Distortion—does it matter ? " Further discussion by I.E.E. Radio Section, W.W. 55.1 (Jan. 1949) 11.
A36. " Quality appreciation " W.W. 55.2 (Feb. 1949) 52. Review of paper by E. A. Vetter " Some psychological factors in quality appreciation."
A37. Licklider, J. C. R. " Effects of amplitude distortion upon the intelligibility of speech " J. Acous. Soc. Am. 18.2 (Oct. 1946) 429.

A38. Nixon, G. M. " Higher fidelity in sound transmission and reproduction " J. Acous. Soc. Am. 17.2 (Oct. 1945) 132.
A39. What constitutes high fidelity reproduction ? " (Review of papers presented at Acoustical Society meeting) Audio Eng. 32.12 (Dec. 1948) 8.
A40. Wiener, F. M. " Phase distortion in electro-acoustic systems " J. Acous. Soc. Am. 13.2 (Oct. 1941) 115.
A41. Powell, T. (letter) " Phase bandwidth " Audio Eng. 33.6 (June 1949) 6.
A42. Hilliard, J. K. " Phase distortion " being chapter 13 of book " Motion picture sound engineering " (D. Van Nostrand Co. New York, 1938).
A43. Institute of Radio Engineers, U.S.A. " Standards on radio receivers—methods of testing amplitude modulation broadcast receivers " (1948).
A44. Information kindly supplied by the Radio Corporation of America.
A45. Shorter, D. E. L. " The influence of high-order products in non-linear distortion " Electronic Eng. 22.266 (April 1950) 152. Correspondence 22.272 (Oct. 1950) 443 ; 23.281 (July 1951) 278-279.
A46. Salmon, V. (letter) " Types of reproduced sound " J. Acous. Soc. Am. 21.5 (Sept. 1949) 552.
A47. Canby, E. T. " Record revue—better audio " Audio Eng. 34.7 (July 1950) 30.
A48. Salmon, V. " Imagery for describing reproduced sound " Audio Eng. (i) 34.8 (Aug. 1950) 14 ; (ii) 34.9 (Sept. 1950) 14.
A49. Voigt, P. G. A. H. " A controversial idea from England " Audio Eng. 34.10 (Oct. 1950) 40.
A50. Canby, E. T. " The other side of the wall " Audio Eng. 35.1 (Jan. 1951) 24.
A51. Olson, H. F., and A. R. Morgan " A high-quality sound system for the home " Radio and T.V. News 44.5 (Nov. 1950) 59.
A52. Souther, H. T. " Design elements for improved bass response in loudspeaker systems " Audio Eng. 35.5 (May 1951) 16.
A53. Fletcher, H. " Auditory patterns " Review of Modern Physics 12 (Jan. 1940) 47.
A54. F.L.D. " Approach to high fidelity " W.W. 57.7 (July 1951) 289.
A55. Snyder, R. H. " Towards a more realistic audio " Audio Eng. 35.8 (Aug. 1951) 24.
Additional references will be found in the Supplement commencing on page 1475.

(B) REFERENCES TO INTERMODULATION DISTORTION
B1. Harries, J. H. O. " Amplitude distortion " W.E. 14.161 (Feb. 1937) 63.
B2. Hilliard, J. K. " Distortion tests by the intermodulation method " Proc. I.R.E. 29.12 (Dec. 1941) 614 ; Discussion 30.9 (Sept. 1942) 429.
B3. Scott, H. H. " Audible audio distortion " Elect. 18.1 (Jan. 1945) 126.
B4. Minter, J. " Audio distortion in radio reception " F.M. and T. 6.3 (March 1946) 24.
B5. Hilliard, J. K. " Intermodulation testing " Elect. 19.7 (July 1946) 123.
B6. Roys, H. E. " Intermodulation distortion analysis as applied to disk recording and reproducing equipment " Proc. I.R.E. 35.10 (Oct. 1947) 1149.
B7. Frayne, J. G., and R. R. Scoville, " Variable density recording" Jour. S.M.P.E.32 (June 1939) 648.
B8. Sturley, K. R. (book) " Radio Receiver Design " Part 2 (Chapman and Hall, 1945) pp. 79-84.
B9. McProud, C. G. " Simplified intermodulation measurement " Audio Eng. 31.3 (May 1947) 21.
B10. Instruction Book " T.1 401 Signal Generator and T.1 402 Intermodulation Analyzer " Altec Lansing Corporation, New York.
B11. Warren, W. J., and W. R. Hewlett " An analysis of the intermodulation method of distortion measurement " Proc I.R.E. 36.4 (April 1948) 457.
B12. Avins, J. " Intermodulation and harmonic distortion measurements " Audio Eng. 32.10 (Oct. 1948) 17.
B13. " A report on the 1949 IRE National Convention—highlights of papers presented by R. H. Tanner and A. Peterson " Comm. 29.4 (April 1949) 10.
B14. Aston, R. H. " Intermodulation measurements on Radiotron Amplifier A515 " Radiotronics No. 130 (March/April 1948) 34.
B15. Read, G. W., and R. R. Scoville " An improved intermodulation measuring system " Jour. S.M.P.E. 50.2 (Feb. 1948) 162.
B16. Fine, R. S. " An intermodulation analyzer for audio systems " Audio Eng. 34.7 (July 1950) 11.
B17. van Beuren, J. M. " Simplified intermodulation measurements " Audio Eng. 34.11 (Nov. 1950) 24.
B18. Berth-Jones, E. W. " Intermodulation testing " W.W. 57.6 (June 1951) 233.
B19. Le Bel, C. J. " A new method of measuring and analyzing intermodulation " Audio Eng. 35.7 (July 1951) 18.
B20. Roddam, T. " Intermodulation distortion " W.W. 46.4 (Apr. 1950) 122.
B21. Callendar, M. V., and S. Matthews " Relations between amplitudes of harmonics and intermodulation frequencies " Electronic Eng. 23.280 (June, 1951) 230.
B22. Peterson, A. P. G. " Intermodulation distortion " G.R. Exp. 25.10 (Mar. 1951) 1.

(C) REFERENCES TO FREQUENCY RANGE AND SOUND-INTENSITY PREFERENCES
C1. Chinn, H. A., and P. Eisenberg " Tonal-range and sound intensity preferences of broadcast listeners " Proc. I.R.E. 33.9 (Sept. 1945) 571.
C2. Chinn, H. A., and P. Eisenberg " Tonal-range and sound intensity preferences of broadcast listeners " Discussion, Proc. I R.E. 34.10 (Oct. 1946) 757.
C3. Moir, J. " Perfect v. pleasing reproduction " Electronic Eng. 19.227 (Jan. 1947) 23.
C4. Wells, L. V. (letter) " Acoustic preferences of listeners " Proc. I.R.E. 35.4 (April 1947) 378.
C5. Powell, T. (letter) " Tonal-range and sound-intensity preferences of broadcast listeners " Proc. I.R.E. 35.4 (April 1947) 378.
C6. Moir, J. (letter) " Distortion and acoustic preferences " Proc. I.R.E. 35.5 (May 1947) 495.
C7. Le Bel, C. J. " Psycho-acoustical aspects of listener preference tests " Audio Eng. 31.7 (Aug. 1947) 9.
C8. " High audio frequencies—are they necessary, are they nice ? " W.W. 53.11 (Nov. 1947) 415.
C9. Haynes, N. M. " Factors influencing studies of audio reproduction quality," Audio Eng. 31.8 (Oct. 1947) 15.
C10. Olson. H. F. " Frequency range preference for speech and music " J. Acous. Soc. Am. 19.4 Part 1 (July 1947) 549.
C11. Olson, H. F. " Frequency range preference for speech and music " Elect. 20.8 (Aug. 1947) 80.
C12. Somerville, T., and S. F. Brownless " Listeners' sound-level preferences " B.B.C. Quarterly 3.4 (Jan. 1949) 245.
C13. Chinn, H. A. and P. Eisenberg " Influence of reproducing system on tonal-range preferences " Proc. I.R.E. 36.5 (May 1948) 572. Discussion Proc. I.R.E. 37.4 (April 1949) 401.
C14. Stuntz, S. E. " The effect of sound intensity level on judgment of tonal range and volume level" Audio Eng. 35.6 (June 1951) 17.

(D) REFERENCES TO LIMITED RANGE, SPEECH AND NOISE
D1. " Radio Club hears de Rosa on synthetic high fidelity " Elect. 15.12 (Dec. 1942) 126.
D2. " The effective reproduction of speech " Jensen Technical Monograph No. 4 (1944).
D3. Beranek, L. L. " The design of speech communication systems" Proc. I.R.E. 35.9 (Sept. 1947) 880.
D4. Sanial, A. J. " Acoustic considerations in 2-way loudspeaker communications " Comm. 24.6 (June 1944) 33.
D5. Dunn, H. K. and S. D. White, " Statistical measurements on conversational speech " J. Acous. Soc. Am. 11.3 (Jan. 1940) 278.
D6. Seacord, D. F. " Room noise at subscribers' telephone locations," B.S.T.J. 12.1 (July 1940) 183.
D7. Hoth, D. F. " Room noise spectra at subscribers' telephone locations " B.S.T J. 12.4 (April 1941) 499.
D8. Fletcher, H., and J. C. Steinberg " Articulation testing methods," B.S.T.J. 8.4 (Oct. 1929) 806.
D9. Sivian, L. J. " Speech power and its measurement " B.S.T.J. 8.4 (Oct. 1929) 646.
D10. French, N. R., and J. C. Steinberg " Factors governing the intelligibility of speech sounds " J. Acous. Soc. Am. 19.1 (Jan. 1947) 90.
D11. Pollack, I. " Effect of high pass and low pass filtering on the intelligibility of speech in noise " J. Acous. Soc. Am. 20.3 (May 1948) 259.
D12. Egan, J. P., and F. M. Wiener " On the intelligibility of bands of speech in noise" J. Acous. Soc. Am. 18.2 (Oct. 1946) 435.
D13. Licklider, J. C. R., and I. Pollack " Effects of differentiation, integration and infinite peak clipping upon the intelligibility of speech " J. Acous. Soc. Am. 20.1 (Jan. 1948) 42.
D14. Schafer, T. H., and R. S. Gales " Auditory masking of multiple tones by random noise " J. Acous. Soc. Am. 21.4 (July 1949) 393.
D15. Hirsh, I. J., and F. A. Webster " Some determinants of interaural phase effects " J. Acous. Soc. Am. 21.5 (Sept. 1949) 496.
D16. Bolt, R. H., and A. D. MacDonald " Theory of speech masking by reverberation " J. Acous. Soc. Am. 21.6 (Nov. 1949) 577.
D17. Cook, E. " White-noise testing methods " Audio Eng. 34.3 (March 1950) 13.
D18. Gottschalk, J. M. " A white-noise generator for audio frequencies " Audio Eng. 34.5 (May 1950) 16. Also A3, E4, E5, E6.
D19. Beranek, L. L., W. H. Radford, J. A. Kessler and J. B. Wiesner " Speech-reinforcement system evaluation " Proc. I.R.E. 39.11 (Nov. 1951) 1401.
Additional references will be found in the Supplement commencing on page 1475.

(E) MUSIC AND ACOUSTICS
E1. Tillson, B. F. " Musical acoustics " Audio Eng. (1) 31.5 (June 1947) 34 ; (2) 31.6 (July 1947) 25 ; (3) 31.7 (Aug. 1947) 34 ; (4) 31.8 (Sept. 1948) 30 ; (5) 31.9 (Oct. 1947) 25 ; (6) 31.10 (Nov. 1947) 31. Correction 32.6 (June 1948) 8.
E2. Wood, A. (book) " The Physics of Music " (Methuen and Co. Ltd., London, 3rd edit. 1945).
E3 Olson, H. F. (book) " Elements of Acoustical Engineering " (D. Van Nostrand Co. Inc., New York, 2nd edit. 1947).
E4. Fletcher, H. " Speech and hearing " (MacMillan and Co. Ltd. London, D. Van Nostrand Co. Inc., New York, 1929).
E5. Fletcher, H. " Some physical characteristics of speech and music," B.S.T.J. 10.3 (July 1931) 349.
E6. Snow, W. B. " Audible frequency ranges of music, speech and noise," B.S.T.J. 3.1 (July 1931) 155.
E7. Sivian, L. J., H. K. Dunn and S. D. White " Absolute amplitudes and spectra of musical instruments and orchestras," J. Acous. Soc. Am. 2.3 (Jan. 1931) 330.
E8. Fletcher, H., and W. A. Munson " Loudness, its definition, measurement and calculation " J. Acous. Soc. Am. 5.2 (Oct. 1933) 82.
E9. Fletcher, H., and W. A. Munson " Relation between loudness and masking " J. Acous. Soc. Am. 9.1 (July 1937) 1.
E10. Steinberg, J. C., H. C. Montgomery and M. B. Gardner, " Results of the World's Fair hearing tests," B.S.T.J. 19.4 (Oct. 1940) 533.
E11. Trimmer, J. D., and F. A. Firestone " An investigation of subjective tones by means of the steady tone phase effect " J. Acous. Soc. Am. 9.1 (July 1937) 24.
E12. Goodfriend, L. S. " Problems in audio engineering " Audio Eng. (1) 33.5 (May 1949) 22 ; (2) 33.6 (June 1949) 15 ; (3) 33.7 (July 1949) 20.
E13. Bell Laboratories Record, June 1934 page 315.
E14. Ladner, A. W. " The analysis and synthesis of musical sounds " Electronic Eng. 21.260 (Oct. 1949) 379.
E15. Canby, E. T. " Record revue " Audio Eng. 34.5 (May 1950) 24.
E16. Wolf, S. K., and W. J. Sette " Acoustic power levels in sound picture reproduction " J. Acous. Soc. Am. 2.3 (Jan. 1931) 384.

CHAPTER 15

TONE COMPENSATION AND TONE CONTROL

BY F. LANGFORD-SMITH, B.Sc., B.E.

SECTION 1 : INTRODUCTION

(i) The purpose of tone compensation   (ii) Tone control   (iii) General considerations (iv) Distortion due to tone control   (v) Calculations involving decibels per octave   (vi) Attenuation expressed as a time constant   (vii) The elements of tone control filters (viii) Fundamental circuit incorporating R and C   (ix) Damping of tuned circuits (x) Tolerances.

(i) The purpose of tone compensation

An " ideal " audio frequency amplifier is one having a response which is linear and level (i.e. " flat ") over the whole a-f range. An amplifier which has a drooping characteristic at the extremes of its frequency range may be made practically flat by the incorporation of a suitable degree of bass and treble boosting. This device is adopted in studio amplifiers where rigid tolerances are imposed on the frequency response of each amplifier or unit, and also in video and wide-band amplifiers where the frequency range is so great that conventional a-f designs are unsatisfactory. (Refs. 35, 38, 53 are typical.)

Studio microphones and pickups are " equalized " by a suitable filter to give a flat response. Thus any input source may be connected to any amplifier or amplifiers, and the overall result will be flat, since each unit in the chain is flat.

A different approach is usually adopted in complete units such as home gramophones where the whole is always used as one unit, and where we are only concerned with the overall performance.

In such a case it is possible to vary the frequency response of the amplifier so as to compensate for certain components such as pickups or loudspeakers which do not have a flat response.

All these are examples of tone compensation, the purpose of which is to give a flat overall frequency characteristic.

(ii) Tone control

A tone control is a variable filter (or one in which at least one element is adjustable) by means of which the user may vary the frequency response of an amplifier to suit his own taste.

Tonal balance is covered in Chapter 14 Sect. 4(ii).

(iii) General considerations

It is usual to regard the middle range of audio frequencies from say 500 to 2000 c/s as the " body " of musical reproduction, with 1000 c/s as the reference frequency.

Thus a lower or higher frequency is said to be attenuated if it is reproduced at a lower level than 1000 c/s or boosted if it is at a higher level than 1000 c/s.

An equalizer for a pickup or microphone is usually placed either between the source and the first amplifier valve or, if this is at a low level, between the first and second amplifier valves.

All sources of input voltage should have a frequency range which is more limited than the range which the amplifier and loudspeaker are capable of handling without noticeable distortion. In other words, no voltage should be applied to the input terminals of the amplifier which has a frequency lower than the low frequency limit of the amplifier and loudspeaker or a frequency higher than the high frequency limit of the amplifier and loudspeaker. If this does not hold, it is highly desirable to limit the frequency range of the input voltage, either at one or both ends of the range as required, by means of a suitable filter (see Sects. 3 and 6 below), inserted before the input terminals of the main amplifier. If there is a pre-amplifier followed by a main amplifier, the filter may be inserted between the two.

If a tone control is to be fitted to an amplifier, its position is of considerable importance, the only exception being when the amplifier has low distortion at all frequencies with a loudspeaker load. In an amplifier using a pentode or beam power amplifier without negative feedback, it is desirable to fit a r.c. filter across the primary of the output transformer. This may be fixed, with a supplementary tone control elsewhere, or it may form the tone control itself. In most amplifiers it is desirable to connect a small capacitance (say 0.001 μF for a 5000 ohm load) directly from the plate of the output valve to earth, or from the plate of each output valve in the case of push-pull operation. This assists in by-passing any radio or ultrasonic audio-frequencies which may be present, without having much effect on the frequency response. Its capacitance might be increased with advantage if some attenuation of the highest frequencies is permissible.

In a typical radio receiver without negative feedback, the tone control may be in the plate circuit of the power valve, in the coupling circuit between the first a-f amplifier valve and the power valve, or in the grid circuit of the first a-f valve.

If an amplifier incorporates negative feedback, the tone control may be incorporated in the feedback network (see Sect. 9 below) or else should be connected to a part of the amplifier which is external to the feedback loop.

Tone controls may be continuously variable or stepped. The more complicated types are stepped but, like a studio attenuator, the steps may be made barely perceptible.

(iv) Distortion due to tone control

Tone control frequently has a pronounced effect on the effective distortion due to stages in the amplifier preceding the tone control filter. **Treble attenuation** results in a reduction in the amplitude of harmonics compared with that of the fundamental —that is a reduction in the harmonic distortion—provided that the predominant harmonics are within the attenuation range.   In general, intermodulation distortion is also reduced by treble attenuation, subject to the same limitation.

On the other hand, any form of **treble boosting** will increase the effective distortion at any frequency so long as the harmonics are accentuated by the treble boosting with respect to the fundamental ; the same also applies to intermodulation distortion. This is one of the reasons why treble boosting is not widely used, or is limited to a very slight rate of boosting.

If an amplifier naturally has some treble attenuation, and is equalized by the correct amount of treble boosting, the distortion will be the same as though the amplifier had a naturally flat characteristic.   However, in the case of a radio receiver, effectively having a treble attenuation through side-band cutting, the distortion at the output terminals will be greater than that at the second detector if treble boosting is used, even if the a-f amplifier distortion could be zero.

Bass boosting gives a reduction in harmonic distortion provided that the fundamental frequency is amplified more than the harmonics, which condition holds over a limited frequency range.

Bass attenuation gives increased distortion over a limited frequency range where the harmonics are amplified more than the fundamental.

References to distortion : 3, 61.

(v) Calculations involving decibels per octave

The rate of attenuation, or of boosting, is usually given in the form of so many decibels per octave.   For example, a simple shunt condenser across a resistive network produces an ultimate treble attenuation of 6 db/octave—see Chapter 4 Sect. 8(ii) and Fig. 4.38.   Ultimate attenuation is normally in multiples of 6 db/octave—e.g. 6, 12, 18 and 24 db/octave.   Actual values of attenuation may differ from the ultimate values, particularly in the range from 0 to 7 db.

In some cases the frequencies at which readings are taken do not conveniently cover an exact number of octaves.   In such a case the following procedure may be adopted.

(A) To convert db/specified frequency ratio to db/octave

Frequency ratio	Multiply db/specified frequency ratio by factor		
1.2 : 1	6.02	to give	db/octave
1.25 : 1	3.10	,, ,,	,,
1.33 : 1	2.43	,, ,,	,,
1.5 : 1	1.71	,, ,,	,,
2 : 1	1.00	,, ,,	,,
3 : 1	0.63	,, ,,	,,
4 : 1	0.50	,, ,,	,,
5 : 1	0.43	,, ,,	,,
6 : 1	0.39	,, ,,	,,
7 : 1	0.36	,, ,,	,,
8 : 1	0.33	,, ,,	,,
10 : 1	0.30	,, ,,	,,

Example—A change of 0.7 db occurs with an increase of frequency from 1000 to 1250 c/s.   What is the rate of change in db/octave ?

Rate of change = 0.7 × 3.10 = 2.17 db/octave.

Note.   A table relating frequency ratio, octaves and decades is given on page 368.

(B) To convert db/octave to db/specified frequency ratio

Frequency ratio	Multiply db/octave by factor				
1.2 : 1	0.263	to give db/specified frequency ratio			
1.25 : 1	0.322	,,	,,	,,	,,
1.33 : 1	0.412	,,	,,	,,	,,
1.5 : 1	0.585	,,	,,	,,	,,
2 : 1	1.00	,,	,,	,,	,,
3 : 1	1.59	,,	,,	,,	,,
4 : 1	2.00	,,	,,	,,	,,
5 : 1	2.33	,,	,,	,,	,,
6 : 1	2.59	,,	,,	,,	,,
7 : 1	2.81	,,	,,	,,	,,
8 : 1	3.00	,,	,,	,,	,,
10 : 1	3.33	,,	,,	,,	,,

Example—What is the change in level for a frequency ratio of 1.5 to 1 when the rate of change is 6 db per octave?

Change in level $= 0.585 \times 6 = 3.51$ db.

(vi) Attenuation expressed as a time constant

In F-M receiver design it is common practice to express the degree of pre-emphasis (treble boosting) in the transmitter as a time constant of so many microseconds, and the degree of de-emphasis (treble attenuation) in the receiver in the same form. The two methods are fundamentally related because the time constant in seconds is equal to RC where R is the resistance in ohms and C is the capacitance in farads. In the general case

attenuation in db $= - 10 \log_{10} (1 + \omega^2 T^2)$

where $\omega = 2\pi f$

$\qquad T = CR =$ time constant in seconds

and $R =$ total effective resistance of supply network.

In the particular case where $T = 75$ microseconds and f is expressed in Kc/s, this becomes

attenuation in db $= - 10 \log_{10} (1 + 0.222 f^2)$.

A de-emphasis curve for a time constant of 75 microseconds is given in Fig. 15.1. This curve, like all other curves of this class, has an ultimate rate of attenuation of 6 db/octave.

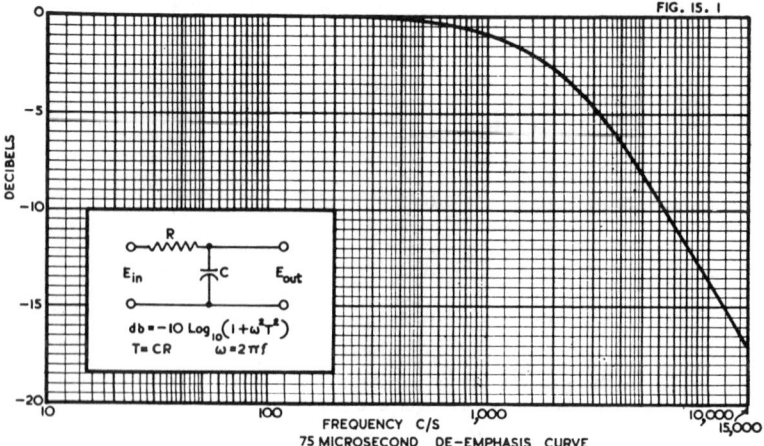

Fig. 15.1. *De-emphasis curve with time constant of 75 microseconds, as used in F-M receivers.*

(vii) The elements of tone control filters

Tone control filters are networks including at least two of the basic elements—resistance, capacitance and inductance.   In adjustable tone controls it is simplest to use the resistance as the variable element, since continuously variable resistors are available.   To vary capacitance or inductance values, it is necessary to use separate components or tappings, with a step switch.

In radio receivers and home amplifiers it is usual, wherever possible, to avoid using inductors for tone control purposes.   This is firstly because inductors are generally more expensive than condensers, and secondly inductors are very prone to pick up hum unless elaborate and often expensive precautions are taken.   An additional reason is that inductors with the desired values of inductance and tolerances are not usually stock lines.

The inductance of an inductor with any form of iron core is not constant, but is affected by the applied alternating voltage, the frequency, and by any direct current.

When a composition " potentiometer " is used with the moving arm in the grid return circuit of a valve, there will be additional noise caused by the movement of the arm.   In such a case, the subsequent amplifier gain should be low.

(viii) Fundamental circuit incorporating R and C

The circuit shown in Fig. 15.2 may be used for the purpose of obtaining many forms of tone compensation.   The values of the six components may be varied as desired to provide bass boosting or attenuation as well as treble boosting or attenuation.   It is to be understood that the choice of values extends from zero to infinity, this being equivalent to the optional short-circuiting or open-circuiting of any one or more resistors or condensers.   This form of filter is intended for use with a constant input voltage (E_i).   This holds approximately when a triode valve is used.

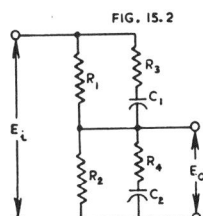

FIG. 15.2

Fig. 15.2.   Fundamental circuit incorporating R and C for use with a constant input voltage.

(ix) Damping of tuned circuits

When tuned circuits are used for purposes of tone control, the damping should be sufficient to reduce the " overshoot " to a small value when a unit step (pulse) input is applied to the amplifier.   This overshoot may be reduced to a small value if Q does not exceed 0.7.   Under no circumstances should Q exceed 1.0 by more than a small margin.   Insufficient damping results in distortion of transients.

(x) Tolerances

In all cases where a tone control, tone compensating network or filter is required to have specified frequency response characteristics, the elements of the network must be specified within narrow tolerances, and should preferably also be tested after delivery by the component manufacturer.   Even in cases where the response characteristic is not specified, the effect of normal tolerances with capacitors, resistors and inductors is to cause a wide " spread " of response characteristics since the tolerances are sometimes cumulative.

When a single item of equipment is being built, it is important to check either the values of the critical components or the overall response characteristic, or preferably both.

SECTION 2 : BASS BOOSTING

(i) General remarks (ii) Circuits not involving resonance or negative feedback (iii) Methods incorporating resonant circuits (iv) Circuits involving feedback (v) Regeneration due to negative resistance characteristic.

(i) General remarks

Bass boosting, either for tone compensation or tone control, implies that the gain of the amplifier at 1000 c/s must be reduced sufficiently to permit the full gain to be available for amplifying bass frequencies. For example, if the maximum amplifier gain is 56 times (35 dbvg*), and a maximum bass boost of 15 db is required, then the gain at 1000 c/s cannot be more than 20 dbvg*.

Bass boosting for tone control purposes may be used to provide better tonal balance at acoustical levels lower than the original sound, and is generally controlled manually—for automatic control see Section 10   Bass boosting increases the power output from the power valve and loudspeaker at low audio frequencies, and therefore tends to cause overloading if used at maximum volume ; this effect does not occur at lower volume levels. When using bass boosting, the apparent loudness at which overloading occurs is less than that without bass boosting. For this reason, bass boosting should be used with discretion except in amplifiers having ample reserve of power.

For general tone control purposes in typical radio receivers and amplifiers, a bass boost variable from zero to $+ 6$ db is a good compromise. This involves, at maximum bass boosting, four times the power without bass boosting. Any increase beyond 6 db would involve a careful analysis of the whole design.

The bass boost should reach its maximum value at some suitable frequency, say 75 c/s, and should fall fairly rapidly below this frequency. There should be zero boost or even attenuation at the minimum frequency which the amplifier and loudspeaker are capable of handling without distortion.

Bass boosting should not extend appreciably above a frequency of 250 c/s at maximum volume, or the male voice will sound unnatural. When the boosting is only used at lower levels this limitation does not hold.

Some of these limitations are removed by the use of automatic frequency-compensated tone control (Sect. 10).

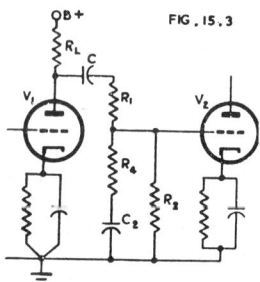

Fig. 15.3. *Conventional bass boosting circuit (plate shunt compensation).*

(ii) Circuits not involving resonance or negative feedback

(A) **Conventional bass boosting circuit** (Fig. 15.3)

(also known as plate shunt compensation)

The bass boosting is due to the increasing impedance of C_2 as the frequency is lowered. It is assumed here that the reactance of C is negligibly small compared with R_4 ; in practice the value of C may be selected to give attenuation below a specified frequency. R_2 should be at least 20 times R_4 and preferably higher. The top limit is the maximum grid circuit resistance permitted for V_2 ; if V_2 is a resistance coupled stage, see Chapter 12 Sect. 2(iii)E and Sect. 2(iv) for triodes, or Sect. 3(iv) and Sect. 3(v) for pentodes. If V_2 is a power valve, R_2 will usually be limited to 0.5 megohm for cathode bias.

*Decibels of voltage gain—see Chapter 19 Sect. 1(vi)A.

The resistor R_1 is intended to reduce distortion through low a.c. load impedances connected to the plate circuit of V_1 ; it may be omitted if the stage is operating at a low level or if R_4 is not less than twice the plate resistance of V_1. A suitable value is not less than $2r_p$ in the case of a triode or not less than $2R_L$ in the case of a pentode. The resistor R_1 also has the effect of decreasing the gain at high frequencies more than at low frequencies, thereby increasing the ratio of gains at low and high frequencies.

It may be shown that

$$\text{Gain at high frequencies} = \frac{\mu}{(1 + r_p/R_L)(1 + R_1/R') + r_p/R'} \tag{1}$$

where $R' = R_2 R_4/(R_2 + R_4)$

$$\text{Gain at zero frequency} = \frac{\mu}{(1 + r_p/R_L)(1 + R_1/R_2) + r_p/R_2} \tag{2}$$

$$B = \text{ratio of gains} = 1 + \frac{R_2}{R_4}\left\{\frac{(1/r_p + 1/R_L)R_1 + 1 - R_4/R_2}{(1/r_p + 1/R_L)(R_1 + R_2) + 1}\right\} \tag{3}$$

As a sufficiently close approximation, R_4/R_2 may be taken to have a value as indicated below :

Boost	6	10	15	20	db
R_4/R_2	0.1	0.05	0.02	0.01	approx.

Using this approximation, we may determine the value of R_4 to provide any desired ratio of gains, that is the bass boost expressed as a ratio (B) :

$$R_4 \approx \frac{R_2}{B - 1}\left\{\frac{(1/r_p + 1/R_L)R_1 + 1 - R_4/R_2}{(1/r_p + 1/R_L)(R_1 + R_2) + 1}\right\} \tag{4}$$

where $B =$ ratio of amplification at zero frequency to amplification at high frequencies.

Typical values of resistors are given below :

	General case	Valve type 6J5	Valve type 6AV6
Plate resistance	r_p	7 700*	62 500*ohms
R_L	5 r_p	50 000	220 000 ohms
R_1	2 r_p	20 000	120 000 ohms
R_2	20 r_p	500 000	1 000 000 ohms
R_4 for 6 db boost	2.4 r_p	25 000	140 000 ohms
R_4 for 10 db boost	1.1 r_p	11 600	67 000 ohms
R_4 for 15 db boost	0.53 r_p	5 500	30 000 ohms
Gain at high frequencies :			
6 db boost	0.36 μ	7.6	30 times
10 db boost	0.22 μ	4.7	19 times
15 db boost	0.13 μ	2.6	10 times

A typical example incorporating a pentode is Fig. 15.37A (Sect. 8).

The procedure in design is

(1) to determine the desired total bass boost in db from high frequencies to zero frequency.

(2) to select a suitable valve type.

(3) to assume suitable values for R_L, R_1 and R_2.

(4) to calculate the value of R_4 to give the desired boost, using eqn. (4).

(5) to determine a suitable value for C (this may be done by following Chapter 12 Sect. 2(ii), assuming $R_1 + R_2$ to be the effective value of R_{g2}).

(6) to determine the value of C_2 to give the required position of the boosting curve on the frequency characteristic, as described below.

Frequency characteristics

The shape of the frequency characteristic is determined solely by the amount of the total boost in db (Fig. 15.4). It is convenient to consider the frequency at which each characteristic reaches half the total boost in decibels, and to call this the " half-

---

*As normal Class A_1 amplifier. Higher values are to be anticipated for resistance-coupled conditions.

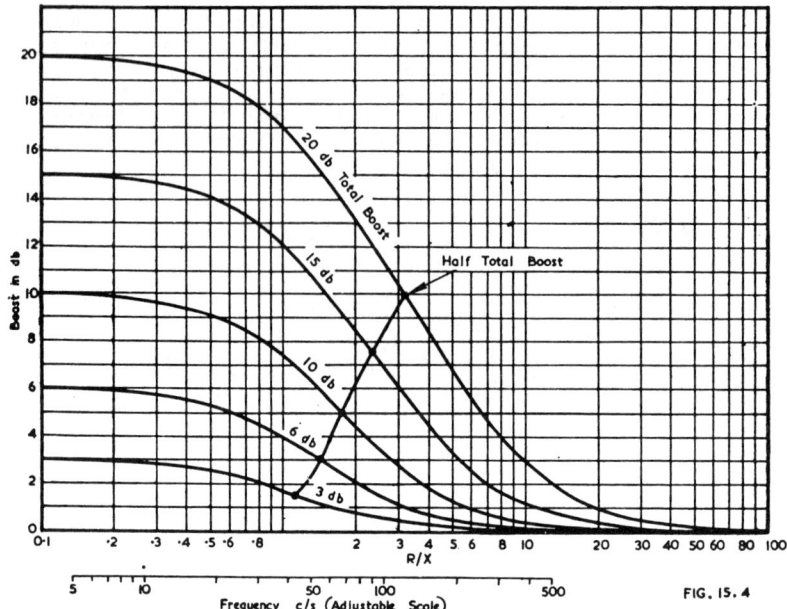

Fig. 15.4.   Bass boosting frequency characteristics. These are quite general and may be applied to any r.c. boosting provided that the value of the total boost in db is known. See text for value R/X (Ref. 10).

boost point." It will be seen that the frequency of the " half boost point " (f_0) increases as the total boost increases.

Total boost	20	15	10	6	3	db
Half-boost	10	7.5	5	3	1.5	db
Boost ratio (B)	10	5.62	3.16	2.0	1.41	
R/X for half-boost*	3.16	2.37	1.78	1.41	1.19	
C_2†	$3.16/\omega R$	$2.37/\omega R$	$1.78/\omega R$	$1.41/\omega R$	$1.19/\omega R$ F

* $(R/X) = \sqrt{B}$ at half-boost point.
† $C_2 = 1/(\omega X) = (R/X)/(\omega R)$ and $\omega = 2\pi f_0$
where $R = [r_p R_L/(r_p + R_L)] + R_1 + R_4$ (Fig. 15.3)
and f_0 = frequency at half-boost point.

The slope of the frequency characteristic at the half-boost point, which is very nearly the point of maximum slope, is approximately :

Total boost	20	15	10	6	3 db
Slope at half-boost point	4.9	4.1	3.0	2.0	1.0 db/octave

If a tangent is drawn to the curve at the half-boost point, it will be seen that the slope of the frequency characteristic does not fall off to any appreciable extent from the half-boost point to the three-quarters boost point (i.e. 75% of the total boost in db). The falling off in slope does not cause a difference of more than 1 db between the curve and the tangent up to 90% of the total boost in db, the reading of 90% being taken on the tangent.

This circuit may be applied to continuously variable bass boosting by using a variable resistor in place of R_4, but this has the effect of varying the amplifier gain at the middle audio frequencies and hence varying the apparent loudness. One possible alternative which avoids this defect, is to put a high variable resistor (say 0.5 megohm logarithmic taper) across C_2. This method gives a variation in total boost without much change in level of the middle frequencies, but the shapes of the frequency characteristics are not the most desirable for tone control purposes—Fig. 15.5 ; see also (D) below.

Several values of C_2 may be selected by means of a tapping switch, leaving the total boost unchanged, but this merely moves the frequency characteristic horizontally and is not satisfactory, on its own, for tone control purposes.

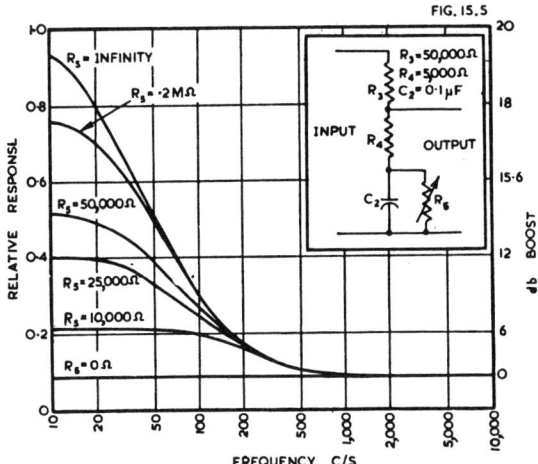

Fig. 15.5.   *Frequency characteristics with conventional bass boosting circuit, having variable resistor across C_2 (Ref. 9).*

A modification which has some advantages is shown in Fig. 15.6.   Here R_3 has been moved from the grid circuit of V_2 to reduce the shunting effect on R_4 and C_2.   The total grid circuit resistance of V_2 is $(R_1 + R_2)$. Ref. 11.

References to conventional bass boosting circuit—9, 10, 11, 20, 23, 38, 51, 55.

(B) Plate series compensation (Fig. 15.7)

This is a simple method of providing a fixed amount of bass boosting which uses the plate decoupling circuit.   It is generally limited to use with r.c.c. pentodes. With the values of components shown, the frequency response curves for two values of C are given in Fig. 15.8.

References to plate series compensation : 38, 52, 53.

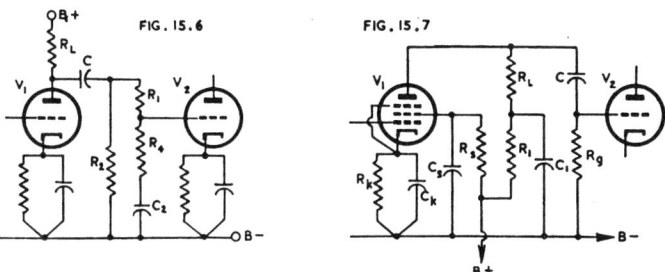

Fig. 15.6.   *Modified   form   of   Fig. 15.3.*

Fig. 15.7.   *Bass boosting—plate series compensation.*

(C) Grid series compensation

This is the same in principle as plate series compensation (Fig. 15.7) except that C_1 is taken to a tapping point on R_g instead of on R_L.   This is often preferred in wide-band amplifiers as C_1 will be smaller for the same frequency characteristic than with plate series compensation.

Reference to grid series compensation : 38.

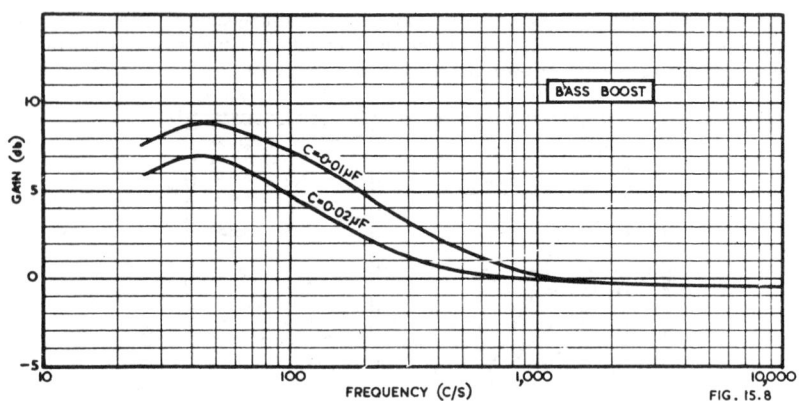

Fig. 15.8. *Frequency characteristics with the circuit of Fig.* 15.7, $V_1 = 6J7$,
$R_L = 0.05\ M\Omega$, $R_1 = 0.2\ M\Omega$, $R_k = 2000\ \Omega$, $C_k = 25\ \mu F$, $R_s = 1.5\ M\Omega$,
$C_s = 0.1\ \mu F$, $C = 0.02\ \mu F$, $R_g = 1\ M\Omega$.

(D) Improved variable bass boost

A method for obtaining continuously variable bass boost having improved shape of
the frequency characteristics, is incorporated in the bass and treble boost circuit of
Sect. 8(x)K and Figs. 15.55 and 15.56A.

(iii) Methods incorporating resonant circuits

Resonant circuits give greater flexibility than those incorporating only capacitance
and resistance.   They are limited, however, to values of Q not greater than 1 (Ref. 54)
and preferably not greater than 0.7—see Sect. 1(ix).

(A) Parallel resonant circuits

A parallel resonant circuit, which may be connected in the plate circuit of a r.c.c.
pentode (Fig. 15.9) provides boosting in the vicinity of its resonant frequency, with
maximum boost at its resonant frequency.   For bass boosting the resonant frequency
is often between 50 and 120 c/s, although higher and lower frequencies are sometimes
adopted.

The inductor L_1 has to carry the greater part of the plate current of V_1 (several
milliamperes in a typical case) and should have a butt joint or air-gap to reduce the
effect of the plate current on the inductance.

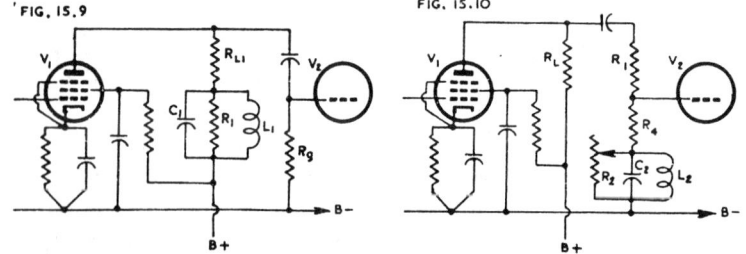

Fig. 15.9. *Bass boosting with parallel resonant tuned circuit in plate circuit.*
Fig. 15.10. *Bass boosting with parallel resonant tuned circuit in grid circuit.*

In all circuits of this type, it is advisable to select a high L/C ratio in order to give
the highest gain for a fixed amount of boosting.

$$Q = \omega_0 C R_s \qquad\qquad (5)$$

For the limiting value of $Q = 1$,

$$R_s \not> 1/\omega_0 C \qquad\qquad (6)$$

where R_s = equivalent total shunt resistance in ohms
$\omega_0 = 2\pi f_0$
f_0 = frequency of resonance in c/s
$LC\omega_0^2 = 1$
L = inductance in henrys
and C = capacitance in farads.

Typical values for a resonant frequency of 70 c/s are : $L_1 = 51.5$ H, $C_1 = 0.1$ μF, $R_1 = 22\,000$ Ω, $R_{L1} = 3\,900$ Ω, $R_g = 0.47$ MΩ, total bass boost 15 db at 70 c/s, $Q = 0.93$. The exact value of R_1 will be influenced by the winding resistance of L_1—the value given above is on the assumption that the winding resistance is zero.

An alternative arrangement which has the advantage of avoiding direct current flow through the inductor is Fig. 15.10—this has the resonant circuit in the grid circuit of V_2. As a result, it is possible to use a higher value of L and to shunt the tuned circuit by a variable resistor as a control. Typical values for $f_0 = 70$ c/s are : $R_L = 0.1$ MΩ, $R_1 = 50\,000$ Ω, $R_4 = 50\,000$ Ω, $L_2 = 250$ H, $C_2 = 0.02$ μF, and $R_2 = 0.5$ MΩ maximum. In this case the winding resistance of L_2 must be taken into account.

Let r_2 = winding (series) resistance of L_2
R_e = equivalent shunt resistance corresponding to r_2
then $R_e = L_2/C_2 r_2$ $\qquad\qquad (7)$

The total shunting on the tuned circuit is therefore R_e in parallel with R_s in parallel with $(R_L + R_1 + R_4)$.

If r_2 is 10 000 ohms*, then
$R_e = 250/(0.02 \times 10^{-6} \times 10\,000) = 1.25$ megohms
$R_s = 0.5$ megohm (max.)
$(R_L + R_1 + R_4) = 100\,000 + 50\,000 + 50\,000 = 0.2$ megohm.

Then $\dfrac{1}{R} = \dfrac{1}{1.25 \text{ M}\Omega} + \dfrac{1}{0.5 \text{ M}\Omega} + \dfrac{1}{0.2 \text{ M}\Omega}$

Thus $R = 0.13$ megohm = total shunting resistance
and $Q = 1.15$ at maximum setting of R_2 and would normally be below 1.0.

(B) Series resonant circuits

A series resonant circuit has a low impedance at the resonant frequency and a gradually increasing impedance at frequencies off resonance. These are used in combined bass and treble controls (Sect. 8) and in circuits incorporating negative feedback (Sect. 9).

(C) Transformer primary resonance

When parallel feed is used with an a-f transformer, the coupling capacitance C may be made to resonate with the primary inductance of the transformer. By this means a limited degree of bass boosting is provided. See Chapter 12 Sect. 4(xii) and Fig. 12.24. The Q should not exceed unity.

The same principle holds when an inductor (choke) is used in the grid circuit, with parallel feed.

References to methods incorporating resonant circuits : 51, 54.

(iv) Circuits involving feedback

(A) Amplifiers with feedback over several stages
This is dealt with in Section 9.

(B) Negative current feedback
Negative current feedback may be applied by omitting the cathode capacitor on the power amplifier valve. This increases the output resistance and provides a close approach to a constant current source for the loudspeaker. It has the effects of peaking the loudspeaker at the bass resonant frequency, reducing the damping on the

---

*This value is typical for a silicon steel core. Very much lower values are typical for mu-metal cores.

loudspeaker, and decreasing the stage gain. This device is usually avoided on account of its short-comings.

(C) Conventional circuits with decreased negative feedback at bass frequencies

One very popular circuit is Fig. 7.33 [Chapter 7 Sect. 2(vi)] which gives feedback over two stages—this has been adapted to provide bass boosting in Fig. 15.11 (Ref. 58). As explained in Chapter 7, this circuit inherently tends to give a slight degree of bass boosting—the boosting is continuously variable from 1.3 db to 12 db using 1000 c/s as the reference frequency, Fig. 15.12. As a modification, the feedback may be taken from the secondary of the transformer—it is advisable to check experimentally to see that the phase rotation due to the transformer does not adversely affect the frequency characteristics.

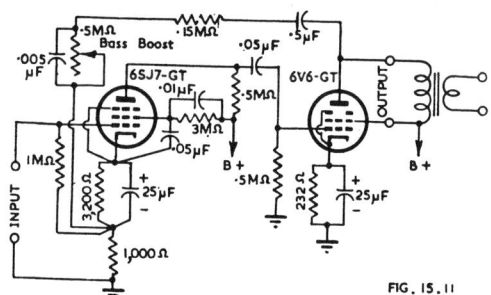

FIG. 15.11

Fig. 15.11. Two stage amplifier with negative feedback over both stages providing bass-boost tone control (typical application).

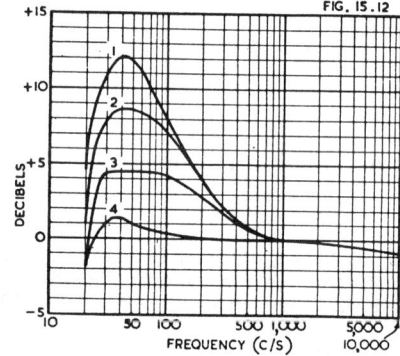

FIG. 15.12

FREQUENCY (C/S)

Fig. 15.12. Frequency characteristics of amplifier in Fig. 15.11. Condition (1) tone control max. resistance (2) three quarters max. (3) half max. (4) one quarter max. resistance. The curves have been superimposed to coincide at 1000 c/s.

Another possible circuit is Fig. 15.13 (Ref. 36) in which the feedback is taken from the plate to the grid through C and R_2. The response is given by

$$(1 - \alpha)(1 + j\omega_0/\omega) \tag{8}$$

where $\omega_0 = \alpha/[CR_2(1 - \alpha)]$

and $\quad \alpha$ = proportion added to the original signal by the setting of VR_1.

Typical values are : $R_1 = R_2 = 1$ MΩ, $R = 0.22$ MΩ, $C = 200$ $\mu\mu$F, giving $\omega_0/2\pi = 250$ c/s when $\alpha = 0.25$. The resistance of VR_1 and VR_2 should be high, preferably with VR_2 greater than VR_1 and $R_3 = VR_1$.

Still another possible circuit is Fig. 15.14A in which feedback is taken from the cathode of the second stage to a resistor at the earthy end of the volume control feeding the grid of the first stage (Ref. 65). It is stated that this gives 10 db of negative feedback at middle and high frequencies and 3 db of positive feedback at low frequencies, equivalent to 13 db bass boost.

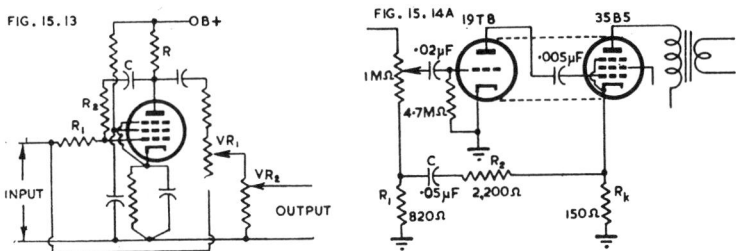

Fig. 15.13.   *Single stage amplifier giving bass boosting with adjustable frequency characteristic (Ref. 36).*
Fig. 15.14A.   *Two stage amplifier giving 13 db effective bass boost (Ref. 65).*

In Fig. 15.14B a capacitance is inserted in the feedback network from the plate of V_2 to the plate of V_1, with r.c. coupling between the stages. This capacitance provides a fixed amount of bass boosting which is generally limited to about 6 db maximum (Ref. 68).

Fig. 15.14B.   Amplifier with 0.005 µF condenser in feedback network to provide bass boosting. V_2 is a power amplifier valve (Ref. 68).

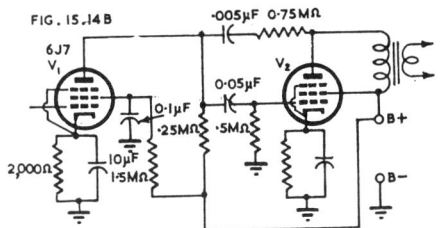

A circuit using an inductor in the feedback network is Fig. 15.54—this may be used for bass boosting only by omitting C_2 and R_4.

A particularly interesting circuit is Fig. 15.15 which incorporates a parallel-T network in the feedback loop (Ref. 57).   In the figure the parallel-T network is tuned to 80 c/s, but it may be tuned to any other desired frequency.   The frequency characteristic is shown in Fig. 15.16.

There have been many other applications of negative feedback to provide bass boosting which have appeared in articles and patents, too numerous to give in detail. Some are given in the References (Sect. 15).

See also Sects. 4, 8, 9, 10 of this chapter, and Chapter 17, Sect. 5.

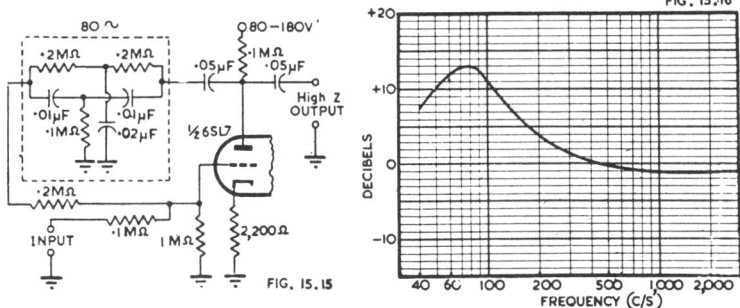

Fig. 15.15.   *Single stage amplifier incorporating a parallel-T network in the feedback loop (Ref. 57).*
Fig. 15.16.   *Frequency characteristics of circuit shown in Fig. 15.15 (Ref. 57).*

FIG. 15.17

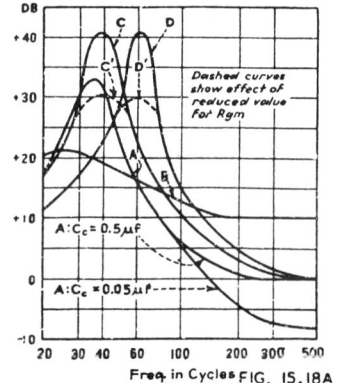

Fig. 15.17. Circuit providing bass boosting due to a negative resistance characteristic of the valve (Ref. 2).

(v) Regeneration due to negative resistance characteristic

A circuit which has interesting possibilities is Fig. 15.17 (Ref. 2). It is capable of very high degrees of bass boosting in the bass without having any appreciable effect at frequencies above 300 c/s (Figs. 15, 18A and B). Since the circuit depends on the negative resistance characteristics of valves, it is subject to larger variations than usual between valves, and is not recommended for quantity production without careful investigation in conjunction with the valve manufacturer.

The frequency of the peak response is determined largely by the values of C_{sg} and C_{gm}. The greater the capacitance of C_t, the sharper will be the bend of the knee of the response curve and the lower the gain between 100 and 500 c/s. Coupling condenser C_{bi} may be considered as regulating the amount of coupling between C_{sg} and C_{gm} with some effect on the shape of the lower end of the curve.

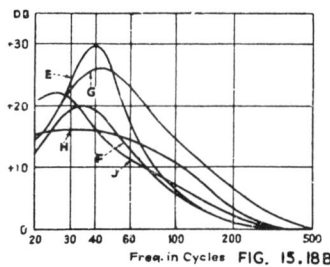

Curve	C_{sg} μF	C_{bt} μF	C_{gm} μF	C_t μF	R_{gm} ohms	R_t ohms
A	0.45	0.1	—	0.1	100 000	7 000
B	0.45	0.1	—	0.1	100 000	100 000
C	0.45	0.1	—	0.2	100 000	7 000
D	0.25	0.02	—	0.2	100 000	7 000
E	0.25	0.1	0.05	0.2	40 000	7 000
F	0.25	0.02	0.05	0.2	100 000	7 000
G	0.45	0.1	—	0.1	100 000	7 000
H	0.45	0.02	—	0.1	100 000	7 000
J	0.25	0.1	0.05	0.1	100 000	7 000

Fig. 15.18 (A and B). Bass boosting characteristics produced by circuit of Fig. 15.17 (Ref. 2).

SECTION 3 : BASS ATTENUATION

(i) General remarks   (ii) Bass attenuation by grid coupling condensers   (iii) Bass attenuation by cathode resistor by-passing   (iv) Bass attenuation by screen by-passing (v) Bass attenuation by reactance shunting   (vi) Bass attenuation by negative feedback (vii) Bass attenuation by Parallel-T network   (viii) Bass attenuation using Constant k filters   (ix) Bass attenuation using M-derived filters.

(i) General remarks

It is desirable for any amplifier to have bass attenuation, although the frequency at which attenuation commences, and the rate of attenuation, should be carefully determined during the design.

Shortwave receivers are often fitted with about 6 db/octave attenuation below about 150 c/s.   In multi-band receivers the attenuation may be incorporated only on the bands where it is desired.   A preferable arrangement is to incorporate a choice of two or three bass cut-off frequencies, say 150, 250 and 400 c/s—this is usually only practicable in communication type receivers.   This extreme attenuation is only usable under bad conditions for barely intelligible speech.   A slight degree of attenuation is helpful in eliminating acoustical feedback to gang condensers and valves and other minor receiver troubles, as well as giving better listening under average shortwave conditions.

Conventional table and mantel model receivers may have an ultimate attenuation of 12 db/octave below say 100 and 150 c/s respectively.   Conventional consoles may have a similar attenuation below 80 c/s, but this may be extended down to say 60 c/s in the case of those having loud-speakers capable of handling the lower frequencies. Fidelity amplifiers should have an ultimate attenuation of 18 db/octave below the lowest frequency which the loudspeaker and output transformer can handle at maximum power output without distortion.   Fidelity amplifiers incorporating negative feedback should connect the filter prior to the main amplifier input terminals.

(ii) Bass attenuation by grid coupling condensers

Every grid coupling condenser introduces some bass attenuation, and it is a matter of good design to select a value to provide the best overall performance of the equipment.   The choice of a capacitance to give a known attenuation at a certain frequency is covered in Chapter 12 Sect. 2(ii) and Fig. 12.9A.   Each coupling condenser gives an ultimate 6 db/octave attenuation.   If it is desired to eliminate low frequency interference or to reduce hum, the coupling condensers may be designed to have identical frequency characteristics, thus giving 12 or 18 db/octave attenuation for 2 or 3 stages respectively.   Even if the grid resistors differ in resistance, the frequency characteristics may be made the same by maintaining the same value of time constant RC.   This method does not apply to the design of feedback amplifiers (see Chapter 7 Sect. 3).

Communication receivers are usually fitted with a switch giving the choice of 2 or 3 bass attenuation characteristics—this may operate by changing the value of grid coupling condenser, or by any other convenient means.

Typical attenuation/frequency characteristics for 1, 2, and 3 stages having identical time constants are given by curves 1, 2 and 3 in Fig. 15.19.

(iii) Bass attenuation by cathode resistor by-passing

The attenuation/frequency characteristics of a resistance-coupled triode with by-passed cathode resistor are given in Chapter 12 Sect. 2(iii) and Fig. 12.3A for a typical case.   These frequency characteristics differ from those for grid coupling condensers in that they are limited to a maximum attenuation which is usually less than 10 db, typical values for r.c. triodes being from 3.7 to 8 db (Chapter 12).   Frequency characteristics for a typical r.c.c. pentode with by-passed cathode resistor are given in Fig. 12.12, the total attenuation being nearly 8 db in this example.

The shape of a typical attenuation/frequency characteristic is illustrated by Curve 4 in Fig. 15.19.

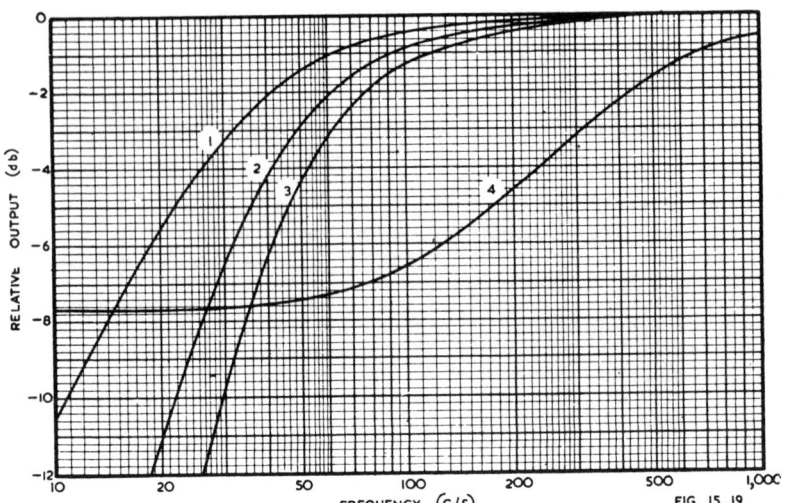

Fig. 15.19. Bass attenuation versus frequency characteristics : (1) Single grid coupling condenser (R = 1 megohm, C = 0.005 μF) ; (2) Two stages each with identical values of time constant RC ; (3) Three stages each with identical values of time constant ; (4) R.c.c. pentode with by-passed cathode resistor (6J7, R_L = 0.25 MΩ, R_k = 2000 Ω, C_k = 0.5 μF) and screen adequately by-passed.

(iv) Bass attenuation by screen by-passing

The attenuation/frequency characteristics of a r.c.c. pentode with a by-passed screen are given in Chapter 12 Sect. 3(iii) and Fig. 12.11A. These characteristics have the same general form as those for cathode resistor by-passing, with a maximum attenuation which is a function of the series screen resistor—with a high resistance series resistor a typical maximum attenuation is 24 db (see Chapter 12). Although the screen by-pass condenser could be used as a step type tone-control, the use of a low capacitance by-pass generally necessitates additional screen filtering for the elimination of hum.

It is possible to design an amplifier so that all grid coupling condensers, cathode and screen by-pass condensers have the same attenuation (say 2 db each) at the same frequency. In such a case, particularly if there are two or more stages, the rate of change of attenuation becomes quite high, although the cathode and screen by-passing tends to introduce steps in the characteristic. However, care is necessary with tolerances—see Sect. 1(x).

(v) Reactance shunting

Any inductance, whether a-f transformer or output transformer, causes ultimate 6 db/octave attenuation below a certain frequency. If the amplifier is to provide good fidelity, it is advisable to design so that this frequency is below the lower frequency limit of the whole amplifier. In other cases, the cut-off frequencies may be made to coincide for more rapid attenuation—however see Sect. 1(x).

If a transformer primary is resonated with the coupling capacitance, in the case of parallel-feed, the ultimate rate of attenuation below the resonant frequency is 12 db/octave.

Owing to the wide tolerances in inductance normally occurring in quantity production, and to the variation caused by the d.c. and signal currents, it is not considered good practice to use the transformer primary resonance for tone control purposes ; a further contributory factor is the increased valve distortion caused by a low impedance load—see Chapter 12 Sect. 4(xii).

If it is desired to vary the cut-off frequency of an a-f transformer with parallel feed, a variable resistor may be inserted to adjust the equivalent source impedance (R_1 in Fig. 15.20). The low frequency attenuation (Ref. 51) is

$$\text{attenuation in db } = 10 \log_{10} [1 + (R/X_1)^2] \tag{1}$$

where $R = R_1 + [r_p R_L/(r_p + R_L)]$
$\quad X_1 = \omega L_1$
and L_1 = primary inductance in henrys.

(vi) Bass attenuation by negative feedback

A network may be inserted in the feedback loop which reduces the feedback voltage at middle and high audio frequencies, thus effectively giving bass attenuation (Fig. 15.21, Ref. 23). Condenser C_2 is merely for blocking the d.c. path, while R_1, R_2 and C_1 form a voltage divider across the output voltage of V_2. The voltage drop across R_2 and C_1 is applied across R_3, thus applying the feedback voltage between V_1 cathode and earth.

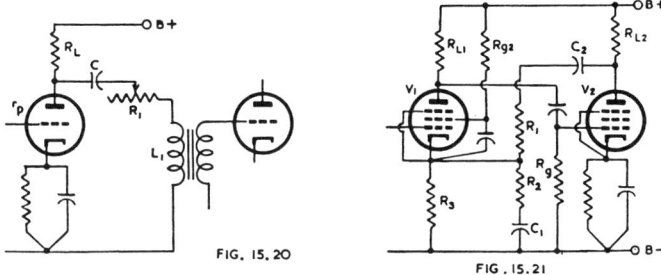

FIG. 15.20   FIG. 15.21

Fig. 15.20. Amplifier, giving variable low-frequency cut-off using a-f transformer and parallel feed.
Fig. 15.21. Amplifier providing bass attenuation by negative feedback (Ref. 23).

(vii) Bass attenuation by Parallel-T network

The parallel-T network—Chapter 4 Sect. 8(iii)—may be used to provide infinite attenuation at one frequency. An example of its application is Fig. 15.22 and the frequency characteristic is given in Fig. 15.23. In its simple form the attenuation is severe even at frequencies five times the frequency giving infinite attenuation. One suggested way of providing a more level characteristic is to incorporate bass boosting around 200 c/s (Ref. 59).

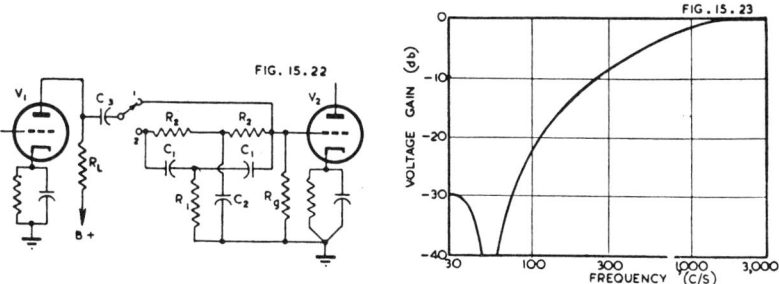

Fig. 15.22. Amplifier providing bass attenuation by Parallel-T network (Ref. 59).
Fig. 15.23. Frequency characteristics of amplifier Fig. 15.22 when V_1 = typical r.c.c. pentode, $C_1 = 0.001$ µF, $C_2 = 0.004$ µF, $C_3 = 0.02$ µF, $R_1 = R_2 = 2$ MΩ, $R_L = 0.2$ MΩ, $R_y = 0.5$ MΩ. The frequency for infinite attenuation is 56 c/s. Switch position (1) Fidelity (2) Bass attenuation (Ref. 59).

(viii) **Bass attenuation using Constant k filters**

If it is desired to achieve a rapid rate of attenuation and a sharp cut-off, it is necessary to use some form of correctly designed filter incorporating both inductors and capacitors. The constant k filter [Chapter 4 Sect. 8(vii)] in its simplest form only incorporates 2 condensers and 1 inductor. An example of its incorporation in an amplifier is Fig. 15.24 (Ref. 60). This filter is fairly non-critical as regards the inductance of L, a 2 : 1 variation still giving a reasonable characteristic with, of course, a change in cut-off frequency. The frequency characteristic obtained with balanced loading is Fig. 15.25 Curve 1 ; when the source impedance is 50 000 ohms and load 0.5 megohm the frequency characteristic is curve 2, showing the typical rise above 0 db due to unbalanced conditions ; curve 3 shows the effect of increasing the resistance of the inductor L.

If sharper attenuation is required, two or more filter sections may be incorporated.

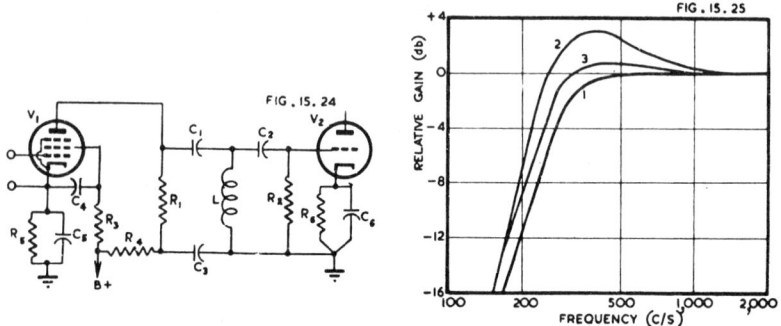

Fig. 15.24. *Amplifier providing bass attenuation by constant k filter (Ref. 60).*
Fig. 15.25. *Frequency characteristics of amplifier Fig. 15.24 when $V_1 = typical$*
r.c.c. pentode, $C_1 = C_2 = 0.002$ $\mu F \pm 5\%$, $C_3 = 1$ μF, $L = 70$ H with $Q = 6$;
Curve (1) $R_1 = R_2 = 0.25$ $M\Omega$; Curve (2) $R_1 = 50\,000$ Ω, $R_2 = 0.5$ $M\Omega$
(unbalanced) ; Curve (3) as Curve 2 with additional resistance in inductor to give
$Q = 3$ *(Ref. 60).*

(ix) **Bass attenuation using M-derived filters**

One T section of an M-derived filter may be used to provide a more rapid attenuation than that of a Constant k filter—see Chapter 4 Sect. 8(viii). The filter itself is shown in Fig. 15.26 and the frequency characteristics for balanced and unbalanced impedances in Fig. 15.27. Here, also, a considerable tolerance is permissible with the value of L and in the degree of mismatching. With mismatching, the value of R may be taken as $\sqrt{R_1 R_2}$ (Ref. 60).

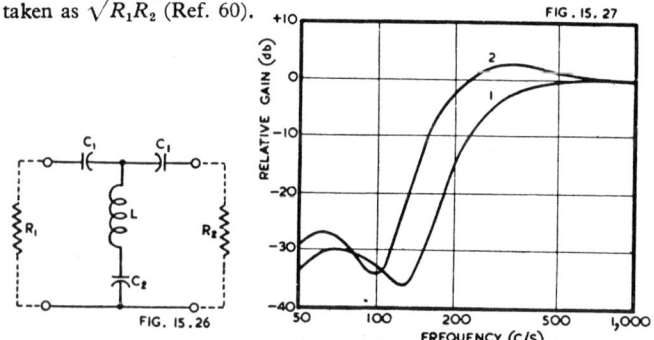

Fig. 15.26. *One T section of a series M-derived filter.*
Fig. 15.27. *Frequency characteristics of filter Fig. 15.26 ; Curve (1) $R_1 = R_2 =$*
0.25 $M\Omega$, $C_1 = 0.0023$ μF, $C_2 = 0.024$ μF, $L = 74$ H ; *Curve (2) $R_1 = 50\,000$ Ω,*
$R_2 = 0.5$ $M\Omega$ *(unbalanced), $C_1 = 0.0058$ μF, $C_2 = 0.035$ μF, $L = 73$ H (Ref. 60).*

SECTION 4 : COMBINED BASS TONE CONTROLS

(i) Stepped controls   (ii) Continuously variable controls.

(i) Stepped controls

A typical stepped control (Fig. 15.28) has 2 positions giving bass boost, 1 giving "flat" response, and 2 giving bass attenuation. The values of capacitance may be selected to give the desired frequency characteristics.

See also Sect. 8 for combined bass and treble tone controls, particularly Figs. 15.37A, 15.37B, 15.39, 15.40.

References 17, 50, 54, 55, 64.

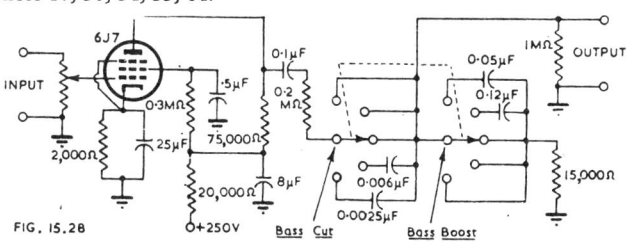

Fig. 15.28.   *Step-control giving bass boosting and bass attenuation.*

(ii) Continuously variable controls

Circuits providing only bass boosting or attenuation with a single continuously variable control are not widely used.

SECTION 5 : TREBLE BOOSTING

(i) General remarks   (ii) Circuits not involving resonance or negative feedback (iii) Methods incorporating resonant circuits   (iv) Circuits involving feedback.

(i) General remarks

Treble boosting, mainly on account of the increased distortion which it causes, is usually avoided altogether or else used only for the purpose of equalizing an unavoidable attenuation in the amplifier at the maximum frequency limit.

The subject of distortion has been covered in Section 1(iv).

(ii) Circuits not involving resonance or negative feedback

(A) Conventional treble boosting circuit (Fig. 15.29)

This is an adaptation of the fundamental circuit Fig. 15.2. It may be shown that

$$\text{Gain at low frequencies} = \frac{\mu R R_2}{(R_1 + R_2)(R + r_p) + R r_p} \tag{1}$$

$$\text{Gain at very high frequencies} = \frac{\mu R'}{R' + r_p} \tag{2}$$

where $R = R_I(R_1 + R_2)/(R_L + R_1 + R_2)$

and $R' = R_2 R_L/(R_2 + R_L)$.

Then $B =$ ratio of amplification at very high frequencies to amplification at low frequencies

$$= 1 + \frac{R_1}{R_2 + R''} \tag{3}$$

where $R'' = r_p R_L/(r_p + R_I)$.

We may determine the value of R_1 for any desired value of B from the equation

$$R_1 = (B - 1)(R_2 + R'') \qquad (4)$$

In the case of pentodes, R'' approximates to R_L.

The frequency characteristics for treble boosting have been plotted (Ref. 51 Fig. 9.18 ; also Refs. 9,55) but for most purposes it is sufficient to work on the " half-boost " points as for bass boosting, the curves being approximately symmetrical S curves about these points. The values of R/X for the half-boost point are identical to those for bass boosting, and not the inverse as might be expected.

Total boost	20	15	10	6	3	db
Half boost	10	7.5	5	3	1.5	db
Boost ratio (B)	10	5.62	3.16	2.0	1.41	
(R_1/X_1) for half boost*	3.16	2.37	1.78	1.41	1.19	
C_1†	$3.16/\omega R_1$	$2.37/\omega R_1$	$1.78/\omega R_1$	$1.41/\omega R_1$	$1.19/\omega R_1$ F

$*(R_1/X_1) = \sqrt{B}$ at half boost point. $X_1 = 1/\omega C_1$

$†C_1 = 1/\omega X_1 = (R_1/X_1)/\omega R_1$ $\omega = 2\pi f_0$

where R_1 and C_1 are as shown in Fig. 15.29

and f_0 = frequency at half-boost point.

A variable control may be achieved in various ways (Ref. 9) :

(a) Varying C_1—This can only be done in steps. The result is to change the frequency of the half-boost point (see table).

(b) Varying R_1—This varies the amplification for middle frequencies, and is unsatisfactory for tone control purposes.

(c) Adding a variable resistor in series with C_1—This is fairly effective but the shapes of the frequency characteristics are not ideal for tone control. It is used in a slightly modified form in the combined bass and treble boost circuit of Fig. 15.46.

References to conventional treble boosting circuit : 9, 17, 48, 51, 55.

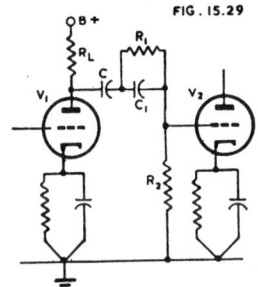

Fig. 15.29. Conventional treble boosting circuit.

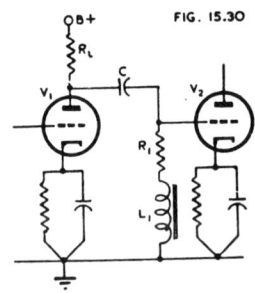

Fig. 15.30. Treble boosting with inductance in grid circuit.

(B) Using inductance in grid circuit (Fig. 15.30)

This gives the same shape of frequency characteristics as the conventional treble boosting circuit. The same expressions may be used except that in this case

$$X_1 = \omega L_1 \text{ and } B = R/R_1$$

where $R = [r_p R_L/(r_p + R_L)] + R_1$.

If R_1 is made variable, it has a very pronounced effect on the gain at middle frequencies. A preferred method of tone control is to use a step switch and tappings on the inductance (see Sect. 8 and Fig. 15.40).

References 17, 51.

(C) Correction for side-band cutting

The methods which may be used for the equalization of side-band cutting in a receiver are described in Refs. 3 and 62.

(iii) Methods incorporating resonant circuits

Resonant circuits are rarely used for treble boosting tone control since there are other preferable methods which give the required performance (Ref. 17).

(iv) Circuits involving feedback

(A) Cathode resistor by-passing

If a suitable small capacitance is used as a by-pass condenser across any cathode resistor, a limited degree of treble boosting will be achieved.   In the case of resistance-coupled amplifiers the maximum boost is not more than 6 or 8 db.   Greater boosting may be achieved by increasing the cathode resistor and returning the grid resistor to a suitable tapping point in the cathode circuit.

(B) Network in feedback loop

The same method may be used as for bass attenuation—see Sect. 3(vi) and Fig. 15.21.

SECTION 6 : TREBLE ATTENUATION

(i) General remarks   (ii) Attenuation by shunt capacitance   (iii) Treble attenuation by filter networks   (iv) Treble attenuation in negative feedback amplifiers.

(i) General remarks

Every amplifier should have a maximum frequency of response, beyond which attenuation should be at a rate of not less than a nominal 12 db per octave for conventional amplifiers or 18 db per octave for fidelity amplifiers.   This attenuation may be partly prior to the input terminals of the main amplifier or in the early stages of the amplifier, and partly (at least 6 db/octave) at the output end.

Because the simplest methods of treble attenuation have a gradual commencement, there tends to occur a rounding of the response curve which may be appreciable at the maximum frequency of response.   It is suggested that 2 db attenuation at this frequency is a reasonable compromise, or even higher attenuation for other than fidelity amplifiers, but where an amplifier must be designed within narrow tolerances it is generally necessary to use one of the more elaborate methods of attenuation.

In the case of amplifiers incorporating negative feedback, the filter to provide the attenuation must be outside the feedback loop.

(ii) Attenuation by shunt capacitance

The theory has been covered in Chapter 4 Sect. 8(ii) under the heading " r.c. low-pass filter " with the circuit of Fig. 4.37 ; the attenuation characteristic is given in Chapter 4 eqn. (4) and plotted in Fig. 4.38.

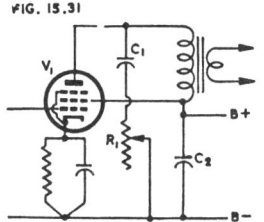

FIG. 15.31

Fig. 15.31.   Typical " tone control" used in radio receivers.   In a typical case $C_1 = 0.05$ μF, $R_1 = 50\,000\Omega$ (max.), $V_1 = $ power pentode with load resistance about 5000 ohms.

This forms part of the usual " tone-control " in many radio receivers of the less expensive class, as illustrated in Fig. 15.31.   The control is in the form of a variable resistance in series with the capacitance which limits its attenuation.   A suitable maximum value for the resistance is 10 times the load resistance. In some cases a fixed capacitance only is used—typical values are from 0.005 to 0.02 μF for a loud-speaker load resistance of 5000 ohms. In this application the " effective resistance " for the calculation of the attenuation will be the loudspeaker impedance in parallel with the output resistance (R_0) of the amplifier.

Tone control by shunt capacitance, when used with discrimination, is fairly satis-factory with a flat amplifier operating from an equalized source. It is far from satis-factory in selective radio receivers in which the high audio frequencies are already heavily attenuated ; this effect is minimized when variable selectivity i-f amplifiers are used. One reason for the popularity of shunt capacitance tone control is that the intermodulation frequencies produced by the distortion in the pentode or beam power amplifier valve are much reduced, and listening thereby made less fatiguing. If methods are taken to reduce the distortion, e.g. by negative feedback, the tone control may preferably take the form of bass boosting.

In order to achieve the desired treble attenuation, a shunt capacitance may be placed across the source (radio receiver, pickup or microphone), from any grid to earth, from any plate to earth or across the output terminals of the amplifier. In each case the effective resistance (R in eqn. 4 of Chapter 4 and Fig. 4.38) is the resultant a.c. re-sistance between the points across which the shunt capacitance is connected. In the case of a r.c.c. pentode it is approximately the load resistance in parallel with the following grid resistor, and it does not make any appreciable difference whether the capacitance is shunted across the plate load resistor or the following grid resistor, owing to the coupling through the grid coupling condenser. Even if no shunt con-denser is added, the output capacitance of the valve plus the dynamic* input capacit-ance of the following valve plus wiring capacitances provide appreciable treble attenua-tion—see Chapter 12 Sect. 2(xi). The effect on treble attenuation may be reduced by reducing the plate load resistance ; it may be increased by increasing the plate load resistance or by adding shunt capacitance. A typical r.c.c. pentode followed by a similar stage has only slight attenuation at 10 000 c/s when the load resistance is less than 0.25 megohm. In the case of a r.c.c. triode the effective resistance is the plate resistance of the valve (under r.c.c. conditions) in parallel with the load resist-ance and the following grid resistance.

In the general case the effective resistance is given by

$$R = R_s + R_L'r_p/(R_L' + r_p) \tag{1}$$

where R_s = resistance in series with C as part of the tone control,
  $R_L' = R_L R_g/(R_L + R_g)$
  R_L = load resistance
  R_g = following grid resistance
and r_p = plate resistance of valve.

With a high-mu triode such as type 6SQ7 the effective input capacitance is of the order of 100 $\mu\mu$F. If this capacitance is shunted across an effective resistance of 0.5 megohm, the attenuation is about 10 db at 10 000 c/s.

The **total attenuation** at any frequency will be the attenuation at that frequency prior to the amplifier plus the sum of the attenuations of the stages in the amplifier at that frequency plus the attenuation by the shunt capacitance at that frequency, all expressed in db. It is not sufficient to assume an attenuation of 6 db/octave from each stage in the amplifier unless all the stages have the same cut-off frequency. In the general case it is necessary, for design purposes, to calculate the attenuation of each stage or filter for convenient frequencies, e.g. 10 000, 14 000, 20 000, 28 000, 40 000 c/s and then to add the values in db to determine the overall attenuation characteristic. The latter will, eventually, almost reach a slope of 6 db/octave per stage or per shunt capacitance, but the knee of the curve will be very rounded unless the cut-off fre-quencies are identical.

---

*Under operating conditions, including the " Miller Effect " capacitance from the plate.

(iii) Treble attenuation by filter networks

(A) Constant k low-pass filter

An approximate Constant k low-pass Π section filter to provide a nominal treble attenuation of 24 db/octave with the choice of several cut-off frequencies is shown in Fig. 15.32. This is intended to be connected between a pickup or microphone pre-amplifier and the main amplifier. For theory see Chapter 4 Sect. 8(vii) and Fig. 4.49. In reality only C_2 has been calculated to give a constant k filter for the correct terminating impedances, and C_1 and C_3 have been calculated merely to give the desired cut-off frequencies leaving L unchanged. Thus it is only a constant k filter on tapping 3 and a sufficiently close approximation on the other tappings. A considerable degree of mismatching is permissible between R_1 and R_2 provided that $\sqrt{R_1 R_2} = 0.1$ megohm.

The calculation of the constant k filter is based on the expressions :

$L/C = R^2$ (see Fig. 4.49C for symbols)

$f_0 = 1/\pi\sqrt{LC}$ where $f_0 =$ cut-off frequency.

FIG. 15.32

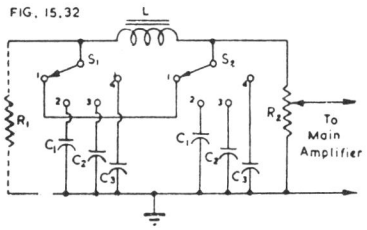

Fig. 15.32. Approximate constant k low-pass Π section filter. $R_1 =$ source impedance. When $R_1 = R_2 = 0.1\,M\Omega$, $L = 5\,H$, $C_1 = 0.0001$, $C_2 = 0.0002$ and $C_3 = 0.0004\ \mu F$, the cut-off frequencies will be (1) none (2) 10 000 c/s (3) 7000 c/s (4) 5000 c/s.

(B) M-derived filters

One T section of an M-derived filter may also be used for treble attenuation. See remarks for bass attenuation.

(C) Parallel-T network

The parallel-T network—Chapter 4 Sect. 8(iii)—may be used to provide infinite attenuation at one frequency but the rate of attenuation begins very gradually. One improvement is described in Chapter 17 Sect. 5(iv) Figs. 17.24B and 24C (Ref. 57). See also Sect. 11(iv).

(iv) Treble attenuation in negative feedback amplifiers

A network may be inserted in the feedback loop to reduce the feedback voltage at low and middle audio frequencies, thus effectively giving treble attenuation. The circuit is effectively the same as for bass boosting (Fig. 15.11) except that the capacitance is selected to give the required frequency characteristics. See also Ref. 23, Fig. 9.

There have been many negative feedback circuits which give a measure of tone control among which are—

(A) Fig. 15.33. This circuit incorporates a very small condenser C (of the order of 10 to 100 $\mu\mu F$) to provide some negative feedback at the higher audio frequencies without seriously affecting the gain at 400 c/s. This is only fully effective when no plate by-pass capacitor is used.

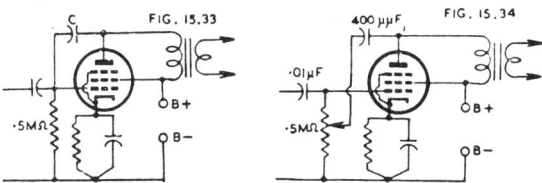

Fig. 15.33. Power amplifier valve with small condenser C giving negative feedback at high audio frequencies (treble attenuation).

Fig. 15.34. Power amplifier valve with variable negative feedback at high audio frequencies (treble attenuation).

(B) Fig. 15.34 is a development of Fig. 15.33 to provide continuously variable tone control. The feedback condenser is larger and a potentiometer is used in the grid circuit to control the tone. At maximum treble cut it gives an attenuation of 28 db at 20 000 c/s (Ref. 39).

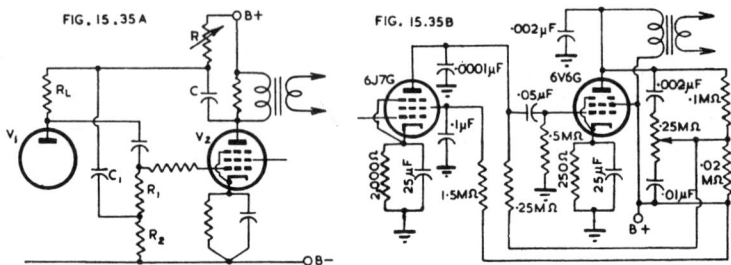

Fig. 15.35A. *Amplifier with variable treble attenuation due to negative feedback.*
Fig. 15.35B. *Amplifier with negative feedback providing treble attenuation or treble boosting (Ref. 67).*

(C) Fig. 15.35A is a further variation in which R and C form a potential divider across the output from V_2. At increasing frequencies a larger negative feedback voltage is applied, the amount being limited by the variable resistance R. C_1 is a blocking condenser (Ref. 63).

(D) For feedback over more than one stage, see Sect. 9.

SECTION 7 : COMBINED TREBLE TONE CONTROLS

Treble controls giving a choice of treble boosting or treble attenuation without also incorporating bass boosting or attenuation are comparatively rare. Attention is directed to Sect. 8 and Figs. 15.37A, 15.37B, 15.39, 15.40, 15.46.

One example in this class is Fig. 15.35B which provides continuously-variable tone control from treble boost to treble attenuation by means of a resistor-capacitor network in the feedback circuit (Ref. 67).

SECTION 8 : COMBINED BASS AND TREBLE TONE CONTROLS

(i) *Stepped controls—general* (ii) *Quality switch* (iii) *Universal step-type tone control not using inductors* (iv) *Universal step-type tone control using inductors* (v) *Fixed bass and treble boosting* (vi) *Step-type tone control using negative feedback* (vii) *Continuously-variable controls—general* (viii) *Single control continuously-variable tone controls* (ix) *Ganged continuously-variable tone controls* (x) *Dual control continuously-variable tone controls.*

(i) Stepped controls—general

Stepped controls are capable of better performance than continuously variable controls, partly because it is possible to vary two or more component values simultaneously and partly because the attenuation at middle frequencies may be adjusted so as to avoid an apparent change in volume as the control is moved. For most radio purposes it is sufficient to have two control knobs, one controlling the bass and the other the treble, with five positions on each—one "flat," two boosting and two attenuating with alternative degrees of each.

(ii) Quality switch

An alternative arrangement which is very effective with the non-technical public is to select suitable values for each type of listening conditions. For example, one manufacturer* has brought out a model with an 11 position tone control as under :

1. for distant stations, extremely noisy—maximum sensitivity (10% feedback), bass normal, treble attenuation severe to counteract static and noise.
2. for distant stations, noisy—maximum sensitivity, bass normal, treble attenuation less severe.
3. for distant stations, less noisy—maximum sensitivity, bass normal, treble attenuation still less.
4. for distant stations, still less noisy—maximum sensitivity, bass normal, treble attenuation slight.
5. for distant stations, slight noise—maximum sensitivity, bass normal, no treble attenuation.
6. for local stations with good fidelity at low volume—reduced sensitivity (25% feedback), bass boost, treble boost.
7. for fidelity at medium volume—slight treble boost, less bass boost.
8. for fidelity at normal volume—no treble boost, no bass boost.
9. for fidelity at normal volume but reducing needle scratch—slight treble attenuation, no bass boost.
10. as (9) but increased treble attenuation.
11. as (10) but increased treble attenuation.

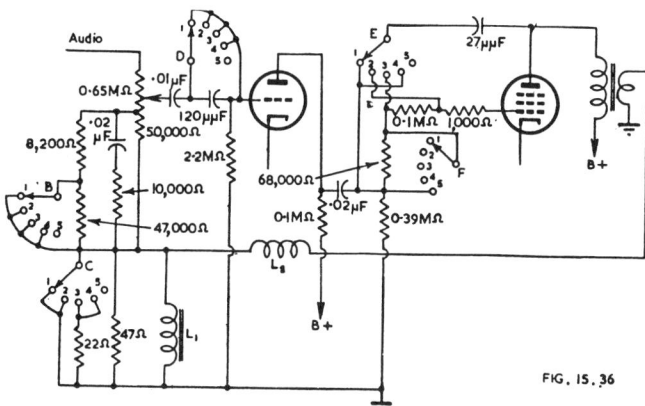

Fig. 15.36.   One form of Quality Switch (Ref. 33).

An alternative form using a 5 position switch is shown in Fig. 15.36 (Ref. 33). The switch positions are

1. heavy bass attenuation and very limited treble attenuation for very distant reception, no feedback.
2. normal bass, heavy treble attenuation, no feedback.
3. bass attenuation, medium treble attenuation, half feedback.
4. normal bass, slight treble attenuation, half feedback.
   Note : Positions 1 to 4 inclusive are for use with narrow i-f bandwidth.
5. normal bass, no treble attenuation, maximum feedback, wide i-f bandwidth, for fidelity reception of local stations.

At low settings of the tapped volume control the capacitance in series with the inductance provide a dip at middle high frequencies, thus effectively giving bass and treble boost.

---

*Columbus Radio (Radio Corporation of New Zealand Limited).

(iii) Universal step-type tone control not using inductors

A simple form is shown in Fig. 15.37A where treble boost is obtained by cathode by-passing (C_1), treble attenuation by shunt capacitance (C_4), bass attenuation by grid coupling condenser (C_3), and bass boosting by the conventional method (C_5). Each of these may have two or three values—C_1 may be 0.01, 0.02, 0.03 μF ; C_3 may be 0.0025, 0.006 μF ; C_4 may be 0.002, 0.005 μF ; C_5 may be 0.02, 0.05, 0.1 μF.

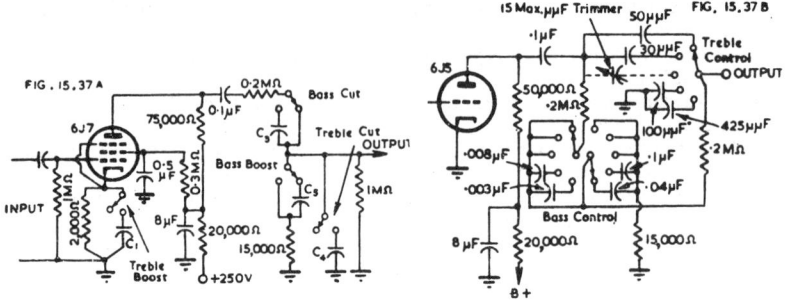

Fig. 15.37A. *Simple universal tone control not using inductors.*   Fig. 15.37B. *Step-type tone control not using inductors (Ref. 70).*

An improved form is given in Fig. 15.37B and its frequency characteristics in Fig. 15.37C (Ref. 70). The 15 $\mu\mu$F max. trimmer capacitor is for the purpose of compensating in the " flat " position for the loss of high audio frequencies caused by stray shunt and Miller Effect capacitances when the following stage is a triode.

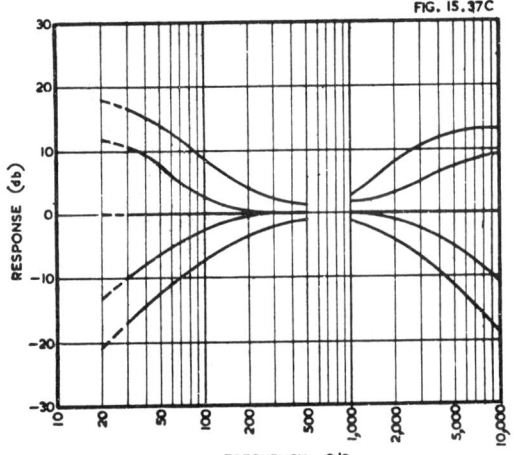

Fig. 15.37C. *Frequency response characteristics of the circuit of Fig. 15.37B (Ref. 70).*

Increased rate of boosting and attenuation is obtainable by connecting two or more r.c. filters in cascade. A step of 6 db/octave (nominal) is too great for a flexible tone control, so that networks giving reduced rates of boosting and attenuation may be devised (Fig. 15.38). These may be combined to give any desired steps. A very comprehensive example is Fig. 15.39 (Ref. 50) which uses two six-pole 11 position control switches. Its overall gain is zero db at 500 c/s at any setting of the control switches.

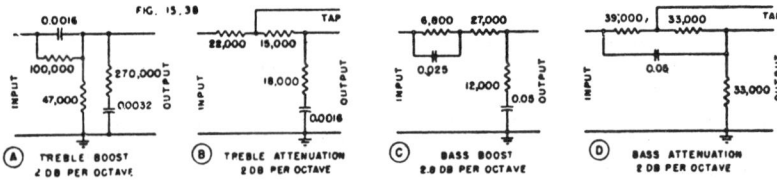

Fig. 15.38. *Networks giving selected values of decibels per octave for boosting and attenuation (Ref. 50).*

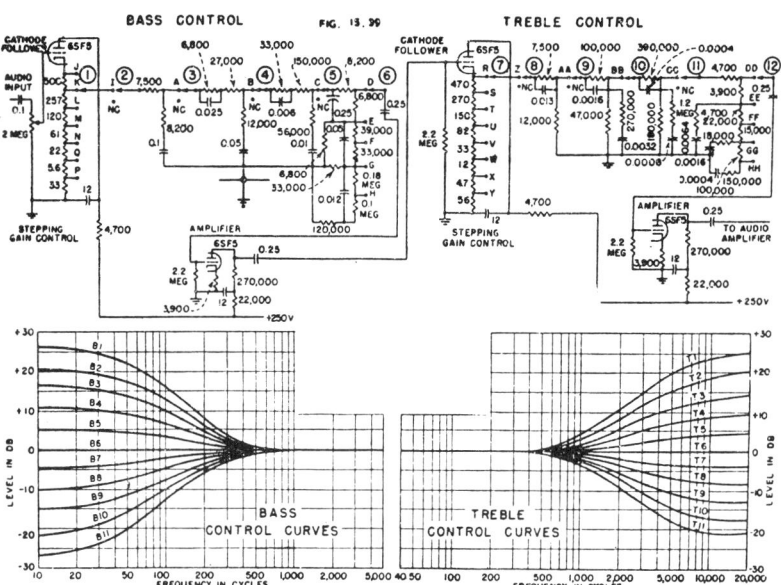

Fig. 15.39.   *Step-type tone control system not using inductors (Ref. 50).*

RESP. CURVE.	7	8	CONTACT. 9	10	11	12
T1	R	Z	AA	BB	NC	CC
T2	S	NC	Z	BB	NC	CC
T3	T	Z	AA	NC	NC	BB
T4	U	NC	Z	NC	NC	BB
T5	V	Z	NC	NC	NC	AA
T6	Y	NC	NC	NC	NC	Z
T7	X	NC	NC	NC	Z	DD
T8	X	NC	NC	NC	Z	EE
T9	X	NC	NC	NC	Z	FF
T10	X	NC	NC	NC	Z	GG
T11	W	NC	NC	NC	Z	HH

RESP. CURVE.	1	2	CONTACT 3	4	5	6
B1	J	I	A	B	NC	C
B2	K	NC	I	B	NC	C
B3	L	I	A	NC	NC	B
B4	M	NC	I	NC	NC	B
B5	N	I	NC	NC	NC	A
B6	P	NC	NC	NC	NC	I
B7	P	NC	NC	NC	I	D
B8	P	NC	NC	NC	I	E
B9	P	NC	NC	NC	I	F
B10	P	NC	NC	NC	I	G
B11	Q	NC	NC	NC	I	H

(iv) Universal step-type tone control using inductors

The treble boosting circuit of Fig. 15.30 may be used, combined with treble attenuation by shunt capacitance and conventional bass boosting and attenuation.

A typical example is Fig. 15.40 (Ref. 64 ; see also Refs. 17, 54, 55). The nominal slope is 6 db/octave for all positions, but the capacitances and inductances may be selected to give any desired frequencies for the commencement of attenuation or

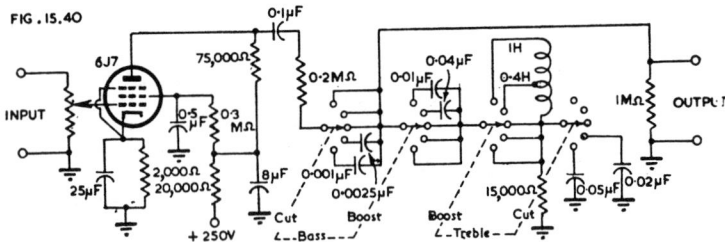

Fig. 15.40. Universal step-type tone control using inductors.

boosting. The inductor may be wound with an air-core as under : Former diameter ¾ in. Length between cheeks ¾ in. Winding wire 40 S.W.G. (or 36 A.W.G) SSE. Total turns 6740, tapped at 4520 turns. Layer wound. Total radial depth of winding say 0.6 in.

One defect of this method of treble boosting is the tendency for the inductor to pick up hum.

(v) Fixed bass and treble boosting

When it is desired to incorporate a fixed amount of bass and treble boosting, as for example equalizing the response of a pair of headphones, the circuit of Fig. 15.41 may be used. For headphone equalizing (Ref. 7) suitable values are $C_1 = 0.0003$, $C_2 = 0.015$ μF, $R_1 = R_4 = 0.1$ MΩ, $R_5 = 0$. For bass and treble boost for tone control suitable values are (Ref. 25) :—$C_1 = C_2 = 0.0002$ μF, $R_1 = 0.2$ MΩ, $R_4 = 0.3$ MΩ, $R_5 = 0$, giving 16 db bass boost (20 c/s) and 7 db treble boost (10 000 c/s) relative to 1000 c/s. If $R_5 = 50\,000$ ohms, other values being unchanged, the bass boost is 16 db and treble boost 12 db.

(vi) Step type tone control using negative feedback

Circuits providing universal step-type tone control using negative feedback are described in Sect. 9 (Fig. 15.58A).

(vii) Continuously-variable controls—general

Continuously-variable controls are in two groups, those which have one control knob and those which have two or more. A single control knob is obviously limited in its capabilities—for example it may be used to provide bass boosting when turned in one direction from the centre point, or treble boosting in the other direction, with flat response at the centre point. On the other hand, with two control knobs it is possible to make any desired combination of bass and treble characteristics.

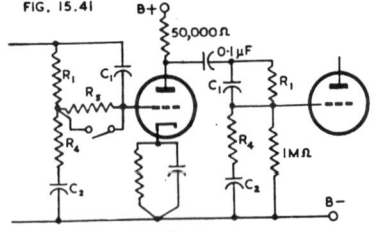

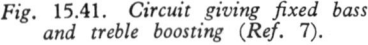

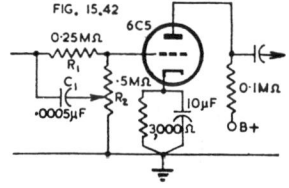

Fig. 15.41. Circuit giving fixed bass and treble boosting (Ref. 7).

Fig. 15.42. Single control continuously variable tone control (Ref. 22).

(viii) Single-control continuously variable tone controls

(A) A typical example is Fig. 15.42 (Ref. 22). When C_1 goes to the grid end of potentiometer R_2, the bass is attenuated by the grid coupling condenser C_1, limited by R_1. When C_1 goes to the earthed of R_2, the treble is attenuated by the shunt capacitance of C_1.

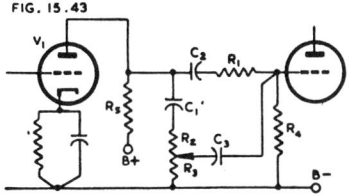

Fig. 15.43. More elaborate single control continuously variable tone control. Typical values are : V_1 = pentode, R_1 = 5 $M\Omega$, $R_2 + R_3$ = 0.25 $M\Omega$, R_4 = 0.3 $M\Omega$, C_1 = 0.0001 μF, C_2 = 0.05 μF, C_3 = 0.001 μF (Ref. 62).

FIG. 15.43

(B) Another example is Fig. 15.43 (Ref. 62) which is capable of giving treble boost 18 db at 10 000 c/s and bass attenuation 6 db at 50 c/s in one extreme position ; bass boost 4 db and treble attenuation 21 db in the other extreme.

(C) An interesting circuit which gives simultaneous bass and treble boosting, linear response, or simultaneous bass and treble attenuation is Fig. 15.43A (Ref. 86). The

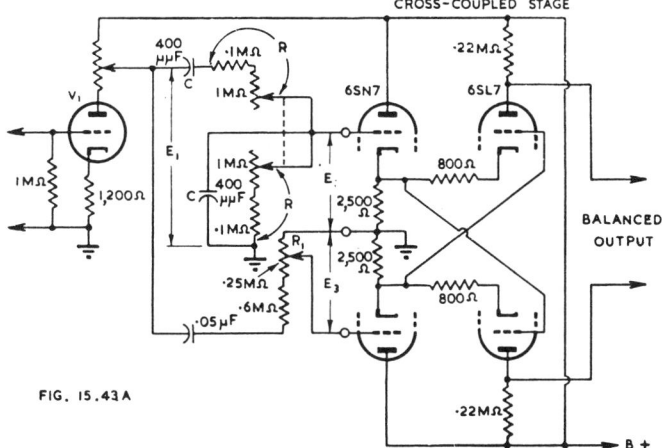

FIG. 15.43A

Fig. 15.43A.   Tone control circuit using cross coupled input stage (Ref. 86).

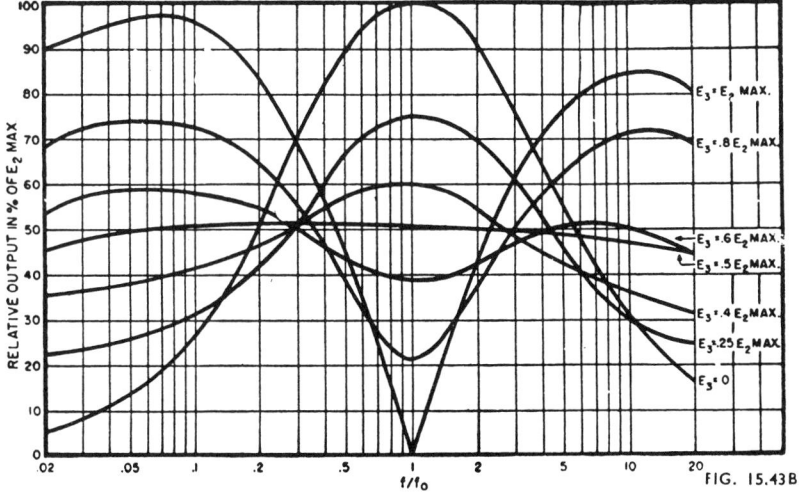

Fig. 15.43B.   Frequency response characteristics of circuit of Fig. 15.43A, the value of R_1 being varied (Ref. 86).

two twin triodes form the " cross-coupled phase inverter "—see Chapter 12 Sect. 6(viii)—which is here used to amplify the difference between the input voltages to the 6SN7 grids. The frequency response curves are shown in Fig. 15.43B for different values of R_1. The maximum output and the point of zero phase shift occur at $f/f_0 = 1$, where $f_0 = 1/2\pi RC$. The frequency of the peak in the curve may be determined by choice of R and C; in the circuit shown, this frequency is adjustable from 360 to 4000 c/s by means of a dual 1 MΩ potentiometer. A choice of $f_0 = 600$ to 800 c/s is pleasing in many cases.

(ix) Ganged continuously-variable tone controls

Two or more controls may be ganged and operated by a single knob. One control may be the tone control and the other a control of gain so that the apparent output level is held approximately constant at all settings.

(x) Dual control continuously-variable tone controls

(A) Simple duo-control circuit, Fig. 15.44A (Ref. 22)

This filter incorporates resistors and condensers only, and R_3 controls the bass and R_4 the treble. A filter having a similar function is also described in Ref. 31.

A slight modification of this circuit is Fig. 15.44B which gives either bass or treble boosting (Ref. 41).

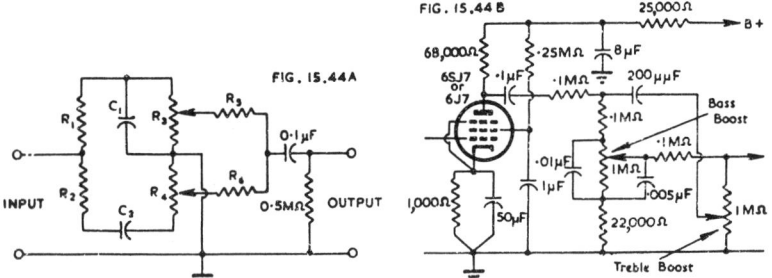

Fig. 15.44A.   *Simple duo-control circuit giving individual control of bass and treble. Typical values are : $R_1 = 0.25$ MΩ; $R_2 = R_3 = 0.5$ MΩ; $R_4 = 0.25$ MΩ; $R_5 = R_6 = 0.5$ MΩ; $C_1 = 0.01$ μF; $C_2 = 0.001$ μF (Ref. 22).*
Fig. 15.44B.   *Simple tone control giving bass and treble boosting (Ref. 41).*

(B) Duo-control circuit incorporating L and C (Fig. 15.45, Ref. 26)

In this filter, R_1 controls the bass and R_3 the treble. Hum may be troublesome on account of L.

(C) Cutler duo-control circuit (Fig. 15.46, Ref. 26)

This filter may only be used to feed into the grid of an amplifier valve ; R_4 controls the bass boost and R_3 the treble boost. The maximum boost is 17 db at 40 c/s and 15 db at 10 000 c/s, relative to 1000 c/s, with the values shown.

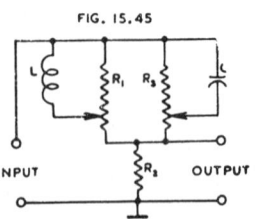

Fig. 15.45.   *Duo-control circuit incorporating L and C (Ref. 26).*

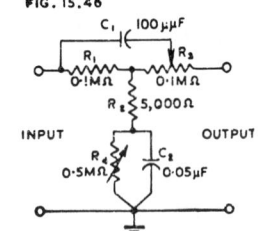

Fig. 15.46.   *Cutler duo-control circuit (Ref. 26).*

(D) Paraphase bass-treble tone control (Figs. 15.47A, Ref. 56)

This is an outstandingly flexible circuit which may be designed for any cross-over frequency—400 c/s in Fig. 15.47A and the curves in Fig. 15.47B. The design procedure is :

1. Choose a cross-over frequency f_c.
2. Choose $R \geqq 10R_L$ where R_L = generator impedance.
3. Make $C = 1/(2\pi f_c R)$.
4. Make $K = 10$; then $KR = 10R$ and $C/K = C/10$.

Example : Choose $f_c = 400$ c/s. Say generator impedance = 1000 ohms, then select

$R = 82\,000$ ohms, from which

$C = 1/(2\pi \times 400 \times 82\,000) = 0.005$ μF ; $C/K = 0.0005$ μF.

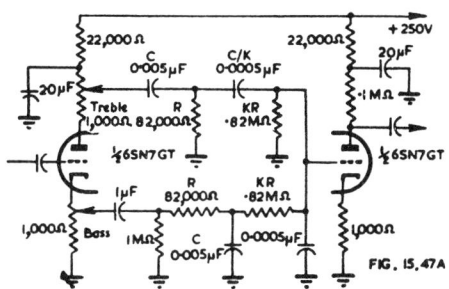

Fig. 15.47A.   *Paraphase bass-treble tone control* (*Ref.* 56).

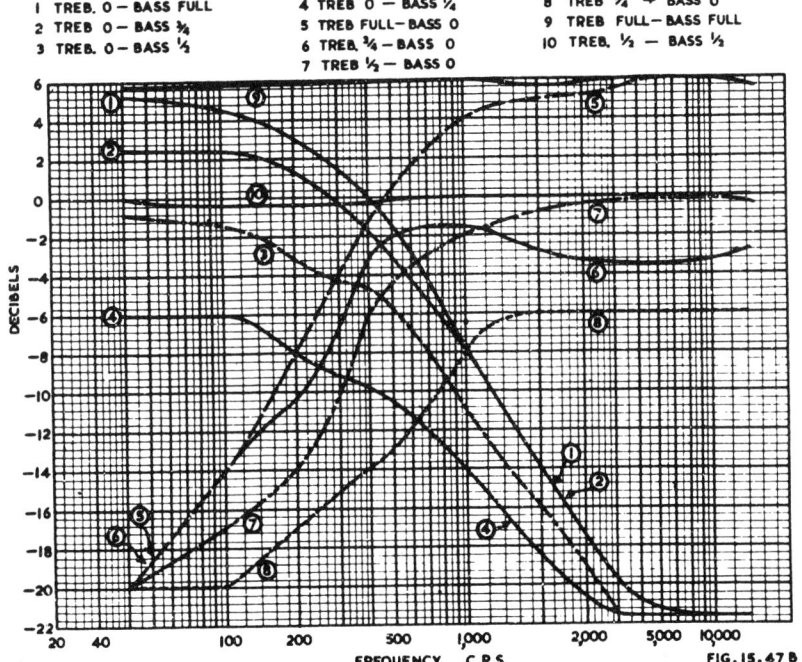

Fig. 15.47B.   *Frequency characteristics of paraphase bass-treble tone control* (*Fig.* 15.47A) (*Ref.* 56).

(E) Two-stage bass and treble tone control (Fig. 15.48, Ref. 41)

This employs a twin triode with the treble boost control in the coupling from V_1 to V_2 and the bass boost control in the coupling from the plate of V_2 to the following stage.

(F) Simple two channel amplifier (Fig. 15.49, Ref. 22)

Owing to the separate amplifier valves, this circuit does not require the series resistors R_5 and R_6 used in Fig. 15.44 and has thereby less attenuation in the filter network.

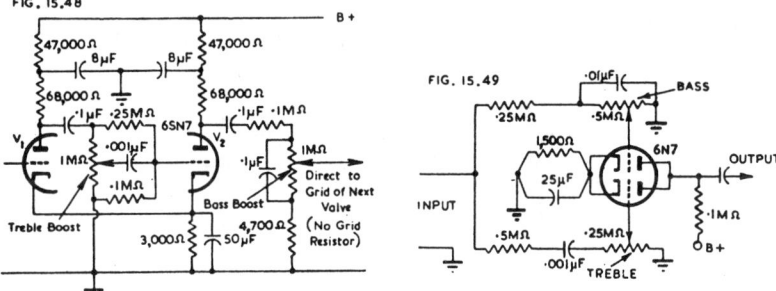

Fig. 15.48. Two-stage bass and treble tone control (Ref. 41).   *Fig. 15.49. Simple two channel amplifier (Ref. 22).*

(G) Wide-range two-channel amplifier (Fig. 15.50, Ref. 31)

In this circuit V_1 is an amplifying valve common to both channels, V_2 is a bass amplifier only, while V_3 is a twin triode having one grid fed from the bass amplifier and the other from V_1 through a filter network which only passes the higher frequencies. However two plates of V_3 are approximately equal in amplitude for all frequencies, owing to the common cathode coupling, and opposite in phase, so that the output may be applied directly to a push-pull power output stage. The bass amplifier V_2 is a true bass booster, so that it is not necessary to attenuate all except bass frequencies. The two controls provide (1) flat response (2) independent treble boost (3) independent bass boost equal to the gain of V_2. If it is desired to add treble attenuation, the filter circuit of Fig. 15.51 may be used. This gives a flat response when the moving arm of the potentiometer is at the centre tap.

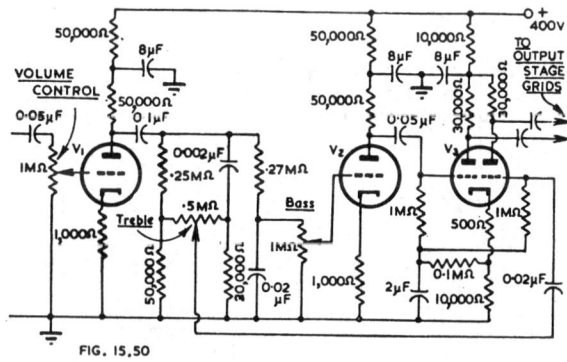

Fig. 15.50. Wide-range two-channel amplifier (Ref. 31).

(H) Resonant plate loading (Fig. 15.52, Refs. 26, 54)

Two parallel tuned circuits, one of which may be tuned to 50-80 c/s and the other to 6000-8000 c/s (or any other desired frequencies) are connected in series with the

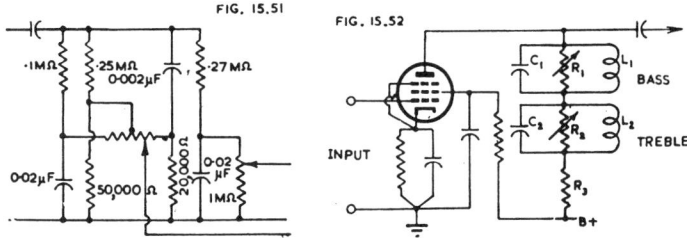

Fig. 15.51. *Modification to Fig. 15.50 to provide treble attenuation (Ref. 31).*
Fig. 15.52. *Resonant plate loading to provide bass and treble boost (Refs. 26, 54).*

plate load resistance R_3. Control of the bass is given by R_1 and of the treble by R_2. The Q of each of these circuits should not exceed 1—see Sect. 1(ix).

(I) Negative feedback incorporating L and C in cathode circuit (Fig. 15.53, Ref. 26)

Feedback provides degeneration across R_1 and R_2, which determines the gain at middle frequencies. With L and C connected directly across R_1 and R_2, the feedback is decreased at low and high frequencies respectively and the stage gain is consequently greater at these frequencies. L may require shielding to reduce hum pickup.

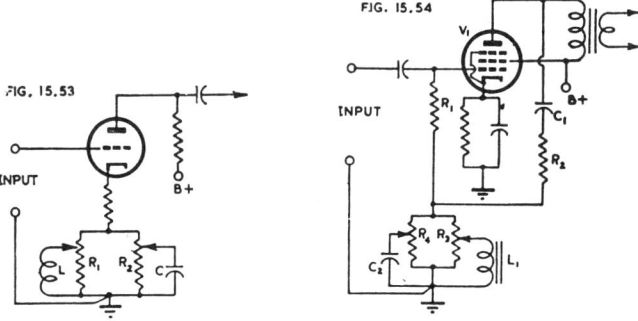

Fig. 15.53. *Bass and treble boosting due to negative feedback incorporating L and C in cathode circuit (Ref. 26).*
Fig. 15.54. *Bass and treble boosting due to negative feedback incorporating L and C (Ref. 15).*

(J) Negative feedback incorporating L and C (Fig. 15.54, Ref. 15)

V_1 is a power amplifier valve with L_1 and C_2 and their control potentiometers in the feedback network. L_1 may have an inductance of about 10 henrys and C_2 may be 0.002 μF ; $R_3 = R_4$ may be 0.1 megohm.

(K) Patchett tone control (Fig. 15.55, Ref. 9)

This circuit has been developed to provide continuously-variable independent bass and treble boost without variation of the middle frequencies and at the same time with frequency characteristics which are close to the ideal for tone control purposes. V_1 is a phase splitter and develops equal voltages across its plate and cathode load resistances. The former feeds directly through an isolating resistor to the output ; the latter feeds the bass and treble filters and potentiometers which in turn feed the grids of the twin triode V_2 whose plates are connected to the output. The output voltage at middle frequencies is about one third of the input voltage. An input of 2.5 volts r.m.s. may be applied with negligible distortion. V_1 may be type 6C5 or 6J5 ; V_2 may be type 6N7 or other twin triode.

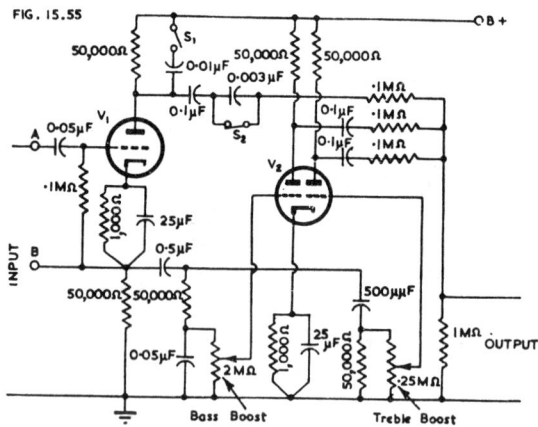

FIG. 15.55

Fig. 15.55.   Patchett tone control circuit (Ref. 9).

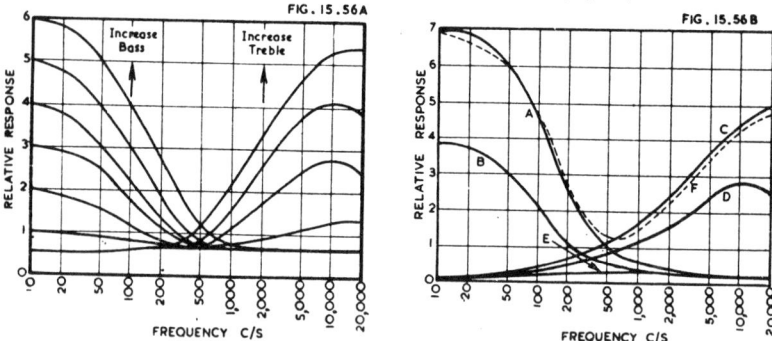

*Fig. 15.56A.   Response characteristics of Patchett tone control circuit with S_1 open
and S_2 closed, and other control at minimum (Ref. 9).*
*Fig. 15.56B.   Response characteristics of Patchett tone control circuit with S_1
closed and S_2 open (Ref. 9).*

Curve A Max. bass, min. treble	Curve D Min. bass, $\frac{1}{2}$ max. treble
,, B $\frac{1}{2}$ Max. bass, min. treble	,, E Min. bass, min. treble
,, C Min. bass, max. treble	,, F Max. bass, max. treble

Fig. 15.56A shows the response characteristics without provision for bass or treble
attenuation ; Fig. 15.56B shows the response with bass and treble attenuation.

If desired, step switches may be arranged to give a wider choice of frequency characteristics by changing the values of the filter condensers.

**(L) Two-stage universal tone
control** (Fig. 15.57A, Ref. 77)
This is a very effective arrangement incorporating a twin triode,
which gives entirely independent
control of both bass and treble
with control extending over a wide
range of boosting and attenuation in
each case (Fig. 15.57B).   The
overall gain of the amplifier of Fig.
15.57A at 800 c/s is only slight.

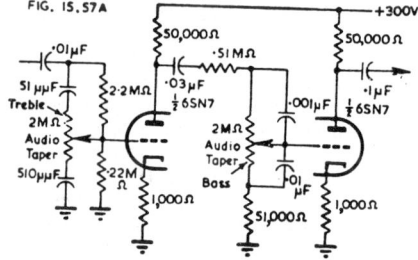

*Fig. 15.57A.   Two stage universal tone control
(Ref. 77).*

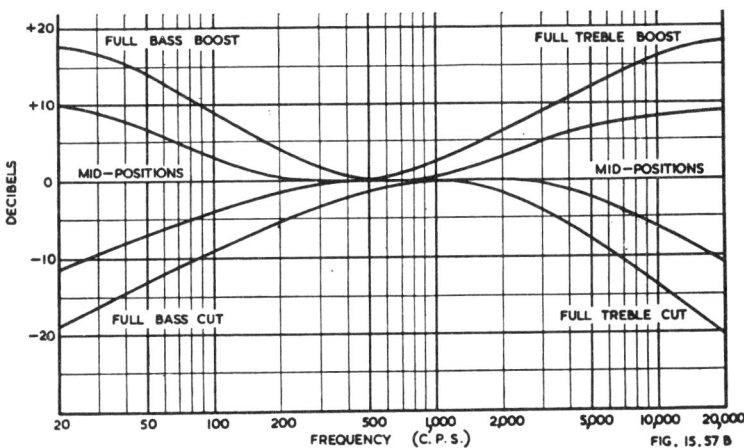

Fig. 15.57B. *Response characteristics of Fig.* 15.57A *(Ref. 77).*

SECTION 9 : FEEDBACK TO PROVIDE TONE CONTROL

(i) Introduction   (ii) Amplifiers with feedback providing tone control   (iii) Whistle filters using feedback.

(i) Introduction

Simple feedback circuits to give tone control have been described above (Figs. 15.11, 13, 14A, 14B, 15, 35B). See also Ref. 80. Tone control may be achieved by incorporating a suitable frequency selective network in the feedback loop, but this involves the question of stability if the feedback is over 2 or more stages—in this case the design should be based on Chapter 7 Sect. 3.

The general principles have been covered very fully in the literature e.g. Refs. 16, 19 (see also References Chapter 7).

In any amplifier in which negative feedback tone control is to be used, maximum freedom in other aspects of the design is desirable. For example it should be possible to alter the feedback factor, and a variation in the feedback over the a-f range should be acceptable. It is also desirable not to be limited to a specified value of output resistance. The feedback should preferably be applied over a small number of stages.

In the case of fidelity amplifiers, there should be sufficient negative feedback at the bass resonant frequency of the loudspeaker to give adequate damping.

Most tone-control amplifiers are only stable for values of β up to a limiting value. One good experimental test is to determine the maximum of β before instability occurs. The working value of β should be less than this limiting value by a comfortable margin (say 8 to 10 db) to allow for all variable factors and for tolerances in the various components.

(ii) Amplifiers with feedback providing tone control

(A) Dual Control continuously variable bass and treble tone control

A simple but effective circuit is shown in Fig. 15.57C and its frequency characteristics in Fig. 15.57D (Ref. 91). Figs. 15.57C and 15.57D will be found on page 1483. The mid-frequency gain is approximately unity, and the output is 4 V r.m.s. (or 1 V r.m.s. with low gain valve) with not more than 0.1% total harmonic distortion up to 5000 c/s. An Americanized form is given in Ref. 98.

A more elaborate circuit is given in Ref. 84.

(B) Dual control step-type bass and treble tone control

Fig. 15.58A (Ref. 42).

This is an example of a well designed amplifier with feedback taken from the secondary of the output transformer over three amplifying stages, having RC networks in the feedback loop to provide bass and/or treble boosting. V_1 is a Schmitt type phase inverter—Chapter 12 Sect. 6(vi)A—V_2 is a similar stage but operating with push-pull input, V_3 and V_4 are push-pull beam power amplifiers. The total voltage gain from the first grid to push-pull output load is 87.7 dbvg, giving full output with a peak input to the first grid of 17 mV without feedback or 0.14 V with feedback ($\beta = -0.0066$). The maximum bass boost is 26 db, and treble boost 11.7 db. The treble boost operates at a low impedance level where the effect of stray capacitances is negligible, while the bass boost, which is less sensitive in this respect, operates at a high impedance level.

V_1 and V_2 are type ECC32 (approx. 6N7) while V_3 and V_4 are type KT61 (high slope output tetrode). T_1 is a miniature screened line-to-grid transformer to match 25-600 ohms on input. T_2 is an output transformer with 22 : 1 turns ratio to match 10 000 ohms plate-to-plate to 20 ohms voice coil. Total primary inductance not less than 60 henrys

(C) Modified "straight" feedback amplifier to provide bass boosting

In many cases a "straight" feedback amplifier may have a capacitive impedance added to the feedback loop to provide bass boosting. This decreases the feedback at bass frequencies, thereby causing the amplifier to lose some of the advantages of feedback; the arrangement is less desirable than a special circuit such as Fig. 15.58A which has been specially designed for tone control.

For example, a capacitor of 0.1 μF may be inserted in series with the 5000 ohm resistor in the feedback circuit in Fig. 7.44 (as in Ref. 69); alternatively a switch giving a choice of several values of capacitance may be used.

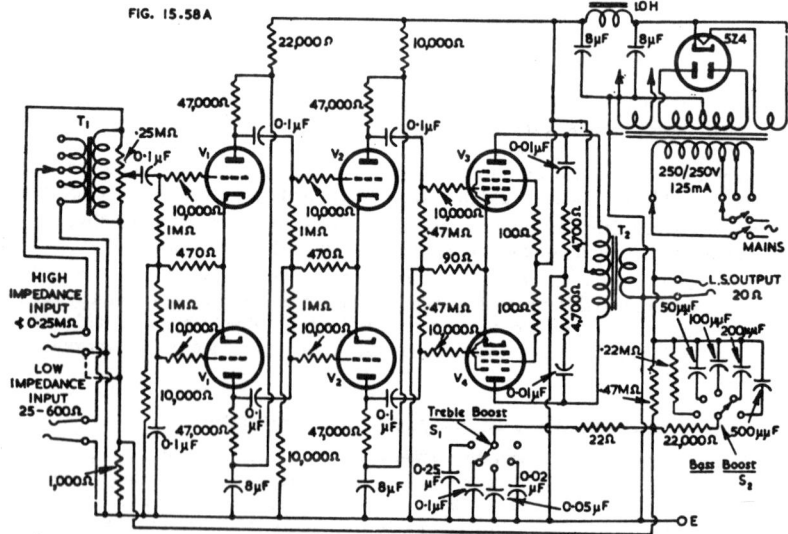

Fig. 15.58A. *High gain amplifier with overall feedback giving bass and treble boosting (Ref. 42).*

(D) The use of feedback to provide special attenuation characteristics at low or high frequencies

If an amplifier is designed with a falling low frequency characteristic, possibly due (in part) to economy in the output transformer design, and negative feedback is applied over the amplifier, its response can readily be made flat or, owing to a phase shift caused by coupling capacitors and by the inductance of the output transformer, the response may rise at those frequencies where it fell without feedback.

If in such an amplifier one of the grid coupling capacitors (included in portion of the amplifier covered by the feedback loop) is replaced by one of smaller capacitance, the feedback will tend to counteract the attenuation produced by this capacitor. The response will be more or less flat down to a certain frequency below which the attenuation will be rapid.

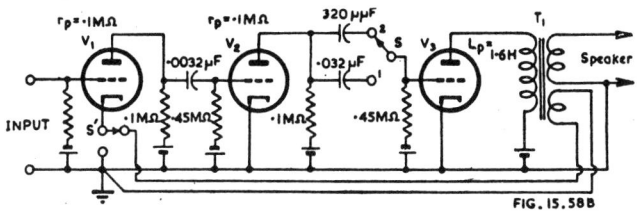

Fig. 15.58B. Use of feedback to provide special attenuation characteristics at low frequencies. The total effective resistance across T_1 primary is taken as 1000 ohms. Switch S_1 normal; S_2 tone control (low frequency attenuation). Switch S' controls feedback.

Some practical response characteristics are described below, based on a theoretical analysis of Fig. 15.58B (after E. Watkinson) :

(a) **To provide rapid low frequency attenuation below a specified frequency** f_0 (Curves are in Fig. 15.58C)

1. Design the amplifier so that the response without feedback and without tone control is only slightly down at f_0, say -6 db (100 c/s in Curve A).

2. Design the tone control to provide a total attenuation, without feedback, of about 3 db at 10 f_0 (Curve D).

3. Apply sufficient negative feedback to flatten the response over the desired frequency range (Curve C). In the example, 20 db of feedback is used, and the combined effect is to give an attenuation of 18 db in the octave from 100 to 50 c/s.

(b) **To provide optional bass boosting**

If now the tone control is switched out of circuit, there will be bass boosting as the result of phase shift in the amplifier (Curve B).

Thus by a single switch S it is possible to change from curve C to curve B, using only one additional component.

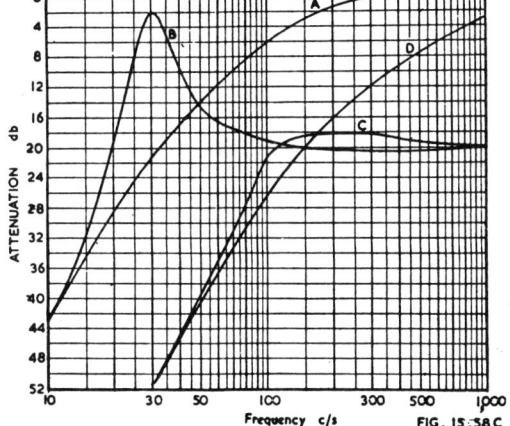

Fig. 15.58C. Frequency characteristics of circuit of Fig. 15.58B. (A) No feedback, no tone control; (B) feedback without tone control; (C) feedback with tone control; (D) no feedback, with tone control.

(c) To provide high frequency attenuation or boosting.
The same principles apply to the high frequency end of the range.

(iii) Whistle filters using feedback
See Figs. 17.24A and 17.24B incorporating a parallel-T network and feedback—also Section 11(iv) of this chapter.

SECTION 10 : AUTOMATIC FREQUENCY-COMPENSATED VOLUME CONTROL

(i) Introduction   (ii) Methods incorporating a tapped potentiometer   (iii) Methods incorporating step-type controls   (iv) Method incorporating inverse volume expansion with multi-channel amplifier.

(i) Introduction
Owing to the special characteristics of the human ear, it is necessary for bass boosting and (to a less extent) treble boosting to be applied to music or speech when reproduced at a lower level than the original sound, if it is desired to retain the full tonal qualities of the original. Provision is therefore often made in the better quality receivers and amplifiers for this to be done automatically as the volume control is adjusted. For this to be fully effective, the volume with the volume control at its maximum setting should be the same as that of the original sound—a condition which it is rarely possible to fulfil in radio reception. One possible way of achieving an approach to the true condition would be to fit two auxiliary volume controls—one with settings for say (1) speech (2) orchestral and (3) solo instrument, and the other to be adjusted to bring the reading on some form of level indicator to a predetermined value. The first of these auxiliary controls is necessary because the various original sounds differ in level, the second control to provide for imperfect a.v.c. in the receiver and variations in percentage modulation between stations. These complications are not likely to be popular. However the intelligent listener should have a second volume control which may be pre-set to give a desired output level for maximum volume—this is particularly important in reproduction from records.

In practice, some form of simple automatic frequency-compensated volume control is found beneficial, in spite of its technical imperfections. One reason for this popularity is that it permits a considerable degree of bass boosting to be used at low volume but not at maximum volume-control setting. It is therefore more fool-proof than manually-operated bass boosting, and less likely to cause overloading of the power amplifier.

(ii) Methods incorporating a tapped potentiometer
A typical application is Fig. 15.59 in which L and C form a series resonant circuit tuned to about 1000 c/s, while R is a limiting resistance to reduce the by-passing at middle frequencies. The tapping point is usually about one fifth of the total resistance, being equivalent to about 14 db below maximum. With this method it is impossible to obtain theoretically correct compensation.

If treble boosting is not desired, L may be omitted leaving C and R to provide bass boosting.
References 71, 72, 89.

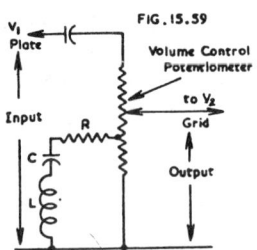

Fig. 15.59. *Tone-compensated volume control with tapped potentiometer.*

A more satisfactory result over a wider range of volume levels can be obtained by using a volume control with two tappings (e.g. at one sixth and one third of the resistance) as in Ref. 88.

A more elaborate type of continuously-variable control has been described (Ref. 79) but requires a special volume control.

Another more elaborate type (Ref. 82) uses three ganged volume controls which are not tapped.

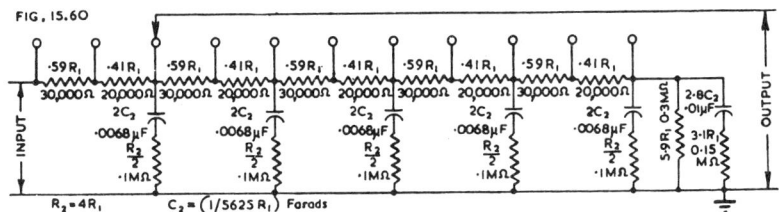

FIG. 15.60

$R_2 = 4R_1$ $C_2 = (1/5625 R_1)$ Farads

Fig. 15.60. Step-type tone-compensated volume control for bass correction only (Ref. 43).

(iii) Methods incorporating step-type controls

All the simpler systems with step-type controls neglect the treble boosting, which is only of secondary importance, but they are capable of a fairly close approach to the ideal for the lower frequencies. One design is shown in Fig. 15.60 (Ref. 43) which uses five 6 db filter sections each divided into two parts, thus giving 11 steps with a total range of 30 db. This is for use with an amplifier having low output resistance (less than 10 000 ohms). The numerical values shown in the figure are for $R_1 = 50\ 000$ ohms.

See also Ref. 78 for a design mounted on a single switch assembly, with a greater number of positions.

(iv) Method incorporating inverse volume expansion with multi-channel amplifier (Ref. 21)

In Fig. 15.61 there are three amplifier channels. The centre one is essentially flat over the whole a-f range. The upper (high-pass) and lower (low-pass) amplifiers may be adjusted to give zero output at any selected output level (generally the maximum) and at any lower level there will be bass and treble boosting which is a function of the output level. Each of the boosting amplifiers is brought to zero output at the desired output level by adjusting three potentiometers (R_1, R_2, R_3 in the bass-boost amplifier). Under these conditions the output voltages of V_7 and V_8 are equal but opposite in phase, hence giving a combined output which is zero.

SECTION 11 : WHISTLE FILTERS

(i) Resonant circuit filters (ii) Narrow band rejection filter (iii) Crystal filters (iv) Parallel-T network (v) Filters incorporating L and C.

A whistle filter is one which is sharply tuned to eliminate a particular frequency, usually 9 or 10 Kc/s, with the least possible effect on adjacent frequencies.

The same principles may also be applied to the elimination of scratch from a pickup when this is peaked, but in this case the maximum attenuation is adjusted to occur at the frequency of pickup resonance.

(i) Resonant circuit filters (Ref. 15 62)

The ratio of normal gain (A_n) to gain (A_r) at the resonant frequency of $L_1 C_1$ is

FIG. 15.61

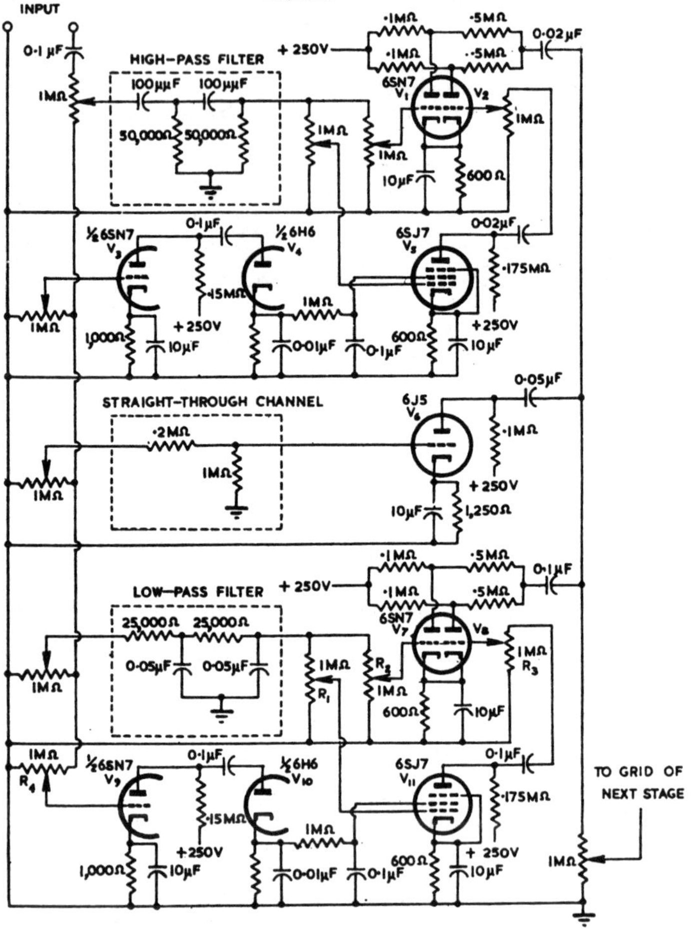

Fig. 15.61. *Inverse volume-expansion circuit with 3 channel amplifier for automatic frequency-compensated volume control (Ref. 21).*

given by
$$A_n/A_r = (R' + R_1)/R_1$$
where $R' = r_p R_L'/(r_p + R_L')$
and $R_L' = R_L R_a/(R_L + R_a)$.
For effective filtering, the maximum attenuation may have to be 20 db or more. The sharpness of the attenuation curve depends upon the Q of the tuned circuit, where $Q = \omega_r L_1/R_1$. A value of $Q = 10$ is probably the highest practicable value, and under these conditions the filter will have a minimum attenuation of 4 db over a bandwidth of 0.64 f_r, or a minimum attenuation of 8 db over a bandwidth of 0.38 f_r where f_r is the frequency of resonance. This is equivalent to an attenuation of at

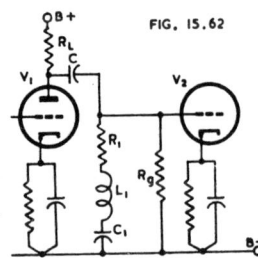

Fig. 15.62. *Whistle filter using resonant circuit.*

least 8 db over a frequency range from approximately 8000 to 12 000 c/s when $f_r =$ 10 000 c/s. Thus the filter cuts such a serious hole in the amplifier frequency characteristics that it is not a satisfactory solution. It is reasonably satisfactory for a maximum attenuation of 6 db, but this may not be sufficient to eliminate the whistle. For theory and curves see Refs. 17, 51, pp. 52-54.

(ii) Narrow band rejection filter

A more complicated and more effective circuit is Fig. 15.62A (Ref. 81) which, when tuned to 9000 c/s, gives an attenuation of 2 db at 8000 c/s, over 40 db at 9000 c/s, 8 db at 10 000 c/s and 5 db at 20 000 c/s relatively to the level at low frequencies. The frequency of resonance may be changed either by adjusting the inductance of L_1, or by varying the capacitance of both C_1 and C_2. The attenuation at low frequencies (insertion loss) is 2 db.

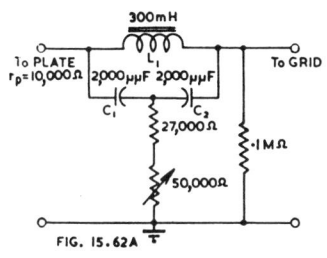

FIG. 15.62A

Fig. 15.62A. Narrow-band rejection filter (Ref. 81).

A narrow band rejection filter is described in Ref. 75 which has the disadvantage of reducing the maximum a-f power output. An improved form for application to a linear reflex detector is Fig. 15.63 in which an attenuation of more than 40 db is obtained at 9000 c/s while the attenuation is only 3 db at 8400 c/s. The principle may also be applied to a cathode-loaded amplifier by omitting C_1 and providing the correct bias for amplification.

The resonant circuit LC_3 is tuned approximately to the whistle frequency, a vernier control for C_3 being desirable. The value of C_1 should be such that its reactance is very much smaller than R_1 at the lowest signal frequency and should be very much larger than R_1 at 9000 c/s. The value of C_2 should be such that its reactance is approximately equal to $R_2 + R_3$ at the low frequency limit of the amplifier, say 50 c/s. If its reactance is higher, a slight amount of bass boosting will occur. The potentiometer R_2R_3 is adjusted to give zero output at the whistle frequency (Ref. 74).

An elaboration of this principle, using two valves, is given in Ref. 83.

(iii) Crystal filters

A crystal filter is one of the few really satisfactory methods of eliminating a whistle, but the cost precludes its use in all except the most expensive communication receivers.

(iv) Parallel-T network

See also Sect. 6(iii)C.

The circuit of Fig. 4.40 may be inserted between two r.c.c. stages with a load resistance of say 0.25 megohm and following grid resistor say 1 megohm. Suitable values for a 10 000 c/s whistle filter are :—$C_1 = C_2 = 0.0001$ μF, $C_0 = 0.0002$ μF, $R_1 = R_2 = 160\,000$ ohms, $R_0 = 80\,000$ ohms.

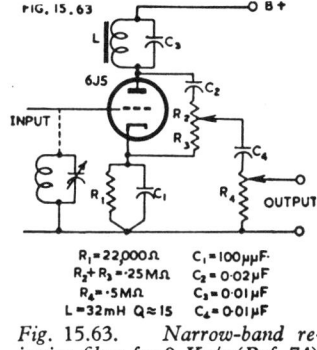

FIG. 15.63

$R_1 = 22{,}000\,\Omega$	$C_1 = 100\mu\mu F$
$R_2 + R_3 = \cdot25\,M\Omega$	$C_2 = 0\cdot02\mu F$
$R_4 = \cdot5\,M\Omega$	$C_3 = 0\cdot01\mu F$
$L = 32mH$ $Q \approx 15$	$C_4 = 0\cdot01\mu F$

Fig. 15.63. Narrow-band rejection filter for 9 Kc/s (Ref. 74).

A special application of the parallel-T network is described in Chapter 17 Sect. 5 and Figs. 17.24B and 24C where an attenuation of 34 db at 10 000 c/s with almost flat response up to 6 Kc/s is obtained by means of a parallel-T feedback loop which gives minimum feedback at 6 Kc/s, together with a parallel-T filter across the output of the amplifier to give the desired attenuation at 10 Kc/s (Ref. 57).

The parallel-T network is also used in the equalizer circuit of Fig. 17.35B and in the filter unit of Fig. 17.35C.

The frequency of maximum attenuation may be varied by varying one element in each of the component T's (see Ref. 85).

(v) Filters incorporating *L* and *C*

Constant *k* and M-derived filters may be used to attenuate a very narrow band of frequencies, but the cost precludes their general use.

SECTION 12 : OTHER METHODS OF TONE CONTROL

(i) Multiple-channel amplifiers   (ii) Synthetic bass.

(i) Multiple-channel amplifiers

With a three-channel amplifier, or one having more than three channels, it is possible to exercise a certain amount of tone control by controlling the volume of the bass and treble amplifiers. For this to be satisfactory for tone control, the cross-over frequencies should be :

Bass—not higher than 200 c/s.

Treble—not lower than 3000 c/s.

In each case the cross-over network may produce an attenuation of 6 or 12 db/octave nominal.

An amplifier in this class is described in Ref. 5.   See also Sect. 10(iv) and Ref. 21.

(ii) Synthetic bass

In small receivers, in which the loudspeaker is incapable of reproducing the bass at all adequately, a device is sometimes used to introduce distortion of the bass frequencies. The fundamental bass frequency is heavily attenuated to avoid overloading, but the harmonics are reproduced and provide " synthetic bass." One such circuit is Fig. 15.64 (Ref. 73, based on Sonora Model RCU-208). Positive feedback at low frequencies is applied from the cathode of the second valve to the cathode of the first by means of the network R_3, R_2, R_1 and C_1.

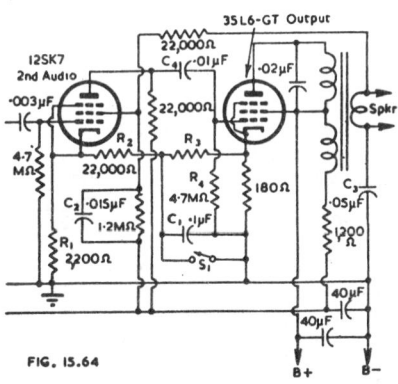

FIG. 15.64

Fig. 15.64.   *Amplifier providing synthetic bass* (*Ref.* 73).

Another circuit (Ref. 87) has a supplementary channel with a distorting valve (triode 6SF5) functioning effectively only below about 100 c/s. Grid distortion is produced through zero bias operation. A plate load resistor of 0.5 MΩ is used, and the plate coupled to the grid of the output pentode. It is stated that the loudspeaker bass resonance should preferably be less than 60 c/s, and that the results sound unnatural to the ear on music when the loudspeaker resonance is about 150 c/s.

The principles of synthetic bass are given in Chapter 14 Sect. 3(viii).

SECTION 13 : THE LISTENER AND TONE CONTROL

Tone control is a controversial subject with very strong conflicting views held by many competent authorities. It is the author's opinion that, in view of the differences of opinion, it is only reasonable to provide the listener with some degree of tone control to permit him to derive the maximum degree of satisfaction while listening. This freedom of choice should not be unlimited, otherwise some listeners will become lost in the possible variations.

1. The cheapest type of set will probably be fitted with a fixed shunt capacitance tone control.

2. Sets which are somewhat higher-priced will be fitted with a single control. This may cover any one of the following :

    (a) Variable resistance and fixed shunt capacitance (this has obvious shortcomings).

    (b) Bass boosting only (this is quite satisfactory in its class provided that the bass boosting is limited to about 6 db total).

    (c) Continuously-variable control giving, say, bass boosting in one direction and treble boosting in the other.

    (d) Step-type control in the form of a " quality switch."

3. The most elaborate radio receivers and amplifiers may be fitted with two controls, one for bass and the other for treble. In addition, many equipments in this class will be fitted with automatic frequency-compensated volume control, thereby limiting the degree of manual bass boosting which is necessary (say 6 db maximum).

Anything which can be done to assist the listener to obtain the best results with the least trouble is to be commended.

SECTION 14 : EQUALIZER NETWORKS

Equalizer networks for pickups are covered in Chapter 17 Sect. 5 while those for microphones are covered in Chapter 18 Sect. 1(xii).

An example of a universal equalizer is given in Ref. 49

A helpful general article on design is Ref. 92. See also Ref. 102.

SECTION 15 : REFERENCES

1. Light, G. S. " Frequency compensating attenuators " Electronic Eng. 16.195 (May 1944) 520.
2. Barcus, L. M. " New bass boosting circuit " Elect. 16.6 (June 1943) 216.
3. Colebrook, F. M. " A note on the theory and practice of tone-correction " W.E. 10.112 (Jan. 1933) 4.
4. " Universal equaliser provides a-f amplifier design data " Elect. 16.8 (Aug. 1943) 120.
5. Widlar, W. L. " Compensating audio amplifier for three channels " Comm. 23.8 (Aug. 1943) 28.
7. Amos, S. W. " Improving headphone quality ; simple correction circuits for use with conventional diaphragm-type earpieces " W.W. 50.11 (Nov. 1944) 332.
8. Austin, K. B. " A Bass and Treble Booster " U.S. Patent 2,352,931 Radio Craft 16.4 (Jan.1945) 230.
9. Patchett, G. N. " A new versatile tone control circuit" (1) Basic principles of tone control W.W. 51.3 (Mar. 1945) 71 ; (2) " Bass and treble lift without variation of middle frequencies " 51.4 (Apr. 1945) 106.
10. Sturley, K. R. " L.F. Amplification (7) Increasing l.f. response " Electronic Eng. 17.207 (May 1945) 510.
11. Winget, D. (letter) " A new versatile tone control circuit " W.W. 51.6 (June 1945) 182.
13. Smith, H. " Tone compensation amplifier " Australasian Radio World (July 1945) 20.
14. Shepard " Improved bass for small radios " Elect. 18.7 (July 1945) 224.
15. Sands, L. G. " More highs—more lows " Radio Craft (1) 14.10 (July 1943) 596 ; (2) 14.11 (August 1943) 662.
16. Hendriquez, V. C. " The reproduction of high and low tones for radio receiving sets " Philips Tec. Rev. 5.4 (April 1940) 115.
17. Scroggie, M. G. " Amplifier tone control circuits " W.E. 9.100 (Jan. 1932) 3.
19. Fritzinger, G. H. " Frequency discrimination by inverse feedback " Proc. I.R.E. 26.1 (Jan. 1938) 207.
20. Design Data (5) " Low-frequency correction circuit " W.W. 52.6 (June 1946) 199.
21. Goodell, J. D., and B. M. H. Michel " Auditory perception " Elect. 19.7 (July 1946) 142.
22. Wortman, L. A. " Tone control circuits " Radio Craft (Aug. 1946) 763.
23. Reference Sheet " Fundamental Tone Control Circuits " Electronic Eng. 18.223 (Sept. 1946) 278.

25. Powell, E. O. " Tone control " W.W. 46.14 (Dec. 1940) 491.
26. Cutler, S. " Audio frequency compensating circuits " Elect. 15.9 (Sept. 1942) 63.
27. Haines, F. M. " High fidelity bass compensation for moving coil pickups " Electronic Eng. 18.216 (Feb. 1946) 45.
28. " Cathode Ray " " Working out tone control circuits " W.W. 43 (Sept. 22, 1938) 269.
29. Najork, J. " Why bass boost " Service 17.6 (June 1948) 18.
31. Hill, J. M. " Wide range tone control-circuit suitable for correction at low volume levels " W.W. 52.12 (Dec. 1946) 422.
32. Case, N. " All purpose F-M A-M receiver " Comm. 7.1 (Jan. 1947) 34.
33. Beard, E. G. " A quality switch in lieu of a tone control " Philips Tec. Com. No. 2 (Feb. 1947) 8. Based on Philips C.A.L.M. Report No. 7.
34. Stevens, P. " The control of tone " Australasian Radio World 11.8 (Jan. 1947) 13.
35. Moody, W. " Audio amplifier equalization methods " Service 16.1 (Jan. 1947) 11.
36. Ellis, J. " Bass compensation—a system using negative feedback " W.W. 53.9 (Sept. 1947) 319.
37. " Negative feedback tone control used by R.G.D." extract from review Radiolympia 1947, W.E. 24.290 (Nov. 1947) 335.
38. Schlesinger, K. " Low frequency compensation for amplifiers " Elect. 21.2 (Feb. 1948) 103.
39. Banthorpe, G. H. " A simple tone control circuit " Electronic Eng. 20.241 (March 1948) 91.
40. Winder, N. " A bass correction circuit for moving-coil pickups " Electronic Eng. 20.244 (June 1948) 187.
41. Sugden, A. R. " Pre-amplifier tone control " Australasian Radio World 13.2 (July 1948) 11.
42. Whitehead, C. C. " Quiet high-gain amplifier—tone control by negative feedback " W.W. 54.6 (June 1948) 208.
43. Bomberger, D. C. " Loudness control for reproducing systems " Audio Eng. 32.5 (May 1948) 11.
44. Bonaira-Hunt, N. " Direct-coupled amplifier " W.W. 54.7 (July 1948) 266.
45. Dahl, H. M. " RC circuits as equalizers " Audio Eng. 32.6 (June 1948) 16.
46. Childs, P. A. " A 9 Kc/s whistle filter " Electronic Eng. 20.248 (Oct. 1948) 320.
47. Gerry, C. C. " Feedback and distortion " W.W. 54.9 (Sept. 1948) 347.
48. Savory, W. A. " Design of audio compensation networks " Tele-Tech (1) 7 : 1 (Jan. 1948) 24 ; (2) 7 : 2 (Feb. 1948) 27 ; (3) 7 : 4 (April 1948) 34.
49. Mitchell, I. A. " AF circuit equalizer " FM and T. 8.6 (June 1948) 28.
50. Lurie, W. B. " Versatile tone control " Elect. 21.12 (Dec. 1948) 81.
51. Sturley, K. R. (book) " Radio Receiver Design " (Part 2) (Chapman and Hall, London, 1948) pp. 41-55.
52. " Bass boosting " Radiotronics No. 73 (February 1937) 14.
53. Lynch, W. A. " Video amplifier l-f correction " Comm. 23.4 (April 1943) 16.
54. Scroggie, M. G. " Flexible tone control " W.W. 41.11 (Sept. 10th 1937) 263.
55. Cocking, W. T. " Tone control systems " W.W. 44.23 (June 8th 1939) 532.
56. Jaffe, D. L. " Paraphase bass-treble tone control " Radio 30.3 (March 1946) 17.
57. Rogers, G. L. " Simple rc filters for phonograph amplifiers " Audio Eng. 31.5 (June 1947) 20.
58. " Radiotron receiver RC52," Radiotronics No. 117 (Jan./Feb. 1946) 3.
59. " Applications of a Parallel-T network," A.R.T.S. and P. Bulletin No. 118 (Jan. 1942) 1.
60. " The design of high-pass filters for bass tone control " A.R.T.S. and P. Bulletin No. 117 (Dec. 1941) 1.
61. Callendar, M. V. " Problems in selective reception " Proc. I.R.E. 20.9 (Sept. 1932) 1427.
62. " Tone Compensation and tone control " A.R.T.S. and P. Bulletin Nos. 27 and 28 (April and May 1936).
63. " Tone control," review in W.W. 46.8 (June 1940) 314 ; British Patent No. 516286, June 20th 1938, Mullard Radio Valve Co. Ltd. and R. G. Clark.
64. " Frequency compensation in audio amplifiers " Radiotronics No. 114 (July 1941) 43.
65. Gustafson, G. E., and J. L. Rennick " Low cost FM-AM receiver circuit " Tele-Tech 7.10 (Oct. 1948) 36.
66. Austin, K. B. " A bass and treble booster " U.S. Patent No. 2,352,931 ; Reviewed Radio Craft 16.4 (Jan. 1945) 230.
67. Williams, W. N. " Two high gain amplifier circuits " Radio and Hobbies, Australia (June 1942) 19.
68. Williams, W. N. " An 807 amplifier " Radio and Hobbies, Australia (April 1947) 42.
69. Williams, W. N. " Using a new amplifier " Radio and Hobbies, Australia (April 1948) 49.
70. Williams, W. N. " A new tone control unit " Radio and Hobbies, Australia (March 1949) 25.
71. Wolff, I., and J. I. Cornell " Acoustically compensated volume control " Elect. 6 (Jan. 1933) 50.
72. D'Oris, P. A., and R. de Cola " Bass compensation design chart " Elect. 10.10 (Oct. 1937) 38.
73. " A synthetic bass note circuit " Radio Electronics 20.1 (Oct. 1948) 37.
74. Childs, P. A. " A 9 Kc/s whistle filter " Electronic Eng. 20.248 (Oct. 1948) 320.
75. Krauke, J. E. " A 10 KC suppressor " Radio News 37.3 (March 1947) 46.
76. James, E. J. " Simple tone control circuit " W.W.55.2 (Feb. 1949) 48.
77. Sterling, H. T. " Flexible dual control system " Audio Eng. 33.2 (Feb. 1949) 11 ; letter J. J. Faran 33.6 (June 1949) 31.
78. Winslow, J. " Full-range loudness control " Audio Eng. 33.2 (Feb. 1949) 24.
79. Turner, J. W. " Construction details of a continuously variable loudness control " Audio Eng. 33.10 (Oct. 1949) 17.
80. Edwards, J., and T. J. Parker " An application of frequency selective negative feedback " Tele-Tech 8.12 (Dec. 1949) 30.
81. Osram Valve Technical Publication TP4 " Radio receiver for use with high fidelity amplifiers " (The General Electric Co. Ltd. of England).
82. Johnson, E. E. " A continuously variable loudness control " Audio Eng. 34.12 (Dec. 1950) 18.
83. Nevin, R. B. " An effective frequency rejection circuit " Audio Eng. 35.2 (Feb. 1951) 20.
84. Fling, W. D. " A continuously variable equalizer " Audio Eng. 35.3 (March 1951) 16.
85. Dunn, S. C. (letter) " Twin-T circuits " W.E. 28.332 (May 1951) 162.
86. Van Scoyoc, J. N. " A cross-coupled input and phase inverter circuit " Radio News 40.5 (Nov. 1948) 6.
87. Exley, K. A. " Bass without big baffles " W.W. 57.4 (Apr. 1951) 132. Letters 57.7 (July 1951) 264 ; 57.8 (Aug. 1951) 321 ; 57.9 (Sept. 1951) 369; 57.10 (Oct. 1951) 408.
88. Brooks, W. O. " A two-tap bass and treble compensated volume control " Audio Eng. 35.8 (Aug. 1951) 15.
89. Toth, E. " The design of compensated volume controls " Audio Eng. (1) 36.1 (Jan. 1952) 13 ; (2) 36.2 (Feb. 1952) 14.
90. Lavin, C. J. " A scratch filter with continuously variable cut-off point " Audio Eng. 35.11 (Nov. 1951) 6.

Additional references will be found in the Supplement commencing on page 1475.

CHAPTER 16

VOLUME EXPANSION, COMPRESSION AND LIMITING

By F. LANGFORD-SMITH, B.Sc., B.E.

with assistance from D. G. Lindsay (Fellow I.R.E. Australia)

SECTION 1 : GENERAL PRINCIPLES

(i) *Introduction*   (ii) *An ideal system*   (iii) *Practical problems in volume expansion* (iv) *Distortion*   (v) *General comments*.

(i) Introduction

The maximum volume range of any sound reproducer is the difference in decibels between the maximum sound output and the level of masking by background noise which latter may include hum, random noise, needle scratch or microphone noise.

The volume range transmitted by a broadcast station may vary from a low value up to at least 60 db for a F-M transmitter (F.C.C.), the value depending on the type of programme. In the case of an A-M transmitter the maximum volume range is about 50 db—see Chapter 14 Sect. 7(iv).

If the original sound has a volume range greater than the maximum volume range of the transmitter, it is usual to compress it in some way. The compression may be accomplished manually by the control engineer, or automatically by a device known as a peak limiter or a volume compressor.

A similar case arises with recorded music, where the maximum volume range may be as low as 35 db for shellac lateral-cut disc records.

Many types of programme have a maximum volume range less than 35 db, and therefore do not require compression. However, most broadcast transmitters have an automatic peak limiter permanently operating, while it has even been found beneficial to use volume compression in F-M broadcasting (Ref. 29).

If reproduced music is to have the same volume range as in its original form, some kind of volume expander is required. But, as will be shown later, volume expansion cannot be applied indiscriminately and cannot be perfect.

One advantage of the use of volume expansion is that it reduces the background noise, other than room noise. This is a considerable advantage when the source is a disc record.

A serious shortcoming of automatic volume expansion is that the circuit can never duplicate the original volume range because it has no way of knowing what these original levels were—it will make the loud portions louder and the soft portions softer, but always by the same amount for a given input level. Thus the expander circuit might increase every signal that is 10 db above the average volume level to 15 db above the average, but the level in the studio might have been higher or lower than this.

The amplitude ratios of a soft and a loud note when rendered simultaneously cannot be compressed or expanded by any system. If the loud note causes a certain compression in the transmitter, the soft note will be compressed in the same ratio and may fall below the noise level (Ref. 33).

(ii) An ideal system

In an ideal system the transmitter would broadcast two separate carrier frequencies, one modulated with the music and the other modulated with some signal indicating the degree of compression being used at each instant. In the case of disc recorded music it would be necessary to have two grooves and two pickups. In each case the reproducer would incorporate a volume expander in which the degree of expansion is controlled by the second (indicating) signal.

Alternatively, only a single modulated carrier may be used provided that the expander characteristic in the receiver is the inverse of the compression characteristic in the transmitter. In addition, the time lags of the compressor and the expander should be equal. This is an ideal which cannot be reached at the present time owing to the lack of standardization and the use of manual controls.

There are serious technical problems both in the compression and expansion operation, but these are considered in Sections 2 and 4 respectively.

(iii) Practical problems in volume expansion

At the present time most broadcast stations employ some form of automatic volume compression, while compression is also used in recording on discs. The problem facing the receiver and amplifier designer is how to make the best use of volume expansion under conditions where it is impossible to reach the ideal.

(A) Orchestral reproduction

This subject is dealt with in a general manner in Chapter 14 Sect. 7(iv) and (v).

It is in orchestral reproduction that volume expansion can be employed to its full advantage. Experience indicates that the operation of the volume expander in the upwards direction should be as rapid as practicable (see Sect. 4) but that the fall should be very gradual—up to 1 or 2 seconds or even more, the optimum rate varying with the type of music and with the listener's choice.

Most competent audio-frequency engineers agree that automatic volume expansion is capable of giving more realistic reproduction of recorded orchestral music when used under the best possible conditions, even though it is known to have many technical defects. The usual amount of volume expansion used in such cases is 12 to 20 db maximum, with the maximum expansion variable at the desire of the listener.

(B) Average home listening

The average home radio receiver, used for a variety of purposes including speech, background music, drama and miscellaneous programmes, is usually better without expansion. Expansion might beneficially be used on orchestral programmes provided that the audience is prepared to listen intently, the room noise level is sufficiently low, and the neighbours do not object. These conditions are the exception rather than the rule.

On the other hand, volume expansion with manual control is well worth incorporation into a home gramophone amplifier.

(C) Background music

Music which is intended to form a background should be compressed and not expanded. The maximum dynamic range may be from 25 db down to possibly 10 db (Ref. 45).

(D) Factory music

Factory music must be heard above a very high noise level without becoming inaudible for any appreciable period. It should be compressed and not expanded, with a maximum dynamic range from 20 db down to possibly 5 db in exceptionally noisy locations.

References to General Principles : 3, 20, 29, 33, 34, 45, 48, 59.

(iv) Distortion

Apart from the harmonic and intermodulation distortion and extraneous noises produced by the compressor and expander, which may be kept low by good design, there are some special features which require consideration. Some of these are described in Sect. 2(iv) in connection with volume compression.

(v) General comments

There is an extraordinarily wide variety of devices for both volume compression and volume expansion, but in many cases the information published is too meagre to permit comparisons between the different methods. There is the additional problem that some of the designs may not be suitable for quantity production, on account of unduly critical adjustments and/or critical selection of valves.

The subject is a very complicated one, and the present unsatisfactory state of the " published art " has forced the author to refrain deliberately from giving much comment. The methods and circuits described in the following sections have been compiled mainly from very limited sources, often from a single article, and the information and claims based solely on the articles, to which full references have been given.

SECTION 2 : VOLUME COMPRESSION

(i) Introduction   (ii) Peak limiters   (iii) Volume limiters   (iv) Distortion caused by peak limiters or volume limiters   (v) Volume compression   (vi) Volume compression plus limiting   (vii) Compression of commercial speech.

(i) Introduction

In this Section it is intended to approach the subject from the point of view of the radio receiver or amplifier designer.

Volume compressors may be divided into three groups :

(A) Peak limiters

A peak limiter is an amplifier whose gain will be quickly reduced and slowly restored when the instantaneous peak power of the input exceeds a predetermined value The output for all inputs in excess of this value is substantially constant.

(B) Volume limiters

A volume limiter is an amplifier whose gain is automatically reduced when the average input volume to the amplifier exceeds a predetermined value, so that the output for all inputs in excess of this value is substantially constant. A volume limiter differs from a peak limiter in that it is controlled by the average volume instead of by the instantaneous peaks.

(C) Volume compressors

These are used for the purpose of compression over a substantial part of the entire operating range.

(ii) Peak limiters

Peak limiters are used principally in broadcast transmitters. They are intended to prevent overmodulation and to increase the average level of the programme. The latter result follows automatically if the input is increased and the peak limiter left to look after the peaks.

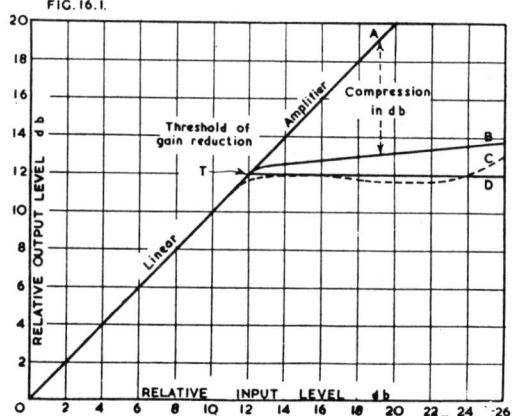

FIG. 16.1.

Fig. 16.1. Compression characteristics (B, C, D) of typical peak limiter or volume limiter.

Some static limiter characteristics are shown in Fig. 16.1. Curve OA is the curve of a linear amplifier. Curve OTD is that of an " ideal " limiter, with point T as the threshold of limiting. It will be seen that the output does not rise above a level of 12 db, no matter how great the input may become.

Curve OTB is that of a practical peak limiter, the curve TB indicating by its slope a compression ratio of 10 to 1 (both being measured in db). The compression in db at any point is as indicated by the arrows.

Curve OC is that of a peak limiter having " over-control " (Refs. 36, 40, 44) which does not exceed 100% modulation (line TD) until the compression reaches 12 db.

With peak limiters a volume expander cannot be used to give correct compensation for the compression.

Most of the earlier and simpler forms of peak limiters have a comparatively slow " attack " time, that is the time taken for the limiter to operate on a sudden increase in input level—usual times are 1 or 2 milliseconds. A slow attack time results in overmodulation and distortion for short intervals following each transient. More recent examples of good design (Refs. 36, 38, 40, 44, 63) claim to have " attack " times less than 100 microseconds, while some (Refs. 36, 40, 44) incorporate a time delay filter in the signal circuit which is claimed to permit almost instantaneous operation. Tests have been carried out under dynamic conditions to demonstrate the performance of various types of limiters (Refs. 4, 40, 63). It is good practice for the " attack " time of a peak limiter to be less than 200 microseconds, while still shorter times (e.g. 100 microseconds or less) are desirable.

The recovery times in most cases are manually controlled, with times up to 2 or even 3 seconds, but automatic control has also been used to lengthen the recovery time so as to avoid " pumping " when several programme peaks occur in rapid succession (Ref. 36). The recovery time is not made longer than necessary, since it causes reduced modulation percentage.

One design (Ref. 86) has a subsidiary series *CR* circuit that has only a slight effect with a single sharp peak, thereby giving a fairly short recovery time (0.33 sec.) but with a sustained peak the recovery time increases to 2 seconds.

If a fixed value of recovery time is used, the most suitable range is from 0.5 to 1.5 seconds, and " pumping " may be avoided by restricting the limiting action to 4 or

6 db. A preferable arrangement incorporates a double value of recovery time, in which a longer recovery time occurs on sustained or repetitive peaks, and a short recovery time on brief, non-repetitive peaks (e.g. Ref. 72).

Careful listening tests with modern peak limiters having very short "attack" time, indicate that differences in quality can only be discerned when the limiting action (compression) reaches 5 db with a good modern type (Ref. 63) or 8 to 10 db with the nearly instantaneous type (Ref. 44).

Some broadcast stations unfortunately abuse the limiter by increasing the percentage modulation to such an extent that the distortion is distinctly audible in any good receiver.

References to peak limiters : 4, 36, 37, 38, 40, 44, 63.

(iii) Volume limiters

These are very similar in most respects to peak limiters, and the same general remarks hold for both. Examples of volume limiters are Refs. 12, 29.

(iv) Distortion caused by peak limiters or volume limiters

(A) Distortion caused by slow attack time

This is definitely audible in some types of programmes when the attack time is greater than 100 microseconds.

(B) Transient waveform distortion

This occurs in some types of limiters during limiting (Ref. 40). It may only affect part of one cycle, or it may continue over many cycles.

(C) Thump

Control current surges caused by rapid changes in input level may or may not be audible as a thump. Careful design is necessary to reduce the trouble. One design is claimed to have a very high signal/thump ratio (Refs. 36, 40, 44).

(D) Non-linear distortion

In the best designs the total harmonic distortion is below 1% and the intermodulation distortion below 3% under all conditions, over the whole a-f range.

(E) " Pumping "

This has already been described above.

(F) Sibilant speech sounds

The high frequency components in speech (sibilant sounds) are normally at a much lower level than the low frequency components. When volume compression is applied to speech, the control voltage derived from sibilants alone is much less than that from vowel sounds. Consequently the amplifier gain is higher for sibilants alone than for other speech sounds, leading to accentuation and distortion of the sibilants. This trouble may be avoided by incorporating a suitable equalizer in the control circuit (Ref. 9).

(G) Effect of pre-emphasis

Pre-emphasis, as used in certain recording systems and in F-M broadcasting, sometimes tends to give overmodulation with high frequency peaks. It is therefore advisable to place the pre-emphasis network ahead of the limiting amplifier. In such cases the limiting should be held to low values such as 2 or 3 db (Refs. 4, 44).

(v) Volume compression

Volume compression is distinguished from limiting in that it extends over the whole, or a substantial part of, the entire operating range. The compression characteristic may either be a straight line (A) or a smooth curve (B) shown typically in Fig. 16.2 ; in both examples an input range of 60 db is compressed to an output range of 40 db.

Either type of compression characteristic may be used satisfactorily in conjunction with an expander provided that the expansion characteristic is the inverse of the compression characteristic.

Some examples of volume compressors are Refs. 2, 31, 66—see also the a.g.c. circuit of Ref. 39. The attack time of a volume compressor should be fairly slow—this will permit the attack time of the expander to be made the same value. A compressor for use with recording is described in Ref. 69.

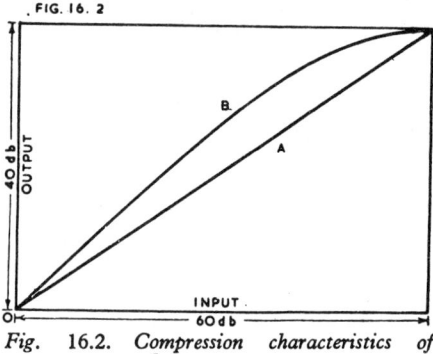

FIG. 16. 2

Fig. 16.2. Compression characteristics of volume compressors.

(vi) Volume compression plus limiting

In broadcast transmitters it is desirable to adopt both volume compression and limiting—volume compression for use with discretion depending on the type of programme ; limiting for all programmes to permit a high percentage modulation to be used without peaks exceeding 100%.

One combined equipment has been designed (Ref. 39) incorporating a memory circuit which holds the gain constant for a predetermined time to preserve the dynamic range of the programme.

References to Section 2 (Volume compression, general) : 2, 4, 5, 9, 12, 29, 31, 32, 36, 37, 38, 39, 40, 44, 63, 66, 69, 85, 89.

(vii) Compression of commercial speech

The comments above apply to good fidelity systems. For commercial speech, some form of " level governing amplifier " (W.E. Co.) or " constant volume amplifier " (British Post Office) is commonly used in complex systems. However, for simple systems, it is difficult to improve on an efficient speech clipper (see Sect. 6).

SECTION 3 : GAIN CONTROL DEVICES

(*i*) *Remote cut-off pentodes* (*ii*) *Pentagrids and triode-hexodes* (*iii*) *Plate resistance control* (*iv*) *Negative feedback* (*v*) *Lamps* (*vi*) *Suppressor-grid control.*

Gain control devices are used in both volume compressors and expanders to provide a gain which is a function of the input signal.

(i) Remote cut-off pentodes

This is an application of audio a.v.c., and remote cut-off pentodes are capable of being used in both applications. The predominant (second) harmonic distortion does not exceed 1% with valve type 78 having a load resistance of 50 000 ohms, with an input of 0.15 volt (Ref. 62). The distortion may be reduced by push-pull operation and by limiting the input voltage to a low level.

A method of testing valves for the preselection of remote cut-off pentodes is described in Ref. 70.

It is the author's opinion that, despite all the troublesome features, the push-pull circuit using suitable selected remote cut-off pentodes, gives the most satisfactory control performance. This control stage must be followed by a " combining circuit " such as a push-pull transformer or centre-tapped-choke, either in the plate circuit or in the following grid circuit—the latter being preferable. See also remarks in Sect. 4(iv).

(ii) Pentagrids and triode-hexodes

Type 6L7 pentagrids are capable of handling higher input voltages than pentodes for the same distortion. The predominant (second) harmonic distortion does not exceed 1% with valve type 6L7 having a load resistance of 50 000 ohms, with an input of 0.5 volt (Ref. 62). The input signal is applied to the remote cut-off grid (No. 1) here operated at fixed bias, and the control voltage is applied to the sharp cut-off grid (No. 3). See Refs. 14, 35.

Type 6A8 pentagrids may also be used, with the signal applied to the oscillator grid (No. 1) and the control voltage applied to the signal grid (No. 4). See Ref. 14.

Type 6K8 triode-hexodes may be used, with the signal applied to No. 1 grid and the control voltage applied to No. 3 grid. See Ref. 14.

Many other types of multi-grid valves may also be used.

(iii) Plate resistance control

The plate resistance of a valve is a function of the grid bias voltage. A triode has a sharp change from a fairly low value at normal bias voltages to a high value when approaching cut-off, and infinity beyond cut-off. This effect has been applied in a volume compressor (Ref. 12).

(iv) Negative feedback

The gain of an amplifier may be controlled by means of a negative feedback network in which one element is the plate resistance of a valve whose grid is connected to the control voltage. One application (Ref. 31) gives less than 1.8% total harmonic distortion under all conditions, with an output of 12 volts.

Another application of negative feedback is Fig. 16.3 which employs a remote cut-off pentode with negative feedback provided by capacitance C_2 from plate to grid.

FIG. 16.3

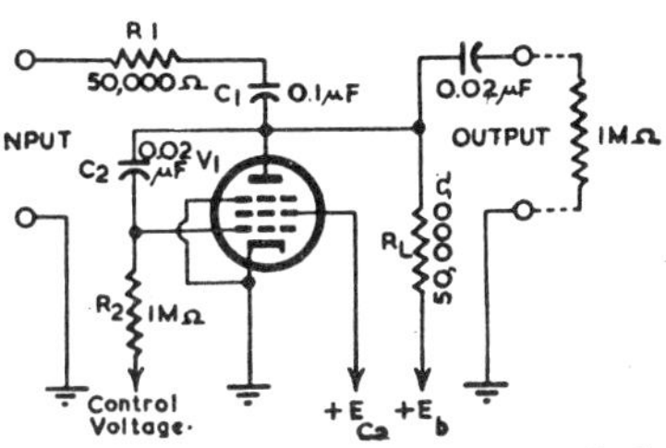

Fig. 16.3. Pentode with negative feedback used as shunt resistance in an attenuator to provide gain control.

By this means the effective plate resistance with feedback becomes approximately $1/g_m$. Thus a valve with a range of mutual conductance from 2000 to 2 micromhos may be used as a resistance varying from 500 to 500 000 ohms. The ratio between output and input voltages is given by

$$R_L/(g_m R_1 R_L + R_1 + R_L).$$

FIG. 16.4

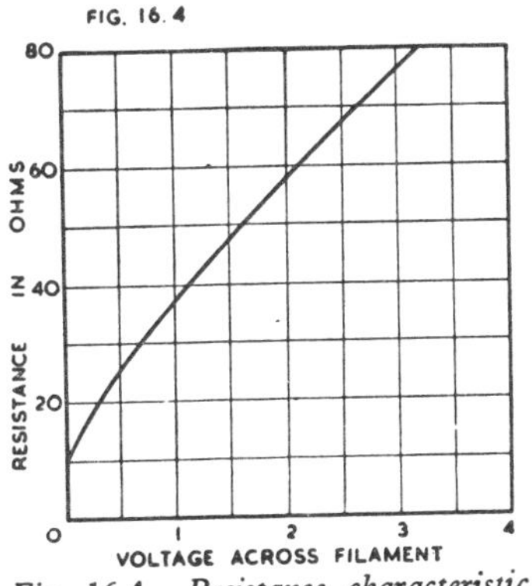

Fig. 16.4. Resistance characteristic of typical 4-6 volt 0.04 A metal filament lamp used in volume expanders (Ref. 45).

With the values given in Fig. 16.3, a variation approaching 33 db is obtainable when V_1 has a maximum g_m of 2000 micromhos (Ref. 62). The dominant (second) harmonic distortion is less than 0.2% for an input voltage of 2 volts.

(v) Lamps

Metal filament (dial) lamps may be used as control devices in volume expanders. A typical tungsten filament lamp has a resistance at maximum brilliancy of about 10 times that at room temperature (Fig. 16.4). See Sect. 4(ii) for applications of metal filament lamps in expanders.

Note : Carbon filament lamps have a resistance which is less when at maximum brilliancy than at room temperature.

(vi) Suppressor-grid control

The control voltage may be applied to the suppressor grid of a suitable pentode valve. This is used in Figs. 16.11 and 16.12.

References to suppressor-grid control : 7, 10, 16, 27, 33, 46.

SECTION 4 : VOLUME EXPANSION

(i) Introduction (ii) Expanders incorporating lamps (iii) Expanders utilizing feedback (iv) Expanders incorporating remote cut-off pentodes (v) Expanders incorporating remote cut-off triodes (vi) Expanders incorporating suppressor-grid controlled pentodes (vii) Expanders incorporating valves with five grids (viii) Expanders incorporating plate resistance control.

(i) Introduction

Volume expansion is very similar to volume compression, and the same control methods are used for both, but the control voltage is of opposite polarity. In fact, an expander/compressor may be designed with a switch to change from one to the other (Fig. 16.14).

The desirable characteristics which a volume expander should have are :

1. Negligible non-linear distortion.

2. The degree of expansion should be under control.

3. The degree and control of expansion should be independent of the volume level at which the amplifier is operated.

4. The expansion should result in the upwards expansion of loud passages and the downwards expansion of soft passages.

5. The attack time should be short—times from 0.2 to 200 milliseconds are in common use, but the shorter times are preferable (say not exceeding 20 milliseconds).

6. The recovery time should be adjustable from a fraction of a second to 1 or 2 seconds.

7. There should be no audible thump or transient distortion with sudden large transients.

8. There should be no appreciable reduction in maximum power output.

9. The overall gain should not be reduced seriously by the expander.

10. The shape of the expansion characteristic should provide some expansion at low output levels, the amount of expansion steadily increasing all the way to maximum power output.

For most purposes an expansion of 10 or 12 db is satisfactory, although some prefer up to 15 or even 20 db.

It is preferable to introduce the expander into the amplifying chain so that minimum amplification follows it. It is therefore desirable to select a type of expander which is capable of a fairly high output voltage. It is preferable for tone controls to precede the expansion unit.

Electronic methods of volume expansion may be divided into two groups, those in which the control voltage is derived from the output voltage, and those in which it is derived from the input voltage, suitably amplified. The former method is cheaper and employs fewer valves, but it has a less desirable shape of expansion characteristic and may be unstable. The latter method is used in Figs. 16.10 to 16.15 inclusive, and is preferable in order to avoid thumps and blocking effects.

When the control voltage is derived from the input voltage, a " side-chain " amplifier is used (e.g. Fig. 16.10) with the final stage transformer-coupled to a suitable full-wave rectifier and load network.

(ii) Expanders incorporating lamps

Small metal filament lamps are sometimes used in volume expanders, the variation of about 10 to 1 in resistance (see Fig. 16.4) giving sufficient range of control.   The simplicity and cheapness of some of these volume expanders make a strong appeal, although there are many shortcomings.   If the lamps are placed in the voice coil circuit there is a loss of something like 50% of the power output.   It is difficult to find lamps with characteristics suitable for all applications.   It is not possible for the listener to control either the amount of expansion or the time constant.   The degree of expansion falls rapidly as the operating level is reduced, and an appreciable amount of expansion can only be achieved at maximum power output.   The more complicated designs overcome a few of these defects.

A lamp with a 40 mA current has an attack time of about 30 milliseconds and a recovery time of about 150 milliseconds (Ref. 45).

(A) Lamps in voice coil circuit

The simplest form is Fig. 16.5 in which the lamp is shunted across the voice coil, but this is only effective with a pentode valve without voltage feedback, and the distortion due to mismatching is severe except at one level.   It is not used in practice.

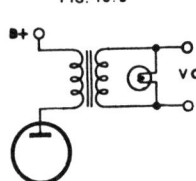

FIG. 16.5

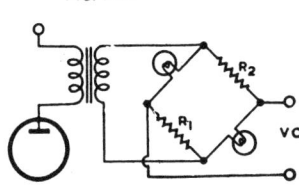

FIG. 16.6

Fig. 16.5.   Simplest form of volume expander using a lamp

Fig. 16.6.   Bridge type of volume expander using lamps.

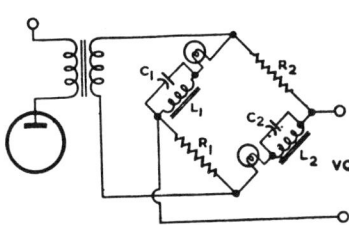

FIG. 16.7

Fig. 16.7.   Bridge type of volume expander using lamps which provides bass boosting at low levels.

The bridge circuit of Fig. 16.6 is quite practical, and may be used with any type of output valve.   At maximum volume this may possibly have an efficiency of 66%. With the addition of two inductors L_1 and L_2 and two capacitors C_1 and C_2 (Fig. 16.7) it may be used to provide bass boosting at low output levels only (Refs. 50, 51, 52, 61). Correct load matching is arranged for maximum output, leaving the low level condition to look after itself.

A suggested application of both carbon and metal filament lamps is Ref. 53.

(B) Volume expander in stage preceding loudspeaker

The loss of power caused by lamps in the voice coil circuit may be avoided by incorporating the expander in an earlier stage.   This necessitates a low power amplifier (driver) stage which provides enough power to operate the expander, say 3 watts, followed by a step-up transformer to the output stage, which may be a high-power push-pull amplifier.   The volume control should follow the expander, or else two volume controls should be used (Ref. 54).

(C) Lamp-controlled feedback

This is undoubtedly the best form of expander using lamps, although it suffers from most of the limitations of the lamp. An expansion of about 10 db is practicable and Fig. 16.8 shows one form which it may take. Refs. 22, 34, 45.

References to expanders incorporating lamps : 22, 33, 34, 45, 50, 51, 52, 53, 54, 55, 61.

FIG 16.8

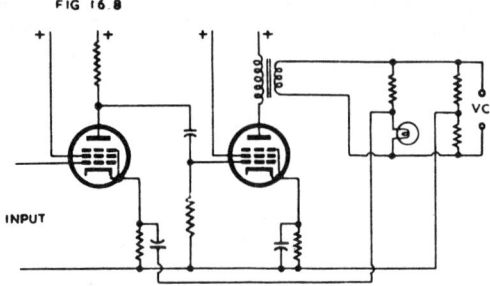

Fig. 16.8. Lamp-oper-ated negative feedback volume expander (Ref. 22).

(iii) Expanders utilizing feedback

Two designs have been developed by Stevens (Ref. 58) and subsequently received attention (Ref. 22) but have certain limitations.

An improved form is shown in Fig. 16.9 which gives an expansion of 29 db, with minimum gain of 200 times. The distortion is not visible on a C.R.O. with an output of 30 volts r.m.s. The attack time is 50 milliseconds and the recovery time about 1 second ; the latter may be adjusted by varying R_{10}. For further details see Refs. 22, 24 and 68.

(iv) Expanders incorporating remote cut-off pentodes

This is the oldest type of volume expander and a very satisfactory one, the valves being types suitable for use as i-f amplifiers or audio a.v.c. stages in receivers. It is possible to obtain a good control characteristic and any reasonable degree of expansion, to provide independent volume and expansion controls, and to arrange the time constants of the circuit to give the desired attack and recovery times. In order to avoid non-linear distortion, the controlled stages should operate with input voltages not greater than 0.15 volt, with a load resistance of 50 000 ohms. The controlled stages are nearly always arranged in push-pull —this eliminates even harmonic distortion and reduces intermodulation distortion. Push-pull operation is particularly valuable in reducing " thump " with large transients ; it also reduces any residual ripple that may come through from the rectifier. A further advantage of push-pull operation is that cathode and screen by-pass capacitors

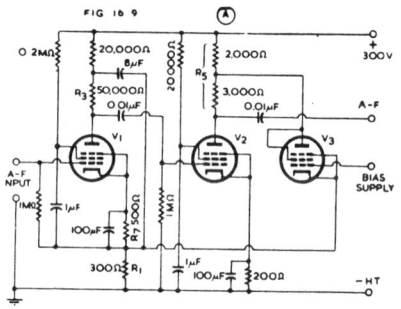

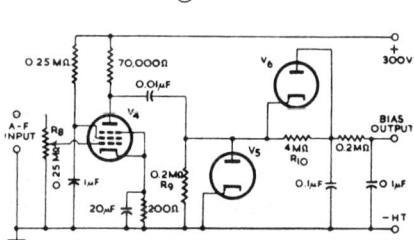

Fig. 16.9. Volume expander using negative feedback ; (A) amplifier (B) circuit for supply-ing control bias to the amplifier (Refs. 22, 24). $V_1 = V_2 = V_3 = V_4 = SP41$; V_5 and V_6 together = 6H6.

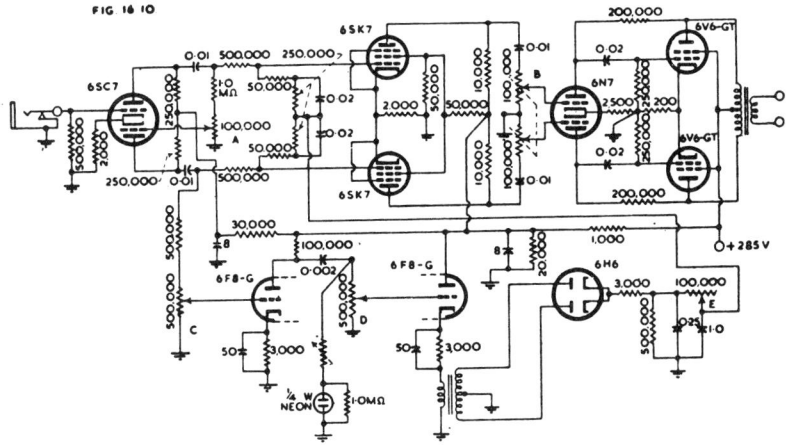

FIG. 16·10

Fig. 16.10. [1] *Volume expander incorporating remote cut-off pentodes (Ref. 23). Controls. A = balance, B = volume, C = input level, D = expansion, E = time constant.*

may be omitted, thus eliminating time delay in the adjustment of electrode voltages. Valves used in each push-pull pair must be very carefully matched for plate current and mutual conductance at several points over the operating range of grid bias voltages. Owing to the limited input-voltage which may be used, there should be one voltage amplifier stage between the controlled stage and the power amplifier.

An example of good design is Fig. 16.10 (Ref. 23). The minimum attack time is about 10 milliseconds and the recovery time of the order of 1 second. The expansion curve (db versus voltage) is approximately linear up to a d.c. control voltage of 20 volts, which gives 10 db expansion. The 6SC7 stage is common to both sections. The 6SK7 push-pull stage is the controlled stage ; this is followed by the 6N7 voltage amplifier stage which is coupled to the 6V6-GT power amplifier. The signal to feed the rectifier is taken from the plate circuit of the 6SC7 stage, amplified in two stages (6F8-G), the second being a cathode follower, and rectified by the 6H6. The direct voltage from the 6H6 filter circuit is applied to the signal grids of the 6SK7 valves.

In any expander of this general type, whatever may be the method of controlling the gain, it is necessary to prevent the transients in the output of the individual expander valves from becoming so large as to cut off the following stage. This can be accomplished by transformer coupling, by a direct-coupled phase inverter (as in Fig. 16.13) or by the use of low values of load resistances and coupling capacitances (as Fig. 16.10). With the third method it is necessary to incorporate an equalizing network to give a flat overall frequency response.

Refs. 22 (Part 1), 23.

(v) Expanders incorporating remote cut-off triodes

Any remote cut-off pentode may be connected as a triode (screen, suppressor and plate tied together) to form a triode which may be used in a similar manner to a pentode, except that the gain will be lower. It has been stated that a single valve has 0.37% total harmonic distortion unexpanded, 0.64% expanded, predominantly second harmonic, when the output level is − 25 db (i e. 0.056 volt). For further details see Refs. 1, 42.

(vi) Expanders incorporating suppressor-grid controlled pentodes

A sharp cut-off pentode of suitable design may be used for controlling the gain— the English Mazda AC/SP1 and the American 6SJ7 are typical examples. One example of this application is Fig. 16.11 (Refs. 7, 10) which has an attack time of about 1 millisecond, and a variable recovery time of about 1 second maximum. Resistors

FIG 16 11

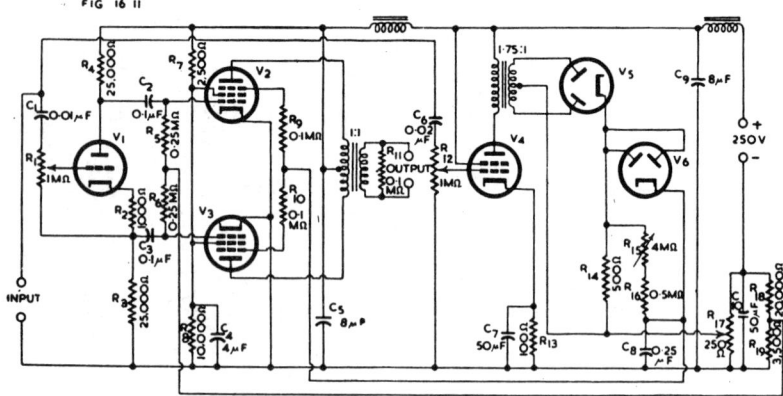

Fig. 16.11. *Volume expander incorporating push-pull suppressor-controlled pentodes (Ref. 7).* $V_1 = MH_4$, $V_2 = V_3 = AC/SP1$, $V_4 = KT41$, $V_5 = V_6 = AZ3$.

R_9 and R_{10} prevent the suppressor grids from being driven positive. With the moving contacts of R_{12} and R_{17} at the chassis ends, R_1 should be adjusted in conjunction with the volume control of the main amplifier so that the latter will just be fully loaded with the loudest signal. R_{17} is then adjusted to give the desired expansion. R_{12} is advanced until the loudest signal just causes the suppressors to be at cathode potential. Any further alteration in volume level should be made by R_1. Some suggested modifications are given in Refs. 10, 16.

A second example is Fig. 16.12 (Ref. 27) in which the need for push-pull operation and for a transformer is avoided by an ingenious device. V_1 is the suppressor-controlled stage while V_2 is a " dummy " designed to balance the plate and screen currents of V_1 so that the current passing through each load resistor is constant. The maximum signal input to the expander stage is about 0.25 volt, and the maximum signal output of the expander is of the order of several volts, for low distortion. The controlled range is about 15 db. The diode rectifier incorporated in the suppressor circuit prevents the suppressor from being driven positive. The time constants are such that 75% of final gain is achieved in the fast position in approximately 20 milliseconds, and in the slow position 60 milliseconds. The recovery times are about 0.5 and 1.2 seconds.

References 27, 46.   FIG. 16.12

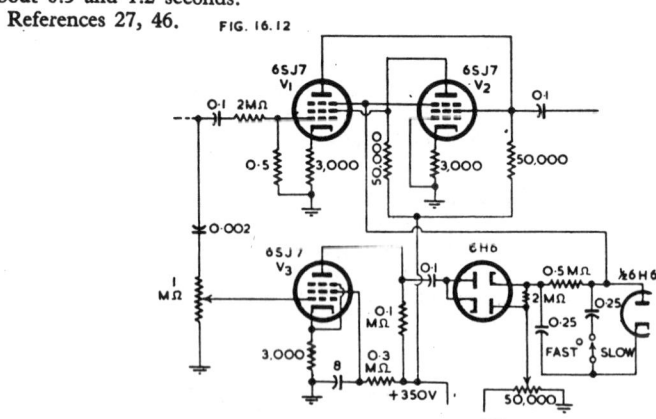

Fig. 16.12. *Single-ended surgeless volume expander incorporating suppressor-grid control (Ref. 27).*

(vii) Expanders incorporating valves with five grids
(A) Incorporating type 6L7

This valve type has two signal input grids, grid No. 1 having a remote cut-off characteristic, and grid No. 3 having a sharp cut-off characteristic. Normally the signal is applied to No. 1 grid which operates at a fixed bias of about -10 volts, grid No. 3 is operated with a static bias of approximately -18 volts, the screen is maintained at 100 volts and the static plate current is from 0.12 to 0.15 mA. The expansion available is about 20 db, and the maximum input is 1 volt, or say 0.5 volt for reasonably low distortion. There may be difficulty with hum, owing to the low signal amplification with small input voltages. There must be a careful compromise between time of " attack " and suppression of the ripple voltage produced by the bias rectifier. Particularly when the input level is low, the transient voltages on No. 3 grid become audible as clicks or thumps. These remarks apply to the old form of single valve design (Refs. 13, 35, 55, 64).

An improved form is shown in Fig. 16.13 (Ref. 21) which makes use of push-pull 6L7 control valves and direct-coupling to the following 6SJ7 phase inverter. The attack time is 12 milliseconds and the recovery time adjustable from 0.07 to 0.9 second. References to 6L7 expanders : 19, 21, 35, 55, 64.

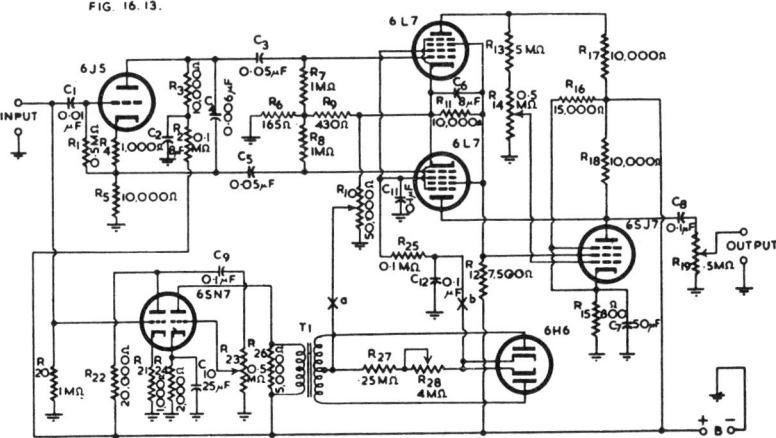

Fig. 16.13.   *Volume expander incorporating push-pull 6L7 control valves (Ref. 21).*

(B) Incorporating 6A8

The input signal is applied to Grid No. 1 (oscillator grid) and the amplified voltage for control purposes is taken from a volume control in the plate circuit of Grid No. 2 (anode grid), rectified by a separate rectifier and then applied to Grid No. 4 (control grid). Ref. 14 ; also Ref. 56 (incorporating MX40).

(C) Incorporating triode-hexode (6K8)

The input signal is applied both to Grid No. 1 of the hexode and to the grid of the triode, and the amplified voltage for control purposes is taken from a volume control in the triode plate circuit, rectified by a separate rectifier and then applied to Grid No. 3 of the hexode. Ref. 14.

(D) Incorporating type 6SA7, 6BE6, or 7Q7

One possible application is Fig. 16.14 (Ref. 25) in which push-pull 7Q7 valves form the controlled stage. The maximum expansion is 18 db, attack time 3 milliseconds and recovery time 0.5 second. The push-pull signal is applied from the phase splitter to the No. 3 grids of the 7Q7 valves, while the control voltage is applied to the No. 1 grids in parallel. In the original article this expansion circuit is incorporated

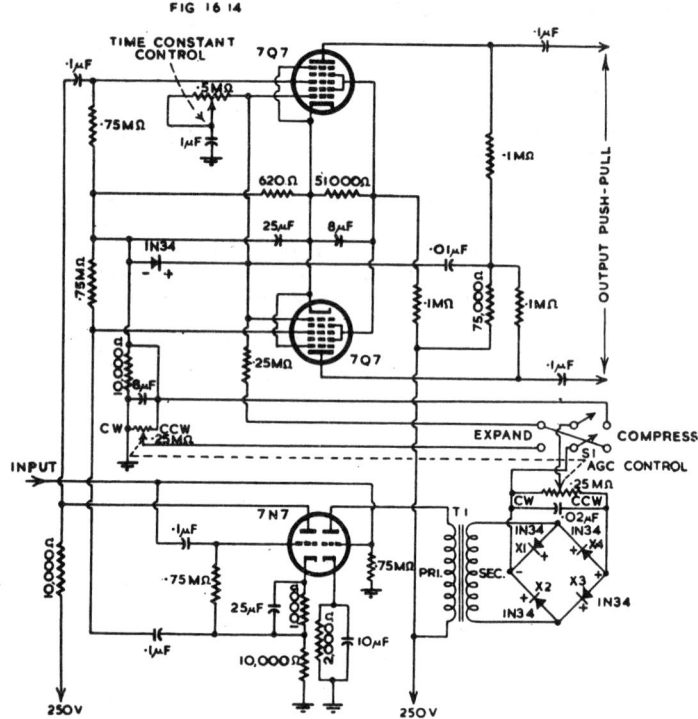

Fig. 16.14.   Volume expander/compressor incorporating push-pull 7Q7 (= 6SA7 = 6BE6) control valves and 7N7 (= 6SN7-GT) as phase splitter and control amplifier with 1N34 germanium crystal diode rectifiers (Ref. 25).

in a complete 30 watt amplifier with 4% total harmonic distortion at zero expansion which is increased to 5.5% with maximum expansion.

(viii) Expanders incorporating plate resistance control

The plate resistance of a valve may be varied either by varying the grid bias or by applying feedback. In either case it may be used to shunt across a network and thereby to control the overall gain.

Some possible applications are described in Refs. 22 (Part 1 Fig. 3), 41, 43 and 55 (Fig. 6). A good design is Fig. 16.15 (Ref. 43) in which a rapid attack is combined with a recovery time which may be varied from 0.5 to 10 seconds. The maximum expansion is 12 db, and the intermodulation distortion is always less than 1.75%. The expansion characteristic (db versus db) is almost linear. The push-pull 6J5 valves form a Class A_1 amplifier with a normal load of 15 000 ohms across each valve. The 6P5 (= 76) valves form the controlling stage and place an additional shunt load from a high value to 10 000 ohms per valve. The maximum input voltage is 3 volts peak, grid to grid. If a high impedance input circuit is desired, transformer T_1 may be replaced by a phase splitter stage (Fig. 6 of Ref. 43). The close approach to linearity in the expansion characteristic is obtained by arranging 6 db more expansion than necessary, and then reducing the overall gain by 6 db through negative feedback.

Some alternative applications are discussed in Ref. 13.

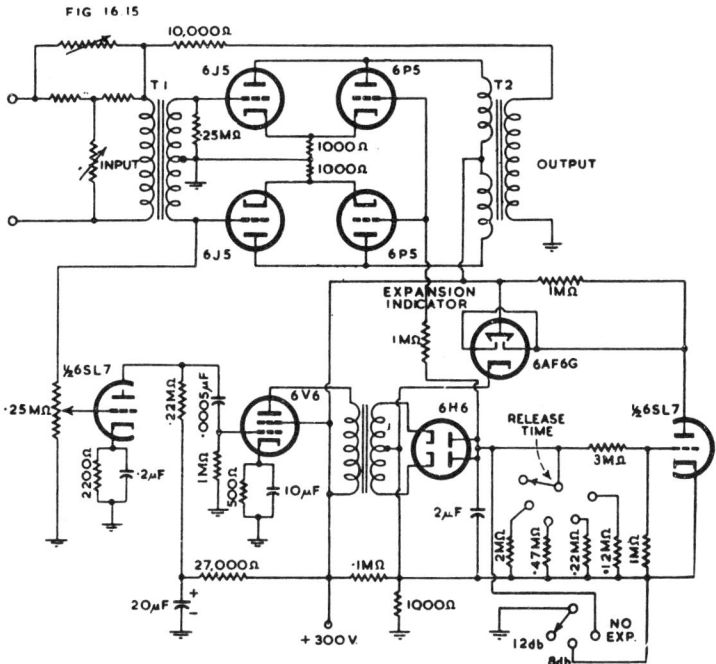

FIG 16.15

Fig. 16.15. *Volume expander incorporating plate resistance control* (*Ref.* 43).

SECTION 5 : PUBLIC ADDRESS A.V.C.

Audio a.v.c. may be applied to an a-f amplifier so as to permit the speaker to move his head without causing a serious drop in sound output level from the loudspeaker. The usual method is to incorporate into the amplifier a control valve, at such a level that the input voltage to the control valve is about 0.5 volt for type 6L7 or 0.15 volt for a remote cut-off pentode.   Other methods of control can, of course, be used. The control voltage may be derived from a suitable point on the main amplifier, preferably through an isolating stage, with a full wave rectifier and filter in the usual way.   A typical example is Ref. 65.

References to public address a.v.c. : 39, 65, 92.

SECTION 6 : SPEECH CLIPPERS

In the case of speech it is practicable to clip the peaks without seriously affecting the tonal qualities of the voice, and with a bearable degree of distortion.   With a suitably designed speech clipper it is possible to operate at a higher average level than with a limiter, owing to the long recovery time which is necessary with the latter. A speech clipper is practically instantaneous in its action.

When peaks have been clipped there should be a minimum of phase distortion in the remainder of the amplifier, at least up to 8000 c/s.   The response should be flat from 200 to 4000 c/s, but there must be treble attenuation above 4000 c/s—at least 25 db at 10 000 c/s.   The treble attenuation may be provided by a single constant *k* section filter, since the harmonics are attenuated sufficiently before they are shifted far enough in phase to increase appreciably the peak amplitude of the wave. At the bass end, satisfactory results are obtained if the response following the clipper stage is attenuated 3 db or less.

The Plex amplifier (Ref. 37, Fig. 3) is capable of 20 db peak clipping before the distortion becomes serious. The average increase in power level is about 12 db.

An alternative design of speech clipper incorporated in a speech amplifier is given in Fig. 2 of Ref. 28.

FIG. 16.16

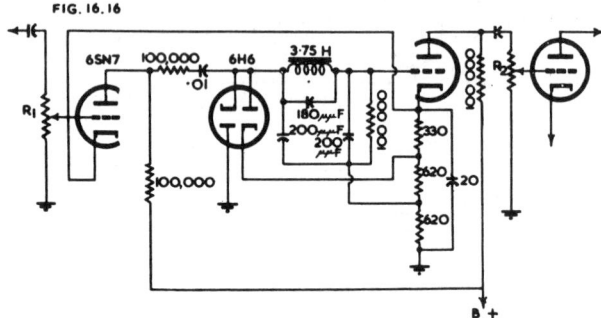

Fig. 16.16. *Simple form of speech clipper suitable for an amateur transmitter. The* 3.75 *H choke must have low d.c. resistance and good a-f characteristics* (*Ref.* 28).

A much simpler form of speech clipper, but one good enough for use with amateur transmitters, is shown in Fig. 16.16 (Ref. 28).

" Infinite " peak clipping has been used successfully, giving articulation from 50 to 90%. If preceded and followed by suitable frequency-tilting filters, the articulation may reach 97% with a quality sounding very much like normal speech. See Chapter 14 Sect. 11(ii), also Ref. 71 of this chapter.

References to speech clippers : 28, 37, 63, 71.

SECTION 7 : NOISE PEAK AND OUTPUT LIMITERS

(*i*) *Introduction* (*ii*) *Instantaneous noise peak limiters* (*iii*) *Output limiters* (*iv*) *General remarks.*

(i) Introduction

Limiters are restrictive devices to mitigate the effects of undesired electrical disturbances of an impulsive nature such as static and ignition noise on the output of an A-M receiver.

(ii) Instantaneous noise-peak limiters

(A) Series noise-peak limiters

These are highly effective in radio receivers. Fig. 16.17 is a simple series limiter which only requires four additional components as shown inside the dash-dash rectangle. Threshold bias is derived from the rectified carrier.

FIG. 16.17

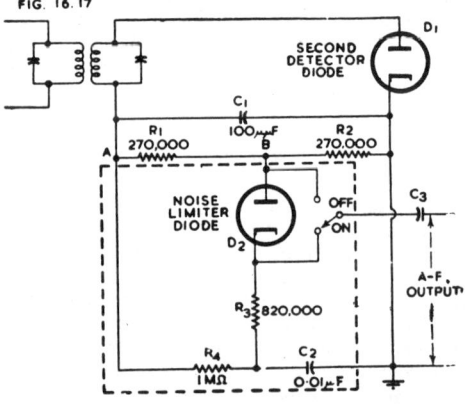

Fig. 16.17. *Simple series limiter* (*Ref.* 30).

When a 6H6 valve is used for both diodes, the hum may be reduced by earthing the end of the heater which is closer to the diode D_2. The hum may still be troublesome even with this precaution.

A circuit giving lower hum is Fig. 16.18 ; if a limiter on/off switch is used, two different values of C_5 will be required.

(B) Series-type noise limiter with threshold adjustment

In this form the limiter threshold can be varied from about 65% modulation down to substantially zero, on half of the modulation cycle (Fig. 16.19). Diode D_3 may be added to buck the thermionic potential of the limiter diode and thereby improve the effectiveness at low carrier levels ; this is at the cost of increased distortion.

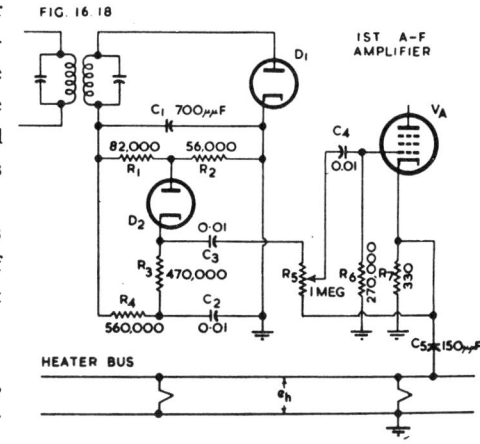

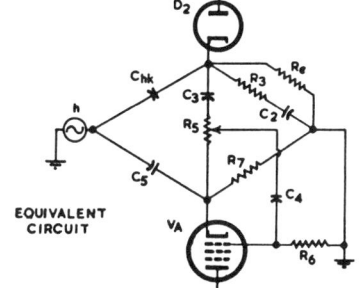

Fig. 16.18. *Hum-reduction version of simple series limiter with equivalent circuit to show bridge configuration (Ref. 30).*

(C) Low-loss series-type noise limiter (Fig. 16.20)

This provides the a-f amplifier with a higher percentage of the a-f voltage across the detector diode load (2 or 3 db improvement).

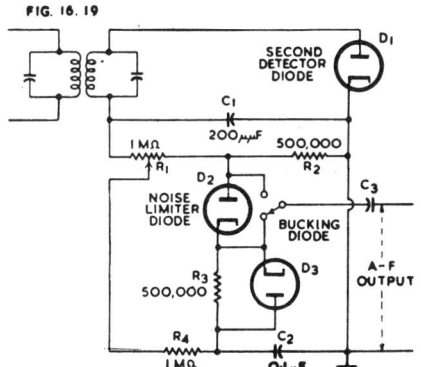

Fig. 16.19. *Series-type noise limiter with threshold adjustment (Ref. 30).*

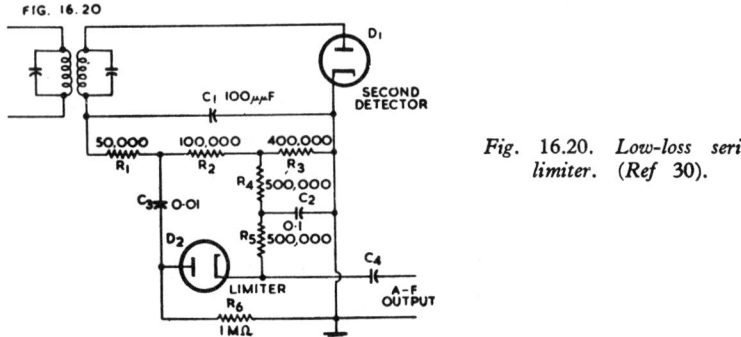

Fig. 16.20. *Low-loss series limiter.* (*Ref* 30).

(D) Balanced-detector noise limiter (Fig. 16.21)

This functions as a balanced bridge arrangement for detector voltages above the limiting threshold, with unbalance at all other times. It must be adjusted manually for each carrier level. This is not a very satisfactory form of limiter.

Fig. 16.21. *Balanced-detector noise limiter* (*Ref.* 30).

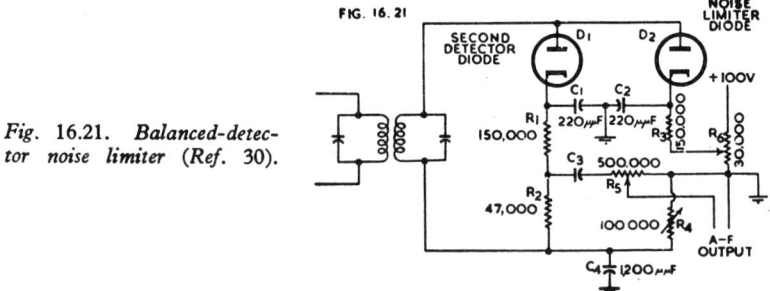

(E) Automatic balanced-detector noise limiter (Fig. 16.22)

The additional diode is operated from a tertiary winding on the final i-f transformer. Limiting does not take place at modulation depths below 100%. This is an improvement on Fig. 16.21 but it also is limited in performance.

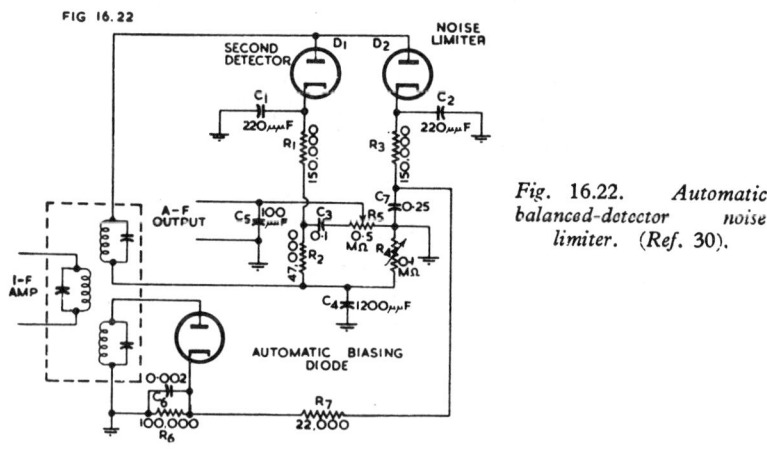

Fig. 16.22. *Automatic balanced-detector noise limiter.* (*Ref.* 30).

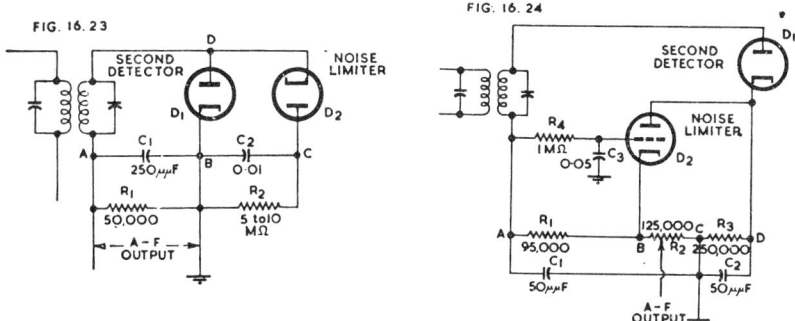

Fig. 16.23. *Balancing-type noise limiter (Ref. 30).*   Fig. 16.24. *Triode shunt-type noise limiter (Ref. 30).*

(F) Balancing-type noise limiter (Fig. 16.23)
This uses a limiter diode with reversed polarity, shunted across the detector diode. The modulation distortion is quite high, even on relatively low modulation depths. This circuit gives effective limiting of noise peaks. It provides about twice the normally obtainable a.v.c. voltage when the direct potential across R_2 is utilized.

(G) Triode shunt-type noise limiter (Fig. 16.24)
This circuit employs the plate resistance of a triode shunted across a portion of detector diode load, the magnitude of the shunt resistance being controlled by the grid and plate voltages, which act in conjunction with differential time-constants.

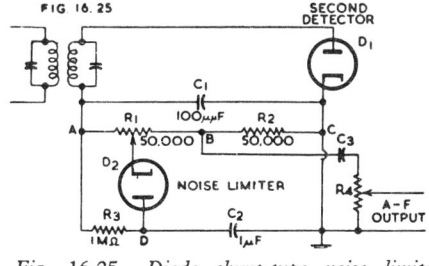

Fig. 16.25. *Diode shunt-type noise limiter (Ref. 30).*

The percentage of modulation at which distortion begins depends on the triode used, the values of R_1, R_2 and R_3, the time-constants involved, and the absolute carrier level. The higher carrier levels produce no distortion and no limiting. Serious distortion has been observed with 10% modulation at low signal levels. Effective limiting action is restricted to a narrow range of carrier input levels, generally above 10 Mc/s in carrier frequency.

(H) Simple diode shunt-type noise limiter (Fig. 16.25)
This is the simplest form of limiter, but the performance is not very good ; some improvement is evident on pulse type interference on signals above 10 Mc/s.

(I) Modified shunt-type noise limiter (Fig. 16.26)

This limiter begins to cause distortion at about 100% modulation for the values shown. This form is much more effective than the simple shunt limiter, although not so good as the simple series-type limiter at the lower carrier frequencies.

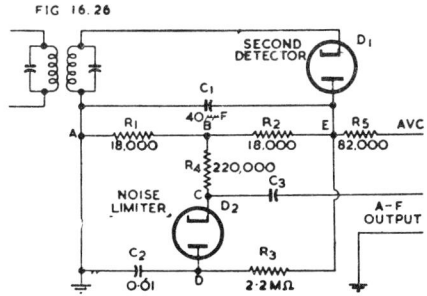

Fig. 16.26. *Modified shunt noise limiter (Ref. 30).*

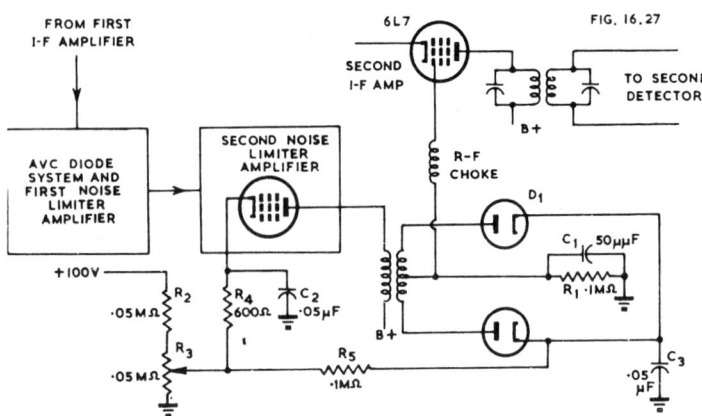

Fig. 16.27. *Degenerative noise limiter, acting between first and second i-f stages*
(*Ref.* 30).

(J) Degenerative noise-limiter (Fig. 16.27)

Degenerative feed to the i-f amplifier prior to final detection is used in this circuit.
A portion of the output from the first i-f amplifier is amplified in the first and second
noise-limiter amplifiers, and the resulting i-f output is coupled into a full-wave rectifier having R_1 as a load resistance. The direct voltage developed across R_1 provides
the bias for the grid of a 6L7 serving as second i-f amplifier. Front panel control
R_3 provides a positive delay voltage for the diode cathodes, to prevent rectification
until the signal or noise peaks exceed this bias. This limiter requires manual adjustment of the threshold of operation, and is useless on fading signals. Modulation
distortion is determined by the delay bias obtained from R_3 or by an accessory a.v.c
bias if provision is made for automatic biasing.

(iii) Output limiters

(A) R-F output limiter

This method, using low voltages on r-f amplifier plates and screens, is mainly used
for telegraphy (Ref. 30).

(B) A-F saturation-type output limiter

This method gives control of the maximum output of the a-f power amplifier by
suitably adjusting the plate, screen and grid voltages. A low-pass filter filters out
distortion produced in the output stage. A similar result may also be achieved in
an a-f voltage amplifier by controlling the screen voltage of a pentode valve. These
are mainly used for modulated c-w (Ref. 30).

(C) Logarithmic compressor (Fig. 16.28)

This circuit is a two-stage a-f amplifier with negative feedback, having two diodes
connected in series with the feedback path in such a way as to provide a path for both
positive and negative half-cycles of the feedback voltage. Potentiometer R_{10} provides
bias to the diodes for setting the limiting level.

Using the diodes in this manner is equivalent to varying the feedback percentage
from a low value to a maximum as the instantaneous feedback voltage rises, with
the reverse effect as the instantaneous feedback voltage falls. This circuit produces
considerable distortion on speech or music but does not destroy intelligibility. Noise
interference is reduced substantially, and almost as effectively in some cases as with a
series noise-peak limiter.

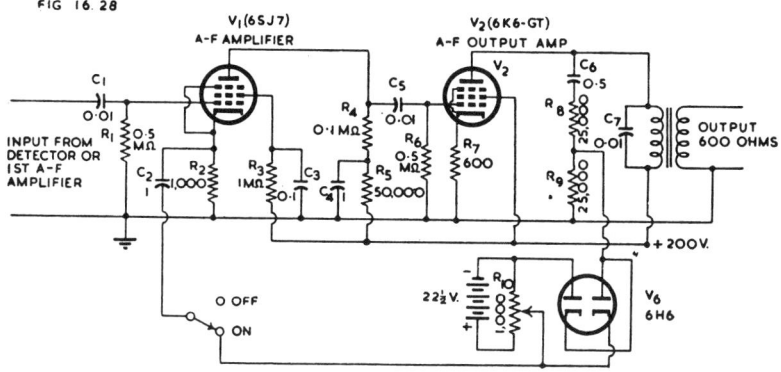

Fig. 16.28.   *Logarithmic compressor (Ref.* 30).

(iv) General remarks

Satisfactory protection against blocking may be provided by arranging for the stage preceding the final i-f amplifier to overload before the final i-f amplifier, then for each preceding stage to overload in turn.

A linear detector (or second detector) is desirable when handling a high noise level.

It is desirable to have control of the a-f gain of the stage following an instantaneous limiter—shunt limiters operate better with low following a-f gain.   Ref. 30.

SECTION 8 : REFERENCES

1. McProud, C. G. " Volume expansion with a triode " Elect. 13.8 (Aug. 1940) 17.
2. Schrader, H. J. U.S. Patent 2,343,207 (1944)—Improved voltage limiter.
3. " Contrast expansion etc." (letters) A. H. King W.W. 48.12 (Dec. 1942) 290 ; J. R. Hughes 49.1 (Jan. 1943) 29 ; J. Moir 49.3 (Mar. 1943) 92 ; D. T. N. Williamson, W.W. 49.5 (May 1943) 151 ; J. R. Hughes 49.8 (Aug. 1943) 240 ; C. E. G. Bailey 49.10 (Oct. 1943) 313 ; J. R. Hughes 49.12 (Dec. 1943) 382 ; A. A. Tomkins 50.5 (May 1944) 152.
4. Hilliard, J. K. " A limiting amplifier with peak control action " Comm. 23.5 (May 1943) 13.
5. Lewis, R. " A peak-limiting amplifier for recording " Q.S.T. 27.9 (Sept. 1943) 26.   Letter 28.8 (Aug. 1944) 55.
7. Williamson, D. T. N. " Contrast expansion unit—design giving unequal pick-up and decline delays " W.W. 49.9 (Sept. 1943) 266.   Corrections W.W. 49.10 (Oct. 1943) 315.
9. Miller, B. F. " Sibilant speech sounds " Electronic Eng. 16.185 (July 1943) 69.
10. Williamson, D. T. N. " Further notes on the contrast expansion unit " W.W. 49.12 (Dec. 1943) 375.
12. Herrick, G. Q. " Volume compressor for radio stations " Elect. 16.12 (Dec. 1943) 135.
13. Felix, M. O. " New contrast expansion circuit—applying the principle of the cathode follower" W.W. 50.3 (March 1944) 92.   Letters W. C. Newman 50.5 (May 1944) 152 ; D. T. N. Williamson 50.6 (June 1944) 187.
14. " Simple volume expander circuits " Australasian Radio World (April 1944) 11—based on R.C.A. (E. W. Herold) Patent 264, 942 (March 30, 1939) U.S.A.
16. Ingham, W. E., and A. Foster " Variable contrast expansion—control of contrast range without change of average level " W.W. 50.8 (Aug. 1944) 243.
17. Roddam, T. " New thoughts on volume expansion—contrast should be proportional to size of room " W.W. 50.9 (Sept. 1944) 286.
18. Cosens, C. R. " RF volume expansion " W.W. 50.12 (Dec. 1944) 381.
19. Hansen, I. C. " Contrast without distortion " Australasian Radio World (April 1945) 15.
20. Crane, R. W. " Suggestions for design of volume expanders " Elect. 18.5 (May 1945) 236.
21. Weidemann, H. K. " A volume expander for audio amplifiers—reducing time constant for more rapid response " Q.S.T. 29.8 (Aug. 1945) 19.
22. White, J. G. " Contrast expansion—the use of negative feedback ; its advantages over earlier methods " W.W. 51.9 (Sept. 1945) 275 ; 51.10 (Oct. 1945) 309 with bibliography.
23. Ehrlich, R. W. " Volume expander design " Elect. 18.12 (Dec. 1945) 124.
24. White, J. G. " Contrast expansion—some practical results using negative feedback " W.W. 52.4 (April 1946) 120.
25. Moses, R. C. " A volume expander compressor preamplifier " Radio News 35.6 (June 1946) 32.
27. Butz, A. N. " Surgeless volume expander " Elect. 19.9 (Sept. 1946) 140.
28. Smith, J. W., and N. H. Hale " Speech clippers for more effective modulation " Comm. 26.10 (Oct. 1946) 20.

29. Phillips, W. E. " Volume compression for FM broadcasting—Raytheon volume limiter " F.M. and T. 6.9 (Sept. 1946) 28.
30. Toth, E. " Noise and output limiters " Elect. (1) 19.11 (Nov. 1946) 114 ; (2) 19.12 (Dec. 1946) 120.
31. Stewart, H. H., and H. S. Pollock, " Compression with feedback " Elect. 13.2 (Feb. 1940) 19
32. Moorhouse, C. W. " A high fidelity peak-limiting amplifier " Q.S.T. 28.5 (May 1944) 19.
33. Philips, " Volume expansion " Philips Tec. Com. 77 (Sept./Oct. 1940).
34. Henriquez, V. C. "Compression and expansion in transmission sound " Philips Tec. Rev. 3. 7 (July 1938) 204.
35. R.C.A. " Application Note on the 6L7 as a volume expander for phonographs " No. 53 (Nov. 27, 1935).
36. Maxwell, D. E. " Automatic gain-adjusting amplifier " Tele-Tech 6.2 (Feb. 1947) 34.
37. Dean, M. H. " The theory and design of speech clipping circuits " Tele-Tech 6.5 (May 1947) 62.
38. " Level-governing audio amplifier " (W.E. model 1126C) Tele-Tech. 6.8 (Aug. 1947) 67.
39. Jurek, W. M., and J. H. Guenther " Automatic Gain Control and Limiting Amplifier " Elect. 20.9. (Sept. 1947) 94.
40. Maxwell, D. E. " Dynamic performance of peak-limiting amplifiers " Proc. I.R.E. 35.11 (Nov. 1947) 1349.
41. McProud, C. G. " Experimental volume expander and scratch suppressor " Audio Eng. 31.7 (Aug. 1947) 13.
42. Johnson, M. P. " Multi-purpose audio amplifier " Audio Eng. 31.7 (Aug. 1947) 20.
43. Pickering, N. C. " High fidelity volume expander " Audio Eng. 31.8 (Sept. 1947) 7.
44. Dean, W. W., and L. M. Leeds " Performance and use of limiting amplifiers " Audio Eng. 31.6 (July 1947) 17.
45. Korn, T. S. " Dynamic sound reproduction " Elect. 21.7 (July 1948) 166.
46. Tomkins, A. A. " Surgeless volume expansion " W.W. 54.6 (June 1948) 234. Correction 54.9 (Sept. 1948) 347.
48. Tillson, B. J. " Musical acoustics," Part 5, Audio Eng. 31.9 (Oct. 1947) 25.
50. Weeden, W. N. " Simplified volume expansion " W.W. 38.17 (24th April 1936) 407.
51. Tanner, R. H., and V. T. Dickins " Inexpensive volume expansion " W.W. 38.21 (22nd May 1936) 507.
52. Tanner, R. H. " Lamps for volume expansion " W.W. 39.7 (14th August 1936) 146.
53. Sayers, G. " Notes on contrast expansion " W.W. 39.12 (18th Sept. 1936) 313.
54. Weeden, W. N. " Improving the simplified volume expander " W.W. 40.3 (15th Jan. 1937) 68.
55. Weeden, W. N. " Contrast amplification ; a new development " W.W. 39.25 (18th Dec. 1936) 636.
56. " Contrast expansion unit " W.W. 41.24 (9th Dec. 1937) 590.
57. " Contrast expansion unit ; a correction " W.W. 41.25 (16th Dec. 1937) 621.
58. Stevens, B. J. " Low distortion volume expansion using negative feedback " W.E. 15.174 (March 1938) 143.
59. Amos, S. W. " Distortion in radio receivers " Electronic Eng. 14.169 (March 1942) 686.
61. " Light-bulb volume expander " Elect. 9.3 (March 1936) 9.
62. Barber, A. W. " Plate resistance control in vacuum tubes as audio gain control means " Comm. 17.10 (Oct. 1937) 23.
63. Black, W. L., and N. C. Norman " Program-operated level-governing amplifier " Proc. I.R.E. 29.11 (Nov. 1941) 573.
64. Sinnett, C. M. " Practical volume expansion " Elect. 8.11 (Nov. 1935) 14.
65. Paro, H. " Public address avc " Elect. 10.7 (July 1937) 24.
66. Cook, E, G. " A low distortion limiting amplifier " Elect. 12.6 (June 1939) 38.
67. Grimwood, W. K. " Volume compressors for sound recording " Jour. S.M.P.E. 52.1 (Jan. 1949) 49.
68. Wheeler, L. J. " Contrast expansion " W.W. 55.6 (June 1949) 211.
69. Singer, K. " High quality recording electronic mixer " Jour. S.M.P.E. 52.6 (June 1949) 676.
70. Singer, K. " Preselection of variable gain tubes for compressors " Jour. S.M.P.E. 52.6 (June 1949) 684.
71. Licklider, J. C. R., and I. Pollack "Effects of differentiation, integration and infinite peak clipping upon the intelligibility of speech " J. Acous. Am. 20.1 (Jan. 1948) 42.
72. A.W.A. Model G51501 (Amalgamated Wireless Australasia Ltd.).
73. Weller, J. A. " A volume limiter for leased-line service " Bell Lab. Rec. 23.3 (March 1945) 72.
74. Hilliard, J. K. " The variable-density film recording system used at MGM studios " Jour. S.M.P.E. 40 (March 1943) 143.
75. Miller, B. F. " Elimination of relative spectral energy distortion in electronic compressors " Jour. S.M.P.E. 39 (Nov. 1942) 317.
76. Smith, W. W. " Premodulation speech clipping and filtering " Q.S.T. 30.2 (Feb. 1946) 46.
77. Smith, J. W., and N. H. Hale, " Let's not overmodulate—it isn't necessary " Q.S.T. 30.11 (Nov. 1946) 23.
78. Smith, W. W. " More on speech clipping " Q.S.T. 31.3 (March 1947) 18.
79. Mather, N. W. " Clipping and clamping circuits " Elect. 20.7 (July 1947) 111.
80. Kryter, K. D., J. C. R. Locklider and S. S. Stevens " Premodulation clipping in AM voice communication " J. Acous. Soc. Am. 19.1 (Jan. 1947) 125.
81. Winkler, M. R. " Instantaneous deviation control " Elect. 22.9 (Sept. 1949) 97.
82. U.S. patent appln. 793,916—H. E. Haynes, Variable gain amplifier (RCV 11688).
83. U.S. patent appln. 794,050—H. J. Woll Variable gain systems (RCV 11504).
84. U.S. patent appln. 768,319 — W. W. H. Dean Peak clipper and indicator therefor (RCA 25944).
85. Hathaway, J. L. " Automatic audio gain controls " Audio Eng. (1) 34.9 (Sept. 1950) 16 ; (2) 34.10 (Oct. 1950) 27.
86. Singer, G. A. " Performance and operation of a new limiting amplifier " Audio Eng. 34.11 (Nov. 1950) 18.
87. Haynes, H. E. " New principle for electronic volume compression " Jour. S.M.P.T.E. 58.2 (Feb. 1952) 137. Reprinted Radiotronics 17.8 (Aug. 1952) 136.

Additional references will be found in the Supplement commencing cn page 1475.

CHAPTER 17

REPRODUCTION FROM RECORDS

BY F. LANGFORD-SMITH, B.SC., B.E.

SECTION 1 :   INTRODUCTION TO DISC RECORDING

(i) Methods used in sound recording   (ii) Principles of lateral recording   (iii) Frequency range   (iv) Surface noise and dynamic range   (v) Processing   (vi) Turntables and driving mechanism   (vii) Automatic record changers.

(i) Methods used in sound recording

There are many methods which have been used for sound recording, but these may be arranged in the following principal groups :

1. **Magnetic recording** includes magnetic wire and magnetic tape. See Refs. 258 (Chapter 29), 272, 316, 318.

2. **Sound on film** finds its principal application in cinema films. The Philips-Miller engraved film system is used to a limited extent for broadcast transcription. Film is also used for embossed lateral recording for some special applications. The high cost of the film precludes the use of this medium in most other fields. See Ref. 258.

3. **Mechanical groove recording** is used in various forms including
   (a) the cylinder (e.g. dictaphone)
   (b) the disc, which is the only type of recording considered in this chapter.

Two methods of recording are used with discs :

1. **Vertical recording** (" hill and dale "), which has had only a limited field of application, and appears unlikely to be used extensively in the future.

2. **Lateral recording** which is the method used most generally, and is the subject of this chapter.

Lateral recording is used with five types of discs—
   (a) The 78 r.p.m. " shellac " pressings,
   (b) The 33-1/3 r.p.m. fine groove, otherwise known as microgroove or long-playing (LP) records,
   (c) The 45 r.p.m. fine groove records,
   (d) Lacquer discs used for direct playback—see Sect. 8,
   (e) Discs used principally for broadcast transcription recording, usually operated at 33-1/3 r.p.m.—see Sect. 9.

The nominal speed of rotation is interpreted for recording as follows :

Supply frequency	50 c/s	60 c/s (R.M.A.)
78 r.p.m. nominal	77.92	78.26 r.p.m. (R.M.A.)
33-1/3 r.p.m. nominal	33-1/3	33-1/3 r.p.m. (R.M.A.)
Tolerance in speed of rotation	± 0.5%	± 0.3% (N.A.B.)

The speed is usually checked by means of a **stroboscope** illuminated by a lamp supplied from a.c. mains. The usual arrangement is :

Supply frequency	50 c/s	60 c/s (N.A.B.)
Number of bars on stroboscope	77	92 for 78 r.p.m.
	180	216 for 33-1/3 r.p.m.

At either 78.26 or 33-1/3 r.p.m. not more than 21 dots per minute in either direction may drift past a reference point (N.A.B. for recording).

See Reference 89 for a " Glossary of disk-recording terms." See Reference 105 for a " Bibliography of disc recording " (1921 to 1947).

See References 2, 87, 237 and 260 for American Recording Standards. See Sect. 2 for current English and American practice.

(ii) Principles of lateral recording

In lateral recording the groove forms a spiral either from the outside to the inside (as with all commercial " home " recordings) or from the inside to the outside (as with some transcription discs). The groove undulates horizontally from side to side of the mean path so as to deflect the stylus (needle) and armature of the pickup in accordance with the recorded sound (Fig. 17.1). The undulation of the groove is called " modulation " and the movement to one side of the mean path at any instant is called the amplitude of the modulation. The stylus is wedged in the groove by the effective vertical pressure due to the weight of the pickup, and moves radially about the pivot of the armature (Fig. 17.2).

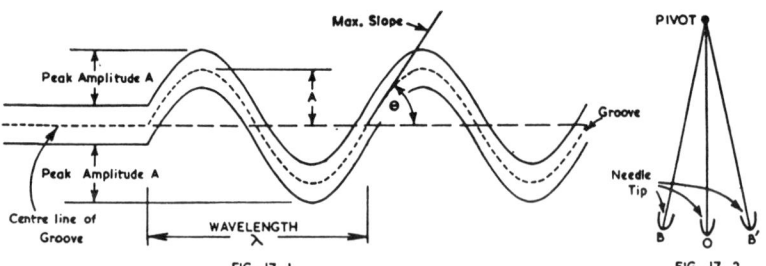

FIG. 17. 1   FIG. 17. 2

Fig. 17.1. Unmodulated (left) and modulated (right) groove of lateral recording.
Fig. 17.2. Motion of stylus tip with lateral recording.

If the recorded sound is of sine-wave form, the maximum transverse velocity of the stylus tip will occur at O and zero velocity will occur at B and B', the two extremities of travel. This is an example of simple harmonic motion (see Chapter 6 Sect. 4). The maximum transverse velocity is $2\pi fA$ where f is the frequency in c/s and A is the peak amplitude. The r.m.s. transverse velocity is given by 1.41 πfA.

The **recorded level** may be specified in terms of r.m.s. velocities at 1000 c/s, or in decibels with a reference level 0 db = 1 cm/sec. lateral r.m.s. stylus velocity. The following tables are based on this usage.

(A) 78 r.p.m.—Cross-over frequency = 500 c/s

Level	Velocity r.m.s.	Peak amplitude*	Comments
+ 10 db	3.16 cm/sec	0.56 mil	R.M.A. Frequency Test Record No. 1(A)
+ 16 db	6.31 cm/sec	1.1 mils	⎱ R.M.A. Frequency Test Record No. 1(B)
+ 18 db	7.94 cm/sec	1.4 mils	⎰ max. velocity (1000 c/s).
+ 22 db	12.6 cm/sec	2.2 mils	
+ 26.8 db	22 cm/sec	3.8 mils	Max. instantaneous programme peak.

* Over constant amplitude portion.

(B) 78 r.p.m.—Cross-over frequency = 250 c/s

Level	Velocity r.m.s.	Peak amplitude*	Comments
+ 10 db	3.16 cm/sec	1.1 mils	
+ 12 db	3.98 cm/sec	1.4 mils	
+ 15 db	5.62 cm/sec	2.0 mils	Normal maximum level
+ 18 db	7.94 cm/sec	2.8 mils	

N.B.   1 mil = 0.001 inch.

(C) 45 r.p.m.

Level	Velocity r.m.s.	Comments
+ 14.3 db	5.2 cm/sec	RCA Test record 12-5-31 (1000 c/s)
+ 22.9 db	14.0 cm/sec	Max. instantaneous programme peak.
+ 25.1 db	18.0 cm/sec	Max. level on R.C.A. Test Record 12-5-37.

(D) 33-1/3 r.p.m. (LP)

Level	Velocity r.m.s.	Comments
+ 7.5 db	2.4 cm/sec	Columbia Test record RD-103 (1000 c/s).
+ 22.9 db	14.0 cm/sec	Max. instantaneous programme peak.

There are two basic methods of recording sounds of different frequencies—constant velocity and constant amplitude. "**Constant velocity**" refers to the maximum transverse velocity of the stylus tip at the zero axis, this being held constant as the frequency changes. A diagrammatic representation of constant velocity recording for two frequencies is given in Fig. 17.3.

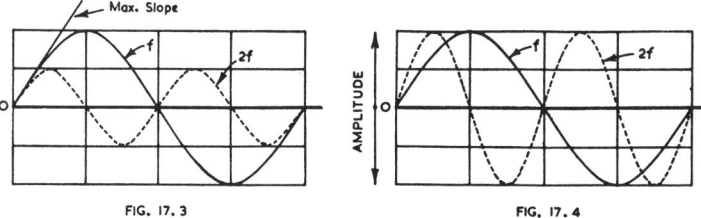

FIG. 17. 3   FIG. 17. 4

Fig. 17.3.   Diagram of constant velocity recording for two frequencies f and 2f.
Fig. 17.4.   Diagram of constant amplitude recording for two frequencies f and 2f.

It may be shown that constant velocity recording has the following characteristics for constant power at all frequencies—

1. The peak amplitude is inversely proportional to the frequency.
2. The maximum slope of the curve is the same for all frequencies.

In the general case when the power is changing, the maximum velocity at any frequency is proportional to the peak amplitude. With an "ideal" electro-magnetic pickup, which inherently follows a constant velocity characteristic, the output voltage for sine waveform is proportional to the maximum velocity at all frequencies.

Constant velocity recording is not suitable for use over a very wide frequency ratio, owing to the large variation in peak amplitudes. For example, over a range of 8 octaves the ratio of maximum to minimum amplitude is 256 to 1.

"**Constant amplitude**" recording indicates that the maximum amplitude is held constant when the frequency changes, for constant power output (Fig. 17.4). It may be shown that the maximum slope is proportional to the frequency.

Constant amplitude recording is very suitable for low frequencies, but is not satisfactory with large amplitudes at the highest audio frequencies because the transverse velocity of the needle tip becomes excessive, leading to distortion in recording and in reproducing. On the other hand, constant velocity recording is satisfactory over a

---

*Over constant amplitude portion.

limited frequency range of medium or high frequencies.   Therefore most recording systems employ an approximation to constant amplitude recording at low frequencies and an approximation to constant velocity recording for at least part of the medium and higher frequency range (see Sect. 5).

It may be shown (Ref. 146) that the minimum radius of curvature at the peak of the curve is given by

$$\rho = 0.025\,\lambda^2/A \qquad (1)$$

where ρ = radius of curvature in inches

λ = wavelength of sine-wave curve in inches

and A = peak amplitude of curve in inches.

Equation (1) may also be put into the form

$$A = 0.025\ \lambda^2/\rho. \qquad (2)$$

Any stylus is unable to fit accurately an undulation in the groove with a radius of curvature at the point of contact less than its own radius of curvature.   We may therefore apply eqn. (2), taking ρ as the radius of curvature of the stylus, to give the maximum amplitude (called the **critical amplitude**) for correct operation,

$$A_{crit} = 0.025\ \lambda^2/\rho \qquad (3)$$

where ρ = radius of curvature of stylus, in inches.

But since the r.m.s. transverse velocity is equal to 1.41 $\pi f A$ we can determine the critical velocity in cm/sec,

Critical velocity r m.s. = 11.3 $f A_{crit}$ cm/sec $\qquad (4)$

where A_{crit} = critical amplitude in inches.

(iii) Frequency range

The maximum recorded frequency in commercial shellac records manufactured prior to 1940 was of the order of 6000 c/s.   At the present time most new recordings are recorded up to at least 8000 c/s, while many manufacturers record up to 10 000 or 15 000 c/s.   Recordings up to ultrasonic frequencies (20 000 c/s) have been made, and are claimed to provide improved fidelity even though the higher frequencies are inaudible (Refs. 195, 271).

The lowest recorded frequency is of the order of 30 c/s, although some record manufacturers do not publish their recording characteristics below 50 c/s.

(iv) Surface noise and dynamic range

Surface noise (needle scratch) is unavoidable with any method of disc recording, but the maximum signal-to-noise ratio may be made quite high.   The maximum dynamic range will be greater than the measured signal-to-noise ratio on account of the nature of the noise, which approaches the characteristics of random noise if there are no resonances anywhere in the equipment—see Chapter 14 Sect. 7(v).

Surface noise is not really objectionable if it is reasonably low in level and if it can be described as " silky."   This latter criterion can only be attained when the pickup and loudspeaker both have smooth response throughout the whole frequency range. If the pickup has a prominent resonance within the audible frequency range—even if correctly equalized—the noise will not be " silky " and will tend to become objectionable.

In all cases the high frequency range of the equipment should be limited by a suitable filter giving a choice of cut-off frequencies.   Filters with very rapid cut-off characteristics are undesirable on account of distortion effects, but a combination of a " roll-off " characteristic and a filter with an ultimate attenuation of at least 12 db/octave appears to be satisfactory.   Any other method of achieving a " rounded " cut-off characteristic would probably also be satisfactory.

The maximum signal to noise ratio of **shellac pressings** varies considerably.   The older shellac pressings had a maximum signal to noise ratio* of the order of 30 or 35 db.   The most recent English shellac pressings have a maximum signal to noise ratio* of 50 db (Ref. 88) for a new record.   The best of recent American 78 r.p.m. shellac recordings have a maximum signal to noise ratio* of 45 or 46 db when new

---

*Measured with flat frequency response.

(Refs. 255, 254). The average post-war shellac record, taking into account all the various types of records on the market, both English and American, has a maximum signal to noise ratio* of about 38 db (Ref. 256).

A new vinyl fine groove record can have an average surface noise, measured on a system whose response is flat on a velocity basis from 500 to 10 000 c/s, about 56 db below the peak recording level ; if measured on a system whose response is the inverse of the recording characteristic, the surface noise is approximately 62 db below peak recording level. The B.B.C " D Channel " recording has a weighted signal to noise ratio greater than 60 db (Ref. 297).

The surface noise increases with the use of the record. After 100 playings with a 2 ounce pickup some shellac records showed negligible increase in noise level while others showed 5 db increase, the average being about 2.5 db (Ref. 181). After 200 playings of a fine groove vinyl record (LP) the surface noise increased by 2 db (Ref. 308).

The cost factor limits the use of **polyethylene** (Ref. 181) which has a scratch level even lower than that of vinyl. See also Sect. 9(ii) for the measurement of noise in accordance with NAB standards for transcription records.

(v) Processing

The processing of shellac pressings is described in Refs. 7, 10 and 25. The processing of vinyl (LP) records is described in Ref. 308.

(vi) Turntables and driving mechanism

Turntables should be strongly constructed, free from flexing, and should run true. The typical home-type turntable and motor is unsatisfactory for good fidelity on account of insufficient motor power, insufficient flywheel action and insufficient mechanical rigidity ; it suffers from rumble, wow and (with some pickups) hum due to the proximity of the motor. The motor should be as far as possible from an electro-magnetic pickup. Motor and turntable vibrations are transmitted through the mounting board and pickup arm to the pickup head and are finally reproduced in the loud-speaker as rumble. Steel turntables may be used without bad effects with crystal pickups and high level moving iron types, except that in the latter case the turntable should be solid, without perforations. Steel turntables are unsatisfactory with some low-level pickups, particularly moving coil types.

Either rim or centre drive may be used, and first class products using both methods are available. In the moderate price class, it seems that (for the same price) the rim drive is generally the more satisfactory arrangement.

The specification for a motor and turntable unit should include—

1. Type of motor.
2. Voltage and frequency range.
3. Power consumption.
4. Torque to brake the turntable from the nominal speed to a lower speed (e.g. 78 to 77 r.p.m.), the torque being expressed in oz-ins and the applied voltage and frequency to be stated.
5. Turntable diameter.
6. Form of automatic stop.

(vii) Automatic record changers

With most record changers it is important to adjust the vertical angle between the stylus and the record so that the deviation from normal is the same (although opposite in direction) at both the top and bottom of a stack of records.

Care should be taken to minimize the impact of the stylus on to the record surface when the arm is released by the changer mechanism.

There are three types of trip mechanisms, one being the velocity trip in which the changer mechanism is triggered by a sudden change in groove pitch. This system has

*Measured with flat frequency response.

the advantage of not requiring adjustments for different inside diameters. It has the disadvantage of requiring, in most designs, that the stylus, cartridge and arm operate a spring tension as they move inwards. This presses the stylus against the outside wall of the grooves and tends to produce differential wear as well as other undesirable effects.

The second type of trip mechanism operates when the stylus reaches a specific inside diameter, and the eccentricity of the inside groove has no effect. The disadvantage of this arrangement is that there are large differences in the inside diameters of the various types of records, so that an adjustable (preferably continuously-variable) control is required. This method is used in 45 r.p.m. record changers, where there is no eccentric inside groove.

The third type is the eccentric groove.

Difficulty is often experienced in an attempt to incorporate automatic record changers in high fidelity equipment owing to rumble and, in some cases, hum pickup. In addition, some high fidelity pickups are unsuitable for use on any automatic record changers. The choice of a pickup for use with an automatic record changer is neces- sarily a compromise, and many high fidelity enthusiasts prefer to use a good quality turntable with manual operation.

Automatic record changers for 45 r.p.m.

The R.C.A. record changer is described in Ref. 263. The tripping mechanism requires an extremely small lateral force from the pickup arm, since the work of put- ting the mechanism into cycle is supplied by the moving turntable.

SECTION 2 : DISCS AND STYLI

(*i*) *General information on discs*   (*ii*) *Dimensions of records and grooves*   (*iii*) *Styli* (*iv*) *Pinch effect*   (*v*) *Radius compensation*   (*vi*) *Record and stylus wear.*

(i) General information on dics

· **Shellac** discs are used only with 78 r.p.m. standard groove records.   The material is hard, and has no appreciable elastic deformation under pressure.

Vinyl* is used for all forms of fine groove discs, and has appreciable elastic de- formation under pressure. As a consequence, high peak accelerations are reduced as compared with the same recording in shellac. For the effects of elastic deformation see Ref. 212. Vinyl has the defect that it becomes electrified and collects a consider- able amount of dust. The effect of dust collection on vinyl records through electro- static attraction may be avoided by the use of a liquid anti-static agent which may be sprayed or wiped on to the surface of the record.

(ii) Dimensions of records and grooves

(A) 78 r.p.m.

The usual nominal sizes of shellac pressings are 10 in. and 12 in. diameter. Dimen- sions and tolerances as recommended by the American Radio Manufacturers' Associa- tion and as used by English and American manufacturers are tabulated opposite.

European continental records do not differ in any significant respect from English practice. The outside diameters are 25 cm (9.85 in.) and 30 cm. (11.8 in.) and the hole diameter is 0.284 inch.

On the basis of these dimensions, the ratio of maximum to minimum R.M.A. groove diameter is 3.07 to 1 for a 12 inch record, providing a groove speed from approximately 3.9 to 1.3 feet per second (47 to 15.3 inches per second).

---

*The name vinyl is a general name covering unfilled vinyl co-polymer resins, which may vary somewhat in their physical characteristics.

DIMENSIONS OF RECORDS AND GROOVES (78 r.p.m.).

	U.S.A. (in inches)	English*	
		Present range	Recommended
Outer diameter : 12 in.	11⅞ ± 1/32†	11-15/16 — 11-27/32	11⅞ — 11-27/32
10 in.	9⅞ ± 1/32†	9-15/16 — 9-27/32	9⅞ — 9-27/32
Thickness : 12 in.	0.090 ± 0.010†	0.092 — .082	0.090 — .080
10 in.	0.080 ± 0.010†	0.080 — .070	0.080 — .070
Centre hole diameter	0.286 + 0.001† / − 0.002	0.292 — 0.284	0.2845 — .2835
Centre pin diameter	.02835 ± .0005	0.282 — 0.280	0.2825 — .2820
Outermost groove diameter : 12 in.	11¼ ± 0.02†	11.500 — 11.395	11.500 ± 0.015
10 in.	9¼ ± 0.02†	9.515 — 9.485	9.500 ± 0.015
Innermost groove diameter	not less than 3¾†	3.875 — 3.750	3.875
Grooves per inch	88, 96, 104, 112, 120‡	103 — 72 (84 to 96 usual on classical recordings)	104 — 72
Width of groove at top	6.5 — 8 mils‡	6.5 — 8 mils	6.5 — 6.7 mils
Depth of groove	not standardized	—	2.9 mils
Radius at bottom of groove	2.2 max. mils‡	0.8 — 2.5 mils	1.5 — 1.8 mils
Angle of groove	87° — 97°‡	82° — 98°	85° — 90°

N.B. 1 mil = 0.001 inch.

*Based on Ref. 235 for most dimensions; Refs. 131, 174 for recommended groove cross-sectional dimensions.
†R.M.A. REC-103 (Ref. 260).   ‡R.C.A. Victor practice

Additional data (REC-103)
Eccentric stopping groove diameter 3⅜ in.; run-out relative centre hole 0.250 ± 0.015 in.; groove minimum depth 0.003 in.
Lead-in spiral: at least 1 complete turn between outer edge of record and recording pitch.
Lead-out spiral: Nominal pitch ⅛ in. minimum.

The wavelength for 78 r.p.m. is given by

$$\lambda = (1.3\pi d)/f \qquad (5)$$

where λ = wavelength in inches for 78 r.p.m.
$\quad d$ = groove diameter in inches
and $\quad f$ = frequency in c/s.

Equation (5) gives the following values—

Frequency	50	250	500	10 000 c/s
Wavelength—outermost groove	0.94	0.19	0.094	0.0047 in.
Wavelength—innermost groove	0.31	0.061	0.031	0.0015 in.

The effective width of the stylus with dimensions as in Fig. 17.5B is about 4 mils, giving a radius of curvature of 2 mils. Using eqn. (3) of Sect. 1,

$$A_{crit} = 0.025\ \lambda^2/0.002$$

and for 10 000 c/s the critical amplitude is—

$A_{crit} = 0.000\ 28$ in. for outermost groove
$A_{crit} = 0.000\ 028$ in for innermost groove.

The critical velocity is given by eqn. (4) of Sect. 1,

Critical velocity r.m.s. = 31.6 cm/sec at outermost groove
        = 3.16 cm/sec at innermost groove.

The critical velocity at the outermost groove is more than sufficient for the highest level of recording, but as the groove diameter decreases a point will be reached where the peak recorded velocity is greater than the critical velocity—thus leading to distortion and loss of high frequency response.

A slightly higher critical velocity could be obtained by the use of a somewhat smaller stylus radius, since the two values are inversely proportional.

The maximum instantaneous peak recorded velocity is about 15 cm/sec (American practice) so that there is every likelihood that the critical velocity will sometimes be exceeded on the smaller diameter grooves. The saving feature in practice is that maximum amplitude does not normally occur at 10 000 c/s with either speech or music, except when cymbals are recorded at maximum amplitude.

(B) 45 r.p.m.*

The R.C.A. Victor 45 r.p.m. fine groove records are only made with a nominal diameter of 7 inches, and are primarily intended for ready use in record changers. The inner portion of the record forms a collar which is thicker than the playing area, thus preventing the playing surfaces from touching. The groove dimensions are

Width across shoulders 3.0 mils + 0 − 0.5
Angle 92° ± 3°
Radius at bottom of groove not greater than 0.25 mil.

The recommended vertical stylus force is 5 ± 1 grams.

The grooves per inch vary from 178 to 274. The maximum instantaneous programme peak recording velocity is 14 cm/second (about 4 db below 78 r.p.m. records).

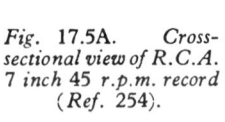

Fig. 17.5A. Cross-sectional view of R.C.A. 7 inch 45 r.p.m. record (Ref. 254).

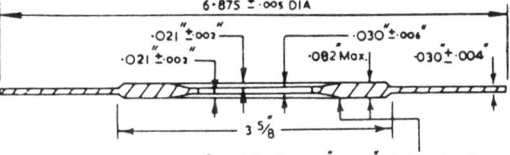

FIG. 17.5A

The cross-sectional view of a record is shown in Fig. 17.5A. The outermost music groove is 6⅝ inch diameter, while the innermost music groove is 4.875 inches minimum diameter—the latter gives 10% intermodulation distortion and a groove velocity of 11.5 inches per second.

The maximum playing time is 5¼ minutes. A lead-in groove extends from the edge of the record and makes from 1 to 1¼ turns before entering the first music pitch. There are from 1 to 1¼ turns of unmodulated music pitch before the start of music. There is from 0 to ½ turn of unmodulated music pitch at the end of music. The

*For R.C.A. 45 r.p.m. Extended Play records see Supplement.

lead-out groove makes from $\frac{3}{4}$ to $1\frac{1}{4}$ turns from the music pitch to 1-15/16 inch radius, beyond which it makes from 1 to 2 turns to the concentric circle (3-13/16 inch diameter).

The central hole is 1.504 $\pm$ 0.002 inches in diameter.

References 254, 263, 264, 265.

(C) 33-1/3 r.p.m. (LP)

The long-playing microgroove records revolve at 33-1/3 r.p.m. and are made with outside diameters of 12, 10 and 7* inches.

The R.C.A. Victor records are identical to the 45 r.p.m. records with respect to groove shape, recording pitch, recorded level and recommended playback stylus.

Dimension	Columbia	R.C.A. Victor
Groove angle	87° ± 3°	92° ± 3°
Radius at bottom of groove less than	0.2 mil	0.25 mil
Width across shoulders of groove	—	0.0025 in.
Maximum amplitude	0.9 mil	—
Max. instantaneous programme peak velocity	—	14 cm/sec
Lead-in groove pitch	—	35 to 40 lines/in.
Blank grooves (normal pitch)	—	1 to 2
Grooves per inch (normal pitch)	200 to 300	178 to 274
Grooves per inch (usual values)	224, 260	—
Lead-out groove	—	4 grooves/in.
Thickness	—	0.070 to 0.085 in.
Thickness depressed label area	—	0.060 in. min.
Maximum diameter (12 in.)	—	11-7/8 ± 1/32 in.
(10 in.)	—	9-7/8 ± 1/32 in.
Outer music groove dia. (12 in.)	—	11-15/32 $^{+\ 0}_{-\ 3/64}$ in.
(10 in.)	—	9-15/32 $^{+\ 0}_{-\ 3/64}$ in.
Inner music groove dia.	4-3/4	4-3/4 in.
Nomin. dia of eccentric groove	—	4-7/16 in.
Eccentric groove off-centre	—	0.125 ± 0.008 in.
Centre hole	—	0.286 in.
Label diameter	—	4-1/16 in.
Outer edge, included angle	—	80°
edge radius	—	1/64 in.
Maximum playing time 12 in.	22-1/2	— mins.
10 in.	15	— mins.
7 in.	5-1/2	— mins.
Optimum stylus force	{ 6 0.21	5 ± 1 grams. 0.18 ounce

References to L-P microgroove : 159, 166, 178, 179, 232, 236, 255, 262.

(iii) Styli

(A) Styli used with shellac pressings are in five principal groups :

(1) **Ordinary steel needles** which are ground to shape by the abrasive in the record.   The exact shape of the needle point when new is not so important as with permanent or semi-permanent styli.   For examples of record wear see Ref. 10 Part 4.

(2) **Semi-permanent needles** such as chromium plated steel. The needle shape is important and " shadow-graph " needles are recommended.   For the best results these should not be used for more than one playing of a 12 inch shellac record (for photograph of wear see Ref. 274).   Alloys such as osmium are also used.

*Columbia only.

(3) "**Permanent**" styli employ a jewel such as a diamond, sapphire or ruby, usually in the form of a jewel tip. The tips of permanent needles should be accurately ground to shape, and highly polished. Diamond tipped needles have a life of several thousands of playings of shellac records for good fidelity. Tungsten carbide is also used. For stylus wear see Sect. 2(vi).

(4) **Fibre needles** are used by some enthusiasts in the belief that they reduce scratch and record wear. While it is true that they reduce scratch they do so by attenuating all the higher frequencies, and they do this no more efficiently than an electrical attenuator. The attenuation of a fibre needle at 4000 c/s is 19 db down as compared with a loud tone steel needle in a typical pickup (Ref. 7). After the first few grooves they wear sufficiently to occupy the whole of the groove and thereby spread the weight of a heavy, stiff pickup of old design. If such pickups are used, contrary to all good advice, then fibre needles are possibly the best compromise because a soft needle damps down the pronounced high-frequency resonance of the pickup. There is evidence to indicate that abrasive particles from the record become embedded in the fibre and thereby cause wear, even though the fibre itself is softer than the record. With modern pickups having light weight and high lateral and vertical compliance the wear from a well polished sapphire point is undoubtedly less than that from a fibre needle.

For photographs of new and worn fibre needles see Ref. 274.

(5) **Thorn needles** are very much harder than fibre, but the amount of needle wear is very much dependent upon the pickup compliance and stylus force. With suitable lightweight pickups, the needle wear is reasonable up to 6 playings (Ref 309), but some loss of high frequency response is inevitable   If sharpened with the use of very fine glasspaper, there is risk of glass dust becoming embedded in the needle—this danger may be avoided by the use of a very fine rotary cutting wheel (Ref. 309, April 1951).

Since thorn needles are normally in contact with the whole of the bottom of the groove, they will tend to remove all dust from the groove. It therefore seems probable that the noise from a record which has been played exclusively with thorn needles will be no greater at the bottom of the groove than on the sides.

Fine point thorn needles have also been used with fine groove (LP) records (Ref. 309).

The shape of the stylus tip to give optimum results with most English and American recordings may be taken as (RTMA Standard REC-126-A, Ref. 260) :

	Radius of tip	Angle
Metal point	2.7 (+ 0.2 - 0.3) mils	40° to 50°
Sapphire	3.0 (+ 0.2 - 0.3) mils	40° to 50°

A cross-sectional view of a correct size of sapphire stylus point in a typical groove is given in Fig. 17.5B. It will be seen that the stylus does not reach to the bottom of the groove ; this is very important because the abrasive dust collects at the bottom of the groove. In addition, there should be some allowance for the stylus to wear on the sides without the bottom of the stylus coming too close to the bottom of the groove.

FIG. 17.5B

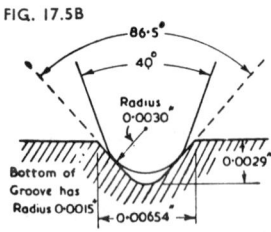

Fig. 17.5B. Cross-sectional view of typical record groove and needle tip.

In some modern records, as well as in transcription discs, a tip radius of 2 mils may be used, but in other (and particularly older) recordings it is likely to scrape the groove bottom and suffer from " groove skating " and single point contact.

One interesting suggestion is to combine a 2 mil radius with a " flattened " extremity (actually a 3.5 mil radius). Although more expensive to produce it seems to have distinct advantages (Ref. 244).

Oval sapphire styli are available with a minor axis of 1.5 to 2.0 mils and a major axis of 2.5 to 3.5 mils. These should be used with the major axis at right angles to the centre-line of the groove, and will permit better response to high-amplitude high-frequency recordings.

The N.A.B. " secondary standard " stylus (permanent point) has an angle of 40° to 55° and a bottom radius of 2.5 ± 0.1 mils. It provides a compromise suitable for the reproduction of both transcriptions and shellac discs.

The use of larger styli (radius 4 mils) is advocated by some engineers (Refs. 223, 225) but it does not appear to be the best all-round compromise.

Needles are made in various sizes. The ordinary steel needles used in mechanical reproducers are used in the stiff heavy weight pickups whose frequency response does not exceed about 6000 c/s ; they are also used in some " needle armature " pickups. For the best high frequency response the " loud tone " or " full tone " needles should be used, because of their smaller size and weight. " Soft tone " and " trailing " needles (as used with acetate discs) cause severe treble attenuation when used with typical pickups. Most light weight pickups with a frequency response extending to 10 000 c/s or over use miniature needles to enable the armature resonance to be at a high frequency. A typical example is the H.M.V. Silent Stylus or Columbia 99 (chromium-plated long-playing). Thorn needles are also made of similar size.

Efforts have been made to develop a stylus tip suitable for both 78 r.p.m and fine groove (see Refs. 279, 281) but a considerable degree of compromise is necessary and the best results are not obtainable from either type of recording.

(B) Styli for fine groove records

With a few exceptions, all styli for fine groove records are permanent types—usually jewels. Diamond tips are the only really satisfactory ones for long life, although expensive to purchase in the first case. Sapphire tips are very common, but wear rather rapidly—see Sect. 2(vi) for stylus wear. Other materials used are tungsten carbide, osmium and other metal alloys.

Dimensions of styli for fine groove records

(RTMA Standard REC-126-A, Ref. 260, for home phonographs)

Radius of tip   0.001 + 0.0001 − 0.0002 inch

Included angle of tip   40° to 50°.

Note : With a 90° groove angle, a tip with a radius of 0.001 inch has an effective radius of 0.0007 at the point of contact.

(C) Colour codes for styli

1. RTMA REC-126-A for home phonographs (Ref. 260). Needles with a 0.001 inch radius shall be colour coded red.

2. English Gramophone Equipment Panel of the Radio and Electronic Component Manufacturers' Federation (Jan. 1951). Red—0.0010 in. ; lemon—0.0020 in. ; green—0.0025 in. ; french blue—0.0030 in. ; orange—0.0035 in ; violet—universal ; sky blue —oval tip. Material (marked by band on shaft) black—hard metal ; white—diamond ; no colour—sapphire

(iv) Pinch effect

Owing to the fact that the cutting is done by a stylus having an effectively flat cutting face, the width of the groove measured at right angles to the groove is narrower at two places in each cycle (see Sect. 6). As a result of this effect, the stylus tip should rise and fall twice in each cycle —only a limited number of pickups, however, make adequate provision for this movement.

The flexibility of the stylus, if mounted at an acute angle to the record, together with that of the record itself, tends to prevent the needle from riding merely the peaks of the vertical undulations. The pinch effect undoubtedly increases needle wear and possibly also record wear (Ref. 212).

Various forms of " bent shank " needles have been developed to provide some vertical compliance. All " bent shank " needles tend to give a drooping high-frequency response the amount of which varies considerably from one make to another ; it may, of course, be compensated if desired.

(v) Radius compensation

Radius compensation does not appear to be used with shellac discs, although it is used with LP discs (R.C A. Victor) and in transcription discs.

(vi) Record and stylus wear

Record wear is a complicated effect depending upon many conditions of operation including :

1. Vertical stylus force.
2. Lateral mechanical compliance at stylus tip.
3. Vertical mechanical compliance at stylus tip.
4. Lateral and vertical pivot friction in pickup arm, or lateral spring tension.
5. Mechanical resonances of pickup and arm.
6. Tracking error.
7. Shape, material and polish of stylus tip.
8. Maximum transverse velocity of record grooves.
9. Record material.
10. Dust.

It is understood that the turntable is free from vibration and wobble and that the record is neither eccentric nor warped ; if not, these are additional causes of record and stylus wear.

The **vertical stylus force** for use on shellac records should preferably not exceed about 1 ounce weight (28.35 grams) and it should be as light as possible to reduce record and stylus wear to the minimum. However care should be taken to provide sufficient force to ensure satisfactory tracking, otherwise the wear will increase. Increased force, say up to 2 ounces (about 60 grams), causes very little additional record wear provided that a well-polished jewel point is used and the pickup has high lateral compliance.

The stylus force with fine groove records is quite critical, but is a function of the pickup. Forces approximating 6 grams weight are common, but it is safer to err slightly on the high side than on the low side. Some manufacturers give a " minimum tracking weight " and a " normal " value—the latter is the correct one to use.

If the required vertical pressure is not known, it may be determined by using an intermodulation test record, decreasing the pressure until the distortion commences to rise—and then increasing by a margin of, say, 50%.

High lateral compliance at the stylus tip is one of the most important requirements for low record and stylus wear. In general, high lateral compliance and light weight go together.'

Vertical compliance at the stylus tip permits the tip to maintain contact with the groove in spite of the " pinch effect." Bent shank styli reduce record wear in cases where the pickup has insufficient vertical compliance.

Mechanical resonances of the pickup are serious causes of record and stylus wear. Both the " arm resonance " (if it comes within the recorded frequency range) and the " armature resonance " affect the record and stylus wear, even if electrically equalized.

Tracking error is discussed in Sect. 4 ; if reduced to the usual low values, record and stylus wear from this cause are very slight.

The **shape and polish** of a jewel stylus tip are of the utmost importance in regard to record wear.

Wear of shellac records

One example is a record which had been played 1000 times with a 1.35 ounce (38 grams) pickup and a sapphire stylus and was then demonstrated as giving fidelity virtually unaffected (Refs. 128, 131). It may be stated as a rough guide that the limit of life of an average shellac record is about 4000 playings under good conditions.

Unmodulated grooves on shellac records will wear indefinitely with a jewel stylus on which the vertical force does not exceed 1 ounce (30 grams). By far the most significant cause of wear is the dynamic action of the stylus in the grooves. If the dynamic force at any moment exceeds the elastic limit of the record material, the shellac crumbles. This permits the stylus to drop down a little deeper in the groove, eventually increasing the bearing area of the needle by contact with the bottom of the groove. At this point with a pickup having low stylus force, the wear apparently stops. This wear does not produce any apparent change in quality of the reproduced sound, although it can be seen as a light streak on the surface of the record. Both lateral and vertical forces must be considered in producing the wear (Ref. 155). Wherever there is excessive stylus wear, it is always accompanied by excessive record wear.

It is important to distinguish between visible wear and audible wear. Visible wear does not necessarily cause any audible defect. It is possible that visible wear may occur earlier with a diamond tip, but audible wear, which is caused by actual widening of the groove at corners, will occur far more rapidly with a worn sapphire or metal tip. The flats on the sides act as scrapers in attempting to negotiate sharp bends in the record groove. A well polished spherical diamond or sapphire tip is believed to cause less record wear than a metal alloy tip.

Wear of vinyl records

Wear of vinyl fine groove records is very slight, provided that the stylus force does not exceed about 7 grams, that a pickup of ample lateral and vertical compliance is used, that the jewel tip is well polished, that there are no marked mechanical resonances, and that dust is excluded. **Dust** is the most important cause of record and stylus wear with vinyl records in the home, and care should be taken to reduce it to a minimum—see Sect. 2(i). Wear in vinyl records shows itself principally in widening of the groove thus leading eventually to distortion and rattles (Ref. 155).

Wear of sapphire styli

The wear of sapphire styli is a function of the material of which the pressings are made, the lateral and vertical compliance of the pickup at the stylus tip, the dynamic mass of the pickup at the stylus tip, the characteristics of the armature (high frequency) resonance and the pickup arm (low frequency) resonance, and the stylus force. Additional wear may be caused by " skating " (due to the use of a tip radius which is too small for the groove), by insufficient vertical force to maintain the stylus always in contact with the groove, and by warped or eccentric records or turntables. When using a pickup with high lateral and vertical compliance and low dynamic mass at the stylus tip, an increase of stylus force has only a minor effect in increasing the wear on the stylus tip. It has been shown (Ref. 282) that with such a pickup (GP20) on a heavily modulated **shellac record,** an increase from 7.5 to 14.5 grams in stylus force causes increased flats on the stylus tip resulting in a drop of only 3 db at 10 000 c/s on a test record. Such a stylus and pickup may be used for 800 playings of a 12 inch shellac record with reasonably good fidelity, or 2000 playings with fair fidelity ($-$ 5 db at 10 000 c/s on a test record).

On the other hand a typical light-weight pickup with a stylus force of 1 ounce (28 grams) has been shown to produce wear on sapphire styli on shellac records as follows (Ref. 59).

Wear just noticeable after 50 playings of 12 inch disc.

1 mil flats after 200 playings   ⎫
2 mil flats after 750 playings   ⎬ The effect of flats on distortion is covered
2.5 mil flats after 1500 playings ⎭ in Sect. 6(iv).

It is obvious that this pickup produces much greater wear than the one with high lateral and vertical compliance (GP20) described earlier. It is probable that the greater wear is due principally to the lower lateral and vertical compliance, although the increased stylus force would also be a contributing factor.

Some excellent photographs of stylus wear are given in Refs. 274 and 290.

The use of a test record at, say, 10 000 c/s to indicate wear of the stylus tip requires careful interpretation of the results. The level of recording of the test record is lower than that which may be reached with music, while the effect of wear of the stylus tip is shown less prominently towards the outer diameter of the record than it would be with a smaller groove diameter.

It is wise to avoid using sapphire or other permanent stylus tips on records which have been played previously with steel needles. For a photograph of wear after 20 playings under such conditions see Ref. 274.

Stylus wear with fine groove records
Apparently stylus wear is far more rapid on microgroove than on standard groove vinyl records ; a rough estimate is about three times faster (Ref. 290). For this reason, diamond styli are much to be preferred to sapphire or osmium, even for home use, where high fidelity is required.

In one case, when tested on 12 inch long playing records, with a GE RPX-041 pickup having 8 grams pressure, a diamond stylus showed a slight flat after 37 hours playing time, whereas a sapphire under similar conditions was badly worn after 5¾ hours (15 playings)—see photographs Ref. 290.

There appear to be considerable variations on the hardness of the sapphires, which are practically all made of synthetic sapphire (Ref. 302), so that it is impossible to quote any figures of stylus wear which can be regarded as typical. In all cases the stylus life is very much affected by the pickup.

Osmium tipped styli have a life only two fifths (Ref. 290) or one sixth (Shure) that of sapphire. This makes the effective life with an osmium tipped stylus, for high fidelity, extremely short.

Method for giving positive indication of stylus condition
Obtain a lacquer disc with unmodulated grooves cut on both sides of it—one side for a 3 mil stylus and the other for a 1 mil stylus. Both types of grooves may be on the one side of the disc. Whenever in doubt, play two or three grooves on this disc If the stylus leaves the grooves unchanged in lustre and smoothness, it is in good condition. If, however, the groove walls show score marks or any other difference when compared with the unplayed grooves, the stylus needs replacement. An ordinary magnifying glass will be of assistance (Ref. 298).

SECTION 3 : PICKUPS

(i) *General survey* (ii) *Electro-magnetic (moving iron) pickups* (iii) *Dynamic (moving coil) pickups* (iv) *Piezo-electric (crystal) pickups* (v) *Magnetostriction pickups* (vi) *Strain-sensitive pickups* (vii) *Ribbon pickups* (viii) *Capacitance pickups* (ix) *Eddy-current pickups.*

(i) General survey
The performance expected from good quality pickups includes—
1. Wide frequency response with the minimum of equalizing. No sharp peaks over + 2 db. For most applications a slight droop in the high frequency response is not undesirable as it may be used to compensate for the pre-emphasis in most recordings.
2. Low stylus pressure.
3. High lateral compliance.
4. High vertical compliance at stylus tip.
5. Low effective vibrating mass of armature at high frequencies.
6. Fairly high output level.
7. Low distortion.
8. Freedom from major resonances over the useful frequency range (see below). The low frequency (arm) resonance should be below 30 c/s so that it will have a negligible effect on record wear. The high frequency (armature) resonance should be as high as possible—certainly over 8000 c/s—and preferably above the useful frequency range. Even so, the resonance should be effectively damped.
9. Minimum " needle talk " (direct acoustical radiation).
10. High signal to hum ratio.
11. Negligibly small voltage generation from vertical movement of the stylus tip.
12. General ruggedness.
13. Freedom from the effects of excessive humidity. Pickups having a restricted frequency range (say from 70 to 6000 c/s) should meet the same requirements as good quality pickups, apart from frequency.

Pickups which have sufficient vertical compliance are sometimes fitted with needle guards to reduce or prevent damage to the pickup from accidentally dropping it.

Pickup manufacturers sometimes quote the minimum vertical stylus force to provide perfect tracking. It is generally regarding as sound practice to adopt a stylus force at least 1.5 times this minimum value, thus making some provision for warped and eccentric records.

For styli, see Sect. 2(iii).

Single purpose pickups for home use often incorporate a fixed spring counterbalance. Dual purpose pickups usually apply a higher stylus pressure for 78 r.p.m. than for fine groove records. In some cases the pickup head (or cartridge) includes the additional weight for 78 r.p.m., in other cases this is provided by an adjustment of spring tension or by a counterbalancing weight.

It is desirable in all cases to keep the moment of inertia about both vertical and horizontal pivots to a minimum.

Where a spring counterbalance is used, it should be checked periodically for pressure ; during weighing, the height of the stylus from the baseboard must be exactly the same as the top of the record.

Up and down movements of the stylus tip, due to the pinch effect and small record surface irregularities should be absorbed by the vertical compliance of the stylus, without generating any output noise.

Pickups for 78 r.p.m.

A stylus force not more than 1 ounce (say 30 grams) is desirable, with 2 ounces as the upper limit.

One of the difficult problems with permanent tipped styli is provision for changing needles without adding seriously to the armature mass. Certain pickups are designed for use with miniature needles pushed into place and gripped by a wedge against which they are held by the pressure from the record. There are two troubles which may occur. The first is that the needle may rotate about its axis, thus placing its " flats " at some unknown angle to the groove—such a case is known to the author. This may be eliminated by tightening the clutch or by using a little cement. The second trouble is " buzz," which is more serious because there appears to be no cure ; however it only seems to occur on deeply modulated recordings of complex waveform, and is only apparent when the frequency range is extended.

Nearly all high fidelity pickups have the stylus permanently fixed to the armature— this appears to be essential for the high frequency resonance to occur above 10 000 c/s.

Pickups for fine groove records

The maximum desirable needle pressure is 6 grams (0.21 ounce) and high lateral compliance is essential. Pickups have been produced by many manufacturers, mainly using diamond, sapphire or osmium tips.

Careful pickup arm design, specifically for fine groove recording, is of paramount importance. It is inadvisable to use a fine groove pickup head or cartridge on an arm other than that designed for it.

A very simple and effective test for the tracking efficiency of a pickup and arm for fine groove records is known as the **McProud Test** (Ref. 314). A 45 r.p.m. record is placed on a standard turntable running at 45 r.p.m. with the maximum possible eccentricity (1¼ inches swing), and the pickup is required to track the record satisfactorily.

Dual purpose pickups have been designed which have two opposed needles 180° apart, suitable for playing either standard 78 r.p.m. or fine groove records.

Pickups employ many principles of operation, but most types are basically either constant velocity or constant amplitude. For example both electro-magnetic (moving iron) and dynamic (moving coil) pickups are basically constant velocity, while piezoelectric (crystal) types are basically constant amplitude. Since the recording characteristic includes both constant velocity and constant amplitude sections, either group is satisfactory although requiring different equalization.

Resonances in pickups

Most pickups have two major resonances. The **arm resonance** occurs at some frequency below 100 c/s, while the armature resonance occurs at some frequency about 3000 c/s. Both resonances have deleterious effects in increasing record wear. The arm resonance tends to cause the pickup to jump out of the groove when a high amplitude is recorded at the frequency of resonance—the frequency of resonance (if undamped) should be well below the lowest frequency for which acceptable tracking is required ; say from 15 to 25 c/s.

A recent development is the use of pickup arms with high viscosity oil or other viscous fluid to damp arm resonance. The damping may be applied as a viscous film in a ball and socket (hemispherical) joint between the pickup arm and the mounting socket, and gives damping in both horizontal and vertical directions. The arm is pivoted at a point which is the centre of both ball and socket, and above the centre of gravity of the arm. The clearance between the two surfaces is about 0.006 inch. By this means the resonant force is greatly reduced, improved resistance to external shock is obtained and protection against damage from accidental dropping of the pickup head is achieved. The amount of mechanical resistance is not a critical value the upper limit is reached when it interferes with the tracking of records having reasonably small values of eccentricity or warpage. With the usual values of suspension compliance and mass, this would occur at several times the amount of resistance necessary to give critical damping of the arm resonance. Even if the damping is considerably below critical, it still has quite a beneficial effect. The viscosity of the fluid is a function of temperature, but the variation does not seriously affect the performance over a reasonable temperature range. Ref. 311.

The **armature resonance** has a serious effect on needle scratch (see Sect. 1). " During operation the needle is subjected to a continuous shower of blows, and although highly damped by rubber buffers, it is in a state of perpetual oscillation at its own resonance frequency " (Ref. 229). In any quality pickup, the armature resonance with the recommended needle should be over 8000 c/s ; many light-weight pickups have the resonance at a frequency over 12 000 c/s—in some cases over 18 000 c/s.

Pickups tend to be rich in harmonics at frequencies in the region of half the armature resonance frequency. In all cases the armature resonance should be effectively damped. Damping material used to obtain smooth response characteristics may adversely affect the tracking capabilities of the pickup, and therefore should be investigated carefully and used judiciously (Ref. 285).

The frequency of the armature resonance is affected by the record material—for example, a change from a lacquer disc to a shellac pressing causes in one case an increase in the resonant frequency in the ratio 1: 2 approximately (Ref. 276). A pickup with an armature resonance well outside the a-f range on shellac discs, shows a much lower resonance frequency on vinyl—usually inside the a-f range. This is an unfortunate characteristic of vinyl. The only pickups known to the author with armature resonances above 15 000 c/s on vinyl fine groove recordings are of the ribbon armature type.

There is a third resonance caused by the pickup head and arm rotating about the arm. With heavy pickups this resonance occurs at frequencies between 100 and 400 c/s, but with light-weight pickups the frequency would be higher. This resonance does not have any effect on the general tonal balance, and in any case should be very slight with good mechanical design.

Crystal pickups have a fourth resonance which occurs at a high frequency—the resonance of the crystal itself.

References to resonances in pickups : 10 (Part 2), 17 (second letter), 276, 283, 285, 311, 331.

The testing of pickups

The testing of pickups for frequency response characteristics is covered by R.M.A. REC-125-A which states that the test record shall be R.M.A. Frequency Test Record

No. 1 when available. This is recorded on side A at a r.m.s. velocity of 3.16 cm/sec (+ 10 db). Crystal pickups are terminated by a load resistance of 1 megohm (or 5 megohms for ammonium phosphate crystals) shunted by a capacitance of 100 $\mu\mu$F. Other types of pickups should be terminated as required. The response should be stated in decibels (0 db = 1 volt).

Pickups should be tested under the same conditions with which they will operate in normal service. Pickups intended to be used on shellac records should be tested on shellac discs ; pickups intended for use on vinyl records should be tested on vinyl discs, while those intended for use on either type of disc should be tested separately on each. Pickups which are satisfactory on vinyl discs give, in some cases, quite poor results on shellac discs owing to the increased stiffness of the material and the reduction in damping by the record. This also applies to pickups with very limited vertical compliance—while passable on vinyl, they are poor on shellac.

The variable speed turntable is sometimes used as an alternative to the standard frequency records for the calibration of pickups—see Refs. 273, 295.

In the design and production of pickups it is usual to supply some form of electro-mechanical calibrator such as in Ref. 284.

For distortion in pickups and the procedure for determining the " tracking " capabilities of a pickup see Sect. 6(vi).

(ii) Electro-magnetic (moving iron) pickups

All " moving iron " pickups have a steady field supplied by a permanent magnet, a coil wound over the magnetic circuit, and an iron armature in the magnet gap. The old type of heavy pickup of this type used with shellac records weighed about $3\frac{1}{2}$ ounces (say 100 grams) and had low lateral compliance. It will not be further described.

The more recent light-weight models are capable of good fidelity, and some models are among the best pickups available. The increased fidelity and higher compliance are obtained at the cost of lower output level—very much lower in some cases.

Electro-magnetic pickups may arbitrarily be divided into three groups, based on the output voltage developed across the pickup with recorded velocity 3.16 cm/sec (+ 10 db).

(a) High level—output voltage above 100 millivolts.

(b) Medium level—output voltage from 20 to 100 millivolts.

(c) Low level—output voltage below 20 millivolts.

Some typical representatives in each group are described below. In most cases the descriptions are based on published information supplied by the manufacturer. The inclusion of certain models should not be taken as indicating their superiority over types not included.

(a) High level pickups
Decca ffrr pickup type D (English) — standard groove

This pickup is of the needle armature type, having a hollow armature fitted with a sapphire, tungsten carbide or diamond stylus. The armature suspension and damping are of rubber ; the stylus, armature and rubber suspension are replaceable as one unit. The low frequency resonance is 25 c/s, while the armature resonance is about 17 000 c/s. The impedance of the pickup is 4200 ohms at 1000 c/s and the output 0.2 volt at the same frequency ; lower impedances are also available. The frequency response corresponds to the velocity characteristic of the ffrr record within $\pm$ 1 db. The whole armature including stylus and damping blocks is replaceable by the user. The stylus tip is available in both round and oval shapes.

Lateral compliance 1.1×10^{-6} cm/dyne

Impedance at arm resonance frequency (approx.)   17 000 dynes/sec./cm.
Distortion in pickup   less than 2%
Stylus force   22 to 24 grams (0.78 to
                                                   0.85 ounce).
Load resistance   100 000 ohms

(b) Medium and low level pickups

Connoisseur Super Lightweight pickup (English)—standard and microgroove

This is a high-fidelity pickup with interchangeable heads. Heads are available with 1 mil, 2.5 mil and 3 mil radii sapphire tips. The frequency response is level ± 2 db from 25 to 15 000 c/s. The 25 ohm model gives 10 millivolts output, or 300 millivolts from a 1 : 50 transformer. The 400 ohm model gives 40 millivolts, or 200 millivolts from a 1 : 6 transformer. The armature mass is 20 milligrams, and the dynamic mass* is 0.8 milligram. The stylus force is 8-10 grams (standard) or 4-6 grams (microgroove). The pressure is automatically corrected by weights in the plug-in head.

Goldring Headmaster (English)—standard and microgroove

This is a high-fidelity pickup with interchangeable heads. Heads are available with 1 mil, 2 mil, 2.5 mil and 3.5 mil radii sapphire tips.   The frequency response with 2.5 mil needle is level to 7000 c/s, − 1 db at 10 000 c/s, + 1 db at 16 000 c/s, 0 db at 17 000 c/s and − 5 db at 20 000 c/s ; at the other end it rises smoothly from 100 to 30 c/s where it is + 3 db.   The output is 40 millivolts (from the pickup itself) and stylus force 20 grams (0.7 ounce). It has high lateral stylus compliance, low inertia, and high vertical compliance. This is a good-all-round pickup for home gramophones.

Pickering cartridges (American) standard and microgroove

Model S-120M (sapphire stylus), or D-120M (diamond), is for standard recordings, and has a 2.7 mil radius tip. A stylus force of 15 grams is sufficient for tracking.

Model S-140S (sapphire), or D-140S (diamond), is for microgroove recordings, and has a 1.0 mil radius tip. A stylus force of 6 grams is recommended.

Both models give an output of 70 millivolts with a recorded velocity of 10 cm/sec. The frequency response with a load resistance of 27 000 ohms is level from 30 to 9000 c/s, + 2 db at 20 c/s, + 2.5 db at 12 000 c/s and falling above that frequency. A lower load resistance attenuates the higher frequencies, while a higher load resistance reduces the damping on the pickup, decreases the signal-to-noise ratio, and is generally undesirable.

Constants of moving system

Lateral moment of inertia	11 mg cm^2
Lateral compliance of stylus	1.0 × 10^{-6} cm/dyne
Vertical compliance of stylus	0.2 × 10^{-6} cm/dyne

Pickering Turn-over pickup (American)--all records

This model 260 pickup may be turned over for either 78 r.p.m. or microgroove operation. It has an output of 30 millivolts at a recorded velocity of 10 cm/sec. It is available with diamond stylus only.

---

* Dynamic mass (also known as equivalent mass or effective mass) is that mass which, if concentrated at the stylus point, would possess the same inertance as that of the moving system.
See also Refs. 248, 283.

G.E. variable reluctance pickup (American)—standard and fine groove
This is a magnetic pickup having almost ideal characteristics except that the output voltage is low (10 millivolts at 4.8 cm/sec.). It has no " needle talk " and has considerable resistance to shock. It has a large degree of vertical compliance, while vertical movement does not cause any electrical output. The suspension is free from any deleterious effects such as standing waves or cross-modulation, and the construction is designed to reduce hum pickup. The frequency response is from 30 to 15,000 c/s. $\pm$ 2 db. The inductance is 520 mH for the home type, and 250 mH for the broadcast types (RPX-046 and RPX-047) which also have a lower output (8 mV). The normal stylus force is from $\frac{3}{4}$ to $1\frac{1}{4}$ oz. (1 oz. on home type) with 78 r.p.m., or 6-8 grams on microgroove.

Dynamic mass of stylus	8 milligrams
Suspension compliance (lateral)	0.87×10^{-6} cm/dyne
Load impedance, normal	6800 ohms.

The cartridge is fitted with interchangeable styli, including 1 mil, both diamond and sapphire. For a suitable pre-amplifier see Fig. 17.26.
References 93, 106, 187, 234, Catalogues.

R.C.A. Light weight pickup—transcription and fine groove
Model MI-11874 pickup, MI-11875 arm. See Refs. 296, 313 and R.C.A. catalogues.

A plug-in pickup head is used, permitting immediate change from 1.0 to 2.5 mils without necessitating any adjustment of arm balance. Stylus force is 8 grams for 1.0 mil and 12 grams for 2.5 mil. Diamond styli are used. Tone arm resonances are outside of the operating frequency range   The vertical and horizontal pivots have very low friction. Frequency response (1.0 mil) at output from filter 50 to 12 000 c/s $\pm$ 1 db, 40 c/s at $-$ 4 db. Voltage output at 1000 c/s on open circuit $=$ 11 mV with 6.1 cm/sec. test record. Output from filter $-$ 64 dbm. Test record 460625-6. Hum level $-$ 139 dbm with magnetic flux density 1 milligauss. Output pickup impedance 135 ohms at 1000 c/s. Filter output should be connected to unloaded input transformer of amplifier designed to operate from 250 ohm source. Intermodulation distortion is low to the point where it is not possible to determine accurately whether the distortion is in the record or the pickup.

E.M.I. Unipivot transcription pickup (English)—standard and microgroove
This Model 17 pickup has interchangeable styli for standard and microgroove, and the arm has an oil damping system. Its impedance is 1 ohm at 1000 c/s, and the output at the secondary of the transformer is 30 millivolts (high ratio), or 4 millivolts (600 ohms), or 2.25 millivolts (200 ohms). The frequency response is sensibly level from 30 to 12 000 c/s with standard stylus (armature resonance above 15 000 c/s). Total harmonic distortion at 400 c/s is less than 5% for a recorded level of $+$ 20 db (0 db $\doteq$ 1 cm/sec.). The stylus force is 6 grams.

Audak Polyphase (American)—two styli
This has two separate styli, each with its own vibrating armature. One stylus force (6-8 grams) is used for all records. Either sapphire or diamond styli may be used, and may readily be replaced. Output is about 20 millivolts, and frequency response 30 to 10 000 c/s, level $\pm$ 1 db from 80 c/s upwards, with gradual rise to $+$ 2 db at 30 c/s. Armature resonance is at 9000 c/s.

(iii) Dynamic (moving-coil) pickups
A dynamic pickup has a coil, with a stylus attached to cause movement of the coil, mounted in a magnetic field and operating as a generator.

With good design this construction is virtually distortionless at low frequencies, but the output voltage is very low and a high-ratio transformer is always used in conjunction with it. One possible difficulty is hum which may arise either through induction from the turntable motor to the moving coil, or through hum picked up by the transformer ; special mu-metal shielded transformers may be necessary to obviate the latter trouble.

Fairchild dynamic pickup (American)—transcription

The low frequency resonance is at 18 c/s and the armature resonance is over 12 000 c/s with only 2 db rise at the latter frequency. Tracking may be obtained on a flat and true record with a needle pressure of only 5 grams (say 0.2 ounce) but the normal needle pressure has been increased to 25 grams (say 0.9 ounce) for best performance under all conditions. The moving coil is supported by two plastic vanes, and a diamond point is used. Ref. 57.

Leak dynamic pickup (English)—home type

A diamond stylus is standard in the latest moving coil pickup developed by H. J. Leak. The height of the pickup is adjustable. The turntable should be non-magnetic. Spring loading is used without any counterweight. Gimbal mounting is used, with hardened steel pivots. This pickup cannot be used with automatic record changers. The frequency response is level $\pm$ 1 db 40-20,000 c/s, the high frequency resonance being above 27,000 c/s (21,000 $\pm$ 2000 for LP). The low frequency resonance is 20 $\pm$ 5 c/s. The output is 11 mV per cm/sec. recorded r.m.s. velocity. The playing weights are 5-6 gm. on 78 r.p.m. and 2-3 gm. on long-playing records.

(iv) **Piezo-electric** (**crystal**) **pickups**

A piezo-electric crystal is one which, when strained, produces electric charges on certain of its faces, the magnitude of these charges being directly proportional to the strain. In practice, two slabs of a suitable piezo-electric material are usually cemented together with one electrode between the slabs and another in contact with both outer faces. The slabs are cut in such a manner that a torque (in the case of a " twister " crystal) or a flexure (in the case of a " bender " crystal) will produce a potential difference across the electrodes. This assembly is known as a " bimorph," the torsional variety being used in most crystal pickups. The output from a crystal pickup is directly proportional to the amplitude of the stylus displacement.

The most commonly used crystal is sodium potassium tartrate (Rochelle salt). This is seriously affected by both high temperatures (over 125°F) and high humidity ; some units are protected from humidity by a water-tight casing.

Other materials include ammonium dihydrogen phosphate (Ref. 158) which is not affected by high temperature or humidity, and ceramic piezo materials (barium titanate etc.). The latter is described on page 721.

The equivalent circuit of a crystal pickup is a capacitance in series with a generator of zero impedance. At 1000 c/s the impedance of many typical pickups is about 0.5 megohm, indicating a capacitance of about 0.0003 μF. One manufacturer (W.E.) produces pickups with impedances from 0.08 to 0.2 megohm, indicating capacitances from 0.002 to 0.0007 μF respectively.

Some crystal pickups are fitted with bent styli to provide vertical compliance for the " pinch effect " and to reduce " needle talk," while others make provision for vertical compliance inside the pickup.

Crystal pickups may be divided into general-purpose types and light-weight high-fidelity types, although there are some with intermediate characteristics. General-purpose crystal pickups have fairly high output, about 1 volt at 1000 c/s, with both arm and armature resonances within the working frequency range. A simple equalizer circuit is Fig. 17.29, although these pickups are frequently used without any equalizer. The stylus force of modern types is usually less than 2.5 ounces (70 grams). Some crystal pickups are available—both standard and fine groove—with a highly compliant drive and capable of tracking with a stylus force of about 8 grams (e.g. Shure " vertical drive ").

The output from a crystal pickup, which may be 1 volt at 1000 c/s, may rise to more than 3 volts r.m.s. at frequencies around 100 to 250 c/s—if no equalizing is used, there is danger of overloading the first valve in the amplifier. A resistance attenuator may be used to reduce the voltage to any desired value.

One manufacturer (Electro-Voice) produces a torque-drive pickup in which vertical movement of the needle produces no output voltage, thereby reducing noise and rumble (Refs. 142, 158).

High fidelity crystal pickups have outputs between 0.5 and 1 volt at 1000 c/s and stylus forces between 0.5 and 1.5 ounces (14-42 grams). The frequency response, when equalized, is approximately flat from a low limit of 25, 30, 40 or 50 c/s to a top limit of 8000, 10 000 or 12 000 c/s. Suitable equalizer circuits are given in Figs. 17.31, 17.32, 17.33. The lateral needle tip impedance of a typical model (Acos GP12) is 1300 grams per centimetre (compliance $= 0.8 \times 10^{-6}$ cm. per dyne).

References to crystal pickups : 81, 106, 120, 142, 158, 301.

The ceramic pickup described in Ref. 178 uses barium titanate in the form of a ceramic. It is free from any appreciable effects from temperature from $-70°$ to $+70°$C, and is also independent of humidity effects. The output obtained from the pickup is 0.75 volt at 1000 c/s and the stylus force is 22 grams (0.8 ounce). The dynamic mass at the stylus point is 4 milligrams at 10 000 c/s. The lateral compliance at the stylus point is 0.5×10^{-6} cm. per dyne or better. The equivalent capacitance of the pickup is about 900 $\mu\mu$F and the optimum load impedance is one megohm. The unequalized output voltage curve is free from sharp peaks and extends from 50 to 10 000 c/s, maximum output occurring at 200 to 400 c/s and falling by 6 db at 50 c/s and 15 db at 10 000 c/s.

A ceramic pickup has been developed for fine groove reproduction with a needle pressure of 6 grams, an output voltage of 0.25 volt at 1000 c/s on a standard test record, and a lateral compliance of 0.75×10^{-6} cm/dyne (Ref. 178).

(v) Magnetostriction pickups

Magnetostriction is that property of certain ferro-magnetic metals, such as nickel, iron, cobalt and manganese alloys which causes them to shrink or expand when placed in a magnetic field. Conversely, if subjected to compression or tension, the magnetic reluctance changes, thus making it possible for a magnetostrictive wire or rod to vary a magnetic field in which it may be placed. This is true for lateral as well as for longitudinal strains, and on this principle the magnetostriction pickup works.

In the pickup a permanent stylus is fastened at right angles to the centre of a piece of nickel wire. Two pickup coils are placed over the magnetostriction wire on each side of the stylus, and the wire is given a slight twist and placed between the poles of a permanent magnet. This effectively gives a push-pull output.

The TM pickup described (Ref. 82) has an output of 0.086 volt from the secondary of the step-up transformer loaded by 0.1 megohm. The stylus force is 0.7 ounce (20 grams). The frequency response (Ref. 76) is level within $\pm$ 2 db from 200 to 8000 c/s and -8 db at 15 000 c/s ; it rises to $+5$ db at 100 c/s and $+6$ db at 50 c/s. The design is such that magnetic fields cause very little hum pickup.

References 76, 82.

(vi) Strain-sensitive pickups

This pickup is based on the principle that the resistance of a conductor changes when the conductor is strained. Direct current is passed through the conductor while in operation. Some low impedance designs are described in Ref. 234 ; these all require a step-up transformer.

A recent high-impedance commercial type is the Pfanstiehl which has a resistance of about 250 000 ohms, giving an output of about 10 to 15 millivolts and a noise level of about 5 micro-

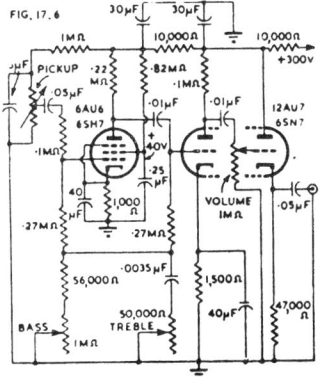

Fig. 17.6. *Pre-amplifier for use with strain-sensitive pickup (Ref.* 336).

volts. Since the pickup is a constant-amplitude device, special methods are required for frequency compensation. The pre-amplifier of Fig. 17.6 gives the required frequency compensation with continuously-variable controls (see Ref. 336 for settings for some American records) and an output of about 2 volts with types 6AU6 and 12AU7. The final stage is a cathode follower. Refs. 96, 303, 336.

A later model of the pickup has a resistance of 125,000 ohms, an output of 5 to 10 millivolts, and a compliance of 1.2×10^{-6} cm/dyne.

(vii) Ribbon pickups

Ribbon pickups operate on the same general principle as the ribbon microphone, and the output is necessarily very low. For this reason a high ratio step-up transformer, magnetically shielded, and a high gain pre-amplifier are required. One design is described below.

Brierley ribbon pickup (English)—home type (Type 4)

This pickup has two gold ribbons backed by a material selected for its self-damping properties. There is, in consequence, no measurable high frequency resonance and this is largely responsible for an extremely low level of buzz. For vibrations in the direction of the groove the damping is independently controlled and very high, but for lateral movements the restoring force is extremely low—lower than that of any other pickup known to the author. The vertical compliance is high, although only half that of the Pickering transcription pickup. The pickup is supplied with a transformer in a mu-metal case, the secondary being loaded by a 0.1 or 0.25 megohm resistance. This pickup may be used on any type of record, including microgroove, without changing tracking force. However the force on microgroove can be halved.

Normal tracking pressure	1/8 ounce (3.5 grams)
Lateral compliance	13×10^{-6} cm/dyne
Vertical compliance	0.09×10^{-6} cm/dyne
Arm resonance	5 c/s
Armature resonance not less than 32 000 c/s.	
Frequency response flat from 30 to 20 000 c/s, rising to $+ 2$ db at 20 c/s.	
Voltage output (across secondary)	10 millivolts
Dynamic mass	about 2 milligrams.

Stylus tips are either diamond or tungsten carbide, the life of the latter being claimed to be much longer than that of sapphire under similar conditions (Refs. 157, 277). Stylus tip radii are 1 mil and 2.5 mils.

It seems likely that the **signal to hum ratio** of the pickup itself is approximately the same as that of a multi-turn moving coil pickup—however see Ref. 312. The hum introduced in the pickup leads and the hum introduced by the transformer are more significant with the ribbon pickup, but may be reduced to very low values by good design.

(viii) Capacitance pickups

A capacitance pickup, in which the movement of the stylus causes a change of capacitance, provides such a small output voltage that it normally requires a very high gain pre-amplifier. However, when it is used to change the frequency of an oscillator, and the output from the oscillator is applied to some form of F-M detector, the output voltage is ample to operate the usual 2-stage amplifier in a radio receiver. In the case of a F-M receiver, the output from the oscillator may be coupled directly to the aerial terminal or to some convenient point in the circuit. Care must be taken to avoid radiation on any band used for radio communication.

A simplified arrangement is shown in Fig. 17.7 in which V_1 is used as an oscillator tuned to resonance by $L_1 C_1$. The inductor L_2 in the grid circuit is made self-resonant at a frequency near that of the plate circuit. Conventional grid-resistor biasing is

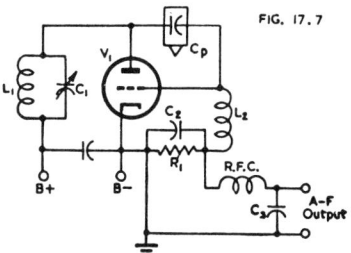

FIG. 17.7

Fig. 17.7. Using a capacitance pickup to provide audio frequency output from an oscillator (Ref. 119).

provided by R_1C_2. The capacitance pickup is connected between grid and plate, and any variation in capacitance will affect the amount of feedback and therefore also the amplitude of oscillation and the voltage drop across R_1. The latter, filtered by RFC and C_3, provides the a-f output which is of the order of 1 volt (Ref. 119).
References to capacitance pickups : 14, 45, 50, 119.

(ix) Eddy-current pickups

In this type of pickup the stylus moves a high-resistance vane in proximity to the inductor of a resonant circuit in an oscillator. The motion of the vane changes the resistance reflected into the tuned circuit and thereby produces amplitude modulation of the oscillator by varying the losses. The amplitude modulation is detected to provide the a-f output voltage. In the Zenith " Cobra " the total harmonic distortion does not exceed 2% and the frequency range extends from 50 to nearly 4000 c/s, with a sharp cut-off at 4000. The level from 1500 to 3500 c/s is 8 db below that at 500 c/s (Ref. 65).

SECTION 4 : TRACKING

(i) General survey of the problem   (ii) How to design for minimum distortion   (iii) The influence of stylus friction.

(i) General survey of the problem

With any pivoted pickup arm it is obvious that the angle between the axis of the pickup and the tangent to the unmodulated groove must change as the pickup moves across the record. This results in (1) harmonic and inharmonic spurious frequencies caused by frequency modulation ; (2) side thrust on the record grooves ; (3) increased record wear (in extreme cases only).

It is therefore advisable to take steps to reduce these effects to satisfactorily small values. The condition for minimum angular tracking error does not provide minimum distortion, because the tracking angle is most critical at the innermost groove. As a matter of interest, it is possible to design a pickup with an angular tracking error not exceeding $2\frac{1}{2}°$ with a length of about $7\frac{1}{2}$ inches, but this is not the optimum design.

References to methods for minimizing angular tracking error : 54 (correction in Ref. 55), 56, 85.

The following summary is based on the detailed analysis by Baerwald (Ref. 228). The subject has also been dealt with in more popular form by Bauer (Ref. 52). The frequency modulation effect is a frequency modulation of the signal by itself, which produces " side band " frequencies. When the signal is a pure sine-wave of frequency f, the dominant " side band " frequency is $2f$, which is the same as the second harmonic. The distortion may therefore be treated as second harmonic distortion, the value being given by :

$$\text{Percentage 2nd harmonic} = (\omega A\alpha/V) \times 100 \qquad (1)$$
$$= (v_0\alpha/V) \times 100 \qquad (2)$$

where $\omega = 2\pi f$

A = maximum groove amplitude in inches

α = tracking error expressed in radians (= angle in degrees divided by 57.3)

V = longitudinal groove velocity in inches per second.

and v_o = maximum transverse velocity in inches per second = ωA.

Equations (1) and (2) indicate that

(1) Distortion is directly proportional to the maximum transverse velocity.

(2) Distortion is directly proportional to the maximum groove amplitude at any one frequency.

(3) Distortion is directly proportional to the tracking angle.

(4) Distortion is inversely proportional to the revolutions per minute. For equal distortion, more careful tracking angle correction is required with 33-1/3 r.p.m. than with 78 r.p.m., other conditions being the same

(5) Distortion is constant over the " constant velocity " portion of a recording characteristic, other conditions being unaltered

(6) The angular tracking error may be increased in the same proportion that the radius is increased, for the same distortion (roughly 3 : 1 ratio over a 12 inch record).

Example of distortion

Consider the distortion with a 12 inch 78 r.p.m. record having 2° angular tracking at the innermost groove, recorded with a sine wave having a peak amplitude of 0.002 inch at a frequency of 250 c/s (this being the cross-over frequency), the r.m.s. velocity being 6.31 cm/sec.

Here $\alpha = 2/57.3$ radians ; $A = 0.002$; $\omega = 2\pi \times 250$; $r = 1.875$ inches ; $V = 2\pi r(78/60) = 15.3$ inches/second.

From equation (1)

$$\text{Percentage 2nd harmonic} = \frac{2\pi \times 250 \times 0.002 \times 2 \times 100}{57.3 \times 15.3} = 0.72\% .$$

If the cross-over frequency had been 500 c/s, with the same recorded velocity (i.e. half the amplitude), the distortion would have been the same. Some frequency test records are recorded with a velocity less than 6.31 cm/sec, so that the distortion due to tracking error would be less than the value stated above. The harmonic distortion as determined for sinusoidal signals gives a fair estimate of the relative **tracking distortion produced by complex signals** over the " constant velocity " portion of the characteristic.

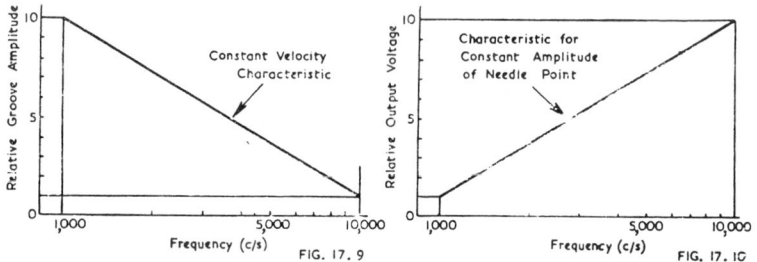

Fig. 17.9. *Constant velocity recording characteristics showing relative variation of groove amplitude with frequency.*

Fig. 17.10. *Frequency characteristic of ideal constant velocity pickup, for constant amplitude of needle point.*

The constant velocity recording characteristic provides a groove amplitude which is inversely proportional to frequency (Fig. 17.9). The ideal constant velocity pickup (a high-fidelity electro-magnetic type is a close approximation) provides an output voltage which is proportional to the frequency, for constant amplitude at the stylus point. It follows that with tracking distortion (or any other form of distortion caused by the groove, needle, or pickup) the harmonics are accentuated in proportion to their frequencies. This effect has been taken into account in the derivation of equations (1) and (2).

It has been proposed (Ref. 228) that a suitable upper limit for tracking distortion is 2% with 0.001 inch peak amplitude as for transcriptions or 4% with 0.002 inch peak amplitude as for shellac discs ; it seems that 1% is a preferable limiting value for good fidelity. On this latter basis the maximum tracking error for 12 inch discs will be 2.8° for 250 c/s cross-over frequency, or 1.4° for 500 c/s crossover frequency, both measured at the innermost groove, or three times these values at the outermost groove.

No additional record wear occurs due to tracking error with permanent needles having spherical tips. Additional record wear may occur with steel needles or with worn sapphire tips if the tracking error is serious.

(ii) How to design for minimum distortion

A straight-arm pickup (not offset) is shown in Fig. 17.11 where D represents the distance from the centre of the record to the pivot of the pickup arm, L represents the length of the arm measured from the pivot to the needle point while r_1 and r_2 represent the radii of the outermost and innermost grooves   In this example the pickup is mounted so that the needle point will pass over the centre of the record, but this is not the position giving the best results.   It is obvious that the axis of the pickup is not a tangent to the groove.

Minimum distortion is always obtained from a straight-arm pickup when it is " underhung,'' that is when the needle point comes short of the centre of the record by a small distance d, the optimum value of which is given below.

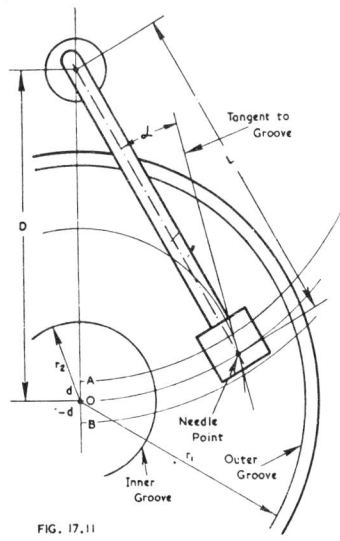

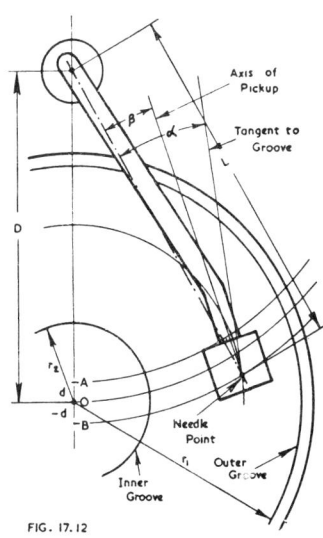

Fig. 17.11.   *Straight-arm pickup on record.*   *Fig.* 17.12.   *Offset-arm pickup on record.*

Straight-arm—not offset (Fig 17.11)

$$\text{Optimum underhang} = d_{opt} = \frac{r_1{}^2 r_2{}^2}{L(r_1{}^2 + r_2{}^2)} \tag{3}$$

For 10 inch discs d_{opt} = 3.04/L.
For 12 inch discs d_{opt} = 3.18/L.

These are based on the values :
For 10 inch discs r_1 = 4.75 r_2 = 1.875 inches.
For 12 inch discs r_1 = 5.75 r_2 = 1.875 inches.
With L = 8 inches, d_{opt} = 0.380 inch for 10 inch discs
= 0.397 inch for 12 inch discs.

These values for d_{opt} are critical, and should be measured accurately. It is safer to keep below than to go above the optimum overhang.

The angular tracking error α at any position on the record is the angle between the axis of the pickup and the tangent to the groove at the needle point. (Figs. 17.11 and 17.12).

With a straight arm, using any value of underhang (Fig. 17.11),

$$\alpha \approx 57.3\left(\frac{r}{2L} + \frac{d}{r}\right) \tag{4}$$

with an error less than 1°.

The optimum value of underhang is given by eqn. (3); using this value and also the values of r_1 and r_2 as for 12 inch discs, with $r = r_2$ (innermost groove) and $L = 8$ inches, the corresponding value of tracking error is approximately 18.6°. The distortion is therefore approximately 6.7% at this point, on the constant velocity characteristic.

Fig. 17.12 shows a pickup with an **offset arm** for the purpose of reducing distortion due to tracking error. This differs from the straight arm in that there is an offset angle β between the axis of the pickup and the straight line joining the pivot to the needle point. Note that this angle β is not equal to the angle of the bend in the arm. In this case minimum distortion is always obtained when the pickup is " overhung," that is when the needle point passes beyond the centre of the record by a distance $- d$, the optimum value of which is given below, on the assumption that the optimum value of offset angle is used.

When a pickup is mounted to provide minimum tracking distortion when used with ordinary needles, the use of bent-shank or trailing type needles will seriously affect the tracking distortion. In cases where either type of needle may be used, a compromise may be necessary. When bent-shank needles only will be used, the increased length of arm should be allowed for.

Offset arm (optimum offset angle) (Fig. 17.12)

$$\text{Optimum overhang} = -d_{opt} = \frac{r_1^2 r_2^2}{L[\frac{1}{4}(r_1 + r_2)^2 + r_1 r_2]} \tag{5}$$

For 10 inch discs $-d_{opt} = 3.99/L$.
For 12 inch discs $-d_{opt} = 4.60/L$.
When $L = 8$ inches :
For 10 inch discs $-d_{opt} = 0.499$ inch.
For 12 inch discs $-d_{opt} = 0.575$ inch.

The optimum offset angle (β in Fig. 17.12) is given by

$$\sin \beta_{opt} = \frac{r_1 r_2 (r_1 + r_2)}{L[\frac{1}{4}(r_1 + r_2)^2 + r_1 r_2]} \tag{6}$$

For 10 inch discs $\sin \beta_{opt} = 2.96/L$.
For 12 inch discs $\sin \beta_{opt} = 3.26/L$.
When $L = 8$ inches :
For 10 inch discs $\beta_{opt} = 21°41'$.
For 12 inch discs $\beta_{opt} = 24°3'$.

The harmonic distortion with optimum overhang and offset angle is given for 12 inch discs approximately by

$$\text{Percentage 2nd harmonic distortion} \approx \frac{5.5}{\sqrt{L^2 - 11.6}}$$

If $L = 8$ inches, 2nd harmonic distortion $\approx 0.76\%$.

(iii) The influence of stylus friction

The stylus friction gives rise to an undesirable excess pressure on the inner groove wall when the optimum offset angle is used. A moderate amount of side-thrust is not detrimental, because it helps to overcome pivot bearing friction, but unduly large values of side-thrust tend to pull the pickup and needle out of the groove towards the centre of the record; this effect is most pronounced towards the inner groove of the record. In automatic record changers it is sometimes necessary to use an offset angle smaller than the optimum value, accompanied by a reduction in " overhang," and to put up with the increased distortion. With light-weight pickups, manually operated, the side-thrust is harmless when the design is based on minimum distortion.

References to tracking : 52, 53, 54, 55, 56, 85, 228, 249.

SECTION 5 : RECORDING CHARACTERISTICS, EQUALIZERS AND AMPLIFIERS

(i) Recording characteristics (ii) Pre-amplifiers for use with pickups (iii) Introduction to equalizers (iv) High-frequency attenuation (scratch filter) (v) Equalizers for electro-magnetic pickups (vi) Equalizers for crystal pickups (vii) Equalizers applying negative feedback to the pickup (viii) Miscellaneous details regarding equalizing amplifiers (ix) Complete amplifiers (x) Pickups for connection to radio receivers (xi) Frequency test records.

(i) Recording characteristics

If a constant input voltage is applied to a recording amplifier, the curve relating frequency and the r.m.s. velocities recorded is known as the recording characteristic Some ideal theoretical recording characteristics are shown in Fig. 17.13.

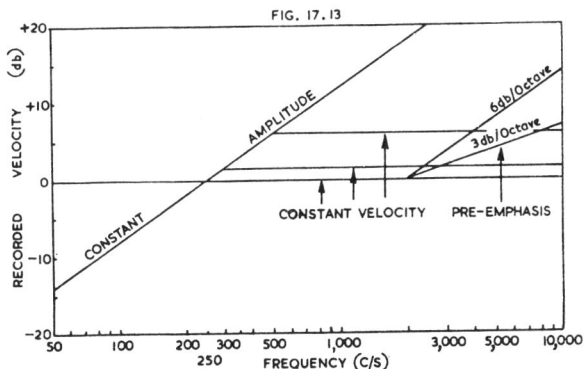

Fig. 17.13. Ideal recording characteristics.

The recording characteristic is usually regarded as having three sections—

1. A constant amplitude characteristic from the lowest recorded frequencies to the cross-over frequency. This has a slope of 6 db per octave.

2. A flat characteristic (constant velocity) from the cross-over frequency to the frequency at which high-frequency pre-emphasis commences.

3. A high-frequency pre-emphasis section.

When examining recording characteristics it is helpful to be able to convert readily from db per octave to db for a frequency ratio of 10* :

6 db/octave = 20 db/10 times frequency ratio.
3 db/octave = 10 db/10 times frequency ratio.

The recorded velocity is proportional to the output voltage from an " ideal " electro-magnetic pickup. When using such a pickup it is necessary to boost the output voltage at frequencies below the cross-over frequency, and to attenuate at frequencies covered by the pre-emphasis. This frequency selective amplification is called **equalizing.**

*A table of frequency ratios, octaves and decades is given in Chapter 7 Sect. 3(v), page 368.

Cross-over frequency

The most popular cross-over frequencies are 250, 300, 350 and 500 c/s, and these are compared in Fig. 17.13 on the basis of equal peak amplitude in the constant amplitude sections. For convenience the 250 c/s cross-over point has been taken as 0 db The constant velocity section of the characteristic with 300 c/s cross-over frequency is at a level of + 1.5 db, while that of the characteristic with 500 c/s cross-over frequency is at a level of + 6 db relative to the characteristic with a 250 c/s cross-over frequency.

Thus it is possible to increase the level of recording of the constant velocity section of the characteristic without increasing the maximum amplitude, provided that the cross-over frequency is increased. This appears very attractive at first sight since it increases the signal-to-noise ratio, but for an increase of 6 db above the cross-over frequency it involves

    (a) Twice the recorded velocity,
    (b) Twice the tracking distortion,
    (c) At least twice the harmonic distortion from a given pickup,
    (d) An additional 6 db of " equalizing " amplification of low frequencies, tending
        to cause trouble with hum and rumble.

There is no " optimum " cross-over frequency because the choice is necessarily a compromise. Where distortion is the principal criterion, a low cross-over frequency from 250 to 350 c/s will be adopted for standard groove 78 r.p.m. Where needle scratch is troublesome with 78 r.p.m. a high cross-over frequency of say 500 c/s may be adopted. With fine groove recordings a cross-over frequency of 400 or 500 c/s is about the optimum choice.

In practice the sharp knee between sections 1 and 2 is rounded off ; the cross-over frequency is then defined as the frequency of intersection of the two asymptotes, provided that the recording characteristic at this frequency is not more than 3 db below the point of intersection of the asymptotes.

If the recording characteristic includes high-frequency pre-emphasis, it is necessary to re-draw the curve for the condition without high-frequency pre-emphasis before determining the cross-over frequency as described above.

High-frequency pre-emphasis

The third section of the recording characteristic is the high-frequency pre-emphasis curve. This may, in the ideal case, be taken as a straight line with a slope of, say, 3 or 6 db/octave as in Fig. 17.13. Of course in practice the sharp junction is rounded off and the recording characteristic gradually approaches the " ideal " straight line. Alternatively the pre-emphasis may be specified in the form of a time constant such as, say, 100 microsecond pre-emphasis. The latter method is covered in Chapter 15 Sect. 1(vi) ; it is restricted to an asymptotic slope of 6 db/octave, but the time constant determines the frequency at which it becomes effective as tabulated below :

Time constant	40	50	75	100	150 μsecs.
Rise of 3 db at	4000	3220	2150	1610	1075 c/s.

All modern American and some English records use some degree of high-frequency pre-emphasis*, although they differ in degree. Assuming that in each case the amplifier is equalized* to provide a level-play-back characteristic, the greater the pre-emphasis the greater the possible dynamic range and the lower the scratch level.

However there are two factors which place a limit on the amount of pre-emphasis which may be used. The first of these factors is the radius of curvature of the groove which at high frequencies and high amplitudes, especially towards the inner portion of the record, cannot be tracked accurately. The radius of curvature is inversely proportional to the maximum amplitude of recording. If the maximum amplitude is reduced, the high frequency pre-emphasis may be increased and in the extreme case constant amplitude recording may be used throughout (Ref. 240).

*Any equalizing which is necessary to correct deficiencies of any of the components of the recording system is corrective equalization, not pre-equalization. The distortion of the frequency characteristic from flat (from air to record track) to some other characteristic is pre-equalization or pre-emphasis (Ref. 247).
    There is reason to believe that some record manufacturers who nominally adopt 16 db pre-emphasis at 10 000 c/s use part of this for corrective equalization for deficiencies of their equipment at high frequencies, and only the balance for true pre-emphasis (Refs. 247, 256 ; 193 discussion by E. W. Kellogg).

The second factor is the peak frequency spectrum characteristic of the music being recorded. The effect of the decreased amplitudes at high frequencies in speech and music has been investigated (notably Ref. 42) but there is insufficient information available concerning the instantaneous distribution of energy within the recorded spectrum. Curves based on the work of Sivian, Dunn and White indicate that " no large increase in distortion occurs in reproducing the 16 db pre-emphasized continuous spectrum . . . yet the addition of a few prominent tones to this spectrum in the region above 2000 cycles will result in intolerable distortion " (Ref. 42).

Pre-emphasis on shellac discs

If the tracking* difficulty could be overcome, it is possible that the N.A.B. pre-emphasis characteristic would not be excessive (Ref. 193) but unfortunately it is the tracking* problem which is the stumbling block. In one test, three records were cut simultaneously with different values of pre-emphasis and played back on a system with the maximum de-emphasis. The record with the 6 db pre-emphasis " reproduced more highs and cleaner highs than the one with the 15, indicating that the overload on the latter was so bad that it was not being tracked* " (Ref. 193, discussion by J. P. Maxfield).

The use of the full N.A.B. pre-emphasis (16 db at 10 000 c/s) on standard shellac discs has come in for much criticism. It is claimed that the cymbals and certain brass instruments will overload the system at high frequencies, although this could be taken care of by the use of a limiting amplifier which reduces the gain for the short period required (Ref. 197). Another writer states that trouble had been experienced through overmodulation by second and third harmonics of the soprano voice when a rising characteristic between 1000 and 5000 c/s had been introduced (Ref. 174). Another writer refers to the " muddiness " and " smearing " in the high level, full-orchestra passages of many records (Ref. 117). Still another writer states that the present N.A.B. pre-emphasis curve effectively guarantees excessive distortion and he refers to the insufficient attention which is paid to the difference between transient response tests and the steady state (Ref. 257). Another refers to the fact that the N.A.B. characteristic unduly weights the importance of the signal to noise ratio and produces the undesirable condition of signals having high velocity with high amplitude (Ref. 189). Another again states " since the pickup stylus can track* the recorded high frequencies more clearly without excessive pre-emphasis, the high frequency reproduction from such records is notable for its clarity " (Ref. 241). Other criticisms have also been published (Refs. 155, 174).

It is interesting to note that one leading American manufacturer, R. C. A. Victor, limits the high-frequency pre-emphasis on home records to a maximum value of 12.5 db at 10 000 c/s in place of 16 db as with the N.A.B. characteristic, while the AES Standard Playback Curve provides for 12 db pre-emphasis at 10 000 c/s.

Pre-emphasis with fine groove recordings

Owing to the lower recording level, smaller radius stylus tip and more elastic material with fine groove records, a higher value of high frequency pre-emphasis may be used than with 78 r.p.m. shellac discs, for the same distortion in both cases.

Measurement of recorded velocities

The recorded velocities may be measured either by the light-pattern method (see (v) below) or by the use of a special high-fidelity pickup. These two methods do not give the same result, owing partly to the mechanical characteristics of the record material and partly to the finite needle size which causes tracing distortion (see Sect. 6). It is current practice to use the special pickup at low frequencies and the light-pattern method at high frequencies ; the latter may be checked by the pickup after allowing for the loss caused by tracing distortion. It has been stated that the mechanical impedance limits at the reproducing point should be included in any standardization of frequency characteristic (Ref. 131).

---

*It is unfortunate that the word " tracking " should be used both here (meaning that the stylus tip is capable of following the modulations in the groove) and also in " tracking distortion " which is due to the angle between the axis of the pickup and the tangent to the unmodulated groove (see Sect. 4). No alternative nomenclature seems to be in current use.

Practical recording characteristics

There does not appear to be any generally accepted definition of published recording characteristics. In most cases, however, it is fairly safe to assume that the user is expected to provide an equalizer amplifier characteristic which is the inverse of the recording characteristic. Of course, some discretion is required in using full equalization for extreme high and low frequencies—this may be provided in the form of tone controls.

The recording characteristics used by English, Australian and the majority of European manufacturers are shown in Fig. 17.14.   Curve 1 is that used by the E.M.I. group of companies for all normal 78 r.p.m. recordings.   Curve 2 is that used by the Special Recordings Department of E.M.I. Studios Ltd. and is notable in that it extends to 20 000 c/s.   Curve 3 is that used by Decca ffrr (1949).   Curve 4 that used by the B.B.C. for transcriptions (Ref. 294).

The Decca (London) LP characteristic is not shown on the curves, but is − 17.5 db at 30 c/s, − 14 db at 50 c/s, − 9 db at 100 c/s, − 3 db at 300 c/s, 0 at 1000 c/s, + 14 db at 10 000 c/s and + 16 db at 15 000 c/s (Jan. 1951).

The recording characteristics used by most American record manufacturers are shown in Fig. 17.15.   Curve 1 is that used in the Columbia long playing micro-groove

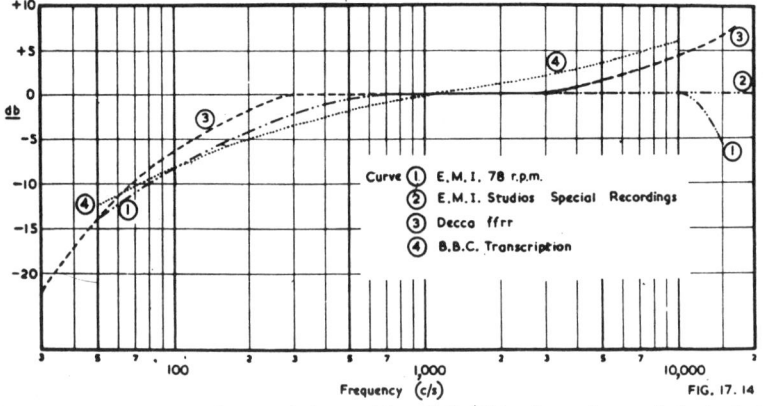

Fig. 17.14.   *Recording characteristics used by all English, Australian and the majority of European record manufacturers for 78 r.p.m., together with B.B.C. transcriptions.*

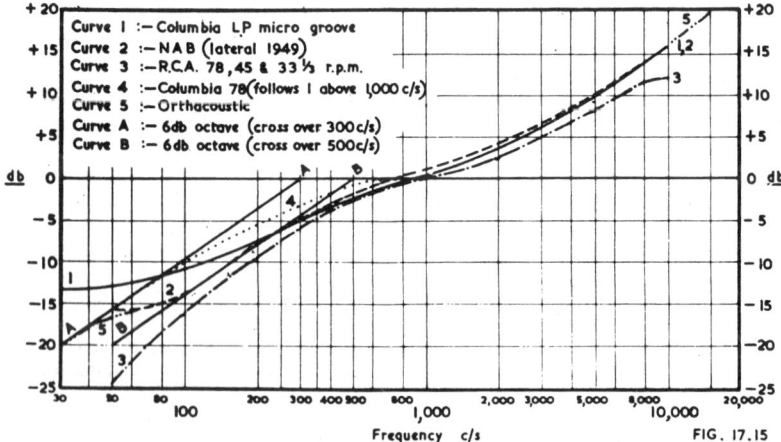

Fig. 17.15.   *Recording characteristics used by most American record manufacturers.*

records. Curve 2 is the N.A.B. (lateral 1949) standard for transcriptions. Curve 3 is that used by R.C.A. Victor for 78 r.p.m. shellac discs, 45 r.p.m. and 33-1/3 r.p.m. fine groove. Curve 4 is that used by Columbia for 78 r.p.m. shellac discs—the cross-over frequency is 300 c/s. Curve 5 is that used for Orthacoustic transcription recording—it follows very closely the N.A.B. curve from 60 to 10 000 c/s but is extended down to 30 c/s (− 20 db) and up to 15 000 c/s.

Standard Playback Curve

For many reasons it has been impossible to achieve a standard recording characteristic, even in a single country and for a specified cross-over frequency. An entirely different approach has been made by the Audio Engineering Society of U.S.A. which has put forward a Standard Playback Curve with the idea of getting this adopted by all designers of equipment for reproduction from records (Ref. 307). The onus would then be on the record manufacturers to produce records which sound well when played with such equipment. The curve (Fig. 17.15A) is based on the frequency of 1000 c/s as a reference point, and the de-emphasis at 10 000 c/s is 12 db, being less than the N.A.B. de-emphasis at this frequency. However the AES playback curve is extended to 15 000 c/s with a de-emphasis of 15.5 db. Both the straight portions of the curve have slopes of 6 db/octave, and the intersections of the extensions of these straight portions with the reference axis occur at 400 c/s (the cross-over frequency) and 2500 c/s.

This playback curve may be duplicated on a flat amplifier with two sections of RC equalization, as shown in Fig. 17.15B, which is one possible arrangement. Alternatively the network of Fig. 17.15C may be used.

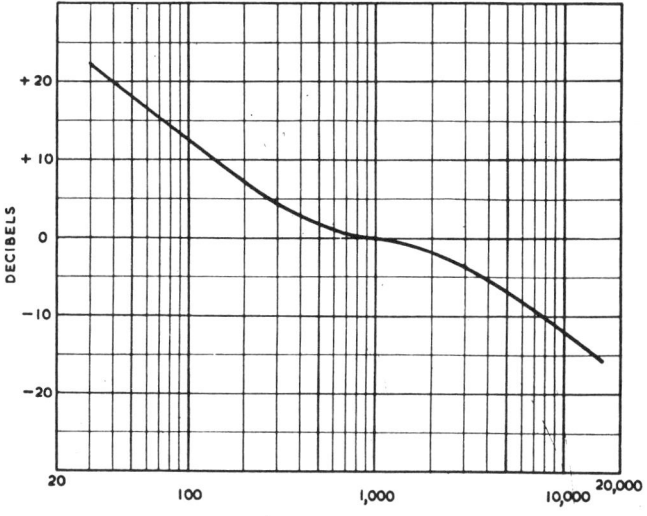

FIG. 17.15A

Fig. 17.15A. *AES Standard Playback Curve (Ref.* 307).
See Supplement for new Standard Playback Curve.

Fig. 17.15B. *High impedance network to provide standard playback curve in grid circuit of amplifier stage (Ref.* 307).

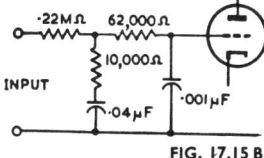

FIG. 17.15 B

It seems that this Standard Playback Curve is what is really needed by the designers of equipment for reproduction from records. Even if record manufacturers do not accept it, the error for any cross-over frequency from 325 to 500 c/s is not more than 2 db, and no problems will be encountered in the reproduction of NAB recording, all fine groove records and most 78 r.p m. discs except those with a cross-over frequency of 250 c/s and without any high frequency pre-emphasis. The latter may be covered by a separate network.

The C.C.I.R. (Geneva, June 1951) proposed a compromise characteristic with a 450 μsec. curve below 1000 c/s and a 50 μsec. curve above 1000 c/s, giving turnover frequencies of 360 and 2800 c/s, for radio programmes for international exchange.

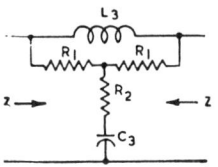

Z (Ohms)	L₃ (H)	C₃ (μF)	R₁ (Ohms)	R₂ (Ohms)
150	0·545	24·08	123°	30
250	0·910	14·44	204	50
500	1·81	7·22	408	101
600	2·18	6·02	491	121

FIG. 17.15C

Fig. 17.15C. Low impedance network to provide close approximation to standard playback curve, with total insertion loss 20 db at 1000 c/s (Ref. 307).

General comments

It has been shown that the choice of recording characteristic is necessarily a compromise between many factors.

By careful attention to all aspects of record manufacture, particularly the material used for the pressings, it is possible to produce records having extremely fine performance.

The best of the fine groove records have demonstrated by results that the performance obtainable is equal to that from the best shellac discs, and superior as regards noise. The fact that there are many mediocre recordings of all kinds proves nothing from the technical angle.

Reference to recording characteristics : 62, 63, 88, 131, 155, 156, 174, 197, 214, 240, 247, 307.

(ii) Pre-amplifiers for use with pickups

Pickups may be divided into three groups, those of the high impedance type (e.g. crystal) normally always used with direct connection to a grid circuit, those of the low impedance type normally always used with a step-up transformer, and those which may be either used with or without a transformer.

In all cases, pickups should be loaded by a substantially resistive load of the value specified by the manufacturer, or determined experimentally—see also Sect. 3. A change in load resistance usually results in a change of frequency characteristic.

The equalizer may be incorporated into the first stage of the pre-amplifier—see Sect. 5 : 2(iii), (v), (vi) and (vii)—or it may follow a normal pre-amplifier stage. In the latter case its design is quite straight forward and follows the general principles of r.c.c. amplifiers—see Chapter 12 Sect. 2 for triodes and Sect. 3 for pentodes ; also Chapter 18 Sect. 2 for microphone pre-amplifiers.

Pre-amplifiers for pickups with high output level (e.g. crystal) may usually be designed without special regard to hum or noise. In other cases the pre-amplifier should be designed as for a microphone pre-amplifier—see Chapter 18 Sect. 2.

A modified cathode follower for use as a low-noise input stage for a crystal microphone or pickup is described in Chapter 18 Sect. 2(vi)D and Fig. 18.6B.

(iii) Introduction to equalizers

(A) Low-frequency equalizers

The amount of bass boosting required by various " ideal " recording characteristics is given below. It is assumed that a constant amplitude characteristic is maintained below the cross-over frequency.

Cross-over frequency	250	300	350	400	500	800 c/s
Bass boost at 70 c/s	11.0	12.6	13.8	15.1	17.1	21.1 db
Bass boost at 50 c/s	14.0	15.6	16 8	18.0	20.0	24.0 db
Bass boost at 30 c/s	18.4	20.0	21.2	22.5	24.5	28.5 db

Cheap pickups usually have a peak in the bass region which reduces the amount of bass boosting required, and in some cases no bass boosting whatever is provided by the amplifier.

Some practical recording characteristics have an approach towards a constant velocity characteristic below some low frequency (e.g. Fig. 17.15 Curves 1, 2 and 5). The equalizing circuit should be designed to suit the individual recording characteristic. The following treatment, however, is based on the 6 db/octave constant amplitude characteristic as approached by Fig. 17.14 Curve 1 and Fig. 17.15 Curves 3 and 4.

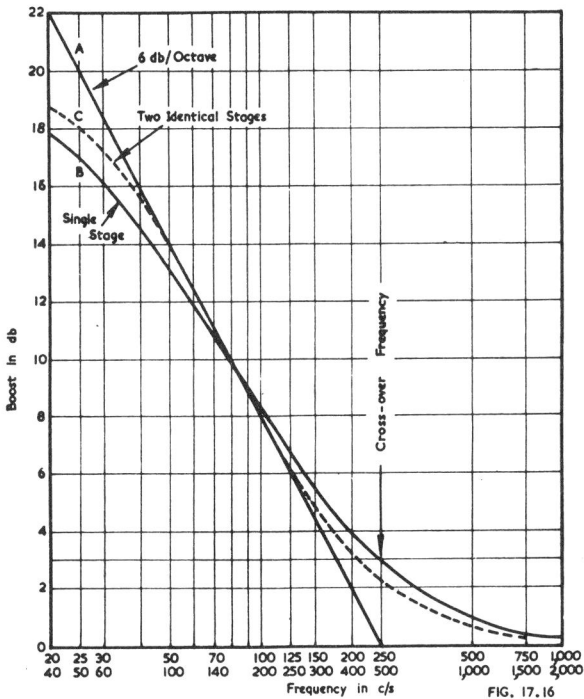

Fig. 17.16. *Bass equalizer characteristics* (*A*) " *ideal* " *characteristic with slope of 6 db/octave* (*B*) *characteristic with single stage r-c equalizer having total boost of 20 db and max. slope 4.9 db/octave* (*C*) *characteristic with two r.c. stages in cascade, each having total boost 10 db and max. slope 3 db/octave* (*circuit as Fig. 17.17*).

Conventional resistance-capacitance equalizing circuits

The form of bass boosting which is normally used results in a " saturation " shape of curve as in Fig. 15.4. Maximum slope is obtained at the point of half total boost, the value being a function of the total boost (Chapter 15 Sect. 2) :

Total boost	20	15	10	6	3	db
Slope at half-boost point	4.9	4.1	3.0	2.0	1.0	db/octave

The desired slope is 6 db/octave, so that there is an appreciable error even with a total boost of 20 db or more. This is plotted in Fig. 17.16 for a total boost of 20 db with alternative frequency scales for 250 and 500 c/s cross-over frequencies. With a cross-over frequency of 250 c/s the error is about − 1 db at 50 c/s and − 2.5 db at 30 c/s, which is generally acceptable, even though there is inevitably some additional loss in the extreme bass, due to coupling condensers.   In the case of a cross-over frequency of 500 c/s, however, the error is − 3 db at 50 c/s and − 4 db at 40 c/s, here again increased by the effect of coupling condensers.   Some designers adopt a total boost of as much as 40 db to provide a satisfactory characteristic for use with high cross-over frequencies such as 800 c/s.   In either case, it is good practice for the amplifier to include a bass tone control which will permit an adjustment to suit the circumstances. If no tone control is fitted, it is usually desirable to attenuate the extremely low frequencies ; a value of coupling capacitance may be chosen to give, for example, an additional 3 db attenuation at 70 c/s and 5 db at 50 c/s.

In the circuit of Fig. 15.3 suitable values of components would be :

V_1 = 6J5 (half 6SN7-GT) ; R_L = 50 000 ohms ; R_1 = 20 000 ohms ; R_2 = 1 megohm ; R_4 = 2800 ohms ; C = 0.25 μF for very good bass response or 0.05 μF for bass attenuation.   Voltage gain at 1000 c/s = 1.5 times = + 3.5 db. Total boost = 20 db. C_2 = 0.2 μF for cross-over 250 c/s or 0 1 μF for 500 c/s.

Alternatively the total boost of 20 db could be obtained in **two separate stages** each having a total boost of 1.1 db at the cross-over frequency.   By this means a maximum slope of 6 db is obtained at a frequency of 100 or 200 c/s with a cross-over frequency of 250 or 500 c/s and more accurate equalizing is possible.   The frequency characteristic obtained is shown with a broken line (C) in Fig. 17.16, and the circuit diagram is given in Fig. 17.17.

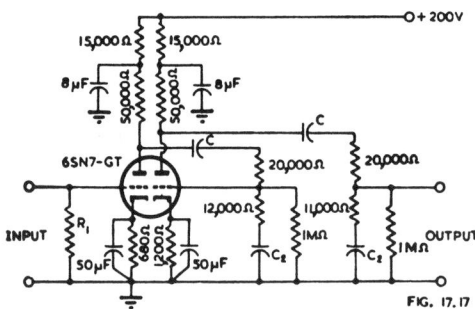

FIG. 17.17

Fig. 17.17.   Circuit diagram of two stage bass equalizing amplifier with bass boosting in each stage.   Pin 4 of the 6SN7-GT should be used for the input circuit.   Total boost is 20 db and max. slope 6 db/octave.   C = 0.25 μF for good bass response or 0.1 μF for some bass attenuation ; C_2 = 0.1μF for cross-over 250 c/s or 0.05 μF for 500 c/s. Voltage gain at 1000 c/s = 22 times = 26.8 dbvg.

Instead of having two networks separated by a valve, the complete two-section equalizer may be incorporated into a single network as in Fig. 17.18.   The maximum slope is 7 db/octave which is adjustable down to 3 db/octave by means of control R_7, which should be tapered ; maximum slope is obtained with maximum value of R_7. The plate load resistors may be increased if it is desired to increase the gain.   This circuit provides 25 db bass boosting with a very close approach to the true 6 db octave slope (when R_7 is correctly adjusted) and has the additional merit of incorporating a tone control providing a total control of 13 db in the region 20 to 50 c/s depending on the cross-over frequency.   With an input voltage of about 50 millivolts, the output will be about 1 volt.

Frequency-selective feedback to provide equalizing

A resistance-capacitance network may be used in the feedback loop of an amplifier to provide bass boosting which gives a close approximation to the correct degree of

Fig. 17.18.   Two section equalizer to provide for 3 values of cross-over frequency and adjustable slope of characteristic (Ref. 127). V_1 + V_2 = 6SL7 or 6SC7 or 7F7 or any high-mu triodes.

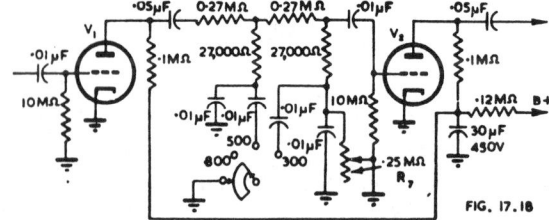

FIG. 17.18

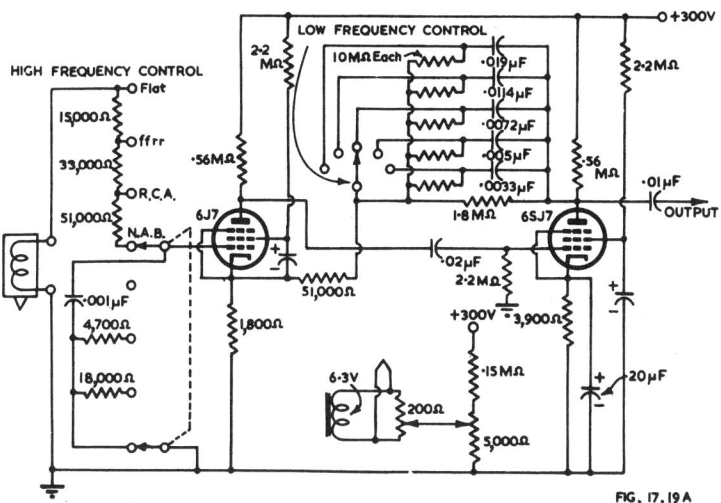

Fig. 17.19A. *Circuit diagram of equalizing amplifier using negative feedback to provide bass equalizing* (1) 250 *c/s* (2) 300 *c/s* (3) 500 *c/s* (4) 750 *c/s* (5) 1200 *c/s cross-over frequency. A choice of four high-frequency de-emphasis circuits is provided* (1) *flat* (2) *ffrr* (3) *R.C.A.* (4) *N.A.B.* (*Ref.* 239).

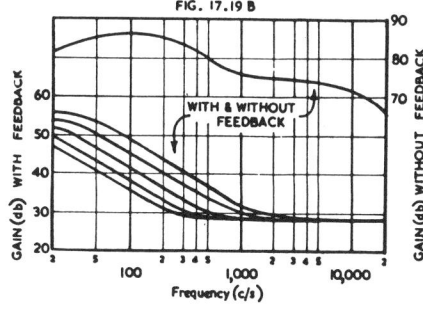

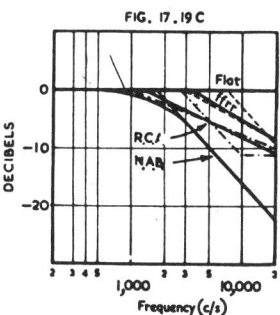

Fig. 17.19B. *Frequency characteristics of circuit of Fig.* 17.19A *with high frequency control in "flat" position (Ref.* 239).

Fig. 17.19C. *High frequency de-emphasis characteristics of the circuit of Fig.* 17.19A *(Ref.* 239).

equalizing. The principle is demonstrated in the circuit of Fig. 15.11 and the frequency characteristics of Fig. 14.12, which are limited to a maximum boost of 12 db. Increased bass boosting could, of course, be obtained by increasing the feedback. With the value of capacitance used in Fig. 15.11 (0.005 μF) a satisfactory compromise is obtained for a cross-over frequency of 250 c/s ; half this capacitance would be suitable for 500 c/s.

This same principle is used in the circuit of Fig. 17.19A to produce the low frequency characteristics of Fig. 17.19B (Ref. 239). The low-frequency control is a shorting type, while the 10 megohm resistors are click suppressors. This circuit also provides four values of high frequency de-emphasis (curves Fig. 17.19C).

Another variation of the same general principle is incorporated in Fig. 17.20 (Ref. 144). It is suitable for use with the GE variable reluctance pickup, or any other low level electro-magnetic pickup which has been equalized to give a constant-velocity characteristic. The values of components are designed for a cross-over frequency of 500 c/s and a high-frequency hinge frequency of 2000 c/s with a slope of 6 db/octave.

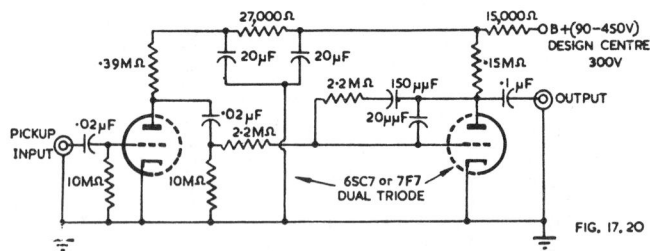

Fig. 17.20. Circuit diagram of equalizing amplifier for low-level electromagnetic pickup using negative feedback over the second stage to accomplish equalization (Ref. 144).

A different form of frequency-selective feedback is used in Fig. 17.21 which provides correct equalizing for English (H.M.V.) records from 25 to 8500 c/s within 2.5 db. V_1 is the first pre-amplifier valve having a series tuned circuit L_1C_1 (with a minimum impedance below 25 c/s) in its cathode circuit so that at higher frequencies an increasing proportion of the input signal is fed back. The condenser C_2 in series with R_3 serves to flatten the response above 1000 c/s ; no provision is made for de-emphasis. The level between 250 and 1000 c/s is flattened by the network $L_2C_3R_4$ in which L_2 and C_3 are tuned to 250 c/s. R_4 and L_2 carry the plate current of the valve, and the value of R_1 primarily determines the degree of feedback at 1000 c/s.

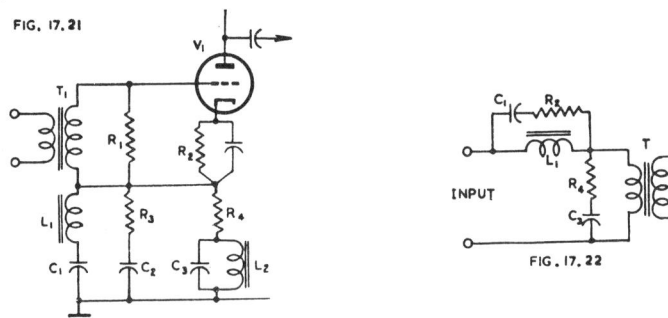

Fig. 17 21. Frequency selective negative feedback used to provide equalizing for English records. (Ref. 199). V_1 = MH4 (μ = 40, r_p = 11 000 ohms), load resistance = 50 000 ohms.

Fig. 17.22. Bass equalizer incorporating L, C and R (Ref. 188).

Bass equalizers incorporating *L*, *C* and *R*

A network incorporating *L*, *C* and *R* may be used to provide a very close approach to the 6 db/octave ideal characteristic, or even higher values if desired. The simplified circuit is shown in Fig. 17.22 in which L_1 is tuned by C_1 to the cross-over frequency, R_2 controls the shape of the band and R_4 controls the slope of the characteristic.

Suggested values for a cross-over frequency of 400 c/s are—L_1 = 2.9 henrys ; C_1 = 0.07 μF ; C_3 = 1.0 μF ; R_2 = 6800 ohms ; R_1 = 2200 ohms.

For a slight change in cross-over frequency only C_1 and R_2 need be altered (Ref. 188).

References to equalizers (general) : 15, 54, 106, 108, 109, 113, 117, 120, 126, 127, 133, 144, 187, 188, 197, 198, 199, 209, 220, 226, 239, 240.

(B) High-frequency equalizers (de-emphasis)

A shunt capacitance provides for attenuation at the nominal rate of 6 db/octave (see Chapter 15 Sect. 6). The attenuation is 3 db when the reactance of the shunt condenser is equal to the resistance of the circuit across which it is connected, that is when

$$C = 1/(2\pi f_o R) \tag{1}$$

or capacitance in microfarads $= 157\,000/(f_oR)$ \hfill (2)

where C = shunt capacitance in farads

$\qquad f_o$ = frequency at which the attenuation is 3 db (it is also the " hinge point " of the ideal 6 db/octave line)

and $\quad R$ = resistance of circuit across which C is connected, in ohms.

If it is desired to limit the attenuation to a specified value, while approaching the nominal slope of 6 db/octave, the circuit of Fig. 17.23A may be used. Here R_1 represents the resistance of the input circuit ; R_3 should be as high as is practicable— it is assumed to be very much greater than R_2. The maximum attenuation in decibels will be given by

$\qquad$ max. attenuation (x db) $= 20 \log_{10}[R_2/(R_1 + R_2)]$ \hfill (3)

Eqns (1) and (2) may be used with reasonable accuracy for maximum attenuations not less than 10 db—at lower values the attenuation at the " hinge point " frequency f_0 will be less than 3 db (the actual values are 2.4 db for 10 db total boost, 2 db for 6 db total boost, 1.2 db for 3 db total boost). The frequency characteristic is indicated in Fig. 17.23B.

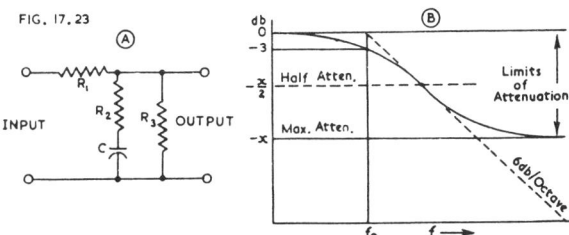

Fig. 17.23. (A) *Circuit providing high-frequency attenuation with a specified limit to the attenuation* (B) *Frequency characteristics.*

If the maximum attenuation is 10 db, the maximum slope (which occurs at half maximum attenuation) is 3 db/octave. This may therefore be used to provide a 3 db/octave de-emphasis characteristic within the limitations of its range.

High frequency equalizers are incorporated in the circuits of Figs. 17.19, 17.20, 17.27.

For general treatment of equalizers see Ref. 258.

(iv) High-frequency attenuation (scratch filter)

In addition to the normal manual tone control it is advisable in good amplifiers to incorporate a filter which provides very rapid attenuation above a certain frequency, say 7000 c/s, so that the upper scratch frequencies on old or noisy records may be rendered inaudible. This is provided by the circuit of Fig. 17.24A (Ref. 15) which has negligible attenuation at 7000 c/s, — 60 db at 8700 c/s and over 36 db attenuation at all frequencies above 8000 c/s. The filter is designed for an impedance of 24 000

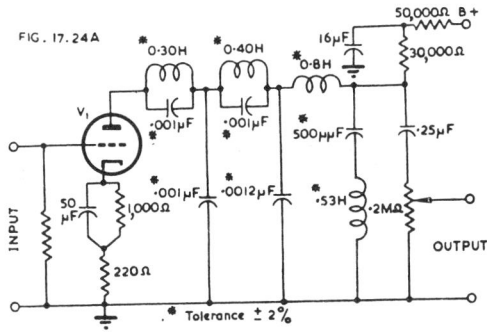

Fig. 17.24A. *Amplifier stage incorporating filter with 7000 c/s pass-band and very rapid attenuation at higher frequencies (Ref. 15).* $V_1 = MH4$, $\mu = 40$, $r_p = 15\,000$ *ohms.*

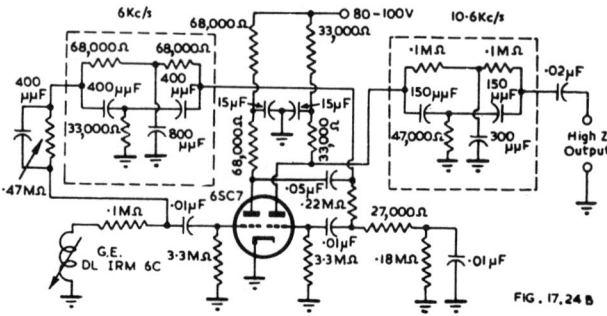

Fig. 17.24B. *Circuit of pre-amplifier and filter circuit with high attenuation at* 10 000 *c/s* (*Ref.* 113).

ohms ; the plate resistance of V_1 is increased to this value by means of a partially un bypassed cathode resistor, while the output end is loaded directly. See reference for coil winding and other details.

An alternative method is used in the pre-amplifier circuit of Fig. 17.24B to produce the response characteristic of Fig. 17.24C. This is suitable for use with low level electro-magnetic pickups. The 10.6 Kc/s parallel-T network provides a maximum attenuation of 34 db, while the 6 Kc/s parallel-T network in the feedback loop removes the attenuation which would otherwise occur at this.frequency and thus give a sharper knee (Ref. 113).

There appears to be substantial evidence that too sharp a knee on the attenuation characteristic gives an unpleasant effect to the listener. If the knee is initially too sharp, it may be slightly rounded by the addition of a "roll-off" characteristic at some suitable point—e.g. in the.preceding stage.

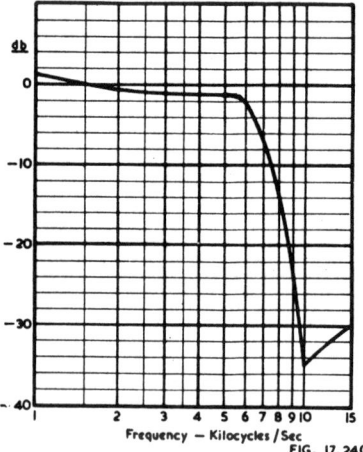

Fig. 17.24C. *Response curve of amplifier and filter shown in Fig.* 17.24B (*Ref.* 113).

(v) Equalizers for electro-magnetic pickups

The general principles of equalizing have been covered in the earlier portion of this section, and may be applied directly to electro-magnetic pickups (Figs. 17.17, 17.18, 17.19, 17.21, 17.22, together with high-frequency de-emphasis). Fig. 17.24B is directly applicable.

With any form of electro-magnetic or dynamic pickup the pickup itself has an impedance which, above about 1000 c/s, is almost entirely inductive reactance. If a capacitance is connected in shunt with the pick-up, or across the secondary of the transformer, this capacitive reactance will resonate with the inductive reactance of the pickup at some frequency determined by L and C. As a result, the output voltage is boosted at the frequency of resonance and falls rapidly at higher frequencies. If, now, a variable resistance is also connected in shunt with the pickup or across the secondary of the transformer, there will be found a value at which the resonance peak just disappears—this value of shunt resistance is the maximum which can be used to permit the condenser to act purely as an attenuator. For an analysis of this effect on the E.M.I. No 12 pickup see Ref. 109 and for that on the Pickering and G.E. variable reluctance pickups see Ref. 117. The inductance of the E.M.I. No. 12 pickup is of the order of 1 mH. The inductance of the Pickering pickup is approximately 100 mH and that of the G.E. pickup 120 mH.* The correct value of shunt-resistance is given by

R (ohms) $= 1.2\ f_c$ for the Pickering pickup

and R (ohms) $= 0.9\ f_c$ for the G.E. pickup

where f_c is the frequency at which the attenuation is 3 db. Beyond this frequency the rate of attenuation is approximately 15 db/octave.

The shunt capacitance required for various values of the cut-off frequency is :

Frequency	4000	5000	6000	7000 c/s
Capacitance Pickering	0.02	.013	.009	.0065 μF
G.E.	0.03	.019	.013	.0095 μF

Fig. 17.25. Equalizer for use with E.M.I. and Marconiphone Model 12A pickup. Output as shown is 200 ohms ; with link closed, output is 600 ohms. For English E.M.I. records join terminals 1 and 2, also 3 and 4 ; for N.A.B. and American records leave unjoined. (E.M.I. Australia).

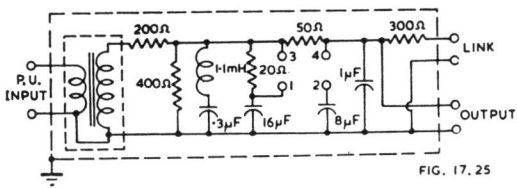

FIG. 17.25

Pre-amplifier for E.M.I. and Marconiphone Model 12A

A suitable equalizer is shown in Fig. 17.25 which gives the following output levels at 1000 c/s with Decca Z718 frequency record.

Termination	Equalizer English	Equalizer N.A.B
200 ohms	— 49 dbm	— 50 dbm
600 ohms	— 54 dbm	— 55 dbm

Pre-amplifier for use with G.E. variable reluctance pickup

The circuit of Fig. 17.26 is suitable for use with the G.E. variable reluctance pickup. This pickup gives an output of about 10 mV (with recorded velocity of 4.8 cm/sec.), the amplifier has a voltage gain of 35 db at 1000 c/s, so that the output voltage is about 0.6 volt. The input shunt resistor of 6800 ohms (which is suitable for the N.A.B. characteristic) may be varied to control the frequency response, a higher value increasing the high frequency response. A low plate voltage is desirable. A slight improvement in response may be effected by shunting the 0.01 μF condenser by a resistance of 180 000 ohms (Ref. 187). See also Fig. 17.34 for input transformer arrangement and Fig. 17.24B as an alternative pre-amplifier.

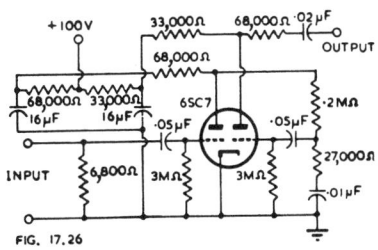

Fig. 17.26. Pre-amplifier and equalizer for use with the G.E. pickup (Ref. 187).

FIG. 17.26

Continuously-variable equalizing pre-amplifier

If a sufficiently flexible equalizer is used, it does not require any additional tone control. Such a circuit is shown in Fig. 17.27A in which the additive method of equalization is employed. This has three transmission channels, one channel having a characteristic which is essentially flat below 1000 c/s, falling off at the rate of 12 db/octave at higher frequencies ; a second channel having 40 db more gain than the basic channel at very low frequencies but with its gain falling off at the rate of 12 db/octave above about 50 c/s ; and a third channel whose gain rises at the rate of 12 db/octave up to 15 000 c/s, above which frequency its gain is 40 db more than the basic

*Later models of G.E. pickups have inductances of 250 mH (Broadcast type RPX-046) and 520 mH (Home type).

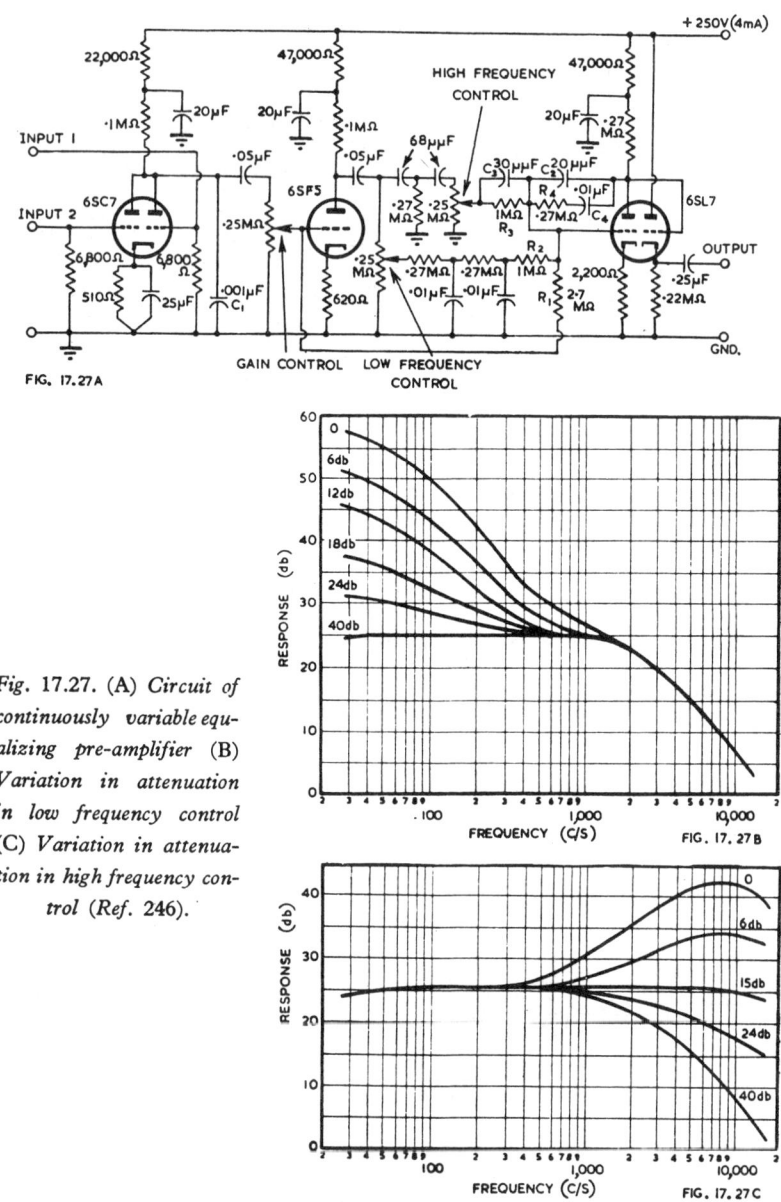

FIG. 17.27A

Fig. 17.27. (A) Circuit of continuously variable equalizing pre-amplifier (B) Variation in attenuation in low frequency control (C) Variation in attenuation in high frequency control (Ref. 246).

channel. Outputs from the three channels are added in a single valve feedback summing amplifier ; potentiometers which add flat loss in the auxiliary channels permit control of the resultant transmission characteristic. The input allows for the connection of two G.E. variable reluctance pick-ups simultaneously. The maximum output is of the order of 1 volt, and the pre-amplifier may be used at a moderately remote location, owing to the cathode follower output.

The frequency characteristics are given in Figs. 17.27B and C (Ref. 246).

Equalizers applying feedback to the first pre-amplifier stage

It has been demonstrated by Ellis (Ref. 209) that the feedback circuit may be made frequency-selective to give a close approach to the required equalizing characteristic for an electro-magnetic pickup, as in Fig. 17.28.   This circuit does not apply any appreciable feedback to the pickup itself, owing to the isolating resistor R_1.   The cross-over frequency may be adjusted by changing either C or R_2, but changing R_2 without making a similar change in R_1 affects the gain.   This arrangement is said to be " the quietest and most stable " (Ref. 287), and a very low hum level is possible if the cathode is well by-passed.   The input impedance of the stage is low, owing to the feedback circuit.   The resistor R_1 should be at least 10 times the nominal impedance of the pickup.   The normal gain, above the cross-over frequency, is approximately (R_2/R_1), so that unity gain is obtained when $R_2 = R_1$.   By making $R_2 > R_1$ this circuit can provide gain as well as bass boosting, but this reduces the available increase for low frequencies, since the upper limit of gain is the normal gain of the valve without feedback.

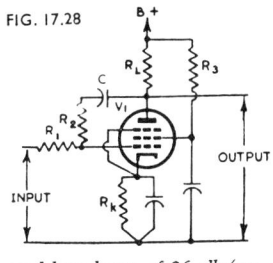

FIG. 17.28

Fig. 17.28.   Equalizer which applies negative feedback to an electro-magnetic pickup (Ref. 209).   Original values were $V_1 = EF36$, $R_1 = R_2 = 1\ M\Omega$, $R_L = 0.22$ $M\Omega$, $R_3 = 2.2\ M\Omega$, $C = 600\ \mu\mu F$, $E_{bb} = 250\ volts$, unity gain.
Suggested adaptation for 19 db gain and bass boost of 26 db is : $V_1 = 6J7$ or 1620, $R_1 = 0.1\ M\Omega$, $R_2 = 1\ M\Omega$, $R_L = 0.47\ M\Omega$, $R_3 = 3.0\ M\Omega$, $C = 330\ \mu\mu F$ for 500 c/s, or 510 $\mu\mu F$ for 300 c/s, or 620 $\mu\mu F$ for 250 c/s. V_1 may also be replaced by a high-mu triode (6AT6) where $R_1 = 0.1\ M\Omega$, $R_2 = 0.33\ M\Omega$, $R_L = 0.22\ M\Omega$, R_3 omitted, $C = 1200\ \mu\mu F$ for 500 c/s, 2000 $\mu\mu F$ for 300 c/s or 2400 $\mu\mu F$ for 250 c/s, giving gain of 8 db and bass boost of 26 db (see Ref. 287).

A modification of the circuit of Fig. 17.28 is described in Ref. 280 which gives both bass boosting (with a " very low frequency roll-off ") and high frequency de-emphasis. Values of components are : $R_1 = 0.33$ megohm, $R_2 = 1.0$ megohm, $R_L = 0.22$ megohm, $C = 600\ \mu\mu F$ approx. (adjustable), $C_2 = 25\ \mu\mu F$ approx. (adjustable) shunted across R_2, and $V_1 = EF37$.

A different modification of this circuit is given in Fig. 15.13 ; see Chapter 15 Sect. 2(iv)C   This modified circuit enables adjustment to be made for any cross-over frequency by means of a continuously variable potentiometer ; unfortunately no measured frequency characteristics have been published but this circuit seems to be close to the ideal for simplicity and to provide a response characteristic as good as that of any other RC feedback amplifier.

(vi) Equalizers for crystal pickups

A crystal pickup, when working into a high load resistance of the order of 2 to 5 megohms gives a practically level response characteristic over the constant amplitude section of a gliding-tone frequency test record (A in Fig. 17.30).   If the load resistance is decreased to 0.5 megohm (Curve B) the output at 50 c/s may fall in a typical case by 6 db, the attenuation ratio at other frequencies being inversely proportional to the frequency.

High temperatures cause a reduction in bass response ; an increase from 68°F to 104°F causing a decrease of about 5 db at 50 c/s in a typical case (Curve B).

The output from a crystal pickup is so high that, if it is not attenuated between the pickup and the first amplifier stage, it may overload the valve.   Unless the equalizer provides heavy attenuation, it is advisable to place the volume control between the pickup and the first grid.

Above the cross-over frequency the output voltage from an unequalized high-fidelity crystal pickup tends to fall at the rate of 6 db/octave.   In popular types there is usually a resonance peak which gives some additional lift to the top end of the characteristic, thus decreasing the amount of equalization required.

A simple equalizer for use with crystal pickups is shown in Fig. 17.29. If $R_2 = 0.5$ megohm, R_1 may be taken as $10R_2$ or 5 megohms. This gives an attenuation at very low frequencies of 20.8 db. The effect of variation of C_1 is shown in curve C Fig. 17.30 ; these response curves are only satisfactory when considerable high-frequency de-emphasis is required.

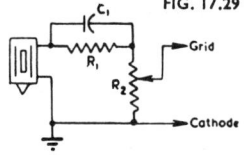

FIG. 17.29

Fig. 17.29. Simple equalizer circuit for a crystal pickup.

Better equalization is obtained when $R_1 = 40R_2$, thus giving a maximum attenuation of 32 db. Suitable values are : $R_1 = 5$ megohms, $R_2 = 0.125$ megohm, $C_1 = 100$ $\mu\mu F$. The resultant response curve is shown in curve D.

A more flexible equalizer suitable for use with a high-fidelity crystal pickup is shown in Fig. 17.31 (Ref. 127). This gives a choice of three load resistors, thus giving some control of the bass, and three values of capacitance suitable for three cross-over frequencies.

An improved equalizing circuit suitable for use with the Acos GP12 high-fidelity pickup is shown in Fig. 17.32 (Ref. 120). This gives correct equalization for H.M.V. records within $+$ 2.5 db, $-$ 3.5 db from 30 to 8000 c/s.

An equalizer suitable for use with the Brush high-fidelity pickup is shown in Fig. 17.33 (Ref. 106). Position (1) is for quiet shellac records with a 500 c/s cross-over

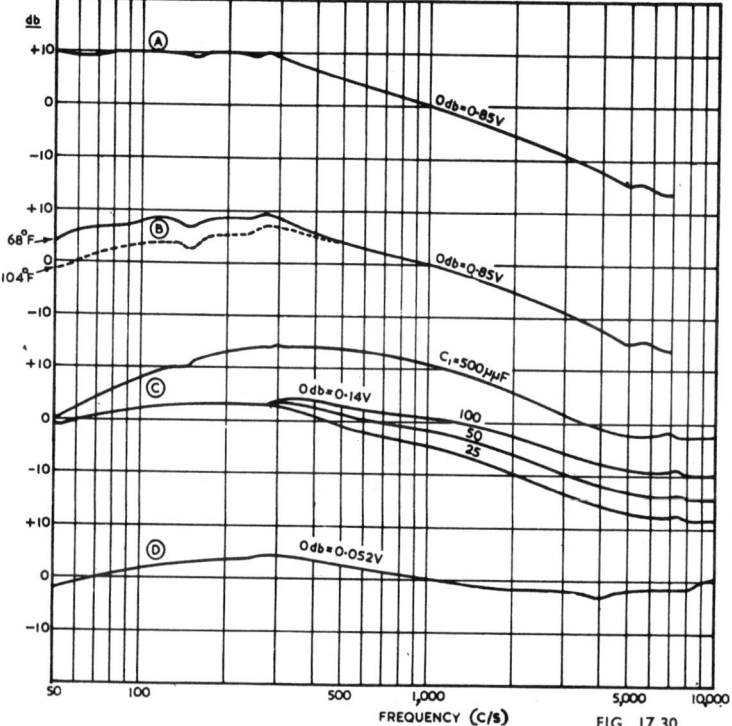

FIG. 17.30

Fig. 17.30. Response curves of typical crystal pickup. (A) load resistance 2 megohms, temperature 68° F ; (B) load resistance 0.5 megohm, temperature 68° and 104°F ; (C) with simple equalizer circuit as Fig. 17.29 having $R_2 = 0.5$ megohm, $R_1 = 5$ megohms ; (D) with $R_2 = 0.125$ megohm, $R_1 = 5$ megohms, $C_1 = 100$ $\mu\mu F$. Test record Audiotone for (A), (B) and (C) ; Columbia M-10 003 for (D). Curves based on data for Astatic FP-18.

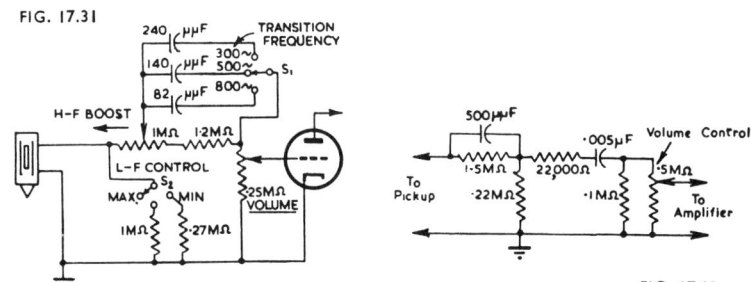

Fig. 17.31. *Flexible equalizer for high-fidelity crystal pickups (Ref.* 127).
Fig. 17.32. *Equalizer for Acos GP12 high fidelity crystal pickup, for 250 c/s cross-over frequency (Ref.* 120).

frequency ; (2) for scratchy and noisy shellac records ; (3) good Orthacoustic transcriptions and (4) noisy transcriptions. The low frequency filter reduces rumble and hum.

See also Sect. 5(x).

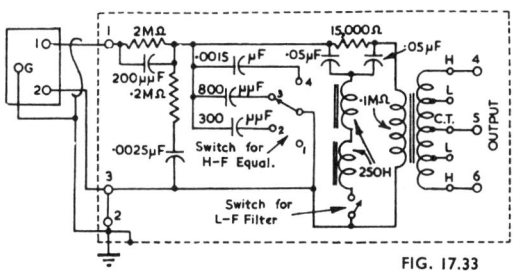

Fig. 17.33. *Equalizer for use with Brush high-fidelity crystal pickup (Ref.* 106).

(vii) Equalizers applying negative feedback to the pickup

It is advantageous in all cases to apply negative feedback to the pickup, whether electro-magnetic or crystal. This may be accomplished in any conventional manner and the feedback reduces non-linear distortion and the effects of mechanical resonances.

Various possible combinations of *CR* and *LCR* in the feedback network of both crystal and electro-magnetic pickups are described in Ref. 220. Most of these apply a very limited amount of feedback, but the same principle could be applied to a high-gain pre-amplifier stage with increased feedback. It is evident that any type of pickup may be equalized by this method, with the additional advantages of increased damping on the pickup. It seems that more detailed investigation of the subject is well merited.

(viii) Miscellaneous details regarding equalizing amplifiers

Position of equalizer

In general it is preferable to place the equalizer following the first stage of the pre-amplifier in the case of low level pickups. With high level pickups it may be possible to incorporate the equalizer in the main amplifier.

Input transformers

Although input transformers are not necessary with electro-magnetic pickups, they have some distinct advantages. The step-up ratio provides useful gain, while they permit the use of a balanced-to-earth input line which gives improved signal-

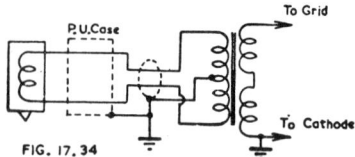

FIG. 17.34

Fig. 17.34. Low impedance pickup with balanced-to-earth input to step-up transformer.

to-noise ratio (Fig. 17.34). With low level pickups the transformer improves the signal-to-hum and signal-to-noise ratios.

Rumble

Rumble may be reduced, if necessary, by a filter which cuts off sharply below, say, 50 c/s. The cut-off frequency and attenuation necessary are dependent upon the motor, turntable and amount of bass boosting in the equalizer and tone control. One example is incorporated in Fig. 17.35B.

(ix) Complete amplifiers

Complete amplifiers for the reproduction of sound from records may include—

1. A suitable pickup.

2. An equalizer to provide constant velocity output (only required for electro-magnetic pickups, and even then only when not provided as part of the main equalizer).

3. A first stage pre-amplifier (only required in the case of a low level pickup).

4. A main equalizer, preferably provided with adjustable cross-over frequencies and variable high-frequency de-emphasis.

5. A tone control, preferably comprising separate bass and treble controls. If the main equalizer has flexible adjustments of its characteristics, it may be used in place of a separate tone control.

6. A rumble filter, preferably with switch to cut in or out (may be omitted if response is attenuated below 70 c/s).

(Continued on page 748)

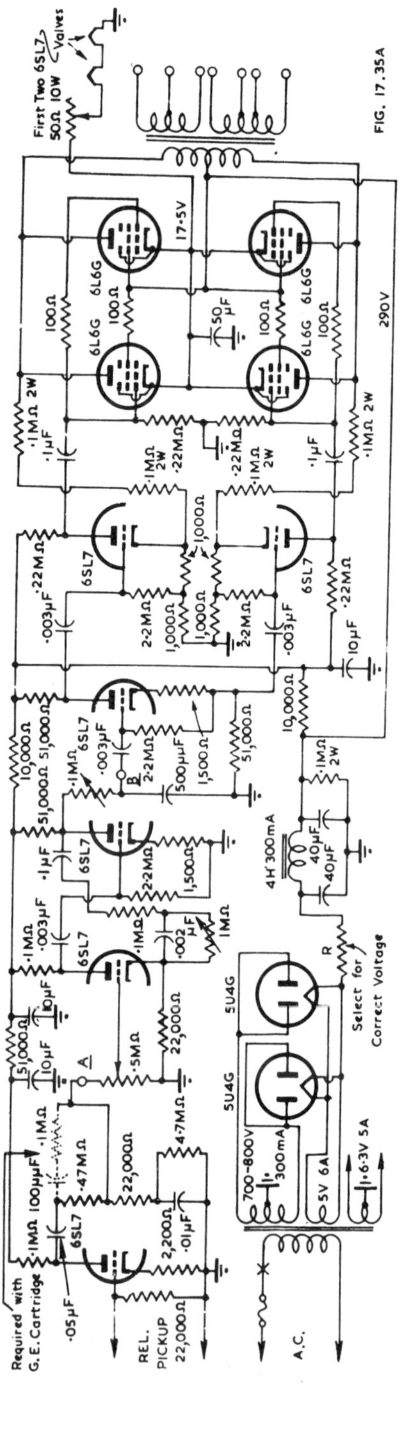

Fig. 17.35A. Circuit of 25 watt amplifier using G.E. variable reluctance or Pickering 120M pickup (Ref. 240).

FIG. 17.35A

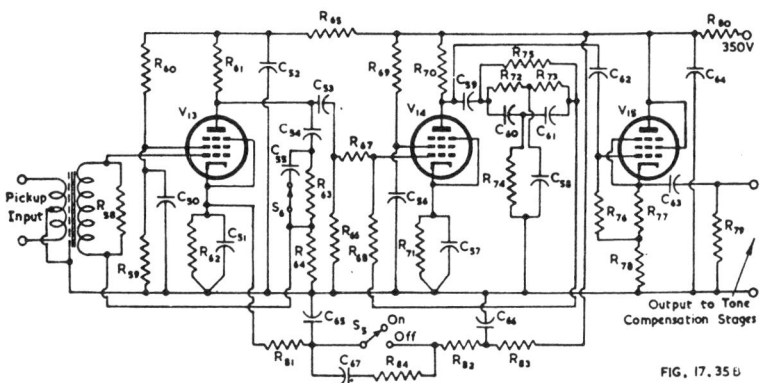

FIG. 17.35 B

Fig. 17.35B. Williamson pre-amplifier including equalizer and high-pass filter (rumble attenuator). Ref. 270. In the original amplifier $V_{13} = V_{14} = V_{15} =$ type EF37. V_{13} and V_{14} may be replaced by type 6J7 or 1620 if $R_{71} = 1600\ \Omega$.

		Type	*Rating*	*Tolerance*
R_{58}	*Value to suit trans-former*	*High-stability carbon*		
R_{59}	0.1 $M\Omega$	*High-stability carbon*	$\frac{1}{2}W$	20%
R_{60}	0·68 $M\Omega$	*High-stability carbon*	$\frac{1}{2}W$	20%
R_{61}	0.22 $M\Omega$	*High-stability carbon*	$\frac{1}{2}W$	20%
R_{62}	4700 Ω	*High-stability carbon*		20%
R_{63}	0.22 $M\Omega$	*Composition*		10%
R_{64}	20 000 Ω	*Composition*		
R_{65}	22 000 Ω	*High-stability carbon*	$\frac{1}{2}W$	20%
R_{66}	0.22 $M\Omega$	*Composition*		10%
R_{67}	0.20 $M\Omega$*	*Composition*		
R_{68}	4.7 $M\Omega$	*Composition*		5%
R_{69}	1.0 $M\Omega$	*Composition*	$\frac{1}{2}W$	20%
R_{70}	0.22 $M\Omega$	*Composition*	$\frac{1}{2}W$	20%
R_{71}	2200 Ω	*Composition*		20%
R_{72}	2.0 $M\Omega$	*Composition*		1% or matched
R_{73}	2.0 $M\Omega$	*Composition*		1% or matched
R_{74}	1.0 $M\Omega$	*Composition*		1% or matched
R_{75}	10 $M\Omega$	*Composition*		5%
R_{76}	47 000 Ω	*Composition*		10%
R_{77}	1000 Ω	*Composition*		20%
R_{78}	47 000 Ω	*Composition*	1W	20%
R_{79}	0.22 $M\Omega$	*Composition*		20%
R_{80}	10 000 Ω	*Composition*	1W	20%
R_{81}	0.22 $M\Omega$		$\frac{1}{2}W$	
R_{82}	0.22 $M\Omega$		$\frac{1}{2}W$	
R_{83}	47 000 Ω			
R_{84}	100 Ω			

All resistors may be $\frac{1}{4}W$ rating, tolerance 20% unless otherwise specified.
*May require adjustment.

			(V d.c. working)	
C_{50}	0.5 μF	*Paper*	250	20%
C_{51}	50 μF	*Electrolytic*	12	

C_{52}	16 μF	Electrolytic	450	
C_{53}	0.02 μF	Paper	350	10%
C_{54}	4000 $\mu\mu F$	Silvered mica	350	10%
C_{55}	100 $\mu\mu F$	Silvered mica	350	10%
C_{56}	50 μF	Electrolytic	12	
C_{57}	50 μF	Electrolytic	12	
C_{58}	0.01 μF	Silvered mica	350	1% or matched
C_{59}	0.25 μF	Paper	500	20%
C_{60}	5000 $\mu\mu F$	Silvered mica	350	1% or matched
C_{61}	5000 $\mu\mu F$	Silvered mica	350	1% or matched
C_{62}	7000 $\mu\mu F$	Silvered mica	350	10%
C_{63}	0.5 μF	Paper	500	20%
C_{64}	16 μF	Electrolytic	450	
C_{65}	4 μF		250	
C_{66}	2 μF		350	
C_{67}	0.1 μF		350	

S_5 single pole single throw
S_6 single pole single throw

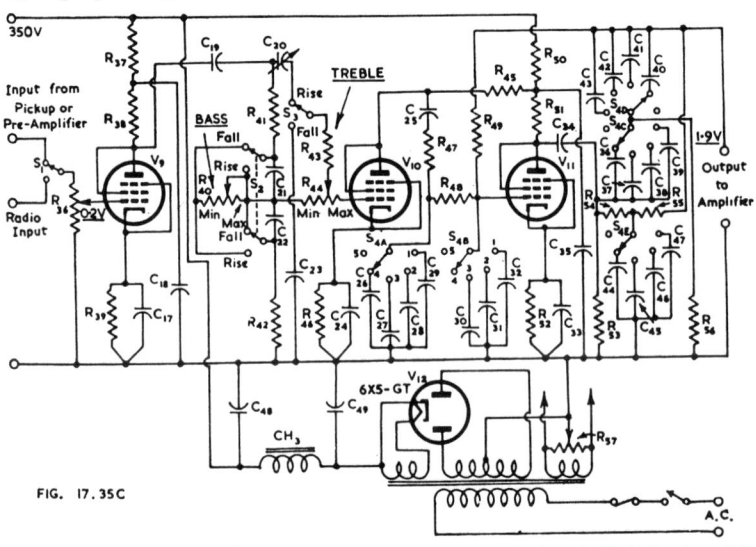

FIG. 17.35C

Fig. 17.35C. *Williamson tone compensation and filter unit (Ref. 270). In the original amplifier* $V_9 = V_{10} = V_{11} = type\ EF37$. *All may be replaced by type 9002, or by type 6AU6 with a slight increase in gain, or by type 6J7 (or 6J5 triode) with a slight decrease in gain.*

		Rating	Tolerance
R_{36}	0.25 $M\Omega$ log.		
R_{37}	47 000 Ω	1W	
R_{38}	47 000 Ω	1W	
R_{39}	3300 Ω		
R_{40}	0.25 $M\Omega$ log.		
R_{41}	100 000 Ω		
R_{42}	6800 Ω		
R_{43}	10 000 Ω		
R_{44}	0.1 $M\Omega$ linear		

R_{45}	100 000 Ω	1W	
R_{46}	2200 Ω		
R_{47}	0.1 $M\Omega$		10%
R_{48}	0.47 $M\Omega$		10%
R_{49}	0.47 $M\Omega$		10%
R_{50}	33 000 Ω	1W	
R_{51}	100 000 Ω	1W	
R_{52}	3300 Ω		
R_{53}	1 $M\Omega$		
R_{54}	0.1 $M\Omega$		1% or matched
R_{55}	0.1 $M\Omega$		1% or matched
R_{56}	50 000 Ω		1% or matched
R_{57}	100 Ω		

All resistors may be $\frac{1}{4}W$ rating, tolerance 20% unless otherwise specified.

		Type	Rating (V d.c. working)	Tolerance
C_{17}	50 μF	Electrolytic	12	
C_{18}	8 μF	Electrolytic	450	
C_{19}	0.25 μF	Paper	500	20%
C_{20}	150 $\mu\mu F$ max.	Preset		
C_{21}	0.01 μF	Paper	250	20%
C_{22}	0.05 μF	Paper	250	20%
C_{23}	1000 $\mu\mu F$	Silvered mica		20%
C_{24}	50 μF	Electrolytic	12	
C_{25}	0.05 μF	Paper	500	20%
C_{26}	100 $\mu\mu F$	Silvered mica		5%
C_{27}	200 $\mu\mu F$	Silvered mica		5%
C_{28}	300 $\mu\mu F$	Silvered mica		5%
C_{29}	500 $\mu\mu F$	Silvered mica		5%
C_{30}	50 $\mu\mu F$	Silvered mica		5%
C_{31}	100 $\mu\mu F$	Silvered mica		5%
C_{32}	250 $\mu\mu F$	Silvered mica		5%
C_{33}	50 μF	Electrolytic	12	20%
C_{34}	0.05 μF	Paper	500	
C_{35}	8 μF	Electrolytic	450	
$C_{36, 40}$	75 $\mu\mu F$	Silvered mica		1% or matched
$C_{37, 41}$	100 $\mu\mu F$	Silvered mica		1% or matched
$C_{38, 42}$	150 $\mu\mu F$	Silvered mica		1% or matched
$C_{39, 43}$	200 $\mu\mu F$	Silvered mica		1% or matched
C_{44}	150 $\mu\mu F$	Silvered mica		1% or matched
C_{45}	200 $\mu\mu F$	Silvered mica		1% or matched
C_{46}	300 $\mu\mu F$	Silvered mica		1% or matched
C_{47}	400 $\mu\mu F$	Silvered mica		
C_{48}	16 μF	Electrolytic	450	
C_{49}	16 μF	Electrolytic	500	

Choke : CH_3 50 H at 20 mA. Resistance about 1500 Ω.
Mains Transformer—
Primary : 200-220-240 V, 50 c/s.
Secondaries : 1. 325-0-325 V, 20 mA d.c.
           2. 6.3 V, 0.6A.
           3. 6.3 V, 1.5A.

Switches—
   S_1 *Single pole single throw.*
   S_2 *Double pole single throw.*
   S_3 *Single pole single throw.*
   S_4 *5 bank, 5 position selector switch.*

7. A high frequency attenuator or filter for worn or noisy records (may be omitted if a sufficiently flexible tone control is fitted). One very effective form of filter is a capacitance connected across the secondary of the output transformer—see page 214.

An example of a complete amplifier is given in Fig. 17.35A. This is a 25 watt amplifier suitable for operation from a G.E. variable reluctance pickup, or Pickering 120M. The total harmonic distortion is under 1%. The pre-amplifier consists of a triode with proper compensation in the output circuit. Bass boosting is obtained by the use of frequency-selective negative feedback in the second and third stages. Hum from the heater is reduced to the vanishing point by making the heaters of the first two twin triodes part of the cathode resistor of the output stage. The output stage draws approximately 300 mA, which is the correct current for the heaters of the 6SL7 valves, and the additional bias required for the output valves is obtained by a resistor (Ref. 240).

Another example of a complete amplifier is that due to D. T. N. Williamson (Ref. 270) shown in Figs. 17.35 B, C, D, E, F and G. The main amplifier (Fig. 17.35F) is a new version of the earlier Williamson amplifier of which one adaptation is Fig. 7.44. The circuit differs in minor details only from Fig. 7.44 ; the balancing adjustment on V_1 and V_2 has been omitted as unnecessary, provided that R_5R_7 R_{11} and R_{13} have the specified values and tolerances, while a transitional phase-shift network $R_{26}C_{10}$ has been added to increase the margin of stability at high frequencies.

Output transformer specifications (3.6 ohm secondaries)—Williamson Amplifier :
   Core : $1\frac{3}{4}$ in. stack of 28A Super Silcor laminations (M. and E.A.). The winding consists of two identical interleaved coils each $1\frac{1}{2}$ in wide on paxolin formers $1\frac{1}{4}$ in. $\times$ $1\frac{3}{4}$ in. inside dimensions. On each former is wound :
      5 primary sections, each consisting of 440 turns (5 layers, 88 turns per layer) of 30 S.W.G. enamelled copper wire interleaved with 2 mil. paper,
alternating with
      4 secondary sections, each consisting of 84 turns (2 layers, 42 turns per layer) of 22 S.W.G. enamelled copper wire interleaved with 2 mil. paper.
Each section is insulated from its neighbours by 3 layers of 5 mil. Empire tape. All connections are brought out on one side of the winding, but the primary sections may be connected in series when winding, two primary connections only per bobbin being brought out. Windings to be assembled on core with one bobbin reversed, and with insulating cheeks and a centre spacer.

Curves showing the loop gain and phase-shift characteristics of the main amplifier are shown in Fig. 17.35G.

The pre-amplifier Fig. 17.35B has a voltage gain of 250 times from the grid of the first valve to the voltage across the output terminals. The input grid voltage must not be less than 0.8 mV ; this minimum voltage can readily be provided by a suitable transformer from any pickup. Care should be taken to avoid overloading the pre-amplifier by too high an input voltage. The overall frequency response, when tested with an ideal " velocity " pickup on an English Decca disc is level within 1 db from 20 to 14 000 c/s ; below 20 c/s there is attenuation at the rate of 30 db/ octave to eliminate rumble. For the English E.M.I. characteristic, switch S_1 should be opened. Other recording characteristics can be handled by the use of the tone compensation unit.

This overall frequency characteristic, which provides bass boosting and high-frequency de-emphasis to suit the recording characteristic, together with a rumble filter, is provided by a careful combination of

(1) bass boosting by V_{13} together with the feedback network C_{54} C_{55} R_{63} R_{64}.

(2) Attenuation produced by the combined inter-valve couplings.

(3) Feedback over V_{14}, through the parallel-T network. This provides a peak at 20 c/s which is used to give a sharp knee to the frequency characteristic instead of the gradual attenuation produced by the inter-valve couplings. (Note that the recording characteristic contributes 6 db/octave to the slope below 20 c/s).

The final stage V_{15} is merely a cathode follower to permit the use of long leads between the pre-amplifier and the tone-compensation unit.

If a high-impedance output is permissible, and an attenuation of 8 db at 10 c/s is sufficient to reduce the rumble, stage V_{13} may be used as a single stage pre-amplifier with a gain of 11 times. Under these circumstances $C_{53} = 0.05\ \mu F$, $C_{54} = 4000\ \mu\mu F$, $R_{64} = 22\ 000$ ohms, $R_{66} = 2.2\ M\Omega$, R_{67} is omitted, and the output is taken from the junction of C_{53} and R_{66}.

In the cathode circuit of V_{13}, closing the switch S_5 reduces the gain to zero in about 1 second for use while changing records.

The noise level, with R_{57} adjusted for minimum hum, is about 3 to 5 μV at V_{13} grid, excluding the noise due to the pickup transformer and auxiliaries. The total harmonic distortion of the pre-amplifier and tone compensation units combined, is considerably less than 0.1%.

The tone compensation and filter unit Fig. 17.35C provides—

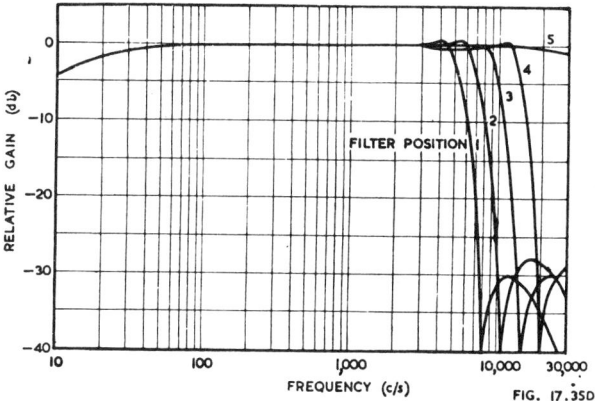

Fig. 17.35D. *Measured overall response of low-pass filter (stage V_{11} in Fig. 17.35C together with network controlled by switch S_4). Ref. 270.*

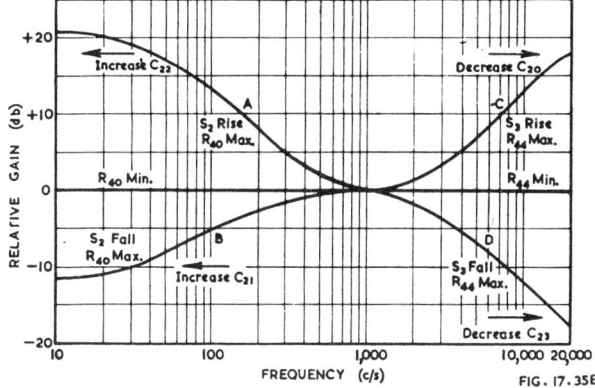

Fig. 17.35E. *Response curves of the tone compensation circuit following V_9 in Fig. 17.35C. Curves are limiting positions for continuously-variable controls (Ref. 270).*

(1) A universal tone control (switches S_2S_3 and controls $R_{40}R_{44}$) with frequency characteristics as Fig. 17.35E. The curves may be shifted bodily along the horizontal axis by modifying the capacitance values shown by the arrows in Fig. 17.35E.

(2) A low-pass filter with nominal cut-off values of 5000, 7000, 10 000 and 13 000 c/s, together with a flat position (switch S_4) providing response characteristics as Fig. 17.35D. This characteristic is provided by a parallel-T network in the feedback loop of V_{11} which gives a symmetrical valley at the frequency of resonance. A capacitor from the grid of V_{11} to earth introduces a lagging phase shift which gives

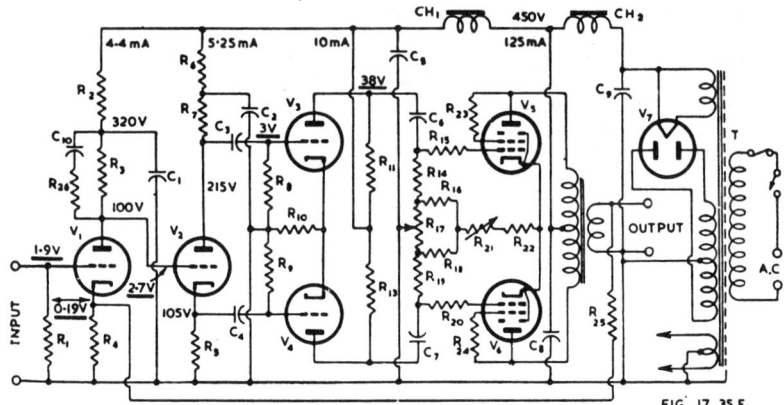

FIG. 17.35 F

Fig. 17.35F. Williamson main amplifier—new version (Ref. 270). Voltages underlined are peak signal voltages at 15 watts output.

R_1	1 $M\Omega$	$\frac{1}{4}$ watt $\pm$ 20%
R_2	33 000 Ω	1 watt $\pm$ 20%
R_3	47 000 Ω	1 watt $\pm$ 20%
R_4	470 Ω	$\frac{1}{4}$ watt $\pm$ 10%
R_5, R_7	22 000 Ω	1 watt $\pm$ 5%
		(or matched)
R_6	22 000 Ω	1 watt $\pm$ 20%
R_8, R_9	0.47 $M\Omega$	$\frac{1}{4}$ watt $\pm$ 20%
R_{10}	390 Ω	$\frac{1}{4}$ watt $\pm$ 10%
R_{11}, R_{13}	47 000 Ω	2 watt $\pm$ 5%
		(or matched)
R_{14}, R_{19}	0.1 $M\Omega$	$\frac{1}{4}$ watt $\pm$ 10%
R_{15}, R_{20}	1000 Ω	$\frac{1}{4}$ watt $\pm$ 20%
R_{16}, R_{18}	100 Ω	1 watt $\pm$ 20%
R_{17}, R_{21}	100 Ω	2 watt wirewound variable
R_{22}	150 Ω	3 watt $\pm$ 20%
R_{25}	1200 $\Omega \times \sqrt{speech\ coil\ impedance}$	
		$\frac{1}{4}$ watt
R_{26}	4700 Ω	$\frac{1}{4}$ watt $\pm$ 20%
C_1, C_2, C_5, C_8	8 μF	500V wkg.
C_3, C_4	0.05 μF	350V wkg.
C_6, C_7	0.25 μF	350V wkg.
C_9	8 μF	600V wkg.
C_{10}	200 $\mu\mu F$	350V wkg.
CH_1	30H at 20 mA	
CH_2	10H at 150 mA	
T	Power transformer	

Secondary 425-0-425V, 150 mA, 5V, 3A, 6.3V, 4A, centre-tapped
V_1, V_2 2 $\times$ L63 or 6J5, 6SN7 or B65
V_3, V_4 do.   do.
V_5, V_6, KT66, V_7 Cossor 53KU, 5V4.

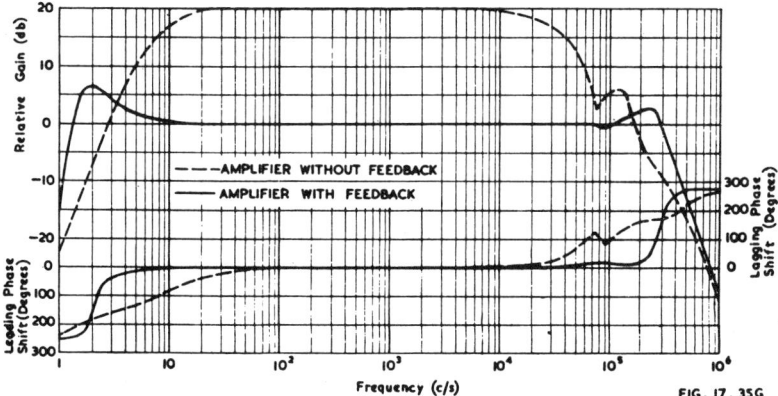

Fig. 17.35G. *Loop gain and phase-shift characteristics of Williamson main amplifier* (*Ref.* 270).

positive feedback below resonance and negative feedback above resonance, thereby unbalancing the characteristic and giving a sharp knee before the attenuation commences. The addition of an independent RC network controlled by S_{4A} gives further treble attenuation and leads to the final result in Fig. 17.35D.

(x) Pickups for connection to radio receivers

When a separate pickup and turntable unit is used in conjunction with a radio receiver, it is often most convenient for the user to operate the device at some distance (say up to 15 feet) from the receiver. The most desirable arrangement is a Record Player in the form of a small cabinet enclosing the motor, turntable, pickup, and (possibly) volume control, pre-amplifier and equalizer, with the top of the cabinet about 20 inches from the floor so as to be accessible from an easy chair.

The design problem hinges on the use or otherwise of a pre-amplifier. **If no pre-amplifier is permissible**, on account of cost or other factors, it is necessary to employ a pickup delivering a high output voltage or whose output voltage is capable of being stepped up by a high ratio transformer.

When pickup terminals are fitted to receivers, they usually provide a resistive load of the order of 0.5 megohm, with some capacitive shunting. The input voltage required to give full power output is usually about 0.25 volt, although some receivers require a higher voltage. If a receiver is being designed specially for good performance with a pickup, full power output should be obtained with an input voltage not exceeding 0.1 volt. This usually demands a pentode first a-f amplifier, especially if negative feedback is used.

If bass equalization of 20 db is required for use with an electro-magnetic pickup, the available voltage must be at least 10 times the input voltage to the amplifier, e.g. 1 volt and 0.1 volt respectively There is a strong temptation to use a popular pickup which provides at least part of the bass boosting by arm resonance, but this is very undesirable on account of record wear, distortion and heavy stylus force required. There are electro-magnetic pickups capable of providing an output of about 1 volt from the secondary of a transformer, which do not employ arm resonance in the vicinity of 100 c/s ; these are much to be preferred to those employing arm resonance. However, the high cost of a good quality transformer, and the hum and poor frequency response with a cheaper transformer, make this a rather unsatisfactory arrangement. The necessary bass boosting may be provided by a resistor and capacitor in series, connected across the secondary of the transformer A choice of cross-over frequency may be provided by a tapping switch and several values of capacitance. High-frequency de-emphasis may be provided by one or more values of shunt capacitance, although there is danger of pickup resonance unless an isolating resistor is used, with consequential loss of gain.   Volume control may be provided by a potentiometer having a total resistance preferably not exceeding 50 000 ohms.   A screened cable may be used to link the record player to the pickup terminals of the receiver.

If a crystal pickup is used, it is quite satisfactory to incorporate the equalizer and volume control in the Record Player.   In this case there is no danger of hum from the pickup or transformer.   With a general-purpose crystal pickup the arrangement of Fig. 17.29 may be used in which $R_1 = 2$ megohms, $R_2 = 0.5$ megohm and $C_1 = 0.000\,25\ \mu$F for a cross-over frequency of 250 c/s or half this value for 500 c/s.   Some measure of tone control may be provided by a variable resistance of 1 megohm in series with C_1.

With a high fidelity crystal pickup, the circuit of Fig. 17.31 or any similar equalizer may be used.

In all cases it is advisable for the listener, if possible, to set the volume control on the record player to about half-way, then to adjust the volume control on the receiver for satisfactory operation under normal conditions, and to make all adjustments (during listening) with the control on the record player.

There will be some inevitable loss of the higher frequencies due to the screened cable.   This may be kept within reason by arranging that the total effective impedance of the line, under operating conditions, is of the order of 50 000 ohms or less.   For example, the equalizer of Fig. 17.32 may be modified by changing the 0.1 megohm fixed resistor into a volume control.

If a pre-amplifier is used, both it and the equalizer will normally be incorporated into the receiver chassis.   A volume control may readily be fitted across the pickup, and shunt capacitances may be used in conjunction with a tapping switch for high frequency de-emphasis, with due precautions against resonance.   With this arrangement it is difficult to make any adjustment for cross-over frequency in the record player.

Alternatively the pre-amplifier and equalizer may be incorporated into the record player, thus giving the maximum flexibility of control.   The final stage in the record player may be either a cathode follower, or a general-purpose triode with a low ratio transformer in its plate circuit.   The record player may incorporate a power transformer, rectifier and filter, or it may obtain its plate supply from the receiver.   Trouble may be experienced with hum owing to the proximity of the motor, transformer and pre-amplifier, particularly if a low level pickup is used.   However, in spite of these design problems, the arrangement is practicable.

(xi) Frequency test records

Frequency test records are of two general types—banded tone and gliding tone. Banded tone records are recorded with a number of constant frequencies which may be played in sequence ; they are generally used when measurements are being made. Gliding tone records are useful when it is desired to check for peaks but may also be used for frequency response measurements by appropriate timing.   These normally commence at a high frequency f_1, maintain constant velocity characteristics down to the cross-over frequency f_0, then follow the recording characteristic down to the final frequency f_2 (approximately at the rate of 6 db/octave).   In some cases the high-frequency response follows the recording characteristic—details are given in the table. The frequency response may be checked to a fair degree of accuracy by the optical method (see below).   Over the " constant velocity " range the width of the light pattern should be constant, while over the " constant amplitude " range the width should taper downwards in proportion to the frequency.

References to frequency test records : Refs. 161, 254, 260 ; record catalogues.

The optical method of testing frequency test records

A frequency test record, either of the stepped-frequency or gliding-tone type, may be checked optically to indicate the amplitude at all parts of the record, with a possible error of about 1 or 2 db.   The lamp should have a concentrated filament and clear bulb ; it should be at least 8 feet from the record and only high enough above the plane of the record to give a brilliant pattern. The eye should be directed vertically downwards on to the side of the record nearer to the lamp, and should be as far above the record as possible.

A pair of dividers may be used to measure the width of the reflected light pattern for each frequency ; the width of the light pattern is proportional to the voltage output from a perfect constant-velocity pickup. This method becomes increasingly less accurate below the cross-over frequency, although it is very satisfactory at higher frequencies.

The theoretical analysis and calculation of errors is given by Ref. 243.

References 25 (Part 5), 99, 219, 243, 258, 293.

TABLE 1 : GLIDING FREQUENCY RECORDS

Record	f_1	f_0 (cross-over)	f_2
English			
Decca ffrr K1802	14 000 c/s		10 c/s
K1803	14 000 c/s	300 c/s	10 c/s
American			
R.C.A. 12 in. (12-5-5)	10 000 c/s	500 c/s (-1.5 db)	30 c/s
London (album LA-32)			
T-4996 12 in. 78 r.p.m.	14 000 c/s ($+8$ db)	250 c/s (-1 db)	100 c/s (-5 db)
T-4997 12 in. 78 r.p.m.	14 000 c/s ($+8$ db)	250 c/s (-1 db)	10 c/s (-33 db)
Universal D61B Band 1	10 000 c/s (0 db)	500 c/s (0 db)	
Band 2		500 c/s (-7 db)	200 c/s (-7 db)
Band 3		200 c/s (-14 db)	50 c/s (-14 db)
Columbia XERD281 12 in.	10 000 c/s	500 c/s	50 c/s

TABLE 2 : COMBINED GLIDING FREQUENCY AND BANDED FREQUENCY RECORDS

	f_1	f_0 (cross-over)	f_2
R.M.A. Standard REC-128 78 r.p.m.			
No. 1 Side A	10 000 c/s	500 c/s	30 c/s

No. 1 Side A: Constant frequencies 1000 ($+$ 10 db), 10 000, 9000, 8000, 7000, 6000, 5000, 4500, 4000, 3500, 3000, 2500, 2000, 1500, 1000, 700, 500, 400, 300, 250, 200, 180, 160, 140, 120, 100, 90, 80, 70, 60, 50, 40, 30.

No. 1 Side B: Voice announcements. Constant frequency 1000 c/s at levels $+10$, $+12$, $+14$, $+16$, $+18$ db. Then alternate gliding and constant frequencies from 3000 to 30 c/s at level $+16$ to $+18$ db (1000 c/s).

TABLE 3 : BANDED FREQUENCY RECORDS

English and Australian (78 r.p.m.) E.M.I. Australia (ED1189)	20 000, 18 000, 16 000, 14 000, 12 000, 10 000, 8000, 6000, 4500, 3500, 2000, (arbitrary 0 db) ; 500 (-1), 160 (-5.5), 70 (-12 db). 19 000, 17 000, 15 000, 13 000, 11 000, 9000, 7000, 5000, 4000, 3000, 1000 (arbitrary 0 db) ; 250 (-3), 100 (-8.5), 50 (-14 db).
E.M.I. Australia (ED1190)	13 000 (-1.2 db), 12 000 (-1.3), 11 000 (-0.4), 10 000 (-0.2), 9000 (-0.2), 8000 (-0.2), 7000 ($+0.4$), 6000 ($+0.2$), 5000 ($+0.2$), 4000 ($+0.7$), 3000 ($+1.2$), 2000 ($+0.6$), 1000 (0 db arbitrary level), 500 ($+0.2$), 250 (-2.9), 140 (-6), 100 (-8.9), 70 (-10.4), 50 (-12.8), 35 c/s (-14 db).
E.M.I. Studios JG449 (1) (2)	20 Kc/s, 18, 16, 14, 12, 10, 8, 6, 4.5, 3.5, 2 Kc/s (0 db) ; 500 c/s (-1), 160 c/s (-5.5), 70 c/s (-12 db). 19 Kc/s, 17, 15, 13, 11, 9, 7, 5, 4, 3, 1 Kc/s (0 db) ; 250 c/s (-3), 100 c/s (-8.5), 50 c/s (-14 db).
Columbia LOX-650	1000 (0 db), 10 000 (-1), 9000 (-1.6), 8000 (-1.6), 7000 (-1), 6000 (-0.8), 5000 (-0.8), 4000 (-0.8), 3000 (-0.8), 2000 (-0.8), 1500 (-1), 800 (-0.8), 500 (-0.8), 300 (-3.4), 200 (-5.2), 150 (-7.1), 100 (-9.4), 70 (-11.2), 50 (-13.6 db).
Decca Z718	$\left\{\begin{array}{l} 50, \quad 70, 100, 160, 250, 500, 1000, 2000, \text{c/s} \\ -14 \quad -11 \quad -8 \quad -4 \quad 0 \quad 0 \quad 0 \quad 0 \text{ db} \end{array}\right.$ $\left\{\begin{array}{l} 3000, 3500, 4000, 4500, 5000, 6000 \text{ c/s} \\ 0 \quad 0 \quad 0 \quad 0 \quad 0 \quad 0 \text{ db} \end{array}\right.$
Decca K1804	14 000 to 400 c/s (constant velocity) 250 to 30 c/s (constant amplitude) ; in steps. Level $+10$ db at 1000 c/s.
American R.M.A. Standard ⎤ No. 2(A) REC-128 ⎬ 78 r.p.m. ⎦ No. 2(B)	Constant frequency bands at same frequencies as No. 1 Side A without announcements. 8000, 6000, 4000, 2000 ($+10$ db) ; alternate gliding and constant frequencies 3000 to 30 c/s ($+16$ to 18 db) ; 8000, 6000, 4000, 2000 (⏐ 10 db).
R.C.A. (12-5-19) 12 in. ⎤ 78 r.p.m. ⎬ Unfilled vinyl, SF ⎦	Approx. const. vel. 10 000 ; 9000 ; 8000 ; 7000 ; 6000 ; 5000 ; 4000 ; 3000 ; 2000, 1000 ; (0 db = 8.6 cm/sec. approx.) 800 c/s ; 500 (-1.5 db) ; Const. ampl. 300 ; 200 ; 100 ; 50 ; 1000 (-2 db) ; 10 000.
R.C.A. (12-5-25) ($=460625$-6) 12 in. 33-1/3 r.p.m. Unfilled vinyl, SF	12 000 (-2.5 db) ; 11 000 (-3) ; 10 000 (-1.5) ; approx. const. vel. down to 700 (-0.5 db) ; 400 (1.2 db below const. ampl.) ; 300 (0.6 db below const. ampl.). Also 400 and 4000 c/s tone for intermodulation testing. Groove has bottom radius less than 0.5 mil. Max. stylus radius 3 mils.
R.C.A. (12-5-31) 7 in. 45 r.p.m. Unfilled vinyl, D.F. (Opposite side of 12-5-29)	Approx. const vel. 10 000 (-0.5 db) ; 9000 ; 8000 ; 7000 (-1) ; 6000 ; 5000 (-1.3) ; 4000 ; 3000 (-1) ; 2000 ; 1000 (0 db = 5.2 cm/sec.) ; 700 (-0.5) ; 400 (-1.5) ; approx. const. ampl. 400 ; 300 ; 200 ; 100 ; 50 ; 1000. Velocity at 1000 c/s = 5.2 cm/sec.

Columbia
TL-1 LP microgroove
33-1/3 r.p.m.

10 000 (+12 db) ; 9000 (+12) ; 8000 (+11.5) ; 7000 (+11) ; 6000 (+11) ; 5000 (+10) ; 4000 (+7.5) ; 3000 (+5.5) ; 2000 (+3) ; 1500 (+2) ; 1000 (0 db = 1 cm/sec. r.m.s.) ; 800 (−1) ; 500 (−2.5) ; 400 (−4) ; 300 (−6) ; 200 (−8.5) ; 150 (−10) ; 100 (−11) ; 70 (−12) ; 50 (−13). This record is cut with the microgroove recording characteristic including high-frequency pre-emphasis.

TL-2 LP microgroove
33-1/3 r.p.m.

10 000 (+5.5 db) ; 9000 (+5.5) ; 8000 (+5) ; 7000 (+5.5) ; 6000 (+5) ; 5000 (+7) ; 4000 (+7.5) ; 3000 (+7) ; 2000 (+4.5) ; 1500 (+6.5) ; 1000 (+7.5) ; 800 (+7.5) ; 500 (+7.5) ; 400 (+6) ; 300 (+3.5) ; 200 (−0.5) ; 150 (−3.0) ; 100 (−6) ; 70 (−7.5) ; 50 (−10.5) ; 30 (−12 db). 0 db = 1 cm/sec. r.m.s.

RD-103 LP microgroove
33-1/3 r.p.m.
12 in. Bands

10 000 (+5 db) ; 9000 ; 8000 ; 7000 ; 6000 (+5) ; 5000 ; 4000 (+7) ; 3000 ; 2000 (+5) ; 1500 ; 1000 (+7.5) ; 800 ; 500 (+7.5) ; 400 ; 300 ; 200 (0 db) ; 150 ; 100 (−6) ; 70 ; 50 ; 30 c/s (−12) ; 0 db = 1 cm/sec. Essentially flat over constant-velocity portion, with cross-over 500 c/s.

RD-103A LP microgroove
33-1/3 r.p.m.
12 in. Bands

10 000 (+12 db) ; 9000 ; 8000 ; 7000 ; 6000 (+10 db) ; 5000 ; 4000 ; 3000 (+5) ; 2000 ; 1000 (0 db) ; 800 ; 500 (−2.5) ; 400 ; 300 (−6) ; 200 ; 100 (−11) ; 70 ; 50 c/s (−13 db). 0 db = 1 cm/sec. Levels arranged to reproduce ± 2 db of " flat " on correctly equalized LP reproducing system.

(10003-M) 78 r.p.m.
12 in. shellac

1000 c/s ; 10 000 (−7 db) ; 9000 ; 8000 ; 7000 ; 6000 ; 5000 ; 4000 ; 3000 ; 2000 ; 1500 ; 1000 (−1 db) ; 800 ; 500 ; 300 ; 200 ; 150 ; 100 ; 70 ; 50 (−17 db). Cross-over 300 c/s. Level approx. 4.8 cm/sec. at 1000 c/s.

(10004-M) 78 r.p.m.
12 in. shellac

As (10003-M) but with cross-over frequency 500 c/s.

Audiotone
78-1 78 r.p.m.

50 to 250 c/s (peak amplitude 0.0017 inch) ; 250 c/s upwards (constant velocity).

London Gramophone
T4998 12 in. 78 r.p.m.

14 000 c/s (0 db) ; 13 000 ; 12 000 ; 11 000 ; 10 000 ; 9000 ; 8000 ; 7000 ; 6000 ; 5000 ; 4000 ; 3000 ; 2000 ; 1000 ; 400 (0 db) ; 250 (−2) ; 100 (−6) ; 55 ; 30 c/s (−16 db).

Clarkstan
(2000S) 78 r.p.m.

50 to 500 (constant amplitude) 500 to 10 000 c/s (constant velocity) in 17 steps.

(20002S) LP microgroove

One side recorded flat, other side with NAB curve.

Cook
Series 10, 10 in. Plastic
DF. Side A 78 r.p.m.

Constant velocity (9 cm/sec.) above 500 c/s. with 3 db knee at crossover. Bands 1000 c/s ; 20 000 ; 17 000 ; 15 000 ; 12 000 ; 10 000 ; 9000 ; 8000 ; 7000 ; 6000 ; 5000 ; 4000 ; 3000 ; 2000 ; 1500 ; 1000 ; 700 ; 500 ; 350 ; 250 ; 125 ; 62.5 ; 40 ; 35 ; 1000 c/s.
Side B : 33-1/3 r.p.m. ; Band 1-LP spot check for standard LP pre-emphasis—100 ; 1000 ; 3000 ; 6000 ; 10 000 c/s. Band 2—100 and 7000 c/s intermodulation test, with 7000 c/s 12 db lower than 100 c/s, on

flat basis. No pre-emphasis. Band 3—slow sweep frequencies from 1000 c/s to 35 c/s with 350 c/s crossover. Both sides cut for use with 1 or 3 mil reproducer stylus.

TABLE 4 : SPECIAL TEST RECORDS

R.C.A. Victor

(12-5-1) 10 in. 78 r.p.m. Shellac DF	Unmodulated grooves at normal recording pitch with lead out and eccentric. Same both sides.
(12-5-3) 12 in. 78 r.p.m. Shellac DF	Unmodulated grooves at normal recording pitch with lead out and eccentric. Same both sides.
(12-5-7) 12 in. 78 and 33-1/3 r.p.m. Shellac DF	On 78 r.p.m. bands 2300 ; 1000 c/s. On 33-1/3 r.p.m. 1000 ; 433 c/s. Frequencies constant within 0.2% instantaneous (12-5-5 on opposite side).
(12-5-9) 12 in. 78 r.p.m. Shellac SF	Band 1—silent ; 2—400 c/s at 5.9 cm/sec. ; 3—1000 c/s at 9.6 cm/sec ; 4—silent. Frequencies constant within 0.2% instantaneous.
(12-5-11) 12 in. 78 r.p.m. Shellac DF	Landing area with no lead-in spiral ; at least one unmodulated normal pitch groove ; steep blank spiral ; normal pitch inside groove ; lead out and eccentric Same both sides.
(12-5-13) 10 in. 78 r.p.m Shellac DF	As for (12-5-11). Same both sides.
(12-5-15) 12 in. 78 r.p.m. Unfilled vinyl DF	Warble frequency bands, sweep rate 5.5 c/s. Band (1) 500—2500 c/s ; (2) 750—1250 c/s ; (3) 1250—1750 c/s ; (4) 1800—2600 c/s. Same both sides.
(12-5-17) 10 in. 78 r.p.m. Shellac DF	As (12-5-15). Same both sides.
(12-5-21) 12 in. 78 r.p.m. Unfilled vinyl DF	For checking automatic record changers. Three modulated bands with large eccentric groove. Bands define standard 10 and 12 in. landing areas without lead in grooves. Inner band 936 c/s (1 minute) for checking wow.
(12-5-23) 12 in. 78 r.p.m. Unfilled vinyl DF	Various modulated bands joined by spiral grooves to indicate limits of standard recording dimensions (12-5-21 on other side).
(12-5-29) 7 in. 45 r.p.m. Unfilled vinyl DF	For checking landing and tripping action of changer mechanisms. Bands 1000 ; 400 ; 1000 c/s.
(12-5-35) 7 in. 45 r.p.m. Unfilled vinyl DF	For checking landing and tripping action of changer mechanisms ; for checking pickup sensitivity, turntable flutter and rumble.
(12-5-37) 7 in. 45 r.p.m. Unfilled vinyl DF (RL-419)	Bands of 400 and 4000 c/s signals combined, the 4000 c/s being 12 db below the 400 c/s level. Peak velocities from 3.8 to 18 cm/sec. in approx. 2 db steps. 0 db = 6 cm/sec. Intermodulation distortion in the record is less than 4%. For testing pickup " tracking " at various levels and stylus forces (Ref. 285).
(12-5-39) 12 in. 78 r.p.m. Unfilled vinyl DF (RL-420)	Bands of 400 and 4000 c/s signals combined, the 4000 c/s being 12 db below the 400 c/s level. Peak recorded velocities run from 27 to 4.4 cm/sec. in approx. 2 db steps. 0 db = 9.1 cm/sec. Intermodulation distortion in the record is less than 3%. Groove has small bottom radius suitable for testing with 1.0 or 3 0 mil styli. Same use as 12-5-37.

(12-5-41) 12 in. 33.3 or 78 r.p.m. Unfilled vinyl DF	For routine testing of record changer operation with 1.0 or 3.0 mil styli. Standard R.M.A. landing areas for 10 and 12 inch records are defined by short interrupted tones. Short bands of 400 and 1000 c/s tones are included for routine pickup sensitivity measurements.
Clarkstan 1000 A 12 in. 78 r.p.m. vinylite SF	Sweep frequency record for oscillographic observation of equipment response. Frequency range 70 to 10 000 c/s, flat within ± 1 db. Sweep frequency rate 20 times/sec. Crossover 500 c/s.
1000 D 12 in .78 r.p.m. vinylite SF.	Sweep frequency record as 1000A but covering range from 5000 to 15 000 c/s.
E.M.I. Studios JH138 10 in. 78 r.p.m. for use with 2.5 mil radius stylus.	Side 1 : 400 c/s (+22.5 db) with approx. 4000 c/s (+10.5 db) superimposed additively for I.M. testing of pickups. Peak lateral velocity of combined wave is equal to that of a sine wave at a level +24.5 db. The succeeding 10 bands have levels of both tones reduced 2 db below those of the foregoing band. Side 2 : 60 c/s (+8.6 db) with 2000 c/s (+10.3 db) superimposed additively. When the 2000 c/s is reduced in the pickup bass correction equalizer by the correct amount relative to 60 c/s (i.e. 13.7 db), its effective level will be −3.4 db, i.e. 12 db below the 60 c/s tone. On a velocity basis, the peak lateral velocity of the combined wave is equal to that of a sine wave at a level +15.5 db. The peak combined amplitude is equivalent to that of a 60 c/s sine wave having a level of +10 db. The succeeding 10 bands each have the level of both tones reduced by 2 db below those of the foregoing band. 0 db = 1 cm/sec. r.m.s. (Ref. 310).

See Supplement for additional Frequency Test Records.

SECTION 6 : DISTORTION * AND UNDESIRABLE EFFECTS

(i) Tracing distortion and pinch effect   (ii) Playback loss   (iii) Wow, and the effects of record warping   (iv) Distortion due to stylus wear   (v) Noise modulation   (vi) Pickup distortion   (vii) Acoustical radiation   (viii) Distortion in recording.

(i) Tracing distortion and pinch effect

Tracing distortion (also known as playback distortion) is non-linearity introduced in the reproduction of records because of the fact that the curve traced by the centre of the tip of the reproducing stylus is not an exact replica of the modulated groove.

Detailed mathematical analyses of tracing distortion have been made (Refs. 41 and 213, 42, 224, 264) and some popular articles have been written (Refs. 40, 146). The distortion arises from the shape of the cutter, which has a chisel edge, so that a spherical tipped needle follows a different path—in other words it introduces harmonic distortion. Odd harmonics are shown up in the lateral movement of the needle, while even harmonics make their presence felt through vertical movement of the needle—the "**pinch effect.**"

* See also Chapter 37 Sect. 1 (vi) N for I.R.E. tests on phonograph combinations.

Harmonic tracing distortion

It had earlier been believed (Ref. 42) that the dominant harmonic was the third, but a more recent analysis (Ref. 264) has proved that the higher harmonics also should be considered. The values of the fundamental and harmonics up to the seventh may be calculated from the equations below. The harmonic distortion, when reproducing with a constant velocity pickup, is given by the ratio of the harmonic to the fundamental, multiplied by the harmonic number (e.g. 3 for third harmonic).

$$\text{Fundamental amplitude} = A - \frac{1}{16}R^2A^3\left\{1 - \tfrac{1}{4}A^2 + \frac{5}{64}A^4 - \ldots\right\} +$$

$$\frac{1}{768}R^4A^5\left\{1 - \tfrac{5}{8}A^2 + \ldots\right\} - \frac{1}{73\,728}R^6A^7\left\{1 - \ldots\right\} + \ldots \quad (1)$$

$$\text{Third harmonic amplitude} = -\frac{1}{16}R^2A^3\left\{1 - \tfrac{3}{8}A^2 + \frac{9}{64}A^4 - \ldots\right\} +$$

$$\frac{9}{512}R^4A^5\left\{1 - \tfrac{3}{4}A^2 + . .\right\} - \frac{81}{40\,960}R^6A^7\left\{1 - \ldots\right\} + \ldots \quad (2)$$

$$\text{Fifth harmonic amplitude} = \frac{1}{128}R^2A^5\left\{1 - \tfrac{5}{8}A^2 + \ldots\right\} +$$

$$\frac{25}{1536}R^4A^5\left\{1 - \frac{5}{4}A^2 + \ldots\right\} - \frac{625}{73\,728}R^6A^7\left\{1 - \ldots\right\} - \ldots \quad (3)$$

$$\text{Seventh harmonic amplitude} = -\frac{1}{1024}R^2A^7\left\{1 - \ldots\right\} -$$

$$\frac{49}{6144}R^4A^7\left\{1 - \ldots\right\} - \frac{2401}{368\,640}R^6A^7\left\{1 - \ldots\right\} - \ldots \quad (4)$$

where $A = 2\pi a/\lambda$
$\quad\quad R = 2\pi r/\lambda$
$\quad\quad a$ = amplitude of lateral modulation of groove, measured in the plane of the record
$\quad\quad r$ = stylus radius
and $\quad \lambda$ = wavelength of sinusoidal modulation measured in the direction of an unmodulated groove.

These formulae only hold when RA does not exceed $\sqrt{2}$.

Example : Groove velocity 10 inches/sec., lateral groove velocity 2 inches/sec for $f = 4000$ c/s,
$\quad r = 2.3$ mils.
Wavelength $= \lambda = 10/4000 = 0.0025$ inch.
Amplitude $= a = 2/2\pi(4000) = 1/4000\pi$ inch.
$A = 2\pi a/\lambda = 0.2$.
$R = 2\pi(0.0023)(400) = 1.84\pi$.

From (1) : Fundamental amplitude $= 0.1839$ (this is 91 9% of the groove amplitude).
From (2) : Third harmonic amplitude $= -0.0113$.
Percentage third harmonic with constant velocity pickup $=$
$\quad 3 \times 0.0113 \times 100/0.1839 = 18.4\%$.
From (3) : Fifth harmonic amplitude $= +0.0035$.
Percentage fifth harmonic with constant velocity pickup $=$
$\quad 5 \times 0.0035 \times 100/0.1839 = 9\%$.
Total r.m.s. distortion $= 20\%$.

Intermodulation tracing distortion

The harmonic distortion so far discussed implies the existence of intermodulation distortion which causes the most distressing effects on the listener and is still audible when the offending harmonics are inaudible. The ratio of intermodulation distortion to total harmonic distortion may be as high as 10 times (Ref. 110).   Intermodulation distortion may, in severe cases, be audible as a buzz.

Since the higher order harmonics are substantial, the most convenient method of comparison for different conditions is given by intermodulation distortion (Ref. 264) and four curves for selected operating conditions are given in Fig. 17.37A.   Under these operating conditions the upper limit for good reproduction is 10% intermodulation—this does not hold for other conditions.

The minimum groove diameters for reasonable fidelity (less than 10% calculated intermodulation distortion as given by Fig. 7.37A) are :

R.P.M.	Stylus radius	Min. groove diameter
78	3.0 mils	6.1 inches
45	1.0	4.9
33-1/3	1.0	6.5
33-1/3	2.3	10.1

Fortunately, owing to the smaller amplitudes which usually occur in both speech and music at the higher frequencies, the actual distortion under the prevalent listening conditions is usually less than the values given above.   In addition, the yield of the record material reduces the distortion (Ref. 265) ; this is appreciable with vinyl, but negligible on shellac records.   A further effect is the translation loss which may be about 6 to 8 db at 10 000 c/s at the inner radius with 78 r.p.m. standard groove, whereby both fundamental and harmonics are reduced.   All these subsidiary effects tend to decrease the actual tracing distortion.

However, the deterioration in both frequency response and fidelity is plainly audible to a critical listener as the groove radius decreases towards the end of a record ; this applies to 78 r.p.m., transcriptions and all types of fine groove recording.

The effect of recording characteristics on tracing distortion

Over the range of constant velocity recording, the tracing distortion is proportional to the square of the frequency

With high-frequency pre-emphasis followed by the correct amount of de-emphasis, at the rate of 6 db/octave above a " hinge frequency " f_1 (P in Fig 17.37B) the third harmonic tracing distortion at a recorded frequency f_1 will be one third of the value calculated by the procedure outlined above provided the constants are the same in both cases. This relationship will, however, not hold at any other point.   For example, at frequency f_2 where $f_2 = 2f_1$, the amplitude at Q' will be twice that at Q and the

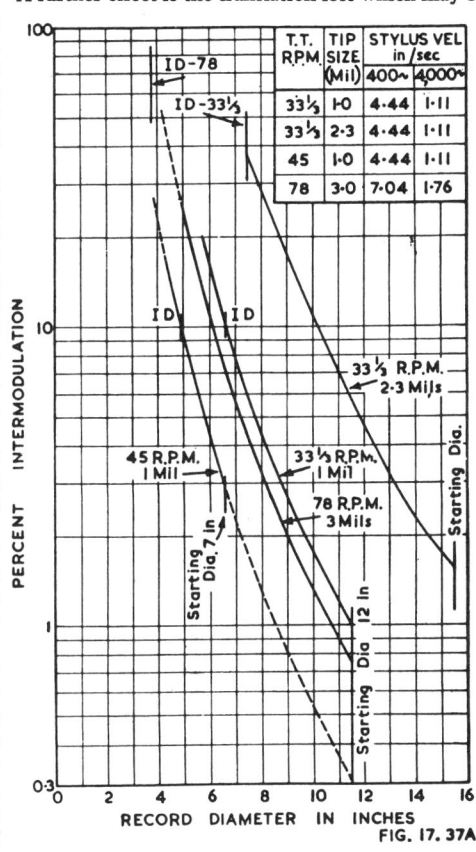

Fig. 17.37A.   Variation of calculated intermodulation tracing distortion with record diameter (based on Ref. 292).

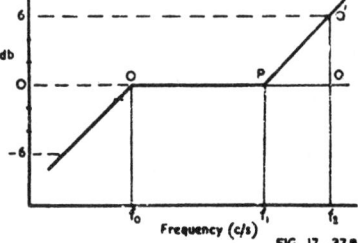

*Fig. 17.37B. Ideal recording char-
acteristic for calculating tracing
distortion.*

distortion will therefore be four times as great with pre-emphasis : de-emphasis
as at point P. This may be expressed in the general form

$$\frac{H_3\% \text{ (constant amplitude)}}{H_3\% \text{ (constant velocity)}} = \frac{m^2}{3} \tag{5}$$

where $m = f_2/f_1$

f_2 = recorded frequency

f_1 = hinge frequency of pre-emphasis

and the pre-emphasis and de-emphasis are both at the rate of 6 db/octave above f_1

If $f_2 = f_1$	$m = 1$	Ratio of $H_3\% = 1/3$
$f_2 = 2f_1$	$m = 2$	Ratio of $H_3\% = 4/3$
$f_2 = 4f_1$	$m = 4$	Ratio of $H_3\% = 16/3$

Other distortion effects

In the foregoing treatment it is assumed that the pickup stylus and armature are
able to follow ideally all the sharp and sudden changes in the modulation of the groove
which occur with complex sounds. In many cases this is obviously not the case, so
that additional causes of spurious intermodulation products (mainly inharmonic)
arise. These are reduced by the use of a really good pickup but in most cases it is
probably true that the best available pickup will introduce distortion additional to
the theoretical tracing distortion.

The distortion arising from overmodulation at high frequencies due to a high value
of pre-emphasis has already been covered in Sect. 5(i). Similar effects may occur
at low frequencies with organ music, but may be avoided by the use of a low frequency
limiter in the recording amplifier.

The effect of finite size of needle tip on the reproduction of the fundamental recorded frequency is covered in Sect. 2(i), also translation loss.

The **pinch effect** has already been mentioned. It causes increased combination
products, forced vibration of the armature at its high resonant frequency, increased
noise in the output voltage and needle talk. Fortunately these bad effects may be
considerably reduced by the introduction of vertical compliance either in the needle
(bent shank or trailer type) or in the pickup itself.

References to tracing distortion, intermodulation distortion and pinch effect :
11, 40, 41, 42, 47, 110, 146, 193, 212, 213, 223, 224, 225, 248, 264, 265, 285.

(ii) Playback loss

Playback loss has been defined as the difference between the recorded and the reproduced level at the very same point of a record. It is due to the physical properties
of the record material, being evident as a loss of the higher frequencies. A comparatively stiff and hard material such as is used with shellac pressings has very little
playback loss ; vinyl and lacquer are very much more flexible and exhibit appreciable
playback loss. It is also a function of the pickup used. Reference 227.

(iii) Wow and the effects of record warping

Wow is a low-frequency modulation effect caused by spurious variations in groove
velocity, either in recording or in reproduction. In recording, the American N.A.B.
recommends that the maximum instantaneous deviation from the mean speed of the
recording turntable shall not exceed ± 0.1% of the mean speed. The B.B.C. " D

channel " limit is ± 0.05% (Ref. 297).   In reproduction, most turntables used in combination sets suffer from an appreciable slowing down on heavily recorded passages.   This effect is reduced by the use of a light-weight pickup with high lateral compliance.   The effect may be noticed, in severe cases, by a movement in the pattern produced on a stroboscopic disc at the commencement and end of a heavily recorded passage.

A warped disc or wobbling turntable will produce greater wear and eventually greater scratch level on the high portions of the record.   This is much less with well-designed light-weight pickups than with older models.   However, it is important that the stylus force should be considerably more than the minimum for correct tracking on a perfectly flat record.   A record with warp in excess of 1/16 inch is likely to cause trouble.

An eccentric disc or turntable will tend to cause wow ; a very small amount of eccentricity is sufficient to be noticeable in recording (max. ± 0.002 inch N.A.B. ; also English—Ref. 235).   This will produce a tonal pitch variation of 0.1% on reproduction (10 or 12 inch records).

Flutter is similar to wow but at a high frequency ; the ear is very sensitive to this form of distortion.   This also may occur both in recording and in reproduction.

References to wow etc. in recording : 84, 90, 193, 231.

Reference to tests on wow in phonograph combinations : 267.

See also Chapter 37 Sect. 1 (vi) N for I.R.E. tests on phonograph combinations.

(iv) Distortion due to stylus wear

All theoretical work on disc reproduction assumes that the needle point is perfectly spherical.   A jewel or other form of permanent tip suffers wear on two opposite faces. The wear is usually measured in terms of the width of the nearly flat portion.   The distortion arising from the worn stylus is independent of amplitude, and consists principally of odd harmonics.   The distortion is a function of d/l

where d = width of flat on needle

and l = wavelength of recorded tone.

The following table has been calculated from published data (Ref. 59) for standard groove constant velocity recording :

d/l	fundamental	H_3	H_5	at innermost groove	
				$f = 1000$ c/s	$f = 10\,000$ c/s
0.16	− 1 db	4.8%	6.6%	$d = 1.5$ mil	$d = 0.15$ mil
0.22	− 2 db	10%	12% .	$d = 2.4$ mil	$d = 0.24$ mil
0.275	− 3 db	20%	17%	$d = 3.3$ mil	$d = 0.33$ mil
0.36	− 5 db	44%	—	$d = 4.1$ mil	$d = 0.41$ mil
0.5	−10 db	—	—	—	—

The distortion becomes very severe before there is any appreciable attenuation of the fundamental caused by the needle wear.   Although the harmonics of the higher frequencies may be outside the frequency range of the equipment, the intermodulation products will be apparent.

If the recording characteristic is constant amplitude (6 db/octave high frequency pre-emphasis followed by de-emphasis) the values of distortion in the table should be divided by 3 and 5 for H_3 and H_5 respectively.   In this case pre-emphasis : de-emphasis shows to considerable advantage over constant velocity, in that styli last longer before requiring replacement, for the same distortion.

It is obvious that, for high fidelity, no observable stylus flat is permissible.   This seriously limits the life of sapphire styli, even when using a good quality pickup with high lateral and vertical compliance—see Sect. 2(vi).

(v) Noise modulation

The noise level is usually measured as the high-frequency noise developed by an unmodulated groove. When a high degree of noise is present and the groove is modulated by a single-frequency signal, it is found that the signal tends to modulate the noise. The noise modulation reaches a maximum twice each cycle, and the peak amplitude may be over ten times the peak amplitude of the measured noise in an unmodulated groove. The use of pre-emphasis : de-emphasis is to reduce the noise itself but to increase noise modulation. This effect is not apparent when the intermodulation is less than 4%.

This effect may be minimized during recording, by special means suggested in the article.

Reference 145. See also Ref. 193 (proposed standards).

(vi) Pickup distortion

Non-linear distortion in a pickup may vary from practically nil in the best high-fidelity types to over 5% total harmonics in some types. The total harmonic distortion with a well designed crystal pickup is well under 2% (Ref. 120).

The usual method **to test harmonic distortion** is to use a suitable frequency test record, check it for distortion with a pickup which is known to have very low distortion, then measure the distortion with a wave analyzer. A simple test is to use a C.R.O., but this is not capable of indicating less than about 2% or 3% total harmonic distortion. The distortion is always a function of frequency, so that it should be checked at all practicable frequencies. " The minimum stylus force for a given pickup to track a given frequency recorded at a given level can be determined by the observation of waveform on an oscilloscope. The waveform produced by failure to track has a characteristic ' spiked ' appearance, and a very small trace of this form of distortion is easily detectable "—Ref. 310(b).

Intermodulation distortion is preferable to total harmonic distortion as a general indication of pickup performance. Recommended test frequencies are 400 and 4000 c/s, with the latter 12 db lower than the former. Test records are available with bands recorded at different levels, from below normal level, increasing by 2 db steps to +10 db (e.g. R. C. A. Victor RL-419, RL-420, E.M.I. JH 138).

In most cases the intermodulation distortion of a pickup increases very gradually up to a certain level, and then rises very rapidly as the level is further increased. This " knee " should occur at a level of about +6 db, on record RL-420, that is at about 18.2 cm/sec. A high fidelity pickup, tested on record RL-420 shows about 2.5% intermodulation distortion at a level of +6 db ; most of this distortion is due to the record (Ref. 291). Other modern lightweight pickups give up to 8% intermodulation at 0 db (9.1 cm/sec.).

The " tracking " capability of a pickup may be checked by plotting a curve of intermodulation distortion against recorded level, for selected values of vertical stylus force (for details see Ref. 285).

Intermodulation distortion may be measured either as the r m.s. sum or as the arithmetical peak sum—see Chapter 14 Sect. 3(ii) and (v). An intermodulation analyser using the peak sum method is described in Ref. 291.

(vii) Acoustical radiation

Acoustical radiation or needle chatter varies considerably among pickups. Some pickups have no audible radiation. In other cases the radiation may be reduced, if desired, by the use of a bent-shank needle or trailer type (Ref. 225).

(viii) Distortion in recording

If the distortion in the cutter head includes, say, 1% third harmonic distortion, this will result in 4% third harmonic distortion if both the fundamental and the harmonic frequencies are recorded and reproduced at constant velocity. On the other hand pre-emphasis followed by de-emphasis will reduce this form of harmonic distortion, although intermodulation products will still be present (Ref. 189).

See also Sect. 9(i) for distortion on original recordings.

Proposed standards for the measurement of distortion in sound recording : Ref. 193.

SECTION 7 : NOISE REDUCTION

(i) Analysis of noise (ii) High-frequency attenuation (iii) High-frequency pre-emphasis and de-emphasis (iv) Volume expansion (v) Olson noise suppressor (vi) Scott dynamic noise suppressor (vii) Price balanced clipper noise suppressor.

(i) Analysis of noise

Noise in the reproduction of sound from records may be divided into two distinct groups, low frequency noise such as hum and rumble, which has already been dealt with, and surface noise (scratch) which is the subject of this section.

Surface noise, when reproduced with a high fidelity pickup, covers the whole audible frequency range and beyond, but is most distressing to the ear at frequencies from about 1500 to 15 000 c/s. The surface noise per 1 c/s increases with increasing frequency.

When a pickup is used which has a pronounced peak in the region of 3000 to 6000 c/s, the noise in the output appears to have a peak at this frequency. This effect is due partly to the increased response of the pickup at the frequency in question (which could be removed by equalization) and partly to shock excitation of the pickup armature. The remedy is to use a better quality pickup.

References 15, 21.

(ii) High-frequency attenuation

The distressing effects of surface noise on the listener may be reduced by any form of high-frequency attenuation. Soft needles, such as fibre, thorn, bent shank or trailer type, introduce attenuation of this kind but it is not controllable and, in the case of fibre and thorn needles, variable.

Electrical attenuation is to be preferred, although it may be combined with the use of, say, a bent shank needle. Best results are obtained with a combination of two attenuation characteristics—

1. A very rapid attenuation above a fixed frequency (say 7000 c/s, or a choice between 4500, 7000 or 10 000 c/s) for use with old or noisy records. A suitable circuit is Fig. 17.24A.

2. A gradual attenuation characteristic with a hinge point at about 1500 c/s with variable rate of attenuation up to at least 6 db/octave, in addition to any de-emphasis as such. Some possible circuits are described in Chapter 15 Sect. 6. See also Figs. 17.19, 17.23 and 17.27. Crystal pickups used without an equalizer give an attenuation approaching 6 db/octave.

If it is not practicable to have two independent attenuation characteristics, a single characteristic may be used provided that it is capable of giving an attenuation of at least 30 db at 7000 c/s, for use with worn records.

(iii) High-frequency pre-emphasis and de-emphasis

This method has already been described, and reduces the noise by approximately half the amount of pre-emphasis at 10 000 c/s (for characteristics which follow the time-constant roll-off).

(iv) Volume expansion

Volume expansion provides an increase in the dynamic range and, effectively, reduces the surface noise level with reference to the maximum power output. Full details are given in Chapter 16. On account of the limitations of available types of volume expanders, a maximum expansion of 8 db is suggested as a satisfactory compromise.

A combined volume expander and scratch suppressor is described in Ref. 211.

(v) Olson noise suppressor

The Olson noise suppressor works on the principle of the threshold effect. If the noise is below a certain low threshold value it is not amplified. The output versus input voltage characteristic is approximately horizontal for an input up to, say, 0.5 volt ; this may be accomplished by means of a network incorporating voltage-delayed diodes (such as 6H6) or germanium crystal diodes. The distortion products (all harmonic and most of the intermodulation) are eliminated by splitting the high frequency part of the amplifier into channels each covering only one octave. Some possible combinations are

1. 0-3000 c/s ; 3000-6000 c/s (2 channels).
2. 0-2000 c/s ; 2000-4000 c/s ; 4000-8000 c/s (3 channels).
3. 0-1500 c/s ; 1500-3000 c/s ; 3000-6000 c/s ; 6000-12 000 c/s (4 channels).

Filters are used which provide attenuation at the rate of approximately 30 db/octave at both ends of each channel, except the low frequency end of the first channel. A filter network for two channels is given in Ref. 115, while one for four channels is given in Ref. 138.

The only known defects are that signals below the threshold value are lost, and that there are some intermodulation products.

(vi) Scott dynamic noise suppressor

The Scott dynamic noise suppressor controls the bandwidth of the amplifier by means of separate high-frequency and low-frequency tone controls which are automatically controlled by the signal. Fig. 17.38 shows one simple application suitable for home use (Ref. 251 based on Ref. 114). V_1 is an amplifying stage which also provides the voltage for application to both high-frequency and low-frequency control circuits which in turn control the grid voltages of V_2 and V_3. The parallel resonance between L_1 and C_1 provides an attenuation at high frequencies and a high attenuation above the normal operating range ; it may also be used as a whistle filter. V_2 is used as a reactance valve providing variable capacitive reactance which, together with C_2 and its companion series condenser, forms a series resonant circuit with L_2. The low-frequency gate including V_3 is quite readily understood. Switch S_1 allows the suppressor to be opened, providing maximum frequency range when no suppression is desired When S_1 is closed, R_3 controls the amount of suppression. For bad records, switch S_2 allows restriction of the maximum frequency range. Switch S_3 closes the high-frequency gate and leaves the low-frequency gate open.

The frequency characteristics of an experimental model built in our Applications Laboratory is given in Fig. 17.39 (Ref. 251). Scott recommends that C_1 should be

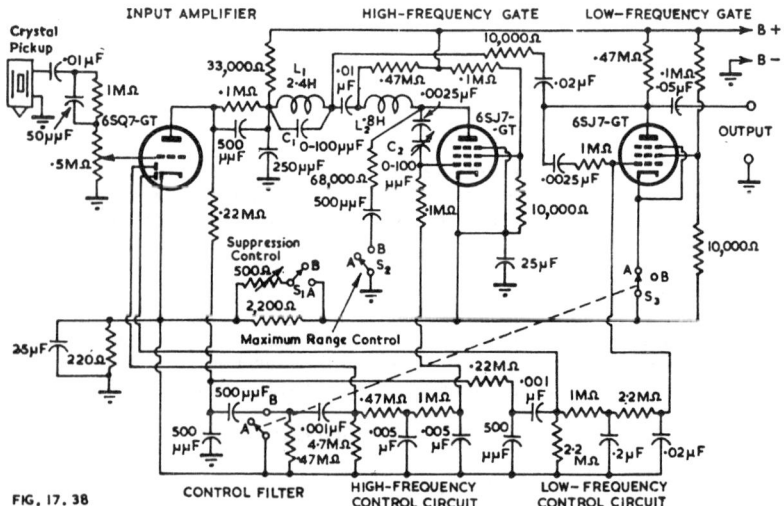

FIG. 17. 38

Fig. 17.38. *Simple Scott dynamic noise suppressor (Ref. 251 based on Ref. 114).*

adjusted for minimum output at 9000 c/s, but it was found preferable to tune C_1 and L_1 for minimum output at 7500 c/s to avoid too great a rise in output between the two frequencies of maximum attenuation.   This simple circuit cannot provide a response extending above about 6000 c/s.

A more recent model with minimum output at 10 000 c/s is described in Ref. 184.

Performance data of a more flexible dynamic noise suppressor built into a complete amplifier are given in Ref. 173.

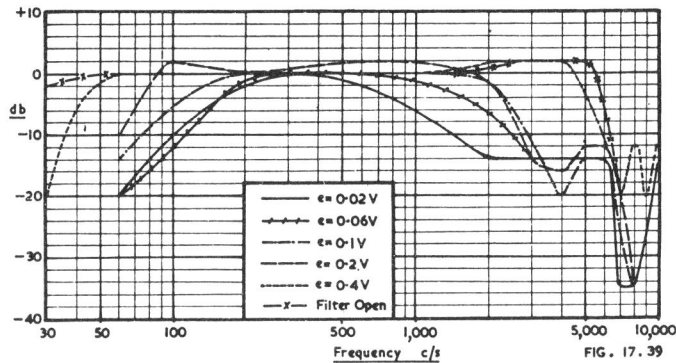

Fig. 17.39.   Frequency characteristics of circuit of Fig. 17.37 *(Ref.* 251).

A modified circuit for home use is that used in the Goodell radio-phonographs (Ref. 133 Fig. 1).   This includes provision for a switch to short-circuit the high-frequency filter ($L_1 C_1$ in Fig. 17.38) ; the sharp cut-off is at a frequency from 10 000 to 16 000 c/s.   This and the Scott amplifier (Ref. 173), are to be preferred to the earlier circuit of Fig. 17.38.

A more elaborate circuit, which also includes a pre-amplifier suitable for use with a low-level electromagnetic pickup, a 3 stage power amplifier, frequency-compensated volume control and bass and treble tone controls, is described in detail in Ref. 250. See also Ref. 188 for further information.

A simpler modification is described in Ref. 165 in which a combined equalizer suitable only for use with low-level electro-magnetic pickups (such as G.E. or Pickering).   This uses one 12SG7 valve and two 12SL7 twin triodes, with d.c. heater supply.

Much more elaborate dynamic noise suppressors for use in broadcast stations are described in Refs , 114, 133, 275.

An analysis of the filter characteristics for the dynamic noise suppressor are given in Ref. 140.

References to Scott dynamic noise suppressor : 83, 114, 133, 165, 173, 184, 188, 250, 275.

To obtain the best results from the dynamic noise suppressor requires intelligent attention from the operator, using the minimum degree of suppression rather than the maximum.   The level should be controlled at some point after the suppressor (see comments, Ref. 149).

A comparison between the Scott, Fisher and Goodell versions of the dynamic noise suppressor is given in Ref. 182.

(vii) Price balanced clipper noise suppressor

This is a modification of the Olson noise suppressor described above.   It differs from the Olson design in that a push-pull clipper is used in order to eliminate second harmonic distortion and thus permit an increase in the frequency coverage of each high-frequency channel.   Possible arrangements are 0-3500 ;   3500-8000 c/s or 0-3000 ;   3000-7000 ;   7000-15 000 c/s (Ref. 150).

The principle of operation is open to the objection that it will respond to certain intermodulation products, more so than the Olson design.

SECTION 8 : LACQUER DISC HOME RECORDING (DIRECT PLAYBACK)

(i) General description (ii) Recording characteristic (iii) Cutting stylus (iv) Cutter head (v) Equalization of cutter (vi) Motor and turntable (vii) Amplifier (viii) Pickups for use on lacquer discs (ix) Recording with embossed groove.

(i) General description

Lacquer discs for home recording usually consist of a thin coating of cellulose nitrate on an aluminium disc, but other coating and disc materials are also used. There are considerable differences in the performance obtained with the best quality discs as compared with others.

The principal defects shown by some lacquer discs are
1. High noise level.
2. Loss of high frequencies on play-back.
3. Distortion, particularly due to drying out after cutting.

Lacquer discs can be used either with a standard stylus or a fine groove stylus. The following details apply to the standard stylus except where otherwise indicated.

With a suitable light-weight pickup, as many as 100 playings are possible. The discs may be processed for making a large number of pressings (for studio use). The signal to noise ratio with the best lacquer discs under the best conditions with a standard stylus may be between 50 and 60 db, the noise level being lowest at the greatest recording diameter and increasing by 2 to 8 db at a recording diameter of 5 inches. With a fine groove stylus the signal-to-noise ratio may be about 58 to 60 db at 10 in. recording diameter, 54 to 58 db at 8 in. and 40 to 50 db at 5 in. diameter. The signal-to-noise ratio may be less than the values quoted above, owing to the use of poor discs and technique ; the difference between a good and a poor disc may exceed 10 db. The overall distortion under the best conditions may be less than 5%.

Discs are available with diameters of 5, 6, 7, 8, 10, 12, 13½ and 16 inches. Discs for home recording are usually limited to 12 inches. If recording is made at 33-1/3 r.p.m. it is wise not to record below a diameter of 8 inches, owing to the loss of high frequencies and increased noise. With 78 r.p.m. it is wise not to record below a diameter of 5 inches, for good fidelity.

The thread (swarf) is highly inflammable, and should be disposed of in water or a closed metal container.

References : 9, 163, 189, 218, 230, 274

Standards for " Disc Home Recording "
(based on R.M.A. REC-105, Aug. 31, 1947)
Drive pin at radius of 1 in. from centre of turntable.
Drive pin diameter 0.180 to 0.185 in.
Drive pin hole 1/4 inch nominal.
Centre hole of rigid base discs, 0.284 in. min. diameter.

Cutting stylus length	5/8 in.
shank diameter	0.0625 in.
face angle	87°
heel angle	50°
shank flat length	3/8 in.
shank flat depth	0.010 in.
max. tip radius (if any)	0.002 in.

Cutting face to be parallel with stylus axis.

(ii) Recording characteristics

The most generally used cross-over frequency is 500 c/s. It is usual to adopt a close approach to the N.A.B. pre-emphasis characteristic ($+16$ db at 10 000 c/s) followed by de-emphasis. The soft coating material causes some loss of higher frequencies during playback, so that a smaller amount of de-emphasis is required than would be used with hard discs, such as shellac.

(iii) Cutting stylus

The usual cutting stylus has a tip radius of about 2 mils with an included angle of about 90°. The depth of cut is usually between 1.5 and 2.5 mils. Steel cutting styli are cheap, but give poor performance all round, and only last for about 15 to 30 minutes of recording. Sapphire cutting styli are very much to be preferred ; they may be reground when necessary Some alloys (e.g. stellite) approach the sapphire performance, and are readily re-ground.

The shape of the cutting stylus is vitally important ; it normally has burnishing facets to produce a noise-free polished groove (Refs. 95, 222).

A hot stylus recording technique has been developed to reduce the noise level by 12 to 18 db while retaining good high-frequency response. No burnishing facets are required (Refs. 288, 289, 315).

(iv) Cutter head

The cutter head requires an available electrical driving power of at least 10 watts, with a higher value preferred. Electro-magnetic cutters with good characteristics (Ref. 233) are preferable but expensive. Crystal cutters with reasonably flat characteristics from about 50 to 9000 c/s are cheaper and satisfactory for home use.

The cutter head is mounted on a feed mechanism which usually provides 96, 112 or 120 grooves per inch.

(v) Equalization of cutter

Equalization is required, firstly to correct any shortcomings in the cutter itself, and secondly to provide the desired recording characteristic. In home recorders using electro-magnetic cutters it is usual to omit the first, and to limit the second to high-frequency pre-emphasis. The cross-over frequency is thus fixed by the cutter design.

High-frequency pre-emphasis (treble boosting) may be provided in a conventional manner to give any desired characteristic. Home recorders usually limit the maximum frequency to 8000 c/s or less, with a maximum boost from 10 to 14 db.

If a crystal cutter is directly connected to the secondary of the output transformer, and if the output resistance of the amplifier (reflected on to the secondary) is less than the capacitive reactance of the crystal at the highest frequency, the recording characteristic will be constant amplitude. The Brush RC-20 crystal cutter has a capacitance of 0.007 μF, with a reactance of about 2500 ohms at 9000 c/s. The maximum signal voltage across the cutter (RC-20) should be about 50 volts ; the step-down ratio of the transformer should be calculated to provide this voltage at maximum power output. The English Acos cutting head requires an input of the order of 150 volts for 1 mil amplitude. This method is only satisfactory with triode valves.

If pentode valves are used it is necessary to connect a suitable shunt dummy load to provide correct matching at about 500 c/s.

The conventional constant velocity characteristic above a specified cross-over frequency may be achieved with a crystal cutter by designing so that the impedance of the driving source is equal to the capacitive reactance of the cutter at the cross-over frequency. With a low impedance source it is necessary to use a series loading resistance.

(vi) Motor and turntable

Motors and turntables for recording purposes must be specially designed for the purpose, as higher power, less vibration and more constant speed are required than with play-back alone. Some equipments are designed for 78 r.p.m. alone, while others are designed for either 33-1/3 or 78 r.p.m.

(vii) Amplifier

The amplifier should be capable of an output of at least 10 watts, with low distortion and good frequency response. The noise and hum level should be at least

45 to 50 db below maximum power output (40 db is an extreme limit for poor recordings). The amplifier should preferably have push-pull triodes with negative feedback, although triodes without feedback or beam power valves with feedback may be used.

(viii) Pickups for use on lacquer discs

The pickup must be a light-weight type, preferably well under 1 ounce (28 grams) needle pressure. The play-back needle should be of the permanent type, usually sapphire, with a point radius of about 3 mils. The needle or pickup must have vertical compliance, and a trailer type of needle is frequently used.

Reference 252.

(ix) Recording with embossed groove

An embossed groove in place of a cut groove has, so far at least, had very limited use. One application (Ref. 103) makes use of $33\text{-}1/3$ r.p.m. discs with 220 grooves per inch which provide 15 minutes playing time on each side of a 7 inch disc. The frequency range is from 150 to 4000 c/s, being suitable for dictation machines. Smaller discs provide $7\frac{1}{2}$ minutes playing time. A special application called for 330 grooves per inch with 30 minutes playing time on each side of a 7 inch disc, working at 22 r.p.m.

See also Refs. 63, 210, 242.

SECTION 9 : REPRODUCTION FROM TRANSCRIPTION DISCS

(i) Introduction   (ii) Characteristics of record material, wear and noise   (iii) Sound track   (iv) Recording characteristics and equalization   (v) Translation loss and radius compensation.

(i) Introduction

Reproduction from transcription discs follows the same general principles as other disc reproduction, and will only be dealt with briefly.   The special features of transcription disc recording and reproduction have been adequately covered in the literature, to which a number of references have been given.

The principal characteristics of N.A.B. (Ref. 237) and B.B.C. (Ref. 214) 16 inch 33-1/3 r.p.m. transcription discs are given below :

Detail	N.A.B. (1949)	B.B.C.
Outer diameter	15-15/16 ± 3/32	16 ins.
Outermost groove diameter	15-1/2 ± 1/16	15-1/2 ins.
	(for inside start,	
	15-9/16 ins max.)	
Innermost groove diameter	7-1/2 ins. min.	
Grooves per inch	96, 104, 120, 128,	120
	136 etc.	
Width of groove at top	< 4.0 mils	—
Radius at bottom of groove	1.5 mils max.	1.5 mils
Angle of groove	88° ± 5°	90°
Turntable speed	33-1/3 ± 0.3%	33-1/3 r.p.m.
Wow factor	≯ ± 0.1%	—
Reproducing stylus : Angle	40° − 55°	—
Bottom radius (primary standard)	2.0 ± 0.1 mils	2.5 mils
(secondary standard*)	2.5 ± 0.1 mils	
Recorded level† (1000 c/s)	Peak velocity	
	7 cm/sec.	
Direction of recording	either	outside-in

Some 10 inch and 12 inch discs are also used, but these have been covered in Sect. 2.

The pickups to be used with transcription discs should have a stylus force not exceeding 1-1/2 ounces (42 grams).   Only permanent points should be used.

The total harmonic distortion on an original recording at 1000 c/s or less, in accordance with good practice, would be less than 2% (Ref. 219).

Intermodulation distortion measured on recordings made by three recording heads, with low frequency peak amplitudes of 2.5 mils and high frequency velocity 12 db lower, were approximately 1.0, 9.2 and 52% (Ref. 269).   The first head is one giving exceptionally low distortion, while the second is representative of good practice. The readings were very little affected by the choice of low or high frequency.   When the level was reduced 6 db, the I.M. distortion was reduced to 23% with the third recording head (say 6% harmonic distortion).

References to transcription recording (general) : 4, 7, 63, (92, 99, 163), (135, 205, 146), 189, 214, 216, 218, 219, 230, 237.

Standards : Refs. 2, 87, 214, 237 :   Specifications : 185 ; Bibliography : 105.

(ii) Characteristics of record material, wear and noise

Most processed transcription disc are made of vinyl.   Vinyl records will reproduce up to 1000 playings with a suitable light-weight pickup and permanent tip.   For the effects of elastic deformation see Ref. 212

---

*Compromise for reproduction of both lateral transcriptions and 78 r.p.m. shellac discs.
†This is the deflection of a standard volume indicator. Programme peaks up to 21 cm/sec. would be anticipated.

Wear of stylus tips is covered in Sect. 2(vi).

The N.A.B. (1949) standard for **signal to noise ratio** states that the noise level measured with a standard volume indicator (ASA Standard C.16.5-1942) when reproducing a record on a flat velocity basis over a frequency range between 500 and 10 000 c/s shall be at least 40 db below the level obtained under the same conditions of reproduction using a tone record of 1000 c/s having a peak velocity of 7 cm/sec. Response of the system at 500 c/s shall be 3 db below the response at 1000 c/s, and the response shall fall at the rate of at least 12 db/octave below 500 c/s. Response of the noise measuring system at 10 000 c/s shall be 3 db below the response at 1000 c/s and the response shall fall at the rate of at least 12 db/octave above 10 000 c/s.

The standard N.A.B. pre-emphasis will increase this value by approximately 8 db, resulting in an effective signal to noise ratio (under minimum conditions) of 48 db. The peak signal to noise ratio will be at least 10 db better than this figure, with normal programme material—say 58 db. Lacquer discs (direct recordings) under similar conditions may have peak signal to noise ratios up to 68 db or even higher (Ref. 63).

(iii) Sound track

The ratio of maximum to minimum groove radius (N.A.B.) is 2.07 : 1. The groove speed varies from 27 to 13.1 inches per second. As the groove diameter approaches 7-1/2 inches there is a progressive loss of high frequencies and increase in harmonic distortion during both recording and play-back. For an analysis of the radius of curvature see Ref. 146. For tables giving groove/land ratios see Ref. 205.

(iv) Recording characteristics and equalization

The recording characteristic standardized by N.A.B. (1949) is given in Fig. 17.15 Curve 2, while that used by the B.B.C. is given in Fig. 17.14 Curve 4. The Orthacoustic recording characteristic is given in Fig. 17.15 Curve 5.

Equalization has been covered in Sect. 5. De-emphasis of the high frequencies (N.A.B.) may be accomplished by a capacitor connected across a resistive network. The value of the capacitance is given by

$$C = 100/R \qquad (1)$$

where C = capacitance in microfarads

R = total circuit impedance (supply resistance and load resistance in parallel) across which C is placed

and 100 = time constant in microseconds.

This provides de-emphasis of approximately 16 db at 10 000 c/s.

(v) Translation loss and radius compensation

Translation loss is defined as the loss in the reproduction of a mechanical recording whereby the amplitude of motion of the reproducing stylus differs from the recorded amplitude in the medium (N.A.B. 1949). The translation loss is a function of the record, the needle tip radius and the pickup. Usual values are from 8 to 16 db at 10 000 c/s at the inner groove.

Radius compensation (diameter equalization)

In order to reduce the loss of high frequencies during play-back some recording organizations boost the higher frequencies during recording, the degree of boosting increasing as the stylus approaches the innermost groove. It is important to remember that the purpose of radius compensation is to give, as nearly as practicable, constant output at all frequencies and at all positions along the groove, when played by a pickup with the recommended needle tip dimensions.

The R.C.A. automatic recording equalizer MI-11100 provides two degrees of radius compensation. The low setting gives 10.8 db at 8 inches diameter and 8 db at 9 inches, both for 10 000 c/s. The high setting gives 13.5 and 9.9 db respectively under the same conditions.

The B.C.C. have a maximum radius compensation of 10 db, but the frequency at which it reaches 10 db is decreased as the diameter becomes smaller (minimum 8-3/4 inches).

There is no doubt that witl certain kinds of music it is impossible to apply the full N.A.B. pre-emphasis of lo db at 10 000 c/s together with the necessary amount of radius compensation to provide a nearly level frequency characteristic without serious over-modulation and distortion. The only answer seems to lie in the use of fine groove recording.

References 4, 99, 189, 214, 227.

SECTION 10 : REFERENCES TO LATERAL DISC RECORDING

2. N.A.B. " Recording Standards " Comm. 22.8 (Aug. 1942) 20.
4. Lynn, R. A. " Factors contributing to good recording " RCA Rev. 6.4 (April 1942) 463.
7. Brooker, V.M. " Sound on disc recording " W.W. 43.26 (Dec. 29th 1938) 588.
8. " Pick-up design " W.W. 43.17 (Oct. 27th, 1938) 364. (Description of H.M.V. " Hyper-sensitive " model).
9. Andrews, H. " Home recording " W.W. (1) 43.2 (July 14, 1938) 24 ; (2) 43.3 (July 21, 1938) 50 ; (3) 43.4 (July 28, 1938) 73 ; (4) 43.5 (Aug. 4, 1938) 95.
10. Voigt, P.G.A.H. " Getting the best from records " W.W. (1) The recording characteristic 46.4 (Feb. 1940) 141 ; (2) " The pick-up " 46.5 (Mar. 1940) 177 ; (3) " More about tone correction circuits " 46.6 (April 1940) 210 ; (4) " The record has the last word " 46.7 (May 1940) 242.
11. Burt, A. D. " The reduction of record noise by pick-up design " Elect. 16.1 (Jan. 1943) 90.
14. Beers, G. L., & C. M. Sinnett " Some recent developments in record reproducing systems " Proc. I.R.E. 31.4 (April 1943) 138.
15. Brierley, J. " Pick-up accessories " W.W. 49.4 (April 1943) 100.
16. Hay, G. A. " Needle armature pick-up ; design giving good frequency response and low amplitude distortion " W.W. 49.5 (May 1943) 137.
17. Brierley, J. " A moving coil pick-up " W.W. 48.7 (July 1942) 154; G. A. Hay (letter) W.W. 48.9 (Sept. 1942) 222. J. H. Mole (letter) W.W. 48.8 (August 1942) 192.
21. Scroggie, M.G. " Gramophone needle scratch " W.W. 46.1 (Nov. 1939) 3.
24. Thrurrell, D. W., G. E. Horn, D. W. Aldous, A. S. Robb (letters) W.W. 49.6 (June 1943) 184.
25. de Soto, C. B. " How recordings are made " Q.S.T. (1) Principles of disc recording 26.7 (July 1942) 30 ; (2) The recording 26.8 (Aug. 1942) 56 ; (3) The amplifier 26.9 (Sept. 1942) 65 ; (4) Playback 26.10 (Oct. 1942) 54 ; (5) Tests and trouble shooting 26.12 (Dec. 1942) 51.
29. W.W. Brains Trust " Future of the Disc " W.W. 50.1 (Jan. 1944) 8.
30. Aldous, D. W. " Defects in direct disk recording " Electronic Eng. 15 (Nov. 1943) 233.
31. " Future of sound recording " W.W. 50.4 (April 1944) 114.
40. Devereux, F. L. " Gramophone needle buzz—an inherent form of distortion : Possible cures " W.W. 50.10 (Oct. 1944) 290.
41. Hunt, F. V., & J. A. Pierce " Distortion in sound reproduction from records " Jour. S.M.P.E. 31.2 (Aug. 1938) 157.
42. Lewis, W. D., & F. V. Hunt " Theory of tracing distortion in record reproduction" J. Acous. Soc. Am. 12.3 (Jan. 1941) 348.
45. Miessner, B. F. " Frequency modulation phonograph pickup " Elect. 17.11 (Nov. 1944) 132.
47. Parvey, J. C. " Vibratory momentum and groove skating in disc reproduction " Comm. 20.10 (Oct. 1940) 22.
50. Miessner, B. F. " Capacity phono pick-up " Radio Craft (Feb. 1945) 282.
52. Bauer, B. B. " Tracking angle in phonograph pickups " Elect. 18.3 (Mar. 1945) 110.
53. Glover, R. P. " A record-saving pick-up " Electronics, 102 (Feb. 1937) 31.
54. Bird, J. R., & C. M. Chorpening " The offset head crystal pickup " Radio Eng. 17.3 (March 1937) 16.
55. Knapp, D. G. " The offset head crystal pickup " Comm. 18.2. (Feb. 1938) 29.
56. Olney, B. " Phonograph pickup tracking error " Elect. 10.11 (Nov. 1937) 19.
57. Lindenberg, T. Jnr. " Moving coil pickup design " Elect. 18.6 (June 1945) 108.
59. Bauer, B. B. " Calculation of distortion due to phonograph needle wear " Elect. 18.77 (July 1945) 250 (based on J. Acous. Soc. Am. 16.4 April 1945 p. 246).
62. Bauer, B. B. " Crystal pickup compensation circuits " Elect. 18.11 (Nov. 1945) 128.
63. Brooker, V. M. " Some aspects of sound recording " Proc. I.R.E. (Aust.) 6.3 (Sept. 1945) 3.
65. Kalmus, H.P. " Pickup with low mechanical impedance " Elect. 19.1 (Jan. 1946) 140.
71. " A new moving coil pickup " (" Lexington ") Electronic Eng. 18.221 (July 1946) 224.
76. Rich, S. R. " Torsional magnetostriction pickup " Elect. 19.6 (June 1946) 107.
82. Scott, R. F. " Magnetostriction pickup " Radio Craft (Nov. 1946) 16.
83. Scott, H. H. " Dynamic suppression of phonograph record noise " Elect. 19.12 (Dec. 1946) 92.
84. Furst, U. R. " Periodic variations of pitch in sound reproduction by phonographs " Proc. I.R.E. 34.11 (Nov. 1946) 887.
85. MacDonald, G. E. " The reduction of tracking error " Comm. 21.1 (Jan. 1941) 5.
87. Smeby, L. C. " Recording and reproducing standards " Proc. I.R.E. 30.8 (Aug. 1942) 355.
88. Southey, R. V. " Things to come in sound recording and reproduction " Proc. I.R.E. (Aust.) 7.12 (Dec. 1946) 4.
89. Chinn, H. A. " Glossary of disc-recording terms " Proc. I.R.E. 33.11 (Nov. 1945) 760.
90. " A stud· of wows " Comm. 22.7 (July 1942) 42.
91. Martinson, B. K. " Phono pickups " Service 15.8 (Aug. 1946) 12.
92. Robinson, W. H. " Lateral Recording " Comm. 27.2 (Feb. 1947) 26. See also Ref. 99.
93. " Two band phono receiver with variable reluctance pickup " (G.E.), Service, 15.11 (Nov. 1946) 26.

94. Chinn, H. A. " Disc recording " Electronic Industries (Nov. 1946) 64.
95. Capps, I.L. " Recording styli " (for lacquer records) Electronic Industries (Nov. 1946) 65.
96. Germeshausen, K. J., & R. S. John " Phonograph pickup using strain gauge " Electronic Industries (Nov. 1946) 78.
97. Leidel, W. F., & N. E. Payne " Tuned ribbon pickup " (Audak) Electronic Industries, 5.10 (Oct. 1946) 67.
99. Robinson, W. H. " Lateral recording " Comm. 27.4 (April 1947) 38. See also Ref. 92.
103. Thompson, L. " Technics of sound methods ; recording with embossed groove " Tele-Tech. 6.5 (May 1947) 48.
105. Jorysz, A. " Bibliography of disc recording " Tele-Tech. 6.6 (June 1947) 73.
106. Queen, I. " Postwar features of phonograph pickups " Radio Craft 18.12 (Sept. 1947) 36.
109. Francis, E. H. " Moving iron pickups " W.W. 53.8 (Aug. 1947) 285.
110. Roys, H. E. " Intermodulation distortion analysis as applied to disk recording and reproducing equipment " Proc. I.R.E. 35.10 (Oct. 1947) 1149.
113. Rogers, G. L. " Simple rc filters for phonograph amplifiers " Audio Eng. 31.5 (June 1947) 28.
114. Scott, H. H. " Dynamic noise suppressor " Elect. 20.12 (Dec. 1947) 96. See also Ref. 275.
115. Olson, H. F. " Audio noise reduction circuits " Elect. 20.12 (Dec. 1947) 119.
116. Gordon, J. F. " A Vacuum-tube-type transducer for use in the reproduction of lateral phonograph recordings " Proc. I.R.E. 35.12 (Dec. 1947) 1571.
117. McProud, C. G. " High Frequency equalization for magnetic pickups " Audio Eng. 31.8 (Sept. 1947) 13.
119. Hayes, A. E. " A new phonograph pickup principle " Audio Eng. 31.9 (Oct. 1947) 14.
120. Wheeler L.J., & K. G. Lockyer " Crystal pickups ; basis of design for high fidelity reproduction " W.W. 53.11 (Nov. 1947) 412.
126. Savory, W. A. " The design of audio compensation networks " Tele-Tech 7.1 (Jan. 1948) 24.
127. McProud, C. G. " Transition frequency compensation " Audio Eng. 31.6 (July 1947) 10.
128. Mittell, B. E. G. " Commercial disk recording and processing in England " (Review) Audio Eng. 32.4 (Apr. 1948) 19.
131. Mittell, B. E. G. " Commercial disc recording and processing " (abstract) W.W. 54.2 (Feb. 1948) 67.
133. Goodell. J. D. " The dynamic noise suppressor " Audio Eng. 31.10 (Nov. 1947) 7.
135. Tillson, B. F. " Musical Acoustics " (5) Recording groove, Audio Eng. 31.9 (Oct. 1947) 25.
136. " Stylus " (letter) " Musical acoustics" (refers to recording frequency ranges) Audio Eng. 31.9 (Oct. 1947) 3.
137. Hull, A. G. " Limelight on quality " Australasian Radio World (May 1948) 5.
138. Cole, C. D. " Experimental noise suppressor " Audio Eng. 32.1 (Jan. 1948) 9.
140. McCracken, L. G. " Filter characteristics for the dynamic noise suppressor " Elect. 21.4 (April 1948) 114.
142. Kahn, A. " Torque-drive phono pickup " Service, 17.6 (June 1948) 32.
144. Burwen, R. S. " Feedback preamplifier for magnetic pickups " Audio Eng. 32.2 (Feb. 1948) 18.
145. Cook, E. G. " Noise modulation in recording " Audio Eng. 31.11 (Dec. 1947) 8.
146. Tillson, B. F. " Musical Acoustics " Part 7, Audio Eng. 31.11 (Dec. 1947) 23.
149. Canby, E. T. " Record Revue " (Scott noise suppressor) Audio Eng. 32.5 (May 1948) 24.
150. Price, S. L. " Balanced clipper noise suppressor " Audio Eng. 32.3 (March 1948) 13.
151. Cook, E. " Increasing volume level in disc recording " Audio Eng. 32.3 (March 1948) 17.
153. Maxfield, J. P., & J. K. Hilliard " Notes on pre-equalization for phonograph records " Audio Eng. 32.4 (April 1948) 15.
155. Pickering, N. C. " Misconceptions about record wear " Audio Eng. 32.6 (June 1948) 11.
157. " Ribbon pickup " (Brierley) W.W. 54.8 (Aug. 1948) 306.
158. " Modern crystal phonopickups " Radio Electronics 20.1 (Oct. 1948) 29.
159. Gernsback, M. H. " Microgroove phonograph records " Radio Electronics 20.1 (Oct. 1948) 30.
161. Dorf, R. H. " Frequency test records " Radio Electronics 20.1 (Oct. 1948) 46.
162. Queen, I. " New magnetic pickups " Radio Electronics 20.1 (Oct. 1948) 55.
163. Robinson, W. H. " Lateral recording " Comm. 28.9 (Sept. 1948) 34.
164. Olson, H. F., & J. Preston " Electron tube phonograph pickup " Audio Eng. 32.8 (Aug. 1948) 17.
165. McProud, C. G. " Simplified dynamic noise suppressor " Audio Eng. 32.8 (Aug. 1948) 22.
166. " Columbia LP microgroove records " Audio Eng. 32.8 (Aug. 1948) 24.
170. Harris, H. " Making phonograph record matrices " Audio Eng. 32.7 (July 1948) 14.
173. Scott, H. H., & E. G. Dyett, " An amplifier and noise suppressor unit " F.M.&.T. 8.3 (March 1948) 28.
174. Southey, R. V. " Modern practice in disc recording " Proc. I.R.E. (Aust.) 9.11 (Nov. 1948) 16.
178. Hector, L. G., & H. W. Koren " Ceramic phonograph pickup " Elect. 21.12 (Dec. 1948) 94.
179. " Design of L-P records " Elect. 21.12 (Dec. 1948) 110.
181. Smith, O. J. M. " Polyethylene phonograph records " Audio Eng. 32.9 (Sept. 1948) 13.
182. Canby, E. T. " Record revue " (3 types of noise suppressors) Audio Eng. 32.9 (Sept. 1948) 33.
183. Baruch, R. M. " The LP microgroove record system " Service 17.11 (Nov. 1948) 14 ; 17.12 (Dec. 1948) 16.
184. Dyett, E. G. " Three tube dynamic noise suppressor " Service 17.11 (Nov. 1948) 26.
185. " Recorder specifications " Tele-Tech. 7.8 (Aug. 1948) 40.
187. Douglas, A. " The G.E. variable reluctance pickup " Electronic Eng. 21.251 (Jan. 1949) 21 (based on G.E. catalogue data).
188. McProud, C. G. " Elements of residence radio systems " (3) Audio Eng. 32.11 (Nov. 1948) 20.
189. Viol, F.O. " Some problems of disc recording for broadcasting purposes " Proc. I.R.E. (Aust.) 10.2 (Feb. 1949) 42. Reprinted Proc. I.R.E. 38.3 (March 1950) 233.
193. " Proposed standards for the measurement of distortion in sound recording " Jour. S.M.P.E. 51. (Nov. 1948) 449.
194. Nicholas, W.R. " Some aspects of phonograph pickup design " Proc. I.R.E. (Aust.) 10.3 (March 1949) 63.
195. " Supersonic high fidelity." W.W. 54.1 (Jan. 1948) 23.
197. Hilliard, J. K. " Factors affecting pre-and post-equalization " Comm. 32.9 (Sept. 1943) 30.
198. Savory, W.A. " The design of audio compensation networks " Tele-Tech 7.2 (Feb. 1948) 27.
199. Haines, F.M. " High fidelity bass compensation " Electronic Eng. 18.216 (Feb. 1946) 45.
205. Tillson, B. F. " Musical acoustics " Part 6 Audio Eng. 31.10 (Nov. 1947) 31.
209. Ellis, J. " Bass compensation " W.W. 53.9 (Sept. 1947) 319. See also Ref. 278.
210. Griffin, E. E. " Embossing at constant groove speed—a new recording technique " Elect. 13.7 (July 1940) 36.
211. McProud, C. G. " Experimental volume expander and scratch suppressor " Audio Eng. 31.7 (Aug. 1947) 13.

212. Begun, S. J., & T. E. Lynch " The correlation between elastic deformation and vertical forces in lateral recording " Jour. Acous. Socy. Am. 13.3 (Jan. 1942) 284. Correction 13.4 (April 1942) 392.
213. Pierce, J. A., & F. V. Hunt " On distortion in sound reproduction from phonograph records " Jour. Acous. Socy. Am. 10.1 (July 1938) 14.
214. Davies, H. " The design of a high-fidelity disc recording equipment " Jour. I.E.E. Vol. 94 Part III No. 30 (July 1947) 275.
215. Scott, H. H. " Dynamic noise suppressors " Service 17.9 (Sept. 1948) 16.
216. Henney, K. (book) " The Radio Engineering Handbook " (McGraw-Hill Book Co., New Yo & London, 3rd edit. 1941) Sec. 21.
218. Aldous, D. W. " Manual of Direct Disk Recording " (Bernards, Publishers, Ltd. London, 1944).
219. " N.B.C. Recording Handbook " National Broadcasting Co. Inc., Engineering Department.
220. O'Brien, E. J. " High fidelity response from phonograph pickups " Elect. 22.3 (March 1949) 118 265.
222. Le Bel, C. J. " Properties of the dulled lacquer cutting stylus " J. Acous. Soc. Am. 13.3 (Jan. 1942) 265.
223. Reid, J. D. " A large radius stylus for the reproduction of lateral cut phonograph records " J. Acous Soc. Am. 13.3 (Jan. 1942) 274. Review W.W. 49.1 (Jan. 1943) 5.
224. Sepmeyer, L. W. " Tracing distortion in the reproduction of constant amplitude recordings " J. Acous. Soc. Am. 13.3 (Jan. 1942) 276.
225. Goldsmith, F. H. " A noise and wear-reducing phonograph reproducer with controlled response " J. Acous. Soc. Am. 13.3 (Jan. 1942) 281.
226. Savory, W. A. " Design of audio compensation networks " Tele-Tech 7.4. (April 1948) 34.
227. Kornei, Q. " On the playback loss in the reproduction of phonograph records " Jour. S.M.P.E. 37 (Nov. 1941) 569.
228. Baerwald, H. G. " Analytic treatment of tracking error and notes on optimal pick-up design " Jour. S.M.P.E. 37 (Nov. 1941) 591.
229. McLachlan, N. W. (book) " The New Acoustics " (Oxford University Press, 1936) p. 88.
230. Dorf, R. H. (book) " Practical Disc Recording " (Radcraft Publications Inc., New York, 1948).
231. Axon, P. E., & H. Davies " A study of frequency fluctuations in sound recording and reproducing systems " Jour. I.E.E. Part III 96.39 (Jan. 1949) 65.
232. " Columbia LP record specifications " Tele-Tech 7.10 (Oct. 1948) 63.
233. Roys, H. E. " An improved lacquer disc recording head " Audio Eng. 33.2 (Feb. 1949) 21.
234. Bachman, " Phonograph reproducer design " E.E. 65.3 (March 1946) 159.
235. Mittell, B. E. G. " Annexure to recorded talk on disc-recording standards " delivered to I.R.E. (Australia) Convention at Sydney, Australia, Nov. 3, 1948.
236. Aldous, O. W. " American microgroove records " W.W. 55.4 (Apr. 1949) 146.
237. " N.A.B. Recording and reproducing standards " Engineering (May 16, 1949)
238. " Stylus ", " Characteristics of the new 45 r.p.m. record " Audio Eng. 33.3 (March 1949) 6.
239. St. George, P. W., & B. B. Drisko " Versatile phonograph preamplifier " Audio Eng. 33.3 (March 1949) 14.
240. Van Beuren, J. M. " Cost vs quality in af circuits " F.M. & T. 9.2 (Feb. 1949) 31.
241. Scott, H. H. " Audio developments : observations on the quality of records made here and overseas " F.M. & T. 9.1 (Jan. 1949) 23.
242. Morse, M. " Sound embossing at the high frequencies " J. Acous. Soc. Am. 19.1 (Jan. 1947) 169.
243. Hornbostel, J. " Improved theory of the light pattern method for the modulation measurement in groove recording " Electronic Eng. 19.1 (Jan. 1947) 165.
244. McClain, E. F. " A distortion reducing stylus for disk reproduction " J. Acous. Soc. Am. 19.2 (March 1947) 326.
245. Duffield, A. W. " Improvements in disc records through constant amplitude recording " Comm. 20.3 (Mar. 1940) 13.
246. Bomberger, D. C. " A continuously variable equalizing pre-amplifier " Audio Eng. 33.4 (Apr. 1949) 14.
247. Maxfield, J. P., & J. K. Hilliard " Notes on pre-equalization for phonograph records " Audio Eng. 32.4 (Apr. 1948) 15.
248. Fleming, L. " Notes on phonograph pickups for lateral-cut-records " J. Acous. Soc. Am. 12.3 (Jan. 1941) 366.
249. Bauer, B. B. " Pickup placement " Elect. 22.6 (June 1949) 87.
250. McProud, C. G. " General purpose 6AS7G amplifier " Audio Eng. 32.6 (June 1948) 24.
251. Aston, R. H. " A dynamic noise suppressor " Radiotronics No. 132 (July/Aug. 1948) 73.
252. Williams, A. L. " Further improvements in light-weight record reproducers " Jour. S.M.P.E. 33 (Aug. 1939) 203.
254. Data kindly supplied by Radio Corporation of America, R.C.A. International Division.
255. Data kindly supplied by William S. Bachman, Director of Research, Columbia Records Inc.
256. Data kindly supplied by R. V. Southey, E.M.I. (Australia) Pty. Ltd.
257. Hunt, F. V. (in discussion on " What constitutes high fidelity reproduction ? ") Audio Eng. 32.12 (Dec. 1948) 8.
258. Frayne, J. G. & H. Wolfe (book) " Elements of Sound Recording " (John Wiley and Sons, New York ; Chapman and Hall, Ltd., London 1949).
259. Langford-Smith, F. (Letter) " Recorded supersonic frequencies " W.W. 55.7 (July 1949) 275
260. R.M.A. (U.S.A.) Standards and other data REC-103 (Oct. 1946) : Dimensional characteristics— phonograph records for home use.
    REC-105 (June, 1947) : Disc home recording.
    REC-125-A (July 1949) : Phonograph pickups.
    REC-126-A (Oct. 1950) : Playback needles for home phonographs.
    REC-128 (May, 1949) : Standard frequency test records.
261. Andrews, D. R. " Importance of groove fit in lateral recordings " Audio Eng. 33.7 (July 1949) 18.
262. Goldmark, P. C., R. Snepvangers & W. S. Bachman " The Columbia long-playing microgroove recording system " Proc. I.R.E. 37.8 (Aug. 1949) 923.
263. Carson, B. R., A. D. Burt & H. I. Reiskind " A record changer and record of complementary design " R.C.A. Rev. 10.2 (June 1949) 173.
264. Corrington, M. S. " Tracing distortion in phonograph records " R.C.A. Rev. 10.2 (June 1949) 241.
265. Roys, H. E. " Analysis by the two-frequency intermodulation method of tracing distortion encountered in phonograph reproduction " R.C.A. Rev. 10.2 (June 1949) 254.
267. Institute of Radio Engineers, U.S.A. "Standards on radio receivers—methods of testing amplitude-modulation broadcast receivers " (1948).
268. Sterling, H. T. " Simplified preamplifier design " Audio Eng. 33.11 (Nov. 1949) 16.

269. Yenzer, G. R. " Lateral feedback disc recorder " Audio Eng. 33.9 (Sept. 1949) 22.
270. Williamson, D. T. N. " High quality amplifier—new version " W.W. 55.8 (Aug. 1949) 282 ; 55.10 (Oct. 1949) 365 ; 55.11 (Nov. 1949) 423.
271. " Disc recording system developments—solutions of many problems to bring about a fidelity range of 20 KC " Tele-Tech. 9.1 (Jan. 1950) 14.
272. Begun, S. J. (book) " Magnetic Recording " (Murray Hill Books Inc. 1949).
273. Haynes, H. E., & H. E. Roys " A variable speed turntable and its use in the calibration of disk reproducing pickups " Proc. I.R.E. 38.3 (March 1950) 239.
274. Briggs, G. A. (book) " Sound Reproduction " Wharfedale Wireless Works, Bradford, Yorks., England, 1949.
275. Scott, H. H. " The design of dynamic noise suppressors " Proceedings of the National Electronics Conference, Chicago, Illinois, Nov. 3-5, 1947, Vol. 3. p. 25.
276. Brierley, J. H. " Reproduction of Records—1 " booklet published by J. H. Brierley (Gramophones and Recordings) Ltd., Liverpool, England.
277. Data supplied by J. H. Brierley of J. H. Brierley (Gramophones and Recordings) Ltd., Liverpool, (England).
278. Fleming, L. " Equalized pre-amplifier using single stage feedback " Audio Eng. (Mar. 1950) 24. See also Ref. 209.
279. Reid, J. D. " Universal phonograph styli " J.A.S.A. 21.6 (Nov. 1949) 590.
280. Warren, H. G. (letter) " Pre-amplifier circuit " W.W. 46.6 (June 1950) 238.
281. Bauer, B.B. " All-purpose phonograph needles " Elect. 23.6 (June 1950) 74.
282. Wood, G. H. H. " Record and stylus wear—advantage of compliant stylus mountings " W.W. 56.7 (July 1950) 245.
283. Mallett, E. S. " The determination of gramophone pick-up tracking weights " Electronic Eng. 22.267 (May 1950) 196.
284. Woodward, J. G. " A feedback-controlled calibrator for phonograph pickups " R.C.A. Rev. 11.2 (June 1950) 301.
285. Roys, H. E. " Determining the tracking capabilities of a pickup " Audio Eng. 34.5 (May 1950) 11.
286. " Playback standards " Audio Eng. 34.7 (July 1950) 4.
287. Fleming, L. " Equalized pre-amplifier using single stage feedback " Audio Eng. 34.3 (March 1950) 24.
288. Bachman, W. S. " The Columbia hot stylus recording technique " Audio Eng. 34.6 (June 1950) 11.
289. Wortman, L. A. (Summary of discussion) " Heated stylus recording technique " Audio Eng. 34.7 (July 1950) 24.
290. Marcus, E. J. & M. V. " The diamond as a phonograph stylus material " Audio Eng. 34.7 (July 1950) 25.
291. Fine, R. S. " An intermodulation analyzer for audio systems " Audio Eng. 34.7 (July 1950) 11.
292. Roys, H. E. " Recording and fine-groove technique " Audio Eng. 34.9 (Sept. 1950) 11.
293. Moyer, R. C., D. R. Andrews & H. E. Roys " Methods of calibrating frequency records " Proc. I.R.E. 38.11 (Nov. 1950) 1306.
294. Godfrey, J. W. " Reproduction of discs and records for broadcasting " B.B.C. Quarterly 4.3 (Autumn 1949) 170.
295. Terry, P. R. " The variable-disc-speed method of measuring the frequency characteristics of pick-ups " B.B.C. Quarterly 4.3 (Autumn 1949) 176.
296. Anderson, L. J. & C. R. Johnson " New broadcast lightweight pickup and tone arm " Audio Eng. 35.3 (March 1951) 18.
297. Moir, J. " Recorders and reproducers " FM-TV 11.1 (Jan. 1951) 14.
298. Weil, M. " Phono facts " Audio Eng. 35.6 (June 1951) 20.
299. Dutton, G. F. " Gramophone turntable speeds " W.W. 57.6 (June 1951) 227.
300. " Frequency test records—Calibration methods discussed by B.S.R.A. " W.W. 57.6 (June 1951) 227.
301. Kelly, S. " Piezo-electric crystal devices " Electronic Eng. 23.278 (April 1951) 134 ; 23.279 (May 1951) 173.
302. Shirley, G. (letter) " Diamond vs sapphire " Audio Eng. 35.8 (Aug. 1951) 6.
303. John, R. S. " A strain-sensitive phonopickup " Radio and TV News 43 (Feb. 1950) 40.
304. Roys, H. E. " Recording and fine groove technique " Broadcast News No. 60 (July-Aug. 1950) 9.
305. West, R. L., & S. Kelly " Pickup input circuits " W.W. 56.11 (Nov. 1950) 386.
306. Goodell, J. D. " Problems in phonograph record reproduction " Radio & TV News 44.5 (Nov. 1950) 39.
307. " AES Standard playback curve " Audio Eng. 35.1 (Jan. 1951) 22.
308. " British long-playing records " W.W. 57.1 (Jan. 1951) S7.
309. Pollock, A. M. " Thorn gramophone needles " W.W. 56.12 (Dec. 1950) 450 ; correspondence 57.3 (March 1951) 121 ; 57.4 (April 1951) 145.
310. Kelly, S. " Intermodulation distortion in gramophone pickups " W.W. 57.7 (July 1951) 256. (b) Letter L. J. Elliott, 57.9 (Sept. 1951) 370.
311. Bachman, W. S. " The application of damping to phonograph reproducer arms " Proc. I.R.E. 40.2 (Feb. 1952) 133.
312. Discussion on hum in moving coil and ribbon pickups : H. J. Leak, W.W. 56.4 (Apr. 1950) 132 ; J. H. Brierley, 56.6 (June 1950) 238 ; 57.2 (Feb. 1951) 81 ; P. J. Baxandall and R. L. West 57.4 (Apr. 1951) 146, 147.
313. Anderson, L. J., & C. R. Johnson " New lightweight pickup and tone arm " Broadcast News No. 64 (May-June 1951) 8.
314. Weil, M. (letter) " Tracking efficiency " Audio Eng. 34.8 (Aug. 1950) 2, also front cover.
315. " Hot stylus technique " W.W. 58.2 (Feb. 1952) 50.
316. Read, O. (book) " The recording and reproduction of sound " (Howard W. Sams & Co. Inc. Indianapolis, Indiana, 2nd ed. 1952).
317. Parchment, E. D. " Microgroove recording and reproduction " Jour. Brit. I.R.E. 12.5 (May 1952) 271.

Additional references will be found in the Supplement commencing on page 1475.

CHAPTER 18

MICROPHONES, PRE-AMPLIFIERS, ATTENUATORS AND MIXERS

By F. LANGFORD-SMITH, B.Sc., B.E.

SECTION 1 : MICROPHONES

(i) General survey (ii) Carbon microphones (iii) Condenser microphones (iv) Crystal and ceramic microphones (v) Moving coil (dynamic) microphones (vi) Pressure ribbon microphones (vii) Velocity ribbon microphones (viii) Throat microphones (ix) Lapel microphones (x) Lip microphones (xi) The directional characteristics of microphones (xii) The equalization of microphones (xiii) Microphone transformers (xiv) Standards for microphones.

(i) General survey

Microphones may be divided into two basic groups, pressure- and velocity-operated. Any microphone that has its diaphragm exposed to sound waves on one side only, is a **pressure-operated type**—that is, the displacement of the diaphragm is proportional to the instantaneous pressure developed in the sound waves. At low frequencies a pressure-operated microphone is non-directional, that is to say it responds uniformly to sounds from all directions. As the frequency increases, however, the response becomes more and more uni-directional and there is a peak in the high frequency response for sound impinging directly on the diaphragm ; this peak may be reduced or eliminated by placing the diaphragm at an angle of 45° to 90° to the direction of the sound.

Examples of pressure-operated microphones are carbon, crystal, moving coil and pressure-ribbon microphones.

A **velocity-operated microphone** is one in which the electrical response corresponds to the particle velocity (or pressure-gradient) resulting from the propagation of a sound wave through the air. Examples are the free-ribbon velocity microphone and pressure-gradient microphones. There are also combinations of pressure-operated and velocity-operated microphones.

For good fidelity a microphone should have a wide frequency response without peaks, low distortion and good transient response. For public address a microphone need not have such a wide frequency response, particularly at the bass end, and a slight peak in the 3000 to 5000 c/s region is not usually considered detrimental, while distortion requirements are less stringent.

The output voltage from a microphone is approximately proportional to the sound pressure.

The peak r.m.s. sound pressure at a distance of 1 foot from a man's mouth is of the order of 10 dynes/cm² with conversational speech. It decreases 6 db each time the distance is doubled. When speaking with the mouth as close as possible to the microphone, the peak r.m.s. sound pressure is about 100 dynes/cm².

General references A5, D3, D4, D7, D11.

Microphone ratings

Microphones may be rated in terms of either voltage or power, as described in detail in Chapter 19 Sect. 1(iv). Ratings in common use include

Voltage ratings

A. Open circuit voltage (0 db = 1 V) for sound pressure 1 dyne/cm², expressed in dbv*.

B. Open circuit voltage (0 db = 1 V) for sound pressure 10 dynes/cm², expressed in dbv.

C. Volume units (as read by a standard Volume Indicator) for sound pressure 1 dyne/cm², expressed in vu.

D. High impedance grid circuit voltage (0 db = 1 V, Z = 40 000 ohms) for sound pressure 1 dyne/cm², expressed in dbv.

Power ratings

E. Output power† (0 db = 1 mW) for sound pressure 1 dyne/cm², expressed in dbm.

F. Output power† (0 db = 1 mW) for sound pressure 10 dynes/cm², expressed in dbm.

G. Output power (0 db = 6 mW) for sound pressure 1 dyne/cm². This rating is now rarely used.

H. Output power (0 db = 6 mW) for sound pressure 10 dynes/cm². This rating is now rarely used.

J. Output power (0 db = 1 mW) for sound pressure 0.0002 dyne/cm². This is the R.M.A. Microphone System Rating G_M (R.M.A. Standard SE-105 ; see Chapter 19 Sect. 1(iv)D).

Voltage ratings—To convert from A to B, add + 20 db.
                  To convert from B to A, add − 20 db.

If a voltage rating is used for a low impedance microphone, the impedance across which the voltage occurs should be specified.

Volume units—This is not an official microphone rating, but is sometimes used as a convenient method of measurement. The Volume Indicator is effectively a voltmeter in which 0 vu = 0.77 volts r.m.s. across 600 ohms. To convert vu (Rating C) to voltage (rating A), add − 2 db. When the output from a low impedance microphone is quoted in terms of vu, it may be inferred that this is equivalent to the power output in milliwatts (Rating E).

Power ratings—The power rating is the power developed in a load of specified value. Power ratings E, F, G and H may be calculated one from the other as under:‡

To convert from	add	To convert from	add	To convert from	add
E to F	+ 20 db	F to J	− 94 db	H to G	− 20 db
E to G	− 8	G to E	+ 8	H to J	− 86
E to H	+ 12	G to F	+ 28	J to E	+ 74
E to J	− 74	G to H	+ 20	J to F	+ 94
F to E	− 20	G to J	− 66	J to G	+ 66
F to G	− 28	H to E	− 12	J to H	+ 86
F to H	− 8	H to F	+ 8		

Relationship between voltage and power ratings‡

Provided that the sound pressure is the same in both cases : Microphone rating in dbv (0 db = 1 volt) = microphone rating in dbm (0 db = 1 milliwatt) + correction factor (db)

---

*dbv indicates a voltage, expressed in decibels, with 0 db = 1 volt.
†Also applies to effective output level.
‡In calculations involving Power Rating J there may be an error, not greater than 2.5 db, due to the Rating Impedance differing from the actual impedance.

where the correction factor (C.F.) is given by.

Z	=	25	50	150	250	600	25 000	40 000 ohms
C.F.	=	−16	−13	−8	−6	−2	+14	+16 db

Crystal and condenser microphones cannot be rated in terms of power, while other types of microphones can only be compared directly with crystal and condenser types on the basis of the voltage on the unloaded grid circuit. To make this comparison it is here assumed that the nominal secondary impedance of the transformer is 40 000 ohms and that the voltage across the secondary is the voltage that would occur across a resistance of 40 000 ohms which was dissipating a power equal to the output rating of the microphone. This voltage is here called the high impedance grid circuit voltage (Rating D).

To convert from	add	To convert from	add
A to D (crystal mics.)	0 db	F to D	− 4 db
B to D (crystal mics.)	−20 db	G to D	+ 24 db
C to D	− 2 db	H to D	+ 4 db
E to D	+16 db	J to D	+ 90 db

Typical microphone ratings (db)

Rating / Type	A 0 db = 1 V 1 dyne/cm²	B 0 db = 1 V 10 dynes/cm²	D 0 db = 1 V $Z = 40\,000\ \Omega$ 1 dyne/cm²	E 0 db = 1 mW 1 dyne/cm²
	dbv	dbv	dbv	dbm
Carbon (S.B.)	−50*	−30*	−23	−39
Condenser	−50 to −60	−30 to −40	−50 to −60	—
Crystal (sound cell)	−54 to −76	−34 to −56	−54 to −76	—
Crystal (diaphragm)	−46 to −65	−26 to −45	−46 to −65	—
Moving coil	—	—	−49 to −70	−65 to −86
Pressure ribbon	—	—	−61 to −65	−77 to −81
Velocity ribbon	—	—	−58 to −67	−74 to −83

*Measured across microphone (approximate resistance 100 ohms).

Type	F 0 db = 1 mW 10 dynes/cm²	G 0 db = 6 mW 1 dyne/cm²	H 0 db = 6 mW 10 dynes/cm²	J 0 db = 1 mW 0.0002 dyne/cm²
	dbm	db	db	db
Carbon (S.B.)	−19	−47	−27	−113
Moving coil	−45 to −66	−73 to −94	−53 to −74	−139 to −160
Pressure ribbon	−57 to −61	−85 to −89	−65 to −69	−151 to −155
Velocity ribbon	−54 to −63	−82 to −91	−62 to −71	−148 to −157

(ii) Carbon microphones

An example of the **single button** type is the modern telephone inset, which gives an output of about 1 mW with close speaking. The frequency response is seriously peaked, while non-linear distortion and noise level are very high.

The **double button** (push-pull) type has a wider frequency range (70 to 7000 c/s in one **example**), and the second harmonic distortion is considerably less, but the output is about 10 db lower than that of the single button type.

Carbon microphones require a source of low voltage to pass a current of from 10 to 50 mA through each button. When the circuit is broken by a switch there is a tendency for the carbon granules to cohere—this may be reduced by connecting two condensers and three inductors as shown in Fig. 18.1.

All carbon microphones have a tendency to blasting, while their other defects including high noise level limit their application. The principal advantage is the high output level.

References A5, A21, A23, D1, D3, D4, D7, D13, D14.

(iii) Condenser microphones

A condenser microphone consists of a stretched diaphragm and a back plate, with a polarizing voltage between them. Owing to the high impedance, the pre-amplifier must be mounted very close to the microphone. A possible circuit arrangement is shown in Fig. 18.2. The response may be made almost flat from 30 to 10 000 c/s with an incident angle of 90°—on the axis there is a high frequency peak of about 8 db as with all pressure microphones. The average response is about − 60 dbv across the microphone itself (0 db = 1 volt per dyne/cm²).

References A5, A7, A23, D1, D3, D4, D7, D11, D13, D14.

The high frequency peak which occurs with microphones of standard size may be reduced by making the diaphragm smaller. A miniature condenser microphone with a diaphragm having the area of a human ear drum has been introduced with non-directional pickup characteristics (Refs. A18, A26). The output level from the pre-amplifier is − 50 dbm in a sound field of 10 dynes/cm².

(iv) Crystal and ceramic· microphones

Crystal microphones are of two types, directly-actuated and diaphragm types. Directly-actuated (sound-cell) microphones, as a class, have lower sensitivity but flatter frequency characteristics and they are almost non-directional. Uniform response up to 17 000 c/s can be obtained. There is a resonance at the high frequency limit causing a rise of response (+ 12 db in a typical case)—this should be equalized or else high frequency attenuation provided.

The frequency response of diaphragm type crystal microphones is less uniform than with the sound-cell type, and varies from 80 to 6000 c/s for speech only (with a pronounced peak in the 3500 c/s region) and from 50 to 10 000 c/s (± 5 db) for improved fidelity. The diaphragm type is more directional at high frequencies than the sound-cell type. There are also cardioid directional crystal microphones—see (xi) below.

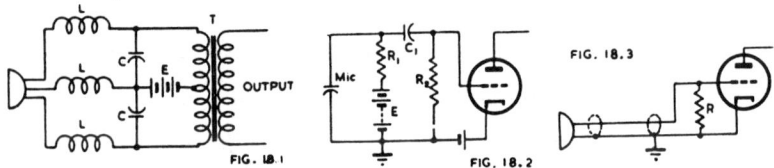

Fig. 18.1.　Circuit diagram of double button carbon microphone with filter (optional) to reduce cohering (L = 0.0014 H, C = 0.02 μF, E = 4.5 to 6 V).
Fig. 18.2.　Circuit diagram of condenser microphone.
Fig. 18.3.　Circuit diagram of crystal microphone.

All crystal microphones using Rochelle salt or similar materials tend to be affected by excessive humidity and temperatures above 125°F. Most of the crystals are now coated to provide considerable protection against humidity, although their use under tropical conditions appears to be risky.

A crystal microphone is effectively a capacitance, of the order of 0.03 μF for a diaphragm type or 0.0004 to 0.015 μF for a sound-cell type. This is effectively in series

with the generated voltage and following grid resistor.   The following grid resistor*
(R in Fig. 18.3) should be from 3 to 5 megohms for a sound cell type, or 1 to 5 megohms
for a diaphragm type.   A low resistance causes attenuation of low frequencies.

A long connecting cable will result in loss of output voltage, but will not affect the
frequency response.   The loss is given by

   db loss = 20 log [1 + ($L \times C_L$)/C_M]

where L = length of cable in feet

         C_L = capacitance of cable in microfarads per foot

and C_M = capacitance of microphone in microfarads.

A few crystal microphones are equipped with step-down transformers for use in
low impedance circuits.

Ceramic piezo-electric microphones have advantages over crystal types as regards
high temperature and humidity.   These have substantially similar performance to
that of crystal microphones.   One model has a nearly flat response from 30 to 10 000
c/s with an output level of − 62 dbv (0 db = 1 V per dyne/cm^2).

References to crystal and ceramic microphones : A5, A21, A23, D1, D3, D4, D7,
D13, D14.

(v) Moving coil (dynamic) microphones

A dynamic microphone is, in essence, a small version of a dynamic loudspeaker,
and some are even used in a dual capacity.   It is pressure-operated and there is the
usual tendency towards a rise in the high-frequency response for sound waves impinging directly on the diaphragm.   It may be mounted vertically to make the microphone non-directional for horizontal sound waves, but this causes some attenuation
of the higher frequencies.

In its simplest form the level frequency range is limited but in the more elaborate
types it may extend from 60 to 10 000 c/s.   It is a general-purpose good-quality
microphone widely used for public-address and indoor and outdoor broadcasts.
References A5, A21, A23, A32, D1, D3, D4, D7, D11, D13, D14.

(vi) Pressure ribbon microphones

The pressure ribbon microphone has a ribbon, suspended in a magnetic field, that
is exposed on one side and terminated in an acoustical resistance on the other.   It is
non-directional but has the characteristic, in common with all pressure types, of a
high frequency response that is a function of the direction of the incident sound.   The
distortion may be around 2% or 3% at low frequencies and 100 dynes/cm^2.

This construction is generally combined with a manual control of the size of an
aperture on the enclosed side which gives control of the directional characteristics
(e.g. R.C.A. 77-D).

A small unobtrusive pressure type ribbon microphone has been produced (R.C.A.
BK-4A) which employs a small angle horn coupled to a cylindrical tube which in turn
is coupled to the front of the ribbon by means of a round-to-rectangular connector
of constant cross-section.   The back of the ribbon is coupled to the damped folded
pipe or labyrinth by means of a rectangular-to-round connector.   This is non-directional, has a frequency response from 70 to 15 000 c/s and an effective output level of
− 61 dbm (sound pressure 10 dynes/cm^2).   (Refs. A25, A21, D3.)

(vii) Velocity ribbon microphones

This is a free ribbon type in which the ribbon resonance is usually below the audible
limit.   With good design it is capable of a very wide frequency response (e.g. 30 to
15 000 c/s for response ± 5 db), while over the most useful part of the range the variations of level are very slight and gradual.   It is strongly bi-directional.   This is undoubtedly the best type of microphone for high fidelity, particularly for transients,
but it is not suitable for general use outdoors.   The distortion in one case is less than
0.33% at 80 c/s and 1000 dynes/cm^2.   When a velocity microphone is placed very

*For maximum values of grid resistance see Sect. 2(vi).

close to the source of sound, the low frequencies are strongly accentuated. For this reason, a good, quiet studio and correct placement of speakers at a reasonable distance is necessary if a normal velocity microphone is used for speech. The bi-directional characteristic is useful where there are two speakers, or two groups, arranged on opposite sides of the microphone.

One model incorporates a switch and compensating reactor to reduce the accentuation of low frequencies with close speaking (R.C.A. KB-2C, Ref. A27).

A modified form incorporates a large amount of acoustical resistance to give a nearly flat response at all frequencies when speaking very close to the microphone (1 to 6 inches away), and noise cancellation for sounds originating at a distance. This is particularly suitable for outdoor public address (R.C.A. KB-3A, Ref. A28).

References A5, A21, A23, A27, A28, D1, D3, D4, D7, D11, D13, D14.

(viii) Throat microphones

A throat microphone is one which is actuated by direct contact of the diaphragm with the throat. The high frequency response must be accentuated to obtain intelligible speech. Both carbon and magnetic types have been used (Refs. A19, D3).

(ix) Lapel microphones

Normal types of carbon, crystal, dynamic and velocity microphones have been used for this application (Refs. A21, D3).

(x) Lip microphones

The velocity ribbon microphone has been used for this application (Ref. A5).

(xi) The directional characteristics of microphones

Pressure microphones are **non-directional** (Curve A Fig. 18.4) for low frequencies but at high frequencies the response is a function of the angle of sound incidence—this effect becomes smaller as the diaphragm diameter is reduced.

Velocity ribbon microphones are **bi-directional** (Curve B) and the sound source should be on one, other or both sides.

There are various types of **uni-directional** microphones, among which the cardioid (Curve C) is most popular. This characteristic may be achieved by a carefully designed combination of a velocity and a pressure microphone. Similar characteristics may be achieved by the use of acoustical delay systems in combination with any pressure microphone such as dynamic or crystal. A reduction of response of about 15 db at the back of the microphone with respect to that at the front is obtainable with a typical cardioid microphone. A single-element ribbon type microphone has been developed which has a reduction of 20 to 25 db at the back of the microphone (Ref. A22). A complete survey of directional microphones is given by Olson (Refs. A17, D3).

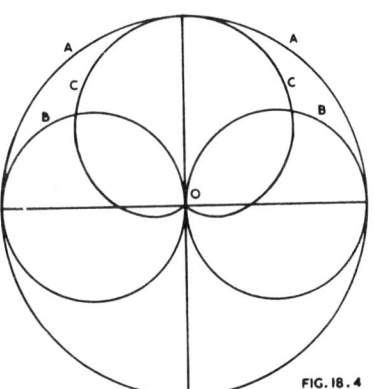

FIG. 18.4

Fig. 18.4. *Polar diagrams showing directional characteristics of microphones (A) Non-directional ; (B) bi-directional ; (C) cardioid.*

Polydirectional microphones provide a choice of two or three directional characteristics or a continuously variable characteristic. A typical example is the R.C.A. 77-D ribbon type.

The correct use of directional characteristics in microphone placement minimizes room reflection and reduces acoustical feedback and background noise.

References to directional microphones A17, A21, A23, D3, D13.

(xii) The equalization of microphones

It is usual, both in broadcasting and recording, to provide frequency equalization in the microphone or in its pre-amplifier to give a flat or any other desired characteristic. In other applications it is advisable to equalize, or at least to attenuate any high peaks in the response characteristics, so as to reduce acoustical feedback and other bad effects.

It is sometimes desirable to reduce the low-frequency response of a microphone (e.g. velocity type) to reduce the effects of accentuated low frequency response due to " close talking " or to reduce " building rumble."

(xiii) Microphone transformers

There is at present a very unhappy state due to the lack of accepted standards for microphone transformer output impedances. The American R.M.A. Standard for broadcasting is 150 ohms (Ref. A16). See also (xiv) below.

Most microphones are designed to work into an unloaded transformer. The best quality microphone transformers have special provision for reducing hum, such as outer steel shields, inner alloy shields and hum-bucking core construction. By this means it is possible to reduce hum to a level far below the level of random noise.

(xiv) Standards for microphones

(A) Microphones for sound equipment

The following summary is based on R.M.A. Standard SE-105 (Ref. A20).

A microphone is defined as an electro-acoustical transducer which converts acoustical energy into electrical energy, the waveform in this conversion remaining substantially unaltered.

The field response of a microphone at a given frequency is defined by

$$20 \log_{10} (E/p)$$

where E is the open circuit voltage generated by the microphone at its accessible terminals and p is the undisturbed sound pressure in dynes/cm^2 at the specified frequency . . . . , the microphone being placed at a specified angle with respect to the wave front.

The electrical impedance of a microphone at a given frequency is equal to the complex quotient at the specified frequency of the alternating voltage applied to the accessible terminals divided by the resulting alternating current when the microphone is placed in a free air field . . . . It shall be expressed in terms of its magnitude and phase angle and plotted as a function of frequency.

The nominal microphone impedance is the electrical impedance at 1000 c/s. The nominal impedance of low-impedance microphones shall lie between 120 and 180 ohms, or between 19 and 75 ohms. The nominal impedance of high impedance magnetic microphones shall lie between 20 000 and 80 000 ohms. The nominal impedance of any other microphone shall be specified.

The microphone system rating G_M (sensitivity) is described in Chapter 19 Sect. 1(iv)D.

The microphone rating impedance (R_{MR}) is a pure resistance to be taken from the table below

Magnitude of nominal microphone impedance	Rating impedance
19 to 75 ohms	38 ohms
75 to 300 ohms	150 ohms
300 to 1 200 ohms	600 ohms
1 200 to 4 800 ohms	2 400 ohms
4 800 to 20 000 ohms	9 600 ohms
20 000 to 80 000 ohms	40 000 ohms
80 000 ohms or more ohms	100 000 ohms

The directional pattern of a microphone at a specified frequency is the variation in the field response at that frequency for different angles of sound incidence measured from a specified zero position. Zero position refers to that microphone orientation in which the direction of the wave propagation lies along the axis of normal usage . . . .

This pattern shall be plotted in the form of a polar diagram as a function of the angle of incidence.   The electrical output at any one angle, E_θ, shall be referred to the output at the zero position, E_0, by the relation 20 $\log_{10}$ (E_θ/E_0).

The directional characteristic of a **bi-directional microphone** shall be at least − 20 db for all frequencies of a specified frequency range for sounds arriving from any point in a plane perpendicular to the axis of normal usage.   For a **uni-directional microphone** the angle of minimum response . . . . shall be on an average of at least − 10 db, the average to be based on minimum values for a number of frequency bands.

Standards for microphones : Refs. A8, A15, A16, A20.

(B) Microphones for radio broadcasting

Extract from R.M.A. Standard TR-105-B (Ref. A16).   The nominal impedance of all microphones intended for broadcasting service shall be 150 ohms ± 10% when measured at a single frequency of 1000 c/s ± 10%.

SECTION 2 : PRE-AMPLIFIERS

(*i*) *Introduction* (*ii*) *Noise* (*iii*) *Hum* (*iv*) *Microphony* (*v*) *Valves for use in pre-amplifiers* (*vi*) *Microphone pre-amplifiers* (*vii*) *Pickup pre-amplifiers* (*viii*) *Gain-controlled pre-amplifiers* (*ix*) *Standard pre-amplifiers for broadcasting* (*x*) *Standard pre-amplifiers for sound equipment.*

(i) Introduction

A pre-amplifier is a voltage amplifier suitable for operation with a low level input and whose output is intended to be connected to another amplifier operating at a higher input level.   Pre-amplifiers are commonly used with low-level microphones and pickups.   When a pre-amplifier has to feed the a-f amplifier in a typical radio receiver, the output voltage is usually taken for design purposes as 1 volt r.m.s. across a high impedance.   This allows for a volume control between the pre-amplifier and the main amplifier.   In some cases it is necessary to incorporate the volume control in the pre-amplifier itself in order to avoid overloading with abnormally high input levels.   In such cases this volume control is usually placed between the first and second stages of the pre-amplifier—with very high input levels it may be necessary to incorporate either a volume control or a fixed attenuator prior to the first grid.

A pre-amplifier follows the same general principles of design as any other voltage amplifier (see Chapter 12) except that particular attention has to be paid to hum, noise and microphony, all of which tend to be troublesome.

(ii) Noise

(A) The characteristics of random noise

Random noise includes both **thermal agitation noise** and **valve noise.**   The energy of such noise is distributed uniformly over the frequency spectrum.   The distribution of amplitude with time has been covered in Ref. B1—occasional peaks exceed four times the r.m.s. value.

Random noise voltages must be added in quadrature—that is to say as the root of the sum of the squares.

(B) Circuit noise

The **thermal agitation noise** of resistors has been covered in Chapter 4 Sect. 9(i)1 and in Chapter 23 Sect. 6.

When two or more resistances are in parallel, the total thermal agitation noise is that corresponding to the resultant resistance.   For example, if a microphone having a d.c. resistance of 30 ohms is loaded by a resistance of 100 ohms, the noise voltage is that corresponding to 23 ohms.

When a resistance R is shunted by a capacitance C, the thermal agitation voltage is decreased, and at 30°C (80°F) is given by

$$e = 1.29 \times 10^{-10}\sqrt{RF_0[\tan^{-1}(F_2/F_0) - \tan^{-1}(F_1/F_0)]} \tag{1}$$

where R = resistance in ohms
C = capacitance in farads
F_0 = $1/(2\pi RC)$
F_1 = upper frequency limit
and F_2 = lower frequency limit.
(See Reference B19, also B8 giving chart and B45 giving theory).

In any network comprising L, C and R it is only the resistive elements which generate noise voltages.

References to circuit noise : A23, B8, B13, B28, B40, B44, B45, B51, D1, D2, D9, D12.

(C) Valve noise

Valve noise has been covered in a general way in Chapter 23 Sect. 6. In this chapter we are only concerned with its effects in pre-amplifiers.

It is convenient to express the **shot-effect noise** of a triode in terms of an **equivalent noise resistance** (R_{eq}) at room temperature, connected from grid to cathode in a noiseless valve. The value of R_{eq} for triodes is given approximately by

$$R_{eq} \approx 2.5/g_m \tag{2}$$

where R_{eq} = equivalent noise resistance in ohms
and g_m = mutual conductance in mhos at operating point.

Pentodes produce additional noise due to the random partition of the cathode current between screen and plate, known as **partition noise**. In general, the noise energy from a pentode will be from 3 to 5 times as great as that from a triode producing the same amplification.

There are other, but minor, sources of noise in valves. The **flicker effect** produces low frequency variations, while any gas in the valve causes **ionization noise.**

References to valve noise : B2, B7, B13, B17, B40, B44, B51, D1, D2, D9.

(D) Methods used in the design of amplifiers to ensure low noise level

Noise in low level amplifiers is due to three principal causes—noise in the plate load resistor of the first stage, valve noise, and noise in the grid (input) circuit. If the stage gain is of the order of 20 or more, the effect of noise in the following stage may usually be neglected. Obviously any unnecessary loss of signal voltage through any form of attenuator between the signal source and the grid should be avoided.

Noise in the plate load resistor (**current noise**) may be avoided by using some form of special low-noise resistor, such as a high-stability cracked carbon resistor described in Chapter 4 Sect. 9(i)m. The plate supply voltage may be from 100 to 180 volts—if, as is usual, a higher supply voltage is available, it is merely necessary to incorporate a RC filter to drop the voltage and by-pass any noise and hum components. Such components are frequently required for decoupling in any case. Composition resistors may be used as screen resistors with pentode valves, because the noise voltage is by-passed to earth. Current noise, however, occurs in the grid circuit due to the negative grid current of the valve. Cathode bias resistors (if used) should be wire wound.

Valve noise may be made low, if necessary, by the use of a r.c.c. **triode valve** in the first stage. The equivalent resistance for shot-effect valve noise is inversely proportional to the mutual conductance (see eqn. 2) so that a high mutual conductance at the operating point is desirable if reduction of shot-effect noise is the principal object. Triodes in common use have published values of mutual conductance from 1000 to 5000 μmhos giving calculated values of noise resistance from 2500 to 500 ohms respectively under published conditions. The noise resistance of a resistance loaded valve is greater than that for the same valve under published conditions. For example type 6J5 operating with a plate current of 1 mA has $g_m = 560\mu$mhos and a calculated noise resistance of 4500 ohms, which is about 4 times the value under published conditions.

The calculated values of noise resistance for three typical **r.c. pentodes** are given below :

Operating conditions : $E_{bb} = 180$ volts ; $R_L = 0.1$ megohm.

Type	6J7 or 1620	6SJ7	6AU6
g_m (working)*	1000	1200	1700 μmhos
Noise resistance	8000	6000	2000 ohms

*Actually the slope of the dynamic characteristic.

There are two other possible sources of valve noise. **Leakage in the valve** from the grid to any other electrode, and particularly to any positive potential electrode, is a possible source of valve noise. **Reverse grid current** is another source of valve noise. It is advisable to use valves which have been tested and selected both for low leakage and reverse grid current, if a very low valve noise level is desired. Under all circumstances the d.c. resistance from grid to cathode should be as low as possible, and no composition resistors should be used in the grid circuit.

If the input source, whether microphone or pickup, is of low impedance and is coupled to the grid by means of a **step-up transformer**, it is usually not necessary to design for minimum valve noise. It is then only necessary to design the input circuit so that the impedance of the microphone reflected on to the grid is at least twice, and preferably four times, the valve noise resistance. With a low impedance microphone and transformer this may readily be accomplished by selecting a transformer secondary impedance of the order of 25 000 or 40 000 ohms. Valve noise will then have a negligibly small effect on the total noise, and pentodes or high-mu triodes may be used satisfactorily.

It is usual to employ as high a ratio in the microphone transformer as is possible, consistent with the frequency response required—very high step-up ratios can be used where the frequency response is not very important. Even in " high fidelity " applications (with a frequency range from 30 to 15 000 c/s) by using nickel-alloy cores, it is practicable to obtain a reflected secondary impedance up to 100 000 ohms, although somewhat lower values are more common.

For the measurement of noise in amplifiers see page 829.

(iii) Hum

The general features of power supply hum filtering and neutralization have been covered in Chapter 31 Sections 4 and 5. Hum due to conditions within the valves and hum due to circuit design and layout have been covered in Chapter 31 Sect. 4. Hum in voltage amplifiers has been covered in Chapter 12 Sect. 10(vi).

In pre-amplifiers the permissible hum voltage on the grid of the first valve depends on the maximum signal level at this point. If a high ratio transformer is used, with a reflected secondary resistance of 25 000 ohms or more, the signal level will be high and an extremely low hum level will not be necessary. On the other hand, if a low level pickup is coupled directly to the grid circuit without a step-up transformer, extreme care and ingenuity will be required to make the hum inaudible.

Some of the precautions which may be necessary in certain cases to reduce hum to a sufficiently low level are given below—

1. Complete electrostatic shielding of first pre-amplifier valve and associated components and wiring. The grid and plate circuits should be separately screened, so that there is negligible capacitive coupling between them—in the case of single-ended valves this requires care in wiring and either a shield or its equivalent between grid and plate.

2. Electromagnetic and electrostatic shielding of microphone transformer (if any). The more elaborately screened microphone transformers may include two or three concentric magnetic shields of permalloy or equivalent, with copper shields between (Ref. D2). For the lowest hum levels, hum-bucking is sometimes also provided in the transformer.

3. It is usual for the pre-amplifier to be built on a separate, preferably non-magnetic, chassis situated a considerable distance from the main amplifier chassis carrying the power transformer and filter chokes. If this is not possible, the power transformer should be designed to have low leakage flux and be mounted independently of the chassis, with the laminations vertical, and all a-f transformers should either be avoided or elaborately screened.

4. If insufficiently screened a-f transformers are used in the pre-amplifier, they should be oriented experimentally to the position giving least hum pickup.

5. With very low level pre-amplifiers it sometimes happens that less hum is obtained with a direct earth connection to a metal stake in damp soil than to a water pipe.

6. The leads from the power transformer to the valve heaters should be twisted, mounted as far as possible from all hum-sensitive components and wiring, and covered with earthed metal braid (for details see Ref. B42).

7. The loop formed by the valve, grid connection, input source and return lead to cathode should include as small an area as possible. Single-ended valves are preferable to double-ended types in this respect.

8. Metal valves are less sensitive to magnetic fields than glass types—they may sometimes be used with advantage if magnetic fields are unavoidable (Ref. B42).

9. If glass valves are used in a magnetic field they should be mounted so as to produce the minimum hum. This occurs when the flux vector is perpendicular to the valve axis and normal to the plane of the grid side rods. The ratio of maximum to minimum hum voltage is of the order of 30 to 40 db. Similar results occur with metal valves but the ratio is only from 10 to 20 db. Glass types may be fitted with a metal shield if desired (Ref. B42).

10. The waveform of the heater current should be closely sine wave, to eliminate capacitance coupling effects with harmonic frequencies.

11. A low leakage socket (e.g. isolantite or polystyrene) may be used to reduce leakage. With single-ended valves, leakage causing hum may occur from the heater pins to the grid or to the plate—the latter is normally the more serious because the grid pin is protected from hum leakage by adjacent pins. With double-ended valves the only possible socket leakage is from the heater pins to the plate. Valves with synthetic resin bases may have appreciable base leakage—for this reason, valves of the all-glass type (e.g. miniature) are preferable to those with separate bases.

One way of avoiding this trouble is to use a pentode valve with the plate, or plate and suppressor, earthed or returned to the cathode. The screen then is used as the anode, and the valve is thus effectively a triode with approximately the same characteristics as with the screen connected to the plate. The screen dissipation should be checked to see that it does not exceed the maximum, but it may be reduced, if necessary, by increasing the grid bias. See (v) below for application to type 6AU6.

12. A low resistance potentiometer may be connected across the heater supply, with the moving arm earthed—this may be adjusted for minimum hum. In some cases less hum may be obtained with the moving arm returned to a point of positive (or sometimes negative) voltage of the order of 5 to 50 volts (see Fig. 17.19A). Minimum hum may be obtained by optimum adjustment of both potentiometer and voltage.

13. A low resistance potentiometer may be used to inject heater voltage in antiphase to the hum at some convenient point such as the suppressor of a pentode valve. In this case a single heater winding is satisfactory with one potentiometer as in (12) above, together with one anti-phase control for each low level stage (Ref. B26).

14. In very low level stages the heaters are sometimes supplied with direct current. However, with good design (including the use of a high step-up ratio transformer) it is possible to achieve, with a.c. heating, a hum level which is entirely inaudible with a frequency range up to 15 000 c/s. Thus a d.c. heater is not really necessary, but rather an alternative method of achieving a similar result. If the frequency band of the amplifier is 5000 c/s or less, the hum usually dominates the noise, even with the best design—in such a case d.c. heater supply may be desirable.

There are many possible methods of providing the d.c. supply for the heaters, including

(a) Connecting the heaters of the low level valves in series at the low voltage end of the plate supply, with a shunt—if necessary—across either the heaters or the plate circuit. This method is most convenient when a heavy plate current is drawn by the amplifier, so that no power is wasted by a shunt across the plate circuit (e.g. Fig. 17.35A). It has the disadvantage that the heater current of the low-level valves is dependent upon the total plate current of the amplifier, which is liable to vary. Ref. B52.

(b) The heaters may be connected in series, and supplied from a separate full-wave rectifier and filter, using a thermionic or selenium rectifier—alternatively a selenium bridge circuit may be used. Refs. B25, B52.

(c) The heaters may be connected in parallel, and supplied from a selenium or other suitable rectifier, using either a full-wave or bridge circuit. Refs. B25, B52.

All methods employing rectified and filtered heater supply have a ripple component which may be large enough to cause audible hum.

15. A r-f oscillator with a frequency of 30 or 40 Kc/s may be used as an alternative to d.c. heater supply (Ref. B25).

16. If a.c. is supplied to the heater of the first stage, the voltage of this stage alone may be decreased to about 70% to 85% of normal. This is only permissible with a limited number of valve types and even so may require selection of valves. A cathode current well under 1 mA is essential (Ref. B20).

17. The cathodes of all pre-amplifier valves should preferably be earthed directly to the chassis. However, in practice this is sometimes inconvenient and it may be found that a minimum capacitance of 100 μF shunted across the bias resistor from cathode to chassis is satisfactory. High resistance grid-leak bias has not been found satisfactory for low level operation.

18. An electrostatic shield in the power transformer is essential, even where the transformer concerned supplies only heaters.

19. The pre-amplifier valves may be de-magnetized in a decreasing a.c. field. This is only useful when the valves have become magnetized by some means.

20. Magnetic shielding of power leads and all conductors carrying a.c. may be accomplished by the use of an iron pipe, flexible conductor or—most effective—a permalloy wrap similar to that used for continuous loading of submarine cable (Ref. B51).

Additional notes on hum with pentode valves

1. Hum due to lack of sufficient filtering in the plate supply may be neutralized by some suitable form of neutralizing circuit (e.g. Fig. 12.57).

2. The hum voltage caused by a magnetic field decreases more rapidly than the gain as the load resistance is decreased. A low load resistance is therefore desirable if there is a strong magnetic field (Ref. B42).

3. Hum caused by heater-to-plate leakage is also reduced more rapidly than the gain as the load resistance is decreased.

4. In pentode valves operating with low input levels there is a further source of hum in that the magnetic field of the heater varies the partition of current between plate and screen and so introduces hum. This hum is worse with remote cut-off valves than with those having sharp cut-off characteristics. This form of hum may be made zero by a suitable adjustment of grid or screen voltage (Ref. B46). See also Refs. B23 (Jan. 1949) and B50.

(iv) Microphony

There are very large variations between valves with regard to microphony, and special low-level types are recommended. Individual selection is, however, desirable in addition if the valves are required to operate at very low levels. In all cases some improvement may be made by a cushion socket and a thick rubber pad around the valve.

(v) Valves for use in pre-amplifiers

It is generally desirable to use valve types manufactured and tested for low level operation (e.g. types 12AY7, 1620 and 5879) but even with any one of these types there are large variations in hum, noise and microphony. If a particular pre-amplifier is required to meet stringent test conditions it is advisable to select valves to meet its requirements, and to use the balance of the valves in other less-stringent pre-amplifiers, or in later sockets of the same pre-amplifier.

A choice is possible in the American range between pentode types 1620 and 5879. The former is very much more expensive than the latter, and its use can generally be avoided by careful design.

If the designer decides to use ordinary radio valves for the first stage in a pre-amplifier, it is advisable to test and select the valves in accordance with the following procedure. A sufficient number of valves should be operated for about 48 hours under the same electrode voltages as in the pre-amplifier. They should then be tested for reverse grid current, and only those with reverse grid currents less than 0.1 μA should be used in very critical positions. If a sensitive microammeter is not available, a 1.0 megohm grid resistor may be cut in and out of circuit and the change in plate current measured—see Chapter 3 Sect. 3(iv)A. Low reverse grid current not only reduces the noise, but also permits a higher grid resistor to be used. The valves with low reverse grid currents should then be tested in the first socket of the pre-amplifier for noise, hum and microphony.

Some single-ended pentodes have one heater pin adjacent to the plate, and very high leakage resistance is required to avoid hum (except when this heater pin is earthed).

Type 6AU6 may be used as a triode with earthed plate and suppressor, using the screen as the anode, to reduce hum from leakage between heater and plate.

This device is only suitable for the first stage in the pre-amplifier owing to the limited screen dissipation rating (0.65 watt). The following operating conditions are suggested with a plate supply voltage of 180 volts (Ref. B48) :

Load resistance	0.05	0.1	0.25 megohm
Cathode bias resistance	450	750	1600 ohms
Plate current	2.4.	1.3	0.59 mA
Stage gain*	21	23	21

*With following grid resistor 0.5 megohm.

Characteristics of some special low-noise valve types

Type 5879 is a 9-pin miniature low-noise pentode with published $g_m = 1000$ μmhos at a plate current of 1.8 mA. The grid resistor (pentode or triode operation) may be up to 2.2 megohms under maximum rated conditions, while the voltage gain with 180 volts supply is 87 with $R_L = 0.22$ megohm and following grid resistor 0.47 megohm. As a pentode, the noise referred to No. 1 grid with a bandwidth of 13 000 c/s is of the order of 7.2 μV with grid resistor* 0.1 megohm, with only 10% exceeding 16 μV. It may also be used as a triode having $\mu = 21$, $g_m = 1530$ μmhos and $I_b = 5.5$ mA at $E_b = 250$, $E_c = -8$ volts. As a triode, the noise under the same conditions as for the pentode is of the order of 6 μV with only 10% exceeding 12 μV.

Conditions : Triode operation	Hum
1. One side of heater earthed, cathode by-passed (40 μF), zero grid resistance—median value	9 μV
2. One side of heater earthed, cathode unbypassed, zero grid resistance —median value	100 μV
3. One side of heater earthed, cathode by-passed (40 μF), grid resistor 0.1 mΩ—median value	13 μV
4. Centre-tapped resistance across heater supply, returned to fixed bias point ($+$ 20 to $+$ 50 volts) will reduce hum on most valves to less than	20 μV
5. Centre-tapped resistance across heater supply, returned to voltage to give minimum hum, cathode by-passed (40 μF), grid resistor 0.1 megohm—median value	3.5 μV
—less than 10% exceeding	8 μV

When adjusted for minimum hum, this feeds a small hum bucking signal into the grid to oppose other minor sources of hum in the amplifier (Ref. B16).

Type 12AY7 is a low noise twin triode (Ref. B17) with $\mu = 40$, $g_m = 1750$ μmhos and $I_b = 3$ mA at $E_b = 250$, $E_c = -4$ volts. The following tests were made with a grid resistor 0.1 megohm and plate load resistor 20 000 ohms. Shot-effect noise (40 to 13 000 c/s) referred to the grid—median valve 4.7 μV, maximum limit 8 μV.

---

*The noise due to the grid resistor alone is about 3.8 μV.

Hum referred to the grid—median valve 3.9 μV, maximum limit 12 μV with cathode resistor by-passed (40 μF).

When used in a typical circuit, placed with the electron stream in a magnetic field* of 1 gauss, the hum voltage on the grid is about 10 μV.

With a grid resistor of 30 000 ohms, approximately 4 μV of hum per volt of heater potential per $\mu\mu$F of capacitance between heater circuit and grid circuit will appear at the grid.   If the heater voltage is 6.3, this will give approximately 25 μV per $\mu\mu$F of coupling.   For applications see Figs. 18.7A, 18.7B.

(vi) Microphone pre-amplifiers

The limit on practicable amplification is placed by the noise level caused by the thermal agitation noise in the grid circuit of the first stage together with some valve noise.   The thermal agitation noise alone is at a level of about $-$ 129 dbm with a matched load (Ref. B28), or $-$ 132 dbm without loading, for a bandwidth of 15 000 c/s.   If the effective microphone level is taken as $-$ 60 dbm, with an unloaded microphone the maximum possible signal-to-noise ratio will be 72 db, or 69 db with a loaded microphone.   In practice with multiple mixers and other contributing factors†, a broadcast station overall noise can be considered satisfactory if within 10 db of the thermal noise.

If the valve noise is at least 3 db below the thermal agitation noise, it may be neglected as an approximation since the resultant is given by the root of the sum of the squares, and the additional noise will be less than 1 db.   This may be put in the alternative form, that the equivalent valve noise resistance should be less than half the effective thermal agitation noise resistance.

In the case of crystal microphones the whole of the microphone noise arises from the grid resistor, but the signal-to-noise ratio is worst for values of about 0.1 megohm, and improves as the grid resistor is made smaller or larger than this value.   For low noise the resistor should be at least 5 megohms, but better signal-to-noise ratio is obtained with 50 megohms and even better with no grid resistor at all (Ref. B19).

However, for ordinary applications, a resistance of 5 megohms is a good compromise, giving a noise voltage of about 4 μV or $-$ 108 dbv (0 db = 1 V).   The noise resistance of a triode or pentode valve is negligibly small in comparison.   This permits a signal-to-noise ratio of at least 60 db with all diaphragm types and the more sensitive sound cell types.   With the less sensitive sound cell types a very high grid resistance may be necessary.

Values of grid resistor greater than 1 megohm may only be used satisfactorily in low-level pre-amplifiers where the valve is one specially manufactured or tested for this class of service, under a specification which ensures that the reverse grid current is very low.   With a maximum reverse grid current of 0.2 μA and with μ not greater than 40, a grid resistor up to 5 megohms may be used provided that the plate load resistor is not less than 0.05 megohm from a plate supply of 150 volts, or proportionately higher than 0.05 megohm for voltages greater than 150 volts.

Hum from the plate supply may be made as small as desired by filtering, neutralizing, and possibly the use of a triode in the first stage.   Hum from other sources, including a.c. heaters, may be made completely inaudible in wide frequency range amplifiers (up to 15 000 c/s) by careful design, using some of the methods in (iii) above.

Pre-amplifiers for use with crystal microphones

The pre-amplifiers described below (A, B and C) have an output level of about 1 volt r.m.s. and are suitable for use with a diaphragm type crystal microphone.

(A) Single stage pentode pre-amplifier (Fig. 18.5)

This is the simplest possible pre-amplifier, with a voltage gain of 118 (41 dbvg‡). The input voltage to provide the specified output level is $-$ 41 dbv (0 db = 1 V). It may therefore be used with an average diaphragm-type crystal microphone with a

---

*Flux densities from 1 to 3 gauss are found quite frequently as far as 7 inches from a typical radio power transformer.
     †The effective noise bandwidth of a studio amplifier is usually wider than 15 Kc/s.
     ‡dbvg = decibels of voltage gain—See Chapter 19 Sect. 1(ii).

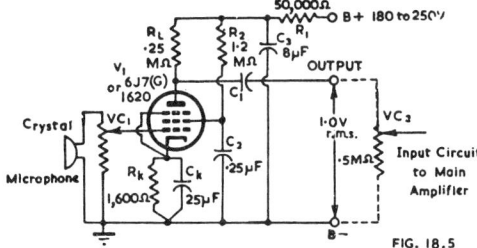

Fig. 18.5. Single stage pentode pre-amplifier for use with diaphragm-type crystal microphone.

FIG. 18.5

sound pressure of 10 dynes/cm², allowing a margin of 5 db for attenuation by the volume controls.

This is reasonably satisfactory for simple public address systems, home recording and amateur transmitters, although the margin of gain is small. It may also be used to feed the pickup terminals of any normal radio receiver.

Higher gain (45 dbvg*) may be obtained by the use of a selected type 6AU6 with $R_L = 0.22$ megohm, $R_2 = 0.43$ megohm and $R_k = 1700$ ohms. Under these conditions VC_2 may be adjusted to give about 10 db attenuation, VC_1 set to provide the desired output level under average conditions, and VC_2 then used as the control— this will prevent overloading of the valve.

The recommended maximum value of VC_1 is 1 megohm. Higher values may be used without damage to the valve, on account of the plate and screen resistors, but the operating point may be seriously shifted by reverse grid current, leading to distortion and loss of gain. Normally VC_1 would be set at maximum, and only reduced for very close working.

This arrangement has many limitations, and the noise level is high, but it is good enough for the simplest and cheapest applications mentioned above.

(B) Single stage pentode pre-amplifier followed by cathode follower (Fig. 18.6)

This circuit has the advantage over Fig. 18.5 that the pre-amplifier may be some distance from the main amplifier, the two being linked by a low impedance line, and that low impedance mixing may readily be applied. The gain and performance are very similar to those of Fig. 18.5 except that the second volume control would normally be incorporated in the low impedance line (reflected impedance about 400 ohms).

Fig. 18.6. Single stage pentode pre-amplifier followed by cathode follower, for use with diaphragm-type crystal microphone.

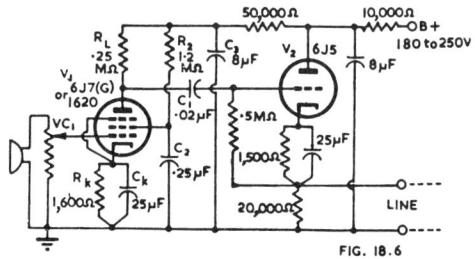

FIG. 18.6

This circuit has the disadvantage that the cathode follower stage may give rise to audible hum. It should therefore be restricted to use with a sensitive type of crystal microphone or to applications where a low hum level is not essential.

(C) A preferred arrangement is a good transformer in the cathode circuit of the 6J5, stepping down from 20 000 ohms primary impedance to any desired line impedance.

(D) Cathode follower as low-noise input stage

Fig. 18.6A shows the conventional cathode follower, and Fig. 18.6B shows the modified circuit for use as a low-noise input stage in connection with crystal microphones

*dbvg = decibels of voltage gain—see Chapter 19 Sect. 1(ii).

or pickups.   It has been shown (Ref. B53) that the modified circuit will have the same stability as the conventional circuit if $R_4 = (1 + G)R_1$, where G is the voltage gain of the valve defined as the ratio of the voltage across R_3 to that across R_1.   If Z_L is high, as it will be if coupled to another valve grid, then G will be approximately constant and a real number.   In a properly designed circuit R_4 can be made considerably higher than R_1.

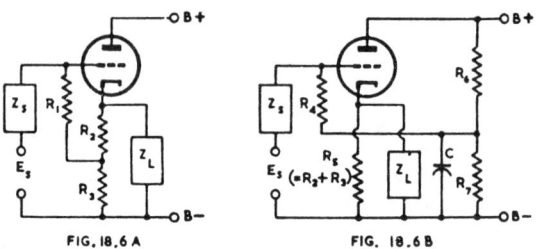

FIG. 18.6 A   FIG. 18.6 B

Fig. 18.6A.   *Conventional cathode follower circuit.*
Fig. 18.6B.   *Modified cathode follower for use as low-noise input stage with crystal microphone or pickup.   (Ref. B53).*

The circuit of Fig. 18.6B shows an improvement in signal to noise ratio compared with that from the conventional circuit, the ratio being $\sqrt{(R_4/R_1)}$ for thermal noise produced by R_1 and R_4.   For noise generated in R_2 and R_3, noise from the power supply, valve noise and hum and microphony voltages other than those due to capacitance coupling or resistance leakage into the grid circuit, the noise ratio is

$$\frac{R_4}{R_1}\left|\frac{Z_s + R_1}{Z_s + R_4}\right|$$

When Z_s is large compared with R_1, this represents an improvement by a factor R_4/R_1.   When $Z_s = R_1$ the ratio is 1/2 or 6 db.   When $Z_s = jR_1$ it is approximately $1/\sqrt{2}$ or 3 db.   When Z_s is small the ratio is negligible.

Thus if Z_s is a crystal microphone or pickup, its impedance will rise as the frequency is reduced, and in the conventional circuit this causes an increase in noise level at low frequencies.   The modified circuit shows a distinct improvement under these conditions.

Great care should be taken with the insulation of the grid circuit.

Pre-amplifiers for use with low-impedance microphones

In broadcast stations it is usual to have a gain of about 30 or 40 db in the pre-amplifier, and for this to be followed by mixing (with a loss of about 18 db for four or

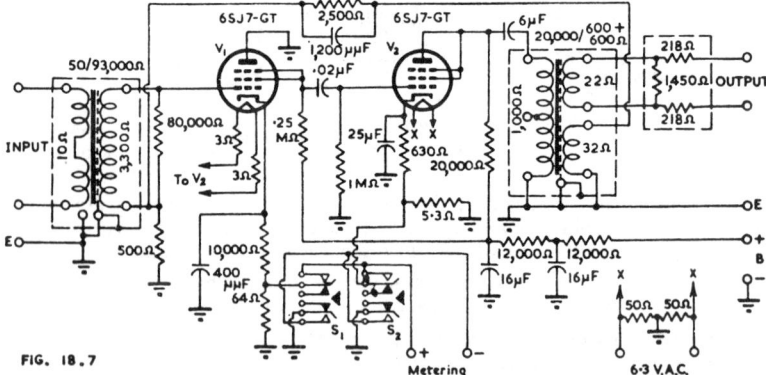

FIG. 18.7

Fig. 18.7.   *Two stage broadcast station microphone pre-amplifier with gain of* 29 *db.*

five channels) and by a second amplifier with a gain of about 40 db followed by a master gain control with a minimum loss of 6 db and a third amplifier with a gain of 40 or 50 db with an output level of + 18 dbm.   The total gain is sufficient for full output to be obtained with the lowest possible input level (say − 70 to − 80 dbm). The minimum signal-to-noise ratio is usually 60 db at − 60 dbm (alternatively 65 db at − 50 dbm).

A pre-amplifier gain of 40 db may result in over-loading of the pre-amplifier and distortion if used with very high microphone input levels—a device may be incorporated to give an optional attenuation of 10 db in the pre-amplifier and thus reduce its total gain to 30 db when there is danger of overloading.

A typical microphone pre-amplifier with a gain of 29 db is shown in Fig. 18.7.   Both valves are connected as triodes, the first stage using the screen as the anode, and the gain without feedback is 47 db.   When operated with an input level of − 60 dbm, the signal to noise ratio is 62 db.   The maximum power output is 10 mW (+ 10 dbm) at which the distortion is less than 1%—for this level to be reached, the input level must be − 29 dbm.   The frequency response is from 30 to 10 000 c/s with less than 0.5 db variation.   This circuit may be criticized on the choice of valve type for the first stage of a low level pre-amplifier.   Type 5879, 1620 or Z729 could be used to advantage   in   the   first   stage.

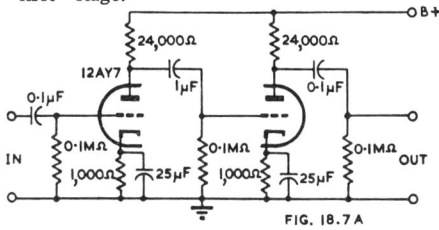

Fig.   18.7A.   *Two-stage pre-amplifier using twin triode type* 12AY7 *(Ref. B17).*

A two-stage pre-amplifier using type 12AY7 twin triode is shown in Fig. 18.7A. This has a total gain of 50 db, and the average noise and hum level referred to the input grid is 11 μV (Ref. B17).

A balanced pre-amplifier employing cross-neutralization and negative feedback is shown in Fig. 18.7B, using a.c. on the heaters.   The frequency response is flat from

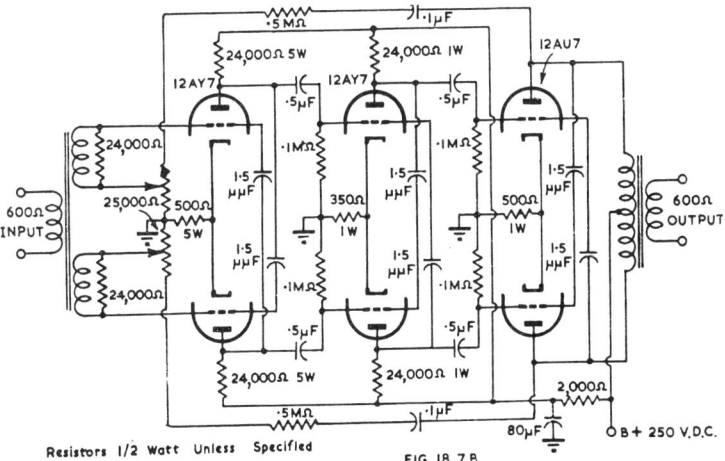

Fig. 18.7B.   *Three-stage balanced cross-neutralized pre-amplifier using two* 12AY7 *valves and* 12AU7 *(Ref. B17).*

30 to 20 000 c/s, the average noise is 87 db down from the + 24 dbm level and the distortion is 0.55% at 24 dbm output.   The combined hum and noise voltages are only 7.6 db above the theoretical value on the poorest valves, and only 2.2 db on the average (Ref. B17).

The essential amplifying and attenuating circuits of the four pre-amplifiers and one " A " amplifier of a single unit equipment for studio use are shown in Fig. 18.7C. Each pre-amplifier has a 3-position key.   In the pre-amplifiers, type 6AU6 is used as a triode with earthed plate and suppressor, and negative feedback is applied from plate circuit to grid.   Each pre-amplifier has a separate attenuator normally " holding " 6 db.   The " A " amplifier uses type 6AU6 as a high gain pentode followed by the master gain control normally " holding " 20 db.

The " B " amplifier of the same equipment is shown in Fig. 18.7D, and has negative feedback from the secondary of the output transformer to the cathode of V_6.

With all controls set at maximum, an input of − 86 dbm gives an output of + 8 dbm.   The distortion is less than 1% at 18 dbm output.   The noise level through any one channel is 60 db or more below output level with input − 60 dbm, output + 8 dbm.

A pre-amplifier circuit used by the B.B.C. is described in Ref. B41.   A special low-noise microphone amplifier for acoustical measurements is described in Ref. B54.

A pre-amplifier for the Western Electric type 640AA condenser microphone, using subminiature valves, is described in Ref. B43.   Extremely high insulation resistance is required in the network between the microphone and the first grid, particularly in the coupling capacitor.

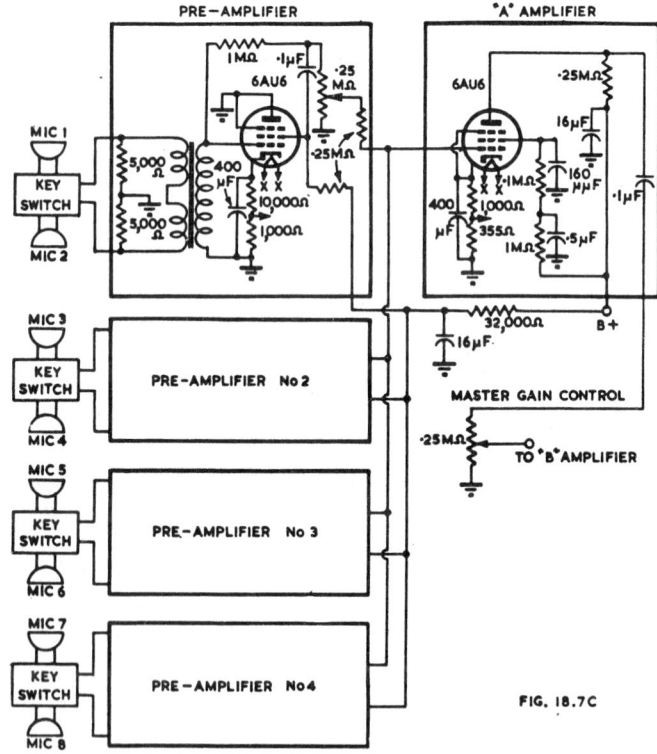

Fig. 18.7C.   *Essential amplifying and attenuating circuits of pre-amplifiers and " A " amplifier of single unit equipment for studio use (Ref. B18).   See also Fig. 18.7D.*

References to microphone pre-amplifiers : A16, A24, B19, B28, B41, B43, B48, B53, B54.

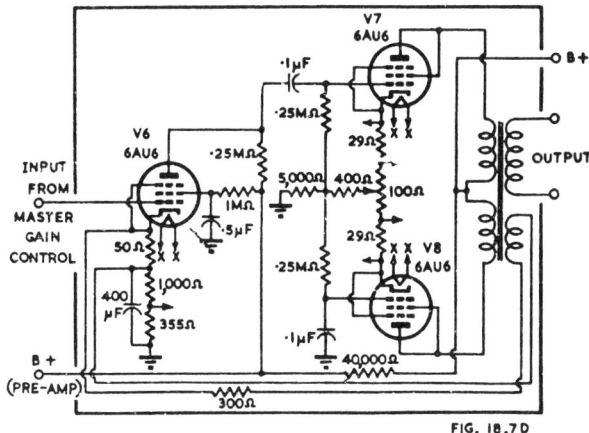

FIG. 18.7D

Fig. 18.7D. " B " amplifier of same equipment as Fig. 18.7C (Ref. B18).

(vii) Pickup pre-amplifiers
This subject has been covered in Chapter 17 Sect. 5.

(viii) Gain-controlled pre-amplifiers
This subject has been covered in Chapter 16 Sect. 5.

(ix) Standard pre-amplifiers for broadcasting
(Extracts from R.M.A. Standard TR-105-B, Ref. A16.)

Standard input signal is 2.45 millivolts r.m.s. in series with 150 ohms (for purposes of calculating insertion gain this corresponds to an input level of − 50 dbm).

Standard output level for feeding telephone lines is + 18 dbm ; for feeding radio transmitters is + 12 dbm. The equivalent " complex wave " level is 10 db lower than these sine wave equivalent testing level values (i.e. + 8 vu and + 2 vu respectively) to provide a margin of 10 db for peaks.

Source and load impedances—600/150 ohms.

Frequency range and harmonic distortion—see Ref. A16.

Signal to noise ratio (noise below standard output)—minimum 65 db (50-15 000 c/s).

(x) Standard pre-amplifiers for sound equipment
(Extracts from R.M.A. Standard SE-101-A, Ref. A30.)

To operate from a **source impedance** of 150 ohms and into a load impedance of 600 and/or 150 ohms.

Noise level—for measurement see Chapter 19 Sect. 6.

Amplifier gain is defined as the ratio, expressed in db, of the power delivered to the load to the power which would be delivered to the same load if the amplifier were replaced by an ideal transformer which matches both the load and source impedances. Frequency response, distortion and power output—see Ref. A30.

SECTION 3 : ATTENUATORS AND MIXERS

(i) Potentiometer type attenuators (volume controls) (ii) Single section attenuators—constant impedance (iii) Single section attenuators—constant impedance in one direction only (iv) Multiple section attenuators (v) Electronic attenuators (vi) Mixers and faders—general (vii) Non-constant impedance mixers and faders (viii) Constant impedance mixers and faders.

An attenuator is a resistance network used for the purpose of reducing voltage, current or power in controlled (and usually known) amounts. An attenuating network is sometimes called a " pad."

In this section only a brief outline is given of the most popular audio frequency applications of attenuators. Additional information is available from the references.

(i) Potentiometer type attenuators (volume controls)

The continuously variable volume control is widely used in radio receivers and a-f amplifiers (Fig. 18.8). The voltage ratio is proportional to the resistance ratio only when the load across the output terminals is very much greater than R. Characteristics commonly used include linear and several types of logarithmic characteristics of resistance versus angular rotation—see Chapter 38 Sect. 3(viii). The input loading is constant (R) only when the output load resistance is infinite. When the output terminals are connected to the grid of a valve which does not draw grid current, the input capacitance together with strays will bring about some change in frequency characteristics as the setting is varied. Direct current through R and through the moving contact should be avoided if a composition type resistor is used, to avoid noise.

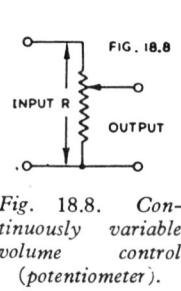

FIG. 18.8

INPUT R

OUTPUT

Fig. 18.8. Continuously variable volume control (potentiometer).

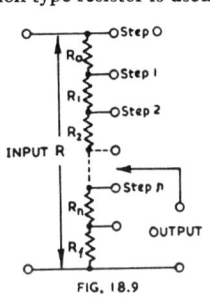

INPUT R

Step 0
Step 1
Step 2

Step n

OUTPUT

FIG. 18.9

Fig. 18.9. Step-type volume control.

The step-type volume control (Fig. 18.9) has some advantages over the continuously variable type—it is more reliable, has lower noise as the contact is being moved, and the degree of attenuation is definitely known. In all the best designs, wire-wound resistors are used. It is usually designed with logarithmic characteristics so that there is a constant step (in decibels) between successive tapping points, e.g. 2 db. The method of calculating R_0, R_1 etc. is given below.

Let R = total resistance of attenuator (Fig. 18.9)

   N = voltage ratio corresponding to the decibel step between successive tapping points (N less than unity)

and $M = (1 - N)$.

Then $R_0 = MR$; $R_1 = MNR$; $R_2 = MN^2R$ etc.

   $R_n = MN^nR$; $R_f = R - R_0 - R_1 - R_2 \ldots - R_n$.

As a practical case take $R = 100\ 000$ ohms with 2 db steps.

Then $N = 0.794$ and $M = 1 - 0.794 = 0.206$. By simple calculation $R_0 = 20\ 600$; $R_1 = 16\ 340$; $R_2 = 12\ 980$; $R_3 = 10\ 300$; $R_4 = 8\ 190$; $R_5 = 6\ 500$; $R_6 = 5\ 170$; $R_7 = 4\ 100$; $R_8 = 3\ 260$; $R_9 = 2\ 590$ and $R_{10} = 10\ 000$ ohms (all values correct to three significant figures).

Slide rule or logarithm accuracy is sufficient for most purposes ; for greater accuracy see Ref. C9.

(ii) Single section attenuators—constant impedance

It is assumed that it is required to maintain constant input and output impedances (Z), that the attenuator is terminated at both ends with resistive impedances Z, and that all impedances are purely resistive. The T type section in Fig. 18.10 may be designed to provide any desired attenuation.

Let $K = E_i/E_0$, this being the voltage or current ratio corresponding to the desired attenuation in db (K being greater than unity).

Then $R_1 = Z\left(\dfrac{K-1}{K+1}\right)$ and $R_2 = \dfrac{2ZK}{K^2-1}$ (1)

It is possible to select any desired values of attenuation in a single T section by varying the values of the three resistors by means of a tapping switch. Values of R_1 and R_2 are given in Table 1 on page 796.

The Π type section in Fig. 18.11 is equivalent to the T type section in Fig. 18.10 provided that the values of the resistances are given by

$$R_3 = Z\left(\frac{K^2-1}{2K}\right) \text{ and } R_4 = Z\left(\frac{K+1}{K-1}\right)$$ (2)

Values of R_3 and R_4 are given in Table 1 on page 796. For formulae giving transformations from T to Π or from Π to T forms see Ref. C12.

Both T and Π sections may be arranged in a " balanced " form for use with an earthed centre-tap—in this case the series resistors R_1 and R_3 are divided into two halves, one on each side of the centre (for table of formulae see Ref. D6). They may also be designed to match unequal source and load impedances (see Refs. C7, C12, D6, D11).

For the derivation of the equations above, see Ref. C13.

If the output impedance is increased beyond the design value, the input impedance will be increased (although to a less extent) and the attenuation will be reduced—see Refs. C6, C12, D6.

There are many other forms of constant impedance attenuators, but none does more than the T or Π section—they may be preferred for practical reasons in certain applications. See Refs. C12, D6, D8, D11.

(iii) Single section attenuators—constant impedance in one direction only

The L section of Fig. 18.12 may be used when it is not necessary for the looking-backwards output impedance of the attenuator to be constant. In this case

$R_5 = Z(K-1)/K$ and $R_6 = Z/(K-1)$ (3)

For further information see Refs. C12, D2.

The values of R_5 and R_6 have been tabulated for steps from 1 to 10 db (Table 1 on page 796).

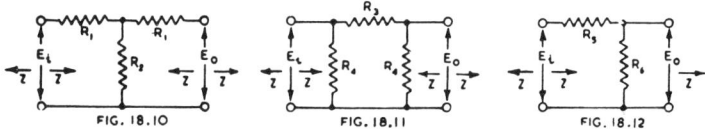

Fig. 18.10. *Constant impedance T section attenuator.*

Fig. 18.11. *Constant impedance Π section attenuator.*

Fig. 18.12. *The L type single section attenuator.*

The values of resistances given in Table 1 are for $Z = 100$ ohms. For any other value of Z, the tabulated values of resistances should be multiplied by $(Z/100)$. See Figs. 18.10, 18.11 and 18.12.

TABLE 1 (SINGLE T, Π AND L SECTIONS)

		T section		Π section		L section	
db	K	R_1	R_2	R_3	R_4	R_5	R_6
1	1.122	5.76	866	11.6	1740	10.9	819
2	1.259	11.46	430	23.2	874	20.6	386
3	1.413	17.10	284	35.2	585	29.3	242
4	1.585	22.6	210	47.6	443	36.9	171
5	1.778	28.0	164	61.0	357	43.8	129
6	1.995	33.2	134	74.6	302	49.9	100.5
7	2.239	38.2	112	89.3	262	55.4	80.7
8	2.512	43.1	94.6	106	232	60.2	66.1
9	2.818	47.6	81.2	123	210	64.5	55.0
10	3.162	51.9	70.3	142	193	68.4	46.2

The following references will be found helpful in supplying further general information on attenuators :

Periodicals C6, C12 (the most comprehensive of all), C13, C17 (the basic treatment). Books D2, D6, D8, D11 (brief treatments are given in several other books).

(iv) Multiple section attenuators

Any convenient number of Π sections may be connected in tandem, and the effective resultant is the " ladder " attenuator of Fig. 18.13 in which the two parallel resistors at the end of each section and the commencement of the next are drawn as a single equivalent resistor $(R_4/2)$. The values of R_3 and R_4 are calculated as for a single Π section (eqn. 2 and Table 1) to provide the desired attenuation per stage. The value of R_8 is given by $R_4 Z/(R_4 + Z)$. The minimum attenuation is 3.5 db on tapping point P_1. The input impedance varies considerably as the tapping point is changed but may be made constant, if desired, by inserting resistors at the points marked X. The output impedance also varies, although less so than the input impedance. This device may also be used when the impedance of the source differs from Z.

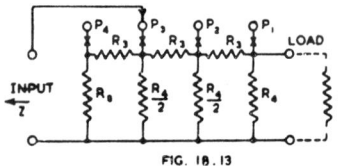

Fig 18.13. *Ladder attenuator (Π section).*

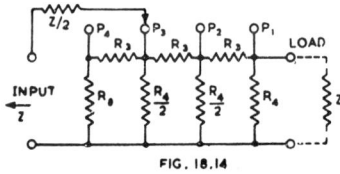

Fig. 18.14. *Modified form of ladder attenuator.*

A modified form of ladder attenuator is shown in Fig. 18.14 which has a minimum attenuation of 6 db but whose input impedance is nearly constant except for high values of attenuation and whose output impedance is nearly constant except for low values of attenuation. This can only be used when the impedance of the source is equal to Z. The values of R_3, R_4 and R_8 are the same as in Fig. 18.13. This is the commonest type for broadcast station " mixer " control equipment.

A form of ladder attenuator suitable for connecting a low impedance line to a high impedance grid circuit is shown in Fig. 18.15. This is built up from three L sections and the input impedance is equal to Z under all conditions but the output impedance varies (this being unimportant). The values of R_5 and R_6 are given by eqn. (3) and Table 1, while R_7 is given by $ZR_6/(Z + R_6)$. As an example take $Z = 600$ ohms and 2 db steps. Then the voltage ratio corresponding to 2 db is $K = 1.259$. From eqn. (3)

$$R_5 = 600 \times 0.259/1.259 = 123.6 \text{ ohms}$$

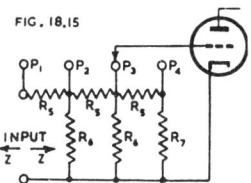

FIG. 18.15

Fig. 18.15. Ladder attenuator for connection to high impedance grid circuit (L section).

$R_6 = 600/0.259 = 2316$ ohms
while $R_7 = 600 \times 2316/2916 = 477$ ohms.

References to multiple section attenuators : C11, C12 (Part 6) ; C14, D2, D6, D11. References to non-uniform sections, C11, C14.

A multiple T section attenuator is shown in Fig. 18.16. This has the advantage of no loss in the zero attenuation position, and is suitable for use in positions where no power loss is permissible, such as an attenuator on a tweeter loudspeaker. The values of resistors for 1 db steps for $Z = 100$ ohms are tabulated below—for any other impedance multiply the resistances by $(Z/100)$.

Loss (db)	Series arm (each)	Shunt arm (total)	Loss (db)	Series arm (each)	Shunt arm (total)
1	5.8 ohms	867 ohms	6	33.2 ohms	134 ohms
2	11.4	430	7	38.2	112
3	17.1	284	8	43.0	94.5
4	22.6	209	9	47.6	81.3
5	28.0	164	10	52.0	70.2

(v) Electronic attenuators

For some purposes an electronic attenuator may be advantageous, and one practicable circuit is a cathode follower in which the screen voltage may be varied, while in addition the desired value of cathode resistor may be selected by means of a tapping switch (Fig. 18.17). Any sharp cut-off pentode valve is suitable, but one with high mutual conductance will limit the maximum attenuation. A valve with $g_m = 2000$ μmhos has an attenuation range from 4 to 74 db with values of cathode resistors from 1000 ohms to 0.1 ohm ; this may be extended by about 10 or 20 db by screen voltage

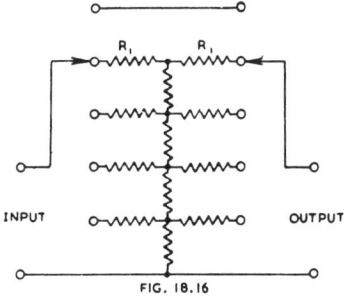

FIG. 18.16

Fig. 18.16. Multiple T section attenuator.

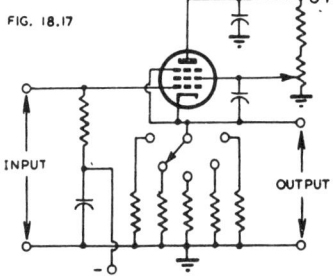

FIG. 18.17

Fig. 18.17. Electronic attenuator using cathode follower (Ref. C15).

adjustment provided that the input level is not too high. Fixed grid bias must be used. A universal design curve is given in Fig. 18.18 (Ref. C15).

A circuit using type 6BE6 as a fader-mixer is shown in Fig. 18.18A. This has a voltage gain of 62 from grid 1 input, and 25 from grid 3 input. The distortion for

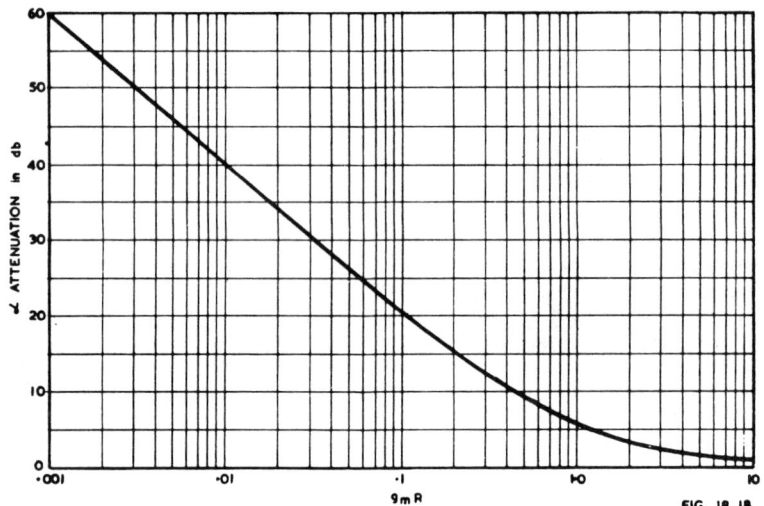

Fig. 18.18. Universal design curve for electronic attenuator (Ref. C15).

an output of 5 volts is 0.19% with input to grid 1, and 1.4% with input to grid 3. A plate load resistor of 50 000 ohms gives lower gain and distortion. For electronic gain control purposes, it is advisable to apply the signal to grid 3 and the control voltage to grid 1 (Ref. C20).

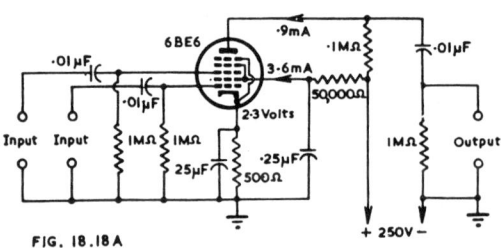

Fig. 18.18A. Fader-mixer circuit using type 6BE6 (Ref. C20).

(vi) Mixers and faders—general

When more than one input source is employed, some form of control is necessary to enable the operator to change from one to the other, or to mix the outputs of two or more sources. A properly designed mixer and fader system enables these objects to be achieved without perceptible jumps.

Mixers (as the complete system will here be called) may be divided into high impedance and low impedance ; constant impedance and non-constant. In addition there are those that have sources at approximately the same level and those with sources having a considerable difference in level. To achieve correct functioning of mixers, it is necessary to provide for the control of the level of individual input sources. Any frequency equalization of input sources is applied prior to the mixer.

(vii) Non-constant impedance mixers and faders

These are commonly used in public address systems, amateur transmitters and other applications which do not require precise adjustment to a predetermined level. The controls are of the continuously variable potentiometer type, and the mixing is usually carried out between a pre-amplifier and a subsequent amplifying stage—under these conditions there is no necessity for maintaining either constant input or output impedance. In some cases the mixing is carried out between the input sources and the grid of the first amplifier stage, and here the load impedance presented to each source must be maintained constant.

The simplest type of fader*, which does not provide mixing, is shown in Fig. 18.19. It is only suitable for high level pickups, or the secondaries of step-up transformers, since the noise from the moving contact is appreciable. The control is a centre-tapped potentiometer with a total resistance $2R$. The value of R should be that recommended for each pickup.

Fig. 18.20 shows a series network mixer which may be used as a fader, but it has serious drawbacks. Both sides of input source A are above earth, and any hum picked up in this channel is fed without appreciable attenuation to the following grid. More-over stray capacitances to earth of channel A tend to by-pass the high frequency signal voltages of channel B. This arrangement is not recommended.

A modified series network mixer is shown in Fig. 18.21 in which one side of each channel may be earthed, but a transformer must be used to couple the outputs to the grid. This may be used for fading and mixing.

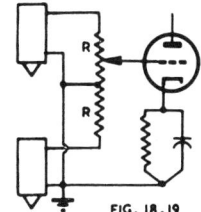
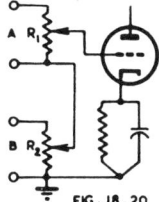
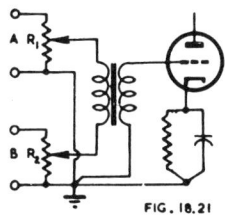

Fig. 18.19. Simple type   Fig. 18.20. Series   Fig. 18.21. Modified
of fader for two pickups.   network mixer.   series network mixer.

The parallel network mixer is shown in Fig. 18.22 where three input sources are provided for, although any number may be used. The value of R_4 may be made equal to or greater than R_1, and similarly with the other channels. If the mixer is con-nected between two amplifying stages, all resistances may be equal (say 0.5 megohm) and the maximum insertion loss will be 6 db for 2 channels, 9.5 db for 3 channels or 12 db for 4 channels. The insertion loss varies when the controls are moved—the maximum variation in insertion loss caused by any one potentiometer is 2.5 db for two channels but less for more channels. This interaction is less if the source im-pedance is considerably smaller than the resistances in the mixer, so that triode valves

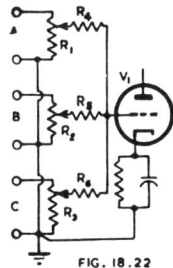

FIG. 18.22

Fig. 18.22. Parallel
network mixer.

are preferable to pentodes in the preceding stage.

The circuit of Fig. 18.22 may also be used when the sources are microphones or pickups, but R_1, R_2 and R_3 should be the correct load resistance in each case. The isolating resistors R_4, R_5 and R_6 may be made equal to the highest of the load re-sistances. This arrangement is only practicable with fairly high source levels, on account of the insertion loss and its effect on the signal-to-noise ratio.

One application of the parallel network mixer is given in Fig. 18.23—this general set-up may be modified to suit any practical case when there is a considerable difference in level between the two sources.

Probably the most popular of all mixing circuits in this class is the method using a common plate load for two or more valves. When two valves are used with a common

---

*The word " fader " is here used in the sense of fading out one input source and fading in another. The fading of a single source is here called attenuation. Mixing is the combination of two (or more) sources so that both are amplified simultaneously ; a mixer may however be used also for fading.

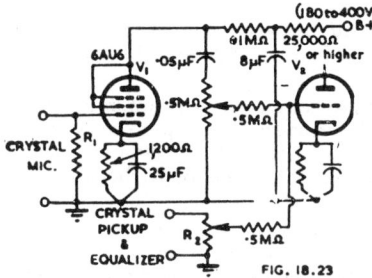

Fig. 18.23. Complete pre-amplifier incorporating parallel network mixer, for use with high level crystal microphone and crystal pickup. $R_1 = 1$ to $5M\Omega$; $R_2 = 0.5M\Omega$ *or higher. Gain in* V_1 *is 30 dbvg.*

FIG. 18.23

plate load resistor, the plate resistance of each valve acts as a shunt load on the other, thereby reducing the gain and the output voltage for a limited distortion. With triodes, the gain is equal to the normal gain for a single valve multiplied by $(r_p + R_L)/(r_p + 2R_L)$. If R_L is considerably greater than r_p, the gain is slightly greater than half that for a normal valve. With pentodes, the loss of gain due to shunting is slight, and may be neglected in most calculations. The effect of the shunting on the maximum output voltage is much more severe than on the gain. The output voltage for limited distortion is approximately equal to the normal output voltage multiplied by $r_p/(r_p + R_L)$. If $R_L = 5r_p$ as for a typical triode, then the output voltage is reduced to one fifth of its normal value. For this reason it is suggested that $R_L = 2r_p$ would be more suitable, giving a reduction to one third of the normal value. Here again, the effect on pentodes is small.

One simple but very effective mixer incorporating common plate load mixing is Fig. 18.24. V_1 may be any twin triode, R_L may be determined as outlined above, while the plate supply voltage may be about 250 volts, or higher if a high output voltage is required with low distortion. If desired, a second twin triode with a further two input channels can be added. Alternatively, two r-f pentodes may be used in place of V_1 to provide increased gain. Various combinations may be devised incorporating one or more common plate load mixers with other types of mixers or faders to meet almost any possible requirements. Fig. 18.25 is one example— V_1 and V_2 may be any suitable twin triodes, R_1 provides fading for the two microphones and R_2 is a combined microphone volume control, R_4 and R_5 provide fading and mixing for the pickups while R_3 is a master volume control. V_2 has a common plate load resistor.

One of the problems with mixers of the common plate load type is the control of volume without introducing noise or circuit complications. This problem may be overcome by the use of 6L7 type valves in which the amplification is controlled by varying the voltage on the third grid (Fig. 18.26). This mixer is intended for use

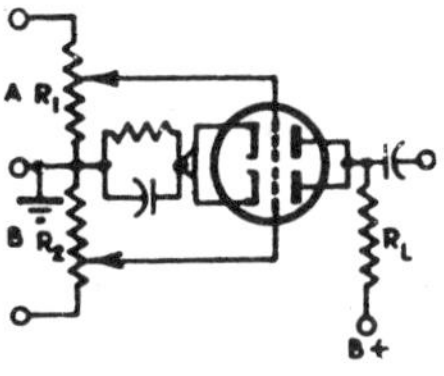

FIG. 18.24

Fig. 18.24. Simple mixer using twin triode with common plate load resistor.

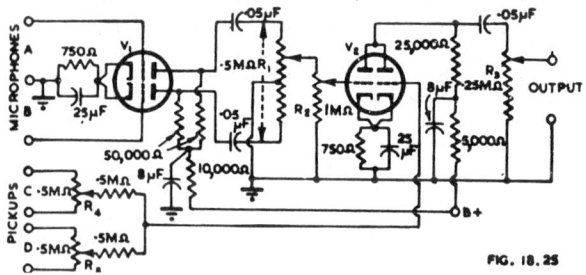

FIG. 18.25

Fig. 18.25. Mixer for two microphones and two pickups. Values of resistors are for $V_1 = V_2 = 6SN7$-GT.

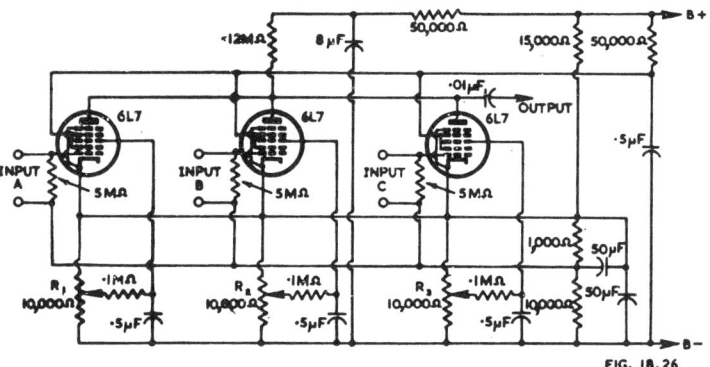

FIG. 18.26

Fig. 18.26. Three channel mixer with electronic volume control (Ref. C4).

with three high-level crystal microphones ; the volume controls operate on d.c. and may therefore be placed some distance from the mixer (Ref. C4).

References to non-constant impedance mixers : C3, C4, D10.

(viii) Constant impedance mixers and faders

Constant impedance* attenuators may be used to provide both attenuation and mixing, and are widely used in studio equipment. In such equipment they usually work out of and into 600 (sometimes 500) or 150 (sometimes 200) ohm circuits.

The constant impedance attenuators may be of the T, bridged T, Π, ladder, bridge, or any other form, together with their balanced equivalents. They may be arranged in parallel, series, series-parallel or bridge circuits.

FIG. 18.27

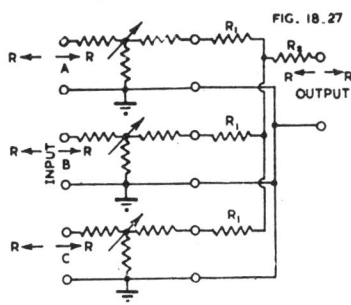

Fig. 18.27. Constant impedance mixer using three T type attenuators in parallel.

(A) Parallel type mixers

Fig. 18.27 shows a three channel mixer using T type attenuators in parallel—the same principle may be applied for any number of channels. Since the attenuators are of the constant impedance type with input and output impedances R, and the mixer output is loaded by the same impedance R, in this case $R_1 = R_2 = \frac{1}{3}R$ and the mixing loss is 9.5 db. In the general case

$R_1 = R_2 = R(n - 1)/(n + 1)$ and Mixing loss $= 20 \log_{10} n$

where n is the number of channels. Values are tabulated below :

$n =$	2	3	4	5	6
$R_1 = R_2 =$	$(1/3)R$	$\frac{1}{2}R$	$(3/5)R$	$(2/3)R$	$(5/7)R$
Mixing loss	6.02	9.54	12.04	14.0	15.56 db.

An alternative form omits R_2 and thereby obtains lower mixing losses, but the output impedance of the mixer and hence the input impedance of the master attenuator will be R_L instead of R_1, the values being given below :

*Constant impedance indicates that both the input impedance and the looking-backwards output resistance are maintained constant under all conditions.

$$R_1 = R(n-1)/n \qquad\qquad R_2 = 0$$
$$R_L = R(2n-1)/n^2 \qquad\qquad \text{Mixing loss} = 10 \log_{10}(2n-1)$$

n	=	2	3	4	5	6
R_1	=	0.50R	0.67R	0.75R	0.80R	0.83R
R_L	=	0.75R	0.56R	0.44R	0.36R	0.31R
Mixing loss	=	4.77	6.99	8.45	9.54	10.41 db.

If it is desired to maintain all input and output impedances constant and equal to R, it will be necessary to insert a matching transformer or to add a matching pad, the latter introducing loss.

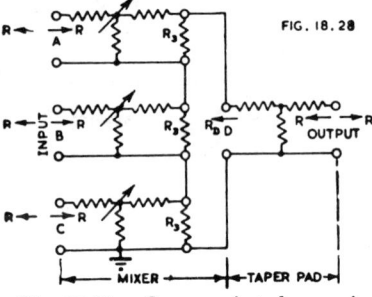

FIG. 18.28

(B) Series type mixers

Fig. 18.28 shows a three channel mixer using T type attenuators in series. Only one of the channels can be earthed, so that there tends to be some cross-talk although this may be kept low by good design. The output impedance at terminals D is given by R_D where $R_D = Rn^2/(2n-1)$, and the mixing loss up to terminals D is $10 \log_{10}(2n-1)$, where n is the number of channels.

Fig. 18.28. *Constant impedance mixer using three T type attenuators in series.*

Values are tabulated below:

n	=	2	3	4	5	6
R_3	=	2.00	1.50	1.33	1.25	1.20
R_D	=	1.33	1.80	2.29	2.78	3.27
Mixing loss	=	4.77	6.99	8.45	9.54	10.41 db.

If this mixer is followed by a transformer there is no further loss, but if it is followed by a taper pad as in Fig. 18.28 there must be added the loss in the taper pad, the value of which is given by

$$N = 20 \log_{10}(R_Z + \sqrt{R_Z^2 - 1}) \text{ decibels}$$

where $R_Z^2 = $ impedance ratio (greater than unity)
$$= n^2/(2n-1).$$

n	=	2	3	4	5	6
impedance ratio	=	1.33	1.80	2.29	2.78	3.27
loss in pad (N)	=	4.77	6.99	8.45	9.54	10.41 db.

(C) Series-parallel type mixers

Fig. 18.29 shows a four channel mixer using T type attenuators in series-parallel. The input and output are both balanced. The values of R_4 and R_0 are given by

$$R_4 = R(n-3)/n \text{ and } R_0 = 4R(2n-3)/n^2$$

where n is the total number of input channels.

If n	=	4	6	8
then R_4	=	0.25R	0.5R	0.625R
R_0	=	1.25R	R	0.75R
and loss in mixer	=	7.0	9.5	11.1 db.

With six channels, the output resistance is equal to R and the taper pad will not be required. In other cases its loss should be added to that in the mixer.

(D) Bridge type mixers

Fig. 18.30 shows a bridge type mixer with four input channels. This can only be earthed at one point, but is otherwise satisfactory for four input sources. The output resistance is equal to R, so that no taper pad or matching transformer is required—the mixing loss is 6 db, which is less than that of any other 4 channel resistance mixer.

(E) Coil mixing (Ref. C18)

Coil mixing refers to the use of a special transformer (Fig. 18.31). The loss is the theoretical minimum and is given by $10 \log_{10} n$ where n is the number of input sources —it is 6 db when $n = 4$. The value of R_6 is given by

$$R_6 = \tfrac{1}{2}R(n-1)$$

and is equal to R when $n = 4$. The secondary transformer impedance is equal to nR.

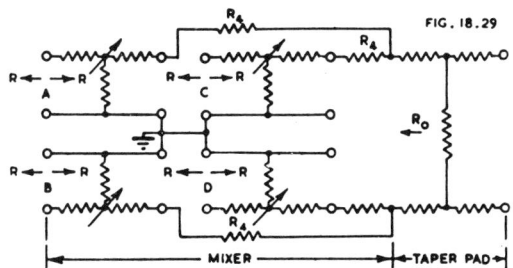

Fig. 18.29.   Constant
impedance mixer using
four T type attenuators
in series-parallel.

Coil mixing appears to be very little used.

(F) Choice of mixer circuit
    The parallel circuit has the very important advantage that one terminal of all (un-balanced) attenuators can be earthed.   Other circuits in which one or more attenuators are above earth nearly always suffer from cross-talk and variation of frequency response at high attenuations.   There are thus two features on which the choice of a mixer circuit must be based—the earthing of all attenuators, and minimum mixing loss. Many engineers insist on the earthing even at the expense of increased mixing loss in some cases.

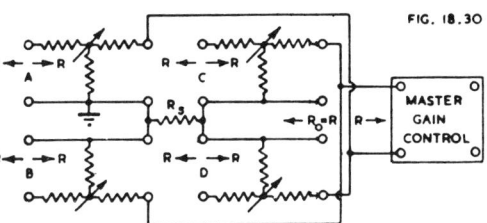

Fig. 18.30.   Constant
impedance mixer using
four T type attenuators
in bridge connection.

    The following comments are based solely on minimum mixing loss, and are there-fore to be applied with discretion—

1. When the impedance of the input circuit is required to be higher than that of the output circuit, a parallel mixer should be used.
2. When the impedance of the output circuit is required to be higher than that of the input circuit, a series or coil type mixer should be used.
3. When the input and output circuits are required to be nearly the same im-pedance, a series-parallel or bridge type mixer should be used.

References to constant impedance mixers :
C1, C2, C3, C12 (parts 7 and 8), C16, C18, D10, D11.

(G) Precautions with studio type mixing
        systems
    Under any possible operating conditions, the level at any stage should be kept at least 6 db and preferably 10 db above the critical pre-amplifier level of — 60 vu.
    With mixer controls turned right off, and a normal level applied to each input in turn, the ratio of " leakage " at 10 000 c/s to normal pro-gramme level should be better than 70 db.
    Cross-talk between two different circuits may be reduced by using twisted leads for all speech circuits with an electrostatic screen around each pair.

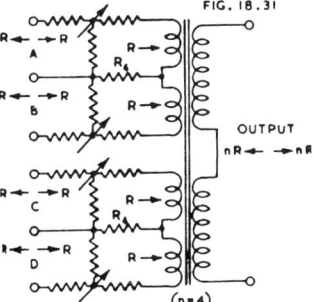

Fig. 18.31.   Constant impedance
mixer using coil mixing with
four input sources.

Where there are long circuits between pre-amplifier outputs and mixer outputs, it is usually considered wise to use balanced transformer inputs to avoid the possibility of cross-talk or noise affecting the programme.

SECTION 4 : REFERENCES

(A) MICROPHONES AND THEIR APPLICATION, STUDIO EQUIPMENT
A2. Queen, I. " Microphones " Radio Craft (Sept. 1944) 723.
A5. Amos, S. W. and F. C. Brooker " Microphones—a detailed review of their design and characteristics " (1) Electronic Eng. 18.218 (April 1946) 109 ; (2) " Sound waves and the physical properties of microphones " 18.219 (May 1946) 136 ; (3) " Pressure operated microphones " 18.220 (June 1946) 190, also 18.221 (July 1946) 221 ; (4) " Ribbon velocity and combined types " 18.222 (Aug. 1946) 255.
A7. Nygren, A. C. " Condenser microphone design—tiny condenser microphone (W.E. 640-AA) combined with bullet shaped amplifier " F.M. and T. 6.8 (Aug. 1946) 38.
A8. Report of the Standards Committee (1933) of the Institute of Radio Engineers, U.S.A.
A15. American Standards Association " American recommended practice for the calibration of microphones " Z24.4—1938.
A16. Radio Manufacturers Association (U.S.A.) " Standard audio facilities for broadcasting systems " TR-105-B (Nov. 1949).
A17. Olson, H. F. " Gradient microphones " J. Acous. Soc. Am. 17.3 (Jan. 1946) 192.
A18. Hilliard, J. K. " An omnidirectional microphone " (Altec-Lansing miniature condenser type) Audio Eng. 33.4 (Apr. 1949) 20.
A19. Martin, D. W. " Magnetic throat microphones of high sensitivity " J. Acous. Soc. Am. 19.1 (Jan. 1947) 43.
A20. Radio Manufacturers Association (U.S.A.) " Microphones for sound equipment " SE-105 (Aug. 1949).
A21. Catalogues of microphones.
A22. Olson, H. F., and J. Preston "Single-element unidirectional microphone" Jour. S.M.P.E. 52.3 (March 1949) 293.
A23. Beranek, L. L. (book) " Acoustic Measurements " (John Wiley and Sons Inc. New York ; Chapman and Hall Ltd. London, 1949).
A24. Telfer, J. E. " Audio-frequency equipment for broadcasting services " Proc. I.R.E. (Aust.) 11.5 (May 1950) 107.
A25. Olson, H. F., and J. Preston " Unobtrusive pressure microphone " Audio Eng. 34.7 (July 1950) 18.
A26. Hilliard, J. K. " Miniature condenser microphone " Jour. S.M.P.E. 54.3 (March 1950) 303.
A27. Anderson, L. J., and L. M. Wigington " The Bantam velocity microphone " Audio Eng. 34.1 (Jan. 1950) 12.
A28. Anderson, L. J., and L. M. Wigington " The KB-3A high-fidelity noise-cancelling microphone " Audio Eng. 34.4 (April 1950) 16.
A29. Beaverson, W. A., and A. M. Wiggins " A second-order gradient noise cancelling microphone using a single diaphragm " J. Acous. Soc. Am. 22.5 (Sept. 1950) 592.
A30. Radio Manufacturers Association " Standard amplifiers for sound equipment " SE-101-A (July 1949).
Additional references will be found in the Supplement commencing on page 1475.

(B) PRE-AMPLIFIERS, NOISE AND HUM
B1. Landon, V. D. " The distribution of amplitude with time in fluctuation noise " Proc. I.R.E. 29.2 (Feb. 1941) 50.
B2. Bell, D. A. " Measurements of shot and thermal noise " W.E. 18.210 (March 1941) 95 ; (letter) W. H. Aldous and E. G. James, W.E. 18.214 (July 1941) 278.
B7. Bell, D. A. (letter) " Shot noise and valve equivalent circuits " W.E. 20.242 (Nov. 1943) 538.
B8. Merchant, C. J. " Thermal noise in a parallel r.c. circuit " Elect. 17.7 (July 1944) 143.
B13. Campbell, N. R., and V. J. Francis " A theory of valve and circuit noise " Jour. I.E.E. Part 3 ; 93.21 (Jan. 1946) 45.
B14. Hooke, A. H. " A method of measuring grid primary emission in thermionic valves " Electronic Eng. 18.217 (March 1946) 75.
B15. Crawford, K. D. E. " H.F. Pentodes in electrometer circuits " Electronic Eng. 20.245 (July 1948) 227.
B16. Heacock, D. P., and R. A. Wissolik " Low-noise miniature pentode for audio amplifier service " Tele-Tech 10.2 (Feb. 1951) 31. See also Ref. B56.
B17. Knight, C. A. and A. P. Haase " New low noise input tube " Radio and T.V. News 12.3 (March 1949) 15.
B18. Amalgamated Wireless Australasia Ltd., Consolette G52107.
B19. Meyer, R. J. " Open-grid tubes in low-level amplifiers " Elect. 17.10 (Oct. 1944) 120.
B20. Shipton, H. W. " Valve and circuit noise in high gain amplifiers " Electronic Eng. 19.229 (March 1947) 81.
B23. Walter, W. G., H. W. Shipton and W. J. Warren " Demagnetising valves—as a cure for residual ripple " Electronic Eng. 20.245 (July 1948) 235 ; Correspondence 20.248 (Oct. 1948) 339 ; 20.250 (Dec. 1948) 406 ; 21.251 (Jan. 1949) 30.
B25. Smith, F. W. " Heater supplies for amplifier hum reduction " Audio Eng. 32.8 (Aug. 1948) 26.
B26. Britton, K. G. " Reducing heater hum " W.W. 54.10 (Oct. 1948) 360.
B28. Chinn, H. A. " Audio system design fundamentals " Audio Eng. 32.11 (Nov. 1948) 11.
B40. Pierce, J. R. " Noise in resistances and in electron streams " B.S.T.J. 27.1 (Jan. 1948) 158.
B41. Ellis, H. D. " Studio equipment—a new design " B.B.C. Quarterly (April 1946).
B42. Dickerson, A. F. " Hum reduction " Elect. 21.12 (Dec. 1948) 112.
B43. Le Bel, C. J. " New developments in preamplifiers " Audio Eng. 33.6 (June 1949) 9.
B44. Thompson, B. J., D. O. North and W. A. Harris " Fluctuations in space-charge-limited currents at moderately high frequencies " R.C.A. Rev. Jan., April, July, Oct 1940, Jan., April, July, 1941.
B45. Williams, F. C. " Thermal fluctuations in complex networks " Jour. I.E.E. Wireless Section 13 (1938) 53.

B46. Zakarias, I. " Reducing hum in pentodes " Elect. 21.11 (Nov. 1948) 170.
B48. " Grounded plate type 6AU6 triode connection for pre-amplifier use " Radiotronics 142 (April 1950) 45.
B49. Data Sheet XIX " Circuit noise due to thermal agitation " Electronic Eng. 14.167 (Jan. 1942) 591.
B50. Zakarias, I. (letter) " Hum in a.c. valves " Electronic Eng. 22.263 (Jan. 1950) 33. See also B23, B46.
B51. Hopper, F. L. " Noise considerations in sound-recording transmission systems," Jour. S.M.P.E. 54.2 (Feb. 1950) 129.
B52. Hedge, L. B. " D.C. heater supply for low-level amplifiers " Audio Eng. 35.6 (June 1951) 13.
B53. Tucker, M. J. " Cathode followers as low-noise input stages " Electronic Eng. 23.281 (July 1951) 270.
B54. Shorter, D. E. L., and D. G. Beadle " Equipment for acoustic measurements—A portable general purpose microphone amplifier using miniature valves " Electronic Eng. 23.283 (Sept. 1951) 326.
See also Chapter 31 Refs. 6, 7, 8, 9, 10. For additional references see Supplement page 1479.

(C) ATTENUATORS AND MIXERS

C1. Wright, P. B. " Mixer and fader control circuit design " Comm. (1) (Nov. 1943) 44 ; (2) (Dec. 1943) 44.
C2. Cooper, M. F. " Audio frequency mixers " W.E. 21.246 (March 1944) 117.
C3. Crane, R. W. " Audio mixer design " Elect. 18.6 (June 1945) 120.
C4. Patchett, G. N. " Mixing crystal microphones " W.W. 52.2 (Feb. 1946) 57.
C6. Espy, D. " Attenuator design " Elect. (Nov. 1941) 51.
C7. Honnell, P. H. " Unsymmetrical Attenuators " Elect. 15.8 (Aug. 1942) 41.
C9. Wright, P. B. " Attenuator design—for amplifier gain controls " Comm. 23.10 (October 1943) 38.
C11. Blackwell, R. F., and T. A. Stranglar " Attenuator design " W.E. 21.246 (March 1944) 122.
C12. Wright, P. B. " Resistive attenuator, pad and network, theory and design " Comm. (1) 24.8 (Aug. 1944) 49 ; (2) 24.10 (Oct. 1944) 62 ; (3) 25.1 (Jan. 1945) 50 ; (4) 25.5 (May 1945) 62 ; (5) 25.6 (June 1945) 68 ; (6) 25.7 (July 1945) 50 ; (7) 25.8 (Aug. 1945) 64 ; (8) 25.8 (Sept. 1945) 68 ; (9) 25.10 (Oct. 1945) 72 ; (10) 25.11 (Nov. 1945) 61.
C13. West, S. S., and E. D. McConnell " Attenuator design " W.W. 46.14 (Dec. 1940) 487.
C14. Baker, W. G. " Notes on the design of attenuating networks " A.W.A. Tec. Rev. 1.2 (June 1935) 19 ; 2.1 (Jan. 1936) 42.
C15. Smith, F. W., and M. C. Thienpont " Electronic attenuators " Comm. 27.5 (May 1947) 20.
C16. Morrical, K. C. " Design and use of mixing networks " Audio Eng. 31.10 (Nov. 1947) 11.
C17. McElroy, P. K. " Designing resistive attenuating networks " Proc. I.R.E. 23.3 (March 1935) 213.
C18. Miller, W. C., and H. R. Kimball " A recording console, associated circuits and constant B equalizers " Jour. S.M.P.E. 43 (Sept. 1944) 186.
C19. Leipert, C. J. " Simplified methods for calculation of H and T attenuation pads " T.V. Eng. 1.7 (July 1950) 7.
C20. Circuit Laboratory Report " Audio frequency applications of type 6BE6 " Radiotronics 16.2 (Feb. 1951) 41.

(D) GENERAL REFERENCES—BOOKS

D1. Terman, F. E. " Radio Engineering " (3rd ed. 1947 McGraw-Hill Book Co., New York and London).
D2. Terman, F. E. " Radio Engineers' Handbook " (1st ed. 1943 McGraw-Hill Book Co., New York and London).
D3. Olson, H. F. " Elements of Acoustical Engineering " (2nd ed. 1947, D. Van Nostrand Company, New York).
D4. Henney, K. (Editor) " The Radio Engineering Handbook " (4th ed. 1950, McGraw-Hill Book Co., New York and London).
D5. Reich, H. J. " Theory and Application of Electron Tubes " (2nd ed. 1944, McGraw-Hill Book Co., New York and London).
D6. " Reference Data for Radio Engineers " (3rd ed. Federal Telephone and Radio Corporation, New York).
D7. Boyce, W. F., and J. J. Roche, " Radio Data Book " (Boland and Boyce Inc., Montclair, N.J. U.S.A.).
D8. " Motion Picture Sound Engineering " (D. Van Nostrand Company Inc. New York, 1938).
D9. Valley, G. E., and H. Wallman (Editors) " Vacuum Tube Amplifiers " (McGraw-Hill Book Company, New York and London, 1948).
D10. Greenless, A. E. " The amplification and distribution of sound " (Chapman and Hall Ltd. London, 2nd edit. 1948).
D11. Frayne, J. G., and H. Wolfe, " Elements of Sound Recording " (John Wiley and Sons, New York ; Chapman and Hall Ltd., London, 1949).
D12. Blackburn, J. F. (Editor) " Components Handbook " (McGraw-Hill Book Company Inc. New York and London, 1949).
D13. Beranek, L. L. " Acoustic Measurements " (John Wiley and Sons, New York ; Chapman and Hall, London, 1949).
D14. Kinsler, L. E., and A. R. Frey " Fundamentals of Acoustics " (John Wiley and Sons, New York ; Chapman and Hall, London, 1950).

CHAPTER 19

UNITS FOR THE MEASUREMENT OF GAIN AND NOISE

By F. Langford-Smith, B.Sc., B.E.

SECTION 1 : BELS AND DECIBELS

(i) Power relationships expressed in bels and decibels   (ii) Voltage and current relationships expressed in decibels   (iii) Absolute power and voltage expressed in decibels   (iv) Microphone output expressed in decibels   (v) Pickup output expressed in decibels   (vi) Amplifier gain expressed in decibels   (vii) Combined microphone and amplifier gain expressed in decibels   (viii) Loudspeaker output expressed in decibels   (ix) Sound system rating   (x) Tables and charts of decibel relationships   (xi) Nomogram for adding decibel-expressed quantities   (xii) Decibels, slide rules and mental arithmetic.

(i) Power relationships expressed in bels and decibels

If a sound is suddenly increased in magnitude, the listener receives an impression of increased loudness which is roughly proportional to the logarithm of the ratio of the two acoustical powers.   In mathematical form

$$\text{loudness} \propto \log (P_2/P_1) \tag{1}$$

This is quite general, and holds for a decrease in power as well as for an increase in power.   Now the ultimate effect of any change of electrical power in a transmitter, receiver or amplifier is to produce a change of acoustical power from the loudspeaker, so that it is convenient to adopt a logarithmic basis for indicating a change of electrical power.

The common logarithm of the ratio of two powers gives their relationship in **bels**—

$$N_b = \log_{10}(P_2/P_1) \tag{2}$$

where P_1 is the reference power

and P_2 is the power which is referred back to P_1.

The more commonly used unit is the **decibel**, which is one tenth of a bel.   Thus, the difference in level between P_1 watts and P_2 watts is given by

$$N_{db} = 10 \log_{10} (P_2/P_1) \text{ decibels} \tag{3}$$

If P_2 is less than P_1 the value of N_{db} becomes negative.   A negative value of N_{db} thus indicates that the power in which we are interested is less than the reference power. Note that these relationships (1, 2 and 3) are independent of any other conditions such as impedance.

Applications and examples

Suppose for example that a power valve driving a loudspeaker is delivering 1 watt which is then increased to 2 watts.   To say that the power has " increased by one watt " is misleading unless it is also stated that the original level was 1 watt. A far more satisfactory way is to state that a rise of 3 db has occurred. This may be calculated quite simply since the gain in decibels is

$$10 \log_{10} (2/1) = 10 \log_{10}2 = 10 \times 0.301 = 3.01 \text{ db}$$

or approximately 3 db.

In a similar manner a decrease from 2 watts to 1 watt is a change of approximately − 3 db.

It has been found that a change in level of 1 db is barely perceptible to the ear, while an increase of 2 db is only a slight apparent increment. For this reason variable attenuators are frequently calibrated in steps of 1 db or slightly less. In a similar manner an increase from 3 watts to 4.75 watts is only a slight audible increment, being an increase of 2 db.

In order to simplify the understanding of **barely perceptible changes** the following table has been prepared, and it will be seen that a move from one column to the nearest on left or right is equivalent to a change of 2 db. In this table 0 db is taken as 3 watts.

db :	−10	−8	−6	−4	−2	0	+2	+4	+6	+8	+10	+12
watts :	0.30	0.47	0.75	1.2	1.9	3.0	4.75	7.5	12	19	30	47.5

In addition to the application of decibels to indicate a change in level at one point, they may also be used to indicate a difference in level between two points such as the input and output terminals of a device such as an amplifier or attenuator. For example, consider an amplifier having an input power of 0.006 watt and an output power of 6 watts. The power gain is 6/0.006 or 1000 times, and reference to the tables shows that this is equivalent to 30 db. The amplifier may therefore be described as having a gain of 30 db, this being irrespective of the input or output impedance. References 1, 8, 9, 16 (Chap. 32).

(ii) Voltage and current relationships expressed in decibels

Since $P_1 = E_1 I_1 = E_1{}^2/R_1 = I_1{}^2 R_1$
and $\qquad P_2 = E_2 I_2 = E_2{}^2/R_2 = I_2{}^2 R_2$
where $\quad R_1 = $ resistance dissipating power P_1
and $\qquad R_2 = $ resistance dissipating power P_2,
it is obvious that the decibel relationship between E_1 and E_2 or between I_1 and I_2 must involve the resistance.

If $R_1 = R_2 = R$, then $P_2/P_1 = E_2{}^2/E_1{}^2 = I_2{}^2/I_1{}^2$ and the difference in level is given by

$$N_{db} = 10 \log_{10}(E_2{}^2/E_1{}^2) = 20 \log_{10}(E_2/E_1) \text{ decibels,} \tag{4}$$
$$\text{or } N_{db} = 10 \log_{10}(I_2{}^2/I_1{}^2) = 20 \log_{10}(I_2/I_1) \text{ decibels,} \tag{5}$$

provided that R remains constant.

If R does not remain constant the difference of level is

$$N_{db} = 20 \log_{10}(E_2/E_1) + 10 \log_{10}(R_1/R_2) \tag{6}$$
$$\text{or } N_{db} = 20 \log_{10}(I_2/I_1) + 10 \log_{10}(R_2/R_1) \tag{7}$$

In the general case with an impedance $\mathbf{Z} = R + jX$ which is the same for both P_1 and P_2, equations (4) and (5) also hold.

When the two impedances are not identical, the difference in level in decibels is

$$N_{db} = 20 \log_{10}(E_2/E_1) + 10 \log_{10}(Z_1/Z_2) + 10 \log_{10}(k_2/k_1) \tag{8}$$
$$\quad = 20 \log_{10}(I_2/I_1) + 10 \log_{10}(Z_2/Z_1) + 10 \log_{10}(k_2/k_1) \tag{9}$$

where $k_1 = $ power factor of $Z_1 = R_1/Z_1 = \cos \phi_1$
and $\quad k_2 = $ power factor of $Z_2 = R_2/Z_2 = \cos \phi_2$.
References 7, 8, 9, 16 (Chap. 32).

(iii) Absolute power and voltage expressed in decibels
(A) Power

Although the decibel is a unit based on the ratio between two powers, it may also be used as an indication of absolute power provided that the reference level (or " zero level ") is known. There have been many so-called " standard " reference levels, including 1, 6, 10, 12.5 and 50 milliwatts, but the 1 milliwatt reference level is very widely used at the present time. As an example, a power of 1 watt may be described as

    30 db (reference level 1 mW)
    or 30 db (0 db = 1 mW).
    or 30 dbm*.

The abbreviation db 6m is sometimes used to indicate a level in decibels with a 6 milliwatt reference level.

To convert from a reference level of 1 mW to 6 mW, add − 7.78 db.
To convert from a reference level of 1 mW to 10 mW, add − 10.00 db.
To convert from a reference level of 1 mW to 12.5 mW, add − 10.97 db.
To convert from a reference level of 6 mW to 1 mW, add + 7.78 db.
To convert from a reference level of 10 mW to 1 mW, add + 10.00 db.
To convert from a reference level of 12.5 mW to 1 mW, add + 10.97 db.

With any reference level, a power with a positive sign in front of the decibel value indicates that this is greater than the reference power, and is spoken of as so many " decibels up." A negative sign indicates less power than the reference power, and is spoken of as so many " decibels down." 0 db indicates that the power is equal to the reference power.

A statement of power expressed in decibels is meaningless unless the reference level is quoted.

References 1, 5, 8, 9, 10.

(B) Voltage

A reference level of 1 volt has been standardized in connection with high impedance microphones. The abbreviation dbv has been standardized (Ref. 38) to indicate decibels referred to 1 volt.

(iv) Microphone output expressed in decibels

The output of a microphone may be expressed either in terms of voltage or power.

(A) In terms of output voltage

The response of a microphone at a given frequency may be stated in decibels with respect to a reference level 0 db = 1 volt (open-circuit) with a sound pressure of 1 dyne per square centimetre (Ref. 36). The abbreviation dbv is used to indicate a voltage expressed in decibels, with reference level 1 volt (Ref. 38).

For example, the output of a microphone may be stated as − 74 dbv with a sound pressure of 1 dyne per square centimetre. This is the open-circuit voltage developed without any loading such as would be provided by the input resistance of the amplifier. Table 1 [Section 1(x)] may be used to determine the corresponding open-circuit voltage, which for the example above is approximately 0.0002 volt r.m.s. (Column 1). If the input resistance of the amplifier is equal to the internal impedance of the microphone (here assumed to be resistive as the worst possible case) the voltage across the input terminals will be only half the generated voltage, giving a loss of 6 db or an effective input voltage of − 80 dbv.

FIG. 19.1

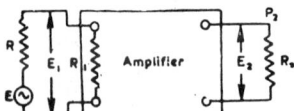

Fig. 19.1. Amplifier with source of input voltage E having internal resistance R.

In the general case, the input voltage to the amplifier will be (Fig 19.1):
$$E_1 = ER_1/(R + R_1)$$
where E = open-circuit voltage developed by the microphone
           R_1 = input resistance of amplifier
and R = internal resistance of microphone (here assumed purely resistive).

(B) In terms of output power.

Alternatively the output of a microphone may be given in terms of output power for a stated sound pressure.

*dbm indicates a power expressed in decibels with a reference level 1 mW.

For example, the output of a microphone may be stated as − 65 dbm into a load of 150 ohms, with an input sound pressure of 1 dyne per square centimetre. From Table 3—Sect. 1(x)—the power is 3.2 × 10⁻⁷ milliwatt. If desired this may be converted into voltage across 150 ohms.

(C) In terms of effective output level
When a microphone is connected to an unloaded input transformer, its output cannot be expressed in terms of power delivered, as no appreciable power is delivered by the microphone. For this reason, microphone output ratings are sometimes given in terms of effective output level, expressed in dbm. The effective output level is so calculated that when the amplifier power gain in db is added to the microphone effective output level in dbm the correct output level from the amplifier (in dbm) will be obtained. The effective output level rating is based upon the assumption that the microphone works into a load impedance equal to its own rated impedance. The voltage corresponding to this effective output level is actually 6 decibels below that which is actually obtained when the microphone is worked into a high impedance pre-amplifier input. This 6 db difference is a function of the pre-amplifier input termination and not of the microphone itself. The " power gain "* ratings of pre-amplifiers take into account this 6 db increase in gain where it occurs, so that it is not necessary to apply any coupling factor.

If the effective output level of a microphone is not known, it may be derived from available data :
(1) The output power may be converted to dbm, or
(2) Where the microphone open-circuit voltage output is known in dbv (0 db = 1 volt), this may be converted to volts (E_G) and the effective output level in milliwatts is given by
    $1000\ E_G{}^2/(4R_M)$ milliwatts
where R_M is the nominal microphone impedance. The power in milliwatts may then be converted into dbm.

(D) R.M.A. microphone system rating (RMA Standard SE-105)
This is particularly useful when it is desired to calculate the combined " system gain " of a microphone, amplifier and loudspeaker. The R.M.A. microphone system rating is defined as the ratio in db relative to 0.001 watt and 0.0002 dyne per square centimetre of the electric power available from the microphone to the square of the undisturbed sound field pressure in a plane progressive wave at the microphone position.

The R.M.A. microphone system rating (Ref. 35) is given by

$$G_M = \left(10\ \log_{10} \frac{E^2/4R_{MR}}{p^2}\right) - 44 \text{ db} \qquad (10)$$

which reduces for practical applications to

$$G_M = (20\ \log_{10} (E/p) - 10\ \log_{10} R_{MR}) - 50 \text{ db} \qquad (11)$$

where G_M = microphone system rating
    E = open-circuit voltage generated by the microphone
    p = sound pressure in dynes per sq. cm.
    R_{MR} = microphone rating impedance—see Chapter 18 Sect. 1(xiv). This may differ from the actual microphone impedance.

The R.M.A. Microphone System Rating is essentially the same as expressing the microphone output in terms of Effective Output Level, except that the acoustical pressure at the microphone is 0.0002 dyne/cm² (the limit of audibility). To convert from R.M.A. Microphone System Rating to the Effective Output Level Rating, it is only necessary to allow for the change of acoustical pressure. For example, if a microphone has an R.M.A. Microphone System Rating of − 154 db, the Effective Output-Level Rating for an acoustical pressure of 10 dynes/cm² will be − 60 dbm, and for 1 dyne/cm² will be − 80 dbm.

---

* An amplifier is normally tested as described in Chapter 37 Sect. 3(ii)C with an input voltage from a generator applied through a constant impedance attenuator, which combination effectively applies a constant voltage through a resistance equal to the rated source impedance of the amplifier as in Fig. 19.3.

Microphone system ratings are most commonly used in a complete sound system—sound, microphone, amplifier, loudspeaker, sound—see (ix) below.

See Chapter 18 Sect. 1 for the relationships between various forms of microphone ratings.

(v) Pickup output expressed in decibels

Although many pickups are rated on the basis of output voltage, some are rated on a power basis with respect to a specified reference level. The procedure is the same as for microphones.

Reference 9.

(vi) Amplifier gain expressed in decibels

Much confusion has been caused by the incorrect or careless use of decibels to indicate the gain of a voltage amplifier. Decibels may be used in various ways to indicate the gain of an amplifier—

(A) In terms of voltage gain (Fig. 19.4)

This is really an arbitrary use of decibels, but it is so convenient that it cannot be suppressed.

$$\text{Gain in decibels of voltage gain} = 20 \log E_2/E_1 \qquad (12)$$

where E_2 = voltage across output terminals of voltage amplifier

and E_1 = voltage across input terminals of amplifier.

It is important to distinguish these decibels from decibels of power, which have an entirely different meaning. *The abbreviation dbvg is suggested as indicating decibels of voltage gain.*

Some engineers express the gain of an amplifier in terms of voltage by taking E_1 (eqn. 12) as the open circuit generator voltage (E in Fig. 19.1). If the load resistance (R_1 in Fig. 19.1) is equal to the generator resistance (R), the indicated gain by this method will be 6 db less than that given by eqn. (12). If R_1 is greater than $10R$, both methods give approximately the same result. If the generator impedance has an appreciable reactive component, the difference between the two methods will be less than indicated above.

(B) Amplifier gain in terms of power

Amplifier gain is defined as the ratio expressed in db of the power delivered to the load, to the power which would be delivered to the same load if the amplifier were replaced by an ideal transformer which matches both the load and source impedances (R.M.A. Standard SE-101-A, Amplifiers for sound equipment—Ref. 35).

In Fig. 19.1 the power delivered to the load is

$$P_2 = E_2{}^2/R_2 \qquad (13)$$

The power which would be delivered to the same load if the amplifier were replaced by an ideal transformer which matches both the load and source impedances is

$$P_1 = E_1{}^2/R_1 \qquad (14)$$

Therefore gain in decibels = $10 \log_{10}(P_2/P_1)$.   (15)

If the input resistance of the amplifier is made equal to the internal resistance of the source,

i.e. $R_1 = R$,

then $E_1 = E/2$ and $P_1 = E^2/4R_1 = E^2/4R$.

Gain in decibels = $10 \log_{10} (4RE_2{}^2/R_2E^2)$

$$= 20 \log_{10} \frac{2E_2}{E}\sqrt{\frac{R}{R_2}} = 6 + 20 \log_{10} \frac{E_2}{E}\sqrt{\frac{R}{R_2}} \qquad (16)$$

If, in addition, $R_2 = R_1 = R$, then

gain in decibels = $6 + 20 \log_{10} (E_2/E)$ (17)

(C) Gain of a bridging amplifier

A bridging amplifier is one whose internal input impedance is such that it may be connected across a circuit without appreciably affecting the circuit performance in any respect. Its function is to operate into programme circuits or similar loads (Ref. 33).

Bridging gain is the ratio, expressed in db, of the power delivered to the bridging amplifier load to the power in the load across which the input of the amplifier is bridged (Ref. 33).

The commonest case is that of the input to an amplifier having a load of 600 ohms, with the input terminals of the bridging·amplifier connected across it. The output load of the bridging amplifier is most commonly 600 ohms.

(vii) Combined microphone and amplifier gain expressed in decibels

(A) **When a microphone, rated in terms of voltage, is connected to a voltage amplifier which is rated in decibels of voltage gain (dbvg)**, the output may be calculated as under:

Output in dbv* = microphone rating in dbv* + coupling factor + amplifier gain in decibels of voltage gain (dbvg).   (18)

The coupling factor = 20 log $[R_1/(R + R_1)]$ (Fig. 19.1).   (19)

Typical values of the coupling factor are tabulated below:

$R_1/(R + R_1)$	=	0.5	0.56	0.63	0.71	0.79	0.89
Coupling factor	=	−6	−5	−4	−3	−2	−1 db

Example: Microphone −54 dbv*
Amplifier + 80 dbvg
$R_1/(R + R_1)$ = 0.5
Output = −54 − 6 + 80 = + 20 dbv* = 10 volts.

It should be noted that the calculated output applies for the rated sound pressure, for example 1 dyne per square centimetre. At other sound pressures the voltage will be proportional to the sound pressure.

(B) **When a microphone, rated in terms of effective output level, is connected to an amplifier having its gain expressed in terms of power in accordance with R.M.A. Standard SE-101-A**:

The amplifier power gain is measured effectively with a constant input voltage in series with a resistance equal to the rated source impedance of the amplifier as in Fig. 19.3. Under these conditions no correction factor is necessary and the output from the amplifier in dbm is equal to the sum of the microphone effective output level in dbm and the amplifier power gain in decibels. This output level will only be attained when the pressure at the microphone is equal to the rated pressure—e.g. 10 dynes/sq. cm.

FIG. 19.3   FIG. 19.4

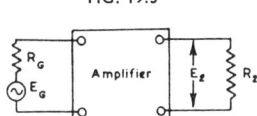

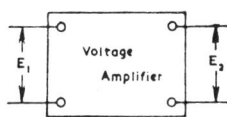

Fig. 19.3. Method of testing amplifier for gain.   *Fig. 19.4. Voltage amplifier showing input and output voltages.*

For example, a ribbon microphone has an effective output level of −59 dbm with an acoustical pressure of 10 dynes/cm². If this is connected to an amplifier with a power gain of 40 db, the output level will be −59 + 40 = −19 dbm with an acoustical pressure of 10 dynes/cm².

(C) **When a microphone, rated in accordance with the R.M.A. microphone system rating is connected to an amplifier having its gain expressed in terms of power in accordance with R.M.A. Standard SE-101-A**:

For general remarks see (B) above.

The output from the amplifier in dbm is equal to the sum of the R.M.A. microphone system rating, the amplifier power gain in decibels, and the sound pressure in decibels.

For example, a ribbon microphone with a R.M.A. system rating (G_M) of −153 db, operating with a sound pressure of 10 dynes/sq. cm. (i.e. +94 db) and connected to

*0 db = 1 volt

an amplifier with a power gain of 40 db will give an output level $-153 +40 +94 =$
-19 dbm. The microphone amplifier and sound pressure in this example are the
same as for (B) above.

(viii) Loudspeaker output expressed in decibels, in terms of acoustical pressure

In accordance with the American R.M.A. Standard SE-103 (Ref. 29) the loudspeaker pressure rating is the difference between the axial sound pressure level (referred to a distance of 30 feet) and the available input power level, and is expressed in
db.

It is expressed by the following forms (equations 20, 21, 22, 23) :

$$G_{SP} = 10 \log_{10} [(p_S/p_0)^2/(W_{AS}/W_0)] \qquad (20)$$
$$G_{SP} = 44 + 20 \log_{10} p_S - 10 \log_{10} W_{AS} \qquad (21)$$
$$G_{SP} = 44 + 20 \log_{10} p_S - 20 \log_{10} E_G + 10 \log_{10} R_{SR}$$
$$+ 20 \log_{10} [1 + (R_{SG}/R_{SR})] \qquad (22)$$

(pressure in db above p_0) $= G_{SP} +$ (power in dbm)   (23)

where G_{SP} = loudspeaker pressure rating in db

p_S = axial, free-space, r.m.s. sound pressure at 30 feet, in dynes/cm²

p_0 = reference r.m.s. sound pressure = 0.0002 dyne/cm²

W_{AS} = electrical power available to the speaker, in watts, and is equal to
$E_G^2 R_{SR}/(R_{SG} + R_{SR})^2$ (Fig. 19.5)

W_0 = reference power = 0.001 watt

E_G = r.m.s. value of the constant voltage of the source, in volts

R_{SR} = loudspeaker rating impedance, in ohms*

R_{SG} = speaker measurement source impedance, in ohms*

and   dbm = power in decibels referred to 1 milliwatt.

FIG. 19.5

*Fig. 19.5.   Loudspeaker
testing conditions (R.M.A.
Standard SE-103).*

The sound pressure (p_d) at any distance d feet may be used to compute the pressure
p_s at 30 feet by the relation

$$p_s = (d/30)p_d \qquad (24)$$

See also Ref. 34.

Example : If the loudspeaker pressure rating (G_{SP}) is 46 db, what is the axial
sound pressure level at 30 feet, with an available power input of 10 watts, using a
standard test signal ?

$G_{SP} = 46$ db $W_{AS} = 10$ watts = 40 dbm

From equation (21),

$20 \log_{10} p_S = 46 - 44 + 10 \log_{10} 10 = 12$ db.

Therefore $p_S = 4$ dynes/cm² (i.e. 86 db above p_0).

The same result may be derived more directly from equation (23), pressure =
$46 + 40 = 86$ db above p_0.

(ix) Sound system rating

The total gain of a system from sound, through microphone, amplifier and loudspeaker to sound again may be calculated by adding the system ratings of the several
sections and coupling factors (if any).

(A) **Using the American R.M.A. system ratings** for microphone, amplifier
and loudspeaker we may put

$$SR = G_M + G + G_{SP} \qquad (25)$$

where SR = sound system rating (gain in db)

G_M = R.M.A. microphone system rating (equations 10 and 11)

G = amplifier power gain (equation 15)

and G_{SP} = loudspeaker pressure rating in db (equations 20, 21 and 22).

*See Chapter 20 Sect. 6(x)B for definitions of R_{SR} and R_{SG}.

A system rating of 0 db indicates that the sound pressure 30 feet from the loudspeaker is the same as that at the microphone. Similarly a system gain of x db indicates that the sound pressure 30 feet from the loudspeaker is x db greater than that at the microphone.

(B) Proposed method of rating microphones and loudspeakers for systems use by Romanow and Hawley (Ref. 11)

This proposed method has not been adopted generally in the precise form expounded, but the article gives a most valuable analysis of the whole subject of system gain. This method is also described in Ref. 34.

(x) Tables and charts of decibel relationships

Table 1 : Decibels expressed as power and voltage or current ratios

Note that the Power Ratio columns give power values in milliwatts when the reference level is 1 mW. The Power Ratio columns also give power values in milliwatts when the centre column represents dbm*.

The Voltage Ratio columns also give values in volts when the centre column represents dbv.†

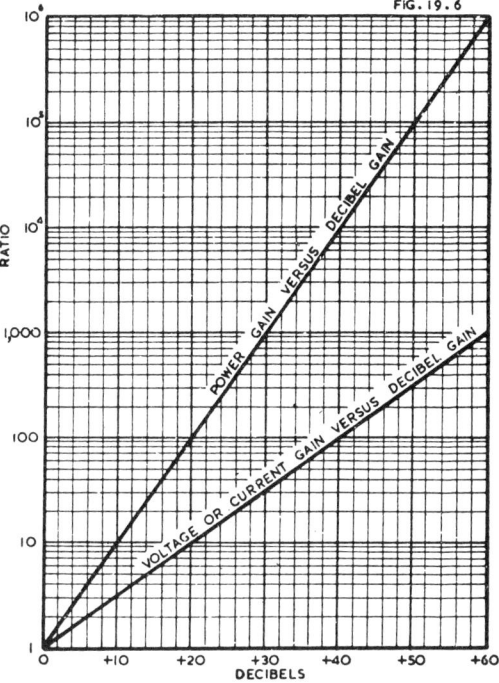

Fig. 19.6. Gain ratio plotted against decibel gain. This chart may also be used for attenuation by inverting the ratio and making the decibels negative.

Interpolation : If it is required to find the power ratio corresponding to 22.5 db, or any other value which is not included in the table, the following procedure may be adopted :—

1. Take the next lowest multiple of 20 db (in this case 20 db), and note the corresponding power ratio (in this case 100).

---

*dbm is unit of power expressed in decibels with 0 db = 1 mW.
†dbv is unit of voltage expressed in decibels with 0 db = 1 volt.

2. Take the difference between the specified level and the multiple of 20 db (in this case 22.5 − 20 = 2.5 db) and note the corresponding power ratio (in this case 1.778).

3. Multiply the two power ratios so determined (in this case 100 × 1.778 = 177.8).

TABLE 1 : DECIBELS EXPRESSED AS POWER AND VOLTAGE RATIOS

Voltage or Current Ratio	Power Ratio (= mW to Reference Level 1 mW)	db	Voltage or Current Ratio	Power Ratio (= mW to Reference Level 1 mW)
1.000 0	1.000 0	−0+	1.000	1.000
.988 6	.977 2	0.1	1.012	1.023
.977 2	.955 0	0.2	1.023	1.047
.966 1	.933 3	0.3	1.035	1.072
.955 0	.912 0	0.4	1.047	1.096
.944 1	.891 3	0.5	1.059	1.122
.933 3	.871 0	0.6	1.072	1.148
.912 0	.831 8	0.8	1.096	1.202
.891 3	.794 3	1.0	1.122	1.259
.841 4	.707 9	1.5	1.189	1.413
.794 3	.631 0	2.0	1.259	1.585
.749 9	.562 3	2.5	1.334	1.778
.707 9	.501 2	3.0	1.413	1.995
.631 0	.398 1	4	1.585	2.512
.562 3	.316 2	5	1.778	3.162
.501 2	.251 2	6	1.995	3.981
.446 7	.199 5	7	2.239	5.012
.398 1	.158 5	8	2.512	6.310
.354 8	.125 9	9	2.818	7.943
.316 2	.100 0	10	3.162	10.000
.281 8	.079 43	11	3.548	12.59
.251 2	.063 10	12	3.981	15.85
.223 9	.050 12	13	4.467	19.95
.199 5	.039 81	14	5.012	25.12
.177 8	.031 62	15	5.623	31.62
.158 5	.025 12	16	6.310	39.81
.141 3	.019 95	17	7.079	50.12
.125 9	.015 85	18	7.943	63.10
.112 2	.012 59	19	8.913	79.43
.100 0	.010 00	20	10.000	100.00
.089 13	.007 943	21	11.22	125.9
.079 43	.006 310	22	12.59	158.5
.070 79	.005 012	23	14.13	199.5
.063 10	.003 981	24	15.85	251.2
.056 23	.003 162	25	17.78	316.2
.050 12	.002 512	26	19.95	398.1
.044 67	.001 995	27	22.39	501.2
.039 81	.001 585	28	25.12	631.0
.035 48	.001 259	29	28.18	794.3
.031 62	.001 000	30	31.62	1 000
.028 18	7.943×10^{-4}	31	35.48	1 259
.025 12	6.310×10^{-4}	32	39.81	1 585
.022 39	5.012×10^{-4}	33	44.67	1 995
.019 95	3.981×10^{-4}	34	50.12	2 512
.017 78	3.162×10^{-4}	35	56.23	3 162

Voltage or Current Ratio	Power Ratio (= mW to Reference Level 1 mW)	db	Voltage or Current Ratio	Power Ratio (= mW to Reference Level 1 mW)
.015 85	2.512×10^{-4}	36	63.10	3 981
.014 13	1.995×10^{-4}	37	70.79	5 012
.012 59	1.585×10^{-4}	38	79.43	6 310
.011 22	1.259×10^{-4}	39	89.13	7 943
.010 000	1.000×10^{-4}	40	100.0	10 000
.008 913	7.943×10^{-5}	41	112.2	12 590
.007 943	6.310×10^{-5}	42	125.9	15 850
.007 079	5.012×10^{-5}	43	141.3	19 950
.006 310	3.981×10^{-5}	44	158.5	25 120
.005 623	3.162×10^{-5}	45	177.8	31 620
.005 012	2.512×10^{-5}	46	199.5	39 810
.004 467	1.995×10^{-5}	47	223.9	50 120
.003 981	1.585×10^{-5}	48	251.2	63 100
.003 548	1.259×10^{-5}	49	281.8	79 430
.003 162	1.000×10^{-5}	50	316.2	100 000
.002 818	7.943×10^{-6}	51	354.8	125 900
.002 512	6.310×10^{-6}	52	398.1	158 500
.002 239	5.012×10^{-6}	53	446.7	199 500
.001 995	3.981×10^{-6}	54	501.2	251 200
.001 778	3.162×10^{-6}	55	562.3	316 200
.001 585	2.512×10^{-6}	56	631.0	398 100
.001 413	1.995×10^{-6}	57	707.9	501 200
.001 259	1.585×10^{-6}	58	794.3	631 000
.001 122	1.259×10^{-6}	59	891.3	794 300
.001 000	1.000×10^{-6}	60	1 000	1 000 000
8.91×10^{-4}	7.943×10^{-7}	61	1 122	1.259×10^{6}
7.94×10^{-4}	6.310×10^{-7}	62	1 259	1.585×10^{6}
7.08×10^{-4}	5.012×10^{-7}	63	1 413	1.995×10^{6}
6.31×10^{-4}	3.981×10^{-7}	64	1 585	2.512×10^{6}
5.62×10^{-4}	3.162×10^{-7}	65	1 778	3.162×10^{6}
5.01×10^{-4}	2.512×10^{-7}	66	1 995	3.981×10^{6}
4.47×10^{-4}	1.995×10^{-7}	67	2 239	5.012×10^{6}
3.98×10^{-4}	1.585×10^{-7}	68	2 512	6.310×10^{6}
3.55×10^{-4}	1.259×10^{-7}	69	2 818	7.943×10^{6}
3.16×10^{-4}	1.000×10^{-7}	70	3 162	1.000×10^{7}
2.82×10^{-4}	7.943×10^{-8}	71	3 548	1.259×10^{7}
2.51×10^{-4}	6.310×10^{-8}	72	3 981	1.585×10^{7}
2.24×10^{-4}	5.012×10^{-8}	73	4 467	1.995×10^{7}
1.99×10^{-4}	3.981×10^{-8}	74	5 012	2.512×10^{7}
1.78×10^{-4}	3.162×10^{-8}	75	5 623	3.162×10^{7}
1.58×10^{-4}	2.512×10^{-8}	76	6 310	3.981×10^{7}
1.41×10^{-4}	1.995×10^{-8}	77	7 079	5.012×10^{7}
1.26×10^{-4}	1.585×10^{-8}	78	7 943	6.310×10^{7}
1.12×10^{-4}	1.259×10^{-8}	79	8 913	7.943×10^{7}
1.00×10^{-4}	1.000×10^{-8}	80	10 000	1.000×10^{8}
8.91×10^{-5}	7.943×10^{-9}	81	11 220	1.259×10^{8}
7.94×10^{-5}	6.310×10^{-9}	82	12 590	1.585×10^{8}
7.08×10^{-5}	5.012×10^{-9}	83	14 130	1.995×10^{8}
6.31×10^{-5}	3.981×10^{-9}	84	15 850	2.512×10^{8}
5.62×10^{-5}	3.162×10^{-9}	85	17 780	3.162×10^{8}
5.01×10^{-5}	2.512×10^{-9}	86	19 950	3.981×10^{8}
4.47×10^{-5}	1.995×10^{-9}	87	22 390	5.012×10^{8}

Voltage or Current Ratio	Power Ratio (= mW to Reference Level 1 mW)	db	Voltage or Current Ratio	Power Ratio (= mW to Reference Level 1 mW)
3.98×10^{-5}	1.585×10^{-9}	88	25 120	6.310×10^{8}
3.55×10^{-5}	1.259×10^{-9}	89	28 180	7.943×10^{8}
3.16×10^{-5}	1.000×10^{-9}	90	31 620	1.000×10^{9}
2.82×10^{-5}	7.943×10^{-10}	91	35 480	1.259×10^{9}
2.51×10^{-5}	6.310×10^{-10}	92	39 810	1.585×10^{9}
2.24×10^{-5}	5.012×10^{-10}	93	44 670	1.995×10^{9}
1.99×10^{-5}	3.981×10^{-10}	94	50 120	2.512×10^{9}
1.78×10^{-5}	3.162×10^{-10}	95	56 230	3.162×10^{9}
1.58×10^{-5}	2.512×10^{-10}	96	63 100	3.981×10^{9}
1.41×10^{-5}	1.995×10^{-10}	97	70 790	5.012×10^{9}
1.26×10^{-5}	1.585×10^{-10}	98	79 430	6.310×10^{9}
1.12×10^{-5}	1.259×10^{-10}	99	89 130	7.943×10^{9}
1.00×10^{-5}	1.000×10^{-10}	100	100 000	1.000×10^{10}

TABLE 2: POWER AND VOLTAGE OR CURRENT RATIOS EXPRESSED
IN DECIBELS

Ratio	db (Power Ratio)	db (Voltage* Ratio)*	Ratio	db (Power Ratio)	db (Voltage* Ratio)
1.0	0	0	5.7	7.559	15.117
1.1	0.414	0.828	5.8	7.634	15.269
1.2	0.792	1.584	5.9	7.709	15.417
1.3	1.139	2.279	6.0	7.782	15.563
1.4	1.461	2.923	6.1	7.853	15.707
1.5	1.761	3.522	6.2	7.924	15.848
1.6	2.041	4.082	6.3	7.993	15.987
1.7	2.304	4.609	6.4	8.062	16.124
1.8	2.553	5.105	6.5	8.129	16.258
1.9	2.788	5.575	6.6	8.195	16.391
2.0	3.010	6.021	6.7	8.261	16.521
2.1	3.222	6.444	6.8	8.325	16.650
2.2	3.424	6.848	6.9	8.388	16.777
2.3	3.617	7.235	7.0	8.451	16.902
2.4	3.802	7.604	7.1	8.513	17.025
2.5	3.979	7.959	7.2	8.573	17.147
2.6	4.150	8.299	7.3	8.633	17.266
2.7	4.314	8.627	7.4	8.692	17.385
2.8	4.472	8.943	7.5	8.751	17.501
2.9	4.624	9.248	7.6	8.808	17.616
3.0	4.771	9.542	7.7	8.865	17.730
3.1	4.914	9.827	7.8	8.921	17.842
3.2	5.051	10.103	7.9	8.976	17.953
3.3	5.185	10.370	8.0	9.031	18.062
3.4	5.315	10.630	8.1	9.085	18.170
3.5	5.441	10.881	8.2	9.138	18.276
3.6	5.563	11.126	8.3	9.191	18.382
3.7	5.682	11.364	8.4	9.243	18.486
3.8	5.798	11.596	8.5	9.294	18.588
3.9	5.911	11.821	8.6	9.345	18.690
4.0	6.021	12.041	8.7	9.395	18.790
4.1	6.128	12.256	8.8	9.445	18.890
4.2	6.232	12.465	8.9	9.494	18.988
4.3	6.335	12.669	9.0	9.542	19.085
4.4	6.435	12.869	9.1	9.590	19.181
4.5	6.532	13.064	9.2	9.638	19.276
4.6	6.628	13.255	9.3	9.685	19.370
4.7	6.721	13.442	9.4	9.731	19.463
4.8	6.812	13.625	9.5	9.777	19.554
4.9	6.902	13.804	9.6	9.823	19.645
5.0	6.990	13.979	9.7	9.868	19.735
5.1	7.076	14.151	9.8	9.912	19.825
5.2	7.160	14.320	9.9	9.956	19.913
5.3	7.243	14.486	10.0	10.000	20.000
5.4	7.324	14.648	100	20	40
5.5	7.404	14.807	1000	30	60
5.6	7.482	14.964	10000	40	80

*Or Current Ratio.
To find the decibels corresponding to ratios above 10, break the ratio into two factors
and add the decibels of each.   For example—
Voltage ratio = 400 = 4 × 100.   Decibels = 12.041 + 40 = 52.041.

TABLE 3 : DECIBELS ABOVE AND BELOW REFERENCE LEVEL 6 mW INTO 500 OHMS

Note that the power in watts holds for any impedance, but the voltage holds only for 500 ohms.

db down		Level	db up	
Volts	Watts	db	Volts	Watts
1.73	6.00×10^{-3}	$-0+$	1.73	.006 00
1.54	4.77×10^{-3}	1	1.94	.007 55
1.38	3.79×10^{-3}	2	2.18	.009 51
1.23	3.01×10^{-3}	3	2.45	.012 0
1.09	2.39×10^{-3}	4	2.75	.015 1
.974	1.90×10^{-3}	5	3.08	.019 0
.868	1.51×10^{-3}	6	3.46	.023 9
.774	1.20×10^{-3}	7	3.88	.030 1
.690	9.51×10^{-4}	8	4.35	.037 9
.615	7.55×10^{-4}	9	4.88	.047 7
.548	6.00×10^{-4}	10	5.48	.060 0
.488	4.77×10^{-4}	11	6.15	.075 5
.435	3.79×10^{-4}	12	6.90	.095 1
.388	3.01×10^{-4}	13	7.74	.120
.346	2.39×10^{-4}	14	8.68	.151
.308	1.90×10^{-4}	15	9.74	.190
.275	1.51×10^{-4}	16	10.93	.239
.245	1.20×10^{-4}	17	12.26	.301
.218	9.51×10^{-5}	18	13.76	.379
.194	7.55×10^{-5}	19	15.44	.477
.173	6.00×10^{-5}	20	17.32	.600
.097 4	1.90×10^{-5}	25	30.8	1.90
.054 8	6.00×10^{-6}	30	54.8	6.0
.030 8	1.90×10^{-6}	35	97.4	19.0
.017 3	6.00×10^{-7}	40	173	60.0
.009 74	1.90×10^{-7}	45	308	190
.005 48	6.00×10^{-8}	50	548	600
.003 08	1.90×10^{-8}	55	974	1 900
.001 73	6.00×10^{-9}	60	1 730	6 000
.000 974	1.90×10^{-9}	65	3 080	19 000
.000 548	6.00×10^{-10}	70	5 480	60 000
.000 308	1.90×10^{-10}	75	9 740	190 000
.000 173	6.00×10^{-11}	80	17 300	600 000

References : Nomographs Refs. 3, 6 ; Tables Ref. 9.

TABLE 4 : DECIBELS ABOVE AND BELOW REFERENCE LEVEL 6 mW
INTO 600 OHMS

Note that the power holds for any impedance, but the voltage holds only for 600 ohms.

db down		Level	db up	
Volts	Watts	db	Volts	Watts
1.90	6.00×10^{-3}	$-0+$	1.90	006 00
1.69	4.77×10^{-3}	1	2.13	.007 55
1.51	3.79×10^{-3}	2	2.39	.009 51
1.34	3.01×10^{-3}	3	2.68	.012 0
1.20	2.39×10^{-3}	4	3.01	.015 1
1.07	1.90×10^{-3}	5	3.37	.019 0
.951	1.51×10^{-3}	6	3.78	.023 9
.847	1.20×10^{-3}	7	4.25	.030 1
.775	9.51×10^{-4}	8	4.77	.037 9
.673	7.55×10^{-4}	9	5.35	.047 7
.600	6.00×10^{-4}	10	6.00	.060 0
.535	4.77×10^{-4}	11	6.73	.075 5
.477	3.79×10^{-4}	12	7.55	.095 1
.425	3.01×10^{-4}	13	8.47	.120
.378	2.39×10^{-4}	14	9.51	.151
.337	1.90×10^{-4}	15	10.7	.190
.301	1.51×10^{-4}	16	12.0	.239
.268	1.20×10^{-4}	17	13.4	.301
.239	9.51×10^{-5}	18	15.1	.379
.213	7.55×10^{-5}	19	16.9	.477
.190	6.00×10^{-5}	20	19.0	.600
.107	1.90×10^{-5}	25	33.7	1.90
.060 0	6.00×10^{-6}	30	60.0	6.0
.033 7	1.90×10^{-6}	35	107	19.0
.019 0	6.00×10^{-7}	40	190	60.0
.010 7	1.90×10^{-7}	45	337	190
.006 00	6.00×10^{-8}	50	600	600
.003 37	1.90×10^{-8}	55	1 070	1 900
.001 90	6.00×10^{-9}	60	1 900	6 000
.001 07	1.90×10^{-9}	65	3 370	19 000
.000 600	6.00×10^{-10}	70	6 000	60 000
.000 337	1.90×10^{-10}	75	10 700	190 000
.000 190	6.00×10^{-11}	80	19 000	600 000

TABLE 5 : DECIBELS ABOVE AND BELOW REFERENCE LEVEL 1 mW
INTO 600 OHMS

Note that the power holds for any impedance, but the voltage holds only for 600
ohms.

db down		Level	db up	
Volts	Milliwatts	dbm	Volts	Milliwatts
0.774 6	1.000	− 0 +	0.774 6	1.000
0.690 5	.794 3	1	0.869 1	1.259
0.616 7	.631 0	2	0.975 2	1.585
0.548 4	.501 2	3	1.094	1.995
0.488 7	.398 1	4	1.228	2.512
0.435 6	.316 2	5	1.377	3.162
0.388 2	.251 2	6	1.546	3.981
0.346 0	.199 5	7	1.734	5.012
0.308 4	.158 5	8	1.946	6.310
0.274 8	.125 9	9	2.183	7.943
0.244 9	.100 0	10	2.449	10.000
0.218 3	.079 43	11	2.748	12.59
0.194 6	.063 10	12	3.084	15.85
0.173 4	.050 12	13	3.460	19.95
0.154 6	.039 81	14	3.882	25.12
0.137 7	.031 62	15	4.356	31.62
0.122 8	.025 12	16	4.887	39.81
0.109 4	.019 95	17	5.484	50.12
0.097 52	.015 85	18	6.153	63.10
0.086 91	.012 59	19	6.905	79.43
0.077 46	.010 00	20	7.746	100.00
0.043 56	.003 16	25	13.77	316.2
0.024 49	.001 00	30	24.49	1.000W
0.013 77	.000 316	35	43.56	3.162W
0.007 746	.000 100	40	77.46	10.00W
0.004 356	3.16×10^{-5}	45	137.7	31.62W
0.002 449	1.00×10^{-5}	50	244.9	100W
0.001 377	3.16×10^{-6}	55	435.6	316.2W
0.000 774 6	1.00×10^{-6}	60	774.6	1 000W
0.000 435 6	3.16×10^{-7}	65	1 377	3 162W
0.000 244 9	1.00×10^{-7}	70	2 449	10 000W
0.000 137 7	3.16×10^{-8}	75	4 356	31 620W
0.000 077 46	1.00×10^{-8}	80	7 746	100 000W

TABLE 6 : WATTS, DBM AND VOLTS ACROSS 5000 OHMS

Watts	dbm	volts across 5000 Ω	milli-watts	dbm	volts across 5000 Ω	micro-watts	dbm	milli-volts across 5000 Ω
20	+43	315	63	+18	17.8	1.0	−30	71
15.8	+42	280	40	+16	14.0	0.63	−32	56
12.6	+41	250	25	+14	11.2	0.40	−34	45
10	+40	213	16	+12	8.9	0.25	−36	35
7.9	+39	200	10	+10	7.1	0.16	−38	28
6.3	+38	178	6.3	+ 8	5.6	0.10	−40	21.3
5.0	+37	158	4.0	+ 6	4.5	0.063	−42	17.8
4.0	+36	140	2.5	+ 4	3.5	0.040	−44	14.0
3.16	+35	126	1.6	+ 2	2.8	0.025	−46	11.2
2.5	+34	112	1.0	0	2.13	0.016	−48	8.9
2.0	+33	100	0.63	− 2	1.78	0.010	−50	7.1
1.59	+32	89	0.40	− 4	1.40	0.006 3	−52	5.6
1.26	+31	79	0.25	− 6	1.12	0.004 0	−54	4.5
1.0	+30	71	0.16	− 8	0.89	0.002 5	−56	3.5
0.79	+29	63	0.10	−10	0.71	0.001 6	−58	2.8
0.63	+28	56	63 μW	−12	0.56	0.001 0	−60	2.13
0.50	+27	50	40 μW	−14	0.45	0.000 63	−62	1.78
0.40	+26	45	25 μW	−16	0.35	0.000 40	−64	1.40
0.32	+25	39	16 μW	−18	0.28	0.000 25	−66	1.12
0.25	+24	35	10 μW	−20	0.213	0.000 16	−68	0.89
0.20	+23	31.5	6.3μW	−22	0.178	0.000 10	−70	0.71
0.16	+22	28	4.0μW	−24	0.140	3.2×10⁻⁵	−75	0.39
0.13	+21	25	2.5μW	−26	0.112	1.00×10⁻ᵇ	−80	0.21
0.10	+20	21.3	1.6μW	−28	0.089	3.2×10⁻⁶	−85	0.13

(xi) Nomogram for adding decibel-expressed quantities

Two or more sounds combine to give a total sound whose acoustical power is the sum of the powers of the individual components. This nomogram* (Fig. 19.6A) may be used·to determine the resultant level in db when the two component sound levels are also expressed in db, or for adding any decibel-expressed quantities.

The difference between the two component values in decibels is first determined by algebraic subtraction, and this difference value is found on Scale A. The corresponding figure on Scale B then indicates the number of decibels to be added to the greater original quantity to give the required resultant.

For example take two sound levels of 35.2 db and 37.0 db. The difference is 1.8 db which, when located on Scale A, indicates 2.2 db on Scale B. This value is then added to 37.0 db to give the resultant sound level of 39.2 db.

When the difference in level exceeds 20 db, disregarding the smaller quantity produces an error of less than 1%.

The nomograph scales are based on the formula :

$$B = \left[10 \log_{10}\left(1 + \log_{10}{}^{-1}\frac{A}{10}\right)\right] - A \qquad (26)$$

where A and B correspond to points on the respective scales.   Ref. 37.

---

* For Nomogram see page 822.

(xii) Decibels, slide rules and mental arithmetic

The following holds for power calculations under all conditions, and for voltage and current calculations when the impedance is constant.

Use of log-log slide rule

For voltage ratios, set the cursor to 10 on the log-log scale, and set the C scale to 20 on the cursor. Then set the cursor to the required voltage ratio on the log-log scale and read the decibels on the C scale. To obtain good accuracy it is advisable to use the section of the log-log scale between 3 and 100, dealing with powers of 10 separately.

For power ratios, set cursor to 10 on the log-log scale, set the C scale to 10 on the cursor and then proceed as before.

Decibel calculations by mental arithmetic

For occasions when a rule is not available, it is useful to be able to perform conversions mentally. It is possible to memorize a conversion table, but this is unnecessary because, starting with the knowledge that an increase of two times in a voltage ratio is equivalent to 6 db, and 10 times to 20 db, it is comparatively simple to build up such a table mentally as required. The backbone of the table is as follows—

db	0	2	6	8	12	14	18	20
factor	1	1.25	2	2.5	4	5	8	10

This is built up simply by adding 6 db for each two times increase from a ratio of unity, i.e. 2, 4 and 8 times represent 6, 12 and 18 db, and by subtracting 6 db for each halving of a multiplying factor of 10 times i.e. 10, 5, $2\frac{1}{2}$ and $1\frac{1}{4}$ times are represented by 20, 14, 8 and 2 db.

The next step in the table can be built up from noticing that a 2 db decrease represents a multiplying factor of 0.8 times. From this fact we can add the values for 4, 10 and 16 db—

db	4	10	16
factor	1.6 (2 × 0.8)	3.2 (4 × 0.8)	6.4 (8 × 0.8)

To complete the table we need values for each of the odd numbers of db. Noticing that 2 db down is equivalent to multiplying by 0.8 we can assume that the error in taking 1 db down as 0.9 is small, and the table can then be written in full.

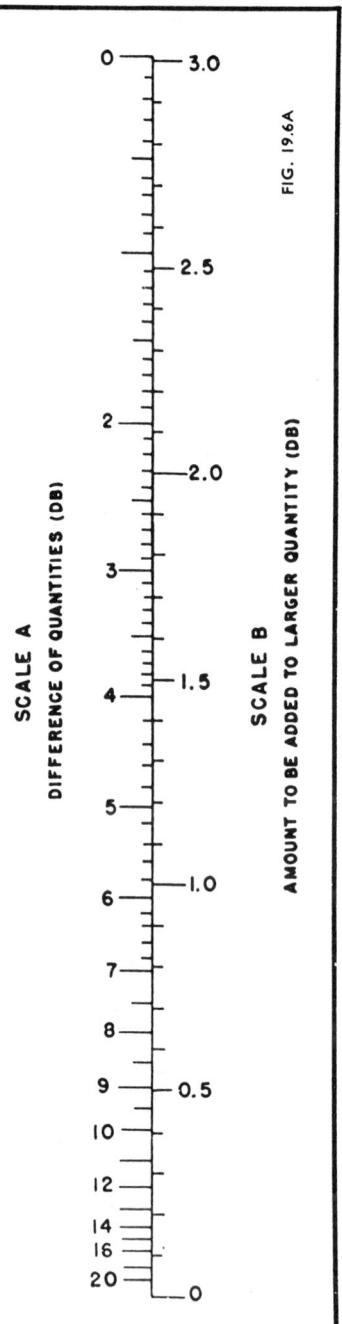

Fig. 19.6A.

Accurate values are recorded in the third line for comparison.

db—										
0	1	2	3	4	5	6	7	8	9	10
factor—										
1	1.12	1.25	1.4	1.6	1.8	2	2.3	2.5	2.9	3.2
true factor—										
1.00	1.12	1.26	1.41	1.59	1.78	2.00	2.24	2.51	2.82	3.16

db—									
11	12	13	14	15	16	17	18	19	20
factor—									
3.6	4	4.5	5	5.8	6.4	7.2	8	9	10
true factor—									
3.55	3.98	4.47	5.01	5.62	6.31	7.08	7.94	8.91	10.0

Comparing the multiplying factors computed in this manner with the true factors it will be seen that the greatest error (at 15 db) is just over 3% which is negligible for the type of calculation intended to be performed by this method.

When working with " db down," the number of db can be subtracted from 20 and the multiplying factor from the table can be divided by 10, e.g. 7 db (20-13) down is equivalent to a multiplying factor of 0.45. A small amount of practice at such mental conversions soon results in many of the factors being memorized (or else becoming immediately obvious) without any conscious effort in this direction.

SECTION 2 : VOLUME INDICATORS AND VOLUME UNITS

(i) Volume indicators   (ii) Volume units.

(i) Volume indicators

When an instrument is required merely to measure power under steady conditions, no particular difficulties are encountered and the scale may be calibrated in decibels with respect to any desired reference level. In studio programmes, however, the power is constantly fluctuating and the indication of the instrument will depend upon its speed of response and damping.

Volume is here applied to the indications of a device known as a Volume Indicator, which is calibrated and read in a prescribed manner.

The **Volume Indicator** is a standardized instrument (Ref. 19) which has been developed primarily for the control and monitoring of radio programmes. The Volume Indicator is a root-mean-square type of instrument with a full-wave copper-oxide type of rectifier. The rectifier law is intermediate between linear and square-law, having an exponential of 1.2 ± 0.2. Its dynamic characteristics are such that, if a sinusoidal voltage of frequency between 35 and 10 000 c/s of such amplitude as to give reference deflection under steady-state conditions is suddenly applied, the pointer will reach 99% of reference deflection in 0.3 second (± 10%) and should then overswing reference deflection by at least 1.0% and not more than 1.5% (Ref. 19). It will give a reading of 80% on an impulse of sine wave form as short as 0.025 second (Ref. 18).

It is fitted with two scales, a vu scale marked 0 (" reference deflection ") at about 71% maximum scale reading, extending to + 3 (maximum) and − 20 (minimum) and a percentage voltage scale with 100% corresponding to 0 vu reading, calibrated downwards to 0%.

The instrument is available with the scales marked differently : With " Scale A " the vu scale is made more prominent, being situated above the percentage scale—

this is for use in measuring instruments etc. where the reading of the meter together with the reading of an associated attenuator (where necessary) give the actual power level. With " Scale B " the percentage scale is made more prominent being situated above the vu scale—this is for use in studio controls etc. where the operator is non-technical and is not interested in actual levels.

The sensitivity shall not depart from that at 1000 c/s by more than 0.2 db between 35 and 10 000 c/s at an input level of 0 vu, nor more than 0.5 db between 25 and 16 000 c/s (Ref. 19).

The instrument is calibrated by connecting it in shunt with a resistance of 600 ohms through which is flowing 1 milliwatt of sine-wave power at 1000 c/s, when a reading shall be 0 vu (or n vu when the calibrating power is n db above 1 milliwatt).

If the instrument is connected across any impedance other than 600 ohms, the volume indicated must be corrected by adding $10 \log_{10} (600/Z)$, where Z is the actual impedance in ohms.

The total impedance of the volume indicator is usually about 7500 ohms, of which about 3600 ohms is external to the instrument.

The Volume Indicator is intended to be read as deviations from the reference volume (0 vu), after making allowance for the sensitivity control (attenuator) which is also calibrated in vu. The reading is determined by the greatest deflections occurring in a period of about a minute for programme waves, or a shorter period (e.g. 5 to 10 seconds) for message telephone speech waves, excluding not more than one or two occasional deflections of unusual amplitude.

References 18, 19, 39, 49.

(ii) Volume units

The **volume unit** (abbreviated vu and pronounced " vee-you ") is a unit to express the level of a complex wave in terms of decibels above or below a reference volume as defined below. A level referred to as x vu means a complex wave power reading on a standard vu meter. Volume units should never be used to indicate the level of a sine-wave signal—the latter should always be referred to as so many dbm. Even if a vu meter is used to read the level of a steady single-frequency sine-wave signal, which is quite permissible, the reading should be referred to as so many dbm.

A volume unit implies a complex wave—a programme waveform with high peaks. The usual convention is to assume that the peak value is 10 db above the sine-wave peak.* For this reason an amplifier for radio broadcasting systems is tested with sine-wave input at a level 10 db above the maximum vu level at which it is intended to be used. For example, a system working at a level of + 12 vu would be tested for distortion at a level of + 22 dbm sine-wave.

Reference volume is defined as that strength of electrical speech or programme waves which gives a reading of 0 vu on a volume indicator as described above, and which is calibrated to read 0 vu on a steady 1000 c/s wave whose power is 1 milliwatt in 600 ohms. Care should be taken to distinguish between this definition of reference volume, which is arbitrary and not definable in fundamental terms, and a reference level of 1 milliwatt used for power measurements under steady conditions with a single frequency.

References 18, 19, 39, 49.

---

*Occasional peaks under certain conditions may be somewhat higher.

SECTION 3 : INDICATING INSTRUMENTS

(i) Decibel meters   (ii) Power output meters   (iii) Volume indicators   (iv) Acoustical instruments.

(i) Decibel meters

A decibel meter is usually a rectifier type of instrument calibrated in decibels, with 0 db situated somewhere near the centre of the scale. The scale is usually marked to indicate the reference level and the line impedance, e.g. " 0 db = 6 mW ; calibrated for 500 ohm lines."

These instruments are frequently provided with attenuators to extend their range and are then sometimes called **power-level indicators.** They are actually voltmeters intended to be connected across a line of known impedance.

If a decibel meter is connected across a resistance other than the correct value the error will be

$$\text{error in db} = 10 \log_{10} (R_x/R) \tag{1}$$

where R_x = resistance across which the meter is connected
and R = correct resistance for which the meter is calibrated.

A chart of corrections based on equation (1) is given in Ref. 4.

(ii) Power output meters

These differ from decibel meters in that the instrument dissipates and measures power. The input impedance is usually variable by means of a selector switch, and the scale of the instrument may be calibrated both in milliwatts and decibels.

(iii) Volume indicators

See Section 2 above.

(iv) Acoustical instruments

Sound level meters—see Section 5 below.
Noise meters—see Section 6 below.
References to indicating instruments—1, 4, 9, 17, 18, 19, 20, 21, 23, 25, 26, 27, 39.

SECTION 4 : NEPERS AND TRANSMISSION UNITS

(i) Nepers   (ii) Transmission units.

(i) Nepers

Just as there are two systems of logarithms in general use, so there are two logarithmic units for the measurement of difference of power levels. The bel and the decibel are based on the system of Common Logarithms (to the base 10).

The neper is based on the system of Naperian Logarithms (to the base ϵ).

Nepers are not commonly used in English-speaking countries, but are used by some European countries.

Equation (1) below should be compared with Section 1, equation (2) above :

$$N_n = \tfrac{1}{2} \log_\epsilon (P_2/P_1) \tag{1}$$

where N_n is the ratio of two powers expressed in nepers
    P_1 = reference power
and P_2 = power which is referred back to P_1.

When the impedances relating to P_1 and P_2 are the same,

$$N_n = \log_\epsilon (E_2/E_1) = \log_\epsilon (I_2/I_1) \tag{2}$$

Equation (2) should be compared with equations (4) and (5) in Section 1 above.

R.D.H.—27

Relationship between decibels and nepers
      1 neper = 8.686 db
      1 db   = 0.1151 neper

Power ratio	1	1.259	1.585	3.162	10	100	10^5
db	0	1	2	5	10	20	50
nepers	0	0.1151	0.2303	0.5757	1.151	2.303	5.757

References 1, 9.

(ii) Transmission units

A transmission unit was the early name of the decibel, but is no longer used.
      1 TU = 1 db

SECTION 5 : LOUDNESS

(i) Introduction to loudness   (ii) The phon   (iii) Loudness units.

(i) Introduction to loudness

The loudness of any tone is a function, not only of its intensity, but also of its frequency.   This is indicated by the contour curves of equal loudness as shown in Fig.
19.7.

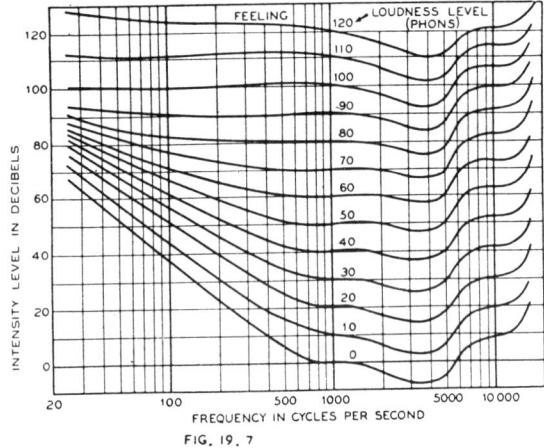

FIG. 19. 7

*Fig. 19.7.   Contours of equal loudness level (0 db = 10^{-16} watt/cm²)—Ref. 28, after
Fletcher and Munson (Ref. 40).*

(ii) The Phon

The phon is the unit of loudness level.   The loudness level, in phons, of a sound
is numerically equal to the intensity level in decibels of the 1000 c/s pure tone which is
judged by the listeners to be equally loud.   The reference intensity is 10^{-16} watt
per square centimetre*, which is near the value of the threshold of audibility for a
1000 c/s pure tone (Ref. 22).

When listening to a 1000 c/s pure tone, the loudness level in phons is equal to the
number of decibels above the reference intensity, but with any other frequency the
loudness level in phons will normally differ from the intensity in decibels (Fig. 19.7).

---

*The equivalent reference pressure for sound pressure measurements is 0.0002 dyne per square centi-
metre.   The equivalent reference velocity for sound velocity measurements is 0.000 005 centimetre
per second.

(iii) Loudness units

For purposes of noise measurement, the **loudness unit** has been standardized (Ref. 28), and is based on the principle that doubling the number of loudness units is equivalent to a sensation of twice the loudness. The relation between loudness level in phons and loudness in loudness units is given in Fig. 19.8.

References 40, 41, 28.

See also Supplement.

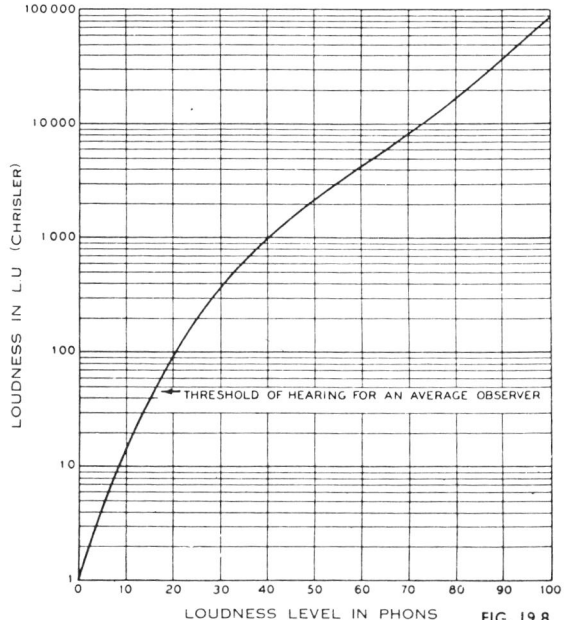

LOUDNESS LEVEL IN PHONS   FIG. 19.8

Fig. 19.8.   Relation between loudness and loudness level ; reference frequency 1000 c/s—
Ref. 28, after Fletcher and Munson (Ref. 41).

SECTION 6 : THE MEASUREMENT OF SOUND LEVEL AND NOISE

(i) Introduction   (ii) The sound level meter   (iii) The measurement of noise in amplifiers   (iv) The measurement of radio noise.

(i) Introduction

Sound includes wanted sound—music or speech—and also unwanted sound—noise. Noise may be measured acoustically, as for example by a sound level meter, or electrically. Acoustical methods of measuring noise and other sounds are described in (ii) below. Electrical noise may be measured either with or without a weighting network. If a weighting network is used, there will be appreciable attenuation of the lower frequencies, including hum frequencies. It is normally assumed in good engineering practice that the hum components have negligible effect on the noise reading ; if this is not so, they should be filtered out before the noise voltage is applied to the measuring equipment.

(ii) The sound level meter

Sound measurements, when made with an American standard sound level meter (Ref. 23) are determinations of sound intensity levels and are expressed in decibels—the reference sound pressure level is 0.0002 dyne/cm^2 at 1000 c/s. The standard sound level meter has a scale calibrated in decibels, and the readings should be referred to as e.g. " 60 decibels sound level " or " sound level of 60 decibels." A sound level meter may provide a choice of frequency response characteristics :

(A) The 40 decibel equal loudness contour*
(B) The 70 decibel equal loudness contour*
(C) Flat frequency response.

In all cases the frequency response should be stated in connection with any measurements. If only one frequency response is provided, this should be (A) above. Curve (A) is recommended for measurements up to 55 db, curve (B) for measurements from 55 to 85 db, and curve (C) for very loud sounds (85 to 140 db).

Within certain tolerances the sound level meter will indicate the sum of the equivalent 1000 c/s intensities of the different single frequency components in a complex wave—that is the power indicated for a complex wave will be the sum of the powers which would be indicated for each of the single frequency components of the complex wave acting alone.

The dynamic characteristics of the indicating instrument are specified in detail (Ref. 23).

The American sound level meter is a very simple instrument giving only approximate indications of loudness levels. However the ear does not judge the loudness of wide-band noises in the same way that it judges the loudness of pure tones. Consequently the use of the equal-loudness contours based on pure tones introduces errors which may be as great as 22 phons between the readings of an American sound level meter and subjectively measured loudness levels in phons (Ref. 42).

British objective noise meter

A more accurate acoustical noise meter has been designed by King and associates (Ref. 42) which includes a phase-shifting network, a peak-indicating meter and a larger number of weighting networks. See also Ref. 34 (pp. 894-896).

References to sound level meters : 12, 23, 24 (pp. 392-393), 34 (pp. 888-896), 42.

TABLE 7 : TYPICAL SOUND LEVELS

measured with American Standard Sound Level Meter

(after H. F. Olson, Ref. 24, by kind permission of the author and publishers—copyright by D. Van Nostrand Company, Inc., New York, U.S.A.)

Source or Description of Noise		Noise Level in Decibels
Threshold of Pain		130
Hammer Blows on Steel Plate	2 ft.	114
Riveter	35 ft.	97
7-passenger sedan car†		87
Factory		78
Busy street traffic		68
Large office		65
Ordinary conversation	3 ft.	65
Large store		63
Factory office		63
Medium store		62
Restaurant		60
Residential street		58
Medium office		58
Garage		55

*Each modified by the differences between random and normal free field thresholds.
†Data from G. L. Bonvallet " Levels and spectra of transportation vehicle noise " J. Acous. Soc. Am. 22.2 (March 1950) 201.

Small store		52
Theatre (with audience)		42
Hotel		42
Apartment		42
House, large city		40
House, country		30
Average whisper	4 ft.	20
Quiet whisper	5 ft.	10
Rustle of leaves in gentle breeze		10
Threshold of hearing		0

(iii) The measurement of noise in amplifiers*

(A) Amplifiers for sound equipment

(based on R.M.A. Standard SE-101-A—Ref. 35).

This is an interim standard for the measurement of steady-state noise, and of pulse noise waves having a peak factor (ratio of peak to r.m.s.) approximating the maximum obtainable in speech (20 db).

Noise level is the level of any noise signals appearing at the output terminals with no signal applied to the input.

The weighted noise level is the noise level weighted in accordance with the 70 decibel equal-loudness contour of the human ear and expressed in dbm. The weighted noise level shall be measured under test conditions specified in Ref. 35. The measuring amplifier shall be one whose a-f response is weighted with the 70 db equal-loudness contour in accordance with Curve B of A.S.A. Specification Z24.3-1944 (Ref. 23) with a standard vu meter as defined by A.S.A. Specification C16.5-1942 (Ref. 19) as an indicator.

A properly weighted amplifier may be obtained by applying a r.c. network with a 1 millisecond time constant to an amplifier having a frequency response of $\pm$ 1 db from 50 to 15 000 c/s. This will give an attenuation of 1 db at 300 c/s, 5.7 db at 100 c/s and 9 db at 60 c/s.

(B) Audio facilities for radio broadcasting systems

(Based on R.M.A. Standard TR-105-B—Ref. 33).

Measurement of **steady state noise** shall be made with a device having an a-f response flat within $\pm$ 2 db from 50 to 15 000 c/s and having the ballistic characteristics of the standard vu meter (Ref. 19) but reading the r.m.s. value of a complex wave. Measurement of pulse noise conditions has not been included because of lack of definition and equipment.

Measurement made on those instruments incorporating average rather than r.m.s. rectifiers will give indications differing in general less than 1 db from the latter due to this difference. This is within the overall accuracy ordinarily obtained on this type of measurement.

Audio frequency signal to noise ratio—definition

The a-f signal to noise ratio is the numerical ratio between the sine-wave signal power required for standard output and the noise power measured with zero applied signal, received by the rated load impedance from the output of the equipment under test, expressed as a power ratio in decibels (Ref. 33).

(iv) The measurement of radio noise

Radio noise is any electrical disturbance which excites a radio receiver in such a way as to produce acoustical noise.

" A major objective of radio-noise-meter design is to provide an instrument which will give, for all kinds of radio noise, indications which are proportional to the annoyance factor or nuisance capability of the noise." Instruments complying with the specification of the Joint Co-ordination Committee (Ref. 25) are stated to be very satisfactory in this regard (Ref. 21). A useful review of the progress made up to 1941

---

* For noise audibility test for radio receivers see Chapter 37 Sec. 1 (vi) K.

is given in Reference 26 with an extensive bibliography. The equivalent British Standard is described in Ref. 27.

References to noise—2, 9, 10, 12, 15 (Sect. 10-35), 20, 21, 24 (pp. 419-420,483-487), 25, 26, 27, 28, 33, 34, 35, 42.

Bibliography : " The measurement of noise " Jour. Acous. Soc. Am. 11.1 (July 1939) 117.

SECTION 7 : REFERENCES

1. Wright, P. B. " Evolution of the db and the vu " Comm. (1) 24.4 (April 1944) 54 ; (2) 24.5 (May 1944) 44 ; erratum and additional data 24.6 (June 1944) 68.
2. Harrison, C. E. " Electrical and acoustical equivalents " Comm. 25.6 (June 1945) 44.
3. Rhita, N. " A decibel nomograph " Radio Craft (Sept. 1945) 769.
4. Hudson, P. K. " Calibration of decibel meters " Comm. 25.7 (July 1945) 58.
5. Scott, F. S. G. " Absolute bels " W.E. 23.272 (May 1946) 132.
6. Miedke, R. C. " Decibel conversion chart " Proc. I.R.E. 34.2 (Feb. 1946) 76W.
7. Haefner, S. J. " Amplifier gain formulas and measurements " Proc. I.R.E. 34.7 (July 1946) 500.
8. Perry, S. V. " The decibel scale " R.M.A. Tech. Bull. U.S.A. (Nov. 1940).
9. Rao, V. V. L. (book) " The Decibel Notation and its Applications to Radio Engineering and Acoustics " (Addison and Co. Ltd. Madras, India, 1944).
10. Fowler, N. B. " Measurements in communications " E.E. 66.2 (Feb. 1947) 135.
11. Romanow, F. F., and M. S. Hawley " Proposed method of rating microphones and loudspeakers for systems use " Proc. I.R.E. 35.9 (Sept. 1947) 953.
12. Morgan, R. L. " Noise measurements " Comm. 28.2 (Feb. 1948) 28 ; 28.7 (July 1948) 21.
13. Chinn, H. A. " dbm vs. vu " Audio Eng. 32.3 (March 1948) 28.
14. Shea, T. E. (book) " Transmission Networks and Wave Filters " (D. Van Nostrand Company Inc., New York, 1929).
15. Pender, H. and K. McIlwain " Electrical Engineers Handbook—Vol. 5, Electric Communication and Electronics " (John Wiley and Sons Inc. New York ; Chapman and Hall Ltd., London, 3rd edit. 1936).
16. Research Council, Academy of Motion Picture Arts and Sciences (book) "Motion Picture Sound Engineering " (D. Van Nostrand Co. New York, 1938).
17. Affel, H. A., H. A. Chinn and R. M. Morris " A standard VI and reference level " Comm. 19.4 (April 1939) 10.
18. Chinn, H. A., D. K. Gannett and R. M. Morris " A new standard volume indicator and reference level." Proc. I.R.E. 28.1 (Jan. 1940) 1.
19. " American recommended practice for volume measurements of electrical speech and program waves " C16.5—1942, American Standards Association. Superseded by Ref. 49.
20. McLachlan, N. W. (book) " Noise " (Oxford University Press, 1935).
21. Burrill, C. M. " An evaluation of radio-noise-meter performance in terms of listening experience " Proc. I.R.E. 30.10 (Oct. 1942) 473.
22. " American standard acoustical terminology " Z24.1—1942. American Standards Association.
23. " American Standards for sound level meters " Z24.3—1944, American Standards Association.
24. Olson, H. F. (book) " Elements of Acoustical Engineering " (D. Van Nostrand Co. Inc. New York, 2nd edit. 1947).
25. " Methods of Measuring radio noise," Report of Joint Co-ordination Committee on Radio Reception of the Edison Electric Institute, National Electrical Manufacturers Association and Radio Manufacturers Association—R.M.A. Engineering Bulletin No. 32 (Feb. 1940).
26. Burrill, C. M. " Progress in the development of instruments for measuring radio noise," Proc. I.R.E. 29.8 (Aug. 1941) 433.
27. " The characteristics and performance of apparatus for the measurement of radio interference " B.S.S. No. 727, British Standards Institution, London (1937).
28. " American standard for noise measurement " Z24.2—1942, American Standards Association.
29. American R.M.A. Standard SE-103 " Speakers for sound equipment " (April 1949).
30. " Reference Data for Engineers " (book) (Federal Telephone and Radio Corporation, New York ; 3rd ed. 1949).
31. Black, W. L., and H. H. Scott, " Audio frequency measurements " Proc. I.R.E. 37.10 (Oct. 1949) 1108.
32. American R.M.A. Standard, SE-105 " Microphones for sound equipment " (Aug. 1949).
33. American R.M.A. Standard TR-105-B " Audio facilities for radio broadcasting systems " (Nov. 1949).
34. Beranek, L. L. (book) " Acoustic Measurements " (John Wiley and Sons Inc. New York ; Chapman and Hall Ltd., London, 1949).
35. American R.M.A. Standard SE-101-A " Amplifiers for sound equipment " (July 1949).
36. American R.M.A. Standard Proposal 198 " Microphones " (adopted Nov. 1938).
37. Di Mattia, A. L., and L. R. Jones " Adding decibel-expressed quantities " Audio Eng. 35.7 (July 1951) 15.
38. I.R.E. " Standards on abbreviations of radio-electronic terms 1951 " Standard 51 I.R.E. 21 S1 ; Proc. I.R.E. 39.4 (April 1951) 397.
39. Chinn, H. A. " The measurement of audio volume " Audio Eng. 35.9 (Sept. 1951) 26.
40. Fletcher, H., and W. A. Munson " Loudness, its definition, measurement and calculation " J. Acous. Soc. Am. 5.2 (Oct. 1933) 82.
41. Fletcher, H., and W. A. Munson " Loudness and masking " J. Acous. Soc. Am. 9.1 (July 1937) 1.
42. King, A. J., R. W. Guelke, C. R. Maguire and R. A. Scott " An objective noise-meter reading in phons for sustained noises, with special reference to engineering plant " Jour. I.E.E. 88, Part II (1941) 163.
43. Chinn, H.A. " The measurement of audio volume " Audio Eng. 35.10 (Oct. 1951) 24.

SECTION 1 : INTRODUCTION

(i) Types of loudspeakers (ii) Direct-radiator loudspeakers (iii) Horn loudspeakers
(iv) Headphones (v) Loudspeaker characteristics (vi) Amplitude of cone movement
(vii) Good qualities of loudspeakers (viii) Loudspeaker grilles.

(i) Types of loudspeakers
In this chapter the point of view is that of the receiver or amplifier designer, not that of the loudspeaker designer.
Loudspeakers are in three principal groups :
    1. Direct radiators—where the cone or diaphragm is directly coupled to the air.
    2. Horn loudspeakers—where the diaphragm is coupled to the air by means of a horn.
    3. Ionic loudspeakers—in which no diaphragm is used. See Refs. 212, 227.

(ii) Direct radiator loudspeakers
(A) **The moving coil or electro-dynamic** type is by far the most popular for radio receivers. It has a voice-coil mounted in a strong magnetic field ; the a-f output power from the amplifier is passed through a suitable matching transformer (see Chapter 21) to the voice coil, and the a-f currents in the coil cause a force to be exerted in the direction of the axis of the coil. If the voice coil is free, it will move in the direction of the force. The voice coil is loaded by the cone which is (normally) firmly fixed to it.
The magnet may be either an electromagnet or a permanent magnet. In all cases, the higher the flux density the greater the efficiency and electro-magnetic damping. See Sect. 2 for detailed description.

(B) The electromagnetic type has two forms, the **reed armature** and the **balanced armature.** In the former there is a steel reed armature at a short distance from the pole piece wound with insulated wire carrying the a-f current, and supplied with steady magnetic flux by means of a permanent magnet. In the balanced armature type, only the a-f flux flows longitudinally through the armature. Both types require relatively stiff springs to return the armature to the position of equilibrium, so that the bass resonance usually occurs at a frequency above 100 c/s.

Refs. 13, 14, 15.

(C) The inductor dynamic type differs from the older electromagnetic type in that the two iron armatures are balanced so that they lie at rest without a strong spring tension.   The natural resonance frequence may be about 70 c/s, and the amplitude of movement is about the same as that of a moving-coil type.   Ref. 15.

(D) The piezo-electric (" crystal ") type generally uses Rochelle salt for the bimorph elements.   It takes two forms, in which the elements either bend or twist with the voltage applied between the two electrodes.   The crystal has a predominately capacitive impedance, the capacitance being of the order of 0.02 μF.

The impedance varies considerably with frequency in both scalar value and phase angle, so that it is difficult to provide a correct load impedance for the output valve. This is not very important with a triode or if another loudspeaker is connected in parallel.

The crystal loudspeaker naturally responds to the higher frequencies, and is sometimes used as a " tweeter " in a dual or triple speaker system (Sect. 5).   In this case the response may be maintained roughly constant over the desired band of frequencies by a small inductance in series with the primary of the transformer.   Crystal loudspeakers are not normally used alone.

Refs. 13, 14, 24, 60.

(E) The condenser type of loudspeaker usually takes the form of one large solid electrode, and a thin movable electrode which is mounted in such a way that it can vibrate (usually in sections) without touching the solid electrode.

Either the solid or the movable electrode may be corrugated ; the movable electrode may be of insulating material with a metal foil coating.   An alternative form has both electrodes flexible (Ref. 22).

The sensitivity is dependent upon the total area ; with 300 to 500 square inches the sensitivity may approximate that of an electro-dynamic type.

The Kyle speaker (Ref. 58) has a capacitance of 0.004 μF for an area of 96 square inches.   The impedance is predominately capacitive.   A high polarizing voltage is required for high efficiency and good performance—500 to 600 volts is usual.

Condenser loudspeakers are little used at the present time.

Refs. 13, 14, 15, 22, 58.

(F) Throttled air-flow loudspeaker—this consists of a mechanical valve, actuated by an electrical system, which controls a steady air stream, the air flow being made proportional to the a-f input current.   It is normally limited to very large sizes.

Ref. 14.

(iii) Horn loudspeakers

Any type of driving unit actuating a cone or diaphragm may be given greater acoustical loading by means of a horn, thereby improving the efficiency and power output. The horn ceases to have any beneficial effect below a frequency whose value depends on the law of expansion and the size of the horn.   Horn loudspeakers have considerably greater efficiency than direct radiators.   For further information see Sect. 4.

(iv) Headphones

Headphones (telephone receivers) make use of a diaphragm which is effectively sealed to the ear by means of a cap with a central opening.   The pressure of the small quantity of air enclosed between the diaphragm and the ear drum varies in accordance with the displacement of the diaphragm.

The driving mechanisms of headphones resemble those for loudspeakers.

(A) Magnetic diaphragm : The a-f force caused by the a-f current in the electromagnet operates directly upon a steel diaphragm.   The bipolar type is the most popular of all headphones at the present time.   A permanent magnet supplies the steady flux.   At frequencies above the second diaphragm resonance the response falls off rapidly.   Refs. 13, 14, 59, 62.

(B) **Moving armature type** : The principles are the same as for loudspeakers. Ref. 14.

(C) **Moving-coil type** : This follows the same principles as the loudspeaker, and with careful design is capable of greater fidelity and wider frequency range than the magnetic types. One model has a response from 10 to 9000 c/s ± 2 db except for a dip of 6 db at 4000 c/s. (Ref. 63). Refs. 13, 14, 63.

(D) **Crystal type** : The impedance of one model is 80 000 ohms (predominately capacitive) at 10 000 c/s. A high resistance may be connected in series to raise the low frequency response relatively to the high frequency response. Refs. 13, 64.

(E) **Ribbon type** : The principle is the same as for a ribbon microphone. It is only used for high-fidelity reproduction. Refs. 13, 14, 65.

(F) **Inductor type** : This has been developed to deliver practically constant sound pressure to the ear cavity from 50 to 7000 c/s. It has a V shaped diaphragm driven by a straight conductor located in the bottom of the V. Refs. 13, 14, 66, 67.

Correction circuits for magnetic diaphragm headphones

A typical magnetic diaphragm unit has a pronounced peak slightly below 1000 c/s. A circuit for attenuating this peak and thereby giving reasonably uniform response is Fig. 20.1 (Ref. 25).

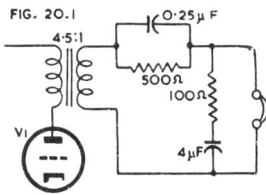

Fig. 20.1. Tone correction circuit for magnetic diaphragm headphones (d.c. resistance 4000 ohms) giving 15 db bass and treble boost. Secondary impedance roughly 500 ohms, primary load resistance 10 000 ohms (Ref. 25).

(v) Loudspeaker characteristics

The amplifier designer is mainly interested in a limited number of characteristics.

Impedance : See Sect. 2(iv) for moving-coil type, and Sect. 4(v)G for horn type.

All types of electro-magnetic loudspeakers have a somewhat similar form of impedance characteristic, although the relative values vary considerably.

Phase angle : See Sect. 2(iv) for moving-coil-type.

Frequency response characteristic

This only applies under the strict conditions of the test, particularly generator impedance, and is very often misleading. The most serious defect is the existence of sharp upward peaks extending higher than say 5 db above the smooth curve, particularly in the region from 1500 to 4000 c/s. These are not always shown by the published curves, which frequently show signs of having been smoothed.

Reflections in rooms may result in changes of 15 db or more in level at certain frequencies, and the shape of the frequency characteristic may be altered out of all recognition by variations in placement of loudspeaker and microphone, or by a change of room.

It is important to know the source impedance (i.e. the output resistance of the amplifier) used in the test. A loudspeaker should always be used with the correct source impedance—too low a source impedance may result in loss of low frequency response, while too high a source impedance will usually result in over-accentuation of a narrow band of frequencies in the vicinity of the loudspeaker bass resonance. See Sect. 9(ii) for R.M.A. Standard loudspeaker measurement source impedance. Refs. 13, 14, 15, 26 (No. 1), 42, 70.

Damping

The source impedance (i.e. the output resistance of the amplifier) also affects the damping of the loudspeaker at the bass resonant frequency. One value of source impedance will normally (in loudspeakers of high flux density) give critical damping—this being the value for no overshoot—while a slightly lower degree of damping may be considered as the optimum. Too high a resistance will give insufficient damping for good transient response. For further information see Sect. 2(x).

Directional characteristics

The usual frequency response characteristic is measured with the microphone on the axis of the loudspeaker, this being the position of maximum high frequency response. There is a serious loss of higher frequencies (4000 c/s and above) even with the microphone 30° off the axis.

See Sect. 2(vii) for further information on moving-coil types, and Sect. 4(vii) for horn types.

Refs. 13, 14, 15, 26 (No. 1), 61, 70, 133.

Non-linear distortion : see Sect. 7(i) and (v).

Efficiency versus frequency characteristic : see Sect. 2(vi).

Transient response : see Sect. 7(iii).

Matching : see Chapter 21.

Characteristics of cones : see Sect. 2(ii).

(vi) Amplitude of cone movement

The peak amplitude of a rigid disc to give an acoustical power output of 0.2 watt (equivalent to 6.7 watts amplifier power with a loudspeaker efficiency of 3%) is given by the following table. Radiation from one side only is considered (derived from Chart No. 61, Ref. 69).

Diameter of cone	Frequency (c/s)					
	1000	400	200	100	50	30
13 in.	0.0003	0.002	0.008	0.03	0.12	0.34 in.
10 in.	0.0005	0.003	0.013	0.05	0.21	0.58 in.
8 in.	0.0008	0.005	0.02	0.08	0.32	0.90 in.
5 in.	0.002	0.013	0.051	0.21	0.82	2.26 in.

(vii) Good qualities of loudspeakers

A loudspeaker should have the following good qualities—

Satisfactory sensitivity
Broad directivity
Low distortion over the whole frequency range
Smooth frequency response
Balanced response
Good transient response
Sufficient damping at the bass resonant frequency
Adequate power-handling capacity

Any loudspeaker is necessarily a compromise. It has been demonstrated by Olson (Ref. 133) that with cone type loudspeakers, increased sensitivity brought about by reducing the thickness of the cone material is accompanied by increased distortion, ragged frequency response, bad transient response, and increased " break up " effects in the cone. On the other hand, the power output for the same distortion is many times greater for the loudspeaker having the more massive vibrating system.

If increased sensitivity is required, this may be achieved, within limits, by increasing the flux density. High flux density has the additional valuable property of increasing the damping.

A prominent peak in the response characteristic at the bass resonant frequency will increase the apparent sensitivity of the loudspeaker, particularly with small speakers having the bass resonance above 150 c/s. Similarly, prominent peaks in the response characteristic above 1000 c/s will also increase the apparent sensitivity, although both are features to be avoided for good fidelity.

When loudspeakers are compared in listening tests, the sound outputs should be adjusted to the same level before a comparison is made, because sensitivity is usually of secondary importance. Refs. 150, 155, 183.

(viii) Loudspeaker grilles

If the grille cloth is not to reduce sound pressure, its resistance must be small compared with the radiation resistance of the loudspeaker. The most suitable type of cloth is one which is very loosely woven and which has hard threads—cotton or plastic. Fuzzy threads increase the resistance of the cloth. Ref. 180.

SECTION 2 : CHARACTERISTICS OF MOVING-COIL CONE LOUDSPEAKERS

(i) Rigid (piston) cone in an infinite flat baffle   (ii) Practical cones   (iii) Special constructions for wide frequency range   (iv) Impedance and phase angle   (v) Frequency response   (vi) Efficiency   (vii) Directional characteristics   (viii) Field magnet (ix) Hum bucking coil   (x) Damping.

(i) Rigid (piston) cone in an infinite flat baffle

A rigid cone is one which moves forwards and backwards like a piston—at low frequencies a loudspeaker cone approaches this ideal. The air acts as a load on the piston at all frequencies when it is set in an infinite flat baffle. The **radiation resistance per unit area** of the piston is approximately constant for frequencies above

$$f_1 = 8120/\text{diameter of piston in inches} \tag{1}$$

In tabular form :

Diameter of piston	12	10	8	6	4	inches
Frequency f_1	677	812	1015	1353	2030	c/s.

Below frequency f_1 the radiation resistance is approximately proportional to the square of the frequency.

For amplitude of cone movement see page 834.

(ii) Practical cones

In practical cones the voice coil is mounted towards the apex of the cone and the driving force is transmitted through the cone. The cone must be a compromise between strength and lightness. A cone for reproducing the low audio frequencies should be fairly rigid, and should have a large diameter. If such a loudspeaker is tested at high audio frequencies above 1000 c/s it will be found to have a poor response owing to :

1. Large mass, causing poor efficiency at the higher frequencies.

2. The vibrations tend to take the form of waves radiating from the voice-coil outwards through the cone. At the higher frequencies the outer portion of the cone may be out of phase with the inner portion, or there may be a phase difference greater or less than 180°.

Cones for reproducing a wide frequency range often are designed with corrugations to reduce the effective vibrating area at the higher frequencies. Other designers achieve a somewhat similar result by a suitable choice of cone material together with changes in the thickness or compliance of the material. It is desirable that the material for the cone be no harder than is necessary to maintain response up to the highest frequency required, in order to give good reproduction of complex tones and transients.

The cone and voice-coil assembly is usually mounted at two points—an annular " surround " and a " spider " or centre mounting. Each of these introduces a restoring force which should increase linearly with the displacement. Actual mounting systems introduce a non-linear restoring force which usually increases rapidly when the displacement is fairly large. The suspension has two characteristics directly affecting the performance of the loudspeaker—compliance and mechanical resistance. Compliance is the inverse of stiffness (e.g. the stiffness of a spring), and controls the frequency of the bass resonance.   The mechanical resistance of the suspension affects the Q of the vibrating system, but usually at the bass resonant frequency the damping due to the suspension is much less than that due to the output resistance of the amplifier, except when a high output resistance is used.   There appears to be an optimum relationship between the stiffness of the cone and the compliance of the rim to minimize peaks and troughs in the 1500 to 3500 c/s region.

Non-linear suspensions are sometimes used with the object of preventing damage to the cone through overloading.   This procedure causes serious distortion at high amplitudes, and the quoted value of " maximum power output " is that capable of being handled without the loudspeaker suffering damage.   Loudspeakers for good fidelity should be capable of handling their full power without exceeding the limits of a fairly linear restoring force.   These should be rated on the maximum power (within certain frequency limits) which they are capable of handling without the distortion exceeding a predetermined value.

As one result of the suspension system, the whole cone assembly tends to resonate at the **bass resonant frequency.**   Below this frequency the cone is stiffness-controlled by the suspension and the output waveform tends to have flattened peaks resulting from the non-linear suspension.

Most loudspeaker manufacturers supply at least two alternative types of cone for use with each model—one with a nearly flat response up to 6000 c/s or higher (for broadly tuned or F-M receivers or amplifiers), the other with a response peaked in the 2000 to 4000 c/s region to compensate for sideband cutting in the receiver.   It is desirable, in each case, to have a choice between a cone for pentode output (without feedback), and one for triode output (or pentode with negative voltage feedback).

Cones are also available with differing bass resonant frequencies (from below 50 c/s to about 225 c/s depending on cone size and application) and differing high frequency limits (say 4500 to 8500 c/s for single cone units).   Most manufacturers also supply special cones to order.   It is of the utmost importance for the receiver or amplifier designer to investigate the complete range of available models and cones, and to make comparative listening tests between different makes, models and cones.

A loudspeaker using a metal cone is described in Ref. 225.

Amplitude and phase measurements on cones are described in Ref. 179.

References to cones : 1, 13, 15, 16, 19, 23, 40, 54 (Part 4), 84, 133, 179, 225.

(iii) Special constructions for wide frequency range

As mentioned above, a simple cone loudspeaker is not capable of giving satisfactory performance over a wide frequency range.   One solution is to use two or more loudspeakers with each designed to cover a limited frequency range or dual or triple integral units (see Sect. 5).   Another solution is to use some device to extend the effective frequency range of a single unit, as described below.

(A) **Multiple single coil, single cone** : It is possible to employ two or more identical fairly light loudspeakers with fairly small cones, all connected in parallel, series or series-parallel and correctly phased.   The number to be so connected depends on the maximum power to be handled at the lowest frequency.   All loudspeakers operate at all frequencies, and no filter is used.

(B) **Single coil, double cone** : This consists of an ordinary voice coil with two cones, the smaller being firmly fixed to the voice coil, and the larger cone being flexibly connected by means of a compliance corrugation.   At low frequencies both cones move together as a whole, but at high frequencies the small cone moves while the large cone remains stationary.

Refs. 13A, 14, 71.

(C) **Double coil, single cone** : This consists of a voice coil in two parts separated by a compliance, the smaller being fixed to a single corrugated cone.   The larger part of the voice coil is shunted by a capacitance to by-pass the higher frequencies.   At

low frequencies the whole assembly moves together, but at high frequencies the larger part of the voice coil remains stationary, and the smaller part drives the cone. The corrugations in the cone are designed to decrease the effective cone area at the higher frequencies.
Refs. 13A, 14, 57.

(D) **Double coil, double cone** : This consists of a light coil coupled to a small cone, the light coil being connected by a compliance to a heavy coil which is firmly fixed to a large cone. The heavy voice coil is shunted by a capacitance to by-pass the higher frequencies. At low frequencies both voice coils are operative, and the whole assembly moves together, but at high frequencies the small cone is driven by the light coil, and the large cone and heavy coil remain stationary. This is really equivalent to two separate loudspeakers.
Refs. 13A, 71.

(E) **Duode** : This has a light aluminium sleeve as the voice coil former covered on the outside by a layer of rubber, over which is wound the voice coil. At high frequencies, it is claimed, the relatively heavy voice coil winding remains stationary, the aluminium sleeve acting as a one turn coil receiving its power by transformer action and driving the cone. Refs. 242, 243.

(iv) **Impedance and phase angle**

The impedance characteristic (e.g. Fig. 20.2) is drawn from measurements made on an electrical impedance bridge, or equivalent method, with the loudspeaker mounted on a flat baffle. The nominal value is usually taken as the impedance at 400 c/s and is reasonably constant from about 200 c/s to about 600 c/s. At lower frequencies it rises to a peak at the bass resonant frequency and then falls rapidly to the resistance of the voice coil at still lower frequencies ; at higher frequencies it rises steadily throughout.

The impedance at the bass resonant frequency is a function of the damping of the cone suspension, the height of the peak decreasing as the damping is made heavier (see Ref. 13A, Fig. 6.31). It is also a function of the applied voltage, owing to non-linearity of the cone suspension.

If the loudspeaker is mounted other than on a flat baffle, the impedance characteristic at frequencies in the vicinity of the bass resonant frequency may be modified. For example with a vented baffle there are normally two impedance peaks —— Sect. 3(iv).

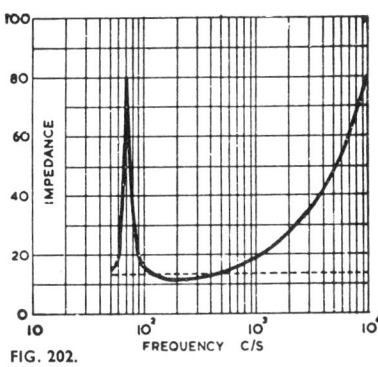

FIG. 202.

Fig. 20.2. Impedance versus frequency characteristic of a typical popular loudspeaker.

The impedance at 10 000 c/s is always greater than that at 400 c/s, the ratio varying between 1.1 and 10 times. A high ratio is most undesirable with pentode operation, with or without feedback, but is of no great consequence with triodes. The impedance at 10 000 c/s is increased by any leakage inductance in the transformer.

The equivalent electrical circuit which gives approximately the same impedance characteristic up to 400 c/s is shown in Fig. 20.3 (Refs. 15, 72). If adjustments are made in the values of L_0, R_0, R_1 and C_1, the circuit may be extended to higher frequencies. The bass resonant frequency is the parallel resonance of L_1 and C_1.

Fig. 20.3. Equivalent electrical circuit providing the same impedance characteristic up to 400 c/s as a moving-coil loudspeaker. Typical values are $L_0 =$ 0.0024 H, R_0 = 10.4 *ohms,* L_1 = 0.018 H, R_1 = 71.6 *ohms,* C_1 = 282 μF.

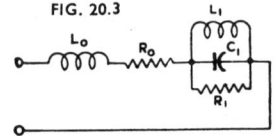

FIG. 20.3

The impedance so far discussed is called the " free " impedance ; there is also the " blocked " impedance which is measured with the voice coil prevented from moving. The " motional " impedance is found by a vector subtraction of the " blocked " from the " free " impedance. The difference between the free and blocked impedances is small except in the vicinity of the bass resonant frequency, where the motional impedance becomes the major portion of the free impedance (Refs 13, 41). Refs. 13, 13A, 14, 15, 26 (No. 2), 36, 37, 41, 47, 70.

Phase Angle : The phase angle versus frequency characteristic of a typical moving-coil loudspeaker is given in Fig. 20.4. The impedance is resistive at two frequencies only, being capacitive between the two points and inductive at lower and higher frequencies. The resistive and reactive components are shown in Fig. 20.5. Refs. 13, 13A, 37, 41, 61.

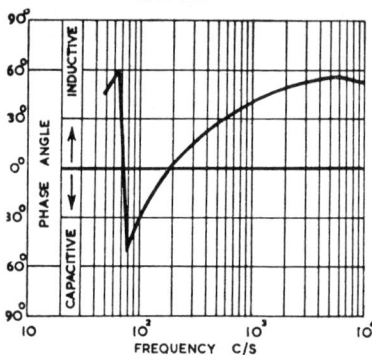

FIG. 20.4

Fig. 20.4. Phase angle versus frequency characteristic of a typical popular loudspeaker (same as Fig. 20.2).

(v) Frequency response

When used on an adequate flat baffle, a moving-coil loudspeaker only operates satisfactorily at and above the bass resonant frequency. The upper limit of frequency response rarely extends much beyond 6000 c/s for popular 10 or 12 inch single coil, single cone loudspeakers. Any further extension adds appreciably to the cost. If wide frequency range is required, there is a choice between a complex single unit— see (iii) above—and a multiple unit (Sect. 5). Refs. 13, 13A, 14, 15, 26 (No. 1), 42, 61, 70.

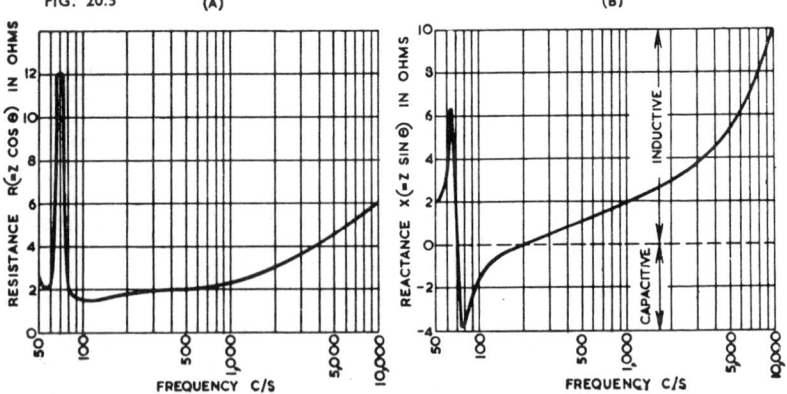

Fig. 20.5. (A) Resistive and (B) Reactive components of the impedance of a typical loudspeaker. Ref. 37.

(vi) Efficiency

Loudspeaker efficiency may be regarded either as the efficiency of the loudspeaker itself, or in the form of available power efficiency.

The efficiency of the loudspeaker itself is the ratio of the radiated acoustical power to the electrical input power, and is not influenced by the regulation of the power source. The efficiency of moving-coil cone loudspeakers on large flat baffles varies from about 2% to 10% at 400 c/s, depending on the design and on the flux density, 3% being a typical average. The maximum possible electro-acoustical efficiency is given by (Fig. 20.7)

$$\eta_{max} = R_2/(R_0 + R_2 + R_3) \qquad (2)$$

and the electro-mechanical efficiency is given by

$$\eta_{em} = (R_2 + R_3)/(R_0 + R_2 + R_3) \qquad (3)$$

where R_0 = blocked (d.c.) resistance of voice coil

R_2 = equivalent radiation resistance

and R_3 = equivalent frictional and eddy current loss resistance.

The electro-acoustical efficiency is always less than the electro-mechanical efficiency owing to the losses represented by R_3. It is possible to measure the electro-mechanical efficiency fairly readily because $(R_2 + R_3)$ is the motional resistance, and

$$(R_2 + R_3) = \text{free resistance } -R_0 \qquad (4)$$

but the accuracy is very poor except at frequencies in the region of the bass resonant frequency.

The available power efficiency is usually defined as the electrical power available to the load when the loudspeaker is replaced by a resistance equal to the rated load impedance with a constant voltage applied in series with the loudspeaker measurement source impedance (see pp. 812, 874-876). The electrical power W_{AS} available to the loudspeaker is

$$W_{AS} = E_G{}^2 R_{SR}/(R_{SG} + R_{SR})^2$$

where E_G = r.m.s. value of constant source voltage

R_{SR} = loudspeaker rating impedance

and R_{SG} = loudspeaker measurement source impedance.

If $R_{SG} = 40\%$ of R_{SR} (as in SE-103, see page 874), then the power available to the load will be half that when $R_{SG} = 0$. Maximum available power efficiency occurs at the bass resonant frequency if the baffle is sufficiently large and if the amplifier output resistance is fairly high, but if the output resistance is low, the available efficiency may be less at this frequency than at some higher frequency. The available power efficiency/frequency characteristic at low frequencies is affected by the type of baffle—see Sect. 3(iii) and (iv).

FIG. 20.7

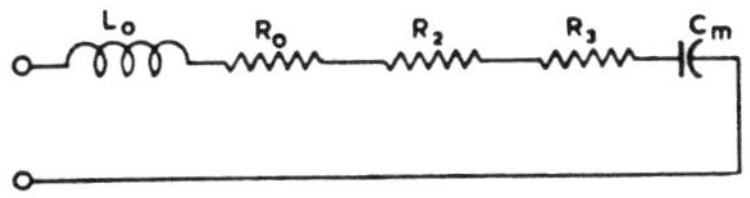

Fig. 20.7. *Equivalent circuit for power considerations, at frequencies above the bass resonance. The values of R_2, R_3 and C_m are not constant with change of frequency.*

Refs. to efficiency, 13, 13A, 14, 15, 36, 43, 54 (Part 3), 56, 70, 71, 74, 95, 121, 156.
References to equivalent circuits, 13, 13A, 15 (page 135), 41, 43, 54 (Part 3), 59, 75, 166, 188.

(vii) Directional characteristics

At low frequencies (up to 400 c/s for a 10 inch cone) a cone has only very slight directional characteristics. The angle of radiation from a flat disk 10 inches in diameter for a decrease of 4 db in sound pressure as compared with that on the axis is :

Frequency	678	1355	2710	5420	c/s
Angle (approx.)	180°	70°	35°	18°	

In practice it is possible to improve the angle of radiation by various expedients. One is to fit a deflector or diffusing lens (Ref. 120) so as to spread the higher frequency waves. Another is to use corrugations (compliances) in the cone so that only a small central area is vibrating at the higher frequencies. A further expedient is to use two co-axial loudspeakers and a frequency dividing network, the high frequency unit having some form of deflector. The ordinary popular loudspeaker is highly directional at high frequencies, but the best of the expensive designs are capable of 120° radiation up to 15 000 c/s (Refs. 36, 61). General references : 13, 14, 15, 54 (Part 2), 70.

(viii) Field magnet

Field magnets may be either electro-magnets or permanent magnets. In either case the requirement is to provide the greatest possible flux density in the air gap. Fidelity loudspeakers should have a uniform field up to the limits of movement of the voice-coil ; cheaper models are usually lacking in this respect. Alternatively the voice coil may be longer than the gap length (Ref. 15).

Permanent magnets have advantages in that they have lower heat dissipation, leading to a lower ambient temperature for the voice coil, and the voltage drop across a suitable choke is less than that across the field coil.

Increasing the flux density increases the efficiency, the output and the damping ; it reduces the rise of impedance at the bass resonant frequency. There is an optimum value of flux density to give the most nearly uniform frequency response (Ref. 56, Fig. 3).

(ix) Hum-bucking coil

Some electro-magnetic fields are fitted with a hum-bucking coil connected in series with the voice coil such that the field coil induces equal hum voltages in the voice coil and hum-bucking coil. By connecting the hum-bucking coil and the voice coil in series opposition, the hum due to ripple in the field is minimized.

(x) Damping

Damping on a loudspeaker is partly acoustical, partly frictional, and partly electro-magnetic. The electro-magnetic damping is a function of the effective plate resistance of the power amplifier.

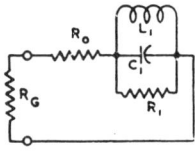

FIG. 20.8

Fig. 20.8. Equivalent circuit of loudspeaker at bass resonant frequency, showing damping effect of amplifier output resistance R_G.

With a direct radiator loudspeaker on a flat baffle the damping is slight, except in the vicinity of the bass resonant frequency, where the electro-magnetic damping may become quite large. The equivalent circuit at the bass resonant frequency is given by Fig. 20.8, which is identical with Fig. 20.3 except that L_0 has been omitted as being negligibly small, and R_G has been added to allow for the damping effect of the amplifier output resistance. R_1 is the damping resistance equivalent to the combined effect of friction and acoustical loading on the cone. R_0 is equal to the resistance of the voice coil plus that of the secondary of the transformer and leads plus that of the primary referred to the secondary. R_G is the effective plate resistance of the valve referred to the secondary. Any reduction in R_G below about one fifth of R_0 has only a very small effect on the Q of the tuned circuit.

At the bass resonant frequency, the Q of the tuned circuit is given by—

Case 1—R_G is infinite $Q = \omega_0 C_1 R_1$

Case 2—$R_G = 0$ $Q = \omega_0 C_1 \dfrac{R_1 R_0}{R_1 + R_0}$

Case 3—general $Q = \omega_0 C_1 \dfrac{R_1(R_0 + R_G)}{R_1 + R_0 + R_G}$

where ω_0 is angular velocity at the bass resonant frequency.

Critical damping is defined* as given by $Q = 0.5$.

Hence for critical damping,

$$R_G = \frac{R_1 + R_0 - 2\omega_0 C_1 R_1 R_0}{2\omega_0 C_1 R_1 - 1} \qquad (5)$$

*See " American Standard Definitions of Electrical Terms " A.S.A. C42—1941 (05.05.365) : also C. E. Crede " Vibration and Shock Isolation " John Wiley & Sons, Chapman & Hall Ltd. 1951, p. 171, and other textbooks on vibration. However some engineers take $Q = 1$ as critical damping.

From Fig. 20.8 it will be seen that critical damping is only possible with a positive value of R_G when R_0 is not too high, that is to say with loudspeakers having fairly high efficiencies, since the values of L_1 and C_1 are functions of the flux density. The damping of any loudspeaker may be increased by increasing the flux density.

In general, there is no advantage in using greater than critical damping.

With enclosed cabinet loudspeakers as a result of the effect on frequency response, the value of Q should not fall below 1, and the most desirable all-round condition appears to be with Q from 1.0 to 1.5—see Sect. 3(iii).

Alternatively, if the impedance is carefully measured and the impedance curve plotted over the region of the bass resonant frequency, the frequencies of the two half-power points may be noted, and the Q of the loudspeaker may be computed from the relation

$$Q \approx \left(\frac{f_0}{2\Delta f}\right)\left(\frac{R_0 + R_G}{Z_m + R_G}\right) \qquad (6)$$

where f_0 = bass resonant frequency
  $2\Delta f$ = band width at half power point (3 db below maximum impedance)
  R_0 = voice coil resistance
  Z_m = maximum value of the impedance (i.e. at f_0)
and R_G = effective plate resistance, referred to the secondary.

The measurements of frequency during this test are somewhat critical. The article by Preisman (Ref. 146) gives further details, although it does not derive eqn. (6) in this form.

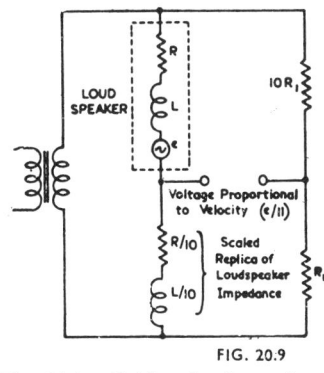

FIG. 20.9

Fig. 20.9. Bridge circuit to give a voltage proportional to the velocity of the cone, to be fed back to the input. (Ref. 184).

The majority of direct radiator loudspeakers have values of Q between 8 and 18 with a high valve plate resistance, although this may be decreased very considerably with a low effective plate resistance. Tests have been carried out to demonstrate the effect of a low plate resistance on the damping of a loudspeaker at the bass resonant frequency (Refs. 13A Fig. 6.30, 126, 132).

While there is no doubt that it is desirable to achieve a near approach to critical damping, the results are very much masked by the effects of the listening room which behaves as a resonant structure. In one case a room showed 8 resonances below 100 c/s, with Q values of the order of 12 to 15, and all above 8 (Ref. 155).

There are some loudspeakers in which the efficiency is too low to permit sufficient damping by reduction of the effective plate resistance. In these cases improvement in the electrical damping is possible by the use of positive current feedback combined with negative voltage feedback which can make the effective plate resistance zero or negative (Refs. 157, 158, 200, 201).

Alternatively, the e.m.f. generated by the movement of the cone may be fed back degeneratively to the input of the amplifier. One method of achieving this result is to arrange a special bridge circuit (Fig. 20.9) to provide the voltage to be fed back. The amplifier must have good characteristics with very low phase shift. This has been used satisfactorily over a frequency range from 10 to 1000 c/s, using a standard 12 inch loudspeaker, whose coil velocity was made proportional to input amplitude and completely independent of frequency (Refs. 184, 186, 148B, 191, 198).

Another method is to wind a separate feedback coil of very fine wire over the existing voice coil in a conventional loudspeaker. The voltage induced in this coil by the motion of the voice coil is a pure motional voltage at most frequencies. This voltage may be used as a feedback voltage to increase the damping and to reduce distortion arising from non-linearity of the cone suspension and from fringing of the magnetic field.

At very high audio frequencies the mutual inductance between the driving voice coil and the feedback coil produces in the feedback voltage a component which is dependent on the induction between the coils rather than on the motion alone. This difficulty is overcome by incorporating additional mutual inductance equal in value but of opposite sign, between the voice coil and feedback circuits at a point external to the magnetic field (Ref. 148).

SECTION 3 : BAFFLES AND ENCLOSURES FOR DIRECT-RADIATOR LOUDSPEAKERS

(i) Flat baffles (ii) Open back cabinets (iii) Enclosed cabinet loudspeakers (iv) Acoustical phase inverter (" vented baffle ") (v) Acoustical labyrinth loudspeakers (vi) The R-J loudspeaker (vii) Design of exterior of cabinet.

(i) Flat baffles

The baffle is intended to prevent the escape of air pressure from the front to the back of the cone, which is out of phase at low frequencies. In many theoretical calculations an infinite flat baffle is assumed, but in practice a baffle is only made just large enough to produce the desired results. When the air distance around the baffle from front to back of the cone is equal to the wavelength, there is a dip in the response. For this reason, baffles are frequently made of irregular shape or the loudspeaker is not mounted in the centre, the object being to produce air distances from front to back of the cone which vary in a ratio at least 5 : 3. For example, with a 4 ft. square baffle and a 10 in. cone, the centre of the cone may be 18 inches from each of two adjacent sides.

The size of a square baffle, with the loudspeaker mounted off-centre, for desired bass response is given approximately by the following table for a 10 inch diameter cone (Ref. 13A) :

| Minimum frequency | 300 | 170 | 80 | c/s |
| Side of square | 2 | 4 | 8 | feet |

Below this minimum frequency, and below the loudspeaker bass resonant frequency, the response falls off at the rate of 18 db/octave.

The best flat baffle is provided by mounting the loudspeaker in a hole in the wall of a room, with the rear of the cone radiating into another room.

In a very large flat baffle, a loudspeaker which has a sufficiently high Q will have a peak in the response characteristic at, or somewhat above, the bass resonant frequency ; below the peak frequency the attenuation is at the rate of 12 db/octave. Refs. 13, 13A, 14, 15 (Part 2), 75, 147, 151, 177.

(ii) Open back cabinets

An open back cabinet, as commonly used with radio receivers of the console, table or mantel type, has a resonance in the enclosure to the rear of the cone. This resonance causes a peak in the response characteristic at a frequency which is mainly a function of the cabinet although also influenced by the loudspeaker characteristics. For example, a cabinet 2 feet × 2 feet × 8 inches depth gave peak response at 180 c/s with a loudspeaker having a bass resonant frequency of 20 c/s, while the same cabinet gave a peak response at 110 c/s with a loudspeaker having a bass resonant frequency of 100 c/s. (Ref. 13A, Figs. 6.19B and 6.20B). The height of the peak is about 3 to 6 db for shallow cabinets or about 6 to 10 db for deep cabinets.

Open back cabinets are undesirable for good fidelity. If unavoidable, they should be as shallow as possible, with the minimum of acoustical obstruction, particularly at the back of the cabinet. Open back cabinets should be placed at least 6 inches out from the wall.

References to open back cabinets : 13A, 14, 147, 151, 188.

(iii) Enclosed cabinet loudspeakers

An enclosed loudspeaker is one which is totally enclosed so that there can be no interference between the front and back of the cone. There is no critical value for the volume of the enclosure but a large volume is desirable because it reduces the rise in resonant frequency above that on a large flat baffle. The increase in resonant frequency for Jensen speakers is shown in Fig. 20.10 and may be taken as fairly typical. There is practically no non-linear distortion caused by the suspension, since this only contributes a small part of the total stiffness.

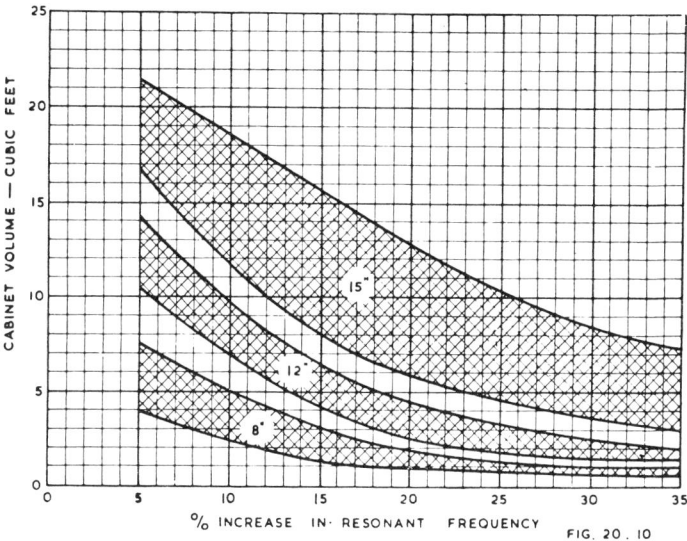

Fig. 20.10.

Fig. 20.10. Increase in resonant frequency of totally enclosed cabinet loudspeaker versus cabinet volume. For each nominal speaker size, the upper limit of the shaded area corresponds to speakers with highest compliance, while the lower limit corresponds to those with least compliance. (Ref. 151).

The enclosure should be adequately damped by at least ½ inch thickness of damping material all over (or double this thickness on one of each pair of parallel surfaces), although this is insufficient to eliminate standing waves at low frequencies.

The ineffectiveness of sound absorbent linings at low frequencies is due to the fact that all the absorbent material lies within a small fraction of a wavelength of the inner wall, which is a velocity node. Without motion of the air particles, or of the material itself, acoustical power cannot be absorbed and dissipated. Shorter (Ref. 135) has examined the problem of standing waves, and puts forward the method of damping by partitions, concentrating the absorbing material (¼ inch carpet felt) into one or two partitions strategically placed across the cabinet. Those sections of the cabinet which are separated from the cone by one or more sound absorbing partitions receive little sound at the high frequencies and therefore require very little acoustical treatment. The loudspeaker back e.m.f. may be measured by a bridge circuit and should show a single peak without subsidiary peaks over the range from 30 to 200 c/s.

It is desirable for the enclosure to have appreciably different values of linear dimensions—typical ratios are 1 : 1.5 : 2.7. However, the longest dimension should be less than $\frac{1}{4}$ of the wavelength at the lowest working frequency. The enclosure should have rigid walls—timber at least $\frac{1}{2}$ inch thick, and braced where necessary.

One limitation of the enclosed cabinet loudspeaker is the bass attenuation which occurs with critical damping, although this only happens with high flux density and low amplifier output resistance—see below under Equivalent Circuit.

Some loudspeakers have been designed specially for use with enclosed cabinets. These usually have very low bass resonant frequencies, when measured on a flat baffle, so that when loaded by the enclosure, the resonant frequency is still sufficiently low.

The volume of the cabinet is given below for some examples :

Western Electric	755A	8 in.	70 c/s	2	cubic feet
,,	,,	756A	10 in.	65 c/s	2½	,, ,,
,,	,,	754A	12 in.	60 c/s	3	,, ,,
Goodmans "Axiom 80"		9½ in.*		3½	,, ,,
R.C.A. (Ref. 84)		peak at 80 c/s		1.5	,, ,,
Stromberg-Carlson		8 in.		1.7	,, ,,
		12 in.		3.9	,, ,,

Equivalent circuit

Fig. 20.11 is a simplified electrical equivalent circuit of the acoustical system of the loudspeaker (Ref. 135).

L_U represents cone mass + effect of radiation reactance.

R_U represents radiation resistance (which varies with frequency)†.

C_C represents acoustical capacitance of cabinet volume.

C_U represents equivalent capacitance of cone suspension.

R_S represents effect of electrical circuit of loudspeaker and driving amplifier reflected into acoustical circuit. The mechanical resistance of the cone suspension may be taken as being included with R_S.

E_S = constant voltage generator.

I_U = alternating air current produced by cone, which is proportional to cone velocity.

Now $R_S \propto \dfrac{B^2}{R_G + R_0}$ at low frequencies

where B = flux density in gap,

R_G = output impedance of amplifier referred to voice coil circuit,

and R_0 = resistance of voice coil.

The acoustical response of the loudspeaker is proportional to $I_U f$.

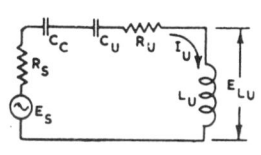

FIG. 20.11

Fig. 20.11. Equivalent electrical circuit of acoustical elements in an enclosed cabinet loudspeaker at low frequencies. (Ref. 135).

The voltage $E_{LU} = 2\pi f L_U I_U \propto I_U f \propto$ acoustical response. Therefore the variation with frequency of E_{LU}, for any one value of L_U, gives the frequency response of the loudspeaker. Over the frequency range for which the equivalent circuit is valid, and the wavelength is large compared with the size of the cone, the loudspeaker can be reduced—so far as frequency response is concerned—to a half-section high-pass filter working into open circuit. Small values of R_S, resulting from low flux density or high amplifier output impedance, give a resonance peak and bad transient response, while large values of R_S, corresponding to high flux density and low amplifier output impedance, can give a serious loss of bass. Fig. 20.12 shows curves for three values of R_S and hence for three values of Q, for a resonant frequency of 45 c/s. This indicates that values of Q less than about 1 result in serious attenuation of low frequencies.

---

*Resonance 17 c s on flat baffle.

†Note that R_U is small compared with other impedances in the circuit.

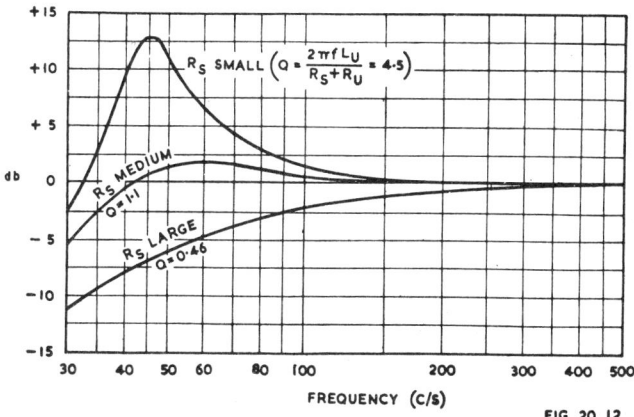

FIG.20.12

Fig. 20.12. Theoretical response of enclosed cabinet loudspeaker for various values of the effective damping resistance R_s. Resonance frequency 45 c/s. (Ref. 135).

References to enclosed cabinets : 13A, 29, 36, 80, 81, 84, 116, 116A, 135, 147, 151, 166, 168, 188.

(iv) Acoustical phase inverter (" vented baffle ")

Also known as a bass reflex baffle.

This has a vent* or duct in the front of the cabinet which augments the direct

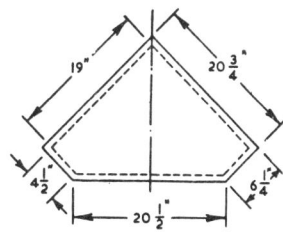

radiation from the cone at low frequencies (Fig. 20.13). The box should be at least partially lined with sound absorbent material to absorb the higher frequencies, but should not be too heavily damped at frequencies below about 150 c/s. The cabinet should be strongly made of heavy timber with adequate bracing to prevent vibration. Many cabinets which have been built for this purpose are partially ineffective as the

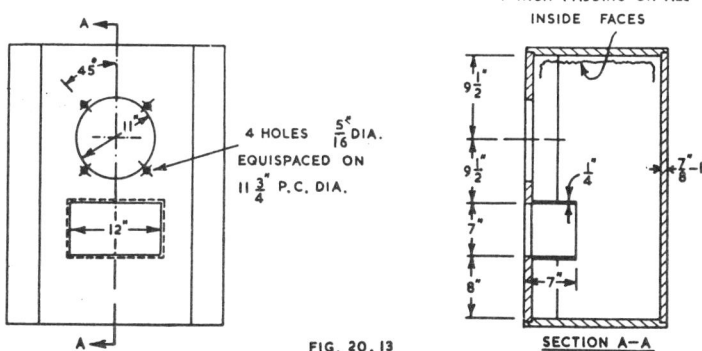

FIG. 20.13   SECTION A–A

Fig. 20.13. Typical corner cabinet with vented baffle, designed for a Goodmans Axiom 12 inch loudspeaker with a bass resonant frequency of 55 c/s, which can be used with a speaker resonant at 75 c/s by removing the tunnel. Cubic capacity about 8000 cu. ins. = 4.6 cu. ft. (Ref. 182).

*An alternative form uses several tuned resonators from the inside to the outside of the cabinet (Ref. 18).

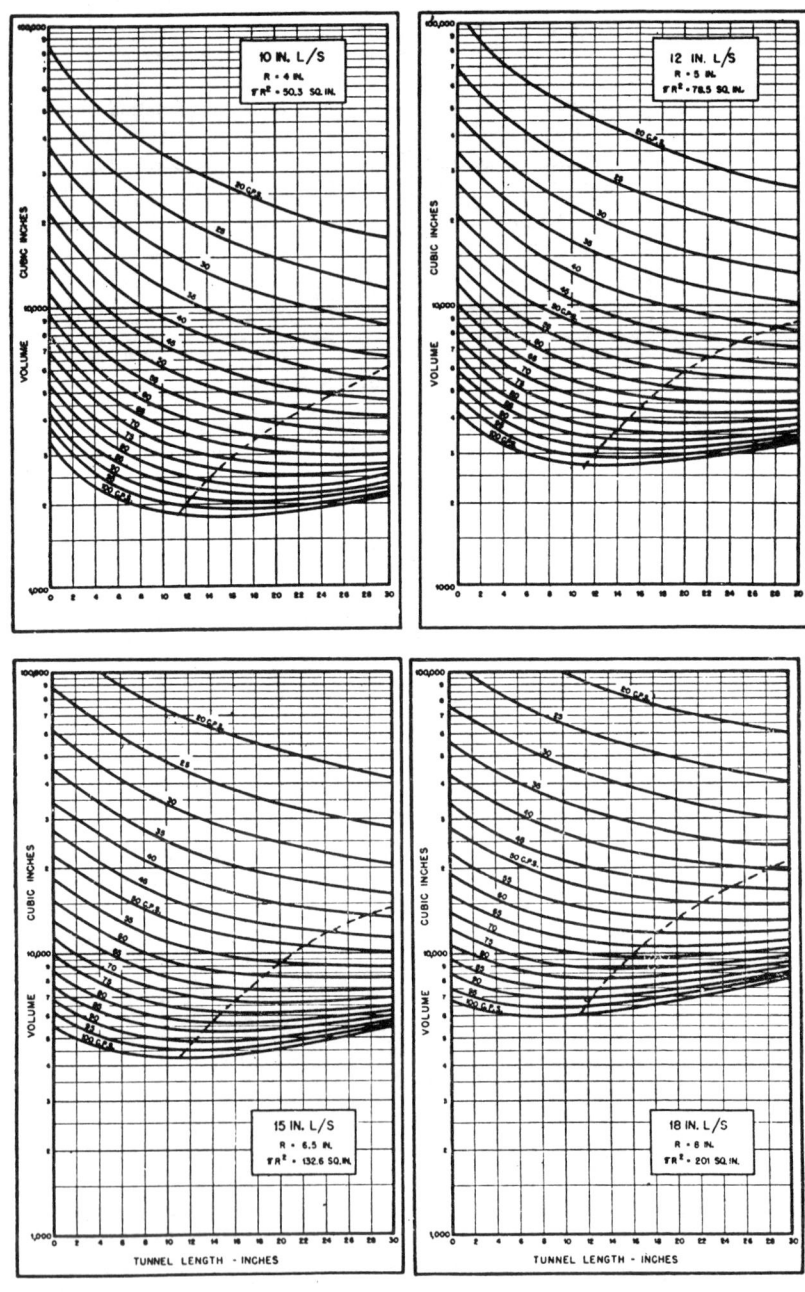

FIG. 20.14

Fig. 20.14. Curves showing tunnel length for 10, 12, 15 and 18 inch Goodmans loudspeakers with bass resonant frequencies from 20 to 100 c/s. The tunnel length should not exceed 1/12 of the wavelength of the speaker bass resonant frequency ; points to the right of the broken lines should not be used (Ref. 182).

result of wall vibration under the extremely arduous conditions of operation.   Other materials which have been used satisfactorily include brick and concrete.

Vented baffles are normally designed so that the loudspeaker bass resonant frequency is matched to the acoustical resonant frequency of the cabinet ; unmatched combinations are, however, also used.

(A) Matched vented baffles

The merits of the vented baffle, when correctly applied, include

(1) improved bass response,

(2) much reduced amplitude of movement of the cone at the resonant frequency, and hence reduced distortion at the same power output or increased power handling capacity for the same distortion,

(3) decreased peak electrical impedance,

(4) increased acoustical damping over a limited range of low frequencies (not including the vent resonance),

(5) increased radiation resistance and decreased reactance of the loudspeaker at low frequencies.

The vented baffle appears to be preferred by most engineers to the acoustical labyrinth, even apart from cost.

The vent area should normally be the same as the effective radiating area of the cone, and the curves in Fig. 20.14 are based on this relationship.   Thus the cross-sectional area of the tunnel is given by πR^2 where R is the radius of the piston equivalent to the speaker cone at low frequencies.   The tunnel may have any length from the thickness of the timber (i.e. when no tunnel, as such, is constructed) to a maximum of approximately one twelfth wavelength, indicated by the broken curves in Fig. 20.14 ; consequently the area to the right of the broken curves should not be used.

A reasonable length of tunnel enables a smaller enclosure to be used, but the space between the end of the tunnel and the rear of the cabinet should not be less than R. The volume occupied by the loudspeaker must be added to the volume derived from the curves to obtain the volume of the enclosure.   If the volume of the loudspeaker is not known, the following approximation may be used :

Volume displaced by loudspeaker :

Loudspeaker dia.	8	10	12	14	16	18	ins.
Volume (approx.)	250	400	650	1000	1600	2300	cu. ins.

Also add volume displaced by internal timber bracing.

Note that no allowance should be made for the thickness of the damping material.

The curves for the volume of the cabinet are derived from the relationship

$$V = \pi R^2 \left[\frac{1.84 \times 10^8}{\omega^2} \times \frac{1}{1.7R + L} + L \right] \tag{1}$$

where V = volume in cubic inches,

R = radius of equivalent piston in inches,

L = length of tunnel in inches

and ω = $2\pi \times$ frequency of vent resonance.

When correctly designed a vented baffle loudspeaker has two impedance peaks, one above and one below the vent resonance, while the same loudspeaker on a flat baffle has only one impedance peak—see Fig. 20.15.   In all cases where optimum performance is desired, the electrical impedance characteristic below 150 c/s should be measured using a constant current source and a voltmeter.   The ratio of the two peak frequencies should not exceed about 2.4 : 1 nor be less than about 1.5 : 1 and any subsidiary peaks caused by standing waves should be removed by the addition of damping material (as for the enclosed cabinet above ; see also Fig. 20.17).

Tests for acoustical output should not be made until the cabinet dimensions and damping have been adjusted to give the correct impedance characteristic.   In general, with a vented baffle the bass range over an octave or more is increased by several decibels over either a very large baffle or an enclosed cabinet loudspeaker.

The power gain may be greater than 3 db since the air loading of the cone may be increased appreciably over the lower octave of the frequency range.   For about one third of an octave above and below the resonant frequency of the system, the greater

FIG. 20.15

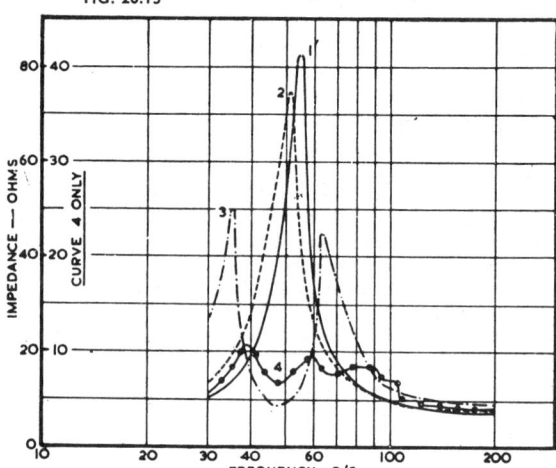

Fig. 20.15. *Impedance versus frequency characteristics of a* 12 *inch loudspeaker*
(1) *in free air* (2) *on a flat baffle and* (3) *in a vented baffle.* *Curve* (4) *shows two*
" *staggered*" *vented baffle loudspeakers in parallel* (*measured*).

part of the energy is radiated by the port.   Phase shift occurs suddenly at the resonant frequency, so that the radiation below this frequency is reduced, being the
vector difference between the two sources.

At the resonant frequency, the air in the vent or tunnel moves vigorously while the
acoustical impedance on the rear of the cone is resistive and reaches a maximum.

At the frequency of the vent resonance, the motion of the speech coil is so small that
no appreciable electrical damping can take place and there is no danger of bass attenuation such as may occur with the enclosed cabinet or flat baffle.   The energy dissipated
by mechanical resistances and by the radiation of sound is generally small, so that the
vent resonance on which the maintenance of the bass response depends, is quite lightly
damped.   For this reason, when good fidelity is required, the vent resonance frequency should not be greater than 60 c/s. (Ref. 135).

A **corner speaker cabinet** for 15 inch cones is described in Ref. 100, while one
for 12 inch cones is described in Ref. 105.   These both employ a vented baffle arrangement with the vents on both sides of the cabinet, between the cabinet and the walls.

Equivalent circuit (Ref. 135)

The Fig. 20.16 is a simplified electrical equivalent circuit
of the acoustical system and is based on Fig. 20.11 with the
addition of R_V representing the radiation resistance of the vent,
and L_V, the acoustical inductance of the vent plus the effects of
its radiation reactance.   Let f_V be the frequency of the vent
resonance (i.e. the resonance of L_V and C_C) and assume that
the cone resonant frequency is the same.   Below f_V, the series
combination $L_U C_U$ appears as a capacitance and the parallel
combination $L_V C_C$ as an inductance, and the overall effect is to
produce a series resonance at some frequency f_1 in this region.
At frequencies above f_V, $L_U C_U$ appears inductive and $L_V C_C$
capacitive, and a second series resonance appears at some frequency f_2.   The resonances at f_1 and f_2 are responsible for the
characteristic double hump in the electrical impedance frequency characteristic of a vented baffle and are subject to the
circuit damping represented by R_S.

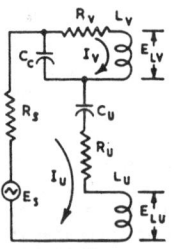

FIG. 20.16

Fig. 20.16. *Equiva-*
lent electrical circuit
of vented baffle
loud-speaker at low
frequencies. (*Ref.*
135).

I_U and I_V represent the alternating air current flow in the cone and vent respectively. I_U originates at the back of the cone and I_V must therefore be reversed in phase with respect to I_U if the acoustical output of the vent is to reinforce that from the front of the cone. At f_V, the frequency of the vent resonance, I_U and I_V are approximately in quadrature ; above f_V the desired reinforcing condition is approached, but below f_V the two air currents oppose one another.

The performance of the system is determined by the value of f_V, which is normally made equal to the loudspeaker bass resonant frequency, and also by the ratio L_V/C_C. With a large cabinet and large vent (C_C large, L_V small) the response can rise to a peak in the bass, while a small cabinet and a small vent (C_C small, L_V large) will give much the same effect as a completely closed cabinet, i.e. there may be bass loss.

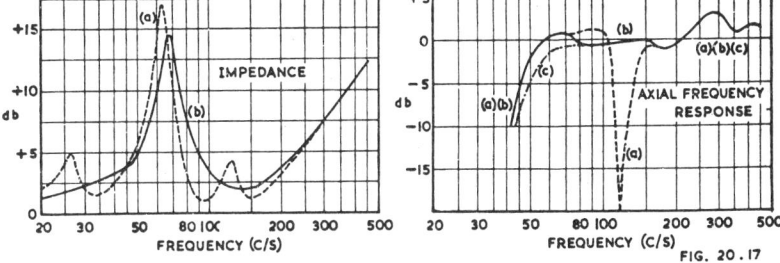

Fig. 20.17. *Free-air axial response of a vented baffle loudspeaker and corresponding impedance characteristics (a) conventional damping (b) optimum damping (c) excessive damping of partition. (Ref. 135.)*

Between these two extremes, there will be a pair of values for L_V and C_C which gives the best approximation to a flat frequency response. The cabinet volume required for this last condition depends on the efficiency of the loudspeaker unit and the output impedance of the amplifier. Fig. 20.17 shows the response and impedance characteristics of a 15 inch unit in a 10.5 cu. ft. cabinet. Curves (a) are for the cabinet lined with kapok quilt and carpet felt, with internal space left free ; a serious internal resonance shows at about 120 c/s. Curves (b) show how this resonance was suppressed by introducing into the cabinet a partition having a window of area about 170 sq. ins. covered with three layers of $\frac{1}{4}$ in. carpet felt. This damping does not seriously affect the frequency response but it suppresses the lower frequency impedance hump. Curve (c) shows the effect of too many layers of felt on the window. In this case the design aimed at a nearly level frequency characteristic down to 50 c/s.

See also Ref. 211 for other form of equivalent circuit.

(B) Unmatched vented baffles

Unmatched vented baffles preferably incorporate a loudspeaker having a very low bass resonant frequency. The cabinet is designed quite independently of the loudspeaker and may, if desired, be fitted with an adjustable vent. In the latter case the maximum vent area may be somewhat greater than the loudspeaker cone area, while in the other extreme the vent may be completely closed. Using a cabinet 2 ft. × 2 ft. × 18 ins. (volume 6 cu. ft.) and a loudspeaker with a bass resonant frequency of 30 c/s, the frequency of maximum acoustical pressure increases from 40 to 60 thence to 75 c/s as the vent area is varied from small to medium to large respectively. The height of the maximum response above that at 400 c/s rises from 4 db (small vent) to 6 db (medium vent) thence to 8 db (large vent) (Ref. 13A).

In other cases a loudspeaker is used which has a bass resonant frequency of conventional value, but here also the vent frequency is higher than the loudspeaker bass resonant frequency. This permits the use of a smaller cabinet while still retaining some of the good features of a matched vented baffle, although it is not so good on transients. In all cases an adjustable vent area is desirable. Tunnels are not generally used, since they give no flexibility of adjustment.

The vent frequency should not exceed (say) 90 c/s, otherwise the tone will be poor. This limits the reduction in cabinet volume to about $2\frac{1}{2}$ cubic feet.

The shape of the cabinet, as well as its volume, affect its frequency response and impedance characteristics. For example, a change from 30 ins. × 15 ins. × 8 ins. to an equivalent volume 15 ins. × 15 ins. × 16 ins. caused an increase from 45 to 58 c/s in the lower impedance peak and from 75 to 110 c/s in the higher peak (Ref. 117). In the cubic shape of cabinet the addition of an internal partition between loudspeaker and vent, extending approximately half-way from front to back, causes a reduction in the frequencies of both impedance peaks, and is claimed to give improved results on both speech and music (Ref. 117).

Careful attention to damping, along the lines given above for both enclosed cabinets and matched vented baffles, would be well repaid.

(c) **Special types of vented baffle loudspeakers**—see Supplement.

Refs. to vented baffles : 4, 9, 12, 13, 13A, 29, 36, 76 (Part 2), 85, 96, 113, 116, 116A, 117, 118, 135, 143, 144, 147, 150, 151, 166, 168, 175, 182, 188, 202, 209, 211, 225, 229, 230.

(v) Acoustical labyrinth loudspeaker

The acoustical labyrinth gives a performance somewhat similar to that of a vented baffle. The rear of the cone drives a long folded tube, lined with sound absorbing materials, the mouth of which opens in front of the cabinet (Fig. 20.18). The length of the tube is approximately 7 feet (measured on the centre-line) for nearly linear response down to 70, c/s. The loudspeaker bass resonance loaded by the labyrinth is preferably at a frequency at which the wavelength is four times the length of the tube ; in the example this is 40 c/s. If this latter condition is not fulfilled, the frequency response will not be linear. The loudspeaker resonance frequency in the example was reduced from 50 to 40 c/s by the loading of the labyrinth. This is the only form of baffle which reduces the bass resonant frequency of a loudspeaker. The rise of impedance from 400 c/s down to the bass resonant frequency is reduced considerably by the acoustical labyrinth—in one case the ratio was reduced from 10 : 1 to 4.3 : 1 (Ref. 27).

References 13, 14, 27, 28, 188, 204, 225 Part 2.

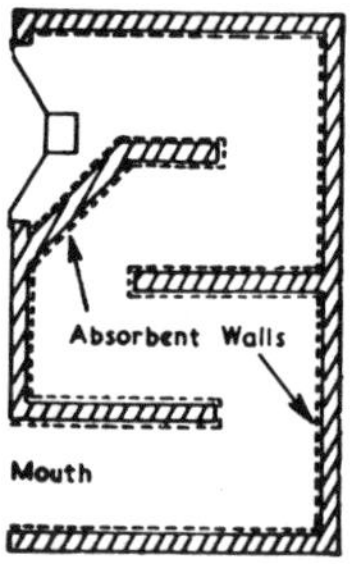

Fig. 20.18. Sectional view of acoustical labyrinth loudspeaker.

(vi) The R-J loudspeaker

The R-J loudspeaker has a particularly compact bass unit, to which any desired tweeter can be added. In one design a 15 inch woofer unit is mounted in a cube with 18 inch sides and fundamental bass reproduction is claimed down to 20 c/s. Only the forward radiation is used, but both sides of the cone are loaded. The back of the speaker is completely enclosed within a small stiff cavity. The front of the woofer works into a carefully designed rectangular duct, and the sound issues from a slot extending across the base of the enclosure.

Refs. 189, 190, 219, 234. See also Supplement.

(vii) Design of exterior of cabinet

The sharp corners on the usual more-or-less box shaped cabinet, particularly those of the side in which the loudspeaker is mounted, produce diffraction effects causing a sequence of peaks and valleys up to ± 5 db in the response characteristic. The best shaped enclosure is a complete sphere, while the worst is a cube with the loud-speaker in the centre of one side. A rectangular parallelepiped (box shaped) with the loudspeaker closer to one of the short sides than to the other is an im-

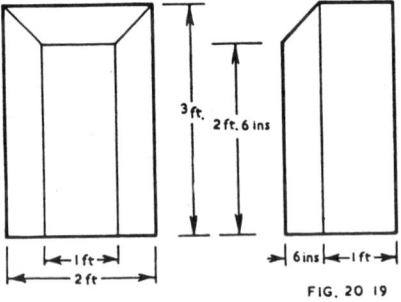

Fig. 20.19. Loudspeaker enclosure to minimize diffraction effects (after Ref. 193).

provement over the symmetrical cube, but still far from the ideal. A very close approach to the ideal is given by a rectangular truncated pyramid mounted on a rectangular parallelepiped (Fig. 20.19) Ref. 193.

In all cases it is desirable that the edge of the cone should be flush with the front of the cabinet.

SECTION 4 : HORN LOUDSPEAKERS

(i) Introduction (ii) Conical horns (iii) Exponential horns (iv) Hyperbolic exponential horns (v) Horn loudspeakers—general (vi) Folded horn loudspeakers (vii) High frequency horns (viii) Combination horn and phase inverter loudspeakers for personal radio receivers (ix) Materials for making horns.

(i) Introduction

A horn is used in conjunction with a diaphragm or cone loudspeaker for the purposes of increasing the acoustical loading on the diaphragm (over a limited frequency range) and thereby increasing the efficiency and reducing non-linear distortion. With a horn, inside the useful frequency limits, the movement of the diaphragm is much less and the acoustical damping is much greater than on a flat baffle, for the same acoustical power output. Thus with a horn, a smaller diaphragm can radiate a given acoustical power.

A horn is essentially a device which transforms acoustical energy at high pressure and low velocity to energy at low pressure and high velocity.

Horns are of various shapes—conical, parabolic, hypex and exponential, but the exponential is most widely used.

A well designed and well executed exponential or hypex horn loudspeaker is capable of giving a flatter frequency output characteristic with less distortion than any other form of loudspeaker.

References 13A, 14, 15, 59, 82, 142, 177, 188.

(ii) Conical horns

A conical horn is one having straight sides ; it functions in a manner similar to an exponential horn but its throat resistance is less than that of an equivalent exponential horn, except at high frequencies. Conical horns are sometimes used with cone type loudspeakers to give directional characteristics ; the angle of propagation is then approximately equal to the angle of the horn.

Conical horns are sometimes used, for economy, as the first section at the throat end, with an exponential horn forming a second section at the mouth end. This device is frequently used with folded horns. Sometimes an approach to an exponential horn is made with a number of conical sections, each with a different degree of taper. All such compromises lead to inferior performance.

References 13A, 14, 15, 142, 175.

(iii) Exponential horns

In an exponential horn, the cross-sectional area (S) at any point distant x feet along the axis is given by

$$S = S_0 \epsilon^{mz} \tag{1}$$

where S_0 = cross-sectional area at throat (sq. feet)
ϵ = 2.71828 (Naperian base)
and m = flaring constant.

Equation (1) may be put into the form
$$S/S_0 = \epsilon^{mz}$$
Therefore $mx = \log_\epsilon (S/S_0) = \log_{10} (S/S_0) \times 2.3026$.
If $(S/S_0) = 2$ then $mx = 0.3010 \times 2.3026 = 0.6931$.

If $(S/S_0) = 4$ then $mx = 0.6020 \times 2.3026 = 2 \times 0.6931$.

We may therefore deduce that

(1) the cross-sectional area doubles itself each time the distance along the axis is increased by $0.6931/m$.

(2) the value of m (in inverse feet) is equal to 0.6931 divided by the distance along the axis in feet for the cross-sectional area to double itself.

This is shown more clearly below where $l' = 0.6931/m$:

Distance along axis	Cross-sectional area
$x = 0$ (at throat)	S_0
$x = l'$	$S = 2S_0$
$x = 2l'$	$S = 4S_0$
$x = 3l'$	$S = 8S_0$ etc.

The **corresponding diameter and side of circular and square cross sections** are :

Distance along axis	Diameter	Side of square
$x = 0$ (at throat)	d_0	a_0
$x = l'$	$1.414d_0$	$1.414a_0$
$x = 2l'$	$2d_0$	$2a_0$
$x = 3l'$	$2.818d_0$	$2.818a_0$
$x = 4l'$	$4d_0$	$4a_0$

The flare cut-off frequency for an infinite exponential horn is given by

$$f_0 = m \times 89.5 \qquad (2a)$$

while the corresponding cut-off wavelength is

$$\lambda_0 = 12.6/m \qquad (2b)$$

where m is expressed in inverse feet.

FIG. 20. 20

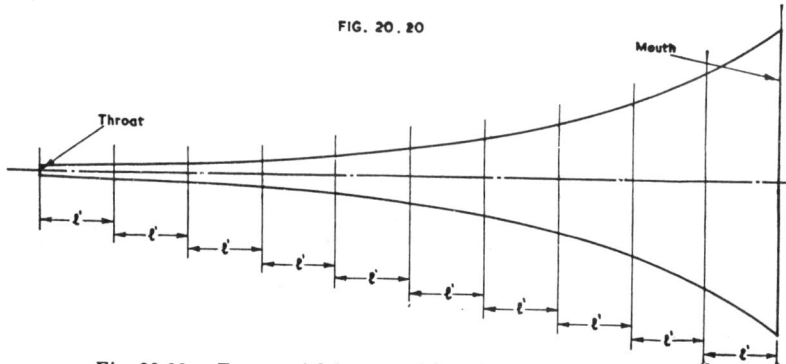

Fig. 20.20. *Exponential horn of either circular or square cross-section.*

Design on basis of minimum useful frequency

The characteristics below are given as a function of f', the lower frequency limit for satisfactory horn loading. The value of f' is arbitrarily taken as 1.2 times the flare cut-off frequency for an infinite exponential horn ; it is, of course, influenced also by the dimensions of the mouth and the length of the horn.

The value of m, the **flaring constant**, should then be equal to or less than $10.5/\lambda'$

$$(2c)$$

where λ' = wavelength (in feet) of the minimum useful frequency (f') for satisfactory horn loading.

Minimum useful frequency (f')	50	100	150	200	c/s
Maximum useful wavelength (λ')	22.6	11.3	7.52	5.65	feet
Maximum value of m	0.47	0.93	1.40	1.85	inverse feet

The **diameter of the mouth** should preferably not be less than one third of the maximum useful wavelength, while the effect of resonances at the low frequencies may be made negligible by increasing the mouth diameter to two thirds of the maximum useful wavelength (λ'). Taking the smaller mouth diameter, and referring to the minimum useful frequency :

$$\text{Minimum circumference} = 1.05\lambda' = 1200/f' \qquad (3)$$

Equation (3) may be applied to any shape of mouth—see table below :

Circumference (min.)	24	12	8	6	feet
Maximum useful wavelength (λ')	22.6	11.3	7.52	5.65	feet
Minimum useful frequency (f')	50	100	150	200	c/s

Horns with circular cross-section

In this case the equation is

$$d = d_0 \epsilon^{mx/2} \qquad (4)$$

where d = diameter in feet
and d_0 = throat diameter in feet.

Diameter at mouth : $\mathrm{d}_m = d_0 \epsilon^{ml/2}$ (5)
or $\mathrm{d}_m = d_0$ antilog$_{10}$ ($ml/4.60$ (6)

Length : $l = (4.605/m) \log_{10} (\mathrm{d}_m/d_0$ (7)

For chart see Ref. 39 Chart III.

Flaring constant : $m = (4.605/l) \log_{10} (\mathrm{d}_m/d_0)$ (8)

If the minimum useful frequency is taken as f' where $f' = 1.2f_0$, and if the minimum diameter of mouth is taken as one third of the wavelength at frequency f', then m may have its maximum value of $10.5/\lambda'$ and

minimum diameter of mouth $= \lambda'/3 = 380/f'$ (9)

minimum length $= (500/f') \log_{10} (380/f'd_0)$ (10)

where λ' = wavelength in feet at frequency f'
$f' = 1.2f_0 = 1.2 \times$ flare cut-off frequency
d_0 = diameter at throat in feet

and all dimensions are in feet.

Horns with square cross-section

These follow the same laws as horns of circular cross-section having diameters equal to the sides of the square, but the length and mouth dimensions are slightly less for the same minimum useful frequency f' :—

minimum side of mouth $= 300/f'$ (11)

minimum length* $= (500/f_0) \log_{10} (300/f'a_0)$ (12)

where a_0 = length of side at throat.

For cutting the side of a square horn from a sheet, see Ref. 77.

General references to horn dimensions : 39, 46, 82.

References to theory of exponential horns : 13A, 142, 175, 188.

(iv) Hyperbolic exponential horns (" hypex ")

A hyperbolic exponential horn is one which follows the law

$$S = S_0[\cosh mx + T \sinh mx]^2 \qquad (13)$$

where S = cross-sectional area at distance x along the axis in sq. ft. (or in any other convenient units)
S_0 = cross-sectional area at throat in sq. ft.
m = flaring constant
x = distance along the axis in feet
and T = shape parameter (T may have any value from zero to infinity).

*For chart see Ref. 46 Figs. 1 and 2

The cut-off frequency is the same as for an exponential horn having the same flaring constant—see eqn. (2a). When $T = 1$, the horn is exponential. When $T = 1/(mx_0)$ and m is allowed to go to zero, the horn is conical. When $T = 0$ the horn is hyperbolic cosine or catenoidal (Refs. 13A, 123, 140, 142, 188). The name " hypex " is usually applied when T is greater than zero and less than unity.

The value of T in " hypex " horns is usually between 0.5 and 0.7, and within these limits the throat resistance of an infinite horn is more nearly constant than that of an exponential horn, at frequencies slightly above the cut-off frequency. These comparisons are for constant throat, mouth and length of horn ; under these conditions a " hypex " horn with T of the order of 0.6 has improved low frequency characteristics as compared with those of an exponential horn. Consequently, for equivalent performance, a " hypex " horn may be made more compact than an exponential horn. A further useful feature of the " hypex " characteristic is that it makes possible a gradual transition from conical, via " hypex " with varying T, to exponential (Ref. 123).

An analysis of the response peaks in finite hyperbolic horns, and design procedure for horns to have peaks at pre-determined frequencies, is given in Ref. 139.

(v) Horn loudspeakers—general
(A) Frequency limitations
The resistive component of the throat impedance drops rapidly below about 1.2 times the flare cut-off frequency. Some output is obtained with finite horns at, and even below, the flare cut-off frequency (Ref. 45), but at frequencies below the flare cut-off frequency there is a strong tendency towards the production of harmonics. Care should therefore be taken to eliminate from the amplifier any appreciable output power below about 1.2 times the flare cut-off frequency, unless the loudspeaker is known to be capable of handling such frequencies without damage or distortion.

The resonant frequency of the diaphragm should not be less than the flare cut-off frequency, and preferably not less than 1.2 times this value, in order to ensure sufficient loading at the resonant frequency.

The only known method for handling frequencies below the flare cut-off frequency of an exponential horn, with good fidelity, is the use of an enclosed air-chamber behind the diaphragm, resonant at a frequency in the vicinity of the flare cut-off frequency, as used with the Klipsch loudspeaker. This air-chamber is designed to provide a capacitive reactance approximately equal to the inductive reactance (inertance) of the horn at the flare cut-off frequency. The volume of the air-chamber is given by eqn. (15) in Sect. 4(vi) below—see also Ref. 31.

(B) Diaphragm and throat.
Some horns are designed with the throat area the same as the diaphragm area. Greater acoustical loading is obtained by the use of a larger diaphragm and sound chamber. The maximum size is limited by the high frequency response—the distances from any parts of the diaphragm to the throat opening should vary by less than one quarter of the wavelength of the highest frequency to be reproduced. The simplest form of sound chamber is Fig. 20.21A, with a single hole in the centre. Better high frequency performance is obtained with the more complex forms of Figs. 20.21 B, C.

Relatively large throats are necessary for high efficiency at low frequencies, and relatively small throats are necessary for high efficiency at high frequencies. Consequently any loudspeaker must be a compromise, and **the highest efficiencies are obtained with limited frequency ranges.**

Second harmonic **distortion** is generated in the throat and follows approximately the theoretical relationship

$$\text{Percentage second harmonic} = (\sqrt{W}/81)\,(f_1/f_0) \times 100 \qquad (14)$$

where W = acoustical watts per square centimetre of throat area
$\quad\quad f_1$ = frequency being radiated
and $\quad f_0$ = cut-off frequency due to flaring (eqn. 2).

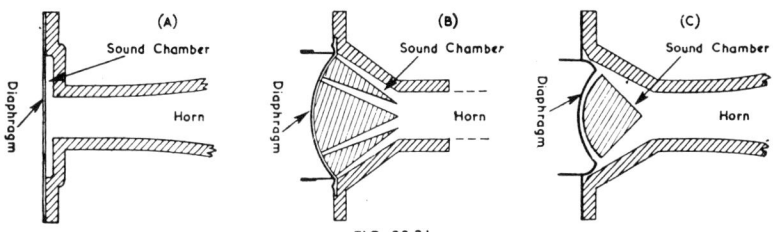

FIG. 20.21

Fig. 20.21.   Different forms of sound chambers in horn loudspeakers.

For example if a horn with a 40 c/s cut-off is to reproduce a 4000 c/s note, there will theoretically be 8% second harmonic distortion for an acoustical power of 0.01 watt per square inch of horn throat.   The actual distortion appears to be about half the theoretical value.

Thus for reasons of both efficiency and distortion, a horn should only be designed to cover a limited frequency range.

(C) Driving units

The simplest driving mechanism is the electro-magnetic type with an iron diaphragm as the armature (Fig. 20.21A).

All modern driving units are of the moving-coil type and may have a diaphragm of aluminium alloy or some form of paper or a cloth base impregnated with synthetic resin.

Well designed direct-radiator loudspeakers may be used as the driving mechanisms for horns.   The most suitable size is from 8 to 12 inches diameter for medium power requirements.   Some loudspeakers are designed specially for this application.

Where high power output at very low frequencies is required, a 15 inch unit may be used, as in the Klipsch corner horn described below.

(D) Distortion

In addition to the distortion caused by the throat, there is also **distortion due to the sound chamber.**   The acoustical capacitance is a function of the position of the diaphragm, and the effect is most apparent at low frequencies where the amplitude is large.   This distortion may be reduced by the use of a large sound chamber, thus limiting this unit to low frequencies only.   A separate high frequency unit with a small sound chamber may be used, since it will not be required to handle large amplitudes.

Distortion is also caused by a non-linear suspension ; this is only serious at low frequencies (Ref. 14, Fig. 8.21).   **Distortion at low frequencies due to non-linear suspension** may be reduced by

(1) the use of a large dynamic driver,
(2) increased compliance in the suspension,
(3) an enclosed air-chamber at the rear of the diaphragm as in the Klipsch corner horn.

Another cause of distortion is **frequency modulation** through the Doppler Effect—see Sect. 7(ii)—which can be reduced to small proportions by the use of separate high and low frequency units.

It is impossible to design a horn loudspeaker that covers a wide frequency band and is simultaneously free from non-linear distortion.   Thus two separate units for low and high frequencies are essential for fidelity.   This is a limitation peculiar to horn loudspeakers.

(E) Efficiency

With horn loudspeakers efficiencies as high as 80% can be achieved over a limited frequency range.   A typical horn of the type used in cinema theatres has an efficiency from 30% to 45% over its useful range.   The Klipsch corner horn has an efficiency around 50% over its useful range.

(F) Directional characteristics

For wavelengths larger than the mouth diameter, the directional characteristics are approximately the same as for a cone the same size as the mouth. At frequencies above 3000 c/s the directional characteristics are only slightly affected by the flare or length. At intermediate frequencies the directional characteristics are broader than those obtained from a piston (cone) the size of the mouth.

When a number of horns are arranged in a line, a sharp beam will be obtained when the horns are parallel to one another, while a broader beam will be obtained when they are arranged radially.

(G) Electrical impedance

A rise in electrical impedance normally occurs at the diaphragm resonance, although less than that with a direct radiator, with a rise at the higher frequencies as with a moving-coil direct radiator.

On account of the variation in impedance, triode power valves (preferably with negative voltage feedback) are highly desirable for fidelity.

(H) Damping and transients

With a well designed horn of sufficient mouth area, or with a smaller horn in which the resistive loading has a peak in the vicinity of the bass resonant frequency, the acoustical damping at that frequency will be high. The damping due to the output resistance of the amplifier will be additive.

At higher frequencies, experimental results indicate that spurious transients are much lower in horn loudspeakers than with direct radiators. For example (Ref. 155), a 12 inch direct radiator loudspeaker, when loaded by a short horn, gave 12 to 15 db less transient level than when used without the horn, using Shorter's method—see Sect. 7(iii).

(I) Horns and rooms

Horns require larger rooms than direct radiator loudspeakers, even apart from the space occupied by the speaker itself, owing to the roughness of the overlap between the low and high frequency units. The distance from loudspeaker to listener must be greater than that with cone type direct radiators (Ref. 150).

Horn loudspeakers are frequently mounted in the corner of the room, so that the angle between the walls acts as sort of continuation of the horn. Some ingenious corner horns have been developed, for example one in which the difference between the solid angle between the walls and the desired exponential flare is taken up by a suitably shaped " plug " (Ref. 130).

(vi) Folded horn loudspeakers

Owing to the large space required by a horn, much attention has been paid to folded horns. These all involve some loss of the higher frequencies, due to reflections and differences in length of path causing cancellation of some frequencies. The number of folds should be kept to the minimum—most folded horns have only one or two folds. Any increased number of folds, if unavoidable, must be accompanied by a reduction in the maximum frequency.

Concentric folded horns

In these, the high frequency loss may be made fairly small, and it is possible to cover a range from below 200 up to 8000 c/s. One popular form is the three section directional reflex (Fig. 20.22A) in which the total length of the air column is nearly three times the overall length ; this gives useful radiation over an angle of 60°. A modified form is the radial type which may be suspended from the ceiling to give an angle of 360° ; this gives almost the same radiation upwards as downwards. Refs. 2, 82.

Low frequency folded horns

When a folded horn is used as the low frequency unit in a dual system, the loss of higher frequencies is actually an advantage, and the inner surface of the horn is frequently covered with sound absorbent material to eliminate high frequency reflections. Some of the forms which such a horn may take are given in Figs. 20.23 (Ref. 51) and 20.24 (Ref. 52).

(A)   FIG. 20.22   (B)

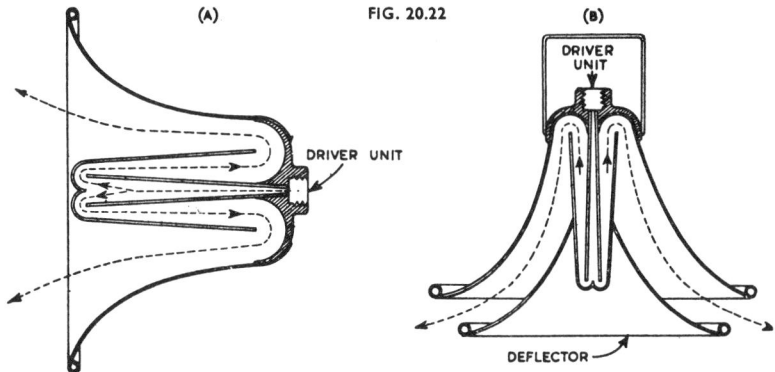

Fig. 20.22. Concentric folded horns (A) *three section reflex* (B) *three section reflex radial type.*

One outstanding design is the Klipsch corner type speaker which includes a folded low frequency horn.

This is notable mainly on account of the comparatively small space which it occupies —20 cubic feet in all—compared with other dual horn systems having equivalent performance. The efficiency of the low frequency woofer unit (K-3-D) is not less than 50% down to 36 c/s, and still fairly high at 32.7 c/s ; it is capable of radiating a clean fundamental at reduced power down to 27 c/s.   Maximum electrical power input is 15 watts, so that the maximum acoustical power output is over 7 watts.

The low frequency unit has a 15 inch direct radiator loudspeaker with an enclosed cabinet baffle at the rear and a folded horn in front.   The recommended driver for ordinary home and small theatre power levels is the Stephens P-52-LX-2 woofer motor ; this is specially treated to increase its compliance before installation.   The enclosed cabinet baffle at the rear is designed to offset the mass reactance of the throat impedance at low frequencies.   The volume of the enclosed baffle is given theoretically by

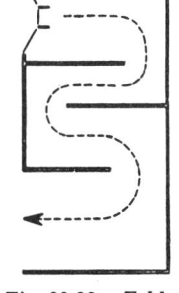

FIG. 20.23

Fig. 20.23. Folded horns for low frequencies actuated by rear of cone with area increasing in steps.

$$V = 2.9 \; AR \qquad (15)$$

where V = volume in cubic inches
  A = throat area in square inches
and R = length of horn in inches within which the horn area doubles.

FIG. 20.24

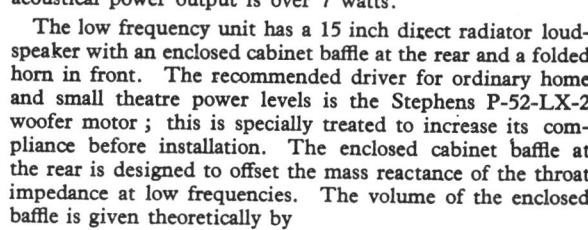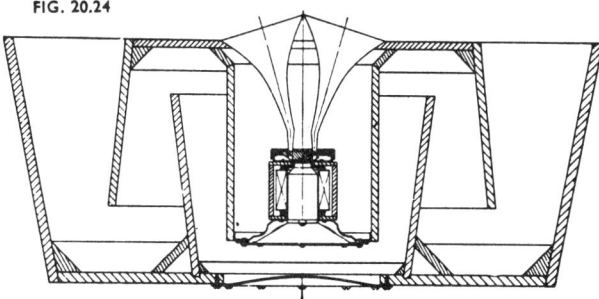

Fig. 20.24. Folded horns for low frequencies—as part of dual system, with conical individual sections. (Ref. 52.)

For the K-3 woofers, $A = 88$ sq. ins., $R = 16$ and the calculated volume is 5270 cubic inches. This equation presumes the suspension compliance to be infinite, which is not the case ; some experimental adjustment is therefore necessary.

The general construction is shown in Figs. 20.25 and 20.26. The voice coil impedance of two designs, the second with increased cone compliance, is shown in Fig. 20.27.

The cross-over frequency is 500 c/s, so that the **high frequency horn** has to be of unusual design, capable of handling 15 watts input down to 500 c/s. The recommended driver for home use is the Stephens P-15 HF motor, with which the response is good to 15 000 c/s. Other suitable drivers are the Jensen XP-101, Western Electric 713A, or Stephens P-30, P-40. The cell structure is contained in the region of small cross-section of the horn, so that the individual mouths of the cells are small. The construction of the high frequency horn is shown in Fig. 20.28.

References to Klipsch horn loudspeakers : 30, 31, 32, 33, 34, 48, 99, 115, 119, 122.

(vii) High frequency horns

High frequency horns are frequently used in dual or triple systems in conjunction

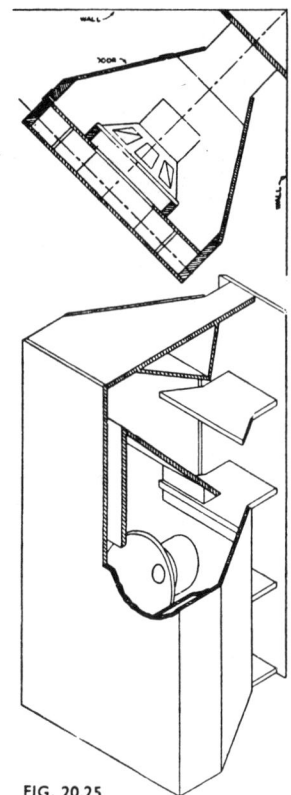

FIG. 20.25

Fig. 20.25. Top and isometric view of the Klipsch bass reproducer Model K-3. (Ref. 48).

with either horn or direct-radiator low frequency units. They are only capable of satisfactory radiation over an angle of from 20° to 40° for a single horn, but up to six similar horns may be built into one unit operated by a single driver to give good coverage over a wide horizontal angle (Refs. 76 Part 1 ; 52, 136). Alternatively the multi-cellular construction may be adopted (Ref. 33). Diffusing lenses may also be used with high frequency horns as an alternative to multi-cellular construction, to spread the sound over a wider angle. In one example the angle was increased from about 20° to over 50° at 8000 c/s by the use of such a diffusing lens. The construction appears cheaper than a multi-cellular

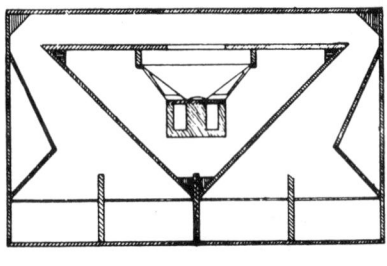

FIG. 20.26

Fig. 20.26. Side and front views of the Klipsch bass reproducer Model K-3. (Ref. 48).

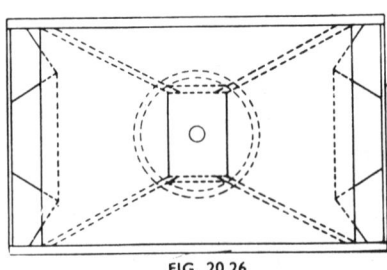

FIG. 20.27

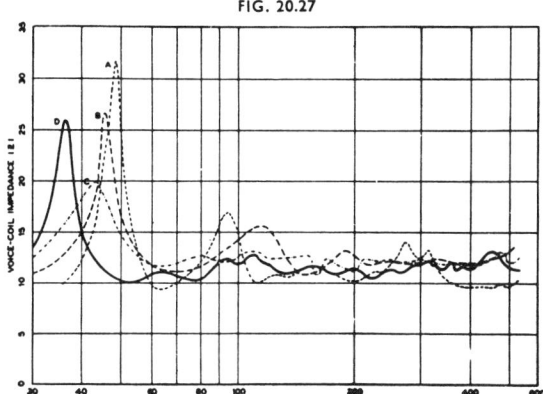

Fig. 20.27.   Voice coil impedance of two designs of Klipsch bass reproducer (A) Model K-3-B as Ref. 48 *(B) Model K-3-C (cone with increased compliance). (Ref. 99).*

horn and permits greater flexibility in application (Refs. 120, 141).

References to corner ribbon high frequency horn : 229, 230.

Reference to reverse flare principle in high frequency horn : 232.

See also the Klipsch high-frequency horn (Fig. 20.28) and page 858.

References to high frequency horns : 33, 76 (Part 1), 80, 137, 165.

General references to horns : 2, 13A, 14, 15, 30, 31, 32, 33, 34, 38 (Parts 1 and 2), 39, 45, 46, 47, 48, 51, 52, 59, 69, 73, 76, 77, 82, 99, 115, 119, 128, 130, 137, 142, 147, 165, 175, 177, 206, 209, 211, 217, 232.

(viii) Combination horn and phase inverter loudspeakers for personal radio receivers

A combination horn and phase inverter loudspeaker for personal radio receivers has been described (Ref. 124) which has an efficiency of about 25% and a frequency range from 300 to 4000 c/s.   A sound level of 84 db is obtained at 3 feet from the loudspeaker with an input of 10mW—this level is somewhat higher than conversational speech.   Sub-miniature valves and very small B batteries may therefore be used to give an acceptable sound output from a receiver having a cubic capacity of 41 cubic inches.

(ix) Material for making horns

The best materials are concrete, brick and masonry.   The most practical material for horn construction, with many good features, is untempered 3/16 inch Masonite. It should be reasonably strutted and backed with absorbent material (Ref. 191).

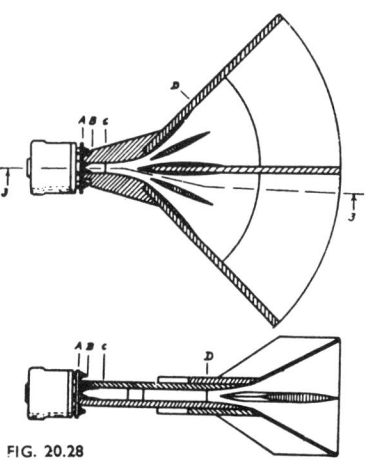

FIG. 20.28

Fig. 20.28.   Klipsch high frequency horn loudspeaker :   Upper view—sectional view from top ; lower view—sectional view from the side through line 3-3 of upper view.   (Ref. 33).

SECTION 5 : DUAL AND TRIPLE SYSTEM LOUDSPEAKERS

(*i*) *Introduction* (*ii*) *Choice of the cross-over frequency* (*iii*) *The overlap region* (*iv*) *Compromise arrangements.*

(i) Introduction

When a wide frequency range has to be covered there is a choice between

1. A single unit employing some special construction to extend the frequency range—see Sect. 2(iii),
2. Two or three separate loudspeakers, each covering a limited frequency range,
3. An integral dual or triple system with two or three loudspeakers mounted (usually co-axially) in one equipment, and
4. Some compromise arrangement—see (iv) below.

If separate loudspeakers are used, they should be mounted as closely together as possible, with co-axial mounting as the ideal ; this is not so important when the cross-over frequency is below 500 c/s. In addition, they should be co-planar, with the plane of horn loudspeakers taken as the plane of the diaphragm. The loudspeakers should be correctly phased, so as to be additive in the overlap region.

A frequency dividing network is used to split the output between two loudspeakers so that neither unit is called upon to handle large amplitudes of frequencies beyond its range. This has the advantage that Doppler Effect distortion (see Sect. 7) is much reduced. Other advantages are that, on account of limitations in the frequency range of each unit, the system efficiency is increased, while the directivity characteristic is improved due to the smaller diaphragm (or horn mouth) for the high frequency unit. In addition, the transient response is improved, and there is less intermodulation and reduced frequency modulation.

The loudspeakers should preferably have the same efficiency, otherwise one will have to be attenuated. The directional characteristics of the high frequency unit should receive careful attention.

Integral dual systems

One excellent arrangement employs two cones, co-axial and co-planar, with the small high-frequency cone mounted near the apex of the large cone. Both cones vibrate in unison in the overlap region. (Refs. 61, 116, 116A).

Another arrangement employs a large low-frequency cone and a co-axial high-frequency horn. This has the disadvantage that the sound sources are not co-planar, since the diaphragm of the high frequency unit is mounted to the rear of the cone.

Integral triple systems

In one design there is a large low-frequency cone, a mid-frequency horn using the flared low-frequency cone for its mouth, and a high-frequency horn mounted in front of the mid-frequency horn.

(ii) Choice of the cross-over frequency

In a dual system, that is one having a low-frequency and a high-frequency loudspeaker, the cross-over frequency is usually between 400 and 1200 c/s. The following points must be satisfied.

1. The low frequency unit (" woofer ") must be capable of handling at least half an octave above the cross-over frequency, at full power. See also (4) below.
2. The high frequency unit must be capable of handling at least half an octave below the cross-over frequency at full power. See also (4) below.
3. Provided that point (2) can be satisfied, the cross-over frequency should be as low as possible, say 400 to 500 c/s.
4. If a 6 db/octave frequency dividing network is used, each unit should be capable of handling one octave beyond the cross-over frequency at full power, and about 3 octaves beyond the cross-over frequency at reduced power.

Some systems have cross-over frequencies from 1200 to 2000 c/s, in which case the high frequency unit approaches more closely to a true " tweeter," since the latter is generally limited to frequencies above 2000 c/s (Refs. 11, 45).

In a triple system there is no necessity for such a compromise regarding the cross-over frequency. The " woofer " may handle up to between 300 and 600 c/s, the middle unit up to between 2000 and 5000 c/s, and the " tweeter " will then look after the higher frequencies (e.g. Ref. 134).

The design of frequency dividing networks is covered in Chapter 21, Sect. 3.

(iii) The overlap region

Serious distortion often occurs in the overlap region when both units are contributing to the total acoustical output.   In a dual direct-radiator system in which the distances from each cone to the listener are not equal, the response characteristic will have pronounced peaks and valleys at frequencies where the two sources are in and out of phase.   In the case of a dual horn system, particularly when one horn is folded, the acoustical paths may differ sufficiently to cause the same effect.   Similar effects occur with other combinations, but the trouble is minimized in all cases by a low cross-over frequency (less than 600 c/s) and, in the case of dual horns, by a fairly steep attenuation characteristic (12 or 18 db/octave nominal).   Most loudspeakers give serious distortion below the frequency of minimum rated response, even though the level is attenuated by the cross-over network—hence the desirability of an extended frequency range.

There tends to be a peak of distortion at the cross-over frequency owing to incorrect impedance matching and the partially reactive load.   This is most serious with pentodes and beam power amplifiers (see Chapter 21).

(iv) Compromise arrangements

One possibility is to use a single moving-coil loudspeaker with the front acting as a direct radiator, and the rear as bass horn (Ref. 51).   Another arrangement uses a single driver with a high-frequency horn facing forwards and a folded bass horn driven by the rear of the diaphragm (Ref. 52).   Owing to the use of a single cone or diaphragm, the high frequency response does not extend as far as with two separate units, each specially designed.

General references to dual and triple systems : 11, 30, 33, 45, 48, 51, 52, 61, 76, 130, 134, 150, 165, 167, 194, 215.

SECTION 6 : LOUDSPEAKERS IN OPERATION

(i) Loudness   (ii) Power required   (iii) Acoustics of rooms   (iv) Loudspeaker placement   (v) Stereophonic reproduction   (vi) Sound reinforcing systems   (vii) Open air Public Address   (viii) Inter-communicating systems   (ix) Background music in factories.

(i) Loudness

Loudness may be measured in loudness units—see page 827 and Fig. 19.8.

(ii) Power required

(A) Direct radiation (outdoor)

Some loudspeaker manufacturers quote the intensity level on the axis at a distance of, say, 30* feet with a stated electrical input power, usually with a warble frequency (from 300 to 3300*, 500 to 2500 or 500 to 1500 c/s).   The intensity level decreases by 6 db each time the distance is doubled, or increases by 6 db each time the distance is halved.

The intensity level increases by 3 db when the power input to the loudspeaker is doubled, by 6 db when the power is quadrupled, and so on.

*R.M.A. SE-103.   See Sect. 6(x).

Example : At a distance of 30 feet on the axis, a certain loudspeaker is stated to produce a level of 81.5 db above 10^{-16} watt per square centimeter at 8 watts input with a warble frequency from 500 to 2500 c/s. Find the intensity level on the axis at a distance of 6 feet with 3 watts input.

Effect of change of distance $= 20 \log_{10} (30/6) = + 14$ db.

Effect of change of power input $= 10 \log_{10} (3/8) = - 4.3$ db.

Net change $= + 9.7$ db.

Intensity level at 6 feet with 3 watts input $= 81.5 + 9.7 = 91.2$ db.

In other cases loudspeaker manufacturers (such as R.C.A.) publish the curves of sound pressure versus frequency, from which it is easy to estimate the average over the frequency range of interest. In this case the reference level is 0 db = 10 dynes per square centimetre, and the microphone distance is 24 inches. In a typical case the output level is $- 1$ db for an input of 0.1 watt to the loudspeaker. The level at a distance of 30 feet for a power input of 8 watts will therefore be

$- 1 + 19.03 - 23.52 = - 5.49$ db (0 db = 10 dynes/cm.^{2})

or converting to the basis of 0 db = 0.0002 dyne/cm^2,

$- 5.49 + 93.98 = + 88.5$ db (0 db = 0.0002 dyne/cm^2).

Alternatively the R.M.A. loudspeaker pressure rating may be quoted. For calculation and example see page 812.

FIG. 20.29

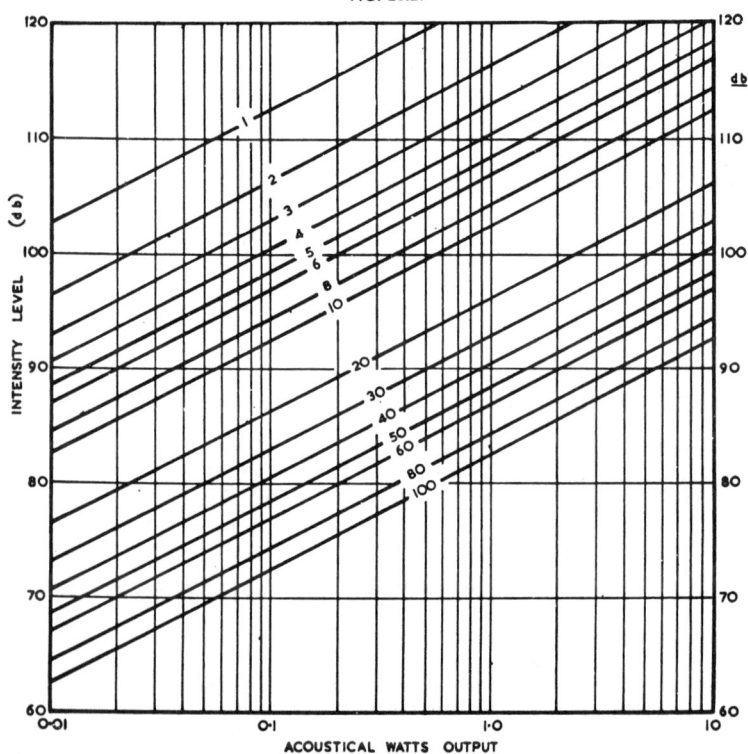

Fig. 20.29. *Intensity in db versus acoustical watts output at specified distances along the axis of a loudspeaker which is considered as a small source in an infinite baffle. This holds fairly accurately at low frequencies, with an error of 1 db at 500 c/s with a cone 10 inches in diameter. The numbers on the sloping lines indicate the distance in feet from the loudspeaker (based on F. Massa, Ref. 69, but extended and modified).*

Again, the loudspeaker percentage efficiency may be known, say 3%, and it is desired to calculate the intensity at a distance of, say, 30 feet with an input of 8 watts. The unknown factor is the angle of radiation, but at frequencies up to 500 c/s this may be taken as approximately 180° (assuming a direct radiator with a large flat baffle) and we may therefore make use of Fig. 20.29.   In this case the acoustical output will be 8 × 0.03 = 0.24 watt and the intensity will be 87 db at 30 feet (Fig. 20.29).

At higher frequencies there will be an increasingly greater focusing effect, and the intensity on the axis will be somewhat greater than the value calculated above.

(B) Power required indoors

When a loudspeaker is operated indoors the direct radiation is supplemented by the reflected sound.

The **reverberation time** is the time in seconds for a sound to fall to one millionth of its original intensity (− 60 db) after stopping the source.   Eyring* gives

$$T = \frac{0.05 \ V}{- \ S \ \log_\epsilon \ (1 \ - \ \alpha)}$$

where T = reverberation time in seconds,
  V = volume of room in cubic feet,
  S = surface area of walls, ceiling and floor in square feet,
and α = average absorption coefficient per square foot, in sabins (values are less than unity).

The sabin is an absorption unit representing a surface capable of absorbing sound at the same rate as does 1 sq. ft. of perfectly absorbing surface, such as an open window.

In a typical living room, the reverberation time is about 0.5 second at 500 c/s and the absorption coefficient is about 0.25. The reverberation time falls to possibly 0.3 second at 5000 c/s and rises to possibly 0.75 second at 200 c/s.(Ref. 14, Fig. 13.26).

In a very large living room, the reverberation time would probably be about 0.8 second at 1000 c/s.

FIG. 20.30

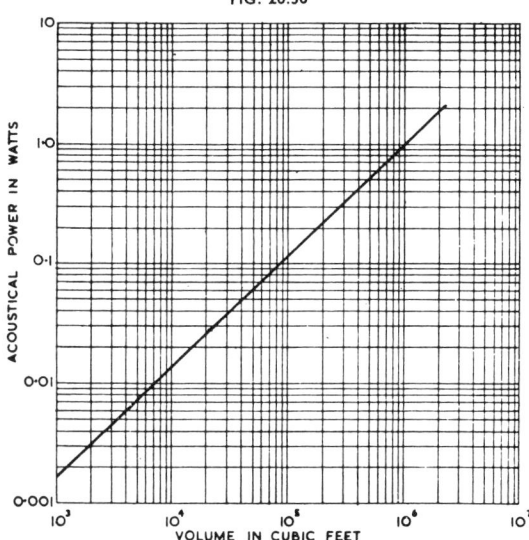

Fig. 20.30.   Acoustical power required to produce an intensity level of 80 db as a function of the volume of the auditorium for optimum reverberation times (after Olson, Ref. 13, by kind permission of the author and of the publishers and copyright holders, Messrs. D. Van Nostrand Company Inc.).

*See Ref. 13A page 399.

The " optimum reverberation time " at 1000 c/s depends on the volume of the room (Ref. 13A) :

Volume	1000	2000	4000	20 000	100 000	500 000	cu. ft.
T_{opt}	0.7	0.75	0.82	1.01	1.28	1.62	secs.

In a typical room 20 ft. $\times$ 15 ft. $\times$ 10 ft., with a reverberation time of 0.5 second, and with the listener 6 feet from the loudspeaker, the reflected sound adds 3 db to the direct radiation at 500 c/s under steady state conditions, and 1 db when the sound only continues for 0.02 second as for speech (Ref. 14, Fig. 13.26). At higher frequencies the reflected sound becomes rapidly smaller, and may be neglected. At 200 c/s the reflected sound adds 5.5 db under steady state conditions or 2 db when the sound continues for 0.02 second. At 100 c/s the reflected sound adds 9.5 db under steady state conditions, or 6 db when the sound continues for 0.02 second. The effect of the reflected sound becomes more important as the listener moves further away from the loudspeaker.

References 13A, 14, 15, 50, 55, 175, 178.

The **acoustical power** in watts to produce an intensity level of 80 db (0 db = 0.0002 dyne/cm²) is plotted in Fig. 20.30 against the volume of the enclosure in cubic feet (after Olson, Ref. 13A). For living rooms the following information may be derived—

Volume	1000	2000	3000	4000	cu. ft.
Intensity level	Acoustical power in watts				
55 db	0.000 0051	.000 0098	.000 014	.000 019	watt
60 db	0.000 016	.000 031	.000 045	.000 059	watt
65 db	0.000 051	.000 098	.000 14	.000 19	watt
70 db	0.000 16	.000 31	.000 45	.000 59	watt
75 db	0.000 51	.000 98	.001 4	.001 9	watt
80 db	0.001 6	.003 1	.004 5	.005 9	watt
90 db	0.016	031	.045	.059	watt
100 db	0.16	.31	.45	.59	watt
110 db	1.6	3.1	4.5	5.9	watts

These values apply to enclosures with optimum reverberation times. Living rooms may have reverberation times which are lower than the optimum values, and the acoustical power required will then be greater than indicated above.

The intensity levels for home listening under various conditions are set out in Chapter 14 Sect. 7(iii).

(iii) Acoustics of rooms

When a musical item is produced in the studio and reproduced by a loudspeaker in a living room, there are two reverberation times—that of the studio and that of the room. Good listening conditions are usually achieved by controlling the reverberation time of the studio so that the music will sound well in the average medium or large living room. Large concert halls with long reverberation times are not ideal, but there is nothing which can be done in the living room to improve matters.

The desirable and usual rise of reverberation time at low frequencies with rigid walls may be replaced by a general fall, due to the **vibration** of walls, floor and ceiling. This is one prevalent cause of loss of bass, below 150 c/s ; if an attempt is made to counteract it by bass boosting, the acoustical effect is not quite natural (Ref. 53).

It is preferable to mount the loudspeaker on sound absorbing pads or thick carpet to prevent the direct transmission of vibration to the floor.

At low frequencies it is doubtful whether the reverberation time of a small or even medium sized room means much, because of **room resonances** producing standing waves. In a typical case with a room 16 ft. 9 in. × 11 ft. 6 in. × 8 ft. 6 in. the first nine resonances are 36.8, 51.1, 62.9, 68.6, 73.6, 77.6, 85.5, 102.3, 137.2 (Ref. 53). The most desirable shape for a listening room is approximately in the ratio height : width : length = 1 : 1.27 : 1.62, so as to distribute the resonances fairly uniformly. The peaks due to these resonances are of the order of 20 db and completely mask less pronounced effects. The bad effects may be reduced by

(1) increased acoustical damping at low frequencies, especially near the corners of the room where it is twice as effective as elsewhere (Ref. 175), and

(2) an open door or window, or preferably both.

In addition to these room resonances, which occur over the whole room, there are other peaks of response due to focusing or interference between different paths, at middle and high frequencies, which vary from point to point. Thanks to our two ears, these sharp peaks and valleys are much less evident than they otherwise would be. References 13A, 14, 53, 175, 192, 208, 220, 221.

(iv) Loudspeaker placement

The ideal position for a loudspeaker is in one corner of the room, either low down near the floor or in the corner between two walls and the ceiling. In either case the boundaries of the room form a triangular conical horn, and the angle of radiation is $\pi/2$ steradians. The axis of the loudspeaker may point directly at the head of one listener, say seated with his head 42 inches above the floor, but every other listener will then lose something of the high frequencies. If a ceiling-corner position is used, the maximum angle of elevation between a listener and the loudspeaker should not normally exceed 30° ; this position is therefore limited to large rooms or those with low ceilings. If a special loudspeaker is used having wide angle radiation at all frequencies, the problem disappears. In other cases the trouble may be minimized by

(1) accentuating the high frequencies and pointing the axis of the loudspeaker away from the listeners, who should be seated at approximately the same angle to the axis, or

(2) keeping the listeners within an angle of 15° from the axis.

Another possible position for the loudspeaker is in the corner of a room, half-way between floor and ceiling—this gives an angle of radiation of π steradians.

In all cases where a loudspeaker is mounted in the corner of a room, the adjacent walls and floor or ceiling should have reflecting surfaces so as to act as a horn. The rest of the room should have a considerable amount of damping material such as upholstered furniture, heavy drapings, carpets, books or acoustical tiles.

In all cases, no listener should be closer than 4 feet from the loudspeaker, with a greater distance for dual or triple systems and considerably greater distance for horns.

Loudspeakers with open-backed cabinets are very difficult to place satisfactorily. They should be about 9 inches out from the wall, and the wall behind them should preferably be damped by a heavy curtain or sheet of acoustical absorbent material. A corner position is desirable.

If a loudspeaker must be placed along one of the walls, it should preferably be along the shorter side, about the centre. This gives an angle of radiation of 2π steradians.

If two separate loudspeakers are used, each handling the full frequency range, they may, if desired, be placed in positions wide apart, such as the two ends of a room. In a fairly large living room some approach to a third dimensional effect may be obtained with the listener slightly nearer to one loudspeaker—this is the pseudo-stereophonic effect, see (v) below.

(v) Stereophonic reproduction

Reproduced sound normally comes from a single source—the loudspeaker—so that there can be no indication of the direction from which the sound originally came. In stereophonic reproduction with three or more channels there can be a 3 dimensional effect which enables the listener to determine the position of the source. Stereophonic reproduction in the strict sense can only be applied in large concert halls or cinemas where the audience is situated further from the loudspeakers than the distance between any two loudspeakers.   An illusion which gives a sense of spatial reproduction can be applied to a room by means of two loudspeakers each fed from a separate microphone through separate amplifiers.   In radio broadcasting at least two separate transmitters and receivers are required—three channels are desirable.   Some methods of recording are capable of carrying two or three channels, such as sound on film and magnetic tape (Refs. 86, 101, 226).

There are methods of obtaining an approach to stereophonic reproduction with a single channel (Ref 87)

When there are two loudspeakers so placed with respect to the listener as to give a transit time difference from 1 to 30 milliseconds, the impression is that of more "liveliness" and "body" (Ref. 125).   This is the **pseudo-stereophonic effect,** and to be fully effective requires the listener to be at least 4 feet—and preferably a greater distance—nearer to one loudspeaker than the other.   With suitable positioning and relative volumes, the effect is quite surprising.

A more elaborate method for heightening sound perspective utilizes volume expansion in one of the two channels (Ref. 49).

References 6, 13, 14, 36, 49, 68, 86, 87, 97, 101, 112, 160, 161, 162, 163, 169, 195, 203, 216 (giving additional references), 226, also Supplement.

(vi) Sound-reinforcing systems

With this system there are two sources of sound, the original sound and the augmented sound from the loudspeakers.   Usually the intensity of the original sound issuing from the stage will be sufficient over a limited area, but the sound energy from the loudspeakers must progressively increase towards the rear of the auditorium.   The problem is to select a loudspeaker or loudspeakers with suitable directional characteristics, and to adjust the power output and orientation so that the total intensity level is constant for all parts of the listening area.

The acoustical power required to produce an intensity level of at least 80 db for ordinary speech and music is as follows (Ref. 13A) :

Volume (cu. ft.)	10 000	50 000	100 000	500 000
Acoustical power (watts)	0.014	0.06	0.11	0.5

For full orchestral dynamic range, a maximum intensity level of 100 db is required, and the acoustical powers should be multiplied by 100.   In noisy situations the sound level should be 10 to 16 db above steady noise or 20 to 30 db above the noise in relatively quiet intervals.

One of the problems in design is the avoidance of acoustical feedback.   It is usual to adopt directional loudspeakers and sometimes also directional microphones.   The microphone is mounted as close to the sound source as possible and in a position of low radiated sound intensity.

For the best illusion to the listener the loudspeaker(s) should be mounted so that the sound appears to come from the direction of the original source.   The difference in path length between the original sound and the sound from the loudspeaker should not exceed 60 to 80 feet.

The overall response of the whole system should be free from peaks.   When feedback occurs, it does so at the frequency at which the system has a peak.

If at all possible, low frequencies should be attenuated so as to reduce the size of the loudspeaker and of the amplifier, and to reduce acoustical feedback.   A minimum frequency range for speech is 400 to 4000 c/s (Ref. 196).   If orchestral music is to be amplified the whole frequency range should be provided for, although here also some compromise may be necessary.

In small or medium halls a single loudspeaker or bunch of loudspeakers may be mounted 20 to 30 feet above the microphone and somewhat more forward than the microphone.   The loudspeakers should be pointed downwards to the audience (Ref. 196).

In larger halls one or more **line-source loudspeakers** may be used.   A line-source consists of a number of loudspeakers mounted close together in a straight line.   A vertical line-source gives directionality in the vertical plane but not in the horizontal plane.   If the line-source is "tapered" in strength so that the sound from each element varies linearly from a maximum at the centre to zero at either end, the directionality is improved. The line-source should be tilted slightly forwards so that the central loudspeaker points directly to the centre of the audience. Better results may be obtained by using two line-sources, a long one for the low frequencies and a shorter one (one quarter the length of the long one) for the high frequencies, with a cross-over network for $f_c = 1000$ c/s (Ref. 197).

In large installations, time delay may be incorporated (Ref. 145, 197).

Multiple loudspeakers

Haas (Ref. 125) has investigated the effects of a single echo on the audibility of speech. He has shown that when the time delay of the echo is from 5 to 30 milliseconds, the echo is not detectable as a separate source if the level of the echo is less than 10 db above that of the direct sound, and that the listener only has the sensation of hearing from the nearer loudspeaker. At greater time delays he determined how loud such an echo could be before interfering with speech (e.g. echo 0 db, 50 msecs, path difference 56 ft.). For echoes at lower levels see Ref. 308. See also Refs. 308, 326, 327.

(vii) Open air public address

A typical reflex horn, with an input of 1 watt to the voice coil, gives the following sound levels on the axis (Ref. 73) :

Distance	10	20	40	80	160	320	feet
Sound level	100	94	88	82	76	70	db

With a single horn, the axis should point directly to the farthest part of the audience. Horns may be arranged radially to cover a larger audience, allowing 30° for each horn.

For speech reproduction only, a 10 watt amplifier with two horn loudspeakers is sufficient for a crowd of 5000 people in quiet surroundings. For musical reproduction a power of 40 watts with four loudspeakers will be required. It is usual to allow from 5 to 10 watts for each horn loudspeaker. Alternatively a large number of any convenient type of loudspeaker may be used, with the spacing between loudspeakers not greater than 70 feet. When a wider frequency range is required, some dual horn system is frequently used, particularly with open-air orchestral sound reinforcing.

For large crowds, several line-source loudspeakers—see (vi) above—may be used at the centre of the crowd.

References 13A (pp. 292-296), 14 (pp. 406-409), 73, 104, 125, 176.

(viii) Intercommunicating systems

Only speech is to be reproduced, and the usual requirement is merely to have sufficient articulation to be understood. A reduced frequency range is almost universal, while a very restricted range is used in noisy surroundings (see Chapter 14 Sect. 11).

References 1, 13 (pp. 297, 299), 13A (p. 426).

(ix) Background music in factories

If the noise level is comparatively low, the system may be quite conventional, with a frequency range from say 100 to at least 6000 c/s. Loudspeakers should be placed fairly close together to give good coverage of high frequencies. The spacing should be adjusted so that, with the prevailing noise level, the range of each is slightly over half the distance between them.

If the noise level is high, the highest noise intensity is often limited in frequency range at both extremes. In such a case, as an alternative to over-riding the noise, the full frequency range up to 8000 c/s may be used for the music, which may be at the same level as the mid-frequency noise. The low frequency music range may either be used or attenuated as desired. Careful choice of the source of music is required to give the full frequency range without distortion. A conventional A-M receiver is unsuitable, owing to sideband cutting. Many shellac discs, especially the older ones, are unsuitable on account of limited frequency range.

In all cases some form of volume compressor or limiter is required to reduce the volume range, and the music should be selected to avoid sudden large changes in volume.

SECTION 7 : DISTORTION IN LOUDSPEAKERS

(i) Non-linearity (ii) Frequency-modulation distortion in loudspeakers (iii) Transient distortion (iv) Sub-harmonics and sub-frequencies (v) Intermodulation distortion.

(i) Non-linearity

Non-linearity in a cone occurs when the force versus displacement characteristic deviates from a straight line. The principal causes of non-linearity are—
1. Insufficient rigidity in the cone,
2. Non-linear suspension, and
3. Non-uniform flux density.

Lack of rigidity in the cone is usually the result of reduction in the mass of the cone to achieve high sensitivity. In addition to the other defects—ragged response and poor reproduction of transients—this has a marked effect in increasing the harmonic distortion over the whole frequency range (Ref. 15 Fig. 3).

All cone or diaphragm suspensions are non-linear to a greater or less extent—their stiffness usually increases with larger amplitudes. The total harmonic distortion of any such loudspeaker is fairly low (of the order of 1%) at frequencies of about 300 c/s and above, and is not appreciably affected by non-linearity in the suspension. As the frequency is decreased, however, the distortion rapidly rises in loudspeakers having non-linear suspension. For example, one 10 inch dynamic loudspeaker with a non-linear suspension gives 10% total harmonic distortion with an input of 2 watts at 60 c/s, and 30% distortion with an input of 10 watts at the same frequency (Ref. 13A Fig. 6.34). On the other hand a good quality loudspeaker is capable of handling an input of 10 watts at 60 c/s with only 3% total harmonic distortion, or an input of 2 watts at 60 c/s with less than 1% distortion (Ref. 36 Fig. 25).

When a loudspeaker is used in a well designed vented baffle, or in a suitable horn, the maximum amplitude of the cone will be decreased and the distortion arising from a non-linear suspension will also be decreased.

Non-uniform flux density up to the maximum amplitude of operation is another source of harmonic distortion. This distortion is usually less than 1% so long as the amplitude of movement is small, consisting of odd harmonics only unless the field is not symmetrical about the voice coil. However at high input levels the distortion is usually severe. In one case total distortion at 50 c/s with an input of 5 watts, was reduced from 9% to less than 5%, by careful design (Ref. 84).

At frequencies above 3000 c/s, the harmonic distortion with single large cone loudspeakers is negligibly small compared with the frequency-modulation distortion— see (ii) below.

With dual loudspeakers, or a dual cone loudspeaker, the total harmonic distortion tends to reach a peak in the vicinity of the cross-over frequency—see Sect 5(iii), also Refs. 36, 116. In one example the total harmonic distortion is over 3% from 500 to 1500 c/s for a 15 inch duo-cone loudspeaker with an input of 5 watts.

The cross-over distortion, and the other forms of distortion described above, all increase rapidly as the input power is increased.

The harmonic distortion below the bass resonant frequency is selective (Fig. 13.54). When the load imposes a greater impedance to the harmonic than to the fundamental, the measured harmonic distortion increases. At a frequency equal to one-half of the bass resonant frequency the second harmonic rises to a peak since the second harmonic frequency is equal to that of the bass resonance. Similarly at a frequency equal to one-third the frequency of the bass resonance, the third harmonic rises to a peak, and so with higher harmonics.

At frequencies above about 1000 c/s all harmonics tend to increase since the impedance of the load to the harmonics is greater than the impedance to the fundamental. This is offset to some extent by the fact that with a triode valve or with most of the commonly used negative feedback circuits, as the load impedance is increased, so the distortion decreases. The nett effect is found by the combination of these separate effects.

A system has been developed (Ref. 152) for automatically recording the total harmonic distortion of a loudspeaker over the a-f range. This is particularly helpful in detecting narrow frequency bands where the distortion is high.

One method of recording the performance of a loudspeaker with regard to non-linear distortion is to measure on the axis the maximum sound level before a certain amount of harmonic distortion is produced. This may be plotted as a curve of maximum undistorted sound pressure as a function of frequency (Ref. 153). Alternatively the distortion may be plotted as a function of frequency for constant electrical input. In all cases the loudspeaker must be tested in the enclosure for which it has been designed.

References 13A (pp. 163-173), 20, 21, 23, 36, 61, 106, 108, 109, 133, 152.

(ii) Frequency-modulation distortion in loudspeakers

The origin of this form of distortion is the **Doppler effect** whereby the pitch rises when the source of sound is advancing towards the listener and falls when it is receding (Refs. 3, 13A p. 18). The effect in loudspeakers is entirely independent of non-linearity. If a loudspeaker has two input frequencies, say 50 and 5000 c/s, the acoustical output can be resolved, like a frequency-modulated wave, into a carrier frequency (the 5000 c/s input) and sidebands (intermodulation frequencies) plus the 50 c/s input. The distortion may be measured in terms of the distortion factor, and so made comparable with the total harmonic distortion when a single input frequency is applied. The distortion factor on the axis is given by

$$d.f. = 0.33\ A_1 f_2 \text{ per cent} \tag{1}$$

where A_1 = amplitude of cone motion (each side of the mean position) in inches at the modulating frequency—say 50 or 100 c/s

     f_2 = modulated frequency (variable frequency),

and $d.f.$ = distortion factor in per cent, defined as the square root of the ratio of the power in the sidebands to the total power in the wave.

The distortion factor is proportional to the variable modulated frequency f_2, and is quite small below 1000 c/s. As a typical example take a 12 inch loudspeaker with an equivalent single frequency power input of 0.125 watt to the voice coil—

f_2	1000	2000	5000	10 000	c/s
$d.f.$	0.65	1.3	3.2	6.4	%

This condition applies to an input of 4.2 watts with a loudspeaker efficiency of 3%.

Tests have indicated that the greater part of the distortion above 3000 c/s with single loudspeakers is due to frequency modulation distortion ; this is reduced very considerably by the use of separate high and low frequency loudspeakers.

References 3, 7, 13A.

(iii) Transient distortion

A transient is a waveform, usually with a steep wavefront, which does not repeat at periodic intervals. Any sudden commencement or cessation of a periodic wave has a transient component. For reproduction without distortion the acoustical waveform must be the same as the input waveform.

It is known from the theory of linear dynamic systems of minimum phase shift type that the amplitude response, the phase characteristic and the transient response to various applied waveforms are merely equivalent ways of observing the same inherent performance of the circuit (Ref. 131). Poor transient response leads to fuzzy reproduction with poor definition. Some requirements for good reproduction of transients are

(1) the loudspeaker must respond to the highest audible frequencies,

(2) the loudspeaker frequency response characteristic must be smooth and uniform, and free from sharp peaks and dips,

(3) the loudspeaker must be sufficiently damped, particularly near the bass resonant frequency,

(4) the bass resonant frequency should be as low as possible, and

(5) the phase-shift characteristics of the loudspeaker should be good.

Reasonably good response to transients is obtainable with loudspeakers having a **frequency response** to 10 000 c/s, but better response is achieved with an increase to 15 000 c/s. Generally speaking it is safe to say that a loudspeaker with a smooth response characteristic has a better transient response than one having sharp resonances, but considerable skill is required for an accurate interpretation of the response characteristic (see Ref. 131 for assistance in this direction).

Loudspeaker damping at the bass resonant frequency has been dealt with in Sect. 2(x). With a suitable choice of loudspeaker and enclosure, and with a sufficiently low amplifier output resistance, critical damping or any desired lesser degree is obtainable. There are differences of opinion as to the amount of damping to be used for good fidelity. Some prefer critical damping without any overshoot, others prefer a slight overshoot with its more rapid uptake (e.g. Ref. 174).

At frequencies above 400 c/s, the cone of a direct radiator loudspeaker ceases to act as a piston, and the effective damping at any point on the cone, at any frequency within this range, is likely to be less, and may be very much less than that with a true piston cone.

Horns provide acoustical damping over a wide frequency range, and well designed horn loudspeakers have better transient response than direct radiators.

The **phase-shift characteristics** of loudspeakers are not readily evaluated. Above 1000 c/s the smoothness of the response characteristic appears to be the best available guide to good phase-shift characteristics ; below 1000 c/s the " envelope delay " may be used to supplement the information which can be derived from the response characteristic (Ref. 138).

When a tone burst* is applied, some of the frequencies generated are not harmonically related to the applied signal, and they cause an annoying type of distortion somewhat similar to intermodulation. The ear is very sensitive to this type of inharmonic distortion (Ref. 131).

Ringing at the frequency of the bass resonance becomes progressively less pronounced as the frequency of the applied tone burst is increased away from the resonant frequency.

Testing loudspeakers for transients
There are four methods of testing loudspeakers for transient response—
1. Applied unit impulse,
2. Suddenly applied square wave,
3. Suddenly applied tone burst, and
4. Shorter's method.

Applied unit impulse gives all the information at one test, but is hard to interpret because it causes all the peaks to ring simultaneously (Ref. 131).

The suddenly applied square wave is more selective, since it emphasizes the ringing of the peaks of nearly the same frequency as the applied wave (Refs. 13A, 15, 36, 126, 131).

The suddenly applied tone burst method is capable of giving much valuable information. Ref. 131 gives a theory of ringing and interpretation of results. It seems that those loudspeakers that sound best generally reproduce tone bursts well, although this is better substantiated for the high frequencies than for the low (Refs. 131, 138, 166).

References to tone burst methods : 44, 88, 121, 131, 133, 138, 158, 166, 213.

It has been demonstrated (Refs. 44, 88) that a smooth frequency response curve means a rapid build-up of a transient. The decay characteristic, after the input voltage has ceased, requires further consideration. Each lightly-damped resonant element, when shock-excited by the sudden cessation of the applied voltage, will dissipate its stored energy by radiating sound at its own natural frequency. The method of testing by Shorter (Refs. 44, 117, 121) is to measure, at all frequencies, the response at time intervals $t. = 0$, 10, 20, 30 and 40 milliseconds after the cessation of the applied voltage. Good transient response appears to be indicated by

---

*A tone burst is a wave-train pulse which contains a number of waves of a certain frequency.

1. Attenuation of 35 db at 1000 c/s, increasing steadily to 50 db at 3000 c/s and higher frequencies, for $t = 10$ milliseconds.
2. Further attenuation of about 10 db for $t = 20$ milliseconds.
3. Further substantial attenuations at $t = 30$ and $t = 40$ milliseconds. These should never be negative, even over narrow frequency bands. A negative value may indicate a spurious ripple frequency.

In double-unit loudspeakers, the important frequency region from the standpoint of transient response is near or below the overlap frequency band (Ref. 36).

References 5 (Part 1), 13A (pp. 159-163, 375), 15 (pp. 138-141, 202-212, 332-339), 36, 40, 44, 54 (Part 4), 88 (pp. 59-64), 117, 121, 131, 132, 133, 135, 138, 150, 155, 158, 166, 172, 174, 179.

(iv) Sub-harmonics and sub-frequencies

In addition to harmonic frequencies, a direct radiator loudspeaker produces sub-harmonics, that is frequencies one half, one third, one quarter, etc., of the applied frequency. Of these, the one half frequency is the only one of consequence. It occurs only in limited frequency regions, and it does not occur below a moderate, critical, level. However when this level is reached, it increases rapidly at first, and is accompanied by frequencies of 3/2, 5/2, 7/2, and 9/2 of the applied frequency (Ref. 188). Sub-harmonics are more objectionable to the listener than harmonics of the same percentage, but they require a relatively long time to " build up " and are generally assumed to be not very obvious in ordinary sound reproduction. However one authority states that a good correlation has been found between the number of sub-harmonics produced by the speaker and the quality rating of the speaker as determined by listening tests (Ref. 153).

References 13A (pp. 167-168), 19, 21, 133, 153, 171, 188 (pp. 751-752).

(v) Intermodulation distortion

Tests on the intermodulation distortion of loudspeakers have been described but no general conclusions can yet be derived (Refs. 108, 109, 171).

SECTION 8 : SUMMARY OF ACOUSTICAL DATA

(i) Definitions in acoustics (ii) Electrical, mechanical and acoustical equivalent (iii) Velocity and wavelength of sound (iv) Musical scales.

(i) Definitions in acoustics

Sound energy density is the sound energy per unit volume. The unit is the erg per cubic centimetre.

Sound energy flux is the average rate of flow of sound energy through any specified area. The unit is the erg per second.

The sound intensity (or sound energy flux density) in a specified direction at a point, is the sound energy transmitted per second in the specified direction through unit area normal to this direction at the point. It may be expressed either in ergs per second per square centimetre or in watts per square centimetre.

Sound pressure is exerted by sound waves on any surface area. It is measured in dynes per square centimetre as the r.m.s. value over one cycle. The sound pressure is proportional to the square root of the sound energy density.

The pressure level, in decibels, of a sound is 20 times the logarithm to the base 10 of the ratio of the pressure P of this sound to the reference pressure P_0. Unless otherwise specified, the reference pressure is understood to be 0.0002 dyne per square centimetre.

The velocity level, in decibels, of a sound is 20 times the logarithm to the base 10 of the ratio of the particle velocity of the sound to the reference particle velocity. Unless otherwise specified, the reference particle velocity is understood to be 5×10^{-8} centimetre per second effective value.

The intensity level, in decibels, of a sound is 10 times the logarithm to the base 10 of the ratio of the intensity I of this sound to the reference intensity I_0. Unless otherwise specified, the reference intensity I_0 shall be 10^{-16} watt per square centimetre.

Compliance is ease in bending.   It is the reciprocal of stiffness.

See Chapter 14 Sect. 7 for volume range and peak dynamic range.

See Chapter 19 Sect. 5 for loudness and Sect. 7 for sound level and noise.

See also I.R.E. Standards on Acoustics : Definitions of Terms, 1951, Proc. I.R.E. 39.5 (May 1951) 509; " American Standard of Acoustical Terminology ", A.S.A. Z24.1—1951.

(ii) Electrical, mechanical and acoustical equivalents

Electrical	Acoustical	Acoustical unit
Capacitance	acoustical capacitance	cm^5/dyne
Inductance	inertance	$grams/cm^2/cm^2$
Resistance	acoustical resistance	acoustical ohms
E.M.F.	pressure	$dynes/cm^2$
Impedance	acoustical impedance	acoustical ohms

Electrical	Mechanical	Mechanical unit
Capacitance	Compliance	cm/dyne

(iii) Velocity and wavelength of sound

The velocity of sound in a medium varies according to the relation $V = \sqrt{E/\rho}$ where E is the elasticity and ρ the density of the medium.   For any particular medium, the velocity depends also on the temperature (because of its effect on the density of the medium) and the intensity of the sound.   In the audible range, velocity decreases with decreasing intensity ; but in the ultrasonic range, velocity increases with decreasing intensity.

Ref. Kaye, G. W. C., and T. H. Laby, " Physical and Chemical Constants " (ninth edition, Longmans, Green and Co. London).

Velocity of Sound in Some Common Media

Medium	metres/sec.	Medium	metres/sec.
Air (dry)	342 (at 18° C)	Steel	4700—5200
Hydrogen	130	Nickel	4970
Water vapour (sat'd)	413	Glass	4000—5300
Water (sea)	1540	Brass	3650
Alcohol	1260	Wood	3300—5000
Aluminium	5100	Rubber	50—70
Copper	3970		

Sound transmission in air at 20°C and 760 m.m.

Frequency	30	50	100	200	400	1000	4000	c/s
Wavelength	452	271	136	67.7	33.9	13.6	3.39	inches
	37.7	22.6	11.3	5.65	2.82	1.13	0.282	feet

Velocity of sound in air = 13 550 inches per second
                        = 1129 feet per second
                        = 344 metres per second.

$f \times \lambda = 1129$

where f = frequency in cycles per second

and λ = wavelength in feet.

(iv) Musical scales

Every musical tone has a frequency which is measured in cycles per second.

A **scale** is a series of tones ascending or descending in frequency by definite intervals.

The **octave** is the most important interval ; two tones are separated by one octave when the frequencies are in the ratio 2 : 1.

Each octave is subdivided into a number of smaller intervals. In the **equally tempered scale** the octave is divided into twelve equal intervals (1.0595 : 1) to allow a change of key without retuning. Thus in the equally tempered scale, if any frequency is multiplied by 1.0595 it is raised by one semi-tone ; if multiplied by 1.1225 it is raised by one tone. All normal musical instruments follow the equally tempered scale.

In addition, there are several systems of what is called the **natural scale,** or **just intonation,** in which for example C♯ has a different frequency from D♭ (Ref. 90 Part 1).

Table 1 gives one version of " just intonation."

TABLE 1

Tone	Interval	Frequency ratio		Tempered scale
		Natural (just) scale		
C	Unison	1	= 1.000	1.000
C♯	Semitone	25/24	= 1.042 ⎫	
				1.0595
D♭	Minor second	27/25	= 1.080 ⎭	
D	Major second (= tone)	9/8	= 1.125	1.1225
D♯	Augmented second	75/64	= 1.172 ⎫	
				1.189
E♭	Minor third	6/5	= 1.200 ⎭	
E	Major third	5/4	= 1.250 ⎫	
				1.260
F♭	Diminished fourth	32/25	= 1.280 ⎭	
E♯	Augmented third	125/96	= 1.302 ⎫	
				1.335
F	Perfect fourth	4/3	= 1.333 ⎭	
F♯	Augmented fourth	25/18	= 1.389 ⎫	
				1.414
G♭	Diminished fifth	36/25	= 1.440 ⎭	
G	Perfect fifth	3/2	= 1.500	1.498
G♯	Augmented fifth	25/16	= 1.563 ⎫	
				1.587
A♭	Minor sixth	8/5	= 1.600 ⎭	
A	Major sixth	5/3	= 1.667	1.682
A♯	Augmented sixth	125/72	= 1.736 ⎫	
				1.782
B♭	Minor seventh	9/5	= 1.800 ⎭	
B	Major seventh	15/8	= 1.875 ⎫	
				1.888
C♭	Diminished octave	48/25	= 1.920 ⎭	
B♯	Augmented seventh	125/64	= 1.953	2.000
C′	Octave	2	= 2.000	2.000

The frequency of all tones is determined by the "pitch." The international standard pitch is a frequency of 440 c/s for tone A (equivalent to 261.63 for C) whereas the "physical pitch" is a frequency of 256 for C.

The octaves of C in the two examples above are :

CCCC	CCC	CC	C	c^i	c^{ii}	c^{iii}	c^{iv}	c^v	
16.4	32.7	65.4	130.8	261.6	523.3	1046.5	2093	4186	c/s
16	32	64	128	256	512	1024	2048	4096	c/s

The frequency of any tone may be calculated by multiplying the value of C next below it by the frequency ratio of the tempered scale as given by the appropriate row in Table 1.

References 13 (pp. 334-335), 89, 90, 91, 92, 93, 127.

SECTION 9 : STANDARDS FOR LOUDSPEAKERS

(i) Voice coil impedance for radio receivers   (ii) Loudspeaker standard ratings for sound equipment.

(i) Voice coil impedance for radio receivers

The American R.M.A. Standard REC-104 (Jan. 1947) "Moving coil loudspeakers for radio receivers" specifies a voice coil impedance of 3.2 ohms $\pm$ 10% measured at the first frequency above resonance giving a minimum value tested without a baffle. This impedance differs from that specified for sound equipment (see below).

This standard (REC-104) is applied to loudspeakers having a maximum pole piece diameter not over 1 inch.

(ii) Loudspeaker standard ratings for sound equipment

The American R.M.A. Standard SE-103 "Speakers for sound equipment" (Ref. 95) lays down certain definitions and ratings which are summarized below.

The **loudspeaker impedance** (Z_S) is the complex value of the electrical impedance given as a function of frequency, and is measured at the accessible signal terminals of the speaker.

Published data to give magnitude and phase angle as function of frequency.

The **loudspeaker rating impedance** (R_{SR}) is the value of a pure resistance, specified by the manufacturer, in which the electrical power available to the speaker is measured. The value shall be 4, 8 or 16 ohms.

Loudspeakers may also be rated in terms of **power drawn from standard distribution lines.** In this case it is necessary to specify the power and the line voltage.

The **loudspeaker measurement source impedance** (R_{SG}) is the value of a pure resistance to be connected in series with the speaker and a constant voltage source to measure the speaker performance. The value of R_{SG} shall be 40% of the value of the loudspeaker rating impedance R_{SR}.* This is equivalent to operating the loudspeaker from a source having voltage regulation of 3 db.

The **loudspeaker pressure rating** (or pressure "efficiency") G_{SP} is the difference between the axial sound pressure level (referred to a distance of 30 feet) and the available input power level, and is expressed in db. For further details see Chapter 19 Sect. 1(viii).

Standard Test Signal No. 1, as indicated by G_{SP1}, is a loudness weighted signal covering the frequency band from 300 to 3300 c/s. (Refs. 95, 50).

*Some manufacturers (e.g. R.C.A.) apply constant voltage to the loudspeaker, so that R_{SG} is made zero.

The **loudspeaker pressure-frequency response** (L_S) is the variation of the 30 foot axial free-space pressure level as a function of frequency, and is expressed in db. The pressure level L_d at distance d feet may be used to compute the pressure level L_S at 30 feet by the relation

$$L_S = (L_d + 20 \log_{10} d) - 29.5 \text{ db} \tag{1}$$

- The **loudspeaker directivity index** (K_S) is the ratio, expressed in db, of the power which would be radiated if the free-space axial sound pressure were constant over a sphere, to the actual power radiated. It is expressed in the form :

$$K_S = 10 \log_{10} \left[4\pi / \int_0^\pi \int_0^{2\pi} (p/p_a)^2 \sin \theta \ d\theta \ d\phi \right] \tag{2}$$

where p_a = axial free space sound pressure, dynes/cm²
   p = general free-space sound pressure at the same distance, dynes/cm²
and θ and ϕ are the angular polar co-ordinates of the system, and the speaker axis is at $\theta = 0$.
(When standard test signal No. 1 is used, K_{S1} is the speaker loudness directivity index).

The value of the directivity index (K_S) may be calculated, over the frequency range which determines loudness, by using the following table for a rigid piston vibrating axially in an infinite baffle (based on curves in Ref. 50) :

df =	500	2000	5000	10 000	15 000	25 000	35 000	45 000
K_S =	3.0	3.0	4.0	7.0	10.5	15.0	18.2	20 db

where f = frequency in cycles per second
and d = piston diameter in inches.
   When using standard test signal No. 1, the value of the **loudness directivity index** K_{S1} is tabulated for similar conditions to those above :

d =	0	4	8	12	16	20	inches
K_{S1} =	3	4.4	5.7	7.0	8.0	9.0	db

Values of K_{S1} for various types of baffles and horns are given by Ref. 50.
   The loudness directivity index K_{S1} is used to evaluate the **loudness efficiency rating,** but the Standard refers to the article by Hopkins and Stryker (Ref. 50) for detailed information. Loudness efficiency rating is defined as the ratio of the total " effective " acoustical power produced by the loudspeaker to the available electrical power. The loudness efficiency rating is equal to

$$LR = 100e = 100 \times 10^{L_e/10} \tag{3}$$

where e = electroacoustical efficiency
   $L_e = 20 \log_{10} p_{ax} - 16 - k - K_{S1}$
   p_{ax} = effective sound pressure on axis of loudspeaker at distance 30 ft., in dynes/cm²,
   $k = 10 \log_{10} W_R$
   W_R = maximum available electrical power in watts
and K_{S1} = loudness directivity index (see above).
   See Refs. 50, 121 (pp. 697-706).

Loudspeaker efficiency in terms of acoustical power
(Based on SE-103—Ref. 95).
   The loudspeaker efficiency (G_{SW}) in terms of acoustical power is the difference between the output acoustical power level and the available input electrical power level, and is expressed in db.
   The symbol G_{SW1} indicates the loudspeaker efficiency when standard test signal No. 1 is used.

$$G_{SW} = 10 \log_{10} (W_S/W_0)/(W_{AS}/W_0) \tag{4}$$
$$= 10 \log_{10} (W_S/W_{AS}) \tag{5}$$
$$= 20 \log_{10} p_S - K_S - 20 \log_{10} E_G + 10 \log_{10} R_{SR}$$
$$+ 20 \log_{10} (1 + R_{SG}/R_{SR}) - 16 \tag{6}$$

where G_{SW} = loudspeaker efficiency in db
   W_S = total radiated acoustical power, in watts
   W_{AS} = electrical power available to the speaker, in watts

W_0 = reference power = 0.001 watt
E_G = r.m.s. value of constant voltage of source, in volts
R_{SR} = loudspeaker rating impedance, in ohms
R_{SG} = loudspeaker measurement source impedance, in ohms
p_S = 30 ft. axial free space sound pressure, in dynes/cm²
and K_S = loudspeaker directivity index, in db.

It is obvious that $(W_S/W_{AS}) \times 100$ is the loudspeaker percentage efficiency, and hence

Efficiency	2	4	6	10	20	30	100	%
G_{SW}	−17	−14	−12.2	−10	−7	−5.2	0	db

The total radiated acoustical power in dbm is given by the electrical power available to the speaker in dbm plus the speaker efficiency in db. For example, if the available electrical power is 10 watts = 40 dbm, and the speaker efficiency is −13 db, then the total radiated acoustical power will be (40 − 13) dbm = 27 dbm = 0.5 watt.

See also " American Recommended Practice for Loudness Testing " C16.4—1942 (Ref. 70).

For loudspeaker testing see also Refs. 13A (pp. 353-376), 121 (pp. 607-609, 661-706), 175 (pp. 292-295), 188 (pp. 768-773).

For Standards for multiple loudspeakers in sound systems see Chapter 21 Sect. 2(ii).

SECTION 10 : REFERENCES TO LOUDSPEAKERS

1. Sanial, A. J. " Acoustic considerations in 2-way loudspeaker communications " Comm. 24.6 (June, 1944) 33.
2. Sanial, A. J. " Concentric folded horn design " Elect. 12.1 (Jan. 1939) 16.
3. Beers, G. L., and H. Belar, " Frequency modulation—distortion in loudspeakers " Proc. I.R.E. 31.4 (April, 1943) 132, also S.M.P.E. 40 (April, 1943) 207.
4. Carson, B. R., K. A. Chittick, D. D. Cole and S. V. Perry, " New features in broadcast receiver designs " R.C.A. Rev. 2.1 (July, 1937) 45. Section " Cabinet acoustics " p. 53.
5. Ebel, A. J. " High Fidelity Systems " Comm. 23.4 (Apr. 1943) 38. 23.5 (May, 1943) 24.
6. Goodell, J. D. and B. M. H. Michel, " Auditory perception " Elect. 19.7 (July, 1946) 142.
7. " FM Loudspeaker distortion—its cause, magnitude and cure " (based on Ref. 3 above) W.W. 49.8 (Aug. 1943) 248.
8. Phelps, W. D. " Acoustic line loudspeakers " Elect. 13.3 (March, 1940) 30.
9. Hoekstra, C. E. " Vented speaker enclosure " Elect. 13.3 (March, 1940) 34.
10. Olney, B. " The co-axial loudspeaker " Elect. 13.4 (April, 1940) 32.
11. Lansing, J. B. " The duplex loudspeaker " Comm. 23.12 (Dec. 1943) 22.
12. Gilbert, R. M. " Baffling the loudspeaker " Radio No. 264 (Dec. 1941) 32.
13. Olson, H. F. (Book) " Elements of acoustical engineering " (D. Van Nostrand Co. N.Y., 1st edit. 1940).
13A. Olson, H. F. (Book) " Elements of acoustical engineering " (D. Van Nostrand Co. N.Y., 2nd edit. 1947).
14. Olson, H. F., and F. Massa (Book) " Applied Acoustics " (P. Blakiston's Son & Co. Philadelphia, 1939).
15. McLachlan, N. W. (Book) " Loud Speakers " (Clarendon Press, Oxford, 1934).
16. Bozak, R. T. " Design of a 27 inch loudspeaker " Elect. 13.6 (June, 1940) 22.
17. Rettenmeyer, R. D. " Improving low-frequency response " Radio Eng. 16.6 (June, 1936) 14.
18. Caulton, C. O., E. T. Dickey and S. V. Perry " The Magic Voice " Radio Eng., 16.10 (Oct., 1936) 8.
19. Tiedje, T. Q. " Speaker design—recent developments in the loudspeaker industry " Radio Eng. 16.1 (Jan. 1936) 11.
20. Olson, H. F. " Action of a direct radiator loudspeaker with non-linear cone suspension system " Jour. Acous. Socy. Am. 16.1 (July, 1944) 1.
21. Hartley, R. V. L. " The production of inharmonic sub-frequencies by a loudspeaker " Jour. Acous. Socy. 16.3 (Jan. 1945) 206.
22. Edelman, P. E. " Condensed loudspeaker with flexible electrodes " Proc. I.R.E. 19.2 (Feb. 1931) 256.
23. Strutt, M. J. O. " On the amplitude of driven loudspeaker cones " Proc. I.R.E. 19.5 (May, 1931) 839. Discussion 19.11 (Nov. 1931) 2030 ; 21.2 (Feb. 1933) 312.
24. Ballantine, S. " A piezo-electric loudspeaker for the higher audio frequencies " Proc. I.R.E. 21.10 (Oct., 1933) 1399.
25. Amos, S. W. " Improving headphone quality—simple correction circuits for use with conventional diaphragm-type earpieces " W.W. 50.11 (Nov. 1944) 332.
26. Jensen Technical Monographs. (1) " Loudspeaker frequency-response measurements." (2) " Impedance matching and power distribution in loudspeaker systems." (3) " Frequency range and power consideration in music reproduction." (4) " The effective reproduction of speech."
27. Olney, B. " A method of eliminating cavity resonance, extending low frequency response, and increasing acoustic damping in cabinet-type loudspeakers " Jour. Acous. Socy. Am. 8.2 (Oct. 1936) 104.
28. Olney, B. J. " The acoustical labyrinth " Elect. 10.4 (April, 1937) 24.
29. Smith, F. W. " Resonant loudspeaker enclosure design " (vented baffle) Comm. 25.8 (Aug. 1945) 35.
30. Klipsch, P. W. " Design of compact two-horn loudspeaker " Elect. 19.2 (Feb. 1946) 156.
31. Klipsch, P. W. " A low frequency horn of small dimensions " Jour. Acous. Socy. Am. 13 (Oct. 1941) 137.
32. Klipsch, P. W. " Improved low frequency horn " Jour. Acous. Socy. Am. 14 (Jan. 1943) 179.

33. Klipsch, P. W. " A high quality loudspeaker of small dimensions " Jour. Acous. Socy. Am. 17.3 (Jan. 1946) 254. With extensive bibliography on horn loudspeakers.
34. Klipsch, P. W. " A note on acoustic horns " Proc. I.R.E. 33.7 (July, 1945) 447.
35. Fletcher, H. " Hearing, the determining factor for high-fidelity transmission," Proc. I.R.E. 30.6 (June, 1942) 266.
36. Olson, H. F., and J. Preston " Wide range loudspeaker developments " R.C.A. Rev. 7.2 (June, 1946) 155.
37. Stanley, A. W. " The output stage-effect of matching on frequency response " W.W. 52.8 (Aug. 1946) 256.
38. Olson, H. F. " Horn loud speakers " R.C.A. Rev. (1) 1.4 (April, 1937) 68 ; (2) Efficiency and distortion, 2.2 (Oct. 1937) 265.
39. Sanial, A. J. " Graphs for exponential horn design," R.C.A. Rev. 3.1 (July, 1938) 97.
40. Wait, E. V. " Loudspeaker cones " A.W.A. Tec. Rev. 1.1 (March, 1935) 14.
41. Rudd, J. B. " The impedance of the voice-coil of a loudspeaker " A.W.A. Tec. Rev. 3.3 (Jan. 1938) 100.
42. Rudd, J. B. " Equipment for the measurement of loudspeaker response " A.W.A. Tec. Rev. 3.4 (April, 1938) 143.
43. de Boer, J. " The efficiency of loudspeakers " Philips Tec. Rev. 4.10 (Oct. 1939) 301.
44. Shorter, D. E. L. " Loudspeaker transient response," B.B.C. Quarterly 1.3 (Oct. 1946) 121 ; summary " Loudspeaker transients " W.W. 52.12 (Dec. 1946) 424. See also Ref. 117.
45. Hilliard, J. K. " High fidelity reproduction " F-M & T 6.1 (Jan. 1946) 28.
46. " The design of acoustic exponential horns " Electronic Eng. 19.235 (Sept. 1947) 286.
47. Sanial, A. J. " Loudspeaker design by electro-mechanical analogy " Tele-Tech 6.10 (Oct. 1947) 38.
48. Klipsch, P. W. " The Klipsch sound reproducer—a corner type speaker which uses the walls as part of the acoustical system," FM & T 7.9 (Sept. 1947) 25.
49. " Enhancing sound reproduction " (Review of patent No. 2, 420, 204 by C. M. Sinnett) Audio Eng. 31.6 (July, 1947) 43.
50. Hopkins, H. F., and N. R. Stryker " A proposed loudness-efficiency rating for loudspeakers and the determination of system power requirements for enclosures " Proc. I.R.E. 36.3 (March, 1948) 315.
51. Olson, H. F., and R. A. Hackley " Combination horn and direct radiator loud-speaker " Proc. I.R.E. 24.12 (Dec. 1936) 1557.
52. Massa, F. " Horn-type loud speakers—a quantitative discussion of some fundamental requirements in their design," Proc. I.R.E. 26.6 (June, 1938) 720.
53. Moir, J. " Acoustics of small rooms," W.W. 50.11 (Nov. 1944) 322.
54. Roder, H. " Theory of the loudspeaker and of mechanical oscillatory systems " Part 1, Radio Engineering 16.7 (July 1936) 10 ; Part 2, 16.8 (Aug. 1936) 21 ; Part 3, 16.9 (Sept. 1936) 24 ; Part 4, 16.10 (Oct. 1936) 19.
55. Morison, G. E. " Power and realism—estimating the watts required for a given sound intensity " W.W. 48.6 (June, 1942) 130.
56. Seabert, J. D. " Electrodynamic speaker design considerations " Proc. I.R.E. 22.6 (June, 1934) 738.
57. Olson, H. F. " A new cone loud speaker for high fidelity sound reproduction " Proc. I.R.E. 22.1 (Jan. 1934) 33.
58. Greaves, V. F., F. W. Kranz and W. D. Crozier " The Kyle condenser loud speaker " Proc. I.R.E. 17.7 (July, 1929) 1142.
59. Terman, F. E. (book) " Radio Engineering " (McGraw-Hill Book Co., New York and London, 3rd edit. 1947).
60. Sawyer, C. B. " The use of rochelle salt crystals for electrical reproducers and microphones " Proc. I.R.E. 19.11 (Nov., 1931) 2020.
61. Rand, G. E. " An F-M quality speaker " (R.C.A. LC-1A) Broadcast News, No. 46 (Sept. 1947) 24.
62. Kennelly, A. E. " Electrical vibration instruments " (The Macmillan Company, New York).
63. Wente, E. C., and A. L. Thuras " Moving-coil telephone receivers and microphones," Jour. Acous. Socy. Am. 3.1 (July, 1931) 44.
64. Williams, A. L. " New piezo-electric devices of interest to the motion picture industry " Jour. S.M.P.E. 32.5 (1939) 552.
65. Olson, H. F., and F. Massa " A high-quality ribbon receiver," Proc. I.R.E. 21.5 (May, 1933) 673.
66. Olson, H. F., and F. Massa " Performance of telephone receivers as affected by the ear " Jour. Acous. Socy. Am. 6.4 (1935) 250.
67. Olson, H. F., Jour. S.M.P.E. 27.5 (1936) 537.
68. Fletcher, H. and others " Auditory perspective " (6 papers) E.E. 53.1 (Jan., 1934) 9.
69. Massa, F. (Book) " Acoustic design charts " (The Blakiston Company, Philadelphia, 1942).
70. " American recommended practice for loudspeaker testing " C16.4-1942 (American Standards Association, Nov. 6, 1942).
71. Olson, H. F. " Multiple coil, multiple cone loudspeakers " Jour. Acous. Socy. Am. 10.4 (April, 1939) 305.
72. Langford-Smith, F. " The relationship between the power output stage and the loudspeaker " Proceedings World Radio Convention, Australia, April 1938 (reprinted W.W. 44.6 (Feb. 9, 1939) 133 ; 44.7 (Feb. 16, 1939)) 167.
73. Edwards, E. M. " Conversion efficiency of loudspeakers " Service 17.3 (March, 1948) 24.
74. Cook, E. D. " The efficiency of the Rice-Kellogg loudspeaker " G.E. Review 33.9 (Sept. 1930) 505.
75. Massa, F. " Loud speaker design " Elect. 9.2 (Feb. 1936) 20.
76. McProud, C. G. " Two-way speaker system " Audio Eng. (1) 31.11 (Nov. 1947) 18 ; (2) 31.12 (Dec. 1947) 19 ; (3) 32.2 (Feb. 1948) 21.
77. Logan, G. H. " Design for exponential horns of square cross section " Elect. 12.2 (Feb. 1939) 33. Reprinted in " Electronics for Engineers " (McGraw-Hill, 1945).
78. Barker, A. C. " Single diaphragm speakers—obtaining wide frequency response and effective damping " W.W. 54.6 (June, 1948) 217.
79. Lanier, R. S. " What makes a good loudspeaker ? " Western Electric Oscillator No. 8 (July, 1947) 11.
80. Nickel, F. " Quality loudspeakers for every use " Western Electric Oscillator No. 8 (July, 1947) 7.
81. Davies, L. A. " Electro-acoustic coupling—how to get high fidelity from ordinary speakers," Australasian Radio World, 11.10 (March, 1947) 7.
82. White, S. J. " Horn type loudspeakers " Audio Eng. 32.5 (May, 1948) 25.
83. Olson, H. F. and Radio Corporation of America " Improvement in folded suspension loudspeaker " U.S. Patent 545,672 (Div. 5) July 19, 1944.
84. Langford-Smith, F. " The design of a high-fidelity amplifier (1) The power valve and the loudspeaker " Radiotronics No. 124 (March-April, 1947) 25.
85. " Stereophonic sound," Elect. 21.8 (August, 1948) 88.

87. Beard, E. G. (Book) " Philips Manual of radio practice for servicemen " (Philips Electrical Industries of Australia Pty. Ltd., 1947) pp. 257-259 describing the Philips Miller system.
88. Meyer, E. (Book) " Electro-acoustics " (G. Bell & Sons Ltd. London 1939).
89. Wood, Alexander (Book) " The physics of music " (Methuen & Co. Ltd., London, 3rd edit. 1945).
90. Tillson, B. F. " Musical acoustics " Audio Eng. Part 1, 31.5 (June, 1947) 34 ; Part 2, 31.6 (July, 1947) 25 ; Correction 32.6 (June, 1948) 8.
91. Vermeulen, R. " Octaves and decibels," Philips Tec. Rev. 2.2 (Feb., 1937) 47.
92. Jeans, Sir James (Book) " Science and Music " (MacMillan Company and Cambridge University Press, 1938).
93. Mills, J. (Book) " A Fugue in cycles and bels " (Chapman & Hall, London, 1936).
94. Souther, H. " Coaxial and separate two-way speaker system design " Comm. 28.7 (July, 1948) 22.
95. " Speakers for sound equipment " American R.M.A. Standard SE-103 (April, 1949).
96. Drisko, B. B. " Getting the most out of a reflex-type speaker " Audio Eng. 32.7 (July, 1948) 24.
97. Angevine, O. L., and R. S. Anderson " The problem of sound distribution " Audio Eng. (1) 32.6 (June, 1948) 18 ; (2) 32.9 (Sept. 1948) 16 ; (3) 32.10 (Oct. 1948) 28.
98. Kendall, E. L. " Electric power required for sound systems " Service 17.9 (Sept. 1948) 12.
99. Klipsch, P. W. " Progress in Klipsch speakers " FM-TV 8.11 (Nov. 1948) 36.
100. McProud, C. G. " A new corner speaker design " Audio Eng. 33.1 (Jan. 1949) 14 ; 33.2 (Feb. 1949) 13.
101. Camras, M. " A stereophonic magnetic recorder " Proc. I.R.E. 37.4 (April 1949) 442.
102. Seashore, C. E. (Book) " Psychology of music " (McGraw-Hill Book Co., New York and London, 1938).
103. Briggs, G. A. (Book) " Loudspeakers—the why and how of good reproduction " (Wharfedale Wireless Works, Bradford, Yorkshire, England, 2nd edit. 1948).
104. Greenlees, A. E. (Book) " The amplification and distribution of sound " (Chapman and Hall Ltd., London, 2nd edit. 1948).
105. McProud, C. G. " Corner speaker cabinet for 12 inch cones " Audio Eng. 35.5 (May 1949) 14.
106. Cunningham, W. J. " Non-linear distortion in dynamic loudspeakers due to magnetic fields " Jour. Acous. Socy. Am. 21.3 (May 1949) 202.
107. Hartley, H. A. " New notes in radio " (booklet published by H. A. Hartley Co. Ltd., London).
109. Veneklasen, P. S. " Physical measurements of loudspeaker performance " Jour. S.M.P.E. 52.6 (June 1949) 641.
110. Maxfield, J. P. " Auditorium acoustics " Jour. S.M.P.E. 51.2 (Aug. 1948) 169.
112. Grignon, L. D. " Experiment in stereophonic sound " Jour. S.M.P.E. 52.3 (March 1949) 280.
113. Chapman, C. T. " Vented loudspeaker cabinets—basis of design to match existing loudspeaker units," W.W. 55.10 (Oct. 1949) 398.
114. American R.M.A. Standard REC-104 " Moving coil loudspeakers for radio receivers " (Jan. 1947).
115. Klipsch, P. W. " New notes on corner speakers " FM-TV 9.8 (Aug. 1949) 25.
116. Olson, H. F., J. Preston and D. H. Cunningham " New 15-inch duo-cone londspeaker " Audio Eng. 33.10 (Oct. 1949) 20.
116A. Olson, H. F., J. Preston and D. H. Cunningham " Duo-cone loudspeaker " R.C.A. Rev. 10.4 (Dec. 1949) 490.
117. Briggs, E. A. (Book) " Sound reproduction " Wharfedale Wireless Works, Bradford, Yorks. England, 1949.
118. Planer, F. E., and I. I. Boswell " Vented loudspeaker enclosures " Audio Eng. 32.5 (May 1948) 29.
119. Klipsch, P. W. " Woofer-tweeter crossover network " Elect. 18.11 (Nov. 1945) 144.
120. Harvey, F. K. " Focussing sound with microwave lenses " Bell Lab. Record 27.10 (Oct. 1949) 349.
121. Beranek, L. L. (Book) " Acoustic measurements " (John Wiley & Sons, Inc. New York ; Chapman & Hall Ltd. London, 1949).
122. U.S. Patents 2,310,243 and 2,373,692.
123. Salmon, V. " A new family of horns " Jour. Acous. Soc. Am. 17.3 (Jan. 1946) 212.
124. Olson, H. F., J. C. Bleazey, J. Preston and R. A. Hackley " High efficiency loud speakers for personal radio receivers " R.C.A. Rev. 11.1 (Mar. 1950) 80.
125. Haas, H. (report) " The influence of a single echo on the audibility of speech " Translated into English and available from Director, Building Research Station, Garston, Watford, Herts, England.
126. Moir, J. " Transients and loudspeaker damping " W.W. 56.5 (May 1950) 166.
127. Barbour, J. M. " Musical scales and their classification " J.A.S.A. 21.6 (Nov. 1949) 586.
128. Mawardi, O. K. " Generalized solutions of Webster's horn theory " J.A.S.A. 21.4 (July 1949) 323.
129. Goodfriend, L. W. " Simplified reverberation time calculation " Audio Eng. 34.5 (May, 1950) 20.
130. Gilson, W. E., J. J. Andrea " A symmetrical corner speaker " Audio Eng. 34.3 (March 1950) 17.
131. Corrington, M. S. " Transient testing of loudspeakers " Audio Eng. 34.8 (Aug. 1950) 9.
132. Moir, J. " Developments in England—new loudspeaker measurement technique are being used to correlate subjective and instrumental performance evaluations." FM-TV 10.9 (Sept. 1950) 23.
133. Olson, H. F. " Sensitivity, directivity and linearity of direct radiator loudspeakers " Audio Eng. 34.10 (Oct. 1950) 15.
134. Plach, D. J. and P. B. Williams " A new loudspeaker of advanced design " Audio Eng. 34.10 (Oct. 1950) 22.
135. Shorter, D. E. L. " Sidelights on loudspeaker cabinet design " W.W. (1) 56.11 (Nov. 1950) 382 ; (2) 56.12 (Dec. 1950) 436. British Patent Application No. 24528/49.
136. Smith, B. H. " A distributed-source horn " Audio Eng. 35.1 (Jan. 1951) 16.
137. Taylor, P. L. " Ribbon loudspeaker " W.W. 57.1 (Jan. 1951) 7.
138. Ewaskio, C. A., and O. K. Mawardi " Electroacoustic phase shift in loudspeakers " J. Acous. Soc. Am. 22.4 (July 1950) 444.
139. Molloy, C. T. " Response peaks in finite horns " J. Acous. Soc. Am. 22.5 (Sept. 1950) 551.
140. Thiessen, G. J. " Resonance characteristics of a finite catenoidal horn " J. Acous. Soc. Am. 22.5 (Sept. 1950) 558.
141. Kock, W. E., and F. K. Harvey " Refracting sound waves " J. Acous. Soc. Am. 21.5 (Sept. 1949) 471.
142. Morse, P. M. (Book) " Vibration and Sound " (McGraw-Hill Book Co. Inc. 2nd ed. 1948).
143. Worden, D. W. " Design, construction and adjustment of reflexed loudspeaker enclosures " Audio Eng. 34.12 (Dec. 1950) 15.
144. Smith, B. H. " Resonant loudspeaker enclosures " Audio Eng. 34.12 (Dec. 1950) 22.
145. Parkin, P. H., and W. E. Scholes " Recent developments in speech reinforcement systems " W.W. 57.2 (Feb. 1951) 44.
146. Preisman, A. " Loudspeaker damping " Audio Eng. 35.3 (March 1951) 22 ; 35.4 (April 1951) 24.
147. Souther, H. T. " Design elements for improved bass response in loudspeaker systems " Audio Eng. 35.5 (May 1951) 16.

148. (A) Tanner, R. L. " Improving loudspeaker response with motional feedback " Elect. 24.3 (Mar. 1951) 142.   (B) Letter R. B. McGregor 24.6 (June 1951) 298.
149. Gardner, M. F., and J. L. Barnes (Book) " Transients in Linear Systems " (John Wiley & Sons, New York 1942) Chapter 2.
150. Moir, J. " Loudspeaker performance—achieving optimum reproduction in small rooms—comparative advantages of open-cone and horn-loaded speakers " FM-TV 11.4 (Apr. 1951) 32.
151. Plach, D. J., and P. B. Williams " Loudspeaker enclosures " Audio Eng. 35.7 (July 1951) 12.
152. Olson, H. F., and D. F. Pennie " An automatic non-linear distortion analyzer " R.C.A. Rev. 12.1 (Mar. 1951) 35.
153. Hall, H. H., and H. C. Hardy " Measurements for aiding in the evaluation of the quality of loudspeakers " Abstract J. Acous. Soc. Am. 20.4 (July 1948) 596.
154. Peterson, A. P. G. " Intermodulation distortion " G. R. Exp. 25.10 (March 1951) 1.
155. Moir, J. " Loudspeaker diaphragm control " W.W. 57.7 (July 1951) 252.
156. Salmon, V. " Efficiency of direct radiator loudspeakers " Audio Eng. 35.8 (Aug. 1951) 13.
157. Clements, W. " A new approach to loudspeaker damping " Audio Eng. 35.8 (Aug. 1951) 20.
158. Hardy, H. C. and H. H. Hall " The transient response of loudspeakers " (summary) J. Acous. Soc. Am. 20.4 (July 1948) 596.
159. Sleeper, M. B. and C. Fowler " The FAS audio system " FM-TV 10.10 (Oct. 1950) 22 ; 10.11 (Nov. 1950) 31 ; 10.12 (Dec. 1950) 24 ; 11.2 (Feb. 1951) 32 ; 11.3 (Mar. 1951) 46 ; 11.5 (May 1951) 34.
    Allison, R. F. " Improving the air-coupler " FM-TV (1) 11.10 (Oct. 1951) 26 ; (2) 11.12 (Dec. 1951) 32.
160. Fletcher, H. " The stereophonic sound film systems—general theory " J. Acous. Soc. Am. 13.2 (Oct. 1941) 89.
161. Maxfield, J. P., A. W. Colledge and R. T. Freibus " Pickup for sound motion pictures (including stereophonic) " J. Acous. Soc. Am. 30.6 (June 1938) 666.
162. Koenig, W. " Subjective effects in binaural hearing " J. Acous. Soc. Am. 22.1 (Jan. 1950) 61.
163. Goodfriend, L. S. " High fidelity " Audio Eng. 34.11 (Nov. 1950) 32.
164. Meeker, W. F., F. H. Slaymaker and L. L. Merrill " The acoustical impedance of closed rectangular loudspeaker housings " J. Acous. Soc. Am. 22.2 (March 1950) 206.
165. Smith, B. H., and W. T. Selsted " A loudspeaker for the range 5 to 20 Kc " Audio Eng. 34.1 (Jan. 1950) 3.
166. Beranek, L. L. " Enclosures and amplifiers for direct radiator loudspeakers " Paper presented at the National Electronics Conference, Chicago, Illinois, Sept. 26, 1950.
167. Plach, D. J. and P. B. Williams " A 3-channel unitary loudspeaker " Radio and T.V. News 44.5 (Nov. 1950) 66.
168. Olson, H. F. and A. R. Morgan " A high-quality sound system for the home." Radio and T.V. News 44.5 (Nov. 1950) 59.
169. Moir, J. " Stereophonic sound " W.W. 57.3 (March 1951) 84.
170. Pipe, D. W. " Sound reinforcing " W.W. 57.3 (March 1951) 117.
171. Peterson, A. P. G. " Intermodulation distortion " G.E. Exp. 25.10 (Mar. 1951) 1.
172. Shorter, D. E. L. " Loudspeaker transient response " B.B.C. Quarterly 1 : 3 (Oct. 1946) 121.
173. Bordoni, P. G. " Asymmetrical vibrations of cones " J. Acous. Soc. Am. 19.1 (Jan. 1947) 146.
174. Powell, T. (letter) " Transients " Audio Eng. 35.9 (Sept. 1951) 12.
175. Kinsler, L. E. and A. R. Frey (Book) " Fundamentals of Acoustics " John Wiley & Sons Inc., New York ; Chapman & Hall Ltd., London, 1950.
176. " Reference Data for Radio Engineers " Federal Telephone and Radio Corporation, 3rd ed. 1949.
177. Frayne, J. G. and H. Wolfe (Book) " Elements of Sound Recording " John Wiley & Sons Inc., New York ; Chapman & Hall, London, 1949.
178. Sabine, H. J. " A review of the absorption coefficient problem " J. Acous. Soc. Am. 22.3 (May 1950) 387.
179. Corrington, M. S. and M. C. Kidd " Amplitude and phase measurements on loudspeaker cones " Proc. I.R.E. 39.9 (Sept. 1951) 1021.
180. Beranek, L. L. " Design of loudspeaker grilles " Newsletter of I.R.E. Professional Group on Audio, 2.4 (July 1951) ; Corr. 2.5 (Sept. 1951) 5.   Reprirt, Radiotronics 17.2 (Feb. 1951).
181. Schneider, A. W. " Sound reinforcing systems " Audio Eng. 34.11 (Nov. 1950) 27.
182. Youngmark, J. A. " Design data for a bass-reflex cabinet " Audio Eng. 35.9 (Sept. 1951) 18.
183. Moir, J. " Developments in England—A discussion of the latest trends in the design of English loudspeakers for high fidelity applications " FM-TV, 10.7 (July 1950) 26.
184. Williamson, D. T. N. (letter) " More views on loudspeaker damping " W.W. 53.10 (Oct. 1947) 401.
185. Langford-Smith, F. (letter) " Loudspeaker damping " W.W. 53.8 (Aug. 1947) 309 ; Replies 53.9 (Sept. 1947) 343-4 ; 53.10 (Oct. 1947) 401-2 ; also Ref. 186.
186. Voigt, P. G. A. H. (letter) " Loudspeaker damping " W.W. 53.12 (Dec. 1947) 487.
187. Geppert, D. V. " Loudspeaker damping as a function of the plate resistance of the power output tube " Audio Eng. 34.11 (Nov. 1950) 30.
188. Knowles, H. S. " Loud-speakers and room acoustics " being Chapter 16 of " Radio Engineering Handbook " edited by K. Henney, McGraw-Hill Book Co., 4th ed., 1950.
189. Canby, E. T. " Record Revue—The RJ speaker " Audio Eng. 35.10 (Oct. 1951) 23.
190. Joseph, W., and F. Robbins " The R-J speaker enclosure " Audio Eng. 35.12 (Dec. 1951) 17.
191. Wentworth, J. P. " Loudspeaker damping by the use of inverse feedback " Audio Eng. 35.12 (Dec. 1951) 21.
192. Yeich, V. " Listening room design " Audio Eng. 35.11 (Nov. 1951) 28.
193. Olson, H. F. " Direct radiator loudspeaker enclosures " Audio Eng. 35.11 (Nov. 1951) 34.
194. Kramer, K. " A three-channel loudspeaker " FM-TV 11.9 (Sept. 1951) 40.
195. Kobrak, H. G. " Auditory perspective—a study of the biological factors related to directional hearing " J.S.M.P.T.E. 57.4 (Oct. 1951) 328.
196. Beranek, L. L., W. H. Radford, J. A. Kessler and J. B. Wiesner " Speech-reinforcement system evaluation " Proc. I.R.E. 39.11 (Nov. 1951) 1401.
197. Parkin, P. H., and J. H. Taylor " Speech reinforcement in St. Paul's Cathedral " W.W. 58.2 (Feb. 1952) 54 ; 58.3 (Mar. 1952) 109.
198. Lowell, H. H. " Motional feedback " Elect. 24.12 (Dec. 1951) 334.
199. Childs, U. J. " Loudspeaker damping with dynamic negative feedback " Audio Eng. 36.2 (Feb. 1952) 11.
200. Clements, W. " It's positive feedback " Audio Eng. 36.5 (May 1952) 20.
201. Childs, U. J. " Further discussion on positive current feedback " Audio Eng. 36.5 (May 1952) 21.·

Additional references will be found in the Supplement commencing on page 1475.

CHAPTER 21

THE NETWORK BETWEEN THE POWER VALVE AND THE LOUDSPEAKER

By F. Langford-Smith, B.Sc., B.E.

SECTION 1 : LOUDSPEAKER "MATCHING"

(i) Loudspeaker characteristics and matching (ii) Optimum plate resistance (iii) Procedure for " matching " loudspeakers to various types of amplifiers.

(i) Loudspeaker characteristics and matching

If the impedance of a loudspeaker were a constant resistance, the problem would be simple. In reality, the magnitude of the impedance may vary in a ratio up to 10 : 1 and the phase angle may be anything between 50° leading and 60° lagging—see Chapter 20 Sect. 1(v).

A loudspeaker, or output, transformer is an impedance changing device, with an impedance ratio approximately equal to the square of the turns ratio (see Chapter 5). The transformer by itself has no "impedance" in this sense—it merely reflects across the primary the load impedance placed across the secondary terminals, multiplied by its impedance ratio. Example : A transformer has a turns ratio 10 : 1 from primary to secondary. Its impedance ratio is therefore 100 : 1 primary to secondary. If the secondary load is 8 ohms, the reflected " primary impedance " will be 800 ohms. The only limitations on the wider use of the same transformer are the primary inductance, which will reduce the low frequency response, the maximum d.c. plate current and the maximum power output.

(ii) Optimum plate resistance

The power output of a loudspeaker is equal to the power input multiplied by the efficiency. The power input to the primary of the loudspeaker transformer, under the usual operating conditions, is equal to the square of the signal plate current multiplied by the reflected load resistance—

Power input in watts $= I_p{}^2 R_L$ (1)

where I_p = signal plate current in amperes, r.m.s.
and R_L = load resistance reflected on to the primary of the transformer.

The signal plate current is given by

$I_p = \mu E_g/(r_p + R_L)$ (2)

where μ = amplification factor of power output valve
       E_g = signal grid voltage, r.m.s.
and r_p = effective plate resistance of power output valve.

Consequently, at the bass resonant frequency where the resistance R_L may rise to a high value, it is possible to have maximum plate circuit efficiency and yet to have

less than the maximum electrical power output, because the signal plate current is reduced by the high load resistance.   This effect is most pronounced with low values of plate resistance, and with triode valves or negative feedback the loudspeaker acoustical output may actually fall at the bass resonant frequency.   With pentodes without feedback the reverse occurs, since a pentode approximates to a constant current source, and there is a high peak of output at the bass resonant frequency.
   A similar effect occurs at the higher frequencies, the highest level of response occurring with pentodes, an intermediate value with triodes having $R_L = 2r_p$, and the lowest with constant voltage at all frequencies (a condition approached when a high value of negative feedback is used).
   For any one loudspeaker, there is a value of plate resistance which provides most nearly constant response at all frequencies.
   Reference 5.

(iii) Procedure for " matching " loudspeakers to various types of amplifiers

   When the load resistance is constant, the only procedure necessary is to select a transformer ratio so that the resistance reflected into the primary is the correct load for the amplifier.   When the load is a loudspeaker, the procedure is outlined below.

   Class A Triodes, either singly or in push-pull, may be treated very simply by arranging for the loudspeaker impedance at 400 c/s to equal the correct load for the amplifier.   Thanks to the shape of the triode characteristics, the rise of impedance at the bass resonant frequency decreases the distortion, and although the power output from the valve is lower than at 400 c/s this is counterbalanced by the rise of loudspeaker efficiency at this frequency.   The rise of impedance above 400 c/s results in a tendency towards a falling response, but loudspeakers specially designed for use with triodes are capable of giving fairly uniform response up to their limiting frequency.   As a result, the designer of an amplifier with a triode output stage need not consider the impedance versus frequency characteristics of the loudspeaker, but only the response.
   The ratio of the nominal load resistance R_L to the output resistance R_0 of the amplifier is not unimportant.   If R_L/R_0 is very high, the loudspeaker is being operated with nearly constant voltage at all frequencies.   If R_L/R_0 is around 2 or 3, the voltage applied to the loudspeaker is slightly greater at frequencies where the loudspeaker impedance rises—this is generally an advantage.   See also Chapter 13 Sect. 2(iv).
   Pentodes without feedback are very critical with regard to load impedance [Chapter 13 Sect. 3(viii)].   Steps which may be taken to minimize the serious distortion include :
   (1) The choice of a loudspeaker with a smaller variation of impedance with frequency, at least over the low frequency range [Chapter 20 Sect. 1(v)].
   (2) Shunting the load by a capacitance, or a capacitance in series with a resistance, to reduce the impedance somewhat at high audio frequencies [Chapter 13 Sect. 3(viii)]. However, if the capacitance is sufficient to reduce the impedance to a more or less horizontal curve, there will inevitably be attenuation of the higher frequencies.
   (3) Shunting the load by an inductance, through the use of a fairly low inductance of the transformer primary [Chapter 5 Sect. 3(iii)c ; Chapter 13 Sect. 3(vii), (viii)]. This only affects the low frequency peak, and is undesirable because of the resultant distortion.
   (4) The use of a vented baffle to reduce the low frequency impedance peak [Chapter 20 Sect. 3(iv)].
   (5) The selection of a loudspeaker impedance at 400 c/s rather lower than the nominal value, so that there is a variation on both sides of the nominal value instead of only on the upper side [Chapter 13 Sect. 3(viii)].
   (6) The selection of an output valve capable of giving considerably more—say 3 times—the desired power output, so as to reduce the distortion with an incorrect load impedance.
   (7) The use of negative voltage feedback (see below).

Fortunately, owing to the characteristics of music, the input voltage applied to the grid of the power amplifier at high frequencies will normally be less than the maximum. This somewhat reduces the distortion at high frequencies.

Pentodes with negative voltage feedback, when operated well below maximum power output, are less affected by the loudspeaker impedance variations than pentodes without feedback.   If the feedback is large, the output voltage remains nearly constant irrespective of the loudspeaker.   There is an optimum value of amplifier output resistance to give most nearly uniform response at all frequencies with any one loudspeaker.

When operated at maximum power output with the small degree of feedback usual in radio receivers, the load resistance is critical.   Satisfactory operation with low distortion can only be obtained by reducing the grid input voltage.

With a large amount of feedback, as for a cathode follower, the impedance at 400 c/s should be the nominal value for the valve.   The increase of impedance at other frequencies will not then cause distortion.   See also Chapter 7 Sect. 5(iv).

SECTION 2 :   MULTIPLE AND EXTENSION LOUDSPEAKERS

  (*i*) *Multiple loudspeakers—general*   (*ii*) *Sound systems*   (*iii*) *Extension loudspeakers* (*iv*) *Operation of loudspeakers at long distances from amplifier.*

(i) Multiple loudspeakers—general

  Two or more loudspeakers may be connected to a single amplifier either to reinforce the sound in a large space, or to give a pseudo-stereophonic effect.   In all normal cases the loudspeakers will each handle the whole frequency range and operate at approximately the same level.   The usual arrangement is to connect the loudspeakers in parallel either on the primary or secondary side.   In the former case an a-f choke may be used in the plate circuit with capacitive coupling to the transformer primaries (Fig. 21.1).   If the correct load resistance is R_L, the nominal impedance of T_1 and T_2 should each be $2R_L$ If there are N loudspeakers in parallel, each should have a nominal impedance of NR_L.   The value of C_1 may be calculated as a coupling condenser [Chapter 12 Sect. 2(ii)] into a resistance R_L.   If $R_L = 5000$ ohms, C_1 may be 1μF for about 1 db loss at 60 c/s.   The value of L_1 may be calculated as for an a-f transformer [Chapter 5 Sect. 3(iii)c and Table 2 p. 213].   The total attenuation is the sum of that due to L_1, the coupling condenser C_1 and the transformer T_1.   A suitable value for L_1 is normally from 10 to 30 henrys.

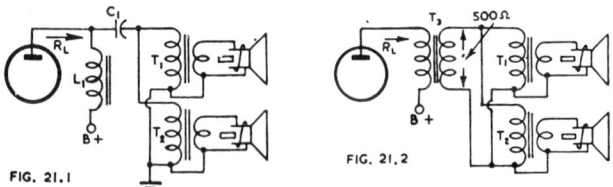

    FIG. 21.1   FIG. 21.2

  Fig. 21.1.   *Choke-capacitance coupling to two or more loudspeakers in parallel.*
  Fig. 21.2.   *Transformer coupling to 500 ohm line, across which two or more loudspeakers are connected in parallel.*

  Alternatively the parallel loudspeakers may be connected across a 500 ohm (or any other convenient value) line from the secondary of transformer T_3 as in Fig. 21.2. Here T_3 will have a 500 ohm secondary and will reflect an impedance R_L into the primary. T_1 and T_2 will have nominal impedances each 1000 ohms.   If there are N loudspeakers in parallel, each will have a nominal impedance of $N \times 500$ ohms.

The simplest arrangement is to connect the secondaries in parallel as for extension loudspeakers (Fig. 21.5).

Loudspeakers may also be connected in series or in series-parallel, provided that the units are identical, with identical impedances.   If loudspeakers having different impedances are connected in parallel, the power output from each will be inversely proportional to its impedance—

Loudspeaker No. 1.   Impedance R_1 Power output P_1
Loudspeaker No. 2   Impedance R_2 Power output P_2

Connected in parallel with applied voltage E :

$P_1 = E^2/R_1$; $P_2 = E^2/R_2$ etc. for any number.

If it is desired to operate two loudspeakers with power outputs in the ratio A to 1, where A is greater than 1 :

$P_1/P_2 = A$; $R_2/R_1 = A$

The impedance of the two in parallel will be

$R = R_1R_2/(R_1 + R_2)$.

Example : A monitor loudspeaker is to operate with an input power of 1 watt and is to be connected across a combined loudspeaker load of 30 watts with a total (line) impedance of 500 ohms.   The ratio of power is 30 : 1 so that the impedance of the monitor loudspeaker (measured across the primary of its transformer) should be 30 $\times$ 500 = 15 000 ohms.   In this case the monitor will have negligible effect on the combined impedance.

(ii) Sound systems

In medium and large installations it is convenient to design amplifiers for a constant output voltage.   The usual output voltage in U.S.A. is 70.7 volts but higher voltages 100, 141, 200 etc. are also used (R.M.A. SE-101-A, July 1949, SE-106, July 1949). Each amplifier normally has its own distribution line suitably loaded by loudspeakers. Each loudspeaker may be arranged—by the use of transformer tappings—to take from the line any desired power, while any loudspeaker may be moved from one line to another without affecting its output level.   If the loudspeaker transformers are correctly matched to the voice coil impedance and the line voltage, the matching of the whole distribution line to the amplifier will be correct when the line is loaded to the full capabilities of the amplifier.

The American Standard (R.M.A. SE-106) specifies transformer secondary impedances of 4, 8 or 16 ohms ; this does not apply to transformers which are furnished only as part of a loudspeaker.   The primaries are tapped to provide output power levels of 1 watt and proceeding upwards and downwards in 3 db steps (i.e. 1, 2, 4 watts etc., 0.5, 0.25 etc.) when the standard input voltage is applied.

(iii) Extension loudspeakers

An extension loudspeaker is one which may be added to the existing loudspeaker in a radio receiver, for which provision may or may not be made by the set manufacturer.

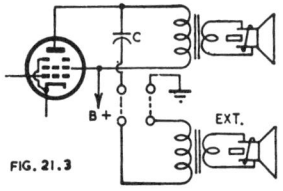

FIG. 21.3

The most common provision for an extension loudspeaker is shown in Fig. 21.3 in which the extension loudspeaker (marked " EXT ") is fed through a blocking condenser connected to the plate of the power valve.   It is necessary for the user to arrange for flexible leads to be taken from two terminals on the chassis, across the primary of the stepdown transformer, to the permagnetic extension speaker.   If the extension speaker has the same reflected load impedance as the transformer of the original speaker,

Fig. 21.3.   Common form of extension loudspeaker.

the power output will be shared equally between the two, but the impedance of the two speaker primaries in parallel will be only half what it should be for maximum power output.   If no provision is made to provide correct matching, the arrangement will be quite practical, except that the power output of both speakers together will be less than with correct matching.

An alternative arrangement, which has much to commend it, is the use of an extension speaker with an impedance of about twice that of the speaker in the set. This means that the extension speaker will operate at a lower sound level than the one in the set, but the mis-matching will be less severe and the maximum volume obtainable from the set will not be seriously affected.

If negative feedback is used in the receiver, the effect of the connection of the external speaker on the volume level of the loudspeaker in the set will not be very noticeable at low or medium levels, although the same problem arises in regard to the overload level.

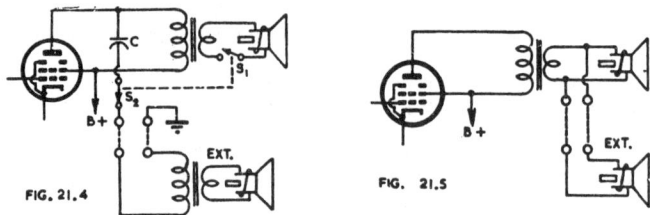

Fig. 21.4.   *Extension loudspeaker with provision for switching either loudspeaker alone.*
Fig. 21.5.   *Extension loudspeaker connected to the voice coil circuit.*

It is possible to modify the previous arrangement by means of switches which can open-circuit the voice coil of the first speaker and at the same time close a switch in the primary circuit of the second speaker. This is illustrated in Fig. 21.4 from which it will be seen that only one speaker will be operating at the one time and there is therefore no problem with correct matching. With this arrangement, the extension speaker should have the same impedance as the one in the set and the power input to both speakers will then be equal. The two switches S_1 and S_2 could, of course, be combined into a single wafer switch.

The previous arrangements have all adopted an extension from the primary of the loud speaker transformer and therefore at a high impedance. There are advantages to be gained in using the voice-coil circuit for the extension, as illustrated in Fig. 21.5. This avoids the necessity for a step-down transformer on the extension speaker and for a blocking condenser. The set manufacturer may fit two terminals on the loudspeaker housing, connected to the voice coil, as an alternative to Fig. 21.3. To obtain the same power from both speakers, it is necessary for the voice-coil impedances to be equal. If it is required to have one speaker operating at a higher level of sound than the other, the impedance of the second (lower output) speaker should be made higher than that of the first speaker. In such a case it is possible to obtain correct matching by calculating the impedance of both voice-coils in parallel and selecting a step-down transformer to suit. Very few voice-coils have an impedance less than about 2 ohms, so that it is possible to use quite ordinary wiring in the connections of the extension speaker. If the second speaker is to be situated more than say 10 feet from the first speaker, it may be desirable to use heavy wire, such as power flex (twin plastic is very convenient). In some cases it may be desired to operate the extension speaker at a rather lower level than the first speaker, in which case losses in the extension line may be desirable.

A single-pole double-throw switch may be used to change over from one to the other voice-coil as shown in Fig. 21.6 (A). Here switch S is used to open-circuit the first voice-coil and at the same time to close the circuit to the second voice-coil. The further refinement of a series volume control R is shown in the extension speaker circuit, so that the volume may be reduced below that of the speaker in the receiver. The resistance R should have a maximum value about 20 times that of the voice-coil impedance, but even so this arrangement cannot be used to reduce the volume to zero.

In order to have a complete control over the volume from the extension loudspeaker, the arrangement of (B) may be used in which R is a potentiometer with the moving contact taken to the extension speaker.

The series resistor volume control shown in Fig. 21.6 (A) increases the effective impedance of the extension loudspeaker circuit at low volumes, but this is not a serious detriment since the volume will be low and the mis-matching of only minor importance, particularly if negative feedback is used.   The potentiometer method of volume control in Fig. 21.6 (B) has to be a compromise, and is incapable of giving satisfactory matching under a wide range of conditions.   A reasonable compromise for the resistance R would be about five times the impedance of the voice-coil, but this will result in appreciable loss of power even at maximum volume.   For perfect matching, resistance R should be taken into account, but for many purposes the effect

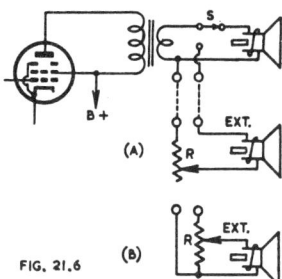

Fig. 21.6. Extension loud-speaker circuits incorporating volume control of the extension speaker.

FIG. 21.6

may be neglected provided that R is not less than five times the voice-coil impedance. At low levels this arrangement has a high impedance but here again the effect will not be serious.

Another form of volume control is a tapped secondary winding, with the extension speaker connected through a rotary switch to a selected tap.   Taps may be made with each one 70% of the turns of the one above it, giving roughly 3 db steps, or 80% for 2 db steps.   This method causes serious mis-matching, and is not advisable unless there are several main loudspeakers and only one " extension " with volume control.

The best method of volume control is a L type level control (Fig. 21.11) which provides constant input resistance and may be continuously variable (see Chapter 18).

All the preceding arrangements are limited to the use of either of two speakers or have given a choice between one and both.   **The ideal arrangement is to permit the use of one or other or both.**   This is particularly helpful when the extension speaker is used in a different room and one may wish to operate the extension speaker alone.   Tuning-in may be done by switching over to both speakers, adjusting the volume level to suit the extension speaker, and then switching over so that only the extension is in operation.   Then, if at any time the speaker in the receiver is required to operate, this may be done simply by moving the switch without causing any interference to the extension speaker.

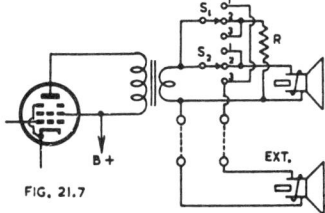

Fig. 21.7. Extension loudspeaker circuit incorporating switching to operate either or both speakers.

FIG. 21.7

If it is desired to use a single secondary winding on the transformer it is possible to arrange two speakers as shown in Fig. 21.7 so that either or both may be operated. With this arrangement the resistor R (which should have a resistance equal to the voice-coil impedance of one of the speakers) has to be used to provide correct matching when one speaker only is operating.   Only half the full power output is available when one speaker is operating alone ; the full power output is, however, available when both speakers are operating together (switch position 1).   This has the result that the switching in of the second speaker does not affect the volume level of the first speaker, the resistor R really being a dummy load to take the power which would otherwise be supplied to the second speaker.   The two switches S_1 and S_2 may be made from a single wafer wave-change switch.

A preferred arrangement, which does not result in any power loss under any circumstances, is the use of a tapped secondary winding on the transformer.   The full winding on the secondary is used for one or other loudspeaker, while the tap is used for both operating together ; this involves a more elaborate switching arrangement but one which may be justified in certain circumstances.

The use of an appreciable amount of negative voltage feedback results in nearly constant voltage across the load under all conditions, and minimizes changes in volume level with any form of switching.

When the loudspeaker is some distance from the transformer secondary, the power loss in the connecting wiring should not exceed 15%.

Line length for 15% line power loss in low impedance lines :

Wire size		Load Impedance (Ohms)					
A.W.G.	S.W.G.	2	4	6	8	10	15
14	16	60 ft.	120 ft.	180 ft.	240 ft.	300 ft.	450 ft.
16	18	38 ft.	75 ft.	113 ft.	150 ft.	190 ft.	285 ft.
18	19	23 ft.	47 ft.	70 ft.	95 ft.	118 ft.	177 ft.
20	21	15 ft.	30 ft.	45 ft.	60 ft.	75 ft.	112 ft.
22	23	9 ft.	18 ft.	28 ft.	37 ft.	47 ft.	70 ft.

Reference 20.

(iv) Operation of loudspeakers at long distances from amplifier

When one or more loudspeakers are to be operated at a considerable distance from the amplifier, it is usual to have at the amplifier a transformer to step-down, for example, to a 500 ohm " line " which may be in the form of ordinary electric power wiring for distances up to several thousand feet.   At the far end there must be another step-down transformer from 500 ohms to the correct impedance to match the voice coil circuit impedance.   In such a case the " line " does not itself impose any appreciable load ; the loudspeaker impedance is reflected back through the transformer to load the line.

Line length for 15% line power loss in 500 ohm lines :

Wire size (A.W.G.)	19	21	23	25
Wire size (S.W.G.)	20	22	24	26
Length of line (feet)	4750	2880	1780	1500

Although 500 ohms is a popular value, any lower or higher value may be used provided that the resistance of the line is not too great, and that the capacitance across the line does not seriously affect the high frequency response.

A very popular line impedance is 600 ohms.

SECTION 3 : LOUDSPEAKER DIVIDER NETWORKS

When two loudspeakers are used in a 2 way system, it is necessary to have a frequency dividing network. The attenuation characteristic may have an ultimate slope of 6, 12 or 18 db per octave, but the most generally satisfactory compromise is 12 db per octave. A slope of 6 db per octave is usually insufficient to prevent overloading of the high frequency unit at frequencies below the cross-over frequency.

The theory of frequency dividing networks is covered in Chapter 4 Sect. 8(x). The simplest possible arrangement is Fig. 4.53A for which the values of L_0 and C_0 are given on page 184 for known values of R_0 and the cross-over frequency f_c. In order to obtain reasonably satisfactory performance, the high-frequency unit should be a fairly substantial 5 in. or larger loudspeaker, and the cross-over frequency should preferably be between 800 and 1200 c/s. The high frequency response will obviously be restricted. The low frequency unit should have reasonable response up to about 2000 c/s. One of the most popular arrangements for good fidelity is the series-connected constant-resistance type of Fig. 4.53B giving an ultimate attenuation of 12 db. Both inductors and both capacitors have identical values: Here $L_1 = R_0/(2\sqrt{2}\pi f_c)$ and $C_1 = 1/(\sqrt{2}\pi f_c R_0)$. For example, let the voice coil impedance R_0 be 10 ohms, and the cross-over frequency 800 c/s.

Then $L_1 = 10/(2 \times 1.41 \times \pi \times 800) = 1.41$ millihenrys
and $C_1 = 1/(1.41 \times \pi \times 800 \times 10) = 28.2$ μF.

The parallel-connected constant resistance type is also used (Refs. 21, 24).

Air cored multilayer solenoids may be used for inductances up to 8 millihenrys. Suitable sizes of formers are $1\frac{1}{4}$ in. dia., with axial length $\frac{3}{4}$ in. for inductances from 0.5 to 2.0 mH and $1\frac{1}{4}$ in. for inductances from 2.0 to 8.0 mH. The winding wire may be 17 A.W.G. (18 S.W.G.) double cotton enamelled copper. The outside diameter will be less than 4 in. for the values quoted above.

FIG. 21.8

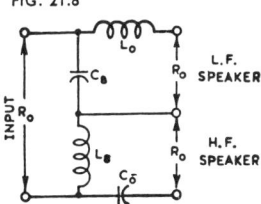

L.F. SPEAKER

H.F. SPEAKER

Fig. 21.8.   Filter type frequency dividing network, series connection.

The number of turns may be calculated from the formulae of Chapter 10 Sect. 2(iv)A or alternatively from curves (Ref. 11 for 17 A.W.G. D.C.E.).

Iron cores introduce distortion, and are therefore undesirable. However if iron-cored inductors are used, an air gap of 0.008 in. or 0.010 in. should be provided (Ref. 6 p. 101). Ferrite cores present possibilities for this application.

In the filter type arrangement which is also popular, the two inductors and the two capacitors have different values, and the nominal attenuation is 12 db per octave (Fig. 21.8). Taking the design constant $m = 0.6$, we have

$L_0 = R_0/(2\pi f_c)$ $C_0 = 1/(2\pi f_c R_0)$
$L_8 = R_0/(3.2\pi f_c)$ $C_8 = 0.8/(\pi f_c R_0)$

where L is in henrys and C is in farads (Refs. 6, 19).

In all cases the series connection is preferable to the parallel connection. Resistance in the inductors has a slight effect on the attenuation at the cross-over frequency, while it also introduces insertion loss (0.3 to 1 db). Care should be taken to keep the resistance as low as practicable. The usual position is between a single power amplifier valve (or two in push-pull or parallel) and the loudspeakers. The insertion loss is therefore a loss of maximum output power. The divider network may be connected either on the primary or secondary side of the output transformer, the latter being more usual— in this case the transformer must be suitable for the total frequency range of both units.

If the dividing network is placed on the primary side, each of the two output transformers is only called upon to handle a limited frequency range, and may be of cheaper construction. One interesting application is Fig. 21.9 (Refs. 9, 16). For a plate-to-plate load of 5000 ohms and a cross-over frequency of 400 c/s the component values are :

L_1 = 2.0 henrys (series aiding) ; inductance T_1 primary = 50 henrys min. ; L_2 = 1.0 henry (with air gap) ; T_2 primary inductance = 2.0 henrys (with air gap) ; C_1 = 0.16 μF : C_2 = 0.04 μF : leakage inductance of T_1 not over 1 henry : leakage inductance of T_2 not over 0.05 henry.

FIG. 21.9

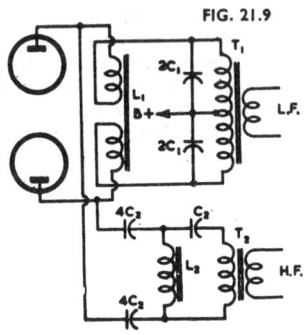

Fig. 21.9. Alternative form of divider network using two stepdown transformers (Ref. 9).

It has been shown that the arrangement of Fig. 21.9 produces less distortion than the conventional circuit : this is due to two features. Firstly, the low-pass filter L_1, $2C_1$ attenuates high frequencies propagating in either direction ; as a result the harmonic components of magnetizing current in T_1 have less effect on the high-pass channel. Secondly, the low-pass filter attenuates high frequencies on their way to T_1 so that the transformer is only called upon to handle a limited range of frequencies, thereby reducing the distortion (Ref. 9).

In the case of separate amplifiers for low and high audio frequencies, the divider network preferably precedes the amplifiers. This permits separate attenuators to be used for each channel ; these may be used for balancing the two units, or for tone control.

If the high frequency unit is more sensitive than the low frequency one, a simple form of fixed attenuator may be incorporated (Fig. 21.10). Here

$R_1 + R_3 = R_0$ and $R_1/R_3 = (E_1/E_2) - 1$
where $R_3 = R_2 R_0/(R_2 + R_0)$.

For example, to give $(E_1/E_2) = 2$ (i.e. an attenuation of 6 db) :
$R_1 = \frac{1}{2}R_0$ and $R_2 = R_0$.

FIG. 21.10

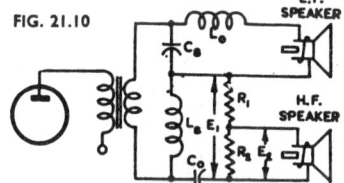

Fig. 21.10. Complete circuit from valve plate to loudspeakers, incorporating the filter of Fig. 21.8 with an attenuator on the h. f. speaker.

If it is desired to use this as a variable attenuator to give tone control, it should preferably be of the constant input impedance type such as the L pad [Fig. 21.11, also see Chapter 18 Sect. 3(iii)]. If a simpler type must be used, it may be designed as for Fig. 21.10 to give the correct input impedance at the normal operating position, but with a sliding contact. It will then have an incorrect impedance at any other setting ; the effect is only slight if the sensitivity of this unit is very much greater than the other, as happens with horn and direct-radiator assemblies.

If one loudspeaker unit has a lower impedance than the other, an auto-transformer may be used to provide correct matching. This may not be necessary if an attenuator is used on one unit.

FIG. 21.11

Fig. 21.11. L type level control providing constant input resistance.

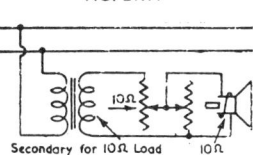

Secondary for 10Ω Load   10Ω

It is important to check for the phasing of the two loudspeakers ; in the vicinity of the cross-over frequency both units should be in phase (aiding one another). This may be checked by the use of a single dry cell connected in series with the secondary of the transformer ; in the case of 2 ohm voice coils there may be a 5 ohm limiting resistor connected in series with the cell.

References 1, 2, 6, 9, 11, 12, 16, 19, 21, 22, 24, 25, 26.

SECTION 4 : REFERENCES

1. Sowerby, J. McG. " Radio Data Charts 10—loudspeaker dividing networks " W.W. 49.8 (Aug. 1943) 238.
2. Sieder, E. N. " Design of crossover networks for loudspeaker units " Q.S.T. 28.12 (Dec. 1944) 35.
4. " Matching transformers for loudspeakers " Philips Tec. Com. 78 (Nov.-Dec. 1940) and 79 (Jan.-Feb. 1941).
5. Stanley, A. W. " The output stage—effect of matching on frequency response " W.W. 52.8 (Aug. 1946) 256.
6. Terman, F. E. " Radio Engineers' Handbook " (McGraw-Hill Book Company, New York and London, First edition 1943) pp. 249-251.
7. Langford-Smith, F. " The relationship between the power-output stage and the loudspeaker " A.W.A. Tec. Rev. 4.4 (Feb. 1940) 199. Originally published in Proceedings of World Radio Convention, Sydney, April 1938 (Institution of Radio Engineers, Australia). See also Ref. 7a.
7a. Langford-Smith, F. " The output stage and the loudspeaker " W.W. 44.6 (Feb. 9, 1939) 133 ; 44.7 (Feb. 16, 1939) 167.
8. R.C.A. " Application Note on receiver design—output transformer " (variation of impedance with current) No. 75 (May 28, 1937).
9. Klipsch, P. W. " Low distortion cross-over network " Elect. 21.11 (Nov. 1948) 98.
10. Langford-Smith, F. " The design of a high fidelity amplifier—(1) The power valve and the loudspeaker," Radiotronics 124 (March-April 1947) 25.   (2) " Negative feedback beam power amplifiers and the loudspeaker," Radiotronics 125 (May-June 1947) 53.
11. McProud, C. G. " Design and construction of practical dividing networks " Audio Eng. 31.5 (June 1947) 15.
12. Schuler, E. R. " Design of loudspeaker dividing networks " Elect. 21.2 (Feb. 1948) 124.
13. McProud, C. G. " Two-way speaker system " Part 3 Audio Eng. 32.2 (Feb. 1948) 21.
14. Angevine, O. L. " Impedance matching " Audio Eng. 31.11 (Dec. 1947) 20.
15. Jonker, J. L. H. " Pentode and tetrode output valves " Philips Tec. Com. 75 (July 1940) 1.
16. Klipsch, P. W. " Woofer-tweeter crossover network " Elect. 18.11 (Nov. 1945) 144.
18. Amos, S. W. " Feedback and the loudspeaker " W.W. 50.12 (Dec. 1944) 354.
19. Hilliard, J. K. " Loudspeaker dividing networks " Elect. 14.1 (Jan. 1941) 26.
20. " Impedance matching and power distribution in loud speaker systems " Technical Monograph No. 2, Jensen Radio Mfg. Co.
21. Smith, B. H. " Constant-resistance dividing networks " Audio Eng. 35.8 (Aug. 1951) 18.
22. White, S. " Design of crossover networks " FM-TV 12.1 (Jan. 1952) 42.

Additional references will be found in the Supplement commencing on page 1475.

CHAPTER 22

AERIALS AND TRANSMISSION LINES

By W. N. Christiansen, M.Sc.

SECTION 1 : INTRODUCTION

Aerials and transmission lines differ from simple electrical networks in that their inductance, capacitance and resistance are not lumped but are distributed over distances such that the time required for electrical energy to travel from one part to another has to be taken into account. In a single chapter it is, of course, impossible to attempt to give a theoretical treatment of these devices. What will be done will be to present some useful formulae, a few physical pictures of the processes which occur and some results of practical experience with aerials and transmission lines.

SECTION 2 : THE TRANSMISSION LINE

(i) Introduction   (ii) The correct termination for a transmission line (iii) Impedance-transforming action of a transmission line.

(i) Introduction

A transmission line consists of an arrangement of electrical conductors by means of which electromagnetic energy is conveyed, over distances comparable with the wavelength of the electromagnetic waves, from one place to another. The theory of transmission lines provides a link between circuit theory and the field theory of electromagnetic waves inasmuch as the properties of such lines may be determined either from the picture of a transmission line as a filter network with an infinite number of elements, or from the picture of electromagnetic waves guided between (usually) a pair of conducting surfaces.

Some of the properties of transmission lines will now be given.

(ii) The "correct" termination for a transmission line

A uniform transmission line has what is called a " characteristic impedance ". This is the impedance that would be measured at the end of such a line if it were infinitely long. The importance of this characteristic impedance lies in the fact that if any length of line is terminated in an impedance of this value, then all the energy flowing along the line is absorbed at the termination and none is reflected back along the line. A result of this is that the input impedance of any length of transmission line terminated in its characteristic impedance is equal to the characteristic impedance. At radio frequencies, the characteristic impedance of all normally-used types of transmission line is almost purely resistive.

The value of the characteristic impedance for a low-loss line is

$$Z_0 = \sqrt{L/C} \tag{1}$$

where L and C are the distributed inductance and capacitance per unit length of the line. The velocity of propagation of electromagnetic waves along such a line is

$$v = 1/\sqrt{LC} \tag{2}$$

With air as the dielectric, $v = 3 \times 10^8$ metres per second, which is the velocity of light*. For other dielectrics

$$v = 3 \times 10^8/\sqrt{K} \text{ metres per second} \tag{3}$$

where K = dielectric constant.

In cables with polythene dielectric, for example, $K = 2.2$ and v is therefore 0.67 of the velocity of light. This means that the wavelength of the waves in the cable is only 67% of the wavelength in air.

For a low-loss transmission-line the characteristic impedance may be found if C and L are determined. Actually it is not necessary to determine both—one is sufficient.

If we combine (1) and (2) we obtain

$$Z_0 = 1/vC = vL. \tag{4}$$

For a transmission line consisting of a pair of parallel conductors

$$Z_0 = \frac{276}{\sqrt{K}} \log_{10} \frac{d}{r} \text{ ohms} \tag{5}$$

where d is the distance between the centres of the conductors and r is the radius of each conductor.

For a co-axial line

$$Z_0 = \frac{138}{\sqrt{K}} \log_{10} \frac{r_1}{r_2} \tag{6}$$

where r_1 is the inside radius of the outer conductor and r_2 is the outside radius of the inner conductor.

(iii) Impedance-transforming action of a transmission line

If the line is terminated in an impedance Z_L which is not equal to the characteristic impedance Z_0 of the line (or, in other words, the load is not matched to the line), then the energy is not completely absorbed at the termination, but some is reflected back along the line. As a result of this, **standing waves** are formed on the line, and the value of the input impedance to the line depends on the length of the line. The ratio of the voltage (or current) in the backward wave to that in the forward wave is called the **reflection coefficient** and has a value

$$k = \frac{Z_L - Z_0}{Z_L + Z_0} \tag{7}$$

The **standing-wave ratio** in the line is the ratio of the maximum to the minimum voltage (or current) that appears at points along the line. (The distance between a maximum and an adjacent minimum is one quarter wavelength).

The standing-wave ratio has a value—

$$\rho = \frac{1 + |k|}{1 - |k|} \tag{8}$$

If Z_L is resistive and has a value R_L, then (8) becomes

$$\rho = R_L/Z_0 \text{ or } Z_0/R_L \tag{9}$$

depending on whether R_L is greater or less than Z_0.

---

*A more precise value is 2.9979×10^8 metres/sec.—see Chapter 9 Sect. 11(iii) and Ref. E1.

For a transmission-line having no dissipation, the **input impedance** Z_i is given by

$$Z_i = Z_0 \frac{Z_L \cos \beta l + j Z_0 \sin \beta l}{Z_0 \cos \beta l + j Z_L \sin \beta l} \qquad (10)$$

where $\beta = 2\pi/\lambda$, λ being the wavelength of the waves in the transmission line and l the length of the transmission line.

If l is equal to any integral number of half waves, then

$$\cos \beta l = \cos n\pi = \pm 1$$
$$\sin \beta l = 0$$
and $Z_i = Z_L$, $\qquad (11)$

i.e. the input impedance is equal to the load impedance irrespective of the value of Z_0.

On the other hand, if the line has a length equal to an odd integral number of quarter-waves, then

$$Z_i = Z_0^2/Z_L \qquad (12)$$

This relation is interesting, because it shows the basis of the **quarter-wave transformer**, which is used extensively. The section of line will match a generator to a load if Z_0 is chosen to be equal to the square-root of the product of the generator and load impedances.

SECTION 3 : AERIALS AND POWER TRANSFER

(i) *Introduction*   (ii) *Power transfer.*

(i) Introduction

The function of a receiving aerial is to collect electromagnetic energy which is passing through the space surrounding the aerial, and to pass this energy into a radio receiver.

(ii) Power transfer

We consider first the simplest case of energy transfer between a transmitter and a receiver, where **two straight dipole aerials** are placed a large number of wavelengths apart and are far removed from any reflecting surface, such as the earth. Suppose that one aerial is energized by a transmitter, and that the other is connected to a receiver. If the power of the transmitter remains fixed, then we find that the power appearing at the receiver is dependent on a number of factors—

(a) The received power is greatest when the aerials are parallel to each other (for short aerials, the maximum occurs when the axis of each aerial is perpendicular to the line joining the centres of the two aerials ; for aerials longer than one half-wave length, this is not necessarily so).

(b) The received power is inversely proportional to the square of the number of wave-lengths between transmitting and receiving aerials.

(c) For aerials less than one half-wave in length, the power received is independent of the length of either aerial, provided the aerials can be matched **without loss** to the transmitter or receiver. In practice, however, aerials very short compared with a wave-length are characterised by high ohmic losses.

Paragraphs (b) and (c) may be summarized quantitatively as follows.

If P_T is the power emitted from a dipole aerial (less than half-wave in length) and P_R is the power available for transfer from the receiving aerial to a receiver, and n is the number of wavelengths between transmitting and receiving aerials, then

$$\frac{P_R}{P_T} = \left(\frac{0.119}{n}\right)^2 \qquad (13)$$

Although this formula applies only when the aerials are situated in free space (or, as is shown later, when the transmitting and receiving aerials are of the short vertical type, situated close to a flat perfectly conducting earth), the formula is useful in illustrating a very important fact, which is, that **in the transfer of power between two aerials, the number of wave-lengths, rather than the distance, is the significant parameter.** As an illustration of this we may compare the energy available at a receiver when the frequency of the emitted wave is (a) 600 Kc/s and (b) 150 Mc/s. We assume that the energy radiated in each case is 1000 watts and the distance between transmitter and receiver is 20 kilometres.

On applying the formula (13) we find that in the medium-frequency case, the power available at the receiver is 8.8 milliwatts, while in the v-h-f case the power available is only 0.142 microwatts.

This means that if both signals are to be amplified to produce, say 1 watt at the detector then the amplification required in the first case is approximately 20 decibels while in the second, it is nearly 70 decibels.

These figures partly explain why a crystal receiver is very effective for long-wave reception, whereas a receiver required for v-h-f work usually has a large number of stages of amplification.

SECTION 4 : CHARACTERISTICS OF AERIALS

(i) *Effective area of a receiving aerial*   (ii) *The power gain of an aerial*   (iii) *The beam-width of an aerial.*

(i) Effective area of a receiving aerial

The radio-frequency power passing through a unit area placed at right angles to the direction from a short dipole transmitting aerial does not (in free space) depend on the frequency. Its value is given by

$$\Phi = \frac{3P_T}{8\pi d^2} \text{ watts per square metre} \tag{14}$$

if the distance d from the transmitting aerial is given in metres, and P_T is in watts. Why is the power available at the receiver dependent on the frequency ? The reason is that the **effective area** for capture of energy by the receiving aerial depends on the frequency.

The power available at the receiver is

$$P_R = \Phi A \tag{15}$$

where A is defined as the " effective area " of the receiving aerial.

If we substitute values of P_R and Φ from (13) and (14) we find that the effective area for a short dipole is $0.119\lambda^2$. In the example given above, A is approximately 30 000 square metres in one case and is less than 0.5 square metre in the other. The reason for the enormous difference in the energy available at the receiver in the two cases is now clear.

In the following table, the effective areas of various types of aerial are given :

TABLE 1

Type	Effective Area
Short dipole	$3\lambda^2/8\pi$ or approx. $0.119\lambda^2$
Half-wave dipole	$0.130\lambda^2$
Half-wave dipole with reflector	$0.25\lambda^2$ to $0.50\lambda^2$ (depending on spacing)
Broadside array with reflectors	Physical area, approximately
Broadside array, without reflectors	Physical area/2, approximately
Short vertical aerial near earth	$3\lambda^2/16\pi$

It will be seen from the table that the effective area for capture of electromagnetic energy by a short dipole is approximately equal to an area bounded by a circle which is at a radial distance of $\lambda/1.6\pi$ from the dipole. The physical significance of this is that such a circle very roughly forms the boundary between the region in which the local induction field of the aerial predominates and the region in which the radiation field is the major component.   Hence we can picture the aerial as capturing the energy which falls within the region in which the induction field of the aerial is of significant magnitude.

A broadside array of dipoles with a reflecting curtain can absorb all the energy that falls on it.   Without the reflectors, it absorbs only half the incident energy and radiates one quarter back towards the transmitter and one quarter in the opposite direction. A similar effect occurs with a dipole.   If it absorbed all the energy available to it and reradiated none, then its effective area would be double the value given in Table 1. A reflector is required to achieve this.

(ii) The power-gain of an aerial

The power-gain of a receiving aerial is the ratio of the power appearing at the input terminals of a receiver which is attached to the aerial to the power that would appear at the receiver if the aerial were replaced by a simple type of aerial (usually a half-wave dipole).

The **power gain,** G, of an aerial, therefore, must be equal to the ratio of the effective area A of the aerial to that of the comparison aerial.

With the half-wave dipole as the comparison aerial, we have

$$G = A/0.130\lambda^2 \qquad (16)$$

As an example we may consider the case of an array of N half-wave elements spaced one half-wave length and arranged in the form of a rectangular curtain.   Then from Table 1 we find

$$A \approx \text{Physical area}/2 \approx (N/2)(\lambda/2)(\lambda/2) = N\lambda^2/8 \qquad (17)$$

Therefore $G \approx N\lambda^2/1.04\lambda^2 \approx N$ (18)
i.e. the power-gain of a broadside array of half-wave elements is approximately equal to the number of elements in the array.

(iii) The beam-width of an aerial

For a directional aerial having one major lobe in its directivity pattern there is a rough but useful rule to determine the angular width of this lobe.   If the aerial has a breadth of n wave-lengths in any particular direction and the beam-width θ of the main lobe is measured in the same plane, then

$$\theta \approx 60/n \text{ degrees} \qquad (19)$$

SECTION 5 : EFFECTS OF THE EARTH ON THE PERFORMANCE OF AN AERIAL

(i) Introduction   (ii) A perfectly-conducting earth   (iii) An imperfectly-conducting earth   (iv) The attenuation of radio waves in the presence of an imperfectly-reflecting earth.

(i) Introduction

In the previous section the effect of reflecting surfaces has been disregarded.   In practice, these effects must be taken into account, except in rare cases.   The two " surfaces " to be considered are the earth and the ionosphere.   We deal here with the effect of the earth.

(ii) A perfectly-conducting earth

The effect of the earth on the propagation of radio-waves is complicated and no attempt will be made here to treat it at all fully.   As a first approximation one can consider the earth as a perfectly conducting flat surface.   This enables one to use a device which simplifies the treatment considerably.   One imagines an image of the aerial in the reflecting surface, and then the problem may be treated as one in which the aerial and its image are situated in free space.   (The sign of the instantaneous potentials in the image are reversed from those in the aerial).

In Fig. 22.1 are shown the effects of the earth on the field pattern of a short vertical aerial and of a short horizontal dipole. It will be seen that the earth has no effect on the field pattern of the vertical aerial apart from removing the lower half of the pattern. With the horizontal aerial, however, there is complete cancellation of the waves in the horizontal plane and the resultant effect is zero.

The conclusion may be drawn from the diagram that where waves are arriving at the aerial in directions parallel to the plane of the earth, then one would expect a vertical aerial to be very effective while a horizontal aerial would be ineffective. (This applies to a transmitting as well as to a receiving aerial). It will be seen later that this conclusion does not fit the facts, in all cases, because the earth is not a perfect conductor. At medium and low radio-frequencies, however, it is in agreement with the facts—horizontal aerials are of no use in communicating between two points both close to the earth.

When waves are arriving at the aerial in directions inclined to the horizontal (as, for example, in the case of short-wave communication via the ionosphere) then either horizontal or vertical aerials may be used. If the distance between transmitter and receiver is not great, however, then the waves will be arriving at nearly vertical incidence and the horizontal aerial will be superior, provided that it is placed at a suitable height above the earth. The height at which the aerial should be placed above the earth depends on the angle of elevation of the waves which are to be received. The height is different for horizontal and vertical aerials.

In Fig. 22.1, for an angle of elevation θ, the distance d between the path lengths of waves to the aerial and to its image should be equal to an even number of half-waves in the case of a vertical aerial, and an odd number of half-waves in the case of a horizontal aerial. Simple trigonometry shows that this corresponds to a height h given by

$$h = n\lambda/(4 \sin \theta) \qquad (20)$$

where n is an even integer for a vertical aerial and an odd integer for a horizontal aerial.

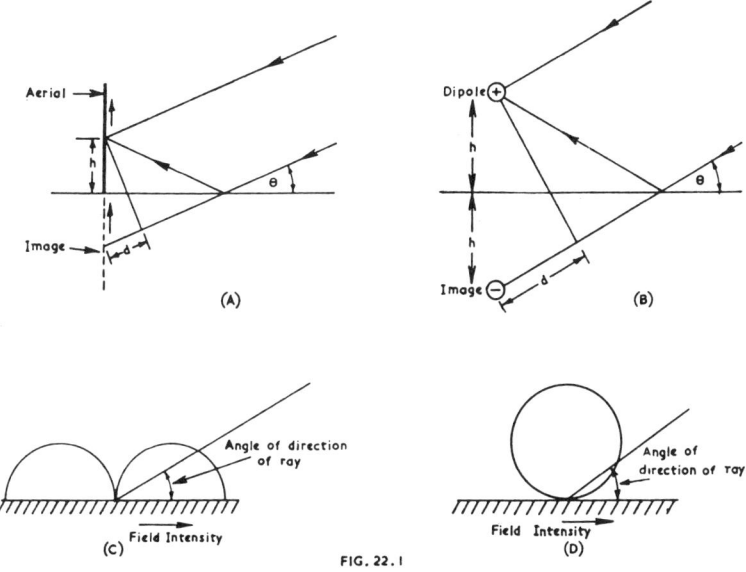

(C) Field Intensity   (D) Field Intensity

FIG. 22.1

Fig. 22.1. Effects of the earth on the field pattern of a short vertical aerial (A) and of a short horizontal dipole (B). Both (A) and (B) are cross-sectional views, and in (B) the sign + indicates direction of current into the paper while the sign − indicates the opposite direction. Polar diagrams of a short vertical aerial and a short horizontal dipole, both close to earth, are shown in (C) and (D) respectively.

(iii) An imperfectly-conducting earth

The propagation of radio-waves close to the surface of a finitely conducting earth has been studied theoretically for many years. Zenneck first produced a solution to the problem, and later Sommerfeld gave a more accurate analysis. Unfortunately, the incorrectness of Zenneck's analysis combined with a small error in Sommerfeld's, created some confusion that has persisted until the present day. As a result of the errors, it appeared that a special type of wave was propagated in the vicinity of the ground-plane. This wave was called a " surface-wave." It is known now, that this Zenneck " surface-wave " does not exist, but the name still appears in the literature on wave propagation. It is applied to the waves that travel between transmitting and receiving aerials when both are close (in terms of a wavelength) to the surface of the earth, as for example, in medium-wave broadcasting. When the aerials are raised several wave-lengths above earth, then the waves travelling between them are called " space-waves." These terms are artificial, but they serve some purpose in that they indicate that the conditions of propagation between two aerials placed close to the earth, are markedly different from those between two aerials elevated several wave-lengths above the earth's surface (as in v-h-f broadcasting).

In the conditions of propagation to which the term " surface-wave " is applied, vertical aerials are superior to horizontal aerials. Thus vertical aerials are always used in medium-frequency broadcasting. (The horizontal portion of a medium-frequency receiving or transmitting aerial plays no useful part, except to increase the efficiency of coupling to the receiver or transmitter).

When the aerials are raised several wave-lengths above the earth, as is usually the case in v-h-f communication, then horizontal and vertical aerials have roughly equal effectiveness.

When waves are not arriving at the aerial in a horizontal direction, then the difference between an imperfectly conducting earth and a perfectly conducting one is not so marked. Hence for short-wave communication one can treat the earth as a perfect conductor without introducing serious error in the calculations.

(iv) The attenuation of radio-waves in the presence of an imperfectly-reflecting earth

At short distances from a transmitting aerial the resistivity of the earth does not have a major effect on the energy reaching the vicinity of the receiving aerial. The principal effect of the earth is that the energy flowing through unit area placed at right angles to the line of propagation is doubled. The reason for this is that the transmitting aerial is now emitting its energy into a hemisphere instead of into a full sphere as was the case in free-space propagation. Instead of the relation (14) we have now

$$\Phi = 3P_T/4\pi d^2 \tag{21}$$

where the symbols have the same meaning as in formula (14).

For a similar reason the effective area for capture of radiation by the receiving aerial is half that of the free-space case. For a short vertical aerial the effective area is

$$A = 3\lambda^2/16\pi. \tag{22}$$

The effects at the transmitting and receiving aerials, however, are such that the power attenuation between the two is the same as in the free-space case, so (13) is still applicable.

As the distance between the two aerials is increased, it is found that the attenuation actually found between transmitter and receiver becomes increasingly larger than that given by (13). This is the result of both the finite conductivity and of the curvature of the earth.

For the range of frequencies used in medium-wave broadcasting, and for earth of average conductivity ($\sigma = 7 \times 10^{-14}$ e.m.u.), values of the power-flux per unit area at the receiver are shown in Fig. 22.2.

Such curves are frequently given in terms of the strength ϵ of the electric field of the wave. To convert this into power-flux per unit area, one may use the relation

$$\epsilon = \sqrt{377\,\Phi} \text{ volts per metre} \tag{23}$$

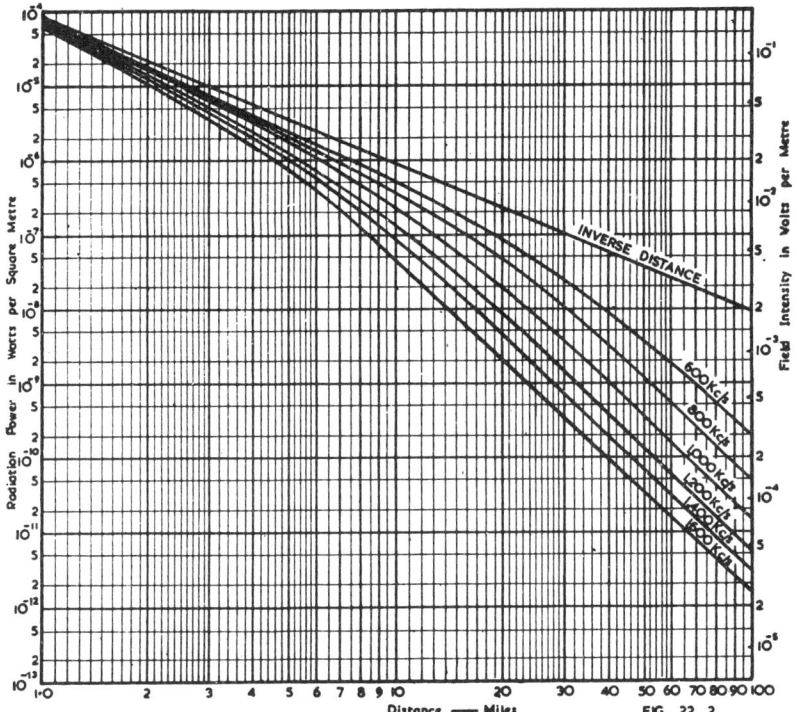

*Fig. 22.2. Attenuation at medium frequencies of the radiation from a transmitter
(1 KW radiated, short aerial, average earth conductivity, $\sigma = 7 \times 10^{-14}$ e.m.u., $\epsilon = 15$).*

where Φ is expressed in watts/metre².

Sometimes it is required to know the strength H of the magnetic field of the wave.
This is related to the electric field by the expression

$$H = (\epsilon/377) \text{ ampere-turn/metre*} \qquad (24)$$

As a matter of interest, it may be noted that ϵ and H are analogous to voltage and
current in electrical circuit theory. On this analogy the number 377 appears to denote
a resistance, and is called the **intrinsic resistance of free space.**†

At very-high-frequencies, the attenuation of waves between two aerials close to
the ground is rapid. This corresponds to the case of the so-called " surface waves "
referred to in the previous section. Fortunately it is not difficult, at these frequencies,
to raise the transmitting and receiving aerials several wave lengths above ground and
achieve the conditions for " space-wave " propagation, where the attenuation is very
much less.

For such conditions, a rough but useful formula for the attenuation is given by

$$P_R/P_T = (1.5 \ h_T h_R/n^2)^2 \qquad (25)$$

where h_T and h_R are the heights (in wavelengths) above earth of the transmitting and receiving aerials, and the other symbols are the same as in (13).

More accurate values may be found by using Figs. 22.3, 22.4 and 22.5.

---

*The ampere-turn/metre is the unit of magnetic field strength in the Rationalized Meter Kilogram
Second (M.K.S.) system of units.
    1 ampere-turn/metre = 4×10^{-3} e.m.u. (oersted).
† Note : ϵ/H in e.m.u. = c ; ϵ/H in e.s.u. = $1/c$.
    ϵ expressed in e.s.u. = H expressed in e.m.u. Refer to Chapter 38 Sect. 1(ii) on electrical and
magnetic units.

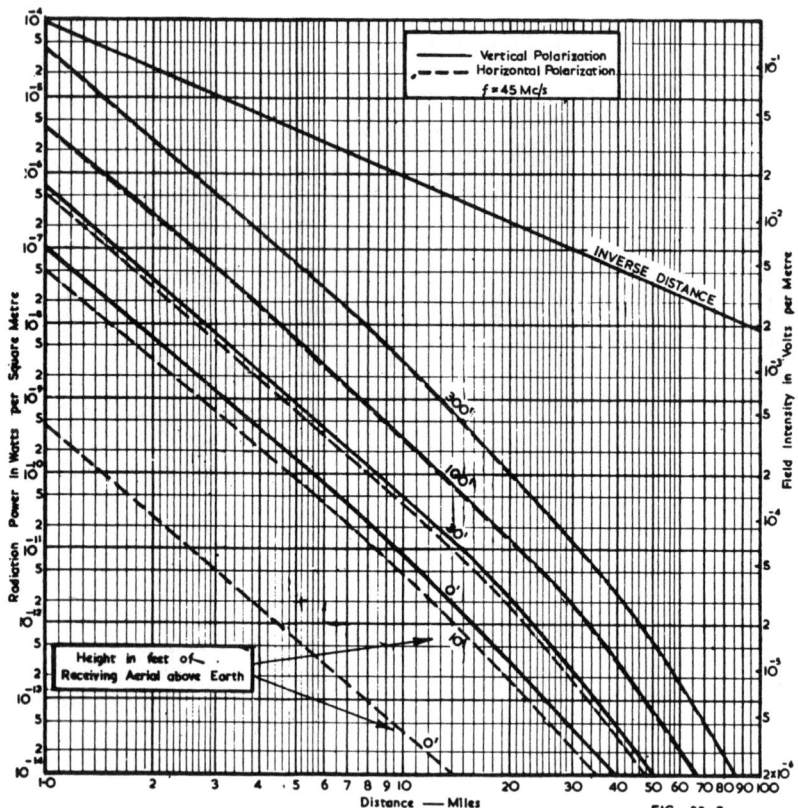

FIG. 22.3

Fig. 22.3. Effect of raising the receiving aerial above the earth (1 KW at 45 Mc/s radiated over earth with conductivity $\sigma = 10^{-13}$ e.m.u. and transmitting aerial 30 feet above the earth). The same curves also apply when the transmitting and receiving aerials are transposed.

Fig. 22.3 shows the effect of raising the aerial above the earth* for a frequency of 45 Mc/s, and indicates that unless the aerials are close to the earth, horizontal and vertical polarizations have practically equal effectiveness.

Fig. 22.4 and 22.5 for frequencies of 75 and 150 Mc/s were derived by T. L. Eckersley for vertically polarized waves, but may be used also for horizontally polarized ·waves with little error when the aerials are at heights above earth corresponding to normal practice.

---

*An infinitely short dipole has been assumed.

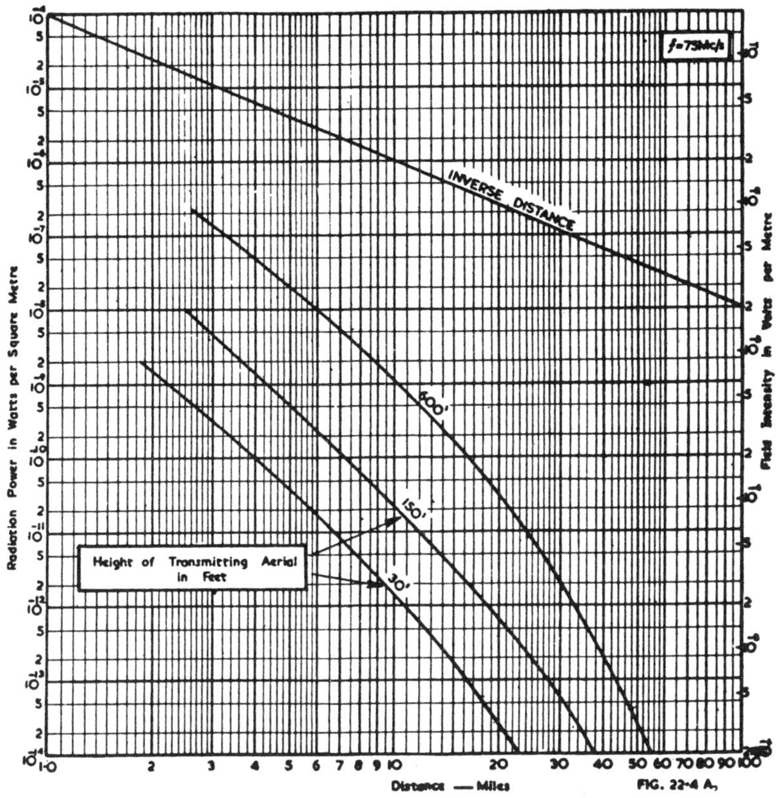

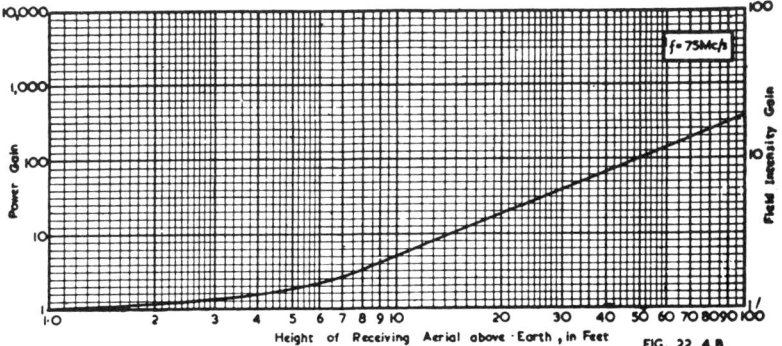

Fig. 22.4 (A) *Attenuation at 75 Mc/s of the radiation from a transmitter* (1 *KW radiated over earth with* $\sigma = 10^{-13}$ *e.m.u.,* $\epsilon = 5$, *with receiving aerial at zero height*) ; (B) *Multiplying factor for height of receiving aerial above earth.*

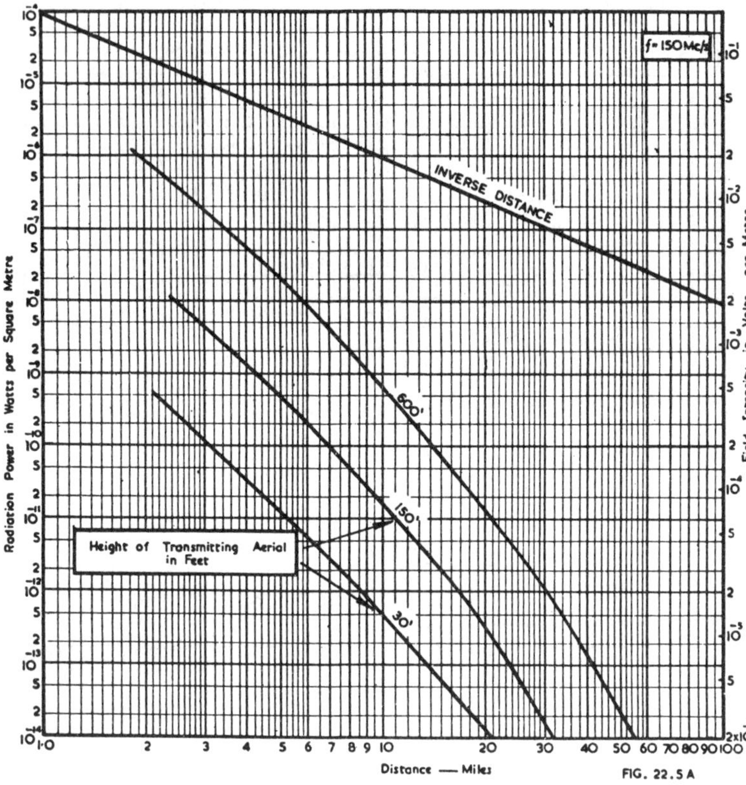

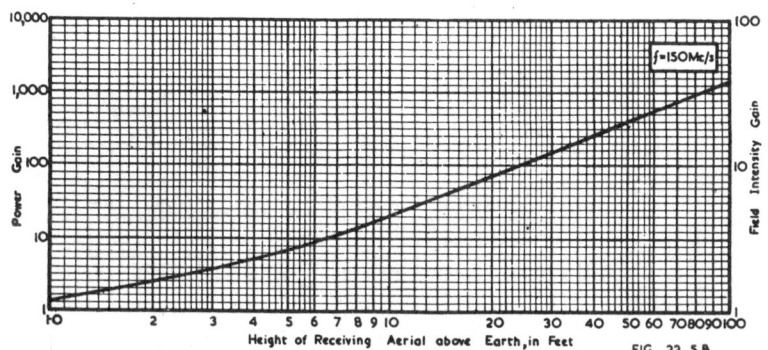

Fig. 22.5.   (A) *Attenuation at 150 Mc/s of the radiation from a transmitter* (1 KW *radiated over earth with* $\sigma = 10^{-13}$ *e.m.u.*, $\epsilon = 5$, *with receiving aerial at zero height*) ; (B) *Multiplying factor for height of receiving aerial above earth.*

SECTION 6 : THE EFFECT OF THE IONOSPHERE ON THE RECEPTION OF RADIO SIGNALS

Radio propagation over long distances on the earth is possible because of the presence of layers of ionized gas in the higher atmosphere of the earth. Signals reflected from the ionosphere are used exclusively in short-wave reception.

The principal reflecting layers in the ionosphere are the E and F_2 layers, situated at roughly 60 miles and 200 miles, respectively, above the surface of the earth. The E layer is a poorly reflecting surface and hence short-wave communication almost exclusively takes place via the F_2 layer. In designing a receiving aerial for short-wave communication, one must calculate the angle of elevation at which signals will arrive at the aerial, and then use the relation (20) to determine the height at which the aerial should be placed above earth. This calculation involves a knowledge of the height of the ionospheric layer, the distance between transmitting and receiving aerials, and the curvature of the earth.

The Australian reader is referred to the " Radio Propagation Handbook " and the " Monthly Ionospheric Predictions " produced by the Ionospheric Prediction Service of the Commonwealth Observatory, for information on communication by way of the ionosphere. In America the equivalent publications are the N.B.S. Circular 465 " Instructions for the use of basic radio propagation prediction " and CRPL-D publication " Basic radio propagation predictions " obtainable from the Central Radio Propagation Laboratory, National Bureau of Standards, Washington, D.C. The English reference is Bulletin A, published by Radio Research Station, National Physical Laboratory.

See also Chapter 8 References 5, 7 and 10.

In medium-wave broadcasting the ionosphere does not reflect any appreciable amount of energy during daylight hours. At night, however, waves are reflected back to the earth from what remains of the E layer (unlike the F layer, the E layer practically disappears at night) and these waves provide a secondary, or fading, service during the night hours over distances of hundreds of miles. Unfortunately, these reflected waves reach the earth also at relatively short distances (50 to 100 miles) from the transmitter and combine with the ground waves to produce a very objectionable type of fading. Hence the primary or non-fading service area of a medium-wave broadcasting station may be less at night than during the day. Aerials that favour signals which arrive horizontally and discriminate against those that arrive from high angles will diminish this undesirable effect. This normally cannot be done very effectively because the height required for the aerial is neither possible nor economical in a receiving installation. The best that can be done is to avoid types of aerials that accentuate this effect. Where night-time fading of signals is encountered, straight vertical aerials should be used. Loop aerials and inverted-L aerials with long horizontal sections should be avoided, and care should be taken that r-f coupling between the power mains and the receiver is kept to a minimum.

SECTION 7 : THE IMPEDANCE OF AN AERIAL

(i) Introduction   (ii) Resistive component of impedance   (iii) Reactive component of impedance   (iv) Characteristic impedance of aerial   (v) Examples of calculations (vi) Dipoles   (vii) Loop aerials.

(i) Introduction

In the previous sections, the energy ideally available at the receiver terminals has been calculated. This energy is available, however, only if the receiver is matched to the aerial, and if there are no resistive losses associated with the aerial or aerial-earth system.

If the input circuit of a receiver is to be designed to extract the maximum energy from the radio-waves that are passing the aerial, then a knowledge of the component parts of the impedance of the aerial is required.

(ii) Resistive component of impedance

The first part of the aerial impedance to be considered is the resistive part concerned with the coupling between the aerial and the space around it. This is called the **Radiation Resistance** of the aerial and is the resistance that would be measured at the point of maximum current in a tuned aerial, in the absence of resistive losses.

If a load resistance of this value is placed in the aerial at the point considered, then the aerial would extract from a passing wave an amount of power equal to the " available power " calculated in the previous sections.

The calculation of the radiation resistance of an aerial is not easy, and in most cases the calculations provide only approximate answers. The reader is referred to standard texts for these approximate methods of calculation.

If the impedance is measured at the base of an ideal vertical aerial, then the resistive component is called the **Base Radiation Resistance.** Values of this for the types of aerial commonly used in medium-frequency reception are given in Fig. 22.6.

If the r-f resistance at the base of an aerial is measured, it will be found to be greater, in general, than the calculated radiation resistance. The difference represents the loss resistance associated with the aerial-earth system. For a short aerial with small radiation resistance, the loss resistance may be many times greater than the radiation resistance. With such an aerial one can extract only a fraction of the power ideally available at the terminals of the aerial.

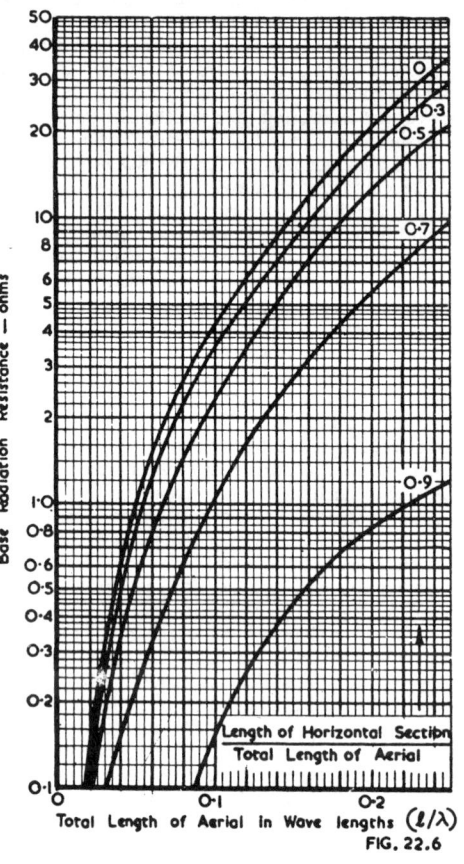

Fig. 22.6. Base radiation resistance of an inverted L aerial.

If we wish to calculate the power that may be extracted from the aerial in the presence of resistive losses we proceed as follows.

The power P_R ideally available at the aerial may be calculated and we can use information such as that given in Fig. 22.6 to find the radiation resistance R_R of the aerial. By measuring the base resistance and subtracting R_R from this we are left with R_L, the equivalent series loss resistance of the aerial-earth system. The equivalent circuit is shown in Fig. 22.7. It may be noted that we have not yet calculated e, the equivalent e.m.f. induced in the aerial by the radio-wave. This can be done easily however, because we know the resistance R_R and the power P_R that would be generated in a load resistance of this value if R_L were absent. Across such a load we would have a voltage of $e/2$.

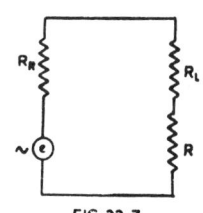

FIG. 22.7

Fig. 22.7. Equivalent circuit of aerial-earth system.

Hence $e = 2\sqrt{P_R R_R}$ (25A)

When R_L is not negligible, then the load resistance R must be increased to $R_R + R_L$ for maximum power transfer. In this case, the power in the load is

$$P' = \left(\frac{e}{2}\right)^2 \Big/ (R_R + R_L)$$ (26)

Hence the ratio of power actually extracted from the aerial to that ideally available (or the **radiation efficiency of the aerial**) is given by

$$P'/P_R = R_R/(R_R + R_L).$$ (27)

To take a simple example, suppose that we have a vertical aerial of 1/16 wavelength, and an earth resistance of 20 ohms and that we wish to calculate the radiation efficiency of the aerial.

From Fig. 22.6 we find $R_R = 1.6$ ohms, and we are given $R_L = 20$ ohms. Then $P'/P = 0.08$, which is the radiation efficiency of the aerial.

(iii) Reactive component of impedance

So far, only the resistive component of the base impedance of the aerial has been mentioned and we have assumed that the reactive part has been tuned out. It is most desirable that we should be able to calculate this reactive component, at least roughly, from the aspects of both the design of the input circuit of the receiver and the estimation of losses introduced by the tuning reactances.

A simple way to make an approximate calculation of the reactance of an aerial, is to treat it as a section of transmission line. We can simplify this treatment still further by neglecting the resistive component of the aerial impedance and by assuming that the transmission line has no loss. (This treatment breaks down when the base of the aerial is close to the point of minimum current of an aerial, but this case is very seldom met in an aerial designed for reception). Hence we can use the simple transmission line formula (10).

If the aerial is a straight wire, then this may be taken as an open circuited length of line. The terminating impedance Z_L is then infinite and (10) becomes

$$Z_i = X_i = -jZ_0 \cot \beta l$$ (28)

This formula can be applied also to an inverted L aerial, by neglecting the bend in the aerial and by taking l as the total length of wire in the aerial. If the aerial has a more elaborate type of capacitance top, then one can proceed as follows. Imagine the capacitance top broken at the point at which it joins the downlead of the aerial. Use (28) to calculate Z_i for each section of the capacitance top, then combine in parallel the values of Z_i. The resultant is then the terminating impedance Z_L of the aerial downlead. The impedance of this is then obtained from (10).

(iv) Characteristic impedance of aerial

As yet, we have given no indication of the value of characteristic impedance Z_0 to be assigned to the aerial. Various approximate formulae are available for this, but that of Steinmetz is possibly the best. This is

$$Z_0 = 138 \log_{10} (\lambda/d) - 104 \text{ ohms}$$ (29)

where λ is the wave-length
and d is the diameter of the aerial conductor (or cage) measured in the same units as λ.

(v) Example of calculations

The following worked example uses much of the work covered so far in this chapter.

Example

A receiving aerial of the inverted-L type has a horizontal top section of 50 ft. length and a 30 ft. vertical downlead. The aerial conductor is composed of 7/.029 in. copper wire. (a) What is the input impedance of the aerial at a frequency of 1000 Kc/s, if the equivalent series earth resistance is 20 ohms ? (b) What voltage will appear at the input terminals of a receiver, matched to the aerial, if the receiving aerial is at a distance of 20 miles from a transmitter ? The transmitter has a short aerial (less than quarter-wave) and radiates 500 watts. The ground between transmitting and receiving aerials has a conductivity of 7×10^{-14} e.m.u.

(a) **Calculation of input impedance**

The length of conductor is 80 feet or 24.4 metres, therefore $l/\lambda = 24.4/300 = 0.0813$. From (28), $X_i = - jZ_0 \cot 2\pi l/\lambda$.

We obtain Z_0 from (29) :

$Z_i = 138 \log_{10} 1.36 \times 10^5 - 104 = 603$ ohms
and $2\pi l/\lambda = 29.3°$.

Therefore $X_i = - j\ 603 \cot 29.3° = - j\ 1070$ ohms.

Next, to obtain R_R we use Fig. 22.6. For this aerial the ratio of the horizontal portion of the aerial to the total length is 0.625. By interpolation between plotted values we find that R_R is approximately 1.0 ohm. The series loss resistance is given as 20 ohms, hence the total resistance at the base is 21 ohms.

The base impedance of the aerial is then

$Z_i = 21 - j\ 1070$ ohms.

(b) **Calculation of voltage at receiver terminals**

From Fig. 22.2 we see that at the given frequency and distance the power flux per unit area Φ is 2.0×10^{-8} watts/m^2 for a radiated power of 1 KW. Hence Φ is 1.0×10^{-8} for 500 watts radiated. The effective area of the aerial is given in Table 1 as $3\lambda^2/16\pi$. Hence the power ideally available is

$P_R = 1.0 \times 10^{-8} \times 3 \times 300^2/16\pi$
     $= 0.535 \times 10^{-4}$ watts.

From (25A) we find that the e.m.f. induced in the aerial (referred to the base radiation resistance) is $e = 2\sqrt{P_R R_R} = 1.46 \times 10^{-2}$ volts or 14.6 millivolts.

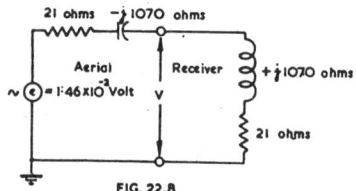

FIG. 22.8.

Fig. 22.8. Equivalent circuit of aerial and input circuit of the receiver.

The voltage appearing at the aerial terminals will be greater than the induced e.m.f. because of the highly reactive impedance of the aerial. The equivalent circuit of the aerial and input circuit of the receiver are shown in Fig. 22.8. We see that the aerial current is

$i \doteq 1.46 \times 10^{-2}/42$ and the voltage at the receiver terminals is approximately
$V = i\ 1070 = 0.373$ volt.

(vi) **Dipoles**

The second type of aerial, the impedance of which we may wish to calculate is the dipole aerial. Dipole aerials well removed from the earth may be treated in a manner similar to that of a vertical aerial, because, at least approximately, a vertical aerial with its image forms a dipole aerial. With the vertical aerial we measure the impedance with respect to the neutral point (earth) of the equivalent dipole, so the impedance is only half that of the equivalent dipole. Hence in calculating the impedance of a dipole we can use the results obtained previously, but multiplied by a factor of two. Because a dipole aerial (remote from the earth) has no earth loss, it is in general more efficient than the equivalent vertical aerial with earth return.

The most commonly used variety of dipole aerial is the resonant, or half-wave variety. When remote from earth it has twice the impedance of a quarter-wave aerial, i.e. its impedance is resistive and has a magnitude of 73 ohms. When in the vicinity of the earth, the impedance depends on the height above earth. The variation is shown in Fig. 22.9.

Reflectors are sometimes used with half-wave aerials at very high frequencies.

In Fig. 22.10 is shown the effect, on the radiation resistance of a half-wave dipole, of the spacing between aerial and reflector, when the latter is tuned to provide maximum forward radiation. (For further information on the effects of reflectors the reader is referred to Ref. 14).

(vii) Loop aerials

Another common type of aerial used in broadcast reception is the loop aerial. With such aerials, the radiation resistance at medium frequencies is normally negligible compared with the loss resistance of the conductors. The aerial may be treated as a lumped inductance, the electrical constants of which can be calculated from well known formulae.

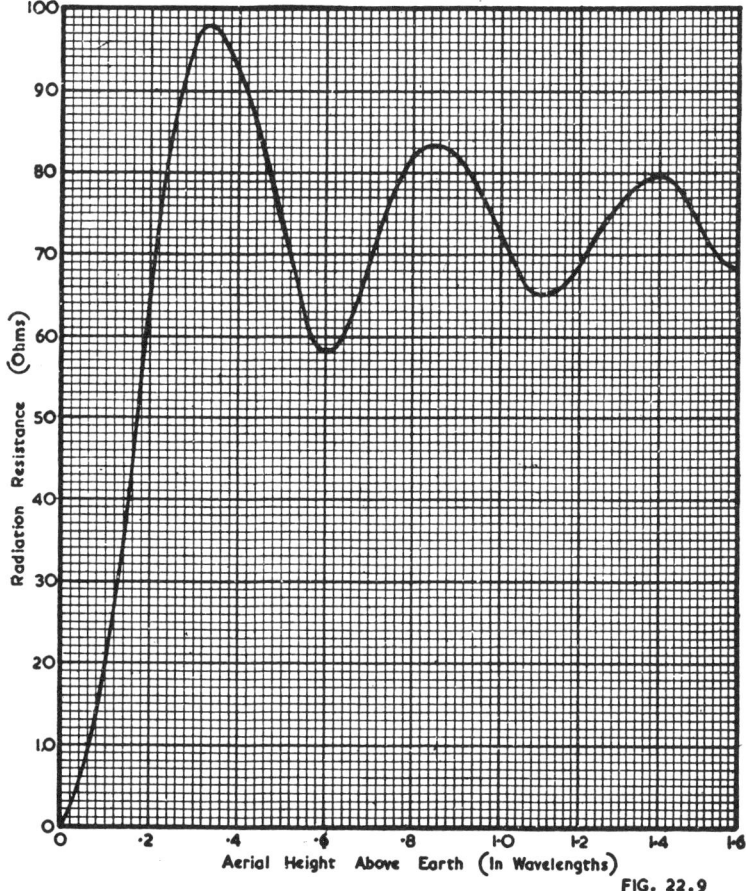

FIG. 22.9

Fig. 22.9. Radiation resistance of horizontal half-wave dipole aerial plotted against height of aerial above the earth.

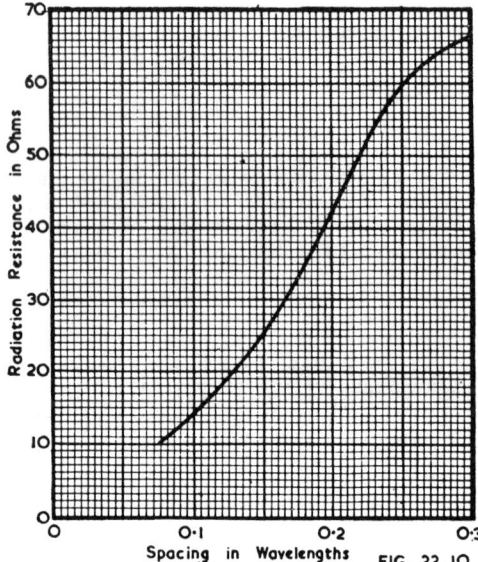

FIG. 22.10

Fig. 22.10. Radiation resistance of half-wave dipole aerial plotted against spacing between aerial and reflector (tuned to provide maximum forward radiation).

SECTION 8 : DUMMY AERIALS

When a signal generator is used in adjusting a receiver and in measuring its performance, it is necessary that the generator should present an impedance at the input terminals of the receiver which is equal to that of the aerial. At a single frequency this presents no difficulty ; one may calculate the aerial impedance in the manner shown in the previous section and then add appropriate amounts of resistance and reactance in series with the generator impedance to produce an impedance equal to that of the aerial.

When the receiver is required to be used over a range of frequencies, the production of a dummy aerial that will simulate the aerial is not quite so easy. If the aerial is very short and has a radiation resistance less than the aerial-earth loss resistance, then we can assume the latter to remain constant in value. We have to produce, therefore, a reactance which changes in the same way as that of the aerial. With a very short aerial, the term $\cot \beta l$ in (28) is approximately equal to $1/\beta l$ and we have

$$X_i \approx - jZ_0/\beta l \qquad (30)$$

which may be reduced to

$$X_i = \frac{-j}{2\pi f C} \qquad (31)$$

where C is the total capacitance of the aerial.

Hence, for such an aerial, the dummy consists of a fixed resistor in series with a fixed capacitor. The impedance of a longer aerial may be represented moderately well over a range of frequencies by a resistor, inductor and capacitor in series, all shunted by a capacitor. For such networks we refer the reader to sources listed in the bibliography.

For standard dummy aerials to be used in receiver testing, see Chapter 37 Sections 1 and 2.

With a receiver designed to operate with a loop aerial, it is not usual to use a dummy aerial when testing the receiver. Instead the aerial is left connected to the receiver and a known e.m.f. is induced in the loop. This is effected by connecting a second loop to a signal generator, and placing the two loops at a suitable distance from each other. The e.m.f. induced in the receiving loop can be calculated if, in addition to the distance, the dimensions of the loops, the number of turns, and the current in the transmitting loop are known.

SECTION 9 : TYPES OF AERIAL USED FOR BROADCAST RECEPTION

(i) Introduction (ii) Medium-frequency receiving aerials (iii) Short-wave receiving aerials (iv) V-H-F aerials.

(i) Introduction

Only simple types of aerials, such as are employed in broadcast reception, will be described here ; no mention will be made of the more complicated aerials used in point-to-point communication services.

There are three important factors to be considered in the design of an aerial for reception.

(a) Its effective area for the capture of radio energy and its efficiency should be sufficiently great to provide a signal that will override the internal noise of the receiver.

(b) It should be placed as far as possible from sources of noise-interference and, in particular, it should be situated outside the induction field of such generators.

(c) If the radio signal should arrive at the receiver by two different paths (e.g. ground-wave and sky-wave) so that distortion of the received signal results, then the aerial should have sufficient directivity to favour the wanted signal and reject the unwanted one.

At very-high frequencies, the effective area for capture of radio-energy by an aerial becomes so small that the first factor is of great importance. In medium-frequency reception, on the other hand, the second factor is the most important. One reason for this is that at such wavelengths it is not easy to place the aerial outside the induction field of noise sources. This may be seen from the following discussion.

At a distance of several wave-lengths from an aerial the power flowing through unit area falls off as the inverse square of the distance from the aerial. This implies that the field strength of the signal falls inversely as the distance [see (23)] from the aerial. This field is called the radiation field of the aerial. An aerial has also another component of its field which is the predominating one at points close to the aerial, but falls off rapidly and becomes negligible compared with the radiation field at a distance of a few wavelengths from the aerial. This is called the "induction field." At a distance of approximately one sixth of a wavelength ($\lambda/2\pi$) from a short aerial, the components of the induction and radiation field are equal. At shorter distances the intensity of the magnetic field of the aerial increases roughly as the inverse square of the distance from the aerial while the intensity of the electric field increases roughly as the inverse cube of the distance. It is obviously desirable that a receiving aerial should be situated outside this zone around a noise source in which the field intensity increases very rapidly as the distance between receiving aerial and the generator is decreased. This is comparatively easy at very-high frequencies where one sixth of a wavelength may be only one or two feet. At medium frequencies, the corresponding distance is of the order of 50 yards and it may be difficult to place the aerial at such a distance from sources of electrical interference.

(ii) Medium-frequency receiving aerials

Aerials short compared with one half-wave length are generally used. These may be of the straight-vertical, inverted-L or T types. The characteristics of these have been discussed in previous sections.

Very small aerials of this type are sometimes used **indoors.** Such aerials are useful only when the required signals are of large intensity, because these indoor aerials do not fulfil any of the three conditions listed above, except when close to broadcast stations. Besides being inefficient because of their small size, they are also partially shielded by earthed conductors in the house-wiring. They are also very liable to pick up r-f noise carried along the power mains.

Loop aerials are sometimes used in broadcast receivers. When used indoors they have the same drawbacks as the indoor capacitance type aerial, except that if shielded, or balanced with respect to earth, they are less sensitive to inductive interference from nearby sources of r-f noise. The reason for this is that a balanced or shielded loop responds to the magnetic component of the field of an aerial, whereas the capacitance type aerial responds to the electric component. In the radiation field, these two components are equal and are mutually dependent. In the immediate vicinity of an aerial they are not equal, and the electric component increases more rapidly than does the magnetic field as one approaches the aerial. Hence in the vicinity of a radiation source of r-f noise, a loop aerial will pick up less energy than will a vertical aerial.

A loop aerial is more sensitive to waves arriving at steep angles to the plane of the earth than is a vertical aerial. Hence it should not be used where interference between ground and skywaves is experienced. In such conditions a straight vertical aerial is normally the best type of aerial that can be used. In special circumstances, such as occur sometimes in country areas, it is possible to use another type of aerial, called a wave-antenna, to reduce ground-wave - sky-wave interference.

The wave antenna consists of a horizontal wire several wave lengths long suspended a few feet from the ground and directed towards the required broadcast station.

It functions because of an effect that has not been mentioned previously, called wave-tilt. This is the production of a radial component of the electromagnetic field, when the waves are passing over imperfectly conducting ground. This component is picked up by the horizontal wire which, when a wavelength or more long, has maximum directivity along its axis.

(iii) Short-wave receiving aerials

Since short-wave reception usually is concerned with signals of low field intensity, efficient aerials placed well away from sources of interference are required. In an earlier section it was shown that the height of the aerial above ground is also important. Indoor aerials are most unsuitable for short-wave reception. A half-wave dipole placed at the correct height above earth and connected to the receiver by a transmission line provides an efficient receiving system over a restricted range of frequencies. In short-wave reception, however, one is concerned with a range of approximately 3 to 1 in frequency. If the aerial is to provide a reasonably good match to the transmission line or receiver over such a frequency range, then a more elaborate aerial than the simple half-wave type is required.

One way of doing this is to arrange a number of half-wave dipole aerials in the form of a fan and connect them in parallel. Each dipole is tuned to one of the frequency bands allotted for short-wave broadcasting. At resonance the half-wave dipole matches a 70 ohm transmission line. The dipoles that are off resonance provide high and predominantly reactive impedances of both signs in parallel with the 70 ohms of the resonant dipole, and have small overall effect.

Such an aerial is connected to the receiver by a balanced transmission line. The receiver must be provided with an input circuit that is also balanced with respect to earth. Alternatively an aperiodic balance/unbalance transformer must be connected between the transmission line and the receiver.

An aerial that has similar characteristics to the one described above, but is not of the balanced type, is shown in Fig. 22.11. It is essentially half of the fan arrangement of dipoles. A common type of co-axial transmission line, having a characteristic impedance of 50 ohms, is suitable for use with this aerial. A few buried radial wires roughly half-wave (at the mid-frequency) in length provide a good earthing system for this aerial.

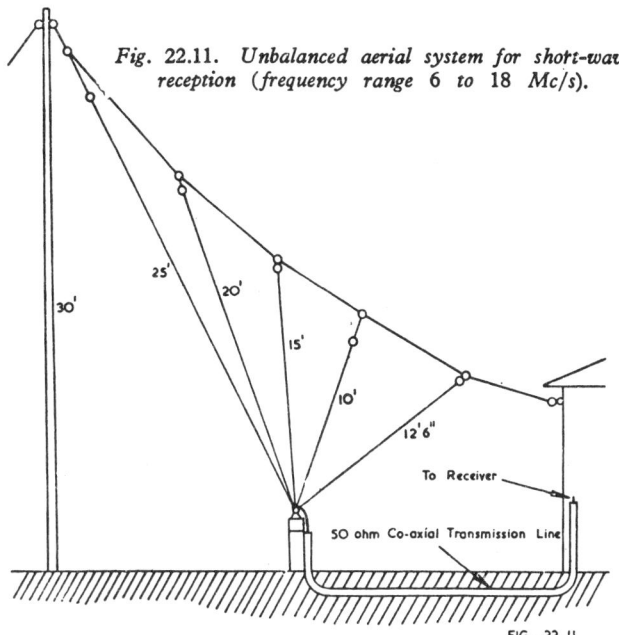

Fig. 22.11. Unbalanced aerial system for short-wave reception (frequency range 6 to 18 Mc/s).

To Receiver

50 ohm Co-axial Transmission Line

FIG. 22.11

(iv) V-H-F aerials

Because the effective area of an aerial at very high frequencies is small and the amount of energy that it can supply to the receiver is correspondingly small, it is essential that the efficiency of the aerial system should be high. At these frequencies the signal power per unit area at the aerial increases as the square of the height above ground, so the aerial should be placed at the maximum available height.

At very high frequencies the dimensions of the aerial are so small that rigid rods, rather than flexible wires, are used as aerial conductors. This simplifies the problem of the mechanical support for the aerial. Half-wave dipoles and quarter-wave vertical aerials with a counterpoise are the most common arrangements. The provision of a reflector is not difficult and is sometimes advantageous.

Matching the aerial to a transmission line is effected, when necessary, by means of sections of transmission line, rather than by combinations of lumped inductors and capacitors. Space does not permit a treatment to be given here of the great variety of line-sections, stubs and other impedance transforming devices that are in common use at very high frequencies. Nor can we deal with the very large number of varieties of aerial in use. A few of the common types are shown in Fig. 22.12. Of these, (a), (b), (c) and (d) are designed for reception of vertically polarized waves, and (e) and (f) are for horizontally polarized waves. The aerial (a) is the simplest type. It consists of a quarter wave radiator connected to the inner conductor of a co-axial line. Connected to the end of the outer conductor of the coaxial line is a quarter-wave " skirt," which forms, in effect, the second half of a vertical dipole. The impedance of the aerial is roughly 70-75 ohms and matches directly a common type of co-axial transmission-line. The disadvantage of this aerial is that the " skirt " is coupled to the outer conductor of the transmission-line and will pick up signals or noise that are present on this.

Type (b) consists of a vertical quarter-wave aerial with four horizontal quarter wave elements which form an artificial earth plane. The impedance of this aerial is lower than that of the normal quarter-wave aerial above a full earth plane, and it requires a matching stub, or a quarter-wave transformer to transform its impedance to that of a usual type of co-axial transmission line. The impedance of this aerial may be increased by a factor of four, by the device shown at (c), and the necessity of providing an impedance transformer is eliminated. Type (d) is a discone aerial, useful over a wide range of frequencies.

(e) Is the most commonly used type of horizontal aerial. It consists of two quarter-wave arms, one connected to the inner, and one to the outer conductor of a co-axial line. An outer quarter-wave sheath is connected at its lower end to the outer conductor of the co-axial line. The effect of this is to isolate the latter from earth, and hence preserve the balance of the aerial with respect to earth.

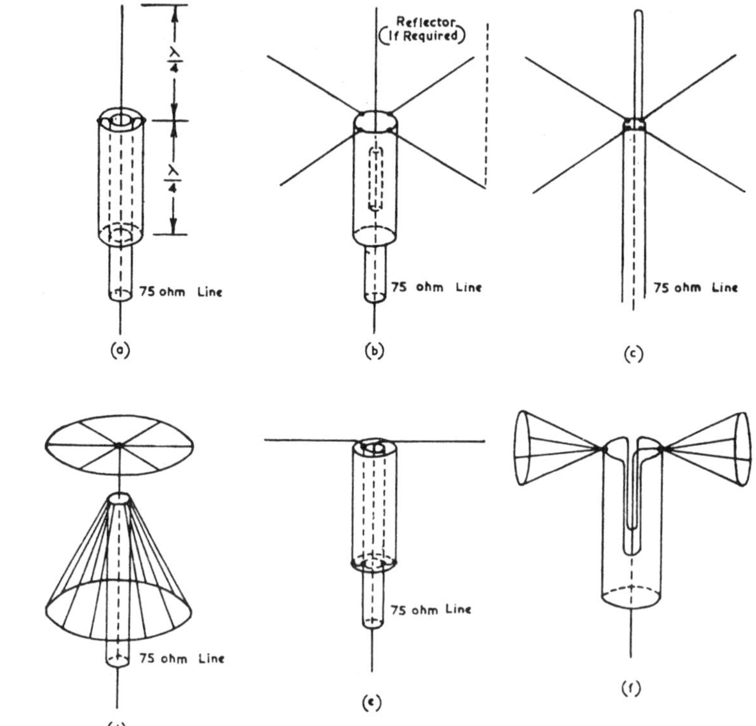

Fig. 22.12. Some varieties of v-h-f aerials.

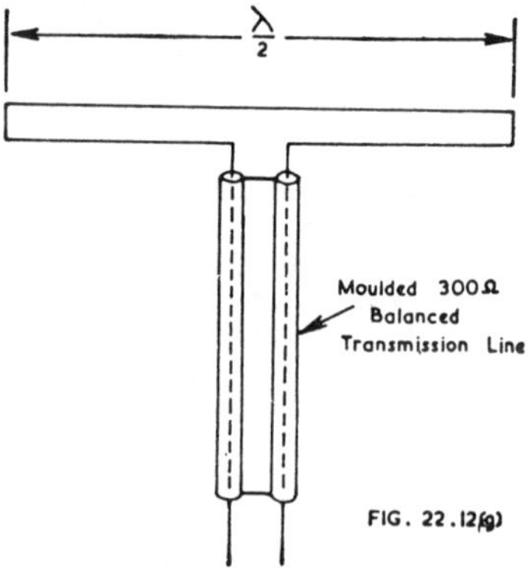

(f) Is a modification of (e).　The aerial conductors are made in the form of a conical cage to increase the frequency-range over which the aerial is effective.　In this aerial another method of providing the transformation from the unbalanced transmission line to the balanced dipole is shown.　It consists of splitting the outer conductor of the co-axial line into halves, for a distance of one quarter-wave.　At the top of this " split " the halves of the dipole are connected directly to the outer conductor, while the inner conductor of the transmission line is connected to one of them.

One of the most easily constructed aerials is the folded half-wave dipole, which is shown in Fig. 22.12 (g).　Whilst the radiation characteristic is the same as that of a conventional half-wave dipole, the folding produces an impedance transformation. When the conductors are of the same diameter throughout, the input impedance of the aerial is approximately 300 ohms.　This enables the aerial to be matched directly to a 300 ohms moulded transmission line, which is commercially available.　The spacing between the parallel conductors is not critical but, as in the case of a transmission line, it must be very small compared with a wavelength.

SECTION 10 : REFERENCES

1. Southworth, " Certain factors effecting the gain of directive antennas " Proc. I.R.E. 18.9 (Sept. 1930) 1502.
2. Burgess, R. E. " Aerial characteristics—relation between transmission and reception," W.E. 21.247 (April 1944) 154.
3. Schelkunoff, S. A. " Theory of antennas of arbitrary shape and size " Proc. I.R.E. 29 (1941) 493.
4. Pistolkors, A. " The radiation resistance of beam antennas " Proc. I.R.E. 17 (1929) 562.
5. Colebrook, F. M. " An experimental and analytical investigation of earthed receiving aerials " Jour. I.E.E. 71 (1932) 235.
6. Sturley, K. R. " Receiver aerial coupling circuits " W.E. 18.211 (April 1941) 137.
7. Burgess, R. E. " Receiver input circuits—design considerations for optimum signal/noise ratio " W.E. 20.233 (Feb. 1943) 66.
8. Wells, N. " Shortwave dipole aerials " W.E. 20.236 (May 1943) 219, also letter J. S. McPetrie 20.237 (June 1943) 303.
9. Stewart, H. E. " Notes on transmission lines—use of general equations in determining line properties " Q.S.T. (August 1943) 25.
10. Laport, E. A. " Open wire radio frequency transmission lines " Proc. I.R.E. 31.6 (June 1943) 271.
11. Wheeler, H. A., and V. E. Whitman, " Design of doublet antenna systems " Proc. I.R.E. 24.10 (Oct. 1936) 1257.
12. Carlson, W. L., and V. D. Landon, " New antenna kit design " R.C.A. Rev. 2.1 (July 1937) 60.
13. Carter, P. S. " Circuit relations in radiating systems " Proc. I.R.E. 20.6 (June 1932) 1004.
14. Brown, G. H. " Directional antennas " Proc. I.R.E. 25.1 (Jan. 1937) 78.
15. Data Sheet 53 " Aerial characteristics II—radiation resistance and polar characteristics " Electronic Eng. 15.188 (Oct. 1943) 197.
16. King, R. " Coupled antennas and transmission lines " Proc. I.R.E. 31.11 (Nov. 1943) 626.
17. King, R., and C. W. Harrison " The receiving antenna," Proc. I.R.E. 32.1 (Jan. 1944) 18.
18. Burgess, R. E. " Reactance and effective height of screened loop aerials " W.E. 21.248 (May 1944) 210.
19. Burgess, R. E. " The screened loop aerial " W.E. 16.193 (Oct. 1939) 492.
20. Colebrook, F. M. " The application of transmission line theory to closed aerials " Jour. I.E.E. 83 (1938) 403.
21. Burgess, R. E. " Aerial characteristics " W.E. 21 (1944) 154.
22. Kronenberg, M. H. " A multiple antenna coupling system—a method of operating several receivers on a single antenna " Q.S.T. (Aug. 1944) 9.
23. " Loops and small doublet antennas for F.M. receivers " A.R.T.S. and P. Bulletin No. 136 (9th Oct. 1944).
24. Carter, P. S. " Simple television antennas " R.C.A. Rev. 4.2 (Oct. 1939) 168.
25. Sherman, J. B. " Circular loop antennas at u-h-f's," Proc. I.R.E. 32.9 (Sept. 1944) 534.
26. Kendell, E. M. " Receiver loop antenna design factors " Comm. 25.11 (Nov. 1945) 62.
27. Moody, W. " F-M antennas " Service 14.1 (Jan. 1945) 24.
28. Burgess, R. E. " Aerial-to-line couplings—cathode follower and constant-resistance network " W.E. 23.275 (Aug. 1946) 217.
29. Vladimir, L. O. " Low impedance loop antenna for broadcast receivers " Elect. 19.6 (Sept. 1946) 100.
30. Sturley, K. R. (book) " Radio Receiver Design " Part 1 Chapter 3 (Chapman and Hall, London 1943).
31. " The A.R.R.L. Antenna Book " (American Radio Relay League, 1949).
32. McPherson, W. L. " Electrical properties of aerials for medium and long-wave broadcasting " Elect. Comm. 17.1 (July 1938) 44.
33. Bruce, E. " Developments in s.w. directive antennas " Proc. I.R.E. 19.8 (Aug. 1931) 1406.
34. " Capacity aerials for mains receivers " Philips Tec. Com. 3 (March 1947), 24, based on C.A.L.M. Report No. 38.
35. Design Data (11) " Aerial coupling in wide-band amplifiers " W.W. 53.2 (Feb. 1947) 50.
36. Browder, J. E., and V. J. Young " Design values for loop antenna input circuits " Proc. I.R.E. 35.5 (May 1947) 519.
37. Kobilsky, M. J. " A note on coupling transformers for loop antennas " Proc. I.R.E. 35.9 (Sept. 1947) 969.
38. Scroggie, M. G. " Is a big aerial worth while ? " W.W. 53.9 (Sept. 1947) 314.

Additional references will be found in the Supplement commencing on page 1475.

CHAPTER 23

RADIO FREQUENCY AMPLIFIERS

By B. Sandel, A.S.T.C.

SECTION 1 : INTRODUCTION

(i) Aerial coupling (ii) Tuning methods (iii) R-F amplifiers (iv) Design considerations.

(i) Aerial coupling

In any radio receiving system it is necessary to have some means of effectively transferring the modulated carrier voltage from the aerial system to the grid of the first radio frequency voltage amplifier. The usual method of achieving this result is to use some form of aerial coupling transformer which will give the desired voltage a larger amplitude than that of any other radio frequency voltages which may also be present at the receiver input terminals.

There are many possible forms of aerial coupling arrangements, but the most common type is the inductively coupled transformer, consisting of a primary and a secondary winding. The tapped inductance, or auto-transformer, is also extensively used, particularly with receivers tuning the broadcast F-M band and in portable battery receivers using single turn loops.

The aerial coupling unit is equally applicable to either tuned radio frequency or superheterodyne receivers, since the fundamental problem of effective voltage transfer is the same in each case.

Fixed tuned and untuned aerial stages are sometimes employed. The fixed tuned arrangement is more often used where the desired frequency range to be covered is only a small percentage of the operating frequency, e.g. in a band-spread range of a multi-range receiver, or at v-h-f. The more usual practice is to tune the secondary of the transformer by means of a variable capacitor, or alternatively to make the inductance variable, since this allows the optimum operating conditions, or a close approach to them, to be obtained over the required tuning ranges. It is not generally of great importance whether the tuning is accomplished by using a variable capacitor or a variable inductor. Because of mechanical difficulties with inductance tuners the variable capacitor has been more widely used up to the present time. The inductance tuner is rapidly gaining favour, however, at frequencies around 100 Mc/s, and higher, because of the difficulties due to common metal shafts in ganged capacitors and the inductance of the metal parts making up the capacitor. With the ganged capacitor arrangement considerable trouble is experienced due to coupling from one circuit to another, because it is not an easy matter to earth the shaft effectively between sections. The earthing wipers or leads have appreciable reactance and resistance at v-h-f and these factors are no longer negligible in comparison with the tuned circuit impedances. Insulated rotors and stators are of considerable assistance in overcoming difficulties due to common shaft coupling.

(ii) Tuning methods

Apart from ganged capacitors, many types of tuners have been used at v-h-f, including vane, guillotine, permeability, resonant line arrangements, etc., but all appear to have some disadvantages and it cannot be said that any one method is greatly superior to all others. Permeability tuning has also achieved some success on the 540 to 1600 Kc/s A-M, broadcast band, but it does not appear likely, in its present form, to supersede the variable capacitor.

(iii) R-F amplifiers

In conjunction with the aerial stage it often becomes necessary to incorporate a further stage to obtain additional radio frequency amplification and to increase further the discrimination against undesired signals, such as those from other broadcasting stations, and in the case of a superheterodyne receiver to limit the effects due to image and other spurious responses. This additional stage usually takes the form of a pentode voltage amplifier operating under class A_1 conditions and using a fixed or variable tuned circuit to act as a load impedance ; an untuned stage may be used in some cases. The valve is connected between the aerial transformer and the r-f tuned circuit which is in turn coupled to a converter valve or another r-f voltage amplifier.

Whether the r-f stage carries out its various functions efficiently depends on a number of factors. Firstly it is necessary that the input resistance of the valve should not be too low at the operating frequency, as this will adversely affect the performance of the input circuit—the aerial circuit in the case of one r-f stage. In addition, the noise generated by the valve should be low, and in a superheterodyne receiver the noise voltage should be very much less than that developed by the converter valve if there is to be appreciable improvement in the signal-to-noise ratio. Also the r-f stage gain should not be too low, otherwise the converter noise will still have an appreciable effect on the signal-to-noise ratio. This last requirement calls for valves operating at v-h-f to have fairly high values of mutual conductance, since the tuned circuit impedances are low and the stage gain is the product of these two factors. Grid-to-plate capacitance should be small in conventional voltage amplifiers to avoid the necessity for neutralization. Input and output capacitances are also important as they set a limit on the possible tuning range. Further considerations of operating conditions for valves used as r-f voltage amplifiers are given in Chapter 2 Sect. 4, and in Sects. 5, 6, 7 and 8 of this chapter.

Since the fundamental principles of operation of the aerial and r-f circuits are very similar whether we use a t-r-f or superheterodyne receiver, it can be taken that the coupling methods and the stage gain and selectivity calculations will apply equally well in both cases. Each type of receiver presents its own particular problems but it is not intended to enter into a discussion of these matters here. Moreover the arguments relating to the relative merits and de-merits of the two systems are well known and details can be found in the references listed at the end of the chapter.

A question which sometimes arises is the desirability of having more than one r-f stage in a superheterodyne receiver. It should be quite clear that, provided the gain of a single r-f stage is sufficient (say 10 to 15 times) to over-ride the effects of converter noise, no appreciable improvement in signal-to-noise ratio is obtained by adding further r-f stages of the same type. The advantage of additional r-f stages lies in the improved selectivity which is possible against image and other undesired frequencies. It is generally preferable to obtain any additional amplification which may be required in the i-f and a-f stages, depending on the purpose for which the additional gain is required. Further, a number of tuned r-f stages using ganged capacitors presents a rather difficult tracking problem, as well as giving rise to the increased possibility of unstable operation. Normally, only the transformer secondary is tuned and the overall selectivity characteristic is not as good as can be obtained with fixed-tuned i-f transformers. The gain and selectivity of the tuned r-f amplifier varies with the signal frequency, and a fixed intermediate frequency will allow more constant gain and higher gain per stage, for a given range of signal frequencies. The higher gain per stage is due to the higher dynamic impedance of the i-f transformer when the intermediate frequency is lower than the signal frequency. Even for an intermediate frequency higher than the signal frequency, larger stage gains are possible because of the better L/C ratio which can generally be obtained.

Even in cases where signal-to-noise ratio and r-f selectivity are not a problem, it is still sometimes advantageous to include a r-f stage. The r-f stage is used as a buffer to prevent the radiation of undesired signals from the receiver. A typical case of radiation occurs with a v.h.f. F-M receiver operated in conjunction with an efficient aerial system. If no r-f stage is used, radiation can occur at the oscillator frequency due to voltages at this frequency appearing at the input circuit of the frequency changer. This is serious because it opens up the possibility of interference between radio receivers which are operating in close proximity to one another.

It is not proposed, in this chapter, to deal in any great detail with grounded grid and other special types of r-f voltage amplifiers, as these are rather specialized applications which are not normally used in the design of narrow-band receivers covering limited bands of frequencies in the range from say 50 Kc/s to 150 Mc/s. However, at frequencies in excess of about 50 Mc/s, input stages using a grounded-grid triode are generally capable of giving better signal-to-noise ratios than circuits using the conventional grounded-cathode pentode voltage amplifier, but only at the expense of r-f selectivity. At the lower frequencies the input resistance of the pentode is much higher than that of the grounded grid triode, and so improved gain in the aerial coupling circuit can be used to offset the deterioration in signal-to-noise ratio due to valve noise. Clearly a better input circuit is possible using a pentode having a lower equivalent noise than with a pentode having the same input resistance but higher equivalent noise, and valve types should be selected with this in mind.

In cases where a wide range of frequencies has to be covered, and where the pass band of the receiver must be very wide (e.g. in television or multi-channel F-M link receivers), the grounded-grid triode offers considerable improvement in signal-to-noise ratio even at frequencies well below 100 Mc/s. Other applications occur in circuits where neutralization cannot be obtained by normal methods or where the circuit requirements are such that the grid must be earthed.

A special circuit (Cascode) which is of interest (see Ref. A10 Chapter 13) uses a grounded-cathode triode voltage amplifier, which is neutralized, followed by a grounded-grid triode. This circuit has high input resistance and low equivalent noise. It offers appreciable improvement in signal-to-noise ratio over the other two types of circuits, particularly at frequencies in excess of about 70 Mc/s. The combined circuit can be considered as equivalent to a single grounded-cathode triode having zero plate-grid capacitance, and a mutual conductance and noise resistance equal to that of the first triode valve (the valves used can be types 6J4 or 6J6 etc.).

Triode circuits of the types mentioned should not be used blindly, particularly at the lower frequencies, because the plate resistances of triodes are extremely low when compared with those of pentodes, and so the r-f stage gain may be insufficient, because the damping of the tuned circuits will be heavy, and also the r-f stage gain may be insufficient to make the effects of converter noise negligible. Detailed information on grounded-grid and triode r-f amplifiers can be found in Refs. B8, B18, A10 and A4.

(iv) Design considerations

The design methods to be discussed are equally applicable over any part of the range of frequencies mentioned above, but their practical application is rather more difficult as the frequency becomes greater, mainly because of the physical size of the components (including valves) and because the circuit layout is increasingly important if satisfactory results are to be obtained. Electron transit time effects in valves become of greater importance as the frequency is raised because of the changes which occur in the valve input and output admittances*. The valve input admittance must be considered when designing an aerial or r-f circuit as it will affect the dynamic impedance of the circuit, and consequently the selectivity and gain. Valve sockets and other components will add further damping to the input circuits. The effects of valve noise will also be governed by the dynamic impedances of the input circuits and so the various valve effects must be considered in conjunction with the external circuits. This becomes of greater importance as the frequency increases.

A design difficulty with aerial and r-f stages is in obtaining satisfactory performance over a large tuning range. The three to one frequency coverage normally employed on dual wave receivers for the medium (540—1600 Kc/s) and shortwave (e.g. 6-18 Mc/s) bands presents some problems in regard to tracking, and constancy of gain and selectivity. The L/C ratios obtainable with a variable capacitor giving an incremental capacitance range of at least 9 : 1 are satisfactory on the medium wave range but are rather poor on the shortwave band. For these reasons it is common practice to limit the coverage on shortwave bands, in the better class of receiver, to enable improved all round performance to be obtained ; a frequency ratio of about 1.3 : 1 is usual in the region of 10 to 20 Mc/s, with larger ratios at the lower frequencies. An obvious advantage of multi-band receivers is the greatly improved ease of tuning, apart altogether from other considerations of better performance.

To obtain the reduced frequency coverage with variable ganged capacitors having a capacitance range of about 10-400 $\mu\mu$F, various arrangements of series and parallel fixed capacitors are used. If the circuits are arranged to use a combination of series and parallel capacitors for bandspreading, it is possible to make the tuning follow practically any desired law. Further, if the tuning range is small, the oscillator frequency in a superheterodyne receiver can be above or below the signal frequency and three point tracking can still be obtained using series padders in both the signal and oscillator circuits. These matters will receive further consideration in Chapter 25 Sect. 3 on superheterodyne tracking.

Methods for measuring the value of k, the coefficient of coupling, are given in Chapter 26 Sect. 4(ii)E and (iii)B, C. Methods for measuring the primary resonant frequency of an aerial or r-f transformer are covered in Chapter 37 Sect. 4(ii).

SECTION 2 : AERIAL STAGES

(i) Difficulties involved   (ii) Generalized coupling networks   (iii) Mutual inductance coupling   (iv) Tapped inductance   (v) Capacitance coupling   (vi) General summary.

(i) Difficulties involved

The principal difficulty in the design of an aerial stage is that the type of aerial with which the receiver will eventually be used is, in general, not known. If the aerial impedance were known it would usually be a fairly simple matter to draw an equivalent circuit of the aerial and aerial input stage and to determine the complete performance by solving for the conditions existing in the circuit. When the conditions are not known, however, it is usually best to adopt constants for an aerial system which will have the most adverse effect on the receiver performance and to solve for these conditions. A compromise is then made in the circuit arrangement to minimize the effects of changes of gain, selectivity, and tracking with other circuits in the receiver.

(ii) Generalized coupling networks

Fig. 23.1 shows the generalized form for an aerial coupling unit. E_1 is the voltage induced in the aerial. E_2 is the voltage applied to the grid of the first amplifier stage.

---

*Valve admittances are covered in Chapter 2 Sect. 8.

Z_1 Z_2 and Z_3 are intended to represent any possible arrangement of components for coupling the aerial to the receiver, where Z_1 includes all impedances external to the receiver input terminals and Z_L is the load impedance. This generalized T section can be made to represent all possible coupling arrangements, since it is well known from circuit theory that any complex network composed of linear bilateral impedances can be transformed, at any one frequency, into an equivalent T network

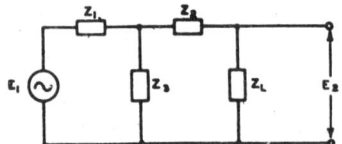

FIG. 23-1 GENERALIZED AERIAL COUPLING UNIT.

From the generalized network

$$\frac{E_2}{E_1} = \frac{Z_3 Z_L}{Z_1 Z_3 + (Z_1 + Z_3)(Z_2 + Z_L)} \tag{1}$$

This expression is of little assistance unless it is applied to the specialized networks used in radio receivers. Additional generalized expressions for this network are developed in K. R. Sturley's "Radio Receiver Design" Part 1, Chapter 3. It is proposed to deal specifically with a few of the more common types of coupling circuits.

(iii) Mutual inductance coupling

Fig. 23.2A shows one of the most widely used aerial coupling circuits. E_1 is the voltage induced in the aerial and Z_A represents the aerial constants, which may be any combination of inductance, capacitance and resistance. L_1 is the aerial coil primary winding which has a r-f resistance R_1. L_2 is the secondary winding coupled by mutual inductance to L_1. R_2 is the r-f resistance of L_2. C_2 is used to tune the secondary circuit to resonance; it should be noted that the setting for C_2 when the aerial circuit is connected across the primary is different from the setting obtained when L_1 is on open circuit. Fig. 23.2B shows the T section equivalent of the aerial transformer, and Fig. 23.2C shows an alternative equivalent secondary

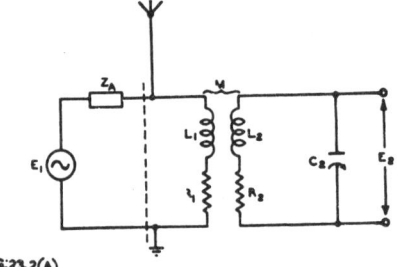

FIG: 23.2(A)

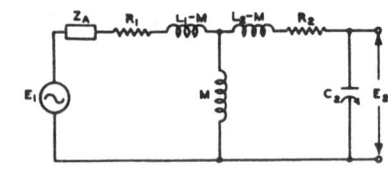

FIG. 23.2(B)

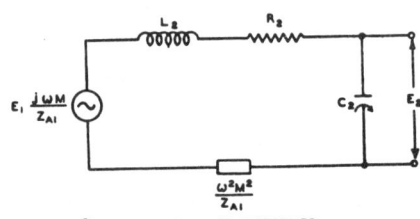

FIG. 23-2(C)   AERIAL TRANSFORMER.

circuit which is very helpful in understanding the idea of "reflected" impedances. For the present, the effects of valve loading will not be considered. When the secondary is tuned to resonance, the aerial circuit gain is given by

$$\frac{E_2}{E_1} = \frac{M/C_2}{Z_{A1}\left[R_2 + \frac{\omega^2 M^2 R_{A1}}{|Z_{A1}|^2}\right]} \tag{2}$$

where $Z_A = R_A + jX_A$
$Z_1 = R_1 + j\omega L_1$
$M = k\sqrt{L_1 L_2}$

$$Z_{A1} = Z_A + Z_1 = R_{A1} + jX_{A1}$$

and $|Z_{A1}| = \sqrt{R_{A1}{}^2 + X_{A1}{}^2}$.

It should be noted that the condition for secondary circuit resonance (partial resonance " S ") is that

$$X_2 = \frac{\omega^2 M^2 X_{A1}}{|Z_{A1}|^2} \tag{3}$$

and the " reflected " resistance and reactance in the secondary are given respectively by

$$\frac{\omega^2 M^2 R_{A1}}{|Z_{A1}|^2} \tag{4}$$

and

$$\frac{-\omega^2 M^2 X_{A1}}{|Z_{A1}|^2} \tag{5}$$

If the mutual inductance M is also made variable, the maximum value of secondary current (and of E_2/E_1 when the primary circuit Q is low) is obtained when

$$M = \frac{|Z_{A1}|}{\omega} \sqrt{\frac{R_2}{R_{A1}}} \tag{6}$$

and

$$\frac{E_2}{E_1} = \frac{1}{2\omega C_2 \sqrt{R_2 R_{A1}}} \tag{7a}$$

$$= \tfrac{1}{2} \sqrt{\frac{R_D}{R_{A1}}}. \tag{7b}$$

where R_D is the dynamic impedance of the secondary circuit. The approximations involved in these equations are discussed in Ref. B12.

From this it can be shown, with M and C_2 adjusted for maximum secondary current, that the selectivity of the tuned circuit is half the value it would have if considered in the absence of the primary. This condition of optimum coupling is not usual in broadcast receivers, except where matching to a transmission line, since the coupling between primary and secondary is made loose to prevent serious tracking errors due to the variable reactance " reflected " into the secondary circuit over a band of frequencies. The expression most generally required is that of eqn. (2) and the calculation of voltage gain need only be made at, say, two or three points across the tuning range.

It should be noted, however, that in some cases loose coupling can lead to a considerable reduction in signal-to-noise ratio if site noise is not predominant, e.g. at the higher frequencies on the short-wave band it is advantageous to use optimum coupling, or slightly greater, to improve the signal-to-noise ratio (see Ref. A2 Chapter 26).

The effects of valve input admittance can be included by considering an additional resistance and a capacitance shunted across C_2. The value of the resistance component may be negative or positive depending on the type of valve and the operating conditions, e.g. some types of converter valves have a negative input resistance.

Having obtained the stage gain it is next important to find the **selectivity** of the circuit. As an approximation for ordinary types of aerial coils it is usually convenient to determine the selectivity of the secondary as for a single tuned circuit. The simplest approach is to consider the resultant Q of the secondary circuit at the resonant frequency when all the secondary resistance, including that due to reflection from the primary and valve loading, is included. The value of Q is given by $\omega L_{eff}/R_{eff}$, where L_{eff} is the resultant secondary inductance. The selectivity is then found as for a single tuned circuit, resonant at the frequency under consideration, using the universal curve given in Chapter 9 (Fig. 9.17) in conjunction with the methods detailed in Chapter 26 Sect. 4(iv)A. This method is satisfactory provided that the amount of mistuning is not too large a proportion of the central reference frequency, and that the aerial coil follows the usual practice with the primary resonated above or below the tuning range or is designed for connection to a transmission line (or aerial) of which the characteristic impedance is largely resistive ; otherwise the variations in R_{eff} and L_{eff} would have to be considered. Further, in a number of practical cases it is permissible to calculate R_{eff} and use L_2 when finding the value for Q, since the reflected reactance may not be very large.   It is often sufficient to calculate selectivity at, say, the mid-frequency of the tuning range, although it should be noted that selectivity is a function of frequency and dependent on the circuit operating conditions.

It is now worth while considering three special cases for the aerial coupling transformer :

(1) Where the aerial can be represented by a capacitive reactance in series with a resistance.

(2) Where the aerial behaves like an inductive reactance in series with a resistance.

(3) Where the aerial is coupled to a transmission line which " looks " like a pure resistance at the aerial terminals ; or the aerial itself, coupled to the receiver, looks like a pure resistance.

Any aerial can be made to appear resistive if the reactance components are tuned out by a suitable arrangement of the aerial coupling network ; this is a common procedure with fixed frequency installations.

In **case** (1) it is usual to make the primary circuit resonate below the lowest tuning frequency, since this offers a good compromise with regard to constancy of gain, tracking, etc.   This arrangement is also advantageous because the effect of the aerial is merely to move the primary resonant frequency still further away from the tuning range, thereby minimizing the effect of the aerial on the secondary without losing too much gain.   In general the high impedance primary winding is resonated at 0.6 to 0.8 of the lowest tuning frequency, either with the aerial capacitance alone, or with an added capacitor.   In either case the resonant frequency must not be close to the intermediate frequency in a superheterodyne receiver, and normally a broadcast band primary is resonated at a lower frequency than the i-f so that no aerial which is likely to be used will resonate the aerial primary at the intermediate frequency.

Since the gain obtained from an aerial transformer with its primary resonant outside the low frequency end of the tuning range decreases towards the high frequency end of the range, a small capacitance is often added across the top of the transformer. This capacitance increases the gain at higher frequencies and allows a substantially flat gain and noise characteristic to be obtained from the transformer over the desired range.   This type of aerial transformer is fairly representative of those likely to be encountered on the medium and long wavebands.   Practical values for the coefficient of coupling between primary and secondary are a compromise between various factors such as gain, tracking error and signal-to-noise ratios ; usual values are from 0.15 to 0.3, with most aerial coils using values of about 0.2 (this should include any added top capacitance coupling).   The value for primary Q is of the order of 50, and the secondary Q is about 90 to 130 in typical aerial coils using the range of coupling coefficients stated, although Q values greater than 200 can readily be obtained by using special iron cores.   Values of secondary Q above 100 generally call for smaller values of k than 0.2 if good tracking is to be achieved.

If for some particular reason it is necessary to use a higher coefficient of coupling than say 0.4, then the primary inductance should be small and arranged to resonate with the aerial at about $\sqrt{2}$ of the highest signal frequency.   This improves the signal-to-noise ratio but may give rise to tracking difficulties.

Case (2) normally applies with untuned loop aerials but the condition can occur with some types of aerials on the shortwave bands.   In this case it is usual to make the coefficient of coupling fairly high, of the order of 0.5 or more, and a useful rule is to make the value of primary inductance approximately equal to 0.4 of the loop (or aerial) inductance (see Refs. B17 and A2, p. 44).

For use with domestic type receivers operating on the short wave band, the standard dummy antenna is taken as being 400 ohms resistive, even though the aerial reactance will not be zero across the tuning range, as this gives an indication of average operating conditions.   The aerial coupling transformer for this type of receiver can be designed using the procedure given in Case (3) below.

The untuned loop antenna (which should be balanced to ground for best results) finds application with portable receivers, and in some cases with ordinary A-M broadcast receivers.   Loops have also been used in F-M receivers operating on the 40-50

Mc/s band, but not in receivers for the 88-108 Mc/s band because of the ease with which a half-wave dipole can be arranged inside the receiver cabinet. With loops it is necessary to keep stray capacitances to a minimum so as to avoid resonance effects within or close to the band of frequencies to be received ; this also assists in achieving good tracking because a loop that behaves like a small fixed inductance can be compensated for quite readily in the aerial coil secondary.

The number of turns used in untuned loop aerials is not very critical, since a loss in effective height by using fewer turns can be offset by the increased voltage step-up possible with the coupling transformer. To keep stray capacitances low the fewer turns used the better. The best compromise for receivers operating on the medium or short-wave bands is a single turn loop, or perhaps a few turns (say not more than four) of wire wrapped around the carrying case for medium wave portable receivers (this wire can be litz or copper braid, and even rubber covered hook-up wire is often satisfactory). The best type of single turn loop is one made out of copper tubing (or any other good conductor such as aluminium or one of its alloys) with the tube diameter not less than about 0.25 in. and preferably considerably larger than this ; the loop area should always be as large as possible. If the loop is to be used on the long-wave band it is preferable to use a few turns (about ten) of litz wire rather than a single turn loop. This is necessary because of the drop in Q which occurs with the single turn loop at the lower frequencies. However, satisfactory results have been achieved with a 0.25 in. diameter copper tube loop at a frequency of 500 Kc/s.

In many cases a two winding transformer is not used and the loop is made part of the tuned circuit (being in series with the main tuning inductance and earthed at one side). An alternative arrangement is to tap the loop across part of the main tuning inductance ; this can be made to give very satisfactory results.

Because the inductance of loops (of the type being considered) is usually very small, it is often difficult to measure the Q at the working frequencies. This difficulty arises with standard types of Q meters because the maximum capacitance is limited to about 400 $\mu\mu$F. A useful method (suggested by J. B. Rudd) is as follows :

Couple the loop fairly tightly to an inductance which is sufficiently large to be tuned to the working frequency. Measure the Q of the inductance with the loop open circuited (call this Q_1) and then the Q (call this Q') with the loop short circuited at its terminals. The Q of the loop (Q_L) is given by

$$Q_L = \frac{yQ_1Q'}{Q_1(1-y) - Q'}$$

where Q_1 = magnification factor for the primary inductance with the loop open circuited.

Q' = magnification factor for the primary inductance with the loop terminals short circuited.

$y = 1 - (L'/L_1)$

L' = inductance of primary with loop short circuited.

and L_1 = inductance of primary with loop open circuited.

See (vi) below for some discussion of directly tuned loops.

Case (3) permits of a fairly simple solution. Normally k is about 0.2 or less, and the primary winding reactance (at the geometrical mean frequency of the tuning range) is made equal to the characteristic impedance of the transmission line or aerial. A particularly simple solution has been made by Rudd (Ref. B12) to give conditions for two point matching in the tuning range. The procedure is as follows :

Geometrical mean angular frequency = $\omega_0 = \sqrt{\omega_1\omega_2}$ where ω_1 and ω_2 are the lower and upper angular frequency band limits. The primary inductance = L_1 = Z_0/ω_0 where Z_0 is the characteristic impedance of the transmission line.

The required coefficient of coupling is then determined from

$$k = \sqrt{\frac{1}{Q_s}\left[\frac{1+\alpha}{\alpha^{\frac{1}{2}}}\right]} \tag{8}$$

where $\alpha = \omega_2/\omega_1 = f_2/f_1$.

and Q_s = secondary circuit Q in the absence of the primary. The value for Q_s should include the effect of valve loading, and so

$$Q_s = \frac{QR}{Q\omega_0 L_2 + R} \qquad (9)$$

where Q is for the unloaded secondary
and R is the valve input resistance.

A useful approximate formula for estimating the aerial coil gain under this condition is

$$\frac{E_2}{E_1} = \frac{1}{2}\sqrt{\frac{Q_s \omega L_2}{Z_0}}. \qquad (10)$$

This expression gives the maximum possible value of gain and corresponds to eqn. (7a) given previously.

Although elaborate methods are available for calculating optimum signal-to-noise ratios etc., the details above should give a fairly close approximation to the required practical conditions where the aerial coupling circuit is a compromise between the various conflicting factors discussed previously. These factors of gain, selectivity etc. should be carefully considered as in nearly every case the performance of a receiver is largely governed by the aerial coupling circuit.

(iv) Tapped inductance

Another common type of aerial coupling arrangement is shown in Fig. 23.3. This arrangement has found wide application in receivers tuning the 88-108 Mc/s F-M band because of its simplicity and the ease with which the correct aerial tapping point may be found. This can also be considered as a particular case of the generalized aerial circuit. This circuit can be treated in a simplified manner due to Zepler (Ref. A2). The required tapping point on the total coil, which is usually a solenoid, can be found approximately from the turns ratio

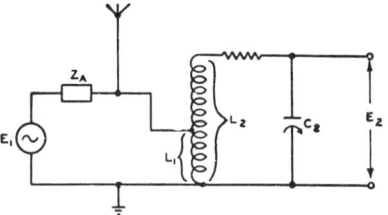

FIG. 23-3   TAPPED INDUCTANCE AERIAL TRANSFORMER

$$\frac{n_2}{n_1} = \sqrt{\frac{Q_s \omega L_2}{Z_0}} \qquad (11)$$

where n_2 = total number of turns
n_1 = number of turns across which aerial is connected
ω = $2\pi \times$ operating frequency
Q_s = magnification factor for complete circuit
and Z_0 = impedance of the aerial (or the characteristic impedance of a transmission line).

The gain is found from equation (10),

$$\frac{E_2}{E_1} = \frac{1}{2}\sqrt{\frac{Q_s \omega L_2}{Z_0}}$$

The selectivity is calculated using the same methods as before, from the equivalent circuit. Whether this equivalent circuit takes the form of a series or parallel tuned circuit is usually not important, as the results obtained are substantially the same, and provided the circuit Q exceeds about 10 the Universal Selectivity Curves can be applied directly in either case.

It should be noted that the treatment is only approximate, as the exact tapping point is dependent on the coupling between the two sections of the aerial coil winding. However, for a wide variation in coupling, the correct tapping point is not very different from that obtained by the above procedure, which assumes unity coupling, and the loss in secondary voltage as compared with optimum coupling is quite small.

Valve input loading is usually a problem at v-h-f and a procedure to minimize this effect is to tap down across L_2.   This will allow a higher value of dynamic impedance for the tuned circuit with improved selectivity, and under some conditions the increase in impedance can more than offset the loss in voltage gain caused by tapping down.

(v) Capacitance coupling

A further type of coupling arrangement fairly commonly used is shown in Fig. 23.4.

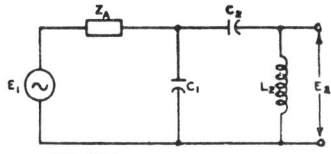

Fig. 23.4.   Capacitance Coupled Aerial Transformer.

FIG. 23.4

This shows a common disadvantage with the circuit of Fig. 23.3 in that the variation in gain across the tuning range is quite appreciable.   The treatment is quite straightforward and is left to the reader ; a fairly complete analysis is given in the references.

(vi) General summary

From the considerations detailed above, it can be seen that the coupled circuit arrangement of Fig. 23.2 is generally most satisfactory since it readily lends itself, with minor modifications, to applications using balanced or unbalanced aerial systems. Under some conditions of receiver operation other methods are used because of the practical consideration of ease of adjustment for best operation, or for use with a particular type of aerial system.

Fixed tuned aerial stages are sometimes used with receivers covering a limited range of frequencies.   A typical case would be in the range of 88-108 Mc/s where the loaded Q of the aerial coil may be less than 20.   In such a case a simple arrangement is for the secondary to be resonated near the centre frequency in the band ; this introduces gain variations of about 2 : 1 across the tuning range.   The gain variations can be offset by resonating the secondary at a point in the band such that the gain variations offset those in r-f or mixer stages.   The overall gain variations for the band under consideration can be limited to less than 1.5 : 1 with a superheterodyne receiver using a tuned r-f stage, and as the loss in gain and signal-to-noise ratio is not very serious, the saving in the cost of a gang section is often well worth while when weighted against the other factors.   If the aerial circuit loaded Q is above about 20, this method is not recommended because of the deterioration in circuit performance.   In some cases it is preferable to tune the aerial circuit and use an untuned r-f stage (when this is practicable).

One type of coupling circuit which has not been mentioned previously is the directly tuned loop.   In this case the loop provides the tuned circuit inductance as well as the direct signal pick-up.   As for any other loop, the area and the Q should be as large as possible.   This loop will probably be used, in most cases, with portable receivers. If the loop can be placed in such a position that its Q is not materially affected by the presence of the receiver and batteries, then its performance will usually be superior to that of the single turn loop (an improvement of about 6 db can be expected).   However, this arrangement is seldom convenient and in many cases better results are possible with the single turn loop (or one with a few turns as discussed previously) because any effect on its Q can be offset to a large extent by the use of an aerial coupling transformer having a high Q winding in the tuned circuit.   The improvement can be seen most readily by considering a low Q loop, having a small inductance, connected directly in series with an aerial coil which has a high Q and whose inductance is practically the whole of the tuned circuit inductance.

In some cases it is required to use an additional external aerial with the loop to increase signal pick-up.   There are many possible arrangements but amongst the simplest is the use of a small capacitance (say 10 to 20 $\mu\mu F$) connected between the external aerial.and the top of the tuned circuit (i.e. to the grid connection) or alternatively a tap on the aerial coil secondary a few turns up from the earthy end.   For a directly tuned loop the aerial is tapped into the loop in the same way as for the other coil arrangement ; alternatively the series capacitor arrangement is satisfactory.   The main disadvantage of both these simple arrangements is the large change in gain which occurs across the tuning range, but for medium-wave commercial portable receivers additional circuit complications are seldom justified.

The subject of aerial-to-receiver coupling is a large one and only a few of the more important design factors have been considered.   For more complete information on particular systems it is necessary to consult the text books and references.

SECTION 3 : R-F AMPLIFIERS

(i) Reasons for using r-f stage   (ii) Mutual-inductance-coupled stage   (iii) Parallel tuned circuit   (iv) Choke-capacitance coupling   (v) Untuned and pre-tuned stages (vi) Grounded grid stages.

(i) Reasons for using r-f stage
The necessity for adding a r-f amplifier in a superheterodyne receiver arises from the need for

    (1) greater gain

    (2) improved rejection against undesired signals

    (3) improved image frequency rejection

    (4) reduction of the effects of spurious frequency combinations,

    (5) improved signal-to-noise ratio,

    (6) the prevention of radiation at the local oscillator frequency (under some conditions).

In the case of the t-r-f receiver, additional r-f stages are usually added for reasons (1) and (2) above.

The design of a r-f stage is quite similar for both types of receiver and so they will not be considered separately.   The only point which will be mentioned is that, in the case of the t-r-f receiver the final r-f amplifier feeds into a detector stage which, if a diode is used, heavily loads the tuned circuit unless a tapping point is used.   For this reason it is common practice to incorporate anode bend or linear reflex detectors which do not appreciably load the input circuit.   However, a diode is convenient as a means of supplying an a.v.c. bias voltage of the correct polarity, a condition not fulfilled by the other two types of detectors mentioned without additional circuit arrangements.

(ii) Mutual-inductance-coupled stage
The most common arrangement for coupling a r-f amplifier valve to the following stage (Fig. 23.5) is by means of a transformer using a tuned secondary with the primary winding resonated either above or below the tuning range.   The primary winding resonates with the valve output capacitance, winding capacitance, stray wiring capacitances and sometimes an added capacitor, these being lumped together and shown as C_p.   If the winding is of the low impedance type it is made to resonate at approximately 1.2 to 1.5 times the highest tuning frequency.   Under this condition the load presented to the valve is inductive and thus will give rise to regeneration by introducing a negative resistance component plus a capacitance into the r-f valve input circuit.

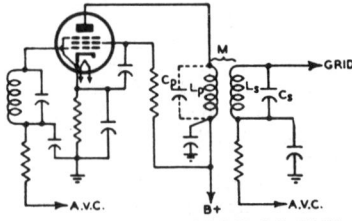

FIG.23·5   TRANSFORMER COUPLED R-F STAGE.

If the winding is of the high impedance type, then it is usual to resonate the primary at approximately 0.6 to 0.8 of the lowest tuning frequency, but the resonant frequency should not be close to the intermediate frequency in superheterodyne receivers. Under this condition the valve has a capacitive load which causes degeneration by introducing a positive resistance component plus a capacitance into the r-f valve input circuit due to the " Miller Effect." It is not usual to resonate the primary winding within the tuning range because of the very large changes in gain and selectivity which would be introduced. In addition, the problem of tracking the various tuned circuits becomes almost hopeless.

When the aerial coil primary is of the low impedance type it is usual, although not essential, for the r-f stage to be of this type also and vice-versa. This simplifies tracking problems.

The methods to be used for calculating the gain and selectivity of this arrangement are set out in detail in Chapter 9 Sect. 6(ii) under the heading " Coupled Circuits— Tuned Secondary." For the usual practical arrangement the stage gain is given approximately by $g_m Q_2 \omega M$,

where g_m = mutual conductance of r-f amplifier valve,

ω = $2\pi \times$ operating frequency,

$M = k\sqrt{L_1 L_2}$ = mutual inductance,

and Q_2 = secondary circuit Q including effect of valve input resistance, but not " reflected " resistance from the primary.

The selectivity can also be calculated, approximately, by the method suggested for aerial coils.

When designing a stage of this type it is necessary, as in the case of the aerial coil, to arrive at a compromise between gain, selectivity and tracking errors. This usually leads to a value for the coefficient of coupling of approximately 0.15 to 0.3, with the lower values preferred for good tracking between aerial, r-f, and, in the case of the superhet., the oscillator stage.

As with aerial coils, the primary Q is about 50 (or perhaps less) and the secondary Q from say 90 to 130 in typical coils using this range of k values. Higher Q generally calls for the lower values of k.

In the case of receivers working at say 50–400 Kc/s it is often inconvenient to resonate the primary winding in the manner suggested. Under these circumstances it is sometimes necessary to damp the primary winding heavily so that no pronounced resonance occurs. It is then permissible to resonate the winding within the tuning range. This procedure can lead to a severe loss in gain but is sometimes justified under practical conditions where the loss of gain is outweighed by other factors.

The addition of top capacitance coupling, i.e. a small capacitance from the plate connection to the grid connection of the following stage, is often employed to equalize gain variations across the tuning range. The value of the capacitance is usually of the order of a few micro-micro-farads and the exact value is best determined experimentally. It should be noted that the presence of this capacitance will alter the co-efficient of coupling, stage gain, selectivity and the primary resonant frequency, as well as circuit tracking. There is always some capacitance coupling present with any practical transformer and this largely determines the relative physical arrangement of the primary and secondary windings, which are connected (in nearly all practical cases) so that the capacitance coupling adds to the mutual inductance coupling.

(iii) Parallel tuned circuit

A simple parallel tuned circuit is sometimes used in the r-f stage, and can be made to give higher gains than the transformer coupled arrangement. The difficulties encountered are that

(1) large gain variations occur across a band of frequencies
(2) the skirt selectivity is rather poor
(3) tracking with conventional aerial circuits is difficult
(4) the tuning capacitor has to be isolated from B+.

For these and other reasons this circuit is not in common use in broadcast receivers, but it might be incorporated in a receiver working at high frequencies covering a

restricted tuning range with some possible advantage.   The gain of the stage is given
by $g_m Q_{eff} \omega L$
where g_m = mutual conductance of r-f amplifier valve,
    Q_{eff} = Q of r-f coil when loaded by plate resistance of r-f amplifier and input
        resistance of following stage,
    ω = 2π × frequency,
and L = inductance of tuned circuit.
Selectivity is calculated as for any other single tuned stage, using the methods of
Chapter 9 or Chapter 26, Sect. 4(iv)A.

(iv) Choke-capacitance coupling

Fig. 23.6 shows a common r-f coupling circuit for use at v-h-f such as on the 88-108
Mc/s F-M broadcast band.   The usual arrangement takes the form of a r-f choke,
resonated well below the lowest tuning frequency, coupled to a tuned circuit $L_2 C_2$
by means of a capacitor C.   The choice of a suitable value for C will allow some reduction in variations of stage gain across the tuning range as it can be considered as
part of a voltage divider formed with the tuned grid circuit.   A further useful function
of the arrangement is that when a suitable resonance frequency has been chosen for
the choke, the total stray capacitance across the tuned circuit is effectively reduced,
thereby reducing limitations on the tuning range ; a small value for C assists
in this latter regard.   However, the application of the circuit is largely one of practical
convenience and, as always, is a compromise.

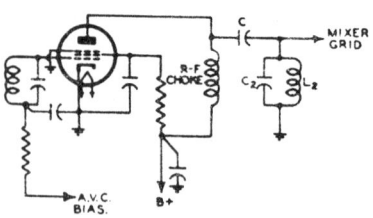

FIG. 23-6   CAPACITANCE COUPLED R-F STAGE.

With a choke resonated at $1/\sqrt{2}$ of the
lowest tuning frequency in a receiver with
a low frequency limit of 87.5 Mc/s and an
upper limit of 108.5 Mc/s, the values
obtained for the additional capacitance
shunted across the tuned circuit were 3.6
$\mu\mu$F at 88 Mc/s and 4 $\mu\mu$F at 108 Mc/s,
excluding the input capacitance of the following stage.   The value of C was 10 $\mu\mu$F.
Because of the low value of dynamic impedance for this circuit at v-h-f, the effect
of the plate resistance of the r-f valve can usually be neglected (but see Ref. A8).
In this case the selectivity for the circuit can readily be calculated as for a single tuned
circuit, including the effects of the input resistance of the following stage, without
introducing very large errors.   The gain may be calculated with sufficient accuracy
as for the case of the simple tuned circuit acting as the anode load and multiplying
the result by a suitable factor to allow for the voltage division due to the capacitive
reactances at the frequency being considered.   This factor is readily found from a
knowledge of the equivalent capacitance due to the choke circuit, at the working
frequency, and the value selected for C.

(v) Untuned and pre-tuned stages

Untuned, or in some cases pre-tuned, r-f stages are sometimes used in receivers
where economy and/or limited space are important factors, or where only a small
frequency range is to be covered.   Unfortunately in the case of portable battery receivers, where the application could be valuable, the use of untuned r-f stages is made
difficult by the fairly low values of mutual conductance for the usual r-f amplifier
valves.   The gain obtained is limited and it may be preferable to use an additional
i-f stage if the r-f stage cannot be tuned and gain is a prime requirement.   With
mains operated receivers the problem is different, as many suitable valve types are
available and quite good stage gains (about 5 times or more, on the medium and short
wave bands) can be obtained over a wide range of frequencies.   Suitable   design
methods using untuned stages for the medium and short wave bands have been given
in R.C.A. Application Note No. 116 (reprinted in Radiotronics No. 124) to which the
reader is referred for full details   A typical coupling circuit of the type discussed in

the Application Note is shown in Fig.
23.7.

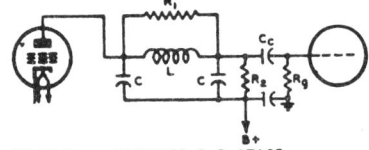

FIG. 23·7   UNTUNED R-F STAGE.

(vi) Grounded-grid stages

As previously mentioned this type of
amplifier will not be discussed in detail.
However, the following equations should prove to be of assistance, and are generally
a sufficiently good indication since experimental techniques usually provide the most
suitable method for determining the best operating conditions for this type of amplifier.

$$\text{Voltage gain} = \frac{R_L}{r_p + R_L} (\mu + 1)$$

Resistance loading across input circuit $\approx \dfrac{r_p + R_L}{\mu + 1}$

(loading across output circuit $\approx r_p$).

For $\mu \gg 1$ the optimum load impedance is given by

$$R_L = r_p \sqrt{1 + R_i g_m}$$

Noise factor $= N = \sqrt{2 + \dfrac{R_1}{R_0 A_1{}^2} + \dfrac{R_2}{R_0 A_1{}^2 A_2{}^2}}$

where R_L = load impedance connected between plate and B+

r_p = valve plate resistance

μ = valve amplification factor

g_m = valve mutual conductance

R_i = valve input resistance at working frequency

R_1 = equivalent noise resistance of first valve

R_2 = equivalent noise resistance of second valve

R_0 = generator impedance e.g. aerial resistance

A_1 = voltage gain of input circuit

and A_2 = voltage gain of first stage, excluding the voltage gain of the input
circuit.

It should be noted that it is always advisable to operate the heater and cathode of
the valve at the same r-f potential. This is readily achieved by using suitable chokes
in the heater leads. For more detailed information on grounded-grid amplifiers
see References B8, B18, A10 and A4.

SECTION 4 : IMAGE REJECTION

*(i) Meaning of image rejection   (ii) Image rejection due to aerial stage   (iii) Other
considerations.*

(i) Meaning of image rejection

With the superheterodyne receiver it is necessary to consider not only adjacent
channel selectivity due to the aerial and r-f stages but also whether the selectivity
of these stages is sufficient to prevent signals on image frequencies from passing into
the i-f amplifier. If there were no tuned circuits ahead of the converter valve a signal
lower in frequency than the oscillator frequency, and which differs from the oscillator
frequency by the value of the intermediate frequency, would pass into the i-f amplifier.
Similarly a signal higher in frequency than the oscillator frequency, and which differs
from the oscillator frequency by the value of the intermediate frequency, would also
pass into the i-f amplifier.

When one of these frequencies is the desired one (it may be either the lower or the higher), the other is referred to as its image frequency.   Clearly the desired and un-desired signal frequencies are separated by twice the intermediate frequency.   Putting this statement in symbolic form :

$$f_{image} = f_{osc} + f_{i-f} \text{ when } f_{osc} > f_{sig}$$
$$f_{image} = f_{osc} - f_{i-f} \text{ when } f_{osc} < f_{sig}$$

where f_{sig} = desired signal frequency.

As an example : On the 540-1600 Kc/s A-M broadcast band the oscillator fre-quency is normally above the signal frequency.   If the reasons for this are not obvious, consideration of the oscillator tuning range and the values of variable capacitance required will make it so.   The usual i-f is 455 Kc/s and if a signal frequency of 1000 Kc/s is taken, then f_{image} = (1000 + 455) + 455 = 1910 Kc/s.

Since the converter cannot discriminate between the two signals of 1000 Kc/s and 1910 Kc/s, it is necessary for the aerial circuit, and r-f circuit if this is used, to make the magnitude of the 1000 Kc/s voltage much greater than the magnitude of the 1910 Kc/s voltage.   Otherwise severe interference can result when signals are being trans-mitted at both these frequencies.

When making calculations for image rejection the worst conditions are usually found at the high frequency end of the tuning range, provided that the selectivity of the tuned circuits does not vary appreciably with change in frequency.   This is because the separation between the two frequencies is a smaller proportion of the fre-quency to which the receiver is tuned.   It should also be clear that higher values of i-f will materially assist in reducing image effects because of the wider frequency separation between the desired and undesired signals ; it is for this reason that some v-h-f receivers use intermediate frequencies of the order of 10.7 Mc/s.

(ii) Image rejection due to aerial stage

The image rejection due to an aerial stage alone is calculated from

$$\sqrt{1 + Q^2\left(\frac{f}{f_0} - \frac{f_0}{f}\right)^2}$$

where $Q = \omega L_{eff}/R_{eff}$ (i.e. loaded value for secondary circuit),

$L_{eff} \approx L_2$, the secondary circuit inductance,

R_{eff} = total secondary circuit series resistance including that due to re-flection from primary circuit and valve input resistance (Valve input admittance is discussed in detail in Chapter 2 Sect. 8, and further comments are given in Sect. 5 of this chapter),

f = image frequency when oscillator frequency is higher than signal frequency,

and f_0 = signal frequency under oscillator conditions as for f.

If the oscillator is lower in frequency than the signal frequency then f and f_0 should be interchanged.

Exactly the same expression and procedure are used to calculate the image protection afforded by a r-f stage.   Since the calculated image rejection is a voltage ratio, the ratios obtained for the various stages are multiplied together, or the ratios expressed in decibels can be added to give the total image protection provided by the input cir-cuits to the converter.

(iii) Other considerations

The image protection due to the r-f stage may be greater than that due to the aerial stage.   A typical example occurs when the loading on the aerial stage due to the r-f amplifier valve has a positive value of input resistance and the loading on the r-f stage due to a pentagrid converter, using inner grid oscillator voltage injection, has a nega-tive value of input resistance.   The resultant Q, simply from these considerations, may then be higher in the case of the r-f coil.   Also, since the resultant Q of the aerial coil is halved, under some conditions of operation, by using two point matching to an aerial transmission line, the greater part of the image protection may here again be due to the r-f stage. These considerations are particularly important at the higher frequencies. On the long and medium wave bands sufficient image protection can often be obtained from the discrimination afforded by the aerial coil alone, but on the shortwave bands deterioration in image rejection is serious (and rapid) unless the intermediate frequency is increased above the usual value of about 455 Kc/s.

It is important to note that a good image rejection ratio is desirable since this also indicates the selectivity of the signal circuits and the degree of rejection against spurious frequency combinations. A high degree of selectivity preceding the first r-f amplifier valve has obvious advantages in reducing cross-modulation effects.

SECTION 5 : EFFECTS OF VALVE INPUT ADMITTANCE

(i) *Important general considerations* (ii) *Input loading of receiving valves at radio frequencies* (A) *Input conductance* (B) *Cold input conductance* (C) *Hot input conductance* (D) *Change in input capacitance* (E) *Reduction of detuning effect.*

(i) Important general considerations

A fairly complete discussion of some of the factors relating to valve input admittance has been given in Chapter 2 Sect. 8. These considerations are additional to the following details which take into account electron transit time effects at frequencies up to about 100-150 Mc/s, where conventional receiver design techniques begin to fail.

From the receiver designer's point of view the main factors (if we neglect changes in valve output admittance—but see Ref. A8) are

(1) A knowledge of the actual values of input resistance and capacitance (neither of which is constant with changes in valve operating conditions) with which he will have to contend when these are shunted across the tuned circuit (this must include the effects of feedback etc.), and

(2) A knowledge of the effects on input capacitance, in particular, when the signal voltage changes, giving rise to an alteration in a.v.c. bias. From this information the amount of detuning of the circuits can be determined. As the input components are in parallel with the tuned circuits, it is often more convenient to discuss the values of conductance, susceptance, and admittance rather than resistance, reactance, and impedance. Several extremely useful lists of values of conductance and capacitance for some common types of receiving valves are given in Chapter 2 Sect. 8, and further details of some additional types are set out later in this section. In cases where the value of short circuit input conductance at a required frequency is not given, then a close approximation can be found by multiplying the conductance value by $\left(\dfrac{\text{Frequency required}}{\text{Frequency stated}}\right)^2$; but extrapolation should not be carried too far e.g. a conductance value at 100 Mc/s should not be used to find a conductance at 455 Kc/s, but will give a reasonable approximation at, say, 10 Mc/s.

A **figure of merit,** which forms a useful basis for comparing various types of r-f voltage amplifier valves, is given by

$$\frac{g_m}{\sqrt{g_i}} \text{ (or } g_m\sqrt{R_i}\text{)}$$

where g_m = control grid to plate transconductance (mutual conductance)

g_i = short circuit grid input conductance

and R_i = short circuit grid input resistance.

At the higher frequencies cathode lead inductance is important.

If the valve types on which information regarding short circuit input admittance is required are not listed, experimental techniques can always be resorted to and reasonably accurate results are possible using a high frequency Q meter to determine the capacitive and resistive loading effects. The usual precautions as to length of leads, earthing etc. must be carefully observed when making these measurements, if the results are to be of any practical value. Something of this nature is usually required when considering the effects of the converter valve on its input circuit, since the published information is rather meagre. In this regard it is well to remember that some types of converter valves, such as those using inner grid injection, e.g. types 6A8, 6K8, 6SA7, 6BE6 etc., give negative loading whilst other types such as X61M, X79, 6J8-G, 6L7, etc. (which use outer grid oscillator injection) give positive loading. Increasing the negative bias on the signal grid will reduce the loading effect with all types, but at very high values of bias the negative loading may reverse its polarity and become positive. [The reasons for these effects are discussed later in this section in connection with input loading of receiving valves.] To reduce input conductance some valve types have more than one cathode connection (see Ref. A8), and typical examples are types 6AK5 and 6AG5, each having two cathode terminals. Even when the cathode terminals are directly connected to ground the length of the cathode lead is still sufficient to provide appreciable inductive reactance at frequencies of the order of 100 Mc/s. For this reason all plate and screen by-pass capacitors should be returned to one cathode lead and the grid returns to the other. The alternating voltages developed across the cathode lead inductance due to currents from the plate and screen circuits are in this way prevented from being directly impressed in series with the grid circuit, since it is the grid to cathode voltages which are of major importance. Direct current divides between the two available cathode paths, but this is not important since the path taken by this current normally does not affect the grid input admittance. To obtain grid bias the " grid " cathode lead employs the usual resistor and capacitor combination. The r-f by-pass capacitors from the plate and screen circuits connect directly to the other cathode terminal which is not directly connected to ground in this arrangement. A typical circuit is shown in Fig. 23.8.

In pentodes working at v-h-f it is advisable to connect the suppressor grid terminal directly to ground, when a separate terminal is available for this electrode, rather than to the cathode, because the suppressor lead inductance would then be connected in series with any external cathode lead inductance. If this precaution is not taken the coupling between control grid and plate is increased, because of the capacitance from control grid to suppressor and from suppressor to plate, the junction of these two

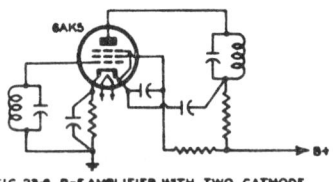

FIG. 23-8 R-F AMPLIFIER WITH TWO CATHODE CONNECTIONS.

capacitances having an impedance to ground depending on the total lead inductance.

Similar considerations also apply to screen grid circuits, particularly when the valve has a single cathode lead, and the shielding action of this grid can be seriously affected if proper precautions are not taken, such as directly earthing by-pass capacitors. These circuit arrangements are not very important at low frequencies where degenerative effects may be more serious, necessitating the connection of the suppressor grid and by-pass capacitors directly to the cathode.

(ii) Input loading of receiving valves at radio frequencies*

(See also Chapter 2, Sect. 8 and R.C.A. Application Note No. 118—Ref. B25).

The input resistance of r-f amplifier valves may become low enough at high radio frequencies to have appreciable effect on the gain and selectivity of a preceding stage. Also, the input capacitance of a valve may change enough with change in a.v.c. bias to cause appreciable detuning of the grid circuit. It is the purpose of this Note to discuss these two effects and to show how the change in input capacitance can be reduced.

---

*From R.C.A. Application Note No. 101.

(A) Input conductance

It is convenient to discuss the input loading of a valve in terms of the valve's input conductance rather than input resistance. The input conductance g_i of commercial receiving valves can be represented approximately by the equation

$$g_i = k_c f + k_\lambda f^2 \qquad (1)$$

where f is the frequency of the input voltage. A table of values of k_c and k_λ for several r-f valve types is shown below. The approximate value of a valve's input conductance in micromhos at all frequencies up to those in the order of 100 megacycles can be obtained by substituting in Eq. (1) values of k_c and k_λ from the table. In some cases, input conductance can be computed for conditions other than those specified in the table. For example, when all the electrode voltages are changed by a factor n, k_λ changes by a factor which is approximately $n^{-1/2}$. The value of k_c is

TABLE OF APPROXIMATE VALUES OF k_c AND k_λ FOR SEVERAL VALVE TYPES

Valve Type	Description	Plate volts	Screen volts	Signal-Grid bias volts	Supress-or volts	k_c micro—mhos/Mc/s	k_λ micro—mhos/Mc/s²
6A8	Pentagrid	250	100	−3	—	0.3	−0.05*
6J7	Pentode	250	100	−3	0	0.3	0.05
6K7	Pentode	250	100	−3	0	0.3	0.05
6K8	Triode-hexode	250	100	−3	—	0.3	−0.08†
6L7	Mixer	250	100	−3	—.	0.3	0.15‡
6SA7¹¶	Pentagrid	250	100	0	—	0.3	−0.03§
6SA7²¶	Pentagrid	250	100	−2	—	0.3	−0.03§
6SJ7¶	Pentode	250	100	−3	0	0.3	0.05
6SK7¶	Pentode	250	100	−3	0	0.3	0.05
954	Pentode	250	100	−3	0	0.3	0.005
1851 6AC7/1852¶	Pentode	250	150	−2	0	0.3	0.13
6AB7/1853¶	Pentode	250	200	−3	0	0.3	0.065

*For oscillator-grid current of 0.3 mA through 50 000 ohms.
†For oscillator-grid current of 0.15 mA through 50 000 ohms.
‡For wide range of oscillator currents.
§For grid No. 1 current of 0.5 mA through 20 000 ohms.
¹Self-excited. ²Separately excited.
¶Denotes single ended valve type.

practically constant for all operating conditions. Also, when the transconductance of a valve is changed by a change in signal-grid bias, k_λ varies directly with transconductance over a wide range. In the case of converter types, the value of k_λ depends on oscillator-grid bias and oscillator voltage amplitude. In converter and mixer types, k_λ is practically independent of oscillator frequency.

In eqn. (1), the term $k_c f$ is a conductance which exists when the cathode current is zero. The term $k_\lambda f^2$ is the additional conductance which exists when cathode current flows. These two terms can be explained by a simple analysis of the input circuit of a valve.

(B) Cold input conductance

The input impedance of a valve when there is no cathode current is referred to as the cold input impedance. The principal components of this cold impedance are a resistance due to dielectric hysteresis, and a reactance due to input capacitance and cathode-lead inductance. Because these components are in a parallel combination, it is convenient to use the terms admittance, the reciprocal of impedance, and susceptance, the reciprocal of reactance.   For most purposes, the effect of cathode-lead inductance is negligible when cathode current is very low.   The cold input admittance is, therefore, a conductance in parallel with a capacitive susceptance.   The conductance due to dielectric hysteresis increases linearly with frequency.   Hence, the cold input conductance can be written as $k_c f$, where k_c is proportional to the power factor of grid insulation and is the k_c of eqn. (1).

(C) Hot input conductance

The term $k_h f^2$, the input conductance due to the flow of electron current in a valve, has two principal components, one due to electron transit time and the other due to inductance in the cathode lead.   These two components can be analysed with the aid of Fig. 23.9.   In this circuit, C_h is the capacitance between grid and cathode when cathode current flows, C_g is the input capacitance due to capacitance between grid and all other electrodes, except cathode, g_t is the conductance due to electron transit time, and L is the cathode-lead inductance.   Inductance L represents the inductance of the lead between the cathode and its base pin, together with the effect of mutual inductances between the cathode lead and other leads near it.   Analysis of the circuit of Fig. 23.9 shows that, with L small as it generally is, the input conductance, g_h, due to the presence of cathode current in the valve, is approximately

$$g_h = g_m \omega^2 L C_h + g_t \qquad (2)$$

where $\omega = 2\pi f$.   The term $g_m \omega^2 L C_h$ is the conductance due to cathode-lead inductance.   It can be seen that this term varies with the square of the frequency. In this term, g_m is the grid-cathode transconductance because the term is concerned with the effect of cathode current flowing through L.   In a pentode, and in the 6L7, this transconductance is approximately equal to the signal-grid-to-plate transconductance multiplied by the ratio of direct cathode current to direct plate current. In the converter types 6A8, 6K8 and 6SA7, the signal-grid-to-cathode transconductance is small.   Cathode circuit impedance, therefore, has little effect on input conductance in these types.

For an explanation of the conductance, g_t, due to electron transit time, it is helpful to consider the concept of current flow to an electrode in a valve.   It is customary to consider that the electron current flows to an electrode only when electrons strike the surface of the electrode.   This concept, while valid for static conditions, fails to account for observed high-frequency phenomena.   A better concept is that, in a diode for example, plate current starts to flow as soon as electrons leave the cathode. Every electron in the space between cathode and plate of a diode induces a charge on the plate ; the magnitude of the charge induced by each electron depends on the proximity of the electron to the plate.   Because the proximity changes with electron motion, there is a current flow to the plate through the external circuit due to the motion of electrons in the space between cathode and plate.

Consider the action of a conventional space-charge-limited triode as shown in Fig. 23.10.   In this triode, the plate is positive with respect to cathode and the grid is negatively biased.   Due to the motion of electrons between cathode and grid, there is a current I_a flowing into the grid.   In addition, there is another current I_b flowing out of the grid due to the motion of electrons between grid and plate receding from the grid.   When no alternating voltage is applied to the grid, I_a and I_b are equal and the net grid current I_g is zero.

Suppose, now, that a small alternating voltage (e_g) is applied to the grid.   Because the cathode has a plentiful supply of electrons, the charge represented by the number of electrons released by the cathode (Q_k) is in phase with the grid voltage, as shown in Figs. 23.11(a) and 23.11(b).   The charge induced on the grid (Q_g) by these electrons would also be in phase with the grid voltage if the charges released by the cathode were to reach the plane of the grid in zero time, as shown in Fig. 23.11(c).   In this hypothetical case, the grid current due to this induced charge (Fig. 23.11(d)) leads the grid voltage by 90 degrees, because by definition, current is the time rate at which charge passes a given point.   However, the charge released by the cathode actually propagates towards the plate with finite velocity ; therefore, maximum charge is induced on the grid at a time later than that corresponding to maximum grid voltage, as shown in Fig. 23.11(e).   This condition corresponds to a shift in phase by an angle θ of Q_g with respect to e_g ; hence, the grid current lags behind the capacitive current of Fig. 23.11(d) by an angle θ, as shown in Fig. 23.11(f).   Clearly, the angle θ increases with frequency and with the time of transit τ.   Expressed in radians, $\theta = \omega\tau$.

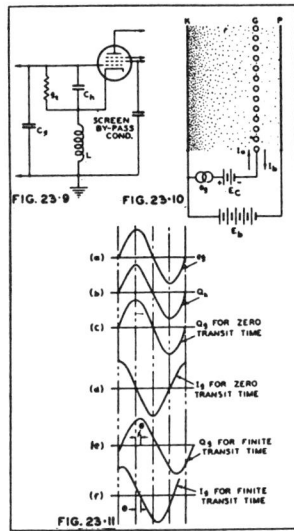

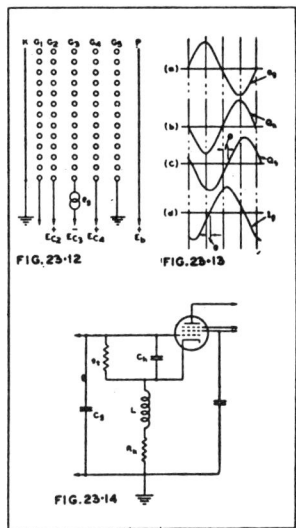

FIGURES 23·9 TO 23·14 INCLUSIVE

The amplitude of Q_g is proportional to the amplitude of the grid voltage ; the grid current, which is the time rate of change Q_g, is thus proportional to the time rate of change of grid voltage.   For a sinusoidal grid voltage, $e_g = E_g \sin \omega t$, the time rate of change of grid voltage is $\omega E_g \cos \omega t$.   Therefore, for a given valve type and operating point, the amplitude of grid current is
$$I_g = KE_g \omega$$
and the absolute value of grid-cathode admittance due to induced charge on the grid is
$$Y_t = I_g/E_g = K\omega \qquad\qquad (3)$$
The conductive component (g_t) of this admittance is
$$g_t = Y_t \sin \theta = Y_t \theta = K\omega\theta \text{ (for small values of } \theta).$$
Because $\theta = \omega\tau$, this conductance becomes, for a given operating point,
$$g_t = K\omega^2\tau. \qquad\qquad (4)$$
Thus, the conductance due to electron transit time also varies with the square of the frequency.   This conductance and the input conductance, $g_m \omega^2 LC_h$, due to cathode-lead inductance, are the principal components of the term $k_h f^2$ of eqn. (1).

This explanation of input admittance due to induced grid charge is based on a space-charge-limited valve, and shows how a positive input admittance can result from the induced charge.   The input admittance due to induced grid charge is negative in a valve which operates as a temperature-limited valve, that is, as a valve where cathode emission does not increase when the potential of other electrodes in the valve is increased.   The emission of a valve operating with reduced filament voltage is temperature limited ; a valve with a screen interposed between cathode and grid acts as a temperature-limited valve when the screen potential is reasonably high.   The existence of a negative input admittance in such a valve can be explained with the aid of Fig. 23.12.

When the value of E_{c2} in Fig. 23.12 is sufficiently high, the current drawn from the cathode divides between g_2 and plate ; any change in one branch of this current is accompanied by an opposite change in the other.   As a first approximation, therefore, it is assumed that the current entering the space between g_2 and g_3 is constant and equal to ρv, where ρ is the density of electrons and v is their velocity. g_2 may now be considered as the source of all electrons passing to subsequent electrodes.

Suppose now, that a small alternating voltage is connected in series with grid g_3, as shown in Fig. 23.12. During the part of the cycle when e_g is increasing, the electrons in the space between g_2 and g_3 are accelerated and their velocities are increased. Because the current ρv is a constant, the density of electrons (ρ) must decrease. In this case, therefore, the charge at g_2 is 180 degrees out of phase with the grid voltage, as shown at a and b of Fig. 23.13. This diminution in charge propagates toward the plate with finite velocity and induces a decreasing charge on the grid. Because of the finite velocity of propagation, the maximum decrease in grid charge occurs at a time later than that corresponding to the maximum positive value of e_g, as shown in Fig. 23.13(c). The current, which is the derivative of Q_g with respect to time, is shown in Fig. 23.13(d). If there were no phase displacement ($\theta = 0$), this current would correspond to a negative capacitance ; the existence of a transit angle θ, therefore, corresponds to a negative conductance. By reasoning similar to that used in the derivation of eqns. 3 and 4, it can be shown that the absolute value of negative admittance due to induced grid charge is proportional to ω, and that the negative conductance is proportional to ω^2. These relations are the same as those shown in eqns. 3 and 4 for the positive admittance and positive conductance of the space-charge-limited case.

A negative value of input conductance due to transit time signifies that the input circuit is receiving energy from the " B " supply. This negative value may increase the gain and selectivity of a preceding stage. If this negative value becomes too large, it can cause oscillation. A positive value of input conductance due to transit time signifies that the signal source is supplying energy to the grid. This energy is used in accelerating electrons toward the plate and manifests itself as additional heating of the plate. A positive input conductance can decrease the gain and selectivity of a preceding stage.

It should be noted that, in this discussion of admittance due to induced grid charge, no mention has been made of input admittance due to electrons between grid and plate. The effect of these electrons is similar to that of electrons between grid and cathode. The admittance due to electrons between grid and plate, therefore, can be considered as being included in eqn. (3).

(D) Change in input capacitance

The hot grid-cathode capacitance of a valve is the sum of two components, the cold grid-cathode capacitance, C_c, which exists when no cathode current flows, and a capacitance, C_t, due to the charge induced on the grid by electrons from the cathode. The capacitance C_t can be derived from eqn. (3), where it is shown that the grid-cathode admittance due to induced grid-charge is

$Y_t = K\omega$.

The susceptive part of this admittance is $Y_t \cos \theta$. Since this susceptance is equal to ωC_t, the capacitance C_t is

$C_t = K \cos \theta = K$ (for small values of θ).

Hence, the hot grid-cathode capacitance C_h is

$C_h = C_c + K$.

The total input capacitance of the circuit of Fig. 23.9 when the valve is in operation, includes the capacitance C_h and a term due to inductance in the cathode lead. This total input capacitance, C_i, can be shown to be approximately

$$C_i = C_g + C_h - g_m g_t L \qquad (5)$$

where the last term shows the effect of cathode-lead inductance. This last term is usually very small. It can be seen that if this last term were made equal in magnitude to $C_g + C_h$, the total input capacitance would be made zero. However, the practical application of this fact is limited because g_m and g_t change with change in electrode voltages, and g_t changes with change in frequency.

When cathode current is zero, the total input capacitance is practically equal to $C_g + C_c$. Subtracting this cold input capacitance from the hot input capacitance given by eqn. (5), we obtain the difference, which is $K - g_m g_t L$. In general, K is greater than $g_m g_t L$. Therefore, in a space-charge-limited valve, where K is positive, the hot input capacitance is greater than the cold input capacitance. In a temperature-limited valve, where K is negative, the hot input capacitance is less than the cold input capacitance. In both valves K changes with change in transconductance. Because of this change, the input capacitance changes somewhat with change in a.v.c. bias. In many receivers, this change in input capacitance is negligible because it is small compared with the tuning capacitances connected in the grid circuits of the high-frequency stages. However, in high-frequency stages where the tuning capacitance is small, and the resonance peak of the tuned circuit is sharp, change in a.v.c. bias can cause appreciable detuning effect.

(E) Reduction of detuning effect

The difference between the hot and the cold input admittances of a space-charge-limited valve can be reduced by means of an unbypassed cathode resistor, R_k in Fig.

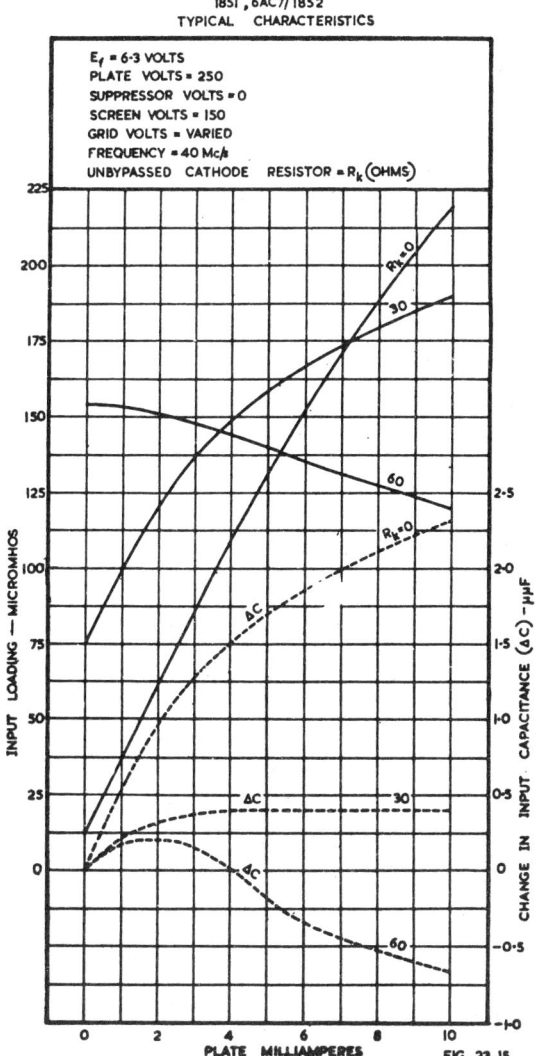

1851 , 6AC7/ 1852
TYPICAL CHARACTERISTICS

E_f = 6·3 VOLTS
PLATE VOLTS = 250
SUPPRESSOR VOLTS = 0
SCREEN VOLTS = 150
GRID VOLTS = VARIED
FREQUENCY = 40 Mc/s
UNBYPASSED CATHODE RESISTOR = R_k (OHMS)

INPUT LOADING — MICROMHOS

CHANGE IN INPUT CAPACITANCE (ΔC) — $\mu\mu$F

PLATE MILLIAMPERES

FIG. 23.15

23.14.   The total hot input admittance of this circuit is made up of a conductance and a capacitive susceptance C_i'.   Analysis of Fig. 23.14 shows that, if cathode-lead inductance is neglected, the total hot input capacitance, C_i', is approximately

$$C_i' = C_s + C_e \frac{1 + K/C_e}{1 + g_m R_k}. \tag{6}$$

Inspection of this equation shows that if K is positive and varies in proportion with g_m, the use of the proper value of R_k will make C_i independent of g_m.   In a space-charge-limited valve, K is positive and is found by experiment to be approximately proportional to g_m.   It follows that the proper value of R_k will minimize the detuning effect of a.v.c. in a space-charge-limited valve.   Eqn. (6) is useful for illustrating the effect of R_k but is not sufficiently precise for computation of the proper value of R_k to use in practice.   This value can be determined by experiment.   It will be found that this value, in addition to minimizing capacitance change, also reduces the change in input conductance caused by change in a.v.c. bias.   The effect of unbypassed cathode resistance on the change in input capacitance and input conductance of types 6AC7/1852 and 6AB7/1853 is shown in Figs. 23.15 and 23.16.   These curves were taken at a frequency of 40 megacycles.   The curves for the 6AC7/1852 also hold good for the 1851.

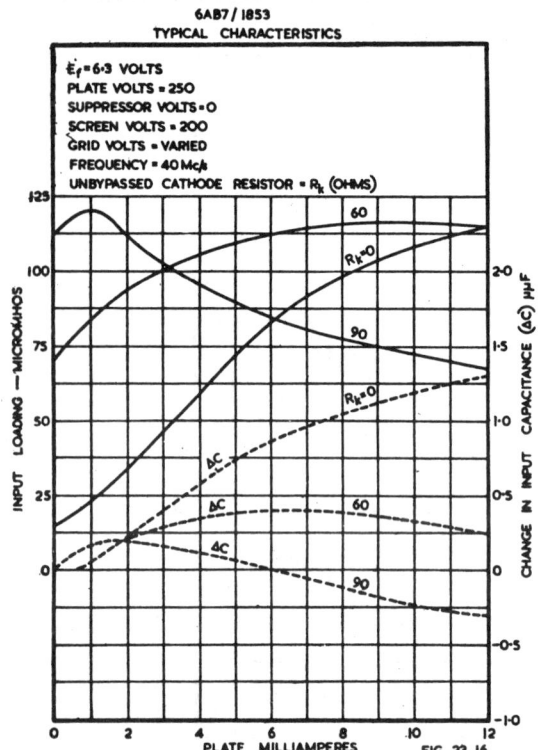

6AB7 / 1853
TYPICAL CHARACTERISTICS

E_f = 6·3 VOLTS
PLATE VOLTS = 250
SUPPRESSOR VOLTS = 0
SCREEN VOLTS = 200
GRID VOLTS = VARIED
FREQUENCY = 40 Mc/s
UNBYPASSED CATHODE RESISTOR = R_k (OHMS)

INPUT LOADING — MICROMHOS
CHANGE IN INPUT CAPACITANCE (ΔC) $\mu\mu F$
PLATE MILLIAMPERES   FIG. 23.16

It should be noted that, because of degeneration in an unbypassed cathode resistor, the use of the resistor reduces gain.   The reduced gain is $1/(1 + g_m R_k)$ times the gain with the same electrode voltages but with no unbypassed cathode resistance.

The hot input conductance of a valve with an unbypassed cathode resistor can be determined by modification of the values of k_h in the table.   The value of k_h in the table should be multiplied by $g_m/(1 + g_m R_k)$.   The resultant value of k_h, when substituted in eqn. (1), with k_c from the table, gives the input conductance of a valve with an unbypassed cathode resistor.   In the factor $(1 + g_m R_k)$, g_m is the grid-cathode transconductance when R_k is by-passed.

When an unbypassed cathode resistor is used, circuit parts should be so arranged that grid-cathode and plate-cathode capacitances are as small as possible.   These capacitances form a feedback path between plate and grid when there is appreciable impedance between cathode and ground.   To minimize plate-cathode capacitance, the suppressor and the screen by-pass condenser should be connected to ground rather than to cathode.

SECTION 6 : VALVE AND CIRCUIT NOISE*

(i) Thermal agitation noise (ii) Shot noise (iii) Induced grid noise (iv) Total noise calculations (v) Sample circuit calculations (vi) Conclusions.

Maximum receiver sensitivity is not, in most cases, determined by the gain of the particular receiver but by the magnitude of the input circuit noise, which is generated by the antenna, the tuned input circuit, and the first tube.   This is true of A-M, F-M and television except that in F-M and television the random noise effect assumes a far greater degree of importance than in the standard broadcast band.   The reason for this is twofold :

(1) At the frequencies where these two services operate, 50 to 250 Mc/s, the relative values of the several different noise sources assume entirely new proportions and the heretofore unimportant and little known **induced grid noise** becomes one of the predominant components of the total.

(2) Most random input and tube noise is proportional to the square root of the bandwidth.   Both television, with a 4 Mc/s band, and F-M, with a 200 Kc/s band, occupy much wider sections of the frequency spectrum than anything previously encountered by the commercial receiver engineer.

(i) Thermal agitation noise

When an alternating electric current flows through a conductor, electrons do not actually move along the conductor but they are displaced, an infinitesimal amount, first in one direction and then in the other.   A voltage is built up across the conductor equal to the magnitude of the current times its resistance.   Applying heat to the conducting material agitates the molecules of the conductor and, consequently, varies the instantaneous position in space of the electrons.   This random electron motion is, in a sense, a minute noise current flowing through the material and is known as **thermal agitation noise**.   That is, the application of heat agitates the electron distribution of the substance thereby creating the noise.

The magnitude of the short-circuit noise current is given by

$$i_n{}^2 = \frac{4KT\varDelta F}{R} \tag{1}$$

where $i_n{}^2$ = mean squared noise current (amperes²)
  K = Boltzmann's Constant (Joules per degree Kelvin), 1.38×10^{-23}
  T = temperature (degrees Kelvin)
  $\varDelta F$ = bandwidth (c/s)
and R = resistance (ohms).

All noise currents and voltages are random fluctuations and occupy an infinite frequency band.   Because of the random effect, the most convenient terminology to use in expressing their magnitude is average noise-power output.   Mean-squared noise current or mean-squared noise voltage, either of which is proportional to average power, is generally used.

In the expression for various noise components the term $\varDelta F$ refers to the effective bandwidth of the circuit.   This is determined from a curve of power output versus frequency by dividing the area under the curve by the amplitude of the power at the noise frequency in question.   For most calculations, however, where only approximate values are desired, the bandwidth between half power points, or 0.707 voltage points, will give sufficient accuracy.

The equation below expresses thermal agitation noise as a voltage in series with a given resistor ;

$$e_n{}^2 = 4KT\varDelta FR. \tag{2}$$

Both the above forms are true of all resistive circuit elements or combination of elements including parallel and series-tuned circuits.

---

*This section is taken directly from an article " Input Circuit Noise Calculations for FM and Television Receivers " by W. J. Stolze, published in Communications, Feb. 1947, and reprinted by special permission.

Referring to Fig. 23.17(a), let us suppose a resistance of 10 000 ohms were connected to the input of an amplifier with a 5 Kc/s bandwidth, i.e., 5 Kc/s between half power points or an audio band of 2.5 Kc/s. At room temperature, 20°C or 293°K, the term $4KT$ in the expressions for noise simplifies to 1.6×10^{-20}, which may be used in most receiver calculations. The noise in Fig. 23.17(a) is therefore :

$$e_n{}^2 = 1.6 \times 10^{-20} \Delta FR$$
$$e_n = \sqrt{1.6 \times 10^{-20} \times 5\,000 \times 10\,000}$$
$$e_n = 0.89 \text{ microvolt.}$$

The noise bandwidth is generally determined by the narrowest element in the entire circuit under consideration. In the example for Fig. 23.17(b) the bandwidth of the amplifier is narrower than the tuned circuit and therefore its ΔF is used in the calculations.

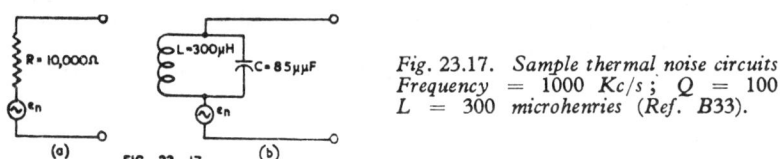

Fig. 23.17. Sample thermal noise circuits :
Frequency = 1000 Kc/s ; Q = 100 ;
L = 300 microhenries (Ref. B33).

FIG. 23.17

Fig. 23.17(b) is a simple parallel-tuned circuit where the noise generating resistance is equal to the tuned circuit impedance. Again let us assume the bandwidth to be five Kc per second.

$$R = Q(\omega L) = 100 \times 1900 = 190\,000 \text{ ohms}$$
$$e_n{}^2 = 1.6 \times 10^{-20} \Delta FR$$
$$e_n = \sqrt{1.6 \times 10^{-20} \times 5\,000 \times 190\,000}$$
$$e_n = 3.9 \text{ microvolts.}$$

Thermal agitation noise voltage may be calculated easily with eqn. (2) but by using the graph shown in Fig. 23.18 the room temperature values may be found directly.

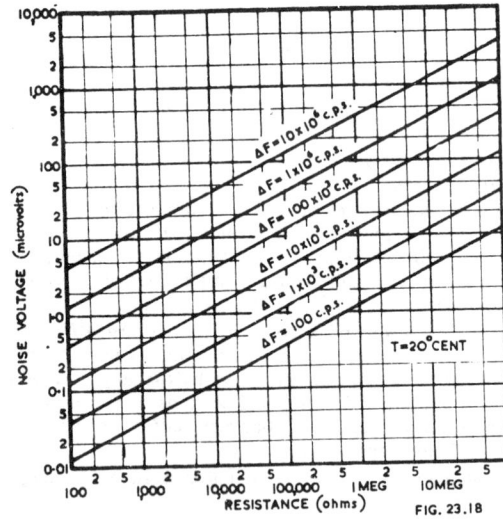

FIG. 23.18

Fig. 23.18. Thermal agitation noise voltage versus resistance and band-width (Ref B33).

(ii) Shot noise

Another important component of the total receiver noise is shot noise. This noise is generated inside the vacuum tube and is due to the random fluctuations in the plate current of the tube, or, to state it in another manner, random variations in the rate of arrival of electrons at the plate. When amplified, this noise sounds as if the plate were being bombarded with pebbles or as if a shower of shot were falling upon a metal surface, hence the name shot noise.

Although generated essentially in the plate circuit of the tube, which is not a convenient reference point for sensitivity or signal-to-noise ratio calculations, the shot noise is nearly always referred to as a noise voltage in series with the grid. Since the following equation is true,

$$e_s = i_p/g_m \qquad (3)$$

where e_s = a.c. grid voltage

i_p = a.c. plate current,

and g_m = transconductance,

by simply dividing the noise current in the plate circuit by the transconductance of the tube, the shot noise may be referred to the grid and expressed in terms of grid voltage.

Another step is taken, however, to simplify the noise nomenclature. Suppose a given tube has a shot noise equal to e_n microvolts in series with its grid. It is perfectly valid to imagine that this voltage could be replaced by a resistance whose thermal agitation noise is equal to e_n (the shot noise) and to consider the tube to be free of noise. This imaginary resistance, which when placed in the grid of the tube generates a voltage equal to the shot noise of the tube, is known as the shot noise equivalent resistance or just as the equivalent noise of the tube. The advantage of this terminology is that when the equivalent noise resistance of the particular tube is known, the noise voltage may be calculated directly for any given bandwidth by substituting values in the following formula :

$$e_n^2 = 4KT\Delta F R_{eq} \qquad (4)$$

where R_{eq} = equivalent noise resistance,

or at room temperature

$$e_n^2 = 1.6 \times 10^{-20}\Delta F R_{eq} \qquad (5)$$

If the noise were expressed as a voltage or current its value would be correct only for one particular bandwidth.

TRIODE AMPLIFIER	$R_{EQ} = \dfrac{2\cdot5}{G_M}$
PENTODE AMPLIFIER	$R_{EQ} = \dfrac{I_B}{I_B+I_{G_2}}\left(\dfrac{2\cdot5}{G_M} + \dfrac{20I_{G_2}}{G_M{}^2}\right)$
TRIODE MIXER	$R_{EQ} = \dfrac{4}{G_C} \qquad G_C = \dfrac{G_M}{4}$
PENTODE MIXER	$R_{EQ} = \dfrac{I_B}{I_B+I_{G_2}}\left(\dfrac{4}{G_C} + \dfrac{20I_{G_2}}{G_C{}^2}\right)$
MULTIGRID CONVERTER OR MIXER	$R_{EQ} = 20 \ \dfrac{I_B(I_K-I_B)}{I_K \ G_C{}^2}$

R_{EQ}= EQUIVALENT SHOT NOISE RESISTANCE	I_{G_2}=AVERAGE SCREEN CURRENT
G_M= GRID PLATE TRANSCONDUCTANCE	G_C=CONVERSION TRANSCONDUCTANCE
I_B= AVERAGE PLATE CURRENT	I_K=AVERAGE CATHODE CURRENT

FIG. 23.19

Fig. 23.19. *Approximate calculated equivalent noise resistance of various receiving-type tubes (Ref. B33).*

By knowing the R_{eq} of any given tubes their relative shot noise merit is also known regardless of what bandwidth they are to operate at, while if the noise voltages were given alone the operating bandwidth at which the calculation was made would also have to be noted if the relative merits of the two tubes were to be defined.

Noise-equivalent resistance values for a number of different tube types (triodes, pentodes, and converters) and for various circuit applications (amplifiers and mixers) can be calculated by applying the expressions presented in the chart, Fig. 23.19*.

When the term converter is used it refers to a tube that is used for frequency conversion where the single tube acts as the local oscillator and the mixer (6SA7) ; the term mixer where two tubes are used, one as the mixer (6SG7), and one as the local oscillator (6C4).

*W. A. Harris " Fluctuations in vacuum tube amplifiers and input systems," R.C.A. Review, April 1941.

TUBE TYPE	APPLICATION	PLATE VOLTS[a]	SCREEN VOLTS	TRANSCONDUCTANCE MICROMHOS	EQUIVALENT NOISE RESISTANCE OHMS
6AC7	PENTODE AMPLIFIER	300	150	9,000	720
6AC7	PENTODE MIXER	300	150	2,200	2,800
6AG5	PENTODE AMPLIFIER	250	150	5,000	1,650
6AG5	PENTODE MIXER	250	150	1,250	6,600
6AG7	PENTODE AMPLIFIER	300	150	11,000	1,540
6AK5	PENTODE AMPLIFIER	180	120	5,100	1,880
6AK5	PENTODE MIXER	180	120	1,280	7,520
6AK6	PENTODE AMPLIFIER	180	180	2,300	8,800
6AT6	TRIODE AMPLIFIER	250	—	1,200	2,100
6AU6	PENTODE AMPLIFIER	250	150	5,200	2,660
6BA6	PENTODE AMPLIFIER	250	100	4,400	3,520
6BA6	PENTODE MIXER	250	100	1,100	14,080
6BE6	CONVERTER	250	100	475[*]	190,000
6C4	TRIODE AMPLIFIER	100	—	3,100	810
6C4	TRIODE MIXER	100	—	770	3,240
6C5	TRIODE AMPLIFIER	250	—	2,000	1,250
6C5	TRIODE MIXER	250	—	500	5,000
6J5	TRIODE AMPLIFIER	250	—	2,600	960
6J5	TRIODE MIXER	250	—	650	3,840
½6J6	TRIODE AMPLIFIER	100	—	5,300	470
½6J6	TRIODE MIXER	100	—	1,320	1,880
6K8	CONVERTER	250	100	350[*]	290,000
6SA7	CONVERTER	250	100	450[*]	240,000
6SB7-Y	CONVERTER	250	100	950[*]	62,000
6SC7	TRIODE · AMPLIFIER	250	—	1,325	1,890
6SG7	PENTODE AMPLIFIER	250	125	4,700	3,100
6SG7	PENTODE MIXER	250	125	1,180	12,400
6SJ7	PENTODE AMPLIFIER	250	100	1,650	6,100
6SK7	PENTODE AMPLIFIER	250	100	2,000	11,000
6SL7	TRIODE AMPLIFIER	250	—	1,600	1,560
6SQ7	TRIODE AMPLIFIER	250	—	1,100	2,300

(a) VALUES OF PLATE VOLTAGE AND CURRENT AND SCREEN VOLTAGE AND CURRENT ARE FOR TYPICAL OPERATING CONDITIONS.

(*) CONVERSION TRANSCONDUCTANCE - MICROMHOS

FIG. 23.20

After the equivalent noise resistance is known the value of r.m.s. noise voltage at the grid of this tube can be calculated by applying the same expression that is used for thermal agitation noise,

$$e_n{}^2 = 1.6 \times 10^{-20} \Delta FR$$

or, by using the graph of Fig. 23.18.

Fig. 23.20 presents calculated equivalent noise resistance values for a number of commonly used tubes acting as various types of circuit elements. These are, of course, approximate figures.

It can be seen from Figs. 23.19 and 23.20 that the noise resistance or voltage is at a minimum for a triode, increasing for the pentode and the multigrid tube, following in that order.

Shot noise is unique among the noise sources in the sense that the shot-noise voltage should be considered to exist in series with the grid inside the tube. The reason for this is that nothing can be done to the external grid circuit that will alter the magnitude of this component. Even though the shot noise must be tolerated, its effect can be minimized by designing the input circuit for maximum- signal at the grid. This does not reduce the magnitude of the noise but does improve the signal-to-noise-ratio of the receiver.

(iii) Induced grid noise

Also present in the receiving tube is a third source of noise which is generated internally in the tube but whose magnitude and effect are determined partially by the external input circuit. Known as induced grid noise, this minute current is induced in the grid wires of the tube by random fluctuations in the plate current. It is known that a varying electron beam will induce a current in any nearby conductor. Therefore, the fluctuating plate current which is in a sense a varying electron beam, will induce a noise current in the nearby grid conductors.

The input impedance of a vacuum tube has a reactive and a resistive component. At relatively low frequencies the resistive component is very high (below about 30 Mc/s) ; as the frequency is increased the resistive component decreases and its magnitude eventually becomes comparable to or even lower than the external grid circuit impedance. The resistive component is composed of two parts, the portion due to transit time effect, and the portion due to the inductance of the cathode lead.

An expression for induced-grid-noise* for tubes

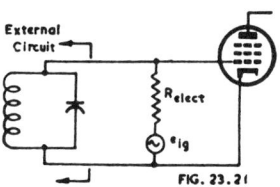

FIG. 23.21

Fig. 23.21. Position of induced grid noise in vacuum-tube circuit (Ref. B33).

with control grid adjacent to the cathode follows :

$$i^2_{i.g.} = 1.4 \times 4KT_k \Delta FG_{elect}$$

or when expressed in the form of a voltage generator,

$$e^2_{i.g.} = 1.4 \times 4KT_k \Delta FR_{elect} \quad (6)$$

where : T_k = cathode temperature (degrees Kelvin)

G_{elect} = electronic (transit time) component of input conductance

and R_{elect} = electronic component of input resistance.

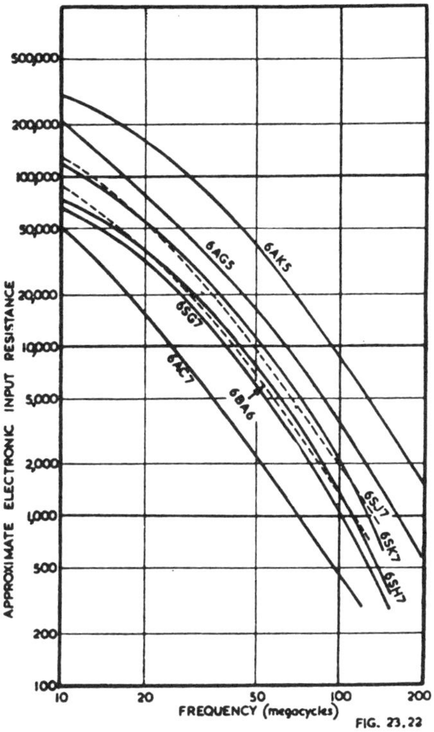

From eqn. (6) it can be seen that the induced grid noise is proportional to the electronic or transit time component of the input resistance. Measurement of the total input resistance is a comparatively simple matter with the use of a high frequency Q meter, but the separation of the electronic and the cathode inductance components, which are essentially two resistances in parallel between the grid and ground, is a very difficult matter. Since most high-frequency tubes are constructed with either two cathode leads or one very short lead, assuming the total measured input resistance to be electronic would not introduce too great an error. Another factor in favour of this approximation is that it would be the case for maximum induced grid noise and any error introduced would more than likely be on the safe side.

Fig. 23.22. Approximate electronic input resistance versus frequency (Ref. B33).

[The left side of the page contains a graph: APPROXIMATE ELECTRONIC INPUT RESISTANCE (vertical axis, values 100, 200, 500, 1,000, 2,000, 5,000, 10,000, 20,000, 50,000, 100,000, 200,000, 500,000) versus FREQUENCY (megacycles) (horizontal axis, 10, 20, 50, 100, 200). Curves labelled 6AG5, 6AK5, 6SG7, 6AC7, 6AB6, 6SJ7, 6SK7, 6SH7. FIG. 23.22]

*D. O. North " Fluctuations induced in vacuum-tube grids at high frequencies " Proc. I.R.E. Feb. 1941.

Cathode temperature in most receiving tubes, which almost exclusively use oxide-coated cathodes, is approximately 3.6 times the normal room temperature in degrees K. Eqn. (6) can be rewritten therefore as

$$e^2_{i.g.} = 5 \times 4KT\Delta FR_{elect} \qquad (7)$$

where T = room temperature (degrees Kelvin),
or, when $T = 300$ degrees Kelvin,

$$e^2_{i.g.} = 8 \times 10^{-20}\Delta FR_{elect} \qquad (8)$$

In circuit calculations this noise is essentially in series with a resistance equal to R_{elect} located between the grid and ground—Fig. 23.21.

The approximate input resistance for a number of common receiving tubes in the frequency range of F-M and television is given in Fig. 23.22. This chart can be used to find approximate input resistance values for induced grid-noise calculations.

(iv) Total noise calculations

Calculations of total input noise are made by using the grid of the input tube as a reference point. There are many sources of noise and each must be calculated and referred to the grid reference point before a summation is made. Since noise is a random effect and calculated on a power basis, the separate components cannot be added directly but as the square root of the sum of the squares.

$$\text{Total Noise} = \sqrt{e_1^2 + e_2^2 + e_3^2 + \text{etc.}} \qquad (9)$$

The various noise voltages that must be referred to the first grid are :
(1) Thermal agitation noise of the antenna radiation resistance.
(2) Thermal agitation noise of the tuned grid circuit.
(3) Shot noise of the input tube.
(4) Induced grid noise of the input tube.
(5) Grid circuit noise of the following stages referred back to the first grid.

In Fig. 23.23(a) appears a diagram of a practical input circuit and the location of all the circuit parameters and noise voltages. Fig. 23.23(b) is essentially the same except that the antenna circuit is reflected through the transformer and considered to exist at the grid. This is the diagram that is most useful in calculating the total input circuit noise.

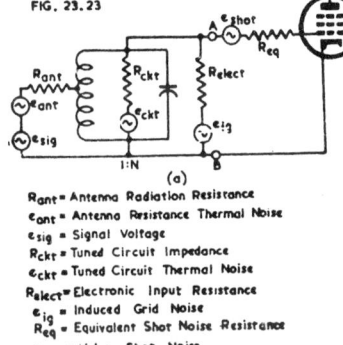

FIG. 23.23

(a)

Rant = Antenna Radiation Resistance
eont = Antenna Resistance Thermal Noise
esig = Signal Voltage
Rckt = Tuned Circuit Impedance
eckt = Tuned Circuit Thermal Noise
Relect = Electronic Input Resistance
eig = Induced Grid Noise
Req = Equivalent Shot Noise Resistance
eshot = Valve Shot Noise
N = Turns Ratio of Coupling Transformer

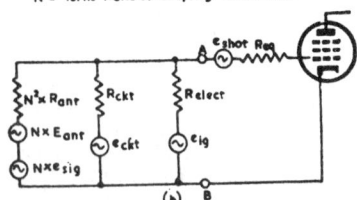

(b)

Fig. 23.23. Position of various noise sources in input circuit (Ref. B33).

The steps necessary to find specific values for each of these factors are shown in Fig. 23.24. Antenna radiation resistance varies widely with the type of antenna chosen, but for F-M and television work it is generally in the order of 75 to 300 ohms. When the noise is known in terms of an equivalent resistance, as is the case here for antenna, tuned circuit, and shot noise, the equivalent voltage can be either calculated or obtained directly from Fig. 23.18.

In order to add the antenna, tuned circuit, and induced grid noise to the shot noise the effective voltage of these three components at the grid, or between the points A and B, must be known. Each must go through what is essentially a resistive divider and may be calculated as shown in Fig. 23.25.

After knowing the magnitude of the separate sources that exist between A-B, the total noise voltage is

(1) R_{ANT} - DEPENDS UPON SPECIFIC ANTENNA

(2) $E_{ANT} = \sqrt{1{\cdot}6 \times 10^{-20} R_{ANT} \Delta F}$ - OR DIRECTLY FROM FIG. 23·18

(3) $R_{CKT} = Q\omega L = \frac{Q}{\omega C}$

(4) $E_{CKT} = \sqrt{1{\cdot}6 \times 10^{-20} R_{CKT} \Delta F}$ - OR DIRECTLY FROM FIG. 23·18

(5) R_{ELECT} - FROM ACCOMPANYING CHART, FIG. 23·22

(6) $E_{IG} = \sqrt{8 \times 10^{-20} R_{ELECT} \Delta F}$

(7) R_{EQ} - FROM ACCOMPANYING CHART, FIG. 23·20

(8) $E_{SHOT} = \sqrt{1{\cdot}6 \times 10^{-20} \Delta F\, R_{EQ}}$ - OR DIRECTLY FROM FIG. 23·18

FIG. 23.24

Fig. 23.24. Procedure for calculating various noise voltages (Ref. B33).

$$e_{total} = \sqrt{(e_{shot})^2 + (e_{ant} \text{ at A-B})^2 + (e_{i-g} \text{ at A-B})^2 + (e_{ckt} \text{ at A-B})^2} \qquad (10)$$

One other factor may affect this total, however. If the total noise of the following stages, which is calculated similarly, ignoring the antenna of course, is appreciable, it must be added to the constants of Fig. 23.25. In reflecting it to the first grid the second stage noise should be divided by the gain of the first tube. When the gain is about ten or more this factor may usually be neglected.

Effective signal voltage across A-B is calculated in the same way as the antenna noise in Fig. 23.25. The signal-to-noise ratio is now also known.

Since the signal-to-noise ratio is determined by the signal strength and the total noise at the grid of the input tube, for a receiver that has a mixer, such as 6SK7, for the input tube, the signal-to-noise ratio may be considerably improved by the addition of an r-f tube, such as a 6SG7, which has considerably less total noise. By adding additional r-f tubes (6SG7's), however, since the total noise and signal at the grid will be the same, the signal-to-noise ratio will not be improved.

FIG. 23. 25

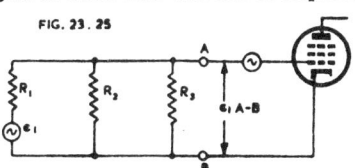

Fig. 23.25. Circuit for reflecting various voltages to the grid (Ref. B33). To find the effective voltage of the antenna, the tuned circuit, and the induced grid noise at the grid of the tube let R_1 equal one of the above noise resistances and e_1 its generated voltage. If R_2 and R_3 equal the other two noise resistances the effective voltage at the grid is

$$e_{1A-B} = \frac{e_1}{R_1 + \dfrac{R_2R_3}{R_3 + R_3}} \times \frac{R_2R_3}{R_2 + R_3}$$

This calculation must be performed for the three components in turn.

(v) Sample circuit calculations

For a sample problem let us calculate the total noise at the grid of an F-M receiver r-f amplifier stage, assuming the circuit in Fig. 23.26(a) to be under consideration.

As a simplification of procedure the steps in the calculations will be numbered.

(1) N^2R_{ant} = 1200 ohms (calculated)

(2) R_{elect} = 1200 ohms (Figure 23.22)

(3) R_{ckt} = $Q\omega L$ = 8000 ohms (calculated)

(4) R_{eq} = 3100 ohms (Figure 23.20)

At this point it will be convenient to redraw the circuit as shown in Fig. 23.26(b).

(5) Ne_{ant} = 2 microvolts (Fig. 23.18)

(6) $e_{i.g.} = \sqrt{8 \times 10^{-20} \times 200 \times 10^3 \times 1200}$

= 4.4 microvolts (equation (8))

(7) e_{ckt} = 6 micróvolts (Fig. 23.18).
(8) e_{shot} = 3.5 microvolts (Fig. 23.18).
The next step is to find the effective voltage of each source between the grid and ground (or A-B) as shown in Fig. 23.25.

(9) e_{ant} A-B = $\dfrac{2}{1200 + 1040}$ × 1040 =· 0.93 microvolt.

(10) $e_{i.g.}$ A-B = $\dfrac{4.4}{1200 + 1040}$ × 1040 = 2.0 microvolts.

(11) e_{ckt} A-B =· $\dfrac{6}{8000 + 600}$ × 600 = 0.42 microvolts.

and the total noise is therefore

(12) e_{total} = $\sqrt{3.5^2 + 0.93^2 + 2.0^2 + 0.42^2}$
       = 4.3 microvolts [equation (10)].

(vi) Conclusions

Selection of an input tube for a television or F-M receiver is dependent upon many varying circuit conditions and individual requirements. The choice of using balanced or unbalanced input, permeability or capacitor tuning, noisy pentodes or quiet triodes that possibly require neutralization, among others, lies entirely with the design engineer. Considering these reasons and various engineering and economic compromises no particular tube can be chosen and defined as **the input tube.** Complete noise information about the circuits involved is necessary, however, as this is one of the determining factors for good sensitivity and signal-to-noise ratio

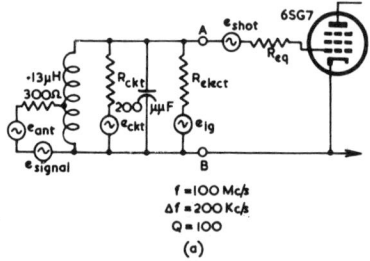

(a)

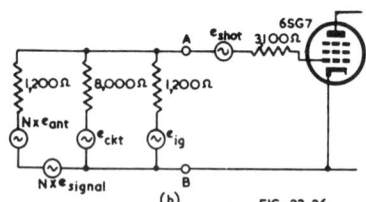

(b)   FIG. 23.26

Fig. 23.26.   *Typical F-M receiver input circuit (Ref. B33).*

SECTION 7 : INSTABILITY IN R-F AMPLIFIERS

(i) *Causes of instability*   (ii) *Inter-electrode capacitance coupling*   (iii) *Summary.*

(i) Causes of instability

Instability in r-f amplifiers can be due to many causes and some of these are listed below.

(1) Inter-electrode coupling due to capacitances within the r-f amplifier valve. This coupling may be augmented by additional capacitance external to the valve, due to wiring etc.

(2) Coupling between r-f and aerial coils, and leads, due to lack of adequate shielding or care in placement of the coils and leads relative to one another.

(3) Impedances common to several stages such as the metal shaft of a variable capacitor, power supply impedance (including heater leads), capacitances between switch contacts, a.v.c. line etc.

(4) Feedback at the intermediate frequency or on harmonics of the i-f.
Feedback at i-f can be serious when there is only an aerial stage preceding the converter. In some cases a simple expedient to overcome this difficulty is to connect a series resonant i-f trap between the aerial and earth terminals.

(5) Overall feedback from the a-f section into the aerial stage. This can be checked by bringing the speaker leads into close proximity with the aerial terminal, and the feedback usually manifests itself as a characteristic a-f howl. Practically all of the above factors are under the receiver designer's control even though the elimination of undesired oscillations is often a very difficult practical problem. Further discussion is given in Chapter 35, Sect. 3(v).

(ii) Inter-electrode capacitance coupling

Coupling due to inter-electrode capacitances calls for special consideration since these are an irreducible minimum when due care has been taken with the external circuits. A description of the effects which may be expected has been given in some detail in Chapter 2 Sect. 8 (see also Chapter 26, Sects. 7 and 8), where the nature of the impedance reflected into the grid circuit has been discussed. The connection with circuit instability is largely bound up in the magnitude of the input resistance component of the reflected impedance appearing across the valve grid circuit, and as to whether this input resistance is positive or negative. If the resistance is positive the dynamic impedance of the grid input circuit is lowered and there is a loss in gain and a broadening of the tuning characteristic. For a negative input resistance component the opposite effects hold with the dynamic impedance of the tuned circuit increased and the tuning becoming sharper. When the negative resistance, due to coupling by inter-electrode capacitances from other circuits associated with the valve, equals or is less than the positive resistance of the grid input circuit, oscillation will occur.

In the usual case for a r-f amplifier the primary winding of the r-f coil is resonated outside the tuning range. For a high impedance primary resonated below the lowest tuning frequency the plate circuit of the valve acts as a capacitance and the grid circuit is affected as though a positive resistance and a capacitance were connected in parallel across it. This effect is due to the total grid-to-plate capacitance. When the primary of the r-f coil is resonated above the highest tuning frequency, the valve sees an inductive load and a capacitance and a negative resistance component appear in parallel across the grid input circuit. (This circuit is the aerial coil when only one r-f stage is used).

For the case of the tuned anode load circuit i.e. a single tuned circuit connected directly between plate and B+, the valve sees a resistive load and there is only a capacitance effectively reflected in parallel with the valve grid input circuit. In this latter case, however, because of circuit mistracking, the valve may only see a resistive load at a few points in the tuning range and at other settings of the tuning dial the load may appear as either a capacitance or an inductance depending on whether the resonant frequency is below or above the required signal frequency.

The above effects can be neutralized by suitable circuit arrangements but usually the additional trouble and expense involved are avoided whenever possible.

To obtain a quantitive idea of the permissible values of grid-to-plate capacitance which would just put a circuit on the verge of instability, an investigation was made by Thompson (Ref. B41) who gave the following results.

$$C_{gp} = \frac{A}{\omega g_m R_D^2}$$

where C_{gp} = total capacitance grid to plate (valve internal and external)

$$A = \begin{cases} 2 & \text{for 1 r-f stage} \\ 1 & \text{for 2 r-f stages} \\ 0.764 & \text{for 3 r-f stages} \end{cases}$$

$\omega = 2\pi \times$ operating frequency

g_m = mutual conductance of r-f amplifier valve (assumed same type and operating conditions in each case),

and R_D = dynamic impedance of the input and output circuits (assumed identical in all cases for simplicity).

Obviously, even for values of C_{gp} less than those obtained from the above expression, there would be a marked effect on the gain and selectivity of the circuits before oscillation actually started.

(iii) Summary

Possible sources of feedback giving rise to instability can often be predicted from the circuit and component layout diagrams of a receiver. An estimate can be made of the possible magnitudes of many of the undesired voltages involved.

It should be evident that it is far better to avoid possible feedback and instability by good electrical and mechanical design rather than spend many fruitless hours tracking down an oscillation which could have been avoided.

A most helpful discussion of instability problems is given in E. E. Zepler's "Technique of Radio Design" (in particular Chapter 9) and the reader is recommended to consult this book as an excellent practical guide.

Some further considerations will be given to circuit instability in Chapter 26, in connection with i-f amplifiers. These circuits, being fixed tuned, are usually more amenable to calculation of possible instability than r-f stages and the results obtained more closely approximate to the practical set-up.

It is necessary to mention that at very high frequencies (say above 50 Mc/s or so) the inductance of the screen-grid lead in screen-grid tetrodes and pentodes can cause an apparent change in plate to control-grid capacitance which is often sufficient to cause instability. If the screen is earthed by means of a capacitor, it is often possible to select a capacitance value which will be series resonant with the screen lead inductance, at the working frequency, and so form a low impedance path to ground. This arrangement is often sufficient to prevent instability from this cause, even over a range of frequencies. For further discussion see Ref. A8.

It is also important to note that at the higher frequencies a capacitor does not behave as a pure capacitance, and its effective inductance and resistance become increasingly important as the frequency increases. With many types of capacitors it is possible that the inductive reactance will exceed the capacitive reactance even at frequencies as low as 30 Mc/s (with electrolytics an additional r-f by-pass should always be used). This effect is sometimes used to make the capacitor series resonant at the working frequency. Even if the capacitor behaves as a small inductance it will be appreciated that in some cases the reactance can be very low and effective by-passing is still possible. In tuned circuits the Q can be materially affected by the increase in r-f resistance of the capacitors and both this and the previous effects should be checked.

The effectiveness of by-pass capacitors and their Q at the working frequency can be readily checked with the aid of a Q meter by placing the capacitor in series with a coil of known inductance and Q. See Ref. B45. Finally, r-f chokes do not always behave as such, and their resonant frequencies should always be checked (see also Chapter 11 Sect. 6).

SECTION 8 : DISTORTION

(*i*) *Modulation envelope distortion*   (*ii*) *Cross modulation distortion.*

Modulation envelope distortion introduced by r-f voltage amplifiers is usually small when compared with that introduced by the later stages in a radio receiver e.g. the i-f amplifier, the detector and, perhaps, the frequency converter. This is particularly true when the magnitude of the signal voltage is small. Amplitude modulation distortion in r-f amplifiers is usually caused by the curvature of the valve characteristic relating control grid voltage to plate current. The usual explanation for the introduction of this distortion is to consider that the portion of the characteristic curve being used can be represented by a power series relating the grid voltage (in this case the modulated carrier wave) and the plate current. It can be shown from this treatment that second order terms (terms containing squares) in the series do not introduce distortion because of the selectivity of the tuned circuits. Third order terms (those containing cubes) do introduce distortion of the modulation envelope.

R-F amplifier valves are usually designed so that the third order curvature of their characteristics is minimized as far as possible. This is achieved by using a variable pitch for the control grid winding and results in the well known variable-mu (or remote cut-off) characteristic. The variable-mu (or more exactly " variable g_m " characteristic) and the resultant shape of the $e_g - i_p$ curve, have a large bearing on cross modulation distortion as will be discussed presently.

(i) Modulation envelope distortion

Because of the presence of the higher order terms in the power series representation, the magnitude of the r-f signal which can be handled by a particular type of r-f valve is limited if the distortion of the modulation envelope is not to be serious. Methods of measuring and calculating the signal handling capabilities of r-f amplifiers have been discussed in the literature and the reader is referred to K. R. Sturley's " Radio Receiver Design " Part 1 Chapter 4 Sect. 7 for a description of these methods, and to the other references listed at the end of this chapter.

(ii) Cross modulation distortion

Cross modulation distortion is an effect well known to receiver designers. It occurs when two signals are applied to any non-linear element such as a radio valve, under certain operating conditions, having appreciable third order curvature of the grid voltage-plate current characteristic. The effect is most noticeable when a strong local signal is present and the receiver is tuned to a weak signal. Insufficient r-f selectivity is often a major contributing factor in this regard. It is quite useless to have good selectivity in, say, the i-f amplifier once cross modulation has occurred in the r-f or converter stage, as no amount of subsequent selectivity can remove the undesired signal which is now superimposed on the same frequency band as the desired signal, and the modulation on both signals will be heard in the received output.

The r-f amplifier valve must also be designed so that the higher order modulation terms are kept as small as possible. A variable-mu valve will materially assist in handling a wide range of signal voltages without the grid voltage-plate current curvature being such as to allow serious cross modulation to occur.

A careful choice of the bias voltage for the r-f valve, and making the cathode by-pass capacitor of sufficiently small value to be effective at radio frequencies only, will often assist in reducing cross modulation.

When the local interfering station is a very powerful one it often becomes necessary to use special rejector circuits as well as high r-f selectivity.

Effects such as external cross modulation are not discussed here but reference can be made to F. E. Terman's " Radio Engineers' Handbook " page 647 for some details.

SECTION 9 : BIBLIOGRAPHY

(A) BOOKS DEALING WITH RADIO FREQUENCY AMPLIFIERS, AERIAL COUPLING, NOISE, ETC.
A1. Sturley, K. R. " Radio Receiver Design " Parts 1 and 2 (Chapman and Hall, London, Part 1, 1943 ; Part 2, 1945).
A2. Zepler, E. E. " The Technique of Radio Design " (Chapman and Hall, London, 1943, John Wiley and Sons, New York, 1943).
A3. Everitt, W. L. " Communication Engineering " (2nd edit., McGraw-Hill, New York and London, 1937).
A4. Terman, F. E. " Radio Engineers' Handbook " (1st edit. McGraw-Hill, New York and London, 1943).
A5. Terman, F. E. " Radio Engineering " (2nd edit. McGraw-Hill, New York and London, 1937).
A6. Glasgow, R. S. " Principles of Radio Engineering " (McGraw-Hill, New York and London, 1936).
A7. Moullin, E. B. " Spontaneous Fluctuations of Voltage " (Oxford University Press, London, 1938).
A8. Harvey, A. F. " High Frequency Thermionic Tubes " (Chapter 2) John Wiley and Sons, New York, 1943.
A9. Kiver, M. S. " F-M Simplified " (Chapter 12—R. F. Tuners for F-M Receivers). D. Van Nostrand Co. Inc., Toronto, New York, London, 1947.
A10. Valley, G. E., and H. Wallman, " Vacuum Tube Amplifiers " (McGraw-Hill, New York, Toronto and London, 1940).

A11. Thomas, H. A., and R. E. Burgess " Survey of existing information and data on radio noise over the frequency range 1-30 MC (His Majesty's Stationery Office, London, 1947).
A12. Moxon, L. A. " Recent Advances in Radio Receivers " (Cambridge University Press, London, 1949).
A13. Goldman, A. " Frequency Analysis, Modulation and Noise " (McGraw-Hill Book Company, 1948).
A14. " Valve and Circuit Noise—survey of existing knowledge and outstanding problems " Radio Research Special Report No. 20 (1951)—Dept. of Scientific and Industrial Research, London.

(B) REFERENCES TO PERIODICALS
B1. Polydoroff, W. J. " Permeability tuning " Elect. 18.11 (Nov. 1945) 155.
B2. Benin, Z. " Modern home receiver design " Elect. 19.8 (Aug. 1946) 94.
B3. Vladimir, L. O. " Permeability tuning of broadcast receivers " Elect. 20.8 (Aug. 1947) 94.
B4. Miner, C. R. " FM receiver front end design " (Guillotine Tuner) Tele-Tech 6.5 (May 1947) 34.
B5. Miccioli, A. R., and C. R. Pollack " Design of FM receiver front ends " (Vane Tuner) Tele-Tech 6.7 (July 1947) 40.
B6. Hershey, L. M. " Design of tuners for AM and FM " Tele-Tech 6.8 (Aug. 1947) 58.
B7. Berkley, F. E. " Tuning without condensers " (resonant lines) FM & T 7.8 (Aug. 1947) 35.
B8. Jones, M. C. " Grounded-grid radio-frequency-voltage amplifiers " Proc. I.R.E. 32.7 (July 1944) 423.
B9. Kiver, M. S. " FM and Television Design—Part 1 (Tubes and circuits as used in high frequency r.f. amplifiers) " Radio Craft 19.2 (Nov. 1947) 32.
B10. Wheeler, H. A., and W. A. Macdonald " Theory and operation of tuned r.f. coupling systems " Proc. I.R.E. 19.5 (May 1931) 738.
B11. R.C.A. " Properties of untuned r.f. amplifier stages " R.C.A. Application Note 116 (July 2, 1946) ; Reprinted Radiotronics No. 124.
B12. Rudd, J. B. " Theory and design of radio-frequency transformers " A.W.A. Tec. Rev. 6.4 (March 1944) 193.
B13. Bachman, W. S. " Loop-antenna coupling—transformer design " Proc. I.R.E. 33.12 (Dec. 1945) 865.
B14. Maurice, D., and R. H. Minns " Very-wide band radio-frequency transformers " W.E. 24.285 (June 1947) 168 ; 24.286 (July 1947) 209.
B15. Vladimir, L. O. " Low impedance loop antenna for broadcast receivers " Elect. 19.9 (Sept. 1946) 100.
B16. Kobilsky, M. J. " A note on coupling transformers for loop antennas " Proc. I.R.E. 35.9 (Sept. 1947) 969.
B17. Burgess, R. E. " Receiver input circuits—design considerations for optimum signal/noise ratio " W.E. 20.233 (Feb. 1943) 66.
B18. Burgess, R. E. " Signal noise characteristics of triode input circuits " W.E. 22.257 (Feb. 1945) 56.
B19. " R.F. amplifying valves and shortwave performance " Philips Tec. Com. No. 48.
B20. " Performance of r.f. amplifying valves " (tests between 1.5 and 60 Mc/s) Philips Tec. Com. No. 51.
B21. Ferris, W. R. " The input resistance of vacuum tubes at ultra high frequencies " Proc. I.R.E. 24.1 (Jan. 1936) 82.
B22. North, D. O. " The analysis of the effects of space charge on grid impedance " Proc. I.R.E. 24.1 (Jan. 1936) 108.
B23. Strutt, M. J. O. " Electron transit time effects in multigrid valves " W.E. 25.177 (June 1938) 315.
B24. " Application Note on input loading of receiving tubes at radio frequencies " R.C.A. Application Note No. 101 (Jan. 25, 1939) see Text Chapt. 23, Sect. 5.
B25. " Input admittance of receiving tubes " R.C.A. Application Note No. 118 (April 15, 1947) ; Reprinted Radiotronics No. 126.
B26. North, D. O., and W. R. Ferris " Fluctuations induced in vacuum tube grids at high frequencies " Proc. I.R.E. 29.1 (Feb. 1941) 49.
B27. Harris, W. A. " Fluctuations in vacuum tube amplifiers and input systems " R.C.A. Review 5.4 (April 1941) 505.
B28. " A discussion of noise in portable receivers " R.C.A. Application Note No. 115 (June 2, 1941) Reprinted Radiotronics No. 123.
B29. Herold, E. W. " An analysis of the signal-to-noise ratio of ultra-high-frequency receivers " R.C.A. Rev. 6.3 (Jan. 1942) 302.
B30. Moxon, L. A. " Noise Factor 1—A new conception of receiver sensitivity and signal/noise ratio " W.W. 52.12 (Dec. 1946) 391.
B31. Moxon, L. A. " Noise Factor 2—Methods of measurement : sources of test signals " W.W. 53.1 (Jan. 1947) 11.
B32. Moxon, L. A. " Noise Factor 3—Design of receivers or amplifiers for minimum noise factor " W.W. 53.5 (May 1947) 171.
B33. Stolze, W. J. " Input circuit noise calculations for FM and television receivers " Comm. 27.2 (Feb. 1947) 12. See Text Chapter 23 Sect. 6.
B34. Campbell, N. R., V. J. Francis and E. G. James " Valve noise and transit time " W.E. 25.296 (May 1948) 148 ; letters 26.311 (Aug. 1949) 277.
B35. Bell, R. L. " Negative grid partition noise " W.E. 25.300 (Sept. 1948) 294.
B36. van der Ziel, A., and A. Versnel " Measurements of noise factors of pentodes at 7.25 M wavelength " Philips Research Reports 3.2 (April 1948) 121.
B37. Goldberg, H. " Some notes on noise figures " Proc. I.R.E. 36.10 (Oct. 1948) 1205.   Correction 37.1 (Jan. 1949) 40.
B38. Houlding, N. (letter) " Valve noise and transit time " W.E. 25.302 (Nov. 1948) 372.
B39. Callendar, M. V. " Thermal noise output in A.M. receivers—effect of wide pre-detector bandwidth " W.E. 25.303 (Dec. 1948) 395.
B40. Harris, W. A. " Some notes on noise theory and its application to input circuit design " R.C.A. Rev. 9.3 (Sept. 1948) 406.
B41. Thompson, B. J. " Oscillation in tuned r.f. amplifiers " Proc. I.R.E. 19.3 (March 1931) 421 ; Discussion 19.7 (July 1931) 1281.
B42. Carter, R. O. " Distortion in screen-grid valves " W.E. 9.102 (March 1932) 123.
B43. Carter, R. O. " The theory of distortion in screen-grid valves " W.E. 9.107 (Aug. 1932) 429.
B44. Sloane, R. W. " The signal handling capacity of H.F. valves " W.E. 26.194 (Nov. 1939) 543.
B45. Price, J. F. " Effectiveness of by-pass capacitors at v.h.f." Comm. 28.2 (Feb. 1948) 18.
B46. Shone, A. B. " Variable filter coupling " W.W. (1) 56.10 (Oct. 1950), 355 ; (2) 56.11 (Nov. 1950) 393.

CHAPTER 24

OSCILLATORS

BY B. SANDEL, A.S.T.C.

SECTION 1 : INTRODUCTION

In this chapter it is proposed to consider, briefly, some of the fundamental types of oscillator circuits (used in conjunction with valves) which are commonly employed in radio receivers. Of course there are numerous variations of the fundamental circuits, but generally these changes are only a practical convenience for obtaining some required special condition of operation from the basic circuit.

The types of circuits to be considered are :

(a) The tuned-plate oscillator
(b) The tuned-grid oscillator
(c) The Hartley oscillator
(d) The Colpitts oscillator
(e) The electron-coupled oscillator
(f) The negative transconductance oscillator.

For those interested in the general theory of oscillation, and in the many special circuits available, a list of suitable references is given at the end of the chapter which will serve as a starting point, at least, for a more complete survey of this field of knowledge. Crystal oscillator circuits are not discussed here but a number of useful circuits are given in Refs. 27 (page 164), 26 (page 97) and 31 ; the latter giving a discussion of overtone (harmonic mode) crystal oscillators.

Before proceeding to a discussion of particular types of oscillator circuits, a few of the fundamental principles will be briefly reviewed.

The simplest form of electrical oscillator consists of a combination of inductance (L) and capacitance (C) connected together (as shown in Fig. 24.1), to which has been

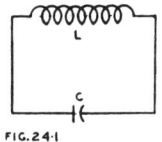

FIG. 24·1

added, initially, electric or magnetic energy. Suppose that the capacitor C has been charged by some means. The energy stored in the capacitor is then $\frac{1}{2}CE^2$, where E is the maximum potential difference between the plates. At this instant, when E has its maximum value, the current in the circuit is zero. The presence of the inductor will allow the energy stored in the electric field of the capacitor to be transferred, and to form a magnetic field around the inductor. The capacitor discharges until finally E becomes zero and the current I becomes a maximum. At the instant at which I is a maximum the energy in the magnetic field is $\frac{1}{2}LI^2$, all the available energy is stored in the magnetic field and there is no electric field. The process now reverses, the magnetic field collapses and energy is transferred back to the electric field of the capacitor. This process repeats itself indefinitely if there is no loss of energy in the circuit (radiation is not considered here).

Since the total energy which is stored in each field in turn, is the same, it is permissible to write

$$\frac{1}{2}CE^2 = \frac{1}{2}LI^2 \tag{1}$$

where E and I have their maximum values.

Also $\qquad E = 1/\omega C \tag{2}$

and so $1/\omega^2 C = L \tag{3}$

or $\qquad f = 1/2\pi\sqrt{LC} \tag{4}$

where f is natural resonant frequency of the oscillations occurring in the circuit.

Since there is always some resistance (R) present with practical circuit elements, the amplitude of each successive oscillation will decrease until eventually all of the energy is dissipated—mainly in the form of heat in the resistance—and the oscillations will cease. (This is discussed further in Chapter 9, Sect. 1). The addition of extra energy to the circuit from some external source, such that the added energy equals that being lost, would allow the oscillations to continue indefinitely. In the circuits to be considered the power supply is the external source of energy, and the valve is the device controlling the energy which is added to the $L\ C\ R$ circuit, in the correct phase and amplitude, to maintain oscillations.

With any valve oscillator an exact analysis of the method of operation is very difficult, if not impossible, and it is usual to treat the circuits as being linear (at least for simple design procedure) although they depend on conditions of non-linearity for their operation. This simplification is valuable because the mathematical analysis which can be carried out yields a great deal of useful information concerning the behaviour of the circuits. That the circuit operation is non-linear can be readily appreciated by considering the fact that the amplitude of the oscillations, once started, does not continue to build up indefinitely. The energy gain of the system reaches a certain amplitude and then progressively falls until equilibrium is established. The limits are usually set by the valve—plate current cut-off occurs beyond some value of the negative grid voltage swing, and plate current saturation or grid current damping will limit the amplitude of the grid swing in the positive direction. For a discussion of the factors governing oscillation amplitude, the reader should consult Refs. 1 and 2.

In the sections to follow, typical circuits applicable to radio receivers will be discussed. It is not proposed to give a mathematical analysis or detailed physical explanations of the operation of the circuits, as this has been more or less adequately done in many text books and periodicals. Suitable references are 1, 3, 5, 7, 10, 11, 24.

At the outset it may be mentioned that, since high oscillator efficiency is not usually as important a factor as some other requirements in radio receivers, the design is generally a combination of empirical and experimental techniques. A large number of the circuit component values are pre-determined by such considerations as tracking the tuned circuit with signal circuits over a band of frequencies and maintaining constant oscillator amplitude over the tuning range. Other factors will be discussed in Sect. 9 of this chapter.

Design procedures such as those used for class " C " power oscillators in transmitters are not carried out in detail for receiver oscillators, although obviously the basic principles are the same in both cases ; this will receive further consideration in Sect. 9

SECTION 2 : TYPES OF OSCILLATOR CIRCUITS

(i) Tuned-plate (ii) Tuned-grid (iii) Hartley (iv) Colpitts (v) Electron-coupled (vi) Negative transconductance oscillators.

(i) Tuned-plate oscillator

Fig. 24.2 shows three arrangements for the tuned plate oscillator. Circuits (A) and (B) use series feed, and circuit (C) uses shunt feed. Circuit (B) is the usual practical arrangement of (A) ; P is a padding and C a blocking capacitor.

The shunt feed circuit (C) is generally preferred in receivers for the following reasons :

(1) The rotor plates of the tuning capacitor can be directly earthed without using additional series blocking capacitors directly in the tuned circuit.

(2) The resistor R_1 can often be selected to assist in maintaining a constant amplitude of oscillation over the tuning range, since it will give greater damping of the tuned circuit as the frequency increases.

Either arrangement of the grid-leak resistor (R_g) shown in circuits (B) and (C) is satisfactory, but in circuit (C) it gives increased damping on the feedback winding (L_f) which may be advantageous in maintaining constant grid voltage amplitude over a range of frequencies.

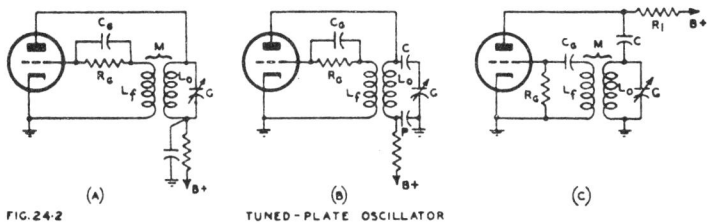

FIG.24-2   TUNED-PLATE OSCILLATOR

Tight coupling and low values of mutual inductance (M) are helpful in maintaining frequency stability with mains voltage variations, as these factors tend to make the circuit relatively independent of valve constants other than interelectrode capacitances (see Ref. 21). L_f is made as small as possible, to reduce M, and also to keep its natural resonant frequency well above the tuning range ; this latter factor means that stray capacitance across the feedback winding must be kept as small as possible. The high natural resonant frequency for the feedback circuit is helpful in reducing variations in the tuned circuit (due to reflected impedance), this being particularly important since it reduces tracking error and limitations on the maximum possible tuning range.

The resonant frequency (f) of the tuned-plate oscillator is given approximately by

$$f = \frac{1}{2\pi\sqrt{L_0 G}}\sqrt{1 + \frac{R}{r_p}} = f_0\sqrt{1 + \frac{R}{r_p}} \tag{5}$$

where L_0 = inductance in tuned circuit

G = capacitance in tuned circuit (this strictly includes any additional stray capacitances shunted across G, e.g. in circuit (C) some capacitive reactance due to C would be present, depending on the value of R and the frequency of operation)

R = total series r-f resistance in tuned circuit [including in circuit (C) the equivalent series resistance due to C and R_1 being shunted across the circuit]

r_p = plate resistance of valve

and f_0 = natural resonant frequency of tuned circuit alone.

For our purposes it is sufficiently close to take

$$f = f_0 = \frac{1}{2\pi\sqrt{L_0 G}} \tag{6}$$

where G is the actual capacitance (including trimmers, padders etc.) tuning L_0.

The important point to observe is that other components, including the valve, affect the frequency of oscillation. As these components are capable of variation with voltage or temperature fluctuations they will affect oscillator stability, and their effects should be minimized as far as possible. Some improvement in stability is possible by making the grid current small, which agrees with the condition of a low value for L_f, but this is also governed by the permissible value of the grid resistor (R_g).

The condition for the maintenance of oscillation is given by

$$M = -\left[\frac{L_0}{\mu} + \frac{GR}{g_m}\right] \qquad (7)$$

where L_0, G and R have the same meanings as previously,

M = mutual inductance = $k\sqrt{L_f L_0}$ (in the absence of additional capacitance coupling)
k = coefficient of coupling
μ = amplification factor of valve

and g_m = mutual conductance of valve.

The negative sign for M indicates that oscillation will only occur for one connection of the feedback winding.

A suitable factor of merit for an oscillator valve is the product μg_m, which should be as high as possible. This is also a suitable factor for power output valves, and explains why types such as the 6V6 beam power valve work well in feedback oscillator circuits.

The advantages claimed for the tuned-plate oscillator are its relative freedom from frequency changes due to mains voltage variations and, when the circuit is used with converter valves, freedom from signal-grid bias voltage changes.

The tuned plate oscillator in its standard form is not very satisfactory for use at frequencies above about 50 Mc/s.

(ii) The tuned-grid oscillator

The general circuit arrangement for a tuned-grid oscillator is shown in Fig. 24.3. Much of the general discussion given in connection with the tuned-plate oscillator is applicable to this circuit.

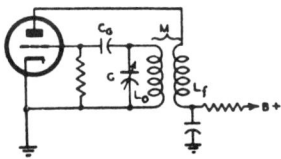

The tuned-grid oscillator is probably the most widely used in receivers for the standard long-, medium- and short-wave bands. One of its features is the ease with which oscillation can be obtained. The arrangement of the components offers little difficulty, and one particular advantage is that the tuned circuit is completely isolated from the plate supply voltage.

FIG. 24.3 TUNED-GRID OSCILLATOR

The approximate frequency of oscillation (f) for this circuit is given by

$$f = \frac{f_0}{\sqrt{1 + \dfrac{L_f R}{L_0 r_p}}} \qquad (8)$$

where $f_0 = 1/2\pi\sqrt{L_0 G}$ = natural resonant frequency of tuned circuit
L_0 = tuned circuit inductance
r_p = plate resistance of valve
L_f = feedback winding inductance
R = series r-f resistance (total) in tuned circuit

and G = total tuning capacitance.

The minimum value of M required to maintain oscillation is given approximately by

$$M = -\left[\frac{L_f A}{\mu(1 + A)} + \frac{GR}{g_m}\right] \qquad (9)$$

where $A = L_f R / L_0 r_p$,
  μ = amplification factor of valve
  g_m = mutual conductance of valve
  r_p = plate resistance of valve
and L_f, R and G are as previously.

If r_p is large and R is small, frequency variation from the natural resonant frequency is reduced, as can be seen from an examination of the equation. Also, as a point of interest, it is seen that f is less than f_0 for the tuned-grid oscillator, and greater than f_0 for the tuned-plate circuit ; so that the tuned circuit acts as an inductive reactance (neglecting resistance effects) in the first case, and as a capacitive reactance for the tuned-plate oscillator. These points are further discussed in Ref. 5. The useful upper frequency limit for the tuned-grid circuit is about 50 Mc/s, as the value of the feedback-winding inductance and the resonant frequency of the feedback circuit become troublesome. This applies also to the tuned-plate arrangement.

(iii) Hartley oscillator

Fig. 24.4 shows two possible arrangements for the Hartley oscillator circuit. The circuit (A) is the conventional series-fed circuit, but is not very convenient for use in radio receivers for several reasons.

(1) The tuning capacitor G cannot be earthed without using blocking capacitors.
(2) The presence of blocking capacitors presents tracking difficulties.
(3) The coil L_0 has H.T. applied which may be awkward when making adjustments, particularly during developmental work.
(4) Untracked stray capacitances are likely to be high.

The alternative arrangement of Fig. 24.4 (B) is the circuit most commonly employed in receivers. The tuned circuit is connected directly to ground. This arrangement avoids the difficulties of circuit (A) ; furthermore it is particularly convenient for use with converter valves of the 6SA7, 6BE6 pentagrid type.

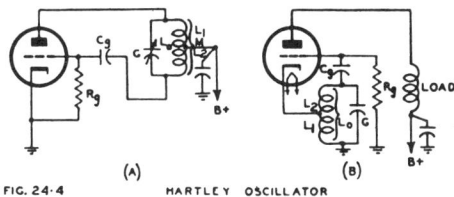

(A)   (B)

FIG. 24.4   HARTLEY OSCILLATOR

The angular frequency of oscillation for the Hartley circuit of Fig. 24.4(A) (see Ref. 5) is given approximately by

$$\omega = \omega_0 \sqrt{1 + R_1/r_p} \qquad (10)$$

where $\omega_0 = 1/\sqrt{L_0 G}$
  $\omega_0 = 2\pi \times f_0$ (f_0 is natural resonant frequency of tuned circuit)
  R_1 = r-f resistance of L_1
  r_p = plate resistance of valve
  L_0 = oscillator coil inductance ($= L_1 + L_2 + 2M$)
and G = tuning capacitance.

For the maintenance of oscillation

$$\mu = \frac{L_1 + M}{L_2 + M} + \frac{G r_p (R_1 + R_2) L_0}{(L_1 + M)(L_2 + M)} \qquad (11)$$

where μ = amplification factor of valve.

L_1 and L_2 are the two sections of L_0 coupled by mutual inductance M and R_1 and R_2 are the r-f resistances of L_1 and L_2 respectively.

The frequency equation shows that the oscillation frequency f is higher than f_0, and to reduce this difference r_p should be high and R_1 small. Also, for the circuit to oscillate readily, the mutual conductance of the valve should be high.

A Hartley oscillator offers advantages over the tuned-plate and tuned-grid circuits at frequencies in excess of about 40 Mc/s, the greatest advantage being that the feedback winding, being a part of the tuned circuit, does not offer the same difficulties as the other cases, in which the natural resonant frequency approaches the operating frequency.

In the case of the other two types of oscillators mentioned, if the feedback winding has greater inductance than the tuned winding, the oscillator can easily change over from one type to the other (i.e. tuned-plate becomes tuned-grid and vice-versa) and satisfactory tuning is obviously impossible. This trouble is not as unlikely as may be thought since it is often very difficult to obtain sufficient amplitude of oscillation as the frequency increases.

The Hartley circuit can be made to give satisfactory operation to frequencies at least as high as 150 Mc/s, even with multi-grid converter valves (depending, of course, on the valve type used), but its chief disadvantages are

(1) Its liability to parasitic oscillations as the frequency increases,
(2) The possibility of the circuit acting like a modified Colpitts oscillator because of stray and valve interelectrode capacitances,
(3) That it is rather awkward to find the optimum tapping point for the best conditions of oscillation, particularly when the frequency becomes fairly high. This is particularly so with valve types such as the 6BE6 on the F-M broadcast band since the tapping point on the oscillator coil has a large effect on conversion gain.

(1) and (2) generally manifest themselves as sudden changes in oscillation frequency when tuning over the working range, or as " dead " spots, or through the oscillator stopping altogether when the gang capacitance becomes less than a certain value.

It may be mentioned finally that the Hartley circuit has been extensively used on all broadcast bands including the F-M 88-108 Mc/s band ; its application in receivers has been largely confined to use with pentagrid converters such as types 6SA7, 6BE6, 6SB7-Y and 6BA7.

(iv) Colpitts oscillator

A circuit for a shunt fed Colpitts oscillator is shown in Fig. 24.5(A). A suitable arrangement (Ref. 9) for use in radio receivers operating at frequencies of the order of 100 Mc/s is that of Fig. 24.5(B). In this latter circuit C_B is the self-capacitance (plus added capacitance if required) of the choke L in the cathode circuit, and C_A is the grid-to-cathode capacitance ; L_0 is the oscillator coil inductance and G the variable tuning capacitor. The cathode impedance can readily be controlled by connecting a variable trimmer across L.

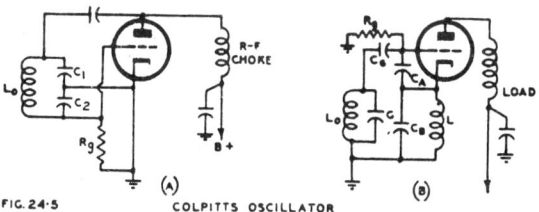

FIG. 24·5   COLPITTS OSCILLATOR

The circuit of Fig. 24.5(B) has been successfully used with type 6BE6 converters tuning the 88-108 Mc/s F-M broadcast band and avoids the necessity for tapping the oscillator coil as in the Hartley circuit.

The angular frequency of oscillation of the Colpitts oscillator, from the solution of the circuit of Fig. 24.5(A), is given by

$$\omega = \omega_0 \sqrt{1 + \frac{R}{r_p}\left(\frac{C_2}{C_1 + C_2}\right)} \tag{12}$$

The condition for oscillation maintenance is that

$$\mu = \frac{C_2}{C_1} + \frac{r_p R(C_1 + C_2)}{L_0} \tag{13}$$

where $\omega_0 = 2\pi \times f_0$ (f_0 is the natural resonant frequency of the tuned circuit alone)

R = r-f resistance of L_0

r_p = plate resistance of valve

μ = amplification factor of valve

L_0 = oscillator coil inductance

and C_1 and C_2 tune L_0 ; they also form a capacitive voltage divider across L_0 ; and the excitation voltage on the grid is proportional to $C_1/(C_1 + C_2)$.

This circuit is most convenient for use at v-h-f, it is very easy to make oscillate, and is not so liable to parasitic oscillations as the Hartley circuit. Its use on the medium and short-wave bands has been rather limited because of difficulties in circuit arrangement when covering frequency bands of about 3 : 1, and also because of the very simple and satisfactory manner in which the tuned-grid and tuned-plate circuits can be made to operate over these ranges. The Colpitts circuit readily lends itself to inductance tuning and with this arrangment has been used satisfactorily on all of the frequency ranges encountered in normal broadcast receiver design.

(v) Electron-coupled oscillator

Electron-coupled oscillators (Refs. 1, 22) can use a large variety of fundamental circuits, such as the Hartley or Colpitts, for generating oscillations. The circuit usually employs a tetrode or pentode valve, or, most often in receivers, is used with a pentagrid converter. Valves having suppressor grids will generally give better frequency stability.

The fundamental principle is that the actual oscillating circuit is connected to the load circuit only by means of the electron stream within the valve. In this way changes in load conditions and high tension voltages have a reduced effect on the actual frequency of oscillation. The improvement in frequency stability with voltage variations appears to be closely bound up with compensating effects due to simultaneous voltage changes on the plate and screen. Variation in load conditions would, ideally, leave the oscillation frequency unchanged since the two circuits are only connected unilaterally through the electron stream. Actually, of course, interelectrode capacitances and stray coupling prevent this condition from being completely fulfilled.

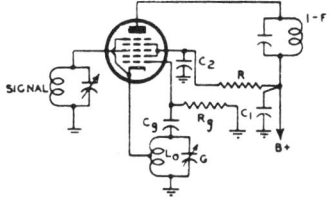

FIG.24.6 PENTAGRID CONVERTER USING ELECTRON-COUPLED HARTLEY OSCILLATOR

Those pentagrid converters which commonly use the Hartley oscillator circuit arrangement are typical examples of the application of electron-coupling. One such arrangement is shown in Fig. 24.6. It is often advantageous in this type of circuit to connect the screen to the rectifier output through R as this reduces " flutter " ; the high tension for the plate is obtained from the power supply in the usual manner.

With this latter arrangement C_2 generally consists of an electrolytic capacitor and a mica capacitor (to act as a r-f by-pass) connected in parallel.

(vi) Negative transconductance oscillators

Several types of negative transconductance oscillators have been suggested using r-f pentode valves. These circuits have a negative bias on the suppressor grid and rely for their operation on the fact that, over a particular range of negative voltage on this grid, the suppressor-screen transconductance is negative. With suitable operating conditions a positive increment in the negative suppressor grid voltage will allow the plate current to increase and the screen current to decrease even when the screen voltage is increased ; the changes in screen and suppressor voltage being approximately equal. The screen and suppressor are coupled together by means of a capacitor, and the tuned circuit is connected in a suitable manner between plate and screen. Detailed descriptions of this type of oscillator can be found in Refs. 7 and 23. Circuits using this arrangement with pentode valves are not very convenient since the negative transconductance is only of the order of -250 micromhos.

Since this oscillator is a two terminal type it is often a very convenient arrangement ; one particular example is its use as a beat-frequency oscillator in a radio receiver. A particular form of the negative transconductance oscillator (Ref. 13)—which employs, also, the principle of electron coupling—using a pentagrid converter valve (e.g. type 6A8, but **not** 6SA7, 6BE6 etc.) is shown in Fig. 24.7. The negative transconductance is brought about as follows. Electrons moving towards the plate are turned back to the inner screen (G_3) and the oscillator anode (G_2) when the control grid (G_4) has a more negative voltage applied to it. The net effect of an increase of negative voltage on the signal grid is to increase the current to the oscillator-anode and to grid G_3.

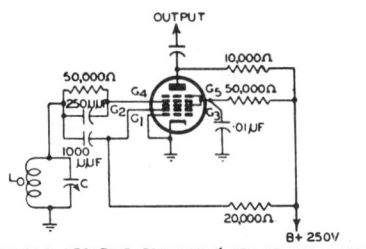

FIG. 24·7 NEGATIVE TRANSCONDUCTANCE OSCILLATOR

Any increase in the current to G_3 is practically offset by a decrease in the current to G_5, the outer part of the screen grid, and the result is that the screen current remains fairly constant for wide variations in signal-grid voltage. The variation in oscillator-anode current, however, is equivalent to a negative transconductance between the control grid (G_4) and the oscillator-anode (G_2). In type 6A8 this amounts to about -400 micromhos. Because of this negative transconductance it is possible to create an oscillatory condition by coupling G_4 to G_2, provided that the rest of the circuit is suitably arranged.

The circuit of Fig. 24.7 has been used at frequencies up to 18 Mc/s. It has also been employed in several types of communications receivers (having intermediate frequencies of from 255 Kc/s to 3 Mc/s) as a beat-frequency oscillator and has given very good results as regards stability of operation, particularly when temperature compensation has been applied to the tuned circuit.

It is near enough, for practical purposes, to take the frequency of oscillation as being

$$f_0 = 1/(2\pi\sqrt{L_0 C}) \tag{14}$$

SECTION 3 : CLASS A, B AND C OSCILLATORS

An oscillator can be made to work under a variety of conditions. The impulses applied to the tuned circuit, to maintain oscillation, can be such that the valve plate current may be flowing from something less than 180° of the electrical cycle to almost the full 360°. Depending on the period for which the impulses are applied, the oscillator may be classified as Class A, B or C.

High quality laboratory oscillators often use the Class A condition, while high power oscillators, such as those sometimes used in transmitters (where efficiency might occasionally be more important than obtaining an output voltage relatively free from harmonics) might possibly use Class C operation ; of course even in high power transmitters a master oscillator is the usual arrangement and efficiency is not an important factor in its operation. Oscillators in superheterodyne receivers do not need to have very high efficiency, but the output must be relatively free from harmonics if whistle interference is not to be a serious problem ; this generally leads to something approaching Class B operation (Ref. 5).

It should be noted that, for all usual operating conditions, grid current flows for at least part of the input voltage cycle.

SECTION 4 : CAUSES OF OSCILLATOR FREQUENCY VARIATION

(*i*) General   (*ii*) Changes in supply voltage   (*iii*) Temperature and humidity changes (*iv*) Oscillator harmonics.

(i) General

Frequency drift in the oscillator section of superheterodyne receivers (F-M or A-M) is important for a number of reasons. The most obvious is that as a result of such drift the actual intermediate frequency output from the frequency changer will not be that to which the i-f amplifier is tuned, giving a loss in amplification and the added possibility of considerable frequency distortion in A-M receivers, and non-linear distortion in F-M receivers. Other special effects will be discussed separately.

Oscillator frequency variations which can be offset in the original design may be due to changes in
(a) Supply voltage
(b) Temperature
(c) Humidity
(d) A.V.C. bias on the signal grid of multielectrode converter valves. This will be discussed in Chapter 25 Sect. 2.

Also, it should be noted that a high percentage of oscillator harmonics can lead to oscillator instability. Mechanical stability of the circuit is a prime requirement.

Detailed discussion regarding oscillator frequency stability can be found in Refs. 1, 4, 5, 10, 11, 15, 19, 20, 21 and 28.

(ii) Changes due to supply voltage

Some change in oscillator frequency always occurs when there is a change in the voltages on the valve electrodes, because of changes in valve " constants."

For the simple triode oscillator, plate voltage changes may be minimized by using a supply voltage having good regulation ; the regulation requirement is often determined by the variations in the total current drawn by the complete receiver when the signal voltages are changing. This may necessitate a separate voltage supply for the oscillator valve, and where extremely high oscillator stability is required a separate voltage regulator valve (e.g. OD3/VR150) may be incorporated. A common manifestation of poor regulation of the H.T. is the " flutter " experienced on the short-wave bands.

For screen grid valves the same precautions are taken as for a triode, but often very appreciable improvement is possible, with some valve types, by offsetting screen and plate current changes one against the other. Although the principle of offsetting changes due to various electrodes against each other is of assistance, it should be noted that with some converter valves in which voltage changes are additive (e.g. 6J8-G), better stability is possible than with other valve types (e.g. 6A8, 6K8) in which the plate and screen voltage variations are subtractive. Pentagrid converter valves of the 6SA7 class are particularly critical as to screen voltage changes and a separate screen supply is often essential.

(iii) Temperature and humidity changes

After a receiver has been switched on, it is found that the oscillator frequency tends to alter for some considerable time, up to say an hour or more. Usually it is found that the internal effects of the valve heating up become negligible after a period of about ten minutes. The base of the valve is responsible for a certain amount of drift, but this can be minimized by the use of high-quality porcelain or micanol bases. With the miniature valve types the electrode connections come straight through the button stem instead of through a glass pinch, and the temperature variations due to the complete valve assembly are kept to an absolute minimum.

The most serious frequency drift through temperature changes is generally caused by capacitance variations due to the heating of the various dielectrics in the circuit.

This is minimized by the use of high quality dielectrics such as special grades of porcelain.

Inductance changes due to temperature rise also cause frequency variation, and low-loss dielectric formers are essential.   If sufficient mechanical stability is possible air cored coils are often preferable, but not always.

Good circuit layout is an essential requirement and all sources of heat should, be kept away from tuning capacitors and coils.

When all the requirements as to dielectrics, circuit layout etc. have been met as far as is possible, final temperature compensation is made by using capacitors having negative temperature coefficients.   These capacitors are available commercially in a wide range of capacitance values, and with various negative temperature co-efficients. The manufacture of this type of capacitor is possible because of the availability of ceramic materials having dielectric constants which decrease with an increase in temperature*.   It should be realized that exact compensation by simple capacitance adjustment is only feasible, in general, at one point in a given tuning range, but a very appreciable improvement over a band of frequencies is possible without exact compensation at any one point.   A method for obtaining exact compensation at two points (although here again a compromise adjustment may be preferable) in a given tuning range, using the principles of superheterodyne tracking, has been given by Bushby (Ref. 19) ; it is also shown that a compromise adjustment is possible, although exact compensation is not practicable, at three points in the tuning range.   Methods for carrying out long period frequency-drift compensation, due to temperature changes, are outlined in Ref. 19.

Changes in oscillator frequency with humidity can be minimized by using coils, capacitors and insulating materials which are properly baked to remove moisture and then impregnated (with suitable waxes or varnishes) to prevent the absorption of moisture.†

It may be worth noting here that carbon resistors have negative temperature coefficients.   Wire-wound resistors have positive or negative temperature co-efficients depending on the type of resistance wire used.‡

(iv) Oscillator harmonics

There is an effect, supposed to be due to cross modulation between the fundamental frequency and its harmonics and between the harmonics themselves, which results in the production of currents at the fundamental frequency.   These new currents may be out of phase with the original current at the fundamental frequency and so tend to shift the frequency of oscillation.   Most of the evidence regarding this effect is experimental and a complete explanation has not been given (see Ref. 1, p. 82). However, it does provide an additional reason for having a minimum of oscillator excitation.

---

*See also Chapter 4 Sect. 9(ii)f and Chapter 38 Sect. 3(vi).
†See Chapter 11 Sect. 7.
‡See Chapter 4 Sect. 9(i).

SECTION 5 : METHODS OF FREQUENCY STABILIZATION

Some of the causes of oscillator frequency variations have already been discussed, and some general methods of reducing the frequency changes have been outlined. A large number of methods have been detailed in the literature for obtaining frequency stability, such as the use of negative feedback, resistance stabilization and the various methods using reactances as given by Llewellyn (Ref. 21). Since this subject is extensive, we shall content ourselves here with stating a few of the main factors which should be observed when designing an oscillator for a superheterodyne receiver. Suitable references are listed at the end of this chapter.

(A) **Mechanical stability** of all components and wiring should be as good as possible. Care is necessary to avoid mechanical vibration (such as that due to sound waves from the loudspeaker).

(B) **Circuit layout** should be such that sources of heat (e.g. valves) are kept as far away as possible from frequency determining elements. In some cases it may be feasible to place the oscillator coil and tuning capacitor under the chassis, with the oscillator valve mounted on the top of the chassis.

(C) For feedback type oscillators, such as tuned-plate and tuned grid circuits, **the windings should be closely coupled.** The feedback winding should be as small as possible and stray capacitances across it should be low.

(D) **Grid current** should be kept as low as possible consistent with stable operation.

(E) All **supply voltages** should be as constant as possible. With pentodes it may be possible to offset plate and screen current changes.

(F) All **dielectrics** in the circuit should have low temperature coefficients and be non-hygroscopic. This applies particularly to the valve socket which becomes hotter during operation than most other sections of the oscillator circuit.

(G) Efforts should be made to stabilize the electrical constants of all components by suitable **heat treatment and impregnation.**

(H) When designing the oscillator and signal circuits it will be found from the data on superheterodyne tracking given in Chapter 25 Sect. 3, that there is some choice of inductance and capacitance values (due to the distribution of trimming capacitances) even though the frequency coverage and type of tuning capacitor are pre-determined by other conditions. Where it is permissible, **the L/C ratio of the oscillator tuned circuit** should be arranged so that it is as small as possible.

(I) The **Q of the oscillator tuned circuit** should be as large as possible consistent with other circuit requirements.

(J) Better stability is possible if the valve can be connected across only part of the tuned circuit. Often this is not convenient in radio receivers because of the reduction in available oscillator voltage, but considerable improvement in stability is possible in those cases where **tapping down** can be applied.

(K) When all other precautions have been taken, **temperature compensation** should be carried out, using capacitors having negative temperature co-efficients [see Chapter 4 Sect. 9(ii)f also Chapter 38 Sect. 3(vi)].

(L) When possible causes of frequency drift are being considered it should not be overlooked that **valve interelectrode capacitances** will change during the initial warm-up period. A typical example occurs in the Hartley circuit of Fig. 24.4(B) in which the heater-cathode capacitance is directly across part of L_0 ; in this case connecting one side of the heater directly to cathode and adding a suitable r-f choke in series with the other heater lead will minimize the trouble. This arrangement is also helpful in reducing microphonics caused by heater-cathode capacitance variations.

(M) Judicious **location of negative temperature co-efficient capacitors** is helpful. For example if part of the compensating capacitance is located at the valve socket terminals, short-time drift due to the valve and socket warm-up period can often be very appreciably reduced.

(N) The values of R_G and C_G should be selected carefully to avoid " **squegging** " (see Sect. 6). This should be checked with a large number of representative valves of the same type.

(O) Selection of the most suitable **valve type** has received some discussion in previous sections. Usually the choice is rather limited in receiver applications.

(P) **Electron coupling** to the load circuit is often helpful, and is used in many applications. Most frequency changers employ this principle.

(Q) **Harmonic operation of the oscillator** may be advantageous in some cases, since better frequency stability is sometimes possible when the fundamental frequency of operation is reduced. The disadvantage of this arrangement is the greatly increased possibility of spurious frequency combination.

The majority of receivers use the oscillator fundamental frequency, and this frequency is generally (but not always) higher than the received signal frequency. This includes all broadcast bands up to 108 Mc/s.

(R) **Increased spacing between the plates of variable capacitors** is helpful in obtaining good oscillator stability, as the effects of small variations in the position of plates, during operation, is reduced. The opportunity to make use of this occurs, for example, with the variable capacitors used in F-M receivers, where the capacitance range may be approximately 5-20 $\mu\mu$F. In this case it is often possible to use double spacing between the plates of the oscillator tuning capacitor.

SECTION 6 : UNSTABLE OSCILLATION

Unstable oscillation can be due to a number of causes, some of which have been mentioned in Sections 4 and 5.

Flutter has received some consideration and the cure is generally to use a separate series resistor from the power supply to provide the B+ for the oscillator plate (or screen in some converters), by-passed by a large capacitor (say an 8 μF electrolytic).

Squegging (see Refs. 5 and 8) is caused by excessive oscillator-grid voltage amplitude and incorrect proportioning of the values for the grid resistance and capacitance. The effect manifests itself as a variation in oscillator output which changes at an audio or ultrasonic rate. The only precautions usually required are to select suitable values for R_g and C_g and to adjust the oscillator grid current to the lowest suitable value. Most valve manuals suggest that values for C_g should lie between 20 to 100 $\mu\mu$F depending on the operating frequency. Small values of the product $R_g C_g$ are called for at the higher frequencies. The effect should be carefully checked with several valves of the same type. In some cases " squegging " occurs at the high frequency end of the tuning range only, and this often calls for the use of a series grid " stopping " resistor, or a parallel resistor shunted across the oscillator tuned circuit ; these resistors provide greater damping at the higher frequencies and so serve to equalize the oscillator grid voltage across the tuning range.

Other miscellaneous effects which are of interest are :

(A) Hum causing amplitude or frequency modulation of the local oscillator in a receiver.

In A-M receivers the selectivity of the i-f amplifier can cause the frequency modulation to give rise to amplitude modulation (superimposed on the desired carrier) and the hum, after detection, appears in the receiver output, in addition to the hum due to amplitude modulation of the oscillator. With F-M receivers the hum provides additional frequency modulation of the carrier. The cure is generally to improve the power supply filtering.

(B) Frequency modulation, at an audio frequency rate, caused by vibration of the oscillator tuning capacitor plates, or some other part of the oscillator circuit, by sound waves from the loud speaker (often called " microphonics "). This effect is particularly troublesome at the higher carrier frequencies. The natural period of the tuning capacitor is important and any method of altering this often provides a cure ;

this frequency should not coincide with the bass resonant frequency of the loud speaker. Often it is necessary to mount the capacitor on rubber and provide flexible leads to the dial, etc.

(C) Microphonics caused by heater-cathode capacitance variations. This effect is minimized by connecting one side of the heater directly to cathode, where this is possible, and using a r-f choke in the other heater lead to avoid shunting the tuned circuit with the heater circuit, e.g. see Fig. 24.6 where the result of omitting a r-f choke would be to short out part of the tuning inductance.

Microphony in superhet. oscillators is covered in Ref. 32.

SECTION 7 : PARASITIC OSCILLATIONS

" Parasitic oscillation " is the name given to any undesired oscillation in a circuit. Tuned circuits will always have at least one additional resonance point determined by the leads and stray capacitances. Usually in radio-frequency circuits these parasitic oscillations have very high frequencies, but they become troublesome and lead to " dead " spots and large fluctuations in oscillator amplitude if sufficient care is not taken with the arrangement of leads and components. Audio frequency circuits are also liable to this form of oscillation which in this case is generally apparent as distortion.

The usual cure for these troubles is the use of " stopping " resistors (or neutralization in some converter circuits) in series with one or more of the valve electrodes concerned. Care is necessary, however, as the presence of the resistors may adversely affect the performance of the stage concerned.

Carelessness in shunting large capacitors with small ones may result in an undesired resonance. The leads provide the inductance and the circuit may be series resonant in the working range of frequencies. Band-switching in all-wave receivers needs to be carried out carefully if an undesired resonance is not to appear in a tuning range.

For a further discussion of parasitic oscillations the reader is referred to Ref. 8 (Chapter 12, p. 264) and Ref. 5 (Chapter 6, p. 269). Some useful data are also given in Ref. 9, p. 9.

SECTION 8 : METHODS USED IN PRACTICAL DESIGN

In the design of oscillator circuits for radio receivers many of the component values required are specified by other circuit considerations. This is readily seen from the procedure given in Chapter 25 Sect. 3 (in connection with tracking) for determining the value of inductance, and the tuning, trimming and padding capacitance values for the oscillator circuit.

In most cases, once a suitable valve has been selected (the choice here is usually restricted), details of the required grid-leak resistance, oscillator grid current (which in conjunction with R_g determines the grid voltage) electrode voltages etc. are directly available from the valve data sheets. The type of circuit is usually restricted to the tuned-plate, tuned-grid, or the Hartley for frequencies up to about 50 Mc/s and to the Hartley and Colpitts for frequencies above this. It is required of the receiver designer to reproduce, if possible, the specified operating conditions (although this is often a rather tricky job) together with such modifications as he deems necessary.

For the feedback type of oscillator a useful " rule of thumb " is to make the number of feedback turns about half to one third of the total on the main winding ; correct amplitude is obtained by selecting the number of turns to give the required oscillator-grid current. The tapping point on Hartley and Colpitts circuits also approximates, very roughly, to this rule, which provides a convenient starting point for the necessary experimental work.

With the feedback type of circuit, provided that the coupling between the windings is sufficiently high, the number of turns on the feedback winding suggested by the above procedure will usually be somewhat greater than is necessary, and it is only necessary to strip off the excess turns until the required conditions are approached. Of course, this does not immediately ensure that the circuit will be satisfactory in all respects, since the oscillator normally tunes over a band of frequencies. Owing to the tolerances permitted in the manufacture of all valve types, it is necessary to check the circuit with a number of valves. These, preferably, should include samples which fall on the upper and lower limits of the permissible range of g_m allowed by the particular manufacturer, otherwise satisfactory results may not be obtained in the mass production of receivers.

Where the available operating data are given for voltages somewhat different from those required, valve conversion factors (see Chapter 2, Sect. 6) can be applied to obtain the new conditions.

Occasionally a receiver designer is faced with the problem of using a valve for which some data are available, but for which conditions for oscillator operation are not given. Several procedures are available. Exact or approximate methods of analysis can be used, just as for power oscillators (see Refs. 10, 11 and 24), but often it is possible to arrive at a suitable set of operating conditions on the basis of past experience (or by comparison with similar valve types for which the required data are given). The only important factor, as far as the use of the valve is concerned, is that the maximum ratings—including the peak plate current—should not be exceeded. Other factors previously discussed, such as keeping the oscillator grid current as low as possible and operating the valve so that harmonic generation is not excessive, suggest that conditions more nearly approaching Class B operation are called for, particularly as high efficiency with receiving type valves is not often (if ever) a design requirement in a radio receiver.

Once a suitable set of operating conditions has been obtained, experimental techniques are resorted to, exactly as before. Preliminary calculations other than the empirical methods suggested, as to tapping points, number of feedback turns etc., can be made if so desired, but the extra trouble involved is hardly worth the effort.

Methods for calculating the inductance required for feedback windings have been suggested (see Ref. 8, page 117) but generally the results give values which are too low. For this reason the experimental approach is generally the best procedure since it must be resorted to finally, in any case, before a design is completed.

SECTION 9 : BEAT FREQUENCY OSCILLATORS

The use of a beat-frequency (or heterodyne) oscillator is called for in the communications type of receiver ; for example when receiving Morse signals from a c.w. transmitter. The principles involved are fully explained in the usual text books (see Ref. 10, page 446 and Ref. 25, pages 63, 67-70). The output from the local beat oscillator is injected at a suitable point in the i-f amplifier, and when the resultant voltage is applied to the detector circuit together with the received carrier voltage, one of the currents at the detector output has a frequency which is equal to the difference between the frequency of the two applied voltages ; this difference frequency is the one heard in the receiver output.

The amplitude of the local oscillator voltage introduced into the i-f amplifier should be sufficiently large so that it is always considerably greater than the voltage, at the same point, due to the strongest signal likely to be received. If this condition is not observed there will be appreciable loss in signal-to-noise ratio, and so one of the main advantages of this method of reception will be lost. Because the amplitude of the oscillator voltage should be large, very careful screening is necessary if spurious frequency responses are to be avoided, or at least reduced to negligible proportions.

If locking between the local beat-oscillator and a very strong signal is to be avoided, very loose coupling into the i-f amplifier is called for. This often suggests an oscillator which can be electron-coupled to the load. The circuit of Fig. 24.7 is a particularly useful one for this application, and only requires a single winding for the tuning coil.

However, almost any oscillator circuit is suitable. Variation of the beat-frequency is best accomplished by variation of the beat-oscillator tuning (say a range of $\pm$ 3 Kc/s) since it should be clear that the best conditions for reception normally occur when the receiver is tuned exactly to the required carrier frequency.

The a.v.c. should not be operated by the beat oscillator as this would cause a serious loss in receiver sensitivity. For this reason a.v.c. is often disconnected when heterodyne reception of this type is being carried out.

Some receivers have the beat-oscillator voltage injected directly into the diode circuit, but it does not necessarily follow that this is the most suitable point. At the detector the incoming carrier voltage has its largest amplitude, and so the beat oscillator voltage also has to be correspondingly high. A simple solution which is helpful in some cases is to inject the oscillator output voltage into the grid circuit of the last i-f amplifier valve, which can be operated with fixed bias to provide constant gain.

Harmonic operation of the beat oscillator is sometimes helpful in obtaining improved stability and in reducing undesired interference effects.

Considerations as to stability of operation are of prime importance, and the factors previously discussed should be kept in mind when designing this circuit.

Useful data on beat oscillator circuits are given in Ref. 8 (page 128), and also in Refs. 26 and 27.

SECTION 10 : BIBLIOGRAPHY

(1) Thomas, H. A. (book) " Theory and Design of Valve Oscillators " (Bibliography of 172 references) Chapman and Hall, London, 2nd. ed. 1951.
(2) Pol van der, B. " The nonlinear theory of electric oscillations " Proc. I.R.E. 22.9 (Sept. 1934) 1051. (Includes a bibliography of 87 items).
(3) Rider, J. F. (book) " The Oscillator at Work " John F. Rider Publisher Inc., New York, 1940.
(4) Colebrook, F. M. " Valve Oscillators of Stable Frequency—A critical survey of present knowledge " H.M. Stationery Office, London (1934).
(5) Sturley, K. R. (book) " Radio Receiver Design "—Part 1 (Chapter 6) Chapman and Hall, London ; Wiley and Sons, New York (1943).
(6) Schlesinger, K. " Cathode-follower circuits " Proc. I.R.E. 33.12 (Dec. 1945) 843 (Oscillating cathode-follower).
(7) Reich, H. J. (book) " Theory and Applications of Electron Tubes " (Chapter 10) McGraw-Hill Book Co. Inc., New York and London (1944).
(8) Zepler, E. E. (book) " The Technique of Radio Design " (Chapter 4) Chapman and Hall, London ; Wiley and Sons, New York (1943).
(9) Beard, E. G. " Some notes on oscillating valve circuits " Philips Tec. Com. No. 9 (Oct. 1947) 6.
(10) Terman, F. E. (book) " Radio Engineers' Handbook " McGraw-Hill Book Co. Inc., New York and London (1st edit. 1943).
(11) Terman, F. E. (book) " Radio Engineering " McGraw-Hill Book Co. Inc., New York and London (2nd edit. 1937).
(12) Brunetti, C. " The Transitron oscillator " Proc. I.R.E. 27.2 (Feb. 1939) 88.
(13) " Negative transconductance oscillator—a useful circuit " Radiotronics No. 114 (July 1941) P. 48.
(14) Dedman, E. A. " Transitron oscillators " (letter) W.W. 49.5 (May 1943) 152 (gives circuit using 6A7. freq. limit 30-40 Mc/s.).
(15) White, S. Y. " V-H-F receiver oscillator design " Elect. 16.7 (July 1943) 96.
(16) Tucker, D. G. " The synchronisation of oscillators " Electronic Eng. (Part 1) 15.181 (March 1943) 412.
(17) Butler, F. " Cathode coupled oscillators " W.E. 21.254 (Nov. 1944) 521. Letter, M. Felix 22.256 (Jan. 1945) 14.
(18) Adler, R. " A study of locking phenomena in oscillators " Proc. I.R.E. 34.6 (June 1946) 351.
(19) Bushby, T. R. W. " Thermal frequency drift compensation " Proc. I.R.E. (Dec. 1942) 546. [Discussion 31.7 (July 1943) 385]. Also, A.W.A. Tec. Rev. 6.3 (1943) 14.
(20) " Effect of Temperature on Frequency of 6J5 Oscillator " R.C.A. Application Note No. 108 (Nov. 13th, 1940). Reprinted Radiotronics No. 110, p. 15.
(21) Llewellyn, F. B. " Constant-frequency oscillators " Proc. I.R.E. 19.12 (Dec. 1931) 2063.
(22) Dow, J. B. " A recent development in vacuum tube oscillator circuits " Proc. I.R.E. 19.12 (Dec. 1931) 2095.
(23) Herold, E. W. " Negative resistance " (gives a bibliography of 55 items) Proc. I.R.E. 23.10 (Oct. 1935) 1201.
(24) Everitt, W. L. (book) " Communication Engineering " (Chapters 17 and 18) 2nd edit. McGraw-Hill Book Co. Inc., New York and London (1937).
(25) Ratcliffe, J. A. (book) " The Physical Principles of Wireless " (5th edition) Methuen and Co. Ltd., London, 1941.
(26) " The Radio Amateur's Handbook " 24th edit. A.R.R.L., Connecticut, 1947, and later editions.
(27) " The Radio Handbook " 10th edit. Editors and Engineers, Los Angeles, 1946.
(28) Miller, J. M. " Thermal drift in superheterodyne receivers " Elect. 10.11 (Nov. 1937) 24.
(29) Keen, A. W. " Negative resistance characteristics—graphical analysis " W.E. 27.321 (June 1950) 175.
(30) Tombs, D. M. and M. F. McKenna " Amplifier with negative resistance load—measurement of stage gain." W.E. 27.321 (June 1950) 189.
(31) " Overtone crystal oscillator design " Elect. 23.11 (Nov. 1950) 88.

Additional references will be found in the Supplement commencing on page 1475.

CHAPTER 25

FREQUENCY CONVERSION AND TRACKING

SECTION 1 : THE OPERATION OF FREQUENCY CONVERTERS AND MIXERS

(i) Introduction (ii) General analysis of operation common to all types (iii) The oscillator section of converter tubes (iv) The detailed operation of the modulator or mixer section of the converter stage (v) Conclusion (vi) Appendix.

Reprinted by special permission from the article by E. W. Herold entitled " The operation of frequency converters and mixers for superheterodyne reception " in the Proceedings of the I.R.E. Vol. 30 No. 2 (February 1942) page 84.

(i) Introduction

The better modern radio receivers are almost universally designed to use the super-heterodyne circuit.   In such a circuit, the received signal frequency is heterodyned with the frequency of a local oscillator to produce a difference frequency known as the intermediate frequency.   The resultant signal is amplified by a selective, fixed-tuned amplifier before detection.   Since the heterodyne action is usually accomplished by means of a suitable vacuum tube, it is the purpose of this paper to discuss the chief similarities and differences among the tubes which might be used, as well as to explain their behaviour.

The combination of signal and local-oscillator frequencies to produce an inter-mediate frequency is a process of modulation in which one of the applied frequencies causes the amplitude of the other to vary.   Although this process was originally called heterodyne detection and, later, first detection, it is now called frequency con-version.   The portion of the radio receiver which produces conversion may, therefore, be identified as the converter.   If conversion is accomplished in a single vacuum tube which combines the functions of oscillator and modulator, this tube may logically be termed a converter tube.   When separate tubes are used for the oscillator and the modulator portions of the converter, respectively, the tube for the latter purpose is conveniently called a modulator or mixer tube.   This terminology will be used in this paper.

Although in some of the earliest superheterodynes, frequency conversion was accomplished by a triode oscillator and triode modulator (Ref. 1) other circuits used a single triode which served as both modulator and oscillator (Ref. 2).   A triode used in the latter way could, therefore, be called a converter tube.   The introduction of two-grid tubes (i.e. tetrodes) permitted a wide variety of modulator and converter arrangements which frequently gave superior performance to that possible with triodes (Ref. 3-7).

When indirectly heated cathodes became more common, conversion circuits in which the oscillator voltage was injected in the cathode circuit were used.   These circuits reduced considerably the interaction between oscillator and signal circuits which would otherwise be present. (Ref. 8). When tetrodes and pentodes became available, the use of the triode was dropped except as the local oscillator. It was not long, however, before the desirability of more complete separation of oscillator and signal circuits became evident. Multigrid converter tubes were, therefore, devised to permit this separation in a satisfactory manner, at least for the frequencies then in common use. (Refs. 9-14). In some of these it was also possible to control the conversion gain by an automatic-volume-control voltage, a decided advantage. The most satisfactory of the earlier multigrid tubes was known as the pentagrid converter, a type still widely used. A similar tube having an additional suppressor grid is used in Europe and is known as the octode.

When it became desirable to add high-frequency bands to superheterodyne receivers which also had to cover the low broadcast frequencies, the converter problem became more difficult. The highest practicable intermediate frequency appeared to be about 450 to 460 kilocycles, a value which was only about 2 per cent of the highest frequency to be received. Its use meant that the oscillator frequency separated from the signal frequency by only 2 per cent and the signal circuit, therefore, offered appreciable impedance at the oscillator frequency. A phenomenon known as " space-charge coupling," found in the pentagrid converter, indicated that signal and oscillator circuits were not separated as completely as would be desirable. (Ref. 15). In addition, the permissible frequency variations of the oscillator had to be held to less than the intermediate-frequency bandwidth, namely, 5 to 10 kilocycles ; at the highest frequency to be received, the oscillator frequency was required therefore to remain stable within 0.05 per cent. In the pentagrid converter, the most serious change in oscillator frequency occurred when the automatic-volume-control voltage was changed, and was sometimes as much as 50 kilocycles. Economic considerations have led to the use of at least a three-to-one frequency coverage for each band in the receiver. With capacitance tuning, the circuit impedance is very low at the low-frequency end of the high-frequency band so that failure to oscillate was occasionally observed in the pentagrid converter.

In Europe, where converter problems were similar, a tube known as the triode-hexode (Ref. 16) was developed to overcome some of the disadvantages of the pentagrid converter. In the pentagrid tube, the oscillator voltage is generated by, and therefore applied to, the electrodes of the assembly closest to the cathode (i.e., the *inner* electrodes). In the European form of triode-hexode, the oscillator voltage is generated by a separate small triode section mounted on a cathode common to a hexode-modulator section. The triode grid is connected internally to the third grid of the hexode section. In this way, by the application of the oscillator voltage to an *outer* grid and the signal to the inner grid of the modulator, space-charge coupling was greatly reduced and automatic-volume-control voltage could be applied to the modulator section of the tube without seriously changing the oscillator frequency. In some European types, a suppressor grid has been added so that such tubes should be called triode-heptodes.

The first American commercial development to provide improved performance over that of the pentagrid converter also utilized oscillator voltage injection on an outer grid but required a separate tube for oscillator (Ref. 17). This development, therefore, resulted in a modulator or mixer tube rather than a converter. There were many advantages accompanying the use of separate oscillator tube so that such a solution of the problem appeared to be reasonably satisfactory.

The demand arose shortly, however, for a one-tube converter system with better performance than the original pentagrid type for use in the standard all-wave receiver. A tube, the 6K8, in which one side of a rectangular cathode was used for the oscillator and the other side was used for the mixer section, was developed and made available. (Ref. 18). This tube used inner-grid oscillator injection, as with the pentagrid converter, but had greatly improved oscillator stability. Another solution, also introduced in the United States, was a triode-heptode which is an adaptation of the European triode-hexode. This type used outer-grid injection of the oscillator voltage generated by a small auxiliary triode oscillator section. A recent converter (the SA7 type) for broadcast use is designed to operate with oscillator voltage on both cathode and first-grid electrodes. (Ref. 19). This tube, in addition to having excellent performance, requires one less connecting terminal than previous converter tubes.

This paper will present an integrated picture of the operation of converter and modulator tubes. It will be shown that the general principles of modulating or mixing by placing the signal on one grid and the oscillator voltage on another, or by placing both voltages on the same grid, are the same for all types of tubes. The differences in performance among the various types particularly at high frequencies are due to a number of important secondary effects. In this paper, some of the effects such as signal-grid current at high frequencies, input impedance, space-charge coupling, feedback through interelectrode capacitances, and oscillator-frequency shift will be discussed.

(ii) General analysis of operation common to all types

A. Conversion transconductance of modulator or mixer tubes

The basic characteristic of the converter stage is its conversion transconductance, i.e., the quotient of the intermediate-frequency output current to the signal input voltage. The conversion transconductance is easily obtained by considering the modulation of the local-oscillator frequency by the signal in the tube and, as shown in another paper (Ref. 17) is determined by the transconductance of the signal electrode to the output electrode. The general analysis of a modulator, or mixer tube, is applicable to all mixers no matter how or on what electrodes the oscillator and signal voltages are introduced.

Under the assumption that the signal voltage is very small and the local oscillator voltage large, the signal-electrode transconductance may be considered as a function of the oscillator voltage only. The signal-electrode-to-plate transconductance g_m may, therefore, be considered as periodically varying at the oscillator frequency Such a periodic variation may be written as a Fourier series

$$g_m = a_0 + a_1 \cos \omega_0 t + a_2 \cos 2\omega_0 t + \ldots$$

where ω_0 is the angular frequency of the local oscillator. Use of the cosine series implies that the transconductance is a single-valued function of the oscillator electrode voltage which varies as $\cos \omega_0 t$. When a small signal, $e_s \sin \omega_s t$, is applied to the tube, the resulting alternating plate current to the first order in e_s may be written

$$i_p = g_m e_s \sin \omega_s t$$

$$= a_0 e_s \sin \omega_s t + e_s \sum_{n=1}^{\infty} a_n \sin \omega_s t \cos n\omega_0 t$$

$$= a_0 e_s \sin \omega_s t + \tfrac{1}{2} e_s \sum_{n=1}^{\infty} a_n \sin (\omega_s + n\omega_0)t$$

$$+ \tfrac{1}{2} e_s \sum_{n=1}^{\infty} a_n \sin (\omega_s - n\omega_0)t.$$

If a circuit tuned to the frequency $(\omega_s - n\omega_0)$ is inserted in the plate, the modulator tube converts the incoming signal frequency ω_s to a useful output at an angular frequency $(\omega_s - n\omega_0)$ which is called the intermediate frequency. Since n is an integer, it is evident that the intermediate frequency, in general, may be chosen to be the difference between the signal frequency and any integral multiple of the local-oscillator frequency ; this is true even though a pure sine-wave local oscillation is applied to the tube. The harmonics of the local-oscillator frequency need only be present in the time variation of the signal-electrode transconductance. The ordinary conversion transconductance is simply a special case when $n = 1$. The conversion transconductance at the nth harmonic of the local oscillator is given by

$$g_{cn} = \frac{i_{\omega_s - n\omega_0}}{e_s} = \frac{a_n}{2}.$$

Substituting the value of the Fourier coefficient a_n it is found that

$$g_{cn} = \frac{1}{2\pi} \int_0^{2\pi} g_m \cos n\omega_0 t \, d(\omega_0 t).$$

When n is set equal to unity, this expression becomes identical with the one previously derived. (Ref. 17).

Thus, the conversion transconductance is obtained by a simple Fourier analysis of the signal-electrode-to-output-electrode transconductance as a function of time.

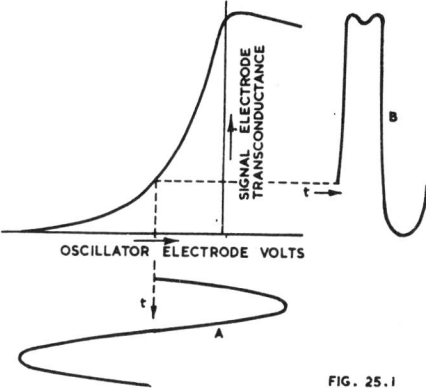

Fig. 25.1. *Signal - electrode transconductance versus oscillator-electrode voltage for a typical mixer tube. The applied oscillator voltage is shown at A and B is the resulting time variation of transconductance.*

Such an analysis is readily made from the tube characteristics directly by examination of the curve of signal-electrode transconductance versus oscillator-electrode voltage. The calculation of the conversion transconductance at the nth harmonic of the oscillator is made from this curve by assuming an applied oscillator voltage and making a Fourier analysis of the resulting curve of transconductance versus time for its nth harmonic component. The analysis is exactly similar to the one made of power output tubes, except that, in the latter case, the plate-current-versus-control-electrode-voltage curve is used. Fig. 25.1 shows a curve of signal-grid transconductance versus oscillator-electrode voltage for a typical modulator or mixer tube. In the usual case, the oscillator voltage is applied from a tuned circuit and so is closely sinusoidal in shape as at A in the figure. The resulting curve of transconductance versus time is shown at B. Any of the usual Fourier analysis methods may be used to determine the desired component of curve B. Half of this value is the conversion transconductance at the harmonic considered. Convenient formulas of sufficient accuracy for many purposes follow. Referring to Fig. 25.2a, a sine-wave oscillator voltage is assumed and a seven-point analysis is made (i.e., 30-degree intervals). The conversion transconductances g_{cn} are

$$g_{c1} = \frac{1}{12}[(g_7 - g_1) + (g_5 - g_3) + 1.73(g_6 - g_2)]$$

$$g_{c2} = \frac{1}{12}[2g_4 + \tfrac{3}{4}(g_3 + g_5 - g_6 - g_2) - (g_7 + g_1)]$$

$$g_{c3} = \frac{1}{12}[(g_7 - g_1) - 2(g_5 - g_3)].$$

The values g_1, g_2, etc., are chosen from the transconductance characteristic as indicated in Fig. 25.2a. The values computed from the above formulas are, of course, most accurate for g_{c1} and of less accuracy for g_{c2} while a value computed from the formula for g_{c3} is a very rough approximation.

Simple inspection of the formula for g_{c1}, the conversion transconductance used for conversion at the fundamental, is somewhat instructive. It is evident that highest conversion transconductance, barring negative values, as given by this formula, occurs when g_1, g_2 and g_3 are all equal to zero, and g_5, g_6, and g_7 are high. These requirements mean that sufficient oscillator voltage should be applied at the proper point to cut off the transconductance over slightly less than the cycle as pictured in Fig. 25.2b. For small oscillator voltages optimum operation requires the differences $(g_7 - g_1)$, $(g_5 - g_3)$ and $(g_6 - g_2)$ to be as large as possible ; this is equivalent to operation at the point of maximum slope. It should be noted that the minimum peak oscillator voltage required for good operation is approximately equal to one half the difference between the oscillator-electrode voltage needed for maximum signal-grid transconductance and that needed to cut off this transconductance. Thus, inspection of the curve of transconductance versus oscillator-electrode voltage gives both a measure of the fundamental conversion transconductance which will be obtained and the amount of oscillator excitation required. Conversion at a harmonic, in general, requires considerably greater oscillator excitation for maximum conversion transconductance.

In practical cases using grid-controlled tubes of the usual kind, the maximum fundamental conversion transconductance which a given tube will give can quickly be determined within 10 per cent or so by simply taking 28 per cent of the maximum signal - grid - to - plate transconductance which can be attained. For conversion at second harmonic, optimum oscillator excitation gives a conversion transconductance of half this value, while for third-harmonic conversion the value is divided by three.

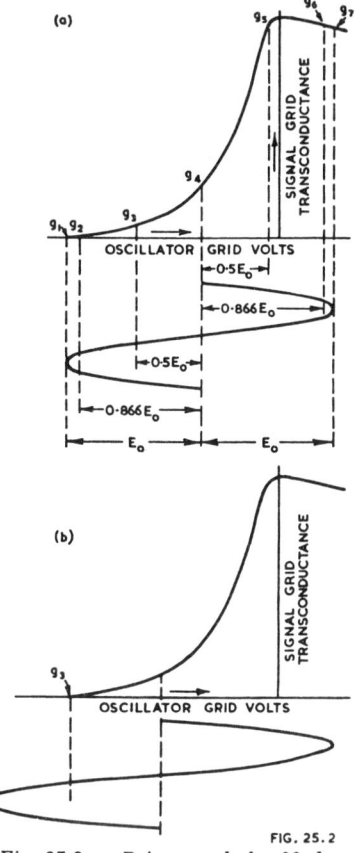

FIG. 25.2

Fig. 25.2a. Points used for 30-degree analysis of conversion transconductance. Fig. 25.2b. Oscillator amplitude and bias adjusted for high conversion transconductance at oscillator fundamental, i.e., $g_1 = g_2 = g_3 = 0$.

Although the same characteristic of all modulator or mixer tubes is used to determine the conversion transconductance, the shape of this characteristic varies between different types of mixers. This variation will be more clearly brought out in the later sections of the paper.

B. Conversion transconductance of converter tubes

In converter tubes with oscillator sections of the usual kind, the oscillator voltage is usually present on more than one electrode. Furthermore, the phase of the oscillator-control-grid voltage is opposite to that of the oscillator-anode alternating voltage, so that the two would be expected partially to demodulate each other. The transconductance curve which should be used in this case is the one in which the oscillator electrode voltages are simultaneously varied in opposite directions.

Fortunately, with most of the commonly used converter tubes such as the pentagrid, octode, triode-hexode, etc., the effect of small variations of oscillator-anode voltage on the electrode currents is so small that usually it may be neglected. Thus, the conversion transconductance of these converter tubes may be found exactly as if the tube were a modulator or mixer, only.

With the circuit of Fig. 25.3 (Refs. 19-21) a Hartley oscillator arrangement is used and oscillator-frequency voltage is present on the cathode. The effect of such a voltage is also to demodulate the electron stream through the action of the alternating cathode potential on the screen-to-cathode and signal-grid-to-cathode voltages. When a relatively high-transconductance signal grid is present, as in the figure, this demodulation is considerably greater than in the normal cathode-at-ground circuit.

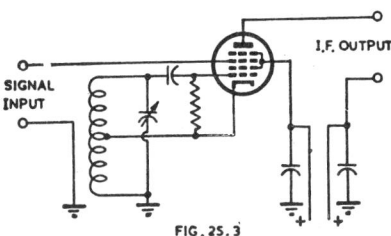

FIG. 25.3

Fig. 25.3. Converter circuit with oscillator voltage on both grid No. 1 and cathode.

In order to determine the conversion transconductance of a tube to be used in this circuit, a signal-grid transconductance curve is needed. Such a curve, however, must be taken with cathode and oscillator-grid potential varied simultaneously and in their correct ratio as determined by the ratio of cathode turns to total turns of the coil which is to be used. However, because the conversion transconductance is approximately proportional to the peak value of signal-grid transconductance, it is often sufficiently accurate to disregard the alternating-current variation of cathode potential and simply shift the signal-grid in the negative direction by the peak value of the alternating cathode voltage. If the resulting signal-grid-transconductance-versus-oscillator-grid-voltage curve is used for an analysis of conversion transconductance, the data obtained will not be far different from the actual values obtained in the circuit of Fig. 25.3 where normal (unshifted) signal-grid bias values are used.

C. Fluctuation noise

The fluctuation noise of a converter stage is frequently of considerable importance in determining the over-all noise. The magnitude of the fluctuation noise in the output of a converter or mixer tube may be found either by direct measurement using a known substitution noise source such as a saturated diode or by making use of the noise of the same tube used as an amplifier and finding the average mean-squared noise over an oscillator cycle. (Refs. 22, 23). Since these methods give values which are substantially in accord, and since the noise of many of the usual tube types under amplifier conditions is readily derived from theory (Ref. 24), the latter procedure is convenient. Thus, if $\overline{i_{p n}{}^2}$ is the mean-squared noise current in the output of the converter or mixer tube considered as an amplifier (i.e. steady direct voltages applied) the mean-squared intermediate-frequency noise is

$$\overline{i_{i-f}{}^2} = \frac{1}{2\pi} \int_0^{2\pi} \overline{i_{y n}{}^2} d(\omega t)$$

or the average of $\overline{i_{p n}{}^2}$ over an oscillator cycle. The values of $\overline{i_{p}{}^2}$ obtained from theory usually require a knowledge of the currents and transconductance of the tube and are usually proportional to these quantities. Thus, the converter-stage output noise, which is the average of $\overline{i_{p n}{}^2}$ over the oscillator cycle, is usually proportional to the average electrode currents and average transconductance when the oscillator is applied. Specific examples will be given in following sections of this paper treating typical modes of converter operation.

Tube noise is conveniently treated by use of an equivalent grid-noise-resistance concept whereby the tube noise is referred to the signal grid. The equivalent noise resistance of a converter or mixer tube is

$$R_{e q} = \frac{\overline{i_{i-f}{}^2}}{(4 k T_R \varDelta f) g_{c n}{}^2}$$

where $k = 1.37 \times 10^{-23}$, T_R is room temperature in degrees Kelvin, and Δf is the effective over-all bandwidth for noise purposes. Since Δf is invariably associated with $\overline{i_{t-f}{}^2}$, the bandwidth cancels in the determination of R_{eq} which is one of the advantages of the equivalent-resistance concept. For $T_R = 20$ degrees centigrade,

$$R_{eq} = 0.625 \times 10^{20} \frac{1}{g_{cn}{}^2} \frac{\overline{i_{t-f}{}^2}}{\Delta f}.$$

A summary of values of R_{eq} for common types of converter will be found in a preceding paper. (Ref 23).

The equivalent noise resistance R_{eq} alone does not tell the entire story as regards signal-to-noise ratio, particularly at high frequencies. For example, if the converter stage is the first stage of a receiver, and bandwidth is not a consideration, the signal energy which must be supplied by the antenna to drive it will be inversely proportional to the converter-stage input resistance. On the other hand, the noise energy of the converter or mixer tube is proportional to its equivalent noise resistance. The signal-to-noise ratio therefore, will vary with the ratio of input resistance to equivalent noise resistance, and this quantity should be as high as possible. When bandwidth is important, the input resistance should be replaced by the reciprocal of the input capacitance if it is desired to compare various converter systems for signal-to-noise ratio.

(iii) The oscillator section of converter tubes

The oscillator section of converters is often required to maintain oscillation over frequency ranges greater than three to one for circuits using capacitance tuning. Although this requirement is easily met at the lower broadcast frequencies, the effect of lower circuit impedances, transit-time phenomena in the tube, and high lead reactances combine to make the short-wave band a difficult oscillator problem. Ability to oscillate has, in the past, been measured by the oscillator transconductance at normal oscillator-anode voltage and zero bias on the oscillator grid. Recent data have shown that, in the case of pentagrid and some octode converters, an additional factor which must be considered is the phase shift of oscillator transconductance (i.e. transadmittance) due to transit-time effects*. (Ref. 26).

The ability of a converter to operate satisfactorily at high frequencies depends largely on the undesirable oscillator frequency variations produced when electrode voltages are altered. The frequency changes are mainly caused by the dependence on electrode voltages of oscillator-electrode capacitances, oscillator transconductance, and transit-time effects. There are many other causes of somewhat lesser importance. Because of the complex nature of the problem no satisfactory quantitative analysis is possible. In the case of the pentagrid and the earlier forms of octode converters there are indications that the larger part of the observed frequency shift is due to a transit-time effect. It is found that the phase of the oscillator trans-admittance and, therefore, the magnitude of the susceptive part of this transadmittance varies markedly with screen and signal-grid-bias voltages. Since the susceptive part of the transadmittance contributes to the total susceptance, the oscillation frequency is directly affected by any changes.

(iv) The detailed operation of the modulator or mixer section of the converter stage

This section will be devoted to a consideration of the modulator or mixer portion of the converter stage. This portion may be either a separate mixer tube or the modulator portion of a converter tube. Since with most of the widely used converter tubes in the more conventional circuits the alternating oscillator-anode voltage has a negligible effect on the operation of the modulator portion, only the effect of oscillator control grid need be considered. Thus the analysis of the operation of most converter tubes is substantially the same as the analysis of the same tubes used as a mixer or modulator only, just as in the treatment of conversion transconductance.

---

*M. J. O. Strutt (see Ref. 26) has published data on this phase shift in octodes. It was measured to be as high as 60 degrees at 33 megacycles.

There are three methods of operation of mixer or modulator tubes. The oscillator voltage may be put on the same grid as the signal voltage, it may be put on the inner grid (the signal applied to an outer grid), or it may be impressed on an outer grid (with the signal on the inner grid). Each of these modes of operation has characteristics which depend on the mode rather than on the tube used in it. Tubes which may be used in any one mode differ from one another mainly in the degree in which they affect these characteristics. The treatment to follow, therefore, will not necessarily deal with specific tube types : instead, the phenomenon encountered will be illustrated by the use of data taken on one or more typical tubes for each of the modes of operation.

A. Tubes with oscillator and signal voltages applied to same grid

Typical tubes used for this type of operation are triodes and pentodes. The oscillator voltage may be introduced in series with the signal voltage, coupled to the signal input circuit inductively, capacitively, and/or conductively, or it may be coupled into the cathode circuit. In all but the last case, by operating below the grid-current point, the oscillator circuit is not loaded directly by the mixer tube. When cathode injection is used, however, an effective load equal to the mean cathode conductance (slightly greater than the mean transconductance) is imposed on the oscillator circuit. The cathode injection circuit has the advantage that oscillator-frequency voltage between the signal input circuit and ground is minimized, thus reducing radiation when the converter stage is also the first stage of the receiver.

A typical transconductance-versus-bias curve for a variable-μ radio-frequency pentode is shown in Fig. 25.4. The use of the Fourier analysis for conversion transconductance at oscillator fundamental indicates that a value of approximately a quarter of the peak transconductance can be attained. Because of the tailing off of the lower end of the curve, highest conversion transconductance requires a large oscillator swing. Very nearly the maximum value is obtained, however, at an operating bias shown by the dotted line, with an oscillator peak amplitude approximately equal to the bias. With lower oscillator amplitudes, and the same fixed bias, the fundamental conversion transconductance drops in approximate proportion to the oscillator amplitude

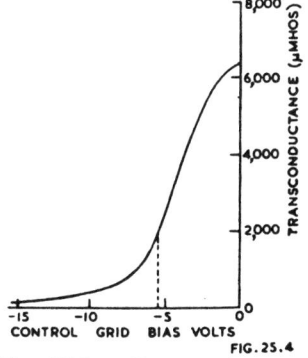

FIG.25.4

Fig. 25.4. Transconductance characteristic of a typical variable-μ, radio-frequency pentode.

Strictly speaking, when the cathode injection type of operation is used the effect of the oscillator voltage which is impressed between screen and cathode, and plate and cathode should be considered. Practically, however, there is little difference over the simpler circuit in which the oscillator voltage is impressed on the signal grid only. It is for this reason that the cathode-injection circuit is placed in the same category as those in which the oscillator voltage is actually impressed on the same electrode as the signal.

In a practical circuit the effective oscillator voltage is, of course, the oscillator voltage actually existent between grid and cathode of the tube. When the oscillator voltage is impressed in series with the signal circuit or on the cathode, this effective voltage is different from the applied oscillator voltage by the drop across the signal circuit. In the usual case, with the oscillator frequency higher than the signal frequency, the signal circuit appears capacitive at oscillator frequency. This capacitance and the grid-to-cathode capacitance, being in series, form a capacitance divider and reduce the effective oscillator voltage. The reduction would not be a serious matter if it remained a constant quantity ; but in receivers which must be tuned over an appreciable frequency range this is not the case. The result is a variation in conversion gain over the band. A number of neutralizing circuits have been described in the patent literature which are designed to reduce the oscillator-frequency voltage across the signal circuit and thus minimize the variations. (Refs. 27, 28).

Coupling of the oscillator voltage into, or across, the signal circuit is also accompanied by changes in effective oscillator voltage when the tuning is varied. These changes are not so great with pure inductive coupling as with pure capacitance coupling. In many practical cases, both couplings are present.

A method of reducing the variation of conversion gain with effective oscillator voltage in tubes in which oscillator voltage and signal are placed on the same grid, employs automatic bias. Automatic bias may be obtained either by a cathode self-bias resistor (by-passed to radio frequency) or by a high-resistance grid leak, or both. An illustration of the improvement which may be obtained in this way is shown in Fig. 25.5. Three curves of conversion transconductance, at oscillator fundamental, against effective peak oscillator volts are shown for the typical variable-μ pentode of Fig. 25.4 used as a mixer. For the curve a, a fixed bias was used at approximately an optimum point. The curve is stopped at the grid-current point because operation beyond this point is not practicable in a receiver. Curve b shows the same tube operated with a cathode self-bias resistor. This curve is also stopped at the grid-current point. Curve c shows operation with a high-resistance grid leak. It is evident that, above an oscillator voltage of about 3, curve b is somewhat flatter, and c is considerably flatter than the fixed-bias curve a. The high-resistance grid leak used for c may be made a part of the automatic-volume-control filter but care must be taken that its value is considerably higher than the resistance in the automatic-volume-control circuit which is common to other tubes in the receiver. If this is not done, all the tubes will be biased down with large oscillator swings. When a high-resistance leak is used, the automatic-volume-control action does not begin in the mixer tube until the automatic-volume-control bias has exceeded the peak oscillator voltage. Because of the high resistance of the leak, the signal circuit is not loaded appreciably by the mixer tube. In a practical case, precautions must be taken that a pentode in the converter stage is not operated at excessive currents when accidental failure of the oscillator reduces the bias. A series dropping resistor in the screen-grid supply will prevent such overload. When a series screen resistor is used, the curve of conversion transconductance versus oscillator voltage is even flatter than the best of the curves shown in Fig. 25.5. Series screen operation, therefore, is highly desirable (Ref. 23).

One of the effects of feedback through interelectrode capacitance in vacuum tubes is a severe loading of the input circuit when an inductance is present in the cathode circuit. Thus, in mixers using cathode injection, the signal circuit is frequently heavily damped since the oscillator circuit is inductive at signal frequency in the usual case. The feedback occurs through the grid-to-cathode capacitance and can be neutralized to some extent by a split cathode coil with a neutralizing capacitance. (Ref. 28). Such neutralization also minimizes the voltage drop of oscillator frequency across the signal circuit.

Loading of the signal circuit by feedback from the plate circuit of modulators or mixers may also be serious when the signal-grid-to-plate capacitance is appreciable. This is especially true when a low capacitance intermediate-frequency circuit, which presents a comparatively high capacitive reactance at signal frequency, is used, as in wide-band intermediate-frequency circuits. The grid-plate capacitance of radio-frequency pentodes is usually small enough so that the effect is negligible in these tubes. In triodes, however, feedback from the intermediate-frequency circuit may be serious and the grid-plate capacitance should be minimized in tube and circuit design. Although neutralization is a possible solution to the plate feedback, a more promising solution is the use of a specially designed intermediate-frequency circuit which offers a low impedance at signal frequency by the equivalent of series tuning and yet causes little or no sacrifice in intermediate-frequency performance.

At high frequencies, the converter stage exhibits phenomena not usually observable at low frequencies. One group of phenomena is caused not by the high operating frequency, per se, but rather by a high ratio of operating frequency to intermediate

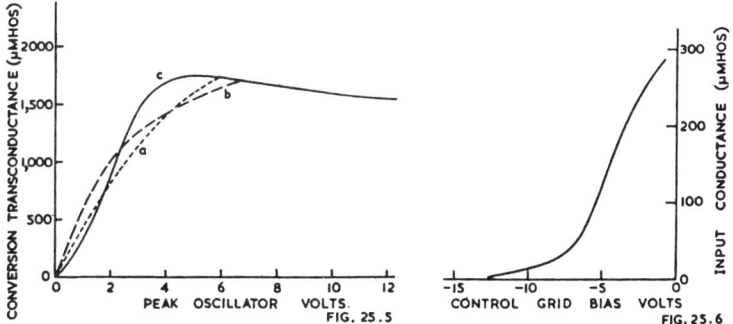

Fig. 25.5. *Conversion transconductance of a typical variable-μ, radio-frequency pentode. Oscillator and signal voltages both applied to grid No. 1.a, fixed-bias operation ; b, cathode resistor used to obtain bias ; c, bias obtained by means of a high-resistance grid leak.*
Fig. 25.6. *Input conductance of a typical variable-μ, radio-frequency pentode, at 60 megacycles.*

frequency (i.e., a small separation between signal and oscillator frequencies). Among these phenomena may be listed pull-in and inter-locking between oscillator and signal circuits and poor image response. In mixers in which oscillator and signal are impressed on the same grid, the first of these effects is usually pronounced because of the close coupling between the oscillator and signal circuits. It can be reduced by special coupling from the local oscillator at an increase in the complexity of the circuit.

Other phenomena, which are due to the high operating frequency, occur in mixers irrespective of the intermediate-frequency. The most important of these are those caused by transit-time effects in the tube and by finite inductances and mutual inductances in the leads to the tube. When the oscillator and signal are impressed on the grid of a mixer, the effects are not dissimilar to those in the same tube used as an amplifier. So far as the signal is concerned, the operation is similar to that of an amplifier whose plate current and transconductance are periodically varied at another frequency (that of the oscillator). The effects at signal frequency must, therefore, be integrated or averaged over the oscillator cycle. The input conductance at 60 megacycles of the typical radio-frequency pentode used for Figs. 25.4 and Fig. 25.5 as a function of control-grid bias is shown in Fig. 25.6. The integrated or net loading as a function of oscillator amplitude, when the tube is used as a mixer at this frequency, is given in Fig. 25.7, both with fixed-bias operation and with the bias obtained by a grid leak and condenser. The conductance for all other frequencies may be calculated by remembering that the input conductance increases with the square of the frequency. The data given do not hold for cathode injection because of the loading added by feedback, as previously discussed.

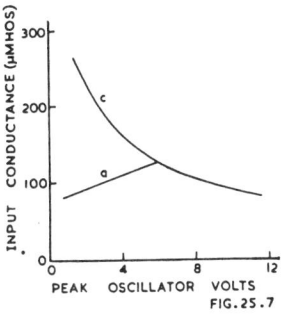

Fig. 25.7. *Input conductance of a typical pentode when used as a mixer at 60 megacycles, a, fixed-bias operation ; c, bias obtained by means of a high-resistance grid leak.*

When automatic volume control is used on the modulator tube, an important effect in some circuits is the change in input capacitance and input loading with bias. This is especially true when low-capacitance circuits are in use, as with a wideband amplifier. With tubes having oscillator and signal voltages on the same grid, because of the integrating action of the oscillator voltage, the changes are not so pronounced as with the same tube used as amplifier. A small, unbypassed cathode resistor may be used with an amplifier tube (Refs. 29, 30) to reduce the variations ; it should give a similar improvement with the modulator.

The question of tube noise (i.e., shot-effect fluctuations) is important in a mixer, or modulator, especially when this tube is the first tube in a receiver.   There is little doubt that triode or pentode mixers, in which signal and oscillator voltages are impressed on the control grid, give the highest signal-to-noise ratio of any of the commonly used types of mixers.   The reason for this has been made clear by recent studies of tube noise. (Ref. 24).   It is now well established that tube noise is the combined result of shot noise in the cathode current which is damped by space charge to a low value and additional fluctuations in the plate current caused by random variations in primary current distribution between the various positive electrodes. Thus, in general, tubes with the smallest current to positive electrodes other than the plate have the lowest noise.   It is seen that the tetrode or pentode modulator, with a primary screen current of 25 per cent or less of the total current, is inherently lower in noise than the more complex modulators in which the current to positive electrodes other than the plate usually exceeds 60 per cent of the total current.   The triode, of course, has the lowest noise assuming an equivalent tube structure.   The conversion transconductance of triode, tetrode, or pentode mixers is usually higher than that of multielectrode tubes using a similar cathode and first-grid structure.   That this is so is again largely due to the lower value of wasted current to other electrodes.

The noise of triodes and pentodes used as mixers in the converter stage is conveniently expressed in terms of an equivalent noise resistance R_{eq} as mentioned in Sect. 1(ii)C.   The noise as a mixer, of both the triode and the pentode, may be expressed in one formula based on the now well-understood amplifier noise relations. (Ref. 24).   The equivalent noise resistance of the triode is obtained simply by equating the screen current to zero.   An approximate formula for equivalent noise resistance of oxide-coated-cathode tubes is

$$R_{eq} \text{ (of triode and pentode mixers)} = \frac{2.2\,\overline{g_m} + 20\,\overline{I_{c_2}}}{g_c{}^2} \cdot \frac{1}{1 + \alpha}$$

where $\overline{g_m}$ is the average control-grid-to-plate transconductance (averaged over an oscillator cycle), $\overline{I_{c_2}}$ is the average screen current, g_c is the conversion transconductance, and α is the ratio of the screen current to plate current.   Valuable additions to the above relation are given by formulas which enable a simple calculation of noise resistance from amplifier data found in any tube handbook.   These additional relations are approximations derived from typical curve shapes and are based on the maximum peak cathode current I_0 and the maximum peak cathode transconductance g_0.   The data are given in Table 1.   It has been assumed that oscillator excitation is approximately optimum.   In this table, E_{c0} is the control-grid voltage needed to cut off the plate current of the tube with the plate and screen voltages applied, and α is the ratio of screen to plate current.

TABLE 1
Mixer Noise of Triodes and Pentodes
(Oscillator and Signal both Applied to Control Grid)

Operation	Approximate Oscillator Peak Volts	Average Transconductance $\overline{g_m}$	Average Cathode Current $\overline{I_k}$	Conversion Transconductance g_c	Equivalent Noise Resistance R_{eq}
At Oscillator Fundamental	$0.7\,E_{c0}$	$\frac{0.47}{1+\alpha}g_0$	$0.35\,I_0$	$\frac{0.28}{1+\alpha}g_0$	$\frac{13}{g_0} + 90\frac{I_0}{g_0{}^2}\alpha$
At Oscillator 2nd Harmonic	$1.5\,E_{c0}$	$\frac{0.25}{1+\alpha}g_0$	$0.20\,I_0$	$\frac{0.13}{1+\alpha}g_0$	$\frac{31}{g_0} + 220\frac{I_0}{g_0{}^2}\alpha$
At Oscillator 3rd Harmonic	$4.3\,E_{c0}$	$\frac{0.15}{1+\alpha}g_0$	$0.11\,I_0$	$\frac{0.09}{1+\alpha}g_0$	$\frac{38}{g_0} + 260\frac{I_0}{g_0{}^2}\alpha$

As an example of the use of the table, suppose it is desired to find the equivalent noise resistance of a particular triode operated as a converter at the oscillator second harmonic. The local oscillator can be permitted to swing the triode mixer grid to zero bias. With a plate voltage of 180 volts and zero bias, the tube data sheet shows a transconductance, $g_0 = 2.6 \times 10^{-3}$ mho. Thus the equivalent noise resistance is $31/g_0$ or 12 000 ohms and the conversion transconductance at second harmonic is 0.13 g_0, or 340 micromhos. Since, with this plate voltage the tube cuts off at about 8 volts, a peak oscillator voltage of around 12 volts will be required.

The above table may also be used to obtain a rough estimate of the input loading of pentode or triode mixers, since the high-frequency input conductance is roughly proportional to the average transconductance $\overline{g_m}$ and to the square of the frequency. Thus, if the loading at any transconductance and frequency is known, the loading as a mixer under the conditions of the table may quickly be computed.

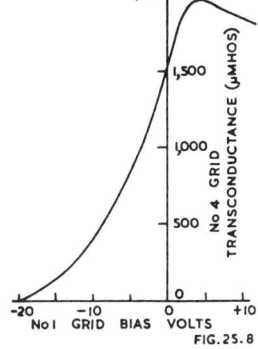

Fig. 25.8.   Signal-grid-(grid No. 4) to-plate transconductance versus oscillator-grid (grid No. 1) voltage curve of a typical mixer designed for inner-grid injection. Signal-grid bias = − 3 volts.

FIG. 25.8

B. Tubes with oscillator voltage on an inner grid, signal voltage on an outer grid

When the oscillator voltage is impressed on the grid nearest the cathode of a mixer or converter, the cathode current is varied at oscillator frequency. The signal grid, on the other hand, may be placed later in the electron stream to serve only to change the distribution of the current between the output anode and the other positive electrodes. When the two control grids are separated by a screen grid, the undesirable coupling between oscillator and signal circuits is reduced much below the value which otherwise would be found.

The signal-grid-to-plate transconductance of the inner-grid injection mixer is a function of the total current reaching the signal grid ; this current, and hence the signal-grid transconductance, will vary at oscillator frequency so that mixing becomes possible. The signal-grid transconductance as a function of oscillator-grid potential of a typical modulator of this kind is shown in Fig. 25.8. It will be observed that this characteristic is different in shape from the corresponding curve of Fig. 25.4 for the tube with oscillator and signal voltages on the same grid. The chief point of difference is that a definite peak in transconductance is found. The plate current of the tube shows a saturation at approximately the same bias as that at which the peak in transconductance occurs, indicating the formation of a partial virtual cathode. The signal grid, over the whole of these curves, is biased negatively and so draws no current. The oscillator inner grid (No. 1 grid) however, draws current at positive values of bias. This separation of signal and oscillator grids is advantageous, inasmuch as the signal circuit is not loaded even though the oscillator amplitude is sufficient to draw grid current. In fact, in the usual circuit, the oscillator grid is self-biased with a low-resistance leak and condenser and swings sufficiently far positive to attain the peak signal-grid transconductance.

The conversion transconductance of such a tube has a maximum with an oscillator swing which exceeds the point of maximum signal-grid transconductance in the one direction and which cuts off this transconductance over slightly less than half the cycle, in the other. Curves of conversion transconductance against peak oscillator voltage are shown in Fig. 25.9. Curve *a* is for fixed-bias operation of the oscillator grid, curve *b* is with a high-resistance (i.e. several megohms) grid leak and condenser for bias, and curve *c* is with the recommended value of grid leak (50 000 ohms) for this type of tube. It is seen that best operation is obtained with the lower resistance value of grid leak. With this value, the negative bias produced by rectification in the grid circuit is reduced enough to allow the oscillator grid to swing appreciably positive over part of the cycle. An incidental advantage to the use of the low-resistance leak when the tube is self-oscillating (i.e., a converter) is that undesirable relaxation oscillations are minimized.

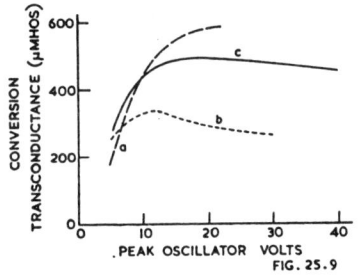

Fig. 25.9. *Conversion transconductance of a typical mixer designed for inner-grid injection of oscillator. Signal-grid bias = −3 volts. a, fixed-bias operation of oscillator grid ; b, oscillator-grid bias obtained through high-resistance grid leak ; c, oscillator-grid bias obtained through a 50,000-ohm grid leak.*

In mixers or converters in which the oscillator voltage is present on both the cathode and the oscillator grid in the same phase (e.g. Fig. 25.3) it is usually necessary to utilize a relatively sharp cut off in the design of the oscillator grid so as to cut off the cathode current when the signal grid is positive (Ref. 19). By this means, the signal grid is prevented from drawing current. At the same time, however, the high currents needed for a high peak value of signal-grid transconductance cannot be obtained without a greater positive swing of the oscillator grid than with a more open oscillator grid structure. Thus, it is clear that it is desirable to have a negative bias on the oscillator electrode which is considerably smaller than the peak oscillator voltage. For this reason, optimum results are obtained on these tubes with very low values of oscillator grid leak (e.g. 10 000 to 20 000 ohms).

The effects of feedback through the interelectrode capacitance are small in well-designed multigrid mixers and converters of the kind covered in this section. The signal-grid-to-plate capacitance is usually small enough to play no part in the operation ; even with a high *L*-to-*C* ratio in the intermediate-frequency transformer, the capacitive reactance of the intermediate-frequency circuit at signal frequency is only a very small fraction of the feedback reactance. The other interelectrode capacitance which plays some part in determining circuit performance (excluding, of course, the input and output capacitances) is the capacitance from the oscillator electrode or electrodes to the signal grid. This capacitance is a source of coupling between these two circuits. In well-designed converter or modulator tubes of the type discussed in this section, however, the coupling through the capacitance may be made small compared with another form of internal coupling known as " space-charge coupling " which will be treated later in this discussion.

Coupling between oscillator and signal circuits is of no great consequence except when an appreciable voltage of oscillator frequency is built up across the signal-grid circuit. This is not usually possible unless the signal circuit is nearly in tune with the oscillator as it is when a low ratio of intermediate frequency to signal frequency is used. The effect of oscillator-frequency voltage induced across the signal circuit depends on its phase ; the effect is usually either to increase or to decrease the relative modulation of the plate current at oscillator frequency and so to change the conversion transconductance. This action is a disadvantage, particularly when the amount of induced voltage changes when the tuning is varied, as usually occurs.

In some cases, another effect is a flow of grid current to the signal grid; this may happen when the oscillator-frequency voltage across the signal-grid circuit exceeds the bias. Grid current caused by this effect can usually be distinguished from grid current due to other causes. By-passing or short-circuiting the signal-grid circuit reduces the oscillator-frequency voltage across the signal-grid circuit to zero. Any remaining grid current must, therefore, be due to other causes.

Current to a negative signal grid of a tube operated with inner-grid oscillator injection is sometimes observed at high frequencies (e.g. over 20 megacycles) even when no impedance is present in the signal-grid circuit. This current is caused by electrons whose effective initial velocity has been increased by their finite transit time in the high-frequency alternating field around the oscillator grid. These electrons are then able to strike a signal grid which is several volts negative. The magnitude of the signal-grid current is not usually as great as with tubes applying the oscillator voltage to an outer grid* although it may prevent the use of an automatic-volume-control voltage on the tube.

An investigation of coupling effects in the pentagrid converter showed that the coupling was much larger than could be explained by interelectrode capacitance. It was furthermore discovered that the apparent coupling induced a voltage on the signal circuit in opposite phase to that induced by a capacitance from oscillator to signal grid. (Ref. 15). The coupling which occurred was due to variations in space charge in front of the signal grid at oscillator frequency. A qualitative explanation for the observed behaviour is that, when the oscillator-grid voltage is increased, the electron charge density adjacent to the signal grid is increased and electrons are repelled from the signal grid. A capacitance between the oscillator grid and the signal grid would have the opposite effect. The coupling, therefore, may be said to be approximately equivalent to a negative capacitance from the oscillator grid to the signal grid. The effect is not reversible because an increase of potential on the signal grid does not increase the electron charge density around the oscillator grid. If anything, it decreases the charge density. The equivalence to a negative capacitance must be restricted to a one-way negative capacitance and, as will be shown later, is restricted also to low-frequency operation.

In general, the use of an equivalent impedance from oscillator grid to signal grid to explain the behaviour of " space-charge coupling " is somewhat artificial. A better point of view is simply that a current is induced in the signal grid which depends on the oscillator-grid voltage. Thus, a transadmittance exists between the two electrodes analogous to the transconductance of an ordinary amplifier tube. Indeed, the effect has been used for amplification in a very similar manner to the use of the transconductance of the conventional tube. (Ref. 32, 33).

It is found that the transadmittance from the oscillator to the signal electrode Y_{mo-s} is of the form

$$Y_{mo-s} = k_1 \omega^2 + j k_2 \omega.$$

At low frequencies (i.e., $k_1 \omega^2 \ll k_2 \omega$) the transadmittance is mainly a transusceptance but, as the frequency rises, the transconductance component $k_1 \omega^2$ becomes of more and more importance, eventually exceeding the transusceptance in magnitude. The early work on " space-charge coupling " indicated that the effect was opposite to that of a capacitance connected from oscillator to signal grid and could be cancelled by the connection of such a capacitance of the correct value (Refs. 15, 34). The effect of such cancellation could be only partial, however, since only the transusceptance was balanced out by this arrangement. For complete cancellation it is also necessary to connect a conductance, the required value of which increases as the square of the frequency, between the oscillator grid and the signal grid so that the transconductance term is also balanced out. (Ref. 35, 36).

The cancellation of " space-charge coupling " may be viewed in another way. A well-known method of measuring the transadmittance of a vacuum tube is to connect an admittance from control grid to output electrode and to vary this admittance until no alternating-current output is found with a signal applied to the control grid.

*The next part of this section contains a more detailed discussion of signal-grid current in outer-grid oscillator injection tubes.

(Ref. 37).   The external admittance is then equal to the transadmittance.   In exactly the same way, the transadmittance which results from the space-charge coupling may be measured.   As a step further, if an admittance can be found which substantially equals the transadmittance at all frequencies or over the band of frequencies to be used, this admittance may be permanently connected so as to cancel the effects of space-charge coupling.   As has been previously stated, the admittance which is required is a capacitance and a conductance whose value varies as the square of the frequency.   Such an admittance is given to a first approximation by the series connection of a capacitance C and a resistance R.   Up to an angular frequency $\omega = 0.3/CR$ the admittance of this combination is substantially as desired.   At higher values of frequency, the conductance and susceptance fail to rise rapidly enough and the cancellation is less complete.   Other circuits are a better approximation to the desired admittance.   For example, the connection of a small inductance, having the value $L = 1/2CR^2$, gives a good approximation up to an angular frequency $\omega = 0.6/CR$.   The latter circuit is, therefore, effective to a frequency twice as high as the simple series arrangement of capacitance and resistance.   Inasmuch as in some cases the value of inductance needed is only a fraction of a microhenry, the inductance may conveniently be derived from proper proportioning and configuration of the circuit leads.

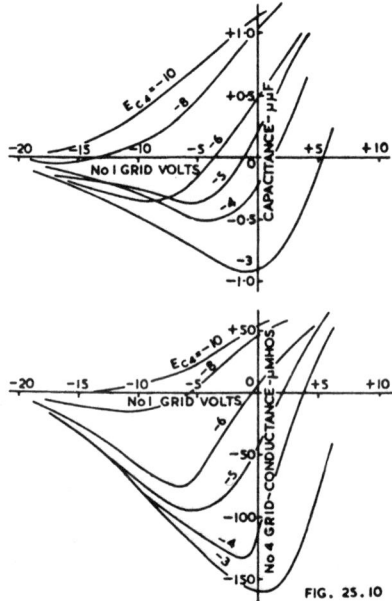

Fig. 25.10.   *Signal grid (grid No. 4) admittance of a typical mixer designed for inner-grid injection of oscillator at 31.5 megacycles.   Curves taken with no oscillator voltage applied.   Data represents electronic admittance only (i.e. " cold " values were subtracted from measured values before plotting.)*

It is of interest to note the order of magnitude of the transadmittance which is measured in the usual converter and mixer tubes. (Refs. 26, 35, 36, 38).   In the formula for Y_{mo-s} given above, k_1 is in the neighborhood of 10^{-21} and k_2 is around 10^{-12}.   Cancellation is effected by a capacitance of the order of one or two micro-microfarads and a series resistance of 500 to 1000 ohms.

The correct value of the cancelling admittance may be found experimentally by adjustment so that no oscillator voltage is present across the signal-grid circuit when the latter is tuned to the oscillator frequency.   Another method which may be used is to observe either the mixer or converter plate current or the oscillator grid current as the tuning of the signal is varied through the oscillator frequency. With proper adjustment of the cancelling admittance there will be no reaction of the signal-circuit tuning on either of these currents.

There are two disadvantages which accompany the cancellation of space-charge coupling as outlined. In the first place, the signal-grid input admittance is increased by the cancelling admittance. This point will be brought up again after discussing the input admittance. The second disadvantage is that the oscillator frequency shift with voltage changes in converter tubes may be somewhat increased by the use of this cancelling admittance. When separate oscillator and mixer tubes are used, the latter effect may be made less serious.

The next point to be considered is the input admittance of the signal grid. Signal-grid admittance curves of a typical modulator designed for use with the oscillator voltage impressed on the first grid are shown under direct-current conditions (i.e., as a function of oscillator-grid bias for several values of signal-grid bias) in Fig. 25.10. The admittance is separated into conductive and susceptive components, the latter being plotted in terms of equivalent capacitance. The admittance components of the " cold " tube (no electrons present) have been subtracted from the measured value so that the plotted results represent the admittance due to the presence of electrons only. The data shown were taken at 31.5 megacycles with a measuring signal which did not exceed 1.0 volt peak at any time. A modified Boonton Q meter was used to take the data. It should be noted that the presence of a marked conductive component of admittance is to be expected at frequencies as high as those used.

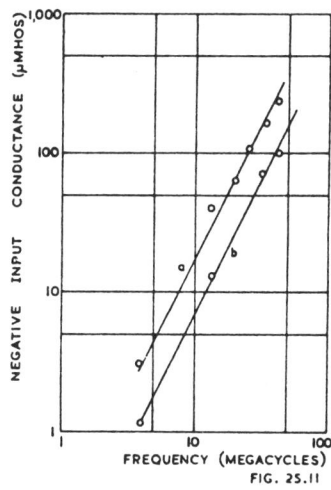

Fig. 25.11. Signal-grid (grid No. 4) conductance of a typical mixer designed for inner-grid injection of oscillator. Lines are drawn with slope of 2. Curve a taken with $E_{c_1} = 0$, $E_{c_4} = -3$ volts. Curve b taken with $E_{c_1} = -6$, $E_{c_4} = -6$ volts.

The most striking feature of the data of Fig. 25.10 is that both susceptive and conductive components are negative over a large portion of the characteristic. The Appendix discusses this feature in somewhat more detail. The measurements show that the susceptive component is analogous to a capacitance. The capacitance curves given are independent of frequency up to the highest frequency used (approximately 50 megacycles). The conductive component, on the other hand, increases as the square of the frequency also up to this frequency. The conductance is, therefore, negative even at very low frequencies although its magnitude is then very small. Thus, the conductance curves of Fig. 25.11 are valid for any frequency by multiplication of the conductance axis by the square of the ratio of the frequency considered, to the frequency used for the data (i.e., 31.5 megacycles). Data taken at various frequencies for two particular values of grid bias voltage E_{c_1} are plotted in Fig. 25.11. The square-law relation is shown to check very closely.

Fig. 25.10 should be considered remembering that the oscillator voltage is applied along the axis of abscissas. Considering an applied oscillator voltage, the admittance curves must be integrated over the oscillator cycle to find the admittance to the signal frequency. The operation is just as if the tube were an amplifier whose input admittance is periodically varied over the curve of Fig. 25.10 which corresponds to the signal-grid bias which is used. Curves of the modulator input conductance at 31.5 megacycles for various applied oscillator voltages are shown in Fig. 25.12. The oscillator-grid bias is obtained by means of the recommended value of grid leak for the tube (50 000 ohms). Curves are shown for two values of signal-grid bias voltage E_{c4}. As before, data for other frequencies are obtained by multiplying the conductance by the square of the frequency ratio.

The practical effect of the negative input admittance in a circuit is due to the conductive portion only, inasmuch as the total input capacitance remains positive in general* An improved image ratio, and somewhat greater gain to the converter signal grid over other types of modulator is to be expected when this type of oscillator injection is used. At high frequencies, when a comparatively low intermediate frequency is used, it is usually desirable to cancel the space-charge coupling of the tube in the manner previously discussed. When this cancellation is made reasonably complete by the use of a condenser and resistor combination connected from the oscillator grid to the signal grid, the losses in this admittance at signal frequency are usually sufficient to wipe out the negative input admittance. The net positive input conductance however is often less than that found with other types of mixer.

The change in signal-grid input capacitance with automatic-volume-control is small in this type of modulator, particularly with the larger values of oscillator swing because of the integrating action of the oscillator voltage.

The fluctuation noise which is found in the output of inner-grid oscillator-injection mixers and converters is not readily evaluated quantitatively. The fluctuation noise is primarily due to current-distribution fluctuations but is complicated by the possibility of a virtual cathode ahead of the signal grid. Data have been taken, however, which indicate some degree of proportionality between the mean-square noise current and the plate current. The signal-to-noise ratio for this type of modulator is, therefore, approximately proportional to the ratio of conversion transconductance to the square root of the plate current. It is considerably less than for the pentode modulator with both signal and oscillator voltages on the control grid.

The noise of the converter or mixer with oscillator on an inner grid may be expressed in terms of an equivalent grid resistance as

$$R_{eq} = \frac{20\bar{I}_b}{g_c^2} F^2$$

where $\bar{I}_b$ is the operating plate current, g_c is the conversion transconductance, and F^2 is a factor which is about 0.5 for tubes with suppressor grids and at full gain. For tubes without suppressor or for tubes whose gain is reduced by signal-grid bias, F^2 is somewhat larger and approaches unity as a maximum. With this mode of operation there is not so much value in expressions for R_{eq} based on maximum transconductance and maximum plate current because these quantities are neither available nor are they easily measured. For operation at second or third harmonics of the oscillator (assuming optimum oscillator excitation) the plate current $\bar{I}_b$ and the conversion transconductance g_c are roughly $\frac{1}{2}$ or 1/3, respectively, of their values with fundamental operation so that the equivalent noise resistance for second-harmonic and third-harmonic operation is around two and three times, respectively, of its value for fundamental operation.

C. Mixers with oscillator voltage on an outer grid, signal voltage on inner grid

With this type of mixer, the cathode current is modulated by the relatively small signal voltage which is impressed on the control grid adjacent to the cathode. The oscillator voltage, on the other hand, is impressed on a later control grid so that it

---

*It should not be forgotten that the data given do not include the " cold " susceptance and conductance of the tube. The latter is a relatively small quantity, however.

periodically alters the current distribution between anode and screen grid. The connections of signal and oscillator voltages to this type of modulator are just the reverse, therefore, of the mixer treated in the preceding section. The behaviours of the two types are also quite different although they both include internal separation of signal and oscillator electrodes through a shielding screen grid.

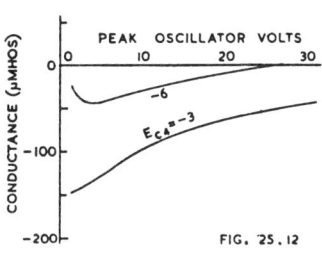 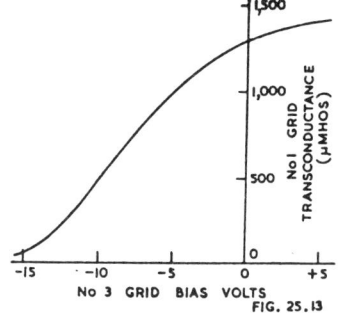

Fig. 25.12. Signal-grid (grid No. 4) conductance of a typical mixer designed for inner-grid injection of oscillator, at 31.5 megacycles. Oscillator voltage applied. Oscillator-grid bias obtained through 50,000-ohm grid leak. Electronic portion of conductance, only, plotted.

Fig. 25.13. Signal-grid (grid No. 1) transconductance versus oscillator-grid (grid No. 3) voltage of a typical mixer designed for use with outer-grid injection of oscillator. Signal-grid bias = −3 volts.

The signal-grid transconductance curve as a function of oscillator-grid voltage of a typical mixer designed for use with the oscillator on an outer grid is shown in Fig. 25.13. It differs in shape from similar curves for the other two classes of modulator in that an approximate saturation is reached around zero bias on the oscillator grid. The conversion transconductance for such a tube is, therefore, more accurately predicted from normal amplifier transconductance. In fact, in the manufacture of this type of mixer, a test of signal-grid transconductance at somewhere near the saturation point (e.g., zero bias) on the oscillator grid has been found to correlate almost exactly with the conversion transconductance. The cut off point of the curve must remain approximately fixed, of course, since this point affects the oscillator amplitude which is necessary.

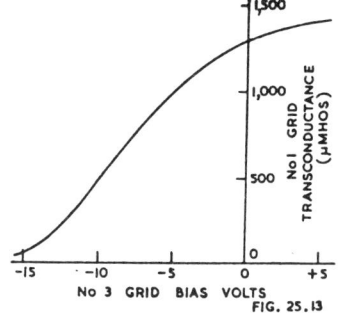

Fig. 25.14. Conversion transconduct-ance of a typical mixer designed for outer-grid injection of oscillator. Signal-grid bias, E_{c_1} = −3 volts. Curve a corresponds to fixed No. 3 grid bias, E_{c_3} = − 8 volts. Curve b corresponds to bias obtained through a 50 000-ohm grid leak.

The conversion transconductance of the typical outer-grid injection mixer tube which was used for Fig. 25.13 is shown in Fig. 25.14. Curve a which is for fixed bias on the oscillator grid is seen to be higher than curve b for which bias is obtained by a 50 000-ohm grid leak and condenser. The latter connection is most widely used, however, because of its convenience. A compromise using fixed bias together with a grid leak is most satisfactory of all. (Ref. 40). When this combination is used, the curve of conversion transconductance follows curve a of Fig. 25.14 to the intersection with curve b and then follows along the flat top of curve b.

In a well-designed mixer with the signal voltage on the grid adjacent to the cathode and the oscillator voltage on an outer grid, effects due to feedback through the inter-electrode capacitance may usually be neglected. The only effect which might be of importance in some cases is coupling of the oscillator to the signal circuit through the signal-grid-to-oscillator-grid capacitance.   In many tubes a small amount of space-charge coupling between these grids is also present and adds to the capacitance coupling (contrary to the space-charge coupling discussed in Section B which opposes the capacitance coupling in that case).   Measurements of the magnitude of the space-charge coupling for this type of modulator show that it is of the order of 1/5 to 1/10 of that present in inner-grid-injection modulators.   Coupling between oscillator and signal circuits causes a voltage of oscillator frequency to be built up across the signal input circuit.   This oscillator-frequency voltage, depending on its phase, aids or opposes the effect of the normal oscillator-grid alternating voltage.   The action is additive when the signal circuit has capacitative reactance to the oscillator frequency, as in the usual case.   When the oscillator-frequency voltage across the signal input circuit exceeds the bias, grid current is drawn to the signal grid, an undesirable occurrence.   This grid current may be distinguished from signal-grid current due to other causes by short-circuiting the signal-input circuit and noting the change in grid current.   With the majority of tubes, another cause of signal-grid current far exceeds this one in importance.   This other cause will now be discussed.

The most prominent high-frequency effect which was observed in mixers of the kind under discussion, was a direct current to the negative signal grid even when no impedance was present in this grid circuit.   This effect was investigated and found to be due to the finite time of transit of the electrons which pass through the signal grid and are repelled at the oscillator grid, returning to pass near the signal grid again. (Refs. 17, 41, 42).   When the oscillator frequency is high, the oscillator-grid potential varies an appreciable amount during the time that such electrons are in the space between screen grid and oscillator grid.   These electrons may, therefore, be accelerated in their return path more than they were decelerated in their forward path.   Thus, they may arrive at the signal grid with an additional velocity sufficient to allow them to strike a slightly negative electrode.   Some electrons may make many such trips before being collected ; moreover, in each trip their velocity is increased so that they may receive a total increase in velocity equivalent to several volts.   A rough estimate of the grid current to be expected from a given tube is given by the semi-empirical equation

$$I_{c1} = A I_k E_{osc} \omega \tau_{2-3} \epsilon^{B E_{c1}}.$$

Where A and B depend on electrode voltages and configuration, I_{c1} is the signal-grid current, E_{c1} is the signal-grid bias, I_k is the cathode current, E_{osc} is the impressed oscillator voltage on the oscillator grid, ω is the angular frequency of the oscillator, and τ_{2-3} is the electron transit time in the space between screen grid and oscillator grid.

Data on the signal-grid current of a typical mixer at 20 megacycles are shown in Fig. 25.15 where a semi-logarithmic plot is used to indicate the origin of the above equation.

The reduction of signal-grid current by operation at more negative signal-grid bias values is an obvious remedy.   When this is done, in order to prevent a reduction in conversion transconductance, the screen voltage must be raised.   A better method of reducing the undesired grid current lies in a change of tube design.   It will be shown in a later part of this discussion that the constant A and/or the transit time τ_{2-3} of the above formula may be reduced considerably by proper electrode configuration.

Another high-frequency phenomenon which is particularly noticed in outer-grid-injection mixers is the high input conductance due to transit-time effects.   The cause for this was first made evident when the change of signal-grid admittance with oscillator-grid potential was observed.   Fig. 25.16 gives data on the susceptive and conductive components of the signal-grid admittance of this type of modulator as a function of oscillator-grid bias (no oscillator voltage applied).   The data were taken at 31.5 megacycles and, as in the other input admittance curves, show the admittance components due to the presence of electrons only.   It is seen that when the No. 3 grid is made sufficiently negative the input admittance is greatly increased.   This

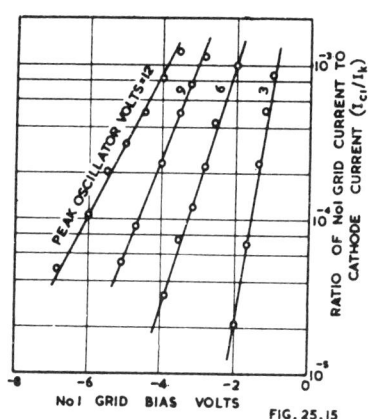

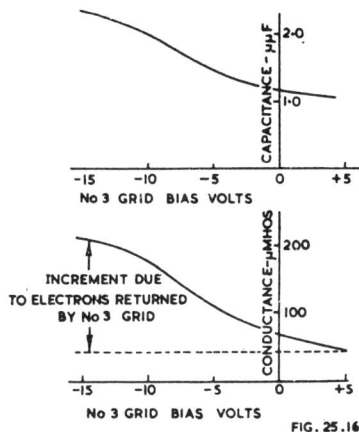

FIG. 25.15

FIG. 25.16

Fig. 25.15. Signal-grid (grid No. 1) current in a typical mixer with a 20-megacycle oscillator voltage applied to grid No. 3. $E_{c_3} = -10$ volts, $E_{c_{2\ and\ 4}} = 100$ volts, $E_b = 250$ volts.

Fig. 25.16. Signal-grid (grid No. 1) admittance of typical mixer designed for outer-grid injection of oscillator. Data taken at 31.5 megacycles with no oscillator voltage applied. $E_{c_1} = -3$ volts, $E_{c_{2\ and\ 4}} = 100$ volts, $E_b = 250$ volts.

behaviour coincides, of course, with plate-current cut off. It seems clear that the electrons which are turned back at the No. 3 grid and which again reach the signal grid are the cause of the increased admittance. · Calculations based on this explanation have been published by M. J. O. Strutt (Ref. 43) and show reasonable quantitative agreement with experiment. As in the other cases above, the upper curve of Fig. 25.16 is approximately independent of frequency while the lower one may be converted to any other frequency by multiplying the ordinates by the square of the frequency ratio.

When an oscillator voltage is applied, the No. 3 grid bias is periodically varied at oscillator frequency The net input admittance is then the average value over the oscillator cycle. Such net values of the conductance component are shown in Fig. 25.17. The frequency for these curves is 31.5 megacycles. Values for other frequencies are obtained by multiplying the ordinates by the square of the frequency ratio. Curve *a* coincides with the fixed bias condition of curve *a* of Fig. 25.14 while curve *b* corresponds to the grid-leak-and-condenser bias as in *b* of Fig. 25.14. The conductance is approximately twice as high when the tube is used as a mixer as when it is used as an amplifier. This is a serious disadvantage, particularly at very high frequencies.

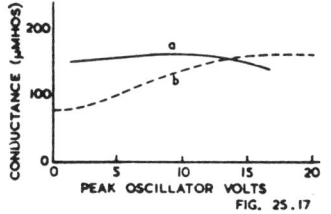

FIG. 25.17

Fig. 25.17. Signal-grid (grid No. 1) conductance of typical mixer designed for outer-grid injection of oscillator. Frequency, 31.5 megacycles, signal-grid bias, $E_{c_1} = -3$ volts. *Curve a corresponds to fixed No. 3 grid bias,* $E_{c_3} = -8$ volts. *Curve b corresponds to bias through a 50 000-ohm grid leak.*

It is thus seen that two serious disadvantages of the outer-grid-injection mixer are both due to the electrons returned by the oscillator grid which pass again to the signal-grid region. It was found possible to prevent this in a practical tube structure by causing the returning electrons to traverse a different path from the one which

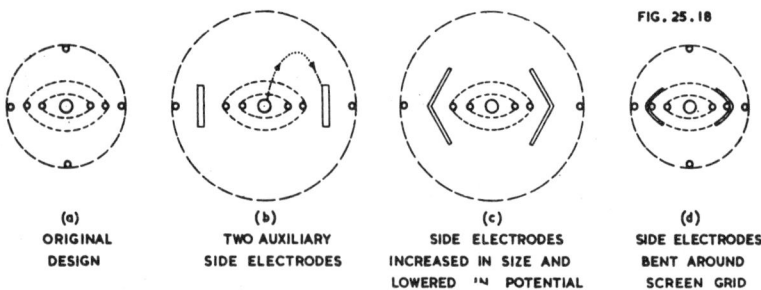

FIG. 25.18

(a)	(b)	(c)	(d)
ORIGINAL DESIGN	TWO AUXILIARY SIDE ELECTRODES	SIDE ELECTRODES INCREASED IN SIZE AND LOWERED IN POTENTIAL	SIDE ELECTRODES BENT AROUND SCREEN GRID

Fig. 25.18. Cross-sectional views of mixer designed for outer-grid-injection of oscillator. The views show only the portions of the tube inside of and including the oscillator injection grid.

they travelled in the forward direction*. (Ref. 45). The progressive steps towards an improvement of this kind are illustrated in Fig. 25.18 where cross-sectional views of the portion inside the oscillator grid of various developmental modulators are shown. The drawing (a) shows the original design, data on which have been given in Figs. 25.15, 25.16 and 25.17. Drawing (b) of Fig. 25.18 shows a tube in which two side electrodes operated at a high positive potential were added. In a tube of this kind many of the electrons returned by the No. 3 grid (oscillator grid) travel paths similar to the dotted one shown ; they are then collected by the auxiliary electrodes and thus do not re-enter the signal grid space. Tubes constructed similarly to (b) showed a considerable improvement in the signal-grid admittance increment due to returned electrons. Construction (c) shows the next step in which the side electrodes are increased in size and operated at somewhat lower potential. Because of the undesirability of an additional electrode and lead in the tube, the construction shown at (d) was tried. In this case the auxiliary electrodes are bent over and connected electrically and mechanically to the screen grid. Curves showing the progressive reduction in the signal-grid conductance increment due to returned electrons are shown in Fig. 25.19. The curves are labelled to correspond with the drawings of Fig. 25.18.

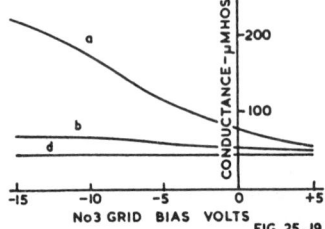

CONDUCTANCE—μMHOS

-200

-100

-15   -10   -5   0   +5
No3 GRID BIAS VOLTS
FIG. 25.19

Fig. 25.19. Signal-grid (grid No. 1) conductance of the outer-grid-injection mixers shown in Fig. 25.18. Data taken at 31.5 megacycles with no oscillator voltage applied.

It should be noted that the use of the oscillator-grid support rods in the centre of the electron streams as shown in Fig. 25.18 (d) was found to improve the performance. No change in signal-grid conductance with oscillator-grid potential could be observed with this construction.† The conductance of the tube as a modulator, therefore, was reduced to less than half of that of construction (a). At the same time, a check of signal-grid current with a high-frequency oscillator applied to the No. 3 grid showed that this current was reduced to 1/20 of that of the original construction (a). The change in construction may be looked upon as dividing the constant A in the grid-current formula previously given, by a factor of more than 20.

---

*The same principles have now been applied to inner-grid-injection mixers and converters. See references 19 and 45.
†It should be mentioned that it is also possible to construct tubes in which the signal-grid conductance decreases somewhat with increasingly negative No. 3 grid bias. This effect is caused by the inductance of the inner screen-grid lead which causes a negative conductance in the input circuit when the inner screen current is high, as at negative No. 3 grid bias values. This negative conductance cancels part of the positive conductance of the signal grid.

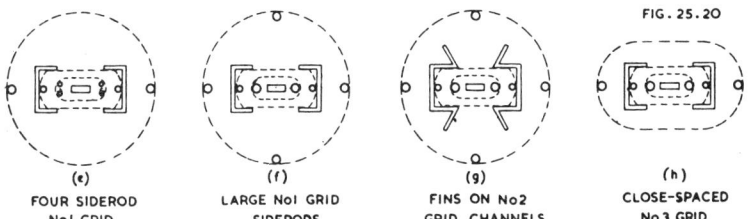

FIG. 25.20

| (e) | (f) | (g) | (h) |
| FOUR SIDEROD No1 GRID | LARGE No1 GRID SIDERODS | FINS ON No2 GRID CHANNELS | CLOSE-SPACED No3 GRID |

Fig. 25.20.   Cross-sectional views of improved mixers designed for outer-grid injection of oscillator.   The views show only the portions of the tube inside of and including the oscillator injection grid.

Another method of reducing the effect of electrons returned by the oscillator grid is to reduce the effect of electron transit time in the tube.   This may be done by reducing the spacings, particularly the screen-grid-to-oscillator-grid spacing.   This method of improving modulator performance has two disadvantages compared with the one discussed in connection with Fig. 25.18.   The reduction in spacing is accompanied by a more sloping (i.e. less steep) signal-grid transconductance versus oscillator-grid voltage curve.   This change in construction requires an increase in applied oscillator voltage to attain the same conversion transconductance.   The second disadvantage is that such a method reduces the transit time and hence, the undesirable high-frequency effects only by an amount bearing some relation to the reduction in spacing.   Since this reduction is limited in a given size of tube, the method whereby electron paths are changed is much more effective.   The method of reducing spacing, on the other hand, is extremely simple to adopt.   A combination of both methods may be most desirable from the point of view of best performance with least complexity in the tube structure.

Fig. 25.21.   Signal-grid (grid No. 1) current of the outer-grid injection mixers shown in Fig. 25.20.   Curve a corresponds to the original design (a) of Fig. 25.18 and is shown for comparison.   Data taken with a 20-megacycle oscillator voltage of 12 volts peak amplitude applied to the oscillator-grid. $E_{c_3} = -10$ volts, $E_{c_{2\ and\ 4}} = 100$ volts, $E_b = 250$ volts.

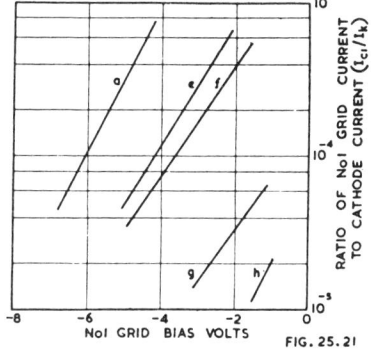

RATIO OF No1 GRID CURRENT TO CATHODE CURRENT (I_{c1}/I_k)

No1 GRID BIAS VOLTS

FIG. 25.21

In a mixer which must operate at high frequencies, it is not usually sufficient to eliminate the effects of returned electrons in order to assure adequate performance. For this reason the development of the principles shown in Fig. 25.18 was carried on simultaneously with a general programme of improving the tube.   To this end, tubes were made with somewhat reduced spacing and with a rectangular cathode and a beam-forming signal grid (i.e., one with comparatively large supports).   A number of developmental constructions are shown in Fig. 25.20.   Construction (g), it will be noted, has finlike projections on the screen-grid channel members*.   In construction (h) a reduction of spacing between screen and oscillator grids was combined with the channel construction.   The relative performance of these constructions, so far as signal-grid current is concerned is shown in Fig. 25.21.   The frequency used was 20 megacycles.   The curve for the original design (taken from Fig. 25.15)

*This construction was devised by Miss Ruth J. Erichsen who was associated with the writer during part of the development work herein described.

is included and is drawn as *a*. All four of the constructions of Fig. 25.20 were satisfactory as regards signal-grid conductance ; in every case the change in conductance as the oscillator grid was made negative was a negligible factor. Construction (h) required approximately 20 per cent more oscillator voltage than (e), (f), or (g) because of the reduction in slope of the transconductance versus No. 3 grid voltage curve which accompanied the reduced spacing between the screen and the No. 3 grid.

Outer-grid-injection mixers have the same or slightly greater signal-grid capacitance changes with automatic volume control as are found in amplifier tubes. In this respect they are inferior to inner-grid-injection converters or mixers. The use of a small unbypassed cathode resistance (Ref. 29, 30) is a help, however.

In closing this section, the subject of fluctuation noise will be considered. Experimental evidence indicates that the major portion of the noise in mixers with oscillator voltage on an outer grid is due to current-distribution fluctuations. (Ref. 24). The oscillator voltage changes the current distribution from plate to screen so that the mixer noise is given by the average of the distribution fluctuations over the oscillator cycle. In terms of the equivalent noise resistance the average has been found to be (Ref. 23),

$$R_{eq} = \frac{20\left[\overline{I_b} - \dfrac{\overline{I_b}^2}{I_a}\right]}{g_c^2}$$

where $\overline{I_b}$ is the average (i.e. the operating) plate current and $\overline{I_b}^2$ is the average of the square of the plate current over an oscillator cycle. I_a is the cathode current of the mixer section and is substantially constant over the oscillator cycle. This relation is not very useful in the form given. It is usually sufficiently accurate for most purposes to use an expression identical with that which applies to tubes with inner-grid oscillator injection, namely,

$$R_{eq} = \frac{20\overline{I_b}}{g_c^2}\, F^2,$$

where F^2 is about 0.5 for tubes with suppressor grids and somewhat higher for others. By assuming a typical tube characteristic, the noise resistance may be expressed in terms of the cathode current I_a of the mixer section and the maximum signal-grid-to-plate transconductance g_{max} as

$$R_{eq} = 120\, \frac{I_a}{(g_{max})^2}$$

for operation at oscillator fundamental. For operation at second or third harmonic of the oscillator, the noise resistance will be approximately doubled, or tripled, respectively.

(v) Conclusion

It has been shown that the principle of frequency conversion in all types of tubes and with all methods of operation may be considered as the same (i.e. as a small-percentage amplitude modulation). The differences in other characteristics between various tubes and methods of operation are so marked, however, that each application must be considered as a separate problem. The type of tube and method of operation must be intelligently chosen to meet the most important needs of the application. In making such a choice, it is frequently of assistance to prepare a table comparing types of tubes and methods of operation on the basis of performance data. An attempt has been made to draw such a comparison in a qualitative way for general cases and for a few of the important characteristics. Table 2 is the result. It must be understood, of course, that the appraisals are largely a matter of opinion based on experience and the present state of knowledge. Furthermore, in particular circuits and with particular tubes, the relative standings may sometimes be quite different. A study of the fundamentals brought out in the previous sections of this paper should help in evaluating such exceptions.

TABLE 2

Approximate Comparative Appraisals of Methods of Frequency Conversion

Desirable Characteristic	Oscillator and Signal Voltages on No. 1 Grid		Oscillator Voltage on No. 3 Grid, Signal on No. 1 Grid	Oscillator Voltage on No. 1 Grid Signal on No. 3 Grid
	Triode	Pentode	Pentode	Hexode or Heptode	Hexode or Heptode
High conversion trans-conductance	Good	Good	Fair	Fair	Fair
High plate resistance	Poor	Good	Poor	Good	Good
High signal-to-noise ratio	Good	Good	Poor	Poor	Poor
Low oscillator-signal circuit interaction and radiation	Poor	Poor	Good	Good	Fair
Low input conductance at high frequencies	Poor[1]	Fair	Poor	Poor[2]	Good
Low signal-grid current at high frequencies	Good	Good	Poor	Poor[2]	Fair
Low cost of complete converter system	Good	Fair	Fair	Poor	Good

1. Due to feedback ; may be increased to Fair by proper circuit design.
2. May be increased to Fair by special constructions as described in text.

(vi) Appendix
Discussion of negative admittance of current-limited grids

In Figs. 25.10, 11 and 12 it was seen that the electronic signal-grid (i.e. input) admittance components (i.e. the admittance due to the presence of electrons) of a mixer designed for No. 1 grid injection of the oscillator are negative over a considerable portion of the normal operating range. Figs. 25.10 and 11, however, were taken with static voltages applied and so indicate that the phenomenon is not caused by an alternating oscillator voltage but is associated with the characteristics of the tube itself.

The input admittance of negative grids in vacuum tubes is the sum of three factors : (1) the "cold" admittance, or the admittance of the tube with the electron current cut off ; (2) the admittance due to feedback from other electrodes through tube and external capacitance, etc. ; and (3) the admittance due to the presence of the electrons in the tube. The first two factors have been well known for many years although certain aspects of the second have only recently received attention. (Refs. 29, 30 and 48). The third factor, however is not so well understood although the excellent work done during the last ten years has paved the way for a complete understanding of the subject. (Ref. 49). The present discussion is concerned only with this last point, namely the admittance of negative grids due to the presence of electrons in the tube.

Early work on transit-time effects in diodes and negative-grid triodes had indicated that, at very high frequencies, the conductance became negative in certain discrete bands (i.e., at large transit angles). It was not, at first, appreciated that conditions were possible with negative-grid triodes in which the input conductance could become negative even at low frequencies (i.e., at small transit angles). Data taken on the input (No. 4 grid) conductance of pentagrid converters by W. R. Ferris of this laboratory during 1934 showed that these tubes had a negative input conductance which varied as the square of the frequency and which remained negative at low frequencies. The conductance appeared, therefore, to behave in the same way as the positive input conductance of ordinary negative-grid tubes, except for a reversal in sign. The data on the pentagrid were taken with an external oscillator voltage applied to the No. 1 grid. The work of Bakker and de Vries (Ref. 50) disclosed the possibility of a negative input conductance at small transit angles in a triode operated under current-limited conditions. They gave an experimental confirmation for a triode operated at reduced filament temperature. Data taken by the writer during 1936 on a pentagrid converter showed that the negative conductance was present in this tube even when direct voltages, only, were applied and that it was accompanied by a reduction in capacitance. A fairly complete theory of the effect was developed in unpublished work by Bernard Salzberg, formerly of this laboratory, who extended the theory of Bakker and de Vries to the more general case of multigrid tubes with negative control grids in a current-limited region. Other experimental work was done on the effect during 1936 by J. M. Miller and during the first half of 1937 by the writer. In the meantime, the papers of H. Rothe (Ref. 51), I. Runge (Refs. 52, 53), and L. C. Peterson (Ref. 54) showed that independent experimental and theoretical work had been done on the negative-admittance effect in other laboratories.

In a rough way, the negative admittance found under current-limited conditions may be explained as follows : The electron current in a tube is equal to the product of the charge density and the electron velocity. If this current is held constant, a rise in effective potential of the control electrode raises the velocity and so lowers the charge density. A reduction in charge density with increase in potential, however, results in a reduction in capacitance, provided no electrons are caught by the grid. Thus, the susceptive component of the part of the admittance due to the current through the grid, is negative. Because of the time lag due to the finite time of transit of the electrons, there is an additional component of admittance lagging the negative susceptance by 90 degrees, i.e., a negative conductance. The value of the negative conductance will be proportional to both the transit angle and to the value of the susceptance. Since both of these quantities are proportional to frequency, the negative conductance is proportional to the square of the frequency.

The general shape of the curves of Fig. 25.10 may be explained as follows : At a No. 1 grid bias of about −20 volts, the cathode current is cut off and the electronic admittance is zero. At slightly less negative values of No. 1 grid bias, the electron current is too small to build up an appreciable space charge ahead of the signal grid (No. 4 grid). The latter grid, although it exhibits some control of the plate current does not control the major portion of the current reaching it and is thus in a substantially current-limited region. Its susceptance and conductance are, therefore, negative. Higher currents increase the negative admittance until, at some value of No. 1 grid bias, the electron current is increased to the point at which a virtual cathode is formed in front of some parts of the signal grid. At these parts, the current which reaches the grid is no longer independent of this grid potential, and as a result, a positive susceptance and conductance begin to counteract the negative admittance of other portions of the grid. The admittance curves reach a minimum and for still higher currents approach and attain a positive value. The current necessary to attain the minimum admittance point is less when the signal-grid bias is made more negative so that the minima for increasingly negative No. 4 grid-bias values occur at increasingly negative No. 1 grid-bias values.

It may be noted that the signal-grid-to-plate transconductance is at a maximum in the region just to the right of the admittance minima of Fig. 25.10 (compare Fig. 25.8). The admittance of such a tube used as an amplifier remains negative, therefore, at the maximum amplification point.

SECTION 2 : CONVERTER APPLICATIONS
By E. Watkinson, A.S.T.C., A.M.I.E. (Aust.), S.M.I.R.E. (Aust.).

(*i*) *Broadcast frequencies*   (*ii*) *Short waves*   (*iii*) *Types of converters.*

(i) Broadcast frequencies
A. Spurious responses
If a sinusoidal signal is applied to the signal grid of a converter and a sinusoidal oscillator voltage to the oscillator grid, harmonics of each signal appear in the mixer plate circuit, together with sum and difference frequencies between each of the applied voltages and their harmonics. There is a component of plate current at each of these spurious response frequencies and each could be selected by a suitably tuned circuit.

Alternatively when the plate circuit is fixed tuned to the intermediate frequency, undesired combinations of signal and oscillator harmonics can produce components of plate current at or near the intermediate frequency. These components heterodyne the desired difference frequency between signal and oscillator voltages.

The most important spurious response is at the image frequency of the desired signal, i.e. removed from the oscillator frequency by an amount equal to the intermediate frequency but on the side of the oscillator frequency remote from the desired signal. Such a signal mixes with the oscillator to produce the intermediate frequency in the same way as the desired signal, and interference from it is not dependent on the characteristics of the converter. With other responses, produced by a combination of signal and oscillator harmonics, the amount of interference is dependent on the operating conditions of the converter. Thus it is desirable to reduce the oscillator amplitude as far as possible without affecting sensitivity or low-voltage operation, with the object of decreasing the magnitude of the higher order components in the transconductance-time curve (B in Fig. 25.1) of the mixer.

Spurious responses are also possible at the harmonics of the intermediate frequency but these are due to feedback from the output of the i-f amplifier rather than to the converter.

For a discussion and chart of the various possible combinations of the signal, harmonics of the signal, the oscillator voltage and harmonics of the oscillator voltage see Ref. 78.

B. The signal-frequency circuit
1. Signal-grid loading
The reason for positive and negative loading effects produced respectively by outer-grid and inner-grid oscillator injection converters is explained in Sect. 1 of this chapter.

Converters with high capacitance between signal-grid and plate may also load the input circuit by feedback from the plate circuit, particularly at the low-frequency end of the broadcast band where the signal frequency approaches the intermediate frequency.

The input loading has resistive and reactive components, and the value of the resistive component (Ref. 78) is

$$R_{g} = \frac{(G_{p} + G_{0})^{2} + B_{0}^{2}}{g_{m}B_{gp}B_{0}}$$

where $G_{p} = 1/r_{p}$ = anode slope conductance,

$\quad G_{0}$ = conductance of the external anode-load admittance at the signal frequency,

$\quad B_{0}$ = susceptance of the external anode-load admittance at the signal frequency,

$\quad g_{m}$ = mutual conductance of the signal grid with oscillator operating,

and B_{gp} = susceptance of the grid-plate capacitance at the signal frequency.

Since the resonant frequency of the plate circuit is so far from the resonant frequency of the grid circuit when tuned to any point in the broadcast band, the expression can be simplified to

$$R_{g} = \frac{B_{0}}{g_{m}B_{gp}}$$

The grid-circuit loading is thus directly proportional to the grid-plate capacitance of the converter, and it is only in types in which this is high that input loading becomes appreciable. For example the 6A8-G has a plate to signal grid capacitance of 0.26 $\mu\mu$F and with an i-f transformer tuned by 85 $\mu\mu$F to 455 Kc/s the resistive component of the loading due to feedback across the valve is about 0.25 megohm at 600 Kc/s rising to 0.45 megohm at 1000 Kc/s and 0.55 megohm at 1400 Kc/s.

The resistive component of the loading is also proportional to the reactance of the capacitor tuning the plate circuit of the converter. This capacitor should be given as large a value as possible if grid loading is the main consideration.

The reactive component of the input loading, which is always capacitive, appears in parallel with the tuning capacitor and can usually be ignored.

Some converters have much lower plate-grid capacitance, e.g. the 6J8-G has a maximum of 0.01 $\mu\mu$F, and in such cases the signal-grid circuit loading due to feedback from the plate circuit is negligible.

External coupling between oscillator and signal frequency circuits can cause another effect which may be mistaken for input circuit loading. This effect occurs in triode-hexode and similar types of converters when a tuned-plate oscillator circuit is used, the loss of gain being most noticeably at the high-frequency end of the broadcast band. In such a case, conversion sensitivity over the broadcast band is reasonably flat, but aerial (or r-f grid) sensitivity shows poor gain at the high-frequency end even although the coil, checked separately, gives constant gain over the band.

The coupling (capacitive or inductive) between the oscillator plate and the signal-grid causes an oscillator voltage to be applied to the signal grid which opposes the effects of the correctly injected oscillator voltage—see Sect. 1(iv)A of this chapter. In severe cases the coupling may even be sufficient to give an oscillator voltage on the signal grid in excess of the bias, leading to severe damping of the input circuit. Tuned-plate oscillator circuits are most likely to produce this effect because the required oscillator grid voltage is a constant whether plate tuning or grid tuning is used and with a tuned plate circuit this voltage must be developed across the smaller untuned grid winding of the oscillator coil. At the same time the oscillator plate voltage, instead of being appreciably smaller than the oscillator grid voltage as in a tuned-grid circuit, is appreciably greater.

A mechanical re-arrangement of the layout to isolate the oscillator-plate circuit from the signal-grid circuit will usually cure the trouble. In a difficult case it may be possible to arrange the wiring so that an opposite coupling between oscillator-grid and signal-grid will produce the required result.

2. Signal-circuit regeneration

It is possible to use controlled regeneration in a converter stage to give improved gain, image ratio and signal-to-noise ratio. Various possibilities are given in Ref. 65.

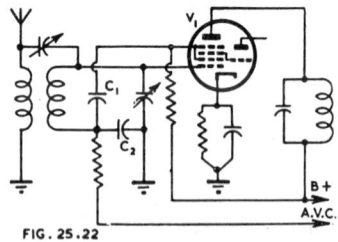

FIG. 25.22

Fig. 25.22. Screen regeneration in converter stage.

One practical and economical circuit is shown in Fig. 25.22. V1 is a triode-hexode and its 0.01 μF screen by-pass capacitor $C1$ is returned to ground through the 0.05 μF a.v.c. by-pass capacitor $C2$. The signal-frequency voltage from the screen circuit developed across $C2$ is injected into the grid circuit in the correct phase to give positive feedback at signal frequencies.

The regeneration obtained with this circuit is proportional to the reactance of $C2$ and is therefore inversely proportional to frequency.   An improvement of about 4 db in sensitivity and image ratio and 2 db in signal-to-noise ratio can be obtained at the low-frequency end of the broadcast band with sufficient stability margin for production purposes.   The regeneration is negligible at 1400 Kc/s.

In a small 3/4 valve receiver, in which the aerial coil trimmer is connected from aerial to grid for maximum gain, sensitivity is greater at 1400 Kc/s than at 600 Kc/s, thus the circuit of Fig. 25.22 is useful in minimizing the sensitivity difference between the ends of the band.

Regeneration always has the disadvantage that variations in gain e.g. with different valves, are emphasized, but this need not be serious so long as moderate amounts of regeneration are used—sufficient for instance to give an improvement of the order outlined above.   The circuit is applicable to other outer-grid injection converters, although component values may need modification.

C. Operating conditions

Recommended conditions for all types of converters are published by valve manufacturers and these are satisfactory for normal applications.   However variations are often required, perhaps to use common voltage supplies for electrodes in the converter and other valves, or perhaps to obtain say maximum signal-to-noise ratio even at the expense of sensitivity.

The usual design procedure in such cases is to supply screen, oscillator plate and bias voltages by means of variable resistor boxes and to control the oscillator amplitude by means of resistors shunted across the untuned primary.   In some cases, to be mentioned later, the effect of varying the primary impedance may need investigation.

The valve operating conditions are set approximately to those recommended and then each voltage in turn is adjusted for maximum performance.   This may need to be done more than once as variations in, say, bias voltage will affect the required screen voltage.

In a typical case, the requirement from a converter might be maximum sensitivity and as the bias voltage is reduced the gain might rise to a maximum and then fall. This is due to an increase in conversion transconductance as the bias is reduced, followed by a decrease in plate resistance which more than offsets the increasing transconductance.   However, with the bias voltage set for maximum sensitivity the screen dissipation may be excessive.   A decrease in screen voltage followed by a further decrease in bias may give a similar sensitivity with satisfactory screen dissipation.   It is important that the bias should not be reduced to a point at which some valves may draw grid current.   The contact potential of indirectly-heated types normally does not exceed -1.0 volt and the applied bias in such cases can be safely reduced to between -1.5 to -2.0 volts provided that electrode ratings are not exceeded.   In the case of battery types, the contact potential is normally zero or positive, so that it is not necessary to apply a negative bias voltage to avoid grid damping.

Increased oscillator amplitude is likely to improve sensitivity to a broad maximum followed by a slight fall.   However, as spurious responses increase rapidly with oscillator drive, the amplitude should not be any greater than necessary.   It is advisable to carry out checks at least at the limits of the tuning range to be covered.

To obtain maximum signal-to-noise ratio a similar procedure would be adopted but the final operating conditions would probably differ, e.g. in the case of a triode-hexode the screen voltage might be reduced.   On the other hand, if the converter noise is of the same order as the signal-grid circuit noise, as is possible on the broadcast band, the effect of variations in converter conditions on the signal-to-noise ratio of the receiver would not be great.

When operating voltages have been determined, resistor boxes may still be useful for obtaining a suitable screen voltage-divider circuit if a.v.c. design is at all critical. If a minimum amount of control is required on the converter, for example to avoid a rapid increase of noise as a.v.c. is applied, a single series resistor from B+ will allow the screen voltage to rise in a triode-hexode type of converter and so decrease the amount of control for a given a.v.c. voltage.   Such a circuit will also minimize cross-modulation.   However in converters without a suppressor grid the screen voltage must not approach the plate voltage because secondary emission from the plate will cause a large reduction in plate resistance and lead to damping of the first i-f transformer, so that some bleed from screen to ground may be needed.   The best compromise can readily be determined with resistor boxes.

If maximum a.v.c. control is required from a triode-hexode converter, for example when it is followed by a reflexed amplifier, and play-through must be reduced to a minimum, a large screen bleed may be needed.

Maximum a.v.c. control can be obtained in pentagrid converters by reducing the oscillator plate voltage to a minimum consistent with satisfactory oscillator performance.   For example with 20 volts bias on the signal grid, the conversion transconductance of a 6A8-G is reduced to less than one half of the original figure when the oscillator plate voltage is changed from 200 to 100 volts, although at minimum signal-grid bias the oscillator plate voltage has little effect on the conversion transconductance.

Final operating conditions should not be decided upon until the tests outlined above have been repeated with a number of valves, some of which should preferably have characteristics near the upper and lower acceptance limits for the type.

(ii) Short waves
A. Alignment
Short wave alignment of the converter stage is complicated by two factors, firstly by " pulling " of the oscillator frequency due to adjustments to the signal-grid circuit, and secondly by the fact that in many cases the range of the oscillator circuit trimmer is sufficient for it to be adjusted either to the correct response or to the image response.

1. **Pulling** : Pulling is due to coupling, either in the converter valve or externally, between the signal and oscillator circuits, and unless suitable precautions are taken it can result in faulty alignment at the high-frequency trimming point on the short-wave band.   The effect is that as the signal-frequency trimmer is adjusted towards its correct setting, thus increasing the output, the oscillator is simultaneously detuned, which decreases the output when the signal is detuned from the peak of the i-f response.   Thus the output rises to a maximum and then falls again, as is normal, but the maximum, instead of being the correct setting, is the point at which the detuning of the oscillator causes the gain to fall more rapidly than the increase due to the approach towards resonance of the signal-grid trimmer.

The true maximum can be found by alternately peaking the signal circuit and retuning the oscillator circuit to resonance (by means of the main tuning control on the receiver) or by carrying out the two operations simultaneously—i.e. by slowly moving the signal-grid trimmer and continually tuning through the signal, noting the amount of output on each occasion and continuing until a maximum is reached.   This procedure is known as " rocking " the tuning control.

If there is a noise source available with constant output over a band of frequencies at the alignment point (e.g. valve noise in a sensitive receiver, or a noise diode) it is possible to trim the signal-frequency circuit without rocking as the wide-band signal source avoids the effects of detuning.   The signal input should not be large enough to operate the a.v.c.

2. **Images** : Some of the difficulty experienced in short wave alignment at the end of the band is due to confusion between the correct response and the image.   This results from the fact that the correct signal appears to differ depending on whether the signal generator is tuned or the receiver is tuned.   For example, if a receiver with a 450 Kc/s intermediate frequency and with its oscillator on the high-frequency side of the signal frequency is correctly tuned to an 18 Mc/s signal the receiver oscillator is set to 18.45 Mc/s.   The generator can then be tuned to a frequency 450 Kc/s **higher** than the oscillator i.e. 18.9 Mc/s and the image response will be found, or alternatively with the generator still set to 18 Mc/s the receiver can be tuned to the **lower** frequency of 17.1 Mc/s when the local oscillator will be at 17.55 Mc/s, 450 Kc/s from the signal, and an image will be heard again.

Whenever there is any doubt as to which is the correct response, the frequencies should be worked out—conditions will be reversed from the example above if the oscillator is at a lower frequency than the signal—but a simple rule is that if the receiver is being tuned and the oscillator is on the **high** frequency side of the signal, then the **higher** frequency signal is the correct one.

A similar effect to an image response can also occur during signal-frequency trimmer adjustment. With the oscillator on the high-frequency side of the signal it may be found that as the signal-frequency trimming capacitance is increased, the output, with rocking, rises to a maximum, falls and rises to a second maximum, which is the correct one. On the other hand if the receiver is left aligned at the first maximum it is found that the adjustment is approximately correct for the image frequency.

This effect has no direct connection with the image response, but is due to the decrease in sensitivity which can occur when the signal-grid circuit is tuned to the oscillator frequency—owing to grid damping in a severe case, or out-of-phase oscillator injection in other cases. Thus as the signal-circuit tuning initially approaches the correct setting the gain rises, decreases again when the separation from the oscillator frequency is small, and then rises to the correct peak. In a sensitive receiver the illusion of correct alignment on the wrong peak is heightened by an increase in the level of the background noise due to the increased receiver sensitivity near the image frequency.

B. Operating conditions

Converters can be set up for operation on the short-wave band in a manner similar to that described for the broadcast band. However additional steps are also worthwhile. Particularly when harmonic mixing—see C(4) below—is being used, it is advisable to investigate the effects of different values of oscillator-grid resistor. To avoid varying two quantities at the same time the oscillator voltage should be kept as constant as possible while the value of the grid leak is altered, and readings should be taken at least at each end of the short-wave band. After the optimum grid leak value has been chosen, previously selected operating potentials should be rechecked.

When the converter is a type with inner-grid oscillator voltage injection, improved sensitivity may be obtained with the oscillator operating at a lower frequency than the signal instead of under the more usual higher frequency condition.

In Sect. 1 of this chapter it was explained that the space charge in front of the signal grid develops a voltage at oscillator frequency on the signal grid and that this voltage, depending upon its phase, increases or decreases the oscillator-frequency modulation of the electron stream. In a particular case with the oscillator at a higher frequency than the signal, the tuned circuit connected to the signal grid would present capacitive reactance to the oscillator voltage. A positive increment of oscillator-grid voltage increases the current flow and thus increases the space charge in front of the signal grid, i.e. the space charge becomes more negative. Since there is capacitive reactance between the space charge and signal grid and between signal grid and ground, a negative increment of space charge potential results in a (smaller) negative increment of signal-grid potential. Thus a positive change in potential on the oscillator grid results in a negative change in potential on the signal grid, so that the effective oscillator modulation is reduced with a consequent reduction in sensitivity.

However when the oscillator is operated at a lower frequency than that of the signal, the signal-grid circuit presents an inductive reactance to the oscillator voltage and the phase of the oscillator voltage on the signal grid is reversed. Thus the effective modulation of the electron stream is increased and sensitivity is improved.

In practice, if neutralizing is not used, an improvement in sensitivity of two or three times is sometimes obtainable by changing from high-frequency to low-frequency oscillator operation, with an increase of the image ratio in the same proportion. The improvement in sensitivity and image ratio decreases towards the low-frequency end of the short-wave band as the separation between signal and oscillator circuits (i.e the intermediate frequency) becomes a larger fraction of the oscillator frequency

For correct tracking with the oscillator on the low-frequency side of the signal, the padder is moved from the oscillator to the signal-frequency circuit(s) but this does not introduce serious complications. However, a large band coverage is more difficult with the oscillator on the low side because the oscillator circuit covers a greater frequency ratio than the signal circuit, whereas on the high side it covers a smaller ratio.

It is possible for external coupling between oscillator and signal circuits to exceed that which occurs due to internal coupling in the converter itself and it is advisable to check that the external coupling is negligible before investigating the effects of low-side and high-side oscillator operation. A valve voltmeter can be used across the circuit and the variation of the indication as the signal circuit is tuned through the oscillator frequency, with the circuit connected to and disconnected from the signal grid, shows the relative amount of coupling due to internal and external sources.

A further desirable step in the investigation of short-wave performance is the determination of the best sensitivity that can be obtained by adjustment of the magnitude and phase of the oscillator voltage on the signal grid. This can be done by using a very small capacitor, variable in order to adjust the amplitude of the oscillator voltage, and connected between either signal-grid and oscillator grid or signal-grid and oscillator plate to vary the phase of the oscillator voltage. In the absence of appreciable coupling within the receiver, some increase in sensitivity is usually obtained with a particular value of neutralizing capacitor, because maximum sensitivity is obtained not when the oscillator voltage on the signal grid is a minimum but when it is the correctly-phased maximum that can be present without signal-grid current flowing.

Thus, when using a plate tuned triode-hexode converter, it may be found experimentally that sensitivity is improved at 18 Mc/s with a 0.5 $\mu\mu$F capacitor connected between oscillator grid and signal grid. In such a case it is usually possible to obtain a capacitance of approximately the correct value without using a separate component by a suitable arrangement of the wave-change switch wiring.

However, when neutralizing is used, a measurement should be made of the effect of the neutralizing capacitance on the frequency stability of the receiver at 18 Mc/s. In general, sensitivity improvements due to oscillator voltage on the signal-grid must be offset against decreased frequency stability. The frequency shift in pentagrid converters of the 6SA7 type increases so rapidly with increased capacitance between signal-grid and oscillator grids that a neutralizing capacitance cannot be used in a normal short-wave receiver if a.v.c. is applied to the control grid.

Unless special precautions are taken, grid current variations over the short-wave band are greater than desired. The increase at the high-frequency end of the band occurs because, although the coupling between primary and secondary of the oscillator coil which is required to maintain oscillation is reduced, the actual coupling remains unchanged. A small resistor (say 25 ohm $\frac{1}{4}$ watt) wired between the tuned circuit and the oscillator grid is the usual method of obtaining reasonably uniform grid current over the band. The effective parallel circuit damping of the resistor in series with the capacitance of the oscillator grid is inversely proportional to the square of the frequency so that a large reduction in grid current can be obtained at the high-frequency end of the band without noticeably affecting the grid current at the low-frequency end.

Such a resistor is sometimes also required to prevent squegging of the local oscillator when the amplitude of oscillation is high and the time constant of the oscillator grid circuit is much greater than the time required for one cycle of the oscillator frequency. For further information on squegging see Chapter 35 Sect. 3(vi)C and Ref. 73.

C. Frequency stability

All the problems of frequency stability experienced with a separate oscillator (see Chapter 24) occur when the oscillator section of a converter is used as the local oscillator in a superheterodyne receiver and in addition there are further causes of instability due to interaction between the oscillator and mixer sections of the valve.

1. Frequency variations due to the local oscillator

Local oscillators are always self-biased triodes so that the only potential applied to the circuit from external sources is the plate-supply voltage. Two common types of frequency instability are due to this link with the rest of the receiver. The first is " flutter " due to fluctuations in the B supply voltage as a result of the output-valve current varying at audio frequencies. These fluctuations vary the oscillator frequency so that the signal is continually detuned, giving a-f signals which make the " flutter " self-sustaining. The mechanism of " flutter " is more fully described in Chapter 35 Sect. 3(vi)B and methods of overcoming it are given.

The second detuning effect for which the B supply is responsible is due to the application of a.v.c. to the controlled stages of a receiver. This reduces the B current, causes the B voltage to rise, and thus produces detuning which will vary, for instance, with the instantaneous level of a fading station.

Although in a particular case the detuning may not be sufficient to produce the symptoms described, it may still give unpleasant tuning at the high-frequency end of the short-wave band, a typical effect being that the receiver does not tune smoothly but jumps from one side of a signal to the other.

When testing for these effects, modulated and unmodulated signals of magnitudes varying from maximum to minimum should be used at the highest tuning frequency of the receiver and the volume control setting should also be varied. If equipment for measuring frequency shift is not available a useful test is to tune the receiver as accurately as possible to a large signal (perhaps 0.1 volt) with the volume control suitably retarded and then reduce the input to the smallest usable signal and turn up the volume control. If the signal is still tuned, frequency shift may be considered satisfactory. The test should be repeated with different signal inputs in case positive and negative frequency shifts should cancel over the range of inputs first selected

If flutter occurs on a large unmodulated signal there are circumstances under which it can be ignored. When the modulation is switched on it may be found that the output valve is severely overloaded, and that when the volume control is turned down to give no more than full a-f output the flutter may not occur. In such a case the flutter would not be noticed in normal use of the receiver and if a cure cannot be effected cheaply it may not be warranted.

In practice it is found that frequency stability can often be improved by the simple expedient of increasing the oscillator grid current. In one particular triode-hexode the 18 Mc/s frequency shift for a given change in control grid bias was reduced to one tenth by increasing the oscillator grid current from 200 to 600 μA. Frequency shift due to other causes, and with other types of converters can be improved in the same way, although in varying degrees. The improvement may of course have to be offset against a decrease in sensitivity and an increase in noise.

A third type of oscillator frequency instability due to the B supply is experienced in F-M receivers. If there is a hum voltage superimposed on the B supply to the oscillator plate this will produce frequency modulation of the local oscillator, and the F-M detector will convert the resultant hum modulation of the signal carrier into a-f hum.

2. Frequency variations due to the mixer

The space charge adjacent to oscillator electrodes in a converter valve can be varied by altering the potentials on mixer electrodes. This, of course, gives rise to oscillator-frequency variations. The mixer screen voltage is particularly important because one of the two grids comprising the screen is adjacent to an oscillator electrode, so that it is desirable to keep this voltage as constant as possible. This frequently necessitates a separate decoupling resistor and capacitor for the converter screen because with a common supply to converter and i-f amplifier screens the application of a.v.c to the i-f amplifier (even if the converter stage is not controlled) alters the mixer screen voltage and thus causes frequency shift. On rare occasions this shift may be used to offset another in the opposite direction due to a different effect.

When frequency shift is troublesome it is desirable to isolate each of the causes and determine its direction, as cancellation is sometimes possible even if a complete cure is not.

The effects upon oscillator frequency of varying a.v.c. voltages applied to a signal-grid electrode are more complex. With either outer-grid or inner-grid oscillator injection a space charge adjacent to the oscillator grid can be varied by altering the direct potential applied to the signal grid.   That this is not always the main source of frequency drift can be seen in some receivers with this type of instability by measuring the frequency shift for a given change in converter a.v.c. voltage with the signal generator connected firstly to the aerial terminal and then to the converter grid.   It may be found that the frequency shift in the former case is many times that with the signal generator connected to the converter-grid, and it may even be in the opposite direction.

This type of instability is due to coupling between oscillator and control-grid circuits.   In Sect. 1(iv)B of this chapter it is explained that at high frequencies it is usual for oscillator voltage to appear at the signal-grid of the converter.   Coupling between signal- and oscillator-grid circuits results in some of this voltage being returned to the oscillator circuit and, in a manner similar to the operation of a reactance-tube modulator, this results in a modification of the effective capacitance at the oscillator grid.   A constant variation of this capacitance would be of no significance, but any alteration of converter operating conditions which resulted in a variation of oscillator voltage on the signal grid, e.g. the application of a.v.c. voltages, would alter the effective reflected capacitance at the oscillator grid and thus cause oscillator frequency shift.   (Ref. 87).

With a given set of conditions the   simplest method of curing this type of frequency shift is usually to minimize the coupling between oscillator and signal grid circuits.   In general, coupling is capacitive, partly within the valve and partly external to it.   In most modern valves the internal capacitance is not troublesome and to effect a cure it may be necessary to separate short-wave aerial and oscillator coils from each other, to separate the wiring from the two grids or from the two sections of the gang condenser, and to use wave-change switch contacts on the opposite side of the wafer.   In the particular case of the 6BE6, it is desirable to reduce the external capacitance between the two grids of the valve to 1 $\mu\mu$F or less.

Another possible cure for this type of frequency shift is the use of harmonic mixing, in which the frequency difference between signal and oscillator circuits is so great that no significant amount of oscillator frequency voltage appears on the signal grid.

3. Tuned-plate operation

An analysis of tuned-grid and tuned-plate oscillators (Chapter 24 Sect. 2) shows the differences between the two types of operation.   However, it is mainly because of the structure of some types of converters that tuned-plate operation is preferable. When the oscillator section of a converter has a common electron stream with the mixer section, the oscillator-plate electrode is designed to modulate the electron stream as little as possible since modulation by the oscillator plate cancels that from the oscillator grid.   It is also desirable in equipment design to keep the oscillator plate alternating voltage as small as possible.   With a tuned-grid oscillator, the plate voltage is smaller than the grid voltage and the reverse is the case with a tuned-plate oscillator.   Accordingly, tuned-grid oscillators are always used for converters having a common electron stream for the oscillator and mixer sections.

Triode-hexode and triode-heptode converters have a separate triode for use as a local oscillator, and oscillator injection into the mixer section is carried out by a special injector grid which is in most cases internally connected to the oscillator grid.   The capacitance and coupling effects of the injector grid are thus introduced into the oscillator circuit, and if a tuned-grid oscillator is used they appear directly across the tuned circuit.

However with a tuned plate oscillator these effects are coupled to the tuned circuit from the oscillator primary winding and their effect is thus minimized.   An improvement in oscillator frequency stability, with respect to the effect of a.v.c. application to the signal grid, of five times has been measured with a typical triode-hexode on converting from a tuned-grid to a tuned-plate oscillator, while frequency shift for a given change in B supply voltage was halved under the same conditions.

4. Harmonic mixing

Many of the difficulties experienced in high-frequency applications of converters are due to the small percentage of frequency difference between the oscillator and signal-frequency circuits. It was shown in Sect. 1 of this chapter that a sinusoidal oscillator voltage applied to the injector grid of the mixer section can also give mixing at harmonics of the oscillator frequency. By taking advantage of this inherent converter characteristic it is possible to separate widely the resonant frequencies of the oscillator and signal frequency tuned circuits and thus eliminate many difficulties in high-frequency converter applications.

There are disadvantages to harmonic mixing but in some cases they are outweighed by the advantages to be obtained. Firstly there is some loss in sensitivity over a part, at least, of the tuning range and secondly there are spurious responses due to incoming signals mixing with the oscillator fundamental.

With careful design the loss in sensitivity is small. In the normal 6 to 18 Mc/s short-wave range it will probably not be greater and may be less than 2 db from 10 to 18 Mc/s. The maximum loss is usually at 6 Mc/s, and should not exceed 3 db, about 2 db or less being normal. Even when the converter is the first valve in the receiver no serious decrease in signal-to-noise ratio of the complete receiver should be experienced.

Because of a number of secondary effects which occur with harmonic mixing, this performance is better than would be expected from an investigation of maximum second harmonic conversion transconductance alone, although it presupposes that the best conditions for harmonic mixing are used. It will be found that higher grid current is necessary, particularly at the low-frequency end of the band, and a lower value of grid leak than normal is often useful in obtaining maximum sensitivity. In general, it is desirable to investigate the converter conditions carefully, as described previously, and for satisfactory performance inner-grid mixers should operate with the second harmonic of the oscillator on the low-frequency side of the signal.

The reason for this is as follows. The modulation of the electron stream by grid 1 gives a transconductance-time curve similar to Fig. 25.1B, and the space charge in front of the signal grid fluctuates at the same rate. This fluctuation, although of the same frequency as the local oscillator, has large harmonic components and since the space charge is coupled capacitively to the signal-grid circuit and this is tuned approximately to the second harmonic frequency, most of the voltage induced into the signal-grid circuit is at the frequency of the second harmonic of the oscillator. This second harmonic voltage remodulates the electron stream with a phase which depends on whether the signal frequency circuit has capacitive or inductive reactance at the second harmonic frequency so that the re-modulation aids or opposes, depending on its phase, the original second-harmonic modulation by grid 1. For inner-grid injection converters, the oscillator second-harmonic frequency should be lower than the signal frequency for aiding re-modulation.

It will be found that with harmonic mixing the same padder can be used as that required for the same frequency coverage with fundamental mixing.

Although the spurious responses which arise due to fundamental mixing are a disadvantage of harmonic operation, a tuned r-f stage ahead of the converter will reduce them to negligible proportions, not greater than −60 db in a typical case. Moreover, even when fundamental mixing is used there are responses, although smaller ones, due to signals mixing with the oscillator harmonics. Fundamental spurious responses 30 db below the signal could be taken as representative of receivers, without a r-f stage, using second-harmonic mixing so that these responses cause appreciably less interference than the normal image.

The benefits to be obtained from harmonic mixing are evident when operating, for example, the 1R5 converter on the 6 to 18 Mc/s short-wave band. Because the 1R5 has no separate oscillator-plate electrode, an oscillator circuit is often used in which the screen is connected to the oscillator primary winding to operate as the oscillator plate. A neutralizing capacitor is then needed between signal grid and oscillator grid and the determining of a suitable capacitance and its tolerances, the minimizing of coupling between signal and oscillator circuits and the stabilizing of the oscillator with reference to supply-voltage variations involve a considerable amount of work, as does the duplicating of the results in production receivers.

However, with harmonic mixing neutralizing is not required. This is because a neutralizing capacitor is used with a 1R5 to balance-out the grid 3 oscillator voltage due to internal capacitive coupling from grids 2 and 4 and not due to space charge coupling. Since the oscillator frequency voltage on grids 2 and 4 is developed across a tuned circuit the harmonic content is low, and thus with harmonic mixing the second harmonic voltage developed on grid 3 is low. Moreover, no other voltages of any magnitude appear on grid 3 because, at frequencies other than those in the vicinity of the oscillator second harmonic, the impedance between grid 3 and ground is small.

Thus a major difficulty experienced with fundamental oscillator frequency voltage on the control grid, is eliminated by harmonic operation. On the other hand the small amount of second-harmonic voltage that does appear on the grid 3 cannot be neutralized because no source consisting only of second harmonic voltages is available.

Another feature of harmonic operation of the 1R5, and similar types of converters, is that sensitivity is increased by negative input impedance at high frequencies due to a negative transconductance between the signal grid and screen and to large capacitance between these electrodes. The oscillator primary winding acts as a capacitive load at signal frequencies and feedback through the inter-electrode capacitance produces negative resistance across the signal-grid circuit.

In a normal case, the amount of regeneration is small, but if there is an unusually large impedance in the screen circuit, e.g. if a very large oscillator primary winding is used, the regeneration may become excessive.

The regeneration is not usually noticed when the oscillator fundamental is used for mixing because, as the screen circuit impedance is increased, the feedback of oscillator frequency voltage from screen to signal grid becomes excessive before noticeable signal-frequency regeneration occurs.

Another aspect of harmonic mixing, common to all types of converters, is that because the signal and oscillator circuits are tuned to widely differing frequencies, coupling between the two circuits can be ignored and variations of coupling within the valve due to changing electrode potentials have little effect on the frequency of the local oscillator. The improvement in stability during alignment is very noticeable and in most cases the need for rocking the tuning control while aligning the short-wave aerial trimmer disappears.

Frequency stability due to effects such as varying oscillator-input capacitance is not improved by harmonic oscillator operation because in general the oscillator tuning capacitance is unchanged at a given signal-frequency setting so that for a given capacitance change the frequency change is proportional to frequency. Although the oscillator operating on half the normal frequency has half the normal shift, the second harmonic will be no more stable than an oscillator fundamental on the same frequency.

(iii) Types of converters
A. Outer-grid oscillator injection
Converters using outer-grid oscillator injection always have a separate oscillator section and are characterized by relatively good oscillator-frequency stability, relative freedom from interaction between oscillator and signal circuits on high frequencies and positive input loading of the signal-grid circuit, which increases in proportion to the square of the frequency.

In the absence of undesired coupling between signal and oscillator-grid circuits, improved sensitivity will be obtained on short waves with these types by operating the oscillator at a higher frequency than the incoming signal.

(1) **Type 6J8-G** : The electrode arrangement of the triode-heptode type 6J8-G is shown in Fig. 25.23. This converter is very stable in operation which makes it useful in dual-wave receivers in spite of its low conversion transconductance. It has a high plate resistance (4 megohms) under recommended operation conditions so that the use of a high impedance first i-f transformer is more effective in increasing conversion gain than in the case of other converter types, for example the 6A8-G.

It is desirable not to apply a.v.c. to the 6J8-G on short waves if this is possible, but if a.v.c. is used a tuned-plate oscillator circuit will greatly improve the frequency stability.

(2) **Types 6AE8, X79 and X61M** : The electrode arrangements of these triode-hexodes is similar to that of Fig. 25.23 but there is no suppressor grid adjacent to the plate.   Because of this it is necessary in circuit design to avoid conditions which may lead to secondary emission from the hexode plate to the screen.

Frequency stability with respect to B voltage variations is good, although it is desirable to use plate tuning of the oscillator if a.v.c. is applied to the hexode section on short waves.   The conversion transconductances are high (approximately 750 micromhos with 2 volts bias and a screen voltage of 85) and few special precautions are required to obtain non-critical operation and good performance on the broadcast and short-wave bands.

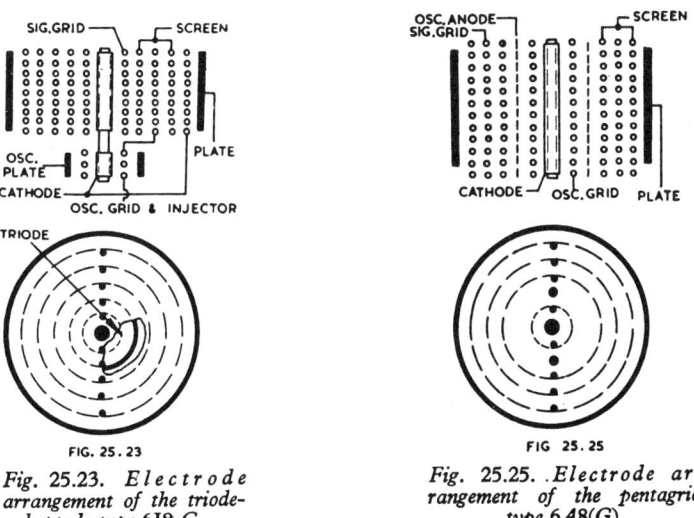

Fig. 25.23. *Electrode arrangement of the triode-heptode type 6J8-G.*

Fig. 25.25. *Electrode arrangement of the pentagrid type 6A8(G).*

The operating characteristics of most outer-grid injection triode-hexodes are similar to those of the above types.

B. Inner-grid oscillator injection

Converters using inner-grid oscillator injection, with the exception of the 6K8(G), have combined oscillator and mixer sections using a common cathode stream.   In general the oscillator frequency stability is poorer than that of outer-grid injection types and coupling between control and signal grids is greater, although modern valves are much improved in these respects.   Under maximum gain conditions the input loading of the signal grid is negative but becomes less negative and ultimately positive as the signal grid bias is increased.

To obtain maximum short-wave sensitivity without neutralizing, the oscillator should operate on the low-frequency side of the signal.

(1) **Type 6A8-G** : The structure of the pentagrid (or heptode) type 6A8-G is shown in Fig. 25.25.   It will be noticed that the oscillator plate consists only of two side rods without a normal grid winding, and that there is no suppressor grid, which leads to a low plate resistance (0.36 megohm under typical operating conditions). Nevertheless the conversion transconductance of 550 micromhos is sufficient to give reasonable conversion gain.

When used with high impedance i-f transformers the 6A8-G may introduce some grid loading at the low-frequency end of the broadcast band due to feedback (Miller

Effect) from the plate circuit to the control grid through the relatively high (0.26 $\mu\mu$F) plate-to-grid capacitance. This effect is normally only just noticeable but may be aggravated by locating the 6A8-G against the back of the signal-frequency section of the tuning condenser so that appreciable capacitance is present between the plate of the valve and the stator assembly of the gang. The effect shows up as apparently poor aerial coil gain so that the cause may not be suspected. It is not of course peculiar to the 6A8-G but the already high plate-to-grid capacitance, the size of the valve and the lack of internal or external shielding make the effect more likely to occur with this type. The effect may be minimized by external shielding of the valve.

Part of the oscillator plate (grid 2) current is due to electrons which have passed from the cathode through grids 1 and 3 and then been repelled by grid 4 back through grid 3 to grid 2 again. The oscillator characteristics and thus the oscillator frequency stability are therefore very dependent on the potentials applied to grids 2, 3 and 4 and, for short wave operation, a.v.c. should not be applied to the valve. In addition it may be necessary to decouple the oscillator-plate voltage supply with an electrolytic capacitor and even to obtain it directly from the rectifier output—suitably decoupled to eliminate hum—in order to avoid feeding a-f variations from the output valve back to the oscillator plate and thus causing flutter.

To obtain consistent performance it is desirable to keep the oscillator-plate voltage higher than the screen voltage.

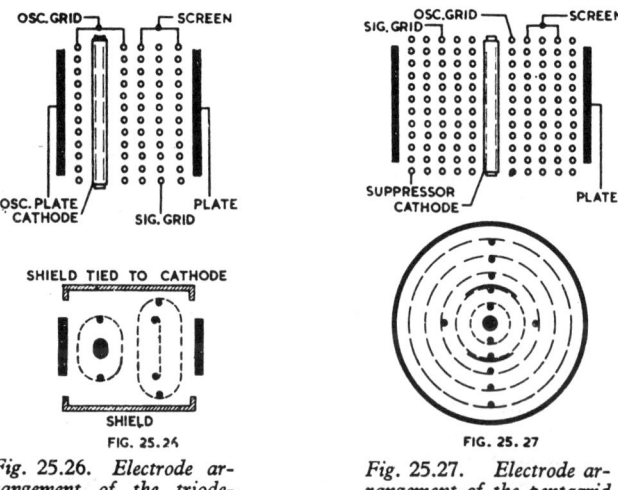

Fig. 25.26. Electrode arrangement of the triode-hexode type 6K8(G).

Fig. 25.27. Electrode arrangement of the pentagrid type 6SA7(GT).

(2) **Type 6K8-G** : The structure of the inner-grid injection triode-hexode type 6K8-G is shown in Fig 25.26. It was designed to provide more stable short-wave performance than the 6A8-G and has a higher oscillator transconductance and operates with lower oscillator excitation. Its lower conversion transconductance (350 micromhos) gives a lower conversion gain than that obtainable from the 6A8-G although the plate resistance (0.6 megohm) is higher.

A.V.C. can be applied to the 6K8-G on short-waves but the frequency stability with respect to changes in other electrode voltages is only fair.

(3) **Types 6BE6 and 6SA7(GT)** : The main point of interest in the structure of the pentagrid types 6BE6 and 6SA7(GT), Figs. 25.27, and 25.29 is that there is no electrode which functions only as an oscillator plate.

By the omission of this electrode, it is possible to obtain a relatively high oscillator transconductance, without greatly increasing the total cathode current.

The oscillator circuits employed with these types have certain unconventional features and Fig. 25.28 may be taken as typical. The lack of a separate oscillator plate results in certain disadvantages in short-wave performance but because a voltage supply is required for one less electrode than usual the cost of components required is a minimum.   The main difference between the 6BE6 and the 6SA7(GT) is the considerably higher oscillator transconductance of the 6BE6, due to the use of a formed No. 1 grid in this type.

When the circuit of Fig. 25.28 is used, the oscillator provides peak plate current at the instant when the oscillating voltage (E_k) on the cathode (with respect to ground)

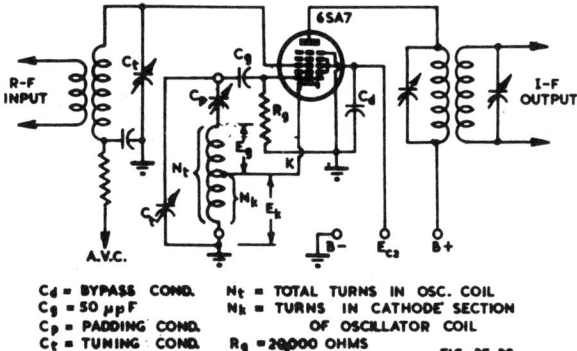

C_d = BYPASS COND. N_t = TOTAL TURNS IN OSC. COIL
C_g = 50 $\mu\mu$F N_k = TURNS IN CATHODE SECTION
C_p = PADDING COND.   OF OSCILLATOR COIL
C_t = TUNING COND. R_g = 20,000 OHMS   FIG. 25.28

Fig. 25.28.   Typical self-excited converter circuit for type 6SA7(GT).

and the oscillating voltage (E_g) on the No. 1 grid are at their peak positive values. For maximum conversion transconductance this peak value of plate current should be as large as possible.   The effect on plate current of the positive voltage on the cathode is approximately the same as would be produced by an equal voltage of negative sign applied to the signal grid.   Hence the amplitude of oscillator voltage on the cathode limits the peak plate current.   This amplitude should therefore be small.

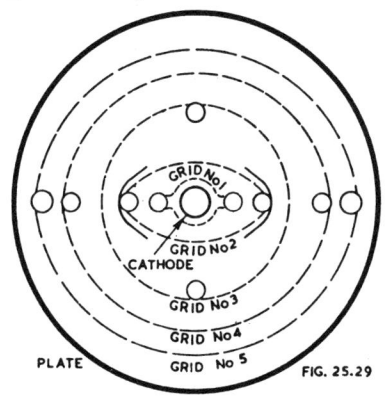

Fig. 25.29.   Electrode arrangement of pentagrid type 6BE6.

During the negative portion of an oscillation cycle the cathode may swing more negative than the signal grid.   If this occurs, positive signal-grid current will flow unless the oscillator grid is sufficiently negative to cut-off the cathode current.   This signal-grid current flowing through the signal-grid circuit resistance will develop a negative bias on the signal grid and may also cause a negative bias to be applied to the i-f and r-f stages through the a.v.c. system.   As a result, sensitivity will be decreased. In order to prevent signal-grid current, the d.c. bias developed by the oscillator grid should be not less than its cut-off value.

Because the peak plate current depends on how far positive the oscillator grid swings with respect to the cathode, it is desirable that this positive swing be as large as possible. It follows that the oscillator grid-leak resistance should be low, but not so low as to cause excessive damping of the tank circuit. It has been found, for operation in frequency bands lower than approximately 6 Mc/s, that all these requirements are generally best satisfied when the oscillator circuit is adjusted to provide, with recommended values of plate and screen voltage, a value of E_k of approximately 2 volts peak, and a d.c. oscillator-grid current of 0.5 mA through a grid-leak resistance of 20 000 ohms. This will give a peak positive voltage of the oscillator grid with respect to cathode of about 4 volts

On the normal short-wave band of 6 to 18 Mc/s, minimum grid current occurs at the low-frequency end of the band and the design procedure consists in adjusting the oscillator circuit so that sufficient grid current (200 μA minimum) is obtained at 6 Mc/s without developing excessive cathode voltage (approximately 2.5 volts r.m.s. maximum) at 18 Mc/s. The oscillator-grid bias is then somewhat less than cut-off at 6 Mc/s, but the signal-grid current should not be so high as to cause trouble. Oscillation at the high-frequency end of the band however, may be greater than optimum unless a grid stopper is used, but over-excitation will improve frequency stability.

If, for manufacturing reasons, the use of a tapped coil is not desirable, a normal primary winding can be used for the cathode connection, although it will be necessary to reverse the connections from those for a plate-tickler oscillator. This coil arrangement allows one side of the padder to be grounded.

Another method of connection which is satisfactory for the broadcast band is to use a plate-tickler oscillator circuit with the screen electrode as the oscillator plate and the cathode grounded. This connection can also be used on the short-wave band but the interaction between signal and oscillator circuits causes severe pulling and other associated troubles. A neutralizing capacitor between signal and oscillator grids will minimize these effects—neutralizing is not recommended with a cathode-coupled oscillator—and an alternative system is the use of harmonic mixing.

Whenever the screen is connected to an oscillator primary winding the oscillator voltage on this electrode must be kept to a minimum, i.e primaries must have as few turns as possible with maximum coupling to secondaries. In addition, it is often desirable to use a small carbon resistor in series with the oscillator grid electrode to prevent the oscillator voltage on the screen from rising excessively at the high-frequency end of the band. The reasons are, firstly, that modulation by the oscillator voltage on the screen is out of phase with the modulation due to the oscillator grid, which decreases conversion sensitivity and, secondly, that as the signal grid is between the two electrodes forming the screen a large oscillator voltage on the screen results, through capacitive coupling, in a comparatively large oscillator voltage on the signal grid, particularly on short waves. This voltage can be neutralized, but normal neutralization is effective for only one frequency and set of operating conditions and when a large amount of neutralization is used and a balance between two large equal and opposite voltages is obtained, any small change in conditions results in a large oscillator voltage reappearing on the signal grid.

The arrangement of shields on grid 2 and of the siderods of grid 3 make the oscillator section of these types much more stable with respect to electrode voltage variations than the 6A8-G, and it can be used satisfactorily with a.v.c. applied on short waves.

Nevertheless in any short-wave application care must be taken to reduce to a minimum any coupling between signal and oscillator grids if frequency stability is to be satisfactory. In the case of the 6BE6, the external capacitive coupling between oscillator and signal grids should be limited to 1 $\mu\mu$F for satisfactory short wave performance.

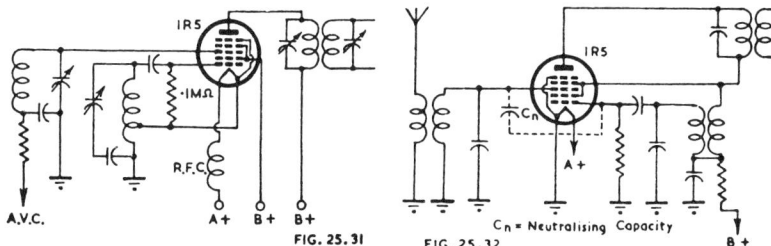

FIG. 25.31   FIG. 25.32   Cn = Neutralising Capacity   B +

Fig. 25.31. Self-excited 1R5 converter circuit.   *Fig. 25.32. Circuit for short-wave operation of 1R5 converter.*

(4) **Type 1R5** : The electrode structure of the 1R5 is similar to that of the 6BE6 and 6SA7(GT) without the shields on grid 2, and with a filament instead of the heater and cathode.   Circuits for use with the 1R5 are complicated by the lack of a separate cathode electrode but this may be overcome by using a choke as shown in Fig. 25.31 Alternatively the choke may be omitted and a filament lead returned to A+ through a separate winding of the same number of turns as, and wound over the tapped section of, the oscillator coil.

However, neither of these circuits is satisfactory on short waves if the full range— 6 to 18 Mc/s—is to be covered, and even on the broadcast band there may be difficulty in obtaining sufficient grid current.   The circuit of Fig. 25.32, with or without the padder feedback shown, is more suitable.

Neutralizing is desirable on short waves if conversion is carried out with the oscillator fundamental.   The neutralizing capacitor should be connected across the short-wave coils only, because a different value is normally required on the broadcast band, and oscillator voltage on grid 3 is usually less with no neutralizing than with the capacitance (from $2\frac{1}{2}$ to 5 $\mu\mu$F) required on short waves.

The purpose of the neutralizing capacitor in a circuit such as Fig. 25.32 is not so much the balancing out of oscillator voltage due to the space charge in front of grid 3, but the neutralizing of a voltage due to capacitive coupling between grid 3 and grids 2 and 4.   Therefore, as previously mentioned, it is essential to reduce the oscillator voltage on the screen to the lowest practicable value.

If there is a tendency for the 1R5 to squeg at the high-frequency end of the short-wave band it is often possible to save the small carbon resistor in the oscillator grid lead, which is the usual cure, by reducing the value of oscillator grid leak to say 30 000 ohms or slightly lower.   This frequently increases sensitivity, other conditions being unchanged, and it is worth trying even if there are no signs of squegging.   In most receivers the oscillator grid circuit damping caused by the low value of oscillator grid leak or by the grid stopper is desirable to reduce the oscillator voltage on the screen of the converter at the high-frequency end of the band with a consequent improvement in sensitivity.

A very satisfactory way to operate the 1R5 is to use harmonic mixing as described previously.   The increased freedom in layout, the increased stability in all respects— which simplifies production alignment—and the removal of the need for neutralizing more than compensate for the slight disadvantages which in any case are of little importance if the receiver has a r-f stage.

SECTION 3 : SUPERHETERODYNE TRACKING
By B. Sandel, A.S.T.C.

(i) General (ii) (A) Formulae and charts for superheterodyne oscillator design (B) Worked examples (iii) (A) Padded signal circuits (B) Worked example.

(i) General

The problem of tracking (whether it be in a straight or a superheterodyne receiver) is to set, simultaneously, to some desired resonant frequency, each of a series of tuned circuits which are mechanically coupled together and operated from a single control.

In a superheterodyne receiver the problem is one of maintaining a constant frequency difference (equal to the intermediate frequency) between the signal circuits (such as the aerial and r-f stages) and the oscillator circuit. It is a relatively simple matter to make the difference between the signal and oscillator frequency equal to the i-f at two points in the tuning range. This condition is called two point tracking, and is applied whenever the error in frequency difference between the circuits does not become a large percentage of the total pass band. Where the tuning error is likely to become excessive, it can be reduced by the addition of another component, in the form of a capacitor or inductor, into the oscillator or signal circuits. In this case it is possible to proportion the circuit components so that zero frequency error exists between the intermediate frequency and the difference between the signal and oscillator circuits at three frequencies in the tuning range, instead of two ; furthermore the error at frequencies between the tracking points is appreciably reduced.

The most usual application of three point tracking has been in receivers covering the standard long, medium and short wave bands, and where the tuning element is a variable capacitor. For this reason attention will be confined to this system of tuning and a suitable design method will be detailed. Those interested in tracking permeability tuned circuits are referred to the articles of Refs. 107 and 111. These give a method of three point tracking using identical variable inductors in the signal and oscillator circuits.

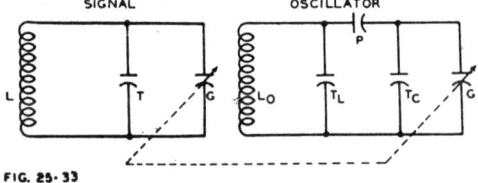

FIG. 25.33

Fig. 25.33. Circuits used for tracking analysis.

It is necessary to point out that the theoretical solutions, so far published, of the three point tracking problem are all idealized in so far that they ignore in part, or completely, the effects of primary windings on the signal and oscillator coils. The best approach under these conditions appears to be to select a method of determining values for circuit components which

(a) Does not involve an excessive number of arithmetical operations and will allow the use of a slide rule (or four figure logs) for all calculations, except in special cases.

(b) Gives values for the components that fall within a few per cent of those actually required in the circuits.

(c) Allows the change required in other circuit components to be rapidly estimated, when the value of one component in the circuit is changed by an amount which falls within a previously determined range of values.

(d) Lends itself to graphical methods of estimating the component values.

After the component values have been calculated it is usually advisable to build a pilot model receiver in which the values of all the elements (padder, coils and trimmers) can be varied over a small range (say ± 10%). In this way it is a fairly simple matter to secure tracking at the three points required, and at the same time to determine what tolerances are permissible in the component values before mistracking becomes excessive. Fixed padders and coils can then be used for further models of the same receiver ; although production difficulties can be reduced by using " slug " inductance variation in the oscillator coils as well as the usual variable parallel capacitance trimmers in the oscillator and signal circuits (many receiver manufacturers also use " slug " tuning of the signal circuits).

A few points which may be of interest are :

(1) The tracking error on the broadcast bands need not exceed a few Kc/s e.g. the error between tracking points need not exceed about 3 Kc/s on the 540-1600 Kc/s medium wave band using tracking frequencies of 600, 1000 and 1400 Kc/s. This error is negligible, in most cases, since the oscillator tuning takes charge and the lack of alignment affects the aerial and r-f circuits which are relatively unselective.

A typical curve of tracking errors is shown in Fig. 25.34.

(2) High impedance primaries, on the aerial and r-f coils, are practically always used in modern receivers covering the medium waveband. The coefficient of coupling in aerial and r-f coils, and the location of the primary resonant frequencies have very important effects on tracking. Suitable values for these factors are discussed in Chapter 23, Sects. 2 and 3.

(3) When a series capacitance (padder) is used in the oscillator circuit only, the operating frequency of the oscillator must be higher than the signal frequency if three point tracking is to be secured.

Three point tracking is obtainable if both the signal and oscillator circuits use series padders, whether the oscillator frequency is higher or lower than the signal frequency provided, of course, that the component values are correctly proportioned. If only the signal circuits are padded the oscillator frequency must be lower than the signal frequency to obtain three point tracking.

A typical example of padding of the signal and oscillator circuits occurs in band-spread receivers, where the short wave band is covered in a number of steps (e.g. 6 bands may be used to tune from 6-18 Mc/s), and it is required to have approximately straight line frequency tuning with a standard variable capacitor (i.e one whose capacitance versus rotation approximates to a straight-line frequency law). The tuning law need

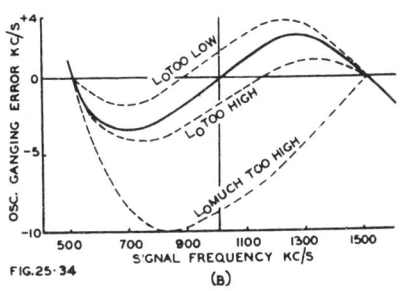

FIG.25·34

Fig. 25.34.   *Typical tracking curves.*

not be linear, however, and the shape of the calibration curve is largely under the control of the receiver designer.

(4) Care is necessary in the placement of components so as to minimize untracked stray capacitances. For example, the padder may be placed at the earthy end of the oscillator coil to assist in this regard.

(5) The selection of the best tracking points to give minimum error over a band of frequencies has received considerable attention from a number of designers. It has been generally accepted that the tracking error is reduced by setting the two outer tracking points somewhat in from the band limits, and locating the third tracking point at the geometrical-mean frequency (or some frequency near this value) of the outer tracking frequencies. However, it has been shown by Green (Ref. 105) that the **maximum** tracking error is reduced by bringing the high frequency point in from the band limit, setting the low frequency tracking point at or very close to the band limit, and making the third frequency the geometrical-mean of the two outer tracking frequencies. It has also been shown that the tracking errors are independent of the manner in which the total trimming capacitance is distributed across the oscillator coil and the variable capacitor ; although it should be noted that the manner in which these capacitances are distributed can have an adverse effect on the L/C ratio, particularly at low frequencies.

(6) Although the details set out in (5) are of interest they result in additional complication in the initial receiver design. A more suitable arrangement, for preliminary calculations, is to take the band limits as coinciding with the two outer tracking fre-

quencies, and make the third point either the arithmetical-mean or geometrical-mean of the outer frequencies. This leads to fairly simple design calculations, and gives values for the components which are reasonably close to the optimum. The small differences in component values required to set to the optimum tracking frequencies (or whatever the designer chooses) can then be made experimentally. It should be observed that it has been usual to align receivers to tracking points on various bands which have been established by long practice as giving sufficiently satisfactory results e.g. on the medium wave band 600, 1000 and 1400 Kc/s are in common use, but these are a compromise, due partly to a lack of more exact knowledge, and also because of the frequency allocations of the main broadcasting stations. If a station were to be located at, say, 540 Kc/s then the loss in receiver sensitivity, due to mistracking, would be serious if 600 Kc/s were to be taken as the low frequency tracking point.

(7) The feedback winding on the oscillator coil should be as small as possible, consistent with correct oscillator operation, if good tracking is desired. For minimum error the oscillator feedback winding should have its natural resonant frequency well above the highest oscillator tuning frequency ; this means that stray capacitances across the winding should be kept as low as possible. The natural resonant frequency is important because the coefficient of coupling between the two oscillator coil windings is usually large, and the amount of reactance reflected into the tuned circuit not only varies with frequency, but also sets a limit to the maximum possible tuning range. The limitation in tuning range is a particularly serious factor in receivers covering the short wave range in one band.

An additional and important reason for making the feedback winding as small as possible, is that for good oscillator stability the highest possible coefficient of coupling should be used, consistent with the smallest possible value of mutual inductance $(M = k\sqrt{L_p L_s})$.

(8) When tracking a superheterodyne receiver at three points in the tuning range (we will assume that the oscillator frequency is higher than the signal frequency) it often happens that the centre tracking (crossover) frequency does not fall at the frequency required. The question then arises as to how the oscillator series padder value should be altered.

First the signal circuit trimmer capacitance (or inductance) is altered so as to give maximum output. If the capacitance was increased then the oscillator series padder should be increased. For a decrease in signal trimmer capacitance (or inductance) the oscillator series padder should be decreased.

Complete re-alignment and re-checking is necessary after the padder value has been altered and it is important to remember that the oscillator inductance and trimmer capacitance values will also require alteration. This is a simple process when the oscillator coil is " slug " tuned, and the trimmer capacitance is adjustable. Fig. 25.34 is also helpful in this regard, and the tracking error can be considered in terms of L_0 if so desired. It should be clear that if the value of L_0 is increased the padder value should be decreased (and vice versa) to retain the desired oscillator frequency coverage.

A good alignment procedure is to set first the signal and oscillator circuits at the band limits. Tracking is then obtained at the required points by setting the signal generator to the required tracking frequency and rocking the receiver dial while altering the oscillator trimmer capacitance (at the high frequency tracking point) and the oscillator inductance (at the low frequency tracking point) until maximum signal output is obtained. Tracking at the centre frequency can be checked either by alteration of the signal circuit trimmers, which have not been altered after the initial adjustment at the band limits, or, alternatively, by leaving the signal circuit trimmers untouched and rocking the receiver dial while adjusting the oscillator inductance (or capacitance) for maximum output. It is preferable to track the oscillator to the signal circuits since these give the required tuning law. A complete re-check is always necessary after the initial alignment procedure is completed Of course correct tracking can be obtained by a number of methods, but the procedure suggested above has proved very satisfactory for receiver development.

(ii) (A) Formulae and charts for superheterodyne oscillator design

The equations quoted below are due to Payne-Scott and Green (Ref. 102) and Green (Refs. 103, 105).   Design charts are available in the references for the cases of arithmetical-mean and geometrical-mean tracking.   The charts shown in Figs. 25.35 to 25.39 are for geometrical-mean tracking only.

For the circuit arrangements used in the derivation of these equations see Fig. 25.33.

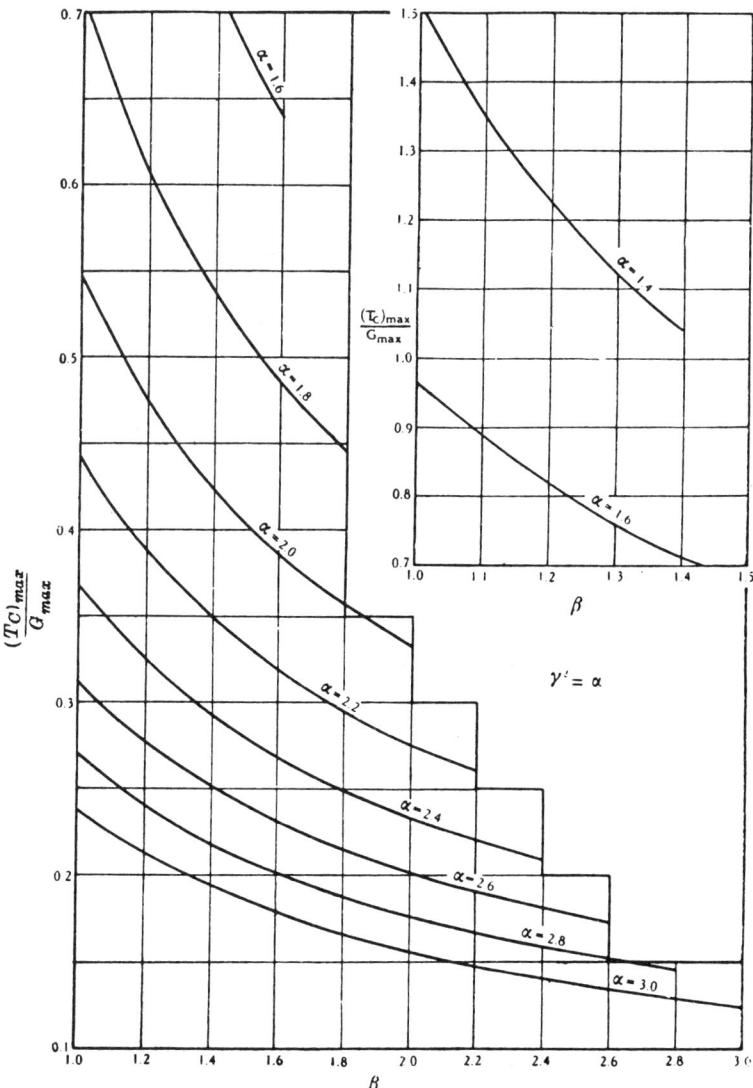

FIG. 25.35

Fig. 25.35.   Charts giving $T_{c_{max}}$ *for geometrical-mean tracking.*

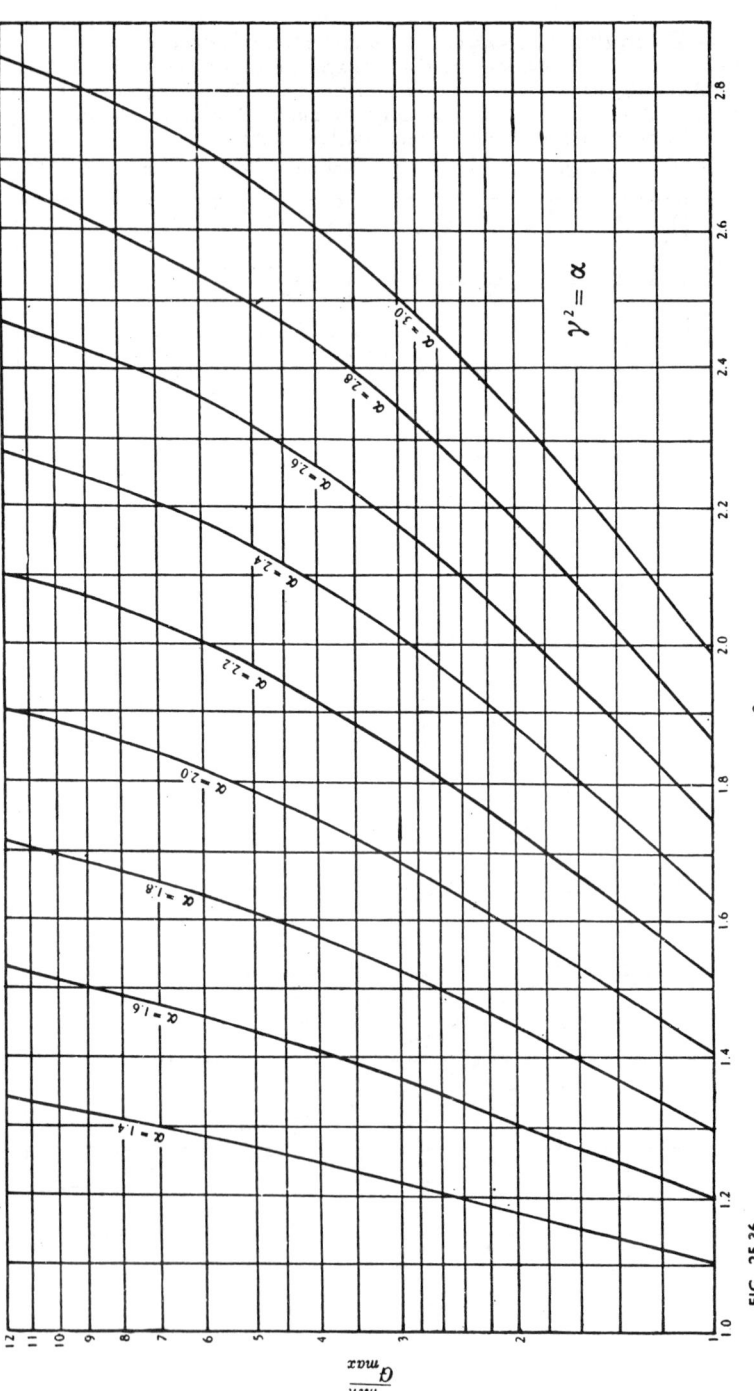

FIG. 25.36

Fig. 25.36.   *Charts giving high values of* P_{min} *for geometrical-mean tracking.*

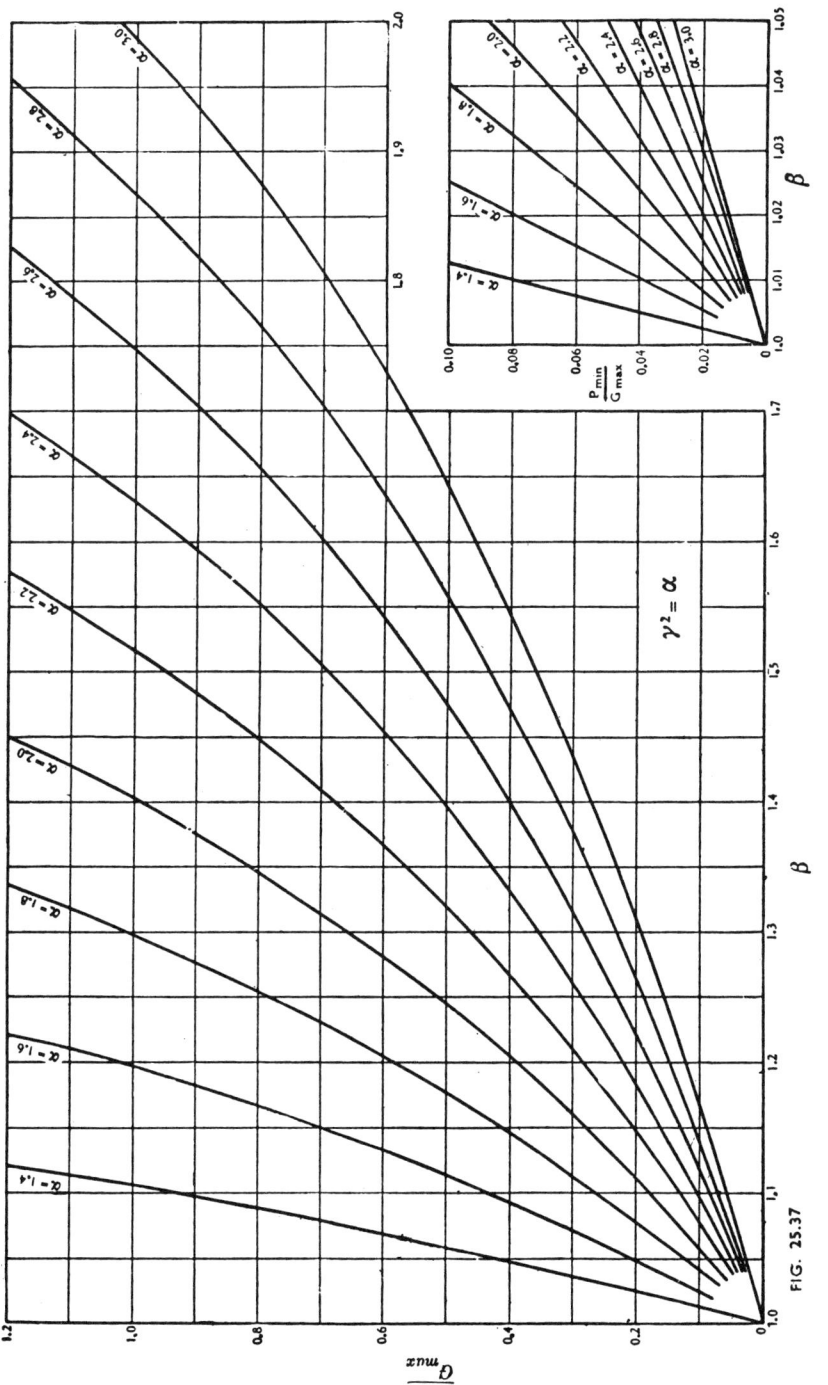

FIG. 25.37

Fig. 25.37. Chart giving low values of P'_{min} for geometrical-mean tracking.

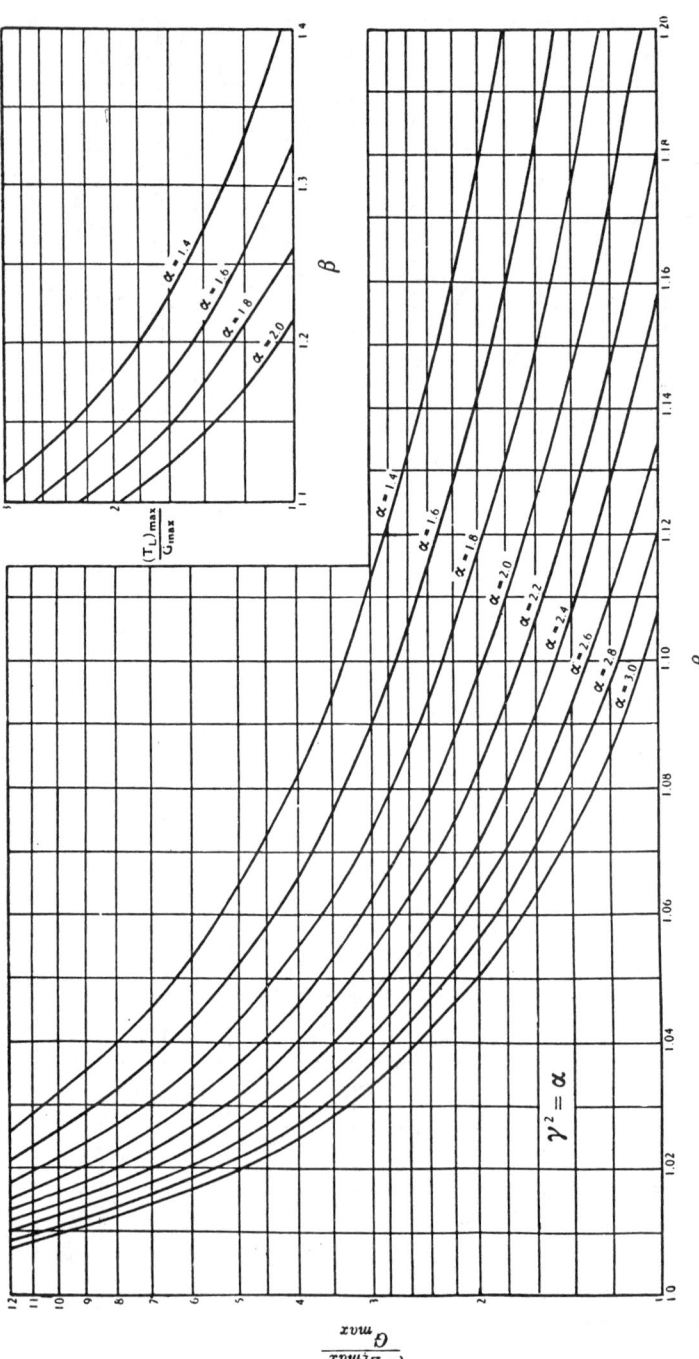

FIG. 25.38

Fig. 25.38. *Chart giving high values of $T_{L_{max}}$ for geometrical-mean tracking.*

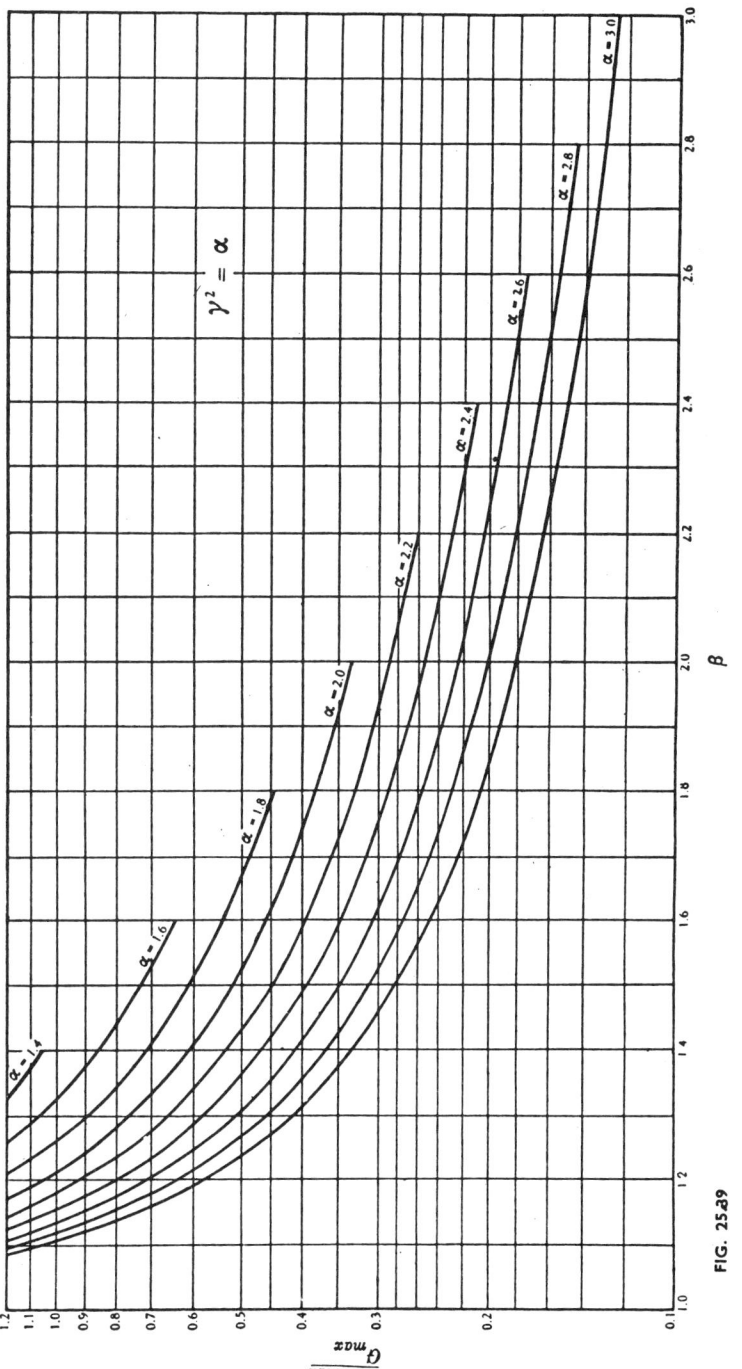

FIG. 25.39

Fig. 25.39. Chart giving low values of $T_{L_{max}}$ for geometrical-mean tracking.

Notation for the equations

$\omega_1 = 2\pi \times$ low-frequency tracking point of the signal circuit (f_1)

$\omega_2 = 2\pi \times$ high-frequency tracking point of the signal circuit (f_2)

$\omega_3 = 2\pi \times$ third tracking frequency for the signal circuit (f_3)

$\omega_i = 2\pi \times$ intermediate frequency (f_i)

$$\alpha = \frac{\omega_2}{\omega_1} = \frac{f_2}{f_1}$$

$$\beta = \frac{\omega_2 + \omega_i}{\omega_1 + \omega_i} = \frac{f_2 + f_i}{f_1 + f_i}$$

$$\gamma = \frac{\omega_3}{\omega_1} = \frac{f_3}{f_1}$$

$L, L_0 = $ signal and oscillator inductances

$G_{max} = $ incremental capacitance of each section of the ganged capacitors (i.e. difference between the maximum and minimum values of the capacitor)

$T = $ capacitance in signal circuit at signal frequency f_2 (includes gang min. cap.)

$T_L = $ fixed capacitance in parallel with L_0

$T_c = $ fixed capacitance in parallel with G in the oscillator·circuit

$P = $ oscillator padding capacitance.

When the extreme tracking points do not coincide with the band limits, the following additional notation is required.

$\omega_1' = 2\pi \times$ low-frequency limit of signal band (f_1')

$\omega_2' = 2\pi \times$ high-frequency limit of signal band (f_2')

$$\alpha' = \frac{\omega_2'}{\omega_1'} = \frac{f_2'}{f_1'}$$

$G'_{max} = $ total incremental capacitance of each section of the ganged capacitors between signal frequencies f_1' and f_2'

$T' = $ T less the incremental capacitance of each section of the ganged tuning capacitors between signal frequencies f_2 and f_2'

$T_c' = $ T_c less the incremental capacitance of each section of the ganged tuning capacitors between signal frequencies f_2 and f_2'.

Signal circuits :

For the tracking points coincident with the band limits.

$$T = \frac{G_{max}}{\alpha^2 - 1} \; ;$$

$$L = \frac{1}{T\omega_2^2} \text{ (or } L = \frac{25\,330}{Tf_2^2} \text{ } \mu H \text{ where } T \text{ is in } \mu\mu F \text{ and } f \text{ is in Mc/s)}$$

Oscillator circuit :

For the tracking points coincident with the band limits.

$$P_{max} = \frac{G_{max}}{r - 1}$$

$$T_{c\,max} = \frac{G_{max}}{r\beta^2 - 1}$$

and $$r = \frac{\alpha^2}{\beta^2} \cdot \frac{1 + \dfrac{2(\alpha - \beta)}{(\beta - 1)(\alpha + \gamma)}}{1 + \dfrac{2(\alpha - \beta)}{(\beta - 1)(1 + \gamma)}}$$

For the case of arithmetical-mean tracking $\left(\text{i.e. } f_3 = \dfrac{f_1 + f_2}{2}\right)$

$$r = \rho = \frac{\alpha^2}{\beta^2} \cdot \frac{3 + \alpha}{3 + \beta} \cdot \frac{1 + 3\beta}{1 + 3\alpha}$$

For geometrical-mean tracking (i.e. $f_3 = \sqrt{f_1 f_2}$)

$$r = R = \frac{\alpha^2}{\beta^2} \cdot \frac{2\beta + (1 + \beta)\alpha^{\frac{1}{2}}}{2\alpha + (1 + \beta)\alpha^{\frac{1}{2}}}$$

(a) General formulae for padding and trimming capacitances.

(1) T_c given.

$$P = P_{max} - T_c = P_{min} + T_{c_{max}} - T_c$$

$$T_L = \frac{P}{P_{min}}\left[T_{c_{max}} - T_c\right]$$

(2) T_L given.

$$P = \frac{P_{min}}{2}\left(1 + \sqrt{\frac{P_{min} + 4T_L}{P_{min}}}\right)$$

$$T_c = P_{max} - P = P_{min} + T_{c_{max}} - P.$$

Formulae when $T_L \ll P$

$$P = P_{min} + T_L$$

$$T_c = T_{c_{max}} - T_L$$

error in P is given by $-\left(\dfrac{T_L}{P}\right)^2 \times 100$ per cent.

error in T_c is given by $\dfrac{T_L^2}{P \times T_c} \times 100$ per cent.

(b) Formulae for inductance of oscillator coil

$$L_0(\omega_2 + \omega_i)^2 = \left(\frac{P_{max}}{P}\right)^2 \cdot \frac{1}{T_{L_{max}}} = \frac{P_{min}P_{max}}{T_{c_{max}}} \cdot \frac{1}{P^2}$$

or $L_0 = \dfrac{25\,330 \times P_{min}P_{max}}{T_{c_{max}}P^2(f_2 + f_i)^2}$

(capacitances in $\mu\mu$F, frequencies in Mc/s)

(c) Relations between P_{max}, P_{min}, $T_{c_{max}}$, and $T_{L_{max}}$

$$P_{max} = P_{min} + T_{c_{max}}$$

$$\frac{P_{max}}{P_{min}} = \frac{T_{L_{max}}}{T_{c_{max}}}$$

(d) Relations between oscillator-circuit components.

The effect of a change in value of one component of the oscillator circuit on the values required for the other components can be found from the following expressions :

$$\delta T_c = -\delta P$$

$$\delta T_L = \left(1 + \frac{2T_L}{P}\right)\delta P$$

$$\delta L_0 = -\frac{2L_0}{P}\delta P$$

where δL_0, etc. means a small change in L_0, etc. These changes must give values of P, T_c etc. which lie between the limits of the maximum and minimum values (P_{max}, P_{min}, etc.) of the components considered.

(e) Auxiliary formulae when the outer tracking frequencies do not coincide with the band limits

$$G_{max} = \left(\frac{\omega_2'}{\omega_2}\right)^2 \cdot \frac{\alpha^2 - 1}{(\alpha')^2 - 1} \cdot G'_{max}$$

$$\frac{T'}{T} = \left(\frac{\omega_2}{\omega_2'}\right)^2$$

$$T_c - T_c' = T - T'$$

(B) Worked examples

To illustrate the application of the equations and charts, worked examples are appended.

A tuning capacitor is available having a capacitance range of 12-432 $\mu\mu$F. It is required to tune over the range of frequencies 530 to 1620 Kc/s, and the i-f of the receiver is 455 Kc/s. Determine the component values foi the signal and oscillator circuits.

For ease of working the outer tracking frequencies will be made coincident with the band limits, and the third tracking frequency (f_3) will be taken as the geometrical-mean of the outer frequencies ($\sqrt{f_1 f_2} = 926.6$ Kc/s).

Signal circuits

$$G_{max} = 432 - 12 = 420 \ \mu\mu F$$

$$\alpha^2 = \left(\frac{1620}{530}\right)^2 = 9.342$$

$$T = \frac{420}{9.342 - 1} = 50.24 \ \mu\mu F \text{ (includes 12 } \mu\mu F \text{ gang min. cap. so that actual trimmer would be 38.24 } \mu\mu F)$$

$$L = \frac{25\,330}{50.24 \times 1.62^2} = 192.2 \ \mu H$$

Oscillator circuit

$$\alpha^2 = \left(\frac{1620}{530}\right)^2 = 9.342 \ ; \ \alpha^{\frac{1}{2}} = 1.748 \ ;$$

$$\beta = \frac{1620 + 455}{530 + 455} = 2.107 \ ; \ \beta^2 = 4.440 \ ;$$

$$(1 + \beta)\alpha^{\frac{1}{2}} = 5.431$$

$$r = R = \frac{9.342}{4.44}\left[\frac{(2 \times 2.107) + 5.431}{(2 \times 3.056) + 5.431}\right] = 1.758$$

$$P_{max} = \frac{432 - 12}{1.758 - 1} = 554 \ \mu\mu F$$

$$T_{c_{max}} = \frac{(432 - 12)}{(1.758 \times 4.44) - 1} = 61.7 \ \mu\mu F$$

(a) A value for T_L (about 8 $\mu\mu F$) will be assumed, after preliminary calculation, as this is the usual case.

As obviously $T_L \ll P$ (Compare 8 $\mu\mu F$ with P_{max})

$$P_{min} = P_{max} - T_{c_{max}} = 492.3 \ \mu\mu F$$

$$T_{L_{max}} = \frac{554 \times 61.7}{492.3} = 69.42 \ \mu\mu F.$$

From this it is permissible to estimate T_L as 8 $\mu\mu F$ since this is about the order of stray capacitances likely to be across L_0. Any small error here is unimportant, as long as the value chosen for T_L is less than $T_{L_{max}}$, since this is taken up during circuit alignment.

Hence $P = 492.3 + 8 = 500.3 \ \mu\mu F$

$T_c = 61.7 - 8 = 53.7 \ \mu\mu F$ (includes gang min. cap. of 12 $\mu\mu F$. Actual trimmer would be 41.7 $\mu\mu F$)

(b) $L_0 = \dfrac{25\,330 \times 492.3 \times 554}{61.7 \times 500.3^2 \times (1.62 + 0.455)^2} = 103.9 \ \mu H$

(c) and (d)

Suppose now that it is required to make a change in the value of one of the oscillator components. What should be the new values of the other components?

A good example is the padder; a more suitable value may be 510 $\mu\mu F$ (which lies between the maximum and minimum values of 554 $\mu\mu F$ and 492.3 $\mu\mu F$).

Then

$$\delta T_c = -9.7 \ \mu\mu F$$

$$\delta T_L = \left(1 + \frac{2 \times 8}{500.3}\right) 9.7 = 10 \ \mu\mu F$$

$$\delta L_0 = - \frac{2 \times 103.9 \times 9.7}{500.3} = -4.03 \ \mu H.$$

So that the new values for the oscillator components are $P = 510 \ \mu\mu F$

$T_c = 53.7 - 9.7 = 44 \ \mu\mu F$

$T_L = 8 + 10 = 18 \ \mu\mu F$

$L_0 = 103.9 - 4.03 = 99.87\ \mu H$
and these values all lie within the permissible range of values, set by the maximum and minimum values calculated for P, T_L and T_c; denoted by P_{max}, P_{min} etc.

A table showing a series of values between the calculated maxima and minima is often very useful. In this way the most suitable component values can be selected. Also, three point tracking is still maintained at the selected frequencies with the range of component values determined, as the change in values is independent of the tracking and intermediate frequencies provided these remain unaltered.

The same design problem, previously solved algebraically, will now be carried out using the tracking charts.

Signal circuits
Procedure exactly as before.

Oscillator circuit

$$\alpha = \frac{1620}{530} = 3.056$$

$$\beta = \frac{1620 + 455}{530 + 455} = 2.107$$

$G_{max} = 432 - 12 = 420\ \mu\mu F$

(a) From the charts :

$$\frac{T_{c\ max}}{G_{max}} = 0.15\ ;\ \text{therefore}\ T_{c\ max} = 63\ \mu\mu F$$

$$\frac{P_{min}}{G_{max}} = 1.2,\ \text{therefore}\ P_{min} = 504\ \mu\mu F$$

$$\frac{T_{L\ max}}{G_{max}} = 0.17,\ \text{therefore}\ T_{L\ max} = 71\ \mu\mu F$$

$P_{max} = 504 + 63 = 567\ \mu\mu F$

Using $T_L = 8\ \mu\mu F$ (as previously)
$P = 504 + 8 = 512\ \mu\mu F$
$T_c = 63 - 8 = 55\ \mu\mu F$

(b) $L_0 = \dfrac{25\,330 \times 504 \times 567}{63 \times 512^2 \times (1.62 + 0.455)^2} = 102\ \mu H$

(c) and (d) Any circuit component changes are made exactly as before. Suitable practical values are then selected after allowing for strays.

(iii) (A) Padded signal circuits

The case of the multi-band all-wave receiver is of interest to designers. It can hardly be said that the receiver covering the complete short wave band (6 − 18 Mc/s) in one step gives very satisfactory performance, particularly in the hands of an unskilled operator.

Consider first the case where bandspreading is obtained merely by loading additional parallel capacitance across the tuned circuits. The scale calibration will become non-linear with frequency and the scale will be crowded towards the low-frequency end of the band. If, instead of using parallel capacitance, a small capacitance is inserted in series with the tuning capacitor so as to restrict the tuning range, the scale will be crowded towards the high frequency end. From this it should be clear that a combination of the two methods can probably be made to give very much improved scale linearity. The method is not restricted, however, to producing a linear scale calibration.

The circuit arrangement for the padded signal circuit takes the same form as the oscillator circuit. The analysis of these circuits, due to Green (Ref. 103), is based on tracking the padded signal and oscillator circuits to a (non-existent) pilot circuit (which also determines the form of scale calibration) at three points, and thereby making the padded circuits track with one another at the same three points. The error at frequencies between tracking points is usually not serious as the range of frequencies to be covered is generally fairly small.

The design problem is to select values for the padder P and the two trimming capacitances T_L and T_c such that simultaneously the following conditions are fulfilled :—

(a) The desired frequency ratio is attained.

(b) The scale calibration is linear, or, more generally, corresponds to a desired form.

(c) The L/C ratio is maintained at a high value in order to achieve adequate gain in the signal circuits.

To simplify notation the virtual pilot circuit takes the same form as the original signal circuit shown in Fig. 25.33 and the same notation is retained, for both the padded signal and oscillator circuits, as was previously used for the oscillator circuit alone. The design equations are limited to the cases of arithmetical-mean and geometrical-mean tracking. Charts are available (in Ref. 103) covering the arithmetical-mean case.

Summary of formulae

Pilot circuit :

$$T = \frac{G_{max}}{\alpha^2 - 1}$$

$$\left(\frac{\omega_2}{\omega}\right)^2 = 1 + (\alpha^2 - 1)\frac{G}{G_{max}}$$

Padded circuit :

$$P + T_c = \frac{G_{max}}{r - 1}$$

$$T_{c_{max}} = \frac{G_{max}}{r\beta^2 - 1}$$

$$T_{L_{max}} = \frac{G_{max}}{r(\beta^2 - 1)}$$

$$P + T_c = P_{max} = P_{min} + T_{c_{max}}$$

where r is limited to the values given by ρ and R.

T_c specified :

$$P = P_{max} - T_c$$

$$T_L = \frac{P}{P_{min}} T_{c_{max}} - T_c$$

where P is first evaluated.

T_L specified :

$$P = \frac{P_{min}}{2}\left[1 + \sqrt{1 + \frac{4T_L}{P_{min}}}\right]$$

$$T_c = T_{c_{max}} - T_L \cdot \frac{P_{min}}{P}$$

where P is first evaluated.

When $T_L \ll P$

$$T_c = T_{c_{max}} - T_L$$

$$P = P_{min} + T_L$$

Gain is proportional to the dynamic impedance Z of the circuit and

$$Z = \frac{Q}{\omega C};$$

$$C = \frac{P}{P_{min}} \cdot \frac{G + T_{c_{max}}}{G + P_{max}} \quad \text{(effective capacitance across inductance).}$$

When $G = G_{max}$ then $C = C_{max}$. The value of P is obtained from the appropriate equations, according to whether T_c or T_L is specified.

Inductance

$$L_0 = \frac{1}{(\omega_1 + \omega_j)^2} \cdot \frac{P_{min}}{P^2} \cdot \frac{1 + \dfrac{P_{max}}{G_{max}}}{1 + \dfrac{T_{c\,max}}{G_{max}}}$$

or for numerical computation $L_0 = \dfrac{25\,330\,P_{min}(G_{max} + P_{max})}{(f_1 + f_j)^2 P^2 (G_{max} + T_{c\,max})}$

using μH, $\mu\mu$F, Mc/s.

in which $(\omega_1 + \omega_j)$ is the known low frequency tracking point of the padded signal circuit. For a padded oscillator circuit replace $(\omega_1 + \omega_j)$ by $(\omega_1 + \omega_j \pm \omega_i)$. The value for P is derived from the appropriate equation according as to whether T_c or T_L is specified.

For an unpadded signal circuit

$$L = \frac{\alpha^2 - 1}{\alpha^2 \omega_1{}^2\, G_{max}}$$

The notation is identical with that given previously, with the addition of

    ω = $2\pi \times$ any frequency in the tuning range

    ω_1 = $2\pi \times$ low frequency tracking point of pilot circuit

    ω_2 = $2\pi \times$ high frequency tracking point of pilot circuit

    ω_3 = arithmetical- or geometrical-mean tracking point of pilot circuit

    ω_i = $2\pi \times$ true intermediate frequency of receiver

    ω_j = $2\pi \times$ virtual intermediate frequency for the combination of a padded signal circuit with a virtual pilot circuit

    α = $\dfrac{\omega_2}{\omega_1} = \dfrac{f_2}{f_1}$ = frequency ratio of pilot circuit

    β = $\dfrac{\omega_2 + \omega_j}{\omega_1 + \omega_j}$ = frequency ratio of padded signal circuit

    G = incremental capacitance of each section of the ganged tuning capacitors, measured from its value at ω_2

    G_{max} = value of G at ω_1

    C = total effective capacitance across L_0

    C_{max} = value of C when $G = G_{max}$

    Q = $\omega L_0/R$ where L_0 and R apply to the signal circuit, and this R is the r-f resistance of the circuit.

(B) **Worked example**

The method will be applied to the design of a receiver tuning from 6 to 9 Mc/s, using an i-f of 455 Kc/s, with a capacitor having a straight line frequency characteristic and a capacitance range of 12-432 $\mu\mu$F. Arithmetical-mean tracking will be used.

Pilot circuit

For the pilot circuit assume that the required scale calibration is given by

    α = 2 .   (A sufficient range of values for most receiver calibration curves is from about 3 to 1.5)

    α^2 = 4

Signal circuits

    β = $9/6$ = 1.5

    β^2 = 2.25

    $r = \rho = \dfrac{4}{2.25} \cdot \dfrac{3 + 2}{3 + 1.5} \cdot \dfrac{1 + (3 \times 1.5)}{1 + (3 \times 2)}$ = 1.552

    $P_{max} = P + T_c = \dfrac{432 - 12}{1.552 - 1}$ = 760.8 $\mu\mu$F

    $T_{c\,max} = \dfrac{432 - 12}{(1.552 \times 2.25) - 1}$ = 301.7 $\mu\mu$F

    $T_{L\,max} = \dfrac{432 - 12}{1.552\,(2.25 - 1)}$ = 216.5 $\mu\mu$F

$P_{min} = 760.8 - 301.7 = 459.1 \ \mu\mu\text{F}$

Take the case where T_c is specified. Draw up a table of component values, and select the most suitable.

T_c	T_L	P	L_0
50		710.8	
100		660.8	
150		610.8	
200		560.8	
250	57.53	510.8	2.024
300		460.8	

From this it seems that a convenient padder value is 510.8 $\mu\mu$F. Then find T_L and L_0 to complete the table. If the values so found, using the first selection for P, are considered as being suitable, only those need be computed.

$$T_L = \frac{510.8}{459.1} [301.7 - 250] = 57.3 \ \mu\mu\text{F}$$

$$L_0 = \frac{25\,330 \times 459.1 \times (420 + 760.8)}{6^2 \times 510.8^2 \times (420 + 301.7)} = 2.024 \ \mu\text{H}$$

So that the signal circuit components are

$L_0 = 2.024 \ \mu\text{H}$
$T_L = 57.3 \ \mu\mu\text{F}$
$P = 510.8 \ \mu\mu\text{F}$
$T_c = 250 \ \mu\mu\text{F}$ (which includes 12 $\mu\mu$F gang min. cap.)

(Suitable practical values are selected after due allowance for strays).

An estimate can now be made of the circuit dynamic impedance (Z), as this will serve as a guide to the possible circuit gain. Improvement can often be effected by a redistribution of the component values, and the most favourable L/C ratios can be determined by completing the table set out earlier.

However the L/C ratio and the component values selected are usually a compromise forced on the designer by considerations of practical convenience.

Oscillator circuit

It will be taken that the oscillator frequency is higher than the signal frequency.

$$\beta = \frac{9 + 0.455}{6 + 0.455} = 1.464$$

$$\beta^2 = 2.145$$

$$r = \rho = \frac{4}{2.145} \ \frac{3 + 2}{3 + 1.464} \ \frac{1 + (3 \times 1.464)}{1 + (3 \times 2)} = 1.609$$

$$P_{max} = \frac{432 - 12}{1.609 - 1} = 689.6 \ \mu\mu\text{F}$$

$$T_{c\,max} = \frac{432 - 12}{(1.609 \times 2.145) - 1} = 171.4 \ \mu\mu\text{F}$$

$$T_{L\,max} = \frac{432 - 12}{1.609 \ (2.145 - 1)} = 228 \ \mu\mu\text{F}$$

$$P_{min} = 689.6 - 171.4 = 518.2 \ \mu\mu\text{F}.$$

Take the case, this time, where T_L is specified and draw up a table of component values. T_L is not necessarily very much smaller than P.

T_L	T_c	P	L_0
50		564.1	
100	85.6	604	1.619
150		639.7	
200		1072	

Then for the tabulated values of T_L we have

$$P = \frac{518.2}{2}\left[1 + \sqrt{1 + \frac{4T_L}{518.2}}\right]$$

These values are listed above.

Taking the padder of 604 $\mu\mu$F as being suitable, then T_L is 100 $\mu\mu$F, and now find T_c and L_0.

$$T_c = 171.4 - 100\left(\frac{518.2}{604}\right) = 85.6 \ \mu\mu\text{F}$$

$$L_0 = \frac{25\,330 \times 518.2\,(420 + 689.6)}{(6 + 0.455)^2 \times 604^2\,(420 + 171.4)} = 1.619 \ \mu\text{H}$$

So that one suitable set of oscillator component values would be
$T_L = 100 \ \mu\mu\text{F}$, $P = 604 \ \mu\mu\text{F}$, $L_0 = 1.619 \ \mu\text{H}$
and $T_c = 85.6 \ \mu\mu\text{F}$ (including 12 $\mu\mu$F gang min. cap.).

The signal and oscillator circuits should track at 6, 9 and 7.5 Mc/s.

References to superheterodyne tracking
Refs. 101-114, 121, 123, 134, 135.

SECTION 4 : REFERENCES

1. Armstrong, E. H. " A New System of Short-Wave Amplification " Proc. I.R.E. 9.1 (Feb. 1921) 3.
2. German Patent No. 324, 515.1918.
3. J. Scott-Taggart, German Patent No. 383,449.1919.
4. de Mare, J., R. Barthelemy, H. de Bellescize and L. Levy " Use of double-grid valves in frequency-changing circuits " L'Onde Electrique 5 (April 1926) 150.
5. " A four-electrode valve supersonic circuit " Experimental Wireless, 3.37 (Oct. 1926) 650.
6. Barthelemy, R. " Valve frequency-changers " Rev. Gen. d'Elec. 19 (April 1926) 663.
7. See also : M. Gausner, French Patent No. 639,028 : G. Thebault, French Patent 655,738 : H.J.J.M. de Regauld de Bellescize, United States Patent No. 1,872,634.
8. Whitman, V. E., United States Patent No. 1,893,813 : H. A. Wheeler, United States Patent No. 1,931,338.
9. Llewellyn, F. B., United States Patent No. 1,896,780.
10. Wheeler, H. A. " The hexode vacuum tube " Radio Eng. 13.4 (April 1933) 12.
11. Hassenberg, W. " The hexode " Funk. Tech. Monatshefte, (May 1933) 165.
12. Application Note No. 3 R.C.A. Radiotron Co., Inc.
13. Robinson, E. Y. British Patent No. 408,256.
14. Smith, J. C., Discussion on H. A. Wheeler paper " Image suppression and oscillator-modulators in superheterodyne receivers " Proc. I.R.E. 23.6 (June 1935) 576.
15. Harris, W. A. " The application of superheterodyne frequency conversion systems to multirange receivers " Proc. I.R.E. 23.4 (April 1935) 279.
16. Shelton, E. E. " A new frequency changer " W.W. 35.21 (Oct. 5, 1934) 283.
17. Nesslage, C. F., E. W. Herold and W. A. Harris " A new tube for use in superheterodyne frequency conversion systems " Proc. I.R.E. 24.2 (Feb. 1936) 207.
18. Herold, E. W., W. A. Harris and T. J. Henry " A new converter tube for all-wave receivers " R.C.A. Rev. 3.1 (July 1938) 67.
19. Harris, W. A. " A single-ended pentagrid converter " Presented, Rochester Fall Meeting, Rochester, N.Y., Nov. 15, 1938. See Application Note No. 100, R.C.A. Manufacturing Co., Inc., Radiotron Division, Harrison, N.J.
20. Klipsch, P. W. " Suppression of interlocking in first detector circuits " Proc. I.R.E. 22.6 (June 1934) 699.
21. " Suppressor grid circuit novelties " Radio World 22.15 (Dec. 24, 1932) 13.
22. Lukacs, E., F. Preisach and Z. Szepcsi " Noise in frequency changer valves " (Letter to Editor) W.E. 15.182 (Nov. 1938) 611.
23. Herold, E. W. " Superheterodyne converter system considerations in television receivers " R.C.A. Rev. 4.3 (Jan. 1940) 324.
24. Thompson, B. J., D. O. North and W. A. Harris " Fluctuations in space-charge-limited currents at moderately high frequencies " R.C.A. Rev. 4.3 (Jan. 1940) 269 ; 4.4 (April 1940) 441 ; 5.1 (July 1940) 106 ; 5.2 (Oct. 1940) 244 ; 5.3 (Jan. 1941) 371 ; 5.4 (April 1941) 505 ; 6.1 (July 1941) 114.
26. Strutt, M. J. O. " Electron transit-time effects in multigrid valves " W.E. 15.177 (June 1938) 315.
27. de Bellescize, H., United States Patent No. 1,872,634 ; Gausner, M., French Patent No. 639,028.
28. Whitman, V. E. United States Patent No. 1,893,813.
29. Strutt, M. J. O. and A. van der Ziel " Simple circuit means for improving short-wave performance of amplifier tubes " Elek. Nach. Tech. 13 (Aug. 1936) 260.

30. Freeman, R. L. " Use of feedback to compensate for vacuum-tube input-capacitance variations with grid bias " Proc. I.R.E. 26.11 (Nov. 1938) 1360.
32. Bakker, C. J. and G. de Vries " Amplification of small alternating tensions by an inductive action of the electrons in a radio valve " Physica 1 (Oct. 1934) 1045.
33. Nicolson, A. M. United States Patent No. 1,255,211 (applied for in 1915).
34. Strutt, M. J. O. " Frequency changers in all-wave receivers " W.E. 14.163 (April 1937) 184.
35. Herold, E. W. " Frequency changers in all-wave receivers " (Letter to Editor) W.E. 14.168 (Sept. 1937) 488.
36. Herold, E. W. United States Patent No. 2,141,750.
37. Llewellyn, F. B. " Phase angle of vacuum tube transconductance at very high frequencies " Proc I.R.E. 22.8 (Aug. 1934) 947.
38. Strutt, M. J. O. " Frequency changers in all-wave receivers " (Letter to Editor) W.E. 14.170 (Nov. 1937) 606.
40. Herold, E. W. United States Patent No. 2,066,038.
41. Steimel, K. " The influence of inertia and transit time of electrons in broadcast receiving tubes " Telefunken-Rohre 5 (Nov. 1935) 213.
42. Knol, K. S., M. J. O. Strutt and A. van der Ziel " On the motion of electrons in an alternating electric field " Physica 5 (May 1938) 325.
43. Strutt, M. J. O. and A. van der Ziel " Dynamic measurements of electron motion in multigrid tubes " Elek. Nach. Tech. 15 (Sept. 1938) 277.
45. Jonker, J. L. H. and A. J. W. M. van Overbeek " A new converter valve " W.E. 15.179 (August 1938) 423.
48. Strutt, M. J. O. and A. van der Ziel " The causes for the increase of admittances of modern high-frequency amplifier tubes on short waves " Proc. I.R.E. 26.8 (Aug. 1938) 1011.
49. An excellent historical summary of this work is found in W. E. Benham's " A contribution to tube and amplifier theory " Proc. I.R.E. 26.7 (July 1935) 683.
50. Bakker, C. H. and C. de Vries " On vacuum tube electronics " Physica 2 (July 1935) 683.
51. Rothe, H. " The operation of electron tubes at high frequencies " Telefunken-Rohre 9 (April 1937) 33 : Proc. I.R.E. 28.7 (July 1940) 325.
52. Runge, I. " Transit-time effects in electron rubes " Zeit fur Tech. Phys. 18 (1937) 438.
53. Runge, I. " Multigrid tubes at high frequencies." Telefunken-Rohre 10 (Aug. 1937) 128.
54. Peterson, L. C. " Impedance properties of electron streams " B.S.T.J. 18.7 (July 1939) 465.
55. Strutt, M. J. O. " On conversion detectors " Proc. I.R.E. 22.8 (Aug. 1934) 981.
56. Howe, G. W. O. " Second channel and harmonic reception in superheterodynes " W.E. 11.132 (Sept. 1934) 461.
57. Wey, R. J. " Heptode frequency changers " W.E. 11.135 (Dec. 1934) 642.
58. Strutt, M. J. O. " Mixing valves " W.E. 12.137 (Feb. 1935) 59.
59. Stewart, J. " The operation of superheterodyne first detector valves " J.I.E.E. 76 (Feb. 1935) 227.
60. Strutt, M. J. O. " Whistling notes in superheterodyne receivers " W.E. 12.139 (April 1935) 194.
61. Strutt, M. J. O. " Diode frequency changers " W.E. 13.149 (Feb. 1936) 73.
62. Bell, D. A. " The diode as rectifier and frequency changer " W.E. 18.217 (Oct. 1941) 395.
63. Mitchell, C. J. " Heterodyning and modulation : do additive and multiplicative mixing amount to the same thing ? " W.W. 53.10 (Oct. 1947) 359.
64. Whitehead, J. W. " Additive and multiplicative mixing " W.W. 53.12 (Dec. 1947) 486.
65. Strutt, M. J. O. " Noise figure reduction in mixer stages " Proc. I.R.E. 34.12 (Dec. 1946) 942.
66. Jonker, J. L. H. and A. J. W. M. van Overbeek " A new frequency-changing valve " Philips Tec. Rev. 3.9 (Sept. 1938) 266.
67. " Application Note on Operation of the 6SA7 " R.C.A. Manufacturing Company, Inc. Radiotron Division, Application Note No. 100 (Dec. 2, 1938).
68. " Characteristics of pentodes and triodes in mixer service " Radio Corporation of America, Tube Dept. Application Note 139 (March 15, 1949).
69. Goddard, N. E. " Contour analysis of mixer valves " W.E. 26.314 (Nov. 1949) 350.
70. Herold, E. W. and L. Malter " Some aspects of radio reception at ultra-high frequency " Proc. I.R.E. 31.10 (Oct. 1943) 567.
71. Tapp, C. E. " The application of super-regeneration to F-M receiver design " Proc. I.R.E. (Aust.) 8.4 (April 1947) 4.
72. " Use of the 6BA6 and 6BE6 miniature tubes in F-M receivers " Radio Corporation of America, Tube Dept., Application Note 121. Reprinted Radiotronics 129 (Jan. 1948) 15.
73. Gladwin, A. S. " Oscillation amplitude in simple valve oscillators " Part 1, W.E. 26.308 (May 1949) 159 ; Part 2 (contains section on squegging) W.E. 26.309 (June 1949) 201.
74. Stockman, H. " Frequency conversion circuit development " Comm. (Part 1) 25.4 (April 1945) 46 ; (Part 2) 25.5 (May 1945) 58.
75. Bedford, L. H. " The impulsive theory of the hexode frequency changer " W.E. 15.182 (Nov. 1938) 596.
76. " Compensation of frequency drift " (Particular reference to use of 6BE6 at 110 Mc/s) Radio Corporation of America, Tube Dept., Application Note AN-122. Reprinted Radiotronics 130 (March 1948) 36.
77. " Receiver microphonics caused by heater-cathode capacitance variations " Particular reference to use of 6BE6 in F-M band, Radio Corporation of America, Tube Dept., Application Note AN-123. Reprinted Radiotronics 130 (March 1948) 38.
78. Sturley, K. R. " Radio Receiver Design " Part 1 (Chapman and Hall, 1943).
79. Zepler, E. E. " The Technique of Radio Design " (Chapman and Hall, 2nd edit. 1949) 151.
80. Erskine-Maconochie, H. C. C. " Regeneration in the Superheterodyne—where and how it may be applied " W.W. 45.4 (July 27, 1939) 77.
81. Lukacs, E., and J. A. Sargrove (1) " Frequency changing problems : further development in mixer valves " W.W. 44.4 (Jan. 26, 1939) 81. (2) " Frequency changing problems : The latest octodes " W.W. 44.5 (Feb. 2, 1939) 119.
82. Cocking, W. T. " Short-wave oscillator problems : avoiding parasitic oscillation " W.W. 44.6 (Feb. 9, 1939) 127.
83. Sargrove, J. A. " Parasitic oscillation in frequency-changers : kinetic grid current in the anode-heptode and triode hexode " W.W. 45.6 (Aug. 10, 1939) 121.
84. James, E. G. " Frequency changer valves : Little known effects on short waves " W.W. 44.13 (March 30, 1939) 295.
85. " The modern receiver stage by stage : (iv) The frequency changer " W.W. 44.12 (March 23, 1939) 281.
86. Aske, V. H. " Gain-doubling frequency converters " Elect. 24.1 (Jan. 1951) 92.

87. Dammers, B. G., J. Otte " Comments on interelectrode feedback in frequency changers " Philips Tec. Com. No. 5 (1950) 13 ; No. 6 (1950) 11 ; reprinted from Philips Electronic Application Bulletin 11, 3 (March 1950).

101. Green, A. L. " Superheterodyne tracking charts " A.W.A. Tec. Rev. 5.3 (Feb. 1941) 77, Reprinted W.E. 19.225 (June 1942) 243. (Introductory, gives extensive bibliography).

102. Payne-Scott, Ruby, and Green, A. L. " Superheterodyne tracking charts—II " A.W.A. Tec. Rev. 5.6 (Dec. 1941) 251. Reprinted W.E. 19.226 (July 1942) 290 (Derives general design equations and arithmetical-mean tracking charts).

103. Green, A. L. " Superheterodyne tracking charts III " A.W.A. Tec. Rev. 6.2 (Feb. 1943) 97 : Reprinted W.E. 20.243 (Dec. 1943) 581 (Padded signal circuits and design charts).

104. Green, A. L. " Superheterodyne tracking charts—IV " A.W.A. Tec. Rev. 6.7 (July 1945) 423 (additional data on padded signal circuit but mainly confined to tracking R-C oscillators).

105. Green, A. L. " Superheterodyne tracking charts—V " A.W.A. Tec. Rev. 7.3 (April 1947) 295 (Tracking errors and design charts for geometrical-mean tracking).

106. Ross, H. A. and Miller P. M. " Measurement of Superheterodyne Tracking Errors " A.W.A. Tec. Rev. 7.3 (April 1947) 323 (Experimental method of checking magnitude and distribution of tracking errors).

107. Simon, A. W. " Tracking permeability tuned circuits " Elect. 19.9 (Sept. 1946) 138 (Method of tracking identical inductors in signal and oscillator circuits).

108. Sowerby, A. L. " Ganging and tuning controls of a superheterodyne receiver " W.E. 9.101 (Feb. 1932) 70.

109. Cocking, W. T. " Wireless Servicing Manual " Text book (6th edit. 1944) Iliffe and Sons, London (Gives general information on practical alignment of receivers etc.).

110. Sandel, B. " Tracking " Radiotronics No. 121 Sept.-Oct. 1946 page 101 (Gives practical design details for typical receiver using arithmetical mean tracking. Method based on that of Payne-Scott and Green (1941).

111. Yu, Y. P. " Superheterodyne tracking charts " Tele. Tech. 6.9 (Sept. 1947) 46. (Tracking Charts for permeability and capacitively tuned circuits).

112. Simon, A. W. " Tracking the permeability-tuned circuits " Elect. 20.11 (Nov. 1947) 142. (Method of neutralizing effect of loop antenna reactance on tracking).

113. Simon, A. W. " Three point tracking method " Radio 30.11 (Nov. 1946) 20 (Simplified eqns. for inductance or capacitance tuning ; broadcast band only).

114. Menzies, E. B. " Method of plotting tracking error " Elect. 20.1 (Jan. 1947) 128.

115. Strutt, M. J. O. " Noise-figure reduction in mixer stages " Proc. I.R.E. 34.12 (Dec. 1946) 942.

116. Stolze, W. J. " Input circuit noise calculations for F-M and television receivers " Comm. 27.2 (Feb. 1947) 12.

117. Strutt, M. J. O. " Properties of gain and noise figures at v.h.f. and u.h.f." Philips Tec. Com. 3 (March 1947) 3. Precis by E. G. Beard, Philips Tec. Com. 3 (March 1947) 17.

118. Miller, P. H. " Noise spectrum of crystal rectifiers " Proc. I.R.E. 35.3 (March 1947) 252.

119. Roberts, S. " Some considerations governing noise measurements on crystal mixers " Proc. I.R.E. 35.3 (March 1947) 257.

120. Tapp, C. E. " The application of super-regeneration to frequency-modulation receiver design " Proc. I.R.E. (Australia) 8.4 (April 1947) 4.

121. Fairman, H. E. " Oscillator tracking methods in permeability tuning " Tele-Tech. 6.3 (March 1947) 48.

122. Middleton, D. " Detector noise " (review of article in Journal of Applied Physics, Oct. 1946) Radio 31.1 (Jan. 1947) 4.

123. Kirkpatrick, C. B. " Three point tracking formulae " Proc. I.R.E. (Aust.) 8.7 (July 1947) 18.

124. Mueller, W. P. (extract) " H.F. Conversion with the 7F8 " Service 16.5 (May 1947) 28.

125. Edwards, C. F. " Microwave converters " Proc. I.R.E. 35.11 (Nov. 1947) 1181.

126. Whitehead, J. W. " Additive and multiplicative mixing " W.W. 53.121 (Dec. 1947) 486.

127. Stockman, H. " Superregenerative circuit applications " Elect. 21.2 (Feb. 1948) 81.

129. Haantjes, J. and B. D. H. Tellegen " The diode as converter and as detector " Philips Research Reports, 2.6 (Dec. 1947) 401.

130. Bradley, W. E. " Superregenerative detection theory " Elect. 21.9 (Sept. 1948) 96.

131. Hazeltine, A., D. Richman and B. D. Loughlin " Superregenerator design " Elect. 21.9 (Sept. 1948) 99.

133. " Characteristics of pentodes and triodes in mixer service " R.C.A. Application Note AN-139 (March 15, 1949).

134. Coppin, K. J. " The tracking of superheterodyne receivers " Jour. Brit. I.R.E. 7.6 (Nov.-Dec. 1948) 265.

135. de Koe, H. S. " Tracking of superheterodyne receivers " W.E. 28.337 (Oct. 1951) 305.

136. Dammers, B. G., and L. J. Cock " Comments on frequency changers for 30 Mc/s to 120 Mc/s for T.V. and F.M. receivers " Philips Tec. Com. Nos. 5,6,7 (1951). Reprinted from Electronic Application Bulletin June/July 1950, Vol. 11 Nos. 6,7.

CHAPTER 26

INTERMEDIATE FREQUENCY AMPLIFIERS

By B. Sandel, A.S.T.C.

SECTION 1 : CHOICE OF FREQUENCY

(i) *Reasons for selection of different frequencies* (ii) *Commonly accepted intermediate frequencies.*

(i) Reasons for selection of different frequencies

The choice of a particular intermediate frequency in a receiver is governed by a number of factors.

(a) If the frequency is very low the circuits will be generally too selective, resulting in side-band cutting. Very high selectivity will also make a receiver more difficult to tune, and imposes severe requirements on oscillator, and other circuit, stability.

(b) The lower the intermediate frequency the more difficult it becomes to eliminate image interference. This difficulty becomes greater as the carrier frequency is increased.

(c) The intermediate frequency should not fall within the tuning range of the receiver, as this would lead to instability and severe heterodyne interference. Also lower harmonics (principally second and third) of the i-f should not fall within the tuning range if this can be avoided. This requirement is not always easy to meet because of the other factors discussed in this section.

(d) Too high a value of intermediate frequency should be avoided as it generally leads to a serious reduction in selectivity and, usually, gain.

(e) The intermediate frequency should not be such that it approaches the range of frequencies over which the receiver is required to tune.

(f) If the intermediate frequency is made too high, tracking difficulties between the signal and oscillator circuits may be experienced.

(g) In some cases, where the intermediate frequency is higher than the highest received signal frequency (as in single span receivers), it is possible to tune the oscillator only and use low-pass filters for the signal circuits. However, it is generally preferable to use tunable signal circuits.

(ii) Commonly accepted intermediate frequencies

As a result of the experience gained over a number of years in addition to the considerations stated previously the values selected for the intermediate frequencies of most commercial receivers have become fairly well standardized. For the majority of broadcast receivers tuning the bands 540-1600 Kc/s and 6-18 Mc/s, an i-f of about 455 Kc/s is usual. A frequency of 110 Kc/s has been extensively used in Europe where the long wave band of 150-350 Kc/s is in operation. Receivers for use only on the short wave bands commonly employ an i-f of 1600 Kc/s or higher. Frequency modulation receivers covering the 40-50 Mc/s band generally use a 4.3 Mc/s i-f, and for the 88-108 Mc/s band they use 10.7 Mc/s. This latter value has been adopted as standard in U.S.A., and some other countries, for v-h-f receivers.

SECTION 2; NUMBER OF STAGES

The number of stages required in the i-f amplifier is generally a compromise between the factors of selectivity, gain, and cost. For the usual broadcast receiver having a 455 Kc/s i-f, one stage employing two transformers is generally considered as being adequate (when using an i-f valve with a g_m of about 2000 micromhos). The transformers in common use are two parallel tuned circuits coupled by mutual inductance and stray capacitance. The windings are arranged so that any capacitive coupling aids the mutual inductance. Care should be taken to keep capacitive coupling small as it alters the symmetry of the overall response curve of frequency versus attenuation.

Common values of capacitance tuning the primary and secondary windings in commercial transformers are from about 50 $\mu\mu$F to 120 $\mu\mu$F to which must be added valve and circuit stray capacitances. Values as high as 800 $\mu\mu$F are used in special circumstances, however. This point will receive further discussion in Section 7 of this chapter. Unloaded values for the primary and secondary Q's are from about 70 to 130. The transformer windings are in most cases coupled from 0.8 of critical to critical. The amount of coupling when the transformer is wired into the receiver is the important factor as stray coupling (including regeneration and alteration of tuning slug positions) will often alter any value of k or conditions for critical k measured external to the receiver.

It should be noted particularly that the second i-f transformer usually feeds into a diode detector and this will appreciably affect the selectivity and gain of the preceding stage because of the loading across the transformer secondary. This loading is commonly taken as half the d.c. diode load resistance, which in most cases does not exceed 0.5 megohm because of the detector circuit requirements. If improved selectivity is required then the secondary of the i-f transformer may be tapped and the diode circuit fed from this tapping point to reduce loading on the tuned circuit ; but this generally involves a loss in stage gain.

For small battery receivers of the portable type it may be advantageous to employ three i-f transformers (i.e. two stages) rather than a r-f stage and one i-f stage. Although there is some reduction in signal-to-noise ratio with this arrangement, additional gain is often the main requirement. Further, it is possible to operate the two i-f valves in a very economical condition, e.g. by using reduced screen voltage, as the full gain available can seldom be utilized without difficulties arising from instability. The Q's of the i-f's need not be more than 60 or 70 with this arrangement as the selectivity is more than adequate for ordinary reception.

With receivers using higher intermediate frequencies at least two stages (three transformers) are often required to give improved selectivity. Another important consideration, however, is that the dynamic impedance of the load presented by the transformer to the valve is lower, and stage gain is decreased. Valves used in the i-f stages of receivers having a high value of i-f generally have a mutual conductance

(g_m) of the order of 4000 micromhos to allow additional gain to be obtained to offset the loss due to the lower circuit dynamic impedance. These conditions apply particularly in F-M receivers using an i-f of 10.7 Mc/s. Short wave receivers using 1600 Kc/s i-f transformers commonly employ two stages (3 transformers) although one stage is often used and generally, but not always, the valves are similar to those used at lower frequencies and have a g_m of about 2000 micromhos or less. The passband and gain requirements will largely determine the values of the constants for the transformers.

A 10.7 Mc/s transformer in a F-M receiver has to pass a band of frequencies about 240 Kc/s wide and at the same time must not introduce such an appreciable amount of amplitude modulation (due to the selectivity of the transformers) that the limiter (or whatever device is used that is insensitive to amplitude variations) cannot give substantially constant output. To secure these results, some designers use combinations of overcoupled and critically-coupled transformers, while others prefer to use only critically-coupled transformers because of the simplification in the alignment procedure. For the type of F-M receiver using a ratio detector, two i-f stages are generally considered as being adequate (although this may lead to difficulties because of insufficient adjacent channel selectivity) with the second i-f valve feeding into a discriminator transformer. For most F-M receivers using limiters, three 10.7 Mc/s i-f transformers are used in cases where one limiter stage is included ; or for two limiter stages an additional very wide band i-f transformer may be incorporated. The purpose of the additional i-f transformer is to provide interstage coupling, but at the same time the design must be such as to introduce no appreciable amplitude modulation of the signal after it has passed through the first limiter stage. If a locked oscillator type of F-M detector (such as the Bradley circuit) is employed, then it is common practice to use three normal i-f transformers plus a fourth transformer giving a voltage step down of about 6 to 1 to provide a low impedance voltage source for driving the detector. An additional i-f stage is included in receivers using limiters for the purpose of giving the greater gain required to obtain satisfactory amplitude limiting with very weak signals ; the additional i-f stage also provides improved selectivity outside the required passband.

The number, and type, of stages used in any i-f amplifier is decided, of course, by the various requirements of the particular receiver and the selectivity and gain are under the control of the designer. Discussion as to the number of i-f stages used in typical A-M and F-M broadcast receivers is meant only to serve as a guide to common practice. However, it should not be overlooked that although gain and pass band requirements can sometimes be met with fewer stages than suggested above, adjacent channel selectivity will generally call for additional transformers. It is important that both pass band and adjacent channel selectivity be considered during the design, and if this is done there is usually little difficulty in selecting the number of i-f stages required.

Special i-f requirements can be fulfilled by making preliminary calculations for any number of stages which a designer may consider necessary. In Section 4 it is proposed to carry out the complete design of several transformers showing the various factors to which attention must be paid to meet a given set of conditions.

SECTION 3 : COMMONLY USED CIRCUITS

(i) *Mutual inductance coupling* (ii) *Shunt capacitance coupling* (iii) *Composite i-f transformers.*

In this section it is proposed to confine attention, particularly, to the most commonly used i-f transformer arrangement which uses mutual inductance coupling. Shunt capacitance coupling is not very widely used, but is included here as being of some interest because of its occasional application in F-M receivers. Details of various coupled circuit arrangements are given in Chapter 9 ; the references should be consulted for a more comprehensive survey of circuit arrangements, and analysis of their properties. Of course, almost any type of coupling can be used in i-f circuits but transformers other than those using two windings coupled by mutual inductance are the exception rather than the rule in ordinary broadcast and communications receivers.

(i) Mutual inductance coupling

Fig. 26.1 shows the most widely used circuit arrangement for an i-f coupling transformer ; C_1, L_1 and C_2, L_2 are the primary and secondary capacitances and inductances respectively. The two windings are coupled together by mutual inductance. The total capacitances tuning the circuits are due to valve input and output, wiring and coil capacitances as well as C_1 and C_2.

The transformer is set to the required i-f by using powdered iron " slugs " which are moved inside the primary and secondary windings to vary the inductance values. Variable capacitance trimmers are often used as an alternative, and in some cases this may be advantageous since the inductance values can be pre-set fairly accurately and the capacitance, which is not known accurately because of additional strays, can be set to give the required resonant frequency ; this arrangement also allows close control on the coefficient of coupling. However, capacitance trimmers are not always completely reliable and it is often of greater practical convenience to use variable iron cores in transformer windings. Further, the fixed capacitors C_1 and C_2 can be of high quality (e.g. silvered mica) to improve the circuit stability.

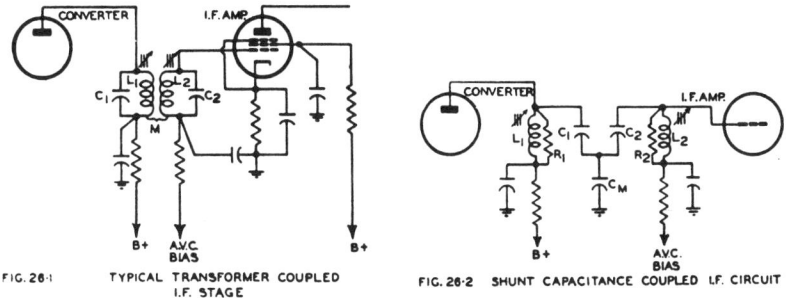

FIG. 26·1   TYPICAL TRANSFORMER COUPLED
I.F. STAGE

FIG. 26·2   SHUNT CAPACITANCE COUPLED I.F. CIRCUIT

The type of circuit being considered is discussed also in Chapter 9 Sect. 6(iii), (iv), (v). Graphical methods of determining selectivity are given in Chapter 9 Sect. 10.

Section 4 below gives detailed design methods and examples for several transformers of this type.

(ii) Shunt capacitance coupling

This type of coupling arrangement is illustrated in Fig. 26.2. Circuits of this type can be conveniently arranged to give a fairly wide pass band and are sometimes used in F-M receivers, in particular, to couple two cascaded limiter stages. Some discussion of this type of circuit is given also in Chapter 9 Sect. 8(ii).

The design of circuits of this type is carried out in a similar manner to those using mutual inductance coupling. Usually the values selected for C_1 and C_2 are about 50 $\mu\mu$F and the value for C_M is determined by the required coefficient of coupling However, C_M is generally fairly large, being of the order of 1000 to 2000 $\mu\mu$F in typical cases. L_1 and L_2 (which usually have equal values) are determined from :

$$L_1 = L_2 = 25\,330/f^2 C$$

where L_1 is in microhenrys

C is in $\mu\mu$F and is equal to $\dfrac{C_1 C_M}{C_1 + C_M}\left(\text{or } \dfrac{C_2 C_M}{C_2 + C_M}\right)$ + stray capacitance across the primary circuit

and f is in Mc/s, and is taken as the intermediate frequency.

The resistors R_1 and R_2 are for the purpose of lowering the Q of the tuned circuits to the values required for the bandwidth desired.

The coefficient of coupling is given by

$$k = \sqrt{\frac{C_1 C_2}{(C_1 + C_M)(C_2 + C_M)}}$$

$$\approx \frac{\sqrt{C_1 C_2}}{C_M} \qquad \text{when } C_M \gg (C_1, C_2).$$

(iii) Composite i-f transformers

Receivers used for F-M and A-M reception, on the 88-108 Mc/s and the 540-1600 Kc/s bands respectively, generally have the i-f amplifiers arranged so that the intermediate frequency is 10.7 Mc/s for the F-M band and 455 Kc/s for the A-M range. For reasons of economy many manufacturers use the same amplifier valves for both i-f's, and so it is necessary to find a solution to this problem which does not require elaborate circuit arrangements.

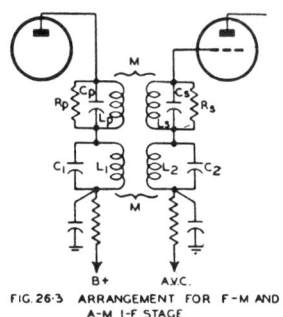

FIG. 26·3   ARRANGEMENT FOR F-M AND A-M I-F STAGE

One solution is the circuit shown in Fig. 26.3. The 10.7 Mc/s transformer is represented by C_p, L_p etc. and the 455 Kc/s transformers by L_1, C_1 etc. When the carrier output from the converter stage is 10.7 Mc/s, the primary and secondary circuits of the 455 Kc/s i-f act as capacitances which allow the lower ends of the 10.7 Mc/s i-f to be effectively earthed. At 455 Kc/s the 10.7 Mc/s i-f transformer tuned circuits act as inductances in series with the transformers in use. The effective inductance, in series with the 455 Kc/s transformer, tends to offset any loss in gain due to the voltage drop across it because it reduces the loading on the valve input circuit. Any possible loss in gain is not important, however, as the g_m of the amplifier valves is about 4000 micromhos, and there are usually two i-f stages. The problem is generally one of getting rid of excess gain on the lower frequency channel ; it is for this reason that very low L/C ratios are used in the 455 Kc/s i-f transformers, in receivers for this purpose, and the primary and secondary capacitances are often of the order of 750 $\mu\mu$F. This system of gain reduction is sometimes preferred to tapping down on 455 Kc/s i-f transformers using the more conventional component values as discussed in Sect. 2.

In receivers using a common converter, for F-M and A-M, it is usual to short out whichever transformer is not in use. It is not considered necessary, in most cases, to use switching with all i-f stages. Receivers adopting the method of using separate converters for improved oscillator stability on the v-h-f band, and a common i-f channel, do not require any switching in the i-f amplifier ; the converter not in use having the high tension removed. Switching of detector circuits is also unnecessary ; the last 455 Kc/s A-M transformer can be connected directly to a diode in the first a-f voltage amplifier valve, and a separate double diode, connected to the discriminator transformer, is used for detection in the F-M channel after F-M to A-M conversion has been carried out. Some valves, such as type 6T8, incorporate three diodes, a triode, and two separate cathodes, and so will allow all the detection and first a-f voltage amplification functions to be performed with one valve without switching being required. An alternative arrangement, common in receivers using ratio detectors, is to use the grid-cathode circuit of the last i-f valve (the driver stage) as a diode for A-M detection. When the receiver has a.v.c. applied to the driver stage, it is possible to select suitable values for the a.v.c. load resistor and the i-f by-pass capacitor to provide a satisfactory R-C combination for A-M detection.

The two transformers may have all the windings in one can, or two cans may be used for each i-f, one for the 10.7 Mc/s i-f, and the other for the 455 Kc/s i-f. This latter arrangement has the advantage that it reduces interaction between the windings, and also permits both transformers to be " slug " tuned without using elaborate mechanical arrangements. In some cases where the transformers are both mounted in one can, combinations of " slug " and capacitance adjustment are provided ; an example of this is illustrated in Ref 12.

Many arrangements have been suggested for combined i-f transformers, and some of these are described in Ref. 22. However, the circuit shown, or modifications of it, has received fairly wide acceptance as providing quite a good solution to the problem.

SECTION 4 : DESIGN METHODS

(*i*) *General*
(*ii*) *Critically-coupled transformers*
    (*A*) *Design equations and table*   (*B*) *Example*
    (*C*) *Design extension*   (*D*) *Conclusions*
    (*E*) *k measurement*
(*iii*) *Over-coupled transformers*
    (*A*) *Design equations and table*   (*B*) *Example*
    (*C*) *k measurement (when k is high)*
(*iv*) *Under-coupled transformers and single tuned circuits*
    (*A*) *Single tuned circuit equations*   (*B*) *Example*
    (*C*) *Under-coupled transformer equations*
    (*D*) *Example*
(*v*) *F-M i-f transformers*
    (*A*) *Design data*   (*B*) *Example*
(*vi*) *I-F transformer construction*
(*vii*) *Appendix : Calculation of coupling coefficients.*

(i) General

The design procedure for i-f transformers can be greatly simplified by the use of charts and tables. If certain assumptions are made, which approximate to practical conditions, the design procedure can be reduced to a few routine operations. Here we will consider only the two winding transformer using mutual inductance coupling ; the added capacitance coupling, which is always present, does not seriously affect the results particularly as its presence is taken into account when setting the coefficient of coupling (k).

The methods given below can be applied to inductive or capacitive coupling provided that the coefficient of coupling (k) is suitably interpreted [see appendix (vii) below for methods of calculating coupling coefficients]. The practical two winding transformer, as previously mentioned, has mixed coupling but this does not cause any difficulty when the two forms of coupling assist each other [suitable connections are given in (vi) below]. However, if the two forms of coupling are in opposition a rejection frequency is possible which can occur within the pass band of the transformer. This effect is well known to receiver designers who have accidentally reversed the connections of one of the i-f transformer windings, and found that the i-f stage gives practically no gain. This fact can also be made the basis of a useful method for measuring k, the capacitive coupling being increased (using a calibrated capacitor) until it equals the inductive coupling and zero voltage transfer then occurs at the working frequency. For further data on mixed coupling see Ref. 4.

The initial assumptions will be that the primary and secondary inductances L_1 and L_2 are given by $L = \sqrt{L_1 L_2}$. Also, it will be assumed that the values of Q do not alter appreciably over the range in which the selectivity curves are taken. We will **not** always take the primary and secondary Q's as being equal, and the advantages to be gained will become clear as we proceed.   The magnification factor, or Q, will be defined as

$$Q = \sqrt{Q_1 Q_2}$$

and provided the ratio of Q_1/Q_2 or Q_2/Q_1 is not greater than 2, the error in the usual design equations is negligible for most practical purposes.

As long as $\sqrt{Q_1 Q_2} \approx (Q_1 + Q_2)/2$ the error in any of the usual approximate design equations will be small.   If Q_1 and Q_2 differ by large amounts then the exact design equations are necessary and can be obtained from Refs. 2, 3, 6 and 8, or the design can be modified by using the universal selectivity curves to obtain the required results. It is of interest to note that the simplified equations given by Kelly Johnson (Ref. 1). Ross (Ref. 2) and Maynard (Ref. 6 and Figs. 9.17 and 9.18 of this book) are identical when it is assumed that $Q = Q_1 = Q_2$ and the various notations are made the same.

It may be thought that writing $Q = \sqrt{Q_1 Q_2}$ and $L = \sqrt{L_1 L_2}$ will be inconvenient since the i-f transformer, as constructed, will have its primary and secondary inductances and Q's equal.   However, in the majority of cases $L = L_1 = L_2$ is applied, and the method is extended to fulfil the condition that the unloaded primary and secondary Q's should be equal while allowing the required $Q = \sqrt{Q_1 Q_2}$ to be obtained in the receiver, without further adjustment.

Critical-coupling, or a close approach to it, is most often employed in i-f transformers but there is little difficulty in designing transformers for almost any degree of coupling.   All cases will be treated.

Universal selectivity and phase shift curves are given in Chapter 9, Sect. 10 (Figs. 9.17 and 9.18).   Additional charts and tables are given, to be used as described in the appropriate sections.

The design procedure generally consists of finding values of Q, k and stage gain for given bandwidths at some value of i-f ; or of finding the required bandwidth for values of Q and k previously determined.   For clarity the cases of critical-, over-, and under-coupled transformers will be dealt with separately.   Single tuned circuits are also included as they are sometimes required in i-f amplifiers.   Additional data for the design of F-M transformers will be given in Sect. 4(v).

Stagger tuning (Refs. 8, 13, 17 and 38) of i-f transformers (e.g. tuning primary and secondary to different frequencies) to give substantially the same bandwidths as over-coupled transformers, does not have a very wide application in F-M and A-M receivers (except in cases where variable selectivity is to be used) and will not be discussed in detail.

Stagger tuning of single and double tuned circuits is widely used in television receivers, but this is a different application from the case where it is applied to the comparatively narrow bandwidths of ordinary sound receivers.   In this case the transformer primary and secondary are tuned to the same frequency, but this is not necessarily the intermediate frequency.

Since i-f transformers for television receivers involve special problems they are not treated here (see Refs. 27, 28, 94, 95, 96 and 97)   Also, triple tuned transformers are not discussed, but the article of Ref. 26 gives an excellent treatment.

Finally, the design methods do not make allowance for regenerative effects, nor should they be applied for finding the shape of resonance curves at frequencies far removed from resonance.   The selectivity curve shapes are assumed to be perfectly symmetrical although in practice it will be found that this is seldom true.

(ii) Critically-coupled transformers

(A) Design equations and table

In general the procedure given is similar to that due to Kelly Johnson (Ref. 1) and Ross (Ref. 2).   However, these procedures assume that $Q = Q_1 = Q_2$; we shall take $Q = \sqrt{Q_1 Q_2}$.   For most cases it is not advisable to use the equations given below for Q_1/Q_2 or Q_2/Q_1 greater than about 2, unless some additional adjustment is made from the universal selectivity and phase shift curves.

For N critically-coupled transformers,

$$\rho = \left(1 + \frac{X^4}{4}\right)^{N/2} \tag{1}$$

$$X = \sqrt{2}\,(\rho^{2/N} - 1)^{\frac{1}{2}} \tag{2}$$

$$Q = \frac{Xf_0}{2\Delta f} \tag{3}$$

and $\quad \theta = \tan^{-1}\dfrac{2X}{2 - X^2}$ $\qquad\qquad$ (4)

where ρ = attenuation at Δf c/s off resonance
$\quad\quad N$ = number of identical transformers used ($N = 1$ for one transformer)
$\quad\quad f_0$ = resonant frequency (the i-f in our case)
$\quad\quad Q$ = $\sqrt{Q_1 Q_2}$ = $1/k_c$ (in which Q_1 and Q_2 are actual primary and second-
$\quad\quad\quad$ ary Q's ; k_c is critical-coupling coefficient)
$\quad 2\Delta f$ = total bandwidth for a given attenuation (ρ)
and $\quad\;\;\theta$ = phase shift between secondary current at resonance and secondary
$\quad\quad\quad$ current at Δf c/s off resonance.

The required design information is given in equations (1) to (4). Usually either ρ is stated for a given bandwidth and a known i-f, or X can be found to allow ρ to be determined. Once these two factors have been found, the determination of k and/or Q is a simple matter.

Table 1 lists various values of ρ and X. Suppose ρ is known, then X is read from the table and used in equation (3) to find Q (since $2\Delta f/f_0$ is already known). The coefficient of coupling is then $k_c = 1/Q$.

If complete resonance and/or phase shift curves are required, then the universal curves of Figs. 9.17 and 9.18 (Chapter 9, Sect. 10) are used. These curves apply to one transformer only. For N identical transformers the attenuation in decibels is multiplied by N ; for transformers which are not identical, the individual attenuations (in db) are added. Resonance and phase shift curves can also be determined directly, by using table 1 and eqns. (1) to (4).

In the application of the universal curves take $D = X = Q2\Delta f/f_0$ (this Q being $\sqrt{Q_1 Q_2}$ as determined) and $b = k/k_c = Qk$ (for critical coupling $b = 1$, in this case) ; which are the same as for $Q_1 = Q_2$. If the values of Q_1 and Q_2 differ by more than about 2 to 1, then the more exact expressions for D and b^2 are applied to check how closely the required conditions are approached, it being carefully noted that in all expressions on the curves involving a that the Q shown is Q_2.

TABLE 1. CRITICALLY-COUPLED TRANSFORMERS
For use with equations (1), (2) and (3)
N = Number of Transformers

Attenuation (ρ)		$N = 1$	$N = 2$	$N = 3$
Times Down	db Down	X	X	X
$\sqrt{2}$	3	1.41	1.14	1.01
2	6	1.86	1.41	1.25
4	12	2.76	1.86	1.57
7	17	3.73	2.21	1.81
10	20	4.46	2.46	1.95
20	26	6.32	2.96	2.26
40	32	8.96	3.54	2.56
70	37	11.9	4.08	2.82
100	40	14.1	4.46	3.02
1000	60	—	7.96	4.46
10 000	80	—	14.1	6.66

The maximum stage gain is given by

$$\text{Gain} = g_m Q \omega_0 L/2 \tag{5}$$

where g_m = mutual conductance of i-f valve (if conversion gain is required, conversion conductance (g_c) is substituted for g_m)

$\omega_0 = 2\pi \times$ resonant frequency (f_0) i.e. $f_0 = i\text{-}f$

$L = \sqrt{L_1 L_2}$; L_1 and L_2 are primary and secondary inductances

and $Q = \sqrt{Q_1 Q_2}$; Q_1 and Q_2 are primary and secondary magnification factors.

A condition, not specifically stated in the equations, is that $L_1 C_1 = L_2 C_2$ in all cases.

The maximum gain is usually converted to decibels, so that the gain at any point on the resonance curve can be found by subtraction of the attenuation, also expressed in decibels.

(B) Example and additional design extension

A 455 Kc/s i-f transformer, using critical-coupling, is required to give a total bandwidth of 20 Kc/s for an attenuation of 20 db (10 times).

(a) $f_0/2\Delta f = 455/20 = 22.75$.

(b) From table 1 we have $X = 4.46$ (since $N = 1$).

(c) From eqn. (3), $Q = 4.46 \times 22.75 = 101$.

(d) $k_c = 1/Q = 0.0099$.

(e) Select a suitable value for C_1 ($= C_2$) ; a capacitance of 100 $\mu\mu$F is satisfactory (made up of fixed + stray capacitances).

Then $L = \dfrac{25.33}{f^2 C} = \dfrac{25.33}{0.455^2 \times 100} = 1.22$ mH

(f in Mc/s ; C in $\mu\mu$F)

and take $L = L_1 = L_2$ since this is convenient in this case.

(f) The i-f valve (e.g. type 6SK7) has $g_m = 2$ mA/volt ($= 2000$ μmhos). From equation (5),

Max. stage gain $= \dfrac{2 \times 10^{-3} \times 101 \times 2\pi \times 455 \times 10^3 \times 1.22 \times 10^{-3}}{2}$

$= 352$ times (or 51 db).

(g) Some designs might stop here and the magnification factor would be taken as $Q = Q_1 = Q_2 = 101$. It would be realized that valve loading would have an effect although possibly nothing more would be done (or else some attempt would be made to allow for the plate and grid resistances by finding new values of Q_1 and Q_2).

Let us proceed further and ask whether the transformer as it stands fulfils the design conditions in a radio receiver. The answer is that obviously it does not, since it would be connected in most cases between two i-f valves, a converter and i-f valve or between an i-f valve and a diode detector. Suppose the connection between two i-f amplifier valves (type 6SK7 would be representative) is considered since this appears a fairly innocuous case. The plate resistance (r_p) of a type 6SK7 under the usual conditions of operation is 0.8 megohm. The short circuit input resistance, also under one set of operating conditions, is 6.8 megohms (this is calculated from the data given in Chapter 23, Sect. 5) ; other effects, which would alter this value, will be ignored for simplicity, although they may not be negligible. It is first required to determine what values of Q_1 and Q_2 are required to give $Q = Q_1 = Q_2 = 101$.

This is found from

$$Q_u = \frac{QR}{R - Q\omega_0 L} \tag{6}$$

where Q_u = unloaded Q

Q = loaded Q

R = parallel resistance across winding

$\omega_0 = 2\pi f_0$; (resonant frequency $= f_0$)

and L = inductance.

For the primary
$$Q = 101 \, ; \, R = r_p = 0.8 \text{ M}\Omega \, ;$$
$$\omega_0 = 2\pi \times 455 \times 10^3 \, ; \, L = L_1 = 1.22 \text{ mH}$$
and $Q\omega_0 L = 101 \times 2\pi \times 455 \times 10^3 \times 1.22 \times 10^{-3} = 0.352 \text{ M}\Omega.$

Then $Q_u = \dfrac{101 \times 0.8}{0.8 - 0.352} = 180.$

For the secondary
$$Q_u = \frac{101 \times 6.8}{6.8 - 0.352} = 106.8.$$

(C) Design extension

The value $Q_u = 180$ could not be obtained very easily, if at all, with a normal type of i-f transformer. In addition, the disadvantage of unequal primary and secondary Q's should be apparent. For values of Q only about 10% higher than that given, or where the transformer is coupled to a diode detector, the situation becomes so much worse that it is clear that a revised approach is necessary What is actually needed is

(1) A transformer with equal values of primary and secondary Q's when unloaded. These will be denoted by $Q_u = Q_{u1} = Q_{u2}$.

(2) The values of Q_{u1} and Q_{u2} to be such that when the transformer is connected into the i-f amplifier, and loaded by the valve output and input resistances, the desired value of $Q = \sqrt{Q_1 Q_2}$ will be obtained.

(3) The required coefficient of coupling (k) (critical for this particular example) to be unchanged. It will be described later how k can be pre-set to any desired value for any two circuits coupled together.

(4) Excessive values of Q_{u1} and Q_{u2} are to be avoided (see the previous method of determining Q_{u1}) as far as possible, because of the practical difficulties involved.

(5) The response curve of frequency versus attenuation to be that specified (or very close to it).

All of these conditions can be fulfilled very closely, provided the approximations made in deriving the design equations hold. A simple analysis of the circuits involved, and including the required conditions, gives

$$Q_u = \frac{\alpha + \sqrt{\alpha^2 + 2Q^2 R_1 R_2 \beta}}{\beta} \tag{7}$$

where $Q_u = Q_{u1} = Q_{u2}$ (unloaded primary and secondary Q)

$Q = \sqrt{Q_1 Q_2}$ (in which Q_1 and Q_2 are loaded primary and secondary Q's)

$R_1 =$ parallel resistance (r_p in our case) shunted across trans. primary

$R_2 =$ parallel resistance (grid input in our case) shunted across trans. secondary

$\alpha = Q(Q\omega_0 L)(R_1 + R_2)$; in all cases it will be taken that $L = L_1 = L_2$

and $\beta = 2[R_1 R_2 - (Q\omega_0 L)^2]$.

For our example :
$$Q = 101 \, ; \, Q^2 = 1.02 \times 10^4 \, ; \, R_1 = 0.8 \text{ M}\Omega \, ; \, R_2 = 6.8 \text{ M}\Omega$$
$$Q\omega_0 L = 0.352 \text{ M}\Omega \text{ (found previously)}$$
$$\alpha = 101 \times 0.352 \times 7.6 = 270 \, ; \, \alpha^2 = 7.29 \times 10^4$$
$$\beta = 2[5.44 - 0.124] = 10.63 \text{ (M}\Omega)^2$$
$$Q_u = \frac{270 + \sqrt{7.29 \times 10^4 + 2 \times 1.02 \times 10^4 \times 5.44 \times 10.63}}{10.63} = 131$$

so that $Q_{u1} = Q_{u2} = 131.$

To check that the transformer, when placed in the receiver, gives the desired value of $Q = \sqrt{Q_1 Q_2}$ use

$$Q = \frac{Q_u R}{Q_u \omega_0 L + R} \tag{8}$$

from which
$$Q_1 = \frac{131 \times 0.8}{0.458 + 0.8} = 83.4$$

$$Q_2 = \frac{131 \times 6.8}{0.458 + 6.8} = 123$$

and so

$Q = \sqrt{Q_1 Q_2} = \sqrt{83.4 \times 123} = 101$, which is the desired value (as determined previously).

(D) Conclusions

All that is required to design the specified transformer is to go through the simple steps (a) to (e) and, knowing R_1 and R_2, apply eqn. (7). Overall response and phase shift are determined from the universal curves, as explained previously.

It should be obvious that eqn. (7) will not hold under all practical conditions, but it is not limited by the ratio of Q_1/Q_2 or Q_2/Q_1, and failing cases can be checked by the condition for $\beta = 0$. It has been assumed for simplicity that $L = L_1 = L_2$ but this is not essential, and the design equation could be extended to the case of $L = \sqrt{L_1 L_2}$. In some failing case, if it is essential to fulfil the specified conditions, Q_1 and Q_2 (and if necessary L_1 and L_2) can be selected to give the desired values of Q and L; this will be illustrated in the section on the design of variable bandwidth crystal filters.

(E) k Measurement

The coefficient of coupling, k, for two circuits resonant at the same frequency, can be set on a Q meter (provided Q_b lies within the useful working range) using the relationship

$$Q_b = \frac{Q_{u1}}{1 + Q_{u1} Q_{u2} k^2} \tag{9}$$

If $\quad Q_{u1} = Q_{u2} = Q_u$

then $\quad Q_b = \dfrac{Q_u}{1 + (Q_u k)^2} \tag{9A}$

where Q_{u1} = primary Q (sec. o/c or detuned by large amount)

$\qquad Q_{u2}$ = secondary Q (pri. o/c or detuned by large amount)

$\qquad Q_b$ = Q to be obtained when primary and secondary are coupled and the secondary tuned to make the primary Q a minimum.

(Usual precautions as to can and earthy side of secondary winding being grounded to be observed—it is preferable to use the same order of connections for measurement as those to be used in the receiver.)

When the transformer has different primary and secondary Q's, it is often advantageous to use the higher Q winding as the primary when setting the coefficient of coupling; this applies particularly when the coupling is very loose.

The actual capacitance values tuning the primary and secondary for Q meter measurements should include the allowance made for stray capacitance otherwise incorrect slug positions (i.e. incorrect inductance values) will give rise to an error which can be avoided.

In our example we desire a value for $k = 0.0099$ (for critical-coupling when $Q = \sqrt{Q_1 Q_2} = 101$); $Q_u = 131$.

Then from (9A)

$$Q_b = \frac{131}{1 + (1.297)^2} = 48.9.$$

All that is required is to adjust the spacing between the two resonant circuits until the Q meter reads 48.9. The desired co-efficient of coupling has then been obtained. Alternatively, by transposing terms in the equation, k is given for any values of Q_b, Q_{u1}, and Q_{u2},

so that $\quad k = \sqrt{\dfrac{Q_{u1} - Q_b}{Q_{u1} Q_{u2} Q_b}} \tag{9B}$

The method applies directly to under-, over-, or critically-coupled transformers and is useful within the limits set by the usable range of the Q meter. For over-coupled transformers additional methods are sometimes required, and the procedure will be indicated in Sect. 5.

It is sometimes required to measure k in terms of critical or transitional coupling. In this case the circuits are loaded to give the values of Q_1 and Q_2 required when the transformer is connected into the receiver, and the following expressions can be applied :

$$\frac{k}{k_c} = \sqrt{\frac{Q_1 - Q_b}{Q_b}} \qquad (9C)$$

$$\text{and} \quad \frac{k}{k_t} = \sqrt{\frac{2Q_1 Q_2 (Q_1 - Q_b)}{Q_b (Q_1{}^2 + Q_2{}^2)}} \qquad (9D)$$

It might be noted that when $Q_1 = Q_2$ the expressions (9C) and (9D) are identical, as would be expected.

(iii) Over-coupled transformers

(A) Design equations and table
Here the method to be followed is based on that due to Everitt (Ref. 3).
Fig. 26.5 illustrates the terms used regarding bandwidth.
It should be noted that when the primary and secondary Q's differ appreciably, two peaks of secondary output voltage do not appear immediately critical-coupling is exceeded. The actual value of k, which corresponds to the condition for two peaks of secondary voltage, is called the **transitional-coupling factor** [see Chapter 9, Sect. 6(v)], and Ref. 8.

In what follows we shall use $Q = \sqrt{Q_1 Q_2}$ and $L = \sqrt{L_1 L_2}$, as was done for critically-coupled transformers, but this is not an approximation in the derivation of the design equations (10), (11) and (12) provided that $L_1 C_1 = L_2 C_2$. It will also be assumed, for simplicity, that the peaks of the response curve are of equal height and symmetrically placed in regard to f_0.

$$Q_k = A + \sqrt{A^2 - 1} \qquad (10)$$

$$A = \frac{(Qk)^2 + 1}{2Qk} \qquad (11)$$

$$\frac{2 \Delta f_p}{f_0} = k \sqrt{1 - \frac{1}{(Qk)^2}} \qquad (12)$$

$$\theta = \tan^{-1} \frac{2X}{1 - X^2 + (Qk)^2} \qquad (13)$$

where $Q = \sqrt{Q_1 Q_2} = \dfrac{1}{k_c}$ (in which Q_1 and Q_2 are primary and secondary Q's respectively ; k_c is coefficient of critical-coupling)

k = any coefficient of coupling equal to or greater than critical

A = gain variation from peak to trough (i.e. difference in transmission level)

$2 \Delta f_p$ = bandwidth between peaks ; $\sqrt{2} (2 \Delta f_p)$ is the total bandwidth for two other points on the resonance curve with the same amplitude as at f_0

θ = phase shift between the secondary current at resonance and the secondary current at Δf c/s off resonance

$X = (2 \Delta f / f_0) Q$

and f_0 = resonant frequency of transformer (i-f).

The universal resonance and phase shift curves of Figs. 9.17 and 9.18 (Chapter 9 Sect. 10) are directly applicable, using the exact expressions if desired and taking Q as Q_2 for all terms involving a. It is more convenient, and sufficiently accurate, to use the conditions for $Q_1 = Q_2$ when Q_1/Q_2 or $Q_2/Q_1 \not> 2$; in this case $b = Qk$ (or k/k_c), $D = (2 \Delta f / f_0) Q$ and since these expressions do not involve a, the value $Q = \sqrt{Q_1 Q_2}$ as determined in the design problem is used. A check will reveal that it is difficult to read any difference from the curves whichever method is used.

It should be observed that five points on the resonance curve are given directly from the design equations.

To find the maximum stage gain which occurs at the peaks, the equation (5) as given for critically-coupled transformers, is applied directly. Generally it is the average gain in the pass band ($\sqrt{2} \times 2\varDelta f$) which is required and this is given by multiplying eqn. (5) by

$$\frac{(Qk + 1)^2}{2[(Qk)^2 + 1]} \tag{14}$$

If the gain at f_0 (i.e. at the trough of the curve) is required, equation (5) is multiplied by

$$\frac{2Qk}{(Qk)^2 + 1} \tag{15}$$

which is the same as multiplying equation (5) by $1/A$, since equation (15) is equivalent to $1/A$.

Equations (14) and (15) can be evaluated directly from Fig. 26.4 when Qk is known ; the dotted line being for eqn. (14) and the solid line for eqn. (15). Equation 5 is multiplied by the gain reduction factor so found. The gain reduction indicated by eqn. (15) can also be read directly from the $1/A$ column in table 2.

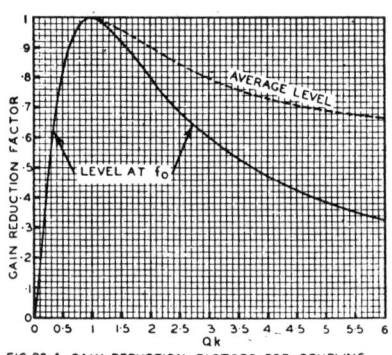

FIG.26·4 GAIN REDUCTION FACTORS FOR COUPLING
OTHER THAN CRITICAL

TABLE 2.   OVER-COUPLED TRANSFORMERS
For use with equations (10), (11), (12), (14), (15)

A (= peak to trough gain variation)		Qk	$\sqrt{1 - \dfrac{1}{(Qk)^2}}$	$(Qk)^2$	$\dfrac{1}{A}$
db	Times Down				
0	1.00	1.00	0.000	1.00	1.00
0.25	1.03	1.27	0.616	1.61	0.971
0.50	1.06	1.41	0.707	2.00	0.943
1	1.12	1.73	0.817	3.00	0.893
1.9	1.25	2.00	0.866	4.00	0.800
2	1.26	2.02	0.869	4.08	0.794
3	1.41	2.41	0.910	5.81	0.709
3.1	1.43	2.45	0.913	6.00	0.670
4	1.59	2.81	0.935	7.90	0.629
4.4	1.67	3.00	0.943	9.00	0.599
5	1.78	3.25	0.952	10.56	0.562
6	2.00	3.72	0.963	13.84	0.500
6.6	2.13	4.00	0.968	16.00	0.469
7	2.24	4.24	0.972	17.98	0.446

(B) Example

A 455 Kc/s i-f transformer is required to pass a band of frequencies 16 Kc/s wide (i.e. $\pm$ 8 Kc/s). The variation in gain across the pass band is not to exceed 0.5 db.

(a) From Fig. 26.5 it is reasonable to take the **total** bandwidth as 16 Kc/s and so the peak separation is $16/\sqrt{2} = 11.3$ Kc/s.

(b) $2\Delta f/f_0 = 11.3/455 = 0.0248.$

(c) From Table 2, $\sqrt{1 - 1/(Qk)^2} = 0.707.$

(d) From eqn. (12), $k = 0.0248/0.707 = 0.035.$

(e) From Table 2, $Qk = 1.41$
$$Q = 1.41/0.035 = 40.2.$$

(f) Assuming a value for $C_1 (= C_2)$ of 80 $\mu\mu$F (including strays)
$$L = 25.33/0.455^2 \times 80 = 1.53 \text{ mH}$$
and take $L = L_1 = L_2 = 1.53$ mH.

(g) To determine the average stage gain in the pass band. From Fig. 26.4 (or eqn. 14), with $Qk = 1.41$, reading from the dotted curve, gain reduction factor equals 0.97. Assuming we use a type 6J8-G converter valve having a conversion conductance of 290 μmhos (0.29 mA/volt), then from eqn. (5) and the gain reduction factor,
Average stage gain $= 0.97 \times \pi \times 0.29 \times 40.2 \times 0.455 \times 1.53$
$$= 24.7 \text{ times (or } 27.9 \text{ db)}.$$

(h) Assume that the transformer is connected between a type 6J8-G converter and a type 6SK7 voltage amplifier and that both valves are working under a particular set of operating conditions.

For type 6J8-G the conversion plate resistance $r_p = 4\text{M}\Omega = R_1$ and for the type 6SK7 the short circuit input resistance $= 6.8$ M$\Omega = R_2$ (as determined from Chapter 23, Sect. 5).

From Equation (7)
$\alpha = 40.2 \times 0.17 \times 10.8 = 73.8$; $\alpha^2 = 0.544 \times 10^4$
$\beta = 2[27.2 - 0.17^2] = 54.4$ approximately
$$Q_u = \frac{73.8 + \sqrt{0.544 \times 10^4 + 4.76 \times 10^6}}{54.4} = 41.4,$$

so that $Q_u = Q_{u1} = Q_{u2} = 41.4$ which is the unloaded value for primary and secondary Q's before the transformer is connected between the two valves ; the additional refinement in design is hardly necessary here, and it would be sufficient to make $Q = Q_1 = Q_2 = 40$ (approx.).

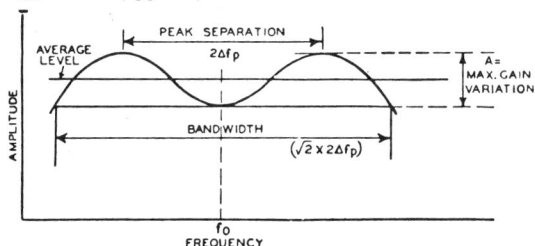

FIG. 26·5 ILLUSTRATION OF TERMS USED FOR OVER-COUPLED TRANSFORMERS

It can easily be checked, using the procedure set out for critically-coupled transformers, that the geometrical-mean of the loaded values Q_1 and Q_2 is 40.2 as required. Complete resonance and phase shift curves are plotted as previously explained. The values for k can be set in exactly the same way as explained in the section on critically-coupled transformers, if this is convenient. If k is high it is preferable to apply one of the following methods.

(C) k measurement (when k is high)

When k is high, one of the following methods may be used

The first of these uses the relationship
$$\Delta C = C_1 k^2/(1 - k^2)$$

or $k = \sqrt{\dfrac{\Delta C}{\Delta C + C_1}}$

where C_1 = capacitance required to tune the primary to resonance with the second-
                        ary open circuited.
          k = coefficient of coupling required (say k greater than 0.1 or so)
and ΔC = increment in capacitance required to tune the primary to resonance
                        when the secondary is short circuited.

As an illustration suppose $k = 0.2$ and $C_1 = 200 \ \mu\mu$F (the exact working frequency
may not always be convenient) then $\Delta C = 8.34 \ \mu\mu$F.   Using a " Q " meter, the spacing between primary and secondary is adjusted until this increment in capacitance is
obtained ; this gives the required value of k.

With some transformers, neither of the two methods given previously will be convenient, and a third method is required.   In this case a " Q " meter is again used,
and the two transformer windings are connected firstly " parallel aiding " and then
" parallel opposing."   Any convenient resonance frequency may be selected, and
the two capacitance values (C_1 and C_2) required to resonate the coils with the two
different connections are noted.   The coefficient of coupling is then given by

$$k = \left(\frac{1 + a^2}{2a}\right)\left(\frac{C_1 - C_2}{C_1 + C_2}\right)$$

where $a^2 = L_2/L_1$.   (This result was determined by J. B. Rudd).

For a pre-determined value of k the relationship

$$\Delta C = \frac{4akC_2}{1 + a(a - 2k)}$$

is applied ; where ΔC = change in capacitance = $C_1 - C_2$.

As an example, suppose $L_1 = L_2$ (i.e. $a^2 = 1$) the capacitance $C_2 = 100 \ \mu\mu$F and
$k = 0.1$ is required.   Then

$$\Delta C = \frac{4 \times 1 \times 0.1 \times 100}{1 + 1(1 - 2 \times 0.1)} = 22.2 \ \mu\mu\text{F}.$$

This method should not be used for small values of k (say below about 0.02) as it
does not take into account capacitive coupling.   For k equal to 0.02 or less the method
of Sect. (ii)E should always be used.   For k greater than about 0.2 the short-circuit/
open-circuit method, given previously, is usually the most convenient.   All measurements must be made with the transformer in its can, and the can should be earthed.

In cases where a " Q " meter is not available, the operation of the transformer can
be checked using a single stage amplifier and measuring the selectivity curve with a
signal generator and a valve voltmeter.   Alternatively, measurements can be made
in the receiver, and a typical case is illustrated in Chapter 27 Sect. 2(iv) in connection
with measurements on ratio detectors (see also Chapter 14 of Ref. 17).

(iv) Under-coupled transformers and single tuned circuits

It is sometimes necessary to use combinations of single tuned circuits or under-
coupled transformers in conjunction with over-coupled circuits to give a substantially
level response over the pass band.   Another application for the under-coupled transformer often arises when an improvement in selectivity is needed, without an excessive
loss in stage gain.   It has been shown by Adams (Ref. 9) that the optimum conditions
for selectivity and gain for a given Q, are obtained when the coefficient of coupling
is approximately 0.82 of critical.   However, it is readily shown for a transformer
having k equal to 0.5 k_c that the loss in gain is only about 2 db (approx. 0.8 of maximum gain).

The design equations (Ref. 1) that follow are applicable to single tuned circuits,
which will be considered first :

(A) Single Tuned Circuits

$\rho = (1 + X^2)^{N/2}$ (16)

$X = (\rho^{2/N} - 1)^{\frac{1}{2}}$ (17)

$Q = \dfrac{X f_0}{2 \Delta f}$ (18)

and $\qquad \theta = \tan^{-1} \dfrac{2\Delta f}{f_0} Q = \tan^{-1} X$ $\qquad\qquad$ (19)

where $\quad \rho =$ attenuation at Δf c/s off resonance
$\qquad N =$ number of identical tuned circuits (for a single tuned circuit $N = 1$)
$\qquad f_0 =$ resonant frequency
and $\quad \theta =$ phase shift between current at resonance and the current at Δf c/s off resonance.

The design of the single tuned circuit presents no difficulty, and it is only necessary to use Table 3 in conjunction with eqns. (16), (17) and (18). Selectivity and phase shift can be found from the universal curves of Figs. 9.17 and 9.18 (Chapter 9 Sect. 10), or selectivity can be evaluated directly from eqns. (16), (17) and (18) used in conjunction with Table 3.

TABLE 3. SINGLE TUNED CIRCUITS
For use with equations (16), (17) and (18)
$N =$ number of identical tuned circuits

Attenuation (ρ)		$N = 1$	$N = 2$	$N = 3$
Times Down	db Down	X	X	X
$\sqrt{2}$	3	1.00	0.644	0.509
2	6	1.73	1.00	0.767
4	12	3.87	1.73	1.23
7	17	6.93	2.45	1.63
10	20	9.95	3.00	1.91
20	26		4.36	2.52
40	32		6.25	3.27
70	37		8.31	4.00
100	40		9.95	4.53
1000	60			9.95

(B) Example

Two single tuned circuits are required to give an attenuation of 17 db for a total bandwidth of 10 Kc/s. The i-f is 455 Kc/s.

(a) From Table 3, $X = 2.45$ (since $N = 2$).

(b) $Q = 2.45 \times 455/10 = 111.5$ from eqn. (18).

(c) Assuming $C = 200 \ \mu\mu\text{F}$ (including all strays)
$\qquad L = 25.33/0.455^2 \times 200 = 0.611$ mH.

(d) The unloaded Q required depends on the combined effects of valve input and output resistance. Take the loading for the two transformers as being the same, for simplicity.
$\qquad$ Then suppose $r_p = 0.8$ MΩ and grid input resistance $= 6.8$ MΩ the effective shunt resistance is 0.715 MΩ.
$\qquad$ From eqn. (6)
$\qquad\qquad Q_u = 111.5 \times 0.715/(0.715 - 0.194) = 153$.

(e) The gain of each stage (twice that for a critically-coupled transformer) is $g_m Q \omega_0 L$, so that taking $g_m = 2$ mA/volt (2000 μmhos) in each case, stage gain $= 2\pi \times 2 \times 111.5 \times 0.455 \times 0.611 = 390$ times (or 51.8 db).

(f) Suppose we have the loaded Q given as 111.5 (as in our previous problem using $N = 2$), and we require the bandwidth for 6 db attenuation ; from table 3 obtain $X = 1.0$ and from eqn. (18) the total bandwidth ($2\Delta f$) is 4.08 Kc/s. In a similar manner the attenuation can be found when the bandwidth is stated. The resonance curves could also be used to find X ($= D$ in Chapter 9, in this case) and ρ.

(C) Under-coupled transformers

If the transformers use very loose coupling, the methods for single tuned circuits could be applied ($N = 2$ for each transformer) in conjunction with eqn. (7). This approach does not lead to very accurate results since the values of coupling are seldom less than 0.1 of critical and more often are of the order of 0.5 to 0.8 of critical.

General design equations applicable to transformers having any degree of coupling are given below, but it will be seen that they are not quite as tractable as in previous cases unless an additional factor such as Q, k, or Qk (i.e. a given proportion of critical-coupling) is specified. However, this will offer little difficulty.

$$X = [\{(1 - \alpha^2)^2 + (\rho^{2/N} - 1)(1 + \alpha^2)^2\}^{\frac{1}{2}} - (1 - \alpha^2)]^{\frac{1}{2}} \qquad (20)$$

$$\rho = \left[\frac{(1 + \alpha^2 - X^2)^2 + 4X^2}{(1 + \alpha^2)^2}\right]^{N/2} \qquad (21)$$

$$X = (2\Delta f/f_0)Q \qquad (22)$$

$$Qk = \alpha = \left[\frac{X\{X^2 + (X^2 + 4)(\rho^{2/N} - 1)\}^{\frac{1}{2}} - (X^2 + \rho^{2/N} - 1)}{\rho^{2/N} - 1}\right]^{\frac{1}{2}} \qquad (23)$$

$$\theta = \tan^{-1} 2X/\{1 - X^2 + \alpha^2] \qquad (13)$$

where ρ = attenuation at Δf c/s off resonance

$\quad \alpha = Qk = k/k_c$

$\quad Q = \sqrt{Q_1 Q_2} = 1/k_c$ (in which Q_1 and Q_2 are the primary and secondary Q's and k_c is critical coupling coefficient)

$\quad k$ = any coefficient of coupling

$\quad N$ = number of identical transformers

$\quad f_0$ = resonant frequency

and $\quad \theta$ = phase shift between secondary current at and off resonance.

The restriction of Q_1/Q_2 or $Q_2/Q_1 \not> 2$ is applied, as previously explained.

It may be observed that these equations are the most general ones, e.g. if $Qk = 1$ the equations reduce to those for critical-coupling.

Stage gain is given by evaluating eqn. (5) and multiplying by eqn. (15) (or reading the gain reduction factor from Fig. 26.4).

The coefficient of coupling can be set as described for critically-coupled transformers.

(D) Example

A 455 Kc/s i-f transformer is required to give a total bandwidth of 20 Kc/s for an attenuation of 4 times. The transformer is to be connected between a voltage amplifier, having a plate resistance of 0.8 MΩ (e.g. type 6SK7), and a diode detector having a load resistance of 0.5 MΩ.

In this case other loading effects due to the a.v.c. system etc. will be neglected. For a typical case where the a.v.c. diode plate is connected by a fixed capacitor to the i-f transformer primary, there will be appreciable damping of the primary circuit due to the diode circuit (approx. $R_L/3$ when diode is conducting). This damping will not be constant for all signal input voltages, particularly if delayed a.v.c. is used.

(a) The difficulty first arises in evaluating X. If we select a suitable value for Q the problem becomes quite straightforward.

To select a value for Q it is necessary to realize that an unloaded value of 150 would be about the absolute maximum with normal types of construction, and even this figure is well on the high side unless an " iron pot " or a fairly large can and former are used. Assume $Q = 150$ for this problem (so far as the procedure is concerned it is unimportant if a lower value is selected).

The next point is that circuit loading will set a limit to the value of $Q \omega_0 L$ ($= Q/\omega_0 C$). Now L will be set, normally, by the minimum permissible value of C. Suppose $C = 85$ $\mu\mu$F including strays, then $L = 1.44$ mH ; and we will take $L = L_1 = L_2$ and $C = C_1 = C_2$ for practical convenience Better performance would be possible by making $L_1 > L_2$ but the improvement is only small, and hardly worthwhile unless the secondary load is very small.

(b) For our problem

$$\omega_0 L = 1/\omega_0 C = 4120 \ \Omega \ ; \ \text{also} \ R_1 = 0.8 \ \text{M}\Omega \ \text{and}$$
$$R_2 = 0.5/2 = 0.25 \ \text{M}\Omega \ \text{(half the d.c. diode load resistance)}.$$

Then applying eqn. (8)

$$Q_2 = \frac{150 \times 0.25}{0.25 + (150 \times 4.12 \times 10^{-3})} = 43.3$$

and $Q_1 = \dfrac{150 \times 0.8}{(0.8 + 0.618)} = 84.6$

so that

$$Q = \sqrt{Q_1 Q_2} = \sqrt{84.6 \times 43.3} = 60.5.$$

(c) From eqn. (22)

$$X = \frac{20 \times 60.5}{455} = 2.66.$$

(d) From eqn. (23)

$$Qk = \left[\frac{2.66\{2.66^2 + (2.66^2 + 4)(4^2 + 1)\}^{\frac{1}{2}} - (2.66^2 + 4^2 - 1)}{4^2 - 1} \right]^{\frac{1}{2}} = 0.93$$

and $k = 0.93/60.5 = 0.0154$.

(e) From eqns. (5) and (15) (the solid curve of Fig. 26.4)

$$\text{Stage gain} = \frac{(2 \times 60.5 \times 4.12)}{2}\, 0.996$$

$$= 249 \text{ times or } 47.9 \text{ db.}$$

(f) The completed transformer has primary and secondary Q's of 150 (before connection into the receiver), a coefficient of coupling equal to 0.0154 (which is 0.93 of critical-coupling when the transformer windings are loaded) primary and secondary inductances of 1.44 mH and tuning capacitances of 85 $\mu\mu$F (including strays). The stray capacitances across the primary would be valve output (7 $\mu\mu$F for type 6SK7) plus distributed capacitance of winding, plus capacitances due to wiring and presence of shield can ; across the secondary there would be diode input capacitance (about 4 $\mu\mu$F for a typical case), plus distributed capacitance of secondary winding plus wiring and shield can capacitances. The total capacitances can be measured in the receiver or estimated using previous experience as a guide ; typical values would be 10-20 $\mu\mu$F depending on the type of i-f transformer, valves etc. If the second valve is not a diode, the input capacitance should also include that due to space charge, Miller effect etc. as discussed in Chapter 2 Sect. 8 ; Chapter 23 Sect. 5 and Sect. 7 of this chapter. However, in most practical cases the total capacitance across the secondary is estimated by adding a suitable value to the valve input capacitance. Changes in input resistance which would affect the loading across the i-f transformer secondary, are also discussed in these same sections. Input capacitance changes with a.v.c. are considered in Sect. 7 of this chapter.

(g) A complete resonance curve can be obtained from Fig. 9.17 (Chapter 9 Sect. 10), by taking

$$D = Q2\Delta f/f_0 \; (= X) \text{ and } b = Qk = k/k_c,$$

or directly from the design equations (20), (21) and (22).

(v) F-M i-f transformers
(A) Design data

The design methods given so far are applicable to both F-M and A-M transformers, but there are additional data available which will be of assistance (see also Sect. 9(ii) of this chapter for an alternative design procedure based on permissible non-linear distortion).

Bandwidth requirements are of importance, and it is fairly generally accepted that the i-f amplifier should be capable of passing all significant sideband frequencies of the frequency modulated wave ; where significant sidebands are taken as those having amplitudes which are greater than about 1% of the unmodulated carrier amplitude. The bandwidths for this condition can be found from Table 4 (see also Ref. 19) for commonly occurring values of modulation index.

TABLE 4. BANDWIDTHS FOR USE WITH F-M TRANSFORMERS

$$\text{Modulation Index} = \frac{\Delta F}{f} = \frac{\text{carrier frequency deviation}}{\text{audio modulating frequency}}$$

Values for $\dfrac{\Delta F}{f}$ may be interpolated with sufficient accuracy

$\Delta F/f$ =	0.01-0.4	0.5	1.0	2.0	3.0	4.0	5.0	6.0
Bandwidth =	$2f$	$4f$	$6f$	$8f$	$12f$	$14f$	$16f$	$18f$

$\Delta F/f$ =	7.0	8.0	9.0	10.0	12.0	15.0	18.0	21.0
Bandwidth =	$22f$	$24f$	$26f$	$28f$	$32f$	$38f$	$46f$	$52f$

As an example, for the, F-M broadcast band $\Delta F = \pm 75$ Kc/s and the highest audio frequency is 15 Kc/s, then $\Delta F/f = 75/15 = 5$.

Then the required bandwidth is $16 \times 15 = 240$ Kc/s.

The highest audio frequency is chosen because this imposes the most severe requirements on bandwidth ; e.g. suppose we had taken $f = 7.5$ Kc/s, then the modulation index would be 10, and the bandwidth $= 28f = 28 \times 7.5 = 210$ Kc/s.

The bandwidths actually employed in a receiver should also make allowance for possible drift in the oscillator frequency. A reasonably good oscillator should not drift by more than about ± 20 Kc/s when operating around 110 Mc/s ; so that an additional 40 Kc/s should be added to the bandwidth. Of course, this is only a rough approximation, since the oscillator frequency variation is random and would introduce additional frequency modulation ; the determination of the true bandwidth would be quite a difficult problem, unless several simplifying assumptions are made.

From what has been said, it appears that the receiver total bandwidth should be about 280 Kc/s to fulfil the most severe requirements. However, most practical receivers limit the total bandwidth to about 200 Kc/s, which is not unreasonable since the average frequency deviation is about ± 50 Kc/s.

For such large bands of frequencies to be passed through tuned circuits which do not give transmission (for the bandwidth desired) without attenuation it is necessary to have some criterion which will allow the permissible amount of attenuation to be estimated. To eliminate non-linear distortion the circuits should provide a uniform amplitude and a linear phase characteristic over the operating range. Curvature of the phase characteristic of the tuned circuits will cause non-linear a-f distortion, while curvature of the amplitude characteristic may cause additional distortion if the amplitude happens to drop below the operating voltage range of the amplitude limiting device incorporated in the receiver. A suitable criterion can be determined from the phase angle/frequency characteristics of the tuned circuits (the phase angle being that between the secondary current at resonance to that at Δf c/s off resonance). Inspection of universal phase shift curves will show that the greatest range of linearity of phase shift versus frequency change, is given by critically-coupled transformers (Ross, Ref. 2) ; but slight overcoupling does not lead to excessive non-linearity. Overcoupling has some advantages, in particular slightly greater adjacent channel selectivity can be obtained ; but there is the disadvantage of more difficult circuit alignment. As an extension of this work, Ross (Ref. 2) has also shown that for a critically-coupled transformer a suitable criterion of permissible non-linearity is that

$$X \not> 2 \tag{24}$$

where we will take

$X = (2\Delta f/f_0)Q$ (this is the same X as previously)
$2\Delta f$ = total bandwidth
f_0 = central carrier frequency

and $Q = \sqrt{Q_1 Q_2}$ (where Q_1 and Q_2 are the primary and secondary magnification factors).

The amount of introduced amplitude modulation can be estimated from

$$m = \frac{\rho - 1}{\rho + 1} \qquad (25)$$

where m = amplitude modulation factor
and ρ = attenuation at bandwidth of twice deviation frequency (i.e. $2\Delta f = 2\Delta F$).
Fig. 26.6 shows directly values of m for various values of X. Values of m are of importance since they allow an estimate to be made of the amplitude limiting requirements demanded from whatever device is incorporated in the receiver to "iron out" amplitude variations.

Many F-M receiver designs allow anything from 20% to 50% of introduced amplitude modulation, but these figures should always be considered in connection with the amount of non-linear distortion introduced by the tuned circuits [see Sect. 9(8)]. Good designs often allow considerably less than 20% of introduced amplitude modulation.

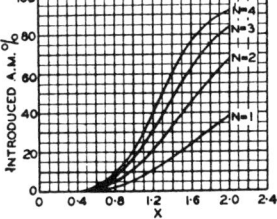

FIG.26·6 PERCENTAGE OF INTRODUCED A-M IN AN F-M CURRENT FOR N CRITICALLY COUPLED TRANSFORMERS

It is also worth noting, before leaving this section, that the carrier frequency should be regarded as a reference point only, since, unlike amplitude modulation, its amplitude varies and becomes zero under some conditions of modulation. This is the basis of a method due to Crosby (Ref. 25) used for measuring frequency deviation.

(B) Example
A F-M i-f amplifier is required using three critically-coupled transformers. The i-f is 10.7 Mc/s and the frequency deviation ± 75 Kc/s ; the highest a-f modulating frequency is 15 Kc/s (critical-coupling has been selected in this case but a combination of critical and over-coupled transformers might lead to a better solution).
The converter valve to be used has a conversion r_p of 1.5 MΩ and a conversion conductance of 475 μmhos ; and the two i-f valves each have plate resistances of 2 MΩ and g_m = 5000 μmhos.
  (a) For simplicity it will be taken that any additional selectivity, due to the other tuned circuits in the receiver, is negligible. Also, as will be illustrated, the dynamic impedances of the i-f transformers will be so low as to render additional damping due to plate and grid input resistances negligible ; this is not true, however, if the final transformer is connected to a limiter stage because of grid current damping—some consideration will be given to this later.
  (b) From previous considerations regarding bandwidth, in connection with Table 4, we will adopt 220 Kc/s as a compromise.
    If we design on the limit of $X = 2$, then from eqn. (3)
$$Q = \frac{2 \times 10.7 \times 10^6}{220 \times 10^3} = 97.4$$
    and $k_c = 1/Q = 0.0103$.
  (c) Taking $2\Delta f = 2\Delta F = 150$ Kc/s, for maximum frequency deviation, since $Q = 97.4$ and $N = 3$, then, from eqn. (3), $X = 1.365$ and, from equation (1), $\rho = 2.55$, whence from eqn. (25) or Fig. 26.6,
$$m = \frac{2.55 - 1}{2.55 + 1} = 0.436 \text{ or } 43.6\% \text{ amplitude modulation.}$$
  (d) This is a severe additional requirement for limiters etc., and could also lead to high distortion, although the condition for linearity of phase shift with frequency is fulfilled. The distortion can be found from eqn. (54) for each of the transformers.
  (e) Select suitable values for C_1 and C_2. To obtain the highest possible dynamic impedance these are usually made rather small. Take
$$C_1 = C_2 = 60 \ \mu\mu\text{F (including strays)}$$
$$L(= L_1 = L_2) = \frac{25\,330}{10.7^2 \times 60} = 3.68 \ \mu\text{H.}$$

There would be an advantage in making $C_1 < C_2$ and $L_1 > L_2$ but this is awkward for winding i-f's on a machine.

(f) The dynamic impedance of each winding (considered uncoupled from one another) is
$$R_0 = Q\omega_0 L = 97.4 \times 2\pi \times 10.7 \times 3.68 = 24\,000\ \Omega$$
which is very much less than the valve plate or input resistances in typical stages.

(g) To find the overall gain of the i-f stages :—For the first stage, connected to the converter, we are concerned with conversion gain.
From eqn. (5) and step (f),
Conversion gain $= 475 \times 10^{-6} \times 24\,000/2 = 5.7$ times (15.1 db). For the second and third stages in each case, gain $= 5.7 \times 5/0.475 = 60$ times (35.6 db). Hence the overall gain is $15.1 + (2 \times 35.6) = 86.3$ db.

(h) If the third transformer connects to a limiter its design should be modified for best results ; however, this is not done in many receivers. Grid current will alter the effective input capacitance of the limiter valve and cause very appreciable detuning of the transformer secondary. To overcome this detuning it may be necessary to make the pass band of this transformer somewhat greater than for the other two, or else to use a very large capacitance (of the order of 600 $\mu\mu$F) to tune the secondary (tapping down is also effective). It has also been suggested that Q_1 should equal Q_2 in this case (Ref. 21). The important point, apart from possible distortion, is that the susceptibility to certain types of impulse noise is increased if the tuned circuits are not accurately aligned to the centre frequency of the discriminator. Further, when the i-f circuit is detuned, additional amplitude modulation will be introduced, and this increases the difficulty of obtaining effective limiting. Grid current damping, of course, would tend to offset the effect to some extent (see also Refs. 21, 24). Since this transformer would probably be non-standard in any case, it would be advantageous to make L_1 as large as possible, to assist in keeping the stage gain high. This follows as a result of the stage gain being directly proportional to $\sqrt{L_1 L_2 Q_1 Q_2} = LQ$.

(i) It is important to check whether the i-f transformers will provide sufficient adjacent channel selectivity. This will depend on the frequency allocations of the various transmitters. In the U.S.A. local F-M transmissions are usually spaced 400 Kc/s apart, and the past experience of some designers has indicated that adjacent selectivity is adequate if the receiver bandwidth is not more than about 800 Kc/s for a relative attenuation of 60 db.
In the example above there are three critically-coupled transformers each having a nominal Q of 97 :
From Table 1, $X = 4.46$ for 60 db attenuation.
From eqn. (3)
$$2\Delta f = \frac{Xf_0}{Q} = \frac{4.46 \times 10.7 \times 10^6}{97} = 494 \text{ Kc/s.}$$
However, commercial F-M receivers have given reasonably satisfactory performance with two i-f transformers with Q's of about 75 leading to a bandwidth of 1140 Kc/s for 60 db attenuation. It might also be noted that larger commercial receivers generally use three transformers with Q's of about 70 to 75 and not 97 as found in the worked example, which is not to be taken as indicating good design practice.
Whether the selectivity performance is entirely adequate will depend very largely on the relative field strengths of the desired and undesired transmissions at the point of reception, and it is to be expected that more exact figures will only be decided on after a number of years' experience with practical operating conditions. The problem is similar to that of the A-M broadcast receivers (using 455 Kc/s i-f transformers) where it is usually taken that adjacent channel selectivity will be adequate if the overall band-

width does not exceed about 40 Kc/s for 60 db attenuation relative to the response at the centre frequency, although many commercial receivers have bandwidths of less than 30 Kc/s for the same attenuation.

(vi) I-F transformer construction

The methods of (i) to (v) in this section will allow the necessary design data for an i-f amplifier to be collected together. The final step is to determine the winding details and physical arrangement of the transformers. It is not proposed to discuss the merits of various types of windings, but merely to give a few details which have proved helpful in practice.

For transformers working at the higher frequencies, the windings are quite often solenoids and the determination of the number of turns required is a simple matter. Satisfactory results can be obtained by applying Hayman's modification of Wheeler's formula as given in Chapter 10. Methods are also set out for determining the number of turns per inch, and suitable wire diameters for obtaining optimum values of Q. Usually wire gauges between 18 and 28 s.w.g. are suitable as they are not so heavy as to be awkward to bend and they do not tear the usual type of coil former when the construction requires the leads to be passed through the inside of the former down to the base connections. Some error in the calculated number of turns will be apparent unless allowance is made for the inductance of leads. The number of turns required is finally determined experimentally in any case so that the calculated number of turns provides a good starting point.

Measurements must always be made with the coils in the cans because the effects of the can, brass mounting bosses, slugs etc. on inductance and Q are quite large.

Powdered iron " slugs " are commonly used for setting the inductance values and, for 10.7 Mc/s in particular, the iron must be very finely divided if the coil Q is not to be seriously changed as the cores are moved through the windings. Sufficient inductance variation also requires that the " slugs " have a certain minimum size.

Silvered mica fixed capacitors are to be preferred for good frequency stability with temperature changes, particularly at high frequencies such as 10.7 Mc/s, and temperature compensation using negative temperature coefficient capacitors is essential if the best results are to be obtained. Cheaper mica types are very often used at the lower frequencies around 455 Kc/s. If capacitance trimmers are to be used, care must be taken in their choice as pressure types are often mechanically unstable and sometimes have very low Q values. Suitable wax or other treatment should be applied to fixed capacitors to offset the effects of changes in humidity.

Coils should be baked* to remove moisture and then given suitable wax or varnish treatment to prevent humidity changes altering their properties. The electrical characteristics of the coils will be altered by wax etc. and it is essential to check the final values for Q, k etc. after the treatment is complete.

Typical former diameters for i-f transformers range from about $\frac{3}{8}$ in. to $\frac{3}{4}$ in., and the grade of material used for the former will affect the stability of the transformer directly when variations in temperature and humidity occur.

The design of transformers with pie windings (a larger number of pies generally reduces distributed capacitance and increases Q) is not as simple as for solenoids since most equations require a knowledge of coil dimensions which are not always available. There are also optimum sizes of winding to give the highest possible Q—see Chapter 10. A method which has proved satisfactory in practice is to make measurements on various types of coils which are likely to be used fairly often and apply the relationship

$$N = B\sqrt{L} \tag{26}$$

where N = turns per pie
       L = inductance in μH
and B = experimentally determined constant.

As an example : A winding is to be made to have an inductance of 1.44 mH. Previous experiments have shown that, in the frequency range of 300-900 Kc/s and for

*See also Chapter 11 Sect. 7.

an inductance of about 0.5-2 mH, a two pie winding on a 9/16 in. former (each pie 5/32 in. wide, with 3/32 in. spacing between the pies and using 5/44 A.W.G. Litz wire) has the factor $B = 4.33$.   Then from eqn. (26)

$$N = 4.33\sqrt{1440} = 165 \text{ turns per pie};$$

the two pies each, of 165 turns, being connected series aiding.   The same method can be applied in cases where it is convenient, to any type of winding.

Since the presence of the iron " slug " will affect inductance (and r-f resistance) it is necessary to determine its effect and also to calculate the variation in inductance which can be made.   The turns required are found for the condition with the " slug " in the winding and in the position giving the mean inductance value.   This means that the value of L used in eqn. (26) will be less than the calculated value by the increase due to the iron.

The value of the inductance (L) using a powdered iron core (e.g. magnetite) can be found from

$$L = L_0\left[1 + a\left(\frac{r_1}{r_2}\right)^2\left(\frac{l_1}{l_2}\right)(\mu_{eff} - 1)\right] \qquad (27)$$

where L_0 = inductance of air cored coil
 r_1 = radius of iron core
 r_2 = mean radius of coil
 l_1 = length of core
 l_2 = length of coil
 μ_{eff} = effective permeability of iron core (Refs. 88 and 89 list values for μ and μ_{eff} for various types of iron powders ; typical values for μ_{eff} are from 1.5 to 3, depending on the type of iron)
 $a = 0.8$ when $l_1 < l_2$
and $a = 1$ when $l_2 < l_1$.

The iron cores in common use are about $\frac{3}{8}$ in. to $\frac{3}{4}$ in. in diameter and range in length from about $\frac{1}{4}$ in. to 1 in.   An inductance change of about $\pm 10\%$ when the " slug " is moved through the winding, is generally sufficient for most requirements.   The dimensions required for solving eqn. (27) are available if the experimental procedure previously suggested has been carried out on air cored coils ; or the whole procedure can be carried out experimentally.

Some manufacturers make up 455 Kc/s i-f transformers completely enclosed in powdered iron pots.   There is often little difficulty with this construction in obtaining Q's in the order of 150.   Stray capacitances are often large, however, and sometimes lead to very unsymmetrical resonance curves.

At the lower frequencies (up to about 1 or 2 Mc/s) Litz wire is advantageous for obtaining high Q values and 3, 5, 7 and 9 strands of about 44 A.W.G. (or near S.W.G. or A.W.G. gauges) wire are common ; Q values greater than about 120 will require some care in the transformer construction, and the size of can selected will materially affect the value obtainable (see Ref. 91 for illustration).

The required values of L_1, L_2, k, etc. for developmental purposes are conveniently found using a Q meter.   Methods for setting k, using a Q meter, have already been outlined in this section.   For an experimental transformer it is of assistance to place the windings on strips of gummed paper (sticky side outwards) wrapped around the former.   In this way the windings can be moved quite readily along the former.

It should be noted that some variation in k can be expected when the transformer is connected in the receiver because of alteration of " slug " position, added top capacitance coupling (this occurs for example because of capacitance between the a.v.c. and detector diodes) etc.   Regeneration is also troublesome as it alters the effective Q values, and hence the conditions for critical k.   To avoid overcoupling it is sometimes desirable to make the value of k somewhat less than is actually required (often about 0.8 to 0.95 of the critical value depending on the receiver construction and i-f).   If all added coupling is accounted for and the Q values are those specified then no difficulty arises.   Some slight increase in coupling, in the receiver, is not necessarily serious, because the loaded primary and secondary Q's are not always equal.   In this case a double hump in the secondary voltage does not appear until transitional coupling has been exceeded, and the k required for this to occur is always higher than k critical ; transitional and critical k are the same, of course, when primary and secondary Q's are equal.

For the capacitance and mutual inductance coupling to be aiding, the primary and secondary windings are arranged so that if the plate connects to the start of the primary, then the grid (or diode plate) of the next stage connects to the finish of the secondary winding ; both coils being wound in the same direction. This method of connection also assists in keeping the undesired capacitance coupling to a minimum. The order of base connections is also important in reducing capacitance coupling and the grid and plate connections should be as far from one another as possible.

The cans to be used with i-f transformers should be as large as is practicable. They are generally made from aluminium, although copper was extensively used at one time. Cans should preferably be round and seamless. Perfect screening is not obtained, in general, and care is necessary in the layout of the various stages to ensure that the transformers are not in close proximity to one another. When mounting the transformer into a can, if there is a choice as to the position of the leads (although this is largely determined by the valve type available) it is always preferable to bring the connections for each winding out to opposite ends, as this reduces stray capacitance coupling. The effects of the coil shield on inductance and r-f resistance can be calculated (see Chapter 10 and Ref. 4, p. 134) and the results serve as a useful guide, but direct measurement on the complete transformer is the usual procedure. Mechanical considerations generally ensure that the can is thick enough to provide adequate shielding (for considerations of minimum thickness see Ref. 4, p. 135).

Methods for determining gear ratios, winding pitch and so on, for use with coil winding machines are discussed in the literature (Refs. 29, 30, 31, 32) ; see also Chapter 11, Sect. 3(iv) in particular.

Finally, the measure of the success of any i-f transformer design will be how closely the predicted performance approaches the actual results obtained when the transformer is connected into the receiver.

(vii) Appendix : Calculation of Coupling Coefficients

For the calculation of the coefficient of coupling (see also Chapter 9) it is helpful to use the following rules of procedure :

(a) If the coupling circuit is drawn as a T section, the coefficient of coupling is given by the ratio of the reactance of the shunt arm divided by the square root of the product of the sum of the reactances of the same kind in each arm including the shunt arm,

i.e. $k = \dfrac{X_m}{\sqrt{(X_1 + X_m)(X_2 + X_m)}}$

(b) If the coupling circuit is drawn as a $\varPi$ section, the coefficient of coupling is given by the ratio of the susceptance of the series arm divided by the square root of the product of the sum of the susceptances of the same kind in each arm including the series arm,

i.e. $k = \dfrac{B_m}{\sqrt{(B_1 + B_m)(B_2 + B_m)}}$

A few examples will now be given to illustrate the procedure. From rule (a applied to Fig. 9.15,

$$k = \frac{X_m}{\sqrt{(X_1 + X_m)(X_2 + X_m)}} = \frac{L_m}{\sqrt{(L_1 + L_m)(L_2 + L_m)}}$$

For Fig. 9.12 using rule (b)

$$k = \frac{B_m}{\sqrt{(B_1 + B_m)(B_2 + B_m)}} = \frac{C_m}{\sqrt{(C_1 + C_m)(C_2 + C_m)}}$$

For Fig. 9.14 using rule (a)

$$k = \frac{1/C_m}{\sqrt{\left(\frac{1}{C_1} + \frac{1}{C_m}\right)\left(\frac{1}{C_2} + \frac{1}{C_m}\right)}} = \sqrt{\frac{C_1 C_2}{(C_1 + C_m)(C_2 + C_m)}}$$

For Fig. 9.13 using rule (b)

$$k = \frac{1/L_m}{\sqrt{\left(\frac{1}{L_1} + \frac{1}{L_m}\right)\left(\frac{1}{L_2} + \frac{1}{L_m}\right)}} = \sqrt{\frac{L_1 L_2}{(L_1 + L_m)(L_2 + L_m)}}$$

It is sometimes helpful to make use of the obvious relationship in the form of k for the two types of circuit, by drawing dual networks and writing down the value of k for the duals as though capacitances were inductances and vice versa, e.g. the inductances of Fig. 9.13 are the duals of the capacitances of Fig. 9.14, and so the value of k is written down for, say, Fig. 9.14 and if k is required for Fig. 9.13 we merely substitute the symbols $L_1 \, L_2 \, L_m$ for $C_1 \, C_2 \, C_m$.

As a more difficult example the coefficient of coupling for Fig. 9 16(A) will be calculated. Firstly redraw the two coupled circuits as two T sections and then combine the two directly connected series arms. The complete coupling circuit now comprises a series arm $L_1 - M_1$; a Π section made up from M_1, $L_1' - M_1 + L_2' - M_2$, M_2; a series arm $L_2 - M_2$. Transforming the Π section to a T section and adding the two series arms of the previous circuit to the new series arms, we have a single T section in which the first series arm is

$$\frac{M_1(L_1' - M_1 + L_2' - M_2)}{L_1' + L_2'} + L_1 - M_1 = \frac{L_1(L_1' + L_2') - M_1(M_1 + M_2)}{L_1' + L_2'}$$

the second series arm is

$$\frac{M_2(L_1' - M_1 + L_2' - M_2)}{L_1' + L_2'} + L_2 - M_2 = \frac{L_2(L_1' + L_2') - M_2(M_1 + M_2)}{L_1' + L_2'}$$

and the shunt arm is $\dfrac{M_1 M_2}{L_1' + L_2'}$

Applying rule (a), and writing $L_m = L_1' + L_2'$,

$$k = \frac{M_1 M_2}{\sqrt{[L_1 L_m - M_1(M_1 + M_2) + M_1 M_2][L_2 L_m - M_2(M_1 + M_2) + M_1 M_2]}}$$

$$= \frac{M_1 M_2}{\sqrt{(L_1 L_m - M_1^2)(L_2 L_m - M_2^2)}}$$

$$= \frac{k_1 k_2}{\sqrt{(1 - k_1^2)(1 - k_2^2)}}$$

where $k_1 = \dfrac{M_1}{\sqrt{L_1 L_m}}$ and $k_2 = \dfrac{M_2}{\sqrt{L_2 L_m}}$

As a further example a bridged T coupling section will be considered, made up from inductances bridged by a capacitance. This corresponds to the familiar case of a mutual inductance coupled transformer with added top capacitance coupling (e.g Fig. 9.16D) and also refers to any practical i-f transformer because of the presence of stray capacitance affecting the coupling between the primary and secondary windings.

The procedure in this case is to use rule (a) and determine the coupling coefficient due to mutual inductance (the inductances are considered as forming a T network); call this k_m. Next use rule (b) to determine the coupling coefficient due to capacitance coupling (the capacitances form a Π network); call this k_c. Then if the connections are such that the two forms of coupling aid each other, the resultant coupling coefficient is

$$k_T = \frac{k_m + k_c}{1 + k_m k_c}$$

and if $k_m k_c \ll 1$ then

$k_T = k_m + k_c$ (approximately).

For the case of the two forms of coupling being in opposition

$$k_T = \frac{k_m - k_c}{1 - k_m k_c}$$

and if $k_m k_c \ll 1$ then

$k_T = k_m - k_c$ (approximately).

The above expressions for k_T can be derived quite simply from data given in Ref. 101, or directly using a similar method to that set out below. The results are most helpful in determining rapidly the effect of additional top capacitance coupling e.g. an i-f transformer is connected between the last i-f amplifier valve and a diode detector, and the a.v.c. diode is connected to the primary of the transformer in the usual manner. Then clearly there is top capacitance coupling added across the i-f transformer due to the direct capacitance between the two diodes, and a knowledge of the value of this capacitance will allow the added coupling to be taken into account when designing the transformer. If the capacitance value is not known its effect can be estimated when a response curve is taken.

This method for finding k_T can be applied in exactly the same way to a bridged T coupling network in which the two main series arms are inductances with mutual inductance coupling between them, the shunt arm is a capacitance and the T is bridged by a capacitance.

An alternative procedure, and a most important one, for determining k will now be discussed. This method can be applied quite generally to determining coupling coefficients for Π and T networks which have $L_1 C_1 = L_2 C_2$. The method is due to Howe and the reader should consult Refs. 99 and 100 for a more detailed explanation. This procedure uses the relationship

$$k = \frac{\omega_1{}^2 - \omega_2{}^2}{\omega_1{}^2 + \omega_2{}^2}$$

where ω_1 is the higher angular frequency of free oscillation in the circuit and ω_2 is the lower angular frequency of free oscillation.

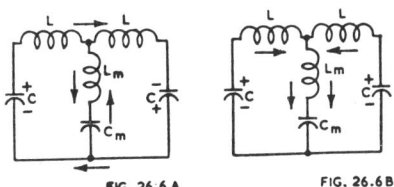

FIG. 26.6 A   **FIG. 26.6 B**

ω_1 is calculated from Fig. 26.6A, which shows a symmetrical T network with mixed coupling, by considering the capacitors C to be charged as shown. During the discharge of the capacitors it is clear that $L_m C_m$ have no effect on the resonant frequency in this case, and

$\omega_1{}^2 = 1/LC$.

ω_2 is calculated from Fig. 26.6B in which the capacitors C are charged as shown. In this case, when the capacitors are discharging, it is convenient to consider the shunt arm as being made up of two equal parts each carrying the primary and secondary current alone ; the reactances of L_m and C_m being doubled to allow for the single current. Then

$$\omega_2{}^2 = \frac{1}{(L + 2L_m)\left(\dfrac{CC_m/2}{C + C_m/2}\right)}$$

so that $k = \dfrac{L_m C_m - LC}{L_m C_m + LC + LC_m}$.

A somewhat similar procedure can be applied to a Π network, or an equivalent T network can be found and the above procedure applied.

For cases where the sections are unsymmetrical some modification in the procedure is necessary. Taking the case of a T section using inductance coupling (with $L_1C_1 = L_2C_2$) the shunt arm is replaced by two inductances (x_1 and x_2) in parallel such that their combined inductance equals that of the shunt arm (L_m) and such that the values selected will give equal resonant frequencies for the two circuits

i.e. $\dfrac{L_1 + x_1}{L_2 + x_2} = \dfrac{C_2}{C_1}$.

The case of capacitance coupling is treated in a similar way. The procedure is then exactly as before for the symmetrical coupling network. There would be little difficulty in extending this procedure to the case of the unsymmetrical T section using mixed coupling, but it is worth noting that numerical solutions are much easier to handle than a general algebraic solution even in the case of simple coupling.

For the two mesh network just considered, using mixed coupling and in which $L_1C_1 = L_2C_2$ it is usually simpler to use the relationship

$$k_T = \frac{k_m \pm k_c}{1 \pm k_m k_c}$$

as was done for the bridged T network. Taking the example just considered,

$$k_m = \frac{L_m}{L + L_m}$$

$$k_c = \frac{C}{C + C_m}$$

and $k_T = \dfrac{\dfrac{L_m}{L + L_m} - \dfrac{C}{C + C_m}}{1 - \dfrac{L_m C}{(L + L_m)(C + C_m)}}$

$$= \frac{L_m C_m - LC}{L_m C_m + LC + LC_m}$$

exactly as before.

It should be carefully noted that in all cases of mixed coupling the sign of L_m is most important. For the example just given L_m was taken as being positive ; this is the case of the two forms of coupling being in opposition, and zero voltage transfer occurs when the shunt arm is series resonant. With L_m negative the two forms of coupling aid each other and there is no series resonant frequency. The aiding condition, which is the one usually required, is possible only when the coupling between the two inductances is due to mutual induction, as a physical coupling inductance would act so as to oppose the capacitive coupling.

The methods given so far, for determining the coupling coefficient, are useful for the types of circuit considered. However, it should be observed that they can be applied directly only to the following cases :

(1) Circuits having coupling elements all of the one kind i.e. all capacitances or all inductances (simple coupling).

(2) Circuits having mixed coupling in which the primary and secondary circuits are tuned to the same frequency. However, it is only necessary that $L_1C_1 = L_2C_2$ and the primary and secondary circuit elements need not be identical.

Determination of k for the cases just stated usually presents little difficulty and is simplified by choosing one or other of the procedures outlined ; these cases cover most practical requirements. A real difficulty does arise when the primary and secondary are tuned to different frequencies and the coupling is mixed.

It may now be of interest to consider the procedure given below which makes use of the result obtained by Howe (Ref. 98) and follows from the definition that the coupling between two circuits is the relation between the possible rate of transfer of energy and the stored energy of the circuits ; by the possible rate of energy transfer is meant the rate of energy transfer in the absence of all resistance other than that utilized for coupling.

The definition leads to the relationship

$$k = \frac{1}{\sqrt{\omega_1 \omega_2}} \sqrt{\frac{E_{12}}{W_1} \frac{E_{21}}{W_2}}$$

where we have taken for our purposes that

$\omega_1 = 2\pi \times$ resonant frequency of primary circuit with sec. o/c
$\omega_2 = 2\pi \times$ resonant frequency of secondary circuit with pri. o/c
$E_{12} =$ voltage transferred from primary to secondary
$\quad = Z_{m_{12}}$ for a primary alternating current of 1 amp. and angular frequency ω_1

$E_{21} =$ voltage transferred from secondary to primary
$\quad = Z_{m_{21}}$ for a secondary alternating current of 1 amp. and angular frequency ω_2

$W_1 =$ maximum energy stored in primary circuit for an alternating current of 1 amp.
$W_2 =$ maximum energy stored in secondary circuit for an alternating current of 1 amp.

This method can be applied to any two oscillatory circuits coupled together in any manner, when all the elements are linear and bilateral.

As a simple example consider the case covered previously of an unsymmetrical T section of inductances. Then

$$Z_{m_{12}} = \omega_1 L_m; \ Z_{m_{21}} = \omega_2 L_m; \ W_1 = L_1 + L_m; \ W_2 = L_2 + L_m$$

and $k = \dfrac{1}{\sqrt{\omega_1 \omega_2}} \sqrt{\dfrac{\omega_1 L_m \omega_2 L_m}{(L_1 + L_m)(L_2 + L_m)}}$

$\quad = \dfrac{L_m}{\sqrt{(L_1 + L_m)(L_2 + L_m)}}$ exactly as before.

It can be seen that the primary and secondary frequencies need not be considered in cases of this type involving simple coupling only.

Now consider an unsymmetrical T section with mixed coupling, and different primary and secondary resonant frequencies:

$$Z_{m_{12}} = \omega_1 L_m - \frac{1}{\omega_1 C_m} = \omega_1 \left(L_m - \frac{1}{\omega_1^2 C_m} \right)$$

$$Z_{m_{21}} = \omega_2 L_m - \frac{1}{\omega_2 C_m} = \omega_2 \left(L_m - \frac{1}{\omega_2^2 C_m} \right)$$

$$W_1 = L_1 + L_m$$
$$W_2 = L_2 + L_m$$

$$k = \frac{1}{\sqrt{\omega_1 \omega_2}} \sqrt{\frac{\omega_1 \omega_2 \left(L_m - \frac{1}{\omega_1^2 C_m} \right)\left(L_m - \frac{1}{\omega_2^2 C_m} \right)}{(L_1 + L_m)(L_2 + L_m)}}$$

$$= \sqrt{\frac{\left(L_m - \frac{1}{\omega_1^2 C_m} \right)\left(L_m - \frac{1}{\omega_2^2 C_m} \right)}{(L_1 + L_m)(L_2 + L_m)}}$$

and since

$$\omega_1^2 = \frac{C_1 + C_m}{(L_1 + L_m)C_1 C_m}$$

$$\omega_2^2 = \frac{C_2 + C_m}{(L_2 + L_m)C_2 C_m}$$

the expression for k can be rewritten as

$$k = \sqrt{\frac{(L_m C_m - L_1 C_1)(L_m C_m - L_2 C_2)}{(L_1 + L_m)(L_2 + L_m)(C_1 + C_m)(C_2 + C_m)}}$$

It should be carefully noted that the ω_1 and ω_2 used here are not the same as found previously for the natural resonant frequencies of the coupled circuits.

If now we take $L_1 = L_2 = L$ and $C_1 = C_2 = C$,

$$k = \frac{L_m C_m - LC}{(L + L_m)(C + C_m)}.$$

Since this case corresponds to that used in connection with Figs. 26.6A and 26.6B it would be expected that the results would correspond. However, it is seen that there is an additional term $L_m C$ in the denominator of the second solution. This is explained by the method used for determining ω_1 and ω_2 in the latter case, and so the method as given here cannot be considered exact. The approximation involved in this particular example is that the result is equivalent to

$$k_T = k_m \pm k_c \text{ and not } k_T = \frac{k_m \pm k_c}{1 \pm k_m k_c}.$$

This is seen quite readily because

$$k_m = \frac{L_m}{L + L_m} \; ; \; k_c = \frac{C}{C + C_m}$$

and so $k_T = \dfrac{L_m(C + C_m) - C(L + L_m)}{(L + L_m)(C + C_m)} = \dfrac{L_m C_m - LC}{(L + L_m)(C + C_m)}$

which is the result just obtained.

A rather similar difficulty occurs when determining the central reference frequency for the transformer of Fig. 26.2. The angular frequency (ω) is determined exactly as above by considering the secondary on open circuit, and for this case

$$\omega = \sqrt{\frac{\omega_1{}^2 + \omega_2{}^2}{2}}$$

where ω_1 and ω_2 are now determined in the same way as for Figs. 26.6A and 26.6B.

SECTION 5 : VARIABLE SELECTIVITY

(i) General considerations   (ii) Automatic variable selectivity.

(i) General considerations

Any method giving variable-coupling may be used to provide variable selectivity.

Variable-coupling by means of pure mutual inductance is the only system in which mistuning of the transformer does not occur without introducing compensation of other component values. As the coupling is increased above the critical value, the trough of the resonance curve remains at the intermediate frequency (Ref. 8).

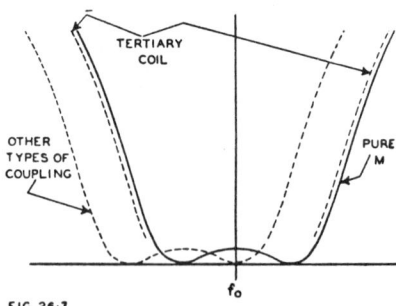

FIG. 26.7

With any other type of coupling which has no mutual inductance component, one of the two humps remains at the intermediate frequency. The mistuning is then half the frequency bandwidth between humps.

The method of switching tertiary coils approximates far more closely to the curve for pure mutual inductance (M in Fig. 26.7) than it does to the other curve. The tertiary coil may have no more than 5% of the turns on the main tuning coil and less than 0.5% ratio of inductance. The symmetry of the overall selectivity curves is usually good.

Variable capacitance coupling may be used and the coupling capacitor may be either a small capacitor linking the top end of the primary to the top end of the secondary or it may be a common capacitor in series with both primary and secondary circuits. This latter arrangement is commonly known as " bottom coupling."

For " top coupling " very small capacitances are required and the effect of stray capacitances is inclined to be serious, particularly in obtaining low minimum coupling. It may be used, however, with a differential capacitor arrangement whereby continuously variable selectivity is obtained (Refs. 41, 42). The differential capacitance in this case adds to or subtracts from the capacitance in the primary and secondary circuits to give the requisite tuning. The disadvantage of this arrangement is that sufficiently low minimum coupling is very difficult to obtain and the capacitor is a non-standard type.

" Bottom coupling " has results similar to " top coupling," but is easier to handle for switching and also has advantages for low coefficients of coupling. A two or three step tapping switch may be used to give corresponding degrees of bandwidth provided that simultaneously other switch contacts insert the necessary capacitances in the primary and secondary circuits, for each switch position, to give correct tuning.

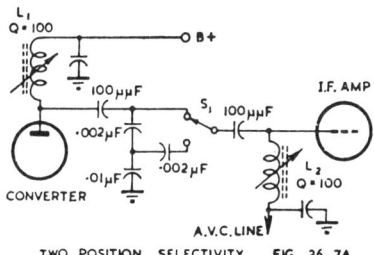

TWO POSITION SELECTIVITY   FIG. 26.7A

One such bottom coupling method which requires the minimum of switching for two degrees of coupling is shown in Fig. 26.7A (Ref. 42a). The two coils L_1 and L_2 should each be in a separate shield can. With switch S_1 in the position shown, the selectivity is broad ; in the lower position the selectivity is normal. Capacitances as shown are only illustrative. The same principle can be used for three values of coupling, using seven capacitors.

In most cases, the second i-f transformer is slightly under-coupled and its peak is used to fill in the " trough " between the peaks of the first transformer in its broad position.

A wide variety of circuit arrangements for obtaining variable selectivity has been described in the literature. Design methods and practical constructional details for many of these can be found in the references listed at the end of this chapter. Further discussion is also given in Chapter 11, Sect. 3(iii).

(ii) Automatic variable selectivity

Automatic variable selectivity allows the bandwidth of the i-f amplifier to be varied with signal strength, the pass band being a maximum for the strongest signal.

Many types of circuits have been developed using variable circuit damping, varying coupling reactance and circuit detuning. A number of the circuits are not very practical arrangements since they involve the addition of a number of valves whose sole function is to provide variable bandwidth.

One interesting circuit arrangement (Ref. 46), which does not require any additional valves, is shown in Fig. 26.8.

To understand the operation of the circuit it is first necessary to consider what happens in an ordinary mutual inductance coupled transformer. It is well known, when two circuits are coupled together, that there is a back e.m.f. induced in the primary winding from the secondary which will have a marked effect on the magnitude and phase characteristics of the primary current as a function of frequency. The primary current response curve exhibits a double peak at values of k even less than critical, because the back e.m.f. represents a voltage drop that decreases rapidly on either side of the resonant frequency. Double peaking of the primary current is masked in its effect on the secondary current by the selectivity of the secondary circuit. When the coupling exceeds critical (actually transitional) coupling the secondary current exhibits two peaks, but to a lesser extent than the primary current.

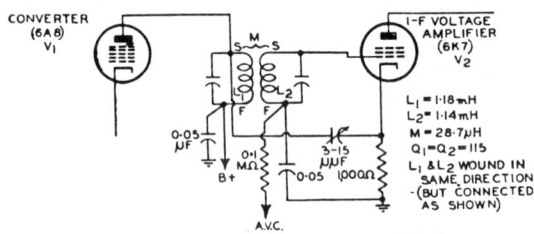

FIG. 26·8   AUTOMATIC SELECTIVITY ARRANGEMENT

If the magnitude of the counter e.m.f. is increased by increasing the mutual inductance, and at the same time a voltage of almost equal magnitude and opposite phase is added in series with the primary circuit, a single peak of output current can be obtained in the transformer secondary. The additional series voltage, so introduced, will increase the stage gain and selectivity from the condition which existed prior to its introduction ; in our case the prior condition is that the stage gain is less and the secondary current has two peaks, i.e. the transformer is initially over-coupled.

In the circuit of Fig. 26.8 the valve V_2 performs the dual functions of amplification and production of the series voltage to be added to the primary circuit. A resistance in the cathode of V_2 produces an a.c. voltage drop, which is proportional to the g_m of the valve and the secondary current of the transformer ; this voltage drop across the cathode resistor is the one applied back to the transformer primary. When a strong signal is being received a.v.c. will increase the bias on the grid of V_2 and so its g_m is reduced. When the g_m (and consequently the voltage drop across the cathode resistor) is sufficiently small, peak separation is obtained ; the width between the peaks depending on the value of mutual inductance. Weak signals reverse the process and narrow down the overall response curve.

A quantitive analysis (setting out design data and instability conditions) is given in the Ref. (46). For the circuit of Fig. 26.8 the peak separation is given as 8 Kc/s for an i-f of 450 Kc/s. Other designs are shown for peak separations of 15 Kc/s and also for obtaining two position automatic selectivity ; the latter arrangement necessitates the use of a relay.

Other types of automatic selectivity are discussed in the references listed.

SECTION 6 : VARIABLE BANDWIDTH CRYSTAL FILTERS

(i) Behaviour of equivalent circuit   (ii) Variable bandwidth crystal filters   (iii) Design of variable bandwidth i-f crystal filter circuits (A) Simplifying assumptions (B) Gain (C) Gain variation with bandwidth change (D) Selectivity (E) Crystal constants (F) Position of filter in circuit (G) Other types of crystal filters   (iv) Design example.

(i) Behaviour of equivalent circuit

Fig. 26.9 shows the generally accepted equivalent electrical circuit for a quartz crystal of the type used in the i-f stage of a communications receiver.

In this network suppose we consider that R is zero, then between terminals 1 and 2 there is a reactance

$$X = \frac{\omega L - 1/\omega C}{\omega C_0 \left[\omega L - \frac{1}{\omega}\left(\frac{1}{C} + \frac{1}{C_0}\right)\right]}.$$ (28)

When $\omega L = 1/\omega C$ the reactance is zero and the circuit is series resonant.

When $\omega L = \frac{1}{\omega}\left(\frac{1}{C} + \frac{1}{C_0}\right)$ the reactance is infinitely large, and the circuit is parallel resonant (anti-resonant).

The difference in frequency between f_p and f_r is given approximately by

$$\Delta f = \frac{C}{2C_0}\, f_r.$$

From the values of the 455 Kc/s crystal constants which are given in (iii)E below, it will be seen that Δf is about 250 c/s in a typical case.

The presence of R will slightly modify the conditions for which parallel resonance occurs. For our purposes the conditions given are near enough.

It should be clear from the circuit that the parallel resonance frequency (f_p) will be higher than the series resonance frequency (f_r). For frequencies above f_r the resultant reactance due to L and C in series is inductive, and this is connected in parallel with a capacitive reactance due to C_0. A typical curve of output voltage versus frequency change, for this type of circuit, is shown in Fig. 26.10.

If the value of C_0 could be altered as required, then the position of f_p would be variable. Suppose by some means we are able to connect a negative capacitance C_N (or a parallel inductance will give somewhat similar results) across C_0 ; then the value of the capacitance shunted across the series circuit will be

$$C' = C_0 - C_N.$$ (29)

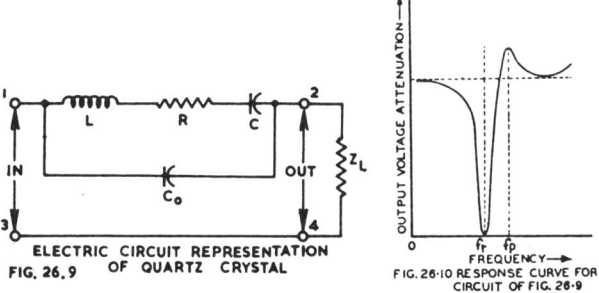

ELECTRIC CIRCUIT REPRESENTATION
FIG. 26.9   OF QUARTZ CRYSTAL

FIG. 26·10 RESPONSE CURVE FOR
CIRCUIT OF FIG. 26·9

From this, it follows that C' can be reduced from its initial value of C_0 (when C_N is zero) until it becomes zero, and C_N has then exactly neutralized C_0. The response curve would now be that for the series circuit (made up from L, R and C) alone, which behaves as a pure series resistance at f_r. If the magnitude of C_N is further increased, then C' becomes negative and the series circuit is shunted by a negative capacitance ; which is equivalent to shunting an inductance across the series circuit (which behaves like a capacitive reactance for frequencies below f_r). This means that the parallel resonance frequency (f_p) will now be lower than the series resonance frequency (f_r). Any frequency above or below the series resonance frequency can now be chosen as the rejection, or parallel resonance, frequency and the current through the circuit, at this point, will be reduced ; the values for f_p and the current depending on the circuit constants.

To achieve the variation in C' the circuit of Fig 26.9 (i.e. the crystal) is incorporated in the bridge circuit of Fig. 26.11, in which C_N is made variable to achieve the results discussed. In this circuit Z_L is the load impedance ; Z_s is the impedance of the voltage source ; Z_1 and Z_2 are any two impedances used to make up the resultant bridge circuit. It will be realized that the presence of Z_s and Z_L would have some modifying effect on our previous discussion, but the general principles remain unchanged ; these impedances will be taken into account when the design of the crystal filter stage is carried out.

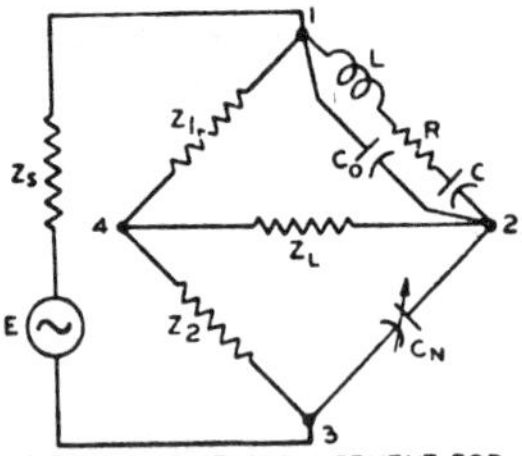

The arrangement of any practical circuit using a single crystal can be reduced to the general form of Fig. 26.11.

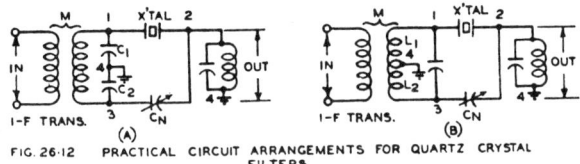

FIG. 26·11 CIRCUIT ARRANGEMENT FOR VARYING $C'(=C_0-C_N)$

Two typical circuit arrangements are shown in Fig. 26.12. Fig. 26.12(A) shows that Z_1 and Z_2 are obtained by using two capacitances C_1 and C_2 for the bridge arms. Fig. 12(B) uses a tap on the i-f transformer secondary (L_1 and L_2) to obtain the bridge arms Z_1 and Z_2.

In any circuit arrangement such as those shown, the best overall selectivity is obtained when C' is zero (i.e. $C_N = C_0$). Under this condition, if the circuit is designed with some care, the response curve of output voltage with frequency change will be reasonably symmetrical. If C_N is varied to place f_p above or below f_r, although we may improve the selectivity (and so the rejection) against an unwanted signal at f_p, the resonance curve is no longer symmetrical and the selectivity is decreased on the opposite side of the curve.

FIG. 26·12   PRACTICAL CIRCUIT ARRANGEMENTS FOR QUARTZ CRYSTAL FILTERS

(ii) Variable bandwidth crystal filters

It is often necessary, in practical receivers, to have means available for varying the bandwidth of the crystal filter circuit. This can be achieved in a number of ways, such as detuning the input and (or) output circuits or varying their resonant impedances. Also, as well as having variable bandwidth it is desirable that there should not be appreciable change in stage gain as the bandwidth is altered. Further, the overall response curve of voltage output versus frequency should remain as symmetrical as possible with changes in bandwidth when the bridge circuit is arranged so that C_0 is neutralized by C_N.

To fulfil all of the conditions above, a simple solution is to use a tuned circuit as the load for the filter circuit, and to alter its dynamic impedance by switching-in either series or parallel resistors. Simple detuning of the input circuit, or the output circuit, is not very satisfactory and the best results can only be achieved by detuning both the input and output circuits in opposite directions by an amount depending on their relative Q's. (For a description of circuits using the latter method see Refs. 48 and 63). To obtain constancy of gain it is necessary that the voltage source impedance be low, and this again suggests the switched output circuit when the bandwidth is to be varied. The choice of series or shunt resistors to alter bandwidths is largely bound up in stray capacitances across switch contacts and the resistors themselves ; careful consideration here suggests that resistors in series with the inductor, or the capacitor, of the tuned load will probably give the least detuning effects. Unless the values of the tuning capacitances for the input and load circuits are fairly large, alteration of C_N will have some appreciable effect on the resonant frequencies of these circuits, giving rise to asymmetry of the response curve and loss in sensitivity. This effect can sometimes be overcome, partly, by the use of a differential type of neutralizing capacitor. However, circuits using this arrangement should be carefully examined as usually there are still some detuning effects. For most practical cases it is sufficient to choose suitable values of tuning capacitance, particularly when operation is confined to an i-f of 455 Kc/s, and to tap down on the circuit when this is possible.

Two other points are worth mentioning. The first is that the crystal stage gives high selectivity around resonance, but the " skirt " selectivity may be quite poor ; for this reason the other stages in the receiver must provide the additional " skirt " selectivity required. In addition, good " skirt " selectivity is a requirement of the i-f amplifier to minimize any possible undesirable effects which may arise because of crystal subsidiary resonances. The other point is that having C_N as a variable control is not necessarily a great advantage, and simpler operation is obtained when C_N can be pre-set to neutralize C_0. This cannot be done with all types of variable bandwidth circuits, since the conditions for neutralization may be altered as the bandwidth is changed.

(iii) Design of variable bandwidth i-f crystal filter circuits
(A) Simplifying assumptions

From Fig. 26.11, when a capacitance balance is obtained as far as C_0 is concerned,
$$C_N \approx C_0(Z_1/Z_2). \tag{30}$$
Since C_0 is generally about 17 $\mu\mu$F it is convenient to make $Z_1 = Z_2$. Also, C_N must have very low losses if the attenuation at the rejection frequencies is to be high. The general design will call for all capacitors to be of the low loss type. When the condition of eqn. (30) is fulfilled, the equivalent circuit reduces to that of Fig. 26.13(A).

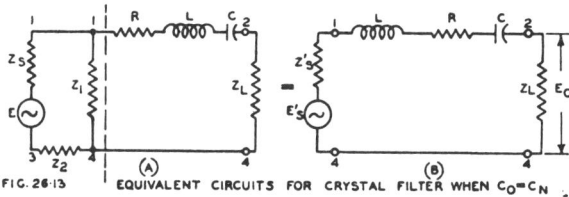

FIG. 26.13   EQUIVALENT CIRCUITS FOR CRYSTAL FILTER WHEN $C_0 = C_N$.

For purposes of analysis it is convenient to rearrange Fig. 26.13(A) as shown in Fig. 26.13(B), and, if we take $Z_1 = Z_2$, the value of Z'_s will be a quarter of the total dynamic impedance of the secondary circuit of the i-f input transformer. E'_s will be half the voltage developed across the two series capacitors tuning the i-f transformer secondary.

(B) Gain

First we will derive an expression for the overall gain of the i-f stage.
From Fig. 26.13(B),
$$\frac{E_0}{E'_s} = \frac{Z_L}{Z_L + Z'_s + Z_x} \tag{31}$$
where E_0 = output voltage across load circuit applied to grid of following i-f amplifier valve

Z_L = load impedance

Z'_s = voltage source impedance

Z_x = crystal impedance (C_0 neutralized)

$\quad = R + j\left(\omega L - \dfrac{1}{\omega C}\right)$

$\quad = R$ at resonance (i.e. i-f)

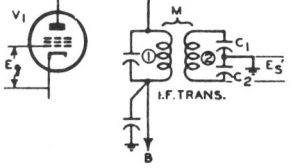

FIG. 26.14   CIRCUIT FOR DETERMINING E_s'

and E'_s = available voltage developed at tap on input i-f transformer secondary.

To compute the complete stage gain we must now consider Fig. 26.14. Here, since we require maximum gain, a critically-coupled transformer is connected to a voltage amplifier valve (V_1) whose grid to cathode input voltage is E_g. Then since E'_s is half the total voltage appearing across C_1 and C_2
$$E'_s = \tfrac{1}{2}\left(\frac{g_m E_g Q \omega_0 L}{2}\right) \tag{32}$$

where g_m = mutual conductance of V_1

$$Q =\sqrt{Q_1 Q_2} = 1/k_c$$
$$\omega_0 = 2\pi \times i\text{-}f$$

and $L =\sqrt{L_1 L_2}$ (always provided that $L_1 C_1 = L_2 C_2$).

Combining eqns. (31) and (32), the gain from the grid of V_1 to the grid of the next amplifier valve (V_2) is

$$\text{Stage gain} = \frac{E_0}{E'_s} \times \frac{E'_s}{E_g} = \frac{E_0}{E_g} = \left(\frac{Z_L}{Z_L + Z'_s + Z_x}\right)\left(\frac{g_m Q \omega_0 L}{4}\right) \qquad (33)$$

Eqn. (33) gives a great deal of information about the circuit. For constancy of gain, it follows that Z_L should be very much greater than ($Z'_s + Z_x$). Since Z_x (or R at resonance) cannot be altered by the receiver designer, it is necessary to make Z'_s as small as possible. However, as Z'_s is made smaller the overall gain will be affected since it is a centre tap on the output of the input i-f transformer. The maximum value for Z_L will be limited by the maximum bandwidth requirements, and the permissible values of circuit constants. The other factors affecting gain are g_m, Q and L. For a given valve under a fixed set of operating conditions, g_m is practically outside the designer's control ; the value of $L(=\sqrt{L_1 L_2})$ is made as large as possible consistent with the requirements of minimum permissible tuning capacitance ; $Q =\sqrt{Q_1 Q_2}$ is adjusted so that Q_1 is made as high as possible, and since we desire Z'_s to be low a suitable value is selected for Q_2. The method of determining Q has a large effect on the stage gain which can be obtained, as will be seen later in the illustrative example.

(C) Gain variation with bandwidth change

For constancy of maximum gain at f_0 (i.e. the i-f) it has been suggested that Z'_s should be small. There is a limit, however, to the minimum gain variation that can be obtained for a given change from maximum to minimum bandwidth. From the preceding equations it may be deduced that the maximum stage gain (A_{max}) is obtained when the bandwidth is greatest, and the minimum gain (A_{min}) when the bandwidth is least. This limit in gain variation (since R for the crystal is fixed) is the condition for which the voltage source impedance (Z'_s) becomes zero, which occurs when

$$\alpha_0 = \frac{A_{min}}{A_{max}} = \frac{R_{T_2}(R_{T_1} - R)}{R_{T_1}(R_{T_2} - R)} \qquad (34)$$

where R_{T_1} = total series resistance (at f_0) when the bandwidth is least

$$(= R + Z'_s + Z_L \text{ ; in which } Z_L \text{ has its smallest value})$$

and R_{T_2} = total series resistance (at f_0) when the bandwidth is greatest

$$(= R + Z'_s + Z_L \text{ ; in which } Z_L \text{ now takes up its largest value}).$$

To determine the voltage source impedance (Z'_s) for the usual condition where Z'_s is not zero, we use

$$Z'_s = \frac{R_{T_2}(R_{T_1} - R) - \alpha R_{T_1}(R_{T_2} - R)}{R_{T_2} - \alpha R_{T_1}} \qquad (35)$$

which, for convenience, is rewritten in terms of eqn. (34) as

$$Z'_s = \frac{\alpha_0 - \alpha}{\dfrac{\alpha_0}{R_{T_1} - R} - \dfrac{\alpha}{R_{T_2} - R}} \qquad (35A)$$

where R_{T_1} and R_{T_2} have the values given immediately above,

R = equivalent series resistance of crystal

and $\alpha = \dfrac{A_{min}}{A_{max}}$ = gain variation desired, and must always be less than one.

If a value is selected for Z'_s then

$$\alpha = \frac{R_{T_2}(R_{T_1} - R) - Z'_s R_{T_2}}{R_{T_1}(R_{T_2} - R) - R_{T_1} Z'_s}. \qquad (36)$$

(D) Selectivity

Selectivity around resonance can be calculated by the methods to be outlined, but for most purposes it is sufficient to determine the bandwidth for a given attenuation at the half power points (3 db attenuation or $1/\sqrt{2}$ of the maximum voltage output) on each of the required selectivity curves.

Considerable simplification in the design procedure is possible, if the bandwidths for 1 db attenuation (or less) are known. In this case it would be sufficiently accurate to take Z_L and Z_s equal to their dynamic resistances at resonance, at least for most practical conditions, without introducing appreciable error. The advantage obtained being that, in what follows, $R_{T_1} = R'_{T_1}$ and $R_{T_2} = R'_{T_2}$ for all conditions. The value of Q_x would be given by $Q_x = f_0/4\Delta f$ for 1 db attenuation, and this expression would be used in place of eqn. (38). However, the procedure given is more general and there should be little difficulty in applying the simplified procedure, if necessary. Further the method given illustrates (in reverse) how the bandwidths near resonance can be calculated for various amounts of attenuation.

To determine the bandwidth at the half power points proceed as follows : first it may be taken that the equivalent inductive reactance of the crystal is very much greater than the inductive reactance of the load and source impedances. Also, provided the variations in gain are not allowed to become excessve, at and near resonance the value of Z'_s is very closely R'_s. For very narrow bandwidths, near resonance, it is also sufficiently close to take $Z_L = R_L$; for large bandwidths the resistive component (R_L) of Z_L at the actual working frequency will have to be found. From these conditions, we have

$$Q_x = \frac{\omega_0 L}{R_L + R'_s + R} = \frac{\omega_0 L}{R'_T} = \frac{1}{\omega_0 C R'_T} \tag{37}$$

where Q_x = the equivalent Q of the circuit of Fig. 26.13(B)
  L = equivalent inductance of the crystal
  C = equivalent series capacitance of the crystal
  R = equivalent series resistance of the crystal
  R_L = resistive component of Z_L, the load impedance, at the frequency being considered ($= Z_L$ at resonance)
  R'_s = resistive component of Z'_s the voltage source impedance
  R'_T = $R_L + R'_s + R$ = total series resistance at frequencies away from f_0 (for very narrow bandwidths $R_L \approx Z_L$ and $R'_T \approx R_{T_1}$)

and $\omega_0 = 2\pi \times f_0$ (where f_0 is the i-f).

From the principles of series resonant circuits we know that the total bandwidth $(2\Delta f)$ for the half power points (3 db atten.) is given by

$$2\Delta f = f_0/Q_x. \tag{38}$$

It follows from eqns. (38) and (37) that for large bandwidths Q_x should be small and so R'_T should be large. For narrow bandwidths the reverse is true, and the limiting case for the narrowest bandwidth would be when $R'_T = R$ (i.e for the crystal alone) ; a condition impossible to achieve in practice because of circuit requirements.

Since $Z_L \gg (Z'_s + R)$ for constancy of maximum gain at f_0, under conditions of varying bandwidth, it also follows that Q_x will mainly be determined by the magnitude of the resistive component of the dynamic impedance of Z_L.

Some additional selectivity is given by the input i-f transformer, and for more exact results the attenuation for a particular bandwidth would be added to that found for the crystal circuit. This additional selectivity is usually negligible around the " nose " of the resonance curve, but can be determined from the universal resonance curves of Chap 9 Sect 10, evaluating D and b for unequal primary and secondary Q's (Q_1 and Q_2) and noting that Q in these expressions is Q_2

The resonance curve for the complete circuit is seldom necessary for a preliminary design. The procedure is rather lengthy but not very difficult. First it is necessary to find the resistive component of Z_L (i.e. R_L) at frequencies off resonance. This is carried out from a knowledge of $|Z_L|(= Z_L)$ at resonance, and by determining the reduction factors from the universal selectivity curves. Multiplying $|Z_L|$ by the

indicated attenuation factors gives the required magnitudes of load impedance $|Z'_L|$.
To find the resistive component, values of θ corresponding to the various bandwidths
(and values of $|Z'_L|$) are read from the universal phase-shift curves of Chapt. 9 Sect. 10.
Then

$$R_L = |Z'_L| \cos \theta. \tag{39}$$

The resistive component of Z'_s can be found in a similar way from

$$R'_s = Z'_s \cos \phi$$

but since the impedance of Z'_s is generally small, and the circuit relatively unselective,
R'_s can often be taken as equal to Z'_s for a limited range of frequencies near resonance.
When Z'_s is very low, it can be neglected in comparison with $R_L + R$.

Then, since the resistance (R) and the inductance (L) of the crystal are known,
eqn. (37) can be applied to determine the values of Q_x corresponding to the various
bandwidths. (Strictly f_0 would be replaced by the actual operating frequency off
resonance, but this is hardly necessary).

Knowing the different values of Q_x, and since the bandwidths corresponding to
each Q_x value are known, the various points for the complete selectivity curve can be
found from the universal resonance curve for a single tuned circuit (it is unimportant
that this is for a parallel tuned circuit rather than a series tuned circuit, since the
conditions for parallel and series resonance are practically the same provided the Q
is not less than about 10 ; there is, however, a reversal in sign of the phase angles
for the two cases). Additional selectivity due to the input transformer can be taken
into account if this degree of accuracy is thought to be necessary.

(E) Crystal constants

Before a complete design can be carried out, some data on the equivalent electrical
constants of the quartz crystal must be available. In most cases details can be obtained from the crystal manufacturer.

Typical values for the electrical constants of 455 Kc/s quartz crystals widely used
in Australian communications receivers are :

$$R = 1500\Omega \; ; \; C = 0.018 \; \mu\mu F \; ; \; C_0 = 17 \; \mu\mu F.$$

The crystals are a special type of X cut bar and have no subsidiary resonances for a
range of at least $=. 30$ Kc/s from f_r. They are approximately 20 mils (0.02 in.) thick,
$\frac{1}{4}$ in. wide, $\frac{3}{8}$ in. long and are mounted between flat electrodes with an air gap not
greater than 1 mil. (0.001 in.).

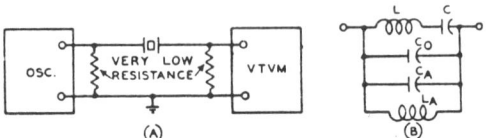

FIG. 26·15 APPROXIMATE METHODS FOR DETERMINING EQUIVALENT
ELECTRICAL CONSTANTS OF QUARTZ CRYSTALS

If the required data cannot be obtained, then details of methods for measuring,
firstly, the value of Q for the crystal, can be found in Ref. 65. The experimental
set-up is shown in Fig. 26.15(A) and Q is found from

$$Q = \frac{f_r}{2(f_p - f_r)} \sqrt{\frac{E_r}{E_p}} \tag{40}$$

where f_r = series resonance frequency
$\quad\quad f_p$ = parallel resonance (or antiresonance) frequency
$\quad\quad E_r$ = voltage across terminating resistance at series resonance
and $\quad E_p$ = voltage across terminating resistance at parallel resonance.

A knowledge of Q will allow the filter circuit to be designed, but if values for C
and C_0 are required these can be readily measured (Ref. 64) by using the arrangement
of Fig. 26.15(B) and the relations

$$\frac{C}{C_A - C_0} = \left(\frac{f_{p_2} - f_{p_1}}{f_r}\right)^2 \tag{41}$$

and $\qquad f_{pA} = \dfrac{1}{2\pi\sqrt{L_A C_A}}$ $\qquad\qquad$ (42)

where $\qquad C$ = series capacitance of crystal

$\qquad\qquad C_A$ = parallel capacitance of circuit $L_A C_A$

$\qquad\qquad L_A$ = parallel inductance of circuit $L_A C_A$

$\qquad\qquad C_0$ = shunt capacitance across crystal due to holder etc.

$\qquad f_{p_1}$ and f_{p_2} = parallel resonance frequencies of combined circuits. The combination has two parallel resonance frequencies, one above and one below f_r.

$\qquad\qquad f_r$ = series resonance frequency, measured with combined circuit.

and $\qquad f_{pA}$ = parallel resonance frequency of circuit $L_A C_A$ alone.

C_0 is determined by disconnecting the crystal and adding capacitance to $L_A C_A$ to retune the circuit to the frequency f_r.

(F) Position of filter in receiver

In most receivers using crystal filters at least two i-f stages are included, since the gain of the crystal stage is usually well below that for a normal i-f stage. Low stage gain immediately after the converter valve may have an adverse effect on signal-to-noise ratio, and it is sometimes preferable to place the filter between the first and second i-f voltage amplifier valves ; however, it is fairly common practice to place the selective crystal stage immediately after the converter to reduce the effects of spurious responses.

It would not be satisfactory to place the filter between the last i-f valve and the detector because of the low impedance of the load in this case.

Since it is common practice to incorporate at least one r-f stage in receivers of the type mentioned, it is unlikely that effects such as possible deterioration in signal-to-noise ratio would be of any consequence. However, good design suggests at least two i-f stages plus one or more r-f stages, depending on performance requirements.

(G) Other types of crystal filters

There are several types of crystal filters suitable for use in radio receivers other than the simple bridge circuit, to which attention has been confined. Typical examples are double-crystal and bridged T filters. These circuits are characterized by having two rejection frequencies, usually placed with geometric symmetry about f_r. Details can be found in Refs. 49, 54, 55 and 61.

Resistance balancing is sometimes applied to the complex types of crystal filters to increase attenuation at the rejection frequencies. This method has also been applied to wave filters using LC circuits only. Details are given in Refs. 59, 60 and 61.

(iv) Design example

Using the constants for the 455 Kc/s crystal given in (E) above, it is required to design a crystal filter circuit to have a total bandwidth $(2\Delta f)$ for 3 db attenuation,

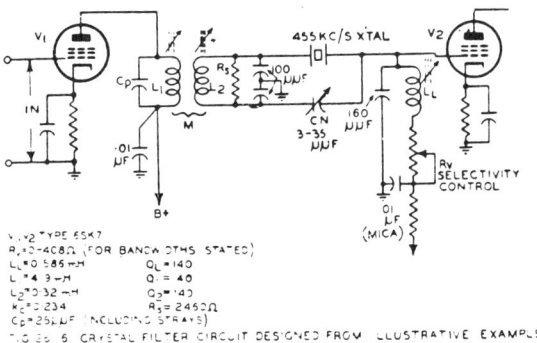

FIG. 26.6 CRYSTAL FILTER CIRCUIT DESIGNED FROM ILLUSTRATIVE EXAMPLE

which can be varied from 200 c/s to 3 Kc/s. The variation in maximum stage gain should not exceed about 2 db ($\pm$ 1 db about the average gain) but it is desirable to keep the stage gain as high as possible, consistent with stable operation with varying signal input voltages (i.e. large detuning of the i-f circuits should not occur when the signal voltages vary over a wide range).

To make the problem complete, it will be assumed that the filter is connected between two type 6SK7 pentode voltage amplifier valves. The complete circuit is shown in Fig. 26.16.

(1') Since $Z_1 = Z_2$ and $C_0 \approx 17\ \mu\mu F$, let us select a suitable capacitance range for C_N. The smallest residual capacitance for C_N will be about 3 $\mu\mu F$. From this, we have to increase C_N a further 14 $\mu\mu F$ to neutralize C_0. In addition, it is desired to move the rejection frequency (f_r) below f_r, so that it is reasonable to allow C_N to increase at least a further 14 $\mu\mu F$. The total increment in C_N is thus 28 $\mu\mu F$; which gives a range of 3 to 31 $\mu\mu F$; for convenience this is made, say, 3 to 35 $\mu\mu F$ (or whatever is the nearest standard capacitance range).

(2') From eqn. (38):
for 200 c/s total bandwidth, and 3 db attenuation,
$Q_{x_1} = 455/0.2 = 2275$;
for 3 Kc/s,
$Q_{x_2} = 455/3 = 151.6$.

(3') From eqn. (37) (and C = 0.018 $\mu\mu F$)

$$R'_{T_1} = \frac{10^{12}}{2275 \times 2\pi \times 455 \times 10^2 \times 0.018} = 8550\ \Omega$$

and $R'_{T_2} = \dfrac{8550 \times 2275}{151.6} = 0.128\ M\Omega.$

So that (since $R = 1500\ \Omega$)
$R_{L_1} + Z'_s = 8550 - 1500 = 7050\ \Omega$
$R_{L_2} + Z'_s = 0.128 - 15 \times 10^{-3} = 0.126\ M\Omega.$

For narrow bands of frequencies, $R_{L_1} \approx Z_{L_1}$, and we may write
$R'_{T_1} = R_{T_1}$ and $Z_{L_1} + Z'_s \approx 7050\ \Omega.$

(4') To find Z_{L_2}. It should be clear, from the values just given (Z'_s remains unchanged) that for fairly large bandwidths
$R_{L_2} \gg Z'_s.$

Assume that, for the load circuit, $Q_L = 140$. Then by calculation, or from the universal resonance curves, we have for a single tuned circuit and a frequency of 455 Kc/s,
$$\frac{Z_{L_2}\ \text{at resonance}}{|Z'_{L_2}|\ \text{at 3 Kc/s bandwidth}} = 1.37.$$
Also, the phase shift $\theta = 42° 43'$.
Using these factors in conjunction with eqn. (39),
$$|Z'_{L_2}| = \frac{R_L}{\cos\theta} = \frac{0.126}{0.735}$$
and so $Z_{L_2} = \dfrac{1.37 \times 0.126}{0.735} = 0.235\ M\Omega.$

(If the maximum bandwidth required is too large it will be found that Z_{L_2} cannot be obtained with ordinary circuit components).
From this, since $Z_{L_2} = Q_L \omega_0 L_L$,
$$L_L = \frac{0.235 \times 10^3}{140 \times 2\pi \times 0.455} = 0.586\ mH$$
and $C_L = \dfrac{25\,330}{0.455^2 \times 586} = 208\ \mu\mu F$ (see below).

(5') If it is possible to make Z_L higher than 0.235 megohm then a voltage step-up is possible using the arrangement of Fig. 26.17, which also reduces detuning effects when C_N is varied.   Good circuit stability (which requires a large value for C_L) is most important since the circuit bandwidth and gain are very critical to detuning. The main causes of detuning are capacitance changes at the input of V_2 due to grid bias variations, and resetting of C_N; this latter capacitance change is offset, to some extent, in the alternative circuit since it appears across only part of the tuned circuit capacitance.   Any form of tapping down is helpful in reducing detuning variations due to C_N, but this is limited by the impedance required for the filter load circuit.

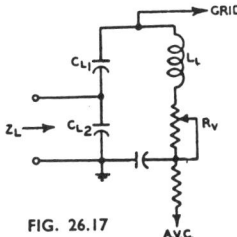

Fig. 26.17. *Alternative load circuit for crystal filter stage.*

FIG. 26.17

Methods of reducing detuning effects are discussed in Sect. 7(ii), and could be applied here to offset changes in valve input-capacitance.   Connecting the grid to a tap on the tuned load is an obvious means of reducing the effects of valve input-capacitance variation.   This method involves a loss in gain, but this is not necessarily serious as in some receivers the total gain is more than can usefully be employed.

The actual fixed capacitance for C_L is found approximately as follows :

Valve input capacitance $\qquad\qquad = \quad 6.5 \ \mu\mu\text{F}$
Strays across coil + wiring etc. $\qquad = \quad 8.5 \ \mu\mu\text{F}$
$C_N + C_0 = 2C_0 = 2 \times 17 \qquad = \underline{\ 34 \ \mu\mu\text{F}}$ (when C_0 is neutralized by

$\qquad\qquad\qquad\qquad\qquad\qquad\qquad\qquad\qquad\qquad C_N)$

$\qquad\qquad\qquad$ Total strays $\quad = \ 49 \ \mu\mu\text{F}$
Fixed capacitance req. $= 208 - 49 = 159 \ \mu\mu\text{F}$ (say 160 $\mu\mu\text{F}$).
(6′) From eqn. (34) ; and since $R_T = 8550 \ \Omega$ and
$\qquad R_{T_2} = 0.237 \ \text{M}\Omega$ (approximately ; i.e. $Z_{L_2} + R$),

$$\alpha_0 = \frac{0.237 \ (7050)}{8550 \ (0.235)} = 0.83$$

(which shows that the gain variation is practicable).

Then, since $\alpha = 2$ db down $= 0.794$, we have from eqn. (35A)

$$Z'_2 = \frac{0.83 \quad - 0.794}{\dfrac{0.83}{7050} - \dfrac{0.794}{0.235 \times 10^6}} \approx \frac{0.036 \times 7050}{0.83} = 306 \ \Omega,$$

so that $Z_2 = 4Z'_2 = 1224 \ \Omega$.

(7′) To determine the approximate range of the selectivity control R_v.   For maximum bandwidth $R_v = 0$.   For minimum bandwidth, since $Z_{L_1} (= R_{L_1}) = 7050 - 306 = 6744 \ \Omega$,

then $Q_{L_1} = \dfrac{Z_{L_1}}{\omega_0 L_L} = \dfrac{6744}{1680} = 4$ [This is an approximation only ; see end of step (8′)]

and $\quad R'_v = \dfrac{\omega_0 L_L}{Q_L} = \dfrac{1680}{4} = 420 \ \Omega$.

Resistance already in circuit when Q is 140, is
$\qquad R_L = 1680/140 = 12 \ \Omega$.
Range of R_v required is from $0 - 408 \ \Omega$.
This is the range if R_v is connected in series with L_L.   If R_v is to be connected in series with the capacitive arm of Z_L it should be realized that it would actually be in

series with the fixed capacitance of 160 $\mu\mu$F only, and not the total capacitance, so that new values for the range of R_v would have to be calculated in this case.

(8') The design of the input transformer is the next step. For maximum gain the transformer will be critically-coupled (this term is hardly a correct one in the design which follows and for this particular case it might be preferable to use other methods).

Consider first the secondary circuit. The total secondary capacitance, if we select 100 $\mu\mu$F capacitors (connected in series) for the ratio arms of the bridge circuit, will be $100/2 + 14.5 = 64.5$ $\mu\mu$F. The 14.5 $\mu\mu$F represents the approximate total for

$$\frac{C_N}{4} + \frac{C_0}{4} = \frac{C_0}{2} \text{ (since } C_N = C_0) = 8.5 \ \mu\mu\text{F plus an allowance of 6 } \mu\mu\text{F}$$

for coil and circuit strays.

Then the apparent secondary inductance required is

$$L_2 = \frac{25.33}{0.455^2 \times 64.5} = 1.9 \text{ mH.}$$

From step (6') above, $Z_s = 1224$ Ω. Since critical-coupling (see remarks above) will halve the actual value of Q_2, we must use an uncoupled value for Z_s of 2448 Ω. This allows us to determine the uncoupled secondary magnification factor

$$Q_2 = \frac{2448}{\omega_0 L_2} = \frac{2448}{5440} = 0.45.$$

Because of the low value for Q_2, the condition of $\omega_0^2 L_2 C_2 = 1$ is no longer sufficiently accurate. For cases such as this, where Q is less than about 10, proceed exactly as before but modify the value of L_2 by a factor $Q^2/(1 + Q^2)$. This gives the true condition for resonance (unity power factor) if Q is assumed constant and L_2 is variable as it will be in most i-f transformers of the type being considered. The actual value required for L_2 is now

$$L_2 \text{ (actual)} = 1.9\left[\frac{0.45^2}{1 + 0.45^2}\right] = 0.32 \text{ mH.}$$

(9') Since the secondary Q is very low, and L_2 is fixed by other considerations, it should be clear that if we require reasonably high stage gain the primary $Q(= Q_1)$ and the primary inductance L_1 should be as high as possible; since this will allow $QL = \sqrt{Q_1 Q_2 L_1 L_2}$ to be increased.

As the minimum capacitance across L_1 will be about 25 $\mu\mu$F (valve output + strays, which will be fairly high for a large winding), then it is possible to make

$$L_1 = \frac{25.33}{0.455^2 \times 25} = 4.9 \text{ mH.}$$

Of course, it may not be advisable to resonate the primary with stray capacitances only, but there are practically no detuning effects present (except those due to temperature and humidity variations). The possibility of instability must not be overlooked, however, when the grid to plate gain is high, and for the case given it would probably be necessary to neutralize the i-f stage or to reduce the gain by increasing the capacitance value above 25 $\mu\mu$F (and so reducing L_1). However, L_1 will be made large here to illustrate the design procedure to be adopted in cases such as this, where L_1 is not equal to L_2 and also to bring out an additional useful point which could be overlooked in connection with the design of a.v.c. systems; see Chapter 27 Sect. 3(xv).

(10') Assume that unloaded Q values of 140 for the primary and secondary can be obtained. The plate resistance (r_p) for type 6SK7 is 0.8 MΩ for a set of typical operating conditions. Then the actual value of Q_1 is

$$Q_1 = \frac{Q_u r_p}{Q_u \omega_0 L_1 + r_p} = \frac{140 \times 0.8}{1.96 + 0.8} = 40.6.$$

(11') In order that the actual value of Z_s shall be 1224 Ω (and so $Z'_s = 306$ Ω) the secondary circuit must be loaded with a resistance R_s given by

$$R_s = \frac{Q_u Q_2 \omega_0 L_2}{Q_{u_s} - Q_2} = \frac{140 \times 2448}{140 - 0.45} = 2460 \text{ }\Omega$$

where L_2 has the apparent value of 1.9 mH, and not its actual value, so as to make allowance for the use of approximate expressions which are normally only applicable when Q exceeds about 10.

(12') $Q = \sqrt{Q_1 Q_2} = \sqrt{40.6 \times 0.45} = 4.28$.

The coefficient of coupling $= k_c = 1/4.28 = 0.234$.

$L = \sqrt{L_1 L_2} = \sqrt{4.9 \times 0.32} = 3.06$ mH.

The actual value of L_2 is required here.

(13') It is now proposed to estimate the gain. To find the true gain for the conditions selected it would be necessary to solve the complete equivalent circuit. It should also be noted that the condition for secondary circuit series resonance does not coincide with the condition for parallel resonance. Correct circuit matching (i.e. that the value of Z'_s is the required one) can be readily checked from the equivalent secondary circuit. This has been done to ensure that the previous modifications to the transformer constants are quite satisfactory. Here the original gain equation (33) will be used, and then a suitable correction factor (namely X_{c_2}/X_{L_2}) will be applied for this special case where $X_{c_2} \neq X_{L_2}$. The results show an appreciable error, about $+ 20\%$ in the worst practical cases, but the additional labour involved in a more exact analysis is hardly justified.

For the valve type selected $g_m = 2000$ μmhos (2 mA/volt). Gain from grid of V_1 to grid of V_2, using eqn. (33), is

$$\text{Gain (max. bandwidth)} = \left[\frac{0.235}{0.235 + 306 \times 10^{-6} + 1500 \times 10^{-6}}\right] \times$$

$$\left[\frac{2 \times 4.28 \times 2\pi \times 0.455 \times 1.25}{4}\right]$$

$$= 0.992 \times 7.7 = 7.6 \text{ times (17.6 db)}.$$

Gain (min. bandwidth) $= 6744/8550 \times 7.7 = 6.07$ times (15.6 db).

The factor $X_{C2}/X_{L2} = 5470/917 = 5.95$ and so the maximum gain is about 45 times (33 db) and the minimum gain about 36 times (31 db). The calculated value from the equivalent circuit is 38 times.

Thus the maximum total gain variation (including all approximations) is about 2 db as specified i.e. ± 1 db about the average gain. Larger variations in gain, if permissible, would also allow increased overall gain ; the disadvantages have been discussed previously.

(14') For some purposes a standard i-f transformer using fixed capacitances of about 100 $\mu\mu$F, and the coupling increased to critical when the secondary and primary are correctly loaded, would give satisfactory results. The gain variation, and maximum gain, are largely controlled by the value of the damping resistor R_s connected across the secondary of the transformer.

To control bandwidth it is necessary to increase Z_L to increase the maximum bandwidth, R_v is increased to decrease the minimum bandwidth. If switched steps are required for bandwidth control, R_v can be calculated for each step ; the remainder of the design is exactly as before.

The arrangement used for the input transformer is only one of many possible circuits. For cases such as the one given here (where the total gain is not a prime requirement) it may be preferable to leave the secondary circuit untuned and to use a tapped resistor to provide the ratio arms for the bridge circuit. The design procedure is readily developed from the usual coupled circuit theory.

SECTION 7 : DETUNING DUE TO A.V.C.

(i) Causes of detuning   (ii) Reduction of detuning effects (A) General (B) Circuits.

(i) Causes of detuning

Detuning of i-f (and r-f) circuits is largely due to changes in valve input capacitance. The capacitance change is brought about when the grid bias on the valve is altered, the bias change being due to the a.v.c. voltage altering with different signal inputs.   This alteration in bias also leads to variation in the valve input conductance which will have to be considered in some circuits, since the conductance increases by about 2 : 1, from cut-off to normal operating conditions, with typical valve types operating at 100 Mc/s (see Ref. 68 ; also Chapter 2 Sect. 8 and Chapter 23, Sect. 5).

If all the factors which cause valve input capacitance and conductance changes are considered the problem is rather involved as can be seen from the discussions given in Chapter 2, Sect. 8 and Chapter 23, Sect. 5.· We will confine attention here, mainly, to input capacitance changes which are most troublesome in receiver design.

One of the principal causes of change in valve input capacitance is the variation in position and density of the space charge distribution between grid and cathode, brought about by the change in grid bias.

Variation in grid-plate transconductance (g_m) is another principal cause of input capacitance change as will be seen later.

Short-circuit input capacitance changes are not very much affected by frequency, although the change in capacitance with transconductance is slightly greater at low frequencies than at high.   The slight variation with frequency will not usually be important when compared with the changes due to alteration of transconductance, so that the data given in Chapter 2, Sect. 8 can be used directly.   From this data, it is seen that the short-circuit input capacitance alters by about 1 to 2 $\mu\mu$F in typical cases.   It is interesting to compare these changes with that due to grid-to-plate coupling i.e. Miller effect (see Chapter 2, Sect. 8).

Suppose, for simplicity, that the plate load acts as a pure resistance.   Then the input capacitance change is given by

$$\varDelta C = C_g - C_{gp} = C_{gp}\left(\frac{g_m r_p R_L}{r_p + R_L}\right) \tag{43}$$

Taking a type 6SK7, with a plate load of 0.25 megohm, as a typical example.   Then since $C_{gp} = 0.003 \ \mu\mu$F, $r_p = 0.8$ MΩ (this value is not constant, but it has been assumed so for simplicity) and g_m changes from 0-2000 μmhos, the input capacitance change is

$$\varDelta C = 0.003 \ \frac{2 \times 0.8 \times 0.25 \times 10^3}{1.05} = 1.14 \ \mu\mu\text{F}.$$

If C_{gp} is augmented by stray capacitances, such as those inevitably present due to the socket, wiring etc., this figure for $\varDelta C$ will be very appreciably increased.   The corresponding change in short-circuit input capacitance is 1.18 $\mu\mu$F (see table Chapter 2 Sect. 8).

The effects are additive and cause an increase in input capacitance as the grid bias voltage becomes more positive.

(ii) Reduction of detuning effects
(A) General

There are a number of methods available for reducing the effects of valve input capacitance variations, and these methods (in most cases) also reduce input conductance variations.   Usually the reduction of these effects will lead to a loss in stage gain unless additional circuit changes are made.

In the case of the ordinary broadcast receiver special circuits are not often used. The only precautions taken are :

(1) The total capacitance tuning the i-f transformer secondary, in particular, is made as large as is practicable consistent with other circuit requirements.   Values of 200 $\mu\mu$F, or so, should be used if possible, although values of the order of 100 $\mu\mu$F, are generally used, since it is unlikely that the change in input signal will be such as to cause the maximum input capacitance change.   Large values of tuning capacitance for the transformer are also helpful in increasing the attenuation at frequencies far removed from resonance.   Also, with some types of converter valves (e.g. 6A8-G) a large tuning capacitance for the primary of the first i-f transformer can be of assistance in reducing degeneration in the signal input circuit connected to the converter ; this applies particularly to signal frequencies which approach the intermediate frequency.

(2) The receiver is aligned on small input signal voltages, since detuning effects will be less serious when the input signal is large.

(3) The circuit layout is such as to minimize stray grid-to-plate capacitance.

The degree of circuit mistuning in terms of frequency shift can be determined by comparing the input capacitance change with the total capacitance across the valve input circuit. A number of helpful practical examples of circuit mistuning are given in Ref. 69.

There are several additional factors which are of importance.

(1) Detuning will cause sideband asymmetry and so lead to the possibility of increased distortion. This is of particular importance in high fidelity receivers.

When the carrier is received on the side of the i-f selectivity curve, one set of sidebands is almost completely eliminated (or at least substantially reduced in amplitude), and the amplitudes of the carrier and low frequency components of the other set of sideband frequencies are reduced. The tuned circuits will introduce phase and frequency distortion and the diode detector will give rise to non-linear distortion because of the absence of one set of sidebands. Experiment has shown that with the usual type of receiver the amount of detuning that can be tolerated before distortion becomes noticeable to a critical listener is about $\pm$ 1 Kc/s.

(2) Detuning of the r-f and converter input circuits can lead to a reduction in signal-to-noise ratio. This effect is not often very serious, when compared with the deterioration in signal-to-noise ratio caused by the reduction in gain of the r-f stages as the a.v.c. bias is increased ; further discussion of this point will be given in connection with the design of a.v.c. systems in Chapter 27.

(3) Detuning can result in considerable loss in adjacent channel selectivity. This is not particularly serious if the receiver has been aligned on small signal input, since the deterioration occurs when the desired signal is large.

(4) The i-f stages lend themselves more readily to compensation methods than the r-f circuits ; in the case of the r-f circuits exact compensation can usually be obtained at one frequency only. Since the r-f stages are often relatively unselective, detuning is not often a very serious factor and it is usual to compensate only in the i-f circuits.

(5) Even although exact compensation for input capacitance changes can be obtained at the resonant frequency, the compensation is not complete at frequencies removed from the i-f centre frequency, and there will always be some departure from symmetry on the " skirts " of the overall response curve. However, since compensation is required mainly at the centre frequency, this is not serious.

(6) The neutralization methods to be described in Section 8, in connection with stability, will affect input capacitance variations which are due to grid-to-plate coupling. If the amplifier is neutralized, then the input capacitance changes with grid bias will be due, mainly, to space charge effects.

(B) Circuits

Attention will be confined to the circuits of Fig. 26.18 (some discussion of which has already been given in Chapter 23, Sect. 5), since these allow satisfactory results to be obtained with a small number of components. Other types of compensating circuits can be found in Refs. 67, 71, 72.

In Figs. 26.18A and B, C_b is a by-pass capacitor and may be neglected in the discussion which follows. It is possible to show that, for complete compensation of the input capacitance variation, the condition required is

$$R_k g_k = \Delta C / C_{gk} \qquad (44)$$

where R_k = unbypassed cathode resistance

$\quad\quad g_k$ = grid-to-cathode transconductance ; which is given by $g_m(I_k/I_b)$ in which g_m is the mutual conductance ; I_k is the total d.c. cathode current (for a pentode this is usually the sum of the plate and screen-grid currents) ; I_b is the d.c. plate current.

$\quad\Delta C$ = valve input capacitance change

and $\quad C_{gk}$ = grid-to-cathode capacitance.

For a typical case suppose $\Delta C = 3 \ \mu\mu\text{F}$, $C_{gk} = 6 \ \mu\mu\text{F}$ and $g_k = 2000 \ (11.8/9.2)$ = 2560 μmhos, so that

$R_k = (3/6) \times (10^6/2560) = 195 \; \Omega.$

Since $\Delta C/C_{gk}$ determines the value of R_k for a particular valve, it is sometimes convenient to increase C_{gk} artificially to obtain a more suitable value for R_k. This is done by adding a small capacitance between the grid and cathode terminals. However, too large a total value for C_{gk} can have an undesirable effect on the valve input conductance, particularly at high frequencies, and so it is not advisable to increase C_{gk} to more than about twice its usual value or even less, at high frequencies (say 10 Mc/s or so). For typical results see Ref. 68. For 455 Kc/s i-f circuits it is often permissible to connect the whole of C_2 in parallel with C_{gk}.

Because of the presence of the unbypassed cathode resistor there will be a change in the effective mutual conductance (and consequently a loss in gain) given by

$$g_{m(effective)} = \frac{g_m}{1 + R_k g_k}. \tag{45}$$

In our example :

$$g_{m(effective)} = \frac{2000}{1 + 195 \times 2.56 \times 10^{-3}} = 1330 \; \mu mhos.$$

If C_{gk} is increased to 20 $\mu\mu$F then $R_k = 58.5 \; \Omega$ and the effective $g_m = 1740 \; \mu$mhos. The actual cathode bias resistor required is approximately 260 Ω so that about 200 Ω ($= R'_k$) would be used and this is then by-passed in the usual manner by C_k, as shown in Fig. 26.18B. The increased value of C_{gk} would need some corresponding reduction in C_2 since the total capacitance tuning the i-f transformer secondary includes the resultant capacitive reactance due to C_{gk} and R_k in series if the reactances of C_b and C_k are neglected.

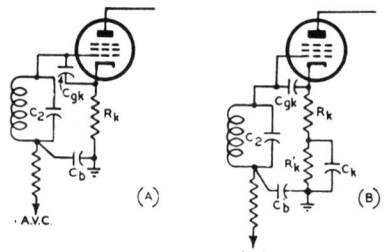

FIG. 26-18 METHOD FOR REDUCING VARIATIONS OF INPUT ADMITTANCE

The use of an unbypassed cathode resistor results in an increase of the short-circuit feedback admittance, and this effect can be used to increase the stage gain so as to offset the loss in g_m and, also, the reduction in Q of the input transformer due to the cathode resistor forming part of the tuned circuit. Two methods are available for increasing the feedback admittance ; the first is to connect the plate and screen by-pass capacitors back to the cathode ; the second is to add a small capacitance between plate and cathode, the plate and screen by-pass capacitors being connected to ground in the usual manner. These methods of increasing gain need to be treated with some care as they can easily lead to oscillation and particular attention is needed if instability is to be avoided. With some valve types (e.g. 6SG7, 6SH7, 9001, 9003, 6AG5, 6AK5) the plate-to-cathode capacitance is already relatively high because of the internal connections between the suppressor grid, beam confining electrodes, and the cathode ; this can lead to oscillation difficulties without external capacitance being added.

In the majority of cases it is preferable to return plate and screen by-pass capacitors (also the suppressor grid) to ground, and add a capacitance from plate to cathode if this is thought to be desirable.

With any circuit the capacitances can be measured and the value for R_k accurately calculated, but in a practical case it is generally much simpler to determine an approximate value for R_k and then find the exact value experimentally.

SECTION 8 : STABILITY

(i) Design data   (ii) Neutralizing circuits.

(i) Design data

The discussion of Chapter 23, Sect. 7 should be used in conjunction with the data to be given here ; also, the detuning effects discussed in the previous section are closely bound up with the data which follow.

For stages using single tuned circuits the expression given in Chapter 23 is more conveniently arranged as

$$\text{Max. stable impedance} = \sqrt{\frac{2}{g_m \omega_0 C_{gp}}} \qquad (46)$$

from which it immediately follows that the

$$\text{Max. stable gain} = \sqrt{\frac{2g_m}{\omega_0 C_{gp}}} \qquad (47)$$

For double tuned circuits which are critically-coupled—the usual case—Jaffe (Ref. 74) has shown that

$$\text{Max. stable impedance} = \sqrt{\frac{0.79}{g_m \omega_0 C_{gp}}} \qquad (48)$$

$$\text{and max. stable gain} = \sqrt{\frac{0.79 \, g_m}{\omega_0 C_{gp}}} \qquad (49)$$

where g_m = mutual conductance of amplifier valve
      ω_0 = $2\pi \times f_0$ (and f_0 is operating frequency ; i-f in this case)
      C_{gp} = total grid-to-plate capacitance made up from that due to the valve, wiring and valve socket etc.

Additional data are given in Ref. 75 for cases where the input and output circuits are not identical. This reference also shows that greater gain is possible when the impedance of the input circuit is less than that of the output circuit ; a condition which is fulfilled only in special cases in practical i-f circuits.

The equations given above, and in the references, are useful in forming an estimate of the possible maximum gain obtainable and in indicating some of the causes of instability ; but detailed calculations for various circuit possibilities hardly seem to be justified.

It may be helpful to note that when selecting a particular type of valve for maximum gain combined with stability, the most suitable is that having the greatest value of g_m/C_{gp} consistent with other circuit requirements (and cost). For an i-f valve, g_m/C_{gp} is of considerable importance but for a r-f valve operating at v-h-f it is also very important to consider the factor $g_m\sqrt{R_i}$, as pointed out in Chapter 23, Sect. 5.

(ii) Neutralizing circuits

A simple neutralizing circuit, which requires only the addition of one capacitor, is shown in Fig. 26.19(A). The equivalent capacitance bridge is shown in Fig. 26.19(B). It is assumed that the cathode by-pass capacitance is large.

The condition for the effects of the grid-to-plate capacitance to be neutralized is

$$C_N = C \frac{C_{gp}}{C_{gk}} \qquad (50)$$

where C_N = neutralizing capacitance required
      C = a.v.c. by-pass capacitance (usually about 0.01 μF)
      C_{gp} = total grid-to-plate capacitance including strays
and C_{gk} = total grid-to-cathode capacitance including strays.

For a typical case $C = 0.01$ μF, $C_{gp} = 0.01$ $\mu\mu$F and $C_{gk} = 15$ $\mu\mu$F.
Then $C_N = 0.01 \times 10^{-6} (0.01/15) = 6.7$ $\mu\mu$F.
The exact value for C_N is determined experimentally. For receiver production it is generally sufficient to use a fixed value for C_N which is reasonably close to the

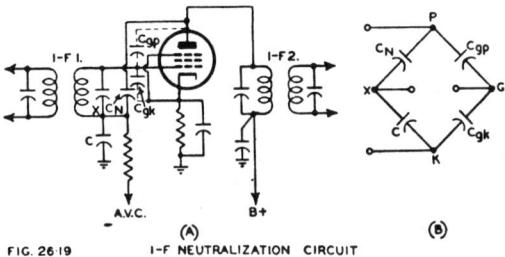

FIG. 26·19   I-F NEUTRALIZATION CIRCUIT

value required. The neutralization is generally not sufficiently critical to require different capacitance values in different receivers of the same type, as even partial neutralization is helpful and often sufficient.

Other circuits using various arrangements of inductance and capacitance for neutralization and stabilization can be found in the references. Two interesting alternatives (Refs. 79 and 80) to that of Fig. 26.19 are shown in Fig. 26.20.

Circuit (1) will be briefly discussed ; circuit (2) should be self-explanatory. In both cases all the components have their usual values except C_3 in circuit (1), and L_a is an added inductance (in i-f and some r-f circuits) in circuit (2) ; the latter arrangement is of interest when the cathode is grounded for d.c. To obtain complete neutralization in circuit (1) it is required to make

$$C_3 = C_1(C_{pk}/C_{gp}) \tag{51}$$

where C_1 = i-f tuning capacitor

C_{pk} = total plate-cathode capacitance including all strays

and C_{gp} = total grid-plate capacitance including all strays.

In a typical case for a 455 Kc/s i-f transformer $C_1 = 100 \ \mu\mu F$, $C_{pk} = 10 \ \mu\mu F$, $C_{gp} = 0.01 \ \mu\mu F$.

Then $C_3 = 100 \times (10/0.01) = 0.1 \ \mu F$, which is used as the by-pass capacitor across the cathode resistor. Also, remember that exact neutralization may not be necessary.

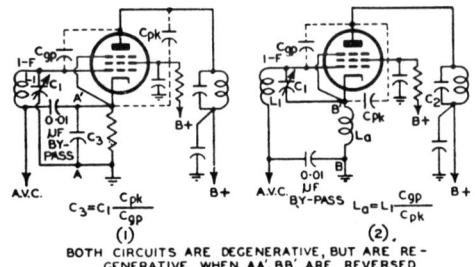

BOTH CIRCUITS ARE DEGENERATIVE, BUT ARE RE-GENERATIVE WHEN AA', BB' ARE REVERSED

FIG. 26·20   METHODS FOR STABILISING R-F AND I-F AMPLIFIERS

This circuit requires no additional components and C_3 is set to the value required for neutralization. It will be noted that the capacitance of C_3 is somewhat larger than is generally used for a by-pass capacitor. The only disadvantage is that an additional lug is required on the i-f transformer base and care is necessary to keep critical leads as short as possible. The advantage of the circuit in a r-f stage should be apparent.

Receivers which combine the detector diode in the i-f amplifier valve sometimes require additional neutralization because of feedback (which may be either regenerative or degenerative) caused by the grid-to-diode capacitance. A typical circuit is shown in Fig. 26.21. This requires two capacitors, C_{N1} and C_{N2}, in addition to the usual circuit components.

Analysis of this circuit (Ref. 77) shows that sufficient accuracy can be obtained if two bridge circuits are used to find the values of C_{N1} and C_{N2}. Although there is interaction between the two circuits, which affects the values selected for C_{N1} and C_{N2}, the calculated values are sufficiently close to allow the practical circuit to be satisfactorily neutralized. The values of C_{N1} and C_{N2} are adjusted in the circuit until complete neutralization is obtained. Fixed values can then be selected, since the circuit is not very critical to small changes in the values of the neutralizing capacitors.

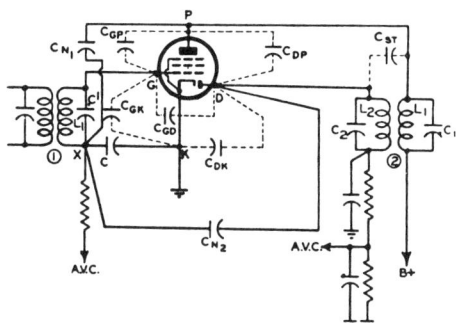

FIG. 26·21   NEUTRALIZATION IN DIODE PENTODE

The equations required are (50) above, which is used to determine C_{N1}, just as previously, and

$$C_{N2} = C(C_{gd}/C_{gk}) \tag{52}$$

where C = a.v.c. by-pass capacitance (as before)
    C_{gd} = grid-to-diode capacitance including strays
and C_{gk} = grid to cathode capacitance (as before).
    For a typical case :
    C_{gp} = 0.007 $\mu\mu$F (including strays), C_{gd} = 0.009 $\mu\mu$F (including strays),
    C = 0.01 μF, C_{gk} = 15 $\mu\mu$F (including strays),
    C_{N1} = 0.01 × 10⁻⁶ × (0.007/15) = 3.3 $\mu\mu$F,
    C_{N2} = 0.01 × 10⁻⁶ × (0.009/15) = 6 $\mu\mu$F.
    In an actual receiver, using a type 6SF7, the values required were C_{N1} = 4 $\mu\mu$F ; C_{N2} = 7 $\mu\mu$F. With wiring re-arrangement the values required were C_{N1} = 3 $\mu\mu$F and C_{N2} = 5 $\mu\mu$F.
    For methods of alignment and a suitable neutralizing procedure see Ref. 77.

SECTION 9 : DISTORTION

(*i*) *Amplitude modulation i-f stages*   (*ii*) *Frequency modulation i-f stages.*

(i) Amplitude modulation i-f stages
    In Chapter 23, Sect. 8(i) the causes of modulation envelope distortion were briefly discussed. These same factors apply, also, to i-f amplifier stages, but become more serious because of the larger signal voltages which must be handled by the valves.
    The most severe conditions for non-linear distortion will be obtained in the voltage amplifier preceding the detector stage, and the worst condition is given when the grid bias, due to a.v.c., is large. To reduce the distortion due to non-linearity of the valve characteristics, the final amplifier valve is sometimes operated with fixed bias, but this also affects the a.v.c. characteristic (see Chapter 27 Sect. 3). As an alternative, partial a.v.c. may be applied to the last stage. If the screen voltage is supplied by means of a series resistor (as is common practice) instead of a voltage divider, appreciable reduction in distortion is possible. The screen voltage rises as the grid bias voltage is increased and, although there is some loss of efficiency in the a.v.c.

system, the distortion is materially reduced ; this can be confirmed experimentally. As a compromise between distortion and a.v.c. action, the screen supply voltage can be obtained by means of a high resistance voltage divider arrangement if so desired. Detailed measurements of distortion have been made on typical i-f voltage amplifier valves for various methods of screen voltage supply and some of the results can be found in Ref. 83. In cases where the range of bias voltage is limited, a high resistance voltage divider can be arranged to give less distortion than that given by the series feed arrangement. A total increase in non-linear distortion of the order of 2% for 90% amplitude modulation is to be expected in typical cases using a series screen resistor. Because of the internal construction of remote cut-off valves, such as those used in i-f and r-f amplifiers, the screen current is often rather variable and this suggests a voltage divider in cases where variations in screen voltage are important (see Ref. 84).

For methods of measuring the signal handling capabilities of i-f amplifier valves and the non-linear distortion from this cause, the reader is referred to Ref. 4 (page 335) and Ref. 83.

Frequency distortion, caused by sideband cutting due to the selectivity of the tuned circuits, is always present to some extent ; this should be obvious from the discussion on the design of i-f transformers and it is not proposed to treat the matter further here.

(ii) Frequency modulation i-f stages

With frequency modulation receivers, distortion of the amplitude (in itself) of the frequency modulated wave is generally not of great importance, provided the amplitude limiting device is capable of smoothing out the variations. Non-linearity in i-f and r-f valve characteristics are of secondary importance, with this system of reception, and the non-linear distortion appearing in the receiver output from this cause is usually negligible ; this is the reverse of the case for amplitude modulation.

The amount of distortion caused by the non-linearity of phase shift/frequency characteristics of the tuned circuits is of considerable importance, however, and poorly designed i-f transformers will lead to non-linear distortion in the receiver output. If the criterion for non-linearity given in eqn. 24 [Sect. 4(v) of this chapter] is fulfilled, then the distortion in the receiver output due to the i-f circuits preceding the discriminator (or whatever detection system is used) will be quite small, and in most cases less than that introduced by the detection and audio systems.

When a F-M receiver is accurately tuned to the centre frequency of the carrier, the non-linear distortion of the modulating signal introduced by the tuned circuits will consist mainly of odd order harmonics. Of these the third harmonic (H_3) is the largest and a method of estimating its magnitude is helpful when i-f transformers are being designed. The required expression is (Ref. 82),

$$H_3 = \frac{4A^3}{\beta} (1 + A^2)^{\frac{1}{2}} \left[\frac{3Q^2k^2 - 4A^2 - 1}{(1 + Q^2k^2)^3} \right].$$ (53)

For a critically-coupled transformer $Qk = 1$ and so

$$H_3 = \frac{A^3}{\beta} (1 + A^2)^{\frac{1}{2}}(1 - 2A^2)$$ (54)

where $A = \dfrac{\Delta F}{f_0} Q$

ΔF = frequency deviation
f_0 = central carrier frequency
Q = magnification factor ($Q_1 = Q_2$)
β = modulation index = $\Delta F/f$
f = a-f modulating frequency
and k = coefficient of coupling (the coupling may be inductive or capacitive and either shunt or series).

For 1 db attenuation it is known that, for a critically-coupled transformer with identical primary and secondary circuits, $Q = f_0/$total bandwidth. From this we may write $A = \Delta F$/total bandwidth. If it is taken that the total frequency deviation $(2\Delta F)$ is equal to the total bandwidth for 1 db attenuation, then we obtain

$$A = \tfrac{1}{2} \text{ and } H_3 = \frac{\sqrt{5}}{32\beta} = \frac{7}{\beta}\%. \tag{55}$$

The Q required for this condition is given from the previously stated relationship $Q = f_0$/total bandwidth.

Suppose $Q = 71$ is obtained, for a critically-coupled transformer, with $f_0 = 10.7$ Mc/s, $2\Delta F = 150$ Kc/s and $\beta = 5$. Then the third harmonic audio distortion introduced by each transformer is 1.4% (approx.) ; using either eqns. (54) or (55).

If $Q = 80$ is obtained from the procedure of Sect. 4, then eqn. (54) gives $H_3 = 1.57\%$; eqn. (55) is not applicable in this case.

It should be clear that a simple approximate design procedure can be developed from the conditions leading to eqn. (55).

The distortion given in the examples is somewhat higher than that experienced under normal operating conditions. The third harmonic of 15 Kc/s will hardly trouble the listener, but harmonics of the lower audio frequencies would be serious if their amplitudes are appreciable, and the possibility of intermodulation effects should not be neglected.

Applying eqn. (54) for $Q = 71$, $\Delta F = 50$ Kc/s (more nearly the usual operating condition) and a modulation frequency of 5 Kc/s, then $H_3 = 0.074\%$. (See also Chapter 27, Sect. 2 under heading " Non-linear distortion ").

It should be carefully noted that eqn. (55) applies only for the special conditions under which it was derived. Distortion is normally calculated from eqns. (53) and (54).

Only a few of the significant factors have been mentioned here, and for more detailed information the reader is referred to Refs. 82, 85, 86 and 87.

SECTION 10 : REFERENCES

(A) I-F AMPLIFIER DESIGN
(1) Kelly Johnson, J. " Selectivity of superheterodyne receivers using high intermediate frequencies " A.R.T.S. and P. Bulletin No. 12 (Reprint from Hazeltine Service Corp. Lab. Bulletin) April 1935.
(2) Ross, H. A. " The theory and design of intermediate-frequency transformers for frequency-modulated signals " A.W.A. Tec. Rev. 6.8 (March 1946) 447.
(3) Everitt, W. L. (book) " Communication Engineering " 2nd ed., 1937, Chap. 16, pages 496-504. McGraw-Hill Book Co. Inc., New York and London.
(4) Sturley, K. R. (book) " Radio Receiver Design " Part 1, Chap. 7, Chapman and Hall, London, 1943.
(5) Espy, D. " Double-tuned transformer design " Elect. 17.10 (Oct. 1944) 142.
(6) Maynard, J. E. " Universal performance curves for tuned transformers " Elect. 10.2 (Feb. 1937) 15.
(7) Maynard, J. E. and P. C. Gardiner, " Aids in the design of intermediate-frequency systems " Proc. I.R.E. 32.11 (Nov. 1944) 674.
(8) Aiken, C. B. " Two-mesh tuned coupled circuit filters " Proc. I.R.E. 25.2 (Feb. 1937) 230.
(9) Adams, J. J. " Undercoupling in tuned coupled circuits to realize optimum gain and selectivity " Proc. I.R.E. 29.5 (May 1941) 277.
(10) Spaulding, F. E. " Design of superheterodyne intermediate Frequency Amplifiers " R.C.A. Rev. 6.4 (April 1940) 485.
(11a) Stern, E. " Measurements on intermediate-frequency transformers " Proc. I.R.E. (Aust.) 9.1 (Jan. 1948) 4 ; Reprinted Jour. Brit. I.R.E. 9.4 (April 1949) 157.
(11b) Rudd, J. B. " A design for double-tuned transformers " Proc. I.R.E. (Aust.) 10.1 (Jan. 1949) 3 ; A.W.A. Tec. Rev. 8.2 (April 1949) 147 ; Reprinted Jour. Brit. I.R.E. 9.8 (Aug. 1949) 306.
(11c) See also bibliography Chapter 9.

General I-F Amplifier Data
(12) Benin, Z. " Modern home receiver design " Elect. 19.8 (Aug. 1946) 94.
(13) Terman, F. E. (book) " Radio Engineers' Handbook " 1st edit. McGraw-Hill Book Co. Inc., New York and London, 1943.
(14) Kees, H. " Receiver with 2 Mc i-f " Elect. 18.4 (April 1945) 129.
(15) Adams, J. J. " Intermediate-frequency amplifiers for frequency-modulation receivers " Proc. I.R.E. 35.9 (Sept. 1947) 960.
(16) Rust, Keall, Ramsay and Sturley " Broadcast receivers : A review " Jour. I.E.E. 88.2 Part 3 (June 1941) 59.
(17) Zepler, E. E. (book) " The Technique of Radio Design " Chapman and Hall, London, 1943. Wiley and Sons, New York, 1943.
(17a) Ridgers, C. E. S. " Practical aspects of the design of intermediate frequency transformers " Jour. Brit. I.R.E. 10.3 (March 1950) 97.

Additional data for F-M i-f Amplifiers
(18) See (2), (12) and (15) above.
(19) Hund, A. (book) " Frequency Modulation " (1st edit.) McGraw-Hill Book Co. Inc., New York and London, 1942.
(20) Foster, D. E. and J. A. Rankin " Intermediate frequency values for frequency-modulated-wave receivers " Proc. I.R.E. 29.10 (Oct. 1941) 546.
(21) Landon, V. D. " Impulse noise in F-M reception " Elect. 14.2 (Feb. 1941) 26.
(22) Beard, E. G. " Intermediate frequency transformers for A-M/F-M receivers " Philips Tec. Com. No. 6 (Nov. 1946) 12.
(23) Sturley, K. R. (book) " Radio Receiver Design " Part 2 Chapman and Hall, London, 1945.
(24) Tibbs, C. E. (book) " Frequency Modulation Engineering " Chapman and Hall, London, 1947.
(25) Crosby, M. G. " A method of measuring frequency deviation " R.C.A. Rev. 6.4 (April 1940) 473.

Special I-F Transformers
See also references 94, 95, 96 and 97 below.
(26) Dishal, M. " Exact design and analysis of triple tuned band-pass amplifiers " Proc. I.R.E. 35.6 (June 1947) 606. (Discussion Proc. I.R.E. 35.12 (Dec. 1947) 1507.
(27) Weighton, D. " Performance of coupled and staggered circuits in wide band amplifiers " W.E. 21.253 (Oct. 1944) 468.
(28) Cocking, W. T. (book) " Television Receiving Equipment " 2nd edit., Iliffe and Sons, London, 1947.

I-F Transformer Construction
(29) " The design of the universal winding " A.R.T.S. and P. Bulletin No. 131 (Reprint Hazeltine Corp. Bulletin) 13th Sept. 1943.
(30) Simon, A. W. " Universal coil design " Radio 31.2 (Feb.-Mch. 1947) 16.
(31) Simon, A. W. " On the theory of the progressive universal winding " Proc. I.R.E. 33.12 (Dec. 1945) 868.
(32) Kantor, M. " Theory and design of progressive and ordinary universal windings " Proc. I.R.E. 35.12 (Dec. 1947) 156.
(33) Scheer, F. H. " Notes on intermediate frequency transformer design " Proc. I.R.E. 23.12 (Dec. 1935) 1483.
(34) Callender, M. V. " Q of solenoid coils " (letter) W.E. 24.285 (June 1947) 185.
(35) Amos, S. W. " Calculating coupling co-efficients—useful formulae for finding the optimum spacing of I-F transformer windings " W.W. 49.9 (Sept. 1943) 272.
(36a) Vergara, W. C. " Determining form factors of I-F transformers " Elect. 22.7 (July 1949) 168.
(36b) See also Refs. 88 to 93 below ; Chapters 10 and 11 and their bibliographies.

(B) VARIABLE SELECTIVITY
(37) See (4), (16) and (17) above.
(38) Wheeler, H. A., and J. Kelly Johnson, " High fidelity receivers with expanding selectors " Proc. I.R.E. 23.6 (June 1935) 594.
(39) Varrall, J. E. " Variable selectivity I-F amplifiers " W.W. 48.9 (Sept. 1942) 202.
(40) Cocking, W. T. " Variable selectivity and the I-F amplifier " (3 parts) W.E. 13.150 (Mch. 1936) 119 ; 13.151 (Apr. 1936) 179 ; 13.152 (May 1936) 237.
(41) " The problem of variable selectivity " Radiotronics No. 82 (15th Dec. 1937) 89.
(42) " The Radiotron fidelity tuner—Continuously variable selectivity " Radiotronics No. 84 (15th March 1937) 105 ; No. 85 (16th May, 1937) 120.
(42a) " A variable-selectivity i-f amplifier system " Radio and Electronics (New Zealand) 5.1 (March 1950) 9 ; also British Patent No. 598662. Review, W.W. 56.1 (Jan. 1949) 40.

Automatic Variable Selectivity
(43) See (4) above.
(44) Beers, G. T. " Automatic selectivity control " Proc. I.R.E. 23.12 (Dec. 1935) 1425.
(45) " Model 8-Tube AXPS receiver " A.R.T.S. and P. Bulletin No. 44 (Reprint Hazeltine Service Corp. Bulletin) 19th July, 1937.
(46) " AXPS circuits without extra tubes " A.R.T.S. and P. Bulletin No. 70 (Reprint Hazeltine Service Corp. Bulletin) 6th March, 1939.
(47) " American A.S.C. circuits " W.W. 40.13 (26th March 1937) 296.

(C) CRYSTAL FILTERS
See (17) above.
(48) Sturley, K. R. " Single crystal filters " W.E. 22.262 (July 1945) 322.
(49) " The Amateur Radio Handbook " (2nd edit.) The Incorp. Radio Society of Gt. Britain, London.
(50) " The Radio Amateur's Handbook " (24th edit.) American Radio Relay League Inc. 1947.
(51) Oram, D. R. " Full-range selectivity with 455 Kc/s quartz crystal filters " Q.S.T. 22.12 (Dec. 1938) 33.
(52) Erlich, R. W. " The crystal filter " Radio Craft 17.6 (March 1946) 398 ; 17.7 (April 1946) 476.
(53) Colebrook, F. M. " High selectivity tone-corrected receiving circuits " Brit. Radio Research Board Special Report No. 12, 1932 (Reprinted 1934).
(54) Builder, G., and J. E. Benson, " Simple quartz crystal filters of variable bandwidth " A.W.A. Tec. Rev. 5.3 (Feb. 1941) 93.
(55) Mason, W. P. " Electrical wave filters employing quartz crystals as elements " B.S.T.J. 13.3 (July 1934) 405.
(56) Robinson, J. " The stenode radiostat " Radio News 12.8 (Feb. 1931) 682.
(57) Stanesby, H. " A simple narrow band crystal filter " British P.O.E.E.J. Vol. 28 Pt 1 (April 1942) 4.
(58) Stanesby, Broad and Corke " Calculation of insertion loss and phase change of 4-terminal re-actance networks " British P.O.E.E.J. Vol. 25 Pt. 3 (Oct. 1942) 88 ; Vol. 25 Pt 4 (Jan. 1943) 105 (Gives transfer impedance etc. for equivalent bridge circuit)
(59) Mason, W. P. " Resistance compensated band-pass filters for use in unbalanced circuits " B.S.T.J. 16.4 (Oct. 1937) 423.
(60) Builder, G. " Resistance balancing in wave filters " A.W.A. Tec. Rev. 3.3 (Jan. 1938) 83.
(61) Starr, A T. (book) " Electric Circuits and Wave Filters " 2nd edit. Sir Isaac Pitman and Sons, London, 1940.

(62) Booth, C. F. " The application and use of quartz crystals in telecommunications " Jour. I.E.E. Part 3, 88.2 (June 1941) 97.
(63) Grisdale, G. L., and R. B. Armstrong " Tendencies in the design of the communication type of receiver " Jour. I.E.E. Part 3, 93.25 (Sept. 1946) 365.

Measurements on Crystals
(64) Builder, G. " A note on the determination of the equivalent electrical constants of a quartz-crystal resonator " A.W.A. Tec. Rev. 5.1. (April 1940) 41.
(65) Mason, W. P., and I. E. Fair " A new direct crystal-controlled oscillator for ultra-short-wave frequencies " Proc. I.R.E. 30.10 (Oct. 1942) 464.

(D) DETUNING EFFECTS AND COMPENSATION
(66) See above for Ref. 17, p. 171 : Ref. 4, p. 37 ; Chapter 2 Sect. 8 ; Chapter 23, Sect. 5.
(67) Ref. 13 above ; page 472.
(68) " Input Admittance of Receiving Tubes " R.C.A. Application Note No. 118 (April 25th, 1947) ; Reprinted Radiotronics No. 126 (July-Aug., 1947) 62.
(69) " Compensation for amplifier tube input capacitance variation " A.R.T.S. and P. Bulletin No. 67 (Dec. 22nd 1938) 5 (Reprint from Hazeltine Service Corp. Lab. Bulletin).
(70) Harvey, A. F. (book) " High Frequency Thermionic Tubes " (Chapter 2 and Bibliography). John Wiley and Sons, New York, 1943 ; Chapman and Hall, London, 1943.
(71) " Television topics—valve input capacity and resistance " W.W. 43.15 (Oct. 13th, 1938) 340.
(72) Freeman R. L. " Input conductance neutralization " Elect. 12.10; (Oct. 1939) 22.

(E) STABILITY
(73) Chapter 23, Sect. 7 and Ref. B41 for that chapter.
(74) Jaffe, D. L. " Intermediate frequency amplifier stability factors " Radio 30.4 (April 1946) 26.
(75) " Effect of transconductance and grid-plate capacitance on the stability of i-f amplifiers " A.R.T.S. and P. Bulletin No. 93 (July 8th, 1940) (Reprint of Hazeltine Service Corp. Bulletin).
(76) " I-F neutralization " Radiotronics No. 115 (July 1941) 54.
(77) " Neutralization in circuits employing a valve as a combined i-f amplifier and diode detector " Radiotronics No. 118 (Mar./April 1946) 33.
(78) Ref. 17 above (Chapter 9). Ref. 13 above (p. 469).
(79) Root, C. S. " Method for stabilising R-F and I-F amplifiers " G. E. Docket 63495.
(80) " A neutralisation circuit for I.F. amplifiers " No. 122 (18th May, 1942) A.R.T.S. and P. Bulletin.
(81) Hultberg, C. A. " Neutralization of screen grid tubes to improve the stability of intermediate-frequency amplifiers " Proc. I.R.E. 31.12 (Dec. 1943) 663.

(F) DISTORTION
Ref. 4 above (pages 154 and 335) Refs. B34, B35, B36, bibliography Chapter 23.
(82) Guttinger, P. " The distortion of frequency modulated signals during transmission by band-pass filters " The Brown Boveri Review 23.8 (Aug. 1946) 185.
(83) " Screen supply for the final I.F. amplifier " A.R.T.S. and P. Bulletin No. 88 (21st March 1940).
(84) " Screen dropping resistors—Restrictions with super-control valves " Radiotronics No. 96 (15th March 1939).
(85) Jaffe, D. L. " A theoretical and experimental investigation of tuned-circuit distortion in F.M. systems " Proc. I.R.E. 33.5 (May 1945) 318.
(86) Roder, H. " Effects of tuned circuits upon a frequency modulated signal " Proc. I.R.E. 25.12 (Dec. 1937) 1617.
(87) Cherry, E. C., and R. S. Rivlin, " Non-linear distortion with particular reference to the theory of F.M. waves " Phil. Mag. (i) 32.213 (Oct. 1941) 265 ; (ii) 32.219 (April 1942) 272. Pt (i) covers non-linear distortion due to valves ; Pt. (ii) tuned circuits.

(G) POWDERED IRON CORES
(88) Shea, H. G. " Magnetic powders " Electronic Industries 4.8 (Aug. 1945) 86 (Gives useful lists of μ_{eff}, Q_{eff} and standard test data).
(89) " Magnetic materials in broadcast receivers—radio dust cores—Permalloy " A.R.T.S. and P. Bulletin No. 19 (22nd Nov. 1935) (Reprint of data from Standard Telephones and Cables (A'sia) Ltd.).
(90) Foster, D. E., and A. E. Newlon, " Measurement of iron cores at radio frequencies " Proc. I.R.E. 29.5 (May 1941) 266.
(91) Starr, A. T. (book) " Electric Circuits and Wave Filters " Chapter 4 (2nd edit.) Sir Isaac Pitman and Sons, London, 1940. (Gives useful data on general coil construction ; also see references to W.E. etc.).
(92) " Ferromagnetic Dust Cores " Loose-leaf folder issued by Kingsley Radio Pty., Ltd., Aust.
(93) " Carbonyl Iron Powders " Catalogue issued by General Aniline Works, New York (also includes useful list of references).

(H) ADDITIONAL REFERENCES ON SPECIAL I-F TRANSFORMERS
(94) Wallman, H. " Stagger-tuned transformers Elect. 21.5 (May 1948) 100 (corrections 21.7 (July 1948) 224).
(95) " Use of Miniature Tubes in Stagger-Tuned Video Intermediate Frequency Systems " R.C.A. Application Note AN-126 ; Reprinted in Radiotronics No. 134 (Nov.-Dec. 1948) 109.
(96) Valley, G. E., and H. Wallman (book) " Vacuum Tube Amplifiers " McGraw-Hill Book Co. Inc New York, Toronto and London ; 1948.
(97) Baum, R. F. " Design of broad band i.f. amplifiers " Journal of Applied Physics Pt i., 17.6 (June 1946) 519 ; Pt ii 17.9 (Sept. 1946) 721

(I) COUPLING COEFFICIENTS
(98) Editorial " Coupling and coupling coefficients " W.E. 9.108 (Sept. 1932) 485.
(99) Editorial " Natural and resonant frequencies of coupled circuits " W.E. 18.213 (June 1941) 221.
(100) Editorial " Coupled circuits " W.E. 21.245 (Feb. 1944) 53.
(101) Editorial " Effect of stray capacitance on coupling coefficient " W.E. 21.251 (Aug. 1944) 357.
(102) Glasgow, R. S. (book) " Principles of Radio Engineering " McGraw-Hill Book Co. Inc., New York and London, 1936. For additional references see Supplement page 1480.

CHAPTER 27

DETECTION AND AUTOMATIC VOLUME CONTROL

BY B. SANDEL, A.S.T.C.

SECTION 1 : A-M DETECTORS

(i) Diodes (A) General (B) Diode curves (C) Quantitative design data (D) Miscellaneous data (ii) Other forms of detectors (A) Grid detection (B) Power grid detection (C) Plate detection (D) Reflex detection (E) Regenerative detectors (F) Superregenerative detectors.

(i) Diodes

(A) General

A diode has two electrodes namely plate and cathode. It is therefore identical in structure with a power rectifier but the term is generally restricted to valves which are used for detection or a.v.c. as distinct from rectifiers which are used for power supply. The operation of diodes with a.v.c. is considered in detail in Sect. 3 of this chapter. The operation of a diode on a modulated wave is rather different from the operation of a power rectifier, and it is necessary to consider the characteristic curves of a diode valve if a full understanding of the operation is to be obtained. The operation of diodes on a modulated input is considered in (B) below. Fig. 27.1 shows typical distortion curves for a diode operating firstly under ideal conditions with no a.c. shunting (curve B) and secondly the distortion resulting when a load of 1 megohm is shunted across a diode load resistance of 0.5 megohm (curve A). The respective percentages of harmonic distortion at 100% modulation are approximately 6% and 12% so that the presence of such shunting has a very marked effect on performance ; these curves apply only for the particular conditions under which they were derived.

The design of a diode detector for low distortion is based on the following requirements :

(1) That the input voltage should not be less than 10 volts peak.

(2) That no appreciable a.c. shunting should be present.

The first of these two requirements is easily met for local stations and a voltage from 10 to 20 volts is quite common in receivers fitted with a.v.c. The second requirement is one which is difficult to satisfy. Shunting of the diode load may be due to :

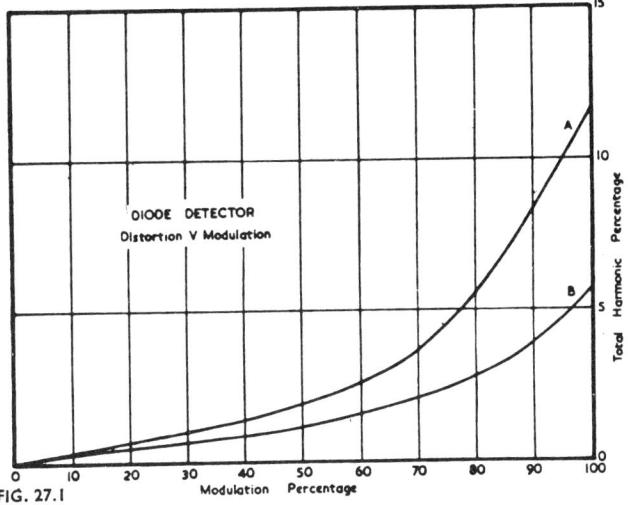

FIG. 27.1

DIODE DETECTOR
Distortion V Modulation

Modulation Percentage

Total Harmonic Percentage

(1) The a.v.c. system.

(2) The following grid resistor.

(3) An electron ray tuning indicator.

The circuit of a typical diode detector is shown in Fig. 27.2 in which the diode load resistance is R_2 together with R_1; the latter, in conjunction with C_1 and C_2, form a r-f filter so that the r-f voltage passed on to the a-f system may be a minimum. R_1 is generally made about 10% of R_2 and typical values are 50 000 ohms and 0.5 megohm. The capacitances of C_1 and C_2 depend upon the frequency of the carrier ; for an intermediate frequency of 455 Kc/s they may both be 100 $\mu\mu$F.

If the volume control (R_2) is turned to maximum the shunting effect due to R_3 will be appreciable, since R_3 cannot exceed 1 or 2 megohms with most types of valves. If, however, grid leak bias is used on a high-mu triode valve, R_3 may be approximately 10 megohms and the input resistance of the valve will then be of the order of 5 megohms. This is sufficiently high to be unimportant but for lower values of R_3 the distortion with the control set near maximum may be severe. It is found in most conventional receivers that the a-f gain is considerably higher than that required for strong carrier voltages and under these conditions the control will be turned to a low setting. The a.c. shunting effect due to R_3 is practically negligible provided the control is below one-fifth of the maximum position. A good method of overcoming the a.c. shunting due to the audio amplifier is shown in Fig. 27.3. The diode may be in the same envelope as the i-f amplifier valve. Negative feedback is applied across part of the cathode bias resistor for the a-f voltage amplifier valve. For the component values shown, the measured input resistance between points A and B is in excess of 10 megohms for frequencies up to about 10 Kc/s. A possible disadvantage is the reduction in overall gain because of the negative feedback, although this is not generally serious. Increasing the feedback resistance R increases the overall gain, but reduces the input resistance. Intermediate values of gain and input resistance may be selected as desired. For further details of receivers using this arrangement see Refs. 6 and 7.

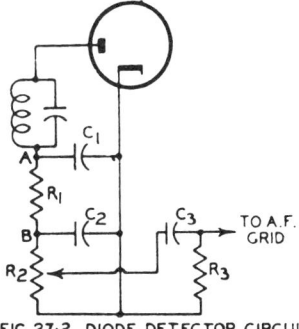

C_1
R_1
C_2
C_3
TO A.F. GRID
R_2
R_3

FIG. 27-2 DIODE DETECTOR CIRCUIT

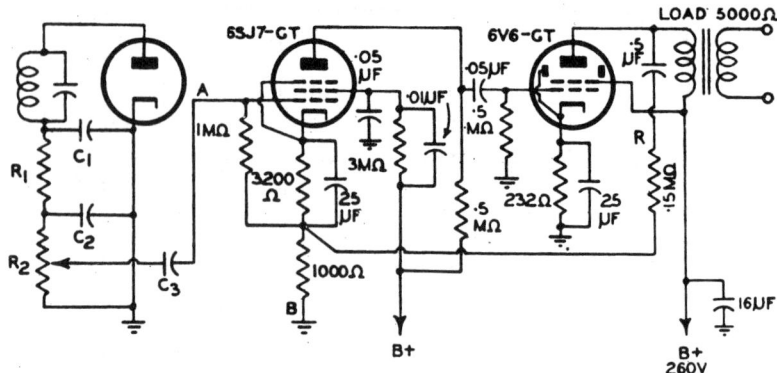

FIG.27·3   AUDIO AMPLIFIER GIVING LOW A.C. SHUNTING ACROSS DIODE LOAD.

Another useful method of reducing a.c. shunting effects is to use a cathode follower stage between the detector and the a-f voltage amplifier. This is often conveniently accomplished by employing a double triode valve, one section serving as the cathode follower and the other as an a-f voltage amplifier.

Distortion due to the a.v.c. system will be discussed in detail in Sect. 3 below. One type of distortion often encountered is caused by delayed a.v.c. systems at the point where the a.v.c. diode just starts to conduct. This form of distortion is called differential distortion and may be kept to low values by making the delay voltage small. The conventional arrangement for obtaining a.v.c. voltage from the primary of the last i-f transformer is generally preferred, as it reduces the a.c. shunting effect across the detector diode circuit.

The a.c. shunting due to the addition of an electron ray tuning indicator to the diode detector circuit is serious and difficult to avoid. In order to reduce the distortion to a minimum the resistor feeding the grid of the tuning indicator may be made 2 megohms and the effect will only then be apparent at high percentages of modulation. If the tuning indicator is connected to the a.v.c. system it will not operate at low carrier levels unless the delay voltage is extremely small. One possible method, where the utmost fidelity is required, is to use the same circuit as for delayed a.v.c. but with a delay voltage of zero, and to connect the tuning indicator to this a.v.c. circuit. With this arrangement a.c shunting due to the a.v.c. circuit and the tuning indicator is eliminated, while differential loading no longer occurs.

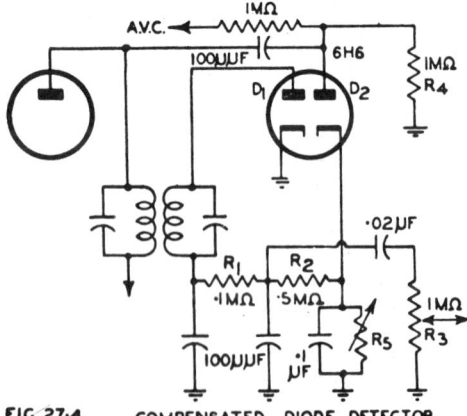

FIG.27·4   COMPENSATED DIODE DETECTOR.

An interesting arrangement (Ref. 8) for counteracting the effect of a.c. loading, and so increasing the maximum percentage of modulation which can be handled without excessive distortion is accomplished by the use of the circuit of Fig. 27.4. In this arrangement a positive bias is applied to the diode plate in such a way as to be proportional to the carrier input. A fixed positive bias would not be satisfactory since it would only give low distortion at one carrier level.

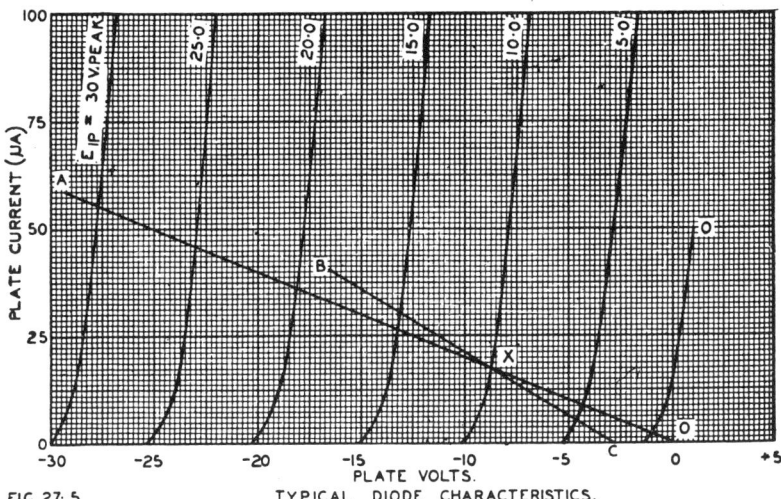

FIG.27·5   TYPICAL DIODE CHARACTERISTICS.

(B) Diode curves
Typical diode characteristics are shown in Fig. 27.5. Each curve corresponds to
the peak voltage of a constant unmodulated carrier voltage. On this graph may be
drawn loadlines corresponding to diode load resistances in a similar manner as for
triodes. The intersection of the applied loadline (OA) with the peak input voltage
curve indicates the d.c. voltage·developed by the diode and available for a.v.c. As
the load resistance increases, so the d.c. voltage approaches the peak input voltage.
For example, if the carrier input is 10 volts peak and the diode load resistor 0.5
megohm, the diode current will be 17 µA, and the d.c. voltage —8.7 volts. If 100%
modulation is applied to the carrier, the operating point will move at audio frequency
along the loadline from the intersection with the " 0 " curve, through X to the intersection with the 20 volt curve. The distortion over this excursion is small (about
5% second harmonic) and may be reduced still further by operating with a higher
carrier input voltage.
Typical diode curves with loadlines already drawn are shown in valve data books.
Average curves are shown in Fig. 27.6 and these are applicable to the diodes incorporated in standard types of Radiotron diode-triode and diode-pentode valves. It has
become the usual practice to·show the signal input voltages in r.m.s. values rather than
the peak voltages indicated in Fig. 27.5.
If the d.c. load resistance of 0.5 megohm, represented by OA in Fig. 27.5, is shunted
by an a.c. load (such as would occur due to the grid resistor of the following valve)
the dynamic loadline will be similar to BC, which passes through the static point X
but which has a slope corresponding to the total effective a.c. load resistance. This
loadline (BC) reaches cut-off at about 75% modulation and the distortion at higher
percentages of modulation will consequently be severe. For any combination of
d.c. and a.c. loads it is possible to draw the loadlines and determine the limiting
percentage of modulation before distortion becomes excessive. It should not be
overlooked that there will be some distortion present even before the limiting percentage of modulation is reached.   The evaluation of this distortion is discussed below.

(C) Quantitative design data
Design formulae will be set out below so that the performance of the diode detector
circuit can be assessed. These expressions are used in conjunction with the diode
curves of Fig. 27.6. The design procedures are necessarily a series of compromises
but the results form a useful practical guide. For a more detailed discussion the
reader is referred particularly to Ref. 1 (p. 339), Ref. 2 (p. 413) and Ref. 5 (p. 553).

(a) Diode detection efficiency

This can be found from curves such as those shown in Fig. 27.6.   The efficiency (assuming a sine wave input) is

$$\eta = \frac{E_{dc}}{\sqrt{2}E_{rms}} \qquad (1)$$

Suppose the d.c. load is 0.5 megohm and the signal input (E_{rms}) is 15 volts (r.m.s.). Then from the curves $E_{dc} = 16.8$ volts.

$$\eta = \frac{16.8}{\sqrt{2} \times 15} = 0.793 \text{ or } 79.3\%.$$

Actually the a.c. load should be determined and used for all calculations.   The a.c. load is, from Fig. 27.2,

$$R_{ac} = R_1 + \frac{R_2 R_3}{R_2 + R_3} \qquad (2)$$

The d.c. load for the diode detector is

$$R_{dc} = R_1 + R_2. \qquad (3)$$

If simple a.v.c. is used the effect of additional a.c. shunting due to this circuit should be taken into account.   In this case the effects of additional capacitive reactance are usually neglected for simplicity.

(b) Critical modulation ratio

The highest percentage of modulation which can be handled by the detector, before serious distortion of the modulation envelope occurs, is given by

$$m = 1 - \eta F \frac{R_{dc}}{R_{dc} + R_3/F} \qquad (4)$$

where F = fraction of R_{dc} across which R_3 is tapped

$$= \frac{R_2}{R_1 + R_2} = \frac{R_2}{R_{dc}}.$$

If $R_{dc} = 0.5$ MΩ, made up from $R_1 = 50\ 000$ Ω and $R_2 = 0.45$ MΩ, then $F = 0.45/0.5 = 0.9$. Taking $\eta = 0.793$ as before, and $R_3 = 1$ MΩ we have,

$$m = 1 - 0.793 \times 0.9 \frac{0.5}{0.5 + 1/0.9} = 0.78 \text{ or } 78\%.$$

It should be noted that the smaller F is made, the higher the critical modulation ratio.   If F is fixed, then R_3 is made as large as possible but is usually limited to about 1 or 2 megohms unless special circuit arrangements, such as those discussed previously, are made.

The critical modulation ratio can also be increased by applying a positive d.c.

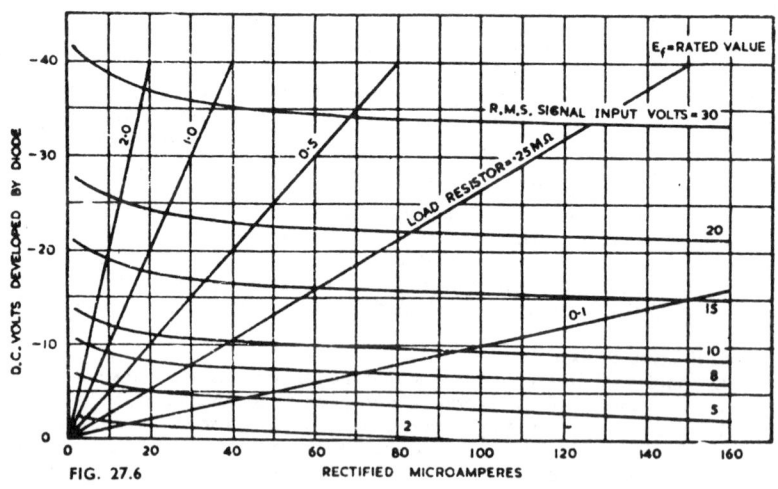

FIG. 27.6   RECTIFIED MICROAMPERES

bias voltage to the diode plate.   This arrangement is only satisfactory if the bias can be changed for each i-f input voltage (E_{rms}) see Ref. 1 (page 348).
For this case

$$m = \left(\frac{\sqrt{2}E_{rms} + E_{bias}}{\sqrt{2}E_{rms}}\right)\left(1 - \eta F\frac{R_{dc}}{R_{dc} + R_3/F}\right) \tag{5}$$

The circuit of Fig. 27.4 shows a practical arrangement for increasing the critical modulation ratio by applying positive bias to the diode plate, as mentioned previously in Sect. (A) above.

(c) Equivalent damping across i-f transformer

The secondary circuit damping is found approximately from

$$R_E = \frac{R_{ac}(\sqrt{1 - \eta^2} - \eta \cos^{-1}\eta)}{\eta(\cos^{-1}\eta - \eta\sqrt{1 - \eta^2})}. \tag{6}$$

R_E/R_{ac} can be read directly from Fig. 27.7 for various values of η.
Suppose $R_{ac} = 0.36$ MΩ and $\eta = 0.793$ then, since $R_E/R_{ac} = 0.7$, the equivalent damping resistance is $R_E = 0.7 \times 0.36 = 0.25$ MΩ.
It is often taken in practice (as a " rule of thumb ") that the equivalent damping resistance is half the d.c. diode load resistance.   This gives an equivalent damping resistance of $R_{dc}/2 = 0.5/2 = 0.25$ MΩ in this case.   However, it should be realized that if the diode efficiency is taken as unity, the equivalent damping resistance which appears across the tuned circuit is half the a.c. load resistance ; this is indicated by Fig. 27.7.

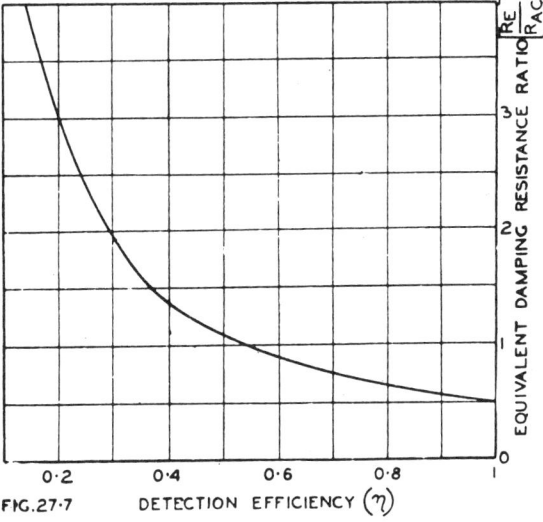

FIG.27.7   DETECTION EFFICIENCY (η)

Damping of the i-f transformer primary is serious when the a.v.c. voltage is derived from a diode detector connected to this voltage source.   The usual arrangement is a capacitor (100 $\mu\mu$F or less) taken from the plate connection of the transformer primary to the diode plate.   The additional primary damping from this source is approximately one third of the d.c. load resistance for the a.v.c. diode.   This also follows fairly readily if it is considered that R_{dc} shunts the primary of the transformer, and in addition the diode conduction current also adds damping equivalent to a parallel resistance of approximately $R_{dc}/2$.   From this

$$R_E = \frac{R_{dc}(R_{dc}/2)}{R_{dc} + R_{dc}/2} = \frac{R_{dc}}{3}.$$

For a typical case, the a.v.c. resistance would be 1 megohm, and so the equivalent damping resistance would be approximately 0.33 megohm.   If the i-f amplifier valve has a plate resistance of 0.8 megohm the primary damping resistance would be 0.23 megohm.

This means that, with detector and a.v.c. diodes conducting, both the secondary and primary circuits of the i-f transformer are heavily damped and this damping must be taken into account when the i-f transformer, connected between the last i-f voltage amplifier valve and the diode detector, is being designed. Delayed a.v.c. will, of course, affect the primary circuit damping and three conditions arise : the diode not conducting, the diode just starting to conduct, and the diode conducting when the applied voltage is reasonably large. When conduction just starts the damping depends very largely on the diode detection efficiency and will vary quite appreciably for a small range of input voltages. It is usually sufficient, however, to know the tuned circuit damping for the two conditions, diode not conducting, and diode conducting with a reasonably large input voltage. For the diode not conducting the added damping is approximately R_{ac} ; R_{dc} should not be used here. With the diode conducting the added damping can be taken as $R_{dc}/3$ with sufficient accuracy in some cases since in this case the fact that the diode detection efficiency is not unity offsets to some extent the increase in damping which is obtained when R_{ac} is used in place of R_{dc}.

Damping of the transformer tuned circuits as determined above assumes linear diode plate-voltage—plate-current characteristics. For parabolic diode detection the damping differs by only a small amount from that given by the linear characteristics and so, in practice, it is seldom necessary to treat the two cases separately. In any case the order of accuracy to be expected in the final results from the procedures set out, would hardly justify any additional refinements in the design calculations.

(d) The actual degree of modulation

The degree of modulation applied to the detector is always less than the modulation of the incoming carrier. The actual value of m which the detector will be required to handle is given approximately by

$$m = m' \frac{R_3(R_{dc} + 2Z_L)}{R_{dc}R_3 + 2Z_L(R_{dc} + R_3)} \qquad (7)$$

where m' = original modulation percentage
$\quad\quad R_{dc} = R_1 + R_2$ (see Fig. 27.2)
$\quad\quad R_3$ = grid resistor of following stage (see Fig. 27.2)
and $\quad Z_L$ = load impedance presented to the i-f voltage amplifier preceding the detector stage ; including the effects of all damping on the primary and secondary of the i-f transformer.

In addition, at higher audio frequencies the reduction of side band amplitude by the selectivity of the earlier stages in the receiver results in a further decrease in depth of modulation.

(e) Audio frequency output voltage and response (frequency distortion)

The approximate r.m.s. audio frequency output voltage can be determined for a particular arrangement by using the equivalent circuit of Fig. 27.8. It is often assumed that all capacitive reactances are negligible, for simplicity in carrying out gain calculations. However, the circuit is readily modified to include C_1, C_2 and C_3 as

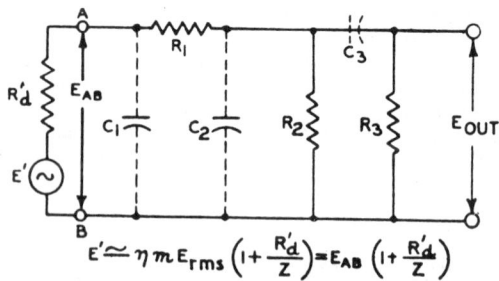

$$E' \simeq \eta\, m\, E_{rms} \left(1 + \frac{R'_d}{Z}\right) = E_{AB}\left(1 + \frac{R'_d}{Z}\right)$$

FIG. 27·8 EQUIVALENT DIODE CIRCUIT AT AUDIO FREQUENCIES.

shown by the dotted lines, and any additional capacitances in the circuit can be included if necessary. For this diagram

m = modulation factor
E_{rms} = i-f input voltage
η = detection efficiency determined from curves of Fig. 27.6
Z = total load impedance connected between the terminals AB. (This is equal to R_{ac} over the middle audio frequency range)

$$R_{ac} = R_1 + \frac{R_2 R_3}{R_2 + R_3}$$

and R'_d = internal equivalent resistance of diode at audio frequencies. This is given by the inverse slope of the curves of Fig. 27.6.

R_1, R_2 and R_3 are as shown in Fig. 27.2.

It should be noted that R'_d is not the diode conduction resistance (R_d) but is related to it by

$$R'_d = \frac{\pi R_d}{\cos^{-1}\eta} \tag{8}$$

Also R'_d and R_d are not constant for all operating conditions. For our previous example we see from the 15 volt (r.m.s.) curve of Fig. 27.6, and taking the diode load as being

$R_1 + R_2 = 0.5$ MΩ, for simplicity,

$$R'_d = \frac{(17 - 16.5) \times 10^6}{(42 - 26)} = 31\,200\ \Omega.$$

It will be seen that R'_d is determined in a rather similar manner to that used for finding the plate resistance of multi-element valves. The diode conduction resistance in this case is (approximately) 6500 ohms.

It is sometimes required to find the audio frequency response and this can readily be determined from the complete circuit of Fig. 27.8. A suitable procedure would be as for r.c. amplifiers in which the capacitances are ignored over the middle frequency range (say around 400 to 3000 c/s), and the audio voltage output is calculated for the resistance network only. At low frequencies the effects of C_1 and C_2 can be ignored since they are usually about 50 to 100 $\mu\mu$F each, and only the effect of C_3 would be considered. At high frequencies C_3 would be neglected and the effects of C_1 and C_2 considered. Often only the maximum output voltage is required plus the total output variation from say 50 c/s to 10 Kc/s, and this simplifies the calculations. If Fig. 27.8 is inspected it will be seen that it is a simple matter to apply the generalized low frequency response curves given in Chapter 12. The high frequency curves can also be applied directly if R_1 can be neglected (this is usually permissible) and C_1 and C_2 are considered as being in parallel.

When using the circuit of Fig. 27.8 it is more convenient to determine the voltage E_{AB} since this is given directly by

$E_{AB} = \eta m E_{rms}.$

It follows that, over the middle frequency range,

$$E_{out} = E_{AB}\left(\frac{R_{ac} - R_1}{R_{ac}}\right)$$

and $$\frac{E_{out}}{E'} = \frac{E_{out}}{E_{AB}} \times \frac{E_{AB}}{E'} = \frac{R_{ac} - R_1}{R_{ac} + R_d}$$

If R_1 can be neglected in comparison with R_2 and R_3 in parallel, then $E_{out} = E_{AB}$ over the middle frequency range.

It is more accurate to determine E_{AB} from the curves of Fig. 27.6, since η is not constant for a modulated carrier voltage and the effects of contact potential are included in the measured data. The procedure is to draw the a.c. loadline on the graph and note the d.c. voltages E_{dc_1} and E_{dc_2} given by the maximum and minimum values of E_{rms} during modulation. From this

$$E_{AB} = \frac{1}{2\sqrt{2}}(E_{dc_1} - E_{dc_2}).$$

The simplified graphical procedure is only suitable when the amplitudes of harmonics higher than the second are negligible [see also (h) below]. More accurate graphical methods are required if this is not true.

To compare the results obtained by the two methods, assume $E_{rms} = 15$ volts and the percentage modulation is 50% (i.e. $m = 0.5$). As determined previously $\eta = 0.793$, and it will be taken that $R_{ac} = R_{dc} = 0.5$ megohm. Then

$$E_{AB} = \eta m E_{rms} = 0.793 \times 0.5 \times 15 = 5.95 \text{ volts (r.m.s.).}$$

From Fig. 27.6, $E_{dc1} = 26.5$ volts and $E_{dc2} = 7.5$ volts so that

$$E_{AB} = \frac{1}{2\sqrt{2}} (26.5 - 7.5) = 6.7 \text{ volts.}$$

The discrepancy between these two results is easily explained, as an examination of typical diode circuits giving the d.c. voltage developed by the diode in the absence of signal input shows that there is about -0.5 to -1 volt developed across a diode load of 0.5 megohm. This difference is also illustrated by the curve of Fig. 27.5, which shows that -1 volt is developed for the conditions being considered. It is also seen that the variation in detection efficiency (η) over the range of voltage used does not have a large effect on the final result. The presence of d.c. voltage across the load resistor, in the absence of signal input voltage, is due to contact potential in the diode.

(f) Effect of shunt capacitance on detection efficiency

Too low a value for C_1 and C_2 will affect the detection efficiency. Provided the **total** shunt capacitance across the diode load is not less than $\dfrac{12.72}{fR_{ac}}$ $\mu\mu$F the effect can be neglected where

f = frequency (expressed in Mc/s)

$$R_{ac} = R_1 + \frac{R_2 R_3}{R_2 + R_3} \text{ (expressed in M}\Omega\text{).}$$

For our previous example of $R_{ac} = 0.36$ MΩ, and i-f of 455 Kc/s, the capacitance due to C_1 and C_2 should not be less than about 78 $\mu\mu$F ; this includes all stray capacitances. In the circuit of Fig. 27.2 typical values of C_1 and C_2 are 100 $\mu\mu$F each and if R_1 is neglected it is seen that the total capacitance is more than twice the value required even when strays are neglected. It would be feasible to reduce C_1 and C_2 to 50 $\mu\mu$F respectively with a reduction in attenuation at the higher audio frequencies.

(g) Non-linear distortion

If the value of total capacitance across the diode load resistance is too large, the discharge time constant will be too long and the voltage across the diode load will not follow the modulation envelope. This will give rise to non-linear distortion and suggests that the rate of discharge of the load circuit should not be less than the maximum rate of change of the modulation envelope. It can be shown that non-linear a-f distortion due to this cause can be almost completely avoided provided that

$$\frac{1}{\omega C_t R_{dc}} \gg m. \tag{9}$$

Suppose we neglect the decoupling resistance (R_1) in Fig. 27.2, as this unnecessarily complicates calculations when its value is about $1/10$ of R_2 (the usual case).

What is required, is to determine a value for $C_t R_{dc}$ which will allow $m = 1$, i.e. the incoming signal to be 100% modulated. However, it should be clear from previous discussion that the modulation percentage is generally a good deal lower than this figure, and it seldom happens that very high modulation percentages occur at high audio frequencies.

Take R_{dc} as 0.5 megohm, then $1/\omega C_t = m R_{dc} = 1 \times 0.5 \times 10^6$.

Assume that the highest audio frequency is 10 Kc/s, then $\omega = 2\pi \times 10^4$ and so

$$C_t = \frac{10^{12}}{2\pi \times 10^4 \times 0.5 \times 10^6} = 31.8 \ \mu\mu\text{F.}$$

This is the largest value C_t should have if the detector is to be capable of handling audio frequencies of 10 Kc/s and 100% amplitude modulation.

It will be noticed from (e) above that the detector efficiency will be reduced if C_t is given the value determined. A suitable practical compromise would be to make $C_1 = C_2 = 50$ $\mu\mu$F and retain the values of R_1, R_2 and R_3 given in the example. If improved decoupling is required then R_1 could be increased to say 0.1 megohm. This also permits a higher critical modulation ratio at the expense of some reduction in available audio output. The difficulty with non-linear distortion is not encountered when C_t is charging, since the diode is conducting and the charging time constant is approximately $C_tR'_d$ which in our example is, taking $C_t = 100$ $\mu\mu$F, $100 \times 10^{-12} \times 3.12 \times 10^4 = 3\,12$ μ secs. The discharge time constant is approximately $100 \times 10^{-12} \times 0.5 \times 10^6 = 50$ μ secs.

(h) Estimate of magnitude of non-linear distortion

The distortion can be found from the curves of Fig. 27.6 in the same way as for power amplifiers (Ref. 3, p. 100). The operating point corresponding to a given signal input is marked on the a.c. loadline. The maximum and minimum excursions of the modulation envelope are now marked on this same line. If only second harmonic distortion is required (and higher order harmonic distortion is very small) these points are sufficient. For distortion calculations which involve harmonics higher than the second the methods detailed in Chapter 13 should be used. The expression to be used here is

$$\text{2nd harmonic percentage} = \frac{1}{2}\left(\frac{A-1}{A+1}\right) \times 100\%, \tag{10}$$

where $A = \dfrac{\text{positive current swing}}{\text{negative current swing}}$.

Voltage swing would give the same results, since the load is taken as being a pure resistance.

Suppose we take $R_{ac} = R_{dc}$ and use the 0.5 MΩ loadline. The carrier input voltage is 15 volts (r.m.s.) and is modulated 50%. The voltage swing is thus $\pm$ 7.5 volts (r.m.s.) about 15 volts.

From Fig. 27.6 this gives (approximately)

$$A = \frac{53-33.5}{33.5-16} = 1.11$$

$$\text{2nd harmonic percentage} = \frac{1}{2}\left(\frac{0.11}{2.11}\right) \times 100 = 2.6.$$

The true a.c. loadline would indicate somewhat more distortion than is given by this simple example.

(D) Miscellaneous data

(a) With radio receivers using diode-triode or diode-pentode valves as combined detectors and a-f or i-f amplifiers several effects require consideration. Where the combined valve is used in the a-f application, difficulty is often experienced with **residual volume effect** (" play-through "). This effect is quite distinct from the minimum volume effect experienced with receivers using a reflexed amplifier. The residual volume is heard in the receiver output when the audio volume control is turned to zero (the grid may also be earthed as a further check). The cause of the effect is capacitive and electronic coupling between the detector diode and the plate of the a-f voltage amplifier.

In some cases there may be direct coupling at audio frequencies, but usually the important factor is coupling of modulated i-f currents which are detected in the audio amplifier.

A complete cure for the trouble is to combine the detector diode with the i-f voltage amplifier. However, this can sometimes lead to difficulties with regeneration or degeneration at i-f due to coupling between the detector diode and signal grid. When the valve is retained as an a-f amplifier it is necessary to keep all stray coupling to a minimum by careful layout and wiring. Diode pentodes are often helpful, particularly when series screen feed is used, as adequate screen by-passing gives a marked improvement. Adequate cathode circuit by-passing is essential in all cases. Neutralization to reduce the effect is almost useless unless rather elaborate circuits are used. In some cases different types of sockets offer an improvement. If simple a.v.c. is used, earthing, or even leaving disconnected, the second diode will often effect an almost complete cure. Obviously the diode connection nearest the plate of the voltage amplifier is the one which should be earthed.

A second effect which is less frequently encountered is residual volume due to capacitance between a diode and the control grid of a combined detector and a-f amplifier. Normally this capacitance is unimportant because with the volume control turned right down the control grid is grounded through the grid coupling capacitor (see Fig. 27.2). However if the audio amplifier is grid leak biased it may have a 10 megohm grid leak, and in this case a 0.001 μF grid coupling capacitor would give adequate bass response for a small receiver. Under these conditions a voltage of the order of one thousandth of the i-f input to the diode could appear on the control grid and with a high gain receiver this is ample to give annoying minimum volume. The remedy, of course, is to increase the size of the grid coupling capacitor.

(b) In general, diodes do not start to conduct at precisely the point where the plate voltage exceeds zero. **Contact potentials** and other effects will sometimes allow the valve to conduct when the diode plate voltage is slightly negative (this is usual with valves having indirectly heated cathodes), but in other cases (e.g. some battery valves) conduction does not occur until the plate voltage is appreciably positive. This point can be appreciated by an examination of typical diode curves. The effects of the contact potentials will be further considered in Sect. (3) below when automatic volume control is being discussed.

For a discussion of the effects of positive and negative start of plate current (which may be deliberately introduced as mentioned in (C)(b) above) in diode detectors the reader is referred to Ref. 1, Chapter 8.

(c) There are distinct advantages in having the **a-f volume control as the diode load.** For the usual operating conditions on local signals the setting of the control is fairly well down, and so the effects of a.c. loading are very much reduced. This advantage is lost when the control is in the grid circuit of the a-f voltage amplifier. The disadvantage is that many controls become noisy, usually after a fairly short period, when they have the diode current passing through them. A compromise arrangement may be best with the control in the grid circuit (R_s in Fig. 27.2) and R_1 about 0.1 megohm and R_2 say 0.5 megohm.

(d) As a summary of the characteristics of the diode detector, it may be stated that its performance as regards frequency and non-linear distortion is excellent provided the input voltage is high and the factors discussed above regarding a.c. shunting etc. are incorporated in the detector design. All forms of detectors suffer from distortion at low input levels, but the diode has the particular advantage that the input may be increased to a very high level with consequent reduction of distortion, without any overloading effect such as occurs with other forms of detectors.

(e) **Crystal diodes** are described in Section 7.

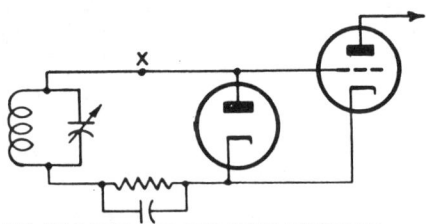

FIG. 27·9 FORM OF GRID LEAK DETECTOR.

(ii) Other forms of detectors
(A) Grid detection

Leaky grid or "cumulative detection" has been used for many years and is still widely used for certain applications. The theory of its operation is essentially the same as that of the diode except that a triode is also used for amplification. The derivation of a leaky grid detector from the combination of a diode and triode is shown in Fig. 27.9. Whether the grid capacitor and resistor are inserted as shown (as is usual with the diode) or at the point X is immaterial from the viewpoint of operation. The diode is directly coupled to the triode and therefore the audio frequency voltages developed in the diode detector are passed on to the triode grid, but at the same time this grid is given a d.c. bias through the d.c. voltage developed in a similar way to that by which a.v.c. is obtained. Consequently the operating point of the triode varies along the e_g-i_p curve from zero towards more negative grid bias voltages as the carrier voltage is increased. This is the same effect as that obtained when the diode is omitted (Fig. 27.10) since the grid and cathode of the triode act as a diode and produce the same results. The illustration given was purely to demonstrate the derivation of the one from the other and not to be a practical form of detector since no advantage is gained by retaining the diode in the circuit.

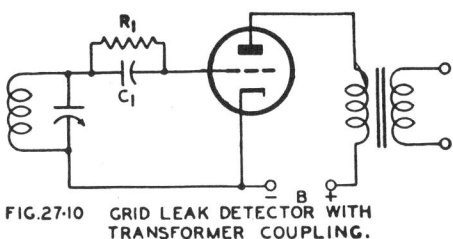

FIG.27·10 GRID LEAK DETECTOR WITH TRANSFORMER COUPLING.

It will be seen that the operating point varies along the e_p-i_p characteristic curve between zero bias and the cut-off point (Fig. 27.11). There will be a certain strength of carrier at which the detection will be most satisfactory, and at lower or higher levels detection will not be so satisfactory on account of improper operating conditions. If with a certain carrier input voltage the d.c. bias on the grid is OA, then the point corresponding to peak modulation is B where OB equals twice OA. If the point B is on the curved part of the characteristic, or in the extreme case actually beyond the cut-off, the distortion will be severe. A valve having low μ and low g_m is capable of operating with a higher carrier voltage than a valve with improved characteristics, but the gain in the detector stage will be less. There is a further difficulty in that the plate current at no signal, or at very weak signal, may be excessively high. If transformer coupling is used this may, in extreme cases, damage the valve, or pass too much direct-current through the transformer, unless the plate supply voltage is reduced. If resistance coupling or parallel-feed is used the efficiency of the detector is decreased. As with diode detection there is distortion at low levels due to the " diode characteristics " but as distinct from the diode, the overload point occurs at quite a low carrier voltage. This method of detection is therefore very much limited in application.

With battery type valves used as cumulative grid detectors it is often advantageous to connect R_1 (Fig. 27.10) to filament positive ; C_1 remains as before. The advantage obtained is that larger modulation percentages can be handled and detection will start with smaller input voltages. The arrangement is equivalent to supplying the grid, which is acting as the plate of a diode, with a positive bias voltage (see Ref. 1, p. 357) and moving the position of the start of grid current.

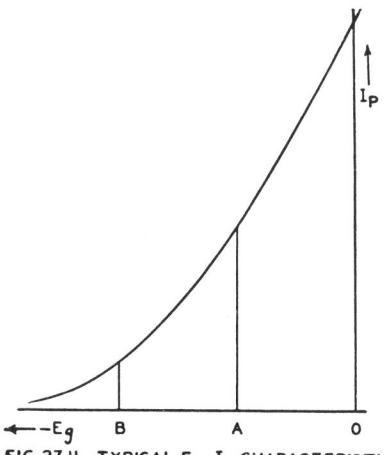

FIG.27·11 TYPICAL E_g-I_p CHARACTERISTIC.

A typical circuit for a grid leak detector as used in radio receivers is shown in Fig. 27.12.

Damping of the input circuit occurs in the same manner as for a diode but additional damping occurs because of grid-plate coupling. The arrangement shown in Fig. 27.12 for the grid resistor further increases the loading on the input circuit. Detailed discussions of this type of circuit can be found in Refs. 2 (p. 414), 1 (p. 377) and 10.

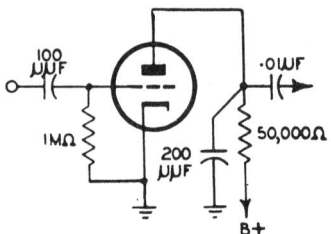

FIG. 27·12· TYPICAL GRID LEAK DETECTOR
WITH R–C COUPLING.

(B) Power grid detection

Power grid detection is a modification of leaky grid or cumulative detection and the circuit is identical in form, but the operating conditions are so chosen that the valve will operate on higher carrier voltages without overloading. In order to obtain a short time constant from the grid capacitor and resistor combination the capacitance and resistance are reduced, thereby improving the high audio frequency response. Under optimum conditions the distortion is at least as high as that of a diode together with increased distortion due to the curvature of the e_g-i_p characteristic. The overload point, even though higher than that of ordinary leaky grid detection is at a much lower level than that with diode detection.

All forms of grid detection, particularly " power grid detection," involve damping of the grid circuit due to grid current, and this damping causes loss of sensitivity and selectivity. Grid detection is thus similar to diode detection in that it damps the input circuit. It has the advantage over diode detection in that gain is obtained in the detector which can be still further increased if transformer coupling is used between it and the following stage. Transformer coupling can, of course, only be used when the valve has a low plate resistance.

The foregoing comparison between a diode and a grid detector is on the basis of the detector alone. In modern practice the diode detector is frequently in the same envelope with a voltage amplifier and the total gain is quite high.

(C) Plate detection

Plate detection or " anode bend detection " involves operation towards the point of plate current cut-off so that non-linearity occurs, thereby giving rectification. Owing to the slow rate of curvature the detection efficiency is small, but there is an advantage in that the amplification which is obtained makes up, to a certain extent, for the poor detection efficiency. Due to the gradual curvature the distortion is very great with low input voltages, and even with the maximum input before overload occurs the distortion is rather high with high percentages of modulation. An important advantage of plate detection is, however, that the grid input circuit is not damped to any great extent, and the detector is therefore sometimes spoken of as being of infinite impedance, although this term is not strictly correct.

With pentode valves it is possible to use either " Bottom Bend Rectification " as with triodes or " Top Bend Rectification " peculiar to pentodes. This " top bend " in resistance coupled pentode characteristics can be seen on valve data sheets e.g. type 6J7 dynamic $e_g - i_p$ characteristics provide a good illustration (Fig. 12.14A).

A similar effect occurs with triodes, but only in the positive grid region, and for this reason it is incapable of being used for plate rectification. For top bend rectification with a pentode valve it is desirable to operate the valve with a plate current in the region of 0.95 (E_B/R_L).   The exact operating point for optimum conditions depends upon the input voltage.

Pentode valves are particularly valuable as plate detectors since the gain is of such a high order. If resistance coupling is used the gain is reduced very considerably, and in order to eliminate this loss it is usual to adopt choke coupling using a very high inductance choke in the plate circuit, shunted by a resistor to give a more uniform frequency response. If the shunt resistor were omitted the high frequencies would be much greater than the low frequencies in relative level.

With all forms of plate detectors the bias is critical and since different valves of the same type require slightly different values of bias the use of fixed bias is not recommended. A very high value of cathode resistance is usually adopted to bias the valve very nearly to cut-off, and in such a way that if valves are changed, or vary during life, the operating point maintains itself near optimum (Fig. 27.13).

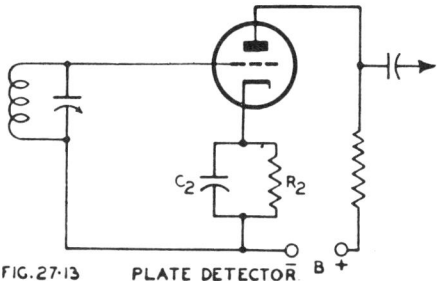

FIG. 27·13   PLATE DETECTOR   B +

Screen grid and pentode valves with self-bias have been used as plate detectors very satisfactorily for a number of years, although the distortion with the usual arrangement is too high for them to be used in any but the cheapest radio receivers at the present time. Such a detector is, however, permissible for some types of short-wave reception and for amateur communication work where its high grid input impedance results in higher sensitivity and selectivity.

With the usual plate detector the cathode bypass capacitor (C_2, Fig. 27.13) has customarily been made sufficiently large to by-pass both audio and radio frequencies. Often C_2 is a 25 μF electrolytic in parallel with a 500 $\mu\mu$F mica capacitor. This arrangement, although widely used in the past, is not a correct one and leads to distortion when the modulation factor is at all high. The effect is similar to that described in connection with diode detectors having the a.c. loadline appreciably different in slope from the d.c. loadline. The correct procedure calls for the resistor R_2 to be by-passed for radio frequencies only, and although this results in a very considerable loss in gain because of negative feedback, the results obtainable are excellent. In a typical case R_2 might be taken as 10 000 to 100 000 ohms and C_2 as 500 to 100 $\mu\mu$F respectively, the plate load resistor would be about 0.25 megohm ; additional r-f decoupling is usual in the plate circuit, although this is not shown in Fig. 27.13. This circuit is often called a reflex detector, although this term is sometimes reserved for cases where 100% negative feedback is used with a plate detector, and the audio output is then taken from the resistor in the cathode circuit ; the plate is earthed for a.c. in this case. The results obtainable from the modified plate detector are excellent, and are comparable with those given by the diode arrangement. The reflex detector is further discussed below.

(D) **Reflex detector**

The reflex detector is essentially a plate detector with negative feedback. Any amount of feedback may be applied from zero to 100%, and as the feedback increases, so the distortion decreases and the stage gain decreases until in the final condition with 100% feedback the gain is less than unity. The reflex detector has an even higher input impedance than the usual type of plate detector, and is therefore valuable in certain applications. Under certain conditions the input resistance is negative and can lead to instability troubles. The increased selectivity can also result in sideband cutting. The degree of feedback may be adjusted to give any required gain (within reasonable limits) by altering the size of cathode by-pass capacitor and the relationship between the plate and cathode load resistance, but the distortion increases with gain and if low distortion is required the maximum gain is limited to about three or four times even with a pentode valve. With maximum degeneration and stable operating conditions, the distortion (see Ref. 11) is about the same as for a diode operating under similar input voltage conditions,·while the reflex detector has the distinct advantage of high input impedance. One application which appears to be of importance is in high fidelity t.r.f. receivers, but even here the reflex detector has not shown any very marked improvement over the diode from a consideration of distortion alone. Input voltages in the order of 10 volts (r.m.s.) are usually most satisfactory when low distortion is desired and in a typical case using 100% feedback about 3% total harmonic distortion can be expected when the modulation depth approaches 100%. This assumes that the a.c. shunting is high (say 10 : 1 or so) compared with the d.c. load. Some useful discussion is given in Refs. 14 (pages 51-55) and 15. These performance figures can be compared with those stated previously for the diode detector under similar operating conditions. Reflex detectors do not provide a.v.c. and so are not used in normal broadcast receivers. They could be used in combination with amplified a.v.c. to provide a receiver with good characteristics.

Typical circuit component values have been discussed in (C) above in connection with modified plate detectors. For the case of 100% negative feedback (i.e. the load resistance is in the cathode circuit and the plate is earthed for a.c.) the cathode resistance is made large compared with $1/g_m$ and a typical value is 25 000 ohms. The by-pass capacitance across this resistance can be about 500 $\mu\mu$F. There is seldom any difficulty with a.c /d.c. ratios as the grid resistance of the following a-f stage is generally about 0.5 megohm. Additional components required are a blocking capacitor, to prevent the d.c. cathode voltage from being applied to the grid of the a-f amplifier, and a series resistor and shunt capacitor to provide additional r-f decoupling between the two circuits.

A limitation of the.reflex detector is that there is a definite maximum for the input signal voltage for freedom from grid current. A further increase of input causes rectification at the grid, with added damping of the grid input circuit, and a steady increase in distortion. An increase in the supply voltage raises the threshold point for grid current. Further data are available in Refs. 11, 12, 13, 14 and 15.

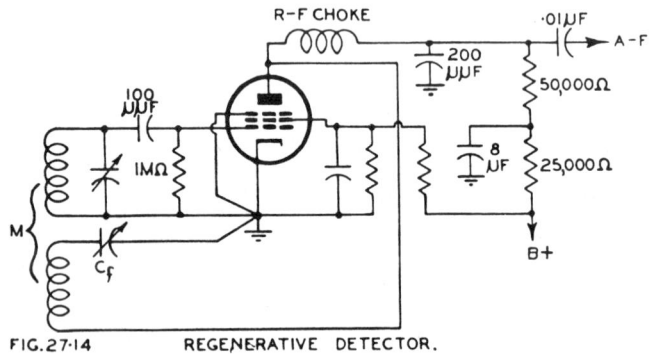

FIG.27·14   REGENERATIVE DETECTOR.

(E) Regenerative detectors

A common arrangement for this type of detector is shown in Fig. 27.14. The amount of positive feedback is adjustable, by means of C_f, to allow increased sensitivity with various signal input voltages. Oscillation will occur when the degree of feedback is sufficiently large, and the circuit can then be used for the detection of CW telegraph signals. For battery operated valves it is usual to return the grid resistor to filament positive.

Best results are generally obtained using pentode valves and resistance-capacitance coupling to the following audio stage. This arrangement minimizes " threshold howl."

Feedback control is possible with a variety of circuits but least difficulty is usually experienced with the variable capacitor arrangement, or the use of a variable resistor in the screen circuit. The latter arrangement is advantageous as regards the reduction of detuning effects.

Further details of this type of detector can be found in Refs. 1, 5, 16, 17, 18, 19 and 20.

(F) Superregenerative detectors

The superregenerative detector is a regenerative circuit in which the detector is automatically switched in and out of oscillation at a very low radio frequency rate (usually about 15 to 100 Kc/s). This switching frequency is called the " quenching " frequency. In general the quenching frequency is increased as the carrier frequency becomes greater, but sensitivity and selectivity are improved by using the lowest permissible value of quenching frequency. For many cases it is usual to make the quench frequency about twice the highest audio frequency contained in the modulation envelope, and it is not considered good practice for the quench frequency to be lower than this value. The amplitude of the quench voltage is also important and it will have a very appreciable effect on·the selectivity characteristic. In general if the quench voltage amplitude is increased in a separately quenched circuit, the selectivity is reduced.

The advantage of this circuit is the extremely high sensitivity which is possible using a single valve. The disadvantages are the high noise level in the absence of a signal, the poor selectivity, and the high distortion.

The circuit of Fig. 27.14 can be made to operate as a superregenerative detector by increasing the time constant of the grid resistance and capacitance combination, and making the amount of regeneration very large. The quenching frequency can be simply adjusted by altering the value of the grid resistance.

A separate quenching oscillator is often used with the regenerative detector arrangement, but as this requires additional circuit components and sometimes an additional valve, it is not so popular as the simple arrangement previously mentioned. The development of suitable types of double triode and converter valves largely overcomes this objection, however, and many modern circuits use separate quenching.

Several other points of interest arise with this detector. Amplitude limiting occurs, so there is less interference from car ignition and similar noises than when other detectors are used ; the output on strong signals is not much greater than for weak signals. Noise quieting and limiting are improved however, by using low quench frequencies. A r-f stage should be incorporated in receivers using this type of detector to reduce radiation. This additional stage will not materially alter the signal-to-noise ratio obtainable with the detector alone. Circuits incorporating the principle of superregeneration have been used in cheap F-M receivers ; this point will receive some further consideration in the section on F-M receivers in Chapter 36.

The method to be adopted for measuring the selectivity of a superregenerative receiver requires some consideration. Conventional methods are usually inadequate since in most cases the problem is similar to that of taking the selectivity curve of a receiver having a.v.c., but in this case the a.v.c. cannot be disconnected. A suitable procedure (Ref. 26) is as follows.

With no input signal applied the audio noise output is measured. A signal is applied at the resonance frequency and its amplitude adjusted until the noise is suppressed by about 10 to 20 db (or any convenient amount). The selectivity is then found by tuning the signal generator to various frequencies around resonance, in the usual manner, and the input voltage is adjusted until the same degree of quieting is obtained as at resonance. The difference between the two signal inputs off and at resonance for the same degree of quieting, gives the attenuation at the particular frequency being considered.

Further details of this detection system are available in Refs. 5 (page 662), 20 (page 148), 21, 22, 23, 24, 25 and 26.

SECTION 2 : F-M DETECTORS

(i) Types of detectors in general use   (ii) General principles   (iii) Phase discriminators (A) General (B) Design data (C) Design example   (iv) Ratio detectors (A) General (B) Operation (C) Types of circuit (D) Design considerations (E) Practical circuits (F) Measurements on ratio detectors.

(i) Types of detectors in general use

Although many ingenious methods have been suggested for the detection of frequency modulated signals, only a few circuits have found general acceptance by receiver manufacturers.   Of these detectors a form of locked oscillator (the Bradley detector) has been used by one large manufacturer, but does not appear to have been employed to any extent outside of this organization.   Amplitude discriminators have been used to a very limited extent with duo-diode triode valves having a separate cathode for the two diodes.   However, the generally accepted method of F-M detection has been **the phase discriminator**.   A modification of the basic phase discriminator circuit has been used for **the ratio detector**, and this arrangement has achieved wide popularity because it allows a satisfactory F-M receiver to be constructed without the use of an additional amplitude limiting stage.   Simple detuning of the signal circuits has been employed in one type of receiver (the **Fremodyne**) to give frequency to amplitude conversion, but this is only satisfactory in a very cheap receiver where cost is more important than quality.

In what follows attention will be confined, mainly, to the design procedures to be adopted for phase discriminators and ratio detectors.   References are listed at the end of this chapter, and can be consulted for details of a number of the alternative detection systems available.   The amplitude and phase discriminators used in connection with Automatic Frequency Control Systems are further considered in Chapter 29.

(ii) General principles

The circuits which will be considered here utilize tuned circuits to convert frequency changes to amplitude changes.   The amplitude modulated carrier is then applied to detectors (usually diodes) to recover the intelligence contained in the received signals.

Although the amplitude modulated carrier applied to the detectors is also frequency modulated, the detectors are only sensitive to amplitude changes, and so it is only the resultant amplitude variations which appear at the output of the detector stage.   It is because the detectors are sensitive to amplitude changes, that some method of limiting is required to overcome undesired amplitude variations.

The tuned circuits to be described are called discriminators, although this name is often taken to include the diode detectors as well.   If distortion is not to be introduced by the discriminator it is essential that the amplitude variations produced be directly proportional to the frequency variations i.e. the circuit must be linear over the full range of applied frequency deviation.   The usual precautions for reducing distortion in the detector circuits must also be applied, just as for any other A-M detector. These precautions have been discussed previously, and those given for diode detectors should be carefully observed here.   One big advantage does appear, however, the percentage amplitude modulation is likely to be quite small, depending on the frequency-amplitude conversion efficiency of the discriminator circuit, and so the possibility of non-linear distortion is considerably reduced.

(iii) Phase discriminators
(A) General

The basic circuit arrangement, together with the voltage distribution at the resonant frequency, for a phase discriminator is shown in Fig. 27.15.   The name arises because the operation is dependent on the 90° phase shift which occurs at resonance between the primary and secondary voltages of a tuned transformer.   When the frequency of the applied primary voltage E_1 (the magnitude of this voltage will be

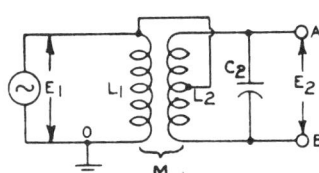
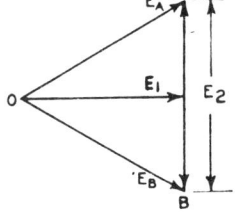

FIG.27.15   BASIC PHASE DISCRIMINATOR CIRCUIT.

assumed constant for the moment) alters, the phase angle between E_1 and E_2 changes from that at resonance.   This leads to a change in the relative magnitudes of E_A and E_B.

The vector relationship between the primary and half secondary voltages, for the phase discriminator, are shown in Fig. 27.16.   This assumes constant primary voltage E_1.

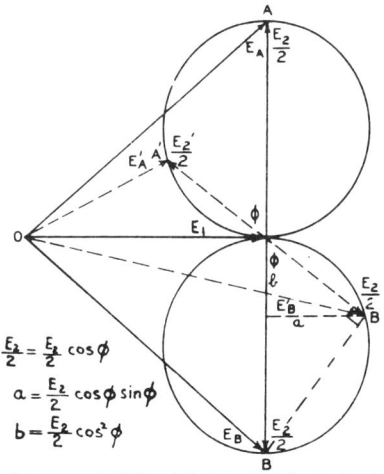

$$\frac{E_A}{2} = \frac{E_2}{2} \cos\phi$$

$$a = \frac{E_2}{2} \cos\phi \sin\phi$$

$$b = \frac{E_2}{2} \cos^2\phi$$

FIG.27.16 VOLTAGE RELATIONSHIPS IN PHASE DISCRIMINATOR WITH CONSTANT PRIMARY VOLTAGE.

If the primary voltage E_1 has its amplitude varied in a suitable manner (e.g. by setting the coupling between the transformer windings, for given primary and secondary Q's, so that two primary voltage humps of the required amplitude appear as the frequency is varied) the linearity and sensitivity of the discriminator can be very appreciably improved.

This statement, regarding linearity and sensitivity, refers, of course, to the relationship between voltage output and frequency-deviation from the central reference frequency (i.e. the nominal intermediate frequency).

Analysis of the phase discriminator for F-M applications has been made by K. R. Sturley (Refs. 27 and 28), and the results of this analysis will be used below   It might be mentioned that a number of discriminator circuits for a wide variety of applications have been designed using the data derived by Sturley and very satisfactory results have been obtained in practice. Before proceeding to set out the design data, the circuit of Fig. 27.17 will be briefly discussed.

The voltages applied to the plates of the double diode valve V_2 will be E'_A and E'_B respectively.   The output voltages from the diodes are developed across R_3 and R_4, and the circuit is arranged so that the available output voltage is equal to the **difference** between the two separate voltages.   This means that when the frequency of the carrier voltage is exactly equal to the intermediate frequency, no output will be obtained from the detector.   Reference to Fig. 27.16 should help to make this point quite clear, since the rectified voltages across R_3 and R_4 in Fig. 27.17 are given by the peak voltages applied to the diode plates multiplied by the detection efficiency of the diodes.   As the signal deviates from the central reference frequency (i.e. the intermediate frequency), a voltage (E_{out}) will appear between the points P and N (Fig. 27.17), and its polarity will depend on the relative magnitudes of the voltages across R_3 and R_4; e.g. point P will be negative with respect to point N when the voltage across R_3 is less than that across R_4.   It follows that it should be possible to calculate

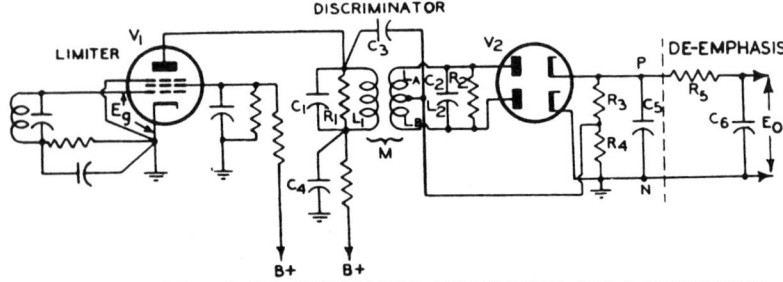

the relationship between output voltage and frequency change from

$$E_{out} = \eta(E'_A - E'_B) \tag{11}$$

where η = the diode detection efficiency, and is assumed to be the same for both diodes of V_2.

(B) Design data

For most designs it is usual to take the total secondary voltage E_2 as being twice the primary voltage E_1 i.e.

$$E_2/E_1 = 2. \tag{12}$$

Primary and secondary Q's are made equal, so that

$$Q = Q_1 = Q_2. \tag{13}$$

The value for Q is determined from

$$Q = f_r/(2\Delta f) \tag{14}$$

where f_r = intermediate frequency

and $2\Delta f$ = total frequency range for which substantially linear operation is required.

The coefficient of coupling for the transformer is found from

$$Qk = 1.5 \tag{15}$$

when a good compromise between sensitivity and linearity is required (the usual case).

For special cases where linearity is the main requirement it is suggested that

$$Qk = 2 \tag{16}$$

be used. The loss in sensitivity in this case is about 1.54 times as can be seen by comparing eqns. (20) and (21).

From the principles of coupled circuits it is possible to show that

$$\frac{E_2}{E_1} = Q_2 k \sqrt{\frac{L_2}{L_1}} \tag{17}$$

and for our particular cases with $E_2/E_1 = 2$ and $Qk = 1.5$ or 2, the relationship between L_2 and L_1 is

$$L_2/L_1 = 1.77 \text{ (for } Qk = 1.5) \tag{18}$$

and $$L_2/L_1 = 1 \text{ (for } Qk = 2). \tag{19}$$

The discriminator sensitivity at f_r is given by

$$S_{(Qk = 1.5)} = 5.465 \times 10^3 g_m Q^2 L_1 \eta E_g \text{ volts per kilocycle deviation,} \tag{20}$$

$$S_{(Qk = 2)} = 3.554 \times 10^3 g_m Q^2 L_1 \eta E_g \text{ volts per kilocycle deviation,} \tag{21}$$

where g_m = mutual conductance of V_1

Q = magnification factor determined from eqn. (14)

L_1 = primary inductance of transformer

η = diode detection efficiency

and E_g = peak voltage between grid and cathode of V_1.

It should be apparent that the sensitivity, given by eqns. (20) and (21), is directly proportional to E_g and so the output from the discriminator is dependent on the input voltage. To overcome this difficulty the valve V_1 is arranged as a limiter in such a way that the product $g_m E_g$ is kept almost constant, and thus the output voltage is no longer directly dependent on the magnitude of the signal input voltage.

Curves of discriminator output voltage versus frequency deviation can be found from vector diagrams similar to that of Fig. 27.16, and then introducing suitable correction factors to allow for variations in E_1 deliberately introduced by the transformer design.   However, it is possible to determine two generalised curves for the design conditions previously imposed, one curve for $Qk = 1.5$ and the other for $Qk = 2$. The curves are shown in Fig. 27.18A, and relate the quantity X (which is a function of bandwidth) to the relative output voltage E_{rel}.   Only the positive halves of the curves are shown since the negative halves have substantially the same shape.   To determine the output voltage E_{out} against frequency, it is only necessary to multiply the horizontal scale by a factor determined by eqn. (14), and the vertical scale by a factor determined by eqn. (20) or eqn. (21) (depending on whether $Qk = 1.5$ or 2 is used).

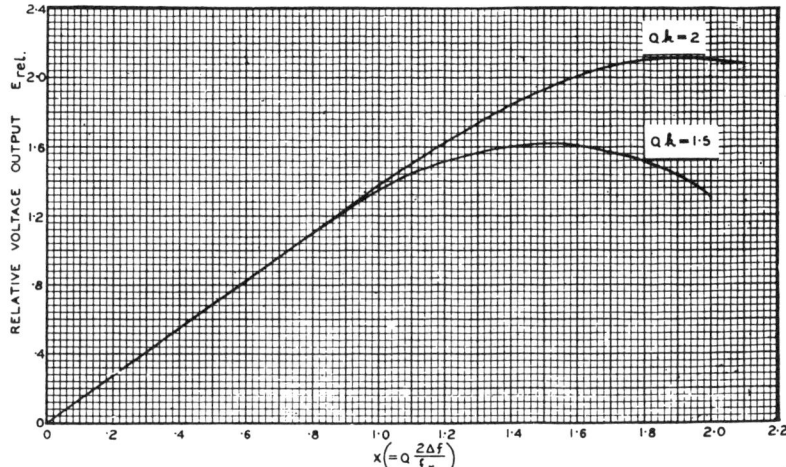

FIG. 27.18(A) GENERALISED PHASE DISCRIMINATOR CURVES OF RELATIVE VOLTAGE OUTPUT $\left(E_{rel.}\right)$ VERSUS X
WITH $E_2/E_1 = 2$; $Qk = 1.5$ AND 2.

The two curves shown in Fig. 27.18(A) are calculated from the following equations.
For $Qk = 1.5$, the relative voltage output is

$$E_{rel} = \frac{3.25A}{\sqrt{(3.25 - X^2)^2 + 4X^2}} \tag{22}$$

and for $Qk = 2$,

$$E_{rel} = \frac{5A}{\sqrt{(5 - X^2)^2 + 4X^2}} \tag{23}$$

where $A = \sqrt{1 + X^2}\left[\sqrt{\left(1 + \frac{X}{1 + X^2}\right)^2 + \left(\frac{1}{1 + X^2}\right)^2} - \sqrt{\left(1 - \frac{X}{1 + X^2}\right)^2 + \left(\frac{1}{1 + X^2}\right)^2}\right]$

$X = Q\, 2\Delta f/f_r$

Q = magnification factor determined from eqn. (14).   Equation (14) is the condition for $X = 1$

Δf = frequency deviation from f_r

and f_r = intermediate frequency.

To illustrate the design procedure, and to bring out additional points, a worked example is appended.

(C) **Design example**

A phase discriminator is required for use with a 10.7 Mc/s F-M i-f channel.   The maximum frequency deviation of the carrier is ± 75 Kc/s.   To make the problem complete it will be taken that the highest audio frequency is 15 Kc/s and the de-emphasis time constant is 75 micro-seconds.

(a) The majority of applications call for reasonable discriminator sensitivity and so the condition $Qk = 1.5$ (see eqn. 15) is practically always used. Also, high primary dynamic impedance will increase the sensitivity of the discriminator. However, since Q is fixed by other considerations (see eqn. 14) L_1 should be large, but L_1 is limited by L_2 which in turn is limited by the permissible minimum secondary capacit-ance C_2. A value is selected for C_2 and the design can then proceed.

(b) A suitable value for C_2, including all strays, is 50 $\mu\mu$F. Then

$$L_2 = \frac{25\,330}{10.7^2 \times 50} = 4.41 \ \mu H.$$

From eqn. (18),

$$L_1 = \frac{L_2}{1.77} = \frac{4.41}{1.77} = 2.5 \ \mu H.$$

From this

$$C_1 = \frac{25\,330}{10.7^2 \times 2.5} = 88.5 \ \mu\mu F \text{ (including all strays).}$$

(c) If the discriminator frequency—voltage characteristic were exactly linear then it would be sufficient to make the total frequency range $2\Delta f = 2 \cdot 75 = 150$ Kc/s. To this would be added an allowance for frequency drift due to the oscillator and the discriminator tuned circuits. Since the discriminator characteristic is not exactly linear (it is linear for about 80% of the total curve using the data given as can be seen from Fig. 27.18) and the frequency drift is not always small it has become common practice in broadcast F-M circuits (of the type being considered) to make $2\Delta f$ from 200 to 400 Kc/s.

As a practical compromise we will take the total bandwidth ($2\Delta f$) as 250 Kc/s, but the design procedure is the same irrespective of what bandwidth is selected.

With $2\Delta f = 250$ Kc/s we have from eqn. (14)

$$Q = 10.7/0.25 = 42.8.$$

From eqn. (15), $Qk = 1.5$
and so $k = 1.5/42.8 = 0.035$.

(d) Summarizing, for the discriminator transformer (see also Fig. 27.17)
$L_1 = 2.5 \ \mu H$; $L_2 = 4.41 \ \mu H$; $C_1 = 88.5 \ \mu\mu F$; $C_2 = 50 \ \mu\mu F$; $Q = Q_1 = Q_2$
42.8 ; $k = 0.035$; $M = 0.116 \ \mu H$. Secondary winding to be centre-tapped. The capacitance values include all strays.

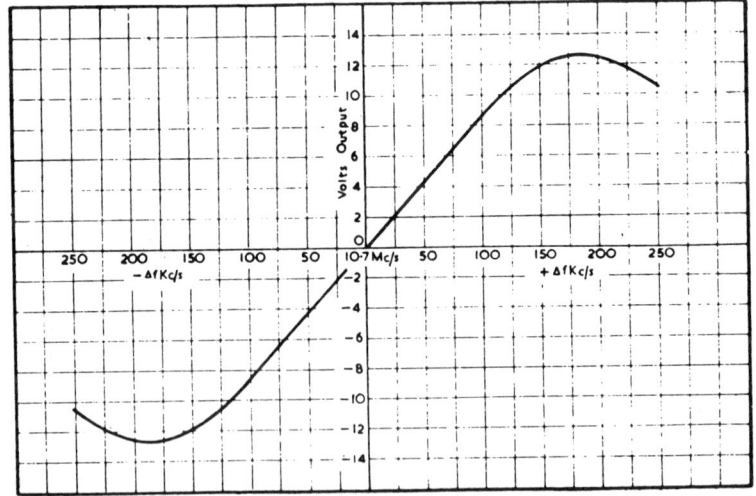

FIG. 27.18(B)   Discriminator. Characteristic Obtained From Design Example

(e) The discriminator sensitivity at f_r is given by eqn. (20).   Assume $Eg = 1$ volt (peak), $g_m = 5$ mA/volt and $\eta = 0.7$.   Then

$$S_{(Qk = 1.5)} = 5.465 \times 10^3 \times 5 \times 10^{-3} \times 42.8^2 \times 2.5 \times 10^{-6} \times 0.7 \times 1$$
$$= 0.0875 \text{ volts per kilocycle deviation.}$$

This also allows the vertical scale factor to be determined for the complete discriminator curve as shown on Fig. 27.18(B).   The horizontal scale factor is found from

$$X = Q\frac{2\Delta f}{f_r} = 1 \text{ [i.e. eqn. (14)].}$$

Therefore $\Delta f = \dfrac{Xf_r}{2Q} = \dfrac{10.7 \times 10^3}{2 \times 42.8} = 125$ Kc/s.

The procedure for finding the plotted points on the curve of Fig. 27.18(B) is as follows.   Draw up a table, as shown below, with values of X corresponding to values of E_{rel} read from the curve of Fig. 27.18(A).   Since it has just been determined that $\Delta f = 125$ Kc/s when $X = 1$, the column for Δf can be filled in (e.g. when $X = 0.8$ then $\Delta f = 125 \times 0.8 = 100$ Kc/s, and so on).   The sensitivity has been determined as 0.0875 volts per kilocycle deviation, and so for 25 Kc/s deviation $E_{out} = 0.0875 \times 25 = 2.19$ volts (the lowest deviation frequency in the table should be used to find E_{out} in this case).   It is now seen that the value of $E_{out} = 2.19$ volts corresponds to $E_{rel} = 0.28$, and so the scale factor is $2.19/0.28 = 7.82$.   If 7.82 is now multiplied by E_{rel} in each case, the column for E_{out} can be filled in.   The complete curve for E_{out} versus Δf can now be plotted.

X	Δf	E_{rel}	E_{out}
0.2	25	0.28	2.19
0.4	50	0.56	4.38
0.6	75	0.84	6.56
0.8	100	1.12	8.76
1.0	125	1.36	10.61
1.2	150	1.52	11.9
1.4	175	1.6	12.5
1.5	187.5	1.62	12.7
1.6	200	1.6	12.5
1.8	225	1.52	11.9
2.0	250	1.31	10.5

(f) It now remains to determine suitable values for the other components in the circuit of Fig. 27.17.

C_3 is for the purpose of connecting the transformer primary to the secondary centre-tap, and providing isolation of the secondary from the h.t. on the plate of V_1. A suitable value is 100 $\mu\mu$F.   The capacitor should have high insulation resistance.

C_4 is a by-pass capacitor ; a value of about 0.01 μF is usual.

R_3 and R_4 are made equal, and are generally about 100 000 ohms each.

C_5 acts as a by-pass for i-f but must not appreciably affect the audio frequency response.   The usual value is about 100 $\mu\mu$F.

The de-emphasis network consists of R_5 and C_6.   For 75 microsecond de-emphasis the nominal values would be 75 000 ohms and 0.001 μF respectively.   If an improved a.c./d.c. ratio is thought to be necessary then suitable values would be say, 0.25 MΩ and 300 $\mu\mu$F.   The grid resistor in the following a-f stage should not be less than 1 MΩ if excessive loss in gain is to be avoided.   However, the actual values used in a receiver will generally deviate from these nominal figures if the overall audio frequency response of the receiver is made to follow the 75 microsecond de-emphasis curve to 15 Kc/s, because of the presence of stray capacitances etc. and in some cases to help compensate for the overall a-f response.

To determine R_1 :—the transformer primary damping due to a single diode circuit, connected as shown, is $R_{dc}/3$.   For this circuit $R_{dc} = R_3 = R_4 = 100\,000$ ohms.

Also, because of the circuit arrangement, R_3 and R_4 are in parallel. The total damping, due to diode conduction currents in addition to R_3 and R_4, is then

$$\tfrac{1}{2}\left(\frac{R_{dc}}{3}\right) = \frac{100\,000}{6} = 16\,600\ \Omega.$$

If R_3 is not equal to R_4, the damping resistance is given by $R_3/3$ and $R_4/3$ in parallel i.e.

$R_3 R_4/3(R_3 + R_4)$. (When $R_3 = R_4 = R_{dc}$ the value $R_{dc}/6$ is obtained as above).

Take the undamped primary Q (written as Q_u) as being 100, and neglect the additional damping due to the plate resistance of V_1. Then the total damping resistance (R) required to obtain a primary Q of 42.8 is

$$R = \frac{Q_u Q \omega L_1}{Q_u - Q} = \frac{100 \times 42.8 \times 2\pi \times 10.7 \times 2.5}{100 - 42.8}$$
$$= 12{,}500\ \Omega.$$

From this

$$R_1 = \frac{12\,500 \times 16\,600}{(16\,600 - 12\,500)} = 50\,600\ \Omega.$$

To determine R_2: The transformer secondary damping is given by $R_3 = R_4 = R_{dc} = 100\,000\ \Omega.$

(This follows because the damping across each half of the secondary is $R_3/2$ and $R_4/2$ respectively, and in each case there is a step up, due to the transformer being centre tapped, of 4 times. From this, across the whole of the transformer secondary there are two resistances $4R_3/2$ and $4R_4/2$ in parallel, and since $R_3 = R_4 = R_{dc}$ the above result is obtained immediately).

The total damping resistance (R') required to make $Q_2 = Q_1 = 42.8$ is

$R' = 12\,500 \times 1.77 = 22\,100\ \Omega.$

From this

$$R_2 = \frac{22\,100 \times 100\,000}{(100\,000 - 22\,100)} = 28\,400\ \Omega.$$

The design of the limiter stage will be discussed in Chapter 29.

(g) Some causes of discriminator unbalance will be mentioned, before leaving this section, and it is helpful to consider the circuit as a bridge in which unbalance has to be eliminated. Even when the transformer secondary is centre-tapped accurately, unbalance can occur because the capacitive coupling between the two halves of the secondary winding and the primary winding is not necessarily equal. This calls for care in the method of arranging the windings. Two methods are in common use. The first uses a bifilar method with the two halves of the secondary wound side by side. The second method arranges the secondary into two halves, placed on either side of the primary winding, and the coupling of both sections, including that due to stray capacitances, is made equal.

Capacitive unbalance will also occur when the input capacitances of the diodes are not equal. Suitable arrangement of stray capacitances can often be used to help in offsetting this effect.

The use of a small capacitance connected across one of the diodes is helpful in reducing capacitive unbalance effects.

Balancing of the conduction resistances of the two diode units is largely outside the control of the receiver designer, but variation in the value of R_3 and R_4 can be made to assist in cases where a very high degree of balance is thought to be necessary. Additional precautions of this nature are seldom carried out in commercial receivers.

It should be noted in Fig. 27.17 that, if the resistors R_3 and R_4 are shunted by two separate capacitors, an additional series resistor or r-f choke will be required between the junction of R_3 and R_4 and the centre tap on L_2. Connecting the junction of R_3 and R_4 through a by-pass capacitor to ground would also effectively short-circuit the i-f primary voltage to ground, if the additional component is not used. For most circuits the arrangement shown is applicable, since it requires a minimum of components consistent with satisfactory performance. The alternative arrangement

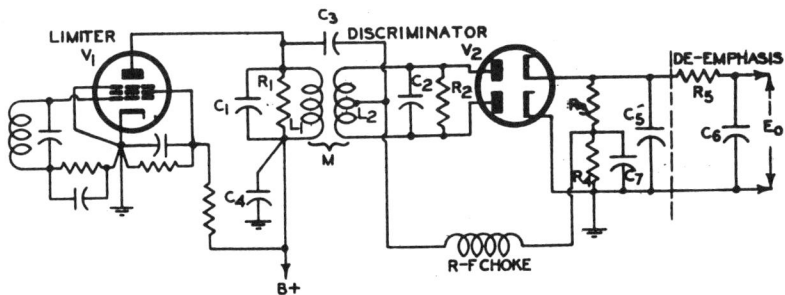

FIG.27·19   ALTERNATIVE ARRANGEMENT FOR PHASE DISCRIMINATOR.

is shown in Fig. 27.19.   The design procedure to be used is exactly as before, except
that the primary circuit damping due to the diode circuits (the damping is actually
across the r-f choke) will be approximately $R_{dc}/4$ instead of $R_{dc}/6$; the secondary
circuit damping is R_{dc} as previously, where $R_{dc} = R_3 = R_4$.   If a resistor is substituted for the r-f choke then the circuit damping is again modified depending on
the value of resistance used in the circuit.   Considerations leading to the choice of
a suitable value for the inductance of the r-f choke will be given in detail in Chapter 29
Sect. 2, in connection with the discussion on a.f.c. discriminators.

It is perhaps worth mentioning that hum due to heater-cathode leakage is sometimes troublesome in discriminator circuits, particularly with miniature diodes, and
a simple and effective cure for this trouble is to make the cathode positive with respect
to the heater.   This can be readily effected, for example in Fig. 27.19, by connecting
a 10 000 Ω resistor by-passed by, say, a 0.01 μF mica capacitor between the earthy
end of R_4 and ground ; the junction of R_4 and the 10 000 Ω resistor is then connected
to B+ via a series resistor whose value is selected so that about 10 to 15 volts appears
across the 10 000 Ω resistor (which now forms one arm of a voltage divider).   In a
typical case the series resistor would be about 0.16 MΩ for a B+ of 250 volts.

(iv) Ratio detectors
(A) General
Many of the details given below, regarding ratio detectors, have been taken from
Refs. 34, 35 and 36.   Practical experience with several of the arrangements shown
has confirmed much of the data given in the design sections.   However, a number of
additional factors will warrant discussion. ·

The principle underlying many circuits for F-M detection is the peak rectification
of two i-f voltages, the relative amplitudes of which are a function of frequency, together with means for combining the rectified voltages in reversed polarity.   The
output is then equal to the difference between the two rectified voltages.   This statement is directly applicable to the phase discriminator circuits of Figs. 27.17 and 27.19,
the two i-f voltages being those applied to the diode plates, and the rectified voltages
combined in reversed polarity being those which appear across the load resistors
R_3 and R_4.   The way in which the two i-f voltages, applied to the diode plates, are
dependent on the instantaneous intermediate frequency has been discussed in connection with Figs. 27.15 and 27.16.   It was mentioned previously that, with the phase
discriminator, changes in the magnitude of the input signal will give rise to amplitude
changes in the resultant output voltage, and the need for some form of amplitude
limiting was emphasized.

In an attempt to eliminate the necessity for a limiter, a ratio type of detector has been
developed from the basic phase discriminator circuit.   In this modified circuit the
rectified voltages are split into two parts in such a way that their ratio is proportional
to the ratio of the instantaneous i-f voltages applied to the detector diodes, and the
sum of the two rectified voltages is kept constant.   It has been found in this type of
circuit that the useful output voltage, which is proportional to the difference between
the two rectified voltages developed by the diode detectors, tends to be independent

of amplitude variations superimposed on a frequency modulated voltage applied to the discriminator circuit input terminals.

A basic ratio detector circuit is shown in Fig. 27.20(A). It can be seen that this is similar to the conventional phase discriminator but one of the diodes is reversed, and so the total voltage between the points P and N is equal to the **sum** of E_1 and E_2. The sum of E_1 and E_2 is held constant by means of a battery or a large capacitance. This point will be further discussed as we proceed. The audio output voltage is taken from the junction of R_3, R_4 and C_5, C_7 and is equal to $(E_1 - E_2)/2$. The voltage output with frequency change is seen from Fig. 27.20(B) to be similar to that for a conventional discriminator circuit.

The sum voltage $(E_1 + E_2)$ can be stabilized by using either a battery or by shunting a large capacitance across the load resistors R_3 and R_4. A battery would limit the operation in such a way that the input signal would need to be at least strong enough to overcome the fixed bias due to the battery voltage. A better solution is to use a capacitor, since the voltage across it will vary in proportion to the average signal amplitude and thus automatically adjust itself to the optimum operating level. This allows amplitude rejection to be secured for a wide range of input signal voltages, the lowest useful signal being determined by the ability of the diode rectifiers to conduct with small input voltages.

When a capacitor is used to stabilize the rectified output voltage its capacitance must be sufficiently large so that the sum of $E_1 + E_2$ cannot vary at an audio frequency rate. This calls for a time constant in the circuit made up from R_3, R_4 and the additional capacitance C of about 0.2 seconds. The effects of C_5 and C_7 on the time constant can be neglected, since the value of the capacitance C will be of the order of microfarads (usually about $8\mu F$).

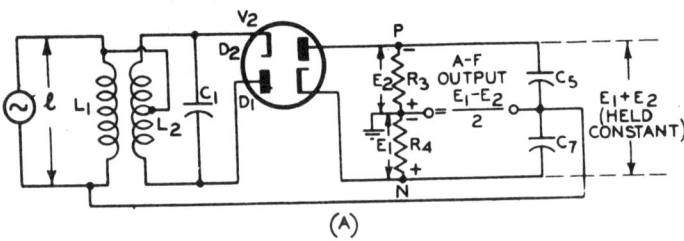

(A)

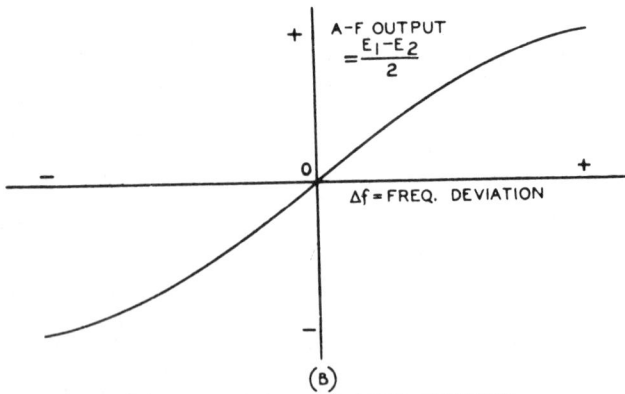

(B)

FIG.27·20   (A) BASIC CIRCUIT OF RATIO DETECTOR.
               (B) TYPICAL OUTPUT CURVE.

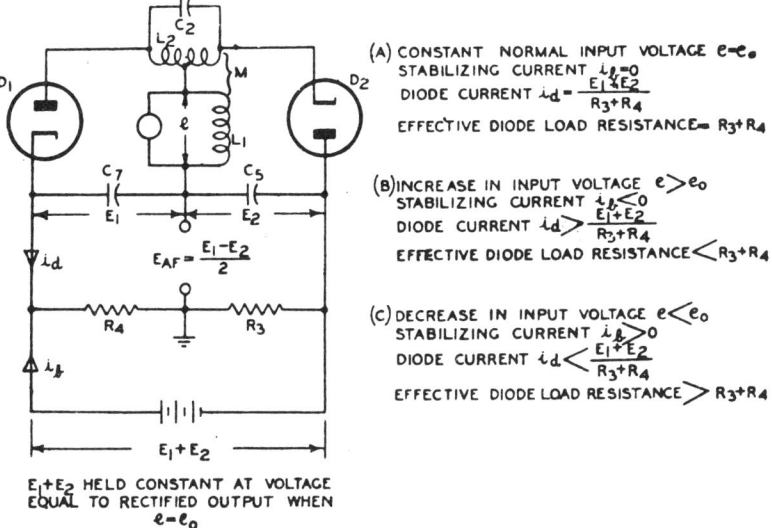

(A) CONSTANT NORMAL INPUT VOLTAGE $e = e_o$
STABILIZING CURRENT $i_s = 0$
DIODE CURRENT $i_d = \dfrac{E_1 + E_2}{R_3 + R_4}$
EFFECTIVE DIODE LOAD RESISTANCE$= R_3 + R_4$

(B) INCREASE IN INPUT VOLTAGE $e > e_o$
STABILIZING CURRENT $i_s < 0$
DIODE CURRENT $i_d > \dfrac{E_1 + E_2}{R_3 + R_4}$
EFFECTIVE DIODE LOAD RESISTANCE $< R_3 + R_4$

(C) DECREASE IN INPUT VOLTAGE $e < e_o$
STABILIZING CURRENT $i_s > 0$
DIODE CURRENT $i_d < \dfrac{E_1 + E_2}{R_3 + R_4}$
EFFECTIVE DIODE LOAD RESISTANCE $> R_3 + R_4$

$E_1 + E_2$ HELD CONSTANT AT VOLTAGE
EQUAL TO RECTIFIED OUTPUT WHEN
$e = e_o$

FIG.27-21 RATIO DETECTOR CIRCUIT SHOWING CONDITIONS FOR DIFFERENT SIGNAL INPUT VOLTAGES.

It is worth noting that the a.v.c. voltages available from ratio detector circuits have values which are not always directly suitable for application to controlled stages. Some form of voltage divider arrangement is often necessary to obtain suitable voltages for securing the desired a.v.c. characteristic.

(B) Operation

A brief qualitative description of the operation of the ratio detector will be given here. A complete quantitative analysis has not been made, although many of the significant factors have been investigated (see Refs. 34, 35 and 36).

Fig. 27.21 shows the same basic circuit as Fig. 27.20(A), rearranged in a more convenient form for discussion. The main details of operation are also indicated, and show what happens in the circuit when the amplitude of the signal input voltage changes. For the case (A) where the amplitude of the signal input voltage is constant, the stabilizing current is zero and the circuit has essentially the same output characteristic as a phase discriminator. When the signal amplitude increases (B), the average diode current also increases, and resultant direct current flows into the stabilizing voltage source (i.e. the battery is being charged). From this since $E_1 + E_2$ remains constant, it is seen that the **effective** diode load resistance must decrease and so the primary and secondary circuits are more heavily damped than for the case (A) where the input voltage has a fixed value. The increased damping on the transformer circuits will tend to offset the increase in the amplitude of the input voltage, and so the output voltage will also tend to be constant.

Similarly, when the signal amplitude decreases—case (C)—the stabilizing voltage source supplies additional current (i.e. the battery is being discharged) to offset the reduction in diode current, so as to maintain $E_1 + E_2$ constant ; the effective diode load resistance is now increased, and so the circuit damping is reduced.

From this it is seen that stabilizing the rectified voltage results in the equivalent load resistance varying in such a way as to offset changes in the amplitude of the signal input voltage. It also follows that there must be optimum circuit conditions which will allow the undesired amplitude variations to be offset most effectively (although these conditions are not necessarily the same for all input voltages). The conditions for obtaining the most satisfactory operation will be summarized below when discussing the design of the ratio detector.

(C) Types of circuit

There are two types of ratio detector circuit in common use. The first of these is the so called balanced circuit and is illustrated by Fig. 27.22. At first glance this does not seem to be the same as the basic circuit previously discussed. However, if the tertiary winding is considered as L_1 for the moment (its presence will be explained presently), and it is taken that C_8 has a high reactance at audio frequencies and negligible reactance at i-f, then a simple re-arrangement of the circuit will show that it is the same as that of Figs. 27.20 and 27.21. Whether the junction of C_5 and C_7 is connected to ground, as shown in Fig. 27.22, or C_8 is connected from the junction of C_5 and C_7 to the junction of R_3 and R_4 as would be expected (since C_8 is part of the audio load circuit) is unimportant, as can be seen by redrawing the circuit. C is the stabilizing capacitor.

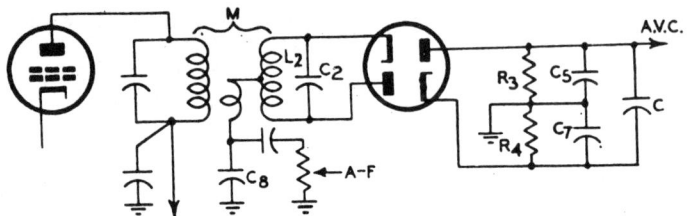

FIG. 27·22   BALANCED TYPE` OF RATIO DETECTOR CIRCUIT.

The second type of circuit in common use is the unbalanced arrangement of Fig. 27.23. The operation is similar to the balanced arrangement but it obviously cannot be redrawn directly into a balanced arrangement. The A-M rejection is the same whether the centre-tap of R_3 and R_4 or any point along the load is grounded, provided that the stabilizing capacitor can hold the voltage across its terminals constant. This condition is not easily met, however, at very low audio frequencies because very long time constants for the diode load circuit have undesirable effects when the receiver is being tuned.

In general the balanced circuit will give better amplitude rejection at low frequencies than the unbalanced arrangement, when the time constants of the two circuits are the same.

Additional advantages of the balanced ratio detector are that the ratio of a.v.c. voltage to audio voltage is comparable with that obtained in conventional A-M detector circuits, and also a voltage is available which is zero at the centre frequency and can be used for a.f.c. purposes if so desired.

The presence of the tertiary winding in the discriminator transformer is readily understood when it is considered that it provides a simple method of tapping down on the transformer primary without the use of blocking capacitors and a choke, and at the same time allows the overall sensitivity of the detector to be increased by permitting a high primary circuit dynamic impedance. This is particularly helpful since the value of the total diode load resistance is generally small (usually less than 50 000 ohms total).

Because of the low value of load resistance it should be clear that high detection efficiencies are not easy to obtain, and so high perveance diodes are of considerable assistance in this regard because their internal impedance will be comparatively small. However, it is often found that high perveance diodes are more susceptible to hum pick-up than the medium perveance types and so are not necessarily the best choice in all circuits.

Before leaving this section an additional point of interest will be discussed. In Fig. 27.24 are shown typical curves of output voltage against signal input voltage for a ratio detector. It can be seen from these curves that the output is not constant with input voltage for a given frequency deviation. This effect can be offset to a large extent by using the valve preceding the ratio detector as a partial limiter. For weak signals the driver valve can be used as a straight voltage amplifier, but if a grid resistor and capacitor are included which have a fairly short time constant (say 10 microseconds or so) they can be used to derive a bias voltage for the grid of the valve when the signal exceeds some pre-determined level, and the valve can then be used to provide a degree of limiting.   In this way the use of an additional valve is still avoided, since most ratio detectors operate satisfactorily with as little as 10 millivolts (or less) applied to the grid of the driver valve, but the advantages of additional amplitude limiting can be obtained on strong signals.   Of course, satisfactory results are obtainable with a ratio detector alone and the application of a.v.c. bias to the earlier stages is helpful but improvement is possible with the arrangement suggested, and with very little additional trouble.   With the partial limiter arrangement a.v.c. is sometimes applied to the first i-f voltage amplifier, the a.v.c. bias being derived from the ratio detector circuit as indicated in Figs. 27.22 and 27.23.   Some receiver manufacturers do not use a.v.c. at all, but use suitable grid resistor-capacitor arrangements with both i-f valves to provide additional bias when the signal is sufficiently large. Some care is necessary with the limiter arrangements, since detuning of the i-f transformer can have adverse effects particularly with some types of noise interference. This point received consideration in Chapter 26 when the design of the F-M i-f transformers was being discussed.

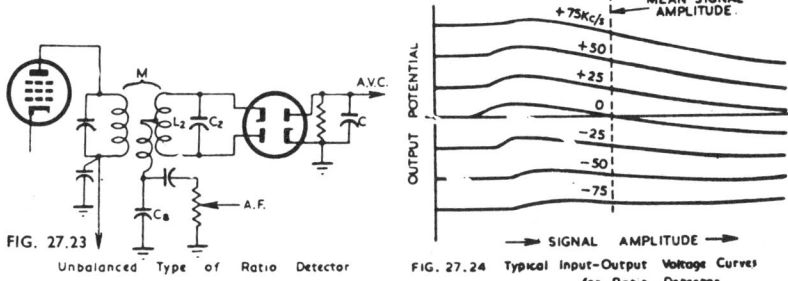

FIG. 27.23
Unbalanced Type of Ratio Detector

FIG. 27.24 Typical Input-Output Voltage Curves for Ratio Detector

(D) Design considerations

The data given below are quite general, but the circuit constants suggested are based on a centre frequency of about 10.7 Mc/s and a peak separation of approximately 350 Kc/s. Most of the information which follows (with certain modifications) is taken from Ref. 34, where a more detailed explanation is available.

Diode characteristics

Good ratio detector performance can be obtained with either high perveance diodes such as type 6AL5, or medium-perveance diodes such as type 6H6.

The circuits used will differ with high and medium perveance diodes in the extent to which the rectified output voltage is held constant.   They will also differ with respect to the compensation used to minimize the residual unbalanced component of amplitude modulation in the output.   The unbalance is brought about by secondary transformer detuning effects caused by input reactance variations due to the diodes. The magnitude of the effect is generally less for diodes having lower perveance.

Because the diode rectification efficiency, and hence the diode circuit loading, varies with signal level, optimum A-M rejection is obtained for a particular input voltage.   The level at which optimum A-M rejection is obtained can be altered by varying the circuit constants, particularly the fraction of the total rectified voltage which is stabilized, or the ratio of secondary and tertiary voltages.

Secondary inductance

The secondary L/C ratio should be as high as possible consistent with circuit stability.   This suggests capacitance values from 25-70 $\mu\mu$F.   Also, the secondary Q should be as high as possible, depending on the required peak separation.   Suitable Q values are 75 to 150, with the higher values giving improved sensitivity, for a given A-M rejection, but less peak separation.

Load resistors

The value R of the diode load resistors is generally selected to reduce the operating secondary circuit Q to a value of approximately one-fourth or less of its unloaded value. Smaller values for R will increase the downward modulation handling capabilities of the detector, but will also reduce its sensitivity.

Primary L/C ratio and Q

The primary L/C ratio should be as large as possible to increase the sensitivity. The limiting factor is the maximum stable gain between the grid and plate of the ratio detector driver valve (i.e. the last i-f valve preceding the detector). When determining the maximum gain it should be remembered that the gain may increase during downward amplitude modulation, particularly on either side of the centre frequency. If the primary Q is high enough so that the operating Q is determined mainly by the diode loading, the grid-plate gain will rise during the A-M cycle. This is advantageous since it increases the ability of the detector to reject downward modulation. The primary Q is made as high as possible consistent with peak separation and stable gain requirements.

Coupling

The value of coupling used, together with the number of turns in the tertiary winding, is the principal factor in determining the ratio of the tertiary voltage to the half secondary voltage. This ratio should always be close to unity. The coupling is generally adjusted to half critical in the actual circuit, since the degree of coupling is dependent on the other circuit component values and these are selected before the coupling is finally set to the desired amount.

Tertiary inductance

The number of turns on the tertiary winding is adjusted so that the required ratio of tertiary voltage to half secondary voltage is obtained (see under Coupling).

Reducing unbalanced A-M component

There will always be some residual amplitude modulation in the detector output because of the variation in the effective diode input capacitance during the A-M cycle, and because of the unbalance introduced by the transformer and other circuit components. Several methods are useful in overcoming this effect. One is to vary the effective centre-tap on the secondary winding (this method is shown in Fig. 27.26) and another is to make the resistors R_3 and R_4 shown in Fig. 27.25 unequal. In addition a resistor R_5 is used in series with the tertiary winding (see Fig. 27.25) to modify the peak diode currents, which has the effect of appreciably reducing the unbalanced A-M component particularly at high input voltages. The methods using resistors to reduce unbalance also reduce the detector sensitivity.

The reduction of unbalance effects is carried out by observing the condition for minimum A-M output when a signal which is simultaneously amplitude and frequency modulated is applied to the detector.

Time constant of stabilizing voltage

The discharge time constant of the stabilizing capacitance and the load resistance should be about 0.2 second. Larger time constants will give better amplitude rejection when the undesired modulating frequency is low, but have undesirable effects on the tuning. The tuning effect is similar to that given by an ordinary A-M receiver when the a.v.c. time constant is too long.

Experiments with the time constant of this circuit give some interesting effects. If the stabilizing capacitance is made very large, say 100 μF or so, and the receiver is tuned rapidly across a signal, it will be found that the point of maximum output is very easily determined, and there is no effect from the usual side responses. When the receiver is detuned the noise level rises, as the capacitor discharges, and the side responses again become evident until the receiver has been tuned once through the point of maximum output.

Additional details

It is not particularly easy to achieve a good balance with the ratio detector circuit, and usually more care is required in this respect than with the conventional phase discriminator. However, the better the balance obtained the better will be the rejection of undesired amplitude modulation.

Although the side responses are normally well down on the response at the main tuning point, with this type of circuit, considerable improvement has been noticed

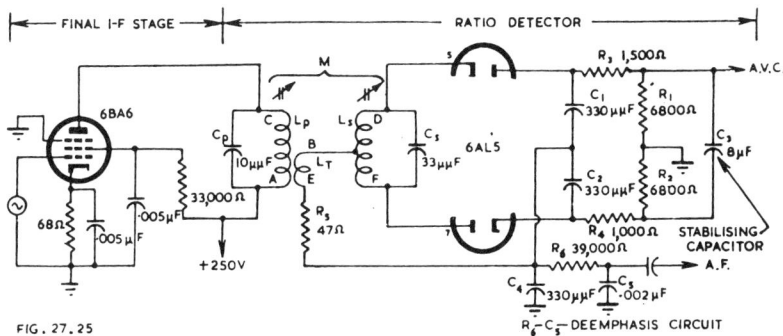

FIG. 27.25 R_6-C_5—DEEMPHASIS CIRCUIT

when the bandwidth of the discriminator is fairly large (say 300 to 400 Kc s) and the i-f response is such as to give steep sides to the overall selectivity curve. With receivers using two i-f transformers (each critically-coupled and both having primary and secondary Q's of about 75) the side responses are still sufficiently large to be noticeable, although they are 12-15 db down on the main tuning position. Receivers using three transformers (each critically-coupled and having Q's of about 70) give a very marked improvement, and the two undesired side responses can only be found after careful searching. If the transformers are made more selective the side responses are reduced, but the additional non-linear distortion introduced by the tuned circuits may no longer be negligible, as can be determined from the data given in Chapter 26. Further, the amount of amplitude modulation on a carrier goes up as the selectivity of the transformers is increased, and this undesired A-M has to be removed by the ratio detector thereby reducing its effectiveness to other undesired external noise. Even in the case of the receiver with only two i-f transformers, the undesired side responses are very much less noticeable than with receivers using the usual limiter and discriminator combination

(E) Practical circuits

Two circuits which have been constructed and tested under working conditions are shown in Figs. 27.25 and 27.26. The intermediate frequency is 10.7 Mc s Constructional details and performance data can be found in Refs. 34 and 37.

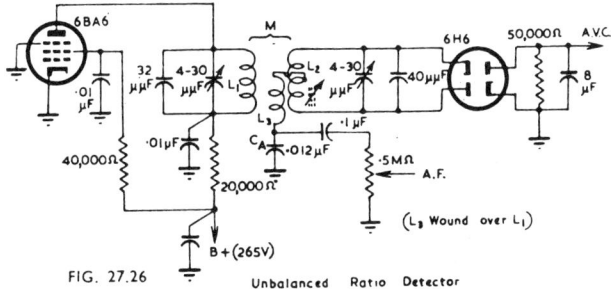

FIG. 27.26   Unbalanced   Ratio   Detector

De-emphasis is obtained in the circuit of Fig. 27.26 by adjusting the value of the capacitor marked C_1. The time constant (0.4 second) of the diode load circuit is about twice as long as that previously recommended for typical cases, but the performance is quite good, and, as suggested in Ref 37, the individual designer can set the constants to suit his own requirements.

The secondary winding for the circuit of Fig. 27.25 is a bifilar arrangement, while that of Fig. 27.26 has the secondary split into two sections placed on either side of the primary. In the latter case slug tuning is used for one half of the secondary so that the effective centre tap can be set to any desired position.

(F) Measurements on ratio detectors
Primary and secondary Q's

To measure the unloaded primary Q, the double diode valve is removed from its socket and the secondary is detuned. The primary Q is then determined by the selectivity of the primary tuned circuit. The loaded primary Q is measured in the same way but with the double-diode valve in the circuit.

To measure the secondary Q the primary is heavily loaded by a shunt resistor. The secondary Q is then determined from the selectivity of the secondary winding as indicated by the rectified voltage variation with frequency. When measuring the unloaded secondary Q, the diode load resistance should be replaced by a resistance of about one megohm. When measuring the loaded secondary Q, the normal load resistors should be used. For these measurements the centre-tap of the secondary winding should be disconnected from the tertiary winding, to prevent the tertiary voltage from contributing to the rectified voltage.

Coupling

The percentage of critical-coupling can be measured by noting the change in primary voltage as the secondary is varied from a tuned to a detuned condition. With all circuit conditions normal, and the signal input voltage set to the intermediate frequency, the primary voltage (E_1) is noted. With the same input signal level, the secondary is detuned so that the primary voltage rises to E_2. The percentage of critical-coupling can then be expressed in terms of the ratio between the primary voltage with the secondary detuned and tuned using the relationship

$$\frac{k}{k_{crit}} = \sqrt{\frac{E_2}{E_1} - 1}.$$

For example, if the signal voltage at the plate rises 25 per cent. (i.e. $E_2 = 1.25\ E_1$) when the secondary is detuned, the coupling is 50 per cent of critical.

The ratio between the secondary and tertiary voltages can be measured indirectly in terms of the rectified output voltage which is obtained (a) with the secondary tuned and (b) with the secondary detuned. In both cases the input signal is adjusted so that the primary voltage remains constant. If the ratio between the voltages read in (a) and (b) is r, that is if

$$r = \frac{\text{Rectified voltage (tuned secondary)}}{\text{Rectified voltage (detuned secondary)}}$$

then $\dfrac{S}{P} = \sqrt{r^2 - 1}$

where S = half secondary voltage at centre frequency
P = " primary " (i.e. tertiary) voltage effective in the diode circuit at the centre frequency.

The ratio S/P is usually made equal to unity.

Signal generator

It is convenient to have available a signal generator of good quality capable of being simultaneously frequency and amplitude modulated. Undesired amplitude and frequency modulation should not be present in the generator output.

Alignment procedure

The ratio detector may be aligned by using either an unmodulated signal set to the centre frequency and a d.c. vacuum-tube voltmeter, or by using a F-M signal generator and an oscillograph.

If an unmodulated signal generator is used, the procedure is to set the signal to the intermediate frequency, and with the d.c. vacuum-tube voltmeter connected to measure the rectified output voltage (this is usually the a.v.c. take-off point), the primary tuning being adjusted for maximum voltage output.

The procedure used for adjusting the secondary tuning depends upon whether the centre-tap of the stabilizing voltage is earthed or whether one end is earthed. If the centre-tap is earthed, the secondary tuning is adjusted so that the d.c. voltage at the audio take-off point is equal to zero ; the d.c. vacuum-tube voltmeter is shifted to the audio take-off point for this measurement. If the detector is of the unbalanced type the secondary tuning is adjusted so that the d.c. voltage at the audio take-off point is equal to half the total rectified voltage. As an alternative an amplitude modulated signal can be used, the primary trimmer is adjusted for maximum d.c. output voltage and the secondary trimmer adjusted to give minimum audio output ; the residual output is measured in the conventional manner. Circuit unbalance is indicated if the two methods do not give the same secondary trimmer setting.

If sweep alignment is used, the primary can be accurately aligned by using a comparatively low deviation, and adjusting the primary trimmer for the maximum amplitude of output voltage. The secondary may be adjusted by using a deviation such that the total frequency swing (twice the deviation) is equal to the peak separation. This procedure makes it possible to adjust the secondary tuning so that a symmetrical detector characteristic is obtained.

Peak separation

The separation between the peaks on the F-M output characteristic of a ratio detector may be measured by applying a frequency modulated signal and increasing the deviation until the response is just observed to flatten at the peaks. When this is done, the peak separation is equal to twice the deviation.

If an attempt is made to measure the peak separation by plotting the output characteristic point by point, the peak separation obtained will usually be considerably less than that obtained under dynamic conditions with the output voltage stabilized. The F-M detector characteristic may be plotted point by point provided a battery of the proper voltage is connected across the stabilizing capacitor. The voltage of this battery must be equal to the rectified voltage which exists at the centre frequency. In practice it is convenient to use a 7.5 volt " C " battery and to adjust the signal input so that the capacitor-stabilized voltage at the centre frequency is equal to the battery voltage.

It is worth noting that the peak separation is not constant for all values of input voltage, and measurements should be made for several representative voltages of the magnitude likely to be encountered under practical conditions of reception. The peak separation is wider for larger input voltages, as a result of the increased diode loading on the tuned circuits. Similarly the peak separation is less for smaller input voltages because of reduced circuit damping.

Measurement of A-M rejection

The measurement of A-M rejection can be carried out using a signal generator which can be simultaneously frequency and amplitude modulated.

Ref. 34 shows a number of the typical wedge-shaped patterns which are obtained when visual methods are used for determining the amplitude modulation present in the audio output. A pattern of the type to be expected is shown in Fig. 27.27.

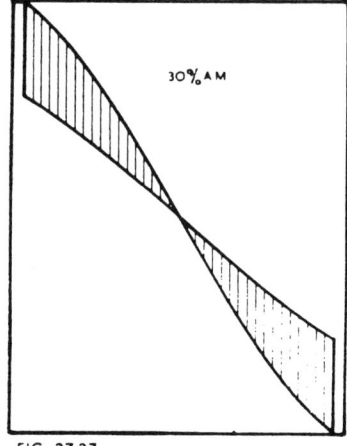

30% A M

The measurement of the A-M rejection of a ratio detector can be described in terms of the pattern obtained for a given frequency deviation and a given percentage of amplitude modulation.

The measurements should be made for several different values of input voltage. Ideally the pattern obtained should be a diagonal line regardless of the presence of amplitude modulation. The amount by which the pattern departs from a straight line indicates the extent to which the detector fails to reject amplitude modulation.

To measure A-M rejection with a generator which can only be frequency modulated the procedure is as follows. A frequency modulated signal fully deviated ($\pm$ 75 Kc/s

FIG. 27.27

in the usual case) is applied, and a battery of which the voltage is equal to the rectified output is shunted across the stabilizing capacitor. The magnitude of the input signal is now reduced until the output becomes distorted as a result of the diodes being biased by the stabilizing voltage, and finally the F-M output drops to zero. The ratio between the initial input signal and the minimum input signal for which the F-M output becomes distorted is then a measure of the amount of downward modulation that the detector can reject. For example if the voltage ratio is r, then the percentage of downward amplitude modulation which can be handled is $100 (r - 1)/r$. To determine the rejection of the detector for upward amplitude modulation the same set-up is used and the change in output is noted as the input signal is increased. Since the amplitude of the input signal is not varied dynamically, this method will not indicate any unbalanced component which may be present in the output. The latter is more conveniently measured using simultaneous F-M and A-M as discussed previously.

Another method which can be applied to measure the amplitude rejection properties of any type of detector for different signal input voltages may be worth discussing. The input signal is simultaneously frequency and amplitude modulated. Any audio modulating frequencies can be selected which are not harmonically related, and suitable values are say 400 c/s for F-M and 30 c/s for A-M. The percentage A-M can be taken as 30% (or any value desired) and the deviation as $\pm$ 22.5 Kc/s (or any other value). The ratio of the desired and undesired audio outputs can then be measured on a frequency selective voltmeter such as a wave analyser. Measurements can be made at the centre frequency to indicate residual A-M, and for various frequencies off the centre frequency to determine the rejection capabilities of the detection system under operating conditions ; e.g. those which occur when oscillator and other frequency drift occurs. Low audio frequencies are preferable for this test (and also for the previous methods), particularly with ratio detectors, since the amplitude rejection is least for this condition. Tests on a number of ratio detectors have indicated similar performance when high audio modulating frequencies are used, but considerable variation occurs in the amount of amplitude rejection obtained for the lower audio frequencies using balanced and unbalanced circuits. This is readily understood when the function of the stabilizing capacitor is considered.

Standard test procedures have been devised for F-M receivers, and standard methods for evaluating the degree of A-M rejection, downward modulation handling capability, and the effects of mistuning are described in Chapter 37.

Non-linear distortion

Distortion measurements are usually quite difficult to carry out with any degree of accuracy because of the inherent non-linear distortion present in the signal source and test equipment. Tests on typical F-M receivers using limiter-discriminator combinations and ratio detectors have been made, using the low power stages of a F-M broadcast transmitter as the signal source. The non-linear distortion in the transmitter signal was less than 0.5% for frequencies from 100 c/s to 15 Kc/s, and less than 1% from 30 to 100 c/s ; the frequency deviation was set to $\pm$ 75 Kc/s in all cases.

Both types of receiver indicated a measured overall distortion of the order of 2 to 2.5% in the frequency range 100 c/s to 7.5 Kc/s, 3 to 5% at 50 c/s and 5% at 15 Kc/s although this latter case was a visual check only. Above 10 Kc/s the non-linear distortion with ratio detector receivers is generally somewhat higher than with limiter-discriminator receivers. The input in all cases was approximately 100 μV and the output was adjusted to 0.5 watt. The test frequency was 97.2 Mc s and the i-f of all receivers was 10.7 Mc/s. The measured distortion for the a-f amplifier alone was 0.5% for the range 100 c/s to 7.5 Kc/s and 2% at 50 c/s ; above 7.5 Kc/s, observation was made using an oscillograph and as there was no visible distortion it was assumed that distortion was less than 3%.

The non-linear distortion is increased with ratio detector receivers when the magnitude of the input voltage is made very large. With the other type of receiver the input signal must be sufficiently large to ensure satisfactory limiter operation. Receivers using a combination of partial limiting and a ratio detector give excellent results for a wide range of signal input voltages, and do not require extra valves.

The test conditions used are very severe, as it is unlikely that a 15 Kc/s audio signal will cause 75 Kc/s deviation ; although it should not be overlooked that pre-emphasis in the transmitter will cause a 15 Kc/s modulating signal to be increased in magnitude by about 17 db. Listening tests indicate that excellent results are obtainable with any of the receiver arrangements. It will be appreciated that the possibility of errors in the distortion measurements are very large, and the results are given to serve as an indication of the magnitude of the distortion to be expected with typical F-M receivers which do not include elaborate design features, but conform to good engineering practice.

SECTION 3 : AUTOMATIC VOLUME CONTROL

(i) Introduction   (ii) Simple a.v.c.   (iii) Delayed a.v.c.   (iv) Methods of feed (a) Series feed (b) Parallel feed   (v) Typical circuits   (vi) A.V.C. application   (vii) Amplified a.v.c.   (viii) Audio a.v.c.   (ix) Modulation rise   (x) A.V.C. with battery valves   (xi) Special case with simple a.v.c.   (xii) The a.v.c. filter and its time constants (xiii) A.V.C. characteristics   (xiv) An improved form of a.v.c. characteristic   (xv) Design methods.

(i) Introduction

Automatic volume control is a device which automatically varies the total amplification of the signal in a radio receiver with changing strength of the received signal carrier wave. In practice the usual arrangement is to employ valves having " remote cut-off " or " variable-mu " grids and to apply to them a bias which is a function of the strength of the carrier.

From the definition given, it would be correctly inferred that a more exact term to describe the system would be automatic gain control (a.g.c.). The older term a.v.c. has been retained here as the name is not likely to lead to any confusion, and is widely used throughout the radio industry.

(ii) Simple a.v.c.

In order to obtain simple a.v.c. it is only necessary to add resistor R_4 and capacitor C_4 to the ordinary diode detector circuit shown in Fig. 27.28. In any diode detector there is developed across the load resistor (R_1 and R_2 in series) a voltage which is proportional to the strength of the carrier voltage at the diode. The diode plate end of the load resistor is negative with respect to earth (Fig. 27.28) and therefore a negative a.v.c voltage is applied to the controlled grids. R_4 and C_4 form a filter to remove the a-f component and any residual i-f and leave only the direct component of voltage.

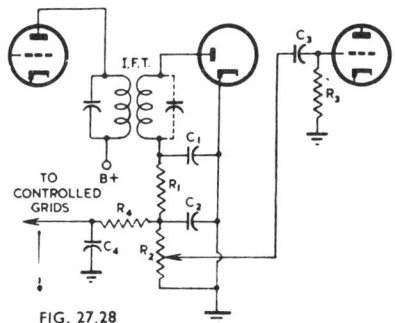

FIG. 27.28

Since a capacitor in series with a resistor takes a finite time to charge or discharge, the a.v.c. filter circuit has a " time constant." The time constant of R_4 and C_4 is equal to R_4C_4. For example, if R_4 is 1 megohm and C_4 is 0.25 microfarad the time constant will be 0.25 second. In this circuit the time constant is also influenced by R_1 during charge and by R_2 during discharge, but since R_1 is small compared with R_4 it will be sufficiently close for practical purposes to regard R_4C_4 as giving the time constant during charge, and $C_4(R_4 + R_2)$ as the time constant during discharge. It is seen that the charge time constant is shorter than the discharge time constant. Time constants in a.v.c. circuits will be discussed in detail in (xii) below.

For 100% rectification efficiency the a.v.c. voltage would equal the peak i-f voltage applied to the diode. Because of losses the a.v.c. voltage is always somewhat less than the peak carrier voltage (this is readily seen from Figs. 27.5 and 27.6), and since, in general, a higher voltage is required for the a.v.c. than for detection, this will be a limitation to the use of simple a.v.c.

The design of the i-f transformer to couple the i-f voltage into the diode circuit is quite straightforward. The damping to be expected across the windings due to the diode circuits has already been discussed in Sect. 1(i)Cc of this chapter. Methods of i-f transformer design are detailed in Chapter 26.

In the circuit of Fig. 27.28 the d.c. diode load consists of R_1 and R_2 in series, R_1 being used in conjunction with capacitors C_1 and C_2 to form a filter to prevent the major part of the undesired i-f voltages from appearing across R_2. The volume control is R_2 in this circuit, and the advantages and disadvantages of using the control in this position were discussed in Sect. 1(i)Dc above. At audio frequencies R_2 is shunted by R_4 and by R_3 when the setting for R_2 is at the maximum. It follows that the diode load for a.c. is considerably different from that for d.c., and, as previously discussed in Sect. 1(i), considerable non-linear distortion of the a-f output will result when the modulation percentage is high. R_4 and C_4 will substantially remove all audio frequency variations from the a.v.c. bias applied to the controlled stages, as well as any residual i-f voltages which may be present across R_2.

Fig. 27.28 shows R_4 connected to the junction of R_1 and R_2, but in some cases R_4 is connected to the top of R_1. This latter arrangement will give a slightly higher a.v.c. voltage, the amount depending on the size of R_1, but the amount of i-f filtering for the controlled stages is reduced. With the usual component values it is not very important which arrangement is used.

Owing to the effects of contact potential in the diode, together with rectification due to unavoidable noise voltages, there is a voltage developed across R_2 even with no carrier input. Even with a weak carrier input this voltage is increased. Consequently it will be seen that with the weakest carrier likely to be received there is an appreciable negative voltage applied to the controlled grids. If no means were taken to compensate for this, the overall sensitivity of the receiver would be decreased. Compensation for the initial standing bias, before a carrier is received, can be carried out by applying a lower minimum negative bias to the controlled stages. However, a very real disadvantage of simple a.v.c. is now apparent in that the sensitivity of the receiver may decrease as soon as a carrier is received. To overcome the loss in sensitivity on weak signals a delayed a.v.c. system is used and this gives very much improved control.

A special case of the effects of contact potential with zero bias valves is worth noting before leaving this section. Suppose a negative bias voltage exists between the grids and cathodes of the controlled valves due to positive grid current. This current can be sufficiently large to cause an appreciable negative bias voltage to appear across the resistance common to the controlled stages and the detector, and so give rise to a negative bias voltage between plate and cathode of the diode. Under this condition, if simple a.v.c. is being used, there will be no output from the receiver until the signal is sufficiently large to overcome this delay bias on the detector. If a separate diode is used for a.v.c. the effect will be to increase the delay bias. Typical examples of the effect have occurred in practice with type 1P5-GT voltage amplifiers used in conjunction with the diode section of a type 1H5-GT. An obvious remedy is to apply a small positive bias voltage to the diode plate, but the grid current can often be appreciably reduced by increasing the screen voltage.

(iii) Delayed a.v.c.

The " delay " in delayed a.v.c. refers to voltage delay, not time delay. A delayed a.v.c. system is one which does not come into operation (i.e. it is delayed) until the carrier strength reaches a pre-determined level. The result is that no a.v.c. voltage is applied to the grids of the controlled stages until a certain carrier strength is reached, and the receiver will have its maximum sensitivity for signals with an amplitude below this pre-determined level.

The circuit used for delayed a.v.c. also makes possible improved rectification efficiency in the a.v.c. circuit, thus producing slightly greater a.v.c. voltage for the same peak diode voltage. This is due to the higher value of a.v.c. diode load resistance (R_7 in Fig. 27.29) permissible in this circuit. With the a.v.c. arrangement shown the a.c. loading on the detector circuit is reduced when compared with the arrangement of Fig. 27.28 and this reduces the distortion of the audio output voltage. Some non-linear distortion of the modulation envelope does occur at the primary of the transformer but this is considerably less than that introduced when the a.v.c. voltage is obtained from the secondary circuit. In addition, the overall selectivity up to the primary of the i-f transformer is less than that at the secondary, and the a.v.c. will start to operate further from the carrier frequency than if fed from the secondary. The advantage here is that the tendency to give shrill reproduction is reduced as the receiver is detuned from a carrier.

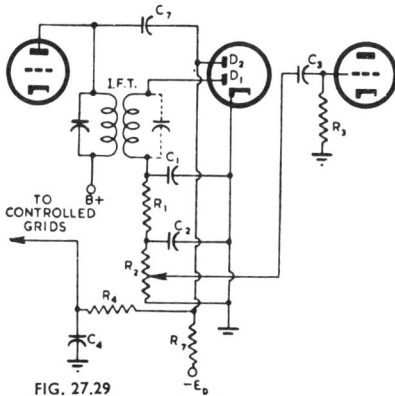

FIG. 27.29

Differential distortion, caused by variable damping of the a.v.c. diode input circuit, is frequently mentioned as an inherent disadvantage of delayed a.v.c. because of the distortion which is introduced when the diode bias line cuts the modulation envelope. Careful measurements have shown that the additional distortion occurring just as the a.v.c. diode starts to conduct is quite small provided that the delay voltage is small and the average depth of modulation is not large (say 50% or so). With a delay voltage (E_D, Fig. 27.29) of 3 volts the total harmonic distortion was found to increase from an average level of about 2.5% to a peak of 4%. Not only is the amount of distortion fairly small, with the usual depth of modulation, but it only occurs over a limited range of input signals, which, with a small delay voltage such as 3 volts or less, occurs at such weak signal strengths as to be relatively unimportant. However with 100% modulation this type of distortion will always occur. Of course, when the input voltage is too small for the a.v.c. diode to conduct there is no distortion. From the evidence available, it appears that the effect of differential distortion is a comparatively minor one when the circuit is correctly designed, and a small delay voltage is used. Detailed discussion of the distortion produced by delayed a.v.c. can be found in Refs. 44 (p. 189) and 46.

If it is desired to apply full a.v.c. voltage to certain controlled stages and a fractional part only to another stage, this may be done by tapping R_7 at a suitable point and by adding a filter circuit similar to that made up from R_4 and C_4.

Staggering* of the a.v.c. is sometimes advantageous, as a considerable improvement in signal-to-noise ratio is often possible by controlling mainly the i-f amplifier valves.

* By staggering of the a.v.c. is meant the application of different values of a.v.c. voltage to various stages.

However, in some cases this leads to difficulties'with cross modulation. When cross modulation is a serious problem it is preferable to control the r-f amplifier valves fully. Often no great care is taken with broadcast receivers in the method of applying a.v.c. With high quality communications receivers precautions are necessary, and it is helpful if the r-f and i-f valves to be controlled by a.v.c. can be selected by the operator. In this way optimum performance can be obtained under all conditions. The usual arrangement, however, is to design an a.v.c. system with compromise characteristics and fit manual switching for the selection of suitable time constants. The signal-to-noise ratio of the receiver should be measured when the a.v.c. characteristic is being determined, to ensure that the performance of the receiver is not seriously impaired when the signal input voltage is increased. Cross modulation tests are also necessary when the receiver performance is being determined. Further discussion of these problems can be found in Ref. 43 (pages 169 and 179). A dual a.v.c. system which has interesting possibilities is described in Ref. 50. Further discussion on a.v.c. and noise is given in Chapter 35 Sect. 3(i).

When a duo-diode triode (or pentode) valve is used with cathode bias, the value of the bias is usually between 2 and 3 volts. Such a voltage is suitable for use as a.v.c. delay bias ; a very simple arrangement is possible by returning R_7 to earth as shown in Fig. 27.34. Alteration of the delay bias because of the presence of contact potentials must be considered, and this effect has already been discussed in (ii) above.

A circuit which eliminates differential distortion, even with large delay voltages and high modulation levels, is shown in Fig. 27.38B. Signal detection and a.v.c. detection are carried out as before, but a diode, D_3, and the three resistors R_8, R_9 and R_{10} are also used. R_9 and R_{10} form a voltage divider from B+ to ground, the junction of these resistors and R_8 having a potential of, say, +50 volts. Typical values for R_4 and R_8 could be 1 megohm and 5 megohms respectively, and it will be seen that R_4 and R_8 form a voltage divider between +50 volts and the source of a.v.c. potential.

The operation of the circuit can be explained most simply if the diode D_3 is ignored initially. Under these conditions, with no signal input to the receiver, point X will be practically at ground potential since there will be no i-f input to the diode D_2 and the voltage at Y will be approximately

$$+ 50 \times \frac{R_4}{R_4 + R_8} = + 8\text{-}1/3 \text{ volts.}$$

Now as an increasing signal is applied to the input of the receiver, the available a.v.c. voltage at X will increase, and when this voltage reaches -10 volts, the potential across the voltage divider $R_4 R_8$ will be 50 $-$ (-10) volts, and the potential at Y will be

$$-10 + (50 + 10) \frac{R_4}{R_4 + R_8} = 0 \text{ volts.}$$

A still larger voltage of -40 volts at X will give at Y

$$-40 + (50 + 40) \frac{R_4}{R_4 + R_8} = -25 \text{ volts}$$

Thus in the absence of the diode D_3, point Y would have a potential with respect to ground which was positive in the absence of a signal and became less positive, zero and then more negative progressively as the signal input increased.

However the diode D_3 modifies this voltage variation by conducting whenever point Y tends to become positive. The result is that under no signal and small signal conditions point Y, from which a.v.c. is applied, is substantially at ground potential but as the signal input increases above some critical value, Y becomes progressively more negative. Thus the requirements of delayed a.v.c. have been met, although the loading of the diode D_2 on the primary tuned circuit is near enough to being constant whether a.v.c. is being applied or not.

After the delay has been overcome, the increase in negative potential at Y is not so rapid as the increase at X, the ratio being the same as that of the voltage divider formed by R_4 and R_8. However by using a large value of resistance for R_8 and a high positive voltage the ratio can be made as close to unity as required by any practical considerations.

The cost of the resistors R_9 and R_{10} can often be avoided by using a source such as the screen of a valve which has no a.v.c. applied, or a voltage divider used for some other purpose, for the positive potential. To save the cost of a valve having an additional diode for D_3 it is quite possible to use the signal diode D_1 as the source of a.v.c. voltages, and D_2 as the " sinking " diode. Alternatively D_1 and D_2 can be used as in Fig. 27.38B and the suppressor grid of say the i-f amplifier can be used as D_3, although in this case with no signal applied the a.v.c. line will take up the potential of the cathode of the i-f valve.

This method of providing delay either to the first stages or to all controlled stages of a receiver is of course equally applicable whether manual or automatic volume control is used

(iv) Methods of feed

The a.v.c. voltage may be fed through the secondary of the r.f. transformer to the grid of the valve (sometimes called " Series Feed ") or directly to the grid (sometimes called " Shunt or Parallel Feed "). In the latter case it is necessary to use a blocking capacitor between the top of the tuned circuit and the grid to avoid short-circuiting the a.v.c. voltage to ground through the comparatively low resistance of the tuned-circuit inductor. Actually the second term is a misnomer, since either circuit arrangement leads to the r-f and d.c. bias voltages being applied, effectively, in series with one another between grid and cathode of the controlled valve. This may not be immediately obvious in the case of so called " parallel feed," but can be readily seen by drawing an equivalent circuit consisting of a r-f voltage generator in series with a capacitor and in parallel with these a resistor in series with a battery. The capacitor charges to some value q because of the battery, and the r-f voltage due to the generator then acts in series with the resultant voltage across the capacitor, varying the total charge and consequently the voltage across the output terminals of the circuit.

It should be noted that the terms " Series Feed " and " Parallel Feed " have no bearing on the type of a.v.c. filter circuit, which may be series, parallel or a combination of both. The two types of a.v.c. feed circuits will now be considered in greater detail.

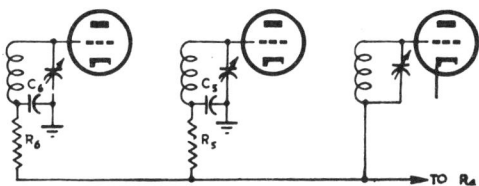

FIG. 27.30

(a) Series feed

One arrangement of series feed is shown in Fig. 27.30. It will be seen that in each r-f tuned circuit a blocking capacitor (C_5, C_6) is used so that the rotor of the ganged tuning capacitors may be earthed and the a.v.c. voltage fed to the lower ends of the coils. In the r-f stages the use of this blocking capacitor will reduce the frequency coverage, and may also affect the tracking of the tuned circuits. To reduce these effects the same capacitance value should be used in each of the 1-f stages, and the value selected should be fairly large consistent with other considerations. Also, if the capacitors are not to affect the Q of the tuned circuits adversely, they must be of a low loss type. A disadvantage of a high capacitance is that it increases the time constant of the a.v.c. circuit.

For typical cases C_5 and C_6 would be about 0.05 μF if the maximum capacitance of the tuning capacitor is about 400 $\mu\mu$F. The effect on the tuning range is easily determined by considering the reduction in total capacitance due to the capacitors being connected in series. Resistors R_5 and R_6 are used in conjunction with the capacitors to provide decoupling between the various stages ; typical resistance values are 100 000 ohms.

Since the tuning capacitors in i-f transformers need not be connected directly to earth, blocking capacitors are not necessary. In these circuits the capacitor C_4 (Fig. 27.29) serves to by-pass all high-frequency components of current from the earthy sides of the secondaries of the tuned circuits of the i.f stages to earth, as well as being part of the a.v.c. filter.

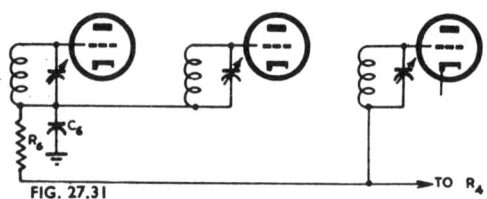

FIG. 27.31

With the r.f. stages an alternative arrangement is to insulate the rotor of the ganged capacitors, and to by-pass it to earth by a single capacitor (C_6, Fig. 27.31). This enables the a.v.c. voltage to be applied without using any blocking capacitors in the r-f tuned circuits, and has the further advantage that the time constant of the a.v.c. circuit may be made small. The arrangement also has obvious disadvantages and is little used.

(b) Parallel feed

The " parallel (or shunt) feed " circuit is shown in Fig. 27.32. In this arrangement a blocking capacitor is necessary to prevent a low resistance d.c. path being formed from grid to earth by the tuning inductor. The resistors R_8, R_9 and R_{10} provide part of the d.c. path from the valve grids back to the cathodes. Since the individual resistors are shunted across their corresponding tuned circuits, the values selected must not be too low as otherwise appreciable damping occurs. A 0.5 megohm resistance in parallel with a typical r-f tuned circuit generally is not serious, but could appreciably reduce the dynamic resistance of an i-f circuit. For these reasons " parallel feed " is sometimes used with r-f stages, but it is seldom applied to i-f stages, and receivers are often designed to use both systems of a.v.c. feed.

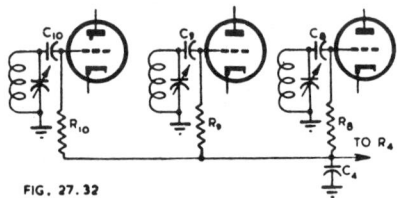

FIG. 27.32

" Parallel feed " for the r-f stages is sometimes more convenient than " series feed," and appears to be satisfactory in most respects, although it has been found that grid blocking is more likely to occur with " parallel feed " than with " series feed."

With any method of feed it is important that the total resistance in the grid circuit should not exceed the maximum for which the valves are rated. Depending on the characteristics of the particular valve type and the effects on electrode dissipations and total cathode current, as determined by equation (6) on page 82, the following maximum values of grid resistor may be used as a general guide.

For one controlled stage   3 megohms
For two controlled stages   2.5 megohms
For three controlled stages   2 megohms

These resistances are to be measured between the grid of any valve and its cathode.
The values above assume that the receiver is normally tuned to a station and that the controlled valve or valves are operating at reduced cathode currents and transconductance, as determined by the a.v.c. bias.

(v) Typical circuits

A typical circuit of a simple a.v.c. system, with three controlled stages, is shown in Fig. 27.33. In order to provide the simplest arrangement the cathode of the duo-diode valve is earthed and grid bias is obtained for the triode section by the grid leak method using a resistor of 10 megohms. This method of biasing reduces the a.c. shunting across the diode load resistor. The cathodes of the controlled stages are normally returned to a point of positive voltage to ensure that the recommended negative grid bias is obtained for conditions where the input signal is very small ; cathode bias is the usual arrangement.

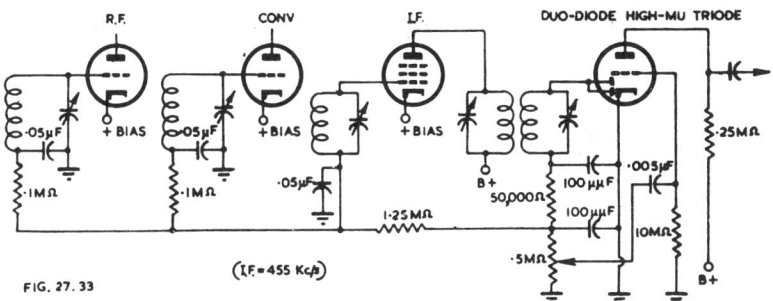

FIG. 27.33   (I.F. = 455 Kc/s)

The circuit values shown are typical for an intermediate frequency of 455 Kc/s. An a.v.c. resistor of 1.25 megohms is used so that the total resistance to earth from any grid does not exceed 2 megohms. If there were only two controlled stages this could be increased to about 1.75 megohms with a consequent decrease in the a.c. shunting.

A typical delayed a.v.c. circuit is shown in Figs. 27.34. Cathode biasing is used for the pentode section of the duo-diode pentode valve, and since the bias will usually be about two or three volts this also provides a suitable a.v.c. delay voltage without any further complication. With circuits using delay voltages on the a.v.c. diode the bias due to automatic volume control on the controlled valves is zero until the peak voltage on the diode exceeds the delay voltage. The controlled stages are arranged to have a self-bias voltage equal to the recommended minimum grid-bias voltage.

(vi) A.V.C. application

Automatic volume control is normally applied to the converter on the A-M broadcast band, irrespective of valve type. On the short-wave band some types of converters give very satisfactory operation with an a.v.c. bias voltage applied to the signal grid, while others introduce difficulties because of the appreciable shift in oscillator frequency caused by the changing signal grid bias. When no r-f stage is used it is often necessary to apply a.v.c. to the converter, but when a r-f stage is incorporated

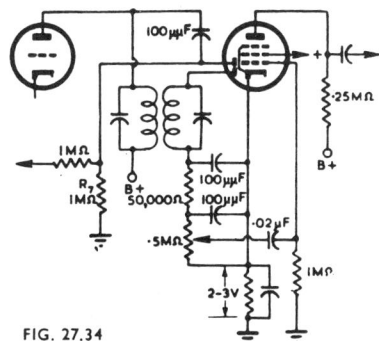

FIG. 27.34

it is frequently advantageous to operate valve types 1A7-GT, 1C7-G, 6A8, 6D8-G on fixed bias. Valve types 6J8-G, 6K8, 6BE6, 6SA7 and all triode hexodes may be used with a.v.c. on A-M broadcast and short wave bands*. For receivers operating in the 88-108 Mc/s F-M broadcast band, a.v.c. bias is not applied to the signal grid of the converter valve ; very often on this band fixed bias is used for the r-f stage, and a.v.c. is applied to the i-f stages only. Whether a.v.c. is used at all with these receivers often depends largely on the type of detector, but it is advisable to avoid overloading of the early stages so as to reduce the possibility of the generation of undesired spurious responses ; this often calls for a.v.c. bias to be applied to the r-f stage.

Some oscillator frequency shift occurs with all types of converters when the signal grid bias is altered. With valve types such as the 6A8 operating on the short wave band the reception of a fading signal is difficult since the variations caused by a.v.c. bias may cause the signal to swing in and out of the pass band of the receiver. An even greater difficulty occurs when the receiver is being tuned to a strong signal, since the magnitude of the output will be different when tuning in from either the high frequency or the low frequency side of the signal. Very careful adjustment is required to obtain the best tuning position. A rather similar effect has been noticed on both the broadcast and short-wave bands, but in this case the effect is generally due to faulty gang wipers. When this is the cause it can occur with any type of receiver (t.r.f. or superheterodyne), but with superheterodynes the oscillator section of the ganged capacitor should be checked first, as this is generally the one causing most of the difficulty.

With ordinary broadcast receivers having a r-f stage, converter, and a single i-f stage, automatic volume control is normally applied to all three stages. If decreased modulation rise is required it is preferable to operate the i-f stage with about one-half of the full a.v.c. bias or, alternatively, to supply the screen voltage by means of a series resistor ; the latter arrangement is the more usual one in practice. Negligible modulation rise is possible, with effective a.v.c. action, if all control is omitted from the i-f stage, but difficulties such as those due to decreased signal-to-noise ratio must not to be overlooked when this system of a.v.c. is used on a number of tuning ranges (see (iii) above).

In a receiver without a r-f stage it is difficult to avoid overloading with large input signals, and a.v.c. is applied to both stages even though modulation rise may be objectionable with very high inputs. In order to obtain maximum control the screen voltage for the i-f valve should be obtained from a voltage divider. The degree of control must, however, be weighed against the possibility of non-linear distortion ; this factor has already been discussed in Chapter 26 in connection with distortion in i-f amplifiers.

For a receiver having two i-f stages the second stage should preferably be operated at fixed bias in order to prevent modulation rise. It is possible that better overall performance at high input signal levels will be obtained when the second i-f stage is operated at a negative bias somewhat greater than the minimum bias. Removal of control from the last i-f valve will often reduce the possibility of overloading this stage, although this is not always a good solution as will be discussed in (xv) below.

(vii) Amplified a.v.c.

There are a number of methods available whereby the a.v.c. voltage to be applied to the controlled stages can be amplified, either before or after rectification.

(a) One or more stages of amplification may be used to form an a.v.c. amplifier channel operating in parallel with the signal channel, and having a separate diode rectifier. If the total gain of the a.v.c. amplifier channel is greater than the total gain of the equivalent section of the signal channel, there is effectively a system of amplified a.v.c. This is more effective than the usual arrangement with a single channel, since it retains the full amplification of the a.v.c. channel under all conditions. It also has advantages in flexibility due to the isolation of the two channels, and the a.v.c. channel may be designed to have any desired selectivity characteristics.

*See also Chapter 25 Sect. 2 pages 990-1001.

(b) If a common i-f channel is used, it is possible to add a further i-f stage with fixed bias for a.v.c. only, followed by a separate a.v.c. rectifier. By this means it is possible to avoid the distortion due to shunting of the diode load resistor, or to " differential loading " at a' point where the a.v.c. delay is just being overcome.

(c) A d.c. amplifier may be used to amplify the voltages developed at. the rectifier. A typical circuit arrangement giving d.c. amplified a.v.c. is shown in Fig. 27.35. Design methods are given in Refs. 44 and 45.

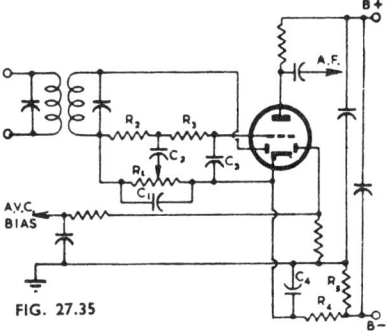

FIG. 27.35

Circuit Arrangement for Obtaining D.C. Amplified A.V.C.

See also Fig. 27.43 and the description given in Sect. 4(ii) later in this chapter.

(viii) Audio a.v.c.

Audio a.v.c. is sometimes used in radio receivers in conjunction with a.v.c. applied to the r-f and i-f stages to flatten out the overall a.v.c. characteristic. Whole or part of the a.v.c. bias voltage is applied to an audio frequency voltage amplifier valve having a grid with a variable-mu characteristic.

It should be clear that conventional a.v.c. systems cannot give a perfectly constant a-f output with a varying input voltage, because conventional a.v.c. is a " back-acting " device in which the effect on the controlled stages must follow the change occurring at a later stage in the receiver. Audio a.v.c. is " forward-acting " and so a constant or even a drooping characteristic is possible for increasing signal input voltages. Simple audio a.v.c. systems tend to introduce a considerable amount of non-linear distortion into the audio output voltage when the signal input voltage is large This is due to the curvature of the $g_m - e_g$ characteristic of the controlled valve.

The amount of distortion can be reduced to negligible proportions by the use of elaborate circuits, but as these systems generally require an additional a-f amplifier valve they are not used to any extent in commercial radio receivers.

If it is decided that audio a.v.c. is desirable, the usual procedure is to obtain the best possible a.v.c. characteristic, apart from the a-f amplifier, and then to add just

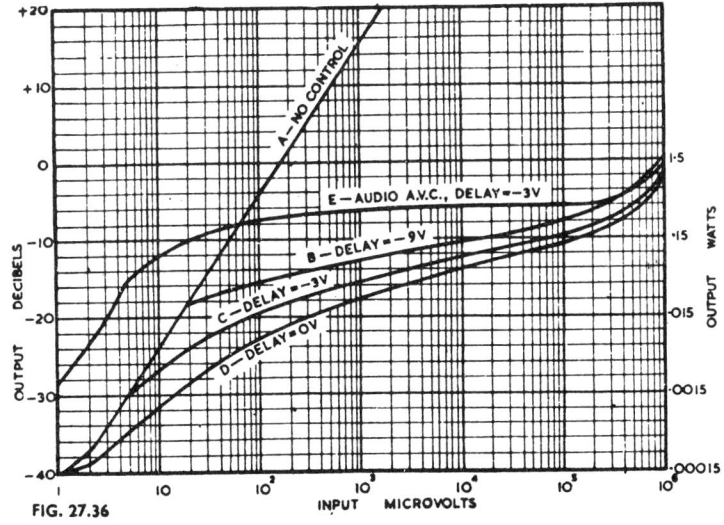

FIG. 27.36

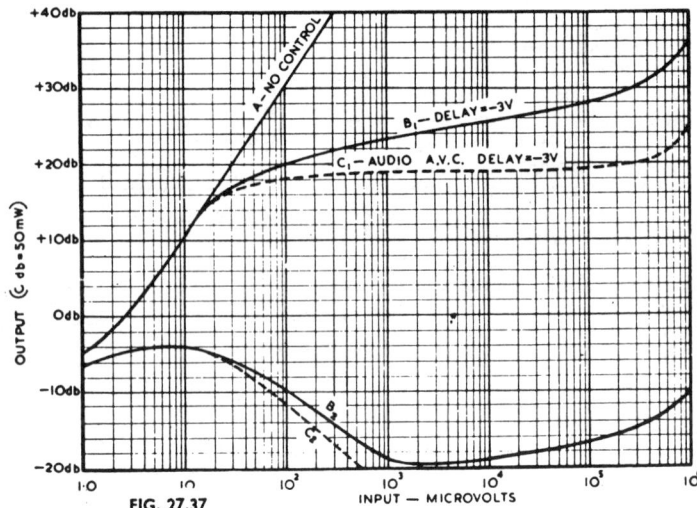

FIG. 27.37

enough a.v.c. bias voltage to the a-f amplifier to give the required overall results. The improvement in the overall a.v.c. characteristics, when audio a.v.c. is used, can be seen from Figs. 27.36 and 27.37.

An example of combined a.v.c. occurs in receivers using a reflexed voltage amplifier stage (see Chapter 28). In this case the a.v.c. bias voltage applied to the valve control grid is effective in controlling both the i-f and a-f gain, and if properly used the system can be made to assist very materially in obtaining good a.v.c. characteristics from a receiver having only one other controlled stage, namely the converter stage in the usual receiver. Another example of a circuit using audio a.v.c. is shown in Fig. 27.45.

(ix) Modulation rise

When a modulated carrier voltage is applied to a voltage amplifier stage in which the valve has a curved $g_m - e_g$ characteristic, the modulation percentage will increase. This modulation rise is noted in the output as audio frequency harmonic distortion (non-linear distortion), mainly second harmonic. Practically all of the modulation rise occurs in the final i-f stage. 20% modulation rise is equivalent approximately to 5% second harmonic. Modulation rise with fixed bias is extremely small even with remote cut-off valves, but there is a slight advantage in using a valve having a sharp cut-off characteristic. Modulation rise may be decreased by operating the i-f stage on a fraction of the a.v.c. voltage, but this adds to the expense of the receiver. Alternatively, a noticeable improvement may be made by supplying the screen from a series dropping resistor connected to B+. This is recommended for circuits having one r-f and one i-f stage. Modulation rise can sometimes be reduced by increasing the gain from the grid of the final i-f valve to the a.v.c. diode. Improving the amount of control on the earlier stages by reducing the screen voltage (this provides cut-off with lower grid voltages) will help in reducing modulation rise, but is likely to lead to difficulties with cross modulation when the receiver is used in close proximity to a powerful transmitter.

(x) A.V.C. with battery valves

When battery valves operating at zero bias are used, it is possible to obtain delayed a.v.c. by incorporating a duo-diode-triode or duo-diode-pentode valve having a diode plate situated at each end of the filament. A delay of between 1 and 2 volts (depending on the filament voltage) is obtainable by this means, and makes a very simple and satisfactory arrangement. The diode at the positive end of the filament is used for a.v.c. and its return is taken to filament negative.   The diode at the negative end of the filament is used for detection and its return is taken to filament positive.

It sometimes happens that the filaments of battery valves are operated in series or series-parallel arrangements.   Under these conditions it becomes more difficult to design an efficient a.v.c. system than when parallel filament operation is used.   In the usual type of a.v.c. circuit the grid returns from the several controlled stages are brought to a common point, and the zero signal grid voltage on these stages is the same. With series or series-parallel operation, however, the filament voltages differ with the result that the zero signal bias on one or more stages may differ from zero by a multiple of the filament voltage.   Circuit arrangements which allow the bias on the grids to be zero for no signal input generally allow only a reduced proportion of the a.v.c voltage developed by the diode to be applied to some of the controlled stages ; full a v.c. bias can only be applied to one or two stages in most cases.   Fig. 27.38A shows a typical circuit arrangement for the case where the filaments are connected in series and operated from the H.T. supply.   The a.v.c. voltage divider is made up from the resistors R_8, R_9, R_{10} and R_{11}.   A value of about 2 megohms is usual for R_{10} while R_{11} is made about 2 megohms per 1.3 volt drop, in the case of 1.4 volt valves (which are operated with 1.3 volts as a recommended value for the series filament connection).   In the circuit shown a higher proportion of the a.v.c. control voltage is applied to valve V_2 than to valve V_3.   The purpose of the resistors R_7, R_2, R_3, R_4 and R_5 is to allow the correct filament voltage to be applied to each of the valves, since it should be obvious that the valve V_1, at the negative end of the chain, carries the total cathode current for all the valves, and each of the other valves carries the cathode currents for all the other valves which are nearer to the positive end of the chain than the valve being considered.   This statement is modified when R_5 is used, because this provides an alternative path for the cathode currents.   From this it follows that no shunting resistor is required for V_5

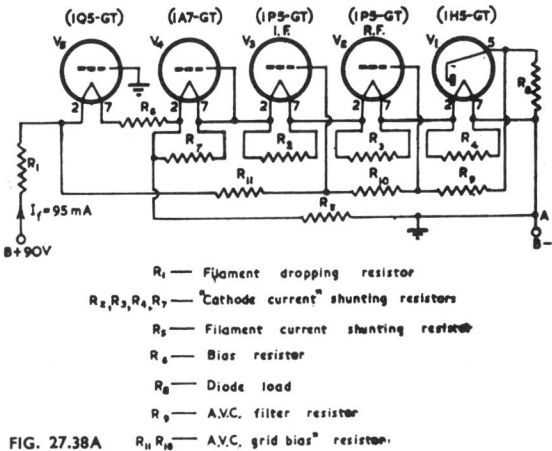

FIG. 27.38A

R₁ — Filament dropping resistor

R₂, R₃, R₄, R₇ — "Cathode current" shunting resistors

R₅ — Filament current shunting resistor

R₆ — Bias resistor

R₈ — Diode load

R₉ — A.V.C. filter resistor

R₁₀ R₁₁ — A.V.C. grid bias resistors

(xi) Special case with simple a.v.c.
If a simple a.v.c. circuit, such as that shown in Fig. 27.28, is used in conjunction with a diode-pentode (or triode) operating with cathode bias, the diode load return will be two or three volts above earth.   It is necessary, in this case, to provide additional negative bias voltage on the controlled stages if the correct minimum bias voltage is to appear between grid and cathode in each case.

(xii) The a.v.c. filter and its time constants
There are several reasons why filter circuits, such as those shown in Fig. 27.33, are required.   One important reason is to prevent r-f and a-f voltages which appear

across the a.v.c. diode load from being applied back to the grids of the controlled stages. If r-f or i-f voltages are applied back to the grids then instability troubles are certain to arise.   A-F applied to the grids can lead to a reduction in the percentage of modulation present on the carrier.   Inter-stage filtering is also necessary because undesired coupling can lead to instability and spurious whistle responses.   From this it can be seen that the prime function of the filter circuits is to allow only the d.c. voltage developed across the diode load to be applied as additional bias on the controlled valves. The components making up the filter also serve the necessary function of completing the r-f and i-f circuits through low impedance paths and, furthermore, the resistors serve to complete the d.c. return path between grid and cathode of each controlled valve so that the correct negative bias voltage can be applied.

It is not permissible to choose values of resistance and capacitance at random if satisfactory circuit operation is to be obtained.   Several conflicting factors must be considered when the component values are being selected.   The time constant (defined below) of the filter network, must be low enough so that the a.v.c. bias voltage can follow the changes in signal input voltage with sufficient rapidity to offset the effects of fading.   Limiting values for the total circuit resistance have been stated in (iv) above, and the choice of suitable values for R and C will be discussed as we proceed, but first the expression " time constant " will be defined.

The " time constant " of a resistance-capacitance network is the time in seconds required for the capacitor to acquire sufficient charge for the voltage between its plates to be equal to 63.2% of the total voltage applied to the circuit.   Alternatively, it is the time taken for a charged capacitor to lose sufficient charge for the voltage across its plates to fall to 36.8% of the initial voltage existing between the plates in the fully charged condition.   It is a simple matter to show that, for the conditions stated, the time constant (T) is given by

$$T = RC \text{ seconds}$$

where R = circuit resistance in ohms
and C = circuit capacitance in farads.

(It is usually more convenient to write

$$T = \text{resistance in megohms} \times \text{capacitance in microfarads} = \text{seconds}).$$

If the total resistance in the circuit is altered in any way from its value during the charging operation to a different value during the discharge of the capacitor, then it is helpful to use the terms " charge time constant " and " discharge time constant." In a.v.c. circuits there is always a difference between the charge and discharge time constants, because during charge the diode is conducting and its conduction resistance effectively short-circuits the d.c. diode load resistor.   This also means that the " charge time constant " is always more rapid than the " discharge time constant " in these circuits.

Suitable values for " charge time constants " are

Broadcast good fidelity receivers	0.25 to 0.5 second
Broadcast receivers	0.1 to 0.3 second
Dual wave or multi-band receivers	0.1 to 0.2 second

For the reception of telegraphy longer time constants are often required to ensure silence between signals, and a value of about 1 second is often used.   In any good quality communications receiver it is usual to provide facilities for selecting any one of a number of a.v.c. time constants.   This enables the operator to select the most suitable condition to offset the particular type of fading being encountered.   Too rapid a time constant is not selected for high quality broadcast reception as rapid fading would cause bass-frequency anti-modulation and so reduce the audio frequency bass response.   See also Chapter 35 Sect. 3(i)B2, page 1233.

In broadcast receivers the charge and discharge time constants are often very nearly equal   However, for some types of reception it may be preferable for the charge time constant to be rapid to prevent the beginning of a signal from being unduly loud, but the discharge time constant is made comparatively slow to prevent a rapid rise in noise output during intervals between signals (see also Ref. 43, p. 181).

The procedure for calculating charge and discharge time constants is as follows. Consider Fig. 27.30 in conjunction with Fig. 27.28, then

For 2 stages
Charge time constant =
$R_1(C_2 + C_4 + C_5) + R_4(C_4 + C_5) + R_5 C_5$
Discharge time constant =
$C_5(R_5 + R_4 + R_2) + C_4(R_4 + R_2) + C_1(R_1 + R_2) + C_2 R_2$.
These follow readily when it is considered that the diode is conducting during charge, and $R_1 \ll R_2$; the diode is non-conducting during discharge.

For 3 stages
Charge time constant =
$R_1(C_2 + C_4 + C_5 + C_6) + R_4(C_4 + C_5 + C_6) + R_5 C_5 + R_6 C_6$.
Discharge time constant =
$C_6(R_6 + R_4 + R_2) + C_5(R_5 + R_4 + R_2) + C_4(R_4 + R_2) + C_1(R_1 + R_2) + C_2 R_2$.
If the circuit of Fig. 27.28 were considered alone the charge time constant would be $R_1(C_2 + C_4) + R_4 C_4$ (assuming $R_2 \gg R_1$), and the discharge time constant $C_1(R_1 + R_2) + C_2 R_2 + C_4(R_4 + R_2)$. The way in which the total time constant is built up for the more complicated cases will be readily seen, and the addition of extra controlled stages should offer little difficulty when the new time constants are to be determined.

For a more detailed discussion of time constants in a.v.c. circuits the reader is referred particularly to Ref. 47.

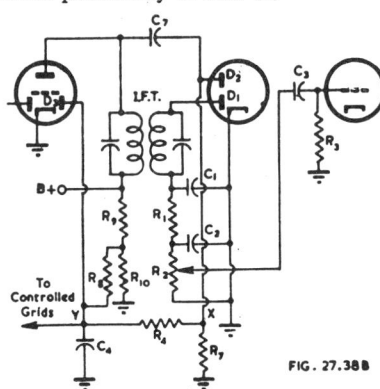

Fig. 27.38B. Circuit for " sinking diode " arrangement which reduces differential distortion.

FIG. 27.38B

(xiii) A.V.C. characteristics

A.V.C. Characteristic Curves may be plotted on 6 cycle log-linear graph paper as shown in Fig. 27.36. The input is usually taken from 1 μV to 1 volt. The output is usually shown in decibels, with an arbitrary zero reference level. The complete curves are useful not only for demonstrating the effectiveness of the a.v.c., but also to indicate modulation rise.

To facilitate the developmental work on the complete receiver it is helpful to draw on the same graph

(a) A curve of distortion against input voltage for 30% modulation at 400 c/s.
(b) A curve of the developed a.v.c. voltage against input voltage.
(c) Curves of the total bias voltages on the controlled stages against input voltage.
(d) A curve of noise output against signal input voltage*. This allows the signal-to-noise ratio, for any input voltage to be determined.

If fixed minimum bias is used, curves (b) and (c) will differ only by the bias voltage. If self-bias is used they will differ by the minimum bias voltage present with no input voltage, and will tend to run together at high input voltages. For methods of conducting the experimental measurements see Chapter 37.

*See also Chapter 35 Sect. 3(i) pages 1229-1234.

Fig. 27.36 shows several a.v.c. curves, each corresponding to a particular condition. In taking these curves two separate diodes were used to maintain constant transformer loading and other conditions. Contact potential in the diode results in a slight increase in the standing bias voltage on the controlled stages ; this effect has been discussed in (ii) above.

Curve A is the " no control " characteristic and is the curve which would be followed, with the a.v.c. removed from the receiver, up to the point at which overloading commences. This characteristic is a straight line and the slope is such that an increase of ten times in the input voltage gives a 20 db increase in output.

Curve B is the a.v.c. characteristic for a delay of −9 volts. For inputs of 3 to 18 μV the experimental curve follows the no control line exactly, and then deviates sharply for inputs above 18 μV. From 18 to 500 000 μV the slope of the curve is fairly constant, the output increasing by about 3.25 db for each 10 times increase in the input voltage. Above 500 000 μV (i.e. 0.5 V) input the curve tends sharply upward, indicating severe modulation rise.

Curve C is the a.v.c. characteristic for a delay of −3 volts. The a.v.c. comes into operation at a lower input voltage, as would be expected, and the average slope is steeper than for the higher delay voltage. In both cases, however, the " knee " of the curve as it leaves the no control line is very clearly defined.

Curve D is the a.v.c. characteristic for a delay of zero voltage, with due compensation for the effect of contact potential on the standing bias of the controlled valves.

Curve E is typical of the characteristics obtained when audio a.v.c. is added to a receiver. Over the range of inputs from 100 to 500 000 μV the total rise in output is only 3 db.

Curve B has been drawn according to the conventional method whereby the output is adjusted to half the maximum undistorted output of the receiver for an input signal of 1 volt. Curves C and D were then taken directly, without any further adjustment to the volume control. Owing to a slight effect on the gain of the receiver when the delay voltage is varied, Curves C and D fall slightly below the datum line at an input of 1 volt. Curve E has been drawn to correspond to Curve C, since both have the same delay voltage. The volume control, however, was advanced considerably for Curve E. It should be noted that no conclusions should be drawn from the relative vertical positions of a.v.c. characteristics drawn according to the conventional method since the volume control settings are unknown.

(xiv) An improved form of a.v.c. characteristic

The conventional method of obtaining a.v.c. characteristics does not give all the information which is available, and an improved method yielding additional data has been suggested by M. G. Scroggie (see Ref. 51). This method will now be described.

Instead of commencing at an input of 1 volt and adjusting the volume control to give one-quarter or one-half of the maximum undistorted power output at 30% modulation, Scroggie's method is to commence at a low input voltage with the volume control set at maximum. The input is increased until the output is approximately one-quarter of maximum, and the volume control is then set back to reduce the power output to one-tenth of the reading. This process is repeated until an input of 1 volt (or whatever is the maximum available from the signal generator if this is less than 1 volt) is reached.

This method is illustrated by the curves shown in Fig. 27.37. The scale of power output represents the output (at 400 c/s) which would be obtained with the volume control at maximum and 30% modulation, provided that no overloading occurred in the audio amplifier. A number of interesting facts may be obtained from an examination of these curves.

1. The residual noise level of the receiver may be shown.

2. The sensitivity of the receiver in microvolts input for any selected output level (e.g. 50 mW or 0.5 W) may be read directly from the curves. It should be noted that the output includes noise.

3. The power output corresponding to any selected input voltage and any position of the volume control may be obtained.   For example, with a delay of −3 volts (curve B1) the delay voltage is overcome at an output level of slightly less than 2 watts with the volume control set at maximum.   As a further example (again assuming 400 c/s and 30% modulation) take the same curve at an input of 1000 μV, where the output is shown as approximately 10 watts with the volume control at maximum. The setting of the volume control to give 4 watts output and using 100% modulation instead of 30% is

$$\sqrt{(30/100)^2 \times (4/10)} = 0.11$$

of the maximum resistance of the control.   The actual amount of rotation will depend on whether the volume control resistance is related to rotation by a linear or a logarithmic law.

4. The voltage at the detector may be calculated from a knowledge of the a-f gain and the detection efficiency.   The a.v.c. bias voltage can also be determined indirectly by calculation from the data obtained.

5. The slope of the initial part of the curve can show up an under-biased valve in the receiver.   Cheap receivers frequently use a common source of bias for two or more valves, and under these conditions it can happen that one of the valves has too little bias and gives less than its maximum gain when there is no signal input to the receiver.   As the input is increased, a.v.c. is applied to each controlled valve and this gives a comparatively rapid increase in gain from the under-biased valve.   Such an increase shows up as an early section of the a.v.c. curve with a slope more vertical than it would otherwise be—usually with a slope in excess of 6 db per octave—and with a noticeable bend towards the horizontal at the point at which the valve receives the bias required for maximum sensitivity.

To operate a valve under these conditions is of course undesirable, and in use the receiver will emphasize the fading of any signals which fade through the range of signal inputs over which the effect operates.

6. The signal-to-noise ratio for any input can be read directly from the graph, as mentioned previously.   In addition, the increments of signal-to-noise ratio with increasing input can be checked   An ideal receiver would give a 20 db improvement of signal-to-noise ratio for each 20 db increase in input, and it is possible for a normal receiver to approach this value, at least over the first decades of the a.v.c. graph. Any significant departure from the ideal can usually be traced to the application of a.v.c. to the first valve in the receiver [see Chapter 35 Sect. 3(i)] and care should be taken to keep the signal-to-noise ratio improving as rapidly as possible until it has reached a value of at least 45 db.

Scroggie's method makes possible the measurement of ratios as great as this, or much greater, because although one reading, the power output, may be beyond the range of the output meter used, the volume control of the receiver is in effect used as a multiplier.

7. The noise curve of the receiver might be expected to decrease indefinitely with increasing a.v.c. voltage, but there are two reasons why it may not do so.   The first is that although the object of the test is to measure the signal-to-noise ratio of the receiver, in practice the signal-to-noise ratio of both the signal generator and the receiver are being measured.   A 50 db ratio between 30% modulation and residual noise with no modulation is a representative figure for a signal generator of reasonable quality, and it is generator noise which causes the noise curve in Fig. 27.37 to become approximately horizontal shortly after 1000 μV input.

The second reason for the noise curve not decreasing indefinitely is that in most receivers in which power supply filtering has been kept to a minimum, for reasons of economy, there is some modulation hum with high inputs.   This hum shows up as an increase in the level of the noise curve, and by plotting the curve over the full range of inputs expected to be applied to the receiver the designer can assure himself that the level of modulation hum does not become objectionable.   An increase in modulation hum can be seen in Fig. 27.37 as the input approaches 1 volt.

The a.v.c. characteristics in Fig. 27.37 can be taken as typical of curves taken by this method. B_1 and B_2 show the signal and noise curves respectively for a 5 valve receiver without audio a.v.c. and C_1 and C_2 show the results for the same receiver with audio a.v.c. It will be seen that the delay is overcome at about 15 μV input and that at say 0.1 volt input there is $8\frac{1}{2}$ db of audio control by the a.v.c. The noise ratio $(B_1 - B_2)$ should be identical with the ratio $(C_1 - C_2)$ since the only difference between the B curve and the C curves is that due to modified audio gain. Thus the curve C_2 would become approximately horizontal at 2000 μV input and at -25 db.

This method enables greater accuracy to be obtained for very small input signals since the power output reading will be well up on the scale of a typical output meter. With the conventional method the output is too small to measure accurately with a standard type of output meter.

At extremely low input voltage levels the noise fluctuations make accurate output measurements very difficult to carry out.

(xv) Design methods

One of the difficulties faced by a receiver designer in making a preliminary calculation of the a.v.c. performance of a receiver is often the lack of complete data on some of the valve types he is expected to use. If a completely satisfactory preliminary design is to be made it is necessary to have available, firstly, data on the manner in which mutual or conversion conductance varies with grid bias voltage ; this data must be for conditions of fixed screen voltage, and also for the case where the screen voltage is applied by means of a series dropping resistor. The second requirement is to know for a given percentage of non-linear distortion the signal handling capabilities (input and output) for various grid bias voltages, of the valves to be used in the controlled stages ; this knowledge is required particularly in the case of the last (and sometimes an earlier stage, as will be discussed below) i-f voltage amplifier valve when a large delay bias is used on the a.v.c. diode. Diode characteristics are also required, but these are usually available. The requirements of valve data are partly fulfilled in some cases, and for later valve types some additional information is available.

Because of the time required to carry out measurements to obtain the additional data, receiver designers, in the majority of cases, make only a few preliminary calculations to determine the desired time constants etc., and then proceed to achieve suitable a.v.c. characteristics by experimental procedures. As an example, with broadcast receivers it is usual to take the cathode bias voltage used for control grid bias on the multi-element valve containing the a.v.c. diode as providing sufficient a.v.c. delay bias (this bias is generally about -3 volts) and with the usual series of valves (e.g. 6SK7, 6J8-G etc.) and transformers a reasonably satisfactory a.v.c. characteristic is obtained ; this assumes that component values are used which are of the order of those discussed in previous sections in connection with Figs. 27.33, 27.34 etc. For the type of receiver where improved a.v.c. characteristics are called for, a fairly large delay bias may be used on the diode. In general, the higher the delay voltage the flatter will be the a.v.c. characteristic (over the range of a.v.c. operation) and this is illustrated by Fig. 27.36. The limitation on the allowable amount of delay bias is set, however, by the magnitude of the signal voltage appearing at the grid and plate of the last i-f amplifier valve, and it is for this reason that, when large delay bias voltage is used, the i-f valve is sometimes operated with reduced or zero a.v.c. bias. However, as this reduces the number of controlled stages it is not necessarily a satisfactory solution, and a better arrangement is to use an additional stage to obtain the necessary control.

Design examples

Attention will be confined mainly to simple delayed a.v.c. systems of the more common types. For additional design information and discussion the reader is referred to Refs. 45, 44, 43 and 53. A good practical approach to the problem is given in Ref. 43.

(1) A receiver is to have one r-f and two i-f stages controlled by a.v.c. ; the converter operates with fixed bias. It is required to estimate, roughly, the diode delay

bias required if the detector output voltage is not to change by more than 6 db for a change in signal input to the aerial stage of 50 db.

Assume first that the a.v.c. and signal detectors are linear, that equal signal voltages are applied to the detector and a.v.c. diodes, and that the detection efficiency (η) is unity. Since stage gain varies with change in mutual conductance (g_m) it can be taken, with typical voltage amplifier valves, that the gain alters by about 1.5 db for each additional volt of negative grid bias (this can be determined with greater accuracy from similar methods to those given later). Since there are three controlled stages using similar valves, the total gain decrease is 4.5 db per volt of bias. (Suppose similar valves are not used, one valve giving a gain change of 2 db per volt of bias and two others giving 1.5 db per volt, then obviously 1 volt of a.v.c. bias change alters the receiver gain by 5 db).

The total decrease in receiver gain which is required equals 50-6 = 44 db. To achieve this the additional a.v.c. bias needed is 44/4.5 = 9.8 volts. That this is also equal to the delay bias required for the stated conditions, can be appreciated because, at the threshold of a.v.c. operation, the voltage on the detector diode will be 9.8 volts peak, and for a further 9.8 volts (peak) increase in signal input the total voltage applied to the detector is 19.6 volts giving an increase in a-f output voltage of 6 db. The effect of 19.6 volts (peak) signal applied to the a.v.c. diode is to produce the required 9.8 volts d.c. bias since half the peak input voltage is required to overcome the delay voltage.

To illustrate a further point, suppose it had been taken that an 80 db input change should only give 6 db output voltage change. Then the delay bias required would be (80-6)/4.5 = 16.4 volts. This also means that the stage preceding the detector must deliver 32.8 volts peak when the carrier is constant. During modulation of the carrier the peak voltage delivered by the last i-f stage can reach 65.6 volts, and if distortion is to be avoided the valve preceding the detector must be capable of delivering this amount of output voltage when the total bias is 16.4 volts plus the standing bias. Whether this is possible can be determined from curves of the signal handling capability of the valve in question, or by direct measurements using a modulated carrier. It will be found, usually, that the maximum output voltage which can be handled without severe distortion is very much reduced when the grid bias voltage is large. This then suggests that reducing the a.v.c. bias on the last i-f amplifier valve will allow the permissible plate voltage swing to be increased. Reducing the bias voltage also results in the stage gain being increased, and so the same a.v.c. bias voltage is produced with a smaller signal input voltage. However, the amount of total control is reduced, and so partial or zero a.v.c. bias on the last i-f valve is not necessarily a satisfactory solution in every case.

It should be observed that the last i-f stage is not necessarily the one in which overloading will occur first. In Chapter 26, when discussing the crystal filter stage, it was pointed out that the grid to plate gain of the valve preceding the filter was very high with the arrangement shown. Suppose the grid to plate gain of the first valve is 400 times, but the gain from the grid of the first to the grid of the second valve (into which the filter feeds) is 5 times, then if the grid to plate gain of the second valve is less than 80 times it becomes immediately apparent that the first i-f valve will be overloaded before the second one. Extreme cases of this type are not likely to occur, but they present possibilities which should not be overlooked.

(2) A detailed design for a single controlled stage is to be carried out from valve data. Simple a.v.c. is to be used.

The valve type selected for this example is the 6SK7 remote cut-off pentode. The average characteristics are shown in Fig. 27.39. The standing bias is −3 volts and the mutual conductance (g_m) for this condition is 2000 micromhos. For a total grid bias of −35 volts the g_m is 10 micromhos. If cathode bias were used some additional error would be introduced, although this is usually neglected.

The first step is to convert the $g_m - E_g$ curve of Fig. 27.39 to a curve showing the change in gain of the controlled stage with variation in grid bias. The actual value of stage gain is not required since it is assumed to be directly proportional to g_m.

The change in g_m is expressed in decibels, taking as the reference level (0 db) the value of g_m for maximum gain, i.e. the g_m with only the standing bias (-3 V) applied to the grid. In this case the g_m used for zero reference level is 2000 micromhos. Then using

$$\text{Change in gain (db)} = -20 \log_{10} \frac{g_m(E_{c1} = -3 \text{ V})}{g_m} = -20 \log_{10} \frac{2000}{g_m}$$

the following values are obtained (the values for g_m are taken from Fig. 27.39).

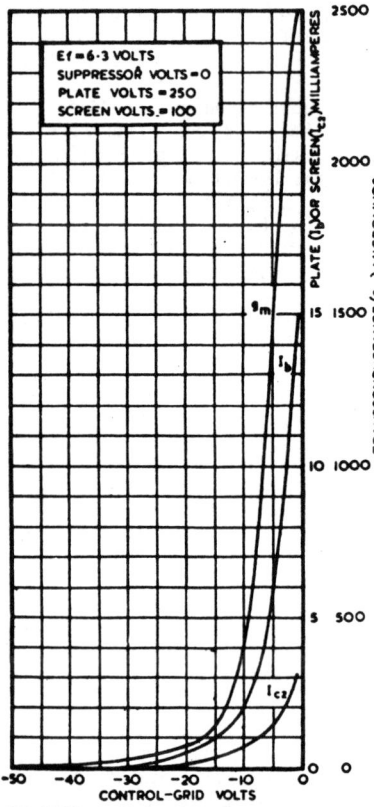

Total Grid Bias*	g_m	Change in Gain (db)
-3	2000	0
-4	1750	$-$ 1.15
-5	1438	$-$ 2.86
-6	1200	$-$ 4.44
-7	925	$-$ 6.69
-8	700	$-$ 9.13
-9	520	-11.7
-10	383	-14.4
-12.5	225	-19
-15	138	-23.2
-17.5	90	-26.9
-20	70	-29.1
-22.5	50	-32
-25	40	-34
-27.5	30	-36.5
-30	25	-38
-32.5	18	-41
-35	10 .	-46

$* = - (3 + \text{a.v.c. bias})$.

FIG. 27.39 6SK7 Mutual Characteristics

These results are plotted in Fig. 27.40 as curve 1. It might be noted that the average slope is 1.4 to 1.5 db per volt.

It is next required to determine the amount of a.v.c. bias voltage developed by the diode circuit, for various input voltages applied to the diode. In a typical case the d.c. diode load resistance would be 1 megohm, and the d.c. voltage developed for various signal input voltages can be taken directly from Fig. 27.6. The developed d.c. voltages are tabulated in column (2) below against r.m.s. signal ($E_{r.m.s.}$) applied to the diode.

$E_{r.m.s.}$ (= signal input to diode)	A.V.C. Bias (volts)	Total bias (= A.V.C. + standing bias)	Diode Signal Input Change (db)
2	$-$ 2.2	$-$ 5.2	0
5	$-$ 6.2	$-$ 9.2	8
8	$-$ 9.1	-12.1	12
10	-11.8	-14.8	14
15	-17.8	-20.8	17.5
20	-23.9	-26.9	20
30	-35.6	-38.6	23.5

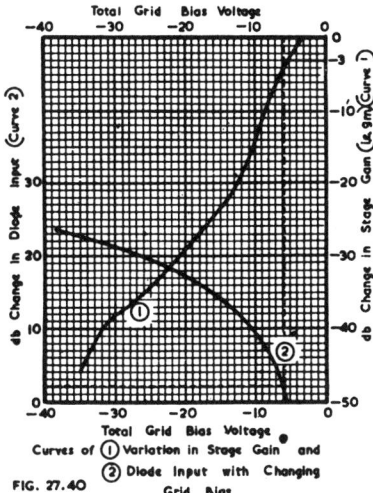

FIG. 27.40

Curves of (1) Variation in Stage Gain and (2) Diode Input with Changing Grid Bias

It is now necessary to relate the total bias (a.v.c. + standing bias) to the **change** in diode input voltage ; the change is conveniently expressed in decibels. Any signal voltage applied to the diode may be taken as zero reference level (0 db), and in our case $E_{rms} = 2$ volts will be convenient. Then the change in diode input voltage is given by,

$$\text{db change in diode input} = 20 \log_{10} \frac{E_{rms}}{E_{rms} \text{ (for total bias } = -5.2 \text{ V)}}$$

$$= 20 \log_{10} \frac{E_{rms}}{2}.$$

The results are tabulated in column (4) of the table given above, and the total bias versus diode signal input change is plotted as curve (2) in Fig. 27.40.

Finally, it is required to find the way in which the output changes with changes in signal input to the amplifier. If we assume various values for the change in signal voltage applied to the diode, it is a simple matter to determine the total grid bias, and from this the change in stage gain, by using curves (2) and (1) Fig. 27.40. Also from the curves, it is seen that, with zero change in diode input (0 db), the amplifier stage gain is 3 db down on the maximum possible gain. (This is because we selected $E_{rms} = 2$ volts as being the zero reference level). If now we assume, say, 2 db change in diode input, the actual change in stage gain is $-4.3 - (-3) = -1.3$ db. Because voltage input applied to the diode has increased 2 db, and the amplifier stage gain has decreased 1.3 db, it follows that the signal voltage applied to the amplifier grid must have increased $2 + 1.3 = 3.3$ db.

The direct output voltage change across the detector load, for 3.3 db change in signal input, is from -2.2 volts to -2.9 (i.e. from curve (1), Fig. 27.40, the total bias voltage change is from -5.2 to -5.9 volts, which correspond respectively to -2.2 volts and -2.9 volts across the diode load, when the standing bias is -3 volts). Expressing the direct output voltage change in decibels, we have

$$\text{Output change (db)} = 20 \log_{10} \frac{\text{a.v.c. bias}}{\text{a.v.c. bias for } E_{rms} = 2 \text{ V}}$$

$$= 20 \log_{10} \frac{\text{a.v.c. bias}}{-2.2}$$

From this we have for 3.3 db change in signal input voltage (2 db change in diode input) a corresponding output change of 2.4 db. The complete a.v.c. characteristic is tabulated below and plotted in Fig. 27.41.

Change in diode input (db)	A.V.C. Bias Voltage	Change in Signal Input (db)	Change in diode output (db)
0	− 2.2	0	0
2	− 2.9	3.3	2.4
4	− 3.5	6.5	4.0
6	− 4.5	10.6	6.2
8	− 6	16.8	8.7
10	− 7.5	22	10.6
12	− 9.1	27.2	12.3
14	−11.8	33.9	14.6
16	−15	40	16.7
18	−18.5	46	18.5
20	−23.9	52.8	20.7
22	−29.5	60	22.5
22.8	−32	65.8	23.2

It is seen that a change of 60 db in signal input gives an increase of 22.5 db in output. It should also be observed that in most cases it can be taken that the diode input is directly proportional to diode output, and the additional set of calculations need not be made—e.g. 60 db change in signal input gives 22 db change in diode input or 22.5 db change in diode output. For three similar controlled stages the output change would be reduced to 7.5 db. If the valves to be controlled are not of the same type, then individual input-output curves must be drawn and the results combined to give the complete a.v.c. characteristic.

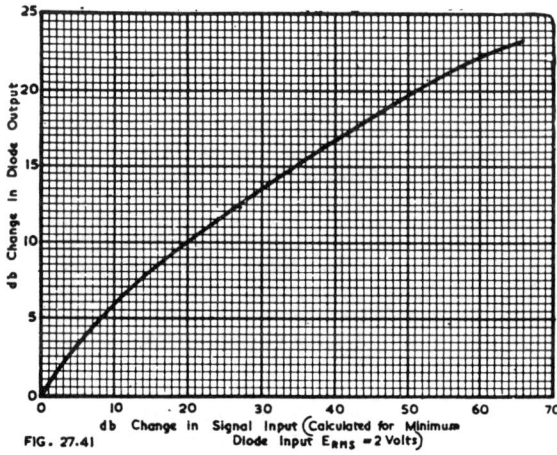

FIG. 27.41

Fig. 27.41. Calculated a.v.c. characteristic for a single controlled stage using type 6SK7.

The a.v.c. characteristic for diode input voltages smaller than $E_{rms} = 2$ volts can be readily plotted, but as the useful range of signal input voltages, where control is mainly required, is usually greater than this value in a reasonably sensitive receiver (and because of the limitations of the diode curves) this value has been taken as affording a satisfactory example.

The variation in gain given by simple a.v.c. is limited, and for very strong signals a local-distance switch or some other method to prevent overloading of the receiver is required.

(3) The voltage amplifier and diode detector used for the previous example are to be used in a delayed a.v.c. circuit, in which the delay bias applied to the diode is −10 volts.

If they are not available, the diode characteristics relating r.m.s. input voltage to d.c. output voltage, for various delay bias voltages, can be measured directly. The rectified voltage, used for a.v.c. bias, is not completely independent of modulation when a delay bias is used, and there is an increase in the available d.c. when the percentage modulation is increased. Usually the a.v.c. comes into operation for lower carrier input voltages than would normally be expected, particularly when the percentage of modulation approaches 100%.

The complete design procedure now follows that given previously for simple a.v.c. using the data derived by the method given in the last paragraph. It should be clear that curve (1) in Fig. 27.40 remains unchanged, but curve (2) must be replotted. The resultant data are then available for plotting the a.v.c. characteristic. For signal input voltages of which the peak value does not exceed the diode delay bias, the receiver operation is the same as for one which does not incorporate a.v.c. (this is clearly shown in Fig. 27.36).

The slope of the no control line is readily determined since it is taken that the output is directly proportional to the input.

SECTION 4 : MUTING (Q.A.V.C.)

(i) General operation (ii) Typical circuits (iii) Circuits used with F-M receivers

(i) General operation

When a sensitive receiver which incorporates automatic volume control is being tuned from one signal to another there is an objectionable increase in noise output. This effect can be overcome by using a circuit arrangement which will make the receiver silent in the absence of a signal. Arrangements of this nature are known by various names such as muting, interstation-noise suppression, quiet automatic volume control (Q.A.V.C.), or squelch systems.

It is desirable, with these circuits, for the receiver to come into operation with the weakest possible useful signal voltage. Since the useful signal voltage may be variable for different conditions of reception, communications receivers are usually fitted with a variable threshold control which can be adjusted to give the best results. For broadcast receivers variable external controls are not used, the muting range being pre-set by the manufacturer.

Most systems of muting depend for their operation on the application of a large bias voltage to an i-f amplifier valve, the detector or an a-f amplifier valve. A wide variety of circuits has been used by receiver designers, but attention will be confined here to a few typical arrangements, since the general principles are much the same in all cases.

(ii) Typical circuits

(a) Biased diode detector

A simple type of biased detector muting system is shown in Fig. 27.42. It is seen that the circuit arrangement for the detector is quite conventional except that the diode load is returned to a tap on the cathode resistor. This makes the diode plate negative with respect to cathode, by the voltage developed between the cathode and the tapping point on the bias resistor R_k and so detection cannot occur until the peak signal input voltage exceeds this negative bias voltage. The components are marked to correspond to those in Sect. 1 of this Chapter.

It should be apparent that there will be a range of signal input voltages around the threshold point over which the audio distortion will be severe.

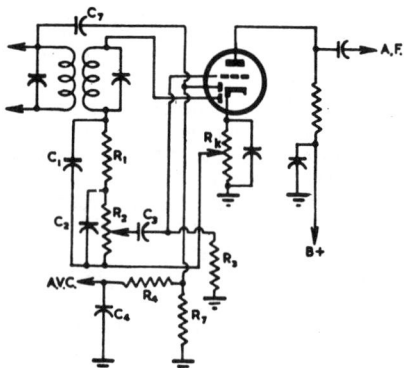

FIG. 27.42 Biased Detector Muting System

A more elaborate muting system, giving improved results over the simple system of Fig. 27.42 has been described by K. R. Sturley (Ref. 44).   This uses a negatively biased diode and at the same time the circuit is arranged to provide d.c. amplified delayed a.v.c.   The operation of this circuit can be readily understood from Fig. 27.43. The diode marked D_3 acts as the automatic volume control detector, and the d.c. voltage developed across the load resistor R_2 is applied to the grid of the triode section of V_3 and amplified.   The diode D_4 will only come into operation when the cathode of V_3 becomes negative with respect to the diode plate, and the a.v.c. bias is delayed until this condition occurs.   Considering now the double diode V_2, it is seen that D_1 acts as the detection diode, but it can only detect when the signal input voltage exceeds the negative bias voltage developed across the cathode circuit R_k, C_k.   This negative

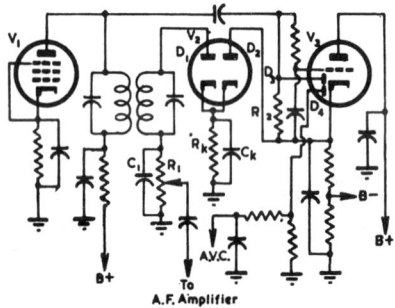

FIG. 27.43 Biased Detector Muting System
with
D C Amplified & Delayed A.V.C.

bias voltage only exists when the diode D_2 is conducting, which occurs when the cathode of V_3 is positive.   As soon as the cathode of V_3 becomes negative, the a.v.c. comes into operation, the diode D_2 no longer conducts, and the negative bias is removed from the plate of the detection diode D_1 which then begins to function in the usual manner.   Part of the detected voltage output will appear across R_k, C_k but this is not serious if the value of R_k is very much less than the resistance of R_1 (say about one tenth or less).   It is recommended that $C_k R_k$ be made equal to $C_1 R_1$.

(b) Inoperative audio amplifier

A typical arrangement of this type (Ref. 57) is shown in Fig. 27.44. V_1 is any double diode, such as type 6H6, and V_2 is a pentagrid mixer, such as the type 6L7, in which grid 1 has a remote cut-off and grid 3 a sharp cut-off characteristic. Some systems use two valves in place of V_2 to effect muting and audio frequency amplification. Improved systems are also available (Ref. 54) which use a duo-diode pentode to eliminate the separate double diode valve, the pentode section being used as the intermediate frequency voltage amplifier. Audio a.v.c. is readily applied in this latter case.

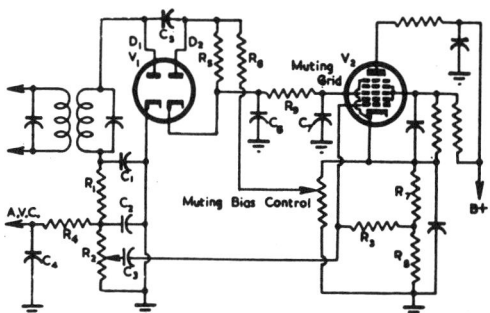

FIG. 27.44   Inoperative Audio Amplifier Muting System

From Fig. 27.44 it is readily seen that the diode detector and simple a.v.c. arrangements are quite conventional. The audio output from the detector is applied to grid 1 of the valve V_2. Only part of the cathode bias is applied to grid 1 by using a tap on the cathode resistor. In the absence of conduction through the diode D_2 there is a large negative bias applied to grid 3 due to the total cathode bias voltage developed across R_7 and R_8 in series. The d.c. path to grid 3 is via R_6, R_5 and R_9. R_6 has a high value of resistance to prevent the plate of D_2 from being connected to ground through the comparatively low impedance of the cathode circuit of V_2 (which includes the muting control as shown). The circuit made up from R_9, C_6 and C_7 is merely a filter to allow the application of direct bias voltage only to the muting grid (3) of V_2; however, R_9 also performs the useful function of stabilizing the voltage applied to grid 3 in the event of the anti-muting bias being sufficiently large to cause grid current. R_5 is the diode load resistor for D_2, and C_5 allows the application of i-f voltage to the plate of the diode D_2. When a signal is received the voltage developed across R_5 is sufficiently large, and of the correct polarity, to reduce the total negative bias on grid 3 and so allow the valve V_2 to perform the function of voltage amplifier for a-f voltages applied to grid 1.

An interesting arrangement (Ref. 54) is shown in Fig. 27.45, in which a duplex-diode pentode and its associated circuit combine the functions of muting, detection, conventional and audio a.v.c., and audio frequency amplification. The valve V_1 is a duo-diode pentode in which grid 1 has a remote cut-off characteristic (the circuit was developed around the Australian-made type 6G8-G). The operation of this circuit is somewhat different from those previously discussed. The maximum gain obtainable from a resistance-capacitance coupled audio frequency amplifier occurs when the negative grid bias has a particular value. For a negative bias less than the optimum value (and for a particular screen voltage, which is generally made fairly low), g_m will drop and the stage will effectively be muted. When the negative bias is made larger than the optimum value the gain of the stage will decrease at a comparatively slow rate. This principle is used in the circuit of Fig. 27.45 and the signal, muting, and audio a.v.c. voltages are all applied simultaneously to the control grid. The cathode resistors R_{10} and R_{11} are used to set the bias voltage to give the optimum value for maximum gain. R_{11} is then adjusted to a lower value to reduce the total bias voltage, and to cause the gain to drop to such a low value that muting has effectively occurred. It is necessary to select the values of bias and screen voltage so that the maximum screen dissipation is not exceeded. When a signal voltage is applied to the a.v.c. diode, the rectified voltage developed across R_7 increases the grid bias in a negative direction which increases g_m (and so the stage gain) and the valve un-mutes. For further increases in signal voltage the a.v.c. comes into operation, and the gain of the audio stage (and also the other controlled stages) begins to fall off in the usual manner. It is seen that only part of the total a.v.c. bias is applied to the grid of V_1.

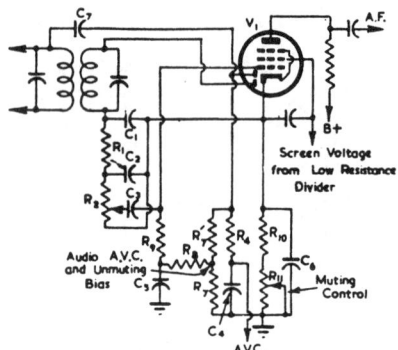

FIG. 27.45 Single Valve Arrangement to Provide Muting, Audio A.V.C., Detection and A.F. Amplification

(iii) Circuits used with F-M receivers

The general principles for muting circuits used in F-M receivers are much the same as for the circuits previously discussed. However, many of the circuits are novel and a few typical arrangements will be discussed.

One typical arrangement (Ref. 59) is shown in Fig. 27.46(A). Valve types V_1, V_2 and V_3 are those normally employed for limiting and detection. Valve V_4 is a double triode with separate cathodes such as the type 6SN7-GT. Section (1) is used as a d.c. amplifier for the muting control bias voltage which is applied to the grid of Section (2). The triode of Section (2) is used as an audio frequency voltage amplifier in the usual manner.

The operation of the circuit is as follows. In the absence of a signal the limiter valve V_2 draws full plate current, and the positive voltage applied to the cathode of triode (1) of V_4 (from the voltage divider R_1, R_2, R_3) is a minimum. At the same time there is a positive voltage on the grid of (1) which is sufficiently large to ensure that there is a resultant positive voltage on the grid with respect to cathode. This allows the plate of (1) to draw full current and makes point A negative with respect to point B across the resistor R_5, and so a negative bias is applied between grid and cathode of triode (2). This bias is made sufficiently large to cut off the plate current of (2) and so prevents any a-f output from being obtained.

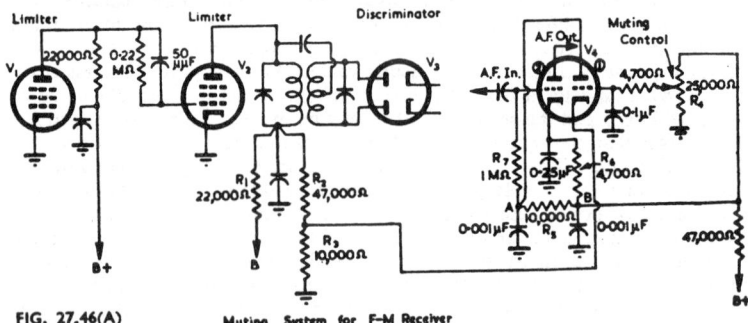

FIG. 27.46(A)   Muting System for F-M Receiver

When a signal voltage is impressed on the limiter V_2 the plate current is reduced, and so the positive voltage applied to the cathode of triode (1) (V_4) is increased.   When the signal input is sufficiently large the positive cathode voltage will exceed the positive voltage on grid (1) by an amount which is sufficient to cut off the plate current.   This puts the points A and B at the same potential and so removes the negative bias from the grid (2).   Triode (2) of V_4 then acts as an ordinary a-f voltage amplifier, and its normal bias is obtained from the voltage drop across the cathode resistor R_6.

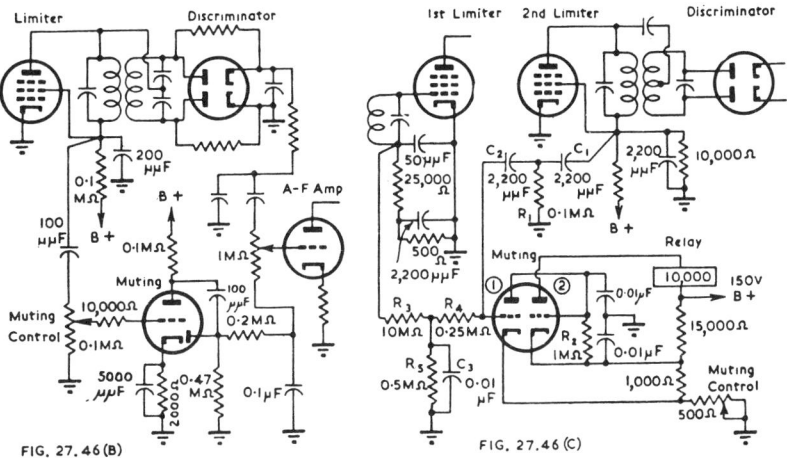

FIG. 27.46(B)   FIG. 27.46 (C)

Fig. 27.46(B) shows a simple but effective arrangement for muting.   Noise voltages, in the absence of a signal, which appear in the screen circuit of the second limiter valve, are applied to the control grid of the triode section of the muting valve (e.g. type 6AV6) and the amplified noise voltage is applied to the diode detector circuit. The d.c. voltage across the diode load is applied as additional negative bias to the grid of the a-f voltage amplifier valve and cuts off this stage.   When a signal is received the noise voltages are reduced and so the additional bias on the a-f valve falls and the stage begins to function in the normal manner.

Another useful arrangement is shown in Fig. 27.46(C).   In this circuit noise voltages from the screen circuit of the second limiter are applied to the grid of the muting valve (which could be a type 12AT7) via the high pass filter $C_1\,C_2\,R_1$.   Triode section (1) operates as an anode bend detector and the direct voltage developed across R_2 cuts off triode section (2) ;   the relay is then in the unoperated condition.   When a signal is received the noise voltages are reduced due to the receiver quieting, and at the same time additional bias is applied to triode (1) by the rectified voltage appearing at the first limiter grid.   The voltage across R_2 is now insufficient to cut off triode (2) and the relay is operated by the plate current.   The relay can operate a contact to short the grid of the output valve to earth : additional contacts can also be utilized should the receiver be used for special purposes.   The circuit can readily be re-arranged so that the relay is not required, and an additional control bias is made available to cut off the a-f voltage amplifier by increasing the voltage in the cathode circuit in the absence of a signal.   The coupling network $R_3\,R_4\,R_5\,C_3$ has a suitably selected time constant, and the circuit arrangement is such as to prevent amplitude modulation or over deviation of the frequency modulated signal from muting the receiver.

Both of the circuits of Fig. 27.46(B) and (C), or variations of them, have been widely used in f-m mobile and V.H.F. link receivers.   In some cases additional negative bias is applied to both the a-f voltage amplifier valve and the a-f output valve ; this has the advantage in the case of mobile receivers that the battery drain is reduced very appreciably during stand-by periods since a large proportion of the battery drain is due to the H.T. current drawn by the output valve.

If muting ON-OFF is required it is usual to switch an additional resistor into the cathode circuit of the muting valve. This resistor is made sufficiently large to cut off the valve and so render the muting inoperative.

A number of other useful circuit arrangements can be obtained from Refs. 58 and 59. Many circuits have appeared in the patent literature, but have not been generally published. One of particular interest uses the two limiters (which are resistance-capacitance coupled) as a multivibrator unit, the output of the second valve being coupled back to the input of the first by means of resistance and capacitance connected in series. In the presence of signal input voltages the relatively low amplification between the valves prevents the feedback network from being effective. In the absence of a signal the grid biases are such that noise voltages will be amplified sufficiently until the circuit operates as a multivibrator. One of the two valves is then always in a non-conducting condition and there is no audio output. This system is described by J. A. Worcester in G.E. Patent Docket 68743. Another useful circuit is given by R. A. Peterson in R.C.A. Patent Docket 21,998. This incorporates d.c. amplification and also requires the use of an additional double diode with separate cathodes. The diodes rectify noise output (which contains A-M) from the limiter. The d.c. voltage developed across the diode load is amplified and applied as a muting bias on the a-f amplifier valve. Unmuting occurs in the presence of a signal by using a negative voltage from the limiter grid to oppose the rectified voltage appearing across the diode load.

Muting circuits are generally applied only to F-M receivers using the limiter and discriminator combination. The noise level between stations with receivers using ratio detectors or locked oscillator arrangements is generally fairly low, and muting is not usually considered to be so necessary in these cases. However, it is a relatively simple matter to utilize the rectified d.c. voltage available at the output of a ratio detector (or also in a conventional discriminator, see Ref. 62 p. 120) to operate a muting system, since the rectified d.c. voltage is proportional to the strength of the incoming signal.

SECTION 5 : NOISE LIMITING

(*i*) *General*   (*ii*) *Typical circuit arrangements.*

(i) General

The purpose of a noise (or crash) limiter in a radio receiver is to assist in the reduction of noise pulses, such as those due to ignition interference and crashes of static, so that their effects may be minimized at the receiver output. The difference between muting systems and noise limiters should be carefully distinguished, as they perform quite different functions in connection with the type of noise 'which they are meant to eliminate.

Noise limiters in general follow two trends. One group " punches a hole " in the signal so that the receiver output is momentarily cut off. The other main group functions by limiting the maximum output to a value which is not appreciably greater than some pre-determined level.

It can be shown fairly readily (Ref. 63) that the best results for noise reduction can be obtained by placing the limiter in the receiver at a point of low selectivity. The Lamb silencer (Ref. 65) for example, is usually placed after the first i-f stage and its operation is such as to cut off the plate current in the second i-f amplifier valve, momentarily.

For many purposes elaborate noise limiting circuits (Refs. 65, 71) are not essential, and the tendency in the usual communications receiver is to use simple diode limiting to remove noise peaks. For amateur work sufficiently good results are often possible using audio output limiters (see Refs. 66, 67, 68). With the A-M receivers used in mobile communications systems a very elaborate noise limiter may be incorporated.

See also Chapter 16 Sect. 6 for speech clippers and Sect. 7 for noise peak and output limiters.

The threshold of operation for noise limiters can be arranged for either automatic or manual setting for different signal input voltages.   Good results are possible using automatic setting, but the more elaborate circuits practically always make some provision for manual control.

(ii) Typical circuit arrangements

The Lamb silencer can be made to give excellent results, but because it is rather elaborate it is not extensively used in commercial communications receivers.   For a description of its operation, together with complete circuit data, the reader should consult Ref. 65.

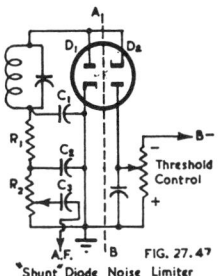

Shunt Diode Noise Limiter

FIG. 27.47

Fig. 27.47 shows a typical example of the " shunt " type of diode limiter.   It is seen that the circuit to the left of the dotted line AB is the usual diode detector circuit and the component values are identical with those normally selected.   The negative direct bias voltage is set to prevent the diode D_2 from conducting until the peak input voltage exceeds the bias.   When this occurs the diode D_2 acts as a virtual short circuit, and there is practically no output from the detector circuit.

A " series " type of diode limiter is shown in Fig. 27.48. The diode D_1 is used for detection in the usual manner. Diode D_2 is biased to be conducting with normal signal voltages applied to the detector circuit.   For bursts of noise there is an instantaneous negative voltage applied to the plate of D_2, and when this negative voltage exceeds the positive bias the diode stops conducting and so opens the audio output circuit.   A large amount of audio gain is required when this type of circuit is used, because of the low audio voltage available.

These simple circuits have several disadvantages.   Firstly they are both susceptible to hum because of heater-cathode leakage when the cathode of D_2 is above earth. For example this limits the size of the resistor R which can be used in the circuit of Fig. 27.48, and so results in a loss in audio output.   (An increase on the value shown would be permissible however).   A further disadvantage is that stray capacitances will allow some of the high frequency noise components to appear in the receiver output, even though the diodes are operating in the prescribed manner ; this can be largely overcome with care in layout and wiring to minimize stray capacitances.

Automatic threshold control may be arranged with simple " series " and " shunt " limiters.   Suitable methods are given in the references (particularly Refs. 62A, 64 and

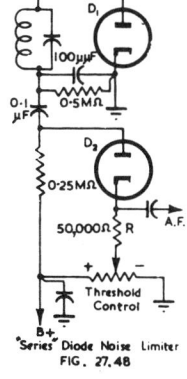

Series Diode Noise Limiter
FIG. 27.48

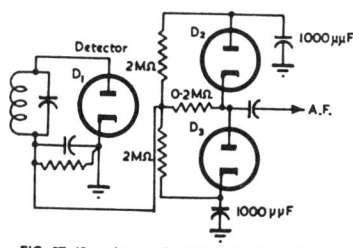

FIG. 27.49   Automatic Noise Limiting Circuit

70). However, a much more satisfactory circuit (Ref. 44) is shown in Fig. 27.49. The diode D_1 is incorporated in the usual detector arrangement and this circuit provides a variable bias voltage for the noise limiting diodes D_2 and D_3. A short burst of noise voltage having a positive polarity will cause the diode D_3 to conduct and so flatten out the audio waveform. A noise voltage having negative polarity causes D_2 to conduct which then flattens out the negative peak of the audio output. In the absence of noise the positive and negative half cycles are equally damped and so distortion is reduced.

Additional methods used for noise limiting in radio receivers are given in Chapter 16 Sect. 7 ; methods of noise reduction for use with reproduction from records are described in Chapter 17 Sect. 7.

SECTION 6 : TUNING INDICATORS

(i) Miscellaneous (ii) Electron ray tuning indicators (iii) Null point indicator using Electron Ray tube (iv) Indicators for F-M receivers.

(i) Miscellaneous

A tuning indicator is a device which indicates, usually by means of a maximum or minimum deflection, when a receiver is correctly tuned. These indicators have taken a number of forms, particularly before the advent of the Electron Ray tuning indicator, originally called the " Magic Eye," and a few of these will be briefly discussed before considering the Electron Ray Tube in detail.

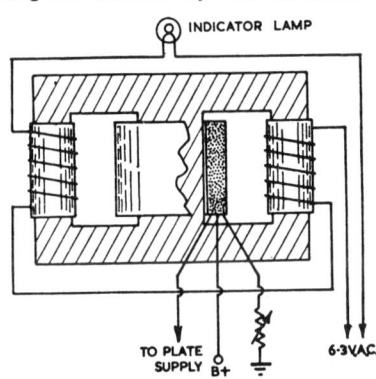

FIG. 27.50   Saturated Reactor Tuning Indicator

One arrangement is to use a milliammeter connected in the plate circuit of a voltage amplifier valve which is controlled by a.v.c. Circuits of this type can also be arranged to indicate relative signal strength, and the meter scale may be calibrated in arbitrary " S " units. Another form is the saturated reactor or dimming lamp indicator. This uses a pilot lamp, in series with a special form of iron-cored inductance, which is excited from a suitable winding on the power transformer. A second winding on the inductance (or transformer) carries the plate current of one or more valves which are controlled by a.v.c. In this simple form the maximum plate current is sufficient to saturate the core and so reduce the impedance in series with the pilot lamp, which is then at full brilliance. When a signal is being tuned the plate current is reduced and the lamp dims. In order to make the pilot lamp reach full brilliance when the receiver is tuned to a station instead of dimming, a " bucking " current can be passed through another winding so that saturation occurs when the plate current is very small, instead of it occurring when the plate current is large. Both arrangements can be understood by studying Fig. 27.50.

A further arrangement is to use a neon tube in which the length of the illuminated column is proportional to the d.c. voltage obtained from a resistance connected in series with the plate of an amplifier valve controlled by a.v.c.

(ii) Electron Ray tuning indicators

Electron Ray tuning indicators are the most popular of all the arrangements being considered. These indicators operate from either the signal diode or the a.v.c. diode circuit. The Electron Ray tube is not limited to use in receivers with a.v.c., since it operates to indicate a change of voltage across any part of the circuit. For example it may be connected to the signal diode circuit irrespective of the presence of a.v.c., or it may be connected across the cathode bias resistor of an anode bend detector. The correct type of Electron Ray tube to be selected for any position depends on the controlling voltage available.

The most popular types of Electron Ray tube (e.g. 6U5/6G5) have a triode amplifier incorporated in the same envelope, so that the voltage necessary to obtain full control is decreased. In most types this amplifier has a remote cut-off characteristic so that the sensitivity may be high for weak signals and yet not cause " overlapping " on strong signals. Type 6E5 has a linear characteristic and is occasionally used for special applications. Type 6AF6-G has two independent Ray-Control Electrodes, one of which may be used for strong signals and the other for weak signals, but has no amplifier incorporated in the same envelope.

In a typical receiver with delayed a.v.c. the Electron Ray tube may be connected to either the signal or the a.v.c. circuit, but since it will give no indication until the diode starts to conduct, it will not operate on weak signals when connected to a delayed a.v.c. circuit. Consequently most receivers using delayed a.v.c. employ the signal diode circuit for operating the tuning indicator. The disadvantage of this latter arrangement is that it introduces additional a.c. shunting effects on the detector diode load resistance and so leads to increased distortion at high modulation levels (see Sect. 1(i) above). The a.c. shunting effects are minimized by using a high resistance in the grid circuit of the tuning indicator, a value of about four times the diode load resistance being typical. In order to prevent flicker due to modulation, adequate decoupling between the diode load and the indicator grid is necessary. A capacitance of about 0.05 μF is usually connected from grid to ground, and this, in conjunction with the grid resistor previously mentioned, should lead to satisfactory operation (see Fig. 27.51). If simple a.v.c. is used (i.e. no delay voltage) it is preferable to operate the tuning indicator from the a.v.c. line to reduce a.c. shunting (see again Fig. 27.51). From the diagram it can be seen that with the switch in the " Det." position the grid of the tuning indicator is connected through a 2 megohm resistor to the detector diode circuit. With the switch in the " a.v.c. " position the indicator is connected directly to the a.v.c. line. This shows the two alternative connections just discussed. The cathode of the tuning indicator is returned to a suitable tapping point on the cathode bias resistor of the power valve as discussed below.

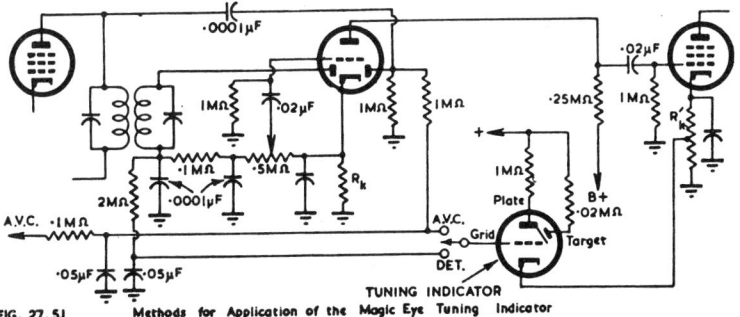

FIG. 27.51   Methods for Application of the Magic Eye Tuning Indicator

The cathode of the Electron Ray tube should be, as closely as possible, at the same potential as the cathode of the diode.   If its cathode is more negative than that of the diode, grid current may occur thereby increasing the initial bias on the controlled stages and reducing the sensitivity of the receiver.   Consequently if the diode cathode is earthed, the indicator tube cathode should also be earthed, but if the diode cathode is positive then the indicator tube cathode should also be positive by an approximately equal amount.   One satisfactory method of obtaining this positive voltage, which however may only be used with a Class A power valve, is to connect the cathode of the Magic Eye to a tapping on the cathode bias resistor of the power valve.   With this arrangement it is advisable for the tapping to be adjusted to make the cathode of the Electron Ray tube about 0.5 volt less positive than that of the diode in order to allow for contact potential in the indicator tube, the " delay " due to this small voltage being negligible.

Alternatively, the cathode return of the Magic Eye may be taken to a suitable tapping point on a voltage divider across the " B " supply.   Due to the fairly heavy and variable cathode currents drawn by the older types of indicator valves it is essential that the voltage obtained from any voltage divider, or other source of voltage, should not be affected appreciably by a current drain of from 0 to 8 mA.   It is for this reason that it is not satisfactory to tie the cathode of the indicator tube to that of the diode. With the newer " space charge grid " construction the cathode currents remain more nearly constant throughout life, and this allowance for change of current need not be made.   However, it is not advisable to base calculations for cathode bias resistors on the published values of cathode currents since in some cases with valves of the newer construction, these are higher than the average currents.

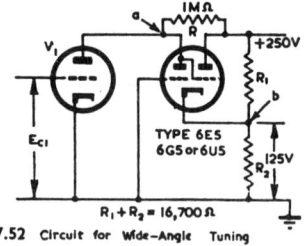

FIG. 27.52 Circuit for Wide-Angle Tuning

Overlapping of the two images is possible on very strong signals, whatever type of indicator tube is selected, but it is generally found with a remote cut-off type (such as the 6U5/6G5) that this is not often experienced under field conditions.   Certain arrangements have been devised to reduce the tendency to overlapping, but none is free from criticism.   Desensitization of the tuning indicator is readily applied, but affects indications on weak signals.   The use of two separate tuning indicators, or a single type 6AF6-G with two separate amplifiers, one for weak and one for strong signals, is excellent but expensive.   If the grid of the indicator tube is excited from the moving contact of the volume control the deflection will depend upon the setting of the control, and " silent tuning " will not be possible.

Wide Angle Tuning with a maximum angle of 180° is practicable if an external triode amplifier is added (see Ref. 73).   With this circuit (Fig. 27.52) the edges of the pattern are sharp from 0° to about 150° to 180°.

(iii) Null point indicator using Electron Ray tube

A Magic Eye tuning indicator may be used in many applications as an indicating device, one of these being as a null point indicator for use with a.c. bridge circuits. Such an arrangement is preferable to the use of head phones or sensitive instruments, since it may be used without disturbance from external noises and is capable of withstanding considerable overload without damage.   The sensitivity of the type 6E5 is 0.1 volt (r.m.s.) for a very clearly marked indication.   When used as a null point indicator the 6E5 grid is biased approximately 4 volts negative, and the a.c. voltage is applied between grid and a cathode.   Any suitable pre-amplifying stage may be used to increase the sensitivity of the device if desired.   When an a.c. voltage is applied the sharp image will change to a blurred half-tone and as the null point is reached the image will again become sharp.   A heavy overload may cause overlapping of the pattern, but this is not detrimental to the tube.

(iv) Indicators for F-M receivers

The usual types of electron ray tuning indicator such as the 6U5/6G5 and 6E5, have been adapted for use as tuning aids in F-M receivers, and several systems using these tubes will be described   A special electron ray indicator designed for use in F-M receivers has been described by F. M. Bailey (Ref. 76), and has been commercially released as type 6AL7-GT.   Data on the application of the type 6AL7-GT can be found in Ref. 77 and in the R.C.A. HB-3 Tube Handbook.

A method that immediately suggests itself for connecting a tuning indicator to a F-M receiver using a limiter is shown in Fig. 27.53.   This method of obtaining a control voltage, from the limiter grid circuit, for operating the indicator tube is by no means the best arrangement available.   The disadvantage of the circuit is that, when a strong signal is being tuned-in, the maximum voltage at the limiter grid is not sharply defined, and exact tuning is almost impossible under this condition. The resistor R and capacitor C are used to provide decoupling, as the presence of a.c. components of voltage at the indicator grid will prevent the fluorescent pattern from being sharply defined.

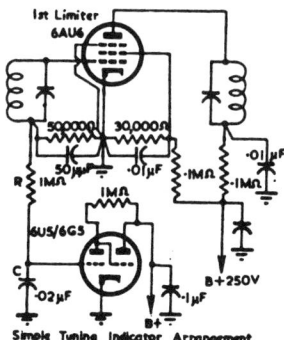

FIG. 27.53   Simple Tuning Indicator Arrangement for F-M Receiver

A much more satisfactory tuning arrangement is shown in Fig. 27.54.   This is similar to the Philco Circuit shown in Ref. 75 but the two components R and C have been added.   The additional resistor appears to be necessary if the d.c. voltages applied to the diode plates of the balanced rectifier are to be equal when the receiver is detuned by an equal amount above and below the centre frequency.   Actual adjustment of R could be made to achieve more exact symmetry, but it is probable that in most cases the tuning operation would not be seriously impaired by leaving R and C out altogether.   The operation of the circuit is straight forward.   Assume firstly that the receiver is tuned to the centre frequency.   The balanced rectifier has no voltage output, and the negative voltage developed across the resistor in the limiter grid is applied to the control grid of the tuning indicator, this voltage causing the pattern to close.   When the receiver is detuned there is a voltage developed across R_L, of the polarity shown, due to the operation of the balanced rectifier.   This voltage plus that developed in the limiter grid are added algebraically and the resultant voltage is applied between grid and cathode of the tuning indicator.   Since the net bias voltage will always be less than that obtained when the limiter voltage alone is applied (the correct tuning position) the pattern on the tuning indicator opens.   The indications given by this circuit are sharply defined, and in addition the indicator opens and closes in the same manner as for indicator circuits used in conventional A-M receivers.

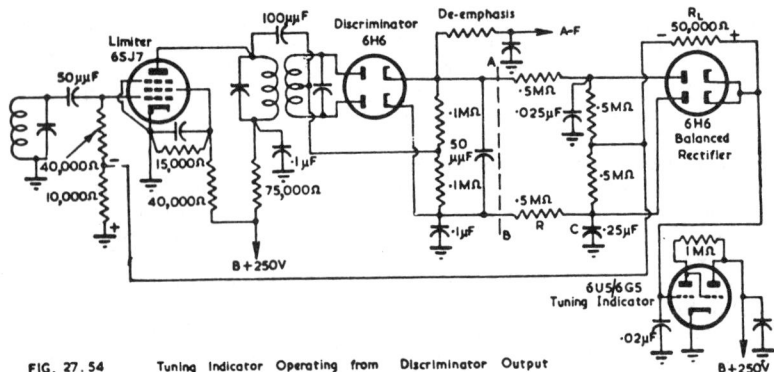

FIG. 27. 54   Tuning Indicator Operating from Discriminator Output

This latter feature is not available with similar arrangements operating from the discriminator output alone.

Tuning aids are sometimes used in receivers with ratio detectors. The simplest arrangement in this case is to take the control bias voltage from the a.v.c. take-off point, or to use whatever fraction of this total voltage that is thought to be necessary for operating the tuning indicator.

SECTION 7 : CRYSTAL DETECTORS

(*i*) *Old type crystal detectors* (*ii*) *Fixed germanium crystal detectors* (*iii*) *Fixed silicon crystal detectors* (*iv*) *Theory of crystal rectification* (*v*) *Transistors*.

(i) Old type crystal detectors

Crystal detectors of the " catswhisker " type have been used for many years, but are very touchy, require frequent adjustment, and are affected by even slight vibration. However, when correctly adjusted, they make quite efficient detectors. Crystal sets are briefly mentioned in Chapter 34 Sect. 1(ii).

(ii) Fixed germanium crystal detectors

These have the advantages of small size and of needing no heater supply, thereby avoiding the possibility of introducing hum into high impedance circuits. Their main disadvantages are that the impedance presented to negative voltages is comparatively low and dependent upon the applied voltage, and that the characteristics from unit to unit show comparatively large variations.

Reverse resistance in a typical case varies from 2 megohms with 20 volts applied to 0.2 megohm at about 100 volts. In some circuits, e.g. a F-M ratio detector, this reverse conductance affects the operation and modifications are required to obtain satisfactory performance. The forward resistance is very low e.g. 200 ohms with one volt applied, falling below 100 ohms at somewhat higher voltages for a typical unit—and this feature can be valuable in some applications.

In this type of rectifier (see Ref. 79 from which this description is taken) as distinct from silicon rectifiers, the outstanding advantage lies in an ability to handle large voltages in the reverse direction, while in the forward direction, the slope (measured at +1 volt) may be of the order of 100 ohms (10 mA at 1 volt).

The slope in the reverse direction reaches a maximum resistance value at the order of −2 volts, beyond which the resistance falls slowly until, at a particular voltage known as the " turnover voltage," the slope resistance falls to zero and then becomes negative.

Measurements of rectification efficiency at various frequencies up to 100 Mc/s indicate that the efficiency falls with frequency by an amount which depends upon the " turnover voltage," the higher the " turnover voltage " the lower the rectification efficiency. In view of this, crystals intended for use at very high frequencies have a maximum as well as a mimimum " turnover voltage " rating.

A typical current/voltage characteristic is shown in Fig. 27.55.

Rectification occurs at the junction between a metallic point and the surface of crystalline germanium. During manufacture this junction is treated to obtain optimum impedance characteristic and time stability. After assembly, the metal point is cemented to the germanium to prevent dislodgment by vibration, and the complete assembly is sealed to prevent ingress of moisture.

A characteristic of fixed germanium crystal detectors is their remarkable property of withstanding severe mechanical shock or vibration. The estimated life of the rectifier is indefinite—in excess of 10 000 hours.

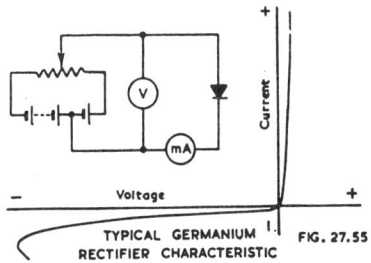

TYPICAL GERMANIUM   FIG. 27.55
RECTIFIER CHARACTERISTIC

There is little change in rectifier characteristics from 15°C to 50°C, but above this temperature up to 100°C, both forward and back slope resistances decrease slowly.

The characteristics of germanium crystal diodes vary considerably, depending on the application and the manufacturer, but the following data are representative of the majority of types.

Allowable direct current	20 to 60 mA.
Allowable surge current	100 to 600 mA.
Allowable reverse voltage	25 to 250 volts
Turnover voltage (reverse voltage for zero dynamic resistance)	40 to 275 volts
Minimum current with +1 volt	2.5 to 15 mA (some are less than 2.5 mA).
Reverse current with −50 volts applied (this only applies to types having an allowable reverse voltage of 50 volts)	40 to 1660 μA
Capacitance	0.8 to 3.0 μμF.
Maximum ambient temperature	+70° to +85°C.
Minimum ambient temperature	−40° to −55°C.
Maximum frequency	up to several hundred Mc/s.

References to germanium detectors : 78, 79, 80, 81, 83, 84, 85, 85A, 85B.

(iii) Fixed silicon crystal detectors

The fixed silicon crystal detector is principally used as a frequency converter at frequencies above 100 Mc/s. Some designs go up to 1000 Mc/s, while others go as high as 5000 or 10 000 Mc/s. The silicon detector is liable to damage by transient voltage overloads in the reverse direction. The conversion loss is about 6 to 8 db.

References to silicon detectors : 82, 83.

(iv) Theory of crystal rectification

The modern theory of crystal rectification is given in Ref. 83 : See also Ref. 79.

(v) Transistors

Transistors are crystal devices with three or more electrodes, which are capable of amplifying. A good introductory article is Ref. 86. See also Refs. 87, 88, 89, 90, 91, 92, 93, 94, 95, 96, 97.

SECTION 8 : REFERENCES

(A) A-M DETECTORS
(1) Sturley, K. R. (book) " Radio Receiver Design " Part 1 (Chapter 8), Chapman and Hall, London, 1943 ; Wiley and Sons, New York, 1943.
(2) Everitt, W. L. (book) " Communication Engineering " (Chapters 13 and 14) McGraw-Hill, New York and London, 1937.
(3) Zepler, E. E. (book) " Technique of Radio Design " (Chapters 4 and 11) Chapman and Hall, London, 1943 ; Wiley and Sons, New York, 1943.
(4) Preisman, A. (book) " Graphical Construction for Vacuum Tube Circuits " (Chapter 6) McGraw-Hill, New York and London, 1943.
(5) Terman, F. E. " Radio Engineer's Handbook " (Section 7) McGraw-Hill, New York and London, 1943.
(6) " Radiotron Receiver RC52 " Radiotronics No. 117 Jan./Feb. 1946.
(7) " Radiotron Receivers RC41 and RC42 " Radiotronics No. 118 Mar./April 1946.
(8) " Distortion in diode detectors " Radiotronics No. 73 (24th Feb. 1937) 11 ; Radiotronics No. 74 (31st March 1937) 20.
(9) Varrell, J. E. " Distortionless detection " W.W. 45.5 (3rd Aug., 1939) 94.
(10) Amos, S. W. " The mechanism of leaky grid detection " Electronic Eng. (1) 17.198 (Aug. 1944) 104 ; (2) 17.199 (Sept. 1944) 158.
(11) " New linear detector with high input impedance " A.R.T.S. and P. Bulletin No. 30 (22nd July 1936).
(12) " Reflex detector application " A.R.T.S. and P. Bulletin No. 32 (21st Sept. 1936).
(13) Healey, C. P., and H. A. Ross, " The linear reflex detector " A.W.A. Tec. Rev. 3.1 (July 1937) 1.
(14) Corfield, D. N. and R. V. Cundy " Valve Technique " Incorporated Radio Society of Great Britain, London, April 1948.
(15) Weeden, W. N. " New detector circuit " W.W. 40.1 (1st Jan. 1937) 6.
(16) Robinson, H. A. " Regenerative detectors " Q.S.T. 17.2 (Feb. 1933) 26.
(17) " Reaction circuits " W.W. 46.2 (Dec. 1939) 40.
(18) Alder, L. S. B. " Threshold howl in reaction receivers " Exp. Wireless and Wireless Engineer 7.79 (April 1930) 197.
(19) Zepler, E. E. " Oscillation hysteresis in grid detectors " W.E. 23.275 (Aug. 1946) 222.
(20) " Radio Amateur's Handbook " (Chapter 7, p. 146) 1947 Edition, A.R.R.L., Connecticut.
(21) " Cathode Ray," " Super-regenerative receivers " W.W. 52.6 (June 1946) 176.
(22) Stockman, H. " Superregenerative circuit applications " Elect. 21.2 (Feb. 1948) 81.
(23) Frink, F. W. " The basic principles of superregenerative reception " Proc. I.R.E. 26.1 (Jan. 1938) 76.
(24) Easton, A. " Superregenerative detector selectivity " Elect. 19.3 (March 1946) 154.
(25) Bradley, W. E. " Superregenerative detection theory " Elect. 21.9 (Sept. 1948) 96.
(26) Hazeltine, Richman and Loughlin " Superregenerator design " Elect. 21.9 (Sept. 1948) 99

(B) F-M DETECTORS
See also bibliography covering A.F.C. given in Chapter 29 for further discriminator circuit data.
(27) Sturley, K. R. " The Phase Discriminator—its use as a frequency-amplitude converter for F-M reception " W.E. 21.245 (Feb. 1944) 72.
(28) Sturley, K. R. (book) " Radio Receiver Design " Part 2, Chapters 13 and 15, Chapman and Hall, London, 1945 ; Wiley and Sons, New York, 1945.
(29) Hund, A. (book) " Frequency Modulation " Chapters 2 and 4, McGraw-Hill Book Co., New York and London, 1942.
(30) Lampitt, R. A. " The frequency discriminator " Elect. Eng. 17.200 (Oct. 1944) 196.
(31) Tibbs, C. " Frequency Modulation 5—demodulation : theory of the discriminator " W.W. 49.5 (May 1943) 140.
(32) Maynard, J. E. " Coupled circuit design " (covers phase discriminator circuit) Comm. 25.1 (Jan. 1945) 38.
(33) Arguimbau, L. B. " Discriminator linearity " Elect. 18.3 (March 1945) 142.
(34) Seeley, S. W., and J. Avins, " The Ratio Detector " R.C.A. Rev. 8.2 (June 1947) 201.
(35) " Ratio Detectors for F-M receivers " Radiotronics No. 120 (July/Aug. 1946) 79.
(36) " Balanced phase shift discriminators " Radiotronics No. 120 (July/Aug. 1946) 83.
(37) Sandel, B. " F-M receiver for the 88-108 Mc/s Band " Radiotronics 125 (May/June 1947) 39 ; Radiotronics 127 (Sept./Oct. 1947) 79. (Correction, Radiotronics 132 (July/Aug. 1948).)
(38) Maurice, D. and R. J. H. Slaughter " F.M. reception-comparison tests between phase discriminator and ratio detectors " W.W. 54.3 (March 1948) 103.
(39) Kiver, M. S. (book) " F-M Simplified " Chapters 7 and 13, D. Van Nostrand Co. Inc., Toronto, New York and London, 1947.
(40) Bradley, W. E. " Single-stage F-M detector " Elect. 19.10 (Oct. 1946) 88.
(41) Loughlin, B. D. " Performance characteristics of F-M detector systems " Tele-Tech, 7.1 (Jan. 1948) 30.
(42) " F-M detector systems " A.R.T.S. and P. Bulletin No. 139 (Reprint Hazeltine Bulletin and covers practically the same data as Ref. 41.)
(42A) Loughlin, B. D. " The theory of amplitude - modulation rejection in the ratio detector " Proc. I.R.E. 40.3 (March 1952) 289.

(C) AUTOMATIC VOLUME CONTROL
(43) Zepler, E. E. (book) " Technique of Radio Design " (Chapter 7) Chapman and Hall, London, 1943 ; John Wiley and Sons, New York, 1943.
(44) Sturley, K. R. (book) " Radio Receiver Design " Part 2 (Chapter 12) Chapman and Hall, London, 1943 ; John Wiley and Sons, New York, 1943.
(45) Cocking, W. T. " The design of a.v.c. systems " W.E. 11.131 (Aug. 1934) 406 ; 11.132 (Sept. 1934) 476 ; 11.133 (Oct. 1934) 542.
(46) Sturley, K. R. " Distortion produced by delayed diode a.v.c." W.E. 14.160 (Jan. 1937) 15.
(47) Sturley, K. R. " Time constants for a.v.c. filter circuits " W.E. 15.180 (Sept. 1938) 480.
(48) James, E. G. and A. J. Biggs " A.V.C. characteristics and distortion " W.E. 16.192 (Sept. 1939) 435.
(49) Sturley, K. R. and F. Duerden " D.C. amplified a.v.c. circuit time constants " W.E. 18.216 (Sept. 1941) 353.
(50) Cocking, W. T. " A.V.C. developments " W.W. 46.2 (Dec. 1939) 51.
(51) Scroggie, M. G. " The a.v.c. characteristic " W.W. 44.18 (4th May 1939) 427.
(52) Moore, J. B. " AGC—noise considerations in receiver design " Elect. 18.5 (May 1945) 116.
(53) Amos, S. W. " A.V.C. calculations " W.W. 53.2 (Feb. 1947) 46.
(54) Watson, S. J. " Some circuits for interstation muting and audio a.v.c." A.W.A. Tec. Rev. 4.4 (1939 but issued Feb. 1940) 139.

(D) MUTING (Q.A.V.C.)
References 54 and 44 above.
(55) Reyner, J. H. " Practical QAVC circuits " W.W. 37.13 (27th Sept. 1935) 348.
(56) Reyner, J. H. " Delayed detector operation " W.W. 28.15 (10th April 1936) 364.
(57) Healy, C. P. and A. L. Green " A muting system for receivers " A.W.A. Tec. Rev. 3.2 (Oct. 1937) 62.
(58) Delanoy, F. " Notes on the design of squelch circuits " Radio 30.10 (Oct. 1946) 12.
(59) Carnahan, C. W. " Squelch circuits for F-M receivers " Elect. 21.4 (April 1948) 98.
(60) Worcestèr, J. A., G. E. Patent Docket 68743 (30th Jan. 1943).
(61) Peterson, R. A., R.C.A. Patent Docket 21,998 (19th March 1943).
(62) Kiver, M. S. (book) " F-M Simplified " (pages 117-120) D. Van Nostrand Co., Inc., Toronto, New York, and London, 1947.
(62A) Toth, E. " Noise and output limiters " Elect. 19.11 (Nov. 1946) 114 ; 19.12 (Dec. 1946) 120.

(E) NOISE LIMITING
Ref. 44, p. 206.
(63) Wald, M. " Noise suppression by means of amplitude limiters " W.E. 17.205 (Oct. 1940) 432.
(64) " Noise suppression " A.R.T.S. and P. Bulletin No. 120 (16th March 1942).
(65) Lamb, J. J. " A noise silencing i.f. circuit for superhet. receivers " Q.S.T. 20.2 (Feb. 1936) 11.
(66) Robinson, H. A. " Audio output limiters for improving the signal-to-noise ratio in reception" Q.S.T. 20.2 (Feb. 1936) 27.
(67) Dawley, R. L. (Editor) " Radio Handbook " (11th edition) Editors and Engineers Ltd., California, 1947 and later editions.
(68) " The Radio Amateur's Handbook " (25th edit.) American Radio Relay League Inc. 1948 and later editions.
(69) Dickert, J. E. " New automatic noise limiter " Q.S.T. 22.11 (Nov. 1938) 19.
(70) Grisdale, G. L. and R. B. Armstrong " Tendencies in the design of the communication type of receiver " Jour. I.E.E. Part 3, 93.25 (Sept. 1946) 365.
(71) Nicholson, M. G. " Comparison of amplitude and frequency modulation " W.E. 24.286 (July 1947) 197.
(72) Weighton, D. " Impulsive interference in amplitude modulation receivers " Jour. I.E.E. Pt. 3, 95.34 (March 1948) 69.

(F) TUNING INDICATORS
(73) R.C.A. Application Note No. 82. Reprinted in Radiotronics No. 86, p. 128, May 1938.
(74) R.C.A. Receiving Tube Manual RC15.
(75) Kiver, M. S. (book) " F-M Simplified " (pages 109-117) D. Van Nostrand Co. Inc., Toronto, New York and London, 1947.
(76) Bailey, F. M. " An electron-ray tuning indicator for frequency modulation " Proc. I.R.E. 35.10 (Oct. 1947) 1158.
(77) " Tube News " Service 16.3 (March 1947) 14.

(G) CRYSTAL DIODES
(78) R.M.A. releases 1N34 etc.
(79) " Germanium crystal diodes " (leaflet published by Sylvania Electric).
(80) Engineering News Letters Nos. 1 to 7 inclusive (Sylvania Electric, Electronics Division, 500 Fifth Avenue, New York, N.Y.).
(81) " Germanium crystal rectifiers " (leaflet published by The General Electric Co., Ltd., London).
(82) " Silicon crystal rectifiers type CS " (leaflet published by British Thomson-Houston, Rugby, England).
(83) Torrey, H. C., and C. A. Whitmer (book) " Crystal Rectifiers " M.I.T. Radiation Laboratory Series, McGraw-Hill Book Co. 1948.
(84) Lovelock, R. T., and J. H. Jupe " Germanium diodes " W.W. 57.2 (Feb. 1951) 57.
(85) Osbahr, B. F. " Characteristics of germanium diodes " Tele Tech 9.12 (Dec. 1950) 33.

(H) TRANSISTORS
(86) Scott, T. R. " Crystal triodes " Proc. I.E.E. Part III 98 (May 1951) 169 ; reprinted in Elect. Comm. 28.3 (Sept. 1951) 195. Gives extensive bibliography.
(87) James, E. G., and G. M. Wells "Crystal triodes" Jour. Brit. I.R.E. 12.5 (May 1952) 285.

Additional references will be found in the Supplement commencing on page 1475.

CHAPTER 28

REFLEX AMPLIFIERS

By F. Langford-Smith, B.Sc., B.E.

SECTION 1 : GENERAL DESCRIPTION

(i) Description   (ii) Advantages and disadvantages of reflex receivers.

(i) Description

A reflex amplifier is one which is used to amplify at two frequencies—usually intermediate and audio frequencies.

Reflex radio and audio frequency stages have also been used in very small T.R.F. receivers (see Sect. 5).

Reflex receivers were fairly common in U.S.A. and Australia in the period 1934 to 1937, but suffered from serious distortion and high play-through. These receivers usually had high a-f plate load resistors giving comparatively high a-f gain, low i-f gain and low operating plate voltages and currents. A.V.C. was not applied to the reflex stage which was frequently a sharp-cut-off valve such as the 6B7 or 6B8. A considerable advance was made in Australia by the use of the remote cut-off type 6G8-G with a.v.c., and a further advance was made in the adoption of a comparatively low a-f plate load resistor. The latter enabled nearly full i-f gain to be obtained with normal voltages on the electrodes, thus increasing the maximum plate voltage swing for a limited distortion. The most common application of a reflex amplifier in Australia* at the present time is in a 3/4 valve receiver comprising converter, reflex stage, power amplifier and rectifier. The reflex stage in such a receiver amplifies at intermediate frequency, provides detection and a.v.c. from its diode or diodes, and then amplifies at audio frequency. It may be compared with a straight receiver in which the second valve is used as i-f amplifier and detector, the output from the detector being used to excite the power stage. Using the same valves and components in both cases, the reflex receiver may have a sensitivity up to 10 times that of the straight set.

(ii) Advantages and disadvantages of reflex receivers
Advantages of the reflex receiver

1. Higher sensitivity.

2. Greater flexibility in design, permitting use of negative feedback either with or without tone compensation (bass boosting) for small cabinets.

---

*Reflex receivers do not appear to be in commercial use in either U.S.A. or Great Britain at the time of writing.

Disadvantages of the reflex receiver

1. Increased cost due to additional components.
2. Increased tendency to overload on strong signals.
3. More complicated design, although once a satisfactory design has been evolved there are no outstanding production difficulties.
4. Somewhat increased distortion at high modulation percentages at medium and high input levels than for a well-designed straight receiver, although both are still in the same class.
5. Play-through—that is the occurrence of a-f output with the volume control at its zero setting. This leads to a further defect known as the "minimum volume effect" whereby minimum volume from the receiver is obtained with the volume control at some setting slightly above zero setting At the point of minimum volume the signal is very badly distorted since there is a balancing-out of the fundamentals between the normal signal and the out-of-phase play-through signal. Both play-through and the minimum volume effect may be reduced by good design so as not to be objectionable. See Sects. 2, 3 and 4. Perhaps a fairer comparison is between a 3/4* valve reflex receiver and a 4/5 valve straight set. In this case the reflex receiver will be more economical to produce, through the saving of one valve, while there will be a slight saving in space, heat dissipation and current consumption. The 4/5 receiver is capable of being designed to have negligible play-through and may have slightly greater sensitivity.

The design of reflex receivers must necessarily be a compromise, with juggling of inter-acting characteristics. However, by reflexing it is possible to achieve an acceptable performance which would not otherwise be obtainable with the same number of valves. The fact that reflex receivers form a substantial percentage of the Australian market for 3/4 valve sets indicates that their performance is acceptable.

On account of the complicated inter-action which occurs in a reflex receiver, it is impossible to design the reflex stage alone ; the whole receiver must be treated as a single design unit. If any change is subsequently made in any other stage, it is usually necessary to make consequential modifications to the reflex stage for the best results.

So far as sensitivity is concerned, a straight 3/4 receiver using modern valves can give 15 μV broadcast sensitivity†, and 30 μV short-wave sensitivity†, which for most applications is adequate. Nevertheless the higher a-f gain of the reflex receiver removes the problem, which is experienced with straight 3/4 sets, of preventing the a.v.c. from limiting the a-f output before the output valve is fully loaded.

The advantages to be obtained at audio frequencies from reflexing a 3/4 valve receiver are considerable. A straight 3/4 valve set without an a-f stage almost invariably suffers from "bubbling" at high a-f outputs on strong signals, and to prevent this the low frequencies are reduced as much as possible without making the reproduction too thin. With the additional stage of a-f amplification provided by reflexing, the regenerative effect which causes bubbling becomes degenerative, and the bass can be boosted to any desired extent without bubbling being experienced.

In addition it is not possible in straight 3/4 valve sets to use negative feedback for any purpose other than mild frequency correction because any reduction in the mid-frequency gain of the output stage makes i-f overloading probable before the output valve is driven to full output. With reflexing an appreciable amount of the added a-f gain can be used for any desired form of negative feedback.

Reflex superhet. receivers may be divided into two principal groups, those employing plate reflexing (see Sect. 3), and those employing screen reflexing (see Sect. 4).

---

*A 3/4 valve receiver is one having three amplifying valves and a rectifier.
†For 50 milliwatts output.

SECTION 2 : SOME CHARACTERISTICS OF REFLEX SUPERHET. RECEIVERS

(i) Play through (residual volume effect)   (ii) Over-loading   (iii) Automatic volume control   (iv) Reduction in percentage modulation   (v) Negative feedback   (vi) Operating conditions of reflex stage.

(i) Play-through (residual volume effect)

This effect has been briefly described in Sect. 1.   Play-through in a reflex receiver is due to the rectification caused by the curvature of the valve dynamic characteristics. Play-through increases as the input signal is increased, and may be measured with a large signal input—preferably with a high modulation percentage and with the volume control at zero.   During design it may be found that play-through increases rapidly when the signal input to the reflex stage passes some critical level.   In such a case the a.v.c. system must be designed to prevent signals of this magnitude from appearing at the grid of the reflex stage.   Since play-through is a function of rectification and therefore of the curvature of the characteristic, it is a variable depending on the bias.   By plotting play-through for varying bias voltage on the reflex stage with a constant input signal to the reflex stage, it is possible to determine the range of a.v.c. voltages which may be applied to the stage without resulting in serious play-through. This may entail using only a small fraction of the developed a.v.c. voltage on the reflex stage.

(ii) Overloading

It is practicable to design a reflex receiver to handle input voltages up to 1 volt without serious distortion (of the order of 10% at 100% modulation).

(iii) Automatic volume control

The a.v.c. system should be designed so that high peak i-f plus a-f voltages are not built-up in the plate circuit of the reflex stage.   There are three principal a.v.c. systems in use :

1. Full a.v.c. applied to both converter and reflex stages.

2. Fractional a.v.c. applied to both stages, although the two fractions may be different.

3. Full a.v.c. is applied to the converter and fractional a.v.c. to the reflex stage.

The choice of converter valve has considerable bearing on the a.v.c. design.   The use of a converter with a not-too-remote cut-off (e.g. 6BE6) assists in the reduction of play-through by limiting the maximum signal voltage applied to the grid of the reflex stage.

If the fraction of the a.v.c. applied to the reflex stage is reduced too much, or omitted entirely, the reflex stage may run into grid current and cause " bubbling " with the volume control at maximum because the rectified a-f signal returned to the grid from the plate of the reflex stage may exceed the bias on the valve.   If the fraction of a.v.c. applied to the reflex stage is increased too much, the a.v.c. characteristic will tend to reach a maximum output and then to fall with increasing input voltages.   The worst effect of excessive control is the inability of a receiver to give full a-f output on strong stations even with the volume control at maximum.   Also, in such a case the effect of tuning to a powerful signal is to produce less output when the receiver is tuned directly to the carrier than when it is tuned to one side, so that there are two adjacent tuning positions of maximum volume.   A further effect of too large a fraction of a.v.c. voltage is the increase in play-through referred to above.

As a criterion of good a.v.c. design, the volume control should be approaching its maximum position for maximum undistorted a-f output at any signal level.   The minimum volume effect may be reduced by using a tone control which gives severe treble attenuation.

(iv) Reduction in percentage modulation

With a large signal input and with the volume control at maximum, a reduction in the percentage of modulation occurs due to the curvature of the characteristics. Under laboratory conditions this can be quite considerable, for example 30% modulation may be reduced to 10%, but it is usually unimportant with ordinary listening because it only occurs under conditions which would also' cause overloading of the power stage. However, where a Scroggie-type of a.v.c. characteristic indicates serious a-f overloading with the volume control at maximum, the actual overloading with a strong input signal may be appreciably less than that indicated, owing to the reduction in percentage modulation.

(v) Negative feedback

If full a.v.c. is applied to the reflex valve, a strong signal will then reduce the a-f gain, which in turn will reduce the a-f feedback, if any. Under these conditions feedback is available on weak signals where it is not needed, but it is very much reduced on strong signals where it is most needed. As a consequence, negative feedback is only practicable with a small fraction of the a.v.c. voltage on the reflex stage.

Negative feedback, in addition to its use for the reduction of distortion and for tone compensation, also assists very considerably in the reduction of play-through (e.g. Fig. 28.3).

(vi) Operating conditions of reflex stage

The reflex stage should be biased, with a very small signal, to give maximum gain unless this bias is insufficient to prevent the valve from drawing grid current due to a-f signals in excess of the bias voltage being applied to the control grid at input signals such that the a.v.c. has not yet become fully effective.

Adequate r-f filtering of the demodulated signal is necessary before returning it to the grid of the same valve again, to avoid regeneration or actual i-f oscillation.

SECTION 3 : DESIGN OF PLATE REFLEX SUPERHET. RECEIVERS

(i) General considerations (ii) Full a.v.c. applied to both stages (iii) Fractional a.v.c. applied to both stages (iv) Full a v.c. on converter, fractional a.v.c. on reflex stage.

(i) General considerations

The majority of reflex receivers are in this class. Plate reflexing is less critical in relation to valve operating point and less critical so far as design is concerned. However, under some conditions, screen reflexing will give less play-through. With high i-f gain in any receiver it is difficult to eliminate regeneration entirely without some filtering, but in a plate reflex receiver i-f decoupling is an inherent feature due to the plate circuit components, and no other i-f filtering is necessary.

(ii) Full a.v.c. applied to both stages

With suitable valve types, this very simple system is capable of handling an input of 1 volt (30% modulated) with distortion as low as can be obtained from any other a.v.c. arrangement. Current designs of reflex receivers employing this a.v.c. arrangement have fairly high play-through but this may be due to factors other than the a.v.c. system.

(iii) Fractional a.v.c. applied to both stages

A typical example is Fig. 28.1 in which a large fraction of the a.v.c. voltage is applied to the converter stage and a small fraction to the reflex stage. As a result of reduced a.v.c. voltage on the reflex stage, the play-through is considerably reduced.

(iv) Full a.v.c. on converter, fractional a.v.c. on reflex stage

An example of a receiver in this class is Fig. 28.2 in which one ninth of the a.v.c. voltage is applied to the reflex stage. The receiver has a sensitivity of 40 μV (for 50 mW output), the comparatively low sensitivity being due principally to the use of low Q i-f transformers. It is capable of handling an input up to 1 volt with complete stability under all conditions, and has several interesting features. The plate load

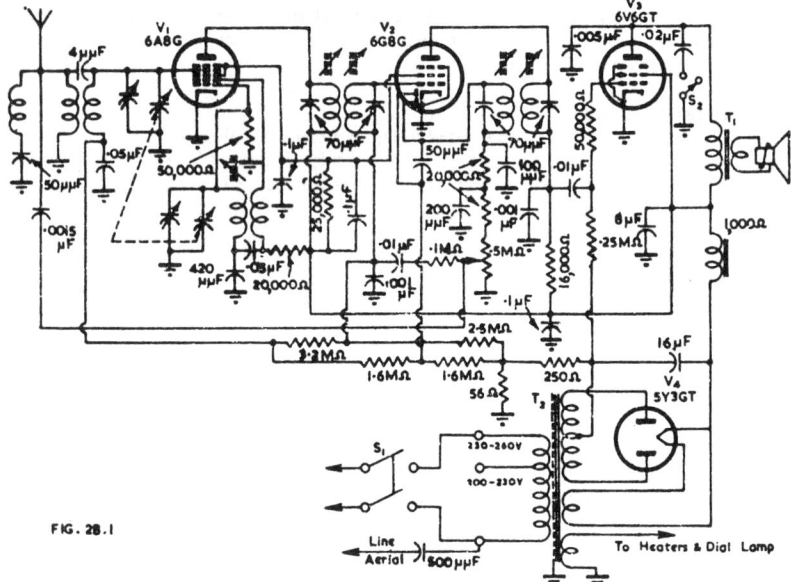

Fig. 28.1. *Reflex superhet. receiver with plate reflexing and fractional a.v.c. on both controlled stages (A.W.A. Radiola Model 517-M).*

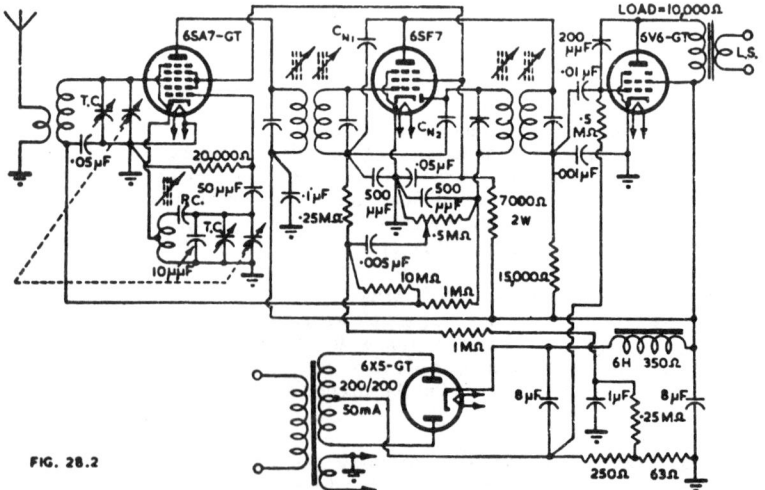

Fig. 28.2. *Reflex superhet. receiver with plate reflexing, full a.v.c. on converter and fractional a.v.c. on reflex stage (Ref. 1).*

resistor was selected to provide optimum operation together with minimum hum ; a certain amount of hum neutralization is possible between the reflex (a-f) and power amplifier stages. Feedback in the reflex stage is neutralized by the very small fixed capacitances C_{N1} and C_{N2}, while a small degree of negative feedback is provided on the output stage.

Considerably improved performance is obtainable by the use of type 6AR7-GT in the reflex stage, together with high Q i-f transformers. This avoids the necessity for neutralization and gives higher gain. Still higher gain is obtainable with a high-slope output valve, although the play-through is increased thereby. This circuit arrangement has considerable merits and there are prospects that, by suitable choice of valve types and a.v.c. design, the play-through may be reduced to a low level.

A possible alternative form of the circuit would be to use type 6BE6 converter, 6BA6 reflex amplifier and a duo-diode-output pentode.

SECTION 4 : DESIGN OF SCREEN REFLEX SUPERHET. RECEIVERS

(*i*) *Screen reflex receivers*   (*ii*) *Comparison between plate and screen reflexing.*

(i) Screen reflex receivers

In a screen reflex receiver the screen of the reflex stage is by-passed to earth for intermediate frequencies only, and the screen is coupled to the grid of the power stage for audio frequencies. For the same number of components and the same economy, screen reflexing permits the use of separate screen dropping resistors for the converter and reflex stages, thus giving greater flexibility in design and less interaction. The play-through is less than with plate reflexing and may be made quite small with the application of negative feedback. It is necessary for the diode-screen capacitance to be low, and in this respect the Australian type 6AR7-GT is satisfactory. The screen dropping resistor and grid bias must be designed for optimum i-f gain ; in the 6AR7-GT this also provides satisfactory a-f gain. Screen reflexing requires good hum filtering.

With screen reflexing, careful design is necessary for optimum results to be obtained. An example of a well designed screen reflex receiver is Fig. 28.3. This has been designed for high sensitivity with reasonably low distortion and low play-

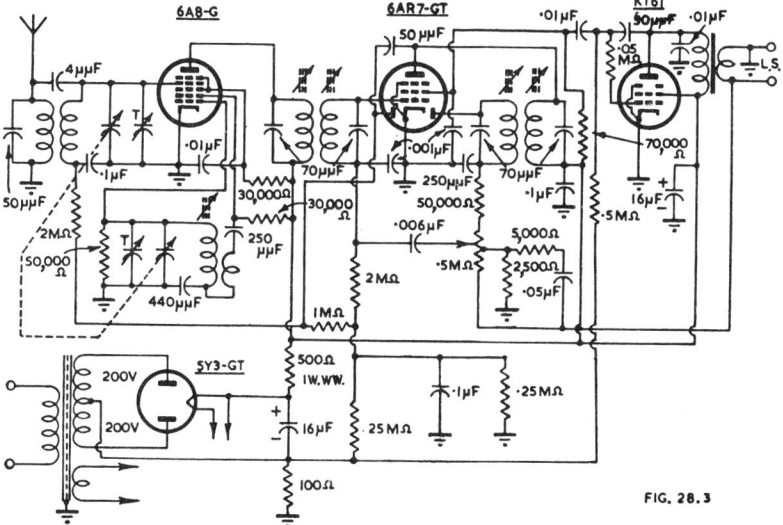

FIG. 28.3

Fig. 28.3.   Reflex superhet. receiver with screen reflexing (Ref. 4).

through. The aerial sensitivity is approximately 5 μV for 50 mW output, and the signal to noise ratio is 10 to 13.5 db at 5 μV input. The distortion is 6.9% with an input of 1 volt, 30% modulated, but 12% with an input of 0.5 volt, 100% modulated. An output of 0.5 watt is obtained with 20 μV input, 30% modulated. With an input of 1 millivolt to the aerial terminal, 30% modulated at 400 c/s, an output of 2 watts is obtained with 10% overall distortion.

Feedback is applied by returning the lower end of the volume control to the voice coil. Maximum bass boosting is provided when the volume control is at the tap, which is adjusted for the lowest listening level. The high frequency peak and bass boosting are reduced as the volume control setting is increased towards maximum. The response above 5000 c/s is cut sharply by a combination of shunt capacitances together with negative feedback. The overall result is an automatic high frequency tone control which gives a " mellow " tone with weak signals and a normal radio tone on strong stations. The gain reduction due to feedback is 14 db with 1 mV signal input and with the volume control at its minimum position. The gain reduction decreases as the volume control is advanced.

The conventional i-f filter in a reflex circuit is a series resistor from the volume control slider, connected to a by-pass capacitor at the cold end of the first i-f transformer secondary. These two components give phase shift at some high frequency with the volume control at its minimum setting, additional to the phase shifts already in the circuit and may thus cause oscillation when heavy feedback is applied with the volume control at zero. However if the resistor is wired in series with the hot end of the volume control, as in Fig. 28.3, with the volume control at its zero setting there is no series resistor in the feedback path, phase shift is reduced and high frequency peaks (or oscillation) are avoided.

Resistance-capacitance filtering is used here for economy. The plate supply voltage is 185 volts. One ninth of the a.v.c. voltage is applied to the reflex stage. The play-through is as low as in some commercial receivers that do not use reflexing.

(ii) Comparison between plate and screen reflexing

Plate reflexing is capable of higher a-f gain than screen reflexing, since the latter is limited by the " triode mu " of the reflex valve. For this reason, types 6AR7-GT and 6BA6 with grid-to-screen mu factors of 18 and 20 respectively are quite suitable for use in screen reflex receivers. However, plate reflexing gives more play-through than screen reflexing for similar conditions in both cases.

Type 6AR7-GT as a screen reflex amplifier allows a higher gain in the output stage than a lower gain valve (e.g. 6G8-G) with plate reflexing, for the same play-through performance in both cases, and similar converter conditions.

SECTION 5 : DESIGN OF T.R.F. REFLEX RECEIVERS

In such a receiver the r-f amplifier is reflexed and used also as an a-f amplifier. The only application at the time of writing is to 2/3 valve receivers having a diode-pentode as a reflexed r-f, a-f amplifier and diode detector, followed by a power pentode (e.g. 6AR7, KT61). A sensitivity of the order of 1000 μV (for 50 mW output) is practicable. A limited degree of regeneration assists gain and selectivity.

SECTION 6 : REFERENCES TO REFLEX AMPLIFIERS AND REFLEX RECEIVERS

1. " Radiotron Receiver RD31 : Four valve a.c. reflex circuit " Radiotronics No. 120 (July/Aug. 1946) 71.
2. A.R.T.S. and P. Bulletins Nos. 56, 57, 60, 63, 65, 66, 78, 84, 89.
3. Beard, E. G. " Some dangers in the use of negative feedback in radio receivers " Philips Tec. Com. 2/3 (1949) 3.
4. Design by Amalgamated Wireless Valve Co. Pty. Ltd. Applications Laboratory.
5. Watson, S. J. " Reflexed amplifiers " A.W.A. Tec. Rev. 4.1 (Jan. 1939) 35.
6. R.C.A. " Application Note on the operation of the 2B7 or 6B7 as a reflex amplifier " No. 16 (July 7, 1933).
7. Holmes, R. G. D. " Designing a modern superheterodyne—three valve reflex circuit with forward and backward a.v.c." W.W. 47.9 (Sept. 1941) 224 ; 47.10 (Oct. 1941) 261.

CHAPTER 29

LIMITERS AND AUTOMATIC FREQUENCY CONTROL

By B. Sandel, A.S.T.C.

SECTION 1 : LIMITERS

(i) General (ii) Typical circuits for F-M receivers (a) Single stage limiter (b) Cascaded limiters.

(i) General

In this section it is proposed to discuss only the conventional type of amplitude limiter using a pentode valve operating as a saturated amplifier. Other circuits, which combine the dual functions of detection and the removal of amplitude variations, such as the ratio detector and locked oscillator, will not be treated here. The ratio detector has been discussed in detail in Chapter 27 Sect. 2. Details of several other alternative systems can be found in Refs. 2, 5, 6, 7 and those at the end of Chapter 27.

The need for some form of amplitude limiting was stressed in Chapter 27 Sect. 2, when the phase discriminator was being discussed. It was pointed out that the diodes in a phase discriminator are amplitude modulation detectors, and if undesired amplitude variations are not to appear in the receiver output the amplitude of the voltage applied to the discriminator should be constant. This was emphasised by reference to eqns. (20) and (21), where it was seen that the product $g_m E_g$ should be held constant if the discriminator sensitivity is to remain fixed for a given frequency deviation.

Circuit arrangements for limiters in F-M broadcast receivers have become quite stereotyped. Sharp cut-off pentodes having fairly high values of g_m are the usual choice, and typical of these are types 6SJ7, 6SH7 and 6AU6. The limiting action is brought about by a combination of grid-leak bias and low values of plate and screen voltage. Grid-bias limiting is obtained by adding a resistor and capacitor, of suitable value, to the grid circuit and using zero or very small values of cathode bias. The grid circuit arrangement is the same as for a grid-leak detector, and the operation is almost identical since the average negative bias on the grid is determined by the $e_g - i_g$ characteristics of the valve in conjunction with the associated circuit. The value of the average grid bias, together with the low screen voltage, determine the condition for which plate current cut-off occurs. Any signal input voltage whose amplitude is sufficiently large will cause the average negative bias to increase and so tend to hold the output voltage constant. Because of the low values of plate and screen voltage the plate voltage swing is limited to a comparatively small value and, for an input signal of sufficient magnitude, there will be practically no corresponding increase in output voltage when the signal input voltage is increased.

The added damping on the i-f transformer secondary, which connects to the limiter grid, is given approximately by $R_g/2$ or $R_g/3$, depending on the circuit arrangement, in exactly the same way as discussed previously for a diode detector ; R_g being the value of resistance selected for the grid leak. This should be taken into account (as well as circuit detuning caused by changes in the valve input capacitance) when the i-f transformer is being designed.

The necessity for high gain in the receiver stages preceding the limiter stage will be appreciated when it is realized that a minimum of about 2 volts peak is required at the limiter grid to obtain satisfactory operation with a typical circuit arrangement ; it is preferable to have voltages of the order of 10 to 20 volts peak for best results under adverse condition of reception. Since this limiter input voltage must be obtained with the smallest signal input voltage likely to be met in the field, the gain of the preceding stages in the receiver should be sufficient to give satisfactory limiter operation with signal voltages as low as 2 or 3 μV ; in F-M mobile communications applications the limiter should saturate with signals of less than 1 μV. Also, because the conversion from F-M to A-M in the discriminator usually results in a low equivalent value of percentage amplitude modulation, it is necessary for the limiter output voltage to be large if the detected audio voltage is to be sufficient to drive the audio amplifier to full output ; this means that something in excess of 10 volts peak is desirable at the plate circuit of the limiter.

For completely satisfactory amplitude limitation two limiter stages are necessary. However, because of the cost factor, commercial domestic type receivers seldom use more than one limiter stage. When two stages of limiting are used, it is essential that the coupling circuit between them should not introduce any appreciable amplitude modulation due to its selectivity characteristic. The selectivity should be sufficient to attenuate harmonics of the intermediate frequency generated by the limiter, although this is also accomplished by the primary of the discriminator transformer ; with single stage limiters the discriminator transformer is relied upon to give the necessary attenuation of the i-f harmonics. Transformer coupling between the two limiter valves is the most satisfactory arrangement, but single tuned circuits are often used.

The choice of the time constant for the grid resistance-capacitance combination is important. It must be sufficiently short for the grid bias to be proportional to changes in amplitude, but not so short as to prevent the bias change from being sufficiently large to control the amplification of the limiter stage, so as to offset any change in signal input voltage. For a single stage, limiter time constants of 2.5 microseconds are usual, although 10 to 20 microseconds and even higher have been used in some receivers. For two stages the first limiter grid circuit uses a time constant of 1.25 to 5 microseconds and 2.5 to 10 microseconds or longer in the second stage, in typical cases. The longer time constants of 10 to 20 microseconds are suitable for most types of noise impulses, but some forms of motor car ignition noise are more completely suppressed when the shorter time constants are used. A careful choice of the time constants is necessary if the noise is not to be heard in the receiver output because the bias on the limiter valve must be able to follow the changes in the amplitude of the input voltage.

No matter how effective the amplitude limiters may be and how carefully their time constants are chosen it will often be found that the F-M receiver will not effectively suppress bursts of noise such as those emitted by car ignition systems. To obtain the best results it is most important that the pass band of the receiver be symmetrical and that the centre frequency of the i-f amplifier coincides exactly with the centre frequency of the discriminator. A useful test is to align the receiver on an unmodulated carrier at the signal frequency and then to switch off the carrier ; if the alignment is correct, and the circuits symmetrical, a d.c. vacuum tube voltmeter connected across the discriminator output will give a reading of approximately zero (of course the usual noise will be heard from the receiver output). Small inaccuracies often arise when the receiver is aligned so as to give maximum grid current at the limiter stage, or stages, even though the signal (unmodulated carrier) frequency is such that zero d.c. output voltage is obtained from the discriminator. The grid current reading is usually rather broad, and it will be found that the i-f and limiter circuits, in particular, can often be realigned to give zero d.c. output voltage at the discriminator on noise without reducing the limiter grid current on signals. As a further check on alignment and symmetry the receiver is tuned very carefully to an unmodulated carrier so that the a.c. voltage at the output transformer is a minimum (i.e. for maximum quieting) and for this condition the reading of the d.c. voltmeter at the discriminator is noted ; then it will usually be found that this latter reading is the same as that obtained from noise alone. (A casual reading of the text may not bring out the full significance of this test, but a practical trial will lead to a better appreciation of its possibilities). If minimum noise output, zero d.c. discriminator output voltage, and maximum limiter grid current do not all occur at the same carrier frequency, then in general the optimum conditions for impulse noise rejection have not been obtained.

For applications other than domestic receivers the above factors require very careful attention.

(ii) Typical circuits for F-M receivers
(a) Single stage limiter
A typical single stage limiter is shown in Fig. 29.1, and an alternative arrangement is shown for the input circuit in Fig. 29.2. The time constant for $R_g C_g$ is 2.5 microseconds in each case.

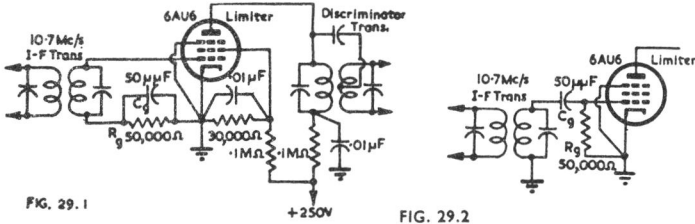

FIG. 29.1   FIG. 29.2

The operation of both circuits is identical, but the damping of the i-f transformer due to grid current is less with the arrangement of Fig. 29.1, being approximately 50 000/2 = 25 000 ohms. The damping in the alternative arrangement is 50 000/3 = 16 600 ohms. Whichever circuit is used will depend, largely, on the additional damping required on the transformer to achieve the required bandwidth or, perhaps, upon considerations of practical convenience. The general method of operation has been discussed in (i) above.

If complete valve characteristics are available then a preliminary design can be made, but the work involved is hardly worth the effort because of the ease with which the practical circuit can be made to give satisfactory results using experimental procedures. For a calculation of gain, or overall discriminator sensitivity, it is necessary to determine the mutual conductance under the actual operating conditions. This can usually be determined, with sufficient accuracy for a preliminary design, from the valve data sheets by ignoring the change in plate voltage (provided the change in plate current with plate voltage does not fall too far down on the knee of the plate characteristics). For example, a type 6AU6 is to be operated with 50 volts on the plate and screen, and zero grid bias. From the average characteristics relating grid No. 1 volts to transconductance, the g_m is 4000 micromhos for zero bias, 50 volts on the screen and 250 volts on the plate. For typical cases of the type being considered, the g_m so found is usually about 10% high, and so 3600 micromhos would be a closer approximation. Alternatively, the valve can be set up as a straight amplifier with the appropriate d.c. voltages applied, and a 1000 ohm resistor connected as the plate load. Then with 1 volt of a-f input (from a source of low d.c. resistance) the output voltage will equal the g_m in mA/volt ; the actual input voltage is selected so that the stage just starts to saturate. However, overall measurements of actual circuit performance are preferable.

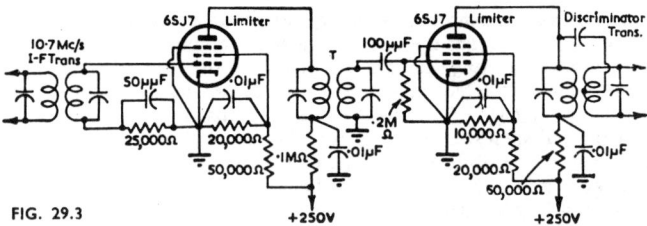

FIG. 29.3

(b) Cascaded limiters

A typical circuit for a two stage (cascaded) limiter is shown in Fig. 29.3. In this case the time constant of the grid resistance-capacitance combination is 1.25 microseconds for the first stage, and 20 microseconds for the second stage.

It should be noted that the limiters not only have to remove peaks of noise, but they must also remove the amplitude modulation introduced onto the frequency modulated signal by the receiver circuits which precede the limiters. Methods for estimating the percentage amplitude modulation introduced by tuned circuits are given in Chapter 26 Sect. 4. It will be realized that A-M introduced by the receiver itself makes it more difficult to effectively remove A-M introduced by external noise voltage sources.

For the transformer T, shown in Fig. 29.3, coupling the two limiter stages, it is essential that the amplitude modulation introduced by its selectivity characteristic should be as small as possible. However, its selectivity characteristic is helpful in removing harmonics of the i-f generated by the limiter, and so a practical compromise between the two conflicting factors is necessary. A critically-coupled transformer is recommended, and the primary and secondary Q's (uncoupled) can be about 30 for an i-f of 10.7 Mc/s. The stage gain will be roughly 6 times depending on the circuits constants selected, and the magnitude of the input voltage.

From the circuit arrangements of Figs. 29.1 and 29.3 it will be noticed that the screen voltage is supplied from a voltage divider in each case. This arrangement is recommended in all cases, but in mobile communications receivers it will be found that many circuits use only a series screen resistor ; the main advantage of this latter arrangement is economy in the total H.T. current drawn by the receiver.

SECTION 2 : AUTOMATIC FREQUENCY CONTROL

(i) General principles   (ii) Discriminators for a.f.c.   (iii) Electronic reactances.

(i) General principles

Although the description given below of automatic frequency control (a.f.c.) systems is confined to simple applications in broadcast receivers, it should not be overlooked that the same general principles are applied in many other types of equipment and are used, for example, to obtain frequency stabilization in F-M transmitters, micro-wave radar receivers and transmitters etc.

Automatic frequency control in a superheterodyne radio receiver is an arrangement for controlling the local oscillator frequency in such a way that, when a signal is being received the correct intermediate frequency will be produced. For example, if a receiver is manually tuned so that it is, say, 3 Kc/s away from a desired signal the oscillator frequency will be varied by the a.f.c. system so as to produce the correct intermediate frequency (or at least within 100 c/s or so of the correct value, since exact compensation is not possible). Alternatively, the oscillator frequency may drift because of temperature or humidity variations affecting the values of the circuit components, and a.f.c. will tend to compensate for this frequency variation. However, the frequency stability of the receiver should be made as good as possible without relying on the a.f.c. system.

The most useful application of the system in broadcast receivers is with those receivers having automatic tuning e.g. push-button station selection, and cam or motor driven variable capacitors. In cases of this type the tuning may not be accurate over extended periods of time, and a.f c. may be used effectively to carry out the final adjustment when the respective capacitances and/or inductances have been selected by the automatic tuning system.

There are two devices necessary for any a.f.c. system. These are :—

(1) A frequency discriminator, which must be capable of changing a frequency variation into a suitable direct voltage change which can be used for control purposes.

(2) A variable reactance, whose value can be controlled by the direct voltage changes due to the frequency discriminator. The variable reactance is connected to the oscillator circuit in such a way as to control its frequency.

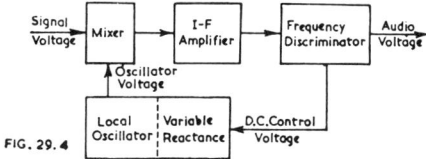

FIG. 29.4

The complete arrangement can be visualized with the aid of the block schematic of Fig. 29.4. It is seen that the additional elements to those normally found in a superheterodyne receiver are the frequency discriminator and the variable reactance. The variable reactance in the discussion to follow will be of the electronic type consisting of a valve (which can be a pentode, a hexode or a heptode) and its associated circuits. It will also be inferred from Fig. 29.4 that the frequency discriminator can be used for normal detection, since there are suitable audio voltages developed in this circuit by the applied modulated i.f. voltage.

Suitable voltages for a.v.c. are also available from the discriminator output, but in this regard it is necessary to point out that a very efficient a.v.c. system is helpful in obtaining satisfactory operation from the a.f.c. system. If reasonably constant input voltage to the discriminator is not maintained, there will be a variation in the "pull-in" and "throw-out" frequencies. Because of the stringent a.v.c. requirements it is fairly common practice to employ a separate diode coupled to the transformer primary, in the usual manner, to provide the a.v.c. bias voltage. There is also a disadvantage in taking the a-f voltage from the discriminator output as the distortion tends to be fairly high. However, in most commercial receivers the cost factor leads to some arrangement such as that of Fig. 29.8.

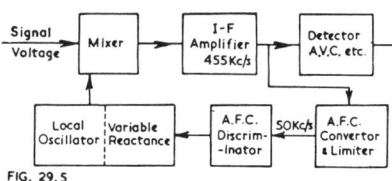

FIG. 29.5

A much more satisfactory method (although considerably more expensive) of obtaining a.f.c. is illustrated by the block schematic of Fig. 29.5. In this case the valve used as an a.f.c. converter and limiter is loosely coupled to the primary of the last i-f transformer. The 455 Kc/s signal is converted to 50 Kc/s to operate the a.f.c. discriminator, and the d.c. output voltage then controls the variable reactance in the usual manner. The use of limiting is helpful in maintaining a constant amplitude for the voltage to be applied to the discriminator. It will also be noticed that the functions of detection and a.v.c. have been separated from the a.f.c. system. A complete circuit using this arrangement can be found in Ref. 15 (page 103), together with a number of other commercial a.f.c. circuits of various types.

(ii) Discriminators for a.f.c.

The function of the frequency discriminator in an automatic frequency control system is to provide a suitable direct controlling voltage for application to the electronic reactance. When the receiver is tuned exactly to the signal frequency, the voltage output from the discriminator should be zero, or else have the same value as that provided in the absence of a signal, so that the controlled reactance will have its normal value. At frequencies above and below the correct frequency the controlling voltage should be appropriately above and below the mean voltage. The operation in connection with a phase discriminator has been discussed in Chapter 27 Sect. 2. The manner in which the output voltage will vary with the frequency change for either the amplitude or the phase discriminator can be seen from Fig. 29.6.

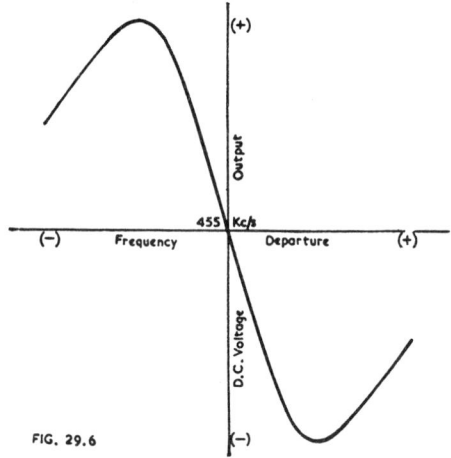

FIG. 29.6

The polarity of the output voltage with frequency change is all important, and, as the electronic reactance is practically always inductive in a.f.c. systems used in tunable broadcast radio receivers, the polarities of the direct output voltage with frequency change as indicated on Fig. 29.6 will be correct. One purpose in making the reactance inductive is so as not to limit the frequency coverage of the receiver. This follows because it is often difficult to keep the minimum capacitance of the tuning unit to a sufficiently low value, and it is hardly wise to deliberately increase this capacitance unnecessarily. However, if carefully designed, the undesired added capacitance due to the capacitive reactance unit can be made very small.

The effect of the added parallel inductance can usually be offset fairly readily by increasing the inductance of the oscillator coil. With receivers using push-button tuning where the circuits are of the preset type either the inductive or capacitive form of electronic reactance is satisfactory. For receivers with inductance tuning a capacitive electronic reactance is usually preferable. The degree of frequency correction is not constant over a band of frequencies, and this again suggests an inductive electronic reactance when the circuits are tuned by a variable capacitor, since the least error is obtained at the low frequency end of the tuning range where the receiver is most selective.

It may be helpful to follow through the steps leading to the polarities shown for the direct voltage with frequency change. Suppose the receiver is tuned to a signal of 1500 Kc/s. Then the oscillator will be set at 1955 Kc/s to give an i-f of 455 Kc/s, and there will be no direct voltage output from the discriminator. If now the oscillator drifts to 1957 Kc/s the i-f produced will be 457 Kc/s, and a direct voltage of negative polarity will be produced by the discriminator and applied to the reactance valve. The equivalent inductance of the electronic reactance is inversely proportional to mutual conductance (g_m) and so, as a more negative bias voltage reduces g_m, the shunt inductance across the oscillator coil is increased, the **total** inductance in the oscillator circuit is now increased, thus lowering the oscillator frequency as required.

The two most commonly used types of discriminator circuits are shown in Figs. 29.7 and 29.8. The first is generally known as the Round-Travis circuit (see Ref. 10), and is a typical example of an amplitude discriminator. The secondary circuits A and B are so tuned that A has its resonant frequency slightly above the intermediate frequency and B is set slightly below the i-f (say +5 Kc/s in one case and −5 Kc/s in the other). Each secondary circuit has its own diode detector and the diode loads are connected in d.c. opposition, so that when the i-f is greater than the required value of 455 Kc/s (and since the voltages developed across R_1 and R_2 have the polarities shown) the voltage across R_1 is greater than that across R_2 and the a.f.c. bias voltage is negative as is required for correction of the oscillator frequency. For example, the voltage across R_1 may be 4 volts when that across R_2 is 3 volts, then the available a.f.c. bias is −1 volt.

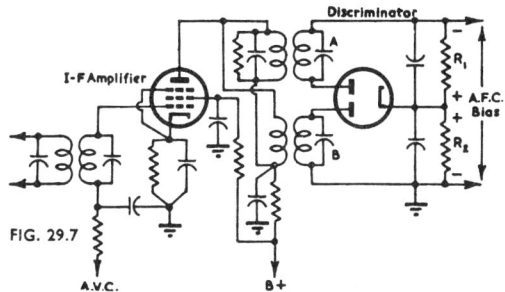

FIG. 29.7

The design procedure for an amplitude discriminator for a.f.c. is set out in detail in Ref. 8. Since this type of circuit is seldom used in modern a.f.c. systems the details of its operation will not be discussed further. There are a number of alternative arrangements for the amplitude discriminator, and some of these are discussed in Refs. 8, 9, 11, 15 and 13. It will be appreciated that a circuit of this type could be used for F-M detection as an alternative to the phase discriminator, and several manufacturers have produced F-M receivers using modified amplitude discriminators (see Ref. 39 given at end of Chapter 27, for typical examples).

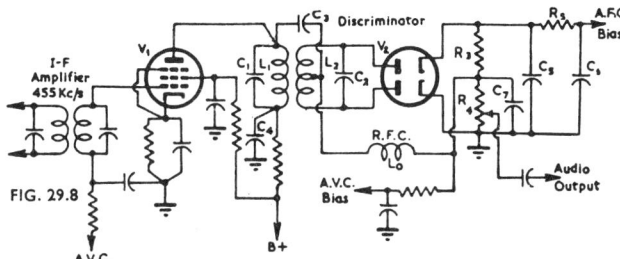

FIG. 29.8

Fig. 29.8 shows the most commonly used circuit, the phase (or Foster-Seeley) discriminator. In this circuit both primary and secondary circuits are tuned to the intermediate frequency. It will be seen that the circuit arrangement is identical with that of the F-M phase discriminator shown in Fig. 27.19, and the general discussion of its operation and the loading effects on the transformer apply equally well here. The de-emphasis circuit is not required here, of course, but the same general arrangements can be retained to provide r-f and a-f filtering, since only a direct control voltage is required. Suitable values for the filter circuit are R_5 equal to 0.5 − 2 MΩ, and C_6 is, say, 0.05 − 0.1 μF. The considerations governing the choice of the component values can be compared with those for selecting filter components for automatic volume control circuits, and a time constant for $R_5 C_6$ of about 0.1 second is usual. The circuit arrangement of Fig. 27.17 could also be used here, and, if the primary circuit is too heavily damped, an additional series resistor can be inserted between the transformer centre-tap and the junction of the diode load resistors R_3 and R_4; however, this will lead to some loss in available output voltage, depending on the value of series resistance selected. The difference in the by-pass capacitor arrangements (i.e. C_5 and C_7) for Figs. 27.17 and 27.19 (or Fig. 29.8) should be observed, and the discussion in Chapter 27 Sect. 2, in connection with Fig. 27.19 will be helpful, if the reasons for the arrangements are not immediately obvious. The considerations leading to the choice of actual circuit values will differ because the a.f.c. discriminator is usually designed for high sensitivity rather than for a very high degree of linearity.

The theory of the circuit for a.f.c. use has been discussed by Roder (Ref. 14) and design procedures have been treated in detail in Ref. 8. A brief discussion, which is quite helpful, is given in Ref. 13.

The method of obtaining a-f output and a.v.c. bias from the a.f.c. discriminator circuit is also shown in Fig. 29.8. To retain the same degree of selectivity as that obtained with a similar receiver not incorporating a.f.c. one extra tuned circuit is required. This is necessary because the discriminator provides very little selectivity as a result of the heavy loading effects produced by the diodes. There is some increase in gain over the usual arrangement where the a-f detector is connected across the i-f transformer secondary only.

The frequency at which the a.f.c. will come into operation when tuning a signal is called the "pull-in frequency." When tuning away from a carrier the frequency at which the a.f.c. loses control is called the "throw-out frequency," and it is always greater than the "pull-in frequency." It is desirable to make these two frequencies as close as possible, because stations which are received when tuning-in may be passed over when tuning-out. Many receivers incorporate arrangements for disconnecting the a.f.c. until a carrier has been approximately tuned-in (see Refs. 8, 15).

Some of the general details of the discriminator design will be set out here, and the reader is referred to Refs. 8, 9 and 14, in particular, for further information. Consider Fig. 29.8. Typical values for R_3 and R_4 are 0.5 MΩ. Capacitors C_5 and C_7 must give adequate by-passing at the intermediate frequency of 455 Kc/s, and suitable values would be 100 to 200 $\mu\mu$F. C_3 is usually 100 $\mu\mu$F. Values for R_5 and C_6 have been discussed previously. The valve V_1 is any of the usual voltage amplifier pentodes such as type 6SK7 etc., and V_2 is a double diode such as the types 6H6 or 6AL5. The next step is to select a suitable value for the inductance of the r-f choke.

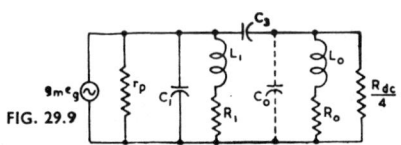

FIG. 29.9

As the inductance and capacitance of the choke will affect the resonant frequency of the transformer primary ($L_1 C_1$) and its Q, care is necessary. The choke, in association with the transformer primary circuit, can be represented by the equivalent circuit of Fig. 29.9 in which $g_m e_g$ represents the equivalent constant current generator for the pentode voltage amplifier; r_p is the plate resistance; C_1 is the total primary capacitance including strays; L_1 is the primary inductance and R_1 its r-f resistance; C_3 is the capacitor connecting the plate circuit to the choke L_0 (whose r-f resistance is R_0); C_0 represents any distributed or stray capacitances across L_0; $R_{dc} = R_3 = R_4$ the diode load resistance. It should be apparent that three cases can arise with the choke (Ref. 14) :—

(1) The capacitance C_0 negligibly small.
(2) The capacitance C_0 small enough to still allow the choke to be self resonant well above the intermediate frequency, but the value of C_0 to be such as to appreciably affect the resonant frequency of the complete primary circuit.

(3) The choke resonant at the intermediate frequency.

It follows from (1) and (2) that L_1 and C_1 should have a resultant capacitive reactance at the intermediate frequency, and in case (3) L_1 and C_1 should be resonant at the intermediate frequency.

Considering the above three cases in conjunction with Figs. 29.9 and 29.8, it can be seen that the derivation of the conditions for discriminator transformer primary resonance, voltage step down between the transformer primary and the choke, and the resultant Q's for the complete discriminator transformer, will follow fairly readily. The derivations are available in Ref. 14, and the results, which are good approximations, are summarized below (with some modifications, as well as changes in notation).

Case (1) (Choke capacitance C_0 negligibly small).
Condition for primary circuit resonance,

$$\frac{X_{L_0} - X_{C_3}}{X_{L_1}} = \frac{X_{C_1}}{X_{L_1} - X_{C_1}} \tag{1}$$

Voltage step down,

$$\alpha = 1 - \frac{X_{C_3}}{X_{L_0}} \tag{2}$$

Reciprocal of equivalent primary circuit Q (this leads to simpler numerical evaluation),

$$\frac{1}{Q_p} = \frac{1}{Q_0} + \frac{X_{L_0}}{R_{dc}/4} + \alpha^2 X_{L_0}\left(\frac{1}{Q_1 X_{L_1}} + \frac{1}{r_p}\right) \tag{3}$$

Secondary circuit Q (since the diode loading is $R_{d\,c}$) is

$$Q_s = \frac{Q_2 R_{dc}}{Q_2 X_2 + R_{dc}} \tag{4}$$

and this equation applies for the three cases—

where
$X_{L_0} = \omega_0 L_0 = $ inductive reactance of choke L_0
$X_{C_3} = 1/\omega_0 C_3 = $ capacitive reactance of C_3
$X_{L_1} = \omega_0 L_1 = $ inductive reactance of L_1
$X_{C_1} = 1/\omega_0 C_1 = $ capacitive reactance of C_1
$\omega_0 = 2\pi \times $ intermediate frequency
$Q_0 = $ magnification factor of choke L_0
$Q_1 = $ magnification factor of L_1
$R_{dc} = R_3 = R_4 = $ diode load resistance value
$r_p = $ plate resistance of voltage amplifier valve
$Q_2 = $ magnification factor of L_2
and $X_2 = \omega_0 L_2 = $ inductive reactance of L_2.

Case (2) (Choke capacitance small but not negligible).
The angular resonant frequency of the choke is

$$\omega_r = 1/\sqrt{L_0 C_0} \tag{5}$$

$$X'_{L_0} = X_{L_0}\left(\frac{\omega_r^2}{\omega_r^2 - \omega_0^2}\right) \tag{6}$$

$$Q'_0 = Q_0\left(1 - \frac{\omega_0}{\omega_r}\right)^2 \tag{7}$$

where $C_0 = $ distributed capacitance of the choke, and all the other symbols are exactly as before.

Equations (6) and (7) are used directly with eqns. (1), (2) and (3) substituting X'_{L_0} for X_{L_0} and Q'_0 for Q_0.

Case (3) (Choke resonant at intermediate frequency).
The inductances L_1 and L_2 are connected in parallel, and α can be taken as unity (since X_{C_3} will be small, compared with the dynamic resistance of $L_0 C_0$), so that the equivalent primary circuit reactance required to give parallel resonance is

$$X_p = \frac{X_{L_1} X_{L_0}}{X_{L_1} + X_{L_0}} . \tag{8}$$

The reciprocal of the equivalent loaded primary circuit Q_1 is

$$\frac{1}{Q_p} = X_p \left[\frac{1}{Q_0 X_{L_0}} + \frac{1}{Q_1 X_{L_1}} + \frac{1}{r_p} + \frac{1}{R_{dc}/4} \right] \tag{9}$$

All notation exactly as for case (1).

It should be noted in all cases that the loaded Q's refer to the case where the primary and secondary are uncoupled from one another. This is the usual definition.

We are in a position to proceed with the determination of suitable values for the primary and secondary circuits of the discriminator transformer, since all the external effects can now be taken into account. What is required next are methods for determining optimum values for Q_p, Q_s, L_p and $L_s (= L_2)$ where these factors have the meanings given by eqns. (1), (3), (4) etc. and $L_s (= L_2)$ is the secondary inductance (since its value is not changed by the presence of the choke). If the choke is not used $X_{L_1} = X_{C_1} = X_p$ and so $L_1 = L_p$; the procedure is then as for any other phase discriminator. The design factors, as well as the bandwidth and sensitivity calculations given below, apply equally well for all phase discriminators provided the values of Q, L and k (coefficient of coupling) so determined are those actually obtained in the receiver.

Optimum values for $k\sqrt{Q_p Q_s}$ for various values of L_s/L_p [see Refs. (8) and (9)] can be found from

$$k(Q_p Q_s)^{\frac{1}{2}} = \left[\frac{(Q_p{}^2 Q_s{}^2 + 2 Q_p Q_s{}^3 L_s/L_p)^{\frac{1}{2}} - Q_p Q_s}{Q_s{}^2 L_s/L_p} \right]^{\frac{1}{2}} \tag{10}$$

Several values are listed below,

$L_s/L_p =$	1	2	4	6	
$Q_p/Q_s = 1/2$	0.786	0.707	0.625	0.578	
$Q_p/Q_s = 1$	0.856	0.785	0.707	0.657	$k(Q_p Q_s)^{\frac{1}{2}}$.
$Q_p/Q_s = 2$	0.909	0.855	0.786	0.740	

The sensitivity of the phase discriminator at the intermediate frequency (f_0) has been derived in Refs. (14) and (8). It is given by,

$$S = \frac{2 g_m Q_p X_p \eta Q_s{}^2 (L_s/L_p)^{\frac{1}{2}}}{f_0} \left[\frac{k}{(1 + Q_p Q_s k)^2 \left(1 + \frac{Q_s{}^2 k^2 L_s}{4 L_p} \right)^{\frac{1}{2}}} \right] \tag{11}$$

and is expressed in direct volts output per Kc/s off tune, for each 1 volt (peak) input to the i-f amplifier valve V_1 (Fig. 29.8).

The symbols are the same as those used previously, with the addition of g_m the mutual conductance of V_1, and η is the detection efficiency of the diodes V_2.

It is necessary to estimate the peak separation for the discriminator characteristic of Fig. 29.6. This can be found (Ref. 8) from,

$$2\Delta f = \frac{f_0}{Q_s} \tan \phi \tag{12}$$

where $\phi = \cos^{-1} \left[\frac{8}{16 + (Q_s k)^2 L_s/L_p} \right]^{\frac{1}{2}}$

For many practical cases it will be found sufficiently accurate to take $\tan \phi = 1$ in eqn. (12).

(iii) Electronic reactances

There are a number of forms which electronic reactance circuits may take and a few of these are given below.

(1) **Resistance in series with a capacitance.** This arrangement is shown in Fig. 29.10, but since the circuit imposes severe resistive loading on the oscillator circuit its use is not advised. (See Ref. 10).

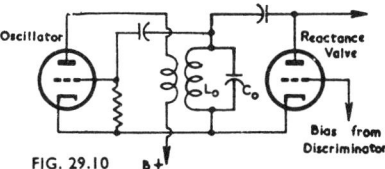

FIG. 29.10

(2) **Miller effect circuits.** These circuits (Fig. 29.11) rely on the change of input capacitance which occurs when the gain is varied. If the plate load can be tuned so that it behaves as a pure resistance, then the valve input resistance can be made very large. For the circuit shown there will be a resistive input component due to Miller effect because of stray capacitance across the load resistor. The application of the circuit is largely confined to use with fixed tuned oscillator circuits.

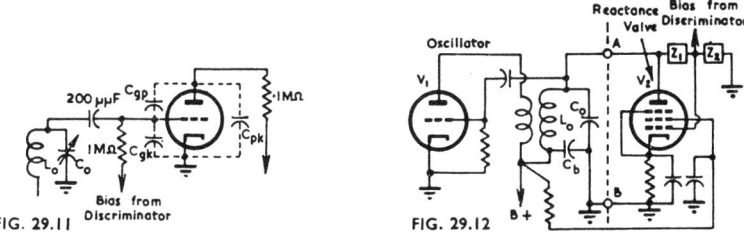

FIG. 29.11

FIG. 29.12

(3) **Quadrature circuits** (Figs. 29.12, 29.13). The grid is fed from a resistance reactance network connected between plate and cathode. This provides a voltage between the grid and cathode which is almost 90° out of phase with the plate to cathode voltage. The source of the alternating plate to cathode voltage is the voltage developed across the tank circuit of the oscillator (see Fig. 29.12). Since the plate current is in phase with the grid voltage (for valves having high plate resistance) the plate voltage and plate current will be approximately 90° out of phase. To the external circuit (oscillator tank circuit in this case) connected between plate and cathode of the reactance valve the behaviour is as though an additional reactance and resistance had been connected in parallel. Whether the valve circuit behaves like an inductive, or a capacitive reactance depends on the resistance-reactance network arrangement, as can be seen from Fig. 29.13.

The value of the apparent reactance and resistance, due to the electronic reactance (more correctly electronic impedance) depends on the mutual conductance (g_m) of the valve, and as the g_m can be controlled by alteration of the grid voltage the equivalent reactance and resistance can also be varied (the resistance variation is usually undesirable). The required grid voltage variation is obtained by utilizing the direct voltage changes at the discriminator output, when the reactance valve is used in a.f.c. circuits. Frequency modulation, using this method, is obtained by applying the audio frequency modulating voltage to the grid of the reactance valve, in the same way as the direct voltage changes are applied for obtaining a.f.c. The magnitude of the a-f voltage determines the change in the equivalent reactance shunted across the oscillator tank circuit, and so determines the frequency deviation from the nominal oscillator centre frequency. The number of times the frequency deviates around the central reference frequency will be determined by the frequency of the a-f modulating voltage. Variation in the value of the shunt resistance, due to the electronic reactance, across the oscillator circuit causes undesired amplitude modulation.

The quadrature circuits are the most widely used for a.f.c. and other purposes, and attention will be confined to discussing some of the possible arrangements. Fig. 29.12 shows the general circuit arrangement using a pentode valve. A hexode or heptode valve can also be used, with the phase shifting network connected to the signal grid and the control voltage to the oscillator grid. The impedances Z_1 and Z_2

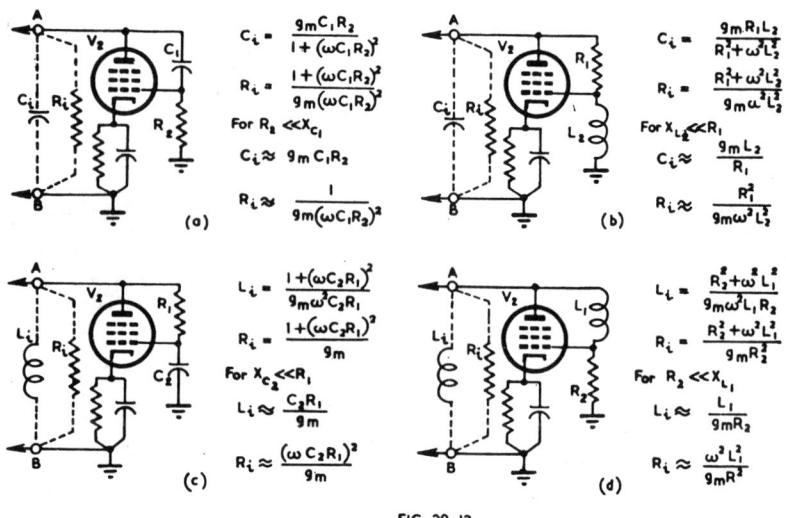

FIG. 29.13

in Fig. 29.12 comprise the phase shifting network. Depending on the form these impedances take it will be clear that an additional blocking capacitor may be required between plate and grid of the reactance valve, and also, since there must be a d.c. path between grid and cathode, an additional grid resistor may be necessary. The values for these additional components should be such as to have negligible effect on the performance of the circuit. Fig. 29.13 shows four possible arrangements for quadrature circuits, together with the equations for the additional resistance (R_i) and inductance (L_i), or capacitance (C_i), connected in parallel with the oscillator tuned circuit. Exact and approximate equations are given, but in most practical circuits the approximate conditions will hold. The equations apply equally well to quadrature circuits using pentode, hexode or heptode valves. The angular frequency ω is that at which the oscillator circuit is meant to operate e.g $2\pi \times 1455$ Kc/s etc. (in the case of F-M this would be the nominal reference frequency). The circuit of Fig. 29.13(c) is the one most commonly used in practical a.f.c. circuits in which the oscillator is tuned by a variable capacitor. Care is necessary when using circuits (b) and (d) as self resonance effects, due to stray capacitances across the inductances L_2 and L_1, often lead to difficulties. Further, the r-f resistances of L_2 and L_1 must be low if the circuits are to behave as relatively pure reactances. Stray capacitances across R_1 and R_2 can also affect performance, and should be kept small.

To carry out the design for an electronic reactance circuit it is necessary to know the manner in which g_m varies with grid bias. This information is generally available on valve data sheets, for a given set of operating conditions. If other operating conditions are required, then direct measurement of the $g_m - e_g$ characteristic is the usual procedure.

It is helpful when designing reactance valve circuits to be able to determine, directly, values for $C_1 R_2$, L_2/R_1, $R_1 C_2$ and L_1/R_2 in terms of the operating frequency, the frequency change required, and the oscillator tank circuit component values. The necessary conditions have been determined (Ref. 18) for the circuits (b) and (c) of Fig. 29.13, using the approximate relationships for C_i and L_i and assuming that the frequency variation is linear (or very nearly so). For circuit (b) the approximate expression is

$$\frac{L_2}{R_1} = \frac{2C_0 f_2 S}{f_1{}^2}$$

For circuit (c) the approximate expression is

$$C_2R_1 = \frac{L_0 f_1{}^2}{2f_2 S} \qquad (14)$$

where C_0 = capacitance tuning the oscillator circuit in the absence of the reactance valve

L_0 = inductance of the oscillator tuned circuit in the absence of the reactance valve

S = sensitivity = $(f_2 - f_1)$/corresponding change in mutual conductance (g_m)

f_2 = high frequency limit of frequency

and f_1 = low frequency limit of frequency

The design procedure is as follows :

(1) Select f_2 and f_1. Suppose the nominal oscillator centre frequency is 1455 Kc/s, and the oscillator frequency is to vary $\pm$ 5 Kc/s. Then $f_2 = 1460$ Kc/s and $f_1 = 1450$ Kc/s.

(2) Select a suitable working range on the $g_m - e_g$ characteristics, for the valve type to be used, which is as nearly linear as is possible. With a valve type 6U7-G (250 volts on plate, 100 volts on screen) a suitable operating range is from -3 to -10 volts, giving a g_m change of $1600 - 275 = 1325$ μmhos.

(3) Choose a suitable value for L_0 or C_0 if these are not already fixed by other circuit considerations. Suppose $L_0 = 110$ μH is required in a typical case ; this value would be modified slightly in the final circuit, to take care of the additional parallel inductive reactance due to the normal value of the electronic reactance in the absence of additional bias from the discriminator. It is not necessary, usually, to take this inductance change into account during the preliminary design.

(4) Compute the sensitivity (S) in cycles per mho. For our example $S = (10$ Kc/s)/1325 μmhos = 7.54×10^6 cycles/mho.

(5) Determine L_2/R_1 from eqn. (13) or C_2R_1 from eqn. (14). Using eqn. (14),

$$C_2R_1 = \frac{110 \times 10^{-6} \times 1450^2 \times 10^6}{2 \times 1460 \times 10^3 \times 7.54 \times 10^6} = 10.5 \ \mu\mu F \times M\Omega$$

(6) Select particular values for L_2, R_1 or C_2 to conform to the circuit requirements ; remembering the previous restrictions of $X_{L_2} \ll R_1$ and $X_{C_2} \ll R_1$.

This should lead to R_1 being at least 5 times X_{L_2} or X_{C_2}, but larger ratios are preferable (see below).

The most convenient procedure is to tabulate various values of R_1, L_2 or C_2 and to select the most suitable combination giving the product found in step (5) e.g. $R_1 = 50\,000\ \Omega = 0.05$ MΩ, then $C_2 = 10.5/0.05 = 210$ $\mu\mu$F, and so $X_{C_2} = 520\ \Omega$. This makes $R_1 \gg X_{C_2}$ as required.

Of course, the actual values of L_i and R_i can be determined directly from the expressions given in Fig. 29.13. R_i should always be so determined for the condition of maximum g_m, after the circuit values have been found, to ensure that the loading on the oscillator circuit is not excessive. The value of g_m to be used in these equations corresponds to the actual bias voltage for a particular operating frequency ; e.g. in the above example, with no additional external bias applied, the operating frequency is 1455 Kc/s, and the standing bias voltage can be taken as -6.5 volts, corresponding to a mutual conductance of 925 μmhos. The total parallel inductance and resistance changes can be found, using the g_m values corresponding to -3 and -10 volts bias, which are the values required when the operating frequencies are 1450 Kc/s and 1460 Kc/s respectively.

Before completing a design it is necessary to check the amplitude of the oscillator voltage applied to the grid of the reactance valve by the phase shifting network. This check is necessary as the possibility of grid current might be overlooked. In our example the minimum bias is -3 volts and so the peak r-f grid voltage should not exceed about 2 volts if grid current is to be avoided. The proportion of the r-f voltage developed across the oscillator tank circuit (and applied between plate and cathode of the reactance valve) which appears between grid and cathode of the reactance valve is for Fig. 29.13(c) $X_{C_2}/\sqrt{X^2_{C_2} + R_1^2}$. For our example, the voltage step down is 0.0104, and so no possibility of grid current exists, as 2 volts peak r-f at the grid corresponds to 196 volts peak across the oscillator tank circuit. This is considerably in excess of the voltage likely to be encountered in a receiver oscillator circuit, where 60 volts peak is about the maximum to be expected with any of the usual arrangements (it is usually considerably less than this value, depending on the type of circuit used).

Further details of the design of reactance valve circuits for a.f.c. can be found in Refs. 8, 9 and 10. A graphical method for determining the " throw-out " and " pull-in " frequencies of an a.f.c. system is given in Ref. 8 (p. 260). The difference in these two frequencies is reduced by using an electronic reactance which gives correction for a limited range of discriminator voltages only ; outside the correction range the added reactance should remain practically constant.

SECTION 3 : REFERENCES

(A) **LIMITERS**
See also Refs. 40, 41, 42 Chapter 27.
1. Tibbs, C. E. " Frequency Modulation 3—Interference suppression, the limiter, and the capture effect " W.W. 49.3 (March 1943) 82 (also gives bibliography).
2. Carnahan, C. W., and H. P. Kalmus " Synchronized oscillators as F.M. receiver limiters " Elect. 17.8 (Aug. 1944) 108 (extensive bibliography).
3. Tibbs, C. E. (book) " Frequency Modulation Engineering " Chapman and Hall, London, 1947.
4. Hund, A. (book) " Frequency Modulation " Chapter 2, p. 209, McGraw-Hill Book Co. Inc., New York and London, 1942.
5. Sturley, K. R. (book) " Radio Receiver Design " Part 2 Chapter 15, p. 328 Chapman and Hall, London, 1945.
6. Kiver, M. S. (book) " F-M Simplified " Chapter 6 D. Van Nostrand Co. Inc., Toronto, New York and London, 1947.
7. " Cascade dynamic limiter in F-M detector systems " A.R.T.S. and P. Bulletin No. 143 (Reprint Hazeltine Electronics Corp. Bulletin).

(B) **AUTOMATIC FREQUENCY CONTROL**
(Discriminators and Electronic Reactances)
8. Sturley, K. R. (book) " Radio Receiver Design " Part 2 Chapter 13, p. 224, Chapman and Hall, London, 1945.
9. Foster, D. E. and S. W. Seeley " Automatic tuning " Proc. I.R.E. 25.3 (Mar. 1937) 289.
10. Travis, C. " Automatic frequency control " Proc. I.R.E. 23.10 (Oct. 1935) 1125.
11. " A.F.C. survey " A.R.T.S. and P. Bulletin No. 55 (11th April, 1938).
12. White, S. Y. " A.F.C. design considerations " Elect. 9.9 (Sept. 1936) 28.
13. Zepler, E. E. (book) " The Technique of Radio Design " (Chapter 4, p. 121), Chapman and Hall, London, 1943 ; John Wiley and Sons, New York, 1943.
14. Roder, H. " Theory of the discriminator circuit for automatic frequency control " Proc. I.R.E. 26.5 (May 1938) 590.
15. Rider, J. F. (book) " Automatic Frequency Control Systems " John F. Rider, Publisher, New York 1937.
16. Rust, Keall, Ramsay and Sturley, " Broadcast receivers : a review " Jour. I.E.E. Part III, 88.2 (June 1941) 59.
17. Cocking, W. T. " Wireless Servicing Manual " 6th edit. Chapter 21 Iliffe and Sons, London.
18. Ross, H. A. and B. Sandel " Design of electronic reactance networks " A.W.A. Tec. Rev. 6.2 (Feb. 1943) 59.
19. Hund, A. (book) " Frequency Modulation " Chapter 2, pages 155-174, McGraw-Hill Book Co. Inc. New York and London, 1942.
20. Butler, F. " Reactance modulator theory " W.E. 25.294 (March 1948) 69.
21. Helfrich, H. D. " Wide deviation reactance modulator " Elect. 21.4 (April 1948) 120.
22. Brunner, F. " Extending linear range of reactance modulators " Elect. 21.5 (May 1948) 134.
(Refs. 18-22 are of general interest in the design of F-M test equipment etc.)

CHAPTER 30

RECTIFICATION

By R. J. RAWLINGS, Grad. I.E.E., Associate Brit. I.R.E.
and F. LANGFORD-SMITH, B.Sc., B.E.

SECTION 1 : INTRODUCTION TO RECTIFICATION

(i) Principles of rectification   *(ii) Rectifier valves and types of service*   *(iii) The use of the published curves*   *(iv) Selenium and copper-oxide rectifiers.*

(i) Principles of rectification

Most electronic equipment requires some form of plate voltage supply which has in the majority of cases to be derived from single-phase a.c. mains. It is the purpose of this chapter to outline the principles and calculations involved, with particular reference to those types of supply required for radio receivers and amplifiers.

The most general and accepted method of a.c. to d.c. conversion, where very large amounts of power are not required, is by valve rectifiers of either the high vacuum or mercury vapour type. Selenium and copper-oxide rectifiers are also used—see (iv) below.

Diagram A in Fig. 30.1 shows a sine wave voltage of which the peak and r.m.s. values are shown as $\hat{E}_\sim$ and $|E_\sim|$. With ideal half-wave rectification and a resistive load with no filter, the positive or upper peaks would also represent the load current, while the negative or lower peaks would be suppressed ; the average voltage would be shown by E_{dc} in the half-wave case. With full-wave rectification the current through the load resistance would be similar* each half cycle, the lower peak being replaced by the dashed line in A. The direct voltage would be the average voltage, i.e., 0.9 of the r.m.s. voltage for a sine wave. For half-wave rectification the average direct voltage over a period would be one half that for full-wave rectification under the same conditions.

Fig. 30.1B illustrates **ideal full-wave† rectification with a condenser input filter** (circuit as Fig. 30.1F in which choke L is assumed to have very high inductance and zero resistance). The voltage at the first filter condenser C_1 follows the line ABA′B′, the condenser charging between A and B but discharging between B and A′.

---

*In practice there is always some lack of symmetry caused by a combination of small factors such as the use of one end of the filament as cathode return in place of a centre-tap, variations in characteristics between the two units in the rectifier and variations in transformer secondary voltages and impedances on both sides of the centre-tap. As a result there is usually a substantial amount of mains frequency ripple with full-wave rectification, although the twice-mains-frequency ripple voltage predominates.
†Also known as biphase half-wave.

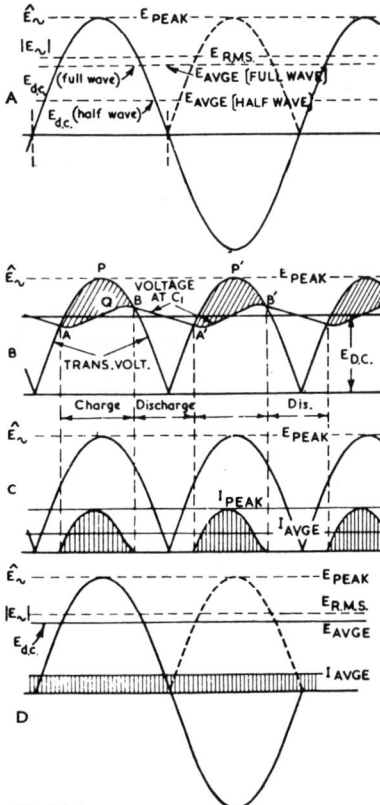

Fig. 30.1. (A, B, C & D). Voltage and current waveforms of condenser input and choke input rectifier systems. The symbols on the left are those used in Section 2.

FIG. 30.1

The mean level of ABA′B′ is the effective direct voltage. The shaded area above the curve AQB represents the voltage by which the transformer voltage exceeds that of C_1. The current through the plate circuit of the rectifier only flows for the interval between A and B and between A′ and B′ because at other parts of the cycle the transformer voltage is below the voltage of C_1. The current through the rectifier, shown in Fig. 30.1C, is similar in form to the difference in voltage between the curves APB and AQB in Fig. 30.1B.

The ripple voltage may be determined from the ABA′B′ curve, and the values of the fundamental and harmonics may be determined by a Fourier analysis.

As the load resistance (R_L in Fig. 30.1F) is increased, BA′ becomes more nearly horizontal and the area APB becomes smaller until in the extreme (theoretical) case when the load resistance is infinite the direct voltage is equal to the peak voltage. This graphical method may be applied to any rectifier with a condenser input filter followed by a high inductance choke. The assumption is made that the current through the inductance remains constant, that is to say that the lines BA′, B′A″ etc. are straight.

With a (full wave) **choke input filter*** the conditions are as shown in Fig. 30.1D assuming a very high inductance choke (L_1 in Fig. 30.1G), although with practical chokes there will necessarily be a certain amount of ripple in the load current.

---

* The term "choke input" is used for convenience in this Handbook to indicate a series inductance followed by a capacitance shunted across the load resistance.

Fig. 30.1E. Basic circuit diagram of half-wave rectifier with condenser input filter.

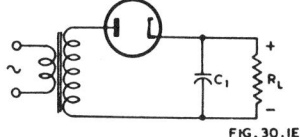

Fig. 30.1F. Basic circuit diagram of full-wave rectifier with condenser input filter.

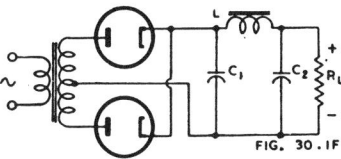

Fig. 30.1G. Basic circuit diagram of full-wave rectifier with choke input filter.

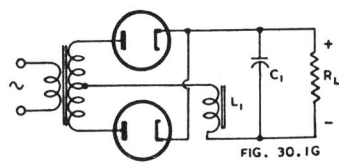

Maximum ratings

Rectifier valves are usually rated for maximum direct-current per plate, maximum peak current per plate, maximum peak inverse plate voltage and maximum r.m.s. supply voltage per plate. In some cases a maximum rated hot-switching transient (or surge) plate current per plate (for a specified maximum time, e.g. 0.2 second) is also given. It is important to ensure that no one of these ratings is exceeded under the conditions of operation. If the ratings are design centre values—see pages 77 and 78—they apply to nominal mains voltages.

Maximum direct current per plate

This may be measured by a d.c. milliammeter in series with R_L (Figs. 30.1 E, F or G).

Maximum peak current per plate

This may either be measured by means of a C.R.O. or calculated by the method described in Section 2 or 3 for condenser or choke input respectively. It is largely influenced by the total effective plate supply impedance per plate, and in any case where the peak current is too high, it may be reduced by adding a resistor in series with each plate or by increasing the effective impedance of the transformer to give the same result. The method of calculating the plate supply impedance per plate for a transformer is given on page 99. In cases where no transformer is used, as in a.c./d.c. receivers, a resistor should be connected in the plate circuit to limit the peak current to a safe value. It is good practice in all cases to limit the peak current to a value below the maximum rated value, to give a margin for safety and longer life.

Maximum peak inverse plate voltage

This is 1.41 times the r.m.s. voltage of the whole secondary winding of the transformer in Fig. 30.1F or G, and twice this value in Fig. 30.1E.

Maximum r.m.s. supply voltage per plate

A higher r.m.s. supply voltage is usually permitted with a choke input to the filter than with condenser input to the filter. In some cases, with a condenser input to the filter, a higher r.m.s. supply voltage per plate is permitted provided that the direct plate current is reduced below its maximum rating and that the total effective plate supply impedance per plate is increased for the higher voltage conditions (e.g. Fig. 30.2A). With a choke input to the filter, the maximum r.m.s. supply voltage per plate is sometimes only permissible provided that the direct plate current is reduced below its maximum rating (e.g. Fig. 30.2C). With a choke input to the filter, it is essential to have a **choke inductance** not less than the critical value for the particular operating condition in question. The critical inductance is a function of the load resistance and the frequency of the supply, as given by eqns. (1), (2) and (3) in Sect. 3.

For any value of inductance, with constant r.m.s. supply voltage, there is a value of current below which operation is not permitted. This is shown in Fig. 30.2C where boundary lines for choke sizes are included.

Equivalent circuit of high vacuum rectifier

The high vacuum rectifier can be considered as being an ideal switch in series with a non-linear resistance and a source of potential which is connected by the switch to the load when the polarity is that required by the load (Ref. 7). As the switching gives rise to pulsating currents (and voltages) it is necessary to assume a linear resistance which is equivalent to the non-linear effective resistance of the rectifier during this pulsating or conduction period. The conduction period (ϕ), and therefore also the magnitude of the current pulse, will depend on the loading and the type of filter connected to the rectified supply. Certain approximations which must be made for the first calculation should be readjusted when the results are known, in order that a second and more accurate calculation can be made.

Mercury vapour rectifiers

In the case of mercury vapour rectifiers the voltage drop in the valve is a constant value of the order of 10 to 15 volts over a wide range of currents. These rectifiers are generally used with choke input filters to provide good regulation for class B amplifiers.

The direct voltage output of such a system is equal to 0.9 times the r.m.s. value of the input voltage minus the valve voltage drop—

e.g. Output voltage $= (0.9 \, E_{rms} - 15)$ volts.

(ii) Rectifier valves and types of service

Rectifier valves may be subdivided into the following groups :—

(1) High vacuum (a) High impedance (e.g. 5Y3-GT)

              (b) Medium impedance (e.g. 6X4, 5R4-GY)

              (c) Low impedance (e.g. 5V4-G, 35Z5-GT)

(2) Mercury vapour—(e.g. 82, 83).

The choice of a rectifier valve for a particular service must take into account the maximum permissible ratings for peak current, average current, and peak inverse voltage. The design of the following filter will influence these last two factors particularly ; the type of filter, either choke or condenser input, will be determined partly by the demands of power supply regulation. In supplies feeding Class A output stages the choice will probably be a condenser input filter, but where Class AB$_1$ and AB$_2$ output stages are to be supplied, the regulation of the power supply becomes a significant feature and choke input filters with low impedance rectifiers must be used.

The following examples represent typical practice—

A.C. radio receivers with Class A power stage :—
High vacuum full wave (e.g. 6X4, 5Y3-GT, 5U4-G).

A.C. radio receivers with Class AB₁ power stage :—
With self bias—high vacuum full wave (e.g. 5Y3-GT, 5U4-G, 5R4-GY, 5V4-G)
With fixed bias—low impedance high vacuum full wave (e.g. 5V4-G).

A.C./D.C. radio receivers :—
Indirectly-heated low impedance high vacuum half-wave types with heaters operating at 0.3 A or 0.15 A (e.g. 25Z6-GT or 35Z5-GT).
In England, heaters operating at 100 mA are widely used.

Battery operated radio receivers with non-synchronous vibrators :—
Indirectly-heated low or medium impedance high vacuum full-wave types (e.g. 6X4).

Amplifiers :—
As for radio receivers except that mercury vapour types may also be used.

In general for radio receiver and small amplifier design high vacuum rectifiers are to be preferred to mercury vapour types because of—

(1) long and trouble-free service ;

(2) the lower transformer voltage which can be used for the same d.c. output voltage when a condenser input filter may be used ;

(3) self protection against accidental over-load due to the fairly high internal impedance of the rectifier. Use can only be made of this last point when the supply is for use with a Class A output stage, when good regulation is not a major consideration and a high impedance rectifier may be used.

With directly-heated rectifiers it is generally found preferable to connect the positive supply lead to one side of the filament rather than to add the further complication of a centre-tap on the filament circuit.

Parallel operation of similar types of vacuum rectifiers is possible but it is preferable to connect together the two sections of a single full wave rectifier and to use a second similar valve as the other half-rectifier if full-wave rectification is required. With low impedance rectifiers as used in a.c./d.c. receivers (e.g. 25Z6-GT) it is desirable to limit the peak current by some series resistance. When two units are connected in parallel it is also desirable to obtain equal sharing, and in such cases a resistance of 50 or 100 ohms should be connected in series with each plate, then the two units are connected in parallel.

Mercury vapour rectifiers may only be connected in parallel if a resistance sufficient to give a voltage drop of about 25 volts is connected in series with each plate, in order to secure equal sharing of the load current.

(iii) The use of the published curves

From published curves on rectifier valves it is possible to predict the output voltage of a rectifier system when provided with the knowledge of the transformer voltage. For this purpose, use may be made of either the constant voltage curves or the constant current curves.

A family of curves is normally published both for condenser-input and for choke-input filters. In the former case the source impedance must be known ; it is usually published as the total effective plate supply impedance per plate.

An example of constant voltage curves, in this case applying to a condenser-input filter, is given in Fig. 30.2A. Each curve is for a specified constant supply voltage per plate ; for intermediate voltages it is possible to interpolate with sufficient accuracy. In this particular case there is the rather unusual feature of one value of total effective plate supply impedance per plate for the lower voltage curves (1 to 5) and a higher value

for the higher voltage curves (6 to 8).   It is always permissible to adopt a higher value
of total effective plate supply voltage per plate than that shown on the curves, but the
direct voltage output will thereby be decreased somewhat.   Operation is only permissible on, and below, the line formed by the highest curve and the " current and
voltage boundary line " ADK.   These curves only apply to one specified value of
capacitance input to the filter, in this case 10 μF.

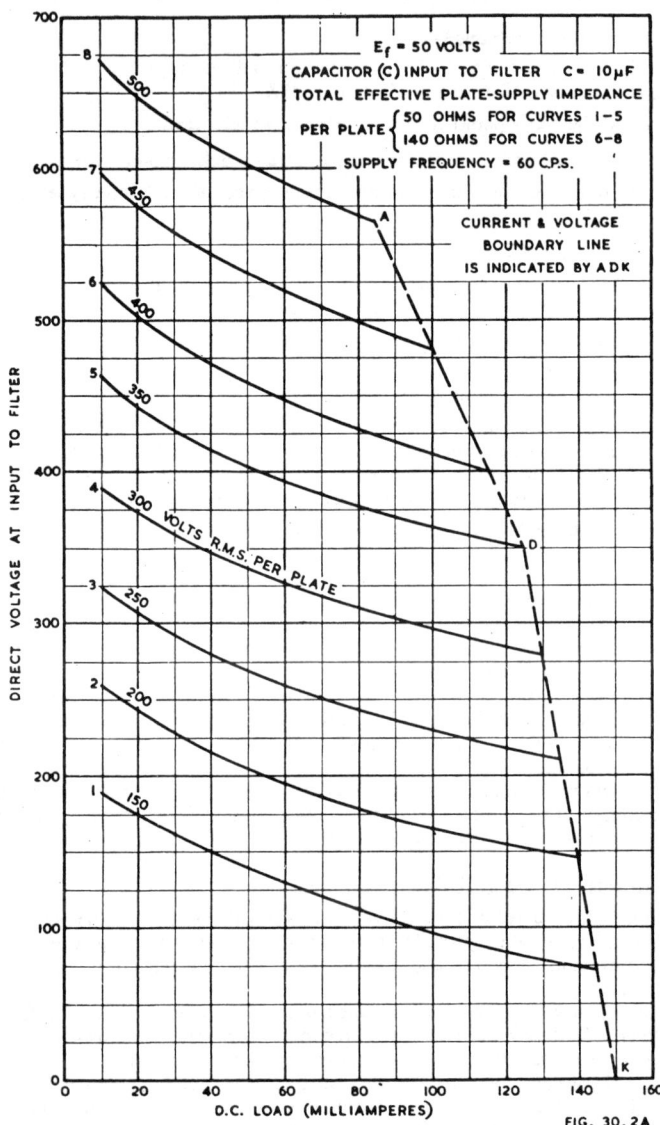

Fig. 30.2A. *Operation characteristics for a typical full-wave rectifier (5Y3-GT) with
condenser input filter.*

The effect of change in the value of capacitance input to the filter is indicated typically by Fig. 30.2B, where curves for 3 values of capacitance are drawn. A higher capacitance gives better regulation and a higher output voltage, but increase in capacitance beyond a certain value (here about 16 μF) has only a very slight effect. If curves are only drawn for one specified value of input capacitance, operation with a higher value is not permissible unless this has been demonstrated by measurement or calculation to be within the peak current rating of the rectifier.

Curves for a typical full wave rectifier with choke input are given in Fig. 30.2C. Operation is only permissible in the area to the right of the boundary line corresponding to the proposed choke size, to the left of the current and voltage boundary line CEK, and below the highest curve for a choke of infinite inductance.   If the direct current varies between two limits, it is important to select a value of inductance at least equal to, or preferably higher than, the value required for the lower limit of direct current ; the inductance should be measured at the lower limit of direct current.

The constant current curves (of which an example is given in Fig. 30.3) are very helpful for deriving certain information.   If it is required to find the input voltage

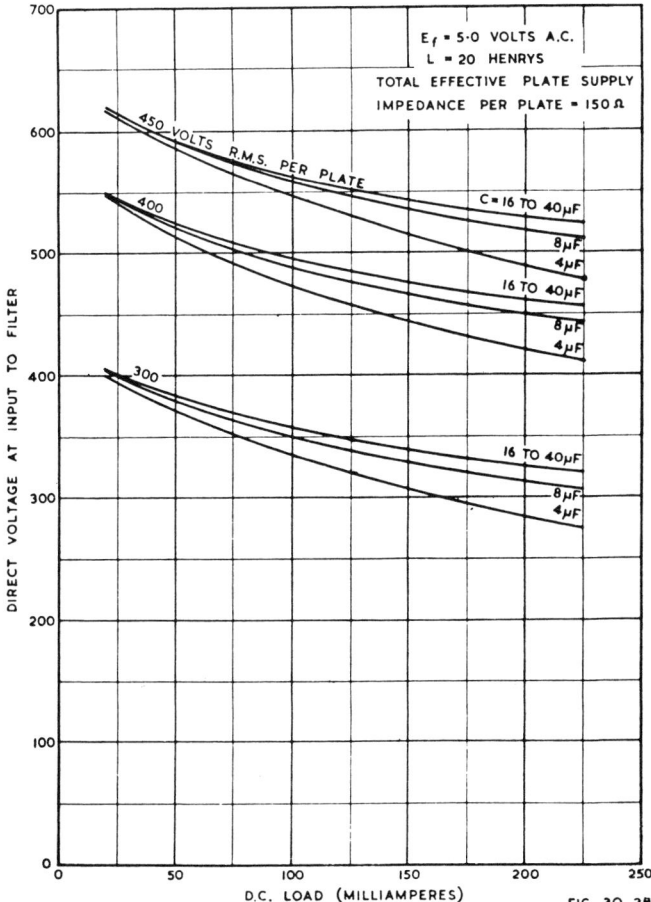

Fig. 30.2B. Operation characteristics for a typical full-wave rectifier (5T4) with condenser input filter, showing effect of input capacitance (C).

to give 250 volts 100 mA direct current, a vertical is drawn upwards from an output voltage of 250 volts to point P on the 100 mA cúrve. From here a horizontal line is drawn which intersects the vertical axis at 277 volts, which is the desired value.

The line OD is for equal input and output voltages. Above and to the left of this line the output voltage is less than the input voltage ; below and to the right of this line the output voltage is greater than the input voltage.

Each of the points A, B, C etc. at which the several current lines cut the vertical axis indicates the combined effective voltage drop in the valve and the transformer ;

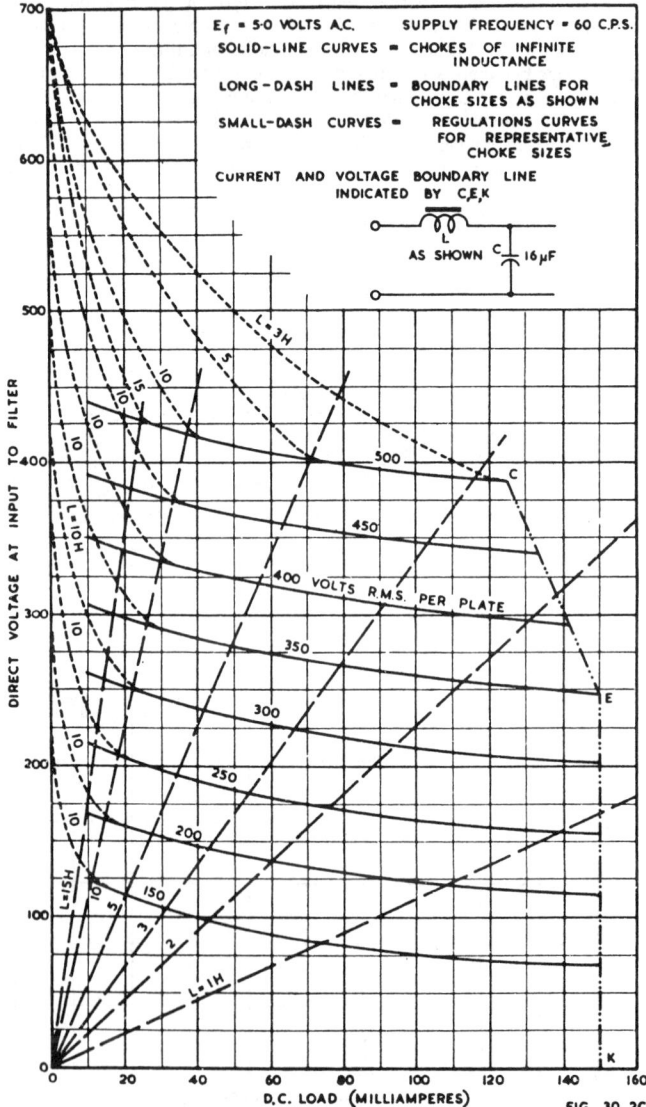

Fig. 30.2C.   *Operation characteristics for a typical full wave rectifier* (5Y3-GT) *with choke input filter.*

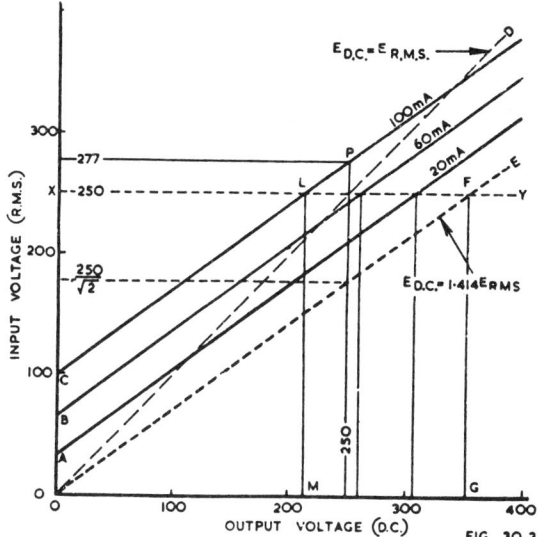

Fig. 30.3. Constant current curves for a 5Y3-GT rectifier ; condenser-input filter with capitance 8 µF and effective plate supply impedance 80 ohms per plate.

in other words it is the input voltage required to maintain the specified load current with the load terminals short-circuited.

The line OE is the theoretical limit of output voltage with no load-current, and is drawn to correspond to an output voltage of 1.414 times the r.m.s. input voltage.

The regulation of the output voltage with varying load currents is indicated by Fig. 30.3. If the input voltage is 250 volts, a horizontal line may be drawn as in XY and vertical lines may be drawn at each intersection with the constant current curves. Line FG is for no load-current, while LM is for 100 mA load current. For this example the output voltage will be seen to drop from 354 volts at no load-current to 213 volts at 100 mA load-current.

It should be noted that the results given by any form of valve curves are only correct for the set of conditions for which the curves were derived.

(iv) Selenium and copper oxide rectifiers

Selenium and copper oxide rectifiers have been used for miscellaneous applications such as for grid bias supplies and instrument rectifiers. Recently, selenium rectifiers have been widely used as plate supply rectifiers in radio receivers.

Both types differ from thermionic rectifiers in that they have appreciable reverse current. When used within their ratings, selenium rectifiers normally have a long life, although a small percentage of breakdowns occurs throughout life.

Copper oxide rectifiers are limited to a temperature rise of about 15 °C while selenium rectifiers may be operated at higher temperatures. Selenium rectifiers are smaller and lighter than copper oxide types, for the same operating conditions.

For further information see Refs. 14, 17, 21, 23, 24, 25, 26, 33.

SECTION 2 : RECTIFICATION WITH CONDENSER INPUT FILTER

(i) Symbols (ii) Rectification with condenser input filter (iii) To determine peak and average diode currents (iv) To determine ripple percentage (v) To determine the transformer secondary r.m.s. current (vi) Procedure when complete published data are not available (vii) Approximations when the capacitance is large (viii) Peak hot-switching transient plate current (ix) The effect of ripple.

(i) Symbols and definitions

$\hat{r}_d$	Effective peak resistance of diode, defined as the anode voltage at the conduction peak divided by the anode current at that time.
$\bar{r}_d$	Effective average resistance of diode, defined as the average anode voltage during the conduction period divided by the average anode current during that time.
$\lvert r_d \rvert$	Effective r.m.s. resistance of diode, defined as the diode anode dissipation divided by the square of the r.m.s. anode current.
R_s	Total resistance in series with diode including transformer winding resistance and any series resistance added to limit the diode peak current.
$\hat{R}_s$	Equal to $\hat{r}_d + R_s$
$\bar{R}_s$	Equal to $\bar{r}_d + R_s$
$\hat{i}_d$	Peak diode current (one anode)
$\bar{i}_d$	Average diode current (one anode)
$\lvert i_d \rvert$	r.m.s. diode current (one anode)
$\hat{e}_d$	Diode anode voltage at peak of conduction period
R_L	Load resistance presented to rectified supply
I_L	Load current from rectified supply
E_{dc}	Rectified direct output voltage across load resistance
$\hat{E}_\sim$	Peak value of alternating input voltage to rectifier
$\lvert E_\sim \rvert$	R.M.S. value of alternating input voltage to rectifier
C	Capacitance of first filter capacitor in farads
ω	$2\pi \times$ supply frequency in c/s
$\lvert E_R \rvert$	R.M.S. value of ripple voltage existing across condenser C
E_p	Peak-inverse voltage across diode
$\lvert I_R \rvert$	R.M.S. value of ripple current through condenser C
$\dfrac{\lvert E_R \rvert}{E_{dc}}$	$= \dfrac{\text{r.m.s. ripple voltage}}{\text{direct voltage across load}} =$ ripple factor.
$\lvert R_s \rvert$	$= \lvert r_d \rvert + R_s.$

(ii) Rectification with condenser input filter

It has been stated in Sect. 1(i) that the nature of the rectified current is pulsating and that it is necessary for the purpose of this simple equivalent circuit to convert the non-linear resistance of the diode to an equivalent linear resistance. For condenser input filters it can be shown (Ref. 6) that the relationship :—

$$\hat{r}_d = 0.88\ \bar{r}_d = 0.93 \lvert r_d \rvert$$

is correct within 5% for all circuits of this type and from the following graphs it is possible to assess the characteristics of a condenser input rectifier system.

Fig. 30.4 gives curves for a number of high vacuum rectifiers from which values of their peak resistance $\hat{r}_d$ can be found. Figs. 30.5—30.7 (based on Schade, Ref. 6)

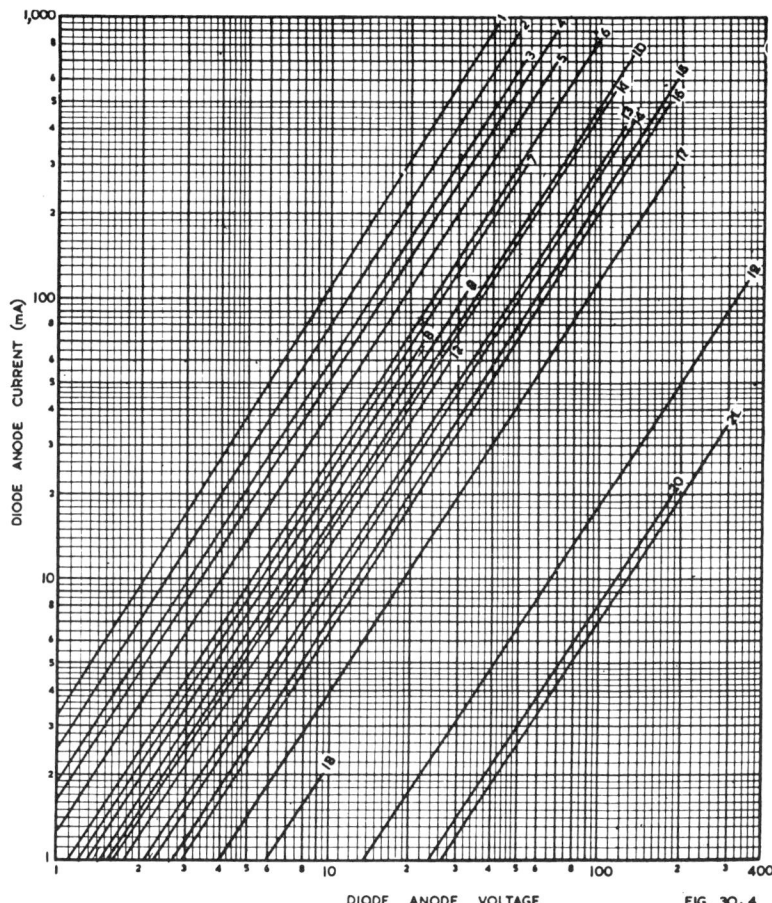

Fig. 30.4. Average anode characteristics of some rectifier valves (based on Ref. 6, with additions). The value of r_d for any rectifier at any diode current may be determined by dividing the diode anode voltage by the diode current.

Valve	Curve	Valve	Curve	Valve	Curve	Valve	Curve
1A3	10	6AQ6	19	6W4-GT	2	35Z5-GT	2
1B3-GT/8016	21	6AR7-GT	18	6X4	8	45Z3	17
1-V	7	6AT6	19	6X5*	7	45Z5-GT	1
2X2-A	19	6AV6	18	6ZY5-G	9	50Y6-GT	4
5R4-GY	11	6B6-G	18	7B6	18	80	14
5T4	6	6B8*	18	7C6	18	81	16
5U4-G	10	6G8-G	18	7E6	18	83I'	4
5V4-G	4	6H6*	12	7E7	18	84/6Z4	7
5W4*	13	6Q7*	18	12Z3	4	117N7-GT	2
5X4-G	10	6R7*	18	25Y5	9	117Z3	3
5Y3-GT	14	6SF7*	18	25Z5	4	117Z6-GT	3
5Y4-G	14	6SQ7*	18	25Z6*	4	217-C	16
5Z3	10	6SR7*	18	35W4	2	836	6
5Z4*	5	6ST7*	18	35Z3	15	878	20
6AL5	3	6SZ7*	18	35Z4-GT	2		

Includes G or GT equivalents.

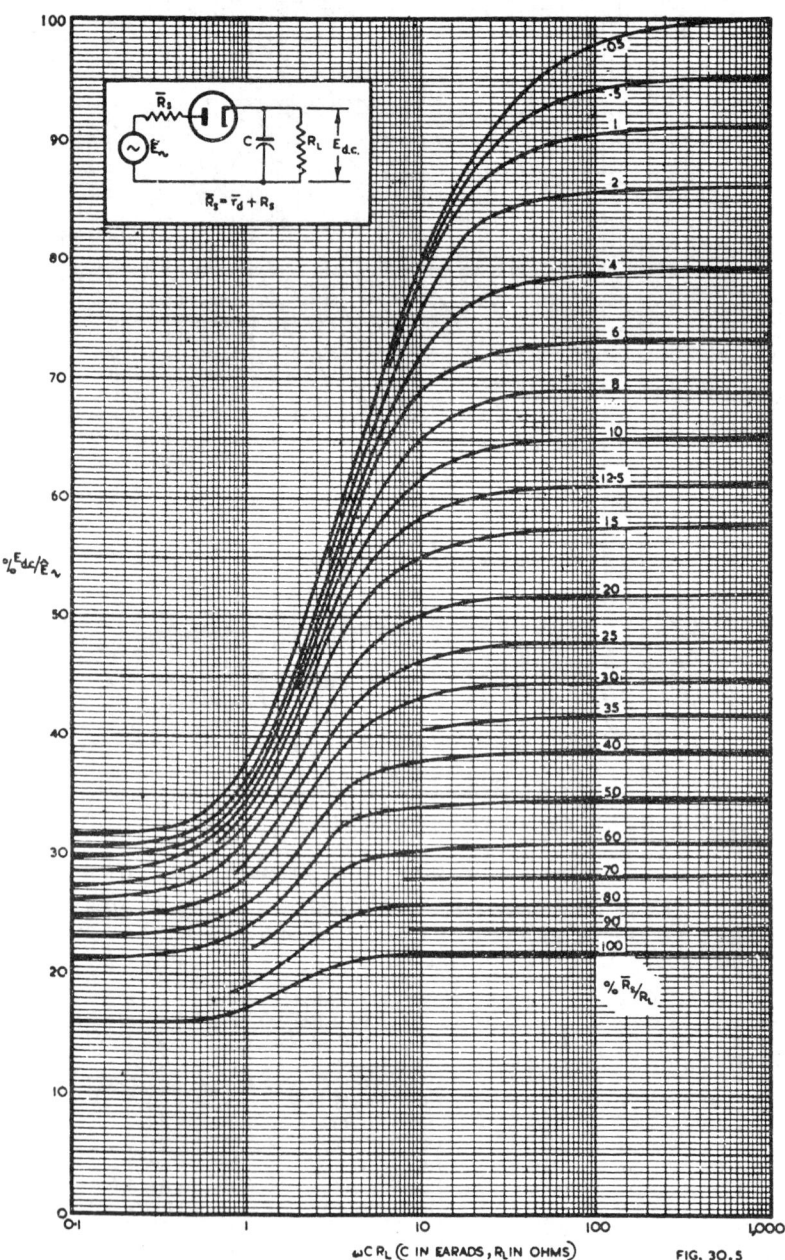

Fig. 30.5.   *Ratio of rectified (direct) output voltage to peak a.c. rectifier input voltage expressed as a percentage, as a function of* ωCR_L *for a half-wave rectifier with a condenser input filter (Ref. 6).*

are curves from which can be found the relationship between E_{dc} and $\hat{E}_{\sim}$ for half-wave, full-wave and full-wave voltage doubler circuits in terms of other circuit parameters. Curves of the ratios of effective $|i_a|$ and peak $\hat{i}_a$ diode currents to the direct current per anode $\bar{i}_a$ are given in Fig. 30.8 ; Fig. 30.9 gives details of the ripple factor and Fig. 30.10, the peak inverse voltage (all based on Ref. 6).

The design considerations to be borne in mind when using these curves are :—

(1) The value of the capacitance C is usually chosen with regard to the maximum permissible ripple in the output (see below) but if $|E_{\sim}|$ is limited to a certain value and the maximum E_{dc} is to be achieved, C may be increased above this value. In doing this due regard must be given to the maximum permissible peak current of the rectifier and, if necessary, limiting resistors placed in series with the anodes of the rectifier.

(2) In order that the direct voltage should not be closely dependent upon the value of C, the value of ωCR_L must be on or to the right of the knee of the appropriate curve in Fig. 30.5, 30.6 or 30.7 as required by the type of rectification.

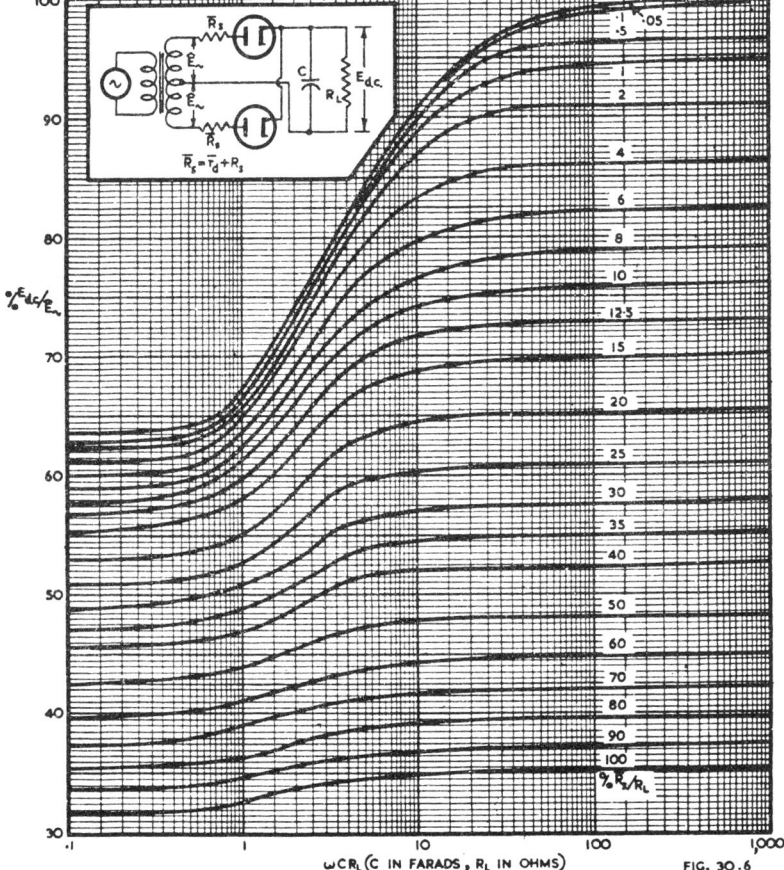

Fig. 30.6. *Ratio of rectified (direct) output voltage to peak a.c. rectifier input voltage, expressed as a percentage, as a function of ωCR_L for a full-wave rectifier with a condenser input filter (Ref. 6).*

(iii) To determine peak and average diode currents

This method is for use when complete published data are available such as are usually supplied by the operation characteristics or equivalent published data.

The procedure is illustrated by an example based on type 5Y3-GT as a full wave rectifier under the following conditions :

r.m.s. voltage $= |E_\sim| = 350$ volts,

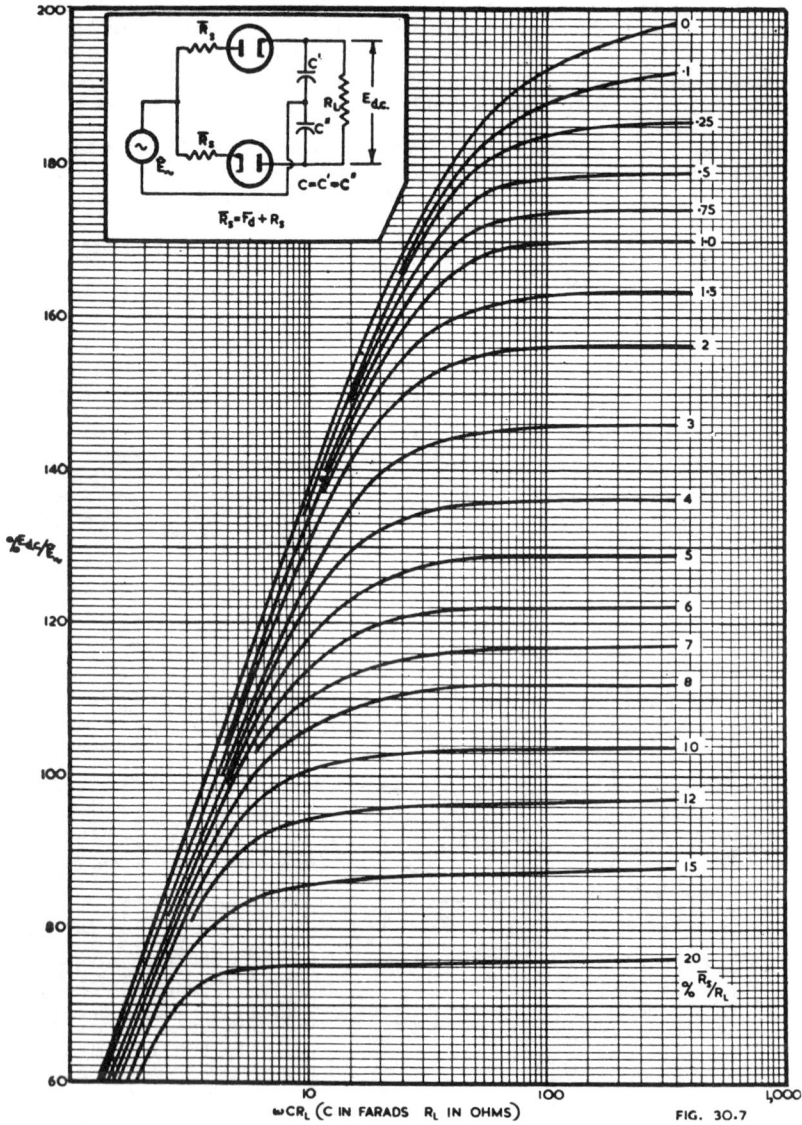

Fig. 30.7. Ratio of rectified (direct) output voltage to peak a.c. rectifier input voltage, expressed as a percentage, as a function of ωCR_L for a condenser-input voltage doubler (Ref. 6).

load current $= I_L = 125$ mA
voltage across load $= E_{dc} = 350$ volts (from Fig. 30.2A),
load resistance $= R_L = 350/0.125 = 2800$ ohms,
total effective plate supply impedance per plate $= R_s = 50$ ohms (from Fig. 30.2A),
$C = 10$ μF (from Fig. 30.2A),
$f = 60$ c/s (from Fig. 30.2A), whence $\omega = 378$.
$\omega CR_L = 378 \times 10 \times 10^{-6} \times 2800 = 10.6$.

Step 1. Determine $\dfrac{E_{dc}}{\overset{\wedge}{E_\sim}}\% = \dfrac{350}{350\sqrt{2}} = 70.7\%$.

Step 2. From Fig. 30.6, knowing $\omega CR_L = 10.6$, we obtain
$$\overline{R}_s/R_L = 13.5\%,$$
whence $\overline{R}_s = 13.5 \times 2800/100 = 378$ ohms.

Step 3. $\overline{r}_d = \overline{R}_s - R_s = 378 - 50 = 328$ ohms.

Step 4. $\hat{r}_d = 0.88\ \overline{r}_d = 0.88 \times 328 = 288$ ohms.

Step 5. $\hat{R}_s = R_s + \hat{r}_d = 50 + 288 = 338$ ohms.

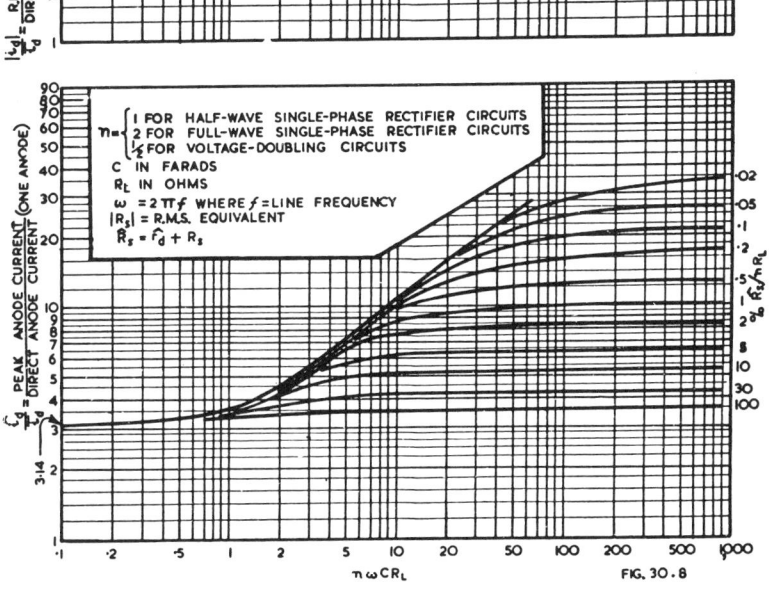

Fig. 30.8 (above). Ratio of r.m.s. diode current to average diode current for one anode, expressed as a function of.ωCR_L; (below) Ratio of peak diode current to average diode current for one anode, expressed as a function of ωCR_L (Ref. 6).

Step 6. $\dfrac{\hat{R}_s}{R_L}\% = \dfrac{338}{2800} \times 100 \doteq 12.1\%.$

Step 7.   Knowing that $n = 2$ for full wave rectification,

$$\frac{\hat{R}_s}{nR_L} = \frac{12.1}{2}\% = 6.05\%.$$

Step 8.   From Fig. 30.8 (lower) where $n\omega CR_L = 21.2$ we may obtain

$$\frac{\hat{i}_d}{\bar{i}_d} = 6.$$

Step 9.   The **average diode current** $\bar{i}_d$:
$\bar{i}_d = I_L$ for half-wave circuits and full-wave voltage doubler circuits,
$\bar{i}_d = \tfrac{1}{2}I_L$ for full-wave circuits.
In this example $\bar{i}_d = 125/2 = 62.5$ mA.

Step 10.   The **peak diode current**, is obtained by substituting the value of i_d given by Step 9 in the result of Step 8.   In the example

$$\hat{i}_d = 6\,\bar{i}_d = 6 \times 62.5 = 375 \text{ mA}.$$

(N.B. the maximum rating is 400 mA).

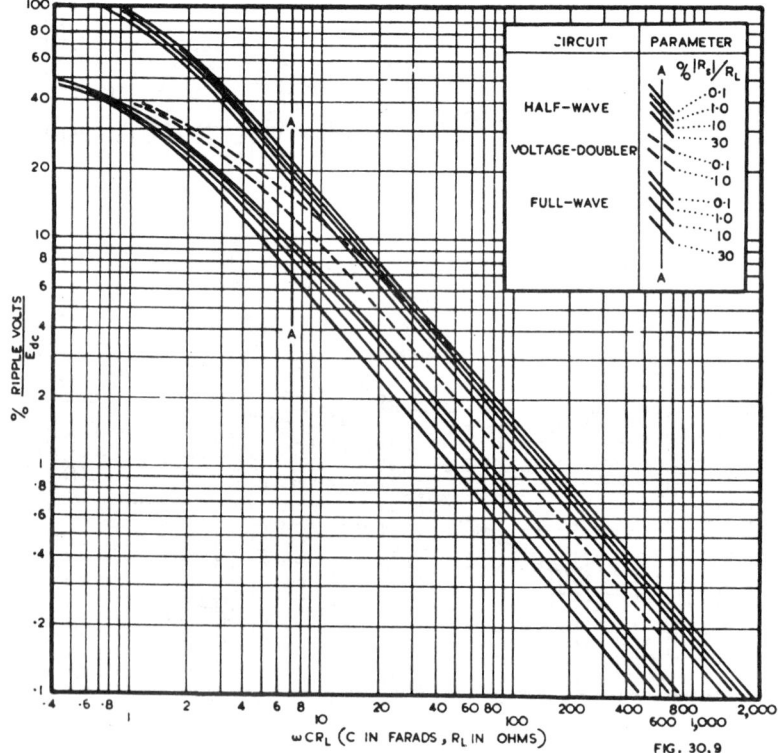

Fig. 30.9.   *Curves for the determination of the ripple factor of condenser input filter rectifier circuits (Ref. 6).*

(iv) To determine ripple percentage

Having determined the peak diode current, we may then proceed to calculate the other unknowns. The same example as in (iii) above is also used here.

Step 1. $|r_d| = \hat{r}_d/0.93 = 288/0.93 = 310$ ohms.

Step 2. $|R_s| = R_s + |r_d| = 50 + 310 = 360$ ohms.

Step 3. $\dfrac{|R_s|}{R_L}\% = \dfrac{360}{2800} \times 100 = 12.8\%.$

Step 4.   Applying this value to Fig. 30.9 (note the values shown in the inset, applying to the various curves), and using the value of ωCR_L determined above, the percentage of ripple voltage to direct voltage is given.

In the example, $\dfrac{|R_s|}{R_L} = 12.8\%$ (full wave) and $\omega CR_L = 10.6$ giving

$\dfrac{\text{ripple voltage}}{\text{direct voltage}} = 5.5\%.$

The ripple voltage $|E_R| =$ ripple percentage $\times$ direct voltage.

(v) To determine the transformer secondary r.m.s. current $|i_d|$

Knowing the value of $\bar{i}_d$ as determined in (iii) Step 9 above, also the values of n, ωCR_L and $|R_s|/R_L$, Fig. 30.8 (upper curves) will give the value of $|i_d|$.

In the same example $\bar{i}_d = 62.5$ mA ; $n = 2$; $\omega CR_L = 10.6$ and $|R_s|/R_L = 12.8\%$, so that $|i_d|/\bar{i}_d = 2.25$ and $|i_d| = 62.5 \times 2.25 = 140$ mA.

(vi) Procedure when complete published data are not available

Step 1.   To determine $\bar{i}_d$.

$\bar{i}_d = I_L$ for half-wave circuits and full-wave voltage doubler circuits,

$\bar{i}_d = \frac{1}{2}I_L$ for full-wave circuits.

Step 2.   The diode peak current $\hat{i}_d$ is tentatively assumed to be $6\,\bar{i}_d$. Alternatively, if the output voltage is known, the current ratio may be derived from Fig. 30.10A.

Step 3.   From Fig. 30.4, and knowledge of the valve type, the diode peak plate voltage $\hat{e}_d$ corresponding to $\hat{i}_d$ can be found.   Therefore $\hat{r}_d = \hat{e}_d/\hat{i}_d$ can be evaluated.

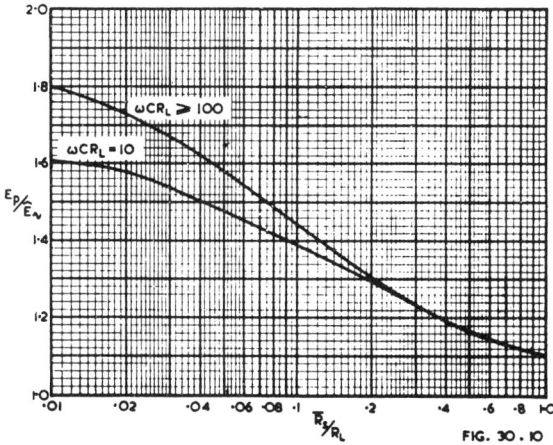

Fig. 30.10.   Ratio of operating peak inverse voltage to peak applied a.c. for rectifiers used in condenser input filter circuits.

Step 4. Calculate $\hat{R}_s = \hat{r}_d + R_s$.

Step 5. Using Fig. 30.8 (lower curves), and knowing $n\omega C R_L$, $\hat{R}_s$ and nR_L, determine $\hat{i}_d/\bar{i}_d$. If this differs appreciably from the assumed value, repeat steps 3, 4 and 5.

Step 6. Calculate $|r_d| = \hat{r}_d/0\ 93$.

Step 7. Calculate $|R_s| = R_s + |r_d|$.

Step 8. Calculate the percentage $|R_s|/R_L$ and apply to Fig. 30.9 to determine the ripple percentage.

Step 9. Calculate $|R_s|/nR_L$ and apply to Fig. 30.8 (upper curves) to determine $|i_d|/\bar{i}_d$, and thence $|i_d|$ which is the transformer secondary r.m.s. current.

Step 10. Calculate $\bar{r}_d = \hat{r}_d/0.88$.

Step 11. Calculate $\bar{R}_s = R_s + \bar{r}_d$.

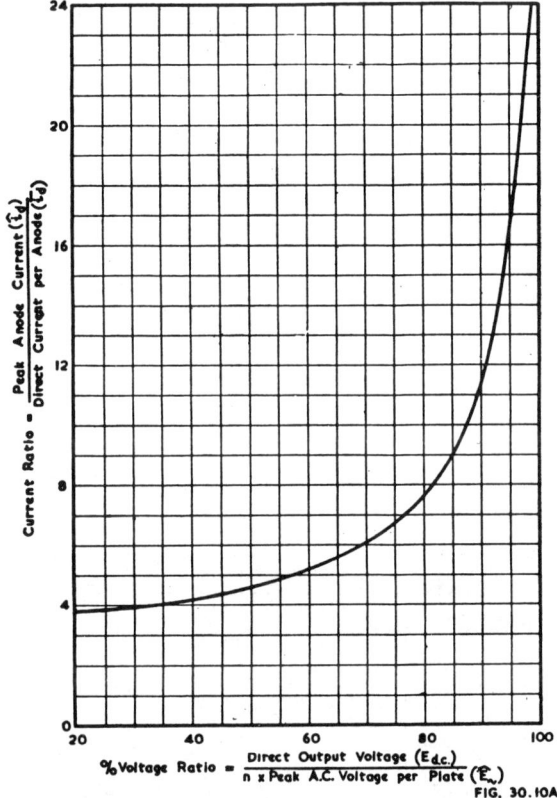

FIG. 30.10A

Fig. 30.10A. *This curve is for full-wave, half-wave and voltage-doubler rectifiers with condenser-input filters. It applies for any size of condenser so long as the condenser is large enough to give maximum output voltage for the given output current and r.m.s. voltage input. For half-wave and voltage-doubler rectifiers the direct current per anode* $\bar{i}_d = I_L$; *for full wave rectifiers* $i_d = \frac{1}{2}I_L$.
  $n = 1$ for full-wave and half-wave rectifiers.
  $n = 2$ for voltage doubler rectifiers.

Step 12.   Calculate the percentage $\bar{R}_s/R_L$.
Step 13.   Using Fig. 30.5 for half wave rectification,
or Fig. 30.6 for full wave rectification,
or Fig. 30.7 for voltage doubler circuits,

determine the percentage $E_{dc}/\hat{E}_\sim$.
Step 14.   The transformer secondary r.m.s. voltage per plate is given by

$$|E_\sim| = \frac{70.7\ E_{dc}}{\text{percentage } E_{dc}/\hat{E}_\sim}$$

Step 15.   The peak inverse voltage under operating conditions is given by Fig.
30.10, but the peak inverse voltage under no-load conditions is equal to $2\hat{E}_\sim$ and
the latter must not exceed the valve rating.
Step 16.   It is then necessary to confirm that none of the maximum valve ratings
has been exceeded.

Example of procedure when complete published data are not available.
Assume type 5Y3-GT as full wave rectifier with $E_{dc} = 350$ volts, $I_L = 125$ mA,
$R_s = 50$ ohms, $f = 60$ c/s, $C = 10$ μF. We may derive directly $R_L = 2800$ ohms,
$n = 2$, $\omega = 378$, $\omega C R_L = 10.6$.

Step 1. $\bar{i}_d = \frac{1}{2} I_L = 62.5$ mA.

Step 2.   Assume $\hat{i}_d = 6\ \bar{i}_d = 375$ mA.

Step 3.   From Fig. 30.4, curve 14, $\hat{e}_d = 123$ volts.

Therefore $\hat{r}_d = \hat{e}_d/\hat{i}_d = 123/0.375 = 328$ ohms.

Step 4. $\hat{R}_s = \hat{r}_d + R_s = 328 + 50 = 378$ ohms.
Step 5.   Fig. 30.8 (lower curves), where $n\omega C R_L = 21.2$

and $\dfrac{\hat{R}_s}{nR_L}\% = \dfrac{378 \times 100}{2 \times 2800} = 6.75\%$,

gives $\hat{i}_d/\bar{i}_d = 5.8$, which differs so slightly from the assumed value of 6 that the calculated values of effective diode resistance may be taken as sufficiently accurate.

If the value of $\hat{i}_d/\bar{i}_d$, as calculated above, differed appreciably from the assumed value,
it would be necessary to repeat steps 3, 4 and 5.

Peak current $\hat{i}_d = 5.8 \times 62.5 = 362$ mA.

Step 6. $|r_d| = \hat{r}_d/0.93 = 328/0.93 = 352$ ohms.
Step 7. $|R_s| = R_s + |r_d| = 50 + 352 = 402$ ohms.
Step 8. $(|R_s|/R_L) \times 100 = 402 \times 100/2800 = 14.4\%$.
Applying to Fig. 30.9 with $\omega C R_L = 10.6$ gives ripple percentage $= 5.5\%$.
Step 9. $(|R_s|/nR_L) \times 100 = 7.2\%$. Applying this to Fig. 30.8 (upper curves)
gives $|i_d|/\bar{i}_d = 2.2$.
Therefore transformer secondary r.m.s. current $= 2.2 \times 62.5 = 138$ mA.

Step 10. $\bar{r}_d = \hat{r}_d/0.88 = 328/0.88 = 373$ ohms.
Step 11. $\bar{R}_s = R_s + \bar{r}_d = 50 + 373 = 423$ ohms.
Step 12. $(\bar{R}_s/R_L) \times 100 = 423 \times 100/2800 = 15.1\%$.

Step 13.   Using Fig. 30.6, $E_{dc}/\hat{E}_\sim = 68.8\%$.
Step 14.   The transformer secondary r.m.s. voltage per plate is equal to
$\dfrac{70.7 \times 350}{68.8} = 360$ volts.

It will be noticed that these values agree within 3% with those determined by the
other procedure.

1180 (vii) APPROXIMATIONS WHEN THE CAPACITANCE IS LARGE 30.2

(vii) Approximations when the capacitance is large

Figs. 30.5, 6 and 7 indicate that, when a certain value of ωCR_L has been reached, all curves flatten out. In other words, if we increase the value of the input capacitance C, the output voltage and hence the output current remain constant above a certain value of C. Similarly with the peak current, as indicated by Fig. 30.8.

It is therefore possible to adopt a simplification when the input capacitance is sufficiently large so that any further increase in C does not have much effect on the direct voltage output. Fig. 30.10A enables the ratio of peak to average (direct) currents to be calculated from a knowledge of the voltage ratio and the type of circuit. It may be used as a fair approximation for most typical radio receivers in which $C \geq 16$ μF for full-wave or 32 μF for half-wave operation. It should not be used in cases where the circuit impedance is very low, such as a half-wave rectifier in transformerless receivers with no added resistance.

A further approximation, which holds under the same conditions as outlined above, may be used when it is desired to reach the maximum rated direct current per plate and the maximum rated peak current per plate simultaneously, when the latter is six* times the former. Under these conditions.

$$\hat{R}_s \approx 0.06 \; nR_L$$

where $\hat{R}_s = \hat{r}_d + R_s$

$\hat{r}_d$ = effective peak resistance of diode
R_s = total effective plate supply impedance per plate
n = 1 for half-wave rectification
    = 2 for full-wave rectification.
and R_L = load resistance.

Also under the same conditions
$E_{dc} = 0.69 \times$ peak supply voltage
       $= 0.975 \times$ r.m.s. supply voltage.
N.B. These relationships are derived from Figs. 30.8 and 30.10A.

(viii) Peak hot-switching transient plate current

The peak hot-switching transient plate current is the current which the diode must carry if the load resistance R_L is short-circuited. This occurs in a practical case when a diode is " hot-switched." The peak hot-switching transient plate current is given by (Ref. 15) :

$$\hat{I}_{mas} = \frac{\hat{E}_{\sim}}{R_s + \hat{r}_{ds}}$$

where $\hat{I}_{mas}$ = peak hot-switching transient plate current in amperes
$\hat{E}_{\sim}$ = peak alternating voltage per plate
R_s = total effective plate supply resistance per plate
and $\hat{r}_{ds}$ = diode resistance when hot-switching current is at its maximum.

The value of $\hat{r}_{ds}$ may be derived from Fig. 30.4, by extending the curves upwards if necessary. If the hot-switching current is greater than 1 ampere, but less than 10 amperes, the resistance may be read from the curves at one tenth of the current value, and the resistance value so derived must then be multiplied by 0.47 (this has an accuracy within about 2% for curves 1 to 17 inclusive). For example, type 5Y3-GT has a rated maximum hot-switching transient plate current of 2.2 amperes per plate. The diode resistance at a plate current of 220 mA is given by Fig. 30.4 curve 14 as 85/0.22 = 386 ohms. The diode resistance $\hat{r}_{ds}$ at a plate current of 2.2 amperes is therefore 386 × 0.47 = 182 ohms. Continuing with the same example, if the peak

---
*This ratio is very commonly used in diode ratings.

alternating voltage per plate is $350\sqrt{2}$, and the peak hot-switching transient current is not to exceed 2.2 amperes, then

$$R_s + \hat{r}_{as} = 350 \times 1.41/2.2 = 225 \text{ ohms.}$$

But $\hat{r}_{as} = 182$ ohms.

Therefore $R_s = 43$ ohms minimum.

(ix) The effect of ripple

The filter condenser C is required to carry a substantial ripple current, the value of which is given approximately by

$$|I_R| = |E_R|\omega C \text{ for half-wave rectification}$$

or $\quad |I_R| = |E_R|2\omega C$ for full-wave rectification

where $\omega = 2\pi f$

$\quad f$ = supply frequency

$\quad |E_R|$ = ripple voltage r.m.s.

$\quad |I_R|$ = ripple current in amperes r.m.s.

and C is measured in farads.

For example, with a supply frequency of 60 c/s, full-wave rectification and $C = 10\ \mu\text{F}$, a ripple voltage of 20 volts r.m.s. will cause a ripple current of 150 mA. This is the maximum permissible for a dry electrolytic condenser with 450 V working voltage, under JAN-C-62 specification [Chapter 38 Sect. 3(x)]. The maximum permissible ripple currents vary with the capacitance, the working voltage, manufacturer and type. Ripple current ratings of a typical English manufacturer are given on page 194 ; they differ considerably from JAN-C-62.

For any predetermined choice of condenser, temperature voltage and value of R_s/R_L there is a maximum load current which can be drawn from the rectifier without exceeding the ripple current ratings. Based on JAN-C-62, with $(R_s/R_L) = 10\%$ and 60 c/s full wave rectification we have :

Capacit-ance	350 volt (d.c.) working			450 volt (d.c.) working		
	max ripple current	max. load current	load curr. / ripple curr.	max. ripple current	max. load current	load curr. / ripple curr.
10 μF	140 mA	120 mA	86%	150 mA	130 mA	86%
20 μF	180 mA	154 mA	85%	180 mA	150 mA	83%
30 μF	200 mA	168 mA	84%	200 mA	168 mA	84%

Based on English T.C.C. condensers, 450 volt (d.c.) working, ambient temperature 40°C, $(R_s/R_L) = 10\%$ and 50 c/s full wave rectification we have :

Capacit-ance	Plain foil			Etched foil		
	max. ripple current	max. load current	load curr. / ripple curr.	max. ripple current	max. load current	load curr. / ripple curr.
8 μF	—	—	—	67 mA	57 mA	85%
16 μF	260 mA	220 mA	85%	122 mA	103 mA	85%
32 μF	405 mA	344 mA	85%	—	—	—

In the cases listed above, the load current is approximately 85% of the maximum ripple current ; this only applies for $(R_s/R_L) = 10\%$. Values for three conditions are given below :

(R_s/R_L)	1%	10%	30%
$\dfrac{\text{load current}}{\text{ripple current}}$	73%	85%	102% approx.

In practice, in radio receivers and a-f amplifiers, (R_s/R_L) is usually well within the extreme limits 1% and 30%.

If the ripple current for any desired condition is greater than the permissible limit, the capacitance of a single unit condenser may be increased and/or the value of R_s may be increased either by selecting a different valve type or adding resistance in series with each plate. Either method of increasing R_s will require a higher transformer voltage. Alternatively two condensers may be connected in parallel, with the total capacitance unchanged, each of which will carry part of the ripple current ; however, equal sharing of current cannot be guaranteed and a large safety margin is desirable. The parallel arrangement confers no appreciable benefits with the T.C.C. ratings, as compared with an increase in capacitance of a single unit.

In general, load currents up to 120 mA may safely be employed with plain foil or 70 mA with etched foil without any investigation.

SECTION 3 : RECTIFICATION WITH CHOKE INPUT FILTER

(*i*) *Rectification with choke input filter*   (*ii*) *Initial transient current.*

(i) Rectification with choke input filter

Where good voltage regulation is required, choke input filters are to be preferred. In this type of circuit, providing the first choke L_1 (Fig. 30.1G) is above a certain critical value L_c, the rectifier valve works under conditions of continuous current current flow and in the ideal case where L_1 is of infinite inductance there would be no fluctuations in this current.

It has been shown (Ref. 6) that L_c should be equal to or greater than

$$\frac{R_s + R_L}{6\pi f} \text{ for full-wave operation} \tag{1}$$

where R_s = total resistance in series with diode
      R_L = load resistance presented to rectified supply
and f = supply frequency.

As an approximation, if R_s is small compared with R_L,

$$L_c \geqq \frac{R_L}{940} \text{ for a 50 c/s supply} \tag{2}$$

and $$L_c \geqq \frac{R_L}{1130} \text{ for a 60 c/s supply.} \tag{3}$$

The above formulae are only stated for full wave rectification as it is not normal practice to use a choke input circuit with a half-wave rectifier owing to the low output voltage which would result.

In applications where the load resistance varies considerably, for example in Class B amplifiers, it is usual to place a bleeder across the supply, thereby reducing the initial value required for L_1. As the required critical value of L_1 decreases with increased load current, the choke can be made with a smaller air gap than necessary for constant inductance at all loads ; the drop in inductance due to d.c. polarization is permissible providing the inductance does not drop below L_c at any load.

This is known as a **swinging choke** ; its design is covered on pages 249 and 250.

The peak diode current is given approximately by

$$\hat{i}_d = I_L + \frac{\hat{E}_\sim \times 0.425}{2\omega L_1 - 1/(2\omega C_1)} \tag{4}$$

or, if $1/(2\omega C_1)$ is very much less than $2\omega L_1$, then

$$\hat{i}_d \approx I_L + \frac{\hat{E}_\sim \times 0.212}{\omega L_1} \tag{5}$$

If $L_1 = L_c = R_L/6\pi f$, then $\hat{i}_d \approx 2I_L$. $\tag{6}$

If $L_1 = 2L_c = R_L/3\pi f$, then $\hat{i}_d \approx 1.5\ I_L$. $\tag{7}$

The approximation in equation (4) is due to the neglect of ripple frequencies higher than twice the supply frequency.

If the rectifier valve has a peak current rating equal to or greater than twice the maximum direct load current rating, and if the value of L_1 never falls below L_c, it is not necessary to calculate peak diode currents. The same holds when the peak current rating is equal to or greater than twice the maximum direct current to be drawn from the rectifier. Owing to the desire to limit the rectifier peak current, it is preferable to make the inductance at least $2L_c$ at the highest load current. The drop in inductance at maximum load current will result in a reduction of filtering, but this is not likely to cause any trouble. As the choke will normally be followed by a filter capacitance (C_1 in Fig. 30.1G), any reduction in the inductance of L_1 below L_c at any value of load will cause the rectifier system to take on the characteristics of a condenser input filter and the output voltage will rise.

It is important to remember, when measurements are being made on a filter choke to determine its suitability for use in choke input circuits, that due consideration should be given to the large value of a.c. potential which will exist across it under working conditions. This potential will increase the inductance at low values of d.c. polarization provided that the sum of the a.c. and d.c. fluxes does not cause saturation of the core.

For the accurate calculation of voltage output and regulation of a choke input type filter, the voltage drop due to the resistance of the choke, rectifier and supply must be taken into consideration. The choke resistance can be easily ascertained and the rectifier resistance may be derived from the curves of Fig. 30.4 using the method outlined below. The supply resistance in series with the anode, in a.c. operated equipments, will be equal to the transformer winding resistance [for calculation see pages 99 and 100] plus any added series resistance. In a.c./d.c. equipment it will be equal to the value of the limiting resistor in series with the anode. It is assumed that the rectifier will be operating under conditions of continuous current flow ; it can be shown (Ref. 6) that $\hat{r}_d \approx \bar{r}_d \approx |r_d|$ also that the average anode current (one anode) $\bar{i}_d = I_L/2$. The procedure is best illustrated by an example.

Example :—It is desired to design a power supply with choke input filter to deliver 0 to 200 mA at 350 volts using a 5U4-G rectifier at 50 c/s. The choke is assumed to have 100 ohms resistance and the effective supply resistance per anode is 75 ohms. In order to reduce the initial value of L_1, a bleeder to take 20 mA is assumed.

At 20 mA : $R_L = 350/0.02 = 17\,500$ ohms

$\qquad L_1 = L_c = 17\,500/940 = 18.6$ H (minimum).

$\qquad \bar{i}_d = 20/2 = 10$ mA.

$\qquad \hat{i}_d = 2 \times 20 = 40$ mA.

Average anode current during conduction $= 20$ mA.

Referring to Fig. 30.4 (curve 10) : anode voltage corresponding to 20 mA is 11.7 volts.

$\qquad \bar{r}_d = 11.7/0.02 = 585$ ohms.

At 220 mA : $R_L = 350/0.22 = 1590$ ohms.

$$L_1 = 2L_c = 2 \times 1590/940 = 3.4 \text{ H (minimum)}.$$
$$\bar{i}_d = 220/2 = 110 \text{ mA}.$$
$$\hat{i}_d = 1.5 \times 220 = 330 \text{ mA}.$$

Average anode current during conduction $= 220$ mA.
Referring to Fig. 30.4 (curve 10) : anode voltage corresponding to 220 mA is 59 volts.
$$\bar{r}_d = 59/0.22 = 268 \text{ ohms}.$$

Voltage drop due to resistance of supply, valve and choke
At 220 mA : $= 0.22 (75 + 268 + 100) = 98$ volts

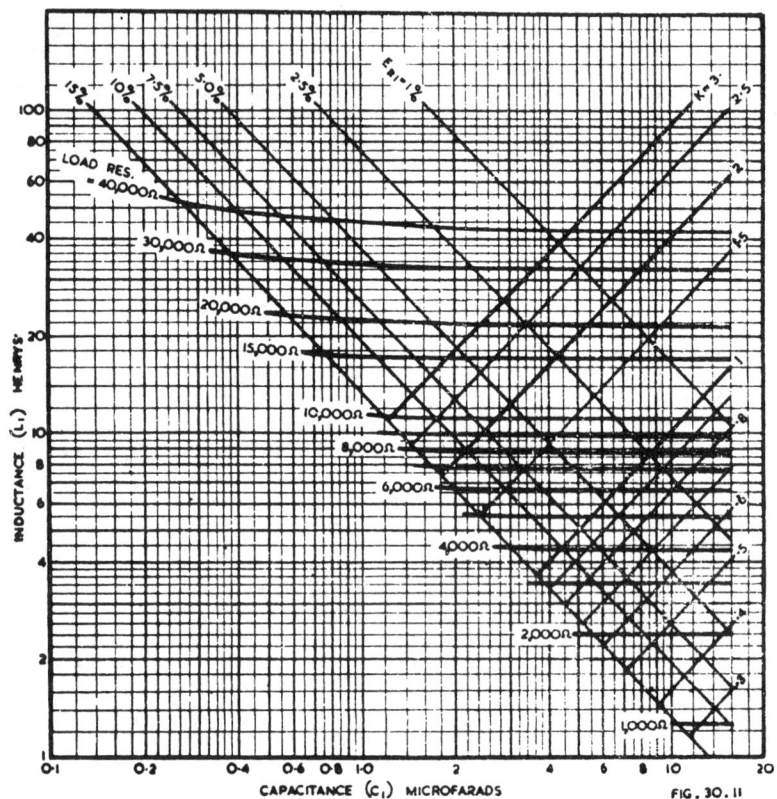

Fig. 30.11. Curves for the determination of the ripple factor of choke input rectifier circuits (based on Ref. 27), applying to the circuit of Fig. 30.1G for full wave rectification on a 50 c/s supply. The curves may be applied to any other supply frequency by multiplying values of inductance and capacitance by 2 for 25 c/s, 1.25 for 40 c/s, or 0.83 for 60 c/s. The ripple curves may be used independently of the K curves and load resistance curves to derive the ripple factor (i.e. ripple voltage E_{R1} expressed as a percentage of the direct load voltage). The operating point at any value of load resistance which occurs in practice should be above (preferably well above) the corresponding load resistance curve, thus determining the inductance L_1. In order to limit the initial (starting) transient current to the maximum peak current rating of the rectifier valve, the operating point should be above and to the left of the corresponding K curve, where

$$K = \frac{\text{r.m.s. voltage per anode } |E_\sim|}{1110 \times \text{peak plate current rating of diode}}$$

At 20 mA : = 0.02 (75 + 585 + 100) = 15 volts.

R.M.S. transformer voltage = (350 + 98)/0.9 = 498 volts.

Voltage across load at 20 mA = (498 × 0.9) − 15 = 449 volts.

Value of C_1 : Assume 10% ripple at maximum current, and referring to Fig. 30.11 for L_1 = 3.4 H, we obtain C_1 = 5.7 μF (8 μF would probably be used).

These results, which show a change of output of 350 to 449 volts when the load is changed from 220 to 20 mA, appear greater than the figures given by the valve curves ; this is because the choke voltage drop has also been taken into consideration.

From the example above it can be seen that the rectifier voltage drop is neither negligible nor constant ; in the case of a 5U4-G it varies from 11.7 volts at 20 mA to 59 volts at 220 mA.

If a lower impedance rectifier such as a 5V4-G had been used, this would have been reduced to 5.4 volts at 20 mA and 27 volts at 220 mA. For practical purposes of calculating the voltage output, a constant value of 16 volts could then be assumed, which would give an error of not more than 3.2%.

Also from the example it can be seen that the effect on output voltage produced by the 75 ohms supply resistance is fairly small, reducing the output by 1.5 volts at 20 mA and 16.5 volts at 220 mA.

When calculations have been completed, a check should be made to see that none of the maximum ratings given in the published data has been exceeded. This will include the value of peak inverse voltage E_p which in a choke input rectifier system will not be greater than 1.65 $\hat{E}_{\sim}$ provided that L_1 does not drop below L_c at any point. If L_1 is lowered below its critical value for any reason, E_p will approach the value for a condenser input rectifier system (see Sect. 2).

(ii) Initial transient current

When initially switching the anode circuit, with the cathode hot, there is a transient current in excess of the steady direct load current. It may be limited to a value equal to the peak anode current rating of the rectifier if the value of the inductance in henrys is equal to, or greater than

$$\left(\frac{|E_{\sim}|}{1110 \times I_{max}} \right)^2 C_1$$

where I_{max} = peak anode current rating of .the valve, in amperes
and C_1 = capacitance in microfarads.

This requirement may be met by ensuring that the L_1 and C_1 values applied to Fig. 30.11 meet at a point on or above the corresponding K curve.

SECTION 4 : TRANSFORMER HEATING

For purposes of calculating transformer heating it is necessary to know the equivalent r.m.s. current in the winding supplying the rectifier.

In the case of **condenser input filters** the r.m.s. value of the anode current can be obtained from Fig. 30.8 (upper curves).

For **choke input filters** in which the inductance L_1 is constant, the r.m.s. value of each anode current is given approximately by (Ref. 31)

$$|i_d| = 0.707\left(1 + \frac{\alpha^2}{2}\right)^{\frac{1}{2}} I_L \qquad (1)$$

where $\alpha = \dfrac{R_L}{3\omega L_1} = \dfrac{L_c}{L_1}$

The following table has been calculated from eqn. (1) :

L_1	α	i_d
$= L_c$	1	$0.87\ I_L$
$= 2L_c$	0.5	$0.75\ I_L$
$= 4L_c$	0.25	$0.72\ I_L$
$=$ infinity	0	$0.707\ I_L$

For design purposes it seems reasonable to calculate on the basis of conditions at maximum current. If $L_1 = 2L_c$ at maximum current, as recommended in Sect. 3, the heating current in the transformer may be taken as $0.75\ I_L$.

For further information on transformer heating, see Chapter 5 Sect. 5 pages 236-237.

SECTION 5 : VOLTAGE MULTIPLYING RECTIFIERS

(*i*) *General* (*ii*) *Voltage doublers* (*iii*) *Voltage tripler* (*iv*) *Voltage quadruplers.*

(i) General

Where it is required to obtain a higher direct voltage from a given a.c. input than is possible with normal rectifier circuits and where for reasons of weight, economy or other factors it is not desired to use a transformer, voltage multiplying rectifier circuits may be used.

These circuits involve the principle of charging condensers in parallel from the input and adding them in series for the output, the switching being accomplished by the rectifier valves.

(ii) Voltage doublers

The voltage doubler can take one of two forms, half- or full-wave.

Half-Wave. In the half-wave circuit (Fig. 30.12) on one half of the cycle the condenser C_1 is charged through V_1 ; this voltage is then added in series on the next half cycle to the voltage of the condenser C_2 charged through V_2. A voltage of approximately $2\hat{E}_\sim$ will appear across R_L depending upon the rectifier type, load resistance and values of C_1 and C_2. The ripple frequency, as in all half-wave circuits, will be the same as the supply frequency. C_1 must be rated at the value of $\hat{E}_\sim$ and C_2 at $2\hat{E}_\sim$.

Full-Wave. In this circuit (Fig. 30.13) C_2' and C_2'' are charged on alternate half cycles, approximately $2\hat{E}_\sim$ appearing across the two in series. The ripple frequency will be equal to twice the supply frequency and the condenser ratings should be each equal to $\hat{E}_\sim$.

Comparing the two circuits

The voltage regulation is better for the full wave circuit at low values of ωCR_L and the rating of both condensers need only be $\hat{E}_\sim$ but the circuit suffers the disadvantage of not having a common input and output terminal. The filtering is easier with the full wave circuit as the ripple frequency is twice the supply frequency, the ripple percentages being approximately equal.

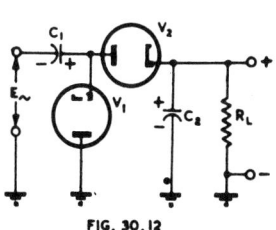

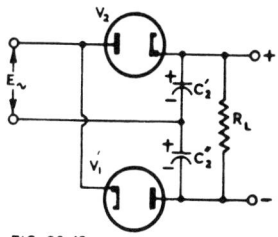

FIG. 30.12

FIG. 30.13

Fig. 30.12. Half wave voltage doubler rectifier circuit.

Fig. 30.13. Full wave voltage doubler rectifier circuit.

In both circuits, the larger the capacitance of the condensers the nearer the output voltage will be to $2\hat{E}_\sim$ and the better the voltage regulation, but care must be taken that the peak current ratings of the rectifiers are not exceeded.

It can be shown (Ref. 10) that at values of ωCR_L greater than 10, the values of $i_d/\bar{i}_d$, $E_{dc}/\hat{E}_\sim$, ripple etc. for a half-wave voltage doubler are for all practical purposes the same as the values for the full-wave voltage doubler and calculations of both types can therefore be made by means of the graphs in Sect. 2.

From these graphs it can be seen that if a voltage multiplication of 1.6 or greater is required, the value of ωCR_L should not be less than 100, also that $\bar{R}_s/R_L$ should not be greater than 1.5%. This means that if a voltage doubler is to give a high output and to be of good regulation, the maximum output current is strictly limited.

(iii) Voltage tripler

This circuit (Fig. 30.14) combines in series the outputs of a half-wave doubler and an ordinary half-wave rectifier, giving an output approximately three times $\hat{E}_\sim$. The ripple frequency will be equal to the supply frequency and the condenser ratings will be as for the individual circuits,

i.e. C_1'' rating $= \hat{E}_\sim$; C_2' rating $= \hat{E}_\sim$; C_2'' rating $= 2\hat{E}_\sim$.

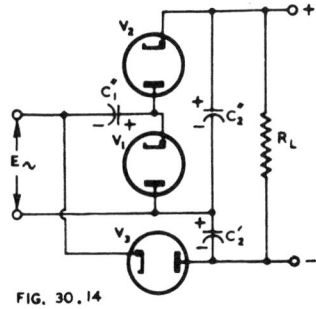

FIG. 30.14

Fig. 30.14. Voltage tripler rectifier circuit.

(iv) Voltage quadruplers

There are two suitable circuits as given in Figs. 30.15A and B. (B) is essentially the same as (A) except for the connection of one of the input leads. This alteration results in a supply which has a common input and output lead, the only other alteration being that C_1'' must now withstand $3\hat{E}_\sim$, while C_2' and C_2'' are (as in Fig. 30.12)

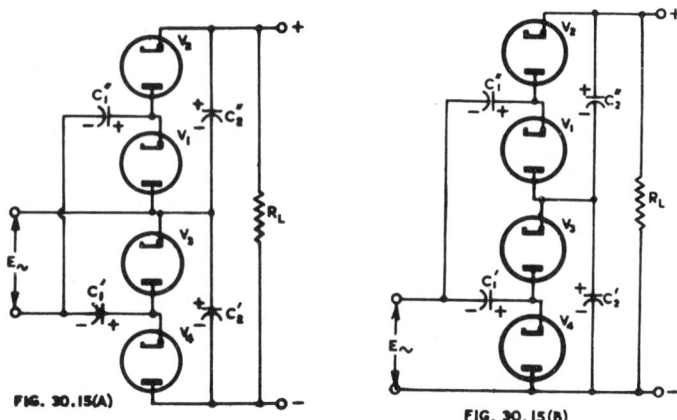

Figs. 30.15(A) and (B). Voltage quadrupler rectifier circuits.

rated at $2\hat{E}_{\sim}$ and C_1' at $\hat{E}_{\sim}$. Another advantage of (B) is that it lends itself to increasing further the number of times the voltage can be multiplied.

For further details on voltage multiplying rectifiers, see References in Sect. 7.

In all the cases polarity of the output can be reversed by reversing the polarity of the rectifiers and condensers.

SECTION 6 : SHUNT DIODE BIAS SUPPLIES

Where a supply of negative bias is required in an amplifier or receiver, as for example the fixed bias operation of an output stage, this can be obtained by the use of a shunt diode without the addition of many components.

If it is desired to make use of a high voltage winding on the power transformer for negative bias, the voltage so obtained is generally much greater than is required

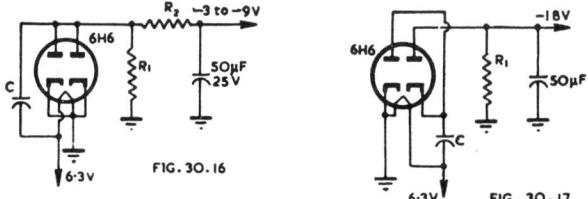

Fig. 30.16. Shunt diode bias supply suitable for the bias of r-f; i-f and a-f amplifier valves.

Fig. 30.17. Voltage doubler bias supply suitable for the bias of output stages.

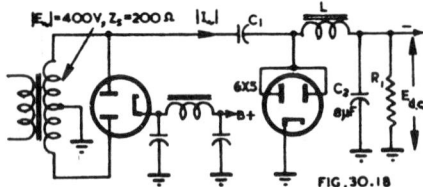

Fig. 30.18. Shunt diode bias supply fed from the transformer winding supplying the main rectifiers.

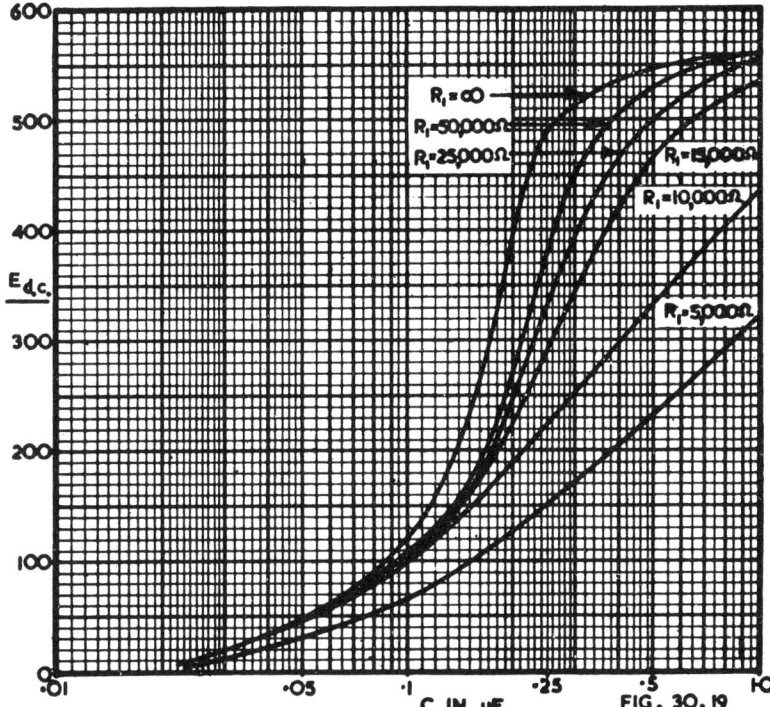

Fig. 30.19. Output voltage of the circuit of Fig. 30.18 for various values of C_1 and R_1.

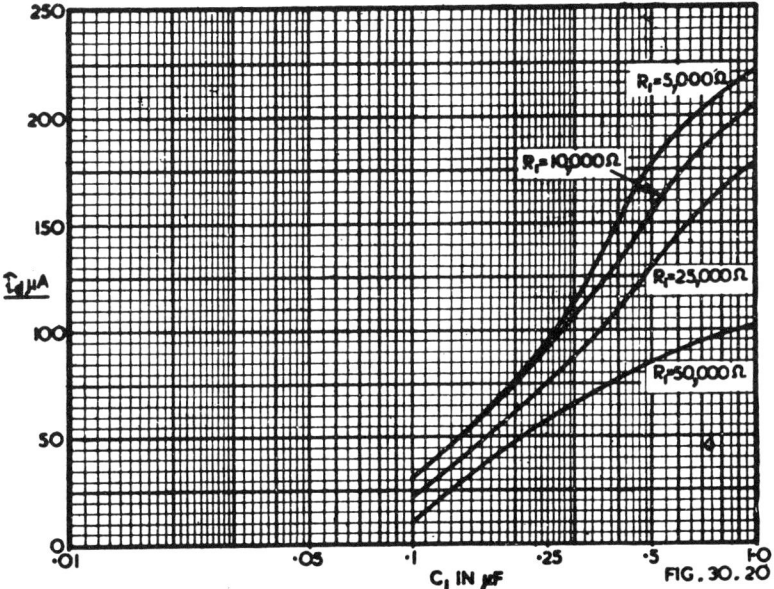

Fig. 30.20. Rectifier peak current values for the circuit of Fig. 30.18 for various values of C_1 and R_1.

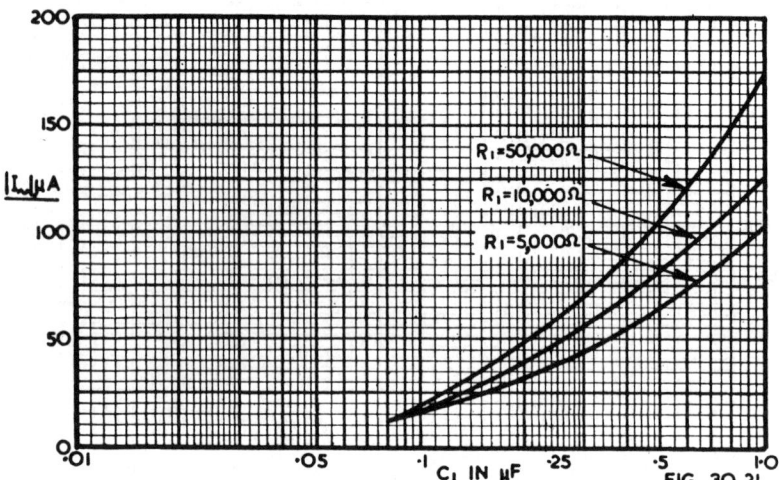

Fig. 30.21. *R.M.S. transformer secondary current in the winding supplying the shunt diode circuit of Fig. 30.18 for various values of C_1 and R_1.*

for bias. In such a case power would be wasted in voltage dividers. The shunt diode system allows the use of a 6H6 type rectifier since the cathode is at earth potential. It is often possible to use a spare diode in one of the valves already in the equipment, thus doing away with the need for an extra valve.

Two versions of the circuit are shown in Figs. 30.16 and 30.17, Fig. 30.16 being suitable for the bias on r-f, i-f and a-f voltage amplifier valves. Fig. 30.17 is a voltage doubling circuit useful for the bias voltage of output stages. It has a maximum output voltage of 18 volts when supplied from a 6.3 volt heater line.

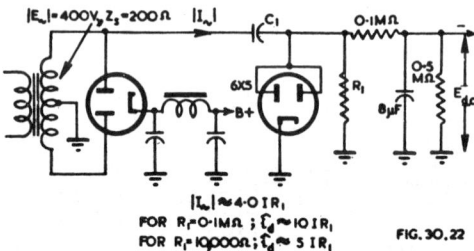

Fig. 30.22.   *Shunt diode bias supply with r.c. filtering.*

When more than 18 volts are required the condenser C_1 may be fed from the transformer winding supplying the main rectifier in the equipment. This is shown in Fig. 30.18 and typical values are given together with the measured performance (Figs. 30.19-21) in order that the magnitude of the various quantities may be assessed. The irregularities in the curves of output voltage for various values of C_1 are due to the resonance of C_1 and L, but apart from modifying the shape of rectifier current pulses no other undesirable effects are evident. Care must be taken that the filter condenser C_2 is isolated from C_1 by a high value of impedance or large currents will be drawn from the transformer windings. In cases where only low output currents are required, L may be replaced by a resistor of not less than 100 000 ohms. This circuit arrangement is shown in Fig. 30.22, the output voltage for various values of capacitance being given in Fig. 30.23.

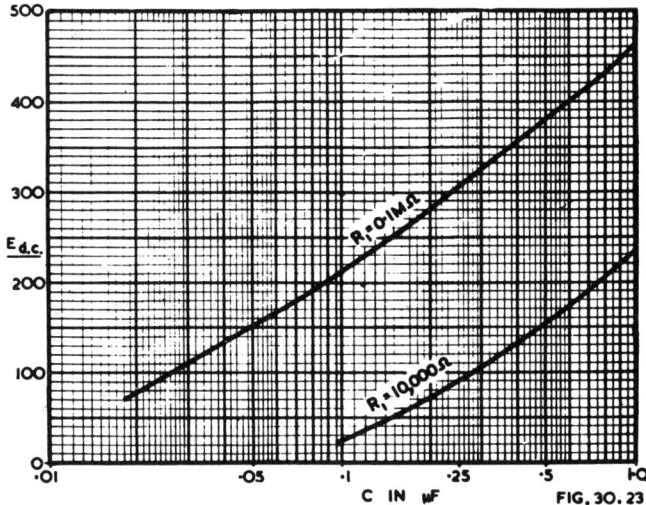

Fig. 30.23. Output voltage of the circuit of Fig. 30.22 for various values of C_1 and R_1.

SECTION 7 : REFERENCES

1. Cockcroft, J. D., and E. T. S. Walton, Proc. Roy. Soc. (London) 136.619 (1932).
2. Dellenbauch, F. S., Jr., and R. S. Quimby, " The important first choke in high voltage rectifier circuits," Q.S.T. 16.2 (Feb. 32) 15.
3. Dellenbauch, S. F., Jr., and R. S. Quimby " The first filter choke—its effect on regulation and smoothing," Q.S.T. 16.3 (March 1932) 26.
4. Dellenbauch, F. S. Jr., and R. S. Quimby, " The economical design of smoothing filters," Q.S.T. 16.4 (April 1932) 33.
5. Reich, H. J. (book) " Theory and Applications of Electron Tubes " (McGraw-Hill, 2nd Edit. 1944) 564.
6. Schade, O. H. " Analysis of rectifier operation," Proc. I.R.E. 31.7 (July 1943) 341.
7. Waidelich, D. L. " Voltage multiplier circuits," Elect. 14.5 (May 1941) 28.
8. Waidelich, D. L. " Full-wave voltage doubling rectifier circuit " Proc. I.R.E. 29.10 (Oct. 1941) 554.
9. Waidelich, D. L., and C. H. Gleason " The half-wave voltage doubling rectifier circuit " Proc. I.R.E. 30.12 (Dec. 1942) 535.
10. Waidelich, D. L., and C. L. Shackelford, " Characteristics of voltage multiplying rectifiers " Proc. I.R.E. 32.8 (Aug. 1944) 470.
11. Waidelich, D. L., and H. A. K. Taskin " Analysis of the voltage tripling and quadrupling rectifier circuits," Proc. I.R.E. 33.7 (July 1945) 449.
12. " Use of the shunt diode for supplying bias voltage " Elect. 18.9 (Sept. 1945) 218.
13. Chadwick, E. W. " Voltage multiplier circuits with selenium rectifiers " Comm. 27.1 (Jan. 1947) 14.
14. " Selenium rectifiers in home receivers " Tele-Tech 6.1 (Jan. 1947) 73.
15. Kauzmann, A. P. " Determination of current and dissipation values for high-vacuum rectifier tubes " R.C.A. Rev. 8.1 (March 1947) 82.
16. Waidelich, D. L. " Analysis of full-wave rectifier and capacitive-input filter " Elect. 20.9 (Sept. 1947) 120.
17. Berkman, R. " Using selenium rectifiers " Q.S.T. 31.10 (Oct. 1947) 50.
18. Tucker, D. G. " Rectifier resistance laws " W.E. 25.295 (April 1948) 117.
19. Rowe, E. G., R. E. B. Wyke and W. Macrae " Power diodes " Part 1 " The design of high-vacuum oxide-coated rectifiers " Electronic Eng. 20.245 (July 1948) 214.
20. Dishington, R. H. " Diode circuit analysis " E.E. 67.11 (Nov. 1948) 1043.
21. " The copper oxide rectifier " E.E. 67.11 (Nov. 1948) 1051.
22. Ramsay, H. T. " The rating of small- and medium-power thermionic rectifiers " Jour. I.E.E. 94. Part III 30 (July 1947) 260.
23. Clarke, C. A. " Selenium rectifier characteristics " Elect. Comm. 20.1 (1941) 47.
24. Yarmack, J. E. " Selenium rectifiers and their design " Trans. A.I.E.E. 61 (July 1942) 488.
25. " Selenium rectifiers for closely regulated voltages " Elect. Comm. 20.2 (1941) 124.
26. Richards, E. A. " The characteristics and applications of the selenium rectifier " Jour. I.E.E. 88 Part III (Dec. 1941) 238 ; 89 Part III (March 1942) 73.
27. " Air Cooled Transmitting Tubes " Technical Manual TT3, Radio Corporation of America (1938).
28. Terman, F. E. (book) " Radio Engineers' Handbook " (McGraw-Hill Book Company, 1st Edit. 1943) p. 601.
29. Lee, R. " Choke-input filter chart " Elect. (Sept. 1949) 112.
30. Lee, R. " Additional uses for rectifier filter chart " Elect. 22.9 (Sept. 1949) 174.
31. Builder, G. " A note on full wave rectifiers " (unpublished paper).
32. Dunham, C. R. " Some considerations in the design of hot-cathode mercury-vapour rectifier circuits " 75.453 (Sept. 1934) 278.

Additional references will be found in the Supplement commencing on page 1475.

CHAPTER 31

FILTERING AND HUM

By R. J. RAWLINGS, Grad.I.E.E., Associate Brit. I.R.E.

SECTION 1 : INDUCTANCE-CAPACITANCE FILTERS

Consideration has already been given in Chapter 30 to the value of ripple voltage E_R in the output of condenser- and choke-input rectifier systems. This chapter takes into consideration the filter components required to reduce the value of ripple given by the graphs in Chapter 30, to the value that is allowable for the equipment under consideration. The latter value will be designated as E_{R1} for one additional filter section, as in Fig. 31.1A, or E_{R2} for two additional filter sections.

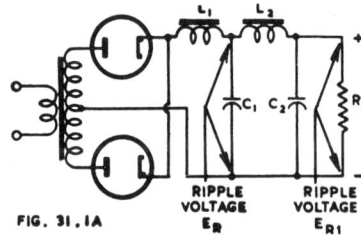

FIG. 31.1A

RIPPLE VOLTAGE E_R RIPPLE VOLTAGE E_{R1}

Fig. 31.1A. Circuit diagram of full-wave choke-input rectifier system, followed by single section filter L_2C_2 to give reduced ripple voltage E_{R1}.

It can easily be shown that for a given total value of LC, maximum filtering is obtained when all the filter sections are similar. Increasing the number of sections for a given total value of LC also increases the filtering, but there is little gained in breaking up the filter into more sections than that number which makes $X_L = 10X_c$. This is only appropriate if there is a definite limitation in the maximum value of LC. In general, better economy will be achieved by using not more than two sections, and increasing the value of C to give the required filtering.

The filter factor $\alpha = E_R/E_{Rn} = (\omega_R{}^2 LC - 1)^n$
where n is the number of additional similar sections.

LC is the value of LC for one section (henrys and farads)
and $\omega_R = 2\pi f_R$
where f_R = ripple frequency.
This can be expressed as :—
$$LC = (^n\sqrt{\alpha} + 1)/\omega_R{}^2 = 0.0254 \, (^n\sqrt{\alpha} + 1)/f_R{}^2.$$
The formula above assumes that $X_L \geqq 20X_c$, the usual practical case.

For quick reference, a graph has been prepared (Fig. 31.1B) for a single section filter to give the relationships between the values of E_R, E_{R1} and L_2C_2 based on the

1192

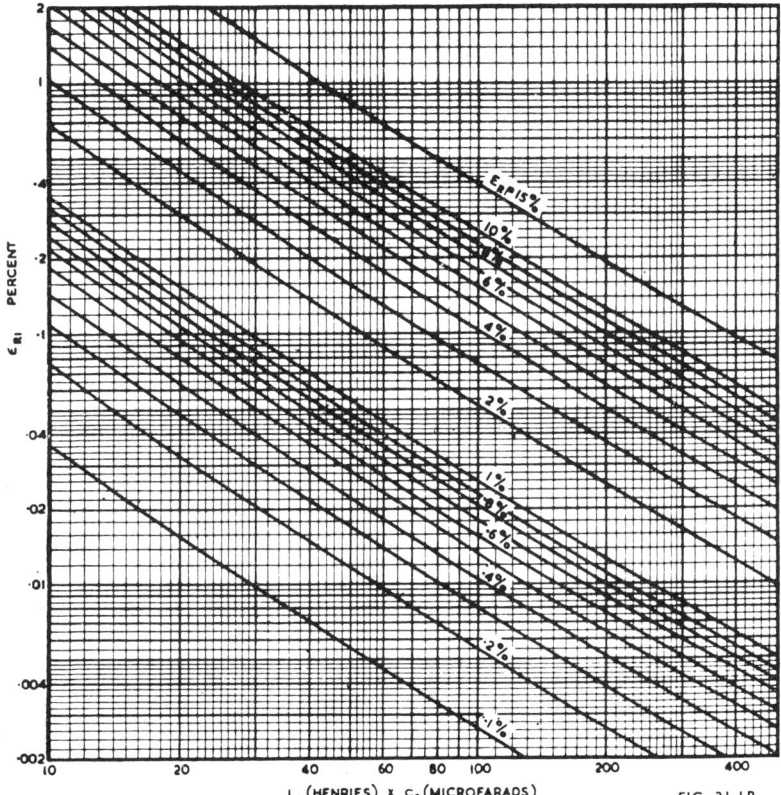

L₂ (HENRIES) x C₂ (MICROFARADS)

FIG. 31.1B

Fig. 31.1B.   Approximate relationships between ER, ER₁ and L₂C₂ for 50 c/s full-wave rectification (i.e. 100 c/s ripple frequency) and single section filter.   For other ripple frequencies, the value of the LC·scale should be multiplied by the appropriate factor as follows.
    40 c/s ripple—6.25 ; 50 c/s ripple—4 ; 60 c/s ripple—2.78 ; 80 c/s ripple—1.56 ; 120 c/s ripple—0.695.

approximate formula $\alpha = \omega_R{}^2 LC$ which is sufficiently accurate for most applications. The values of L_2C_2 in Fig. 31.1B only hold for full-wave 50 c/s, but may be adapted to other conditions by multiplying by a factor (see title).

With either a condenser-input (Fig. 30.1F) or choke-input filter having a second filter section (Fig. 31.1A) it is important to avoid resonance between the choke L (or L_2) and condensers C_1 and C_2 in series.   The inductance should be sufficiently large to avoid resonance, with a comfortable margin, over the whole frequency range of the amplifier and at least down to the ripple frequency (e.g. 100 c/s for 50 c/s mains and full wave rectification).   This may be accomplished by ensuring that in Fig. 31.1A

$$L_2 \geqq \frac{C_1 + C_2}{C_1 C_2} \cdot \frac{25\,000}{f^2} \text{ henrys}$$

where f = minimum rated frequency of amplifier in c/s, and C_1 and C_2 are in microfarads.

When $f = 100$ c/s, $L_2 \geqq \dfrac{C_1 + C_2}{C_1 C_2} \times 3.5$.

When $f = 50$ c/s, $L_2 \geqq \dfrac{C_1 + C_2}{C_1 C_2} \times 14.$

When $f = 25$ c/s, $L_2 \geqq \dfrac{C_1 + C_2}{C_1 C_2} \times 56.$

For example, when $C_1 = C_2 = 8$ μF and $f = 25$ c/s, $L = 14$ henrys minimum.

SECTION 2 : RESISTANCE-CAPACITANCE FILTERS

In cases where filtering of a low-current supply is required, as in the case of the early stages of an audio-frequency amplifier or for certain applications in electronic instruments, resistance-capacitance filters can be used. The filter factor $\alpha = E_R/E_{R1} = \omega C R + 1$ and for ripple frequencies of 50, 60, 100 and 120 c/s this can be read directly from Fig. 31.2, for a single section filter.
See also Chapter 4 Sect. 8(ii) and (iii).

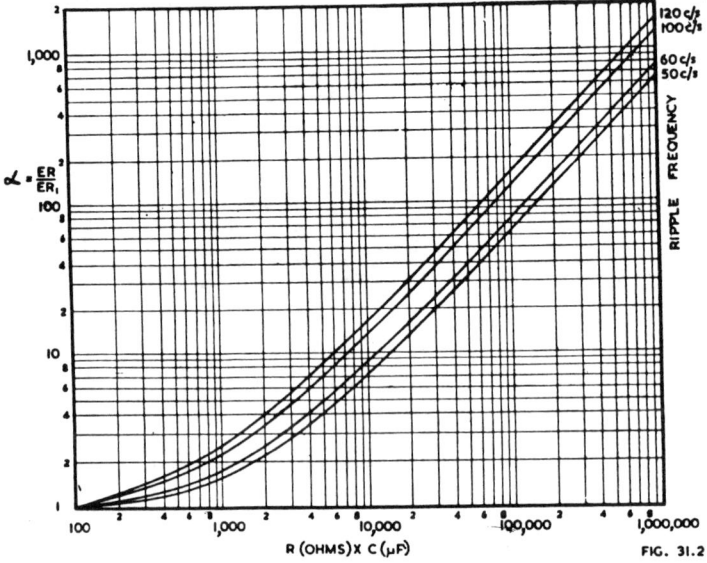

FIG. 31.2

Fig. 31.2. Curve for determining the value of $\alpha = E_R/E_{R1}$ of a resistance-capacitance type single section filter for frequencies of 50, 60, 100 and 120 c/s. This curve is based on the formula, $E_R/E_{R1} = \omega C R + 1$.

SECTION 3 : PARALLEL T FILTER NETWORKS

A useful filter network is that shown in Fig. 31.3(a). This can be transposed, for any one frequency, to its equivalent Π network as in (b) and, by correct choice of the values of circuit components, the attenuation can be made theoretically infinite at any one frequency (f_∞). Conditions for infinite attenuation are—

$1/C_1 \omega = K R_2$
$2 R_1 = K^2 R_2$
$2/C_2 \omega = R_2/K$ where K is a constant.

When the parallel T network is used for power supply filtering, the value of K to be used depends upon a number of considerations :—

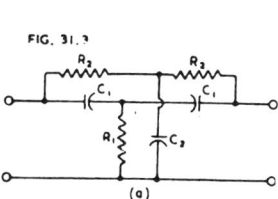 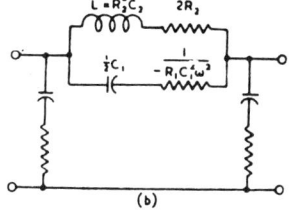

Fig. 31.3. *Resistance-capacitance parallel T network and its equivalent Π network.*
1. The d.c. resistance of the filter (i.e. $2R_2$) should be as low as possible.
2. The total capacitance should be as low as possible.
3. Standard value components may be used.

Table 1 : Showing the relationships between R_1, R_2, C_1 and C_2 for various K values.

K	Resistances	Capacitances	f_∞
$1/2$	$R_2 = 8R_1$	$C_2 = 0.5C_1$	$1/(\pi R_2 C_1)$
$1/\sqrt{2}$	$R_2 = 4R_1$	$C_2 = C_1$	$1/(\sqrt{2}\pi R_2 C_1)$
1	$R_2 = 2R_1$	$C_2 = 2C_1$	$1/(2\pi R_2 C_1)$
$\sqrt{2}$	$R_2 = R_1$	$C_2 = 4C_1$	$1/(2\sqrt{2}\pi R_2 C_1)$

Table 2 : Showing the total capacitance values required for various values of K with $f_\infty = 100$ c/s, $R_2 = 200$ ohms.

K	Capacitance in μF		
	C_1	C_2	$2C_1 + C_2$
$1/2$	15.9	7.95	39.75
$1/\sqrt{2}$	11.25	11.25	33.75
1	7.95	15.9	31.8
$\sqrt{2}$	5.625	22.5	33.75

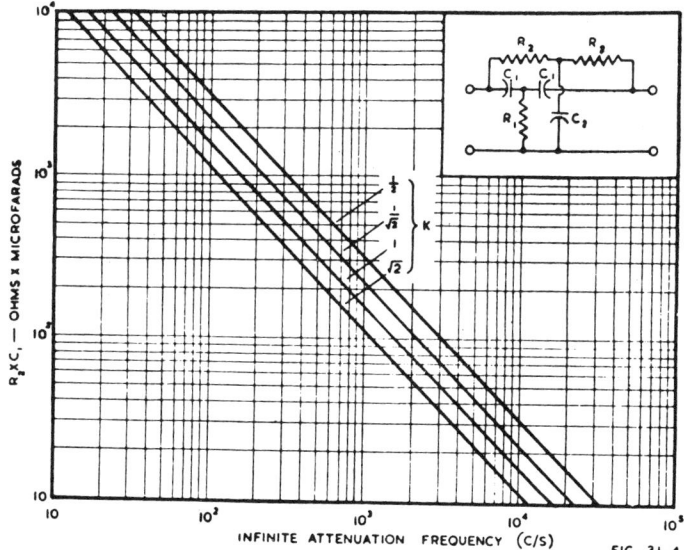

Fig. 31.4. *Values of $R_2 C_1$ required for various values of K to obtain infinite attenuation at frequency f_∞.*

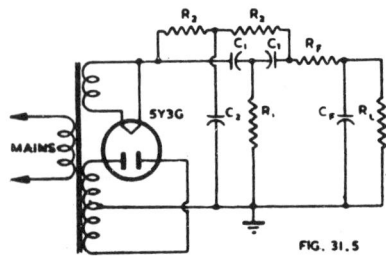

Fig. 31.5. Typical complete filter circuit using a parallel T filter network.

By reference to Fig. 31.4, it will be seen that the minimum value of R_2C_1 will be required for $K = \sqrt{2}$. When $f_\infty = 100$ this gives a value of R_2C_1 not easily obtainable with standard values of components, and $K = 1$ will give more easily obtained values. Also when $K = 1$, the total capacitance required is a minimum. Where use is made of this filter, some attenuation must be provided for the ripple frequency harmonics, and this may be accomplished by means of a resistance-capacitance filter (see Sect. 2). A typical complete circuit is shown in Fig. 31.5.

For the theory of parallel T filter networks reference should be made to Chapter 4, Sect. 8(iii). See also Ref. 1.

SECTION 4 : HUM—GENERAL

(i) Hum due to conditions within the valves   (ii) Hum due to circuit design and layout   (iii) Hum levels in receivers and amplifiers.

Hum in the output of an a.c. operated receiver or amplifier may be due to many causes. These may be sub-divided into two main groups, hum due to conditions within the valves, and hum due to circuit design and layout.

(i) Hum due to conditions within the valves

This is caused by the use of a.c. for the heater supply and can give rise to hum in the output in several ways :

(1) capacitive coupling from the heater to any electrode
(2) heater-to-cathode conduction
(3) heater-to-cathode emission (or vice versa)
(4) modulation of the electron stream by the magnetic field of the filament
(5) conduction (i.e. leakage) from any electrode, with ripple voltage applied, to another electrode
(6) hum due to operating valve in magnetic field from external source.

Explaining these in more detail :—

(1) The amount of hum appearing at the plate of a valve due to capacitive coupling between the valve heater and an electrode, will depend upon the electrode to which this coupling takes place. The most serious case is that between heater and control grid as the voltage transferred will be subject to the full gain of the valve. In fairly high gain amplifiers it is advisable to use a potentiometer of 100 to 500 ohms resistance across the heater supply close to the valve socket ; the slider is earthed and its position adjusted for minimum hum output. In this way the hum voltages fed from each side of the heater to the control grid can be made equal and opposite.

In lower gain amplifiers a centre-tapped heater winding may be used or, more usually, a centre-tapped resistor across the heater supply. In the latter case the resistor may have a resistance of 50 + 50 ohms for 6.3 volt heaters. In some cases it is practicable to omit any form of tapping, and in these cases one of the heater terminals should be earthed. In the case of types in which the control grid or plate pin is adjacent to one heater pin, the latter should be earthed. Where the grid pin is separated from one heater pin only by one other pin, the latter should preferably be earthed directly, or else by-passed to earth by a large condenser.   In other cases the heater terminal to be earthed should be chosen to provide minimum hum with regard to the position of un-earthed or unbypassed terminals.

Valves with the grid connection brought out to a top cap have less capacitance coupling between heater and grid circuits, but there may be appreciable electromagnetic coupling into the loop formed by the cathode-to-grid path and the return path through the valve.   For this reason it is sometimes found that a double-ended valve gives more hum in a low level amplifier than a well designed single-ended stage.

When single-ended valves are used, the all-glass (e.g. miniature) construction gives less capacitance coupling from heater to grid or plate than an equivalent single-ended valve with a " pinch " (stem press) construction.

Reduction of the grid resistor value will also reduce the hum due to capacitive coupling, as the grid resistor and the heater-to-grid capacitance form a potential divider which determines the proportion of the heater voltage fed to the grid, unless there is a large grid coupling condenser and a low impedance to earth from the other side of it.

Diode-heater capacitance can also cause high hum levels in diode-triodes and diode-pentodes as the diode is effectively coupled to the control grid.   For this reason it is recommended that the diode used for signal detection should, in the case of double-diode valves, be the diode further from the heater pins on the valve base.

(2) Conduction from heater to cathode will cause a current to flow from the heater to the cathode, thence through the impedance between cathode and earth.   The resultant hum may be minimized by reducing the effective voltage applied to the circuit, and by reducing the impedance between cathode and earth.   The voltage may be reduced by a potentiometer connected across the heater supply, with the slider earthed and its position adjusted for minimum hum output—as for hum due to capacitive coupling.   The impedance between cathode and earth may be reduced by the use of a large capacitance by-pass condenser across the cathode bias resistor, or may be made zero by earthing the cathode and obtaining negative bias from some hum-free source.

See also Ref. 7.

(3) Emission from-heater to cathode is caused by impurities in the heater insulation material or from the deposit, during valve manufacture, of cathode material on to the heater.   To overcome this fault the heater should be biased positively with respect to the cathode.   As the voltage required may be of the order of 50 volts, the minimum required to give satisfactory hum reduction should be used to avoid exceeding the maximum permissible heater-cathode voltage.

Cathode-to-heater emission can also take place and cause hum, although it is not common.   The cure for this is the biasing of the heater negatively with respect to cathode.

In the general case hum may be due to two or more causes, and minimum hum may be obtained in any particular case by connecting the moving arm of the potentiometer to a point of positive (or sometimes negative) voltage of the order of 5 to 50 volts and by experimentally obtaining optimum adjustment of both potentiometer and voltage.

(4) Modulation of the electron stream by the magnetic field will depend on the type of heater and the electrode construction.   If the centre-point of the heater circuit is earthed, the double helix gives noticeably less hum than the folded heater, but when one side of the heater is earthed the reverse may occur, depending on the electrode arrangement—see Ref. 6, Fig. 47.

Heater-induced hum in a-f amplifiers—summary*

The amplifier designer has to make allowance for the total hum, effective at the grid of the valve, due to the use of a.c. supply to the heater.   There are very few valve types for which information is available in this regard—see Chapter 18 Sect. 2(v) for hum data on types 12AY7 and 5879.   Tables have been published (Ref. 10) giving values of the 60 c/s component of the hum voltage of a fairly wide range of types with

*Contributed by the Editor.

cathode by-passed and unbypassed, grid resistor zero and 0.5 megohm, and with different heater earthing arrangements but this does not give any indication of the total hum and is likely to be misleading since the ratio of total hum to 60 c/s component varies widely from type to type.

By the choice of a suitable valve type (e.g. 5879, triode operation) it is possible to achieve a median hum voltage effective on the grid of less than 4 microvolts under optimum conditions, with only a small percentage exceeding 10 microvolts, with grid resistor 0.1 megohm. On the other hand, with other types of valves and with other than optimum conditions, the hum may exceed 500 microvolts.

See also Chapter 12, Sect. 10(vi) and Chapter 18, Sect. 2(iii).

Refs. 5, 6, 7, 9, 10.

(ii) Hum due to circuit design and layout

The most common of hum troubles due to circuit design is caused by insufficient smoothing in the power supply. Where hum is introduced in the early stages of an amplifier, a simple RC filter can often be used (see Sect. 2). RC filters can also be used to supplement conventional filtering when lack of filtering in r-f and converter stages causes modulation hum.

Modulation hum can also be caused by capacitive coupling between primary and secondary windings of the mains transformer. The cure is the addition of an earthed electrostatic screen between the windings, or in some cases filters can be fitted in the mains leads as in a.c./d.c. receivers. Every care should be taken to eliminate r-f pickup from all sources other than the aerial, and receivers should be earthed directly rather than connected to the mains conduit which may have a high resistance earth connection. Capacitive coupling between the mains lead and aerial can also cause a mains frequency modulated carrier to be passed through the receiver.

In a.c./d.c. receivers care must be taken in the series connection of heaters. Usual practice is to place the a-f valve heater right at the earthy end, with the converter (which is particularly susceptible to modulation hum) next to it. In a.c./d.c. audio frequency amplifiers the earliest stages should be at the earthy end.

Electrostatic pick up of hum voltages by high impedance circuits from nearby leads carrying alternating currents may be avoided by the fitting of electrostatic shields or, where the capacitance of screened lead is permissible (as in the case of filament wiring) the use of screened a.c. leads is recommended. The use of twisted filament leads tends to cancel out the magnetic field around the leads.

The magnetic fields surrounding power transformers can give rise to induced currents in filter chokes and in a steel chassis. For this reason the use of a separate chassis in the case of high gain amplifiers is to be preferred. Alternatively the chassis may be made of non-magnetic metal. The placing of the power transformer laminations vertically usually gives rise to less hum than when they are placed in the same plane as the chassis, particularly when the chassis is of a magnetic material. Thorough screening of intervalve transformers (and more particularly microphone input transformers) from the magnetic field of the power transformer is also necessary, and high permeability shields or astatic type windings may be necessary in extreme cases to reduce hum to the required level. By orientating transformers and chokes with respect to each other a position of minimum coupling can be found which may avoid the necessity for taking more elaborate measures. If the position of minimum coupling is at all critical, it is recommended that an individual adjustment be made for each amplifier. The filter choke and transformer fields must also be kept away from high gain valves to prevent modulation of the electron stream.

The reason for high hum-level output in a receiver or amplifier can often be determined by the pitch of the note. If it is due to poor filtering, the 50 or 60 c/s in the case of half-wave rectification or the 100 or 120 c/s note with full-wave rectification will predominate. If the hum is induced from the power transformer the note will be 50 or 60 c/s while with capacitive hum pick up the note will be rough owing to the accentuation of the harmonics of the supply frequency.

When tracing hum it may be found that the reduction of hum in one part of the circuit may increase the overall hum level as, when the hum is due to more than one cause, a certain amount of cancellation may be taking place.

See also Chapter 18, Sect. 2(iii) on pre-amplifiers.

Hum caused by heater-cathode leakage with an unbypassed cathode resistor is dealt with in Chapter 7, Se   2(ix)B.

The effect of the output valve on hum originating in the plate supply voltage*

(See Chapter 7, Sect. 2(ix) for the effect of feedback).

Case (1)—Triode with transformer-coupled output

The hum output voltage across the primary of the output transformer is

$$E_{h0} = E_h R_L/(R_L + r_p)$$

where E_h = hum voltage in plate supply

R_L = load resistance reflected on to primary of output transformer

and r_p = plate resistance of valve.

If $R_L = 2r_p$ (as is typical)

then $E_{h0} = 0.67 E_h$.

Case (2)—Triode with parallel-feed (series inductor)

This is an excellent circuit for low hum provided that the choke L_1 has a reactance at the hum frequency at least several times the impedance of r_p and R_L in parallel.

See Chapter 7, Sect. 2 equation (55).

$$E_{h0} = \cos \theta . E_h$$

where $\theta = \tan^{-1} L_1/R$

and $R = r_p R_L/(r_p + R_L)$.

Case (3)—Pentode with transformer-coupled output

See Chapter 7, Sect. 2, equation (53).

$$E_{h0} = E_h[R_L/(R_L + r_p) + R_L \mu_{g2p}/(R_L + r_p)].$$

The screen and plate effects are additive. The second term would become zero if the screen were perfectly filtered.

Case (4)—Pentode with parallel-feed (series inductor)

The hum due to the plate circuit is low, but that due to the screen is high. Screen filtering is required for low total hum.

See Chapter 7, Sect. 2, equation (57).

$$E_{h0} \approx E_h \cos \theta - R_L \mu_{g2p}/(R_L + r_p)$$

where θ and R are as in Case (2).

References : Chapter 7 Refs. B3, B4, B5, B7.

Typical values (all cases)

Typical values for types 2A3 and 6V6 are given in Chapter 19, Sect. 10(v).

(iii) Hum levels in receivers and amplifiers*

It is convenient to express hum levels in dbm, rather than as so many db below maximum power output. This hum is measured with zero signal output.

Hum should, ideally, be completely inaudible. This ideal is capable of achievement, although at a price, so far as the hum introduced by the receiver or amplifier is concerned. Audible hum is a defect which, while it may not prevent the sale of the equipment, will certainly be an annoyance to a critical user.

Hum is objectionable firstly because of its direct effect with zero signal, and secondly because of its indirect effect in modulating the signal.

The maximum acceptable hum level may be determined by a listening test in a quiet residential area, or under actual conditions for a custom-built equipment. The listener for the test should be one having at least average, and preferably higher than average, acuity of hearing for the lower frequencies.

Alternatively, the maximum permissible hum level may be calculated. For example if the predominant hum frequency is 120 c/s, the loudspeaker efficiency at 120 c/s is 3%, the angle of radiation at 120 c/s is 180°, and the listener is 1 foot from the

---

*Contributed by the Editor.

loudspeaker, then the limit of audibility is about -35 dbm for an average listener, and -45 dbm for a very critical listener.   A good quality receiver or amplifier should therefore have a hum level less than -35 dbm for 3% loudspeaker efficiency at the hum frequency.   This quoted loudspeaker efficiency of 3% is a fairly typical average value over the a-f frequency band, but the effective efficiency at the hum frequency may be higher or lower than the average value.   The efficiency at the hum frequency will be higher than the average value if bass boosting is used, or if the cabinet/loudspeaker combination accentuates the hum frequency.   The efficiency at the hum frequency will be lower than the average value if the hum frequency is lower than the loudspeaker bass resonance or if the cabinet attenuates the hum frequency.   In cases where the loudspeaker efficiency is higher or lower than 3%, the permissible hum level should be adjusted accordingly.

This calculation for hum is somewhat unrealistic since it is based on the supposition that the hum is predominantly a single frequency.   Any higher frequency components will give a higher acoustical hum level without appreciably affecting the measured value.

Summing up

For critical home listening with low room noise and with loudspeakers of normal sensitivity, the hum level should not be above -40 dbm, with -50 dbm as a preferred limit.

Under other conditions, with higher background noise or a less critical audience, a somewhat higher hum level may not be found objectionable.

SECTION 5 : HUM NEUTRALIZING

(A) One form of hum neutralizing is that using a hum-bucking coil in series with the voice coil in speakers where the field coil is used as the filter choke.

(B) A valve hum neutralizing system which has certain applications where a substantially constant load is to be supplied is shown in Fig. 31.6.   This circuit has been explained in detail elsewhere (Ref. 3) and neutralization is dependent upon the valve producing a hum voltage of the same value and 180° out of phase with that already existing in the plate supply voltage.   The valve must therefore be capable of producing a gain of unity between grid and plate, the feedback resistance $R_c (= R_b + R_f)$ being adjusted to obtain this balance.

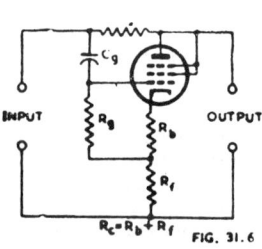

Fig. 31.6.   *Valve hum neutralizing circuit.*

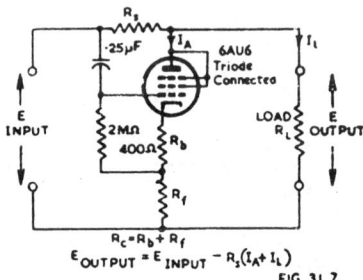

Fig. 31.7.   *Practical valve hum neutralizing circuit.*

A typical circuit is shown in Fig. 31.7 and the value of ripple reduction α obtained together with the values of R_c are plotted in Fig. 31.8 as a function of R_s.

The lowest ripple frequency which this circuit will suppress depends upon the time constant of $R_g C_g$ which for this circuit is given by $T = R_g C_g/(1 - R_g/R_s)$ provided $\mu \gg 1$, and the internal resistance $R_0 = R_s \mu/(\mu + 1)$.

As the condition for neutralization does not involve the load impedance, the relatively high impedance of the regulator can be reduced to a low value for a.c. voltages by shunting the output by a large condenser.

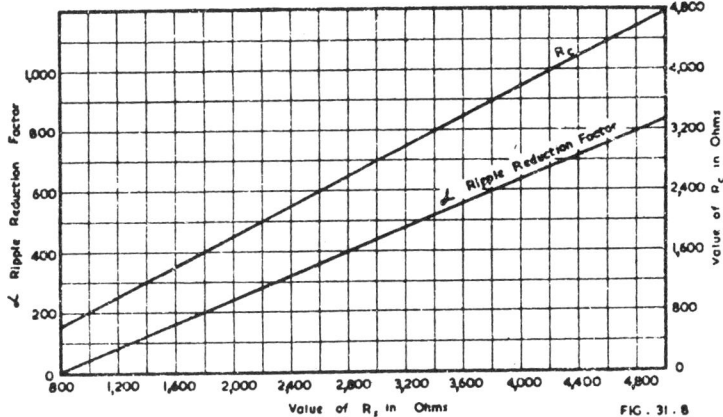

Fig. 31.8.   *Values of* α *and* R_c *as a function of* R_s *for the circuit of Fig.* 31.7.

In this way coupling between stages due to a high common impedance can be reduced.

Care must be taken that the value of input ripple to the circuit does not swing the valve beyond the linear position of its characteristics.

(C) Another form of neutralization, which is limited to pentode voltage amplifiers, is described in Chapter 12 Sect. 10(vi)D and Fig. 12.57.

(D) A form of neutralization suitable for use with multistage amplifiers is described in Ref. 4. Another form is described in Ref. 11.

(E) Various types of neutralizing are commonly used in radio receivers, for example, hum from the back bias supply may be neutralized at the grid of an output valve by hum from the B supply, with a suitable choice of output valve grid leak and plate load resistor for the a-f amplifier. In other cases the balance of these hum components may be adjusted so that the resultant neutralizes hum in the plate circuit of the output valve. More complicated neutralizing circuits even include hum components from the control and screen grids of the a-f amplifier in the balance at the grid of the output valve.

SECTION 6 : REFERENCES

1. " Applications of a parallel ' T ' network " A.R.T.S. and P. Bulletin No. 118 (Jan. 1942).
2. Reich, H. J. " Theory and Applications of Electron tubes " (McGraw-Hill, 2nd Edit. 1944).
3. Sowerby, J. McG. " Shunt voltage stabilizer " W.W. 54.6 (June 1948) 200.
4. Wen-Yuan Pan "'Circuit for neutralizing low-frequency regeneration and power-supply hum " Proc. I.R.E. 30.9 (Sept. 1942) 411.
5. Dickerson, A. F. " Hum reduction " Elect. 21.12 (Dec. 1948) 112.
6. Benjamin, M., C. W. Cosgrove and G. W. Warren " Modern receiving valves : design and manufacture " Jour. I.E.E. 80.484 (April 1937) 401.
7. " Heater-cathode leakage as a source of hum " Elect. 13.2 (Feb. 1940) 48.
8. Terman, F. E., and S. B. Pickles " Note on a cause of residual hum in rectifier-filter systems " Proc. I.R.E. 22.8 (Aug. 1934) 1040.
9. Fleming, L. T. " Controlling hum in audio amplifiers " Radio and TV News 4.5 (Nov. 1950) 55.
10. " Heater-induced hum in audio amplifiers " Tele-Tech 10.11 (Nov. 1951) 58.
11. Hammond, A. L. "Neutralizing hum and regeneration" Audio Eng. 36.5 (May 1952) 22.

CHAPTER 32

VIBRATOR POWER SUPPLIES

By R. J. RAWLINGS, Grad. I.E.E., Associate Brit. I.R.E.

SECTION 1 : VIBRATORS—GENERAL PRINCIPLES

(i) Operation   (ii) Vibrator types   (iii) Choice of vibrator   (iv) Coil energizing   (v) Waveform and time efficiency   (vi) Standards for vibrators for auto-radio.

(i) Operation

The vibrator consists essentially of a vibrating reed upon which are mounted switching contacts.   By the method of connection in a battery circuit, these contacts enable the direct battery voltage to be converted into an approximate square wave. This can then be transformed and rectified to obtain a plate voltage supply.   The standard type vibrators operate at a frequency between 100 and 120 c/s.   High frequency types operating about 250 c/s are available for special applications.

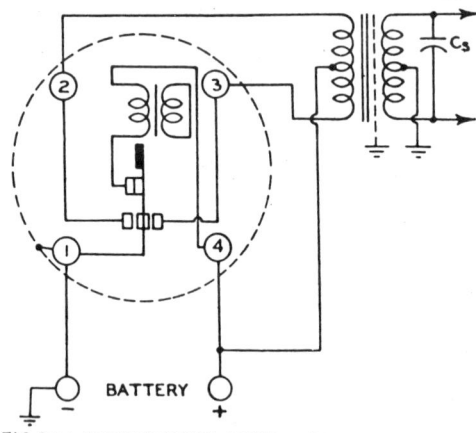

FIG. 32·1 INTERRUPTER TYPE VIBRATOR.

(ii) Vibrator types

There are three basic types of vibrator and these are shown with their circuit connections in Figs. 32.1—32.3.

Fig. 32.1 shows the interrupter (non-synchronous) vibrator which acts as a single pole double throw switch, leaving rectification to be performed by a separate rectifier.

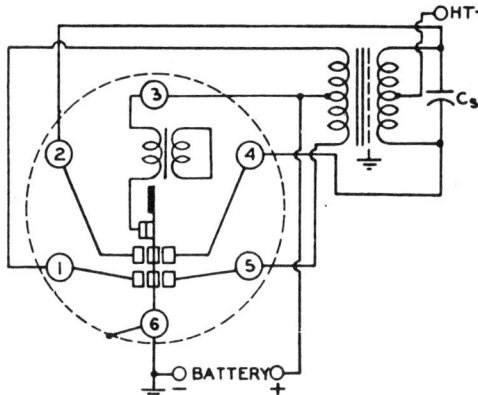

FIG. 32·2   SELF-RECTIFYING   TYPE   VIBRATOR.

The short-circuited secondary is used by some vibrator manufacturers to reduce the inductance of the energizing winding and hence the sparking of the starting contacts.

Fig. 32.2 gives the circuit arrangement for the self-rectifying (synchronous) vibrator which in addition to switching the primary circuit has a further pair of contacts to provide mechanical rectification of the transformer secondary voltage.

The third type, Fig. 32.3, is a modification of the synchronous type in which, by splitting the vibrating reed, the primary and secondary circuits are isolated from each other.

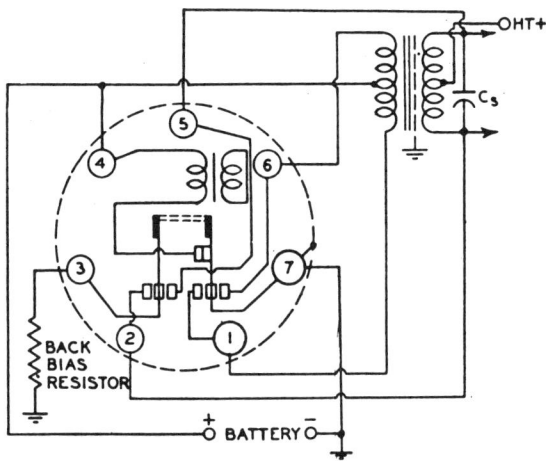

FIG. 32·3   SPLIT-REED   SELF-RECTIFYING TYPE   VIBRATOR.

(iii) Choice of vibrator

The choice of the type of vibrator to be used will depend upon the application for which it is required.

Where vibrator equipment has to operate with either positive or negative earthed systems, either the interrupter type with a valve rectifier or the reversible self-rectifying type must be used. This latter type is a standard self-rectifying vibrator which can be fitted into its socket in two positions, by rotation through 180°. The use of these two positions enables the correct output polarity to be obtained from either positive or negative earthed supplies.

In the self-rectifying type, the elimination of the rectifier valve reduces the overall size, power consumption* and heating of the equipment, although at the expense of increased difficulty in hash elimination and, perhaps, of some of the reliability of the non-synchronous type.

The use of split-reed synchronous vibrators allows greater circuit flexibility, particularly in cases in which one model of radio receiver is to be used with dry-battery or vibrator-operated power supplies.

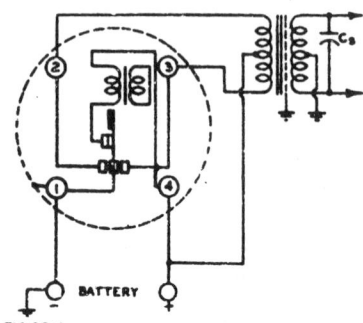

Fig. 32.4. Separate driver system energising.

FIG. 32·4

(iv) Coil energizing

There are three possible methods—Fig. 32.4 shows the separate driver system energizing while Fig. 32.5A shows shunt energizing with separate starting contacts, and Fig. 32.5B is the conventional shunt energizing arrangement. With the separate driver system, the coil is operated by only the battery voltage ; in the conventional shunt connection the coil is operated by greater than the battery voltage due to the auto-transformer action of the transformer primary.

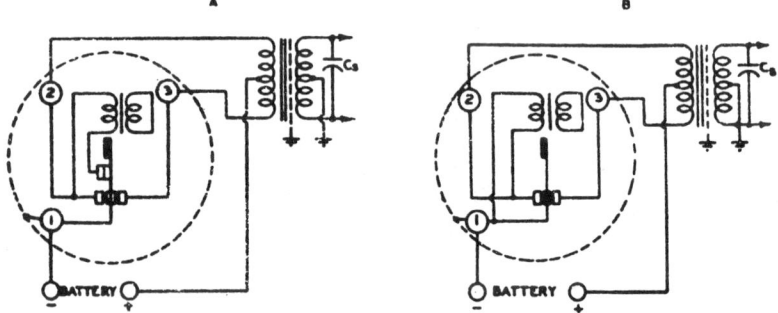

FIG. 32·5

Fig. 32.5.   Two alternative methods of shunt coil energising ;   (A) with separate starting contacts (B) conventional shunt energising.

The advantage claimed for the separate driver system energizing and also for the arrangement shown in Fig. 32.5A, is that as the coil current is not switched by the power contacts, good starting will be obtained even towards the end of life or at low battery voltages. This is not the case in Fig. 32.5B in which the starting performance is dependent upon the condition of the power contacts.

*If a cold-cathode rectifier is used the power consumption of synchronous and non-synchronous types should be the same.

(v) Waveform and time efficiency

The general waveform of a vibrator output voltage is given in Fig. 32.6 ; the periods t_2 and t_4 are referred to as the " off contact " time interval. For reasons of consistent operation with life, these times $t_2 + t_4$ are made from 10% to 30% of the total duration of the cycle. The remaining part of the cycle $t_1 + t_3$, which is the " on contact " time, is therefore from 70% to 90% of the cycle and its value is known as the " time efficiency."

The relationships between the peak, average and r.m.s. values of the vibrator waveform are given below—

Peak value $= E_b$

R.M.S. value $= \sqrt{\omega_t} E_b$

Average value $= \omega_t E_b$

Form factor $= \sqrt{\omega_t}/\omega_t$

where E_b is the battery voltage and ω_t is the time efficiency expressed as a decimal.

The rectangular waveform will be modified to a certain degree by the timing capacitance (see Sect. 3) and will take the approximate form of the dotted curve in Fig. 32.6. This change will not greatly affect the relationships given above.

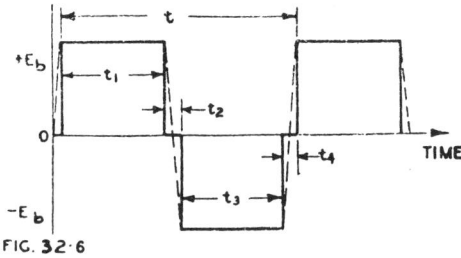

FIG. 32·6

(vi) Standards for vibrators for auto radio

Extract from R.M.A. Standard REC-113—" Vibrating interrupters and rectifiers for auto radio—frequency 115 cycles " (April 1948).

Voltage ratings—nominal 6.3, operating 5.0 to 8.0 volts. Reed frequency at 6.3 volts : 115 ± 7 c/s. See Standard for enclosures, base pins and circuit.

SECTION 2 : VIBRATOR TRANSFORMER DESIGN

(i) *General considerations*   (ii) *Transformer calculations*   (iii) *Standards for vibrator power transformers.*

(i) General considerations

There are many considerations in vibrator transformer design which require particular attention as against a similar transformer used on a sinusoidal supply. However see also Chapter 5 Sect. 5 for power transformer design.

For a given power output the vibrator transformer will be of greater size for a number of reasons. Half the primary is inactive at any instant due to the need for a centre-tapped winding to obtain the necessary flux reversal. The primary winding should be designed to have the minimum possible resistance, in order that as large a portion of the battery voltage as possible is available for transformation. This means the use of a wire of large diameter with a correspondingly poor space factor. Electrostatic shields between the windings, which are sometimes used to assist in the elimination of hash, also add to the size.

An important point in the transformer design which affects the life of the vibrator is the value of primary leakage inductance. This leakage inductance component in the primary circuit can resonate with the reflected value of the timing capacitance to produce an undesired damped oscillation, while possible insulation breakdown and contact arcing may result.   Therefore where practicable a winding of the pancake type should be used, or else the winding shown in Fig. 32.7, the windings being arranged so that the active sections of the primary and secondary are always adjacent. When manufacturing economy requires the use of the cheapest form of winding, best results are obtained by winding the two halves of the primary first and then the two halves of the secondary.   This gives lower leakage reactance than the reverse type of winding.

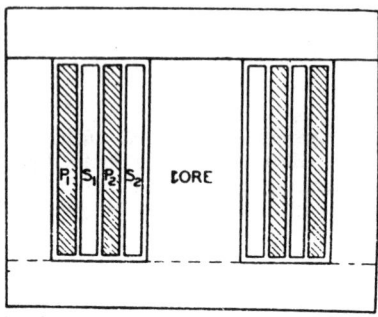

FIG. 32·7

The secondary insulation should be good enough to prevent the risk of a voltage breakdown caused by any irregularity occurring in the vibrator during life which may cause high transient voltages to be developed.

As the magnetizing current of a vibrator transformer has to be supplied by the battery, this represents a power loss and it should be reduced to the minimum possible value—for further details see Sect. 3(iv).   This can be done by keeping the flux density B at a low value.   The value of B will depend upon the primary voltage, all other quantities being fixed.   This value of primary voltage is subject to considerable variations ; e.g. from 5.5 to 7.5 volts on a car radio input when a voltage regulator is fitted to the battery charging system, and this may rise to 9.0 volts without the regulator.   In the general case the transformer may be designed to have satisfactory characteristics at a maximum of 8 volts.

A high value of magnetizing current also limits the maximum output available from the vibrator.   This is because the magnetizing current has to be handled by the primary contacts whose current carrying capacity is limited, thus reducing the current rating available for the load.

(ii) Transformer calculations

The equation for vibrator transformers is

$$N_s = \frac{E.\omega_t.10^8}{4f.B.A} \qquad (1)$$

where E = highest battery voltage at which satisfactory operation is required
  ω_t = vibrator time efficiency expressed as a decimal
  f = vibrator frequency in cycles per second
  A = effective cross sectional area (i.e. actual cross sectional area × the stacking factor) in sq. inches
  B = flux density in lines per sq. inch
and N_s = " effective " number of primary turns (i.e. ½ total primary turns).
The value of A can be estimated from the relationship

$$A = \sqrt{W}/5.58 \text{ sq. ins.} \qquad (2)$$

where W = primary watts under full load.
The r.m.s. value of the transformer primary current is given by

$$|I_b| = (\sqrt{\omega_t}/\omega_t)\bar{I}_b \qquad (3)$$

where $\bar{I}_b$ = average battery current in amperes.

For values of $\omega_t \geqq 0.8$, $\sqrt{\omega_t}/\omega_t \approx 1$ and $|I_b| = \bar{I}_b$.

This value of current can be used in the calculation of the required wire diameter.

As the primary is of heavy gauge wire, and has to be centre tapped, it is of advantage to have an even number of primary layers ; this brings the centre tap to the outside of the winding.   Therefore when commencing the design, where possible it is of advantage to choose the total number of turns ($2N_p$) to give an even number of layers.   The required effective cross sectional area of the core is then given by

$$A = E\omega_t \, 10^8/(4fBN_p).\tag{4}$$

(iii) Standards for vibrator power transformers

R.M.A. Standard REC-119 " Vibrator power transformers " (Sept. 1948) gives purchase specifications, performance specifications, test equipment and procedure.

SECTION 3 : TIMING CAPACITANCE

(i) *The use of the timing capacitance*   (ii) *Calculation of timing capacitance value*
(iii) *Percentage closure*   (iv) *Effect of flux density on timing capacitance value.*

(i) The use of the timing capacitance

The usual simplified form of the vibrator voltage wave is shown in Fig. 32.8.   The dotted curve shows the damped oscillation obtained by the addition of the timing capacitance.   In actual operating this oscillation will be suppressed at C when the other contacts connect the battery to the other half of the transformer.   The oscillation is shown in full in Fig. 32.8 to illustrate how the first part of the cycle is operative in automatically reversing the battery voltage, so that there is zero potential across the contacts when they make.

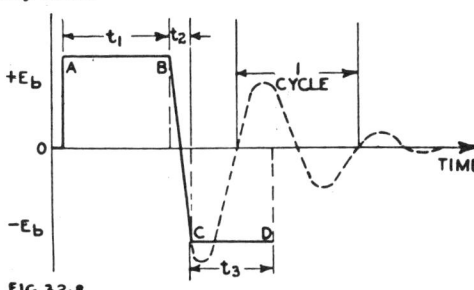

FIG. 32.8

(ii) Calculation of timing capacitance value

The value of the timing capacitance, to give the correct frequency of oscillation for the conditions above, is given by

$$C_p = \frac{H.l_m.(1 - \omega_t)10^6}{8N_p.f.E} \, \mu F \tag{5}$$

where l_m = length of magnetic path in inches
and H = the value given by the BH curves for the transformer iron being used, at the value of B corresponding to the voltage E in the equation above.

The equation will give the required value of timing capacitance C_p to be used across the primary.   For reasons of size and economy, it is usual for the timing capacitance to be placed across the secondary.   The value of the equivalent secondary capacitance C_s will be $C_p/(N_s/N_p)^2$ microfarad.

While these equations will give an approximation to the required value of timing capacitance, some modification of this value will be required in practice.   The final value should not be decided, until after examination of the voltage waveform obtained. This test should be carried out with a number of samples of the vibrator which it is intended to use.

(iii) Percentage closure

When the waveform of Fig. 32.8 is obtained, by the correct choice of timing capacitance, there is said to be 100% closure.   Figs. 32.9 and 32.10 illustrate the effects of over- and under-closure respectively.

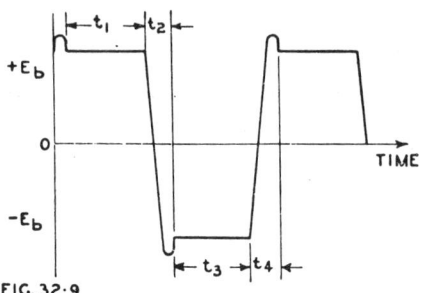

FIG. 32·9

As the effect of overclosure is to produce high transient voltages, and as during life the tendency is for t_1 and t_3 to become shorter, the vibrator will rapidly deteriorate. By aiming at slight under-closure the effect during life is for conditions to approach 100% closure.

For this reason a value of approximately 65% closure is chosen for an average case, as shown in Fig. 32.10A—this means a larger value of timing capacitance than indicated by eqn. (5).

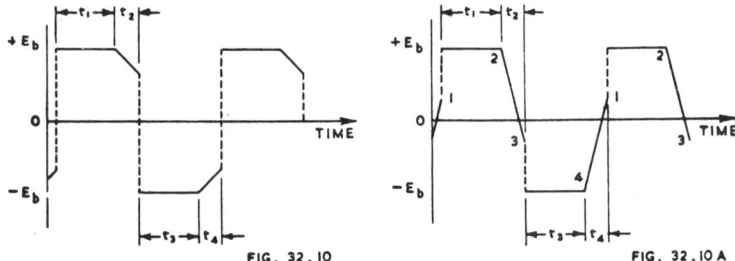

FIG. 32.10   FIG. 32.10 A

(iv) Effect of flux density on timing capacitance value

The equation for the timing capacitance shows the need for the careful choice of the value of flux density used in the transformer design.   For it will be seen that a change in the ratio of H/E in eqn. (5), such as will take place when working on the curved characteristic of the BH curve, will result in different values of timing capacitance being required at different loads.   If the value of H/E is to be kept constant, B must be directly proportional to H over the working range.   Examination of Fig. 32.11 will show that this is not the case, even on the straight portion of the BH curve, as a line drawn through the straight portion of the curve would not originate at the junction of the X and Y axes.   A more exact approximation to the desired condition can be obtained by working around the knee of the curve ; operation at this point will result in greater economy of iron.

In Fig. 32.11 the line OY has been drawn so that the deviation of the BH curve from the ideal curve OY is such that the values AA′, BB′ and CC′ when transferred to the H scale, represent equal values—in this case $7\frac{1}{2}$% of the nominal value.   Thus the timing capacitance does not vary by more than $7\frac{1}{2}$% for a variation in flux from

53 to 81 kilolines/sq. inch. Point A at 71 kilolines/sq. inch would appear to be, in this particular case, the optimum point where a variation of plus 14% and minus 25.5% can be allowed in the value of B. If this value of B results in an excessive value of magnetizing current, point X corresponding to 61 kilolines/sq. inch could be used.

Likewise any other point could be selected providing other design considerations are satisfied, but the point A represents the optimum as regards the change of timing capacitance with input voltage. The point A can be found for any BH curve by suitable positioning of the line OY.

It should be noted that having decided on the position of OY, and from this deriving the operating centre point, the values of B and H required for the timing capacitance calculation should be read from the line OY.

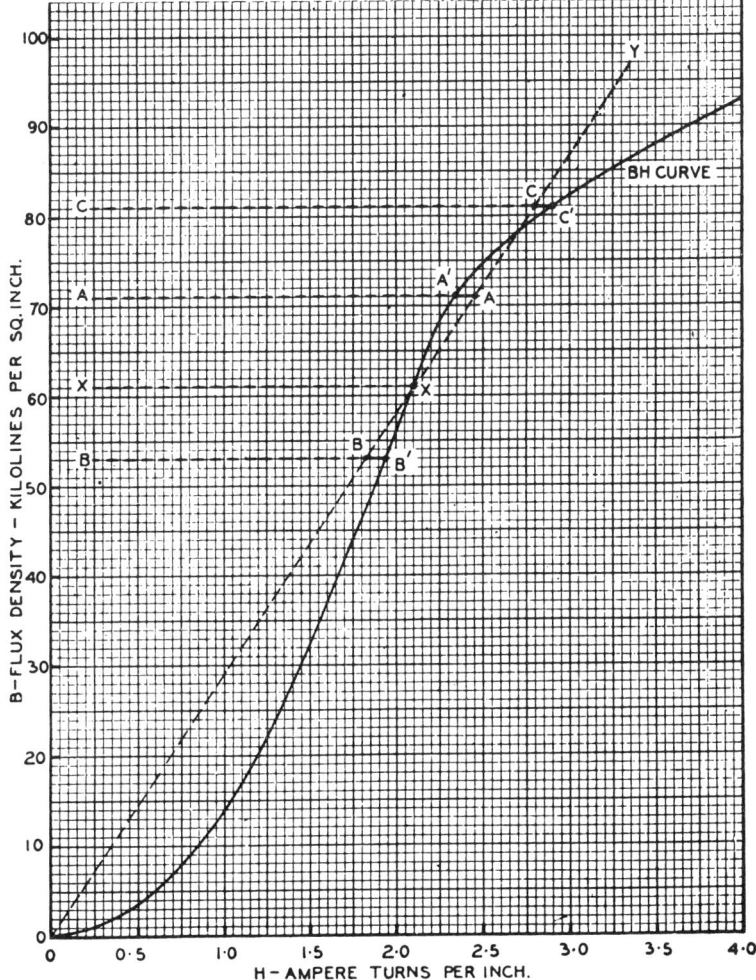

FIG. 32·11

SECTION 4 : ELIMINATION OF VIBRATOR INTERFERENCE

The methods of eliminating vibrator interference can only be stated in general terms and the final arrangement will depend upon the layout and characteristics of the particular power supply.   Where the vibrator power supply is not used with a receiver or similar sensitive equipment, the requirements of ' hash ' elimination are less stringent provided that there is no danger of interference being introduced into adjacent equipment.

With sensitive receivers, particularly those covering short wave ranges, it is desirable to use a separate shielded chassis for the vibrator power supply as it may otherwise be impossible to eliminate hash at all points in the tuning range with the receiver aerial wrapped around the battery cable and battery, and with the receiver at maximum sensitivity.   Such a chassis may be mounted on the main receiver chassis if desired, and in any case a flexible mounting (such as rubber grommets) is useful in reducing mechanical vibration and hum.

When a separate chassis is used a satisfactory method of construction is to have all hot components, vibrator, vibrator transformer and timing capacitors, in a shielded compartment with the leads from this compartment—normally either two or three— by-passed at or near the point of exit and connected to a r-f choke immediately after leaving the compartment.   R-F chokes are less useful on the other side of the shield. A common earthing point can still be used for all components by having a solder lug accessible from both sides of the chassis.

It is sometimes found that better hash reduction is obtained with one or more components earthed in isolated places.   However the improved results at one frequency are usually at the expense of increased hash at some other frequency.   If this is not the case, the earthing point is liable to be critical and perhaps variable from unit to unit, i.e. the improvement is due to cancellation of hash and not to elimination.   This is undesirable in prototypes of equipment which is later to be mass-produced.

R-F chokes are liable to have an appreciable field around them and layout must be such as to avoid introducing hash from such fields into circuits which have already been filtered.   Special types of by-pass capacitors are available with braided leads to reduce their r-f impedance.   These can give a considerable improvement over standard types with wire pigtails.   " Spark plates " are also useful.   These are by-pass capacitors using the chassis as one plate, a thin sheet of mica as the dielectric and a sheet of say brass as the " hot " plate, the assembly being held together with insulated eyelets.

Vibrator design is rarely a matter for ingenious circuits.   Once the transformer and timing capacitors are correctly specified, layout and shielding are the most important considerations and a minimum cost circuit is usually adequate.   If trouble is experienced from hash it is not usual for a cure to result from a mere addition of by-pass capacitors to a circuit which already has a normal complement.   For instance a unit in which all outgoing leads are " hot " is more likely (assuming reasonable filtering) to suffer from shielding troubles than from insufficient filtering of the leads in question.

A convenient method for investigating the presence or absence of interference at a particular point in a receiver is to use a probe of shielded wire with the end bared and connected through a capacitor to the aerial terminal of the receiver used with the vibrator unit.

Before hash elimination from a prototype vibrator unit can be considered complete it is necessary to carry out the standard tests with a number of vibrator cartridges, some of which have been in service for long periods.   Units in which hash suppression is only sufficient to silence the hash from new vibrator cartridges in good condition may give poor apparent cartridge life owing to the need for replacement as soon as sparking and the consequent interference occur due to slight wear.

In Fig. 32.12 are shown the more usual arrangements and typical circuit values are given as a guide to the choice of suitable components.   The timing capacitance is centre-tapped and its centre point is earthed.   The two 100 ohm resistors, R_1 and R_2,

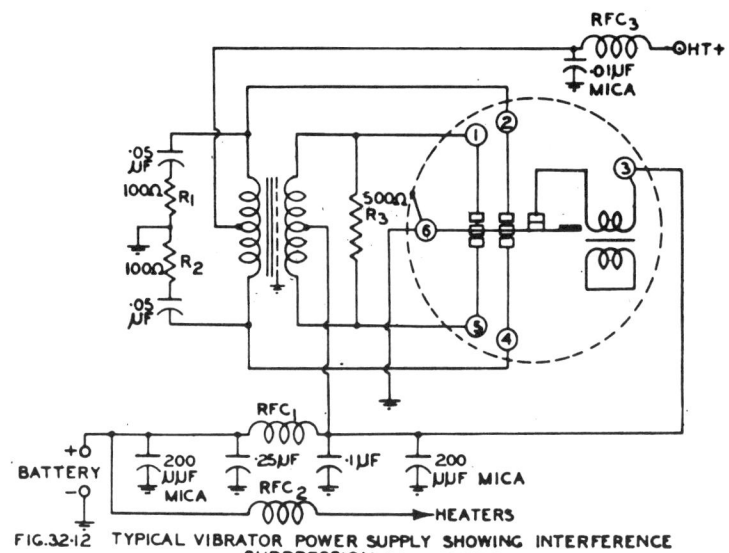

FIG.32·12   TYPICAL VIBRATOR POWER SUPPLY SHOWING INTERFERENCE
SUPPRESSION COMPONENTS

are not for hash elimination but to protect the vibrator transformer in the event of a timing capacitance short-circuiting. RFC_1 and RFC_2 are low resistance r-f chokes made of heavy gauge wire. RFC_1 prevents interference being fed back along the battery leads which could result in these leads radiating interference to the aerial. RFC_2 prevents interference entering the r-f section of the receiver through the heater wiring. RFC_3, a standard receiver r-f choke, is to eliminate interference frequencies in the plate voltage supply. The resistor R_3 is useful in eliminating the strong impulse interference which usually exists, although at the expense of some battery current. Alternatively, two separate resistors in series may be used across the primary contacts, the centre point of these being earthed.

In 12, 24 and 32 volt systems where part of the timing capacitance is usually placed across the primary circuit to aid starting, this capacitance has been found to be effective in reducing interference.. This method can be used for 6 volt systems to reduce the interference, but if the transformer has considerable leakage inductance increased trouble may be experienced with spurious resonances—see Sect. 2(i).

SECTION 5 : 12, 24 and 32 VOLT VIBRATOR SUPPLIES

While the higher voltage of the 12, 24 and 32 volt systems has some advantages over the 6 volt systems, special precautions are necessary if successful vibrator operation is to be obtained.

The higher voltage and consequently lower current, for the same power input, reduce the percentage voltage drop in the battery leads. The larger number of primary turns which are required results in an improved space factor and the centre tapping of the primary winding is made easier because of the larger number of layers and smaller wire diameter.

Against these advantages are the disadvantages of the additional circuit arrangements required for limiting the current during the starting period. Owing to the higher voltages existing across the contacts, ionization of the air between them will more readily take pláce. This condition is particularly serious on starting where the correct conditions for the vibrator have not stabilized and heavy currents may flow.

Fig. 32.13 shows a method of overcoming this difficulty, a three position switch being used to introduce some resistance into the battery lead for the starting condition.

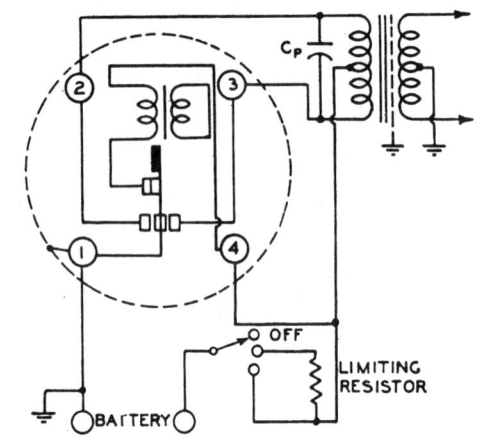

FIG. 32·13   STARTING ARRANGEMENTS FOR
24 AND 32 VOLT VIBRATORS.

Condensers across the contacts to suppress the arc are not to be recommended unless the minimum value that just suppresses the arc is used.   By splitting the timing capacitance and placing part of it across the primary, arcing can be reduced.   If this is done, consideration should be given to possible resonances with the primary leakage inductance due to imperfect addition of the split timing capacitance—see Sect. 2(i).

We wish to give acknowledgment to P. R. Mallory and Co., Inc., for much of the information and some of the diagrams in this Chapter, which have been adapted from their publication " Fundamental Principles of Vibrator Power Supply Design."

SECTION 6 : REFERENCES

1. " Fundamental Principles of Vibrator Power Supply Design " P. R. Mallory and Co.
2. " RMA Standard vibrator power transformers " REC—119 Sept. 1948 (U.S.A.).
3. Williams, M. R. " Heavy duty vibrator type power supplies " Radio News 35.6 (June 1946) 46.
4. Bell, D. A. " Vibrator power packs " W.W. 54.8 (Aug. 1948) 272.

Additional references will be found in the Supplement commencing on page 1475.

CHAPTER 33

CURRENT AND VOLTAGE REGULATORS

By R. J. RAWLINGS, Grad. I.E.E., Associate Brit. I.R.E.

SECTION 1.: CURRENT REGULATORS

(i) *Barretters* (ii) *Negative temperature coefficient resistors* (*Thermistors*).

(i) Barretters

In a.c./d.c. receivers, the valve heaters are wired in series and connected across the mains supply. If the total of the heater voltages is less than the mains voltage a current-limiting fixed resistor or a barretter is used in series with the heater circuit to absorb the extra voltage.

The disadvantage of the fixed resistor is that the current variations through the heater circuit are greater than if the circuit consisted of valve heaters alone.

See Chapter 35 Sect. 6 for the application of series resistors and barretters to a.c./d.c. receivers.

To maintain a constant current through the series heater circuit, with widely varying mains voltages, a barretter is generally employed. This consists of a hydrogen-filled tube containing an iron filament and has the property of nearly constant current flow for wide variations in voltage across it. The actual voltage limits between which satisfactory current regulation is obtained are stated by the manufacturers, and the barretters should be chosen with regard to these values.

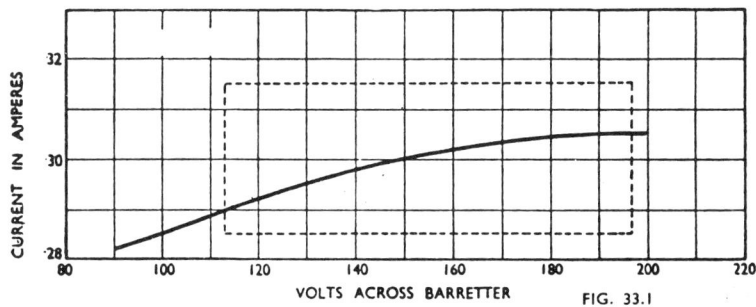

Fig. 33.1. The characteristic of a typical barretter.

A typical barretter characteristic is shown in Fig. 33.1 ; the portion of the curve within the dotted rectangle showing the useful operating voltage range and the tolerances on the current regulated by the barretter.

It should be noted that the barretter is subject to a large surge current when switching on as the resistance of the valve heaters is then only about one seventh or one tenth of their hot (running) resistance.

1213

In certain cases, a maximum total of heater voltage to be connected in series with the barretter is specified on the manufacturer's data sheets.   This then protects the regulator against excessive surge currents which would shorten its life.

Owing to the iron filament, barretters should be kept well clear of any magnetic field which could damage them and for this reason a magnetic screen is sometimes used around them.   As considerable power is dissipated in the barretter, the magnetic shield should be perforated and good ventilation provided.   The barretter should always be operated in a vertical position, base down.

Occasionally it is desired to increase, by a small amount, the current controlled by a barretter and this may be done, within certain limits, by operating it in parallel with a fixed resistance.   This is not normally recommended since the effectiveness of the barretter is reduced.   It may be used fairly satisfactorily for an increase of current up to about 10% of the current in the barretter.

If the barretter is required to control a current which is smaller than its current rating, this may be accomplished by shunting the current-regulated device by a resistor to bring the total load current up to the normal barretter current.

As with any other device, a barretter has manufacturing tolerances, but these are such that if a design is based on an average sample then no damage is likely to be done to the valves with any individual barretter.

(ii) Negative temperature coefficient resistors (Thermistors)
These are described in Chapter 4 Sect. 9(i)n.

SECTION 2 : VOLTAGE REGULATORS

(i) Gaseous tube voltage regulators   (ii) Valve voltage regulators.

In cases where stability of plate voltage supply is essential, some form of voltage regulation must be provided to prevent changes due either to mains input or load current variations.

(i) Gaseous tube voltage regulators
The electronic voltage regulator is shown in its simplest form in Fig. 33.2.   In this circuit a gas-filled two electrode tube is used as the voltage regulator.   The characteristics of this tube are such that quite large variations of current through the tube do not greatly alter the potential drop across it.

The circuit of Fig. 33.2 regulates the output voltage within certain interdependent limits of input voltage and load resistance.   If the input voltage increases for constant load resistance or the load resistance increases for constant input voltage then in either case the voltage across R_L tends to increase.   As E_R tends to rise, the regulator tube takes more current and increases the voltage drop across R_1, maintaining the voltage E_R at nearly its original value.   The resistor R_1 is essential to obtain voltage regulation.   Furthermore, as the striking voltage of the regulator tube is higher than its operating voltage, R_1 is necessary to prevent excessive tube current.   Table 1 gives values of minimum supply voltages required for starting the regulator tube, together with starting and operating voltages of typical tubes.

Referring to Fig. 33.2, if E_{in} = input voltage and I_t = regulator tube current, the following equation holds

$$E_R = E_{in} - (I_L + I_t) R_1 \qquad\qquad (1)$$

The range of regulation or the maximum change in E_{in} or R_L within the region of regulation can be determined accurately for any conditions of operation by substituting in eqn. (1) the limits of the variables E_{in}, I_L, I_t, R_L.   The regulator tube must operate between the published limits of current rating.   The change in one of the variables E_{in}, or R_L, can be much greater than in the other and when one of these changes is small the approximate range of regulation can be simply obtained by considering either E_{in} or R_L constant.   If these conditions apply and E_R is constant it follows by equating the partial differentiations of eqn. (1) to zero

that $\dfrac{\triangle E_{in}}{E_{in}} = \dfrac{R_1 \triangle I_t}{E_{in}}$ $\qquad\qquad (R_L$ constant$)$ $\qquad\qquad (2)$

and $\dfrac{\triangle I_L}{I_L} = -\dfrac{\triangle I_t}{I_L}$ $\qquad\qquad (E_{in}$ constant$)$ $\qquad\qquad (3)$

where $\triangle E_{in}$, $\triangle I_L$, $\triangle I_t$ are the small increments in E_{in}, I_L, I_t. The range of regulation is obtained by inserting in these equations the maximum value of $\triangle I_t$.

TABLE 1

Regulator Tube Type	Approximate Operating Voltage	Approximate Starting Voltage	Minimum Supply Voltage
OA2 (miniature)	150	160	185
OA3	75	100	105
OB2 (miniature)	108	115	133
OC3	105	115	133
OD3	150	160	185

Operation about the midpoint of the type OD3 characteristic gives $\triangle I_t$ (max.) = $(40 - 5)/2 = 17.5$ mA. Consider the regulator subject to small changes in load. If for example $I_L = 175$ mA, from eqn. (3) the range of regulation $\triangle I_L / I_L = 17.5/175 = 10\%$. Equation (2) shows that a large range of regulations with variable E_{in} is obtained when R_1 is large, i.e. when a large part of E_{in} appears across R_1.

When small load currents are required and variations in E_{in} and R_L are small, satisfactory regulation can be obtained by operating the regulator tube at low average current.

Regulation is frequently desired when R_L varies over the very large range from no load to full load. In this case R_1 is adjusted so that at no load the maximum tube current at maximum input voltage does not exceed the tube's rating, i.e., 40 mA for type OD3. Load voltage regulation is then obtained from zero load current to nearly 35 mA if changes in E_{in} are small.

The tube voltage of types OD3, OC3 can change over their operating range by approximately 3% and this represents their limit of regulation under wide range conditions.

If higher regulated voltages are required two or more tubes can be connected in series as in Fig. 33.3. Additional stabilized voltages may be taken off the individual tubes as shown. Tapping points may be provided on R_{L2} to give lower voltages, however if the load varies, good regulation will be obtained only if the current drawn from the tapping point is very small compared with the current through R_{L2}. Grid bias supplies for class A amplifiers may be effectively stabilized by the circuit of Fig. 33.2 with tappings as required—see chapter 13 Sect. 10(ii).

The use of gaseous regulator tubes in parallel is not to be recommended as resistors must be put in series with each tube so that all tubes will start and share current equally. These resistors impair voltage regulation. Therefore when large currents have to be handled, grid controlled valve voltage regulators must be used.

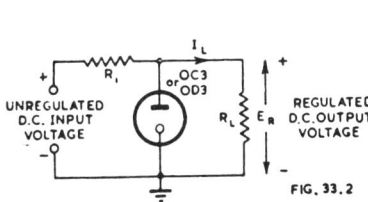

FIG. 33.2

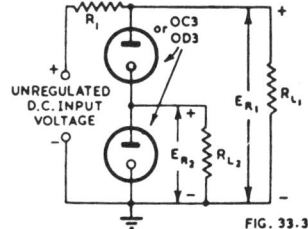

FIG. 33.3

Fig. 33.2. Simplest form of gaseous
tube voltage regulator.

Fig. 33.3. Higher regulated voltage
provided by two gaseous tubes in series.

(ii) Valve voltage regulators

A typical valve voltage regulator is shown in Fig. 33.4. The valve V_1 acts as a variable resistance whose value is varied by a change in bias. When the output voltage tends to decrease, the grid voltage of V_2 becomes less positive. As the cathode voltage of V_2 is kept constant with respect to earth by the gaseous voltage regulator tube V_3, and positive with respect to the grid, the fall in the positive value of the grid will increase the negative bias on the valve, and I_b will decrease. This decrease of plate current will cause an increase in the voltage at the plate of V_2. The plate of V_2 being directly connected to the grid of V_1 will cause a drop in the negative bias on V_1 and the plate-cathode voltage will decrease, tending to restore the original voltage drop across E_R.

The complete mathematical analysis of voltage regulator performance is fairly complex and for detailed information reference should be made to the bibliography at the end of this chapter.

For practical purposes of voltage regulator design, the following method may be used.

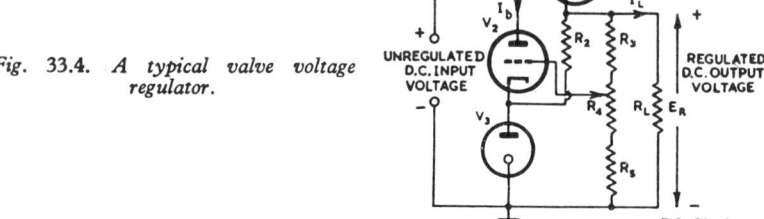

Fig. 33.4. A typical valve voltage
regulator.

Referring again to Fig. 33.4. If constant output voltage is to be maintained across the load, the valve V_1 must be capable of adjusting its voltage drop to compensate for voltage input changes. A similar condition exists if the input voltage is constant and variable output voltage is required. The valve has therefore to satisfy conditions at four points as set out in Table 2.

TABLE 2

Point	Output	Valve V_1	
		Plate Voltage	Plate Current
1	Maximum Voltage Maximum I_L	Supply Voltage Maximum Output Voltage	Maximum I_L
2	Maximum Voltage Minimum I_L	Supply Voltage Maximum Output Voltage	Minimum I_L
3	Minimum Voltage Maximum I_L	Supply Voltage Minimum Output Voltage	Maximum I_L
4	Minimum Voltage Minimum I_L	Supply Voltage Minimum Output Voltage	Minimum I_L

Each of these four points has its particular limitations as regards the operation of the valve to maintain the correct operating conditions without exceeding the maximum ratings. Reference should be made to Fig. 33.5 in which these four points are plotted for a triode-connected type 807 valve.

Point 1—this point must be kept in the negative grid-bias region of the curve and a bias of not less than -3 volts may be taken as the design minimum. This point must also be kept below the maximum plate or screen dissipation curve (whichever is first reached), and below the recommended value of current for a steady d.c. oper-

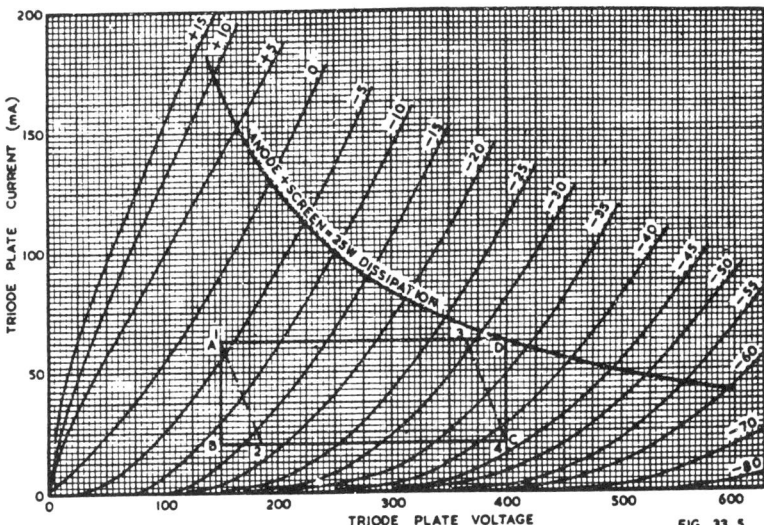

Fig. 33.5.   Plate characteristics of triode-connected 807 valve to illustrate operation of the valve as a voltage regulator.

ting condition.   It is important that point 1 be kept above a certain minimum value of plate voltage if linear characteristics are to be obtained from V_1, a value of not less than 125 volts on the plate being recommended.

Point 2—In order that this point should not come in the non-linear portion of the plate characteristics, a bleed should be arranged across the supply to limit the minimum current.   This also aids power supply regulation (see Chapter 30 Sect. 3).

Point 3—This point should be kept below the maximum dissipation curve and below the maximum rated plate voltage for the valve being used.

Point 4—This point should be kept above the non-linear portion of the characteristics by means of a bleed as in (2) and also kept below the maximum plate voltage value for the valve being used.

The maximum recommended usable section of the valve characteristic for type 807 is shown enclosed by ABCD in Fig. 33.5.

While it is possible to use higher plate current values at low plate voltages than appear in the diagram before the plate dissipation is exceeded, this is not to be recommended as reduced valve life may result.

From the graphical construction can be seen the effect of large variations of input voltage to the regulator with varying output currents (i.e. poor regulation of the power supply).   This will cause the point 2 to move along the plate voltage axis, which results in a reduction of the maximum variation of output voltage obtainable.   When the power supply has good regulation and the input voltage is substantially constant, an output variation of $\pm$ 125 volts is obtainable about a nominal value.   With a constant output voltage the same tolerances may be allowed in the input voltage i.e. $\pm$ 125 volts.   In each case it is assumed that the circuit conditions are set so that the valve is operating at its centre point for the nominal input or output voltage.

It will be seen from this description that a choke input power supply is advisable. Reference should be made to Chapter 30 Sect. 3 for further information on choke input power supply design.

For minimum voltage loss across V_1 a valve with a high g_m should be used.   If the load current required from the regulated supply is greater than that obtainable with one series valve, then two or more valves may be placed in parallel.

TABLE 3
Valves suitable for use as series regulators.

Valve type	Current (mA)
6F6-G*	45
6V6-GT*	50
6L6-G*	80
6Y6-G*	80
807*	80
6AS7-G	250

*Screen connected to plate through 500 ohm 1 watt resistor.

Having now determined the operating conditions for V_1, the design may be extended to consider the operating conditions for the amplifier valve V_2.

From the knowledge of the voltage drop in V_1 required for the four operating points, the corresponding bias values can be determined. The voltage drop across the resistor R_1 is then given by the voltage drop across V_1 plus the required grid bias for V_1.

Having determined these voltages, the resistance R_1 can be chosen to have a value which will give plate current values for the amplifier valve which are on the linear portion of its plate characteristics.

The higher the gain of V_2, the better will be the regulation, as a smaller change in output will be effective in adjusting the voltage drop across V_2 to correct the output voltage change. For this reason a pentode is recommended, the additional complication of adding a screen supply being small for the improvement obtained.

In general, the plate voltage for V_2 should be derived, through its plate load resistance R_1 from the input voltage to the regulator, so that even at low output voltages not less than the minimum voltage drop of V_1 plus its grid bias voltage will exist across R_1. The exception to this connection will be when the output voltage of the regulator is to be constant at a value of 200 or greater and the input voltage is to be varying over wide limits ; the effect of connecting R_1 to the input in this case would be to reduce the effective gain of the valve.

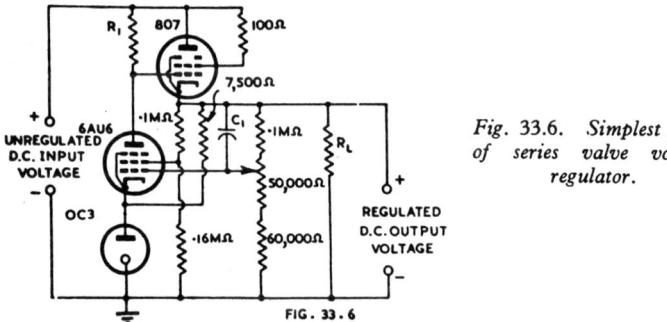

Fig. 33.6. Simplest form of series valve voltage regulator.

FIG. 33.6

The connection of the gaseous voltage regulator tube V_3 also requires some consideration. It has been mentioned in Sect. 2(i) that changing the current through a voltage regulator tube will produce a small change of voltage across it. As this tube is used as the reference voltage for the regulator, any change in potential across it will have an adverse effect upon the regulator performance. For this reason, if a constant output voltage regulator is being designed, the series feed resistor for this tube should be supplied from the output side of the regulator, the extra current drawn by this tube being taken into account when considering the current rating for V_1. In cases where the regulator has to supply varying output voltages, V_3 may be supplied from the input if this is sufficiently constant ; preferably, V_3 should be supplied from a separate source.

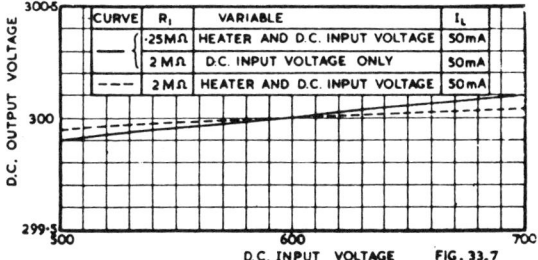

Fig. 33.7. Operation characteristics of circuit Fig. 33.6—output versus input voltage.

Circuits of typical voltage regulators and their performance figures are given in Figs. 33.6 to 33.14.

In Fig. 33.6 is shown the simplest form of series valve voltage regulator, which, however, with the circuit components and values shown, is capable of good performance. The results obtained with this regulator are shown in Figs. 33.7 and 33.8, from which several points should be noted. The variation of heater voltage as well as the input

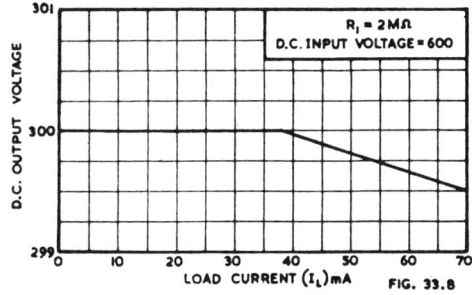

Fig. 33.8. Operation characteristics of circuit Fig. 33.6—output voltage versus load current.

direct voltage, the usual practical case, results in better regulation than that obtained with only direct voltage changes. This is due to a slight change of plate current in V_2 together with other minor circuit changes caused by the variation of cathode temperature in the valves. For rapid input voltage changes the regulation characteristic will be as for direct input voltage changes only since the cathode temperature cannot change rapidly. A value of 250 000 ohms for R_1 will be found to be sufficient for most purposes as little increase in gain is obtained by increasing R_L above this value.

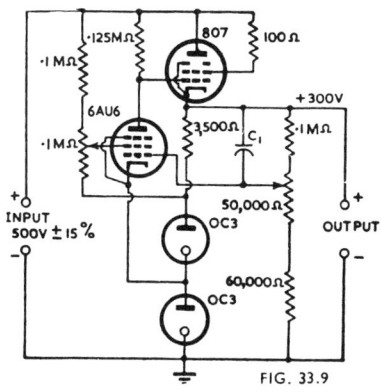

Fig. 33.9. Valve voltage regulator in which the screen is used to correct for input voltage changes.

Such a value will allow this resistor as well as all others to be wire wound if particularly stable operation is required.

No advantage is gained in the circuit of Fig. 33.6 in stabilizing the screen supply, this already being supplied from a substantially constant source. However screen stabilization is necessary when the screen is used to correct for input direct voltage changes A circuit of this type is shown in Fig. 33.9 and a constant or overcorrected output is obtainable.

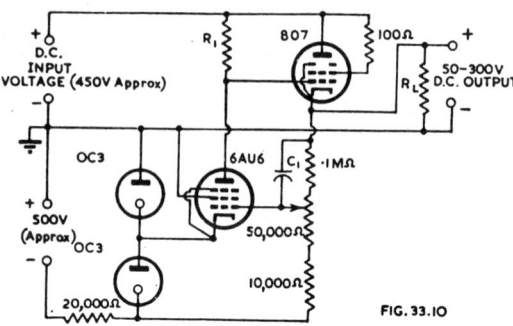

FIG. 33.10

Fig. 33.10. Valve voltage regulator with a controllable output voltage.

The circuit of Fig. 33.10 is useful where a controllable output voltage is required and is suitable for output voltages from 50 to 300. Regulation characteristics for this regulator are shown in Figs. 33.11 and 33.12.

Fig. 33.11. Regulation characteristics for circuit of Fig. 33.10—output versus input voltage.

Fig. 33.12. Regulation characteristics for circuit of Fig. 33.10—output voltage versus load current.

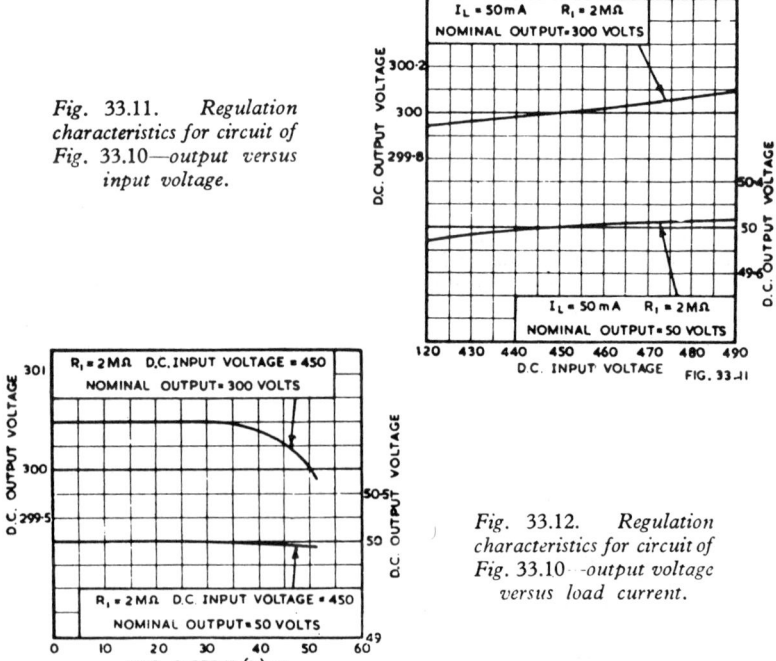

To draw a constant value of bleed current from the variable voltage supply, in order to prevent soaring of the output voltage on no load, some form of constant current device must be used. Refer to explanation of points 2 and 4 under Table 2. A suitable circuit is shown in Fig. 33.13 in which a 6V6-GT type valve is used as a constant current pentode ; with plate voltages from 50 to 300 the plate current will only vary from 18.5 to 19.0 mA.

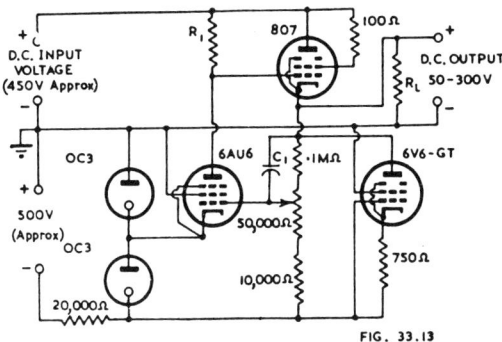

FIG. 33.13

Fig. 33.13. Valve voltage regulator with a controllable output voltage as Fig. 33.10 but including constant current pentode to draw constant bleed current for varying output voltages.

The condenser C_1 in all circuits assists in the reduction of ripple from the regulated supply as the grid of V_2 then obtains the full ripple voltage from the supply instead of the fraction determined by the values R_3, R_4 and R_5 (Fig. 33.4). This condenser should have a value not greater than that required for satisfactory ripple suppression as large values will cause a time lag in the operation of the output voltage control, a value of .01 to .1 μF being satisfactory for most purposes.

The need for negative voltage supplies for several of the given circuits need not require the addition of a separate power supply ; reference should be made to Chapter 30 Sect. 6 which gives details of shunt diode bias supplies which would be suitable for this purpose. The regulator tube current in typical circuits, except Fig. 33.14,

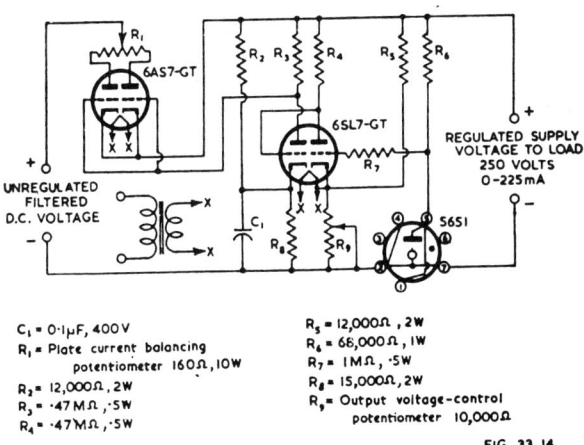

C_1 = 0·1μF, 400 V	R_5 = 12,000Ω , 2W
R_1 = Plate current balancing	R_6 = 66,000Ω , 1W
potentiometer 160Ω,10W	R_7 = 1MΩ , ·5W
R_2 = 12,000Ω , 2W	R_8 = 15,000Ω, 2W
R_3 = ·47MΩ ,·5W	R_9 = Output voltage-control
R_4 = ·47MΩ ,·5W	potentiometer 10,000Ω

FIG. 33.14

Fig. 33.14. Voltage regulator utilizing a special voltage reference tube (type 5651) to give extremely good voltage regulation.

may be adjusted to have a value of approx. 10 mA at the normal operating point of the regulator.

For most applications these circuits will give adequate regulation but for details on compensated and the more complicated regulators reference should be made to the bibliography at the end of this chapter.

The main limitation of all these regulators is the difficulty of obtaining a completely stable reference voltage, as gaseous voltage regulator tubes may stabilize at slightly different voltages when the unit is switched off and on again. This means that, although nearly perfect voltage regulation may be obtained, the voltage about which this regulation takes place is not always a constant value.

This defect may be minimized by the use of a special voltage reference tube such as R.C.A. type 5651 as, for example, in the circuit of Fig. 33.14. The unregulated input is approximately 375 volts at zero load current and 325 volts at 225 mA load current. The variation of output voltage is less than 0.1 volt for a variation of ± 10% in input voltage when operated at maximum load current. When adjusted to an output of 250 volts, the variation in output voltage is less than 0.2 volt over the current range from 0 to 225 mA.

SECTION 3 : REFERENCES

1. Anthony, A. " Basic theory and design of electronically regulated power supplies " Proc. I.R.E. 33.7 (July 1945) 478.
2. Ashworth, J. A., and J. C. Mouzon, " A voltage stabilizer circuit " 8.4 Rev. Sci. Inst. 127.
3. Bereskin, A. B. " Voltage regulated power supplies " Proc. I.R.E. 31.2 (Feb. 1943) 47.
4. Bousquet, A. G. " Improving regulator performance " Elect. 11.7 (July 1938) 26.
5. Cherry, L. B., and R. F. Wild, " Electronic alternating current power regulator " Proc. I.R.E. 33.4 (April 1945) 262.
6. Hamilton, G. E., and J. Maiman, " Voltage-regulated power supplies " Comm. 25.11 (Nov. 1945) 44 ; 25.12 (Dec. 1945) 70 ; 26.1 (Jan. 1946) 48.
7. Hogg, F. L. " Electronic voltage regulators " W.W. 49.11 (Nov. 1942) 327 ; 49.12 (Dec. 1943) 371.
8. Harris, E. J. " A note on stabilizing power supplies " Electronic Eng. 20.241 (March 1948) 96.
9. Hill, W. R. Jnr. " Analysis of voltage regulator operation " Proc. I.R.E. 33.1 (Jan. 1945) 38.
10. Hoyle, W. G. " Circuit design for gas-discharge regulator tubes " Tele-Tech. (Feb. 1948) 46.
11. Hunt, F. W., and R. W. Hickman " On electronic voltage stabilizers " Rev. Sci. Inst. 10.1 (Jan. 1939) 6.
12. Ledward, T. A. " A.C. voltage stabilizer " W.W. 49.6 (June 1943) 166.
13. Miller, S. E. " Resistance coupled d.c. amplifier with a.c. operation " Elect. 14.11 (Nov. 1941) 27.
14. Neher, H. V., and W. H. Pickering " Two voltage regulators " Rev. Sci. Inst. 10.2 (Feb. 1939) 53.
15. Sowerby, J. M. " Shunt voltage stabilizer " W.W. (June 1948) 200.
16. Penners, B. A., and W. Davis " Design of electronically regulated power supplies " Radio 31.2 (Feb.-March 1947) 9.
17. Helterline, L. " Diode-controlled voltage regulators " Elect. 20.6 (June 1947) 96.
18. Koontz, P., and E. Dilatush " Voltage regulated power supplies " Elect. 20.7 (July 1947) 119.
19. Helterline, L. L. " Design of regulated power source " Tele-Tech. 6.7 (July 1947) 63.
20. Hughes, J. W. " Stabilizing direct-voltage supplies " W.E. 24.287 (Aug. 1947) 224.
21. Berg, W. R. " Optimum parameter for gas tube voltage regulators " Elect. 20.10 (Oct. 1947) 136.
22. Helterline, L. L. " A-C voltage stabilizers " Audio Eng. 31.8 (Sept. 1947) 23.
23. " High potential voltage regulator " (Review of patent No. 2416922 by A. J. Irish and D. G. Lindsay) Audio Eng. 31.6 (July 1947) 44.
24. Luo, Y. P. " A negative-current voltage-stabilization circuit " Proc. I.R.E. 36.5 (May 1948) 583.
25. Miller, W. B. " Versatile power supply " Elect. 21.6 (June 1948) 126.
26. Scroggie, M. G. " Stabilized power supplies" (1) W.W. 54.10 (Oct. 1948) 373 ; (2) 54.11 (Nov. 1948) 415 ; (3) 54.12 (Dec. 1948) 453.
27. Scroggie, M. G. " Low-impedance variable-voltage tappings " W.W. 55.1 (Jan. 1949) 2.
28. Benson, F. A. " Voltage stabilisers ".Electronic Eng. (1) 21.255 (May 1949) 155 ; (2) 21.256 (June 1949) 200 ; (3) 21.257 (July 1949) 243 ; (4) 21.258 (Aug. 1949) 300.
29. Yu, Y. P. " Novel regulator circuit " Elect. 22.5 (May 1949) 370.
30. Engineering Dept., Aerovox Corp. " Regulated Power Supply Design " Aerovox Research Worker 20.9 (Sept. 1950).
31. Houle, J. (letter) " Wide-range voltage regulators " Elect. 24.8 (Aug. 1951) 202.
32. Baruch, J. J. " Zero-impedance power supply termination " Elect. 24.8 (Aug. 1951) 240. See also correction, Stockman, H. " Zero impedance " Elect. 25.5 (May 1952) 358.
33. Benson, F. A. (book) " Voltage Stabilizers " Electronic Engineering, London, 1950.
34. Patchett, G. N. " Precision a.c. voltage stabilizers " Electronic Eng. 22.271 (Sept. 1950) 371 ; 22.272 (Oct. 1950) 424 ; 22.273 (Nov. 1950) 470 ; 22.274 (Dec. 1950) 499 ; letters 23.276 (Feb. 1951) 70.
35. Willmore, A. P. " The cathode follower as a voltage regulator " Electronic Eng. 22.271 (Sept. 1950) 399.

Additional references will be found in the Supplement commencing on page 1475.

CHAPTER 34

TYPES OF A-M RECEIVERS

By I. C. HANSEN, Member I.R.E. (U.S.A.)

SECTION 1 : INTRODUCTION AND SIMPLE RECEIVERS

(i) Types of receivers (ii) Crystal sets (iii) Regenerative receivers (iv) Superregenerative receivers (v) Tuned radio-frequency receivers.

(i) Types of receivers

Radio receivers may be divided into several categories as the following tabulation shows :—

(a) crystal

(b) regenerative

(c) superregenerative

(d) tuned-radio-frequency

(e) superheterodyne

(f) synchrodyne.

(ii) Crystal sets

The simplest type of receiver employs a crystal such as galena, plus a "catswhisker," for a detector. This, together with a suitable tuned circuit and a pair of headphones, forms a satisfactory local station receiving set. Its disadvantages are poor sensitivity and selectivity, together with low output. Modern developments have made available " fixed " germanium crystals which are being used satisfactorily as detectors. The transistor, or three element crystal, recently announced, offers the advantage of amplification as well as detection in the one unit.

(iii) Regenerative receivers

Higher sensitivity can be obtained by using a valve as a grid or plate circuit detector. Additional amplification can be obtained by means of reaction ; the feeding back of signal from plate to grid in the correct phase to aid the existing circuit gain. Oscillation will occur if too much feedback is used, hence the necessity for judicious use of the reaction control, particularly for speech modulated signals, which would otherwise be rendered unintelligible. When operated near the point of oscillation, i.e. at maximum sensitivity, the selectivity is such that serious side-band cutting occurs and distortion of the a-f signal results ; this is usually the limiting factor in the amount of feedback that can be successfully employed. As the input circuit is coupled directly to the aerial, should the circuit oscillate, radiation will occur with the risk of interference to other receivers operating nearby—hence the limited use of this circuit. An r-f stage between the aerial and detector assists in minimizing this trouble.

(iv) Superregenerative receivers (Fig. 34.1)

The superregenerative receiver is basically similar to the simpler detector with reaction. It, however, is adjusted to the threshold of oscillation so that an incoming signal will cause the circuit to oscillate. At this instant, a local " quench " oscillator, operating at a low radio frequency, damps out the oscillation, thus ensuring that the receiver is operating continuously at maximum sensitivity. The oscillator can be combined with the detector or it can be a separate valve. Although very sensitive, the superregenerative receiver has poor selectivity due to circuit loading. In addition, interference is also caused by radiation from the " quench " oscillator. It is used mainly for higher frequency work where fidelity is unimportant and high sensitivity from simple apparatus is required.

See also Chapter 27 Sect. 1(ii)F page 1087 for further information.

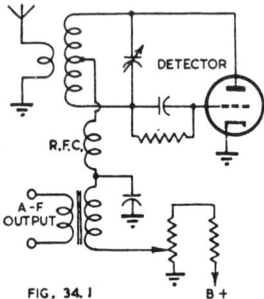

Fig. 34.1.   Self quenched superregenerative receiver incorporating linear reflex detector.

(v) Tuned-radio-frequency receivers (Fig. 34.2)

The simple receivers described above can give improved selectivity, sensitivity and output by the addition of radio frequency and audio frequency amplifier stages. Two or three r-f stages are commonly used ahead of the detector, which is usually of the plate detection or power-grid type. One or more a-f stages follow the detector depending upon the power output required. Such receivers are simple to design and construct for broadcast frequencies but present difficulties at higher frequencies. This is due to the risk of instability resulting from all the gain being achieved at the signal frequency. Tracking also becomes a problem as the frequency increases. In general, the defects of the tuned r-f receiver are variation of sensitivity and selectivity with frequency over the tuning range.

At broadcast frequencies, the circuit simplicity and ease of alignment are the main advantages.

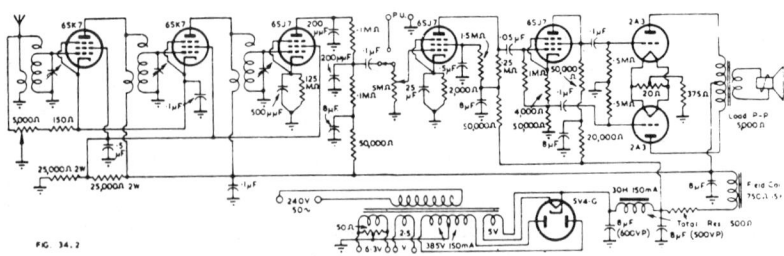

Fig. 34.2.   Typical t.r.f. receiver

SECTION 2 : THE SUPERHETERODYNE

The supe.heterodyne receiver as illustrated in Figs. 34.3 and 34.4, has several important advantages over other types of receivers.   In the superheterodyne circuit, the incoming signal frequency is changed to a lower frequency, known as the intermediate frequency.   The major part of the amplification then takes place at this frequency before detection in the normal manner.   The typical superheterodyne tuner consists of several distinct sections.   These are :—

    (a) Preselector or r-f amplifier

    (b) 1st detector or mixer

    (c) Oscillator

    (d) I-F amplifier

    (e) 2nd detector.

Section (a) may or may not employ one or more r-f amplifiers.   Its main functions are to improve the signal-to-noise ratio and to provide a sufficient degree of selectivity to avoid " double-spotting."   This latter is the term applied to the reception of one station transmitting on a certain definite frequency, at more than one point on the tuning dial.

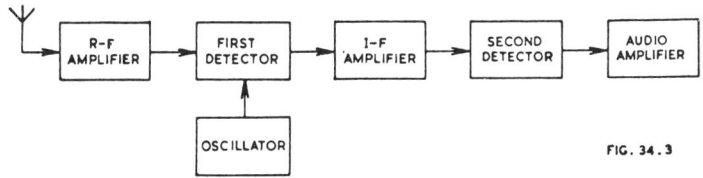

FIG. 34.3

Fig. 34.3.   Superheterodyne receiver—block diagram.

The first detector stage receives the modulated signal from the preselector and also an unmodulated signal from a local oscillator.   These two r-f signals are arranged to differ by a constant frequency by suitable design of the preselector and oscillator tuning circuit constants.   The output of the first detector is tuned to this difference frequency and as a result the original signals and their sum frequency are suppressed.

Various types of first detectors have been used, the most popular being the pentagrid and triode hexode converters (Refer Chapter 25).   The tickler-feedback oscillator is very commonly used and optimum adjustment of this circuit minimizes harmonic production with its attendant whistles and spurious responses.   It also ensures maximum conversion conductance from the first detector circuit and optimum signal-to-noise ratio.

The fixed frequency amplifier, called the intermediate frequency or i-f amplifier, usually consists of 1 stage, i.e. a valve and two i-f transformers, input and output.

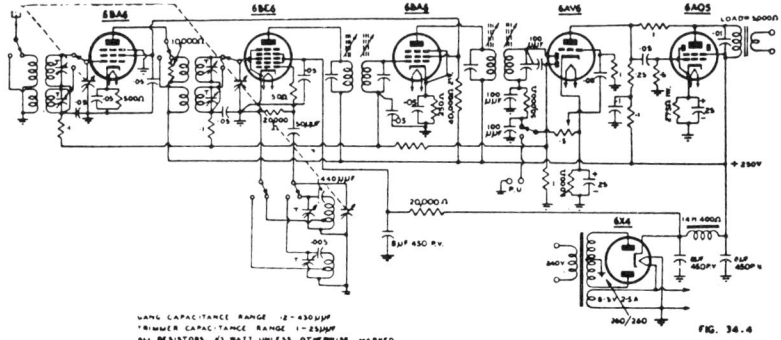

Fig. 34.4.   Typical dual-wave superheterodyne receiver.

In wide band and communication receivers, two or more stages are commonly used. The intermediate frequency in general use is 455 Kc/s. Earlier receivers used 175 Kc/s but with the appearance of powdered iron cores and the development of high slope amplifier valves, the previous objection to the use of higher intermediate frequencies, i.e. lower gain, was nullified.

The higher i-f now in use considerably reduces the incidence of " double-spotting " i.e. the reception of the same station at two points on the receiving dial, one removed from the other by twice the intermediate frequency. Thus with an intermediate frequency of 455 Kc/s the first " double-spot " would be at 1460 Kc/s, which is near the upper limit of the broadcast band if this is taken as from 550 to 1600 Kc/s.

As most of the receiver amplification occurs at the intermediate frequency, which is fixed in frequency, the overall gain does not change appreciably with signal tuning. Similarly the selectivity remains approximately constant as this is largely predetermined by the i-f channel.

From the i-f amplifier the signal passes into a second detector stage where the audio frequency modulation is separated from the carrier and then amplified in the normal manner by a conventional a-f amplifier. If, as is usual, a diode detector is employed, the rectified carrier provides a direct voltage which can be used for various control purposes. The principal use is for automatic volume control, or, more exactly, automatic gain control. In this circuit the negative direct voltage, which increases with an increase in signal, is applied through suitable decoupling networks to the r-f, i-f and first detector grids as required. This overcomes, to a considerable degree, the effect known as " fading."

For the reception of continuous wave code signals another oscillator is necessary. The output from this is fed into the second detector and adjusted to give a suitable audio difference frequency, say 1000 c/s, when beating with the incoming code signals.

The advantages of the superheterodyne over the tuned-radio-frequency receiver are :—

    (a) more uniform sensitivity and selectivity over the tuning range,
    (b) stability is greater, as the major part of the amplification takes place at a low radio frequency,
    (c) the i-f amplifier can be designed for minimum sideband cutting, while preserving reasonable gain,
    (d) greater selectivity.

It should be noted that the superheterodyne contains more tuned circuits than the t.r.f. receiver, but fewer tuned circuits are continuously variable.

See References (B).

SECTION 3 : THE SYNCHRODYNE

A newer type of receiver is the " synchrodyne " (Fig. 34.5). In this design, selectivity is obtained without resort to tuned circuits or band-pass filters. The block diagram of Fig. 34.5 will facilitate the understanding of the operation of this circuit which requires no tuning circuits other than that of the oscillator. Here, the desired incoming modulated signal is heterodyned with a local unmodulated signal of the same carrier frequency. The output from the detector, consisting of the required modulation plus unwanted higher frequency components from stations operating on adjacent channels, is then fed to an a-f amplifier through a low-pass filter.

This simple filter can readily be designed to cut off sharply at any requisite point whereas the usual superheterodyne band-pass filters require many elements and generally reduce the response at a considerably lower frequency than that desirable for optimum results.

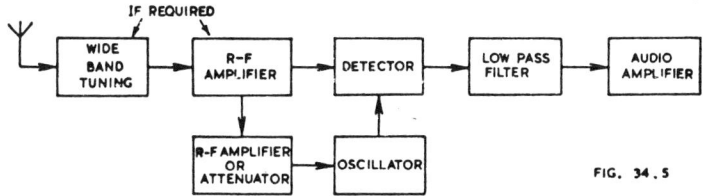

Fig. 34.5.   Synchrodyne receiver—block diagram.

The selectivity of the synchrodyne depends upon the oscillator circuit and a restriction of the frequency band there does not affect the a-f response.

To avoid beats in the output signal the local oscllator must be synchronized or " locked " with the wanted carrier.   One advantage of this receiver is that it is either correctly tuned or not tuned at all.   Distorted output due to mistuning is thus impossible.   The only effect of altering the oscillator tuning within the synchronizing range is to change the volume level.

If necessary, a broad-band r-f amplifier can precede the detector to avoid strong stations overloading the receiver.   As in the typical superhet, a.v c. voltage derived from the detector output can be used for gain control.

The main disadvantage of this receiver is that loud heterodyne whistles are heard when tuning in a station.   This defect can be readily overcome by the use of push-button tuning, and hence the synchrodyne is likely to be most popular for high-quality local-station reception.

See References (C).

SECTION 4 : REFERENCES

(A) GENERAL
Van Dyck, A., and D. E. Foster, " Characteristics of American Broadcast receivers as related to the power and frequency of transmitters " Proc. I.R.E. 25.4 (April 1937) 387.
Foster, D. E. " Receiver characteristics of special significance to broadcasters " Comm. 19.5 (May 1939) 9.
Dean, C. E. " Receiving Systems " Chapter 17 of " Radio Engineering Handbook " by K. Henney (McGraw-Hill Book Co. Inc. N.Y. 4th ed. 1950).
Jordon, E. C. and others (book) " Fundamentals of Radio " (Prentice-Hall, Inc. N.Y. 1943) Chapter 13.

(B) SUPERREGENERATIVE RECEIVERS
Riebman, L. " Theory of the superregenerative amplifier " Proc. I.R.E. 37.1 (Jan. 1949) 29.
Cathode Ray " Super-regenerative receivers " W.W. 52.6 (June 1946) 182.
Fox, L. S. " Super-regeneration—its theory of operation " Q.S.T. (Dec. 1943) 17.
Grimes, D., and W. S. Barden, " Super-regeneration and its application to high frequency reception " Report, R.C.A. Laboratories Industry Service Division.
Russell, O. J. " Super regeneration " W.W. 50.12 (Dec. 1944) 361 (good bibliography).
Easton, A. " Superregenerative detector selectivity " Elect. 19.3 (Mar. 1946) 154.
Riebman, L. " Theory of the superregenerative amplifier " Proc. I.R.E. 37.1 (Jan. 1949) 29.
Hazeltine, A., D. Richman and B. D. Loughlin " Superregenerator design " Elect. 21.9 (Sept. 1948) 99.
Bradley, W. E. " Superregenerative detection theory " Elect. 21.9 (Sept. 1948) 96.
Stockman, H. " Superregenerative circuit applications " Elect. 21.2 (Feb. 1948) 81.
Tapp, C. E. " The Application of super-regeneration to frequency-modulation receiver design " Proc. I.R.E. (Aust.) 8.4 (April 1947) 4.
Whitehead, J. R. (book) " Super-Regenerative Receivers " (Cambridge University Press, London and U.S.A. 1950).

(C) SYNCHRODYNE RECEIVERS
Tucker, D. G. " The Synchrodyne—a new type of radio receiver for A.M. Signals," Electronic Eng. 19.229 (March 1947) 75.
Tucker, D. G. " The design of a synchrodyne receiver " Electronic Eng. Part 1—Design Principles 19.234 (Aug. 1947) 241 ; Part 2—Some suitable designs 19.235 (Sept. 1947) 276.
Tucker, D. G. " The Synchrodyne—further notes," Electronic Eng. 19.237 (Nov. 1947) 366.
The Synchrodyne—correspondence ; Electronic Eng. 19.233 (July 1947) ; 19.237 (Nov. 1947) 368.
Tucker, D. G., and J. Garlick, " The synchrodyne : refinements and extensions," Electronic Eng. 20.240 (Feb. 1948) 49.
Burnup, J. E. (letter) " The synchrodyne " (simpler oscillator) Electronic Eng. 20.240 (Feb. 1948) 63.   Correction April 1948 p. 132.
Langberg, E. " The Synchrodyne " (letter) Electronic Eng. 20.242 (April 1948) 132.
" Cathode Ray " " The synchrodyne " W.W. 54.8 (Aug. 1948) 277.

Additional references will be found in the Supplement commencing on page 1475.

CHAPTER 35

DESIGN OF SUPERHETERODYNE A-M RECEIVERS

By E. WATKINSON, A.S.T.C., A.M.I.E. (Aust.), S.M.I.R.E. (Aust.)

SECTION 1 : INTRODUCTION

The design of A-M receivers as discussed in this chapter is taken as design for quantity production, since it is normally only under these conditions that a signal generator, a wave analyzer, a low-distortion beat frequency oscillator and suitable valve voltmeter and other meters are available.

The design of the individual stages in a receiver has been covered in the earlier chapters of this Handbook.   It is assumed here that the receiver designer has already studied these earlier chapters or has equivalent knowledge.   The present chapter covers the design procedure and certain general features which affect more than a single stage.

The first stage in a design is the drawing of the circuit of the receiver.   This statement contrasts with the views of those who believe the circuit to be the last stage, but whereas most engineers can readily draw a usable circuit for any normal type of receiver, there are few conscientious engineers who could build such a receiver and not find that after alignment and adjustment they did not wish to make some modifications to remove faults or to improve the performance in some way.   Accordingly the circuit is the first stage, modifications to layout and perhaps to the circuit as a result of measurements and operating tests are the second stage, the measurement of all performance figures and checking of all ratings and tolerances the third stage and the making of a final sample the last stage.

The fact that after the final sample is finished it frequently becomes necessary to revert to stage two again, should not be taken as a reflection on the design engineer but rather as an illustration that there is far more to a receiver than the circuit (Ref. 29). What are (apparently) minor changes in layout may introduce unforeseen difficulties with, say, the symmetry of the i-f amplifier, a tendency to instability due to feedback from the i-f amplifier to the aerial terminal or any one of a dozen other possible sources of trouble.

In a good design preliminary calculation will have been carried far enough to ensure that the circuit as first drawn will be such as to allow the specifications to be met with a minimum of components and the model will have been built with an adequate knowledge of the practical troubles likely to be encountered so that they may be a minimum.

SECTION 2 : SPECIFICATIONS AND REQUIREMENTS

The designer of a commercial radio receiver has in general three sets of specifications to meet.   Firstly there are the specifications of the relevant authorities in the countries concerned.

Secondly a brief technical specification is usually supplied to the designer.   This will include the more obvious electrical features of a receiver such as

1. Power supply details ; a.c., a.c./d.c., accumulator or dry battery, with the required range of operating voltages in the first two cases.

2. Frequency coverage of various wave ranges, including any bandspread ranges.

3. Sensitivity at three points on each wave range covered.

4. Noise ratio, usually at one input only, say 5 μV.

5. I-F selectivity for an attenuation of 2 times, 10 times, and 1000 times, and perhaps figures for similar conditions on the broadcast band.

6. Image ratio at three points on each wave range covered.

7. Battery consumption for battery receivers and perhaps power consumption for a.c. receivers.

Thirdly, there is usually a large number of requirements which are unwritten and taken for granted as being part of a good design, but which are all-important from the point of view of the ultimate buyer and user of the receiver.   These requirements include such details as

8. Suitable a.v.c. and noise performance of the receiver, as discussed later in this chapter.

9. A-F fidelity, including the response of the loudspeaker as mounted in the cabinet.

10. A satisfactory tuning response, i.e., a minimum of unpleasant effects as the receiver is tuned to a strong or weak station or even when tuned between stations.

11. An absence of objectionable hum under all conditions, such as, for instance, with a strong unmodulated carrier tuned in and the volume control well advanced.

12. A low volume level when the volume control has been turned to its minimum position.

13. An absence of microphonic effects under normal conditions of use.

14. An absence of unnecessarily objectionable effects when the volume control is turned up to or past the a-f overload point.

15. A satisfactorily low heat rise in the power transformer and other components after long periods of continuous operation, and operation of all components within their maximum ratings under all conditions.

16. Satisfactory performance from battery operated receivers even when the battery voltage under load has fallen by at least one-third of the original voltage.

SECTION 3 : GENERAL DESIGN

(i) A.V.C. and noise   (ii) Audio-frequency response   (iii) Hum   (iv) Microphony (v) Instability   (vi) The local oscillator   (vii) Cabinet design   (viii) Ratings   (ix) Field testing.

(i) A.V.C. and noise

(A) Noise measurements as such can conveniently be made by the e.n.s.i. method as described in Chapter 37 ; however a.v.c. and noise are grouped together in this section as they are plotted on the same a.v.c. curve, around which a large part of the design of a receiver may take place.   Such curves are preferably taken by Scroggie's method (Ref. 50) as described in Chapter 27 Sect. 3(xiv) ; Fig. 35.1 shows typical curves which might have been taken during the development of a receiver.

Before the curves are studied in detail it is proposed to discuss briefly the way in which overall receiver noise may vary with the application of a.v.c. bias to different valves in a receiver. Shot noise is generated essentially in the plate circuit of a valve. The expression connecting noise current in the anode circuit, and direct anode current is

$$I_n = A\sqrt{I_d \Delta F} \tag{1}$$

where I_n = noise component of anode current,
   I_d = direct anode current
   ΔF = bandwidth
and A = a factor which varies with different valve types, triodes having the lowest values, and converters or mixers the highest.

Since the multiplication by g_m of the signal at the grid of a valve gives the plate current due to the signal, so the division of a plate current component by g_m gives the magnitude of an equivalent signal at the grid. Thus the equivalent shot noise at the grid is

$$E_{shot} = \frac{A\sqrt{I_d \Delta F}}{g_m} \tag{2}$$

from which it can be seen that for an equivalent valve type the lower the ratio of $\sqrt{I_d}$ to g_m the lower will be the shot noise of the valve.

It will be found with remote cut-off valves that as the grid bias is increased above the value used to obtain maximum gain the shot noise is also progressively increased. A convenient demonstration of this can be found in Fig. 27.39 which gives characteristics of the 6SK7, a type frequently used as a r-f amplifier. With 5 volts bias the numerical value of the factor $\sqrt{I_d}/g_m$ is 1.6 whilst with 10 volts bias the value is 3.5. Thus the shot noise of the valve has more than doubled when the bias is increased from 5 to 10 volts.

In addition, the gain of the stage is of course also decreased, and if the following stage is contributing to the total noise of the receiver it will add a larger amount to the total equivalent noise as increased negative bias is applied. This is because the noise voltage of the second stage is divided by the gain of the first stage to refer it to the grid of the first valve.

It is worth noting that, if noise calculations are carried out in terms of noise resistance, then the noise resistance of the second stage will be divided by the square of the voltage gain when being referred to the first stage. This follows from eqn. (2) in Chapter 23 Sect. 6 which expresses thermal agitation as a voltage e_n in series with a resistor R where

$$e_n{}^2 = 4KT\Delta FR.$$

Thus as noise voltages are directly multiplied, or divided, by stage gains, noise resistances must be multiplied, or divided, by the square of the stage gain to give the same result.

With an ideal receiver a ten times increase of signal input would give a ten times increase of signal-to-noise ratio. In addition the output of the receiver would be unchanged (because of the ideal a.v.c. curve) so that the gain of the receiver must decrease ten times and the noise must also decrease ten times. This could be accomplished by decreasing the a-f gain ten times, or by decreasing the gain of any stage which made no contribution to the equivalent noise at the first grid. Such a stage would be one with a large amount of gain between its own grid and the grid of the first stage. If however the gain of the input stage were decreased by a.v.c. bias, even in conjunction with a reduction in gain of other stages, then the noise of the input stage could increase and the improvement in signal to noise ratio be less than the maximum possible.

This applies particularly when the input valve is a converter. From Fig. 23.20 it will be seen that the noise resistance of a 6SA7 is 240 000 ohms whereas the impedance of an aerial coil secondary of 200 microhenries and with an effective Q of 50 is approximately 60 000 ohms when resonated at 1000 Kc/s. With such an input valve and coil, receiver noise would be determined by the input valve noise alone—

see Chapter 23 Sect. 6(iv) for method of adding noise voltages—so that any increase in valve noise would decrease the signal-to-noise ratio.

However if the input valve were a 6SK7 r-f amplifier its noise resistance of 11 000 ohms would be appreciably less than the tuned circuit impedance (at 1000 Kc/s) and an increase in valve noise resistance of 14 db due to a.v.c. application would make less than 3 db difference to the receiver's signal-to-noise ratio. At the low frequency end of the band the tuned circuit impedance would be only about 30 000 ohms and a correspondingly smaller increase in valve noise resistance would be permissible.

On the short wave band, where tuned circuit impedances are of the order of 5000 ohms in the middle of the tuning range, noise from valves must always be considered.

Thus from the point of view of signal-to-noise ratio, the best point of a.v.c. application is the last i-f valve. However with a.v.c. applied to the last i-f valve alone, severe overloading of this valve would occur with quite small inputs. The result therefore must be a compromise and a solution is to delay the application of a.v.c. to the first stage of a receiver until the desired signal-to-noise ratio has been achieved for the smallest possible input, and to apply as much a.v.c. as possible to the stage as the input increases above this point.

The importance of this is not always realized, but money and time spent on improving aerial coils to obtain chiefly a good signal-to-noise ratio at, say, 5 μV input, can be largely wasted by poor a.v.c. circuit design which at larger inputs wastes the advantage gained. Consider two receivers, A and B. A has a good aerial coil giving a 1000 Kc/s signal-to-noise ratio of 15 db at 5 μV input and its a.v.c. characteristic is such that a 20 db increase in input gives a 14 db increase in signal-to-noise ratio. B has a poorer input circuit giving a 9 db signal-to-noise ratio at 5 μV and an a.v.c. characteristic giving 19 db increase in signal-to-noise ratio for each 20 db increase in input. The signal-to-noise ratios of the two receivers for various inputs are tabulated below.

SIGNAL TO NOISE RATIO (db)

Input (μV)	Receiver A	Receiver B
5	15	9
50	29	28
500	43	47
1000	47	53

It will be seen that although the noise ratio of A is twice as good (6 db difference) as that of B at 5 μV, yet for an input of 1000 μV B is twice as good as A. Far more use is made of a receiver with inputs above 80 μV—where B is superior—than with inputs below 80 μV where A is superior. In receivers with average a-f characteristics the signal-to-noise ratio cannot be ignored until it is in excess of 40, and preferably 45 db.

A typical a.v.c. design problem is illustrated in Fig. 35.1 where curves A1 and B1 represent the output of a receiver with an input modulated 30% at 400 c/s, and curves A2 and B2 represent the noise output with unmodulated input.

Curves A1 and A2 could be taken on a receiver with high i-f gain, low a-f gain and with a.v.c. obtained from the primary of the 2nd i-f transformer and applied to the converter and i-f amplifier without any delay. The curves show the following faults in the receiver.

(a) The sensitivity of the receiver (for 50 mW output) could readily be improved. As the output is 6 db below 50 mW for 1 μV input, it could be increased to 50 mW by doubling the input, if the effects of noise were neglected. As some of the output from the 1 μV input is noise, somewhat more than 2 μV input would be needed, but the existing sensitivity of the receiver (6.5 μV) could at least be doubled.

(b) An output of even two watts cannot be obtained from the receiver for a 30% modulated signal less than 1000 μV. Although all normal signals have a maximum

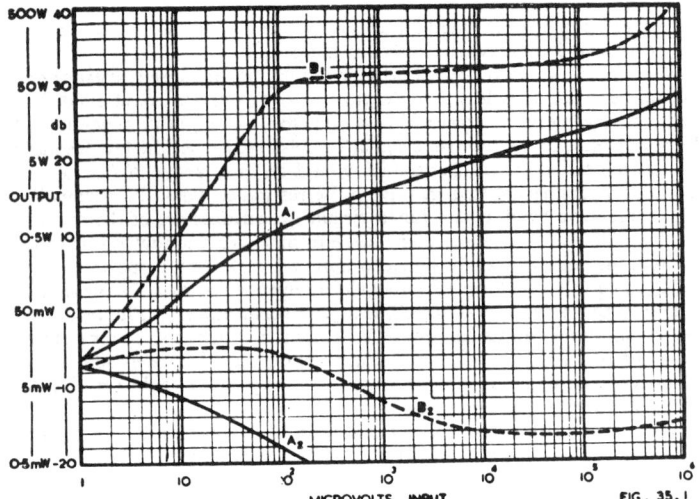

Fig. 35.1. A.V.C. and noise characteristics.

modulation depth greater than this, experience shows that if the impression of insufficient output on a weak signal is to be avoided, a receiver should deliver its full a-f output from a 30% modulated signal.

(c) The signal-to-noise ratio of the receiver could be improved at any input between 5 μV and 1000 μV with the greatest improvement at the higher inputs.

Faults (a) and (b) could be removed by increased a-f gain, perhaps by reduction of negative feedback, or by substituting a pentode a-f amplifier for a triode. For instance the curve A1 shows that with another 6 db of a-f gain the receiver would have a sensitivity of 1 μV and would give a power output of 2 watts with 80 μV input. A-F gain is specified since additional i-f gain would alter the a.v.c. characteristic.

However a different modification to the receiver would simultaneously correct faults (a) (b) and (c). Fault (c) is due to the fact that at all inputs above 1 μV, a.v.c. is applied to input valve of the receiver. This is known because the rate of increase of signal-to-noise ratio is appreciably worse than the ideal. Accordingly an obvious remedy is to use delayed a.v.c., and with the object of obtaining the maximum possible signal-to-noise ratio up to 100 μV, the delay might be removed at this figure, with results as shown in B1 and B2. From these curves the following information is obtained.

(a) The sensitivity of the receiver (for 50 mW output) is now slightly better than 3 μV.

(b) An input of 20 μV will give an output of 2 watts.

(c) The noise ratios of the receiver before and after modification are as tabulated below.

SIGNAL-TO-NOISE RATIO (db)

Input μV	Original receiver	Modified receiver
5	9	10
10	14	15½
10^2	29	34½
10^3	—	43

Correction of the three previous faults has however introduced other troubles.

(d) Although the a.v.c. characteristic of the receiver is now flat within $\pm$ 2db for inputs varying between 100 and 100 000 μV, severe modulation rise occurs with inputs of the order of 1 volt. This is because the much larger signal at the plate of the last i-f valve (3 times larger with 0.1 volt input) is causing overloading.

(e) With any input from 100 μV upwards, the a-f system will be severely overloaded if the volume control is turned to maximum. In other words the useful range of the control is decreased and the receiver will· behave unpleasantly if carelessly handled.

Depending on the amount of design time and component expenditure which can be afforded, a compromise between the various faults might take the form of

⁄ (a) No delay to the a.v.c. and only a fraction (say one half) of the developed a.v.c. voltage applied to the controlled grids. This would allow curve A1 to rise more rapidly, and would minimize the increase in noise in the first valve with a.v.c. application.

(b) A smaller amount of delay than that used for curve B1. An appropriate amount would allow the receiver output to reach its maximum just as the delay disappeared. This would give maximum possible signal-to-noise ratio up to about 30 μV only.

(c) A better solution might be to use a fraction of simple a.v.c. on the i-f valve and full delayed a.v.c. on the first valve. By this means the signal-to-noise ratio could increase at the maximum rate until say 500 μV input, without the maximum output rising excessively, and for larger inputs the first valve could be heavily controlled.

For further information which can be obtained from curves such as those shown in Fig. 35.1 see Chapter 27 Sect. 3.

(B) Miscellaneous matters

1. Consideration of a typical a.v.c. curve will show that it is quite possible to have too much a-f gain (Refs. 26, 31) or too much i-f gain in a normal receiver. Too much a-f gain will merely move the whole a.v.c. curve upwards, and after full output from a receiver can be obtained for some reasonably small input, any further increase in a-f gain will allow severe overloading to occur with the volume control at maximum, and increase any hum troubles.

On the other hand as i-f gain is increased, so developed a.v.c. voltages are increased and an extremely sensitive i-f channel could develop sufficient a.v.c to bias the valves back appreciably for 1 μV input. As a result the signal handling capabilities of the valves would be seriously reduced at large inputs due to their having unnecessarily high bias.

It is even possible to have too much r-f gain, at least on a particular band. If the gain of a r-f stage is 50 on the broadcast band, and 10 at the low frequency end of the short wave band, then the a.v.c. curves for the two bands can only be a compromise, and if there is sufficient sensitivity on the short wave band, excessive a.v.c. voltages will be developed with large inputs on the broadcast band, leading to i-f overloading.

A convenient method of reducing broadcast band r-f gain is to tune the high impedance r-f transformer primary to a very low frequency by using an additional capacitance, say 80 $\mu\mu$F, across a 4 mH primary. This method has the advantage of reducing the pulling of the primary on the tuned secondary circuit which may minimize tracking problems. Also it needs no additional switching as the wave ranges are changed.

2. Mention has previously been made of the necessity for having a suitable time constant for the a.v.c. system [Chapter 27 Sect. 3(xii)]. One type of distortion which can be very distressing is caused by a time constant which is too small. Under these conditions the potential on the a.v.c. line follows the a-f modulation on the diode load, although because the filter used is a resistance capacitance type, only low frequencies are present and the lower the frequency the less the attenuation by the a.v.c. filter. As a result the i-f gain of the receiver is varied at an a-f rate and while the effect on a single low frequency is to reduce its amplitude, the effect on a musical programme is to modulate the whole a-f output with its lower frequency components. The resulting reproduction sounds very rough, broken up and unpleasant. A time constant of 0.1 second must be regarded as a minimum if this effect is to be avoided and larger values are preferable.

3. Unpleasant tuning effects are usually due to the a-f frequency characteristics of a receiver, the a.v.c. system or a combination of the two (assuming that the receiver is stable under all conditions). So far as the a.v.c. is concerned it can be appreciated that with the a.v.c. voltage derived from the signal diode it is quite possible to overload the first valve in a receiver while a strong station is being tuned in. Signals of 0.5 volt are quite frequently met, and as a receiver having a resonant aerial coil gain of 10 is tuned towards such a station the signal voltage on the grid of the first valve where the selectivity is very poor compared with the selectivity of the i-f channel might reach 4 volts before any appreciable a.v.c. voltage is developed. Since the first valve might be operating with only 3, or even 2, volts of standing bias severe overloading and distortion would occur, although with the receiver tuned in correctly the valve would probably receive enough bias to allow it to handle the signal without distortion. To minimize this trouble a.v.c. is best developed from a less selective source, the primary of the last i-f transformer, even at the expense of another few components. Some advantage in smoothness of tuning is obtained on all stations, apart from extreme cases such as mentioned above, since whenever the receiver is detuned from a station, a larger a.v.c. voltage is obtainable from the less selective source, so that the volume of sound distorted by the mistuning of the receiver is reduced.

4. Although the ideal often aimed at is a flat a.v.c. curve, it must be realised that too close an approach to such an ideal has disadvantages, and is in fact undesirable for a normal A-M commercial receiver. Many receivers are used in situations with high noise levels, although in most cases the noise is of lower amplitude than the local stations. Under these conditions an a.v.c. curve flat from say 10 μV to 1 volt input will be responsible for the receiver making far more noise than is necessary when being tuned between stations.

For the purpose of providing minimum noise between stations, an a.v.c. curve would be designed to be reasonably flat over the range of inputs covering most of the local stations, and at lower inputs would decrease as rapidly as possible. Of course the minimum tuning noise requirement for an a.v.c. curve is not the only one, or even the main one, but it should not be overlooked.

It will be found that the flattest a.v.c. curve that is desirable is one which rises perhaps 20 db between 1 μV and say 250 μV input and increases between 250 μV and the maximum input at the rate of 2 or 3 db for each 10 times increase in input. The noise and static heard on stations below 250 μV give the illusion of a curve much flatter than is actually the case.

(ii) Audio-frequency response

(A) The design of the a-f end of a radio receiver is of particular importance, as its ' tone ' is the one feature of a receiver's performance which is continually evident to the user. In spite of flat-frequency-response ideals, the object to be aimed at is always to provide pleasing performance, and consideration of the intended use of a receiver is necessary before it can be known what constitutes pleasing performance.

For the majority of its working life an A-M broadcast receiver provides background music for people whose main attention is concentrated on something else, and the chief purpose of a receiver is consequently to sound pleasant under all conditions.

As a secondary requirement, the frequency range of the receiver should be as great as possible, so long as this does not introduce any undesirable effects. For instance a restricted frequency range is preferable to an extended one with added distortion.

If a receiver were designed with a very flat frequency response the listener's reaction would probably be that it was lacking in bass. This is readily explainable from the effects of scale distortion as discussed in Chapter 14 Sect. 8, and the designer's first modification to the response of the a-f end of a receiver is to increase the bass. This can be done in three ways ; an increase which is not affected by any of the receiver's controls ; an increase which can be altered by means of a tone control ; and an increase which is dependent on the setting of the volume control so that at maximum volume there is little or no bass boost. A circuit of the last type is shown in Fig. 35.2, or negative feedback can be used as in Fig. 35.3. In each case the values of components shown are typical, but final sizes are governed by the acoustical qualities of the cabinet and loudspeaker, and the response of the remainder of the receiver. See also Chapter 15 Sects. 3 and 10.

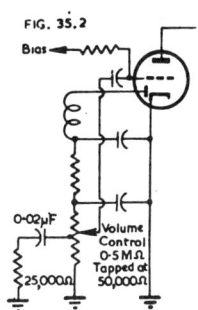

FIG. 35.2

There is a limit both in frequency and amplitude to the amount of bass boosting that can be used. Considering frequency first, if the response is too high in the region of 150 c/s, reproduction becomes boomy and is very annoying to listen to for a period of time. Even worse is excessive gain at frequencies lower than the bass resonant frequency of the loudspeaker. These frequencies cannot be reproduced as fundamentals and any output is due to frequency doubling.

In addition the primary impedance of the output transformer falls to such a low value, due to insufficient primary inductance in a normal cheap transformer, that the output valve can only provide a small fraction of its mid-frequency undistorted output before distortion becomes serious at frequencies appreciably below 100 c/s.

Even if these objections are overcome it is found that a

Fig. 35.2. Bass boost varied with volume control setting.

good response below 50 c/s in the great majority of radio receivers is a liability rather than an asset, as hum, turntable rumble, wow and extraneous noises from broadcasting stations become objectionable. Within the receiver itself microphonic and hum troubles are increased.

From the aspect of the amplitude of the bass boost it must be remembered that a receiver, with say 4 watts maximum undistorted output and a 12 db bass boost at some frequency below 100 c/s will overload very readily at the bass boost frequency. Assuming power output at low frequencies equal to that at mid-frequencies [which is certainly not the case in an average radio receiver owing to the reduced load impedance and the elliptical load line of the output valve (see Chapter 2 Sect. 4)] then an input which will give 300 milliwatts of output at the mid-frequency will overload the amplifier at the bass boost frequency. Although the overloading is due to low frequency excitation, intermodulation products will occur throughout the a-f spectrum when music is being reproduced.

On the other hand excessive high frequency response can also have undesirable effects. The frequency range from 2500 c/s to 3500 c/s is very unpleasant if over-

FIG. 35.3

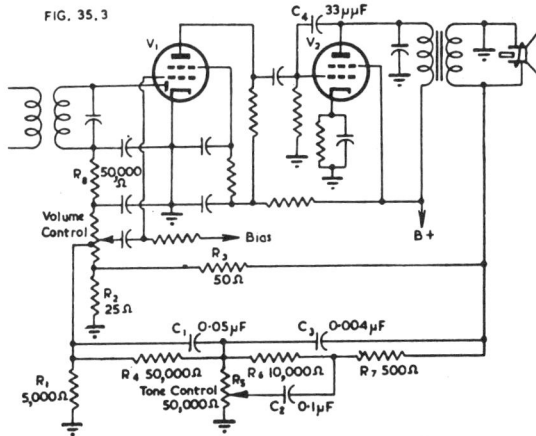

Fig. 35.3. A-F end of receiver with compensated negative feedback and feedback tone control.

accentuated, and where higher frequencies are to be emphasized it is essential that distortion throughout the system be kept to a minimum (Ref. 32).

Particular care must be paid to the tuning characteristics of the receiver if the a-f response is to be extended above 5000 c/s. When listening to a correctly tuned station the selectivity of the receiver normally limits the high frequency input to the a-f end, but during tuning both distortion and emphasis of high frequencies occur, and these effects are exaggerated with an extended high frequency response.

Since in a normal superheterodyne receiver the selectivity is such as to give an attenuation of 6 db at about 3500 c/s and 20 db at about 7500 c/s, it is obvious that the upper end of an a-f response which is flat to 7500 c/s is useless. If any serious attempt is to be made to reproduce frequencies above 5000 c/s the first step to be taken is to broaden the i-f amplifier—Chapter 11 Sect. 3(iii).

It was mentioned above that under conditions of reasonably extended a-f response the distortion throughout the system must be kept to a minimum. The system, of course, can include a recording, a pick-up, and a transmitter, in addition to the receiver, and as only one link in the chain needs to cut the high frequencies from the reproduction—or add distortion—to make extended high frequency response in the receiver either useless or a liability, it is seldom that it is desirable to extend the response of the a-f end of a typical radio receiver beyond about 5000 c/s.

(B) It sometimes becomes necessary to design the frequency response of a receiver to do more than provide the most pleasing tone. For instance, in small battery receivers the a-f output is always restricted because of the small output valve used and the need to economize on battery current, and it is usually desirable to modify the frequency response to make the most of the output that is available. This can be done by providing a comparatively high level between 1000 c/s and 2500 c/s compared with the 400 c/s output. This emphasis should not of course be carried to the stage of making the receiver sound unpleasant either when playing normally, or when overloaded, as receivers with small output are often operated in this latter condition.

(C) In addition, the frequency and damping of the loudspeaker low frequency resonance and the frequency and prominence of the high frequency resonance can have a considerable effect on the apparent maximum output when no electrical frequency compensation is used. If the low frequency resonance is too heavily damped, there will be loss of bass response. In mantel battery-operated receivers the damping is often reduced to the minimum by the use of a power pentode either without feedback or with negative current feedback (e.g. Fig. 35.4). In the latter case there will also be reduction of distortion. Care should be taken when using this circuit that the high frequency response does not extend to frequencies where it is undesirable. A suitable value of resistance for R would be 0.1 or 0.2 ohm (depending on the gain of the a-f amplifier) so that the loss in output power is more than offset by the advantages.

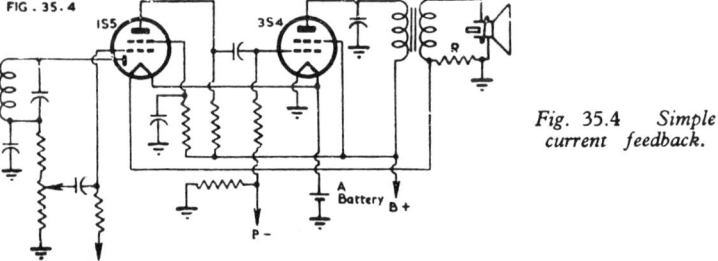

FIG . 35.4

Fig. 35.4   Simple current feedback.

Where frequency compensation is used in a receiver, the speaker resonances are chosen with a different object in view. The electrical response can, owing to the compensation, be raised to any desired level at any frequency within reason, but the loudspeaker will not reproduce satisfactorily frequencies below its main resonance.

Accordingly the frequency of the main resonance is chosen to be as low as possible consistent with the type of receiver—and speaker—being used. For instance the resonance for a 12 inch speaker to operate in a large console model might possibly be as low as 55 to 60 c/s, for an 8 inch speaker in a table model receiver 65 to 70 c/s, and a 6 inch speaker in a small mantel model perhaps 110 c/s.

The high frequency resonance is made as high as possible although for different reasons. When a receiver's high frequency response is increased, a distressing screech occurs if the response between 2500 and 3500 c/s is too high. Many loudspeakers have a peak of sound output in this range of frequencies and when this is the case only a very small amount of high frequency boosting can be used. However if the frequency of the peak can be increased to say 4500 c/s, preferably with a dip between 2500 and 3500 c/s, more high frequency boosting up to say 5000 c/s can be used, and improved results obtained.

(D) The correct application of negative feedback to a radio receiver can reduce the distortion, permit the boosting of appropriate frequencies and provide damping of the loudspeaker bass resonance, but it must be realised that the greater the number of features expected from the feedback, the more critical will be the design. For example bass boosting is usually required at about the same frequency as the loudspeaker main resonance, and unless care is exercised it will be found that when the required bass boosting has been obtained, the feedback may be positive, in the vicinity of the resonant frequency of the loudspeaker, so that the damping required for high quality reproduction will be reduced rather than increased.

Negative feedback is used in the circuit of Fig. 35.3 to give distortion reduction, increased speaker damping, bass boosting, treble boosting with sharp cut-off, hum and minimum volume reduction, a decrease in the low and high frequency boosting as the volume control is advanced, and a tone control. The tone control gives treble boosting at one end, a comparatively flat response in the middle and a treble cut at the other end.

It will be seen that feedback is taken from the voice coil of the loudspeaker to the volume control by two separate paths. The simpler path through the 50 ohm resistor R_3 provides a large amount of feedback, independent of frequency within reasonable limits, to the bottom of the volume control. This serves to reduce hum and play-through with the volume control in its minimum position as explained in Sect 3(iii) of this chapter. Resistors R_3 and R_2 are designed to provide the desired amount of feedback, which must be limited to avoid instability.

The second feedback path is more complicated but can readily be explained in steps. With the variable tone control R_5 set to the position of maximum treble boost (slider connected to ground), R_7 and C_2 form a voltage divider, and the higher the frequency the smaller the voltage that appears across C_2. It is this voltage which is the main feedback voltage applied to the amplifier, and the decreasing negative feedback as the frequency increases gives increasing output from the amplifier, thus providing treble boost.

The next component in the main feedback path is the resistor R_6 which isolates the treble and bass boosting sections of the network. In the absence of R_6, C_1 and C_2 would give 0.033 μF to earth from the tap on the volume control. R_6 also serves a purpose in the operation of the tone control to be described below.

The next part of the feedback network is another frequency discriminating voltage divider composed of C_1 and R_1. The effect of this divider is the reverse of the previous one in that the lower the frequency the smaller the voltage appearing across R_1, owing to the increasing reactance of C_1, and as feedback is reduced at low frequencies so the overall gain of the amplifier is increased. In this way the bass boost is obtained. However as mentioned earlier there is a limit to the amount of bass boosting which can be used and to prevent the feedback from increasing indefinitely as the frequency is reduced the resistor R_4 is used.

Excessive high frequency response is also undesirable and capacitor C_3 provides a sharp cut-off above the desired range. The reactance of C_3 is so high at mid-frequencies that it has no effect on the response, but as the frequency becomes higher it provides a progressively lower impedance path between the source of the feedback voltage and the tap on the volume control. Above the useful range of frequencies it becomes the main feedback path.

To prevent instability at very high frequencies where the fraction of the output voltage fed back to the input is very high owing to the small reactance of C_3, the gain of the amplifier itself is made to decrease rapidly at frequencies above the desired a-f range.   This is the purpose of the 33 $\mu\mu$F capacitor C_4, and it has the added advantage of giving a sharper overall cut-off than could be provided by C_3 alone.

The resistor R_1 has been mentioned as one element of the bass boosting voltage divider, but its size is governed by another consideration, the adjusting of the volume at which maximum bass and treble boost are provided.   The feedback current flowing into the tap on the volume control returns to ground by two main paths, the volume control and the resistors R_2 and R_3 in parallel providing one, and the volume control, the decoupling resistor R_8 and the diode-cathode path of V_1 being the other. Thus as the slider of the control is turned up from its minimum setting the amount of frequency-compensated feedback applied to the amplifier is increased until the slider reaches the tap and thereafter decreases.   With the slider at the tap, the volume should therefore be the lowest normal listening level so that as the volume is increased, the amount of compensation is reduced.   The normal type of tapped volume control has its tap at about 50 000 ohms for a 0.5 megohm control and this gives too much volume for maximum compensation, so the value of R_1 is chosen to reduce it to the required level.

The operation of the tone control is comparatively straightforward.   With the slider at the earthed end of the resistance element, R_7 and C_2 remove most of the high frequencies from the feedback path to provide a treble boost.   However as the slider is turned away from the earthed end the resistance in the capacitive arm of the voltage divider is increased and its frequency discriminating properties are reduced. For example when the resistance between slider and ground is 5000 ohms the amount of feedback used varies only between 10/11 and 11/11 of the total amount as the frequency is increased from zero to infinity.   Thus the treble boost is removed over the first part of the rotation.

When the control is set to the other extreme, C_2 is in parallel with R_6 and the sizes of these components are adjusted so that their impedances are the same in the lower mid-frequency range.   Under these conditions C_2 offers an increasingly lower impedance path for feedback currents as the frequency increases above say 400 c/s, and the larger amount of feedback results in a falling high frequency response for the amplifier as a whole.

It must be realised that while the foregoing explanation of the feedback circuit is correct as far as it goes, a complete picture of its operation cannot be obtained without a knowledge of the phase shifts introduced by the various coupling elements both in the feedback circuit and in the main amplifier circuit. Feedback which is 180° out of phase with the input at the mid-frequency is no longer negative when a phase shift of 90° has taken place.   In Fig. 35.3 at low frequencies the coupling condensers to the grids of V_1 and V_2, and the primary of the output transformer will each give a maximum phase shift of 90° while capacitors C_2, C_1, the screen by-pass of V_1, the cathode by-pass of V_2 and the decoupling by-pass in the B supply of V_1 will each give a phase shift with a maximum varying between perhaps 20° and 50°, so it is obvious that the phases of the feedback voltages become far more important than their magnitudes.   By far the most difficult part of the design of such a circuit is to maintain the feedback negative over the usable a-f range, rather than to obtain the desired frequency response.

Because of the variation in frequency response with the phase of the feedback voltage it is possible to obtain bass and treble boosting without frequency discriminating elements in the feedback path.   Such a circuit would be the same as Fig. 35.3 with all feedback components removed except R_2 and R_3.   In such a case circuit elements such as the grid coupling condensers, which would normally be as large as possible to avoid phase shifts in the working range, would be decreased in size to cause phase shifts and thus develop a bass boost.   This is the reverse of what would be expected from such a change in an amplifier without feedback.   Similarly a high frequency boost could be obtained by adding a by-pass in the plate circuit of V_1 or elsewhere.

As the amount of negative feedback applied to an amplifier is increased, the peaks of response at low and high frequencies are removed further from the mid-frequency and become greater in· amplitude, and with the fewer components available in the simplified circuit it usually becomes necessary to vary the amount of feedback to assist in obtaining the desired response.

(iii) Hum

The a-f gain in a radio receiver is seldom high enough to make hum elimination difficult except in so far as space or economy considerations make it so.   Nevertheless economy is such an ever-present need in commercial design that the majority of receivers have hum reduced to a barely acceptable minimum.

The amount of hum permissible at any stage in a receiver can be expressed as the maximum tolerable hum at the output of the receiver divided by the gain at the appropriate frequency between the output and the stage in question.   Thus the filtering required by the plate of the output valve is less than that for any other part of the a-f system.   Advantage can be taken of this effect in small receivers, in which the plate of the output valve can be fed directly from the capacitor across the rectifier output.   Since the output valve plate draws a large proportion of the total B current in a small receiver, the filter used for the supply for the rest of the receiver can be decreased considerably in size, and it may be possible to use a resistive filter instead of a choke without an excessive voltage drop.   It should be noted however that since there is gain between the screen of a norm⸱l tetrode or pentode and its plate the screen of the output valve in such a case will probably need to be supplied from the filtered source.

Filtering becomes more important when the grid circuit of the output valve and the plate circuit of the a-f amplifier valve are considered.   Designs using back-biasing for the output valve normally have the grid of the output valve decoupled for hum reduction, and similarly the plate supply to the first a-f amplifier is often decoupled. It is worth noting that the hum introduced from these two sources is out of phase, and whilst if one decoupling is used as a matter of course, or if the output valve is cathode biased, the other may become essential, yet the results obtained with both omitted are often satisfactory.

Hum introduced between grid and cathode of the a-f amplifier is usually the most difficult to eliminate since the following amplification is high, and the circuit is usually a high impedance one which makes the effects of small capacitive couplings appreciable.   The impedance of course is variable, depending upon the volume control setting, and when tracing hum introduced into this stage it is best to turn the volume control to maximum.   If noise from the r-f end of the receiver then drowns the hum, the last i-f valve can be removed.

Hum can be introduced into a receiver before the second detector, although under these circumstances (assuming transformer coupling to the detection diode) it can only be troublesome if it modulates the signal.   **Modulation hum** can usually be traced to two general sources, firstly amplitude or frequency modulation of the local oscillator in a superheterodyne receiver by hum from the power supply, and secondly amplitude modulation of the signal in a valve working on a non-linear part of its characteristic and with a hum voltage impressed on one electrode.

Modulation hum due to the local oscillator is as a rule only met when the oscillator plate has a separate filter from the rectifier output to eliminate flutter—see Sect. 3(vi) of this chapter.   The filter is usually resistive and arranged to give the proper voltages so that an increase in the size of the filter capacitor is the cure.

Modulation hum occurring because of a non-linear characteristic in a r-f or i-f valve is greatest at the point of maximum non-linearity and this as a rule is at some intermediate value of bias, so that the hum may increase up to some particular input and decrease thereafter.   This can readily be checked on the a.v.c. characteristic, plotted as described earlier in this section. A common reason for this type of hum is insufficient filtering for the screen of a r-f amplifier. If the a.v.c. line carries bias generated across a back bias resistor it is also possible for hum from this source to modulate incoming signals in any of the remote cut-off valves to which a.v.c. is applied.

A third source of modulation hum is sometimes met in a.c./d.c. receivers. Owing to insufficient filtering in the mains leads or perhaps to insufficient shielding, the rectifier becomes a part of the circuit in which signal currents circulate. The signal in this path is modulated by the mains frequency since the rectifier is conductive for only half of each mains frequency cycle. A cure for this type of hum is to use a r-f by-pass across the rectifier, a typical capacitor being a 0.001 μF mica type.

A receiver with high a-f gain can give trouble with hum due to potentials developed between different soldering lugs on the chassis, and because of this it is usually best to return directly to the cathode of the a-f valve any leads coming from the ' earthy ' side of components in the grid circuit of the valve. In addition, the lead between cathode and ground must be separate from the lead earthing one side of the valve's heater. When building the pilot model of a receiver which is likely to give hum trouble it is best to take all possible precautions initially, trying any economy measures one by one after the model is in operation. The alternative of building first for maximum economy may lead to delay, since when several sources of hum are present cancellation usually occurs, and the removal of one source may lead to an increase in total hum.

One aspect of hum which can be put to good use during design is that a hum level which is objectionable at very low listening levels may be satisfactory at higher levels when it is masked by the programme. A circuit which at very small cost reduces hum at the minimum volume setting of the volume control is shown in Fig. 35.5. The connection from the voice coil is in the direction to give negative feedback, and this feedback is a maximum with the volume control turned to zero, decreasing as the volume control is turned up, until at maximum volume the gain of the receiver is reduced only by one or two db. The feedback is also valuable in reducing play-through, and this aspect has a particular application in reflex receivers.

There is a maximum value of feedback which can be used without instability, and in a high gain amplifier use of the full amount of feedback available from the voice coil may give trouble (see Chapter 7 Sect. 3). In addition, the feedback into the bottom of the diode load results in an apparent reduction of the a.c./d.c. impedance ratio.

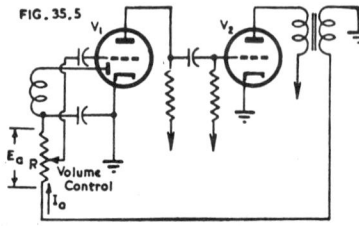

FIG. 35.5.

Fig. 35.5. Negative voltage feedback varied by volume control setting.

The reason for this latter effect can readily be visualized from Fig. 35.5 as follows. Neglecting for the moment the negative feedback, the diode load is the 0.5 megohm resistor R, and an a-f voltage E_a due to the modulation of an incoming carrier produces in the diode load an a-f current I_a of a magnitude given by $I_a = E_a/R$.

Now when feedback is applied, an a-f voltage of magnitude—in a typical case— equal and opposite to E_a with the audio volume control at maximum, is applied to the bottom of the diode load. The a-f voltage across the resistor R is now $2E_a$ and

$$I_a' = 2E_a/R$$

where $I_a' =$ a-f current in diode load with feedback applied.

Thus the a-f current flowing in the diode load has been doubled by the application of feedback for the same demodulated a-f signal E_a. An alternative way of consider-ing this effect is that the a-f voltage E_a is generated across an impedance equal to $R/2$. This is the usual conception of the effect, and the explanation for the statement that the a.c./d.c. ratio is reduced by such a feedback application.

Of course as the volume control is turned down from its maximum setting the feedback voltage applied to the bottom of the volume control is reduced and the a.c./d.c. ratio is improved, even although the amount of a-f gain reduction is increased.

To overcome the trouble of possible instability and increased a.c. loading on the diode load when the full voice coil voltage is used for feedback, the circuit of Fig. 35.6A can be used. Resistors R_1 and R_2 may be of equal size, say 25 ohms each, and the phase of the feedback is such that the feedback at the bottom of the diode load is still negative, so that the feedback into the top of the diode load is positive. The values of the resistors can readily be arranged so that the gain reduction with the volume control at maximum is zero, when the a.c./d.c. ratio is no worse than in the no-feedback condition.

It is quite possible of course to have positive feedback predominating, giving an a.c./d.c. ratio greater than unity, but the reduced detection distortion in this case may be more than offset by increased a-f distortion.

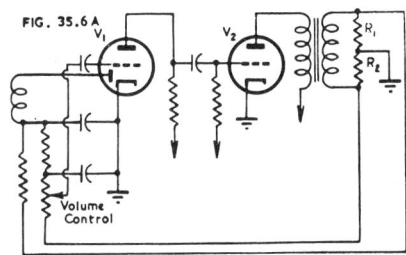

Fig. 35.6A. Combination of positive and negative voltage feedback.

When tracing hum it is essential to be in a location quiet enough for the character of the hum to be recognisable, and a meter reading hum amplitude and an oscilloscope showing hum waveform can each be useful on occasions.

The amount of hum which can be tolerated is very dependent on the frequency of the hum, and pure 50 c/s hum could be of much greater amplitude than say a 150 c/s hum for two reasons. Firstly, the sensitivity of the ear is greatly reduced at 50 c/s, and secondly the sensitivity of the reproducing equipment also decreases rapidly below the resonance of the loudspeaker. Nevertheless a receiver with a small speaker which cannot reproduce 100 c/s satisfactorily can suffer badly from 100 c/s modulation hum, which, although not audible in itself, modulates any other audio frequencies which are present and thus gives a very harsh, broken-up type of reproduction.

Since the effects of hum are to such a large extent subjective it is natural that final approval in a doubtful case must come from a listening test. It is essential that such a test be carried out in a home under the quietest conditions possible, as the masking effects of noise in a normal factory, or even laboratory, make an evaluation of the hum level very difficult.

See also Chapter 31 for a general treatment of hum, Chapter 7 Sect. 2(ix) for hum with feedback, Chapter 12 Sect. 10(vi) for hum in voltage amplifiers, and Chapter 18 Sect. 2(iii) for hum in pre-amplifiers

(iv) Microphony

Microphonic effects in electronic equipment are those in which mechanical movement produces undesired electrical output from the equipment. Many such effects are self-sustaining in that the microphonic noise produced reinforces the original mechanical movement and a continuous output builds up at the frequency of the mechanical vibration.

There are two main sources of microphony in an A-M receiver, the local oscillator and its circuit, and valves. Since all tuned circuits and all valves are microphonic to some extent, the object of a designer when laying out a receiver must be to isolate from the output of the loudspeaker those components which are most likely to give trouble.

Tuning condensers probably give more microphonic trouble than any other component, although frequently this is due to the mounting method used or to the dial drum. Some receivers have the complete tuning unit (condenser, coils, wave-change switch and valves) mounted on a subchassis which is floated on rubber. This precaution, although comparatively expensive, is usually sufficient to remove any microphony, and even the cheaper solution of floating the gang condenser alone will usually give a substantial improvement. A drawback to the use of floating tuning units is that as a rule the dial and tuning knob must float too, or flexible couplings must be provided to them, in which case backlash between the tuning knob and the gang condenser or the dial usually becomes a problem.

The type of gang condenser microphony which can be cured by floating the condenser is usually due to movement of masses of metal such as the complete gang body, or the whole rotor shaft, and an alternative to floating the gang is to make it more rigid. The first step is to stiffen the body by soldering with large fillets of solder the junctions between the cross bracing strips and bars, and the end and dividing plates in the gang. For extreme cases it may be necessary to solder a plate to the back of the body of the oscillator section. An alternative is to mount the gang normally on its feet and then add an additional bracing from the top of the gang to some nearby solid mounting. Such a bracing may be better either rigid or rubber mounted, in the latter case acting more as a vibration damper. A gang condenser with its framework stressed by the mounting is more likely to be microphonic than one which is not stressed. Thus a suitably designed three point mounting may be an improvement on the conventional four point suspension, although in each case care must be taken to see that it is not possible for the condenser to rock.

It does sometimes happen that a complete stator packet vibrates and causes microphony. In such cases it may be possible to reinforce the brackets supporting the stators sufficiently at least to remove the resonance to a frequency at which it will not be troublesome.

Some microphonic effects are due to a large dial drum mounted on the gang spindle. Such a drum may be self resonant if made of metal or it may because of its weight induce flexing effects in the condenser shaft. In other cases its position may give trouble ; for instance the drum may be spaced only slightly from the front panel of a mantel model receiver, with the loudspeaker mounted on the same panel. Very considerable acoustical coupling will then exist between the loudspeaker and the gang rotors. Should the drum itself be resonant it may be possible to alter the frequency sufficiently to avoid microphony by punching suitably placed holes in it. Spraying with flock or attaching a piece of felt may damp the resonance adequately.

Another type of gang microphony is due to the vibration of the plates of the condenser. This is more often due to sound waves impinging directly on the plates themselves and acoustical shielding between speaker and gang condenser should be tried as a cure.

Should a valve be causing microphony it can be mounted on a floating socket and shielded as far as possible from the noise from the loudspeaker. Sometimes a rubber sleeve around the glass envelope may effect a cure.

With battery valves, microphony can be affected by circuit design. If the plate current of a remote cut-off valve has been reduced almost to cut-off by a.v.c. action, movement of the filament under the influence of noise from the loudspeaker may cause some sections of the valve which have been completely cut off to conduct again. The resulting change in plate current creates noise in the loudspeaker which maintains the vibration of the filament and the microphony is sustained. An alteration to the circuit which will result in smaller a.v.c. potentials being applied to the valve in question should be tried.

Microphony is also encountered between pickups and loudspeakers, although more recent types of combined gramophone motors, turntables and pickups use very flexible mountings which almost completely remove such troubles.

If the remedies suggested above fail to cure a particular case of microphony there are other methods of attack. The simplest is to mount the loudspeaker on rubber or felt with no solid connection between loudspeaker and cabinet.   Even if this should result in a gap of up to one quarter of an inch between the edge of the loud-speaker frame and the baffle, little change of a-f response need be expected.   Such a mounting is, of course, mainly effective in keeping low frequency vibration from the cabinet, and is most likely to cure low frequency microphony.

An alternative method of achieving the same result is to mount the loudspeaker solidly to its baffle, and float loudspeaker and baffle from the cabinet on suitable rubber grommets.

Similar results can be obtained electrically by reducing the bass response of a receiver.   If the response is already satisfactory this is undesirable, but it may be found possible to attenuate severely frequencies below say 70 c/s without appreciably affecting the tone, and this could produce the required effect.   Alternatively, since microphony is usually much worse on short waves, a bass cut can be switched in with the wave change switch as the receiver is turned to short waves.   This possible cure can actually improve short wave reception since fading and noise can be less objection-able with a restricted bass response.

Since some types of microphony are due to a change in receiver output as the os-cillator frequency is varied, it follows that the greater the selectivity of a receiver, the more prone it will be to this type of microphony.   Receivers with a sharp peak to the selectivity curve are particularly susceptible and this is frequently caused by re-generation in the i-f channel which may be reduced by improved layout or by neu-tralization—see Chapter 26 Sect. 8(ii).

(v) Instability

This sub-section is confined to instability problems commonly met in A-M re-ceivers.   For general information on the subject see Ref. 22.

Probably the most common trouble with new models when wired up for the first time is instability, although the reason can usually be found with little trouble by by-passing possible " hot " points with a large capacitor, say 0.5 μF, or by placing an earthed piece of metal between stages across which feedback is likely to occur.

More annoying instability problems are those which do not show up in pilot models but which affect production receivers, a typical instance being instability which occcurs only at the low frequency end of the broadcast band.   The regeneration is due to some of the i-f output being returned to the aerial coil and thus back to the i-f input again.   The effect usually becomes evident only towards the low frequency end of the broadcast band where the aerial coil secondary tuning is approaching the inter-mediate frequency, and in addition the high impedance aerial primary must be resonating at or near the intermediate frequency.

The i-f output is often radiated directly from the second detector valve or from the last i-f transformer, in which case shielding may be needed, or at other times from the loudspeaker leads or frame.   Earthing the frame will cure the latter trouble and a r-f bypass in the audio circuit the former.

Since the instability is due to intermediate frequency fed back to the aerial primary, a reduction in impedance of the primary reduces the feedback.   A resistor, say 10 000 ohms, across the primary may be sufficient damping, but usually gives a greater loss in gain, selectivity and signal-to-noise ratio than a condenser tuning the primary to a lower frequency than the intermediate frequency.   When the latter method is used, additional capacitance due to the aerial can only tune the primary further from the intermediate frequency, but otherwise some critical aerial length may resonate the primary at the intermediate frequency and lead to instability.

Battery receivers are particularly prone to trouble in this respect when operated without an earth, as the aerial capacitance is in series with the capacitance between receiver and ground, so that even a large aerial does not tune the primary much lower in frequency.

A very similar trouble is instability when the receiver is tuned to the second harmonic of the intermediate frequency.   In this case the second harmonic is radiated from the second detector circuit—where it has a high amplitude—picked up by the input circuit and returned to its source. Care in confining the i-f harmonics to appropriate sections of the chassis is sufficient to cure this fault. Sometimes actual instability will not be encountered, but the a-f output of the receiver will decrease as the volume control is turned up over the last few degrees. This is due to rectification by an a-f valve of i-f voltages on its grid, thus increasing the bias on the valve and decreasing its gain. By-passing, or if possible improved layout, will cure this trouble, and it is worth checking with a good oscillograph each new model developed, to see that a-f valves are not handling i-f voltages of the same order of magnitude as the a-f voltages.

A more difficult problem is presented by multistage battery receivers in which coupling occurs in the filament leads. Since some couplings may be regenerative and some degenerative it becomes difficult to determine the true gain. However by filament by-passing and by wiring the valves in a suitable order regeneration can be removed, and if the gain can be spared it is well worth while to introduce degeneration. Under these conditions variations in sensitivity with decreasing battery voltage are minimized.

Battery sets in particular are prone to a type of instability which shows up as a squeal when a strong signal is tuned in rapidly. This is due to interaction between i-f and a-f circuits, and in battery receivers adequate a-f by-passing of the B supply will prevent the trouble from developing when the internal impedance of the battery rises with use. If back biasing is used the resistor may need by-passing for audio frequencies. Filtering of i-f voltages from the a-f end of the receiver is in some cases the only cure necessary for other troubles of this type.

Electrolytic condensers are not always satisfactory r-f by-passes and where more than one i-f stage is returned directly to the B supply an additional paper by-pass is advisable. When two or more circuits have a common by-pass they should be returned to it individually to avoid common impedances.

A tendency to instability at the high frequency end of the short wave band of a receiver is often due to excessive coupling between the signal grid and the oscillator circuit of a receiver and can be cured by coupling from the opposite phase of the oscillator. If a r-f stage is used the trouble may be due to resonance or near resonance in the r-f coil primary and a suitable cure is a small carbon resistor, say 25 ohms, between r-f plate and r-f coil primary.

An indication of the stability of an i-f channel can be obtained from its selectivity curve. While complete symmetry is rarely obtained with the mixture of capacitive and inductive coupling encountered in most receivers, a marked degree of asymmetry at small attenuations can usually be traced to regeneration and it may be difficult to remove the last traces. Apart from coupling due to leads or capacitances inside or outside the amplifying valves—not forgetting capacitances above the chassis where they are apt to be overlooked—magnetic coupling can occur between two transformers either in air, or in the metal of the chassis, the latter trouble being more prevalent with the small spacing occurring with miniature valves.

The two leads from the signal generator should be twisted together when i-f selectivity measurements are made, to keep the loop formed by them as small as possible and so avoid coupling between the first and later stages of the i-f amplifier.

See also Chapter 23 Sect. 7 for instability in r-f amplifiers and Chapter 26 Sect. 8 for instability in i-f amplifiers.

(vi) The local oscillator

(A) Recommended values of **oscillator grid current** are specified for all converter types, and it might be thought that if these values were obtained and the oscillator tuned over the required range, no further design work would be required for the oscillator. However the local oscillator circuit can be responsible in a receiver for flutter, squegging, poor sensitivity with high noise level, unstable short wave tuning, apparently short battery life in battery receivers and other more obscure faults.

(B) The term "**flutter**" covers two different effects, the sound emitted from the loudspeaker being the same in each case. As a strong signal is being tuned in, the voltages in a receiver are varied by the application of a.v.c. to the controlled stages and by the changing plate current of the output valve. If the changes in voltage reach the oscillator valve the oscillator frequency is altered, and a case could be visualized in which the tuning-in of a station generated voltages sufficient to detune the receiver. The additional voltages would then disappear and the receiver would retune the signal, only to be detuned again. This is actually what happens when a receiver flutters, although the station may not be completely tuned and detuned on each cycle. The time taken for a cycle is dependent on the time constants of the circuits involved, but it is usually low enough for each cycle to be heard separately.

The voltage applied to the oscillator may come from the B supply or from the a.v.c In each case a receiver is more prone to the trouble when operating on short waves, since a given percentage of detuning is a larger number of cycles per second than on the medium waveband. If the fault originates in the B supply the effect is that as a carrier is being tuned-in the a-f signal causes a large variation in the current drawn from the B supply by the output valve. This variation in current sets up a voltage across the series impedance of the power supply and this voltage is applied to the oscillator plate and detunes the oscillator from the signal.

The time constants concerned are those of the a-f amplifier and of the power supply, and since the a-f amplifier must amplify the flutter frequency, which is very low, a cure is to decrease the bass response of the a-f amplifier. If this leads to excessive bass cutting—although as in the case of microphony it may be necessary only on the short wave band—the alternative of increasing the time constant of the power supply can be tried. This is not so desirable since a larger electrolytic condenser is required at additional expense.

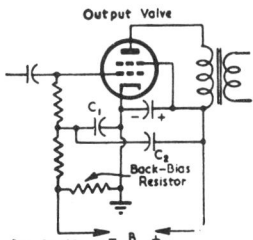

Fig. 35.6B.   *Reduction of "flutter"
by negative feedback.*

A cure which has been used successfully is shown in Fig. 35.6B. The condenser C_2 is connected between B+ and the power valve grid circuit and applies negative feedback to any a.c. potentials (due to flutter or hum) appearing across the B+ filter condenser. The usual decoupling condenser C_1 is also necessary unless the hum level on the B supply is very low (complete cancellation of hum across the B supply is not the same as cancellation of hum across the output transformer primary) and values of 0.1 μF for C_1 and 0.05 μF for C_2 have been used to give appreciable reduction in flutter while leaving the hum level almost unchanged.

To remove the cause of the flutter, rather than curing the symptoms of the trouble, it is necessary to reduce the impedance common to the oscillator and the output valve. A large part of this impedance is the filter in the B supply, and by supplying the oscillator plate directly from the output of the rectifier a big improvement can be effected. A separate filter to remove the hum from the oscillator plate supply will be needed and typical values are a 20 000 ohm resistor from the rectifier cathode and an 8 μF electrolytic condenser. The normal oscillator plate resistor, perhaps 30 000 ohms, is connected from the junction of these components to the oscillator plate. A careful check for modulation hum—see Chapter 37 Sect. 1(vi)H—should be made if this circuit is used.

Flutter due to a.v.c. application to converters is confined to short wave bands, and is due to coupling within the valve between the control grid and the oscillator section. Reference should be made to Chapter 25 Sect. 2 for an explanation of this effect. The simplest cure is to use fixed bias on the converter on the short wave band, and this is not unduly detrimental to receiver performance since the maximum signal input to be expected on short waves is less than on the broadcast band. The a.v.c. curve will not be so flat with one less stage controlled, but this may be an advantage because it is not desirable to hold receiver output too flat when selective fading is experienced as this unduly emphasizes the accompanying distortion. Even when the a.v.c. applied to the converter does not cause flutter, it may make the tuning of strong signals at the high frequency end of the short wave band very difficult, the effect being that no matter how carefully the tuning knob is handled, the receiver tunes just past the station. In addition, the fading of a signal may cause it to be detuned.

(C) **Squegging** of the local oscillator in a receiver is usually confined to the short wave range, but cases have been encountered on the broadcast band when unusual coupling circuits have been used. Squegging is due to high oscillator amplitude in conjunction with a large time constant in the oscillator grid circuit. Oscillation at the desired frequency becomes interrupted at another frequency which is dependent on the time constants in the oscillator circuit. The interruption frequency may be audible or supersonic and depends upon the rate at which the oscillation amplitude at the desired frequency builds up sufficiently to bias the valve to cut-off, and the time required for the charge on the grid condenser to leak away sufficiently for oscillation to start again.

The possible results of a squegging local oscillator are a very high noise level (with or without multiple tuning points and with or without a heterodyne at each point) when the squegging frequency is supersonic, or a continual squeal when the squegging frequency is in the audible range. Such a squeal should not be confused with another which sounds almost identical and which is due to signal grid and oscillator grid circuits being tuned to approximately the same frequency on the short wave band. This second type can be stopped by detuning the signal frequency circuit, but a squeal due to squegging can not.

A third squeal which sometimes occurs at the high frequency end of the broadcast or short-wave band sounds similar but needs a small resistor (say 25 ohms carbon) in series with the control grid of a converter, as close to the grid as possible, for a cure.

The simplest remedy for squegging is to reduce the oscillator grid capacitor and resistor to the lowest values that can be used without reducing unduly the oscillator grid current at the lowest tuning frequency. This may not be sufficient if the same components are used on both the broadcast band and the short wave band (100 $\mu\mu$F and 25 000 ohms are about the smallest combination possible if these components are not to affect the oscillator amplitude excessively at frequencies below 600 Kc/s) and a simple alternative is to switch the grid coupling condenser together with the coil as the wave range is changed.

A third possibility is to connect a small carbon resistor in series with the oscillator grid capacitor or oscillator plate lead. This resistor reduces the amplitude of oscillation at and near the high frequency end of the band, where it is a maximum, without noticeably affecting the low frequency end, and thus removes the tendency to squeg.

The varying effect of the resistor can readily be visualized when it is realised that at the high frequency end of the band the total tuning capacitance may be 30 $\mu\mu$F and the resistor is in series with the oscillator input which amounts to about 10 $\mu\mu$F, or one third of the total. At the low frequency end of the range the 10 $\mu\mu$F with the resistor in series amounts to only about one fiftieth of the total capacitance.

(D) To produce the required **oscillator grid current** at a specified frequency is usually simple, but to obtain the same grid current over a range of frequencies which is usually greater than three to one can be quite difficult. Fortunately considerable variation in grid current is possible with little change in conversion characteristics and full use is made of this in design.

Battery receivers usually present the most difficult local oscillator problems because of the lower initial slope in the oscillator section of the converter and because of the necessity for the oscillator to operate with A and B batteries reduced to (preferably) 2/3 of their initial voltage. For this reason **battery converter problems** **are** considered here.

The first trouble to be experienced is usually insufficient grid current at the low frequency end of the short wave band—assuming that preliminary design is such as to allow the required band to be covered. Figure 35.7 shows a simple circuit (" padder feedback ") for increasing ... g.id current without loss of frequency coverage, the only variation from a conventional circuit being that the cold end of the primary coil is connected to the cold end of the secondary. If this circuit is already in use, attention must be given to coil design as discussed in Chapter 11 Sect. 5.

FIG. 35.7

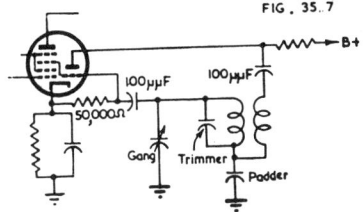

Fig. 35.7.   *Oscillator circuit with padder feedback.*

With valves which do not oscillate readily at high frequencies, it sometimes happens that a frequency range which is impossible when mixing with the oscillator fundamental can be covered with the second harmonic of the oscillator used for mixing. This method of operation also has the advantage of practically eliminating pulling and, in spite of decreased gain at the low frequency end of the short wave band and additional spurious responses arising from unwanted signals beating with the oscillator fundamental, it is often worth trying.

It is important to remember that it is the voltage across a grid resistor which is the essential requirement. The oscillator grid current published by valve manufacturers is used because it is easier to measure, but it only applies to the published value of grid resistance. If a different value of grid resistance is used, the grid current should be inversely proportional to the ratio of the resistances. For example, if the published grid current is 0.4 mA in 50 000 ohms, then the equivalent value would be 0.8 mA in 25 000 ohm resistor.

A choice of series feeding or shunt feeding the oscillator plate is often available (see Figs. 24.2B and C for series and shunt feeding of a tuned plate oscillator). An increase in grid current, dependent in size on the value of the resistor used, can be obtained by using series feed as this removes the damping of the resistor from the plate coil.

With series fed oscillator circuits, an increase in coverage can be obtained by connecting the by-pass capacitor in the oscillator plate supply directly to the cold end of the oscillator primary coil, rather than to some other place in the circuit. All oscillator wiring must be kept as short as possible and removed as far as possible from other components and from the chassis if maximum frequency coverage is to be obtained. As the oscillator grid capacitor is in series with the capacitance of the oscillator valve it should be kept as small as possible, and if the capacitor is switched with the oscillator coil, a substantial improvement in frequency coverage can be obtained by the use of the smallest possible value. At the same time the effect of the change in oscillator input capacitance as the valve warms up is reduced, thereby increasing the frequency stability.

It is unfortunate that in battery receivers, in which frequency coverage with sufficient grid current is difficult to obtain, it is also necessary for the oscillator to provide sufficient output for reasonable conversion conductance when the battery voltage is considerably reduced. As a routine test during oscillator coil design for battery receivers, the sensitivity should be checked from time to time with A and B battery voltages simultaneously reduced to two thirds of their initial values, with a converter valve having an oscillator slope as low as is likely to be encountered in production, and under these conditions the oscillator should not only start readily each time the receiver is switched on, but it should also provide enough grid current to give reasonable conversion conductance.

It should be remembered when laying out any receiver that the characteristics of the valves can be seriously affected by the presence of a **magnetic field**. All valves are likely to fall off in performance if placed too close to a loudspeaker, and oscillator grid current can be seriously reduced in this way.

(E) In a.c. receivers it is a simple matter to obtain sufficient oscillator grid current but excessive values should be avoided. On the broadcast band, where so many signals can be tuned, self generated spurious responses must be reduced to a minimum and it is helpful **to reduce the oscillator amplitude** as far as possible. This should be done experimentally as it is found that the reduction can be carried, without undue loss of sensitivity, further than would be expected from a study of published curves of conversion conductance versus oscillator voltage. Signal-to-noise ratio should also be measured when the grid current is reduced as it may even improve with moderately low values of oscillator voltage.

An additional advantage of using the lowest possible oscillator grid current is that radiation from the local oscillator is reduced.

(F) **With most converters it is found that the shortwave sensitivity is affected when the oscillator frequency is changed from one side of the signal frequency to the other.** The reasons for this are set out in Chapter 25 Sect. 2(ii), but it is not always realised that a useful gain in short-wave sensitivity can be obtained with certain types of converters by determining the better method of operation and using it, even if this necessitates an oscillator frequency lower than the signal frequency. Under these conditions the signal frequency circuit (or circuits) is padded and the oscillator circuit is not, but no other changes are necessary, and an increase in sensitivity of two or three times at the high frequency end of the short wave band is sometimes obtainable. In addition image responses are reduced proportionately.

A disadvantage of operating the oscillator on the low frequency side of the signal is that a given ratio of signal frequencies becomes more difficult to cover.

Under all conditions it is advisable to make sure that the converter is not in need of neutralizing on short waves. Neutralization of oscillator voltages in the signal frequency circuit need not require additional components, as an undesired voltage from the oscillator grid circuit can often be offset by a wiring change which increases the capacitance between the oscillator plate and the control grid and vice versa, the wiring to and the capacitances of the wave change switch being very useful for this purpose. **Plate tuned oscillators** develop high oscillator voltages in the plate circuit and need particular care in layout even on the broadcast band.

(vii) Cabinet design

(A) The subject of cabinet design for direct radiator loudspeakers is covered in Chapter 20 Sect. 3. Open-back cabinets have many shortcomings—see Chapter 20 Sect. 3(ii). Completely enclosed cabinets, even with an enclosed volume as small as 2 cubic feet, are capable of giving very much improved low frequency response— see Chapter 20 Sect. 3(iii). Vented baffles give about 3 db additional response over a limited frequency range equivalent to bass boosting—see Chapter 20 Sect. 3(iv)

The cloth used for loudspeaker grilles should be very light and of suitable texture— see Chapter 20 Sect. 1(viii).

Some form of diffuser is highly desirable to increase the angle of radiation at the higher frequencies—see Chapter 20 Sect. 2(vii). One simple form of diffuser is cone shaped and mounted directly in front of the loudspeaker.

Sharp angles (e.g. 90°) on the exterior front of the cabinet should be avoided because of deleterious diffraction effects—see Chapter 20 Sect. 3(vii).

In console and radiogram cabinets improved reproduction can be obtained if the speaker can be directed upwards, particularly if the controls are on the top of the receiver. The high frequencies produced by a loudspeaker are propagated in a beam, and where the speaker is mounted close to the floor the high frequencies do not reach ear level unless the loudspeaker is directed upwards.

When two loudspeakers are used, both reproducing high frequencies, a desirable effect can be obtained by directing the speakers outwards at an angle of about 30°

with respect to each other and upwards if necessary.   In this way the high frequency response is spread over a much greater angle.

(B) A small amount of time spent on the **ventilation** of a receiver in a cabinet to which a back is fitted can often materially reduce the operating temperature of the components.   Provision must be made for cold air to enter near the bottom of the cabinet and to leave it at the top.   The components which affect the frequency of the local oscillator should have a supply of air from outside the receiver passing over them before it passes over any other heat sources.   Ventilation should also be provided to permit a copious flow of air to pass over the output and rectifier valves and power transformer.

An efficient ventilation system is essential in the case of small a.c./d.c. receivers. For the power—of the order of 60 watts in a typical case—to be dissipated by radiation alone, the cabinet and the components inside it would rise to excessive temperatures. so that the greatest possible amount of heat should be carried away by convection.

This involves a maximum volume of air passing through the receiver at as high a speed as possible, so that restrictions on the flow must be minimized and the air paths provided should be as direct as possible.   At the same time all heat sources within the receiver must be included in the ventilation system.

In the case of wooden cabinets it is essential that some means be employed for preventing hot spots from forming on the timber.   Ventilation is the best method where it is possible, but in extreme cases it may be necessary to use asbestos sheeting to protect the wood.   Failure to avoid hot spots will sooner or later result in the finish of the cabinet being marred where the heat is excessive, although the seriousness of the blemish will depend on the quality of the veneering and on the type of finish.

When investigating ventilation, it is helpful to allow cigarette smoke to be drawn into the back of the receiver, and to watch its course through, and exit from, the receiver, which should be at operating temperature.

See also Chapter 3 page 81.

(viii) Ratings

Standard laboratory tests in most cases cover the requirement that all components be operated within their ratings.   There are some cases however where circumstances arise under which components may be over-run, although the operating conditions under test may be quite satisfactory.   A typical case is that of an i-f valve screen by-pass condenser on which the no-signal operating voltage may be 80 volts, rising to perhaps 180 volts when a strong signal is tuned in, and to 380 volts for some 15 or 20 seconds every time the receiver is switched on.   The last effect would be due to the use of a directly heated rectifier with a complement of indirectly heated valves and of course a condenser with a working voltage of at least 400 must be used.   .

Whilst such a case would rarely be overlooked, it should be remembered that even with an indirectly heated rectifier, the same high voltages may be experienced for a shorter time, and electrolytic and other condenser **working** voltages should be specified accordingly.

Other points which have given trouble are

(A) **The screen dissipation of an output valve** should always be checked with a large signal input to, and a large a-f output from, the receiver.   A large signal input biases the controlled valves, thus causing the B supply voltage to rise and the bias to decrease if back bias is used, whilst the large a-f output increases the screen dissipation. The output valve **plate dissipation maximum** will occur with a large r-f signal and no output, since any output is subtracted from the plate dissipation.

(B) If a transformer is to be operated on **frequencies** of 60 c/s and 50 c/s, or 50 c/s and 40 c/s as is the case with transformers in some Australian receivers, the heat run should always be carried out at the lowest rated frequency.   The flux density on 40 c/s is 25% higher than on 50 c/s, which will seriously affect the temperature rise.

(C) **The dissipation of the resistors** forming screen circuit voltage dividers changes with applied signal, and the resistor from B supply to screen should be checked under no signal conditions, and the resistor from screen to ground at maximum signal input. These dissipations should also be checked when the value of each resistor is on the upper or lower tolerance limit, whichever will give the greater dissipation.

(D) **When 1.4 volt valves are operated in series across a 6 volt wet battery,** a large capacitance (500 μF) electrolytic capacitor is often used to by-pass some of the valves. It frequently happens that every time the receiver is turned on, this capacitor is charged through one or more of the filaments in the string. The resulting flash can readily be seen and can seriously reduce the life of the valve, although the filament voltage may be correct when tested.

Special cases of one sort or another occur in most receivers and emphasize the need to investigate the worst conditions in every case.

(ix) Field testing

Receiver design is, more than anything else, a matter of compromise between various sets of conflicting requirements. The success with which these compromises have been made can only be judged by operation of the receiver over a period of time under normal operating conditions in the field.

Perhaps the most contentious part of a design centres around the frequency response, and final approval can only be given after prolonged listening, so that the widest possible variety of programme material and listening conditions is included. In a receiver in which distortion is kept to a minimum an extension of the a-f response progressively improves such items as a live-artist programme, but also makes progressively more annoying any poor quality recordings. Whether the best fidelity position of the tone control is one which can be consistently used or not is a question which must be decided over a period of time in a home. It may happen too that an extended a-f response sounds pleasant for one item but, probably because of distortion, becomes irritating if left playing for some hours, even on items of good fidelity. A careful check for listener irritation or fatigue after long periods of listening should be made.

Hum may change its apparent volume or character under quiet conditions, and a different level of background noise can also affect the volume control setting at which maximum frequency compensation is needed.

Controls should of course all operate smoothly and it frequently happens that noisy dial drives are heard for the first time in a home, or it may be that non-technical users find a receiver difficult to tune on short waves owing to insufficient drive reduction ratio between tuning and gang spindles.

The number of small points which may be discovered under field testing conditions could be extended indefinitely, but the main point is that no design is complete until it has been used critically and· for a period of time in a home and has been found satisfactory.

SECTION 4 : FREQUENCY RANGES

     (*i*) *Medium frequency receivers*   (*ii*) *Dual wave receivers*   (*iii*) *Multiband receivers* (*iv*) *Bandspread receivers.*

(i) Medium frequency receivers

A large percentage of commercial A-M receivers have two wave ranges, medium frequency (say 540 to 1600 Kc/s) and short wave (say 5.9 to 18.4 Mc/s). The short wave band is largely ignored in actual use and in most cases a more useful receiver could be made to sell more cheaply if the short wave range were omitted.

The obvious advantage would be that the cost of the extra components (wave change switch, two coils, s.w. padder and knob, as a minimum) could be used to provide, for instance, improved fidelity which would be available to the user at all times.

Another factor is that, particularly in cheaper receivers, the inclusion of a short-wave band restricts the design, to the detriment of the broadcast band performance. For example, microphonic trouble is far more prevalent on the short wave band and the bass response of a dual wave receiver is frequently reduced for the sole purpose of avoiding short wave microphony ; a.v.c. application to the various stages may be a compromise between the broadcast band and short-wave band requirements, and in order to save money on switching, the final circuit may lead to increased overloading or reduced signal-to-noise ratio on the broadcast band ; to reduce costs it is also customary to operate the s.w. and broadcast aerial primary coils in series, thus avoiding switching, but somewhat reducing the broadcast aerial coil performance ; if switching is used, the capacitance between the leads to the switch may on the other hand reduce the broadcast band image ratio.

When a receiver is designed only for the broadcast band it is possible to use less i-f amplification owing to the increased aerial coil gain on medium frequencies (average perhaps 8 times) as against that on the short wave band (average not greater than 2 times).   This may result in a more pleasant tuning characteristic as a single i-f stage giving maximum gain usually verges on regeneration.

In an extreme case, say in a 3 valve plus rectifier receiver, the inclusion of a short wave band may even make the difference between a reflex and a straight receiver to obtain the required i-f sensitivity for the short wave band.   This introduces the problems of a remote cut-off valve used as an a-f amplifier, rectification in the i-f amplifier, and high play-through as discussed in Chapter 28 Sect. 2(i), for the sake of increased amplification which is cancelled by turning down the volume control.

(ii) Dual wave receivers

In Australia many country areas are outside the daylight service range of medium frequency broadcasting stations and dual wave receivers are used in such districts to receive daylight programmes from short wave stations provided especially for this purpose.

The international allocations for the world wide broadcasting bands are given in Chapter 38 Sect. 4(ii), and it will be noticed that a tuning range of 4.2 : 1 would be needed to provide coverage for all short-wave bands on one wave range.   Such a tuning range is impossible with standard components and the usual compromise is to omit the two highest frequency bands and to design receivers which tune from about 5.9 to 18.4 Mc/s.   The 13 and 11 metre bands give very variable performance, depending upon ionospheric conditions, and although they can provide some of the best short wave entertainment—owing to the absence of static—when conditions are good they are so often unusable that their omission from a dual-wave receiver is not counted a serious disadvantage.

The signal strengths obtainable from international short wave broadcasting stations are of course less than those from a local medium wave station, and more sensitivity is consequently required from the receiver.   A minimum sensitivity for 50 mW output for international short-wave listening might be taken as 25 μV from 18 to 10 Mc/s, falling off to 50 μV at 6 Mc/s where static drowns the weaker stations.   Local conditions and the aerial used will of course consideably affect the required sensitivity.

Selective fading, i.e. fading accompanied by distortion of the signal, is far more prevalent on short waves than on medium frequencies, and has an effect on the preferred type of a.v.c. curve.   A curve which is too flat results in the signal being held at the same output level throughout each individual fading cycle, so that the unpleasant effect of the distortion is greatly emphasized.   Even in the absence of distortion a flat curve in a sensitive receiver results in the " troughs " of each fading cycle being filled in with noise (receiver or external) which is much less desirable than the silent space in the programme resulting from an a.v.c. curve sloping steeply at low inputs.

To provide the required difference between the broadcast and short wave a.v.c. curves it is often desirable to use a.v.c. on the converter on the broadcast band and not on the higher frequencies.   This has the added advantage of improving the oscillator stability on the short wave band.

The frequency of each fading cycle varies from some seconds for each cycle to many cycles a second on the high frequency bands, and in a good receiver a special tone control position is sometimes fitted to improve short wave intelligibility. Such a control limits the treble response of the a-f section and also reduces considerably the bass response, which removes the " fluttering " effect of rapid fading cycles and some of the more objectionable features of bursts of static and ignition interference from passing cars.

Microphony is an ever-present problem in the design of dual wave receivers, and Sect. 3(iv) of this chapter is devoted to the subject.

A problem peculiar to receivers with more than one wave band is the resonance of unused coils within an operating wave range. It is usually a low frequency coil which, resonated by its own distributed capacitance, interferes with a higher frequency band, and when broadcast and short wave coils are wound on the same former the effect is aggravated. Even in the absence of inductive coupling however, wave change switch wiring often provides enough coupling to give trouble.

A resonance affecting the oscillator coil is easily detected by measuring oscillator grid current. A sharp dip In the grid current (whether to zero or not) is an indication of an unwanted resonance. With an aerial or r-f coil more care is necessary as the resonance will be indicated only by a reduction in sensitivity over a very narrow band. To check for such an effect the receiver must be tuned in step with the signal generator over the whole tuning range using very small increments of frequency.

Whether the resonance is affecting the oscillator, r-f or aerial coil it can be traced by leaving the receiver tuned to the dip (in grid current or sensitivity) and touching with a conductor the " hot " terminals of the disconnected coils. When the resonating coil is touched the grid current (or sensitivity) will return to normal.

The normal cure for such resonances is to use a wave change switch which earths, when necessary, any low frequency coils as the receiver is switched to a higher frequency band.

One type of resonance which is not cured by earthing occurs when a broadcast band coil is wound with two or more pies. Each pie can be self resonant in some other frequency range and as one end, at most, of each pie is available, short circuiting is not possible. It may be possible with a certain value of capacitance to tune such a resonance to some harmless frequency, or it may be necessary to redistribute the turns on the pies, or to remove the coupling between the self resonant coil and the coil in use.

Since the broadcast range of a typical receiver covers 1100 Kc/s and the short wave range 12.5 Mc/s, a movement of the tuning knob sufficient to tune the receiver through 10 Kc/s on the broadcast band will tune more than 100 Kc/s on the short wave band. The difference between the two tuning rates is so great that it is difficult to provide a tuning ratio sufficiently large for the short wave band which does not make the tuning of broadcast stations undesirably slow. One solution is the use of a large ratio and a weighted " flywheel " tuning spindle which will spin when flicked, allowing rapid broadcast tuning.

The requirement for satisfactory tuning seems to be a certain linear movement of the outside of the tuning control for a given frequency change rather than a particular ratio of rotation between tuning spindle and gang spindle. Thus a possible broadcast drive is a very large knob (at least three inches in diameter) mounted directly on to the gang spindle, although this gives quite critical tuning.

For normal knobs of perhaps one inch diameter a twelve to one ratio between tuning and gang spindles is sufficient for the broadcast band but the short wave ratio should be at least eighteen to one for non-technical users of the receiver.

Probably the best mechanical solution is to use a dual ratio tuning control either with concentric knobs or with one knob and the slow speed tuning available for only one revolution of the tuning knob.

(iii) Multi-band receivers

Little expenditure is involved in the provision of additional short-wave bands in a normal dual-wave receiver, and considerable advantages are possible.

Receivers manufactured for export to some tropical countries need to receive frequencies as low as 2.3 Mc/s as well as the normal broadcasting bands and with two short wave ranges and one broadcast range it is possible to provide continuous coverage from 22 Mc/s to 540 Kc/s with standard components.

However the more usual type of multi-range receiver is one in which the coverage of individual ranges is restricted by reducing the effective change in tuning capacitance. From the view point of ease of tuning it is best in such receivers to give each range an equal coverage in megacycles rather than an equal tuning ratio, as the highest frequency band may otherwise still be quite difficult to tune.

Improved oscillator performance is possible because higher (if necessary) and more constant values of oscillator grid current can be obtained, while r-f and aerial coil performance can be improved because higher values of inductance can be used.

Microphony is reduced because part of the oscillator tuned circuit capacitance is a fixed capacitor. Further advantages of using a number of short wave tuning ranges are brought out in the next subsection.

(iv) Band-spread receivers

Although the short wave tuning range of a dual-wave receiver is usually greater than 12 Mc/s, with consequent difficulties in tuning, the majority of short-wave programmes are to be found within the 16, 19, 25, 31 and 49 metre international broadcasting bands, which have a combined coverage of only 1.65 Mc/s. If the bands could be arranged consecutively on one tuning scale the tuning speed would be quite acceptable. However their separation in the spectrum makes this impossible, and many ingenious methods have been used to spread the short-wave broadcasting bands without spreading equally the whole short-wave spectrum.

When the actual tuning-in of a short wave station has been made easy by mechanical or electrical means, other possible refinements to short wave tuning become obvious. As adjacent short-wave stations are, with band-spreading, separated by a greater mechanical movement of the tuning control it may be possible to give short wave calibrations sufficiently accurate for the identification and relocation of short-wave stations. This is most desirable.

On the other hand any frequency drift which occurs during warming up is made very obvious by the much greater tuning control travel required to correct it, and unless the drift is only of the order of 10 Kc/s, useful calibrations are not possible.

The various general types of bandspreading are shown clearly in Fig. 35.8 (Ref. 14). The mechanical types mentioned are not commonly used in mass production because the accuracy needed to provide the necessary mechanical amplification without undue blacklash is not readily achieved. Nevertheless the National H.R.O. dial is one well known example of the precision drive and scale.

A broad distinction is made on the electrical side of the " family tree " shown in Fig. 35.8 between switched inductance and switched capacitance types of band spreading. However it is quite possible for the two types to be used in one receiver, a capacitor being switched in parallel with a coil to tune it to a lower frequency and an inductor to tune it to a higher frequency band. When fixed capacitors are switched for band selection the frequency stability can be at least as good as when inductors are switched, although for draughting convenience this is not indicated in the figure.

Some of the more unusual band-spreading methods include the use of tuning condensers with specially shaped plates to give very rapid tuning between bands, and adequate spreading on the short wave broadcasting bands. Another ingenious system has a normal tuning condenser coupled to an iron core moving continually into and out of the inductor. This increases and decreases the inductance, and the mechanical coupling is arranged so that on the international bands the inductance variation opposes the capacitance variation, giving very slow tuning, while between the bands the two variations are additive, giving very rapid tuning.

The double frequency changing method mentioned on the chart has the advantages of giving the same tuning range on each spread band and of separating the band-spreading and local oscillator circuits. However it is a comparatively expensive system and its use is limited because of this.

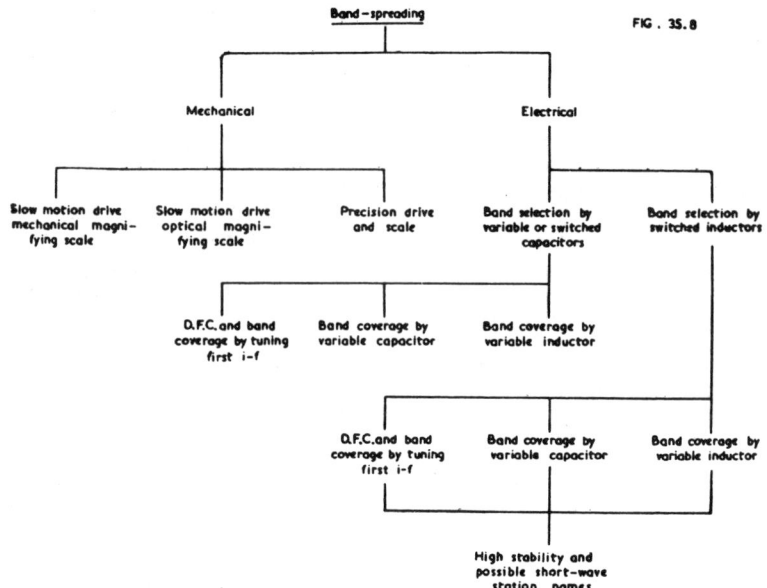

Fig. 35.8.   *Possible types of bandspreading (from Ref. 14).*

Bandspreading by means of moving iron cores has been used in many models, but although this is an electrical type it introduces mechanical problems of core location if the band-spread ranges are calibrated.

The small parallel variable capacitor used for spreading is well known to amateurs but gives a very variable amount of band-spreading, depending upon the capacitance of the main tuning condenser, with which it is in parallel, at different parts of the tuning range.   Moreover unless the band-spreading condenser has as many sections as the main tuning condenser the signal frequency circuits become detuned when band-spreading is used.   This restricts the proper use of the spreading condenser to the passband of the signal frequency circuits between values of about 6 db loss, and when a r-f stage is used this contains very few broadcasting channels.

A simple method of spreading which avoids the use of an additional tuning condenser is the tapped coil system.   When the tuning condenser is tapped down the coil, its effective capacitance is reduced approximately by the square of the tapping ratio.   If distributed capacitance across the whole coil is neglected, the coverage at each tapping point will be a constant proportion of the frequency.   Distributed capacitance reduces the coverage at the higher frequencies, giving a more nearly equal coverage measured in Kc/s.   To restrict the coverage at each tapping point, a capacitor can be connected in parallel (Ref. 20) or in series with the main tuning condenser. Such capacitors can conveniently be brought into circuit by the band-spread switch.

The series capacitor method of bandspreading (Refs. 15, 16) has been widely used. It has the advantage of providing, if desired, continuous coverage of the normal short-wave band (6 to 18 Mc/s) together with tuning on the international broadcasting bands which is comparable with that of the medium wave band.   In addition it requires few additional components, all of which are standard types.

Fig. 35.9 shows the essentials of such a band-spreading circuit as applied to an aerial stage.   Only two wavebands are shown, but as many as can conveniently be switched can be used.   A typical value for the capacitor in series with the gang is 50 $\mu\mu$F and at the high frequency end of the band, with gang capacitances of the order of 15 $\mu\mu$F it has only a small effect on the resonant frequency of the tuned circuit. With the gang approaching its maximum setting of 400 $\mu\mu$F or more, however, C

is the main frequency determining component, as the total capacitance approaches 50 $\mu\mu$F asymptotically.

The value of C can be chosen to give continuous short-wave coverage with one international band at the low frequency end of each range, where the maximum band-spreading is obtained. In this case C will have a capacitance of about 57 $\mu\mu$F for a typical tuning condenser. Alternatively the requirement might be a tuning rate at least as slow as that at the low frequency end of the broadcast band, in which case C will be 30 $\mu\mu$F. Padding for the oscillator circuit is obtained by using a smaller series condenser for the oscillator section of the gang, assuming that the oscillator operates at a higher frequency than the signal circuits.

FIG. 35.10

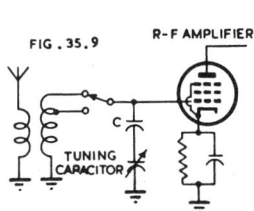
FIG. 35.9

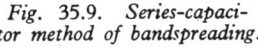

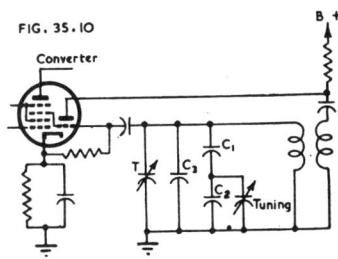

Fig. 35.9. Series-capacitor method of bandspreading.

Fig. 35.10. Bandspread circuit to give linear scale.

The system has the advantage of providing a high L/C ratio, owing to the small maximum tuning capacity, thus giving high gain, and microphony due to the tuning condenser is reduced or eliminated on the spread ranges because capacitance variations in the gang itself are minimized by the series condenser. The main disadvantages are the large number of inductors required—even when full advantage is made of tapped coils—if all short wave bands are to be covered, the unequal spreading within the bands if a large amount of spreading is used, and the small parallel capacitances in the oscillator circuit even on the low frequency ranges, which makes difficult the stabilizing of the oscillator frequency as the valve warms up.

The system just described gives its maximum spreading at the low frequency end of the band. An alternative circuit could be imagined in which no series capacitor was used, but a capacitor in parallel with the gang (say 100 $\mu\mu$F) gave spreading at the high frequency end of the band. A combination of the two circuits (Fig. 35.10) gives an almost linear spread over the whole tuning range when suitable components are used.

For a case in which one short wave band is to be spread over the whole tuning range, typical values for the components in Fig. 35.10 are

Tuning condenser	12 to 420 $\mu\mu$F
C_1	60 $\mu\mu$F
C_2	100 $\mu\mu$F
C_3	80 $\mu\mu$F
T	2 to 25 $\mu\mu$F trimmer.

These values give a tuning range slightly greater than 400 Kc/s on the 31 metre band. Although different values of C_3 will obviously affect the tuning range, different bands can be tuned by altering its capacitance and if the oscillator is operated for example on the low frequency side of the 17 Mc/s band, and on the high frequency side of the adjacent 15 Mc/s band, the difference in coverage is not serious. Higher frequency bands can be tuned by substituting a smaller inductor, or by connecting another inductor in parallel with the tuned circuit.

The unlimited spreading capabilities of the circuit of Fig. 35.10 allow a short-wave band of 200 Kc/s to be tuned over the whole dial scale. This is not advisable however, as the tuning becomes too slow, and a suitable minimum tuning range is about 500 Kc/s. A good reason for restricting the tuning range is that a fixed tuned aerial circuit can be used without any serious loss in gain. The gain of an aerial circuit with an effective Q of 50, tuned to the middle of a band 500 Kc/s wide and including the 15 Mc/s band, falls about 6 db between the centre point and the extremes of the band.

With such a system spreading only the international bands, it may be advisable to provide one short wave band giving continuous coverage from 6 to 18 Mc/s, as many stations of interest are heard outside the bands. This is not a great disadvantage as the components used for continuous coverage can also be used for at least some of the spread ranges.

Apart from the possibility of providing linear spreading up to any reasonable limit (and the consequent need for bandspreading for the oscillator circuit alone in a receiver without a r-f amplifier) the circuit has the advantage of preserving suitable L/C ratios while maintaining at all times a capacitance in parallel with the oscillator input large enough to minimize frequency drift as the valve warms up. Capacitor C_3 can be a temperature compensating condenser to improve the frequency stability.

Microphony from the gang is completely eliminated because, with normal component values, the tuning capacitance is only altered 10% as the gang is turned from maximum to minimum.

An interesting image rejection circuit is given in Ref. 14(1). Its use is confined to bandspread receivers with a r-f stage and covering one short-wave band only on each band spread range, but image ratios from 35 to 50 db are claimed. Few additional components are required.

SECTION 5 : A.C. OPERATED RECEIVERS

(i) Four valve receivers   (ii) Five valve receivers   (iii) Larger receivers   (iv) Communication receivers.

(i) Four valve receivers*
(A) T-R-F receivers

Because it has no local oscillator coil or i-f transformer, the t-r-f receiver is the cheapest to build, but has a performance equivalent to its cost. Using modern valves a sensitivity of a few hundred microvolts can be obtained, or even better if some form of fixed regeneration is used and can be made effective over the whole tuning range, but the selectivity is in general insufficient to separate local stations if they are of high power or situated close to the receiver.

The main application of t-r-f sets is in absolute minimum cost receivers which operate with 3 stages and a rectifier, although using only three or even two actual valve types. In such a case there is a resulting loss in sensitivity, which might then be no better than one millivolt.

Volume controlling is most conveniently carried out by varying the bias of the r-f amplifier, and to ensure low minimum volume on strong local stations the aerial is sometimes wired to the control as in Fig. 35.11 so that it is earthed in the minimum volume position.

(B) Superheterodyne with a-f amplifier

An improvement in selectivity can be obtained by using the superheterodyne principle even when no i-f amplifier is used. Such a receiver would have a converter followed by an i-f transformer, a second detector, an a-f amplifier and an output valve. Detection can be by diode (included in an a-f valve), anode bend, or leaky grid detector with possible regeneration in the last case. Since the input to the detector is fixed tuned, pre-set regeneration can provide a constant and appreciable improvement to both sensitivity and selectivity over the whole tuning range without the care which is needed in a t-r-f receiver.

Selectivity is improved since there are three tuned circuits, two of which are fixed tuned and give constant selectivity over the band.

---

*i.e. three amplifying stages and rectifier.

(C) Superheterodyne with i-f amplifier

The majority of three valve and rectifier receivers are superheterodynes with an i-f amplifier, in which selectivity can be made as good as required. Sensitivity too, while presenting some problems, can usually be made adequate, i.e. less than one hundred microvolts and, in extreme cases, less than fifteen microvolts on the medium wave band.

Three types of second detectors have been used, power grid, anode bend (each carried out by the output valve) and diode, although diodes are now used universally. The power grid detector had the disadvantage of high plate dissipation when no signal was being received, and the anode bend had somewhat higher distortion and needed a very high impedance plate load. The maximum power output when the output valve is detecting is of course seriously reduced. The diodes can be combined with the i-f amplifier or with the output valve, and one receiver has been marketed in which the suppressor grid of the i-f amplifier was used as a diode.

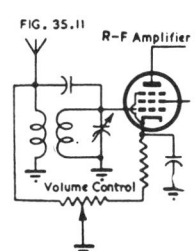

FIG. 35.11. R-F Amplifier

Volume Control

Fig. 35.11. Volume-
control circuit used with
t-r-f receivers.

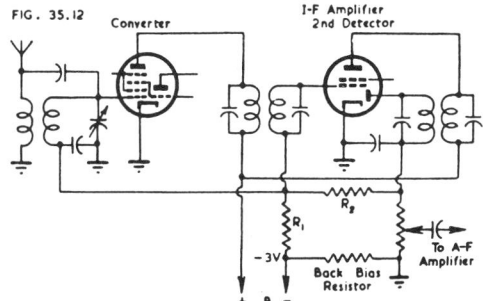

FIG. 35.12. Converter   I-F Amplifier 2nd Detector

R_1 R_2

−3V   To A-F Amplifier

Back Bias Resistor

+ B −

Fig. 35.12. A.V.C. circuit suitable for 3/4 valve
receiver.

Volume controlling is carried out by manual control of converter and i-f bias or by a.v.c. and an a-f volume control. The design of a.v.c. circuits for low gain receivers without an a-f amplifier presents some problems since they must be cheap, and must give full a-f output with the smallest possible input, which prohibits the use of full, simple a.v.c. as commonly used in cheap receivers with an a-f amplifier. On the other hand the low sensitivity means that the required range of a.v.c. control is less than normal and advantage is taken of this in the usual a.v.c. circuit which is shown in Fig. 35.12.

This circuit needs only one more resistor (R_1) than the simplest possible circuit, and R_1 and R_2 form a voltage divider so that only a fraction of the voltage developed across the diode load is applied to the grids of the controlled stages. The most suitable fraction depends on the initial sensitivity and the valves used, but values of one third to one fifth are common. Two details need care, the negative voltage applied to the diode must be kept to a minimum, to reduce distortion and to prevent the diode from being muted in the absence of a signal and the a.c. shunting on the diode load must be as small as possible. Large values for R_1 and R_2 fulfil both requirements and minimum values can be taken as 1 megohm and 2 megohms respectively. With the normal value of 0.5 megohm diode load the bias on the diode is less than $\frac{1}{2}$ volt, and the shunting reduces the a.c./d.c. ratio to 0.8. Better values to give almost the same ratio would be 2 megohms and 5 megohms (0.2 volt bias on diode ; a.c./d.c. ratio = 0.9).

The use of only part of the developed voltage allows the output to rise rapidly with increasing input, which is an advantage for small signals, but there is little flattening of the characteristic with larger inputs.

For a slight increase in cost, delayed a.v.c. can be used and this will give the most rapid rise in output with increasing input, followed by a flat output-input characteristic. A suitable circuit is the " sinking diode " type shown in Fig. 27.38B, and provided that an additional diode is available the increase in cost is slight.

With manual volume control the circuit of Fig. 35.11 is usually employed. However if the B supply voltage is low, the signal handling capabilities of the i-f valve are reduced (owing to the reduced plate-cathode voltage) when a strong signal is tuned in, the volume control is turned back and the cathode voltage of the controlled valves thus increased. This objection can be overcome by using a negative voltage applied to the grids of the controlled stages. One method of obtaining the negative voltage is to connect the power supply filter in the negative lead and use the d.c. voltage generated across it. Some decoupling is necessary, but one stage is sufficient as high value resistors can be used.

A second method of obtaining a suitable negative voltage is to use the volume control resistor as, or in parallel with, the oscillator grid leak (with suitable decoupling). Oscillator grid current must be kept as constant as possible throughout the tuning range when this is done.

Three valve and rectifier receivers with an i-f stage and using i-f transformers with a Q of about 115 and good aerial coils are capable of producing sensitivity figures of the order of 15 μV on the medium wave band when the most suitable commercial valve types are used. This sensitivity is ample for almost any listening conditions providing full volume is not required from very weak signals. In fact, successful dual-wave receivers giving short wave sensitivities not worse than 50 μV have been made with such a circuit.

(D) Superheterodyne with reflexing

Reflexing in a three valve and rectifier superheterodyne receiver provides enough a-f gain to allow full simple a.v.c. to be used, and to allow good short wave performance to be obtained. The additional problems involved are discussed in Chapter 28.

A convenient a.v.c. circuit consists of full a.v.c. voltage applied to both converter and reflexed i-f amplifier, which can give a very flat a.v.c. characteristic. An a.v.c. circuit in which control is applied only to stages before the a.v.c. detector, can never give a completely flat output curve, because some rise is necessary to provide the additional bias needed to reduce the output as the input signal is increased. Such a circuit is a "backward acting" a.v.c. circuit.

A " forward acting " a.v.c. circuit is one in which the developed a.v.c. voltage is applied to a stage after the detector, and with such a circuit the output voltage may even fall with increasing input. The reason is that although the detector output must rise as in the previous case, the gain after the detector will fall, and the net result may be an increase or a decrease or a flat characteristic depending on the constants used.

In a reflex receiver the application of a.v.c. to the reflexed stage controls the mutual conductance of the valve and thus its gain at intermediate and at audio frequencies. As the a-f gain follows the detector the a.v.c. system is a forward acting one.

Alternative a.v.c. designs in a reflex receiver may use full a.v.c. on the converter and a fraction or none on the reflexed stage.

The increased a-f gain makes it possible to use inverse feedback, frequency compensated or otherwise, in the a-f amplifier, and resistive feedback into the bottom of the volume control (see Sect. 3(iii) of this chapter) is useful in reducing minimum volume.

The combination of audio a.v.c. and negative feedback (particularly when frequency compensated) is not always advisable, either in reflex receivers or in larger types in which a.v.c. may be applied to a valve acting only as an a-f amplifier. Since the purpose of the a.v.c. is to vary the gain of the stage, and one of the functions of negative feedback is to minimize any changes within the loop, the two effects oppose each other. This is not undesirable in itself, but the result is that on strong signals, such as local stations, the a-f gain is reduced and the feedback is consequently reduced so that the full effects of distortion reduction are not obtained. When the feedback incorporates frequency compensation the result is that weak stations receive maximum compensation and local stations a reduced amount, which is the reverse of the normal requirement.

Advantage has been taken of this effect in a receiver having one a-f stage with deliberately attenuated high and low frequency response. Audio a.v c. was applied to this stage so that on strong stations no attenuation resulted—slight high and low frequency boosts were incorporated elsewhere in the a-f circuit—whereas on weak stations considerable treble and bass cutting was automatically introduced, giving a very effective automatic tone control. Suitable treble cutting for such a circuit could be by means of a small capacitor between plate and grid of the controlled stage, and bass cutting could be brought about by a small value of screen by-pass in a pentode.

In conjunction with a pick-up of good sensitivity a reflexed three valve receiver can reproduce gramophone records without added circuit complications.

The most common type of reflexing has the a-f and i-f signals applied to the control grid, with the load for each in the plate circuit. Another type which has been used has the two signals applied to the control grid with the i-f load in the plate circuit and using the screen dropping resistor as the a-f load, with an i-f by-pass from screen to ground. The a-f path to the grid of the output valve is provided by means of a normal grid coupling capacitor.

Since the a-f gain available from a reflex stage is comparatively small if the i-f operation is to be satisfactory, the reduced amplification between control grid and screen grid may be more than offset by the increased a-f load resistor which can be used and by the fact that proper i-f by-passing can be used on the cold side of the i-f transformer in the plate circuit.

(ii) Five valve receivers*
General comparison of types

The three valve receiver using a converter, an i-f amplifier and an output valve can be considered to be the smallest conventional superheterodyne receiver, and a valve can be added to it in three ways, as an a-f amplifier, an i-f amplifier or a r-f amplifier. Three very different types of receivers result, and the comparison serves to show the main ways in which the conventional 4 valve receiver with a-f amplifier can be improved.

The table below lists the order of preference for each type of receiver against a variety of headings.

Additional stage.	A-F	I-F	R-F
Cost	1	2	3
Sensitivity	2	1	3
Selectivity	3	1	2
A-F Response	1	2	2
A.V.C.	3	1	2
Stability	2	1	2
Noise	3	2	1
Spurious Responses	2	3	1

The names " a-f receiver," " i-f receiver " or " r-f receiver " are used to identify a receiver with such an additional stage.

Although the performance of the a-f receiver can be improved upon in many ways, its characteristics can be satisfactory, as evidenced by the fact that the great majority of five valve receivers are built in this form. Since most receivers are tuned only to local stations, the advantages of the i-f and r-f types would rarely be displayed, and their additional cost might be more profitably used on a refinement on an a-f receiver such as a tuning indicator, bandspread ranges, or a better loudspeaker, cabinet or a-f system.

*i.e. four amplifying stages and rectifier.

(iii) Larger receivers

The conventional six valve receiver* has r-f, converter, i-f, detector, a-f and output stages, but although the stage usually added to the conventional four valve and rectifier receiver is a r-f one with advantages outlined in the previous section, the customer usually expects improvements in all features of the receiver's performance. To the average listener an improvement in fidelity is more noticeable than, for example, a reduction in image response, and it seems probable that the public would in most cases be better served with a six valve receiver with push-pull output rather than by a single ended receiver with a r-f stage.

Whether or not push-pull is used, some improvement in a-f response is required, for instance an increase in acoustical output, in conjunction with a suitable negative feedback circuit to give a reduction in distortion and a compensated frequency response.

A valve delivering more than four watts into an efficient output transformer and a sensitive loudspeaker can produce a considerable volume of noise but, even without push-pull, impressive reproduction can be obtained from the larger tetrodes and pentodes which draw between 70 and 80 mA of plate current.

The additional selectivity provided by a r-f amplifier with a high Q coil is appreciable and gives even more sideband cutting than is experienced with the usual 4 valve and rectifier receiver, particularly at the low frequency end of the broadcast band. As a result the high frequency response of the larger receiver may be inferior and, in fact, inadequate. The remedy of expanding the selectivity can be quite inexpensive, particularly if a " local broadcast " position is used on the wave change switch, or if one position on a tone control switch is used. Push-button receivers can have one button set aside for expanding the selectivity, which has the additional advantage of minimizing the result of frequency drift of the settings of the other push buttons.

Expanding of the i-f selectivity can be carried out simply and cheaply as described in Chapter 11 Sect. 3(iii), and the wiring is not critical because all leads have a low impedance to ground. It is only necessary to expand the first i-f transformer in a conventional single i-f stage receiver with an untapped second i-f transformer because the heavy damping on the second i-f reduces its selectivity appreciably from that of the less heavily damped first i-f transformer.

After the i-f amplifier has been expanded to give say 6 db attenuation at 8 Kc/s detuning, it may be found that the overall selectivity for small attenuations is still excessive, and the remedy is to expand the r-f stage. The damping of the aerial on the aerial secondary makes it unnecessary to expand the aerial coil. A convenient method of decreasing the r-f coil selectivity is to switch a small resistor (25 ohm carbon) in series with the cold end of the secondary. Gain is not seriously reduced, and the leads involved are low impedance so that they can be taken to the appropriate switch section without giving trouble.

After the high frequency response has been brought up to the required level, the small expense involved in fitting a high frequency diffuser is well repaid. With such a diffuser, the quality of reproduction becomes reasonably independent of the listener's position in front of the receiver, and the usual beam of accentuated high frequency response directly in front of the loudspeaker disappears.

(iv) Communication receivers

(A) Definition

A communication receiver has been defined (Ref. 33) as one which is not designed for limited or specific purposes. In some respects a communication type may be inferior to another receiver (perhaps an interception receiver to operate on a fixed frequency) in a particular application, but in general it has high performance and flexibility. Control of many of the circuits is available to the operator, of whom some technical knowledge is required.

(B) Frequency coverage and calibration

Communication receivers have been designed to provide coverage between 15 Kc/s and 25 Mc/s and between 30 and 300 Mc/s, but many do not tune above

*i.e. five amplifying stages and rectifier.

30 Mc/s or below 540 Kc/s.   Provision for reception of F-M broadcasting on 88-108 Mc/s is included in some models.

Owing to the wide range of frequencies covered, the calibration of a communication receiver is important, and increasing emphasis is being placed on the inclusion of facilities for calibration checking and even on the possibility of the receiver being tuned to any given channel in its range without external aids.

An inbuilt crystal oscillator is the usual method of checking calibrations, and when the oscillator operates at 0.5 or 1.0 Mc/s, convenient checking points occur throughout the tuning range.   Movable cursors have been used with receivers of this kind to allow corrections to be made to the calibration when found necessary.   Figure 35.13 (British patent No. 8706/44) shows a simple circuit to increase the amplitude of the higher harmonics of the oscillator output up to the self-resonant frequency of the choke *L*.   The choke is designed to keep the harmonic output as constant as possible throughout the tuning range of the receiver.   Such an inbuilt calibrator can be very useful in the alignment of the receiver.   A possible refinement is the modulation of the oscillator output by pulsing with a neon tube, or even by the mains frequency, for ease of identification.

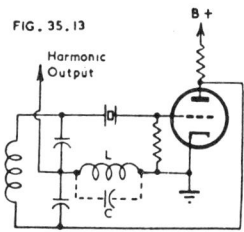

FIG. 35.13

Fig. 35.13.   Circuit for increasing amplitude of high-order oscillator harmonics (from Ref. 33).

For the second requirement of setting the receiver accurately to any desired frequency, a crystal oscillator and its harmonics are used as a part of the mixer input, and tuning is provided by a comparatively low frequency oscillator which bridges the spaces between the harmonics of the crystal.   The variable oscillator can be designed to operate at a low frequency and to have high stability, any residual fluctuations being a much smaller proportion of the total mixing frequency than of the variable oscillator frequency.

To minimize the possibility of incorrect frequency settings, some receivers have sliding dial scales which expose only the calibration actually in use.   Where high setting accuracy is required the frequency may be displayed as the sum of two readings, one in Mc/s and the other in Kc/s.

(C) Bandchanging

Methods of bandchanging can be divided into three main categories, mechanically operated moving coils with stationary contacts, plug-in coils and switched coils. The most common form of moving coil arrangement uses a rotating turret with the coils inside it, but moving platforms which travel along the chassis have also been used.   Particularly when high frequencies are tuned, the contacts for the moving coils present a problem as the inductance of the contacts can form a large part of the required inductance—with consequent loss of gain—and the inductance may vary each time a particular band is brought into circuit or when the receiver is subjected to mechanical movement.

Plug-in coils can be efficient as they do away with the need for a wavechange switch —each plug-in unit can even carry its own calibration— but the necessity for grouping the coils adjacent to an opening in the case restricts the layout somewhat.   Storage must be provided for the unused coil units close to the receiver.

Switched coils are most commonly used and the main associated problem is the disposing of a number of coils adjacent to each switch.   The switches need to be reliable, but modern types using rhodium plating properly applied are satisfactory.

(D) Stability

Many stability problems are automatically eliminated by the use of suitable circuits, e.g. crystal controlled oscillators. Methods of improving electrical stability are outlined in Chapter 24 Sect. 5. However mechanical stability is a first requirement and the tuning units at least should be made as rigid as possible, some receivers even using a cast chassis with this end in view.

(E) Sensitivity and noise

Owing to the number of stages used, the sensitivity is always limited by the noise in the first stages of a receiver. This is taken into account in one method of measurement by defining sensitivity as the lowest intensity of an input r-f carrier modulated at 400 cycles to a depth of 30% such that the total r.m.s. output power (signal and noise) is halved when the modulation is removed from the input carrier. Such a definition of sensitivity gives a good indication of the minimum signal which can be used and figures of the order of one micro-volt and better can be obtained.

Another method of defining the noise ratio is by comparing it with the signal-to-noise ratio of an ideal receiver under the same conditions. The factor by which the noise ratio of an actual receiver is worse than that of an ideal receiver is known as its noise factor. Further details of this method of noise measurement are given in Chapter 37 Sect. 1(vi)G.

(F) Selectivity

The selectivity requirements of a communication receiver are severe and depend on the type of service for which the receiver is being used. For telephony the bandwidth cannot be less than about 5 Kc/s at 10 db attenuation, whereas for telegraphy a bandwidth of a few hundred cycles per second is sufficient while greater bandwidths give increased noise interference.

These requirements can only be met by variable selectivity under the control of the operator, and a good receiver will usually have at least six i-f tuned circuits to provide adequate " skirt " selectivity, with a crystal filter providing varying degrees of " nose " selectivity perhaps together with a variable rejection control for dealing with an interfering station on a nearby frequency. A representative crystal circuit will give an attenuation of 20 db when detuned 200 c/s in its sharpest position and 2000 c/s in its broadest position. With a rejection or " phasing " control in use, an attenuation in excess of 40 db can be obtained within 250 c/s of resonance. The associated i-f amplifier will give an attenuation of 60 db when detuned less than 15 Kc/s. Crystal bandpass filters are becoming increasingly popular and they provide a pass band of say 300 or 3000 c/s with severe symmetrical attenuation on each side. Details on the design of variable selectivity crystal filters are given in Chapter 26 Sect. 6.

An additional means of discriminating between a desired telegraphy signal and an undesired one separated by perhaps only a few hundred cycles per second, is the use of a-f selectivity. A typical selective circuit centred on 1000 c/s gives an attenuation of 15 to 20 db only 100 c/s from resonance and of course much greater reductions with greater separation. Apart from discrimination against undesired signals, such a circuit reduces random noise in the same ratio as it reduces the bandwidth, and the ringing which is experienced with crystals at maximum selectivity when subjected to bursts of noise is eliminated.

Such selective a-f circuits are normally designed with iron-cored inductors because maximum Q is required, but it may be possible to produce equivalent or improved performance with a bridged T negative feedback circuit.

(G) Volume control and a.v.c.

Volume control presents some problems because the range of signal levels which the receiver is expected to use exceeds one million to one, and because different types of control are required for different functions of the receiver. For reception of telegraphy many operators prefer a manual control and some receivers provide separate controls for a-f gain and i-f gain ; even separate r-f gain is sometimes provided. A convenient compromise which obtains good results from the receiver and requires no care on the part of the operator is to gang the a-f and r-f controls and delay the

application of bias to the first stage by means of a diode as discussed in Chapter 27 Sect. 3(iii). The r-f control is rendered ineffective when a.v.c. is switched on, and the a.v.c. voltages can be applied to the r-f control circuit.
   A.V.C. is frequently used for c.w. reception when a suitable time constant is provided. For telephony a normal time constant is 0.2 second but this needs to be increased to at least 1 second for c.w.
   Amplified a.v.c. is used in most large communication receivers to provide maximum control, and results such as a $2\frac{1}{2}$ db increase in output for a 100 db increase in input can be obtained (Ref. 34). Control should begin at the minimum signal level which can be used by the operator.

(H) Beat frequency oscillator
   To produce an a-f output from a c.w. signal it is necessary to beat another signal with it. The beat frequency oscillator operates at, or close to, the intermediate frequency and beats with any signal in the i-f channel. The oscillator frequency can usually be varied a small amount ($\pm 2\frac{1}{2}$ Kc/s is typical) by a control on the front panel to allow adjustment of the beat frequency (perhaps to the peak of the a-f response) without detuning the signal from the peak of the i-f circuits.
   The required amount and method of B.F.O. injection need some consideration. Sensitivity is lost if the B.F.O. amplitude at the second detector is too small, but when the coupling is increased a large input signal pulls the B.F.O. into zero beat from increasing frequency separations. Pulling into zero beat from a difference frequency of the order of 1000 c/s is not uncommon in bad cases. Another effect of too much B.F.O. injection is that the B.F.O. provides a signal at the second detector large enough to generate appreciable a.v.c. voltages and so reduce the sensitivity of the receiver even without an external signal.
   Pulling can be minimized by electron coupling of the B.F.O. into the i-f channel, and with careful control of the amount of beat frequency voltage it is possible to obtain adequate injection without appreciable a.v.c. sensitivity reduction.

(I) Signal strength meter (S meter)
   Many receivers use a calibrated meter to indicate the strength of signals tuned. The calibration is usually arbitrary (S1 to S9 for example) because the variation of receiver sensitivity across the various bands, and on different bands, makes an absolute calibration impossible. If this effect, and the varying efficiency of different aerials on different frequencies are borne in mind, the S meter can be a useful reference.
   Circuits have been devised (Ref. 34) in which the S meter operates in conjunction with the manual volume control to provide signal strength indications on all types of reception, with or without the beat oscillator.
   The usefulness of the meter can be greatly increased by providing a switch which allows appropriate currents and voltages throughout the receiver to be measured.

(J) Aerial input
   To allow maximum efficiency to be obtained from different aerial systems, provision is often made for different types of aerial inputs, such as a single wire with separate earth connection, a balanced two wire feeder or a low impedance concentric cable.
   An aerial trimmer on the front panel is very desirable when a variety of aerial inputs is provided. Even with a single wire aerial the changes of impedance throughout the short wave ranges allow considerable improvements in sensitivity and signal-to-noise ratio to be obtained by the use of an aerial trimmer, and aerial coil design is not restricted so greatly by the possibilities of mistracking.
   An analysis of the effects on signal-to-noise ratio of the ratio of feeder impedance to receiver input impedance with various detuning ratios is given in Ref. 33.

(K) Noise (crash) limiters
   These are commonly fitted to communication receivers. Details of the many varieties will be found in Chapter 16 Sect. 7 and Chapter 27 Sect. 5.

(L) Diversity reception
   Provision for diversity reception is sometimes incorporated in communication receivers by providing facilities for combining the a.v.c. voltages and outputs of the individual receivers.

Another requirement for diversity receivers is that oscillator reradiation should be kept to a minimum. This can be done by using the lowest possible oscillator grid current, by earthing the whole oscillator circuit at one point and treating other amplifier circuits between oscillator and aerial in the same way, by reducing to a minimum capacitance couplings between primaries and secondaries of signal frequency coils (this includes using high impedance externally wound primaries wherever possible) and by minimizing all couplings other than the coils between the signal frequency and oscillator stages. The tuning condenser rotor shaft is the worst of such couplings and a rotor shaft insulated between sections is very desirable. If this is not possible, low impedance earthing should be provided for the rotor shaft and gang framework between each section.

(M) Cross-modulation

Communication receivers are frequently used close to a transmitter and may be required to receive another station while the transmitter is in operation. Under such conditions the cross-modulation characteristics of a receiver may be of more importance than its signal-to-noise ratio or sensitivity (Ref. 33).

A system known as counter-modulation has been used to allow receivers to operate more satisfactorily in the presence of very strong signals. The cathode resistor of the first valve in the receiver is by-passed to radio, but not to audio, frequencies, so that a-f components due to cross-modulation appear across it. The value of the resistor is adjusted so that the a-f voltages developed are sufficient to remodulate the wanted carrier with signals approximately equal and opposite in phase to the original cross-modulation. Correct bias for the valve is obtained by applying a suitable voltage to its grid.

Further improvement can be obtained by operating the first valve solely to give minimum cross-modulation and substantial improvements can be obtained for slight reductions in signal-to-noise ratio.

The possibility of cross-modulation occurring in non-linear conductors close to the receiver should not be overlooked (Ref. 3).

SECTION 6 : A.C./D.C. RECEIVERS

(i) Series-resistor operation (ii) Barretter operation (iii) Dial lamps (iv) Miscellaneous features.

(i) Series-resistor operation

A receiver for use with d.c. supply cannot use a power transformer with low and high voltage windings, and some of the main problems in designing such a receiver are concerned with providing power for the comparatively low voltage valve heaters.

The problems are partly due to the non-linear relationship between voltage and current in a valve heater (owing to the increased resistance of the wire as the heater warms up) and Fig. 35.14 shows this relationship for a normal 6.3 V 0.3 A heater and for a 21 ohm resistor which draws the same current with 6.3 volts applied to it. It will be seen that with the normal heater a 10% variation in applied voltage gives approximately 6% increase in current.

In a.c./d c. receivers, power is supplied to the heaters by connecting them in series across the mains, with or without an additional series impedance. When the sum of the required heater voltages is equal to the mains voltage the method is as satisfactory as when a power transformer is used, but when a series resistor is used, a variation of the mains voltage results in a larger percentage variation of the voltage across the heaters.

This effect is illustrated in Fig. 35.15. Curve ABC represents the current voltage characteristic of a string of heaters used in a radio receiver. The rated operating voltage is 108 for a nominal current of 150 mA, and a series resistor of 880 ohms is used to give the required voltage drop with a 240 volt mains supply. The intersection of the line with a slope of 880 ohms, and the current-voltage characteristic of

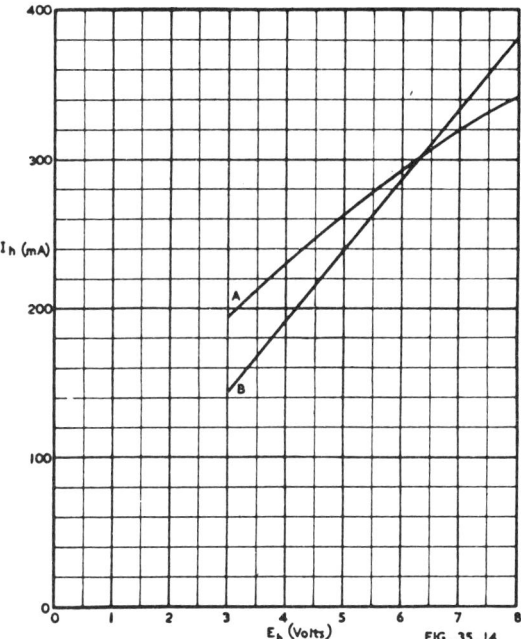

FIG. 35.14

Fig. 35.14 (A).   *Voltage-current characteristic of 6.3 volt 0.3 amp. heater ;* (B) *Voltage-current characteristic of 21 ohm resistor.*

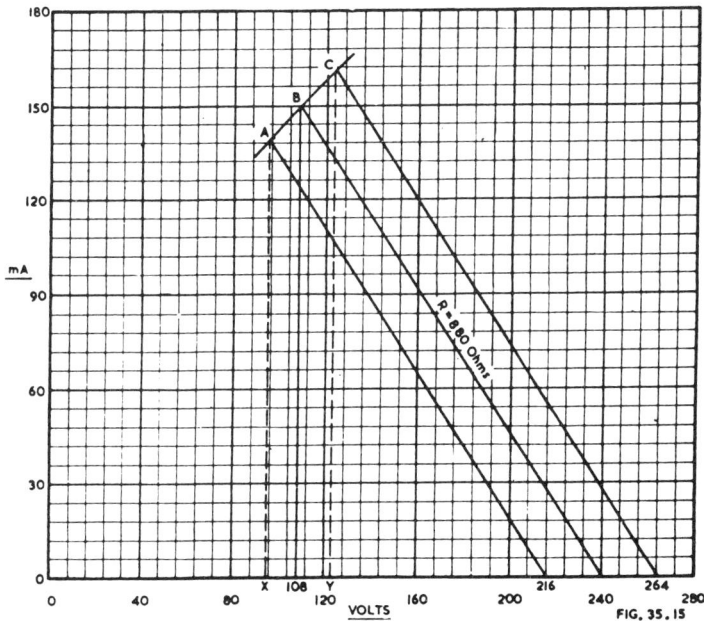

FIG. 35.15

Fig. 35.15.   *Illustration of the effect of mains voltage variation on heater voltage (Method from Ref.* 41).

the heaters, is point B, the operating point under normal conditions. Now if the mains voltage is increased 10% to 264 volts the slope of the resistive line is unchanged and it intercepts the curve at point C. A vertical from C cuts the base line at Y, representing a voltage of 123, and an increase of 15 volts, i.e. 14% in the voltage applied to the heaters. Similarly a decrease of 10% in the mains voltage gives more than 10% decrease in the voltage applied to the heaters.

Because of this fact, it is necessary to provide more tappings on a series dropping resistor for various primary voltages than would otherwise be the case. Variations in supply voltages around the nominal value must be taken into account, in conjunction with the manufacturer's tolerance on the valve heaters. For American types this tolerance is usually taken as $\pm$ 10% of the rated voltage, but other tolerances are also used, and some types have a tolerance on the permissible current variation.

When the dropping resistor is designed for one primary voltage only, it often takes the form of resistance in the power lead. This has the advantage of dissipating the heat outside of the cabinet. However if a receiver was originally designed for 115 volt operation and is converted for 230 volt operation by means of a series resistor, the B+ voltage is liable to be reduced by this method (Ref. 36). The reason is that the value of the series resistor must be adjusted to give the correct average heater voltage (or current) but the rectifier draws its current in pulses and these pulses flowing through the series resistance decrease the applied voltage during the time the rectifier is drawing current. This gives a decrease in the d.c. output of the rectifier, which in a severe case might amount to 30 volts in 110. Methods of providing separate voltage dropping impedances for the heaters and rectifier which are suitable for use on both a.c. and d.c. are given in Ref. 36. Where voltage dropping of this type is required on a.c. only, a series condenser of the correct impedance can be used in series with the power lead (Ref. 42) provided that an adequate voltage rating is specified.

(ii) Barretter operation

A barretter is a device which passes a substantially constant current as the applied voltage varies within the operating range—see Chapter 33 Sect. 1(i) and Fig. 33.1.

The heat dissipation of a barretter is appreciable, and as most of the heat should be carried away by convection, ample ventilation is needed. This is necessary both for the barretter and for the whole receiver, if it is a small one, to prevent the cabinet from becoming unduly hot. The same problem exists, of course, if a series resistor is used.

Apart from eliminating the effects of mains voltage fluctuations on filament current, a barretter does away with the need for changing tappings with different supply voltages over a range covered by the operating characteristics of the barretter. When a receiver is switched on, the initial current surge is limited by the barretter, which has the effect of noticeably increasing the time taken for the receiver to begin playing. The limitation of the surge is of no particular value to the valves themselves but may reduce dial light overloading in some circuits.

Since the barretter contains an iron wire carrying an alternating current it is susceptible to magnetic fields, which cause the filament to vibrate. The vibration leads to mechanical fatigue and breakage and if the barretter cannot be located at some distance from the loudspeaker, a magnetic shield must be used—preferably one which interferes with ventilation as little as possible.

(iii) Dial lamps

The protection of dial lamps from surges when the receiver is first switched on is one of the problems of a.c./d.c. receivers. If a 0.15 A dial lamp were connected in series with an appropriate number of 0.15 A valve heaters across the mains, the high current (several times greater than 0.15 A) drawn by the heaters for some seconds each time the receiver was switched on would drastically reduce the life of the dial lamp. If a series resistor were used the surge would be reduced, but not sufficiently, and even if a barretter replaced the resistor the dial lamp would be overloaded before the barretter filament reached its operating temperature. The use of a dial lamp of higher current rating than the valve heaters is a possible method of operation. Some manufacturers have produced special lamps for series running. Nevertheless in each of these cases the failure of the dial lamp would remove the voltage from the valve heaters and stop the receiver.

One method of avoiding this trouble is to use a negative coefficient resistor (Thermistor) in series with the valve heaters—see page 190 (n).

A system in which both heater current and high tension supply current are used is given in Fig. 35.17. Special rectifier valves (e.g. 35Z5-GT) have been produced for use in such circuits, and the initial surge is offset by the fact that until the rectifier heater reaches its operating temperature the high tension component of the dial lamp current does not flow. No. 40 or No. 47 dial lamps (0.15 amp) should be used, and when the rectified current exceeds 60 mA a shunting resistor (R_s in Fig. 35.17) is required, its value being 300, 150 and 100 ohms for rectified currents of 70, 80 and 90 mA respectively.

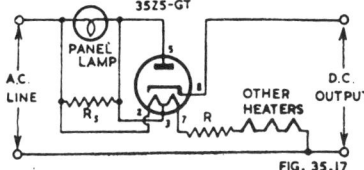

Fig. 35.17. *Circuit using 35Z5-GT and dial lamp. The drop across R and all heaters (with panel lamp) should equal 117 volts at 0.15 ampere. R_S = shunting resistor required when d.c. output current exceeds 60 mA.*

Other rectifiers making similar provision for a pilot lamp are the miniature 35W4 and the 45Z5-GT.

The trend in small a.c./d.c. receivers is to use a simple dial without a lamp, and thus entirely avoid the use of the troublesome light source.

(iv) Miscellaneous features

(A) Rectifier

Since there is no centre-tapped source of high voltage a.c. in an a.c./d.c. receiver it is not possible to use a normal full wave rectifier. Bridge rectifiers or voltage doublers are possible but half wave rectifiers are most commonly used. They introduce some filtering problems since the frequency of the ripple in the output is one half of that from a full wave rectifier. To give the same electrical attenuation, filter capacitors and inductors would need to be increased in size, but speaker inefficiency and the poorer response of the ear at the lower frequency help in minimizing the effects of higher hum level.

When the receiver is operating on d.c., the rectifier acts only as a series impedance but serves a useful purpose in protecting the electrolytic filter condensers from damage if the mains connection is reversed in polarity, although the receiver will not operate under these conditions.

The high tension voltage is likely to be different with a.c. or d.c. mains of the same rated voltage. The reason is that on a.c. the peak voltage applied to the rectifier is $\sqrt{2}$ times the rated r.m.s. voltage and with small to medium loads and a large input condenser for the rectifier the d.c. output voltage will exceed the r.m.s. value of the a.c. input.

Modulation hum is frequently experienced in a.c./d.c. receivers operating on a.c. due to the presence of r-f signals in the mains, and to the direct connection of the rectifier to the mains. A suitable filter in the mains lead will stop the modulation hum and any interference coming from the mains, but a satisfactory cure for the hum alone is a small (1000 $\mu\mu$F) mica condenser connected between plate and cathode of the rectifier.

(B) Valve order

When the valve heaters are wired in series and connected across the mains the potential between heater and cathode of any valve will be dependent on its position in the string.   Each valve must be placed so that the maximum heater-cathode potential specified by the valve manufacturers is not exceeded, and also so that the potential is not great enough to introduce hum into the receiver output.

The position of a valve in the circuit determines its susceptibility to hum pick-up, and an a-f amplifier is generally the most critical.   The usual wiring order, starting from the grounded end is second detector and first a-f amplifier, converter, r-f amplifier, i-f amplifier, output valve, rectifier.   The converter is kept as close to ground as possible to avoid modulation hum.   It will be found that allowable maximum heater-cathode potentials (90 V max. design centre for American amplifying valves) always allow this order, which is the best for modulation hum, to be used.

(C) Earth connection
Two main types of a.c./d.c. receivers are produced, one with one side of the mains directly connected to the chassis, and the other with mains wiring connected to an earth bus which is by-passed to the chassis proper with a suitable capacitor, perhaps shunted by a high value resistor.

The first type of receiver has fewer design difficulties but does not comply with the safety requirements in some countries, e.g. the Radio Code of the Standards Association of Australia (A.S.S. No. C.69—1937) states : V.7 (f) (ii) " Power units and sets of the transformerless type shall have the live parts of the inner structure isolated from the case or frame by an isolating condenser or other approved means, which shall not be capable of passing a current exceeding 5 milliamperes to case or frame when the full rated voltage is applied in the normal manner of operation."

The main trouble encountered with the other type of a.c./d.c. receiver is instability due to impedance between the earthed bus and the chassis proper, the impedance being made up of the reactance of the by-pass capacitor and the inductance of its leads.   This impedance can be reduced to a low value by making up a unit consisting of an inductor wound on the body of the capacitor, the two being connected in series and resonated at the intermediate frequency, at which most amplification occurs.

Many cases of instability need individual treatment and it often becomes necessary to return particular circuits (e.g. the cathode by-pass of the i-f amplifier) to the chassis through a series tuned circuit instead of to the negative bus.

To avoid modulation and other types of hum it may be helpful to use back-bias for amplifying valves so that the cathodes can be directly grounded.   This avoids the possibility of hum voltages appearing between cathode and ground.

Where there is any possibility of interconnection of two or more a.c./d.c. units, such as for instance a radio tuner and a separate public address amplifier, it is essential that the mains be isolated from the chassis.

SECTION 7 : BATTERY OPERATED RECEIVERS
(i) General features (ii) Vibrator-operated receivers (iii) Characteristics of dry batteries.

(i) General features
Most battery operated receivers are used in locations where signal strengths are low, and the main requirements are therefore high sensitivity and low noise.   The cost of power obtained from batteries is high and every effort is made in the design of battery receivers to keep the current drain to a minimum.   This is done by reducing the a-f power output considerably from that available in an a.c. receiver and by using high Q aerial and i-f coils so that high gain can be obtained without operating the valves at their maximum ratings.   Many receivers have an economy switch which reduces battery drain by decreasing the screen voltage on the i-f and output valves or by other means.   The reduced power output and sensitivity are still adequate for most uses.

The mutual conductance of battery valves is lower than that of corresponding a.c. types to such an extent that although the smallest a.c. receivers in general use have three valves and a rectifier, the smallest battery sets need four amplifying valves to obtain similar sensitivities. The four types are a converter, an i-f amplifier, an a-f amplifier and an output valve. The i-f output is severely limited by the low plate voltage on the last i-f valve—about 84 volts with fresh batteries in a typical modern receiver—and an a-f amplifier is needed to obtain sufficient a-f voltage to drive the output valve.

The more general type of battery receiver has five valves and uses the additional valve as a r-f or second i-f amplifier. On the broadcast band sensitivities of the order of 1 μV are readily available in each case, and if a high gain aerial coil is used there is not a great deal of difference in signal-to-noise ratio. On the short-wave band the r-f stage receiver is noticeably quieter but the average sensitivity at 6 Mc/s may be of the order of 25 μV, against perhaps 10 μV for the receiver with an additional i-f stage.

Because of the low i-f plate voltage the signal handling capacity of battery operated receivers is limited, and taken in conjunction with the requirement for high initial sensitivity, this means that a normal battery receiver overloads with a comparatively small r-f signal applied. Nevertheless the conditions of use are such that large inputs are not usual, and a receiver which will not distort seriously with an input of 0.1 volt will give satisfactory service in almost all locations.

There is usually only one diode available in a battery receiver, and this severely restricts the a.v.c. design as it makes delayed a.v.c. impossible. The most satisfactory compromise is probably to apply a fraction (say one third) of the developed a.v.c. voltage to the first and last controlled stages (unless the first valve is a r-f amplifier which may take full a.v.c. without increased signal-to-noise ratio) and full a.v.c. to the intermediate one. The small amount of a.v.c. on the input stage minimizes the increased noise from the first valve as a.v.c. is applied and on the last i-f stage it allows larger signals to be handled without overload. The middle stage has no effect on the noise, handles only small signals, and consequently can do most of the gain controlling.

A.V.C. design affects high tension current consumption and if too small a fraction of the developed a.v.c. voltage is used, current will be wasted when a local station is tuned.

Bias for the output valve is usually obtained from a back-bias resistor in dry battery operated receivers. Receivers have been made in which the voltage developed across the oscillator grid leak has been used for this purpose (with suitable decoupling) but if the A battery voltage falls more quickly than that of the B battery, the output valve is operated in an underbiased condition and draws excessive B current.

Separate C batteries have also been used, but the difficulty of discharging the C battery at the proper rate has led to the popularity of back biasing, in which the voltage remains approximately optimum throughout the B battery life.

When a back-bias resistor is used it should be effectively by-passed to audio frequencies, otherwise a-f voltages developed across the resistor are applied to the plate circuits of the i-f and perhaps oscillator valves. This may cause instability as a receiver is being tuned to a station.

Another advantage of back biasing is that if B+ is short circuited to the valve filaments, the bias resistor restricts the current to an amount which will not damage the valves. With a 500 ohm resistor and a 90 volt supply the current is 180 mA, whereas even a four valve receiver normally draws 250 mA from its filament battery.

When a battery receiver is turned off, the B supply voltage does not disappear as in the case of an a.c. operated receiver. The filament voltage being turned off stops the valves from passing current but small leakage currents (for instance, through condensers) have merely to flow through another series leakage—that of the B supply switch. Particularly in humid conditions, the resistance between switch contacts may be only of the order of tens of megohms and the uninterrupted current which flows can lead to electrolysis of fine wires and consequent open circuits.

Unless precautions are taken, the fine wire of the output transformer primary is particularly susceptible to this trouble, leakage occurring between the primary (connected to B+) and the core (connected to chassis), and interwound coils with one wire earthed and the other connected to the B supply also give trouble.

Output transformer electrolysis can be prevented by connecting the primary winding to the core and isolating the assembly from the chassis, for instance in a pitch filled container.

Interwound coils have a satisfactory life if they are baked to remove moisture and then impregnated and flash-dipped in a moisture resisting wax or varnish.

In designing battery receivers it is important to check power output, sensitivity and oscillator grid current with reduced voltages and with many valves. A large number of valves is required because no tests are made by valve manufacturers at voltages as low as those commonly employed by receiver designers as the " end of life " point.

For battery converter problems see Sect. 3(vi).

(ii) Vibrator-operated receivers

Chapter 32 deals with the design of vibrator operated power supplies. Receivers to operate with such supplies have some associated problems, mostly concerned with the elimination of r-f and a-f interference from the vibrator unit.

Separation of all vibrator circuits from the receiver is essential if hash is to be satisfactorily eliminated. If for instance various parts of the vibrator circuit are earthed at different places in a receiver chassis, interference can occur which no amount of filtering will remove. The correct method is to earth all vibrator by-passes and earth returns at one point and connect the lead from the battery to the same point.

The same principles apply to a-f interference. It is assumed that adequate filtering is used in the B supply so that interference from that source is negligible. Even under these conditions, vibrator noise can be troublesome and the first source is the battery itself. Although the internal impedance of a battery is very low it is not negligible, and the pulses of current drawn by the vibrator set up voltages across it which are applied to the filament of the a-f amplifier and result in noise from the loudspeaker.

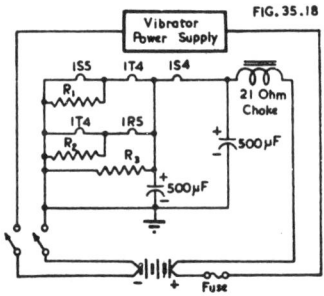

FIG. 35.18

Fig. 35.18.   Filament circuit of vibrator-operated receiver.

Any output valve with 100 mA filament current could be used in place of the 1S4 (e.g. 3Q4, 3S4 or 3V4 with parallel filament connections).

A cure is an iron-cored choke between the positive battery terminal and the filaments of the valves, which are normally connected in series-parallel. A large (500 μF) condenser is usually required in addition to the choke, and receivers with high a-f gain may need two condensers. A typical circuit is shown in Fig. 35.18. With the reduced filament drain of 1.4 volt valves it is possible in some cases to do away with the iron-cored choke and use a resistor. When used with a 6.0 volt battery the filament voltage for each valve should be 1.3 volts so that a dropping resistor of 21 ohms would be required in place of the iron cored choke shown in Fig. 35.18. Where two large condensers must be used in any case the resistor usually has sufficient impedance.

Even when the filtering (due to the electrolytic condensers and choke or resistor) is adequate, serious buzz can still occur. In Fig. 35.18 two leads are taken from each battery terminal, and the vibrator and filament circuits are entirely separate. However it sometimes happens that two switches cannot be used and a common switch section and negative battery lead must be used. This will introduce some buzz although the degree will depend on the a-f sensitivity and the value of the mutual impedance. An undesirable feature is that the interference may vary each time the switch is operated and may become progressively worse with age as the contact resistance of the switch increases.

If the a-f gain is high, even the common impedance of a fuse may cause trouble and in some cases only the vibrator circuit is fused for this reason.

The electrolytic condensers are also useful in preventing coupling between the filament circuits of the various valves. However high gain receivers usually need additional r-f by-passes on the filament string to prevent regeneration at the intermediate frequency or to keep i-f voltages from the a-f end of the receiver.

One undesirable feature of the circuit of Fig. 35.18 is that if any one of the 1R5, 1S5 or 1T4 valves is removed from its socket, or if one filament becomes open circuited, two other valves will have excessive filament voltage applied. The additional voltage is normally not enough to open circuit the other filaments but if the receiver is left unattended in this condition the valves will deteriorate rapidly.

The resistors R_1, R_2 and R_3 equalize the filament voltages on the different valves. Since the filament string requires 100 mA and the cathode current of the 3S4 alone is 8.8 mA under typical operating conditions, some adjustment to the voltage across the filaments (in which the cathode currents flow) is necessary. However calculation of the values required for the various valves is not straightforward since the cathode current of each valve has alternative paths of different impedance. For instance in Fig. 35.18 the 1R5 cathode current will flow partly in the negative filament lead and partly in the positive, this latter current splitting into one component which flows through the 1S5 and another which flows through the 3S4. As a result the resistor values are most readily determined experimentally, and with average valves in a receiver and decade boxes connected where shunting resistors are needed, a few minutes' manipulation of the decade boxes will determine the correct values. A.V.C. application will affect plate currents and hence filament voltages, and the resistor values should be decided with an average input signal. By making slight compromises it is often possible to do without one or more resistors, but if this is done conditions should be checked with freshly charged and discharged batteries as well as at the rated battery voltage.

A.V.C. design is complicated in vibrator receivers with series-parallel filament circuits, but satisfactory solutions can usually be obtained by means of a.v.c. voltage dividers returned to different potentials. For instance in Fig. 35.18, which might be the filament circuit of a receiver with two i-f amplifiers, the diode load would be returned to the 1S5 negative filament, i.e. to ground, and a.v.c. would be taken directly to the grid of the 1T4 acting as the first i-f amplifier. This is possible because the filaments are at the same potential. However the 1R5 and second i-f 1T4 have their filaments more positive and would be biased 1.3 volts negative if connected directly to the diode load. A solution would be to connect a two megohm resistor from the top of the diode load to the 1R5 and 1T4 grids, and another two megohm resistor from the grids to the positive side of the 1R5 filament. This would remove the bias from the 1R5 and 1T4 and reduce the amount of a.v.c. to these valves which would be desirable—see subsection (i) of this section.

The elimination of hash from a vibrator operated receiver usually involves work on the power unit (Chapter 32) but some precautions in the receiver may be necessary. A frequent source of interference which is difficult to trace is hash picked up by the converter valve. A very small amount of interference in the B supply may modulate the oscillator and be fed into the i-f amplifier with the local oscillator injection, or interference on the converter filament may be troublesome. In each case the interference receives the total receiver amplification. A separate r-f choke is often used for the converter filament to eliminate this trouble.

When tracing hash, a convenient method is to connect a shielded wire, with one end bared, to the receiver's aerial terminal through a capacitor. The bared end of the wire is used as a probe and when in contact with various points in the receiver rcuit indicates whether interference is present. The first points to check, of course, are the incoming leads from the vibrator unit to the receiver and in general these should be filtered until they are quite " cold ".

To be sure of trouble-free production runs it is advisable to work on hash elimination until, with an aerial wrapped around the battery leads and battery, no interference can be heard at any frequency with the receiver operated at maximum gain in a screened room.

Mechanical noise and vibration from the vibrator also present a problem. The high frequency noise components can readily be minimized by felt or other sound-absorbing material, and even the metal shielding box in which vibrators are often used may be sufficient. However noise at the fundamental frequency of the vibrator is more difficult to cure and much trouble can be avoided by using a very flexibly mounted vibrator socket. Each new receiver usually presents different problems in this respect.

(iii) Characteristics of dry batteries*

A knowledge of dry battery characteristics is essential in designing receivers which are to use the batteries to the best advantage. It is also helpful in obtaining maximum battery life with a particular receiver.

The decrease in battery life caused by a receiver design which gives unsatisfactory performance—or none—when the battery voltage falls below certain levels is best illustrated in Fig. 35.19. If the average B current of a receiver is taken as 15 mA it will be seen that the useful life of the battery is increased from 1.16 to 1.49 ampere hours, an increase of 28%, by designing the receiver so that it will operate satisfactorily when the output of the 90 volt B battery has fallen to 60 volts, instead of becoming unusable when the output is 68 volts.

It will be noticed that as the current drain in Fig. 35.19 is reduced, the output rises to a maximum and then falls. This is due to the fact that the batteries were discharged for 2 hours each day so that with a 10 mA drain for instance the test lasted for 175 days. Tests with smaller drains take longer and the effects of shelf life become more important than the decreased drain.

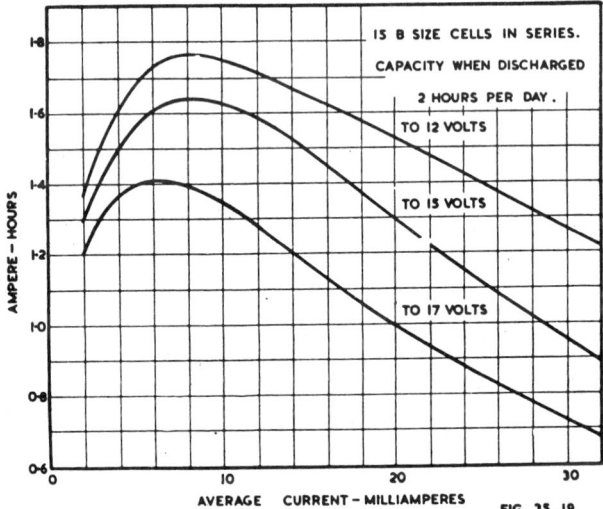

Fig. 35.19. *Ampere-hour output versus discharge current for various end-point voltages (Ref. 25).*

*Information taken from Ref. 25.

Fig. 35.20 shows the importance of reducing battery drain to a minimum or alternatively of specifying batteries of as great a capacity as possible.   As the drain is reduced a progressively greater amount of the battery output is made available at a high voltage, and as an extreme example, a 2.9 mA drain on the cell will give 83% of the battery's output (assuming that the cell can be used until discharged to 0.8 volt) at a voltage greater than 1.3, whereas a 187 mA drain gives only 16% of the battery's output at 1.3 volts or more.   The curves of Fig. 35.20 were taken on continuous discharge, and with operation of say 4 hours a day the effects of shelf life would modify the figures obtained.

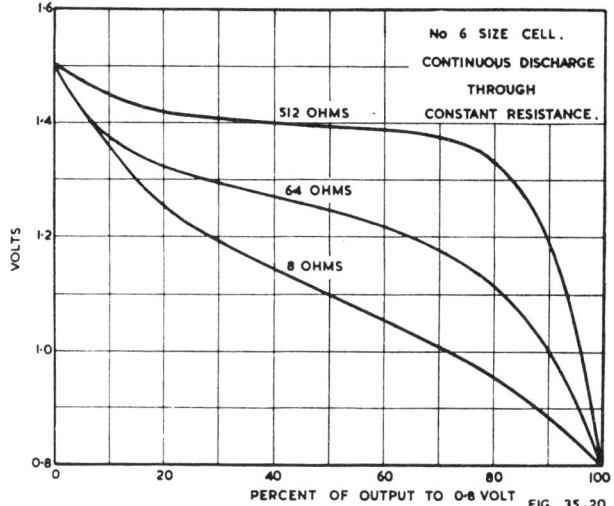

Fig. 35.20.   Working voltage of No. 6 size cells during discharge through various resistances (Ref. 25).

The main interest of Fig. 35.21 is to the user of the receiver, as it shows the effects of heavy battery usage in decreasing battery output.   For a B drain of 15 mA and 8 hours of use each day, each cell will give 1.13 ampere-hours output, but with 2 hours daily usage the output rises to 1.48 ampere-hours.   Thus the batteries will last for about 50 days with 2 hours of use, but less than 10 days with 8 hours of use.   The effects of shelf life in Fig. 35.21 are clearly indicated at low current drain and with decreasing hours of discharge per day.

An aspect of dry battery performance which is of particular importance to receiver designers is the increase of internal resistance of each cell as it becomes discharged. The resistance of a new cell can be ignored for radio purposes, as it varies from about one third of an ohm for small B battery cells, to one fiftieth of an ohm for large A battery cells.   However as the battery ages, an impedance of the order of 20 ohms per cell may develop and for a 90 volt battery this represents a series impedance of 1200 ohms.

Such impedances cause coupling between circuits, and in the case of an A battery may cause instability due to coupling between different i-f stages or between i-f and a-f stages.   The remedy is an adequate by-pass (0.5 μF may be necessary) across the filament circuit, and the necessity for this should be tested with a discharged battery during design.   Impedance in the B supply circuit can cause trouble between almost any pair of stages (including the oscillator) in the receiver.   An electrolytic condenser in parallel with a r-f by-pass (0.1 μF) is commonly used to avoid this trouble.

The only completely satisfactory method of testing a receiver for performance with discharged batteries is to use a set of batteries which has been used normally until the voltage has fallen to the required level.   Accelerated rates of discharge do not

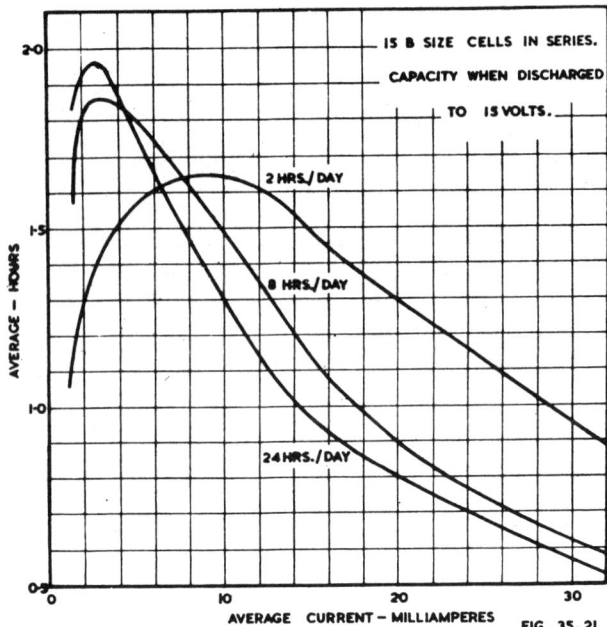

*Fig. 35.21.   Ampere-hours output versus discharge current for various operating cycles
(Ref. 25).*

produce such a high internal resistance.   Noise will sometimes be encountered with
a discharged battery unless a large by-pass is connected in parallel with it.

The effect of temperature on dry cells is very marked.   At high temperatures the
deterioration of unused cells is rapid and tests have shown that batteries stored for
one year at 104°F. have deteriorated more than similar batteries stored for five years
at 48°F.   Small cells have a shorter shelf life than large ones stored under the same
conditions.

The reasons for the increased rate of deterioration at high temperatures are increased moisture loss from the electrolyte, and increased chemical activity in the
battery.   This latter effect is beneficial when the battery is actually in use and a
battery discharged in a comparatively short time at 100°F. has been found to give
140% of the output of a similar battery at 70°F.   At 40°F. the output fell to 48%
of the output at 70°F.

When testing batteries it is important to realise that there is no relationship between the current delivered by a dry cell on a short circuit amperage test and the
service capacity of the cell.   On the other hand the working voltage of a dry cell does
decline progressively as the cell becomes exhausted, so that a voltage test under normal
load is a good indication of a battery's condition.   The voltmeter used should have a
resistance of not less than 100 ohms per volt for single cells or 1000 ohms per volt for
B batteries.   The current drawn by lower sensitivity meters may lead to a lower
voltage indication than actually occurs in use.

In some types of portable receivers which can be used with dry batteries or an a.c.
power supply the dry batteries can be charged during a.c. operation or while the receiver is not in use.   This is claimed to increase the service obtainable from the dry
batteries although the degree of improvement depends on the conditions of use.   It
is important that the charging current should not be excessive.   Further information
is available in Refs. 56, 57, 69, 70, 71.

SECTION 8 : CAR RADIO

(i) Interference suppression   (ii) Circuit considerations   (iii) Valve operating conditions.

(i) Interference suppression

The main problem in car-radio design is the suppression of interference from the car and from the power supply of the receiver. Ignition interference is the more troublesome, and while there are methods of suppression which apply in all cases any particular installation is liable to present problems of its own.

One theory as to the mechanism by which interference is produced is that each spark causes oscillatory currents in the ignition leads covering a wide band of frequencies. The spectrum is propagated from the ignition lead which acts as an aerial, and the receiver amplifies those frequencies to which it is tuned. See Ref. 10 for a comprehensive survey of reports on the nature, measurements and suppression of ignition interference.

Another possible mechanism has been described in Ref. 1, although objections to it are brought forward in Ref. 10C. This conception of interference generation is that the radiated field is an impulse which is short compared with the period of the carrier frequency up to frequencies of hundreds of megacycles. Each burst of interference heard is thus the impulse response of the receiver and its aerial, so that the interference is more dependent on the bandwidth of a receiver than on the frequency to which it is tuned, although of course the signal received will vary in amplitude with large variations in frequency.

In addition to ignition noise, interference is caused by the brushes on the generator but this is readily removed by by-passing. Special suppression capacitors are available of 0.5 μF capacitance, 200 volt working, encased in a metal can to which one side of the capacitor is connected.

Methods of dealing with the majority of straightforward interference suppression problems are given below (Ref. 60).

(a) Cut the high tension lead connecting the ignition coil to the centre of the distributor head as close as possible to the distributor head and screw the two ends of the leads into a resistor suppressor. Suitable suppressors consisting of a carbon resistor of 15 000 ohms with facilities for attaching the two leads are available commercially.

(b) Connect a suppression capacitor to the point where the receiver low tension cable is connected to the car's low tension system, with the metal case clamped to the chassis.

(c) Clamp the metal case of another suppression capacitor to the generator housing and connect the lead to the generator armature-terminal. The capacitor must not be connected to the generator field-terminal. In American cars the correct terminal is usually the larger one.

In most English cars the armature terminal is not accessible on the generator and the suppression capacitor should be connected in such cases to the " D " terminal on the voltage distribution channel.

If trouble is still experienced the following points should be investigated.

(d) See that the distributor contacts and spark plug points are clean, in good condition and correctly adjusted. Replacements may be needed.

(e) If the engine is rubber mounted a bond may be necessary between the engine block and the chassis. This bond should be a piece of flexible copper braid not less than ⅛ in. wide and as short as possible, although allowing for engine movement.

(f) See that the low tension leads do not come close to any of the high tension wiring. In some cars it may be necessary to remove the low tension wires from channels provided for the spark plug leads.

(g) Make sure that there is a good connection between the ignition coil can and the engine block.

(h) Bond all oil pipes, bowden cables, etc., to the bulkhead as they pass through to the driving compartment.

(i) Try the effect of bonding various parts of the bodywork to the bulkhead or engine block with ⅛ in. copper braid. It sometimes happens that a large section of metal, for example a mudguard, is insulated by rubber or fabric beading, or even by paint, from the rest of the body. Bonding of such sections may eliminate interference.

(j) By-pass separately the various electrical components of the car, such as the horn, ignition switch, windscreen wiper motor, petrol pump, head lights, dome light, parking lights, petrol gauge, electric clock, etc.

In fitting the various units of the radio itself it is necessary to remove some paint from the metal to make sure that the unit is electrically earthed.

Occasionally trouble is experienced from wheel and tyre static. Graphite grease in the wheel bearings, may be sufficient to effect a cure, but in some cases spring contacts bearing on the wheel hub have been used.

(ii) Circuit considerations

The main requirements for a car radio are high sensitivity, high signal-to-noise ratio, good a.v.c., high a-f output, compactness, the ability to withstand vibration, and mechanical flexibility for installation purposes.

The first four items have almost standardized the design into the form of a receiver with r-f, converter, i-f, a-f and output stages. Such a combination allows the receiver sensitivity to be limited only by the noise in the first stage—which is desirable. The high over-all sensitivity (1 μV input for 0.5 watt output) is not uncommon) causes high a.v.c. voltages to be developed with comparatively small inputs so that strong a.v.c. action is available even when signals fade to very low levels. The a.v.c. diode load is often returned to a source of bias for the r-f and i-f amplifiers so that a delay of two or three volts is obtained and this flattens the a.v.c. characteristic even more, after the delay has been overcome. The a-f sensitivity is adequate to allow full output.

When full a.v.c. is applied to the r-f stage a check should be made from Scroggie's curves—Sect. 3(i) of this chapter—to see that the noise from the first valve does not become comparable in size with that from the input circuit and so affect the signal-to-noise ratio.

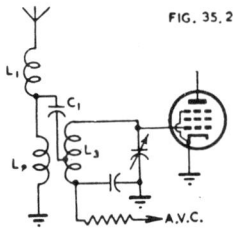

FIG. 35.22

Fig. 35.22.   Car-radio aerial   input   circuit.

The design of the aerial input circuit is very important in a car radio, because of the small effective height and physical size of the aerial itself. Different aerials may be used with the receiver, but telescopic types with a capacitance of about 50 $\mu\mu$F are commonly assumed. One of the most popular circuits is shown in Fig. 35.22. L_1 is a choke consisting of a few spaced turns of thick wire, resonant at about 40 Mc/s. It is used to minimize incoming ignition interference without seriously affecting the desired signals. L_2 is the aerial coil primary, which is arranged to resonate with the aerial and the rest of the circuit outside the low frequency end of the required band. However the resonance is kept as close to the band as possible without introducing too much mistracking between the aerial and r-f circuits, in order to obtain the maximum possible gain at the low frequency end of the band.

The capacitor C_1 is from 50 to 100 $\mu\mu$F in a typical case, and provides capacitance coupling to improve the gain at the high frequency end of the band. L_3 is tapped at from one quarter to one third of the total number of turns from the cold end. Coupling between L_2 and L_3 usually approaches 20%.   The Q of L_3 must be as high as possible and iron cores are commonly used. Q values in air as high as 250 at the maximum point in the band have been used, and a gain of from 8 to 10 is obtainable over the band with such a coil mounted in the chassis.   Gain is measured with the output of the signal generator connected through a 50 $\mu\mu$F capacitor to the aerial end of the shielded input cable.   Noise figures obtainable are of the order of equal noise and power output when the input signal is from 0.3 to 1.0 μV, 30% modulated.

A series trimming condenser (say 50-350 $\mu\mu$F) is sometimes used between the spark coil and the aerial primary so that the aerial coil can still be tracked with the r-f coil even when a high capacitance aerial is connected to the receiver.

Power supplies for car radios are vibrator operated, using either synchronous or non-synchronous types.   The non-synchronous system has the advantage of keeping a constant output polarity when the input polarity is reversed, so that the lack of uniformity between car manufacturers as to positive or negative grounding of the battery becomes unimportant.   Non-synchronous systems also have fewer troubles from vibrator hash in the receiver and are generally assumed to be more reliable in operation but of course they are more expensive, since a rectifier must be provided, and consume more current if a hot cathode type of rectifier is used.

In the elimination of interference either from the vibrator or from ignition noise in a car radio, " spark plates " are often used.   These are by-pass capacitors using the chassis as one plate, a thin sheet of insulating material, preferably mica, as the dielectric and tin-plate or brass as the other plate, the assembly being eyeletted together with suitable insulating washers.   Because components are soldered directly to the top plate and the bottom plate is the chassis, the series inductance of such a capacitor is very small.   It has been recommended that incoming leads should be soldered to one end of a spark plate and outgoing leads to the other end.   Capacitances between 10 and 200 $\mu\mu$F are commonly made up in this way.

Because of the intense ignition interference field it is often necessary to filter all leads coming into the receiver.   Tone control leads and even loud speaker leads may need by-passes or series chokes to eliminate interference.

The effects of vibration must be considered in car radio design.   Reliable air dielectric trimmers are more stable than compression mica types, but need to be sealed after alignment.   Large paper capacitors need to be held by means other than their own leads, and smaller components must be mounted so that they cannot vibrate on their leads.   Older types of valves, such as G types and others of similar size can give trouble through falling out of their sockets under the influence of vibration, particularly when mounted horizontally.   For this reason and for the extreme compactness which they allow, miniature valves are well suited for car radio designs.

(iii) Valve operating conditions

Valves in car radios are subjected to heater voltages outside both the top and bottom limits specified by valve manufacturers.   Extremes of perhaps 8.0 and 4.5 volts may be encountered with the battery on charge and fully charged, and off charge and discharged.

Excessively high heater voltages reduce valve life by evaporating the coating from the cathode, and care should be taken to see that plate or screen dissipations do not exceed valve manufacturers' ratings under any anticipated working condition, particularly as the ambient temperature inside the car radio is likely to be very high at times.

When heaters are operated at excessively high voltages it is possible for control grids to rise to a temperature at which grid emission takes place.   To reduce the effects of this grid emission current, it is advisable to use the lowest convenient values for output valve grid leaks and a.v.c. series resistors in car radios.

When the supply voltage is very low, the receiver loses sensitivity and power output but remains usable so long as the oscillator operates.   To guard against oscillator stoppage a new design of receiver should be checked with a large number of valves including some with low oscillator grid current under normal operating conditions.

Since valves are only tested by manufacturers to the recommended voltage limits, there is no guarantee of uniformity of operation between valves at voltages outside those limits

SECTION 9 : MISCELLANEOUS FEATURES

(i) Spurious responses   (ii) Reduction of interference   (iii) Contact potential biasing (iv) Fuses   (v) Tropic proofing   (vi) Parasitic oscillations   (vii) Printed circuits (viii) Other miscellaneous features.

(i) Spurious responses

With a superheterodyne receiver, several types of signal can be received which do not originate in stations broadcasting on the frequency to which the receiver is tuned.   The signals are not necessarily objectionable of themselves, but owing to the congested condition of the broadcasting band, almost any spurious signal will interfere with a true signal.   These spurious responses are due to several causes which are listed below.

(a) Harmonics of a station operating at a lower frequency.
(b) Two stations broadcasting on frequencies separated by the intermediate frequency of the receiver.
(c) A station broadcasting on the image frequency of a desired station.
(d) A combination of local oscillator harmonics with other signals or harmonics.
(e) Feedback of intermediate frequency harmonics.
(f) A station broadcasting on the intermediate frequency of the receiver.

Interference of type (a) will be received by any receiver of sufficient sensitivity and cannot be minimized in the design of a receiver.

When type (b) interference is experienced, the two stations will be heard together over two bands of frequencies centred on the frequencies of each of the stations. The interference continues even when the local oscillator is stopped.   Since it is necessary for the two signals to be present at the grid of the frequency changer for the interference to be troublesome, selectivity before this point is required to eliminate it.   Alternatively the intermediate frequency of the receiver can be altered to a frequency at least 10 Kc/s removed from the difference frequency.

Type (c) interference can be cured by adequate signal frequency selectivity, and is the main reason for this selectivity.   Particularly in the short-wave range of dual wave receivers is trouble experienced (although not always recognised) from this source, and it is so prevalent that the signal frequency selectivity of a receiver is always expressed as its image ratio.

Methods of calculating image ratio are given in Chapter 23 Sect. 4(ii) for coils without capacitance coupling.   To obtain the full calculated rejection, particularly at the high frequency end of a band, it is necessary to reduce to a minimum all couplings other than that by mutual inductance.   Separation of circuits and earthing at one point for each circuit will be found beneficial when high gain is used.   Capacitance coupling must be particularly avoided.   Ref. 21 (pp. 225-237) gives circuits which increase image rejection.

There are many possibilities of type (d) interference but it is not a major problem in most receiver installations.   Nevertheless it is advisable to restrict oscillator grid current to a minimum, consistent with other requirements, to minimize this source of interference.

Feedback of intermediate frequency harmonics can occur through B supply, a.v.c. or other wiring, or by means of voltages induced into the converter input circuit from the second detector, or any part of the a-f amplifier.   The feedback may be sufficient to cause actual instability or merely an annoying heterodyne on stations adjacent in frequency to the harmonics of the i-f.   Shielding and perhaps decoupling or an r-f by-pass in the a-f circuit readily remove the trouble.

Interference from a signal on or near the intermediate frequency continues when the local oscillator is stopped and becomes progressively worse as the receiver tuning approaches the intermediate frequency.   It is seldom encountered in superheterodynes with a tuned r-f amplifier, but receivers with an untuned r-f amplifier or with none usually have very poor i-f rejection at frequencies lower than 600 Kc/s.

A method of plotting interference signals is given in Ref. 21, Part 1, page 201. An example of the estimation of the interference to be expected in a given set of conditions will be found in Ref. 53.

(ii) Reduction of interference

There are two main methods of minimizing the effects of external noise on radio reception.   The first is by the use of noise limiters in the receiver (Chapter 16 Sect. 7 and Chapter 27 Sect. 5) and this method is equally applicable whether the noise is of man-made or natural origin.   The second method consists of using filters and noise-reducing aerials and is only applicable to noise from man-made sources.

Most mains operated receivers receive appreciable r-f inputs from the supply lines, and when noise is troublesome the ratio of signal to noise is often much lower in the mains than in an aerial.   This occurs when the source of interference is connected to the power supply and induces interference into the mains wiring.   In such a case filtering of the mains leads can give a substantial improvement in signal-to-noise ratio.

An electrostatic shield between primary and secondary windings of the power transformer is the most common method of treatment.   Methods of incorporating such a shield in the power transformer are described in Chapter 5 page 233.   The alternative method (b) of winding the earthed low voltage filament between primary and secondary, due to its inductance, is probably less effective than the normal shield (a).

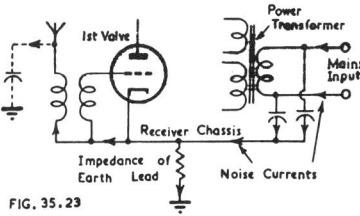

Fig. 35.23.   Interference from power-supply induces noise into input circuit of receiver.

By-passing of the incoming mains leads to the chassis, with or without series inductors, is often used to reduce interference.   Fig. 35.23 shows that this method is of limited use, depending on the impedance between the receiver chassis and ground. Noise currents represented by arrows flow from mains to chassis and from chassis to ground, thus setting up a noise voltage on the chassis relative to ground.   This voltage causes another noise current to flow through the primary of the aerial coil and the capacitance of the aerial to ground, and thus a noise voltage is induced in the aerial coil secondary and applied to the first valve of the receiver.

The fact that interference stops when the aerial is removed from the receiver is not proof that the noise is picked up by the aerial.   If the aerial were removed in Fig. 35.23 the interference would stop because the path to ground through the aerial primary would be removed, but the noise source is the supply mains.

Figure 35.23 indicates how improvements in noise reduction can be made.   Firstly, if the earth lead for the mains by-passes is separated from the receiver earth, the noise potential between chassis and ground is eliminated.   The separation needs to be more than electrical, as if two wires were close to each other the noise currents in the filter earth would induce noise currents in the receiver earth.

Secondly the incoming noise currents can be considerably reduced by using a r-f choke in series with the mains leads before the first by-pass to increase the impedance of the path.   This will reduce the noise voltage between chassis and ground, but increase the noise voltage between mains leads and ground, and the capacitance between mains and aerial leads may lead to increased noise from this source.

A third possibility is the use of a balanced aerial coil primary connected to the ends of a symmetrical lead-in from the aerial. When the two ends of the primary are balanced with respect to capacitance to chassis (a small trimmer can be used on either side of the primary if found necessary to give the greatest noise reduction) then noise currents flowing from receiver chassis to ground through the aerial capacitance do not give rise to noise voltages in the aerial coil secondary or the remainder of the receiver. An electrostatic shield can be used between primary and secondary as an added precaution against electrostatic asymmetry between primary and secondary windings.

The symmetrical lead-in can be terminated at the aerial end in another transformer coupling it to the aerial, in a dipole, or aerial and counterpoise or the lead-in may be merely a two wire feeder with one end extended to form the aerial. In any of these cases additional noise reduction can be expected from the fact that the aerial proper can be placed at some distance from the power supply mains or any other known source of interference, e.g., it can be removed as far as possible from, and placed at right angles to, tram lines.

Before proceeding with elaborate noise reduction methods it is advisable of course to make sure that the receiver itself is not picking up noise directly. In strong fields large voltages are induced into grid leads and under-chassis wiring, and complete shielding of the receiver proper may reduce interference without other assistance, or it may be essential if the full benefit of more elaborate methods is to be obtained.

In-built aerials on mains operated receivers usually work on the principle of providing an earth return for signals coming from the mains, in some cases as shown in Fig. 35.23, and in other cases with a capacitance between mains and aerial terminal, and the receiver chassis providing, through capacitance or direct connection, the return to ground. In the first case the requirement for the " aerial " is merely the largest possible capacitance to ground, and a ground connection to the chassis should reduce the signal input. In the second case an earth connection should improve signal strength as the current through the coil will be increased.

Even when a mains-operated receiver is equipped with an in-built loop aerial, some of the signal input is due to the capacitance effect of the loop as a whole. A mains filter may improve the signal-to-noise ratio in such a case, but with the other two types of in-built aerial a mains filter will reduce the signal and noise by the same amount leaving the ratio unchanged.

(iii) Contact potential biasing

Electrochemical activity between the electrodes of a valve produces potential differences without the application of external e.m.f.'s. The effect is most noticeable on grids and diode plates since their " contact potential " may be of the same order as potentials applied from other sources. The effective bias depends on the surfaces of the two electrodes and the impedance connected between the respective valve pins. The control grid contact potential will vary with different types of valves, with different valves of the same type and with age in a particular valve.

In practice, control grid and diode contact potentials are found in indirectly-heated valves to vary between -0.1 and -1.1 volts with different types of valves, and the contact potential of a particular valve may vary as much as 0.4 volt during life. Because of this, the contact potential of valves has a restricted application as a source of bias, but it can be satisfactory if its limitations are borne in mind.

The two types of contact potential bias are " grid leak bias " of a-f amplifiers and " diode bias " of i-f and other amplifiers. Some valve types which are likely to be used with grid leak bias e.g. Radiotron types, 6B6-G, 6SQ7-GT, and 6AV6, are given a 100% production test for grid leak bias by Amalgamated Wireless Valve Co., under conditions sufficiently severe to ensure that any change of characteristics during valve life will not affect their performance. These types can be recommended for this type of service provided that the plate current is restricted to a maximum of 1 mA, the grid leak is 5 megohms or larger (2 megohms can be used with a small signal input but distortion should be checked) and the output required is not greater than 25 V r.m.s. Under these conditions less distortion can be expected from grid leak bias than from cathode bias (Ref. 43).

Pentode valves have more critical bias requirements when used as low distortion resistance coupled a-f amplifiers and because of this are not so suitable for grid leak biasing. However when used with a series screen resistor of high value and comparatively small a-f inputs they can be satisfactory, and remove some of the hum problems from fairly low-level a-f amplifiers since there is no bias supply to be filtered, and the cathode is at ground potential so that heater-cathode conductance is not troublesome.

Diode biasing finds its main application in minimum cost 5 valve (4 valve plus rectifier) receivers. A.V.C. is applied to the i-f and converter valves directly from the diode plate of the second detector, the diode load and the cathode of the i-f amplifier being returned to ground. Under these conditions the contact potential of the diode is the only bias applied to the i-f amplifier, and its screen voltage is adjusted to keep plate and screen dissipations within tolerances. The converter may need additional bias, obtained from a cathode resistor and by-pass, depending on the type used.

A trouble which occasionally occurs with this circuit is that the diode bias is too small, the i-f valve draws grid current, and its reduced input impedance damps the i-f transformer sufficiently to cause a serious drop in sensitivity with small inputs. As the input is increased, the additional i-f bias from the a.v.c. line restores the sensitivity to normal so that over a small range of input signals the receiver output increases much more rapidly than the input. A receiver with this defect may show poor sensitivity for 50 mW output, but normal sensitivity for the larger signal required to give 0.5 watt output as recommended by the I.R.E. " Standards on Radio Receivers, 1948."

Additional methods of using contact potential for bias are given in Ref. 44.

(iv) Fuses

Commercial A-M receivers rarely incorporate a fuse owing to the difficulties involved in providing a satisfactory one. The peak current in the primary of the power transformer when a receiver is switched on may be twenty times the average current for a time not exceeding 0.01 sec. This occurs during normal operation, but on the other hand a fault resulting in twice the normal primary current over a period of time may lead to transformer breakdown.

Since a normal cartridge fuse will blow within 0.01 sec. with five to ten times its rated current applied, a fuse of four times the rated current of a receiver would be the smallest value which could be used. Such a fuse would probably not fail immediately, even with a complete short circuit on the whole high tension winding of a receiver, and would be almost useless.

Two types of fuse which can be used satisfactorily under such conditions have been described (Ref. 30). The first consists of the normal glass cartridge and ends, containing in series a small spring soldered under tension to a fine manganin resistance wire by means of a blob of low temperature solder (e.g. Wood's metal). A prolonged overload produces enough heat in the resistance wire to melt the solder and allow the fuse to clear, but the thermal inertia of the mass of solder protects the fuse against short duration surges. With a severe overload the resistance wire clears almost instantaneously. An advantage of this type of fuse is the low temperature required for operation—of the order of $100°C$ compared with perhaps $900°C$ for a normal type of fuse.

Another fuse has been made consisting of a high melting point nickel wire with blobs of magnesium powder in a binding varnish supported on it. Once again the thermal inertia of the blob allows short duration surges to occur without damage, but a prolonged overload raises the magnesium powder to ignition point (about 500°C) and the fuse clears.

A different type of protection has been used in other types of receivers. This consists of an insulated strip of metal with high heat conductivity which is included in the power transformer during winding. One end of the strip projects outside the winding and to this a small stirrup is soldered with very low melting point solder. A spring engages in the stirrup under tension and the primary current to the whole receiver passes through the assembly. Any overheating of the power transformer melts the solder and the spring disconnects one side of the mains from the receiver. Such a device might not protect a low impedance rectifier from damage if the input condenser were to break down, but it could save the power transformer and would certainly prevent a fire.

A heat-operated overload cut-out is described in Ref. 65.

See also Chapter 38 Sect. 12 for general information on fuses, and Table 55.

(v) Tropic proofing

Tropic proofing has been the subject of a large amount of investigation (Ref. 40). However many treatments described are not applicable to commercial A-M receivers —even if only because of expense—and in any case are not required for equipment which, although in the tropics, is to operate inside a home.

For " commercial tropic proofing " the following points might be taken as a minimum requirement :—

(a) Good quality resistors, derated to dissipate no more than two thirds of the manufacturer's rating.

(b) Good quality paper capacitors, preferably sealed in glass, but at least moulded in some non-hygroscopic composition, with particular care (non-cracking wax or varnish sealing for instance) against ingress of moisture along leads. Voltage rating to be at least twice that experienced under working conditions. Mica capacitors to be used in positions critical to leakage e.g. a-f coupling capacitors, or a.v.c. by-pass capacitors.

(c) Mica capacitors in tuned circuits to be of silvered mica type, flash-dipped in non-hygroscopic wax or varnish.

(d) Trimmers to have air dielectric.

(e) Power transformer to be moisture proofed, preferably by baking and vacuum impregnation with bitumen, or with a varnish which can be made to set satisfactorily right through the winding (this must be checked). Failing vacuum impregnation, pitch sealing can be used providing good penetration is obtained. Penetration is assisted by baking the transformers, standing them on edge with the top of the winding just above the surface of the pitch immediately after baking for perhaps half an hour, and then lowering the whole transformer below the surface in the same position and leaving for another quarter hour. The transformers can then be drained.

(f) A-F transformers and chokes to be vacuum impregnated and to have the core connected to the winding, the whole assembly being isolated from the chassis.

(g) Coils to be wound on moulded or ceramic formers and baked, impregnated (preferably under vacuum) and flash dipped in non-hygroscopic wax.

(h) Very fine wire to be avoided ; 36 A.W.G. is a suitable minimum size.

(i) Hook-up wire to be covered only with good quality rubber of low sulphur content, or P.V.C.

(j) All corrodible metals to be suitably treated, e.g. steel to be well and heavily plated (and lacquered where possible), aluminium to be anodised if possible or chromate dipped in such places as gang plates, brass to be plated. Die castings and readily corrodible metals, such as zinc, to be avoided.

(k) Plywood to be urea-bonded.

(l) Good quality micarta to be used for lug strips, wavechange switch wafers, etc. Micarta to be baked and vacuum impregnated in non-hygroscopic compound where possible.

(m) All fungus-supporting media such as cloth and other organic materials to be avoided as far as possible and fungus and moisture proofed if used.

(n) Wiring to be arranged as far as possible to minimize the effects of leakage e.g. B+ and a.v.c. not to be wired to adjacent lugs on a lug strip, etc.

(o) All-glass valve types to be used to avoid losses in moulded insulating material, together with high quality valve sockets.

(p) Ceramic insulation to be used in the tuning gang if possible. Micarta—if used —must be baked, vacuum impregnated in varnish and rebaked.

Additional improvements which can be incorporated include the use of ceramic instead of micarta for such items as lug strips, wavechange switch, and valve sockets, hermetic sealing for individual components (particularly transformers), hermetic sealing for the complete equipment with silica gel used for dehydration, and the avoiding of contact potentials in excess of a small fraction of a volt between contacting metals. Platings of appropriate metals can be used to minimize potential differences between adjacent metal surfaces.

See also Chapter 11 Sect. 7 for tropic proofing of coils.

(vi) Parasitic oscillations

The high slope of modern valves makes them particularly liable to generate parasitic oscillations unless suitable precautions are taken. In a radio receiver, a power output stage employing a high slope valve is the most probable cause of trouble, which may show up as a continuous " frying " noise, as an irritating " buzz " perhaps only on loud passages, or merely as reduced output with distortion.

Parasitic oscillations often occur at frequencies of the order of 100 Mc/s with leads from valve sockets forming resonant circuits and with a feedback path provided by a few micromicrofarads of capacitance between plate and grid circuits. In such a case a suitable remedy is a non-inductive resistor (say 50 000 ohms, ½ watt, carbon) wired directly to the grid pin. This increases the losses in the high frequency circuit to such an extent that oscillation is not possible.

Because the placing of wires is liable to vary between receivers, oscillation may occur only in some receivers of a production run, but it is normal practice to provide a grid stopper in each set to avoid rejects. In extreme cases it may also be necessary to use stoppers in screen grid or plate circuits, but the values must be much smaller to avoid excessive d.c. voltage drop.

Valves other than high slope output valves can be troublesome, and in one instance some of a production run of receivers having i-f amplifiers with bare tinned-copper leads a few inches in length connecting the grid and plate pins to the i-f transformers were found to be oscillating at a frequency in excess of 200 Mc/s. The trouble showed up as poor i-f sensitivity with high noise level, and was cured by wiring the hot side of the i-f trimmer directly to the valve socket, thereby including the inductance of the lead to the transformer in the i-f tuned circuit.

The best test for parasitic oscillation in an a-f amplifier is to examine with an oscilloscope the output waveform with the amplifier driven from zero input to overload at various frequencies. Oscillation may be shown as a thickening of the trace over part or all of the cycle or by a higher frequency superimposed on the correct trace. In the absence of an oscilloscope, unusual or varying plate current in an output valve may be taken as an indication of oscillation particularly if the plate current can be varied by moving the hand near the valve.

Additional details on parasitic oscillations are given in Ref. 52.

(vii) Printed circuits*

Printed electronic circuits are no longer in the experimental stage. Introduced into mass production early in 1945 in the tiny radio proximity fuze for mortar shells developed by the National Bureau of Standards, printed circuits are now the subject of intense interest on the part of manufacturers and research laboratories in this country and abroad.

Circuits are defined as being " printed " when they are produced on an insulated surface by any process. The methods of printing circuits fall into six main classifications. (1) Painting. Conductor and resistor paints are applied separately by means of a brush or a stencil bearing the electronic pattern. After drying, tiny capacitors

---

*Reprinted by permission of The Institute of Radio Engineers from Ref. 38.

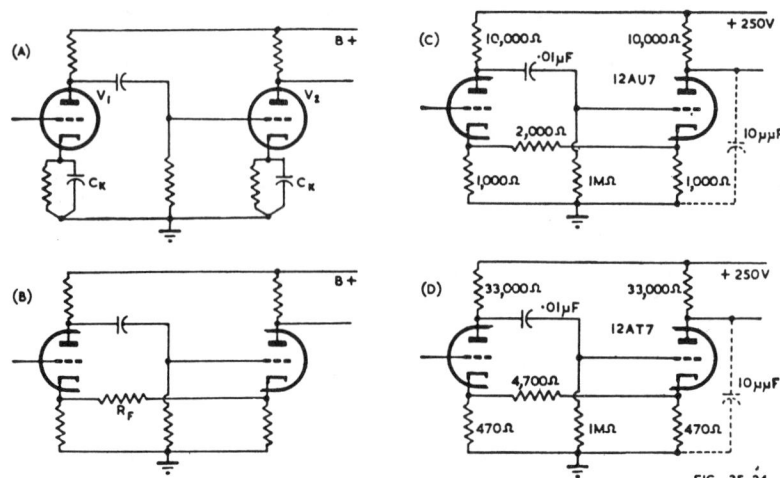

Fig. 35.24. *Positive feedback circuits for eliminating cathode bypass capacitors*
(Ref. 51).

and subminiature tubes are added to complete the unit. (2) Spraying. Molten metal or paint is sprayed on to form the circuit conductors. Resistance paints may also be sprayed. Included in this classification are an abrasive spraying process and a die-casting method. (3) Chemical deposition. Chemical solutions are poured onto a surface originally covered with a stencil A thin metallic film is precipitated on the surface on the form of the desired electronic circuit. For conductors the film is electroplated to increase its conductance. (4) Vacuum processes. Metallic conductors and resistors are distilled onto the surface through a suitable stencil. (5) Die-stamping. Conductors are punched out of metal foil by either hot or cold dies and attached to an insulated panel. Resistors may also be stamped out of a specially coated plastic film. (6) Dusting. Conducting powders are dusted onto a surface through a stencil and fired. Powders are held on either with a binder or by an electrostatic method.

Methods employed have been painting, spraying, and die-stamping. Principal advantages of printed circuits are uniformity of production, and the reduction of size, assembly and inspection time and cost, line rejects, and purchasing and stocking problems.

See Ref. 37 for a description of the production of complete radio receivers by printing techniques. See also Refs. 38, 39, 46, 47, 61, 62, 63, 64.

There are components which it is not practicable to produce by printing, e.g., electrolytic and other high value capacitors. In Ref. 51 details are given of circuit techniques to eliminate such components without seriously affecting performance.

Fig. 35.24A shows a conventional two-stage triode amplifier and Fig. 35.24B a circuit with similar gain which does not use electrolytic capacitors. In Fig. 35.24B the removal of the cathode by-passes decreases gain and results in an a-f voltage appearing across each of the two cathode resistors. These voltages are in opposite phase and by adjusting the positive feedback through R_F from the cathode circuit of the second valve to the cathode circuit of the first valve the gain of the amplifier can be restored to the original level.

Fig. 35.24C shows a practical amplifier with a voltage gain of 80 and a bandwidth (3 db down) of 250 Kc/s. Fig. 35.24D gives a single valve amplifier with a gain of 1000 and a bandwidth of 100 Kc/s. Two such valves can be used in a four stage amplifier with a gain of one million and a volume of six cubic inches.

Fig. 35.25A shows a method of eliminating screen by-passes by using positive feedback from the plate of a following valve, while Fig. 35.25B demonstrates a screen regeneration circuit similar to the cathode regeneration circuit of Fig. 35.24B.

Circuits for high frequency compensation by means of positive feedback used in place of compensating inductors are also given in Ref. 51.

An analysis of a-f circuits with " cathode neutralization " will be found in Ref. 54

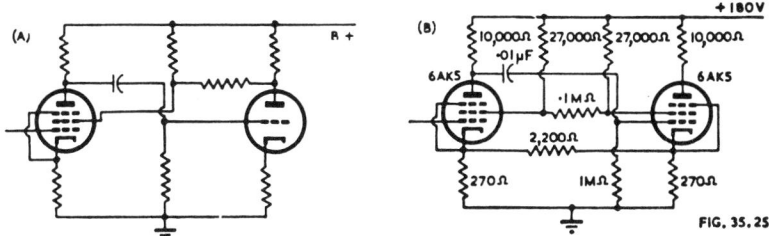

Fig. 35.25. Positive feedback circuits for eliminating screen bypass capacitors (Ref. 51).

(viii) Other miscellaneous features

Synthetic bass is described in Chapter 14 Sect. 3(viii) and its application in receivers in Chapter 15 Sect. 12(ii).

Tone control is covered in Chapter 16.

Whistle filters are covered in Chapter 16 Sect. 11.

SECTION 10 : REFERENCES

REFERENCES DEALING WITH INTERFERENCE AND NOISE
(1) Eaglesfield, C. C. " Motor-car ignition interference " W.E. 23.277 (Oct. 1946) 265.
(2) Blok, L. " Radio interference " Philips Tec. Rev. 3.8 (Aug. 1938) 235.
(3) Ebel, A. J. " A note on the sources of spurious radiations in the field of two strong signals " Proc. I.R.E. 30.2 (Feb. 1942) 81.
(4) Hall, G. G. " Design problems in automobile radio receivers " A.W.A. Tec. Rev. 4.3 (Nov. 1939) 105.
(5) Moore, J. B. " A.G.C.-noise considerations in receiver design " Elect. 18.5 (May 1945) 116.
(6) Herold, E. W. " An analysis of the signal-to-noise ratio of ultra-high-frequency receivers " R.C.A. Rev. 6.3 (Jan. 1942) 302.
(7) North, D. O. " The absolute sensitivity of radio receivers " R.C.A. Rev 6.3 (Jan. 1942) 332.
(8) Friis, H. T. " Noise figures of radio receivers " Proc. I.R.E. 32.7 (July 1944) 419 ; correction 32.12 (Dec. 1944) 729 ; discussion 33.2 (Feb. 1945) 125.
(9) Ledward, T. A. " Transformer screening, suggestions for improvement " W.W. 50.1 (Jan. 1944) 15.
(10) (A) Nethercot, W. " Ignition interference " W.W. 53.10 (Oct. 1947) 352
     (B) Nethercot, W. " Ignition interference " W.W. 53.12 (Dec. 1947) 463.
     (C) Nethercot, W. " Car-ignition interference " W.W. 26.311 (Aug. 1949) 251.
(11) Fabini, E. G., and D. C. Johnson " Signal to noise ratio in A-M receivers " Proc. I.R.E. 36.12 (Dec. 1948) 1461.
(12) Middleton, D. " Rectification of a sinusoidally modulated carrier in the presence of noise " Proc. I.R.E. 36.12 (Dec. 1948) 1467.
(13) (A) Hamburger, G. L. " Interference measurement, effect of receiver bandwidth " W.E. 25.293 (Feb. 1948) 44.
     (B) Hamburger. G. L. " Interference measurement, effect of receiver bandwidth " W.E. 25.294 (Mar. 1948) 89.

REFERENCES RELATING TO BANDSPREADING
(14) Moxon, L. A. " Making the Most o Short Waves " (1 W.W. 47.6 (June 1941) 148 ; (2) W.W. 47.7 (July 1941) 180.
(15) " A simple bandspread system for commercial receivers " A.R.T.S. and P. No. 112 (Aug. 1941) 1.
(16) " An economical bandspread system for communication receivers " A.R.T.S. and P. No. 115 (Oct. 1941) 1.
(17) " The padded signal circuit " A.R.T.S. and P. (1) No. 128 (Mar. 1943) 1 ; (2) No. 129 (May 1943) 1.
(18) Miller, J. M. " Thermal drift in superheterodyne receivers " Elect. 10.11 (Nov. 1937) 24.
(19) Foster, D. E., and G. Mountjoy, " Short-wave spread bands in automobile and home receivers " Proc. I.R.E. 30.5 (May 1942) 222.
(20) Woodbridge, F. H. " Bandspreading, its effect on the tuning rate " W.E. 42.204 (Sept. 1940) 394.

Additional references will be found in the Supplement commencing on page 1475.

GENERAL REFERENCES

(21) Sturley, K. R. " Radio Receiver Design " Parts 1 and 2 (Chapman and Hall, London, Part 1, 1943 ; Part 2, 1945).

(22) Zepler, E. E. " The Technique of Radio Design " (Chapman and Hall, London, 1943, John Wiley and Sons, New York, 1943).

(23) Haller, C. E. " Filament and heater characteristics " Elect. 17.7 (July 1944) 126.

(24) Blatterman, A. S. " Sensitivity limits in radio manufacturing " Elect. 18.11 (Nov. 1945) 141.

(25) Potter, N. M. " Dry battery characteristics and applications " Proc. I.R.E. (Aust.) 7.1 (Jan. 1946) 3.

(26) " Power Sensitivity of A-F Amplifier in Radio Receivers " R.C.A. Application Note No. 89 (March 16, 1938) 2.

(27) " Problems of Series Filament Operation, 1.4 Volt Valves " Radiotronics No. 106 (Sept. 1940) 1.

(28) " Operation of Fifty Milliampere Tubes by the 117N7GT," R.C.A. Application Note No. 109 (Nov. 13, 1940) 1.

(29) " Radio receiver design " Radiotronics (1) No. 117 (Jan. 1946) 13 ; (2) No. 118 (Mar. 1946) 30.

(30) Strafford, F. R. W. " Mains transformer protection " W W. 53.2 (Feb. 1947) 51 ; correction 53.3 (Mar. 1947) 94.

(31) " An Inverse-Feedback Circuit for Resistance Coupled Amplifiers " R.C.A. Application Note No. 93 (June 8, 1938).

(32) Toth, E. " High fidelity reproduction of music " Elect. 20.6 (June 1947) 108.

(33) Grisdale, G. L., and R. B. Armstrong, " Tendencies in the design of the communications type of receiver " Jour. I.E.E. 93 (Pt. 3) 25 (Sept. 1946) 365.

(34) Green, A. L., and J. B Rudd, " A general purpose communication receiver " A.W.A. Tech. Rev. 4.4 (Feb. 1940) 181.

(35) Diallist, " The warning winker " W.W. 55.2 (Feb. 1949) 43.

(36) " A.C./D.C. voltage dropping " W.W. 52.7 (July 1946) 236.

(37) Sargrove, J. A. " New methods of radio production " J. Brit. I.R.E. 7.1 (Jan. 1947) 2.

(38) Brunetti, C., and R. W. Curtis " Printed circuit techniques " Proc. I.R.E. 36.1 (Jan. 1948) 121.

(39) Brunetti, C. " New Advances in Printed Circuits " (National Bureau of Standards, U.S. Dept. of Commerce, Miscellaneous Publication 192, Washington, 1948).

(40) Proskauer, R. and H. E. Smith " Fungus and moisture protection " Elect. 18.5 (May 1945) 119.

(41) " A.C./D.C. heater circuits " Philips Tec. Comm. No. 35, p. 1.

(42) Stanley, A. W. " Series capacitor heater circuits " W.W. 54.9 (Sept. 1948) 332 ; correspondence 54.10 (Oct. 1948) 385.

(43) " Grid leak bias operation of high mu triode valves " Radiotronics 94 (Jan. 1939) 3.

(44) Sterling, H. T. " Diode contact potential for negative bias " Elect. 20.10 (Oct. 1947) 164.

(45) McProud, C. G. " Elements of residence radio systems " Audio Eng. (i) 32.9 (Sept. 1948) 22 ; (ii) 32.10 (Oct. 1948) 21 ; (iii) 32.11 (Nov. 1948) 20 ; (iv) 32.12 (Dec. 1948) 22.

(46) Murray, A. F. " Present status of printed circuit techniques " Tele-Tech, 6.12 (Dec. 1947) 29.

(47) Murray, A. F. " Evolution of printed circuits for miniature tubes " Tele-Tech, 6.6 (June 1947) 58.

(48) McLaughlin, J. L. A. " Heterodyne eliminator " Elect. 22.3 (March 1949).

(49) Jaffe, D. L. " Intermediate frequency amplifier stability factors " Radio 30.4 (April 1946) 26.

(50) Scroggie, M. G. " The a.v.c. characteristic " W.W. 44.18 (May 4, 1949) 427.

(51) Sulzer, P. G. " Circuit techniques for miniaturization " Elect. 22.8 (Aug. 1949) 98.

(52) " Cathode Ray " " Parasitic oscillations " W.W. 55.6 (June 1949) 206

(53) Russell, G. H. " Intermediate frequency and the Copenhagen plan : new and serious interference problem " W.W. 55.9 (Sept. 1949) 322.

(54) Miller, J. M. " Cathode neutralization of video amplifiers " P.I.R.E. 37.9 (Sept. 1949) 1070.

(55) " You Should Know About Batteries " Vesta Battery Company Limited.

(56) " Self-charging battery/a.c. portable " Radio Electrical Weekly 28.26 (April 29, 1949).

(57) " Australian self-charging battery/a.c portable " Radio Electrical Weekly 29.6 (June 10, 1949) 13.

(58) Terlecki, R., and J. W. Whitehead " 28 volts h.t. and l.t." Electronic Eng. 19.231 (May 1947) 157.

(59) Adams, P. H. " Developments in dry batteries and associated electronic equipment " Proc. I.R.E. Aust. 12.6 (June 1951) 167.

(60) Information supplied by Mr. T. M. Chambers of Amalgamated Wireless Australasia Ltd.

(61) Peters, R. G. " Printed circuit production and assembly techniques " TV Eng. 1.11 (Nov. 1950) 20.

(62) Tewell, W. A. " Printed circuit design methods and assembly techniques " TV Eng. 3.2 (Feb. 1952) 10 ; 3.3 (March 1952) 19.

Additional references will be found in the Supplement commencing on page 1475.

CHAPTER 36

DESIGN OF F-M RECEIVERS

By E. WATKINSON, A.S.T.C., A.M.I.E.(Aust.), S.M.I.R.E.(Aust.)

SECTION 1 : F-M RECEIVERS

(i) Comparison with A-M   (ii) Aerial and r-f design   (iii) Local oscillator design (iv) I-F amplifier   (v) F-M detection and·A-M rejection.

(i) Comparison with A-M

Many of the problems encountered in F-M receiver design are new, but the majority are similar to those in A-M receivers and yield to similar treatment to that described in Chapter 35.

The advantages conferred by F-M are not so great that an obviously improved service will necessarily be available to the F-M listener unless suitable programme material is broadcast, and unless the receiver design makes full use of the increased dynamic range, frequency response and noise reduction which are possible with F-M.

To obtain the advantages of the noise reducing properties of frequency modulation it is necessary to design the F-M receiver with considerable care. If this is not done the receiver can destroy much of the improvement obtainable. In many applications F-M does offer undeniable advantages over A-M but this is not true in every case. It should not be assumed that F-M offers a cure-all for every type of noise problem, and for some communication systems A-M may be superior to F-M (see Refs. 9 and 24).

The main differences between A-M and F-M receivers are due to the higher frequencies used for the signal, local oscillator and intermediate frequencies, the need for greater i-f bandwidth, the different type of detection required and the necessity for some form of amplitude rejection. New techniques required for the higher F-M signal frequency would be equally necessary for an A-M system operating on the same frequency.

(ii) Aerial and r-f design

See also Chapter 22 (Aerials) and Chapter 23 (R-F Amplifiers).

(A) The inductance of a straight non-magnetic wire with a length much greater than its diameter can be obtained (Ref. 1) from the approximate formula

$$L = 0.005\ l[2.3\ \log_{10}(4\ l/d - 0.75)] \tag{1}$$

where L = inductance in microhenrys,
l = length of wire in inches,
and d = diameter of wire in inches.

The inductance of one inch of 22 A.W.G. wire is thus 0.025 μH, and its reactance at 100 Mc/s is 16 ohms, so that when laying out circuits for use at v-h-f every lead used must be considered as an inductance. This factor, and the reduced input admittance of valves at v-h-f, are responsible for most of the problems in F-M receiver front-end design.

(B) Since at 100 Mc/s one quarter wavelength is only about 30 inches, it is quite possible to provide a **tuned aerial system** with F-M receivers. A type commonly used with console receivers is made from 300 ohm ribbon type transmission line. The usual arrangement is approximately half a wavelength long with the two wires connected together at the ends and one side cut to take the feeder. The aerial is tacked around the cabinet, on the back or inside, or it may be placed under floor coverings.

Such aerials operate quite satisfactorily at distances up to fifteen or twenty miles from a transmitter in favourable circumstances, but they are too big for mantel models. However a "**mains aerial**" (or "line antenna") consisting of a small capacitor between one side of the mains and the hot end of an unbalanced aerial primary can be used with comparable results.

For improved results in difficult locations an **outside aerial** should be erected as a matter of course, and in all cases it is advisable to investigate the effect of slight changes in position or orientation of the aerial since they may make a large difference to the signal strength. This is due to the fact that the dimensions of the waves are comparable with those of houses, rooms and even people, so that reflections occur, and standing waves exist which vary in amplitude with small changes in location. When a receiver is operated with an inbuilt aerial on low-level signals this is strikingly illustrated by signal strength variations being caused by reflection from people moving in the vicinity of the receiver.

(C) Aerial coil design is greatly influenced by the input conductance of the first valve, but in all cases it is necessary for the layout of the stage to be such as to include as much as possible of the required tuned circuit inductance in the coil. This prohibits the use of switching within the tuned circuit, and since a typical aerial coil secondary inductance might be 0.06 μH (less than the inductance of two inches of 22 A.W.G. wire) a good system is to mount the secondary coil across the tuning condenser (if one is used) when no a.v.c. is applied to the input stage. If a.v.c. is used an assembly can be made of the tuning condenser, tuning coil and a.v.c. by-pass. In the usual case in which a 300 ohm transmission line is used, the main function of the aerial coil is impedance matching and no appreciable voltage gain is obtained. In some cases there may even be a slight loss. See Ref. 14 for calculation of aerial coil constants, although in the example a 75 ohm transmission line is used.

(D) **The valve used in the first stage** has a considerable effect upon the signal-to-noise ratio of a receiver. Although the impedance* between the grid of the input valve and ground might be as low as say 800 ohms in a typical case, reference to Chapter 23 Sect. 6 shows that the equivalent noise resistance of the valve (probably of the order of some few thousands of ohms) is effectively in series with the input circuit and can thus be the most importance source of noise in the first stage.

Also the input impedance of the first valve controls the Q (and thus the gain and selectivity) of the aerial coil. It is possible to increase this input impedance—or even make it negative—by using an unbypassed cathode resistor in the r-f stage, but inductance in the cathode circuit and inter-electrode and electrode-to-ground capacitances, which are not accurately known, make it difficult to calculate the value of resistor required. A convenient experimental procedure is to couple a signal generator to the aerial primary and to measure the voltage on the grid of the r-f valve with the heater open-circuited. When power is applied to the heater, the voltage on the grid of the valve will fall owing to its reduced (hot-cathode) input impedance. A suitable value of unbypassed cathode resistor will reduce this drop to negligible proportions.

The effect of the resistor on changes in input capacitance with changing a.v.c. voltage (see Chapter 26 Sect. 7(ii) also Figs. 23.15 and 23.16) should be checked before the value of the resistor is finalized.

(E) The main problem with r-f stages is **instability**, and apart from the causes experienced at lower frequencies, impedances common to the input and output of the

*See Chapter 2 Sect. 8 for valve admittances.

stage become particularly important. These impedances are usually inductances and at 100 Mc/s the inductance of the rotor shaft of the gang condenser becomes sufficiently important in some cases to determine the maximum gain that can be obtained from a r-f stage. This is one reason for the use of permeability-tuned signal frequency circuits, or of loaded resonant lines with a movable short-circuiting bar (Ref. 2). Other couplings can be avoided to a large extent by returning all leads from the input circuit to one point and leads from the output circuit to a different point. If two cathode leads are available in the r-f valve they should be used for this purpose.

Coupling from output to input of the r-f amplifier can be caused by inadequate internal shielding due to an impedance between the screen grid and the cathode. This is usually due to the inductance of the leads in the screen by-pass capacitor, but this inductance can be put to good effect by series tuning it to the operating frequency. By using a small (100-500 $\mu\mu$F) mica capacitor with leads as short as possible, perhaps $\frac{1}{2}$ in. in all, much better by-passing is obtained than with larger capacitors. The actual frequency of series resonance is not particularly critical. In the case of penta-grid converters of the 6BE6 type satisfactory by-passing is particularly important to prevent undesired impedances from appearing in the oscillator circuit.

Owing to the impedance at signal frequencies of the B+ and filament leads it may happen that more than one B+ by-pass is necessary. Hot spots may occur in the tuning range, and self-resonant by-passes (100-500 $\mu\mu$F mica) should be tried at various points in B+, a.v.c. and heater wiring.

A stage gain of 10 can usually be obtained from a r-f stage without undue trouble, and higher gain (at least up to 15) is possible, particularly when there is no coupling between grid and plate circuits in the rotor shaft of a gang condenser.

(iii) Local oscillator design

See also Chapter 24.

Local oscillator stability is important in a F-M receiver because much of the noise-reducing ability of the receiver depends on the centring of the signal on the discriminator characteristic. Detuning in excess of about 25 Kc/s can noticeably affect noise reduction and audio distortion (Ref. 11). On the other hand even a technical user of a F-M receiver can make tuning errors of this order. It is necessary therefore for the local oscillator stability to be of the order of 25 Kc/s in 100 Mc/s, i.e. 2½ parts in 10 000, if the receiver performance is to be unaffected.

This requirement is not easy to meet and an average figure for an uncompensated oscillator is more like 8 parts in 10 000. By keeping heat sources away from the oscillator circuit and introducing compensation after all possible precautions have been taken, a long and short term stability of 2 parts in 10 000 is possible.

Additional oscillator stability can be obtained by using automatic frequency control if the local oscillator is a separate valve, and if suitable control voltages can be obtained from the detector circuit (Refs. 10 and 21). Improvements in oscillator stability of five to ten times are possible, but care must be taken to see that the a.f.c. will allow the receiver to tune from a strong local station to a weak one in an adjacent channel.

Some designs have used permeability-tuned-oscillator circuits to eliminate the temperature instability of a tuning condenser (Ref. 11). Harmonic operation has also been used and gives a reduction in frequency shift due to capacitance variations when a larger tuning capacitor is used to tune to the lower oscillator frequency.

Harmonic operation has the added advantage of greatly increasing the separation between the oscillator and signal frequencies and it thus improves oscillator frequency stability in cases where pulling is experienced between oscillator and signal frequency circuits.

For other methods of frequency stabilization see Chapter 24 Sect. 5.

Grid-tuned and plate-tuned oscillators are not commonly used in the F-M band because the feedback winding becomes comparable in size with the tuned winding.

This leads to tracking troubles and the possibility of an oscillator changing from grid tuning to plate tuning or vice versa in a particular band.

Both Colpitts and Hartley circuits are used, the Hartley (Fig. 24.4B) probably because it is widely used on the broadcast band with 6SA7 type converters. However unless the receiver has only one waveband, switching of the cathode lead is necessary which introduces undesirable impedance between the cathode and the tap on the coil and may lead to parasitic oscillation. Added to this the Hartley circuit is in any case more prone to parasitic oscillation than the Colpitts, and is also liable to change over to a Colpitts type using valve interelectrode capacitances to give a tapping point in the oscillator circuit.

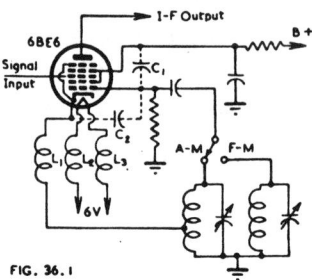

FIG. 36.1

Fig. 36.1. Simplified switching from Colpitts high frequency oscillator to Hartley medium wave oscillator.

Because of this, the Colpitts oscillator is popular in F-M receivers, and a convenient arrangement for a two band receiver is shown in Fig. 36.1 (from Ref. 16). The inductance of the choke L_1 is small enough to be negligible on the broadcast band so that it is only necessary to switch one connection to the oscillator circuit to change from the F-M to the broadcast band. The ratio of internal capacitances between screen grid and cathode and between oscillator grid and cathode determines the amount of feedback, and, if desired, external capacitances can be added to modify the feedback and increase the grid current, as shown (C_1 and C_2) dotted on the circuit. However, these capacitors also have an effect on the signal frequency circuit and sensitivity and signal-to-noise ratio should be checked at the same time as the oscillator grid current. It may be necessary to compromise between maximum sensitivity, maximum signal-to-noise ratio and maximum oscillator grid current.

The chokes L_2 and L_3 comprise the two remaining sections of a trifilar choke made by twisting together three lengths of enamelled 28 A.W.G. copper wire and then winding 15 turns, closely-spaced, on a 9/32 in. former, $\frac{1}{2}$ in. long. The sections L_2 and L_3 introduce impedance between the heater of the oscillator valve and ground so that variations in heater-cathode capacitance do not appear across a part of the tuned circuit. This is a recommended precaution against microphony and may improve frequency stability during the heating period (about five minutes) of the mixer.

Care is needed in the local oscillator design to avoid microphony and modulation hum since any a-f variation in oscillator frequency results in an equivalent output from the frequency discriminator. A check on modulation hum produced by the local oscillator can be obtained by comparing the hum in the receiver output when intermediate frequency and signal frequency inputs are fed into the converter grid.

(iv) I-F amplifier
See also Chapter 26.

When the maximum Q allowable by bandwidth considerations is used in a F-M transformer the gain obtained with the types of valves currently used approaches the maximum gain possible without an undesirable amount of regeneration. Usually there are three valves amplifying at the intermediate frequency and the over-all gain is greater than 10 000 times, so that regeneration problems are common.

A convenient method of tracing regeneration is described in Ref. 1. It consists of feeding the i-f signal to the hot side of the plate circuit of the second last i-f amplifier by means of a very small capacitor (3 $\mu\mu$F or less), retrimming the circuit, and converting the detector to an A-M type by open-circuiting one diode and replacing it with an appropriate capacitor to keep the circuit aligned. When A-M signals are fed into the amplifier the output will be due to the generator and to any regenerative effects from earlier stages which may be present. If the receiver output is decreased by short-circuiting the input of the second last i-f amplifier regeneration exists and various parts of the receiver can be short-circuited to trace it.

The method has the advantage of making it possible to deal with regenerative signals alone in the first stage instead of regenerative plus desired signals as when tracing regeneration by more conventional means. In addition the A-M signal generator can be set to various frequencies in the passband of the i-f amplifier to trace regeneration which may be more severe at some frequency other than the centre of the pass-band.

When regeneration occurs due to plate-to-grid capacitance in a single valve the circuit of Fig. 36.2A (Ref. 5) provides economical neutralization as illustrated in Fig. 36.2B. Phase shifts and stray inductances complicate the circuit sufficiently to make the calculation of C_n difficult but an average value is about 3000 $\mu\mu$F and correct neutralizing is obtained when, using the tracing method just outlined, no change in output occurs as the input transformer is tuned through resonance by C_2.

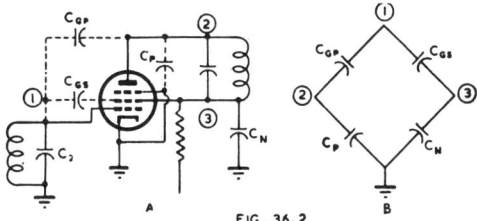

FIG. 36.2

Fig. 36.2. Circuit for i-f amplifier neutralization (A) Actual circuit (B) Capacitance rearranged to show bridge (Ref. 5).

Series resonance can be obtained in i-f by-passes by allowing half-inch leads with typical 0.01 μF paper capacitors. Larger values may give less by-passing and once the self-resonant frequency is passed the effective series impedance increases more rapidly than the frequency at which the by-pass is used.

The i-f amplifier design can be responsible for poor noise rejection on the part of the receiver unless adequate stability is provided against the effects of time, temperature, moisture and varying signal strength. The detuning of one or more i-f tuned circuits from the centre point of the discriminator characteristic will result in decreased noise rejection whether or not a limiter is used (Ref. 3).

Stability versus time characteristics are usually satisfactory when good trimmers are used, but compression type mica trimmers may give trouble, particularly when the effects of vibration in transit between factory and the user's home are considered.

The effects of temperature and humidity can be offset by suitable impregnation methods for coils and the use of low temperature coefficient condensers to tune the transformers. Silvered-mica types are satisfactory as are most air-dielectric trimmers.

Frequency instability with varying signal inputs applies to individual circuits rather than to the i-f amplifier as a whole and can be caused by a decrease in regeneration as gain is reduced by a.v.c. application or by valve input capacitance changes due to a.v.c. applied to i-f amplifiers or to sudden changes in the amplitude of the applied signal in the case of limiters. One method of overcoming the latter trouble is to use resistance coupling between the last i-f amplifier and the limiter (Ref. 21), although this can only be done at the expense of gain and selectivity.

When impulse noise is present the detuning with each pulse may result in phase distortion of the signal which will no longer be centred on a symmetrical pass-band. The noise itself will also be heard and under extreme conditions the effect of a limiter can be to increase noise, as the increase due to mistuning more than offsets any other effect of the limiter (Ref. 3).

In the absence of noise, detuning due to a.v.c. application decreases the amplitude rejection of the receiver and may increase phase distortion.

On the other hand, some detuning due to a.v.c. might be preferable to the amount of detuning caused by overloading in the grid circuits on strong signals. Resistance-capacitance networks having suitable time constants are often used to provide additional self bias when the signal strength is high. This system is usually preferred to normal a.v.c. application.

To minimize detuning effects the i-f amplifier must be non-regenerative and the capacitors on the secondary—at least—of limiter input i-f transformers should be increased as much as the consequent reduction in gain will allow. If the primary can conveniently be wound with a different number of turns from the secondary the primary inductance may be increased when the secondary inductance is reduced and gain will be unchanged if $\sqrt{L_pL_sQ_pQ_s}$ remains constant. This should not be carried to the point where detuning of the primary circuit becomes noticeable either when valves are changed or due to differences in wiring capacitances between different chassis. In addition, an undue increase in L_p can cause the grid-to-plate gain to become excessive, resulting in instability.

The same precautions can be taken with the input circuits of stages to which a.v.c. is applied and in addition an unbypassed cathode resistor can be used to reduce input capacitance changes. Figures 23.15 and 23.16 show the effect of varying values of unbypassed cathode resistors on the change in input capacitance of type 6AC7 and 6AB7 valves. For type 6BA6 a common value of resistor is 68 ohms.

(v) F-M detection and A-M rejection

There are three main methods of performing the functions of F-M detection and A-M rejection in commercial F-M receivers, the limiter-discriminator combination, the ratio detector and the locked oscillator. See Chapter 27 Sect. 2 for design details and Ref. 8 (from which some of the following information is taken) for a consideration of the capabilities of each. · The effect of the type of detector used on over-all receiver performance can be summarized as follows.

(a) Assuming reasonably high a-f sensitivity, the use of a **ratio detector** can save one valve from the number required for a receiver using a limiter and discriminator or a locked oscillator.

(b) The downward A-M handling capability of the ratio detector is less than that of the other two types. For the ratio detector an average figure is about 60% for medium and large inputs whereas the other types rapidly approach 100% once their threshold input is passed. Serious distortion occurs in a ratio detector if the downward amplitude modulation is excessive since this stops the diodes from conducting (see Chapter 27 Sect. 2).

(c) The ratio detector gives some A-M rejection at very small inputs whereas the limiter discriminator combination has a threshold (the beginning of limiting) below which there is no A-M rejection. The locked oscillator also has a threshold—the input required to lock the oscillator. As a result, interstation noise levels on a receiver using a ratio detector are noticeably low for a given sensitivity.

For medium and high inputs all three types reduce amplitude modulation to a small percentage of the original, with the proviso in the case of the ratio detector that the downward amplitude modulation must not exceed about 60%.

An amplitude modulated signal tuned to the mid-point of a balanced discriminator will not give any a-f output, and in such a case the A-M rejection of the discriminator alone is 100%. However, if the signal is not centred on the discriminator characteristic there will be some a-f output.

Even 100% A-M rejection will not remove noise from a F-M carrier. If the carrier is considered as a rotating vector and at a particular instant a noise carrier of twice the amplitude and with a 180° phase difference is added to it, the resulting carrier will be unchanged in amplitude but rotated through 180° by the noise. Thus, phase modulation (and so equivalent frequency modulation) of the desired carrier has occurred and noise must appear in the output of the detector. Since the amplitude is unchanged there will be no limiting. In practice, both amplitude and frequency modulation of a desired carrier occur, and although the receiver may have 100% A-M rejection the effect of the frequency modulation cannot be eliminated.

(d) The output of a ratio detector is proportional to the average amplitude of its input so that some form of a.v.c. is desirable. With the other two types output is almost independent of input above the threshold level, although some form of a.v.c. is often used to prevent overloading of i-f stages and the consequent detuning.

(e) With each type there is a side tuning response when the signal is tuned to the skirt of the i-f selectivity curve. However in the case of the ratio detector this is at a much lower level than the main response and tuning is not a great deal more difficult than with an A-M receiver.

The limiter discriminator circuit gives distorted side responses at a level at least equal to the correct tuning position and tuning is not easy for non-technical users, unless a suitable tuning indicator is incorporated in the receiver.

The locked oscillator circuit may produce severe distortion as the signal crosses the locking threshold in the side tuning positions, or stations may tune with a definite " plop " as the oscillator locks.

If a.f.c. is used it is possible to arrange for the receiver to be pulled rapidly through the side responses by the a.f.c., thus minimizing their effect.

In general, the saving of a valve and the desirable tuning characteristic of the ratio detector make it attractive. However more care is needed in design to obtain correct balance between the two diode circuits for maximum noise rejection, and the limited downward modulation handling capability means that the selectivity of the i-f amplifier must be closely controlled in production. If this is not done, excessive selectivity due to decreased coupling may introduce more amplitude modulation than can be handled by the ratio detector (even without the possible addition of noise) resulting in severe distortion.

The standard de-emphasis characteristic required has a falling characteristic with increasing frequency and is equivalent to that provided by a simple circuit having a time constant of 75 microseconds. The components used to filter i-f voltages from the detector circuit are given suitable values to provide this de-emphasis. Fig. 15.1 gives the standard curve.

Valves designed especially for limiting and F-M detection are available, namely the gated beam discriminator type 6BN6 and the Philips " ϕ-detector " type EQ80.

In the 6BN6 two grids are operated in a quadrature phase relationship at centre frequency and the phase of one grid voltage is caused to vary with respect to the other about the quadrature point as the frequency is varied. The quadrature voltage is developed by space charge coupling to a special grid to which a parallel tuned circuit is connected. Limiting is obtained by controlling the current of an electron beam by an apertured slot located in the beam in a region of high current density.

Typical performance of the 6BN6 (Refs. 25, 26) is that with a F-M intermediate frequency of 10.7 Mc/s the output voltage for 75 Kc/s deviation is about 4.5 volts r.m.s. Maximum A-M rejection with 30% A-M and 30% F-M modulation applied simultaneously is about 35 db. The A-M rejection with respect to signal input varies but at least 20 db of A-M rejection is obtained with 1 volt input and from 15 to 30 db at signals above 1.25 volt r.m.s. An a-f output of 3.7 volts with less than 1% distortion can be obtained using accelerator and plate supply voltages of 60 and 80 volts. Higher voltages and correspondingly higher plate loads will give an a-f output up to 15 volts r.m.s. for 75 Kc/s deviation which is sufficient to drive the output valve directly.

The EQ80 (Ref. 27, from which the following information is taken) contains seven grids of which the second, fourth and sixth grids are screen grids and the seventh is a suppressor grid. To each of the control grids a voltage of at least 8 V r.m.s. is applied from an i-f transformer. The mean value of the anode current is a function of the phase shift between the two control voltages whilst the phase shift is a function of the frequency deviation. Both functions are approximately linear when the phase shift has a sweep between 60° and 120°.

The amplitude of the anode current is not dependent upon the magnitude of the control voltages (provided they are greater than 8 V) so that the valve also acts as a limiter. The ϕ-detector gives an a-f output of about 20 to 25 volts. The first grid can be arranged to cut off the cathode current if the control voltages are not large enough, so that interstation noise is suppressed.

SECTION 2 : F-M/A-M RECEIVERS

(i) R-F section   (ii) I-F amplifier   (iii) General considerations.

(i) R-F section

The valve complement and other major features of a combination F-M/A-M receiver· are determined by the F-M requirements almost exclusively, and the A-M receiver operates with a suitable selection of the parts already in use.

There are of course some additions, and the A-M tuning device is one. The two tuning systems may use separate gang condensers or separate permeability tuning arrangements or a combination of the two. There are two well-known systems of tuning the two bands with one control, one using a tuning condenser with A-M and F-M sections on the same rotor shaft, and the other with a cam on the rotor shaft of a normal A-M type gang, the cam operating slugs in the F-M coils.

On the F-M band, tuning should be accurate within about 20 Kc/s in a band of 20 Mc/s—one part in one thousand—which is comparable with a required A-M setting accuracy of about 1 Kc/s in a band of about 1000 Kc/s. Thus a tuning rate which is satisfactory for the A-M broadcast band should be suitable for the F-M band.

Much of the ingenuity in combined F-M/A-M design is directed towards the elimination of switching, particularly in hot circuits. This is not so much to save the cost of the switching but to avoid the losses in leads to switch contacts and in the additional stray capacitances which are introduced.

One example of this can be seen in Fig. 36.1 where one switch section is saved by converting the oscillator from a Hartley on A-M to a Colpitts on the F-M band. Fig. 36.3 (taken from Ref. 10) shows a system of combining short wave A-M input with F-M band input without switching and without undue loading of one circuit by the other.

Untuned r-f amplification also saves switching, and choke coupling can conveniently be used or a mixture of choke feed for the r-f amplifier on the F-M band and resistance feed on the A-M band. In this case there is of course a considerable loss in image rejection compared with that obtainable with tuned r-f amplification.

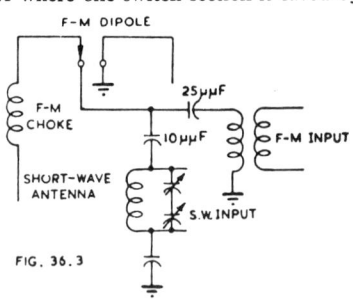

FIG. 36.3

Fig. 36.3. Combined F-M band and A-M short wave band input circuit (Ref. 10).

Other systems which have been used include a circuit in which the r-f stage is used only on the F-M band with an A-M loop switched to the converter grid for A-M reception, and a more complicated one (Ref. 4) in which the F-M signal is converted by a 12BE6 to the 10.7 Mc/s intermediate·frequency and is then amplified by a second 12BE6 operating as an i-f amplifier. On A-M the second 12BE6 becomes the converter by comparatively simple switching, and an A-M loop is connected to its grid while the F-M second i-f amplifier becomes the single A-M i-f amplifier. The first 12BE6 is not used in A-M reception.

(ii) I-F amplifier

As in the r-f section of the receiver, it is important to avoid switching hot circuits in the i-f amplifier, and the fixed tuning and large frequency difference between 10.7 Mc/s and 455 Kc/s fortunately makes a simple series connection of the two i-f's quite satisfactory (see Chapter 26 Sect. 3).

One trouble experienced with this system is that when operating on A-M, spurious responses (e.g. oscillator harmonics) can set up large signals across the 10.7 Mc/s tuned circuits and may overload later i-f amplifiers, interfere with a.v.c. arrangements

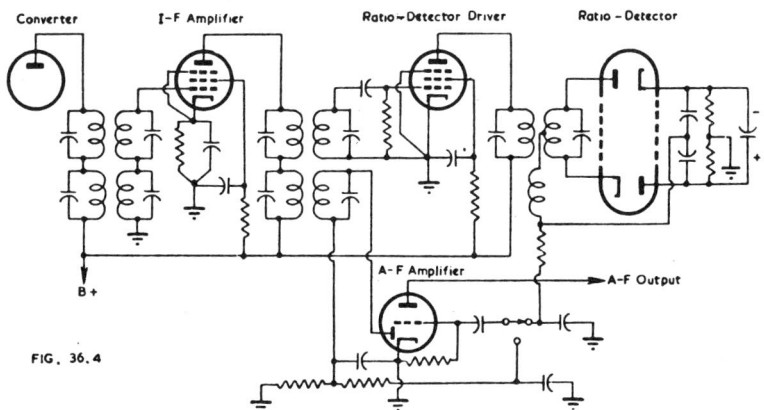

Fig. 36.4. Combined F-M A-M i-f channel using one A-M i-f amplifier and two F-M i-f amplifiers.

or cause other trouble depending on the circuit of the subsequent sections of the receiver.   To minimize this trouble it is common practice to switch the converter output to the appropriate (10.7 Mc/s or 455 Kc/s) winding.   In such a case the secondary and all other i-f windings (except the output windings) would be series connected. Separate converters, each connected only to its own i-f, have also been used.   This eliminates the spurious response trouble, allows very simple switching between bands (only the converter B supply need be switched) and allows more flexibility in the F-M oscillator design.

When using a ratio detector, two i-f amplifiers are needed, and with a limiter-discriminator combination an extra valve is required.   As against this the i-f amplifier for an A-M receiver needs only one valve, particularly if of the high slope type commonly used as F-M i-f amplifiers.

One method of obtaining an extra F-M i-f amplifier was mentioned in sub-section (i) the case in which the A-M converter was switched to carry out this function.   However, the more usual arrangement is shown in Fig. 36.4 for the case of ratio detector F-M receiver.   A combined detector of the triple-diode-triode type could be used to save one valve, but the main point is that the last A-M i-f is not connected in series with the F-M winding, to avoid the necessity for switching a high level i-f circuit.

F-M detection can be carried out by tuning the F-M signal to the skirt of the i-f response and applying the resultant amplitude and frequency modulated wave to a diode.   The amount of A-M introduced can be high (see Fig. 26.6) so that a few times greater detection sensitivity is available than from a conventional discriminator. This is obtained at the expense of distortion, the degree of which is dependent on the degree of non-linearity of the i-f skirt characteristic, and the circuit also gives two tuning points.   Tuning must be for minimum distortion which is not necessarily obtained at the same tuning point as maximum volume or maximum signal-to-noise ratio.   An additional disadvantage is that the circuit has no inherent amplitude rejection properties, as has a discriminator, although a limiter could be used.

(iii) General considerations

The problem of providing comparable a-f output on F-M and A-M has different aspects depending on the type of detection used in each case

With a ratio detector the a-f output is approximately proportional to the r-f input so that some form of a.v.c. is needed, as with an A-M diode detector.   The direct voltage generated by the detector provides an excessive amount of control when applied to the grid of the i-f amplifier unless high a-f gain is provided or unless delay is used.   The circuit of Fig. 27.38B can be used for delay if a diode is available.

However " partial limiting " as discussed in Chapter 27 Sect. 2(iv)C provides a satisfactory a.v.c. characteristic and at the same time removes the need for any other means of A-M detection since if suitable constants are used this is carried out satisfactorily in the control grid circuit of the driver of the ratio detector. The direct voltage available from this circuit can also be used as an a.v.c. voltage for previous stages (on A-M or F-M).

If about one third of the a-f voltage developed in the " partial limiter " grid circuit is used on A-M; and the full ratio detector output on F-M, similar a-f output will be obtained from the receiver over a range of input signals with each system.

A small discriminator bandwidth (some commercial designs have a width of only 200 Kc/s) will increase the a-f output, which would allow a larger fraction of the A-M output to be used, and increase the output available from small signals with each system.

In the case of a limiter-discriminator combination having A-M taken from the input to the limiter stage, the a-f output is high and almost constant on F-M but is much less with small inputs on A-M. A poor A-M a.v.c. characteristic which allows the a-f output to increase rapidly can provide approximately equal F-M and A-M output when reasonably large A-M signals are received.

SECTION 3 : REFERENCES

(1) Freeland, E. C. " F.M. receiver design problems " Elect. 22.1 (Jan. 1949) 104.
(2) Berhley, F. E. " Tuning without condensers " F.M. & T. 7.8 (Aug. 1947) 35.
(3) Landon, V. D. " Impulse noise in F-M reception " Elect. 14.2 (Feb. 1941) 26.
(4) Frankhart, W. F. " Low cost FM-AM receiver design " Tele-Tech 7.4 (April 1948) 40.
(5) Gustafson, G. E., and J. L. Rennick, " Low cost FM-AM receiver circuit " Tele-Tech 7.10 (Oct. 1948) 36.
(6) Beard, E. G. " Intermediate frequency transformers for A.M.-F.M. receivers " Philips Tec. Com. 6 (Nov. 1946) 12.
(7) " Fremodyne F-M receivers " Elect. 21.1 (Jan. 1948) 83.
(8) Loughlin, B. D. " Performance characteristics of FM detector systems " Tele-Tech 7.1 (Jan. 1948) 30.
(9) Nicholson, M. G. " Comparison of amplitude and frequency modulation " W.E. 24. 286 (July 1947) 197.
(10) Aram, N. W., L. M. Hershey and M. Hobbs " F-M reception problems and their solution " Elect. 20.9 (Sept. 1947) 108.
(11) Benin, Z., " Modern home receiver design " Elect. 19.8 (Aug. 1946) 94.
(12) " FM standards of good engineering practice as released by the Federal Communications Commission on September 20, 1945 " FM & T. 5.10 (Oct. 1945) 28.
(13) Bradley, W. E. " Single-stage F-M detector " Elect. 19.10 (Oct. 1946) 88.
(14) Sandel, B. " An F-M receiver for the 88-108 Mc/s band " (i) Radiotronics 125, (May 1947) 39 ; (ii) 127, (Sept. 1947) 79.
(15) " A new approach to F-M/A-M receiver design " Elect. 20.7 (July 1947) 80.
(16) Harris, W. A., and R. F. Dunn, " An a.c.-d.c. receiver for AM and FM " Radio and TV News, 40.5 (Nov. 1948) 62.
(17) Crocker, A. G. " Is discriminator alignment so difficult " W.W. 54.9 (Sept. 1948) 312. Correspondence, 54.10 (Oct. 1948) 386.
(18) " Balanced phase shift discriminators " R.C.A. Lab. Report LB-666. Reprinted Radiotronics 120 (July 1946) 83.
(19) " The ratio detector " R.C.A. Lab. Report LB-645. Reprinted Radiotronics 120 (July 1946) 79.
(20) " Use of the 6BA6 and 6BE6 miniature tubes in FM receivers " R.C.A. Application Note AN-121. Reprinted Radiotronics 129 (Jan. 1948) 15.
(21) Hershey, L. M. " Design of tuners for A-M and F-M " Tele-Tech 6.8 (Aug. 1947) 58.
(22) Plusc, I. " Investigation of frequency modulation signal interference " Proc. I.R.E. 35.10 (Oct. 1947) 1054.
(23) Miner, C. R. " FM receiver front-end design " Tele-Tech 6.5 (May 1947) 34.
(24) Tatt, C. E. " The application of super-regeneration to frequency-modulation receiver design. Proc. I.R.E. (Aust.) 8.4 (April 1947) 4.
(25) Hasse, A. P. " New one-tube limiter-discriminator for F-M " Tele-Tech (1(9.1 (Jan. 1950) 21 ; (2) 9.2 (Feb. 1950) 32.
(26) Adler, R. " A gated beam discriminator " Elect. 23.2 (Feb. 1950) 82.
(27) Jonker, J. L. H., and A. J. W. M. van Overbeek " The φ-detector, a detector valve for frequency modulation " Philips Tec Rev. 2.1 (July 1949) 1.

CHAPTER 37

RECEIVER AND AMPLIFIER TESTS AND MEASUREMENTS

By E. WATKINSON, A.S.T.C., A.M.I.E.(Aust.), S.M.I.R.E.(Aust.).

SECTION 1 : A-M RECEIVERS

(i) Introduction (ii) Definitions (iii) Equipment required (iv) Measurements and operating conditions (v) Measurements (vi) Performance tests.

(i) Introduction

Few of the results of measurements made on radio receivers can be expressed as the difference between an ideal receiver and the receiver under test. For this reason it is desirable to adopt standard methods of measuring the performance characteristics of receivers and of presenting the results of those measurements. When this is done direct comparisons between receivers are possible, whereas if for instance one manufacturer expresses noise as the percentage of output voltage remaining when the 400 c/s 30% modulation is removed from the modulated carrier required to give standard output, and another follows the English R.M.A.* recommendation and expresses it as the unmodulated r-f input carrier for which the output is equal to the output given by a 10% modulation of that carrier at 1500 c/s, then it is not easy to form any opinion of the relative merits of receivers measured under the two systems.

Unfortunately there is no standard international method of measuring receiver performance and the two series of tests due to the American I.R.E. (Ref. 1) and the English R.M.A.* (Ref. 2) differ in important details. The main requirements of the two testing specifications are given in this chapter.

(ii) Definitions

(A) **Standard input voltages** (I.R.E.) " Four standard input voltages are specified for the purpose of certain tests, as follows :

(1) A "distant-signal voltage " is taken as 86 decibels below one volt, or 50 microvolts.

(2) A " mean-signal voltage " is taken as 46 decibels below one volt, or 5000 microvolts.

(3) A " local-signal voltage " is taken as 20 decibels below one volt, or 100 000 microvolts.

(4) A " strong-signal voltage " is taken as one volt.

(B) **Antenna sensitivity-test input** (I.R.E.) " The sensitivity input is the least signal-input voltage of a specified carrier frequency, modulated 30 per cent at 400 cycles and applied to the receiver through a standard dummy antenna, which results

---

*This standard is now obsolete and it is understood that a new R.I.C. standard is in course of preparation. References to the old standard have been included where it appears that they convey useful information or follow current practice.

in normal test output when all controls are adjusted for greatest sensitivity. It is expressed in decibels below 1 volt, or in microvolts."

(C) **Normal test output** (I.R.E.) " For receivers capable of delivering at least 1 watt maximum undistorted output, the normal test output is an audio-frequency power of 0.5 watt delivered to a standard dummy load.

For receivers capable of delivering 0.1 but less than 1 watt maximum undistorted output, the normal test output is 0.05 watt audio-frequency power delivered to a standard dummy load. When this value is. used, it should be so specified. Otherwise, the 0.5-watt value is assumed.

For receivers capable of delivering less than 0.1 watt maximum undistorted output, the normal test output is 0.005 watt audio-frequency power delivered to a standard dummy load. When this value is used, it should be so specified.

For automobile receivers, normal test output is 1.0 watt audio-frequency power delivered to a standard dummy load."

The English R.M.A. recommends a standard output of 50 milliwatts measured across a non-inductive resistance connected in place of the speaker voice-coil and with an impedance equal to the modulus of the voice-coil impedance at 400 c/s.

A level of 1 mW has been suggested (Ref. 3) for communication receivers intended for phone operation or for connection to a land line. An alternative of 0.1 mW is suggested in Ref. 4 for special conditions.

The use of 500 mW for large receivers has the advantage of decreasing the effects of hum and noise on the standard output, but when simple a.v.c. is used the sensitivity may be affected by the a.v.c. before the standard output is reached, so that sensitivity and the a.v.c. characteristic both affect the results obtained.

(D) **Maximum undistorted output** (I.R.E.) " The so-called maximum undistorted output is arbitrarily taken as the least power output which contains, under given operating conditions, a total power at harmonic frequencies equal to 1 per cent of the apparent power at the fundamental frequency. This corresponds to a root-sum-square total voltage at harmonic frequencies equal to 10 per cent of the root-sum-square voltage at the fundamental frequency, if measured across a pure resistance. (The root-sum-square voltage of a complex wave is the square root of the sum of the squares of the component voltages.)"

(E) **Bandwidth** (I.R.E.) " As applied to the selectivity of a radio receiver, the bandwidth is the width of a selectivity graph at a specified level on the scale of ordinates."

(F) **Standard antenna** (I.R.E.) " A standard antenna is taken as an open single-wire antenna (including the lead-in wire) having an effective height of 4 meters."

(iii) Equipment required

(A) **Standard signal generator** (English R.M.A.) " Signal generators shall give readings of microvolts to $\pm$ 10% up to 10 megacycles per sec. and to $\pm$ 25% above 10 megacycles. Their frequency calibration shall be within 1% or within 10 kc. whichever is the greater and incremental frequencies shall be within 0.5 kc ; the modulation shall be accurate to within 1/10 of the nominal percentage."

I.R.E. recommendations include the following three points :—

(1) At broadcast frequencies an output voltage indication accuracy within 10% is usually adequate. At higher frequencies an accuracy of indication within 25% is satisfactory.

(2) A frequency indication accuracy of 1% is generally sufficient, but for selectivity or interfering-signal tests the radio frequency should be adjustable in small increments about the desired frequencies and adjustment and indication to within 0.1 per cent of the carrier frequency are desirable.

(3) Frequency modulation should be kept as low as possible. The maximum permissible frequency modulation in cycles is given by $(50 f_c + 100)m$, in which f_c is the carrier frequency expressed in megacycles and m the modulation factor.

Built-in 400 c/s modulation facilities are usually provided, at least for 30% modulation and sometimes for varying depths. The 400 c/s source should have low distortion (not greater than 1%). It is desirable for modulation depths between 0 and

90% to be available and-modulation distortion should be small.   There should be
provision for the connection of an external a-f generator so that the modulation frequency can be varied between the limits of 30 and 10 000 c/s.

The r-f output should be variable at least between one microvolt and one volt,
but a maximum output of two volts is useful.

A smooth, backlash-free drive is necessary, with a movement slow enough and
scale large enough for small increments of frequency to be accurately recorded when
selectivity curves are being taken.

Fig. 37.1.   Standard
dummy antenna and method
of connection (from Ref. 1).

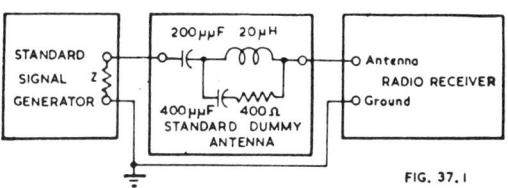

FIG. 37.1

(B) **Standard dummy antenna** (I.R.E.) " The elements of the standard dummy
antenna are capacitors (C_1 and C_2) of 200 and 400 micromicrofarads, respectively,
an inductor L of 20 microhenrys, and a resistor R of 400 ohms, connected as shown
in Fig. 37.1.   The effective values of R, L, and C should be within 10% of the nominal
values.   The stray capacitance between any two points must be so small as to be
negligible at operating frequencies, and the dummy antenna must be so devised as
to avoid coupling to other equipment.   If the output impedance of the attenuator of
the signal generator is not negligible with respect to that of the dummy antenna,
this impedance should be deducted from the respective constants thereof.

The leads used in connecting the standard-signal generator through the dummy
antenna to the receiver should be so short as to introduce negligible voltage drop.
They should be shielded to reduce external fields."

(C) **Output-power-measuring device** (I.R.E.) " The standard dummy load
is a pure resistance which should be of sufficient power capacity to carry the maximum
power output of the receiver without change in its resistance.   If provided with taps,
these should be sufficiently numerous to allow adjustment to within 10% of the proper
value.   The precise value of the resistance of each step should be known.

Dry-rectifier-type voltmeters, vacuum-tube voltmeters, or thermocouple-type
ammeters are suitable for measuring the power delivered to the standard dummy
load.   They may read root-sum-square values and be calibrated in current, voltage,
power, or directly in decibels, the former units being more commonly used.   While
dry-rectifier-type voltmeters are subject to temperature and frequency errors these
are usually not of sufficient magnitude to affect the measurements seriously, so long
as distortion is not high."

(D) **Audio-frequency generator** The output of the a-f generator should be
sufficient to modulate the r-f signal generator to 100% from 30 to 10 000 c/s, and to
drive the output valve of the receiver to full output when connected to its grid circuit.
The maximum harmonic distortion should not exceed 1%.

(E) **Audio-frequency attenuator** An attenuator is desirable which gives known
a-f output voltages continuously variable between 1 V and 1 mV when supplied with
a 1 V (or larger) input.   Commercial types are available with an output range of 1 V
to 1 μV.   The calibration should be accurate (or the correction known) over the
range of frequencies to be used.

(F) **Wave analyser** A wave analyser is required to measure the relative magnitudes of harmonics and inter-modulation products.   The I.R.E. requirements for
such an instrument are " the accuracy of measurement of each frequency should be
within 5% of the indicated harmonic amplitude, and sufficient selectivity should be
available to prevent adjacent harmonics at any measured frequency from influencing
the results.   In this type of analysis it is, of course, necessary that the signal-generator
harmonics be small in comparison with the harmonic distortion being measured."

A distortion-factor meter can be used to measure total distortion* i.e. the receiver output at all frequencies except the fundamental of the signal source. It is worth noting that the English R.M.A. definition of " distortion factor " is

$$\frac{1}{2}\sqrt{\frac{\sum\limits_{n=2}^{n=\infty} n^2 V_n^2}{\sum\limits_{n=1}^{n=\infty} V_n^2}} .$$

where n represents the order of an individual harmonic. This expression† loads the higher harmonics in proportion to their order.

An I.R.E. recommendation is " If an instrument is used which makes a lumped measurement of all distortion, by elimination of the fundamental, a high-pass filter should be used, in order to eliminate hum and similar low-frequency noise from the result."

(G) **Meters.** Meters are required to cover the range of voltages and currents to be measured. Valve voltmeters are useful for measuring small a.c. voltages and currents and for making measurements in which a minimum of circuit loading is necessary (e.g. a.v.c. bias voltages).

(H) **Equipment for acoustical tests** A number of I.R.E. and English R.M.A. tests are of an acoustical nature. For information on the equipment required and the testing conditions, Refs. 1 and 2 should be consulted.

(iv) Measurements and operating conditions

(A) **Output measurements** The I.R.E. requirement is that output measurements are to be made in terms of the power delivered to a standard dummy load, which is a pure resistance of a value equal to the 400 c/s impedance of the loudspeaker to be used with the receiver. Where an output transformer is connected between the radio receiver and the loudspeaker, the output transformer is to be treated as part of the radio receiver.

There are special requirements when power output is to be measured in the presence of background noise. If the background noise power is smaller than the output power being measured, the incremental reading of output power may be used. With a thermocouple output meter the incremental output power is equal to the observed total power minus the observed noise power. If another type of output meter is used, a calibration should be made in terms of incremental power.

If the background noise is greater than the output power being measured, it is desirable to use a band-pass filter tuned to the test audio frequency to remove the background noise wholly or partially from the output meter. This filter should be connected between the load and the output meter.

Alternatively an electrodynamometer with its field synchronously excited from the original source of modulation can be used.

(B) **Summary of additional I.R.E. requirements** " The operating voltage applied to a radio receiver should be held constant at the specified value during measurements of receiver characteristics. Certain receiver characteristics may be desired at other than normal test voltage, or over a range of operating voltages, in which case a statement of the voltages used should be included in the test data. In any case, tests should be made to check whether the receiver operates satisfactorily over the full range of operating voltage that is liable to be encountered in practice.

A.C., A.C./D.C., or D.C. receivers should be tested with 117 volts r.m.s. A.C. or 117 volts D.C. applied, unless some other operating voltage is stated.

Automobile receivers should be tested with 6.6 volts at the receiver-battery terminals. The voltage should be obtained from a battery being charged at the required rate and not by the use of a dropping resistor. 32-volt farm-lighting-plant receivers are tested with 36 volts.

---

*Total harmonic distortion, see Chapter 14 Sect. 2 (iv) p. 609.
†This is the same as the Weighted Distortion Factor described in Chapter 14 Sect. 2(v) p. 610.

Other battery-operated receivers are tested with batteries of the type and voltage specified. The batteries used should not have abnormally high internal resistance. The valves used should be selected to have the rated values of those characteristics which most affect the performance of the receiver."

(C) **Summary of English R.M.A. requirements**   The receiver is to operate in its cabinet for at least half an hour before tests begin.

All batteries are to be at their normal operating voltages. The internal resistance of the H.T. battery is to be increased by adding, in the H.T. negative lead, a resistance equal to 1 ohm per cell. Care should be taken to see that the added resistance does not affect the grid bias.

Where the receiver carries a mains tapping device and a number of marked ranges, the applied voltage should fall in the middle of one of these ranges, the midpoint being calculated as a geometric mean.

(v) **Measurements**

Tests are applied to receivers for various reasons ; as an aid to design, to compare different designs, to compare a mass-produced receiver with a laboratory prototype and as a preliminary to repairs. Appropriate tests should be selected from the following sections to suit the required object, or perhaps the available equipment. An arbitrary distinction has been drawn between measurements (carried out with volt-meters and ammeters) and tests, which in general need more equipment.

For normal **voltage and current measurements** receivers and amplifiers should be operated without input or output, and bias-controlling volume controls should be turned to maximum.

If all voltages are to be checked it is necessary to use a high impedance meter (20 000 ohm/volt or more) but most important checks can be made with 1000 ohm/volt meters. In any case allowance must be made for meter resistance in all high impedance circuits and it is useful to know the actual value of resistance for any particular range in use.

Measurements of grid bias should normally be made either across the cathode bias resistor (if any), or between the cathode and the earthy end of the grid resistor. This will give the grid-cathode voltage unless current is flowing in the grid resistor. The latter may be checked by short-circuiting the grid resistor with a suitable current meter in the plate circuit, and noting any change in reading.

There are two methods which can be used to determine plate and screen voltages when high values of series resistors are used. The first is the direct method, noting the reading on a high impedance voltmeter and making allowance for the meter resistance. If a suitable two range voltmeter is available, the method described by Lafferty may be used to determine the correct voltage—see Chapter 4 Sect. 7(iii) pages 163-164, and eqns. (10) and (11). Alternatively, the series resistor in the voltmeter may be shunted by an external resistor of equal value, with a switch to remove the shunt when required. This gives a scale ratio of 2, and enables eqn. (11) on page 164 to be used. The shunt resistor may well take the form of a decade box, accurately adjusted to the internal voltmeter resistance. Other methods of making allowance for meter resistance may introduce serious errors.

The second method of measuring plate and screen voltages, which may be called the indirect method, is to measure the voltage from B+ to cathode and also the current to the electrode. The value of the series resistor should then be measured, and the voltage drop across it may be calculated by Ohm's Law.

An example of the error which may be introduced by other methods is given below. Suppose that a 20 000 ohm/volt meter on the 25 volt range is used to measure the bias voltage (from grid to cathode) on an output valve which has back bias and a 0.5 megohm grid resistor.

Although the meter resistance is equal to the resistance of the grid leak so that half of the bias voltage appears across the meter, only one half of the generated bias voltage is applied to the grid of the valve, its plate current is increased, and a larger voltage is developed across the back-biasing resistor. The indicated bias voltage is thus more than half the working bias voltage.

When voltages and currents are being measured to check valve electrode dissipation or component ratings it may be necessary to modify receiver operating conditions to obtain the worst conditions for the component in question. For example tuning a strong signal with a back-biased receiver will decrease the bias on the output valve and may cause the plate dissipation to exceed the limit. The effect of component tolerances must also be taken into account.

A measurement of, say, oscillator grid current at low voltages as part of a type test in a battery receiver at low voltages would not be considered finished until many valves had been measured to determine the average and minimum values, or unless valves with bogie and limit characteristics were available.

The maximum **temperature rise** allowed in radio power transformers by R.M.A. specification M4-541 is 65°C (core or windings whichever is greater) when operated continuously with the specified load, at the specified frequency, under standard conditions and at a voltage 10% above that of the standard line voltage.

The standard conditions are set out in M4-546 and require a specially constructed asbestos box. The core temperature is measured with an all-glass mercury bulb thermometer brought into good contact with the surface of the core by means of glazier's putty.

The temperature of the primary and secondary windings are determined by the resistance method using the following formula : (A.I.E.E. Rule 13-207)

t = reference temperature of winding
T = temperature of winding to be calculated
r = resistance at reference temperature t
R = observed resistance at temperature T

$$T = \frac{R}{r} (234.5 + t) - 234.5.$$

The observed resistance R is to be measured within one minute from the completion of the test run, that of the highest voltage secondary winding being measured first and that of the primary winding being measured second.

The temperature rise of the transformer is determined by subtracting the temperature of the ambient air within the enclosure at the end of the test from the temperature of the core or windings (whichever is highest) at the end of the tests.

In determining the ultimate temperature of the core and windings, the operation of the transformer is continued until temperatures within the enclosure show constancy for three successive readings at intervals of at least 15 minutes.

An ambient temperature of 40°C is maintained within the enclosure throughout the test.

(vi) Performance tests
(A) Sensitivity

All measurements, including sensitivity, which are made with the signal generator output applied to the aerial terminal of a receiver require a standard dummy antenna to be connected between the two, except that in the case of a receiver designed for use with a special aerial, a dummy aerial is required with constants giving an impedance characteristic similar to the special aerial. The sensitivity is expressed as the input required to give standard output from the receiver. The input referred to is the open circuit voltage from the signal generator, modulated 30% at 400 c/s.

For purposes of recording results sensitivity is usually measured at or near the three tracking points on each wave range, but when curves are to be drawn additional points are required.

With sensitive receivers noise forms a large part (or all) of the output in sensitivity measurements, and alternative methods of measurement which specify the input required to give a minimum signal-to-noise ratio are becoming increasingly popular. One method is to specify the input required to give equal signal and noise outputs i.e. when the modulation is switched off the output power is halved. A signal-to-noise ratio of 15 db is more commonly used for communication work in England, this ratio being taken as providing satisfactory intelligibility.

When the signal-to-noise ratio method of measurement is used for all types of receivers it is necessary to make provision for insensitive receivers by defining sensitivity as the input required to give a certain signal-to-noise ratio, or the input required to give standard output, whichever is the greater.

Sensitivity measurements on telegraphy receivers are carried out by adjusting the beat note to the peak of the a-f response.

To check the gain of individual stages in a receiver, a measurement is made of the input to successive grids required to give standard output from the receiver. The generator is coupled to the receiver by means of a large capacitor (say 0.1 μF) and the normal connections to the grid are not disturbed.

Measurements made in this way are liable to errors at high frequencies and when regeneration is present. At high frequencies the inductance of the generator output leads and the input capacitance of the stage being measured may seriously affect the results.

During measurement, regenerative effects may be eliminated by the low impedance of the signal generator connected between grid and ground but when the gain from the previous grid is measured the regeneration will re-appear so that the apparent gain of the earlier stage is its true gain multiplied by the gain due to regeneration in the subsequent stage.

The regeneration may be from causes such as feedback due to the plate-grid capacitance of an i-f amplifier, or in the case of a converter it may be due to oscillator voltages on the signal grid. For the latter reason, measurements of conversion gain and r-f gain (or aerial coil gain if no r-f stage is used), particularly at high frequencies, should be treated with reserve unless it is known that these effects were not present.

(B) Selectivity

Selectivity measurements are made with the object of determining the frequency separation necessary to eliminate interference between a desired signal and a stronger undesired signal.

It is not possible to specify one selectivity test the results of which will be a measure of the interference to be expected, because there are several ways in which interference can occur.

These are :—

(1) Because of inadequate selectivity the two modulated carriers may both be present at the second detector, and although no noticeable interaction occurs between the signals, they are both heard.

(2) The desired carrier may be modulated by the modulating frequencies of the undesired carrier (cross-modulation).

(3) A reduction of the a-f output from the desired carrier may be caused by a.v.c. voltages developed by the undesired carrier, or by demodulation by the undesired carrier.

(4) When the frequency separation between adjacent carriers is an audible frequency a continuous heterodyne is heard. This effect is very noticeable in receivers with a good high frequency response as they are capable of reproducing frequencies of 9 or 10 Kc/s by which medium wave broadcasting stations are separated.

(5) Intermodulation may occur between two stations of which the frequency difference or frequency sum is the same (or nearly the same) as the frequency of a desired station. An intermodulation product can then interfere with a desired signal.

(6) Distortion in the receiver may produce harmonics of an undesired signal which interfere with a desired signal.

(7) " Monkey chatter " may occur due to the sidebands of the undesired carrier being detected, with frequency inversion, as sidebands of the desired carrier. For example, if the desired carrier is separated by 20 Kc/s from the undesired carrier, which is lower in frequency and which is modulated by frequencies of 7 and 12 Kc/s, the higher sidebands of the undesired carrier become respectively 13 and 8 Kc/s lower sidebands of the desired signal.

Sturley (Ref. 3) points out a less commonly recognised extension of this effect. The presence of small percentages of a-f harmonics may greatly extend the carrier frequency separation across which monkey chatter may occur. For example, suppose a strong local station on 1000 Kc/s induces 1 volt 50% modulated at 7 Kc/s in the aerial of a receiver tuned to 1030 Kc/s. If there is a fourth harmonic sideband percentage of 0.1% the amplitude of the upper frequency harmonic sideband (1024 Kc/s) is 500 μV. A station on 1030 Kc/s would thus have added to it a 6 Kc/s sideband of ample power to cause over-modulation at low signal levels.

An investigation into the poor agreement between selectivity as measured by the one or two signal generator methods and actual performance on the air is detailed in Ref. 13.

The I.R.E. recommendation for a single signal selectivity test is that the receiver be tuned to the test frequency as in the sensitivity test. The signal generator is then detuned each side of resonance, the radio-frequency input voltage which results in normal test output is observed, and its ratio to the sensitivity-test input is computed. If this test is made at only one frequency in superheterodyne receivers it is recommended that this be 1000 Kc/s.

When the selectivity of the a.v.c. circuit is the same as that of the signal circuit no special precautions are necessary. However if there is any appreciable difference between the two selectivities it is recommended that the a.v.c. voltage be maintained at the value obtained at centre frequency. This method provides an indication of the circuit selectivity rather than of discrimination against interference.

Errors may occur in very selective receivers because the receiver output is the result of applying a band of frequencies (carrier and upper and lower sidebands) to a sloping characteristic. To avoid this effect an unmodulated carrier can be used and diode current in the second detector circuit taken as the output indication.

Another point which may give trouble is that with very sensitive receivers the majority of the standard output at full sensitivity is noise. A different signal-to-noise ratio at standard output will be experienced when tuning towards a strong carrier and a faulty reading will be obtained. To overcome this a reference input is often selected to give a good signal-to-noise ratio while still leaving the a.v.c. inoperative—10 μV is a suitable input in many cases.

Two-signal tests are specified by the I.R.E. for cross-talk, whistle and blocking interference. The signal generators are preferably connected in series but parallel connection is possible with special dummy aerials. The radio receiver is tuned to the desired signal at one of the standard test frequencies and at one of the standard input voltages. The receiver volume control is adjusted to give normal test output when the signal is modulated 30% at 400 c/s, after which the modulation is switched off.

An interfering-signal input voltage is applied to the receiver, in addition to the desired signal which remains unchanged. The interfering signal is modulated 30% at 400 c/s. The interfering signal is tuned through a wide frequency range and the interference-test input voltage which gives interference-test output at 400 c/s (30 db below standard output) is observed wherever its value is less than one volt. Readings are not taken for carrier spacings so close that beat note interference is more severe than cross-talk interference.

Such a test is the only type of selectivity test which shows correctly the selectivity curve, at reduced sensitivity, of a receiver having a.v.c. It is also the only method by which the selectivity of a receiver having automatic selectivity control can be tested directly.

The interfering signal may also reduce the desired-signal output by a.v.c. action or overloading. Blocking interference tests can be carried out for this effect (as specified by the I.R.E.) by modulating the desired signal and measuring the change in output with varying signal strength and frequency of the unmodulated signal. The results of blocking signal tests can be plotted on the same sheet as the results of the two-signal cross-talk tests.

The two-signal whistle interference test specified by the I.R.E. indicates the greatest interference input permitted without the output exceeding the equivalent of approximately 1% modulation of the desired signal. If the desired signal is modulated 30%, the permitted interference output power is 0.001 of the desired-modulation output power.

The signals are the same as for the cross-talk test, except that the interfering signal is unmodulated. The interfering signal is tuned through a wide frequency range, and the interference-test input voltage, which gives interference-test output at 400 c/s, is observed wherever its value is less than one volt. No reading is made with the desired and undesired signals separated by 400 c/s.

The English R.M.A. selectivity test is similar to the I.R.E. cross-talk interference test but a 400 c/s bandpass filter is used between receiver and output meter to eliminate the effect of the heterodyne between carriers. The desired signal is 1 mV, 30% modulated at 400 c/s and the receiver volume control is adjusted to give one quarter of the rated maximum output. Modulation is removed from the desired signal and the input required to give an output 40 db below the previous output is noted for various off-tune frequencies of the 400 c/s 30% modulated undesired carrier.

In Ref. 40 a test for r-f intermodulation is given. A " wanted " signal 30 db above 1 μV is applied to a receiver, the sensitivity control is adjusted for standard output (a.v.c. switched off) and the signal is then removed.

Two interfering signals are then applied to the receiver with a frequency difference or frequency sum equal to the frequency of the wanted signal. The interfering signals are each 110 db above 1 μV and neither is of such a frequency as to give an appreciable output when applied alone. Under these conditions the receiver is not to produce an output exceeding the standard output.

When this test is used care should be taken to see that intermodulation does not occur outside the receiver, for instance in measuring diode circuits of signal generators.

A test for r-f harmonic generation in the receiver is also given in Ref. 40. Standard output is obtained from a wanted signal as above, and then the wanted signal is replaced by an unwanted signal of half the frequency and at a level of 116 db above 1 μV. The output from the receiver should not exceed the standard output. The effect of signal generator harmonics must not be overlooked when this test is being performed.

Sturley (Ref. 3) gives additional tests whereby the effects of heterodyne whistles, monkey chatter and a receiver whistle filter can all be shown on a set of curves.

(C) Electric fidelity (frequency response)

The electric fidelity test specified by the I.R.E. shows the manner in which the electric output of a receiver depends on the frequency of the a-f modulation. The radiation characteristics of the loud speaker are not taken into consideration and this minimizes the value of the test.

30% modulation of a 1000 Kc/s 5000 μV signal is used, and the receiver is tuned to it in the following manner. Firstly the receiver is tuned to the carrier with 400 c/s modulation, and then the modulating frequency is increased until the receiver output has fallen to one fifth. The receiver tuning is then readjusted slightly for **minimum** output.

The receiver output is measured in terms of current or voltage in a standard dummy load, or in terms of the voltage across or of the current in the voice coil of the loud speaker. In the latter case, the loud speaker should be located in its baffle or cabinet, with the receiver chassis in place. The data should include a statement of which load is used. The receiver volume control is adjusted to give normal test output. The modulation frequency is then varied continuously from 30 to 10 000 cycles while maintaining 30% modulation and the output variation is observed.

If the electric-fidelity curve has decided peaks, there is abnormal tendency towards overloading, and the observations may have to be repeated with less output.

It is frequently unnecessary to make or plot these observations below −20 db, but further observations may be desirable, particularly if a large amount of negative feedback is used.

Suitable precautions must be taken to prevent hum or noise, if present, from affecting the accuracy of the results.

If electric-fidelity varies appreciably with the signal frequency, the input level, the volume or tone control settings or any other parameter, additional tests should be made to show the effect of these changes.

(D) Harmonic distortion (I.R.E.)

This test is intended to evaluate the spurious a-f harmonics which appear in the electric output of the radio receiver during normal operation. Care should be taken to avoid appreciable harmonic distortion occurring in any part of the signal-generating equipment, or in the output-measuring circuit. The required harmonic-measuring equipment in the output circuit should not appreciably affect the output load conditions. This equipment may measure each harmonic individually or may measure all harmonics collectively. The proper tuning of the receiver is important in making distortion tests.

No one complete set of conditions can be prescribed for this test, because harmonic distortion depends on so many details of radio receiver design and operating conditions. Harmonic distortion is caused by overloading and many other phenomena, and is present under various operating conditions, especially at high degrees of modulation. The following series of tests is intended to show the effect of operating parameters on distortion.

Variation of output. The receiver is tuned to 1000 kilocycles and a "mean-signal" input, modulated 30% at 400 cycles, is applied. The distortion is noted as the output of the receiver is varied by means of the volume control.

Variation of modulation. With a "mean-signal" 1000-kilocycle, 400-cycle-modulated input, the modulation is varied from 10 to 100% and the distortion observed. The output is maintained at normal test output by the volume control for this test, or as near this value as possible.

Variation of input signal level. With a 1000-kilocycle signal, modulated at 400 cycles, the distortion at normal test output is noted as the input signal voltage is varied. This test is to be performed at both 30% and 80% modulation. It is usually sufficient to take distortion at the standard input levels which fall within the limits of the receiver. When a standard-signal level exceeds the limit of the receiver, a measurement should be made at that limit.

Variation of modulation frequency. To disclose the effect of the modulation frequency on distortion, tests on variation of output and variation of modulation should be repeated at several modulation frequencies throughout the a-f range. The maximum modulation frequency at which harmonic distortion can be measured is one-half the maximum frequency which can produce any appreciable output.

The harmonic distortion is measured across a standard dummy load and may be measured as either root-sum-square total harmonic distortion or each harmonic may be measured separately. It is expressed as the ratio of the harmonic voltage to the fundamental voltage either in per cent or decibels.

Intermodulation measurements. Owing to the lack of signal generators suitable for r-f intermodulation work, these measurements are normally carried out only on the a-f end of radio receivers. See Sect. 3(ii)A of this chapter for details.

(E) Maximum undistorted output (I.R.E.)

" This test is intended to indicate the maximum power output which the receiver will deliver under given conditions, before appreciable overloading or other forms of distortion occur. The maximum undistorted output may be determined under given conditions by observing the total harmonic distortion, and continuously increasing the output from zero up to the least value which contains a total harmonic distortion of 10% (root-sum-square voltage). This value is designated the maximum undistorted output under the given conditions.

The data should include a statement of the operating conditions, including which condition was varied in order to increase the output during this test. It is suggested that the volume control of the radio receiver be varied, the other conditions being unchanged during a single test and being chosen as suggested for the harmonic-distortion test. Freedom from distortion depends on the r-f input voltage and on the frequency and percentage of modulation.

It is understood that there is no sharp dividing line between appreciable and negligible distortion. The figure of 10% has been chosen somewhat arbitrarily as a reasonable basis for the definition of maximum undistorted output as affected by all operations in the receiver."

(F) Automatic volume control

The I.R.E. recommendation is to tune the receiver to a mean-signal input voltage 400 c/s 30% modulated at 1000 Kc/s and to adjust the volume control so that with one volt input to the receiver the output is one half of the maximum power output. The output is then read as the input is varied from one microvolt to one volt.

A disadvantage of this method is that the output readings for low inputs, from which much valuable information can be obtained, are carried out at very low levels (perhaps below 1 mW) and hum in the receiver output may be troublesome.

Scroggie (Ref. 6) suggests beginning the a.v.c. curve with the volume control set at maximum and with the minimum input. Output readings are taken with increasing input until the output reaches one quarter of the maximum. It is then reduced to one tenth of this value by means of the volume control and the input is further increased. This procedure is repeated until the maximum input is applied to the receiver. The sections of the curve are made continuous by the use of appropriate multiplying factors. This method has many advantages—Chapter 27 Sect. 3(xiv).

Useful information can also be obtained from a noise curve plotted on the same sheet. The same procedure is adopted as for drawing the a.v.c. curve but the input carrier is unmodulated.

Other information sometimes added to a.v.c. curves includes graphs of distortion, a.v.c. bias voltage and electrode current and dissipation for the controlled valves vs. input voltage, and it may be useful to draw a horizontal line across the a.v.c. curve (when drawn by Scroggie's method) at a level representing the nominal output of the receiver. Alternatively the maximum output of the receiver may be plotted.

When the converter valve has a.v.c. applied to it, detuning of the signal may occur as the signal input is varied. This should be remedied by retuning the receiver with each input level change if necessary.

(G) Noise

The I.R.E. recommendation is that noise should be expressed as the equivalent-noise-sideband-input (ensi). An unmodulated input carrier of suitable strength, E_s is applied to the receiver, and the noise output power P_n is measured on a R.M.S.-reading instrument. A 400 c/s bandpass filter is connected between receiver and output meter, the carrier is 30% modulated at 400 c/s and the signal power output P_s is measured. Then

$$\text{ensi} = 0.3\ E_s \sqrt{P_n/P_s}.$$

E_s should be at least three times and preferably ten times greater than the computed noise voltage.

Excessive low frequency hum components may usually be filtered out by the use of a 300 cycle high-pass filter without materially affecting the random-noise output.

An alternative method of specifying noise is by comparing the signal-to-noise ratio of a particular receiver with that of an ideal receiver, the difference being called the noise factor of the receiver in question. Ignoring any noise external to the receiving system, the only source of noise which is inseparable from an incoming signal is the thermal agitation noise due to the radiation resistance of the antenna. Thus a signal-to-noise ratio has its highest possible value when the noise due to the antenna radiation resistance is the only noise source in a receiving system.

Where a signal voltage e_{ant} is induced into an antenna of radiation resistance R_{ant}, maximum signal power is delivered to a receiver with an input resistance of R_{ant}. This maximum power available from the antenna is

$$\left(\frac{e_{ant}}{2}\right)^2 \times \frac{1}{R_{ant}} = \frac{e^2_{ant}}{4R_{ant}}.$$

Similarly the noise from the antenna available to the receiver is

$$\left(\frac{e_{noise}}{2}\right)^2 \times \frac{1}{R_{ant}} = \frac{4KT\Delta FR_{ant}}{4R_{ant}} = KT\Delta F,$$

where K = Boltzmann's constant = 1.38×10^{-23} joule/degree Kelvin
    T = absolute temperature in degrees Kelvin
    F = bandwidth of system in cycles per second,
and the signal to noise power ratio of an ideal system is

$$\frac{e^2_{ant}}{4KT\Delta FR_{ant}}.$$

The noise factor of an ideal receiver can thus be defined as the number of times by which the available signal power from the aerial must exceed $KT\Delta F$ in order to give unity ratio of available signal to noise power.

The noise requirement for such an ideal receiver is that it must contain no internal sources of noise. It must also present an infinite impedance to the antenna so that there will be no loss of signal voltage. If for instance the input impedance of the receiver were to match the antenna impedance the voltage available to the receiver would be one half of the signal voltage induced into the antenna and this would be developed across one half of the antenna impedance due to the parallel connection of the antenna and the matching input circuit. Substitution of these factors in the expression above shows that in such a case the maximum noise factor obtainable is 2, i.e. 3 db:

Since the requirement for maximum power transfer is that the input impedance of the receiver should be equal to the antenna impedance, it is evident that the conditions for maximum gain are in general different from those for minimum noise. The best noise figure in practical receivers is obtained when the input impedance is intermediate between an open circuit and an impedance match with the antenna system.

When noise calculations are being made the value $T = 290°K$ is commonly used as this gives KT the convenient value of 4×10^{-21}. However, the effective temperature of the radiation resistance is subject to wide variations, and is affected by the surroundings with which the antenna can exchange energy by radiation. For instance, the noise temperature of a highly directional micro-wave radar antenna pointed at the depths of space may be only slightly above absolute zero. On the other hand a directional 7 metre antenna pointed at parts of the Milky Way may have an effective temperature much higher than ambient.

The noise factor of a receiver can be obtained by using a signal generator or a noise diode (see below). With a signal generator, difficulty is experienced in that the characteristic of the detector in the receiver may affect the ratio of signal power to noise power. If the detector is a true power sensitive device such as a thermocouple no error occurs, but with normal detectors error is introduced to a degree dependent on the variation between the assumed and the actual detection characteristic.

On the other hand, if the signal generator is replaced by a noise diode, detector distortion can be ignored since the noise from the diode has the same spectrum as the noise from the antenna.

The signal generator method is to apply the generator output to the receiver through a resistor R_{ant} and to adjust the output until the indication at the receiver detector is twice that from noise alone. The noise factor is then obtained by calculation from the formula

noise factor $F = \dfrac{e^2_{s.g.}}{4KT\Delta FR_{ant}}$

where $e_{s.g.}$ = voltage output of signal generator.

Alternatively the anode current of a temperature limited diode (noise diode) can be used as a noise source. The diode is connected with suitable d.c. isolation across the receiver input, which is also shunted by a resistor equal in value to R_{ant}. The d.c. anode current of the diode is then increased until the noise output of the receiver alone is doubled, and under these conditions the noise factor of the receiver is

noise factor $F = \dfrac{2eI_b\Delta FR^2_{ant}}{4KT\Delta FR_{ant}}$

where e = charge on an electron = 1.59×10^{-19} coulombs and I_b = diode plate current in amperes.

therefore noise factor $F = 20\ I_b R_{ant}$ at 290°K.

A comparison of the two noise factor formulae shows a second advantage of the noise diode method, viz. : the receiver bandwidth need not be known.

Additional information on noise factors will be found in Refs. 7, 8 and 9.

(H) Hum

Hum is a low pitched composite tone which may include a component at any integral multiple of the a-c power supply frequency. It may be due to the loud-speaker itself, to the a-f stages or to the r-f stages.

When the field winding of a speaker is used for filtering, hum may appear in the voice coil even although some neutralizing is accomplished by connecting a hum-bucking coil in series with the voice coil. The I.R.E. recommendation is that this type of hum " should be measured in terms of the hum current through the loud-speaker voice coil itself, rather than in terms of the voltage across the loudspeaker, and the total hum calculated using this current and the loudspeaker voice coil im-pedance. The loudspeaker is connected in the normal manner to the radio receiver when the observations are made. The current measuring equipment should intro-duce into the voice coil an impedance which is negligible as compared with the voice coil impedance. In the case of a loudspeaker having a field coil carrying hum current, this procedure evaluates the combined effect of hum originating in the radio receiver itself and hum induced in the voice coil from the field coil, with due regard to their phase relations."

" **Hum from the a-f stage** of the receiver should be measured

    (a) with the volume control at minimum,

    (b) with the volume control left at that setting which would produce normal test output with mean-signal input, but with the intermediate-frequency-ampli-fier circuits inactivated, as by-passing the last i-f plate to ground,

    (c) with the volume control at full volume and the i-f system inactivated.

" In the case of phonograph combinations, a-f hum should also be measured with the phonograph pick-up connected to the a-f amplifier under the following conditions :

    (a) with the phonograph motor de-energized, the pick-up on the rest, and the volume control at minimum

    (b) same as (a) but with the volume control at maximum

    (c) with the volume control adjusted, while reproducing the outside 1000 cycle band on R.M.A. Frequency Test Record No. 1 to give normal test output, the needle then lifted not more than ¼ inch above the record, with the motor still running."

" **Hum modulation** is produced by hum sources which modulate a carrier being received, and its intensity generally increases with increasing carrier voltage.

The hum-modulation test is intended to evaluate the hum components introduced in a radio receiver by hum disturbances modulating the received carrier. In order to measure hum modulation, as distinguished from a-f hum, the former is accentuated by the adjustments of the receiver. The receiver is tuned in the normal manner to each of the four standard input voltages at 1000 kilocycles. If the receiver has a tone control, it should be set in the " high " position. The volume control is first adjusted to give the normal test output with the given signal voltage modulated 30% at 400 cycles and the modulation is then reduced to zero. The hum is measured with the load and output-measuring equipment connected as in the electric-fidelity test, and is expressed in terms of reduction in decibels below standard test output."

Hum distortion can also occur. It is identified as sidebands of frequencies differing from the audio frequency of modulation by an amount equal to the frequency of the hum disturbance causing the distortion.

A typical design likely to introduce hum distortion is one in which the plate of the output valve is fed directly from the rectifier output (assuming a condenser-input filter). No specific test procedure for measuring this effect is given by the I.R.E. or R.M.A. but owing to its unpleasant effects it should be eliminated if this is possible.

Acoustical measurement of hum output is desirable but difficult by available methods. The results of the electric measurement must be interpreted with reference to electric- and acoustical-fidelity curves and the characteristics of audition.

(I) Frequency shift (I.R.E.)

" This test is intended to show the variation in the frequency of the oscillator of a superheterodyne receiver. The tests are normally performed with the receiver tuned to a frequency in the middle of each tuning range of the receiver. If observations under the worst conditions are desired, the receiver should be tuned to the highest frequency of each range.

" The variation of frequency is observed with the aid of a beat note obtained between the oscillator under test and another oscillator of constant frequency. For example, the frequency of the beat note may be observed by comparison with a calibrated a-f oscillator.

" (a) The frequency varies with time during the warming-up period of the receiver . . . . The time is measured from switching on the receiver but observations are ordinarily started one minute later.

" (b) The frequency varies with power-supply voltage in a manner that depends on the rate of variation of this voltage. The major change occurs almost instantly following a change of the power voltage. Therefore the test is performed as quickly as possible to minimize other effects. In the case of operation from a 115-volt power line, the line voltage is varied at least between 105 and 130 volts and the resultant frequency shift is observed. The amount of frequency shift is expressed in cycles per 1 per cent change of power voltage, as an average value over the specified range of power-line voltage.

" (c) If the receiver has automatic gain control, the variation of signal-input voltage affects the oscillator frequency indirectly by way of the control circuit. The frequency shift with variation of signal-input voltage is observed after the receiver has been in operation a sufficient length of time to reach temperature stability."

(J) Spurious response (I.R.E.)

" With the radio receiver tuned to each of the test frequencies, the signal generator should be continuously varied over a wide frequency range to discover if the receiver is simultaneously resonant at frequencies other than the test frequency. These other resonant frequencies are called spurious-response frequencies and are most often found in superheterodyne receivers. Each spurious-response frequency is noted and the spurious-response sensitivity-test input is measured as in the sensitivity test, provided it is smaller than 1 volt. Its ratio to the desired-signal sensitivity-test input may be computed, and is called the spurious-response ratio."

" Care should be taken that the harmonic output of the signal generator is attenuated sufficiently not to affect the observation of the spurious response of the receiver."

Image response. " A superheterodyne receiver is generally responsive to two frequencies whose difference from the local-oscillator frequency is equal to the intermediate frequency. One of these (usually the lower) is the desired-signal frequency, and the other is called the image frequency. This is a special case of spurious-response frequency, and is tested as such. Its observed characteristics are referred to as ' image-sensitivity test input ' and ' image ratio.' "

Intermediate-frequency response. " Another special case of a spurious-response frequency in a superheterodyne receiver is that due to the sensitivity to an intermediate-frequency signal input. The test procedure is the same as for other spurious responses, and the observed characteristics are referred to as the intermediate-frequency-response sensitivity and the intermediate-frequency-response ratio."

(K) Noise audibility (I.R.E.)

" The actual audibility of random noise, hum, and miscellaneous noise is best determined by a listening test. Such observations are not capable of precision, but are fundamentally sound, as distinguished from less direct electrical observations. The completely assembled and operating radio receiver is placed in a quiet room, and an experienced observer with normal hearing notes the greatest distance at which the noise is audible under stated conditions. The distance is used to express the audibility of the noise. The room is preferably large, or treated to minimize reverberation. This method takes into account noise produced both by loudspeaker radiation and by mechanical vibration of parts. A brief description of the sound heard is useful, in addition to the audibility observation. Obviously, this method is suited only for observing a small amount of noise, audible for only a short distance. The noise-audibility test is intended to evaluate collectively random noise and hum under operating conditions. The radio receiver is tuned in the normal manner to each of the four standard input voltages at 1000 kilocycles. If the receiver has a tone control, it is set in the ' high ' position. The volume control is adjusted to give normal test output with the signal modulated 30% at 400 cycles, and then the modulation is reduced to zero. The audibility of the remaining noise is then observed. The residual noise audibility is likewise observed, with no signal and with the volume control set at minimum."

(L) Radiation from local oscillator (I.R.E.)
" A local oscillator, such as is employed in a superheterodyne receiver, may radiate sufficient power to cause interference in other radio receivers operating in the same neighbourhood. Such radiation may be caused by coupling to the antenna, power line, or other external leads, or by incomplete shielding of the oscillator and the circuits coupled thereto.
The receiver is connected to its proper antenna, and the electric- and magnetic-field intensity in the neighbourhood is observed by any of the known methods. There is no simple form for expressing the results. (The recommendation for local oscillator radiation in F-M receivers is that results are to be expressed in field intensity as a function of distance from the receiver under test. Observations are made at least at the middle frequency of each tuning band, and preferably at the extreme frequencies of each band."

(M) Microphony
Microphony troubles are of two main types, audio frequency only and those involving radio frequencies. A convenient test to apply for the first type—which might be due to feedback between the speaker and a pick-up mounted in the same cabinet for instance—is to feed the receiver into a dummy load with an output meter across it, and excite the speaker, mounted in its normal position by means of a variable a-f oscillator.
The pick-up is allowed to stand on a stationary record and a tendency to microphony is indicated by appreciable output across the dummy load at any particular frequency. When the voltage across the dummy load equals the voltage across the speaker, microphony would occur (assuming equal speaker and dummy load impedance at the frequency concerned). Another speaker can be used instead of a dummy load if desired, so long as no sound from this speaker reaches the receiver. This test has the advantage of indicating the margin available in a particular model before sustained microphony will be experienced. It can also be used for types of microphony involving radio frequencies, but owing to the increased number of possible variables it loses some of its advantages. Tests outlined in (O) 1 below may be used for r-f microphony.

(N) Phonograph combinations (I.R.E.)
" In addition to the above tests on radio receivers, there are several special tests which reveal useful information on the operation of the phonograph portion of combination receivers . . . In addition to hum, the following characteristics should be measured :
   (1) Electric fidelity
   (2) Rumble
   (3) Maximum output
   (4) " Wow " or flutter.
Definitions of these characteristics, with methods of measurements, are given below.
   (1) **Fidelity.** The fidelity of a phonograph reproducer corresponds to the a-f fidelity described in Sect. 1(vi)C of this chapter except that RMA Frequency Test Record No. 1 is used instead of a modulated carrier. This RMA Frequency Test Record No. 1 contains, on side A, a gliding frequency from 10 000 to 30 cycles for

usual nominal level testing. Side B contains five 1000-cycle bands, recorded at 0, 2, 4, 6 and 8 decibels above the 1000-cycle reference on side A, and a gliding frequency 6 to 8 decibels above side A in the range between 3000 and 30 cycles for checking pickup tracking at various levels.

In making the fidelity test, side A is used, and the output is first set at standard output using the outside 1000-cycle band. Readings of output are then made at all announced frequencies, including the 1000-cycle bands at start and finish.

Suitable correction usually must be made for noise resulting from needle scratch to provide accurate data above 1000 cycles.

(2) **Rumble.** Rumble is a low-frequency tone or series of random pulses generated at the phonograph pickup as a result of vibrations of the record player. It is generally a maximum when the needle is near the outside of a 12-inch record.

Rumble is measured with the load and output-measuring equipment connected as in the electric-fidelity test, except that a low-pass filter is used with a sharp cutoff at 300 cycles. The measuring equipment should have low wave-form error.

Side A of the test record is also used in this test. The volume control is adjusted to give normal test output while reproducing the outside 1000-cycle band and with the low-pass filter disconnected. The filter is then reconnected and measurements made of rumble components while the needle is in the region between the 1000-cycle tone and the 10 000-cycle tone.

In some cases, a particular rumble frequency may predominate, resulting from motor vibration or pickup-arm resonance. Measurements of such a frequency may be accomplished by means of a tuned filter or harmonic analyzer.

(3) **Maximum output.** The maximum audio output on phonograph may be less than that obtained on radio, because of the absence of automatic-gain-control voltage which normally causes a rise in plate and screen voltages in the audio amplifier.

The test is made by reproducing the 1000-cycle tone on side A of the test record and adjusting the volume control to produce maximum output. The measuring equipment used must give an accurate indication of root-sum-square independent of wave form.

(4) **'Wow' or flutter.** 'Wow' or flutter is caused by minute imperfections in the motor and/or transmission means used to drive the phonograph turntable ; and, if present to an appreciable extent, may noticeably impair the quality of reproduction. It is usually evident when a steady note is being reproduced, such as the 1000-cycle band, and evidences itself as a cyclic variation in pitch ; hence the name 'wow.' While there is no great distinction between 'wow' and flutter, the former is usually applied to very low cyclic variation, the latter to the more rapid type.

'Wow' or flutter is measured as a percentage of root-mean-square deviation in frequency of a tone to the average frequency. Special equipment is required to make such a measurement, this equipment being capable of responding uniformly to all flutter rates up to 200 cycles, and of measuring to a precision of 0.02% flutter.

Care should be exercised to insure that the method of measurement permits normal conditions to exist as when playing a commercial pressing."

(O) Miscellaneous

(1) **Tuning tests**

Some receivers which perform well when tuned to a station exhibit undesirable characteristics while being tuned. A test which shows up most of these effects is to tune in unmodulated signals of varying strengths up to the maximum liable to be encountered in service, moving the tuning control slowly and also as rapidly as possible. The volume control should be at its maximum setting, and the test should be repeated at the high and low frequency end of each wave range, and also with modulated signals.

Faults which are liable to be discovered are squeaks as the receiver is tuned rapidly through the sidebands of a signal, flutter on short-waves (and sometimes medium-waves) with slow tuning, microphony, or a rapid variation in the tuning rate near the carrier frequency on short waves due to varying electrode potentials or capacitances affecting the oscillator frequency. This should not be taken as a complete list of possible faults and **any** unusual effect should be investigated.

Other tests should be made with different aerial lengths (including very short aerials, and none) and signals tuned in all sections of the tuning range. Regeneration at the low frequency end of the broadcast band, particularly when a short aerial is placed as closely as possible to the second detector, and oscillation on the short-wave band with no aerial are possibilities in this case.

(2) **Dry battery receiver tests**

A set of batteries which has been discharged by normal use to a required low voltage should be used to test battery receivers at least for sensitivity, power output, stability and oscillator grid current (using a valve on the low limit of oscillator mutual conductance).

The requirement that the batteries should be discharged by normal use is necessary since rapid discharge gives a lower value of internal resistance. To simulate a discharged battery by means of added resistance, fresh heavy duty batteries can be tapped and the following table (Ref. 21) used.

Resistance Required to Simulate Discharged Radio " B " Batteries.

Volts per 22½ volt section	Resistance per 22½ volt section	
	Farm type batteries	Portable type batteries
22½	0 ohms	45 ohms
20	10 ohms	60 ohms
17	50 ohms	110 ohms
15	110 ohms	175 ohms
12	250 ohms	330 ohms

The following points are important. First, only fresh heavy-duty batteries should be used, their resistance being negligible. Second, if intermediate voltage taps are used the added resistance must be distributed between the taps in proportion to their respective voltages. Third, voltage readings should be taken only under load and across the battery cable terminals to include the voltage drop of the load current through the added resistances.

Battery end-of-life tests can be carried out in accordance with R.M.A. (U.S.A.) specifications M4-431 and M4-432 at 1.1 volts per cell for the A battery and 17, 15 and 12 volts per 22½ volt section for the B battery.

The lowest filament voltages used for testing battery valves by valve manufacturers are 1.1 volts for 1.4 volt valves and 1.7 volts for 2 volt valves. Below these voltages, performance may not be consistent.

(3) **Interference tests**

A suitable test for vibrator operation is to wrap a short lead around the battery cable and battery, connect one end of the lead to the aerial terminal of the receiver and check for interference throughout the tuning range of the receiver at maximum sensitivity. In a well designed receiver no interference will be heard.

Car-radio interference tests can be arranged using appropriate parts from the ignition system of a car, but such tests give no guarantee of interference-free reception from the receiver installed in a car.

The R.M.A. Standard REC-11 reads as follows :—

Chassis Pickup of Vehicular Receivers.

Vehicular receivers shall be considered as complying with the principles of good engineering practice if, when installed according to the manufacturers' instructions and using materials supplied by the manufacturer, there is no perceptible chassis pick-up with any setting of the user controls.

" Chassis pick-up " is defined as the interference arriving in the vehicular receiver other than through the antenna.

" Perceptible " is defined as the difference in the noise output of the receiver with the engine running and with it stopped.

Testing for chassis pickup shall be done by replacing the antenna of the installed receiver by an antenna equivalent, adequately shielded and grounded ; the antenna trimmer of the receiver shall be tuned for resonance at the normal aligning frequencies ; and observation made of any perceptible noise output.

(4) Distortion

Causes of distortion may be present in a receiver and remain undetected in spite of reasonably thorough testing with instruments. One such type of distortion occurs with insufficient capacitance by-passing a.v.c. lines, which results in low frequency intermodulation or perhaps hum modulation in a back-biased receiver. Another may be due to simultaneous high modulation levels and high signal strengths, and another may be the hum distortion mentioned in (H) above which is easily missed. Because of this, testing is not completed until critical listening to broadcasting stations over a period of time has shown that such faults are not present.

(P) Acoustical tests

Refs. 1, 2 and 5 should be consulted.

SECTION 2 : F-M RECEIVERS

(i) *Definitions* (ii) *Testing apparatus* (iii) *Test procedures and operating conditions* (iv) *Receiver adjustments* (v) *Performance tests.*

(i) Definitions

The information presented in this section, except where otherwise stated, is taken from the " Methods of Testing Frequency-Modulation Broadcast Receivers, 1947," Standard on Radio Receivers (Ref. 12), published by the Institute of Radio Engineers. Additional F-M information is available from the " Standards of Good Engineering Practice Concerning FM Broadcast Stations " published by the Federal Communications Commission in 1945.

(A) The three **standard test frequencies** are 88, 98 and 108 Mc/s. When only one frequency is used for testing it should be 98 Mc/s.

(B) The five **standard input values** may be expressed in terms of available power or of input voltage. The available power is that delivered by a generator to a matched load and is equal to $E^2/4R$ where E is the equivalent open-circuit voltage of the generator and R is the internal resistance of the generator (including the dummy-antenna resistance).

Five standard input values are specified, the voltage and power figures being equivalent when the receiver has an input impedance of 300 ohms and R is 300 ohms.

(a) Standard Input Powers
    (1) 130 decibels below 1 watt
    (2) 110 decibels below 1 watt
    (3) 90 decibels below 1 watt
    (4) 50 decibels below 1 watt
    (5) 30 decibels below 1 watt

(b) Standard Input Voltages
    (1) 11 microvolts
    (2) 110 microvolts
    (3) 1100 microvolts
    (4) 110 000 microvolts
    (5) 1.1 volts.

(c) The standard mean-signal input is either 90 decibels below 1 watt, or 1100 microvolts.

(C) The **standard test modulation** is a deviation at 400 c/s of 22½ Kc/s (30% of maximum deviation).

(D) The **standard test output** is as set out from the I.R.E. Standards for A-M Receivers in Sect. 1(ii)C of this chapter.

(E) The **maximum undistorted output** is as set out from the I.R.E. Standards for A-M Receivers in Sect. 1(ii)D of this chapter.

(F) The **standard 300-ohm antenna** comprises a pair of resistors, one connected in series with each terminal of the signal generator, of such value that the total impedance between terminals, including the signal generator, is 300 ohms (Fig. 37.2).

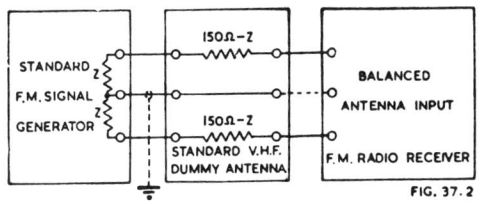

Fig. 37.2. Standard v-h-f dummy antenna and method of connection (from Ref. 12).

(G) The **standard pre-emphasis characteristic** has a rising response with modulating frequency, equivalent to that provided by a single circuit in the modulating source having a time constant of 75 microseconds. The characteristic may be obtained by taking the voltage across an inductor and a resistor connected in series and fed with constant current. The inductance in henries is 0.000 075 times the resistance in ohms.

(H) The **standard de-emphasis characteristic** has a falling response with modulation frequency, the inverse of the standard pre-emphasis characteristic, equivalent to that provided by a simple circuit having a time constant of 75 microseconds. The characteristic may be obtained by taking the voltage across a capacitor and a resistor connected in parallel and fed with constant current. The capacitance in farads is equal to 13 333 divided by the resistance in ohms. The standard de-emphasis characteristic is usually incorporated in the audio circuits of the receiver.

(ii) Testing apparatus

(A) " **A frequency-modulated signal generator** is required for testing frequency-modulation radio receivers.

The signal generator should cover at least the carrier-frequency range from 88 to 108 megacycles. It preferably also covers the intermediate-frequency range and frequency ranges required for spurious-response tests.

The generator output should be controlled by a calibrated attenuator, and the output should be adjustable over a range of at least 1 microvolt to 100 000 microvolts, and preferably from 0.1 microvolt to 1.1 volts. Balanced output terminals should be provided for the radio-frequency ranges, and single-sided output terminals for the intermediate-frequency range. It may be desirable to provide single-sided output terminals for the radio-frequency range also All of these terminals should be provided at the end of a flexible cable.

The output meter of the signal generator should indicate the open-circuit voltage at the terminals, and the internal impedance should be stated.

The generator should be capable of being frequency-modulated at rates from 30 to at least 15 000 cycles per second, and at deviations from zero to at least rated system deviation and preferably to twice that value. It should be provided with a deviation indicator reading from not more than 5 kilocycles up to the maximum deviation.

The modulation circuit of the generator should be provided with a standard pre-emphasis network. A switch should be provided for cutting this pre-emphasis network in or out of the generator circuit at will.

The generator should provide a frequency-modulated signal at 400 cycles up to maximum rated system deviation with less than 2%, and preferably less than 1% (root-sum-square) distortion. Amplitude modulation resulting from the frequency modulation should be kept to a minimum.

The frequency and amplitude modulation of the output voltage due to power-supply ripple should be negligible, in comparison with the effects under observation. The proper connection of a balanced-output signal generator for testing a balanced receiver is shown in Fig. 37.2."

(B) **Audio-output and distortion-measuring devices** are the same as those required for the testing of amplitude modulation receivers. See Sect. 1(iii) of this chapter.

(C) **Two-signal tests** are carried out using two signal generators and dummy antennas of twice the standard 300 ohms value with each generator as shown in Fig. 37.3. The output terminals of the two dummy antennas are then connected in parallel and to the input terminals of the receiver. With this connection the impedance connected across the receiver input terminals is the normal value and the open-circuit signal voltages are half the values indicated by each generator.

(D) **A standard-signal generator for amplitude-suppression testing** should be capable of simultaneous 400 c/s amplitude modulation and 1000 c/s frequency modulation with preferably less than 1% of incidental frequency modulation in the amplitude modulation process. Alternatively two signal generators may be used, connected as described above, but precautions must be taken that the beat note between the two carriers does not influence the undesired output and that the amplitude modulated generator is free of incidental frequency modulation.

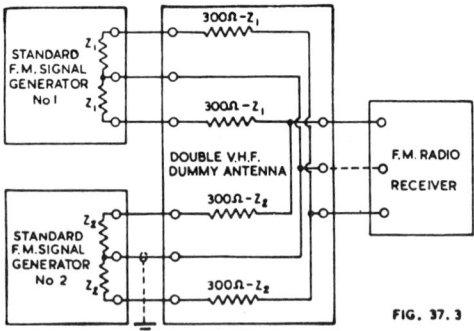

Fig. 37.3. Standard v-h-f dummy antenna for two-signal test and method of connection (from Ref. 12).

(iii) Test procedures and operating conditions

(A) **Input measurements** are made through a standard 300-ohm dummy antenna on receivers designed for a balanced antenna. If only a single-sided signal-generator output is available, one terminal of the input circuit of the receiver is connected to the grounded terminal of the generator output circuit, and the correct value of dummy antenna resistance must be used.

There are two simple tests for unbalance, the balanced receiver input coil can be reversed, or the power-line connection for either the receiver or the signal generator is moved to a different outlet. Any change in the sensitivity observed on alternating these connections indicates an error due to unbalance.

(B) **Output measurements** are made as described in Sect. 1(iv)A of this chapter.

(C) **Operating conditions** are similar to those in Sect. 1(iv)B of this chapter.

(iv) Receiver adjustments

The **tuning control** is adjusted until the desired a-f output is obtained either with the least possible r-f input power or with the lowest possible setting of the volume control. This is an **approximate** adjustment.

A receiver for frequency-modulated waves is tuned **accurately** to a desired signal by first tuning it approximately and then adjusting the tuning controls until either the undesired noise is a minimum or the harmonic distortion of the demodulated desired signal is a minimum. In many receivers these two tuning positions coincide.

When they do not coincide it should be stated whether the tuning is for minimum noise or for minimum distortion.

A simple method of tuning for minimum distortion in many receivers is by observing the audio wave form on a cathode ray oscilloscope while increasing the deviation somewhat beyond 100%.

The **tone control** is adjusted to give maximum modulation-frequency output unless otherwise specified.

(v) Performance tests

(A) Sensitivity

There are three sensitivity values of general interest in relation to frequency-modulation receivers. Each gives information as to the usefulness of the receiver, and in expressing results the type of sensitivity should be specified.

The **maximum sensitivity** is measured with the signal generator connected to the receiver through the dummy antenna. The receiver controls are adjusted for greatest sensitivity, and the output of the generator, which is modulated 30% (22½ Kc/s) at 400 c/s, is adjusted to obtain standard test output from the receiver.

When the tuning for minimum noise does not coincide with that for minimum distortion, the above test and the following test should be repeated with the receiver tuned for minimum noise.

The **maximum-deviation sensitivity** is measured by applying to the receiver a strong signal-input modulated 100% (75 Kc/s) at 400 c/s, tuning the receiver for minimum distortion and adjusting the receiver volume control to give standard output. The output distortion should be observed and the signal input reduced, keeping the indicated receiver output constant by readjusting the volume control if necessary, until the output distortion increases to 10%, or until the input is below that required for standard output. The signal input at which the distortion reaches 10% is the maximum-deviation sensitivity input and is expressed in decibels below 1 watt or in microvolts.

The **deviation sensitivity** is measured with the receiver volume control at maximum and with the standard mean-signal input (1100 μV) applied. The generator is modulated at 400 c/s and the deviation adjusted to the value which gives standard test output. The value of deviation required is the deviation sensitivity, and it is expressed in Kc/s or as a percentage of rated system deviation.

The **quieting-signal sensitivity** is measured by applying a signal of mean value, 30% modulated at 400 c/s, to the receiver with the volume control adjusted to give a convenient output below audio overload. The modulation should then be switched off and the signal intensity reduced to the least value which will produce a 30 db rise in indicated output with standard test modulation as compared with the indicated output with the unmodulated carrier. The results are expressed in decibels below 1 watt, or in microvolts.

(B) Co-channel interference

This test is intended to show the effect of an interfering signal of the same frequency as the desired signal, and includes the inherent effect of the detector, the limiter and the automatic volume control.

Two signal generators are required, only one of which need be capable of frequency modulation. The outputs of both are applied simultaneously to the receiver under test at the mean carrier frequency of 98 Mc/s.

With the desired signal frequency having standard test modulation and an intensity equal to one of the standard input values, the a-f output of the receiver is adjusted by means of the volume control to the standard test output.

The modulation of the desired signal is then removed, keeping the intensity of its carrier unchanged, the interfering signal, frequency-modulated 30% at 400 c/s, is turned on and the output of the receiver read as the level of the interfering signal is increased from zero to 1 volt or more.

The result of the test includes the effects of both the cross-talk and the beat-note components of the interference. If the results are desired for the cross-talk only, a 400 c/s filter is used in the output.

The co-channel interference characteristic may be expressed as the interfering-signal input, in decibels below the desired-signal input, which produces an output 30 decibels below the standard test output

(C) Masking interference

The masking effect of an unmodulated interfering signal is obtained by a test similar to that for the co-channel interference but with the desired signal 30% modulated at 400 c/s and with the interfering signal unmodulated. The output signal is noted as the level of the interfering signal is increased from zero.

(D) Selectivity

Test conditions are similar to those described for the co-channel interference test [(B) above] except that the interfering signal generator is separated in frequency from the desired signal by one standard channel separation.

The desired signal, unmodulated, is applied at the lowest value of standard input and the output of the receiver is recorded as the level of the interfering signal, frequency modulated 30% at 400 c/s, is varied from zero to a value capable of producing standard test output. This procedure is repeated for all values of standard test input. The measurements are then repeated with the interfering signal generator separated from the desired signal by twice the standard channel separation.

The adjacent-channel and second-channel interference may be expressed as the interfering signal input which produces an output 30 decibels below the standard test output.

(E) Amplitude-modulation suppression

This test measures the suppression of amplitude modulation which may be present in a frequency modulated signal. It is carried out at the standard mean carrier frequency. The frequency modulation is at a 1000 c/s rate with a deviation of 30% of maximum system deviation. The standard mean input-signal value having this modulation is applied to the receiver in the usual manner. The volume control is adjusted to produce standard output. The input signal is then amplitude-modulated at 400 c/s and 30% modulation. The intensity of the undesired output of the receiver is measured by filtering out the 1000 c/s frequency.

The amplitude suppression is the ratio of the undesired output to standard test output expressed in decibels. In order to determine the variation of amplitude-modulation suppression with input, the test is repeated with the other standard input-signal values.

(F) Electric fidelity (frequency response)

Methods similar to those for A-M receivers [Sect. 1(vi)C of this chapter] are used at the standard mean carrier frequency and with a standard mean signal input. The standard pre-emphasis characteristic is employed in the standard-signal generator and the modulating voltage is to be maintained constant at that value which provides a modulation of 30% of maximum rated system deviation when the modulation frequency is 400 c/s.

(G) Harmonic distortion

The tests are similar to those used for A-M receivers [Sect. 1(vi)D of this chapter] but variation of input signal testing is recommended with 30% and 100% of maximum rated system deviation. In addition there is a deviation distortion test.

The **maximum-deviation distortion** test is the measurement of the distortion due to inadequate bandwidth and/or inadequate amplitude-modulation rejection. It is measured at 98 Mc/s with the signal generator connected to the receiver through the 300-ohm dummy antenna. Distortion at standard test output is measured at full system deviation over the range from maximum sensitivity levels to 30 decibels below 1 watt.

(H) Maximum undistorted output

As for A-M receivers—Sect. 1(vi)E of this chapter.

(I) Automatic-volume-control characteristic

This test is similar, except for carrier frequency, to the I.R.E. test for A-M receivers — Sect. 1(vi)F of this chapter.

(J) Spurious responses

The testing of F-M receivers for spurious responses is carried out in a similar manner to that specified for the tuning of A-M receivers—Sect. 1(vi)J of this chapter. However spurious response sensitivity test inputs up to 1.1 volts are recorded and an additional intermediate frequency response test is specified.

The procedure is to inject the intermediate-frequency test input signal, amplitude modulated 30%, between the two antenna terminals of the receiver, connected in parallel, and ground, through a standard 300-ohm dummy antenna.   The frequency of the test signal is adjusted for maximum receiver output.

(K) Hum

Hum in F-M receivers has similar characteristics to hum in A-M receivers, and is measured in the same way—See Sect. 1(vi)H of this chapter.

(L) Noise audibility

The noise audibility test for F-M receivers is the same as that for A-M receivers—Sect. 1(vi)K of this chapter.

(M) Tuning characteristic test

The tuning characteristic shows the variation in audio output of the receiver as it is tuned through a signal.   This characteristic is of importance in frequency-modulation receivers since they may have spurious output responses adjacent to the correct tuning point.   The effect is usually more easily measured by variation of the signal-generator frequency than by variation of receiver tuning, since the signal generator usually has better frequency control and calibration than the receiver.

The receiver is tuned to the mean carrier frequency for each standard input signal value with standard test modulation and the volume control adjusted to standard output.   The output is then measured as the signal is detuned to each side of the carrier frequency.   A tuning curve is plotted for each signal input value (Fig. 37.4)

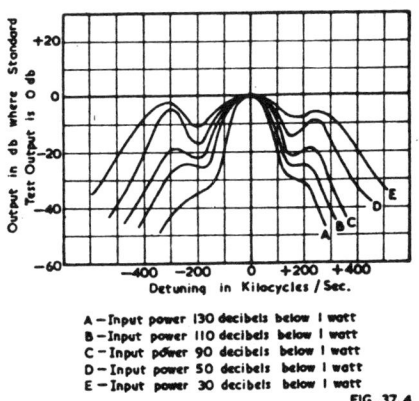

A — Input power 130 decibels below 1 watt
B — Input power 110 decibels below 1 watt
C — Input power 90 decibels below 1 watt
D — Input power 50 decibels below 1 watt
E — Input power 30 decibels below 1 watt

FIG. 37.4

Fig. 37.4.   Tuning characteristics (from Ref. 12).

(N) Frequency drift

This test is carried out in the same way as that for frequency shift in A-M receivers —Sect. 1(vi)I of this chapter—except that it is to be performed at the standard signal carrier frequency.

(O) Low-frequency instability

This test is intended to evaluate the limiting conditions for unstable operation of the receiver as affected by low frequency feed-back which may be electrical or acoustical in nature, and which may involve both carrier frequency and audio frequency and the audio-frequency circuits.   At any given frequency, the variables employed to induce instability are the signal-input power, the tuning control, the manual volume control and tone control, the modulation frequency and degree of modulation, and, in battery receivers, the age and condition of the batteries. All parts of the receiver, including the loudspeaker, are mounted in their normal relations.

The test is performed at the standard test frequencies. The receiver is tuned to a modulated test signal, after which the modulation is switched off. The conditions most conducive to the detection of any tendency to instability are found by trial. It is suggested that the frequency of the signal be varied manually over a range of about 100 Kc/s above and below the normal test frequency as the test input power is varied from zero to a maximum of 0.01 watt. An observation is made of the maximum signal-input power at which any unstable operation appears. The maximum of such input power is also recorded if less than 0.01 watt.

(P) Radiation from local oscillator

Tests are made in a manner similar to that for the A-M receiver tests—Sect. 1(vi)L of this chapter. Observations are made with the receiver tuned to each of the standard test frequencies. Results are expressed in field intensity as a function of distance from the receiver under test.

(Q) Mistuning

The degree of mistuning is represented by the total signal output distortion resulting when the receiver is adjusted to a frequency other than the desired signal frequency. The measurement is made by setting the signal generator to standard input voltages successively, modulating the signal generator to 75 kc deviation at standard test output. The signal generator is then adjusted off tune by successive increments, the volume control is adjusted for standard test output, and the total distortion in per cent (or db) is measured. For each value of input signal a curve is plotted, having as abscissa the frequency difference of detuning, and as ordinate the distortion expressed in per cent, or db. Distortion components will comprise all frequencies present except the fundamental frequency of the modulating tone. In these tests the signal generator is adjusted off tune on each side of the signal frequency.

The standard measurement will comprise setting the signal generator on each side of the signal frequency and noting the amount of mistuning that will produce 10% distortion, expressing the degree of mistuning as the average of the measured plus and minus frequency excursions. The signal input for this test shall be the standard mean signal input (1100 microvolts).

This mistuning test should be correlated with the frequency drift test—Sect. 2(v)N.

(R) Downward modulation

This test will define the ability of the receiver to withstand the effects of downward amplitude modulation. In this test it is assumed that the principal forms of distortion are caused by the downward component of modulation.

The test is made at the standard mean-carrier frequency (98 megacycles). Frequency modulation is impressed at a 400-cycle modulation rate at 30% of maximum rated system deviation and the volume control is adjusted for standard output. The input signal is then simultaneously amplitude modulated at a 100-cycle rate. By means of a band cut-off filter, the 100-cycle modulation is eliminated in the receiver output. The amplitude modulation is then increased until the total distortion reaches 10%. The percentage modulation at this point is the downward modulation capability of the receiver. The test is made at all values of standard input signal voltages.

(S) Open field method of measurement of spurious radiation from frequency
      modulation and television broadcast receivers

See I.R.E. Standard (Ref. 41).

SECTION 3 : AUDIO FREQUENCY AMPLIFIERS

(*i*) *Equipment and measurements* (*ii*) *Tests.*

(i) Equipment and measurements

The equipment already specified for testing the a-f end of a radio receiver is sufficient for the testing of a-f amplifiers. There are however additional tests commonly used in the design of a-f amplifiers and these tests need more instruments.

Intermodulation testing requires at least another a-f signal generator, and equipment is available (Refs. 19, 22 and 27) which is specifically designed for the purpose. The technique of square-wave testing (Refs. 16 and 20) also needs an appropriate generator as do similar techniques using other wave shapes (Ref. 18).

Measurements of voltage, current, dissipation and heating differ if at all only in the higher values found in high-power a-f amplifiers and the appropriate parts of Section 1 of this chapter are applicable.

(ii) Tests

The American R.M.A. Standard SE-101-A on Amplifiers for Sound Equipment gives details of tests for noise level, amplifier gain, amplifier frequency response, distortion and power output. The following conditions of measurement apply to each of the tests mentioned.

" The amplifier shall be operated at its rated power supply voltage and frequency. When a range of line voltage values is indicated for specified power input terminals, the arithmetic mean of the voltages specified for this range shall be applied except when R.M.A. standard M3-217* applies. The frequency of the power supply shall be within ± 2% of the lowest frequency specified. The rms harmonic voltage of the power supply voltage shall not exceed 10%.

Manufacturers of amplifiers designed to work from crystal microphones or other capacitive devices and photocells, may provide additional ratings based on the stated source impedance applicable to condition of use.

" The input and output shall be terminated† in pure resistance equivalent to the rating impedance within ± 5%. Grounding of the circuits and chassis shall be as normally used.

" The gain control shall be adjusted to give a maximum gain " (except when measuring noise and frequency response when it is set to give maximum noise and to reduce the gain 6 db below maximum respectively) " and any other controls shall be adjusted to give the most nearly uniform frequency response.

" The measurements shall be made after the amplifier has been operated for not less than one hour at an output 6 db less than the rated output at 400 cps, and at a room temperature of not less than 20°C."

The vacuum tubes used shall be selected to have the rated values of those characteristics which particularly affect the amplifier characteristic under test.

(A) Amplifier distortion

Definition (R.M.A. standard SE-101-A) :—" Amplifier distortion is the difference between the harmonic content of the voltage at the output terminals and that of the input voltage expressed as a percentage of the total rms output voltage "

Additional test conditions are :—

" Three distortion measurements shall be made with signals having frequencies of 100, 400 and 5000 cps applied to the input terminals. The harmonic content of these signals shall inappreciably affect the distortion of the amplifier.

" A distortion factor meter shall be used that is capable of filtering out or suppressing the fundamental frequency with respect to its harmonics by at least 60 db, without affecting the phase or amplitude of the generated distortion."

---

*R.M.A. Standard M3-217 states " When no means for adapting an electric radio receiver to the line voltage is provided, it shall be standard to adjust the secondary voltages so that the filaments and heaters of tubes are supplied with rated voltage when a line voltage of 117 volts is applied to the receivers."
†The input is effectively a low-impedance generator in series with a resistance equal to the rating source impedance. See Fig. 19.3.

If the amplitude of individual harmonics is measured and the distortion is small, the percentage of total harmonic distortion is given approximately by

$$D \approx \sqrt{(H_2\%)^2 + (H_3\%)^2 + \ldots}$$

where H_2 = second harmonic percentage, etc.
with an error not exceeding 1% if D does not exceed 10%—see Chapter 14 Sect. 2(iv).

A knowledge of the manner in which distortion decreases as output is reduced is desirable. Amplifiers with a large amount of feedback usually have very little distortion until the output is almost a maximum, but a class B amplifier may even give increased distortion with decreased output.

A simple test which can be used as an indication of freedom from distortion at low levels is the plotting of a linearity curve, i.e. a graph of input voltage against output voltage. Over the range of output for which the curve is a straight line the amplifier can be assumed to have a low distortion level. However in many cases the graph is curved almost from the zero point.

Intermodulation measurements are made with two signals applied to an amplifier through a suitable mixing network (e.g. bridge), either from separate a-f oscillators or from a combined unit designed for intermodulation testing. The usual procedure is for one frequency to be low (40, 60, 100, 150 or 400 c/s are used) and for the other to be high (1000, 2000, 4000, 7000 or 12 000 c/s). One test is carried out with the low frequency approximately at the low frequency limit of the amplifier and another at 100 or 150 c/s. The low frequency signal is usually made four times larger than the high frequency signal, although a 1 : 1 ratio is also used. The low frequency test will show distortion due to transformer saturation and the latter will be more representative of normal operation.

Distortion is mainly dependent on the peak voltage output so that an amplifier delivering at the same time two output voltages of 4 units and 1 unit (total output voltage = 5 units) would have distortion equivalent to a power output of 5^2 whereas the true output would be only $4^2 + 1^2$. Because of this, the indicated output power is multiplied by 25/17 when intermodulation tests are carried out under the conditions above.

The intermodulation products from the two signals can be measured separately or collectively. If they are measured separately and the lower frequency is f_1 and the higher f_2, then the intermodulation products to be investigated—considering f_2 as the carrier—are $(f_2 - f_1), (f_2 + f_1), (f_2 - 2f_1), (f_2 + 2f_1), (f_2 - 3f_1), (f_2 + 3f_1), \ldots$
The intermodulation distortion percentage referred to f_2, based on the r.m.s. sum, is

$$\frac{\sqrt{(E_{f_2 - f_1} + E_{f_2 + f_1})^2 + (E_{f_2 - 2f_1} + E_{f_2 + 2f_1})^2 + (E_{f_2 - 3f_1} + E_{f_2 + 3f_1})^2 \ldots}}{E_{f_2}}$$

If the r.m.s. sum of the intermodulation products is measured collectively a high-pass filter can be used to separate the low and high frequencies and their respective intermodulation products. The modulated high frequency is then rectified and passed through a low pass filter, the remainder being low frequency components previously modulating the high frequency signal. Descriptions and circuits of equipment of this type will be found in Refs. 19, 22, 23 and 27.

The **peak sum modulation method** is described in Chapter 14 Sect. 3(v).

Le Bel's Oscillographic method is described in Chapter 14 Sect. 3(vi).

Another method of measurement is to keep a **constant difference frequency** (say 400 c/s) between the outputs of two a-f generators and determine the magnitude of the difference frequency as the generators are tuned through the a-f range of the amplifier (Ref. 28)—see Chapter 14 Sect. 3(iii).

For an indication of the distortion tolerable under different conditions, reference should be made to Chapter 14 Sect. 2(iii). Intermodulation distortion is covered in Chapter 14 Sect. 3 and References (B).

Hum distortion may be overlooked if the individual distortion components are measured separately. It is due to intermodulation between signal and hum frequencies and with a 50 c/s power supply (100 c/s hum) and a 1000 c/s signal, hum distortion products will be found at 900 and 1100 c/s and also as sidebands to the harmonics of the signal. Hum modulation is most likely to be experienced at full output in an amplifier in which the filtering of the plate supply to the output valves is inadequate.

(B) Amplifier Power Output

Definition (R.M.A. standard SE-101-A) :—" Amplifier power output is the maximum rms power output (including distortion) at rated distortion which the amplifier will deliver into its rated load under normal operating conditions."

An additional requirement is that " The power output shall be measured using an indicator which measures rms values, such as thermally actuated meters, simultaneously with the distortion . . . ."

During design it is helpful to measure power output over a range of frequencies as it may decrease at both low and high frequencies. When bass and treble boosts are available to the user it is particularly important for the power output to be maintained over the range of frequencies to which the boosting applies. At high frequencies the response may be falling and the degree of overloading may not be apparent because the amplitude of the harmonics may be appreciably decreased, for instance by a capacitor across the primary of the output transformer. It is in such a case that the value of intermodulation measurements becomes apparent.

When measuring amplifier power output and other characteristics it is advisable to connect an oscilloscope across the output terminals so that the presence of parasitic oscillations can be detected.

(C) Amplifier Gain

Definition (R.M.A. standard SE-101-A) :—" Amplifier gain is the ratio expressed in db of the power delivered to the load, to the power which would be delivered to the same load if the amplifier were replaced by an ideal transformer which matches both the load and source impedances."

Additional test requirements are

" The gain shall be measured at a frequency of 400 cycles per second and at an output 3 db less than the rated power output. The harmonic content of the input signal shall inappreciably affect the gain of the amplifier.

" The gain shall be measured using input and output meters and associated measuring equipment that does not affect the frequency response or gain of the amplifier."

(D) Amplifier Frequency Response

Definition (R.M.A. standard SE-101-A) :—The amplifier frequency response is the variation of gain as a function of frequency over the range specified, expressed in db relative to the gain at 400 cycles per second."

Additional test requirements are

" Any automatic limiting or dynamic range control in the amplifier shall be disconnected during this test.

" The frequencies for testing shall be obtained from a source whose rms harmonic voltage does not exceed 5% of the fundamental voltage at any point in the frequency range of measurement.

" Measurements shall be made with the input voltage adjusted at each frequency to produce output levels of 3 db and 10 db respectively below rated power output.*

" The input and output meters and associated measuring equipment shall not discriminate as to frequency over the frequency range of measurement."

The frequency response of an amplifier is usually measured in terms of the output voltage across a resistive load. If feedback from the output stage is used it is advisable also to check the response with the normal loudspeaker load and with the minimum and maximum loads liable to be used.

In all cases the signal input must be small enough for the input-output characteristic to be linear over the frequency range being investigated. The frequency response may vary with output level at the low frequency end of the range and more than one curve may be necessary for this reason.

The effect of any controls which alter the frequency response should be shown by suitable curves.

---

*Frequency response is also sometimes measured at levels of 30 and 60 db below rated power output.

Square wave testing is used to obtain a rapid indication of frequency response (or phase shift) during design and in production. Two test frequencies can be used, one low enough for some low frequency attenuation or phase shift to be present at the fundamental frequency and the other high enough for some of the harmonics in the square wave to be attenuated. In production testing, limits for the wave shape can be indicated on a blank in front of the screen of the oscilloscope. Peaks in response are indicated (the approximate frequency of the peak being readily determined) as well as the degree of damping of the transient response and parasitic oscillation (Refs.16 and 20).

Details are given in Ref. 18 of a technique in which stepped sine waves are used in a similar manner.

(E) Noise Level

Definition (R.M.A. standard SE-101-A). (This is proposed as an interim standard for the measurement of steady state noise, and of pulse noise waves having a peak factor (ratio of peak to rms) approximating the maximum obtainable in speech) :—
" Noise level is the level of any noise signals appearing at the output terminals with no signal applied to the input.

" The weighted noise level is the noise level weighted in accordance with the 70 decibel equal loudness contour of the human ear and expressed in dbm, i.e. decibels referred to a level of 1 milliwatt."

An additional testing requirement is " The measuring amplifier shall be one whose audio frequency response is weighted with the 70 db equal-loudness contour of the human ear in accordance with Curve B of the ASA Specification Z24.3-1944 (Sound level Meters for Measurement of Noise and Other Sounds) with a standard VU meter as defined by ASA Specification C16.5-1942 (American Recommended Practice for Volume Measurements of Electrical Speech and Program Waves) as an indicator.

" Illustrative : A properly weighted amplifier may be obtained by applying an RC network with a 1 millisecond time constant to an amplifier having frequency response of $\pm$ 1 db from 50 to 15 000 cycles. This will give an attenuation of 1 db at 300 cycles, 5.7 db at 100 cycles and 9 db at 60 cycles."

(F) Microphony

High gain, high output a-f amplifiers may be subject to microphony under some conditions of use. A test similar to the one described in Sect. 1(vi)M of this chapter enables variations between microphonic tendencies between different amplifiers to be checked reliably.

(G) Stability

In amplifiers without feedback the most probable type of instability is parasitic oscillation. This can best be observed on an oscilloscope while the signal input is varied from zero past the overload point at various frequencies (including very low and very high).

Varying load impedances and high mains voltages may also cause trouble and the test should be repeated at least for design purposes with the highest power supply voltage liable to be encountered in the field and with a range of load impedances varying from an open circuit through capacitive and inductive loads to a short circuit. The plate dissipation of the output valves in AB or B class amplifiers should be observed during tests with the load impedance smaller than specified as it may become excessive. On occasions, variation in input impedance can provoke trouble and all possible types of pickup and microphone inputs should be tried. Square wave, instead of sine wave, excitation may induce parasitic oscillation.

(H) Feedback

Probably the most important feedback test is that for stability and the tests under (G) should be adequate, although instability experienced may be due to causes other than parasitic oscillation. For example an amplifier with voltage feedback from the output stage may oscillate when operated with open-circuited output, due to excessive feedback.

The output impedance is liable to be variable with frequency in a feedback amplifier. It can be estimated by varying the output impedance and noting the change in voltage across it (Ref. 26). However a variation in output impedance alters the amount of feedback and thus the impedance being measured. A more accurate determination can be made by connecting the correct load and driving source to the amplifier (it is only necessary to complete the input circuit if feedback is applied to the input stage, see Chapter 7 Sect. 2(vi)A) then connecting an a-f oscillator through a high impedance to the amplifier output and measuring the current flowing into the output circuit and the voltage across it. The impedance of the load in parallel with the output impedance of the amplifier is then obtainable and since the load is known the output impedance can be calculated. If output impedance is important it is desirable to make a measurement at least at the speaker resonant frequency as well as at mid-frequencies.

Another useful feedback test is a measurement of gain reduction vs. frequency. Such a curve is readily plotted and draws attention to any points at which the feedback may become positive. The shape of this curve is similar to an inverted output-impedance curve.

SECTION 4 : MEASUREMENTS ON COILS

(i) Measurement of coefficient of coupling   (ii) Measurement of primary resonant frequencies of aerial and r-f coils   (iii) Measurement of distributed capacitance across coils.

(i) Measurement of coefficient of coupling

The measurement of k, the coefficient of coupling, has been covered in Chapter 26 Sect. 4(ii)E and (iii)B.

(ii) Measurement of primary resonant frequencies of aerial and r-f coils

The usual methods of measuring primary resonant frequencies are either

(a) connecting a valve voltmeter across the secondary circuit, which is tuned to a much higher frequency than that to be measured (or a much lower frequency in the case of low impedance primaries), and feeding in a signal from the signal generator. The primary resonant frequency shows up by a sharp increase in the voltage across the valve voltmeter as the generator is tuned through this frequency. The secondary detuning is readily effected by rotating the rotor of the gang condenser, say, half-way across the band, or

(b) connecting the valve voltmeter straight across the primary winding, finding the resonant frequency, and then finding the true primary frequency by calculation after allowing for the input capacitance of the valve voltmeter, which can be measured on a Q meter or found by other means.

(iii) Measurement of distributed capacitance across coils

The distributed capacitance across a coil can be found with the use of a Q meter, following the methods given in most Q meter instruction books, as summarized below.

Approximate method ($\pm$ 2 $\mu\mu$F)

This is suitable for large inductors only

1. Connect the inductor under test to the coil (L) terminals of the Q meter.

2. Set the tuning capacitor to about 200 $\mu\mu$F (C_1) and adjust the oscillator frequency for resonance.

3. Note the oscillator frequency f_1.

4. Change the oscillator to a frequency exactly twice f_1.

5. Set the tuning capacitor to a new value C_2 for resonance

6. The distributed capacitance C_0 may then be calculated,
$$C_0 = (C_1 - 4C_2)/3 \ \mu\mu\text{F}. \tag{1}$$

If it is not convenient to use a frequency ratio of 2 : 1, the more general expressions for determining distributed capacitance are :
$$C_0 = \frac{(C_1 - C_2) - C_2(\alpha^2 - 1)}{\alpha^2 - 1} \ \mu\mu\text{F} \tag{2}$$

where $\alpha = (f_2/f_1)$.

The true inductance is given by
$$L = \frac{25\,330}{f_2{}^2(C_2 + C_0)} \ \mu\text{H}, \tag{3}$$

f being expressed in Mc/s
C_2 and C_0 in $\mu\mu\text{F}$.

Natural resonant frequency of inductor $= f_n = f_2\sqrt{1 + (C_2/C_0)}$. (4)

Accurate method (accuracy about $\pm$ 4%)

This method involves measuring the natural resonant frequency of the inductor.

1. Connect the inductor to the coil (L) terminals of the Q meter, set the Q circuit tuning capacitor to about 400 $\mu\mu\text{F}$ and adjust the oscillator frequency for resonance. Call the frequency f_1 and the tuning capacitance C_1.

2. Replace the inductor under test by a shielded coil having an inductance about 1/25 of that of the inductor under test.

3. Set the oscillator to a frequency about 10 times f_1, and adjust the Q circuit tuning capacitor for resonance.

4. Connect the inductor under test to the coil (L) terminals of the Q meter, in parallel with the shielded coil, taking care to avoid coupling between the two coils. Then adjust the Q circuit tuning capacitance for resonance, observing whether the capacitance has to be increased or decreased from its previous setting. If the capacitance has to be increased, increase the oscillator frequency by an appreciable amount (10 to 20%). If the capacitance has to be decreased, decrease the oscillator frequency.

5. Disconnect the inductor under test and adjust the Q circuit tuning capacitance to resonance at the new frequency, repeating the procedure of (4) above, changing the oscillator frequency by smaller increments as it approaches the resonant frequency of the inductor under test, until the frequency reaches a value at which the Q tuning capacitance is unchanged when the inductor under test is connected or disconnected. The corresponding oscillator frequency is called f_0.

6. The distributed capacitance, C_0, of the inductor may then be calculated :
$$C_0 = \frac{C_1}{\dfrac{f_0{}^2}{f_1{}^2} - 1} \tag{5}$$

or by the close approximation
$$C_0 \approx \frac{f_1{}^2}{f_0{}^2} \cdot C_1. \tag{6}$$

The accuracy of this measurement depends on the inductance of the coil remaining constant over the two frequencies. This is reasonably accurate for commonly used coils, although it may not hold for coils having iron cores.

This method often requires a considerable amount of trial and error. As a starting point it is helpful to measure L (this is not the true inductance of course) and make a guess at the value of C_0. The natural resonant frequency f_n can then be calculated to give a rough guide for starting more accurate measurements.

SECTION 5 : REFERENCES

1. American I.R.E. Standards :
   48 *IRE 17S1* " *Standards on Radio Receivers* : *Methods of Testing Amplitude-Modulation Broadcast Receivers*, 1948."   47 *IRE 17S1* " *Standards on Radio Receivers* : *Methods of Testing Frequency-Modulation Broadcast Receivers*, 1947 " *adopted by ASA C.16.12—1949.*   49 *IRE 17S1* " *Tests for effects of mistuning and for downward modulation*, 1949 " *being supplement to 47 IRE 17S1 Standards on Radio Receivers* : *Methods of testing Frequency-Modulation broadcast receivers*, 1947 ; *published in Proc. I.R.E.* 37.12 (*Dec.* 1949) 1376.
2. " Specification for Testing and Expressing Overall Performance of Radio Receivers Part 1 : Electrical Tests.   Part 2 : Acoustical Tests " R.M.A. (England) (Dec. 1936).   Reprinted Jour. I.E.E. 81 (1937) 104 and I.E.E. Wireless Proceedings 12.36 (Sept. 1937) 12 with discussion.
3. Sturley, K. R. " Radio Receiver Design " Chapman and Hall (1945) Part 2, Chapter 14.
4. Bray, W. J., and W. H. R. Lowry " The testing of communication-type radio receivers " Jour. I.E.E. 94 Part III A 12 (March 1947) 313.
5. 38 IRE 6 S1 " Standards on Electroacoustics, 1938 " The Institute of Radio Engineers New York.
6. Scroggie, M. G. " The a.v.c. characteristic " W.W. 44.18 (May 4, 1949) 427.
7. Moxon, L. A. " Noise factor " (1) W.W. 52.12 (Dec. 1946) 391 (2) W.W. 53.1 (Jan. 1946) 11.
8. Terman, F. E. " Radio Engineering " 3rd ed. (McGraw-Hill Book Co. 1947) 767.
9. Radio Research Laboratory, Harvard University " Very High-Frequency Techniques " Vol. II (McGraw-Hill Book Co. 1947) 637.
10. Fanker, E. M., and R. A. Ratcliffe " Testing procedures for F.M. V-H-F receivers " Proc. I.R.E. Aust. 8.3 (March 1947) 4.
11. Blatterman, A. S. " Sensitivity limits in radio manufacturing " Elect. 18.11 (Nov. 1945) 141.
12. 47 IRE 17S1 " Standards on Radio Receivers : Methods of Testing Frequency-Modulation Broadcast Receivers" The Institute of Radio Engineers, New York.
13. Rust, N. M., O. E. Keall, J. F. Ramsay and K. R. Sturley " Broadcast receivers : A review " Jour. I.E.E. 88 (Part 3) 2 (June 1941) 59.
14. Smith, N. S. " Performance tests on radio receivers " The Telecommunication Journal of Australia, 7.3 (Feb. 1949) 155.
15. Clack, G. T. " The technique of receiver measurements " Electronic Eng. (1) 14.162 (Aug. 1941) 348 ; (2) 14.163 (Sept. 1941) 404 ; (3) 14.164 (Oct. 1941) 452.
16. Van Duyne, J. P., and M. E. Clark " Square wave analysis at audio frequencies " Audio Eng. 31.3 (May 1947) 27.
17. McProud, C. G. " Simplified intermodulation measurement " Audio Eng. 31.3 (May 1947) 21.
18. Sabaroff, S. " Technique for distortion analysis " Elect. 21.6 (June 1948) 114.
19. Daniel, G. " Instrument for intermodulation measurements " Elect. 21.3 (March 1948) 134.
20. Swift, G. " Amplifier testing by means of square waves " Comm. 19.2 (Feb. 1939) 22.
21. Potter, N. M. " Dry battery characteristics and applications " Proc. I.R.E. Aust. 7.1. (Jan. 1946) 3.
22. " New test equipment circuits " Radio 30.3 (March 1946) 30.
23. Hilliard, J. K. " Intermodulation testing " Elect. 19.7 (July 1946) 123.
24. Avins, J. " Intermodulation and harmonic distortion measurements " Audio Eng. 32.10 (Oct. 1948) 17.
25. Warren, W. J., and W. R. Hewlett " An analysis of distortion methods by the intermodulation method" Proc. I.R.E. 36.4 (April 1948) 457.
26. Richter, W. " Simple method of determining internal resistance " Audio Eng. 32.10 (Oct. 1948) 19.
27. Read, G. W., and R. R. Scoville " An improved intermodulation measuring system " Jour. S.M.P.E. 50.2 (Feb. 1948) 162.
28. Scott, H. H. " Audible audio distortion " Elect. 18.1 (Jan. 1945) 126.
29. Pickering, N. C. " Measuring audio intermodulation " Electronic Industries 5.6 (June 1946) 56.
30. Goldberg, H. " Some notes on noise figures " Proc. I.R.E. 36.10 (Oct. 1948) 1205.
31. Jones, M. C. " Grounded-grid radio-frequency voltage amplifiers " Proc. I.R.E. 32.7 (July 1944) 423. See " Measurement of noise factor " page 426 et. seq.
32. van der Ziel, A. " Method of measurement of noise ratios and noise factors " Philips Research Reports 2.5 (Oct. 1947) 321.
33. Goodman, B. " How sensitive is your receiver ?   The diode noise generator for testing receiver sensitivity " Q.S.T. 31.9 (Sept. 1947) 13.
34. Sulzer, P. G. " Noise generator for receiver measurements " Elect. 21.7 (July 1948) 96.
35. Reynolds, G. D. " Tests for the selection of components for broadcast receivers " Jour. I.E.E. 95 (Part 3) 34 (March 1948) 54.
36. Black, W. L., and H. H. Scott " Audio-frequency measurements " Proc. I.R.E. 37.10 (Oct. 1949) 1108.
37. " Amplifiers for sound equipment " R.M.A. Standard SE-101-A (July 1949) U.S.A.
38. Black, W. L., and H. H. Scott " Audio-frequency measurements " Audio Eng. (1) 33.10 (Oct. 1949) 13 ; (2) 33.11 (Nov. 1949) 18.
39. Allen, M. " Method for determining receiver noise figure " Tele Tech 7.1 (Jan. 1948) 38.
40. " Radio for Merchant Ships : Performance Specifications " His Majesty's Stationery Office (London, 1947).
41. 51 IRE 17S1 " Standards on Radio Receivers : Open Field Method of Measurement of Spurious Radiation from Frequency Modulation and Television Broadcast Receivers, 1951 " Proc. I.R.E. 39.7 (July 1951) 803. Supplement (1952).
42. Aubry, P. J. " Intermodulation testing " Audio Eng. 35.12 (Dec. 1951) 22.

Additional references will be found in the Supplement commencing on page 1475.

CHAPTER 38

TABLES, CHARTS AND SUNDRY DATA

Section		Page

SECTION 1 : UNITS

(i) *General physical units* (ii) *Electrical and magnetic units* (iii) *Photometric units* (iv) *Temperature.*

(i) General physical units

TABLE 1 : UNITS

Quantity	English	Metric

Length	1 mil = 0.001 inch = 0.00254 cm
	1 inch = 1000 mils = 2.54 cm
	1 foot = 12 inches = 30.48 cm
	1 yard = 3 feet = 0.9144 m
	1 mile = 1760 yards = 1.6093 Km
	1.152 miles = 1 nautical mile
	    = 1.853 Km
	60 naut. miles = 1 degree = 111.100
	    Km
		1 mm = 39.37 mils
		1 cm = 0.3937 inches
		    = 0.0328 foot
		1 m = 1.094 yards
		    = 3.272 feet
		1 Km = 0.6214 mile

		1 micron = 10^{-6} metre
		    = 0.0001 cm
		    = 10 000 Angstroms
		1 Angstrom (A^{0}) = 10^{-10} metre
		    = 10^{-8} cm
		    = 0.0001 micron

Area	1 mil² = 6.452×10^{-10} m²
	1 circ. mil* = 0.7854×10^{-6} in²
	    = 5.067×10^{-10} m²
	1 in² = 6.452 cm²†
	1 ft² = 144 in² = 0.0929 m²
	1 yd² = 9 ft² = 0.8361 m²
		1 cm² = 0.1550 in²
		    = 0.001 076 ft²
		1 m² = 10.76 ft²

Volume	1 in³ = 16.39 cm³† = 0.016 39 litres ** 1 ft³ = 1728 in³ = 28.32 litres	1 cm³ = 0.061 02 in³ 1 litre = 61.02 in³

Mass	(Avoirdupois)
	1 grain = 0.0648 grams
	1 dram = 1.772 grams
	1 ounce = 16 drams = 28.35 grams
	1 pound = 16 ounces
	    = 7000 grains
	    = 453.6 grams
	112 pounds = 1 hundredweight
	    = 50.800 Kg
	1 ton = 20 cwt = 1016.1 Kg
		1 gram = 15.432 grains
		    = 0.035 27 ounce
		    = 0.002 205 lb
		1 Kg = 2.205 lb
		1000 Kg = 0.9842 ton
		    = 1 metric ton

(Continued on page 1330)

*1 circular mil is the area of a circle 0.001 inch diameter.

† cm² = sq. cm = square centimetres (similarly for other symbols).
  cm³ = cub. cm = cubic centimetres.

**1 litre = 2.202 lb of fresh water at 62 °F

Quantity	English	Metric
Force	1 pound weight = 4.448 × 10⁵ dynes 1 poundal = 1.382 × 10⁴ dynes	1 dyne = 0.2248 × 10⁻⁵ pound weight = 0.001 0197 gm weight 1 gram weight = 980.62 dynes‡
Intensity of Pressure	1 atmosphere = 760 mm mercury at 0 °C = 1.0132 × 10⁶ dynes/cm² 1 lb/ft² = 478.8 dynes/cm² 1 lb/in² = 0.6894 × 10⁵ dynes/cm² 1 ton/ft² = 1.072 × 10⁶ dynes/cm² 1 inch of mercury at 0 °C = 3.386 × 10⁴ dynes/cm² = 34.53 gms/cm²	1 dyne/cm² = 0.9869 × 10⁻⁶ atmosphere 1 mm mercury at 0 °C = 1.333 × 10³ dynes/cm² = 1.359 gms/cm² = 1.316 × 10⁻³ atmosphere
Angles	1 degree (1 °) = 0.017 4533 radian 1 radian = 57° 17′ 44.806″ = 57° 17.7468′ = 57.295 780° 1 quadrant = 90° = π/2 = 1.571 radians 1 revolution = 360° = 2π = 6.283 radians	

‡The internationally accepted value of gravitational acceleration at latitude 45° and sea level.

$$1 \text{ lb/ft}^2 = 478.8 \text{ dynes/cm}^2$$

(ii) Electrical and magnetic units

There are several systems of units in common use, but they may be divided into three clearly distinguished groups :

1. **Unrationalized systems,** including

    (a) Absolute† c.g.s.* electrostatic system.

    (b) Absolute† c.g.s.* electromagnetic system.

    (c) Absolute† m.k.s. (metre-kilogram-second) system.   Otherwise known as
    the Giorgi system.

2. **Rationalized systems** including

    (a) Rationalized m.k.s. system (Giorgi).

3. **Practical systems**

The common practical system includes the volt, ampere, coulomb, ohm, farad, henry and watt.

All fundamental physical relationships are normally worked out in one of the un-rationalized systems, and the final result may be converted into practical units for general use.

Rationalized systems have been developed to simplify certain calculations. They may be used as alternatives to unrationalized systems.

The m.k.s. system is increasing in popularity because neither the c.g.s. electrostatic system nor the c.g.s. electromagnetic system is convenient for use with all problems, and the combined use of the two systems has been generally adopted in the past. Another reason for its popularity is that it includes many of the practical units, the second, joule, watt, coulomb, ampere, volt, ohm, mho, farad and henry. The rationalized m.k.s. system has been standardized by the American I.R.E. (January 1948).

The Giorgi m.k.s. system absolute system was adopted by the International Electrotechnical Commission (I.E.C.) in Bruxelles, June 1935 (see Proceedings of National Academy of Sciences Vol. 21 No. 10 pp. 579-583, October 1935 ; reprinted by Harvard University, Publications from the Graduate School of Engineering 1935-36 No. 167). See also A. E. Kennelly " I.E.C. adopts MKS System of Units," Electrical Engineering 54.12 (Dec. 1935) 1373. See also References below.

In any one sequence of calculations it is essential to retain the same system throughout. The final result may then, if desired, be converted to any other system. Table 2 should enable any engineer to convert from one quantity in any system to any other of the major systems.

REFERENCES TO MKS SYSTEM

Carr, H. L. A. " The M.K.S. or Giorgi system of units—the case for its adoption " Proc. I.E.E. Part I
    97.107 (Sept. 1950) 235.

Rawcliffe, G. H. " The rationalization of electrical units and its effect on the M.K.S. System " Proc.
    I.E.E. Part I 97.107 (Sept. 1950) 241.

Marriott, H., and A. L. Cullen " The rationalization of electrical theory and units " Proc. I.E.E. Part I
    97.107 (Sept. 1950) 245.

Bradshaw, E. " Rationalized M.K.S. units in electrical engineering education " Proc. I.E.E. Part I
    97.107 (Sept. 1950) 252.

A brief description of all systems including the m.k.s. is given in " Applied Electronics " (Massachusetts Institute of Technology ; John Wiley & Sons Inc. New York ; Chapman & Hall Ltd., London, 1943) Appendix B.

---

*c.g.s. = centimetre-gramme-second.

†An absolute system is one which includes length, mass and time in its fundamental dimensions.

TABLE 2 : THE PRINCIPAL ELECTRICAL AND MAGNETIC UNITS

Quantity	Practical (English)	Giorgi MKS	C.G.S. Electrostatic	C.G.S. Electromagnetic
Length	1 foot, 1 inch	1 metre	1 centimetre	1 centimetre
Mass	1 pound	1 kilogram	1 gram	1 gram
Force	1 pound weight	1 dyne-five = 1 newton	1 dyne	1 dyne
Time	1 second		1 second	1 second
Work, Energy	1 joule		1 erg	1 erg
Power	1 watt		1 erg/second	1 erg/second
Charge	1 coulomb		1 statcoulomb	1 abcoulomb
Current	1 ampere		1 statampere	1 abampere
Electromotive force	1 volt		1 statvolt	1 abvolt
Resistance Resistivity	1 ohm / 1 ohm/cm cube or 1 ohm-centimetre		1 statohm	1 abohm
Conductance Conductivity	1 siemens = 1 mho / 1 mho/cm cube		1 statmho	1 abmho
Capacitance	1 farad		1 statfarad	1 abfarad
Inductance	1 henry		1 stathenry	1 abhenry
		MKS unrationalized	MKS rationalized	
Flux (ϕ)	1 line = 1 maxwell	1 weber	1 weber	1 line = 1 maxwell
Flux density (B)	1 line/sq in	1 weber/sq metre	1 weber/sq metre	1 gauss
Magnetizing force (H)	1 ampere-turn /in	1 praoersted	1 ampere-turn /metre	1 oersted
Magnetomotive force (F)	1 ampere-turn	1 pragilbert	1 ampere-turn	1 gilbert
Reluctance		1 pragilbert/ weber	1 ampere-turn/ weber	1 gilbert/ maxwell
Permeability of free space (μ_o)		10^{-7} henry/ metre (or 10^{-7} weber/sq metre/praoersted)	$4\pi \times 10^{-7}$ henry/metre	1 abhenry/cm (or 1 gauss/ oersted)

RELATIONSHIPS BETWEEN UNITS

1 metre = 100 centimetres	1 centimetre = 1/100 metre
1 kilogram = 1000 grams	1 gram = 1/1000 kilogram
1 newton = 10^5 dynes	1 dyne = 10^{-5} newton

1 joule = 10^7 ergs	1 erg = 10^{-7} joule
1 watt = 10^7 ergs/second	1 erg/second = 10^{-7} watt

1 coulomb = 3×10^9 statcoulombs = 0.1 abcoulomb
1 statcoulomb = 3.33×10^{-11} abcoulomb   1 abcoulomb = 3×10^{10} statcoulombs

1 ampere = 3×10^9 statamperes = 0.1 abampere
1 statampere = 3.33×10^{-11} abampere   1 abampere = 3×10^{10} statamperes

1 volt = 3.33×10^{-3} statvolt = 10^8 abvolts
1 statvolt = 3×10^{10} abvolts = 300 volts   1 abvolt = 3.33×10^{-11} statvolt

1 ohm = 1.11×10^{-12} statohm = 10^9 abohms
1 statohm = 9×10^{20} abohms   1 abohm = 1.11×10^{-21} statohm

1 mho = 9×10^9 statmhos = 10^{-9} abmho
1 statmho = 1.11×10^{-21} abmho   1 abmho = 9×10^{20} statmhos

1 farad = 9×10^{11} statfarads = 10^{-9} abfarad
1 statfarad = 1.11×10^{-21} abfarad   1 abfarad = 9×10^{20} statfarads
  = 1.11×10^{-12} farad   1 $\mu\mu$F = 0.9 statfarad

1 henry = 1.11×10^{-12} stathenry = 10^9 abhenrys
1 stathenry = 9×10^{20} abhenrys   1 abhenry = 1.11×10^{-21} stathenrys

1 weber = 10^8 maxwells = 10^8 lines   1 maxwell = 10^{-8} weber

1 weber/sq metre = 10^4 gauss   1 gauss = 10^{-4} weber/sq metre

1 praoersted = 10^{-3} oersted   1 oersted = 10^3 praoersted
1 ampere-turn/inch = 0.495 oersted = 495 praoersteds
1 ampere-turn/metre = 0.01257 oersted = 12.57 praoersteds

1 ampere-turn = 1.257 gilberts   1 gilbert = 0.796 ampere-turn
1 pragilbert = 0.1 gilbert   1 gilbert = 10 pragilberts
1 ampere-turn = 12.57 pragilberts   1 pragilbert = 0.0796 ampere-turn

1 pragilbert/weber = 10^{-9} gilbert/maxwell = 0.0796 ampere-turn/weber
1 ampere-turn/weber = 1.257×10^{-8} gilbert/maxwell

(iii) Photometric units

TABLE 3 : PHOTOMETRIC UNITS

Quantity	Symbol	Unit	Relationship
Luminous Flux	F	Lumen	1 lumen $= \dfrac{1}{4\pi} \times$ flux emitted by one candle
Light Intensity	I	Candle	Flux emitted by 1 candle $= 4\pi$ lumens
Illumination	E	Footcandle $=$ lumens/ft^2 Phot $=$ cm candle $=$ lumens/cm^2 Lux $=$ metre candle $=$ lumens/m^2	

Relationship between units of illumination

Unit	Foot-candle	Phot	Lux
1 Foot-candle	1	0.001076	10.76
1 Phot	929	1	10^4
1 Lux	0.0929	10^{-4}	1

(iv) Temperature

TABLE 4 : TEMPERATURE

Freezing point of water (normal pressure)	$= 32°$ Fahrenheit $= 0°$ Centigrade* $= 273.16°$ Kelvin (Absolute)
Boiling point of water (normal pressure)	$= 212°$ Fahrenheit $= 100°$ Centigrade* $= 373.16°$ Kelvin (Absolute)
1 Fahrenheit degree	$= \dfrac{5}{9}$ Centigrade* degree $= 0.5556$ Centigrade* degree $= 0.5556$ Kelvin degree
1 Centigrade* degree	$= 1.800$ Fahrenheit degree $= 1.0$ Kelvin degree
Temperature in °C	$= \dfrac{5}{9}$ (°F $-$ 32)
Temperature in °F	$= 1.8$ (°C) $+$ 32
Temperature in °K Absolute temperature	$=$ (°C) $+$ 273.16
Temperature in °C	$=$ (°K) $-$ 273.16
Absolute Zero	$= 0°$K $= -273.16$°C $= -459.68$°F

---

*The Ninth General Conference on Weights and Measures held in Paris in October, 1948, decided to abandon the designation Centigrade and use Celsius (International Temperature Scale) instead. For most practical purposes the two scales may be regarded as identical.

SECTION 2 : COLOUR CODES

(i) Colour code for fixed composition resistors   (ii) Colour code for fixed wire-wound resistors   (iii) Table of R.M.A. colour code markings for resistors   (iv) Colour code for moulded mica capacitors   (v) Colour code for ceramic dielectric capacitors (vi) Colour code for i-f transformers   (vii) Colour code for a-f transformers and output transformers   (viii) Colour code for power transformers   (ix) Colour code for loud-speakers   (x) Colour code for chassis wiring   (xi) Colour code for battery cables   (xii) Colour code for metallized paper capacitors.

(i) Colour code for fixed composition resistors

(A) R.M.A. Standards GEN-101, REC-116, July 1948.

In fixed composition resistors with axial leads the nominal resistance value is indicated by bands of colour of equal width around the body of the resistor (Fig. 38.1).

Band A indicates the first significant figure.

Band B indicates the second significant figure.

Band C indicates the decimal multiplier.

Band D, if any, indicates the tolerance limits about the nominal resistance value.

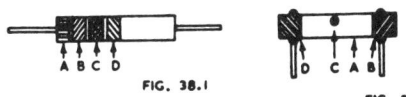

FIG. 38.1

FIG. 38.2

Fig. 38.1.   R.M.A. colour code for fixed composition resistors (axial leads).
Fig. 38.2.   R.M.A. colour code for fixed composition resistors (radial leads).

TABLE 5 : COLOUR CODE

Colour	Significant Figure	Decimal Multiplier	Tolerance %
Black	0	1	
Brown	1	10	
Red	2	10^2	
Orange	3	10^3	
Yellow	4	10^4	
Green	5	10^5	
Blue	6	10^6	
Violet	7	10^7	
Grey	8	10^8	
White	9	10^9	
Gold	—	10^{-1}	±5
Silver	—	10^{-2}	±10
No Colour	—	—	±20

Body colour : Black indicates uninsulated.   Any other colour indicates insulated.

Colour code with radial leads

Resistors with radial leads may use the same colour code as those with axial leads, but alternatively* they may be colour coded as in Fig. 38.2 where

Body A corresponds to Band A above,
End B corresponds to Band B above,
Dot C (or band C) corresponds to Band C above,
Band D corresponds to Band D above.

See also Sect. 3(i)—Standard fixed composition resistors.

*As JAN-R-11 and old R.M.A. specification.

(B) British Standard BS. 1852 : 1952

for fixed resistors for telecommunication purposes.

Interpretation of marking :

Fig. 38.3A	Fig. 38.3B or Fig. 38.3C	Interpretation
1st band (A)	Body (A)	Indicates first significant figure of resistance value.
2nd band (B)	First tip (B)	Indicates second significant figure.
3rd band (C)	Spot (C)	Indicates multiplier.
4th band (D)	Second tip (D)	If present, indicates percentage tolerance on nominal resistance value. If no colour appears in this position the tolerance is ± 20 per cent.
5th band (E)	—	If present, indicates grade of resistor.

Colour values :

As Table 5 with the addition of

Colour	Tolerance %
Brown	± 1
Red	± 2

Colour	Grade
Salmon pink	Grade 1 (" high stability ")

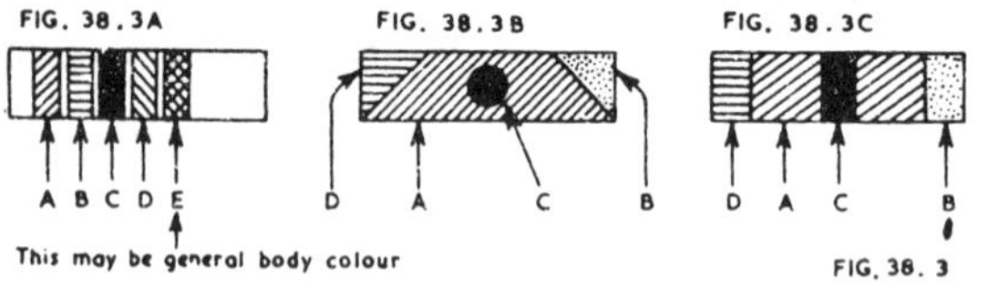

FIG. 38.3A   FIG. 38.3B   FIG. 38.3C

A B C D E   D   A   C   B   D   A   C   B

This may be general body colour   FIG. 38.3

Fig. 38.3A.   British Standard coloured band marking (preferred).
Fig. 38.3B.   British Standard body, tip and spot marking.
Fig. 38.3C.   British Standard body, tip and central band marking.

(ii) Colour code for fixed wire-wound resistors

(A) R.M.A. Standard REC-117, July 1948

In fixed wire-wound resistors with axial leads the nominal resistance value is indicated by bands of colour around the body of the resistor (Fig. 38.3).

A B C D

FIG. 38.3

Fig. 38.3.   R.M.A. colour code for fixed wire wound resistors (axial leads).

Band A is of double width, thereby indicating a wire-wound resistor. The colour of Band A, and those of bands B, C and D, have the same indications regarding resistance and tolerance as for composition resistors. The body colour also has the same indications as for composition resistors.

See also Sect. 3(ii) for standard fixed wire wound resistors.

(B) British Standard colour code for fixed resistors. See (i) (B) above.

(iii) Table of R.M.A. colour code markings for resistors

TABLE 6 : STANDARD COLOUR CODING FOR RESISTORS

Preferred values of resistance (ohms)			Old standard resistance values (ohms)	Resistance designation		
±20% D = no col	±10% D = silver	±5% D = gold		A	B	C
			50	Green	Black	Black
		51		Green	Brown	Black
	56	56		Green	Blue	Black
		62		Blue	Red	Black
68	68	68		Blue	Gray	Black
		75	75	Violet	Green	Black
	82	82		Gray	Red	Black
		91		White	Brown	Black
100	100	100	100	Brown	Black	Brown
		110		Brown	Brown	Brown
	120	120		Brown	Red	Brown
		130		Brown	Orange	Brown
150	150	150	150	Brown	Green	Brown
		160		Brown	Blue	Brown
	180	180		Brown	Gray	Brown
		200	200	Red	Black	Brown
220	220	220		Red	Red	Brown
		240		Red	Yellow	Brown
			250	Red	Green	Brown
	270	270		Red	Violet	Brown
		300	300	Orange	Black	Brown
330	330	330		Orange	Orange	Brown
			350	Orange	Green	Brown
		360		Orange	Blue	Brown
	390	390		Orange	White	Brown
			400	Yellow	Black	Brown
		430		Yellow	Orange	Brown
			450	Yellow	Green	Brown
470	470	470		Yellow	Violet	Brown
			500	Green	Black	Brown
		510		Green	Brown	Brown
	560	560		Green	Blue	Brown
			600	Blue	Black	Brown
		620		Blue	Red	Brown
680	680	680		Blue	Gray	Brown
		750	750	Violet	Green	Brown
	820	820		Gray	Red	Brown
		910		White	Brown	Brown
1,000	1,000	1,000	1,000	Brown	Black	Red
		1,100		Brown	Brown	Red
	1,200	1,200	1,200	Brown	Red	Red
		1,300		Brown	Orange	Red
1,500	1,500	1,500	1,500	Brown	Green	Red
		1,600		Brown	Blue	Red
	1,800	1,800		Brown	Gray	Red
		2,000	2,000	Red	Black	Red
2,200	2,200	2,200		Red	Red	Red
		2,400		Red	Yellow	Red
			2,500	Red	Green	Red
	2,700	2,700		Red	Violet	Red
		3,000	3,000	Orange	Black	Red

±20% D = no col	±10% D = silver	± 5% D = gold	Old standard resistance values (ohms)	A	B	C
Preferred values of resistance (ohms)				Resistance designation		
3,300	3,300	3,300		Orange	Orange	Red
			3,500	Orange	Green	Red
		3,600		Orange	Blue	Red
	3,900	3,900		Orange	White	Red
			4,000	Yellow	Black	Red
		4,300		Yellow	Orange	Red
4,700	4,700	4,700		Yellow	Violet	Red
			5,000	Green	Black	Red
		5,100		Green	Brown	Red
	5,600	5,600		Green	Blue	Red
		6,200		Blue	Red	Red
6,800	6,800	6,800		Blue	Gray	Red
		7,500	7,500	Violet	Green	Red
	8,200	8,200		Gray	Red	Red
		9,100		White	Brown	Red
10,000	10,000	10,000	10,000	Brown	Black	Orange
		11,000		Brown	Brown	Orange
	12,000	12,000	12,000	Brown	Red	Orange
		13,000		Brown	Orange	Orange
15,000	15,000	15,000	15,000	Brown	Green	Orange
		16,000		Brown	Blue	Orange
	18,000	18,000		Brown	Gray	Orange
		20,000	20,000	Red	Black	Orange
22,000	22,000	22,000		Red	Red	Orange
		24,000		Red	Yellow	Orange
			25,000	Red	Green	Orange
	27,000	27,000		Red	Violet	Orange
		30,000	30,000	Orange	Black	Orange
33,000	33,000	33,000		Orange	Orange	Orange
		36,000		Orange	Blue	Orange
	39,000	39,000		Orange	White	Orange
			40,000	Yellow	Black	Orange
		43,000		Yellow	Orange	Orange
47,000	47,000	47,000		Yellow	Violet	Orange
			50,000	Green	Black	Orange
		51,000		Green	Brown	Orange
	56,000	56,000		Green	Blue	Orange
			60,000	Blue	Black	Orange
		62,000		Blue	Red	Orange
68,000	68,000	68,000		Blue	Gray	Orange
		75,000	75,000	Violet	Green	Orange
	82,000	82,000		Gray	Red	Orange
		91,000		White	Brown	Orange
100,000	100,000	100,000	100,000	Brown	Black	Yellow
		110,000		Brown	Brown	Yellow
	120,000	120,000	120,000	Brown	Red	Yellow
		130,000		Brown	Orange	Yellow
150,000	150,000	150,000	150,000	Brown	Green	Yellow
		160,000		Brown	Blue	Yellow
	180,000	180,000		Brown	Gray	Yellow
		200,000	200,000	Red	Black	Yellow

Preferred values of resistance (ohms)			Old standard resistance values (ohms)	Resistance designation		
±20% D = no col	±10% D = silver	± 5% D = gold		A	B	C
220,000	220,000	220,000		Red	Red	Yellow
		240,000		Red	Yellow	Yellow
			250,000	Red	Green	Yellow
	270,000	270,000		Red	Violet	Yellow
		300,000	300,000	Orange	Black	Yellow
330,000	330,000	330,000		Orange	Orange	Yellow
		360,000		Orange	Blue	Yellow
	390,000	390,000		Orange	White	Yellow
			400,000	Yellow	Black	Yellow
		430,000		Yellow	Orange	Yellow
470,000	470,000	470,000		Yellow	Violet	Yellow
			500,000	Green	Black	Yellow
		510,000		Green	Brown	Yellow
	560,000	560,000		Green	Blue	Yellow
			600,000	Blue	Black	Yellow
		620,000		Blue	Red	Yellow
680,000	680,000	680,000		Blue	Gray	Yellow
		750,000	750,000	Violet	Green	Yellow
	820,000	820,000		Gray	Red	Yellow
		910,000		White	Brown	Yellow
1.0 MΩ	1.0 MΩ	1.0 MΩ	1.0 MΩ	Brown	Black	Green
		1.1 MΩ		Brown	Brown	Green
	1.2 MΩ	1.2 MΩ		Brown	Red	Green
		1.3 MΩ		Brown	Orange	Green
1.5 MΩ	1.5 MΩ	1.5 MΩ	1.5 MΩ	Brown	Green	Green
		1.6 MΩ		Brown	Blue	Green
	1.8 MΩ	1.8 MΩ		Brown	Gray	Green
		2.0 MΩ	2.0 MΩ	Red	Black	Green
2.2 MΩ	2.2 MΩ	2.2 MΩ		Red	Red	Green
		2.4 MΩ		Red	Yellow	Green
	2.7 MΩ	2.7 MΩ		Red	Violet	Green
		3.0 MΩ	3.0 MΩ	Orange	Black	Green
3.3 MΩ	3.3 MΩ	3.3 MΩ		Orange	Orange	Green
		3.6 MΩ		Orange	Blue	Green
	3.9 MΩ	3.9 MΩ		Orange	White	Green
			4.0 MΩ	Yellow	Black	Green
		4.3 MΩ		Yellow	Orange	Green
4.7 MΩ	4.7 MΩ	4.7 MΩ		Yellow	Violet	Green
			5.0 MΩ	Green	Black	Green
		5.1 MΩ		Green	Brown	Green
	5.6 MΩ	5.6 MΩ		Green	Blue	Green
			6.0 MΩ	Blue	Black	Green
		6.2 MΩ		Blue	Red	Green
6.8 MΩ	6.8 MΩ	6.8 MΩ		Blue	Gray	Green
			7.0 MΩ	Violet	Black	Green
		7.5 MΩ		Violet	Green	Green
			8.0 MΩ	Gray	Black	Green
	8.2 MΩ	8.2 MΩ		Gray	Red	Green
			9.0 MΩ	White	Black	Green
		9.1 MΩ		White	Brown	Green
10 MΩ	10 MΩ	10 MΩ	10 MΩ	Brown	Black	Blue

(iv) Colour code for moulded mica capacitors

The Standard American R.T.M.A. Colour Marking (R.C.M.) uses six dots for moulded mica capacitors.

(R.T.M.A. Standard REC-115-A, May 1951) (Fig. 38.4).

Dot (1) : White indicates R.T.M.A. moulded mica capacitor colour coding.
           Black indicates JAN colour coding (see below).
           Any other colour indicates old R.M.A. 6 dot system.
Dot (2) : First digit of capacitance (see Table 7 below).
Dot (3) : Second digit of capacitance (see Table 7 below).
Dot (4) : Class of capacitor (see REC-115-A).
Dot (5) : Tolerance (see Table 7 below).
Dot (6) : Multiplier for capacitance (see Table 7 below).

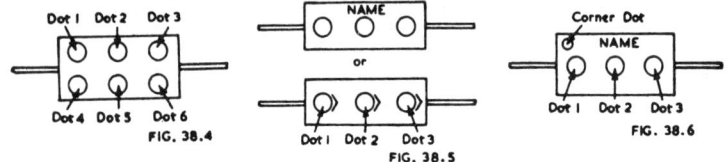

Fig. 38.4.   *Colour code for moulded mica capacitors* (6 *dot*).
Fig. 38.5.   *Colour code for moulded mica capacitors* (3 *dot*).
Fig.. 38.6.   *Colour code for moulded mica capacitors* (4 *dot*).

TABLE 7 : R.T.M.A. MOULDED MICA CAPACITORS

Colour	Numeral	Multiplier	Tolerance	Class
Black	0	1	20%	A
Brown	1	10		B
Red	2	100	2%	C
Orange	3	1000	3%	D
Yellow	4	10 000		E
Green	5		5%	
Blue	6			
Violet	7			
Grey	8			I
White	9			J
Gold		0.1		
Silver		0.01	10%	

JAN-C-5 six dot colour marking (Fig. 38.4)

In the JAN-C-5 code the black Dot 1 signifies a mica dielectric capacitor. Dot 4 gives the capacitor characteristic (for details see JAN specification). The remaining dots have the same significance as in the R.C.M. system.

Old R.M.A. six dot colour marking (Fig. 38.4)

Dots 1, 2 and 3 signify the first three significant figures of the capacitance. Dot 4 signifies the rated direct working voltage. Dot 5 gives the percentage tolerance. Dot 6 gives the decimal multiplier.

Three dot system

This applies only to capacitors with 500 volt ratings and ± 20% tolerances in capacitance (Fig. 38.5).

Dot 1 gives the first significant figure.
Dot 2 gives the second significant figure.
Dot 3 gives the decimal multiplier.

Four dot system

To extend the usefulness of the three dot system, a fourth dot is sometimes added in the top left-hand corner (Fig. 38 6) to indicate the percentage tolerance.

See also Sect. 3(v) for standard moulded mica capacitors.

(v) Colour code for ceramic dielectric capacitors
(A) R.M.A. Standard REC-107, Oct. 1947

The colour markings consist of five colours, one of which unambiguously marks the end of the capacitor bearing the inner-electrode terminal while the remaining four colour markings are successively closely adjacent along the length of the capacitor ; an indicator is provided to avoid ambiguity in the interpretation of the significance of the position of the colour markings.

The end colour indicates the temperature coefficient in accordance with Table 8 below.

The first and second colour markings indicate the first and second digits of the value of the capacitance in micro-microfarads.

The third colour marking indicates the decimal multiplier of the value of the capacitance.

The fourth colour marking indicates the capacitance tolerance in accordance with Table 8 below.

TABLE 8

Position	End	Nos. 1, 2	No. 3	No. 4	
Colour	Temperature coeff. parts/ million/°C	Significant figure	Decimal multiplier	Capacitance tolerance	
				$C > 10 \ \mu\mu\text{F}$	$C < 10 \ \mu\mu\text{F}$
black	0	0	1	±20%	
brown	−30	1	10	± 1%	±0.1 $\mu\mu$F
red	−80	2	10^2	± 2%	
orange	−150	3	10^3	±2.5%	
yellow	−220	4	10^4		
green	−330	5		± 5%	±0.5 $\mu\mu$F
blue	−470	6			
violet	−750	7			
grey	+30	8	0.01		±0.25 $\mu\mu$F
white	General purpose condenser*	9	0.1	±10%	±1.0 $\mu\mu$F

Capacitance change

silver   +25%   (Class 4 capacitors)
silver +25%, − 50% (Class 5 capacitors)

*May have any nominal temperature coefficient between +120 and −750 parts per million per degree Centigrade, at option of the manufacturer.

(B) British R.I.C. colour code for ceramic dielectric capacitors, RIC/133 (Ref. F7)

Colour	End Colour (Temperature Coefficient)*	1st dot, 1st significant figure	2nd dot, 2nd significant figure	3rd dot Multiplier	4th dot Tolerance	
					10 $\mu\mu$F or less	More than 10 $\mu\mu$F
Black	NP0	0	0	1	± 2.0 $\mu\mu$F	$\pm 20\%$
Brown	N030	1	1	10	± 0.1 $\mu\mu$F	$\pm 1\%$
Red	N080	2	2	100		$\pm 2\%$
Orange	N150	3	3	1000		$\pm 2.5\%$
Yellow	N220	4	4	10 000		
Green	N330	5	5		± 0.5 $\mu\mu$F	$\pm 5\%$
Blue	N470	6	6			
Violet	N750	7	7			
Grey	P030	8	8	0.01	± 0.25 $\mu\mu$F	
White	P100	9	9	0.1	± 1.0 $\mu\mu$F	$\pm 10\%$

*N signifies negative, P signifies positive temperature coefficient. Figures give parts per million per °C.

See also Sect. 3(vi) for standard ceramic dielectric capacitors.

(vi) Colour code for i-f transformers
(R.M.A. Standard REC-114, March 1948, for 455 Kc/s)
Blue—plate lead.
Red—B + lead.
Green—grid (or diode) lead.
White—grid (or diode) return.
(For " full-wave " transformer, the second diode lead will be violet).

(vii) Colour code for a-f transformers and output transformers
(R.M.A. Standard U.S.A. S410, M4-507, May 1935)
Blue—plate (finish) lead of primary.
Red—B+ (this applies whether primary is plain or centre-tapped).
Brown—plate (start) lead on C.T. primaries. Blue may be used for this lead if polarity is not important.
Green—grid (finish) lead to secondary (hot end of voice coil).
Black—grid return (this applies whether the secondary is plain or centre tapped).
Yellow—grid (start) lead on centre-tapped secondaries. (Green may be used for this lead if polarity is not important).
Note : These markings apply also to line-to-grid and valve-to-line transformers.

(viii) Colour code for power transformers
(R.M.A. Standard S410, M4-505, May 1935)
1. Primary leads—no tap   Black
    If tapped—Common   Black
        Tap : Black and yellow 50/50 striped design
        Finish : Black and red 50/50 striped design
2. Rectifier—Plate winding   Red
        Centre tap : Red and yellow 50/50 striped design
3. Rectifier—Filament winding   Yellow
        Centre tap : Yellow and blue 50/50 striped design
4. Amplifier—Filament winding No. 1   Green
        Centre tap : Green and yellow 50/50 striped design

5. Amplifier—Filament winding No. 2   **Brown**
   Centre tap : Brown and yellow 50/50 striped design
6. Amplifier—Filament winding No. 3   **Slate**
   Centre tap : Slate and yellow 50/50 striped design

(ix) Colour code for loudspeakers
(R.M.A. Standard U.S.A. M5-181 Nov. 1936)
Loudspeaker field coils
Black and red*—start.
Yellow and red*—finish.
Slate and red—tap (if any).

Note : If two field coils are fitted to the same loudspeaker, the basic colour coding is used for the lower resistance field, and green is substituted for the red in the higher resistance field.

Loudspeaker transformer primaries
Centre-tapped
Blue or brown—start
Blue—finish
Red—centre tap
Untapped
Red†—start
Blue—finish

Loudspeaker transformer secondaries
Black—start
Green--finish

Standard pin arrangement 4A
Pin 1 (large) Yellow and red (field finish).
Pin 2 Blue or brown (transformer start) for push-pull.
    No connection or red (transformer start).
Pin 3 Blue (transformer finish).
Pin 4 (large) Black and red (field start).

Standard pin arrangement 5A
Pin 1 Yellow and red (field finish).
Pin 2 Blue or brown (transformer start).
Pin 3 Red (transformer centre-tap).
Pin 4 Blue (transformer finish).
Pin 5 Black and red (field start).
    The start, tap and finish of all windings occur in clockwise order around the plug pins when the plug is viewed from the socket end.

(x) Colour code for chassis wiring
(A) R.M.A. Standard U.S.A. REC-108-A (December 1949)

Colour	Circuit
Black	Grounds, grounded elements, and returns
Brown	Heaters or filaments, off ground
Red	Power supply B+
Orange	Screen grids
Yellow	Cathodes
Green	Control grids
Blue	Plates
Violet	Not used
Grey	A.C. power lines
White	Above or below ground returns, a.v.c., etc.

*Some manufacturers use a single colour, omitting the red.
†Some manufacturers use brown whether tapped or untapped.

When leads for antenna and ground connections are provided on the receiver, it shall be standard to colour code the antenna lead blue and the ground lead black. Special antenna connection leads shall be coded with combinations of blue and black.

(B) RTMA Standard Colour marking of thermoplastic insulated hook-up wire, GEN-104 (July 1951)
Colour coding for thermoplastic hook-up wire without fabric braids shall be accomplished either by the use of solid coloured insulation or by natural colour or white insulation with coloured helical stripes.   It is intended that these wires be interchangeable by colour with fabric braid wires.   Where, in fabric, braids consist of solid colours with tracer threads of contrasting colours, this standard implies that base colour stripe (wide stripe) is used in lieu of solid coloured fabric braid.   Helical tracer stripes are used in lieu of contrasting colour threads in fabric braid-covered wire.

(xi) Colour code for battery cables
(American R.M.A. Standard S-410, M4-508, April 1939)

A +	Red	B intermediate	White
A −	Black	C +	Brown
B +	Blue	C intermediate	Orange
B −	Yellow	C −	Green

(xii) Colour code for metallized paper capacitors
A 3 dot colour code may be used to indicate capacitance in $\mu\mu$F.   Dot 1 gives the first significant figure.   Dot 2 gives the second significant figure (Table 7 col. 2). Dot 3 gives the multiplier (Table 7 col. 3).

SECTION 3 : STANDARD RESISTORS AND CAPACITORS

(i) Standard fixed composition resistors   (ii) Standard fixed wire wound resistors (iii) Fixed paper dielectric capacitors in tubular non-metallic cases   (iv) Metal encased fixed paper dielectric capacitors for d.c. application   (v) Standard fixed mica dielectric capacitors   (vi) Standard ceramic dielectric capacitors   (vii) Standard variable capacitors   (viii) Standard variable composition resistors   (ix) Standard metallized paper dielectric capacitors   (x) Standard electrolytic capacitors   (xi) References to standard resistors and capacitors.

The information in this Section is in the form of summaries or extracts from certain recognized standards, and is necessarily incomplete.   The purpose of this Section is, in part, to draw the attention of design engineers to the importance of the information obtainable from such standards.   However, all design engineers are urged to secure up-to-date copies of the official standard specifications under which they work.
See also Chapter 4 Sect. 9 for general information.

(i) Standard fixed composition resistors
(A) American R.M.A. Standard REC-116, July 1948
Values of resistance from 1 ohm to 100 megohms.   The preferred values of resistance, over the range from 50 ohms to 10 megohms, are shown in Table 6 in Sect. 2(iii).   The complete list is given in the R.M.A. Standard REC-116.   These preferred values have been adopted in U.S.A. and to some extent in England, but not universally.   The old standard resistance values, shown in bold typeface, are generally used where the preferred values have not been adopted.
The tolerance in resistance is ±5%, ±10% or ±20% as desired.

The **resistance-temperature characteristic** is expressed as the change of resistance between ambient temperatures of $+25°C$ and $+85°C$ in the form of the percentage of the resistance at $25°C$.

The **voltage coefficient** is given by

$$\text{Voltage coefficient} = \frac{100\,(R_1 - R_2)}{R_2\,(E_1 - E_2)} = \frac{111\,(R_1 - R_2)}{E_1 R_2}$$

where E_1 = rated continuous working voltage (to give rated power dissipation, or maximum continuous voltage rating, whichever is the higher)

$E_2 = E_1/10$

R_1 = resistance measured with applied voltage E_1

and R_2 = resistance measured with applied voltage E_2.

The test is carried out firstly with reduced voltage (E_2), and immediately afterwards with voltage E_1.

The **noise** is measured by connecting the resistor in a d.c. circuit and comparing the noise voltage with an audio frequency of 1000 c/s. The measuring equipment includes an amplifier, filter and valve voltmeter (for details see REC-116). The noise is expressed in microvolts per volt (applied direct voltage).

Ambient temperature—effect on maximum working voltage

Temperature	40°	60°	80°	85°	100°	110°C
Max. voltage	100%	77.5%	55%	50%	20%	0

For colour code see Sect. 2(i).

(B) American JAN-R-11 (Ref. E1)

Some commercial resistors are manufactured in accordance with JAN-R-11 specifications while these are compulsory for use in American Army/Navy equipment.

There are two resistance-temperature characteristics ; characteristic F has one half the change in resistance of characteristic E. The de-rating curve is tabulated on page 187. The maximum surface temperature is $135°C$. The maximum continuous working voltage is :

Type RC15	(¼ watt)	200	volts (d.c. or r.m.s.)	
RC10, 16	(¼ watt)	250	,,	,,
RC20, 21, 25	(½ watt)	350	,,	,,
RC30, 31, 35	(1 watt)	500	,,	,,
RC38	(1 watt)	1000	,,	,,
RC40, 41, 45	(2 watts)	500	,,	,,
RC65	(4 watts)	500	,,	,,
RC75, 76	(5 watts)	500	,,	,,

The voltage coefficient shall not exceed 0.035% per volt for ¼ and ½ watt ratings, or 0.02% per volt for larger ratings. Tests are specified for temperature cycling, humidity, vibration, effect of soldering, overloading etc., also salt water immersion and insulation strength where applicable. The root mean-square value of the noise generated shall not exceed 3.0 r.m.s. $\mu V/V$ at the rated continuous working voltage for ¼ and ½ watt ratings, or 1.2 r.m.s. $\mu V/V$ for larger ratings, frequency characteristic —3 db at 70 and 5000 c/s. Marking, resistance values and tolerances are as R.M.A. REC-116.

Amendment No. 3 provides for a characteristic G which has an ambient temperature of $70°C$ with a linear de-rating characteristic which reaches zero at $130°F$.

(C) British R.I.C. Specification No. RIC/112

Grade 1 (high stability). Issue No. 1, May 1950. Resistance values follow the series, 1, 1.1, 1.2, 1.3, 1.5, 1.6, 1.8, 2.0, 2.2, 2.4, 2.7, 3.0, 3.3, 3.6, 3.9, 4.3, 4.7, 5.1, 5.6, 6.2, 6.8, 7.5, 8.2, 9.1, 10 for all tolerances. Range 10 ohms to 10 megohms with some limitations.

Tolerances ± 1%, ±2%, ±5%.

Ratings ⅛, ¼, ½, ¾, 1, 2 watts.

Colour code (five colours)—see Sect. 2(i)B.

The surface temperature shall not exceed $150°C$. The voltage coefficient shall not exceed 0.002% per volt. The noise shall not exceed 0.5 μV per volt (d.c.). Resistance change with soldering shall not exceed ±0.3%.

Temperature coefficient :

Watts	$\frac{1}{8}$	$\frac{1}{4}$	$\frac{1}{2}$	$\frac{3}{4}$	1 and 2	Temp. coeff.
Res up to	100K	250K	0.5M	1M	2M	0 to −0.0004
Res. up to	0.5M	1M	2.5M	5M	10M	0 to −0.0008
Res. up to	2.5M	5M	10M	10M	—	0 to −0.001

Permissible variation in resistance after climatic and durability tests, endurance, tropical exposure, humid atmosphere, salt atmosphere :

Permissible variation	$\frac{1}{8}$W	$\frac{1}{4}$W	$\frac{1}{2}$W	$\frac{3}{4}$W	1 and 2 W
1% up to resistance of	10K	25K	50K	100K	200K
1½% ,, ,, ,, ,,	100K	250K	0.5M	1M	2M
2% ,, ,, ,, ,,	0.5M	1M	2.5M	5M	10M
3% ,, ,, ,, ,,	2.5M	5M	10M	10M	—

Rating curve (dissipation) 100% to 70°C ambient temperature, then decreasing linearly to 0% at 150°C.

(D) British R.I.C. Specification No. RIC/113
Grade 2. Issue No. 1, June 1950.
Resistance values as Table 5 for corresponding tolerances.
Range 10 ohms to 10 megohms with some limitations.
Tolerances ±20%, ±10%, ±5%.
Ratings—axial, non-insulated : $\frac{1}{2}$, 1 watt
        radial, non-insulated 1/10, $\frac{3}{4}$, 1, 2$\frac{1}{2}$ watts.
        axial, insulated : $\frac{1}{4}$, $\frac{1}{2}$, $\frac{3}{4}$, 1 watt.
Colour code—see Sect. 2(i)C.
The surface temperature shall not exceed 120°C. The voltage coefficient shall not exceed 0.025% per volt (below 1 megohm), or 0.05% (1 megohm and above). Temperature coefficient shall not exceed 0.12%. Resistance change with soldering shall not exceed ±2%. Rating curve (dissipation) : 100% at 70°C, decreasing linearly to 0% at 110°C, and increasing to 175% at 40°C ambient temperature.

(ii) Standard fixed wire wound resistors
(A) American R.M.A. Standard REC-117, July 1948
There are three styles :

Style No.	RRU3	RRU4	RRU6
Max. watts dissipation at 40°C ambient	$\frac{1}{2}$	1	2
Minimum resistance, ohms	0.24	0.47	1.0
Maximum resistance, ohms	820	5100	8200
Standard tolerances :			
More than 10 ohms	5% 10% 20%	5% 10% 20%	5% 10% 20%
10 ohms or less	10% 20%	10% 20%	10% 20%

The preferred values of resistance are as for composition resistors, except for the extension to lower values.
Ambient temperature—effect on max. working voltage

Temperature	40°	70°	100°	110°C
Max. voltage	100%	60%	20%	0

Power rating—in still air at 40°C ambient temperature the temperature rise should not exceed 70°C for styles 3 and 4, and 95°C for style 6, for the rated power input. At temperatures higher than 40°C, with reduced working voltage and power dissipation as tabulated above, the hot spot temperature should not exceed 110°C for styles 3 and 4, and 135°C for style 6.

Resistance change with temperature
The change in resistance referred to an ambient temperature of 25°C shall not be more than ±0.025% per 1°C for all resistors over 10 ohms, and not over ±0.15% per 1°C for 10 ohms and lower.

Marking
Typical marking RRU3 511 J.

RRU3   indicates fixed wire wound resistor, style 3.

511   indicates resistance. The first two digits are the significant figures of the value of resistance in ohms, while the last digit indicates the number of zeros which follow the significant figure. The letter R indicates a decimal point e.g. 511 indicates 51 × 10 = 510 ohms. R56 indicates 0.56 ohms.

J   indicates the tolerance :

symbol	J	K	L
tolerance	±5%	±10%	±20%

(B) American JAN-R-26A, JAN-R-19, JAN-R-184 should also be examined. See (xi) below for list.

(C) British R.I.C. Specification No. RIC/111
Resistors, fixed, wirewound, non-insulated. Issue No. 1, July 1950.
Resistance values (all tolerances) as column 3 in Table 6.
Tolerances ±1%, ±2%, ±5%, ±10%.
Colour code as Fig. 38.2B, with Table 5.

(iii) Fixed paper dielectric capacitors in tubular non-metallic cases
(A) American R.M.A. Standard REC-118, Sept. 1948

TABLE 9

Nominal capacitance	Standard tolerance	Voltage Ratings					
1.0 μF	+20%, −10%	100	200	400	—	—	—
0.5	+20%, −10%	100	200	400	600	—	—
0.25	+20%, −10%	100	200	400	600	—	—
0.15	+20%, −10%	100	200	400	600	1000	—
0.10	+20%, −20%	100	200	400	600	1000	—
0.05	+20%, −20%	100	200	400	600	1000	1600
0.03	+20%, −20%	100	200	400	600	1000	1600
0.02	+20%, −20%	100	200	400	600	1000	1600
0.01	+20%, −20%	100	200	400	600	1000	1600
0.005	+40%, −20%	—	—	—	600	1000	1600
0.003	+40%, −20%	—	—	—	600	1000	1600
0.002	+40%, −20%	—	—	—	600	1000	1600
0.001	+60%, −25%	—	—	—	600	1000	1600

Insulation Resistance at 25°C

TABLE 10

Nominal capacitance	Insulation resistance not less than:
1.0 μF	1000 megohms
0.5	2000 megohms
0.25	4000 megohms
0.001 to 0.15	5000 megohms

To determine the minimum insulation resistance at temperatures from 20°C to 40°C, multiply the values above by the factor :

°C	20	25	30	35	40
Factor	1.42	1.00	0.71	0.51	0.35

At still higher temperatures see Table 11.

TABLE 11

Capacitance μF	Insulation resistance at elevated temperatures	
	65°C Class W*	85°C Class M*
1.0	not less than 25 megohms	not less than 10 megohms
0.5	not less than 50 megohms	not less than 20 megohms
0.25	not less than 100 megohms	not less than 40 megohms
0.15	not less than 166 megohms	not less than 66 megohms
0.10	not less than 250 megohms	not less than 100 megohms
0.05	not less than 250 megohms	not less than 200 megohms
0.03 to 0.001	not less than 250 megohms	not less than 250 megohms

*Classification

Class W Operating temperature range −30°C to +65°C. Power factor not greater than 2%. Resistance to humidity as below. Insulation resistance at 65°C as below.

Class M Operating temperature range −30°C to +85°C. Power factor not greater than 1%. Insulation resistance at 85°C as above.

Humidity resistance (Class W only)

After humidity treatment (see REC-118 for details), the insulation resistance shall not be less than :

Nominal capacitance	0.001 to 0.15	0.25	0.5	1.0 μF
Insulation resistance	1750	1400	700	350 MΩ

Power Factor Class M—not more than 1%.

Class W—not more than 2%.

Marking. Typical marking RCP10 W 6 504 K

RCP10 = RMA Standard Capacitor employing Paper Dielectric, with non-metallic cylindrical enclosure equipped with axially positioned wire leads.

W = Class (see above).

6 = Voltage in units of 100 volts.

504 = Indicating capacitance 50 × 10⁴ giving capacitance in micro-microfarads.

K = tolerance.

K indicates +10%, −10%   X indicates +40%, −20%

L indicates +20%, −10%   Y indicates +60%, −25%

M indicates +20%, −20%

(B) American JAN-C-91 should also be examined. See (xi) below for list.

(C) British R.I.C. Specification No. RIC/131

Issue No. 1, July 1950.

Capacitance values conform to the series 1, 2, 5, 10, etc.

Capacitance range 0.001 to 2 μF.

Tolerances ±25% up to and including 0.01 μF.

±20% above 0.01 μF.

±10% (Red Group only).

Maximum voltage ratings (d.c.) 150, 350, 500, 750, 1000 volts at rated temperature

Rated temperature 70°C (Red and Yellow) ; 60°C (Green).

Power factor shall not exceed 0.01.

Insulation resistance (one month after delivery) shall not be less than 2000 ohm-farads or 10 000 MΩ whichever is the less, for Red and Yellow, or 500 ohm-farads or 2500 MΩ, whichever is the less, for Green.

Tests are specified for climatic and durability, humid atmosphere, tropical exposure, endurance etc.

This specification also includes capacitors in metallic and ceramic cases.

(iv) Metal encased fixed paper dielectric capacitors for d.c. application

(A) American R.M.A. Standard TR-113 (April 1949)

TABLE 12

Nominal Capacitance	Standard Tolerance	Voltage ratings* (up to 40°C)						
50 μF	±10%					600		
25	±10%					600		
20	±10%					600		
15	±10%					600	1000	1500
12	±10%					600	1000	1500
10	±10%					600	1000	1500
8	±10%					600	1000	1500
6	±10%					600	1000	1500
4	±10%					600	1000	1500
2	±10%			200		600	1000	1500
1	+20%, −10%			200	400	600	1000	1500
0.5	+20%, −10%			200	400	600	1000	1500
0.25	+20%, −10%			200	400	600	1000	1500
0.1	+20%, −20%		100	200	400	600	1000	1500
0.05	+20%, −20%				400	600	1000	1500
0.02	+20%, −20%		100			600	1000	1500
0.01	+40%, −20%		100	200		600	1000	1500
0.006	+40%, −20%		100			600	1000	1500
0.003	+40%, −20%					600	1000	1500
0.001	+60%, −25%		100					

* Voltage ratings above 1500V are not shown here.

Life : The voltage ratings given above, and the reduced voltage ratings for temperatures above 40°C, are based on an expected life of 1 year continuous operation at these temperatures. Longer life can be expected by operation at still lower voltages, e.g. 5 years at 70% of the values indicated.

Working voltage : The average working voltage over 24 hours should not exceed the adjusted rated voltage by more than 5%. A voltage of 110% of the rated voltage may be applied for not more than 10% of the operating time in any 24 hour period.

Alternating current component : In cases where alternating voltages are present in addition to direct voltages, the capacitor working voltage should be taken as the sum of the direct voltage and the peak value of the alternating voltage, provided that the peak alternating voltage does not exceed 20% of the direct voltage and that the frequency is 60 c/s or less. At higher frequencies the voltage should be adjusted as tabulated below :

Frequency	60	100	200	400	1000	4000	10 000 c/s
Ripple	20	16	12	9	6.2	2.8	1 %

Insulation resistance

Minimum insulation resistance in megohms :

TABLE 13

Temperature	25°C	40°C	55°C	70°C	85°C	
Characteristic A	3000	1000	300	100	30	MΩ
Characteristic B	1000	300	100	30	10	MΩ
Characteristic C	500	150	50	15	5	MΩ

Effect of temperature on capacitance (basis = 25°C)

TABLE 14

Temperature	85°C	50°C	25°C	0°C	−40°C
Maximum tolerances :					
Characteristic A	±5%	±2%	±0	± 2%	± 5%
Characteristic B or C	±5%	±2%	±0	$\left\{ \begin{array}{l} + \ 2\% \\ -30\% \end{array} \right.$	$\begin{array}{l} + \ 5\% \\ -30\% \end{array}$

Effect of ambient temperature on voltage rating

Where condensers are used at ambient temperatures in excess of 40°C, the working voltage shall be reduced as indicated below :

TABLE 15

Temperature	40°C	50°C	60°C	70°C	80°C	85°C
Voltage rating (1)	100%	98%	94%	86%	74%	65%
Voltage rating (2)	100%	97.5%	92%	82.5%	66%	55%

Voltage rating (1) applies to condensers with a watt-second rating from 0 to 5. Voltage rating (2) applies to watt-second ratings over 5 (Voltage ratings over 2000 V are not included here).

Watt-second rating

Watt-second rating = $\frac{1}{2}CE^2$

where C = capacitance in microfarads

and E = nominal rating in kilovolts.

For other details and standard tests see TR-113-A.

Marking—Typical marking 71 B 1 A H 205 K.

71   indicates style (case form)*
B   indicates terminal designation
        (A indicates wire lead, B indicates solder lug, C indicates screw and nuts).
1   indicates schematic circuit and number of terminals*
A   indicates characteristic (A, B or C)—see above.
H   indicates nominal voltage rating, as under :

Code letter	B	C	E	F	G	H	etc.
Voltage	100	200	400	600	1000	1500	etc.

205   indicates capacitance (here 20 × 10⁵ $\mu\mu$F).
K   indicates tolerance :

Code letter	K	L	V	M	W
Tolerance	±10%	±15%	−10%, +20%	±20%	−0, +25%
Code letter	X		Y		
Tolerance	−20%, +40%		−25%, +60%		

*For details see TR-113-A.

The tolerance code may be omitted if the tolerance is standard (see table above).
(B) **American JAN-C-25** should also be examined.   See (xi) below for list.
(C) **British R.I.C. Specification No. RIC/131**
This specification includes both metallic and non-metallic cases—see Sect. 3(iii)C.

(v) Standard fixed mica dielectric capacitors
(A) Standard molded mica capacitors (American R.M.A. Standard REC-115-A, May 1951).

Nominal capacitance ($\mu\mu$F)

TABLE 19

10	36	100	300	820	2400	6800	.020 μF
12	39	110	330	910	2700	7500	.022 μF
15	43	120	360	1000	3000	8200	.024 μF
18	47	130	390	1100	3300	9100	.027 μF
20	51	150	430	1200	3600	.010 μF	.030 μF
22	56	160	470	1300	3900	.011 μF	.033 μF
24	62	180	510	1500	4300	012 μF	.036 μF
27	68	200	560	1600	4700	.013 μF	.039 μF
30	75	220	620	1800	5100	.015 μF	.043 μF
33	82	240	680	2000	5600	.016 μF	.047 μF
	91	270	750	2200	6200	.018 μF	

These are available in 11 styles, but any one value of capacitance is only available in a limited number of styles.

Standard d.c. voltage ratings
At temperature $-20°$C to $+85°$C ; barometric pressure 28 in. to 32 in. of mercury ; relative humidity 10% to 80%.

TABLE 20

Voltage rating	Available in styles
300	20, 25, 30, 35, 40
500	20, 25, 30, 35, 40
600	45, 50, 55 56, 60, 61
1000	40
1200	45, 50, 55, 56, 60, 61
2500	45, 50, 55, 56, 60, 61

Standard Classification

TABLE 21

Class	Q* not less than	Insul. Resist. not less than	Temp. coeff.† not more than	Capacit. drift not more than
A	30%	3000 MΩ	$\pm$ 1000 ppm	$\pm$ (5% + 1) $\mu\mu$F
B	100	6000	$\pm$ 500	$\pm$ (3% + 1)
C	100	6000	$\pm$ 200	$\pm$ (0.5% + 0.5)
I	100	6000	+ 150, -50	$\pm$ (0.3% + 0.2)
D	100	6000	$\pm$ 100	$\pm$ (0.3% +0.1)
J	100	6000	+ 100, -50	$\pm$ (0.2% + 0.2)
E	100	6000	+ 100, -20	$\pm$ (0.1% + 0.1)

*Not less than this percentage of values tabulated below :
Nomin. capacit.   7   10   20   40   60   100   200   1000 $\mu\mu$F
Minimum Q   120   160   280   450   580   760   1000   1000
(Q is measured at approximately 1 Mc/s).
†Temperature coefficient of capacitance, being capacitance change in parts per million per °C.

Insulation resistance after humidity cycle :
Class A capacitors—not less than 1000 megohms.
Other classes—not less than 2000 megohms.

Marking (alternative to colour code for which see Sect. 2(iv)).
Typical marking RCM 20 A 050 M
  RCM = RMA Standard Capacitor employing Mica dielectric.
  20   = style designation (see SP158B)
  A   = class designation (see above).
  050   = 05 × 10° = 5 μμF (capacitance).
  M indicates tolerance as under :

Code letter	G	H	J	K	M
Tolerance	±2%	±3%	±5%	±10%	±20%

(B) **American JAN-C-5** should also be examined.   See (xi) below for list.

(C) **British R.I.C. Specification RIC/132**
Mica dielectric, stacked foil (Ref. F6).
Tolerance ±20%.   Power factor shall not exceed 0.001.   Insulation resistance
shall not be less than 10 000 megohms.   Capacitance values conform to the series
10, 15, 22, 33, 47, 68 and 100.   Range from 100 to 10 000 μμF.   Voltage ratings
350, 750 volts.

(D) **British R.I.C. Specification RIC/137**
Mica dielectric, metallized (Ref. F10).
Tolerances ±20%, ±10%, ±5%, ±2%, subject to a minimum of ±1 μμF.
Power factor shall not exceed 0.001 except on values below 47 μμF where the limit
is 0.002.   Insulation resistance shall not be less than 10 000 megohms.   Capacitance
values conform to the series, according to tolerance, given by the first three columns
of Table 6.   Range from 10 to 10 000 μμF.   Voltage ratings 350, 750 volts (d.c.)
at 70°C.

(vi) Standard ceramic dielectric capacitors
(A) **American R.M.A. Standard REC-107, Oct. 1947**
Nominal capacitances (μμF)

TABLE 22

0.5	0.75	1.0	1.5	2.0	3.0	4.0	5.0	6.0	7.0
8.0	9.0	10	12	13	15	18	20	22	24
27	30	33	36	39	43	47	51	56	62
68	75	82	91	100	110	120	130	150	160
180	200	220	240	270	300	330	360	390	420
470	510	560	620	680	750	820	910	1000	1100
1 200	1 300	1 500	1 600	1 800	2 000	2 200	2 400	2 700	3 000
3 300	3 600	3 900	4 200	4 700	5 100	5 600	6 200	6 800	7 500
8 200	9 100	10 000	11 000	12 000	13 000	15 000			

Tolerances in capacitance

TABLE 23

Classes 1, 2, 3	Classes 4, 5	All classes
±20% ±10% ± 5% ± 2½% ± 2% ± 1%	±20%	(Capacit. 10 μμF or less) ±2.0 μμF ±1.0 μμF ±0.5 μμF ±0.25 μμF ±0.1 μμF

Working voltage (peak)
  Classes 1, 2, 3 :   500 volts (down to 3.4 in. mercury)
  Classes 4, 5 :   350 volts (down to 3.4 in. mercury)

Q and insulation resistance

Class	Q not less than	Insulation resistance not less than
1	1000*	7500 megohms
2	650*	7500
3	335*	7500
4	100	1000
5	40	1000

*For capacitances 30 $\mu\mu$F and over. Lower values for lower capacitances.

Capacitance drift with temperature cycling
Classes 1, 2, 3 : not more than 0.3% or 0.25 $\mu\mu$F whichever is the greater.

Capacitance-temperature characteristic
Classes 1, 2 and 3 :
Standard characteristics—Change of capacitance over the range $-55°C$ to $+85°C$ per unit of capacitance at 25°C per degree change in temperature : $+100, +30$, zero, $-30, -80, -150, -220, -330, -470, -750$ parts per million per °C.
Classes 4 and 5 :
Maximum change in capacitance from its value at 25°C over temperature range from $-55°C$ to $+85°C$:
Class 4 $\pm 25\%$; Class 5 $-50\%, +25\%$.

Tolerances on temperature coefficient
(when based on 2 point measurement, one at 25°C and one at 85°C) $\pm 15, \pm 30, \pm 60, \pm 120, \pm 250, \pm 500$ parts per million per °C.
An alternative method of measurement, with unsymmetrical tolerances, is also given in REC-107.
Humidity tests—see REC-107.

Marking
Inner electrode terminal indicated by dot or depression.
Typical marking : R2 CC 20 CH 100 G
  R2 indicates RMA Class 2
  CC indicates ceramic capacitor
  20 indicates style (see REC-107)
  CH indicates temperature characteristic and capacitance tolerance (see Table 1 below)
  100 indicates capacitance (here $10 \times 10° = 10 \mu\mu$F). For values incorporating decimal fractions, R indicates decimal point, e.g., 1R5 indicates 1.5 $\mu\mu$F, R75 indicates 0.75 $\mu\mu$F
  G indicates tolerance (see Table 24 below)

TABLE 24

Letter symbol	Tolerance for capacitance of	
	10 $\mu\mu$F or less	more than 10 $\mu\mu$F
B	± 0.1 $\mu\mu$F	
C	± 0.25	
D	± 0.50	
F	± 1.0	$\pm 1\%$
G	± 2.0	$\pm 2\%$
H		$\pm 2.5\%$
J		$\pm 5\%$
K		$\pm 10\%$
M		$\pm 20\%$

For alternative colour code see Sect. 2(v).

TABLE 25

First letter symbol	Capacitance-temperature coefficient or per cent decrease* in	Second letter symbol	Capacitance tolerance on temp. coeff. or per cent increase* in
A	+ 100 parts/mln/°C		
B	+ 30 parts/mln/°C	F	15 parts/mln/°C
C	Zero parts/mln/°C	G	30 parts/mln/°C
H	− 30 parts/mln/°C	H	60 parts/mln/°C
L	− 80 parts/mln/°C	J	120 parts/mln/°C
P	− 150 parts/mln/°C	K	250 parts/mln/°C
R	− 220 parts/mln/°C	L**	500 parts/mln/°C
S	− 330 parts/mln/°C	Y	25%
T	− 470 parts/mln/°C	Z	50%
U	− 750 parts/mln/°C		
Y	−25%		
Z	−50%		

*From that at 25°C over temperature range −55°C to 85°C.
**Use only with first letter symbol S (characteristic) to indicate general purpose condenser which may have any nominal temperature coefficient between +120 and −750 parts per million per °C.

TABLE 26

Standard capacitance-temperature characteristics

Capacit. $\mu\mu F$	First letter of temperature characteristic symbol									
	A	B	C	H	L	P	R	S	T	U
0.5 to 2.0	K	K	—	K	K	K	K	KL	K	—
3	JK	JK	JK	JK	JK	JK	JK	JKL	JK	JK
4 to 9	HJK	HJK	HJK	HJK	HJK	HJK	HJK	HJKL	JK	HJK
10 to 91	GHJK	GHJK	GHJK	GHJK	GHJK	GHJK	GHJK	HJKL	HJK	HJK
100 to 1600	GHJK	FGHJK	FGHJK	FGHJK	GHJK	GHJK	GHJK	HJKL	HJK	HJK

(B) **American JAN-C-20A and JAN-C-81** should also be examined. See (xi) below for list.

(C) **British R.I.C. Specification RIC/133**
Ceramic dielectric, Grade 1. (Ref. F7).

Capacitance values conform to the series, according to tolerance, given by the first 3 columns of Table 6. Range from 1 to 1000 $\mu\mu F$. Tolerances ±20%, ±10%, ±5%, ±2%, subject to minimum of ±1 $\mu\mu F$. Voltage ratings 500 and 750 volts (d.c.).

Temperature coefficient (standard) +100, −30, −80, −470, −750 parts per million per °C, with tolerance ±40 ppm or ±15%, whichever is the greater. A special tolerance (available only on values of 47 $\mu\mu F$ and above) shall be ±40 ppm or 7½%, whichever is the greater.

The power factor shall not exceed 0.0015. The insulation resistance shall not be less than 10 000 megohms.

Colour code (optional) : see Sect. 2(v)B.

(vii) Standard variable capacitors
(A) R.M.A. air dielectric, tuning
The standard capacitance range (American R.M.A. REC-106-A Class A) without trimmers (compensators) is given by table 27.

TABLE 27

No. of plates	Difference in capacitance ($\mu\mu$F)	Min. capacitance ($\mu\mu$F)	
		R-F	Osc.
25	530.0	not greater than 15.0	—
23	485*	not greater than 15.0	—
21	441*	not greater than 13.5	13.0
19	397*	not greater than 13.0	12.0
17	353*	not greater than 12.5	11.5
15	309*	—	11.0
13	265*	—	10.5
11	221*	—	10.0

The standard capacitance range (American R.M.A. REC-101 Class B) without trimmers is given by table 28.

TABLE 28

No. of plates	Difference in capacitance ($\mu\mu$F)	Min. Capacitance ($\mu\mu$F)	
		R-F	Osc.
27	420	not greater than 13	9
25	388*	not greater than 12	*
23	355*	not greater than 11	*
21	323*	not greater than 10	*

*Calculated in proportion to the number of dielectrics.

Tolerance in capacitance (Class A and Class B) :
Reference section $\pm(1\ \mu\mu$F $+\ 1\%$ of tabulated value).
Other sections compared with reference section $\pm(1\ \mu\mu$F $+\ \frac{1}{2}\%)$.

Capacitance characteristics with maximum number of plates**
TABLE 29

	REC-106-A Class A		REC-101 Class B	
Dielectrics	24	20	26	18
Rotation	Difference in capacitance ($\mu\mu$F) from that of zero position			
	R-F Section(s)	Oscillator	R-F Section(s)	Oscillator
0*	0.0	0.0	0.0	0.0
10%	9.4	7.3	7.3	5.1
20%	33.7	29.1	22.3	15.2
25%	47.9	41.8	31.8	21.1
30%	63.6	55.1	42.9	27.7
40%	101.9	84.5	71.4	43.3
50%	154.6	119.0	109.9	61.9
60%	222.6	157.5	159.7	82.9
70%	299.9	196.8	219.6	104.9
75%	340.1	215.4	252.0	115.7
80%	380.4	232.4	285.3	126.0
90%	461.3	262.5	354.6	145.5
100%	530.0	285.2	420.0	162.0

**With smaller number of plates, the capacitance values given above should be determined in proportion to the number of dielectrics.
*180° from 100% rotation position (i.e. mechanically maximum position).

Trimmers (compensators) : Each adds a maximum of 2.0 $\mu\mu$F to the minimum capacitance of the section.   The minimum capacitance change of each compensator is 15 $\mu\mu$F.

Rotor torque : Class A : From 2 to 6 inch-ounces.   Class B : From 2 to 5 inch-ounces.

Slotted plates : The outside rotor plates shall be slotted radially.

(B) R.I.C. air dielectric, tuning
RIC/141 (Ref. F11).   This specification was incomplete at the time of going to press.

(C) R.I.C. Capacitors, variable, preset, air dielectric
RIC/142 (Ref. F12).   This specification was incomplete at the time of going to press.

(D) R.I.C. Capacitors, variable, preset, mica dielectric
RIC/143 (Ref. F13).   Single plate : variable between 2 to 15, 3 to 30, or 4 to 40 $\mu\mu$F. Multiple plate types with max. capacitance up to 3000 $\mu\mu$F are included.   The specification was incomplete at the time of going to press.

(viii) Standard variable composition resistors (" potentiometers ")

(A) American RTMA Standard variable control resistors REC-121-A, July 1952.
Resistance values 5 000, 10 000, 25 000, 100 000, 250 0¢0, 500 000 ohms, 1.0, 2.0 megohms (total)

Tapers

Linear :
Midpoint resistance = half total resistance.   Minimum resistance between either terminal and shaft = 0.05% of total resistance for total resistances 100 000 ohms and greater (higher percentages for lower total resistances).

" S " taper :
25% resistance with 25% ± 3% rotation
50% resistance with 50% ± 3% rotation
75% resistance with 75% ± 3% rotation
Minimum resistance as for linear type.

10% clockwise modified logarithmic taper
10% resistance with 50% ± 3% rotation.   Min. resistances 0.02% and 1% of total resistance for 100 000 ohms and greater.

20% clockwise modified logarithmic taper
20% resistance with 50% ± 3% rotation.
Minimum resistances as for 10% logarithmic taper.

10% counterclockwise modified logarithmic taper
20% counterclockwise modified logarithmic taper
(both as for clockwise type with ends reversed).

(B) American JAN-R-94 should also be examined.   See (xi) below for list.

(C) British Radio Industry Council Specification No. RIC/122 for rotary variable resistors (Ref. F1).
Resistance in multiples of 1, 2, 5.   Tolerances ± 20%.
Dissipation range : 0.1 to 1.5 watts.
Resistance range : 1000 ohms to 2 megohms.
Logarithmic taper : (1) resistance between 5% and 15% of actual overall resistance at 45% to 55% effective rotation (2) resistance between 2% and 5% of actual overall resistance at 20% to 30% effective rotation
Linear law : resistance between 35% and 65% of actual overall resistance at 45% to 55% effective rotation.
Hop-on resistance : less than 50 ohms or 0.05% of overall resistance, whichever is the greater.
Hop-off resistance : less than 1% of overall resistance.
Noise : not greater than 50 mV when tested under specified conditions with 20 volts d.c. applied.

(ix) Standard metallized paper dielectric capacitors

RIC/136 (Ref. F9).

Capacitance values conform to series 1, 2, 5, 10 etc. Range of values from 0.0001 to 2 μF. Tolerance ±25%. Voltage ratings 150, 250, 350, 500 volts (d.c), 300 volts r.m.s. (20 to 120 c/s). The power factor shall not exceed 0.015 at 1000 c/s. The insulation resistance shall not be less than (a) for 2-foil types : 200 ohm-farads or 1200 megohms, whichever is the less, (b) for single-foil—castellated—types (below 0.05 μF) : 10 000 megohms, (c) to case : 10 000 megohms, Colour code—see Sect. 2(xii).

(x) Standard electrolytic capacitors

(A) JAN-C-62 specification : **Capacitors, dry-electrolytic, polarized**
Working temperature

Designation	A	B	C	D	E	F
Temp. (°C)	0 to 85	−20 to +85	−40 to +85	0 to 65	−20 to 65	−40 to 65

Voltage limits

TABLE 30

Designation	Working voltage (d.c.)	Surge voltage (d.c.)
E	15	20
F	25	40
G	50	75
H	100	150
J	150	200
K	200	250
M	250	300
N	300	350
P	350	400
Q	400	450
R*	450	500

*Available only with working temperature designations D, E and F.

Direct current leakage (under specified test conditions at rated working voltage) shall not exceed either 10 mA or the value calculated below, whichever is the smaller.

$$I = KC + 0.3$$

where I = d.c. leakage current in milliamperes
C = rated capacitance in microfarads
and K = 0.01 for rated working voltage 15 to 100,
0.02 for rated working voltage 101 to 250,
0.025 for rated working voltage 251 to 350,
0.04 for rated working voltage 351 to 450.

Tolerance in capacitance −10% ; +250%.

Equivalent series resistance (under specified test conditions at maximum rated voltage) shall not exceed the value of P tabulated below, divided by the rated capacitance in microfarads.

Rated working voltage	P	Rated working voltage	P
15	600	250	230
25	500	300	210
50	400	350	200
100	330	400	200
150	300		
200	250	450	200

Test conditions : 20°C to 40°C. Frequency 120 c/s.

Maximum r.m.s. ripple current (extracts from table)

Rated cap.	Working voltage					
	100	150	200	250	350	450 V
10 μF	—	—	130	130	140	150 mA
20 μF	130	170	170	170	180	180 mA
30 μF	160	180	180	200	200	200 mA
40 μF	170	190	190	190	200	200 mA
50 μF	190	200	200	200	200	200 mA

(B) British R.I.C. Specification RIC/134 (Ref. F8)

Range of capacitance 1 to 1000 μF.

Voltage ratings : At 60°C—12, 25, 50, 150, 275, 350, 450, 500.

At 70°C—12, 25, 50, 120, 220, 280, 360, 400.

Tolerances : +50%, −20% for working voltages above 100 V (except etched foil types up to and including 16 μF) ; +100 −20% for working voltages of 100 V and less, and all etched foil types up to and including 16 μF. Ripple rating at 70°C is 0.75 of the ripple rating at 60°C. Power factor at 50 c/s shall not exceed 0.2 for working voltages up to and including 100 V ; 0.15 for working voltages above 100 V.

The leakage current of single winding capacitors shall not exceed $0.5 \ V\sqrt{C} \mu A$ or 100 μA, whichever is the greater, where V is the rated voltage and C the actual capacitance in μF.

(xi) References to standard resistors and capacitors
(A) British Standard for Service Equipment

These standards are not complete, and have been largely superseded by the R.C.S C. Components Book (see below)

A1. BS/RC.G/1. General guide on radio components (Issue 1) Aug. 1944.

A2. BS/RC.G/110 Guide on fixed resistors (Issue 1) Aug. 1944. Superseded by RCG110.

A3. BS/RC.S/110 Group test-specification for fixed resistors (Issue 2) July 1946 ; Amended Aug. 1947, May 1948. Superseded by RCS112.

A4. BS/RC.S/110.1. Test schedule for fixed resistors (Issue 2) July 1946 ; amended Oct. 1946, Feb. 1947, Aug. 1947. Superseded by RCS112.

A5. BS/RC.G/130. Guide on fixed capacitors (Issue 1) Aug. 1945. Superseded by RCG130.

A7. BS/RC.S/130.3. Test schedule for ceramic dielectric fixed capacitors (Issue 1) March 1944 ; amended Jan. 1946, Mar. 1946, Aug. 1947, Dec. 1947.

A8. BS/RC.S/130.4. Test schedule for electrolytic capacitors (Issue 1) March 1944. Superseded by RCS134.

A9. BS/RC.S/130.6m. Test schedule for miniature paper-dielectric capacitors (metallised paper type) (Issue 2) June 1945, amended Jan. 1946, March 1946, Aug. 1947.

A10. BS/RC.S/130.7m. Test schedule for miniature (High K) type ceramic dielectric fixed capacitors (Issue 1) July 1944. Amended Aug. 1947.

A11. BS/RC.S/141. Group test specification for air dielectric rotary variable capacitors (Issue 1) Nov. 1945 ; amended Aug. 1948. Superseded by RCS141.

A12. BS/RC.S/141.1. Test schedule for air dielectric rotary variable capacitors (Issue 1) Nov. 1945 ; amended Aug. 1948. Superseded by RCS141.

A13. BS/RC.S/141.1m. Test and performance specification for miniature variable capacitors (air-spaced ganged type) (Issue 1) July 1944. Superseded by RCS 141.

(B) British R.C.S.C. Components Book

This book is available to manufacturers of British Service Equipment, and is available for reference at Standards Libraries.

38.3 (xi) REFERENCES TO STAND. RESISTORS AND CAPACITORS 1359

New Issues and amendments are made from time to time.
RCS1 General specification for Electronic Components.
RCG4 Guide to approved components.
RCS11 Specification for the climatic and durability testing of service electronic components.
RCG100.9 Guide to the tropic proofing of electrical equipment.

Resistors

RCG110 Guide on fixed resistors.
RCL110.11 Working schedule, fixed resistors.
RCS111 Specifications for wire-wound resistors.
RCS112 Specifications, fixed composition resistors.
RCS121 Specifications for rotary wire-wound resistors.
RCL121 List of standard rotary wire-wound resistors.
RCS122 Specifications for rotary composition resistors.
RCL122 List of standard rotary composition resistors.

Capacitors—general

RCG130 Guide on capacitors, fixed.

Capacitors, fixed, paper dielectric

RCL130.11 Working schedule—Rectangular metal case.
RCL130.12 Working schedule—Tubular type stud mounting and ins' lated.
RCL130.13 Working schedule—Tubular metal case, non-insulated.
RCS131 Specification.

Capacitors, fixed, mica dielectric

RCL130.21 Working schedule—metallized case.
RCL130.22 Working schedule—metallized, wax protected.
RCL130.23 Working schedule—foil, moulded case.
RCL130.24M Working schedule—miniature foil and metallized types.
RCS132 Specification (excluding wax-protected types).
RCS132.1 Specification for wax-protected types.

Capacitors, fixed, ceramic dielectric

RCL130.31 Working schedule—Cup and disc types, and temperature compensating types.
RCL130.71M Working schedule—high K.
RCS133 Specification.

Capacitors, fixed electrolytic

RCS134 Specification.
RCL134 Standard list—tubular, metal case.

Capacitors, fixed, paper dielectric (metallized)

RCS136 Specification.
RCS136.1 Specification, humidity class 3, insulated only.
RCL136 Standard list—tubular, insulated and non-insulated.

Capacitors, variable

BS/RCS141 Group test specification, air dielectric, rotary
BS/RCS141.1 Test schedule, air dielectric, rotary.
BS/RCS141.1m Test specification, air-spaced, ganged, miniature.
RCL141.11m Working schedule, air-spaced, ganged, miniature.
RCL141.12m Working schedule, air dielectric trimmer, miniature.
RCL141.14 Working schedule, air dielectric trimmer, with locking device.
RCL141.15 Working schedule, air dielectric trimmer, concentric.
RCS141.2m Test specification, air-spaced trimmer, vane type, miniature.

(D) American R.M.A. Standards for resistors and capacitors

D1. REC-116 Standard fixed composition resistors (July 1948).
D2. REC-117 Standard fixed wire wound resistors (July 1948).
D3. REC-118 Standard fixed paper dielectric capacitors in tubular non-metallic cases (Sept. 1948).
D4. TR-113-A Metal encased fixed paper dielectric capacitors for d.c. application (May 1951).
D5. REC-115-A Standard molded mica capacitors (May 1951).

D6.   REC-107-A Standard ceramic dielectric capacitors (Aug. 1952).
D7.   REC-106-A Standard variable capacitors Class A (Jan. 1949).
D8.   REC-101 Standard variable capacitors, Class B (Oct. 1946).
D9.   REC-121-A Standard variable composition resistors (July 1952).
D10. S-417 ; M4-571 through 574-Wet electrolytics.
D11., S-418 ; M4-591 through 598-Dry electrolytics.

(E) American Joint Army/Navy Standards for resistors and capacitors.
E1.   JAN-R-11 (31 May 1944) : Resistors, fixed composition. Amendment No. 3
      (22 March 1949).
E2.   JAN-R-19 (31 July 1944) : Resistors, variable wire-wound (low operating
      temperature). Amendment No. 2 (12 Jan. 1949).
E3.   JAN-R-22 (31 July 1944) : Rheostats, wire wound, power-type. Amendment
      No. 4 (26 June 1950).
E4.   JAN-R-26A (17 Sept. 1948) : Resistors, fixed, wire wound, power type. Amend-
      ment No. 1 (28 July 1949).
E4A. JAN-R-93 (16 June 1945) : Resistors, accurate, fixed, wire-wound. Amend-
      ment No. 3 (Jan. 1949).
E5.   JAN-R-94 (4 Oct. 1948) : Resistors, variable, composition. Amendment
      No. 2 (17 Jan. 1949).
E6.   JAN-R-184 (31 July 1945) : Resistors, fixed, wire wound, low power. Amend-
      ment No. 2 (17 Jan. 1949).
E7.   JAN-C-5 (20 April 1944) : Capacitors, mica-dielectric, fixed. Amendment
      No. 2 (6 Jan. 1949).
E8.   JAN-C-20A (4 Dec. 1947) : Capacitors, ceramic-dielectric, fixed (temperature
      compensating). Amendment No. 2 (15th June 1950).
E9.   JAN-C-25 (24 July 1947) : Capacitors, direct-current, paper-dielectric, fixed
      (hermetically sealed in metallic cases). Amendment No. 4 (June 1950).
E10. JAN-C-62 (30 November 1944) : Capacitors, dry-electrolytic, polarized.
      Amendment No. 3 (Jan. 1949).
E11. JAN-C-81 (27 August 1945) : Capacitors, ceramic-dielectric, variable. Amend-
      ment No. 1 (Oct. 1948).
E12. JAN-C-91 (21 Aug. 1947) : Capacitors, paper dielectric, fixed (non-metallic
      cases). Amendment No. 2 (25 Aug. 1950).
E13. JAN-C-92 (30 Dec. 1944) : Capacitors, air-dielectric, variable (trimmer
      capacitors). Amendment No. 4 (25 Aug. 1950).

(F) British Radio Industry Council*
F1.   RIC/111. ⎫
F2.   RIC/112. ⎬Superseded by British Standard BS.1852 : 1952.
F3.   RIC/113. ⎭
F4.   RIC/122. Resistors, rotary, variable, composition track (with or without
      switches).
F5.   RIC/131. Capacitors, fixed, paper dielectric, tubular foil. Issue No. 1,
      July 1950.
F6.   RIC/132. Capacitors, fixed, mica dielectric, stacked foil. Issue No. 1,
      July 1950.
F7.   RIC/133. Capacitors, fixed, ceramic dielectric, Grade 1. Issue No. 1, April
      1951.
F8.   RIC/134. Capacitors, fixed, electrolytic. Issue No. 1, April 1951.
F9.   RIC/136. Capacitors, fixed, paper dielectric, tubular, metallized. Issue
      No. 1, February 1951.
F10. RIC/137. Capacitors, fixed, mica dielectric, metallized. Issue No. 1, April
      1951.
F11. RIC/141. Capacitors, variable, air dielectric, tuning. Issue No. 1, Feb. 1951.
F12. RIC/142. Capacitors, variable, preset, air dielectric. Issue No. 1, Sept. 1951.
F13. RIC/143. Capacitors, variable, preset, mica dielectric. Issue No. 1, Sept. 1951.

---

*It is hoped that these Radio Industry Council Specifications will, in due course, be incorporated into British Standard Specifications.

SECTION 4 : STANDARD FREQUENCIES

(*i*) *Standard frequency ranges*   (*ii*) *Frequency bands for broadcasting*   (*iii*) *Standard intermediate frequencies.*

(i) Standard frequency ranges

The Final Acts of the International Telecommunication and Radio Conferences at Atlantic City, 1947, proposed the following nomenclature of frequencies.

Frequencies shall be expressed in kilocycles per second (Kc/s) at and below 30 000 Kc/s and in megacycles per second (Mc/s) above this frequency.

TABLE 32

Frequency sub-division	Frequency range	Metric sub-division
v-l-f (very low frequency)	below 30 Kc/s	myriametric waves
l-f (low frequency)	30 to 300 Kc/s	kilometric waves
m-f (medium frequency)	300 to 3 000 Kc/s	hectometric waves
h-f (high frequency)	3 000 to 30 000 Kc/s	decametric waves
v-h-f (very high frequency)	30 000 Kc/s to 300 Mc/s	metric waves
u-h-f (ultra high frequency)	300 to 3 000 Mc/s	decimetric waves
s-h-f (super high frequency)	3 000 to 30 000 Mc/s	centimetric waves
e-h-f (extremely high frequency)	30 000 to 300 000 Mc/s	millimetric waves

(ii) Frequency bands for broadcasting

The International Telecommunication and Radio Conference at Atlantic City, 1947, allocated the following frequency bands for broadcasting purposes.

TABLE 33

Band shared with maritime mobile	150—160 Kc/s
Low frequency (not world-wide)	160—285 Kc/s
Medium frequency (Region 1 only)	525—535 Kc/s
Medium frequency (world wide)	535—1605 Kc/s
Short wave frequencies (world wide)	5.95—6.2 Mc/s
	9.5—9.775 Mc/s
	11.7—11.975 Mc/s
	15.1—15.45 Mc/s
	17.7—17.9 Mc/s
	21.45—21.75 Mc/s
	25.6—26.1 Mc/s
Very high frequency (world wide)	88—100 Mc/s
(U.S.A.)	88—108 Mc/s

In Australia, the band 100-108 Mc/s has been allocated for the aeronautical mobile service until required for broadcasting service.

(iii) Standard Intermediate Frequencies

It is recommended that superheterodyne receivers operating in the medium frequency broadcast band use an intermediate frequency of 455 Kc/s. This frequency is reserved as a clear channel for the purpose in most countries of the world.

The European " Copenhagen Frequency Allocations " provide the following two intermediate frequency bands : 415—490 Kc/s and 510—525 Kc/s.

An intermediate frequency of 175 Kc/s is also used.

The American RTMA has standardized the following intermediate frequencies (REC-109-B, March 1950) : Standard broadcast receivers—either 260 or 455 Kc/s. V-H-F broadcast receivers—10.7 Mc/s.

SECTION 5 : WAVELENGTHS AND FREQUENCIES

(i) *Wavelength-frequency conversion tables* (ii) *Wavelengths of electromagnetic radiations.*

(i) Wavelength-frequency conversion table
Convenient points selected for rapid reference.

TABLE 34 : MEDIUM-FREQUENCY BROADCAST BAND

Frequency Kc/s	Wavelength m	Frequency Kc/s	Wavelength m	Frequency Kc/s	Wavelength m	Frequency Kc/s	Wavelength m
540	555.5	810	370.4	1080	277.8	1350	222.2
550	545.5	820	365.9	1090	275.2	1360	220.6
560	535.7	830	361.4	1100	272.7	1370	219.0
570	526.3	840	357.1	1110	270.3	1380	217.4
580	517.2	850	352.9	1120	268.2	1390	215.8
590	508.5	860	348.8	1130	265.5	1400	214.3
600	500.0	870	344.8	1140	263.2	1410	212.8
610	491.8	880	340.9	1150	260.9	1420	211.3
620	483.9	890	337.1	1160	258.6	1430	209.8
630	476.2	900	333.3	1170	256.4	1440	208.3
640	468.8	910	329.7	1180	254.2	1450	206.9
650	461.5	920	326.1	1190	252.1	1460	205.5
660	454.5	930	322.6	1200	250.0	1470	204.1
670	447.8	940	319.1	1210	247.9	1480	202.7
680	441.2	950	315.8	1220	245.9	1490	201.3
690	434.8	960	312.5	1230	243.9	1500	200.0
700	428.6	970	309.3	1240	241.9	1510	198.7
710	422.5	980	306.1	1250	240.0	1520	197.4
720	416.7	990	303.0	1260	238.1	1530	196.1
730	411.0	1000	300.0	1270	236.2	1540	194.8
740	405.4	1010	297.0	1280	234.4	1550	193.5
750	400.0	1020	294.1	1290	232.6	1560	192.3
760	394.7	1030	291.2	1300	230.8	1570	191.1
770	389.6	1040	288.5	1310	229.0	1580	189.9
780	384.6	1050	285.7	1320	227.3	1590	188.7
790	379.7	1060	283.0	1330	225.6	1600	187.5
800	375.0	1070	280.4	1340	223.9		

TABLE 35 : SHORT WAVE BAND

Frequency Mc/s	Wavelength m	Frequency Mc/s	Wavelength m	Frequency Mc/s	Wavelength m	Frequency Mc/s	Wavelength m
1.5	200	11	27.3	21	14.3	65	4.62
2.0	150	12	25.0	22	13.6	70	4.29
3.0	100	13	23.1	23	13.0	75	4.00
4.0	75.0	14	21.4	25	12.0	80	3.75
5.0	60.0	15	20.0	30	10.0	85	3.53
6.0	50.0	16	18.8	35	8.57	88	3.41
7.0	42.9	17	17.6	40	7.50	90	3.33
8.0	37.5	18	16.7	45	6.67	95	3.16
9.0	33.3	19	15.8	50	6.00	100	3.00
10.0	30.0	20	15.0	55	5.45	105	2.86
				60	5.00	108	2.78

(ii) Wavelengths of electromagnetic radiations

TABLE 36

Rays	Wavelengths		
	Angstrom units	Microns	Metres
Cosmic rays	10^{-6}—10^{-2}	10^{-10}—10^{-6}	
Gamma rays	10^{-2}—1	10^{-6}—10^{-4}	
X-Rays	10^{-1}—10^{2}	10^{-5}—10^{-2}	
Ultraviolet rays	100—3900	10^{-2}—0.39	
Visible light rays	3900—7600	0.39—0.76	
Infrared rays	7600—10^{6}	0.76—10^{2}	
Electric or radio rays		10^{2}—10^{10}	10^{-4}—10^{4}
Radio broadcasting rays		10^{7}—10^{9}	10—10^{3}

SECTION 6: STANDARD SYMBOLS AND ABBREVIATIONS

(i) Introduction (ii) Multipliers (iii) Some units and multipliers (iv) Magnitude letter symbols (v) Subscripts for magnitude letter symbols (vi) Magnitude letter symbols with subscripts (vii) Mathematical signs (viii) Abbreviations (ix) Abbreviations of titles of periodicals (x) References to periodicals (xi) References to standard symbols and abbreviations.

(i) Introduction

Owing to the lack of international standardization, the editor has been forced to select suitable symbols for use in this Handbook. The choice which has been made is believed to be a reasonable compromise, and capable of being understood readily throughout the English-speaking world.

(ii) Multipliers

In general, small letters are used for quantities below unity, and capital letters for quantities above unity.

d = deci	= 1/10	= 10^{-1}
c = centi	= 1/100	= 10^{-2}
m = milli	= 1/1000	= 10^{-3}
μ = micro	= 1/1000 000	= 10^{-6}
$\mu\mu$ = micromicro	= 1/1000 000 000 000	= 10^{-12}
K or k = kilo	= 1000	= 10^{3}
M = meg	= 1000 000	= 10^{6}

(iii) Some units and multipliers

A = ampere   mA = milliampere μA = microampere
F = farad μF = microfarad $\mu\mu$F* = micromicrofarad
H = henry   mH = millihenry μH = microhenry
Ω = ohm   MΩ = megohm
V = volt   mV = millivolt   KV or kV = kilovolt
W = watt   mW = milliwatt
m = metre   cm = centimetre   mm = millimetre

*Alternatively, pF = picofarad = $\mu\mu$F = micromicrofarad.

(iv) Magnitude letter symbols†

B = susceptance
C = capacitance
C_r, g = conductance
D = total harmonic distortion
E, e = electromotive force
F_p = power factor
f = frequency
g, C_r = conductance
H_1 = fundamental frequency component of distortion
$H_2, H_3,$ etc. = second (third etc.) harmonic components of distortion
I, i = current
K = dielectric constant
L = inductance
M = mutual inductance
P = power
Q = charge, quantity of electricity
also Q = figure of merit of a reactor
R = resistance
X = reactance
X_L = inductive reactance
X_c = capacitive reactance
Y = admittance
Z = impedance (scalor)
$\mathbf{Z}$ = impedance (vector)

Δ = increment of
ϵ = 2.7182818
η = efficiency
λ = wavelength
π = 3.14159
$\omega = 2\pi f$ = angular velocity.

Magnetic units
B = magnetic flux density
H = magnetic field strength, magnetizing force
ϕ = magnetic flux
μ = magnetic permeability

Operators
d = differential ; δ = partial differential
j = 90° rotational = $\sqrt{-1}$.

(v) Subscripts for magnitude letter symbols

in = input
out = output
max = maximum (reduced to m when combined with another subscript)
min = minimum
av = average
b = plate—steady or total value
bb = plate supply
c = grid—steady or total value
c_1 = grid no. 1 ; similarly c_2, c_3 etc.
cc = grid supply
co = at point of plate current cutoff
f = filament
g = grid—varying component

†A magnitude letter symbol is used to designate the magnitude of a physical quantity in mathematical equations and expressions. Two or more magnitude symbols printed together represent a product.

g_1 = grid no. 1 ; similarly g_2, g_3 etc.
h = heater
k = cathode
o = quiescent (no signal)'
p = plate (anode)—varying component
s = screen or metal shell or other self-shielding envelope
t = triode
N.B. Grids are numbered in order, beginning at the cathode and working outwards towards the plate.

(vi) Magnitude letter symbols with subscripts
Symbols for filament or heater circuits
E_f = filament or heater (terminal) voltage
E_{hk} = voltage of heater with regard to cathode
I_f = filament or heater current.

Symbols for plate circuits
All voltages are taken as being with respect to the cathode unless otherwise indicated.
E_{bb} = plate supply voltage
E_b = average or quiescent value of plate voltage
E_{bo} = quiescent (no signal) value of plate voltage
E_{bm} = maximum value of plate voltage
E_p = r.m.s. value of varying component of plate voltage
E_{pm} = maximum value of varying component of plate voltage
e_b = instantaneous total plate voltage
e_p = instantaneous value of varying component of plate voltage
I_b = average or quiescent value of plate current
I_{bm} = peak total plate current
I_{bo} = quiescent (no signal) value of plate current
I_p = r.m.s. value of varying component of plate current
I_{pm} = maximum value of varying component of plate current
I_s = total electron emission
i_b = instantaneous total plate current
i_p = instantaneous value of varying component of plate current.

Symbols for grid circuits
E_c = average or quiescent value of grid voltage
E_{cm} = maximum value of grid voltage
E_g = r.m.s. value of varying component of grid voltage
E_{gm} = maximum value of varying component of grid voltage
e_c = instantaneous total grid voltage
e_g = instantaneous value of varying component of grid voltage
I_c = average or quiescent value of grid current
I_g = r.m.s. value of varying component of grid current
i_c = instantaneous total grid current
i_g = instantaneous value of varying component of grid current.

Symbols for valve characteristics
(A) Inside the valve
E_{pg} = voltage of plate with regard to grid
μ = amplification factor
μ_t = amplification factor (triode connected)
μ_{g1g2} = mu factor from grid no. 1 to grid no. 2, and similarly for other electrodes
g_m = mutual conductance (= g_{gp}) or " slope "
g_c = conversion conductance
g_d = dynamic conductance (slope of dynamic characteristic)
g_p = plate conductance
g_g = grid conductance
g_n = plate-grid transconductance (inverse mutual conductance) = g_{pg}
g_{jk} = transconductance from electrode j to electrode k

r_p = plate resistance
r_g = grid resistance
r_{g2} = screen (grid no. 2) resistance
C_{gp} = grid-plate capacitance
C_{gk} = grid-plate cathode capacitance
C_{pk} = plate-cathode capacitance
C_{ght} = grid-heater capacitance
C_{ph} = plate-heater capacitance
C_{in} = input capacitance
C_{out} = output capacitance.

(B) Valve and circuit

R_L = plate load resistor
Z_L = plate load impedance
R_h = cathode resistor
C_k = cathode by-pass capacitor
C_c = coupling condenser
R_{g1} = grid resistor
R_{g2} = following grid resistor
r_i = input resistance
r_o = output resistance
R_s = screen series resistor
C_s = screen by-pass capacitor
P_i = power input
P_o = power output
P_p = plate (anode) dissipation
P_{g2} = screen (grid no. 2) dissipation
A = amplification (voltage gain of stage)
A_o = amplification (voltage gain of stage) at mid-frequency
I_{hm1} = peak fundamental current
I_{hm2} = peak second harmonic current.

(vii) Mathematical signs

$+$	plus		$\therefore$	therefore
$-$	minus		$\propto$	varies as
$\pm$	plus or minus		$\sqrt{\ }$	square root
$\times$ or $.$	multiplied by		$\sqrt[n]{\ }$	nth root
$\div$ or $/$	divided by		$!$	factorial
$=$	equal to		F	function
$\neq$	not equal to		$\int$	integration
$\approx$	approximately equal to			
$\geqslant$	equal to or greater than		∞	infinity
$\leqslant$	equal to or less than		$°$	degree
$\equiv$	identical with		$'$	minutes of degree
$>$	greater than		$''$	seconds of degree
$<$	less than		δ or Δ	increment
			Σ	summation

$\log x$ logarithm of x to base 10
$\log_\epsilon x$ logarithm of x to base ϵ
j   square root of minus one (90° angular rotation)
ϵ base of natural logarithms (2.71828)
π ratio of circumference to diameter of circle (3.141593 approx.)
For trigonometrical symbols and differentiation refer to Chapter 6.

(viii) Abbreviations

a.c. = alternating current
d.c. = direct current
a-f = audio frequency

i-f	=	intermediate frequency
r-f	=	radio frequency
v-h-f	=	very-high-frequency
u-h-f	=	ultra-high-frequency
		(For other frequency designations see Sect. 4.)
AWG	=	American Wire Gauge
B & S	=	same as AWG
SWG	=	Standard Wire Gauge
E	=	enamelled
r.m.s	=	root mean square
db	=	decibel
dbm	=	decibels of power referred to 1 milliwatt
c/s	=	cycles per second
Kc/s	=	kilocycles per second (kc/s may also be used)
Mc/s	=	megacycles per second
Hz	=	Hertz = cycles per second
a.v.c.	=	automatic volume control
a.a.v.c.	=	audio automatic volume control
A-M	=	amplitude modulation
F-M	=	frequency modulation
SSC	=	single silk covered
DSC	=	double silk covered
SCC	=	single cotton covered
DCC	=	double cotton covered.

Note : Abbreviations are used to indicate either singular or plural, either as a noun or as an adjective.

(ix) Abbreviations of titles of periodicals

Standard Abbreviations	Title and Publisher
A.R.T.S. & P.	Australian Radio Technical Services and Patents Bulletin (47 York St., Sydney, Australia).
Audio Eng. (formerly Radio)	Audio Engineering (Radio Magazines Inc. 10 McGovern Ave., Lancaster, Pa.).
Australian Radio & Electronics (formerly Australasian Radio World)	Radio & Electronics (Aust.) Pty. Ltd., 17 Bond St., Sydney.
A.W.A. Tec. Rev.	A.W.A. Technical Review (Amalgamated Wireless Australasia Ltd., 47 York St., Sydney, Australia).
B.B.C. Quarterly	British Broadcasting Corporation, 35 Marylebone High St., London, W.1.
B.S.T.J.	Bell System Technical Journal (American Telephone & Telegraph Coy., 195 Broadway, New York 7, N.Y. U.S.A.).
Bell. Lab. Rec.	Bell Laboratories Record (Bell Telephone Laboratories Inc. 463 West Street, New York 14, N.Y.).
British P.O.E.E.J	Post Office Electrical Engineers' Journal, Institution of P.O.E.E. Engr-in-chief's office Alder House, Aldersgate St., London E.C.1.
Brown-Boveri Review	Brown Boveri & Co. Ltd., Baden, Switzerland.
Comm.	Communications (Bryan Davis Publishing Coy., Inc., 52 Vanderbilt Ave., New York 17, N.Y. U.S.A.).
E.E.	Electrical Engineering (American Institute of Electrical Engineers, 33 West 39th St., New York 18 N.Y.).

E.W. & W.E.	Same as Wireless Engineer.
Elect.	Electronics (McGraw-Hill Publishing Co., Inc., West 42nd St., New York 18, N.Y., U.S.A.).
Elect. Comm.	Electrical Communication (International Telephone & Telegraph Corp. 67 Broad St., New York 4, N.Y. U.S.A.).
Electronic Eng.	Electronic Engineering, 28 Essex St., London W.C.2.
Electronic Industries	Caldwell-Clements Inc., 480 Lexington Ave., New York 17, N.Y.
Engineering	J. A. Dixon, 33-36 Bedford St., London W.C.2.
F.M. & T. ⎫ FM-TV ⎭	FM-TV Radio Communication, formerly FM Magazine, FM and Television, FM Radio Electronics (FM Company, 264 Main St., Great Barrington, Mass. U.S.A.).
G.E. Review	General Electric Review (G.E. Co., Schenectady, New York, U.S.A.).
G.R. Exp.	General Radio Experimenter (General Radio Company, 275 Massachusetts Av., Cambridge, 39, Mass., U.S.A.).
J. Acous. Soc. Am.	Journal Acoustical Society of America. Published for A.S.A. by the American Institute of Physics, Prince and Lemon Streets, Lancaster, Pa. U.S.A.
Journal of Applied Physics	American Institute of Physics, 57 East 55th St., New York 22 N.Y.
Jour. I.E.E.	Journal of the Institution of Electrical Engineers (Savoy Place, Victoria Embankment, London, W.C.2., England).
J. Brit. I.R.E.	Journal of British Institution of Radio Engineers (9 Bedford Square, London, W.C.1 England).
Jour. Sci. Instr.	Journal of Scientific Instruments (Institute of Physics, 47 Belgrave Sq., London S.W.1).
Jour. S.M.P.E. ⎫ Jour. S.M.P.T.E. ⎭	Journal of the Society of Motion Picture and Television Engineers 342 Madison Ave., New York 17, U.S.A.
Marconi Review	Marconi's Wireless Telegraph Co. Ltd., Marconi House, Chelmsford.
Phil. Mag.	Philosophical Magazine (Taylor and Francis Ltd., Red Lion Court, Fleet St., London, E.C.4).
Philips Tec. Com.	Philips Technical Communications (Philips Electrical Industries of Australia Pty. Ltd., 69-73 Clarence St., Sydney).
Philips Tec. Rev.	Philips Technical Review (N.V. Philips' Gloeilampen-fabrieken, Eindhoven, Holland).
Phys. Rev.	Physical Review (American Institute of Physics, 57 East 55th Street, New York 22, N.Y.).
Proc. I.R.E.	Proceedings of the Institute of Radio Engineers (I.R.E. Inc. 1 East 79 Street, New York 21, N.Y., U.S.A.).
Proc. I.R.E. Aust.	Proceedings of the Institute of Radio Engineers Australia (Science House, Gloucester St., Sydney, Australia).
Proc. Roy. Soc.	Proceedings of the Royal Society Burlington House London W.1.
Q.S.T.	American Radio Relay League Inc. (38 La Salle Rd., West Hartford 7, Conn. U.S.A.).
R.C.A. Rev.	R.C.A. Review (Radio Corporation of America, R.C.A. Laboratories Division, Princeton, N.J. U.S.A.).
R. & E. Retailer	Radio and Electrical Retailer (Australian Radio Publications Pty. Ltd., 30-32 Carrington St., Sydney).
Radio	Radio Magazines Inc. 10 McGovern Ave., Lancaster, Pa).

Radiotronics	Amalgamated Wireless Valve Co. Pty. Ltd. (47 York St., Sydney, Australia).
Radio Craft ⎱ Radio Electronics ⎰	Radiocraft Publications Inc. (25 West Broadway, New York, 7, N.Y. U.S.A.).
Radio Electrical Weekly	Mingay Publishing Co. 146 Foveaux St., Sydney.
Radio Eng.	Radio Engineering. Same as Communications.
Radio & Hobbies, Australia	Associated Newspapers Ltd., 60-70 Elizabeth-St., Sydney.
Radio Review of Australia	Australian Radio Publications Pty. Ltd., 30-32 Carrington St., Sydney.
Rev. of Sci. Instr.	Review of Scientific Instruments (American Inst. of Physics 57 East 55th St., New York 22 N.Y.).
Radio News ⎱ Radio & TV News ⎰	Ziff Davis Publishing Coy. (185 North Wabash Ave., Chicago 1, Ill. U.S.A.).
Service	Bryan Davis Publishing Coy. Inc., 52 Vanderbilt Ave., New York 17, N.Y. U.S.A.).
TV Eng. (formerly Communications ; formerly Radio Engineering ; formerly Communication and Broadcast Engineering)	Tele Vision Engineering (Bryan Davis Publishing Coy., Inc., 52 Vanderbilt Av., New York 17, N.Y. U.S.A.).
Telecommunications Journal of Australia	Postal Electrical Society of Victoria, G.P.O. Melbourne Australia.
Tele-Tech	Caldwell-Clements Inc. 480 Lexington Ave., New York 17 N.Y.
Trans. A.I.E.E.	Transactions of the American Institute of Electrical Engineers (A.I.E.E., 33 West 39th Street, New York 18 N.Y.).
W.E. ⎱ W.E. & E.W. ⎰	Wireless Engineer (Iliffe & Sons Ltd., Dorset House, Stamford St., London, S.E.1., England).
W.W.	Wireless World (Iliffe & Sons Ltd., Dorset House, Stamford St., London, S.E.1, England).
Western Electric Oscillator	Western Electric Company (Graybar Electric Co, 420 Lexington Ave. New York 17, N.Y.).

(x) References to periodicals

The references to periodicals in this Handbook follow substantially the form recommended in British Standard 1219 : 1945.

Example : Lamson, H.W. " Permeability of dust cores " W.E. 24.288 (Sept. 1947) 267.

Lamson, H. W. indicates the author.

" Permeability of dust cores " indicates the title.

W.E. indicates the periodical (see abbreviations above).

24 indicates the volume number.

288 indicates the number.

Sept. 1947 indicates the date.

267 indicates the first page of the article.

(xi) References to standard symbols and abbreviations

1. " Standards on abbreviations, graphical symbols, letter symbols and mathematical signs." The Institute of Radio Engineers (U.S.A.) 1948.
2. " American standard letter symbols for electrical quantities " A.S.A. Z10.5, 1949.
3. British Standard 1409 : 1950 " Letter symbols for electronic valves " British Standards Institution 28 Victoria St., London, S.W.1.
4. " Standards on designations for electrical, electronic and mechanical parts and their symbols, 1949 " Standard 49 IRE, 21 S1, published Proc. I.R.E. 38.2 (Feb. 1950) 118.
5. " Standards on abbreviations of radio-electronic terms, 1951 " Standard 51 IRE 21 S1, published Proc. I.R.E. 39.4 (April 1951) 397.

SECTION 7 : STANDARD GRAPHICAL SYMBOLS

GRAPHICAL SYMBOLS

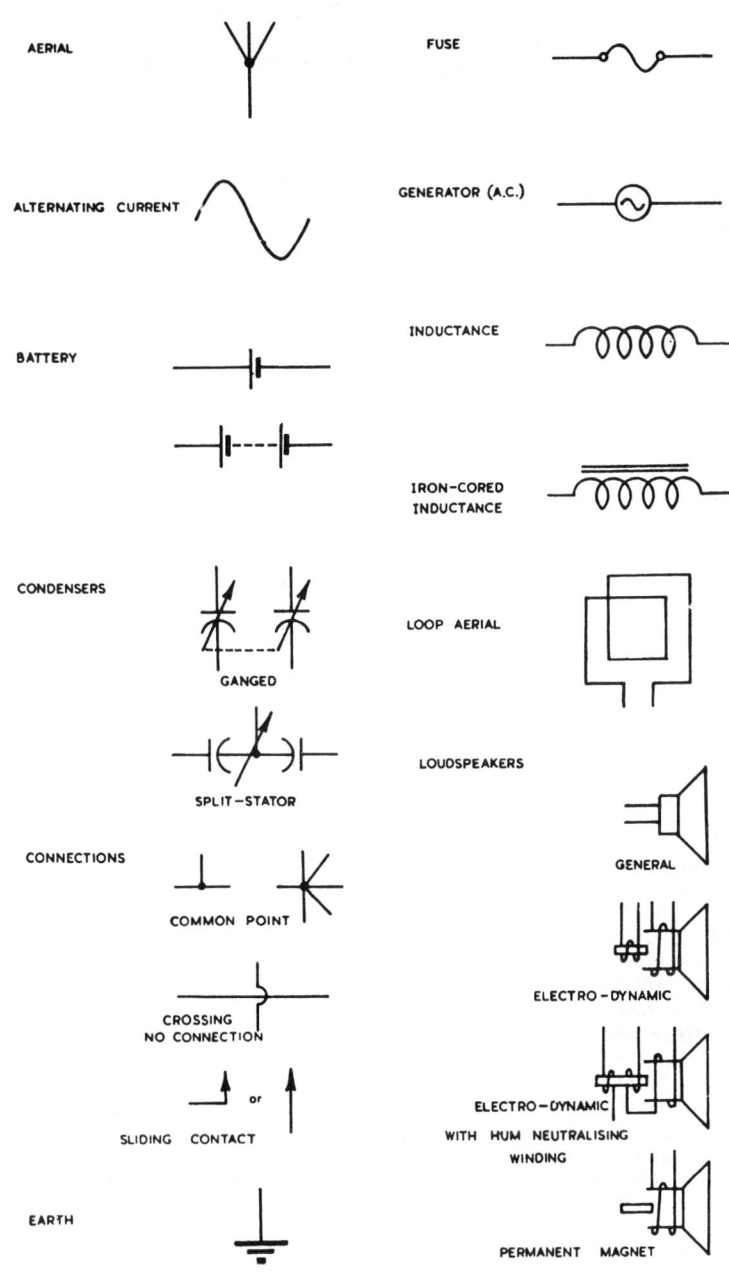

AERIAL

ALTERNATING CURRENT

BATTERY

CONDENSERS

GANGED

SPLIT—STATOR

CONNECTIONS

COMMON POINT

CROSSING
NO CONNECTION

SLIDING   CONTACT

EARTH

FUSE

GENERATOR (A.C.)

INDUCTANCE

IRON—CORED
INDUCTANCE

LOOP AERIAL

LOUDSPEAKERS

GENERAL

ELECTRO—DYNAMIC

ELECTRO—DYNAMIC
WITH HUM NEUTRALISING
WINDING

PERMANENT   MAGNET

GRAPHICAL SYMBOLS

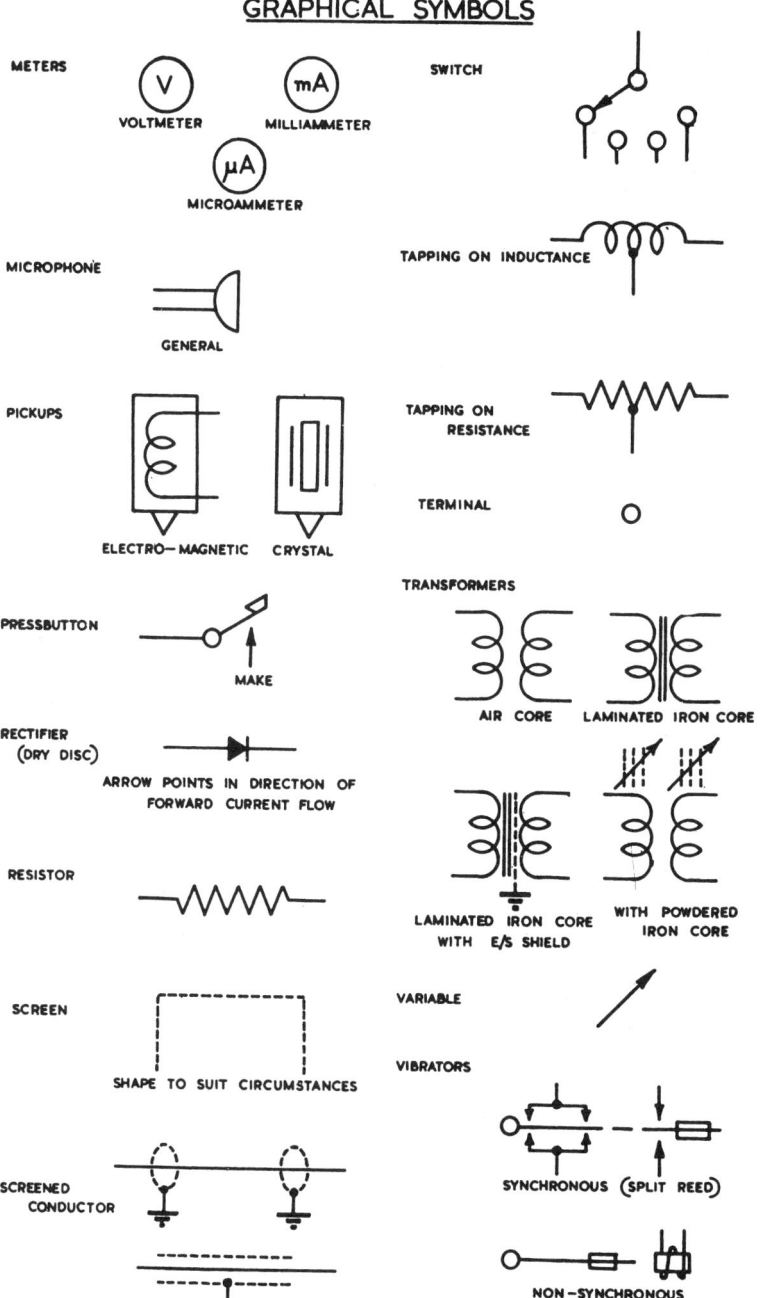

SECTION 8 : PROPERTIES OF MATERIALS AND CHEMICAL AND PHYSICAL CONSTANTS

(i) *Properties of insulating materials* (ii) *Properties of conducting materials* (iii) *Composition of some common plastics* (iv) *Weights of common materials* (v) *Resistance of a conductor at any temperature* (vi) *References to properties of materials* (vii) *Chemical and physical constants.*

(i) Properties of insulating materials

TABLE 37

Insulating Material	Dielectric Constant 60 c/s	Power Factor (per cent)			Dielectric Strength V/mil	Resistivity ohm-cm	Softens at °C	Coeff. of expansion parts in 10^6 per °C
		60 c/s	1 Mc/s	100 Mc/s				
Air-Normal Pressure	1				19.8-22.8	very high		
Amber	2.7-2.9	2.3	0.2-0.6		2300		250	44
Asphalts	2.7-3.1				25-30	Poor		
Casein-Moulded	6.4	6	5.2		400-700		177	80
Cellulose-Acetate	6-8	5-15	10		250-1000	4.5×10^{10}	70	160
Cellulose-Nitrate	4-7		7-10		300-780	$2-30 \times 10^{10}$	85	90-160
Ceresin Wax	2.5-2.6		0.05	0.05			57	
Fibre	2.5-5	6-9	5	5	150-180		130	
Glass-Crown	6.2		1		500	5×10^9		25
Glass-Electrical	4-5		0.5		2000	8×10^{14}	1100	8.9
Glass-Flint	7		0.4					7.9
Glass-Photographic	7.5		0.8-1					
Glass-Plate	6.8-7.6	0.2	0.6-0.8			10^{14}		
Glass-Pyrex	4.5	0.13-1.7	0.2-0.7			$10^{13}-10^{14}$	600	3.2
Halowax	3.4-3.8	3	0.14	0.54			88	
Magnesium silicate	5.9-6.4	0.2	0.037-0.38	10.5	335	$>10^{14}$	1350	6-10.5
Methacrylic Resin	2.8		2	0.03-0.37	200-240		135	70
Mica	2.5-8	0.03-0.05	0.2-6		600-1500	2×10^{17}		
Mica-Clear India	7-7.3	0.64	0.02-0.03	0.03	350	5×10^{13}	1200	3-7
Micalex 364	6-8	1.8	0.21	0.22	305		348	8-9
Nylon	3.6		2	1.8	1250	10^{13}	71	57
Paper	2-2.6				381			
Paraffin Oil	2.2	0.01	0.01	0.04				
Paraffin Wax	2.25	0.02	0.02	0.02	203-305	10^{16}	M.P. 56	710
Phenol-formaldehyde	5	2-8	1-2.8	3.8	400-475	1.5×10^{13}		28
Phenol-Yellow	5.3	2.5	0.7		500			

TABLE 37 (Continued)

Insulating Material	Dielectric Constant 60 c/s	Power Factor (per cent)			Dielectric Strength V/mil	Resistivity ohm-cm	Softens at °C	Coeff. of expansion parts in 10^6 per °C
		60 c/s	1 Mc/s	100 Mc/s				
Phenol-Black Moulded	5.5	8	3.5		400-500	10^{10}-10^{13}		40
Phenol-Paper Base	5.5	6	3.5		650-750			30
Phenol-Cloth Base	5.6	5	5		150-500			20
Polyethylene	2.25	0.03	0.03	0.03	1000	10^{17}	104	Varies
Polystyrene	2.5	0.02	0.02	0.03	508-760	10^{17}	80	70
Polyvinyl Chloride	2.9-3.2	1.2	1.6	0.8	400	10^{14}	54	82
Porcelain-Wet Process	6.5-7	2	0.6		150	5×10^8	1510	4-5
Porcelain-Dry Process	6.2-7.5	2	0.7		40-100	1.3×10^{16}		3-4
Pyrophillite	5.2		0.2-0.7	0.36	500	10^{14}-10^{18}		
Quartz-Fused	3.5-4.2	0.09	0.02	0.02	200	10^{12}-10^{14}	1430	0.45
Rubber-Hard	2-3.5	1	0.5-1		450	10^{16}	70	70-80
Shellac	2.5-4	0.6-2.5	0.9-3.1	3	900		85	6-8
Steatite-Commercial	4.9-6.5	0.02	0.2-0.3	0.5	150-315	10^{14}-10^{15}	1500	
Steatite-Low-loss	4.4	0.02	0.2	0.13	100-210	10^{13}-10^{14}		
Titanium Dioxide	90-170		.02-0.5		300-550	10^{12}-10^{13}	1600	7-8
Urea Formaldehyde	5-7	3-5	2.8	5	450-550		200	26
Varnished cloth	2-2.5		2-3		400-500	10^{14}		
Vinyl Resins	4		1.7					70
Wood-Dry Oak	2.5-6.8		4.2		116			
Wood-Paraffined Maple	4.1							

See also Refs. 1, 2, 3, 4, 5, 6.

(ii) Properties of conducting materials

TABLE 38

Material	Relative Resistance	Temperature Coefficient of Resistivity at 20°C ($\times 10^{-4}$)	Specific Gravity	Thermal Conductivity at 20°C	Coefficient of Linear Expansion ($\times 10^{-6}$)	Melting Point (°C)
Aluminium	1.64	40	2.7	0.48	25.5	660
Brass	3.9	20	8.47	0.26	18.9	920
Cadmium	4.4	38	8.64	0.222	28.8	321
Cobalt	5.6	33	8.71	—	12.3	1480
Constantan*	28.45	±2	8.9	0.054	17.0	1210
Copper	1.00	39.3	8.89	0.918	16.7	1083
Gas Carbon	2900	-5	1.88	0.0004	5.4	3500
Gold	1.416	34	19.32	0.705	13.9	1063
Iron-Cast	5.6	60	7.87	0.18	10.2	1535
Karma	77.1	±0.2	—	—	—	—
Lead	12.78	42	11.37	0.083	29.1	327
Magnesium	2.67	40·	1.74	0.376	25.4	651
Manganin	26	±0.2	8.5	0.053	18.0	910
Mercury	55.6	8.9	13.55	0.0148	—	-38.87
Molybdenum	3.3	45	10.2	0.346	5.0	2622
Monel Metal	27.8	20	8.8	0.06	14	1350
Nichrome	65	1.7	8.25	0.035	12.5	1350
Nickel	5.05	47	8.85	0.142	12.8	1452
Nickel Silver	16	2.6	8.72	0.07	18.36	1110
Phosphor Bronze	5.45	—	8.9	0.15	19.0	1050
Platinum	6.16	38	21.4	0.166	8.9	1773
Silver	0.95	40	10.5	1.006	19.5	960.5
Steel	7.6-12.7	16-42	7.8	0.115	10.5-11.6	1480
Steel—Stainless	52.8	—	7.9	0.069	10-11	1410
Tantalum	9.0	33	16.6	0.130	6.5	2850
Tin	6.7	42	7.3	0.155	21.4	231.9
Tungsten	3.25	45	19.2	0.476	4.44	3370
Zinc	3.4	37	7.14	0.265	26.3	419.5
Zirconium	2.38	44	6.4	—	—	1860

*Also known as Advance, Copel, Eureka, Ideal, etc.   See also Refs. 1, 5, 6.

(iii) Composition of some common plastics

TABLE 39

Trade Name	Composition
Bakelite	Phenol formaldehyde
Bakelite	Urea formaldehyde
Bakelite	Cellulose acetate
Bakelite	Polystyrene
Beetle	Urea formaldehyde
Cellophane	Regenerated cellulose film
Celluloid	Cellulose nitrate
Distrene	Polystyrene
Duperite	Phenol formaldehyde
Erinoid	Casein
Formica	Phenol formaldehyde (lamination)
Glyptal	Glycerol-phthalic anhydride
Lucite	Methyl methacrylate polymer
Micarta	Phenol formaldehyde (lamination)
Mycalex	Mica bonded glass
Neoprene	Chloroprene synthetic rubber
Nylex	Polyvinyl chloride
Nylon	Synthetic polyamides and super polyamides
Paxolin	Phenol formaldehyde
Perspex	Methyl methacrylate polymer
Plexiglass	Methyl methacrylate polymer
Polythene	Polyethylene
PVC	Polyvinyl chloride
Scarab	Urea formaldehyde
Synthane	Phenol formaldehyde
Trolitul	Polystyrene
Vinylite A	Polyvinyl acetate
Vinylite Q	Polyvinyl chloride
Vinylite V	Vinyl chloride-acetate copolymer
Vinylite X	Polyvinyl butyral
Xylonite	Nitrocellulose

(iv) Weights of common materials

The weights of the following materials are given in pounds per cubic inch.

TABLE 40

Aluminium	0.098	Polystyrene	0.038
Armco iron	0.284	Radio metal	0.299
Brass	0.304	Rho metal	0.292
Constantan	0.321	Steel	
Copper	0.310	Cast	0.278
German silver	0.323	2% silicon	0.278
Iron (cast)	0.281	4% silicon	0.271
Mumetal	0.318	Synthetic wax	
Perm-Alloy B	0.299	Seekay	0.056
Perm-Alloy C	0.310	Halowax	0.056
Perspex	0.043	2129 Alloy	0.292
Polyamide (Nylon)	0.041		

See Sect. 10(viii) for weight of metal sheets.

(v) Resistance of a conductor at any temperature

To find the resistance of a conductor at any temperature :

$$R_t = R_{20}[1 + (t - 20)\alpha]$$

where R_t = resistance at temperature $t°C$

R_{20} = resistance at 20°C

t = temperature of conductor °C

and α = temperature coefficient of resistivity at 20°C.

Most conductors increase in resistance when their temperature is increased.

(vi) Reference to properties of materials

1. English readers are referred to the Radio Industry Council Specification No. RIC/1000/A " Choice of materials for radio and other electronic equipment and for components therein " Issue No. 1—July 1949.

2. British Standard 1598 : 1949 " Ceramic materials for telecommunication and allied purposes." Gives summary of electrical, mechanical, thermal and general properties of insulators.

3. British Standard 1540 : 1949 " Moulded electrical insulating materials for use at radio frequencies ".

4. D. W. Thomasson " Silicones and other silicon compounds " Electronic Eng. 22.272 (Oct. 1950) 422.

5. Federal Telephone and Radio Corporation " Reference Data for Radio Engineers " 3rd ed. 31-39, 47-53.

6. " Standard Handbook for Electrical Engineers," McGraw-Hill Book Co. Inc. 8th ed. 1949, Sect. 4.

Extensive treatment ; also gives bibliography of insulating materials Sect. 4-619.

(vii) Chemical and physical constants

Some Chemical and Physical Constants as at 1 Dec. 1950, extracted from Table VII, " A re-evaluation of the fundamental atomic constants " Physical Review 81.1 (Jan. 1, 1951) 73.

Electron mass $\quad m = (9.10710 \pm 0.00022) \times 10^{-28}g$

Electronic charge $\quad e = (4.80217 \pm 0.00006) \times 10^{-10}esu$

$\qquad\qquad\qquad = (1.601844 \pm 0.000021) \times 10^{-20}emu$

Boltzmann's constant $k = (1.38020 \pm 0.00007) \times 10^{-16}erg\ deg^{-1}$

Ratio proton mass to electron mass

$$M/m = (1836.093 \pm 0.044)$$

Velocity of light $\quad c = (299790.0 \pm 0.7)\ km/sec$

Note: Since the first printing of this edition, the International Scientific Radio Union (U.R.S.I.) has recommended that the velocity of light in a vacuum is to be taken as 299792 ± 2 km/sec (see Wireless World, March 1953, page 121).

SECTION 9 : REACTANCE, IMPEDANCE AND RESONANCE

(i) Inductive reactances (ii) Capacitive reactances (iii) Impedance of reactance and resistance in parallel (iv) Impedance of reactance and resistance in series (v) Resonance (vi) Approximations in the calculation of impedance for reactance and resistance in series and parallel (vii) Reactance chart.

(i) Inductive reactances

(Correct to three significant figures)

TABLE 41

AUDIO FREQUENCIES $X_L = \omega L$

Inductance (Henries)	Reactance in Ohms at :—					
	30 c/s	50 c/s	100 c/s	400 c/s	1000 c/s	5000 c/s
250	47 100	78 500	157 000	628 000	1 570 000	7 850 000
100	18 800	31 400	62 800	251 000	628 000	3 140 000
50	9 420	15 700	31 400	126 000	314 000	1 570 000
25	4 710	7 850	15 700	62 800	157 000	785 000
10	1 880	3 140	6 280	25 100	62 800	314 000
5	942	1 570	3 140	12 600	31 400	157 000
1	188	314	628	2 510	6 280	31 400
.1	18.8	31.4	62.8	251	628	3 140
.01	1.88	3.14	6.28	25.1	62.8	314
1000 μH	.188	.314	.628	2.51	6.28	31.4
200 μH	.0376	.0628	.126	.502	1 26	6.28
100 μH	.0188	.0314	.0628	.251	.628	3.14

RADIO FREQUENCIES $X_L = \omega L$

Inductance (Henries)	Reactance in Ohms at :—					
	175 Kc/s	252 Kc/s	465 Kc/s	550 Kc/s	1000 Kc/s	1500 Kc/s
1	1 100 000	1 580 000	2 920 000	3 460 000	6 280 000	9 430 000
.1	110 000	158 000	292 000	346 000	628 000	943 000
.01	11 000	15 800	29 200	34 600	62 800	94 300
1000 μH	1 100	1 580	2 920	3 460	6 280	9 430
200 μH	220	317	584	691	1 260	1 890
100 μH	110	158	292	346	628	943

(ii) Capacitive reactances

(Correct to three significant figures)

TABLE 42

AUDIO FREQUENCIES $X_C = 1/\omega C$

Capacitance Microfarads	Reactance in Ohms at :—					
	30 c/s	50 c/s	100 c/s	400 c/s	1000 c/s	5000 c/s
.00005	—	—	—	—	—	637 000
.0001	—	—	—	—	1 590 000	318 000
.00025	—	—	—	1 590 000	637 000	127 000
.0005	—	—	3 180 000	796 000	318 000	63 700
.001	—	3 180 000	1 590 000	398 000	159 000	31 800
.005	1 060 000	637 000	318 000	79 600	31 800	6 370
.01	531 000	318 000	159 000	39 800	15 900	3 180
.02	263 000	159 000	79 600	19 900	7 960	1 590
.05	106 000	63 700	31 800	7 960	3 180	637
.1	53 100	31 800	15 900	3 980	1 590	318
.25	21 200	12 700	6 370	1 590	637	127
.5	10 600	6 370	3 180	796	318	63.7
1	5 310	3 180	1 590	398	159	31.8
2	2 650	1 590	796	199	79.6	15.9
4	1 310	796	398	99.5	39.8	7.96
8	663	398	199	49.7	19.9	3.98
10	531	318	159	39.8	15.9	3.18
20	265	159	79.6	19.9	7.96	1.59
50	106	63.7	31.8	7.96	3.18	0.637

RADIO FREQUENCIES $X_C = 1/\omega C$

Capacitance Microfarads	Reactance in Ohms at :—					
	175 Kc/s	252 Kc/s	465 Kc/s	550 Kc/s	1000 Kc/s	1500 Kc/s
.00005	18 200	12 600	6 850	5 800	3 180	2 120
.0001	9 100	6 320	3 420	2 900	1 590	1 060
.00025	3 640	2 530	1 370	1 160	637	424
.0005	1 820	1 260	685	579	318	212
.001	910	632	342	290	159	106
.005	182	126	68.5	57.9	31.8	21.2
.01	91.0	63.2	34.2	28.9	15.9	10.6
.02	45.5	31.6	17.1	14.5	7.96	5.31
.05	18.2	12.6	6.85	4.79	3.18	2.12
.1	9.10	6.32	3.42	2.89	1.59	1.06
.25	3.64	2.53	1.37	1.16	.637	.424
.5	1.82	1.26	.685	.579	.318	.212
1	.910	.632	.342	.289	.159	.106
2	.455	.316	.171	.145	.0796	.0531
4	.227	.158	.0856	.0723	.0398	.0265

(Table 42—continued)

RADIO FREQUENCIES $\qquad\qquad X_C = 1/\omega C$

Capacit-ance $\mu\mu F$	Reactance in Ohms at :—					
	6 Mc/s	12 Mc/s	18 Mc/s	25 Mc/s	50 Mc/s	100 Mc/s
10	2650	1330	888	637	318	159
22	1206	603	402	290	145	72.4
47	565	282	188	135	67.7	33.9
100	265	133	88.8	63.7	31.8	15.9
220	120.6	60.3	40.2	29.0	14.5	7.24
470	56.5	28.2	18.8	13.5	6.77	3.39
1000	26.5	13.3	8.88	6.37	3.18	1.59
2200	12.06	6.03	4.02	2.90	1.45	0.724
4700	5.65	2.82	1.88	1.35	0.677	0.339
10 000	2.65	1.33	0.888	0.637	0.318	0.159

(iii) **Impedance of reactance and resistance in parallel**

Table 43 has been prepared to permit the finding of any one of the three quantities X, R or Z when the other two are given. When X and R are given, divide the larger of the two quantities into the smaller one and thus get a ratio less than 1. Find this ratio in the left column and multiply the number obtained in the second column by R or X whichever is the larger and find Z.

Suppose R equals 1000 ohms and X is 200 ohms, which makes $X/R = .20$. Table 43 shows us that Z/R is then 0.1961. Multiplying by R, we have $Z = 0.1961 \times 1000 = 196.1$ ohms.

TABLE 43 : REACTANCE AND RESISTANCE VALUES IN PARALLEL

X/R or R/X	Z/R or Z/X	X/R or R/X	Z/R or Z/X	X/R or R/X	Z/R or Z/X
0.10	0.0995	0.49	0.4400	0.88	0.6606
0.11	0.1093	0.50	0.4472	0.89	0.6648
0.12	0.1191	0.51	0.4543	0.90	0.6690
0.13	0.1289	0.52	0.4613	0.91	0.6730
0.14	0.1386	0.53	0.4683	0.92	0.6771
0.15	0.1483	0.54	0.4751	0.93	0.6810
0.16	0.1580	0.55	0.4819	0.94	0.6849
0.17	0.1676	0.56	0.4886	0.95	0.6888
0.18	0.1771	0.57	0.4952	0.96	0.6925
0.19	0.1867	0.58	0.5017	0.97	0.6963
0.20	0.1961	0.59	0.5082	0.98	0.6999
0.21	0.2055	0.60	0.5145	0.99	0.7036
0.22	0.2149	0.61	0.5208	1.00	0.7071
0.23	0.2242	0.62	0.5269	1.10	0.7400
0.24	0.2334	0.63	0.5330	1.20	0.7682
0.25	0.2425	0.64	0.5390	1.30	0.7926
0.26	0.2516	0.65	0.5450	1.40	0.8137
0.27	0.2607	0.66	0.5508	1.50	0.8320
0.28	0.2696	0.67	0.5566	1.60	0.8480
0.29	0.2785	0.68	0.5623	1.70	0.8619
0.30	0.2874	0.69	0.5679	1.80	0.8742
0.31	0.2961	0.70	0.5735	1.90	0.8850
0.32	0.3048	0.71	0.5789	2.00	0.8944
0.33	0.3134	0.72	0.5843	2.20	0.9104
0.34	0.3219	0.73	0.5895	2.40	0.9231
0.35	0.3304	0.74	0.5948	2.60	0.9333
0.36	0.3387	0.75	0.6000	2.80	0.9418
0.37	0.3470	0.76	0.6051	3.00	0.9487
0.38	0.3552	0.77	0.6101	3.20	0.9545
0.39	0.3634	0.78	0.6150	3.40	0.9594
0.40	0.3714	0.79	0.6199	3.60	0.9635
0.41	0.3793	0.80	0.6246	3.80	0.9671
0.42	0.3872	0.81	0.6289	4.00	0.9702
0.43	0.3950	0.82	0.6341	5.00	0.9807
0.44	0.4027	0.83	0.6387	6.00	0.9864
0.45	0.4103	0.84	0.6432	7.00	0.9902
0.46	0.4179	0.85	0.6477	8.00	0.9921
0.47	0.4254	0.86	0.6520	9.00	0.9939
0.48	0.4327	0.87	0.6564	10.00	0.9950

By means of Chart 38.1 it is possible to determine either the parallel impedance (Z), reactance (X) or resistance (R) knowing the other two values. Expressing the capacitance (C) in microfarads, the inductance (L) in henrys and the frequency (F) in cycles use the bottom scale and read directly on to the chart.

REACTANCE AND RESISTANCE IN PARALLEL
CHART 38.1

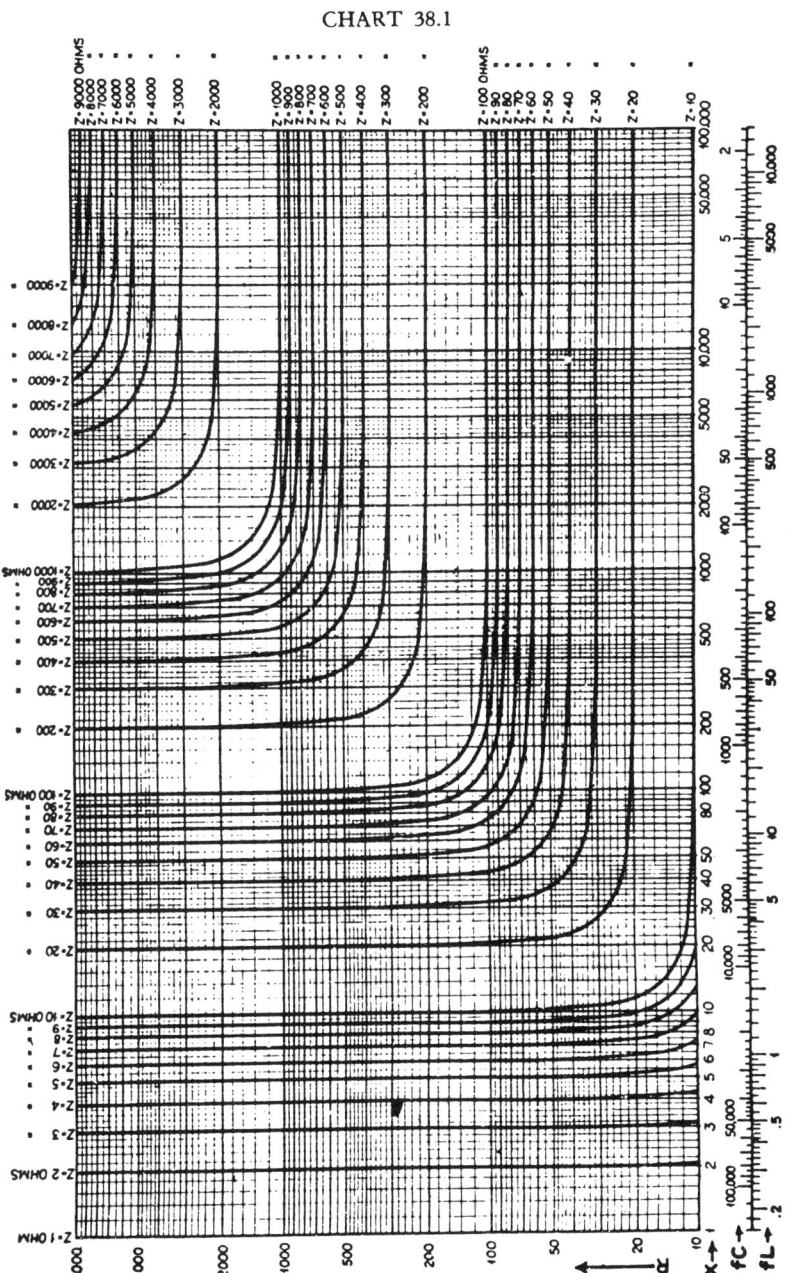

(iv) Impedance of reactance and resistance in series

To use Table 44, find the ratio R/X or X/R, refer to the table and find the corresponding ratio Z/X or Z/R. The table can also be used when Z is given together with one of the other quantities. It was for this reason that the table had to be extended for values of R/X or X/R from 0.1 to 1.0 since otherwise it would have been sufficient to include values from 1 upwards or downwards but not both. Example : suppose $X = 1600$ ohms and $R = 1000$ ohms. Then $X/R = 1.6$; the table shows $Z/R = 1.8868$. Then Z equals 1 8868 R or 1886.8 ohms.

TABLE 44 : REACTANCE AND RESISTANCE VALUES IN SERIES

X/R or R/X	Z/R or Z/X	X/R or R/X	Z/R or Z/X	X/R or R/X	Z/R or Z/X
0.10	1.0050	0.53	1.1318	0.96	1.3862
0.11	1.0060	0.54	1.1365	0.97	1.3932
0.12	1.0072	0.55	1.1413	0.98	1.4001
0.13	1.0084	0.56	1.1461	0.99	1.4071
0.14	1.0097	0.57	1.1510	1.00	1.4141
0.15	1 0112	0.58	1.1560	1.1	1.4866
0.16	1.0127	0.59	1.1611	1.2	1.5621
0.17	1.0144	0.60	1.1662	1.3	1.6401
0.18	1.0161	0.61	1.1714	1.4	1.7205
0.19	1.0179	0.62	1.1765	1.5	1.8028
0.20	1.0198	0.63	1.1819	1.6	1.8868
0.21	1.0218	0.64	1.1873	1.7	1.9723
0.22	1.0239	0.65	1.1927	1.8	2.0591
0.23	1.0261	0.66	1.1981	1.9	2.1471
0.24	1.0284	0.67	1.2037	2.0	2.2361
0.25	1.0308	0.68	1.2093	2.1	2.3259
0.26	1.0333	0.69	1.2149	2.2	2.4166
0.27	1.0358	0.70	1.2207	2.3	2.5080
0.28	1.0384	0.71	1.2264	2.4	2.6000
0.29	1.0412	0.72	1.2322	2.5	2.6926
0.30	1.0440	0.73	1.2381	2.6	2.7857
0.31	1.0469	0.74	1.2440	2.7	2.8792
0.32	1.0499	0.75	1.2500	2.8	2.9732
0.33	1.0530	0.76	1.2560	2.9	3.0676
0.34	1.0562	0.77	1.2621	3.0	3.1623
0.35	1.0595	0.78	1.2682	3.1	3.2573
0.36	1.0628	0.79	1.2744	3.2	3.3526
0.37	1.0662	0.80	1.2806	3.3	3.4482
0.38	1.0698	0.81	1.2869	3.4	3.5440
0.39	1.0733	0.82	1.2932	3.5	3.6400
0.40	1.0770	0.83	1.2996	3.6	3.7362
0.41	1.0808	0.84	1.3060	3.7	3.8327
0.42	1.0846	0.85	1.3125	3.8	3.9293
0.43	1.0885	0.86	1.3190	3.9	4.0262
0.44	1.0925	0.87	1.3255	4.0	4.1231
0.45	1.0966	0.88	1.3321	4.1	4.2202
0.46	1.1007	0.89	1.3387	4.2	4.3174
0.47	1.1049	0.90	1.3454	4.3	4.4147
0.48	1.1092	0.91	1.3521	4.4	4.5122
0.49	1.1136	0.92	1.3588	4.5	4.6098
0.50	1.1180	0.93	1.3656	4.6	4.7074
0.51	1.1225	0.94	1.3724	4.7	4.8052
0.52	1.1271	0.95	1.3793	4.8	4.9030

TABLE 44 (Continued)

X/R or R/X	Z/R or Z/X	X/R or R/X	Z/R or Z/X	X/R or R/X	Z/R or Z/X
4.9	5.0009	6.6	6.6752	8.3	8.3600
5.0	5.0990	6.7	6.7741	8.4	8.4594
5.1	5.1971	6.8	6.8731	8.5	8.5580
5.2	5.2952	6.9	6.9720	8.6	8.6576
5.3	5.3935	7.0	7.0711	8.7	8.7572
5.4	5.4918	7.1	7.1701	8.8	8.8566
5.5	5.5901	7.2	7.2691	8.9	8.9560
5.6	5.6885	7.3	7.3681	9.0	9.0554
5.7	5.7871	7.4	7.4671	9.1	9.1548
5.8	5.8856	7.5	7.5662	9.2	9.2542
5.9	5.9841	7.6	7.6654	9.3	9.3536
6.0	6.0828	7.7	7.7646	9.4	9.4530
6.1	6.1814	7.8	7.8638	9.5	9.5524
6.2	6.2801	7.9	7.9630	9.6	9.6518
6.3	6.3789	8.0	8.0623	9.7	9.7512
6.4	6.4777	8.1	8.1615	9.8	9.8507
6.5	6.5764	8.2	8.2608	9.9	9.9503
				10.0	10.0499

From Chart 38.2 (page 1384) it is possible to determine either the series impedance (Z), reactance (X) or resistance (R) knowing the other two values. Expressing the capacitance (C) in microfarads, the inductance (L) in henrys and the frequency (f) in c/s, use the top scale and read directly on to the chart.

Tables 43, 44 and charts 38.1, 38.2 are reprinted from the Aerovox Research Worker 11, Nos. 1 and 2 (Jan. and Feb. 1939) by courtesy of the Aerovox Corporation.

CHART 38.2

REACTANCE AND RESISTANCE IN SERIES

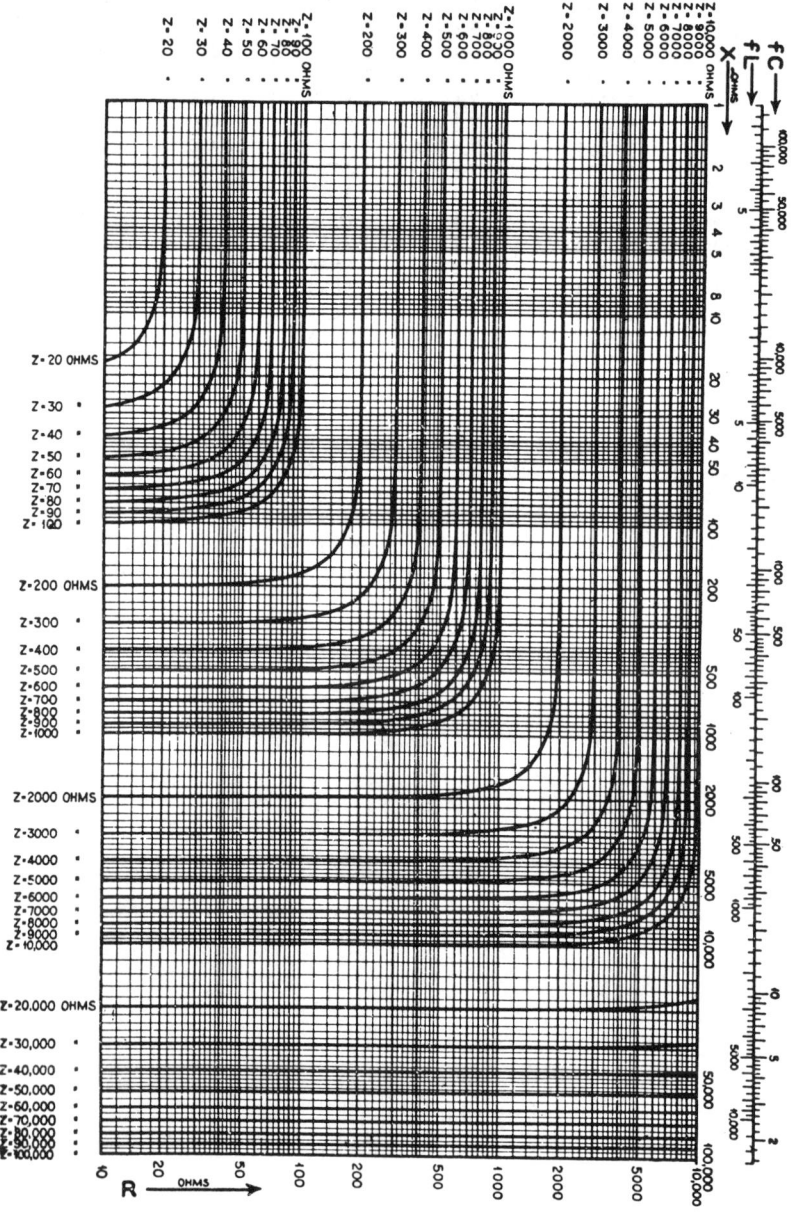

Resistance and reactance in series—Alternative method

Chart 38.3 assists in the evaluation of the impedance Z and its phase angle θ when the values of resistance R and reactance X are known. Firstly read R along the horizontal scale, then read upwards to the value of jX ; the magnitude of Z is given by the radius which may be read by running the eye down the arc to the horizontal axis. The angle of Z is given directly by the angle at the point of intersection. For example, $R = 15$ ohms in series with $jX = 5.6$ ohms gives $R + jX = 16$ ohms, $\angle 20°$ (approx.). When any value of R or X is greater than the limits on the chart, both scales should be multiplied by the same factor. For example, $R = 150$ ohms in series with $jX = 56$ ohms gives $R + jX = 160$ ohms, $\angle 20°$. Negative reactances may be treated as positive when using the chart, but the angle of Z will then be negative.

CHART 38.3

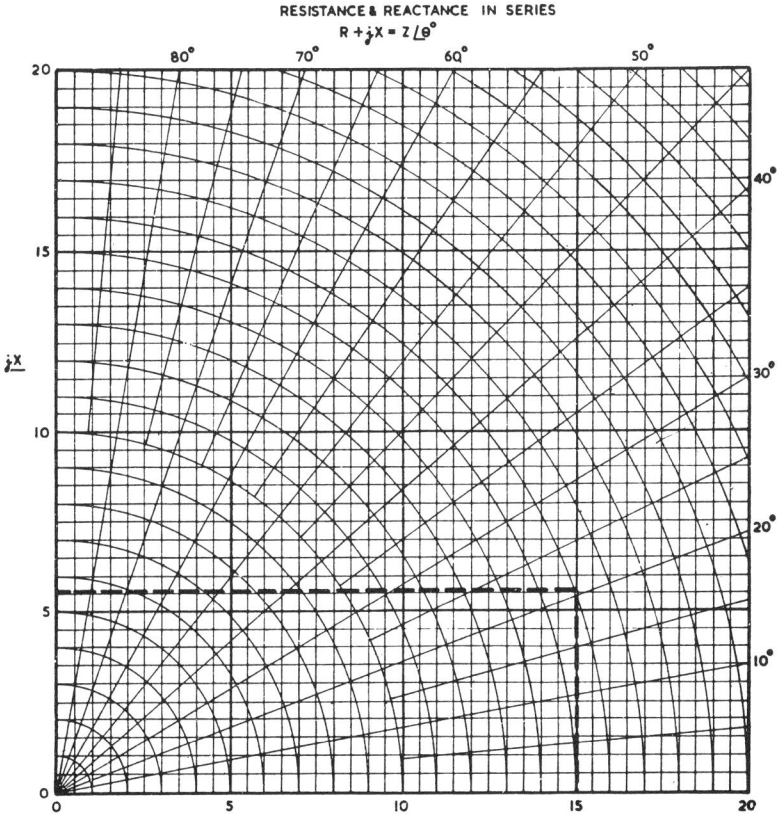

RESISTANCE & REACTANCE IN SERIES
$R + jX = Z\angle\theta°$

(v) Resonance
(A) Resonance frequency table (audio frequencies)

TABLE 45 : RESONANCE FREQUENCY C/S

$C(\mu F)$	50	100	200	400	1000	2000	5000	10 000
1	10.1	2.53	633	158	25.3	6.33	1.01	0.253
0.5	20.3	5.07	1.27	317	50.7	12.7	2.03	0.507
0.25	40.5	10.1	2.53	633	101	25.3	4.05	1.01
0.1	101	25.3	6.33	1.58	253	63.3	10.1	2.53
0.05	203	50.7	12.7	3.17	507	127	20.3	5.07
0.02	507	127	31.7	7.91	1.27	317	50.7	12.7
0.01	1010	253	63.3	15.8	2.53	633	101	25.3
0.005	2030	507	127	31.7	5.07	1.27	203	50.7
0.002	5070	1270	317	79.1	12.7	3.17	507	127
0.001	10 100	2530	633	158	25.3	6.33	1.01	253

Inductance values above the stepped line are in millihenrys and those below in henrys.

(B) R-F solenoid design chart

Chart 38.4 enables the radio engineer to determine the approximate number of turns on a solenoid of specified diameter and length, to resonate at a specified frequency with a specified capacitance. (Chart by P. G. Sulzer in Tele-Tech, May 1951, p. 45 and reproduced by kind permission).

The following example will indicate the method :

It is desired to design a coil for a harmonic generator which is to operate at a frequency of 10 Mc/s with a total capacitance of 50 $\mu\mu$F. Drawing a straight line between 10 Mc/s on the f scale and 50 $\mu\mu$F on the C scale, the inductance L is found to be approximately 5 μH. This value need not be recorded unless it is required for some other purpose. Assuming that the winding is to be one inch in diameter by two inches long, the intersection of the appropriate lines is found in the graph at the right-hand side of the chart. This intersection is projected horizontally to the left as shown, and then a straight line is drawn to the 5 μH point on the L scale. The result is found to be 22 turns, as indicated on the n scale.

(vi) Approximations in the calculation of impedance for reactance and resistance in series and parallel

When a resistance and a reactance are in series, and the reactance is numerically smaller than the resistance, the value of the resistance may be taken as being an approximation to the impedance, with the errors indicated below :

R/X =	1	2	3	4	5	7	10
error =	29%	11%	5%	3%	2%	1%	0.5%

When a resistance and a reactance are in parallel, and the reactance is numerically larger than the resistance, the value of the resistance may be taken as being an approximation to the impedance, with the errors indicated below :

X/R =	1	2	3	4	5	7	10
error =	41%	12%	5.5%	3%	2%	1%	0.5%

(vii) Reactance chart

A very complete form of reactance chart, including reactance, frequency, inductance, capacitance, susceptance, wavelength and time constant, is given by H. A. Wheeler "Reactance Chart" Proc. I.R.E. 38.12 (Dec. 1950) 1392.

CHART 38.4: R-F SOLENOID DESIGN CHART—*see page* 1386.

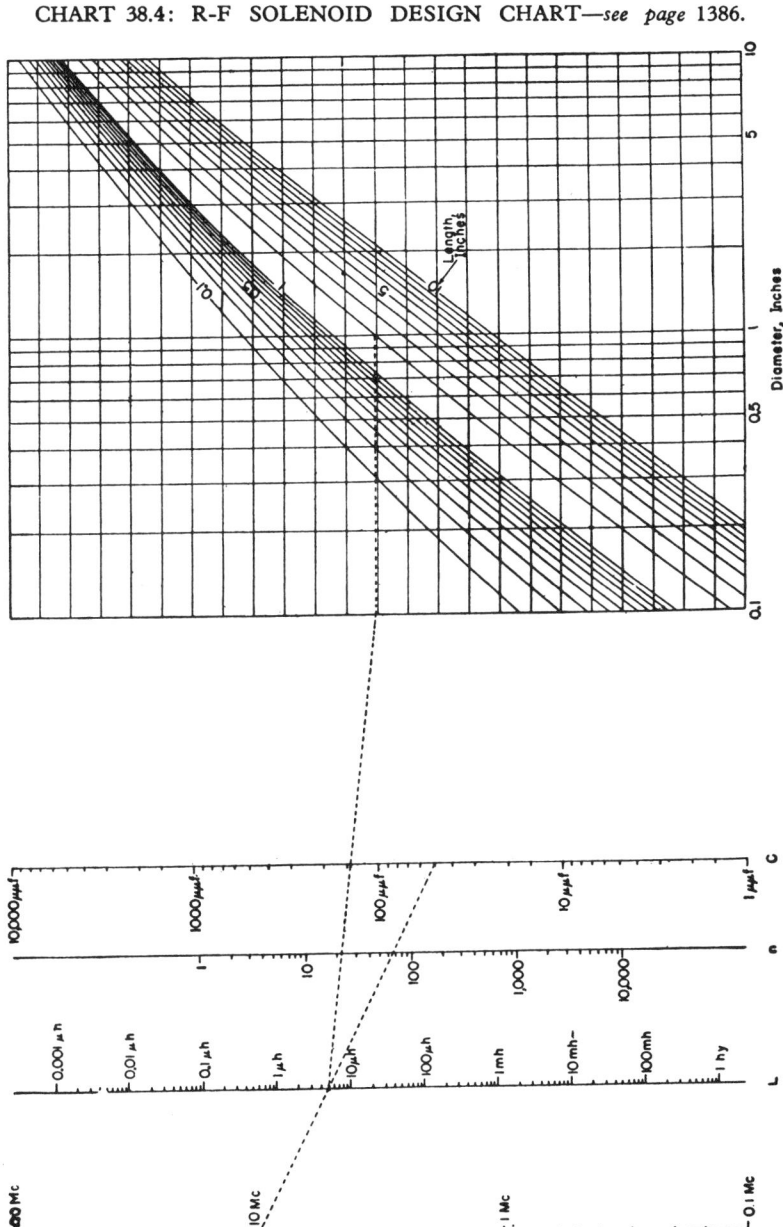

SECTION 10 : SCREW THREADS, TWIST DRILLS AND SHEET GAUGES

(i) Standard American screws used in radio manufacture (ii) B.A. screw threads (iii) Whitworth screw threads (iv) Unified screw threads (v) Drill sizes for self-tapping screws (vi) Wood screws (vii) Twist drill sizes (viii) Sheet steel gauges.

(i) Standard American screws used in radio manufacture
(American Standard B1.1—1935 ; coarse thread).

TABLE 46

Size of screw and T.P.I.	Outside dia. in inches	Pitch dia. in inches	Root dia. in inches	Tap drill steel	Tap drill cast iron	Clearance drill
2-56	.0860	.0744	.0628	No. 49 (.0730)	No. 49 (.0730)	No. 43 (.0890)
3-48	.0990	.0855	.0719	No. 44 (.0860)	No. 44 (.0860)	No. 39 (.0995)
4-40	.1120	.0958	.0795	No. 42 (.0935)	No. 43 (.0890)	No. 33 (.1130)
5-40	.1250	.1088	.0925	No. 34 (.1110)	No. 35 (.1110)	No. 30 (.1285)
6-32	.1380	.1177	.0974	No. 32 (.1160)	No. 33 (.1130)	No. 28 (.1405)
8-32	.1640	.1437	.1234	No. 27 (.1440)	No. 28 (.1405)	No. 19 (.1660)
10-24	.1900	.1629	.1359	No. 21 (.1590)	No. 22 (1.570)	No. 11 (.1910)
10-32*	.1900	.1697	.1494	No. 19 (.1660)	No. 20 (.1610)	No. 11 (.1910)
12-24	.2160	.1889	.1619	No. 16 (.1770)	No. 17 (.1730)	No. 11 (.1910)
¼-20	.2500	.2175	.1850	No. 7 (.2010)	No. 8 (.1990)	No. 2 (.2210), ¼ in.

*Fine thread.

(ii) B.A. screw threads

TABLE 47
Dimensions given are only approximate

B.A. No.	Outside dia.	Core dia.	Turns per in.	Clearing drill	Tapping drill
0	.236	.189	25.4	¼ in. or " B "	10-12
1	.209	.166	28.3	2-3	18-19
2	.185	.147	31.4	10-11	25-26
3	.161	.127	34.8	18-19	30-31
4	.142	.111	38.5	26-27	33-34
5	.126	.098	43.1	29-30	39-40
6	.110	.085	47.9	32-33	44
7	.098	.076	52.9	38-39	48
8	.087	.066	59.2	42-43	51
9	.075	.056	65.1	46-47	53
10	.067	.050	72.5	49-50	55

(iii) Whitworth screw threads

TABLE 48

Outside dia.	Core dia.	Threads per inch	Tapping drill
1/8 in.	.093 in.	40	41
3/16 in.	.134	24	9/64 in.
1/4 in.	.186	20	12
5/16 in.	.241	18	1/4 in.
3/8 in.	.295	16	5/16 in.
1/2 in.	.393	12	13/32 in.
5/8 in.	.509	11	17/32 in.
3/4 in.	.622	10	5/8 in.
1 in.	.840	8	27/32 in.

(iv) Unified screw threads

Provisional British Standard B.S. 1580 : 1949.

American Standard A.S.A. B1.1—1949.

TABLE 49

Designation	Pitch	Turns	Major diameter Nut and bolt	Effective diameter* Nut and bolt	Minor diameter Nut (design size)†	Minor diameter Bolt
Coarse Thread	in.	per in.	in.	in.	in.	in.
1/4-20.UNC	0.05000	20	0.2500	0.2175	0.1959	0.1887
5/16-18.UNC	0.05556	18	0.3125	0.2764	0.2524	0.2443
3/8-16.UNC	0.06250	16	0.3750	0.3344	0.3073	0.2983
7/16-14.UNC	0.07143	14	0.4375	0.3911	0.3602	0.3499
1/2-12.UNC	0.08333	12	0.5000	0.4459	0.4098	0.3978
9/16-12.UNC	0.08333	12	0.5625	0.5084	0.4723	0.4603
5/8-11.UNC	0.09091	11	0.6250	0.5660	0.5266	0.5135
3/4-10.UNC	0.10000	10	0.7500	0.6850	0.6417	0.6273
7/8-9.UNC	0.11111	9	0.8750	0.8028	0.7547	0.7387
1-8.UNC	0.12500	8	1.0000	0.9188	0.8647	0.8466
Fine Thread						
1/4-28.UNF	0.03571	28	0.2500	0.2268	0.2113	0.2062
5/16-24.UNF	0.04167	24	0.3125	0.2854	0.2674	0.2614
3/8-24.UNF	0.04167	24	0.3750	0.3479	0.3299	0.3239
7/16-20.UNF	0.05000	20	0.4375	0.4050	0.3834	0.3762
1/2-20.UNF	0 05000	20	0.5000	0.4675	0.4459	0.4387
9/16-18.UNF	0.05556	18	0.5625	0.5264	0.5024	0.4943
5/8-18.UNF	0.05556	18	0.6250	0.5889	0.5649	0.5568
3/4-16.UNF	0.06250	16	0.7500	0.7094	0.6823	0.6733
7/8-14.UNF	0.07143	14	0.8750	0.8286	0.7977	0.7874
1-12.UNF	0.08333	12	1.0000	0.9459	0.9098	0.8978

*American : Pitch diameter.

†Corresponds to a flat of $p/4$ where p = pitch.

Note : The final decimal point is subject to slight modification when the British Standard has been approved.

(v) Drill sizes for self-tapping screws

TABLE 50

Screw No.	Metal thickness* mils	Drill size	Screw No.	Metal thickness* mils	Drill size
2	16	52	4	50	41
2	20	52	4	62	39
2	25	51	4	78	38
2	31	50	6	16	37
2	39	49	6	20	37
2	50	49	6	25	36
2	62	48	6	31	36
4	16	44	6	39	35
4	20	44	6	50	34
4	25	43	6	62	32
4	31	42	6	78	31
4	39	42	6	99	30

*For steel or brass.   Use a somewhat smaller hole for softer metals.

(vi) Wood screws

TABLE 51

No.	American National Standard		British Practice	
	shank dia. in.	clearance drill No.	shank dia. in.	clearance drill No.
0	0.060	50	0.063	52
1	.073	49	.066	51
2	.086	44	.080	46
3	.099	39	.094	41
4	.112	33	.108	35
5	.125	30	.122	30
6	.138	28	.136	28
7	.151	24	.150	23
8	.164	19	.164	18
9	.177	16	.178	14
10	.190	11	.192	9
11	.203	6	.206	4
12	.216	2	.220	1
14	.242	E	.248	E
16	.268	I	.276	K

(vii) Twist drill sizes

TABLE 52
Number drills

Drill No.	Dia. Inch	Drill No.	Dia. Inch	Drill No.	Dia. Inch
1	.2280	28	.1405	55	.0520
2	.2210	29	.1360	56	.0465
3	.2130	30	.1285	57	.0430
4	.2090	31	.1200	58	.0420
5	.2055	32	.1160	59	.0410
6	.2040	33	.1130	60	.0400
7	.2010	34	.1110	61	.0390
8	.1990	35	.1100	62	.0380
9	.1960	36	.1065	63	.0370
10	.1935	37	.1040	64	.0360
11	.1910	38	.1015	65	.0350
12	.1890	39	.0995	66	.0330
13	.1850	40	.0980	67	.0320
14	.1820	41	.0960	68	.0310
15	.1800	42	.0935	69	.0292
16	.1770	43	.0890	70	.0280
17	.1730	44	.0860	71	.0260
18	.1695	45	.0820	72	.0250
19	.1660	46	.0810	73	.0240
20	.1610	47	.0785	74	.0225
21	.1590	48	.0760	75	.0210
22	.1570	49	.0730	76	.0200
23	.1540	50	.0700	77	.0180
24	.1520	51	.0670	78	.0160
25	.1495	52	.0635	79	.0145
26	.1470	53	.0595	80	.0135
27	.1440	54	.0550		

Letter drills

Letter	Dia. Inch	Letter	Dia. Inch	Letter	Dia. Inch
A	0.2340	J	0.2770	S	0.3480
B	0.2380	K	0.2810	T	0.3580
C	0.2420	L	0.2900	U	0.3680
D	0.2460	M	0.2950	V	0.3770
E	0.2500	N	0.3020	W	0.3860
F	0.2570	O	0.3160	X	0.3970
G	0.2610	P	0.3230	Y	0.4040
H	0.2660	Q	0.3320	Z	0.4130
I	0.2720	R	0.3390		

(viii) Sheet metal gauges

TABLE 53 : Thickness and weight per square foot

Gauge	Birmingham Gauge (B.G.)		American Manufacturers Standard Gauge* for sheet steel		B & S (American) for non-ferrous sheet			
	inch	steel lb/ft²	inch	lb/ft²	inch	lb per square foot copper	brass	aluminium
10	.1250	5.100	1345	5.625	.1019	4.71	4.51	1.44
11	.1113	4.541	.1196	5.000	.0907	4.19	4.02	1.28
12	.0991	4.043	.1046	4.375	.0808	3.74	3.58	1.14
13	.0882	3.599	.0897	3.750	.0720	3.33	3.19	1.01
14	.0785	3.203	.0747	3.125	.0641	2.96	2.84	.903
15	.0699	2.852	.0673	2.8125	.0571	2.64	2.53	.804
16	.0625	2.550	.0598	2.500	.0508	2.35	2.25	.716
17	.0556	2.268	.0538	2.250	.0453	2.10	2.01	.638
18	.0495	2.020	.0478	2.000	.0403	1.86	1.78	.568
19	.0440	1.795	.0418	1.750	.0359	1.66	1.59	.506
20	.0392	1.599	.0359	1.500	.0320	1.48	1.42	.450
21	.0349	1.424	.0329	1.375	.0285	1.32	1.26	.401
22	.03125	1.275	.0299	1.250	.0253	1.17	1.12	.357
23	.02782	1.134	.0269	1.125	.0226	1.05	1.00	.318
24	.02476	1.010	.0239	1.000	.0201	.931	.890	.283
25	.02204	0.898	.0209	.8750	.0179	.829	.793	.252
26	.01961	0.800	.0179	.7500	.0159	.738	.706	.225
27	.01745	0.712	.0164	.6875	.0142	.657	.628	.200
28	.015625	0.636	.0149	.6250	.0126	.589	.560	.178
29	.0139	0.567	.0135	.5625	.0113	.521	.499	.159
30	.0123	0.502	.0120	.5000	.0100	.464	.444	.141
31	.0110	0.449	.0105	.4375	.00893	.413	.395	.126
32	.0098	0.400	.0097	.40625	.00795	.368	.353	.112
33	—	—	.0090	.3750	.00708	.328	.314	.100
34	—	—	.0082	.34375	.00635	.292	.279	.089
35	—	—	.0075	.3125	.00562	.260	.249	.079

*Based on 41.820 lb. per sq. foot per inch thick ; variations in thickness are to be expected.
Other gauges are based directly on thickness ; variations in weight per square inch are to be expected.

American Standard Preferred Thicknesses for uncoated thin flat metals (ASA. B32.1—1941).
This series of thickness is given in inches.

TABLE 54

	0.180	0.090	0.045	0.022	0.011
	0.160	0.080	0.040	0.020	0.010	0.005
	0.140	0.071	0.036	0.018	0.009
	0.125	0.063	0.032	0.016	0.008	0.004
0.224	0.112	0.056	0.028	0.014	0.007
0.200	0.100	0.050	0.025	0.012	0.006

SECTION 11 : TEMPERATURE RISE AND RATINGS

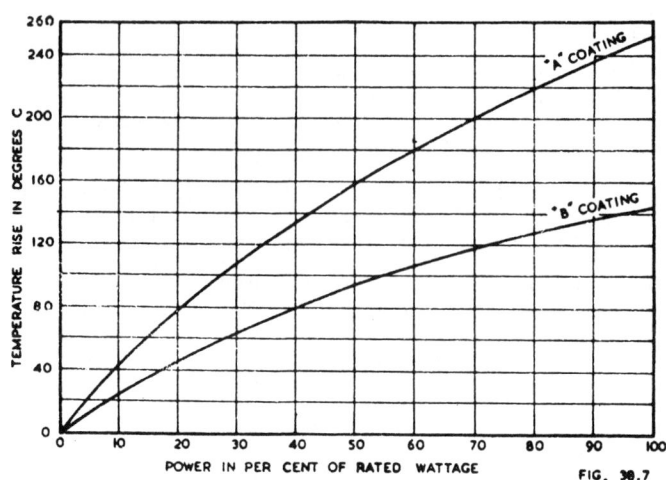

Fig. 38.7. Curves of temperature rise of power wire-wound resistors in free air. The temperature is considerably affected by the radiating properties of the coating. Type A coating refers to a vitreous enamel which is used for " tropic proofed " application and type B coating refers to an organic cement coating.

TABLE 55 : SIZES OF WIRES FUSED BY ELECTRIC CURRENTS

Fusing Current (Amperes)	Copper a = 10 244		Aluminium a = 7585		German Silver a = 5230		Tin a = 1642		Tin-Lead Alloy a = 1318		Lead a = 1379	
	Dia. Inch	S.W.G. Approx.	Dia. inch	S.W.G. Approx.	Dia. inch	S.W.G. Approx.	Dia. inch	S.W.G. Approx.	Dia. inch	S.W.G. Approx.	Dia. inch	S.W.G. Approx.
1	.0021	47	.0026	46	.0033	44	.0072	37	.0083	35	.0081	35
2	.0034	43	.0041	42	.0053	39	.0113	31	.0132	29	.0128	30
3	.0044	41	.0054	39	.0069	37	.0149	28	.0173	27	.0168	27
4	.0053	39	.0065	37	.0084	35	.0181	26	.0210	25	.0203	25
5	.0062	38	.0076	36	.0097	33	.0210	25	.0243	23	.0236	23
10	.0098	33	.0120	30	.0154	28	.0334	21	.0386	19	.0375	20
20	.0156	28	.0191	25	.0245	23	.0529	17	.0613	16	.0595	17
30	.0205	25	.0250	20	.0320	21	.0694	15	.0803	14	.0779	14

Reprinted, with additions, by permission from STANDARD HANDBOOK FOR ELECTRICAL ENGINEERS (8th ed. Sect. 14-148) copyrighted by McGraw Hill Book Company Inc.

SECTION 12 : FUSES

The fusing current of wire depends largely upon external conditions such as atmospheric temperature, method of mounting, proximity of other objects, and time of operation. For sizes up to about 25 amps, a simple construction may be used with metals such as aluminium or special alloys which produce a minimum vapour on fusing.

Copper wire, operated near the fusing current, is particularly subject to corrosion and should therefore be coated with a metal —such as tin—which does not oxidise so readily.

Table 55 applies only to con - ditions with the wire freely suspended in air. The maximum safe working current may be taken as approximately 67% of the fusing current under the same conditions.

Fuses designed for intermediate values of current may be calculated from the formula

$$\text{diameter} = \left(\frac{\text{current}}{a}\right)^{2/3}$$

where a is the factor given in Table 55.

See Chapter 35 Sect. 9(iv) for fuses in radio receivers.

SECTION 13 : CHARACTERISTICS OF LIGHT ; PANEL LAMPS

(i) Visibility curves of human eye, and relative spectral energy curves of sunlight and tungsten lamp   (ii) Velocity of light   (iii) American panel lamp characteristics.

(i) Visibility curves of human eye, and relative spectral energy curves of sunlight and tungsten lamp (Fig. 38.8)

The wavelength of light is measured in Angstrom* Units (A°). One Angstrom Unit is equal to 1/10 000 of a micron, that is 1/10 of a milli-micron. A micron is 1/1 000 000 (10⁻⁶) of a metre.

The wavelengths visible to the human eye extend from about 4000 to 7000 A°. Beyond these extends the region of " invisible light " which, although invisible to the human eye, may be detected by the photo-tube or other means.

The eye, when accustomed to high light intensity, is most sensitive to a wavelength of 5550 A°, in the green-yellow region. The relative visibility curve for these conditions is Curve A, which is taken after H. E. Ives†. As the light intensity is reduced, the wavelength at which the eye is most sensitive decreases until at very low intensity it is approximately as shown in Curve B.

Different light sources have different spectral energy curves. The curve for sunlight (Curve C is for sunlight at the earth's surface at a zenith distance of 25°) is more nearly constant over the range of visible light than that for a tungsten lamp (Curve D is for a 1000 watt gasfilled tungsten lamp, 20 lumens per watt). The relative positions of curves C and D are quite arbitrary.

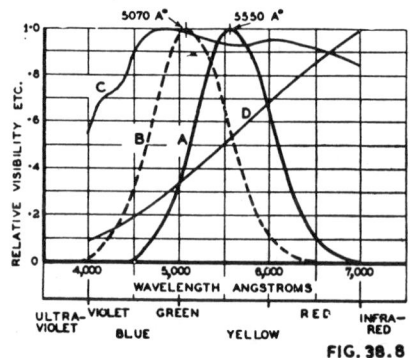

FIG. 38.8

Fig. 38.8.   Relative visibility curves, etc. (for description see text).

(ii) Velocity of light

Velocity of light in a vacuum = 2.99792 × 10⁸ metres/sec.
$$\approx 3 \times 10^8 \text{ metres/sec.}$$

See Editorial, W.E. 28.331 (April 1951) 99.

Also Electronic Eng. (Dec. 1950) 524.

See also page 1376 (vii) Chemical and Physical Constants.

---

*The name Angstrom is Swedish, pronounced " ongstrem."
†See R. A. Houston, " Vision and Colour Vision," Longmans Green and Co. (1932) Chapter 5.

(iii) American panel lamp characteristics

(RTMA Standard REC-137, October 1951).

TABLE 56 : AMERICAN STANDARD PANEL LAMPS

Trade Number	Circuit Volts	Design Volts	Amperes or Watts	Approx. C.P.	Life*	Bulb	Base
40	6.3	6.3	0.15A	$\frac{1}{2}$	3000	T-3$\frac{1}{4}$	(A)
41	2.5	2.5	0.50A	$\frac{1}{2}$	3000	T-3$\frac{1}{4}$	(A)
44	6.3	6.3	0.25A	$\frac{3}{4}$	3000	T-3$\frac{1}{4}$	(B)
46	6.3	6.3	0.25A	$\frac{3}{4}$	3000	T-3$\frac{1}{4}$	(A)
49	2	2	0.06A	—	1000	T-3$\frac{1}{4}$	(B)
48	2	2	0.06A	—	1000	T-3$\frac{1}{4}$	(A)
47	6.3	6.3	0.15A	$\frac{1}{2}$	3000	T-3$\frac{1}{4}$	(B)
—	115-125	120	10W	3	3000+	C7	(C)

A. Miniature screw base. B. Miniature bayonet base, single contact. C. Candelabra bayonet base, double contact.

SECTION 14 : GREEK ALPHABET

TABLE 57 : GREEK ALPHABET

Name	Large	Small	English Equivalent
alpha	A	α	a
beta	B	β	b
gamma	Γ	γ	g
delta	Δ	δ	d
epsilon	E	ϵ	e (short e as in " met ")
zeta	Z	ζ	z
eta	H	η	e (long e as in " meet ")
theta	Θ	θ	th
iota	I	ι	i
kappa	K	$\varkappa$	k
lambda	Λ	λ	l
mu	M	μ	m
nu	N	ν	n
xi	Ξ	ξ	x
omicron	O	o	o (as in " olive ")
pi	Π	π	p
rho	P	ρ	r
sigma	Σ	σ	s
tau	T	τ	t
upsilon	Y	υ	u
phi	Φ	φ	ph
chi	X	χ	ch (as in " school ")
psi	Ψ	ψ	ps
omega	Ω	ω	o (as in " hole ")

* Rated average laboratory life expectancy in hours at design volts.

SECTION 15 : DEFINITIONS

Most words which need explanation are explained or defined in the text and a reference is given in the index. The definitions below are supplementary and not inclusive.

See also other references at end of definitions (pp. 1403-1404).

A

A-battery Battery for supplying power to the valve filaments. (English name L.T. battery).

Absolute value Refers to magnitude without regard to direction or sign. Designated $|A|$.

A.C./D.C. receiver One which operates from either power source.

Acoustical feedback The operation of a microphone by the return to it of sound waves from the loudspeaker in the same sound system, reinforcing the input. It can cause sustained oscillation, or howling. A similar effect can occur from a microphonic valve.

Air-core Having no iron in its magnetic circuit.

Algebraic sum The sum of two or more quantities paying due regard to the positive or negative sign before each quantity.

Alternating current (or voltage) One that alternates regularly in direction, which is periodic and has an average value (over a period) of zero.

Ambient temperature The temperature of the air at a particular point, usually as indicated by a thermometer.

Amplification Increase in signal voltage (or power).

Angular velocity of a rotating body is the angle through which any radius turns in one second, generally expressed in radians per second.

Anode Positive electrode or plate.

Antinode A point on a stationary wave system which has a maximum amplitude.

Asymmetrical Having different characteristics for conduction in the two directions.

Attenuation Decrease in signal voltage or power.

Audio frequency A frequency within the range of frequencies audible to the normal human ear (say 25-16 000 c/s).

Axial leads Leads from the ends of resistors, capacitors, etc.

B

B-battery The battery which supplies power to the plate and screen circuits. (English name is H.T. battery).

B-supply A plate and screen voltage source for a receiver and amplifier.

Band Frequencies between a specific upper and lower limit.

Band-pass Passing a specific band of frequencies only.

Bandspread A means for spreading the coverage, on the dial, of a band of frequencies.

Bass Low audio frequencies.

Bass boost To boost the amplification at bass frequencies relatively to that at higher frequencies.

Beating A periodic variation in the amplitude of an oscillation (or sound) which is the resultant of two or more oscillations (or sounds) of different frequencies.

Bias Voltage applied to the grid to obtain the desired operating point.

Bleeder resistor Resistor connected across a voltage source to provide an additional load. The term is sometimes applied to a voltage divider carrying a relatively heavy current to provide good regulation.

Blocking Cutting off the plate current by a high negative bias on the grid. For Grid Blocking see page 21.

Bogie   Each individual characteristic in the manufacturing specification is normally prescribed as a bogie value with plus and minus tolerances. The bogie value of a characteristic is the exact value specified for that characteristic by the valve manufacturing specifications.

Bucking   Two forces opposing one another.

By-pass condenser   One which allows alternating current to by-pass part of a circuit.

C

C-battery   Battery which supplies voltage to the grid.   Also known as bias battery.

Capacitor   Condenser.

Carrier frequency   The frequency of a component of a modulated wave which is the same as that of the wave before modulation is applied.

Carrier wave   The unmodulated wave radiated by a broadcast station.

Cascade   In an amplifier, when the output of one valve is used to control the grid circuit of another valve, the valves are said to be in cascade.

Cathode   Negative electrode.

Cathode current   The total electronic current passing through the cathode.

Channel   The assigned band of frequencies within which a broadcast station is expected to keep its modulated carrier.

Choke coil   Inductance.

Co-axial   Having a common axis.

Compliance   The ratio of the displacement of a body to the force applied. The inverse of stiffness.

Condenser   Capacitor. See Chapter 4.

Consonance   The agreeable effect produced by certain intervals (in music).

Constant   A quantity which retains its fixed value under all conditions.

Converter   The stage in a receiver which converts the frequency from radio frequency to intermediate frequency.

Co-planar   In the same plane.

Core   The centre of a coil.

Cross modulation   The modulation of the carrier of the desired signal by an undesired signal.

Cut-off frequency   The frequency beyond which the attenuation increases rapidly.

Cycle   A cycle is one complete series of values of a periodic quantity. These values repeat themselves at regular intervals.

D

Delay   Time delay is when an interval of time occurs between cause and effect. Voltage delay, as in an a.v.c. system, is when a specified voltage has to be developed before there is any effect.

Demodulation   A process by which the carrier frequency is removed, and the modulating frequencies retained.

Denominator   The part of a fraction below the fraction bar.

Detection   Demodulation by means of an asymmetrical conducting device.

Dissipation   Loss of electrical energy.

Dissonance   The disagreeable effect produced by certain intervals (in music).

Dividend   Number to be divided by divisor.

Divisor   Number by which the dividend is to be divided.

Drain   To take current from a voltage source.

Driver   A power amplifier stage which drives a Class B or other following stage requiring input power.

Dynamic   (1) Having a moving coil in a magnetic field.

(2) When applied to a valve with signal voltage on the grid, usually with plate load in circuit.

(3) "Dynamic characteristic" is the mutual characteristic when a (usually resistive) load is in the plate circuit.

E

Effective current   The value of alternating or varying current which produces the same heating effect as the same value of direct current.   It is the same as r.m.s. current.

Efficiency   The ratio or percentage of output to input.

Electromotive force   That force which tends to cause an electric current to flow in a circuit.   The practical unit is the volt.

Empirical   Obtained from experimental data.

Envelope   The envelope of a periodic wave is obtained by passing one smooth curve through the maximum, another through the minimum, points or peaks of the wave.

Equation   A mathematical equation is a statement that the terms to the left of the equal sign are equal to those on the right.

Equivalent circuit   A relatively simple circuit which may be used in calculations as having the same effect as the actual circuit.

Excitation   A signal voltage applied to the control electrode of a valve.

Exponential curve   The same as a logarithmic curve.   See Chapter 6 Sect. 2(xvii) for exponential functions.

Expression   A mathematical expression is a combination of terms giving the value of some quantity.

F

Factor   Any of the numbers whose product equals the given number.

Factorial n   Is the multiple product $n(n - 1)(n - 2) \ldots 3 \times 2 \times 1$.

First detector   The same as Converter.

Frequency   The number of cycles per second ; the reciprocal of the period.

Frequency changer   The same as Converter.

Function   A function is a quantity whose value depends on the value of a variable quantity.

Fundamental frequency   The lowest frequency of a number of harmonically related frequencies.

G

Generator   A device which develops electrical voltage either direct or alternating at any frequency.

Gramophone   Also known as a Phonograph.

Ground   Earth, or conductor serving as the earth.

H

Hard valve   A valve which has reverse grid current less than the specified maximum for its type.

Harmonic frequency   A frequency which is a multiple of the fundamental frequency.   Twice fundamental frequency is called the second harmonic, etc.

Hertz   Cycles per second.

High level   A relative term indicating the level of the final stages of an amplifier.

High-pass filter   A filter which passes all frequencies above a critical value.

Honeycomb winding   A coil winding in which spaces are left between turns so as to give the appearance of a honeycomb.

Hunting (of a rotary machine)   An oscillation of angular velocity about a state of uniform rotation.

I

Integer   A whole number, having no fraction or decimal portion

Intensity   The strength of a quantity such as current, voltage or pressure.

Intermediate frequency   A frequency to which the incoming signal is changed in superheterodyne reception.

Intermodulation   The modulation of the components of a complex wave by each other, in a non-linear system.

L

Layer winding A coil winding in which one layer is wound over another.

Lead A connecting wire.

Linear Having an input versus output characteristic which, when plotted on ordinary graph paper, is a straight line. A distortionless amplifier is linear, because its amplification is constant for all values of input voltage.

Linear reflex detector (also called infinite input impedance detector) An anode-bend detector having its negative bias supplied from a high resistance cathode bias resistor which is by-passed for radio frequencies only.

Line voltage The voltage of the power lines. Mains voltage.

Load Any device which absorbs electrical power.

Locus The path followed by a moving point.

Loudness A subjective evaluation which is primarily a function of intensity but is strongly influenced by frequency.

Low level A relative term indicating the level of the early stages of an amplifier (e.g. pre-amplifier).

Low-pass filter A filter which passes all frequencies below a critical value.

M

Manual Adjusted by hand.

Maximum signal Usually applied to the conditions under which the amplifier just reaches the point of " maximum undistorted power output."

Maximum undistorted power output The maximum power output for a specified distortion.

Microphonic A valve or other component which amplifies sound waves or vibration sufficiently to produce a loud sound or continuous howl in the loudspeaker.

Mirror image One curve is a mirror image of another when the first one, seen through a mirror, is identical with the second.

Mixer The stage in a superheterodyne receiver in which the incoming signal is mixed with the voltage from a local oscillator to produce the intermediate frequency signal. Also a control which combines the output from a number of microphones in any desired proportion to the input of the main amplifier.

Modulation The process by which the amplitude, frequency or phase of a carrier wave is modified in accordance with the characteristics of a signal.

Monaural Hearing with one ear only.

Motorboating Regeneration at low audio frequencies causing sounds like those of a motor boat.

N

Network An electrical circuit. See Chapter 4.

Neutralization The process of balancing out an undesirable effect, such as regeneration.

Node A point on a stationary wave system which has zero amplitude.

Non-linear The opposite of linear. Distorted.

Numerator The part of a fraction above the fraction bar.

O

Open circuit A circuit which is not complete and can therefore carry no current.

Order " Of the order of " indicates " that the value is in the general vicinity of." The expression is looser than " approximately equal to."

Output stage The final stage in an amplifier which supplies audio frequency power to a loudspeaker. Also known as power amplifier.

P

Parallel Conductors or components are in parallel when the current flowing in the circuit is divided between them.

Parameter Where there are three variables, any one of them (called the parameter) may be given a series of fixed values, and two dimensional curves may then be drawn to show the relationships between the remaining two variables.

Penultimate Last but one.

Period The time for one complete cycle of values ; the reciprocal of the frequency.

Phonograph Also known as a gramophone.

Pitch A subjective observation of a musical sound, principally dependent on frequency but also affected by intensity.

Polarity (1) electrical—a term applied to electrical apparatus when it is desired to indicate which terminal is positive and which is negative ;
(2) magnetic—that quality of a body by virtue of which certain characteristic properties are manifested over certain regions of its surface, which are known as poles, and whose polarity is indicated by the term north or south pole.

Potential The potential at a point is the potential difference between that point and earth.

Potential difference The potential difference between two points is equal to the work done in transferring unit quantity of positive electricity from one point to the other. The practical unit is the volt.

Power pack The power supply of a radio receiver or amplifier. It converts the available voltage to the values required by the plate, screen, and grid circuits.

Q

Q factor The ratio of reactance to resistance of a coil, condenser or resonant circuit.

Quiescent Stationary. The quiescent operating point is that with instantaneous zero signal voltage.

R

Radial lead A lead at 90° to the axis.

Radio frequency Any frequency at which electromagnetic radiation is used for telecommunication.

Reciprocal of a number is 1 divided by the number.

Rectification The process of converting an alternating current to unidirectional current by means of an asymmetrical conducting device.

Ripple frequency The frequency of the a.c. component of a current when this component is small relative to the d.c. component.

Roll-off A frequency response characteristic is said to have a roll-off when the attenuation at high or low frequencies has the same form as that of a single section low- or high-pass RC filter. The term is also used in a loose sense to cover any characteristic with gradually increasing attenuation at high or low frequencies, in distinction to a sharp cut-off.

S

Scalar Has magnitude and sign but no direction.

Series Conductors or components are in series when traversed by the same current.

Short circuit A low resistance connection, usually accidental, across part of a circuit, resulting in excessive current flow.

Shunt Same as parallel.

Sideband A band of frequencies within which lie the frequencies of the waves produced by modulation.

Side frequency A frequency produced by modulation.

Signal voltage The audio frequency voltage in an a-f amplifier ; the modulating voltage in an r-f or i-f amplifier, also applied to the whole modulated carrier voltage.

Sinusoidal Having the form of a sine (or cosine) wave.

" Small compared with " A vague term indicating that it may be neglected in a rough approximation, or that its square may be neglected with only a small error—say " less than one tenth of."

Soft valve One which has reverse grid current in excess of the specified maximum for its type.

Space current The same as cathode current.

Steradian The solid angle subtended at the centre of a sphere by an area on its surface numerically equal to the square of the radius. The maximum possible value is 2π steradians.

Strain Is the deformation produced by an applied stress, for example change in length per unit length produced by a tensile stress.

Stress is the deforming force per unit area. Examples are tensile stress, compressive stress and bending stress.

Subharmonic A component of a complex wave having a frequency which is an integral sub-multiple of the basic frequency. Example : basic frequency 300 c/s ; subharmonics 150, 100, 75, 60, 50 c/s.

Subscript Printed below the line. For example the letter m in g_m is a subscript.

Superscript Printed above the line. For example the figure 2 in x^2 is a superscript.

N.B. Q' is called " Q dash."

Supersonic Air waves having frequencies above those which may be heard by the human ear. The preferred term is ultrasonic.

Symmetrical Having the same characteristics for conduction in both directions.

T

Tap A connection at some point other than the ends.

Term A portion of an expression which is separated from the other portions by + or — signs. In a sequence, series or progression the terms may be separated by commas.

Terminal A point to which electrical connections may be made.

Tertiary Third.

Tolerance The maximum permissible variation from an assigned value.

Transducer A transducer is a device by means of which energy may flow from one or more transmission systems to one or more other transmission systems. The energy transmitted by these systems may be of any form (for example it may be electrical, mechanical or acoustical), and it may be of the same form or different forms in the various input and output systems (e.g. electro-acoustical).

An **electrical transducer** is an electrical network by means of which energy may flow from one or more transmission systems to one or more other transmission systems ; in most cases it has two input and two output terminals.

An **electron-tube transducer** is an electrical transducer containing one or more electron tubes (valves). Examples are an amplifier, and a frequency converter.

Treble High audio frequencies.

U

Ultrasonic Air waves having frequencies above those which may be heard by the human ear.

Unidirectional current Current which may change in value but never reverses its direction.

Unity The figure 1.

V

Variable A quantity whose value changes.

Variational The variational slope of a curve at a point is the differential slope at that point.

Vector A quantity having both magnitude and direction.

" Very small compared with " A vague term indicating that it may be neglected with only a small error—say " less than one twenty-fifth of."

Voice coil Moving coil of a dynamic loudspeaker.

W

Work function The thermionic work function of a conductor is the definite amount of work which must be applied to release one electron. An electron must possess kinetic energy at least as great as the work function of the conductor to pass out through the surface.

Z

Zero signal Having no signal input voltage.

See also

" American Standard Definitions of Electrical Terms " A.S.A. C42-1941 (American

Institute of Electrical Engineers, New York).

British Standard 204 : 1943 " Glossary of Terms used in Telecommunication " (British Standards Institution).

Also Supplement No. 3 (1949) to BS204 : 1943 " Fundamental Radio Terms."

British Standard 205 : 1943 " Glossary of Terms used in Electrical Engineering " (British Standards Institution).

American I.R.E. Standards

" Standards on electron tubes : Definitions of Terms, 1950," Proc. I.R.E. 38.4 (April 1950) 426.

38 IRE 6 S1 Standards on Electroacoustics 1938.

38 IRE 17 S1 Standards on Radio Receivers 1938.

50 IRE 24 S1 Standards on Wave Propagation, 1950.

(Proc. I.R.E. 38.11 November 1950 p. 1264).

52 IRE 17 S1 Standards on Receivers : Definitions of Terms, 1952.

[Proc. I.R.E. 40.12 (Dec. 1952) 1681].

SECTION 16 : DECIMAL' EQUIVALENTS OF FRACTIONS

TABLE 58 : DECIMAL EQUIVALENTS OF FRACTIONS

Fractions of an inch		Decimal	Fractions of an inch		Decimal
	1/64	0.0156	1/2		0.5000
1/32		0.0313		33/64	0.5156
	3/64	0.0469	17/32		0.5313
1/16		0.0625		35/64	0.5469
	5/64	0.0781	9/16		0.5265
3/32		0.0938		37/64	0.5781
	7/64	0.1094	19/32		0.5938
1/8		0.1250		39/64	0.6094
	9/64	0.1406	5/8		0.6250
5/32		0.1563		41/64	0.6406
	11/64	0.1719	21/32		0.6563
3/16		0.1875		43/64	0.6719
	13/64	0.2031	11/16		0.6875
7/32		0.2188		45/64	0.7031
	15/64	0.2344	23/32		0.7188
1/4		0.2500		47/64	0.7344
	17/64	0.2656	3/4		0.7500
9/32		0.2813		49/64	0.7656
	19/64	0.2969	25/32		0.7813
5/16		0.3125		51/64	0.7969
	21/64	0.3281	13/16		0.8125
11/32		0.3438		53/64	0.8281
	23/64	0.3594	27/32		0.8438
3/8		0.3750		55/64	0.8594
	25/64	0.3906	7/8		0.8750
13/32		0.4063		57/64	0.8906
	27/64	0.4219	29/32		0.9063
7/16		0.4375		59/64	0.9219
	29/64	0.4531	15/16		0.9375
15/32		0.4688		61/64	0.9531
	31/64	0.4844	31/32		0.9688
				63/64	0.9844

SECTION 17 : MULTIPLES AND SUB-MULTIPLES

TABLE 59 : MULTIPLES AND SUB-MULTIPLES

Multiply reading in	by	to obtain reading in
Amperes	1 000 000 000 000	micromicroamperes
Amperes	1 000 000	microamperes
Amperes	1000	milliamperes
Farads	1 000 000 000 000	micromicrofarads
Farads	1 000 000	microfarads
Farads	1000	millifarads
Henrys	1 000 000	microhenrys
Henrys	1000	millihenrys
Volts	1 000 000	microvolts
Volts	1000	millivolts
Mhos	1 000 000	micromhos
Mhos	1000	millimhos
Watts	1000	milliwatts
Cycles per second	0.000 001	megacycles per second
Cycles per second	0.001	kilocycles per second
Microamperes	0.000 001	amperes
Milliamperes	0.001	amperes
Micromicrofarads } Picofarads }	0.000 000 000 001	farads
Microfarads	0.000 001	farads
Millifarads	0.001	farads
Microhenrys	0.000 001	henrys
Microvolts	0.000 001	volts
Millivolts	0.001	volts
Micromhos	0.000 001	mhos
Millimhos	0.001	mhos
Milliwatts	0.001	watts
Kilowatts	1000	watts
Megacycles per second	1 000 000	cycles per second
Kilocycles per second	1000	cycles per second
Megohms	1 000 000	ohms

SECTION 18 : NUMERICAL VALUES AND FACTORIALS

(i) *Numerical values* (ii) *Factorials.*

(i) Numerical values

TABLE 60

	Numeric		Reciprocal	
	Approx.	More accurately	Approx.	More accurately
π	3.1416	3.141 593	0.318 3	0.318 309 9
2π	6.2832	6.283 185	0.159 21	0.159 156
3π	9.4248	9.424 778	0.106 1	0.106 103
4π	12.566	12.566 371	0.0796	0.079 577 4
5π	15.708	15.707 963	0 0637	0.063 662
$\pi/2$	1.5708	1.570 796	0.6366	0.636 620
$\pi/3$	1.0472	1.047 198	0.9549	0.954 930
$\pi/4$	0.7854	0.785 398	1.273	1.273 23
$\pi/6$	0.5236	0.523 598	1.910	1.909 86
π^2	9.8696	9.869 604	0.1013	0.101 321
$(2\pi)^2$	39.4784	39.478 414	0.0253	0.025 330 3
π^3	31.0062	31.006 277	0.0322	0.032 252
$\sqrt{\pi}$	1.7725	1.772 454	0.5642	0.564 190
$\sqrt{\pi/2}$	1.2533	1.253 31	0.7979	0.797 887
$\sqrt[3]{\pi}$	1.4646	1.464 592	0.6818	0.681 784
$\log_{10}\pi$	0.4971	0.497 150	2.011	2.011 46
$\log_{10}\pi/2$	0.1961	0.196 120	5.099	5.098 92
$\log_{10}\pi^2$	0.9943	0.994 300	1.006	1.005 73
$\log_{10}\sqrt{\pi}$	0.2486	0.248 575	4.023	4.022 93
ϵ	2.718	2.718 282	0.3679	0.367 879
ϵ^2	7.389	7.389 057	0.1353	0.135 335
$\sqrt{\epsilon}$	1.649	1.648 721	0.6065	0.606 531
$\log_{10}\epsilon$	0.4343	0.434 294	2.3026	2.302 585
$\sqrt{2}$	1.414	1.4142	0.7071	0.707 113 5
$\sqrt{3}$	1.732	1.7321	0.5773	0.577 334 2
$\sqrt{5}$	2.236	2.2361	0.4472	0.447 206
$\sqrt[3]{2}$	1.260	1.2599	0.7939	0.793 873
$\sqrt[3]{3}$	1.442	1.4422	0.6933	0.693 385
$\sqrt[3]{4}$	1.587	1.5874	0.6300	0.629 960
$\sqrt[3]{5}$	1.710	1.7100	0.5848	0.584 79

(ii) Factorials

TABLE 61

n	$n! = 1.2.3 \ldots n$	$1/n!$
1	1	1
2	2	0.5
3	6	0.166 667
4	24	$0.416\,667 \times 10^{-1}$
5	120	$0.833\,333 \times 10^{-2}$
6	720	$0.138\,889 \times 10^{-2}$
7	5040	$0.198\,413 \times 10^{-3}$
8	40 320	$0.248\,016 \times 10^{-4}$
9	362 880	$0.275\,573 \times 10^{-5}$
10	3 628 800	$0.275\,573 \times 10^{-6}$

SECTION 19 : WIRE TABLES *

TABLE 62

Bare Copper Wire, A.W.G. (20° C. = 68° F.)

AWG No.	Dia-meter Mils	Area Circular Mils	Area Square Inches	Ohms per 1000 Feet	Ohms per Pound	Feet per Pound	Pounds per 1000 Feet
0000	460	211 600	.166 2	.04901	.000 076 52	1.561	640.5
000	410	167 800	.131 8	.06180	.000 121 7	1.968	507.9
00	364.8	133 100	.104 5	.07793	.000 193 5	2.482	402.8
0	324.9	105 500	.082 89	.09827	.000 307 6	3.130	319.5
1	289.3	83 700	.065 73	.1239	.000 489 1	3.947	253.3
2	257.6	66 400	.052 13	.1563	.000 777 8	4.977	200.9
3	229.4	52 600	.041 34	.1970	.001 237	6.276	159.3
4	204.3	41 700	.032 78	.2485	.001 966	7.914	126.4
5	181.9	33 100	.026 00	.3133	.003 127	9.980	100.2
6	162.0	26 250	.020 62	.3951	.004 972	12.58	79.46
7	144.3	20 820	.016 35	.4982	.007 905	15.87	63.02
8	128.5	16 510	.012 97	.6282	.012 57	20.01	49.98
9	114.4	13 090	.010 28	.7921	.019 99	25.23	39.63
10	101.9	10 380	.008 155	.9989	.031 78	31.82	31.43
11	90.7	8 230	.006 467	1.260	.050 53	40.12	24.92
12	80.8	6 530	.005 129	1.588	.080 35	50.59	19.77
13	72.0	5 180	.004 067	2.003	.1278	63.80	15.68
14	64.1	4 110	.003 225	2.525	.2032	80.44	12.43
15	57.1	3 257	.002 558	3.184	.3230	101.4	9.858
16	50.8	2 583	.002 028	4.016	.5136	127.9	7.818
17	45.3	2 048	.001 609	5.064	.8167	161.3	6.200
18	40.3	1 624	.001 276	6.385	1.299	203.4	4.917
19	35.89	1 288	.001 012	8.051	2.065	256.5	3.899
20	31.96	1 022	.000 802 3	10.15	3.283	323.4	3.092
21	28.46	810	.000 636 3	12.80	5.221	407.8	2.452
22	25.35	642	.000 504 6	16.14	8.301	514.2	1.945
23	22.57	509	.000 400 2	20.36	13.20	648.4	1.542
24	20.10	404	.000 317 3	25.67	20.99	817.7	1.223
25	17.90	320.4	.000 251 7	32.37	33.37	1 031.0	0.9699
26	15.94	254.1	.000 199 6	40.81	53.06	1 300	0.7692
27	14.20	201.5	.000 158 3	51.47	84.37	1 639	0.6100
28	12.64	159.8	.000 125 5	64.90	134.2	2 067	0.4837
29	11.26	126.7	.000 099 53	81.83	213.3	2 607	0.3836
30	10.03	100.5	.000 078 94	103.2	339.2	3 287	0.3042
31	8.928	79.70	.000 062 60	130.1	539.3	4 145	0.2413
32	7.950	63.21	.000 049 64	164.1	857.6	5 227	0.1913
33	7.080	50.13	.000 039 37	206.9	1 364.0	6 591	0.1517
34	6.305	39.75	.000 031 22	260.9	2 168	8 310	0.1203
35	5.615	31.52	.000 024 76	329.0	3 448	10 480	0.095 42
36	5.000	25.00	.000 019 64	414.8	5 482	13 210	.075 68
37	4.453	19.83	.000 015 57	532.1	8 717	16 660	.060 01
38	3.965	15.72	.000 012 35	659.6	13 860	21 010	.047 59
39	3.531	12.47	.000 009 793	831.8	22 040	26 500	.037 74
40	3.145	9.888	.000 007 766	1 049.0	35 040	33 410	.029 93
(41)	2.75	7.5625	.000 005 940	1 370	59 990	43 700	.022 89
(42)	2.50	6.2500	.000 004 909	1 660	87 700	52 800	.018 92
(43)	2.25	5.0625	.000 003 976	2 050	133 700	65 300	.015 32
(44)	2.00	4.0000	.000 003 142	2 600	214 000	82 600	.012 11
(45)	1.75	3.0625	.000 002 405	3 390	356 200	107 900	.009 27
(46)	1.50	2.2500	.000 001 767	4 610	676 800	146 800	.006 81

*See also A. F. Maine " Rapid coil calculations for magnetic devices " Jour. Brit. I.R.E. 12.7 (July 1952) 403 for nomographs.

Table 63
Bare Copper Wire S.W.G. (60° F)

SWG No.	Diameter Mils.	Area Circular Mils.	Area Square Inches	Ohms per 1000 Feet	Ohms per Pound	Feet per Pound	Pounds per 1000 Feet
4/0	400	160 000	.125 66	.06368	.000 131 46	2.064	484.4
3/0	372	138 400	.108 69	.0736	.000 175 74	2.390	418.9
2/0	348	121 100	.095 11	.0841	.000 229 5	2.730	366.7
1/0	324	105 000	.082 45	.0971	.000 305 4	3.147	317.8
1	300	90 000	.070 69	.1132	.000 415 5	3.670	272.5
2	276	76 180	.059 83	.1338	.000 580 0	4.338	230.6
3	252	63 500	.049 88	.1605	.000 834 5	5.200	192.3
4	232	53 820	.047 27	.1893	.001 161 7	6.139	162.9
5	212	44 940	.035 30	.2267	.001 666 1	7.348	136.1
6	192	36 860	.028 95	.2764	.002 476	8.961	111.6
7	176	30 980	.024 33	.3289	.003 507	10.66	93.8
8	160	25 600	.020 11	.3980	.005 135	12.90	77.5
9	144	20 740	.016 286	.4914	.007 827	15.93	62.78
10	128	16 380	.012 868	.6219	.012 537	20.16	49.61
11	116	13 460	.010 568	.7570	.018 587	24.55	40.74
12	104	10 820	.008 495	.942	.028 77	30.54	32 75
13	92	8 464	.006 648	1.204	.046 98	39.01	25.63
14	80	6 400	.005 027	1.592	.082 16	51.60	19.38
15	72	5 184	.004 072	1.966	.125 23	63.73	15.69
16	64	4 096	.003 217	2.488	.2006	80.65	12.40
17	56	3 136	.002 463	3.249	.3422	105.4	9.49
18	48	2 304	.001 809 6	4.422	.6340	143.3	6.98
19	40	1 600	.001 256 6	6.368	1.3146	206.4	4.844
20	36	1 296	.001 017 9	7.860	2.004	254.8	3.924
21	32	1 024	.000 804 2	9.950	3.209	322.6	3.100
22	28	784	.000 615 8	12.997	5.475	421.2	2.374
23	24	576	.000 452 4	17.69	10.144	573.4	1.744
24	22	484	.000 380 1	21.05	14.366	682.6	1.465
25	20	400	.000 314 2	25.47	21.03	825.8	1 211
26	18	324	.000 254 5	31.45	32.06	1 019	0.981
27	16.4	269	.000 211 2	37.88	46.52	1 229	0.814
28	14.8	219	.000 172 03	46.52	70.14	1 508	0.6632
29	13.6	185	.000 145 27	55.09	98.37	1 786	0.5600
30	12.4	153.8	.000 120 76	66.27	142.35	2 148	0.4655
31	11.6	134.6	.000 105 68	75.7	185.87	2 455	0.4074
32	10.8	116.6	.000 091 61	87.4	247.4	2 832	0.3531
33	10.0	100.0	.000 078 54	101.9	336.5	3 302	0.3028
34	9.2	84.64	.000 066 48	120.4	469.8	3 901	0.2563
35	8.4	70.56	.000 055 42	144.4	676.0	4 682	0.2136
36	7.6	57.76	.000 045 36	176.4	1 008.7	5 718	0.1749
37	6.8	46.24	.000 036 32	220.4	1 574	7 143	0.1400
38	6.0	36.00	.000 028 27	283.0	2 596	9 174	0.1090
39	5.2	27.04	.000 021 24	376.8	4 603	12 210	0.0819
40	4.8	23.04	.000 018 096	442.2	6 340	14 330	0.0698
41	4.4	19.36	.000 015 205	526.3	8 979	17 060	0.058 62
42	4.0	16.00	.000 012 566	636.8	13 146	20 640	.048 44
43	3.6	12.96	.000 010 179	786.3	20 040	25 480	.039 24
44	3.2	9.734	.000 008 042	995.0	32 090	32 260	.031 00
45	2.8	7.840	.000 006 158	1 299.7	54 750	42 120	.023 74
46	2.4	5.760	.000 004 524	1 769	101 440	57 340	.017 44
47	2.0	4.000	.000 003 142	2 547	210 300	82 580	.012 11
48	1.6	2.560	.000 002 011	3 980	513 500	129 000	.007 75
49	1.2	1.440	.000 001 131	7 077	1 623 000	229 400	.004 36
50	1.0	1.000	.000 000 785 4	10 190	3 365 000	303 000	.003 03

TABLE 64

TURNS PER INCH AND INSULATED WIRE DIAMETER A.W.G.
COPPER WIRE

AWG No.	Diameter (mils)		Turns per inch (exact winding)					
	*Enam.	D.C.C.	Bare	Enam.	S.C.C.	D.C.C.	S.S.C.	D.S.C.
8	130.6	142.5	7.78	7.65	7.32	7.01	—	—
9	116.5	126.4	8.74	8.58	8.23	7.91	—	—
10	104.0	112.9	9.81	9.61	9.26	8.85	—	—
11	92.7	100.2	11.02	10.7	10.4	9.98	—	—
12	82.8	90.3	12.37	12.0	11.6	11.07	—	—
13	74.0	81.5	13.89	13.5	12.9	12 27	—	—
14	66.1	73.6	15.60	15.1	14.4	13.59	—	—
15	59.1	66.6	17.52	16.9	16.1	15.0	—	—
16	52.8	60.3	19.68	18.9	17.9	16.5	18.9	18.2
17	47.1	54.8	22.1	21 2	19.8	18.2	21.1	20.2
18	42.1	49.8	24.8	23.7	22.0	20 0	23.6	22.5
19	37.7	45.4	27.8	26.5	24.4	22.0	26.3	25.0
20	33.8	41.5	31.3	29.5	27.0	24.1	29.4	27.7
21	30.2	38.0	35.1	33.1	29.8	26.3	32.7	30.7
22	27.0	33.8	39.4	37.0	33.5	29.5	36.6	34.1
23	24.1	31.1	44.3	41.4	36.9	32.1	40.6	37.5
24	21.5	28.6	49.7	46.5	40.6	34.9	45.2	41.4
25	19 2	26.4	55.8	52.0	44.6	37.8	50.0	45.6
26	17.1	24.4	62.7	58.4	49.0	40.9	55.8	50.0
27	15.3	22.7	70.4	65.3	53.4	44.0	61.7	54.9
28	13.6	21.1	82.8	73.5	58.4	47.3	68.4	60.2
29	12.2	19.8	88.8	81.9	63.2	50.5	75.1	65.3
30	10.8	18.5	99.7	92.5	68.9	54.0	83.3	71.4
31	9.7	17.4	112.0	103	74.6	57.4	91.7	77.5
32	8.7	16.5	125.8	114	80.0	60.6	100	83.3
33	7.7	15.6	141.2	129	86.2	64.1	109	90.0
34	6.9	14.8	158.6	144	92.5	67.5	120	97.0
35	6.2	14.1	178	161	99.9	70.9	131	104
36	5.5	13.0	200	181	111	76.9	142	111
37	4.9	12.5	224	204	117	80.0	153	117
38	4.4	12.0	252	227	125	83.3	166	125
39	3.9	11.5	283	256	133	86.9	181	133
40	3.5	11.1	318	285	140	90.0	196	140
(41)	3.05	—	363	327	—	—	—	—
(42)	2.64	—	400	378	—	—	—	—
(43)	2.37	—	444	421	—	—	—	—
(44)	2.12	—	500	471	—	—	—	—
(45)	1.91	—	571	523	—	—	—	—
(46)	1.72	—	666	581	—	—	—	—

*Nominal Value.   Actual dimensions vary slightly.

TABLE 65
TURNS PER INCH AND INSULATED WIRE DIAMETER, S.W.G.
COPPER WIRE

SWG No.	Diameter (mils)		Turns per inch (exact winding)					
	*Enam.	D.C.C.	Bare	Enam.	S.C.C.	D.C.C.	S.S.C.	D.S.C.
10	132	142	7.81	7.63	7.35	7.04	—	—
11	120	130	8.62	8.33	8.07	7.69	—	—
12	108	118	9.62	9.26	8.93	8.48	—	—
13	96	106	10.87	10.42	10.00	9.43	—	—
14	84	94	12.50	11.90	11.36	10.64	—	—
15	75.5	84	13.89	13.25	12.66	11.90	—	—
16	67.5	76	15.63	14.81	14.08	13.16	14.93	14.71
17	59	68	17.86	16.95	15.87	14.71	16.95	16.67
18	50.7	59	20.83	19.72	18.18	16.95	20.00	19.61
19	42.6	51	25.00	23.47	21.28	19.61	23.81	23.26
20	38.5	47	27.78	25.97	23.81	21.28	26.32	25.64
21	34.3	43	31.25	29.15	26.32	23.26	29.41	28.57
22	30.0	39	35.71	33.33	29.41	25.64	33.33	32.26
23	25.7	34	41.67	38.91	34.48	29.41	38.46	37.04
24	23.6	32	45.45	42.37	37.04	31.25	42.55	40.00
25	21.5	30	50.00	46.51	40.00	33.33	46.51	43.48
26	19.4	28	55.56	51.55	43.48	35.71	51.81	48.78
27	17.7	26.4	60.98	56.50	46.73	37.88	56.50	52.91
28	16.0	24.8	67.57	62.50	50.51	40.32	62.11	57.80
29	14.8	23.6	73.53	67.57	53.76	42.37	67.11	62.11
30	13.4	22.4	80.65	74.63	57.47	44.64	72.99	67.11
31	12.6	21.6	86.21	79.37	60.24	46.30	77.52	70.92
32	11.7	20.8	92.59	85.47	63.29	48.08	82.64	75.19
33	10.9	20.0	100.00	91.74	66.67	50.00	88.50	80.00
34	10.0	19.2	108.7	100.0	70.42	52.08	95.24	85.47
35	9.1	17.4	119.0	109.9	80.65	57.47	103.1	91.74
36	8.3	16.6	131.6	120.5	86.21	60.24	112.4	99.01
37	7.4	15.8	147.1	135.1	99.21	63.29	123.5	107.5
38	6.6	15.0	166.7	151.5	100.0	66.67	137.0	117.6
39	5.7	14.2	192.3	175.4	108.7	70.42	153.8	129.9
40	5.3	13.8	208.3	188.7	113.6	72.46	163.9	137.0
41	4.8	—	227.3	208.3	—	—	178.6	151.5
42	4.4	—	250.0	227.3	—	—	192.3	161.3
43	3.9	—	277.8	256.4	—	—	208.3	172.4
44	3.5	—	312.5	285.7	—	—	227.3	185.2
45	3.1	—	357.1	322.6	—	—	250.0	200.0
46	2.65	—	416.7	377.4	—	—	277.8	217.4
47	2.25	—	500.0	444.4	—	—	312.5	238.1
48	—	—	—	—	—	—	—	—

*Nominal Value. Actual dimensions vary slightly.

TABLE 66

MULTI-LAYER COIL WINDING AND WEIGHT OF INSULATED WIRE, A.W.G.

COPPER WIRE

AWG No.	Enamelled			D.C.C.		Weight—lbs. per 1000ft.		
	Turns per square inch*	Ohms per cubic inch*	Turns per square inch layer insulated	Turns per square inch*	Ohms per cubic inch*	Enam.	D.C.C.	D.S.C.
8	57	.003 15		48	.002 65	50.55	51.15	
9	72	.004 75		59	.003 88	40.15	40.60	
10	90	.007 48		76	.006 31	31.80	32.18	
11	113	.011 83		93	.009 74	25.25	25.60	
12	141	.018 78		114	.015 19	20.05	20.40	
13	177	.029 5		140	.023 3	15.90	16.20	
14	221	.046 4		171	.035 9	12.60	12.91	
15	277	.073 4		208	.055 1	10.00	10.33	
16	348	.116 2		260	.086 9	7.930	8.210	7.955
17	437	.184 0		316	.133 1	6.275	6.540	6.315
18	548	.291 0		378	.200 8	4.980	5.235	5.015
19	681	.456 0		455	.304 8	3.955	4.220	3.990
20	852	.720 0		545	.460 5	3.135	3.373	3.173
21	1 065	1.134		650	.6920	2.490	2.685	2.520
22	1 340	1.800		865	1.162	1.970	2.168	2.006
23	1 665	2.820		1 030	1.774	1.565	1.727	1.593
24	2 100	4.488	1 420	1 215	2.596	1.245	1.398	1.272
25	2 630	7.080	1 750	1 420	3.822	.988	1.129	1.018
26	3 320	11.27	2 030	1 690	5.740	.7845	.9140	.8100
27	4 145	17.75	2 620	1 945	8.330	.6220	.7560	.6450
28	5 250	28.34	3 250	2 250	12.15	.4940	.6075	.5140
29	6 510	44.32	3 920	2 560	17.30	.3915	.4890	.4130
30	8 175	70.15	4 780	2 930	25.15	.3105	.3955	.3330
31	10 200	110.4	6 780	3 330	36.05	.2465	.3257	.2678
32	12 650	127.6	8 250	3 720	50.76	.1960	.2700	.2170
33	16 200	279.0	10 600	4 140	71.30	.1550	.2270	.1750
34	19 950	433.2	12 400	4 595	99.77	.1230	.1928	.1412
35	25 000	684.5	15 200	5 070	138.7	.0980	.1600	.1130
36	31 700	1 094	21 500	5 550	191.6	.0776	.1361	.0920
37	39 600	1 723	26 300	6 045	263	.0616	.1204	.0740
38	49 100	2 693	32 000	6 510	357	.0488	.1049	.0623
39	62 600	4 332	40 000	6 935	480	.0387	.0937	.0504
40	77 600	6 770	48 400	7 450	650	.0307	.0838	.0429

Note in "Turns per square inch layer insulated" column: "Paper insulated each layer. 20% allowance for waste space at ends of layers." With "4 mil. paper" (rows 24–30), "2 mil. paper" (rows 31–35), "1 mil. paper" (rows 36–40).

*For exact winding with no allowance for space factor.

TABLE 67

MULTI-LAYER COIL WINDING AND WEIGHT OF INSULATED WIRE, S.W.G.
COPPER WIRE

SWG No.	Enamelled		Turns per square inch layer insulated	D.C.C.		Weight—lbs. per 1000ft.		
	Turns per square inch*	Ohms per cubic inch*		Turns per square inch*	Ohms per cubic inch*	Enam.	D.C.C.	D.S.C.
10	58.22	.002 95		49.56	.00256	47.77	50.77	
11	69.39	.004 37		59.14	.003 73	44.40	41.83	
12	85.75	.006 73		71.91	.005 65	33.04	33.71	
13	108.6	.0109		88.92	.008 93	25.89	26.50	
14	141.6	.0189		113.2	.0150	19.60	20.12	
15	175.6	.0287		141.6	.0234	15.87	16.36	
16	219.3	.0456		173.2	.0358	12.56	12.67	12.56
17	287.3	.0776		216.4	.0585	9.607	10.03	9.640
18	388.9	.1431		287.3	.1055	7.060	7.43	7.093
19	550.8	.290		384.6	.204	4.910	5.262	4.945
20	674.4	.440	Paper insulated each layer. 20% allowance for waste space at ends of layers.	452.8	.296	3.987	4.267	4.011
21	849.7	.705		541.0	.448	3.152	3.409	3.181
22	1 109	1.200		657.4	.710	2.419	2.649	2.446
23	1 513	2.23		864.9	1.26	1.779	1.918	1.807
24	1 789	3.14	1 220	976.6	1.71	1.498	1.667	1.305
25	2 162	4.58	1 460	1 109	2.35	1.241	1.392	1.247
26	2 663	6.95	1 760	1 274	3.33	1.008	1.152	1.013
27	3 192	10.05	2 080	1 436	4.51	.836	.977	.844
28	3 906	15.18	2 500	1 624	6.28	.683	.811	.689
29	4 570	20.9	2 870	1 798	8.23	.578	.699	.582
30	5 565	30.7	3 430	1 989	10.98	.478	.604	.486
			4 mil. paper					
31	6 304	39.8	3 840	2 144	13.45	.419	.524	.428
32	7 310	53.1	4 360	2 314	16.83	.364	.467	.371
33	8 409	71.2	5 680	2 500	21.15	.313	.412	.329
34	10 000	100.5	6 670	2 714	27.3	.264	.328	.271
35	12 080	145	7 920	3 306	39.3	.221	.317	.288
			2 mil. paper					
36	14 520	212	9 360	3 624	53.1	.181	.238	.188
37	18 220	336	11 400	4 007	73.5	.146	.207	.151
38	22 950	544	14 100	4 436	104.5	.114	.169	.119
39	30 770	965	20 900	4 956	155.0	.0854	.138	.0868
40	35 610	1 310	24 000	5 256	193.5	.0726	.123	.0791
			1 mil. paper					
41	43 390	1 905	28 800			.0608		.0677
42	51 620	2 740	33 700			.0505		.0571
43	65 740	4 300	41 800			.0408		.0473
44	81 620	6 760	50 800			.0324		.0386
45	104 300	11 250				.0249		.0309
46	142 100	21 000				.0182		.0231
47	197 100	41 700				.0126		.0172
48								

*For exact winding with no allowance for space factor.

TABLE 68

RESISTANCE WIRE TABLE, A.W.G.

20° C. (68° F.)

AWG No.	Dia. mils.	Advance Wire				Nichrome Wire**		
		Ohms per 1,000 feet	Lbs. per 1,000 feet	Feet per Ohm	Current Milli Amps*	Ohms per 1,000 feet	Lbs. per 1,000 feet	Current Milli Amps†
8	128	17.9	50	55.9	—	40.8	45	—
9	114	22.6	39	44.2	—	51.9	36	—
10	102	28.0	32	35.7	—	64.9	29	—
11	91	35.5	25	28.2	—	81.5	23	—
12	81	44.8	20	22.3	—	102	18	—
13	72	56.7	15.7	17.6	—	130	14	—
14	64	71.7	12.4	13.9	—	164	11	—
15	57	90.4	9.8	11.1	—	207	9.2	—
16	51	113	7.8	8.85	—	259	7.2	—
17	45	145	6.2	6.90	—	333	5.6	—
18	40	184	4.9	5.44	800	421	4.42	—
19	36	226	3.9	4.43	650	520	3.58	—
20	32	287	3.1	3.48	522	659	2.83	—
21	28.5	362	2.5	2.76	420	831	2.24	—
22	25.3	460	1.9	2.17	335	1 055	1.77	—
23	22.6	575	1.5	1.74	273	1 321	1.41	—
24	20.1	728	1.2	1.37	220	1 670	1.12	460
25	17.9	919	.97	1.09	178	2 106	.89	390
26	15.9	1 162	.77	.861	144	2 669	.70	330
27	14.2	1 455	.61	.687	117	3 347	.56	278
28	12.6	1 850	.48	.541	95	4 251	.44	228
29	11.3	2 300	.38	.435	78	5 286	.35	196
30	10.0	2 940	.30	.340	63	6 750	.276	165
31	8.9	3 680	.24	.272	52	8 521	.199	158
32	8.0	4 600	.19	.217	43	10 546	.177	117
33	7.1	5 830	.15	.172	36	13 390	.139	97
34	6.3	7 400	.12	.135	29	17 006	.110	82
35	5.6	9 360	.095	.107	24	21 524	.087	69
36	5.0	11 760	.076	.085	20	27 000	.069	58
37	4.5	14 550	.060	.0687	17	33 333	.056	49
38	4.0	18 375	.047	.0544	14.5	42 187	.045	41
39	3.5	24 100	.038	.0415	12	55 102	.034	34
40	3.1	30 593	.028	.0327	10	70 239	.025	28
(41)	2.75	38 888	.0229	.0257	8.5	89 256	.0209	24
(42)	2.5	46 400	.0189	.0215	7.5	108 000	.0173	20.5
(43)	2.25	58 103	.0153	.0172	6.8	133 333	.0140	17.5
(44)	2.0	73 500	.0121	.0136	6.0	168 750	.0110	14.5
(45)	1.75	96 078	.0092	.0104	5.0	220 408	.0084	12.0
(46)	1.5	130 666	.0068	.0076	4.0	300 000	.0062	9.5

**Wire produced by some manufacturers differs considerably from the resistance values given.
*D.S.C. wound on spool.
†Bare wire on slab—well ventilated. Spacing between turns equal to wire diameter.
N.B.—To find current for Advance wire wound on slab, multiply Nichrome Current column by approx. 1.5.

TABLE 69

RESISTANCE WIRE TABLE, S.W.G.

SWG No.	Dia. mils.	Eureka Wire				Nichrome Wire**		
		Ohms per 1,000 feet	Lbs. per 1,000 feet	Feet per Ohm	Current Milli Amps*	Ohms per 1,000 feet	Lbs. per 1,000 feet	Current Milli Amps†
10	128	17.4	49.7	57.5	—	40.8	45	—
11	116	21.2	40.9	47.2	—	50.2	37.3	—
12	104	26.4	32.9	37.9	—	62.4	29.5	—
13	92	33.8	25.7	29.6	—	79.7	23.4	—
14	80	44.6	19.5	22.4	—	105.4	17.7	—
15	72	55.1	15.8	18.15	—	130	14.0	—
16	64	69.8	12.5	14.33	—	164	11.3	—
17	56	91.1	9.5	10.98	—	215	8.7	—
18	48	123.9	7.0	8.07	—	292	6.4	—
19	40	178.5	4.9	5.60	—	421	4.42	—
20	36	220.4	3.9	4.53	650	520	3.58	—
21	32	279.1	3.12	3.58	510	659	2.83	—
22	28	364	2.38	2.75	390	861	2.17	—
23	24	496	1.75	2.02	300	1 170	1.60	—
24	22	590	1.47	1.70	250	1 390	1.33	—
25	20	714	1.21	1.40	210	1 680	1.12	—
26	18	882	0.99	1.134	170	2 080	.897	400
27	16.4	1 062	.82	.942	140	2 510	.746	350
28	14.8	1 305	.67	.766	117	3 080	.607	300
29	13.6	1 545	.56	.647	101	3 650	.513	250
30	12.4	1 858	.47	.538	85	4 390	.427	230
31	11.6	2 123	.41	.471	75	5 010	.373	205
32	10.8	2 450	.35	.408	66	5 780	.324	185
33	10.0	2 857	.304	.350	57	6 750	.276	165
34	9.2	3 376	.257	.296	49	7 970	.235	145
35	8.4	4 049	.215	.247	41	9 560	.195	125
36	7.6	4 947	.175	.202	35	11 690	.160	110
37	6.8	6 179	.140	.1618	29	14 600	.128	91
38	6.0	7 936	.109	.1260	23	18 700	.100	76
39	5.2	10 565	.082	.0947	19	24 900	.075	62
40	4.8	12 395	.070	.0807	16	29 200	.064	55
41	4.4	14 756	.059	.0677	13	34 800	.0536	48
42	4.0	17 855	.049	.0560	11	42 180	.0450	41
43	3.6	22 045	.039	.0454	9.5	52 000	.0358	35
44	3.2	27 888	.031	.0359	8.0	63 900	.0283	30
45	2.8	36 216	.024	.0276	6.5	86 100	.0217	25
46	2.4	49 588	.018	.0202	5.0	117 000	.0160	20
47	2.0	71 428	.012	.0140	4.0	168 000	.0112	15
48	1.6	111 333	.008	.0090	3.0	263 600	.0071	—

**Wire produced by some manufacturers differs considerably from the resistance values given.
*D.S.C. wound on spool.

†Bare wire on slab—well ventilated. Spacing between turns equal to wire diameter.

N.B.—To find current for Eureka wire wound on slab, multiply Nichrome Current column by approx. 1.5.

TABLE 70 : WINDING DATA CHART
FOR ROUND PLAIN ENAMELLED COPPER WIRE, A.W.G.

(By courtesy of Standard Transformer Corporation, Chicago)

Insulation Layer Insulation Thickness (Inches)	Wire A.W.G. No.	Current Based on 700 Cir. Mils per Amp. (Maximum) (Amperes)	Dia. Overall Enamelled Wire (Nominal) (Inches)	Area Bare Copper Wire (Nominal) (Cir. Mils)	Turns per layer — Lamination Size (Dimension of Centre Leg) 0.50"	0.62"	0.75"	0.87"	1.00"	1.12"	1.25"	1.37"	1.50"	Wire A.W.G. No.
.010 Fibre	10	14.90	.1039	10380	—	—	—	9	10	11	12	14	16	10
,,	11	11.80	.0927	8230	—	—	—	10	11	12	14	16	18	11
,,	12	9.35	.0827	6530	—	—	8	11	13	14	16	18	20	12
,,	13	7.42	.0738	5180	—	8	9	13	14	16	18	20	22	13
.007 Fibre	14	5.87	.0659	4110	7	9	11	14	16	18	20	22	25	14
,,	15	4.66	.0588	3260	8	10	12	16	18	20	22	25	28	15
,,	16	3.69	.0524	2580	9	11	13	18	20	22	25	28	31	16
.005 Fibre	17	2.93	.0469	2050	10	12	15	20	22	25	28	31	35	17
,,	18	2.31	.0418	1620	11	14	17	23	25	28	31	35	39	18
.004 Kraft	19	1.84	.0374	1290	12	15	19	25	28	31	35	39	44	19
,,	20	1.46	.0334	1020	14	17	21	28	32	35	39	44	49	20
,,	21	1.16	.0299	810	15	19	23	31	35	39	43	49	55	21
,,	22	0.917	.0267	642	17	22	26	35	39	44	48	55	62	22
,,	23	0.723	.0238	510	20	25	29	39	44	49	54	62	69	23
,,	24	0.578	.0213	404	22	27	33	44	50	55	61	69	77	24
.003 Kraft	25	0.457	.0190	320	25	31	37	49	55	61	68	77	86	25

TABLE 70 : WINDING DATA CHART
FOR ROUND PLAIN ENAMELLED COPPER WIRE, A.W.G.

(By courtesy of Standard Transformer Corporation, Chicago)

Insulation — Layer Insulation Thickness (Inches)	Wire A.W.G. No.	Current Based on 700 Cir. Mils per Amp. (Maximum) (Amperes)	Dia. Overall Enamelled Wire (Nominal) (Inches)	Area Bare Copper Wire (Nominal) (Cir. Mils)	Turns per layer — Lamination Size (Dimension of Centre Leg)									Wire A.W.G. No.
					0.50"	0.62"	0.75"	0.87"	1.00"	1.12"	1.25"	1.37"	1.50"	
.003 Kraft	26	0.363	.0170	254	27	35	41	55	62	68	76	87	98	26
”	27	0.289	.0152	202	31	39	46	62	69	77	85	97	108	27
.002 Kraft	28	0.229	.0136	160	34	43	51	69	77	86	95	108	121	28
”	29	0.181	.0122	127	38	48	57	77	86	95	115	119	133	29
”	30	0.144	.0108	101	43	53	64	86	96	107	118	134	151	30
.0015 Kraft	31	0.114	.0097	79.7	47	60	72	96	107	119	132	150	168	31
”	32	0.0903	.0088	63.2	52	66	78	105	118	130	143	164	183	32
”	33	0.0715	.0078	50.1	59	73	88	118	133	147	162	184	207	33
.001 Kraft	34	0.0568	.0069	39.8	67	83	96	133	150	167	183	209	233	34
”	35	0.0450	.0061	31.5	75	92	111	150	169	186	205	233	265	35
”	36	0.0357	.0055	25.0	83	103	125	166	186	206	227	258	293	36
”	37	0.0283	.0049	19.8	93	115	139	185	209	231	255	290	325	37
.0008 Kraft	38	0.0224	.0044	15.7	102	127	151	205	230	255	284	322	—	38
”	39	0.0179	.0038	12.5	118	148	177	237	266	296	330	—	—	39
”	40	0.0141	.0034	9.89	131	163	197	263	295	329	—	—	—	40
”	41	0.0120	.0032	8.41	139	174	209	279	312	—	—	—	—	41
”	42	0.0089	.0028	6.25	158	198	238	—	—	—	—	—	—	42

SECTION 20 : LOGARITHM TABLES

	0	1	2	3	4	5	6	7	8	9	Differences								
											1	2	3	4	5	6	7	8	9
10	0000	0043	0086	0128	0170	0212	0253	0294	0334	0374	4	8	12	17	21	25	29	33	37
11	0414	0453	0492	0531	0569	0607	0645	0682	0719	0755	4	8	11	15	19	23	26	30	34
12	0792	0828	0864	0899	0934	0969	1004	1038	1072	1106	3	7	10	14	17	21	24	28	31
13	1139	1173	1206	1239	1271	1303	1335	1367	1399	1430	3	6	10	13	16	19	23	26	29
14	1461	1492	1523	1553	1584	1614	1644	1673	1703	1732	3	6	9	12	15	18	21	24	27
15	1761	1790	1818	1847	1875	1903	1931	1959	1987	2014	3	6	8	11	14	17	20	22	25
16	2041	2068	2095	2122	2148	2175	2201	2227	2253	2279	3	5	8	11	13	16	18	21	24
17	2304	2330	2355	2380	2405	2430	2455	2480	2504	2529	2	5	7	10	12	15	17	20	22
18	2553	2577	2601	2625	2648	2672	2695	2718	2742	2765	2	5	7	9	12	14	16	19	21
19	2788	2810	2833	2856	2878	2900	2923	2945	2967	2989	2	4	7	9	11	13	16	18	20
20	3010	3032	3054	3075	3096	3118	3139	3160	3181	3201	2	4	6	8	11	13	15	17	19
21	3222	3243	3263	3284	3304	3324	3345	3365	3385	3404	2	4	6	8	10	12	14	16	18
22	3424	3444	3464	3483	3502	3522	3541	3560	3579	3598	2	4	6	8	10	12	14	15	17
23	3617	3636	3655	3674	3692	3711	3729	3747	3766	3784	2	4	6	7	9	11	13	15	17
24	3802	3820	3838	3856	3874	3892	3909	3927	3945	3962	2	4	5	7	9	11	12	14	16
25	3979	3997	4014	4031	4048	4065	4082	4099	4116	4133	2	3	5	7	9	10	12	14	15
26	4150	4166	4183	4200	4216	4232	4249	4265	4281	4298	2	3	5	7	8	10	11	13	15
27	4314	4330	4346	4362	4378	4393	4409	4425	4440	4456	2	3	5	6	8	9	11	13	14
28	4472	4487	4502	4518	4533	4548	4564	4579	4594	4609	2	3	5	6	8	9	11	12	14
29	4624	4639	4654	4669	4683	4698	4713	4728	4742	4757	1	3	4	6	7	9	10	12	13
30	4771	4786	4800	4814	4829	4843	4857	4871	4886	4900	1	3	4	6	7	9	10	11	13
31	4914	4928	4942	4955	4969	4983	4997	5011	5024	5038	1	3	4	6	7	8	10	11	12
32	5051	5065	5079	5092	5105	5119	5132	5145	5159	5172	1	3	4	5	7	8	9	11	12
33	5185	5198	5211	5224	5237	5250	5263	5276	5289	5302	1	3	4	5	6	8	9	10	12
34	5315	5328	5340	5353	5366	5378	5391	5403	5416	5428	1	3	4	5	6	8	9	10	11
35	5441	5453	5465	5478	5490	5502	5514	5527	5539	5551	1	2	4	5	6	7	9	10	11
36	5563	5575	5587	5599	5611	5623	5635	5647	5658	5670	1	2	4	5	6	7	8	10	11
37	5682	5694	5705	5717	5729	5740	5752	5763	5775	5786	1	2	3	5	6	7	8	9	10
38	5798	5809	5821	5832	5843	5855	5866	5877	5888	5899	1	2	3	5	6	7	8	9	10
39	5911	5922	5933	5944	5955	5966	5977	5988	5999	6010	1	2	3	4	5	7	8	9	10
40	6021	6031	6042	6053	6064	6075	6085	6096	6107	6117	1	2	3	4	5	6	8	9	10
41	6128	6138	6149	6160	6170	6180	6191	6201	6212	6222	1	2	3	4	5	6	7	8	9
42	6232	6243	6253	6263	6274	6284	6294	6304	6314	6325	1	2	3	4	5	6	7	8	9
43	6335	6345	6355	6365	6375	6385	6395	6405	6415	6425	1	2	3	4	5	6	7	8	9
44	6435	6444	6454	6464	6474	6484	6493	6503	6513	6522	1	2	3	4	5	6	7	8	9
45	6532	6542	6551	6561	6571	6580	6590	6599	6609	6618	1	2	3	4	5	6	7	8	9
46	6628	6637	6646	6656	6665	6675	6684	6693	6702	6712	1	2	3	4	5	6	7	7	8
47	6721	6730	6739	6749	6758	6767	6776	6785	6794	6803	1	2	3	4	5	5	6	7	8
48	6812	6821	6830	6839	6848	6857	6866	6875	6884	6893	1	2	3	4	4	5	6	7	8
49	6902	6911	6920	6928	6937	6946	6955	6964	6972	6981	1	2	3	4	4	5	6	7	8
50	6990	6998	7007	7016	7024	7033	7042	7050	7059	7067	1	2	3	3	4	5	6	7	8
51	7076	7084	7093	7101	7110	7118	7126	7135	7143	7152	1	2	3	3	4	5	6	7	8
52	7160	7168	7177	7185	7193	7202	7210	7218	7226	7235	1	2	3	3	4	5	6	7	7
53	7243	7251	7259	7267	7275	7284	7292	7300	7308	7316	1	2	2	3	4	5	6	6	7
54	7324	7332	7340	7348	7356	7364	7372	7380	7388	7396	1	2	2	3	4	5	6	6	7
	0	1	2	3	4	5	6	7	8	9	1	2	3	4	5	6	7	8	9

	0	1	2	3	4	5	6	7	8	9	Differences								
											1	2	3	4	5	6	7	8	9
55	7404	7412	7419	7427	7435	7443	7451	7459	7466	7474	1	2	2	3	4	5	5	6	7
56	7482	7490	7497	7505	7513	7520	7528	7536	7543	7551	1	2	2	3	4	5	5	6	7
57	7559	7566	7574	7582	7589	7597	7604	7612	7619	7627	1	2	2	3	4	5	5	6	7
58	7634	7642	7649	7657	7664	7672	7679	7686	7694	7701	1	1	2	3	4	4	5	6	7
59	7709	7716	7723	7731	7738	7745	7752	7760	7767	7774	1	1	2	3	4	4	5	6	7
60	7782	7789	7796	7803	7810	7818	7825	7832	7839	7846	1	1	2	3	4	4	5	6	6
61	7853	7860	7868	7875	7882	7889	7896	7903	7910	7917	1	1	2	3	4	4	5	6	6
62	7924	7931	7938	7945	7952	7959	7966	7973	7980	7987	1	1	2	3	3	4	5	6	6
63	7993	8000	8007	8014	8021	8028	8035	8041	8048	8055	1	1	2	3	3	4	5	5	6
64	8062	8069	8075	8082	8089	8096	8102	8109	8116	8122	1	1	2	3	3	4	5	5	6
65	8129	8136	8142	8149	8156	8162	8169	8176	8182	8189	1	1	2	3	3	4	5	5	6
66	8195	8202	8209	8215	8222	8228	8235	8241	8248	8254	1	1	2	3	3	4	5	5	6
67	8261	8267	8274	8280	8287	8293	8299	8306	8312	8319	1	1	2	3	3	4	5	5	6
68	8325	8331	8338	8344	8351	8357	8363	8370	8376	8382	1	1	2	3	3	4	4	5	6
69	8388	8395	8401	8407	8414	8420	8426	8432	8439	8445	1	1	2	2	3	4	4	5	6
70	8451	8457	8463	8470	8476	8482	8488	8494	8500	8506	1	1	2	2	3	4	4	5	6
71	8513	8519	8525	8531	8537	8543	8549	8555	8561	8567	1	1	2	2	3	4	4	5	5
72	8573	8579	8585	8591	8597	8603	8609	8615	8621	8627	1	1	2	2	3	4	4	5	5
73	8633	8639	8645	8651	8657	8663	8669	8675	8681	8686	1	1	2	2	3	4	4	5	5
74	8692	8698	8704	8710	8716	8722	8727	8733	8739	8745	1	1	2	2	3	4	4	5	5
75	8751	8756	8762	8768	8774	8779	8785	8791	8797	8802	1	1	2	2	3	3	4	5	5
76	8808	8814	8820	8825	8831	8837	8842	8848	8854	8859	1	1	2	2	3	3	4	5	5
77	8865	8871	8876	8882	8887	8893	8899	8904	8910	8915	1	1	2	2	3	3	4	4	5
78	8921	8927	8932	8938	8943	8949	8954	8960	8965	8971	1	1	2	2	3	3	4	4	5
79	8976	8982	8987	8993	8998	9004	9009	9015	9020	9025	1	1	2	2	3	3	4	4	5
80	9031	9036	9042	9047	9053	9058	9063	9069	9074	9079	1	1	2	2	3	3	4	4	5
81	9085	9090	9096	9101	9106	9112	9117	9122	9128	9133	1	1	2	2	3	3	4	4	5
82	9138	9143	9149	9154	9159	9165	9170	9175	9180	9186	1	1	2	2	3	3	4	4	5
83	9191	9196	9201	9206	9212	9217	9222	9227	9232	9238	1	1	2	2	3	3	4	4	5
84	9243	9248	9253	9258	9263	9269	9274	9279	9284	9289	1	1	2	2	3	3	4	4	5
85	9294	9299	9304	9309	9315	9320	9325	9330	9335	9340	1	1	2	2	3	3	4	4	5
86	9345	9350	9355	9360	9365	9370	9375	9380	9385	9390	1	1	2	2	3	3	4	4	5
87	9395	9400	9405	9410	9415	9420	9425	9430	9435	9440	0	1	1	2	2	3	3	4	4
88	9445	9450	9455	9460	9465	9469	9474	9479	9484	9489	0	1	1	2	2	3	3	4	4
89	9494	9499	9504	9509	9513	9518	9523	9528	9533	9538	0	1	1	2	2	3	3	4	4
90	9542	9547	9552	9557	9562	9566	9571	9576	9581	9586	0	1	1	2	2	3	3	4	4
91	9590	9595	9600	9605	9609	9614	9619	9624	9628	9633	0	1	1	2	2	3	3	4	4
92	9638	9643	9647	9652	9657	9661	9666	9671	9675	9680	0	1	1	2	2	3	3	4	4
93	9685	9689	9694	9699	9703	9708	9713	9717	9722	9727	0	1	1	2	2	3	3	4	4
94	9731	9736	9741	9745	9750	9754	9759	9763	9768	9773	0	1	1	2	2	3	3	4	4
95	9777	9782	9786	9791	9795	9800	9805	9809	9814	9818	0	1	1	2	2	3	3	4	4
96	9823	9827	9832	9836	9841	9845	9850	9854	9859	9863	0	1	1	2	2	3	3	4	4
97	9868	9872	9877	9881	9886	9890	9894	9899	9903	9908	0	1	1	2	2	3	3	4	4
98	9912	9917	9921	9926	9930	9934	9939	9943	9948	9952	0	1	1	2	2	3	3	4	4
99	9956	9961	9965	9969	9974	9978	9983	9987	9991	9996	0	1	1	2	2	3	3	3	4
	0	1	2	3	4	5	6	7	8	9	1	2	3	4	5	6	7	8	9

SECTION 21 : TRIGONOMETRICAL AND HYPERBOLIC TABLES

TABLE 72 : TRIGONOMETRICAL RELATIONSHIPS

Angle	Radians	Sine	Cosine	Tangent	Angle	Radians	Sine	Cosine	Tangent
0°	.0000	.0000	1.000	.0000	45°	.7854	.7071	.7071	1.0000
1	.0175	.0175	.9998	.0175	46	.8029	.7193	.6947	1.0355
2	.0349	.0349	.9994	.0349	47	.8203	.7314	.6820	1.0724
3	.0524	.0523	.9986	.0524	48	.8378	.7431	.6691	1 1106
4	.0698	.0698	.9976	.0699	49	.8552	.7547	.6561	1.1504
5	.0873	.0872	.9962	.0875	50	.8727	.7660	.6428	1.1918
6	.1047	.1045	.9945	.1051	51	.8901	.7771	.6293	1.2349
7	.1222	.1219	.9925	.1228	52	.9076	.7880	.6157	1.2799
8	.1396	.1392	.9903	.1405	53	.9250	.7986	.6018	1.3270
9	.1571	.1564	.9877	.1584	54	.9425	.8090	.5878	1.3764
10	.1745	.1736	.9848	.1763	55	.9599	.8192	.5736	1.4281
11	.1920	.1908	.9816	.1944	56	.9774	.8290	.5592	1.4826
12	.2094	.2079	.9781	.2126	57	.9948	.8387	.5446	1.5399
13	.2269	.2250	.9744	.2309	58	1.0123	.8480	.5299	1.6003
14	.2443	.2419	.9703	.2493	59	1.0297	.8572	.5150	1.6643
15	.2618	.2588	.9659	.2679	60	1.0472	.8660	.5000	1.7321
16	.2793	.2756	.9613	.2867	61	1.0647	.8746	.4848	1.8040
17	.2967	.2924	.9563	.3057	62	1.0821	.8829	.4695	1.8807
18	.3142	.3090	.9511	.3249	63	1.0996	.8910	.4540	1.9626
19	.3316	.3256	.9455	.3443	64	1.1170	.8988	.4384	2.0503
20	.3491	.3420	.9397	.3640	65	1.1345	.9063	.4226	2.1445
21	.3665	.3584	.9336	.3839	66	1.1519	.9135	.4067	2.2460
22	.3840	.3746	.9272	.4040	67	1.1694	.9205	.3907	2.3559
23	.4014	.3907	.9205	.4245	68	1.1868	.9272	.3746	2.4751
24	.4189	.4067	.9135	.4452	69	1.2043	.9336	.3584	2.6051
25	.4363	.4226	.9063	.4663	70	1.2217	.9397	.3420	2.7475
26	.4538	.4384	.8988	.4877	71	1.2392	.9455	.3256	2.9042
27	.4712	.4540	.8910	.5095	72	1.2566	.9511	.3090	3.0777
28	.4887	.4695	.8829	.5317	73	1.2741	.9563	.2924	3.2709
29	.5061	.4848	.8746	.5543	74	1.2915	.9613	.2756	3.4874
30	.5236	.5000	.8660	.5774	75	1.3090	.9659	.2588	3.7321
31	.5411	.5150	.8572	.6009	76	1.3265	.9703	.2419	4.0108
32	.5585	.5299	.8480	.6249	77	1.3439	.9744	.2250	4.3315
33	.5760	.5446	.8387	.6494	78	1.3614	.9781	.2079	4.7046
34	.5934	.5592	.8290	.6745	79	1.3788	.9816	.1908	5.1446
35	.6109	.5736	.8192	.7002	80	1.3963	.9848	.1736	5.6713
36	.6283	.5878	.8090	.7265	81	1.4137	.9877	.1564	6.3138
37	.6458	.6018	.7986	.7536	82	1.4312	.9903	.1392	7.1154
38	.6632	.6157	.7880	.7813	83	1.4486	.9925	.1219	8.1443
39	.6807	.6293	.7771	.8098	84	1.4661	.9945	.1045	9.5144
40	.6981	.6428	.7660	.8391	85	1.4835	.9962	.0872	11.43
41	.7156	.6561	.7547	.8693	86	1.5010	.9976	.0698	14.30
42	.7330	.6691	.7431	.9004	87	1.5184	.9986	.0523	19.08
43	.7505	.6820	.7314	.9325	88	1.5359	.9994	.0349	28.64
44	.7679	.6947	.7193	.9657	89	1.5533	.9998	.0175	57.29

TABLE 73 : HYPERBOLIC SINES, COSINES AND TANGENTS

$$\sinh x = \tfrac{1}{2}(\epsilon^{x} - \epsilon^{-x}) ; \cosh x = \tfrac{1}{2}(\epsilon^{x} + \epsilon^{-x}) ; \tanh x = (\epsilon^{x} - \epsilon^{-x})/(\epsilon^{x} + \epsilon^{-x})$$
$$\epsilon = 2.718 = 1/0.3679$$

x	$\sinh x$	$\cosh x$	$\tanh x$	x	$\sinh x$	$\cosh x$	$\tanh x$
0.00	0.000	1.000	0.000 0	2.50	6.050	6.132	0.986 6
0.10	0.100	1.005	0.099 7	2.60	6.695	6.769	0.989 0
0.20	0.201	1.020	0.197 4	2.70	7.406	7.473	0.991 0
0.30	0.304	1.045	0.291 3	2.80	8.192	8.253	0.992 6
0.40	0.411	1.081	0.380 0	2.90	9.060	9.115	0.994 0
0.50	0.521	1.128	0.462 1	3.00	10.018	10.068	0.995 1
0.60	0.637	1.185	0.537 0	3.10	11.077	11.122	0.996 0
0.70	0.759	1.255	0.604 4	3.20	12.246	12.287	0.996 7
0.80	0.888	1.337	0.664 0	3.30	13.538	13.575	0.997 3
0.90	1.027	1.433	0.716 3	3.40	14.965	14.999	0.997 8
1.00	1.175	1.543	0.761 6	3.50	16.543	16.573	0.998 2
1.10	1.336	1.669	0.800 5	3.60	18.285	18.313	0.998 5
1.20	1.509	1.811	0.833 7	3.70	20.211	20.236	0.998 8
1.30	1.698	1.971	0.861 7	3.80	22.339	22.362	0.999 0
1.40	1.904	2.151	0.885 4	3.90	24.691	24.711	0.999 2
1.50	2.129	2.352	0.905 2	4.00	27.290	27.308	0.999 3
1.60	2.376	2.577	0.921 7	4.10	30.162	30.178	0.999 45
1.70	2.646	2.828	0.935 4	4.20	33.336	33.351	0.999 55
1.80	2.942	3.107	0.946 8	4.30	36.843	36.857	0 999 63
1.90	3.268	3.418	0.956 2	4.40	40.719	40.732	0.999 70
2.00	3.627	3.762	0.964 0	4.50	45.003	45.014	0.999 75
2.10	4.022	4.144	0.970 5	4.60	49.737	49.747	0.999 80
2.20	4.457	4.568	0.975 7	4.70	54.969	54.978	0.999 83
2.30	4.937	5.037	0.980 1	4.80	60.751	60.759	0.999 86
2.40	5.466	5.557	0.983 7	4.90	67.141	67.149	0.999 89
				5.00	74.203	74.210	0.999 91

SECTION 22 : LOG. SCALES AND LOG. SCALE INTERPOLATOR

(i) Log scales   (ii) Log scale interpolator.

(i) Log scales

The exact (to 3 significant figures) relationship between the log. and linear scales are set out in Fig. 38.9. It will be seen that the approximation of Fig. 38.10 is very close and the error is negligible in comparison with ordinary graphical errors. To plot the approximate log. scale, use a decimal rule (inches or centimetres etc.) and mark 0.3, 0.6, 0.9 and 1.0 for each major division, and continue similarly. If linear graph paper is available, select values for 1 and 10 on the log. scale such that the linear scale is 10 or a convenient multiple of 10, then mark log. values as in Fig. 38.10.

See also Chapter 19 Sect. 1(xii) for decibels, slide rules and mental arithmetic.

FIG. 38.9

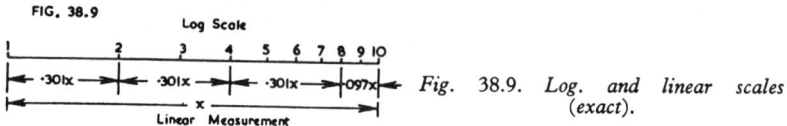

Fig. 38.9. *Log. and linear scales (exact).*

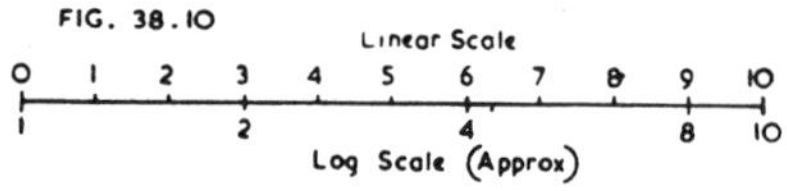

FIG. 38.10

Fig. 38.10. *Log. and linear scales (approx.).*

(ii) Log. scale Interpolator

Often it is desired to know the exact value of a point on a logarithmic scale, when it does not come on one of the scale markings. In such a case, mark on a straight edge of paper the distance from the point to the next lower scale marking which is a multiple of 10, also marking the distance to the next higher marking which is a multiple of 10. The piece of paper with the three pencil marks should then be placed on the Log Scale Interpolator (see opposite page) so that the edge of the paper is roughly horizontal, and the two extreme points come exactly on the outer radial lines, the lower value to the left and the higher value to the right. The value of the point may then be read on the top scale, multiplied by the appropriate multiple of 10.

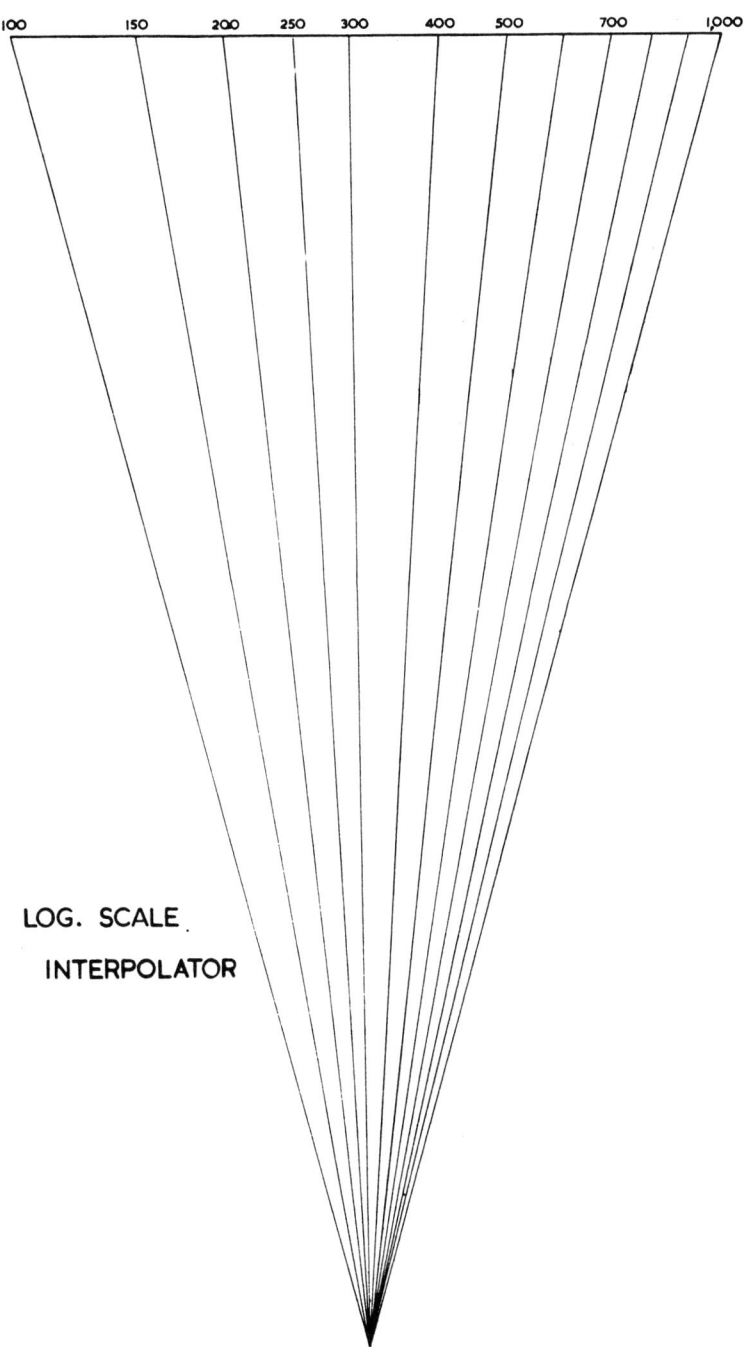

LOG. SCALE
INTERPOLATOR

INDEX

M

SUPPLEMENT

In many cases brief summaries or comments are given, particularly with the later references.

This Supplement completely supersedes the smaller Supplements included in earlier impressions.

CHAPTER 1
INTRODUCTION TO THE RADIO VALVE

ADDITIONAL REFERENCES
11A. Metson, G. H., S. Wagener, M. F. Holmes and M. R. Child " The life of oxide cathodes in modern receiving valves " Proc. I.E.E. 99. Part III 58 (March 1952) 69.
11B. Nergaard, L. S. " Studies of the oxide cathode " R.C.A. Rev. 13.4 (Dec. 1952) 464.
11C. Hallows, R. W., and H. K. Milward " Introduction to valves " (Iliffe and Sons Ltd., London 1953).

CHAPTER 2
VALVE CHARACTERISTICS

ADDITIONAL REFERENCES
B23. Williams, L. E. " Space-charge reactance tube " Elect. 25.6 (June 1952) 166.
B24. Nergaard, L. S. " Studies of the oxide cathode " R.C.A. Rev. 13.4 (Dec. 1952) 464.
B25. Levy, I. E. " The effect of impurity migrations on thermionic emission from oxide cathodes," Proc. I.R.E. 41.3 (March 1953) 365.
B26. Tillman, J. R., J. Butterworth, and R. E. Warren " The independence of mutual conductance on frequency of aged oxide-cathode valves and its influence on their transient response " Proc. I.E.E. 100.5 Part IV (Oct. 1953) p. 8. This variation has been ascribed to an impedance at the interface layer between the oxide coating and the nickel sleeve ; this article shows that the previous representation of the impedance as a single parallel combination of R and C is inadequate. A comprehensive bibliography is given.
B27. Bounds, A. M., and P. N. Hambleton " The nickel-base indirectly heated oxide cathode " E.E. 72.6 (June 1953) 536.
B28. Herrimann, G., and S. Wagener (book) " The oxide coated cathode " 2 vols., Chapman and Hall (1951).
B29. Wright, D. A. " A Survey of present knowledge of thermionic emitters " Proc. I.E.E. 100. Part III 65 (May 1953) 125. A valuable and comprehensive survey of the whole subject, with considerable detail on oxide cathodes and extensive bibliography.
B30. Tillman, J. R., J. Butterworth and R. E. Warren " The independence of mutual conductance on frequency of aged oxide-cathode valves and its influence on their transient response " I.E.E. Monograph, Dec. 1952, Digest, Proc. I.E.E. 100. Part III 65 (May 1953) 175.
B31. Pullen, K. A. " Using conductance curves in electronic circuit design " Proc. National Electronics Conference Vol. 6 (1950) 112.
B32. Pullen, K. A. " G curves " Tele-Tech 1953/4 " UHF oscillator design notes " 12.2 (Feb. 1953) 80 ; " Conductance curves speed triode r-c amplifier design " 12.5 (May 1953) 80 ; " Conductance curves speed pentode r-c amplifier design " 12.7 (July 1953) 44 ; " G curves and impedance amplifiers " 12.9 (Sept. 1953) 71 ; " G curves and degenerative amplifiers," 13.4 (April 1954) 86.

(H) Grid current characteristics
H3. Watkinson, E. " Control grid currents in radio receiving valves " Proc. I.R.E. Australia 15.6 (June 1954) 139. An introduction to the subject, with grid load lines and input resistance.

CHAPTER 3
THE TESTING OF OXIDE-COATED CATHODE HIGH-VACUUM RECEIVING VALVES

ADDITIONAL REFERENCES
94. Kuder, M. L. " Electron tube curve generator " Elect. 25.3 (March 1952) 118.
95. Smith, G. " Two bridges for measuring valve parameters " Electronic Eng. 24.289 (March 1952) 127.
96. Foster, B. C. " A simple valve comparator " Electronic Eng. 24.291 (May 1952) 220. (Simple form of display comparator).
97. Brewer, R. " Radio valve life testing " Proc. I.E.E. 98, Part III 54 (July 1951) 269.
98. Heins, A. J. " Dynamic measurements on receiving valves " J. Britt. I.R.E. 12.1 (Jan. 1952) 63.
99. Metson, G. H., S. Wagener, M. F. Holmes and M. R. Child " The life of oxide cathodes in modern receiving valves " Proc. I.E.E. 99, Part III 58 (March 1952) 69. Discussion 100 Part III 68 (Nov. 1953) 371.
100. Terman, F. E., and J. M. Pettit (book) " Electronic Measurements " McGraw-Hill Book Company, 2nd ed. 1952, pp. 289-310.
101. Flanagan, J. L. " Valve tube testers " Elect. 25.6 (June 1952) 139.
102. Rowe, E. G. " Technique of trustworthy valves " Proc. I.R.E. 40.10 (Oct. 1952) 1166.
103. Acheson, M. A., and E. M. McElwee " Concerning the reliability of electron tubes " Proc. I.R.E. 40.10 (Oct. 1952) 1204.

104. Knight, C. R. and K. C. Harding " General considerations in regard to specifications for reliable tubes " Proc. I.R.E. 40.10 (Oct. 1952) 1207.
105. Koch, D. G. " Increasing tube reliability in industrial circuits " Product Engineering. McGraw-Hill Publishing Co., 23.6 (June 1952) 175. Reprinted in Radiotronics 18.1 (Jan. 1953) 14.
106. Levy, I. E. " Shielding and mounting effects on tube bulb temperatures " Tele-Tech. 12.2 (Feb. 1953) 72.
107. Eaglesfield, C. C. " Vibration test for valves—use of repeated impacts " W.E. 30.3 (March 1953) 57.
108. " Special quality valves—announcement by B.V.A." Electronic Eng. 25. 304 (June 1953) 238.
109. Jones, W. R. " Tube applications for increased reliability " Trans. I.R.E. Professional Group on Quality Control, PGQC-1 (Aug. 1952). Reprinted Proc. I.R.E. (Australia) 14.12 (Dec. 1953) 299. Demonstrates effect of cathode bias and screen resistors on spread of valve characteristics.
110. Wyman, J. H. " Design factors that extend electron tube life " Tele-Tech 12.11 (Nov. 1953) 80. Effect of cathode temperature on life, voltage variations on cathode temperature, bulb temperature on cathode temperature, high altitude and vibration on tube life.
111. " A pulse emission test for field testing of hot-cathode gas tubes " R.C.A. Application Note A.N.-157 reprinted Radiotronics 18.8 (Aug. 1953) 127.
112. Paul, F. A. " Commercial, ruggedized and premium tubes " Elect. 20.10 (Oct. 1953) 212. Gives curves of variation of gm with time, type 6AK5, 5654, 5654 aged 500 hours.
113. Prager, H. J. " Performance evaluation of special red tubes " R.C.A. Rev. 14.3 (Sept. 1953) 413. Gives performance data and curves for types 5691, 5692 and 5693.
114. Niehaus, E. H. " Curve-tracer test set for vacuum tubes " Tele-Tech 13.2 (Feb. 1954) 90. General Electric development, plots smooth curves, 1% accuracy, directly on paper. Chopper provided for use when dissipations would be exceeded.
115. Knight, L. " Valve reliability in digital calculating machines " Electronic Eng. 26.311 (Jan. 1954) 9. Measures to reduce number of valve failures ; also applicable to other electronic equipment.
116. " Special Quality valves " Jour. Brit. I.R.E. 15.5 (May 1953) 274. Based on statement by British Valve Manufacturers' Association.
117. Paul, F. A. " A comparison of the 6AK5 and 5654 tubes " Trans. I.R.E.-PGCP-1 (March 1954) 18. Type 5654 is Premium version of type 6AK5. Comparison on basis of vibration and shock.
118. Handley, P. A., and P. Welch " Valve noise produced by electrode movement " Proc. I.R.E. 42.3 (March 1954) 565. Valve rattle ; resonances ; contribution of various electrodes ; calculation of resonant frequencies of electrodes ; methods of measuring noise ; valve design.
119. Kurshaw, J., R. D. Lohman and G. B. Herzog "Cathode ray tube plots transistor curves" Elect. 26.2 (Feb. 1953) 122. R.C.A. development.
120. Wood, R., and W. H. Hunter " A method of analyzing the microphonic output of a tube and a description of the CK6247 " Proc. National Electronics Conf. 9 (1953) 111.
121. Wohl, R. J., S. Winkler, L. N. Heynick and M. Schnee " Audio frequency impulse noise and microphonism " Proc. Nat. Electronics Conf. 9 (1953) 119.
122. Levy, I. E. " The influence of a vacuum tube component temperatures on characteristics and life " Proc. Nat. Electronics Conf. 9 (1953) 621.

CHAPTER 4

THEORY OF NETWORKS

ADDITIONAL REFERENCES

(A) References to practical resistors
A34. Howes, J. W. " The characteristics and applications of thermally sensitive resistors or thermistors " Proc. I.R.E. Aust. 13.5 (May 1952) 123. Reprinted Jour. Brit. I.R.E. 13.4 (April 1953) 228.
A35. Hooper, C. K. " Stability characteristics of standard composition resistors " Tele-Tech 11.9 (Sept. 1952) 88.
A36. Paul, F. A. " Resistor temperature coefficients " Tele-Tech 12.1 (Jan. 1953) 52.
A37. Keim, L. B. " The deposited carbon resistor, an essential component of good audio design practice " Jour. A.E.S. 1.1 (Jan. 1953) 42.
A38. Gibson, W. T. " Thermistor production " Elect. Comm. 30.4 (Dec. 1953) 263 ; Post Office Electrical Engineers' Journal Part 1.46 (April 1953) 34.
A39. " High temperature carbon film resistors " Elect. 26.1 (Jan. 1953) 148. High stability, deposited carbon film on ceramic.
A40. Langford-Smith, F. " The use of cracked carbon resistors in amplifiers " Radiotronics 19.7 (July 1954) 84.
A41. Bell, D. A., and K. Y. Chong " Current noise in composition resistors " W.E. 31.6 (June 1954) 142. A valuable investigation into several features of current noise.
A42. Forman, A. J. " Thermistors : components for electronic control and measurement " Tele-Tech 13.4 (April 1954) 72.
A43. " Metal film resistors " W.W. 60.7 (July 1954) 318.

(B) References to practical condensers
B18. Maxwell, J. W. " New low-temperature capacitors—electrolytic types for operation to—55° C " Tele-Tech 11.6 (June 1952) 53.
B19. Davidson, R. " R.F. characteristics of capacitors " W.W. 58.8 (Aug. 1952) 301.
B20. Van Buskirk, M. " Electrolytic capacitors, why and when " Jour. A.E.S. 1.1 (Jan. 1953) 46.
B21. Burnham, J., " Breakdown and leakage resistance—investigation of metallized paper capacitors " Trans. I.R.E. PGCP-1 (Mar. 1954) 3.
B22. Podolsky, J. K. Sprague " Some characteristics and limitations of capacitor and resistor components " Trans. I.R.E. PGCP-1 (March 1954) 33.
B23. Altenpohl, D. " Improvements in the field of electrolytic capacitors " Convention Record, I.R.E. Nat. Conv. 1954, Part 3, p. 35.
B24. Geiser, D. T. " An investigation of the lowest resonant frequency in commercially available bypass capacitors " Convention Record, I.R.E. Nat. Conv. 1954, Part 3, p. 43.

ADDITIONAL REFERENCES TO NETWORKS (continued from page 171)

9. Bacon, W., and D. P. Salmon ; " Resistance-capacitance networks with over-unity gain " W.E. 30.1 (Jan. 1953) 20.
10. Langford-Smith, F., " Calculations of impedance with reactance and resistance in series " Radio-tronics " 18.6 (June 1953) 80.   Method applied to slide rule.
11. Brown, D. A. H. " The equivalent Q of RC networks " Electronic Eng. 25.305 (July 1953) 294. Effective Q of RC networks used as frequency determining element in phase shift oscillator. Correspondence 25.307 (Sept. 1953) 394-395 ; 25.310 (Dec. 1953) 534 ; 26.314 (Apr. 1954) 177.
12. Oakes, F. " Simplified calculations—working out resonant circuit constants on the slide rule " W.W. 59.9 (Sept. 1953) 439.
13. Tyler, V. J. " Pocket reactance and resonance calculator " W.W. 59.12 (Dec. 1953) 560.
14. Oakes, F. " Resistances in parallel : calculating effective values on the slide rule " W.W. 60.2 (Feb. 1954) 95.
15. Storch, L. " Rapid parallel-Z calculations " Tele-Tech 12.8 (Aug. 1953) 91.   Uses concentric circle graphical method.

ADDITIONAL REFERENCES TO FILTERS (continued from page 185)

1. Bolle, A. P. " Theory of twin-T RC-networks and their application to oscillators " Jour. Brit. I.R.E. 13.12 (Dec. 1953) 571.   A very thorough and comprehensive analysis, both of the general properties, and the application to oscillators.
2. Linvill, J. G. " RC active filters " Proc. I.R.E. 42.3 (Mar. 1954) 555.   Active transistor low-, high-, and band-pass filters.
3. Mole, J. H. " Filter design data for communication engineers " (Spon, London, 1952).
4. Crowhurst, N. H. " More about filters " Radio Electronics 24.4 (April 1953) 64 ; 24.5 (May 1953) 62.

CHAPTER 5
TRANSFORMERS AND IRON-CORED INDUCTORS

ADDITIONAL REFERENCES

(A) General

A12. " The magnetic properties of the nickel-iron alloys " Mond Nickel Co. Ltd. London, 2nd ed. (June 1950).
A13. MacFadyen, K. A. (book) " Small Transformers and Inductors " (Chapman and Hall, London, 1953).   Covers most forms of iron-cored transformers, but treatment is on impedance basis, using concept of complex permeability.

(C) Audio-frequency transformers

C35. Crowhurst, N. H. " Measuring up an audio transformer " Audio Eng. 36.11 (Nov. 1952) 24.
C36. Ayres, W. R. " Power and voltage amplifiers " Audio Engineering Society Lecture No. 2 (17 January 1952).
C37. Morris, A. L. " Tape wound magnetic cores " Electronic Eng. 24.295 (Sept. 1952) 416.
C38. Crowhurst, N. H. " How good is an audio transformer ? " Audio Eng. 36.3 (March 1952) 20.
C39. Crowhurst, N. H. " Making the best of an audio transformer " Audio Eng. 37.1 (Jan. 1953) 40.
C40. Crowhurst, N. H. " Audio transformer design " Audio Eng. 37.2 (Feb. 1953) 26.
C41. Crowhurst, N. H. "Audio Handbook No. 3 : The use of a.f. transformers" Norman Price (Publishers) Ltd., London, 1953.
C42. Halabi, T. " Audio transformer design charts " Elect. 20-10 (Oct. 1953) 193.   Charts showing primary inductance insertion loss, inductance with 4% silicon steel, leakage inductance, insertion loss due to leakage reactance, phase shift, and effect of primary inductance on reflected impedance.
C43. Howard, L. W. " Review of new materials and techniques in high fidelity transformer design " Jour. A.E.S. 1.3 (July 1953) 265.
C44. Ayres, W. R. " Output transformer design considerations " Audio Eng. 37.4 (April 1953) 14. Very brief survey.
C45. Lehnert, W. E. " Consideration of some factors concerning the use of audio transformers " Jour. A.E.S. 1.1 (Jan. 1953) 105.   Prediction of performance with source and load impedances other than rated ; magnetic distortion; noise reduction ; matching of several impedances simultaneously.

(D) Power transformers

D27. Medina, L. " Prevention of ionization in small power transformers " Proc. I.R.E. Australia 15.5 (May 1954) 114.   Application to transformers with voltages in excess of 2000 volts.

(E) Iron-cored inductors

E20. Crowhurst, N. H. " The design of high Q iron cored inductors " Electronic Eng. 25.309 (Nov. 1953) 478.   Design for max. Q with zero d.c.

CHAPTER 7
NEGATIVE FEEDBACK

SECTION 7: OVERLOADING OF FEEDBACK AMPLIFIERS ON TRANSIENTS

Negative-feedback amplifiers often distort a signal, such as a pulse, which changes rapidly with time, although the amplitude of the signal is less than that required to overload the amplifier when the rate of change of the signal is small.   This is because the feedback voltage changes more slowly than the input voltage, with the result that the voltage applied to the grid of one of the valves becomes large enough to drive it into grid current or beyond cut-off.

The case of a cathode follower with a capacitive load has been covered in Sect. 2(i)(Y) page 327.

The design of single-stage, two- and three-stage resistance-coupled amplifiers is covered in Ref. J3, with curves, of which the following is a summary. In single stage amplifiers the magnitude of signal required to overload the valve decreases as the rise-time of the signal is reduced. In two-stage amplifiers, the voltage applied to the first valve increases as the rise-time of the signal is reduced, but only if the gain of the amplifier is very small is it possible for the first valve to be overloaded by a signal which does not also overload the second valve. The second valve is, therefore, normally the first to overload. If the time-constant of the first stage is sufficiently large compared with that of the second stage, the input signal required to overload the second valve does not decrease as the rise-time of the signal is reduced. In three-stage amplifiers, the voltage applied to the third stage is never greater for a quick change than for a slow change. Therefore, if the first two stages are designed not to overload, the signal required to overload the amplifier is as large when it changes quickly as when it changes slowly.

See also Ref. J10.

ADDITIONAL REFERENCES TO NEGATIVE FEEDBACK

J1.   Shimmins, A. J. " Cathode follower operation—transient and steady-state performance with a capacitive load " W.E. 29.345 (June 1952) 155 ; letter H. H. Adelaar 30.2 (Feb. 1953) 49.

J2.   MacDiarmid, I. F. (letter) " Cathode-coupled amplifier " W.E. 29.345 (June 1952) 169.

J3.   Flood, J. E. " Negative feedback amplifiers, overloading under pulse conditions " W.E. 29.347 (Aug. 1952) 203.

J4.   Thomas, A. B. (letter) " Non-linearity in feedback amplifiers " Proc. I.R.E. 37.5 (May 1949) 531.

J5.   Shimmins, A. J. " Cathode follower performance " W.E. 27.327 (Dec. 1950) 289.

J6.   Mills, B. Y. " Transient response of cathode followers in video circuits " Proc. I.R.E. 37.6 (June 1949) 631.

J7.   Flood, J. E. " Cathode follower input impedance—effect of capacitive load " W.E. 28.335 (Aug. 1951) 231.

J8.   Cooper, V. J. " New amplifier techniques " J. Brit. I.R.E. 12.7 (July 1952) 371. (" Negative feedback amplifiers of desired amplitude frequency characteristics " (p. 384) deals with maximal and optimal flatness, based on Flood's maximal flatness.)

J9.   Baer, R. H. " Cathode follower response—Chart " Elect. 23.10 (Oct. 1950) 114.

J10.   Roddam, T. " Calculating transient response " W.W. 58.8 (Aug. 1952) 292. (Based on thesis by G. F. Floyd of M.I.T. and covers overloading in feedback amplifiers on transients.)

J11.   Bell, D. A. " Amplifier frequency response—effect of feedback " W.E. 29.344 (May 1952) 118 ; 29.349 (Oct. 1952) 281.

J12.   Bell, D. A. " Cathode follower as high-impedance input stage " W.E. 29.351 (Dec. 1952) 313.

J13.   Colls, J. A. " D.C. amplifiers with low-pass feedback " W.E. 29.351 (Dec. 1952) 321.

J14.   Miller, E. J. " A stable, high quality, power amplifier " Electronic Eng. 24.294 (Aug. 1952) 366.

J15.   Garner, L. E. " Improving amplifier response " Elect. 25.9 (Sept. 1952) 213. Letter H. L. Armstrong 25.11 (Nov. 1952) 432.

J16.   Wilson, J. " Design of the complete amplifier system " Audio Engineering Society Lecture No. 4 (1952).

J17.   Crowhurst, N. H. " Audio Handbook No. 2—Feedback " Norman Price (Publishers) Ltd., England, 1952.

J18.   Hekimian, N. C. " Chart speeds design of feedback amplifiers " Elect. 25.9 (Sept. 1952) 153.

J19.   Anspacher, W. B. " Miniaturizing pentode amplifiers by positive feedback " Proc. National Electronics Conference 6 (1950) 103.

J20.   Kean, A. W. " Anode-follower derivatives " W.E. 30.1 (Jan. 1953) 5.

J21.   Dunn, S. C. " RC cathode follower feedback circuits " W.E. 30.1 (Jan. 1953) 10.

J22.   Reeves, R. J. D. " Feedback amplifier design " Monograph No. 51, published Proc. I.E.E. Part IV (Dec. 1952).

J23.   Crowhurst, N. H. " A new approach to negative feedback design " Audio Eng. 37.5 (May 1953) 26.

J24.   Ayres, W. R. " Stability testing of feedback amplifiers " Audio Eng. 37.9 (Sept. 1953) 14.

J25.   Sokal, N. O. " Cathode-follower design charts " Elect. 26.9 (Sept. 1953) 192. Charts show output impedance vs. input voltage for nine tube types, and give required resistor value.

J26.   Crowhurst, N. H. " Why feed back so far ? " Radio Electronics 24.9 (Sept. 1953) 36.

J27.   Diamond, J. M. " Multiple-feedback audio amplifier " Elect. 26.11 (Nov. 1953) 148. (Note by editor : This uses feedback from plates of output stage to plates of preceding RC stage, and this is regarded as undesirable owing to additional distortion produced in preceding stage). See also Ref. J40. Letter by W. B. Bernard, Elect. 27.1 (Jan. 1954) 401.

J28.   Ayres, W. R. " Feedback from output transformer secondary " Audio Eng. 37.7 (July 1953) 34.

J29.   Kuehn, R. L. " Feedback—degenerative and regenerative " Audio Eng. 37.4 (April 1953) 23. Gives condition for oscillation.

J30.   Miller, J. M. " Amplifier with positive and negative feedback " U.S. Patent, 2,652,458 (Bendix). Reviewed by R. H. Dorf, Audio Eng. 37.12 (Dec. 1953) 2. Network in positive feedback circuit to ensure reversal of phase to improve stability.

J31.   Stockman, H. " Inherent feedback in triodes " W.E. 30.4 (April 1953) 94. Treats a triode as a pentode with negative voltage feedback. This transformation makes it possible to obtain practical triode circuit formula from conventional feedback theory.

J32. Rowlands, R. O. " Harmonic distortion and negative feedback " W.E. 30.6 (June 1953) 133. More rigorous method than usually employed. The value of A in the expression giving reduction in distortion is the slope of the output vs. input curve in the vicinity of the distortion. See also correspondence, W.E. 30.9 (Sept. 1953) 232-233 ; 30.10 (Oct. 1953) 262 ; 30.11 (Nov. 1953) 291. See also Refs. J35, J38.

J33. Mason, S. J. " Feedback theory—some properties of signal flow graphs " Proc. I.R.E. 41.9 (Sept. 1953) 1144.

J34. Ayres, W. R. " Feedback from output transformer tertiary " Audio Eng. 38.1 (Jan. 1954) 10.

J35. Roddam, T. " Distortion in negative feedback amplifiers—points at which simple theory breaks down " W.W. 60.4 (April 1954) 169. This is an extension of Ref. J32. See also Ref. J38.

J36. West, J. C., and J. Potts " A simple connection between closed-loop transient response and open-loop frequency response " Proc. I.E.E. 100 Part II. 75 (June 1953) 13. Digest, 100 Part III. 66 (July 1953) 250. Application primarily to servo-mechanisms.

J37. Onder, K. " A tone burst generator " J. Acous. Soc. Am. 25.6 (Nov. 1953) 1154. Frequency 12 kc/s, used to test feedback amplifiers. Correction 26.3 (May 1954) 453.

J38. Zepler, E. E. " Harmonic distortion and negative feedback " W.E. 31.5 (May 1954) 118. An expansion of the treatment in Ref. J32.

J39. Brady, J. W. " Cathode-coupled valves—graphical methods of design " W.E. 31.5 (May 1954) 111.

J40. Knapp, J. Z. " The linear Standard Amplifier " Radio and TV News 51.5 (May 1954) 43. See also Ref. J27 for same amplifier. Also compares transient and square-wave response with Williamson.

J41. Hekimian, N. C. " Feedback amplifiers with stabilized output impedances " Tele-Tech 12.6 (June 1953) 103. This is the same as Bridge Feedback, R.D.H. Pages 313-314.

J42. Whittle, R. L. " Design of cathode followers " Tele-Tech 12.7 (July 1953) 52.

J43. Favors, H. A. " Designing cathode followers for pulse type circuit " Tele-Tech 12.8 (Aug. 1953) 80. Note that the blocks of Figs. 3 and 6 have been interchanged.

CHAPTER 9
TUNED CIRCUITS

ADDITIONAL REFERENCES

(B) Theory of R-F single-tuned circuits
B28. Morris, D. " Q as a mathematical parameter" Electronic Eng. 26.317 (July 1954) 306. Suggested definition of Q for use in circuits with Q less than unity.

(C) Theory of tuned coupled circuits
C48. Polishuk, H. D. " High-Q coupled tuned circuits " W.E. 31.3 (March 1954) 428. Impedance characteristics, resonant frequencies, rate of frequency deviation, input conductance, stored energy, and power dissipation ratios.

CHAPTER 10
CALCULATION OF INDUCTANCE

ADDITIONAL REFERENCES

Approximate formulae for self and mutual inductance
25c. Löfgren, E. " Formulés approchés pour le calcul de l'inductance des bobines circulaires " Revue Générale de L'Electricité (Aug. 1949) 305.

25d. Löfgren, E. " Närmeformler fur induktansen hos runda spolar " Teknisk Tidskrift (Oct. 8, 1949) 711.

25e. Cosens, C. R. " Tapped inductances—calculation of tapping points " W.E. 31.3 (March 1954) 74. Formulae for circular coils of square cross-section with inner diameter twice side of square.

CHAPTER 11
DESIGN OF RADIO FREQUENCY INDUCTORS
ADDITIONAL REFERENCES

(A) References to iron cores
A31. Tucker, J. P. " Powder metal IF cores " TV Eng. 2.10 (Oct. 1951) 22.

A32. " Ferroxcube " Philips Tec. Com. 4 (1952) 3.

A33. Latimer, K. E., and H. B. MacDonald " A survey of the possible applications of ferrites " Communication News (Philips Telecommunication Industries, Hilversum, Holland) 11.3 (Sept. 1950) 76 ; reprinted in Philips Tec. Com. 4 (1952) 13.

A34. Wessels, P. S. " Design of slug-tuned superheterodyne receivers " Elect. 25.11 (Nov. 1952) 176.

A35. Champion, D. F. W., and E. G. Wilkins " Magnetic powder cores—manufacturing techniques and applications in radio and telephony " W.W. 59.2 (Feb. 1953) 83.

A36. Polydoroff, W. J. " Powdered magnetic cores " Tele-Tech 12.2 (Feb. 1953) 69.

A37. Owens, C. D. " Analysis of measurements on magnetic ferrites " Proc. I.R.E. 41.3 (March 1953) 359. Gives useful bibliography.

A38. Hoh, S. R. " Evaluation of high-performance magnetic core materials " Tele-Tech 12.10 (Oct. 1953) ; 12.11 (Nov. 1953) 92. Curves showing core loss, a.c. characteristics and apparent permeability of laminated, powdered and ferrite materials at low flux densities.

A39. Salpeter, J. L. " Developments in sintered magnetic materials " Proc. I.R.E. Aust. 14.5 (May 1953) 105. Reprinted in J. Brit. I.R.E. 13.10 (Oct. 1953) 499 ; Proc. I.R.E. 42.3 (March 1954) 514.

A40. Went, J. J. and E. W. Gorter " The magnetic and electrical properties of Ferroxcube materials " Philips Tec. Rev. 13.7 (Jan. 1952) 181. Abstract Philips Tec. Com. 4 (1953) 18.

A41. Went, J. J., G. W. Rathenau, E. W. Gorter and G. W. van Oosterhaut, " Ferroxdure, a class of new permanent magnetic materials " Philips Tec. Rev. 13.7 (Jan. 1952) 194. Abstract Philips Tec. Com. 4 (1953) 18.
A42. Six, W. " Some applications of Ferroxcube " Philips Tec. Rev. 13.11 (May 1952) 301.
A43. " Applications and properties of Ferroxcube " Philips Tec. Com. 6 (1953) 11 ; 1 (1954) 20. Reprinted from Electronic Application Bulletin 13. 3/4 (March/April 1952).
A44. Richards, C. E., and A. C. Lynch (book) " Soft Magnetic Materials for Telecommunications " Interscience Publishers, New York, 1953. Review Tele-Tech 13.3 (March 1954) 52.
A45. Thomas, L. A. " Modern trends in communication materials " Jour. Brit. I.R.E. 13.7 (July 1953) 356.
A46. Rohan, P. " Notes on permeability tuning for short waves " Proc. I.R.E. Australia 15.5 (May 1954) 111. Discussion and formulae for calculation of components of the resonant circuits for band-changing and band-spreading are derived.
A47. Harvey, R. L. " Ferrites and their properties at radio frequencies " Proc. Nat. Electronics Conf. 9 (1953) 287.
A48. Harvey, R. L. " Ferrite characteristics at radio frequencies " Tele-Tech 13.6 (June 1954) 110.

(B) References to inductance calculation
For curves assisting design of inductors for loudspeaker divider networks see Chapter 21 Refs. 26, 27.

CHAPTER 12
AUDIO FREQUENCY VOLTAGE AMPLIFIERS

ADDITIONAL REFERENCES

(A) Resistance-capacitance-coupled triodes
A17. Ayres, W. R. " Power and voltage amplifiers " Audio Engineering Society Lecture (17 Jan. 1952).
A18. Kruse, O. " Circle diagrams for resistance-capacitance coupled amplifiers " Audio Eng. 37.2 (Feb. 1953) 22. Correction 37.3 (March 1953) 55.
A19. Ayres, W. R. " R-C coupled amplifier charts " Audio Eng. 37.10 (Oct. 1953) 12. Comments on published charts (R.C.A.) and operating conditions of R.C. amplifiers.
A20. Stockman, H. (letter) " Degenerative pentode equivalent circuit " Proc. I.R.E. 41.6 (June 1953) 801. Derives useful formulae.
A21. Goodfriend, L. S. See Ref. B14.
A22. Pullen, K. A. " Conductance curves speed triode r-c amplifier design " Tele-Tech 12.5 (May 1953) 80.

(B) Resistance-capacitance-coupled pentodes
B13. Haycock, J. G. " Pentode gain stabilizing circuit " Elect. 26.11 (Nov. 1953) 200. Stabilizes gain by using voltage-sensitive resistor in low potential section of screen voltage divider of r.c. pentode.
B14. Goodfriend, L. S. " Bypass and decoupling circuits in audio design " Jour. A.E.S. 1.1 (Jan. 1953) 111. Mathematical treatment of partially bypassed cathode and screen resistors, giving gain and phase angle. Also triodes with decoupling network.
B15. Pullen, K. A. " Conductance curves speed r-c amplifier design " Tele-Tech 12.7 (July 1953) 44.

(C) Phase inverters
C28. Bourget, L. R. " Phase splitter " U.S. Patent No. 2,618,711. Described by R. H. Dorf " Audio Patents " Audio Eng. 37.3 (March 1953) 4. Claims perfect balance 20 to 150,000 c/s.
C29. Wen Yuan Pan, " Phase inverter with reduced hum " U.S. Patent No. 2,626,321 (R.C.A.). Described by R. H. Dorf, Audio Eng. 37.6 (June 1953) 4, with circuit diagram. Uses no additional components.
C30. Varkonyi, G. " Cross-coupled inverter " Audio 38.5 (May 1954) 8. Comparison between cross-coupled and split-load—former is deficient at high frequencies and when directly coupled to driver tubes and negative feedback is used, serious trouble is experienced with biasing of drivers, also dynamic balance upset. Split-load type is excellent at high frequencies but gives less low frequency stability. See also letter J. Marshall Audio 38.6 (June 1954) 14.
C31. Boegli, C. P. " Simplified cross-coupled amplifier " Radio and TV News 51.5 (May 1954) 62. Eliminates input tubes of original circuit with some increase in distortion.

(D) Direct-coupled amplifiers
D42. McDonald, D. " Constant current d-c amplifier " Elect. 25.7 (July 1952) 130.

(F) Pulse amplifiers and transients
F2. Boegli, C. P. " Transient and frequency response in audio equipment " Audio Eng. 38.1 (Feb. 1954) 19. Mathematical analysis of the uptake characteristic when unit step input signal is applied to amplifier or pickup.

(H) General
H1. Sodaro, J. F. " The pass band of a transformer-coupled amplifier " Audio Eng. 37.6 (June 1953) 24. Also gives abac for 1 and 3 db attenuation frequencies of a transformer.
H2. Villchur, E. M. " Handbook of sound reproduction—Chapter 13, Voltage amplifiers and phase splitters " Audio Eng. 37.10 (Oct. 1953) 42.
H3. Crowhurst, N. H. (booklet) " Amplifiers " (Norman Price, London, 1951).

CHAPTER 13
AUDIO FREQUENCY POWER AMPLIFIERS

ADDITIONAL REFERENCES

H1. Peterson, A. P. G. " A new push-pull amplifier circuit " G.R. Exp. 26.5 (Oct. 1951) 1. See also Refs. E32, H2, H29, H45.

H2. " Single-ended push-pull amplifier " W.W. 58.5 (May 1952) 203. See also Refs. E32, H1, H29, H45.

H3. Brociner, V., and G. Shirley " The OTL (output-transformer-less) amplifier " Audio Eng. 36.6 (June 1952) 21. Correspondence W. H. and J. R. Coulter ; L. Bourget, 36.9 (Sept. 1952) 10, 14. See also Refs. H36, H43, H45.

H4. Moir, J. " Review of British amplifiers " FM-TV 11.10 (Oct. 1951) 30.

H5. Hafler, D., and H. I. Keroes " Ultra-linear operation of the Williamson amplifier " Audio Eng. 36.6 (June 1952) 26. See also Refs. H6, H11.

H6. Williamson, D. T. N., and P. J. Walker " Amplifiers and superlatives—an examination of American claims for improving linearity and efficiency " W.W. 58.9 (Sept. 1952) 357. See also Ref. H5.

H7. Sarser, D., and M. C. Sprinkle " Musician's amplifier " Audio Eng. 33.11 (Nov. 1949) 11.

H8. Sarser, D., and M. C. Sprinkle " Musician's amplifier senior " Audio Eng. 35.1 (Jan. 1951) 13.

H9. Sarser, D., and M. C. Sprinkle " The Maestro—a POWER amplifier " Audio Eng. 36.11 (Nov. 1952) 19. Gives output 80 watts at 2% intermodulation.

H10. Beaumont, J. H. " Williamson type amplifier using 6A5's " Audio Eng. 34.10 (Oct. 1950) 24.

H11. Hafler, D., and H. I. Keroes " An ultra-linear amplifier " Audio Eng. 35.11 (Nov. 1951) 15. See also Ref. H5.

H12. Kiebert, M. V. " The Williamson type amplifier brought up to date " Audio Eng. 36.8 (Aug. 1952) 18.

H13. Miller, E. J. " A stable, high quality, power amplifier " Electronic Eng. 24.294 (Aug. 1952) 366.

H14. Werner, C. L., and H. Berlin " New medium-cost amplifier of unusual performance." Audio Eng. 36.11 (Nov. 1952) 30.

H15. Williamson, D. T. N. " High quality amplifier modifications " W.W. 58.5 (May 1952) 173.

H16. " Leak Point One Amplifiers " booklet by H. J. Leak and Co. Ltd., Brunel Road, Westway Factory Estate, London, W.3.

H17. Pullen, K. A. " Using conductance curves in electronic circuit design " Proc. National Electronics Conference Vol. 6 (1950) 112. See also Chapter 2 Refs. B14, B22, B32.

H18. Good, E. F. " RC or direct-coupled power stage " W.E. 30.3 (March 1953) 54. Gives conditions for maximum efficiency for ideal triode with parallel feedback.

H19. Bender, W. G. " A power tube figure of merit " Audio Eng. 37.3 (March 1953) 21. Letters A. J. L. Prasil 37.5 (May 1953) 10, G. B. Houck and W. G. Bender 37.6 (June 1953) 10.

H20. Postal, J. " Simplified push-pull theory—a graphical, non-mathematical explanation," Audio Eng. (1) 37.5 (May 1953) 19 ; (2) 37.6 (June 1953).

H21. Bogen, L. H., and A. M. Zuckerman " Loudness contour selector in new amplifier " Audio Eng. 37.5 (May 1953) 31. David Bogen DB20 amplifier, distortion 0.3% total harmonic at 20 watts, 0.25% at 15 watts, 0.2% at 10 watts 1000 c/s. Combined plate and cathode loading. Loudness contour-selection 5 positions. See also Ref. H25.

H22. Werner, C. L., and H. Berlin " Everyman's amplifier—new low cost ten watt unit described as the Ford of the Hi-Fi industry " Audio Eng. 37.10 (Oct. 1953) 40. Distortion 0.25% up to 9 watts output, 1% at 10 watts.

H23. White, S. " The White Powtron Amplifier " Audio Eng. 37.11 (Nov. 1953) 32. Uses ultra-linear amplifiers, with 2 channels, 20 watts main amplifier, 10 watts treble amplifier, with frequency dividing network prior to both amplifiers. No distortion figures quoted. Employs mainly negative voltage feedback, but also small degree of negative current feedback ; this is claimed to eliminate " power distortion " caused by variation in loudspeaker impedance. See comment Chapter 12 Ref. C30.

H24. Langford-Smith, F. " Limiting Class A operation—a useful device for good quality push-pull power amplifiers " Radiotronics 18.10 (Oct. 1953) 177.

H25. Frieborn, J. K. " High quality circuits " Radio Electronics 24.9 (Sept. 1953) 33. Includes Brociner UL-1 (ultra-linear), Bell 2200 (combined plate and cathode loaded), Bogen DB 20 (combined plate and cathode loaded—see also Ref. H21—with distortion curves) and Stromberg-Carlson AR-425 (pentodes with overall feedback—with distortion curves).

H26. Hust, L. B. " Extended Class A amplifier " Radio and TV News 50.3 (Sept. 1953) 40. Two triodes and two pentodes in push-pull parallel, output 50W. See R.D.H. p. 587 and Refs. E31, E13.

H27. Marshall, J. " Junior Golden-Ear amplifier " Radio Electronics 24.11 (Nov. 1953) 55. Modified ultra-linear with push-pull 6V6. See also Ref. H33.

H28. Crowhurst, N. H. (booklet) " Amplifiers " (Norman Price, London, 1951).

H29. Yeh, Chai " Analysis of a single-ended push-pull audio amplifier " Trans. I.R.E.—PGA. AU-12. (March-April 1953) 9. Theoretical analysis and experimental results. See also Refs. E32, H1, H2.

H30. Onder, K. " A new transformerless amplifier circuit " Jour. A.E.S. 1.4 (Oct. 1953) 282.

H31. Corderman, S. A., and F. H. McIntosh " A new 30-watt power amplifier " Jour. A.E.S. 1.4 (Oct. 1953) 292. A Class AB1 McIntosh amplifier. See also Ref. E28.

H32. Pomper, V. H., " The Scott 99A Amplifier " Radio and TV News 51.2 (Feb. 1954) 66. 10 watts output, distortion 0.8%, hum 80 db below full output, pre-amplifier on same chassis. See also Ref. H38.

H33. Marshall, J. " The new Golden-Ear amplifier " Audio Eng. 38.1 (Jan. 1954) 17 ; pre-amplifier and tone control 38.2 (Feb. 1954) 22. Power output 20 watts ; distortion not stated. See also Ref. H27. See comment Chap. 12 Ref. C30.

H34. Macpherson, C. H. " A medium-power tetrode amplifier with stabilized screen supply " Audio Eng. 38.2 (Feb. 1954) 30. Complete with pre-amplifier and tone control on single chassis. Output stage 6V6-GT. Intermodulation distortion less than 0.5% up to 8 watts, 1.2% at 10 watts.

H35. Sterling, H. T., and A. Sobel " Constant current operation of power amplifiers " Jour. A.E.S. 1.1 (Jan. 1953) 16. Also Elect. 26.3 (March 1953) 122. Uses push-pull parallel 5881's in " ultra-linear " Class A2 with high load resistance (12,000 ohms P-P) to give an approach towards constant current operation. Each cathode with its own resistor and bypass ; matching of valves and subsequent adjustments not required. Driver stage P-P 12B4's as triode cathode followers, directly coupled.

H36. Onder, K. " Audio amplifier matches voice-coil impedance " Elect. 27.2 (Feb. 1954) 176. Balanced transformerless amplifiers ; outputs 8 and 18 watts. See also Ref. H3.

H37. Roddam, T. " Grounded-grid A.F. amplifier " W.W. 60.5 (May 1954) 214. Increased power output, using positive feedback and negative overall feedback.

H38. " Tested in the Home : Scott 99-A Amplifier " High Fidelity 4 : 2 (April 1954) 81. Output tubes balanced automatically. See also Ref. H32.

H39. Langford-Smith, F. "Triodes versus pentodes in high-fidelity output stages" Radiotronics 19.7 (July 1954) 73.
H40. "Equipment report—QUAD II" Audio 38.5 (May 1954) 28.
H41. Bereskin, A. B. "A high efficiency-high quality audio frequency power amplifier" Trans. I.R.E. PGA. AU-2.2 (March/April 1954) 49. Push-pull 807's in Class B1 with 24 db feedback from tertiary, output 50W for 0.7% distortion at 400 c/s. Direct coupling to output stage. Does not require matched valves. Permits different plate and screen voltages. Uses special bi-filar output transformers.
H42. Hafler, D. "Ultra-linear operation of 6V6 tubes" Radio and TV News 51.6 (June 1954) 43.
H43. Gilbert, F. H. "The Stephens OTL amplifiers" Radio and TV News 49.3 (March 1953) 45. No output transformer, triode output, distortion 0.1% at 18 watts. See also Ref. H3.
H44. Marshall, J. "The importance of balance in push-pull amplifiers" Radio Electronics 24.7 (July 1953) 28.
H45. Dickie, D. P., and A. Macovski "A transformer-less 25-watt amplifier for conventional loudspeakers" Audio 38.6 (June 1954) 22. Output 20 W into 16 ohms with 0.4% harmonic distortion. Three type 6080 valves with 6 triode units in parallel. See also Refs. H1, H3.

CHAPTER 14
FIDELITY AND DISTORTION

ADDITIONAL REFERENCES

(A) Distortion and fidelity—general
A56. Schjelderup, J. R. "A proposed solution to the loudness control problem" Audio Eng. 36.9 (Sept. 1952) 34.
A57. Robbins, J. G. "The acoustic significance of the amplitude and phase of harmonics present in a source of sound in a room" J. Acous. Soc. Am. 24.4 (July 1952) 380.
A58. DeLange, O. E. "Distortion measurement" U.S. Patent No. 2,618,686 (see R. H. Dorf, Audio Patents, Audio Eng. 37.3 (March 1953) 2. Visual method employing outphasing principle.
A59. Lampard, D. G. "Harmonic and intermodulation distortion in 'power law' devices" Proc. I.E.E. 100.5 Part IV (Oct. 1953) 3. The calculation of the amplitudes of harmonic and intermodulation components produced when two sinusoidal voltages are applied to a device whose transfer characteristic is a simple power law—e.g. in variable density sound-on-film and use of diodes in a.g.c. circuits.
A60. Aerovox, "Phase shift distortion test" reprinted Radio and Hobbies, Australia, 15.11 (Feb. 1954) 37. Combines output with input, phase reversed 180 degrees, to show distortion on CRO.
A61. Tyler, V. J. "Simple distortion meter" W.W. 59.9 (Sept. 1953) 431. Uses filter-amplifier to remove fundamental frequency (letter) G. H. Askew and R. Malchell, W.W. 59.12 (Dec. 1953) 582.
A62. Wigan, E. R. "Diagnosis of distortion—The Difference Diagram and its interpretation" W.W. 59.6 (June 1953) 261.
A63. Peterson, A. G. "The measurement of non-linear distortion" Presented at the I.R.E. Convention, March 1949. Technical Publication B-3, General Radio Company, Cambridge, Mass. Compares various methods of measurement.
A64. Pressey, D. C. "Measuring non-linearity" W.W. 60.2 (Feb. 1954) 60. The fundamental is subtracted by frequency-insensitive element, using adding amplifier with one valve, and test signals need not be pure sine wave. Corrections 60.3 (Mar. 1954) 128.
A65. "High fidelity—what is it? Some suggestions for a high-fidelity yardstick" Jour. A.E.S. 2.1 (Jan. 1954) 56.
A66. Bloch, A. "Measurement of non-linear distortion" Jour. A.E.S. 1.1 (Jan. 1953) 62. The harmonic, heterodyne (CCIF) and intermodulation methods of measurement are examined from a mathematical standpoint.
A67. Maxwell, D. E. "Comparative study of methods for measuring non-linear distortion in broadcasting audio facilities" Jour. A.E.S. 1.1 (Jan. 1953) 68. Compares harmonic, I.M. and C.C.I.F. methods for essentially pure quadratic and cubic distortion, with frequency response uniform; limited to 8,000 c/s; and increasing at low and high frequencies.
A68. Lampard, D. G. "Harmonic and intermodulation distortion in 'Power law' devices". I.E.E. Monograph, Dec. 1952. Digest Proc. I.E.E. 100. Part III 64 (March 1953) 111. Mathematical treatment deriving both intermodulation and harmonic distortion in terms of the index.
A69. I.R.E. Standards on Circuits: Definitions of Terms in the field of Linear Varying Parameter and Non-Linear Circuits 1953. Proc. I.R.E. 42.3 (March 1954) 554.
A70. Dadson, R. S. "The normal threshold of hearing and other aspects of standardisation in audiometry" Acustica 4.1 (1954) 151. NPL determination of normal threshold of hearing. Results differ significantly from those of Sivian and White and given by Fletcher (Ref. A3).
A71. Jones, E. M. "How much distortion can you hear?" Trans. I.R.E. PGA. AU-2.2 (March/April 1954) 42. 35% detected an increase of distortion from 0.3% to 0.9% at 1000 c/s and an increase from 0.8% to 1.7% at 100 c/s; 28% detected an increase in I.M. distortion from 1.5% to 3.7% on tape; 21% detected 1.3% I.M. distortion on a live performance.

(B) Intermodulation distortion
B23. Berth-Jones, E. W. "Intermodulation distortion—its significance and measurement" Jour. Brit. I.R.E. 13.1 (Jan. 1953) 57.
B24. Scott, H. H. "Intermodulation measurements" Jour. A.E.S. 1.1 (Jan. 1953) 56. Describes the two main types of intermodulation measurements—modulation meters and beat tone measurements, with comments. Good bibliography.

(D) Limited range, speech and noise
D20. Dadson on speech analysis, Acoustical Society of America, J. Acous. Soc. Am. 24.6 (Nov. 1952) 581-642 (10 papers).
D21. Hirsh, I. J., and W. D. Bowman "Masking of speech by bands of noise" J. Acous. Soc. Am. 25.6 (Nov. 1953) 1175.

CHAPTER 15
TONE COMPENSATION AND TONE CONTROL
Additional Tone Control Circuit
(see page 669)

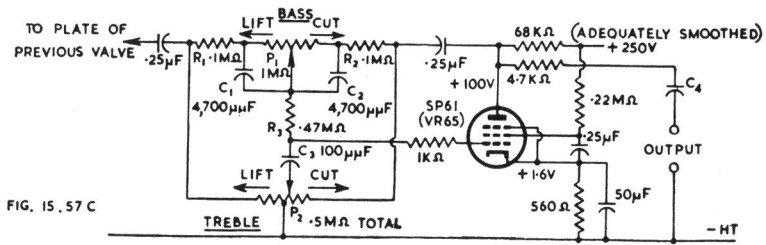

Fig. 15.57C. *Tone control circuit. Tolerances* R_1, R_2, C_1, C_2, R_3, $C_3 \pm 5\%$. P_1 *and* P_2 *are linear potentiometers,* P_2 *having a fixed tapping at* 50% *rotation. The source impedance should not be more than 10,000 ohms.* C_4 *should normally be* 0.05 μF *if following stage has* 0.25 $M\Omega$ *grid leak (Ref. 91). Type 6AU6 could be used as substitute for the type shown, with suitable values of cathode and screen resistors, to give an output of about 2 V r.m.s.*

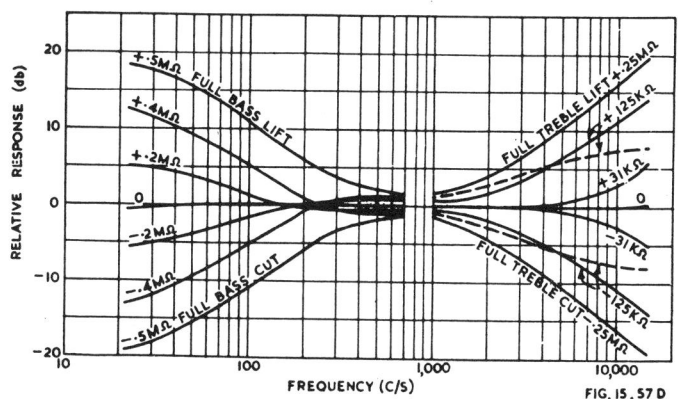

Fig. 15.57D. *Measured frequency response curves of circuit of Fig. 15.57C. Labels on curves are resistance values between potentiometer slider amd centre of element. Dotted curves are with* P_2 *centre-tap disconnected from earth and with one* 0.33 $M\Omega$ *resistor connected from each end of* P_2 *to earth (Ref. 91).*

ADDITIONAL REFERENCES

91. Baxandall, P. J. "Negative feedback tone control—independent variation of bass and treble without switches" W.W. 58.10 (Oct. 1952) 402. Correction 58.11 (Nov. 1952) 444.
92. Douglas, G. A. "Simplified equalizer design—charts and tables to reduce complication and construction hints to ease building" Audio Eng. 36.12 (Dec. 1952) 18.
93. "The 'Vari-Slope' Pre-amplifier" H. J. Leak and Co. Ltd., Brunel Road, Westway Factory Estate, Acton, London, W.3. Uses modified twin-T resistor-capacitor networks in negative feedback loops to give continuously-variable slope of attenuation characteristic (from 5 db to 50 db over the octave immediately following the cut-off frequency), and choice of two cut-off frequencies. See also Ref. 99.
94. Villard, O. G., and D. K. Weaver "The Selectoject" Q.S.T. (Nov. 1949) 11 ; A. Q. Morton "Oscillator/filter unit" W.W. 59.3 (March 1953) 129.
95. Villchur E. M. "The selection of tone control parameters" Audio Eng. 37.3 (March 1953) 22

96. Mountjoy, G., and C. R. Shafer " Tone control circuits " U.S. Patent 2,626,991 (Stromberg-Carlson) described by R. H. Dorf, Audio Patents, Audio Eng. 37.6 (June 1953) 4.   Two circuits shown, with curves, each providing bass boosting, one with potentiometer control.
97. Barber, B. T. " Flexible tone control " Audio Eng. 37.9 (Sept. 1953) 29.   Americanized form of Baxandall circuit, Ref. 91.
98. Villchur, E. M. " Handbook of sound reproduction—Chapter 14.   Tone control and equalization " Audio Eng. 37.11 (Nov. 1953) 25.
99. Crowhurst, N. H. " British audio circuits " Radio Electronics 24.11 (Nov. 1953) 74.   Telrad, Leak " Vari-slope " and QUAD tone control circuits.
100. Dundovic, J. F. " A three-channel tone-control amplifier " Audio Eng. 37.4 (April 1953) 28.   Provides bass and treble boosting of varying slope and fixed hinge-point.   Correction (new diagram) 37.12 (Dec. 1953) 20.
101. Sisson, E. D. " Resistance-capacitance networks in amplifier design " Jour. A.E.S. 1.1 (Jan. 1953) 116.   RC networks reduced to 2 basic types, and attenuation and phase angle characteristics given.
102. Blies, F. R. " Attentuation equalizers " Jour. A.E.S. 1.1 (Jan. 1953) 125.   Comprehensive treatment of equalizers to correct overall gain-frequency characteristic with 11 charts.
103. John, R. S. " Dynamic loudness control " Radio and TV News, R.E.E. Supplement 49.5 (May 1953) 10.   General principles and circuit giving frequency compensation varying with instantaneous level.
104. O'Leary, M. G. " Loudness control : the good and bad features of some popular types " Radio Electronics 24.8 (Aug. 1953) 48.

CHAPTER 16

VOLUME EXPANSION, COMPRESSION AND LIMITING

ADDITIONAL REFERENCES

88. Pope, G. J. " Design for a constant volume amplifier " Electronic Eng. 24.296 (Oct. 1952) 464.   Letters B. D. Corbett and G. J. Pope, 24.298 (Dec. 1952) 580.
89. Roberts, D. E. Volume compressor, U.S. Patent, 2,596,510.   See R. H. Dorf " Audio patents" Audio Eng. 36.12 (Dec. 1952) 2.
90. Scott, R. F. " Volume expanders and compressors " Radio Electronics 24.3 (March 1953) 41.
91. Culicetto, P. J. " Volume expander and compressor " U.S. Patent 2,615,999.   Described by R. H. Dorf Audio Eng. 35.5 (May 1953) 2.   Second harmonic 0.35%, third harmonic 0.15% with 17 db expansion ; or output level ± 3 db for 22 db input level changes.   Tubes 2/6 SNT-GT, 1-6J5, 1-6H6.
92. Schouten, G. H. " A.G.C. by means of miniature NTC resistors with heating element " Philips Tec. Com. 7 (1953) 9.   Reprinted from Electronic Application Bulletin 12.2 (Feb. 1951) 33.   May be applied to public address a.v.c.
93. Roberts, F. W., and R. C. Curtis " Audio automatic volume control systems " Jour. A.E.S. 1.4 (Oct. 1953) 310.   Useful summary of limiting, compression and public address a.v.c. devices.
94. Nigro, J., and J. B. Minter " Concert-hall realism through the use of dynamic level control " Jour. A.E.S. 1.1 (Jan. 1953) 160.   Uses 6SK7 with cathode bias, screen varying from − 2 to + 5 volts, and suppressor from 50 to 150 volts for dynamic control, and frequency response is a function of the output level.

CHAPTER 17

REPRODUCTION FROM RECORDS

DISCS AND STYLI (Continued from page 709)

(D) R.C.A. 45 r.p.m. Extended Play (EP) records

These records have a maximum playing time of 7.9 minutes and the following characteristics :

Grooves per inch, normal max.	300
Peak recording velocity	14 cm/sec.
Diameter innermost music groove	4¼ in. min.
Groove width	2.5 to 3.0 mils
Minimum permissible groove width	2.2 mils

The lead-out groove is reduced in length due to the smaller ending diameters. All other factors (except intermodulation distortion and groove velocity) as shown on pages 708 and 709.   For shorter selections (3 to 4 minutes or so) the number of grooves per inch is selected to bring the last music groove to 4.875 inches, and the lead-out groove is as described on pages 708 and 709.   For longer selections the number of grooves per inch is gradually increased and minimum recording diameter decreased simultaneously to the limiting values stated above.

SECTION 5 (i) Standard Playback Curve
(Continued from pages 731-732)

The original AES Standard Playback Curve (Fig. 17.15A) has now been revised, and the new curve (Fig. 17.15D Curve B) has been adopted by the RIAA, AES, NARTB (transcriptions) and leading American phonograph manufacturers. The original curve is also shown as Curve A in Fig. 17.15D to enable a direct comparison to be made.

This curve may be duplicated on a flat amplifier by the RC network of Fig. 17.15E following a triode, or that of Fig. 17.15F following a pentode.

The history and details of the new curve are given by Moyer (Ref. 346).

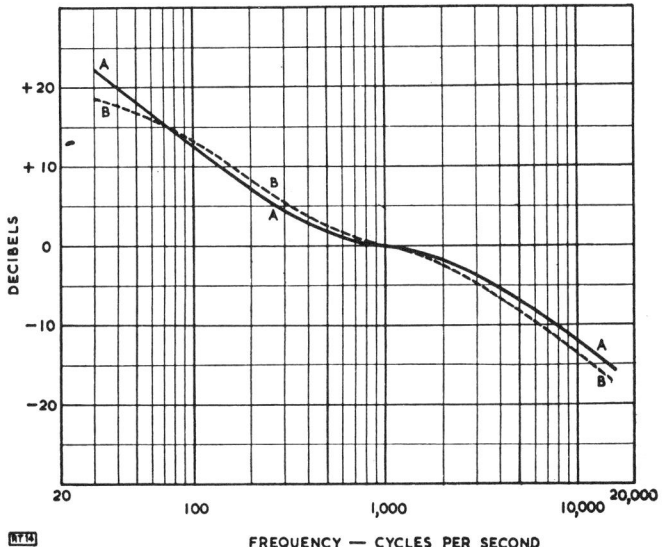

Fig. 17.15D.   (A) Old AES Standard Playback Curve; (B) New RIAA—AES—NARTB—RCA New Orthophonic Standard Playback Curve (RT14).

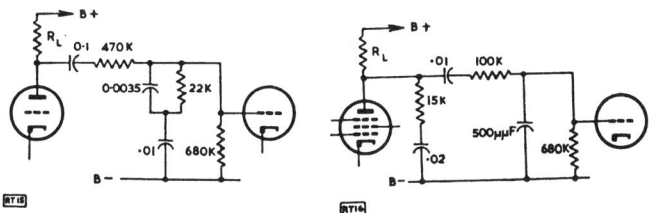

Fig. 17.15E.   Equalizing circuit following a triode (RT15).
Fig. 17.15F.   Equalizing circuit following a pentode (RT16).

ADDITIONAL STANDARD FREQUENCY TEST RECORDS
(Continued from pages 753-757)

LXT 2695   Decca Microgroove Frequency Test Record

This record has been cut at 33-1/3 r.p.m. with the groove width, at the top, of 0.0037 of an inch (0.95mm) with an included angle of $90° + $ or $- 1°$ and a radius at the bottom of the groove of less than 0.0003. The recording is from outside to inside in bands of constant frequency and in the following order :—

15 Kc/s (+ 12.5 db) ; 14 Kc/s (+ 13.1 db) ; 13 Kc/s (+ 12.9 db) ; 12 Kc/s (+ 12.0 db) ; 11 Kc/s (+ 11.5 db) ; 10 Kc/s (+ 10.5 db) ; 9 Kc/s (+ 10.1 db) ; 8 Kc/s (+ 9.2 db) ; 7 Kc/s (+ 8.5 db) ; 6 Kc/s (+ 7.3 db) ; 5 Kc/s (+ 5.9 db) ; 4 Kc/s (+ 4.6 db) ; 3 Kc/s (+ 3.6 db) ; 2 Kc/s (+ 1.9 db) ; 1 Kc/s (0 db) ; 500 c/s (− 2.3 db) ; 250 c/s (− 6.6 db) ; 125 c/s (− 9.0 db) ; 60 c/s (− 11.7 db) ; 40 c/s (− 13.9 db).

The recorded velocity at 1,000 cps is 1.2 cm per sec. r.m.s. These levels are accurate to within + or − 0.5 db.

This record should be played with a pickup using a point radius of 0.001 of an inch and with a vertical force of not greater than 10 grams.

ADDITIONAL REFERENCES : REPRODUCTION FROM RECORDS

318. Godfrey, J. W. and S. W. Amos (book) " Sound Recording and Reproduction " (B.B.C.) Iliffe and Sons Ltd., London, 1952.
319. Williamson, D. T. N. " High quality amplifier modifications " W.W. 58.5 (May 1952) 173.
320. " What is a recording characteristic " W.W. 58.5 (May 1952) 178.
321. Kelly, S. " Further notes on thorn needles " W.W. 58.6 (June 1952) 243.
322. Rabinow, J., and E. Codier " Phonograph needle drag distortion " J. Acous. Soc. Am. 24.2 (March 1952) 216.
323. New method of noise reduction which knows what noise is present—John M. Miller Jr. (assigned to Bendix Aviation) U.S. Pat. 2,589,723. See summary R. H. Dorf. " Audio Patents " Audio Eng. 36.7 (July 1952) 2.
324. Bauer, B. B. " The wear of phonograph needles " Trans. I.R.E. Porfessional Group Audio (Nov. 1951).
325. Reiskind, H. I. " Design interrelations of records and reproducers " Trans. I.R.E. Professional Group Audio, PGA-5 (Feb. 1952) 1.
326. Moir, J. " Review of British amplifiers " FM-TV 11.10 (Oct. 1951) 30. Gives circuits of Acoustical pre-amplifier and QUAD main amplifier.
327. Beggs, G. E. " Precision pre-amplifier " Elect. 25.7 (July 1952) 121.
328. Kiebert, M. V. " A pre-amplifier switching and equalizing unit for critical listening " Audio Eng. 36.9 (Sept. 1952) 21.
329. Voigt, P. G. A. H. " Some pickup design considerations " Audio Eng. 36.10 (Oct. 1952) 64.
330. Shirley, G. " Danger ! worn needles " High Fidelity 1 : 4 (Spring 1952) 28.
331. McLachlan, K. R., and R. Yorke " Objective testing of pickups and loudspeakers " J. Brit. I.R.E. 12.9 (Sept. 1952) 485.
332. Bixler, O. C. " A commercial binaural recorder " Jour. S.M.P.T.E. 59.2 (Aug. 1952) 109.
333. Markow, E. W. " Record improvement with H-F cut-off filters " Audio Eng., 36.11 (Nov. 1952) 27.
334. Cook, E. " Recording binaural sound on discs " Tele-Tech 11.11 (Nov. 1952) 48.
335. Cook, E. " Binaural discs " High Fidelity 2.3 (Nov.-Dec. 1952) 33.
336. John, R. S. " Constant amplitude pickup compensation " Radio Electronic Engineering (April 1951) 10A.
337. " Elements and Practice of Sound Recording." Lectures presented by the New York Chapter of the Audio Engineering Society (Nov. 10, 1949 to Feb. 23, 1950).
     1. Psychoacoustical aspects of the recording problem, (A) H. F. Olson, (B) W. B. Snow.
     2. The recording process—a survey, (A) C. J. Le Bel, (B) C. R. Sawyer.
     3. Disc recording—lathes, recording heads, reproducers, (A) T. Lindberg, (B) N. C. Pickering.
     4. Disc recording—characteristics, distortion, lacquers, styli, (A) H. E. Roys, (B) E. Cook.
     5. Disc recording—test procedures and processing, (A) F. W. Roberts, (B) K. R. Smith. and others on magnetic and film recording etc.
338. Carlson, E. V. " A ceramic vibration pickup " Trans. I.R.E.—PGA PGA-10 (Nov.-Dec. 1952) 2.
339. Pickering, N. " Effect of load impedance on magnetic pickup response " Audio Eng. 37.3 (March 1953) 19.
340. Goldmark, P. C. " The Columbia ' 360 ' " Audio Eng. 37.3 (March 1953) 28. Also E. T. Canby " Record Revue " page 46.
341. Woodward, J. G., and J. B. Halter " The measurement of the lateral mechanical impedance of phonograph pickup " J. Acous. Soc. Am. 25.2 (March 1953) 302. See also Ref. 342.
342. Woodward, J. G., and J. B. Halter " The lateral mechanical impedance of phonograph pickups " Audio Eng. (1) 37.6 (June 1953) 19 ; (2) 37.7 (July 1953) 23. Measures complex mechanical impedance 30-10,000 c/s, with equivalent mechanical system of crystal pickup. Curves of mechanical impedance for various phonographs and pickups. See also Ref. 341.
343. Roys, H. E. " Distortion in phonograph reproduction " Jour. A.E.S. 1 : 1 (Jan. 1953) 78. Re-printed R.C.A. Rev. 14.3 (Sept. 1953) 397. See also Ref. 382.
344. Foster, E. F. " Sharpening thorn needles—simple machine gives well-shaped and burnished points " W.W. 59.12 (Dec. 1953) 564.
345. " The Ferranti pickup " High Fidelity 3.5 (Nov. 1953) 109.
346. Moyer, R. C. " Evolution of a recording curve " Audio Eng. 37.7 (July 1953) 19. Gives early history and R.C.A. " New Orthophonic " curve.
347. Wood, J. F. " A new wide range phonograph cartridge " Audio Eng. 37.12 (Dec. 1953) 22. Correction 38.2 (Feb. 1954) 60. Barium titanate cartridge, damped by silicone, 30-15,000 c/s ± 2.5 db on basis of New Orthophonic curve without equalization, lateral compliance 10⁻⁶ cm/dyne or higher, output 0.6 volt, intermodulation distortion well below 2% throughout entire recorded range. Electro-Voice Model 84. For comments see E. T. Canby, Audio 38.3 (March 1954) 42.
348. Minter, J. B. and A. R. Miccioli " Effect of high-frequency pre-emphasis on groove shape " Jour. A.E.S. 1.4 (Oct. 1953) 321. Analysis of step function input with oscillograms of groove shape, giving values of angle which maximum slope line makes with direction of record travel for various values of pre-emphasis.
349. Williamson, D. T. N. (notes from lecture) " Suppressing gramophone surface noise " W.W. 59.7 (July 1953) 298. Use of gate circuit to remove disturbance from dust particles and clicks.

350. Russell, G. H. " Inexpensive pickups on long-playing records " W.W. 59.7 (July 1953) 299. Simple resonant filter to eliminate peak.
351. Weil, M. " Let's talk about diamonds " High Fidelity 3.1 (March-April 1953) 38.
352. I.R.E. Standard 53 I.R.E. 1951 " Standards on sound recording and reproducing : Methods of measurement of noise " Proc. I.R.E. 41.4 (April 1953) 508.
353. British Standard BS 1928 (1953) " Lateral cut gramophone records and direct recording."
354. McProud, C. G. " Preamp with ' presence ' " Audio Eng. 38.1 (Jan. 1954) 23. Optional 6 db boost—centred on 2700 c/s for solo violin or singer.
355. Boegli, C. P. " Transient and frequency response in audio equipment " Audio Eng. 38.1 (Feb. 1954) 19. Mathematical analysis of the uptake characteristic when unit-step input signal is applied to amplifier or pickup.

356. Marshall, J. " The new Golden Ear amplifier "—Part 2, pre-amplifier and tone control " Audio Eng. 38.2 (Feb. 1954) 22. Uses IU5 valves on rectified and smoothed a.c. in pre-amplifier.
357. Villchur, E. M. " Pickup tracking error " Audio Eng. 34.3 (March 1950) 17.
358. " Piezotronic Technical Data " Brush Electronics Co., Cleveland, 1953.
359. Villchur, E. M. " Handbook of Sound Reproduction " Chapter 16, Pickups and tone arms, Audio Eng. 38.2 (Feb. 1954) 33.
360. Carlson, E. V. " A ceramic vibration pickup " Proc. National Electronics Conference 8 (Sept./ Oct. 1952) 94. Shure Bros. Calibration techniques.
361. Snyder, R. H. " History and development of stereophonic sound recording " Jour. A.E.S. 1.2 (April 1953) 176.
362. " The proposed AES Disk Standard " Jour. A.E.S. 2.1 (Jan. 1954) 3. Proposed standard AES TS-1 (Dec. 1953) intended to supersede earlier AES " Standard playback curve " (1951 Ref. 307), submitted for comments.
362A. Scott, H. H. " The philosophy of amplifier equalization " Jour. A.E.S. 2.1 (Jan. 1954) 45. Divides equalizers into three groups. Curves shown for all.
363. " Measurement of frequency variation in sound recording and reproduction " British Standard BS1988 : 1953. Abstract in W.W. 59.9 (Sept. 1953) 440.
364. Cook, E. " Binaural disc recording " Jour. A.E.S. 1.1 (Jan. 1953) 1.
365. Lindenberg, T. " Analyzing the long-playing pickup problem " Jour. A.E.S. 1.1 (Jan. 1953) 140. Designing pickup with low mass to bring resonance up to 15,000 c/s (Pickering).
366. Kelly, S. " Piezo-electric crystal pick-ups " Jour. Brit. I.R.E. 13.3 (March 1953) 161.
377. Axon, P. E., and W. K. E. Geddes " The calibration of light-pattern measurements " Proc. I.E.E. 100. Part III 66 (July 1953) 217. Analysis of light-pattern ; new apparatus for measurements ; focal-plane measurements found more consistent than on disc surface ; new method gives higher accuracy.
378. Henn-Collins, C. A. " Power supply for multi-speed record players " Electronic Eng. 25.302 (Apr. 1953) 166. Uses stabilized Wien bridge oscillator to excite push-pull amplifier.
379. " Long-playing disc records compared with magnetic tape for sound reproduction in the home " Discussion. Proc. I.E.E. 101. Part III 70 (March 1954) 83.
380. Koren, H. W., H. A. Pearson, H. Klingener and R. W. Sabol " Dual-stylus ceramic phonograph pickup development " Jour. Acous. Soc. Am. 26.1 (Jan. 1954) 15. A Sonotone development, with two styli mounted back-to-back on the same stylus arm.
381. Gayford, M. L. " Distortion and gramophone reproduction—a review " Electronic Eng. 25.299 (Jan. 1953) 24. A useful summary of the various forms of distortion, with references. Letter R. W. Bayliff and author's reply 25.302 (April 1953) 172 on effects of longitudinal tip movement.
382. Roys, H. E. " Distortion in phonograph reproduction " J. Acous. Soc. Am. 25.6 (Nov. 1953) 1140. See also Ref. 343.
383. Parchment, E. D. " Microgroove recording and reproduction " Jour. Brit. I.R.E. 12.5 (May 1952) 271.
384. IRE Standards on Sound Recording and reproducing : Methods for determining flutter content, 1953. Proc. I.R.E. (Mar. 1954) 537.
385. " Mercury disc-charger " High Fidelity 4.2 (April 1954) 83. A small piece of radium-base material is mounted close to the record and discharges static electricity.
386. " Report on the HGP-40 pickup " Radio and Hobbies 16.4 (July 1954) 84. Output curves of Acos pickup with and without equalization.
387. Bachman, W. S. (letter) " Columbia LP and RIAA recording curves " High Fidelity 4.4 (June 1954) 102 (with figure showing curves).
388. " New standard record curve—phonograph record manufacturers agree on standard recording and playback curve " Radio Electronics 25.5 (May 1954) 63. New RIAA, AES, NARTB, RCA, New Orthophonic.
389. Korte, J. W. " R-J type 12 inch speaker enclosure " Radio Electronics 25.5 (May 1954) 68. Gives dimensions, and acoustical damping in box.
390. Kelly, S. " Piezo crystals—survey of physical properties and their practical exploitation " W.W. (1) 60.6 (June 1954) 275 ; (2) 60.7 (July 1954) 345.
391. " Phonograph needle drag distortion " Elect. 26.1 (Jan. 1953) 214. Based on NBS data. Produces even harmonics, with second pre-dominating.
392. Hunt, F. V. " Stylus groove relations in the phonograph playback process " Proceedings of 1st ICA-Congress, Netherlands, 1953 ; Acustica 4.1 (1954) 33. Analysis of effects of elastic deformation of groove walls.
393. Carlson, E. V. " Ceramic vibration pickup " Radio and TV News, Radio Electronic Eng. Supplement, 49.5 (May 1953) 8. Barium titanate transducer by Shure.
394. Mitchell, J. A. " An equalizer for FM pickups " Radio and TV News 51.5 (May 1954) 54. For Weathers pickup.
395. Konins, J. A. " Checking your audio system " Radio and TV News (May 1954) 59. Dubbings Co. test records and test tapes.
396. Canby, E. T. " Audio etc.—On the gadget front " Audio 38.6 (June 1954) 48. Describes some LP records inherently static-free ; also methods of treating and handling existing records to reduce troubles from dust.

References to Pre-amplifiers

Refs. 144, 239, 246, 250, 268, 270, 278, 280, 287, 305, 319, 326, 327, 328, 354, 356.
See also Chapter 18 Refs. A43, A44, B16, B17, B18, B41, B43, B48, B52, B53, B54, B55, B56, B57, B58, B60, B61.

CHAPTER 18

MICROPHONES, PRE-AMPLIFIERS, ATTENUATORS AND MIXERS

ADDITIONAL REFERENCES

(A) Microphone and studio equipment

A31. Staff of B.B.C. " Microphones " Iliffe and Sons Ltd. (1952).

A32. Anderson, L. J. " Pressure microphone for TV and broadcast-service " Tele-Tech 12.1 (Jan. 1953) 58.

A33. Bauch, F. W. O. " New high-grade condenser microphones "—Neumann types, W.W. 59.2 (Feb. 1953) 50 ; 59.3 (March 1953) 111. Reprinted in Jour. A.E.S. 1.3 (July 1953) 232.

A34. Anderson, L. J. " A pressure microphone for television and broadcast service " Broadcast News 71 (Sept.-Oct. 1952) 58. [R.C.A. BK-1A]

A35. Crowhurst, N. H. " Rapid remote microphone control " Radio Electronics 24.3 (March 1953) 34.

A36. Hilliard, J. K. " Microphones measure high-intensity sound " Elect. 26.11 (Nov. 1953) 160. For measurement in range 40-220 db, 3 types of microphones used to cover frequency range 3-30,000 c/s.

A37. Anderson, L. J. " Sensitivity of microphones to stray magnetic fields " Trans. I.R.E.-PGA. AU-1.1 (Jan./Feb. 1953) 1.

A38. Olson, H. F., J. Preston, and J. C. Bleazey " The uniaxial microphone " R.C.A.-Rev. 14.1 (March 1953) 47. See also Ref. A39.

A39. Olson, H. F. " The uniaxial microphone " Trans. I.R.E.-PGA. AU-1 4 (July-Aug. 1953) 12. Uni-directional with sharper directivity pattern than cardioid, and independent of frequency.

A40. Souther, H. T. " An adventure in microphone design " Jour. A.E.S. 1.2 (April 1953) 176. Electro-Voice Model 655 " Slim Trim " pressure-type.

A41. Bauer, B. B. " A miniature microphone for transistorized amplifiers Jour. Acous. Soc. Am. 25.5 (Sept. 1953) ; Trans. I.R.E.-PGA. AU-1.6 (Dec. 1953) 5. A magnetic type for hearing aids etc. —70 db re 1 volt/microbar. Shure Bros.

A42. Medill, J. " A miniature piezo-electric microphone " Jour. Acous. Soc. Am. 25.5 (Sept.1953) 864. Trans. I.R.E. PGA. AU-1.6 (Dec. 1953) 7. Uses acoustical damping of high frequency reasonance. Linear up to 10,000 microbars. Shure Bros. Model 98-99.

A43. Boylan, W. F., and W. E. Goldstandt " A new approach to professional magnetic recording equipment " Jour. A.E.S. 2.1 (Jan. 1954) 25. Uses 6BK7 as cascode microphone pre-amplifier ; input noise figure —127 dbm.

A44. Stewart, W. E. " Basic problems in audio systems practice " Jour. A.E.S. 1.1 (Jan. 1953) 85.

A45. Clark, M. A. " An acoustic lens as a directional microphone " J. Acous. Soc. Am. 25.6 (Nov. 1953) 1152 ; Trans. I.R.E. PGA. AU-2.1 (Jan.-Feb. 1954) 5.

A46. Beaverson, W. A. " Techniques for designing pressure microphones " Tele-Tech 13.5 (May 1954) 84.

A47. Phinney, T. W. " The Vagabond wireless microphone system " Trans. I.R.E., PGA. AU-2.2 (March-April 1954) 44. A cableless system using induction coupling between transmitter and receiver with frequency modulation, carrier frequency 2.1 Mc/s.

A48. Wolfe, B. " Preventing acoustic feedback " Tele-Tech 12.12 (Dec. 1953) 77. System with two pre-amplifiers permits microphone close to loudspeaker with only 3 db loss.

A49. Anderson, L. J. " Sensitivity of microphones to stray magnetic fields " Broadcast News 78 (Mar.-April 1954) 30. Measurements of hum on several R.C.A. microphones under varying magnetic field conditions.

(B) Pre-amplifiers, noise and hum

B55. " The ' Vari-Slope ' Pre-amplifier," data sheet, H. J. Leak and Co. Ltd., Brunel Road, Westway Factory Estate, Acton, London, W.3.

B56. Heacock, D. P., and R. A. Wissolik " Low-noise miniature pentode for audio amplifier service " Proc. National Electronics Conference 6 (1950) 155. See also B16.

B57. Ayres, W. R. " Hum reduction in amplifier development " Audio Eng. 37.6 (June 1953) 14. Gives useful list of precautions to reduce hum level.

B58. Snow, W. B. " Audio-frequency input circuits " Jour. A.E.S. 1.1 (Jan. 1953) 87. A wide and detailed treatment covering noise and frequency response for resistive and reactive generators.

B59. Fremlin, J. H. " Noise in thermionic valves " Proc. I.E.E. 100 Part III 64 (March 1953) 91. A new approach to the relation between statistical and thermodynamic formulae for a temperature-limited diode.

B60. Woll, H. J., and F. L. Putzrath " A note on noise in audio amplifiers " Trans. I.R.E. PGA. AU-2.2 (March/April 1954) 39.

B61. Price, R. L. " The cascode as a low noise and audio amplifier " Trans. I.R.E. PGA. AU-2.2 (March/April 1954) 60.
See also Refs. A43, A44.
See also complete list of References to Pre-amplifiers at end of Chapter 17 References.

(C) Attenuators and mixers

C21. Scott, C. F. " Attenuator types and their application " Jour. A.E.S. 1.1 (Jan. 1953) 95. Covers application to multichannel mixings.

C22. Crowhurst, N. H. " Attenuator design " Electronic Eng. 26.312 (Feb. 1954) 76. Gives chart with wide applicability.

CHAPTER 19

UNITS FOR THE MEASUREMENT OF GAIN AND NOISE

(iii) Loudness Units (Continued from page 827)

The **sone** is commonly used as an alternative to the loudness unit, 1 sone being equal to 1000 loudness units. The solid curve in Fig. 19.9 is the A.S.A. Standard, as in Fig. 19.8. The broken curve is the relationship between subjective loudness and loudness level in phons as determined by Garner (Ref. 46) and is scaled in lambda (λ) units to distinguish it from the present standard system. The difference between the two curves is very great.

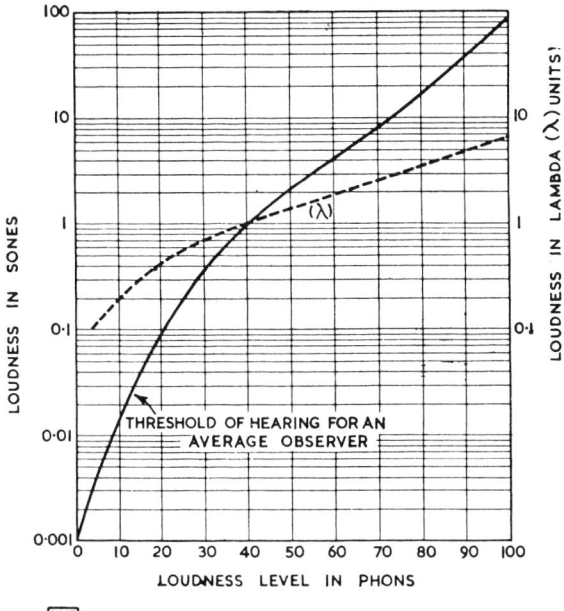

Fig. 19.9. *Relationship between loudness in sones (solid curve) or in lambda units (broken curve) and loudness level in phons (RT17).*

ADDITIONAL REFERENCES

44. I.R.E. Standard 53 IRE 7 S1 " Standards on electron devices : Methods of measuring noise " Proc. I.R.E. 41.7 (July 1953) 890.
45. Garner, W. R. " An equal discriminability scale for loudness judgments " J. Exp. Psychol. 43 (1952) 232.
46. Garner, W. R. " A technique and a scale for loudness measurement " J. Acous. Soc. Am. 26.1 (Jan. 1954) 73. A comprehensive treatment on a new basis, which gives results differing markedly from the A.S.A. Standard (loudness units).
47. Robinson, D. W. " The relation between the sone and phon scales of loudness " Acustica, 3.5 (1953) 344.
48. General Radio Company " Handbook of Noise Measurement " (Cambridge 39, Massachusetts, 1953). Extensive bibliography.
49. I.R.E. Standard 53 IRE 3.S2 " Standards on American recommended practice for volume measurements of electrical speech and program waves 1953 " Proc. I.R.E. 42.5 (May 1954) 815. This will supersede the earlier A.S.A. C16.5-1942 (Ref. 19).

CHAPTER 20
LOUDSPEAKERS
SECTION 3 : BAFFLES AND ENCLOSURES

(Continued from page 850)

(C) Special types of vented baffle loudspeakers

(1) The Baruch and Lang loudspeaker

The Baruch and Lang loudspeaker employs four 5 inch loudspeakers in an enclosure with a volume of only half a cubic foot and is claimed to radiate 0.1 acoustic watt at 3% distortion, with an input of 2 watts and an efficiency about 5%. The response is claimed to be flat ± 3 db from 40 to 12,000 c/s, and the high frequency angular dispersion is 75°. It is a modified acoustical phase inverter (vented baffle) with an array of small holes on one side to provide the requisite port area—15 holes each 15/32 inch diameter spaced 2¼ inches apart. There is an internal baffle with 21 holes spaced 2 inches apart. These holes provide acoustical resistance to damp down the system resonance to the most desirable degree. In addition, it is claimed that, because the holes are distributed over a large area, the radiation impedance of the array is equivalent to that of a 21 inch cone.

The speakers employed are standard low-cost replacement units, modified to meet the requirements of the system. The optimum dimensions of cabinet and holes, as well as the configuration of holes and the speaker array, are determined by the characteristics of the particular speakers used.

This is a most interesting high-fidelity loudspeaker and enclosure which has been designed to bring the cost within reach of those with limited means—a commercial model is now selling in U.S.A. for less than 20 dollars.

Refs. 202, 311, 316.

(2) Additional Notes on the R-J loudspeaker (from page 850)

When one speaker is intended to handle a wide frequency range, the central portion of the cone must not be obstructed by the frontal board ; an oval or lemon-shaped opening may be used. The loudspeaker is mounted on a " speaker board " which is mounted a short distance behind the frontal board. By decreasing the spacing between these two boards, it is possible to lower the Q of the system. Usually apertures are provided between the speaker board and the frontal board on two sides only, the remaining two sides being blocked up. These two apertures should feed in where the frontal board projects furthest over the cone, thus giving maximum front loading on the cone. One effect is to reduce the resonance of the system considerably below the speaker resonant frequency. When properly designed and adjusted, the system is well damped and remarkably free from frequency doubling, even below the system resonance.

Refs. 189, 190, 219, 234, 246, 247.

(3) The Karlson Exponential Slot enclosure

This is totally enclosed except for an exponential form of slot in front. The enclosure is divided into two chambers by a partition on which the loudspeaker is mounted and which includes a port joining the two chambers. The front chamber includes the exponential slot, with the smaller dimension near the top of the cabinet, while the back chamber is enclosed and has some acoustical padding.

Refs. 214, 288 (latter gives dimensions).

ADDITIONAL REFERENCES TO LOUDSPEAKERS

202. " High fidelity at low cost with a new speaker system—inexpensive 5 inch speakers in this small enclosure provide wide-range reproduction " FM-TV 12.6 (June 1952) 26. Design by Baruch and Lang.
203. Moir, J., and J. A. Leslie " The stereophonic reproduction of speech and music " J. Brit. I.R.E. 12.6 (June 1952) 360.
204. Slaymaker, F. H. " An integrated line of high-fidelity equipment " Audio Eng. 36.7 (July 1952) 26.
205. Bartlett, S. C. " Public address systems in generating plants " A.I.E.E. Trans. 70 Part 2 (1951) 1804. Reprinted in Radiotronics 17.10 (Oct. 1952) 159. Covers paging in noisy situations.
206. " Compact back-loading folded horn cabinet for 12 inch and 15 inch loudspeakers " (Jensen Manufacturing Company). Reprinted in Radiotronics 17.10 (Oct. 1952) 172.
207. Olson, H. F. (book) " Musical Engineering " McGraw-Hill Book Company 1952.
208. Moir, J. " Better music-room acoustics " FM-TV 11.8 (Aug. 1951) 32.
209. Plach, D. J., and P. B. Williams " Horn loaded loudspeakers " Trans. I.R.E. Professional Group Audio (Oct. 22, 1951) ; also Proc. National Electronics Conference Vol. 7, p. 108. Reprinted Radiotronics 17.6 (June 1952) 102.
210. Salmon, V. " Coupling the speaker to the output stage " Newsletter I.R.E.—PGA 3.1 (Jan. 1952) 5.
211. Locanthi, B. N. " Application of electric circuit analogies to loudspeaker design problems " Trans. I.R.E.—P.G.A. PGA-6 (March 1952) 15 ; PGA7 (May 1952) 46.
212. Axtell, J. C. " Ionic loudspeakers " I.R.E.—PGA-8 (July 1952) 21. See also Ref. 227.
213. Kidd, M. C. " Tone-burst generator checks a-f transients " Elect. 25.7 (July 1952) 132. Used for loudspeaker and loudspeaker plus enclosure testing.
214. Karlson, J. E. " A new approach in loudspeaker enclosures "—exponential slot—Audio Eng. 36.9 (Sept. 1952) 26. See also High Fidelity 3.3 (July Aug. 1953) 92.
215. Badmaieff, A. " Design considerations of duplex loudspeakers " Audio Eng. 36.9 (Sept. 1952) 28.

216. Moir, J. " Stereophonic reproduction " Audio Eng. 36.10 (Oct. 1952) 26.   Gives useful biblio-graphy.
217. Gately, E. J. " Design for clean bass " Audio Eng. 36.10 (Oct. 1952) 29.
218. Kiebert, M. V. " A corner-mounting infinite baffle " Audio Eng. 36.10 (Oct. 1952) 32.
219. Canby, E. T. " Record revue " Audio Eng. 36.10 (Oct. 1952) 46.   Refers to R-J and EW enclosures.
220. Briggs, G. A. " Room acoustics " High Fidelity 2.1 (Summer 1952) 69.
221. Somerville, T. " Acoustics in broadcasting " Report of Building Research Congress, Division 3, Part 1, Building Research Station, Watford, Herts, England.
222. Briggs, G. A. " Response curves " High Fidelity 1.4 (Spring 1952) 66.
223. Briggs, G. A. " The loudspeaker " High Fidelity 2.2 (Sept.-Oct. 1952) 39.
224. McLachlan, K. R., and R. Yorke " Objective testing of pickups and loudspeakers " J. Brit. I.R.E. 12.9 (Sept. 1952) 485.
225. Brittain, F. H. " Metal cone loudspeaker—principles underlying the design of the G.E.C. High Quality Reproducer " W.W. 58.11 (Nov. 1952) 440 ; 58.12 (Dec. 1952) 490.
226. Bixler, O. C. " A commercial binaural recorder " Jour. S.M.P.T.E. 59.2 (Aug. 1952) 109.
227. " Non-mechanical ' ionic ' loudspeaker " Technicana, Audio Eng. 36.11 (Nov. 1952) 84, being summary of article in TSF and TV (July-Aug. 1952).   See also Ref. 212.
228. Briggs, G. A. " The loudspeaker and the ear " High Fidelity 1.3 (Winter 1951) 17.
229. " Corner ribbon loudspeaker " W.W. 56.1 (Jan. 1950) 11.
230. " The corner ribbon " (booklet) Acoustical Manufacturing Co. Ltd. Huntingdon, Hunts, England, May 1952.
231. Goodwin, J. L. " Sound reinforcement and reproduction " Electrical Review, London, 150.3875 (Feb. 29, 1952) 437.
232. Cohen, A. B. " Wide angle dispersion of high frequency sound " Audio Eng. 36.12 (Dec. 1952) 24.
233. Randall, R. H. (book) " An Introduction to Acoustics " Addison-Wesley Press, Cambridge 42, Mass., 1951.
234. Joseph, W., and F. Robbins " Practical aspects of the R-J speaker enclosure " Audio Eng. 37.1 (Jan. 1953) 19.
235. " Why stereophonic or binaural reproduction ? " letters by J. Versace and T. O. Dixon, Audio Eng. 37.1 (Jan. 1953) 8-12.
236. Sherman, H. T. " Binaural radio broadcasting " Audio Eng. 37.1 (Jan. 1953) 14.
237. Tinkham, R. J. " Binaural or stereophonic ? " Audio Eng. 37.1 (Jan. 1953) 22.
238. Canby, E. T. " Record revue—suspended in space " Audio Eng. 37.1 (Jan. 1953) 46.
239. Brittain, F. H. " The environment of high-quality reproduction " W.W. 59.1 (Jan. 1953) 2.
240. " Friction-driven loudspeaker " W.W. 59.1 (Jan. 1953) 27.
241. Hardy, H. C., H. H. Hall and L. G. Ramer " Direct measurement of the efficiency of loudspeakers by use of a reverberation room " Trans. I.R.E. PGA-10 (Nov.-Dec. 1952) 14, described in " Loud-speaker efficiency " W.W. 59.2 (Feb. 1952) 61.
242. Dempster, B. " The Magnavox duode speaker " W.W. 38 (March 6th 1936) 241 ; also ᵈ New apparatus reviewed " p. 245 same issue.
243. Barker, A. C. " Single diaphragm loudspeakers " W.W. 54.6 (June 1948) 217.
244. Hughes, T. R. " Real theater sound in a small package " Audio Eng. 37.2 (Feb. 1953) 19 ; 37.3 (March 1953) 30 ; 37.4 (April 1953) 24.   Three horn corner reproducer.
245. Adams, C. F. " Binaural public address " Audio Eng. 37.2 (Feb. 1953) 24.
246. " The R-J speaker enclosure " Radio and TV News, 49.4 (April 1953) 53.   Constructional page for 8 inch speaker.
247. Villchur, E. M. " Handbook of sound reproduction " Chapter 10 " Loudspeakers," Audio Eng. 37.3 (March 1953) 32 ; 37.4 (April 1953) 29.   Chapter 11 " Loudspeaker mounting " Audio Eng. 37.5 (May 1953) 34 ; 37.6 (June 1953) 30.   Includes, enclosed cabinet, open back cabinet, bass reflex, R-J enclosure, acoustical labyrinth, horns.
248. Harrison, C. W. " Coupled loudspeakers " Audio Eng. 37.5 (May 1953) 21.   Uses 4 separate enclosures each 22.5° mounted in a corner.   Gives useful information of acoustical damping inside enclosures, using fiberglas etc.
249. Canby, E. T. " Record Revue : True binaural and panoramic " Audio Eng. 37.9 (Sept. 1953) 47.
250. Goldmark, P. C. " The Columbia XD (extra-dimensional) sound system " Audio Eng. 37.10 (Oct. 1953) 36.   Simple means for providing a pseudo-stereophonic effect by using a small loudspeaker for the highs radiating at a point removed from the main loudspeakers in which the frequency range is restricted to the lows.   Reprinted Radio and Hobbies 16.1 (Jan. 1954) 19.
251. McLean, A. " Loudspeaker frequency response curves " Radiotronics 18.7 (July 1953) 103.
252. Veneklasen, P. S. " Power capacity of loudspeakers " Trans. I.R.E., PGA, AU1.5 (Sept.-Oct. 1953) 5.   Method of testing based on departure from linearity of acoustic output versus electrical input curves using octave bands of thermal noise.
253. Olson, H. F. " Subjective loudspeaker testing " Trans. I.R.E., PGA, AU1.5 (Sept.-Oct. 1953) 7.   A useful, brief survey.
254. Allison, R. F. " The junior air coupler " High Fidelity 3.2 (May-June 1953) 80.
255. Tested in the home " Kelton loudspeaker " High Fidelity 3.2 (May-June 1953) 85.   Developed by Henry Lang.   Uses 2 speakers in enclosure 11x 11 x 23 ins., each in separate compartment. 6 in. speaker faces outwards near top ; 8 in. speaker facing downwards into padded cavity with holes for bass outlet. Back areas of both speakers filled with sound absorbing material ; back radiation not used.   Design is critical.   Is designed for maximum performance for limited outlay (49.50).
256. " A sound delay system " Electronic Eng. 25.305 (July 1953) 281.   Acoustic delay line in amplifier chain to produce time delay in PA equipment.   Delay provided by rubber hose cut to appropriate length (B.T.H. Co.).
257. Pickering, N. C., and E. Baender " Two ears in three dimensions " Jour. A.E.S. 1.3 (July 1953) 255.   Reprinted Radiotronics 19.4 (April 1954) 38.
258. Gately, E. J. and T. A. Benham " Super-horn : a folded horn enclosure Radio and TV News 50.3 (Sept. 1953) 38.
259. Gately, E. J. and T. A. Benham " The Purist—a non-corner horn " Radio and TV News 50.6 (Dec. 1953) 56.
260. Briggs, G. A. " Enclosures for loudspeakers " High Fidelity (1) 3.4 (Sept.-Oct. 1953) 98 ; (2) 3.5 (Nov./Dec. 1953) 97 ; (3) 3.6 (Jan. /Feb. 1954) 89 ; (4) 4.1 (March 1954) 86.
261. Baruch, J. J., and H. C. Lang " An analogue for use in loudspeaker design work " Trans. I.R.E.-PGA. AU-1.1 (Jan.-Feb. 1953) 8.   Reprinted Proc. National Electronics Conference, 8 (1952) 92.
262. Bixler, O. C. " A practical binaural recording system " Trans. I.R.E.-PGA. AU-1.1 (Jan./Feb. 1953) 14.

263. Richardson, E. G. (edit) " Technical Aspects of Sound " (Ensevier Publishing Co. 1953) pp. 339-372 loudspeakers ; pp. 425-435 stereophonic reproduction.
264. Fletcher, H. (book) " Speech and Hearing in Communication " (van Nostrand 1953) pp. 210-216 binaural hearing ; pp. 217-299 auditory perspective.
265. Stewart, K., and P. Edwards " Action of conical dome speakers " Service 22.11 (Nov. 1953) 44. Describes R.C.A. LC1A.
266. Klipsch, P. W. " Loudspeaker developments " Trans. I.R.E.-PGA AU-1.3 (May-June, 1953) 16. Gives early history of loudspeakers, especially development of corner horns, and bibliography.
267. Bauer, B. B. " Acoustic damping for loudspeakers " Trans. I.R.E. PGA. AU-1.3 (May-June 1953) 23.
268. Salmon, V. " Loudspeaker impedance " Trans. I.R.E.-PGA. AU-1.4 (July-Aug., 1953) 1. See also Ref. 294.
269. Olson, H. F. " Selecting a loudspeaker " Newsletter I.R.E.—PGA. 2.5 (Sept. 1951) 7. Reprinted Radiotronics 17.2 (Feb. 1952) 37.
270. Snow, W. B. " Foreword—Developments in stereophony " Jour. S.M.P.T.E. 61.3 (Sept. 1953) 353. Special issue. Refs. 271 to 279 below.
271. Fletcher, H. " Stereophonic recording and reproducing system " Jour. S.M.P.T.E. 61.3 (Sept. 1953) 355.
272. Grignon, L. D. " Experiment in stereophonic sound " Jour. S.M.P.T.E. 61.3 (Sept. 1953) 364.
273. Hilliard, J. K. " Loudspeakers and amplifiers for use with stereophonic reproduction " Jour. S.M.P.T.E. 61.3 (Sept. 1953) 364.
274. Singer, K., and M. Rettinger " Multiple track magnetic heads " Jour. S.M.P.T.E. 61.3 (Sept. 1953) 390.
275. Frayne, J. G., and E. W. Templin " Stereophonic recording and reproducing equipment " Jour. S.M.P.T.E. 61.3 (Sept. 1953) 395.
276. Volkmann, J. E., J. F. Byrd and J. D. Phyfe " New theatre sound system for multipurpose use " Jour. S.M.P.T.E. 61.3 (Sept. 1953) 408.
277. Fletcher, H. " Basic requirements for auditory perspective " Jour. S.M.P.T.E. 61.3 (Sept. 1953) 415. Reprinted from B.S.T.J. 13 (April 1934) 239.
278. Steinberg, J. C., and W. E. Snow " Physical factors in auditory perspective " Jour. S.M.P.T.E. 61.3 (Sept. 1953) 420. Reprinted from B.S.T.J. 13 (April 1934) 245.
279. Wente, E. C., and A. L. Thuras " Loudspeakers and microphones for auditory perspective " Jour. S.M.P.T.E. 61.3 (Sept. 1953) 431. Reprinted from B.S.T.J. 13 (April 1934) 259.
280. Mulvey, J. A. " Feedback and loudspeaker damping " Audio Eng. 37.4 (April 1953) 34. Theoretical treatment proposing feedback winding on voice coil, with additional transformer with low impedance primary in series with voice coil and secondary neutralizing the component in the feedback voltage due to inductive coupling between voice coil and feedback coil.
281. Canby, E. T. " Record Revue " Audio Eng. 37.4 (April 1953) 48. Two or more loudspeaker systems or separate channels give placement effect, but do not give any improvement in confused conversation as occurs with binaural. The liveness does not improve.
282. Tinkham, R. J. " Stereophonic recording equipment " E.E. 72.12 (Dec. 1953) 1053.
283. Villchur, E. M. " Handbook of sound reproduction " Chapter 12, Part 2, Audio Eng. 37.8 (Aug. 1953) 26. Loudspeaker damping, equivalent circuits, effects of various types of feedback.
284. Houck, G. B. " Vibration reduction in loudspeaker enclosures—how to brace a speaker cabinet " Audio Eng. 37.12 (Dec. 1953) 24.
285. Lindenberg, T., C. E. Smiley and J. B. Minter " Design of an electrostatic loudspeaker " Jour. A.E.S. 1.4 (Oct. 1953) 273.
286. Plach, D. J. " Design factors in horn-types speakers " Jour. A.E.S. 1.4 (Oct. 1953) 276.
287. Ingerslev, F. (book) " Measurement of linear and non-linear distortion in electro-dynamic loudspeakers " (in Danish, 266 pages, with 14 page English summary). Den polytekniske Laereanstalt, Sologade 83, Kobenhavn, Denmark. See also Ref. 315.
288. Karlson, J. E. " The Karlson speaker enclosure " Radio and TV News 51.1 (Jan. 1954) 58. Gives detailed dimensions, impedance measurements and polar radiation pattern. See also Ref. 214.
289. Haynes, N. M. " Sterophonic nomenclature " Audio Eng. 38.1 (Jan. 1954) 19. Gives suggested nomenclature under headings Sterophonic effects ; Stereophonic equipment ; Deficient stereophonic effects.
290. Olson, H. F., and J. Preston " New line of hi-fi speakers " Radio and TV News 51.2 (Feb. 1954) 69. R.C.A. Models SL-8, SL-12, LC-1A, giving very complete information. Distortion all models less than 0.3% for input 1 watt for all frequencies above 250 c/s. LC-1A (15 inch) distortion less than 0.5% for all frequencies above 200 c/s, and less than 0.3% over most of range above 200 c/s, for 5 watts input.
291. Snitzer, M. S. " Adventures with a bass reflex " Audio Eng. 38.1 (Jan. 1954) 26. Measurements of impedance characteristics under various conditions.
292. Denny, W. B. " A corner horn for the small listening room " Audio Eng. 38.2 (Feb. 1954) 21. Front of speaker is 12 in. direct radiator, back loading enclosure, fiberglas filled, and horn using walls on subtraction principle.
293. Hoodwin, L. S. " The compound diffraction projector " Jour. A.E.S. 2.1 (Jan. 1954) 40. Electro-Voice, single driver with straight h.f. horn and compound diffraction l.f.horn. Response 175-10,000 c/s ± 5 db. Distortion less than 6% at 20 watts input, above 900 c/s. Polar distribution 120° at 5000 c/s.
294. Langford-Smith, F. " Loudspeaker impedance " Radiotronics 19.4 (April 1954) 44. Clarifies the five impedances associated with the operation and testing of loudspeakers. See also Ref. 268.
295. Youngmark, J. A. " Loudspeaker baffles and cabinets " Jour. Brit. I.R.E. 13.2 (Feb. 1953) 89. A valuable treatment of plane baffles, open back cabinets, acoustic labyrinth, bass reflex cabinets, effect of baffle on h.f. response, and multiple speakers close together.
296. " Action of the conical domes in the improved type LC-1A loudspeaker " Broadcast News 76 (Sept.-Oct. 1953) 64. R.C.A.
297. Thurston, W. R. " Testing and adjusting speaker installations with the Sound-Survey Meter " Jour. A.E.S. 1.1 (Jan. 1953) 146. Tests made indoors averaging 10 readings ; also outdoors.
298. Brittain, F. H. " Loudspeakers : relations between subjective and objective tests " Jour. Brit. I.R.E. 13.2 (Feb. 1953) 105.
299. Walker, P. J. " The loudspeaker in the home " Jour. Brit. I.R.E. 13.7 (July 1953) 377. Includes treatment of perspective in depth.
300. Miles, J. W. " Transient loading of a baffled piston " Acous. Soc. Am. 25.2 (March 1953) 200. Application to a loudspeaker shows that a system designed for critical damping on a " steady state "

approximation will be slightly overdamped in its initial motion in a " step " response, but that this time is so small as to be negligible.

301. Smith, B. H. " An investigation of the air chamber of horn type loudspeakers " J. Acous. Soc. Am. 25.2 (March 1953) 305. Gives complete design method for high frequency performance.

302. Mawardi, O. K. " A physical approach to the generalized loudspeaker problem " J. Acous. Soc. Am. 26.1 (Jan. 1954) 1. An analysis solved exactly for a loudspeaker of circular plane shape in an infinite baffle.

303. Veneklasen, P. S. " Power capacity of loudspeakers " J. Acous. Soc. Am. 26.1 (Jan. 1954) 98. A suggested method for specifying the power capacity of loudspeakers, using bands of thermal noise, measuring the output/input linearity characteristic for each band, noting the —1 db point as overload indicator, and plotting max. output level for 1 db deviation against frequency, also max. input level.

304. Vermeulen, R. " Stereophonic reproduction " Audio 38.4 (April 1954) 21. Some salient points on the mechanism of binaural and stereophonic phenomena as they affect the ear and are useful in reproduction.

305. Simonton, T. E. " A new integral ratio chromatic scale " J. Acous. Soc. Am. 25.6 (Nov. 1953) 1167.

306. Hartley, H. A. " High-fidelity loudspeakers " Radio Electronics 25.3 (March 1954) 35 ; 25.4 (April 1954) 60 ; 25.6 (June 1954) 42 (horns and multiple units).

307. Plass, G. " Stereophony from the outside in " High Fidelity 4.2 (April 1954) 78.

308. Muncey, R. W., A. F. B. Nickson and P. DuBout " The acceptability of speech and music with a single artificial echo " Acustica 3.3 (1953) 168.
See also Ref. 125. Results modify findings by Haas for rooms having low reverberation time.

309. Snow, W. B. " Basic principles of stereophonic sound " Jour. S.M.P.T.E. 61.5 (Nov. 1953) 567. Good summing up, with very extensive bibliography.

310. " Tested in the home : Electro-Voice 15TRX speaker " High Fidelity 4.4 (June 1954) 76.

311. Monitor " For golden ears only—Baruch Lang loudspeaker system " Radio Electronics 25.5 (May 1954) 59. Results of listening tests.

312. Lenihan, J. M. A. " The velocity of sound in air " Acustica 2.5 (1952) 205. Appears to be most precise measurement to date ; 331.45 ± 0.04 metres/sec at 13,500 c/s, 0°C, 1013.2 millibars. Very extensive history and bibliography.

312A. Jordan, V. L. " A system for stereophonic reproduction " Proc. 1st ICA-Congress Electro-Acoustics 1953 ; Acustica 4.1 (1954) 36. Relation between angular displacement of the virtual sound source and the difference in intensity level from two loudspeakers.

313. de Miranda, J. R. " The radio set as an instrument for the reproduction of music " Proc. 1st ICA-Congress Electro-Acoustics 1953 ; Acustica 4.1 (1954) 38. General survey ; dip at high frequencies caused by depth of hole in baffle ; attenuation by loudspeaker cloths.

314. Somerville, T. " The establishment of quality standards by subjective assessment " Proc. 1st ICA-Congress Electro-Acoustics 1953 ; Acustica 4.1 (1954) 48. B.B.C. methods for subjective testing. Impossible to make reliable comparisons unless acoustics of originating studio are good. In loudspeaker testing (a) outdoor tests on speech ; (b) recording speech in non-reverberent room, replaying and re-recording several times to accentuate loudspeaker defects ; (c) balance in a studio is obtained for one particular monitor speaker, and no other gives good reproduction.

315. Ingerslev, F. " Measurements of non-linear distortion in loudspeakers " Proc. 1st ICA-Congress, Electo-Acoustics 1953 ; Acustica 4.1 (1954) 74. See also Ref. 287.

316. " Radical new speaker gives Hi-Fi for 30 dollars " Popular Science 161.3 (Sept. 1952) 171. Baruch-Lang speaker, shows interior arrangement and internal baffle with 21 holes spaced 2 inches apart. Four speakers 4 inch holes 6 inches apart.

317. Cohen, A. B. " Horns for the P.A. Technician " Radio and TV News 51.4 (April 1954) 43. Descriptions of Cobreflex wide angle horn.

318. Sodaro, J. H. " Nomograph for bass reflex enclosure design " Audio 38.5 (May 1954) 31.

319. Canby, E. T. " Audio etc.—Small enclosures " Audio 38.5 (May 1954) 36. A general discussion on listening qualities of small enclosures.

320. Stocklin, W. A. " Loudspeaker enclosures " Radio and TV News 49.5 (May 1953) 43. Report on performance of small cabinet Helmholtz resonator-type enclosures.

321. Holzman, J. " EW speaker enclosure " Radio and TV News 50.1 (July 1953) 43. Small modified vented baffle for 8 inch speaker with two air chambers, joined.

322. Souther, H. " Building the EV Regency " Radio and TV News 50.2 (Aug. 1953) 50. Constructional details of the Electro-Voice folded horn enclosure for 3-way system, 15 inch woofer.

323. Gately, E. J., and T. A. Benham " Super Horn ; a folded horn enclosure " Radio and TV News 50.3 (Sept. 1953) 38. Constructional details for 12 or 15 inch speakers.

324. Kantor, F. I. " The Klipsch Rebel IV : a back-loading folded corner horn " Radio and TV News 50.4 (Oct. 1953) 48. For 12 inch speakers.

325. Diefenbach, W. W. " Electrostatic speaker " Radio Electronics 24.4 (April 1953) 66. Isophon and Koerting speakers described.

326. Parkin, P. H. " The application of the Haas effect to speech reinforcement systems " Acustica 4.1 (1954) 98.

327. Parkin, P. H., and W. A. Allen " Acoustic design of auditoria " Nature 172 (July 18, 1953) 98.

References to Stereophonic and Binaural Reproduction (additional to those on page 866) :
245, 249, 250, 262, 263, 264, 270, 271, 272, 273, 274, 275, 276, 277, 278, 279, 281, 282, 289, 304, 307, 309, 312A.

Reference in Supplement to loudspeakers (general) including horns :
204, 206, 209, 210, 212, 222, 223, 225, 227, 229, 230, 232, 239, 240, 243, 244, 247, 251, 258, 259, 265, 266, 268, 269, 280, 283, 285, 286, 290, 292, 293, 294, 296, 299, 300, 301, 303, 306, 310, 313, 317, 322, 323, 324, 325.

References in Supplement to loudspeaker enclosures :
202, 206, 214, 218, 219, 234, 246, 247, 248, 254, 255, 260, 267, 284, 288, 291, 295, 306, 311, 316, 318, 319, 320, 321.

References in Supplement to loudspeaker testing :
213, 224, 241, 252, 253, 287, 297, 298, 314, 315.

References in Supplement to duplex and 3-way systems :
215, 242.

References in Supplement to P.A. systems and sound reinforcement :
205, 231.

(Continued on page 1494)

CHAPTER 20 (continued)

References in Supplement to Acoustics of rooms :
208, 220, 221, 308, 327.

References in Supplement to loudspeaker theoretical design and analogies :
211, 261, 302.

References to echoes and time delay :
125, 256, 308, 326, 327.

References in Supplement to acoustics (General) :
305, 312.

CHAPTER 21
THE NETWORK BETWEEN THE POWER VALVE AND THE LOUDSPEAKER

ADDITIONAL REFERENCES

23. British Standard Code of Practice CP 327.300 (1952) " Sound Distribution Systems " The Council for codes of practice for buildings.
24. Wentworth, J. P. " A discussion of dividing networks " Audio Eng. 36.12 (Dec. 1952) 17. Gives chart for determining L and C in parallel network, constant resistance type, 6 and 12 db/octave.
25. Crowhurst, N. H. " The basic design of constant resistance cross-overs " Audio Eng. 37.10 (Oct. 1953) 21. An analysis of the response and phase characteristics of constant-resistance cross-over networks worked out for filters employing from one to four elements. Letter E. de Boer, 38.1 (Jan. 1954) 8.
26. Meyer, A. " Air-core coil design for crossover networks " Trans. I.R.E.-PGA AU-1.5 (Sept./Oct. 1953) 9. Derives expression for coil shape for max. Q. Then, knowing L and Q, charts give turns, length of wire and resistance. See also Ref. 27.
27. Stewart, J., and F. Langford-Smith " Loudspeaker divider networks " Radiotronics 19.7 (July 1954) 75. A comprehensive treatment incorporating inductor design charts from Ref. 26.
28. Goss, L. C. " Coaxial speaker dividing networks " Radio and TV News 50.1 (July 1953) 36. Brief but useful survey of most types with some attenuation and impedance curves.

CHAPTER 22
AERIALS AND TRANSMISSION LINES

ADDITIONAL REFERENCES

39. Jackson, W. (book) " High Frequency Transmission Lines," Methuen, 1945.
40. Smith, R. A. (book), " Aerials for Metre and Decimetre Wavelengths " Cambridge University Press, 1949.
41. Laport, E. A. (book) " Radio Antenna Engineering " McGraw-Hill Book Co. 1952.
42. Schelkunoff, S. A. and H. T. Friis (book) " Antennas, Theory and Practice " John Wiley and Sons Inc. 1952.
43. Beard, E. G. " Ferroxcube antenna rods as an alternative to loop aerials " Philips Tec. Com. 5 (1953) 14. Based on paper by H. van Suchtelen.
44. IRE Standards on Antennas and Waveguides : Definitions of terms, 1953. See Proc. I.R.E. 41.12 (Dec. 1953) 1721.
45. Kiely, D. G. (book) " Dielectric Aerials " (John Wiley, 1953, 127 pp.).

CHAPTER 23
RADIO FREQUENCY AMPLIFIERS

ADDITIONAL REFERENCES

B47. Houlding, N., and A. E. Glennie " Experimental investigation of grid noise " W.E. 31.2 (Feb. 1954) 35. Excellent bibliography, investigation of triode noise factor. Deduced that correlation of induced grid noise with shot noise is very slight. Although optimum value of noise factor may be calculated from shot noise and optimum source resistance, the latter must be found by experiment, and therefore theory not of major practical importance.
B48. Bell, D. A. " Physical basis of thermal noise " W.E. 31.24 (Feb. 1954) 48.

CHAPTER 24
OSCILLATORS

ADDITIONAL REFERENCES

32. Stibbe, H. " Microphony in superhet. oscillators " W.W. 58.12 (Dec. 1952) 504 ; 59.1 (Jan. 1953) 35.
33. Edson, W. A. (book) " Vacuum Tube Oscillators " John Wiley and Sons Inc. New York, 1953.
34. Gillies, A. W. " Electrical oscillations—a physical approach to the phenomena " W.E. 30.6 (June 1953) 143.
35. Scott, N. R. " Amplitude stability in oscillating systems " Proc. I.R.E. 41.8 (Aug. 1953) 1031.
36. Roddam, T. " ' Chamelion ' Oscillator—versatile modified Hartley circuit giving high frequency stability " W.W. 60.2 (Feb. 1954) 52. May be described as an overbalanced rejector circuit oscillator. See Ref. 37.

37. Roddam, T. " Cathode follower oscillator—using RC networks with a voltage step-up " W.W. 60.3 (March 1954) 106. See also Ref. 36.
38. Dickson, A. W. " Use of servo techniques in the design of amplitude stabilized oscillators " Proc. National Electronics Conference 8 (Sept./Oct. 1952) 166.
39. Bacon, W. " Single stage phase-shift oscillator—method of design " W.E. 31.4 (April 1954) 100.
40. Bolle, A. P. " Theory of twin-T RC-networks and their application to oscillators " Jour. Brit. I.R.E. 13.12 (Dec. 1953) 571. A very thorough and comprehensive analysis.
41. Kretzmer, E. R. " An amplitude stabilized transistor oscillator " Proc. I.R.E. 42.2 (Feb. 1954) 391.
42. Wray, W. J. (letter) " More on the RC oscillator " Proc. I.R.E. 41.6 (June 1953) 801 ; letter B. J. O'Brien, 42.2 (Feb. 1954) 486.
43. Davidson, J. A. B. " A note on a precision decade oscillator " Proc. I.R.E. 40.9 (Sept. 1952) 1124.
44. Holbrook, G. W. " High frequency resistance-capacitance oscillators " Electronic Eng. 25.310 (Dec. 1953) 509.
45. Ward, P. W. " Oscillator feedback networks of minimum attenuation " Electronic Eng. 26.317 (July 1954) 318   Method of determining optimum values of components in RC oscillators.

CHAPTER 26
INTERMEDIATE FREQUENCY AMPLIFIERS

ADDITIONAL REFERENCES
(A) I-F AMPLIFIER DESIGN

11d. Tatan, E. " Simplified i-f amplifier design" Elect. 25.9 (Sept. 1952) 147.
11e. Dougherty, J. J. " Shape factor as a criterion of skirt selectivity " Elect. 26.10 (Oct. 1953) 232.
11f. Hupert, J. J., and A. M. Reslock " A method of bandpass amplifier alignment " Proc. I.R.E. 41.11 (Nov. 1953) 1668. Alignment based on oscillographic display of second order harmonic distortion as function of carrier frequency departure from pass-band centre.
11g. Jelonek, Z., and R. S. Sidorowicz " Bandpass amplifiers—investigation of design and stability " W.E. 31.4 (April 1954) 84.

(E) STABILITY
See also Ref. 11g.

(G) POWDERED IRON CORES
See Chapter 11 References (A).

CHAPTER 27
DETECTION AND AUTOMATIC VOLUME CONTROL

LOW DISTORTION A-M DETECTOR

A low-distortion A-M detector has been developed by W. T. Selsted and B. H. Smith (Fig. 27.56). This consists of a conventional diode rectifier direct-coupled to a cathode follower which is in turn connected to an r-f filter to reduce the carrier signal output. The excellent performance of this circuit is due to two facts :

1. That the load on the diode for normal A-M carrier frequencies is essentially resistive, and the normal effects of excessive shunting capacitance are eliminated.

2. That, since the coupling to the cathode follower is direct, there is no effect of biasing currents which are normally developed in a diode loading circuit using coupling condensers. The distortion for 100% modulation, as shown in Fig. 27.57, is claimed to be 0.3% at a modulating frequency of 420 c/s and 0.8% at 4000 c/s. The carrier input voltage and frequency are not stated.

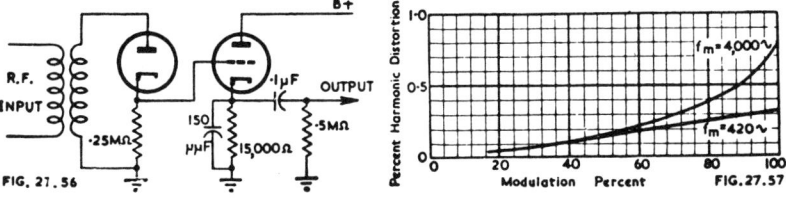

Fig. 27.56.   Circuit of low distortion A-M detector (W. T. Selsted and B. H. Smith).

Fig. 27.57.   Distortion characteristics of low distortion A-M detector (W. T. Selsted and B. H. Smith).

ADDITIONAL REFERENCES

(A) A-M DETECTORS
26A. Costas, J. P. " Synchronous detection of amplitude-modulated signals " Proc. National Electronics Conference 7 (1951) 121.
26B. Schooley, A. H., and S. F. George " Input versus output signal-to-noise characteristics of linear parabolic and semi-cubical detectors " Proc. National Electronics Conference 7 (1951) 151.
26C. Langford-Smith, F. " A low-distortion A-M detector " Radiotronics 18.6 (June 1953) 79. Developed by W. T. Selsted and Bob H. Smith in Radiation Laboratory of University of California. Summary in Electronics 26.11 (Nov. 1953) 214.

(D) MUTING (Q.A.V.C.)
62B. Vilkersom, B. S. " Muting for A-M or F-M " U.S. Patent No. 2,639,375, assigned to R.C.A. Described by R. H. Dorf " Audio Patents " Audio Eng. 38.1 (Jan. 1954) 2. Uses crystal oscillator tuned to centre of pass-band.

(G) CRYSTAL DIODES
85A. Douglas, R. W. and E. G. James " Crystal diodes " Proc. I.E.E. 98 Part III 53 (May 1951) 157, 177.
85B. Jordan, J. P. " The ABC's of germanium " E.E. 71.7 (July 1952) 619.
85C. Lovelock, R. T. " Point contact germanium rectifiers " W.W. 59.11 (Nov. 1953) 511 ; 59.12 (Dec. 1953) 600. Principles of operation and relation to performance and reliability.
85D. Jones, D. D., and B. C. Brodribb " Some high frequency effects in germanium diodes " Electronic Eng. 26.311 (Jan. 1954) 33.

(H) TRANSISTORS
86. Add discussion Proc. I.E.E. 99 Part III 62 (Nov. 1952) 363.
88. Scott, T. R. " Crystal triodes " Proc. I.E.E. 98, Part III 53 (May 1951) 169, 177.
89. Morton, J. A. " New transistors give improved performance " Elect. 25.8 (Aug. 1952) 100.
90. " The new tetrode junction transistor " Tele-Tech 11.11 (Nov. 1952) 38.
91. Rose, G. M., and B. N. Slade " Transistors operate at 300 MC " Elect. 25.11 (Nov. 1952) 116.
92. Oser, E. A., R. O. Enders and R. P. Moore " Transistor oscillators " R.C.A. Rev. 13.3 (Sept. 1952) 369.
93. Transistor issue (48 articles) Proc. I.R.E. 40.11 (Nov. 1952).
94. Shea, R. F. " Transistor power amplifiers " Elect. 25.9 (Sept. 1952) 106.
95. " Development transistorized equipments " Tele-Tech 12.1 (Jan. 1953) 75.
96. Smith, K. D. " Properties of junction transistors " Tele-Tech 12.1 (Jan. 1953) 76.
97. Roddam, T. " Transistors " W.W. (1) 59.2 (Feb. 1953) 70 ; (2) 59.3 (March 1953) 125 ; (3) 59.4 (April 1953) 175 ; (4) 59.5 (May 1953) 205 ; (5) 59.6 (June 1953) 256 ; (6) 59.7 (July 1953) 311 ; (7) 59.8 (Aug. 1953) 359 ; (8) 59.9 (Sept. 1953) 435 ; (9) 59.10 (Oct. 1953) 475 ; (10) 59.11 (Nov. 1953) 543 ; (11) 59.12 (Dec. 1953) 568.

CHAPTER 30
RECTIFICATION

ADDITIONAL REFERENCES
33. Hamann, C. E. " Resistance-capacitance loading of selenium rectifiers " Tele-Tech 11.11 (Nov. 1952) 58.
34. Corbyn, D. B. " Special rectifier circuits—a description of some new high-voltage circuits and consideration of centre-tapped circuits " Electronic Eng. 24.295 (Sept. 1952) 418.

CHAPTER 32
VIBRATOR POWER SUPPLIES

ADDITIONAL REFERENCES
5. Mitchell, J. H. " Recent developments in vibrators and vibrator power packs " Jour. Brit. I.R.E. 12.8 (Aug. 1952) 431.

CHAPTER 33
CURRENT AND VOLTAGE REGULATORS

ADDITIONAL REFERENCES
36. Benson, F. A. (book) " Voltage Stabilizers " Electronic Engineering, London, 1950.
37. Elmore, W. C., and M. Sands (book) " Electronics " McGraw-Hill Book Co. 1949 p. 363.
38. Seely, S. (book) " Electronic Tube Circuits " McGraw-Hill, 1950, pp. 306-316.
39. Benson, F. A. " The design of series-parallel voltage stabilizers " Electronic Eng. 24.289 (March 1952) 118.
40. Armitage, M. D. " Improved stabilization from a voltage regulator tube " Electronic Eng. 24.298 (Dec. 1952) 568. Circuit uses Stabilovolt gas-filled voltage regulator tube, together with a barretter.
41. Benson, F. A. " A study of the characteristics of glow-discharge voltage regulator tubes " Electronic Eng. 24.295 (Sept. 1952) 396 ; 24.296 (Oct. 1952) 456.
42. Kiryluk, W. (letter) " Voltage regulator tubes " Electronic Eng. 25.300 (Feb. 1953) 83. Dynamic test for gas tubes, revealing hysteresis, discontinuities and a.c. resistance of tube.
43. Trigg, R. D. " Voltage stabilization with series valve control " Electronic Eng. 25.304 (June 1953) 254). Gives circuit analysis, including resistor shunted across series valve.
44. " A.C. thermonic voltage regulator using Radiotron AV33 tungsten filament control diode " Radiotronics 18.7 (July 1953) 108.

45. Benson, F. A. " Voltage stabilization " Electronic Eng. 25.302 (April 1953) 160 ; 25.303 (May 1953) 202. With extensive bibliography. For non-linear bridge voltage source see also letter V. H. Attree W.E. 30.8 (Aug. 1953) 208.
46. Thomas, P. A. V. " An alternating current stabilizer for supplying valve heaters " Electronic Eng. 25.310 (Dec. 1953) 522. Review of existing methods and description of transducer controlled system giving voltage ±1%, output 240 watts. Correspondence 26.315 (May 1954) 218.
47. Edwards, P. L. " Relaxation oscillations in voltage regulator tubes " Proc. I.R.E. 41.12 (Dec. 1953) 1756. Oscillations may occur with gas-filled tubes shunted by a condenser ; conditions are given for avoiding oscillation.
48. Patchett, G. N. (book) " Automatic Voltage Regulators and Stabilizers " (Pitman, 1954, 335 pp.).
49. Geer, C. D., and W. C. Brockhuysen " Thermal relays control heater voltage " Elect. 26.1 (Jan. 1954) 166. Use in receivers and amplifiers.
50. Miles, R. C. " How to design VR tube circuits " Elect. 25.10 (Oct. 1952) 135.
51. Jones, W. R. " Voltage regulator tubes " Elect. 26.3 (March 1953) 162.
52. Benson, F. A., and G. Mayo " Impedance-frequency variations of glow-discharge voltage-regulator tubes " Electronic Eng. 26.315 (May 1954) 206. Tubes have resistive and inductive components. Measurements of 4 types of tubes from 20 to 10,000 c/s.
53. Dalton, W. M. " A.C. voltage stabilizers " Electronic Eng. 26.317 (July 1954) 310.
54. Hopkins, E. G. " Self-heating triode for voltage stabilization " W.E. 31.7 (July 1954) 169.

CHAPTER 34
TYPES OF A-M RECEIVERS

ADDITIONAL REFERENCES

1. Smith, R. A. " The relative advantages of coherent and incoherent detectors : a study of their output noise spectra under various conditions " Proc. I.E.E. 89, Part III 55 (Sept. 1951) 401.
2. Tucker, D. G. " The synchrodyne and coherent detectors " W.E. 29.346 (July 1952) 184.
3. Tucker, D. G. " The history of the homodyne and synchrodyne " J. Brit. I.R.E. 14.4 (April 1954) 143. A very complete survey of whole field with exhaustive bibliography.

CHAPTER 35
DESIGN OF SUPERHETERODYNE A-M RECEIVERS

ADDITIONAL REFERENCES

References dealing with interference and noise

13C. Lampard, D. G. " The minimum detectable change in the mean-noise input power to a radio receiver " Proc. I.E.E. 101. Part III 70 (March 1954) 111.

References relating to bandspreading

20A. Parry, C. A. " A method of bandspreading " Proc. I.R.E. Aust. 13.10 (Oct. 1952) 365.
20B. " Fine tuning arrangements " Electronic Eng. 25.305 (July 1953) 298. Trimmer capacitor connected in shunt with main tuning inductor, and other methods.

General references

63. Waverling, E. " Printed circuits for home radio receivers " Elect. 25.11 (Nov. 1952) 140.
64. Davis, B. L. " Printed circuit techniques : an adhesive tape resistor system " N.B.S. Circular 530 (1952) National Bureau of Standards, Washington D.C.
65. Strafford, F. R. W. " Reducing fire risks—a new method of safeguarding receivers " W.W. 58.12 (Dec. 1952) 499.
66. Whitehead, J. C. " The S-P 600 Communication Receivers " (Hammarlund MF-HF-VHF), Comm. Eng. 13.4 (July/Aug. 1953) 32.
67. Eisler, P. " Printed circuits and miniaturization " Electronic Eng. 25.304 (June 1953) 234. Describes foldable 3-dimensional circuits.
68. Kobe, K. A., and R. P. Graham " The effect of applying a counter e.m.f. to a Leclanche cell " paper read before Electro-Chemical Society, U.S.A., 30th April, 1938.
69. Hallows, R. W. " Improving the dry cell " W.W. 59.6 (June 1953) 276.
70. Hallows, R. W. " Reactivating the dry cell " W.W. 59.8 (Aug. 1953) 344.
71. Kelly, A. W. " Review of new printed circuit development and audio frequency applications " Jour. A.E.S. 1.1 (Jan. 1953) 53.
72. Dummer, G. W. A. and D. L. Johnston " Printed and potted electronic circuits " Proc. I.E.E. 100. Part III. 66 (July 1953) 177.
73. Eisler, P. " Printed circuits : some general principles and applications of the foil technique " Jour. Brit. I.R.E. 13.11 (Nov. 1953) 523.
74. Knight, M. B. " Designing trouble-free series tube heater strings " Tele-Tech 12.4 (April 1953) 76. Primarily TV receivers.

CHAPTER 36
DESIGN OF F-M RECEIVERS

ADDITIONAL REFERENCES

28. Willmotte, R. M. " Reception of an F.M. signal in the presence of a stronger signal in the same frequency band, and other associated results " Proc. I.E.E. 101. Part III 70 (March 1954) 69.
29. Medhurst, R. G. " Harmonic distortion of frequency-modulated waves by linear networks " Proc. I.E.E. 101. Part III 71 (May 1954) 171.

CHAPTER 37

RECEIVER AND AMPLIFIER TESTS AND MEASUREMENTS

ADDITIONAL REFERENCES

43. Terman, F. E., and J. M. Pettitt " Electronic Measurements " McGraw-Hill, 2nd ed. 1952.
44. Scroggie, M. G. " Radio Laboratory Handbook " Iliffe, London.
45. Maurice, D., and G. F. Newell and J. G. Spencer " Proposed test procedure for F-M broadcast receivers " Electronic Eng. 24,289 (March 1952) 106 ; reprinted Radiotronics 18.3 (March 1953) 39. Also gives results obtained on three actual receivers.
46. " British Standard Glossary of Terms for the Electrical Characteristics of Radio Receivers " (BS 2065 : 1954). Comments, W.W. 604. (April 1954) 188.
47. " Recommended methods of measurement on receivers for amplitude-modulation broadcast transmissions " (1st ed. 1954), Central office of the International Electrotechnical Commission," Publication 69. Comments W.W. 60.6 (June 1954) 271.

CHAPTER 38

TABLES, CHARTS AND SUNDRY DATA

Sect. 3 (xi) (F) British Radio Industry Council (page 1360) add after F5 :—
F5a.RIC/131/B. Capacitors, fixed, paper dielectric, foil, in rectangular metal cases.

Additional definition

Maximum output (in receivers). The greatest average output power into the rated load regardless of distortion.

Abbreviations of titles of periodicals, pages 1367-1369, add

Acustica	S. Hirzel Verlag, Zurich.
Comm. Eng.	Communication Engineering, The Publishing House, Great Barrington, Mass. U.S.A. Previously known as FM-TV and FM-TV Radio Communication.
High Fidelity	Audiocom Inc., Great Barrington, Mass. U.S.A.
Jour. A.E.S.	Journal of the Audio Engineering Society, Box 12 Old Chelsea Station, New York 11, N.Y.
Proc. I.E.E.	Proceedings of the Institution of Electrical Engineers (Savoy Place, Victoria Embankment, London, W.C.2, England).
Trans. I.R.E.-PGA	Transactions Institute of Radio Engineers, U.S.A. Professional Group Audio.

Additional references to standard symbols and abbreviations page 1369)
6. British Standard BS530 : 1948 " Graphical symbols for tele-communications," supplements 1, 2, 3, 4 (in preparation). See comments, Bainbridge-Bell, L. H. " The standardization of symbols and the arrangement of electronic circuit diagrams " Jour. Brit. I.R.E. 13.7 (July 1953) 339.
7. " I.R.E. Graphical symbols for electrical diagrams " Proc. I.R.E. 42.6 (June 1954) 967. These are the same as the American Graphical Symbols for Electrical Diagrams, Y32.2—1954, American Standards Association.

ADDITIONAL ITEMS

Neutralizing circuits (reference to page 1065)
These circuits in Figs. 26.19 and 26.21 are strictly not neutralizing circuits, but the effect achieved by using C_N is similar to that achieved by true neutralization as it allows the effect of feedback due to grid-to-plate capacitance to be reduced to negligible proportions, although it does not completely eliminate it.